S0-ADN-329

Refrigeration and Air Conditioning

An Introduction to HVAC/R

Fourth Edition

Larry Jeffus
Air Conditioning and Refrigeration Institute

Upper Saddle River, New Jersey
Columbus, Ohio

Library of Congress Cataloging-in-Publication Data

Jeffus, Larry
Refrigeration and air conditioning : an introduction to HVAC/R. —4th ed. / Larry Jeffus [and] Air Conditioning and Refrigeration Institute.
p. cm.
Rev. ed. of: Refrigeration and air conditioning. 3rd ed. c1998.
Includes index.
ISBN 0-13-092571-3
1. Refrigeration and refrigerating machinery. 2. Air conditioning. I. Air-Conditioning and Refrigeration Institute. II. Refrigeration and air conditioning. III. Title.

TP492.R377 2004
621.5'6—dc22

2003063259

Editor in Chief: Stephen Helba
Executive Editor: Ed Francis
Production Editor: Christine Buckendahl
Design Coordinator: Diane Ernsberger
Cover Designer: Jeff Vanik
Cover Photo: Larry Jeffus
Production Manager: Pat Tonneman
Marketing Manager: Mark Marsden

Photo Credits: Photos courtesy of Hampden Engineering Corporation were taken by Larry Jeffus. Photos courtesy of Rheem Manufacturing (except 10-5-12, 11-3-5, and 17-2-6) were taken by Larry Jeffus. Photos courtesy of Taco Bueno were taken by Larry Jeffus.

This book was set in New Century Schoolbook by *The GTS Companies*/York, PA Campus. It was printed and bound by Courier/Kendallville, Inc. The cover was printed by Phoenix Color Corp.

Pearson Education Ltd.
Pearson Education Singapore Pte. Ltd.
Pearson Education Canada, Ltd.
Pearson Education—Japan
Pearson Education Australia Pty. Limited
Pearson Education North Asia Ltd.
Pearson Educación de Mexico, S.A. de C.V.
Pearson Education Malaysia Pte. Ltd.

10 9 8 7 6 5 4 3 2
ISBN: 0-13-092571-3

This book is dedicated in loving memory to my brother, Richard Jeffus.

About the Author

Larry Jeffus is highly respected in both education and the refrigeration and air conditioning field, with more than 30 years of experience. His many years of dedicated service to his students and technical education were influential in leading Texas Governor Rick Perry to appoint Larry to the Texas Workforce Investment Council. Larry's aim in writing this book is to bring together, in a logical order, the data necessary to be successful in the HVAC/R industry. In doing this, he has been guided by his experience as a NATE Certified HVAC/R technician working in the field, as a teacher in the classroom, as a Texas licensed HVAC/R contractor, and as a consultant in the refrigeration and air conditioning field. He provides an excellent blend of theory with job-qualifying skill development and uses color photographs and line drawings that make the principles and service information easy to understand.

Preface

The refrigeration and air conditioning industry today presents a continually growing and changing series of opportunities for skilled technicians. Despite economic ups and downs, there is always a positive job outlook in the HVAC/R market. Because of new technologies and steady growth, the demand for technicians grows ever greater. In addition, our health and comfort depend on new developments in the refrigeration, air conditioning, ventilation, and heating fields. *Refrigeration and Air Conditioning: An Introduction to HVAC/R, Fourth Edition,* is an excellent resource for beginners and apprentices to start and pursue a pleasant and profitable career in the HVAC/R field.

Refrigeration, air conditioning, ventilation, and heating have dramatically expanded in the variety of available equipment and applications. This text contains all the most recent information and the advances necessary to prepare the technician for today's world. It incorporates the latest technical changes and EPA rulings, covers the newer refrigerants, and provides current information on the recovery, reclaiming, and recycling of refrigerants. It also contains basic information on the many certification exams. Material is presented in an easily understood format intended for use in HVAC/R classes in high schools, technical schools, and community colleges, or for those just wanting current information. This book will give students a solid foundation for success in the HVAC/R trades and serve as a great reference source for technicians in the field.

A thorough study of this book in a classroom setting will help students prepare for the many career opportunities in HVAC/R, from installation and servicing to maintaining air conditioning and refrigeration equipment. The extensive descriptions of equipment and supplies with in-depth explanations of their operation and function will familiarize students with the tools of the trade. The text also discusses occupational opportunities in the field and explains the training required. Students preparing for a career in HVAC/R will need not only technical skills, but also need to be able to follow written and verbal instructions, work with or without supervision, read and interpret diagrams and formulas, work well with tools and equipment, communicate effectively both verbally and in writing, be alert to possible problems and safety issues, and be able to keep up with the latest HVAC/R technologies. This book addresses all of these requirements.

ORGANIZATION OF THIS TEXT

The chapters are logically organized into sections that include:

- **Fundamentals** (Introduction; Tools, Meters, and Measuring Devices; HVAC/R Practices)
- **Refrigeration Principles** (Principles of Matter and Thermodynamics; Air Conditioning and Refrigeration System Components; Refrigerants)
- **Electricity** (Basic Electricity; Electrical Diagrams and Controls)
- **Residential Systems** (Central Residential Air Conditioning; Central Residential Gas Warm Air Heating; Central Residential Oil Warm Air Heating; Central Residential Electric Warm Air Heating; Central Residential Heat Pump Systems)
- **Indoor Air Systems** (Air Distribution; Indoor Air Quality; Load Calculation)
- **Commercial Systems** (Packaged Heating/Cooling Systems; Commercial Refrigeration; Central Plant Hydronic Systems)
- **Unitary Systems** (Appliances)
- **Employment Skills** (Job Skills)
- **Appendices** (Temperature Conversions; Properties of Refrigerants; Commonly Used HVAC/R Formulas; and Tables and Figures with Useful HVAC/R Data)

At the beginning of each chapter is a list of objectives. These objectives identify the topics covered and goals to be achieved in the chapter. All the chapters contain boxed features on **Tech Tips, Service Tips, Safety Warnings, Tech Talk,** and facts that are just **Nice to Know.** All chapters end with review questions to determine how well the student is learning the material. There are many tables that contain technical data and troubleshooting information throughout the book. These tables contain information that will help the reader diagnose and repair common problems. A list of these helpful tables is provided in Appendix D so they can be easily referenced once you have finished your studies of this book. Color is used extensively in this edition to enhance understanding and highlight important information. Hundreds of bright and clear color photographs help students grasp the subject material. Finally, both English- and Spanish-Language glossaries are provided at the end of the book.

ANCILLARY MATERIALS

- Student CD with full motion animations of many of the basic HVAC/R concepts.

A comprehensive instructor package is available free when the text is adopted for classroom use from Prentice Hall (call 1-800-526-0485 or visit online at www.prenhall.com). This instructor package includes the following:

- PowerPoint® presentation on topics covered in the text.
- Instructor CD with suggested student activities, a test bank, photo library with hundreds of digital color photos, as well as many other useful elements for the classroom.
- Answers to all questions in the textbook.
- Computerized test generator.

ACKNOWLEDGMENTS

The author would like to acknowledge the following companies for their contributions to this edition:

Air Conditioning Contractors of America (ACCA)
Aeroquip Corporation
Alnor Instrument Company
Amana
Amprobe
American Society of Heating, Refrigeration, and Air-Conditioning Engineers
Anamet Industrial, Inc.
Applied Energy Recovery Systems
A. W. Sperry Instruments
Bacharach, Inc.
Baltimore Aircoil Company
Bard Manufacturing Company
Calmac Manufacturing Corp.
Cooper-Atkins Corp.
Copeland Corp.
Danfoss Inc.
Dole Refrigerating Company
Dunham-Bush, Inc.
DuPont Chemicals
Duro Dyne Corp
Duro Metal Products
Dwyer Instruments Inc.
Elite Software
Evapco
Field Controls Company
Fluke Corporation
Frick Company
Frigidaire
General Electric Appliances
GlowCore A. C. Inc.
Goodway Technologies, Corp.
Heatcraft Refrigeration Products
Honeywell, Inc.
Hussmann Corporation, an Ingersoll Rand business
Imperial
J/B Industries
Johnson Controls/PENN
Kramer Refrigeration
Lennox Industries, Inc.
Manitowoc Company
Mannix
Marsh Instrument Company
Mastercool
Megger®
Mestek, Inc.
Mill-Rose Co.
National Energy Control Corp.
Nicholson
Video General (VGI)
Perfect Sense, Inc. (Nextek Power)
Ranco North America
Reed Manufacturing Company
Research Products Corporation
Rheem Manufacturing Air Conditioning Division
Roberts Gordon Inc.
Robinair SPX/OTC
The Robur Corporation
Scotsman Ice Systems

Selkirk Metalbestos
Skuttle Indoor Air Quality Products
SMACNA
Sporlan Valve Company
SRI International
Suntec Industries, Inc.
Taco Bueno, Inc.
Tecumseh Products
Test Products International (TPI)
Thermo King Corp.
TIF Instruments
Trane
Tunstall, Inc.
Underwriters' Laboratories, Inc.
U.S. Bureau of the Census
Vilter Manufacturing
Weil-McLain
White Industries
Worker's Compensation Board of British Columbia
Yellow Jacket Division, Ritchie Engineering
York International Corp.

I would also like to acknowledge Roger Keils, branch manager, and his staff at Johnstone Supply on Miller Road in Dallas, Texas, for the use of their location, equipment, and supplies for the many photos shot for this book.

Thanks and gratitude go to the following individuals for the information, inspiration, and help they provided in the preparation of the material for this new edition: Jerry Adams, Brian Aga, Donnie Aga, Ronnie Aga, Rex Boynton, Ben Burris, Sam Burris, Richard Carter, Loran Dailey, Bob Grossie, A. M. Jeffus, Amy Jeffus, Johnathan Jeffus, Richard Jeffus, Wendy Jeffus, Josh Kahn, Jeff Lester, Tommy Mills, Joe Moravek, Pat Murphy, Mort Newman, Ron Nolan, Bill Pritchard, Patrice Pruitt, Bill Smith, Mike Stephens, Joe Sutterfield, Wesley Taylor, Robert J. Dohse, Bill D. Smith, Pete Van Lancker, Andy Schoen, Peter Van Lancker, Jeff Rozga, Harvey Caplan, and Paul Wieboldt.

I must also express my heartfelt personal thanks to the dedicated individuals who worked with me for many thousands of hours in the preparation of this textbook: Jeri Bess and Bernice Nolan, for their help with the clerical work; Tina Ivey and Marilyn Burris, my dedicated administrative assistants; Richard Jeffus and Vicki Dolton, for their art and art research; Tommy Mills, for his photographs; and Bill Milam and Ronnie Ashworth, for their technical research and assistance.

Finally, a very special thanks goes out to my family, especially my wife Carol, for all the support and help that made this book possible.

FINAL NOTE

Congratulations on selecting *Refrigeration and Air Conditioning: An Introduction to HVAC/R, Fourth Edition.* This book will guide you to a successful career in the HVAC/R field and provide you with an excellent resource in your new profession.

Features of This Text

To enhance readability and understanding, as well as make study more enjoyable, this text is packed with many special features.

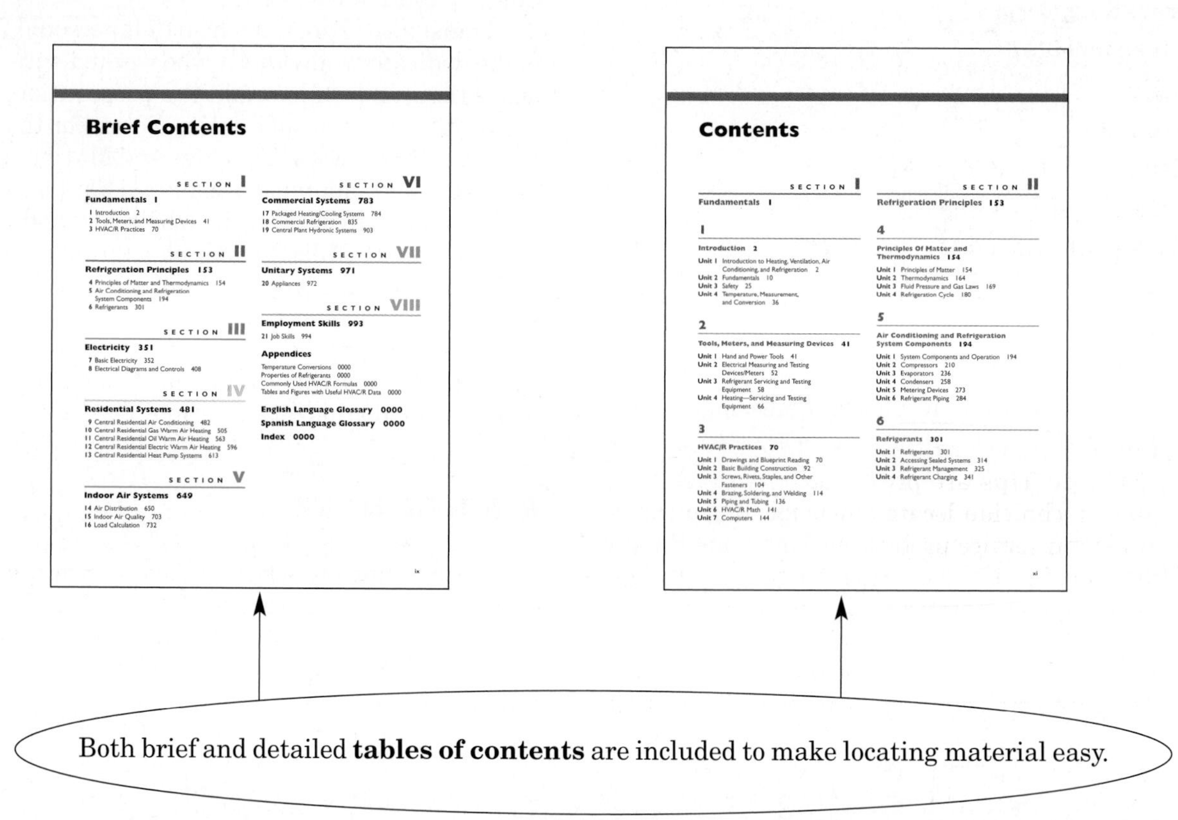

Brief Contents

ix

Contents

xi

Both brief and detailed **tables of contents** are included to make locating material easy.

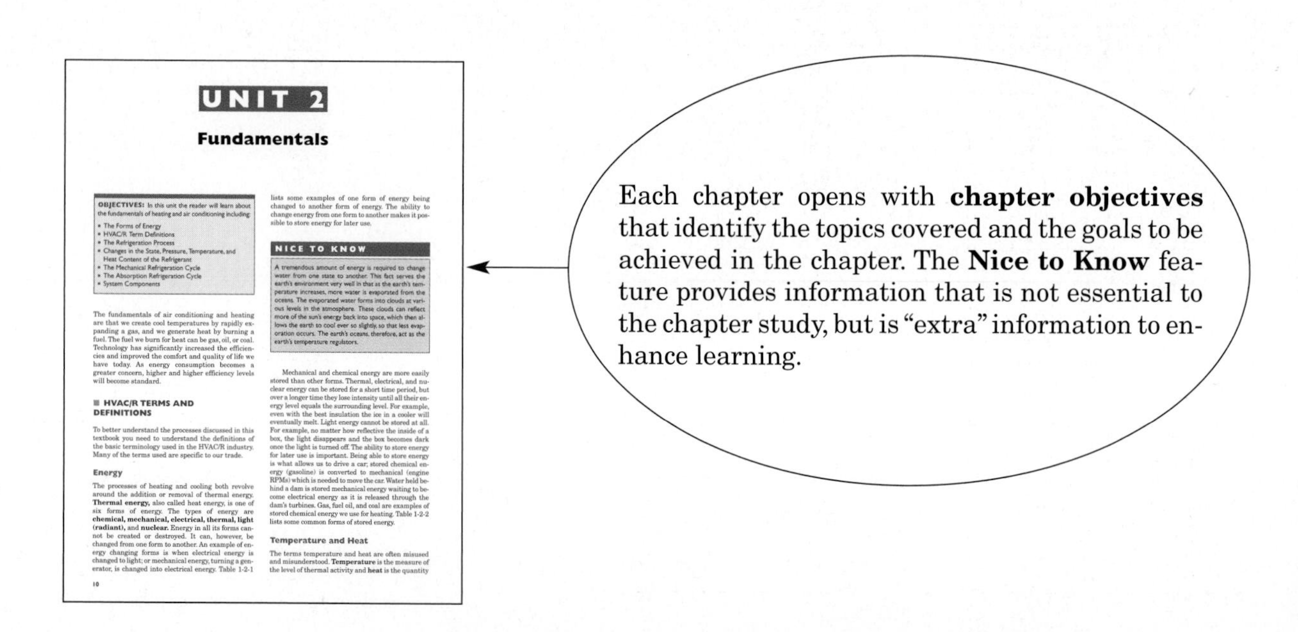

UNIT 2

Fundamentals

OBJECTIVES: In this unit the reader will learn about the fundamentals of heating and air conditioning including:

- The Forms of Energy
- HVAC/R Term Definitions
- The Refrigeration Process
- Changes in the State, Pressure, Temperature, and Heat Content of the Refrigerant
- The Mechanical Refrigeration Cycle
- The Absorption Refrigeration Cycle
- System Components

The fundamentals of air conditioning and heating are that we create cool temperatures by rapidly expanding a gas, and we generate heat by burning a fuel. The fuel we burn for heat can be gas, oil, or coal. Technology has significantly increased the efficiencies and improved the comfort and quality of life we have today. As energy consumption becomes a greater concern, higher and higher efficiency levels will become standard.

HVAC/R TERMS AND DEFINITIONS

To better understand the processes discussed in this textbook you need to understand the definitions of the basic terminology used in the HVAC/R industry. Many of the terms used are specific to our trade.

Energy

The processes of heating and cooling both revolve around the addition or removal of thermal energy. **Thermal energy,** also called heat energy, is one of six forms of energy. The types of energy are **chemical, mechanical, electrical, thermal, light (radiant),** and **nuclear.** Energy in all its forms cannot be created or destroyed. It can, however, be changed from one form to another. An example of energy changing forms is when electrical energy is changed to light; or mechanical energy, turning a generator, is changed into electrical energy. Table 1-2-1 lists some examples of one form of energy being changed to another form of energy. The ability to change energy from one form to another makes it possible to store energy for later use.

NICE TO KNOW

A tremendous amount of energy is required to change water from one state to another. This fact serves the earth's environment very well in that as the earth's temperature increases, more water is evaporated from the oceans. The evaporated water forms into clouds at various levels in the atmosphere. These clouds can reflect more of the sun's energy back into space, which then allows the earth to cool ever so slightly, so that less evaporation occurs. The earth's oceans, therefore, act as the earth's temperature regulators.

Mechanical and chemical energy are more easily stored than other forms. Thermal, electrical, and nuclear energy can be stored for a short time period, but over a longer time they lose intensity until all their energy level equals the surrounding level. For example, even with the best insulation the ice in a cooler will eventually melt. Light energy cannot be stored at all. For example, no matter how reflective the inside of a box, the light disappears and the box becomes dark once the light is turned off. The ability to store energy for later use is important. Being able to store energy is what allows us to drive a car; stored chemical energy (gasoline) is converted to mechanical (engine RPMs) which is needed to move the car. Water held behind a dam is stored mechanical energy waiting to become electrical energy as it is released through the dam's turbines. Gas, fuel oil, and coal are examples of stored chemical energy we use for heating. Table 1-2-2 lists some common forms of stored energy.

Temperature and Heat

The terms temperature and heat are often misused and misunderstood. **Temperature** is the measure of the level of thermal activity and **heat** is the quantity

10

Each chapter opens with **chapter objectives** that identify the topics covered and the goals to be achieved in the chapter. The **Nice to Know** feature provides information that is not essential to the chapter study, but is "extra" information to enhance learning.

6 SECTION I FUNDAMENTALS

Figure 1-1-10 Modern refrigeration display cases provide us with a variety of food products that would not be available without refrigeration. (*Courtesy of Hussman Corporation, an IR [Ingersoll-Rand] business*)

that would not be possible to preserve any other way, Figure 1-1-10.

Before Clarence Birdseye began commercially freezing food, people had allowed food to freeze naturally during the winter months as a way of preserving it. The trick was to come up with a way of freezing food and having it still taste good when it was thawed. Blast freezing, a process of rapidly freezing food, was the key ingredient in improving the quality of thawed frozen foods. The problem with freezing food slowly is that when ice crystals form over time they become much larger. These large sharp ice crystals grow through the cell walls of the food and when the food thaws all of the nutrients in the food are allowed to drain away. Blast freezing causes the ice crystals to be very small and they are less likely to penetrate cell walls. So the food retains nutrients and flavor when it's thawed.

EMPLOYMENT OPPORTUNITIES

The HVAC/R industry represents one of the largest employment occupations in the country. Our industry, for example, is the largest consumer of utilities in the nation. More electricity and natural gas is consumed producing heating and cooling than with any other single use. The size of the industry has been growing steadily since the late 1960s when residential central systems became popular.

There are a variety of occupational specialties offered within the HVAC/R industry. These occupations range from the basic entry level helper to the design engineer.

Entry Level Helper The entry level helper provides the master technician with assistance installing and servicing equipment. Most medium and large mechanical contracting companies use a number of helpers to assist with the installation and service of residential and commercial systems. A helper may be expected to assist in lifting, carrying, or placing equipment or components. They may also run errands to pick up parts and clean up the area following installation or service. Helpers receive basic safety training and must have good driving records.

Rough-in Installer The initial installation process is referred to as rough-in. In this process the technician will install the refrigerant lines, electrical lines, thermostat and control lines, duct boots and duct run, as well as setting the indoor and outdoor units. The rough-in technician must have an understanding of duct layout, blueprint reading, basic hand tools, and good brazing skills.

Start-up Technician Once the system has been installed and all of the components are ready for operation, a start-up technician will go through the manufacturer's recommended procedures to initially start a system. Because much of the HVAC system has been field installed, this checkout procedure is essential to ensure safe and efficient operation. The start-up technician records all of the information requested by the manufacturer's warranty. Start-up technicians must be skilled with electrical troubleshooting, refrigerant charging, and have good reading comprehension skills and writing skills.

Service Technician The service technician is the individual who provides the system owner with repair and maintenance. Service technicians are the people who must be able to diagnose system problems and make the necessary repairs. Service technicians must be skilled in diagnosing electrical problems, refrigerant problems, and air distribution problems.

TECH TIP

Technology has enabled the field tech to stay in close contact with his service manager. This allows the highly experienced service managers to provide assistance to technicians as they come upon new problems. The technician can also call upon the office to research unique problems to determine the best, most efficient way of making the repair.

The **Tech Tip** features give the reader many helpful tips from the pros, such as how to work quicker and more efficiently.

Service Tips are provided to help the field technician locate and troubleshoot common service problems.

54 SECTION I FUNDAMENTALS

SERVICE TIP

Most clamp-on ammeters cannot accurately read low amp draws. A small loop of thermostat wire can be made to enable the meter to accurately read these lower amp draws. To make this loop take a single strand of thermostat wire and make a round loop with a diameter of approximately 1½ in. Make 10 complete wraps of thermostat wire. Use several pieces of black electrical tape to ensure the coil stays tightly packed together so that it is not easily snagged in your tool pouch. To use this amperage multiplier connect the loop wire ends into the circuit so that it is in series where all of the amperage flowing through the circuit will then pass through the 10 turns of the amperage multiplier. Clamp the amperage ammeter jaws through the 10 turn loop. The amperage reading shown on the ampmeter will be 10 times higher than the actual amp draw. For example an amperage reading of 15 would indicate an actual amp draw of 1.5 amps. This amperage multiplier will make it possible to read the very low amp draw of a gas valve so that the thermostat heating anticipator can be accurately set.

Figure 2-2-5 Clamp-on ammeters.

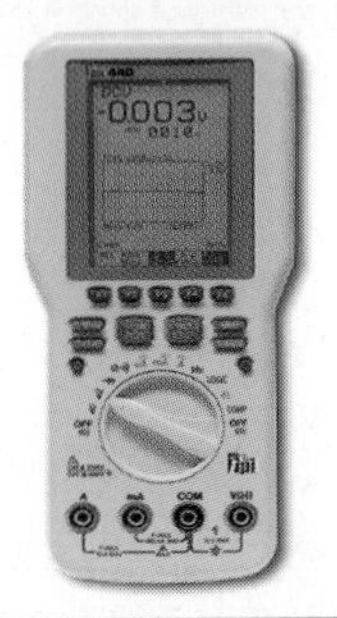

Figure 2-2-4 The safety terminals at the bottom of this multimeter are recessed, so they will not accidentally cause a short circuit. (*Courtesy Test Products International, Inc.*)

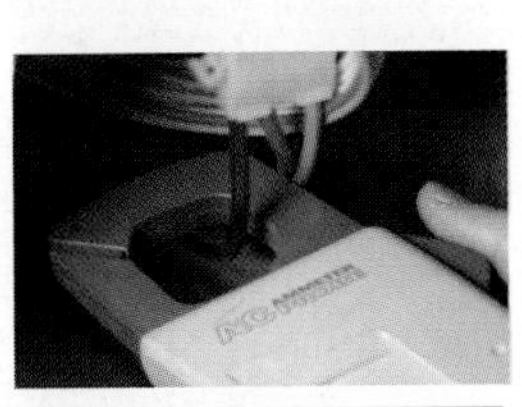

Figure 2-2-6 Measuring amperage with a clamp-on ammeter. (*Courtesy Test Products International, Inc.*)

Wattmeter

This instrument reads true power, including an allowance for the power factor. The meter makes the necessary calculations for power, in accordance with the power formula:

watts = volts * amps * power factor

Chapter 1 Introduction 9

education director and its Education and Training Committee, ARI serves as a resource from the manufacturers to school instructors, department heads, and guidance counselors. In addition to this textbook and its companion materials, ARI produces the *Bibliography of Training Aids*, a career brochure, and a promotional videotape for schools to use to recruit students into HVAC/R programs. Many schools around the country have adopted the ICE competency exams as final exams for their programs. ARI's most recent efforts involve participation in efforts to develop national HVAC/R competency standards.

Having students pass the ICE competency exams and training toward national competency standards will improve the quality of installation and service. New HVAC/R technicians will be better prepared, resulting in three basic advantages:

1. Limited training required for contractors
2. Limited rework or repeat calls due to error
3. Limited warranty/replacement for manufacturers.

The cost of repeat service calls, which is borne by contractors, may be reduced substantially through the use of properly trained technicians. Every new technician receives training and serves as an apprentice for a period of time. That is essentially a period where contractors pay two people to do one job. A properly trained technician will generally require less training time and function sooner than a poorly trained technician.

TECH TALK

Tech 1. I don't understand why the boss won't let me gauge up on a system. I know exactly how to do it because I've been taking classes. What's even more frustrating is that I can do it at school, but he won't let me do it here.

Tech 2. That's right. There is a special provision in the EPA regulations that allow students under the supervision of an instructor to work on refrigeration systems as part of their learning process, but that privilege does not extend to the shop. Before you are going to be able to gauge up out here you are going to have to pass the EPA certification test.

Tech 1. It still would be good practice for me if the boss would just let me do it occasionally. I mean, I could get more efficient and I could help him out.

Tech 2. Well, the real problem would be if someone were to turn you in for working without an EPA license then you *and* the boss *and* the company would wind up being fined for violating the EPA regulations. The regulations are very stringent and very specific. If you choose to intentionally violate these regulations you are placing yourself and the company at significant risk of being fined. The boss is right. Until you get your EPA certification you can't work on any of the systems.

Tech 1. I didn't know that.

Tech 2. That's why you have to take the class and become certified. You will learn all this in the class. It will be part of your test to become certified.

Tech 1. Thanks, I appreciate that. I'll find out when the next class and test will be scheduled.

UNIT 1—REVIEW QUESTIONS

1. According to the text, what was the significance of the discovery of fire?
2. Why did early Romans build channels underneath the floors?
3. Where does it appear that ice production first began?
4. How low does the relative humidity have to be for evaporation to cool water to its freezing point even though outside air temperature is well above freezing?
5. How was the first mobile refrigeration used?
6. What was the primary means of air conditioning for many years?
7. What is Willis Carrier often credited with?
8. When did residential air conditioning start to become popular?
9. Who developed the process of freezing foods?
10. What are the duties of an entry level helper?
11. What are the duties of a start-up technician?
12. What are the requirements to be a service technician?
13. What are the two major categories of HVAC/R sales?
14. What are facilities maintenance personnel responsible for?
15. A service manager is typically a skilled HVAC/R technician with ______.
16. What is ASHRAE?
17. What is ASME composed of?
18. What is RSES dedicated to?
19. What information is covered on the ICE competency exam?
20. What is the reclaimer certification program?
21. Following ARI's goals, new HVAC/R technicians will be better prepared, resulting in what three basic advantages?

The **Tech Talk** feature appears at the end of each unit, along with the unit review questions. Tech Talk presents a dialogue between a new technician and a more experienced one, in which the "old pro" gives the new technician advice.

This page shows an example of the many tables in this edition that provide technical material to be used both in the classroom and as a reference once the student becomes a working professional in the field.

TABLE 1-2-1 Examples of Energy Conversion

Type of Product/Action	Energy Conversion From	Energy Conversion To
Gasoline in a car	Chemical	Mechanical
Electricity going through a light bulb	Electrical	Radiant
Water turning a turbine	Mechanical	Electrical
Heat energy producing steam to turn a turbine	Thermal	Mechanical
Chemicals in a battery	Chemical	Electrical
Burning coal moving the wheels on a steam engine	Chemical	Mechanical
Windmill	Mechanical	Electrical
Electricity in a heater	Electrical	Thermal
Natural gas burning	Chemical	Thermal
Light on a photocell	Radiant	Electrical

of thermal energy. We often relate temperature to hot or cold. Temperature alone does not tell you the amount of thermal energy in a material. A very light object can have a high temperature but because it has little weight it has little thermal energy. The amount of thermal energy present will tell the technician how much refrigerant will be needed.

A **refrigerant** is a fluid that picks up heat by evaporating at a low temperature and pressure and gives up heat by condensing at a higher temperature and pressure. Common types of refrigerants such as R-22, R123, R134a, and R410A are shown with their cylinder colors in Table 1-2-3. As heat is removed from a space, the area becomes cooler. Therefore, **cold** can be defined as the relative absence of heat energy. But remember, there is always some heat energy present down to absolute zero (−460°F, −273°C).

TABLE 1-2-2 Examples of Stored Energy

Material	Type of Energy Stored
Batteries	Electrical
Gasoline	Chemical[1]
Water behind a dam	Mechanical
Glow-in-the-dark strips	Light
Coal	Chemical[1]
Candle	Chemical[1]
Wood	Chemical[1]
Capacitor	Electrical
Ice	Thermal
Food	Chemical

[1]When gasoline, coal, a candle, or wood are burned, the chemical energy is converted to thermal energy and radiant (light) energy.

TABLE 1-2-3 Refrigerant and Refrigerant Cylinder Colors

Refrigerant	Color
11	Orange
12	White
13	Light blue (sky)
13B1	Pinkish-red (coral)
14	Yellow-brown (mustard)
22	Light green
23	Light blue-gray
113	Dark purple (violet)
114	Dark blue (navy)
116	Dark gray (battleship)
123	Light blue-gray
124	Deep green (DOT green)
125	Medium brown (tan)
134a	Light blue (sky)
401A	Pinkish-red (coral)
401B	Yellow-brown (mustard)
401C	Blue-green (aqua)
402A	Light brown (sand)
402B	Green-brown (olive)
404A	Orange
407A	Lime green
407B	Cream
407C	Medium brown (brown)
408A	Medium purple (purple)
409A	Medium brown (tan)
410A	Rose
410B	Maroon
500	Yellow
502	Light purple (lavender)
503	Blue-green (aqua)
507	Blue-green (teal)
717, NH_3	Silver
Any recovered	Yellow/gray

Refrigeration

Refrigeration is the process of removing heat energy from a place where it is not wanted and disposing of it in a place where it is wanted or not objectionable. **Mechanical refrigeration** makes use of mechanical components to produce movement and transfer heat from an area of lower temperature to an area of higher temperature. For example a household refrigerator transfers heat from the interior of the refrigerator cabinet to the surrounding area of the kitchen. Another kind of refrigeration, **absorption refrigeration,** uses heat

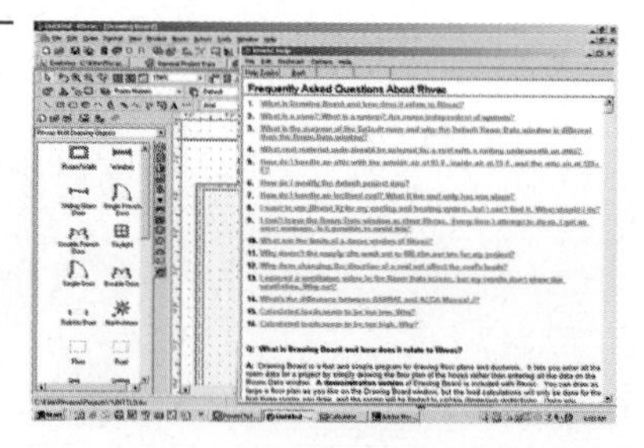

Figure 3-7-2 Help menu for frequently asked questions (FAQ). (*Courtesy Elite Software*)

This page provides an example of the current computer information included in this textbook.

seek help even when working late nights during the busy season. Phone-in service may be limited to standard business hours for the time zone where the company's corporate offices are located. Phone-in technical support may be free, free for a period of time, or fee based. Check with your software manufacturer's guidelines to determine which policy is in effect for that software.

Input Range Often the program will have a range of acceptable input data. This system would give a warning or error message if a number is entered that is outside the expected range for that item. For example, a program would be able to determine that the air changes of the residence were below code or standard requirements if too narrow a duct cross section were entered. These operating parameters

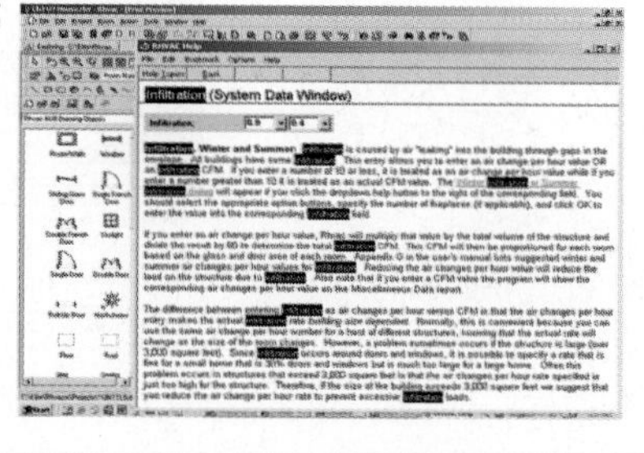

Figure 3-7-3 Help provided by a search for the word "infiltration." (*Courtesy Elite Software*)

This page shows the sequential presentation of the detailed drawings included in this edition.

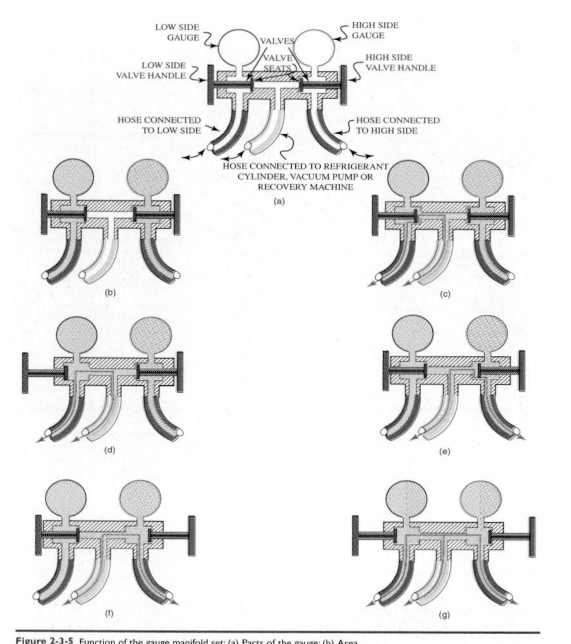

Figure 2-3-5 Function of the gauge manifold set: (a) Parts of the gauge; (b) Area pressurized with both valves closed; (c) Area pressurized with low side valve slightly open; (d) Area having full flow with low side valve open; (e) Area pressurized with high side valve slightly open; (f) Area pressurized with high side valve completely open; (g) Area pressurized with both low and high side valves open.

English Language Glossary

Absolute Humidity The amount of moisture actually in a given volume of air.

Absolute Pressure Gauge pressure plus atmospheric pressure.

Absolute Temperature Temperature measured from absolute zero.

Absolute Zero Temperature at which all molecular motion ceases (−460°F and −273°C).

Absorbent Substance which has the ability to take up or absorb another substance.

Absorber A device containing liquid for absorbing refrigerant vapor or other vapors.

Accumulator A storage vessel located in the suction line ahead of the compressor. Used to limit liquid refrigerant return to the compressor and store excess refrigerant in the heating mode.

Activated Carbon Specially processed carbon commonly used to clean air.

Adiabatic Compression Compressing refrigerant gas without removing or adding heat.

Adsorbent Substance which has property to hold molecules of fluids without causing a chemical or physical change.

Air Binding A condition in which a bubble or other pocket of air is present in a pipeline that prevents the desired flow in the pipeline.

Air Changes The amount of air leakage through a building in terms of the number of building volumes exchanged.

Air Conditioner Device used to control temperature, humidity, cleanliness, and movement of air in conditioned space.

Air Cushion Tank (See Expansion Tank).

Air Shutter An adjustable shutter on the primary air openings of a burner, which is used to control the amount of combustion air.

Air Source Heat Pump Heat pump that transfers heat from outdoor air to an indoor air circulation system.

Air Vent A valve installed at the high points in a hot water system to eliminate air from the system.

Aldehyde A class of compounds that can be produced during incomplete combustion of a fuel gas.

Allen Head Screw Screw with recessed head designed to be turned with a hex shaped wrench.

Alternating Current Abbreviated AC. Current that reverses polarity or direction periodically. It rises from zero to maximum strength and returns to zero in one direction then goes to similar variation in the opposite direction. This is a cycle which is repeated at a fixed frequency. It can be single phase, two phase, three phase, and polyphase. Its advantage over direct or undirectional current is that DC voltage can be stepped up by transformers to a high value which reduces transmission costs.

Alternator A machine that converts mechanical energy into alternating current.

Ambient Temperature Temperature of the air that surrounds an object.

American Wire Gauge Abbreviated AWG. A system of numbers that designate cross-sectional area of wire. As the diameter gets smaller, the number gets larger, e.g., AWG #14 = 0.0641 in, AWG #12 = 0.0808 in.

Ammeter An electric meter used to measure current.

Ampere Unit of electric current equivalent to flow of one coulomb per second.

Ampere Turn Abbreviated AT or NI. Unit of magnetizing force produced by a current flow of one ampere through one turn of wire in a coil.

Amplitude The maximum instantaneous value of alternating current or voltage. It can be in either a positive or negative direction.

Anemometer Instrument for measuring the rate of flow of air.

Angle of Lag or Lead Phase angle difference between two sinusoidal wave forms having the same frequency.

Annealing Process of heat treating metal to obtain desired properties of softness and ductility.

Anticipator A heater used to adjust thermostat operation to produce a closer temperature differential than the mechanical capability of the control.

Armature The moving or rotating component of a motor, generator, relay or other electromagnetic device.

ASME American Society of Mechanical Engineers.

Aspect Ratio Ratio of length to width of rectangular duct.

Aspirating Psychrometer A device that draws in a sample of air to measure for humidity.

Atom Smallest particle of an element.

Atomic Weight The number of protons in an atom of a material.

Atomize Process of changing a liquid to a fine spray.

Attenuate Decrease or lessen in intensity.

Authorized Dealer A dealer authorized to install heat pumps by the manufacturer.

Automatic Defrost System of removing ice and frost from evaporators automatically.

1065

Spanish Language Glossary

Absolute Humidity Humedad Absoluta. La cantidad de vapor de agua en un cierto volumen de aire.

Absolute Pressure Presión Absoluta. Presión de manómetro más presión atmosférica.

Absolute Temperature Temperatura Absoluta. Temperatura medida a partir del cero absoluto.

Absolute Zero Cero Absoluto. Temperatura a la cual cesa todo movimiento molecular (equivalente a −460° Farenheit y −273° Celsius).

Absorbent Absorbente. Substancia que tiene la capacidad de absorber otra substancia.

Absorber Tanque de absorción. Aparato que contiene un líquido para absorber vapor refrigerante u otro tipo de vapor.

Accumulator Acumulador. Depósito que recibe líquido refrigerante del evaporador e impide que fluya hacia el tubo de succión.

Activated Carbon Carbón activado. Carbón procesado industrialmente que se usa comúnmente para limpiar aire.

Adiabatic Compression Compresión adiabática. Compresión de gas refrigerante sin añadir o eliminar calor.

Adsorbent Adsorbente. Substancia que tiene la propiedad de retener moléculas de un fluído sin causar una reacción química o un cambio físico.

Air Binding Atascamiento de aire. Condición en la cual una burbuja o una bolsa de aire en una tubería impide el flujo deseado.

Air Changes Cambios de aire. La cantidad de aire que se escapa de un edificio en términos del número de volúmenes del edificio intercambiados.

Air Conditioner Aire Acondicionado. Aparato utilizado para controlar la temperatura, humedad, limpieza y movimiento de aire en el espacio acondicionado.

Air Cushion Tank Depósito de cojín de aire. Recipiente cerrado que permite la expansión del agua sin que la presión aumente de manera excesiva.

Air Shutter Obturador de aire. Obturador ajustable situado en los orificios primarios de aire de una caldera que se utiliza para controlar la cantidad de aire de la combustión.

Air Source Heat Pump Bomba térmica de alimentación de aire. Bomba térmica que transfiere calor del aire exterior a un sistema de circulación de aire interior.

Air Vent Respiradero. Válvula instalada en los puntos altos de un sistema de agua caliente para eliminar aire de dicho sistema.

Aldehyde Aldehído. Una clase de compuestos que se pueden producir durante la combustión incompleta de un carburante.

Allen Head Screw Tornillo Allen. Tornillo de cabeza empotrada, diseñado para girar con una llave de sección hexagonal.

Alternating Current Corriente Alterna (Abreviatura: AC). Corriente eléctrica que cambia de polaridad o dirección de manera periódica: aumenta de cero hasta un máximo, vuelve a cero y aumenta de nuevo hasta dicho máximo pero en la dirección opuesta, en un ciclo que se repite a una frecuencia fija. La corriente alterna puede ser de una sola fase, dos fases, tres fases y polifásica. La ventaja de la corriente alterna sobre la corriente continua es que el voltaje puede aumentarse a valores muy altos con la ayuda de transformadores, lo cual reduce los costes de transmisión.

Alternator Alternador. Una máquina que convierte la energía mecánica en corriente alterna.

Ambient Temperature Temperatura Ambiente. Temperatura del fluído, generalmente aire, que rodea a un objeto.

American Wire Gauge Sistema Americano de Clasificación de Hilo Conductor (Abreviatura: AWG). Un sistema de numeración utilizado para designar la sección de un hilo conductor. Este número es mayor conforme disminuye el diámetro del hilo; por ejemplo, AWG #14 = 0.0641 pulgadas; AWG #12 = 0.0808 pulgadas)

Ammeter Amperímetro. Un aparato de medición que se utiliza para medir la corriente eléctrica.

Ampere Amperio. Unidad de corriente eléctrica equivalente al flujo de un culombio por segundo.

Ampere Turn Amperio Vuelta (Abreviatura: AT o NI). Unidad de fuerza magnética que se produce cuando una corriente eléctrica de un Amperio circula por una vuelta de una bobina.

Amplitude Amplitud. Valor máximo instantáneo de una corriente o un voltaje alterno, tanto en la dirección positiva como en la negativa.

Anemometer Anemómetro. Instrumento para medir el coeficiente de flujo de aire.

Angle of Lag or Lead Ángulo de Desfase. Diferencia de fase entre dos formas sinusoidales de la misma frecuencia.

Annealing Revenir o destemplar. Proceso de tratamiento térmico de metales para obtener las características deseadas de dureza y ductilidad.

Anticipator Anticipador. Un calentador que se utiliza para ajustar la operación de un termostato con el fin de reducir el diferencial de temperatura por encima de la capacidad mecánica del controlador.

Armature Armadura. El componente móvil o rotatorio de un motor, generador, relé u otro tipo de dispositivo electromagnético.

1080

Both English and Spanish glossaries are included to help every student understand the terms used in the HVAC/R field.

Brief Contents

Contents

SECTION I

Fundamentals

1 Introduction
2 Tools, Meters, and Measuring Devices
3 HVAC/R Practices

Introduction

UNIT 1

Introduction to Heating, Ventilation, Air Conditioning, and Refrigeration

OBJECTIVES: In the following chapter the reader will learn about:

- History of Heating and Refrigeration
- Employment Opportunities
- Trade Associations

HISTORY OF HEATING AND REFRIGERATION

The history of heating a space by burning wood starts at our earliest times and continues to the present. The discovery of fire allowed early mankind to venture beyond the temperate regions, Figure 1-1-1. We think of early humans using fire as a way of preparing food, but more importantly it was a source of heat. Campfires allowed these early settlers to move into the colder climates and still survive. Fire burning in an open fireplace remained the primary source of heat for centuries. Elaborate systems using firewood heated Roman buildings. Early Romans built channels underneath the floors. Heat from the fire could be drawn through so that the building was warmed by the first central heating systems, Figure 1-1-2. In more recent centuries, wood burning stoves provided heat for an entire home, Figure 1-1-3.

The first use of refrigeration used ice harvested from lakes for the preservation of food. Ice harvesting remained a flourishing industry well into the 20th century. Ice production first appeared at about 1000 B.C. in northern Egypt, the Middle East, Pakistan, and India. Archeologists have discovered that people in this region were artificially producing ice by using evaporation. Archeological excavations in these regions have discovered ice-producing fields that covered several acres. The ice was produced in shallow clay plates, about the size of a saucer. The

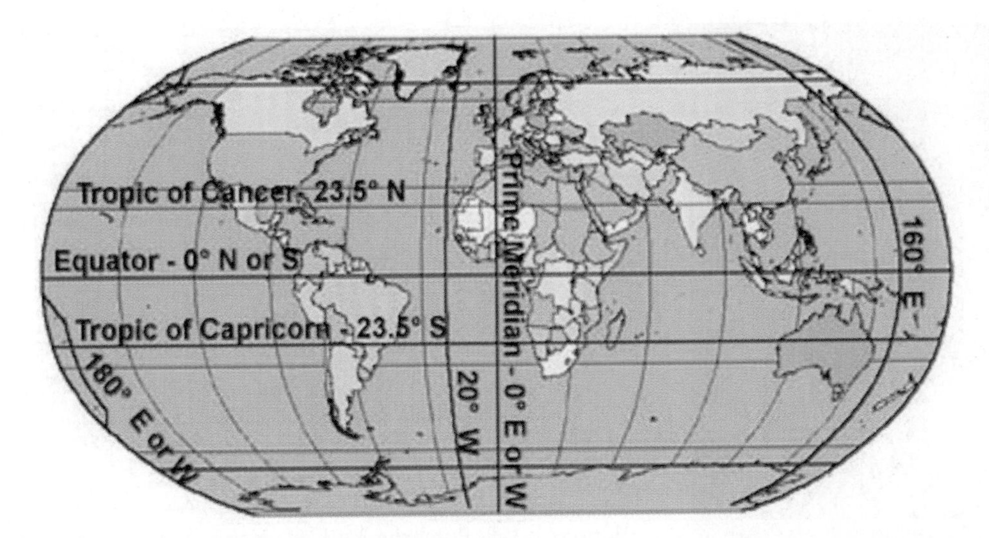

Figure 1-1-1 Before gaining the ability to use fire, early humans were confined to the warmer climates around the equator.

water in these clay plates wept through the clay. This water dampened the small straw mats holding the clay plates in racks a few feet above the ground, Figure 1-1-4. The straw aided with evaporative cooling of the water to below its freezing point. When the relative humidity is low enough, around 5%, evaporation can cool water to its freezing point even though the outside air temperature is well above freezing. Figure 1-1-5 shows a chart of relative humidity at various temperatures to demonstrate this cooling effect. These shallow dishes would produce only thin skins of ice which were harvested early in the morning and placed in straw insulated caves to store food.

NICE TO KNOW

Without our ability to control the environment it would be impossible for us to be exploring space, the bottom of the ocean, or even enjoying the comfort of a transcontinental jet ride at 35,000 ft. So our ability to control our environment has served both to improve the quality of life and to enhance our scientific endeavors.

Evaporative cooling is also the principle behind modern snow making equipment. A snow blower like the one in Figure 1-1-6 can make snow by evaporative cooling even when the temperatures are above freezing. Artificially cooling the air in a living space also dates back to the earliest centuries. In Ancient Greece, large wet woven tapestries were hung in natural drafts so that the air flowing through and around the tapestries was cooled by the evaporating water. Some manufacturers sprayed water in factories for cooling as early as the 1720s. Evaporative cooling is still used extensively in residences and businesses throughout the Southwest United States where typical summer conditions are very hot with very low relative humidities, Figure 1-1-7.

The mechanical refrigeration process was developed as a result of observations made in coal mines in England. Miners used steam-driven or water-powered compressors to force air into the deepest mines so that they could work in a safe atmosphere. Over long hours of operation, miners observed the formation of ice around the air nozzles, Figure 1-1-8. This ice was harvested and used for food preservation. The construction of steam-powered compressed-air ice producing plants soon followed. The first mobile refrigeration units were made by putting steam powered compressors on sailing ships to make it possible for beef to be shipped from Australia to England in the 1800s.

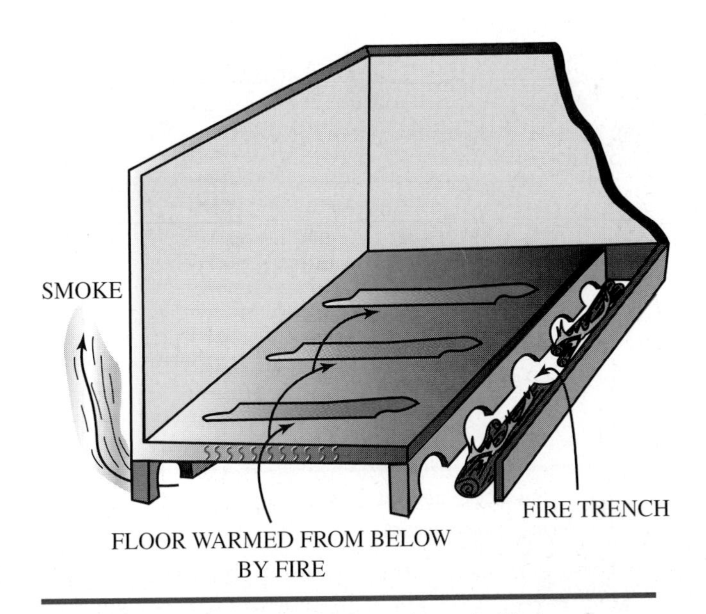

Figure 1-1-2 Romans used fires channeled below floors as early heating systems.

Figure 1-1-3 Wood burning stove used to heat a home.

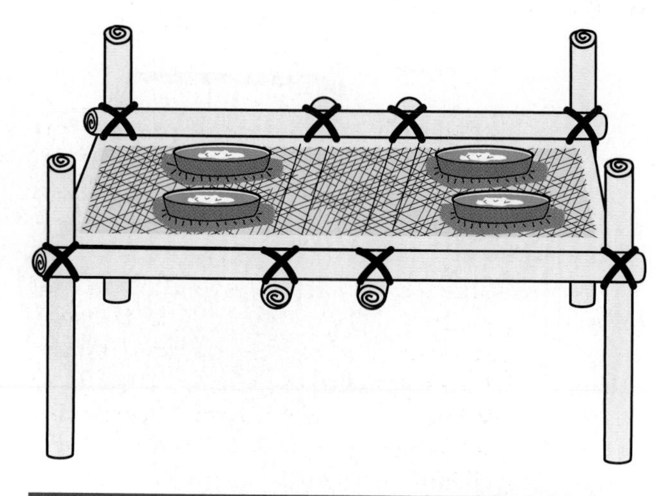

Figure 1-1-4 Ice was first artificially produced to be used for food preservation more than 3000 years ago.

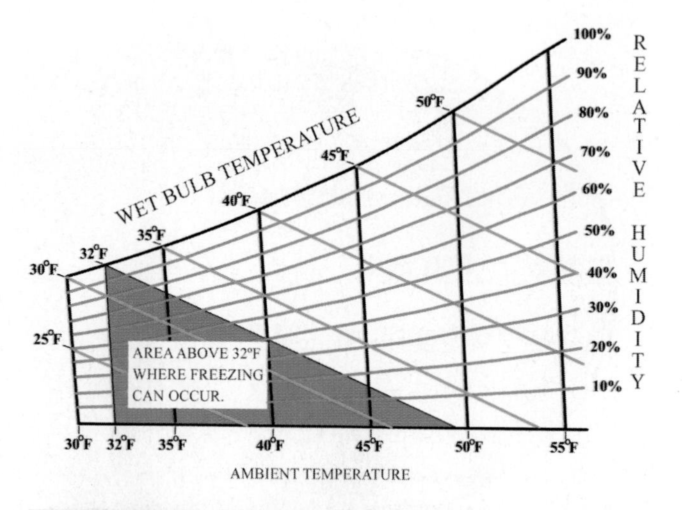

Figure 1-1-5 Psychrometric chart shows that water can be cooled below freezing if the relative humidity is low enough.

Figure 1-1-6 Snow blowers can produce artificial snow by evaporative cooling. (*Courtesy of Red River Ski Area*)

Figure 1-1-7 Evaporative cooler used to cool a store in New Mexico.

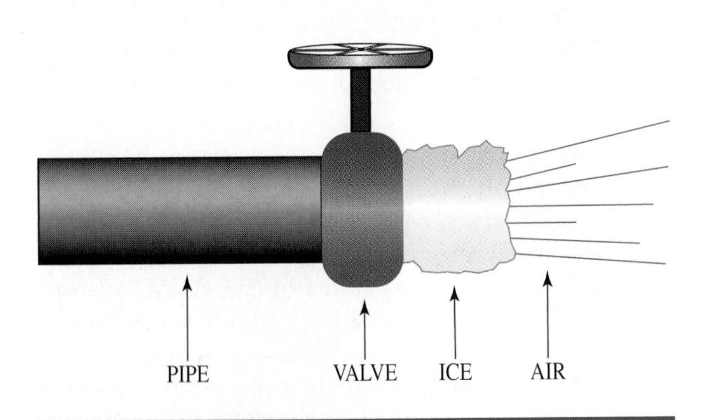

Figure 1-1-8 Ice forming around an air nozzle.

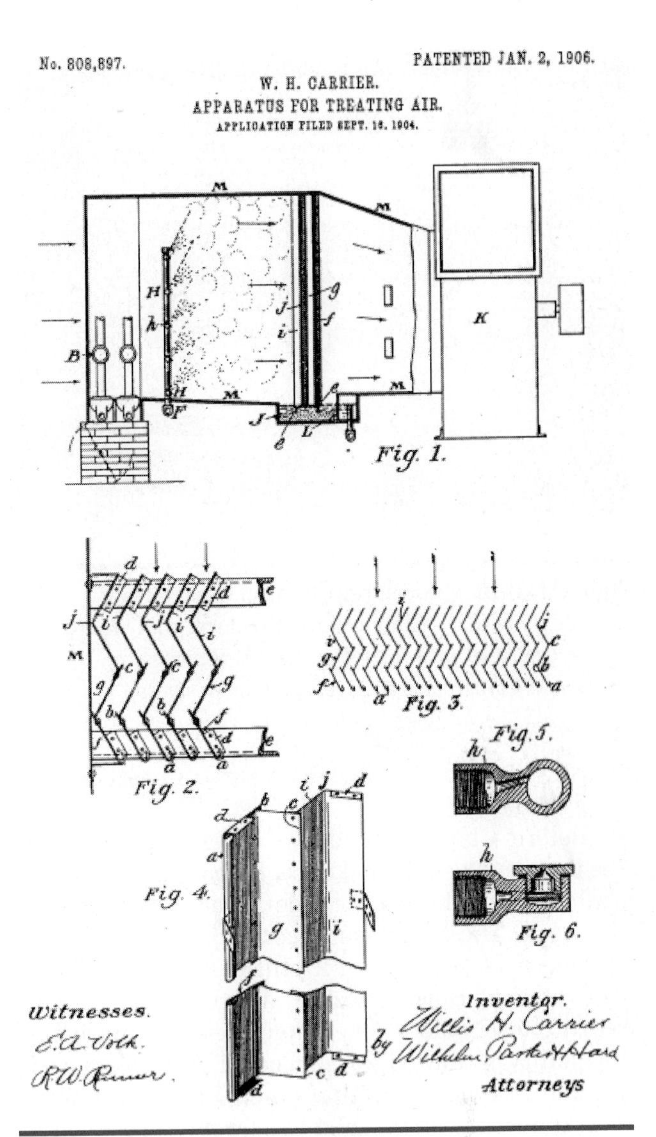

Figure 1-1-9 Patent for the first apparatus for cooling air. Invented by Willis Carrier.

Ice was the primary means of cooling air for many years. The Romans packed ice and snow between double walls in the emperor's palaces. John Gorrie patented the first mechanical air conditioned system in 1844. His system was used to cool sick rooms in hospitals in Florida.

The development of modern air conditioning is often credited to Willis Carrier. Mr. Carrier, an engineer, was confronted with a problem facing printers. As paper was printed with one color, the dampness in the ink caused the paper to stretch slightly, and it was nearly impossible for the second color to be printed without being misaligned. Mr. Carrier determined that a means for controlling the humidity was necessary and developed the first air conditioning system for the printing industry. His invention, called an "Apparatus for Treating Air" was patented in 1906, Figure 1-1-9. His invention quickly found favor not only for dehumidifying, but also for cooling. Through the 1940s and 1950s, businesses would proudly display signs reading "Air Conditioned." Mass air conditioning of homes began in the late 1950s with window air conditioners. Central residential air conditioning started to become popular in the mid 1960s. Today most of us can't imagine living in a home anywhere in the country that didn't have air conditioning.

Before Willis Carrier's invention, buildings were cooled with ice. In fact the phrase "ton(s) of air conditioning" we use today came from this era in history when tons of ice were used for cooling. The United States capitol building in Washington DC was first air conditioned using ice in 1909. Rumor has it that when the legislators got really involved in controversial debates, more ice was required to keep the building cool.

Clarence Birdseye made another major contribution to our industry. He developed the process of freezing foods in 1922. Today supermarket freezer displays provide us with a variety of food products

Figure 1-1-10 Modern refrigeration display cases provide us with a variety of food products that would not be available without refrigeration. (*Courtesy of Hussmann Corporation, an Ingersoll-Rand business*)

that would not be possible to preserve any other way, Figure 1-1-10.

Before Clarence Birdseye began commercially freezing food, people had allowed food to freeze naturally during the winter months as a way of preserving it. The trick was to come up with a way of freezing food and having it still taste good when it was thawed. Blast freezing, a process of rapidly freezing food, was the key ingredient in improving the quality of thawed frozen foods. The problem with freezing food slowly is that when ice crystals form over time they become much larger. These large sharp ice crystals grow through the cell walls of the food and when the food thaws all of the nutrients in the food are allowed to drain away. Blast freezing causes the ice crystals to be very small and they are less likely to penetrate cell walls. So the food retains nutrients and flavor when it's thawed.

EMPLOYMENT OPPORTUNITIES

The HVAC/R industry represents one of the largest employment occupations in the country. Our industry, for example, is the largest consumer of utilities in the nation. More electricity and natural gas is consumed producing heating and cooling than with any other single use. The size of the industry has been growing steadily since the late 1960s when residential central systems became popular.

There are a variety of occupational specialties offered within the HVAC/R industry. These occupations range from the basic entry level helper to the design engineer.

Entry Level Helper The entry level helper provides the master technician with assistance installing and servicing equipment. Most medium and large mechanical contracting companies use a number of helpers to assist with the installation and service of residential and commercial systems. A helper may be expected to assist in lifting, carrying, or placing equipment or components. They may also run errands to pick up parts and clean up the area following installation or service. Helpers receive basic safety training and if they will be driving, they must have good driving records.

Rough-in Installer The initial installation process is referred to as rough-in. In this process the technician will install the refrigerant lines, electrical lines, thermostat and control lines, duct boots and duct run, as well as setting the indoor and outdoor units. The rough-in technician must have an understanding of duct layout, blueprint reading, basic hand tools, and good brazing skills.

Start-up Technician Once the system has been installed and all of the components are ready for operation, a start-up technician will go through the manufacturer's recommended procedures to initially start a system. Because much of the HVAC system has been field installed, this checkout procedure is essential to ensure safe and efficient operation. The start-up technician records all of the information requested by the manufacturer's warranty. Start-up technicians must be skilled with electrical troubleshooting, refrigerant charging, and have good reading comprehension skills and writing skills.

Service Technician The service technician is the individual who provides the system owner with repair and maintenance. Service technicians are the people who must be able to diagnose system problems and make the necessary repairs. Service technicians must be skilled in diagnosing electrical problems, refrigerant problems, and air distribution problems.

TECH TIP

Technology has enabled the field tech to stay in close contact with his service manager. This allows the highly experienced service managers to provide assistance to technicians as they come upon new problems. The technician can also call upon the office to research unique problems to determine the best, most efficient way of making the repair.

Sales HVAC/R sales are divided into two major categories, inside sales and outside sales. Inside sales deal primarily with system sales to other air conditioning contractors. Outside sales may be to both contractors and end users. Working in outside sales or consumer sales requires the technician to have a good understanding of cost and value of equipment so that the owner can make an informed choice.

Equipment Operators Equipment operators are required by local ordinance and state law to be present any time large central heating and air conditioning plants are in operation. Their primary responsibility is to ensure the safe and efficient operation of these large systems. They must have a good working knowledge of the system's mechanical, electrical, and computer control systems in order to carry out their job. Equipment operators generally work by themselves or as part of a small crew. They often are required to have good computer skills when buildings have computerized building management systems.

Facilities Maintenance Personnel Facilities maintenance personnel are responsible for planned maintenance and routine service on systems. They may work at a single location or have responsibilities for multiple locations, such as school systems. Facilities maintenance personnel typically maintain systems and provide planned maintenance. They may work alone or as part of a crew depending on the size of the facility. Maintenance personnel may from time to time have duties and responsibilities outside of the HVAC/R trades, such as doing minor electrical plumbing and carpentry projects for the upkeep of the building.

Service Manager A service manager is typically a skilled HVAC/R technician with several years of experience. This individual oversees the operation of a company or maintenance department. They must have good management skills, communication skills, and technical expertise. Service managers typically assign jobs to other technicians and employees. They must then oversee these individuals' jobs.

TRADE ASSOCIATIONS

With the rapid growth and variety of interests, trade associations naturally evolved to represent specific groups. The list includes manufacturers, wholesalers, contractors, sheet metal workers, and service organizations. Each is important and makes a valuable contribution to the field. Space does not permit a detailed examination of all of these organizations, or all of their activities, but throughout the book many of these associations will be acknowledged as specific subjects are covered.

American Society of Heating, Refrigeration, and Air Conditioning Engineers (ASHRAE)

The American Society of Heating, Refrigeration, and Air Conditioning Engineers is an organization started in 1904 as the American Society of Refrigeration Engineers (ASRE) with 70 members. Today its membership is composed of thousands of professional engineers and technicians from all phases of the HVAC/R industry. ASHRAE also creates equipment standards for the industry. Its most important contribution probably has been a series of books that have become the reference books of the industry. These include the *Guide and Data Books for Equipment, Fundamentals, Applications, and Systems.*

NICE TO KNOW

Becoming an active participating member in a professional trade association will provide you with an opportunity to continue your HVAC education. The HVAC/R field is such a dynamic and evolving industry that in order to stay competitive you must continually attend seminars and take classes.

In cosponsorship with ARI, ASHRAE holds an annual international Air Conditioning Heating Refrigeration Exposition, which may draw 30,000 to 50,000 people in the field. Product exhibits, technical displays, and business seminars highlight the event, Figure 1-1-11.

American Society of Mechanical Engineers (ASME)

The American Society of Mechanical Engineers is an organization composed of engineers in a wide variety of industries. Among other functions, ASME writes standards related to safety aspects of pressure vessels.

Air Conditioning Contractors of America (ACCA)

The Air Conditioning Contractors of America is a service contractor's association concerned with the education of technicians and service managers, and

Figure 1-1-11 Air conditioning and refrigeration is a multimillion dollar industry that puts on trade shows such as the annual International Air Conditioning, Heating, Refrigeration Exposition, sponsored by ARI, ASHRAE, and other trade organizations. (*Courtesy of the American Society of Heating, Refrigerating, and Air Conditioning Engineers*)

with business improvement techniques. In cooperation with Pennsylvania College of Technology, ACCA provides technician EPA certification.

Refrigeration Service Engineers Society (RSES)

The Refrigeration Service Engineers Society is a service association dedicated to education in the HVAC/R industry. RSES also has a technician EPA certification program. RSES chapters conduct classroom training in technical areas and is a source for educational printed material and books.

Air Conditioning and Refrigeration Institute (ARI)

The Air Conditioning and Refrigeration Institute (ARI) is a national trade association representing manufacturers of over 90% of United States-produced central air conditioning and commercial refrigeration equipment. ARI was formed in 1954 through a merger of two related trade associations. Since that time several other trade associations have also merged into ARI. ARI traces its history back to 1903 when it started as the Ice Machine Builders Association of the United States. Today ARI has over 180 companies as members.

Many services are provided by ARI to assist HVAC/R technicians. Some of these services which would supplement this text are listed below:

1. ICE is an industrial competency exam. This test is made available to students of educational institutions to test their knowledge of fundamental and basic skills necessary for entry level HVAC/R technician positions. The information in this text covers the topics in the ARI curriculum guide and would assist the student in taking this examination. A directory of those who pass the examination is published nationally to assist prospective employers in identifying job candidates.

2. Equipment donations to schools participating in the ICE competency exam. ARI contacts industry sources having no-cost or low-cost equipment available to supply a school's laboratory needs.

3. Technician certification program. In accordance with EPA's (the Environmental Protection Agency) enforcement of the Clean Air Act, the sale of refrigerants is made only to those technicians who have been certified. ARI is among those approved by EPA to administer the test for certification. In addition, ARI provides study material to prepare for the test.

4. Reclaimer certification program. EPA also requires certification of any processor of recovered refrigerant for resale. ARI is among those assigned by EPA to carry out a certification program for companies that seek to reclaim refrigerants. Technicians handling reclaimed refrigerant should become familiar with the *Directory of Certified Reclaimed Refrigerants,* published every March and September by ARI.

5. Certification program for equipment used to recover and recycle refrigerant. ARI is one of the companies approved by EPA to certify equipment used to recover and recycle refrigerants. Technicians should become familiar with the *Directory of Certified Refrigerant Recovery / Recycling Equipment,* published every March and September by ARI.

6. HVAC/R equipment certification program. ARI maintains a certification service which tests a wide variety of equipment and products to verify the performance described by the manufacturer. Certified directories for various products are published semiannually and annually.

SERVICE TIP

The ARI list of certified equipment is available to anyone through the world wide web. This material is very helpful when trying to make a determination of the best equipment to recommend for a customer and their specific application needs. On the web, very often all of the various pieces of equipment are available.

ARI has a full program of educational activities geared toward helping the nation's vocational and technical schools improve and expand their education and training programs. Under the direction of ARI's education director and its Education and Training Committee, ARI serves as a resource from the manufacturers to school instructors, department heads, and guidance counselors. In addition to this textbook and its companion materials, ARI produces the *Bibliography of Training Aids,* a career brochure, and a promotional videotape for schools to use to recruit students into HVAC/R programs. Many schools around the country have adopted the ICE competency exams as final exams for their programs. ARI's most recent efforts involve participation in efforts to develop national HVAC/R competency standards.

Having students pass the ICE competency exams and training toward national competency standards will improve the quality of installation and service. New HVAC/R technicians will be better prepared, resulting in three basic advantages:

1. Limited training required for contractors
2. Limited rework or repeat calls due to error
3. Limited warranty/replacement for manufacturers.

The cost of repeat service calls, which is borne by contractors, may be reduced substantially through the use of properly trained technicians. Every new technician receives training and serves as an apprentice for a period of time. That is essentially a period where contractors pay two people to do one job. A properly trained technician will generally require less training time and function sooner than a poorly trained technician.

UNIT 1—REVIEW QUESTIONS

1. According to the text, what was the significance of the discovery of fire?
2. Why did early Romans build channels underneath the floors?
3. Where does it appear that ice production first began?
4. How low does the relative humidity have to be for evaporation to cool water to its freezing point even though outside air temperature is well above freezing?
5. How was the first mobile refrigeration used?
6. What was the primary means of air conditioning for many years?
7. What is Willis Carrier often credited with?
8. When did residential air conditioning start to become popular?
9. Who developed the process of freezing foods?
10. What are the duties of an entry level helper?
11. What are the duties of a start-up technician?
12. What are the requirements to be a service technician?
13. What are the two major categories of HVAC/R sales?

TECH TALK

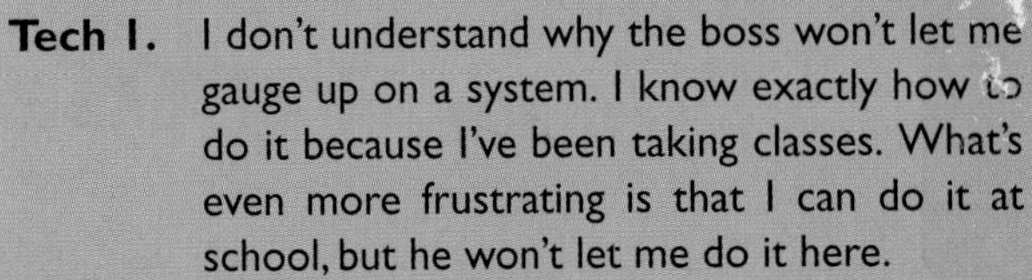

Tech 1. I don't understand why the boss won't let me gauge up on a system. I know exactly how to do it because I've been taking classes. What's even more frustrating is that I can do it at school, but he won't let me do it here.

Tech 2. That's right. There is a special provision in the EPA regulations that allow students under the supervision of an instructor to work on refrigeration systems as part of their learning process, but that privilege does not extend to the shop. Before you are going to be able to gauge up out here you are going to have to pass the EPA certification test.

Tech 1. It still would be good practice for me if the boss would just let me do it occasionally. I mean, I could get more efficient and I could help him out.

Tech 2. Well, the real problem would be if someone were to turn you in for working without an EPA license then you *and* the boss *and* the company would wind up being fined for violating the EPA regulations. The regulations are very stringent and very specific. If you choose to intentionally violate these regulations you are placing yourself and the company at significant risk of being fined. The boss is right. Until you get your EPA certification you can't work on any of the systems.

Tech 1. I didn't know that.

Tech 2. That's why you have to take the class and become certified. You will learn all this in the class. It will be part of your test to become certified.

Tech 1. Thanks, I appreciate that. I'll find out when the next class and test will be scheduled.

14. What are facilities maintenance personnel responsible for?
15. A service manager is typically a skilled HVAC/R technician with ________.
16. What is ASHRAE?
17. What is ASME composed of?
18. What is RSES dedicated to?
19. What information is covered on the ICE competency exam?
20. What is the reclaimer certification program?
21. Following ARI's goals, new HVAC/R technicians will be better prepared, resulting in what three basic advantages?

UNIT 2

Fundamentals

OBJECTIVES: In this unit the reader will learn about the fundamentals of heating and air conditioning including:

- The Forms of Energy
- HVAC/R Term Definitions
- The Refrigeration Process
- Changes in the State, Pressure, Temperature, and Heat Content of the Refrigerant
- The Mechanical Refrigeration Cycle
- The Absorption Refrigeration Cycle
- System Components

The fundamentals of air conditioning and heating are that we create cool temperatures by rapidly expanding a gas, and we generate heat by burning a fuel. The fuel we burn for heat can be gas, oil, or coal. Technology has significantly increased the efficiencies and improved the comfort and quality of life we have today. As energy consumption becomes a greater concern, higher and higher efficiency levels will become standard.

HVAC/R TERMS AND DEFINITIONS

To better understand the processes discussed in this textbook you need to understand the definitions of the basic terminology used in the HVAC/R industry. Many of the terms used are specific to our trade.

Energy

The processes of heating and cooling both revolve around the addition or removal of thermal energy. **Thermal energy,** also called heat energy, is one of six forms of energy. The types of energy are **chemical, mechanical, electrical, thermal, light (radiant),** and **nuclear.** Energy in all its forms cannot be created or destroyed. It can, however, be changed from one form to another. An example of energy changing forms is when electrical energy is changed to light; or mechanical energy, turning a generator, is changed into electrical energy. Table 1-2-1 lists some examples of one form of energy being changed to another form of energy. The ability to change energy from one form to another makes it possible to store energy for later use.

NICE TO KNOW

A tremendous amount of energy is required to change water from one state to another. This fact serves the earth's environment very well in that as the earth's temperature increases, more water is evaporated from the oceans. The evaporated water forms into clouds at various levels in the atmosphere. These clouds can reflect more of the sun's energy back into space, which then allows the earth to cool ever so slightly, so that less evaporation occurs. The earth's oceans, therefore, act as the earth's temperature regulators.

Mechanical and chemical energy are more easily stored than other forms. Thermal, electrical, and nuclear energy can be stored for a short time period, but over a longer time they lose intensity until all their energy level equals the surrounding level. For example, even with the best insulation the ice in a cooler will eventually melt. Light energy cannot be stored at all. For example, no matter how reflective the inside of a box, the light disappears and the box becomes dark once the light is turned off. The ability to store energy for later use is important. Being able to store energy is what allows us to drive a car; stored chemical energy (gasoline) is converted to mechanical (engine RPMs) which is needed to move the car. Water held behind a dam is stored mechanical energy waiting to become electrical energy as it is released through the dam's turbines. Gas, fuel oil, and coal are examples of stored chemical energy we use for heating. Table 1-2-2 lists some common forms of stored energy.

Temperature and Heat

The terms temperature and heat are often misused and misunderstood. **Temperature** is the measure of the level of thermal activity and **heat** is the quantity

TABLE 1-2-1 Examples of Energy Conversion

Type of Product/Action	Energy Conversion From	Energy Conversion To
Gasoline in a car	Chemical	Mechanical
Electricity going through a light bulb	Electrical	Radiant
Water turning a turbine	Mechanical	Electrical
Heat energy producing steam to turn a turbine	Thermal	Mechanical
Chemicals in a battery	Chemical	Electrical
Burning coal moving the wheels on a steam engine	Chemical	Mechanical
Windmill	Mechanical	Electrical
Electricity in a heater	Electrical	Thermal
Natural gas burning	Chemical	Thermal
Light on a photocell	Radiant	Electrical

of thermal energy. We often relate temperature to hot or cold. Temperature alone does not tell you the amount of thermal energy in a material. A very light object can have a high temperature but because it has little weight it has little thermal energy. The amount of thermal energy present will tell the technician how much refrigerant will be needed.

A **refrigerant** is a fluid that picks up heat by evaporating at a low temperature and pressure and gives up heat by condensing at a higher temperature and pressure. Common types of refrigerants such as R-22, R-123, R-134a, and R-410A are shown with their cylinder colors in Table 1-2-3. As heat is removed from a space, the area becomes cooler. Therefore, **cold** can be defined as the relative absence of heat energy. But remember, there is always some heat energy present down to absolute zero (−460°F, −273°C).

TABLE 1-2-2 Examples of Stored Energy

Material	Type of Energy Stored
Batteries	Electrical
Gasoline	Chemical[1]
Water behind a dam	Mechanical
Glow-in-the-dark strips	Light
Coal	Chemical[1]
Candle	Chemical[1]
Wood	Chemical[1]
Capacitor	Electrical
Ice	Thermal
Food	Chemical

[1]When gasoline, coal, a candle, or wood are burned, the chemical energy is converted to thermal energy and radiant (light) energy.

TABLE 1-2-3 Refrigerant and Refrigerant Cylinder Colors

Refrigerant	Color
11	Orange
12	White
13	Light blue (sky)
13B1	Pinkish-red (coral)
14	Yellow-brown (mustard)
22	Light green
23	Light blue-gray
113	Dark purple (violet)
114	Dark blue (navy)
116	Dark gray (battleship)
123	Light blue-gray
124	Deep green (DOT green)
125	Medium brown (tan)
134a	Light blue (sky)
401A	Pinkish-red (coral)
401B	Yellow-brown (mustard)
401C	Blue-green (aqua)
402A	Light brown (sand)
402B	Green-brown (olive)
404A	Orange
407A	Lime green
407B	Cream
407C	Medium brown (brown)
408A	Medium purple (purple)
409A	Medium brown (tan)
410A	Rose
410B	Maroon
500	Yellow
502	Light purple (lavender)
503	Blue-green (aqua)
507	Blue-green (teal)
717, NH_3	Silver
Any recovered	Yellow/gray

Refrigeration

Refrigeration is the process of removing heat energy from a place where it is not wanted and disposing of it in a place where it is wanted or not objectionable. **Mechanical refrigeration** makes use of mechanical components to produce movement and transfer heat from an area of lower temperature to an area of higher temperature. For example a household refrigerator transfers heat from the interior of the refrigerator cabinet to the surrounding area of the kitchen. Another kind of refrigeration, **absorption refrigeration,** uses heat energy instead of mechanical energy to produce the conditions necessary to transfer heat energy from one place to another.

Air Conditioning System

An **air conditioning system** is a refrigeration system used to cool, dehumidify, filter, and/or heat the air of a space. The refrigeration system removes the heat from the air in the space, reducing the temperature of the space. Water vapor contained in the air is collected on the cool surface of the evaporator and drained out, thus air conditioning also dehumidifies air in the space. Mechanical or electrostatic filters can be used to remove dust or other contaminants from the conditioned space.

Refrigeration is measured in tons. **One ton of refrigeration** is the refrigeration produced by melting one ton of ice at a temperature of 32°F (7°C) in 24 hours. The refrigerating effect is expressed in units such as 288,000 Btu/24 hr, 12,000 Btu/hr, or 200 Btu/min. One **Btu,** or British thermal unit, is the quantity of heat required to raise the temperature of 1 lb of water 1°F.

NICE TO KNOW

Early air conditioning in large public buildings used melting ice to provide cooling. An early example of this type of air conditioning is the U.S. Capitol building in Washington, D.C. This system has long since been replaced with more modern, up-to-date, mechanical air conditioning, but the ice banks served our representatives well in helping to keep them cool for a number of years.

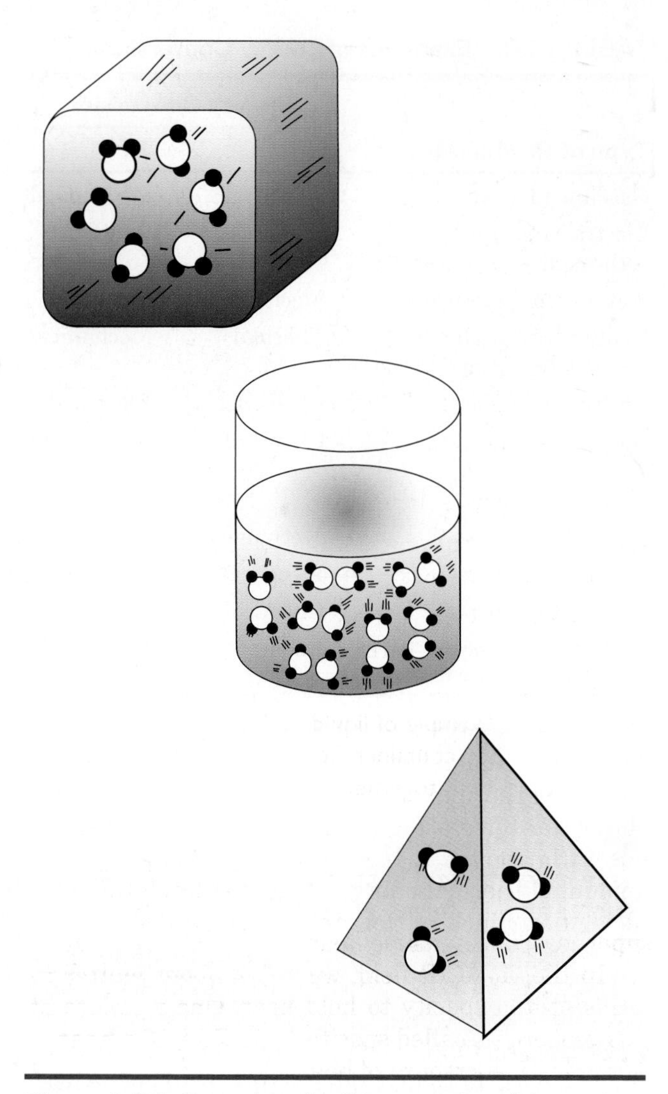

Figure 1-2-1 Examples of solid matter. In solid matter, atoms may vibrate, but do not move about.

Matter

Matter is anything that has weight and takes up space. Matter makes up everything that exists in the universe. All matter is made up of atoms. Matter can exist as a solid, liquid, or gas. **Solid matter** has shape and form and can be supported by one side, Figure 1-2-1. **Liquid matter** takes the shape of its container, Figure 1-2-2. Matter in the form of a **gas** fills its container, Figure 1-2-3. Matter can be changed from one state to another by the increase or decrease of its pressure and/or temperature.

Heat Flow

Heat is a form of energy. Heat does not have weight or occupy space. A quart of water weighs no more or no less whether it is hot or cold. Since heat is energy, it is subject to the **law of conservation of energy.** This means that heat energy cannot be created or destroyed. Heat energy is often the by-product of other forms of energy. The adding to or removal of heat energy from a substance determines the speed or intensity of the molecules. The intensity or speed of molecules determines the physical state of matter: solid, liquid, or gas. The colder matter gets, down to Absolute zero, the less each molecule moves or vibrates. A comparison of Celsius and Fahrenheit scales is shown in Figure 1-2-4.

Heat flow is the movement of heat from a warmer to a cooler body. Heat moves in three ways: conduction, convection, and radiation. **Conduction** is the transfer of heat from molecule to molecule of a substance without movement of the particles themselves. **Convection** is the transfer of heat by a flowing medium. Convection takes place only in liquids and gases, since solids do not flow. **Radiation** is a wave form of heat movement similar to light, except that it cannot be seen. Like light, radiative heat requires no medium to travel.

Figure 1-2-2 Example of liquid matter. Liquids take on the shape of their container. In liquids, atoms may slide about, but still stick together.

In referring to heat, we talk about heat and temperature in terms of specific heat, latent heat, superheat, and sensible heat.

In the HVAC/R field, we think about matter in terms of its capacity to hold heat. One measure of this property is called specific heat. **Specific heat** is defined as the amount of heat energy, in Btu, needed to change 1 lb of the material 1°F. Recall that 1 Btu is the amount of heat required to raise the temperature of 1 lb of water 1°F. The specific heat of water is, therefore, 1.

A **change of state** is the change from a solid to a liquid, a liquid to a gas, a gas to a liquid, or a liquid to a solid. A large amount of heat must be added or removed to cause a change of state. The amount of heat used to cause a change of state is called the latent heat.

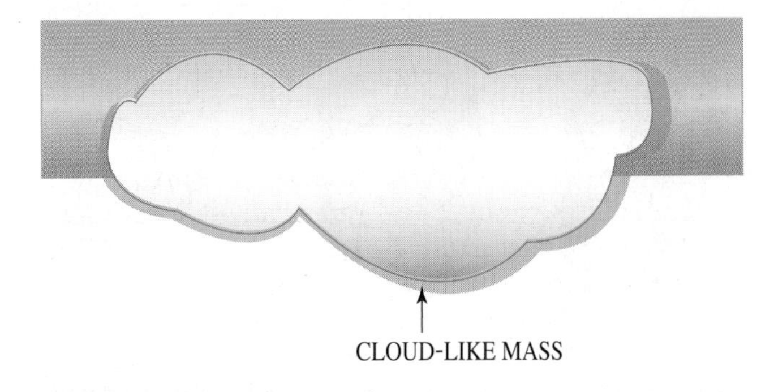

Figure 1-2-3 Example of gaseous matter. In gases, the atoms are completely free to move about. The gas expands to the extent of its container.

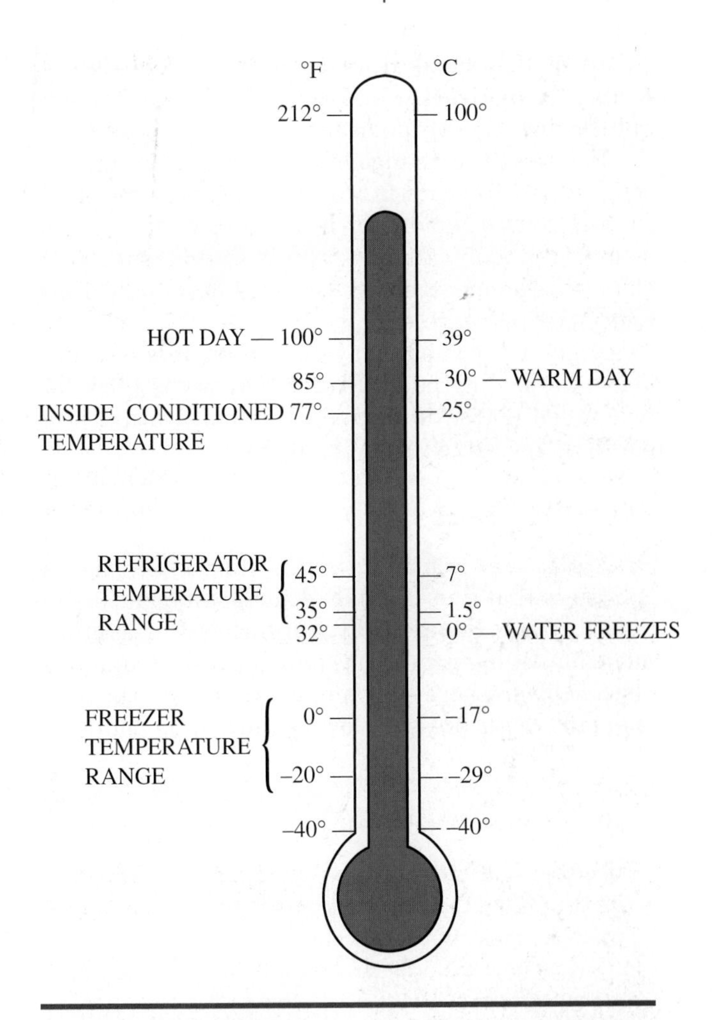

Figure 1-2-4 Fahrenheit and Celsius scale comparison.

Latent heat is heat energy absorbed or released in a change of state with no change in temperature or pressure. Think about this in terms of sweat evaporating. When you sweat, beads of liquid form on your skin. The liquid evaporates, pulling great amounts of latent heat away from your skin. Both the liquid sweat and the evaporated sweat vapor are at the same temperature and pressure. However, the evaporated sweat has a much greater amount of heat energy, the latent heat. Pulling away this latent heat has cooled you down, because heat energy was removed from you. Now you can see how evaporative cooling works because of latent heat.

There are two types of latent heat: latent heat of vaporization/condensation and latent heat of fusion. The **latent heat of vaporization** is the amount of heat energy required to change a substance from a liquid to a vapor without changing its temperature or pressure. The opposite of vaporization is condensation. The **latent heat of condensation** is the heat energy that must be removed from a substance to change the state from a vapor to a liquid without changing its temperature or pressure. The latent heats of vaporization and condensation are equal, but the direction of the heat transfer is opposite.

Fusion is the change of state from a liquid to a solid, called freezing in water. The **latent heat of fusion** is the amount of heat energy required to change a substance from a liquid to a solid without a change in temperature or pressure at standard atmospheric pressure.

A gas or a vapor may be at any temperature above the boiling point. **Superheat** is the term for how much hotter, in degrees, the temperature of a gas is compared to its boiling point.

Sensible Heat

Sensible heat is heat added to or removed from a substance that causes a change in the temperature of the substance. **Saturation temperature** is another name for boiling point. The *boiling point* of a liquid is affected by pressure—*higher* pressure raises the boiling point; *lower* pressure lowers the boiling point.

SERVICE TIP

The efficiency of radiant heating is affected significantly by distance. When the distance from the heat source doubles, the amount of heat received drops four times. So, for radiant heat to be effective, it must be as close as possible to the heat source.

Pressure is a force on a unit area; impact or thrust exerted on a surface, measured in pounds per square inch (psi). Pressure measurements are referred to in one of three ways: atmospheric, gauge, or absolute. **Atmospheric pressure** is stated as 14.7 psia at sea level. **Absolute pressure** (psia) is pressure measured on the absolute scale. The zero point is at zero atmospheric pressure. The total of the gauge and atmospheric pressures is called *absolute pressure*. **Gauge pressure** (psig) is measured on the gauge scale. The zero point is 14.7 psi, or atmospheric pressure at sea level. Gauge pressure is the measured pressure above or below atmospheric. Pressure below atmospheric is measured in inches of mercury.

A vapor at its liquid's boiling point is referred to as *saturated vapor*. A vapor above its liquid's boiling point is referred to as *superheated vapor*. A liquid below its boiling point is referred to as *subcooled liquid*.

REFRIGERATION PROCESS

The transfer of heat in the refrigeration system is performed by a refrigerant operating in a closed system. The refrigeration process has its application in both refrigerated systems and air conditioning systems. Refrigerated systems are chiefly concerned with cooling products, whereas air conditioning systems cool (or heat) people. Air conditioning systems use refrigeration to provide comfort cooling and dehumidification of air.

SERVICE TIP

Removing humidity from air also removes heat from the air. The more humidity there is in air, the more heat the air contains. Reducing the amount of humidity reduces the air's heat. That is why you feel more comfortable at 75° with 40% relative humidity as compared to 70° and 90% relative humidity. So, air conditioning systems must remove both temperature and humidity to be effective.

One of the very useful properties of the refrigerant is the pressure-temperature relationship of the saturated vapor. A refrigerant vapor is said to be saturated whenever both liquid and vapor are present in the same container, in stable equilibrium.

Under these conditions a fixed relationship exists between the temperature of the refrigerant in the container and its pressure.

Components of the Refrigeration Cycle

In the simple vapor compression refrigeration system, Figure 1-2-5, there are four essential parts:

1. The compressor
2. The condenser
3. The metering device
4. The evaporator

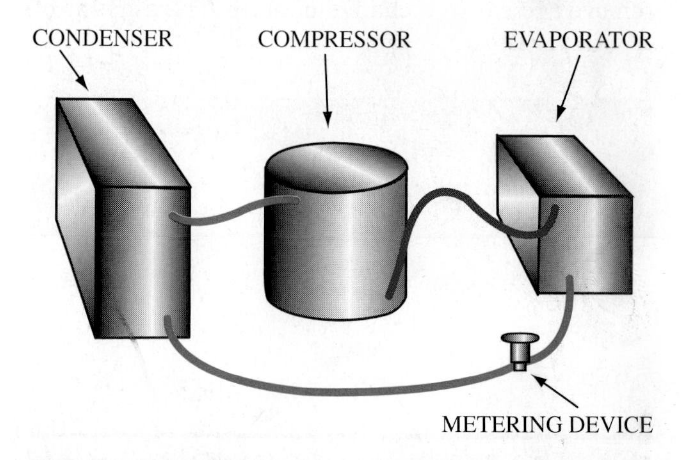

Figure 1-2-5 Basic components of a vapor compression refrigeration system.

The compressor is a mechanical device for pumping refrigerant vapor from a low pressure area (the evaporator) to a high pressure area (the condenser). Since the pressure, temperature, and volume of a gas are related, a change in pressure from low to high causes an increase in temperature and a decrease in volume or a compression of the vapor. The main types of compressors are: reciprocating (piston), Figure 1-2-6a; rotary, centrifugal, screw, and scroll, Figure 1-2-6b.

The different types of compressors get their names from their mechanical parts. In the reciprocating compressor, a piston travels back and forth (reciprocates) in a cylinder. The rotary compressor has a vane that rotates within a cylinder. The centrifugal compressor has a very high speed centrifugal impeller with multiple blades. The impeller rotates within a housing. The screw compressor uses a rotating screw within a tapered housing. The scroll compressor has a stationary and an orbiting scroll that moves within the stationary scroll.

The **condenser** is a device for removing heat from the refrigeration system. In the condenser, the vapor at high temperature and high pressure transfers heat through the condenser tubes to the surrounding medium (usually air or water). When the temperature of the vapor reaches the saturation temperature, the additional latent heat removed causes condensation of the refrigerant, producing liquid refrigerant. There are three types of condensers: air cooled, water cooled, and evaporative, Figure 1-2-7. The air cooled condenser uses air as the condensing medium, the water cooled condenser uses water as the condensing medium, and the evaporative condenser uses both air and water.

Air cooled condensers consist of two types: forced air condensers and natural draft (static) condensers. Two examples of forced air condensers can be seen in Figure 1-2-8. Air cooled condensers can be further classified by their construction: (1) fin and tube, and (2) plate.

There are four types of water cooled condensers: (1) double pipe, (2) open vertical shell and tube, (3) horizontal shell and tube, and (4) shell and coil. An example of a water cooled condenser is shown in Figure 1-2-8b.

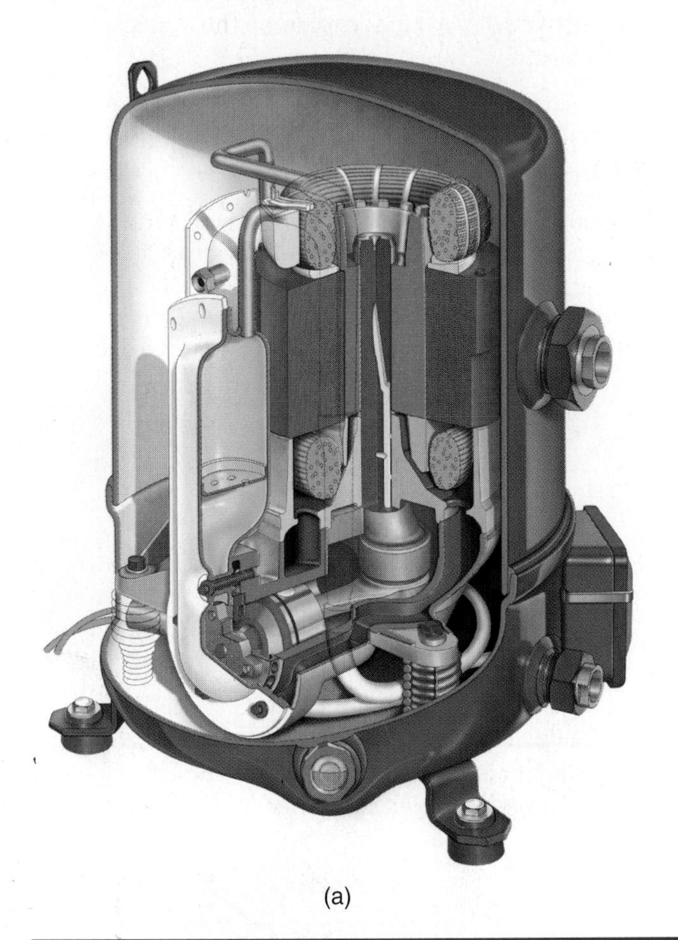

(a)

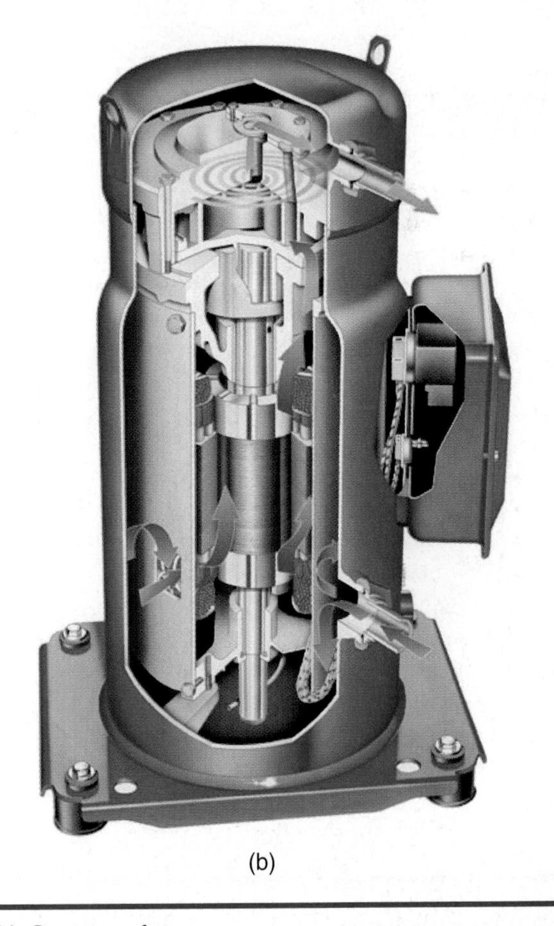

(b)

Figure 1-2-6 (a) Cutaway of a piston (reciprocating) compressor; (b) Cutaway of a scroll compressor. *(Courtesy Danfoss Inc.)*

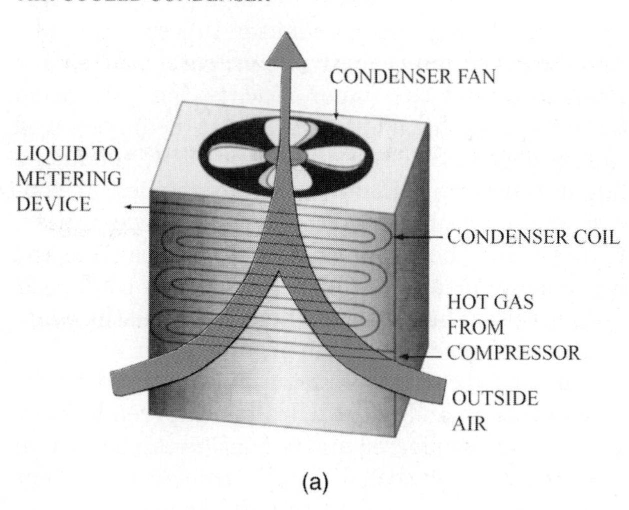

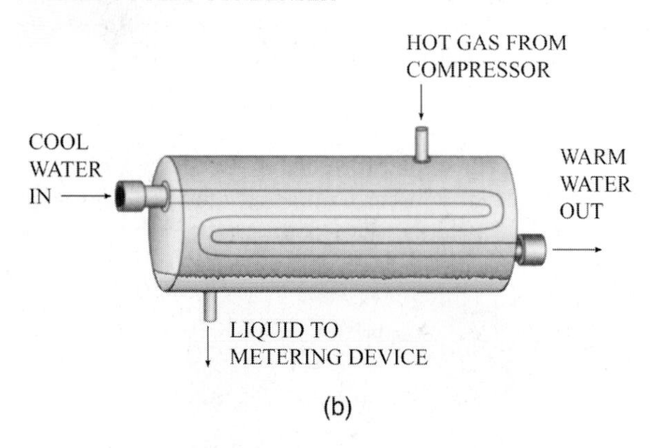

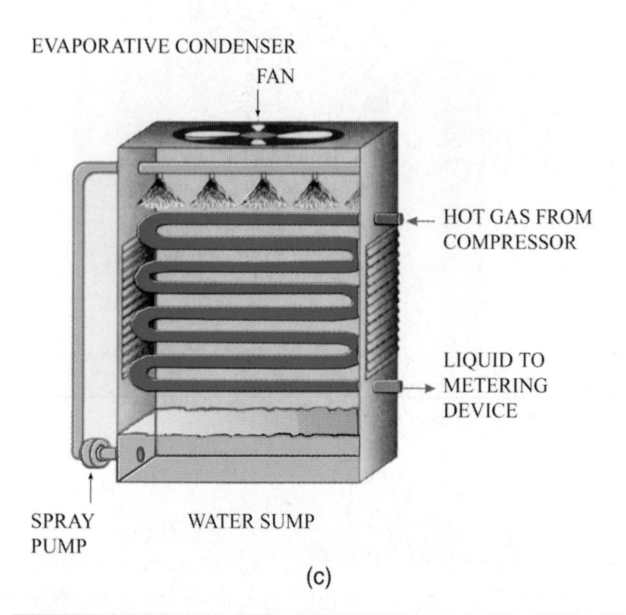

Figure 1-2-7 Condensers: (a) Air cooled; (b) Water cooled; (c) Evaporative condenser, also called a sump.

Figure 1-2-8 Condensing units: (a) Air cooled; (b) Residential air cooled condenser.

A **metering device** controls the flow of refrigerant to the evaporator (described later). It separates the high and low pressure parts of the system. High pressure, high temperature liquid enters the metering device and exits as low pressure, low temperature to enter the evaporator. The pressure is low because the compressor is continuously pumping vapor from the evaporator. Two actions occur in the metering device: (1) the refrigerant liquid is cooled to the evaporator temperature by actual evaporation of some of the liquid refrigerant, and (2) the pressure of the refrigerant is reduced to a pressure corresponding to the evaporator temperature at the saturated condition.

There are six major types of metering devices: (1) hand operated expansion valve, (2) low side float, (3) high side float, (4) automatic expansion valve, (5) thermostatic expansion valve, and (6) capillary tube. Metering devices are selected by application. There are other types of metering devices used for special applications.

NICE TO KNOW

For many years large air conditioning systems used in commercial buildings were controlled by an operator opening and closing a hand valve to control the refrigerant flow in the system. These manually operated valves were the only way to control the refrigerant flow in these large systems for many years. Today's modern, large air conditioning plants for commercial buildings are controlled by computers.

The **evaporator** is a device for absorbing heat into the refrigeration system. In the evaporator, the saturated refrigerant absorbs heat from its surroundings and boils into a low pressure vapor. Liquid refrigerant boiling and vaporizing in a cold evaporator can be compared to water boiling on a stove. The refrigerant is heated by the product load, whereas the water is heated by the stove burner. The action and results are comparable. The difference is that the stove is hot and the refrigerator is cold; however, in both cases heat energy is being transferred.

Some superheating of the vapor takes place before the suction gas reaches the compressor. This is desirable since unevaporated liquid refrigerant could damage the compressor.

Although there are many variations and modifications of evaporators, there are three basic types: bare pipe, finned tube, and plate, Figure 1-2-9. Like condensers, evaporators can be forced air or natural draft (static). Evaporators are designed on the basis of their intended usage. Application determines which type is the best suited.

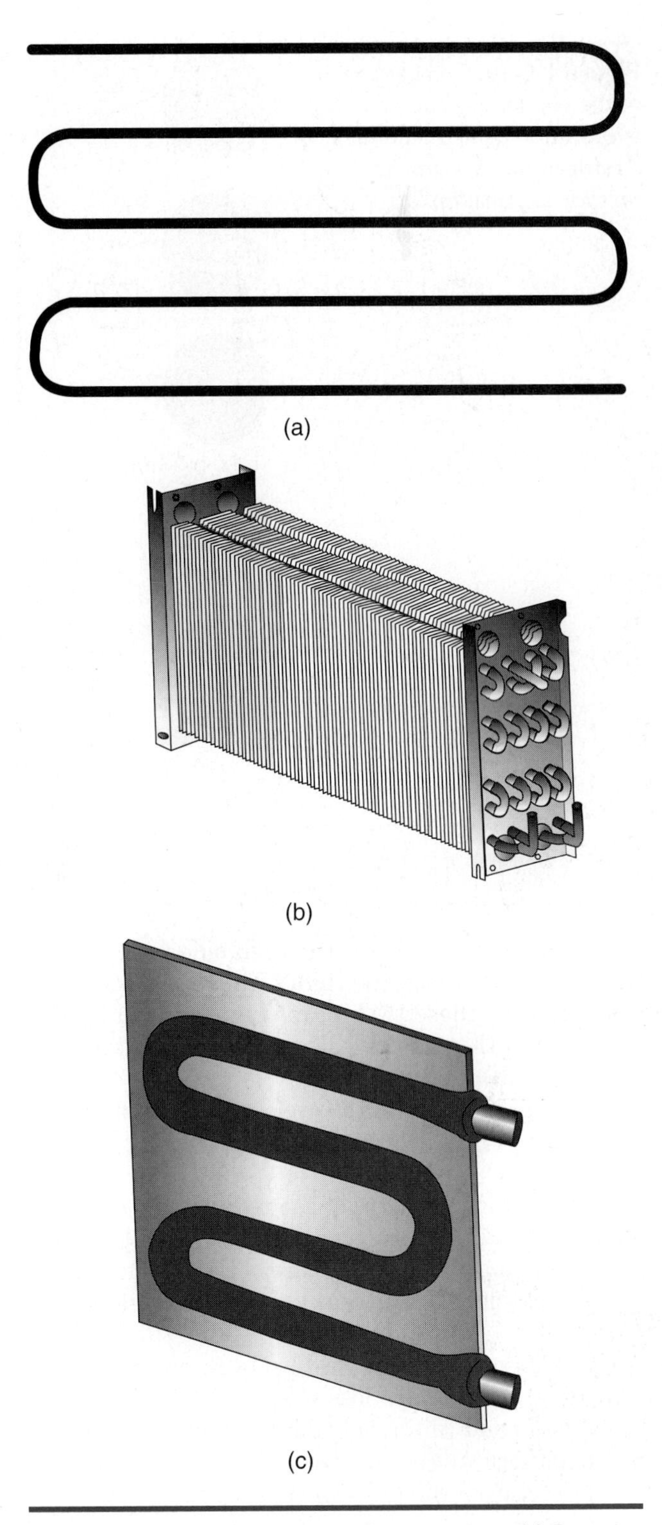

Figure 1-2-9 Different types of evaporators; (a) Bare pipe; (b) Finned tube; (c) Plate.

Action in a Typical Refrigeration Cycle

There are a number of different kinds of refrigeration cycles. The most common type is the vapor compression cycle, Figure 1-2-10.

To review the action that occurs in a typical system, first note the dotted vertical line in the center of

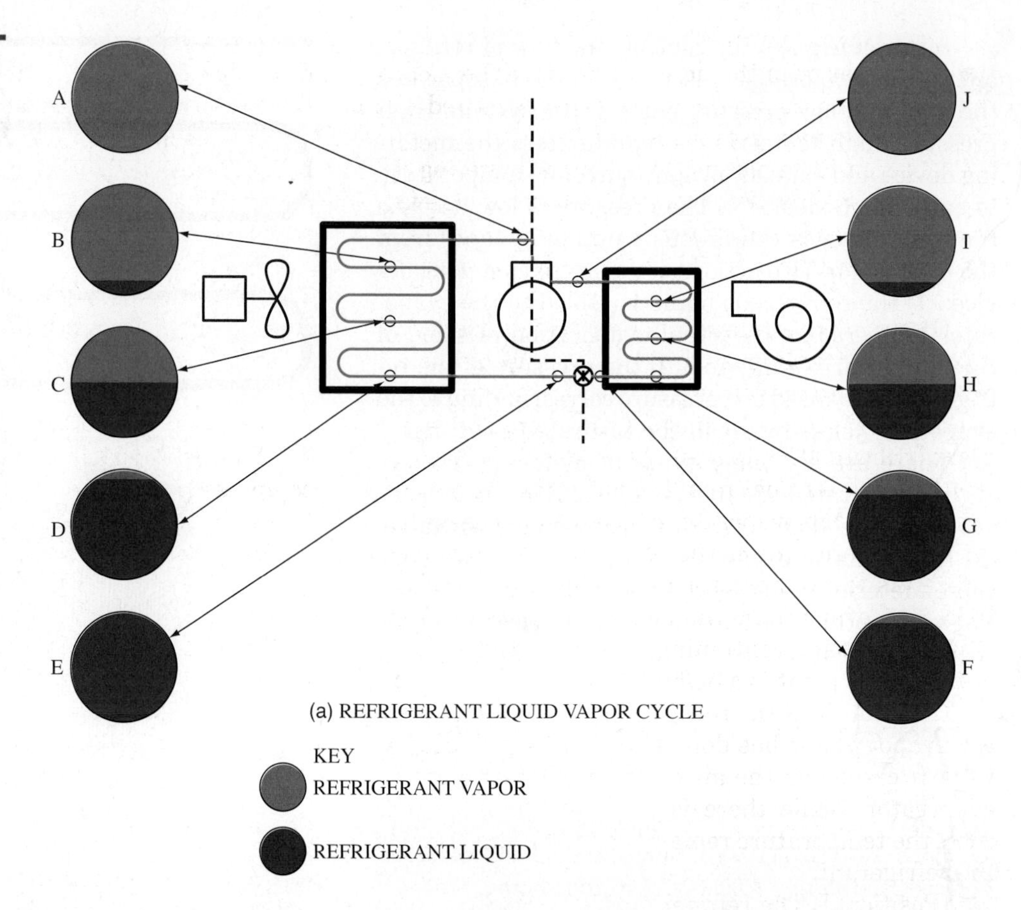

Figure 1-2-10 Comparison of the condensing and evaporation components of a refrigeration system. See text for explanation.

the diagram. It separates the high side (on the left) from the low side (on the right). The pressure difference between these two areas is maintained by the operation of the compressor and the restriction produced by the metering device.

The performance of the cycle in various stages of its operation can be observed in four ways:

1. The changes in state of the refrigerant
2. The changes in pressure of the refrigerant
3. The changes in temperature of the refrigerant
4. The change in heat content of the refrigerant

The following example has been simplified by assuming no pressure loss through the evaporator, condenser, and interconnecting piping. In actual practice there will be some losses; however, this does not detract from the principles explained below.

CHANGE OF STATE, PRESSURE, TEMPERATURE, AND HEAT CONTENT

Referring to Figure 1-2-10 and the numbered positions in the diagram, the changes in state, pressure, temperature, and heat content of the refrigerant are as follows:

(Position A) The refrigerant leaves the compressor as a high pressure, high temperature, superheated vapor. Heat of compression is also absorbed into the refrigerant. The state of the refrigerant is 100% vapor.

(Position B) As the refrigerant enters the condenser, the first portion of the condenser heat is removed and the temperature of the refrigerant falls to the saturation temperature. After one turn through the condenser, the refrigerant is about 10% liquid, 90% vapor.

(Position C) As additional latent heat is removed, the vapor condenses. After two turns through the condenser, the refrigerant is a 50/50 mixture of high pressure saturated liquid and vapor.

(Position D) At the lower portion of the condenser, the refrigerant is 90% liquid. There is still some vapor. The temperature is still the saturation point.

(Position E) After passing through the final portion of the condenser, the refrigerant is a subcooled liquid. **Subcooling** is the process of continuing to remove heat from the refrigerant after all the latent heat has been extracted and the vapor has been changed to a liquid. Subcooling reduces the temperature of the liquid below its boiling point at a particular pressure. Adequate subcooling will prevent the refrigerant from starting to boil as it experiences small pressure drops through the piping or components. Such boiling causes *flash gas* (the flashing of the refrigerant to a gas due to the sudden pressure drop and volume increase at the entrance to the evaporator) and can reduce the system capacity. It is desirable to subcool the liquid refrigerant either in the condenser or in the liquid line before the metering device. Subcooling the liquid refrigerant reduces flash gas, and increases mass flow.

(Position F) In passing through the metering device to the low pressure zone, some refrigerant is evaporated, cooling the remaining liquid. The refrigerant is typically a mixture of 75% liquid and 25% vapor at this point.

(Position G) As the refrigerant enters the evaporator, the air (or liquid, or whatever is being cooled) gives off heat to the refrigerant. The refrigerant takes in this heat as latent heat of vaporization. The refrigerant is about 30% vapor and 70% liquid at this point.

(Position H) Heat from the air or the product being cooled in the evaporator is absorbed by the liquid refrigerant and causes the refrigerant to boil or vaporize. As the compressor draws the vaporized gas from the evaporator, the metering device admits more refrigerant, continuing the process. The refrigerant at this point is a 50/50 mixture.

(Position I) As the refrigerant continues through the evaporator, it has done almost all of the cooling work it can do on the air or product flowing by the evaporator. While there is still 10% liquid in the tube, the temperature remains at the boiling point of the refrigerant.

(Position J) The refrigerant has picked up some superheat in the final evaporator circuit. Superheating is the process continuing to heat the refrigerant after sufficient latent heat has been added to vaporize all the liquid. Superheating ensures no liquid slugs will reach the compressor and cause damage to the valves and pistons. The refrigerant enters the compressor as a low temperature, low pressure, superheated vapor.

TECH TIP

Understanding the relation between liquid and vapor refrigerant and heat temperature and pressure as shown in the refrigerant cycle will enable you to better resolve problems with any air conditioning or refrigeration system.

For all practical purposes there are two pressures in the system: the low side pressure and the high side pressure. From the metering device, evaporator, and suction line up to the compressor inlet represent the low side of the system. The compressor, discharge line, condenser, liquid line, up to the metering device are considered the high side of the system. The compressor is considered to be in the high side and the metering device in the low side of the system. The compressor and the metering device work in partnership to maintain this pressure difference. The metering device controls the flow into the evaporator and the expansion of the refrigerant causes a pressure drop. The compressor pumps the refrigerant out of the evaporator and maintains the pressure.

Heat is added to the refrigerant by absorbing the product load in the evaporator. This constitutes primarily the net refrigeration effect, plus a small gain that occurs in the piping up to where the refrigerant enters the compressor. The compressor adds a sizable quantity of heat to the refrigerant. This is equivalent to the work done in compressing the refrigerant. In a suction gas cooled semi-hermetic or hermetic motor compressor unit, the motor heat is also transferred to the refrigerant.

The heat added in the evaporator and by the compressor is removed in the condenser. Some relatively small additional losses occur in the receiver and the liquid line piping up to the metering device.

THE MECHANICAL REFRIGERATION CYCLE

A mechanical refrigeration system must have several basic components which can circulate the refrigerant and transfer heat. These components are the evaporator (cooling coil) compressor, condenser, metering device (liquid refrigerant control), and interconnecting tubing (suction, discharge, and liquid lines). Auxiliary devices such as accumulators, filter-driers, receivers, etc., may be installed on certain systems but are not considered basic cycle components.

Evaporator

The evaporator or cooling coil is fabricated from metals such as copper or aluminum or both. These metals are selected because of their good thermal conductivity. The evaporator is fabricated to the desired size and shape. The tubing is interconnected by aluminum fins that serve to both direct the air flow through the coil and increase transfer of heat by conduction, Figure 1-2-11.

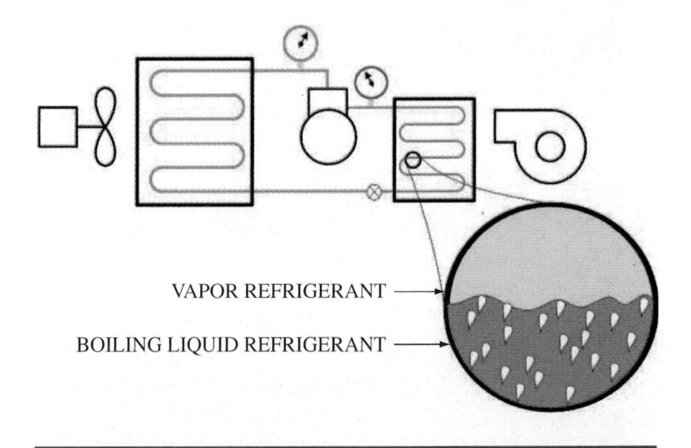

Figure 1-2-11 The boiling of the refrigerant in the evaporator coils as the heat is absorbed.

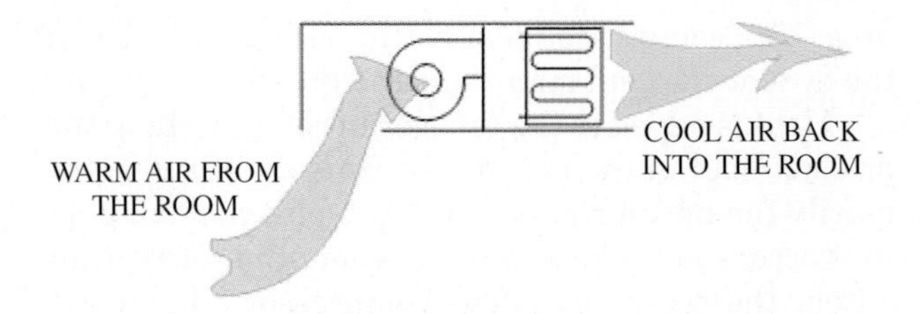

Figure 1-2-12 Air is drawn into the air conditioning evaporator unit by the fan and pushed out through the evaporator coil. The air enters warm and moist and leaves cool and dry.

The evaporator coil is the location for boiling (saturated) liquid refrigerant. It is inside the area to be cooled. Air from this space is drawn by a blower through this coil and redistributed back into the conditioned space, Figure 1-2-12.

A low pressure is maintained inside the evaporator coil, thus lowering the boiling point of the refrigerant. Upon entering the evaporator the refrigerant liquid begins to flash to saturation and boils, as illustrated in Figure 1-2-11. The refrigerant, in order to vaporize, must absorb heat. Heat is transferred from the fins and tubes of the evaporator to the refrigerant circulating inside. The fins and tubing provide a cold surface for the absorption of heat from the surrounding air. As air is pulled into the evaporator coil by the blower, heat is removed from the air and the moisture in the air collects on the cold surface of the evaporator coil. The condensed water vapor drains into a collection pan and is routed to a drain by tubing, Figure 1-2-13.

Figure 1-2-13 The white plastic condensate drain carries the condensed moisture away from the unit.

Compressor

The compressor, Figure 1-2-14, pumps the heat laden vapor away from the evaporator. This causes the low pressure back in the evaporator. The low pressure is therefore maintained by the compressor and by the restricted flow into the evaporator by the metering device at the inlet to the evaporator. The resulting temperature and pressure of the saturated refrigerant is lowered. The refrigerant in the evaporator boils and vaporizes, absorbing latent heat at this low temperature and pressure.

The heat laden vapor is then compressed by the compressor, increasing the pressure and the temperature of the vapor. The temperature of the refrigerant vapor leaving the compressor (discharge line) and entering the condenser must be higher than the temperature of the condensing medium.

The compressor, which is driven by an electric motor lubricated with a special refrigeration oil and mounted in a welded steel shell, is referred to as a welded hermetic compressor. The electrical components associated with the compressor, Figure 1-2-15, are the start relay (current operating) and the bimetal overload. Replacement compressors like the

TECH TIP

Air conditioning and refrigeration compressors have very large motors. Most of these motors are located inside of the refrigerant system so that the refrigerant can keep the motor cooler. Some large compressors actually have the cold suction vapor come into the compressor at the motor end to provide even greater motor cooling.

Figure 1-2-14 The compressor is located inside the condensing unit. It is connected with refrigerant lines and electrical circuits.

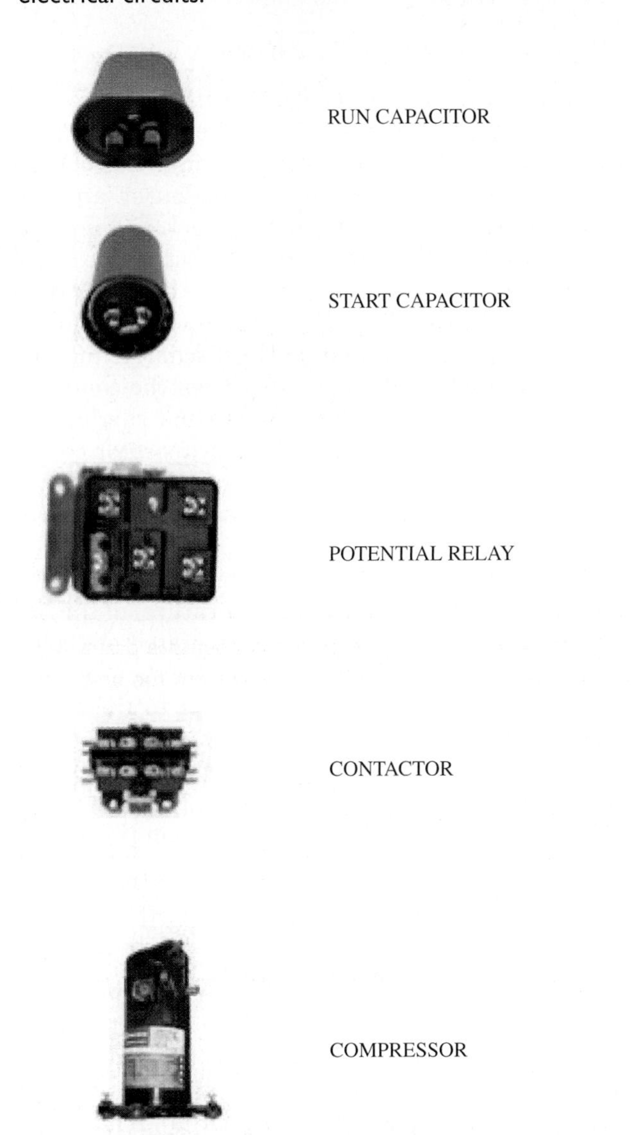

Figure 1-2-15 These are the basic electrical components connected to the compressor.

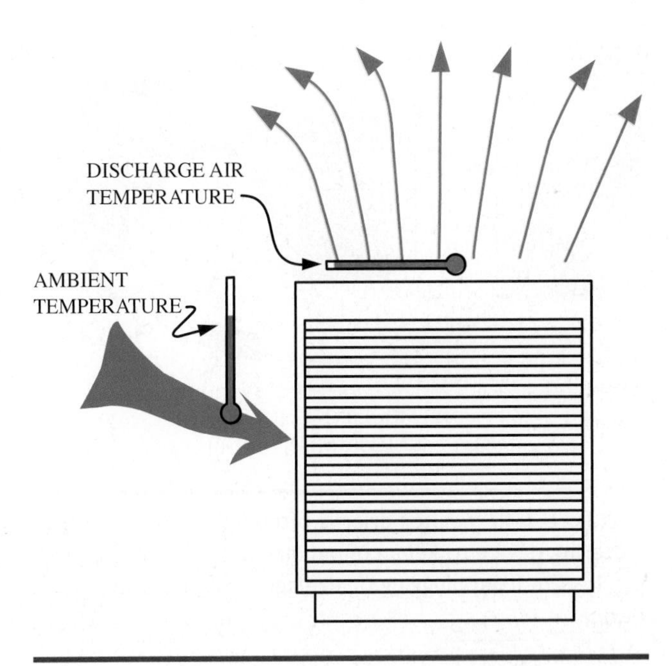

Figure 1-2-16 The condenser takes in ambient air, draws it through the condenser coil, and discharges it out the top of the unit.

one pictured may also have a starting capacitor. The replacement compressor may also come with an installed suction service valve.

Condenser

The condenser, Figure 1-2-16, is similar to the evaporator in construction. It is a series of tubes through which the hot, high pressure vapor passes. Air is forced through the condensing coil by the fan, and heat is given up by the refrigerant to the surrounding air, causing the vapor to condense.

Once the vapor's sensible temperature is reduced to its condensing temperature, latent heat will continue to be removed as the vapor is condensed into a liquid refrigerant. The condenser must be sized in accordance with the size of the compressor and evaporator. Heat that is absorbed in the cooling coil, superheat, plus miscellaneous heat caused by electricity or friction, a by-product of the process of compression, must be removed by the condenser. The suction line, returning to the compressor, is insulated to prevent picking up any excessive superheat through it.

The liquid refrigerant leaving the condenser passes through the filter drier, Figure 1-2-17, before it enters the metering device (capillary tube). The filter drier is a mechanical device with a screen to block foreign particles and a desiccant, or drying agent, to remove moisture from the refrigerant. It is extremely important not to use any antifreeze compounds with the refrigerant in the system, as these will clog the drier.

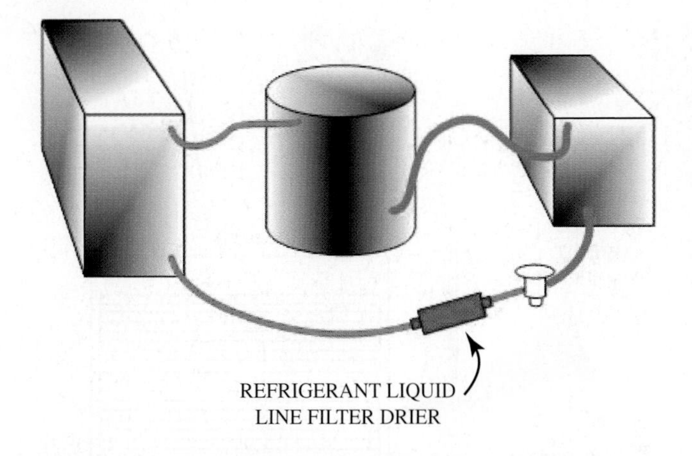

Figure 1-2-17 A filter drier is often installed before the metering device to keep contaminants that might be in the system from plugging or damaging the metering device.

Metering Device

The most common type of metering device used on refrigeration systems is the restrictor type of flow control, commonly known as the capillary tube. The capillary tube, Figure 1-2-18, is located after the filter drier and connects the liquid line and the evaporator. The capillary tube meters the flow of liquid refrigerant from the liquid line and provides a pressure drop from the high pressure side of the system to the low pressure side of the system. The capillary tube's length and inside diameter determine the correct amount of liquid flow into the evaporator and the correct pressure drop.

The accumulator is a storage cylinder located between the evaporator and the suction line to the compressor. The purpose of the accumulator is to store liquid refrigerant. This prevents spillover into the suction line, thus obtaining maximum efficiency from the evaporator coil by permitting it to be fully flooded with refrigerant.

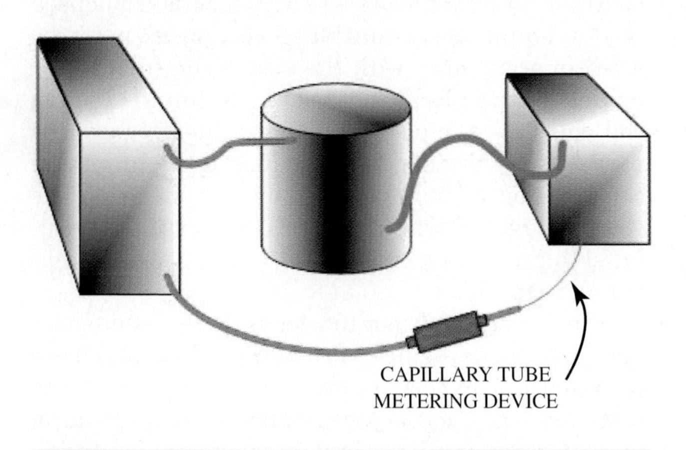

Figure 1-2-18 A capillary tube is a very thin copper tube attached to the liquid line. The capillary tube serves as a metering device.

SERVICE TIP

The compressor is often referred to as the heart of a refrigeration system, and the metering device is often referred to as the brain of the system. The metering device is what controls the flow of refrigerant, and therefore, the actual amount of cooling provided by the system. Some metering devices such as the TXV are self adjusting in order to keep the system functioning efficiently.

Interconnection Tubing

The components of the refrigeration system are connected by three refrigerant lines: the liquid line connecting the condenser and evaporator, the discharge line from the compressor to the condenser, and the suction line, from the evaporator back to the compressor. The liquid line carries the high pressure, warm liquid from the condenser to the capillary tube (metering device). The discharge line is normally smaller than the suction line. The discharge line carries the high-pressure, hot vapor from the compressor to the condenser. The suction line carries the low-pressure, cool vapor from the evaporator coil to the compressor inlet.

The operation of the compressor removes the vapor from the evaporator, reducing the pressure on the liquid refrigerant. Lowering the pressure lowers the boiling point (saturation temperature) of the liquid. Heat from the area to be refrigerated or cooled flows by convection to the evaporator, then by conduction through the fans and tubing to the liquid refrigerant.

This latent heat allows the liquid to vaporize without changing the temperature or pressure of the boiling liquid. The heat-laden vapor from the evaporator is compressed by the compressor. This increases the temperature and pressure of the refrigerant vapor above the condensing medium.

This highly superheated vapor enters the condenser through the discharge line. Condenser heat is removed from the high temperature, high-pressure vapor by conduction to the tubes and fans, then by convection to the air being forced through the condenser by the blower fan. Thus, heat is carried away from the refrigeration system at the condenser.

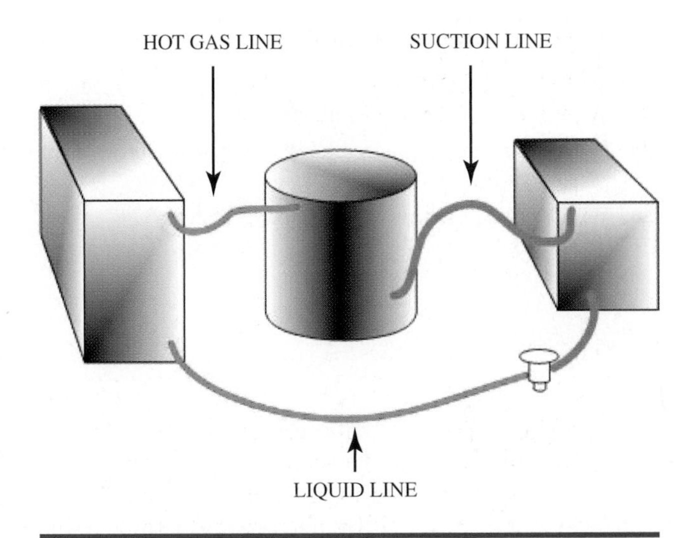

Figure 1-2-19 The three major lines in a refrigeration system are the hot gas line, the liquid line, and the suction line.

As heat is removed from the vapor, the sensible temperature of the conditioned air is reduced. The cool surface of the evaporator also collects water vapor from the conditioned air and drains it.

As latent heat is removed from the vapor at the condenser, the vapor condenses to a liquid. The liquid flows to the outlet from the condenser, through the liquid line to the metering device (capillary tube). The liquid is subcooled as it travels this path. The three major lines in a refrigeration system are shown in Figure 1-2-19.

The compressor maintains a pressure differential between the evaporator coil and the condenser, with high pressure in the condenser and low pressure in the evaporator coil. This pressure differential causes the liquid refrigerant to flow through the filter-drier to the capillary tube restriction to the evaporator coil. The filter-drier removes moisture and contaminants from the refrigerant. The metering device (capillary tube) meters the correct amount of liquid to the evaporator and provides the desired pressure drop.

The cycle of refrigerant flow from the evaporator coil, where heat is absorbed through the refrigeration system and returns to the evaporator after the heat is released at the condenser, ready to absorb more heat, is known as the mechanical refrigeration cycle.

UNIT 2—REVIEW QUESTIONS

1. What do the processes of heating and cooling both revolve around?
2. What are the types of energy?
3. According to the text what is refrigeration?
4. Temperature is measured by a ________?
5. How many Btu's are required to raise the temperature of one pound of water by 1°F?
6. What is matter made up of?
7. Why can't heat energy be created or destroyed?
8. Define change of state.
9. In referring to heat, what are the two kinds of latent heat?
10. What is superheat?
11. List the three different scales of pressure.
12. A liquid at a temperature below its boiling point is referred to as ________.
13. In the refrigeration process, what is one of the very useful properties of the refrigerant?
14. What are the four essential parts in the simple vapor compression refrigeration system?
15. What is the condenser?

TECH TALK

Tech 1. One of the other guys on the crew yesterday told me that with a thermometer he could tell exactly where the last drop of refrigerant disappeared in the evaporator. How can he do that?

Tech 2. Well, as the refrigerant in the evaporator boils, the evaporator tubes will stay at almost exactly the same temperature. Just like a pot of water on the stove stays at 212° as long as there's water in the pot. But just like the pot, temperature rapidly increases once all the water has boiled out. The evaporator coil tubes will have a similar rapid temperature increase once all of the liquid refrigerant has evaporated.

Tech 1. Can you do that on a condenser?

Tech 2. Yes, with an infrared temperature probe it's really easy to see where the last bit of vapor condenses, and the refrigerant begins to become subcooled.

Tech 1. That sounds neat, but what good is it?

Tech 2. By measuring the coil's temperatures you can follow the level of liquid refrigerant in both the evaporator and condenser. This can be a great troubleshooting tool when a system is not functioning properly. With practice you can almost imagine that the pipe is made of glass so you can see exactly what's happening inside by simply taking the pipe's temperature.

Tech 1. I'll have to give that a try. Thanks.

16. What controls the flow of refrigerant to the evaporator?
17. List the six major types of metering devices.
18. What is the most common type of refrigeration cycle?
19. List the four ways the performance cycle in various stages of its operation can be observed.
20. What are the basic components that a mechanical refrigeration system must have to circulate the refrigerant and transfer heat?
21. What is the evaporator coil?
22. What is the compressor?
23. Where is the capillary tube located?
24. In the refrigeration system, what is the purpose of the discharge line?
25. What is the mechanical refrigeration cycle?

UNIT 3

Safety

OBJECTIVES: In this unit the reader will learn about safety in the HVAC/R industry including:

- Personal Safety
- Protective Clothing and Equipment
- Harmful Substances
- Safe Work Practices
- Safely Working with Electricity
- Refrigeration Safety

PERSONAL SAFETY

In every trade, safety is a major concern. Most accidents are caused by carelessness, as well as lack of awareness of proper safety procedures. This chapter deals with some of the safety tips and procedures the refrigeration technician should follow—whether on the job site or at related locations where hazards could exist.

The following are some of the important areas in practicing personal safety that, if followed, will help in preventing accidents:

1. *Protective clothing and equipment* appropriate to activity being performed.
2. Safety procedures for *handling harmful substances.*
3. *Safe work practices,* including proper care and use of tools.
4. Necessary precautions for *preventing electrical shocks.*
5. *Avoiding refrigerant contact* with any part of your body. Always keep the pressure of a confined gas within safe limits.

PROTECTIVE CLOTHING AND EQUIPMENT

The following clothing and equipment should be used by refrigeration technicians:

Head Protection An approved hard hat, Figure 1-3-1, should be worn whenever there is a danger of things dropping on the head or where the head may be bumped. On a construction site, proper safety head gear is a must.

Confine long hair and loose clothing before operating rotating equipment.

Ear Protection Hearing protection devices, Figure 1-3-2a–d, must be worn whenever there is exposure to high noise levels of any duration. These devices are of two types: (1) ear plugs, which are inserted in the ear, and (2) headphones, which cover the ear. Either one must be properly selected on the basis of how much protection is required.

Eye and Face Protection Approved eye or face protectors, Figure 1-3-3, must be worn whenever there is a danger of objects striking the eyes or face. Eye and face protectors have various shapes and sizes, some of them very specialized. If prescription eyeglasses are worn, they must have side shields.

Special eye protectors must be worn when arc welding, spot welding, and burning, in order to cut out harmful light radiation. These special face visors come with various shades of viewing eyepieces which filter out the harmful emissions. Take time to identify the right one for the job. For example, never wear oxyacetylene welding goggles when an arc welding face shield is needed.

NICE TO KNOW

In order to control insurance costs many air conditioning companies have adopted very stringent policies on personal protection equipment. Unlike your shop teacher who may have reminded you each day about safety glasses, ear protection, etc., many employers may terminate you with a single safety infraction. Others may warn you once or twice about wearing proper safety equipment while working. But no HVAC/R companies are going to give you unlimited warnings before your employment with them is terminated. These policies are good for you and the company because they reduce the likelihood of your being injured on the job.

Figure 1-3-1 Hard hats are required to be worn on many job sites.

Figure 1-3-3 Eye protection equipment: (a) These safety goggles can be worn over glasses; (b) Safety glasses with side protection.

Respiratory Protection There are two main types of respirators as shown in Figure 1-3-4: (1) those that purify the air by filtering out harmful dust, mist, metal, fumes, gas, and vapor; and (2) those that supply clean breathing air from a compressed air source. The second type should always be worn when working in a confined space where concentrations of harmful substances are very high or where the concentration is unknown. Remember that most refrigerants are odorless, tasteless, and invisible, and can cause asphyxiation in a very short time.

Respirators must fit tightly against the skin so that there is no leakage from the outside into the face. Workers who are required to use respirators at any time must be instructed in their use, care, maintenance, and limitations.

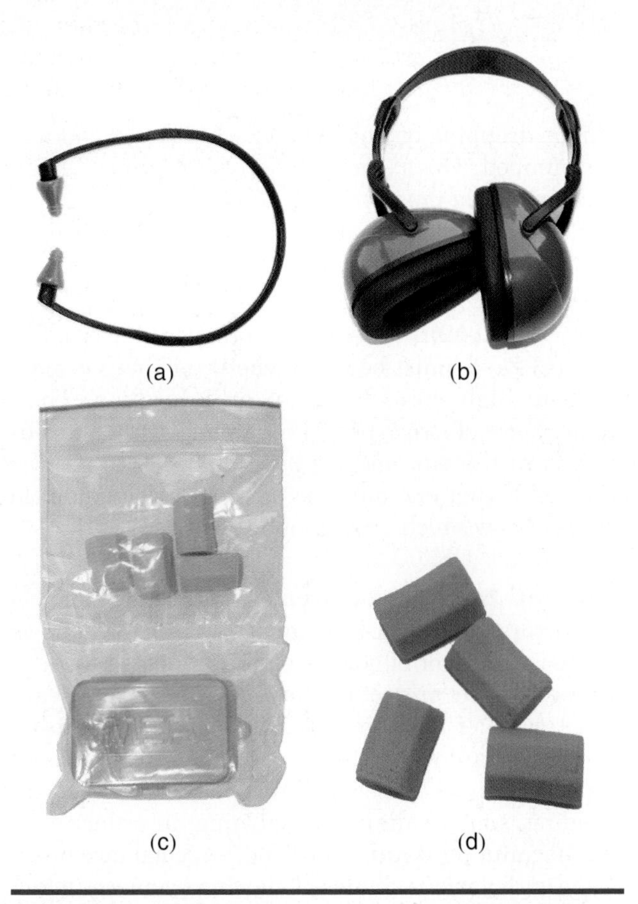

Figure 1-3-2 Ear protection equipment: (a) Disposable earplugs on a lanyard (that fits around the neck) allow for easy removal and reuse; (b) Headphones protect both ears and hearing; (c & d) Disposable, one time use ear plugs.

TECH TIP

Respirator equipment is required to be located in all equipment rooms where that equipment contains large quantities of refrigerant. These respirators are provided in case there is a massive refrigerant leak. If you work in one of these areas, you must familiarize yourself with where the respirators are located and how to quickly put them on. You may have only a matter of seconds once a refrigerant leak alarm is sounded to safely put on this equipment.

Hand and Foot Protection There are many different kinds of gloves used for hand protection, as shown in Figure 1-3-5. Some are made for special usages, such as gloves of steel mesh or Kevlar to protect against cuts and puncture wounds. Different glove materials are needed to protect against a variety of different chemicals. Choose the right kind from a dependable supplier who can supply this information. Discard the damaged ones.

When choosing foot protection as shown in Figure 1-3-6, use the following guidelines:

1. All footwear must be well constructed to support the foot and to provide secure footing.
2. Where there is danger of injury to the toes, top of the foot, or from electrical shock, the proper

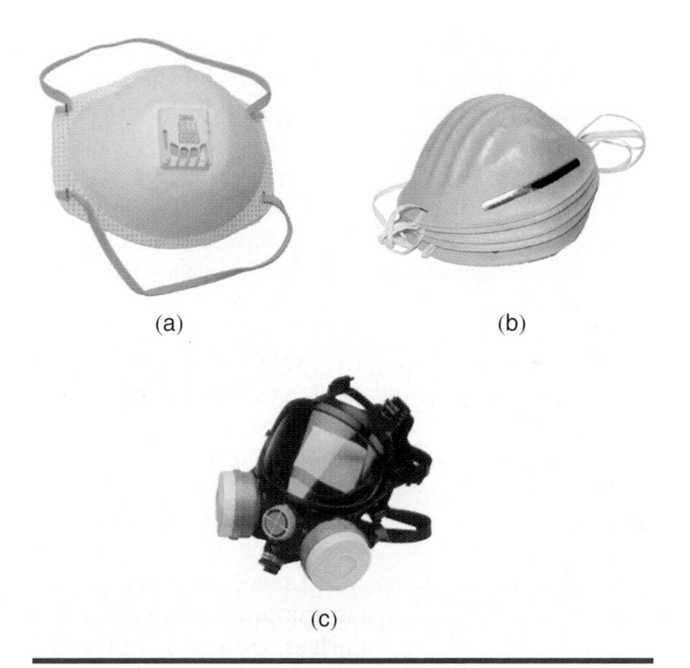

Figure 1-3-4 Filtration masks: (a & b) Light duty filter; (c) Respirator with replaceable filters.

shoe or boot must have Construction Safety Approval (CSA) indicated.

3. Where there is danger of injury to the ankle, footwear must cover the ankle and have a built-in protective element/support.
4. If there is danger of harmful liquids dropping on the foot, the top of the shoe must be completely covered with an impervious material or treated to keep the dripped substance from contacting the skin.

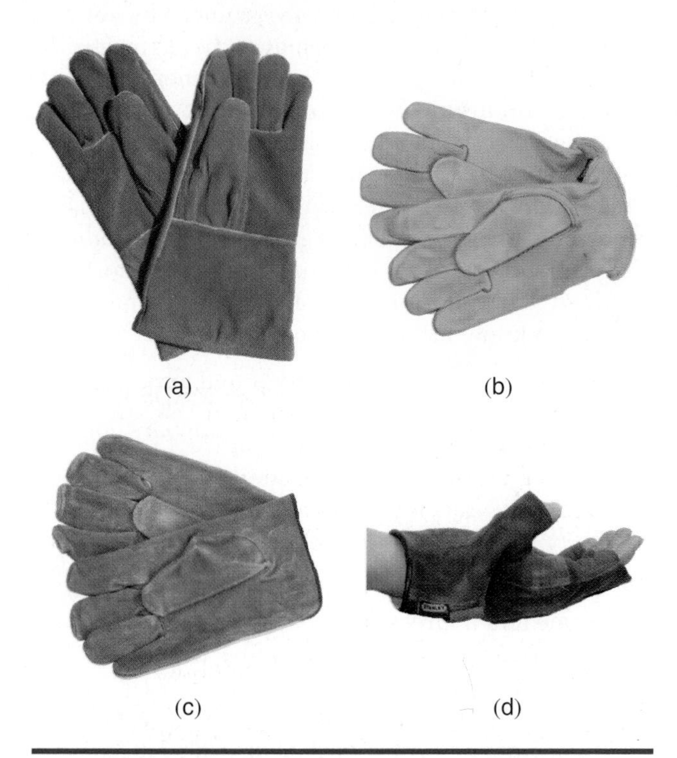

Figure 1-3-5 (a) Gauntlet-type work gloves; (b) Work gloves; (c) Welding gloves; (d) Open-tipped gloves.

Figure 1-3-6 High-top work boots.

Fall Protection Two methods of preventing injury from falling are: (1) fall prevention equipment and (2) fall arresting equipment. Either of these methods is required when working at heights over 10 ft above grade when no other means has been provided for preventing falls. Figure 1-3-7 illustrates a safety belt.

In fall prevention, a worker is prevented from getting into a situation where a fall can occur. For example, a safety belt attached to a securely anchored lanyard will limit the distance a worker can move.

In fall arresting, the worker must wear a safety harness attached to a securely anchored lanyard,

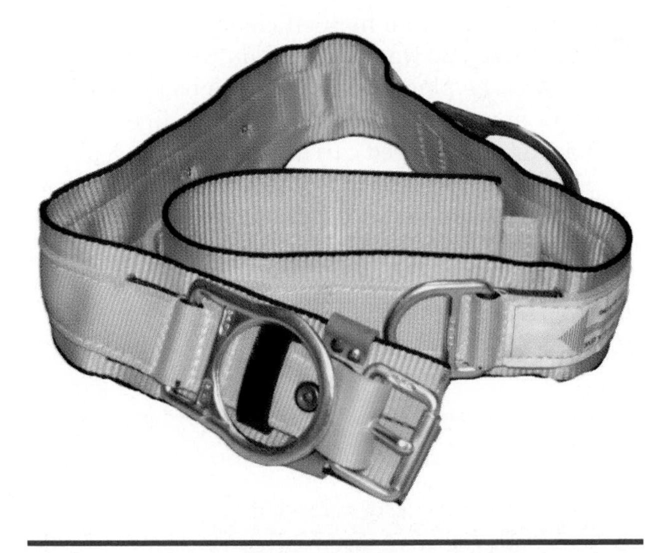

Figure 1-3-7 Safety equipment for heights over 10 ft above grade. A safety lanyard would be attached to this safety belt.

which will limit the fall to a safe distance above impact. The harness helps prevent the worker from suffering internal damage. Belts should not be used to arrest a fall because they do not provide the measure of safety that harnesses do. Where a fall arresting system is not practicable, a safety net should be suspended below the work activity. The worker should be secured separately from the tools and equipment.

CAUTION: Fall protective harnesses are designed to suspend you in a vertical position if you accidentally slip and fall from a height. These harnesses, however, are not designed to suspend you for long periods of time. In recent years workers have survived a fall only to die in the safety harness. The safety harness can constrict blood flow to your legs as you dangle at the end of the safety line. The restriction of blood flow to your legs can cause enough blood to pool in your legs so that you might pass out or even die if allowed to dangle motionless for a long period of time. If you are the victim of a fall and are suspended on your safety line, you should remember to move your legs to help keep the blood flowing until you are rescued.

HARMFUL SUBSTANCES

Workers in the mechanical trades can be exposed to a variety of harmful substances, such as dust, asbestos, carbon monoxide, refrigerants, resins, adhesives, and solvents.

All dust can be harmful. Where dust cannot be controlled by engineering methods, an approved respirator designed to filter out specific dust must be worn.

When asbestos containing material (insulation) is being cut or shaped, the particles must be removed by a ventilation system that discharges the particulate matter through a high efficiency particulate air (HEPA) filter. All waste materials that contain asbestos must be placed in impervious bags for transfer to an approved disposal site. These fibers, when inhaled, are considered carcinogenic.

Engine driven mobile equipment operating in an enclosed area can produce dangerous levels of carbon monoxide (CO). Oil-fired or gas-fired space heaters without suitable vents can also produce carbon monoxide. Areas must be well ventilated while being heated with these devices.

TECH TIP

Read and follow all label safety and use instructions on any materials you use in the HVAC/R field. Some of these materials can be hazardous to your health if you do not follow the label's directions.

Some refrigerants are more dangerous than others. All refrigerants are dangerous if they are allowed to replace the oxygen in the air. Even the so-called safe refrigerants can produce a poisonous phosgene gas when heated to high temperatures. Refrigerants sprayed on any part of the body can quickly freeze tissue. The safe handling of refrigerants will be discussed in detail in a later part of this chapter.

Resins, adhesives, and solvents can be dangerous if not properly handled. Ensure that the workspace is continuously ventilated with large amounts of fresh air.

Never use carbon tetrachloride for any purpose, because it is extremely toxic, either inhaled or on the skin. Even slight encounters with it can cause chronic problems.

To provide workers and health care professionals with specific reactions and treatments for exposure to materials on the job, all manufacturers must provide, on request, Material Safety Data Sheets (MSDS) on all of their products. It is your responsibility to request MSDS's for all of the products you work with or carry on your service vehicle. Some site safety officers, building inspectors, or job managers may even ask to see your MSDS's before you are allowed to start or continue work.

Material Safety Data Sheets (MSDS)

Material Safety Data Sheets are required by law and have specific important information listed in specific areas so that they are easily read by emergency personnel. If an area does not apply to the product, the manufacturer must mark the space as being non-applicable. No blank spaces are allowed on MSDSs. This is done so that there will not be any confusion regarding the safety or reactions to any products. You should read MSDSs on any material before you use it so you know how to use it properly and safely as well as knowing what to do if there is an accident involving the material.

Figure 1-3-8 is an example of an MSDS for a coil cleaner. The following material appears on all MSDSs:

1. *Identity.* This section gives the name of the product.
2. *Section I.* This section gives the manufacturer's name and address. This section contains the emergency contact phone number and company phone number.
3. *Section II: hazardous ingredients / identity information.* A list of the hazardous ingredients by chemical name is given here. If a material has a secret ingredient the company

Material Safety Data Sheet
May be used to comply with
OSHA's Hazard Communication Standard,
29 CFR 1910.1200. Standard must be
consulted for specific requirements.

U.S. Department of Labor
Occupational Safety and Health Administration
(Non-Mandatory Form)
Form Approved
OMB No. 1218-0072

① IDENTITY *(As Used on Label and List)* Coil Master Non Acid Coil Cleaner CMD-2A

Note: Blank spaces are not permitted. If any item is not applicable, or no information is available, the space must be marked to indicate that.

② Section I

Manufacturer's Name	Emergency Telephone Number
Crow Marketing & Distribution, Inc.	800-255-3924
Address *(Number, Street, City, State, and ZIP Code)*	**Telephone Number for Information**
P. O. Box 29171, Dallas, TX 75229	214-241-3049
	Date Prepared 08-23-95 reviewed 12/01/99
	Signature of Preparer *(Optional)*

③ Section II – Hazardous Ingredients/Identity Information

Hazardous Components (Specific Chemical Identity; Common Name(s))	OSHA PEL	ACGIH TLV	Other Limits Recomended	% (optional)
NaOH Sodium Hydroxide 50%	2mg/m^3	2mg/m^3		
CAS#1310-73-2 NA1824 Corrosive				

④ HMIS INFORMATION

Health	2
Reactivity	1
Flammability	0

⑤ Section III – Physical/Chemical Characteristics

Boiling Point	180°F	Specific Gravity (H_2O = 1)	1.01
Vapor Pressure (mm Hg.)	760	Melting Point	N/A
Vapor Density (AIR = 1)	1	Evaporation Rate (Butyl Acetate = 1)	N/A

Solubility in Water
Complete - 100%

Appearance and Odor
Yellow to gold liquid, characteristic odor

⑥ Section IV – Fire and Explosion Hazard Data

Flash Point (Method Used)	Flammable Limits	LEL	UEL
None	None		

Extinguishing Media
N/A

Special Fire Fighting Procedures
Will react with metals to produce flammable hydrogen gas

Unusual Fire and Explosion Hazards
None known. Contents under pressure. Exposure to temperatures above 120°F may cause bursting.

Figure 1-3-8 Sample of a completed Material Safety Data Sheet (MSDS).

must either list it or provide it to a health worker on request. The item listed as OSHA PEL stands for the *personal exposure limit*. This is the maximum safe amount of contact or exposure time that is allowed. The ACGIH TLV is the time weighted average exposure over an 8 hr period of time. It is the maximum amount of contact or exposure allowed during an 8 hr working day. Any other specific limitations are given in this section.

4. *HMIS information.* The information in this section refers to the hazardous material identification system (HMIS). This information tells health care workers a relative number according to how significantly the material will affect health, how reactive it is, and its flammability.
5. *Section III: physical/chemical characteristics.* This section gives the properties of the material, such as its boiling point, vapor pressure, solubility in water, specific gravity, melting point, evaporation rate, and its appearance or odor.
6. *Section IV: fire and explosion hazard data.* Gives the flammability of the material and lists its flash point, flammability limits, extinguishing media, any special firefighting procedures and any unusual fire or explosion hazards.

In addition to the information on the MSDS you must read and follow all manufacturer's listed instructions for proper use and handling of every material you use on the job. Many companies have policies that require that you be fired if you do not follow these instructions. These policies are for both your protection and the protection of the company and their customers. Handle all material properly according to the manufacturer's instructions.

SERVICE TIP

It is a good work practice to request MSDS sheets on all products that you get from the supply house. Some job site safety officials require that you have MSDS sheets on all of the products you will be using on their job site. An easy way of maintaining these documents is to put them all in a large manila envelope that is labeled MSDS and keep this envelope in your service vehicle.

SAFE WORK PRACTICES

The refrigeration and air conditioning technician works in many areas: in the shop, in various types of buildings, in equipment rooms, on rooftops, and on the ground outside buildings. Each location requires different activities where safe performance is essential.

In addition, the worker deals with many potentially dangerous conditions, such as handling pressurized liquids and gases, moving equipment and machines, working with electricity and chemicals, and exposure to heat and cold. It is important, therefore, that the technician practice good safety procedures wherever the work is being done or whatever part of work in which he or she is engaged.

Hand Tools

1. Keep all hand tools sharp, clean, and in safe working order.
2. Defective tools should be repaired or replaced.
3. Use correct, proper fitting wrenches for nuts, bolts, and objects to be turned or held.
4. Do not work in the dark; use plenty of light.
5. Do not leave tools on the floor.

Power Tools

1. Only use power tools that are properly grounded.
2. Stand on dry nonconductive surfaces when using electrical tools.

3. Use only properly sized electrical cords in good condition.
4. Turn on the power only after checking to see that there is no obstruction to proper operation.
5. Disconnect the power from an electrical tool (or motor) before performing any maintenance.
6. Disconnect the power supply when equipment is not in use.

Shop Safety

1. Keep the shop or laboratory floor clear of scraps, litter, and spilled liquid.
2. Store oily shop towels or oily waste in metal containers in an open, airy place.
3. Clean the chips from a machine with a brush; do not use a towel, bare hands, or compressed air.
4. Keep safety glasses and gloves in a prominent location adjacent to machinery used for grinding, buffing, or hammering and where material with sharp edges is handled.
5. Establish cleaning periods. Make sure everyone is clear when using compressed air to clean.

Maintaining an Orderly Shop

1. Arrange machinery and equipment to permit safe, efficient work practices and ease in cleaning.
2. Materials, supplies, tools, and accessories should be safely stored in cabinets, on racks, or in other readily available locations.
3. Working areas and workbenches should be clear. Floors should be clean. Keep aisles, traffic areas, and exits free of materials and obstructions.
4. Combustible materials should be properly disposed of or stored in approved containers.
5. Drinking fountains and wash facilities should be clean and in good working order at all times.

Fire Extinguishers

The danger of fire is always present. Rags soaked in oil, grease, or paint can ignite spontaneously. Keep used rags in tightly closed metal containers.

Sparks, open flames, and hot metal can ignite many materials. Always have a fire extinguisher close at hand when welding or burning.

Extreme caution should be taken with highly flammable and volatile solvents. Due to its low flash point (the temperature at which vapors will ignite), gasoline should never be used as a cleaning solvent.

Fire extinguishers should be readily accessible, properly maintained, regularly inspected, and promptly refilled after use. They are classified according to their capacity for handling specific types of fires, as shown in Figure 1-3-9.

CLASS A EXTINGUISHERS

CLASS B EXTINGUISHERS

CLASS C EXTINGUISHERS

CLASS D EXTINGUISHERS

Figure 1-3-9 Fire extinguisher classification symbols.

1. *Class A extinguishers.* These are used for fires involving ordinary combustible materials such as wood, paper, and textiles, where a quenching, cooling action is required.
2. *Class B extinguishers.* These are for flammable liquid and gas fires involving oil, gas, paint, and grease, where oxygen exclusion or flame interruption is essential.
3. *Class C extinguishers.* These are for fires involving electrical wiring and equipment where nonconductivity of the extinguishing agent is critical. This type of extinguisher should be present whenever functional testing and system energizing take place.
4. *Class D extinguishers.* Some metals, such as magnesium, can actually catch fire. Class D extinguishers are used to put out combustible metal fires.

> **CAUTION:** Having the proper fire extinguisher is not enough if it is not large enough to do the job. Make sure that your fire extinguisher is large enough to handle the size of fire that might occur in your work area. If there are a lot of combustible materials, you need a bigger fire extinguisher. In some cases you may want to have more than one fire extinguisher available.

Material Handling

Use mechanical lifting devices whenever possible. Use a hoist when lifting tools or equipment to a roof. If you are required to lift a heavy object, get help. In order not to strain your back, the following

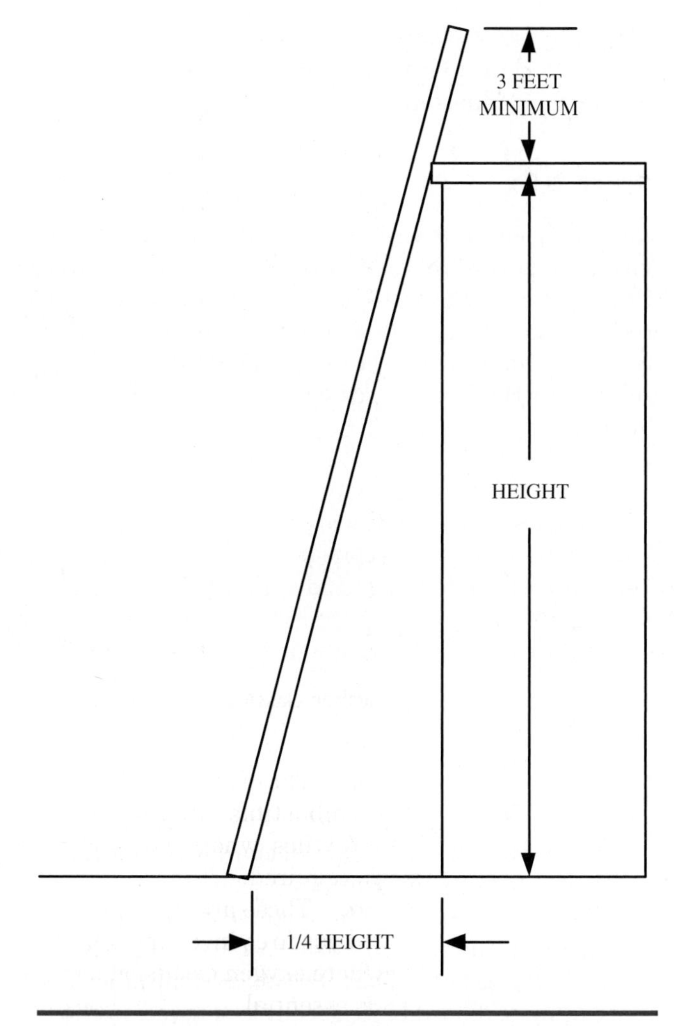

Figure 1-3-10 A ladder must be placed so the top is at least 3 ft above the roof and at an angle in which the distance from the building is one-fourth the height of the building.

procedures should be observed when lifting heavy objects.

1. Bend your knees and pick up the object, keeping your back straight up.
2. Gradually lift the weight using your leg muscles, continuing to keep your back vertical.

Access Equipment

Access equipment refers to ladders and scaffolds that are used to reach locations not accessible by other means. The following precautions should be practiced in the use of ladders.

1. Only use CSA or ANSI-approved ladders. Maintain ladders in good condition. Inspect ladders before each use and discard ladders needing frequent repairs or showing signs of deterioration.
2. All portable ladders must have nonslip feet.
3. Place ladders on a firm footing, no farther out from the wall than one quarter of the height required, as shown in Figure 1-3-10.
4. Ladders must be tied, blocked, or otherwise secured to prevent them from slipping sideways.
5. Never overload a ladder. Follow the maximum carrying capacity of the ladder, including the person and equipment. The American National Standards Institute (ANSI) sets the standard for ladders.
6. Only one person should be on a ladder, unless the ladder is designed to carry more people. Follow maximum load rating.
7. Never use a broken ladder, or a ladder on top of scaffolding.
8. Always face the ladder and use both hands when climbing or descending a ladder.
9. Use fiberglass or wood ladders when doing any work around electrical lines, Figure 1-3-11.
10. Ladders should be long enough so you can perform the work comfortably, without leaning or having to go beyond the two rungs below the top rung safety barrier.
11. Stepladders should only be used in their fully open positions.

CAUTION: Ladders must be inspected from time to time to ensure their safety. Some companies will require that a company safety official inspect the ladders you carry on your service van. Damaged or worn ladders must be repaired or removed from service.

The following recommendations apply to scaffolds.

1. Scaffolds must be supported by solid footings.
2. A scaffold having a height exceeding three times its base dimension must be secured to the structure.

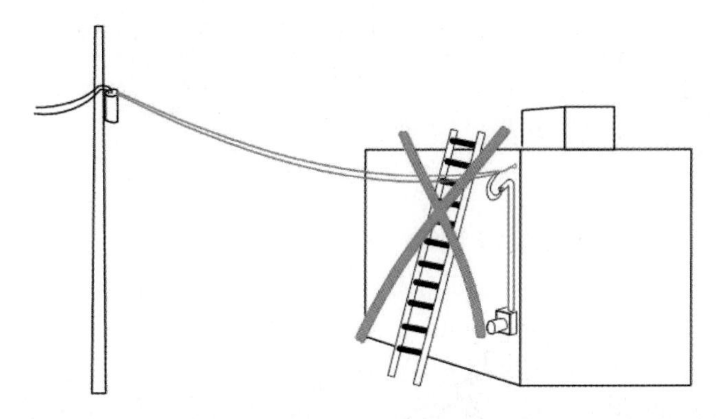

Figure 1-3-11 Never use a metal ladder near electrical wires!

3. When rolling scaffolds are used, the wheels must be locked when there are workers on the scaffold.
4. No worker is to remain on the scaffold while it is being moved. All equipment should also be removed before moving a scaffold.
5. Access to the work platform must be a fixed vertical ladder or other approved means.

Welding and Cutting

Welding and cutting is a specialized skill and requires special training. Many refrigeration and air conditioning technicians require this training due to the need to perform some of these operations as part of their work. It must be recognized, strictly from a safety standpoint, that this work should not be attempted without adequate knowledge and instruction.

Air Acetylene Torches

Air acetylene torches are used by HVAC/R technicians to produce sufficient heat for silver (hard) soldering. These torches use a mixture of air and acetylene as a fuel. Safe use of the air acetylene torch will be covered in detail in Chapter 3. The following safety rules should be practiced when using this equipment.

1. Always use a regulator on the acetylene tank.
2. Always secure the cylinder to something solid to prevent it from being accidentally knocked over.
3. Wear the proper colored safety glasses.
4. Open the valve on the acetylene cylinder only one and one-quarter turns.
5. Light the air acetylene torch with a striker.

CAUTION: All oxygen and acetylene cylinders should be stored and used in secure racks or carts. Federal safety regulations (OSHA) require 55 ft^3 cylinders and larger to be safety chained to carts or structures any time the cylinder is being used with its safety cap removed. There are no exceptions to these OSHA regulations regarding cylinders.

Commercial Job Site Safety

1. Always follow contractors' safety guidelines. Most safety guidelines parallel Occupational Safety and Health Act (OSHA) or state safety rules.
2. Hard hat, steel toe shoes, and safety glasses should be used when required for your own safety.
3. Every person who goes to work in the morning has a right to return home safe after his or her work is done safely.

First Aid

Refrigeration and air conditioning workers are advised to enroll in an approved first aid course. Prompt and correct treatment of injuries not only reduces pain but could also save lives. A classification of accidents that occur to HVAC/R personnel, related to the hazards described, includes the following:

1. Injuries due to mechanical causes.
2. Injuries due to electrical shocks.
3. Injuries due to high pressure.
4. Injuries due to burns and scalds. Injuries due to explosions.
5. Injuries due to breathing toxic gases.

Steps To Be Followed in Case of an Accident

1. All accidents, injuries, and illnesses should be reported to whomever is in charge, no matter how minor the injuries may seem.
2. First aid should be administered, if needed, only by those qualified to do so. Posted emergency procedures should be followed, as applicable. For example, if liquid refrigerant is sprayed on the skin or in the eyes, flush the area with cold water and get treatment.
3. The victim may be sent or taken to receive medical services.
4. An accident report form should be filled out by the person in charge.
5. The area should be cleaned up to remove any contaminants causing the injury, before permitting the area to be used again.
6. An investigation of the accident should be conducted to determine the cause of the accident and to determine ways to prevent similar incidents.

SAFETY WHEN WORKING WITH ELECTRICITY

All possible precautions must be practiced to prevent electrical shock, that is, current passing through the body. Very few realize the damage that can be done by even a small amount of current.

The following information applies to low voltage circuits where current is measured in milliamps (mA). One amp (A) is equal to 1000 mA.

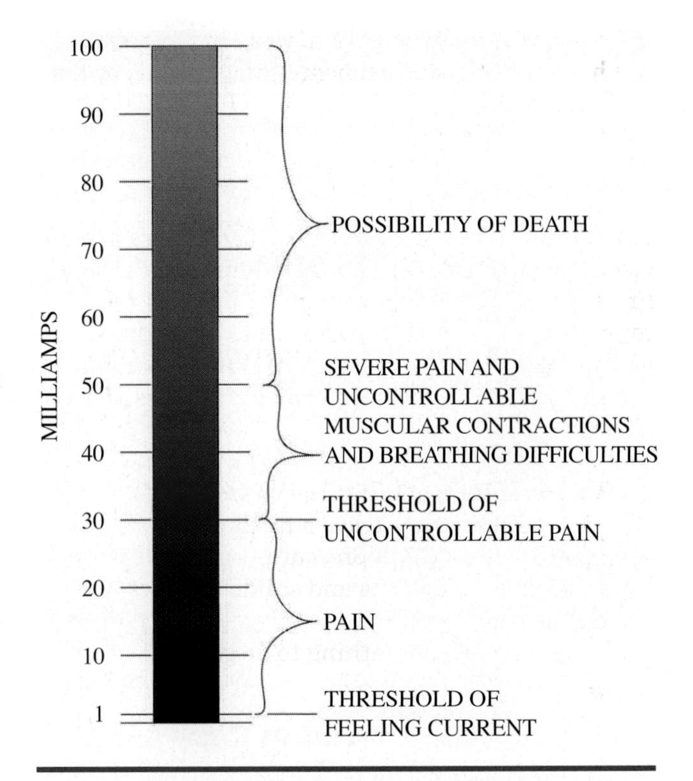

Figure 1-3-12 Amperage rating of electric current creating various shock effects from 1 mA up to 100 mA. *(Courtesy of Workers Compensation Board of British Columbia)*

The illustration in Figure 1-3-12 indicates the effect on the body when various amounts of current pass through the body at 100 mA or less.

Electrical Safety Rules

1. Check all circuits for voltage before doing any service work. **Tag and lock out** all electrical disconnects when working on live circuits.
2. Stand on dry nonconductive surfaces when working on live circuits.
3. Work on live circuits only when absolutely necessary.
4. Never bypass an electrical protective device.
5. Properly fuse all electrical lines.
6. Properly insulate all electrical wiring.

REFRIGERATION SAFETY

The hazards associated with refrigeration service are principally associated with the proper use of refrigerants and their storage in closed containers and systems. A large improvement was made when the HVAC industry started using the so-called safe refrigerants (Class I or fluorocarbons), which were nontoxic and nonflammable. Dangers now relate to the use of pressurized gas or liquid and the fact that these chemicals, when released accidentally, can replace oxygen in a confined space without sensory detection.

CAUTION: From time to time it may be necessary to warm a refrigerant cylinder so that system charging can proceed. Never heat a refrigerant cylinder with an open flame. This could cause two problems. The open flame could cause a rapid rise in cylinder pressure above the rupture point of the cylinder, and the high heat can cause the refrigerant to decompose. Two safe ways of warming cylinders are to place the cylinder in a warm bath of water or to place the cylinder in the warm discharge air from the condenser fan.

The following rules can help decrease the hazards of refrigeration service.

Use of Refrigerants

1. Good ventilation is essential where work is being done with refrigerants, and whenever welding, brazing, or using a cutting torch. Don't use torches in an area where there are high concentrations of refrigerant.
2. Wear safety goggles and gloves when working with refrigerants. Liquid refrigerant can cause frostbite when in contact with eyes and skin.
3. Use low loss hose fitting, or wrap cloth around hose fittings before removing the fittings from a pressurized system or cylinder. Inspect all fittings before attaching hoses or working on them.
4. Ventilate a room containing refrigerants, pressurized gas, or liquid before entering the room to work. Breathing a mixture of air and refrigerant can cause unconsciousness because the refrigerant contains no oxygen.
5. Install oxygen monitors and alarm systems. These are required in machinery rooms with refrigeration equipment. Never enter an area where refrigerant is above exposure limits without proper breathing apparatus.
6. Wear gloves when servicing a system where a compressor has burned out. Refrigerant oil contained in the system can be very acidic. It should never be allowed to touch the skin.
7. Shut off and tag valves before working on refrigerant, steam, and water lines.
8. Follow all codes when making modifications or repairing any system.
9. Never chip ice or frost from refrigeration line, coils, or sight glass.
10. Read all MSDSs.
11. Follow warning and caution signs.
12. Following refrigerant recovery, ensure the system is open to the atmosphere before brazing.

Handling Refrigerant Cylinders

1. Do not fill a cylinder with liquid refrigerant to more than 80% of its volume. Heat can expand

the refrigerant and create a rupture pressure. Space must be available inside the cylinder for proper expansion to take place. In recovering refrigerants, this is particularly important. Special cylinders have been designed for recovery that have an automatic volume limiting device.
2. In using a cylinder or transporting it, the cylinder must be secured with a chain or a rope in an upright position. Do not drop a cylinder.
3. Mixing refrigerants is dangerous. Cylinders are color coded to help identify each refrigerant. Each system has an identifying label. Do not mix refrigerants. Maintain the identification system.
4. Never apply a torch to a system containing refrigerant. If heat is needed to vaporize refrigerant, use hot water at a temperature not to exceed 125°F (52°C).
5. Do not refill disposable refrigerant cylinders.
6. Replace the cylinder cap when not using a cylinder. The cap protects the valve. Do not lift or carry a cylinder by the valve.

SERVICE TIP

Most refrigerant cylinders come in cardboard boxes. These boxes contain important safety information. Some industrial plant safety officers will require that you maintain these cardboard containers with the safety material listed whenever you are working on their job site.

System Safety

1. Never use oxygen or acetylene to pressurize a system. Use dry nitrogen or carbon dioxide from a tank properly fitted with a pressure regulator.
2. When isolating a section of piping or component of a system, exercise caution to prevent damage and potential hazard from liquid expansion.
3. Always charge refrigerant vapor into the low side of the system. Liquid refrigerant entering the compressor could damage the compressor or cause it to burst.
4. Never service a refrigeration system where an open flame is present. The flame must be enclosed and vented outdoors. If a fluorocarbon refrigerant comes in contact with intense heat, it can produce poisonous phosgene gas.
5. Always prevent moisture (water) from entering the refrigeration system. It can cause considerable damage. All parts must be kept dry. Containers of oil must be tightly sealed to prevent contamination from the absorption of moisture.

TECH TALK

Tech 1. Sorry I'm late. The boss made me go take a drug test this morning after the accident I had yesterday in the service van. I don't understand why I had to go get tested for drugs. I was tested when they hired me. I don't use drugs, and besides that the accident wasn't even my fault.

Tech 2. Well, this company like many other HVAC/R requires drug testing of everyone involved in any type of accident whether or not it was your fault.

Tech 1. Why do they do that when it's not even your fault?

Tech 2. Well, the company wouldn't want to ever wind up in court someday over a minor accident and have a lawyer accuse you of having been drinking or taking drugs. By testing you following an accident, the company can prove that you weren't impaired at the time of the accident.

Tech 1. What if it's not an accident in the service van? Do they still test you?

Tech 2. Well, this company does, and a lot of others do also. If you get hurt on the job, and even if you accidentally damage a piece of equipment, you might be tested.

Tech 1. Sounds like this is not the job for someone who has a drinking or drug problem.

Tech 2. That's right. Not being alert or paying attention in this field can kill you.

Tech 1. I didn't think air conditioning and refrigeration was that dangerous.

Tech 2. It's not if you follow all the safety rules. The company just wants to make sure that you are not impaired with drugs or alcohol, which would make you a lot more dangerous to yourself and everyone else you are working with.

Tech 1. I see their point. Thanks.

UNIT 3—REVIEW QUESTIONS

1. List the five important areas in practicing personal safety, as they are found in the text.
2. According to the text, when should head protection be worn?
3. What are the two main types of respirators listed in this chapter?
4. What are the guidelines for choosing foot protection?
5. List the two methods of preventing injury from falling and give an example of each.

6. What must be done when using asbestos-containing material?
7. Under what conditions can the so-called safe refrigerants produce a poisonous phosgene gas?
8. Why should carbon tetrachloride never be used?
9. List the five safety rules for hand tools.
10. List the four classes of fire extinguishers and what type of fire they are used for.
11. List two procedures that should be followed when lifting heavy equipment in order not to strain your back.
12. What does access equipment refer to?
13. When must a scaffold be secured to the structure?
14. List the safety rules that should be practiced when using air acetylene torches.
15. Most safety guidelines parallel _______ or state safety rules.
16. List the five classifications of accidents that occur to HVAC/R personnel.
17. Who should an accident be reported to?
18. One A is equal to _______ mA.
19. When should you work on a live circuit?
20. Since a large improvement was made when the industry started using the so-called safe refrigerant, what do the dangers now relate to?
21. What can happen if moisture enters the refrigeration system?

UNIT 4

Temperature, Measurement, and Conversion

OBJECTIVES: In this unit the reader will learn about the methods of measuring heat energy including:

- Measurement of Temperature
- Temperature Conversion

We are all acquainted with common measurements such as those referring to length, width, volume, etc. We must also acquaint ourselves with the methods of measuring heat energy.

Temperature is the measurement of the speed of molecular movement, or vibration, in a substance, Figure 1-4-1. Most molecules cannot truly move but they can vibrate. The higher the temperature the faster the molecules vibrate. When the vibration caused by higher temperatures becomes fast enough we can see the surface begin to glow. Lower speed vibrations can be seen as a dull red color and higher speed vibrations produce white light. We see the changing frequency of molecular vibration as a change in the color of light produced, Figure 1-4-2. The change in the color of the light corresponds to the change in the temperature. The dull red we see coming from a dying fire tells us how hot the wood's surface is. When materials are not hot to produce visible light they do produce a lower frequency of light. This unseen light is called infrared light. That is the light that an infrared thermometer sees when reading the temperature of a material's surface some feet away, Figure 1-4-3a,b. Temperature is measured by a thermometer using either the Fahrenheit or Celsius scales.

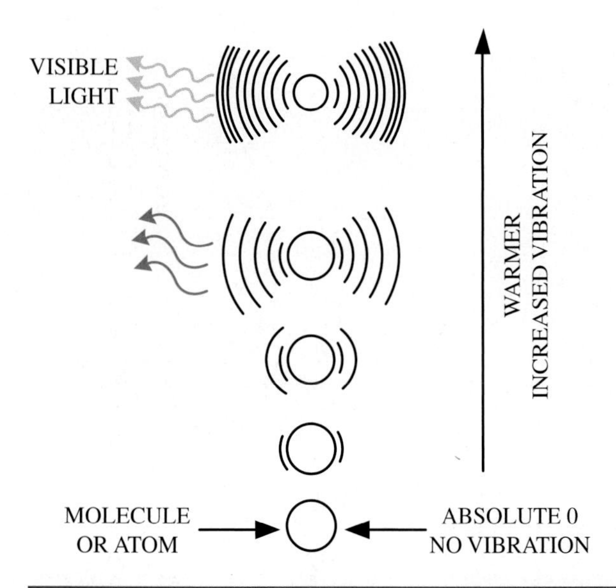

Figure 1-4-1 As the temperature rises above absolute zero, the molecules begin to vibrate. The hotter the material, the faster the vibrations.

TECH TIP

Most of the electronic temperature probes will read in both Fahrenheit and Celsius scales. As you begin working in the HVAC/R field from time to time switch between the scales so that you can begin to get a sense of the relativity from Fahrenheit and Celsius temperature readings. Over time this will help you develop an ability to think both in Fahrenheit and Celsius temperatures, much like learning to speak another language.

The quantity of heat energy in a substance depends on the size of the substance as well as the intensity or level of heat energy in the substance. The level of heat energy is measurable on a comparison basis by means of a thermometer. The thermometer was developed using the principle of the expansion and contraction of a liquid, such as mercury, in a tube of small interior diameter which includes a reservoir for the liquid. Upon being subjected to a temperature (heat intensity) change, the liquid will rise with an increase of temperature, or fall with a decrease of temperature.

The two most common standards of temperature (intensity) measurement are the Fahrenheit and Celsius (formerly centigrade) scales. Figure 1-4-4 shows a direct comparison of the scales of a Fahrenheit and a Celsius thermometer.

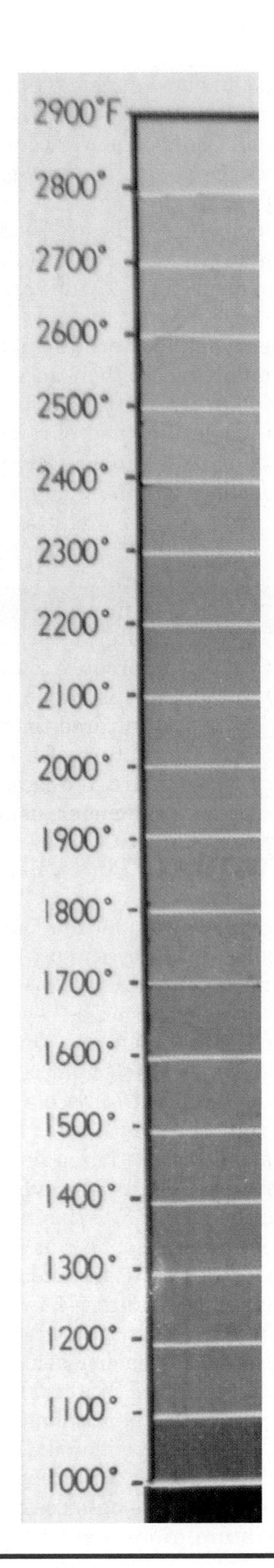

Figure 1-4-2 Above 1000°F, the vibrations are fast enough to begin emitting a dull red visible light. As the temperature continues to rise, the color of the light changes. When it is very hot, it glows white.

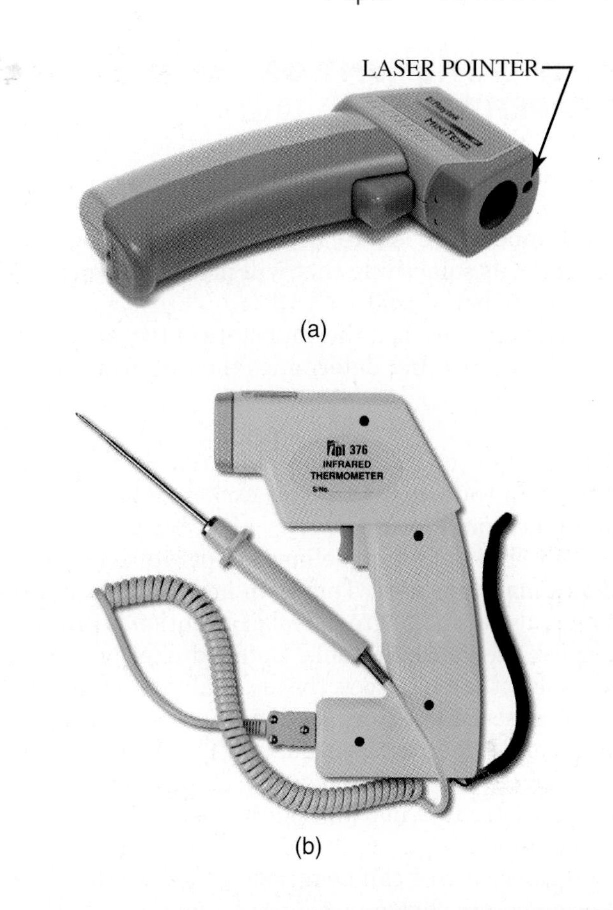

Figure 1-4-3 (a) Infrared thermometer; (b) Infrared thermometer with a temperature contact probe. (*Courtesy Test Products International, Inc.*)

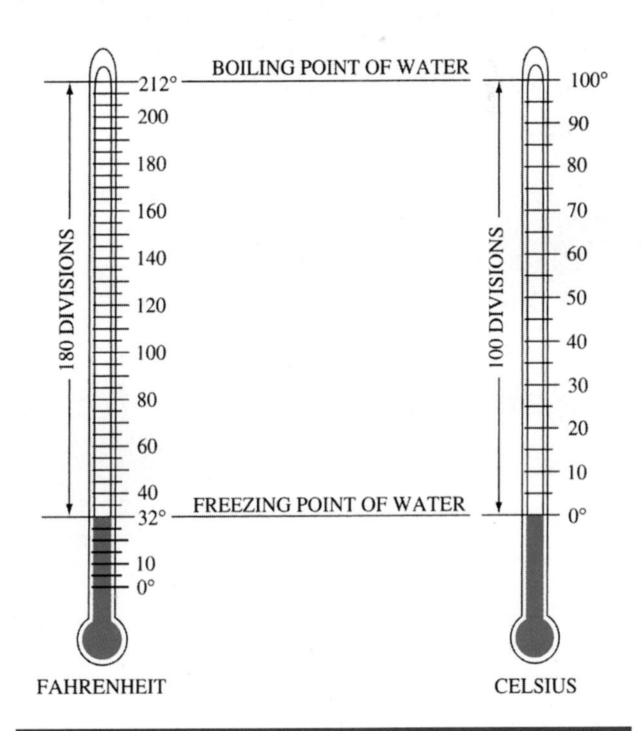

Figure 1-4-4 Comparison of the Fahrenheit and Celsius temperature scales.

MEASUREMENT OF TEMPERATURE

Two definitions of temperature are:

1. Temperature is a relative term that can be defined as something that will produce a sensation of hot or cold.
2. Temperature is a thermal state of two adjacent substances that determines their ability to exchange heat.

A conclusion of this second definition is that substances in contact that do not exchange heat are at the same temperature.

The first definition defines temperature in terms of a human sensation. This is an important aspect of temperature since a great deal of attention is given to keeping people comfortable. Detailed information on this subject is included in the air conditioning section.

The second definition defines temperature in terms of its heat transfer quality. The whole system of refrigeration is involved in moving heat from places where it is not wanted to places where it can be tolerated.

Temperatures can be measured using one of the thermometers shown in Figure 1-4-5a,b.

Glass stem thermometers have been commonly used by refrigeration service people. The temperature is measured by the expansion or contraction of a liquid (mercury or alcohol) with changes in temperature, Figure 1-4-6. In a hollow glass tube the fluid moves up and down a scale that is calibrated to register the ambient temperature. Pocket sizes are available as well as longer, more accurate models for laboratory work. Breakage is a problem since glass thermometers are fragile.

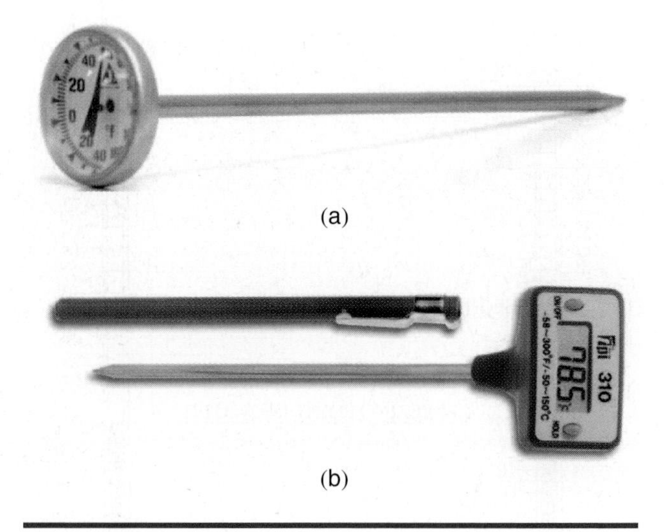

(a)

(b)

Figure 1-4-5 Pocket thermometers: (a) Analog; (b) Digital.

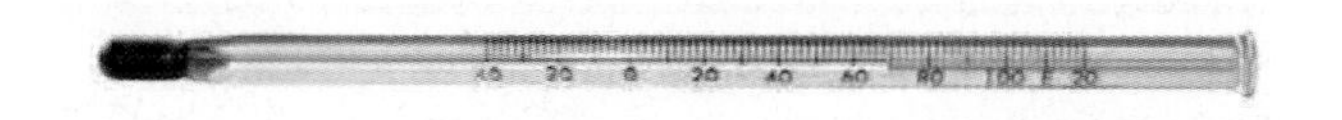

Figure 1-4-6 Glass tube thermometer.

Dial thermometers indicate temperature by a pointer moving over a circular scale. They are more rugged and can be used for a wide variety of applications.

Digital thermometers with compact sensing elements are popular due to their speed in sensing a change in temperature and the ease in reading the alphanumeric display instead of a dial or a scale. They are useful in measuring surface temperatures, such as suction superheat in adjusting an expansion valve.

> **NICE TO KNOW**
>
> The difference between a thermometer and a thermostat is that a thermometer takes the temperature of a room, and a thermostat can do something about the temperature in the room.

TEMPERATURE CONVERSION

Most frequently a conversion from one temperature scale to another is made by the use of a conversion table. But if one is not available, the conversion can be done easily by using formulas based on a definite reference point—absolute zero. This is the point where, it is believed, all molecular action ceases. On the Fahrenheit temperature scale this is about –460°F below zero, or –273°C.

Certain basic laws are based on the use of absolute temperatures. If a Fahrenheit reading is given, the addition of 460°F to this reading will convert it to degrees Rankine or °R; whereas if the reading is from the Celsius scale, the addition of 273° will convert it to Kelvin, K, Figure 1-4-7.

The Fahrenheit scale is associated with the inch-pound (IP) system of measurement, which is still predominant in the United States. The Celsius scale is part of the Standard International (SI) system of measurement, which is considered desirable for universal usage.

The Fahrenheit scale is based on using 32° as the melting temperature of ice and 212° as the boiling temperature of water, at standard atmospheric pressure. The range between these two temperatures is 180°. Scales can go higher or lower depending on the use of the instrument. For example, outdoor thermometers usually have scales from −60°F to 120°F. Indoor thermometers usually range from 50°F to 90°F.

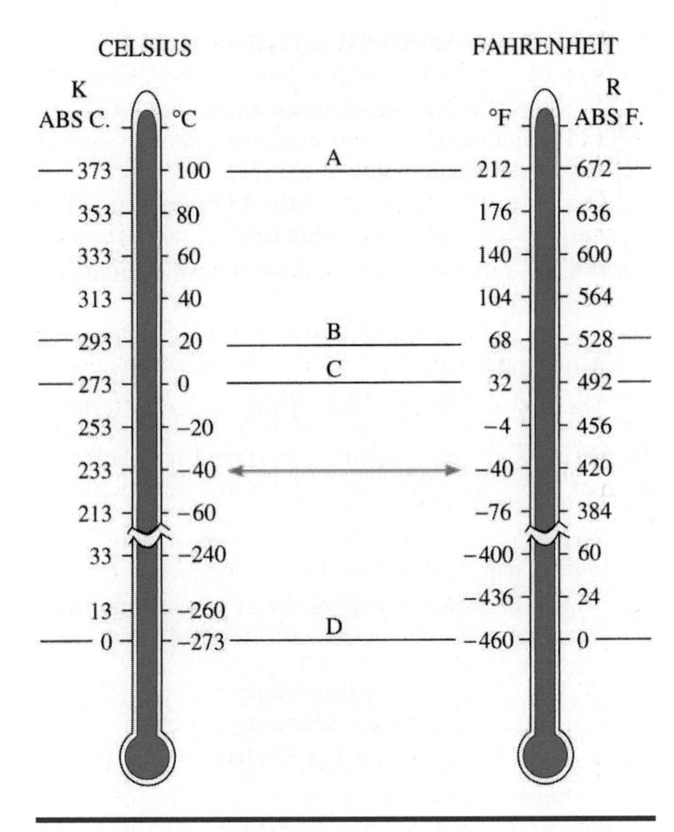

Figure 1-4-7 Fahrenheit, Celsius, Rankine, and Kelvin temperature scales: (a) Boiling temperature of water; (b) Standard temperature; (c) Freezing temperature of water; (d) Absolute zero. Note that −40°C (red arrow) equals 40°F.

On the Celsius scale the melting point of ice is 0° and the boiling point of water is 100°. On an outdoor thermometer the scale is usually from −50°C to +50°C. On an indoor thermometer the scale is usually from 10°C to 30°C.

These reference points (melting temperature of ice and boiling temperature of water) are measured at one atmosphere of pressure, which is standard at sea level. This reference pressure in SI units is 101.325 kPa, which is exactly 1013.25 millibars. In IP units the value is approximately 14.696 psi, or 29.921 inches of mercury at 32°F.

Referring to Figure 1-4-7, it is interesting to note that −40°F is equivalent to −40°C. This is the only place where the two scales coincide.

Two other scales are used for scientific work, the Rankine (IP) scale and the Kelvin scale (SI). Both of these scales start at a theoretical value called absolute zero. This is the lowest hypothetically possible temperature. There is no heat in a substance at this point.

To convert Fahrenheit temperatures to Celsius temperatures or the reverse, the following three methods can be used.

1. Use the conversion tables in Appendix A

TECH TIP

The more metric conversions you do, the easier it will be for you to think in both standard and metric terms. This is much like your recognizing that a football field is both 300 feet and 100 yards long.

2. Calculate the conversion using the following formula:

$$°F = 9/5(°C) + 32$$

or

$$°C = 5/9(°F - 32)$$

3. Estimate the converted value, where precise accuracy is not required, by doubling the Celsius temperature and adding 30° to obtain the Fahrenheit temperature.

EXAMPLE

Suppose you observe a weather report that indicates a temperature in Chicago of 20°C. You are more familiar with Fahrenheit temperatures, so you would like to know the equivalent.

Solution

To convert 20°C to ?°F:

1. Using the tables: Refer to the tables in Appendix A. Note that the table is made up of three columns. The center column gives "Temperature to be converted." The first column gives "Degrees F." The third column gives "Degrees C."

   First, find 20° in the center column. Then project horizontally to the first column. Note 68.0°F. Therefore, 20°C = 68°F.
2. Using the formula: °F = 9/5(°C) + 32°F = 9/5(20°) + 32° = 36° + 32°F = 68°. This is the same answer found using the tables.
3. Using an estimate: It would be done as follows. Double the Celsius degrees and add 30°. Doubling 20° is 40°. Adding 30° is 70°. Although 2° off, it is reasonable that the Chicago temperature is within the comfort range. This value may be close enough for some uses.

Let's convert in the other direction. Suppose you are a Canadian looking at a U.S. weather report. You are familiar with Celsius temperatures. They report the weather in Seattle as 60°F. Would you feel comfortable without a coat? To convert 60°F to °C:

1. Using the table: Find 60° in the center column. Project over to the third column. Read 15.6°C. Therefore, 60°F = 15.6°C.
2. Using the formula: °C = 5/9(°F − 32) °C = 5/9(60° − 32°) = 5/9(28°) °C = 15.6°. This is the same answer found by using the tables.

3. Using the estimate: Subtract 30° from the °F then take half of it. Subtracting 30° from 60° is 30°. Half of 30° is 15°. The temperature in Celsius is approximately 15°. This is a little cool. Better wear a coat. ■

EXAMPLE

What is the freezing point of water in degrees Rankine (°R)?

Solution

Since the freezing point of water is 32°F, adding 460° makes the freezing point of water 492° Rankine (32° + 460° = 492°). ■

TECH TALK

Tech 1. We're learning metric conversions in the night class I'm taking. I don't see any reason to do them. Do you ever really use them in the field?

Tech 2. I don't use metric conversions that often, but there are other guys that use them a lot more. The major thing is that in the HVAC/R field, if you think about it, temperatures are what we deal with all the time. And if you don't learn the metric measurements, it's like not learning a big part of what we do. Besides that, when you get out on a job and all of the manufacturer's literature is in Celsius, it's a little late to start looking for a calculator to try to make the conversions.

Tech 1. Yeah, I can see how that would help. I appreciate it. Thanks.

UNIT 4—REVIEW QUESTIONS

1. In addition to common measurements such as length, width, and volume, you must also acquaint yourself with the measurements of ________.
2. The level of ________ is measurable on a comparison basis by means of a thermometer.
3. What are the two most common measurement of temperature?
4. Give the two definitions of temperature as they are listed in this text.
5. The first definition defines temperature in terms of ________.
6. Which kind of thermometer has been commonly used by refrigeration service people?
7. Which thermometer indicates temperature by a pointer moving over a circular scale?
8. Which thermometers are useful in measuring surface temperatures, such as suction superheat in adjusting an expansion valve?
9. Most frequently a conversion from one temperature scale to the other is made by the use of a ________.
10. The ________ scale is associated with the inch-pound (IP) system of measurement.
11. The ________ scale is part of the Standard International (SI) system of measurement.
12. What is the boiling point of water using the Fahrenheit scale?
13. What is the melting point of ice using the Celsius scale?
14. List the two other scales used for scientific work.
15. What are the three methods that can be used to convert Fahrenheit temperatures to Celsius temperatures?

2

Tools, Meters, and Measuring Devices

UNIT 1

Hand and Power Tools

OBJECTIVES: In this unit the reader will learn about the proper selection and use of hand tools and accessories including:

- Wrenches
- Pliers
- Screwdrivers
- Brushes
- Files
- Vises
- Measuring Tapes and Rulers
- Calipers
- Drills
- Accessory Items

HAND TOOLS AND ACCESSORIES

The care and use of tools are important considerations for the technician. The customer often judges a technician on the appearance of their tools. They often assume that a technician that has a dirty, poorly maintained, and disorderly tool bag will provide the same type of service. To a large extent this is true because it is harder to work with poorly maintained tools, and a tool bag that is not well organized makes it take longer to locate the tools for the job. The way you maintain your tools reflects on your quality of workmanship. In addition, injury can frequently be traced to a lack of, or improper use of, hand tools. Clean sharp tools are better to work with and safer to use.

The common hand tools needed by an HVAC/R technician are described in this chapter.

TECH TIP

Air conditioning and refrigeration hand tools don't get as dirty as most mechanics' tools will. However, it is a good idea to wipe them down with a clean, dry rag after each use. In addition from time to time they should be rubbed with a lightly oiled cloth to protect them from rusting.

WRENCHES

The term wrench is used to describe tools that grip and turn threaded parts such as nuts, bolts, valves, and pipes. Some wrenches are adjustable so they will fit a range of sized nuts, bolts, or pipes, as shown in Figure 2-1-1. Other wrenches have fixed sizes and come as sets to fit common sizes, as shown in Figure 2-1-2. Some wrenches are specifically designed for HVAC/R jobs such as the service valve wrench, as shown in Figure 2-1-3.

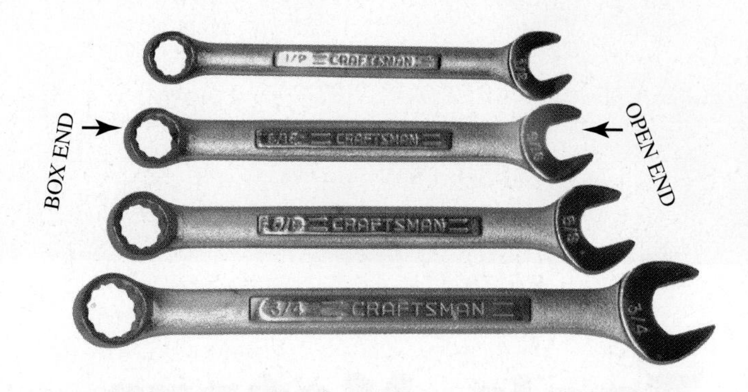

Figure 2-1-2 Combination wrenches. One end is open and the other end is boxed.

Service Valve Wrenches

Service valves on some compressors and condensers are ¼ in (6 mm) square stems and require a special service valve wrench as shown in Figure 2-1-3. Service valve wrenches come in several styles and shapes. Some service valve wrenches use a ratchet mechanism. The ratchet may be fixed in one direction so the wrench has to be removed and turned over to change directions. Other service valve wrenches may have a lever for reversing the rotation of the wrench as required for either opening or closing the valve, Figure 2-1-4. Reversible ratchet service valve wrenches can have up to four fixed sizes of square openings ranging from 3⁄16 in, ¼ in, 5⁄16 in, and ⅜ in (4 mm, 6 mm, 8 mm, and 9 mm), Figure 2-1-5a,b. These service wrenches look similar to the ratchet wrenches used by auto mechanics, but those wrenches have hex openings to fit the heads of standard nuts and bolts.

The handles on service valve wrenches can be flat or offset. The offset gives your hand a little more clearance in tight places, Figure 2-1-6. Most HVAC/R technicians carry more than one type of service valve wrench to meet the varying job needs.

Figure 2-1-3 Refrigerant service valve wrenches have square openings.

Figure 2-1-4 Ratchet service valve wrench.

Socket Wrenches

Socket wrenches, often just referred to as socket(s), are made to slip over the head of bolts and nuts,

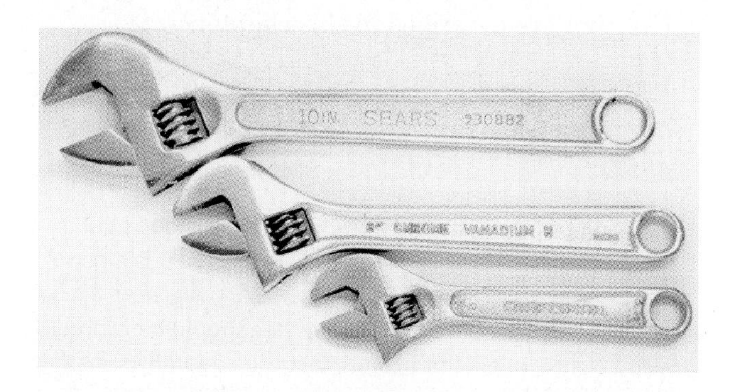

Figure 2-1-1 Adjustable wrenches come in a variety of sizes.

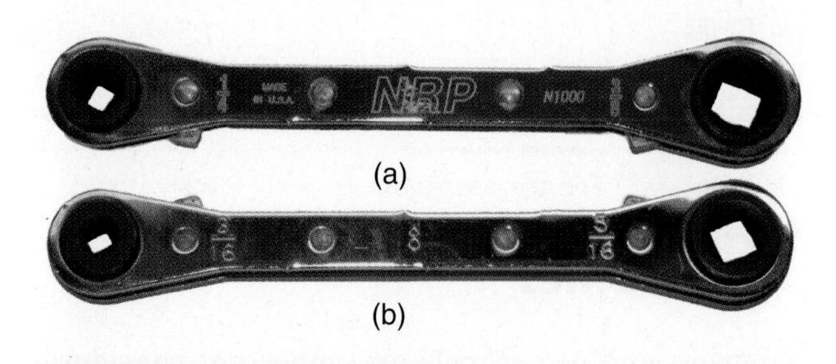

Figure 2-1-5 Some ratcheting refrigerant service valve wrenches provide four different size openings by having two different sizes on each end: (a) Front, ¼ and ⅜ in sizes; (b) Back, 3⁄16 and 5⁄16 in sizes.

SERVICE TIP

From time to time it is necessary to use an extension on a socket to reach way down inside a piece of equipment to remove or install a bolt or a nut. For example the mounting bolts on a compressor inside of a residential condensing unit may require a foot long extension or more. A problem can exist when you try to reach back in this tight space to reinstall the bolts. There may not be room for your hand and wrench. However if you use a small piece of paper possibly folded one or two times and place it over the head of the bolt before pushing the head into the socket, this can be used to hold the bolt or nut in the socket as it is lowered down into the confined space. This trick can save a lot of aggravation. Some manufacturers do provide small magnets that can be placed in the sockets which can serve the same purpose. However the paper trick works when you don't have the magnets.

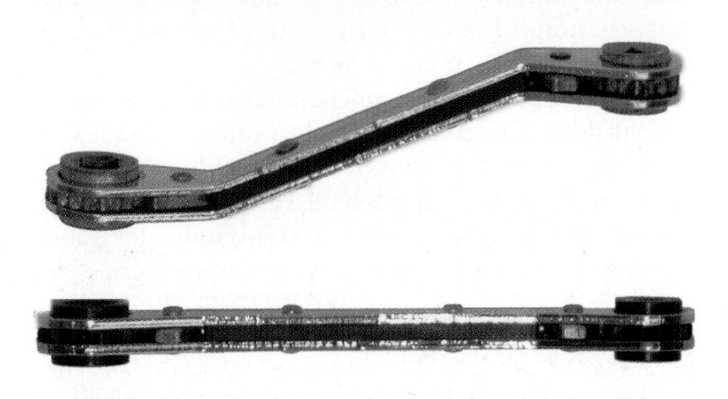

Figure 2-1-6 Refrigerant service valve wrenches are available with offsets to provide some hand clearance when using the wrench.

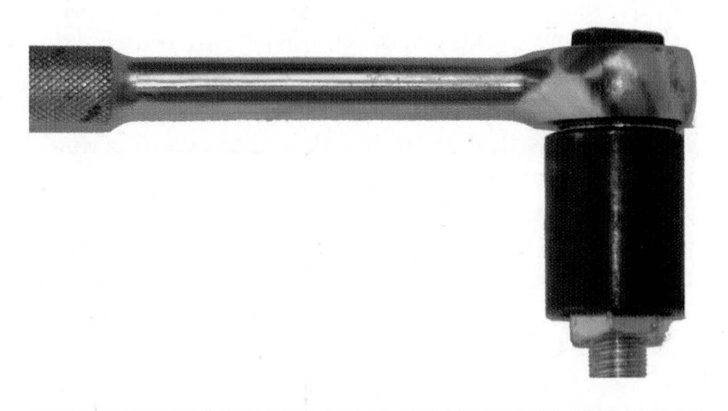

Figure 2-1-7 The socket slides over the heads of nuts and bolts.

Figure 2-1-8 Socket set.

Figure 2-1-7. They are made of tool steel. Sockets usually come as sets in either standard or metric sizes. Although standard sockets are available in sizes ranging from 1/16 in up to several inches in size the most common set would have sockets sizes starting with 1/4 in up to 1 in by 1/8 in or 1/16 in increments, Figure 2-1-8. Common metric socket sets range from 6 mm up to 25 mm.

Both standard and metric socket sets for standard hexagonal nuts or bolts can be either 6 point or 12 point, Figure 2-1-9. The 6 point sockets are stronger; but the 12 point sockets allow easier alignment and shorter swing for tight locations. Swivel sockets and universal joints are also useful for reaching bolts that are hard to get at.

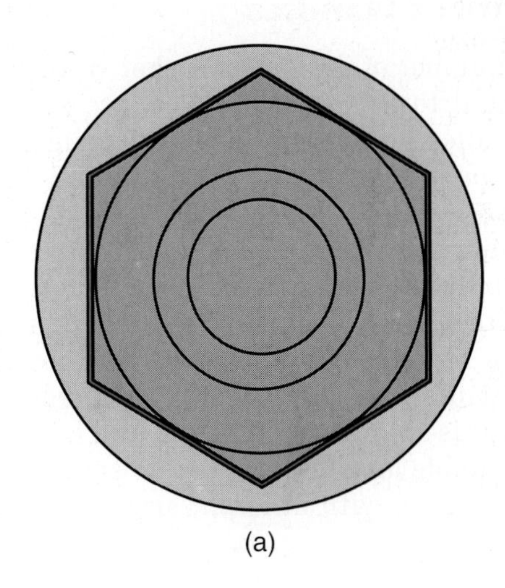

(a)

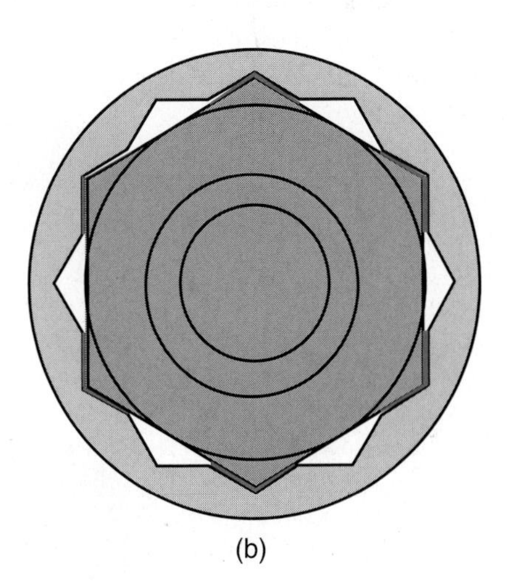

(b)

AREA OF CONTACT BETWEEN SOCKET AND BOLT

Figure 2-1-9 (a) Six point sockets provide greater contact with the head of the nut or bolt than (b) twelve point sockets.

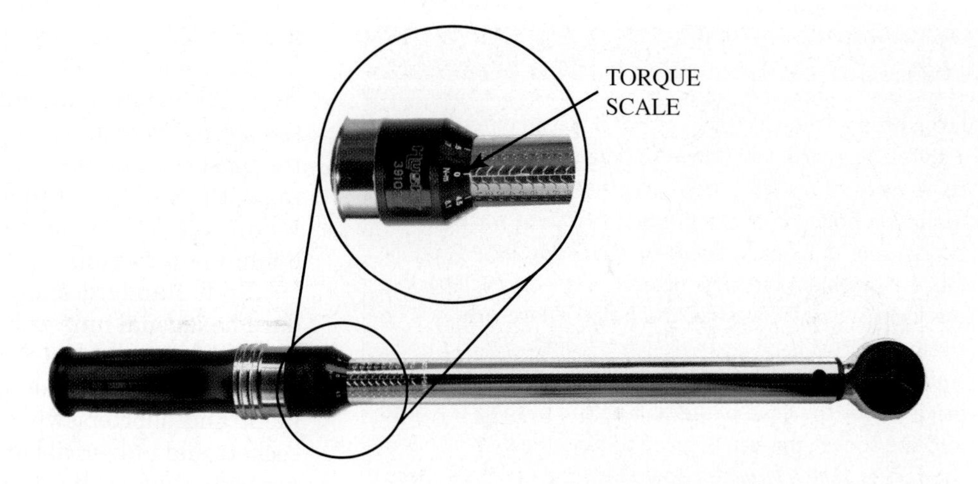

Figure 2-1-10 Torque wrench.

Socket Handles

A number of designs of socket wrench handles, referred to as socket handles, are available. Socket handles range from the most common ratchet type to the breaker bar to specialty designs such as tee driver and torque wrench handle, Figure 2-1-10. There are two dimensions used to refer to socket handles: length and size. The length refers to the distance from the tip of the handle to the head. The size refers to the dimension of the square shank that fits into the socket. The three common sizes for this shank are ⅛ in, ¼ in, and ⅜ in. Adapters are available to allow larger or smaller shank sizes to be used with different sized socket sets, Figure 2-1-11.

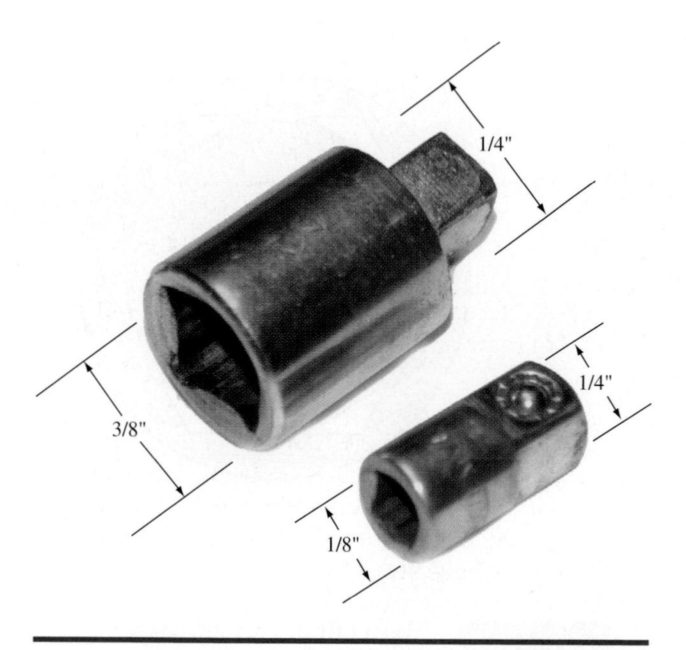

Figure 2-1-11 Socket wrench adaptors are available that permit more than one size socket to be used with different sized wrenches.

Ratchet socket wrench handles are the most common type of handles because they can be used to quickly loosen and remove or install and tighten nuts and bolts. Some ratchets have swivel heads that are helpful for working in tight places, Figure 2-1-12.

Breaker bar socket wrench handles are used when a great deal of force is required to remove a stuck or stubborn bolt or nut. They are designed to take significantly more force than ratchets. However, never use a cheater bar or pipe to extend the handle for more leverage because this could damage the wrench.

Torque Wrenches

Torque wrenches have a gauge to measure the force being applied. Making sure the same force is used on nuts and bolts is important especially when tightening the heads on compressors. Uneven tightening can result in the head being warped or bolts being stripped or broken. Either case can render the compressor unusable or more expensive to repair. Equipment manufacturers have recommended torque specifications that must be followed to ensure a quality assembly, Figure 2-1-13. Typical specifications will give both the torque and the tightening sequence. A good torque practice is to first apply only a light torque to each bolt in the recommended sequence. Then repeat the tightening sequence and each time

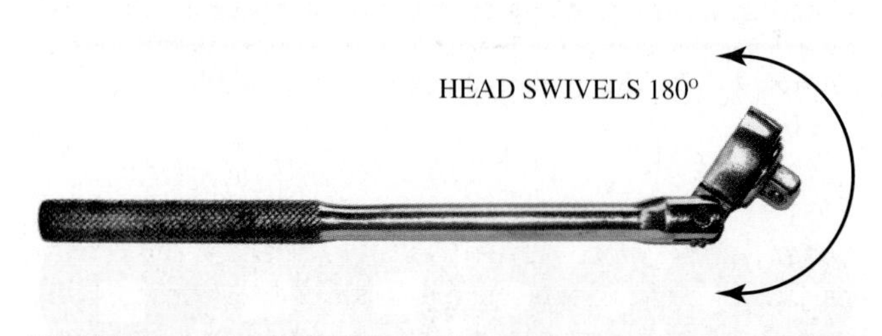

Figure 2-1-12 Swivel head ratchet handle.

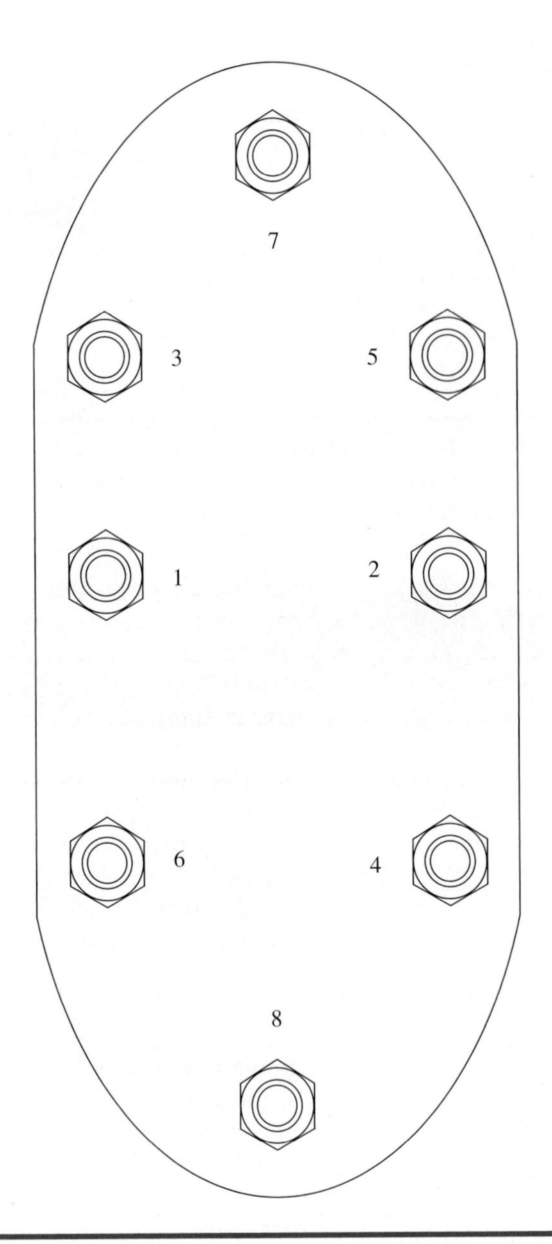

Figure 2-1-13 Typical head torquing sequence. Check with the manufacturer for specific torquing sequences and torque specifications.

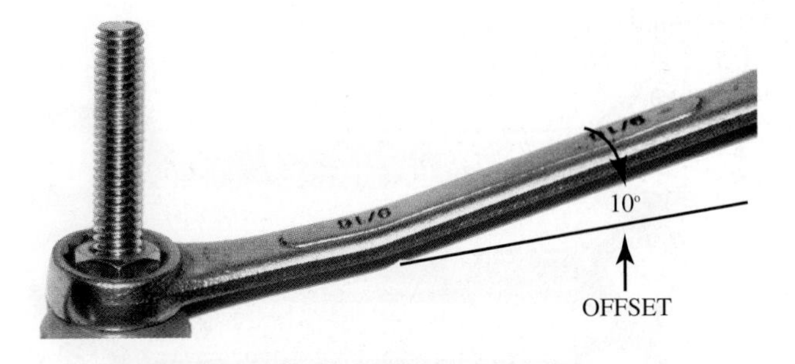

Figure 2-1-14 Offset box end wrench.

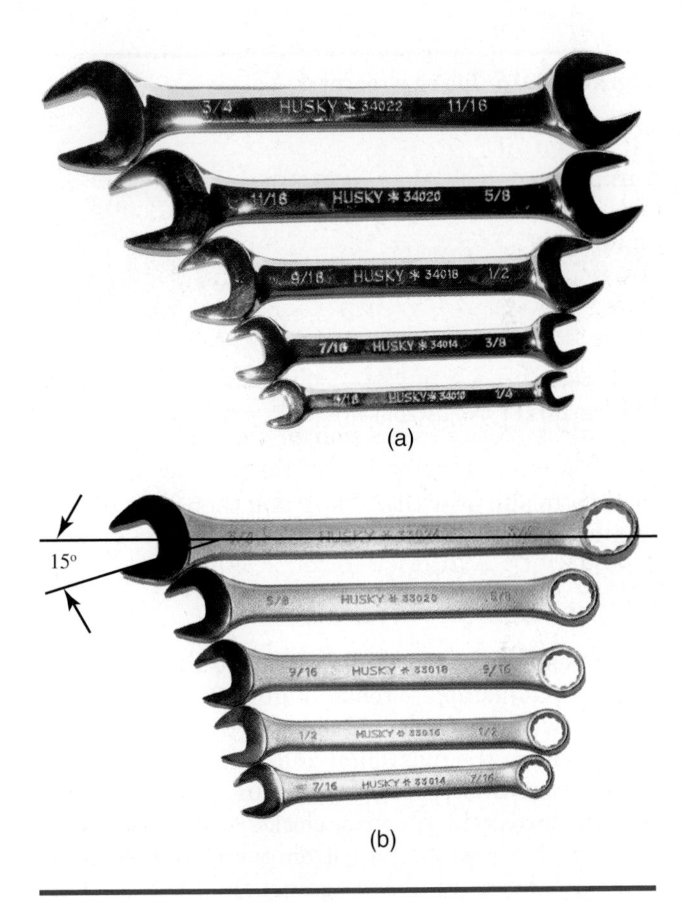

Figure 2-1-15 (a) Open end wrench with 15-degree offset openings, both ends; (b) Open end wrench with 15- and 75-degree openings.

increase the torque slightly until the required torque is reached.

The dial on the wrench indicates the amount of pressure being applied to turn the bolt. Torque is a twisting action. Torque is measured in inch pounds (in lb), foot pounds (ft lb) or Newton meters (Nm) as applied to the wrench handle.

Box End and Open End Wrenches

Box end wrenches, Figure 2-1-14, are strong and are resistant to slipping. They are ideal for working on long shaft bolts where a socket would not fit. The open end wrench is excellent for work on bolts or nuts where access is limited or the end of the bolt cannot be reached, Figure 2-1-15. Some wrench sets, called combination box open end wrenches, have one end open and one box end. All of these types of wrenches may have straight or offset handles.

Flare Nut Wrenches

The flare nut wrench is a special variation of the box wrench in that the heads are slotted to allow the

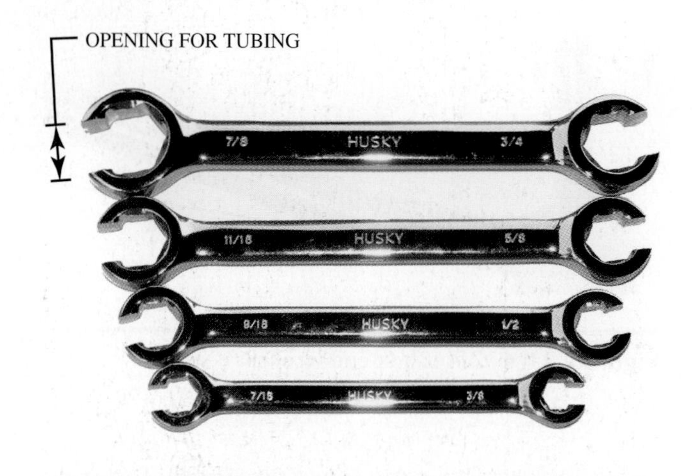

Figure 2-1-16 Tubing wrenches.

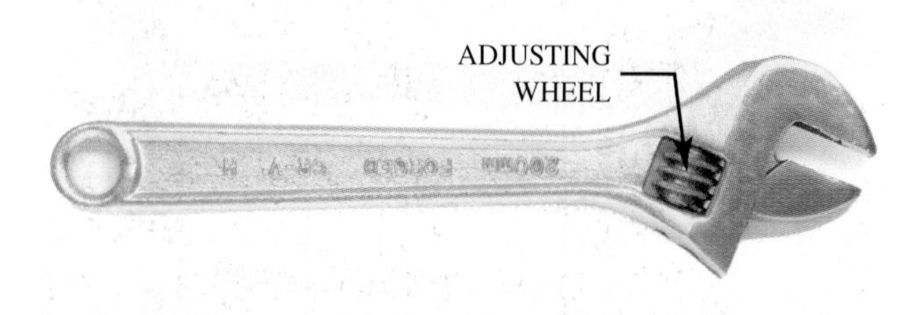

Figure 2-1-17 Adjustable wrench.

wrench to slip over the tubing and then onto a flare nut, Figure 2-1-16.

Adjustable Wrenches

The Crescent Tool Company was the first to introduce the adjustable wrench, Figure 2-1-17. The adjusting wheel permits fitting the flat to any size object within the maximum and minimum opening. Always use this type of wrench in a manner such that the force is in a down or clockwise direction with the movable jaw on the bottom when tightening a bolt. This keeps the force against the fixed head so the wrench is less likely to suddenly slip and injure you, Figure 2-1-18. Adjustable wrenches are available in sizes from 4 in to 12 in (102 mm to 305 mm).

Pipe Wrenches

The pipe wrench is used in refrigeration installation and service work to assemble or disassemble threaded pipe, Figure 2-1-19.

At least two sizes are recommended. An 8 in (203 mm) wrench can handle up to 3 in (76 mm) diameter pipe and a 14 in (356 mm) size can handle up to 8 in (203 mm) diameter pipe. Some have replaceable jaw inserts to extend the life of the tool.

The chain wrench is another form of adjustable pipe wrench. This wrench can make work easier in a confined area or on round, square, or irregular shapes.

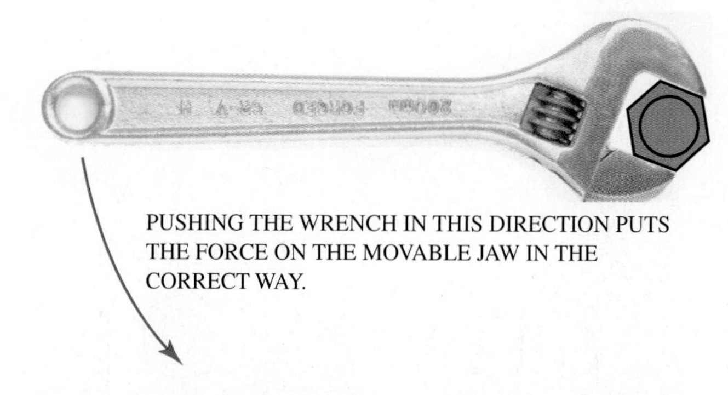

Figure 2-1-18 Always push an adjustable wrench so that the force is correctly applied to the movable jaw and the wrench is less likely to slip.

Figure 2-1-19 Pipe wrench.

Allen Wrenches

Allen wrenches often come in sets. Some are short, referred to as standard Allen wrenches, and others are long. The sets should be marked as to whether they are standard or metric sizes, Figure 2-1-20. The hexagonal shaft of the Allen wrench fits into the Allen screw typically used on fan pulleys, fan blade hubs, and other components. Allen wrenches are also available in a pocket set, Figure 2-1-21. There are specialty Allen wrenches that have balled ends that allow the wrench to be at an angle and still go inside the setscrew. Another specialty Allen wrench is the one used with the refrigerant ratchet wrench to access the service valves on most condensing units, Figure 2-1-22.

SERVICE TIP

Allen wrenches are often used to loosen and tighten pulley setscrews. These screws are often very tight and difficult to remove. Make certain that the Allen wrench tip is fully inserted to the complete depth in the setscrew before trying to tighten it or loosen it. Failure to insert the Allen wrench to its full depth can result in damage to the wrench or to the setscrew.

Figure 2-1-20 (a) Standard size Allen wrenches; (b) Metric Allen wrenches.

Figure 2-1-21 Pocket Allen wrench set.

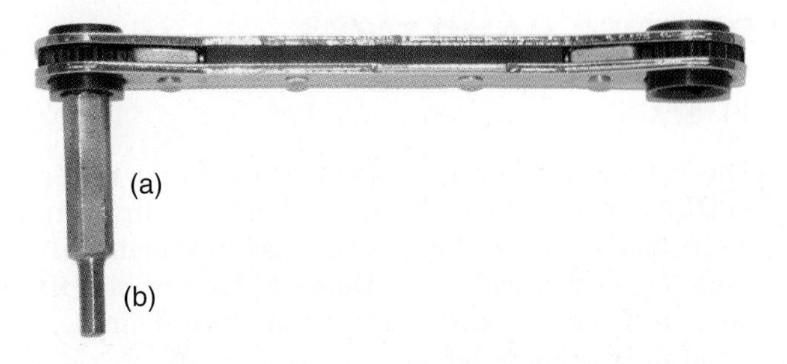

Figure 2-1-22 Allen wrench designed to fit in service ratchet wrench: (a) Part of Allen wrench for accessing most suction line service valves; (b) Section of Allen wrench for accessing liquid line service valves.

Nut Drivers

Nut drivers are like the small socket wrenches Figure 2-1-23. The nut driver consists of a plastic handle that has different drive sockets to fit over the screw or nut head. The nut driver is useful in tightening or removing hex head sheet-metal screws or machine screws that hold equipment panels in place or fasten control box covers. Some nut drivers have magnetic ends that hold the screw head in the nut driver end.

Figure 2-1-23 Nut driver set: (a) Side view; (b) End view. Often the ends of the handles on nut drivers are color coded and identified with the size to make it easier to pick them out of your toolbag.

OTHER HAND TOOLS

Pliers

There are a number of different types of pliers used in HVAC/R work. Some types of pliers are slip joint, multi track and locking pliers, which are primarily used for mechanical work. Diagonal cutters, needle nose, and lineman pliers are primarily used for electrical work, Figure 2-1-24.

Slip joint pliers are handy for general use for holding hot parts or grabbing something that is being removed. Multi track pliers are used like slip joint pliers except they can open wider to hold or grab larger items, such as pipe.

Locking or vise-grip pliers clamp and lock to hold parts securely. Their locking clamp action is beneficial because once the pliers are clamped in place you can let go of them, which frees your hand, Figure 2-1-25.

Electrical pliers, such as diagonal cutters, needle nose, and lineman pliers, can be damaged at the cutting surface if they are used for cutting materials other than copper wire. Cutting a hard material can put a notch in the cutter and/or dull the cutting edge. In either case damaging the cutter can render your pliers useless for HVAC/R work. If you have to cut nails, use a hacksaw or an old pair of pliers, not your good electrical pliers.

CAUTION: Do not use pliers to tighten or loosen brass fittings. No matter how delicately you use pliers on these fittings the fitting surface will be damaged. This damage can prevent the next technician from using the proper wrench on the damaged fitting.

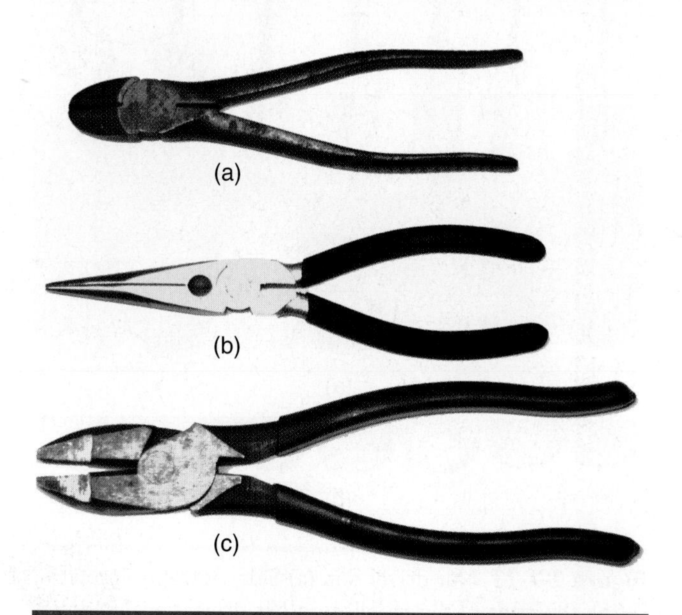

Figure 2-1-24 (a) Wire cutters; (b) Needle nose pliers; (c) Lineman pliers.

Screwdrivers

Flat-blade and Phillips tips are the most common types of screwdrivers, Figure 2-1-26. It is important to have a tight fit between the screw and screwdriver tip, otherwise it can be possible to strip the screw slot. Six-way screwdrivers have two sizes of flat tips and two sizes of Phillips tips as well as two sizes of nut drivers, Figure 2-1-27. These are becoming very popular for HVAC/R service work.

Screwdrivers are versatile tools because, in addition to their primary purpose, they can be used (with discretion) for light duty prying, wedging, or scraping. However, screwdrivers should never be pounded with a hammer.

Brushes

The use of a wire brush is recommended to clean the inside of tubing and fittings. These brushes range in size from ¼ in to 2⅛ in (25 mm to 54 mm) to fit the outside diameter (OD) of the soldered fitting.

A solder flux brush is also recommended in applying paste. A paintbrush is useful to brush dirt or dust out of a control box, for example, or it may be used to apply cleaning solvent to an object.

Files

Files come in several shapes: flat or rectangular, round, half round, triangular, square, etc. In refrigeration

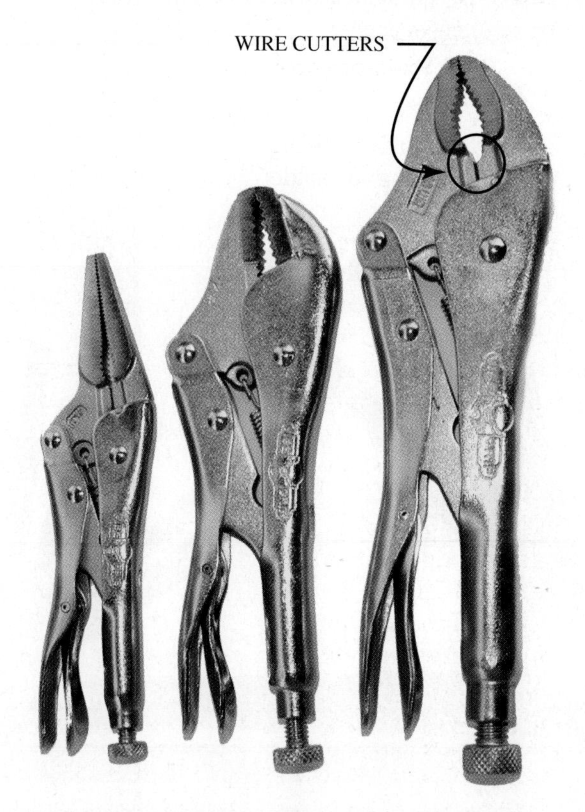

Figure 2-1-25 Locking pliers.

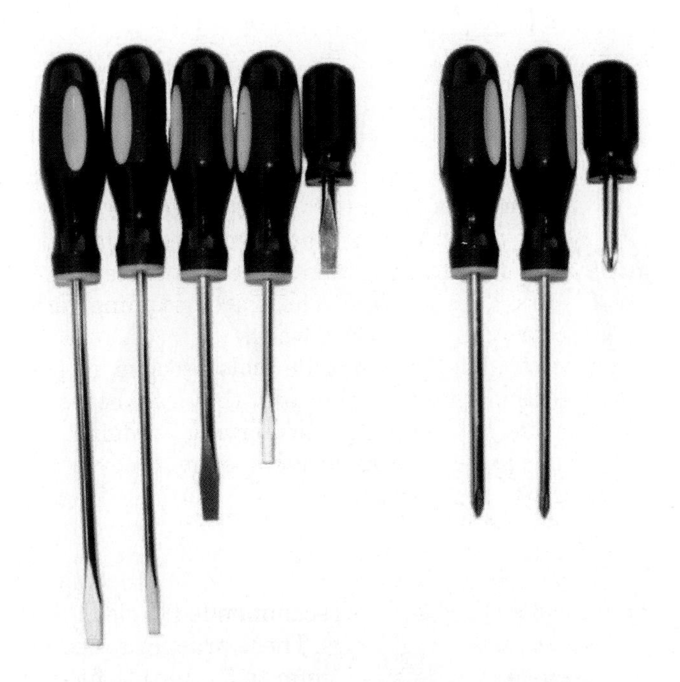

Figure 2-1-26 Screwdrivers come in a variety of sizes and lengths.

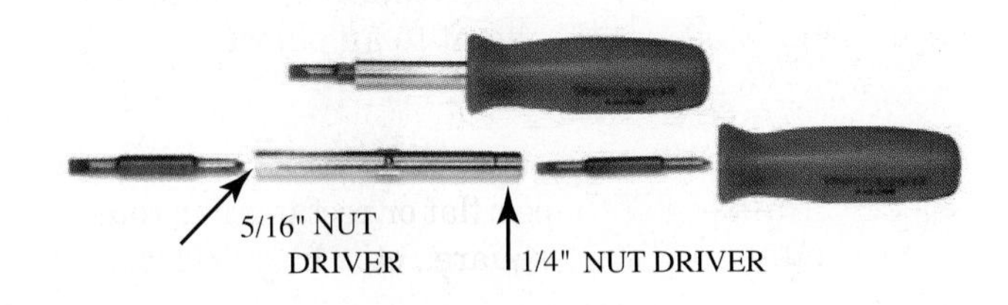

Figure 2-1-27 Six-in-one screwdriver.

work the common flat file and the half round are used in preparing tubing for soldering by squaring the end or for removing burrs.

Files can be either single-cut or cross-cut, Figure 2-1-28. A single-cut file is used for finishing a surface such as in preparing copper pipe for soldering. A cross-cut file is coarser and would be used where deeper and faster metal removal is needed. A rasp is an extremely coarse cross-cut file intended for very rough work.

Vises

A machinist's vise, Figure 2-1-29, can be very useful when mounted in a service van. The pipe vise on a tripod stand, Figure 2-1-30, is a requirement both in the field and in the shop to hold tubing or pipe while cutting or threading operations are taking place.

CAUTION: When holding copper or other soft metal, be sure that the jaws are soft so as not to mar the surface and that the clamping pressure does not squeeze the tubing out of round.

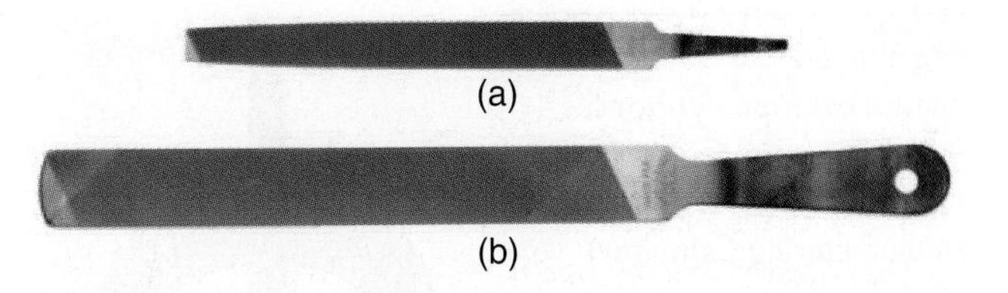

Figure 2-1-28 (a) Single cut file; (b) Cross cut file. Note that file (a) must have a handle attached to the point so it can be used safely. File (b) has a handle formed as part of the file.

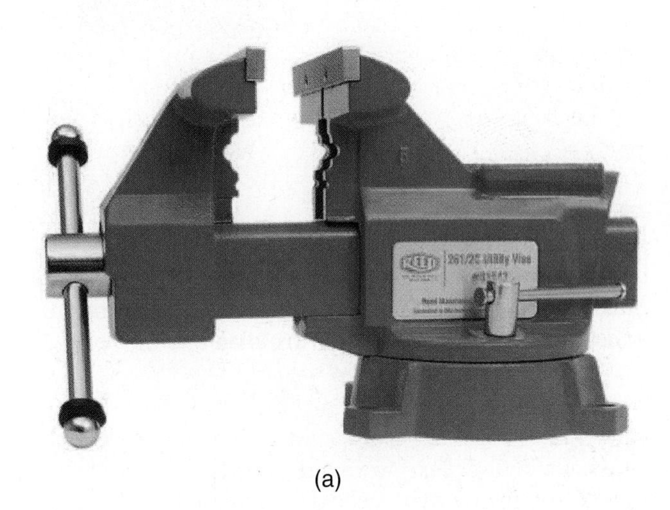

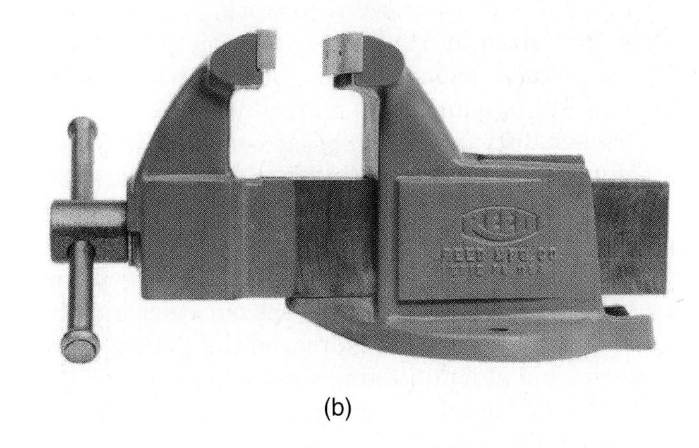

Figure 2-1-29 (a) This vise can be attached permanently to a bench or service van; (b) This vise can be clamped temporarily to any substantial base. (*Courtesy Reed Manufacturing Company*)

Measuring Tapes and Rulers

The familiar 10- or 12-ft (2- or 3-m) flexible steel tape is invaluable for measuring tube diameters, short tube lengths, filter sizes, and duct sizes. A 50- or 100-ft (15- or 30-m) steel or fiberglass tape is essential for measuring long piping runs. Fiberglass tapes are

Figure 2-1-30 (a) Pipe clamp; (b) Pipe support; (c) Universal base with chain pipe clamp. (*Courtesy Reed Manufacturing Company*)

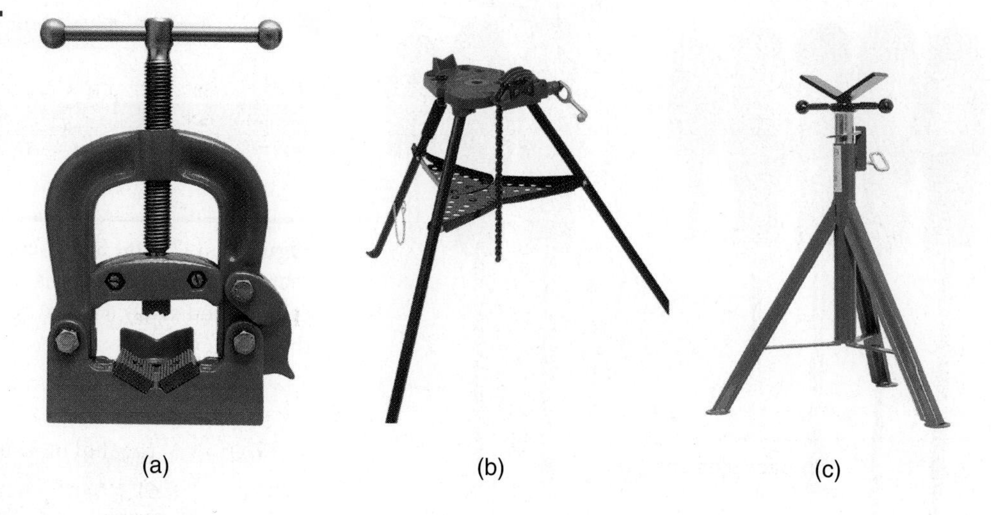

recommended when working near electrical equipment as they are nonconductive. A non-rusting steel machinist's ruler for more precise measurements and a 6 ft folding wood ruler are also useful.

Drills

Drills are frequently used to both drill holes and drive screws. Cordless drills with interchangeable batteries are ideal for HVAC/R work, Figure 2-1-31. Drills are sized by the largest size bit that will fit into the chuck. The most common size drills are the ¼ in and ⅜ in (6 mm or 10 mm) chuck size.

Heavy-duty drilling in brick, concrete, and thick steel is often required during installation work. For such operations a ½ in (12 mm) heavy-duty model drill is recommended.

A useful drill should have a variable speed control plus a reversing switch to back out stuck bits. Bit selection will depend, of course, on the nature of the material, but at a minimum, a set of high-speed alloy steel bits for metal is a must. Such bits can also be used for drilling wood and plastic. However, masonry bits and wood-boring bits may be added for more intensive job requirements.

Rechargeable, battery-operated drilling equipment can be useful time savers where some light drilling or driving of screws is required and no power outlet is handy. These units typically have variable speed controls and a reversing switch.

For more than occasional drilling of concrete or masonry for anchors and shields, nothing beats a rugged rotary hammer drill with the proper carbide masonry bits. The savings in time can pay for the tool in a short time.

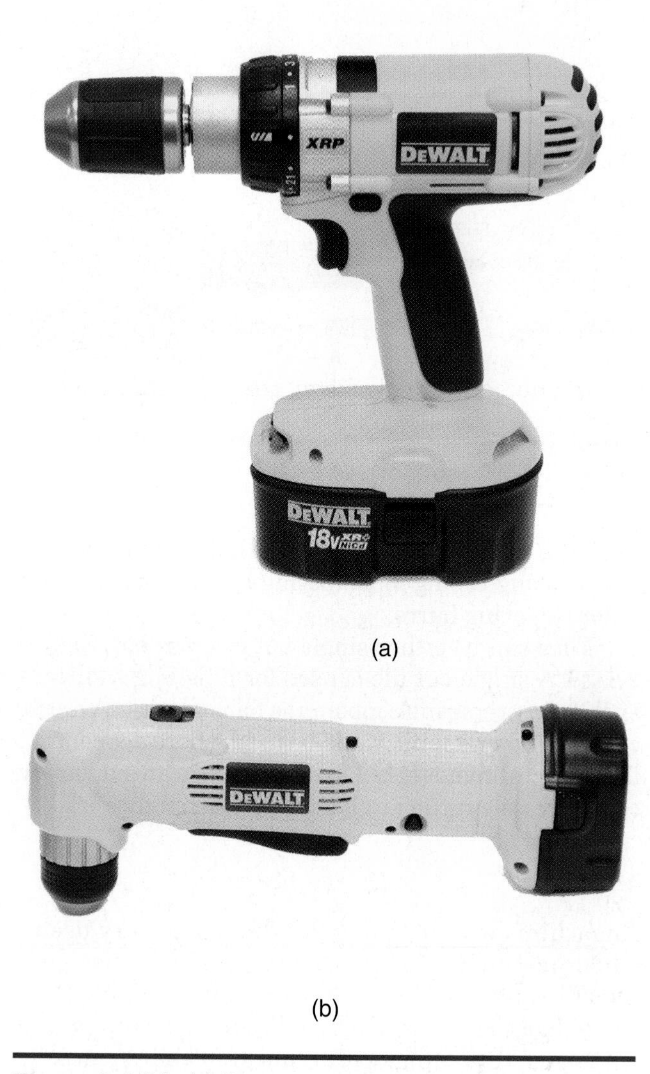

Figure 2-1-31 (a) Battery operated electric drill; (b) Right angle battery operated drill.

Figure 2-1-32 Prepackaged toolkit.

TECH TIP

Power converters are devices that plug into a vehicle's 12 V outlet. They convert 12 V DC to 120 V AC. Power converters are excellent additions to your service vehicle tools because they allow standard portable drill battery chargers to be used during the workday to keep your spare battery charged.

Tool Kits

Tool manufacturers package tools in kits, Figure 2-1-32, for easier storage and greater convenience.

TECH TALK

Tech 1. That list of tools you gave me has some really expensive tools on it. I was able to find some similar tools on the discount tool table down at the hardware store.

Tech 2. Tools are expensive but you get what you pay for. I have had some of my tools for years. In fact most of the tools I have I got years ago when I first got into the HVAC field.

Tech 1. Wow, you mean your tools have lasted that long? One of the screwdrivers I got bent the first time I tried to use it.

Tech 2. I'm not surprised. Quality tools, if you take care of them, will last you as long as you are in the trade. You don't have to buy the most expensive tools, but you do have to buy good quality tools.

Tech 1. I may take some of those tools back that I haven't used and go invest in some better quality tools.

Tech 2. Sounds like a good idea to me.

UNIT 1—REVIEW QUESTIONS

1. The customer often judges the technician on the appearance of ________.
2. The term ________ is used to describe tools that are used to grip and turn threaded parts such as nuts, bolts, valves, and pipes.
3. The handles on service valve wrenches can be ________ or ________.
4. Sockets usually come as sets as either ________ or ________.
5. What are the two dimensions used in referring to socket handles?
6. What does the dial on the torque wrench indicate?
7. The ________ wrench is used in refrigeration installation and service work to assemble or disassemble threaded pipe.
8. The ________ is useful in tightening or removing hex head sheet metal screws or machine screws that hold equipment panels in place or that fasten control box covers.
9. Which pliers are handy for general use for holding hot parts or grabbing something that is being removed?
10. What are the most common types of screwdriver?
11. What is the recommended use for a wire brush?
12. List six shapes that a file comes in.
13. What is a rasp?
14. A ________ can be very useful for determining the outside diameter of tubing or a fan shaft, which would be difficult to measure otherwise.
15. ________ are frequently used by the refrigeration mechanic in the work of installation and repair.

UNIT 2

Electrical Measuring and Testing Devices/Meters

OBJECTIVES: In this unit the reader will learn about various electrical and electronic instruments and how they are best used including:

- Testing Electrical Circuits
- Descriptions of Electrical and Electronic Instruments
- General Principles in the Use of Meters
- Temperature Measuring Instruments
- Leak Detectors

ELECTRICAL AND ELECTRONIC INSTRUMENTS

Testing electrical circuits is an important skill that each technician needs to develop. Although practice and experience are significant, a high degree of success is obtainable by following a proven procedure, such as the following:

1. Know the unit electronically. This means understanding the proper function of each control and the sequence of the control operation.
2. Be able to read schematic wiring diagrams and have them available.
3. Be able to use the proper electrical test instruments. Know the instrument. Read instructions carefully before using.

In the following section, various electrical and electronic instruments will be described along with information on how they are best used. For electrical troubleshooting, the instruments most commonly used are:

1. A voltmeter to measure electrical potential
2. An ammeter to measure electrical current
3. An ohmmeter to measure electrical resistance
4. A megohm meter to measure very high electrical resistance
5. A wattmeter to measure electrical power
6. A capacitor checker to measure electrical capacitance

TECH TIP

Electronic meters are expensive and relatively easily damaged if they receive rough treatment. To ensure that your instruments last as long as possible they should be cleaned after each use and stored in their own individual case or pouch. It is also a good idea to carry your instruments in a separate bag away from your hand tools.

General Principles

Before specific types of meters and their applications are described, a number of general principles should be observed in the use of meters, as follows:

1. Always use the highest scale on the meter first then work down to the appropriate scale. This prevents damaging a meter by applying excessive power. An autoranging voltmeter will do this automatically.
2. Always check the function of a meter before using it. Don't just assume a meter is working. If it is a battery operated meter, the batteries may be run down. The meter could be damaged during transportation. Other things could happen to affect the readings.
3. In using a clamp-on ammeter, be sure the jaws are around only one wire. If it is around two "hot" wires, the reading could be meaningless. Start at the high range and work down as described above.
4. Always have an extra set of meter fuses on hand for replacement. Sometimes the proper fuses are difficult to obtain when you need them. The same applies for batteries.
5. Never use an ohmmeter in a circuit that is powered. An ohmmeter has its own power supply and can be destroyed by connecting to a live power source.
6. Some tests require the use of an adapter for measuring temperature with a thermocouple, thermistor, or remote temperature device (RTD). Be sure an adapter and accessory sensors are included with the testing tools.

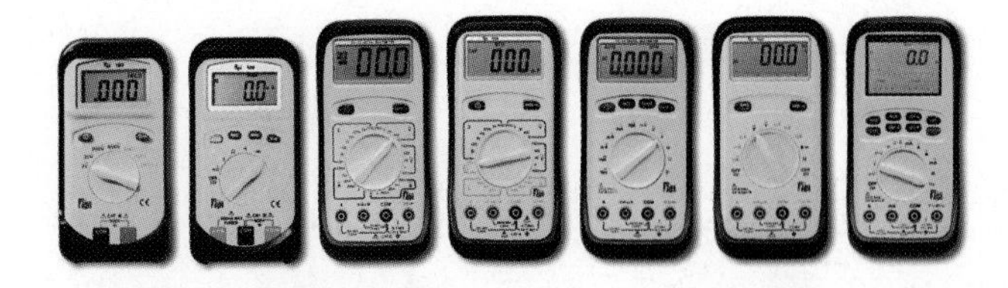

Figure 2-2-1 Multimeters come in a variety of sizes with a variety of functions. (*Courtesy Test Products International, Inc.*)

Multimeters

Many common meters combine a number of functions. This is particularly true for meters measuring volts, amperes, and ohms. These are called multimeters and they are an essential tool for the technician, Figure 2-2-1.

The displays on multimeters can be either digital or analog. Digital meters display a number in a digital display. Many have autoranging so they will automatically select the proper scale to display the voltage or resistance values being tested. They are easily read and the most popular types in use.

Analog meters use a mechanical needle that swings across the dial to a point over a scale, Figure 2-2-2. The meter reading is taken by looking directly at the face of the meter dial and comparing the needle location to the appropriate scale. Analog meters are almost never autoranging. If too high a voltage is being tested for the scale setting, then the meter can be damaged.

The leads that are used with meters should be of the type that has a protective covering over the ends so they do not accidentally short out, Figure 2-2-3. The spring loaded covering will slide back as the lead is used in confined test points. The meter ports should also be recessed so they can not be accidentally contacted by your hand while using the meter, Figure 2-2-4.

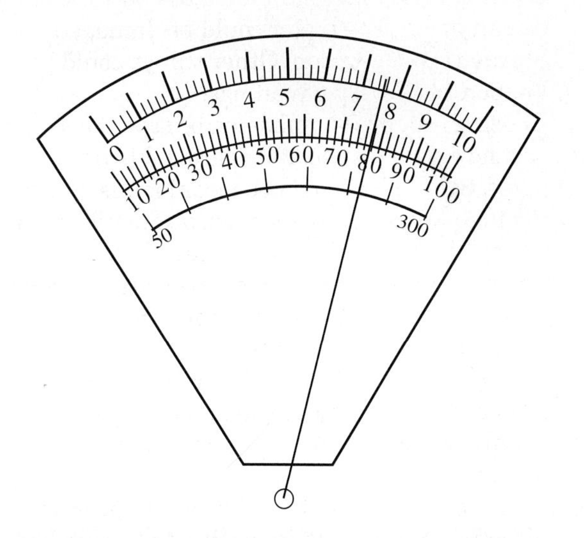

Figure 2-2-2 An analog multimeter has several scales, so you must know which range you are measuring and read the appropriate scale.

Figure 2-2-3 Safety multimeter leads. (*Courtesy Test Products International, Inc.*)

Clamp-on Ammeters

The clamp-on ammeter is designed for measuring current flow through a single wire, Figure 2-2-5. It ranks with the multimeter as an essential tool for the technician.

The clamp jaws can be opened and placed around the wire to be measured, Figure 2-2-6. They close automatically by spring pressure when the trigger is released. The range scale is also positioned by a thumb wheel revolving the scales viewed through the window. Some ammeters also have volt and ohm scales. To use these functions leads are inserted in the instrument.

Selecting the Proper Instrument

The three instruments commonly used for electrical troubleshooting are the voltmeter, ammeter, and ohmmeter. The following guidelines will help the technician decide which one of these instruments to use:

1. If any part of the unit operates, the voltmeter and ammeter are normally used.
2. If no portion of the unit operates, check to make sure power is present at the unit terminals. If present, a short circuit could be the problem and after disconnecting power the ohmmeter is probably the best instrument to use.

SERVICE TIP

Most clamp-on ammeters cannot accurately read low amp draws. A small loop of thermostat wire can be made to enable the meter to accurately read these lower amp draws. To make this loop take a single strand of thermostat wire and make a round loop with a diameter of approximately 1½ in. Make 10 complete wraps of thermostat wire. Use several pieces of black electrical tape to ensure the coil stays tightly packed together so that it is not easily snagged in your tool pouch. To use this amperage multiplier connect the loop wire ends into the circuit so that it is in series where all of the amperage flowing through the circuit will then pass through the 10 turns of the amperage multiplier. Clamp the amperage ammeter jaws through the 10 turn loop. The amperage reading shown on the ampmeter will be 10 times higher than the actual amp draw. For example an amperage reading of 15 would indicate an actual amp draw of 1.5 amps. This amperage multiplier will make it possible to read the very low amp draw of a gas valve so that the thermostat heating anticipator can be accurately set.

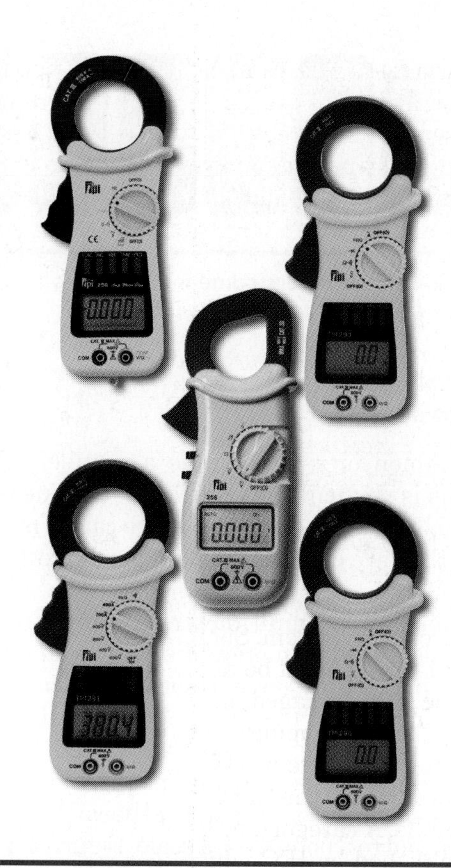

Figure 2-2-5 Clamp-on ammeters.

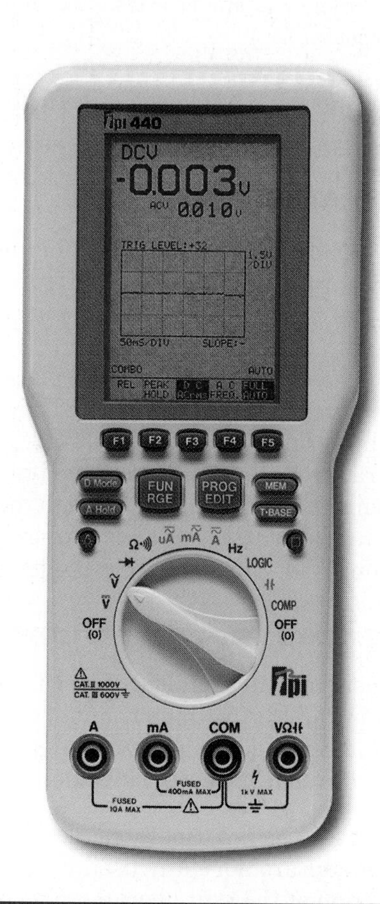

Figure 2-2-4 The safety terminals at the bottom of this multimeter are recessed, so they will not accidentally cause a short circuit. (*Courtesy Test Products International, Inc.*)

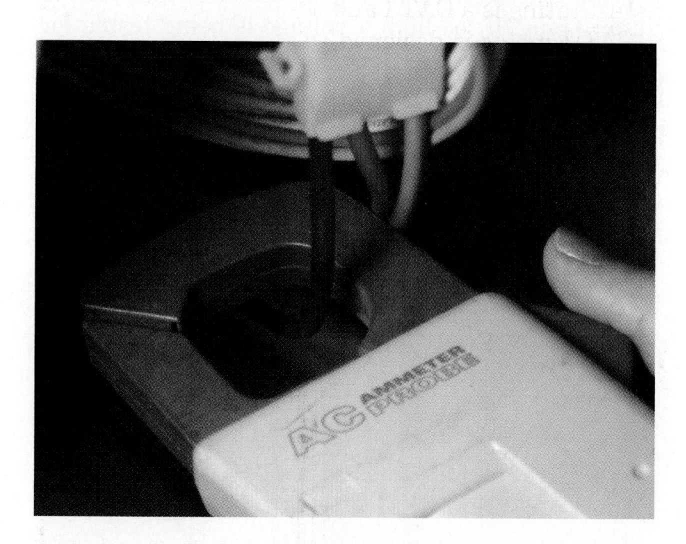

Figure 2-2-6 Measuring amperage with a clamp-on ammeter. (*Courtesy Test Products International, Inc.*)

Wattmeter

This instrument reads true power, including an allowance for the power factor. The meter makes the necessary calculations for power, in accordance with the power formula:

$$\text{watts} = \text{volts} \times \text{amps} \times \text{power factor}$$

The wattmeter zeros itself when not in use and automatically selects the proper range when used. It can be used to measure watts on single, split-phase, and three-phase power sources.

To mathematically calculate watts using the measured volts and amps, it is necessary to know the power factor.

Overvoltage Protection

Any time you are working with electricity there is a possibility of an overvoltage situation occurring. We use surge protectors on our computers to protect them from overvoltages. A momentary surge or spike of voltage can be several times the normal circuit voltage. These surges, called transient voltages, can be caused by electrical storms, even miles away, or they can be caused when a major load is added or dropped from the system. Some of the things causing shifts in the load could be an auto accident where a utility pole was damaged, or a large piece of equipment starting or stopping.

The closer you are working to the power pole or main power box, the more dangerous transient spikes can be. A category, or CAT rating system, has been established by IEC which provides the technician with information about the level of safety provided by their meter. Not all meters are rated. The lowest rating is a CAT I and the highest is a CAT IV. A CAT I meter is designed to be used on protected equipment like a gas furnace in a hallway, equipment room, or attic. CAT II meters can be used on backup heat strips found in the indoor part of a heat pump. CAT III meters can be used on condensers. CAT IV meters can be used on any branch circuit inside or outside of the building. The CAT rating on a meter will list the maximum circuit voltage that the meter can be used on, Figure 2-2-7.

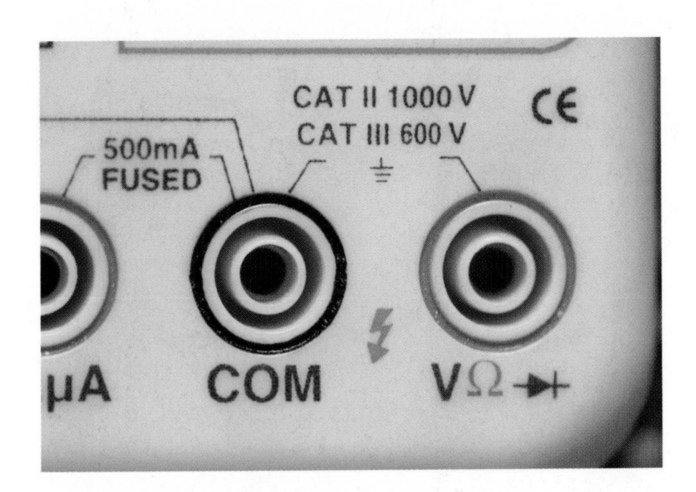

Figure 2-2-7 Category ratings of two voltage levels, CAT II 1000 V, and CAT III 600 V.

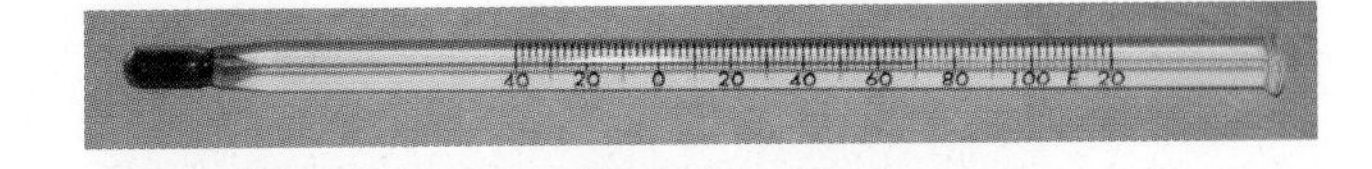

Figure 2-2-8 Glass tube thermometer.

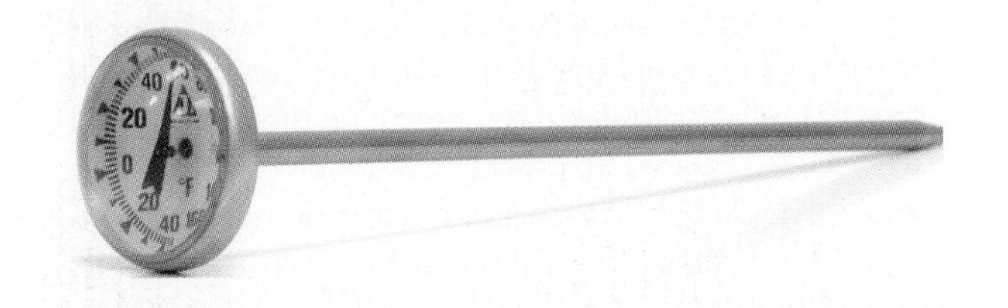

Figure 2-2-9 Pocket thermometer.

REFRIGERATION—SERVICING AND TESTING EQUIPMENT

Digital Thermometer

Temperature measuring instruments can be digital or analog. An example of an analog thermometer is a glass thermometer, Figure 2-2-8. Some pocket thermometers are also analog, Figure 2-2-9.

Digital thermometers come in a wide variety of shapes and with a number of different options, Figure 2-2-10. The units will read temperatures in either Fahrenheit or Celsius degrees. They are battery powered and can may come with a carrying case. Some can take two temperatures and give you the difference, T1 − T2 = TD. Another feature available on some digital thermometers is their ability to record temperatures over time to provide you with the range of high and low temperatures.

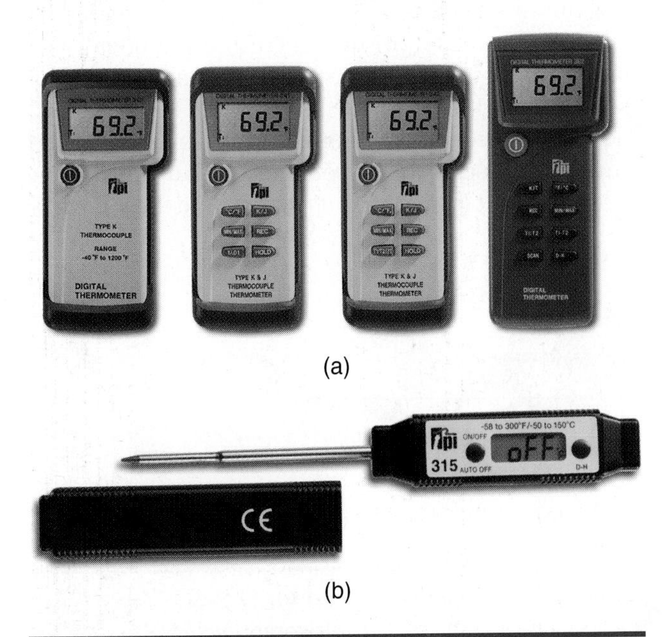

Figure 2-2-10 Digital thermometers. (*Courtesy Test Products International, Inc.*)

TECH TIP

Sometimes the thermocouple leads used with digital temperature gauges may have a slight error in the temperature reading. For that reason it is a good idea to test a new set of temperature probes against each other or against another temperature measuring instrument. This comparison will allow you to determine the accuracy or variance in readings that you may receive from these leads. If you find a variance greater that one degree consistently between any lead and other test instruments you should mark that lead with a small piece of tape with the number of degrees plus or minus that it is off. This will allow you to make corrections in temperature readings when using this probe.

A digital thermometer can be used with a wide variety of probes, three of which are shown in Figure 2-2-11. The different probes allow the single thermometer to be used for more different types of jobs.

Infrared thermometers can tell the temperature of an object some distance away by recording the heat signature they produce, Figure 2-2-12. Every object emits waves of heat energy that a thermal detector can measure. The surface color can affect the temperature reading. Black surfaces give the most accurate infrared readings, and white or reflective surfaces give less accurate readings.

Leak Detectors

Electronic Leak Detector An electronic leak detector is shown in Figure 2-2-13. The electronic leak detector draws air over a platinum diode.

This unit is capable of detecting leaks as low as 0.4 ounce per year. It can be used for HCFC, CFC, and HFC gases. The pump located in the device draws air directly to the sensing tip. No calibration is required. It is battery operated and has both a visual and an audible signal which increase in frequency as the leak source is approached.

Pump-style Leak Detector, Using Corona Discharge Technology This technology creates a high voltage corona between the inner tip and surrounding shell. When refrigerant interrupts the electronic field, the alarm is triggered. The instrument will sense HFC/CFC/HCFC refrigerants by operating a selection switch on the face of the instrument.

Ultrasonic Type Leak Detector An ultrasonic type leak detector instrument, shown in Figure 2-2-14, will detect any gas leaking through an orifice.

The features and specifications for this meter are as follows:

- Detects pressure or vacuum leaks
- Unaffected by windy, roof top conditions
- Unaffected by background noise
- Detects ultrasonic noise from arcing electrical switchgear
- Can be used for finding leaks in ductwork

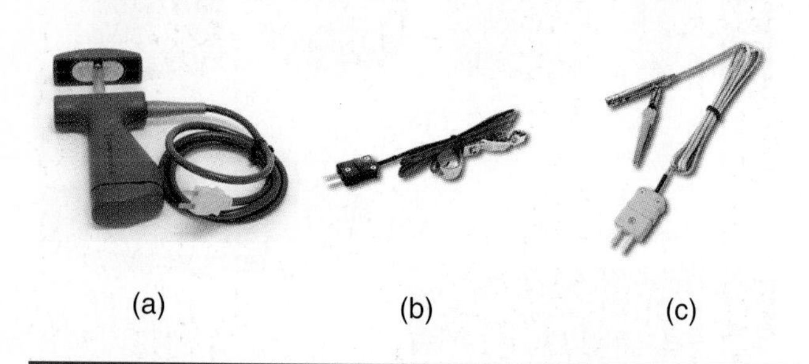

(a) (b) (c)

Figure 2-2-11 Temperature sensor probes: (a) Temperature sensor for refrigerant piping; (b) Ambient temperature sensor with clamp; (c) Flame sensing temperature probe.

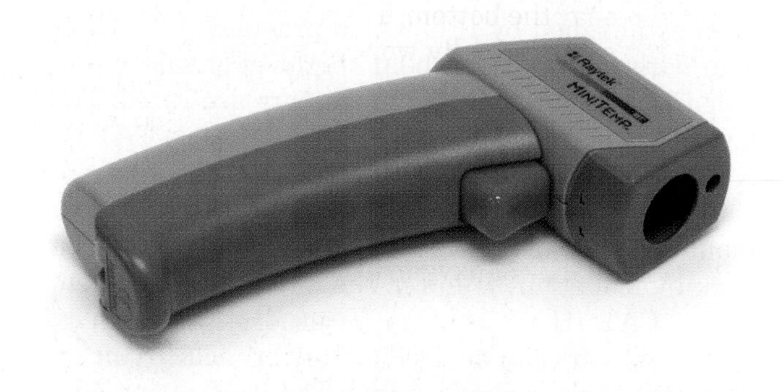

Figure 2-2-12 Infrared temperature measuring instrument with laser sight.

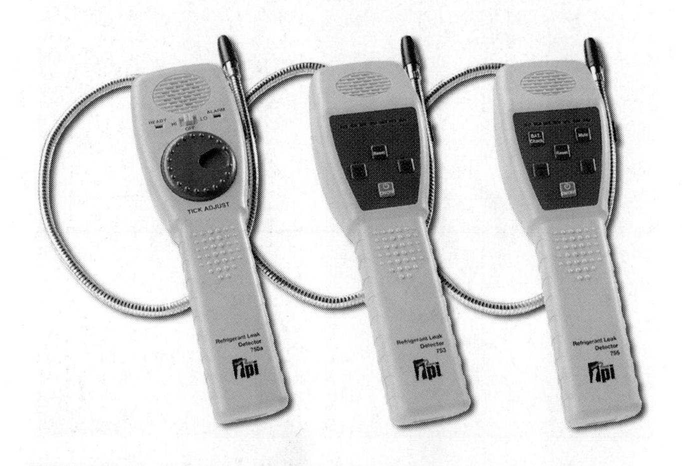

Figure 2-2-13 Electronic leak detectors. *(Courtesy Test Products International, Inc.)*

Figure 2-2-14 Ultrasonic leak detector. (*Courtesy Robinair SPX/OTC*)

Tips for Using a Leak Detector

- In leak-testing a packaged unit, remember that the refrigerant is heavier than air. If there is a leak near the top of the unit, the refrigerant will tend to puddle near the bottom. Therefore, start testing at the top and gradually work your way down.
- If there is a strong wind, such as on the roof of a building, shield the area you are checking to secure a more accurate reading.
- When working with refrigerants, always be sure there is adequate ventilation. Be concerned about the possibility of oxygen depletion.

TECH TALK

Tech 1. I don't know why my multimeter has anything other than the continuity check. That is all I ever use.

Tech 2. Do you use that when you are trying to get an ohm reading?

Tech 1. Sure, I use it for everything. It gives you an ohm reading.

Tech 2. Actually, some meters don't give you an accurate ohm reading when you have it set on continuity. After all it is not trying to give you an accurate ohm reading but simply tell you that the circuit has continuity.

Tech 1. You mean that reading could be off?

Tech 2. Yes, it could be and probably is. Most multimeters are not designed to give you an accurate ohm reading when they are set on continuity.

Tech 1. Gosh, I didn't know that. Thanks, I appreciate that.

UNIT 2—REVIEW QUESTIONS

1. _______ is an important skill that each technician needs to develop.
2. List the instruments most commonly used for electrical troubleshooting.
3. What is a wattmeter used to measure?
4. What is an ohmmeter used to measure?
5. Some tests require the use of an _______ for measuring temperature with a thermocouple, thermistor, or remote temperature device.
6. List the features of the digital multimeter illustrated in Figure 2-2-1b.
7. What is the clamp-on ammeter designed to measure?
8. What are the three instruments commonly used for electrical troubleshooting?
9. To calculate watts, given the volts and amps, it is necessary to know the _______.
10. What category meter can be used to check continuity in fuses, wire resisters, transformer coils, alarm circuits, and relays?
11. An electronic leak detector can be used to detect _______ gases.

UNIT 3

Refrigerant Servicing and Testing Equipment

OBJECTIVES: In this unit the reader will learn about refrigerant servicing and testing equipment including:

- Vacuum Pump
- Gauge Manifold Sets
- Charging Cylinder
- Charging Scale
- Micron Vacuum Gauge

VACUUM PUMP

A typical high-performance vacuum pump is capable of a rapid pull down and thorough evacuation, Figure 2-3-1. Vacuum pumps are rated by the volume of air they can pump each minute. The rating is given in cubic feet per minute (CFM). The higher the CFM rating the faster the pump can pull a vacuum. Pumps that have rating around 4 to 6 CFM are acceptable for central residential AC work. Pumps having a lower rating are very well suited for small appliance work and larger pumps with a higher rating are ideal for commercial air conditioning and refrigeration. Having a vacuum pump that is too large for the job will be excessively expensive and more difficult to carry to the job site. A vacuum pump that is too small for the job will take much more time to evacuate the system.

Figure 2-3-1 Vacuum pump. *(Courtesy Yellow Jacket Division, Ritchie Engineering Company)*

SERVICE TIP

A micron gauge can be used as a very effective leak check for HVAC refrigerant systems. When a system has been evacuated down to 50 microns, isolate the system from the vacuum pump by shutting off the service valve to the vacuum pump. By watching the micron gauge for any release in pressure, leaking systems can be identified. One note, however, is to be certain that all of your connections are tight and leak free. The micron leak check will be adversely affected by any leak, whether it is in the refrigerant system or in your gauges or hoses.

Deep vacuum pumps, sometimes called high vacuum pumps, should be able to pull a vacuum down to 50 microns. However, even the ability of high quality vacuum pumps to pull a deep vacuum will be affected by air and moisture dissolved in the oil. Following the pump manufacturer's recommendations on oil changes will reduce wear and prevent moisture-laden oil from remaining in the pump. Moisture gets into the pump's oil as a normal part of its use. As the air is pumped out of the system, moisture from the system is also pulled out. Over time that moisture will collect in the oil. When a deep vacuum is being pulled, the boiling out of the moisture in the vacuum pump oil can delay or stop the pump from reaching the desired vacuum level.

GAUGE MANIFOLD SETS

Gauge manifold sets are often referred to as AC gauges or just gauges. They are available in a variety of styles as shown in Figure 2-3-2a–c. The combination low side gauge in Figure 2-3-2a reads from 30 in Hg to 350 psi. The high side gauge reads from 0 to 500 psi. Manifold gauge sets are available in both standard and metric scales.

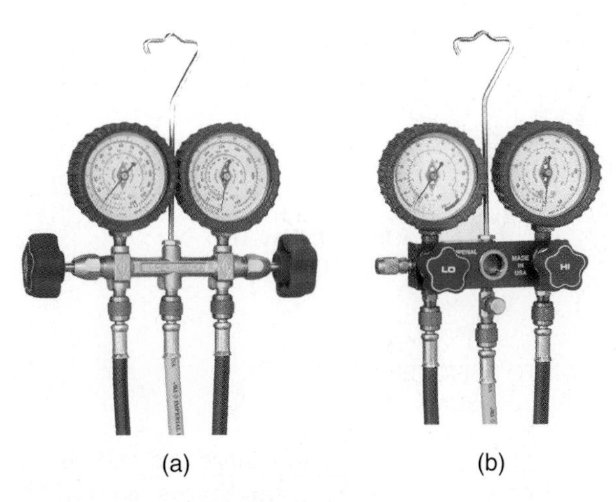

Figure 2-3-2 Gauge manifold sets: (a) Standard gauge manifold set; (b) Six quart manifold set; (c) Six quart, four valve manifold set. *(Courtesy Imperial)*

These liquid filled analog gauges are filled with glycerin to dampen pulsations, lengthen service life, and improve accuracy. The low-side gauge in Figure 2-3-2b reads pressures from 30 in Hg to 120 psi. The four-way block in Figure 2-3-2c is equipped with three refrigerant hoses and one evacuation hose.

The gauge manifold is used for checking operating pressures, adding or removing refrigerant, adding oil, and performing other necessary operations.

The bottom of the manifold has three connections. High vacuum hoses, usually capable of being leak-tight down to 50 microns or less, are attached to these openings. The left hose is connected to the low side of the refrigeration system being serviced. The right hose is connected to the high side of the system. The center hose has a number of uses, all associated with servicing the system.

TECH TIP

Manifold gauge sets have screw on fittings for the system end of the hoses to be attached when the gauges are not in use. It is important that you attach the hose end to these fittings every time you finish a job. These fittings are there to keep dirt out of the hose end because any dirt that gets into the hose end will be pushed into the next system you access.

Refrigerant hoses are available in different lengths and colors, as shown in Figure 2-3-3. The longer hoses give you more working room but are more expensive and can get in the way in small work areas. Hoses are available with check valves to prevent venting of refrigerant, as shown in Figure 2-3-4.

The exteriors of gauge manifolds are often color coded. The compound gauge and low side hose are blue. The high side gauge and high side hose are red. The center utility hose is usually white or yellow. The center hose, Figure 2-3-5a, is useful for connecting

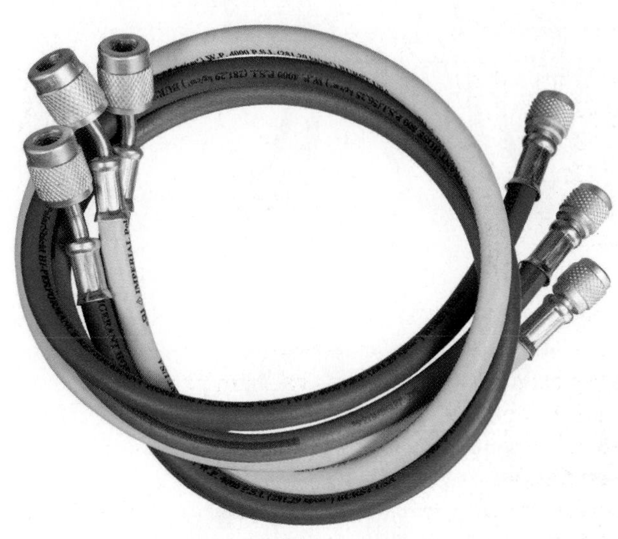

Figure 2-3-3 Refrigerant hoses. *(Courtesy Imperial)*

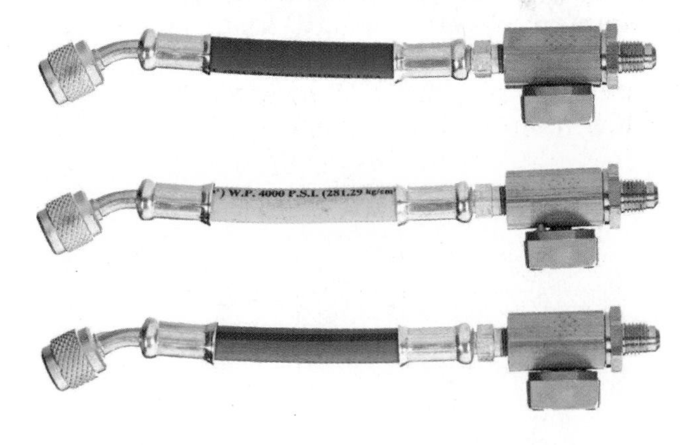

Figure 2-3-4 Refrigerant hose extensions with mechanical shutoff valve. *(Courtesy Imperial)*

to the charging refrigerant cylinder, vacuum pump, or recovery machine.

The hand wheels that control the valves on the end of the manifold can be set in one of three positions; fully closed, fully open, or partially opened. When both hand wheels are closed and the hoses are connected to

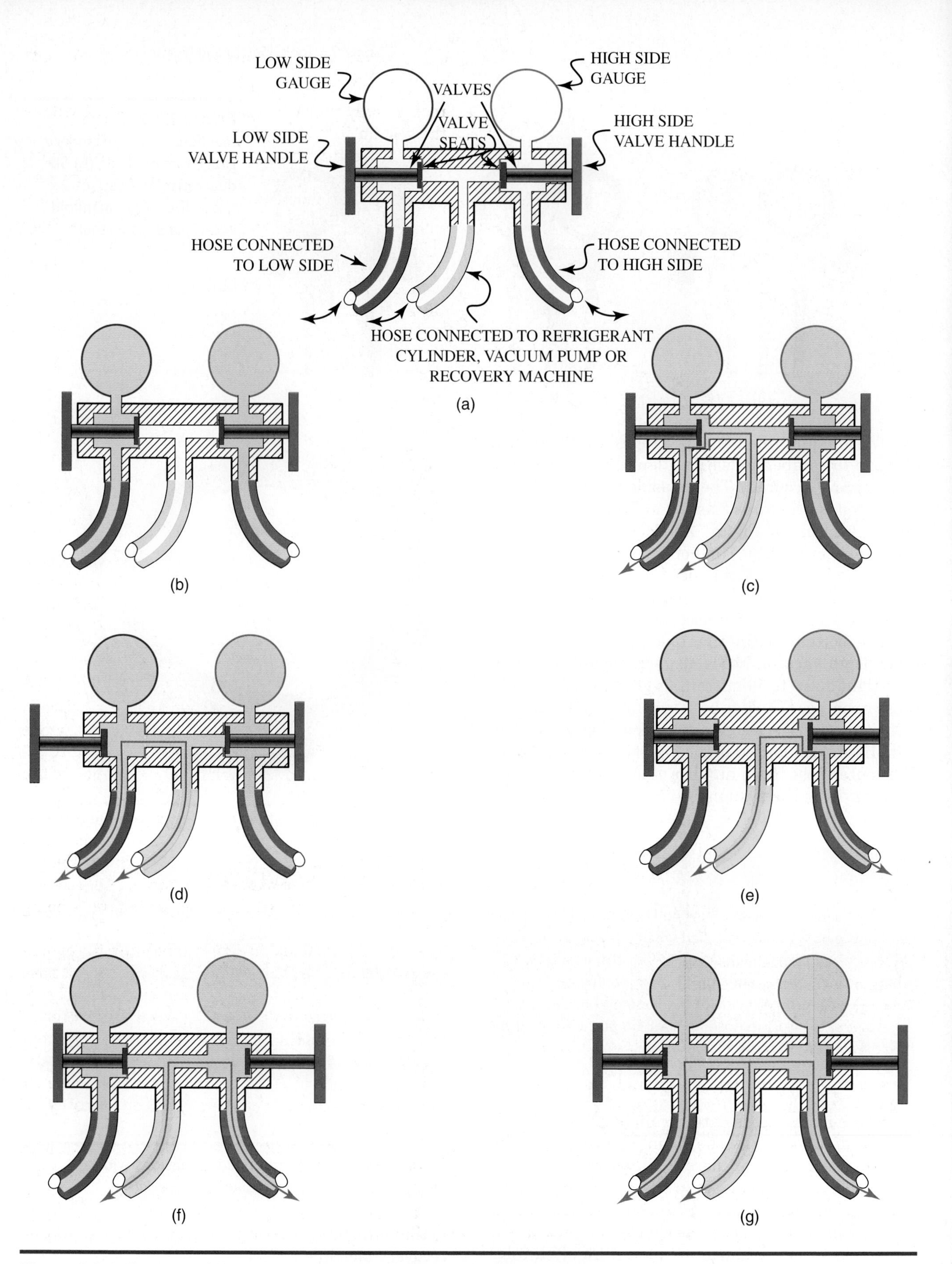

Figure 2-3-5 Function of the gauge manifold set: (a) Parts of the gauge; (b) Area pressurized with both valves closed; (c) Area pressurized with low side valve slightly open; (d) Area having full flow with low side valve open; (e) Area pressurized with high side valve slightly open; (f) Area pressurized with high side valve completely open; (g) Area pressurized with both low and high side valves open.

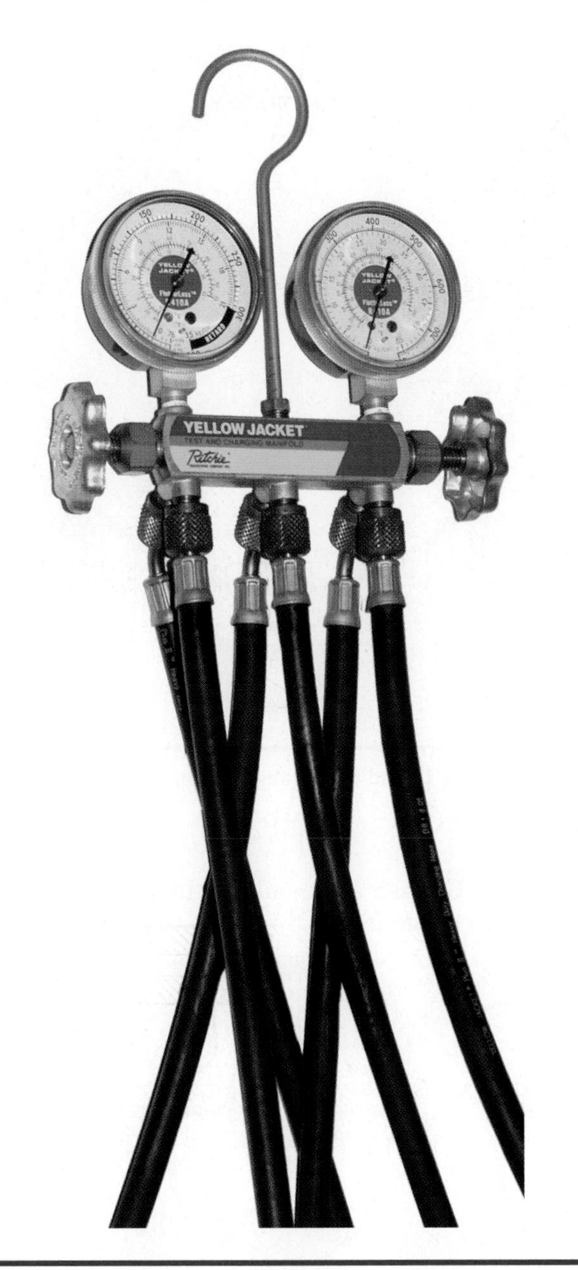

Figure 2-3-6 Gauge manifold set specifically designed for vacuum. *(Courtesy Yellow Jacket Division, Ritchie Engineering Company)*

the system, the low side gauge will indicate the system low side pressure and the high side will indicate the system high side pressure, Figure 2-3-5b.

When the low side hand wheel is turned open slightly a partially open path around the valve and valve seat is opened from the low side to the center hose, Figure 2-3-5c. This is the way a system can be charged with refrigerant. The slight opening allows the technician to control the flow of refrigerant into the system.

As the valve is fully opened, free flow between the low side and the center hose can occur, Figure 2-3-5d. This is the position that would be used to recover only vapor refrigerant from the low side of the system.

When the high side hand wheel is turned open slightly, a partially open path around the valve and valve seat is opened from the high low side to the center hose, Figure 2-3-5e. This is how the valve is operated when a little liquid refrigerant is being removed from the system. The slight opening allows the technician to control the flow of refrigerant out of the system.

TECH TIP

The small O ring in the end of the service valve hose is a replaceable part. These O rings will become damaged over time and will need to be replaced to prevent refrigerant and vacuum leaks from occurring at the service valve connections. It's a good idea to carry a small package of these replacement seals with your tools because from time to time the O rings will need to be replaced and occasionally one may fall out. When the O ring falls out of the hose end the hose cannot be used until a new O ring has been installed.

As the valve is fully opened, free flow between the low side and the center hose can occur, Figure 2-3-5f. This is the position that would be used to recover only liquid refrigerant from the low side of the system.

When both hand wheels are fully opened a free flowing path between the low side, high side, and center port is provided, Figure 2-3-5g. This is the position for the valves when a vacuum is being pulled on the system or when all of the refrigerant is being recovered.

Gauge manifolds designed especially for evacuating and dehydrating a system have larger hose connections and use larger diameter hoses, Figure 2-3-6. The larger, usually ⅜ in, inside diameter of the hoses and valves reduces the pressure drop that occurs when pulling a vacuum. Reducing even the very small pressure drop that occurs in normal gauge manifolds can cut the evacuation time by up to 90%.

Some gauge manifolds have multiple openings on the utility connection to accommodate various devices being used simultaneously. The additional connections allow you to change from evacuation to charging without losing any of the system vacuum or refrigerant.

During operation of an AC or refrigeration system the high side pressure may rapidly pulsate as the compressor valves open and close. This pulsation can cause the high side gauge needle to flutter or vibrate,

Figure 2-3-7 High pressure needle vibrating due to compressor pulsing.

Figure 2-3-8 Liquid filled gauges dampen needle vibration.

Figure 2-3-7. Oil filled gauges are available, and they stop this fluttering from happening, Figure 2-3-8.

SERVICE TIP

The needle does not flutter when the gauges are first attached. That is because the needle flutters when the high side hose and valve get full of liquid refrigerant. Liquid refrigerant is drawn to the cooler hose as vapor condenses in the manifold and hose. Liquids are noncompressible so the pulsing pressure in the high side line is transmitted directly to the gauge when the hose fills with liquid, Figure 2-3-9a. First close the cylinder valve so none of the system refrigerant charge will be lost. Now slightly open the high side valve to let the liquid flow into the center hose, Figure 2-3-9b. This will dampen the needle fluttering for a moment, allowing you to get a pressure reading. After you have the pressure reading, close the high side valve. Next slightly open the low side valve to let the liquid refrigerant that collected in the hose be drawn back into the system through the low side service valve, Figure 2-3-9c. By putting the refrigerant back into the system the refrigerant charge is not changed.

Charging Cylinder

A charging cylinder provides an excellent way to measure a refrigerant charge, Figure 2-3-10. The calibrated shrouds make it simple to compensate for volume fluctuations due to temperature variations. Cylinders with heaters make charging faster and more complete. A pressure relief valve protects all cylinders. Gauges are available in psi, kgcm^2, and kPa. Sizes range from 2½ to 10 lb of refrigerant.

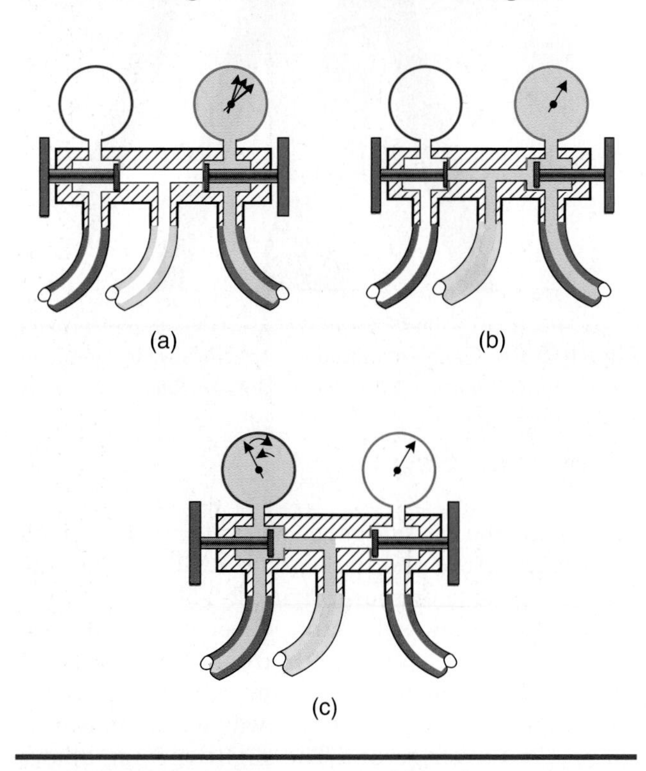

Figure 2-3-9 Illustration of Service Tip: (a) High side gauge needle vibrating due to liquid flooding; (b) Gauge needle dampened as liquid is slowly bled into center hose; (c) Liquid from center hose has bled back into system.

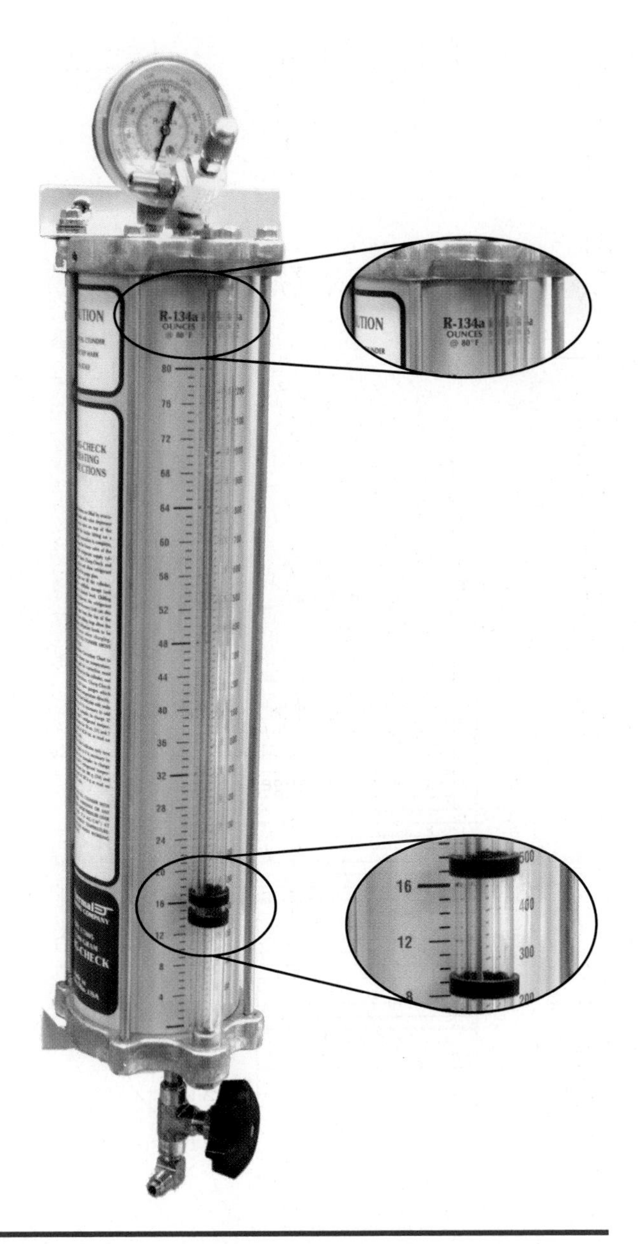

Figure 2-3-10 Charging cylinder. (*Courtesy Robinair SPX/OTC*)

Charging Scale

An electronic charging scale will measure the charge by weight, Figure 2-3-11. It is designed for refrigerant tanks up to 110 lb (50 kg). A solenoid valve stops the charge when the programmed weight has been dispensed.

Figure 2-3-11 Electronic charging scale. (*Courtesy Robinair SPX/OTC*)

Micron Vacuum Gauge

Micron gauges are used to measure deep vacuums below the level that a compound gauge on a manifold gauge set can display. The compound gauge can only show that a vacuum had been pulled down to 29 in Hg. The micron gauge can display vacuums down to around 50 microns, Figure 2-3-12. Microns are very small measurements. There are 25,000 microns in 1 in, which means there are about 400 microns in 1/64 in.

The display on micron gauges can display the reading as a number on a numeric display, as in Figure 2-3-13a, or by lighting an LED next to the micron scale, as in Figure 2-3-13b.

Electronic Sight Glass

An electronic sight glass is shown in Figure 2-3-14. This ultrasonic instrument has both visual and audible bubble detection. The display indicates when actual bubbles passing between the sensor clamps. Transducer clamps fit tubing from 1/8 in to 1 1/4 in (3 mm to 32 mm) in diameter.

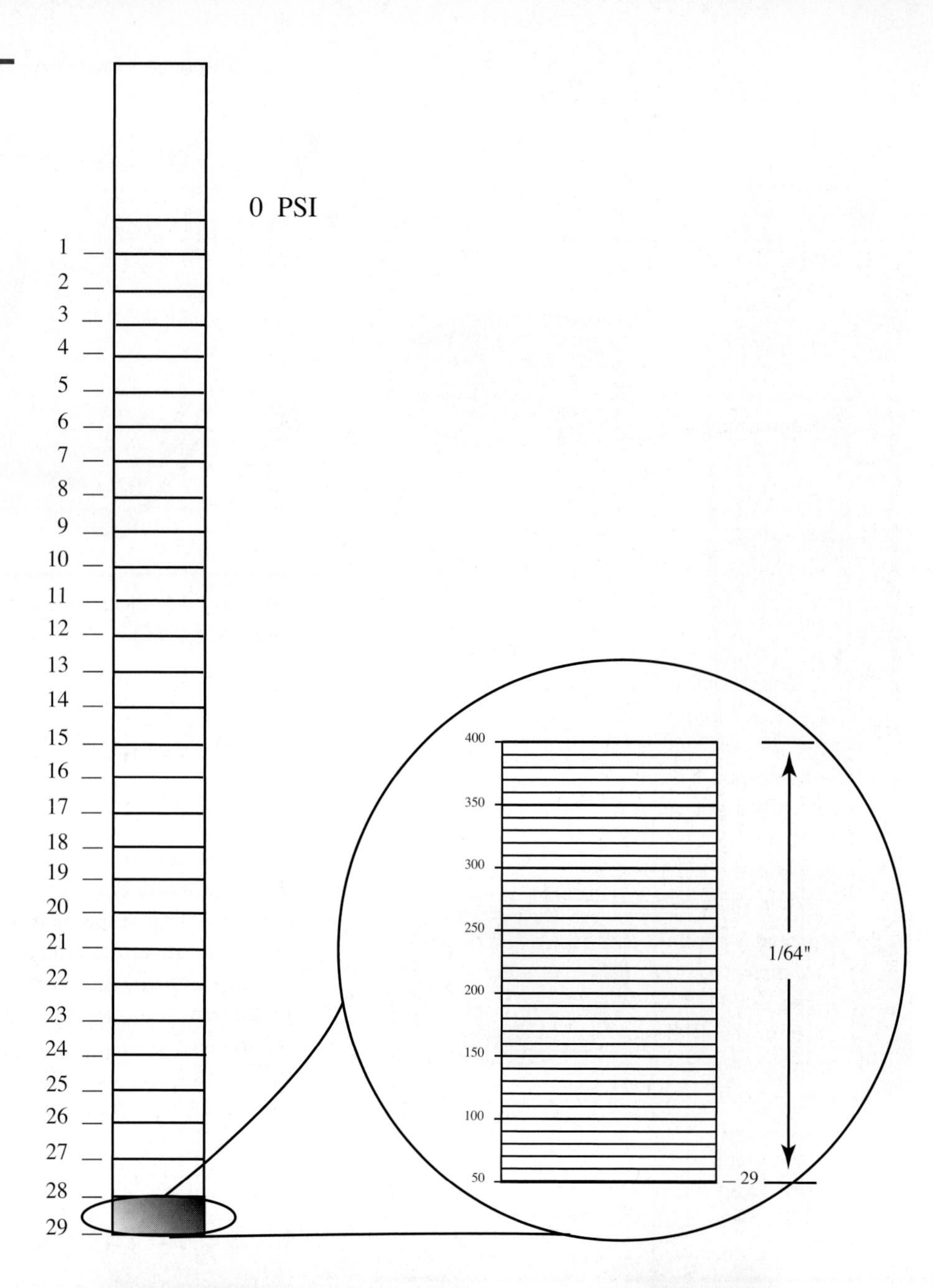

Figure 2-3-12 Micron vacuum gauge.

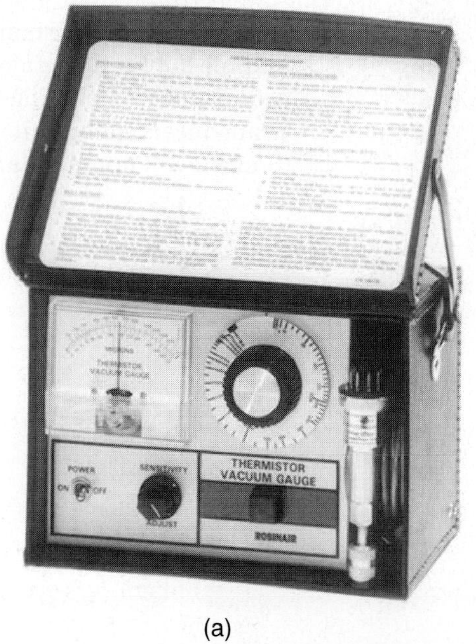

(a)

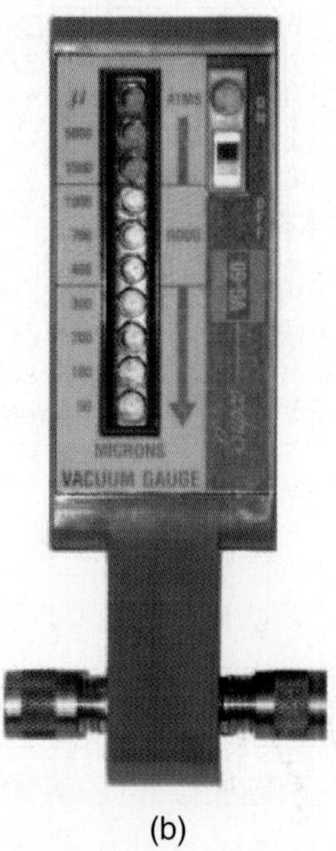

(b)

Figure 2-3-13 (a) Numeric display micron gauge; (b) LED display micron gauge. (*Courtesy Robinair SPX/OTC*)

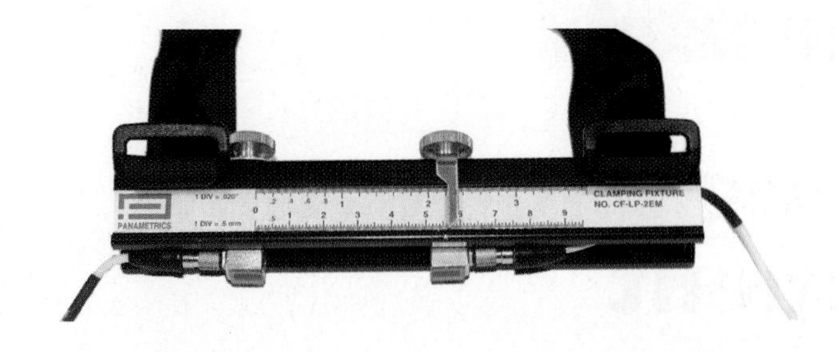

Figure 2-3-14 Electronic sight glass.

TECH TALK

Tech 1. I bought my vacuum pump last year and I thought it was still in good shape but the other day I tried to pull a vacuum on a system and it wouldn't pull below 100 microns. I even connected the micron gauge directly to the vacuum pump but it wouldn't pull it down so the problem has got to be in the vacuum pump. Where can I get it fixed?

Tech 2. When was the last time that you changed the oil in your vacuum pump?

Tech 1. Oh, probably a month ago, more or less. Why? It has plenty of oil, it doesn't need to be changed yet.

Tech 2. Well, it probably does. When you use your vacuum pump to dehydrate a system with moisture, some of that moisture gets trapped in the vacuum pump oil. The next time you use the pump that moisture will boil out of the oil, preventing the pump from pulling a deep vacuum. So you probably only need to change the oil, not rebuild your pump.

Tech 1. You mean it makes that much difference?

Tech 2. Yes, it can. If it has been that long since you changed the oil you need to run the vacuum until it gets good and hot, drain the oil and put in fresh. Run the pump with it connected to your micron gauge and watch what happens. It probably will improve but if it doesn't improve to the point where it can pull a 50 micron vacuum, change the oil again.

Tech 1. That oil is expensive and you want me to keep changing it?

Tech 2. Well, your vacuum pump is a lot more expensive than that vacuum oil and the vacuum pump won't work right without the oil. So you have a choice to either damage your vacuum pump and not be able to pull a deep vacuum or change the oil according to the manufacturer's recommended procedures.

Tech 1. You're right, that pump was very expensive. So I guess I better protect it by changing the oil more often.

Tech 2. That is a good investment. Remember to change the oil when the vacuum pump is good and hot. That will get out most of the moisture and make changing the oil more effective at bringing the pump back so it will pull a deeper vacuum.

Tech 1. Thanks, I will remember that. Now let me go change my oil.

UNIT 3—REVIEW QUESTIONS

1. List the three types of gauge manifold sets illustrated in Figure 2-3-2.
2. The four-way block is equipped with _______ and one evacuation hose.
3. What is a gauge manifold used for?
4. What is the center hose useful for?
5. When both service valves are all the way closed, can the gauge manifold be connected and disconnected?
6. When both service valves are all the way closed, what will the low gauge read?
7. A _______ that has been designed especially for evacuating and dehydrating a system has larger hose connections and uses larger hose.
8. A _______ provides an excellent way to measure a refrigerant charge.
9. An electronic charging scale will measure the charge by _______.
10. A digital micron gauge plus a vacuum indicator instantly recognizes _______.
11. An _______ has both visual and audible bubble detection.

UNIT 4

Heating—Servicing and Testing Equipment

OBJECTIVES: In the unit the student will learn about heating, servicing, and testing equipment including:

- Measurement of CO_2
- Gas Identifiers and Monitors

MEASUREMENT OF CO_2

In the operation of heating systems, it is essential that the CO_2 (carbon dioxide) content of flue gases be maintained as high as practical to improve the efficiency of the heating unit while keeping a low smoke level. In practice, as far as oil furnaces are concerned, 10–12% CO_2 is desirable. For natural gas furnaces, the CO_2 range should be 8.25–9.50%. The CO_2 measurement, plus the stack temperature measurement, can be used with a special slide rule to indicate the efficiency of the furnace.

CAUTION: Many CO_2 (carbon dioxide) detectors used in the HVAC/R industry must be calibrated to ensure their accuracy. Calibration of CO_2 detectors and other similar instruments must be done by a certified calibration company. Check with the instrument's manufacturer, if necessary, to locate an appropriate company to provide this service.

Draft Gauge The draft gauge is supplied with a draft tube to be inserted in the flue pipe with a rubber tubing extension, Figure 2-4-1.

There are two places where the draft is usually taken. One is in the flue pipe at the furnace exit and the other is in the door of the furnace (over the fire). By taking these two measurements, draft problems can be analyzed. The draft tube is inserted in the flue pipe opening and the meter registers the amount of draft, in inches of water column. The probe is then inserted in the door of the furnace to read the draft at this location. If there is less draft at the furnace door, it is an indication that there is a leak in the heat exchanger that needs to be corrected.

Smoke Tester The tube from the smoke tester is inserted in one of the test holes during operation of the furnace and a sample of the flue gas is passed through a filter, Figure 2-4-2. Ten pump strokes of the tester are required to get a proper sample. The color of this filter, as compared to the standard smoke scale, identifies the smoke density in the flue gases (for oil burners this is usually scale number 1 or 2 on a scale of 0–9). Any reading above 2 would require adjusting the burner.

Flue Gas Analyzer The flue gas analyzer measures the CO_2 in the flue gases, Figure 2-4-3. The probe is inserted in the flue pipe during the operation of the furnace. These testers can sometimes measure O_2 (oxygen) in the flue gases. You may also use a CO_2 probe that plugs into a multimeter, Figure 2-4-4.

Figure 2-4-1 Draft gauge. (*Courtesy of Bacharach, Inc.*)

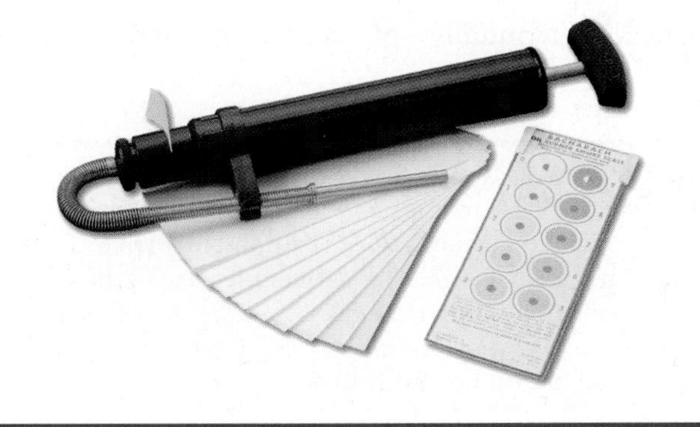

Figure 2-4-2 Smoke tester. (*Courtesy of Bacharach, Inc.*)

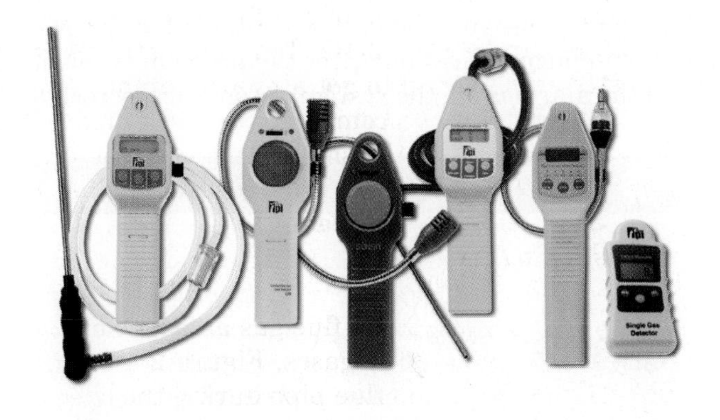

Figure 2-4-3 CO_2 testers. (*Courtesy Test Products International, Inc.*)

Figure 2-4-4 CO_2 tester adapter for multimeters. (*Courtesy Test Products International, Inc.*)

SERVICE TIP

Flue gas analyzers can be used to help the technician set up a furnace so that it is operating at peak energy efficiency. In the past many technicians simply looked in the fire box and made a visual determination that the flame looked OK. With today's rising energy costs and concerns over the environment it is important that a complete and accurate analysis of the flue gas products be taken so that the furnace can be set at its peak performance.

GAS IDENTIFIERS AND MONITORS

There are a number of uses for gas identifiers and monitors:

1. To notify the owner of a refrigerant leak.
2. To protect anyone entering a room where refrigerants are stored or used. Refrigerants can displace oxygen, and therefore become dangerous. Most refrigerants do not have an odor and are not normally detected.
3. To identify the CO content in the air, and in some cases, to automatically control the addition of outside air. There is an increasing amount of concern about indoor air quality and particularly the need for ventilation to prevent the buildup of CO in enclosed spaces.

The refrigerant identifier is a microprocessor-based instrument that extracts a sample of the refrigerant and analyzes it to determine the type of refrigerant, Figure 2-4-5. In some cases systems have been charged with refrigerants without properly

Figure 2-4-5 Electronic analyzer to determine refrigerant types.

tagging the installation with the refrigerant number. The service technician MUST be certain of the refrigerant that has been installed in order to perform any service necessary.

Oxygen Depletion Monitor

The oxygen depletion monitor has a remote oxygen sensor and a controller, which are normally mounted outside the mechanical equipment room door, Figure 2-4-6.

It is designed to continuously monitor the oxygen level in mechanical equipment rooms where class A, group I refrigerants are used. Oxygen is normally about 20% of the atmosphere, and when levels drop to 19.5% an amber warning light goes on, starting a mechanical exhaust. If the oxygen level continues downward to 18.5%, a red warning light goes on and an alarm will sound. Relays provide an option for sending the warning to remote locations. A "purge" switch permits the fan to run continuously or only when signaled by the monitor. The alarm turns off automatically when normal levels of oxygen return, or may be turned off manually—at which time there is automatic reset.

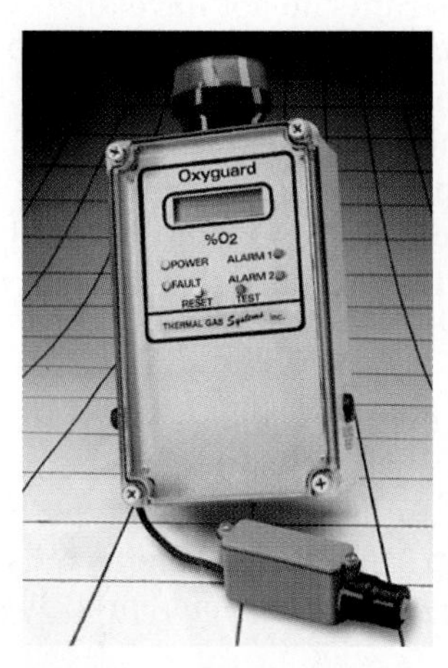

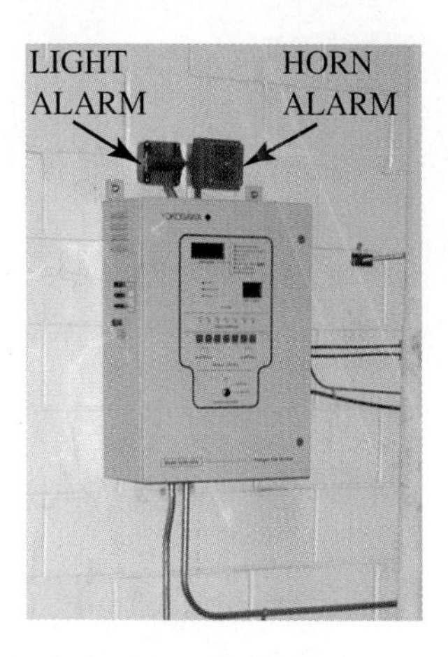

Figure 2-4-6 Oxygen deprivation monitors for equipment rooms. (*Courtesy OI Analytical*)

> **CAUTION:** Oxygen deprivation monitors are mandated by OSHA regulations for most commercial and light commercial mechanical equipment rooms. These monitors are also a good investment for use in those few situations not mandated under the OSHA regulations. Oxygen deprivation can quickly render a technician unconscious and could result in death.

The first stage, or warning stage, activates an amber LED and a set of relay contacts to operate a fan or other mechanical equipment. The second stage, or alarm stage, activates a red LED, an audible alarm, and auxiliary set of contacts for optional remote alarm indication. After the problem is solved and the alarm is off, the system automatically resets.

TECH TALK

Tech 1. I went to buy one of those carbon dioxide testers you sent me after but I decided not to get it after talking to the guy in the supply house.

Tech 2. What are you talking about—some guy in the supply house talked you out of getting a carbon dioxide detector?

Tech 1. Yeah, he said he has been in the business for 35 years and he thinks that they are a total waste of money. He said after you have been in the business for a little while that you can look in the firebox and tell exactly what is going on. There's really no reason to waste your time or the customer's money performing all those tests.

Tech 2. He may have been right 20 or 30 years ago because that is the only way they knew how to do it. But today there is absolutely no reason to do that shoddy of a job. You are taking a chance that a furnace is going to produce too much CO_2, which can cause health problems. In addition, your failing to properly adjust the burner can cause soot build-up in the firebox that can ultimately stop the entire furnace from working properly. Not to mention it can result in flame rollout where you are going to wind up with a lot of CO_2 in the residence. No sir, I would never want to work on a system where I was adjusting the burner and not use one of the latest electronic analyzers.

Tech 1. Well, I guess you are going to have to let me go for a little bit so I can go back down to the supply house and pick up one. Thanks, I appreciate that.

UNIT 4—REVIEW QUESTIONS

1. In practice what is the desirable CO_2 level for oil burners?
2. Where are the two places that a draft sample is usually taken?
3. Any reading above a ________ on a smoke tester would require adjusting the burner.
4. List the uses for gas identifiers and monitors.
5. Where is an oxygen depletion monitor normally mounted?
6. A ________ permits the fan to run either continuously or only when signaled by the monitor.
7. What is activated in the second stage or alarm stage of the oxygen depletion monitor?

3

HVAC/R Practices

UNIT 1

Drawings and Blueprint Reading

OBJECTIVES: The reader will learn about mechanical and architectural drawings and blueprint reading including:

- The Alphabet of Lines
- Mechanical Drawings
- Architectural Drawings
- Architectural Symbols
- Reading Mechanical Drawings
- Reading Architectural Drawings

INTRODUCTION

Mechanical, electrical, and architectural drawings are used extensively in the HVAC/R industry to provide the technician with a graphic representation of equipment, HVAC/R systems, and buildings. Mechanical drawings are used to illustrate equipment, systems, and components such as air handlers, refrigerant coils, component layout, relays, motors, valves, and other parts and pieces of HVAC/R systems. Electrical drawings are used to show how the components in a system are wired together as shown in Figure 3-1-1. Architectural drawings are used to represent buildings, houses, and other structures as well as systems within those structures such as system location for electrical, duct layout, piping layout, and other substantial buildings parts as shown in Figure 3-1-2.

ALPHABET OF LINES

All drawings use a standard set of lines. This set of lines is collectively referred to as the *alphabet of lines*. Each of the lines that make up the alphabet of lines has a specific purpose. Drawings are a universal language as illustrated by these highway signs that can be understood by drivers worldwide as shown in Figure 3-1-3.

Figure 3-1-4 shows how the various lines are used in drawings. The alphabet of lines can also be understood even if the words on the drawings are done in another language.

The basic line types are:

- **Object line** The object line is used to represent the visible lines or edges of an object.

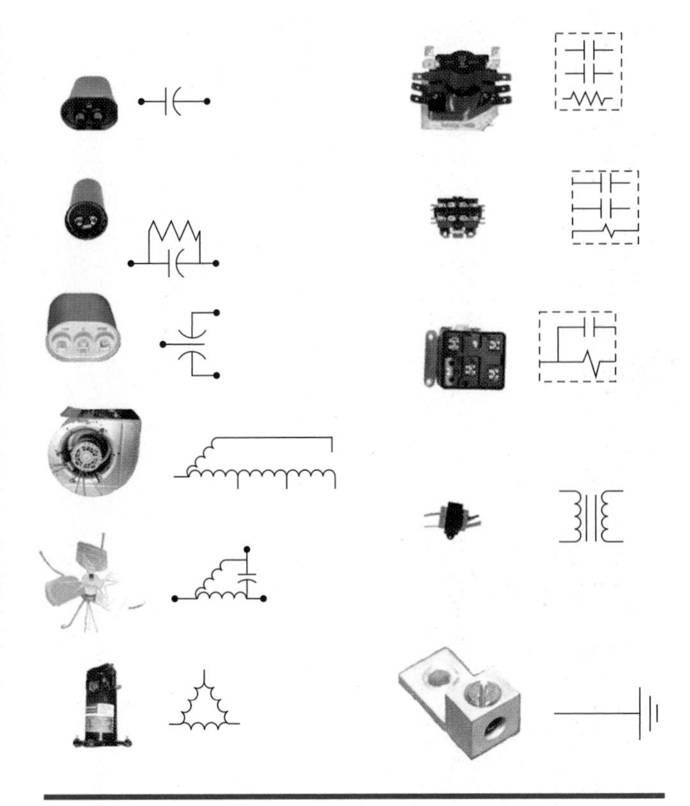

Figure 3-1-1 Standardized symbols are used to represent the various components found in electric circuits including resistors, capacitors, and inductors.

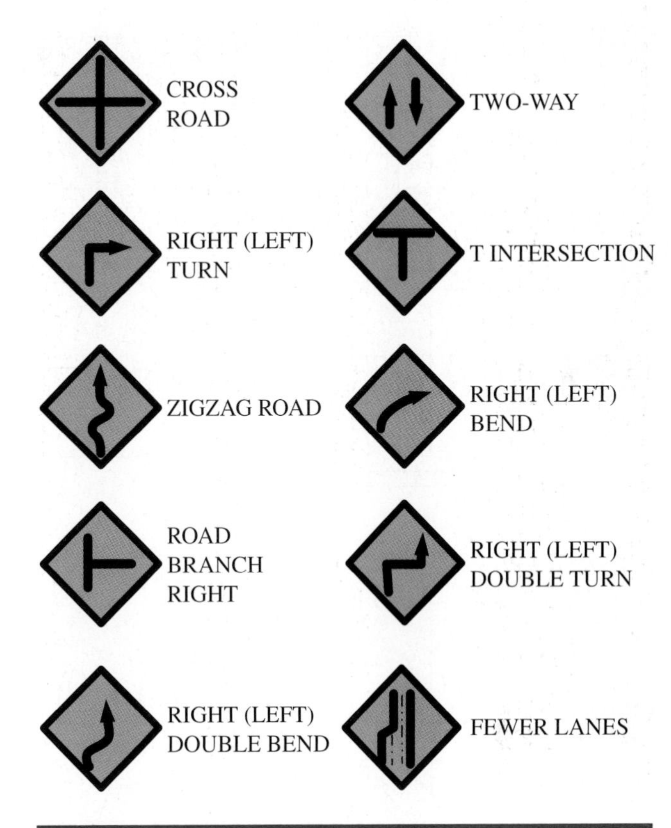

Figure 3-1-3 Road signs are examples of universally understood symbols.

Figure 3-1-2 Architectural drawings are called "blueprints" because of the printing process used to reproduce these large sheets.

- **Hidden lines** Hidden lines are used to represent major components that would be obscured by the surface of the object.
- **Center lines** Center lines are used to represent the center of an object that is round or symmetrical. They are also used to locate the center of holes.
- **Section lines** Section lines are used to represent the edge of a cutaway surface.
- **Dimension lines** Dimension lines are used in conjunction with numbers representing the size or length of an object.
- **Extension lines** Extension lines are lines that extend from a point being dimensioned.
- **Cutting plane lines** Cutting plane lines are used to locate an imaginary cut on a surface to expose the parts inside of an object.
- **Break lines** Break lines are used to show the removal of an area of an object to expose the inside of the object. They are frequently used on pictorial drawings to expose the internal workings of an object.
- **Leaders and arrows** Leaders and arrows are used to locate or identify items on a drawing.
- **Phantom lines** Phantom lines show where a part will be placed or another position where a part can be moved.

MECHANICAL DRAWINGS

Mechanical drawings are divided into two major groups. The first group is called orthographic projections or simply projection drawings. These drawings are typically made up of one or more views of the object, as it would appear if you were looking straight at it. These views are always shown in the orderly

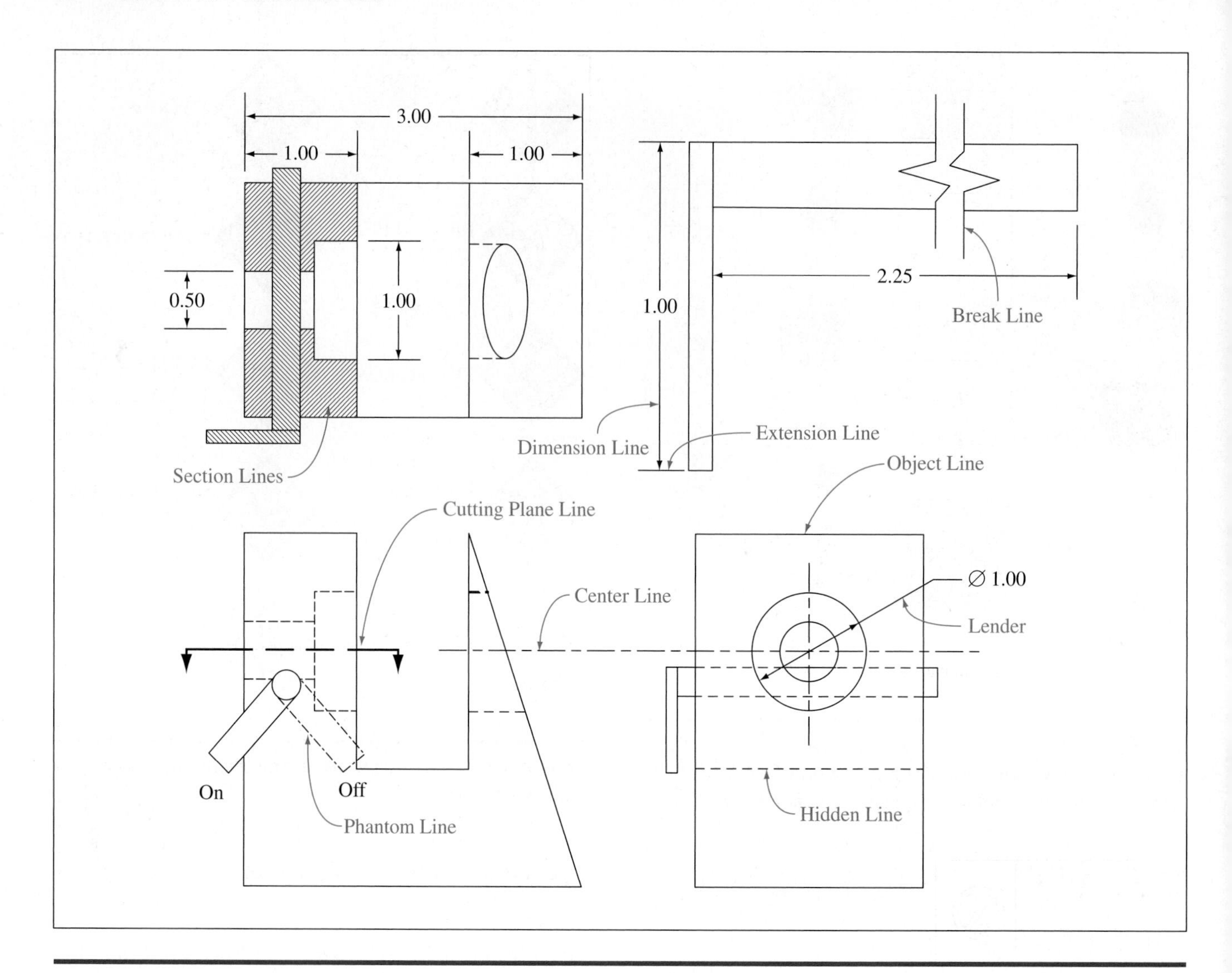

Figure 3-1-4 The alphabet of lines used in an architectural drawing.

arrangement diagrammed in Figure 3-1-5. If an object to be pictured would look the same from the back and front view, you would typically remove or not draw the duplicate view. Most mechanical drawings use the front view, right side, and top views of the object to fully describe the object's shape and appearance. Figure 3-1-6 is an example of the three views used to illustrate one piece of electrical equipment, a grounding lug.

PICTORIAL DRAWINGS

Pictorial drawings are the most common mechanical drawings used in the HVAC industry. Pictorial drawings are used to draw a picture-like representation of the object, such as the connector in Figure 3-1-7. These drawings are made as if the draftsman were standing and looking at the unit. Pictorial drawings are often used to show the technician the location of the various components used to make up an air conditioning or heating unit. Because pictorial drawings are so picture-like they are the easiest drawings to use to understand and identify parts. Because pictorial drawings are so easily understandable they are extensively used in manufacturers' sales literature. A cutaway pictorial drawing like the one in Figure 3-1-8 may even show the arrangement of interior parts you wouldn't be able to see in a photograph.

ARCHITECTURAL DRAWINGS

Architectural drawings are used to show the layout and structure of houses and buildings. These drawings are typically made as if the drafter were in a plane looking straight down at the building. Lines are used on an architectural drawing to represent the building's walls and other major mechanical and

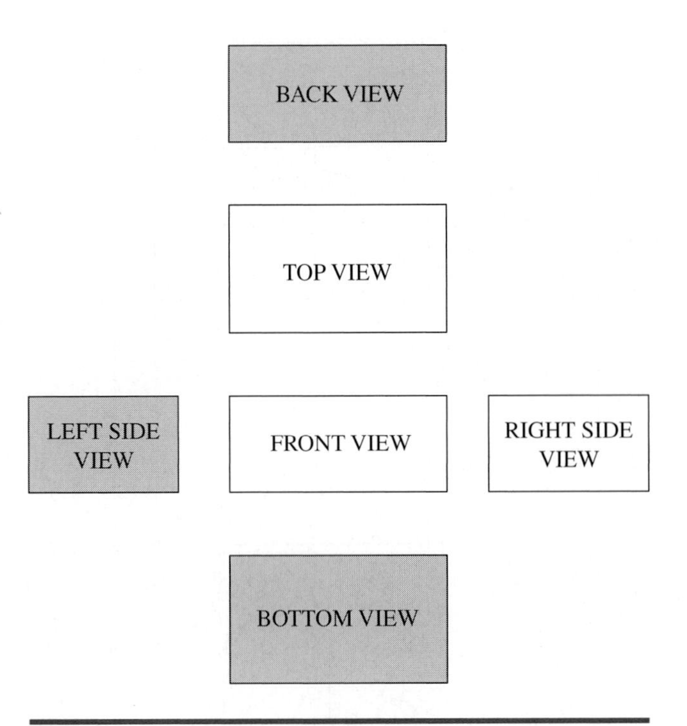

Figure 3-1-5 Orderly arrangement of views for mechanical lines.

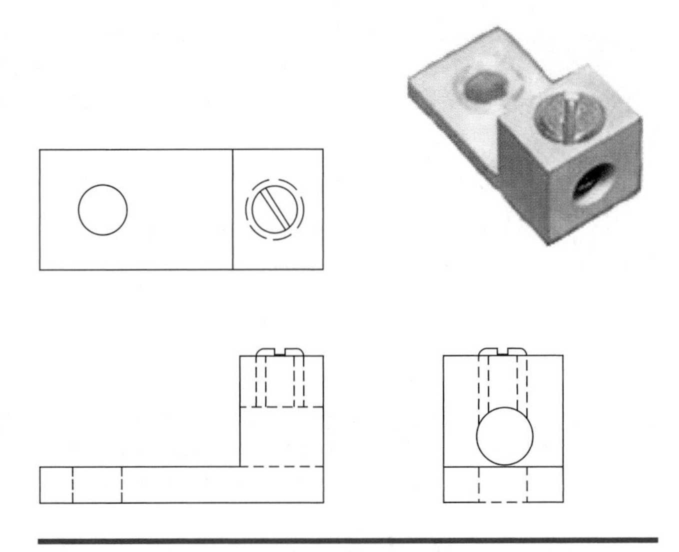

Figure 3-1-6 Three views used to illustrate a grounding lug.

structural components located within the building. Figure 3-1-9 shows a small residential house and you can easily identify the doors, windows, bathroom fixtures, and major kitchen appliances.

In order to completely and clearly illustrate all of the various elements and systems used in the construction of a building, multiple pages of drawings are often used, Figure 3-1-10. These drawings collectively are known as the building plans. In a set of drawings for a single family home or residence some of these drawings pages are combined. But in a set

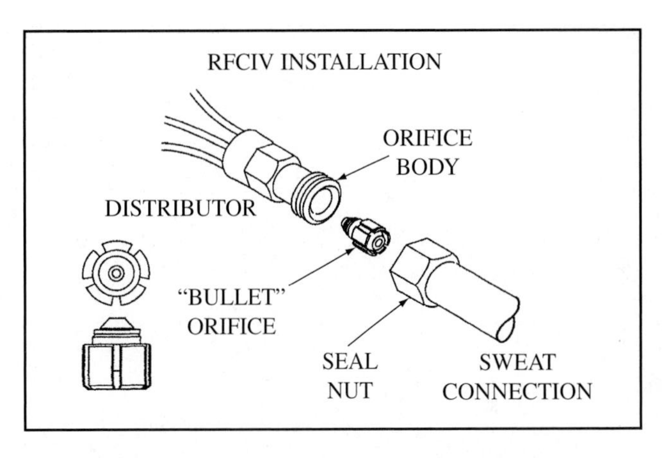

Figure 3-1-7 A pictorial drawing from a manufacturer show how an assembly fits together. This one includes top and side views of a part to help identify it.

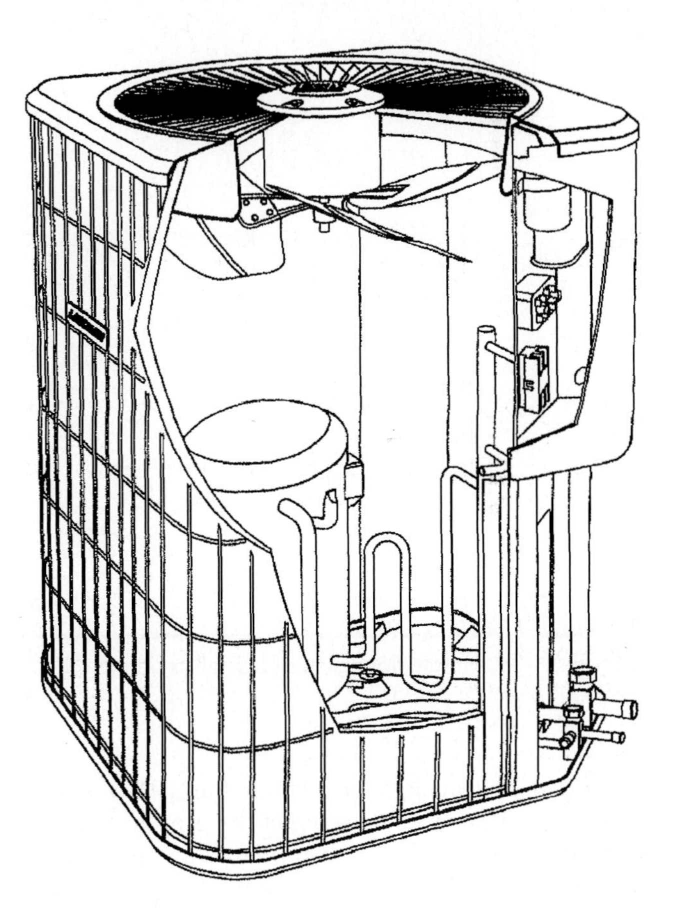

Figure 3-1-8 A cutaway pictorial drawing can show both internal and external appearances. (*Courtesy of Lennox Industries, Inc.*)

of drawings for a commercial building there may be separate drawings for each of the major parts of the building. Figure 3-1-11 shows a comparison of the amount of drawings needed for a commercial building (Figure 3-1-11a) and a residential house (Figure 3-1-11b).

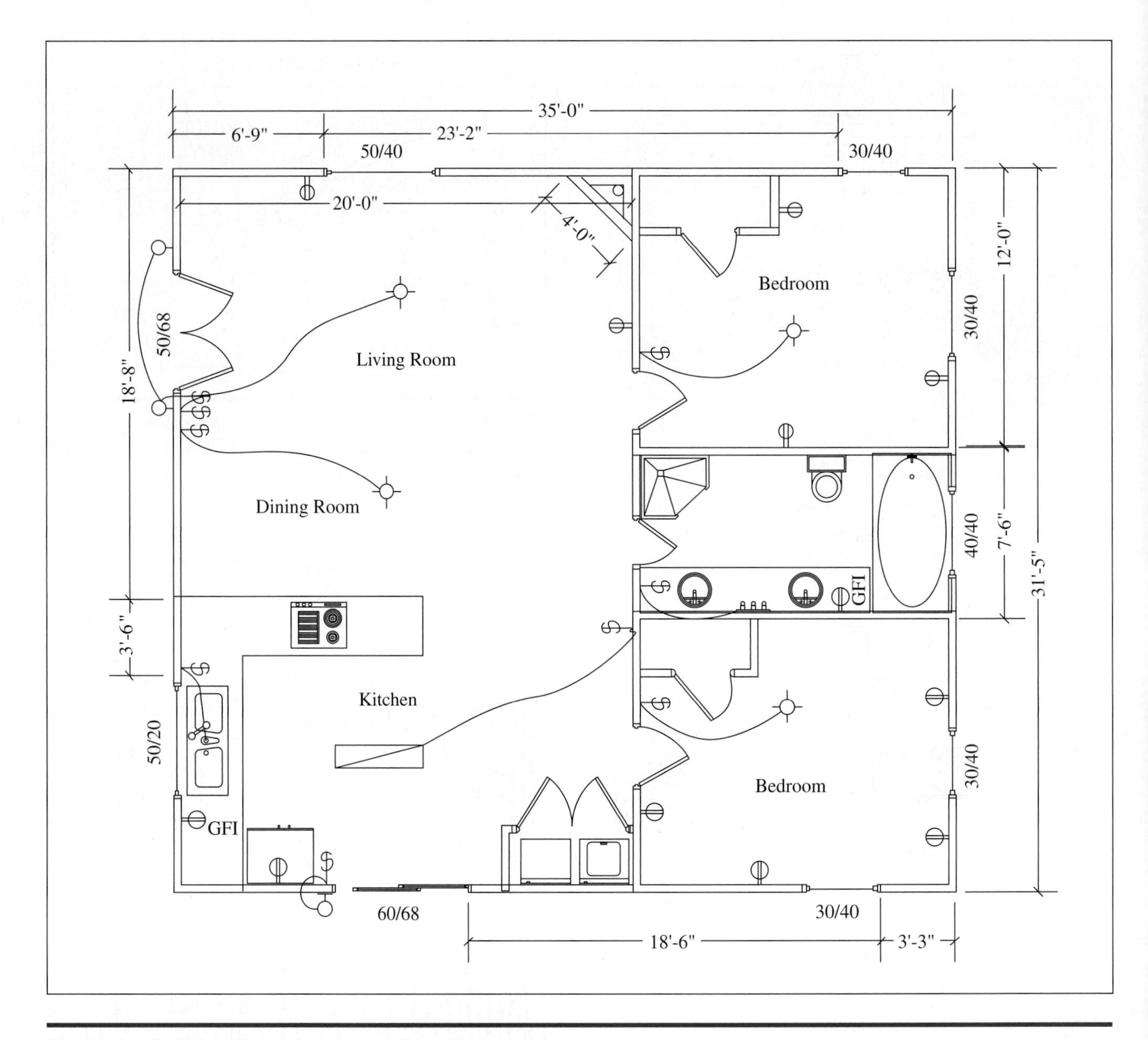

Figure 3-1-9 Simple floor plan of a small residential house.

A set of drawings for a single family home may start with a pictorial drawing of the proposed house, Figure 3-1-12. Not all house plans have a pictorial drawing. A floor plan of the building shows the layout and gives the dimensions, Figure 3-1-13a. Next there may be drawings for the electrical, plumbing, and HVAC, Figure 3-1-13b,c,d. When combined these three are often referred to as the mechanical drawings for a building. Additional drawings may show the various elevations of the building, Figure 3-1-14, such as left, right, front, and back elevations. Other drawings may show a plot or site plan of where and how the building is located on the land, Figure 3-1-15. Detailed drawings of building components, such as roof framing, windows, stairwells, kitchen cabinets, and roof details, may also be included in a set of detail drawings, Figure 3-1-16.

ARCHITECTURAL SYMBOLS

There are many standard symbols used in architectural drawings. Figure 3-1-17 shows the common architectural symbols and gives their description. Some symbols such as the symbol for a ceiling light can be used for a variety of different types of lights, Figure 3-1-18a,b. Variations of the light symbol can be used to represent recessed lights, track lights, a ceiling fan with a light, and a vent fan with a light (Figure 3-1-18c–f). The four common variations in the light

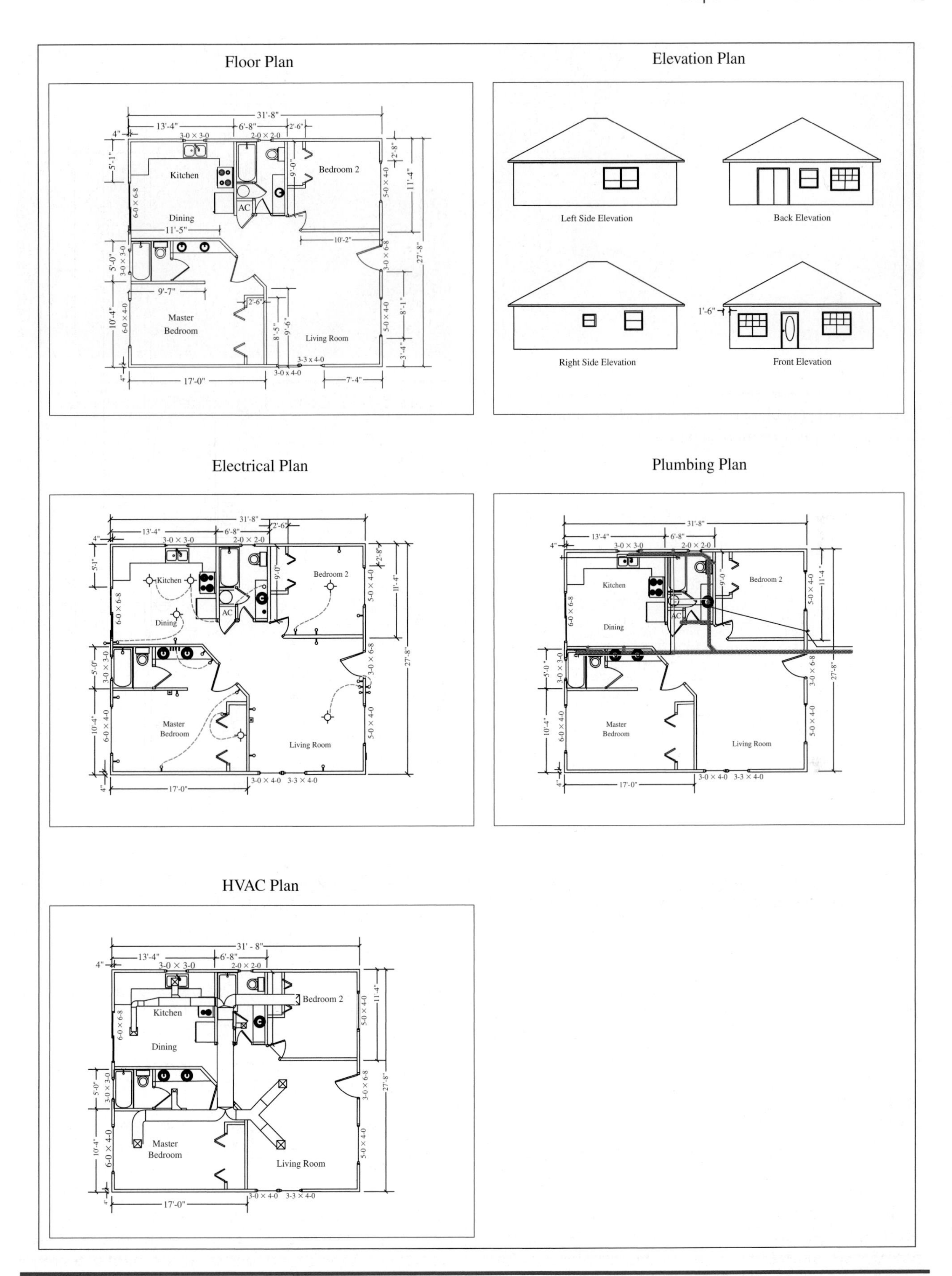

Figure 3-1-10 Typical drawings found in a set of plans used in the construction of a single family home.

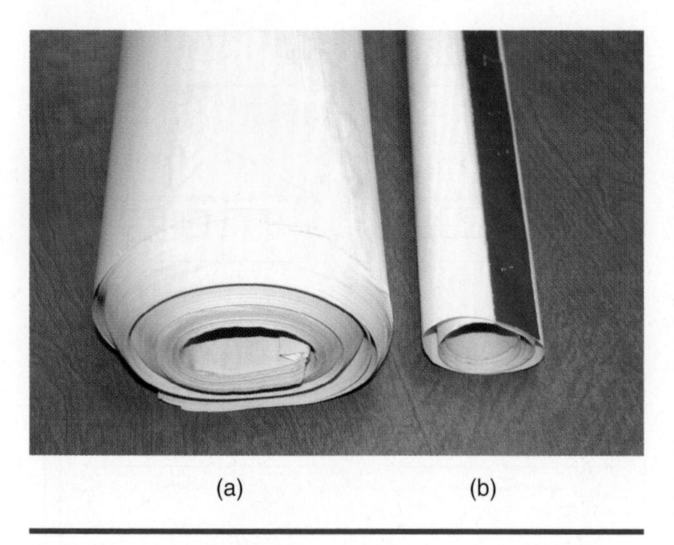

Figure 3-1-11 (a) Blueprints for a commercial building; (b) Blueprints for a residential home.

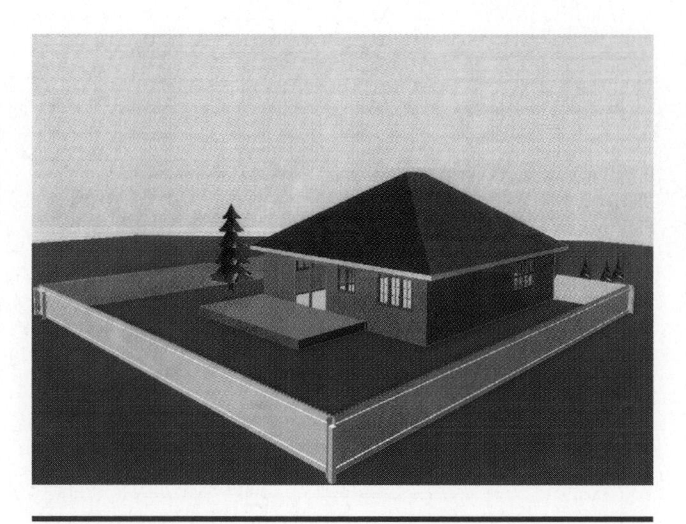

Figure 3-1-12 Computer-generated pictorial view.

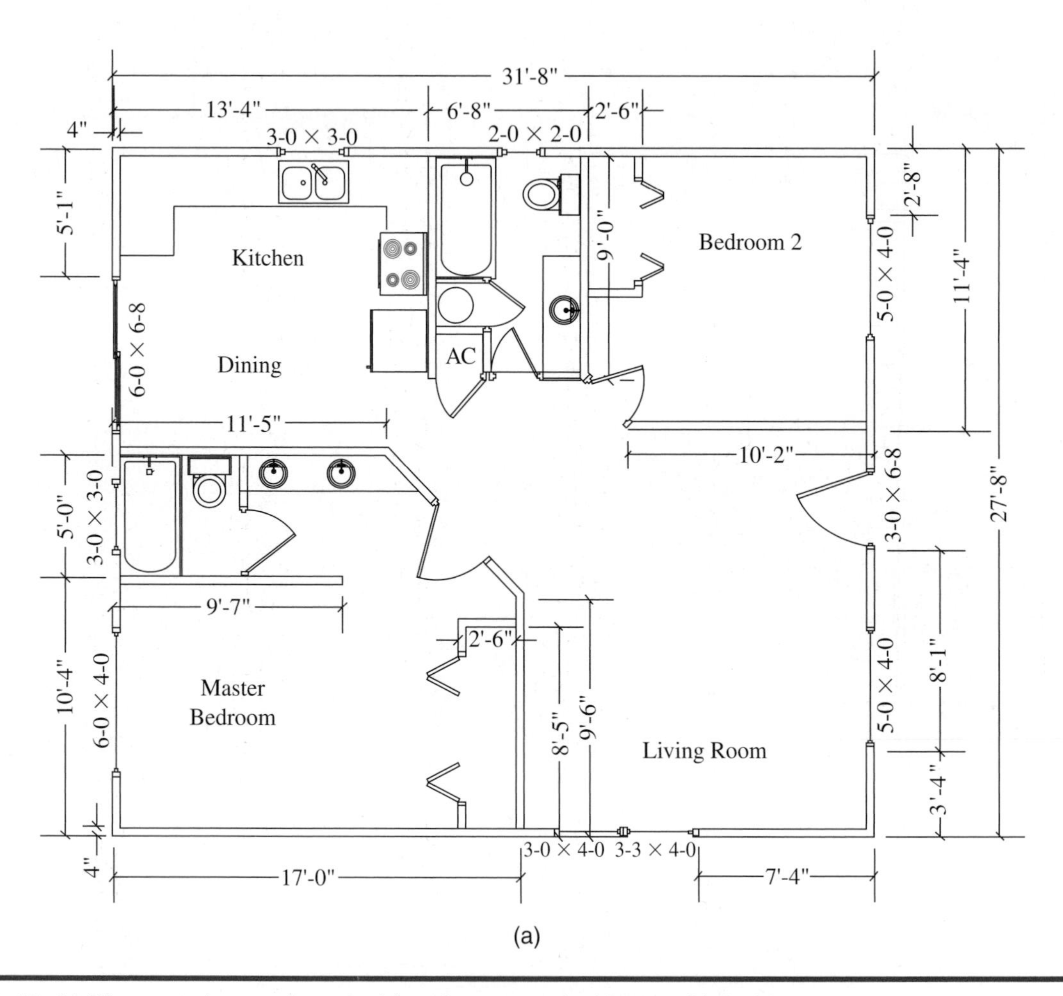

Figure 3-1-13 (a) Blueprint showing basic building dimensions and layout; (b) Building electrical layout; (c) Plumbing layout; (d) HVAC or mechanical duct layout.

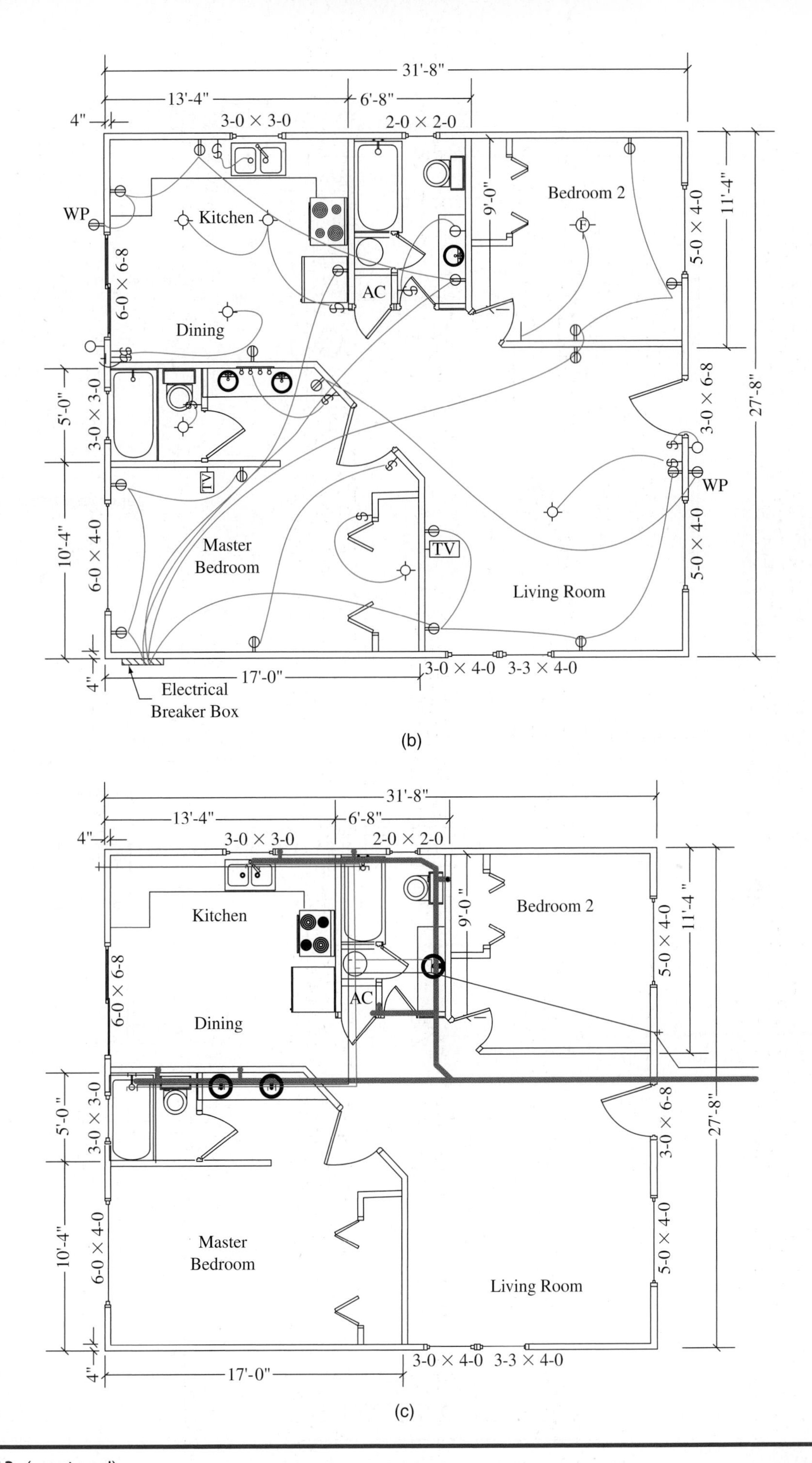

Figure 3-1-13 (continued)

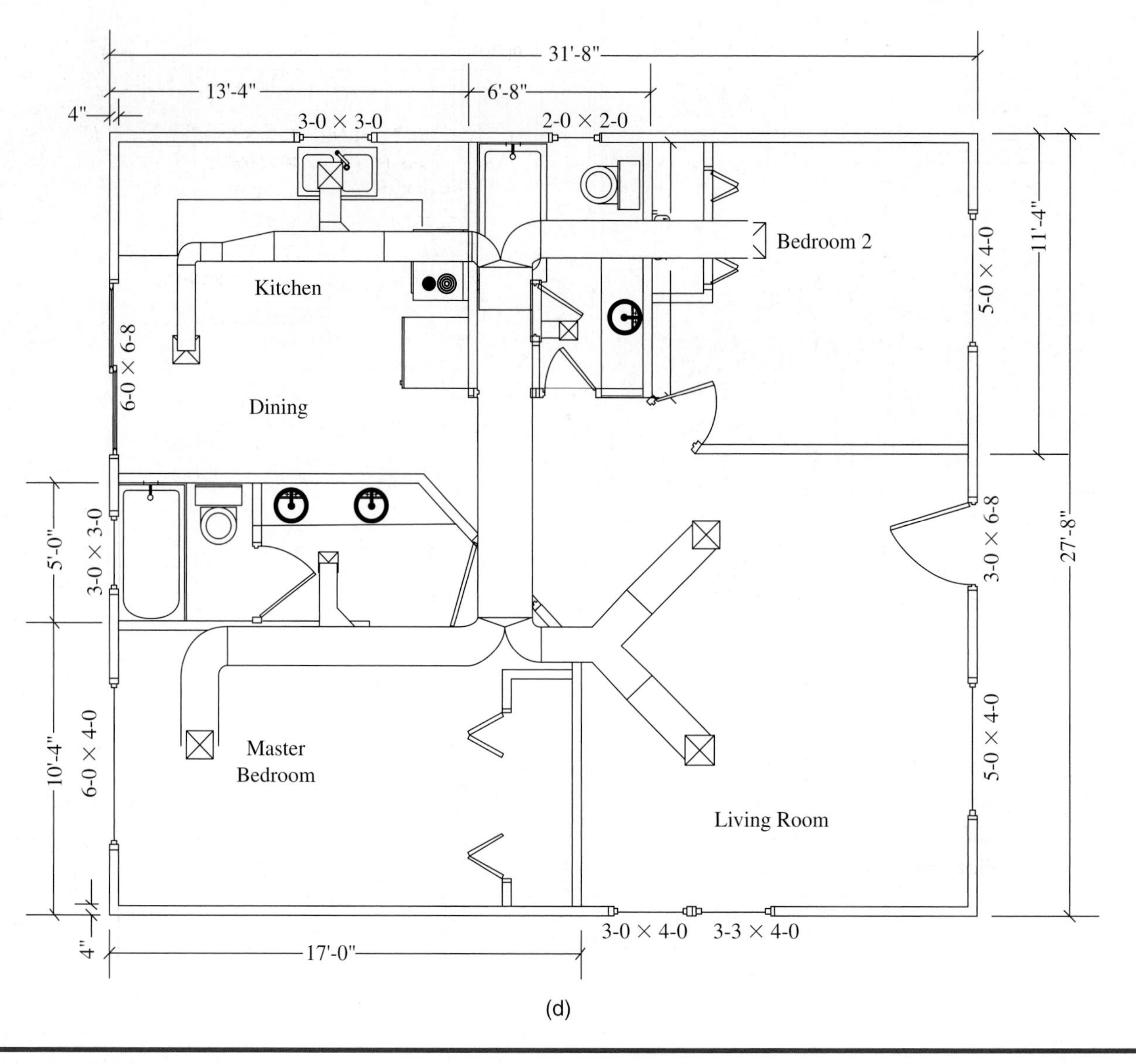

(d)

Figure 3-1-13 (continued)

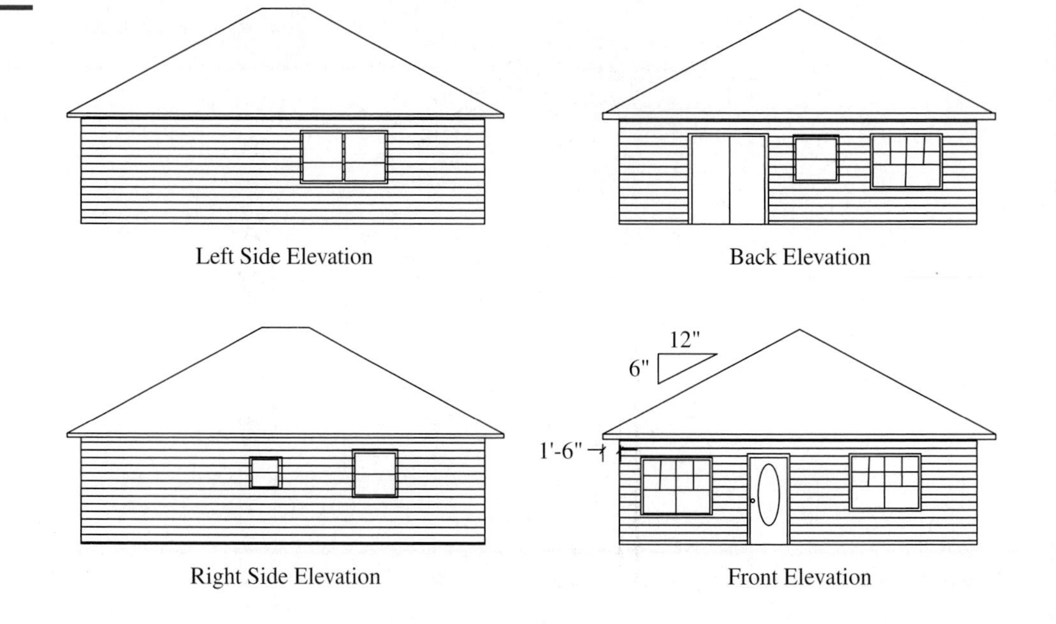

Figure 3-1-14 Elevation views.

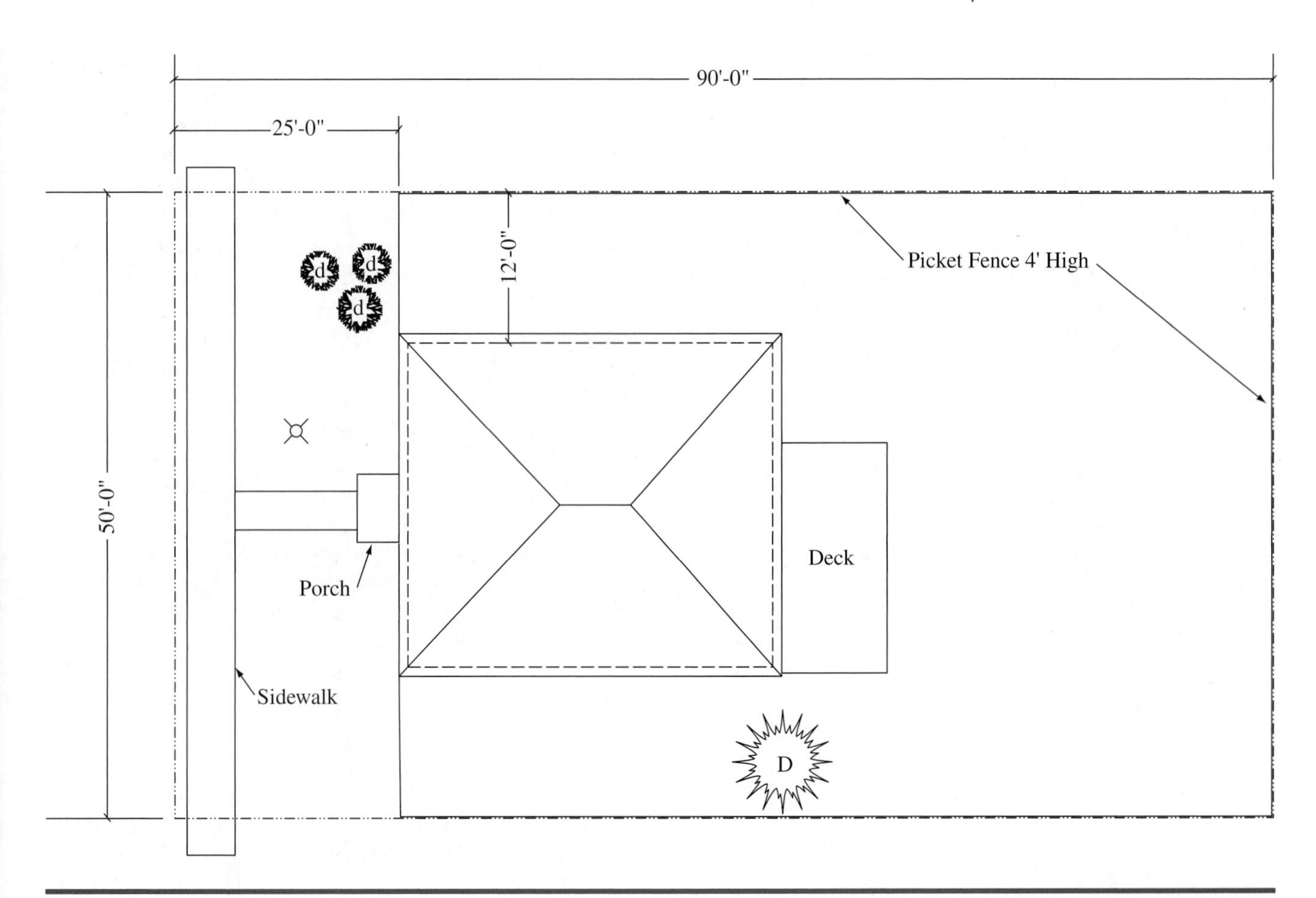

Figure 3-1-15 Plot plan.

Figure 3-1-16 Detail drawings may give specific information about building components.

switch symbol and outlet symbol (Figure 3-1-18g,h) are detailed in Figure 3-1-17. Because many symbols do not resemble the object they are representing, such as the register, vent fan, and shower nozzle (Figure 3-1-18i–k), most drawings will have a table showing the symbols and explaining their use. In addition to the commonly used symbols, architects can invent their own symbols for items located on the drawings. The symbols table should include the standard symbols and any made up symbols the architect will be using for that drawing. For large commercial buildings there may be an entire page dedicated to symbols. Always refer to the symbols table when reading blueprints.

READING MECHANICAL DRAWINGS

Figure 3-1-19 shows a three view drawing of an air conditioning condensing unit. The vapor line or suction line in the top view has been shaded blue, and it can be seen on the right side view of the orthographic projection. The liquid line had been shaded red, and it can be seen on the top and left side views, and just as a sliver in the right side view. The discharge air grill on top of the unit has been shaded yellow, and it too can be seen in all three views. However, in the left side and right side views it only appears as a thin yellow line. Its actual shape can only be seen in the top view. In an orthographic projection not all of the parts of the unit can be seen clearly in each of the views. In the case of the liquid and suction lines, they are only both completely

DESCRIPTION	SYMBOL
SINGLE-POLE SWITCH	
THREE-WAY SWITCH	3
FOUR-WAY SWITCH	4
WATERPROOF SWITCH	WP
DOUBLE OUTLET	
WATERPROOF OUTLET	WP
GROUND FAULT INTERRUPTOR (GFI)	GFI
240 VOLT OUTLET	240v
TRANSFORMER	T
JUNCTION BOX	J
POWER PANEL	
BRANCH CIRCUIT	
CEILING LIGHT	
RECESSED CEILING LIGHT	R
TRACK LIGHTS	
WALL LIGHT	
CEILING FAN WITH LIGHT	F
VENT FAN WITH LIGHT	F
VENT FAN	F
TV CABLE JACK	TV
THERMOSTAT	T

Figure 3-1-17 Standard electrical symbols used in architectural drawings.

visible in the top view. Only the liquid line is shown in the left side view because the suction line is behind it and cannot be seen. However, only the suction line is seen in full in the right side view because in this case the liquid line is mostly hidden, and can only be seen as a thin red line. The unseen portions of the liquid and suction lines could have been drawn as hidden lines, but that would not make this drawing clearer or easier to understand. Often items that are not required for the purpose of the drawing may be left off if their being added would serve to make the drawing less clear. In the actual drawings used in manufacturing this condenser, there would have been far more detail.

Each of the three primary views used in an orthographic projection can contain dimension information as in Figure 3-1-20. In the front view the width and height of the object are shown, in the right side the depth and height of the object are shown, and in the top view the width and depth of the object are shown. In an orthographic projection each view can only contain two dimensions, and this type of drawing is actually referred to as a two-dimensional drawing.

A pictorial drawing can contain all three dimensions, Figure 3-1-21. In this single view you can see the width, height, and depth of the object. For the purposes of installation, service, and identification, usually all of the overall dimensions of the system are given. It is important, however, that the location of parts not dimensioned be kept consistent so that it is easier for the technician to locate that part on a corresponding view. For example in Figure 3-1-19 on page 82 the height of the base is shown in the left side view as ¾ in (19 mm) and the height to the center of the suction and liquid lines is shown as 2¾ in (70 mm) in the right side view. Being able to reference dimensions is important because most manufacturers show one drawing to represent an entire line of different models. This one drawing must reference different model dimensions in a table, such as can be seen at the bottom of Figure 3-1-19 on page 82. The overall height (A) of the nine different models shown in the table is given on the left side view but can be easily transferred to the right side view.

READING ARCHITECTURAL DRAWINGS

It is a common practice to leave off the last dimension from a wall in the architectural drawing. The omission of such a single dimension is often done because it is not possible to make the slab or floor exactly the correct size during construction. Any difference in overall size would simply be added or subtracted from that room's size. However, we need the dimensions for all of the rooms to do a heat load on the house. Knowing all the dimensions will make determining the room size easier.

Any dimension shown that is parallel to the front wall would be the width of the house, room, door, etc., and any dimension shown parallel to the left side wall would be the depth of the house, room, door, etc. The front walls of this house are shown as red and the window and doors are shown as yellow in the plan and elevation drawings of Figure 3-1-22a,b. The

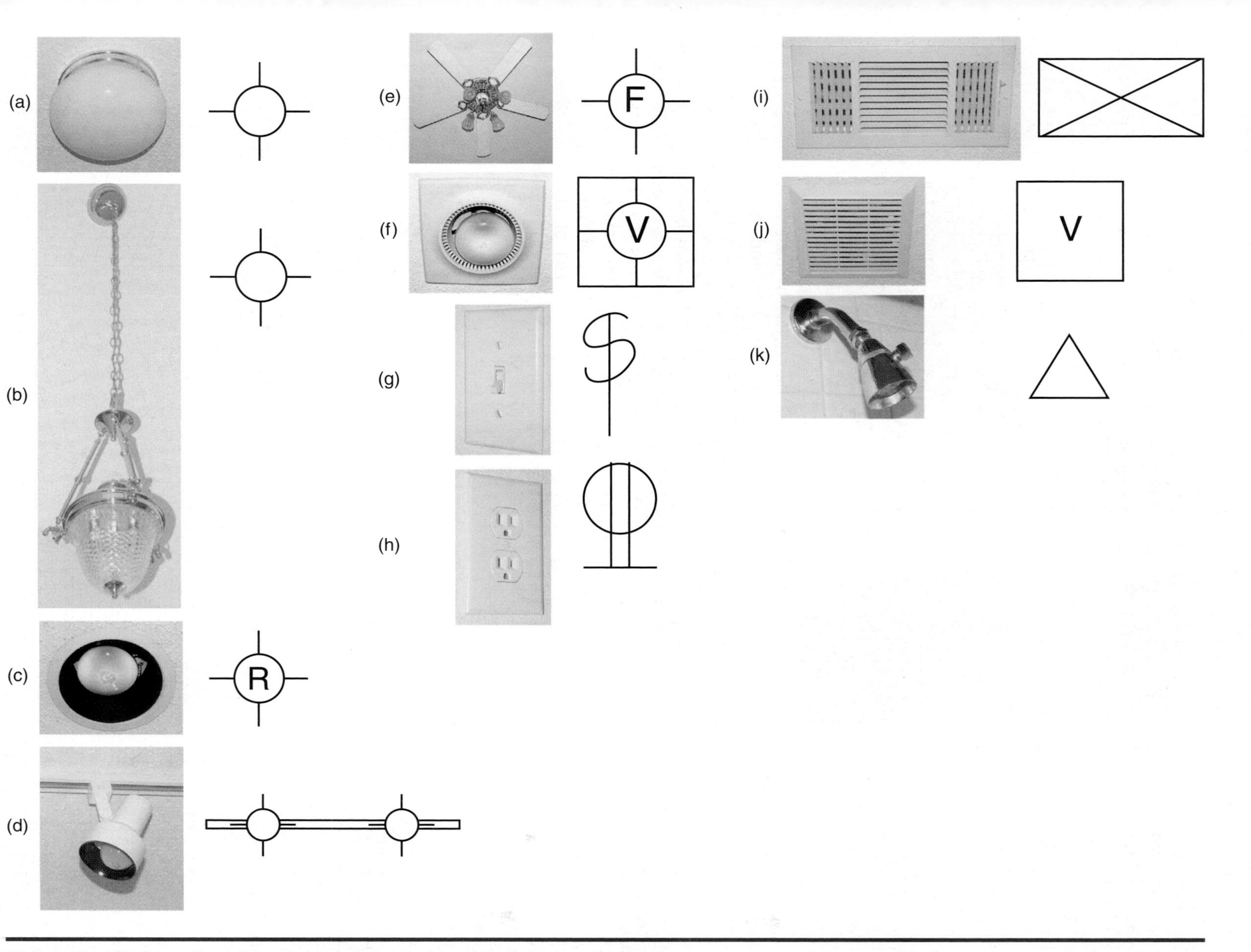

Figure 3-1-18 Examples of standard architectural symbols and the items they represent: (a) Ceiling light; (b) Ceiling light; (c) Recessed light; (d) Track light; (e) Ceiling fan with light; (f) Vent fan with light; (g) Light switch; (h) Electrical outlet; (i) Air register; (j) Vent fan; (k) Shower nozzle.

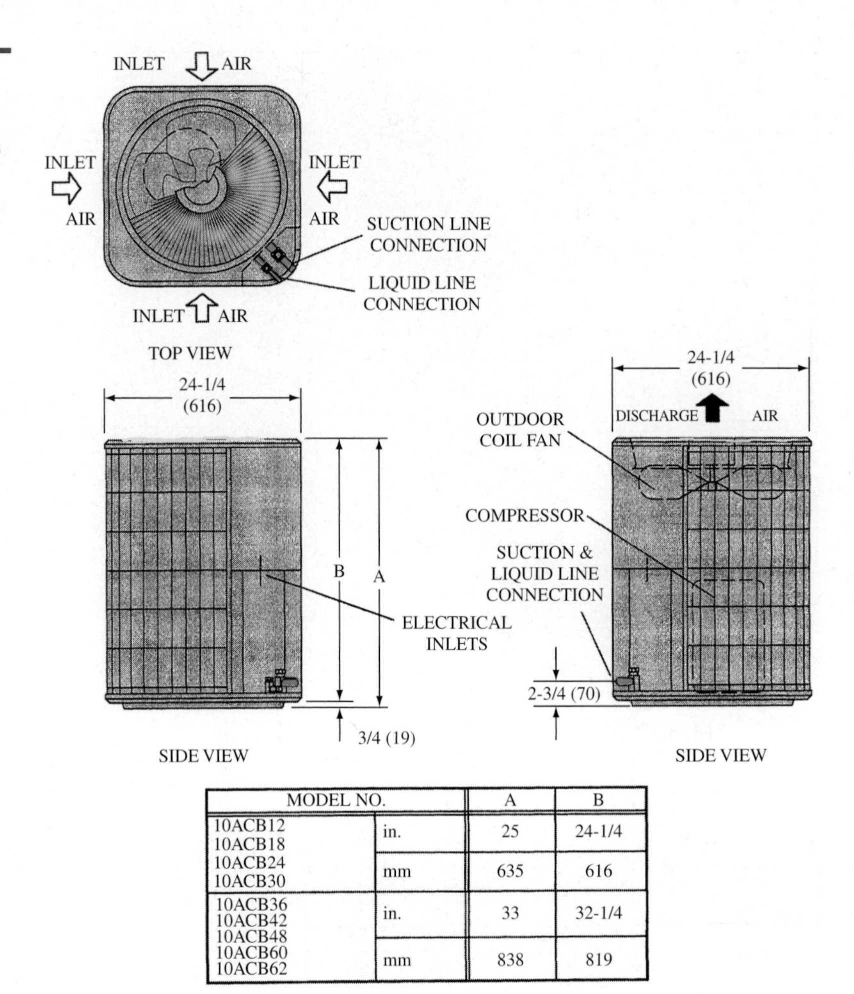

MODEL NO.		A	B
10ACB12 10ACB18 10ACB24 10ACB30	in.	25	24-1/4
	mm	635	616
10ACB36 10ACB42 10ACB48 10ACB60 10ACB62	in.	33	32-1/4
	mm	838	819

Figure 3-1-19 Identifying parts as shown in the three different primary views, top, left side, and right side. In this case, one drawing is enough to give dimensions for nine model numbers. *(Courtesy of Lennox Industries, Inc.)*

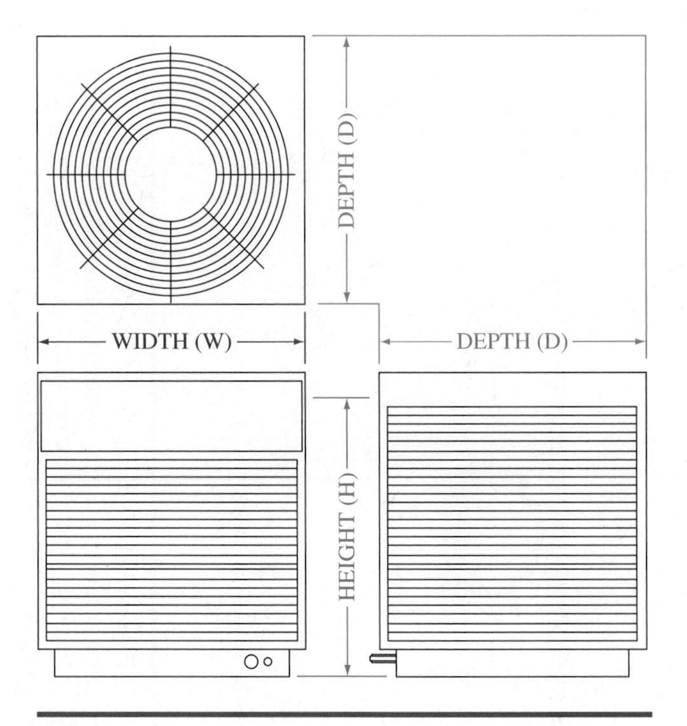

Figure 3-1-20 Dimension layout on orthographic drawing.

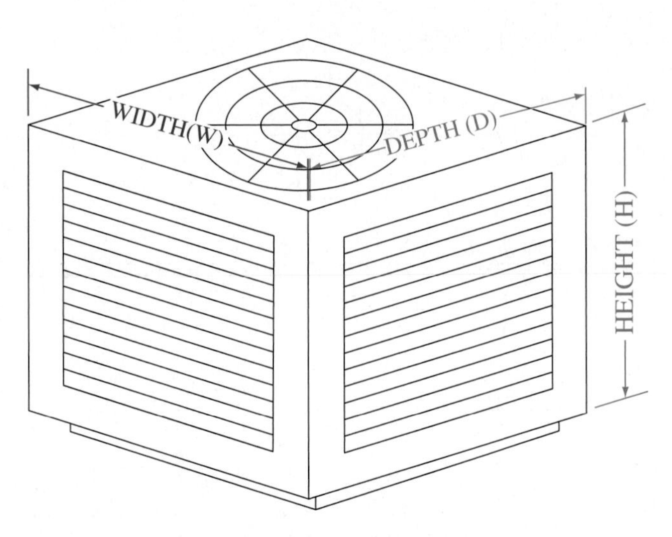

Figure 3-1-21 Dimension layout on a pictorial drawing.

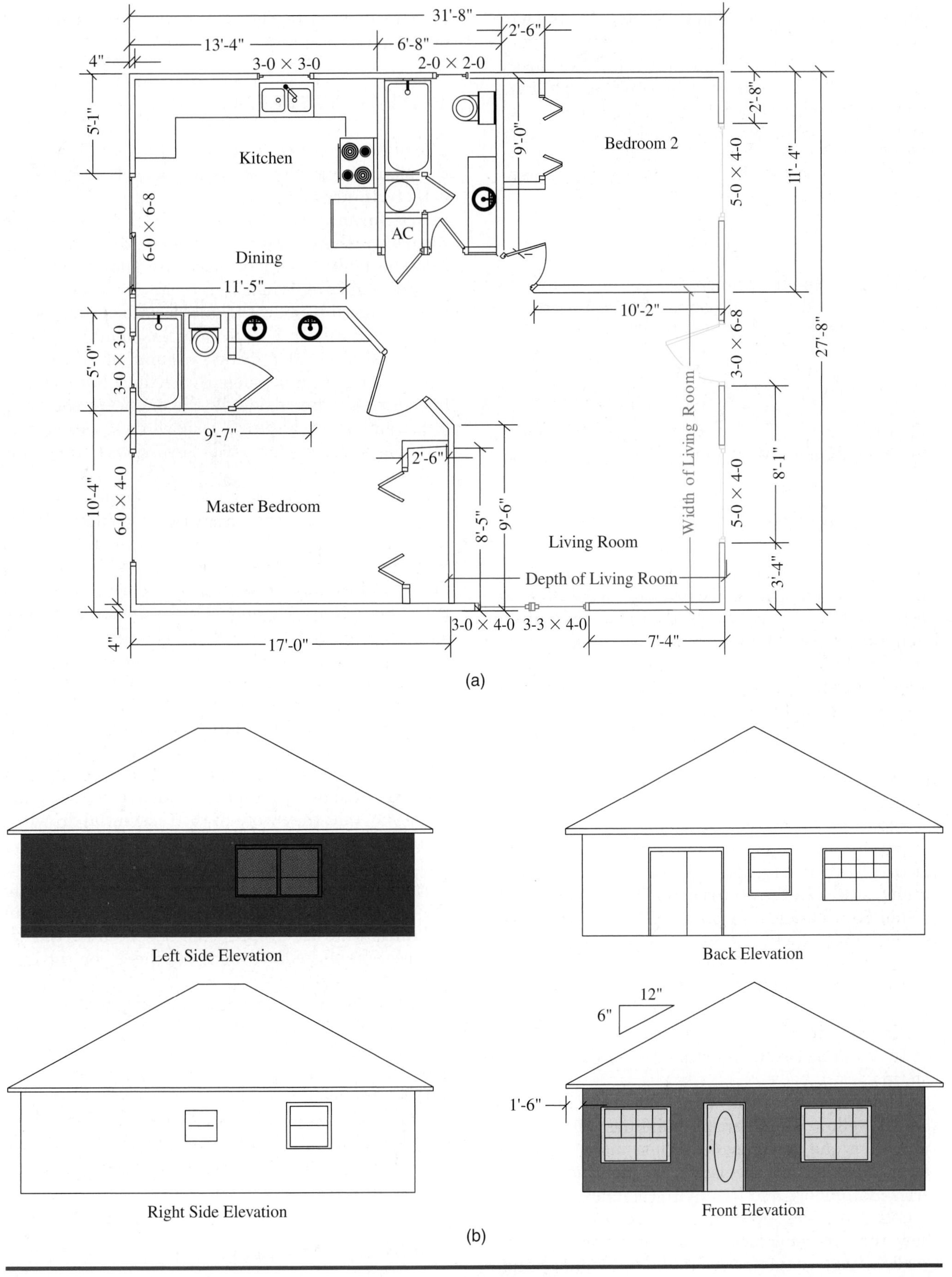

Figure 3-1-22 Walls represented in plan (a) are shown in the same color in elevations (b).

walls on the left side are shown as blue and the windows are green.

The size of the living room is not given on the drawing. First, to determine the depth of the living room you would subtract the size of the master bedroom (17′-0″) from the overall depth of the house (31′-8″).

$$\begin{array}{r} 31'\text{-}8'' \\ -17'\text{-}0'' \\ \hline 14'\text{-}8'' \end{array}$$

The width of the living room is determined by subtracting the width of bedroom 2 (11′-4″) from the total width of the house (27′-8″).

$$\begin{array}{r} 27'\text{-}8'' \\ -11'\text{-}4'' \\ \hline 16'\text{-}4'' \end{array}$$

The living room is 16′-4″ wide and 14′-8″ deep.

ELECTRICAL

Commonly used electrical symbols are shown in Figure 3-1-17 on page 80. On drawings a free flowing line is used to connect switches to lights and outlets in circuits. These branch circuit lines make it easy for the technician to determine which switches control which lights or devices. For example, in Figure 3-1-23, the red light switch next to the front door (on the right side of the plan) controls the red entryway light. In this section, we will examine all of the colored branch circuits in Figure 3-1-23.

All of the blue electrical outlets are connected to the same blue branch circuit. These plugs are located outside, in the kitchen, and in the bathrooms, where electrical shock hazards are highest. All electrical outlets located in these areas must be connected to a ground-fault circuit interpreter (GFCI). The initials GFI are often used to designate an outlet or switch that is attached to one of these safety devices.

In the living room the green switch is connected to the green light in the hall and to the green switch by the back door. These three way switches allow the lights to be turned off and on at more than one location.

The main electrical panel for this house is shaded yellow. Commercial buildings often have more than one power panel. When more than one power panel is used, the panels are identified with a series of numbers and letters. In commercial installations, often, the separate electrical drawings will show the size and location of the various circuit breakers that supply power to the individual building circuits. To make future maintenance easier, many builders will stencil the power panel and breaker number on the surface of every switch and outlet in the building.

HVAC SYSTEMS

The placement of the indoor unit is often located on the plan drawing by a hidden line such as in Figure 3-1-24. An arrow and note are used to identify the indoor furnace, such as the note Vertical Up Gas Furnace with AC shown in the center of the drawing. The sheet metal supply ducts are shown with the supply registers. When flex ducts are used, single lines are used to show the duct locations such as the red lines in Figure 3-1-25. Note the small blue loops with a short leader drawn around the duct run. These mark the dimension of each duct. Round ducts have a single dimension for the diameter, shown in blue numbers on Figure 3-1-25; while square or rectangular ducts will have two dimensions noting their width and height.

On residential drawings the supply registers may be drawn as rectangles. In some cases arrows are used to denote the direction of air distribution from the register. The size and location of residential ducts and supply registers, and even the return grills, are not always noted on the drawings. If they are not provided, you should prepare a drawing showing your proposed sizes and locations and present that drawing to the owner or job site foreman for approval before you begin work.

In commercial installations the unit location will always be shown on the drawing. Also, the drawings will contain the engineered duct layout. Duct sizes are presented as either a note with a leader or in the same way that they were in the residential drawing, using a loop and note. Some commercial rigid sheet metal systems use a short (approximately 10 ft long) section of flexible duct to connect the air diffuser to the end of the rigid duct. This is common practice in buildings that use dropped ceilings, because the exact location of the diffuser is dependent on the ceiling grid.

Commercial air distribution system plans usually will contain the CFM (cubic feet per minute) of air that will be supplied through the room diffuser. This CFM volume of air has been determined to be sufficient to meet that area's heat gain and heat loss requirements. The CFM values listed for each of the diffusers and grills will be used by the air balance technician to determine proper system setup during the initial testing and balancing process.

Detailed Drawings

Some residential drawings and almost all commercial drawings will contain detailed drawings. On

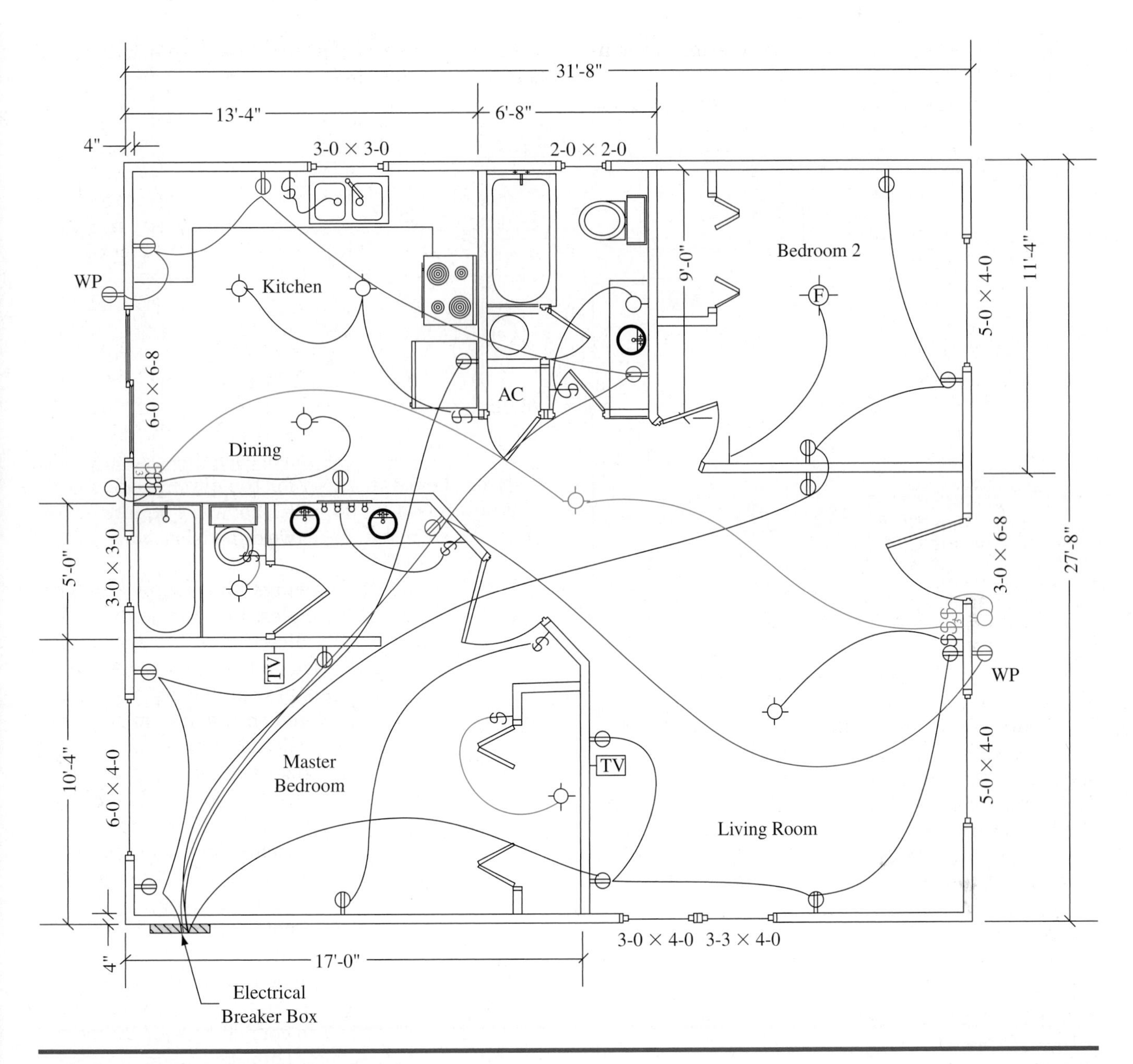

Figure 3-1-23 Electrical branch circuits.

NOTE: As a general rule of thumb 400 CFM is equal to one ton of air conditioning. So a diffuser supplying 800 CFM of air to an area would be providing two tons of air conditioning. A diffuser supplying 600 CFM of air would be producing 1½ tons of air conditioning. This is a quick way of determining the amount of the cooling load requirements for a system.

residential prints details are often used to show the typical wall section insulation in ceiling and walls, Figure 3-1-26. Other detailed drawings are used to show any special millwork for kitchen cabinets, bathroom cabinets, and built-in furniture. On commercial building sets of drawings, more detailed drawings are included. Drawings showing the size and configuration of fittings for natural gas and condensate drain lines are common. Others show roof flashing and the curb for rooftop units on commercial buildings. Many other detailed drawings are included as part of the set of drawings for commercial installations; however most do not relate to HVAC systems.

Drawing Scale

Of course, drawings must be made smaller than the actual building so they will fit on the drawing page.

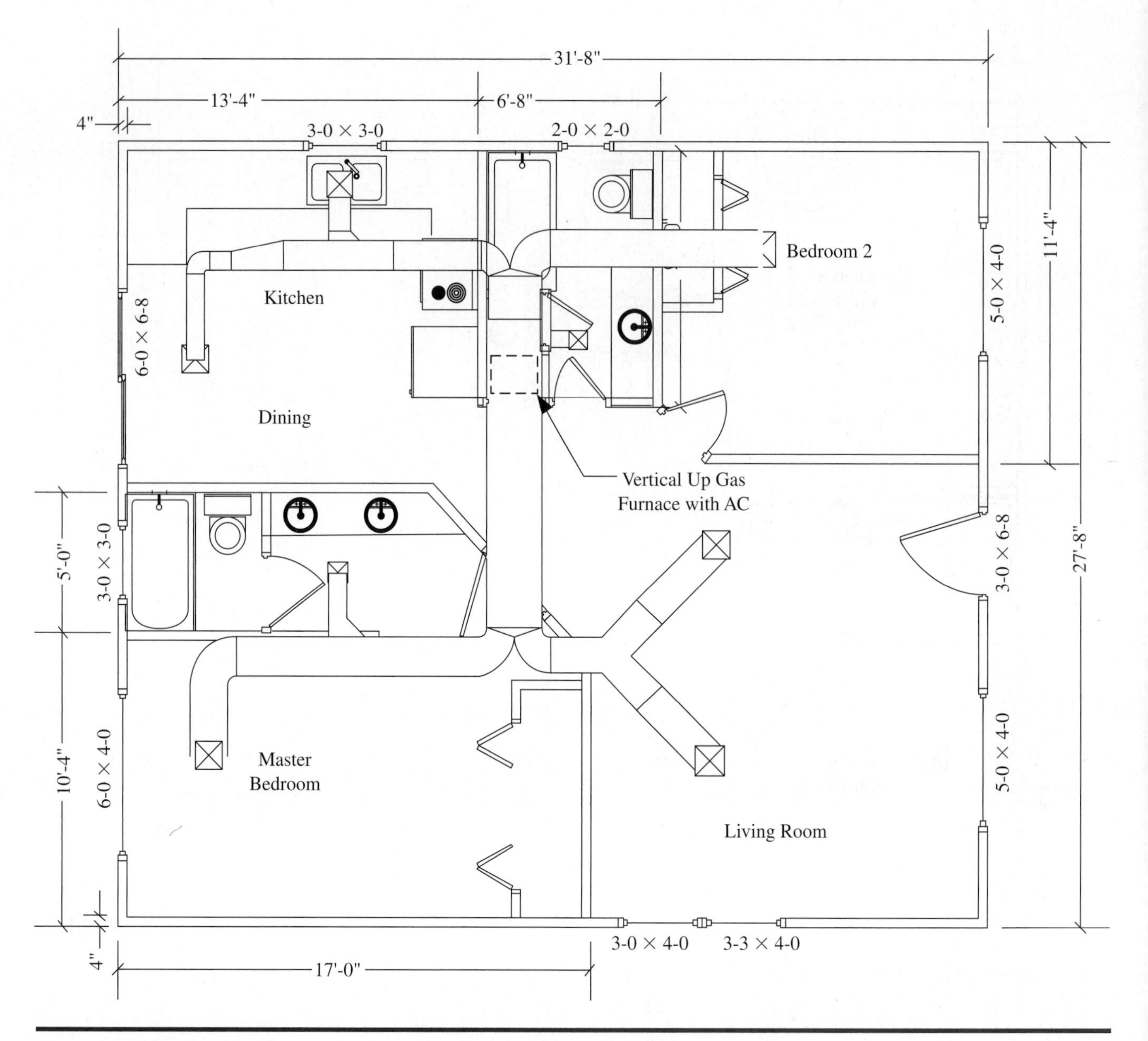

Figure 3-1-24 Mechanical drawing for duct layout for sheet metal duct.

To make the drawing smaller a scale is used. On a scale an inch or foot is represented as shorter than it actually is. Special rulers called scales are marked so they can easily be used to make a drawing of the desired size, Figure 3-1-27a–d. There are two commonly used scales; one is architectural (Figure 3-1-27a,b) and the other is engineering (Figure 3-1-27c,d). Most residential drawings are made using a scale where ¼ in on the drawing equals 1 ft in the building so a large home will fit on a standard size sheet of drafting paper. On this scale 1 ft of length is represented by ¼ in and a 40 ft length would be 40/4 or 10 in long, Figure 3-1-28. Other scales can be used to allow the drawing to fit on the page. The drawing in Figure 3-1-29 was drawn to a 3/16 in. to 1 ft scale. The inset shows a close view of the scale used to check the drawing. Commercial buildings are drawn most frequently at ¼ in equals 1 ft or ⅛ in equals 1 ft scales.

Detailed drawings are not drawn at the same scale as the floor plan. Typically, ¾ in equals 1 ft and ⅜ in equals 1 ft are used for detailed drawings on both residential and commercial drawings.

Because copies of the original drawing are made, the actual size may change slightly. The closeup view of the scale in Figure 3-1-30 shows that the measurement of 5′-1″ shown on the drawing would actually be read as about 5′-3″ if read off the scale. For this reason, it is not usually acceptable for someone to measure directly off of a drawing to

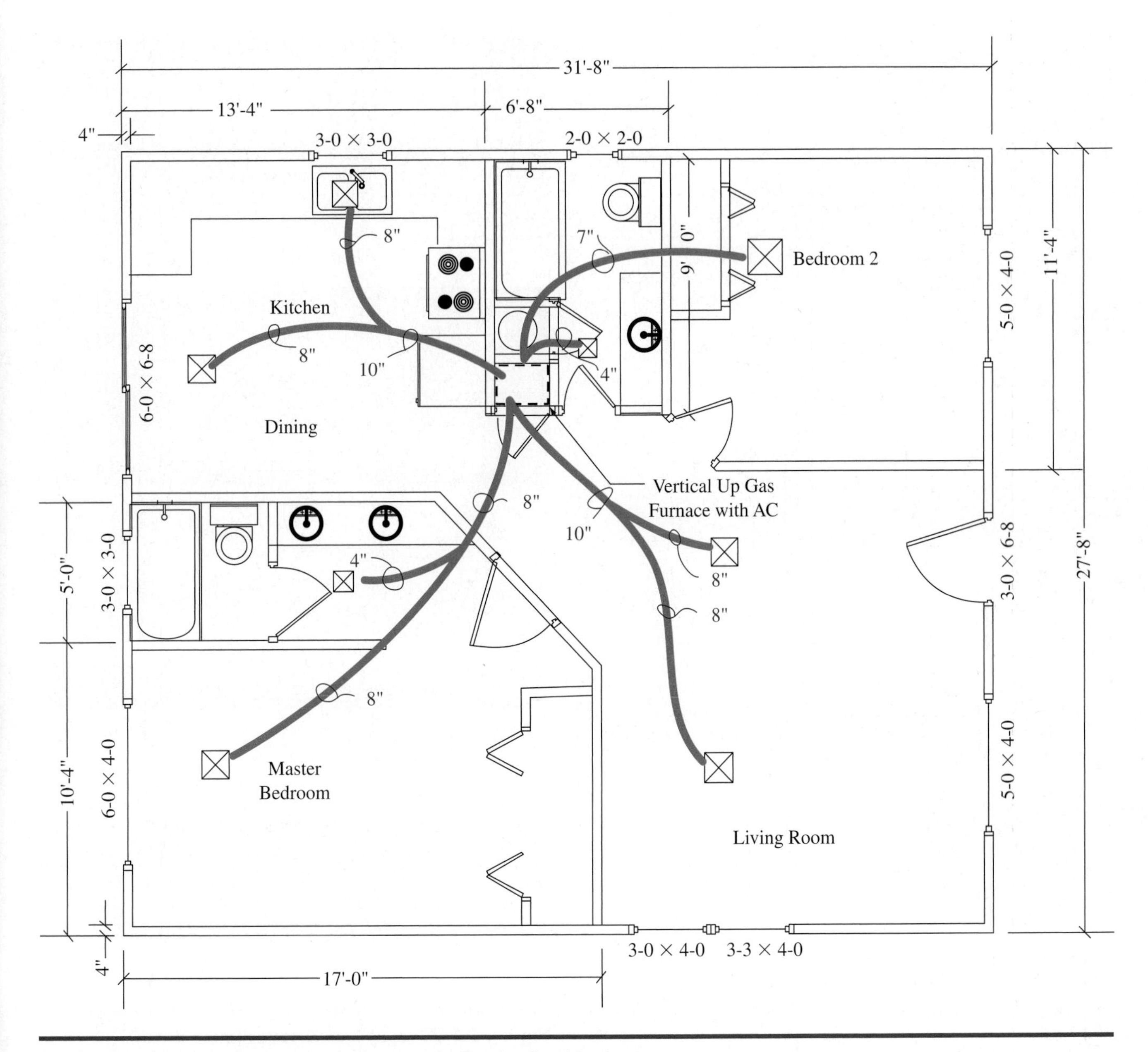

Figure 3-1-25 Mechanical drawing for duct layout for flexible duct.

determine the exact size of a part of the building. In HVAC, however, for the purpose of estimating materials required for a duct system in residential and light commercial applications, direct measurements can be taken from a drawing, as long as the measurements are used for the purpose of estimating only. However, if rigid fabricated sheet metal ducts are to be used, accurate building measurements must be taken before the sheet metal is fabricated. When rigid sheet metal is used, the technician should go to the job site and measure the building as it may differ from the building as drawn. The exact location of walls, supports, or other building systems such as plumbing may require slight changes in the proposed duct system layout so that these obstacles can be considered in the duct design. Unfortunately the HVAC system installation is often one of the last pieces installed before finish construction of the interior space is completed. As a result, we must often work around obstacles such as pipes, drain lines, gas lines, and other previously installed equipment in order for our system to fit and work properly.

Each trade will usually be provided with a set of drawings to work with during construction. The common practice for HVAC technicians is to draw the unit location and duct runs, grills and register locations on these drawings if that information was not already included as part of the set of drawings. Colored felt tip pens are ideal for this purpose. The

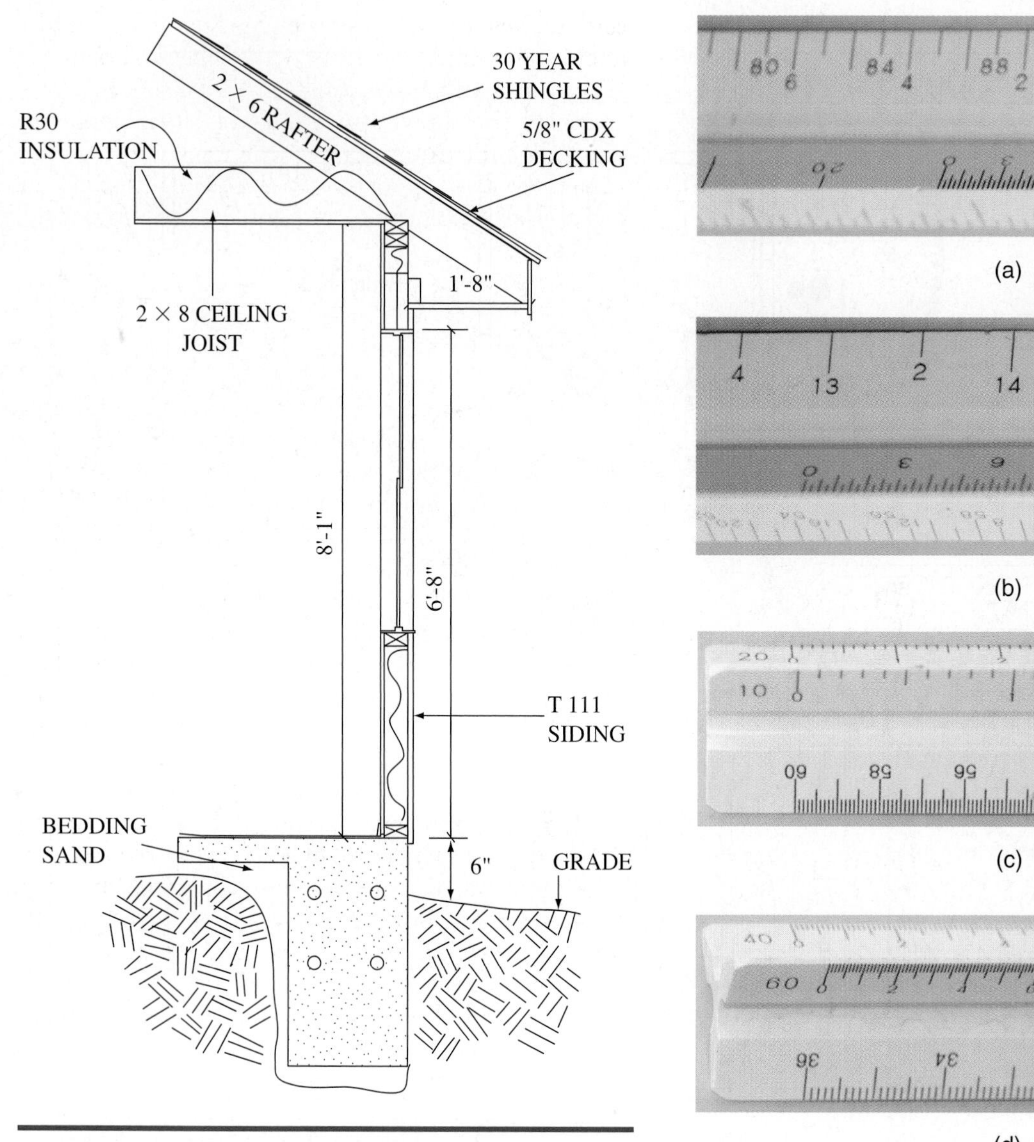

Figure 3-1-26 A detailed drawing of a typical residential wall section, including many details about insulation and shingles.

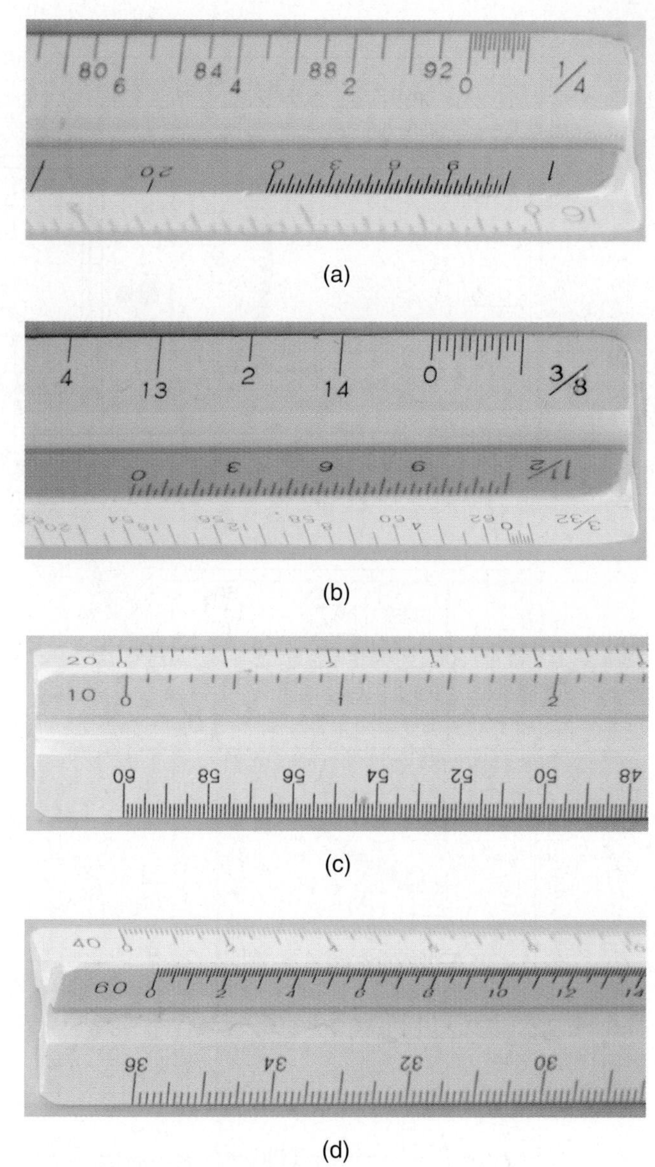

Figure 3-1-27 (a,b) Typical scales found on an architectural scale; (c,d) Typical scales found on an engineering scale.

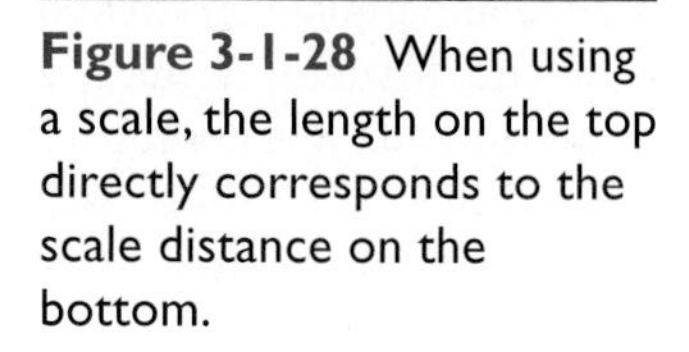

Figure 3-1-28 When using a scale, the length on the top directly corresponds to the scale distance on the bottom.

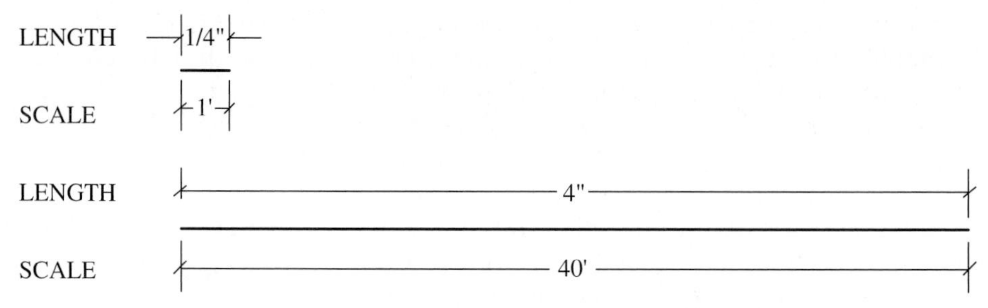

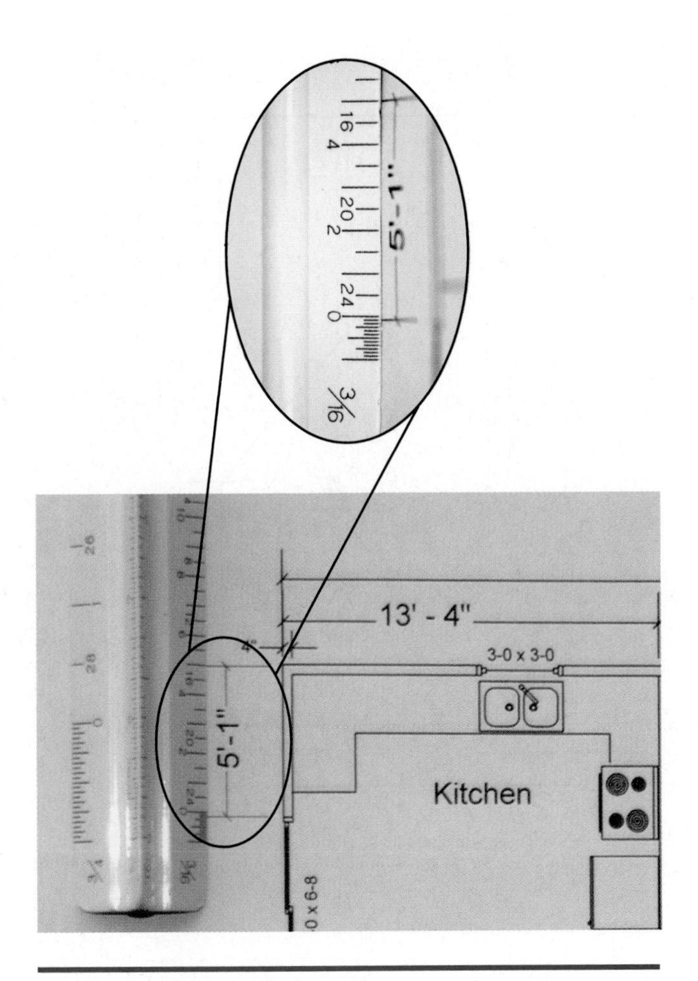

Figure 3-1-29 A drawing at ⅛ in with a 3/16 scale to the left and enlarged above.

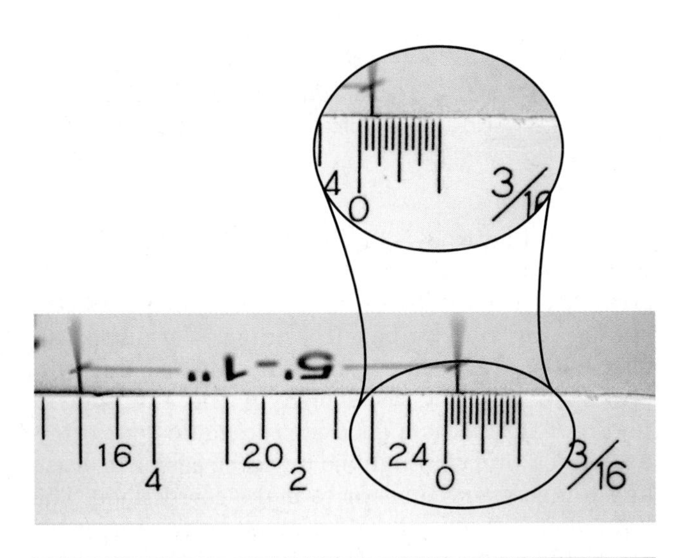

Figure 3-1-30 Close-up views of a reproduction of the drawing in Figure 3-1-29. Note that the scale reading is about 5′-3″, while the drawing states the distance as 5′-1″. This shows the effect of moisture on drawing size.

color makes it much easier for the technician to read and quickly make an important decision about the HVAC mechanical system information that has been added to the drawings. When the job is complete, these as-built drawings should be left with the job site supervisor so they can be included with the final building drawings presented to the building owner at closing.

SKETCHING

The ability to make a sketch of a house or building is not required by most HVAC/R employers. But often there are no drawings available for the building or area where you are going to be installing HVAC or R equipment. So it is nice to know how to make a sketch that can be used to calculate a heat load or how a duct system and/or hydronic piping system can be laid out. Sketches are quick drawings made without the aid of drafting tools. The sketch should look proportional, but because all of the needed dimensions are written on the sketch, it may be made to scale or to approximate scale.

One of the best ways to make a sketch is to use graph paper that has a ¼″ or ⅛″ square spacing, Figure 3-1-31. The scale of graph paper selected is dependent on the size of the house or building you are drawing. A building approximately 32′ by 42′ can be drawn on a standard 8½″ by 11 sheet of ¼″ graph paper; and a building approximately 64′ by 84′ can be drawn on a ⅛″ scale sheet of graph paper.

Sketches can be made as either a single line sketch (Figure 3-1-32a) or a double line sketch (Figure 3-1-32b). The single line sketch is the fastest and easiest to draw and gives enough information for most HVAC/R jobs.

The art of sketching a series of short quickly drawn lines to make up a completed longer line is

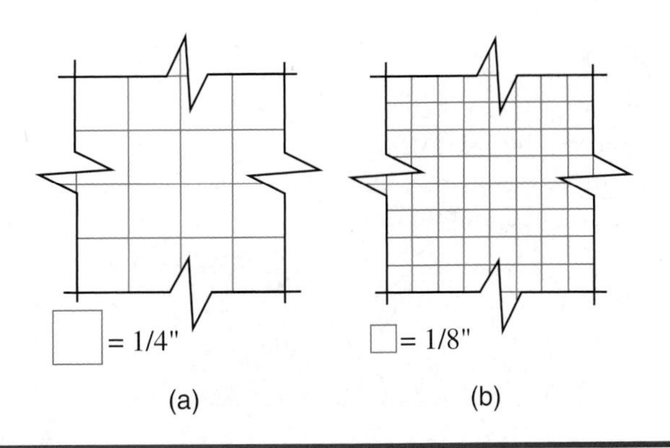

Figure 3-1-31 (a) Quarter inch graph paper; (b) Eighth inch graph paper.

Figure 3-1-32 (a) Single line sketch; (b) Double line sketch.

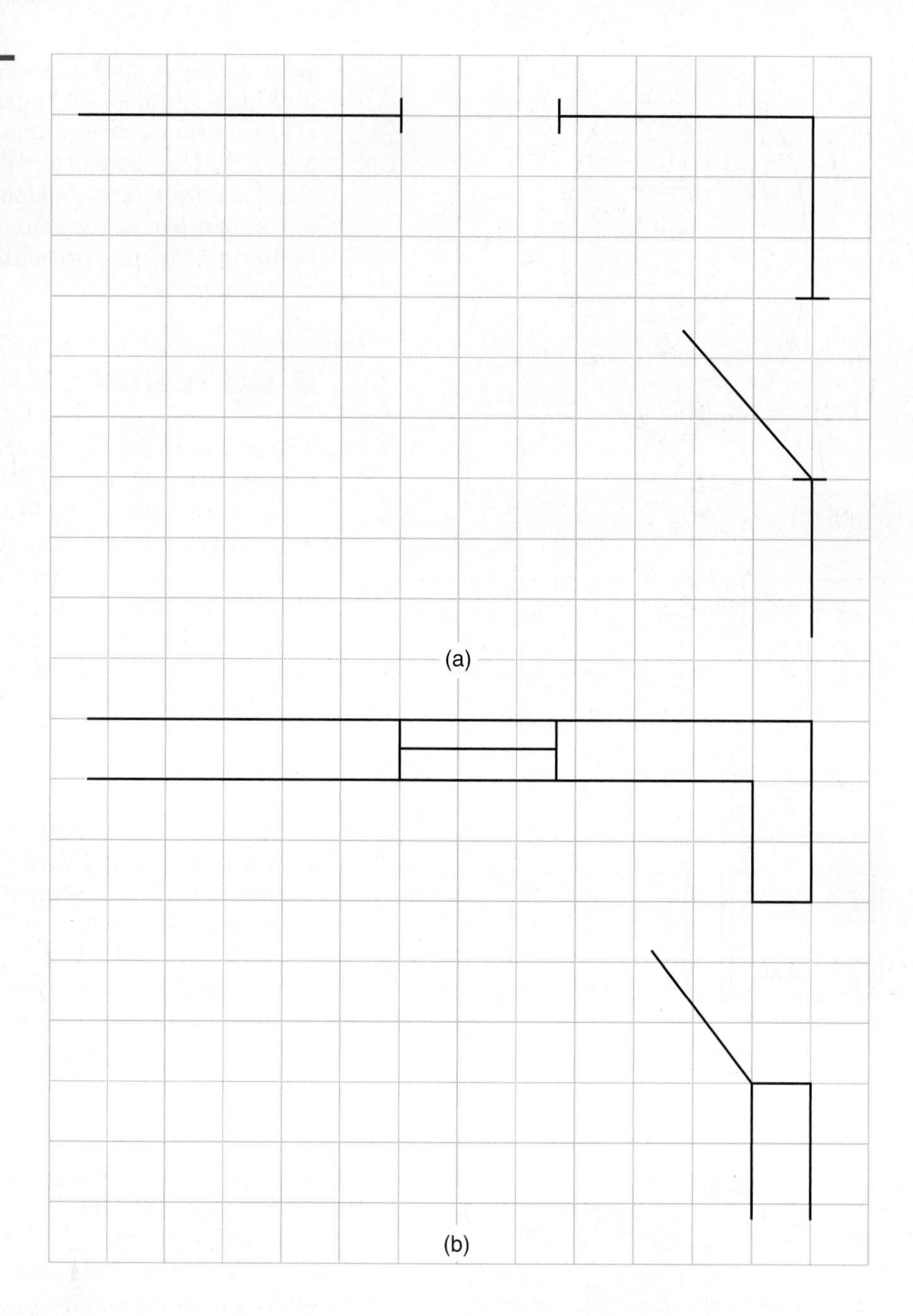

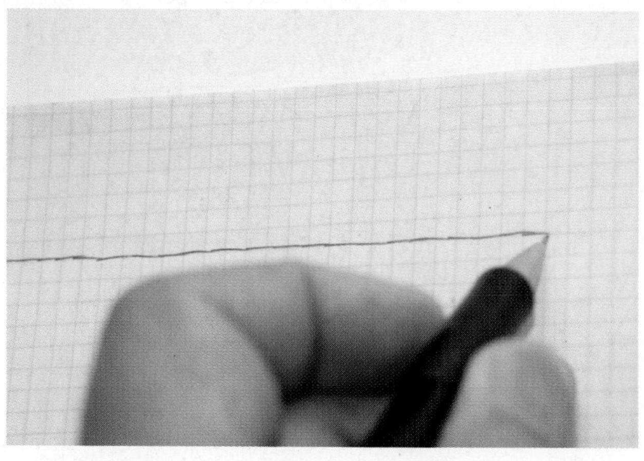

Figure 3-1-33 A sketch line is made up of a series of short lines.

shown in Figure 3-1-33. Because the longer line is made up of a series of short lines, it can be made straighter than if it were drawn continuously without picking up the pencil. The smaller lines can even be slightly askew, and the finished line will still look straight, as shown in the example in Figure 3-1-34. It is a good idea to make your original sketch line very light so it will be easier to erase any unnecessary lines or to make corrections. The light sketch lines can be darkened later as needed.

Using a pencil and graph paper, sketch the floor plan shown in Figure 3-1-35. Change this sketch from a two-line to a one-line sketch. Remember, sketches are made without the use of drafting equipment, including straightedges.

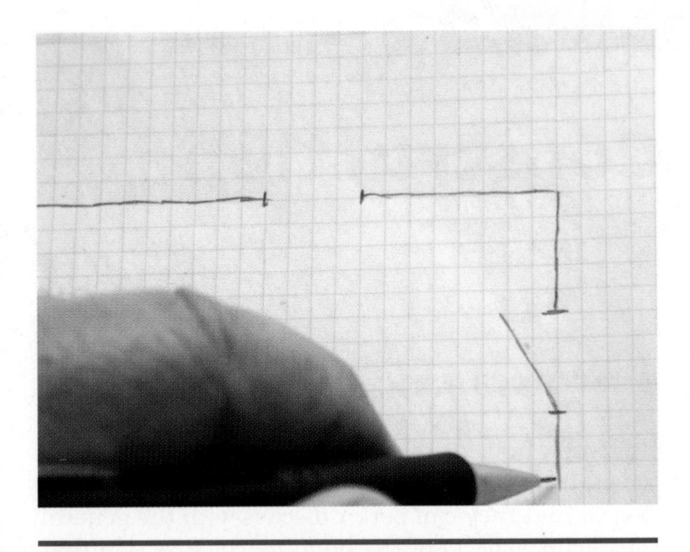

Figure 3-1-34 A sketch line can look straight and even.

TECH TALK

Tech 1. I don't need to be able to make sketches, because I can remember all of the important information about a job.

Tech 2. Sketches are not just for you, they are so that other people can look at the building and be able to do a layout or a heat load. Besides, even the best memory is never as good as a written note.

Tech 1. Well, I guess I have forgotten some things. Thanks.

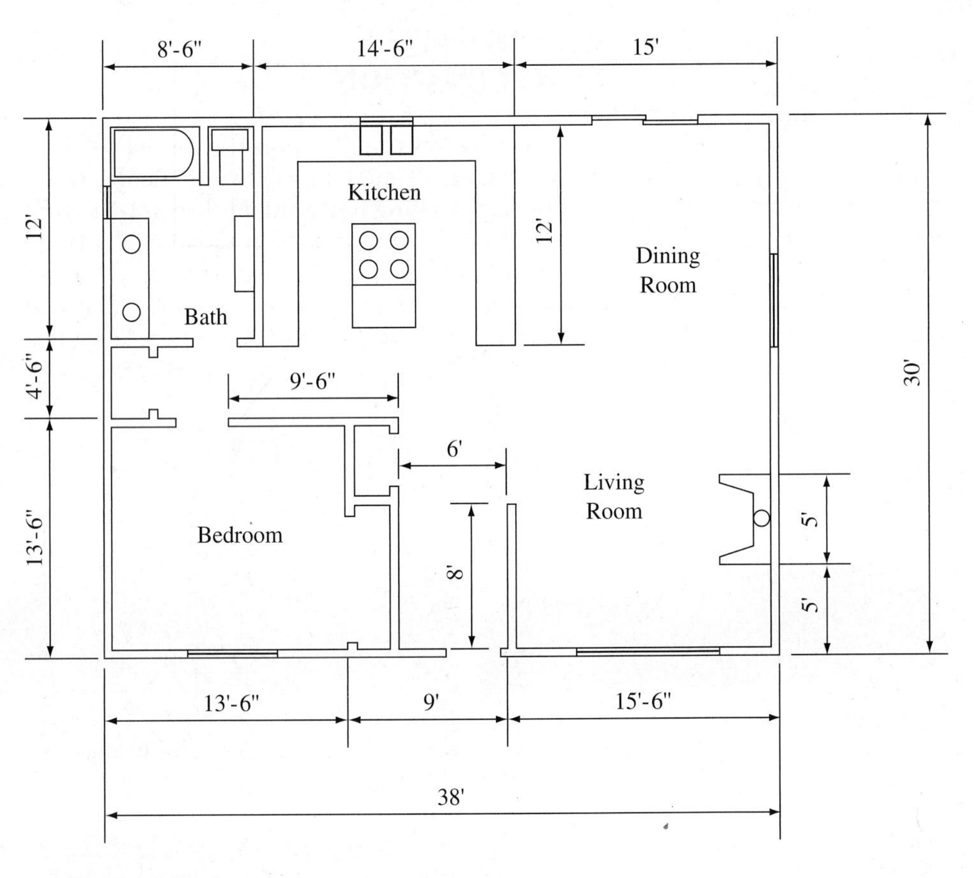

Figure 3-1-35 Floor plan to be sketched.

UNIT 1—REVIEW QUESTIONS

1. What are mechanical drawings used to illustrate?
2. What does the dimension line represent?
3. What is the cutting plane line used to locate?
4. What are break lines?
5. List the two major groups into which mechanical drawings are divided.
6. Architectural drawings are used to show the ________ and ________ of houses and buildings.
7. An orthographic projection is actually referred to ________ dimensional drawing.
8. What are the initials GFI often used to designate o drawing?
9. On drawings the HVAC indoor unit is often located b a ________.
10. List the two commonly used drawing scales.
11. How is a line sketched?
12. Using a pencil, graph paper, and a measuring tape, make a sketch of your home.

UNIT 2

Basic Building Construction

OBJECTIVES: In this unit the reader will learn about basic building construction including:

- Residential Construction
- Basements
- First Floor Construction
- Wall Construction
- Ceiling and Roof Construction
- Insulation
- Walls and Ceilings
- Light Commercial Construction
- Drop Ceilings
- Walls and Partitions

INTRODUCTION

It is important for the air conditioning technician to have an understanding of basic building construction so that they can better discuss with the general contractor aspects of the building construction that will affect the installation of the HVAC system.

RESIDENTIAL CONSTRUCTION

Most residential construction uses wood as the primary structural material for walls, floors, ceilings, rafters, and decking material, Figure 3-2-1a–d. Steel studs and beams can be used in some residential construction but are not in widespread use. Other common materials used in construction include concrete blocks, cinder blocks, and poured concrete walls.

(a)

(b)

(c)

(d)

Figure 3-2-1 Major steps in house construction: (a) Site preparation; (b) Framing; (c) Dried in; (d) Exterior flat work.

Basements

Residential construction may or may not include a basement, as shown in the drawings in Figure 3-2-2. While the standard ceiling height of a normal room in a residence is 8 ft, basement ceilings may be lower. When the basement ceilings are less than 7 ft, laying out and installing ductwork so it does not interfere with the use of the basement can be a problem. Basements are constructed in houses in many states and regions but may not be very common in other areas. Area soil conditions and the level of groundwater both affect the construction of residential basements.

Moisture and poor ventilation are the two most common problems associated with basement areas. Without adequate dehumidification, high levels of moisture can contribute to the growth of mold and mildew. Proper ventilation can help control mold and mildew growth as well as provide better indoor air quality.

Basements are typically constructed with poured concrete or cinderblock walls, Figure 3-2-3 (top, middle). They usually have concrete floors. However, in some arid (dry) climates such as Denver, Colorado, basements may have a wooden joist floor structure, Figure 3-2-3 (bottom). Properly designed and constructed basements provide a relatively low cost per square foot additional space to residences.

First Floor Construction

Houses not having basements may have slab floors (Figure 3-2-4a) or pier and beam floors (Figure 3-2-4b). Concrete slabs are sometimes called concrete pads. They may be constructed using post tension cables, reinforcement rods, or wire mesh as reinforcement materials in the concrete. These materials are used to help prevent cracking of the slab over time. Typical residential slabs are from 4–6 in thick and are poured continuously, incorporating the perimeter beam and cross beams that were used, Figure 3-2-5a,b.

TECH TIP

The plastic put down under a slab is not a moisture barrier. It is not there to keep water vapor out of the residence. It is instead used to keep the newly poured concrete from drying too quickly. The slower concrete dries, the harder it is, and the less likely it is to crack.

If refrigerant lines or condensate plumbing are to be placed below grade, they must be preplaced before the slab is poured, Figure 3-2-6. They should be installed inside a 3 in to 4 in chase or conduit pipe. If they are buried in sand to protect them from sharp objects such as stones.

NOTE: Copper piping, including refrigerant lines, reacts to concrete. Therefore, they must be separated with a nonreactive material, usually foam refrigerant line insulation, as shown in Figure 3-2-7a,b.

When preinstalling refrigerant lines and condensate drain lines before a slab is poured, make certain that your dimensions are taken from the outside of the building to locate these pipes. It is often the case that dimensions are given from the inside wall. If the outside building dimensions are not given, you must look at the wall thickness as provided on the detailed wall section and add that dimension to your interior dimensions so that your pipes will come up in the appropriate location. The condensate drain in the center of Figure 3-2-8 is located from the outside kitchen wall by first adding the wall thickness, 4″, to the 13′-4″ distance to the AC equipment room wall. The center of the condensate pipe is approximately 2″ inside the AC equipment room. The distance to be measured is 13′-10″ from the outside kitchen wall. The distance to the condensate pipe from the outside bathroom wall is 8″–10″. This distance is calculated by adding the outside wall thickness to 9′-0″ and then subtracting the inside wall and pipe distance inside the room. An example of refrigerant lines and condensate line being installed is shown in Figure 3-2-9. Moving pipes after concrete pouring can be done, but it is time consuming and expensive. Always double check your dimensions before leaving the job site after roughing in these lines.

First floor and construction for pier and beam and for basements are similar. Both use joists spaced either 16 in or 2 ft apart. These joists can be solid wood, usually fir, or they may be wood or laminate trusses (Figure 3-2-10a); or wood or laminate materials for beams (Figure 3-2-10b). When solid wood joists are used, your ductwork must be located below the roof. Trussed and glue laminate materials may allow the ductwork to pass through precut or fabricated openings. With permission from the construction manager and the manufacturer of the laminate being used, you may be allowed to cut holes on site, which will allow the ducts to pass through.

SAFETY TIP

Do not cut a hole in any structural member without first obtaining permission from the manufacturer of the product and the job site foreman. Under no circumstances can a beam's structural surface be cut, shown in red in Figure 3-2-11. Cuts or gouges along these surfaces can cause the structure to prematurely fail.

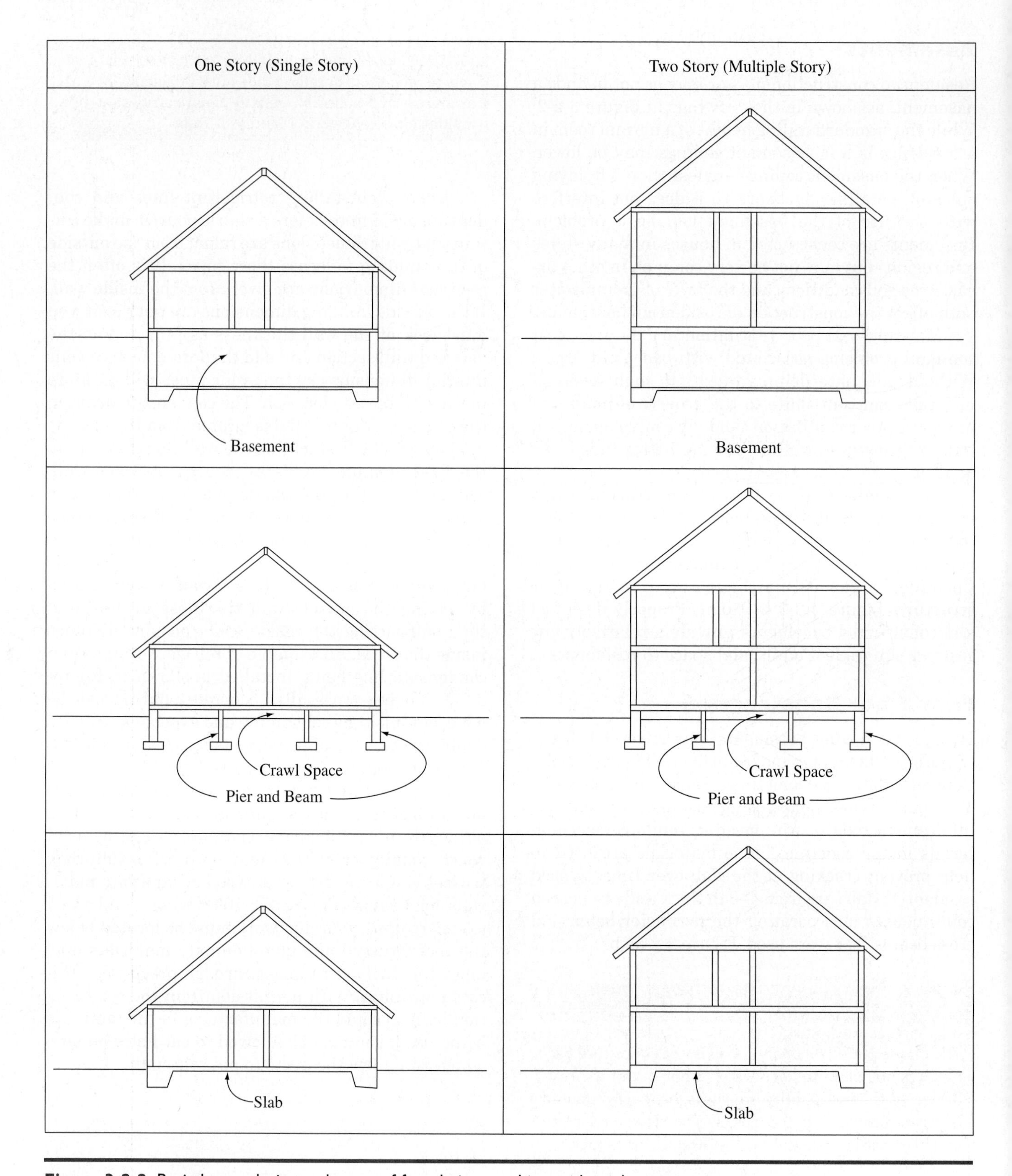

Figure 3-2-2 Basic house design and types of foundation used in residential construction.

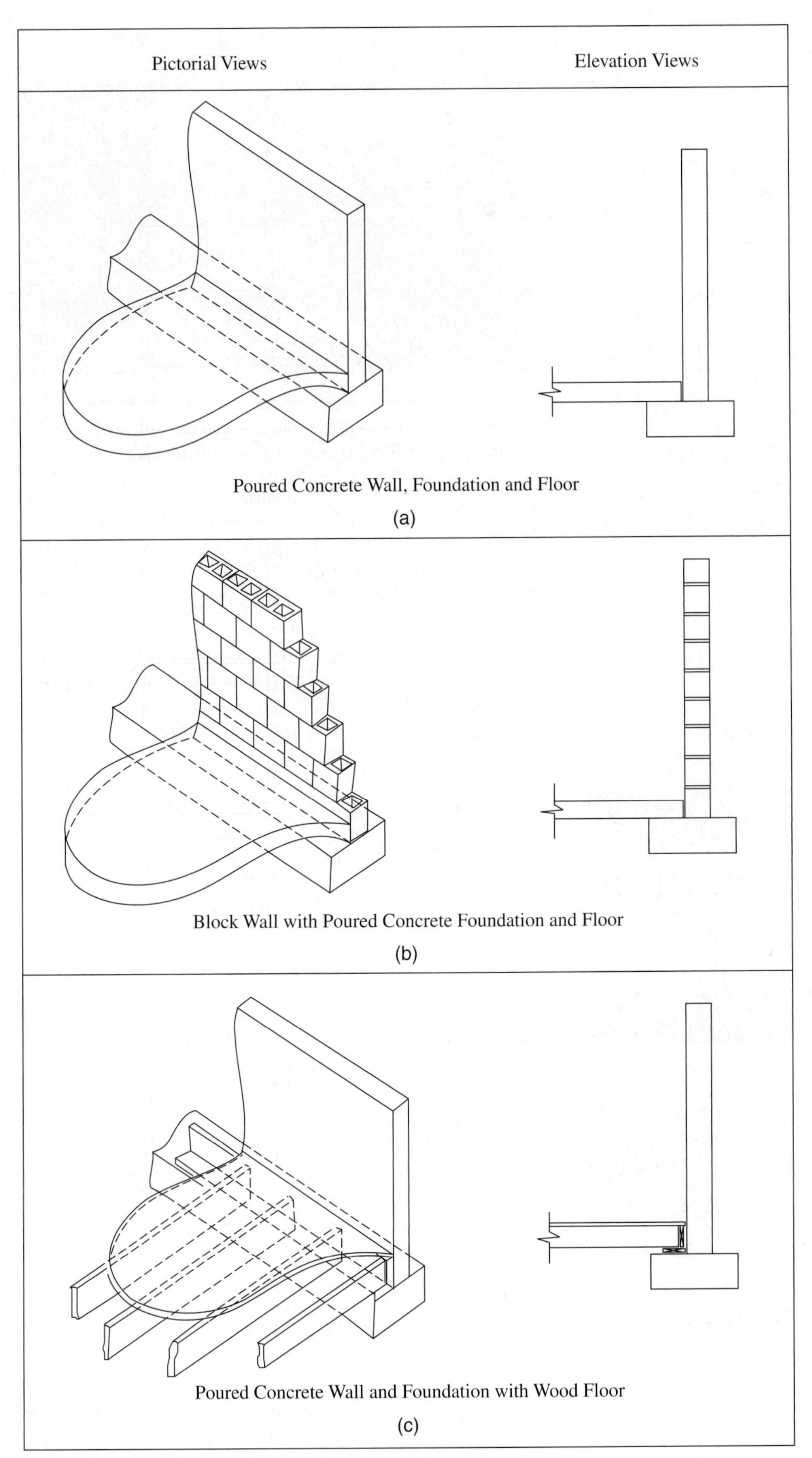

Figure 3-2-3 Types of basement wall and floor construction techniques used for residential homes.

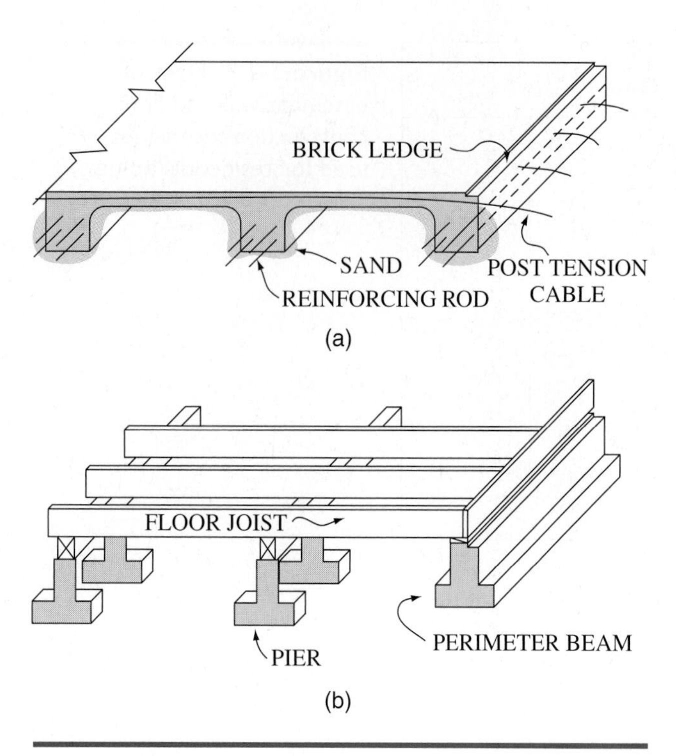

Figure 3-2-4 Floor construction techniques: (a) Slab floor using post tension cable; (b) Pier and beam.

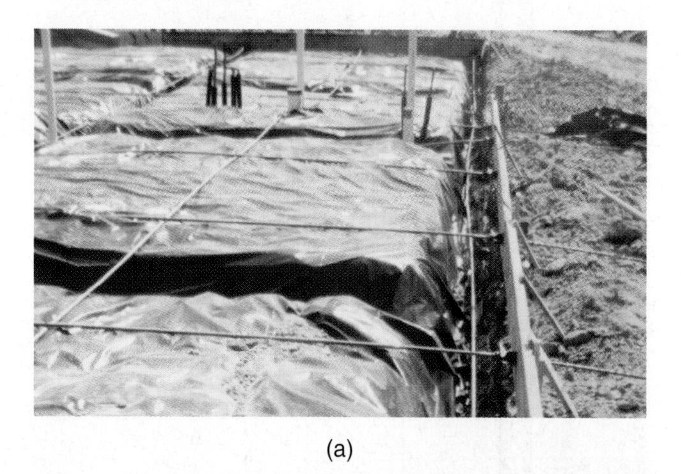

(a)

(b)

Figure 3-2-5 (a) Slab preparation; (b) Slab pouring.

Figure 3-2-6 Plumbing, including refrigerant, lines are buried in sand-filled trenches beneath the slab.

The most common subfloor material used today is 1⅛ tongue and groove plywood decking. Figure 3-2-12a shows 1⅛ plywood decking being installed as the subfloor of a residence. It is often glued and nailed down. Using mastic reduces future floor squeaks. Figure 3-2-12b shows how the plywood decking is offset, that is, each new length starts at a different place from the one next to it. This increases the strength and durability of the subfloor.

Figure 3-2-13a–d shows the typical wall sections for the four most common types of residential wood frame construction. You will note from the first floor construction level that each of the typical wall sections is similar.

Wall Construction

Most residential walls are constructed with two-by-fours located 16 in on center, Figure 3-2-14. In some cases two-by-sixes are used on the exterior walls. This allows for more insulation to both reduce the heat load and the sound level in the dwelling. In some areas when two-by-sixes are used in the perimeter wall, they may be spaced 24 in on center. The two-by-fours or two-by-sixes used for walls are located on a horizontal board called a base plate. If this base plate is located on a concrete floor, it must be pressure treated lumber. This is to reduce both wood rot and insect attacks. Two horizontal boards are placed on top of the stud wall. These are called the top plate. Typical wall heights for residential construction are 8 ft, 9 ft, and 10 ft with 8 ft being the most common. A vapor barrier is placed on the outside of the two-by-four wall in warm climates and on the inside in cold climates before the finished siding or masonry work is performed, as shown in Figure 3-2-15.

(a)

(b)

Figure 3-2-7 All copper lines are protected with foam insulation so that they do not come in direct contact with the slab concrete.

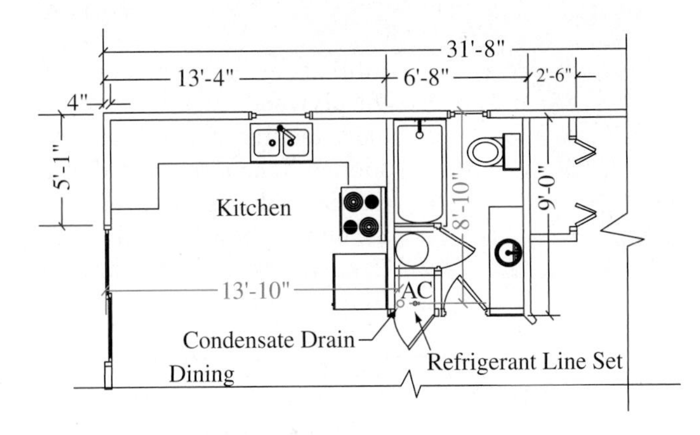

Figure 3-2-8 Locating refrigerant and condensate drain lines.

Figure 3-2-9 When properly located within the slab, the refrigerant lines and the condensate can easily be connected to the indoor unit during rough-in.

(a)

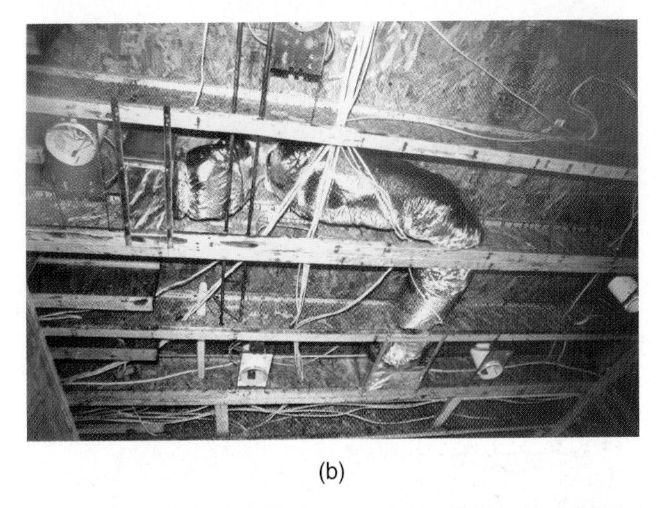

(b)

Figure 3-2-10 Vents going through two types of floor joists: (a) Truss; (b) Beam.

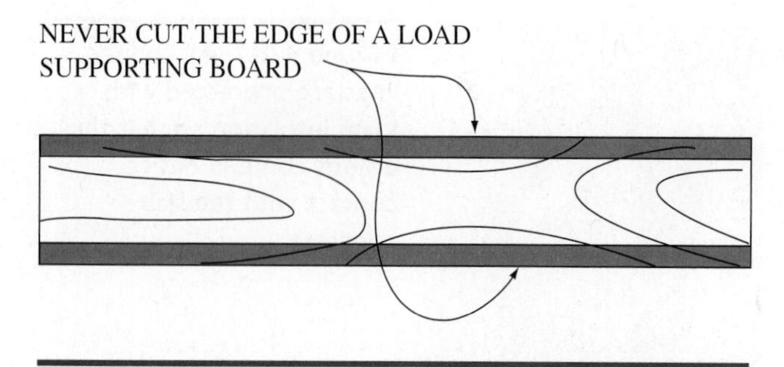

Figure 3-2-11 Cutting the edge of a horizontal supporting beam will significantly reduce the beam's strength.

Ceiling and Roof Construction

Some builders use prefabricated roof trusses, as shown in Figure 3-2-16. These are almost always fabricated out of two-by-four materials with metal cleats connecting the joints. Contractors often favor such trussed roof systems because they are strong, durable, and easily assembled on the job site. They make it slightly more difficult for the air conditioning technician to locate a unit in an attic and make straight duct runs.

(a)

(b)

Figure 3-2-12 (a) Plywood decking is glued and nailed to the floor joist; (b) The plywood joists are offset to increase the strength of the subfloor.

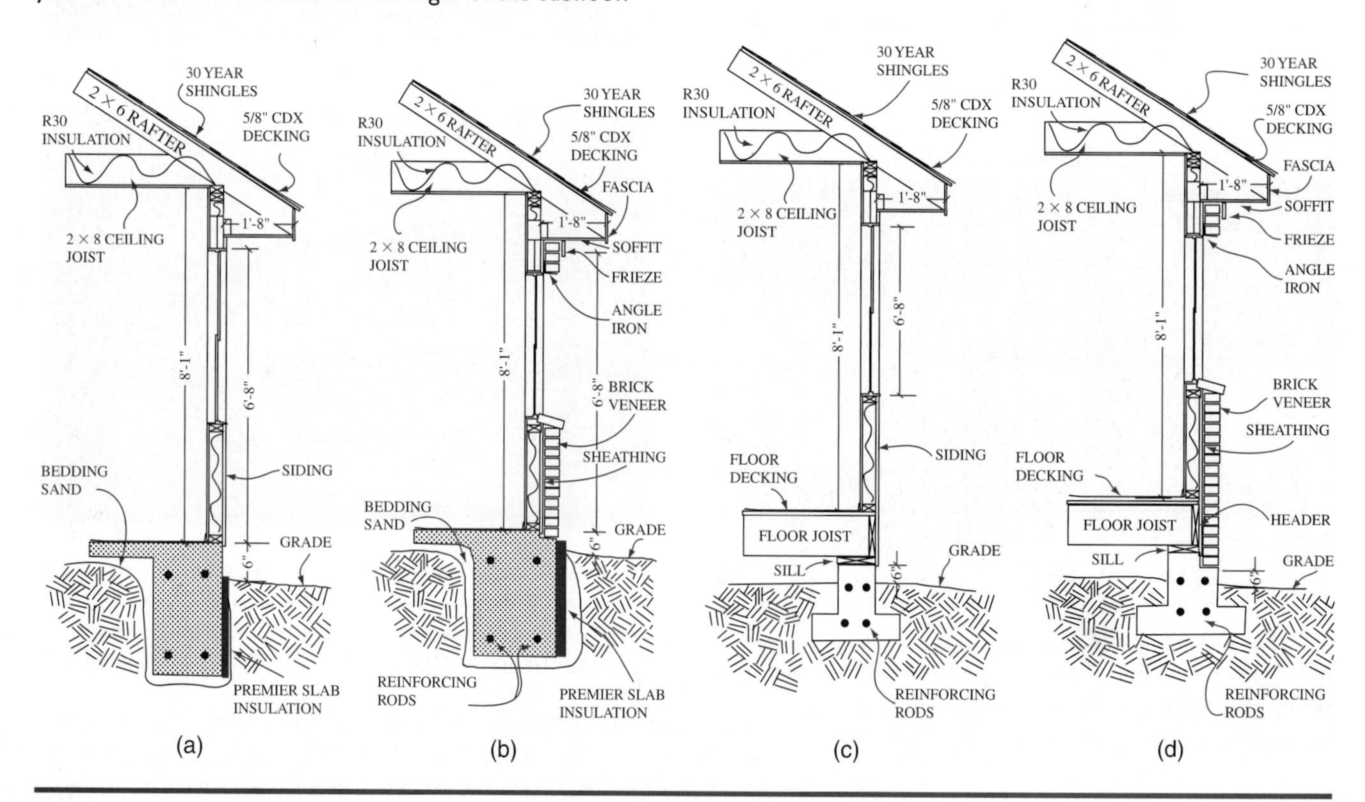

Figure 3-2-13 Typical wall sections: (a) Slab with siding; (b) Slab with brick veneer; (c) Pier and beam with siding; (d) Pier and beam with brick veneer.

Figure 3-2-14 Typical framing is installed on 16 in centers. Blocking is used at wall intersection to provide easier sheetrock installation.

Figure 3-2-15 Vapor barrier is installed before brick veneer is applied.

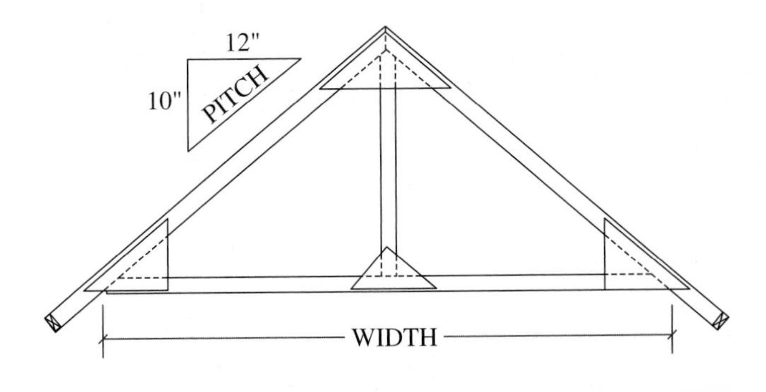

Figure 3-2-16 Prefabricated trusses are available in a variety of widths.

Figure 3-2-17 Ceiling joists can be made from a variety of sizes of conventional lumber, depending on the roof slope, load, and span. The flatter roof section here uses two-by-sixes, but the steeper section uses two-by-twelves.

If trusses are not used, the ceiling rafters are typically two-by-six to two-by-twelve dimensional lumber depending on the room span, as shown in Figure 3-2-17. Roof rafters are typically two-by-sixes; however, depending on the height of the roof, width of the span, and possible snow load or northern climates, two-by-eight (2 × 8), two-by-ten (2 × 10), and even two-by-twelve (2 × 12) may be used for rafters. For composite roofs, asphalt shingles, and sheet metal, the entire roof area will be covered by 7⁄16" thick plywood roof decking material. When wood shingles are used, a lapped system will be installed across the top of the rafters as opposed to roof decking. An advantage of shingle construction is that with this lapped substructure the attic area will vent freely, which reduces the heat load on the residence during summer.

SAFETY TIP

A safety harnesses and lanyard is required under OSHA regulation for construction workers any time they will be working more than 6 ft above the floor.

INSULATION

There are four common types of insulation used in residential construction. They are:

- **Rock wool** Rock wool is a synthetic material fabricated from a mineral.
- **Fiberglass** Fiberglass insulation is a spun glass. Fiberglass is available in rolls and bats with or without an attached vapor barrier. Most fiberglass insulation is pink; however, yellow and

white are also common colors. The color of the insulation has no effect on the capacity of the insulation to resist heat.

- **Cellulose insulation** Cellulose insulation is a synthetically produced insulation product that uses recycled newspaper as its basic material. Cellulose insulation can either be blown loose or else a binder can be added so that it can be blown in wall cavities as a solid fill material.
- **Foam insulation** There are a number of foam insulations that can be used in residential construction as either a sheet or a product that is blown in on site.

SAFETY TIP

When working with insulation, wear a respirator to avoid breathing the fine insulation dust. Avoid touching insulation with unprotected skin. Wearing a long sleeved shirt, gloves, and a cap will reduce your exposure to insulation that can cause irritation to your skin.

For all insulation, the ability to resist the transfer of heat is directly related to its thickness. Some materials may have greater resistance per inch of material than others. But if the insulation material is compressed, it loses its resistance significantly. For example, 4 in of fiberglass insulation may have an insulation value of R11, but if the same insulation is compressed into the 3½ in stud space in a wall, its insulation value drops to R10.25. Further compounding the problem of insulation compression is a standard practice by many insulation installers of pressing the sides of the bat of insulation down rather than stapling the tabs of the vapor barrier to the studs.

The ceiling is the area in a residence that loses and gains most of the heat. Since ceilings are so crucial for heat gain and loss, code requires that a much thicker layer of insulation be installed in the attic. Attic insulation can be blown, loose insulation, or bat insulation. In any case some settling of the insulation will occur over time. As mentioned before, the thickness of the insulation significantly affects this resistance to heat transfer. For that reason residents may want to, from time to time, build the level of insulation back up in the attic to restore its resistance rate values. Table 3-2-1 lists common insulating materials and their R values based on thickness.

WALLS AND CEILINGS

The most common material found on interior walls for residences is sheetrock. Sheetrock is a product that has paper on both sides of a powdered gypsum. Figure 3-2-18 shows a room before (a) and after (b) sheetrock installation. Sheetrock is not a structural material. If it is struck very hard the paper will become damaged and the gypsum can be pulverized. Once sheetrock has been damaged, the damaged portion must be removed and replaced with a new section of sheetrock. Sheetrock is relatively soft and can be easily pierced with a screwdriver.

Sheetrock is attached to the stud using either sheetrock nails or screws. The sheetrock joints are covered with a sheetrock compound and paper tape. The paper tape does not have adhesive. It is held in place by the sheetrock compound. Sheetrock is finished by sanding the seams and joints, and most often has a texture sprayed on as a finishing step.

CAUTION: Sheetrock will not support your weight. If you intentionally or accidentally step on a sheetrock ceiling, you will fall through. Pay very close attention to where you are stepping when working in an attic to avoid a fall.

Often the two-by-four top plates and bottom plate on a wall must have holes drilled in them in order for wires or plumbing to pass through. These holes must be sealed to prevent air infiltration. The holes are

TABLE 3-2-1 R Values of Common Building Materials

Material	R Per Inch
Batt and blanket rock wool	R3.7
Blown cellulose	R3.1–R3.7
Blown mineral wool—horizontal fill	R2.9–R2.2
Blown mineral fiber	R2.2–R2.9
Expanded polystyrene board (white bead board)	R4.0
Fiberglass batt, high density	R3.5–R4.3
Fiberglass batt, standard density	R2.9–R3.7
High density batt and blanket fiberglass	R4.3
Loose fill cellulose	R3.7
Loose fill fiberglass	R2.2
Loose fill rockwool	R2.9
Low density fiberglass batt and blanket	R3.1
Medium density fiberglass batt and blanket	R3.7
Mineral wool batt	R3.14–R3.80
Polystyrene board (blue or pink board)	R6.3–R5.6
Polyurethane board (open cell)	R5.50–R6.25
Rigid foam board	R3.7–R8
Spray-on polyurethane foam	R6.3–R5.6
Spray-on cellulosic fiber	R3.4–R2.9

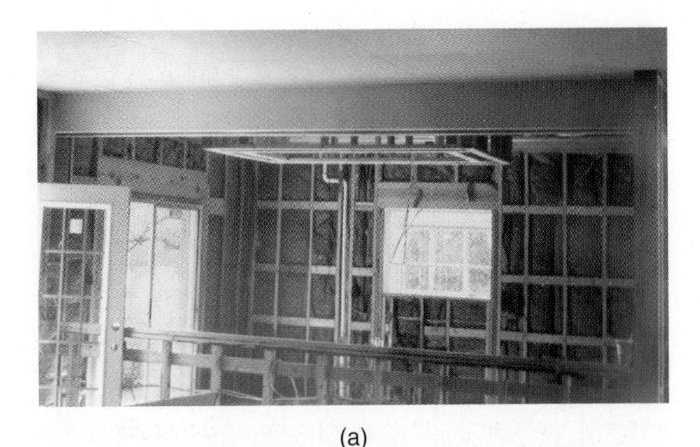

(a)

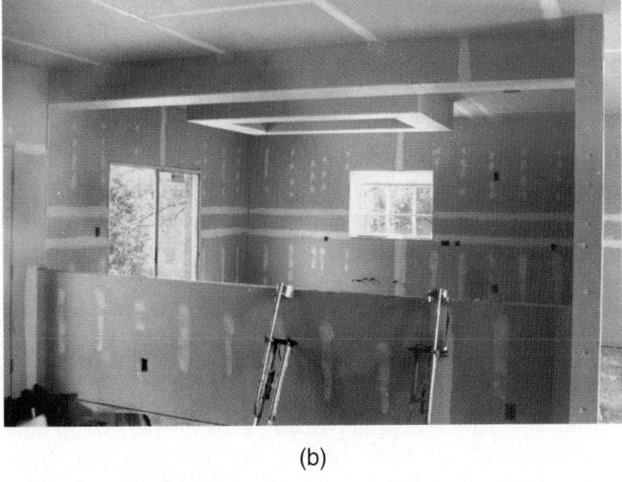

(b)

Figure 3-2-18 Sheetrock installation: (a) Sheetrock installed on ceilings first; (b) Sheetrock installed on walls with joints taped and bedded.

Figure 3-2-19 Wires that go through the top plate of a wall must be either caulked or filled with expanding foam as shown in the photo.

typically filled using expandable foam as shown in white in Figure 3-2-19. Any holes drilled horizontally in two-by-four wall studs are typically not sealed.

Light Commercial Construction

Buildings constructed for light commercial use may either be built in a fashion similar to residential, with all wood construction, or they may be constructed using metal studs and steel structures. When wooden structures are used, the construction of commercial buildings is very similar to residential construction.

Fire codes often require light commercial buildings to be constructed using metal studs, Figure 3-2-20. Metal studs are preformed out of 16, 18, or 20 gauge galvanized sheet metal and are assembled using sheet metal screws.

Flat Roofs

Often commercial buildings have flat roofs. A flat roof is constructed with a roof decking which may be either sheet metal or wood. Sheet metal decking is often covered with a thin layer of lightweight concrete. The concrete or wooden top surface is then sealed using either hot tar, thin layers of roofing paper, or a plastic roof compound layered with sheets of plastic. Usually a thin layer of gravel is applied to the top surface.

When rooftop air conditioning units are used, a curb must be first put in place on the roof and sealed completely. This is typically done by a roofing contractor. Water leaks are a constant potential problem for flat roofs and it is essential that these air conditioning curbs be sealed completely and properly. Water leaks are such an ongoing issue for flat roofs

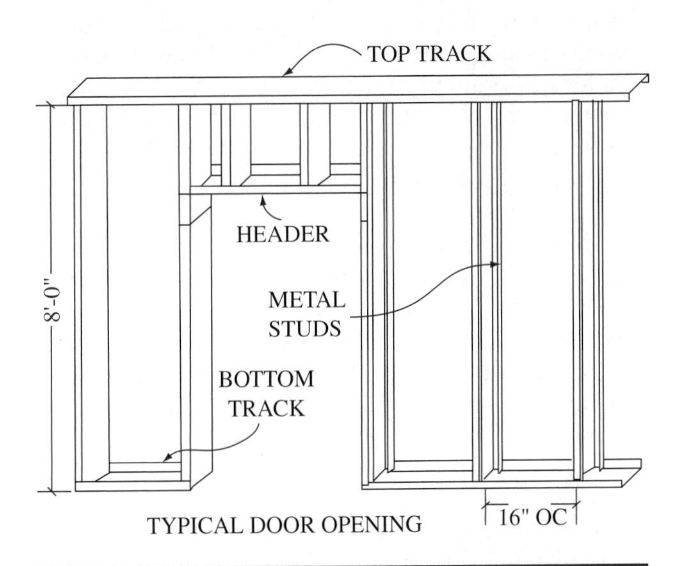

Figure 3-2-20 Typical metal stud wall layout with door opening.

Figure 3-2-21 Rooftop unit on a composite flat roof.

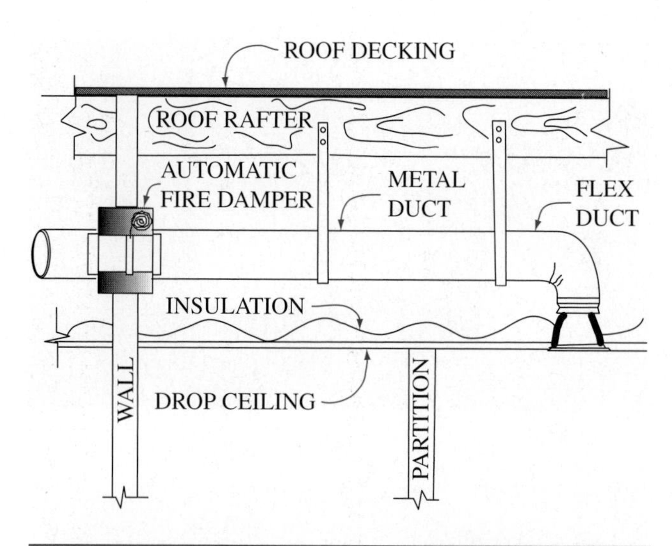

Figure 3-2-22 Partitions in commercial construction are often used as room dividers and do not extend beyond the drop ceiling. Because partitions are not structural, automatic fire dampers are not required when a duct passes over or through partitions. Automatic fire dampers are required anytime a duct passes through an interior wall.

that many, such as the one pictured in Figure 3-2-21, are bonded. Bonding is when the roofing contractor provides the building owner with an insurance bond that would cover water damage if the roof were to leak. Because of this insurance bond provided by the roofing contractor it is essential that any time air conditioning work is being performed that would require a new opening to be cut through the roof the roofing contractor be contacted before the work begins. This must be done whether the hole is for electrical, refrigerant lines, or the installation of a new roof top unit. Failure to contact the roofing contractor may result in your air conditioning company being held liable for any water leaks that occur in the building.

CAUTION: A roof's slope is given as its height in inches as compared to each 12 in of horizontal run. Roofs that have slopes less than 8 in for each 12 in can be worked on relatively safely. Roofs that have slopes greater than 8 in are very hard to impossible to stand on without slipping. If you must work on a sloped roof, you must wear a safety harness.

Drop Ceilings

Drop ceilings are commonly used in light commercial buildings. They allow for easy reconfiguring of the interior space as the business changes. Drop ceilings are constructed by suspending a grid of metal rails. Ceiling tiles are placed in the grid from above. These tiles are easily damaged, so be very careful if you must move them. Do so with clean hands, and be very careful not to scar or scratch the surface of the tile. If the tile becomes damaged, you may have difficulty locating an exact replacement pattern, which can become time consuming and expensive. Often the space above a building's drop ceiling is used as the return air plenum for light commercial buildings.

WALLS AND PARTITIONS

Walls and partitions are similar in commercial construction. Walls, however, extend all the way up through the ceiling to the roof of the building. Partitions only extend up to the ceiling. That makes partitions very easy to move as the business reconfigures as it grows. Some walls extend all the way up to the ceiling and are classified as firewalls. These walls are constructed with fire resistant materials. Building and fire codes regulate the construction of firewalls. Any time supply or return ducts must pass through a firewall, the duct must have a fire damper, Figure 3-2-22. Fire dampers will automatically close, sealing off the duct to prevent the spread of smoke and fire throughout the building.

UNIT 2—REVIEW QUESTIONS

1. What are the two most common problems associated with basement areas?
2. Under what circumstances can a beam's structural surface be cut?

TECH TALK

Tech 1. I have really good balance, so I'm not afraid to work on heights. In fact I don't even need to use a safety harness.

Tech 2. Wearing a safety harness has nothing to do with being afraid of heights or having good balance. It only has to do with working safe.

Tech 1. I work safe!

Tech 2. That may be true, but did you know that if you don't follow safety rules, you could be fired?

Tech 1. You mean the company would fire me for not wearing my safety harness?

Tech 2. Yes.

Tech 1. Thanks.

3. Most residential walls are constructed with ________.

4. A ________ is placed on the outside of the two-by-four wall before the finish siding or masonry work is performed.

5. Why are trussed roof systems favored by contractors?

6. What is the advantage of shingle construction?

7. List the four common types of insulation.

8. What directly affects the ability of all insulation to resist the transfer of heat?

9. What area in a residence loses and gains the most heat?

10. What is sheetrock?

11. Why are many flat roofs bonded?

12. Why are drop ceilings commonly used in light commercial buildings?

UNIT 3

Screws, Rivets, Staples, and Other Fasteners

OBJECTIVES: In this unit the reader will learn about the various kinds of fasteners and their uses including:

- Screws
- Bolts and Nuts
- Rivets
- Staples
- Tape
- Nylon Tie Straps

SCREWS AND BOLTS

Screws and bolts are both types of threaded fastener. Screws are typically smaller in diameter and length than bolts. In addition screws may or may not thread into a nut. Typical types of screws are sheet metal screws, wood screws, and machine screws. Sheet metal screws are the most commonly used threaded fasteners in the HVAC industry. Wood screws are sometimes used in HVAC work to attach hanging straps to wooden rafters. Machine screws have standard threads and screw into nuts or prethreaded holes. Machine screw threads are available with coarse or fine threads in both standard and metric diameters, Table 3-3-1.

SHEET METAL SCREWS

A typical specification for a sheet metal screw might be a ½ in #8 hex sheet metal screw (one-half inch number eight hexagonal head sheet metal screw), Figure 3-3-1. The length of a screw is given from the tip of the screw to the base of the head of the screw or level with the top for flat head screws, Figure 3-3-2. The most commonly used length of screws in HVAC are ½ and ¾ in (13 and 20 mm) in length. Table 3-3-1 lists common sheet metal screw lengths. The diameter of a screw is given as a standard number. Number 8 and number 10 screws are the most commonly used diameter sizes in the HVAC trade.

TABLE 3-3-1 Comparison of Thread Sizes Between Standard and Metric for Both Coarse and Fine Threads

Standard Threads			Metric Threads		
Number or Diameter	Coarse Threads Per Inch	Fine Threads Per Inch	Diameter in mm	Coarse Threads Per mm	Fine Threads Per mm
#6	32	40			
			4	0.7	0.5
#8	32	36			
#10	24	32			
			5	0.8	0.5
#12	24	28			
			6	1	0.5
1/4	20	28			
			7	1	.05
5/16	18	24			
			8	1.25	0.5
			9	1.25	0.75
3/8	16	24			
			10	1.5	0.75
			11	1.5	0.75
7/16	14	20			
			12	1.75	1
1/2	13	20			
			14	2	1
9/16	12	18			
			15	—	1
5/8	11	18			
			16	2	1
			17	—	1
			18	2.5	1
3/4	10	16			
			20	2.5	1
			22	2.5	1
7/8	9	14			
			24	3	1
			25	—	1
1	8	12			

Figure 3-3-1 Number 8, ½ in, hex head sheet metal screw.

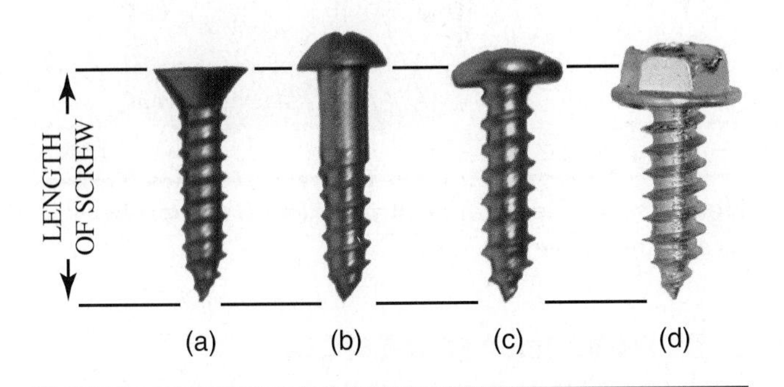

Figure 3-3-2 Screw length.

The hex head sheet metal screw shown in Figure 3-3-3a,b with (a) or without (b) a screwdriver slot are commonly used in the HVAC industry. The ¼ in and 5/16 in (6 to 8 mm) hex head sheet metal screws are the most commonly used sizes in the field.

The shape of the tip of the sheet metal screw may be:

- **Pointed** Pointed sheet metal screws, shown in Figure 3-3-4a, are often referred to as self-starting. This is because they can be spun into sheet metal without having to predrill a hole. Self-starting sheet metal screws can be used on sheet metal ranging from 30 gauge through 18 gauge.
- **Self-threading** Self-threading sheet metal screws must have a pilot hole drilled into the metal before these screws are used. A small notch at the tip of the screw acts as a thread tap to cut the thread into the metal as the screw is being installed. Self-threading sheet metal screws can be used in sheet metal ranging from 16 gauge to ⅛ in thickness. The threads on the self-threading screw of Figure 3-3-4b can tap a hole in 1/16 gauge sheet metal. The threads of the self-threading sheet metal screw in Figure 3-3-4c are more closely spaced, and this screw could thread a hole in ⅛ in sheet metal.
- **Round tipped sheet metal screws** Round tipped sheet metal screws like the one shown in Figure 3-3-4d are often used on electrical panels or within electrical panels to hold components in place. The tip of these screws is rounded and the end is not threaded. This protects against accidental stripping or shorting of wires within the electrical panel.

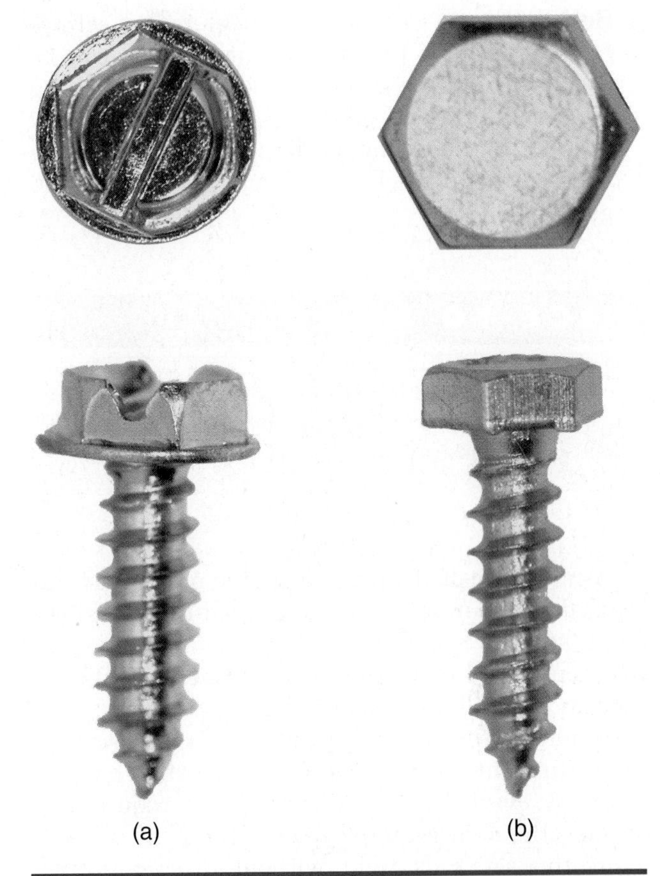

Figure 3-3-3 Sheet metal hex head screws: (a) With flat blade slot; (b) Without flat blade slot.

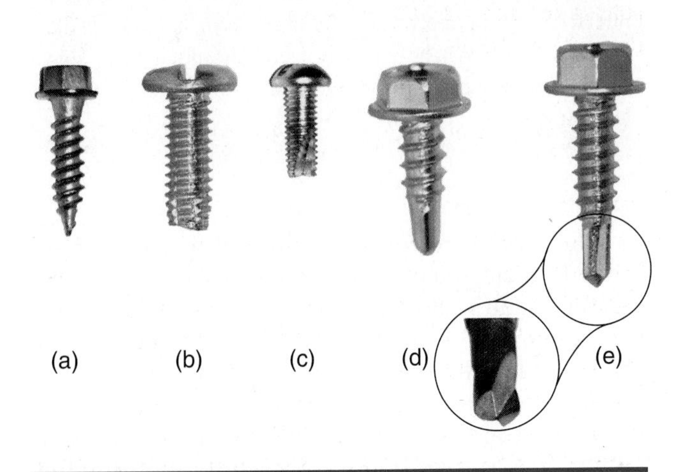

Figure 3-3-4 Common types of screws used with HVAC work: (a) Self-starting; (b) Self-threading; (c) Self-tapping; (d) round tipped; (e) Self-drilling.

- **Self-drilling** Self-drilling sheet metal screws have a tip built like a small drill bit as can be seen in the inset in Figure 3-3-4e. Self-drilling sheet metal screws predrill the hole as the screw goes in. Self-drilling sheet metal screws can be used in metal thicknesses from 30 gauge to 16 gauge.

Figure 3-3-5 Cordless electric drill with adjustable torque chuck.

NICE TO KNOW

The purpose of using round tipped sheet metal screws is to eliminate the possibility of a sharp point damaging electrical wire insulation.

An electric drill with a magnetic nut driver can be used to spin in self-starting and self-tapping sheet metal screws. Self-starting sheet metal screws are commonly used for on-site sheet metal duct fabrication and installation. When using an electric drill with self-starting or self-tapping sheet metal screws, the technician must be careful not to strip out the screw by overtightening it. Practice is required to develop the skill required to stop the drill the moment the screw is tight without overtightening. Some electric drills have torque adjustments, such as the dial on the drill in Figure 3-3-5, that allow the operator to set the drill so that it will apply enough force to sink the screw into the sheet metal but not strip it out.

SERVICE TIP

When using a magnetic nut driver and a drill, be certain that the head of the screw is completely in the end of the nut driver. If the nut is not completely in or is in at a slight angle, the screw head can slip inside the nut driver. This will quickly hollow out the end of the nut driver, rendering it useless.

Flat blade screwdrivers and Phillips head screwdriver tips may be used on some sheet metal screws. The flat blade screwdriver tip works well when manually removing or inserting screws, but it is very difficult to keep the bit aligned in the slot if a drill of power screwdriver is being used. Phillips head screws can be used with a Phillips bit chucked in a drill or power driver more successfully than the flat head type bit. Unfortunately when the screwdriver bit begins to wear it is more difficult to tighten and loosen screws.

SCREW HEAD DESIGN

Screws can have several different head designs, shown in Figure 3-3-6, including hex (a), round (b), pan (c), and flat (d). Hex head screws are the most common type used in HVAC work. They are used for almost everything inside and outside of the units. The round head screw and the pan head screw may be found inside electrical boxes and on electrical terminals. Round and pan head screws can have flat or Phillips slots and some even have combination

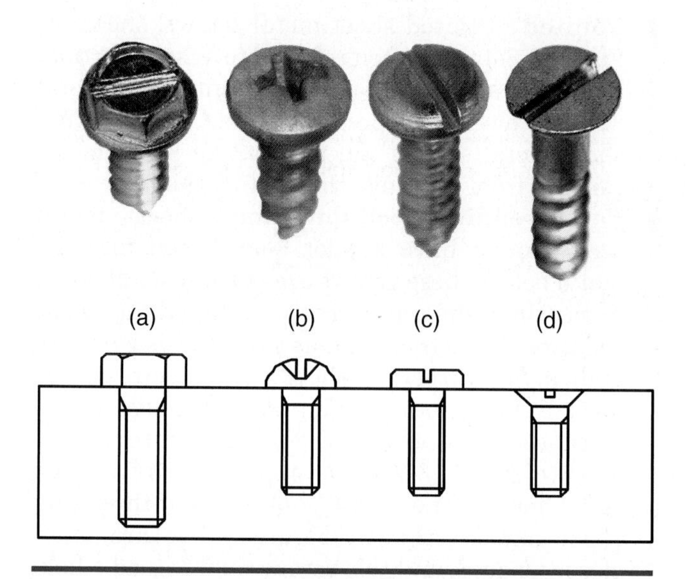

Figure 3-3-6 Common types of screw heads: (a) Hex; (b) Round; (c) Pan; (d) Flat. The bottom half of the figure shows how each screw will be represented in a drawing.

flat/Phillips slots, shown in Figure 3-3-7. Phillips and slot head screws can be used in drills or power screwdrivers with an adaptor, Figure 3-3-8. The adaptor has a magnetic screwdriver bit inside a collar. The collar slides down over the screw so it can be held straight as it is driven in with a power screwdriver or drill. Wood screws are the most common type of flat head screws. They are used to attach strapping to beams and joists.

Other Screws

Wood screws and sheetrock screws, Figure 3-3-9a,b, are often used in the HVAC industry to attach hanging straps to wooden structures. Wood screws for attaching hanging straps (a) are most often flat head screws. Sheetrock screws are almost exclusively Phillips heads. The major difference in appearance between wood screws and sheetrock screws is that the wood screw is shiny and has a portion of the shank that does not have any thread. Sheetrock screws are almost always black and their thread extends the full length of the screw.

Figure 3-3-7 Combination Phillips and slot-headed screw.

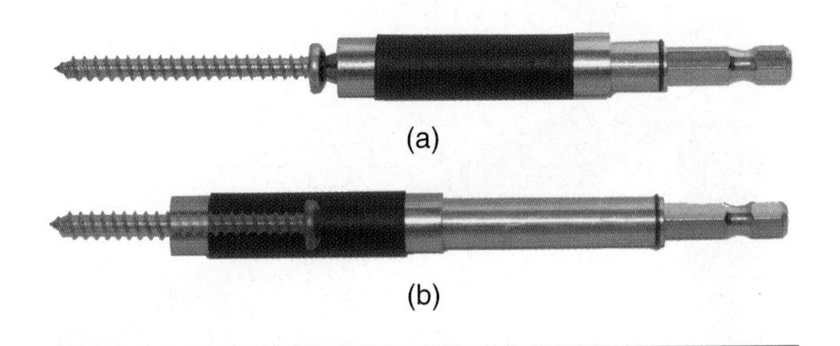

Figure 3-3-8 Screwdriving adapter: (a) With collar retracted to put Phillips bit into screw head; (b) With collar extended to keep contact between Phillips bit and screw head.

Figure 3-3-9 (a) Wood screw; (b) Sheetrock screw.

BOLTS AND NUTS

Bolts and nuts have machined threads. These threads will only work when matched parts are used. A typical specification for a bolt would be 3″ × ½″ NC-13 hex head, which is shown in Figure 3-3-10. The first dimension given, 3″, is the length of the bolt. The second dimension, ½″, is the outside or major

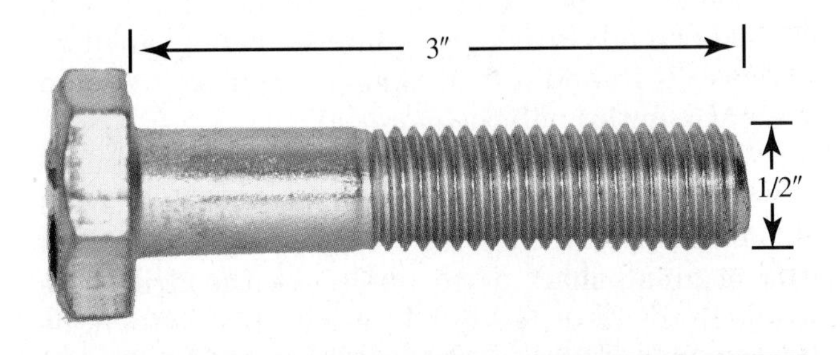

Figure 3-3-10 Bolt, 3″ × ½″ NC-13 hex head.

diameter of the bolt. The letters NC refer to the fact that this a national coarse threaded bolt. It has thirteen threads per inch, and the bolt has a hex head. The length of the bolt is measured from the base of the head to the tip. The diameter of a bolt is measured across the threads. The type of threads is listed as either NC for national coarse or NF for national fine. Each standard diameter bolt has a uniform number of threads per inch depending on whether it is coarse or fine. Although most of the bolts used are hex head bolts, other head designs may include carriage bolt, square, and Allen wrench. Square head and Allen wrench bolts are typically used on fans to hold the blades to the rotating motor shaft.

RIVETS

A rivet is a small unthreaded piece of metal that is installed through a predrilled hole. Once it is through the hole, the end of the rivet is enlarged so it will not fit back through the hole. The process of enlarging the end is called "upsetting the rivet." Over the centuries rivets that had to be upset by hammers were in common use, but today this type is seldom used. Pop rivets have replaced the older style rivets.

Pop rivets are usually made out of aluminum but other metals including copper and stainless are available. A variety of pop rivets is shown in Figure 3-3-11. A pop rivet has a central shaft that is placed in a pop rivet gun, shown in Figure 3-3-12. The sequence showing the installation of a rivet is represented in Figure 3-3-13. The central shaft is withdrawn back through the center of the hollow rivet shaft. As the bead on the end of the center shaft is pulled through the rivet sleeve, the sleeve deforms outward. When the bead at the end of the central shaft is pulled snugly against the back side of the drilled metal it pops off so that the nail-like shaft can be removed and discarded.

Pop rivets come in a variety of sizes ranging from the most popular ⅛ in diameter up to ¼ in diameter. The length of a pop rivet is determined by the distance from the tip to the head of the pop rivet. You should select a pop rivet that will extend through the joint materials by at least 1½ times the diameter of the pop rivet. For example, a ⅛ in diameter pop rivet should extend through the material being joined by at least 3/16 in and a ¼ in diameter pop rivet should extend ⅜ in through the material.

The best way to remove rivets without damaging the base metal is to drill them out. Pop rivets can be easily drilled using a bit slightly larger than the original shaft of the hole. As the drill cuts through the rivet head, the shaft of the rivet will be removed once the head has been cut free. Figure 3-3-14 shows the sequence of the drill bit next to the

Figure 3-3-11 Pop rivets come in a variety of types and sizes.

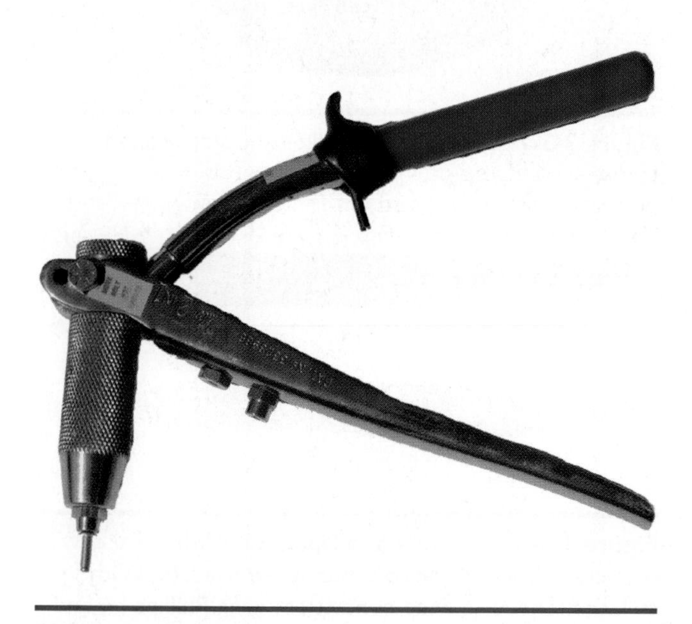

Figure 3-3-12 Pop rivet gun.

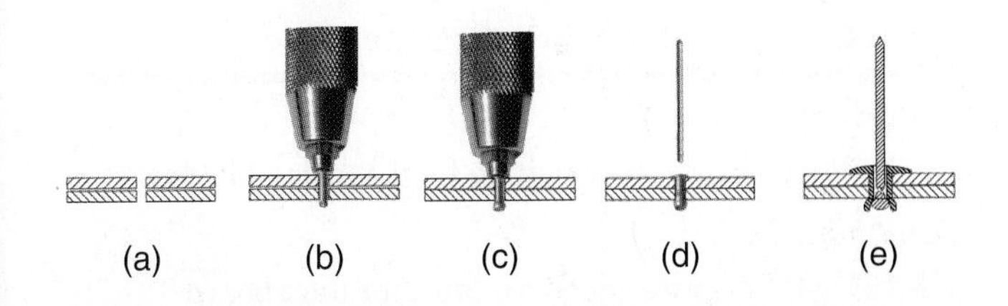

Figure 3-3-13 Steps in pop riveting: (a) Drill a hole; (b) Insert a pop rivet into the hole; (c) Depress the pop rivet gun handle to upset the rivet; (d) Rivet stem pops off when blades are pulled tight and rivet is seated; (e) Cross section of pop rivet.

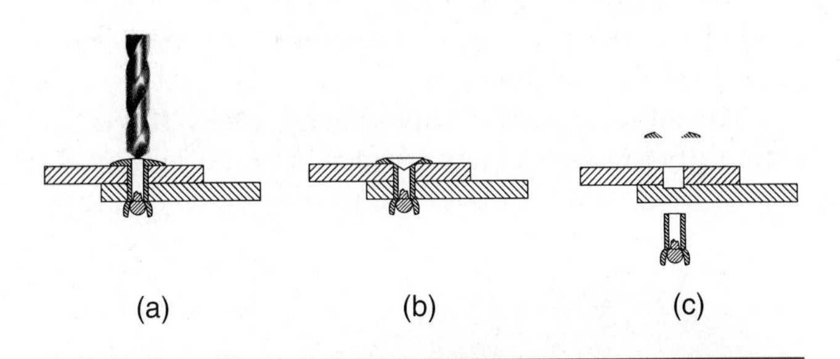

Figure 3-3-14 Steps in removing a pop rivet using a drill: (a) The drill bit should be slightly larger than the original pilot hole; (b) The drill only cuts out the rivet head; (c) The head and the shaft fall away.

SERVICE TIP

Usually rivets have limited field applications in the HVAC industry because of the time required to predrill the holes. However, a ⅛ in pop rivet will fit into a #8 sheet metal screw hole. Placing one pop rivet in the corner of a panel will reduce the possibility of unauthorized people tampering with the system.

rivet (a), drilling out just the rivet head (b), and the pop rivet, with its head drilled out, falling away (c). Take care not to drill through and increase the size of the rivet hole.

SERVICE TIP

Sometimes the drill will grab the rivet and the rivet will spin. If this is happening, you cannot drill out the rivet head. But by angling the drill bit slightly to one side, the bit will begin cutting again. Only angle the bit enough for it to start cutting. Too much angle may cause the bit to slip off of the head, damaging the finish of the part.

STAPLES

Staples hold by either driving their points into a soft material such as wood, or having the legs folded as with a paper stapler. Staples come in one standard width, Figure 3-3-15. Staples can be purchased with a variety of lengths of leg, from ¼ in to 15⁄16 in.

Specially designed staples are used to attach fabric or vinyl backed duct insulation, Figure 3-3-16a. Duct staplers are designed so that the tips of the staple are bent outward as the staple is driven through the fabric, Figure 3-3-16b. Duct staples are the most

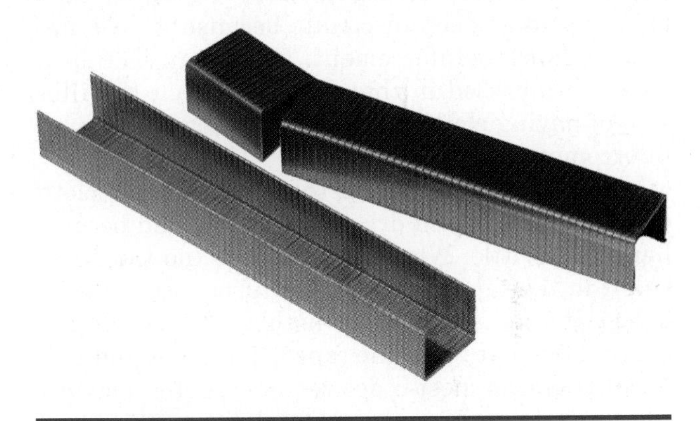

Figure 3-3-15 Staples come as preassembled strips.

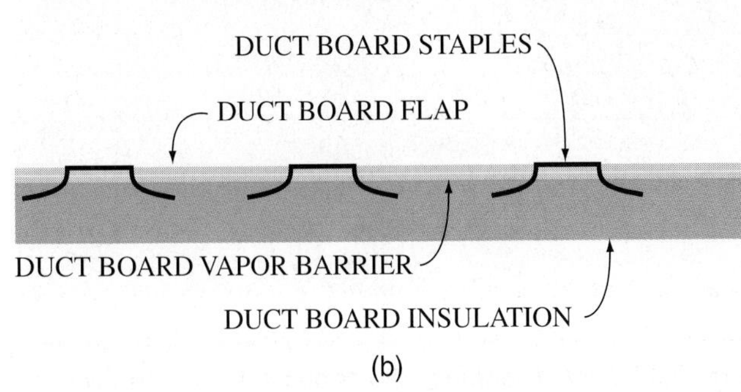

Figure 3-3-16 Staple gun used to (a) Attach duct insulation or (b) Seal duct board flaps.

commonly used method of attaching external insulation to sheet metal ducts.

■ TAPE

Two of the most commonly used types of tape in the HVAC industry are duct tape and electrical tape. Tape is manufactured to a variety of specifications but new codes require that all duct tape meet UL 181 standards. The quality of tape used will directly affect the length of time it stays in place to seal the duct. For example, some vinyl gray duct tape is sold at deep discounts because it does not contain fabric reinforcement. There may even be a pattern embossed in the plastic giving it the illusion of having cloth reinforcement, however, no reinforcement exists.

Because of the heat in the attic, vinyl or cloth duct tape mastic will dry out over time and become hard and brittle. When this happens, no tackiness is left in the tape and it simply turns loose, Figure 3-3-17. It is not a matter of whether the tape will dry out and fail, but when the tape will dry out and fail. For this reason most codes require UL 181 for duct closing systems. There are several different types of tape under the UL 181 listing. Each of these tapes has a different purpose, Table 3-3-2.

Aluminum pressure sensitive tapes are easily formed to the surface, creating a tight fit. They have a rubber or acrylic adhesive that is protected on the roll with a paper backing. The paper backing is removed when the tape is being used, Figure 3-3-18.

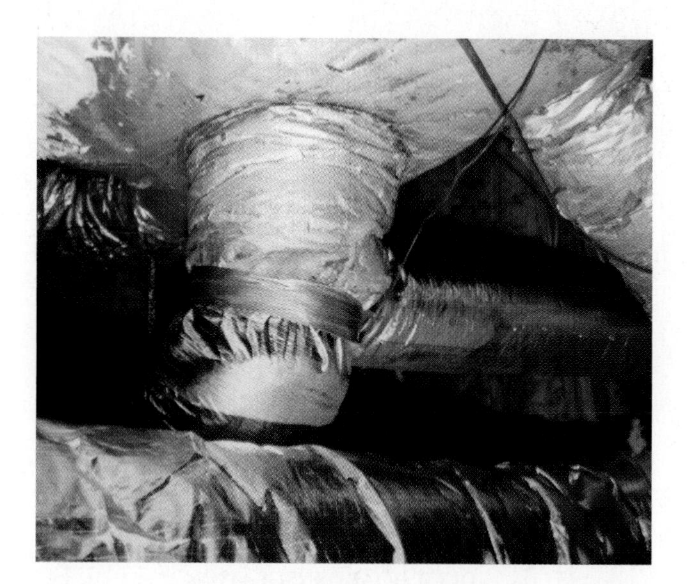

Figure 3-3-17 A number of products have been used in the past to seal duct and duct insulation such as this system that used cloth duct tape.

TABLE 3-3-2 UL Listings for Duct Tape

Flex duct
UL-181B-FX Pressure sensitive tape, for flexible ducts
Duct board
UL 181A-P Pressure sensitive tape, for duct board
UL 181A-H Heat activated tape, for duct board
Metal duct connectors to flex or duct board
UL 181 A-P or UL181 B-FX

Cloth tapes have a mylar coating. They are designed for application on flex ducts, Figure 3-3-19. All approved UL 181 tapes must be identified with lettering on the face of the tape.

The adhesive on foil tapes is very tacky. If it is accidentally touched to any surface before it is in place it is difficult if not impossible to remove it. Foil tape

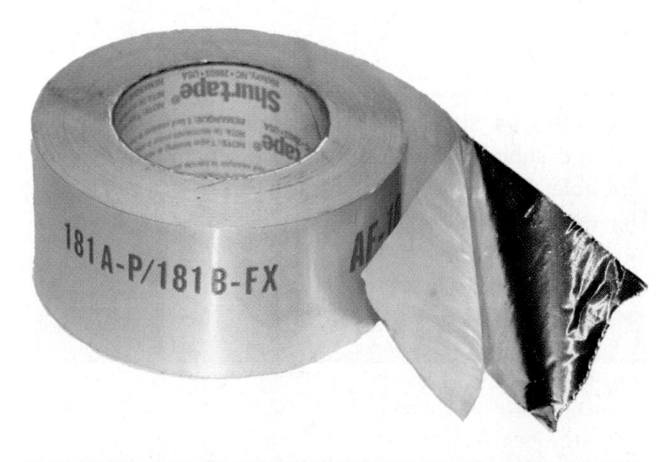

Figure 3-3-18 Aluminum foil pressure sensitive duct tape.

Figure 3-3-19 Mylar coated pressure sensitive duct tape.

comes with a paper backing to keep it from being stuck permanently to itself in the roll. This paper must be peeled back as the tape is applied to the surface. Once the tape is in place it should be smoothed down securely to ensure a proper seal. Failure to wipe the tape down can result in air gaps beneath the tape, which will allow its mastic to dry out. Keeping duct tape securely in place over time is the only way to ensure that conditioned air does not leak out of the duct system.

Vinyl electrical tape is often referred to as electrical tape or simply black tape. As with duct tape there is a range of quality in the tape available. A good vinyl electrical tape should have some stretch so it can form itself around the wires or connection. Too much stretch and the tape cannot be pulled tightly around the parts. The tape must resist heat without loosening or softening.

TIE STRAPS (DRAW BANDS)

Tie straps or draw bands are usually made out of nylon and are widely used in the HVAC industry. Large sizes are used to secure flexible duct to start collars or boots, shown in the installation sequence in Figure 3-3-20a–d. These tie straps must be resistant to ultraviolet light (UV), have a tensile strength of 150 lb and a service temperature of 165°F. When nylon straps are used on duct installations a special tool called a strap tensioner must be used. The strap-tensioning tool pulls the strap tight, reducing the likelihood of air leaks, and securely mechanically connecting the duct to the sheet metal boot or collar. Once the proper tension has been applied, a lever on the strap-tensioning tool extends a razor knife to cut off the unused nylon strap end.

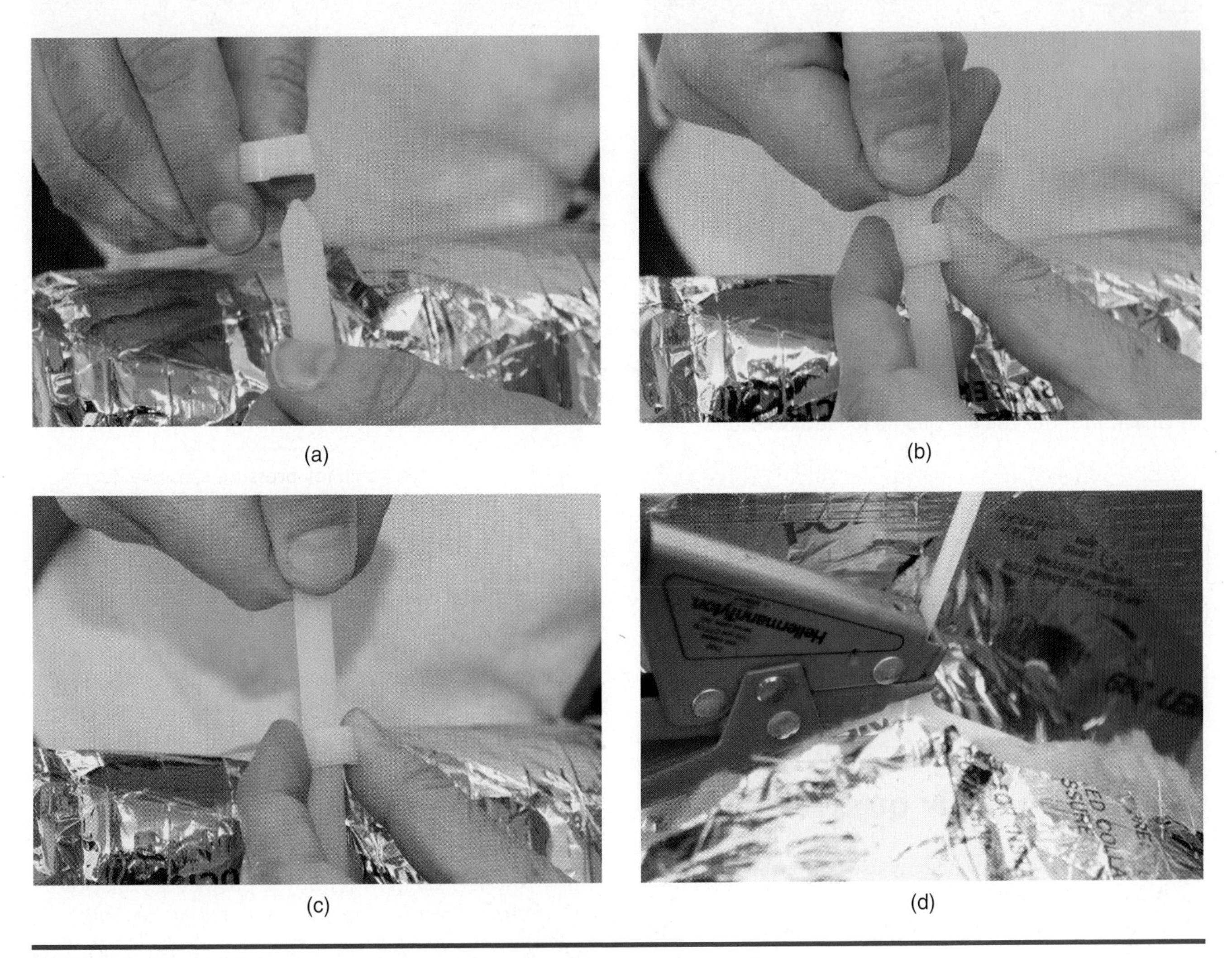

Figure 3-3-20 (a) Insert the end of the tie strap through the ratchet mechanism; (b) Use your hand to pull the strap through; (c) Pull the strap with your hand as tight as possible; (d) Use a duct strap tightening tool to finish tightening and cut the end of the strap.

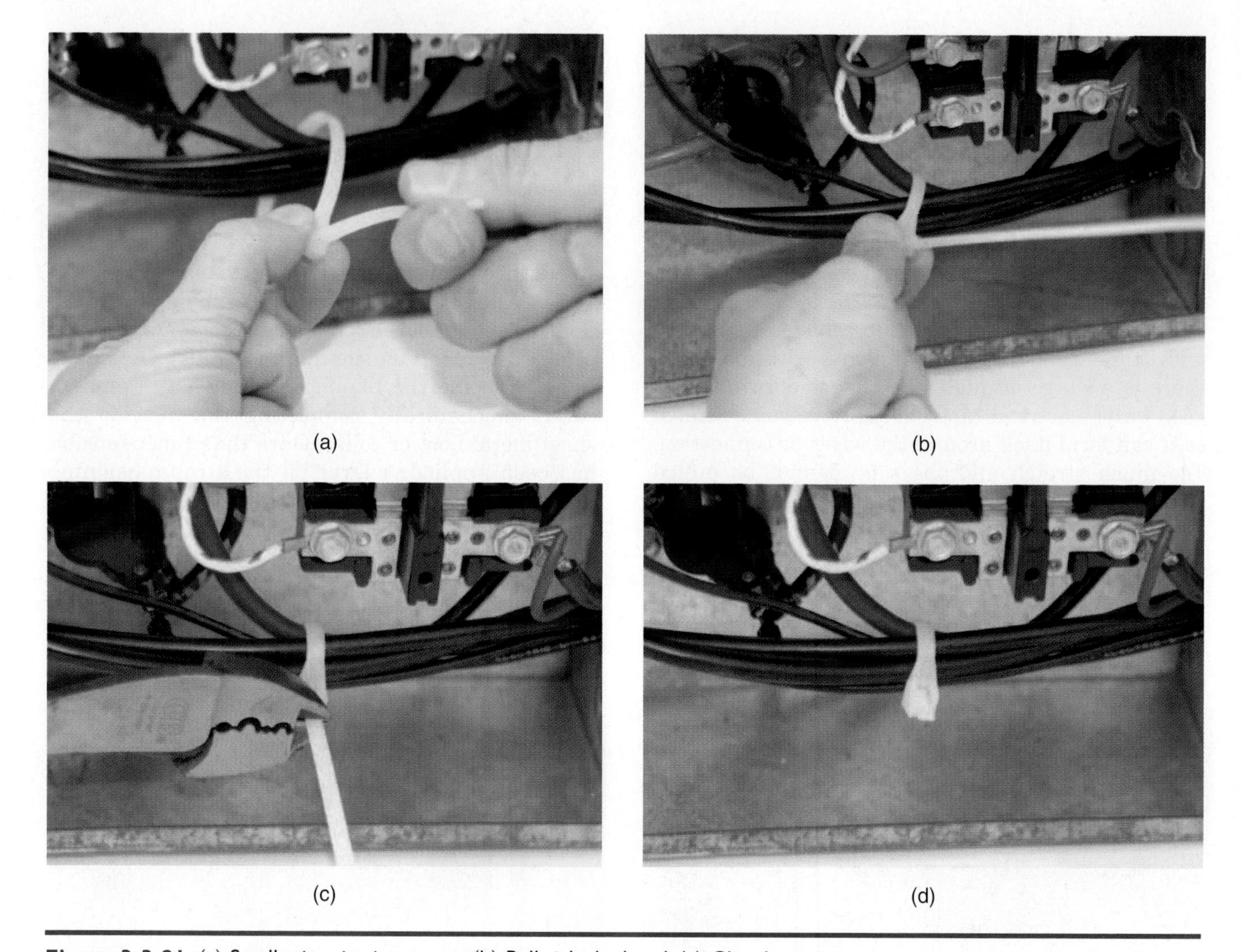

Figure 3-3-21 (a) Small wire tie size straps; (b) Pull tight by hand; (c) Clip the strap; (d) Finally, with the end cut off, the job looks neater.

Smaller ties are used to secure wires into bundles inside of units. When the nylon straps are used around electrical wires, they are simply pulled tight by hand or with a pair of pliers, and the unused portion is cut off, as shown in the installation sequence in Figure 3-3-21a–d.

SERVICE TIP

From time to time a tie strap may be put in place and only later is it found that it has to be removed. Ties may need to be removed when a wire must be added or removed from the bundle, or when a duct needs to be cut shorter or straightened. Tie straps are expensive and must not be wasted. They can be removed easily, if the end has not been cut off, by placing the tip of a small screwdriver under the ratchet mechanism, Figure 3-3-22a–b. By putting just a little force on the corner of the screwdriver blade the tie will be able to slip through the mechanism. Don't force the screwdriver tip under the mechanism. You can damage the ratchet, or the screwdriver could slip and stick your finger. As a precaution you might want to use pliers to hold the strip the first few times you try this. Once the ratchet is released, use your thumb to slide the end up and off to remove the tie strap.

UNIT 3—REVIEW QUESTIONS

1. What are the most commonly used threaded fasteners in the HVAC industry?
2. What is the most commonly used length of screws in HVAC?
3. What is the purpose of using round tipped sheet metal screws?
4. What are self-starting screws commonly used for?

   Flat head screws are used where a ________ is desired.

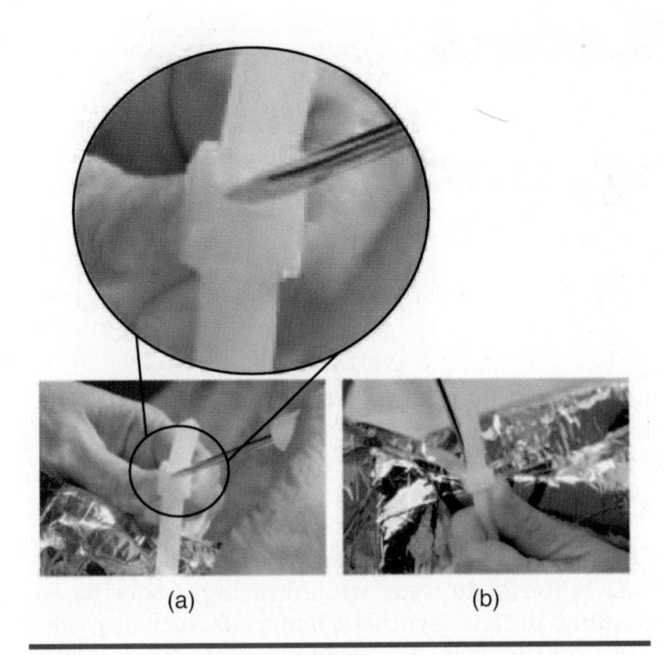

Figure 3-3-22 Use a screwdriver to release the ratchet mechanism so that a duct strap can be removed and reused.

TECH TALK

Tech 1. I use self-drilling screws for all of my sheet metal work.

Tech 2. Well, that is not a good idea on thin sheet metal.

Tech 1. Why not? They go in all right.

Tech 2. That's true, but on thin sheet metal much of the holding power of the screw comes from the little dimple of metal the self-starting screw pushes in.

Tech 1. What do you mean?

Tech 2. Look at this drawing (Figure 3-3-23). See how the self-starting screw pushes a little metal inward? That lets more of the screw threads stay in contact with the sheet metal. The self-drilling screws are designed to be used on thicker sheet metal where, even though they cut out the hole, the metal is thick enough so that the screw has several threads holding in the metal.

Tech 1. Oh, so it is the number of threads that are in contact with the sheet metal that gives sheet metal screws their holding power. I get it now. Thanks.

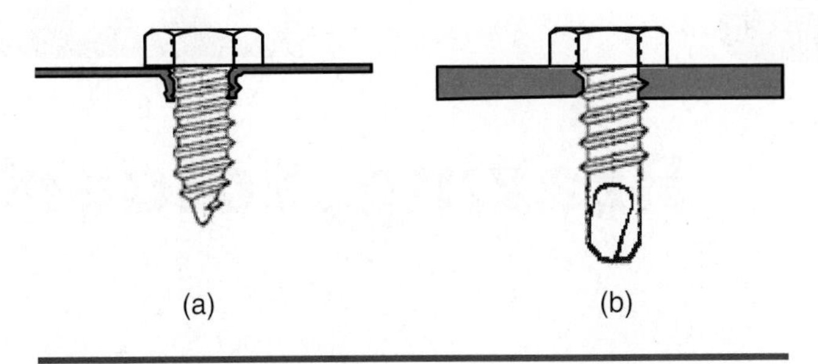

Figure 3-3-23 (a) Self-starting sheet metal screw in thin gauge sheet metal; (b) Self-drilling sheet metal screw in thicker sheet metal.

Why does the pan head screw have limited use in the HVAC industry?

5. What is the major difference in appearance between wood screws and sheetrock screws?
6. In reference to nuts and bolts what do the listings NC and NF mean?
7. What is a rivet?
8. What are the two general types of rivets?
9. What type of staples are commonly used to attach external insulation?
10. List the three most common types of tape used in the HVAC industry.

    What is the purpose of mastic?
11. Keeping ________ securely in place over time is the only way to ensure that conditioned air does not leak out of the duct system.
12. What are nylon tie straps used for?

UNIT 4

Brazing, Soldering, and Welding

OBJECTIVES: In this unit the reader will learn about the brazing, soldering, and welding processes used in the HVAC/R industry including:

- Welding Safety
- Welding Clothing
- Brazing
- Acetylene
- Oxygen
- Lighting and Adjusting the Flame
- Making a Brazed Joint
- Testing Brazed Joints
- Oxyacetylene Torch
- Electrical Resistance Soldering
- Torch Soldering
- Resistance Welding

INTRODUCTION

The processes of brazing, soldering, and welding are often miscategorized as all being welding. In the process of welding the base metal itself is melted as additional metal from the electrode is added to the joint. However, during brazing and soldering, only the metal being added to join the base parts is melted. The difference between brazing and soldering is temperature. Soldering takes place at a lower temperature than brazing, but other than that both processes are similar.

SAFETY

Because of the large quantity of hot sparks produced by most welding processes, proper protective clothing must be worn during welding. In addition, arc welding produces a great deal of ultraviolet (UV) light. That is why a welder's safety shield must be worn over the face at all times when arc welding. UV light can cause skin burns similar to sunburn. However, because of the intensity of the arc light, these burns can occur very quickly to any unprotected skin surface. Properly worn protective clothing will prevent both burns from hot sparks and burns from the welding light.

Clothing

One hundred percent cotton or leather clothing is the best material to wear while brazing, soldering, or welding. Because synthetic materials, such as nylon, are so easily burned by welding sparks, they should never be worn while welding, cutting, soldering, or brazing. Shirts worn during welding should have long sleeves and either have no shirt pocket or else have flaps that cover the pocket to prevent sparks from collecting. They should have a collar and be able to be buttoned securely around the neck. One hundred percent cotton blue jeans are ideal pants for welding. A welder's cap should be worn to protect your head from flying sparks. Earplugs can be used to protect your ears from both the loud noises associated with some welding shops and to keep hot sparks from flying into your ear. Ear protection can be provided by earplugs that fit into the ear or by special headphones that cover the entire ear. Leather gauntlet type welding gloves are preferred, Figure 3-4-1b. The gauntlet keeps both sparks and welding light from burning your exposed wrists. Leather shoes with high tops are required, Figure 3-4-2. Tennis shoes should never be worn in a welding shop. Safety glasses must be worn at all times.

BRAZING AND SOLDERING

Brazing and soldering are similar processes. In both brazing and soldering the base metal must be heated to a temperature within a range where the bonding phase between the liquid filler and the base metal can occur. This process is called tinning. If the metal being joined is too cool or too hot tinning cannot occur. Overheating is a major problem when brazing or soldering copper pipe because the flux will burn (become oxidized), stop working, and become a barrier to tinning. In addition, a heavy oxide can be formed on the pipe itself, preventing a bond forming between the filler metal and the pipe surface.

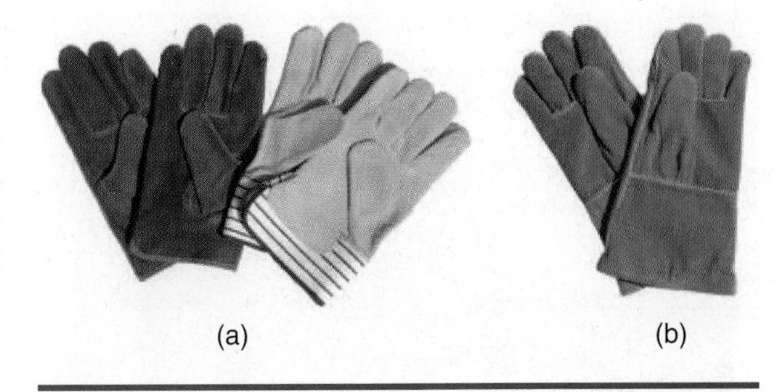

Figure 3-4-1 (a) Work gloves may be cloth, leather palm, or all leather; (b) Welding gloves are gauntlet-type gloves with high cuffs.

Figure 3-4-2 Welding boots should have smooth toes to prevent sparks from being trapped.

SERVICE TIP

When working on copper pipe, once the metal has been overheated you must stop to cool and reclean the parts before the joint can be made correctly. If the parts are not disassembled and recleaned, only the outside of the joint will be able to be joined. This will leave a joint that will someday leak.

The melting temperature of the filler material determines whether the process is brazing or soldering. The brazing process takes place at temperatures above 840°F, and soldering takes place at temperatures below 840°F. Both processes are often referred to in conjunction with the filler material. For example, silver brazing refers to the use of silver alloys commonly used to join copper tubing for HVAC work.

Brazing and soldering use a lot of specialized terms that you should become familiar with. The term soft solder often refers to an alloy of tin and lead. The term hard solder frequently refers to an alloy of tin and antimony. Nowadays tin antimony alloys are the most commonly used for soft soldering because of the potential health and environmental concerns when lead based solders are used. Because the terms soft and hard soldering are not standardized terms, there may be variations of their meanings depending on local usage. For example, sometimes hard soldering is used in reference to alloys containing trace amounts of silver. Table 3-4-1 lists the common brazing and soldering alloys and fluxes.

TABLE 3-4-1 Common Soldering and Brazing Metal and Fluxes Showing Base Metals that Can Be Joined

	Alloy	Flux Type	Base Metal
SOLDER	95-5 Tin antimony solder[1]	C-Flux	Copper pipe, brass, steel
	95-5 Tin antimony solder	Rosin	Copper pipe, copper wiring, brass
	95-5 Tin antimony solder	Acid	Copper pipe, brass, steel, galvanized sheet metal
	98-2 Tin silver solder	Mineral based flux	Copper pipe, brass, steel
	40-60 Cadmium zinc solder	Specific flux from solder manufacturer	Aluminum
BRAZING	Copper phosphorus silver brazing BCuP	1% to 15% Silver no flux required	Copper pipe
	Copper phosphorus silver brazing BCuP	1% to 15% Silver mineral based flux	Copper pipe to brass, brass, steel
	Copper silver brazing BAg	45% Silver mineral flux	Copper pipe to steel, brass, steel

[1]The percentages of the materials in the flux are given in the numbers, for example 95% tin, 5% antimony.

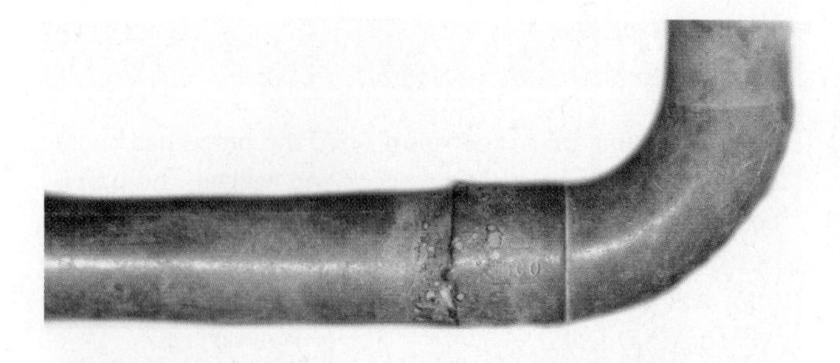

Figure 3-4-3 Excessive oxides on a braze joint.

Figure 3-4-4 Air acetylene torch kit.

In order for the bonding or tinning phase to occur between the filler metal and the base metal, often a flux is required. A flux is a typically active compound that removes light surface oxides and promotes the wetting of the base metal with the liquid filler metal. Not all brazing and soldering requires fluxes. For example, some alloys containing silver, which promotes wetting, are often referred to as flux-less brazing or soldering alloys. Even though these alloys may work without fluxes in most cases, there are some situations where excessive surface oxide may exist and must be removed, Figure 3-4-3.

Some fluxes are neutral at room temperature and only become active when heated to a specific temperature. Other fluxes are active irrespective of the temperature. All fluxes, but particularly active fluxes, must be removed from the completed joint. If flux is allowed to remain on the joint, even though it may be inert at standard temperatures, it can still cause problems. Flux can trap moisture causing pitted corrosion, or it may obscure small pinhole leaks which could open later causing system leaks. Active fluxes can significantly damage metal parts if not completely cleaned away and neutralized.

BRAZING

Codes require that refrigerant line joints be made using silver brazing alloys. The most commonly used torch for this type of brazing is the air acetylene torch, Figure 3-4-4. The air acetylene torch uses acetylene gas from a compressed gas cylinder, and when this gas flows through a venturi in the torch, air is drawn in to the gas where they are mixed. An air acetylene flame burns at a temperature of approximately 4220°F. Air acetylene torches work very well on copper pipe for soldering pipe from ¼ in up to about 3 in and for brazing pipes from ¼ in up to about 2 in in diameter. Although larger diameter copper pipes may be joined using air acetylene torches, the lack of a highly concentrated flame results in a much slower joint completion time. The longer that copper is kept at the higher brazing temperature, the greater the formation of copper oxide, Figure 3-4-5. Copper oxide can become a barrier to the successful completion of the joint as well as becoming a contaminant in the refrigerant circuit. For that reason, technicians usually use an oxyacetylene torch for making braze joints in copper pipes ½ in and larger in diameter. Properly used, the oxyacetylene torch can produce satisfactory brazes in all ranges of copper tubing at diameters from 1/16 in through 6 in. The oxyacetylene torch uses compressed acetylene and oxygen from cylinders.

Figure 3-4-5 Copper oxide formed by excessively heating copper pipes and fittings.

Acetylene

Acetylene gas is provided in two common cylinder sizes for HVAC work. The MC acetylene cylinder contains approximately 10 ft^3 of acetylene, Figure 3-4-6a, and the B acetylene cylinder contains approximately 40 ft^3 of acetylene, Figure 3-4-6b. Acetylene gas under high pressure is unstable. It can explode as the result of being struck or jarred

Figure 3-4-6 (a) The MC acetylene tank holds 10 ft^3; (b) The B acetylene tank holds 40 ft^3.

severely. For this reason, acetylene gas must be stabilized in the cylinder with acetone. The acetone is capable of absorbing 28 times its own weight in acetylene gas. Acetone reacts with acetylene similarly to carbon dioxide and water. Carbon dioxide can be dissolved in water forming a carbonated beverage. As the pressure on a carbonated drink bottle is slowly released, you can observe bubbles being formed within the liquid. These bubbles are CO_2 gas being released. The acetone in an acetylene bottle works the same way. To prevent the acetone from simply being poured out accidentally, the acetylene cylinder is filled with an absorbent coarse material during manufacturing, Figure 3-4-7.

Because of acetylene's instability, it is illegal to operate a torch with an acetylene pressure greater than 15 psi. Most acetylene gauges have a red mark indicating 15 psi, Figure 3-4-8. Torch manufacturers have designed their equipment so that a torch will operate properly with acetylene pressures from 1–5 psi for most air conditioning and refrigeration applications. Turning the acetylene pressure up beyond the recommended pressure value for a specific torch will not make the torch operate better. It does, however, present a greater safety issue. Do not exceed the recommended pressure for any given torch and application.

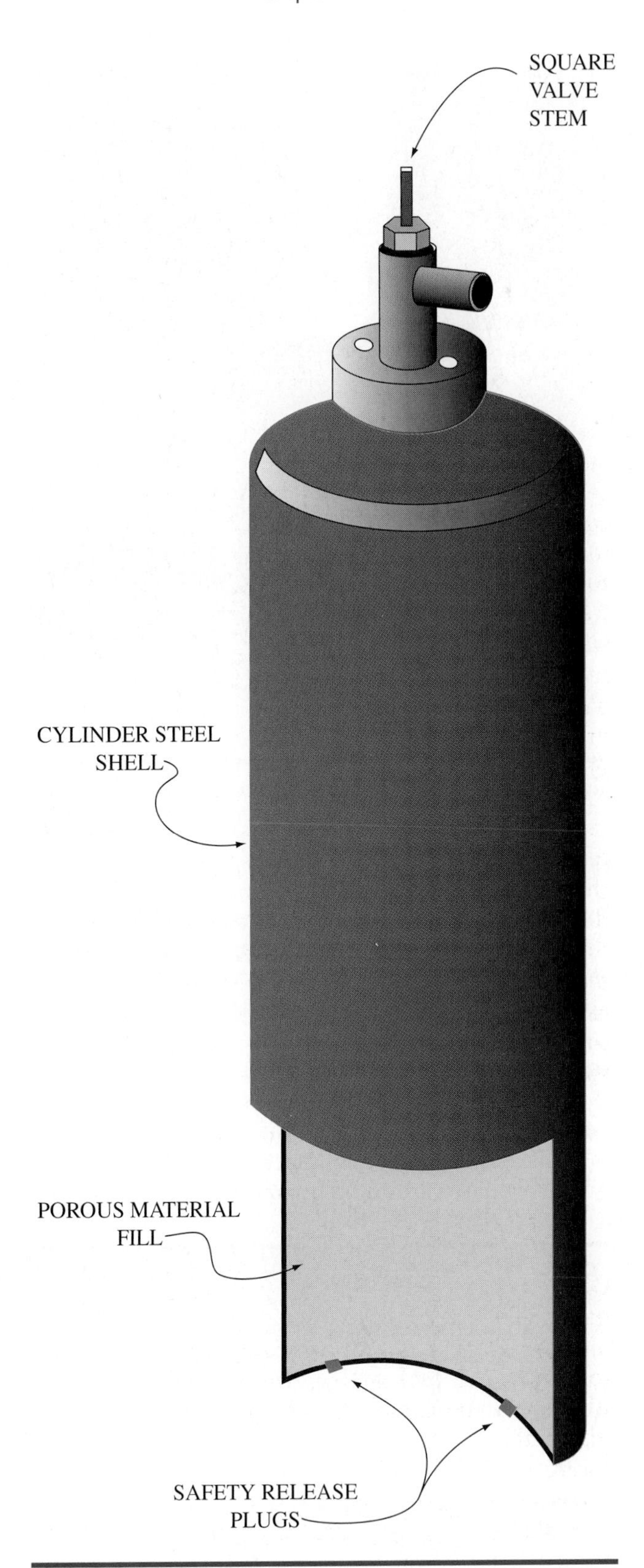

Figure 3-4-7 Diagram of a cutaway section of a B acetylene tank showing the porous absorbent section.

Figure 3-4-8 Acetylene regulator with gauges.

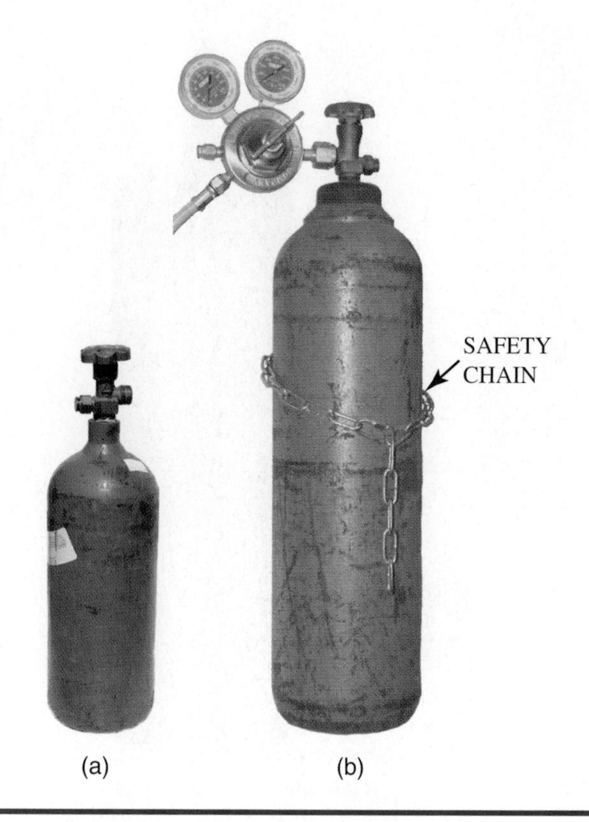

Figure 3-4-9 (a) 22 ft^3 oxygen cylinder; (b) 55 ft^3 oxygen cylinder.

Oxygen

Oxygen is provided in highly pressurized cylinders. A full oxygen cylinder contains approximately 2000 pounds per square inch (psi) of oxygen. The two most common sized cylinders used for the air conditioning and refrigeration trades are the 22 ft^3 (cubic feet) and the 55 ft^3 cylinders, Figure 3-4-9a–b.

Because the oxygen is under such great pressure, care must be taken to ensure that the oxygen cylinder valve is not damaged or knocked off. If the oxygen cylinder valve were to be broken off of a full oxygen cylinder, the cylinder could fly around the room much like a child's balloon. To prevent possible damage to the valves and regulators on oxygen and acetylene cylinders they should be securely attached to a frame or cart when they are in use, Figure 3-4-10. Oxygen is not flammable; however, because of its high pressure and level of purity, even small quantities of oil that may get on the oxygen gauge, valve, or hoses can cause an explosion when subjected to high-pressure oxygen. It is very important that you never allow oil to come in contact with oxygen and acetylene equipment.

The oxygen valve has two seats, one at the bottom and one at the top. The bottom valve shuts the gas off. When the cylinder valve is open all the way tight to the top, that seat prevents leakage of oxygen from around the valve stem. It is recommended that the oxygen cylinder valve be opened all the way, especially when the cylinder valve is going to remain open for any period of time.

CAUTION: Federal safety regulations (OSHA) require 55 ft^3 cylinders and larger to be safety chained to a cart or structure any time the cylinder valve cap is removed. If one of these high-pressure cylinders is knocked over and the valve is broken off, the entire tank will fly around the room like a released balloon.

SAFETY TIP

Always remove regulators from oxygen and acetylene cylinders before you put them back in your service vehicle. In many states and cities it is against the law to transport oxygen and acetylene cylinders in a motor vehicle with the regulators attached. The danger is that in an accident the regulators could be broken off, causing a fire or explosion.

CAUTION: Compressed oxygen should never be used in place of nitrogen or compressed air. If oxygen were used to pressurize a refrigeration system and the system were started, there would be a severe explosion of the compressor and compressor shell. This explosion would be hazardous to anyone within the immediate area.

Figure 3-4-10 Portable oxyacetylene welding and cutting rig. (*Courtesy Victor Equipment Company*)

Figure 3-4-11 Support the cylinder with one hand when turning on the acetylene or oxygen valves.

Figure 3-4-12 Non-adjustable acetylene wrench.

Lighting and Adjusting the Flame

Once the air acetylene torch and regulator have been attached to the acetylene cylinder, the acetylene cylinder valve is opened one quarter turn. OSHA also requires that a non-adjustable wrench be used on any valve stems that are not equipped with a hand wheel, such as the one being used in Figure 3-4-11. A non-adjustable wrench is shown in Figure 3-4-12. It is also a regulation that the wrench be left on the cylinder valve any time the valve is open. Hold the cylinder steady with one hand while opening the valve with the other hand. It is recommended that the acetelene cylinder valve be opened a maximum of 1 1/2 turns. This will allow for a rapid shutdown of the acetylene valve if an accident were to occur.

CAUTION: It is against federal regulations (OSHA) to open any acetylene cylinder valve more than two and one half turns. This is to allow you to turn off the cylinder quickly in an emergency. Best practice is to always open the cylinder one and one half turns.

An air acetylene torch tip is designed to handle a specific flame size. With the pressure adjusted properly according to manufacturer's recommendations for that torch tip, turn on the acetylene valve slightly to light the flame. The only safe devices to use to light any torch are those specially designed for that purpose, such as spark lighters or flint lighters. When using a flint lighter hold it slightly off to one side of the torch tip, as in Figure 3-4-13a. Do not hold it directly over the tip as in Figure 3-4-13b.

SERVICE TIP

Holding the lighter over the end of the tip may cause a pop when the torch is lit. This can be unnerving to your customer as well as doing damage to your torch tip.

CAUTION: Never use a cigarette lighter to light any torch. Butane cigarette lighters will explode with the force of a quarter stick of dynamite if they are used to light an acetylene torch. The torch flame can instantaneously burn through the plastic lighter, causing it to explode.

Open the hand wheel of the torch body completely. Many air acetylene torch tips will whistle loudly when adjusted properly, as in Figure 3-4-14a. Failure to provide the torch tip with the proper gas flow will result in the tip being overheated. With low

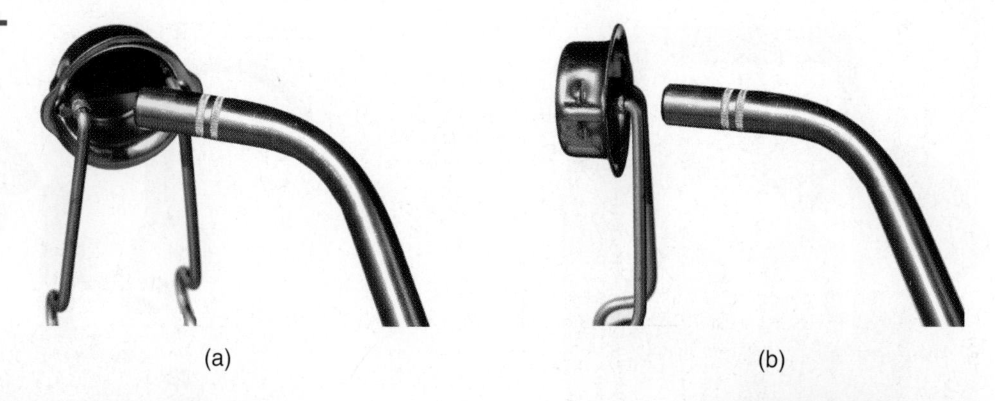

Figure 3-4-13 (a) Correct position for striker, off to the side of the torch tip; (b) Improper position for striker, directly in front of the torch tip.

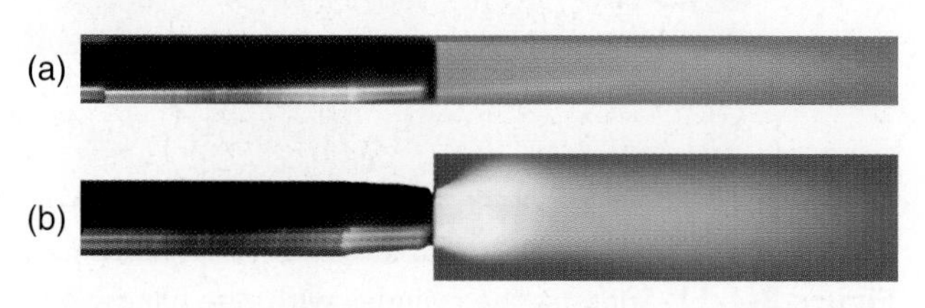

Figure 3-4-14 (a) Correct appearance of oxyacetylene flame for brazing; (b) Flame with too much acetylene brings heat too close to tip, and can damage it.

gas flow rates, the flame gets closer and closer to the tip. If the gas flow rate continues to be too low, the tip will become hot, Figure 3-4-14b. In some severe cases the tip may actually begin to glow red hot. This can damage the tip so severely that it must be replaced.

MAKING A BRAZED JOINT

In order to make a proper brazed joint in copper, the copper fittings must be clean. Clean new fittings might only need to be wiped off; however, pipe and fittings that have been discolored from oxide must be cleaned mechanically or chemically. Mechanical cleaning is easiest performed using sand cloth, Figure 3-4-15a,b, or a wire brush, Figure 3-4-16. Specially designed round wire brushes fit inside of copper fittings to provide better cleaning. These round brushes or small flat brushes can also be used for cleaning the outside of the pipe. Sand cloth or abrasive pads for cleaning copper pipe and fittings are available. Sand cloth should not be confused with Emery cloth. Emery cloth is a very hard abrasive material and if any abrasive grit is left inside the pipe it can cause damage to a piston or other moving parts. AC type sand cloth has a softer grit that will more easily break down if the grit is accidentally left in the system, reducing the possibility of causing damage to moving parts.

The pipe should be cleaned at least ½ in farther back than the pipe fits into the fitting, shown in the series of photos in Figure 3-4-17a–c. Copper oxides form over time. It is best to clean the fittings immediately prior to producing the brazed joint. Once the fittings have been cleaned, caution must be taken not to touch the cleaned surfaces with your hands or tools that may have oil, which can contaminate the cleaned surfaces.

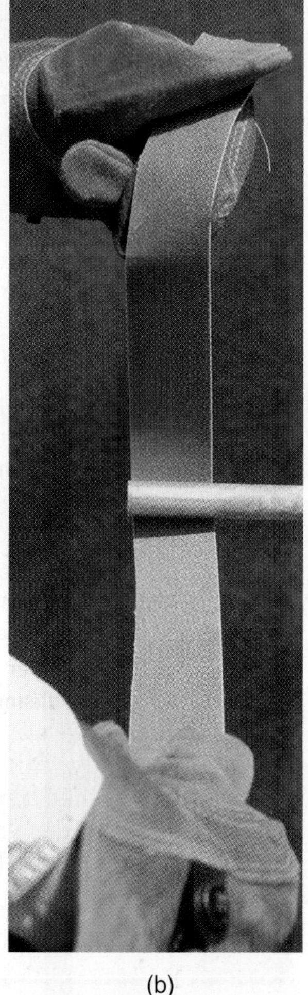

(a) (b)

Figure 3-4-15 (a) Use aluminum oxide sand cloth to clean copper before brazing and soldering; (b) Be sure to clean all the way around the fitting.

Figure 3-4-16 Round tubing brushes can be used to clean the inside of pipe and fittings.

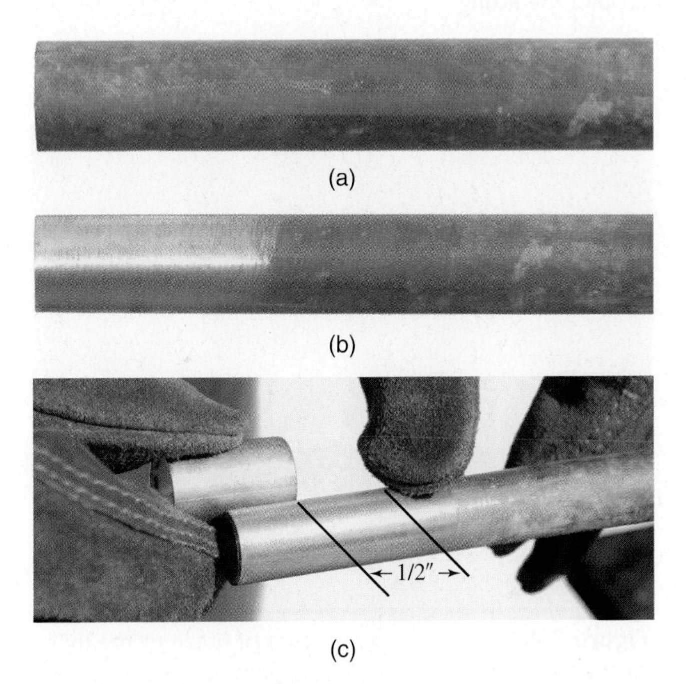

Figure 3-4-17 (a) Copper pipe before cleaning; (b) Copper pipe after cleaning; (c) Clean at least ½ in beyond the fitting.

SERVICE TIP

Push the cleaned copper pipe as deeply as possible into the copper fitting, Figure 3-4-18a. Twisting the pipe once it is completely inserted can help seat it completely to the bottom stop of the joint, Figure 3-4-18b. On fittings that have a stop all the way around the inside or on swaged fittings, this can prevent braze or solder from flowing into the pipe.

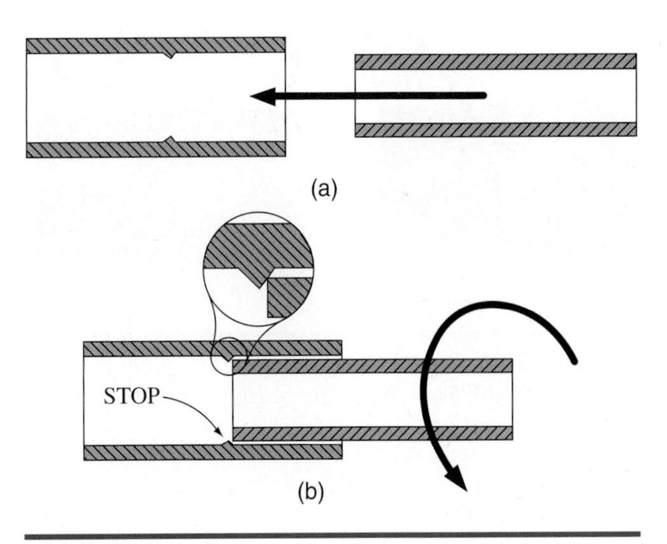

Figure 3-4-18 (a) Insert the copper pipe into the fitting; (b) Twisting the copper pipe against the stop helps seat the fitting.

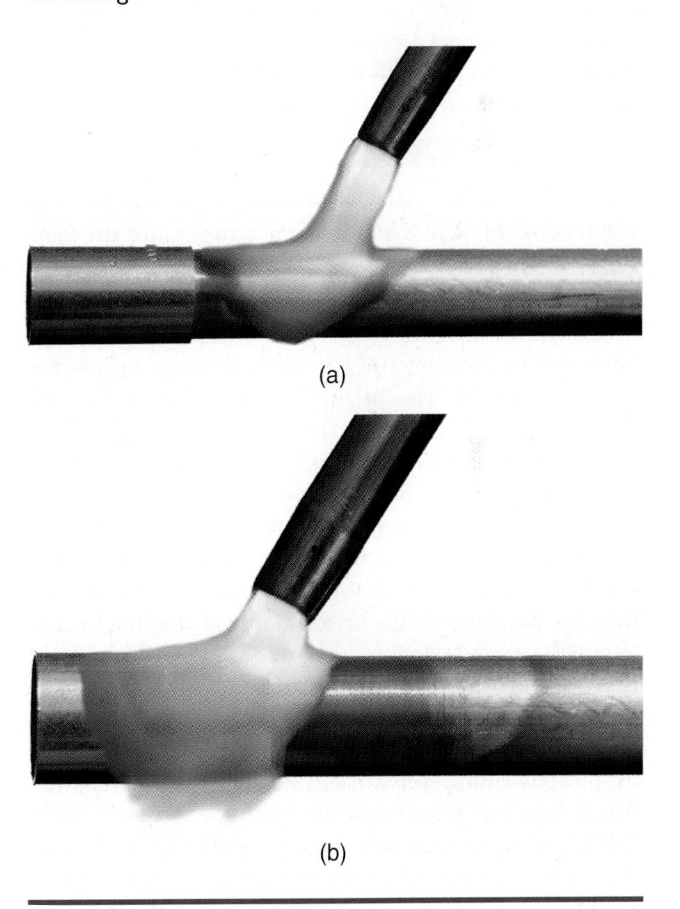

Figure 3-4-19 (a) Begin heating the pipe near the fitting; (b) Once the pipe is hot, move the flame onto the fitting and continue heating.

Heating the Copper Pipe

The order in which you heat the parts is crucial to a quality brazed joint. It is important to heat the pipe first, and then the fitting. As the copper pipe is heated, it will expand, Figure 3-4-19a. This expansion will tighten the joint space between the fitting and the copper pipe. If the fitting were to be heated first, this space would increase. The tighter the space is, the better the heat transfer and the greater the capillary attraction for the liquid braze metal. The capillary attraction is the force that draws molten

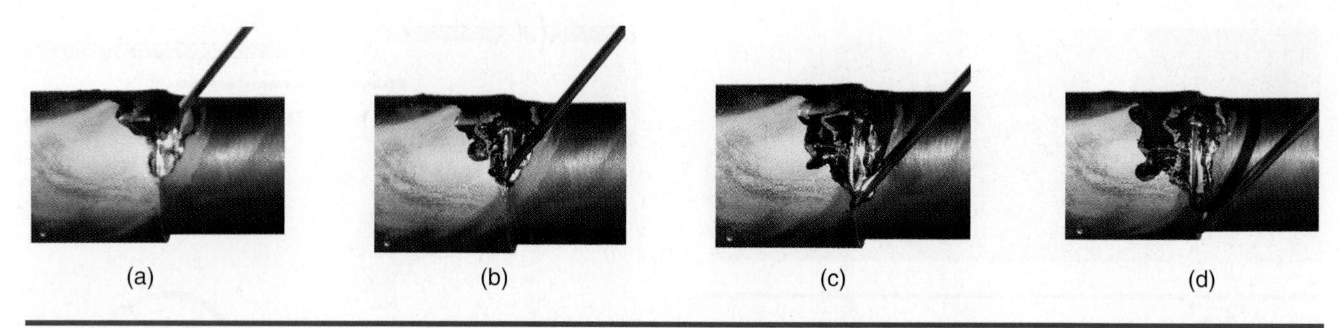

Figure 3-4-20 Once brazing temperatures have been reached, use a slight downward pressure on the filler rod at the braze joint and follow the joint around the fitting.

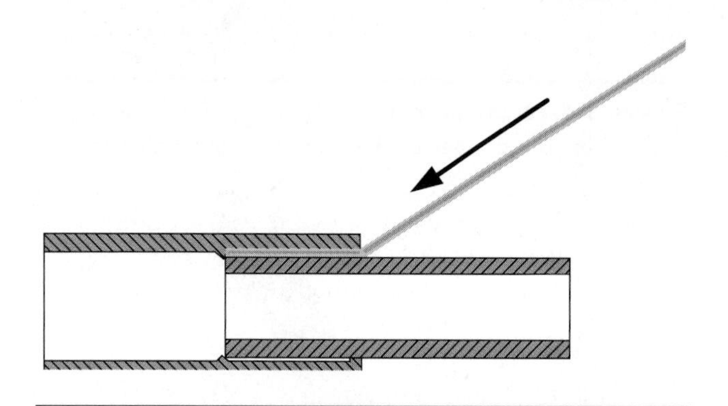

Figure 3-4-21 Add the filler metal straight into the joint gap.

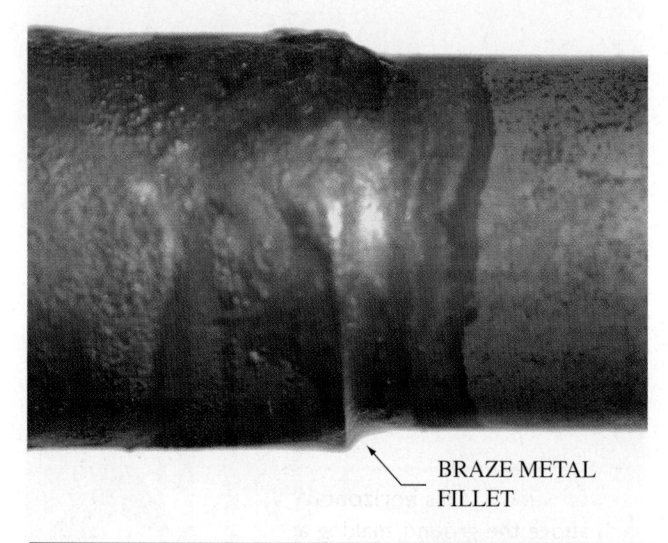

Figure 3-4-22 Add a small amount of braze to the joint so that there is a smooth uniform fillet to ensure a good joint seal.

braze metal deep into the joint. Once the pipe has become hot, approaching a dull red color, move the torch flame onto the fitting so that it envelops both the fitting and the pipe, Figure 3-4-19b. Continue heating the pipe. Occasionally touch the pipe surface with the tip of the brazing rod as a test of temperature readiness. A small drop of braze material may transfer from the rod to the fitting just before the fitting reaches brazing temperature. Continue heating the pipe until the brazing metal begins to flow evenly over the surface, Figure 3-4-20a–d. A slight pressure on the filler metal into the joint will aid in forcing the filler metal more deeply into the joint space, Figure 3-4-21. It is important that the filler metal flow completely to the base of the joint. Any unfilled gaps can result in joint failure, especially when R410A refrigerants are being used. These higher-pressure refrigerants will cause the copper pipe to expand slightly as the pressure inside the pipe increases. As the pressure decreases the pipe contracts. This cyclical expansion can cause a stress fracture if the joint is not completely filled with brazed metal.

If you watch carefully, you can see a slight change in the fitting's color as the filler metal flows into the joint space. This change in color is more obvious for larger diameter fittings. Bringing the torch down on the fitting will help draw the filler metal completely into the joint.

Once the joint gap has been filled, continue adding small amounts of filler braze metal until a fillet of metal surrounds the joint, Figure 3-4-22. This fillet serves two purposes. First, the fillet makes it easier to visually check the joint to see that it has no leaks. Second, it removes any sharp transition between the two pieces of copper, the fitting and the pipe, which will further reduce stress cracking when high pressure refrigerants (i.e., R410A) are being used.

Horizontal Brazed Joint

Horizontal brazed joints are the most common joints experienced in the HVAC field. Horizontal joints are typically found at the junction of the refrigerant line set and the outdoor unit as well as from the refrigerant line set to the evaporative coil. When practicing this joint, it will be beneficial to you later if you will work from just the top side of the joint. Because these joints are typically close to the ground in most installations, you cannot see the bottom of this joint when making it in the field,

Figure 3-4-23 This horizontal refrigerant line is one inch above the ground, making access to the bottom of the joint difficult.

Figure 3-4-24 Use the flame to heat and bend the brazing rod approximately 2 in from the end to make reaching the bottom of the horizontal joint easier.

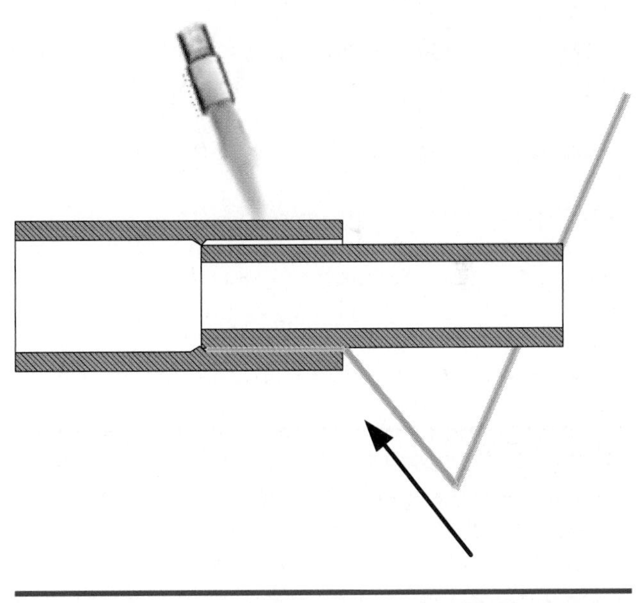

Figure 3-4-25 When the fitting is completely heated, filler metal can be added from the bottom first; then follow the joint around with the tip of the filler metal ending at the top of the joint.

such as the joint shown in Figure 3-4-23. Using the torch, heat the filler metal approximately two inches from the end. When the rod becomes soft, bend it to approximately a 90° angle, Figure 3-4-24. This will allow you to direct the filler metal into the bottom portion of the joint while your hand and torch are working from above. Again, always heat the copper pipe first. Then move your torch to the fitting. When the pipe and fitting are hot enough (indicated by the red color in Figure 3-4-25) add filler metal directly into the groove space.

Some time while you are brazing, you might accidentally touch the tip of the filler metal to the joint. If it becomes stuck, simply move the torch flame over until the rod melts itself free. Watch the filler metal and do not overfill the joint. Overfilling will result in a drip of brazed metal hanging from the bottom of the joint. When the joint has been completely filled, simply wipe the tip of the brazed metal rod across the bottom joint. This will provide a slight fillet around the joint. On larger diameter copper pipe and fittings (1 in or larger) it is usually necessary for you to move the torch around so that the flame is directed on the bottom surface. As you move the torch around, pay particular attention so that you do not inadvertently direct the flame toward yourself, others, or combustible materials in your work area. Learning to be aware of the flame direction will help you in the field to avoid burning the paint off of equipment as braze joints are performed.

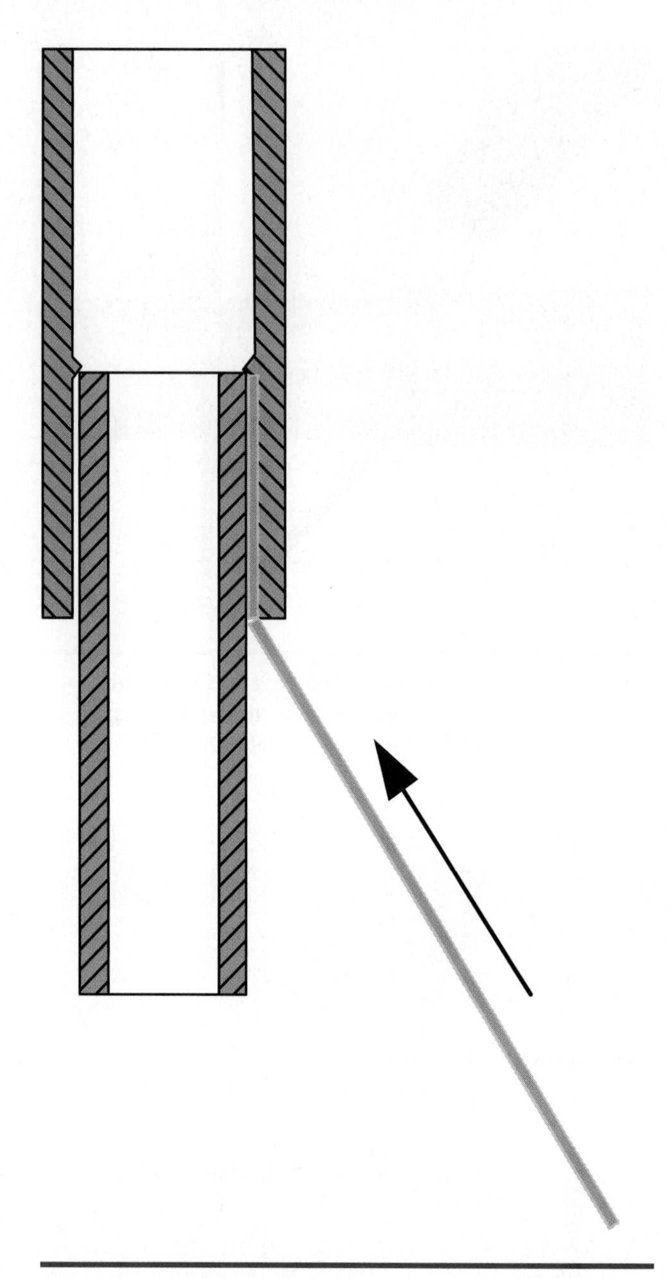

Figure 3-4-26 Inward pressure of the filler metal helps in filling the joint gap completely with brazes in the vertical up position.

Vertical Up Joint

Occasionally in the field, a brazed joint must be made in the vertical up position. These joints are typically found, for example, when a 90° fitting or coupling is joined in refrigerant line sets.

Start by heating the copper pipe just as with the other brazing practices. When the pipe is properly heated, move the torch up on the fitting so that the flame covers both the joint and pipe. Continue heating the copper pipe and fitting until the brazed metal begins to flow. Using a slight inward pressure, hold the tip of the filler metal against the joint and press upward. This slight pressure along with the heat and capillary space will draw the filler metal deeply into the joint,

Figure 3-4-27 The first step in braze joint testing is to cut the fitting just beyond where the pipe is seated.

Figure 3-4-26. When the joint has been completely filled, wipe the tip of the filler metal around the joint groove to produce a soft fillet of brazed metal. It is important not to overfill this joint. Overfilling will result in a drip of brazed metal hanging down on the pipe.

TESTING BRAZE JOINTS

In the field, brazed joints are tested by evacuating and charging the system. In the refrigeration and air conditioning lab, it is possible to test your brazing by destructively testing the joint. To perform this destructive test, first use a hacksaw to saw off the copper and copper fitting at a point just beyond the depth that the pipe was inserted into the joint, Figure 3-4-27. Once the fitting has been cut completely apart, clamp the pipe in a vise and cut straight down through the entire joint into the pipe. Rotate the pipe 90° and repeat this process, cutting as shown in the diagram in Figure 3-4-28. First bend each quarter out with a wrench, Figure 3-4-29a. Using a hammer and anvil, flatten each of the four corners of the pipe and fittings you just made, Figure 3-4-29b. As the joint pieces are flattened, it is easy to see areas where 100% joint penetration did not occur, Figure 3-4-30a,b. Repeat the practice until you can successfully pass each of the four various joints without finding voids during testing.

OXYACETYLENE TORCH

The oxyacetylene torch flame is lit by first opening and adjusting the gas pressures on the cylinders and regulators to the manufacturer's specified pressure. Open only the acetylene hand wheel on the torch slightly. Light the acetylene gas using a spark or flint lighter. Do not cover the end of the torch with the lighter; refer to Figure 3-4-13 on page 120. When the

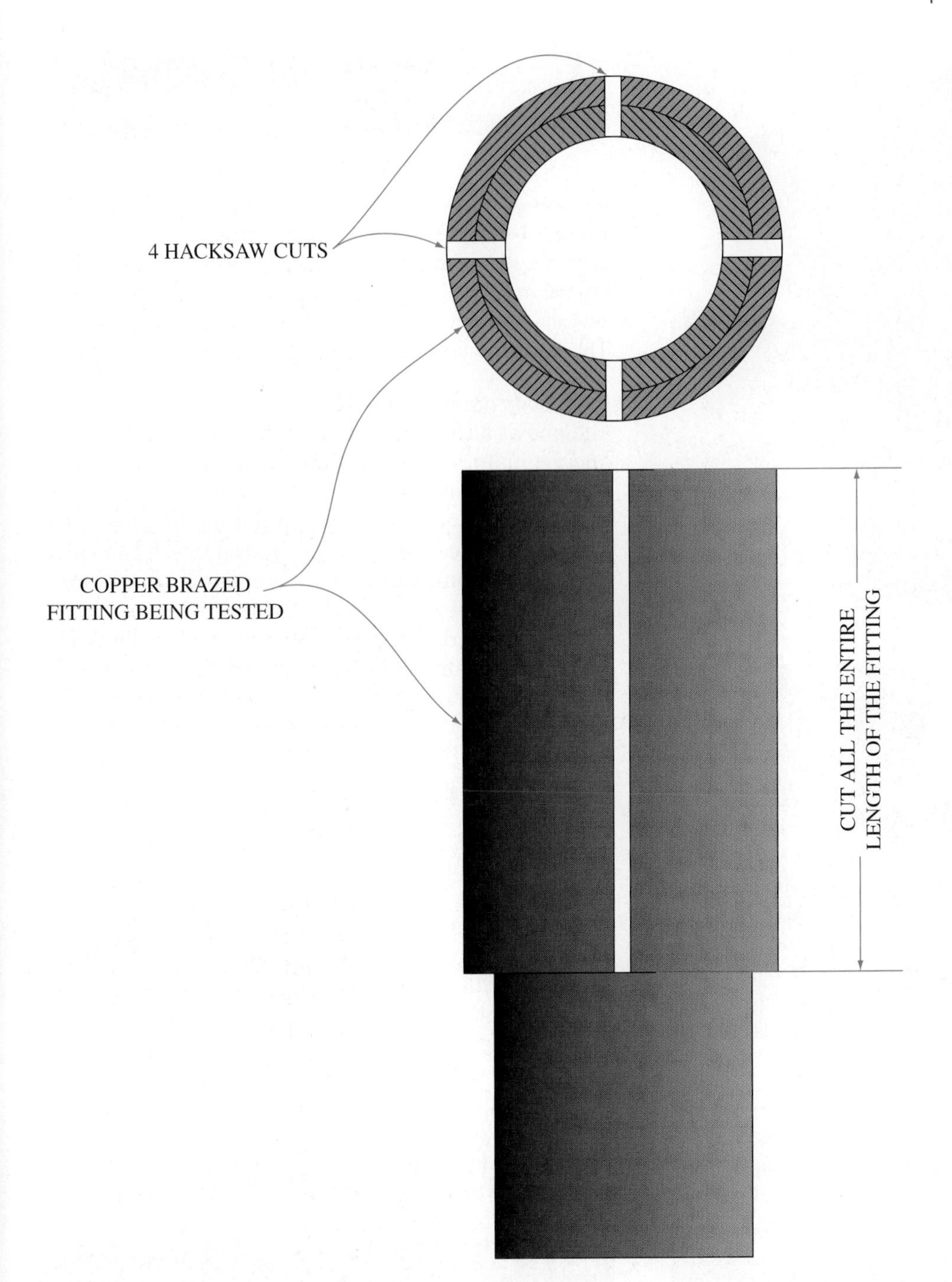

Figure 3-4-28 Use a hacksaw to make two cuts through the entire length of the fitting.

torch lights, Figure 3-4-31a, immediately turn up the flow until all of the smoke in the flame has disappeared, Figure 3-4-31b. Next, slowly open the oxygen torch handle valve, Figure 3-4-31c. Increase the oxygen flow until the feather in the flame completely disappears and joins the center cone, Figure 3-4-31d. Do not increase the oxygen flow rate beyond this point or the flame will have too much oxygen, Figure 3-4-31e. Flames with too much oxygen are called oxidizing flames and cause excessive oxides to form on copper during brazing.

A properly lit and adjusted oxyacetylene flame will burn with a neutral flame. This means there is not an excess of oxygen or acetylene; therefore, the flame is balanced. It is important to use a balanced or neutral flame so that the flame does not introduce contaminants into the braze.

Adjusting the flame flow rate so that the smoke disappears will ensure that you have adequate gas flow to keep the tip cool. If the torch were to be lit and the acetylene still produced smoke, there would be insufficient gas flow once the oxygen is introduced to keep the tip from overheating. As the tip becomes overheated, it will begin to pop. This popping is called a backfire, and backfires can results in damage to the equipment. If backfires become excessive, that can result in a flashback where flame actually races back through the hoses to the cylinders. If this were to occur, the results could be catastrophic. The system could explode.

(a)

(b)

Figure 3-4-29 (a) Use a pair of pliers to bend each quarter section out; (b) Flatten each section using a ball peen hammer and anvil.

SERVICE TIP

It is possible to use an oxyacetylene flame for brazing that is slightly out of adjustment. However, when the flame is adjusted properly, the job will go faster and the results will be better. If the flame is too far out of adjustment, the excessive acetylene or oxygen in the flame can stop the braze metal from wetting the copper surface.

Figure 3-4-32 shows where the torch body hand wheels should be turned off. Note that when welding or brazing, you should always be wearing protective gloves. To properly shut off an oxyacetylene torch, first turn off the acetylene flow and then the oxygen at the torch body hand wheels. Next, close the cylinder valves and one at a time release the pressure from the hoses and regulators. Once the pressure has been relieved from both the oxygen and acetylene hoses and regulators, back out or loosen the regulator-adjusting valve handle. Make certain you do not overloosen the valve handle as in some cases it can completely unscrew and fall out of the regulator body. If this does occur, be very careful not to cross-thread the adjusting handle as you reinstall it. Many manufacturers use nylon threads in the regulator so that oil is never needed for them to operate smoothly. The nylon is easily cross-threaded.

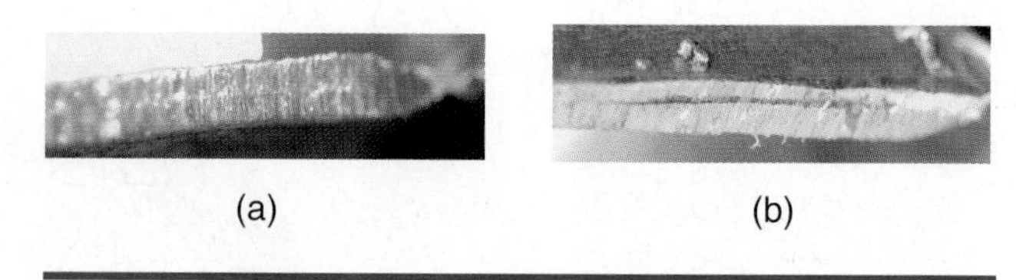

(a) (b)

Figure 3-4-30 (a) A thin layer of brazed metal can be seen between the pipe and fitting; (b) Area not filled with brazed metal will allow the pipe and fitting to slip, slightly revealing a very small shelf between the pipe and fitting. This area was not properly filled with brazing metal.

Tip Cleaning

From time to time the oxyacetylene tip becomes dirty. To clean the tip use a tip cleaner of the proper size and ream out the dirt that has collected in the tip orifice, Figure 3-4-33. When the tip becomes severely damaged, it may be necessary to first file the tip surface flat before cleaning, Figure 3-4-34. Filing the tip is not necessary unless the tip surface has become damaged. Select a size tip cleaner that will easily fit into the hole in the tip. Tip cleaners are actually small round files that are accurately calibrated for each tip size. By sliding the tip cleaner in and out several times you will remove the debris inside the tip as demonstrated by Figure 3-4-35.

NICE TO KNOW

The stubs on a new hermetic compressor that attach to the low and high side refrigerant line look like copper. The copper is actually a very thin coating on a steel pipe. The copper coating is put on the stubs to aid in brazing the refrigerant lines to the compressor. This thin copper coating is easily damaged by overheating if the oxyacetylene torch flame is an oxidizing flame or if too much time is taken during brazing. To keep from burning off the copper coating, be sure to use a neutral oxyacetylene flame, and use a torch tip that is big enough to do the job quickly. If the copper coating is damaged, you will have to stop, disassemble the joint, and clean it before restarting. In that case, it will also be necessary to use a brazing flux.

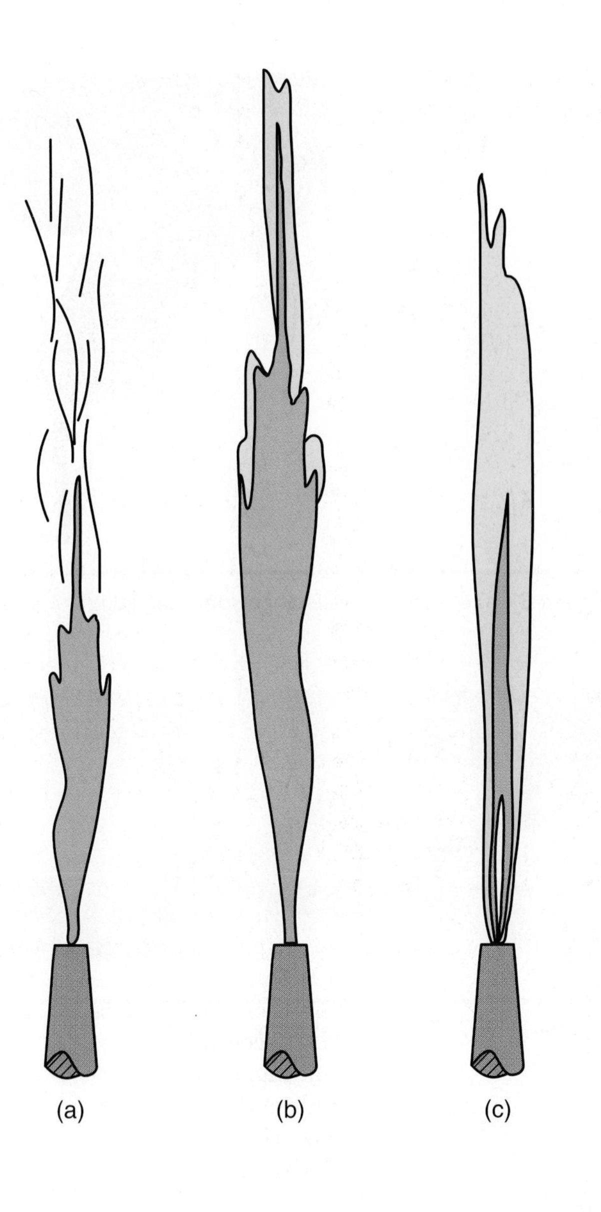
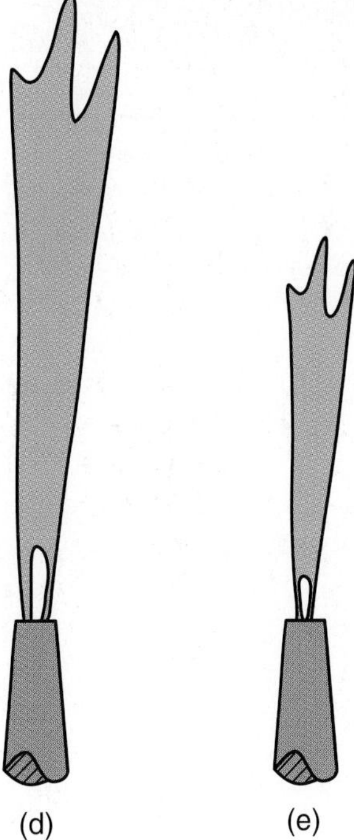

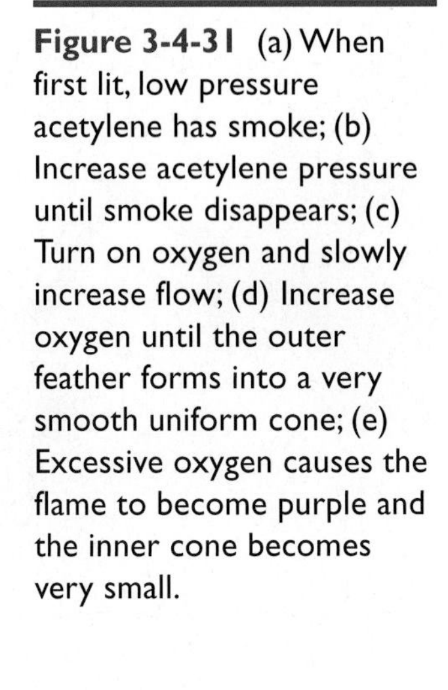

Figure 3-4-31 (a) When first lit, low pressure acetylene has smoke; (b) Increase acetylene pressure until smoke disappears; (c) Turn on oxygen and slowly increase flow; (d) Increase oxygen until the outer feather forms into a very smooth uniform cone; (e) Excessive oxygen causes the flame to become purple and the inner cone becomes very small.

Figure 3-4-32 Turn off the acetylene gas valve first. For clarity, this procedure is shown with bare hands, but welding gloves should always be used.

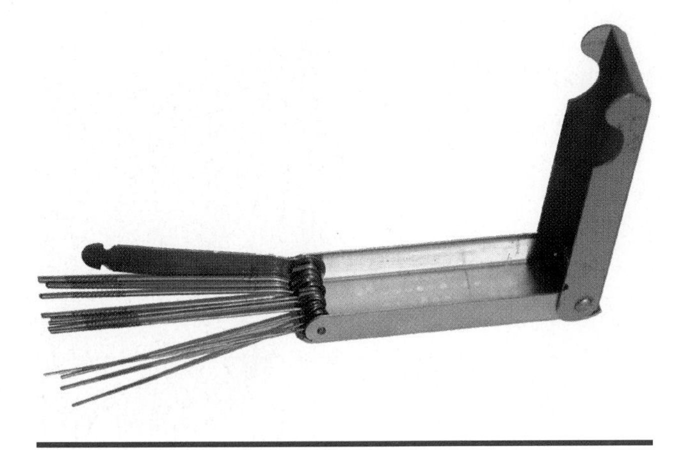

Figure 3-4-33 Tip cleaning set.

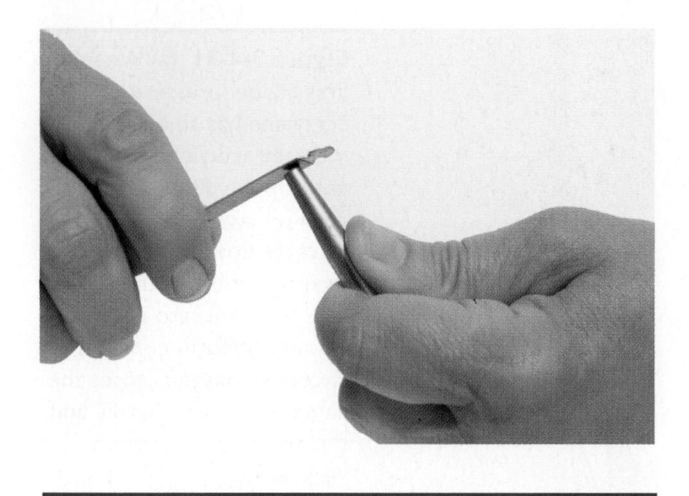

Figure 3-4-34 File the tip flat if it has been damaged.

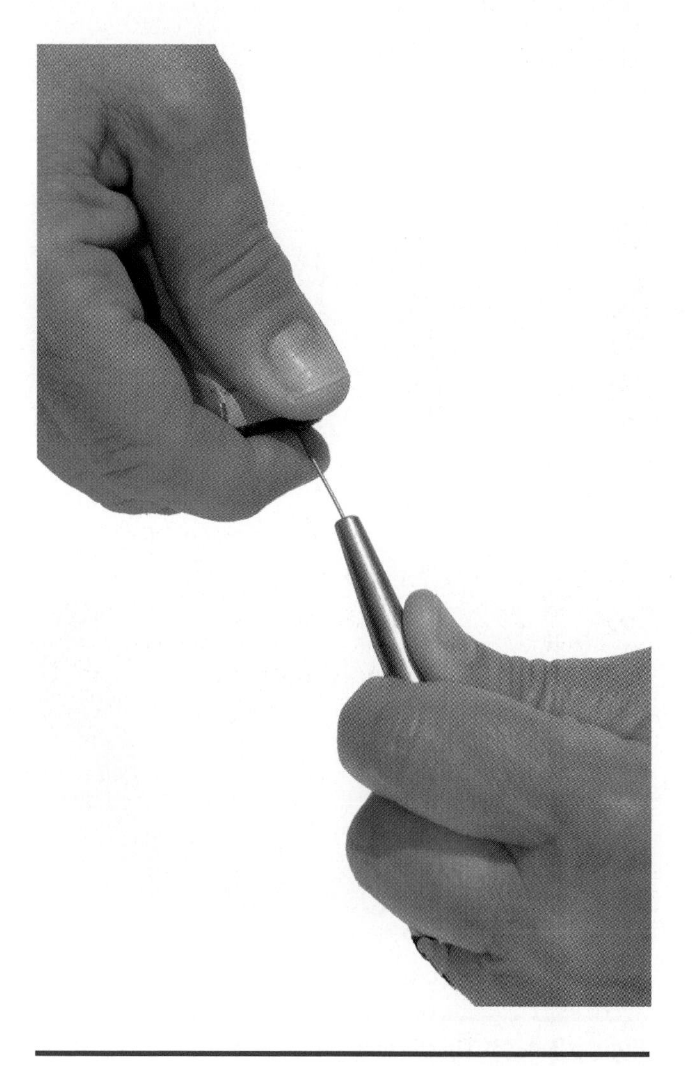

Figure 3-4-35 Use the round tip cleaner to ream the tip orifice.

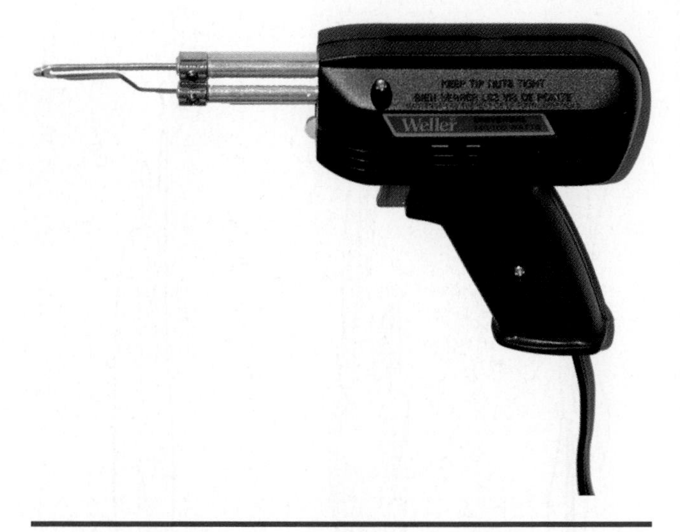

Figure 3-4-36 Electric resistance soldering gun.

Figure 3-4-37 Cleaning soldering gun tip.

SOLDERING

Soldering takes place below 840°F. Soldering is used in the HVAC industry to both join electrical wiring and components and for copper pipes when used for water or condensate drains. Solder is not approved for refrigerant line joints. Electrical soldering is accomplished using an electric soldering iron, Figure 3-4-36. Soldering to join copper pipes uses an air acetylene torch, air Mapp torch, or air propane torch. Oxyacetylene is not recommended for soldering because its flame temperature is so high that it is very easy to overheat the pipe. When overheating occurs the flux breaks down and excessive oxides are formed so that no bonding can occur.

Electrical Resistance Soldering

Electrical resistance soldering irons and guns are used for electrical wiring and component servicing. The tip of an electric soldering iron or gun is made of copper. A new tip or a damaged tip must first be tinned with solder before it is used. To tin a tip first clean the surface with sand cloth or a file, Figure 3-4-37. Apply soldering

Figure 3-4-38 Fluxing soldering gun tip in preparation for tinning.

flux and turn on the gun, Figure 3-4-38. Press the solder against the surface until the solder begins to melt as shown in Figure 3-4-39. Turn off the gun and rub the tip surface with the solder until the entire surface has been tinned with solder.

Figure 3-4-39 Tinning soldering gun tip.

Soldering Electrical Wires

When using solder to join electrical components, first thread the electrical component lead through the eyelet of the part as shown in Figure 3-4-40. Make a solid mechanical connection by bending the wire around in a tight loop.

Turn on the soldering gun and allow the solder of the tip of the gun to melt. Hold this molten solder against the electrical wire ends. Place the solder wire against the opposite side of the wires being joined as shown in Figure 3-4-41. When the wires reach soldering temperature, the solder will melt and flow into the joint. Gently remove the soldering tip and solder wire from the connection. Allow the solder to cool and

Figure 3-4-40 Loop wire through terminal lug to make a mechanical connection.

Figure 3-4-41 Hold the soldering gun against one side of the wire and lug as the solder is introduced to the opposite side.

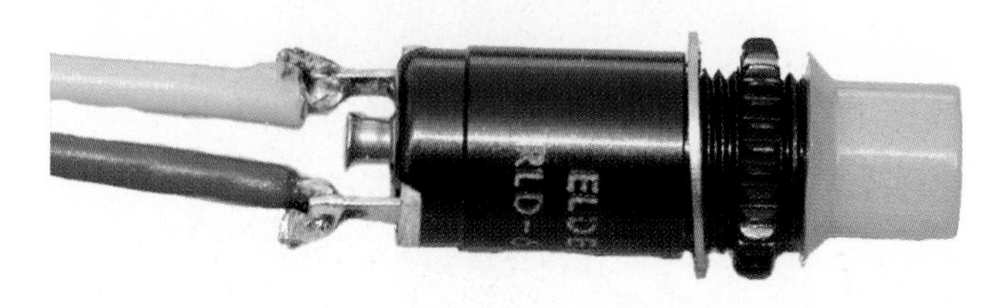

Figure 3-4-42 Clip the ends of the wire as necessary to complete the job. When connecting multiple wires, have the insulation always on the inside so there is less chance of an electrical short.

solidify completely before the solder joint is moved. If the solder is moved before it is completely solidified, it can introduce microscopic cracks into the solder joint. These microscopic cracks can cause resistance, which may result in the joint building up heat and failing.

NICE TO KNOW

When attaching more than one wire in a small space, install the wires so the wire insulation is closest together. You can do this by threading the wire from the inside to the outside through the wire lug, as shown in Figure 3-4-42. This makes soldering easier, and the insulation helps prevent possible shorts.

SERVICE TIP

If the soldering gun is placed against the component being soldered before the gun has heated up, the copper wire will pull heat away from the joint rapidly. This will significantly slow the soldering process and can overheat the part or wire insulation. To prevent this from occurring, turn on the soldering gun and allow the tip solder to melt before it is brought into contact with the part being joined.

Most electrical soldering uses a flux-cored wire. The flux for electrical application is rosin. Rosin is inactive at low temperatures and will not corrode the electrical parts even if the rosin is left on following soldering. Acid and acid core solders are reactive and will damage electrical components. Even after a thorough cleaning of an electrical component joined with acid core solder, the join will often still corrode at some later time. It is difficult if not impossible to remove all the acid from within the fine wires of an electrical connection.

TORCH SOLDERING

Air acetylene or air propane/air Mapp torches are all ideal for soldering copper piping. The flame produces a uniform heating of the joint, Figure 3-4-43. Uniform heating of the joint is important to provide a complete joint fill with the solder alloy.

Begin by cleaning the copper pipe and fitting using abrasive cloths like those shown in Figure 3-4-44. A wire brush or sand cloth may be used for this purpose. When the part is cleaned of all contamination, such as oil, paint, or dirt, it must not be touched with your hand. Just as before, with brazing, oil from your fingers can prevent solder from flowing completely into a joint. The cleaned end should look like the one in Figure 3-4-17b on p. 121.

With the joint clean, use a brush to apply the flux. Do not apply flux to the very end of the pipe. Apply the flux up to approximately $\frac{1}{16}$ in to $\frac{1}{8}$ in from the end of the pipe as shown in Figure 3-4-45. This will help prevent flux contamination into the system. Insert the flux-covered pipe into the copper fitting. Twist the copper pipe inside the copper fitting. This will spread the flux around inside the joint.

Use an air fuel torch. Begin by heating the pipe. As the pipe expands, it will make firm mechanical contact with the inside of the copper fitting, Figure 3-4-46. This mechanical contact will aid in capillary attraction, which pulls the solder into the joint. The mechanical contact also aids in thermal conductivity so that the entire pipe and joint become uniformly heated. As the pipe begins to be heated, move the torch onto the fitting and pipe. Periodically test the back side of the joint with the tip of the solder. Once the solder begins to melt, remove the torch. Continue adding the solder until the joint is filled. Ideally a small fillet of solder should be left at the joint surface, Figure 3-4-47. Three quarters of an inch of solder is adequate to make soldered joints in $\frac{1}{2}$ in through 1 in diameter copper pipe. To prevent overfilling of the joint, before you start, bend the solder approximately three quarters of an inch from the end. Pushing more solder into the joint simply results in solder bb's being formed inside the piping

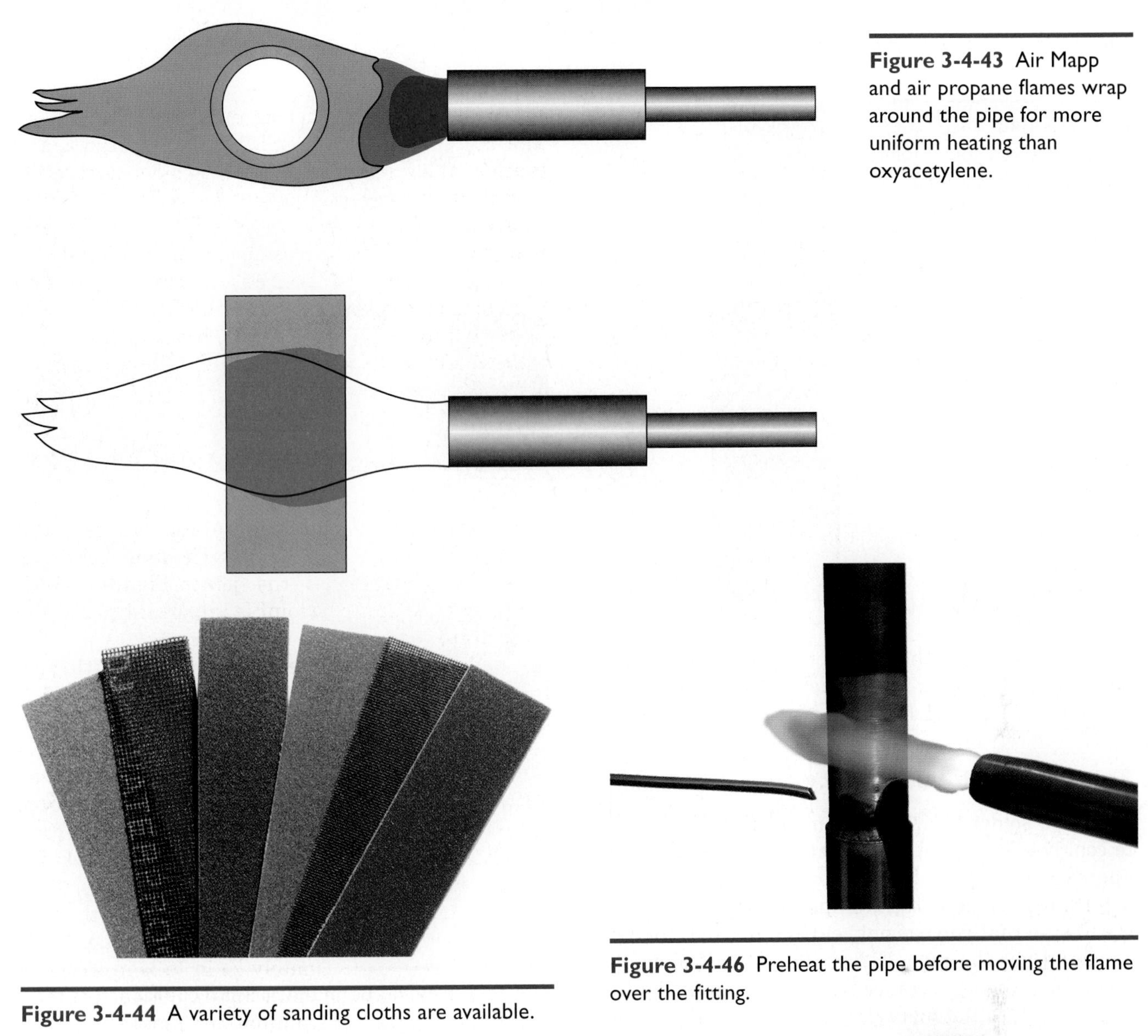

Figure 3-4-43 Air Mapp and air propane flames wrap around the pipe for more uniform heating than oxyacetylene.

Figure 3-4-44 A variety of sanding cloths are available.

Figure 3-4-46 Preheat the pipe before moving the flame over the fitting.

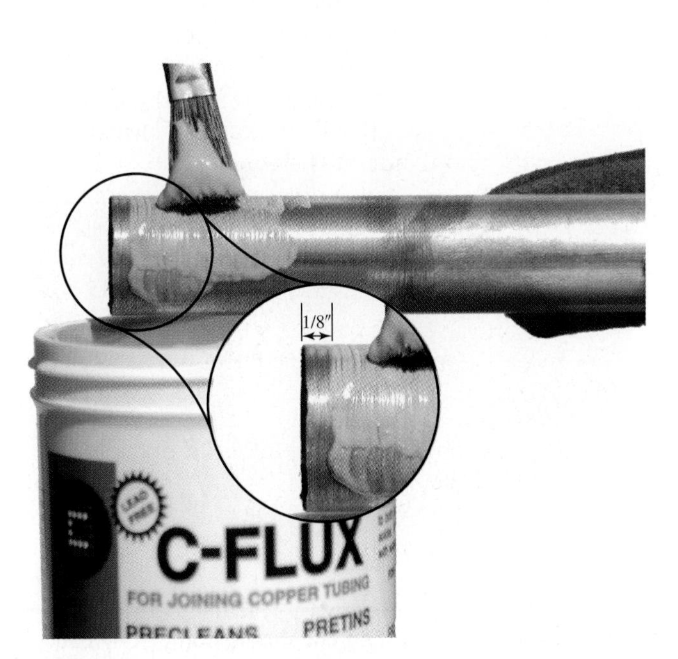

Figure 3-4-45 Apply the flux up to approximately ⅛ in from the end of the copper pipe.

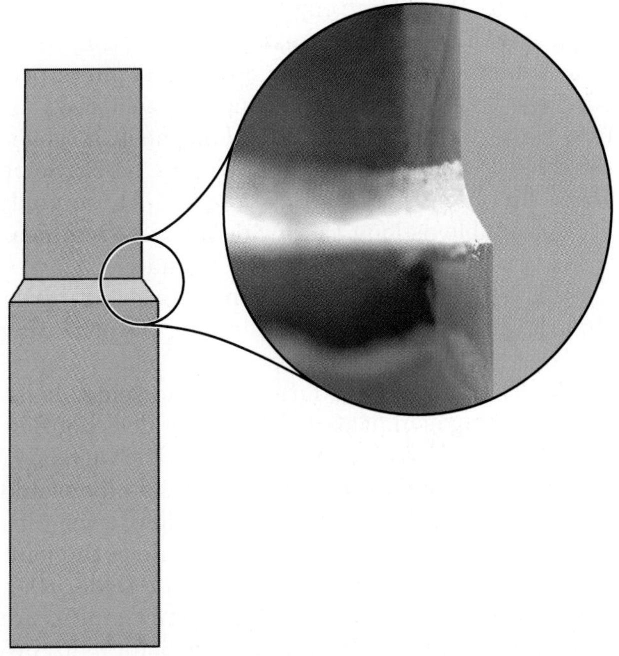

Figure 3-4-47 Add a small amount of additional solder to create a fillet around the joint.

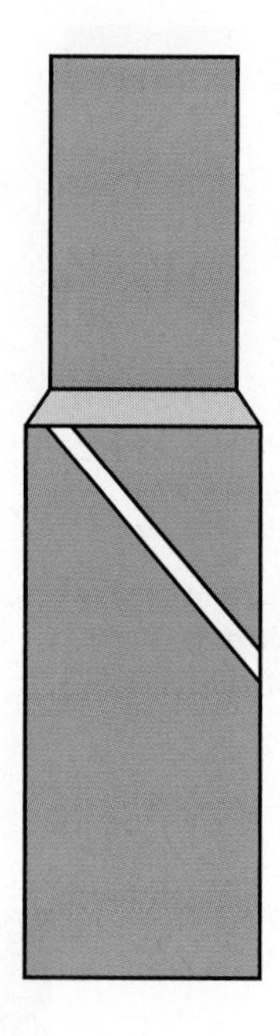

Figure 3-4-48 The first step in testing a solder joint is to make a hacksaw cut diagonally across the fitting.

(a)

(b)

Figure 3-4-49 (a) Begin the hacksaw cut flat across the fitting; (b) Rock the hacksaw as the cut is made.

system. In the field these bb's of solder can circulate back to water pumps where they can cause damage to the pump's impellers.

Testing Soldered Joints

Once the solder joint has been completed, allow it to cool before cutting a slice into the pipe fitting at approximately a 45° angle using a hacksaw, Figure 3-4-48. Use a slight rocking motion as the cut progresses, so that you are only cutting through the fitting, and not the pipe, Figure 3-4-49a,b. The cut is done when the appearance resembles that in Figure 3-4-50. Using a flat blade screwdriver placed into the slot cut in the fitting, apply pressure and twist, Figure 3-4-51. This will cause the pipe fitting to release from the copper pipe. Using a pair of pliers, peel back the copper fitting exposing the soldered surface. If the surface is smooth and has no large voids, the soldering job was successful, Figure 3-4-52. Some small flux-filled pockets may remain. These voids are acceptable as long as they do not run from near the inside of the joint to near the outside.

Some of the common problems with soldering include overheating and underheating. Overheating is indicated by tiny bubbles in the solder, Figure 3-4-53. The bubbles are formed as the solder boils. Underheating is indicated if the solder does not flow into the joint. If the joint were heated first instead of the pipe first, then during soldering the pipe would separate so that heat would not transfer to the pipe from the joint. The solder on the outside of the joint may look great as in Figure 3-4-54. The solder even flowed down the outside of the joint shown in Figure 3-4-54. But looking at the inside of the joint in Figure 3-4-55, we see that the joint was actually very poor. The solder did not flow throughout the joint space

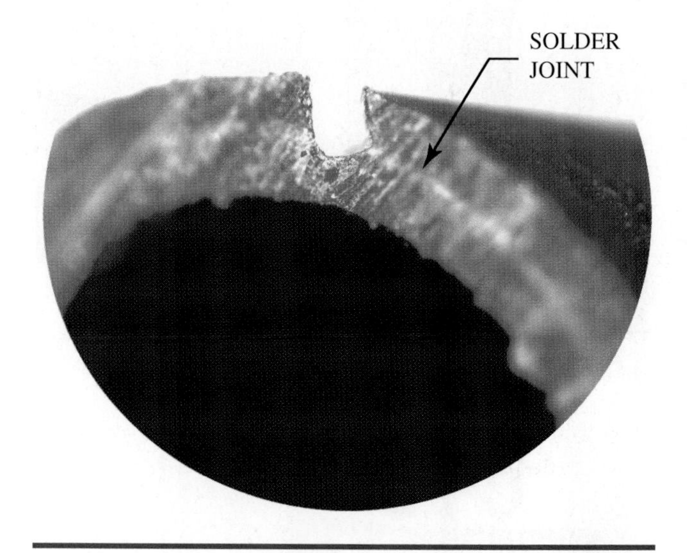

Figure 3-4-50 The finished cut should completely cut through only the outer fitting. A small cut into a portion of the inner fitting is okay.

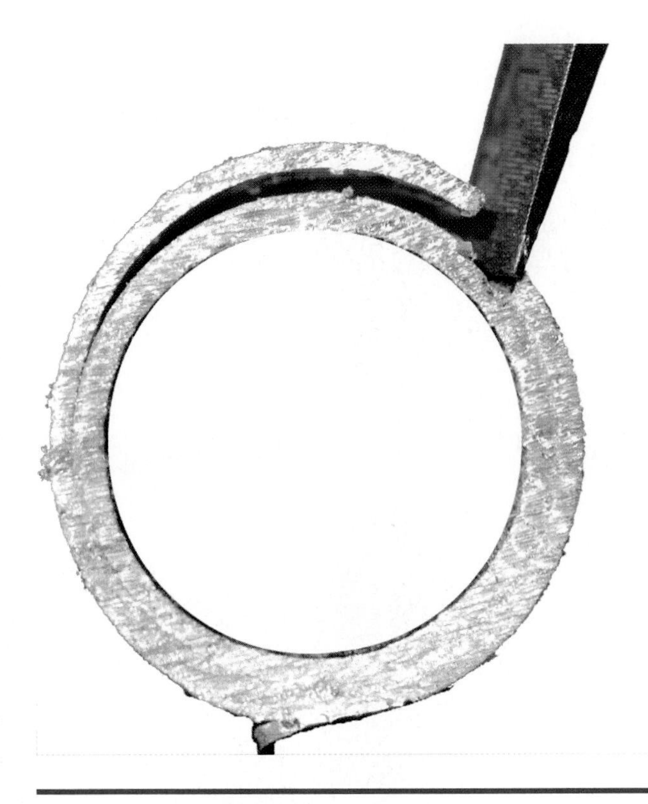

Figure 3-4-51 Place a flat head screwdriver in the hacksaw kerf and twist to begin opening up the joint. Use pliers to finish the job.

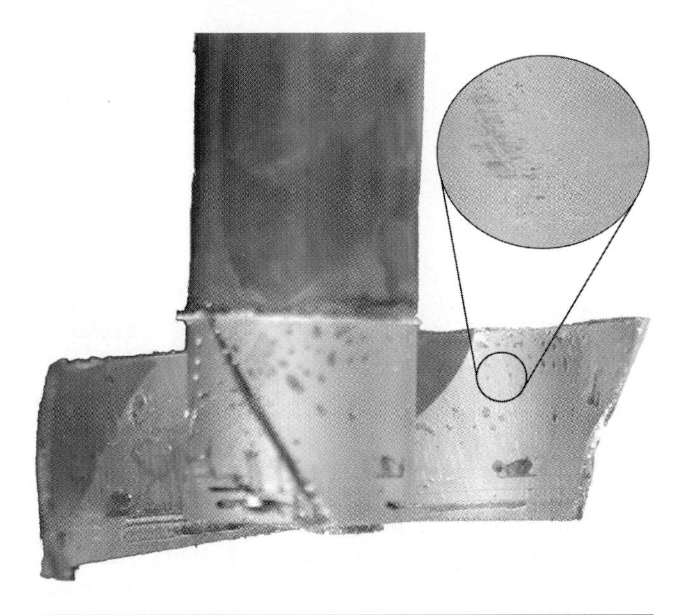

Figure 3-4-52 An acceptable solder joint is one that will have a minimum of small bubbles and flux inclusions.

Figure 3-4-53 Excessive bubbles formed by overheating and boiling solder.

Figure 3-4-54 From the outside, this joint looks very good. Solder flowed completely across the outside face of the joint indicating that the fitting was at the proper soldering temperature.

and left large unfilled voids. The joint temperature was inadequate or it may have been uneven.

WELDING

In the HVAC industry most of the material needing to be welded is thin sheet metal. Of the nearly one hundred different welding processes only a few are commonly used in the HVAC industry. The two most commonly used welding processes in the HVAC industry are resistance spot welding and resistance seam welding. To a lesser extent MIG welding and flux cored arc welding are used. These two latter processes are similar and can be performed using an oxyacetylene rig.

Figure 3-4-55 The inside of the joint from Figure 3-4-54. Solder did not flow through the joint because the pipe had not reached the soldering temperature due to improper heating techniques.

RESISTANCE WELDING

Resistance spot welding is commonly used in sheet welding duct fabrication. The welds produced will securely hold together sheet metal parts. Resistance spot welds are made when a high amperage, low voltage current is passed through the metal causing a rapid increase in temperature at the joint surface, Figure 3-4-56. Force is applied to this hot surface causing it to fuse together, forming a weld. The welding current for spot welding is applied through opposing copper electrodes. The steps in the process are as follows. The opposing copper electrodes are pressed tightly against the outside surfaces of the joint, welding current is applied for a specific period of time, a small weld nugget is formed, pressure from the electrodes is released, and the electrodes are moved to the next location for the process to be repeated.

When making spot welds it is important that the tips on the copper electrodes be clean and flat. As welds are produced these surfaces will become pitted and damaged. Periodically you must stop and refile these tip surfaces, Figure 3-4-57. The higher the welding current and the greater the welding force, the more frequently these tips must be reconditioned. Under normal shop fabrication conditions the welding tip surface should be usable for several hours of spot welding and under ideal conditions may last for several days without the need for reconditioning. To extend the useful life cycle of spot welding tips it is better to use slightly lower

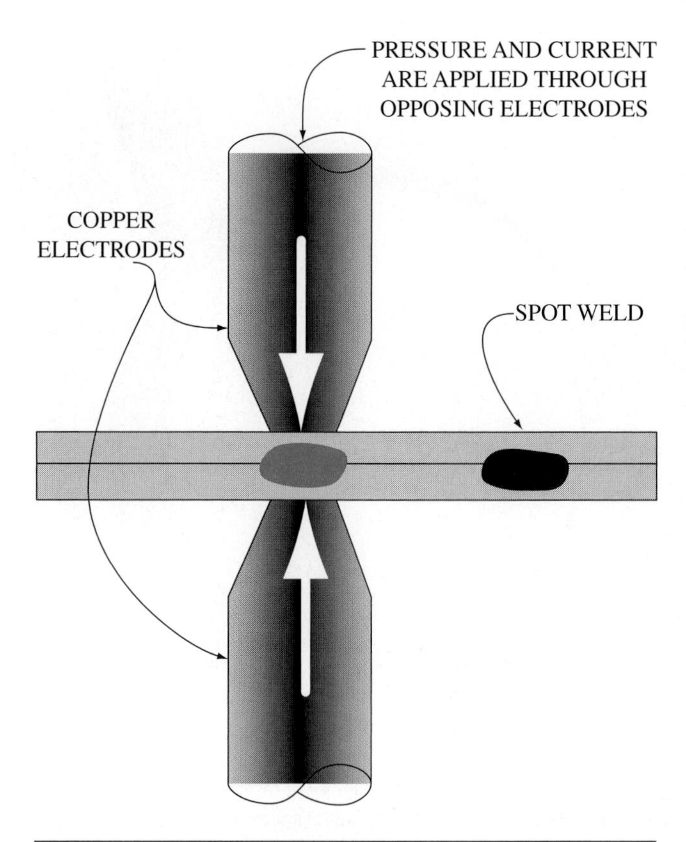

Figure 3-4-56 Resistance spot welding.

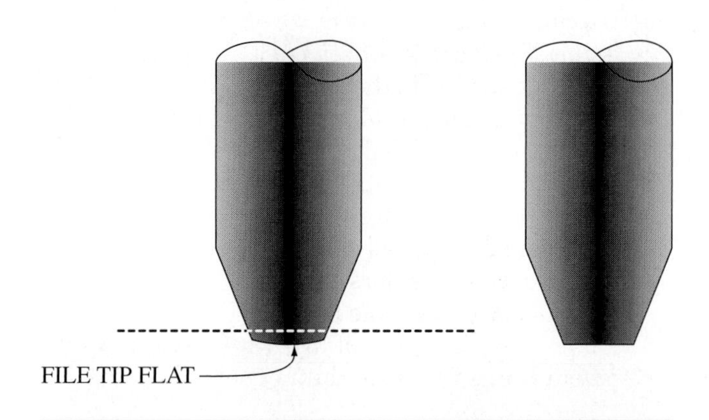

Figure 3-4-57 Resistance spot welding, tip cleanup.

currents and lower pressures with longer welding cycles. Making several practice spot welds will allow you to best set the welding parameters for your job.

Most spot welding equipment found in sheet fabrication shops, such as the equipment in Figure

SAFETY TIP

Often spot welding results in red hot metal sparks being thrown from the spot weld. For this reason safety glasses are required any time you or someone near you is performing spot welds.

Figure 3-4-58 Pedestal spot welder.

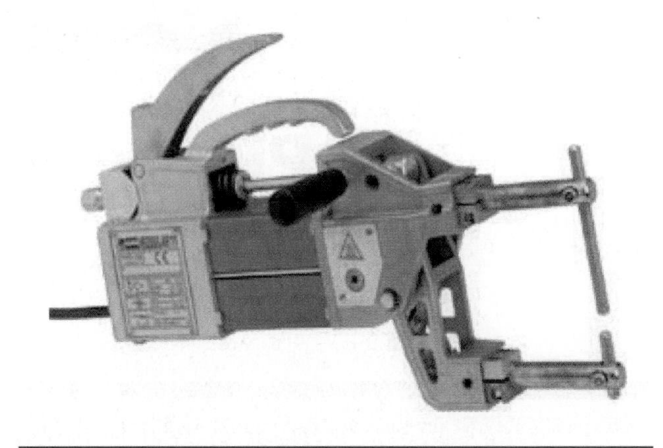

Figure 3-4-59 Portable spot welder.

3-4-58, can produce welds in metal thickness from 32 gauge up to 80 gauge. This range of metal thickness includes most commonly used thicknesses for sheet metal ductwork.

Metal preparation for spot welding simply requires that the surfaces be clean. You must remove any paint, oil, dirt, or other such material from the metal surfaces before spot welding can be performed. Surface contamination may make welding difficult or impossible. Such contamination will significantly reduce the weld strength; therefore, it must be removed prior to welding. Spot welding is most commonly used to join mild steel and galvanized steel sheet metal. Other metals such as stainless steel may be spot welded; however, they are not as commonly used as mild steel and galvanized steel are for HVAC work.

Portable spot welders like the one shown in Figure 3-4-59 are available for some limited field applications. The capacity of portable spot welding equipment limits its use to thinner sections of metal.

SEAM WELDING

Resistance seam welding machines are most commonly used as part of an automatic duct fabrication system. Resistance seam welding can be used to produce the long continuous seams required for this type of automatic duct processing equipment. Because of the size and expense of seam welders, they are not commonly found in HVAC shops.

TECH TALK

Tech 1. Silver brazing is so much stronger. I use it for all of my piping jobs.

Tech 2. Why?

Tech 1. Well it is only a little more expensive, and I know I can get a good joint that is leak free.

Tech 2. The cost is not the main problem, it's that some parts like water valves are easily damaged by too much heat, such as that from brazing. And most water valves are brass, and you would have to use flux for silver brazing.

Tech 1. So I'm better off if I use the right process for the each job, brazing for brazing jobs and soldering for soldering jobs. Thanks.

UNIT 4—REVIEW QUESTIONS

1. What is the difference between brazing and soldering?
2. Why are tin antimony alloys most commonly used for soft soldering?
3. In order for the bonding phase to occur between the filler metal and base metal, often a ________ is required.
4. What are the two most common sizes of oxygen cylinders used for the air conditioning and refrigeration trades?
5. It is against the law to open an acetylene cylinder valve more than ________.
6. What are the most common joints experienced in the HVAC field?
7. Where are vertical up joints typically found?
8. How should you properly shut off an oxyacetylene torch?
9. What is soldering used for in the HVAC industry?
10. Which torches are most ideal for soldering copper piping?
11. What are the most commonly used welding processes in the HVAC trades?
12. ________ are produced by holding the opposing copper electrodes firmly against the metal surface and using high amperage, low voltage current.
13. List the factors that control the time between reconditioning for spot weld electrodes.

UNIT 5

Piping and Tubing

OBJECTIVES: In this unit the reader will learn about refrigerant piping and tubing:

- Basic Principles of Refrigeration Piping
- Refrigerant Piping Materials
- Copper Tubing
- Copper Tubing Classification
- Soft-Drawn Copper Tubing
- Hard-Drawn Copper Tubing
- Tube Sizes for SI Systems

INTRODUCTION

Most air conditioning and refrigeration systems include an outdoor condenser and an indoor evaporator. These two major units are connected with pipes or tubes that allow refrigerant to flow between them. The two pipes or tubes that make up the interconnecting system are referred to as the "line set," shown in Figure 3-5-1a. With air conditioning and refrigeration systems the smaller diameter line is the liquid line and it is high pressure. It carries liquid refrigerant from the outlet of the condenser in to the metering device at the evaporator inlet. The larger diameter line is the vapor or suction line and it is at the lower pressure. It carries vapor refrigerant back to the compressor from the indoor evaporator. Note plastic caps that have been put on the copper pipe ends to keep dirt, moisture, and insects out to the lines before they are installed to the outdoor unit, Figure 3-5-1b.

Refrigerant lines used for residential and commercial air conditioning and refrigeration systems are usually made of copper. They can be made with either copper pipe or copper tubing. All pipe is measured from the inside (ID) and most copper pipe used in HVAC/R is rigid, Figure 3-5-2. All tubing is measured from the outside (OD) and most copper tubing used in HVAC/R is flexible. Depending on the wall thickness, the OD of a ½ in type L copper tube could almost fit inside of a ½ in copper pipe, Figure 3-5-3.

SERVICE TIP

The larger diameter refrigerant line serves two purposes when it is connected to a heat pump. It carries hot high pressure vapor from the compressor to the indoor coil during the heating cycle; and it carries cool low pressure vapor back to the compressor during the air conditioning cycle. Because it is not always the suction line on heat pumps, it is called the vapor line. Heat pumps have a special port to access the suction pressure during the heating cycle. Be aware of what cycle the heat pump is on before you test. You can damage your low side gauge if you connect it to the vapor line during the heating cycle.

REFRIGERANT TUBING AND PIPING MATERIALS

Most tubing used in refrigeration and air conditioning systems is made of copper; however, aluminum is used by some manufacturers for fabrication of the evaporator and condenser internal coil circuits. Aluminum tubing has not become popular for field refrigerant lines, principally because it cannot be easily soldered.

Steel piping is used in some larger factory assembled units as well as in the assembly of the very large refrigeration systems where pipe sizes of 6 in (152 mm) diameter and above are needed. Threaded steel pipe connections are seldom used in modern refrigeration work since they may leak. These systems are welded, and couplings and flanges are then bolted to the equipment, Figure 3-5-4.

Confusion in Copper Sizes

Tubing is specified by outside diameter, whereas pipe is identified by nominal inside diameter. Both tubing and pipe can be soft and ductile or hard and rigid. But sometimes the term tubing is used incorrectly to refer to a soft ductile material and pipe is used incorrectly to refer to a hard rigid material. This misconception has resulted in a certain amount of confusion and is the source of some installation errors.

(a)

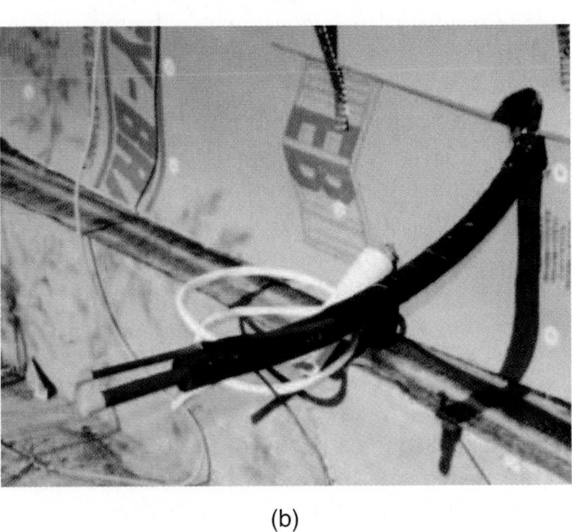

(b)

Figure 3-5-1 (a) The line set, or set of liquid (small) and vapor (large) lines, inside the house; (b) The line set outside the house has yellow plastic end caps until they are ready to be installed.

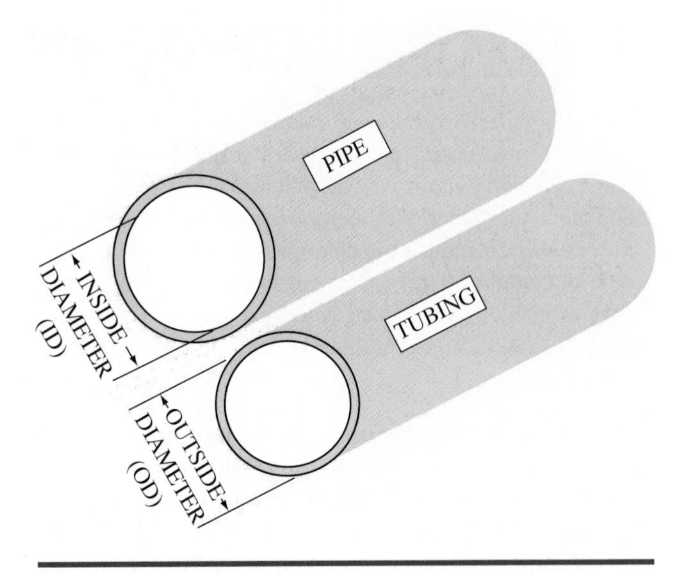

Figure 3-5-2 Pipe (top) is measured by the ID, but tubing (bottom) is measured by the OD.

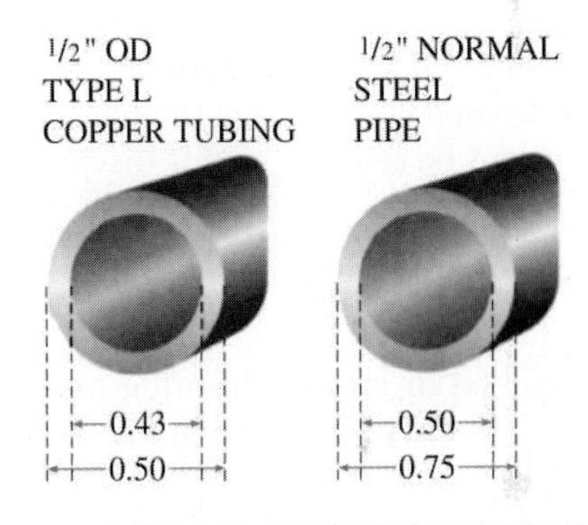

Figure 3-5-3 Method of sizing tubing and pipe.

Figure 3-5-4 In steel piping installations for commercial applications, couplings and flanges are bolted to the system.

SERVICE TIP

The only difference between tubing and pipe is the way they are measured. Manufacturers make soft copper, steel, and aluminum tubing as well as hard copper, steel, and aluminum pipe. With experience you will be able to more easily determine the difference between tubing and pipe but until you get that experience you must make sure which you are working with on a job.

SERVICE TIP

Copper tubing can become work hardened if it bounces around in a service van or truck. To prevent this from happening, copper tubing should never be hung on a hook so it bangs about as the vehicle is driven around. The best way of keeping copper tubing in good shape for the next job is to, when possible, put the unused portion back in the original box.

Pipe sizes in SI units refer to nominal pipe ID. Conversion tables are available to aid in standard to SI sizes. This table indicates the inch equivalent and the metric size in millimeters (mm). Both the inside and outside diameters are given in millimeters. This table also shows the volume of the tube in liters per meter (L/m).

Table 3-5-1 is an example of common pipe and tubing sizes, showing differences in their actual measurement.

Because of the special treatment of aluminum tubing and welded steel piping, the technique of fabrication will not be covered in this discussion.

COPPER TUBING

The tubing used in air conditioning and refrigeration systems is annealed. Annealing is the manufacturing process that helps make a metal as soft as possible. This makes the tubing easier to bend and fit. As copper is bent it will become work hardened. The more it is bent the harder it becomes. If excessive work hardening has occurred it may be impossible to form the tubing into place. Work hardened copper can be re-annealed by heating the copper to a dull red and allowing it to cool. If this is done quickly with a large hot flame a minimum of oxides will form on the copper tubing. Slow heating can cause excessive oxides which can cause problems with the system operation once installation is complete.

TABLE 3-5-1 Tubing and Pipe Size Comparison

Refrigeration Tubing Size: OD (in)	Plumbing Pipe Size: Nominal ID (in)
1/2	3/8
5/8	1/2
7/8	3/4
1 1/8	1

Copper tubing intended for refrigeration and air conditioning work is designated as ACR tubing. This tubing has been cleaned, degreased, and sealed to keep its inside clean. ACR tubing is purged by the manufacturer with nitrogen gas to seal against air, moisture, and dirt, and also to minimize the harmful oxides that are normally formed during brazing. The ends are plugged in the process, and these plugs should be replaced after cutting a length of tubing.

Copper Tubing Classification

Copper tubing has three classifications—K, L, and M, based on the wall thickness:

K Heavy wall; ACR approved

L Medium wall; ACR approved

M Thin wall; plumbing

Type M thin wall tubing is not used on pressurized refrigerant lines, for it does not have the wall thickness to meet the safety codes. It can, however, be used for water lines, condensate drains, and other associated system requirements.

Type K heavy wall tubing is meant for special use where abnormal conditions of corrosion might be expected.

Type L is most frequently used for normal refrigeration applications. Table 3-5-2 provides specifications for both types K and L tubing. Both K and L copper tubing are available as soft-drawn and hard-drawn types.

Soft Copper Tubing

Soft copper tubing, as the name implies, is annealed to make the tubing more flexible and easier to bend and form. It is available in sizes from 1/8″ to 1 5/8″ OD and is usually sold in coils of 25-, 50-, and 100-ft lengths. Soft copper tubing may be soldered, brazed, or used with flared or other mechanical-type fittings. Since it is easily bent or shaped, it must be held by clamps or other hardware to support its weight.

TABLE 3-5-2 Specifications of Common Copper Tubing Sizes

Type	Diameter OD (in)	Diameter ID (in)	Wall Thickness (in)	Weight per foot (lb)
K	1/2	0.402	0.049	0.2691
	5/8	0.527	0.049	0.3437
	3/4	0.652	0.049	0.4183
	7/8	0.745	0.065	0.6411
	1 1/8	0.995	0.065	0.8390
	1 3/8	1.245	0.065	1.037
	1 5/8	1.481	0.072	1.362
	2 1/8	1.959	0.083	2.064
	2 5/8	2.435	0.095	2.927
	3 1/8	2.907	0.109	4.003
	3 5/8	3.385	0.120	5.122
L	1/2	0.430	0.035	0.1982
	5/8	0.545	0.040	0.2849
	3/4	0.666	0.042	0.3621
	7/8	0.785	0.045	0.4518
	1 1/8	1.025	0.050	0.6545
	1 3/8	1.265	0.055	0.8840
	1 5/8	1.505	0.060	1.143
	2 1/8	1.985	0.070	1.752
	2 5/8	2.465	0.080	2.479
	3 1/8	2.945	0.090	3.326
	3 5/8	3.425	0.100	4.292

The more frequent application is for line sizes from 1/4 to 7/8 in OD.

The proper way to unroll soft copper is to roll it out over a flat surface as shown in the sequence in Figure 3-5-5. If the coil were pulled out from the center of the coil it would have lots of little dips along its length, Figure 3-5-6.

Hard Copper Tubing

Hard copper is also used extensively in commercial refrigeration and air conditioning systems. Unlike soft tubing it is hard and rigid and comes in straight lengths. It is intended for use with formed fittings to make the necessary bends or changes in direction. Because of its rigid construction it is more self-supporting and needs fewer supports. Sizes range from 1/4 in OD to over 6 in OD. Hard-drawn tubing comes in standard lengths of 10 ft and 20 ft. It is dehydrated, charged with nitrogen, and plugged at each end to maintain a clean, moisture-free internal condition. The use of hard-drawn tubing is most frequently associated with large line sizes of 7/8 in OD and above or where neat appearance is desired.

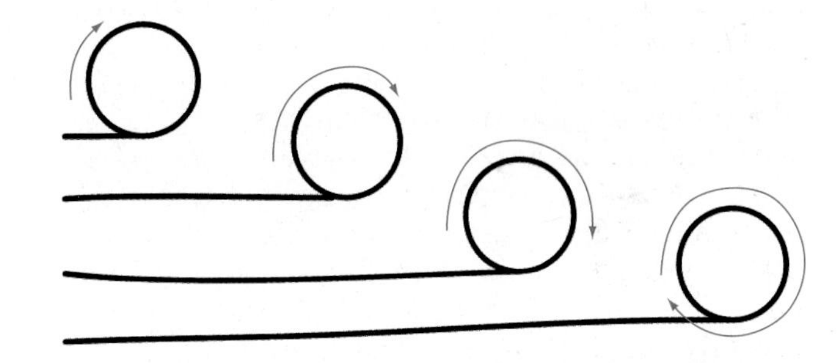

Figure 3-5-5 To use copper tubing, take the whole roll and unroll what you need from the outside edge across a clean, flat surface.

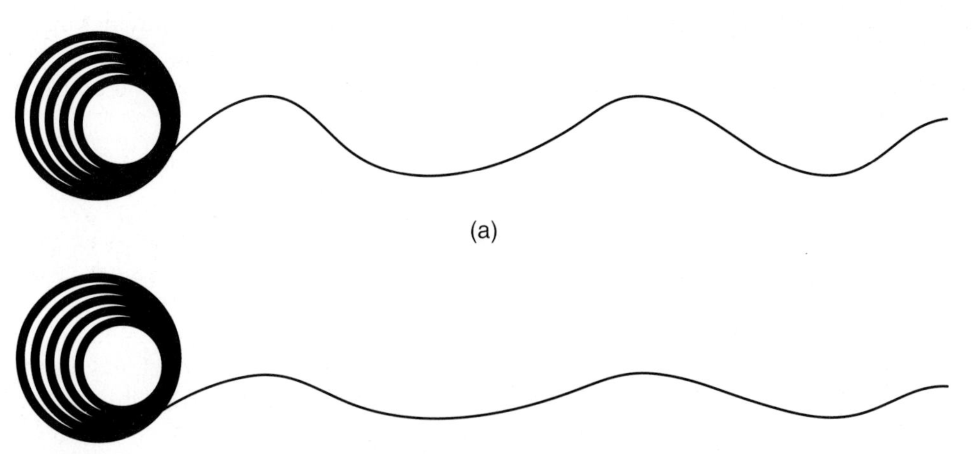

Figure 3-5-6 (a) Pulling copper tubing from the center of the roll will result in curves and dips; (b) These will stay in the metal even when it is pulled taut.

BASIC PRINCIPLES

Certain basic functions of refrigeration piping must be observed when these lines are constructed:

1. The piping size must ensure a proper supply of refrigerant to the evaporator.
2. The sizes selected must be practical; too small will result in an excessive pressure drop and too large will not maintain sufficient velocity to return oil with refrigerant vapor.
3. The piping must prevent refrigerant oil from being trapped in any part of the system.
4. Liquid refrigerant and/or oil slugs must be prevented from entering the compressor.
5. It is important to maintain a sealed system that is leak free.
6. The system must be purged with an inert gas such as nitrogen during brazing.

TECH TALK

Tech 1. I put tape over the end of the refrigerant lines when I finish a rough-in. You don't need to worry about putting the plastic caps back on the refrigerant line set. Tape works OK.

Tech 2. Well not really. The tape can leave sticky adhesive on the copper that can really cause problems later when you're brazing the pipes to the unit. Even if you sand the ends really good, sometimes the sticky will stay. Then you've got a real problem. If you have a leak caused by the tape adhesive, you'll have to disassemble the entire fitting to clean it properly so you can fix the leak.

Tech 1. Oh. What if I didn't keep the plastic tubing caps?

Tech 2. Then you're better off bending the end closed and putting a little braze across the end of the pipe. Really, that's the best way to seal the ends anyway, especially if you know it's going to be several weeks before the outdoor units are installed.

Tech 1. Thanks.

UNIT 5—REVIEW QUESTIONS

1. The ________, if it is air cooled, must be located in an area where a proper air supply is available.
2. The ________ must be protected at all times from loss of lubricating oil.
3. Most tubing used in refrigeration and air conditioning is made of ________.
4. ________ generally applies to thin wall materials that are joined together by means other than threads cut into the tube wall.
5. ________ is the term applied to thick pipe wall material into which threads can be cut and which is joined by fittings that screw onto the pipe.
6. In the refrigeration trades, tubing sizes are expressed in terms of the ________.
7. Pipe sizes are expressed in the plumbing trade as ________.
8. Explain the process of annealing copper tubing.
9. List the three classifications of copper tubing.
10. Which of the three classifications of copper tubing is used the most for normal refrigeration applications?
11. What is the intended use for hard-drawn copper tubing?

UNIT 6

HVAC/R Math

OBJECTIVES: In this unit the reader will learn about the math used in the HVAC/R industry including:

- Basic Math
- Numbers
- Addition and Subtraction
- Multiplication

BASIC MATH

As an HVAC/R technician you will use basic math every day. One of your most important math tasks will be calculating time and charges for your customer's bill. Depending on your company's policies, in addition to time and charges you may have to calculate markups on parts and supplies used on the job. The math covered in this unit is designed to give you the basics that would be required for an entry level technician. Higher, more complex levels of mathematics will be covered in other chapters and units in this textbook. In those chapters and units you will be expected to develop math skills required to calculate heat loads, refrigerant charges, voltage, tolerances, and solve other complex problems.

SERVICE TIP

It is important to be able to do basic math even though calculators are used to do most of the HVAC/R math. If you can do a quick estimate of the math before you enter the numbers into the calculator it will be easier to tell if you have made a mistake while entering the numbers on the keypad. Don't rely on a calculator's answer without having an idea as to whether it is correct or not.

Numbers

Numbers used in HVAC/R can be whole numbers, decimal numbers, fractions, and mixed numbers. Examples of whole numbers would be 2, 6, 22, 55, and 1100. Examples of decimal numbers would be 2.5, 6.11, 22.01, 55.1, and 1100.101. Examples of fractions would be $\frac{1}{2}$, $\frac{3}{4}$, $\frac{5}{8}$, and $\frac{17}{18}$. Finally, examples of mixed numbers would be $2\frac{1}{2}$, $6\frac{3}{4}$, and $11\frac{5}{9}$. There are other numbers used in more advanced mathematics but these are the primary types of numbers used in HVAC/R work. To work with numbers mathematically you will need to have a good understanding of addition, subtraction, multiplication, and division to meet these and other HVAC/R job requirements.

SERVICE TIP

Adding feet and inches and pounds and ounces can be confusing and take a lot of time. It is easier when using calculators that have a fraction key marked

$a\frac{b}{c}$

which allow you to enter fractions. This key can be used to enter feet and inches as well as pounds and ounces. To enter 12′-5″ plus 2′-4″ you would input

$$12\frac{5}{12} + 2\frac{4}{12} =$$

and the answer would be given as $14\frac{9}{12}$ or 14′-9″. Pounds and ounces can be entered the same way by using $\frac{1}{16}$ for the ounces.

Addition and Subtraction

Addition is the mathematical process of finding the total or sum of two or more numbers. Subtraction is the mathematical process of finding the difference between two numbers. It is possible to subtract a series of numbers from one number, but you must first sum the numbers to find the total value to be subtracted. When adding or subtracting numbers using a pencil and paper you should list the numbers in a column. The numbers must be lined up vertically with tens, hundreds, thousands, ten thousands, etc., directly above each other in the column. This is the best way to avoid misreading a number by accidentally shifting it from one column to another.

Multiplication

Multiplication can be thought of as adding the same number over and over again for a specific number of times. We are all familiar with mathematical equations like the ones for finding the area of a square or circle, or for converting degrees Fahrenheit to degrees Celsius. These mathematical equations, like all equations, are used to show the relationship between two or more items. Equations can be used to solve problems when the relationships between the various items are known. An example of an equation is the one we use to find area, $A = L \times W$. In this equation the area is represented by the letter A, the length by the letter L, and the width by the letter W. If we know the length is 5 ft and the width is 10 ft we can find the area by replacing the letters with the numbers. So now we can say that the area $A = 5 \text{ ft} \times 10 \text{ ft}$ or that the area is 50 ft^2.

When the unknown, like the area in $A = 5 \times 10$, is on the left side of an equation the equation is easily solved. The problem for many is when the unknown is not set apart from the known values. For example, if we knew the area A to be 50 ft^2 and the length L to be 10 ft what would our equation have to be to solve for the width W? In math we learned that in order to separate the known information, in this case A and L, from the unknown information, W, we must add, subtract, multiply, or divide both sides of the equation by the same values in order to move the unknown away from the known information. We have learned how to resolve this equation so that the W is on the left side of the equals sign and the A and L are on the right by doing the following:

Step 1 Write the basic equation $A = L \times W$

Step 2 Divide both sides of the equation by L $\frac{A}{L} = \frac{L \times W}{L}$

Step 3 Cancel out the Ls on the right side $\frac{A}{L} = \frac{\not{L} \times W}{\not{L}}$

Step 4 Remove the canceled Ls from the right side $\frac{A}{L} = W$

Step 5 Reverse the equation so the unknown W is on the left side of the equation $W = \frac{A}{L}$

With this revised equation we can now calculate the value of W by dividing A by L. So now we can solve for W by dividing 50 ft^2 by 10 ft so $W = 5$ ft. Remembering all of the steps required to move the knowns and unknowns can be a problem especially if you do not use mathematics very often. There is an easier way to rearrange most equations as long as you remember to do the same operation to both sides of an equation.

In this method you must remember that all math functions have opposites. The opposite of addition is subtraction, the opposite of multiplication is division, and the opposite of squaring is square root. ($+ \Leftrightarrow -$, $- \Leftrightarrow +$, $\times \Leftrightarrow \div$, $\div \Leftrightarrow \times$, $x^2 \Leftrightarrow \sqrt{}$, $\sqrt{} \Leftrightarrow x^2$) First you substitute your own numbers into the equation so you can see how they relate. Calculators have made this process easy to do as long as you do not use 1s or 2s for numbers. The problem with 1s is that $1 \times 1 = 1$ and $1 \div 1 = 1$ and the problem with using 2s is that $2 + 2 = 4$ and $2 \times 2 = 4$. Almost any other numbers work in this process. Let's take the area problem and solve it another way by using our own number first.

Step 1 Write the basic equation $A = L \times W$

Step 2 Write your numbers in the basic equation $A = 3 \times 4$

Step 3 Solve this problem $12 = 3 \times 4$

Next remove the number you used for the length and remembering that division is the opposite of multiplication see what you must divide to get the number 3.

Step 4 Replace the number 3 with the letter L in the basic equation $12 = L \times 4$

Step 5 Using trial and error divide 4 by 12 to see if it equals 3 $\frac{4}{12} \neq 3$

Step 6 Again using trial and error divide 12 by 4 to see if it equals 3 $\frac{12}{4} = 3$

You can see that the Step 5 trial and error did not equal 3 but the Step 6 trial and error did. Use the same process with the real numbers and solve the problem.

Step 7 Using the same process now replace 12 with 50 and 4 with 5 $\frac{50}{5} = L$

Step 8 Divide 50 by 5 to solve for the length $10 = L$

Now use the same trial and error method to solve a more complex problem. What would your hourly rate be if you were offered a job that paid $55,000 a year? First you would write the formula for a yearly salary (YS) at a given hourly pay (HP).

Step 1 Write the basic formula $\text{YS} = \text{HP} \times 40 \times 52$

Step 2 Replace the letters HP with hourly pay of $5 per hour $\text{YS} = \$5 \times 40 \times 52$
$\text{YS} = \$5 \times 2080$

Step 3 Solve the equation to find out the yearly salary at $5 an hour $\text{YS} = \$10{,}400$

Step 4 Replace $5 with HP and solve the equation $\quad \$10{,}400 = HP \times 40 \times 52$

Step 5 Using the trial and error method solve the equation $\quad \frac{40 \times 52}{10{,}400} \neq \$5$

Step 6 If the answer does not equal $5 then retry to solve the equation $\quad \frac{10{,}400}{40 \times 52} = \$5$

Step 7 Using the same process now replace $10,400 with $55,000 and solve the equation one step at a time $\quad \frac{\$55{,}000}{40 \times 52} = HP$

$$\frac{\$55{,}000}{2080} = \$26.44$$

Key to Equation
Yearly Salary = YS
Hourly Pay = HP
hours in a week = 40
weeks in a year = 52

Your hourly rate of pay for a yearly salary of $55,000 would be $26.44 per hour.

UNIT 6—REVIEW QUESTIONS

1. List the most important math task that an HVAC/R technician will use every day.
2. List the four types of numbers used in HVAC/R.
3. The number $6\,^3\!/\!_4$ is an example of a ________.
4. Define addition.
5. ________ can be thought of as adding the same number over and over again for a specific number of times.
6. In the equation $A = L \times W$, ________ is represented by the letter A.
7. Using the equation YS = HP × 40 × 52, what would your hourly rate be if you were offered a job that paid $45,000 a year?

TECH TALK

Tech 1. Did you know that sometimes division is not very accurate?
Tech 2. What do you mean?
Tech 1. Well if you divide 100 by 3 the answer on my calculator says 33.333333333 and if I add those together I wind up with 99.99999999 so when you divide some numbers they don't add back up right.
Tech 2. Oh, I see what you mean. But what difference does it make?
Tech 1. Well, when you are doing a lot of calculations to find an answer, you are better off if you do the multiplication before the division.
Tech 2. Oh, I see.
Tech 1. Yeah, that way any error from the division is not compounded as you do the multiplication.
Tech 2. Does it really make that much of a difference?
Tech 1. Not in most of the work we do. In fact you can actually keep the decimal points to just two places and that is accurate enough for most of the math we need here in air conditioning and refrigeration.
Tech 2. Okay, but it is still a good idea to do the multiplication first then.
Tech 1. Yeah, if you can, if it is okay with the way the formula is written.
Tech 2. Thanks.

UNIT 7

Computers

OBJECTIVES: In this unit the reader will learn about computers including:

- Operating Systems
- Program Functions
- Programs
- Computer Tracking

INTRODUCTION

Computers have a significant impact on the HVAC industry. They are used to do load calculation, duct layout, preparing bills of material, preparing customer statements, and job scheduling and dispatching. The list of HVAC/R computer programs keeps growing, Table 3-7-1. In addition computers can be linked with test instruments, HVAC/R systems, and global positioning to provide assistance with troubleshooting, operation, service, and management of HVAC/R systems. Some test instruments, HVAC/R equipment, and system thermostats have a microcomputer processor built directly into them.

Computers allow building environmental systems to be managed efficiently. These systems can be either managed on site or remotely. The internet allows technicians rapid access to manufacturer data sheets and technical support. Some commercial equipment contains computers that can diagnose their own problems and contact the service company so that repairs can be facilitated.

SERVICE TIP

Major factors affecting the ability of your computer to run software are the amount of RAM your computer has and the processor speed. Check with the software manufacturer before purchasing any program to make sure your system is compatible with the program's requirements.

There are a number of programs available specifically designed for HVAC/R businesses. These programs allow the businesses to computerize service tickets and invoices. Automating this document is important because no matter how careful a technician is when calculating a customer's bill, errors will occur. Each of these errors costs the company. Most companies have a policy that when a mathematical error has resulted in an overcharge, they promptly return the money to the customer. In addition many companies have a policy that when a mathematical error on the service ticket results in an underpayment by the customer no additional funds are requested from the customer. It is generally considered bad business practice to attempt to collect this underpayment from a customer. The net result is that all mathematical errors associated with customer invoices cost the company. Computer programs will automatically perform the math required to determine the customer's bill.

Laptop computers can have wireless connections to the internet. This allows the service technician to transmit a customer's bill back to the office directly from the service vehicle. At the office the bill can be printed and mailed promptly to the customer. In addition, programs are available that allow the technician to fill out a work order requesting additional materials or supplies for completion of a job. Those work orders can be transmitted to the office during regular business hours so that the company can secure those items, making them available to the technician for the next business day. This significantly reduces the time that a technician would have to spend at a parts house picking up parts to return and complete a service call.

SERVICE TIP

Laptop computer batteries are expensive. Reading and following all of the computer and battery manufacturer's instructions can significantly affect battery life and performance.

TABLE 3-7-1 Computer Programs Used in the HVAC Field

Program	Use of Program
Load Calculation	
CHVAC	Commercial HVAC loads
RHVAC	Residential HVAC loads
REFRIG	Refrigeration box loads
HVAC Audit and Analysis	
Audit	Residential energy analysis
HVAC solution	HVAC schematic design
EZDOE	Commercial energy analysis
Ductsize	Duct sizing
HVAC CAD Details Package	
Drawing board	Simple HVAC drawings
ECA	Earth coupled pipe loop sizing
Gasvent	Gas vent sizing
Heavent	Industrial exhaust ventilation
HSYM	Chilled and hot water pipe analysis
HVAC tools	Collection of HVAC utilities
IAQ tools	Indoor air quality tools
Various indoor humidity tools	Humidity measurement and control
Life	Multi-phased life cycle cost analysis
Psychart	Psychrometric analysis
Quote	HVAC estimating and sales
Shadow	Glass shading analysis
Ventilation tools	Ventilation measurement and control

(Courtesy Elite Software)

OPERATING SYSTEMS

Before a computer can be helpful it must have an operating system and a program. Computer programs are often referred to as software. The operating system is the base program that runs all other programs. Although there are a number of different operating systems, the two primary computer operating systems are Microsoft Windows and Macintosh. The operating systems are often referred to as platforms because they are the underlying foundation for other programs. Both major and most minor operating systems use a Windows type user interface. The term Windows comes from the fact that these operating systems "pull down" and "pop up" displays to direct and inform the user.

There are many software programs that can be loaded and run on computers. Programs are a series of complex computer commands that process the inputted information in order to create your desired results. Without a program instructing the computer as to how to process inputted information, the computer would be relatively worthless. These programs are very complex and allow the user to input raw data and receive a comprehensive completed report generated by the software.

Program Functions

Help One feature in most programs is the help function. Help is available in a number of ways. One method to get help is to hover the cursor over an icon. Within moments a phrase or term describing the icon function will appear such as the description General Project Data in Figure 3-7-1. Many programs allow you to right click the mouse on an icon for a short narrative description of its functions. For more significant help, a help button is available that generally brings up three choices of how to receive help.

The three help choices are to type in a question, select from a either list of topics or a complete indexing of the help offerings, or to look at the answers to frequently asked questions, which will look something like the screen shown in Figure 3-7-2. These different help functions will enable the user to more quickly select the source and process needed to resolve their problem, Figure 3-7-3. In addition to help the user through the program itself, program manufacturers may have a web based technical support and phone-in service. The web based service requires that the computer be connected to the World Wide Web. One advantage of the web based service is that it is available 24 hours a day so that a technician can

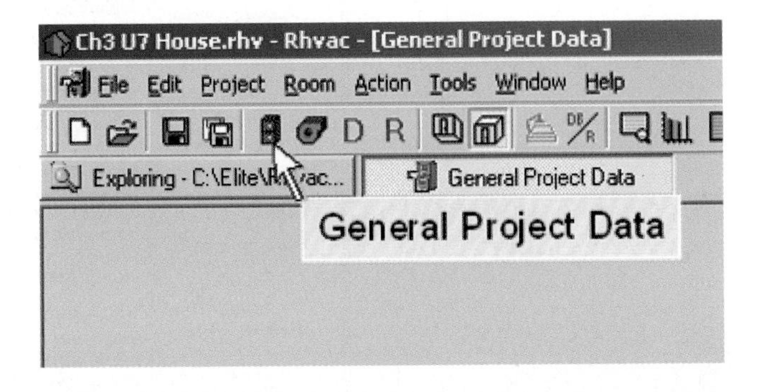

Figure 3-7-1 Icon identification provided by hovering the cursor over an icon. (*Courtesy Elite Software*)

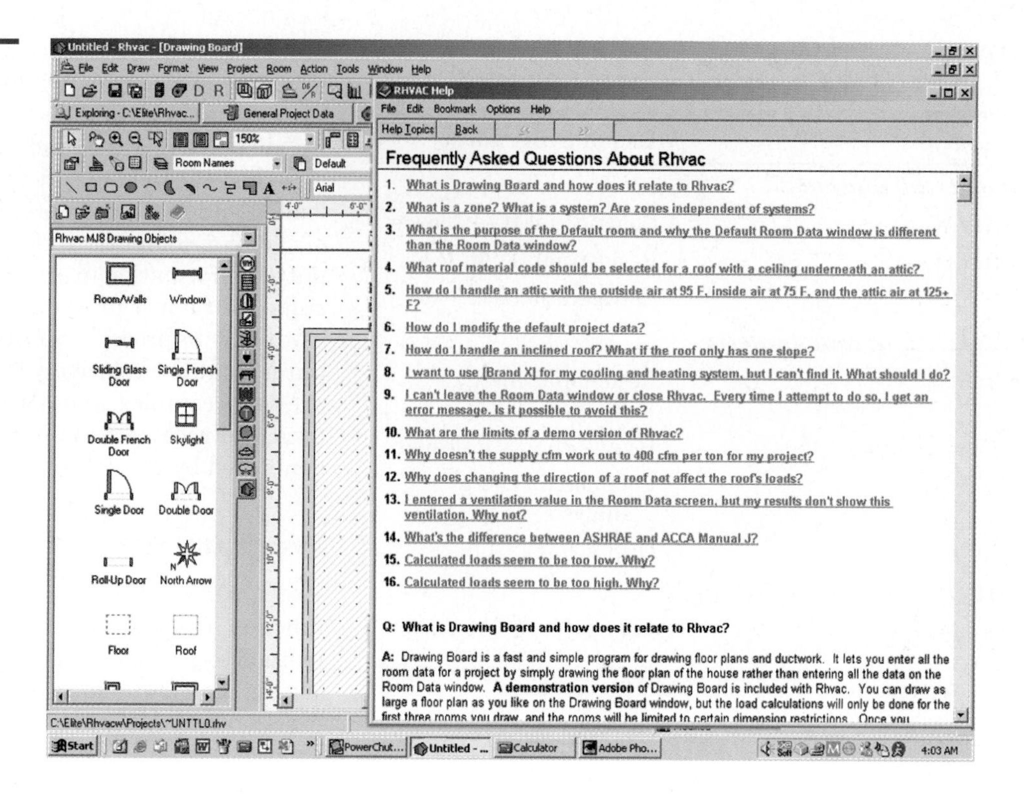

Figure 3-7-2 Help menu for frequently asked questions (FAQ). (*Courtesy Elite Software*)

seek help even when working late nights during the busy season. Phone-in service may be limited to standard business hours for the time zone where the company's corporate offices are located. Phone-in technical support may be free, free for a period of time, or fee based. Check with your software manufacturer's guidelines to determine which policy is in effect for that software.

Input Range Often the program will have a range of acceptable input data. This system would give a warning or error message if a number is entered that is outside the expected range for that item. For example, a program would be able to determine that the air changes of the residence were below code or standard requirements if too narrow a duct cross section were entered. These operating parameters

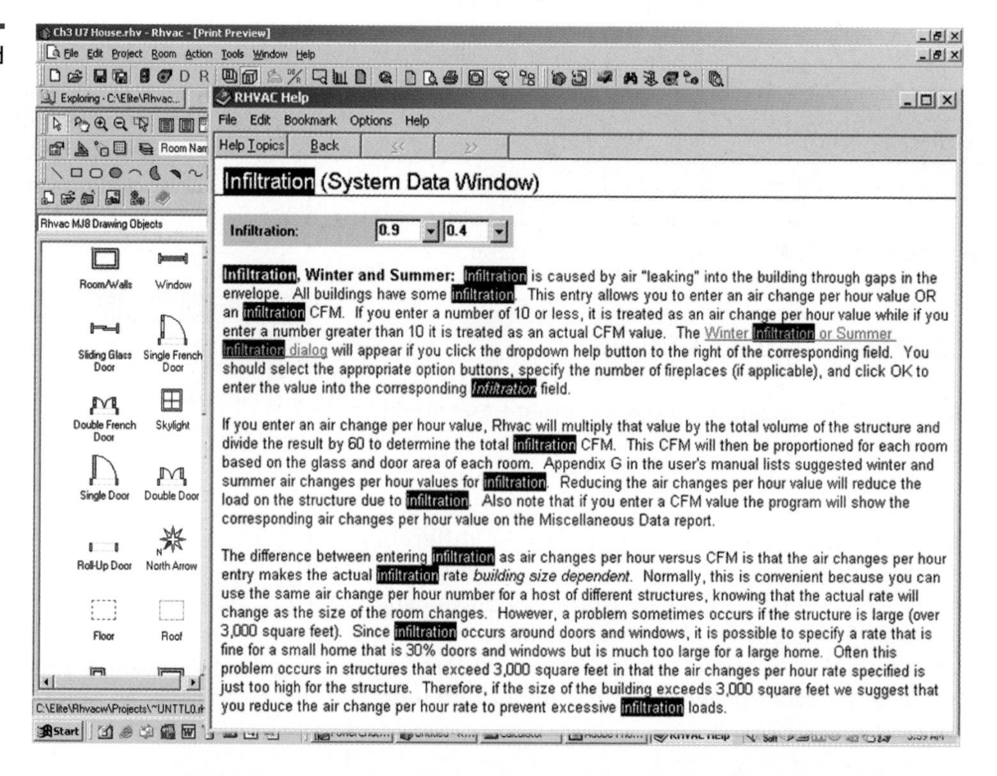

Figure 3-7-3 Help provided by a search for the word "infiltration." (*Courtesy Elite Software*)

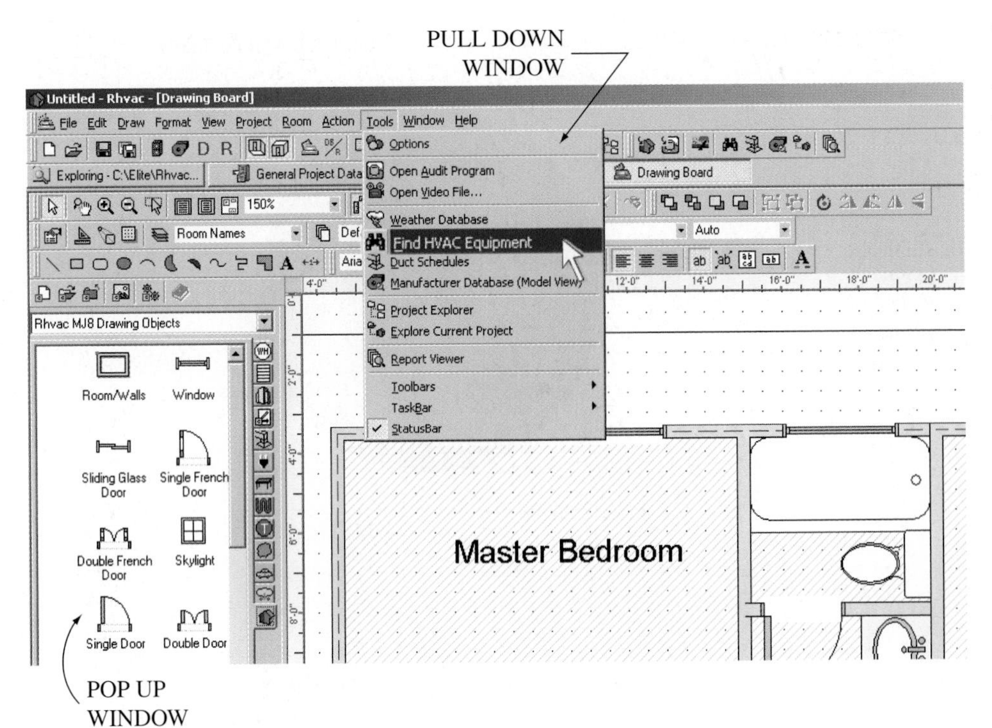

Figure 3-7-4 A pop up window and a pull-down window can include data inputted by the user and data calculated by the program. (*Courtesy Elite Software*)

reduce input errors and enable the technician to more reliably accept program outputs.

Data Input Method Not all of the input to a program must be done from a keyboard. Many programs allow the user to use a mouse to select and input information from a pull down menu or import the information from another program. The program shown in Figure 3-7-4 permits the technician to draw the residence using standard architectural drawing items from a pop up window. It also allows the technician to search for equipment using a pull down window.

Formatting Output There are a variety of ways for the program's data reports to be outputted. A written report consisting of a number of pages detailing the technical specifications can be printed, Figure 3-7-5. Bar graphs or circle graphs, as shown in Figure 3-7-6, are a good way to show the relationship between the various output elements of a program. A drawing can be produced to show the location of parts, ducts, or systems, Figure 3-7-7.

PROGRAMS

The most commonly used programs in the HVAC/R industry are heat load calculation and duct design and layout. A heat load calculation is the only accurate way of determining the correct size of equipment needed. Sizing HVAC equipment improperly can have severe consequences to the system performance and economy and even shorten the equipment service life. Once a proper heat load is determined, a duct layout can be produced. Peak system performance depends on a properly sized and designed air distribution system. An undersized or improperly designed air distribution system can have devastating effects on the ability of the system to heat or cool.

Rhvac - Residential & Light Commercial HVAC Loads
Larry Jeffus
Garland, TX 75043

Elite Software Development, Inc.
House 1
Page 4

Load Preview Report

Scope	Area	Sens Gain	Lat Gain	Net Gain	Sens Loss	Win CFM	Sum CFM	Sys CFM	Duct Size
Building 2.05 Net Tons, 2.46 Recommended Tons, 384 ft.²/Ton, 25.98 MBH Heating									
Building	944	22,100	2,494	24,594	25,975	343	1,022	1,022	
System 1: 2.05 Net Tons, 2.46 Recommended Tons, 384 ft.²/Ton, 25.98 MBH Heating									
System 1	944	22,100	2,494	24,594	25,975	343	1,022	1,022	14x16
Zone 1	944	22,100	2,494	24,594	25,975	343	1,022	1,022	
1-Bedroom 2	128	1,227	65	1,292	4,051	54	57	57	1-4
2-Living Room	240	6,325	899	7,224	7,124	94	293	293	3-6
3-Master Bedroom	176	3,224	481	3,705	5,238	69	149	149	2-5
4-Master Bath	85	994	24	1,018	1,242	16	46	46	1-4
5-Kitchen-Dining	240	9,823	1,001	10,824	7,021	93	454	454	4-6
6-Bath	75	507	25	532	1,299	17	23	23	1-4

Figure 3-7-5 A report printout. (*Courtesy Elite Software*)

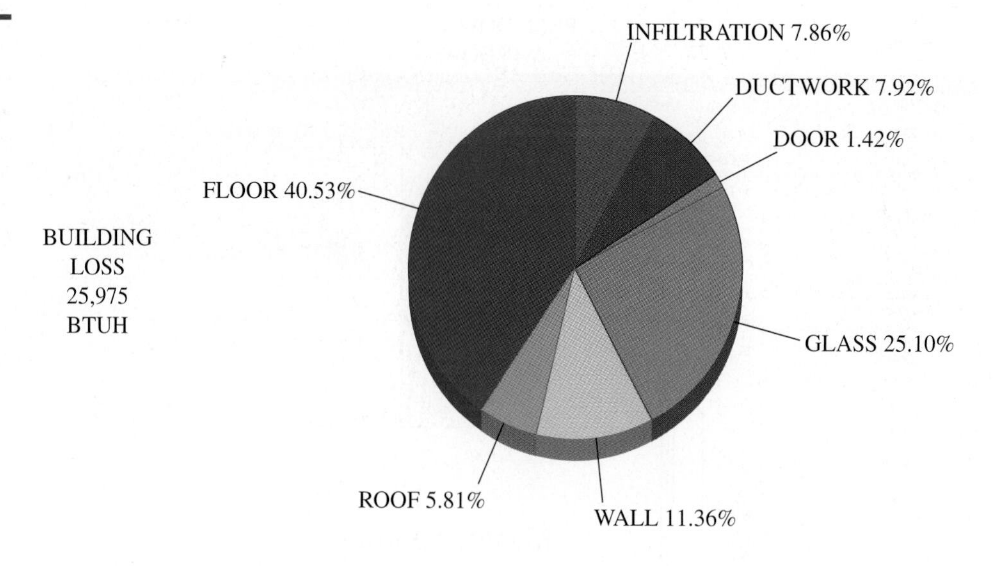

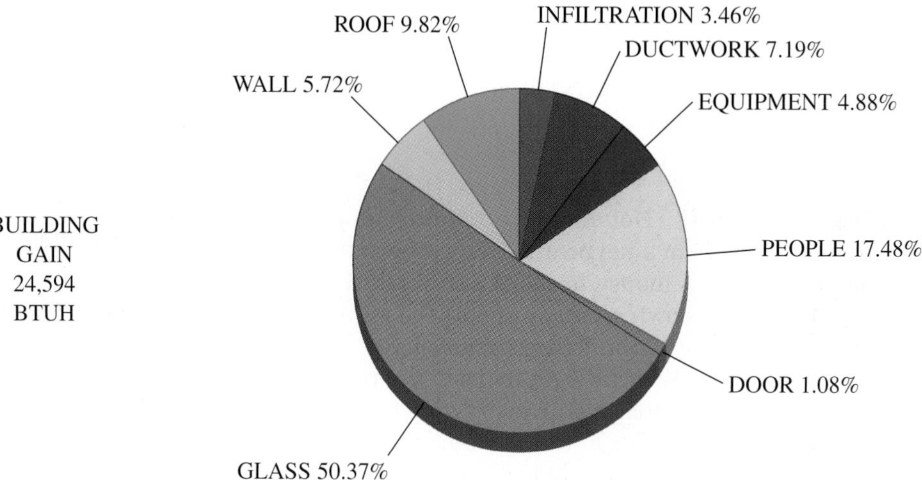

Figure 3-7-6 Report graphs. (*Courtesy Elite Software*)

Heat Load Calculation Programs

In order to properly size heating and cooling systems a load must be calculated on the building. Load calculations take into consideration a myriad of variables such as volume of the building, amount of glass, orientation, size of the wall, insulation in the wall, insulation in the ceiling, type of floor, window treatment, shade, color of the roof, number of occupants, tightness of the construction, and more. Although these calculations can be done by hand, the most efficient way of doing a load calculation is by using a computer program.

A number of software companies have programs available for heat load calculation. Each of these programs uses the same general basic database and the same method of calculation. Some programs go into greater depth than others. Examples of complex programs covering the widest range of variables are Elite Software's RHVAC and CHVAC and R-J8. These programs allow the user to calculate not only the load on the building but also each individual room or zone load as shown in Figure 3-7-8. Room and zone loads can then be used for duct sizing.

There are less complex programs available from manufacturers, supply houses, and on the internet. Because of their cost, they are sometimes referred to as freeware. The accuracy of these programs must be verified before you begin using them. Some inexpensive or freeware programs can be obtained from local suppliers, utility companies, or other organizations. Often these are specifically designed for that region or local weather conditions. That is why it is so important to verify the accuracy of these freeware programs to be sure they are accurate for your local climate conditions.

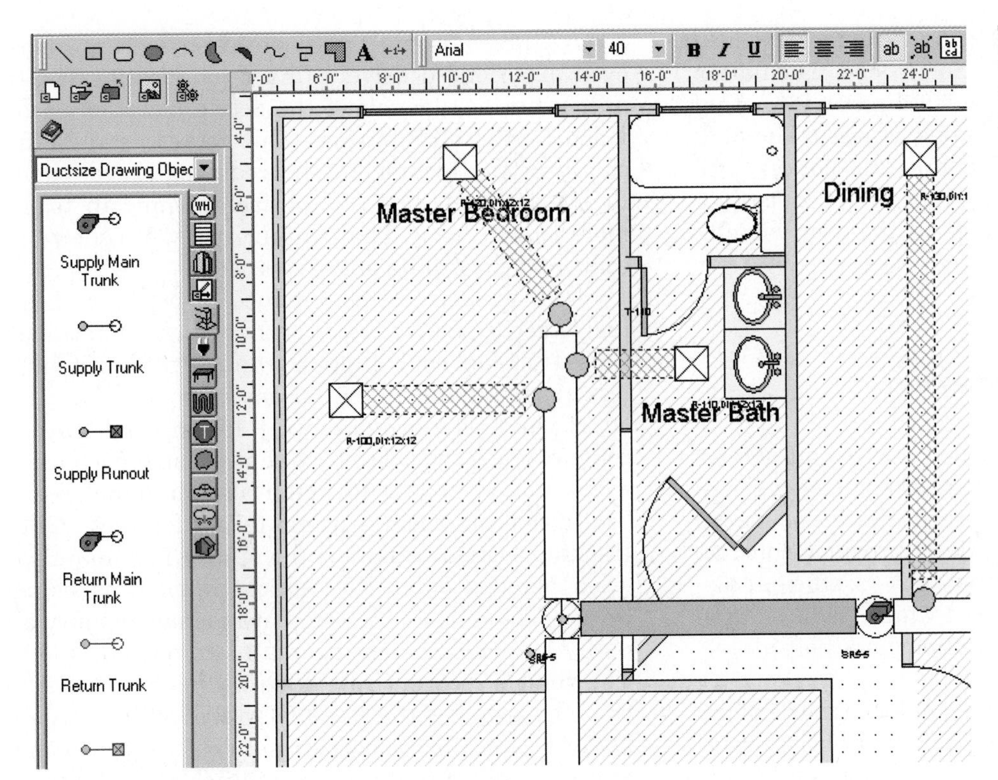

Figure 3-7-7 Report drawings. (*Courtesy Elite Software*)

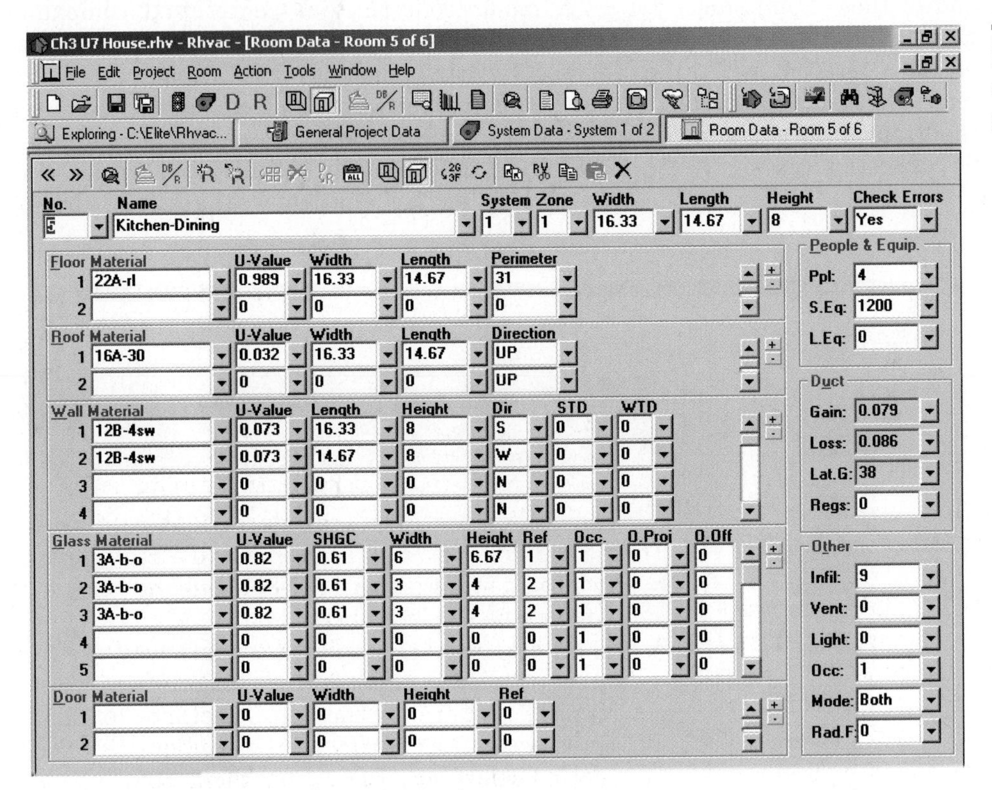

Figure 3-7-8 Typical room data input. (*Courtesy Elite Software*)

Programs for residential and commercial load calculations are available. Some load calculation program results can often be fed directly into other programs designed to calculate duct sizes. The ability to import this material into programs for duct calculation and layout design significantly reduces the possibility of errors and increases the user friendliness of these programs.

COMPUTER TRACKING

Many companies have installed devices on service vehicles that allow them to be tracked using a global positioning system (GPS) to track vehicles as shown in Figure 3-7-9. Vehicle tracking provides contractors with a number of benefits, such as: efficient dispatching of service vehicles, which reduces fuel consumption and vehicle pollution; and theft protection and vehicle recovery.

For larger companies, being able to know exactly where their vehicles are within their service area allows these companies the ability to more efficiently dispatch technicians to new service calls, as shown in Figure 3-7-10. It also allows these companies greater flexibility in dispatching so that it is easier to divert one service vehicle to a location where another technician may need assistance with a problem call. This is very important in larger metropolitan areas, where traffic congestion can significantly increase travel time.

A GPS map and travel direction feature helps even experienced drivers find a job location faster and easier as a result of the active directions provided by the GPS system. This increases the efficiency of the technicians by reducing travel time. Reduced travel time allows each technician to do more jobs in a day. That can increase both their income and the profits of the company.

GPS systems can help reduce air pollution. Less fuel is needed as the result of the reduction in travel time. Less fuel consumption translates directly into less air pollution for a better environment.

Security is a prime concern for air conditioning contractors. A typical service vehicle can have tens of thousands of dollars in materials and supplies on board, making them a prime target for theft. GPS tracking systems can alert the office when unauthorized persons enter the vehicle when it is being driven. This allows the company the opportunity to notify law enforcement agents to an ongoing theft. The vehicle recovery rate where some form of GPS tracking is part of the security system can be as high as 95%. Typically the perpetrator can be stopped and apprehended in a matter of minutes from the time the vehicle is stolen.

A major concern that contracting companies have is individual driver performance. Drivers that "hotdog" by driving fast or erratically can expose the contracting company to more frequent accidents and greater insurance losses. Figure 3-7-11 shows the GPS screen indicating three individual service vehicles. When technicians know that their driving

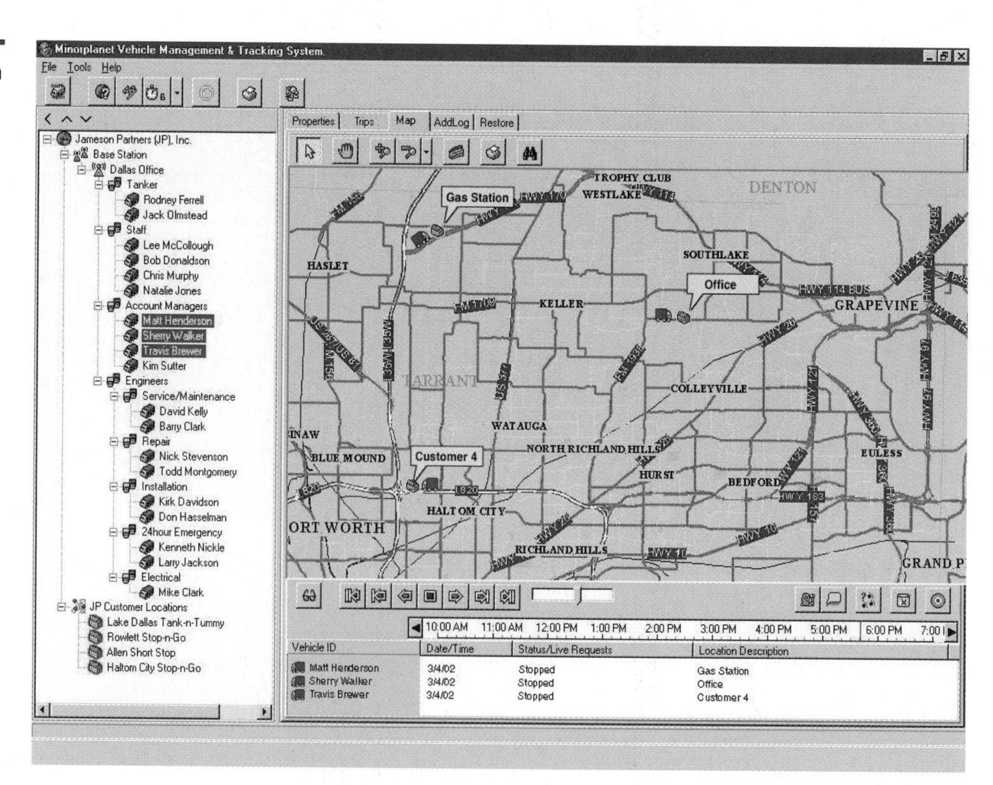

Figure 3-7-9 A GPS system uses satellite and receivers to plot positions of vehicles and show route maps. *(Courtesy of Minorplanet Systems USA)*

Figure 3-7-10 Some GPS systems provide the service company's office with a real-time vehicle tracking map. *(Courtesy of Minorplanet Systems USA)*

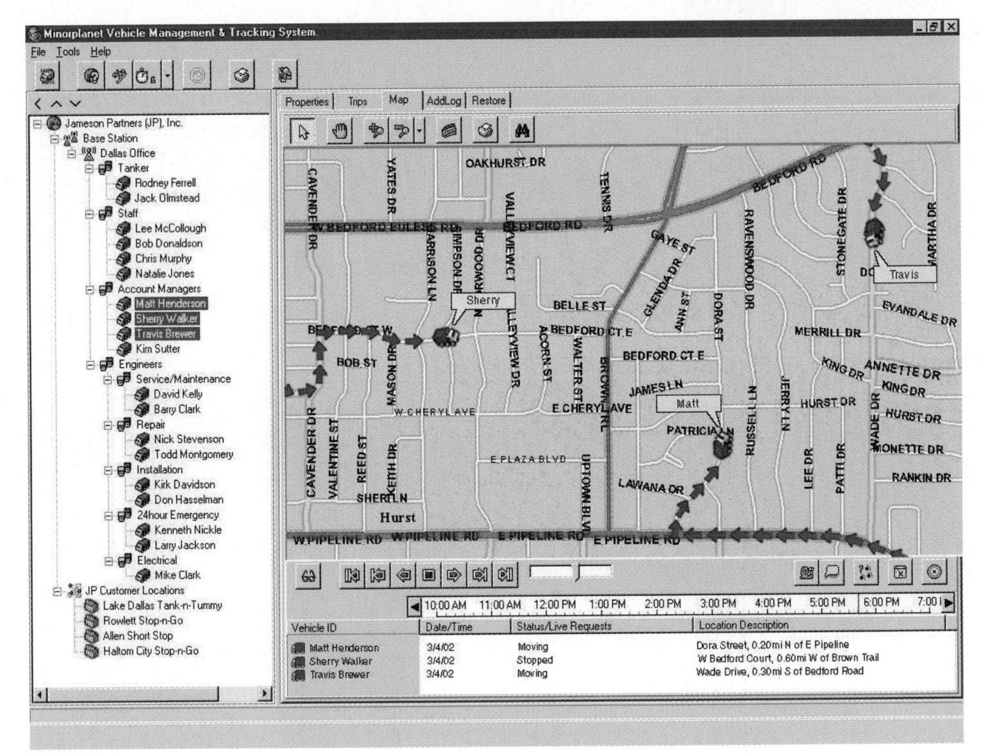

Figure 3-7-11 This GPS program provides the company with the opportunity to replay vehicle trips. *(Courtesy of Minorplanet Systems USA)*

habits can be monitored such habits improve. It also protects the technician from accusations of improper driving by allowing them to confirm or deny the presence of their vehicle at the site of the alleged offense. This backs up the technician saying, "but I wasn't there, it wasn't me," and allows them to prove where they were at any given time.

UNIT 7—REVIEW QUESTIONS

1. What are computers used for in the HVAC/R industry?
2. What is the policy of most companies when a mathematical error has resulted in an overcharge?
3. What is the name used to refer to computer programs?
4. Why are operating systems referred to as platforms?

TECH TALK

Tech 1. I never want to work for one of those companies that have that tracking thing on your truck.

Tech 2. You mean that GPS system?

Tech 1. Yes, I would feel like someone was looking over my shoulder all the time.

Tech 2. Well, you know if you drive responsibly like you would if the boss was in the truck with you, then if wouldn't make a difference if you work for a company that has that tracking device. Also, you know that some people would call in to tell the company that they saw you speeding or you cut them off—things about you that just aren't true. They may be upset with the company for some reason and you wind up with the brunt of the problem because they accuse you and it wasn't you. But, with one of those tracking devices the company can go back and replay the trip and see whether or not you were there. They can also prove you weren't anywhere near the place. If you have an accident they can prove you were not speeding. There are just a lot of advantages to it.

Tech 1. Well I see your point.

Tech 2. In fact I know a technician who works for another company who was in an accident during last year's ice storm. He totaled his vehicle and the insurance company claimed he was at fault because he was driving recklessly.

Tech 1. What happened? Did he lose his job?

Tech 2. No. With the GPS tracking system he was able to prove he was traveling at 15 mph, not recklessly at all.*

Tech 1. Oh, so then maybe it is a good idea to have one of those things on your truck.

Tech 2. It certainly can be.

Tech 1. Thanks. I appreciate that.

*Courtesy of Minorplanet's VMI system.

5. List the two ways the help function is available.
6. Define formatting output.
7. What is the advantage of web based service help?
8. What are the most commonly used programs in the HVAC/R industry?
9. Define load calculation.
10. List five variables that need to be taken into consideration when doing load calculations.
11. Define GPS.
12. What are some advantages of having a GPS tracking system?

SECTION II

Refrigeration Principles

4

Principles of Matter and Thermodynamics

UNIT 1

Principles of Matter

OBJECTIVES: In this unit the reader will learn about principles of matter including:

- States of Matter
- Density
- Specific Volume
- Specific Gravity
- Specific Heat
- Change of State
- Sensible Heat
- Latent Heat

STATES OF MATTER

The physical state of a substance can be controlled by temperature and pressure. For example, many outdoor ice rinks are designed for when the weather goes below freezing temperatures. The weight of a skater on the surface of the ice causes the ice under the hard surface of the skates to melt, making the ice "slippery." In nature, heavy blocks of ice, such as glaciers, pressing on the hard surface of the earth can cause the bottom of the ice to melt. This process lubricates glaciers resting on slopes, allowing them to move.

Matter exists in three states: solid, liquid, and vapor or gas, as shown in Figure 4-1-1. A common example is water, which exists in all three states. Water is in a solid (ice) stage below 32°F (0°C), a liquid state from temperatures of 32°F (0°C) to 212°F (100°C) and a vapor or gas at 212°F (100°C) and above.

The various states of matter have unique characteristics.

A **solid** is a substance that has a definite shape, which it will hold under a certain degree of stress of pressure, depending on the material and the type of disturbance. It must be supported or it will fall to the next level of support. This condition requires the design of an adequate foundation when dealing with solids.

Solids of sufficient density will retain their size and weight. Solids of light density will lose molecular quantity under certain conditions and lose weight and quantity. Carbon dioxide (CO_2) in solid state (known as dry ice) will pass from the solid state into the gaseous state under certain conditions.

IN SOLID FORM, WATER MOLECULES REMAIN RIGIDLY IN PLACE AND FORM IN HOLLOW RINGS, GIVING ICE ITS LOW DENSITY.

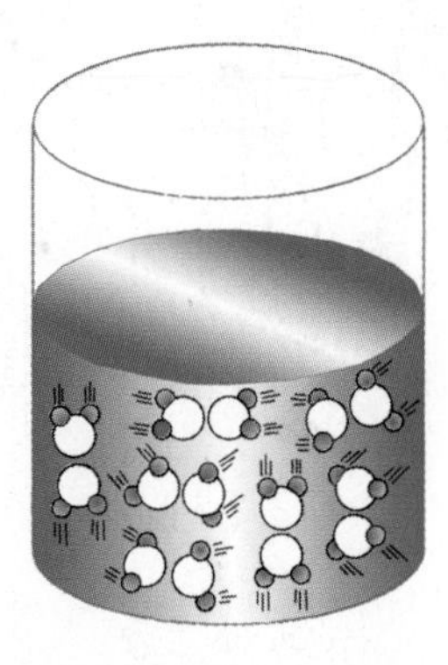

WATER MOLECULES ARE CLOSE TOGETHER YET FREELY SLIP OVER ONE ANOTHER, GIVING LIQUID ITS FLOW.

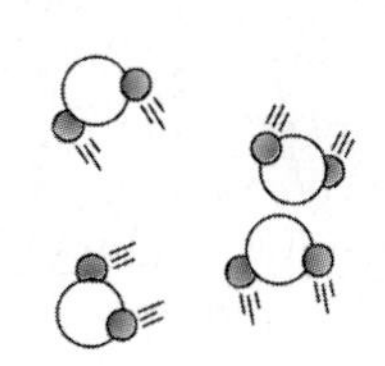

CHAOS! MOLECULES IN GAS (STEAM) ARE WIDELY SPACED, DART RAPIDLY, AND COLLIDE WITH ONE ANOTHER.

Figure 4-1-1 Three states of matter.

A **liquid** is a substance that can take the shape of any enclosure when it is allowed to move freely, but the volume remains constant. It is considered incompressible. When a liquid fills an enclosure it exerts both a horizontal and a vertical pressure on the enclosure. For example, valves located in the basement piping of a hot-water heating system for a multi-story building would need to be capable of withstanding the pressure exerted by the column of water above them.

NICE TO KNOW

A fluid is anything that flows. Most people think of liquid as being the only fluid. However, gases are also fluids in that they flow. The wind blowing outside is air acting as a fluid. Much like water, as the air flows it exerts force from objects. In the same way, refrigerant vapor is a fluid flowing through the system and its inertia applies force. When it makes a turn as a result of a 90° angle in a pipe the inertia of the fluid is affected by the 90. In addition to liquids and gases being fluids some solids may also may be considered as fluid. The most recognizable is a glacier made up of solid ice that over time flows from its origins in the mountains down to the sea. Other solids that are fluids include the toy Silly Putty.

If 1 ft^3 water in a container measuring 1 ft on each side is transferred to a container of different dimensions, the quantity and weight of the water will remain the same, although the dimension will change. In time, liquids of light density, such as water, will gradually lose quantity and weight by molecular loss to the gaseous state.

A **vapor** or **gas** is a substance that has no fixed shape or volume, and therefore must be contained in an enclosure or it will escape into the atmosphere. A good example is the enclosure of a refrigerant in vapor form. Certain refrigerants that are destructive to the environment are being replaced and must not be purged into the atmosphere.

If a 1 ft^3 cylinder containing gaseous state water, called steam, or some other vapor, is connected to a 2 ft^3 cylinder of which, theoretically, a perfect vacuum has been drawn, the vapor will expand to occupy the volume of the larger cylinder as well as the original. Other changes will occur in the vapor which will be discussed later.

Density, Specific Volume, and Specific Gravity

Density is a measurement that can be used to compare the mass of various substances. Density is substance's mass per unit of volume. In the standard system a convenient unit is lb/ft^3. In metric terms the unit is kilograms per cubic meter.

Specific volume (SV) is the volume occupied by a unit mass of a substance, under standard conditions. In the standard system it is expressed in ft^3/lb. The standard conditions are 70°F and 29.92 inches of mercury pressure. In the SI system it is expressed in cubic meters per kilogram. The standard conditions are 20°C and 101.3 kPa pressure. For example, 1 lb of air at standard atmospheric conditions occupies 13.33 ft^3.

SERVICE TIP

The volume of 1 lb of air under standard conditions is a little over 13 ft^3. The equivalent of 1 yd^3 (one cubic yard) is 27 ft^3, so 1 lb of air is a lot smaller than 1 yd^3.

Specific gravity is the ratio of the mass of a given volume of a liquid or solid to the mass of an equal volume of water. Where gases are used, air or hydrogen is the comparing substance. Specific gravity is an absolute term; use density for measurement.

Although these specific differences exist in the three states of matter, quite frequently, under changing conditions of pressure and temperature, the same substance may exist in any one of the three states. For example, water could exist as a solid (ice), a liquid (water) or a gas (steam). Solids always have a definite shape, whereas liquids and gases have no definite shape of their own and so will conform to the shape of their container.

MATTER AND HEAT BEHAVIOR

All matter is composed of small particles known as molecules, and the molecular structure of matter (as studied in chemistry) can be further broken down into atoms.

For the present we concern ourselves only with the molecule, the smallest particle into which any matter or substance can be broken down and still retain its identity. For example, a molecule of water (H_2O) is made up of two atoms of hydrogen and one atom of oxygen. If this molecule of water were divided further into atoms it would no longer be water.

Molecules vary in shape, size, and weight. In physics we learn that molecules have a tendency to cling together. The character of the substance of matter itself is dependent on the shape, size, and weight of the individual molecules of which it is made and also the space or distance between them, for they are, to a large degree, capable of moving about.

NICE TO KNOW

To demonstrate the different sizes of molecules, you can take 50 ml of alcohol and 50 ml of water and pour them together in a 100 ml graduated cylinder. The combined volume of both the alcohol and water most people would assume should be 100 ml, but it is not; it is actually 96 ml. What has happened is very similar to the process of pouring a half bucket of sand into a half bucket of gravel. The sand will fit in the spaces of the gravel so that the combined volume of both buckets will not completely fill one bucket. The molecules of water can fit between the larger molecules of alcohol so that the combined volume of both fluids is less than the sum of each individual volume.

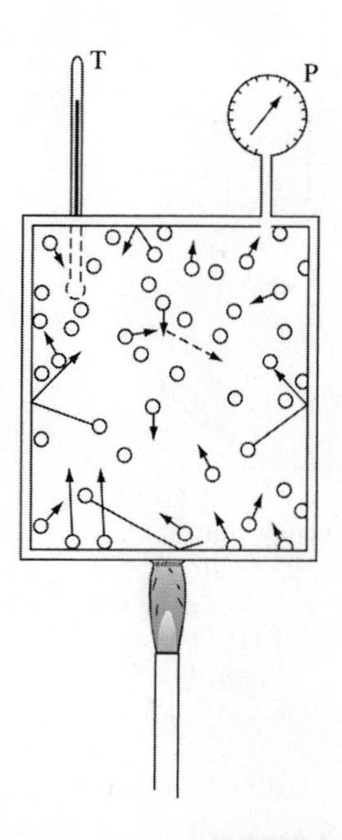

Figure 4-1-2 When heat is applied, the velocity of the molecules will increase.

When heat energy is applied to a substance it increases the internal energy of the molecules, which increase their motion or velocity of movement, such as the gas in Figure 4-1-2. With this increase in movement of the molecules, there is also a rise or increase in the temperature of the substance.

When heat is removed from a substance, the velocity of the molecular movement will decrease and there will also be a decrease or lowering of the internal temperature of this substance.

Heat is a form of energy that causes an increase in temperature when it is added, and a decrease in temperature when it is removed, provided there is no change in state. Recall from Chapter 1, Unit 2, that when heat is used to change the state of a substance no temperature change takes place.

The transfer of heat takes place between two bodies of different temperature. The movement of heat goes from the hot object to the cold object.

Transferring Heat

As we saw in Chapter 1, Unit 2, there are three principal ways that heat is transferred, shown in Figure 4-1-3, convection, conduction, and radiation. Most refrigeration systems utilize all three methods. Further discussion of heat transfer is given later in this chapter.

Recall that the IP unit of heat is the Btu, the amount of heat required to raise 1 lb of water 1°F at sea level. In the SI system, the unit used for heat is the Joule (J), which is also the unit for energy. Since the Joule is small it is common practice to use kJ,

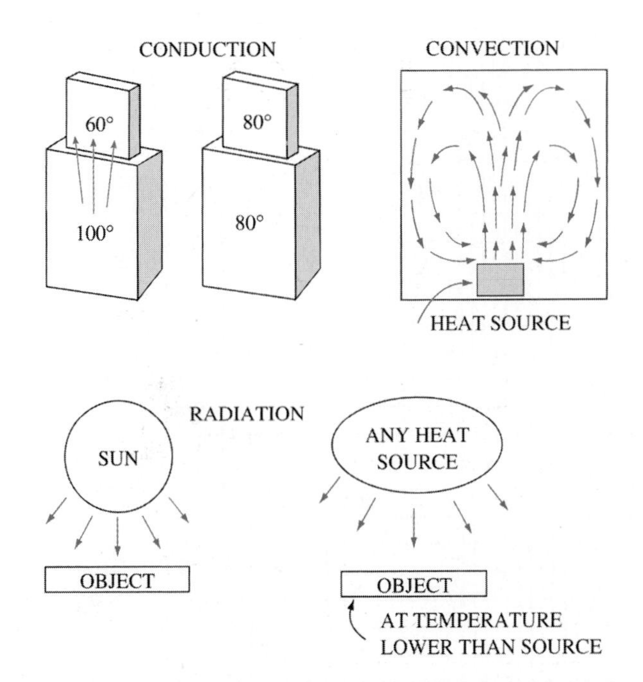

Figure 4-1-3 Three methods for transferring heat.

equivalent to 1000 J. The amount of heat required to raise 1 kg of water 1°C is 4.187 kJ. A comparison of the two units is shown in Figure 4-1-4.

Two useful conversion factors are:

$$1 \text{ J/g} = 0.4299 \text{ Btu/lb}$$
$$1 \text{ Btu/lb} = 2.326 \text{ J/g} \qquad (4\text{-}1\text{-}1)$$

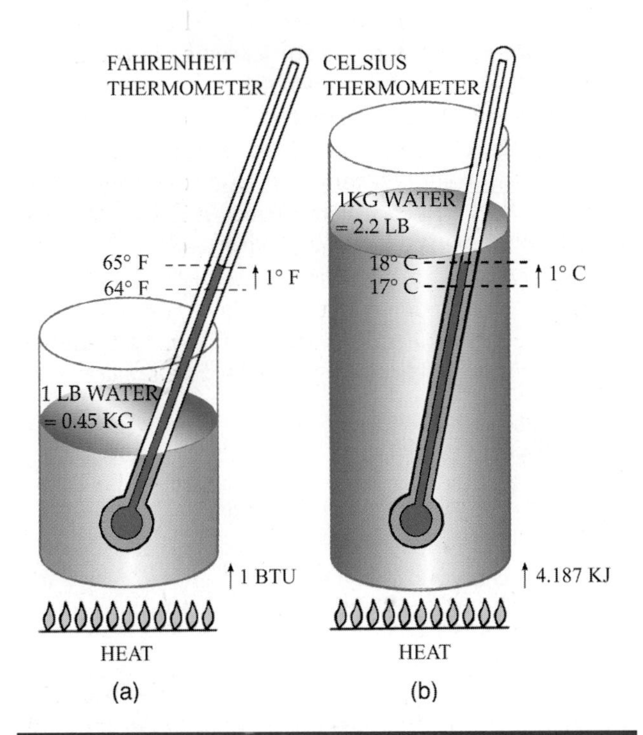

Figure 4-1-4 Comparison of Btu and kJ: (a) raising water 64°F to 65°F requires 1 Btu; (b) 1 kJ of water raised from 17°C to 18°C requires 4.187 kJ.

It is important in measuring heat to note the rate of change, since the amount of heat required to perform an operation is directly affected by the speed of the process. In IP terms we use Btu per hour (Btu/hr) to include time as a factor. In the SI system the element of time is included in the term of Watts (W), 1 W being equivalent to 1 J/s. Either Btu/hr or W, can, therefore, be used to rate the capacity of a refrigeration, heating, or cooling unit. The conversion factors are:

$$1 \text{ W} = 3.412 \text{ Btu/hr}$$
$$1 \text{ Btu/hr} = 0.293 \text{ W} \qquad (4\text{-}1\text{-}2)$$

NICE TO KNOW

In practice you will often see the unit Btu/hr written as Btuh. Do not let this confuse you. Whichever way it is written, it still stands for the rate Btu per hour.

If a building has a heat loss of 100,000 Btu/hr, in SI units this would be equivalent to 100,000/3.412, which is equal to 29,300 W.

NICE TO KNOW

The units Btu/hr and W are also used in the measurement of conductivity, expressed as Btu/hr/ft/°F or W/m/°C.

Sensible and Latent Heat

There are two distinctly different types of heat: *sensible heat*, where the change can be sensed by a thermometer, and *latent heat*, which is not sensed by a thermometer but is required to change the state of a substance.

For an example of latent heat refer to Figure 4-1-5.

The heat added to ice at 32°F (0°C) to change it into water at the same temperature is 144 Btu/lb (335 kJ/kg). This added heat changes the state but not the temperature.

NICE TO KNOW

The capital letter *Q* stands for heat in same way the capital letter *V* stands for voltage and the letter *A* stands for amperage (current). A change in heat will sometimes be shown by the symbols ΔQ where the Greek symbol Δ (Delta) means the change in *Q*. Any time you have a change in heat, pressure, or temperature the change is given as a positive number.

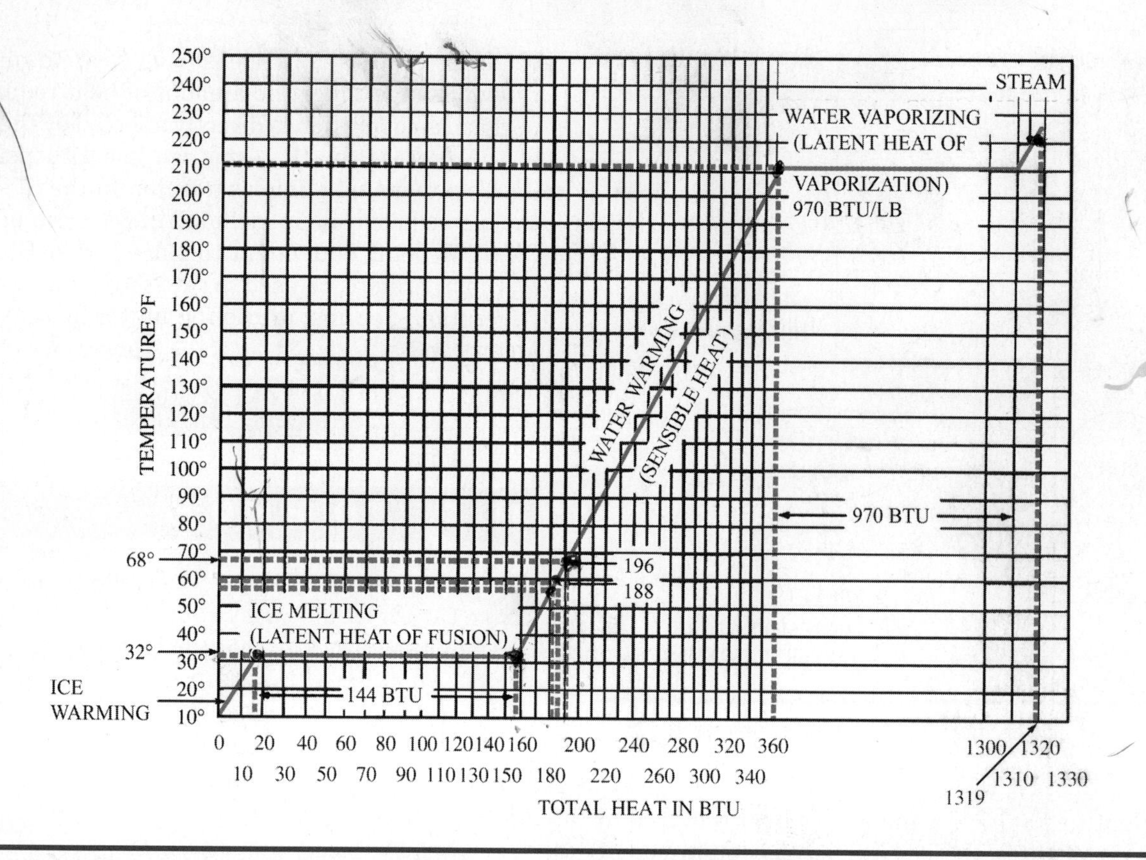

Figure 4-1-5 Chart demonstrating sensible and latent heat relationships in melting ice, changing ice to water and water to steam.

The formula used to determine the sensible heat added or removed from a substance is:

$$Q = W \times \text{SH} \times \Delta T \qquad (4\text{-}1\text{-}3)$$

where

Q = quantity of heat
W = weight of the material
SH = specific heat
ΔT = change in temperature

The formula used to determine the latent heat added or removed from a substance is:

$$Q = W \times \text{LH} \qquad (4\text{-}1\text{-}4)$$

where

LH is the change in latent heat per unit of weight.

For example, if 10 lb (4.5 kg) of water at 212°F (100°C) is changed to steam at 212°F (100°C), the latent heat that is added is 970 Btu per pound of water (IP) or 2260 kJ/kg (SI).

The total heat added in IP units is:

$$\begin{aligned} Q &= W \times \text{LH} \\ &= 10 \text{ lb} \times 970 \text{ Btu} \\ &= 9700 \text{ Btu} \end{aligned}$$

or in SI units

$$\begin{aligned} Q &= \text{Mass} \times \text{LH} \\ &= 4.5 \text{ kg} \times 2260 \text{ kJ/kg} \\ &= 10{,}170 \text{ kJ} \end{aligned}$$

HEAT QUANTITY AND MEASUREMENT

Heat quantity is different from heat intensity, because it takes into consideration not only temperature of the fluid or substance being measured, but also its weight. The unit of heat quantity is the British thermal unit (Btu). As we have seen, a Btu is the amount of heat required to change the temperature of 1 lb of water 1°F at sea level.

TECH TIP

Measuring the quantity of heat is very difficult in the field. The amount of heat in air or in a substance is both its latent and sensible heat and because latent heat is sometimes referred to as hidden heat, it is difficult to determine. It's easier to determine a change in heat because you can look at a change in the humidity or in the change in weight of a material to help determine what the change in heat is.

The addition of 2 Btu will cause a change in temperature of 2°F of 1 pound of water; or it will cause a change in temperature of 1°F of 2 pounds of water;

therefore, when considering a change in temperature of water, the following equation may be utilized:

$$\text{Btu} = W \times \Delta T \qquad (4\text{-}1\text{-}5)$$

where

change in heat (in Btu) = weight (in pounds) × change in temperature (in °F)

EXAMPLE

Calculate the amount of heat necessary to increase 10 lb of water from 50°F to 100°F.

Solution

Since the change in temperature (ΔT) = 50°, then heat = Btu = $W \times \Delta T = 10 \times 50 = 500$ Btu.

In the example above, heat is added to the quantity of water, but the same equation is also used if heat is to be removed. ■

EXAMPLE

Calculate the amount of heat removed if 20 lb of water is cooled from 80°F to 40°F.

Solution

Btu = $W \times \Delta T = 20 \times 40 = 800$ Btu ■

NICE TO KNOW

Here is an illustration of how much heat 1 Btu is; a gallon (gal) of water weighs approximately 8.4 lb and there are 8 pints (pt) in 1 gal. Therefore, 1 pt of water weighs about 1 lb. An ordinary wooden kitchen match produces approximately 1 Btu which is enough heat to raise the 1 lb of water 1°F.

Specific Heat

Recall from Chapter 1 Unit 2 that the specific heat of a substance is the quantity of heat in Btu required to change the temperature of 1 lb of the substance 1°F. The specific heat of water is therefore 1.0; and water is the basis for the specific heat table. You will see that different substances vary in their capacity to absorb or to give up heat. The specific heat values of most substances will vary with a change in temperature; some vary only a slight amount, whereas others can change considerably.

Suppose that two containers are placed on a heating element or burner side by side, with one containing water and the other an equal amount, by weight, of olive oil. You would soon find that the temperature of the olive oil increases at a more rapid rate than that of the water.

If the rate of temperature increase of the olive oil were approximately twice that of the water, it could be said that olive oil required only half as much heat as water to increase its temperature 1°F. Based on the value of 1.0 for the specific heat of water, the specific heat of olive oil must be approximately 0.5, or half that of water. (The table of specific heats of substances shows that olive oil has a value of 0.47.)

TABLE 4-1-1 Specific Heats of Common Substances (Btu/lb/°F)

Substance	Specific Heat
Water	1.00
Ice	0.50
Air (dry)	0.24
Steam	0.48
Aluminum	0.22
Brass	0.09
Lead	0.03
Iron	0.10
Mercury	0.03
Copper	0.09
Alcohol	0.60
Kerosene	0.50
Olive oil	0.47
Glass	0.20
Pine	0.67
Marble	0.21

SERVICE TIP

Specific heat and the weight of a material have some correlation; however, there are some materials, metals for example, that have very low specific heats and are very heavy. Materials such as ceramics have very high specific heats, but are also very heavy. Don't assume because an object is heavy or light that it has a high or low specific heat.

Equation 4-1-5 can now be stated as

$$\text{Btu} = W \times c \times \Delta T \qquad (4\text{-}1\text{-}6)$$

where

c = specific heat of a substance

EXAMPLE

Calculate the amount of heat required to raise the temperature of 1 lb of olive oil from 70°F to 385°F.

Solution

Since ΔT = 315° and c of olive oil = 0.47, then

$$\begin{aligned}\text{Heat} = \text{Btu} &= W \times c \times \Delta T\\ &= 1 \times 0.47 \times 315 = 148 \text{ Btu}\end{aligned}$$ ■

The specific heat of a substance will also change with a change in the state of the substance. Water is a very good example of this variation in specific heat. We have learned that, as a liquid, its specific heat is 1.0, but as a solid (ice) its specific heat approximates 0.5, and this same value is applied to steam (the gaseous state).

Within the refrigeration circuit we will be interested, primarily, with substances in liquid or gaseous form, and their ability to absorb or give up heat. Also, in the distribution of air for the purpose of cooling or heating a given area, we will be interested in the possible changes in the values for specific heat—more about this later.

Air, when heated and free or allowed to expand at a constant pressure, will have a specific heat of 0.24. Refrigerant-22 (R-22) vapor at approximately 70°F and at constant pressure, has a specific heat value of 0.066, whereas the specific heat of R-22 liquid is 0.306 at 86°F. When dealing with metals and the temperature of a mixture, a combination of weights, specific heats, and temperature differences must be considered in the overall calculations of heat transfer. In order to calculate the total heat transfer of a combination of substances, it is necessary to all the individual rates as follows:

$$\begin{aligned}\text{Btu} = {} & (W_1 \times c_1 \times \Delta T_1) \\ & + (W_2 \times c_2 \times \Delta T_2) \\ & + (W_3 \times c_3 \times \Delta T_3) \ldots \text{etc.}\end{aligned} \qquad (4\text{-}1\text{-}7)$$

EXAMPLE

How much heat must be added to a 10 lb copper vessel, holding 30 lb of water at 70°F to reach 185°F if the specific heat of copper is 0.095°F?

Solution

This equation may be used as follows:

$$\text{Btu} = (W_1 \times c_1 \times \Delta T_1) + (W_2 \times c_2 \times \Delta T_2)$$

where

W_1, c_1, and ΔT_1 pertain to the copper vessel, and W_2, c_2, and ΔT_2 pertain to the water. Therefore,

$$\begin{aligned}\text{Btu} &= (10 \times 0.095 \times 115) + (30 \times 1.0 \times 115) \\ &= 109.2 + 3450 \\ &= 3559.2 \text{ Btu}\end{aligned}$$

■

Change of State

We are now ready to look at the five principal changes of state:

Solidification a change from a liquid to a solid
Liquefaction a change from a solid to a liquid
Vaporization a change from a liquid to a vapor
Condensation a change from a vapor to a liquid
Sublimation a change from a solid to a vapor without passing through the liquid state

When a solid substance is heated, the molecular motion is chiefly in the form of rapid motion back and forth, the molecules never moving far from their normal or original position. But at some given temperature for that particular substance, further addition of heat will not necessarily increase the molecular motion within the substance; instead, the additional heat will cause some solids to liquefy (change into a liquid). Thus the additional heat causes a change of state in the material. The temperature at which this change of state in substance takes place is called its melting point. Let us assume that a container of water at 70°F, in which a thermometer has been placed, is left in a freezer for hours, Figure 4-1-6. When it is taken from the freezer, it has become a block of ice—solidification has taken place.

> **NICE TO KNOW**
>
> Water expands as it freezes becoming a solid. Water is one of the few materials that expands during its freezing process. If water didn't expand a pond would freeze from the bottom up killing all the fish. The fact that water expands when it freezes is beneficial to marine life, however, extremely detrimental to plumbing because expansion can break pipes, valves, pumps, etc. Be sure all water lines that are exposed to low temperature are either protected with antifreeze or drained to prevent freeze damage.

Let us further assume that the thermometer in the block indicates a temperature of 20°F. If it is allowed to stand at room temperature, the ice will absorb heat from the room air until the thermometer

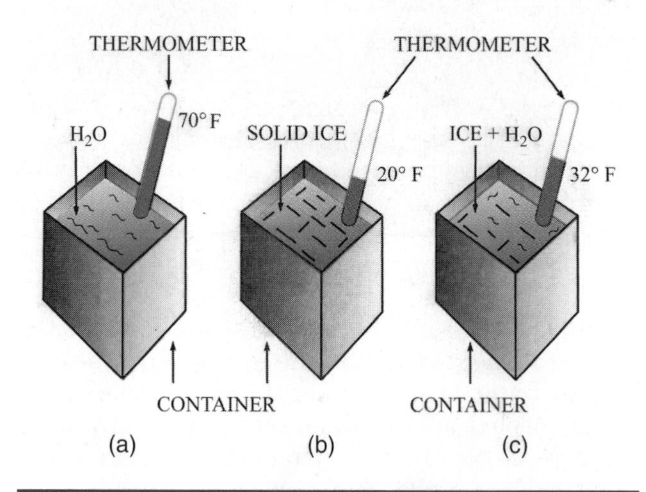

Figure 4-1-6 H_2O changes temperature and changes state as it absorbs sensible and latent heat.

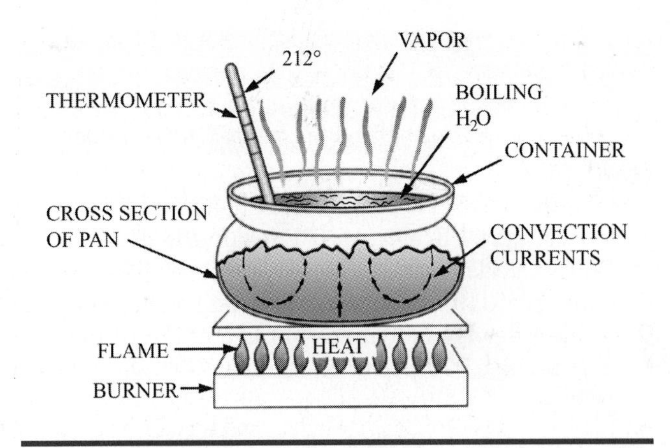

Figure 4-1-7 Convection currents caused by temperature differential.

indicates a temperature of 32°F, when some of the ice will begin to change into water.

With heat continuing to transfer from the room air to the ice, more ice will change back into water; but the thermometer will continue to indicate a temperature of 32°F until all the ice has melted. Liquefaction has now taken place. (This change of state without a change in the temperature will be discussed later.)

As mentioned, when all the ice is melted the thermometer will indicate a temperature of 32°F, but the temperature of the water will continue to rise until it reaches or equals room temperature.

If sufficient heat is added to the container of water through outside means, Figure 4-1-7, such as a burner or a torch, the temperature of the water will increase until it reaches 212°F. At this temperature, and under standard atmospheric pressure, another change of state will take place—vaporization. Some of the water will turn into steam, and with the addition of more heat, all of the water will vaporize into steam, yet the temperature of the water will not increase above 212°F. This change of state—without a change in temperature—will also be discussed later.

If the steam vapor could be contained within a closed vessel, and if the source of heat were removed, the steam would give up heat to the surrounding air, and it would condense back into a liquid form—water. What has now taken place is condensation—the reverse process from vaporization.

SERVICE TIP

If the heat is removed from a sealed pressure vessel containing steam that does not have a relief valve the collapsing steam can produce enough of a vacuum within the vessel to cause it to collapse. In some cases that inward collapse can be as violent as an explosion. The loss of water to a steam boiler can be a catastrophic event if safeguards are not taken.

Oxygen is a gas above –297°F, a liquid between that temperature and –324°F, and a solid below that point. Iron is a solid until it is heated to 2800°F, and becomes iron vapor at a temperature approximating 4950°F.

Thus far we have learned, with some examples, how a solid can change into a liquid, and how a liquid can change into a vapor. But it is possible for a substance to undergo a physical change through which a solid will change directly into a gaseous state without first melting into a liquid. This is known as sublimation.

All of us probably have seen this physical change take place without fully recognizing the process. Damp or wet clothes, hanging outside in the freezing temperature, will speedily become dry through sublimation; just as dry ice (solid carbon dioxide, or CO_2) sublimes into a vapor under normal temperature and pressure.

Sensible Heat

Heat that can be felt or measured is called sensible heat. It is the heat that causes a change in temperature of a substance, but not a change in state. Substances, whether in a solid, liquid, or gaseous state, contain sensible heat to some degree, as long as their temperatures are above absolute zero. Equations used for solutions of heat quantity, and those used in conjunction with specific heats, might be classified as being sensible heat equations, since none of them involve any change in state.

As mentioned earlier, a substance may exist as a solid, liquid, or as a gas or vapor. The substance as a solid will contain some sensible heat, as it will in the other states of matter. The total amount of heat needed to bring it from a solid state to a vapor state is dependent on its initial temperature as a solid, the temperature at which it changes from a solid to a liquid, the temperature at which it changes from a liquid to a vapor, and its final temperature as a vapor. Also included is the heat that is required to effect the two changes in state.

Latent Heat

Under a change of state, most substances will have a melting point at which they will change from a solid to a liquid without any increase in temperature. At this point, if the substance is in a liquid state and heat is removed from it, the substance will solidify without a change in its temperature. The heat involved in either of these processes (changing from a solid to a liquid, or from a liquid to a solid), without a change in temperature, is known as the *latent heat of fusion.*

Figure 4-1-5 on page 158 shows the relationship between temperature in Fahrenheit degrees and both sensible and latent heat in Btu.

As pointed out earlier, the specific heat of water is 1.0 and that of ice is 0.5, which is the reason for the difference in the slopes of the lines denoting the solid (ice) and the liquid (water). To increase the temperature of a pound of ice from 0°F to 32°F requires only 16 Btu of heat; the other line shows that it takes only 8 Btu to increase the temperature of a pound of water 8°F (60°F to 68°F), from 188 to 196 Btu/lb.

Figure 4-1-5 also shows that a total of 52 Btu of sensible heat is involved in the 196 Btu necessary for converting a pound of 0°F ice to 68°F water. This leaves a difference of 144 Btu, which is the latent heat of fusion of water or ice, depending on whether heat is being removed or added.

The derivation of the word *latent* is from the Latin word for hidden. This is hidden heat, which does not register on a thermometer, nor can it be felt. Needless to say, there is no increase or decrease in the molecular motion within the substance, for it would show up in a change in temperature on a thermometer.

$$\text{Btu} = (W_1 \times c_1 \times \Delta T_1) + (W_1 \times \text{latent heat}) + (W_2 \times c_2 \times \Delta T_2) \qquad (4\text{-}1\text{-}8)$$

EXAMPLE

Calculate the amount of heat needed to change 10 lb of ice at 20°F to water at 50°F.

Solution

Utilizing Eq. 4-1-8, we have

$$\Delta T_1 = (32° - 20°) = 12°$$
$$\Delta T_2 = (50° - 32°) = 18°$$
$$\text{Btu} = (10 \times 0.5 \times 12) + (10 \times 144) + (10 \times 1.0 \times 18)$$
$$= 60 + 1440 + 180$$
$$= 1680$$

■

Another type of latent heat that must be taken into consideration when total heat calculations are necessary is called the *latent heat of vaporization*. This is the heat that 1 lb of a liquid absorbs while being changed into the vapor state. It can also be classified as the latent heat of condensation, for when sensible heat is removed from the vapor to the extent that it reaches the condensing point, the vapor condenses back into the liquid form.

The latent heat of vaporization of water as 1 lb is boiled or evaporated into steam at sea level is 970 Btu. That amount also is the heat that 1 lb of steam must release or give up when it condenses into water. Figure 4-1-5 also shows the relationship between temperature and both sensible heat and latent heat of vaporization.

The absorption of the amount of heat necessary for the change of state from a liquid to a vapor by evaporation, and the release of that amount of heat necessary for the change of state from a vapor back to a liquid by condensation are the main principles of the refrigeration process, or cycle. Refrigeration is the transfer of heat by the change in state of the refrigerant.

Figure 4-1-5 shows the total heat in Btu necessary to convert 1 lb of ice at 0°F to superheated steam at 230°F under atmospheric pressure. This total amounts to 1319 Btu. Only 205 Btu is sensible heat; the remainder is made up of 144 Btu of latent heat of fusion and also 970 Btu of latent heat of vaporization.

TECH TALK

Tech 1. I have no idea why the boss wants us to worry so much about Btu's.

Tech 2. Our whole job is involved with either removing Btu's or adding Btu's to a space to make it more comfortable.

Tech 1. I don't understand what you mean. I thought we were involved in heating and air conditioning.

Tech 2. Well, in air conditioning we are taking Btu's or heat out of the inside of the house. And in heating we are putting Btu's back in the house. So, our whole job deals with Btu's.

Tech 1. Hold on a minute. If a heating system isn't working properly, why should Btu's help me figure out what's going on?

Tech 2. Well, if a heating system is not producing enough Btu's then the house will be too cool. The Btu's are produced in the furnace by burning the oil or gas. So not enough Btu's means there is not enough oil or gas being burned, or maybe the fuel is not being burned properly.

Tech 1. Okay, but what about an air conditioning system? How can Btu's help me there?

Tech 2. The same thing is true with air conditioning. If it is not producing enough cooling maybe you're not taking out enough Btu's. Maybe you were taking out latent Btu's, taking out a lot of water, which is going to add load to the system but not really change the room temperature that much.

Tech 1. I thought I only had to worry about latent heat inside the air conditioning system.

Tech 2. If you think about it, taking humidity out of a humid house means condensing the humidity into water that drains out of the system.

Tech 1. That takes a lot of Btu's!

Tech 2. That's right! Having a good understanding of Btu's will help you troubleshoot a system.

Tech 1. So, I guess having an idea of Btu's is a good idea. Thanks, I appreciate that.

UNIT 1—REVIEW QUESTIONS

1. List the three states of matter.
2. A ________ is a substance that can take the shape of any enclosure when it is allowed to move freely.
3. Define density.
4. How much volume does 1 lb of air occupy at standard atmospheric conditions?
5. What is the specific heat of a substance?
6. List the five principal changes of state.
7. Heat that can be felt or measured is called ________.
8. What is the latent heat of fusion?
9. To increase the temperature of 1 lb of ice from 0° to 32°F requires ________ Btu of heat.
10. What is the latent heat of vaporization?

UNIT 2

Thermodynamics

> **OBJECTIVES:** In this unit the reader will learn about thermodynamics including:
>
> - Thermodynamic Principles
> - Heat Transfer
> - Conduction
> - Convection
> - Radiation

INTRODUCTION

In Chapter 1 Unit 2, you learned brief definitions of the three heat transfer methods: conduction, convection, and radiation. In this unit, we cover thermodynamics and then look at heat transfer more deeply.

THERMODYNAMIC PRINCIPLES

Thermodynamics is the branch of science dealing with mechanical action of heat. Two main laws of thermodynamics are of interest in HVAC/R applications.

First Law of Thermodynamics

The first law of thermodynamics states that "energy can neither be created nor destroyed; it can only be converted from one form to another."

Energy itself is defined as the ability to do work, and heat is one form of energy. It is also the final form, as ultimately all forms of energy end up as heat. Other common forms of energy are: mechanical, electrical, and chemical, which may be converted easily from one form to another. The steam-driven turbine generator of a power plant is a device that converts heat energy into electrical energy. Chemical energy may be converted into electrical energy by the use of a battery. Electrical energy is converted into mechanical energy through the use of an electric magnetic coil to produce a push-pull motion or the use of an electric motor to create rotary motions. Electrical energy may be changed directly to heat energy by means of heating resistance wires such as in an electric toaster, grill, or furnace.

Second Law of Thermodynamics

The second law of thermodynamics states that "to cause heat energy to travel, a temperature difference must be established and maintained." Heat energy travels downward on the intensity scale. Heat from a higher temperature (intensity) material will travel to a lower temperature (intensity) material, and this process will continue as long as the temperature difference exists. The rate of travel varies directly with the temperature difference. The higher the temperature difference (commonly called the delta temperature or ΔT), the greater the rate of heat travel. Commonly, the lower the ΔT, the lower the rate of heat travel. Heat and heat transfer are expressed in different forms and terms that are important to the refrigeration industry.

HEAT TRANSFER

The second law of thermodynamics, discussed earlier, states that heat transfers (flows) in one direction only, from a high temperature (intensity) to a lower temperature (intensity). This transfer will take place using one or more of the following basic methods of transfer:

1. Conduction
2. Convection
3. Radiation

> **SERVICE TIP**
>
> Heat is always on the move. It's moving constantly from a warmer body to a cooler body. This movement can be slowed with insulation, but no matter how thick the insulation is heat will continue to move through it until both bodies have reached equilibrium and are at the same temperature. For that reason it is very difficult to store heat for long periods of time without significant loss in the quantity of heat being stored.

Conduction

Conduction is described as the transfer of heat between the closely packed molecules of a substance, or between substances that are in good contact with one another. When the transfer of heat occurs in a single substance, such as a metal rod with one end in a fire or flame, movement of heat continues until there is a temperature balance throughout the length of the rod.

If the rod is immersed in water, the rapidly moving molecules on the surface of the rod will transmit some heat to the molecules of water, and still another transfer of heat by conduction takes place. As the outer surface of the rod cools off, there is still some heat within the rod, and this will continue to transfer to the outer surfaces of the rod and then to the water, until a temperature balance is reached.

The speed with which heat will transfer by means of conduction will vary with different substances or materials if the substances or materials are of the same dimensions. The rate of heat transfer will vary according to the ability of the material or substances to conduct heat. Solids, on the whole, are much better conductors than liquids; in turn, liquids conduct heat better than gases or vapors.

Most metals, such as silver, copper, steel, and iron, conduct heat fairly rapidly, whereas other solids, such as glass, wood, or other building materials, transfer heat at a much slower rate and therefore are used as insulators.

SERVICE TIP

Materials that are excellent electrical conductors tend to be excellent thermal conductors. Conversely, materials that are good electrical insulators tend to be good thermal insulators. There are of course exceptions and in a few cases materials such as ceramics can be designed to be either an insulator or a conductor of electricity or heat. One particular type of ceramic is actually considered to be a superconductor, meaning it has extremely low electrical and heat resistance.

Copper is an excellent conductor of heat, as is aluminum. These substances are ordinarily used in the evaporators, condensers, and refrigerant pipes connecting the various components of a refrigerant system, although iron piping is occasionally used with some refrigerants.

The rate at which heat may be conducted through various materials is dependent on such factors as the thickness of the material, its cross-sectional area, the temperature difference between the two sides of the material, the heat conductivity (k factor) of the material, and the time duration of the heat flow. Table 4-2-1 gives the heat conductivities (k factors) of some common materials.

TABLE 4-2-1 Conductivities for Common Building and Insulating Materials. k Values Expressed in Btu/hr/ft^2/°F/in Thickness of Material

Material	Conductivity k
Plywood	0.80
Glass fiber—organic bonded	0.25
Expanded polystyrene insulation	0.25
Expanded polyurethane insulation	0.16
Cement mortar	5.0
Stucco	5.0
Brick (common)	5.0
Hardwood (maple, oak)	1.10
Softwood (fir, pine)	0.80
Gypsum plaster (sand aggregate)	5.6

Note that the k factors are given in Btu/hr/ft^2/°F/in of thickness of the material. These factors may be used correctly with Eq. 4-2-1.

$$\text{Btu} = \frac{A \times k \times \Delta T}{X} \tag{4-2-1}$$

where

A = cross-sectional area, ft^2
k = heat conductivity, Btu/hr/ft^2/°F/in
ΔT = temperature difference between the two sides, °F
X = thickness of material, in

Metals with a high conductivity are used within the refrigeration system itself because it is desirable that rapid heat transfer occurs in both evaporator and condenser. The evaporator is where heat is removed from the conditioned space or substance or from air that has been in direct contact with the substance; the condenser dissipates this heat to another medium or space.

In the case of the evaporator, the product or air is at a higher temperature than the refrigerant within the tubing and there is a transfer of heat downhill. In the condenser, however, the refrigerant vapor is at a higher temperature than the cooling medium traveling through the condenser, and here again there is a downhill transfer of heat.

Plain tubing, whether copper, aluminum, or another metal, will transfer heat according to its conductivity or k factor, but this heat transfer can be increased through the addition of fins on the tubing.

They will increase the area of heat transfer surface, thereby increasing the overall efficiency of the system. If the addition of fins doubles the surface area, it can be shown by the use of Eq. 4-2-1 that the overall heat transfer will be doubled compared to that of plain tubing.

Convection

Another means of heat transfer is by motion of the heated material itself. Convection is limited to heat transfer within a liquid or a gas. When a material is heated, convection currents are set up within it, and the warmer portions of it rise, since heat brings about the decrease of a fluid's density and an increase in its specific volume.

SERVICE TIP

Homeowners and building operators sometimes like to put up their own thermometers in a room. That thermometer reading sometimes is the one you will hear them talking about when they complain about room temperature. There are several factors that will affect the thermometer reading that may result it being different than the room thermostat. First, the room thermostat is usually close to a height of five feet and should be located on an inside wall. If their thermometer is higher or lower than that it will read a different temperature. Second, if their thermometer is near an outside wall it will read higher or lower depending the season of the year. Certainly one that is in a window can have wide swings in temperature. One common thermometer people use gives indoor and outdoor temperature readings. By its design it must be located on an outside wall so that its probe can go to the outside of the building. Those temperature readings will be significantly different most of the time from the room thermostat.

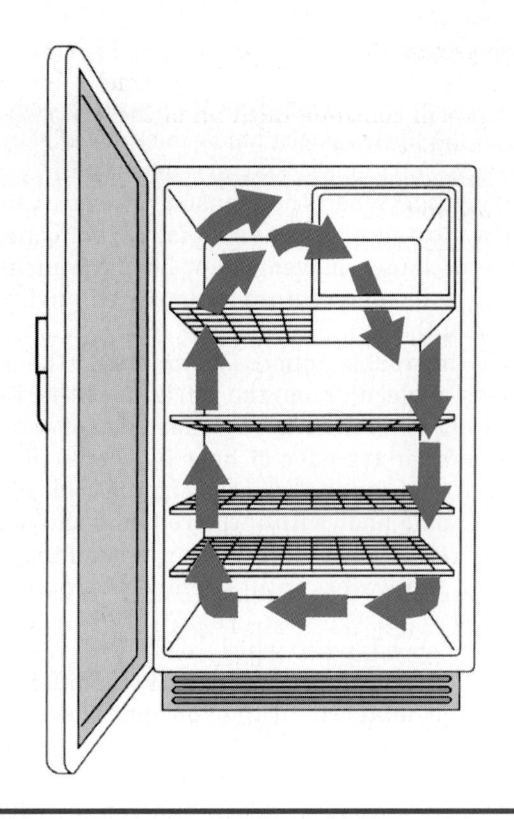

Figure 4-2-1 Convection currents caused by temperature differential.

SERVICE TIP

Convection currents in air or water distribute the temperature only to a certain extent. In a lake for example, during the summer the water near the surface will be very warm and as you dive down you pass through a layer where the temperature change is dramatic, but in the winter the top of the lake will be colder and water at the bottom of the lake warmer. Water and air both stratify even with temperature differences. It is far more noticeable in the summer when much of the heat is collected near the ceiling. Convection currents in both water and air make it difficult to keep a room's temperature consistent from floor to ceiling. When there is a difference of more than 2° from the floor to the ceiling additional air circulation is needed to break up the stratified air temperature.

Air within a refrigerator is a prime example of the results of convection currents, Figure 4-2-1. The air in contact with the cooling coil of a refrigerator becomes cool and therefore more dense, and begins to fall to the bottom of the refrigerator. In doing so, it absorbs heat from the food and the walls of the refrigerator, which, through conduction, has picked up heat from the room. After heat has been absorbed by the air it expands, becoming lighter, and rises until it again reaches the cooling coil where heat is removed from it. The convection cycle repeats as long as there is a temperature difference between the air and the coil. In commercial units, baffles may be constructed within the box in order that the convection currents will be directed to take the desired patterns of airflow around the coil.

Water heated in a pan will be affected by the convection currents set up within it through the application of heat. The water nearest the heat source, in absorbing heat, becomes warmer and expands. As it becomes less dense, it rises and is replaced by the other water, which is cooler and more dense. This process will continue until all of the water is at the same temperature.

Convection currents as explained and shown here are natural (passive), and as in the case of the refrigerator, a natural (passive) flow is a slow flow. In some cases, convection must be increased through the use of fans or blowers. In the case of liquids, pumps are used for forced (active) circulation to transfer heat from one place to another.

Radiation

A third means of heat transfer is through radiation by waves similar to light or sound waves. The sun's rays heat the earth by means of radiant heat waves, which travel in a straight path without heating the intervening matter or air. The heat from a light bulb or hot stove is radiant in nature and felt by those near them, although the air between the source and the object, which the rays pass through, is not heated.

NICE TO KNOW

When you purchase a hot apple pie at a fast food restaurant the package tells you that the contents inside can be very hot. These products are heated with radiant lamps. The lamp heats both the outside of the box and the product at the same time. The reason the box does not feel hot is that its specific heat is very low and cools very rapidly. By the time you pick the box up it is no longer hot, but the product inside is hot because it has a much higher specific heat. Some people feel that because the box is not hot and the product inside is hot that radiant heating does not heat the box but only the product much like a microwave. That's not true. Radiant heating heats the surface and the hot surface conducts the heat to the center.

If you have been relaxing in the shade of a building or a tree on a hot sunny day and move into direct sunlight, the direct impact of the heat waves will hit like a sledgehammer, even though the air temperature in the shade is approximately the same as in the sunlight.

At low temperatures there is only a small amount of radiation, and only minor temperature differences are noticed; therefore, radiation has very little effect in the actual process of refrigeration itself. But the results of radiation from direct solar rays can cause an increased refrigeration load in a building in the path of these rays.

Radiant heat is readily absorbed by dark or dull materials or substances, whereas light-colored surfaces or materials will reflect radiant heat waves, just as they do light rays. Clothing designers and manufacturers make use of this proven fact by supplying light-colored materials for summer clothes.

This principle is also carried over into the summer air-conditioning field, where light-colored roofs and walls allow less of the solar heat to penetrate into the conditioned space, thus reducing the size of the overall cooling equipment required. Radiant heat also readily penetrates clear glass in windows, but will be absorbed by translucent or opaque glass.

When radiant heat or energy (since all heat is energy) is absorbed by a material or substance, it is converted into sensible heat—that which can be felt or measured. Every body or substance absorbs radiant energy to some extent, depending on the temperature difference between the specific body or substance and other substances. Every substance will radiate energy as long as its temperature is above absolute zero and another substance within its proximity is at a lower temperature.

If an automobile has been left out in the hot sun with the windows closed for a long period of time, the temperature inside the car will be much greater than the ambient air temperature surrounding it. This demonstrates that radiant energy absorbed by the materials of which the car is constructed is converted to measurable sensible heat.

TECH TALK

Tech 1. Which heating process is better, conduction, convection, or radiation?

Tech 2. It depends on what you are trying to heat. If you are talking about heating a room then convection is the best way. That's when you heat the air and distribute it through the house. If you are talking about heating a pot of water on the burner then conduction is better. If you want to heat a large open area, radiation is better. The heat travels through the air without having to heat the air so that you feel warmer and the tools you're working with feel warmer.

Tech 1. So really, the type of heat is very much dependent on what you're going to do with it.

Tech 2. That's right, there's not a best way of heating. They are all good and they all have their advantages and disadvantages.

Tech 1. Thanks, I appreciate that.

UNIT 2—REVIEW QUESTIONS

1. What does the first law of thermodynamics state?
2. Define energy.
3. What does the second law of thermodynamics state?
4. _______ is described as the transfer of heat between the closely packed molecules of a substance, or between substances that are in good contact with one another.
5. What is a prime example of the results of convection currents?
6. _______ is readily absorbed by dark or dull materials or substances.
7. What is all matter composed of?
8. What is a molecule of water made up of?
9. In the IP system the unit of heat is the _______.
10. In the SI system the element of time is included in the term of _______.

UNIT 3

Fluid Pressure and Gas Laws

OBJECTIVES: In this unit the reader will learn about fluid pressure and gas laws including:

- Pascal's Law
- Density
- Specific Volume
- Atmospheric Pressure
- Measurement of Pressure
- Absolute Pressure
- Pressure of Gas
- Expansion of Gas
- Boiling Point
- Condensing Temperature
- Fusion Point
- Saturation Temperature
- Superheat
- Subcooling

INTRODUCTION

The dictionary describes a fluid as any substance that can flow: liquid, gas, and in a very few cases solids. A refrigerant may therefore be classified as a fluid, since within the refrigeration cycle, it exists both as a liquid and as a vapor or gas. Although, as previously mentioned, ice—a solid—has also been used in heat removal, its use in refrigeration has been overshadowed by the discovery of the versatility of the chemicals and chemical combinations used as refrigerants today.

FLUID PRESSURE

The weight of a block of wood or any other solid material acts as a force downward on whatever is supporting it. The force of this solid object is the overall weight if the object, and the total weight is distributed over the area on which it lies. The weight of a given volume of water, however, acts not only as a force downward on the bottom of the container holding it, but also as a force laterally on the sides of the container. If a hole is made in the side of the container below the water level, as in Figure 4-3-1, the water above the hole will be forced out because of its force acting downward and sideways.

NOTE: The most commonly overlooked fluid that affects air conditioning performance is air. Many people discount air as a fluid. It however, is a fluid with all of the associated characteristics of weight, density, inertia, and pressure.

Fluid pressure is the force per unit area that is exerted by a gas or a liquid. In IP units, it is given in psi; and in SI units, it is given in kPa. Fluid pressure varies directly with the density and the depth of the liquid. At the same depth below the surface, the pressure is equal in all directions. Notice the difference between the terms used: force and pressure. **Force** means the total weight of the substance; **pressure** means the unit force or pressure per square inch.

If the tank in Figure 4-3-1 measures 1 ft in all dimensions, and it is filled with water, we have 1 ft^3 of water, the weight of which is approximately 62.4 lb. Therefore, we would have a total force of 62.4 lb being exerted on the bottom of the tank, the area of

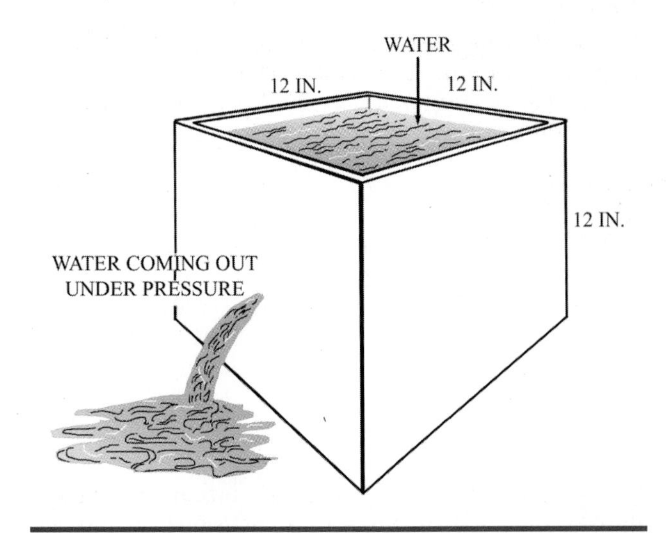

Figure 4-3-1 Water pressure in a container exerts pressure in all directions.

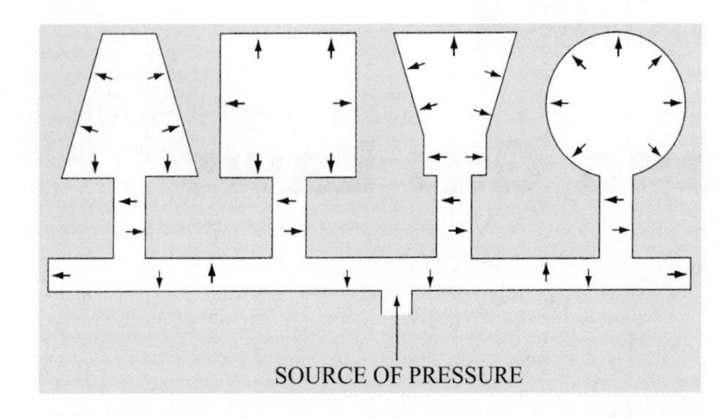

Figure 4-3-2 Pascal's principle of pressure equalization in various shaped containers.

which equals 144 in^2 (12 in × 12 in = 144 in^2). Using the equation:

$$\text{Pressure} = \frac{\text{Force}}{\text{Area}}$$

or

$$P = \frac{F}{A} \qquad (4\text{-}3\text{-}1)$$

the unit pressure of the water will be

$$\text{unit pressure} = \frac{\text{force}}{\text{area}} = \frac{62.4\ \text{lb}}{144\ \text{in}^2} = 0.433\ \text{psi}$$

This pressure of 0.433 psi is exerted downward and also sideways. If a tank is constructed as in Figure 4-3-2, the pressure of water would cause the tube to be filled with water to the same level as in the tank. Fluid pressure is the same on each square inch of the walls of the tank at the same depth, and it will act at right angles to the surface of the tank.

EXAMPLE

A tank has a cross section of 4 ft^2 and is filled with water to its depth of 3 ft. Find (a) the volume of water, (b) the weight of the water, (c) the force on the bottom of the tank, and (d) the pressure on the bottom of the tank.

Solution

a. Volume = 4 ft × 4 ft × 3 ft = 48 ft^3
b. Weight = 48 ft^3 × 62.4 lb/ft^3 = 2995 lb
c. Force = weight = 2995 lb
d. Pressure = weight ÷ area = 2995 ÷ 2304 = 1.3 psi
(area = 4 ft × 4 ft × 144 in^2/ft^2 = 2304 in^2) ■

HEAD PRESSURE

Pressure and depth have a close relationship when a fluid is involved. In hydraulics, a branch of physics that has to do with properties of liquids, the depth of a body of water is called the **head** of water. *Water pressure* varies directly with its depth. As an example, if the tank in Figure 4-3-1 was 2 ft high and filled with water, it would contain a volume of 2 ft^3 of water and would weigh 2 × 62.4 lb, or 124.8 lb. Now the force of the water on the bottom of the tank would still be distributed over 144 in^2 and the unit pressure would be 0.866 psi (124.8 lb ÷ 144 in^2). This is twice the amount of pressure that was exerted when the head of water was only 1 ft. Therefore, in an open top container, the pressure of the water will equal 0.433 psi for each foot of head.

> **SERVICE TIP**
>
> A pump only moves fluid—it does not create pressure. The *resistance to the flow* of a fluid is what results in pressure. A pump that has a fully unrestricted outlet will have zero or nearly zero pressure reading. It is only when a pipe is attached to the outlet and the water directed through the pipe or up some distance that the pressure in the system is created.

If there is a decrease or increase in the head of a body of water, there will be a corresponding decrease or increase in the pressure involved, provided that the other dimensions stay the same. If the head of water is only 6 in ($\frac{1}{2}$ ft), the pressure would equal 0.433 × $\frac{1}{2}$, or 0.217 psi. If there were 10 ft of water in an open tank, the pressure would be 0.433 × 10 or 4.33 psi.

This relationship can be expressed in the equation

$$p = 0.433 \times h \qquad (4\text{-}3\text{-}2)$$

where
p = pressure, psi
h = head, ft of water

The tank in Figure 4-3-1 has an area of 1 ft^2 with 1 ft of head; therefore, the pressure on the bottom of the tank is 0.433 psi. If there is a fish pond covering an area of 50 ft^2, and the depth of water in it is 1 ft, pressure on the bottom of the pond will still be just 0.433 psi, even though there is a larger total volume of water. This points out the relationship between pressure and depth and demonstrates that there is not necessarily a relationship between pressure and volume.

With this relationship between pressure and depth established, we can find the depth of water in a tank if we know the pressure reading at the bottom of the tank.

$$p = 0.433 \times h$$
$$h = p/0.433 \qquad (4\text{-}3\text{-}3)$$

EXAMPLE

If a pressure gauge located at the bottom of a 50 ft high water tower showed a reading of 13 psi, what is the depth of the water in the tank, and what would the gauge indicate if the tower were filled to the top of the 50 ft with water?

Solution

a. $h = \dfrac{p}{0.433} = \dfrac{13}{0.433} = 30$ ft of water

b. $p = 0.433h = 0.433 \times 50 = 21.65$ psi ■

PASCAL'S LAW

In the middle of the seventeenth century, a French mathematician and scientist named Blaise Pascal was experimenting with water and air pressure. His scientific experiments led to the formulation of what is known as Pascal's law, namely that pressures applied to a confined liquid are transmitted equally throughout the liquid, irrespective of the area over which the pressure is applied. The application of this principle enabled Pascal to invent the hydraulic press, which is capable of a large multiplication of force. In Figure 4-3-2, the unit pressure is equal in all vessels regardless of their shape, for pressure is independent of the shape of the container.

Figure 4-3-3, illustrating Pascal's law, shows a vessel containing a fluid such as oil; the vessel has a small and a large cylinder connected by a pipe or tubing, with tight-fitting pistons in each cylinder. If the cross-sectional area of the small piston is 1 in^2 and the area of the large one is 30 in^2, a force of 1 lb when applied to the smaller piston will support a weight of 30 lb on the larger piston, because a pressure of 1 psi throughout the fluid will be exerted.

Since we found earlier that pressure equals force divided by area, a force of 1 lb applied to an area 1 in^2 will create a pressure of 1 psi. By transposition, force equals pressure multiplied by the area; therefore, a weight of 30 lb will be supported when a pressure of 1 psi is applied over an area of 30 in^2.

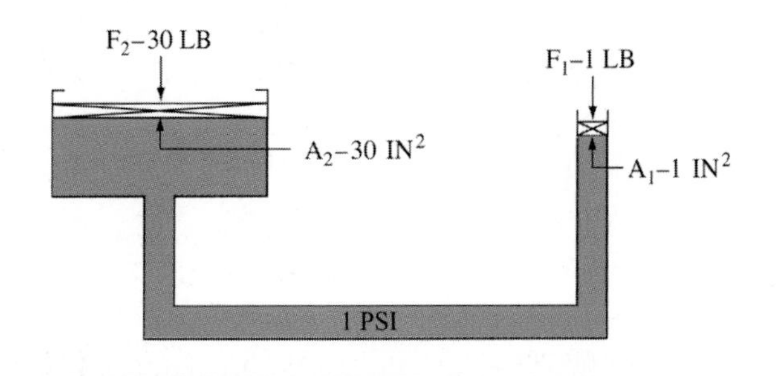

Figure 4-3-3 Pascal's principle of pressure transmission by hydraulic action.

DENSITY

From a scientific or physics viewpoint, **density** is the weight per unit volume of a substance, and it may be expressed in any convenient combination of units of weight and volume used, such as lb/in^3 or lb/ft^3. An equation can be formulated which expresses this relationship.

$$D = \frac{W}{V} \tag{4-3-4}$$

where
D = density
W = weight
V = volume

As mentioned previously, the density of water is approximately 62.4 lb/ft^3, and it can be expressed as 0.0361 lb/in^3 (1 ft^3 contains 1728 in^3, and 62.4 ÷ 1728 = 0.0361). The densities of some other common substances are listed in Table 4-3-1.

The **specific gravity** of any substance is the ratio of the weight of a given volume of the substance to the weight of the same volume of another given substance (where solids or liquids are concerned, water is used as a basis for specific gravity calculations; air or hydrogen is used as a standard for gases).

$$\text{Density (solid liquid)} = \text{specific gravity} \times \text{density of water (lb/ft}^3) \tag{4-3-5}$$

TABLE 4-3-1 Density and Specific Gravity of Some Common Substances

Substance	Density (lb/ft^3)	Specific Gravity
Water (pure)	62.4	1
Aluminum	168	2.7
Ammonia (liquid, 60°F)	38.5	0.62
Brass	530	8.5
Brick (common)	112	1.8
Copper	560	8.98
Cork (average board)	15	0.24
Gasoline	41.2	0.66
Glass (average)	175	2.8
Iron (cast)	448	7.2
Lead	705	11.8
Mercury	848	13.6
Oil (fuel)(average)	48.6	0.78
Steel (average)	486	7.8
Wood		
Oak	50	0.8
Pine	34.2	0.55

The specific gravity of water is given as 1.0 and values for other substances are listed in Table 4-3-1.

Pressure within a fluid is directly proportional to the density of the fluid. Consider the tank in Figure 4-3-1, which, if filled with water weighing 62.4 lb, has a force on the bottom of the tank of 62.4 lb and a unit pressure of 0.433 psi. If this tank were filled instead with gasoline, which has a specific gravity of 0.66, the force on the bottom would be only 66% as great as when water is in the tank, and the pressure of the gasoline would be only 66% as great. Therefore, the relationship can be expressed as

$$\text{Pressure} = \text{head} \times \text{density}$$

$$p = h \times D \tag{4-3-6}$$

where

p = pressure, lb/ft^2
h = head or depth below the surface, ft
D = density, lb/ft^3

or where

p = pressure, psi
h = head or depth below the surface, in
D = density, lb/in^3

SPECIFIC VOLUME

The *specific volume* of a substance is usually expressed as the number of cubic feet occupied by 1 lb of the substance. In the case of liquids, it will vary with temperature and pressure. The volume of a liquid will be affected by a change in its temperature, but since it is practically impossible to compress liquids, the volume is not affected by a change in pressure.

The volume of a gas or vapor is definitely affected by any change in either its temperature or the pressure to which it is subjected. In refrigeration, the volume of the vapor under the varying conditions involved is most important in the selection of the proper refrigerant lines.

The appropriate specific volumes and pressures for refrigerants will be covered later, but as an example of the effect temperature has on a refrigerant vapor, refer to Table 4-3-2 for the properties of Refrigerant-12 (R-12).

Note that at +5°F the specific volume of vapor is 1.46 ft^3/lb, whereas at 86°F it is only 0.38 ft^3/lb. Correspondingly, there is an increase in pressure

TABLE 4-3-2 Properties of R-12 Refrigerant. Note Pressures Corresponding to Standard Evaporating Temperature of 5°F and Condensing Temperature of 86°F

	Pressure		Volume (Vapor)	Density (Liquid)	Heat Content (Btu/lb)	
Temp °F	**Psia**	**Psig***	**(ft^3/lb)**	**(lb/ft^3)**	**Liquid**	**Vapor**
−150	0.154	*29.61*	178.65	104.36	−22.70	60.8
−125	0.516	*28.67*	57.28	102.29	−17.59	83.5
−100	1.428	*27.01*	22.16	100.15	−12.47	66.2
−75	3.388	*23.02*	9.92	97.93	−7.31	69.0
−50	7.117	*15.43*	4.97	95.62	−2.10	71.8
−25	13.556	2.32	2.73	93.20	3.17	74.56
−15	17.141	2.45	2.19	92.20	5.30	75.65
−10	19.189	4.49	1.97	91.70	6.37	76.2
−5	21.422	6.73	1.78	91.18	7.44	76.73
0	23.849	9.15	1.61	90.66	8.52	77.27
5	**26.483**	**11.79**	**1.46**	**90.14**	**9.60**	**77.80**
10	29.335	14.64	1.32	89.61	10.68	78.335
25	39.310	24.61	1.00	87.98	13.96	79.9
50	61.394	46.70	0.66	85.14	19.50	82.43
75	91.682	76.99	0.44	82.09	25.20	84.82
86	**108.04**	**93.34**	**0.38**	**80.67**	**27.77**	**85.82**
100	131.86	117.16	0.31	78.79	31.10	87.63
125	183.76	169.06	0.22	75.15	37.28	88.97
150	249.31	234.61	0.16	71.04	43.85	90.53
175	330.64	315.94	0.11	66.20	51.03	91.48
200	430.09	415.39	0.08	60.03	59.20	91.28

*Pressures in italic are inches of mercury below 1 atm.

(*Source:* The DuPont Company)

(psia) from 26.48 psia to 108.04 psia. Note also the change in density from 90.14 lb/ft^3 to 80.67 lb/ft^3.

Table 4-3-2 gives two different pressure columns: psia for absolute pressure and psig for gauge pressure. The following discussion explains the difference between them.

ATMOSPHERIC PRESSURE

The earth is surrounded by a blanket of air called the atmosphere, which extends 50 or more miles upward from the surface of the earth. Air has weight and also exerts a pressure known as *atmospheric pressure*. It has been shown that a column of air with a cross-sectional area of 1 in^2 and extending from the earth's surface at sea level to the limits of the atmosphere, would weigh approximately 14.7 lb. As was pointed out earlier in the chapter, force also means the weight of a substance, and pressure means unit force per square inch; therefore, standard atmospheric pressure is considered to be 14.7 psi at sea level.

SERVICE TIP

Many refrigerant pressure charts and psychrometric charts are all based on sea level readings. You must use the correct chart when working at altitudes that are significantly above sea level. The pressure temperature chart used by a technician in Denver, Colorado is going to be different from one used by a technician in Dallas, Texas. If you will look at the bottom of the chart there is often a small note stating the barometric pressure, altitude, or other indications as to the chart's intended use.

This pressure is not constant; it will vary with altitude or elevation above sea level, and there will be variations due to changes in temperature as well as water vapor content of the air.

NOTE: Atmospheric pressure was first demonstrated by the type of simple barometer shown in Figure 4-3-4. Early experimenters used a glass tube about 36 in long and closed at one end, an open bowl, and a supply of mercury. They filled the tube with mercury and inverted it in the bowl of mercury, holding a finger at the open end to keep the mercury from spilling out while the tube was inverted. Upon removal of the finger, the level of the mercury in the tube dropped somewhat, leaving a vacuum at the closed top of the tube. The atmospheric pressure bearing down on the open bowl of mercury forced the mercury in the tube to stand up to a height determined by the air pressure, approximately 30 in Hg at sea level. That is how pressure came to be measured in inches of mercury (in Hg).

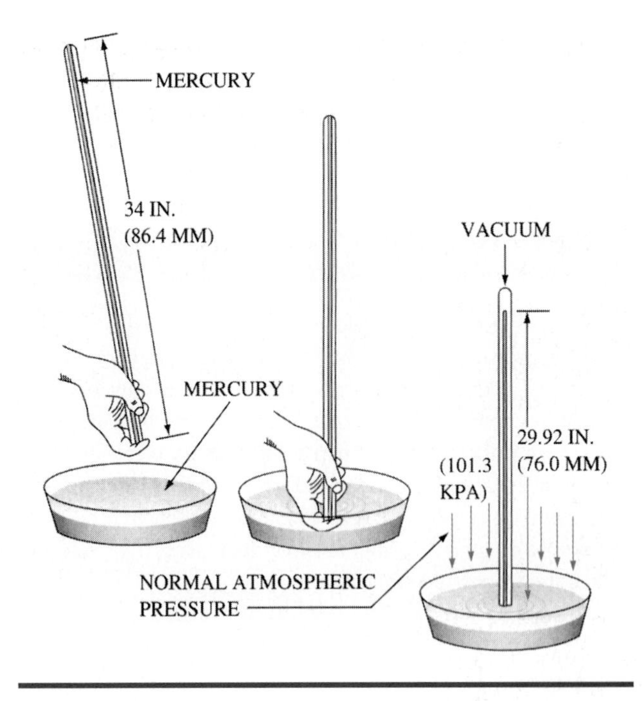

Figure 4-3-4 Column of mercury supported by normal atmospheric pressure.

As previously shown in Table 4-3-1, the specific gravity of mercury is 13.6; that is, mercury weighs and exerts pressure 13.6 times that of an equal volume of water. Since a 1 in high column of water exerts pressure of 0.0361 psi, a similar column of mercury will exert pressure that is $0.0361 \times 13.6 = 0.491$ psi. A 30 in column of mercury will exert pressure of $0.491 \times 30 = 14.7$ psi, or an amount equal to atmospheric pressure. Conversely, atmospheric pressure at sea level (14.7 psi) bearing down on the open dish of mercury divided by the unit pressure that is exerted by a 1 in column of mercury (0.491) should cause the column of mercury in the tube to stand approximately at a height of 30 in (actually, 29.92 in).

$$\frac{14.7}{0.491} = 30 \text{ in Hg}$$

Of course, water could have been used in the closed tube barometer instead of mercury; but the tube would have to be about 34 ft high, and this was not practical.

$$\frac{14.7}{0.036} = 407 \text{ in of water } (33.0 \text{ ft})$$

MEASUREMENT OF PRESSURE

A manometer is one type of device used in the refrigeration and air conditioning field for the measurement of pressure. This type of pressure gauge uses a liquid, usually mercury, water, or gauge oil, as an indicator of the amount of pressure involved. The water manometer or water gauge is customarily used when measuring air pressures, because of the low density of the fluid being measured.

SERVICE TIP

There are electronic devices that can measure air pressure so accurately that they can tell the difference in altitude of a distance as little as 5 ft. These devices use small electronic sensors to determine the difference in pressure. Most of these instruments use a piezo crystal as their sensor. A piezo crystal is a crystal that changes its electrical resistivity as a result of external force. As the cost of piezo crystal sensors decreases, their use in air conditioning will increase.

A simple open arm manometer is shown in Figure 4-3-5a. The U shaped glass tube is partially filled with water and is open at both ends. The water is *at the same level* in both arms of the manometer, because both arms are open to the atmosphere, and no external pressure is being exerted on them.

Figure 4-3-5b shows the manometer in use with one arm connected to a source of positive air pressure that is being measured. The water is at different levels in the arms, and the difference denotes the amount of pressure being applied.

A space that is void, or lacking any pressure, is described as having a *perfect vacuum*. If the space has pressure less than atmospheric pressure, it is defined as being a *partial vacuum*. It is customary to express this partial vacuum in units of in Hg, and not as negative pressure. In some instances it is also referred to as a given amount of absolute pressure, expressed in psia, and this will be covered later in this chapter.

If a partial vacuum has been drawn on the left arm of the manometer by means of a vacuum pump, as shown in Figure 4-3-6, the mercury in the right arm will be lower, and the difference in levels will designate the partial vacuum in units of in Hg.

Pressure gauges most commonly used in the field by service technicians, to determine pressure within the refrigeration system, are of the Bourdon tube type. As is shown in Figure 4-3-7, an internal view, the essential element of this type of gauge is the Bourdon tube. This oval metal tube is curved along its length and forms an almost complete circle. One end of the tube is closed, and the other end is connected to the equipment or component being tested.

As shown in Figure 4-3-8, the gauges are preset at 0 lb, which represents atmospheric pressure of

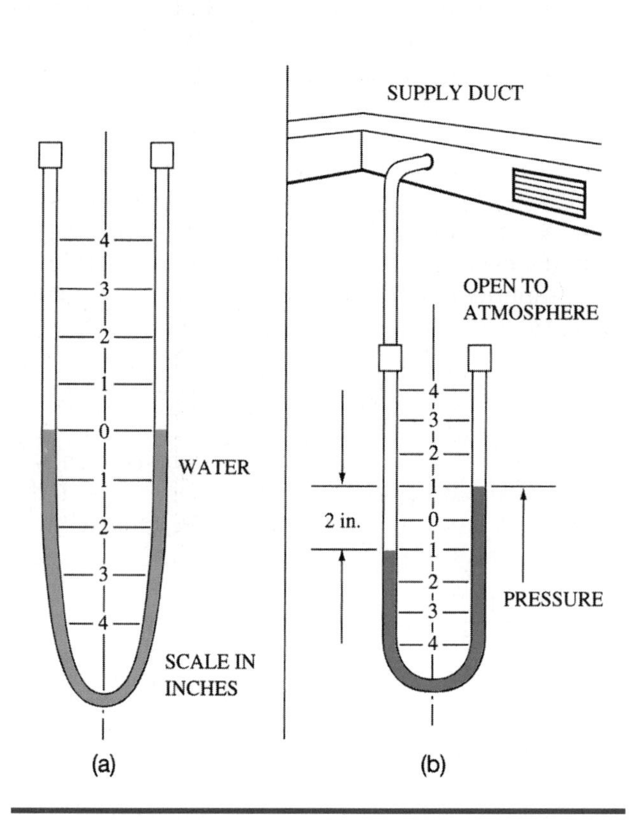

Figure 4-3-5 Water-filled manometer used to measure air pressure.

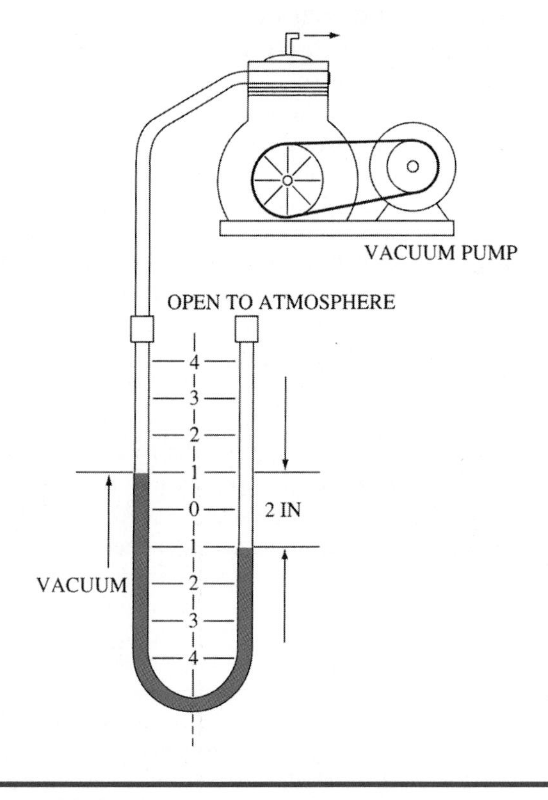

Figure 4-3-6 Mercury-filled manometer measuring vacuum pressure.

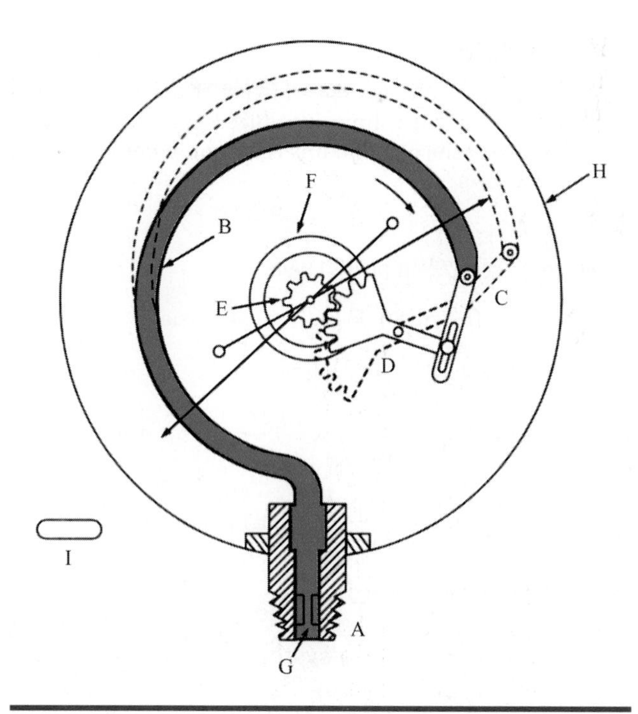

Figure 4-3-7 Internal construction of a pressure gauge: (a) Adapter fitting, usually a ⅛ in pipe thread; (b) Bourdon tube; (c) Link; (d) Gear sector; (e) Pinter shaft gear; (f) Calibrating spring; (g) Restricter; (h) Case; (i) Cross section of the Bourdon tube. The dashed lines indicate how the pressure in the Bourdon tube causes it to straighten and operate the gauge.

14.7 psi. Any additional pressure applied when the gauge is connected to a piece of equipment will tend to straighten out the Bourdon tube, thereby moving the needle or pointer and its mechanical linkage, thus indicating the amount of pressure being applied.

Pressures below atmospheric are customarily expressed in units of in Hg. There is an indication of the range between 0 gauge and 30 in Hg on the compound gauge.

SAFETY TIP

Mercury is considered to be a hazardous chemical and it must be disposed of properly when an instrument containing mercury is to be disposed of. Mercury can contaminate an area resulting in an expensive cleanup operation if it is carelessly discarded on the floor of an office, school, hospital, or any other public area. Mercury vapors have associated with a number of health issues, therefore, any time mercury is spilled you must notify the proper authorities so that its potential threat can be determined and an appropriate method of removing the contamination can be done. Be very careful when instruments are used that contain mercury to avoid any accidental spillage.

Figure 4-3-8 Refrigeration gauge with temperature scales.

Absolute Pressure

Table 4-3-3 shows definite relationship among absolute pressure, atmospheric pressure, and gauge pressure. For many problems, atmospheric pressure does not need to be considered, so the customary pressure gauge is calibrated and graduated to read zero under normal atmospheric conditions. Yet when gases are contained within an enclosure away from the atmosphere, such as in a refrigeration unit, it is necessary to take atmospheric pressure into consideration, and mathematical calculations must be in terms of the absolute pressures involved.

TABLE 4-3-3 Relationship between Absolute and Gauge Pressure

Gauge Pressure	Absolute Pressure
Above Atmospheric Pressure	
40 psig	54.7 psia
30 psig	44.7 psia
20 psig	34.7 psia
10 psig	24.7 psia
Atmospheric Pressure	
0 psig	14.7 psia
Below Atmospheric Pressure	
10 in Hg	9.7 psia
20 in Hg	4.7 psia
30 in Hg	0.0 psia

PRESSURE OF GAS

The volume of a gas is affected by a change in either the pressure or temperature, or both. There are laws that govern the mathematical calculations in computing these variables.

Boyle's law states that the volume of a gas varies inversely as its pressure if the temperature of the gas remains constant. This means that the product of the pressure times the volume remains constant, or that if the pressure of a gas doubles, the new volume will be one-half of the original volume. Or it may be considered that, if the volume is doubled, the absolute pressure will be reduced to one-half of what it was originally.

This concept may be expressed as

$$p_1V_1 = p_2V_2 \qquad (4\text{-}3\text{-}7)$$

where

p_1 = original pressure
V_1 = original volume
p_2 = new pressure
V_2 = new volume

It must be remembered that p_1 and p_2 have to be expressed in the absolute pressure terms for Eq. 4-3-7 to be used correctly.

EXAMPLE

If the gauge pressure on 2 ft^3 of gas is increased from 20 psig to 50 psig while the temperature of the vapor remains constant, what will be the new volume?

Solution

Since

$$p_1V_1 = p_2V_2$$

then

$$V_2 = \frac{p_1V_1}{p_2}$$

First, calculate p_1 and p_2 in terms of absolute pressure (psia).

$$\begin{aligned} p_1 &= 20 + 14.7 \\ &= 34.7 \text{ psia} \\ p_2 &= 50 + 14.7 \\ &= 64.7 \text{ psia} \end{aligned}$$

Next, substitute the values of p_1 and p_2 into the new form of the equation.

$$\begin{aligned} V_1 &= 2 \text{ ft}^3 \\ V_2 &= ? \\ V_2 &= \frac{34.7 \times 2}{64.7} = 1.072 \text{ ft}^3 \end{aligned}$$

The same basic equation may be used to determine the given new pressure. ■

EXAMPLE

If additional pressure is applied to a volume of 2 ft^3 of gas at 20 psig so that the volume is reduced to 1.072 ft^3 and the temperature of the gas remains constant, what is the new pressure in psig?

Solution

First, convert p_1 from psig to psia.

$$\begin{aligned} p_1 &= (20 + 14.7) = 34.7 \text{ psia} \\ p_2 &= \frac{p_1V_1}{V_2} \\ V_1 &= 2 \text{ ft}^3 \\ V_2 &= 1.072 \text{ ft}^3 \\ p_2 &= \frac{34.7 \times 2}{1.072} = 64.7 \text{ psia} \end{aligned}$$

Now convert psia to psig.

$$p_2 = 64.7 - 14.7 = 50 \text{ psig}$$

■

EXPANSION OF GAS

Most gases will expand in volume at practically the same rate with an increase in temperature, provided that the pressure does not change. If the gas is confined so that its volume will remain the same, the pressure in the container will increase at about the same rate as an increase in temperature.

Theoretically, if the pressure remains constant, a gas vapor will expand or contract at the rate of 1/492 for each degree of temperature change. The result of this theory would be a zero volume at a temperature of −460°F, or at 0° absolute.

Charles' law states that the volume of a gas is in direct proportion to its absolute temperature, provided that the pressure is kept constant; and the absolute pressure of a gas is in direct proportion to its absolute temperature, provided that the volume is kept constant. That is,

$$\frac{V_1}{V_2} = \frac{T_1}{T_2} \qquad (4\text{-}3\text{-}8)$$

and

$$\frac{p_1}{p_2} = \frac{T_1}{T_2} \qquad (4\text{-}3\text{-}9)$$

where

T = absolute temperature
p = absolute pressure

To clear the fractions, these may also be expressed as

$$V_1T_2 = V_2T_1 \quad \text{and} \quad p_1T_2 = p_2T_1 \qquad (4\text{-}3\text{-}10)$$

EXAMPLE

If the temperature of 2 ft^3 of gas is increased from 40°F to 120°F, what would be the new volume if there was no change in pressure?

Solution

$$V_2 = \frac{V_1 T_2}{T_1} = \frac{2 \times (120 + 460)}{40 + 460}$$

$$= \frac{1160}{500} = 2.32 \text{ ft}^3$$

EXAMPLE

If a container holds 2 ft^3 of gas at 20 psig, what will be the new pressure in psig if the temperature is increased from 40°F to 120°F?

Solution

$$p_2 = \frac{p_1 T_2}{T_1} = \frac{(20 + 14.7) \times (120 + 460)}{40 + 460}$$

$$= 40.25 \text{ psig}$$

$$40.25 - 14.7 = 25.55 \text{ psig}$$

In numerous cases dealing with refrigerant vapor, none of the three possible variables remains constant, and a combination of these laws must be utilized, namely the general law of perfect gas:

$$\frac{p_1 V_1}{T_1} = \frac{p_2 V_2}{T_2} \qquad (4\text{-}3\text{-}11)$$

or

$$p_1 V_1 T_2 = p_2 V_2 T_1 \qquad (4\text{-}3\text{-}12)$$

in which the units of p and T are always used in the absolute.

EXAMPLE

If a volume of 4 ft^3 of gas at a temperature of 70°F and at atmospheric pressure is compressed to one-half its original volume and increased in temperature to 120°F, what will be its new pressure?

Solution

Transposing Eq. (4-3-12) $p_1 V_1 T_2 = p_2 V_2 T_1$, we have

$$p_2 = \frac{p_1 V_1 T_2}{V_2 T_1}$$

Therefore,

$$p_2 = \frac{14.7 \times 4(120 + 460)}{2 \times (70 + 460)} = \frac{34{,}104}{1{,}060} = 32.17 \text{ psia}$$

$$32.17 - 14.7 = 17.47 \text{ psig}$$

BOILING POINT

The most important point to understand when dealing with the action in a refrigeration system is the boiling point of the liquid (refrigerant) in the system. Lowering the boiling point causes the refrigerant to absorb heat and vaporize or boil. Conversely by raising the boiling point the vapor gives up the heat and condenses. Basically, the refrigeration system operates by control of the boiling point.

Earlier in this text, boiling point was defined as the temperature at which a liquid turns from a liquid to a vapor or condenses from a vapor to a liquid depending on the absorption or rejection of heat energy.

The chart used was based on using water at standard atmospheric pressure of 29.92 in Hg and 70°F. At these conditions water will boil at 212°F or 100°C with the addition of heat energy, or condense at this same temperature with the removal of heat energy.

SERVICE TIP

Not all boiling water is hot. As a vacuum is pulled the boiling temperature drops. In fact the boiling temperature can be dropped so low that ice will actually form as a result of a vacuum being pulled on the container of water. Conversely, the boiling temperature of water can be significantly increased as a result of pressure. Food cooked in a pressure cooker with 15 or 30 lb of pressure will cook much faster than the same food cooked in an open pot at atmospheric pressure. Water at atmospheric pressure cannot get higher than 212°F. But by increasing the pressure, the water in the pressure cooker can get significantly hotter before reaching its new higher boiling point.

When referring to boiling point, the pressure that the liquid is subjected to must also be considered. When referring to the boiling point of water as 212°F or 100°C, the assumption is made that the water is subjected to standard barometric pressure of 29.92 in Hg.

In reality, the boiling point of a liquid will change in the same direction as the pressure to which the liquid is subjected. This is a very important basic law of physics that must be remembered. In a later chapter this law will be applied to various refrigerants, but for our discussion in this chapter water will be used as the refrigerant. Figure 4-3-9 gives examples of the boiling point of water at sea level (212°F) and at 14,100 ft above sea level (167°F). Obviously, this is because the boiling point of water drops as the atmospheric pressure drops. Again, the boiling point of

Figure 4-3-9 Boiling point of water at different elevations.

a liquid will vary in the same direction as the pressure to which the liquid is subjected.

If the boiling point of water is determined at various pressures, both above (pressure) and below atmospheric pressure (vacuum), and the temperatures are plotted on a graph, the results would be as pictured in Figure 4-3-10. Here we can see that the boiling point of water can be raised to 276°F with a pressure of 30 psig or lowered to 40°F at a pressure of 29.67233 in Hg. Therefore, to obtain a desired boiling point of water it is only necessary to maintain an equivalent pressure based on the pressure temperature curve of the liquid.

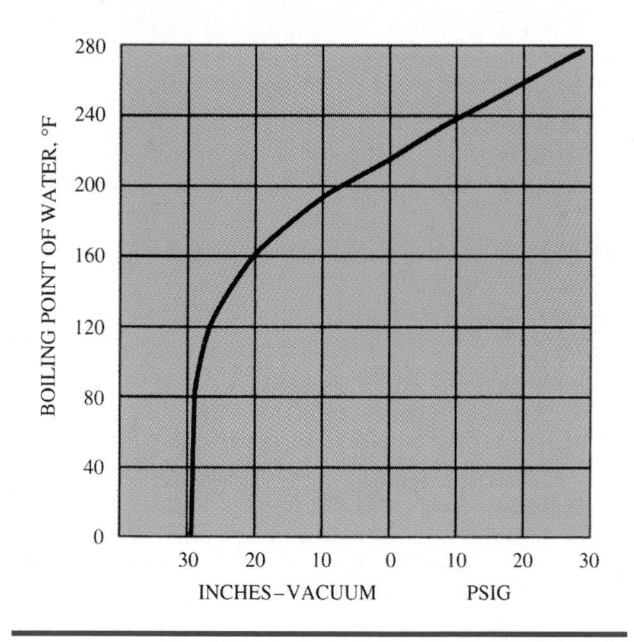

Figure 4-3-10 Pressure-temperature curve for water.

CONDENSING TEMPERATURE

With the liquid in a vapor state, to remain as a vapor, the sensible temperature must be higher than the condensing temperature. If heat energy is removed from the vapor to the point where the sensible temperature would fall below the condensing temperature of the vapor, the vapor will liquefy or condense.

The boiling point and the condensing temperature for liquid are the same temperature. The only difference is in the action taking place. Boiling point implies liquid to vapor; condensing temperature implies vapor to liquid. In each case, the temperature at which the change takes place varies as the pressure to which the liquid is subjected. Lowering the pressure lowers the boiling point or condensing temperature. Raising the pressure raises the boiling point or condensing temperature.

FUSION POINT

The fusion point (temperature at which solidification of liquid or melting of a solid takes place) is also affected by the pressure to which the solid is subjected. If the fusion point of a solid is raised by increasing the pressure on the solid until the fusion point is equal to the sensible temperature of the solid, any further attempt to raise the fusion point will cause the solid to liquefy.

This principle is what allows glaciers to move downhill. When the weight of the snow and ice in the glacier is high enough to raise the fusion or melting point of the base of the ice up to the sensible temperature of the ice, any further rise in pressure will cause the base of the ice to liquefy, and the glacier moves on a layer of water.

Control of the fusion point is of little importance in refrigeration, as it is usually an undesirable result of temperature maintenance. In air conditioning, formation of ice is a definite detriment and must be eliminated.

SATURATION TEMPERATURE

In the section on boiling point, the statement was made that the boiling point and condensing temperature of a liquid at a given pressure are the same. This means that the liquid has reached the point where it contains all the heat energy it can without changing to a vapor. This condition is described by referring to it as a *saturated liquid*. This means that if any more heat energy is added, the liquid will boil.

Commonly, if the vapor is cooled to a point where the vapor is so dense that any further reduction in heat energy causes it to condense to a liquid, the condition is referred to as a *saturated vapor*.

In the sections where sensible heat and latent heat were discussed, sensible heat was said to change temperature and latent heat was said to change state. Therefore, liquid at the boiling point is saturated with sensible heat and any heat added would be latent heat to vaporize the liquid. Vapor at the condensing temperature has been reduced to that temperature by removing the sensible heat until the density of the vapor is to a point where any further removal of heat will cause condensation of the vapor and removal of the latent heat of vaporization. At this point the vapor is said to be a saturated vapor.

SUPERHEAT

Superheat is the heat added to a vapor after it becomes a vapor, a simple rise in temperature of the vapor above the boiling point. If, for example, water were to boil at 212°F and before leaving the passages in the boiler it were to take on more heat and the steam temperature would rise to 220°F, the steam would be superheated 8°F, the difference between the boiling point of the liquid and the actual physical (sensible) temperature of the vapor. This is called *superheated steam*. To remove the heat and condense the steam to a liquid, the first action required is to de-superheat the steam to the saturation point (condensing temperature) and then remove the latent heat of vaporization to produce the liquid.

Superheat is very important in refrigeration systems to produce maximum system capacity. Superheat is also critical in prolonging equipment life, and will be referred to throughout the chapters in *Refrigeration and Air Conditioning*.

SUBCOOLING

When a liquid is at a sensible temperature below its boiling point, it is said to be *subcooled*. For example, water at standard atmospheric conditions with a sensible temperature of 70°F will be subcooled 142°F (212°F − 70°F = 142°F subcooled). The subcooling of the liquid in the refrigeration or air conditioning system is important for maximum capacity and efficiency. This will be discussed further in later chapters.

UNIT 3—REVIEW QUESTIONS

1. What is the definition of a fluid?
2. Fluid pressure is usually expressed in terms of ________.
3. Define Pascal's law.
4. ________ is the weight per unit volume of a substance.
5. How is the specific volume of a substance usually expressed?
6. What is a manometer?
7. What does Boyle's law state?
8. What does Charles' law state?
9. The most important point to understand when dealing with the action in a refrigeration system is the ________.
10. Define fusion point.
11. ________ is the heat added to a vapor after it becomes vapor.
12. When a temperature is at a sensible temperature below its boiling point, it is said to be ________.

TECH TALK

Tech 1. Superheat and subcooling confuse me.
Tech 2. Why don't you start with superheat?
Tech 1. Well, when I feel tubing holding a refrigerant vapor that is superheated it still feels cold.
Tech 2. Oh, you mean the refrigerant tubing suction line. The suction line holds a superheated gas. That doesn't mean the gas is hot. In fact, refrigerant lines can be well below 0°F. What superheated means is that the gas has been heated to the point where there is no liquid at all remaining.
Tech 1. Well, okay, but what about subcooling? When I brush against a hot water pipe that is at a subcooled temperature, it is still hot enough to burn me.
Tech 2. You're right. On subcooling, that line can be hot. What subcooling means is not that it's cold, but there is no vapor remaining in the line. Just liquid.
Tech 1. So why don't they call it something different than superheat and subcooling if it is not hot and cold?
Tech 2. Well, those terms have been around a long time, long before air conditioning. They refer to the physical state of the vapor or liquid.
Tech 1. I see. But it would make it a lot easier for technicians in the field to understand if a line that was subcooled was not so hot or a line that was superheated was not so cold.
Tech 2. Just remember the phrase subcooled liquid, superheated vapor.
Tech 1. I think I can remember it that way.

UNIT 4

Refrigeration Cycle

OBJECTIVES: In this unit the reader will learn about the refrigeration cycle including:

- Refrigeration Effect
- Temperature-Heat Diagrams
- Cycle Diagrams
- Refrigeration Process
- Coefficient of Performance
- Effects on Capacity

REFRIGERATION EFFECT MEASUREMENT

As mentioned in Chapter 1 Unit 2, a ton of refrigeration is the common term used to define and measure refrigeration capacity. It is the amount of heat absorbed in melting a ton of ice (2000 lb) over a 24 hr period.

One ton of refrigeration is equal to 12,000 Btu/hr. This may be calculated by multiplying the weight of ice (2000 lb) by the latent heat of fusion (melting) of ice (144 Btu/lb). Thus

$$2000 \text{ lb} \times 144 \text{ Btu/lb} = 288{,}000 \text{ Btu}$$

in 24 hr or 12,000 Btu/hr (288,000/24). Therefore, 1 ton of refrigeration equals 12,000 Btu/hr.

A 10 ton refrigeration system will have a capacity of $10 \times 12{,}000$ Btu/hr $= 120{,}000$ Btu/hr.

TEMPERATURE HEAT DIAGRAMS

Figure 4-4-1 shows a temperature-heat diagram for water in both the IP and SI units. Referring first to the IP diagram: *A–B* represents the sensible heat added to ice at −40°F to raise it to the melting temperature of 32°F.

$$\text{Btu} = 0.504 \text{ (specific heat)} \times 72°\text{F}(\Delta T) = 36.3$$

B–C represents the latent heat required to melt the ice.

$$\text{Btu} = 144$$

C–D represents the sensible heat added to the melted water at 32°F to raise it to steam temperature at 212°F.

$$\begin{aligned}\text{Btu} &= 1 \text{ (specific heat of water)} \times 180°(\Delta T)\\ &= 180 \text{ Btu}\end{aligned}$$

D–E represents the latent heat required to change 212°F water to steam at 212°F.

$$\text{Btu} = 970$$

E^{+} represents superheated steam.

Referring to the SI system diagram:

A–B represents the heat added to ice at −40°C to raise it to 0°C.

$$\text{kJ} = 2 \text{ kJ/kg (specific heat of ice)} \times 40(\Delta T) = 80$$

B–C is the latent heat added to melt ice at 0°C.

$$\text{kJ} = 335$$

C–D represents the sensible heat required to raise water at 0°C to the steaming temperature of 100°C.

$$\text{kJ} = 4.2 \text{ kJ/kg (specific heat)} \times 100\ (\Delta T) = 420 \text{ kJ}$$

D–E represents the latent heat required to change water to steam at 100°C.

$$\text{kJ} = 2260$$

E^{+} represents superheated steam.

REFRIGERATION EFFECT

If a specific job is to be done in a refrigeration system or cycle, each pound of refrigerant circulating in the system must do its share of the work. It must absorb an amount of heat in the evaporator or cooling coil, and it must dissipate this heat—plus some that is added in the compressor—through the condenser, whether air cooled, water cooled, or evaporative cooled. The work done by each pound of the refrigerant as it goes through the evaporator is reflected by the amount of heat it picks up from the refrigeration load, chiefly when the refrigerant undergoes a change of state from a liquid to a vapor.

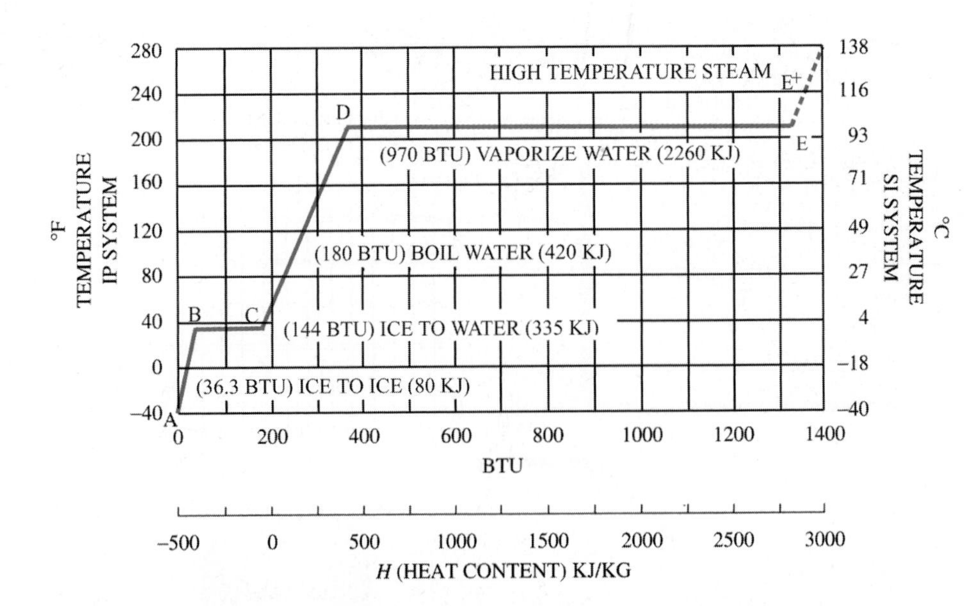

Figure 4-4-1 Temperature/heat diagram for one pound (kg) of water at atmospheric pressure, −40° to vaporization, in both IP and SI systems.

NICE TO KNOW

The refrigerant cycle where the refrigerant absorbs heat in the evaporator and squeezes the heat out in the condenser can be compared to a process of using a sponge to remove water from a bucket. When the sponge is placed in a bucket it absorbs water like the refrigerant absorbs heat in the evaporator. When the sponge is removed and squeezed the water drains out like the refrigerant releasing its heat in the condenser. The sponge then can be returned to the bucket to pick up more water in the same way that the refrigerant returns to the evaporator to pick up more heat as the cycle continues.

As mentioned previously, for a liquid to be able to change to a vapor, heat must be added to or absorbed in it. This is what happens in the cooling coil. The refrigerant enters the metering device as a liquid and passes through the device into the evaporator, where it absorbs heat as it evaporates into a vapor. As a vapor, it makes its way through the suction tube or pipe to the compressor. Here it is compressed from a low temperature, low pressure vapor to a high temperature, high pressure vapor; then it passes through the high pressure or discharge pipe to the condenser, where it undergoes another change of state—from a vapor to a liquid—in which state it flows out into the liquid pipe and again makes its way to the metering device for another trip through the evaporator. A schematic of a simple refrigeration cycle is shown in Figure 4-4-2.

When the refrigerant, as a liquid, leaves the condenser it may go to a receiver until it is needed in the evaporator; or it may go directly into the liquid line to the metering device and then into the evaporator coil. The liquid entering the metering device just ahead of the evaporator coil will have a certain heat content (enthalpy), which is dependent on its temperature when it enters the coil, as shown in the refrigerant tables in the Appendix. The vapor leaving the evaporator will also have a given heat content (enthalpy) according to its temperature, as shown in the refrigerant tables.

The difference between these two amounts of heat content is the amount of work being done by each pound of refrigerant as it passes through the evaporator and picks up heat. The amount of heat absorbed by each pound of refrigerant is known as the refrigerating effect of the system, or of the refrigerant within the system.

This refrigerating effect is rated in Btu per pound of refrigerant (Btu/lb); if the total heat load is known (given in Btu/hr), we can find the total number of pounds of refrigerant that must be circulated each hour of operation of the system. This figure can be broken down further to the amount that must be circulated each minute, by dividing the amount circulated per hour by 60.

EXAMPLE

If the total heat to be removed from the load is 60,000 Btu/hr and the refrigerating effect in the evaporator amounts to 50 Btu/lb, then:

$$\frac{60{,}000\ \text{Btu/hr}}{50\ \text{Btu/lb}} = 1200\ \text{lb/hr} \quad \text{or} \quad 20\ \text{lb/min} \quad ■$$

Since 12,000 Btu/hour equals the rate of 1 ton of refrigeration, the 60,000 Btu/hr in the example above amounts to 5 tons of refrigeration, and the 20 lb of refrigerant that must be circulated each

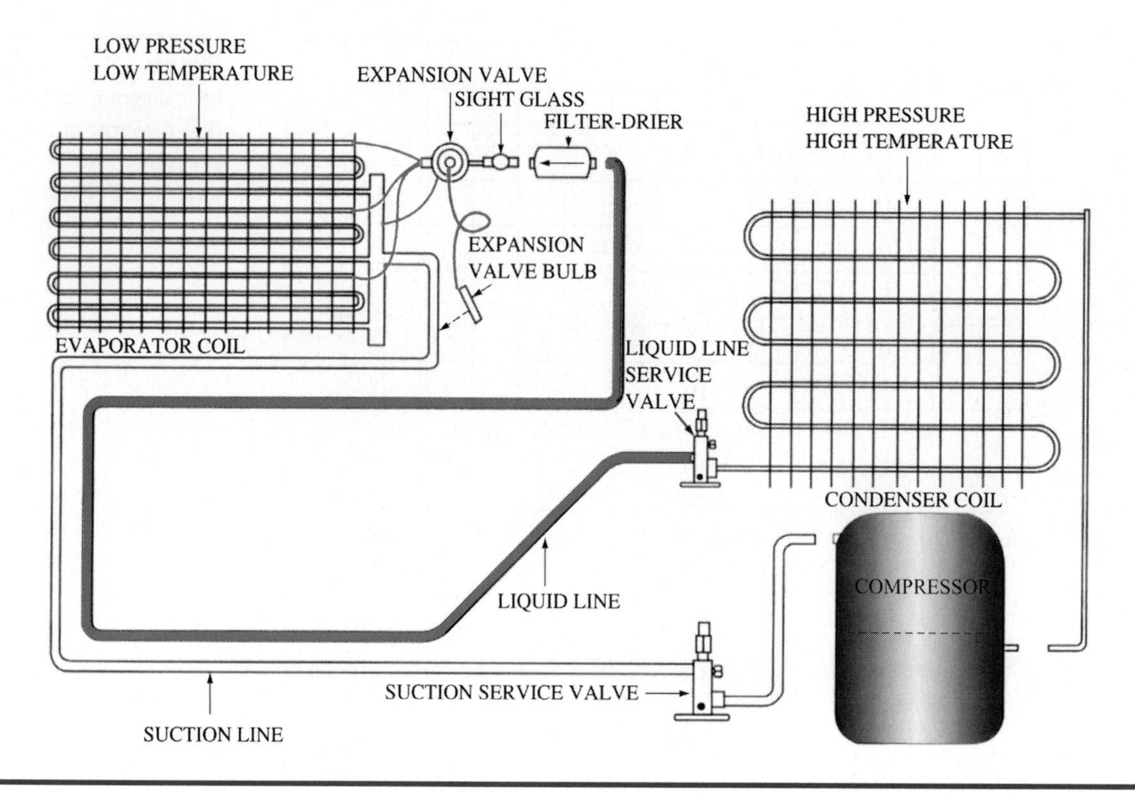

Figure 4-4-2 Schematic diagram of a simple refrigeration cycle.

minute is the equivalent of 4 lb/min/ton of refrigeration. (One ton of refrigeration for 24 hr equals 288,000 Btu).

In this example, where 20 lb of refrigerant having a refrigerating effect of 50 Btu/lb is required to take care of the specified load of 60,000 Btu/hr, the results can also be obtained in another manner. As mentioned previously, it takes 12,000 btu's per hour to equal 1 ton of refrigeration, which is equal to 200 Btu/min/ton.

Therefore, 200 Btu/min, when divided by the refrigerating effect of 50 Btu/lb, amounts to 4 lb/min. This computation can be shown by the equation

$$W = \frac{200}{\text{NRE}} \tag{4-4-1}$$

where

W = weight of refrigerant circulated per minute, lb/min
200 = 200 Btu/min—the equivalent of 1 ton of refrigeration
NRE = net refrigerating effect, Btu/lb of refrigerant

Because of the small orifice in the metering device, a fact that will be discussed more thoroughly in a later chapter, when the compressed refrigerant passes from the smaller opening in the metering device to the larger tubing in the evaporator, a change in pressure occurs together with a change in temperature. This change in temperature occurs because of the vaporization of a small portion of the refrigerant (about 20%) and, in the process of this vaporization, the heat that is involved is taken from the remainder of the refrigerant.

From the table of saturated R-22 in Figure 4-4-3, it can be seen that the heat content of 100°F liquid is 39.27 Btu/lb and that of 40°F liquid is 21.42 Btu/lb; this indicates that 17.85 Btu/lb has to be removed from each pound of refrigerant entering the evaporator. The latent heat of vaporization of 40°F R-22 (from the Appendix tables) is 86.72 Btu/lb, and the difference between this amount and that which is given up by each pound of refrigerant when its liquid temperature is lowered from 100°F to 40°F (17.85 Btu/lb) is 68.87 Btu/lb. This is another method of calculating the refrigerating effect—or work being done—by each pound of refrigerant under the conditions given.

The capacity of the compressor must be such that it will remove from the evaporator that amount of refrigerant which has vaporized in the evaporator and in the metering device in order to get the necessary work done. The compressor must be able to remove and send on to the condenser the same weight of refrigerant vapor, so that it can be condensed back into a liquid and so continue in the refrigeration circuit or cycle to perform additional work.

If the compressor is unable to move this weight, some of the vapor will remain in the evaporator. This, in turn, will cause an increase in pressure inside the

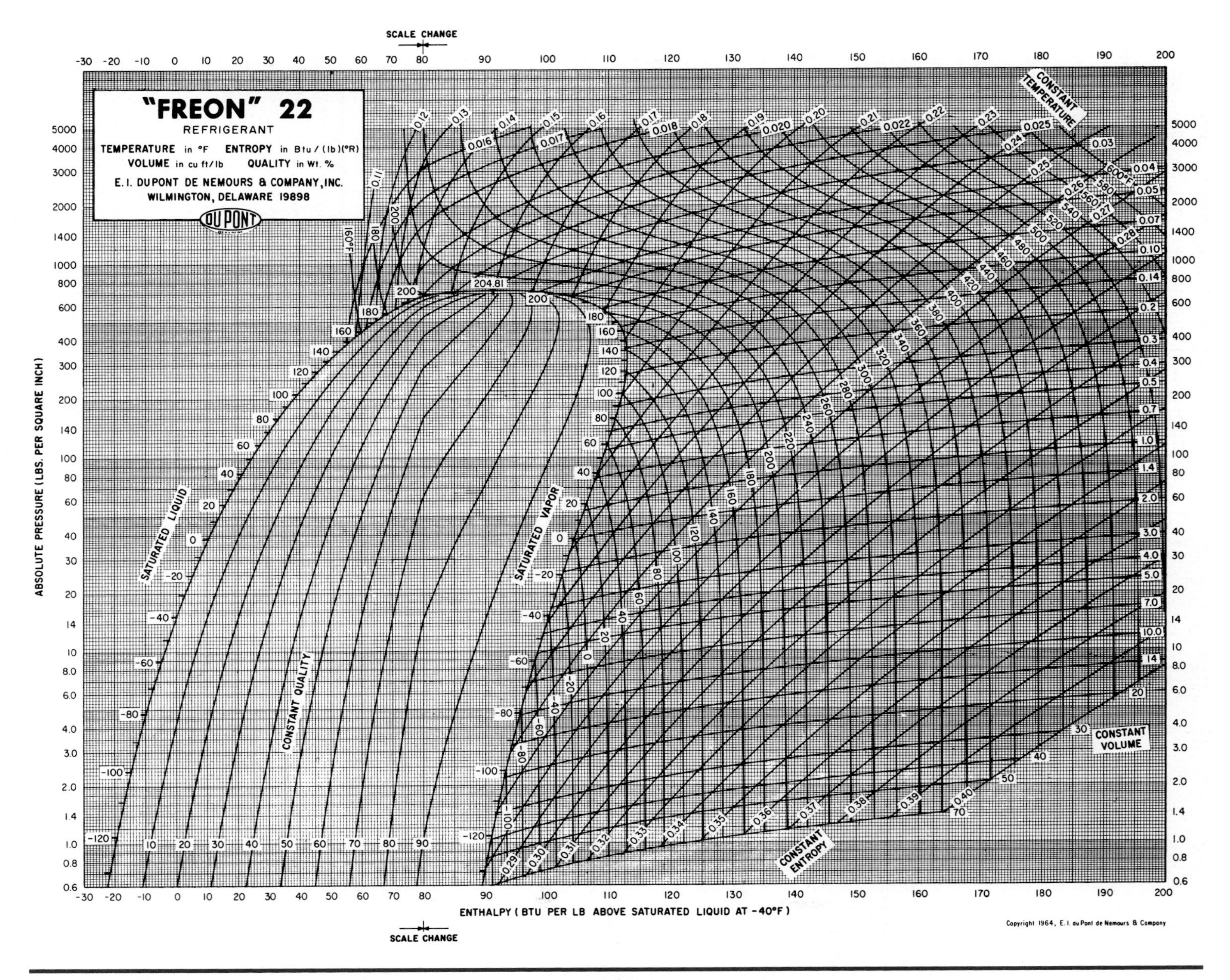

Figure 4-4-3 IP pressure-heat diagram for R-22. (*Courtesy DuPont Chemicals*)

evaporator, accompanied by an increase in temperature and a decrease in the work being done by the refrigerant, and design conditions within the refrigerated space cannot be maintained.

SERVICE TIP

The ability of a compressor to remove heat is based on its ability to move refrigerant. As the evaporator pressure drops, less weight of refrigerant enters the piston and less heat is removed. As the evaporator pressure increases, the refrigerant becomes more dense and a greater weight of refrigerant enters the piston chamber each time, therefore more heat can be removed. The ability to remove heat therefore is significantly affected by the pressure in the evaporator. Likewise as the pressure in the condenser goes up it takes more force to move the refrigerant from the compressor into the condenser. As that pressure increases less refrigerant is moved and less heat is removed. Therefore the most efficient refrigeration system would be one that would have the highest possible evaporator temperature and the lowest possible condenser temperature.

A compressor that is too large will withdraw the refrigerant from the evaporator too rapidly, causing a lowering of the temperature inside the evaporator, so that design conditions will not be maintained in this situation either.

In order for design conditions to be maintained within a refrigeration circuit, there must be a balance between the requirements of the evaporator coil and the capacity of the compressor. This capacity is dependent on its displacement and on its volumetric efficiency. The measured displacement of a compressor depends on the number of cylinders, their bore and stroke, and the speed at which the compressor is turning. Volumetric efficiency depends on the absolute suction and discharge pressures under which the compressor is operating. A thorough and elaborate presentation of these facts concerning displacement, as well as the variables pertaining to volumetric efficiency, will be offered in a later chapter, together with equations and other data.

CYCLE DIAGRAMS

Refer to Figure 4-4-2 for a schematic flow diagram of a basic cycle in refrigeration, denoting changes in phases or processes. First the refrigerant passes from the liquid stage into the vapor stage as it absorbs heat in the evaporator coil. The compression stage, where the refrigerant vapor is increased in temperature and pressure, comes next; then the refrigerant gives off its heat in the condenser to the ambient cooling medium, and the refrigerant vapor condenses back to its liquid state where it is ready for use again in the cycle.

Figure 4-4-3 is a pressure temperature diagram for R-22, which shows the pressure, heat, and temperature characteristics of this refrigerant. Enthalpy is another word for heat content. Diagrams such as Figure 4-4-3 are referred to as pressure-enthalpy diagrams. Pressure-enthalpy diagrams may be utilized for the plotting of the cycle shown in Figure 4-4-3, but a basic or skeleton chart as shown in Figure 4-4-4 might be used as a preliminary illustration of the various phases of the refrigerant circuit. There are three basic areas on the chart denoting changes in state between the saturated liquid line and saturated vapor line in the center of the chart. The area to the left of the saturated liquid line is the subcooled area, where the refrigerant liquid has been cooled below the boiling temperature corresponding to its pressure; whereas the area to the right of the saturated vapor line is the area of superheat, where the refrigerant vapor has been heated beyond the vaporization temperature corresponding to its pressure.

The construction of the diagram, or rather a knowledge and understanding of it, may bring about a clearer idea of what happens to the refrigerant at the various stages within the refrigeration cycle. If the liquid vapor state and any two properties of a refrigerant are known and this point can be located on the chart, the other properties can easily be determined from the chart.

If the point is situated anywhere between the saturated liquid and vapor lines, the refrigerant will be in the form of a mixture of liquid and vapor. If the location is closer to the saturated liquid line, the mixture will be more liquid than vapor, and a point located in the center of the area at a particular pressure would indicate a 50% liquid 50% vapor situation.

NOTE: As the percentage of liquid refrigerant in the evaporator decreases, the evaporator temperature and pressure remain relatively constant. However, the amount of heat entering the refrigerant is significant. The ability of the refrigerant to remain at a relatively constant temperature and pressure as it absorbs the heat helps ensure that the evaporator coil surface remains at a relatively constant temperature. By remaining at a relatively constant temperature the entire coil surface becomes an active part of the heat removal process. If one area of the coil became significantly colder than another area it would then be trying to remove more heat. This would result in a significant decrease in coil efficiency. To better ensure that the coil stays at more uniform temperature, multiple passes are designed into the evaporator coil so that the refrigerant enters the coil in several locations.

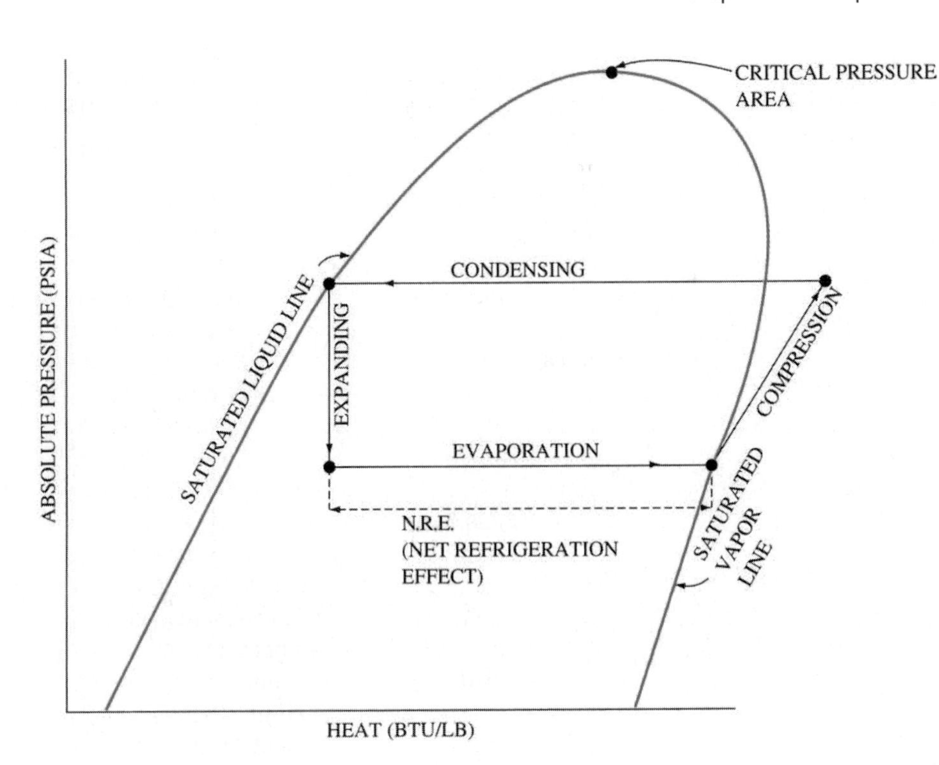

Figure 4-4-4 Pressure-enthalpy changes through a refrigeration cycle.

Referring to Figure 4-4-4, the change in state from a vapor to a liquid—the condensing process—occurs as the path of the cycle develops from right to left; whereas the change in state from a liquid to a vapor—the evaporating process—travels from left to right. Absolute pressure is indicated on the vertical axis at the left, and the horizontal axis indicates heat content, or enthalpy, in Btu/lb.

The distance between the two saturated lines at a given pressure, as indicated on the heat content line, amounts to the latent heat of vaporization of the refrigerant at the given absolute pressure. The distance between the two lines of saturation is not the same at all pressures, for they do not follow parallel curves. Therefore, there are variations in the latent heat of vaporization of the refrigerant, depending on the absolute pressure. There are also variations in pressure-enthalpy charts of different refrigerants and the variations depend on the various properties of the individual refrigerants.

REFRIGERATION PROCESSES

Based on the examples presented earlier in the chapter, it will be assumed that there will be little or no changes in the temperature of the condensed refrigeration liquid after it leaves the condenser and travels through the liquid pipe on its way to the expansion or metering device, or in the temperature of the refrigerant vapor after it leaves the evaporator and passes through the suction pipe to the compressor.

Figure 4-4-5 shows the phases of the simple saturated cycle with appropriate labeling of pressures, temperatures, and heat content or enthalpy. A starting point must be chosen in the refrigerant cycle; we will start on point *A* on the saturated liquid line where all of the refrigerant vapor at 100°F has condensed into liquid at 100°F and is at the inlet to the metering device. What occurs between points *A* and *B* is the expansion process as the refrigerant passes through the metering device; and the refrigerant temperature is lowered from the condensation temperature of 100°F to the evaporating temperature of 40°F.

When the vertical line *A–B* (the expansion process) is extended downward to the bottom axis, a reading of 39.27 Btu/lb is indicated, which is the heat content of 100°F liquid. To the left of point *B* at the saturated liquid line is point *Z*, which is also at the 40°F temperature line. Taking a vertical path downward from point *Z* to the heat content line, a reading of 21.42 Btu/lb is indicated, which is the heat content of 40°F liquid; this area between points *Z* and *B* is covered later in the chapter.

The horizontal line between points *B* and *C* indicates the vaporization process in the evaporator, where the 40°F liquid absorbs enough heat to completely vaporize the refrigerant. Point *C* is at the saturated vapor line, indicating that the refrigerant has completely vaporized and is ready for the compression process. A line drawn vertically downward to where it joins the enthalpy line indicates that the heat content, shown at h_c is 108.14 Btu/lb, and the difference between h_a and h_c is 68.87 Btu/lb, which is the refrigerating effect, as shown in an earlier example.

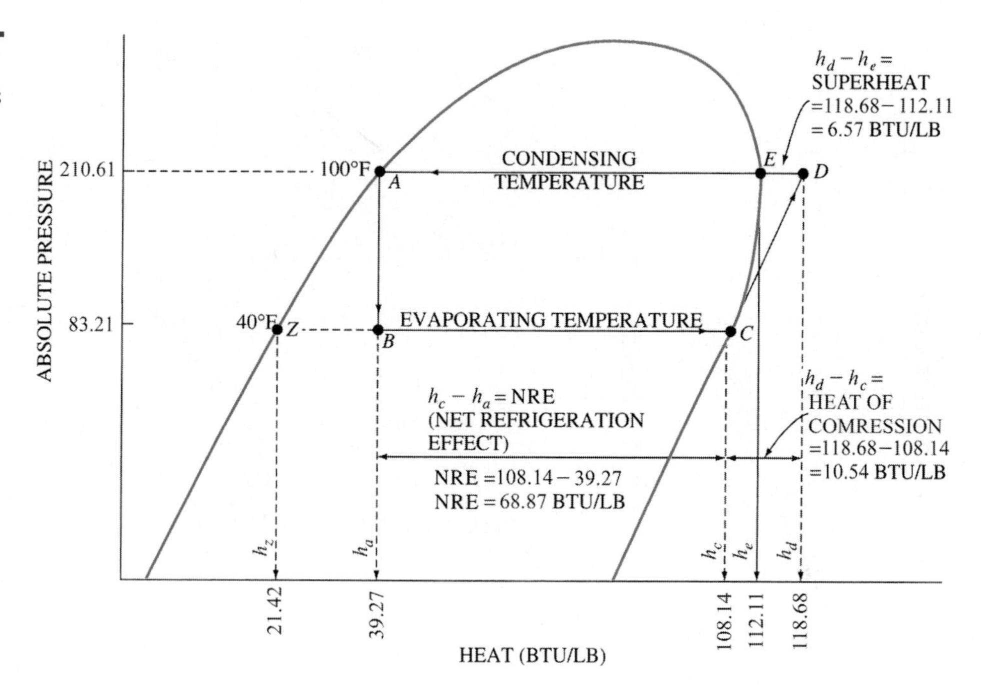

Figure 4-4-5 Pressure, heat, and temperature values for a refrigeration cycle operating with a 40°F evaporator.

The difference between points h_z and h_c on the enthalpy line amounts to 86.72 Btu/lb, which is the *latent heat of vaporization* of 1 lb of R-22 at 40°F. This amount would also exhibit the refrigerating effect, but some of the refrigerant at 100°F must evaporate or vaporize in order that the remaining portion of each pound of R-22 can be lowered in temperature from 100°F to 40°F.

The various properties of refrigerants will be elaborated on before we proceed with the discussion of the compression process. All refrigerants exhibit certain properties when in a gaseous state; some of them are: *volume, temperature, pressure, enthalpy* or *heat content,* and *entropy*. The last property—entropy—is really the most difficult to describe or define. In physics, entropy is defined as the degree of disorder of the molecules that make up. In refrigeration, we define entropy as the ratio of the heat content of the gas to its absolute temperature in degrees Rankine.

The pressure-enthalpy chart plots the line of constant entropy, which stays the same provided that the gas is compressed and no outside heat is added or taken away. When the entropy is constant, the compression process is called *adiabatic,* which means that the gas changes its condition without the absorption or rejection of heat either from or to an external body or source. It is common practice, in the study of cycles of refrigeration, to plot the compression line either along or parallel to a line of constant entropy.

In Figure 4-4-6, line *C–D* denotes the compression process, in which the pressure and temperature of the vapor are increased from that in the evaporator to that in the condenser, with the assumption that there has been no pickup of heat in the suction line between the evaporator and the compressor. For a condensing temperature of 100°F, a pressure gauge would read approximately 196 psig; but the chart is rated in absolute pressure and the atmospheric pressure of 14.7 must be added to the psig, making it actually 210.61 psia.

SERVICE TIP

Understanding the relationship between the various lines on the pressure-enthalpy diagram will help you understand what is happening in the refrigerant cycle. You will then be better able to visualize what is happening with the pressure, temperature, and heat. It is important that you have a good understanding of these concepts. When you are in the field trying to troubleshoot a refrigerant circuit, if you find that you have an unusually low temperature you could therefore check out whether or not the evaporator was clean and was picking up heat or if the system was low in refrigerant. You would not assume that the compressor was not working because from your understanding of the chart you would know that if the compressor is not drawing the vapor out of the evaporator at a fast enough rate the evaporator pressure goes up. Therefore a low suction pressure is not an indication of a bad compressor. In fact the compressor is working so well that the heat can't come into the refrigerant fast enough.

Point *D* on the absolute pressure line is equivalent to the 100°F condensing temperature; it is not on the saturated vapor line, it is to the right in the superheat area, at a junction of the 210.61 psia line, the line of constant entropy of 40°F, and the temperature

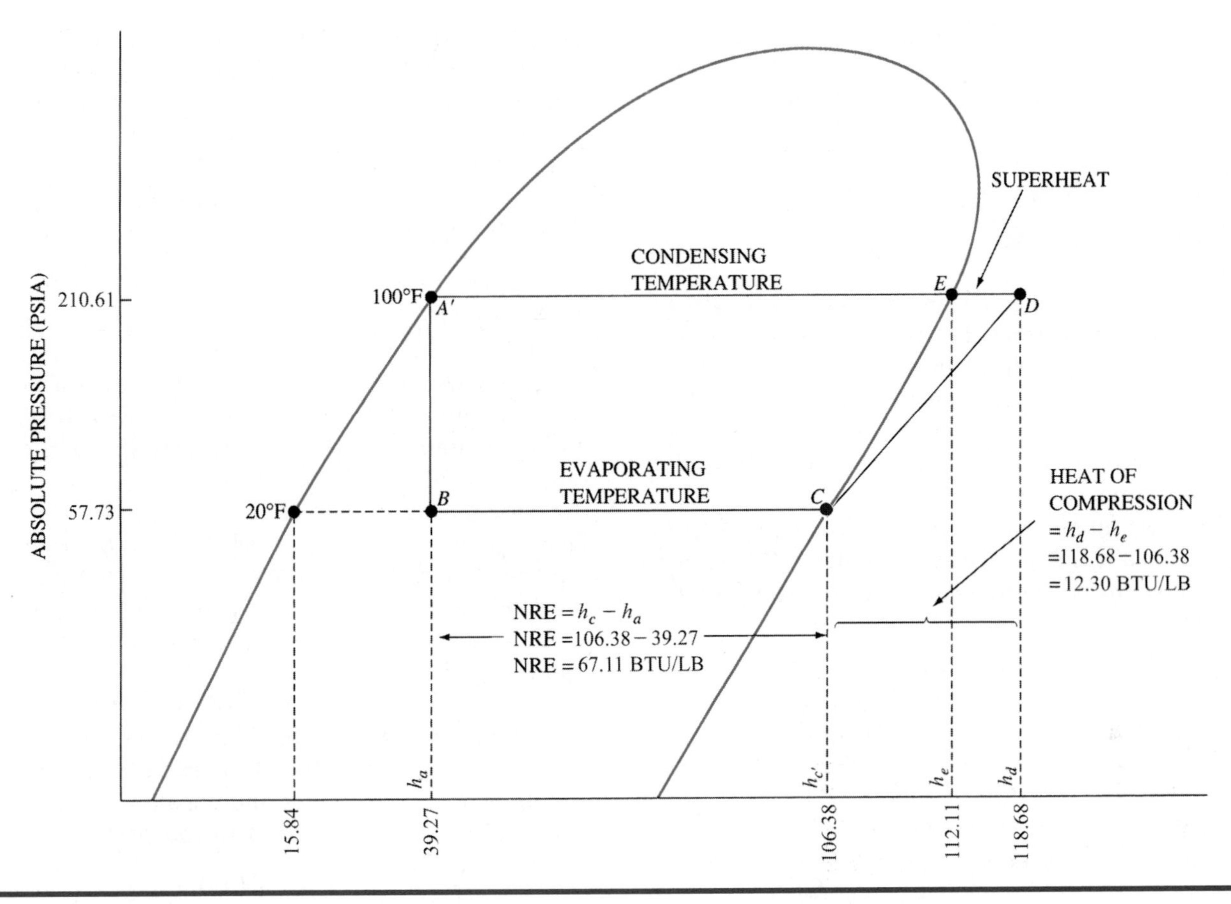

Figure 4-4-6 Pressure, heat, and temperature values for a refrigeration cycle operating with a 20°F evaporator.

line of approximately 128°F. A line drawn vertically downward from point D intersects the heat content line at 118.68 Btu/lb, which is h_d, the difference between h_c and h_d is 10.54 Btu/lb—the heat of compression that has been added to the vapor. This amount of heat is the heat energy equivalent of the work done during the refrigeration compression cycle. This is the theoretical discharge temperature, assuming that saturated vapor enters the cycle; in actual operation, the discharge temperature may be 20° to 35° higher than that predicted theoretically. This can be checked in an operating system by strapping a thermometer or a thermocouple to the outlet of the discharge service valve on the compressor.

During the compression process the heat that is absorbed by the vapor is a result of friction caused by the action of the pistons in the cylinders and by the vapor itself passing through the small openings of the internal suction and discharge valves. Of course, the vapor is also heated by the action of its molecules being pushed or compressed closer together, commonly called heat of compression. Some of this overall additional heat is lost through the walls of the compressor. A lot depends, therefore, on the design of the compressor, the conditions under which it must operate and the balance between the heat gain and heat loss to keep the refrigerant at a constant entropy.

Line D–E denotes the amount of superheat that must be removed from the vapor before it can commence the condensation process. A line drawn vertically downward from point E to point h_e on the heat content line indicates the distance h_d–h_e, or heat amounting to 6.54 Btu/lb, since the heat content of 100°F vapor is 112.11 Btu/lb. This superheat is usually removed in the hot gas discharge line or in the upper portion of the condenser. During this process the temperature of the vapor is lowered to the condensing temperature.

Line E–A represents the condensation process that takes place in the condenser. At point E the refrigerant is a saturated vapor at the condensing temperature of 100°F and an absolute pressure of 210.61 psia; the same temperature and pressure prevail at point A, but the refrigerant is now in a liquid state. At any other point on line E–A the refrigerant is in the phase of a liquid vapor combination; the closer the point is to A, the greater the amount of the refrigerant that has condensed into its liquid stage.

At point A, each pound of refrigerant is ready to go through the refrigerant cycle again as it is needed for heat removal from the load in the evaporator.

COEFFICIENT OF PERFORMANCE

Two factors mentioned earlier in this chapter are of the greatest importance in deciding which refrigerant should be used for a given project of heat removal. Ordinarily, this decision is reached during the design aspect of the refrigeration and air conditioning system, but we will explain it briefly now, and elaborate later.

The two factors that determine the *coefficient of performance* (COP) of a refrigerant are *refrigerating effect* and *heat of compression*. The equation may be written as

$$\text{COP} = \frac{\text{refrigerating effect}}{\text{heat of compression}} \qquad (4\text{-}4\text{-}2)$$

Substituting values, from the pressure-enthalpy diagram of the simple saturated cycle previously presented, the equation would be

$$\text{COP} = \frac{h_c - h_a}{h_d - h_c} = \frac{68.87}{10.54} = 6.53$$

The COP is therefore a rate or a measure of the efficiency of a refrigeration cycle in the utilization of expended energy during the compression process in ratio to the energy that is absorbed in the evaporation process. As can be seen from Eq. 4-4-2, the less energy expended in the compression process, the larger will be the COP of the refrigeration system. Therefore, the refrigerant having the highest COP would probably be selected—provided other qualities and factors are equal.

EFFECTS ON CAPACITY

The pressure-enthalpy diagrams in Figure 4-4-5 and 4-4-6 show a comparison of two simple saturated cycles having different evaporating temperatures, to bring out various differences in other aspects of the cycle. In order that an approximate mathematical calculation comparison may be made, the cycles shown in Figure 4-4-5 and 4-4-6 will have the same condensing temperature, but the evaporating temperature will be lowered 20°F. Data can either be obtained or verified from the table for R-22 in the Appendix; but we will take the values of A, B, C, D, and E from Figure 4-4-5 as the cycle to be compared to that in Figure 4-4-6 (with a 20°F evaporator). The refrigerating effect, heat of compression, and the heat dissipated at the condenser in each of the refrigeration cycles will be compared. The comparison will be based on data about the heat content or enthalpy line, rated in Btu/lb.

For the 20°F evaporating temperature cycle shown in Figure 4-4-6:

$$\text{net refrigerating effect } (h_{c'} - h_a) = 67.11 \text{ Btu/lb}$$
$$\text{heat of compression } (h_{d'} - h_{c'}) = 12.30 \text{ Btu/lb}$$

In comparing the data above with those of the cycle with the 40°F evaporating temperature Figure 4-4-5, we find that there is a decrease in the net refrigeration effect (NRE) of 2.6% and an increase in the heat of compression of 16.7%. There will be some increase in superheat, which should be removed either in the discharge pipe or the upper portion of the condenser. This is the result of a lowering in the suction temperature, the condensing temperature remaining the same.

By utilizing Eq. 4-4-1, it will be found that the weight of refrigerant to be circulated per ton of cooling, in a cycle with a 20°F evaporating temperature and a 100°F condensing temperature, is 2.98 lb/min/ton:

$$W = \frac{200\ (\text{Btu/min})}{\text{NRE}\ (\text{Btu/lb})} = \frac{200 \text{ Btu/min}}{67.11 \text{ Btu/lb}} = 2.98 \text{ lb/min}$$

This of course would also necessitate either a larger compressor, or the same size of compressor operating at a higher rpm.

Figure 4-4-7 shows the original cycle with a 40°F evaporating temperature, but the condensing temperature has been increased to 120°F.

Again taking the specific data from the heat content or enthalpy line, we now find for the 120°F condensing temperature cycle that $h_a = 45.71$, $h_c = 108.14$, $h_d = 122.01$, and $h_e = 112.78$.

$$\text{net refrigerating effect } (h_c - h_{a'}) = 62.43 \text{ Btu/lb}$$
$$\text{heat of compression } (h_{d'} - h_c) = 13.87 \text{ Btu/lb}$$
$$\text{condenser superheat } (h_{d'} - h_{e'}) = 9.23 \text{ Btu/lb}$$

In comparison with the cycle having the 100°F condensing temperature, it can be calculated that by allowing the temperature of the condensing process to increase 20°F, there is a decrease in the NRE of 9.4%, an increase in heat of compression of 31.6%, and an increase of superheat to be removed either in the discharge line or in the upper portion of the condenser of 40.5%.

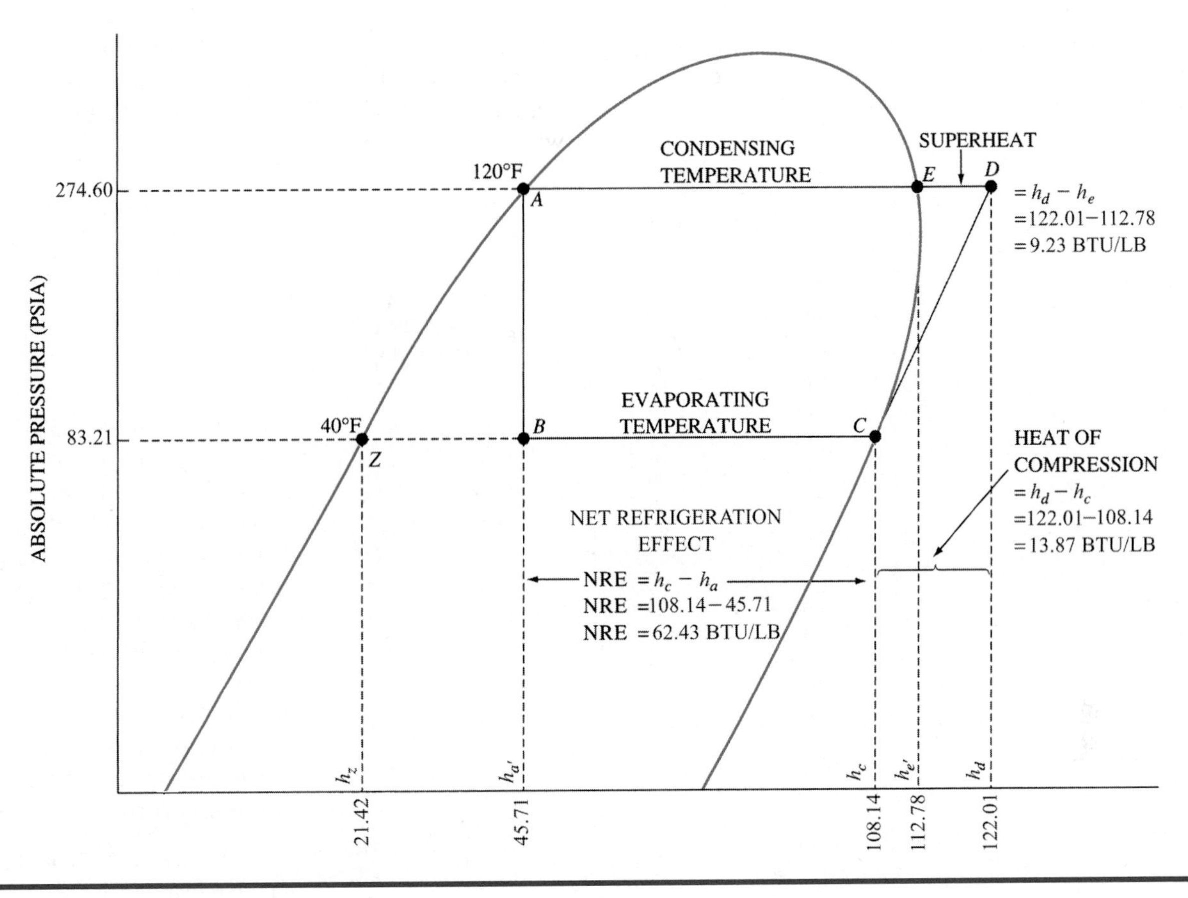

Figure 4-4-7 The C.O.P. is reduced when the system has a condensing temperature of 120°F.

Through the use of the Eq. 4-4-1 it is found that with a 40°F evaporating temperature and a 120°F condensing temperature the weight of refrigerant to be circulated will be 3.2 lb/min/ton. This indicates that approximately 10% more refrigerant must be circulated to do the same amount of work as when the condensing temperature was 100°F.

Both of these examples show that for the best efficiency of a system, the suction temperature should be as high as feasible, and the condensing temperature should be as low as feasible. Of course, there are limitations as to the extremes under which systems may operate satisfactorily, and other means of increasing efficiency must then be considered. Economics of equipment (cost + operating performance) ultimately determine the feasibility range.

Referring to Figure 4-4-8, after the condensing process has been completed and all of the refrigerant vapor at 120°F is in the liquid state, if the liquid can be subcooled to point A' on the 100°F line (a difference of 20°F), the NRE ($h_c - h_a$) will be increased 6.44 Btu/lb. This increase in the amount of heat absorbed in the evaporator without an increase in the heat of compression will increase the COP of the cycle, since there is no increase in the energy input to the compressor.

This subcooling may take place while the liquid is temporarily in storage in the condenser or receiver, or some of the liquid's heat may be dissipated to the ambient temperature as it passes through the liquid pipe on its way to the metering device. Subcooling may also take place in a commercial type water cooled system through the use of a liquid subcooler, which, in a low temperature application, may well pay for itself through the resulting increase in capacity and efficiency of the overall refrigeration system.

Another method of subcooling the liquid is by means of a heat exchanger between the liquid and suction lines, whereby heat from the liquid may be transferred to the cooler suction vapor traveling from the evaporator to the compressor. This type is shown in Figure 4-4-9, a refrigeration cycle flowchart, using a liquid-suction heat exchanger. True, heat cannot be removed from the liquid and then added to the suction vapor without some detrimental effects to the overall refrigeration cycle; for example, the vapor would become superheated, which would in turn cause an increase in the specific

Figure 4-4-8 Subcooling in the condenser improves the refrigerating effect and the C.O.P.

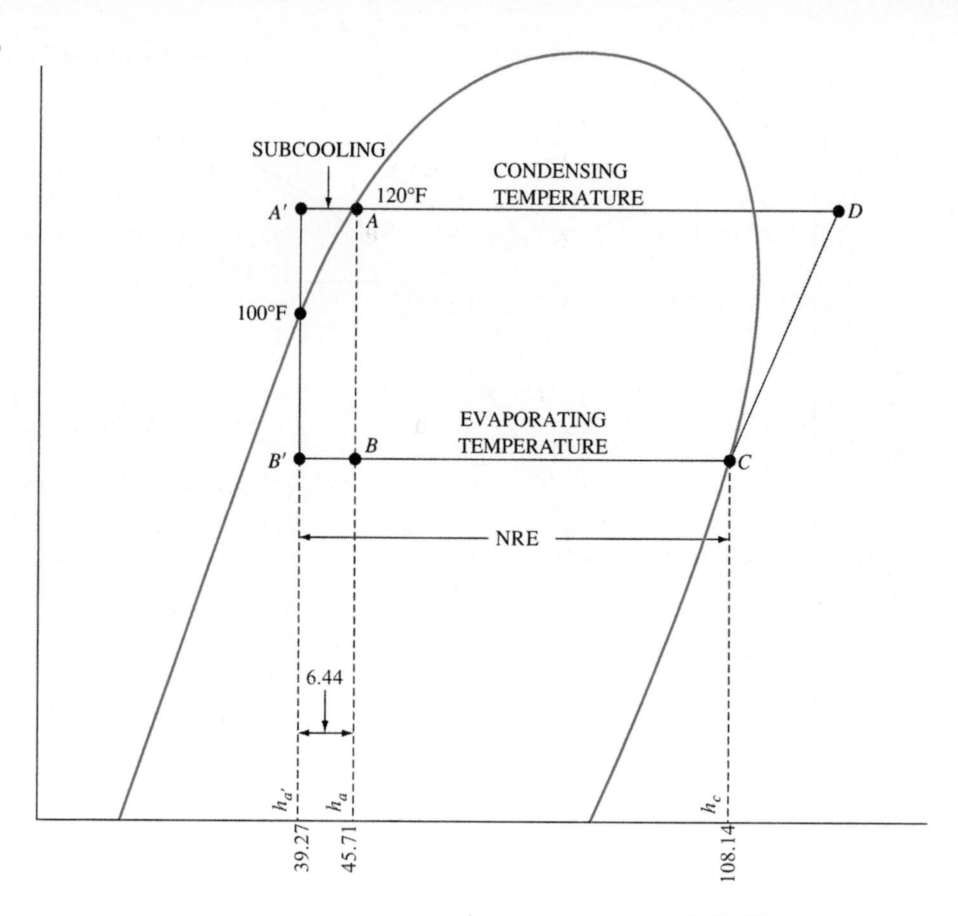

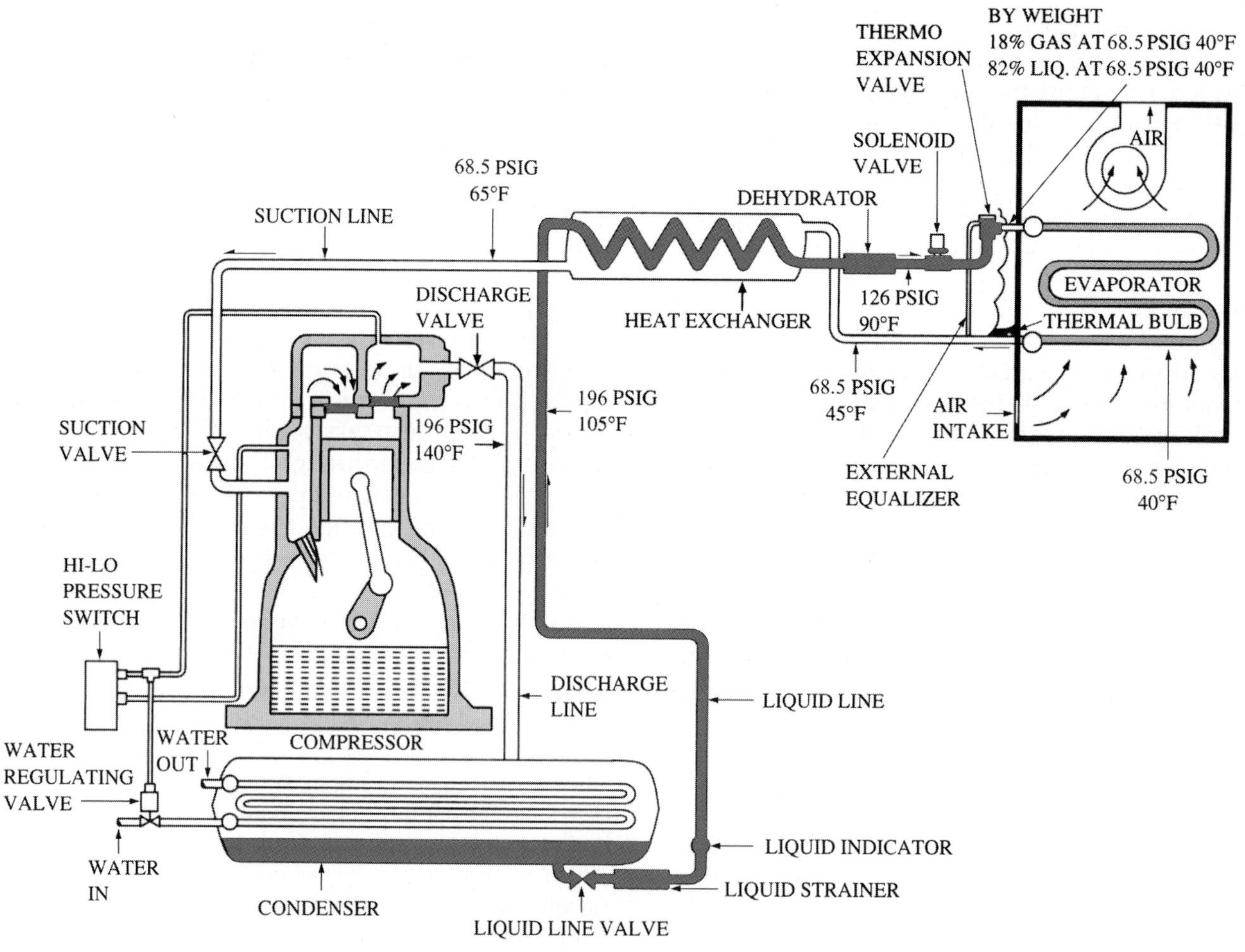

Figure 4-4-9 Flow diagram of the refrigeration system.

volume of each pound of refrigerant vapor and consequently a decrease in its density. Thus, any advantage of subcooling in a saturated cycle would be negated; but, in an actual cycle, the conditions of a simple saturated cycle do not exist.

In any normally operating cycle, the suction vapor does not arrive at the compressor in a saturated condition. Superheat is added to the vapor after the evaporating process has been completed, in the evaporator and/or in the suction line, as well as in the compressor. If this superheat is added only in the evaporator, it is doing some useful cooling; for it too is removing heat from the load or product, in addition to the heat that was removed during the evaporating process. But if the vapor is superheated in the suction line located outside of the conditioned space, no useful cooling is accomplished; yet this is what takes place in the majority—if not all—of refrigeration systems.

Now, were some of this superheating in the suction pipe curtailed through the use of a liquid-suction heat exchanger, this heat added to the vapor would be beneficial, for it would be coming from the process of subcooling the liquid. As an example, suppose that the suction temperature in the evaporator is at 40°F; the superheated vapor coming out of the evaporator may be about 50°F, and the temperature of the vapor reaching the compressor may be 75°F or above, depending on the ambient temperature around the suction. This means that the temperature of the vapor has been increased 25°F, without doing any useful cooling or work, because this heat has been absorbed from the ambient air outside of the space to be cooled.

If some or most of this 25°F increase in the vapor temperature were the result of heat absorbed from the refrigerant liquid, it would be performing useful cooling, since the subcooling of the liquid will result in a refrigerating effect higher than it would be if the refrigerant reached the metering device without any subcooling. It is possible to reach an approximate balance between the amount of heat in Btu/lb removed by subcooling the liquid and the amount of heat added to the refrigerant vapor in the suction pipe without the heat exchanger.

To reinforce your knowledge of the refrigeration cycle, study Figure 4-4-10. The first section, Figure 4-4-10a, repeats the earlier demonstration of the liquid/vapor balance throughout the system. Note that the refrigerant moves counterclockwise through the system. The second section, Figure 4-4-10b, shows the pressure gauge readings along the refrigerant circuit. It should be no surprise that the pressure is high on the high pressure side on the left of the diagram; and low on the low pressure side. The pressure should be uniformly high or low as shown in the figure. The compressor and the metering device maintain the pressure difference. There really shouldn't be any pressure variation within the high side, or within the low side.

The third figure, Figure 4-4-10c, shows the temperature readings throughout the cycle. Note from the liquid/vapor balance that the three middle readings have a combination of liquid and vapor. From

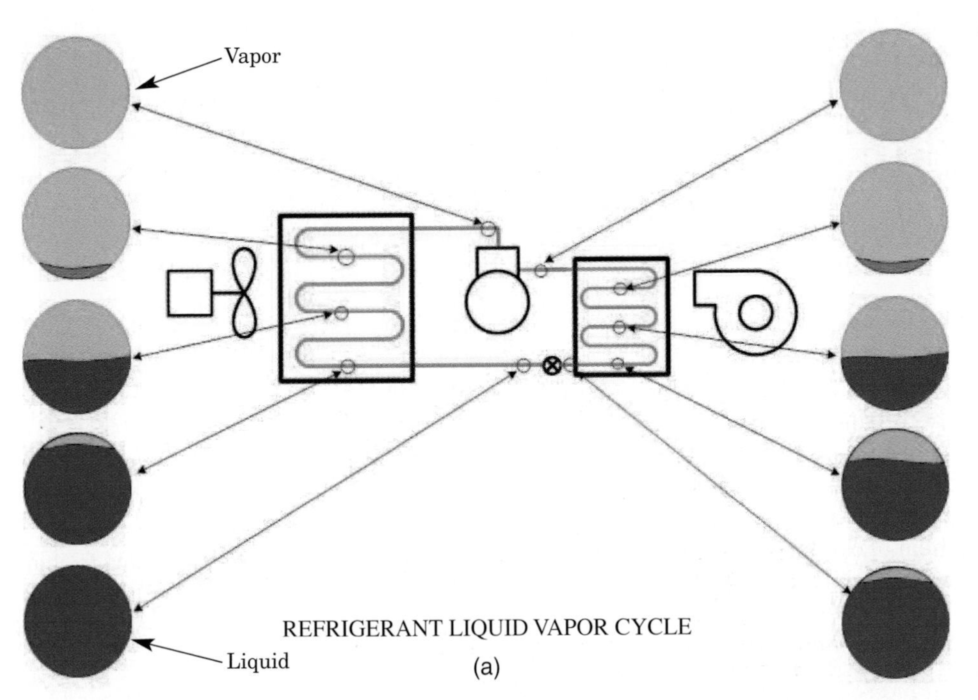

Figure 4-4-10 Refrigeration cycle.

Figure 4-4-10 (continued)

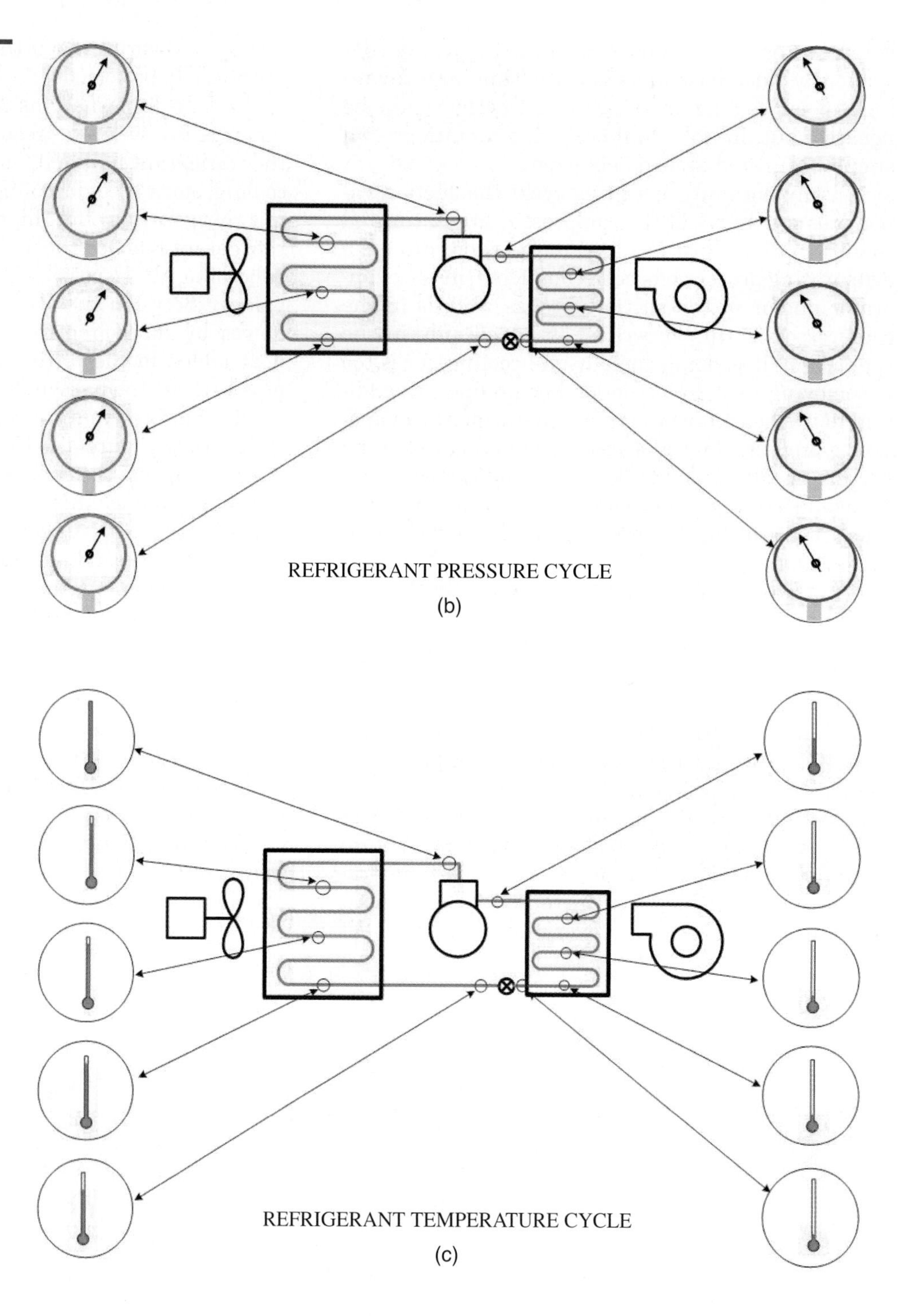

this information, you would expect that there is a change of state occurring in the refrigerant. And indeed, the three middle readings in both sides of Figure 4-4-10c show constant temperature readings. This shows there is no change in sensible temperature while there is still latent heat of condensation or latent heat of vaporization occuring within the condenser or evaporator coils. The temperature readings also show areas of superheat and areas of subcooling.

UNIT 4—REVIEW QUESTIONS

1. What is the basis for a ton of refrigeration?
2. What is the capacity in Btu's of a 10 ton refrigerating system?
3. What must be added for a liquid to be able to change to a vapor?
4. The amount of heat absorbed by each pound of refrigerant is known as the ________ of the system.

TECH TALK

Tech 1. Would a refrigeration system work in space?

Tech 2. Yes, why do you ask?

Tech 1. Well, what about the pressures in the system? Wouldn't they be affected if they were out in the vacuum of space?

Tech 2. No, not at all. In fact we calculate the operating pressures for a pressure enthalpy diagram from atmospheric pressure. That means it's already been corrected back as if it were operating in the vacuum of space.

Tech 1. So even if a refrigerator was taken up in orbit around the earth you are saying it would still work.

Tech 2. That's right. It would work exactly as it does here. No difference.

Tech 1. Well, I still don't understand.

Tech 2. Just think about it. If changing pressure affected the way a refrigerator worked, then what would people do that live in the mountains? Their refrigerators wouldn't work very well either would they? It is certainly like being in space but there is a big pressure difference. Because everything is calculated and works as if it were in a vacuum, if it is in a vacuum it still works.

Tech 1. So, I guess I understand. I do know that the refrigerators up in the mountains work just as well as those down near the coast. So it must be that you're right.

Tech 2. Thanks for giving me credit.

5. The ________ of the compressor must be such that it will remove from the evaporator that amount of refrigerant which has vaporized in the evaporator and in the metering device in order to get the necessary work done.
6. A ________ that is too large will withdraw the refrigerant from the evaporator too rapidly, causing a lowering of the temperature inside the evaporator.
7. What does the pressure-enthalpy diagram show?
8. List the properties exhibited by refrigerants in the gaseous state.
9. Define entropy as used in the refrigeration field.
10. What are the two factors that determine the coefficient of performance of a refrigerant?

5

Air Conditioning and Refrigeration System Components

UNIT 1

System Components and Operation

OBJECTIVES: In this unit the reader will be learn about system components and operation including:

- High Side and Low Side Operations
- Refrigerant Cycle
- Accumulators and Receivers
- Oil Separator
- Filters and Driers

INTRODUCTION

Figure 5-1-1 illustrates the eleven major components that are likely to be found in air conditioning and refrigeration systems. These items include (a) condenser*, (b) hot gas line*, (c) compressor*, (d) accumulator, (e) suction (vapor) line*, (f) evaporator*, (g) operational controls and safeties*, (h) metering device*, (i) liquid line*, (j) receiver, and (k) condensate line. In addition to these eleven major components there are many other items that may be added to a system for the purposes of improved control, energy efficiency, defrost, or other similar function. Of the eleven basic items found on a refrigeration system, eight items marked with an asterisk (*) will be found on every air conditioning and refrigeration system.

A refrigeration system can be divided into the high side and low side portions. The high side contains the high pressure vapor and liquid refrigerant and is the part of the system that rejects heat. The low side contains the low pressure liquid vapor and refrigerant and is the side that absorbs heat. Figure 5-1-2 shows that the compressor and metering device are the dividing lines between the high and low pressure sides of the system. The inlet valve in the compressor is the end of the low side of the system and the discharge valve side of the compressor is the beginning of the high pressure side. The inlet of the metering device is the end of the high pressure side, and the outlet of the metering device is the beginning of the low pressure side.

HIGH SIDE AND LOW SIDE OPERATIONS

Heat is always trying to reach a state of balance by flowing from a warmer object to a cooler object. Heat only flows in one direction, from warmer to cooler.

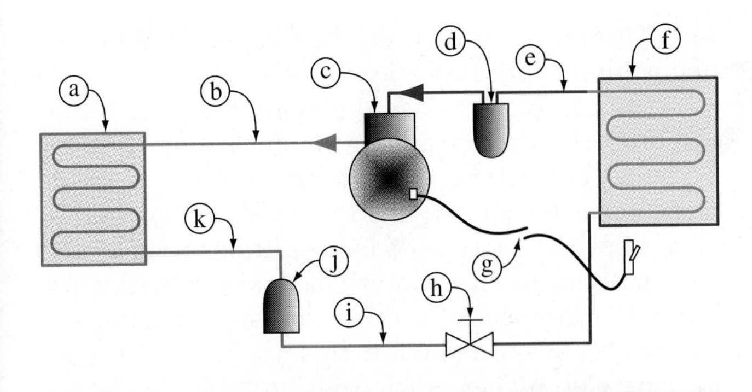

Figure 5-1-1 Major components of air conditioning and refrigeration systems: (a) Condenser; (b) Hot gas line; (c) Compressor*; (d) Accumulator; (e) Suctions (vapor) line*; (f) Evaporator*; (g) Operational controls and safeties*; (h) Metering device*; (i) Liquid line*; (j) Receiver; (k) Condensate line. (Components with asterisks appear on all air conditioning and refrigeration units.)

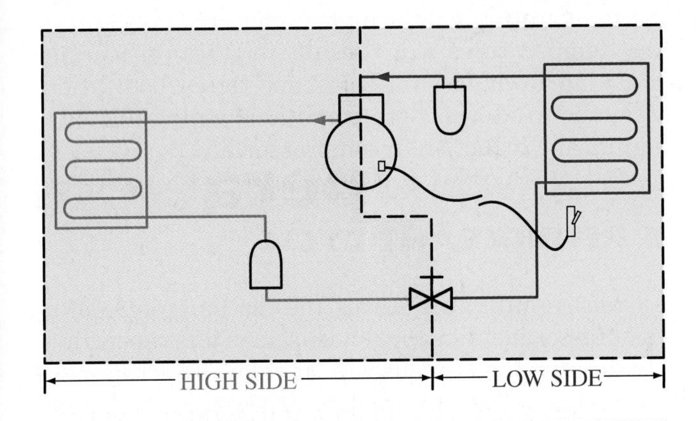

Figure 5-1-2 High pressure, high temperature and low pressure, low temperature sides of a refrigeration system.

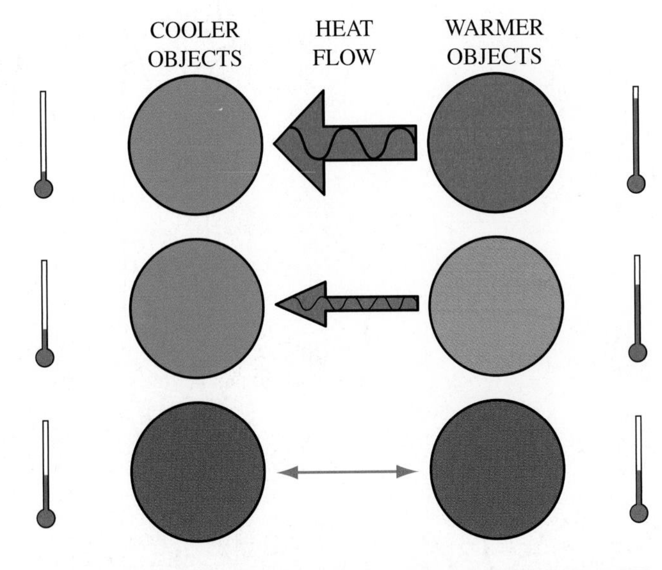

Figure 5-1-3 The greater the temperature difference, the faster the heat flows.

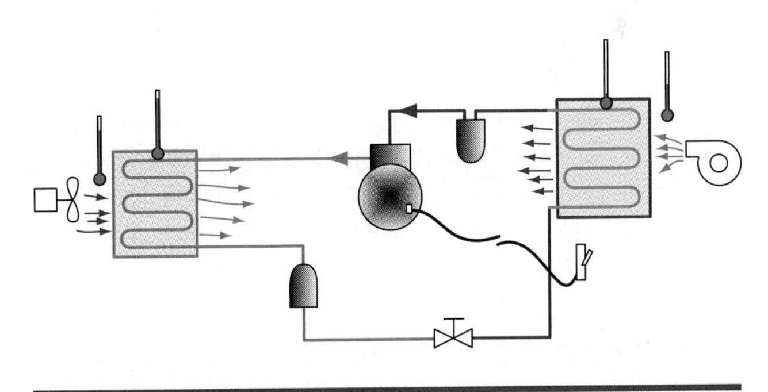

Figure 5-1-4 Heat picked up by the evaporator is rejected by the condenser.

Temperature difference (TD) is what allows heat to flow from one object to another, Figure 5-1-3. The greater the temperature difference the more rapid the heat flow. For the high side of a refrigeration unit to reject heat its temperature must be above the ambient or surrounding temperature. Also for the evaporator to absorb heat, its temperature must be below the surrounding ambient temperature, Figure 5-1-4.

The two factors that affect the quantity of heat transferred between two objects are the temperature difference and the mass of the two objects. The greater the temperature difference between the refrigerant coil and the surrounding air, the more rapid will be the heat transfer. The larger the size of the refrigerant coil, the greater the mass of refrigerant, which also increases the rate of heat transfer. Engineers can either design coils to have high temperature differences or larger areas to increase the heat transfer rate.

To increase energy efficiency systems are designed with larger coils because it is more efficient to have a lower temperature and a larger area to transfer heat. It takes less energy to produce a smaller pressure/temperature difference within a refrigeration system. Manufacturers of new high efficiency air conditioning systems use this principle. That is why the newer outdoor condensing units are significantly larger than older models having the same capacity.

The same principle has been applied to the evaporator coils. Evaporator coil temperature differences between the entering and leaving air are lower than they were on earlier systems. Older lower efficiency air conditioning systems may have evaporative coils that operate at 35°F while newer higher efficiency coils may operate in the 45°F range. Both coils can pick up the same amount of heat provided that the higher temperature, higher efficiency coil has greater area and, therefore, more mass of refrigerant being exposed to the air stream to absorb heat, Figure 5-1-5.

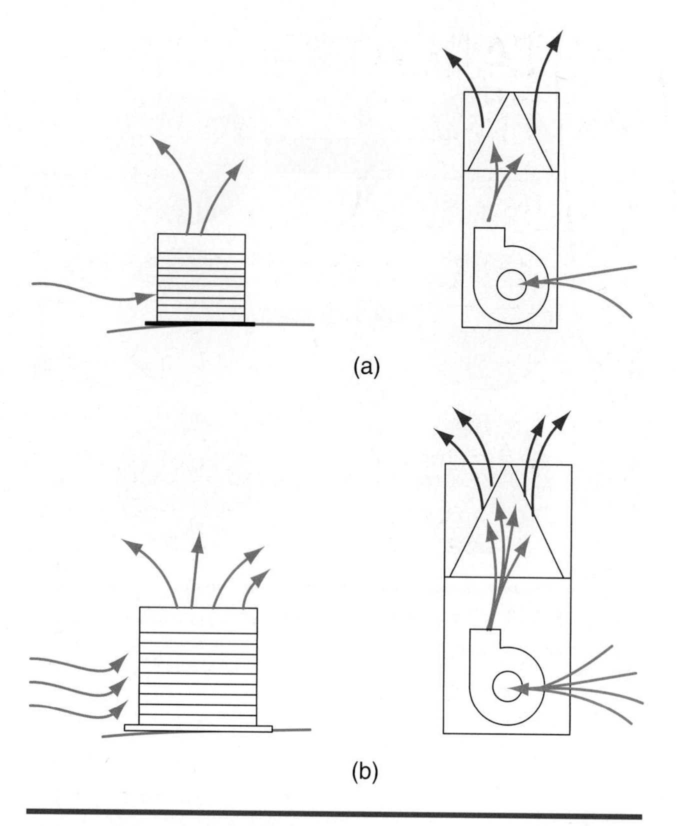

Figure 5-1-5 (a) Low efficiency equipment is smaller than (b) high efficiency equipment. And high efficiency equipment moves more air than low efficiency equipment–with less temperature difference.

TECH TIP

The larger size of the new high efficiency equipment may make impossible to fit it in the same location as an older, smaller, low efficiency system. Check with the manufacturer for the size of new equipment before starting to change out older equipment.

Air conditioning and refrigeration design engineers must take into consideration a variety of factors when designing systems for higher efficiency. For example, the higher evaporative coil temperature may produce less dehumidification. In humid climates dehumidification can be an important part of the total air conditioning. Manufacturers spend thousands of hours and tens of thousands of dollars researching the effective energy efficiency of systems. These tests are carried out in large calibration rooms where the condenser operates at specific temperature and humidity conditions in one area and the evaporator operates under separate conditions in another area.

The results of manufacturing research are incorporated within the manufacturer's technical data sheets provided to the technician during installation. This material may also be found in ARI's certified equipment guides, Figure 5-1-6.

It is important to use the manufacturer's material properly when designing and installing HVAC/R systems. The correct equipment selection is of utmost importance to ensure system operation and to obtain desired energy efficiencies. It has been a common practice in many locations for installers to select an indoor coil of a different tonnage than the outdoor unit capacity. While this practice in the past may provide higher efficiencies, for most of today's more technically designed systems they must be properly matched according to the manufacturer's specifications in order to provide proper operation. Mismatching systems can result in customer complaints of poor humidity control and higher operating costs. In addition to poor energy efficiency and lack of proper humidity control, mismatched system compressors may not receive adequate cooling from returning refrigerant vapor. As a result the compressor temperature will be higher, and higher compressor temperatures can result in the refrigerant breaking down, forming acids, and the carbonization of motor windings. Both conditions will result in a significant reduction in compressor life.

REFRIGERANT CYCLE

As refrigerant vapor leaves the discharge side of a compressor, it enters the condenser. As this vapor travels through the condenser, heat from the refrigerant dissipates to the surrounding air through the piping and fins. As heat is removed, the refrigerant begins to change state from vapor to liquid, Figure 5-1-7. As the mixture of liquid and vapor continues to flow through the condenser, more heat is removed and eventually 100% of the vapor has transformed into liquid. The liquid flows from the outlet of the condenser through the liquid line to the metering device, Figure 5-1-8.

The high pressure, high temperature liquid refrigerant passes through the metering device where its temperature and pressure change. As the pressure and temperature change, some of the liquid refrigerant instantaneously boils off forming flash gas, Figure 5-1-9. As this mixture of refrigerant, liquid, and vapor flow through the evaporator, heat is absorbed, and the remaining liquid refrigerant changes into a vapor. At the outlet of the evaporator 100% vapor flows back through the suction or vapor line to the compressor inlet, Figure 5-1-10.

The compressor draws in this low pressure, low temperature vapor and converts it to a high temperature, high pressure vapor where the cycle begins again, Figure 5-1-11. Figure 5-1-12 shows the refrigerant cycle for the liquid vapor cycle (a), temperature (b), and pressure (c).

INSTALLATION INSTRUCTIONS

Value™ Series 10ACB Condensing Units

CONDENSING UNITS
1–1/2 through 5 ton
503,800M
2/98
Supersedes 503,755M

Technical Publications
Litho U.S.A.

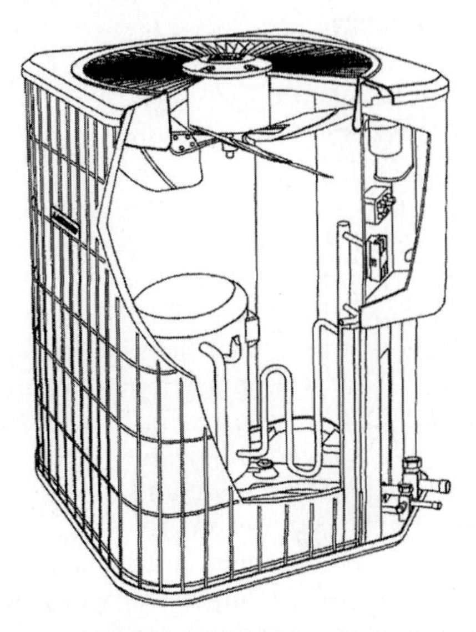

VALUE 10™ CONDENSING UNIT

Value 10 condensing units are designed for expansion valve (TXV) and RFC systems. Refer to Lennox engineering handbook for expansion valve kits which must be ordered separately.

SHIPPING AND PACKING LIST

1– Assembled 10ACB condensing unit

1– 45° copper street elbow

1– RFCIV refrigerant metering device (bullet)

1– Coupling – 5/16 x 3/8″ (18, 24, 30)

Check unit for shipping damage. Consult last carrier immediately if damage is found.

GENERAL INFORMATION

These instructions are intended as a general guide and do not supersede national or local codes in any way. Authorities having jurisdiction should be consulted before installation.

⚠ IMPORTANT

The Clean Air Act of 1990 bans the intentional venting of refrigerant (CFC's and HCFC's) as of July 1, 1992. Approved methods of recovery, recycling or reclaiming must be followed. Fines and/or incarceration may be levied for non–compliance.

TABLE OF CONTENTS

RETAIN THESE INSTRUCTIONS FOR FUTURE REFERENCE

⚠ WARNING

Product contains fiberglass wool.

Disturbing the insulation in this product during installation, maintenance, or repair will expose you to fiberglass wool. Breathing this may cause lung cancer. (Fiberglass wool is known to the State of California to cause cancer.)

Fiberglass wool may also cause respiratory, skin, and eye irritation.

To reduce exposure to this substance or for further information, consult material safety data sheets available from address shown below, or contact your supervisor.

Lennox Industries Inc.
P.O. Box 799900
Dallas, TX 75379–9900

Figure 5-1-6 Manufacturer installation guide. (*Courtesy of Lennox Industries, Inc.*)

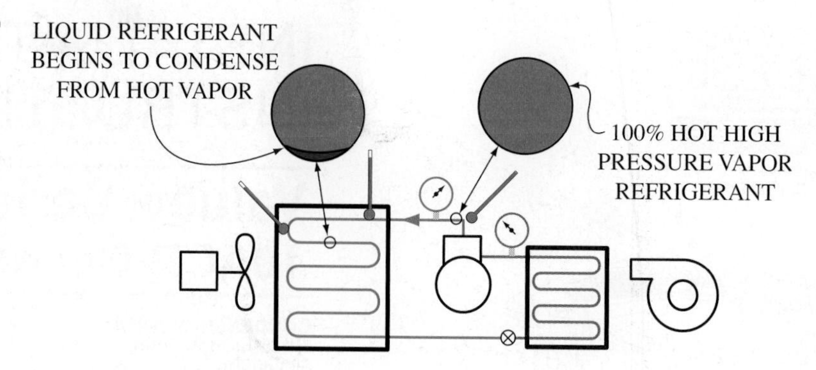

Figure 5-1-7 Liquid refrigerant begins to form in the first part of the condenser.

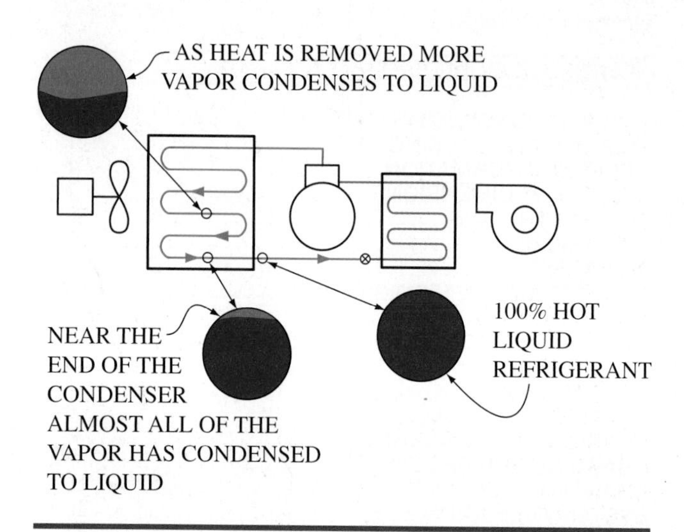

Figure 5-1-8 All of the vapor converts to liquid in the condenser.

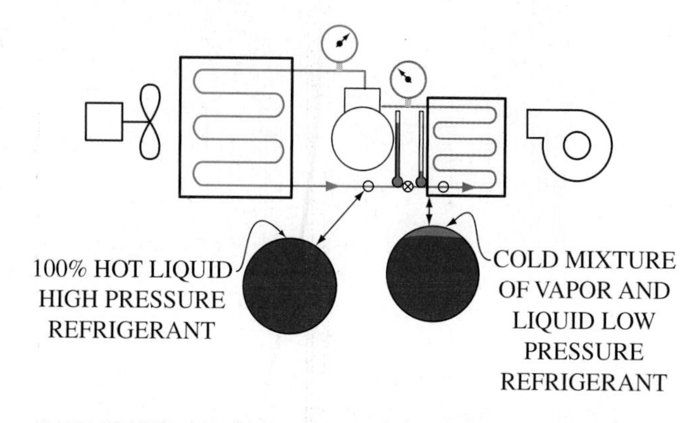

Figure 5-1-9 As the liquid refrigerant passes through the metering device, its pressure drops, its temperature drops, and some of the liquid turns to vapor.

An ideally sized and functioning system is one where the last bit of refrigerant vapor changes into a liquid at the end of the condenser and where the last bit of liquid refrigerant changes into a vapor at the end of the evaporator. However, because it is impossible to have a system operate at this ideal state, units are designed so that we have some additional cooling, called subcooling, of the liquid refrigerant to ensure that no vapor leaves the condenser. Even a small amount of vapor leaving a condenser can significantly reduce the system's efficiency.

On the evaporator side a small amount of additional temperature is added to the refrigerant vapor, called superheat, to ensure that no liquid refrigerant returns to the compressor. Returning liquid refrigerant to the compressor can damage the compressor.

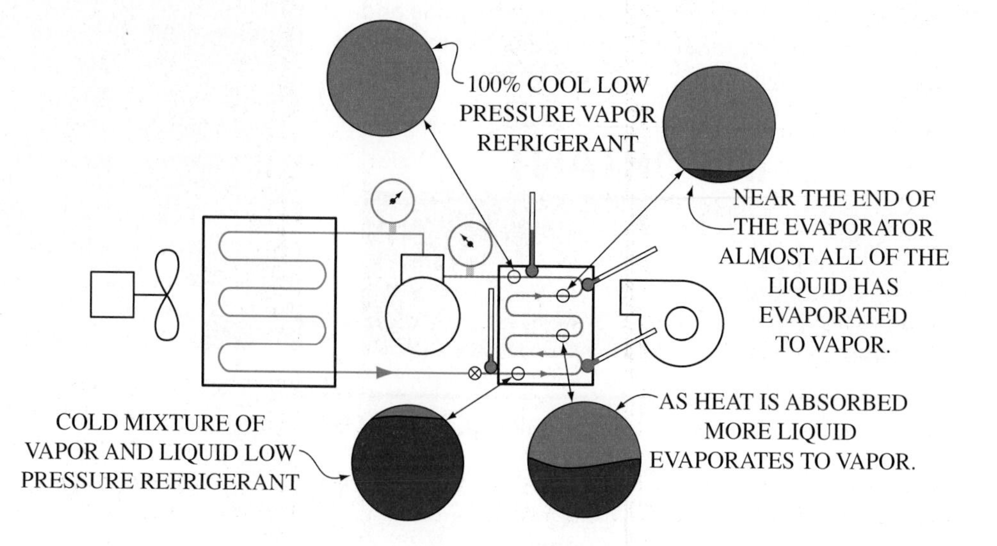

Figure 5-1-10 As the liquid refrigerant passes through the evaporator, it changes from liquid to vapor.

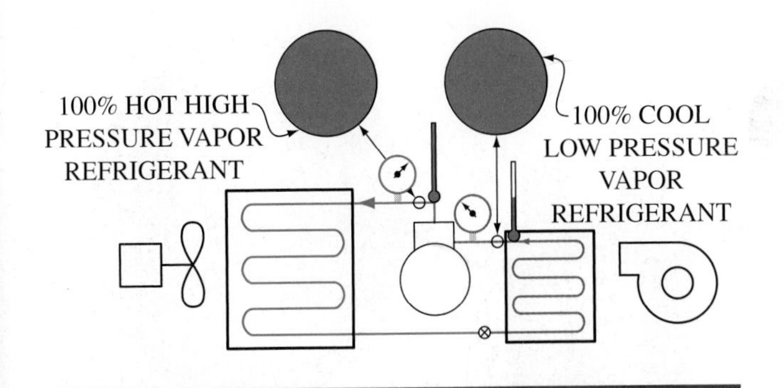

Figure 5-1-11 100% low temperature, low pressure vapor enters the compressor and 100% high pressure, high temperature vapor leaves the compressor.

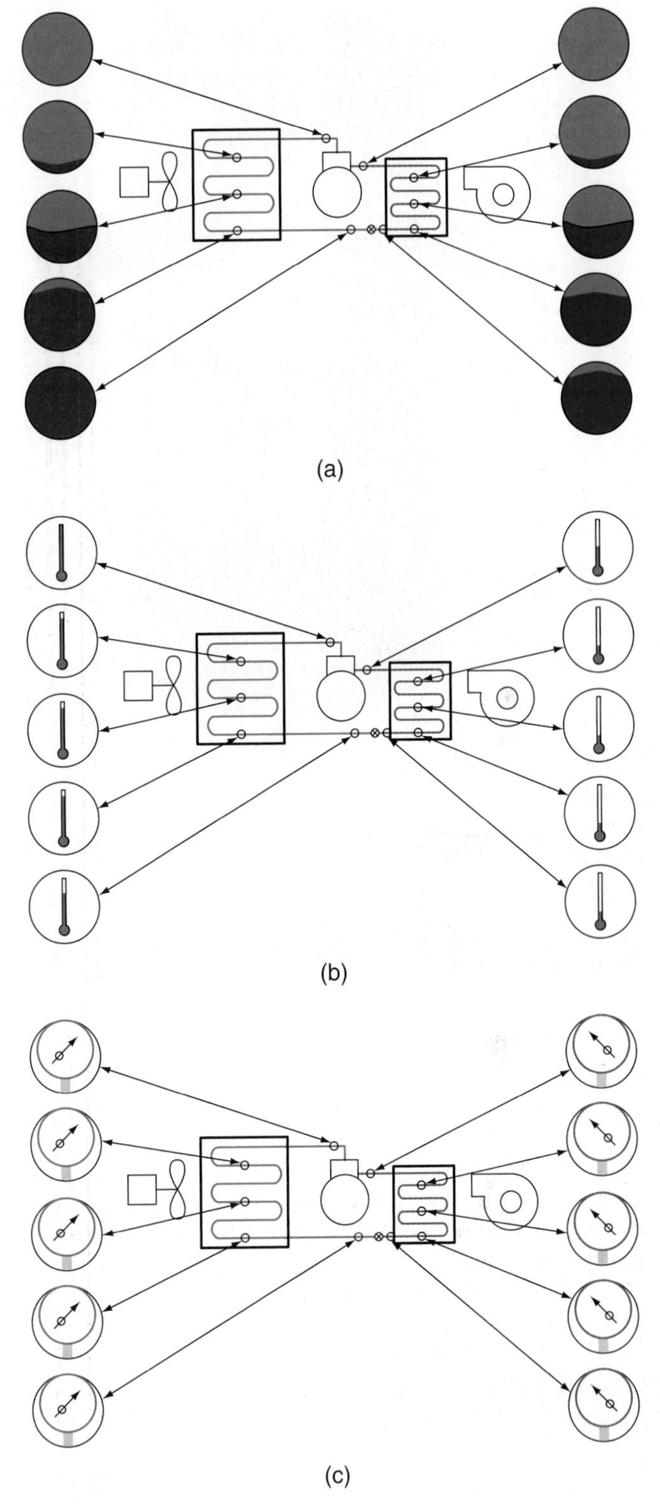

Figure 5-1-12 (a) Refrigerant liquid vapor cycle; (b) Refrigerant temperature cycle; (c) Refrigerant pressure cycle.

ACCUMULATORS AND RECEIVERS

Systems that must operate under a broad range of temperature conditions will have difficulty maintaining the desired level of subcooling and superheat. There are two components that can be used in these systems to enhance the level of efficiency and safety in operation. They are the receiver and the accumulator. The receiver holds a little extra refrigerant so the system has enough for high loads on hot days, Figure 5-1-13. The accumulator holds the liquid refrigerant that would flow back to the compressor on cool days with light loads, Figure 5-1-14.

A liquid receiver can be located at the end of the condenser outlet to collect liquid refrigerant, Figure 5-1-15. The **liquid receiver** allows the liquid to flow into the receiver and any vapor collected in the receiver to flow back into the condenser to be converted back into a liquid. The line connecting the receiver to the condenser is called the condensate line and must be large enough in diameter to allow liquid to flow into the receiver and vapor to flow back into the condenser. The condensate line must also have a slope toward the receiver to allow liquid refrigerant to freely flow from the condenser into the receiver. The outlet side of the receiver is located at the bottom where 100% liquid can flow out of the receiver into the liquid line.

Receivers should be sized so that 100% of the refrigerant charge can be stored in the receiver. When the receiver is properly sized, it can be used to store the refrigerant during a pop down cycle or for some types of approved system service.

Some refrigeration condensing units come with receivers built into the base of the condensing unit, Figure 5-1-16.

Accumulators

An **accumulator** is a device located at the end of the evaporator that allows liquid refrigerant to be collected in the bottom of the accumulator and remain there as the vapor refrigerant is returned to the compressor. The inlet side of the accumulator is connected to the evaporator where any liquid refrigerant and vapor flow in. The outlet of the accumulator

Figure 5-1-13 Receiver.

Figure 5-1-14 Accumulator. *(Courtesy Rheem Manufacturing)*

draws vapor through a U shaped tube or chamber, Figure 5-1-17. There is a small port at the bottom of the U shaped tube or chamber that allows liquid refrigerant and oil to be drawn into the suction line. Without this small port, refrigerant oil would collect in the accumulator and not return to the compressor. The small port does allow some liquid refrigerant to enter the suction line. However, it is such a small amount of liquid refrigerant that it boils off rapidly, so there is no danger of liquid refrigerant flowing into the compressor.

Accumulators are often found on heat pumps. During the changeover cycle, liquid refrigerant can flow back out of the outdoor coil. This liquid refrigerant could cause compressor damage if it were not for the accumulator, which blocks its return.

OIL SEPARATOR

Refrigerant oil flows throughout the refrigerant circuit during normal operation. This oil provides lubrication for some solenoids and metering devices. Some systems may have difficulty returning all of the oil back to the compressor during some conditions of operation. An **oil separator** is a device that collects some of the oil leaving the discharge side of the compressor and returns that oil back to the compressor, Figure 5-1-18. An oil separator uses a fine meshed screen or wire strainer to trap small droplets of refrigerant oil traveling with the refrigerant vapor. The liquid oil collected on the screen drains down into the sump in the oil separator. As the level of oil rises, a small float valve in the base of the sump is automatically opened, Figure 5-1-19. The sump is connected with a capillary tube to the suction side of the compressor crankcase. The pressure difference between the high side pushing down on top of the oil and the low side pulling on the bottom of the oil causes the oil to flow back into the compressor. Once the level of oil drops in the oil separator, the small float valve automatically closes, stopping the flow of oil.

SOLENOIDS

Solenoids are used in air conditioning and refrigeration as refrigerant control devices. They can be used to stop or start flow of refrigerant, water, oil, or other

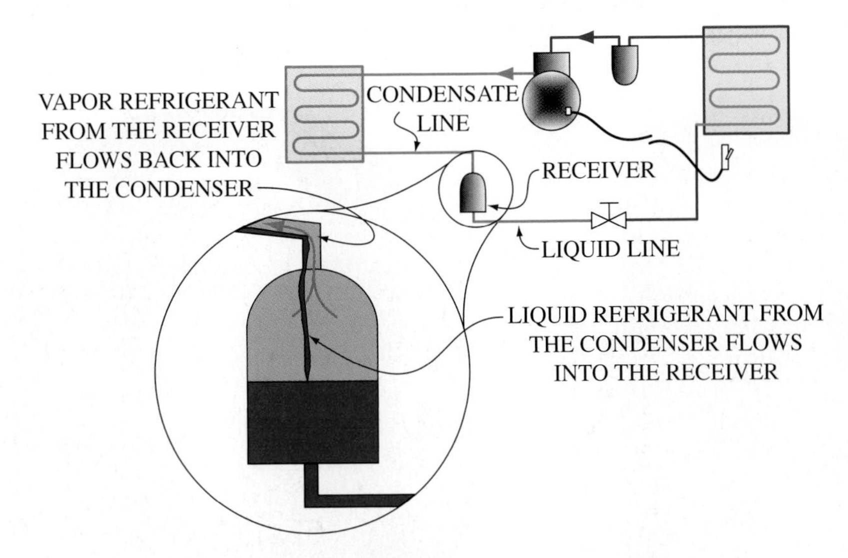

Figure 5-1-15 Liquid refrigerant drains into the receiver, and vapor flows back to the condenser through the condensate line.

Figure 5-1-16 Receiver tank as part of a refrigeration condensing unit. *(Courtesy Danfoss, Inc.)*

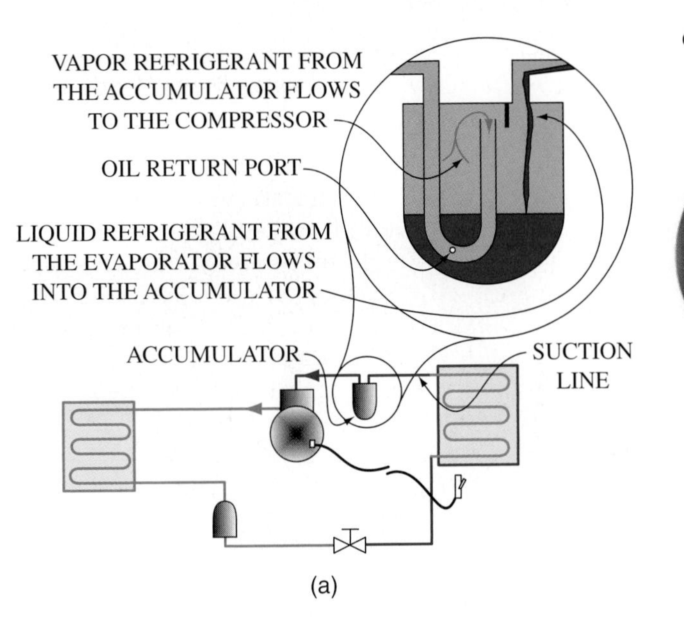

Figure 5-1-17 Unevaporated liquid refrigerant flows into the accumulator. Vapor refrigerant is drawn off the top of the accumulator. A screen covers the oil return port to prevent debris in the system from plugging the port. (a) Diagram; (b) Accumulator.

Figure 5-1-18 Oil separators. (*Courtesy Danfoss, Inc.*)

Figure 5-1-20 Solenoid coil.

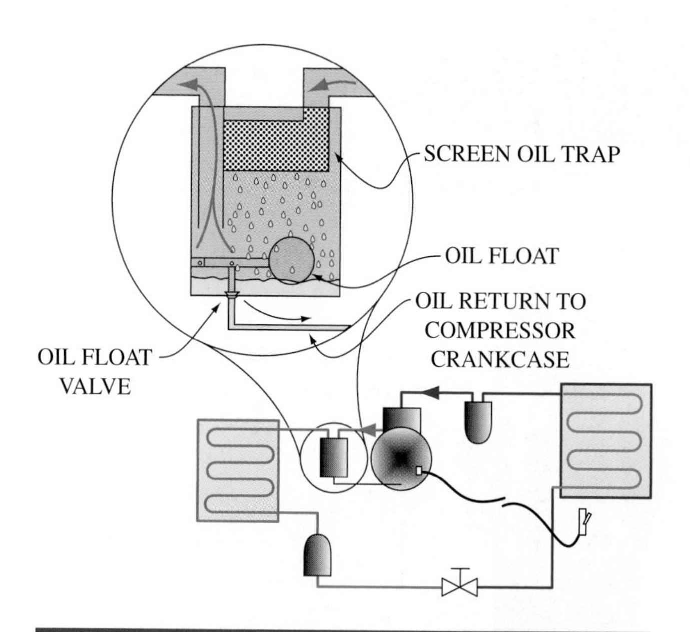

Figure 5-1-19 Oil, which is a very fine mist, collects on the screen at the beginning of the oil separator and drips into the bottom of the separator where it is drawn off into the compressor crankcase.

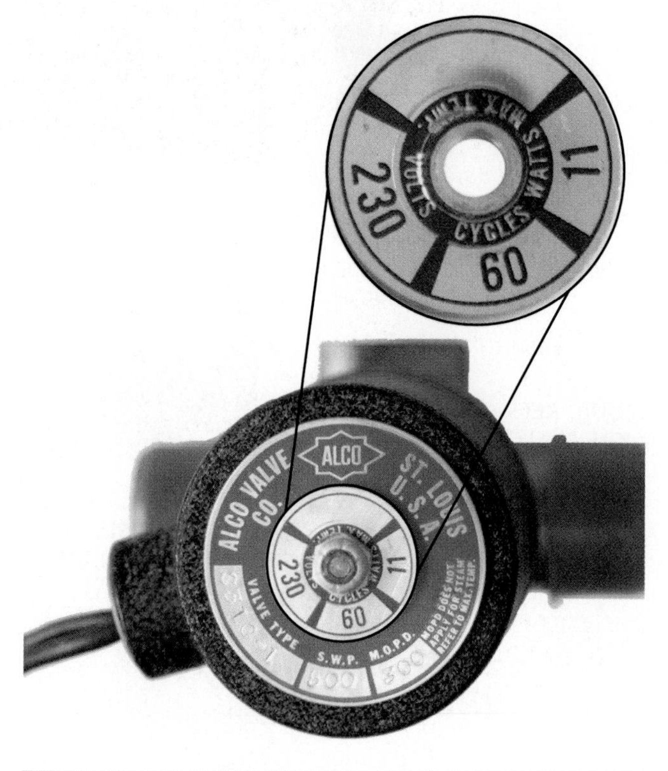

Figure 5-1-21 Typical name and identification plate found on solenoid coils listing the volts, cycles, and lot requirements of the coil. Close up view reads volts: 230, cycles: 60, watts: 11.

fluids. Solenoids may be either normally open (NO) or normally closed (NC). When we use the term normally in conjunction with a control device such as a solenoid, we are talking about its condition as it exists in the box. That is, the condition of the solenoid when it is not energized. Often the term normally is misconstrued because many valves in operation may be energized and in the opposite position. For example, the solenoid used for pop down in a refrigeration system may be energized and open almost continuously in some applications. Therefore, during operation the valve is open, although the valve when de-energized (as it existed in the box before it was installed) is normally closed. The normal condition, then, of a solenoid valve is how it exists in its de-energized position irrespective of how it is during most of its operational life.

A solenoid valve operates as a result of a small magnetic coil, Figure 5-1-20. The coil may be low voltage, usually 24 volts AC, or line voltage, 110 or 220 volts AC, Figure 5-1-21. Figure 5-1-22a shows a pictorial view of the major parts of a solenoid. Figure 5-1-22b shows a diagram version of a solenoid. When a normally closed solenoid is in the off position the small plunger rests on the pilot orifice. The system pressure forces the diaphragm down on the main orifice, stopping all flow as shown in Figure 5-1-23a (pictorial view) and Figure 5-1-23b (diagram version). The pressures on top and on the bottom of the diaphragm are equal to each other in the off position. But there is more area on the top side of the diaphragm; part of the bottom side is covered up by the main orifice, so there is more closing force than opening force, Figure 5-1-24.

When the magnetic coil is energized it lifts the small plunger, Figure 5-1-25. Lifting this small plunger opens up the small pilot orifice in the valve's diaphragm, Figure 5-1-26. With the port open, system pressure on top of the diaphragm escapes through the pilot orifice. At that point the pressure on the diaphragm becomes unbalanced, Figure 5-1-27. Within moments of lifting the plunger the pressure on top of

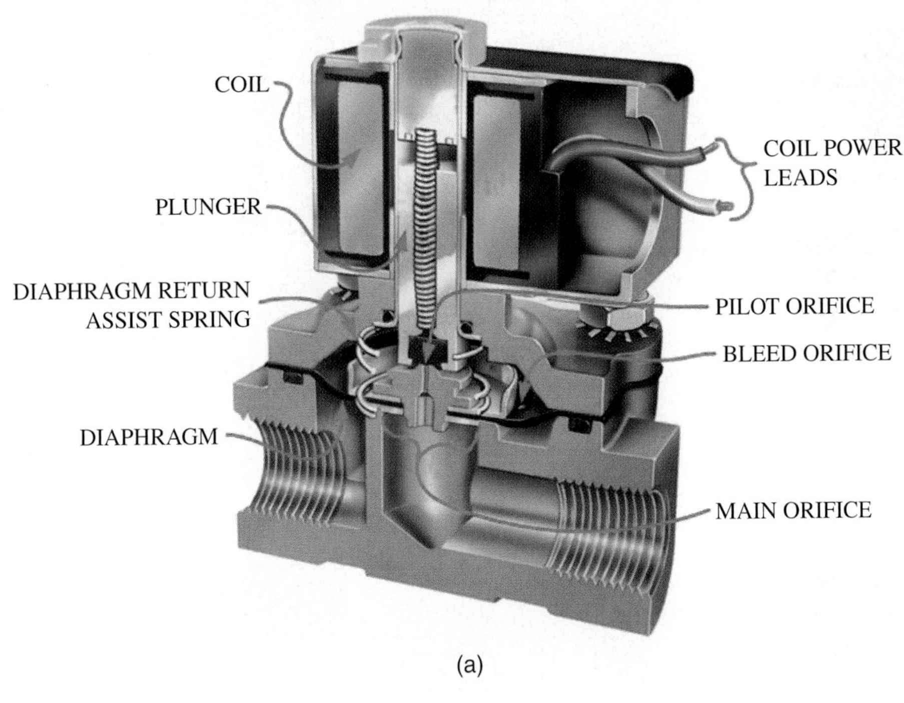

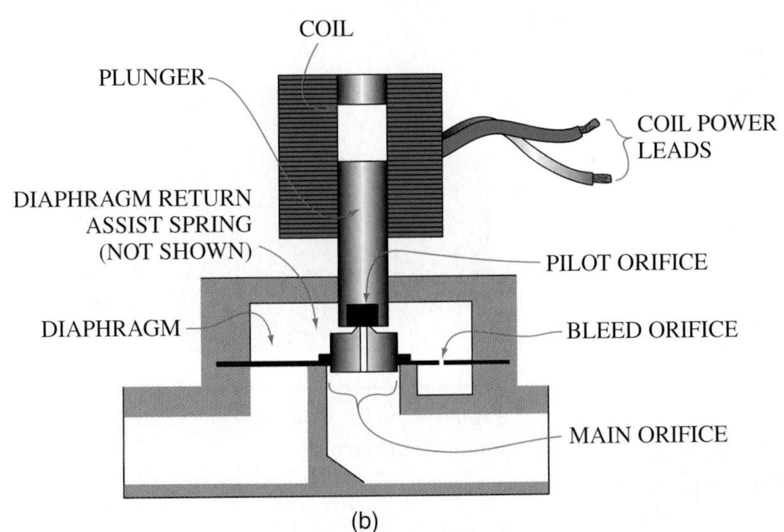

Figure 5-1-22 (a) Pictorial cutaway of solenoid parts (*Courtesy Parker Fluid Control Division [Gold Ring™]*); (b) Schematic of solenoid parts.

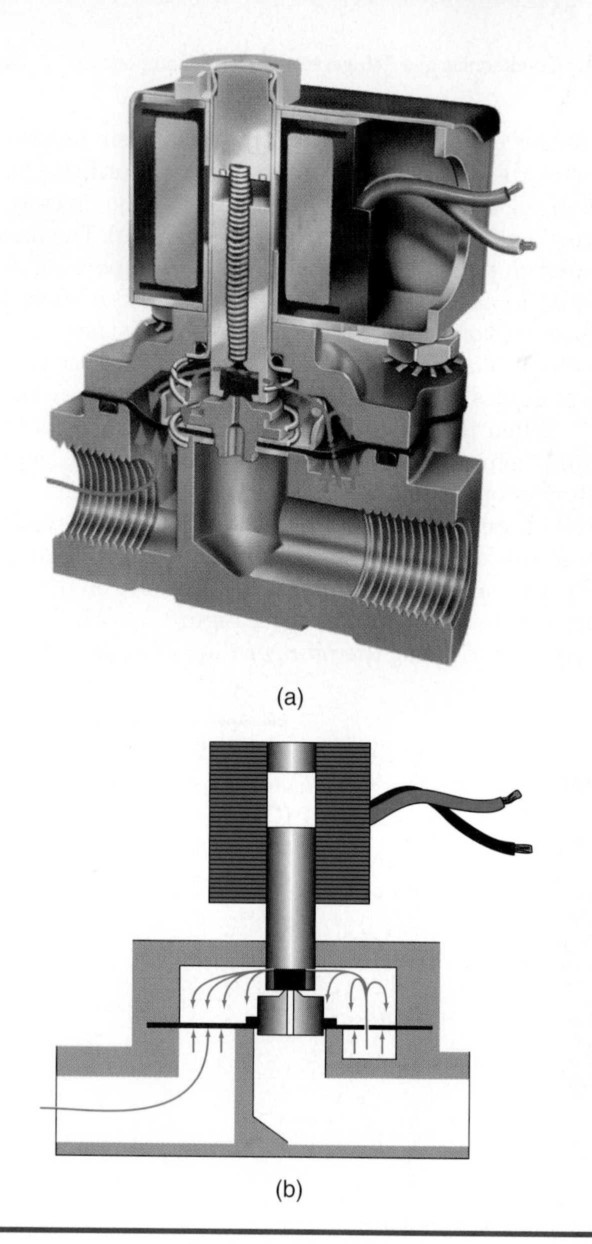

Figure 5-1-23 (a) Path of refrigerant through the solenoid in the OFF position (*Courtesy Parker Fluid Control Division [Gold Ring™]*); (b) Schematic illustration of forces on diaphragm in OFF position.

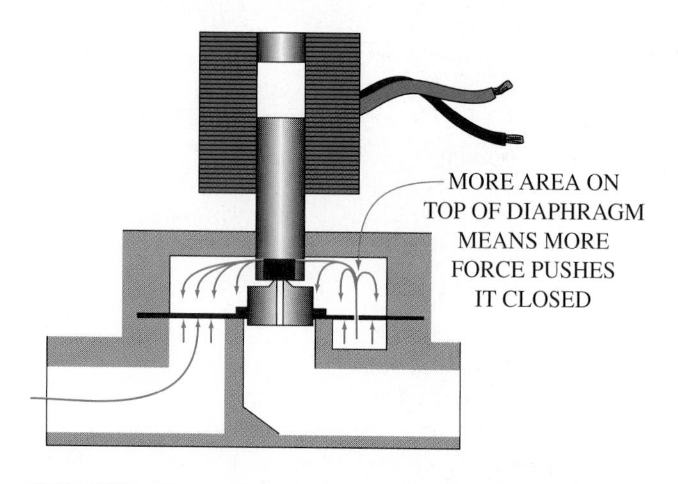

Figure 5-1-24 Pressure on the diaphragm is greatest from the top in the OFF position.

Figure 5-1-25 Major parts of a solenoid pilot orifice in a metal diaphragm.

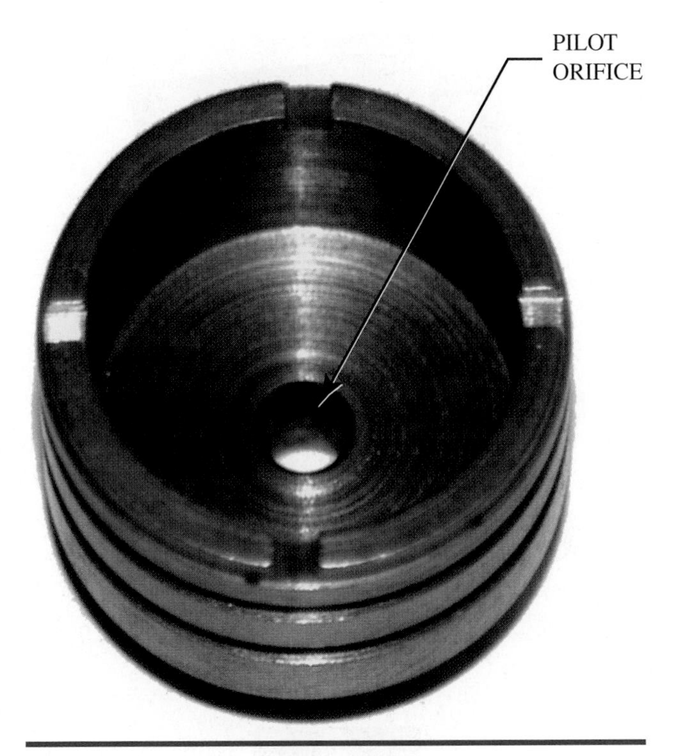

Figure 5-1-26 Metal diaphragm showing the pilot orifice.

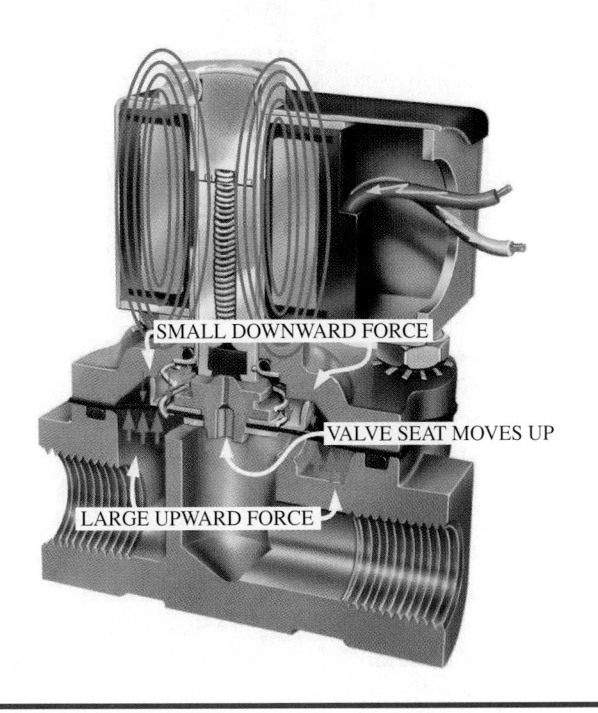

Figure 5-1-27 When the plunger is lifted with the magnetic fields, forces on the top of the diaphragm decrease and forces on the bottom of the diaphragm remain the same. The smaller force on the top results in the valve seat moving upward. *(Courtesy Parker Fluid Control Division [Gold Ring™])*

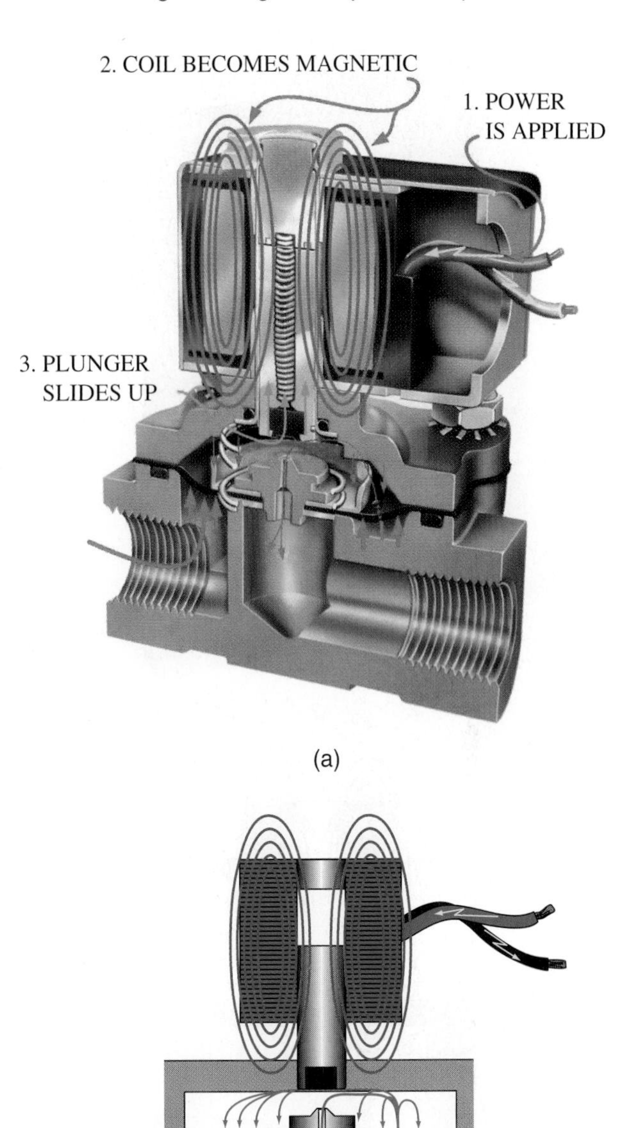

Figure 5-1-28 (a) As the power is applied, the coil becomes magnetic. The plunger lifts, opening the pilot orifice and allowing the pressure on the top of the diaphragm to decrease; (b) Path of refrigerant through pilot orifice. *(Courtesy Parker Fluid Control Division [Gold Ring™])*

the diaphragm drops below the pressure on the bottom side of the diaphragm, Figure 5-1-28a (pictorial drawing) and Figure 5-1-28b (diagram).

The higher pressure on the bottom of the diaphragm forces it up and off of the main seat so flow can begin, Figure 5-1-29a (pictorial drawing) and Figure 5-1-29b (diagram). The system pressure is the force that opens and closes a solenoid valve. There is

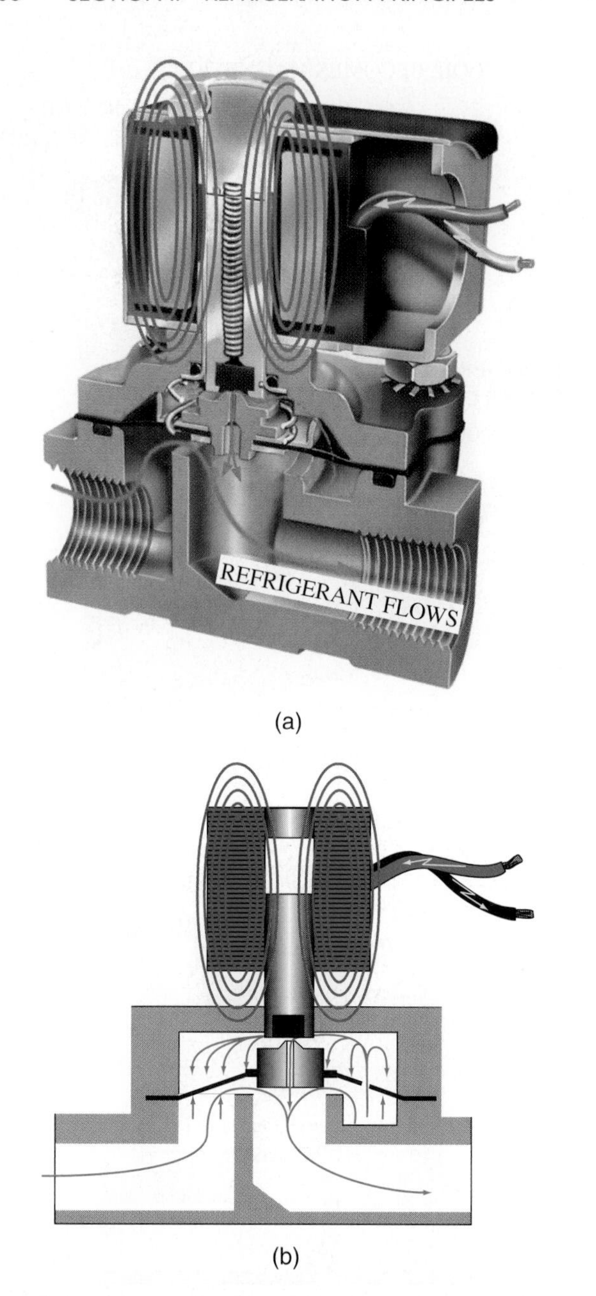

Figure 5-1-29 (a) Solenoid in the full open position with a small amount of refrigerant passing through the bleed orifice in the diaphragm and pilot orifice in the valve seat *(Courtesy Parker Fluid Control Division [Gold Ring™])*; (b) Schematic showing path of refrigerant.

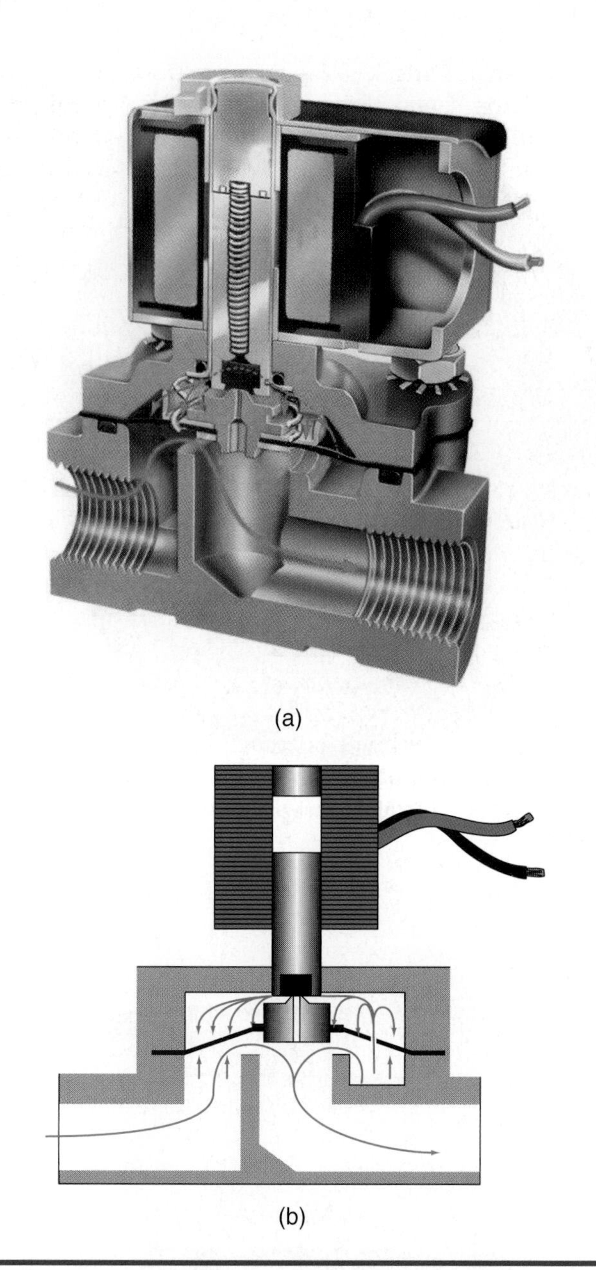

Figure 5-1-30 (a) The coil is de-energized, the pilot orifice is plugged, and pressure builds above the diaphragm as refrigerant flows through the diaphragm bleed orifice *(Courtesy Parker Fluid Control Division [Gold Ring™])*; (b) Path of refrigerant during shutdown.

a slight delay between the time the coil is energized or de-energized, and the time that the valve reacts. This time delay is because of the lag in the time it takes for the pressure on the top side of the diaphragm to change.

When the normally closed valve coil is de-energized, the plunger drops back into the diaphragm orifice, sealing it off. The force applied to the top of the diaphragm becomes greater than the force applied to the bottom side of the diaphragm, Figure 5-1-30a (pictorial drawing) and Figure 5-1-30b (diagram).

TECH TIP

The pressure differential in a system is significant. It is not possible for this diaphragm to stick closed in a manner that rapping on the valve body with a wrench or hammer can free it. Striking the outside body of a solenoid valve will never fix the problem. Unfortunately, we find damaged valve bodies that were obviously struck with tools such as adjustable wrenches or small hammers. Never beat on a valve in an attempt to repair it.

When testing a solenoid coil, if the coil is removed from the central shaft while it is energized it will quickly overheat. The overheating of this coil will damage it. To prevent overheating from occurring place a screwdriver shaft through the core center as it is removed. The magnetism of the coil will pass through the screwdriver shaft and not cause the coil to overheat.

Some solenoid valves have internal diaphragm pistons and springs that are replaceable. When removing these parts for service, the exact manufacturer's replacement parts must be used. Using mismatched or ill fitting parts can cause damage to the system and in some cases even be hazardous.

Solenoid valves can be installed in a system using threaded fittings or sweat fittings. Some manufacturers require that when brazed or sweat fittings are being installed the valve body be disassembled, removing the diaphragm to prevent heat damage that may occur during brazing. Some manufacturers have provided extra long studs on the valve body so that brazing heat can be minimized. When brazing any solenoid valve remove the coil and wrap the valve body with a wet rag to keep it from overheating.

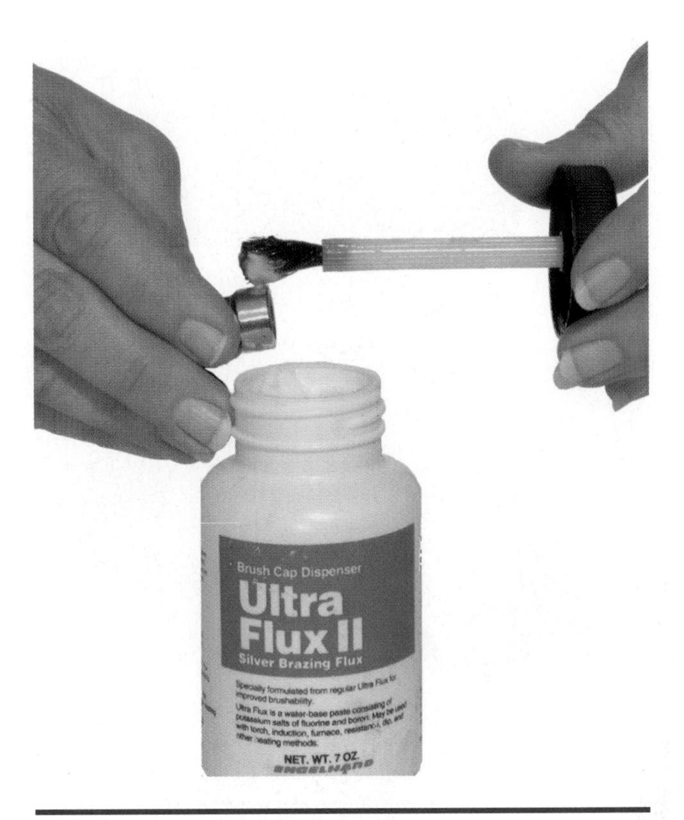

Figure 5-1-31 Silver brazing flux must be applied to glass valve bodies before silver brazing.

TECH TIP

Many solenoid valve bodies are made of brass. It is necessary to use a flux with silver braze material when joining copper to brass. Complete bonding between the silver and brass will not occur without the use of a proper brazing flux such as the type shown in Figure 5-1-31.

SAFETY TIP

Safe work habits require the technician to verify that the solenoid valve is not under pressure before it is disassembled. Some valves have service ports for pressure testing. These service ports can be used to verify system pressure before disassembling the valve. If pressure taps are not located on the valve, then other system pressure taps must be accessed to confirm the system does not have pressure before removing the valve body.

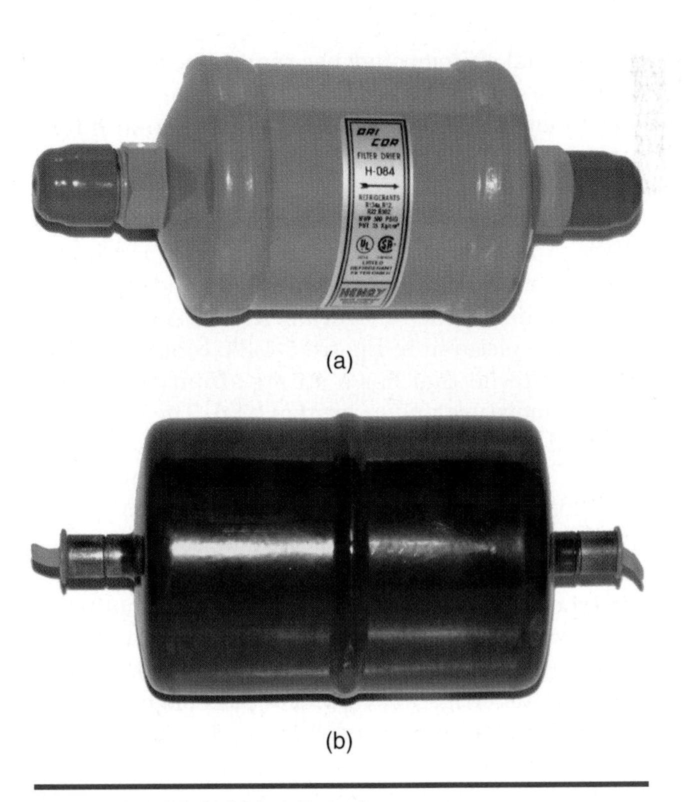

Figure 5-1-32 Welded filter driers.

FILTERS AND DRIERS

Air conditioning and refrigeration systems use both filters and driers. Most residential sized systems use a combination device that performs as both a filter and a drier all contained within a single welded shell, as shown in the examples in Figure 5-1-32a,b.

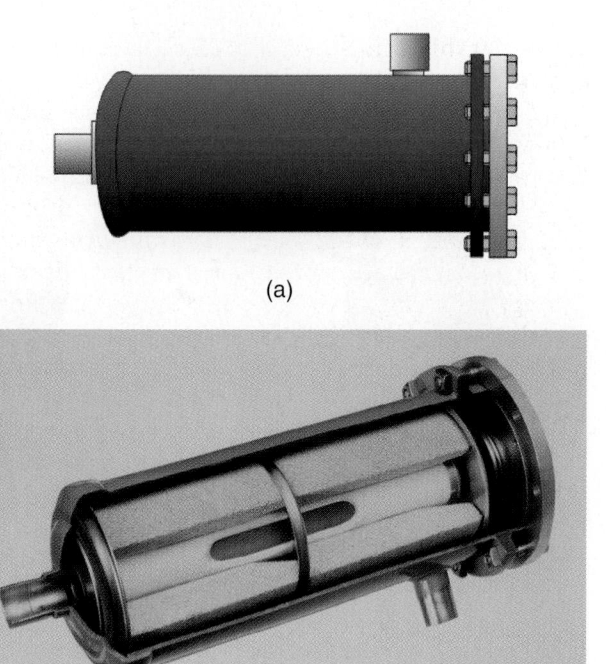

Figure 5-1-33 (a) Schematic of replaceable core filter drier; (b) Pictorial cutaway view of replaceable core filter drier. *(Courtesy Sporlan Valve Company)*

Figure 5-1-34 Solid block dessicant in filter drier. *(Courtesy Rheem Manufacturing)*

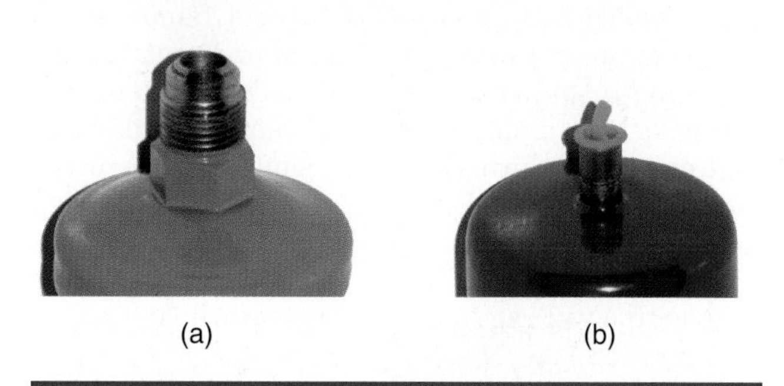

Figure 5-1-35 (a) Flare fitting on filter drier; (b) Braze fitting on filter drier.

Commercial systems may use removable core filters, which are shown in Figure 5-1-33a,b. The advantage of commercial removable core systems is that different core materials can be used depending on the system clean-up needs.

Not all filters are driers, but most driers are also filters. A filter may be a fine wire screen, mesh material, or molecular sieve, all designed to trap solids traveling inside the refrigerant circuit. A drier, such as the one pictured in Figure 5-1-34, contains a desiccant material that has a strong affinity or attraction for moisture and/or system acid. The desiccant material may be formed into a single porous block or shaped into BB sized beads. The most commonly used desiccant material is activated alumina.

Driers are designed to remove system contamination. A properly installed system should contain no such contaminants. It is therefore practical not to install a drier if the system was properly assembled. As a drier becomes clogged with moisture or acid, this will restrict refrigerant flow. It is possible for a filter drier to become so plugged that it acts like a second metering device with a substantial pressure drop.

The ability of all desiccants to hold moisture or acid is temperature dependent. As a desiccant is heated, it will release moisture. As it cools it will absorb moisture. The proper way of removing a sweated filter drier is with a tubing cutter. If a sweated filter drier is removed using a torch, the torch's heat will cause the desiccant to release its trapped contaminants back into the system. If a plugged filter drier is removed with a torch, smoke can be seen wafting out of the just released fitting. That smoke is the revaporized moisture and acidic oil, much of which has just gone back into the refrigerant circuit.

TECH TIP

The end fittings on filter driers are capped with plastic covers to keep ambient moisture from contaminating the core material before they are installed. Do not remove these covers until just before the filter drier is to be installed to keep the desiccant as dry as possible before the filter drier is installed.

Some technicians prefer using threaded filter driers (Figure 5-1-35a) or flare fitted filter driers (Figure 5-1-35b) because they can be removed without heat or cutting. Once the filter drier has been removed, a short piece of tubing with fittings on each end can be installed in its place if a new filter drier is not being reinstalled. When a dirty filter drier is being removed that has flare fittings, it is easily replaced with another filter drier with flare fittings.

Suction line filter driers should be removed according to the manufacturer's recommendations following system cleanup. Low section line pressures have a major impact on system operation and efficiency. As a section line filter drier becomes plugged from picking up acid or moisture, there is an increased pressure drop across the filter drier. Even a small pressure drop of 3 psi can adversely affect system function. Therefore, the technician should return to a system with a section line filter drier within several days to confirm the system's pressure drop. If it has reached the manufacturer's recommended limit, remove it and replace it with a clean filter drier or a section of straight copper pipe.

Under no circumstance should more than one filter drier be placed in a line. Some technicians assume that this is an acceptable practice. It is not, and can cause major restrictions to refrigerant flow.

There are two types of filter driers, one for the liquid line and one for the suction line. There are bi-flow filter driers that are used in heat pumps, such as the one shown in Figure 5-1-34. Bi-flow filter driers have built in check valves that allow the refrigerant to pass through the filter in one direction during the heating cycle and in the opposite direction during the cooling cycles. It is not possible for a heat pump to function properly with a standard filter drier installed.

TECH TALK

Tech 1. Accumulators and receivers look the same. Are they interchangeable?

Tech 2. No. Accumulators are designed to let only vapor refrigerant flow through and receivers are designed to let only liquid refrigerant flow through.

Tech. 1. So they look the same, have the same two fittings but they perform exactly oppositely.

Tech 2. That's right and you need to check the inlet and outlet marking on them before you install them in a system. Get them backwards and they will not work.

Tech 1. Thanks.

UNIT 1—REVIEW QUESTIONS

1. List the ten major components that can be found in air conditioning and refrigeration systems.
2. _______ difference is what allows heat to flow from one body to another.
3. List the two factors controlling heat transfer.
4. _______ should be sized so that 100% of the refrigerant charge can be stored in the receiver.
5. What is an accumulator?
6. _______ are often found on heat pumps where during changeover cycle, liquid refrigerant can flow back out of the outdoor coil.
7. A(n) _______ is a device that collects some of the oil leaving the discharge side of the compressor and returns that oil to the compressor.
8. _______ are used in air conditioning and refrigeration as refrigerant control devices.
9. What will happen if a drier becomes clogged with moisture or acid?
10. What are the two types of filter driers?

UNIT 2

Compressors

OBJECTIVES: In this unit the reader will learn about compressors used in the HVAC/R industry including:

- Compressor Types
- Compressor Lubrication
- Functions of Compressors
- Capacity Control Factors

INTRODUCTION

The compressor is a mechanical device for pumping refrigerant vapor from the low pressure side in the evaporator to the high pressure side in the condenser. Since pressure, temperature, and volume of a gas are related, a change in pressure from low to high causes an increase in temperature and a decrease in volume or a compression of the vapor. The main types of compressors are shown in Figure 5-2-1a–e: reciprocating piston (a), centrifugal (b), rotary (c), screw (d), and scroll (e).

1. **Reciprocating compressors**
   a. Open, direct drive
   b. Open, belt drive
   c. Open, for transporting vehicles
   d. Semihermetic
   e. Hermetically sealed
2. Rotary compressors
3. Scroll compressors
4. Screw compressors
5. Centrifugal compressors

The type of compressor used depends on the application and the size (tonnage) of the project. Figure 5-2-2 indicates the various applications for the different types of compressors.

Compressors are identified by the mechanical part(s) that perform the actual pumping of the refrigerant vapor. In the reciprocating compressor, Figure 5-2-3, a piston travels back and forth in a cylinder. The scroll compressor, Figure 5-2-4, has a

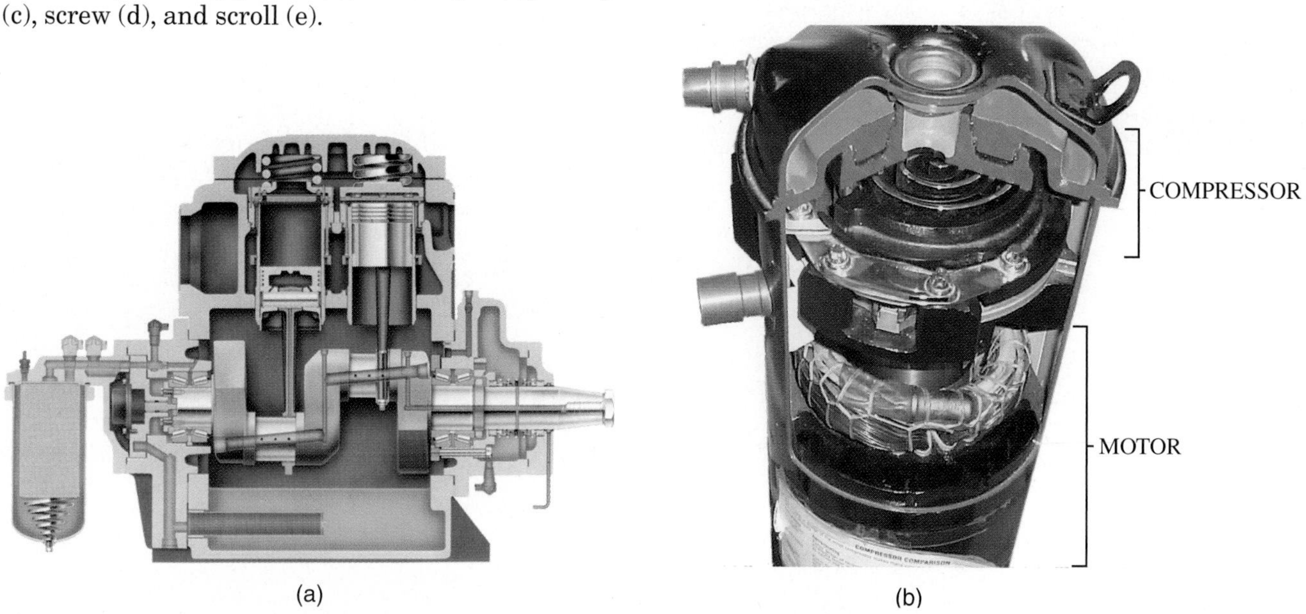

(a)

(b)

Figure 5-2-1 (a) Reciprocating compressor (*Courtesy Vilter Manufacturing*); (b) Centrifugal compressor (*Courtesy of the American Society of Heating, Refrigerating, and Air-Conditioning Engineers, Inc.*); (c) Rotary compressor; (d) Screw compressor (*Courtesy of York International Corp.*); (e) Scroll compressor.

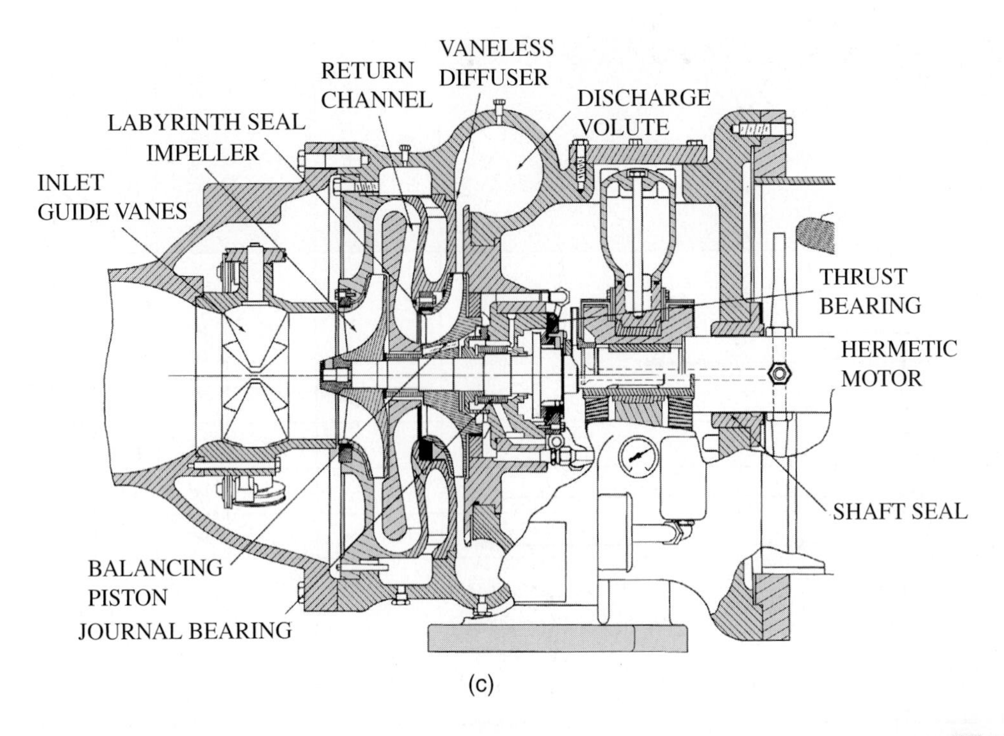

(c)

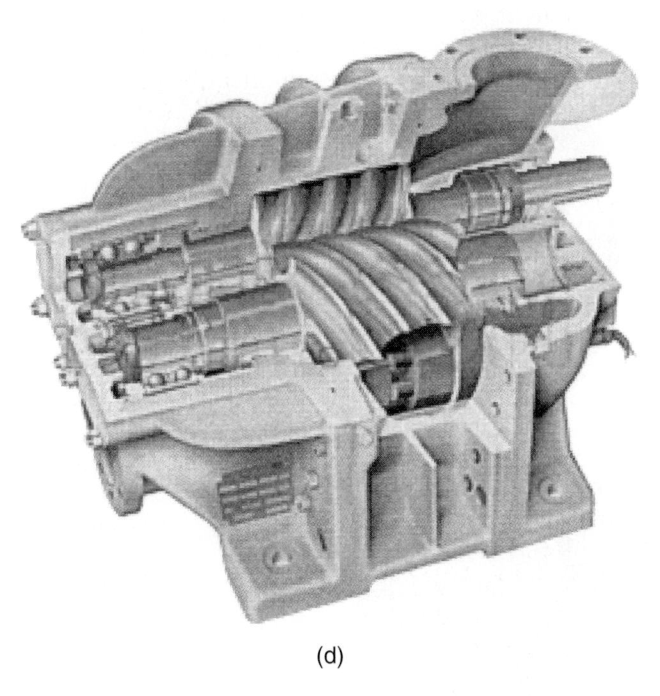

(d)

(e)

Figure 5-2-1 (continued)

stationary scroll and an orbiting scroll that moves within the stationary scroll. The centrifugal compressor has a very high speed centrifugal impeller with multiple blades rotating within a housing, as shown in Figure 5-2-5. The screw compressor, Figure 5-2-6, uses a rotating screw within a tapered housing. The rotary compressor, Figure 5-2-7, has a vane that rotates within a cylinder.

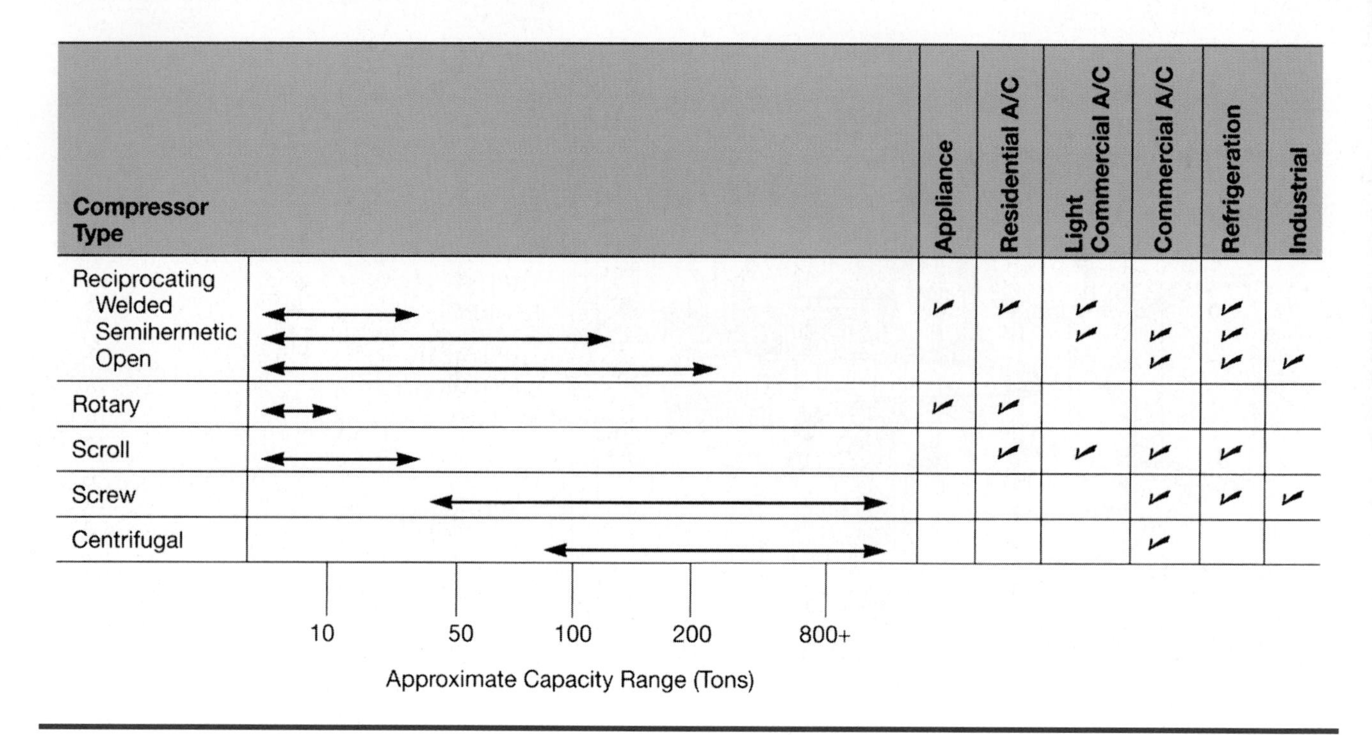

Compressor Type	Appliance	Residential A/C	Light Commercial A/C	Commercial A/C	Refrigeration	Industrial
Reciprocating						
Welded	✓	✓	✓		✓	
Semihermetic			✓	✓	✓	
Open				✓	✓	✓
Rotary	✓	✓				
Scroll		✓	✓	✓	✓	
Screw				✓	✓	✓
Centrifugal				✓		

Figure 5-2-2 Compressor application by types.

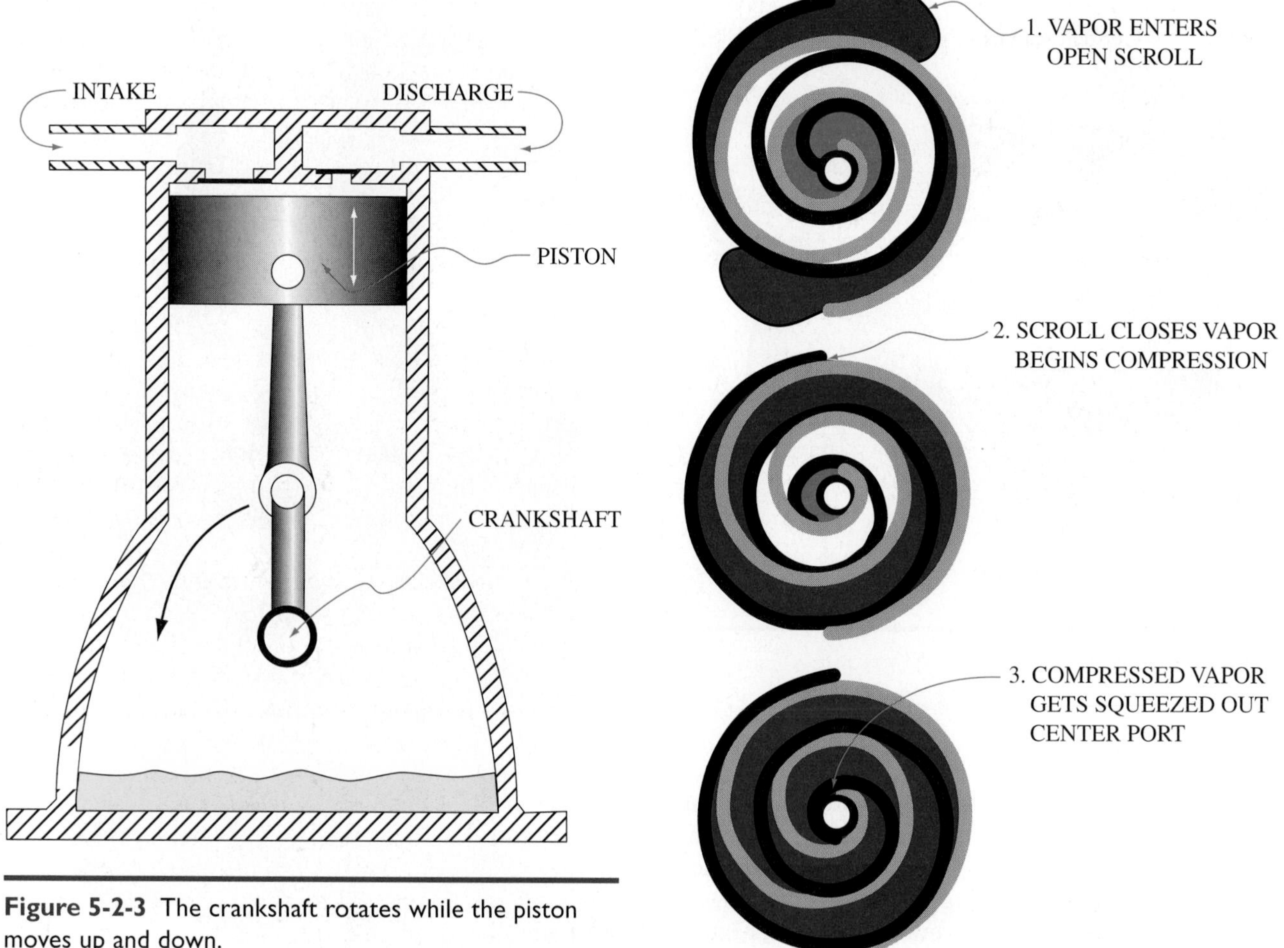

Figure 5-2-3 The crankshaft rotates while the piston moves up and down.

Figure 5-2-4 Scroll compressor parts and movement.

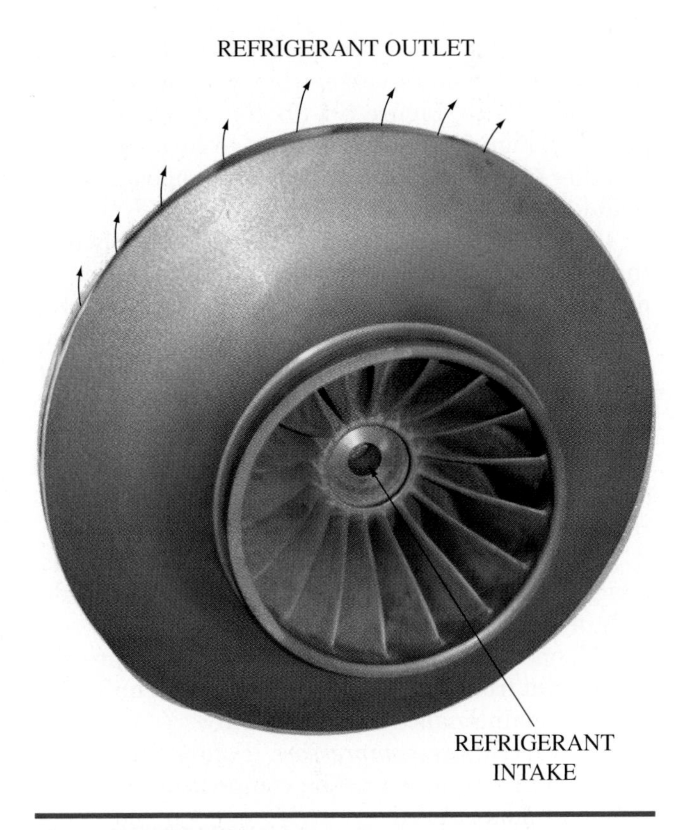

Figure 5-2-5 Centrifugal compressor turbine.

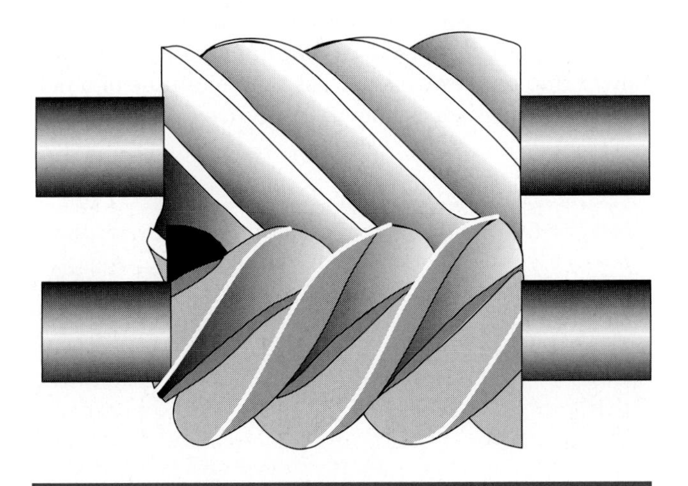

Figure 5-2-6 Screw compressor.

COMPRESSORS

After it has absorbed heat and vaporized in the cooling coil, the refrigerant passes through the suction line to the next major component in the refrigeration circuit, the compressor. This unit, which has two main functions within the cycle, is frequently classified as the heart of the system, for it circulates the refrigerant through the system. The functions it performs are:

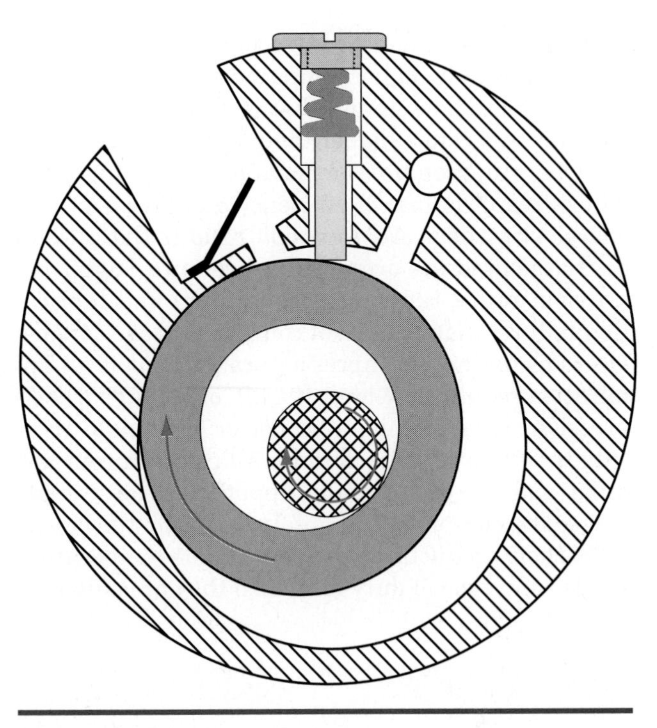

Figure 5-2-7 Rotary compressor.

1. Receiving or removing the refrigerant vapor from the evaporator, so that desired low pressure and low temperature can be maintained.
2. Increasing the pressure of the refrigerant vapor through the process of compression, and simultaneously increasing the temperature of the vapor so that it will give up its heat to the condenser cooling medium.

The compressor is the heart of all air conditioning and refrigeration systems. The compressor pumps refrigerant vapor from the evaporator, lowering the pressure in the evaporator. As the pressure is lowered, so is the corresponding saturation temperature of the refrigerant. As the refrigerant picks up heat in the evaporator, it vaporizes and enters the compressor.

As a comparison, the heat energy to boil water in a kettle on a stove comes from the gas or electric burner. In the same way, the heat to boil liquid refrigerant in an evaporator comes from the air (or fluid) surrounding the evaporator. It may be cold, but it still contains heat energy. As heat flows to the cold refrigerant liquid, more vapor boils away, carrying heat energy with it and keeping the evaporator cold.

The compressor increases the pressure and temperature of the refrigerant to a higher level where the heat can be transferred to air or water flowing through the condenser, thus condensing the high pressure refrigerant vapor to a high pressure liquid and readying the cycle to repeat again. Heat is absorbed in the evaporator at a low temperature and rejected in the condenser to a cooling medium, which is at a noticeably higher temperature. Thus, heat has been made to flow uphill, that is, from a lower temperature space to a higher temperature space.

Operating compressors near or at their high limits can result in loss of efficiency and excessive discharge superheat. *Compression ratio* is the ratio of discharge pressure over suction pressure. High compression ratios can cause overheating and lubrication failures. Some types of compressors are capable of producing high compression ratios. But if an application requires too high of a lift, or too high a compression ratio, it might be better to handle it in stages, with a cascade system. All common refrigeration and air conditioning applications are single stage. The equipment manufacturers rate the capacity of their products for the permissible temperature ranges and type of duty for which they are suited.

NOTE: A refrigeration compressor is much like a pump in a shallow well dug in sand. As the pump lifts water from the well and dumps it out, the level of water in the well is momentarily lowered; however, because the sand is porous and surrounded by water, the well begins to fill. The pump continuously lifts the water out of the well and pours it on the surrounding sand. If the well has a large pump, the water level can be lowered. So the pump can be turned off for short periods of time. It would have to be cycled off and on to maintain the water level at a desired depth. If the pump is too small, it can't keep up with the water, and the level begins to rise. Just like the case of the water pump in sand, the compressor capacity must be sized to match the load.

COMPRESSOR TYPES

Having dealt with the general information on compressors, it is time to look at the various types available for refrigeration and to examine the characteristics of each type—how they work and how they are serviced.

Reciprocating Compressors

Reciprocating compressors have been used in refrigeration service for a long time. The reciprocating compressor is used in the majority of domestic, small commercial, and industrial condensing unit applications. This type of compressor can be further classified according to its construction. The oldest compressor design is the open type compressor shown in Figure 5-2-8a. Open compressors may be driven by means of a pulley on a shaft that extends through a seal. The pulley is driven by a separate motor or engine. Hermetic compressors are enclosed in a welded pressure vessel; they cannot be opened for field service, Figure 5-2-8b. Semihermetic compressors, Figure 5-2-8c, are enclosed inside a bolted together cast iron housing; they can be opened in the field for service.

Open compressors have performed reliably over the decades, but have been largely superseded by newer designs. Open machines (unlike hermetic) can be used with ammonia (NH_3) refrigerant as long as all components and accessories are made of iron or steel (or other suitable materials). Sizes of open compressors range from 5 to 150 tons and larger. They are often used for refrigeration and industrial applications where rugged, serviceable machinery is required.

Figure 5-2-9 shows a cutaway view of an industrial open compressor and Figure 5-2-10 shows details of essential components.

All open compressors use a shaft seal to prevent refrigerant from leaking around the crankshaft. This is also a maintenance consideration.

In *direct drive compressors,* Figure 5-2-11, the motor shaft is coupled to the compressor shaft and driven at motor speed.

Belt driven machines offer the flexibility of selecting a compressor speed to match the load. The belts require additional space and a protective guard. They require increased maintenance and increased power due to operational losses. An example of a belt driven compressor is shown in Figure 5-2-12.

NOTE: One major advantage of an open drive compressor is that the motor and compressor are separate units so that the motor or compressor can be repaired or replaced without having to buy a complete system. Another advantage is that the motor's heat is not absorbed in the refrigerant, so it does not have to be rejected in the condenser like most other compressor types. Open drive compressors are frequently used for ice storage bins located at grocery and convenience stores. The advantage of using the open drive compressor is that these storage bins are typically located outside or where the motor heat can easily be dissipated. This means that the system can operate more efficiently.

The most common use of open compressors is for automobile air conditioning. The main vehicle engine drives the compressor used for air conditioning. Open compressors are also used for refrigeration on trucks, but these are generally driven by a separate engine.

Hermetic compressors were developed in the 1920s and 1930s when refrigerants came into use that

(a)

(b)

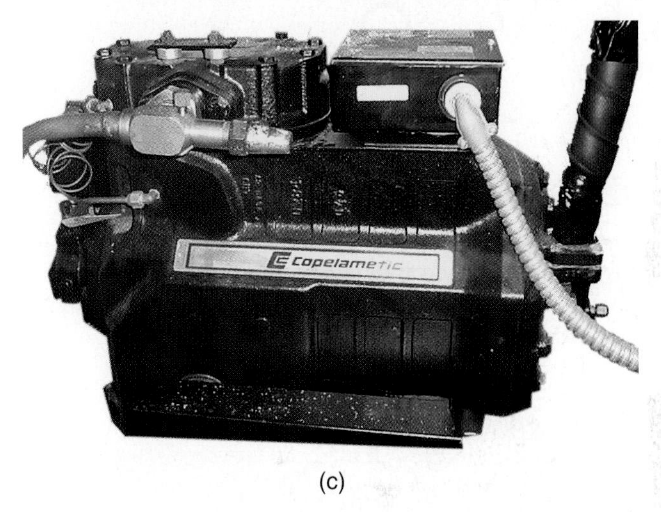

(c)

Figure 5-2-8 (a) Open compressor; (b) Hermetic compressor; (c) Semi-hermetic compressor in cast iron housing.

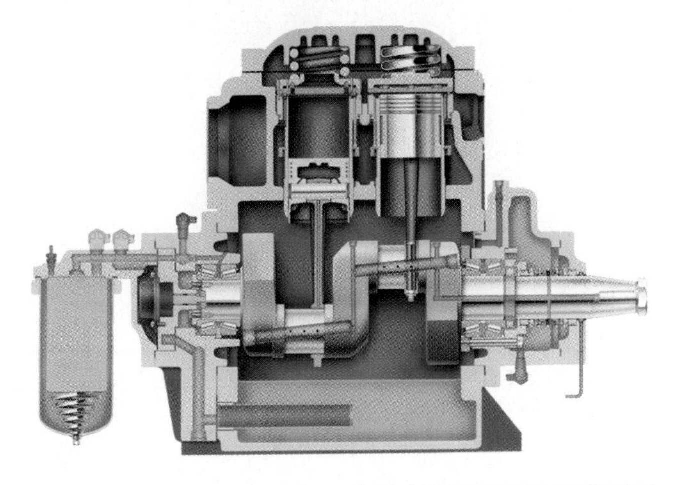

Figure 5-2-9 Cutaway view of open type industrial duty compressor. *(Courtesy Vilter Manufacturing)*

were compatible with electric motor components, especially wire insulation. The benefits for use on a domestic refrigerator were immense and obvious:

- Direct connected, reliable, no belts or couplings to wear;
- Sealed in welded steel shell, no shaft seal to leak;
- Very compact, permitting increased refrigerated storage space;
- Lower sound level, ideal for domestic appliances.

Larger sealed machines were developed for air conditioning applications by the late 1930s and by the mid 1950s the open machine had been largely supplanted in most appliance and packaged HVAC/R applications.

Hermetically sealed electric motor compressor units, Figure 5-2-13, are made in a variety of sizes, from tiny fractional horsepower units meant for small appliances, to larger units up to about 20 tons for air conditioning use. They are sometimes called welded hermetics, full hermetics or sealed hermetic compressors. For the sake of simplicity, they shall simply be referred to hereafter as hermetics.

The welded steel shell prevents any field service access so there are no service procedures to replace damaged internal components such as motors, bearings, valves, etc. If damaged or defective, the entire hermetic compressor is replaced. They are usually

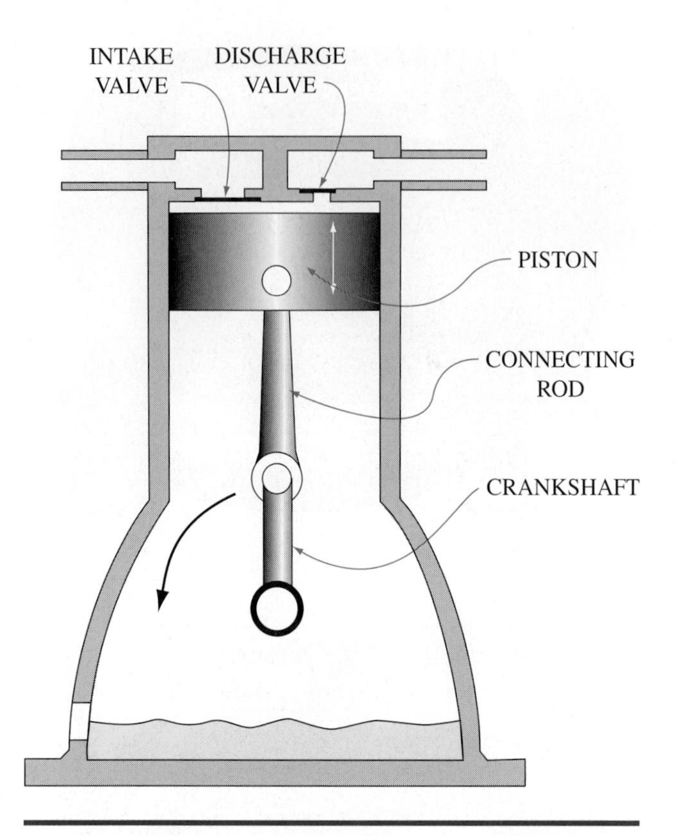

Figure 5-2-10 Industrial duty compressor components: Intake valve, discharge valve, piston, connecting rod, and crankshaft.

Figure 5-2-12 Open type direct drive compressor. *(Courtesy Hampden Engineering Corporation)*

Figure 5-2-11 Direct drive compressor. *(Courtesy Frick, York Refrigeration Systems)*

Figure 5-2-13 Hermetically sealed electric motor compressor units. *(Courtesy Danfoss Inc.)*

internally spring isolated, Figure 5-2-14, to reduce the inherent vibration caused by the reciprocating action of the pistons.

Figure 5-2-15 is a diagram of the piston action in a high efficiency hermetic compressor with the suction valves located in the crown of the piston. Larger sized vertical hermetic compressors have centrifugal oil pumps on the bottom of the motor shaft.

The semihermetic motor compressor unit, which is field serviceable by virtue of its bolted construction, has evolved over the past 50 to 60 years. It may have a bolted cast iron construction or bolted, flanged, drawn steel shell. It is field or factory repairable, and parts such as valve reeds, gaskets, bearing inserts, or motor stators may be replaced on various units. These machines are known by a variety of names: semihermetics, serviceable hermetics, accessible hermetics, and bolted hermetics. For the

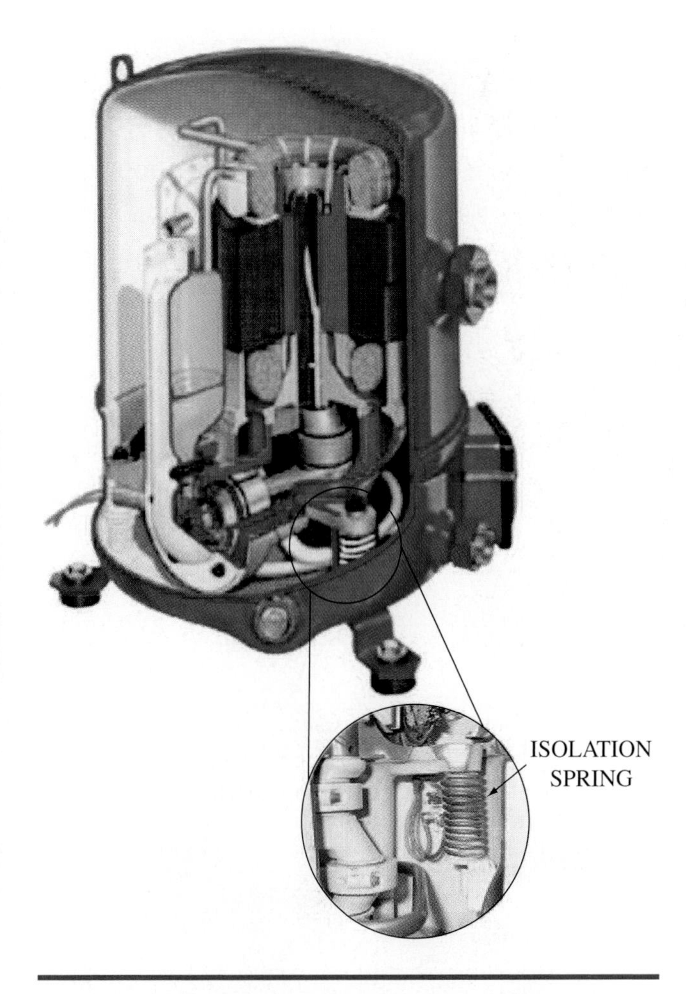

Figure 5-2-14 Cutaway view of hermetic compressor showing internal isolation spring inset. (*Courtesy Danfoss Inc.*)

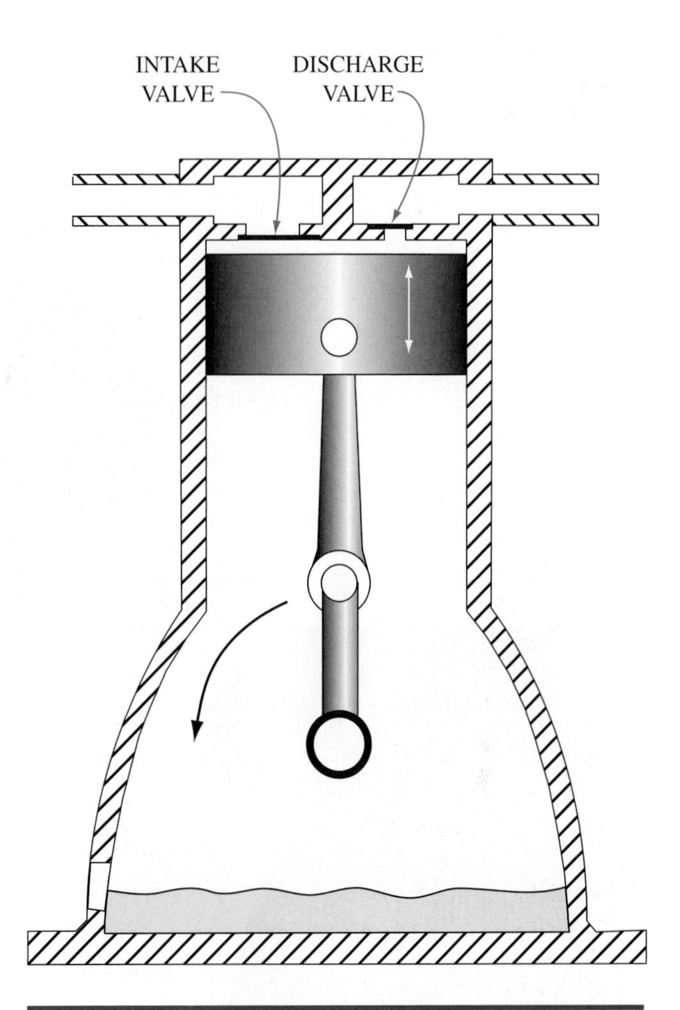

Figure 5-2-15 High efficiency hermetic compressor with suction valves in crown of piston.

sake of consistency, they shall be referred to hereafter as semihermetics. The units are available in sizes that range from approximately 7.5 to 125 tons cooling capacity. Figure 5-2-16 shows a semihermetic compressor used for medium and low temperature refrigeration.

Figure 5-2-17 shows the pressure lubrication system for a refrigeration duty semihermetic compressor. Figure 5-2-18 shows a reversible oil pump that is used in both the larger semihermetic and

SERVICE TIP

Although semihermetic compressors are field serviceable, for cost reasons many are removed and sent to companies that specialize in refurbishing compressors. These facilities are set up to thoroughly clean and inspect the entire compressor and replace those parts that have worn beyond tolerance. Field service work is commonly done for large compressors that would be extremely difficult or expensive to remove.

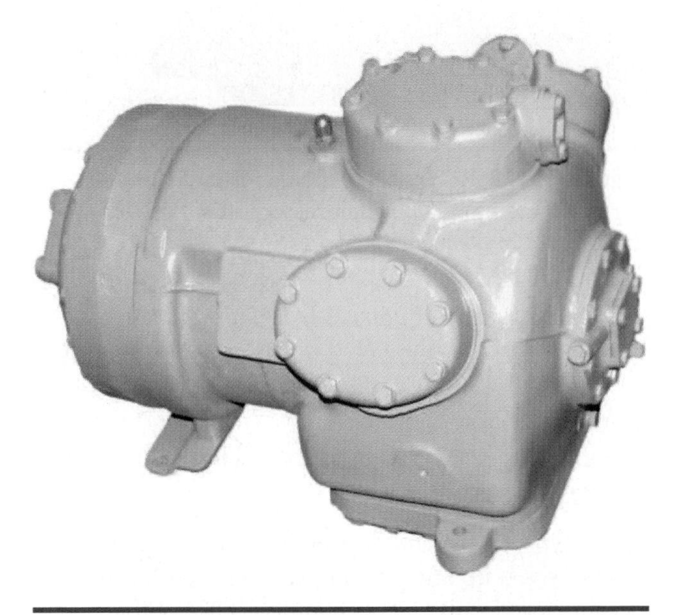

Figure 5-2-16 Semi-hermetic compressor used for AC, medium and low temperature refrigeration.

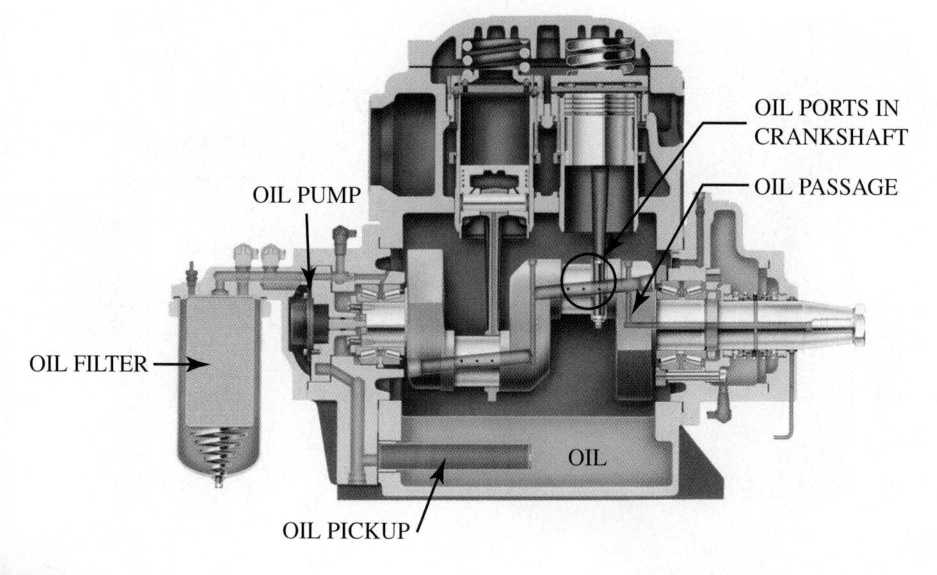

Figure 5-2-17 Section of semi-hermetic compressor showing centrifugal oil pump.

Figure 5-2-18 Reversible oil pump for larger semi-hermetic and open-type compressors. (*Courtesy Hampden Engineering Corporation*)

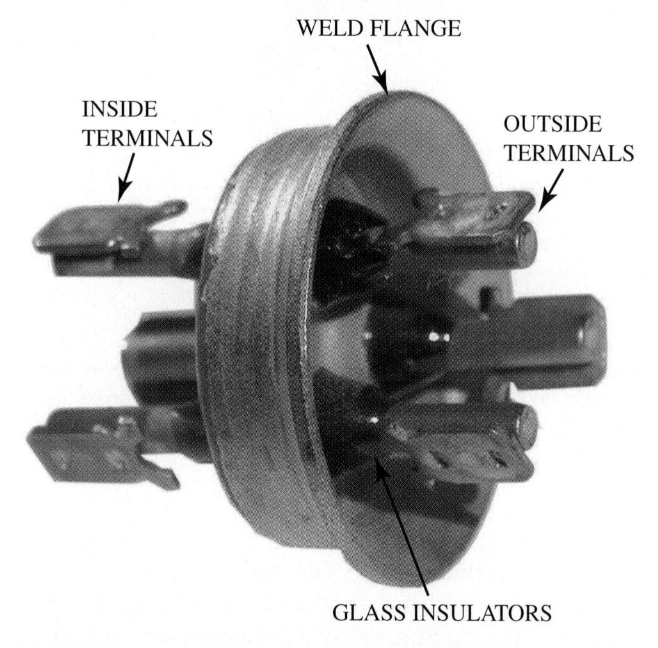

Figure 5-2-19 A glass to metal connection is used to isolate the electrical terminals from the metal compressor shell.

open compressors. Figure 5-2-19 shows the isolation arrangement for the electric power terminals on a semihermetic compressor.

Operation of Reciprocating Compressors

In a reciprocating compressor, Figure 5-2-20, the piston is driven up and down in the cylinder by the connecting rod and crankshaft. The valves are self-opening and closing due to pressure difference. They are sometimes spring loaded, but the principle is the same. Note that this is quite different from an automobile engine where the valves are mechanically opened and closed by a camshaft synchronized to the crankshaft.

Figure 5-2-21 shows the various types of reciprocating cylinder valves: (a) the flexing reed valve; (b) the floating reed valve; (c) the ring valve; and (d) the reduced clearance poppet valve used for the Discus compressor.

Pistons within the compressors may have the suction valve located in the top of the piston; this is classified as a valve-in-head type, or the piston may have a solid head, with the suction and discharge valves located in a valve plate or cylinder head.

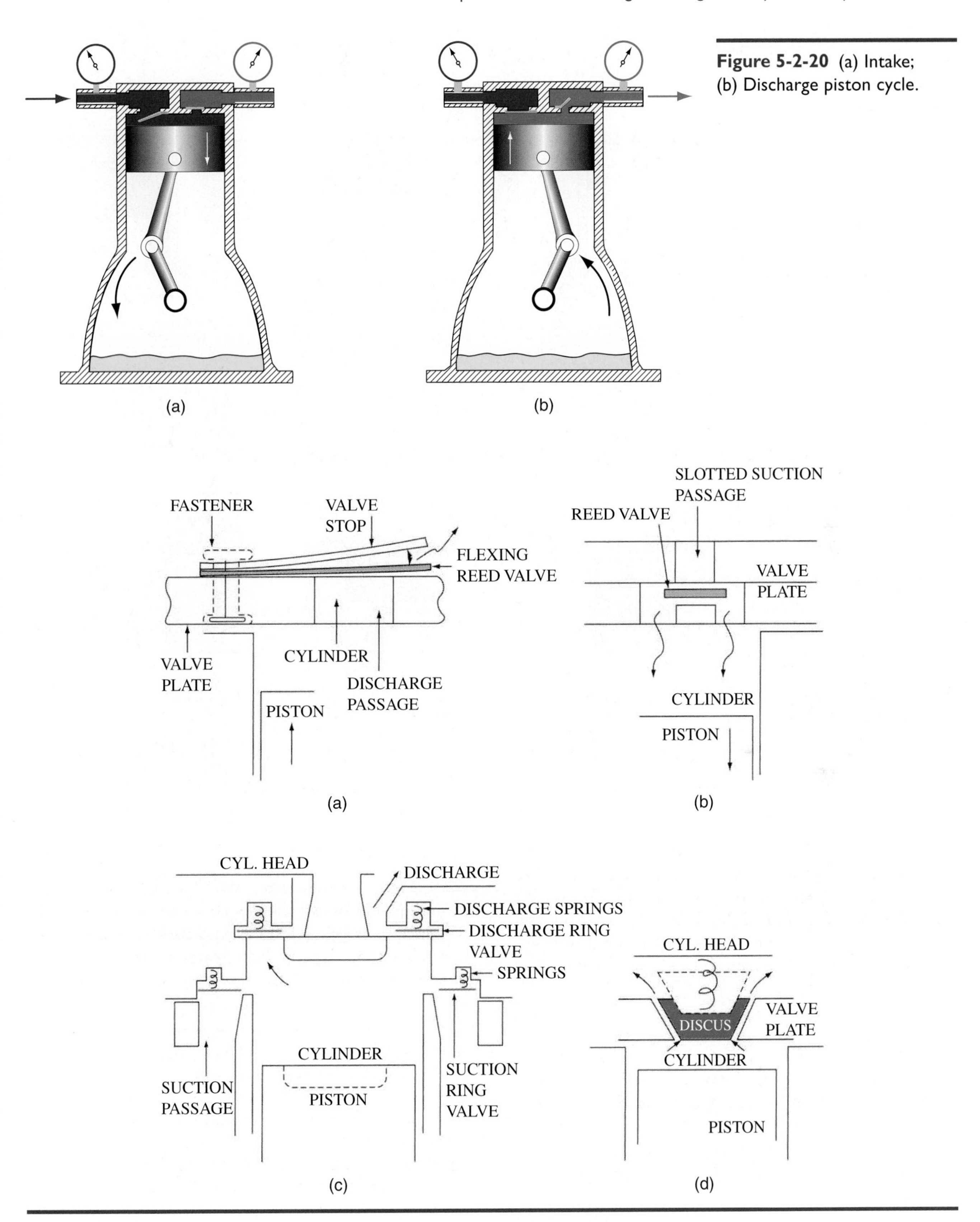

Figure 5-2-20 (a) Intake; (b) Discharge piston cycle.

Figure 5-2-21 Typical reciprocating compressor cylinder valves: (a) Flexing reed valve; (b) Floating reed valve; (c) Ring valves; (d) Reduced clearance poppet valve, discus (TM) compressor.

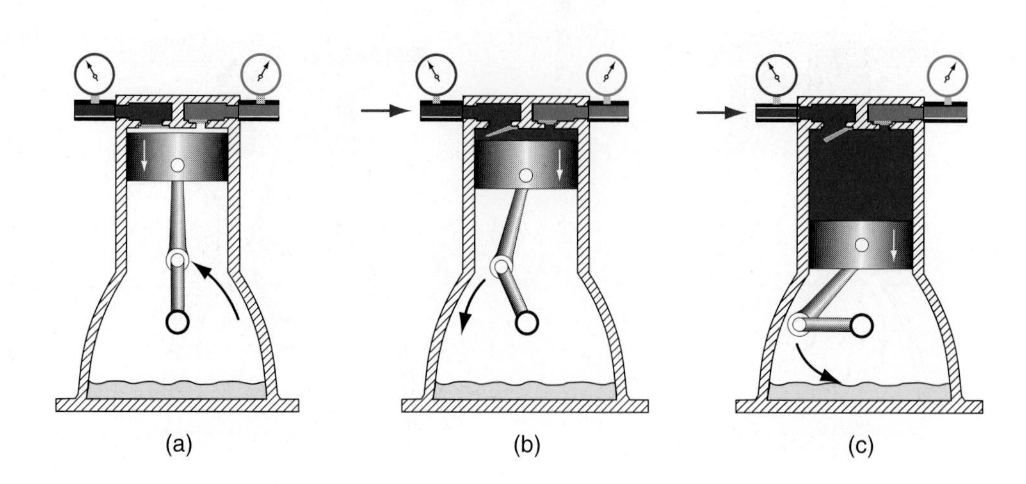

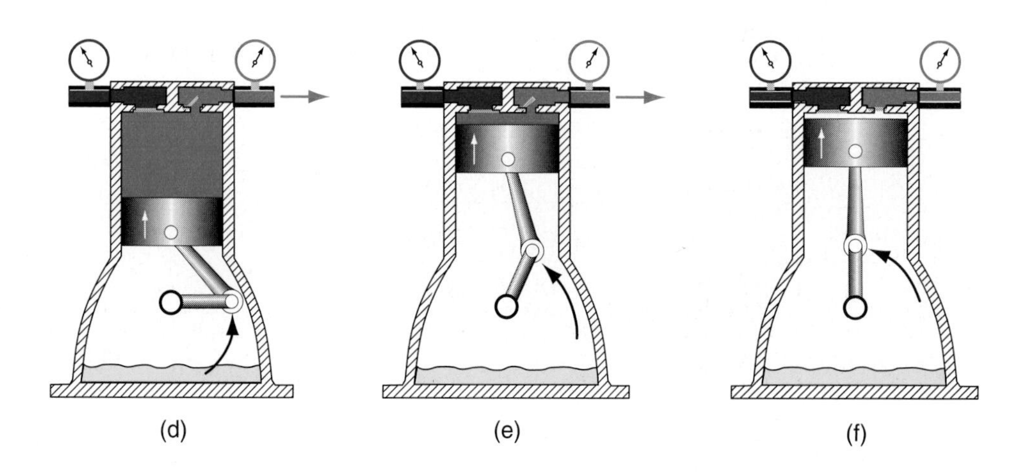

Figure 5-2-22 Complete compression cycle of a reciprocating piston compressor.

As the piston strokes down in the cylinder, Figure 5-2-22, the pressure will drop to a point where it is below the pressure in the suction manifold. At that point the pressure difference will force the suction valve(s) to open, and low pressure, cold (but superheated) suction gas from the evaporator will flow into the cylinder.

Figure 5-2-22a–f shows the complete compression cycle of a recipicating piston compressor. Figure 5-2-22(a) shows the beginning of the intake cycle; the suction valve does not open until the piston has reached a point where the cylinder pressure is below the system suction pressure. In Figure 5-2-22(b), the higher pressure in the suction line forces the suction valve open. In Figure 5-2-22(c), the piston continues downward, drawing in refrigerant vapor. In Figure 5-2-22(d), the refrigerant pressure above the rising piston forces the discharge valve open. In Figure 5-2-22(e) the rising piston pushes out the hot refrigerant vapor. Figure 5-2-22(f) shows that even with a very tightly fitting piston a little hot gas remains at the top of the piston stroke.

Clearance Volume

At the top of the stroke, piston travel will reverse directions again. All the gas that can, passes into the discharge manifold. The **clearance volume** is the space between the top of the piston and the cylinder head is filled with compressed gas that is not going anywhere. This gas will have to expand on the downstroke before the pressure can drop to a point to admit fresh suction gas. This reexpansion gas represents wasted work and lost capacity. The compressor designer strives to minimize the clearance volume while at the same time leaving adequate room for thermal expansion and for unwanted incompressible slugs of oil and liquid refrigerant. These slugs can enter the cylinder, and possess the ability to do great damage to the valves, piston, and cylinder head. Theoretically, 0% clearance volume would be good; practically it would be disastrous.

The horizontal axis in Figure 5-2-23 represents the cylinder volume and the vertical axis represents the cylinder pressure. Point *A* represents top dead center

TECH TIP

Disk valves in compressors make a significantly different sound than other valve mechanisms. Occasionally, technicians who are not familiar with the sound may feel that the compressor has a bad rod or other component that is about to fail. Disk compressors are identified on the nameplate. Check the nameplate before you jump to the wrong conclusion.

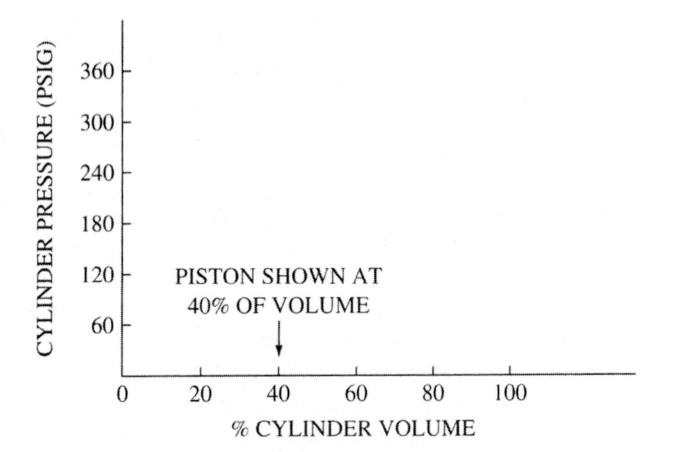

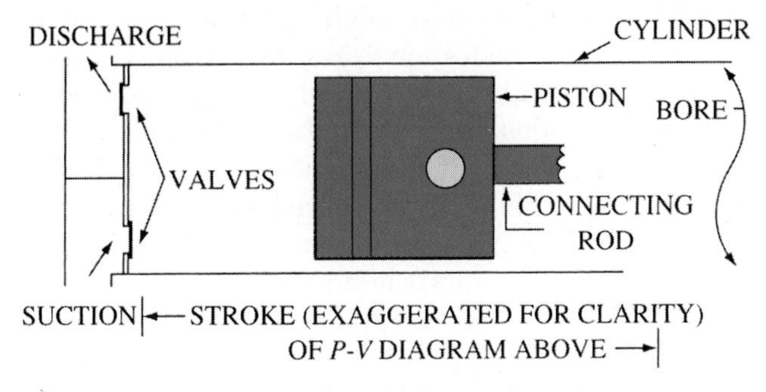

Figure 5-2-23 The two elements of a pressure-volume plot, used in the study of reciprocating compressor performance.

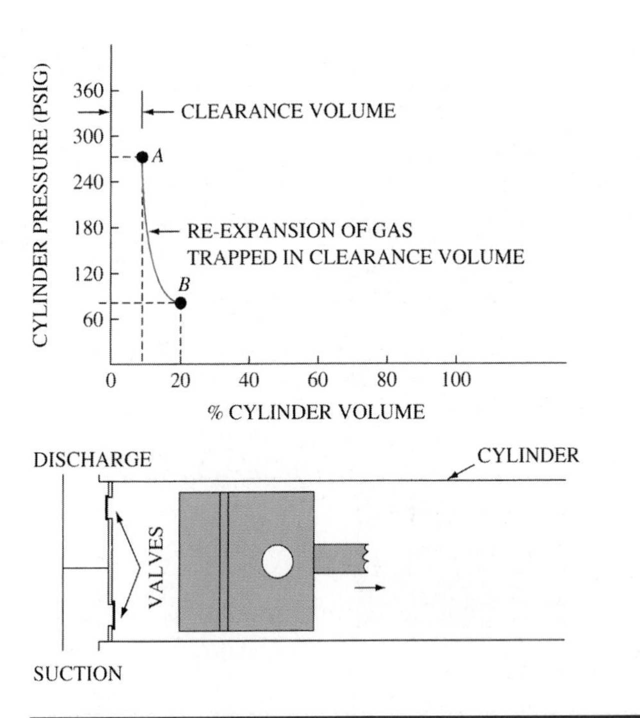

Figure 5-2-24 Beginning of the intake stroke. *A:* top dead center; *B:* suction valve ready to open.

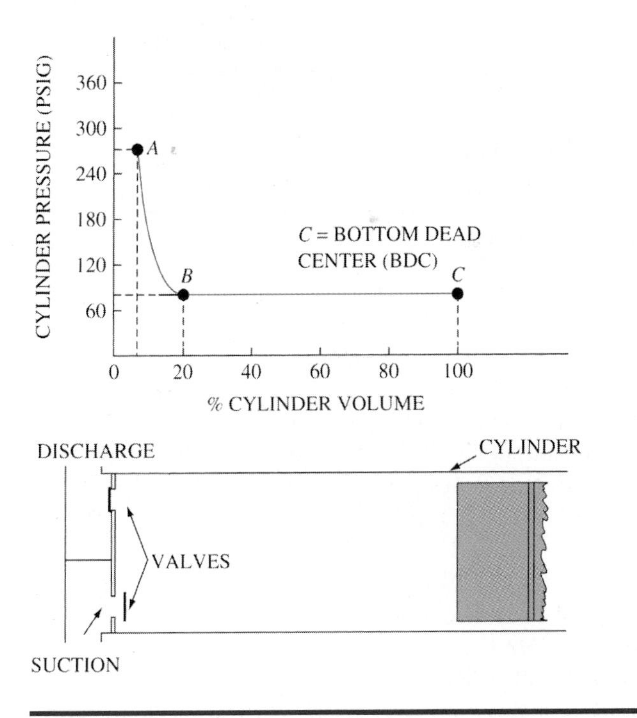

Figure 5-2-25 Useful portion of intake stroke, *B* to *C*.

of the piston stroke and shows the clearance volume and discharge pressure, as shown in Figure 5-2-24. As the piston travels on the downstroke the trapped compressed clearance volume gas has to reexpand until it reaches suction pressure at point *B*. This represents perhaps 20% of the intake stroke and points to the inefficiency involved. The other types of machines (scroll, rotary, etc.) do not have trapped reexpansion gas to deal with and, thus, have enhanced efficiencies.

As the downstroke continues from point *B* to point *C*, the cylinder pressure is below the suction manifold pressure and the pressure difference opens the suction valve(s) and fills the cylinder with new refrigerant vapor, Figure 5-2-25. At the bottom of the stroke (C) the suction valve(s) closes.

The compression stroke from bottom dead center (*C*) to point *D* compresses the gas to a point about equal to discharge manifold pressure, Figure 5-2-26. During the compression stroke, both suction and discharge valves are closed. The piston has traveled from 100% of cylinder volume to a point about 20% of cylinder volume. On air conditioning R-22 applications, the

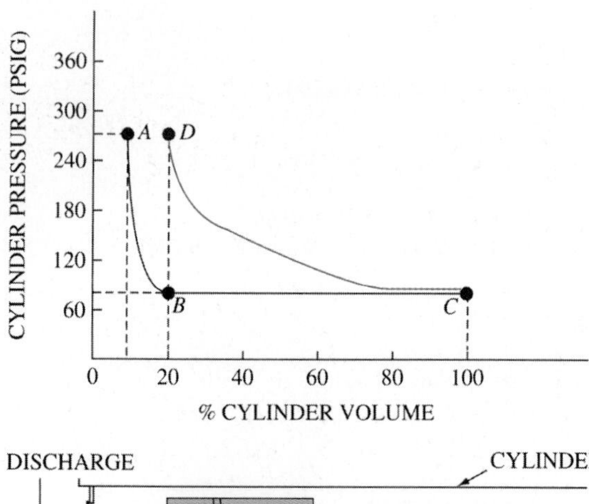

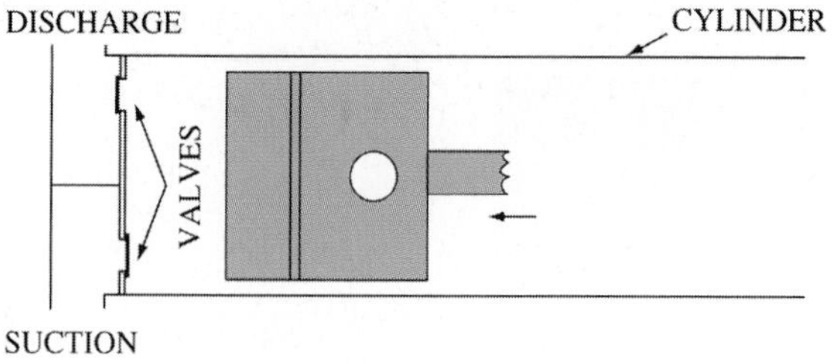

Figure 5-2-26 Compression stroke, *C* to *D*, with cylinder valves closed.

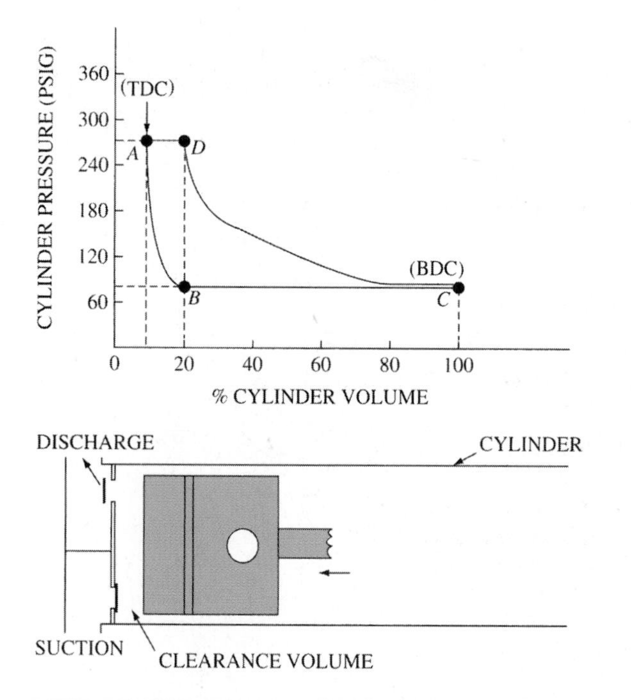

Figure 5-2-27 Discharge portion of compression stroke, *D* to *A*.

suction pressure would be about 69 psi (40°F saturation) and the discharge pressure about 263 psig (120°F saturation).

As the piston completes its compression stroke from point *D* to *A* (top dead center), the cylinder pressure exceeds the discharge manifold pressure and forces the gas charge out with the exception of the gas occupying the clearance volume space (approximately 4% volume), Figure 5-2-27. This trapped gas will remain and reexpand with the next intake stroke.

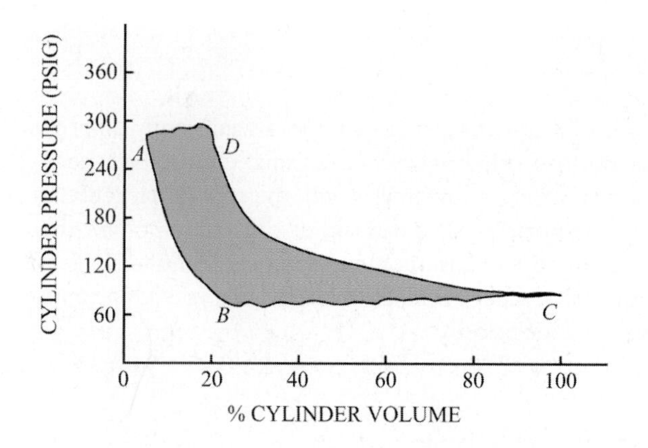

Figure 5-2-28 Complete *P-V* diagram for a one cycle compressor cylinder.

The previous illustrations assume an ideal world with no pressure losses across the valves, minimum turbulence, etc. A more realistic view is found in Figure 5-2-28, which shows pressure losses. The area enclosed by points *A*, *B*, *C*, and *D* on the pressure-volume (*P-V*) plot represents the work done by the compressor on the gas passing through the cylinder.

The wasted effort represented by reexpansion of clearance volume gas from point *A* to *B* can be increased if small amounts of liquid (a wet suction) are compressed, flash, and reexpand. This will reduce the enclosed area *A-B-C-D* and reduce output. Increasing discharge pressure and/or reducing suction pressure likewise reduce the output of the cylinder. Other things being equal, a machine will produce the greatest output by running at the highest suction temperature and the lowest discharge temperature practical: the most work at the lowest energy rate.

Rotary Compressors

Rotary compressors are widely used in small welded hermetic sizes to power small refrigerated appliances, window air conditioners, packaged terminal air conditioners, and heat pumps in sizes up to about 1 ton. They are simple in design, efficient, and run smoothly and quietly. They are described as rolling piston or single valve rotaries, Figure 5-2-29. They are not internally spring isolated.

In addition to the rolling piston type unit is a rotary vane compressor, which includes a multivane type used for higher capacity design.

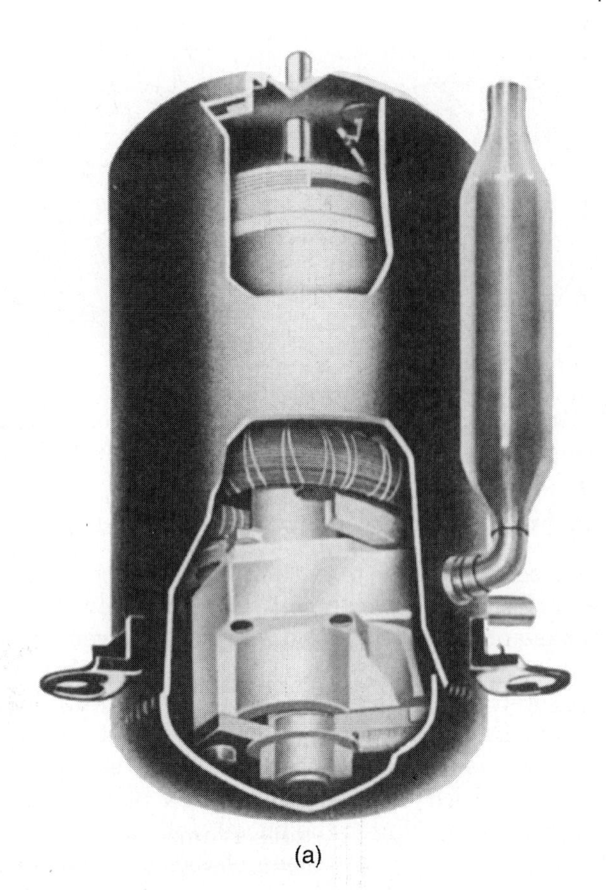

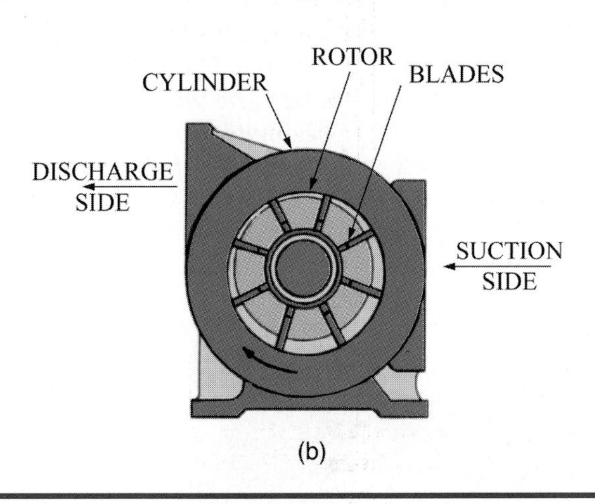

Figure 5-2-29 Rolling piston rotary compressor. *(Courtesy of Rotorex Company, Inc.)*

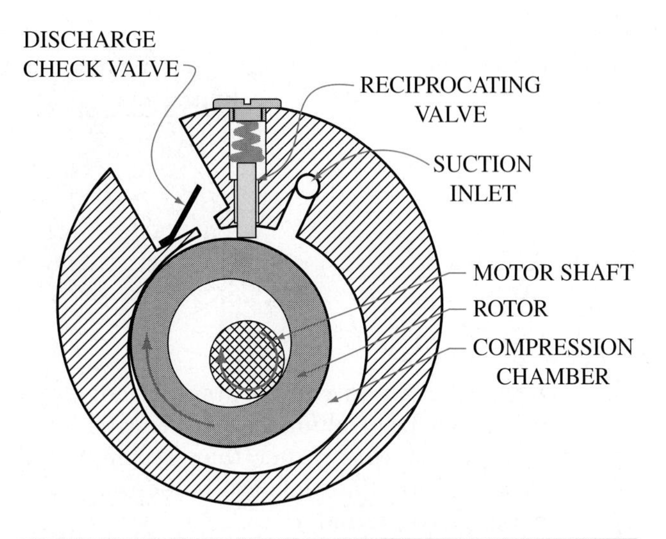

Figure 5-2-30 Diagram of a rolling piston rotary compressor.

NICE TO KNOW

Rotary compressors use a high side shelf meaning that the motor is in a very hot vapor as compared to other compressors. Most other compressors use a low side shelf, where their motors are operating in a much lower temperature environment. Because of that and other design problems, rotary compressors with capacities above a ton had a historic failure rate that was unacceptable. In the late 1980s tens of thousands of refrigerators were recalled as a result of rotary compressor failures. Rotary compressors are a high efficiency compressor for low ton capacities so they are still found in window AC units, window heat pumps, and other small units. Scroll compressors have very few moving parts as compared to reciprocating compressors. Scroll compressor motors are pressed to fit into the compressor housing. As a result, scroll compressors can sometimes produce an unusual high pitched sound. Some manufacturers enclose these compressors in soundproofing boxes to keep the sound level (db) low enough to meet standards. They are particularly noisy on startup, sometimes having a very loud rattle. This rattle is associated with a very slight overcharge in some systems. A hard start kit can be added to correct the rattle if it is not associated with a slight overcharge.

Operation of Rotary Compressors

The rolling piston, Figure 5-2-30, is a rotating off-center cam or lobe which sweeps a path inside a round cylinder. The drive shaft is centered in the cylinder. A close fitting spring loaded sliding vane follows the piston and separates the cylinder openings for suction (inlet) gas and discharge (outlet) gas.

As the rolling piston rotates, gas will flow in and fill the suction cavity due to pressure difference, much the same as in a reciprocating compressor, Figure 5-2-31. The piston continues to rotate, closes off, and passes the suction port and compresses the gas until its pressure is high enough to flow out through the discharge (reed) valve into the discharge

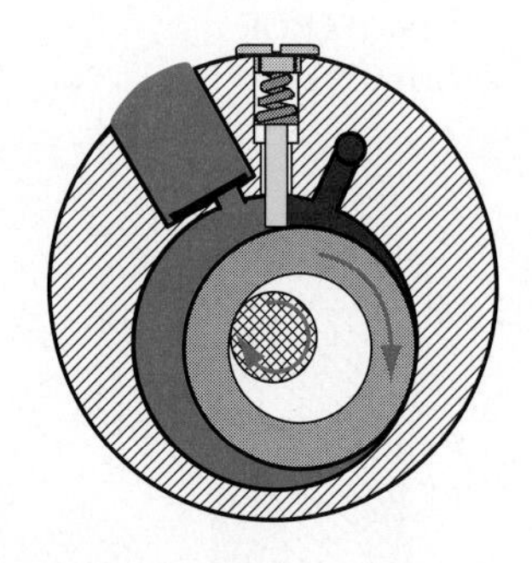
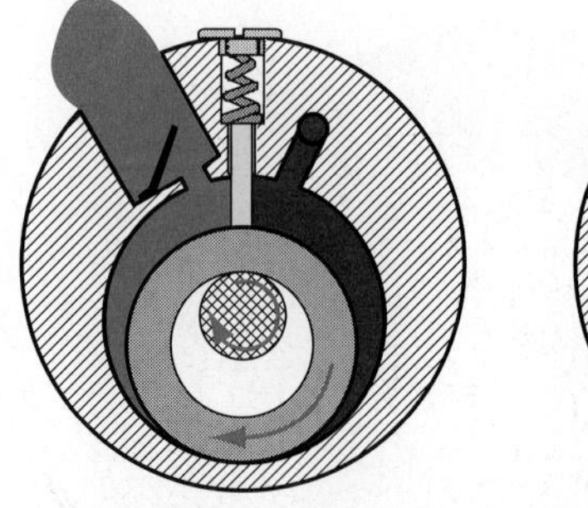
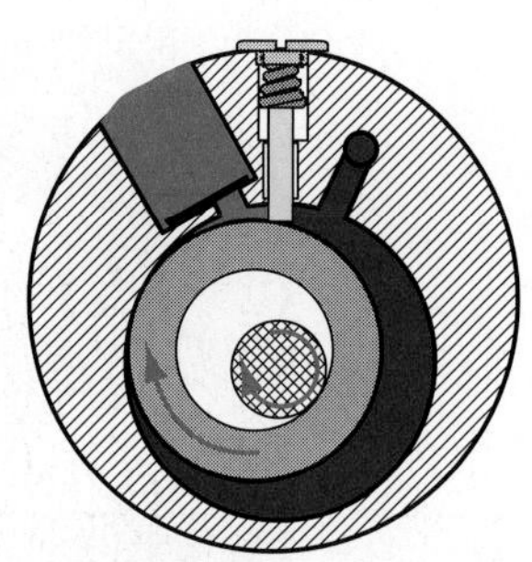

Figure 5-2-31 The compression cycle for a rotary compressor.

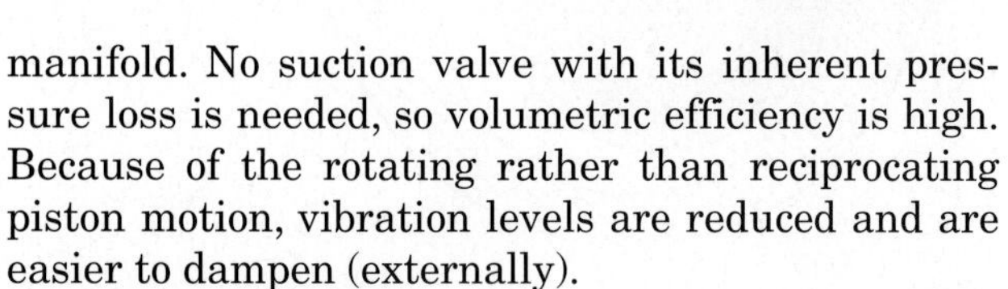

manifold. No suction valve with its inherent pressure loss is needed, so volumetric efficiency is high. Because of the rotating rather than reciprocating piston motion, vibration levels are reduced and are easier to dampen (externally).

Compact hermetic versions are built in sizes from small fractional horsepower up to about 5 Hp. Close production tolerances are required and lubricating oil helps provide a seal between the active surfaces. Rotary compressors are commonly used in window air conditioners, package terminal units, and appliances.

Rotary-vane compressors are sometimes called fixed vane rotary compressors, and are used in larger specialized applications.

A large off-center rotating shaft carries multiple vanes in slots. The vanes have a reciprocating action as they move in and out, following the cylinder walls. The moving cavities (cells) formed between adjacent vanes fill with gas as they sweep by the suction port, then compress the gas as the moving space between the vanes, cylinder wall, and shaft is reduced. Finally, the compressed gas is released through a discharge port and the process repeats. An oil-flooded design provides sealing, and construction is of the open type. Typically, these machines are used as booster compressors for low temperature ammonia systems as well as for single stage compressors for industrial and process applications.

Scroll Compressors

Scroll compressors, Figure 5-2-32, are simple hermetic rotary machines which compress gas between two close fitted spiral scroll members. One is fixed and the other moves (but does not rotate) in an orbital path. Progressively reducing cavities compress the refrigerant gas with little or no vibration.

They are positive displacement machines with high volumetric efficiencies, currently available in sizes from about 1 to 12 tons of air conditioning capacity. Because they are approximately 10% more efficient than a comparable reciprocating compressor, they have experienced extraordinary growth in U.S. residential air conditioning applications where manufacturers must meet increasingly more stringent federal energy efficiency (SEER) regulations. Scroll compressors are built in a welded hermetic design.

Operation of Scroll Compressors

For a scroll machine to work well, it must be finished to very close tolerances so the contact between the flanks and tips of the scroll members are very snug. Suction gas is captured in pockets at the periphery of the scroll members and then as the orbiting scroll motion moves the gas pockets toward the center, their size is progressively reduced. This compresses the gas, which is finally discharged at the center through an opening in the fixed scroll.

Figure 5-2-33 shows one of the matched units that comprise the scroll compressor. Figure 5-2-34 shows the suction inlet and discharge locations and demonstrates that the movable scroll member follows an orbital path, but does not rotate.

Scrolls are classed as compliant or noncompliant designs depending on the method used to accomplish the critical task of gas sealing. This can be accomplished in noncompliant designs with oil flooding and flexible tip seals.

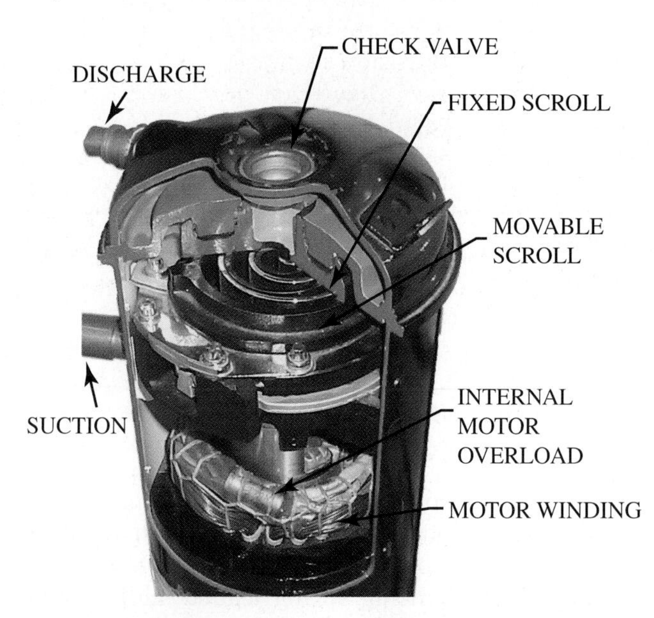

Figure 5-2-32 Scroll compressor parts shown in a cutaway.

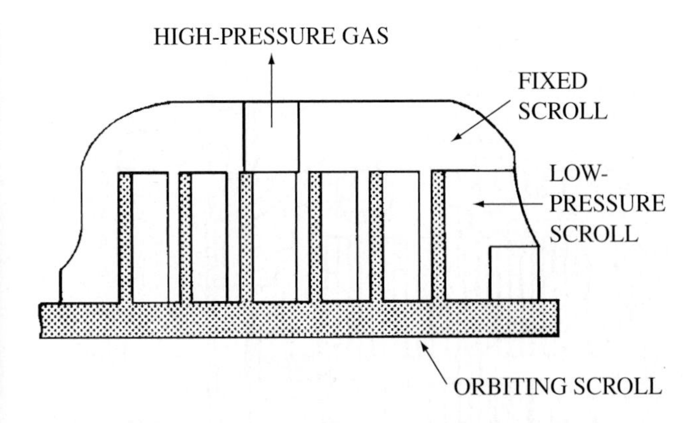

Figure 5-2-33 Scroll compressor inlet and outlet. *(Copyright by the American Society of Heating, Refrigerating, and Air-Conditioning Engineers, Inc.)*

Compliant designs can provide radial compliance by use of an Oldham coupling. This allows the scroll flanks to maintain close contact and yet the orbiting scroll can separate and unload to pass a slug of liquid or foreign material. Axial compliance is a feature whereby an adjustable force is applied to maintain sealing between the scroll tips during compressor operation and is released on shutdown. A compliant design that will enable a compressor to handle liquid accumulations from whatever circumstances should enhance the reliability of system operation, especially under adverse conditions.

Figure 5-2-35 shows a cross section of a scroll compressor. The Oldham coupling is used to drive the orbiting scroll providing radial compliance, permitting the compressor to handle liquid slugs.

Figure 5-2-36 shows the scroll compression sequence in diagram form. One pair of cavities is shown in the compression sequence.

Scroll compressors have low vibration and sound levels. The compressor is not spring isolated inside the welded shell. Vibration isolation is only external. It is sometimes possible to pick up low level frequencies from the compressor motor and scroll pump. An enclosed design and soundproofing can deal with this noise, which is not found in other compressors.

Some scroll compressor observations:

1. Do not run the machine with the suction valve closed. Pulling a high vacuum could damage the machine.
2. Check for proper rotation of three phase machines with gauge manifolds. If normal suction and discharge pressures do not quickly develop, shut off the machine and reverse any two-compressor power leads. Then recheck.
3. Scroll compressors when running backward are extremely noisy. It should be very obvious to the technician when a scroll is running backward.

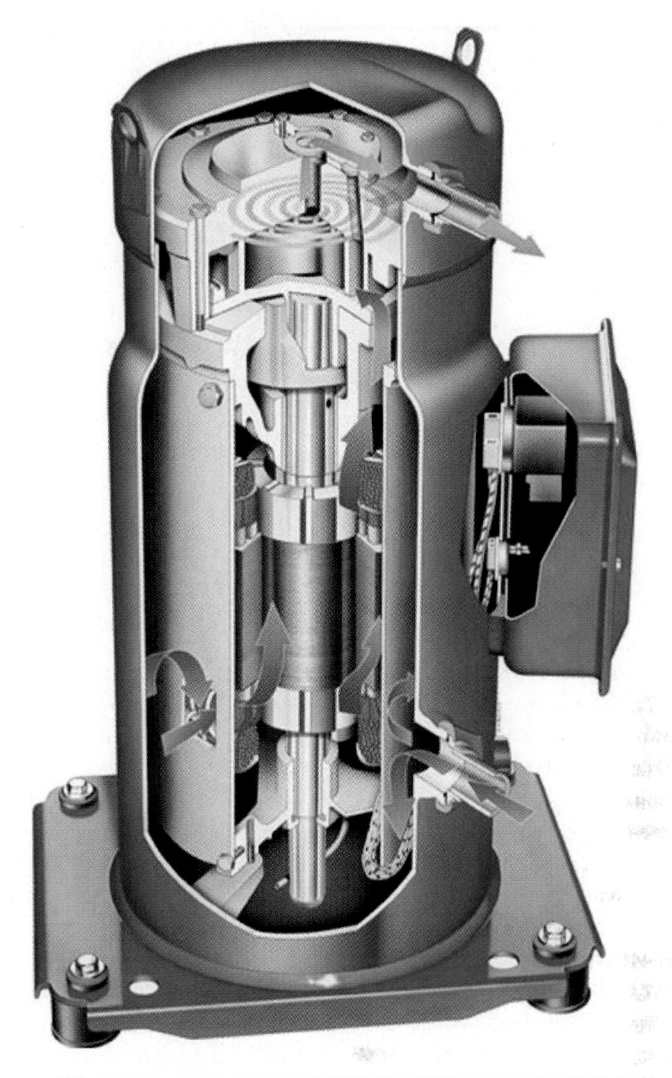

Figure 5-2-34 Partial cutaway of a scroll compressor. *(Courtesy Danfoss Inc.)*

4. A discharge line check valve prevents continued backflow through a scroll when shutting down. A momentary sound will be heard for one or two seconds as the internal pressures equalize backward through the scroll.
5. It is possible for a scroll to seal some refrigerant in the seal. Always be sure to vent both the high and the low sides before working on the lines, especially if brazing or soldering.

Screw Compressors

Helical rotary screw compressors are positive displacement machines made for larger applications. Helical rotary screw compressors are available in both twin rotor and single rotor types. They are relatively simple in concept but, like the scroll, demand advanced manufacturing capability for producing complex shapes to very close tolerances. They are smooth running and can deliver high compression ratios. Their volumetric efficiency is high, and they

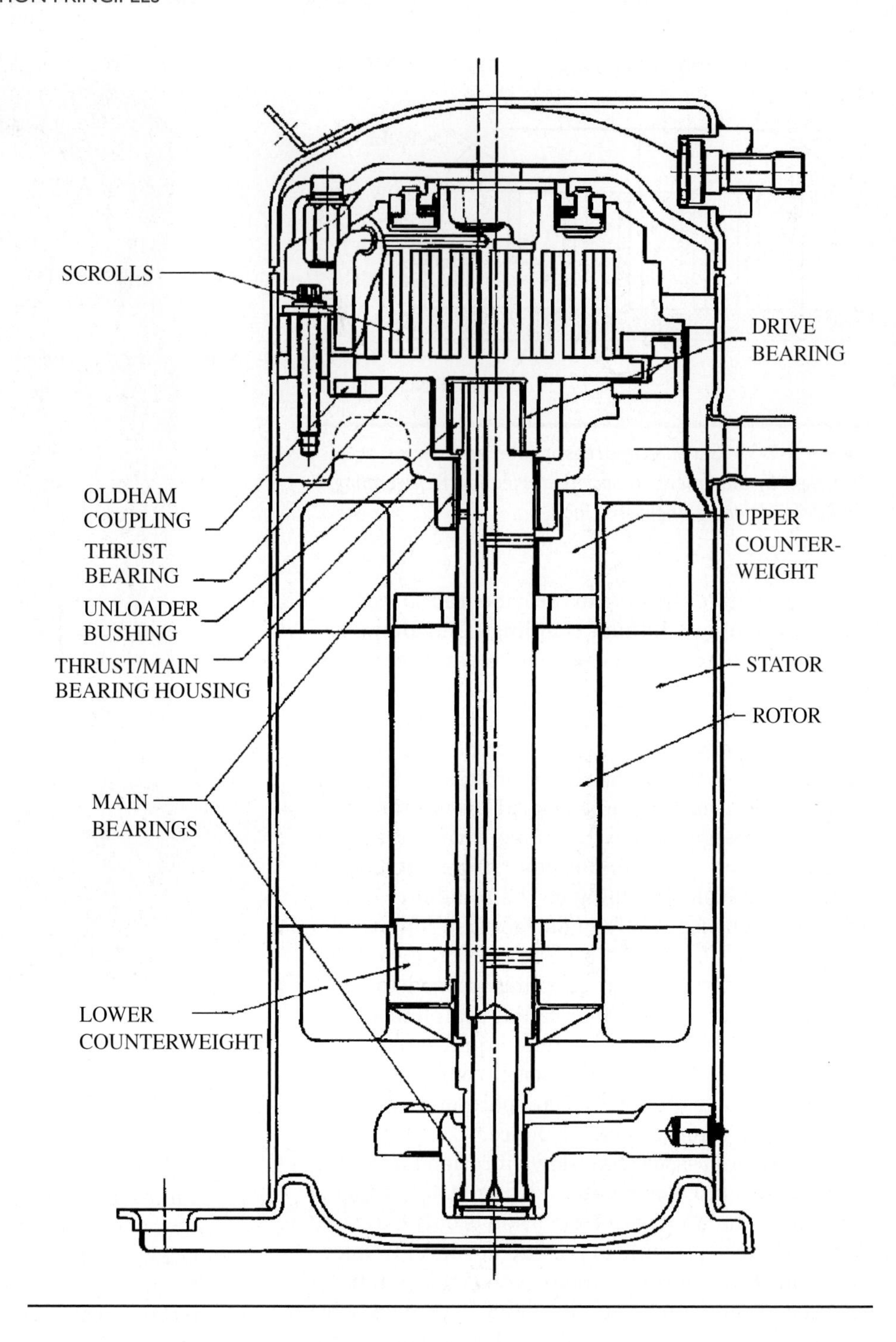

Figure 5-2-35 Cross section of a scroll compressor. *(Copyright by the American Society of Heating, Refrigerating, and Air-Conditioning Engineers, Inc.)*

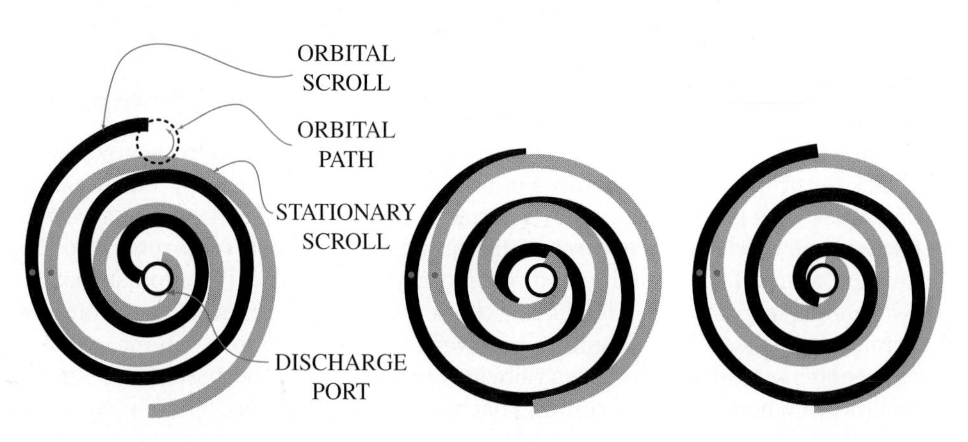

Figure 5-2-36 Scroll compressor movement. Notice that the red dots on the scrolls stay very close together through the compression cycle.

can handle a wide variety of refrigerants and gases. Consequently, they have many applications in refrigeration and air conditioning.

NOTE: Screw compressors are a positive displacement compressor meaning that 100% of the gas that enters the compressor in the low side is discharged on the high side. For comparison, centrifugal compressors are not positive displacement. An advantage of a screw type compressor is it can be used on high pressure refrigerants.

Twin screw compressors are available in open, semihermetic, or hermetic, Figure 5-2-37, construction designs. Hermetic designs are available from 35 to 175 Hp and are often used in multiples for larger capacity equipment.

In addition to the twin screw, a single screw design, Figure 5-2-38, using gate rotor(s) is available. It shares the same attributes as the twin screw and is available in open construction for larger refrigeration applications up to about 1800 CFM displacement (twins go up to about 3500 CFM).

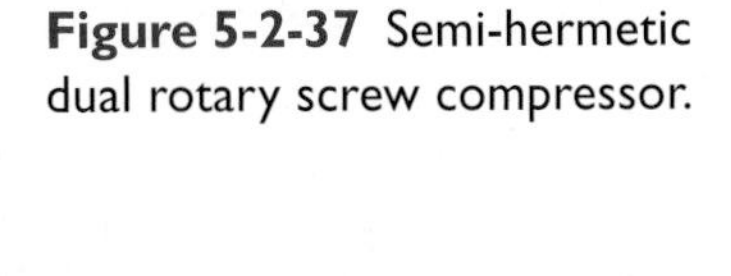

Figure 5-2-37 Semi-hermetic dual rotary screw compressor.

Figure 5-2-38 Single screw open compressor. *(Courtesy of Vilter Manufacturing)*

Operation of Screw Compressors

The twin screw models range in capacity up to 3500 CFM and are used for larger refrigeration and air conditioning (usually water chilling) applications. They often consist of a driven male rotor and mating female rotor carried on bearings and rotating within a stationary housing containing gas inlet and discharge ports. Either rotor can be the driven component.

Sealing and lubrication are usually accomplished by oil flooding, although it is possible to use synchronized timing gears to drive both rotors. Typical combinations of male rotor lobes and female rotor flutes are 4:6, 5:6, and 5:7. The rotors do not run at the same speeds.

The refrigerant gas is drawn through the inlet port into voids created as the rotors turn and unmesh, as shown in Figure 5-2-39. The entire length of the rotor space will fill with gas. As the rotors continue to rotate, they re-engage one another on the suction end. This progressively compresses the trapped gas as they mesh, moving it toward the discharge end of the machine. The compressed gas is released through the discharge port as the rotors turn and uncover the port. The cycle repeats as interlobe spaces are created, fill with gas, diminish (compressing the gas), and vent gas to the discharge. The individual discharges blend into a smooth flow of gas with little pulsation compared to a comparable reciprocating compressor.

Rotor bearings may be sleeve or antifriction type, depending on machine design and size. The oil separator is an important component in oil flooded machines and may be integrated into the design.

While capacity control is possible by speed adjustment, slot, or lift valves, the most commonly used device is the slide valve. It is a simple arrangement that bypasses part of the gas back to the suction and regulates the discharge port. It permits satisfactory unloading over a wide capacity range and the mechanism is uncomplicated and reliable.

Because of the simplicity of the screw machine, one manufacturer is building smaller sizes (35–175 Hp) in fully welded steel enclosures. Larger sizes (up to and above 1000 tons) are available as semihermetic machines for halocarbon refrigerants. Open models are built in a wide range of capacities and are used for ammonia and industrial applications, as well as for air conditioning.

Single screw compressors use a single helical main rotor and either one or two gate rotors as shown in Figure 5-2-40. The principle of operation is shown in Figure 5-2-41.

The compression volume is formed by the rotor grooves, by the cylinder wall, and by the meshing gate rotor tooth. The rotor groove fills with gas, turns, and compresses the gas as the volume is reduced by the meshing action with the gate rotor. Oil flooding is used for sealing as well as lubrication and cooling.

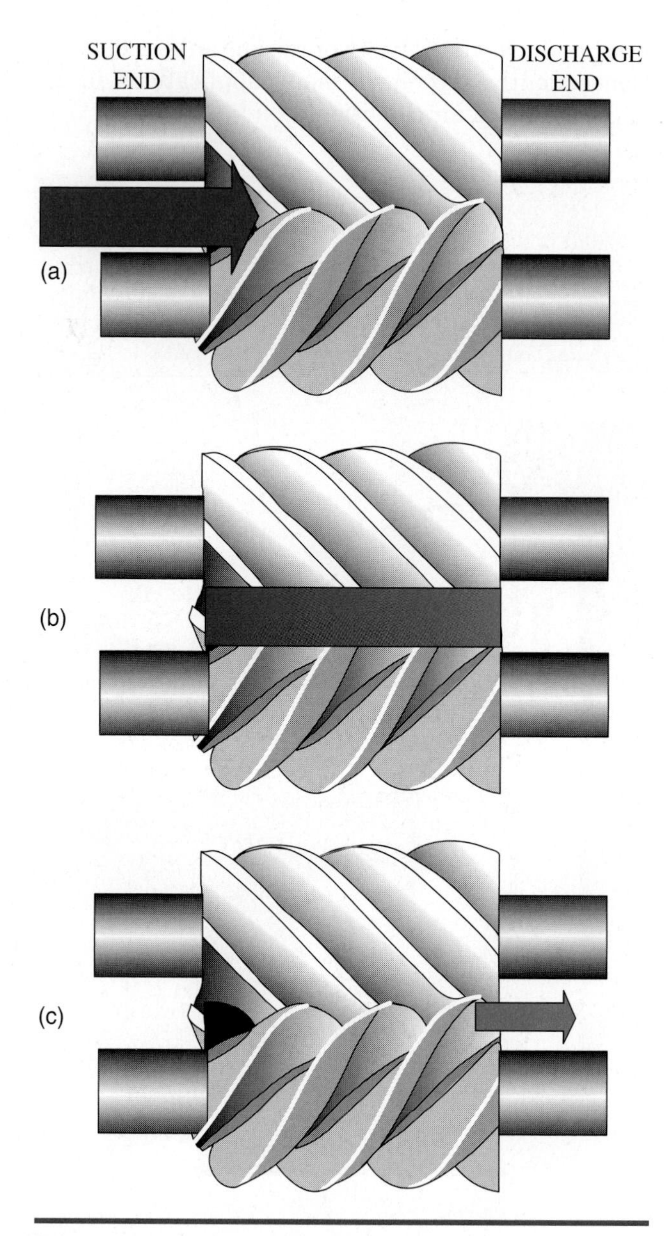

Figure 5-2-39 Cross section view of a twin-screw showing the gas flow through the compressors.

Main rotors are typically cast iron. One manufacturer builds the gate rotors from a reinforced fiberglass engineered plastic material. Sizes range from 150 to 1800 CFM. Capacity is controlled by slide valves in the compressor casing.

Centrifugal Compressors

Centrifugal compressors are widely used in single-stage and multi-stage open and semihermetic configurations to chill water for building air conditioning in capacities from 100 to 8000 tons and higher. One example of a centrifugal compressor is shown in Figure 5-2-42. Until recently, they generally used R-11, a low pressure CFC refrigerant that has been phased out. These compressors handled large volumes of vapor at

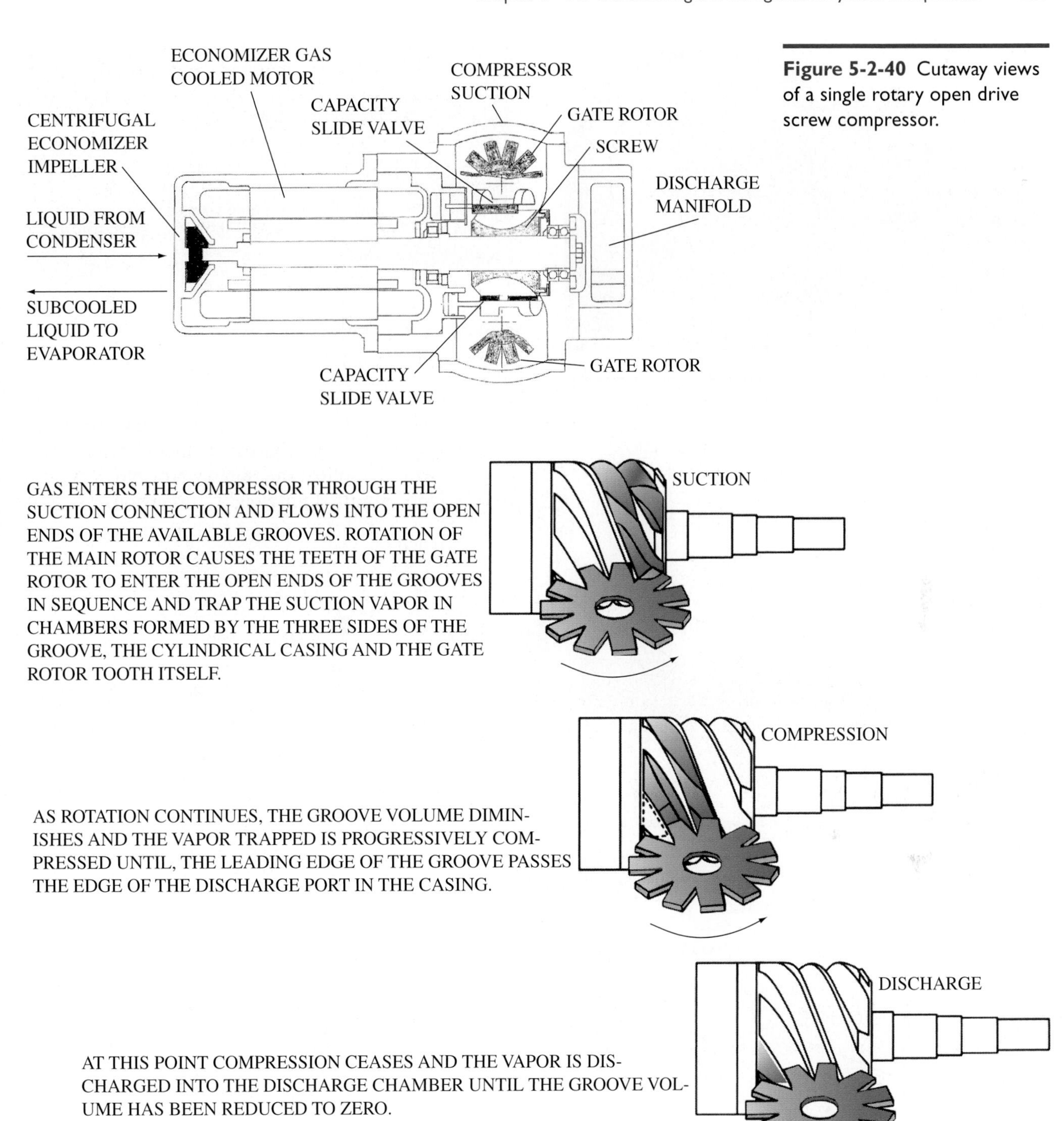

Figure 5-2-40 Cutaway views of a single rotary open drive screw compressor.

Figure 5-2-41 Three views showing the operating principle of a single rotary screw compressor. *(Courtesy of Vilter Manufacturing)*

low pressures (vacuum to 10 psig). An HCFC interim replacement refrigerant, R-123, is available, and a portion of the estimated 80,000 U.S. installations are being retrofitted.

High pressure centrifugal compressors are available in both open and semihermetic configurations. Some were especially designed for CFC R-12. They are now being supplied to run on HFC-134a refrigerant. Other machines are designed to use HCFC R-22. Presently their capacities are limited to less than 1000 tons. Large chiller units utilize multiple compressors. These machines compete with the screw machines for large capacity chilled water cooling plants.

Figure 5-2-42 Multi-stage centrifugal compressor.

NOTE: Most centrifugal systems are extremely large and used as part of a chiller system in large commercial buildings. These systems are so large that they fall under local, state, or federal guidelines for equipment room operators. An individual must be present at all times when these systems are running. Most run 24 hours a day, 7 days a week. The individual who operates these units may be referred to as an equipment room operator or an operating engineer.

Operation of Centrifugal Compressors

Centrifugal compressors, like the one shown in Figure 5-2-43, were once considered low pressure machines until a semihermetic model was introduced for R-12 applications some years ago. They use a relatively small, lightweight, precision die cast

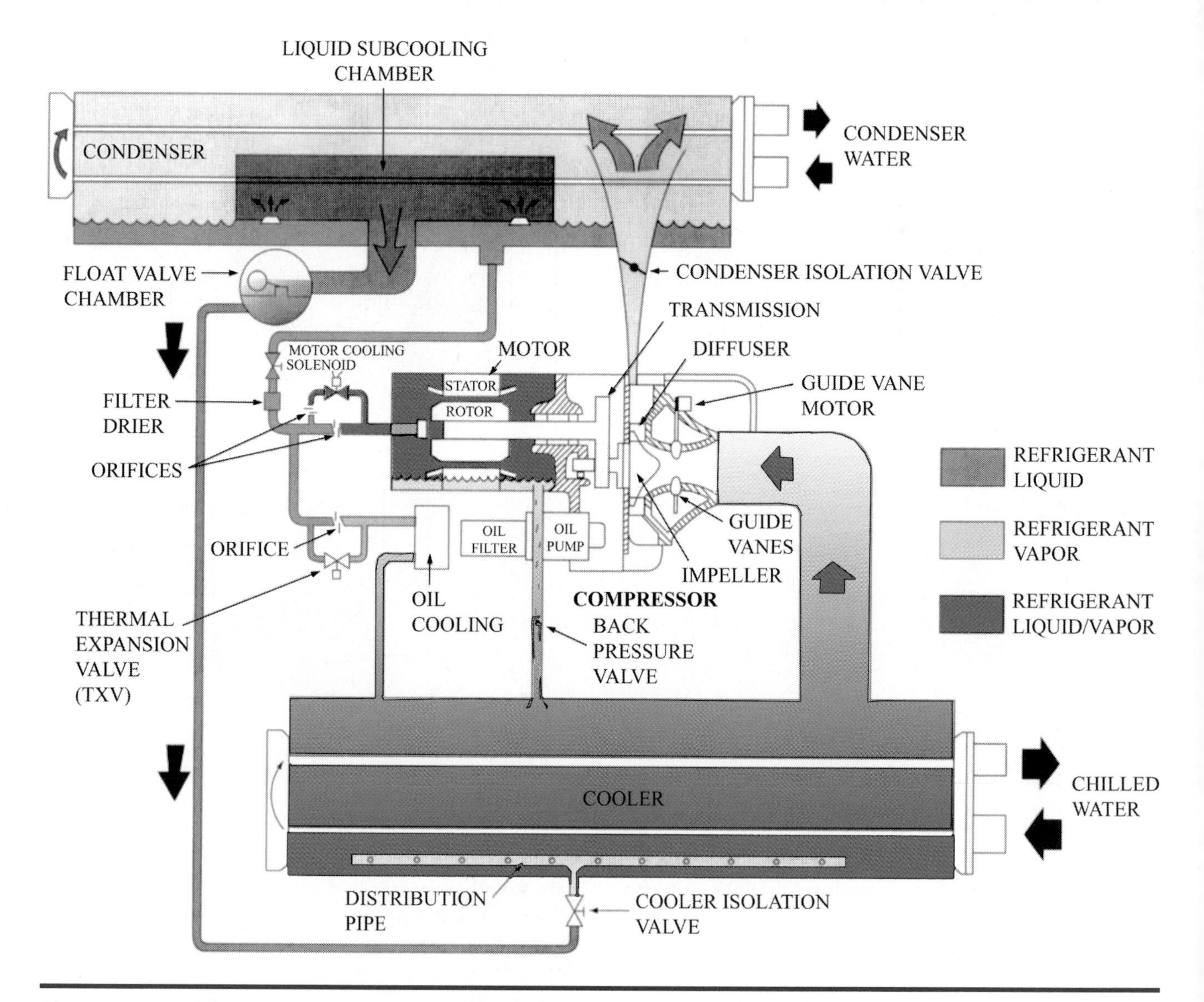

Figure 5-2-43 Centrifugal water chiller using high pressure refrigerant. Refrigerant flow shown.

Figure 5-2-44 Lightweight high-speed centrifugal impeller.

impeller, Figure 5-2-44. A two pole 3600 rpm motor through a gear type speed increaser drove the impeller at speeds of 8000 to 10,000 rpm and higher. To reduce sound levels, the impeller vanes were asymmetrically arranged but fully balanced.

Overall size is reduced and, because the machine always operates in a positive pressure, there is no need for an air purge system. Purge systems and air leakage always have been high maintenance problems on the low pressure machines. Other manufacturers now build high pressure centrifugal compressors. They are available in both open and semihermetic configurations and are being designed for use with HCFC-22 as well as the newer HFC refrigerants.

Capacity control is commonly by inlet vane controls but is also available with variable motor speed control.

Figure 5-2-45 shows a flow diagram for a centrifugal chiller using a high pressure refrigerant. Note the separate oil pump and the use of a float valve metering device.

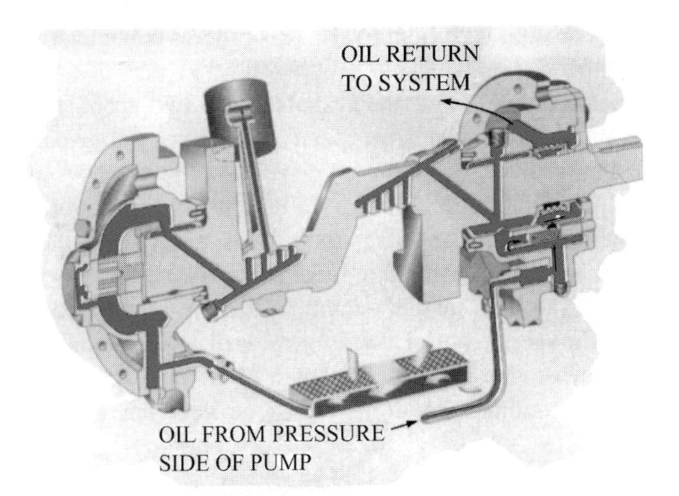

Figure 5-2-45 Lubrication of the compressor bearings using drilled oil passages. (*Courtesy of Frick Company*)

COMPRESSOR LUBRICATION

All refrigeration compressors require lubrication of moving surfaces. Delivery of lubricant can be by simple splash systems, crankshaft, or separate oil pump, Figure 5-2-46. As some oil can be expected to carry over with the discharge gas, the oil should be miscible with the refrigerant and be compatible with the system designed to carry the oil through the condenser and evaporator and return it to the compressor crankcase.

Some machines are designed to have high oil circulation rates and are always provided with discharge oil separators. Oils used in hermetic and semihermetic systems have to be compatible with the electrical components and be good insulators. Mineral oils should be wax free. All oils should have a low moisture content to ensure long corrosion-free system life, and freedom from electrical insulation deterioration and freeze-ups.

CAUTION: Refrigerant oil does not evaporate from a system. If you have a compressor that is operating on low oil, the oil is trapped somewhere in the system. Determine where the oil is and what problem has caused the oil to be entrapped in the system away from the compressor, and resolve that problem. Capacity control is important on any system because it enhances the system's overall operating efficiency. A compressor that runs continuously will draw fewer total watts of power than one that starts and stops. By providing a method of capacity control on large systems, the overall building operation can be significantly improved. Many large buildings have a latent load that must be maintained even when there is a light load on the building. By providing this unloading capacity, the system can run in what is sometimes referred to as *background* so that the air is dehumidified without significantly cooling it. In some applications in light load conditions, a reheat of the air may be done to prevent the occupied space from becoming too cold.

HFC chlorine-free refrigerants require use of special polyol ester (POE) synthetic oils. Older CFC

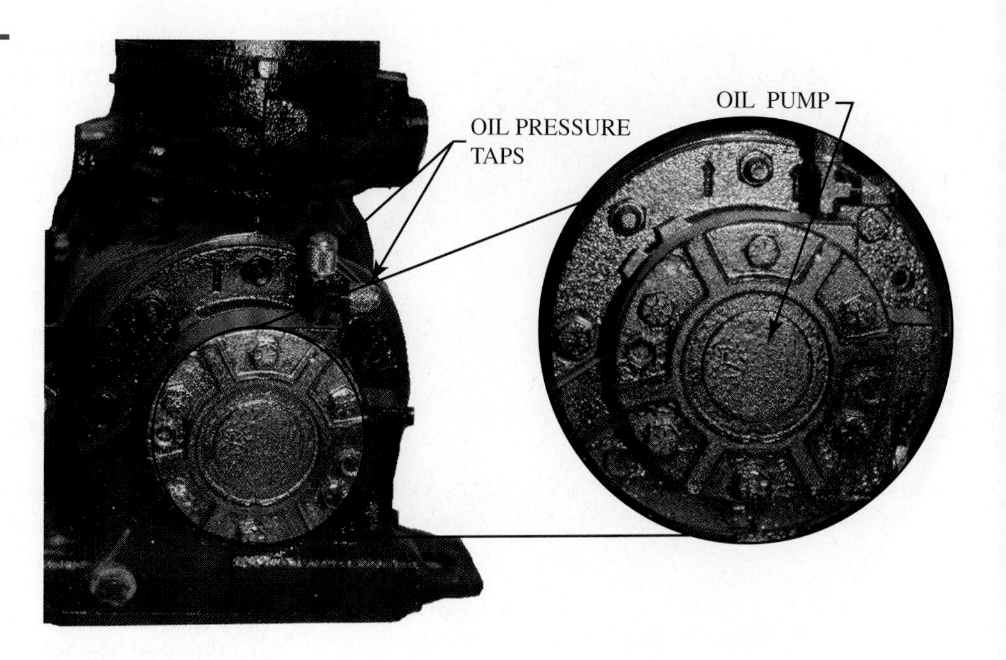

Figure 5-2-46 Compressor with oil pump mounted. (*Courtesy Hampden Engineering Corporation*)

machines used mineral oil based lubricants. Current recommendations for HCFC refrigerant installation and conversions include mineral oil, alkyl-benzene refrigeration oil, or a combination of the two. The choices have become more complicated. Synthetic refrigeration oil is more expensive but can be used with all halocarbon refrigerants. It must be used with HFC refrigerants as specified.

Pressure lubricated compressors usually are equipped with a differential pressure sensor to measure actual oil pressure and a time delay safety lock-out circuit to shut the compressor down should lubrication fail.

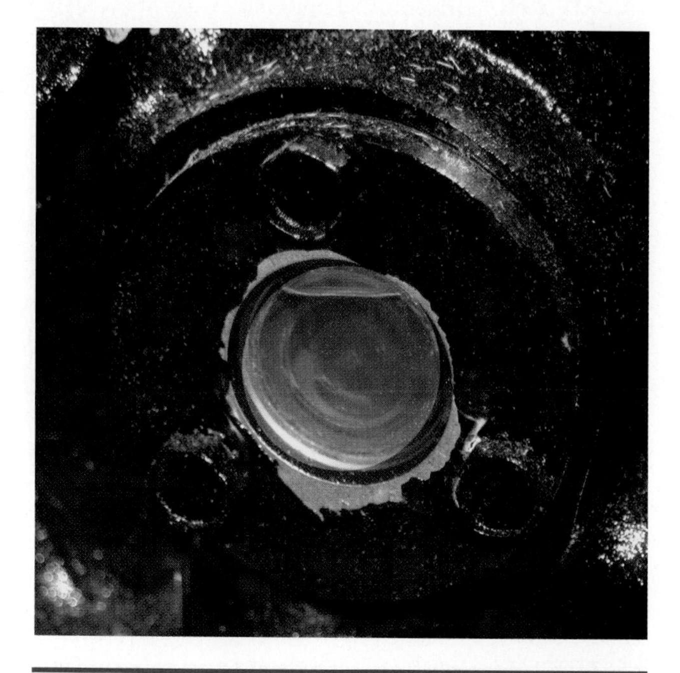

Figure 5-2-47 The oil level must be visible in the sight glass. (*Courtesy Hampden Engineering Corporation*)

Refer to manufacturer data for minimum pressure settings, normal operating pressures, and high relief valve settings. Oil and refrigerant liquid are miscible, so it is possible to have refrigerant migrate during long OFF cycles to a compressor in a cold location, condense, and raise the oil level above the oil sight glass (if so equipped), as shown in Figure 5-2-47.

A crankcase heater can be added to the outside of hermetic compressors to raise the temperature of the oil to prevent this migration from occurring, as shown in Figure 5-2-48. A flat crankcase heater can be attached to the bottom of semihermetic compressors for the same purpose, as shown in Figure 5-2-49.

On startup the crankcase pressure drops, the liquid refrigerant flashes to gas, and the mixture turns to foam which the oil pump cannot handle well and which is a poor bearing lubricant. Where such

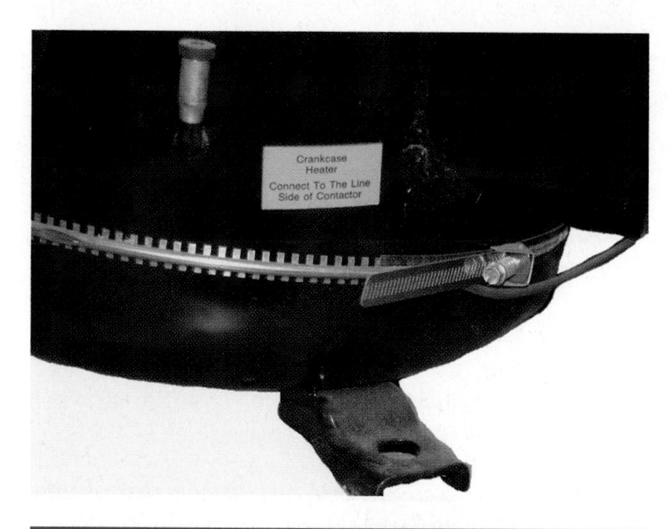

Figure 5-2-48 External crankcase heater for hermetic compressors.

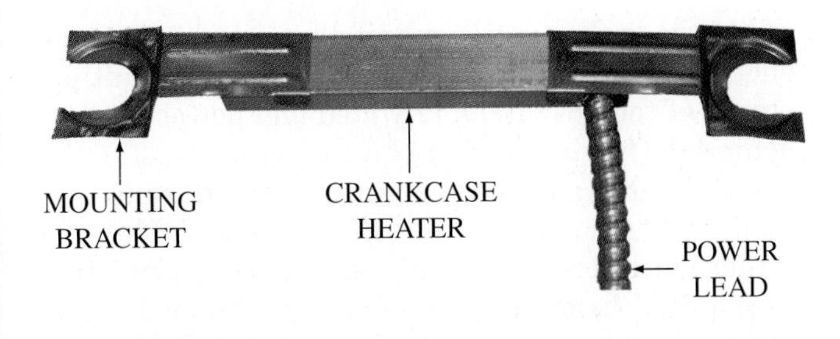

Figure 5-2-49 External crankcase heater for semi-hermetic compressors. (*Courtesy Hampden Engineering Corporation*)

problems are encountered, crankcase heaters can be used as well as timed pumpout cycles and piping changes.

ENGINE DRIVEN COMPRESSORS

Although it is possible to drive a refrigeration compressor by a stationary engine, almost all applications the service technician will encounter will be electric drive. An open compressor with separate motor will require ventilation to keep the motor from overheating, if enclosed in a confined space. If the compressor is a welded hermetic or a semihermetic, the motor cooling is normally provided by the refrigerant suction gas stream, and ventilation usually is not a concern. Engine driven compressors are a special case, need ample ventilation, and should be installed according to manufacturers' recommendations.

CAPACITY CONTROL FACTORS

A number of factors control compressor capacity.

1. The choice of the refrigerant
2. Changes in the compressor displacement

The choice of refrigerant for a given size compressor can produce a wide variation in capacity. This is a design choice and once made is not likely to be changed. It would not be practical on a given system to deal with a fluctuating load by changing the refrigerant.

Changing refrigerants in the field to convert a system may change power requirements dramatically. So if a system was to have the refrigerant changed, in addition to making sure the component materials are compatible, pressure ratings adequate, control devices adequate and properly set, and oil compatible, the motor capacity would have to be checked. A larger motor might be possible for an open machine; it is unlikely that a semihermetic motor could be increased in size and certainly it would not be possible with a welded hermetic.

Displacement offers a variety of ways of changing refrigeration capacity to match system load:

1. Cycling on/off
2. Cycle multiple machines on/off
3. Cylinder unloading
4. Hot-gas bypass
5. Speed control
   a. Two speed
   b. Variable-frequency drives (VFD)
6. Other

On/off cycling is simple and basic: it gives the choice of 0% and 100% capacity. Often it is all that is needed. It is not desirable to excessively short-cycle any equipment, and in some applications the temperature variations of on/off control are too wide to be acceptable. If so, better capacity control is needed.

Multiple machines with either common or independent refrigerant circuits are sometimes used for a variety of reasons, including closer temperature control and better humidity control. In a twin compressor, Figure 5-2-50, installation would give the choice of 0%, 50%, and 100% capacity steps.

Figure 5-2-50 Capacity control by using two compressors on a common refrigerant circuit.

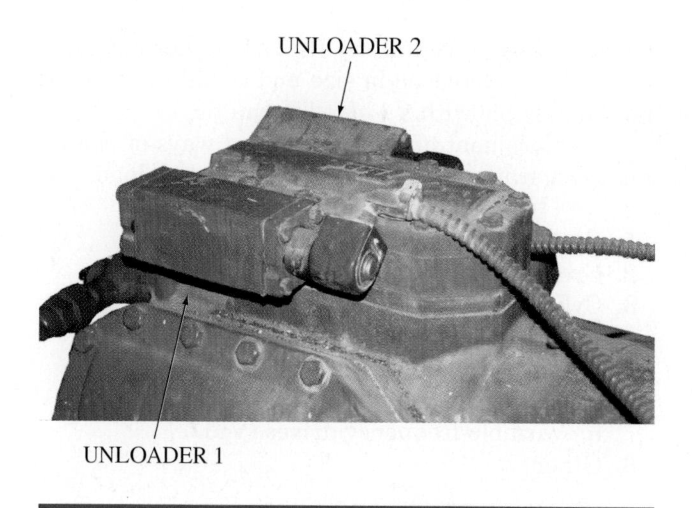

Figure 5-2-51 Capacity control using an unloader.

Cylinder unloading is available on reciprocating compressors of 15–20 ton capacity and larger. Compressor unloaders can be actuated by gas, by oil pressure as shown in Figure 5-2-51, or electrically. They can unload one or more cylinders by holding the suction valves open, bypassing the cylinder discharge to the suction manifold, or blocking off the suction inlet. Depending on the compressor, multiple stages of control can be provided and power savings achieved. Overheating can be a problem.

Hot-gas bypass is sometimes used on small machines where cylinder unloading is unavailable or on larger machines when additional capacity reduction is required below that possible with unloaders, as shown in Figure 5-2-52. Care must be taken not to overheat the compressor. It can be a practical solution in dealing with light loads. This process is not energy efficient.

Some compressors are available with two-speed motors. This technique *(two-speed control)* is not widely used at present.

Variable frequency drives (VFD) enable continuous variable speed (capacity) control on many machines. These have been applied to large centrifugal compressors as well as to smaller comfort cooling applications. Variable speed control works very well and is likely to be increasingly used as the cost of VFDs (also called inverters) continues to drop. It can be very energy efficient.

Slide valves are commonly used on screw compressors to regulate capacity. The valves work well and save energy. They function based on diverting gas to the suction manifold before it is compressed.

Inlet vane capacity controls are commonly used on centrifugal chillers. These vanes can control down to 10–25% capacity. They function by giving the entering gas a prerotation swirl.

Figure 5-2-52 Hot gas bypass. *(Courtesy Hampden Engineering Corporation)*

TECH TALK

Tech 1. Why are there so many different types of compressors?

Tech 2. Each compressor type has its advantages and disadvantages in the way they function.

Tech 1. Can you give me an example?

Tech 2. Sure. Can you imagine how large the pistons would have to be in a compressor that had two or three hundred tons capacity? It would have to be enormous. So, for jobs like that, a centrifugal or screw compressor works better. Because with a piston or reciprocal compressor, the piston has to stop and start at the bottom and top of the stroke. If you made a piston big enough to have a large capacity, can you imagine the type of vibration it would have every time it stopped or started? But a screw compressor or centrifugal compressor gets its compression from spinning parts. They have a whole lot less vibration.

Tech 1. Then why don't they make them small and have just screw and centrifugal compressors for everything?

Tech 2. Well, you can't do that either. On a small piston compressor you can have really tight tolerances and really high efficiencies. And when you're talking about very little refrigerant flowing compared to a two or three hundred ton compressor, every little bit of leakage around the piston is a big deal. Centrifugal compressors have a little leakage back around their compressor parts, but as compared to the whole volume they pump, it's such a small part that it doesn't affect their efficiency like it would in a small compressor.

Tech 1. So there are big compressors and small compressors and you can't use the same design for all of them.

Tech 2. That's right. They all have their own purpose and they all do their jobs very well.

Tech 1. Thanks.

UNIT 2—REVIEW QUESTIONS

1. List the main types of compressors.
2. How do you determine what type of compressor to use?
3. What are the functions performed by the compressor?
4. What is compressor ratio?
5. Where is a reciprocating compressor used?
6. The most commonly used open compressors are used with __________.
7. List the various types of reciprocating compressors.
8. What is another name for rotary vane compressors and where are they used?
9. What determines if a scroll is classed as a compliant or noncompliant design?
10. The __________ is an important component in oil-flooded machines and may be integrated into the design.
11. Where are centrifugal compressors widely used?
12. List the different types of delivery of lubricants.
13. What two factors control compressor capacity?
14. The choice of __________ for a given size compressor can produce a wide variation in capacity.
15. List the ways displacement changes refrigeration capacity to match system load.

UNIT 3

Evaporators

OBJECTIVES: In this unit the reader will learn about evaporators used in the HVAC/R industry including:

- Types of Evaporators
- Evaporator Construction
- Applications
- Capacity Control
- Condensate
- Types of Defrost
- Liquid Coolers
- Servicing and Troubleshooting of Evaporators

INTRODUCTION

The *evaporator* is a device for absorbing heat into the refrigeration system in a vapor compression refrigeration system. See the evaporator diagrammed in Figure 5-3-1. In the evaporator, the liquid refrigerant absorbs heat from its surrounding load and boils into a low pressure vapor. The evaporator is a heat exchanger with the refrigerant contained within tubes, passages, or a vessel. The load can be air, water, brine, or any product to be cooled. The load is separated from the refrigerant by the heat exchanger walls or shell.

The load on evaporators, either air or fluid, can be mechanically blown or pumped or it can be static or natural draft. Evaporators are designed on the basis of their intended usage. The application determines which type is the best suited.

The heat flow occurs because the temperature of the refrigerant is lower than the temperature of the air or water that is being cooled. The refrigerant temperature in the evaporator is maintained at the saturation temperature that corresponds to the evaporator pressure.

When the compressor starts, the pressure in the evaporator drops, causing the refrigerant liquid in the evaporator to evaporate (flash) and cool down, as shown in the diagram in Figure 5-3-2. Liquid refrigerant boiling and vaporizing in a cold evaporator can be compared to water boiling on a stove. The refrigerant is heated by the load, whereas the water is heated by the stove burner. The action and results are comparable. The difference is that the stove is hot and the refrigerator is cold; however, in both cases heat energy is being transferred.

TYPES OF EVAPORATORS

There are three general types of evaporators:

1. Bare pipe
2. Extended surface (fin coils)
3. Plate

In the vapor compression refrigeration cycle, all of the refrigerants have to be confined within the system by components and interconnecting tubing strong enough to withstand the pressures, temperatures, and vibration imposed by the system. Most

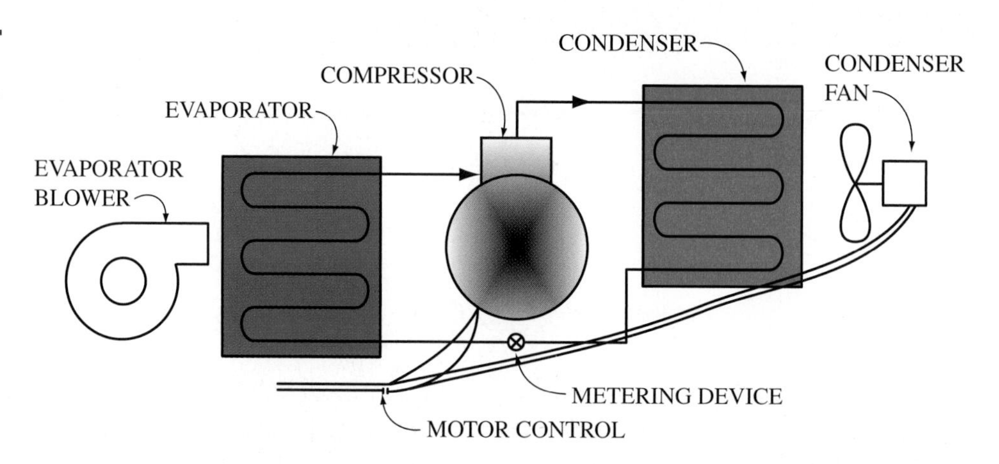

Figure 5-3-1 Location of the evaporator in relation to other components.

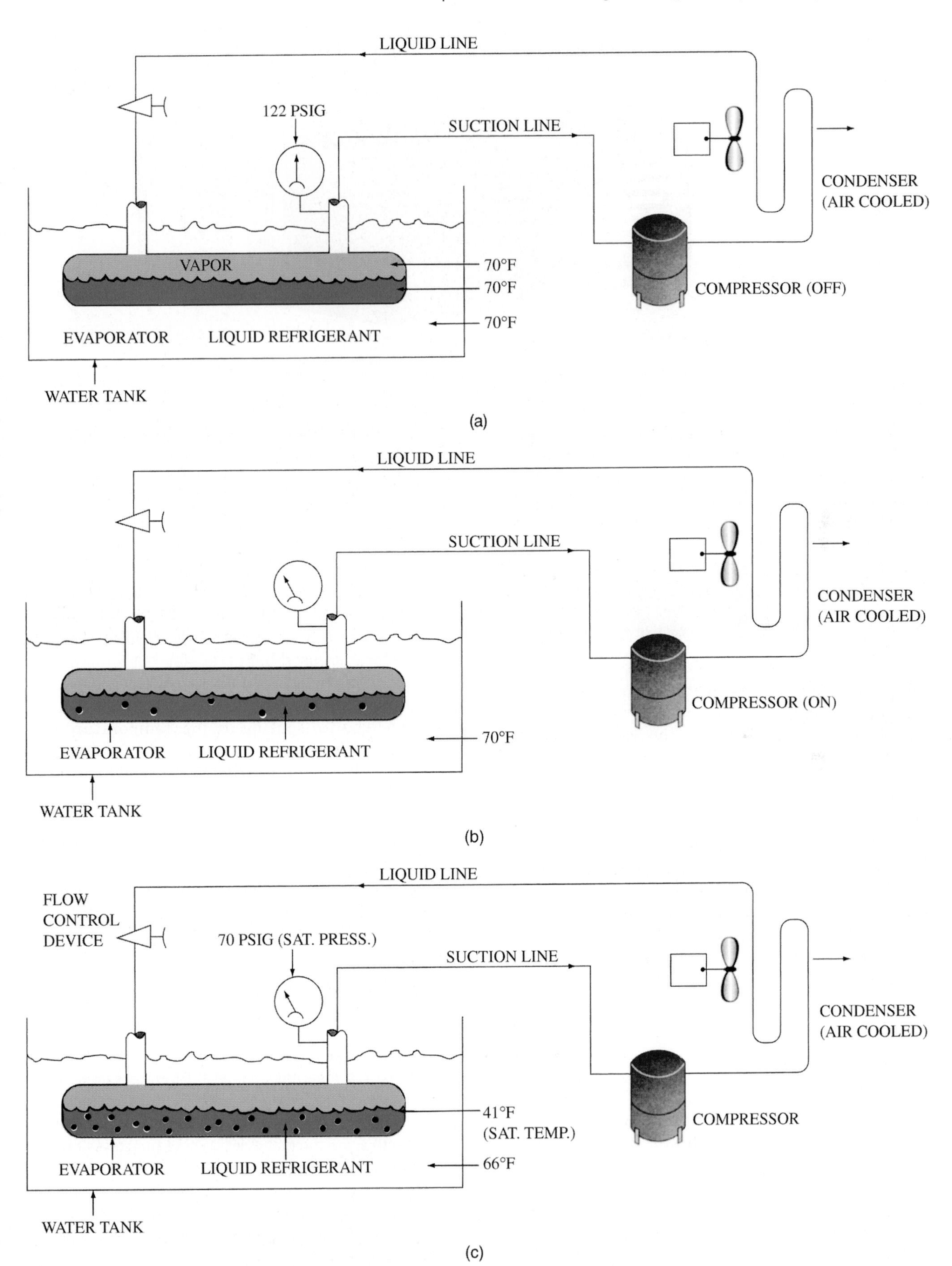

Figure 5-3-2 Diagrams of the refrigeration process (R-22 refrigerant): (a) System off; (b) System startup; (c) System running.

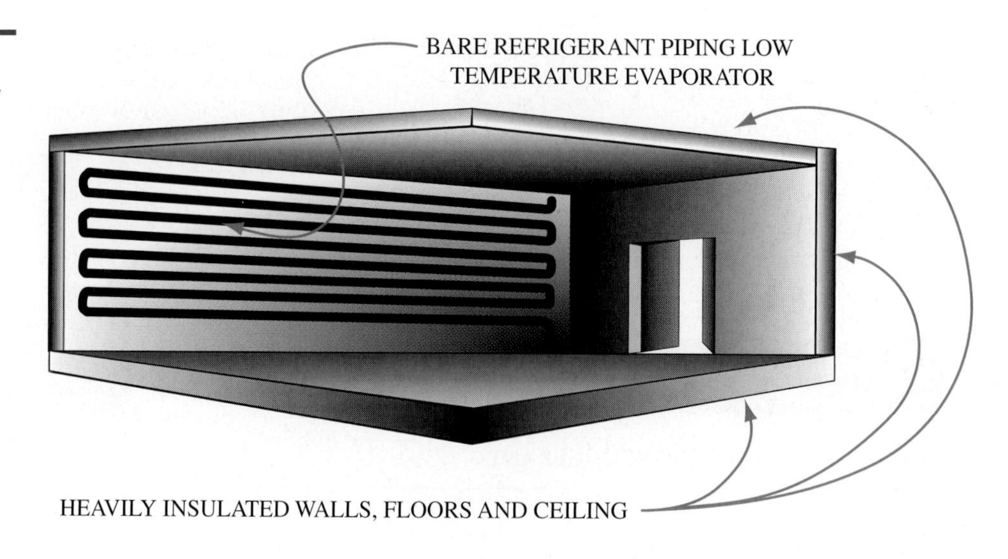

Figure 5-3-3 Bare pipe evaporator coil with gravity air circulation.

refrigerants except ammonia (R-717) use copper or aluminum for most residential and small commercial applications and steel or stainless steel for larger commercial systems. Ammonia refrigerant systems must be all iron, steel, or aluminum.

SAFETY TIP

Ammonia refrigerant, R-717, is reactive with copper and copper alloys (brass or bronze). If any fittings or components made out of copper or copper alloys are used in ammonia systems they will fail. Such failure can be fatal to anyone in the area.

Bare Pipe Evaporators

Bare pipe evaporators such as the one shown in Figure 5-3-3 may be used for low temperature refrigeration, liquid cooling, ice skating rinks, and thermal storage applications. Ice buildup on low temperature bare pipe evaporators will not significantly reduce the evaporator's efficiency unless the ice becomes very thick. However, ice buildup on low temperature refrigeration fin coils is a major problem and defrosting adds heat to the freezer area.

Extended Surface Evaporators (Fin Coils)

Extended surfaces help equalize the heat transfer capabilities of refrigerant evaporator tubes when cooling air. Two examples of extended surfaces are shown in Figure 5-3-4. The heat transfer capacity on the refrigerant side is high due to the turbulent boiling liquid refrigerant. The air side is much more limited. Air is not as good a conductor of heat. By adding tight fitting aluminum fins to the copper tubes, the air side surface and corresponding heat transfer can be greatly increased. In addition, the size and cost of the evaporator may be reduced, compared to a unit without an extended surface. To ensure good heat transfer, the fins must be tightly bonded to the tubes, as shown in Figure 5-3-5a,b. With plate-fin construction, the tubes are mechanically or hydraulically expanded into the tube collar. The manufacturing process of expanding the tubes within the fins is shown in Figure 5-3-5c. Multi-row construction is available and the fins can be shaped or dimpled to provide additional turbulence on the air side for increased heat transfer.

The distance between fins is called "fin spacing" or "fin pitch" and is expressed as the number of fins per inch. The fin spacing on low temperature coils can be 1 to 4 fins per inch to reduce ice bridging. Ice bridging is the term for what happens when the ice on one fin joins up with the ice formed on the adjoining fin. As shown in Figure 5-3-6, it restricts all air flow through the coil. Generally medium temperature coils can have fins spaced from 4 to 14 per inch and air conditioning temperature coils can have fins spaced from 11 to 20 per inch. This measurement will be covered later in Figure 5-3-15 on page 245.

SAFETY TIP

Many plate evaporators are used in small residential refrigerators. Under the counter refrigerators do not defrost automatically. When manually defrosting these units, individuals often puncture the evaporator refrigerant passages with a sharp metal object. These punctures can be repaired using a two-part epoxy specifically designed for repairing holes in plate evaporators. Follow the manufacturer's instructions on using the epoxy and only use those epoxies specifically designed and approved for this application. Some epoxies may have toxic substances that would not be appropriate to have in a domestic refrigerator freezer where food is stored.

(a)

(b)

Figure 5-3-4 Plate fin cooling coils for (a) direct expansion; (b) chilled water. (*Courtesy Hampden Engineering Corporation*)

Plate Evaporators

Plates such as those shown in Figure 5-3-7 are a special form of extended heat transfer surface used in refrigeration and freezer applications. The plate surfaces may be fabricated in a variety of shapes and the refrigerant passages may be integral or attached. That is, the refrigerant passages may be formed between two plates welded together or the refrigerant may flow through tubing welded to the plate surface.

PRESSURE-HEAT DIAGRAM

The pressure-heat diagram of Figure 5-3-8 affords a good view of the cooling process in the evaporator. Initially a high pressure liquid should be subcooled 8–10°F or more if possible.

When subcooled liquid from point A flows through the expansion device, its flow is controlled and its pressure drops to evaporator pressure. Approximately 20% of the liquid boils off to gas, cooling the remaining liquid-gas mixture. Its total heat (enthalpy) at point B is unchanged from A. No external heat energy has been exchanged. From points B to C, the remainder of the liquid boils off, absorbing the heat flowing in from the evaporator load (air, water, or product). At point C, all of the liquid has evaporated and the refrigerant is 100% vapor at the saturation temperature corresponding to the evaporator pressure.

The subcooling increases cycle efficiency and can prevent flash gas due to pressure loss from components, pipe friction, or increase in height.

Flash Gas Problems

Two problems can occur, both causing the formation of flash gas. Flash gas can be formed as a result of:

1. A restricted filter-drier, as shown in Figure 5-3-9.
2. A rise in the liquid line, lowering the pressure, causing some of the refrigerant to boil.

The problem with Figure 5-3-9 can be solved by changing the restricted filter-drier. If the condensing unit had sufficient condenser surface to provide additional liquid subcooling, the likelihood of flash gas problems would greatly diminish.

TECH TIP

Capillary tube metering devices use the formation of flash gas in the end of the capillary tube to help establish the pressure differential. The flash gas bubbles act as a vapor lock in the line to increase the pressure drop through the capillary tube.

Figure 5-3-10 points to the need for caution with systems where the evaporator is significantly above the level of the condenser. The R-22 liquid line pressure will drop 1 psig for every 2 ft increase in vertical rise. A 30 ft rise equals a 15 psig drop in pressure, which is equivalent to a 5°F drop in saturation temperature. This is before any friction losses are considered.

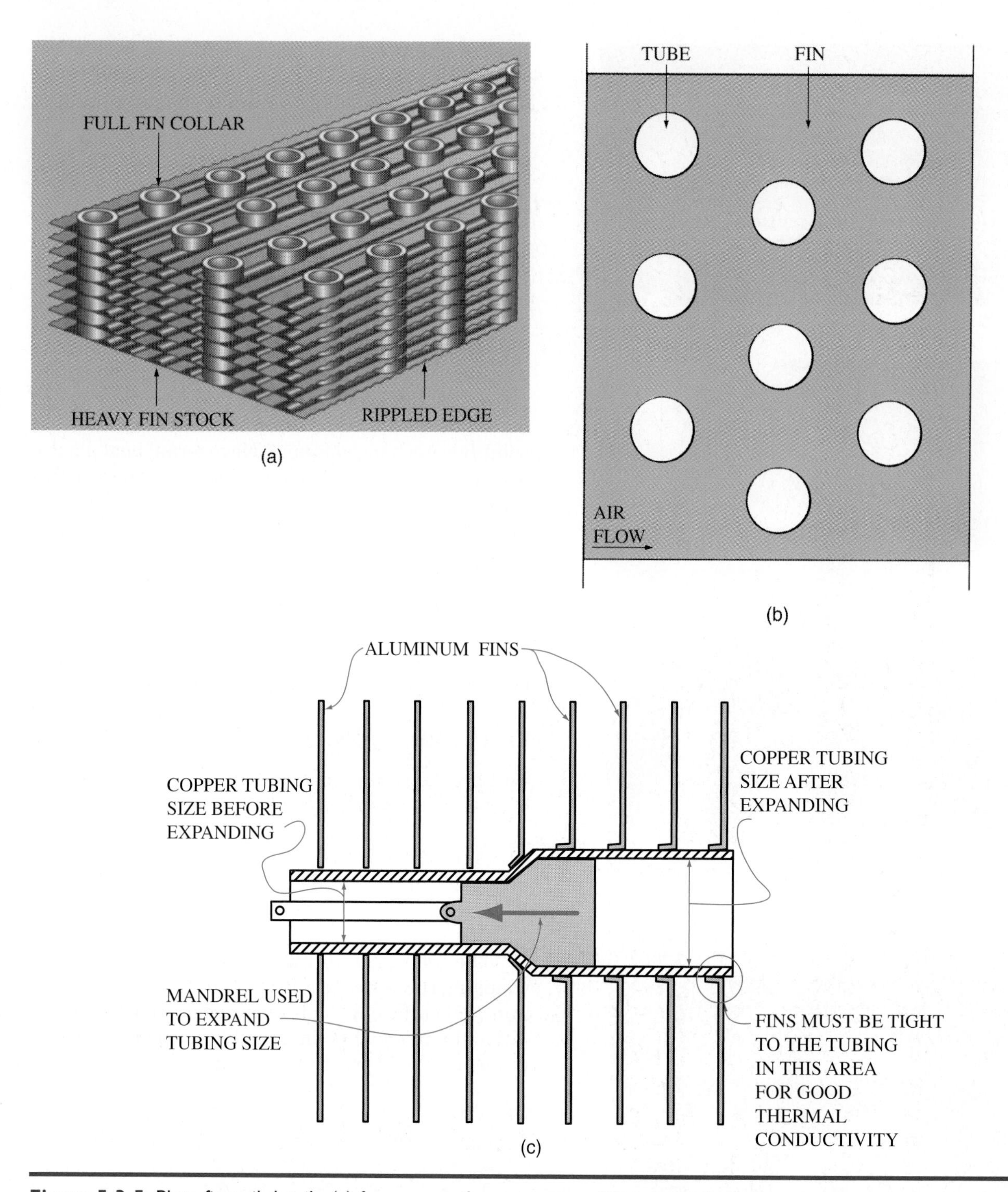

Figure 5-3-5 Plate fin coil details: (a) four row coil construction; (b) staggered tubes; (c) cross section of finned tube assembly.

A possible solution would be the addition of a suction-to-liquid heat exchanger placed near the condensing unit or providing suction-to-liquid heat exchange by running the lines tightly together (soldered every 2 feet and insulated) to achieve the needed liquid subcooling.

The subcooling increases cycle efficiency and can prevent flash gas due to pressure loss from components, pipe friction or from increase in height.

Direct Expansion (DX) Evaporators

Most smaller refrigeration systems are designed to have the expansion device control the refrigerant flow so the evaporator will heat the vapor beyond saturated conditions and ensure no liquid droplets will enter and possibly damage the compressor (liquid slugs). It is assumed here for the sake of

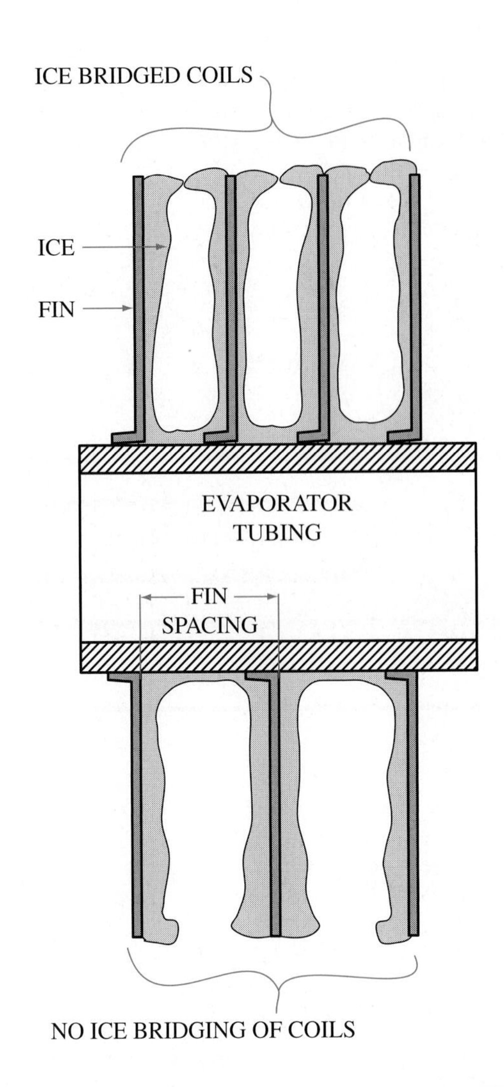

Figure 5-3-6 Ice bridging on a finned coil.

simplicity there is no pressure drop through the evaporator. (In reality there are pressure drops which would slightly shift the evaporating and condensing processes from the constant pressure lines shown.)

This additional heating of the gas (at constant pressure) is called superheating. It simply means heated above saturation temperature. The last pass of the heat exchanger tubes usually provides the superheating Figure 5-3-11, and at somewhat less efficiency because the refrigerant being heated is a gas (not a boiling liquid), and the temperature difference is reduced. A common range of superheat is 8–10°F which occurs between points C' and C. At this point the cold, low pressure, superheated refrigerant vapor is ready to flow to the compressor and complete the closed cycle.

Flooded Evaporators

If an evaporator does not have to superheat refrigerant vapor, it can produce more cooling capacity. On

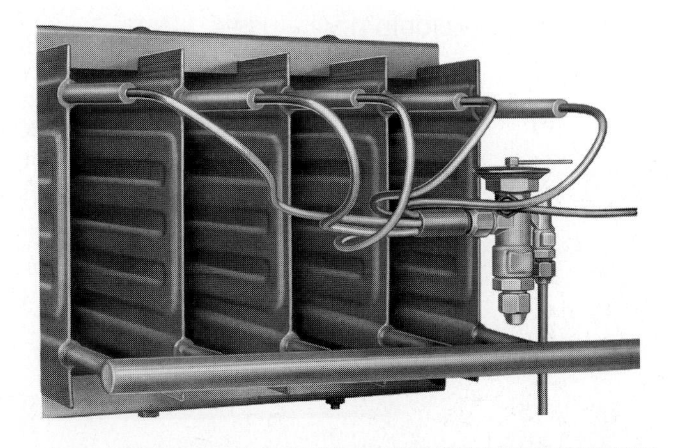

Figure 5-3-7 Plate-type evaporator. (*Courtesy Sporlan Valve Company*)

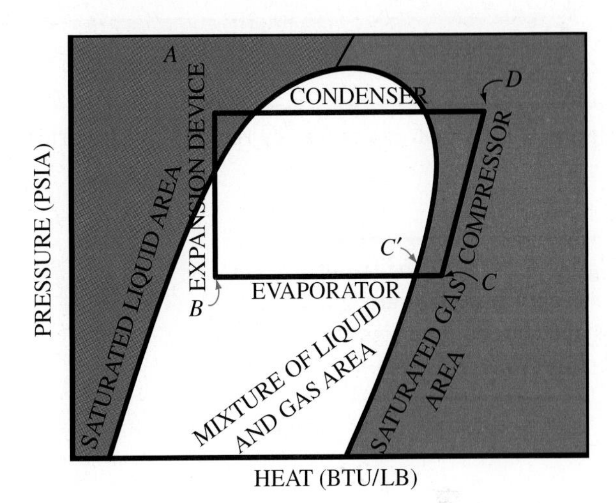

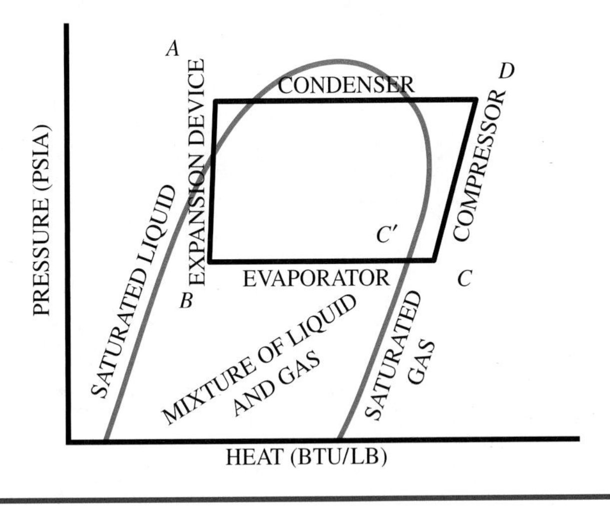

Figure 5-3-8 Pressure-temperature diagram showing refrigerant cycle.

small systems the difference is negligible and it is important to protect the compressor. On very large systems, an increase in evaporator performance can be important. A flooded evaporator absorbs heat from points B to C. It can circulate more pounds of

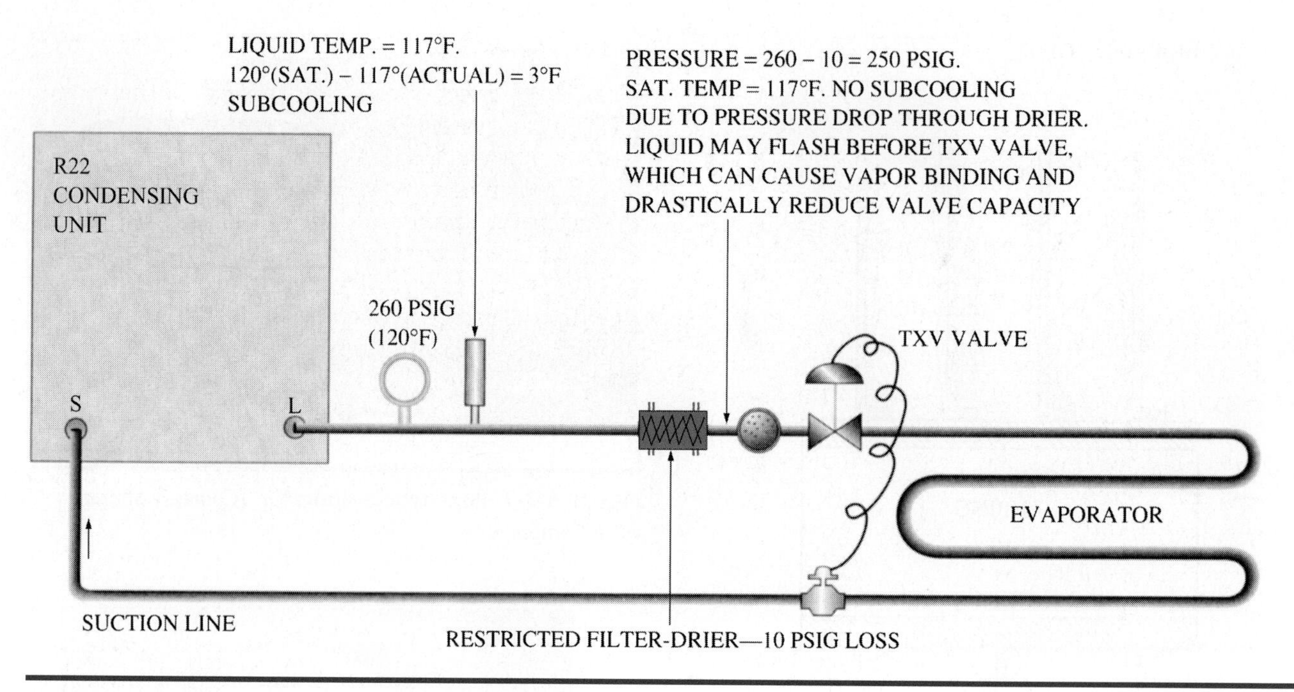

Figure 5-3-9 Flash gas caused by restricted filter drier.

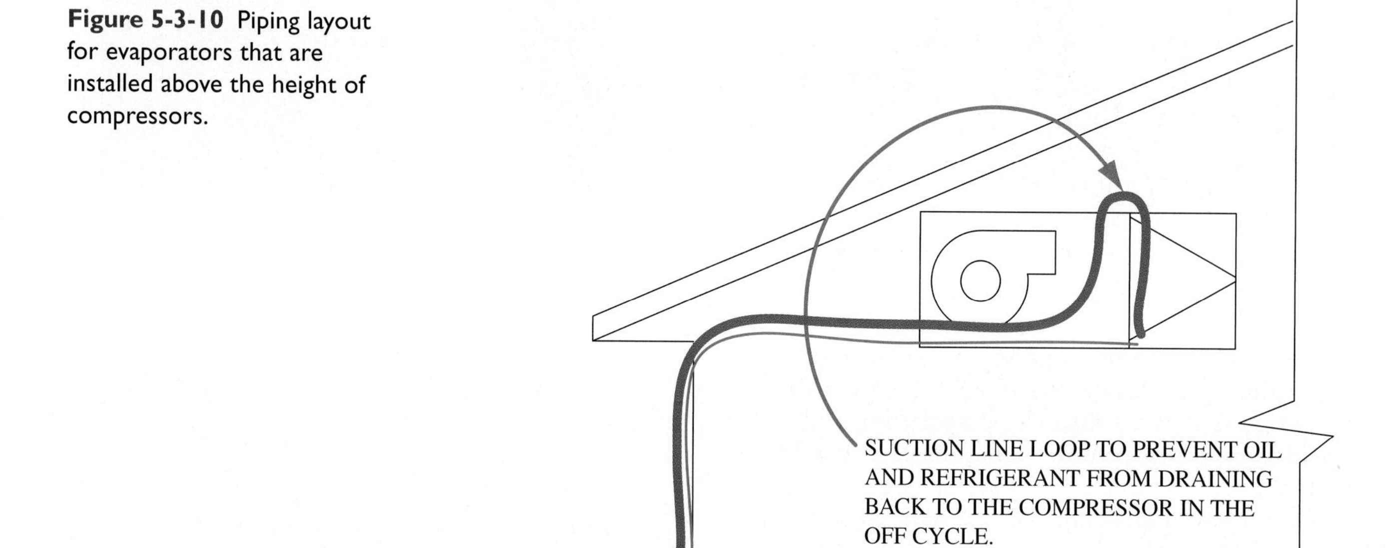

Figure 5-3-10 Piping layout for evaporators that are installed above the height of compressors.

refrigerant (more cooling capacity) per square foot of heat transfer surface. No surface needs to be used to superheat the suction vapor.

Large water chillers use flooded evaporators as shown in the example in Figure 5-3-12. It is very important to ensure that the saturated refrigerant flowing to the compressor does not contain quantities of liquid that could cause mechanical damage.

Load Variation

What happens when the ambient temperature increases? The load on the evaporator increases. When the load on an evaporator increases, the pressures increase. Figure 5-3-13a shows what happens to an air conditioner with an increased heat load. The operating points shift up and to the right, to A_1-B_2-C_2-

NOTE: Flooded evaporators use a low side float mechanism as their metering device. One of the advantages of a flooded evaporator is its ability to maintain a very accurate temperature across the entire evaporator surface. Some large ice rinks use flooded evaporators so that the rink temperature can be maintained very closely. This is particularly important for Olympic skating competitions. Skaters can tell the difference in ice temperature of a few tenths of a degree. Very cold ice sticks to the skate and is more difficult to glide across than slightly warmer ice.

D_2 on the pressure-heat curve. Figure 5-3-13b shows what happens when the load on the evaporator decreases. The dashed line shows that when the load on an evaporator decreases, the pressures decrease. The operating points shift down to A_3-B_3-C_3-D_3.

Undersized Evaporator

What is the effect of reducing the size of the evaporator on the system? A smaller evaporator with less heat transfer surface will not handle the same heat load at the same temperature difference as the correctly sized evaporator, Figure 5-3-14. The new balance point will be reached with a lower suction pressure and temperature. The load will be reduced and the discharge pressure and temperature will also be reduced. Note the similarity between Figure 5-3-13b and Figure 5-3-14. An undersized evaporator (Figure 5-3-14) and a reduced load (Figure 5-3-13b) both have the same effect on the refrigerant cycle because they both are removing less heat from the space.

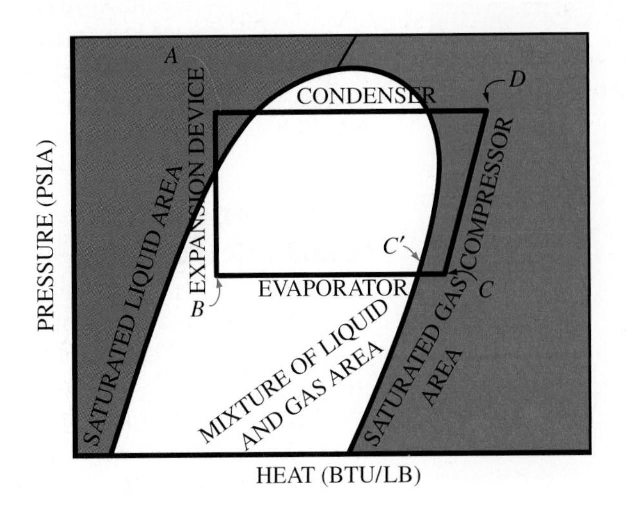

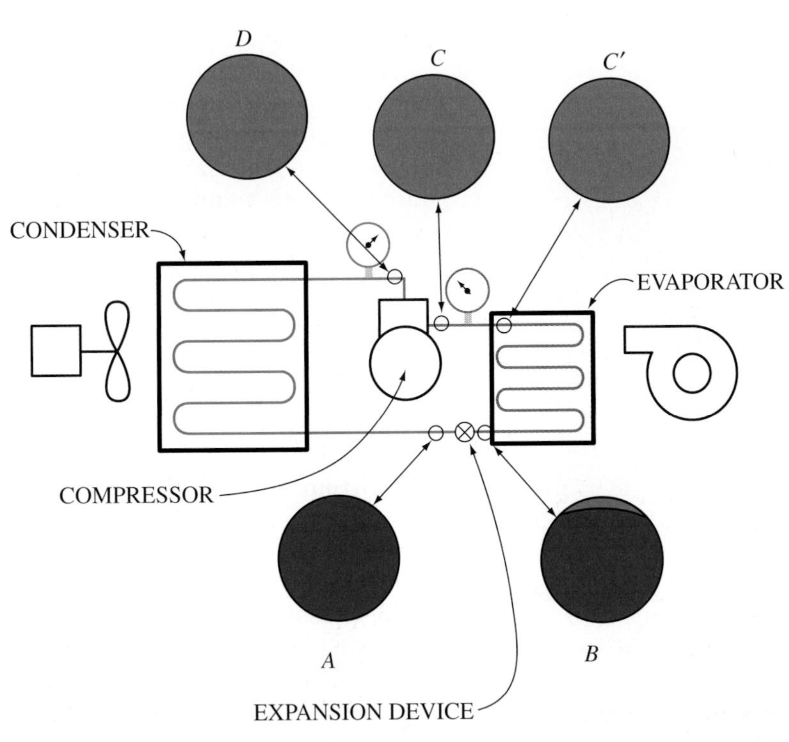

Figure 5-3-11 Superheating of refrigerant.

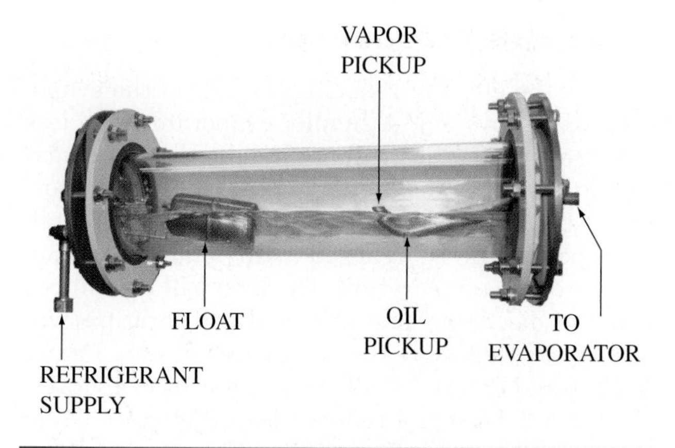

Figure 5-3-12 Flash gas caused by rise of liquid line. *(Courtesy Hampden Engineering Corporation)*

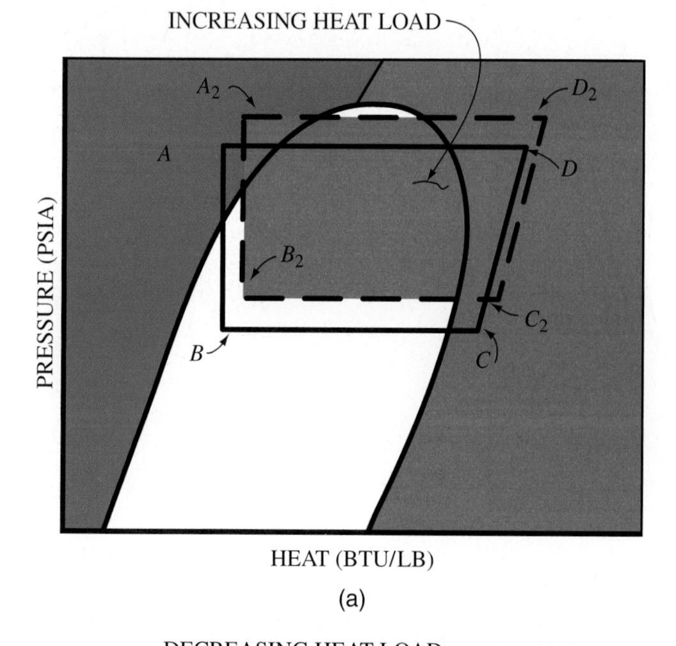

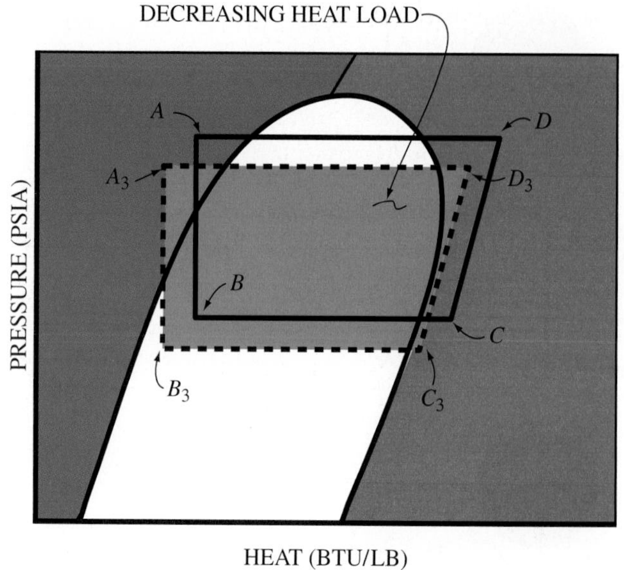

Figure 5-3-13 Superheat and flash gas.

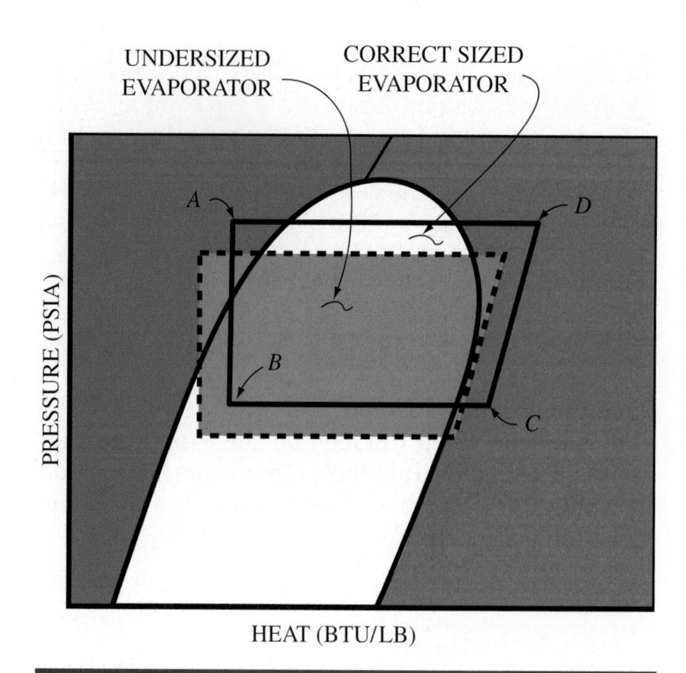

Figure 5-3-14 The effect of undersized evaporator on refrigerant cycle.

EVAPORATOR CONSTRUCTION—AIR COOLING

The most common evaporator for cooling air is the plate fin coil. This applies to small refrigerated reach-in and walk-in boxes, as well as to large comfort air conditioning units. The most common tube material is copper. Aluminum tubes have also been used. Aluminum and steel are suitable for ammonia refrigerant, but copper is not. Aluminum tubes are difficult to patch in the field if they develop a leak.

The most common fin material is aluminum. There are special applications for which other materials are used for fins. For sprayed-coil dehumidifiers, where a water spray on the coil fins is used to maintain close temperature and humidity conditions, copper fins would be recommended. They are costly and not ordinarily used.

It is possible to order coils with special protective coatings for marine applications or industrial use where ordinary fins are attacked and corroded. The coatings may reduce capacity somewhat, but lengthen equipment life.

For air conditioning applications, fin spacing ranges from 8 to 14 fins per inch. Figure 5-3-15 shows how to make this measurement. The closer fin spacing produces increased performance per square foot of coil, but the additional fins produce a higher air-side pressure drop and more energy is required to operate the fan. This type of tradeoff in design is not unusual. Tube rows in the direction of air flow typically range from two to six with three and four most common in direct expansion coils.

Figure 5-3-15 Fourteen fins per inch on this finned-tube coil used for air cooling. (*Courtesy Hampden Engineering Corporation*)

TECH TIP

One of the primary functions of an evaporator is to provide dehumidification. More dehumidification will occur if the fin spacing is closer together. Air only releases its moisture to the evaporator when it comes in contact with the fin material. Moisture vapor in the air that doesn't touch the metal simply passes through the evaporator. Selecting an evaporator with the proper fin spacing can assure that the system provides the correct degree of dehumidification. Coil thickness also affects the dehumidification provided by the coil. The level of dehumidification needed is affected by the local environment. In an arid part of the country a great deal of dehumidification is not needed as compared to coastal regions where humidity is much higher.

Face velocities typically range from 300 to 500 feet per minute (fpm). Most designers limit maximum cooling coil velocity to 500–550 fpm to reduce the possibility of condensate carryover from the coil fins during humid weather. If the condensate carries past the drain pan, it can leak from the air handler cabinet and may cause expensive water damage to the building or its contents.

To ensure maximum coil performance, adjacent rows of tubes are staggered to cause good air mixing and turbulence. The number of tubes in a row is designated as the tube face. This may also be described as how many tubes high the face is, or how many tubes you can see in the front row of the evaporator. The coil face dimensions open to airflow are designated as tube length (or finned length), and the height as finned height.

Counterflow

Heat evaporators generally should be installed and piped with counterflow. Counterflow is when the two fluid streams flow in opposite directions, as in Figure 5-3-16. This produces the greatest mean effective temperature difference (METD), therefore, the greatest heat transfer capacity. In the case of DX coil, the final refrigerant circuits, where the gas must be superheated, are in contact with the warmest entering air.

EVAPORATOR CONSTRUCTION—WATER COOLING

Water chillers (evaporators) are made in a variety of sizes and designs. Small units can be of a tube-in-tube coaxial design, as shown in Figure 5-3-17. A common construction is copper water tube and steel refrigerant tube.

DX water chillers of the shell and tube design range from 5 tons to 350 tons (and up). A shell and tube design is shown in Figure 5-3-18. The shell is usually steel, although brass pipe has been used in some smaller diameters. The refrigerant tubes are copper (some with an integral rolled fin to increase the heat transfer surface). Some designs utilize metal inserts or turbulators in the tubes to enhance heat transfer on the refrigerant side. The shell holds the water and baffles provide for even water flow and loading of all circuits.

Capacity Control

Evaporator coils can be arranged in multiple circuits and the refrigerant flow controlled by solenoid valves in larger systems. Row and face control are two general choices, shown in Figure 5-3-19. Face control is usually preferred for better humidity control (dehumidification).

TECH TIP

Capacity control adds efficiency to systems; however, the fan speed cannot simply be slowed down as a means of capacity control. This is because as the fan speed slows, less heat is picked up by the evaporator. Therefore the evaporator coil temperature can decrease below the freezing point. Capacity control can be achieved on evaporators by using solenoids to shut down part of the refrigerant circuit within a single evaporator or in an evaporator when there are multiple evaporators in the system.

When unloading face control coils, the lower coil section should always be the first on and last off. This ensures that condensate will not drain over an inactive section of coil and re-evaporate.

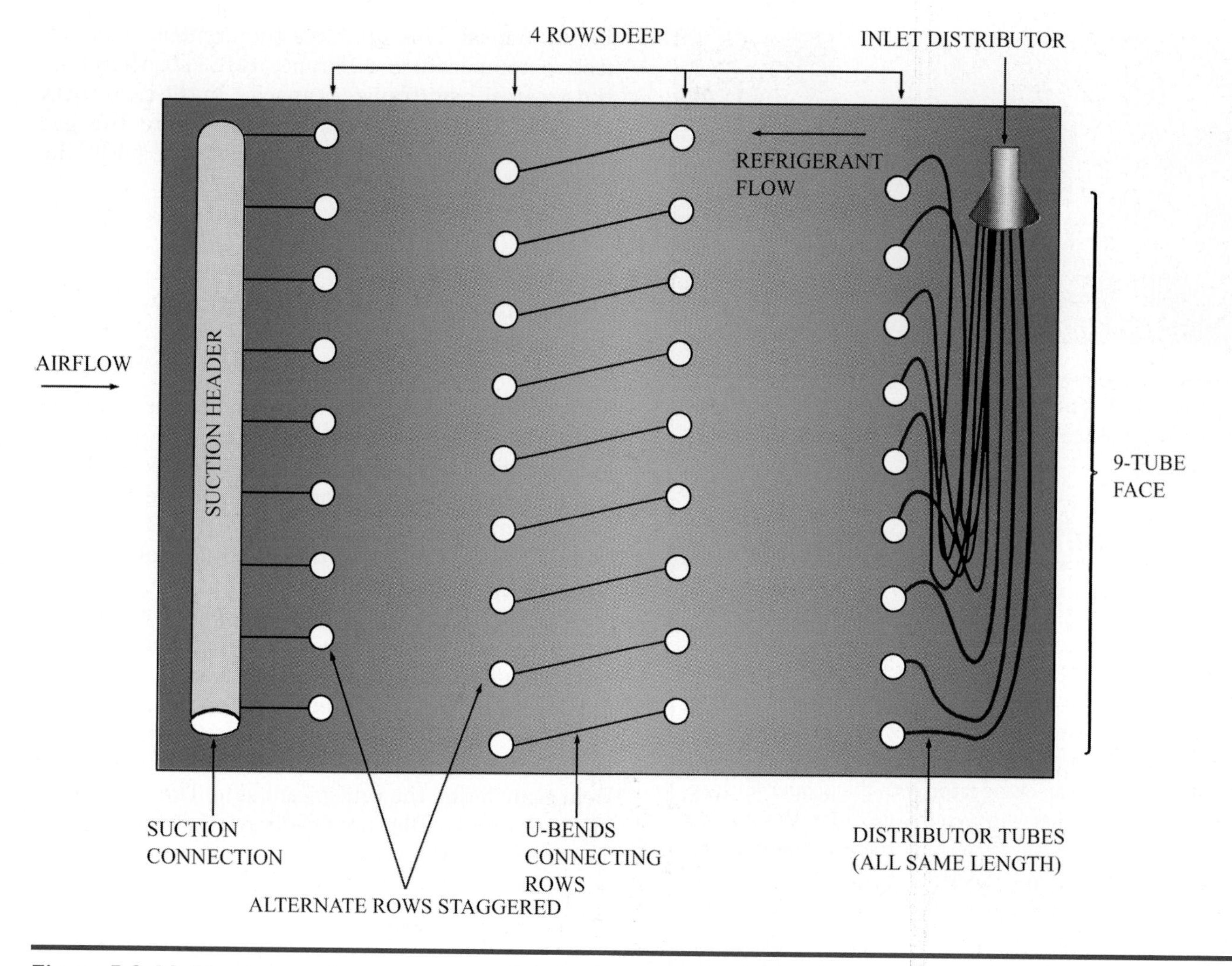

Figure 5-3-16 Direct expansion evaporator coil showing counterflow airflow.

Figure 5-3-17 Coaxial tube heat exchanger.

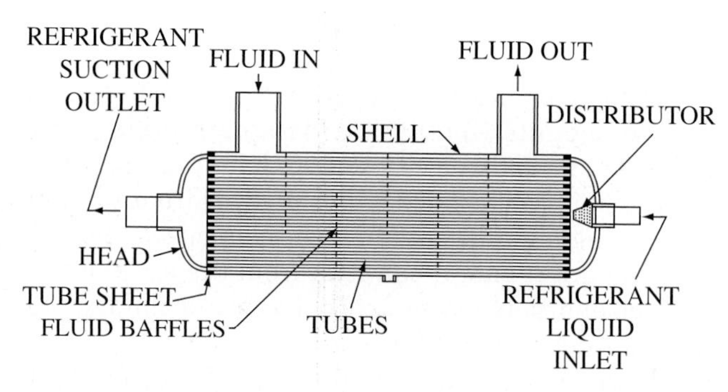

Figure 5-3-18 Direct expansion chiller. Refrigerant in tubes, water in the shell.

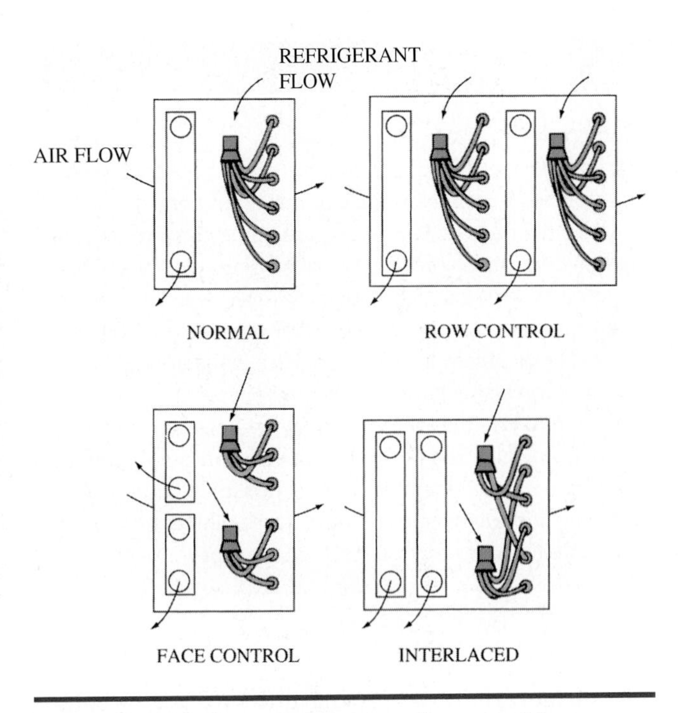

Figure 5-3-19 Direct expansion coil, row, and face control.

Condensate

DX coils operating below the dewpoint of the entering air stream will cool the water vapor (humidity) in the air and condense it as droplets on the coil. The droplets drain by gravity into the drain pan and flow through a condensate drain line to a drain or sink disposal as shown in Figure 5-3-20. The drain line should be provided with a water seal trap and never be connected directly to a sanitary drain.

Condensate on refrigerator-freezers can be drained to a tray under the appliance and evaporated by warm air from the air cooled condenser, as shown in Figure 5-3-21.

Condensate on package terminal air conditioners (PTAC) and window air conditioners is drained to a sump, picked up by a slinger on the condenser fan, and hurled onto the condenser coil to evaporate Figure 5-3-22.

When condensate cannot be drained by gravity, small pumps with a reservoir and float-operated switch can be used to pump the condensate to a nearby drain, as shown in Figure 5-3-23.

Types of Defrost

While air conditioning units remove humidity from the air as water, refrigeration and freezer units run colder and will freeze the condensed water as frost or ice on the coil. To allow for some frost buildup, the fins on a refrigeration coil are spaced much wider apart than in air conditioning.

Although the wide fin spacing allows air to flow with some frost on the fins, the frost will eventually build up sufficiently to block the airflow. Also, the frost acts as an insulator and inhibits heat transfer. The coil will require periodic defrosting to ensure sufficient airflow and cooling. The frost buildup depends on air conditions and evaporating temperature.

Air Defrost Coolers that operate at 28–30°F suction temperature will build up frost on the plate fins. They can be defrosted by simply shutting off

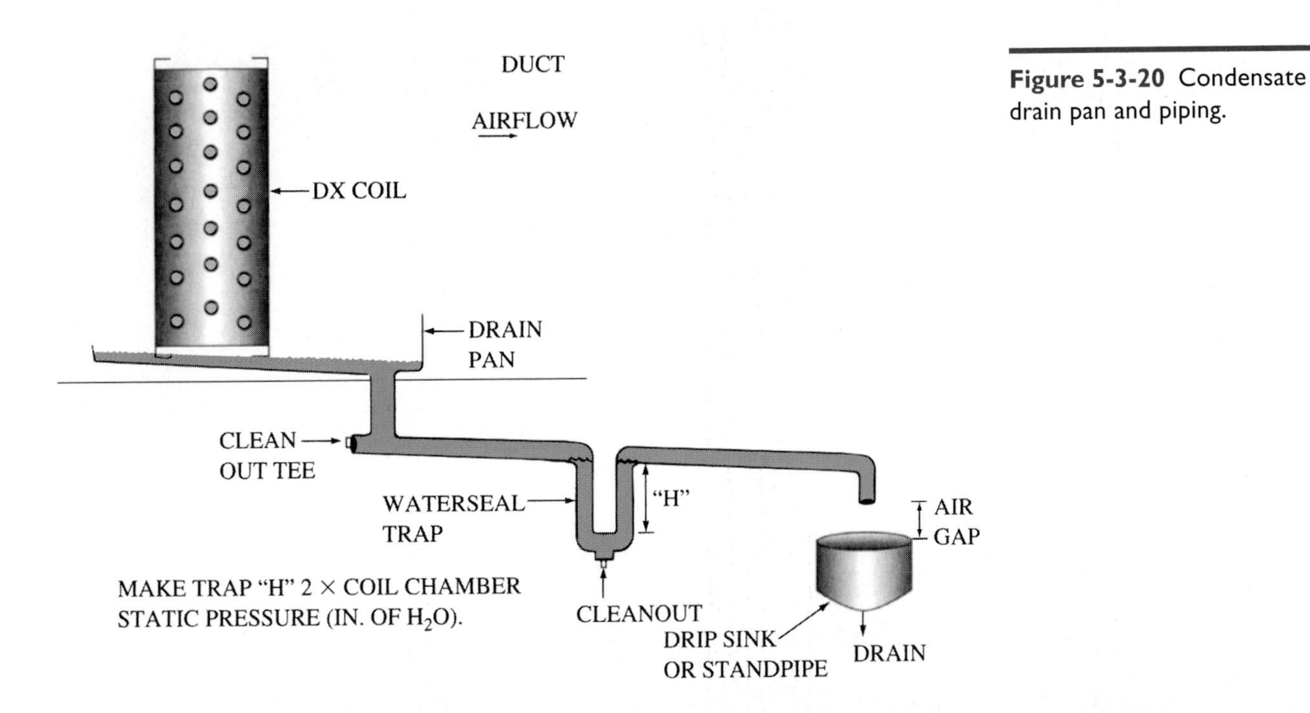

Figure 5-3-20 Condensate drain pan and piping.

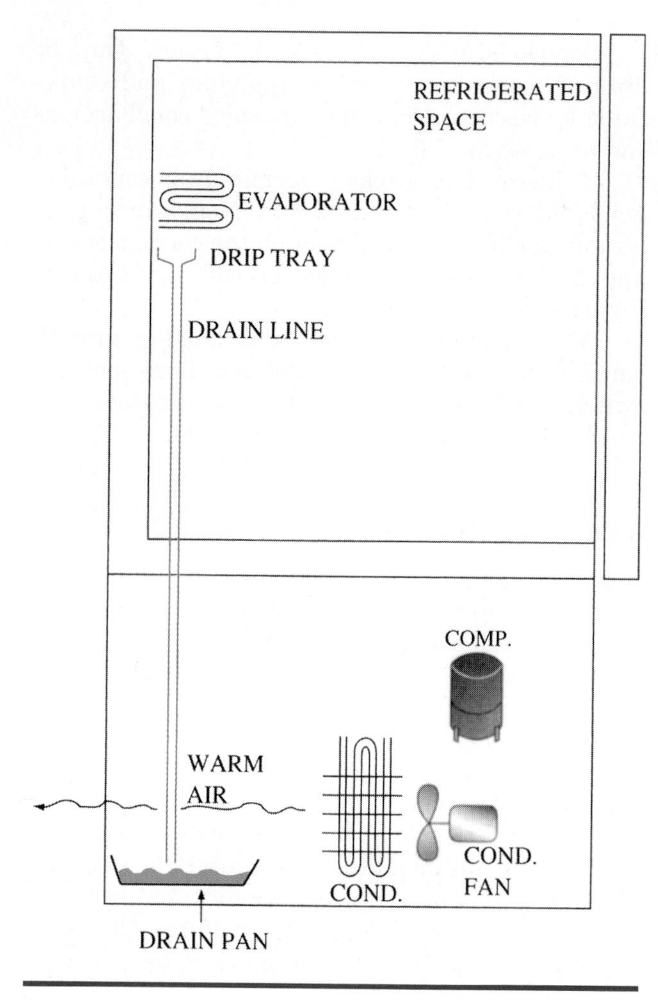

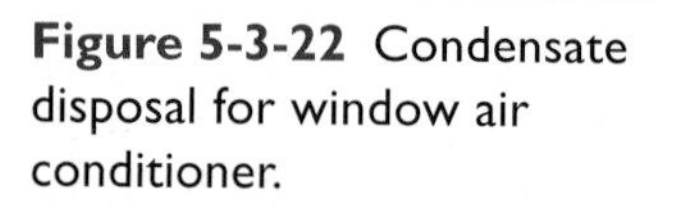

Figure 5-3-21 Condensate disposal for a refrigerator.

SERVICE TIP

Some coils that are used in medium temperature applications where the space temperature is above 32°F may still form ice because the evaporator temperature is below 32°F. These coils may require some form of defrost either using air or other artificial heating systems, such as electric or hot gas. Coils that ice over without defrost systems may result in a compressor tripping on low pressure. The problem for the technician is when they arrive on the job, the ice has melted and there is no evidence of a problem. For that reason, it may be necessary to temporarily install a temperature recorder on the system to ascertain if the system at times is dropping down below freezing and forming ice. Temperature plotters are available as electronic recorders or paper tape.

the refrigeration and letting uncooled air circulate over the coil, as shown in Figure 5-3-24a,b. Air at 35–39°F will melt the frost. Periodic defrosting can be initiated by a time switch. Air defrosting may go slower than other means.

Electric Defrost Refrigeration units running below 28°F with box temperatures below freezing (32°F) will not air-defrost. Electric defrost uses electric resistance heaters to melt ice off the coil, drain

Figure 5-3-22 Condensate disposal for window air conditioner.

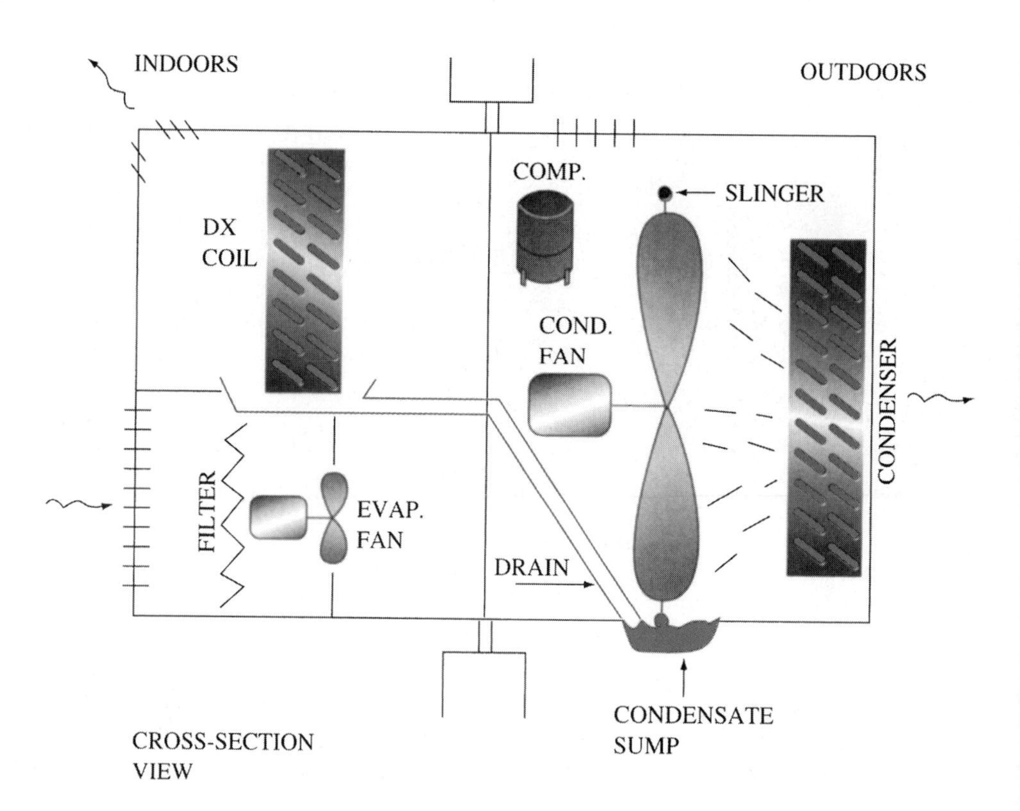

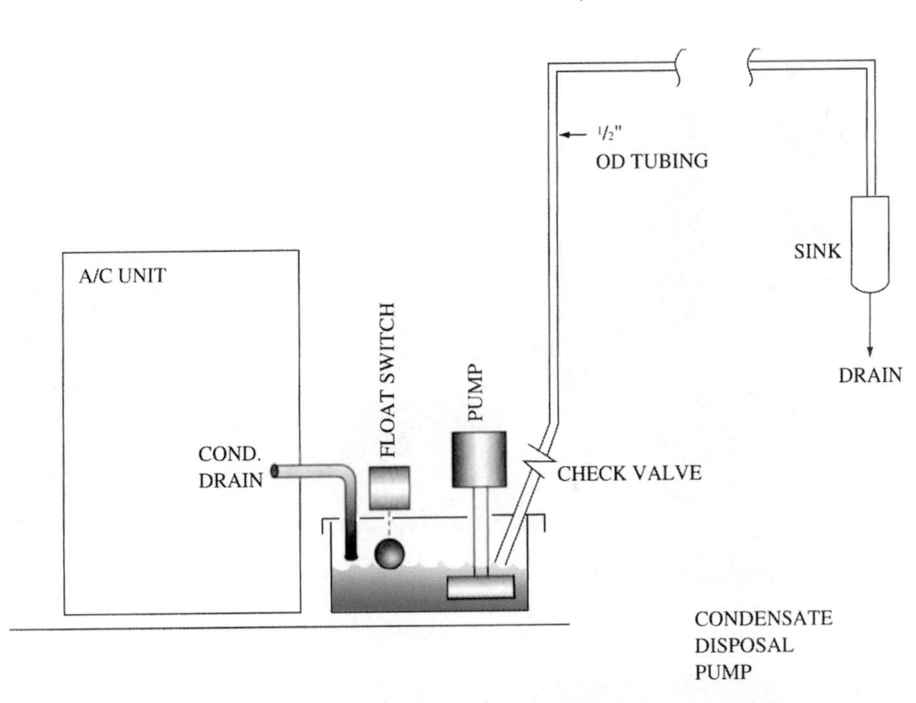

Figure 5-3-23 Condensate disposal using pump.

(a)

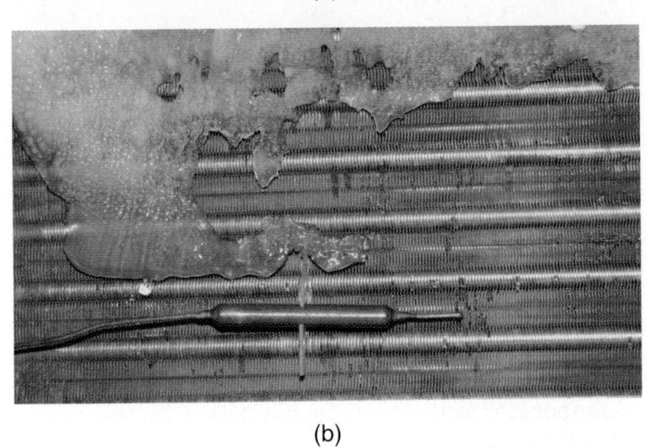

(b)

Figure 5-3-24 (a) Air defrost evaporator unit (*Courtesy Taco Bueno*); (b) The air moved over the coil during the off cycle melts the ice.

pan, and drain line, as shown in Figure 5-3-25a–c. The electric defrost cycle is simple and widely used. Figure 5-3-26a shows a typical wiring diagram for an electric defrost circuit. The circuit is shown in Figure 5-3-26b.

Defrost is usually initiated by a time switch although pressure or temperature sensors also have been used. The operation is as follows. The liquid refrigeration solenoid valve closes and the compressor pumps down. The fan shuts down. The coil and drain pan defrost heaters activate. The cycle is usually temperature terminated with a backup time termination provision to prevent the box from overheating should the temperature control fail for any reason. On cooling cycle restart, there may be a delay period before the fan clicks on.

Hot Gas Defrost Using hot gas available from the refrigeration cycle can be an attractive means of defrost. To successfully implement it in practice requires some care. With a single evaporator, a means is needed to re-evaporate hot gas refrigerant condensed in the evaporator during the defrost cycle. The system shown in Figure 5-3-27 uses a stored heat reservoir to re-evaporate the refrigerant.

In the refrigeration cycle, Figure 5-3-27a, the compressor discharges refrigerant vapor, which passes through water-storage-tank heating coils. Some heat is extracted from the hot gas and stored in the water (A). The refrigerant passes through a normally open (NO) solenoid valve (G) to the condenser (B). Liquid refrigerant drains from the condenser through a check valve into the receiver (C).

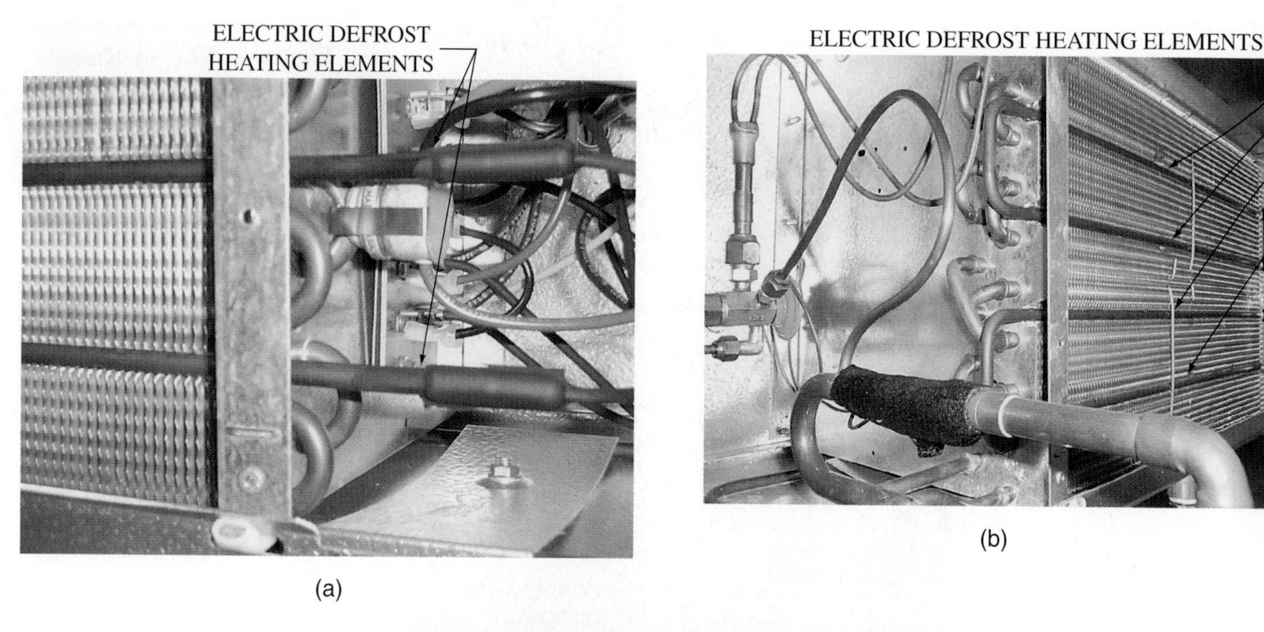

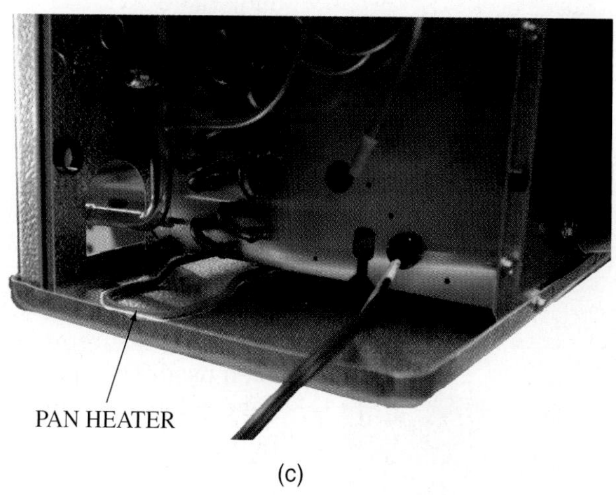

Figure 5-3-25 Electric defrost evaporator unit: (a) End cover removed; (b) Bottom cover removed; (c) Electric resistance pan heater to prevent ice from forming during the defrost cycle. *(Courtesy Hampden Engineering Corporation)*

From there it passes through the filter-drier, the liquid suction heat exchanger (D), the liquid solenoid valve (K), the thermal expansion valve (E), and into the evaporator (F). After vaporizing and picking up the evaporator heat load, the suction gas passes through the liquid suction heat exchanger (D) and through a suction solenoid valve (L) back to the compressor.

When the defrost cycle is initiated, Figure 5-3-27b, the following events occur:

1. Discharge solenoid (G) is energized closed.
2. Evaporator fans (H) are shut down.
3. Evaporator hot gas solenoid (J) is energized open.
4. Liquid solenoid (K) is de-energized closed.
5. Suction solenoid (L) is closed.

Discharge gas is blocked from the condenser and flows through a spring loaded check valve (M) (15 psig), through the receiver (C) and liquid line to the evaporator via the hot gas solenoid (J). The hot gas quickly defrosts the drain pan and evaporator coil and condenses to a high pressure liquid that flows through the suction line toward the compressor. The closed suction solenoid (L) blocks the liquid flow, causing it to be metered through a suction hold-back valve into the re-evaporator coil which is immersed in the water storage tank (A). The hot water provides a load to safely boil off the returning liquid to a low pressure gas that can be returned safely to the compressor to repeat the cycle. The water in the storage tank will cool and then freeze during the re-evaporation process.

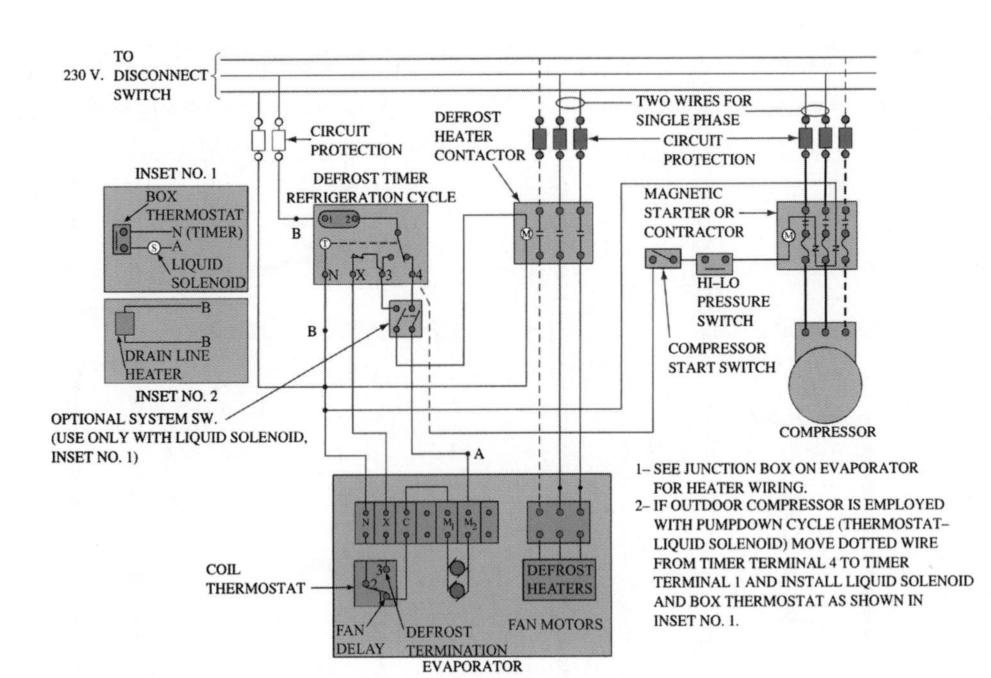

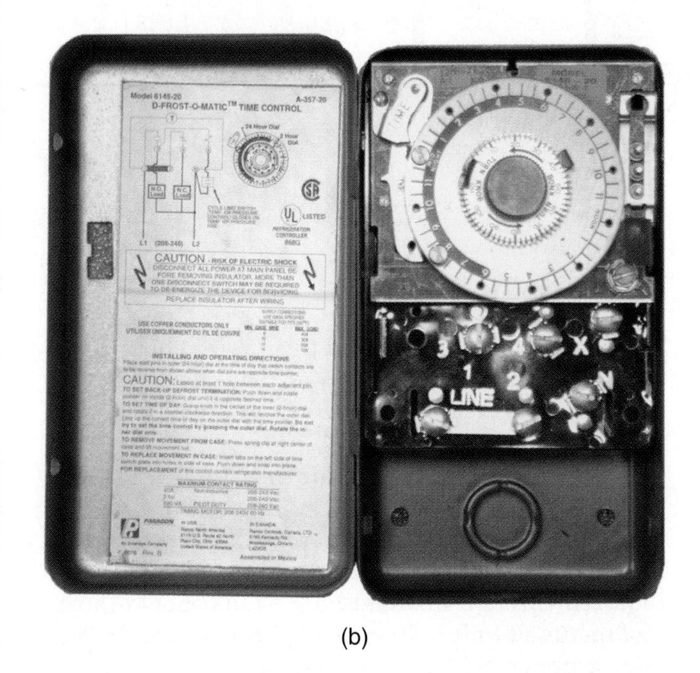

Figure 5-3-26 (a) Typical wiring for electric defrost unit. Timer initiates, thermostat terminates defrost. (*Courtesy of Kramer Refrigeration*) (b) Defrost timer.

The defrost cycle is terminated when the evaporator pressure rises to a set pressure, causing the discharge solenoid to open to the condenser and the evaporator hot gas solenoid to close. Figure 5-3-28 shows the wiring diagram for the hot gas defrost system. The liquid and suction solenoids stay closed and the fans remain off until the evaporator pressure drops to about 20 psig as sensed by the psi-defrost-pressure switch. The cycle is more complicated than electric defrost, but has the advantage of relatively fast defrost times and a savings of electric power.

Other hot gas defrost arrangements are possible. With two evaporators on a single condensing unit it is possible to defrost one evaporator while continuing to utilize the other as an evaporator. Solenoid valves provide the means of rerouting the refrigerant flow. Still another system uses a reversing valve to swap the evaporator and condenser in a fashion similar to a heat pump. The main requirement of any defrost system is that it be quick, reliable, effective, and as simple as possible for ease of service.

All defrost systems must have provisions to allow the melted condensate to drain clear of the coil before the system restarts. This is accomplished by having a time delay between the end of the defrost cycle and the restarting of the compressor and refrigerant cycle.

Drain lines from condensate drain pans in freezers must be free and clear to handle the drainage during the defrost cycle. They can be continuously heated by an electric heater cable, as shown in Figure 5-3-29. They should have a steep pitch (4 in per 12 in) for good drainage and be insulated. Line sizes are typically a minimum of ⅞ in OD tubing, with cleanout tees provided for maintenance.

Heater cable capacity should be sized in accordance with supplier recommendations. A typical capacity is about 6 watts per lineal foot. It is also possible to run a hot gas line strapped to the drain line as a heat source.

Other Means of Defrost In addition to air, electric, and hot gas defrost systems, ice can be melted from evaporator coils by water or water-glycol sprays. These are not as common as the former.

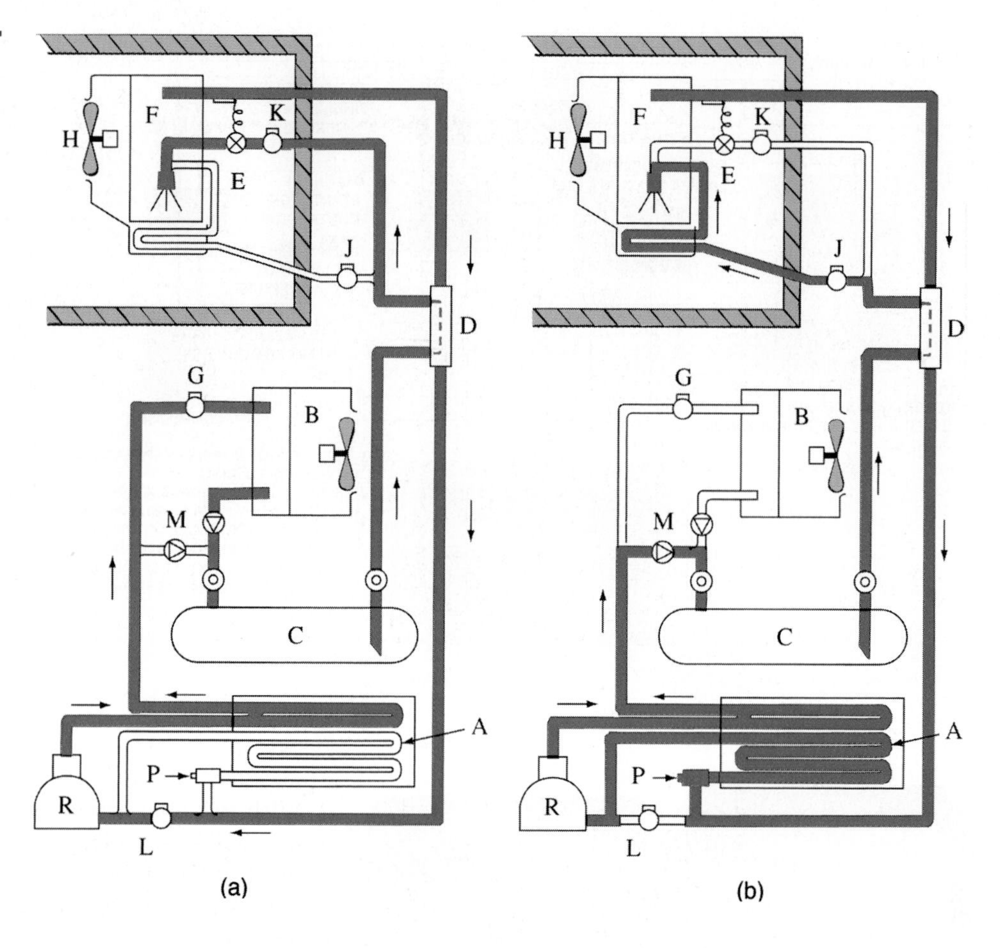

Figure 5-3-27 Thermal bank hot gas defrost system: (a) Normal refrigeration cycle; (b) Defrost cycle. *(Courtesy of Kramer Refrigeration)*

LIQUID COOLERS

Refrigerated liquid coolers can range from small drinking fountains to huge water chillers capable of cooling large buildings.

Drinking Water Fountains

The evaporator consists of a small storage tank cooled by an external coil of tubing wrapped around the tank. Construction can be considered double wall for safety. To enhance capacity, incoming water may be precooled by cold water draining from the fountain.

CAUTION: Any time you are working around systems that have potable water (drinking water) you must keep the system clean. It is a violation of many local and state health codes to return a system to service that has not been thoroughly sanitized. Refer to the manufacturer's guidelines on proper system cleanup before putting the system back in service.

Water Chillers

The construction of shell and tube water chillers was covered earlier in this chapter. The most common application is to provide a secondary coolant, namely chilled water.

Chilled water is used in air conditioning for cooling multiple remote air handling units where it would be impractical to run long multiple refrigerant lines. The chilled water can be pumped wherever required through insulated piping, as shown in the roof mounted unit in Figure 5-3-30. The air handlers have water coils instead of direct expansion refrigerant coils. Water coils are usually several rows deeper than an equivalent DX coil to get an equivalent cooling performance. They are normally the same size in face area.

Flywheel Effect

Control of cooling capacity is important. Water flow to small and large cooling coils can be controlled by modulating valves controlled by room thermostats. The chillers can usually be unloaded down to 25% capacity or less. Beyond that, the water circulating in

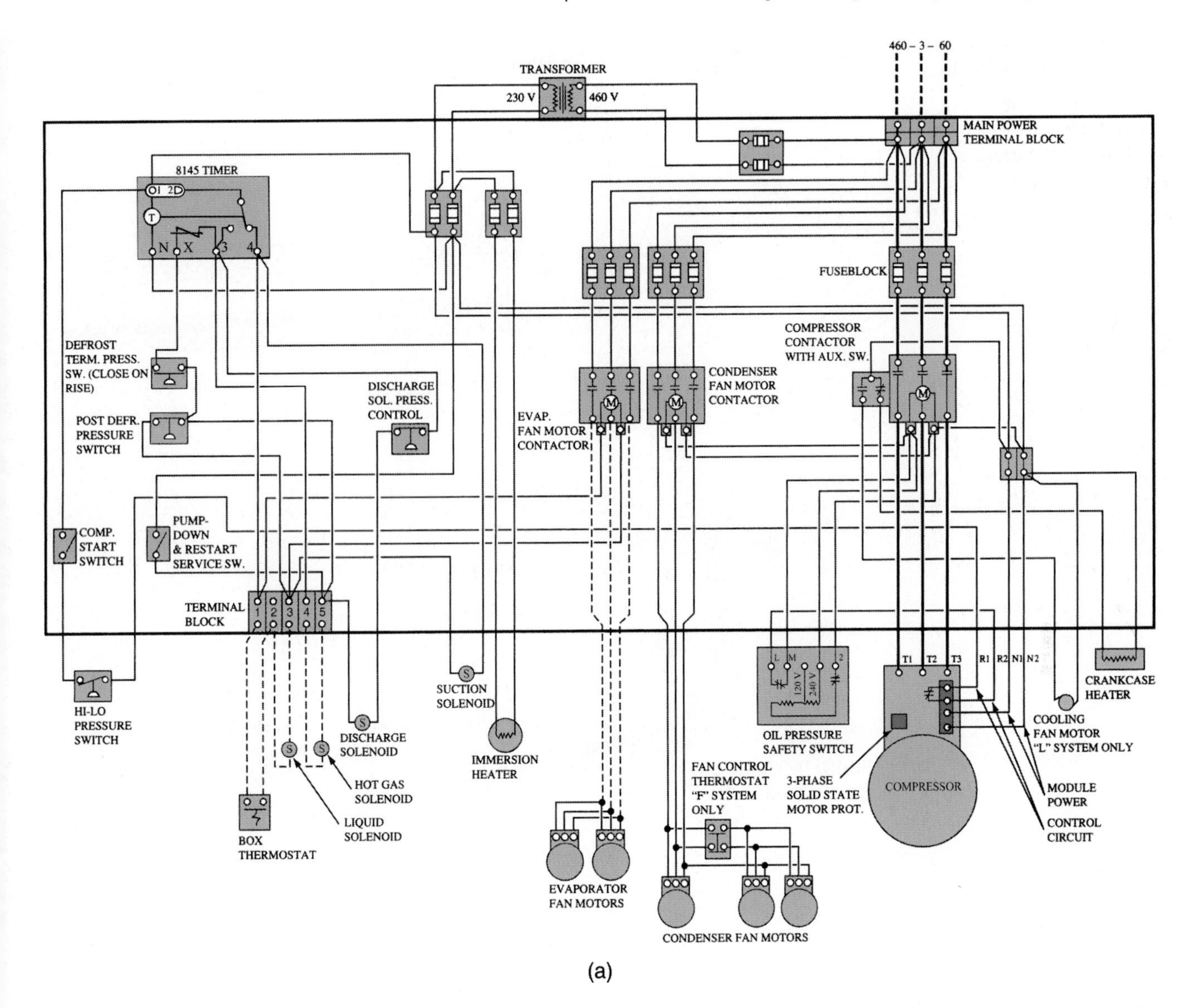

(a)

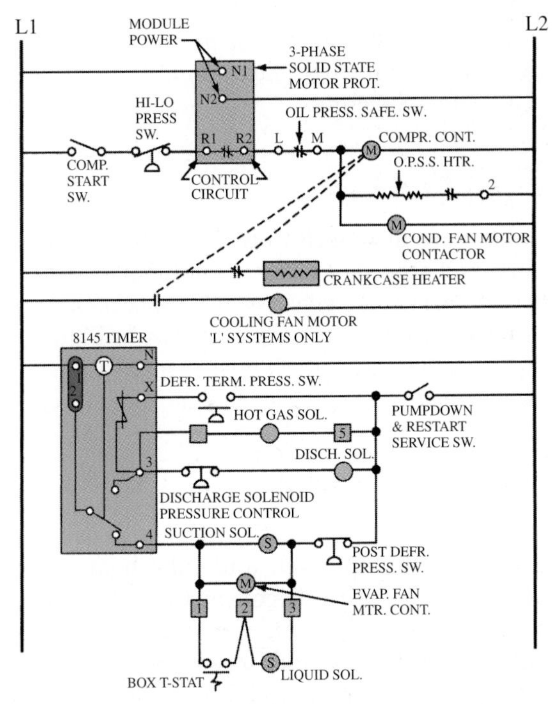

(b)

Figure 5-3-28 Thermal bank wiring diagram: (a) Panel; (b) Schematic. (*Courtesy of Kramer Refrigeration*)

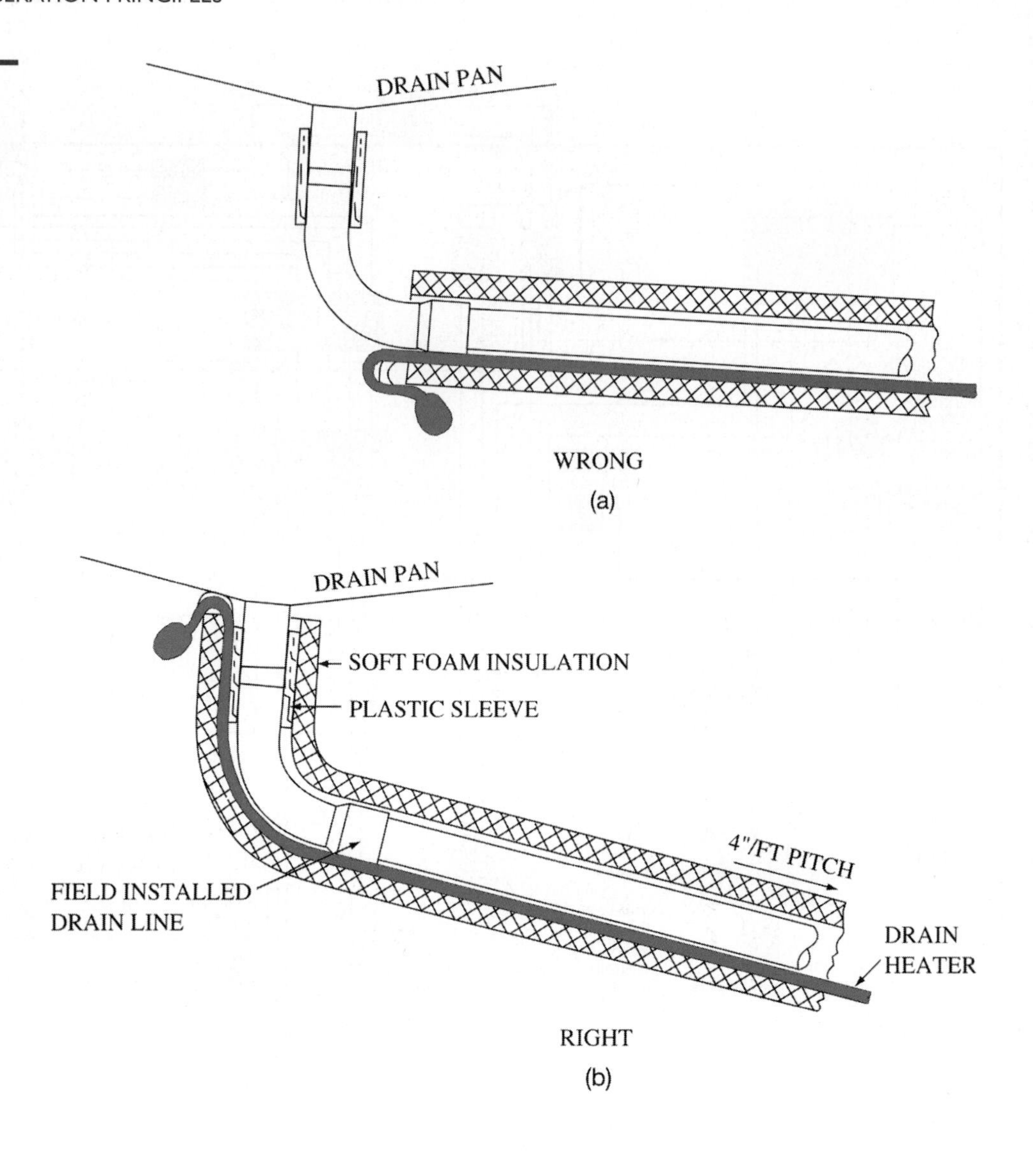

Figure 5-3-29 Drain line heaters: (a) wrong application; (b) correct application. (*Courtesy of Kramer Refrigeration*)

the system provides a reserve load to be cooled and provides a flywheel effect, permitting satisfactory operation during light load conditions without excessive on/off cycling.

Brine Coolers

Chillers are also used to cool brines (saltwater solutions) and other secondary coolants such as ethylene glycol or propylene glycol–water (antifreeze) solutions below 32°F. A brine cooler is shown in Figure 5-3-31. These are found in skating rink applications and can also be used to produce ice for thermal storage.

Chillers—Flooded

Reference has been made to direct expansion (DX) chillers. These are shell and tube heat exchangers (evaporators) with refrigerant in the tubes and water filling the shell. For larger capacity systems (several hundred tons and up) it is common to use flooded chillers, Figure 5-3-32, instead of DX chillers. These reverse the fluids, with the refrigerant filling the shell and water (or brine) passing through the tubes. They use enhanced surface tubing for improved refrigerant heat transfer. The tubes are submerged in liquid refrigerant and fill the lower portion of the shell. Refrigerant is controlled by a float valve or orifice, and the shell can act as a surge chamber, permitting separation of refrigerant vapor from liquid. Closely spaced chevron-type eliminator baffles or mesh screens are normally provided to ensure satisfactory separation of dry vapor from liquid.

Flooded chillers do not superheat suction vapor, and provide very efficient heat transfer.

SERVICING AND TROUBLESHOOTING EVAPORATORS

Evaporator problems can result in unsatisfactory operation and possible compressor failures, so the technician will want to be alert to these problems.

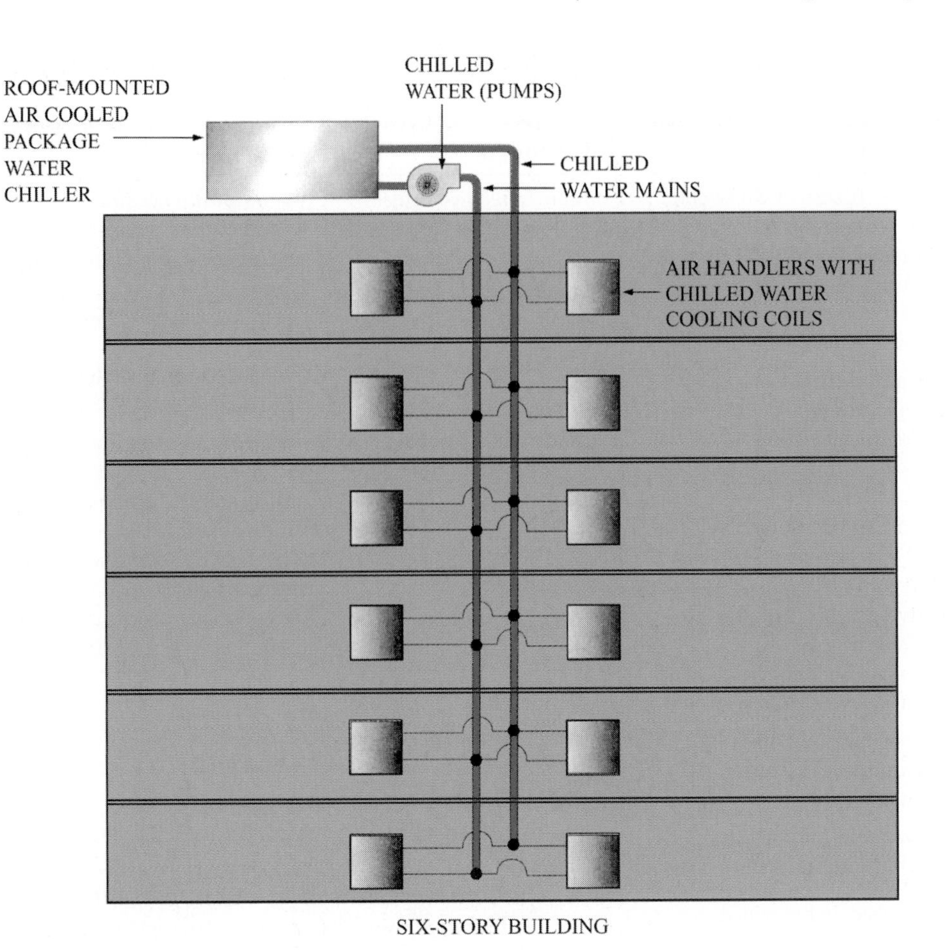

Figure 5-3-30 Roof mounted, air cooled packaged water chiller.

Figure 5-3-31 Packaged brine cooler.

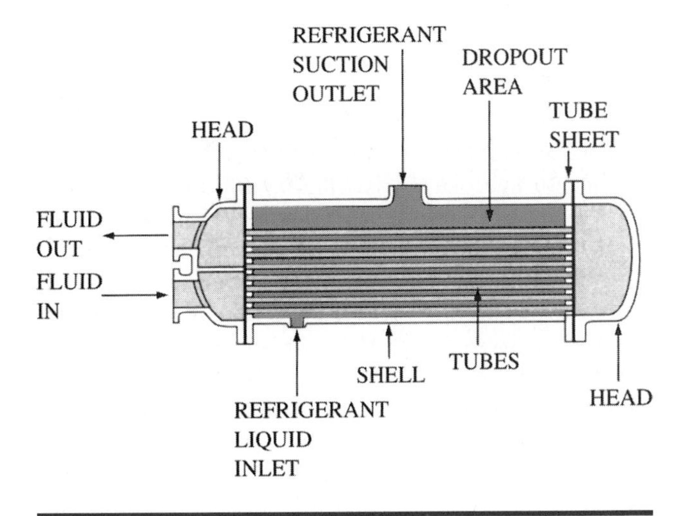

Figure 5-3-32 Flooded chiller; water in tubes, refrigerant in the shell. (*Copyright by the American Society of Heating, Refrigerating, and Air-Conditioning Engineers, Inc.*)

TABLE 5-3-1 Troubleshooting Evaporator Problems

Problems*	Possible Causes
Insufficient cooling (insufficient airflow)	Plugged air filters or dirty coil Loose or broken fan belt Fan blades plugged with dirt Fan running backward Coil iced (from any of above) Slipped or closed damper
Insufficient cooling (airflow normal)	Refrigerant charge low Filter-drier plugged Plugged capillary tube Compressor tripping off on safety Damaged compressor valves Thermal expansion valve, loss of charge TXV valve gas binding due to flash gas in liquid line Frozen water in TXV valve, wet system
Coil flooding liquid to compressor	TXV valve jammed open by foreign material (solder, slag, etc.)
Odors	Check for mold and mildew Clean coil, pan, and drain lines

*All of the problems assume a properly sized and installed system that has performed satisfactorily in the past.

TABLE 5-3-2 Insufficient Evaporator Cooling Problems

Problems	Possible Causes
Insufficient cooling (low water flow)	Plugged pump strainers Plugged chiller, system piping, and/or coils due to scale/muck/rust System low on water or air bound Valve closed or throttled improperly
Insufficient cooling (water flow normal)	Refrigerant charge low Filter-drier plugged Compressor tripping off on safety Damaged compressor valves Thermal expansion valve, loss of charge TXV valve gas binding due to flash gas in liquid line

Table 5-3-1 lists some of the problems and their possible causes.

If a high efficiency (SEER) air conditioning condensing unit has been installed and the old evaporator coil is left in service, the coil may be inadequate. The newer units are designed to run at higher evaporating temperatures and need more evaporator coil surface to achieve a satisfactory balance. Also, the older fixed orifice metering devices are designed to operate with a higher pressure. These metering devices will not operate properly when left installed with newer higher efficiency condensing units as the liquid pressure in these units is lower. Manufacturers recommend changing the DX coil when a new high-efficiency unit is installed, to ensure a proper equipment match.

Another hazard is having someone improperly charge the wrong refrigerant into a system. The technician should be aware of recent repairs made to the system, Table 5-3-2.

When servicing chillers care should be taken to avoid possible damaging freezeups. Make sure:

1. Flow interlocks are properly wired so the chiller cannot operate until water flow is established.
2. Chilled water thermostat is correctly set. Usual minimum exiting water temperature is about 44°F.
3. Chiller freezestats, low suction pressure cutouts, and low refrigerant temperature cutouts are set as recommended by the manufacturer to prevent freezeups.
4. Do not bypass any safety or operating controls.
5. If the chiller is exposed to freezing temperatures:
   a. Make sure electrical heaters and controls are functioning. If possible, protect the secondary coolant by adding antifreeze to depress the freezing point.
   b. At shutdown, drain chiller and piping. Circulate with antifreeze to ensure no unprotected pockets of water remain.

UNIT 3—REVIEW QUESTIONS

1. List the three general types of evaporators.
   What is a halocarbon system?
2. ________ are a special form of extended heat transfer surface used in refrigeration and freezer applications.
3. A rise in the height of the liquid line above the condenser, lowering the pressure and causing some of the refrigerant to boil, can be one of the causes of the formation of ________.
4. What is the most common evaporator for cooling air?
5. What is the most common fin material?
6. What is the difference between a plate fin evaporator and a plate type evaporator?

TECH TALK

Tech 1. I've got a problem on an evaporator. Part of it is freezing up, and the other part is warm. I don't understand why.

Tech 2. Well, have you checked to see if the distributor is feeding refrigerant uniformly to all of the coil?

Tech 1. No. How do you do that?

Tech 2. There are a couple of ways. One is to shut the fan down and let the system run. Coils that are being fed with refrigerant will ice up. Those that are not receiving refrigerant won't. In fact sometimes you can see that some tubes are partially fed because they will only ice up a little while others are icing up completely. Now the other way you can do it is to use a touch thermometer and measure the temperature of each of the circuits at the beginning of the circuit next to the distributor tube. Both methods work well.

Tech 1. Well, how do you fix that?

Tech 2. On large systems you can actually replace the distributor tube. But on small evaporators it isn't cost effective to try to fix it, and the evaporator should be replaced. Certainly on any evaporator you want to look to see the overall evaporator condition. If it has deteriorated as a result of moisture, then you probably want to go ahead and replace it anyway.

Tech 1. Thanks. I appreciate that.

8. When unloading face control coils, why should the lower coil section always be first on and last off?

9. What can be used when condensate cannot be drained by gravity?

10. Frost buildup depends on ________ and ________.

11. What happens when the defrost cycle is initiated?

12. When is the defrost cycle terminated?

13. Why is chilled water used in air conditioning?

14. When troubleshooting and servicing evaporators, list the possible causes if the problem is insufficient cooling.

15. What should be done when servicing a chiller that is exposed to freezing temperatures to avoid possible damage?

UNIT 4

Condensers

OBJECTIVES: In this unit the reader will learn about condensers including:

- Operations of Condensers
- Types of Condensers
- Air Cooled Condensers
- Water Cooled Condensers
- Condenser Capacity

CONDENSER OPERATION

The condenser is located on the discharge side of the compressor, as shown in Figure 5-4-1. The hot refrigerant vapor enters the condenser from the compressor and leaves the condenser as subcooled liquid refrigerant.

The function of the condenser is to transfer heat that has been absorbed by the system to air or water. In an air cooled condenser the outside air passing over the condenser surface dissipates the heat to the atmosphere. Using a water cooled condenser, the water is pumped to a cooling tower where the heat is transferred to the atmosphere by means of evaporation.

The pressure-heat diagram shown in Figure 5-4-2 illustrates the action performed by the condenser. The hot discharge gas from the compressor enters the condenser at point *C*. First the superheat is removed; then the vapor is condensed; then the liquid is subcooled until it reaches point *D*.

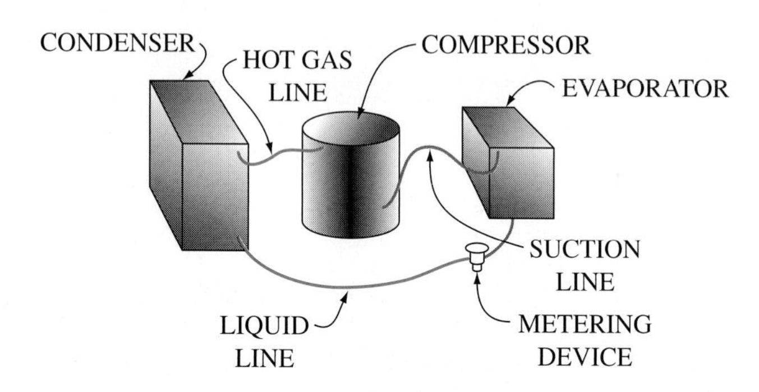

Figure 5-4-1 Schematic of refrigeration cycle.

Figure 5-4-3 summarizes the action that takes place in the condenser. This information is based on using a typical air cooled condenser with R-22 refrigerant and 95°F air passing over the outside surface of the condenser at 165°F and from which the liquid leaves at 105°F. About 15% of the heat is removed as superheat, 80% of the heat is removed by condensation, and 5% of the heat is removed by subcooling. This entire action takes place in the condenser.

Even though the subcooling accounts for only a small amount of the total heat rejection, it is important for two reasons:

1. It ensures that a solid stream of liquid will enter the metering device.
2. It adds to the cooling capacity of the system at a rate of about 0.5% of the total cooling capacity per degree of subcooling. For example: with 10°F of subcooling, 5% (0.5%/°F × 10°F = 5%) additional capacity is added to the system.

TECH TIP

It is very important that air cooled condenser coils be kept clean, as dirt builds up on the condenser. Dirt reduces the airflow and can act as an insulator preventing heat transfer. Condenser coil cleaning should be part of a seasonal startup program. The coil must be cleaned with an appropriate solution to remove the dirt and debris. Failure to use a manufacturer approved solution can result in damage to the coil. Extremely dirty air cooled condenser coils can result in very high head pressures. If the head pressure and temperature get too high, the refrigerant can decompose, forming acids. Dirty coils result in high utility usage and less cooling.

TYPES OF CONDENSERS

Three types of condensers are used in HVAC/R systems:

1. Air cooled
2. Water cooled
3. Evaporative, which is a combination air and water cooled.

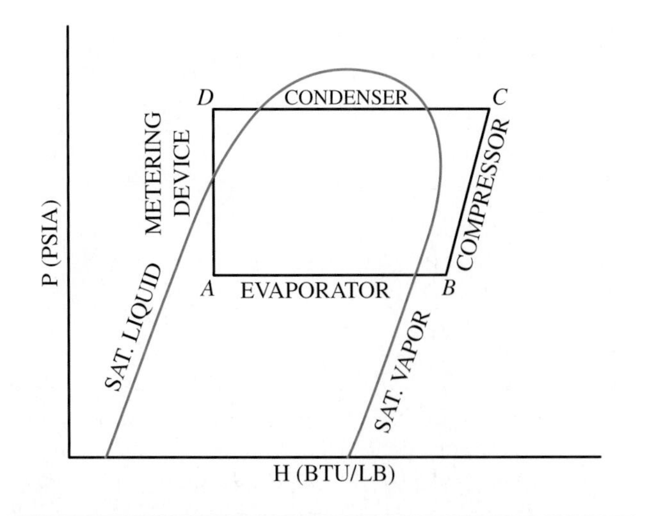

Figure 5-4-2 Pressure-enthalpy (heat) diagram showing the heat removed by the condenser.

Actually, a condenser is a very useful application of a heat exchanger. The heat picked up from various sources within the system is expelled by means of the condenser. Heat exchangers are made of metal to permit fast and efficient heat transfer. The hot refrigerant vapor is in contact with one side of the heat exchanger surface and the transfer medium such as air or water is on the other side.

The application of an air cooled condenser is usually the simplest arrangement, particularly if the condenser is located outside the system or unit. Although the water cooled condenser is more efficient, it is more costly to install. The examples that follow demonstrate some of the common uses of condensers.

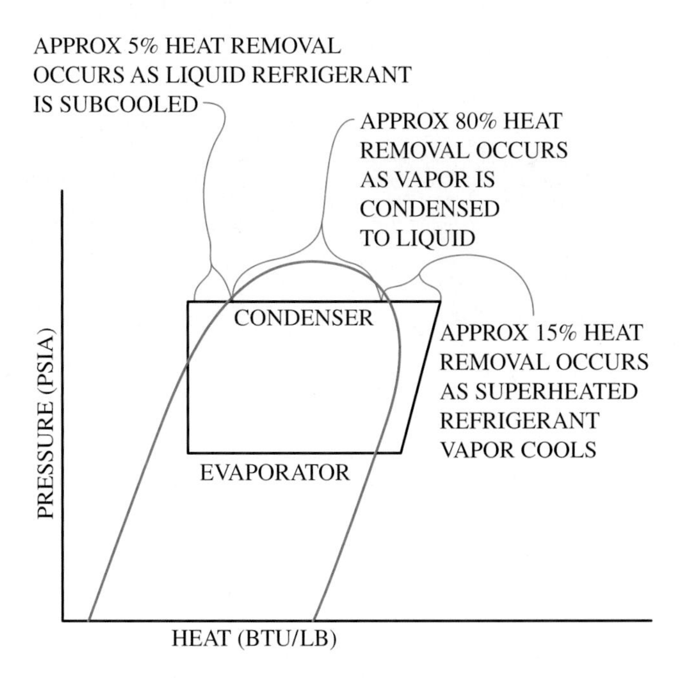

Figure 5-4-3 Heat removed from the refrigerant in the condenser.

AIR COOLED CONDENSERS

Air cooled condensers give off heat to the outdoors. At normal peak load conditions, the temperature of the refrigerant in the condenser is 25° to 30°F higher than ambient temperature. This means that on a 95°F day the condensing temperature is between 120° to 125°F.

With a water cooled condenser, the temperature difference between the medium and the refrigerant is lower. For example, it is common for the water cooled unit to use 85°F water and operate at a 105°F condensing temperature. This makes the water cooled unit more energy efficient but the unit cost is higher.

The air cooled units are simple and easy to install and maintain. They are therefore commonly used for most residential air conditioning systems up to 5 tons and commercial systems up to about 50 tons in capacity.

One common use of the air cooled condenser is for domestic refrigerators and upright freezers, as shown in Figure 5-4-4. The condensers are located in the lower part of the cabinet and a small fan is used to pull room air over the surface of the condenser coil and exhaust it back out into the room. Some refrigerators use a natural draft type condenser coil on the back of the cabinet.

Serious compressor problems can result if the condenser gets dirty. The dirt will reduce the heat transfer rate and the compressor head pressure can rise to damaging levels. The condenser can usually be cleaned with a brush and vacuum cleaner.

SERVICE TIP

The fins on an air cooled condenser can be damaged mechanically or as a result of corrosion. Mechanical damage can be weather related, such as from ice or sleet, and also from vandalism. Corrosion can result from sprinklers, dogs, or discharge from a domestic clothes dryer. Mechanical damage of the fins can be straightened with a fin comb. If more than 20% of the fins have been damaged, the condenser's efficiency will be so low that under heavy summer loads the compressor is likely to fail. Corrosive damage to condenser fins is not repairable, so the condenser coil must be replaced.

Another application of the air cooled condenser is in the residential packaged air conditioning unit, Figure 5-4-5. These are available for central air conditioning in sizes from 1½ to 5 tons of cooling capacity. These units can also include provisions for

Figure 5-4-4 Condenser on a domestic refrigerator.

Figure 5-4-5 Condenser on a residential packaged air-conditioning system.

a system of warm air heating or for cooling—but only where the source of heating is hot water.

Packaged room coolers and through-the-wall air conditioners both use air cooled condensers. Typical units of this type are shown in Figure 5-4-6a,b. These units are manufactured in sizes ranging from ½ to 3½ tons of cooling capacity. They are commonly used in houses, apartments, townhouses, condominiums, offices, schools, and motels.

The residential condensing unit is a popular use of air cooled condensers. This unit includes the compressor, condenser, fan, and outdoor circuitry, as shown in Figure 5-4-7. Refrigerant lines connect to an evaporator coil located on the top of a forced air furnace indoors or to a fan coil unit. Due to the fact that these air conditioning systems consist of two parts, they are called split systems. They use ambient air to remove heat from the condenser.

A commercial field-assembled air conditioning unit will have a remote air cooled condensing unit. The portion of this system inside the building is primarily an air handling unit, including the evaporator, the metering device, and a blower-filter unit. The portion on the outside includes the compressor, condenser, and condenser fan. This type of system can be

(a)

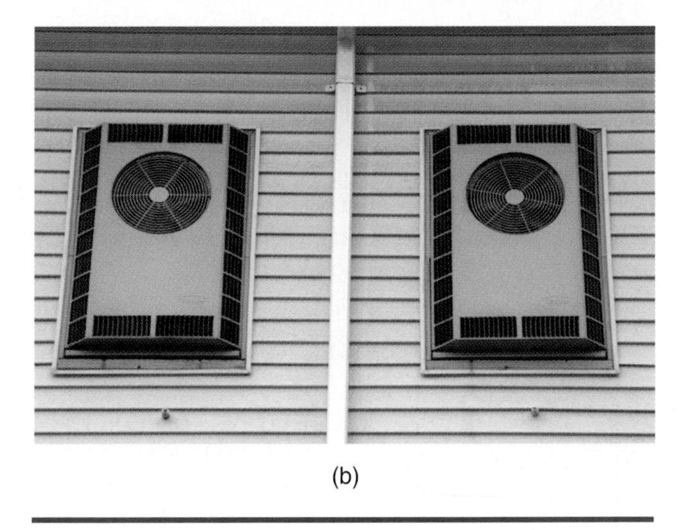

(b)

Figure 5-4-6 (a)(b) Condensers on through-the-wall air conditioners.

labeled as a split system because it is similar to the residential split systems, except larger.

Some air cooled condensers use a centrifugal fan. Condensers of this type can be located within the building and use ductwork to exhaust the heat laden

Figure 5-4-7 Condensers on residential outside condensing units.

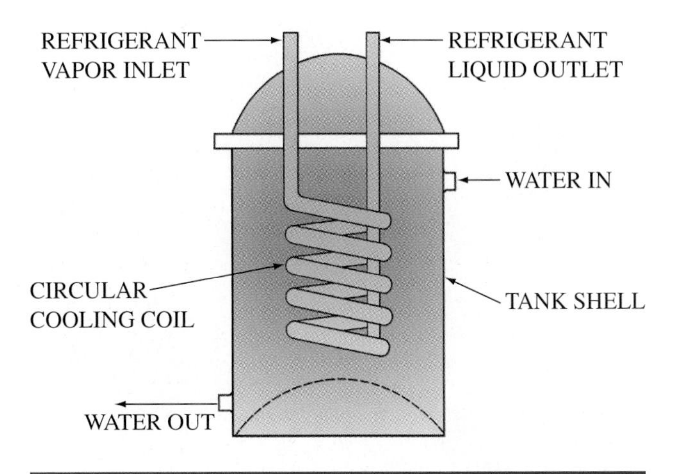

Figure 5-4-8 Water cooled condenser.

Figure 5-4-9 Cooling tower.

air to the outside. This can provide a considerable advantage in keeping the refrigerant line connections to the other parts of the system as short as possible.

Air Cooled vs. Water Cooled Condensers

Air cooled condensers are used on most residential air conditioning systems and on commercial systems up to 20 tons in capacity. They are even used on some commercial systems up to 100 tons. The reason for their popularity is their simplified application and low maintenance requirement. It is necessary to run the compressor discharge line to their location and the refrigerant liquid line back to the evaporator. This presents no problem if the distance is relatively short.

When a water cooled condenser is used, such as the one shown in Figure 5-4-8, a cooling tower, Figure 5-4-9, needs to be added to conserve water. This

TECH TIP

Water cooled condensers are far more efficient than air cooled condensers. This does not mean that a homeowner should put a sprinkler on an air cooled condenser to increase its efficiency. Continuous watering of an air cooled condenser coil will result in mineral buildup on the coil, which will ultimately reduce the coil's efficiency. In addition many minerals are corrosive to the fin material so that over time the aluminum fins will be corroded away.

(a)

(b)

Figure 5-4-10 (a) Remote-mounted air-cooled condenser with all other refrigeration cycle components inside the building; (b) This unit uses R-404A refrigerant. *(Courtesy Taco Bueno)*

requires a constant supply water as well as water treatment to prevent corrosion and scaling, as well as the formation of algae. A tower must include a pump that also requires service and maintenance. Periodically the tower needs to be thoroughly cleaned, and even the condenser tubes may collect deposits that must be removed. There is also the danger of water freezing in the piping during cold weather unless it is drained or receives special treatment.

Air cooled condensers operate at higher condensing temperatures during the day than water cooled condensers; however, the advantage in using air cooled condensers is that they are cost-effective, considering all factors.

The air cooled condenser can be installed as a separate component, as shown in Figure 5-4-10. With this arrangement the compressor is placed inside the building. The alternative is to purchase a packaged condensing unit with the compressor and condenser both in one package as shown in Figure 5-4-11. This is called an air cooled condensing unit.

Figure 5-4-11 Typical commercial type air-cooled condensing unit.

The advantages of air cooled condensing units as compared to separate air cooled condensers are as follows:

1. Compressor noise and vibration is outside of the building.
2. The complete condensing unit can be factory built and tested.
3. Less valuable building space may be used.

Selection of Air Cooled Equipment

To select an air cooled condenser, the following information is needed:

1. The temperature difference between design ambient air temperature and condensing temperature. This should range between 20° and 25°F.

2. The total heat rejection based on published ratings. This allows for the total cooling load plus the heat of compression. The total heat rejection for an air cooled condenser is 14,500 to 18,000/Btu/ton or higher, depending upon whether the system is used for low temperature refrigeration, commercial refrigeration, or air conditioning.

3. The refrigerant being used.

The following example illustrates the method of selecting an air cooled condenser.

EXAMPLE

Select a condenser. Use the following requirements: a nominal 20 ton semihermetic condenser system, using R-22, with a total heat rejection of 293,000 Btu or 290 MBH, when operating at 40°F evaporating temperature and 120°F condensing temperature. Use Table 5-4-1.

Solution

Table 5-4-2 shows a typical rating table for air cooled condensing units. Using the table, the nearest standard unit meeting these requirements would be a Model BRH031 with a capacity of 298 MBH at 20°F TD. This is a direct drive vertical airflow unit. ■

To select an air cooled condensing unit, the following information is needed:

1. The refrigerant being used.
2. The design ambient temperature.
3. The design suction temperature.
4. The tons of refrigeration capacity (at 12,000 Btu/ton) required to match the load.

TABLE 5-4-1 Typical Air Cooled Condenser Rating Table

BRH Model	Fan Config.	R-404A, R-502 and R-507 Total Heat of Rejection, MBH					*R-22 Total Heat of Rejection, MBH					Maximum No. of Circ. Avail.
		1°TD	10°TD	15°TD	20°TD	30°TD	1°TD	10°TD	15°TD	20°TD	30°TD	
Single Row of Fans												
023	1 × 2	11.07	111	166	221	332	11.30	113	170	226	339	14
027	1 × 2	13.00	130	195	260	390	13.27	133	199	265	398	14
031	1 × 2	14.60	146	219	292	438	14.90	149	224	298	447	14
035	1 × 2	17.14	172	257	343	515	17.50	175	263	350	525	14
041	1 × 3	19.53	195	293	391	586	19.93	199	299	399	598	21
045	1 × 3	21.89	219	328	438	657	22.33	223	335	447	670	21
049	1 × 3	24.24	242	364	485	727	24.73	247	371	495	742	28
053	1 × 4	26.04	260	391	521	781	26.57	266	399	531	797	21
061	1 × 4	29.20	292	438	584	876	29.80	298	447	596	894	21
065	1 × 4	32.34	323	485	647	970	33.00	330	495	660	990	28
071	1 × 4	34.30	343	515	686	1029	35.00	350	525	700	1050	28
075	1 × 5	36.95	369	554	739	1108	37.70	377	566	754	1131	21
079	1 × 5	39.36	394	590	787	1181	40.17	402	603	803	1205	28
089	1 × 5	43.48	435	652	870	1304	44.37	444	666	887	1331	28
097	1 × 6	47.24	472	709	945	1417	48.20	482	723	964	1446	28
107	1 × 6	52.14	521	782	1043	1564	53.20	532	798	1064	1596	28
Double Row of Fans												
046	2 × 2	22.15	221	332	443	664	22.60	226	339	452	678	2 @ 14
054	2 × 2	26.04	260	391	521	781	26.57	266	399	531	797	2 @ 14
060	2 × 2	29.20	292	438	584	876	29.80	298	447	596	894	2 @ 14
066	2 × 2	32.34	323	485	647	970	33.00	330	495	660	990	2 @ 14
070	2 × 2	34.30	343	515	686	1029	35.00	350	525	700	1050	2 @ 14
080	2 × 3	39.04	390	586	781	1171	39.83	398	598	797	1159	2 @ 21
086	2 × 3	41.81	418	627	836	1254	42.67	427	640	853	1280	2 @ 21
090	2 × 3	43.77	438	657	875	1313	44.67	447	670	893	1340	2 @ 21
098	2 × 3	48.48	485	727	970	1454	49.47	495	742	989	1484	2 @ 28
106	2 × 4	52.07	521	781	1041	1562	53.13	531	797	1063	1594	2 @ 21
120	2 × 4	58.38	584	876	1168	1751	59.57	596	894	1191	1787	2 @ 21
132	2 × 4	64.65	646	970	1293	1939	65.97	660	990	1319	1979	2 @ 28
140	2 × 4	68.60	686	1029	1372	2058	70.00	700	1050	1400	2100	2 @ 28
152	2 × 5	73.86	739	1108	1477	2216	75.37	754	1131	1507	2261	2 @ 21
162	2 × 5	78.73	787	1181	1575	2362	80.33	803	1205	1607	2410	2 @ 28
168	2 × 5	82.39	824	1236	1648	2472	84.07	841	1261	1681	2522	2 @ 28
178	2 × 5	86.93	869	1304	1739	2608	88.70	887	1331	1774	2661	2 @ 28
194	2 × 6	94.47	945	1417	1889	2834	96.40	964	1446	1928	2892	2 @ 28
202	2 × 6	98.85	988	1483	1977	2965	100.87	1009	1513	2017	3026	2 @ 28
212	2 × 6	104.30	1043	1565	2086	3129	106.43	1064	1597	2129	3193	2 @ 28

*For R-134A capacity, multiply R-22 capacity by 0.95; for 50 HZ capacity multiply by 0.92.

(Courtesy of Heatcraft Refrigeration Products)

TABLE 5-4-2 Typical Condensing Unit Rating Table

Capacity Data* (60 HZ.)** Condensing Units—R22														
		Ambient Temperature °F												
		90°F		95°F		100°F		105°F		110°F		115°F		
Model	Suction Temp °F	TONS	K.W.	TONS	K.W.	TONS	K.W.	TONS	K.W.	TONS	K.W.	TONS	K.W.	EER @ ARI Base Rating Cond
RCU-008S	30	5.8	7.4	5.6	7.5	5.3	7.6	5.1	7.7	4.7	7.8	4.6	7.9	11.2
	35	6.6	7.8	6.5	8.0	6.1	8.1	5.7	8.2	5.5	8.3	5.3	8.5	
	40	7.4	8.2	7.0	8.4	6.8	8.6	6.6	8.7	6.3	8.9	5.9	9.2	
	45	8.0	8.6	7.9	8.9	7.6	9.1	7.4	9.3	7.0	9.6	6.7	9.8	
RCU-008SS†	30	6.4	7.0	6.2	7.1	5.8	7.2	5.6	7.4	5.5	7.6	5.2	7.7	12.4
	35	7.0	7.4	6.8	7.6	6.6	7.8	6.4	7.9	6.2	8.0	5.8	8.2	
	40	7.7	7.7	7.5	7.9	7.0	8.1	6.9	8.3	6.7	8.5	6.5	8.7	
	45	8.4	8.1	8.0	8.4	7.8	8.6	7.6	8.8	7.4	8.9	7.0	9.1	
RCU-010SS†	30	7.7	8.4	7.5	8.6	7.3	8.7	6.9	8.9	6.7	9.1	6.4	9.3	12.2
	35	8.5	8.9	8.3	9.1	8.0	9.2	7.8	9.4	7.5	9.6	7.3	9.8	
	40	9.2	9.3	9.0	9.5	8.8	9.8	8.6	10.0	8.3	10.2	8.0	10.4	
	45	10.0	9.8	9.8	10.1	9.6	10.3	9.4	10.5	9.0	10.8	8.8	11.0	
RCU-010T	30	8.5	11.5	8.2	11.7	7.9	11.9	7.5	12.0	7.2	12.1	7.0	12.2	9.7
	35	9.4	12.4	9.1	12.6	8.7	12.8	8.4	13.0	8.0	13.1	7.8	13.2	
	40	10.3	13.2	10.0	13.4	9.6	13.6	9.3	13.8	8.9	13.9	8.6	14.1	
	45	11.3	14.0	10.9	14.2	10.5	14.6	10.2	14.7	9.9	14.8	9.5	15.1	
RCU-015SS†	30	10.9	12.6	10.6	12.8	10.2	13.6	9.7	13.4	9.5	13.6	9.4	13.8	11.8
	35	12.1	13.2	11.8	13.7	11.4	13.9	11.1	14.2	10.8	14.6	10.5	14.8	
	40	13.3	14.1	12.8	14.2	12.4	14.6	12.0	14.9	11.9	15.4	11.4	15.6	
	45	14.5	14.8	14.5	15.2	13.8	15.4	13.3	16.0	12.9	16.2	12.5	16.4	
RCU-015T	30	11.2	14.8	10.8	15.1	10.2	15.4	9.7	15.5	9.2	15.7	8.8	15.9	10.3
	35	12.7	15.7	12.1	16.1	11.6	16.4	11.1	16.7	10.7	17.0	10.1	17.2	
	40	14.1	16.8	13.5	17.2	13.0	17.7	12.4	17.9	11.9	18.2	11.3	18.6	
	45	15.5	18.0	14.9	18.4	14.3	18.8	13.8	19.2	13.3	19.6	12.8	19.9	
RCU-020T	30	14.5	16.8	14.1	17.1	13.5	17.5	13.0	17.9	12.4	18.3	11.9	18.7	11.6
	35	16.1	17.8	15.6	18.1	15.0	18.5	14.5	19.0	14.0	19.2	13.4	19.6	
	40	17.4	18.5	16.9	19.0	16.3	19.5	15.8	20.0	15.4	20.3	14.6	20.4	
	45	19.0	19.7	18.5	21.0	17.8	20.6	17.3	21.0	16.8	21.4	16.4	21.9	
RCU-020SS†	30	15.7	19.1	15.2	19.5	14.9	19.9	14.3	20.2	14.0	20.5	13.4	21.0	11.1
	35	17.4	20.0	16.8	20.5	16.4	20.9	15.8	21.4	15.5	21.9	14.9	22.4	
	40	19.0	21.3	18.5	21.7	17.9	22.4	17.4	22.8	17.1	23.2	16.3	23.7	
	45	20.7	22.4	20.1	22.8	19.6	23.5	19.0	24.1	18.5	24.5	17.8	25.0	

Notes: ▭ ARI Base rating conditions 90° ambient, 45° suction temperature.

†All models with the suffix 'SS' denote single D/B-Metic accessible Hermetic compressors.

*For capacity ratings at 85°F ambient temperature multiply the ratings of the 90°F ambient by 1.03 x Tons and .97 x KW.

**For 50 hertz capacity ratings, derate above table by .85 multiplier.

(Courtesy of Dunham-Bush, Inc.)

Low Ambient Control

Evaporators for commercial refrigeration have approximate operating ranges from +35°F down to −20°F when matched with appropriate condensers. Outdoor conditions may vary from 115°F down to zero degrees and below, Figure 5-4-12.

Condensing units for comfort air conditioning duty are nominally rated at 95°F outdoor ambient temperature to function with evaporators operating at 40°F with incoming air at 67°F wet bulb temperature (WB). However, there is an increasing need for comfort air conditioning to operate when outside ambients fall below 75°F. High internal heat loads from people, lights, and electronic equipment may demand cooling even when outside temperatures go down to 35°F and below.

Where possible, an economizer providing "free" cooling should be used if outside air at the proper temperature and humidity is available. If low ambient conditions are to be encountered, proper controls need to be provided to enable operation.

TECH TIP

Many air conditioning and refrigeration systems must operate when the outdoor temperature is low because of building heat loads. Restaurants, computer labs, hospitals, and office buildings all have excessive indoor heat loads even in cold weather that require continuous air conditioning for refrigeration. Low ambient controls are essential for these systems. It has been an unacceptable practice to overcharge a condenser during winter months to artificially increase the head pressure by flooding the condenser. This practice is not recommended by any manufacturer because the first warm summer day can result in compressor failure.

In the case of both commercial refrigeration and air conditioning, low ambient operation is an important need. In earlier discussions we learned that in order to have proper refrigerant vaporization in a direct expansion (DX) evaporator coil, it was necessary

Figure 5-4-12 (a) Ambient thermostat used as part of the low ambient control kit; (b) Terminals located inside of the control. (*Courtesy Hampden Engineering Corporation*)

to maintain a reasonable pressure differential across the expansion device. In a normal air cooled condenser operating between 80°F and 115°F ambient, condensing pressures are sufficiently high. But in winter condensing pressures can drop 100 psi or more; thus the pressure across the expansion device may be insufficient to maintain control of liquid flow. Evaporator operation becomes erratic. The thermal expansion valve will alternately open and close, first causing floodback of liquid refrigerant and then starving the coil when the valve closes. The capillary tube (if used) is worse, because it is a fixed metering device, and as the pressure difference falls, the flow of refrigerant is severely reduced. Below outdoor air temperatures of 65°F, a capillary tube system is in real trouble.

The solution to the problem of low ambient operation is to maintain a minimum head pressure in order to ensure proper feeding of refrigerant to the evaporator. There are several different ways of doing this.

Where multiple condenser fans are employed on a single coil, the control system of the condensing unit can be equipped with devices to switch or cycle "off" the fans in stages. These controls are usually sensors that measure outdoor ambient temperature or pressure controls that sense actual head pressure. As fans are turned off the airflow across the coil is reduced. Condensing temperatures therefore rise. All fans can be turned off, and the coil acts as a static condenser with only atmospheric air movement. Its capacity at this point is usually ample to continue operating below freezing conditions. Where only one condenser fan is used, air capacity can be reduced by employing a two speed motor or solid state speed control, which have infinite speed control.

Another technique of restricting airflow across the condenser coil is the use of dampers on the fan discharge. These are used on non-overloading centrifugal type fans, not propeller fans. Dampers modulate from a head pressure controller to some minimum position, at which time the fan motor is shut off.

One common and important characteristic of the airflow restriction methods is that the full charge of refrigerant and entrained oil is in motion at all times, ensuring positive motor cooling and oil lubrication to the compressor.

Another technique of artificially raising the head pressure is to back the liquid refrigerant up into the condenser tubes. The normally free draining condenser has very little liquid in the coil; it is mostly all vapor. But if a control valve is placed in the liquid outlet of the condenser, such as shown in Figure 5-4-13, and is actuated by inlet pressure, some of the discharge gas is allowed to bypass the condenser and enter the liquid drain. This restricts drainage of the liquid refrigerant from the condenser, flooding it exactly enough to maintain the head and receiver pressure. This method is not as common as restricting the airflow, because it is associated with systems that use an external receiver. Critically charged DX systems for air conditioning normally do not employ receivers.

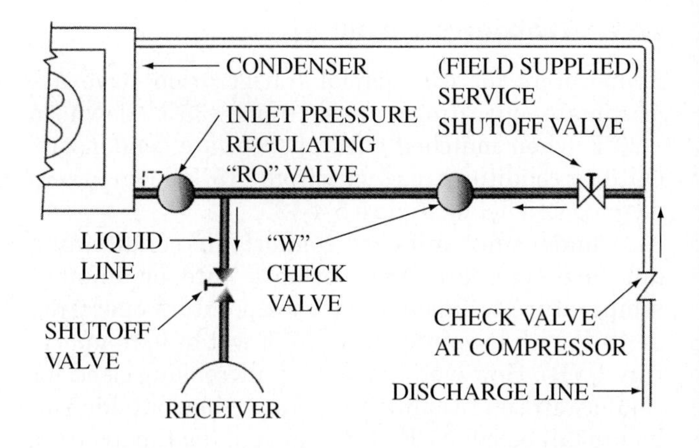

Figure 5-4-13 Head pressure control system for low-ambient operation using the liquid flood-back arrangement.

WATER COOLED CONDENSERS

In earlier applications of water cooled condensers to refrigeration and air conditioning, it was common practice to tap the water supply and then waste the discharge water to a drain connection as shown in Figure 5-4-14. An adjustable automatic water regulating valve was placed in the line and the flow of incoming water was controlled by the condenser operating head pressure through a pressure tap.

The temperature of the incoming water would naturally affect the condenser performance and flow

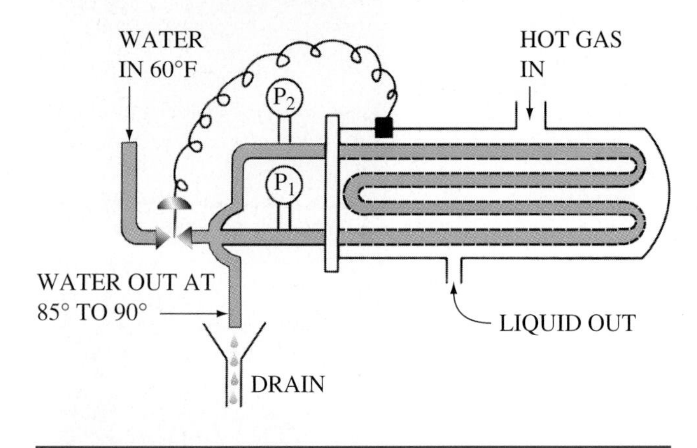

Figure 5-4-14 Series flow condenser for non-recycled water.

rate of any heat load. Depending on the geographic location, water temperatures in the city water mains rarely rise above 60°F in summer and frequently drop to much lower temperatures in winter.

NOTE: Because of water shortages experienced in many areas of the country water cooled condensers that waste water are not approved. Only water cooled condensers that recirculate the water may be installed in these areas of the country. Some water regulations are so restrictive that they may even require existing discharging water cooled systems to be replaced.

Condensers piped for tap water flow were always arranged for series flow, circuited for several water passes to achieve maximum heat rejection to the water, which was then wasted. Condensers drawing on city water could use only 1 or 1½ gallons per minute (gpm) per ton of refrigeration. The multipass circuit created high water pressure drops ($P_1 - P_2$) or 20 psi or more; however, most city pressures were able to supply the minimum pressure requirement (usually 25 psig).

The cost and scarcity of water (unless drawn from a lake or wells and returned) has become prohibitive and is even outlawed for refrigeration and air conditioning use by many local city codes. Ordinances restricted the use of water to the point where such installations were forced to use all air cooled equipment, water saving devices such as the evaporative condenser, or a water tower.

Water Cooled Condensers and Water Towers

The condenser, when used with a recirculated flow such as in a water tower system, Figure 5-4-15, is usually designed for parallel tubes with fewer water passes to accommodate a greater water quantity

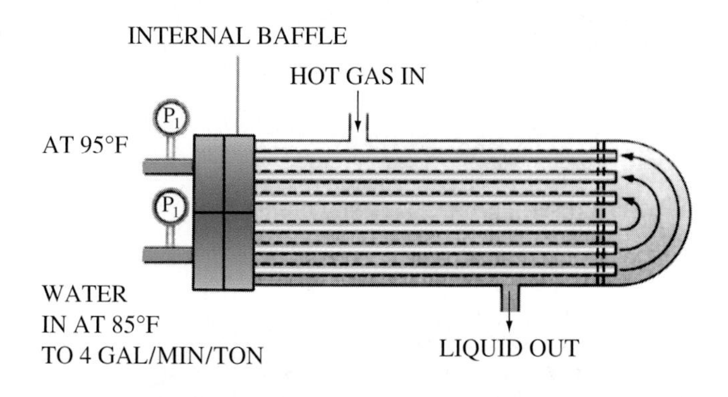

Figure 5-4-15 Parallel flow condenser for cooling-tower operation.

(3 gpm/ton) and lower pressure drop ($P_1 - P_2$) at 8 to 10 psi (the pressure reached with 18 to 23 ft of head) pressure drop. The nominal cooling tower application will involve a water temperature rise of 10°F through the condenser, with a condensing temperature approximately 10°F above the water outlet temperature. Another consideration in using the water cooled condenser for open recirculated flow is the fouling factor, which affects heat transfer and water pressure drop. Fouling is essentially the result of a buildup on the inside of the water tubes, which comes from mineral solids (scale), biological contaminants (algae, slime, etc.) and entrained dirt and dust from the atmosphere. The progressive buildup of these contaminants on the condenser tubes creates an insulating effect that retards heat flow from the refrigerant to the water. As the internal diameter of the pipe is reduced, so is the water flow, unless more pressure is applied. Reduced water flow cannot absorb as much heat and thus condensing temperatures rise—as do operating costs.

SERVICE TIP

The buildup of algae or other growth on water cooled condensers can have the same adverse effect as debris on air cooled condensers. It will reduce the system's operating efficiency. You must use a manufacturer approved cleaning solution to remove these deposits. If the deposits are reoccurring, it may be necessary to use a water treatment to retard their growth.

In selecting condensers, application engineers will usually allow for the results of fouling so that the condenser will have sufficient excess tube surface to maintain satisfactory performance during normal operation, with a reasonable period of service between cleanings. For conditions of extreme fouling and poor maintenance, higher fouling factors are used. Proper maintenance depends on the type of condenser; mechanical or chemical cleaning—or both—may be needed or employed to remove scale deposits.

Water cooled condensers are an efficient type of heat exchanger used to transfer the heat absorbed by the refrigeration cycle to water. In most installations using water cooled condensers, water is supplied to the condenser from a cooling tower and returned to the tower to dissipate the heat picked up from the refrigerant to the air, as shown in Figure 5-4-16. There are four steps in the process.

1. Heat is transferred from the refrigerant to the water in the condenser.
2. The water is pumped from the indoor condenser to the (usually) outdoor cooling tower.

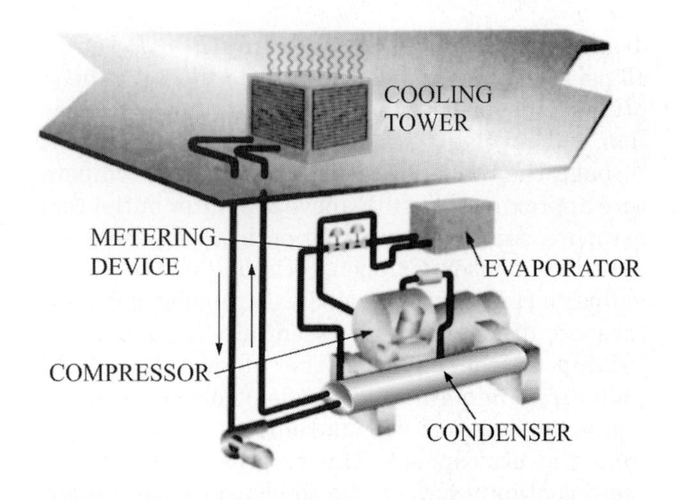

Figure 5-4-16 Refrigeration cycle showing location of water-cooled condenser.

3. At the tower, the heat is rejected to the outdoor air, cooling the water.
4. The cooled water is returned to the condenser to pick up more heat, making a continuous process.

Water cooled condensers permit about 15°F lower compressor discharge temperature than air cooled units. This allows the compressor to operate at a lower discharge pressure, increasing its capacity and lowering the power requirements. These advantages, however, must be considered along with an increase in installation and maintenance costs.

Water cooled systems in North America are most popular in systems over 100 tons of cooling capacity. In other parts of the world greater use is made of them. Sizes manufactured range from 5 to 3000 tons.

A typical water cooled condenser system will work under the following conditions. The refrigerant condensing temperature is about 105°F. The water entering the condenser is about 20°F lower. The water temperature rises about 10°F in passing through the condenser, leaving the condenser at about 95°F, which is only 10°F below the saturated refrigerant temperature.

When the water gets to the cooling tower, it is at 95°F. The warm water drops over the cooling tower's PVC wetted decking, while outside air moves past it in the opposite direction, causing evaporation. Each pound of water that is evaporated removes 1000 Btu from the remaining water, cooling it. A 10°F drop in temperature occurs by the time the water reaches the bottom of the tower. This 85°F water is returned to the condenser to absorb more heat. Water that is lost by evaporation, windage, and blowdown is replaced by make up water, which is regulated by a float valve in the sump of the tower.

Types of Water Cooled Condensers

There are three basic types of water cooled condensers:

- Tube-in-tube (coaxial)
- Shell and coil
- Shell and tube

Each type is constructed to accomplish the same thing, but in a different way. The tube-in-tube or double type is illustrated in Figure 5-4-17. The usual arrangement is for a smaller tube to be placed inside a larger tube, which is sealed at the end. When multiple tubes are used, they connect to the headers. The water flows through the inner tube and the refrigerant flows through the angular space between the tubes. An advantage of this design is that it can be wrapped into a shape to fit the space available. For example, the packaged condensing unit shown in Figure 5-4-18 uses the tube-in-tube condenser.

> **CAUTION:** Care is required when recovering refrigerant from a water cooled condenser. The water must either be continuously circulated or else completely drained before recovering the refrigerant. During the recovery process, liquid refrigerant in the water cooled condenser can vaporize and draw the temperature down to below the freezing temperature of water. Ice formed inside of any sealed container such as a water cooled condenser can rupture the vessel or damage the internal piping.

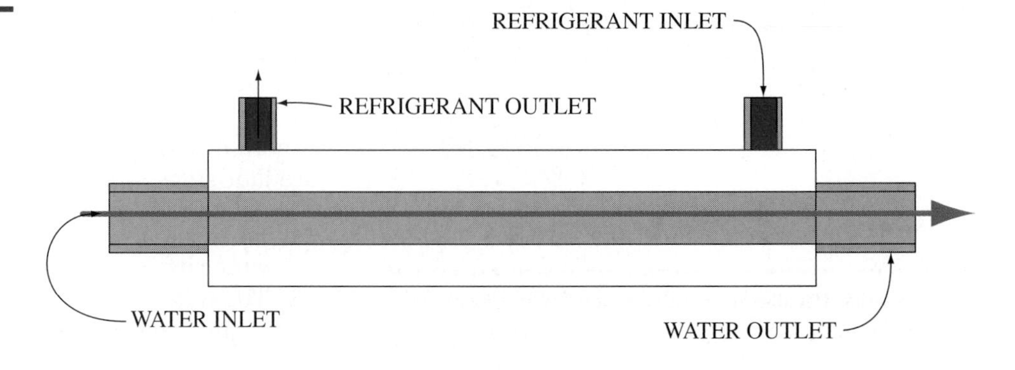

Figure 5-4-17 Typical refrigerant and water flow for tube-in-tube condenser.

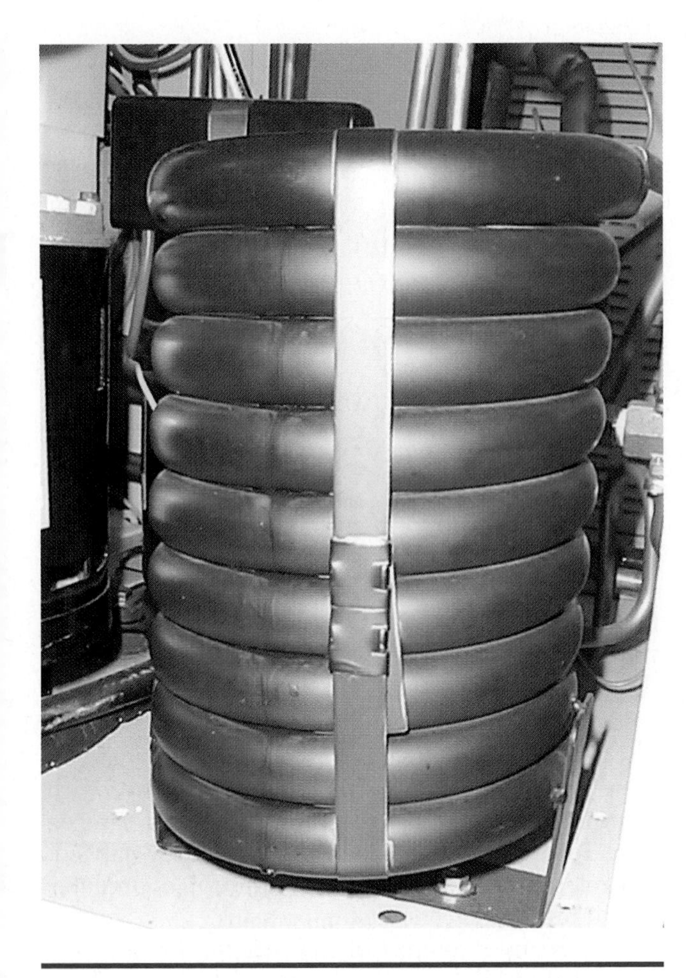

Figure 5-4-18 Typical use of a tube-in-tube condenser designed for in-the-room installation.

The shell and coil design uses a welded steel shell with a coil of continuous finned tubing inside. Water flows in the tubes and refrigerant flows in the shell. The shell can run vertically for better heat transfer through convection, or horizontally to meet possible space requirements.

Shell and coil condensers are sometimes combined with the compressor to form a condensing unit package. These units are usually limited to 20 tons or less. Vertical packaged air conditioning units from 20 to 60 tons use shell and coil condensers due to their compact dimensions.

The shell in tube type condenser, Figure 5-4-19, is used in the largest condensers. Capacities extend to 1,000 tons. As in the previous type, water flows through the tubes (the tube side of the condenser) and refrigerant flows on the outside of the tubes (shell side of the condenser). These condensers have long, straight finned tubes, connected to a steel plate (tube sheet) at each end. At each end, water manifolds, called heads, are bolted to the shell. These heads direct the water to make from one to eight passes, depending on the size and design of the condenser. The heads can be removed to permit cleaning the individual tubes. Rubber gaskets provide a watertight seal. The other two types of water cooled condensers, tube-in-tube and shell and coil, must be chemically cleaned.

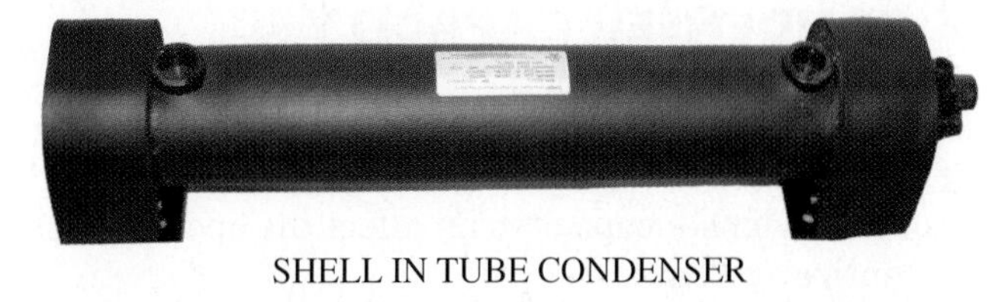

Figure 5-4-19 Typical design of a shell-in-tube condenser.

Figure 5-4-20 Typical water-cooled condensing unit with shell-in-tube condenser.

In a typical water connection, the hot gas inlet and purge valve are on the top, the liquid outlet valve is on the bottom, and the pressure relief valve is on the side. Many of these shell and tube condensers are assembled with compressors to form a water cooled condensing unit, as shown in Figure 5-4-20. They range in size from 5 to 150 tons. A common use of the shell and tube condenser is as a component of a water chiller, as shown in Figure 5-4-21.

Figure 5-4-21 Typical packaged liquid chiller unit with a shell-in-tube condenser.

CONDENSER CAPACITY

Since the condenser is one of the major components of the refrigeration system, any problem in providing proper condenser capacity can affect the operation of the entire system.

CAUTION: If the condenser becomes dirty, its condensing pressure increases. This can be a serious problem when high pressure refrigerants, such as R-410A, are used. R-410A operates at normal head pressure of around 450 psig, but a dirty condenser can cause these pressures to greatly increase. Condensers are designed to withstand these pressures; however, over time, the integrity of the condenser or other system components can result in a catastrophic failure of the refrigerant circuit from these excessively high pressures. It is therefore very important that condensers containing high pressure refrigerant, such as R-410A, be thoroughly cleaned on a regular basis.

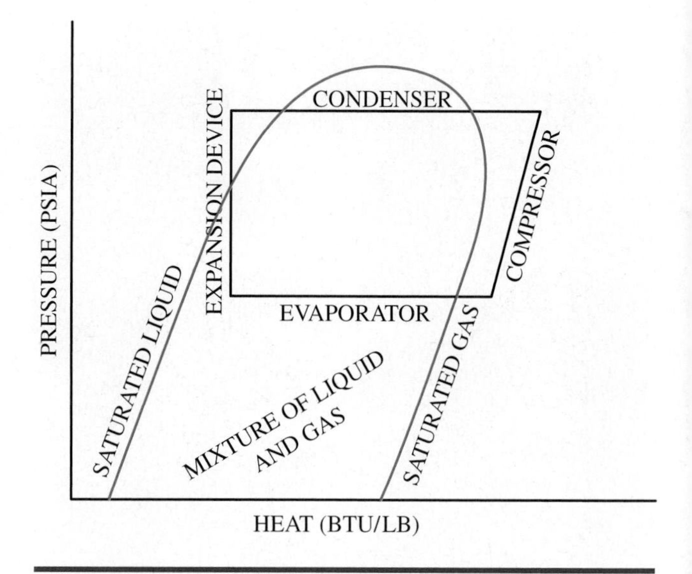

Figure 5-4-22 Drop in head pressure in an air-cooled system caused by decrease in ambient temperature shown on P-H diagram.

A number of factors can cause the condensing temperature to be too high, resulting in reduced capacity:

1. Design error: undersizing of any type.
2. Airflow blockage or recirculation: air cooled condenser, cooling tower, and evaporative condenser.
3. Water flow blockage: water cooled condenser.
4. Dirty condenser coil: any type.

If the system has previously worked properly, the selection is not the problem. High discharge temperature is usually the problem. It can be caused by air or water blockage, or a dirty condenser.

Excessive head pressure is a major cause of compressor failure. Even if compressor failure does not result, high head pressure can produce high bearing loads and poor efficiency. Excessive temperature can increase the chemical reaction between contaminants and cause damage to essential parts.

Controlling Condenser Capacity

All air cooled condensing units that must operate under low ambient conditions require special control equipment for proper operation. The units must be equipped with head pressure control.

Figure 5-4-22 shows how the head pressure drops during low ambient conditions on an air cooled system. Likewise a lower wet bulb temperature reduces the head pressure on systems using cooling towers or evaporative condensers.

There is a benefit from lower head pressure in producing greater compressor efficiency. However, there is a limit to how low the pressure can drop without affecting the proper operation of the metering device. To prevent this condition, condenser capacity control systems are used to maintain a predetermined minimum condensing pressure.

There are a number of methods for doing this on air cooled condensers:

1. Cycle the fans on and off.
2. On a single-fan condenser, use a variable speed fan motor.
3. On a multiple-fan condenser, cycle all but the one fan and modulate the last fan to produce smooth control. This arrangement can control the head pressure for ambient temperature as low as −20°F.
4. Control floodback. This is a control arrangement that will cause liquid refrigerant to fill some of the condenser tubes. This effectively reduces the size of the condenser, since only the unflooded tubes will condense refrigerant.

On water cooled condenser systems, a temperature controlled water valve that bypasses the tower and mixes cooling tower water with condenser water can be used. With this arrangement the desired condenser water temperature can be maintained. The temperature sensing element for the bypass valve is placed at the water entrance to the condenser.

Another simple, inexpensive method is to cycle the cooling tower fan to maintain exiting water temperature in the 75°F to 85°F range.

There are a number of arrangements for controlling evaporative condensers:

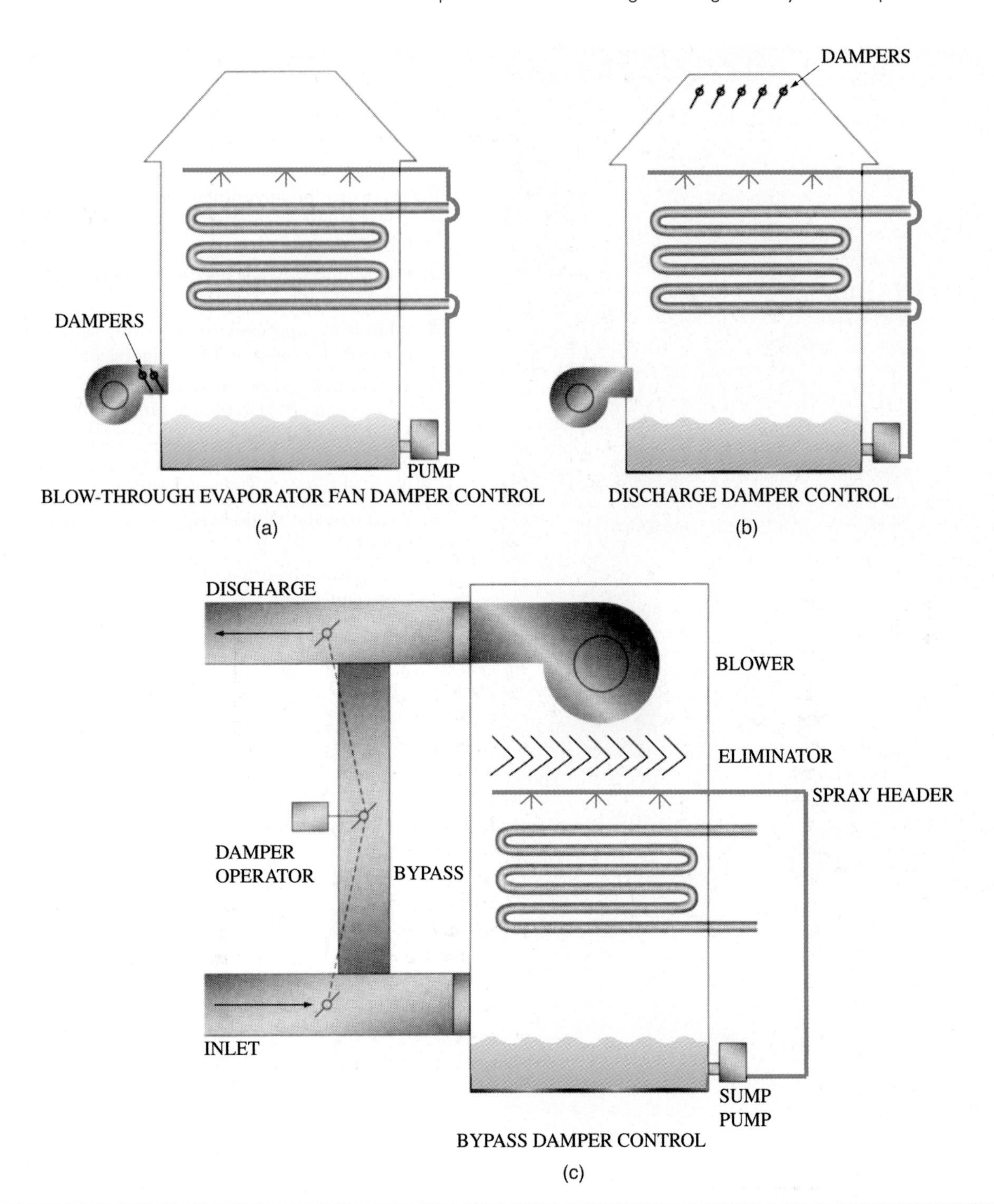

Figure 5-4-23 Three ways to control the airflow on the evaporative condenser.

1. Shut off the water sprays. This reduces the evaporative condenser to an air cooled condenser. The air cooled capacity is about 50 to 60% of the wetted capacity. This, however, is a huge step in capacity reduction and evaporative condensers have a hard time carrying the load in the 35°F to 40°F range. An added disadvantage is that it may scale the coil.

2. The fan can be cycled. This can shorten the belt life. The speed may be modulated.

3. Dampers can be installed in the blower discharge or unit outlet and modulated to produce the desired flow. Three methods of reducing airflow are shown in Figure 5-4-23.

4. Warm humid air from the condenser discharge can be mixed with outdoor air by the use of bypass dampers and duct work. This increases the relative humidity of the air entering the unit, reducing evaporation and, therefore, reducing the head pressure. This is a relatively easy means of control.

TECH TALK

Tech 1. You know that customer's house I went to yesterday? You wouldn't believe it. They had a piece of screen wire around the outside of their condenser.

Tech 2. Don't they live down by the park?

Tech 1. Yeah. What's that got to do with it?

Tech 2. Well, there's a bunch of cottonwoods down there, and all that cottonwood lint those trees put out this time of year will collect on the condenser. That can plug up the coil in a real short period of time. That screen they have out there will keep it out of the condenser coil. If you go back after the cottonwood season, you'll see that they have taken that screen down. It really helps prevent a serious maintenance problem for that particular unit.

Tech 1. What do you mean "for that particular unit"?

Tech 2. Well, that one uses a spine fin coil, and a spine fin coil you can't just go and brush that cottonwood lint off. It's real difficult to remove.

Tech 1. So, just use it on the spine fin coils?

Tech 2. No. Any time you have a unit that's subject to that type of coverage, it's a good idea. And another thing that people don't think of, is they will mow the grass right next to the unit when it's running, blowing the grass right into the condenser.

Tech 1. I've seen people do that too.

Tech 2. They don't realize that the grass gets pulled into the coil, and although they might wash it off after mowing, there is a lot of grass that gets pulled back inside, and it can stop the bottom part of the coil. And the bottom part of the condenser is where the subcooling of the refrigerant takes place. You eliminate that, you eliminate a lot of system efficiency.

Tech 1. Makes good sense. Thanks. I appreciate that.

UNIT 4—REVIEW QUESTIONS

1. What is the function of the condenser?
2. List the three types of condensers that are used in HVAC/R systems.
3. What are the advantages of using an air cooled condensing unit as compared to using separate air cooled condensers?
4. What information is needed to select an air cooled condensing unit?
5. What is the approximate operating range for condensing units for commercial refrigeration?
6. The solution to low ambient operation is to maintain a ________ in order to assure proper feeding of refrigerant to the evaporator.
7. What is one common and important characteristic of the airflow restriction methods?
8. What are the three basic types of water cooled condensers?
9. What is an advantage of the design of the tube-in-tube type condenser?
10. What is the capacity range for a shell and tube type condenser?

UNIT 5

Metering Devices

OBJECTIVES: In this unit the reader will learn about metering devices including:

- Metering Device Operation
- Types of Metering Devices
- Common Metering Device Problems
- Measuring Superheat
- Liquid Distributors
- Low Load Limits

INTRODUCTION

A *metering device* controls the flow of refrigerant to the evaporator. It separates the high pressure side from the low pressure side. The evaporator pressure is low because the compressor is continuously pumping vapor from the evaporator. The metering device helps maintain the low pressure by restricting the flow of refrigerant into the evaporator.

Two actions occur in the metering device: the refrigerant liquid is cooled to the evaporator temperature by actual evaporation of some of the liquid refrigerant, and the pressure of the refrigerant is reduced to a pressure corresponding to the evaporator temperature at the saturated condition.

TECH TIP

There is no reason that any metering device should ever be struck as a means of fixing it. There is never a stuck part in a metering device. Striking a metering device with anything can only damage the device and will prevent it from functioning correctly. If the metering device is not functioning properly it needs to be repaired or replaced, but never struck.

TYPES OF METERING DEVICES

There are eight major types of metering devices: hand operated expansion valve, low side float, high side float, automatic expansion valve, thermostatic expansion valve, electronic expansion valve, fixed orifice, and capillary tube. Metering devices are selected by application. There are other types of metering devices used for special applications.

METERING DEVICE OPERATION

A metering device is a type of restrictor placed in the liquid line between the condenser and the evaporator, to produce a difference in pressure between the high side and the low side of a refrigeration system and to regulate the flow of refrigerant. The amount of the restriction is provided to maintain a condensing temperature high enough above the condensing medium (water or air) to condense the high pressure vapor from the compressor. The restriction is also provided to maintain an evaporating temperature low enough to absorb heat from the product being cooled and to evaporate the liquid refrigerant being supplied to the evaporator.

The location of the metering device in the system is shown in Figure 5-5-1.

The evaporated liquid is called flash gas as shown in Figure 5-5-2. The heat required to lower the evaporating temperature of the refrigerant, in going from the high side of the system to the low side, is absorbed from the liquid refrigerant creating flash gas.

For a better understanding of the normal functioning of a metering device, consider some typical values of an air conditioning unit using R-22 with an air

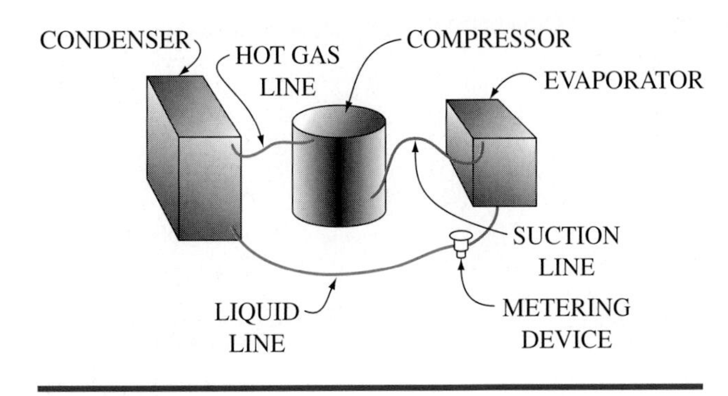

Figure 5-5-1 Location of metering device in refrigeration cycle.

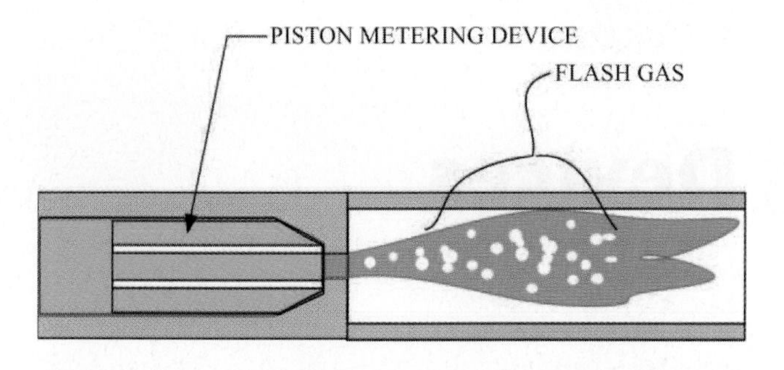

Figure 5-5-2 How flash gases are formed when the liquid refrigerant flows through the metering device.

cooled condenser and an outside ambient temperature of 95°F. The saturation temperature is 120°F before the metering device, and 40°F after. The refrigerant temperature is 105°F at the inlet, and 40°F at the outlet. The typical pressure at the inlet would be 277.7 psia (263.0 psig); and the outlet pressure would be 83.7 psia (69.0 psig). The enthalpy, or heat energy, is unchanged by the metering device. The typical enthalpy is 42.0 Btu/lb.

TYPES OF METERING DEVICES

There are eight different types of metering devices. They are divided into two groups—those that are fixed and those that are adjustable to provide regulation matching the load.

Fixed Metering Devices

There are two types of fixed metering devices: the capillary tube and the fixed orifice.

The *capillary tube* is probably the simplest of all metering devices. The capillary tube is commonly used as a metering device on domestic refrigerators and other small appliances. It is constructed of a single tube with an inside diameter in the size range of .026 in to .090 in. It is normally located at the entrance to the evaporator and sometimes coiled to conserve space, as in Figure 5-5-3. The size and the length of the tube are carefully selected to match the pumping capacity of the compressor at full load.

Figure 5-5-3 Typical location of a capillary tube type metering device.

NICE TO KNOW

Did you know that a fixed metering device will allow some adjustment in system capacity as the load changes? As the outdoor condensing temperature increases, more refrigerant is fed through the metering device into the evaporator, increasing its capacity slightly. Conversely as the heat load goes down, the outdoor condensing temperature goes down and less refrigerant is fed into the evaporator. For a location where the load does not vary widely, fixed metering devices may float with the load quite well. However, for climates where there is a great range in temperature variation, a thermostatic expansion valve (TXV) metering device is superior to a fixed metering device.

Most systems that use the capillary tube type of metering device are packaged units, such as domestic refrigerators and freezers and window air conditioners, where the system is installed at the factory and critically charged with refrigerant. It operates best where the load is more nearly constant and is usually used on small systems not over 3 tons in capacity. It has been applied successfully to room coolers, close-coupled split systems and small heat pumps. Due to the small size of the tube, it can be easily plugged. Many systems use a built-in liquid line filter at the entrance of the tube.

One advantage of the capillary tube is that on shutdown the high side and the low side system pressures are equalized. Not having to start against a pressure differential permits the use of a low starting torque compressor motor. This can also be a disadvantage since on shutdown the evaporator can be filled with liquid and possibly damage the compressor on subsequent startup. To counteract this problem, many systems use a suction line accumulator at the entrance to the compressor, as shown in Figure 5-5-4. This captures the liquid refrigerant and returns it to the system as vapor.

Because liquid refrigerant is evaporated from an accumulator, these devices get very cold and often sweat. This condensation can cause accumulators to rust over time as in Figure 5-5-5. On older units the rust can weaken the accumulator to the point where it can fail.

The *fixed orifice*, shown in Figure 5-5-6, is the second type of fixed metering device. This orifice is

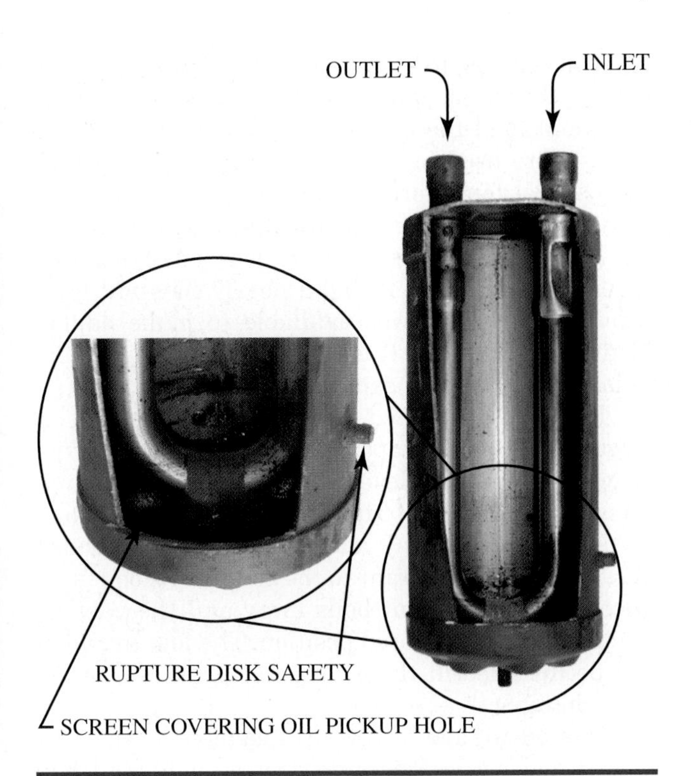

Figure 5-5-4 Location function of the suction line accumulator. *(Courtesy Marilyn Burris)*

Figure 5-5-5 Rust formed on the accumulator due to condensation.

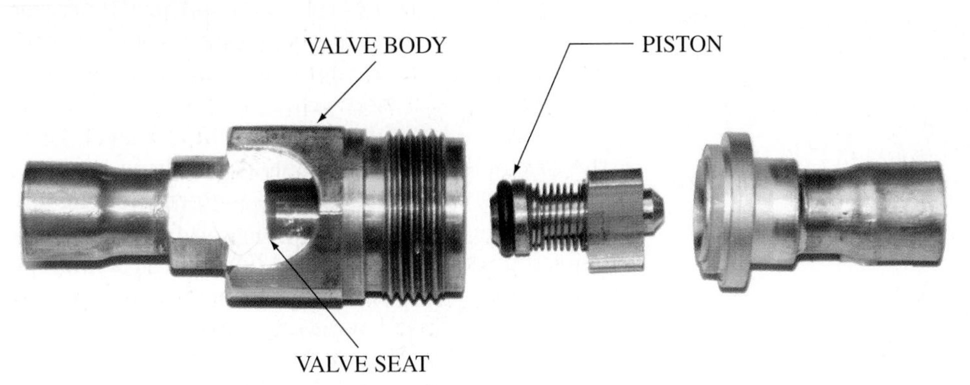

Figure 5-5-6 Construction of a fixed orifice type metering device. *(Courtesy Rheem Manufacturing)*

built into a rugged assembly and has the advantage for heat pump application of including a built-in check valve. During the reverse cycle on the heat pump, this metering device will function with the flow in either direction. It is like the capillary tube in a number of ways:

1. It must be carefully selected to match the load.
2. The system must be critically charged.
3. It permits refrigerant migration into the evaporator during the OFF-cycle and requires all of the same protective accessories as the capillary tube. It is factory selected for the application. For split systems the selected orifice is shipped with the outdoor unit to match the unit's capacity.

Figure 5-5-7 shows the performance of the fixed orifice in the pressure-heat diagram. The orifice works best at full load conditions. For most systems under normal operating conditions it is satisfactory. Problems can occur if the unit is oversized and operates for long periods of time at low load. As a precaution, the unit should not be oversized.

TECH TIP

One of the disadvantages of a fixed orifice metering device is that the high side pressure can bleed off, letting all of the refrigerant collect in the evaporator. A liquid line solenoid placed just before the orifice will stop this bleed-off from occurring. This simple device can significantly increase the system's operating efficiency by keeping the liquid in the condenser so that it reaches operating equilibrium faster. A disadvantage of installing this type of solenoid is that you may have to put a hard-start kit on the compressor.

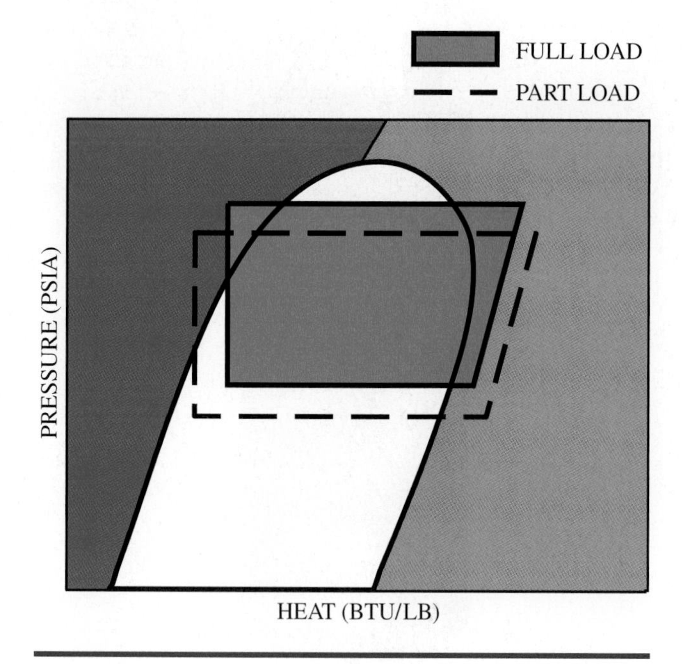

Figure 5-5-7 Function of the fixed orifice in a pressure-enthalpy diagram. Note that there is no adjustment of this type of metering device.

The fixed metering orifice automatically adjusts to normal changes in load. The orifice regulates the flow of refrigerant in a manner described as floating with the load. When the compressor moves less heat, the discharge pressure drops. This reduces the pressure across the orifice, lowering the flow of refrigerant. When the load is increased, the opposite condition takes place.

Adjustable Metering Devices

There are six types of adjustable metering devices:

1. Hand operated expansion valve
2. Low side float (LSF)
3. High side float (HSF)
4. Automatic expansion valve (AEV)
5. Thermostatic expansion valve (TXV)
6. Electric and electronic expansion valve

All of the above mechanically adjust to the changes in load.

The hand operated expansion valve adjusts the rate of flow through the valve as determined by:

1. The size of the valve port opening or orifice.
2. The pressure difference across the orifice.
3. How far the valve is opened.

An increase in any of these conditions will increase the flow. A decrease will reduce the flow.

The obvious disadvantage of the *hand expansion valve* is that it has no automatic arrangement to control the size of the orifice to match the load. These valves were used in the past on applications where the load was fairly constant and an operator was present to manually make adjustments when necessary.

The hand expansion valve was often used on ammonia systems that had a nearly constant load, where an operator was available to make adjustments when needed. It is seldom used today, except in laboratory tests to explore optimum flow rates.

In the *low side float* (LSF) type metering device (Figure 5-5-8), the evaporator is flooded with refrigerant liquid and the level in the evaporator is maintained by a float. If the load is increased, more refrigerant boils away and the valve opens wider, passing more refrigerant to the coil. If the load is reduced, less refrigerant boils away and the valve is moved toward the closed position. The low side float can be installed in the evaporator or in a separate float chamber.

One advantage of this arrangement is that the heat transfer rate from a flooded coil is higher than with a mixture of liquid and vapor. One disadvantage is the possibility of oil collecting in the evaporator under light loads and not returning to the compressor.

The low side float has been rather commonly used on ammonia systems that use flooded evaporators.

The *high side float* (HSF) modulates refrigerant flow to the evaporator based on liquid level. Unlike the low side float, this float assembly is located on the high pressure side of the metering device's orifice. As the load increases, more refrigerant is condensed in the condenser and flows into the high side float assembly. As the liquid level rises in the chamber the valve opens, permitting greater flow into the evaporator. The pressure reducing valve, or weight valve, is used in a long liquid line to prevent evaporation before the liquid reaches the evaporator. If the load is decreased, the float is lowered, reducing the flow of refrigerant.

The most common application of the high side float is in the flooded cooler of a centrifugal chiller.

The *automatic expansion valve* (AEV) consists of a mechanical arrangement for metering the liquid refrigerant into the evaporator to maintain a constant evaporator pressure. The valve has an adjustment at the top for setting the evaporating pressure to produce a desired evaporating temperature. The evaporator pressure exerts a force against the bottom of the diaphragm. An adjustable spring exerts a pressure on the top of the diaphragm. As the evaporator pressure increases, it overcomes the spring pressure and moves the diaphragm up, thus closing the valve. As the evaporator pressure decreases, the spring pressure overcomes the evaporator pressure and pushes the valve open.

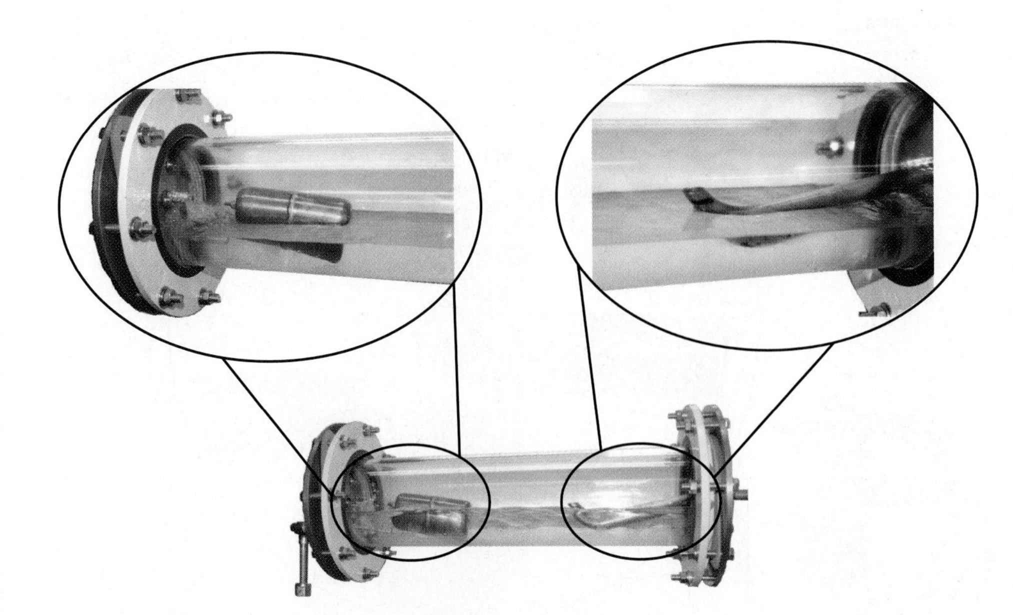

Figure 5-5-8 (Left close-up): Float opens when the level of refrigerant lowers. (Right close-up): The outlet has a small opening that lets some refrigerant and oil flow back to the compressor to ensure proper oil return to the compressor during operation. (*Courtesy Hampden Engineering Corporation*)

NICE TO KNOW

Automatic expansion valves have limited use today. Some of the major valve manufacturers no longer even offer them. They are, however, extremely effective in maintaining an evaporator temperature under very close control. For that reason you may find them on some small precision cooling pieces of equipment used for medical research and scientific activities.

There are some disadvantages to the use of the AEV metering device. If the load is light, less liquid will boil in the evaporator. This causes the evaporator pressure to fall and opens the valve. This excess refrigerant in the evaporator can overflow into the compressor and cause a serious hazard. Then as the load increases, more liquid refrigerant will boil into vapor, and the needle valve will move toward the closed position. This causes the refrigerant vapor going into the compressor to have an increased amount of superheat. This may cause the compressor to overheat, resulting in high discharge temperature and pressure, oil breakdown, carbonizing of the valves, and poor efficiency.

The problem with the valve is that it is not load oriented and therefore has increasingly limited applications. It is usually used on small units where the load is relatively constant.

Applications of this valve include some types of domestic refrigerators and freezers, and small retail freezer cabinets.

The *thermostatic expansion valve* (TXV) is shown in Figure 5-5-9. It supplies the evaporator with enough refrigerant for any or all load conditions. Although it is considered a control, it is not classified as a temperature, suction pressure, humidity, or operating control.

The TXV is constructed in many respects the same as the AEV, except that it has a remote bulb that senses the superheat of the refrigerant leaving the evaporator. The position of the orifice is controlled by three pressures:

1. The evaporator pressure
2. The spring pressure acting on the bottom of the diaphragm
3. The bulb pressure opposing these two pressures and acting on the top of the diaphragm.

Let's examine an R-22 system in equilibrium. The pressures on the bottom of the diaphragm exactly balance the bulb pressure on the top of the diaphragm. The flow of refrigerant through the valve matches the load and produces a satisfactory superheat in the refrigerant leaving the evaporator coil, normally between 8°F to 10°F.

If the evaporating temperature on an R-22 system is 40°F and the pressure is 69.0 psig, then the temperature of the bulb is 50°F, allowing 10°F superheat. The bulb exerts a pressure of 84.0 psig on the top of the diaphragm. The spring, which acts on the bottom of the diaphragm, is set for a pressure of 15.0 psig, equivalent to 10°F of superheat. Thus, the pressure exerted on the bottom of the diaphragm is

$$69 \text{ psig} + 15 \text{ psig} = 84 \text{ psig}$$

This is equal to the pressure exerted on the top and the valve is in equilibrium.

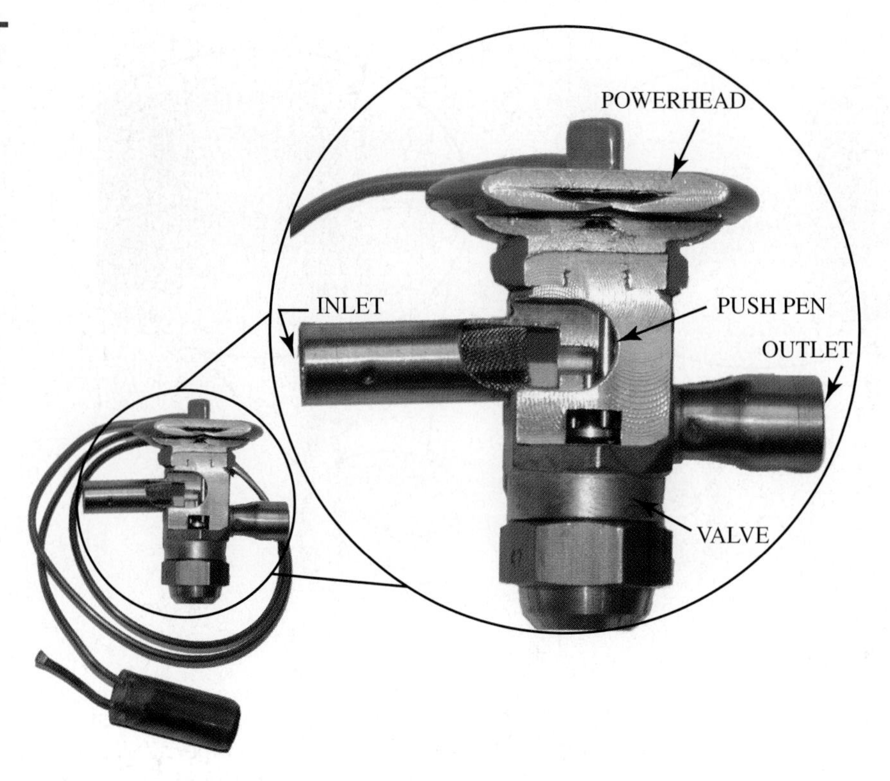

Figure 5-5-9 Construction of thermostatic expansion valve. (*Courtesy Rheem Manufacturing*)

If the load increases, the superheat rises, the valve supplies more refrigerant and gradually returns to an equilibrium condition. If the load decreases, the superheat drops and the valve moves in the direction of the closed position to match the new load, then gradually returns to an equilibrium condition. The advantage of the TXV is its ability to automatically adjust its orifice to match the load.

The TXV is widely used for nearly all types of air conditioning and refrigeration systems, except for small appliances.

For larger applications as shown in Figure 5-5-10a–c, externally equalized TXV valves measure the suction pressure using ¼ in OD tubing at the evaporator outlet for more accurate control. This avoids a pressure drop across the liquid distributor.

TECH TIP

On an externally equalized TXV it is important that the equalizing line not be attached to the bottom of the suction line. If it is attached at the bottom, oil from the refrigerant can collect in the equalizer line and prevent it from working correctly.

The *electronic expansion valve* is activated by an electronically controlled stepper motor, as shown in Figure 5-5-11. The motor shaft moves in and out in tiny steps. When the sleeve attached to the shaft moves upward, it exposes more metering slots. This increases the refrigerant flow to the evaporator. When it moves downward, it covers the slots, thus reducing flow and cooling capacity. Like the TXV, this type of valve is designed to maintain a constant superheat. The stepper motor gets its signal from an electronic control panel that is attached to an electronic sensor that measures refrigerant superheat. The stepper motor valve can have a long stroke and a corresponding low pressure drop when compared to conventional TXV. Since it is controlled independently of pressure, this valve is able to provide safe startup, shutdown, and operation, and high energy efficiency through its full range of operating conditions.

Another electronic control alternative to the TXV is the pulsating solenoid valve, shown in Figure 5-5-12. Unlike the TXV, which closes or opens by pressure, the solenoid valve opens and closes electrically. Pulse modulated valves control flow by increasing or decreasing open time during each cycle. When coupled with the proper electronic control system, a solenoid valve can pulse rapidly, opening and closing quickly in response to the cooling load. The illustration on the left in Figure 5-5-12 shows the valve closed, and the solenoid valve is de-energized; the view on the right shows the energized condition. This valve also serves the function of a liquid line solenoid valve, blocking flow to the evaporator during the OFF-cycle and at other times when refrigerant flow is not required.

One manufacturer uses stepper-motor-type TXV valves on its air cooled packaged chillers. The large

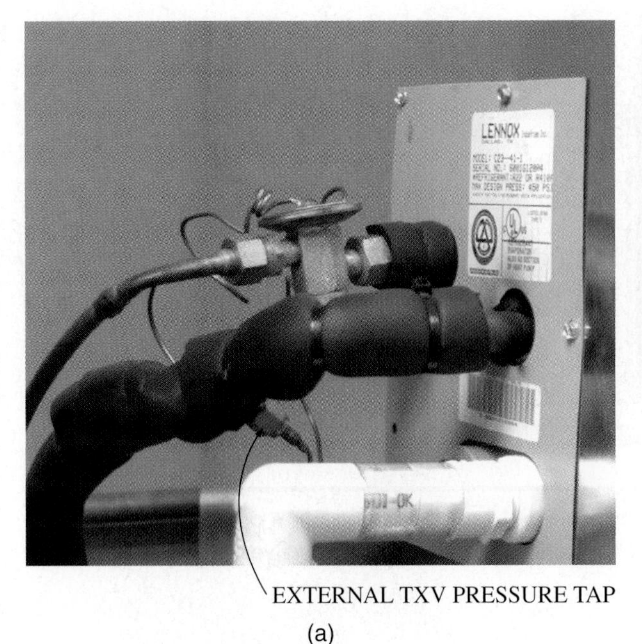

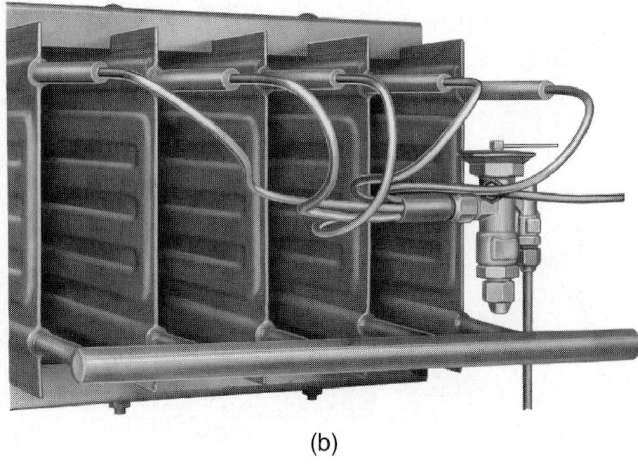

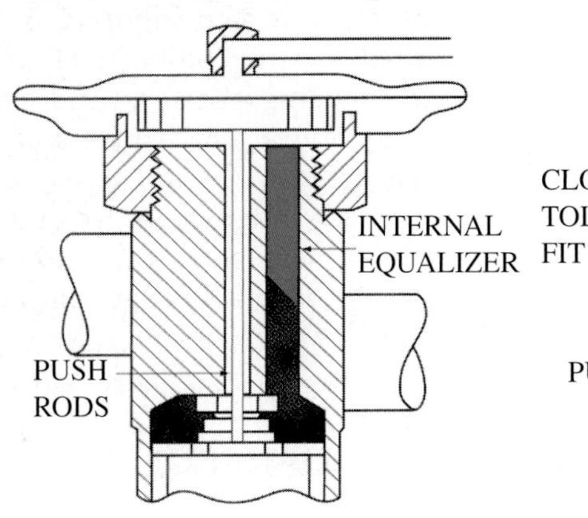

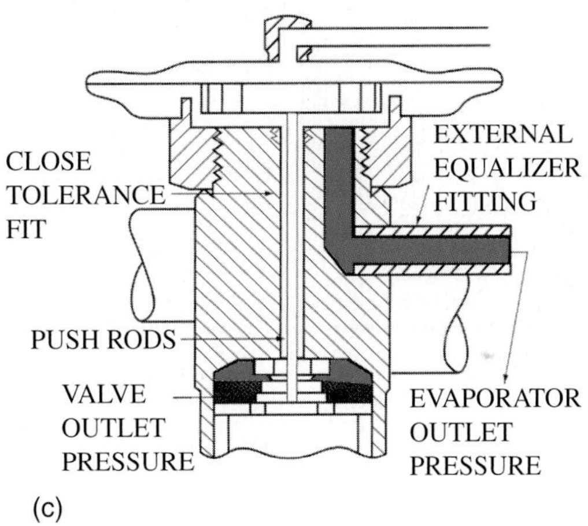

Figure 5-5-10 (a) TXV external; (b) TXV external. (*Courtesy Sporlan Valve Company*); (c) Difference in construction of internal and external equalized valves.

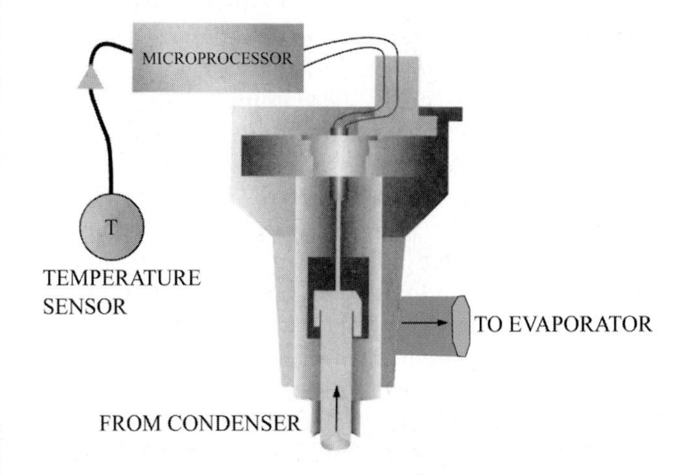

Figure 5-5-11 Construction of an electronic expansion valve.

sizes, using more than four compressors, use a two-circuit chiller with two stepper-motor-type TXV valves.

As electronic control becomes more popular, more electronically controlled valves will be put into use.

Externally Equalized Valves

When the pressure drop in the evaporator (including the distributor) is sizable, in order to obtain an accurate superheat setting, it is necessary to use an externally equalized thermal expansion valve, as shown in Figure 5-5-13. In the valve body itself, the evaporator inlet from the bottom of the valve is sealed off and a piping connection is arranged from above the diaphragm to the end of the evaporator. This connection not only improves the superheat

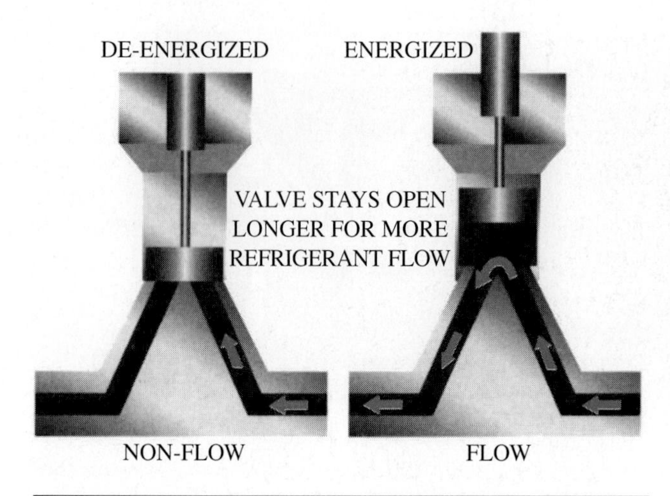

Figure 5-5-12 Pulsating solenoid valve, both energized and de-energized.

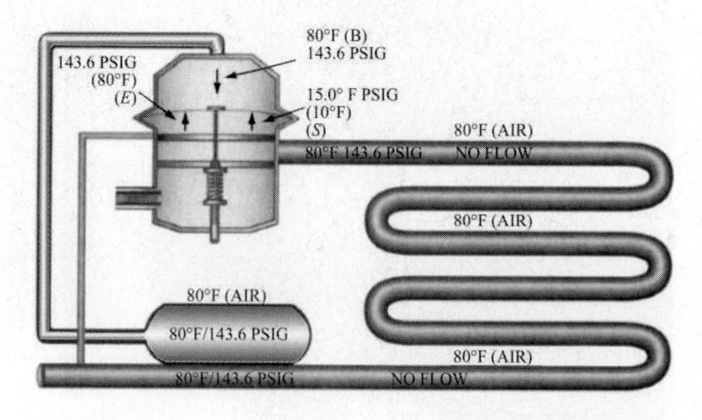

Figure 5-5-14 Use of a maximum operating pressure (MOP) thermostatic expansion valve.

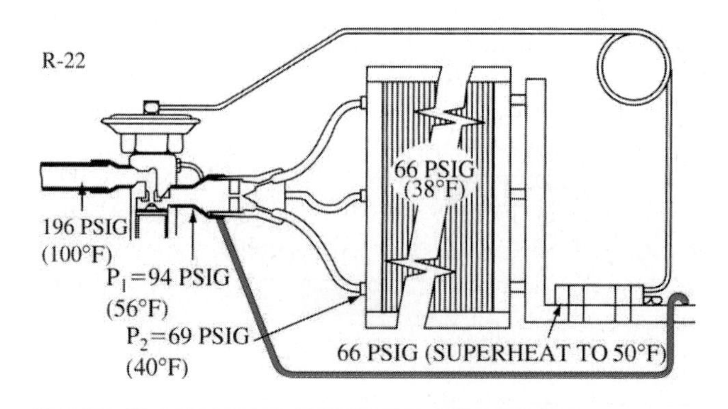

Figure 5-5-13 Effect of an external equalized thermostatic expansion valve on the operation of the system. *(Courtesy Sporlan Valve Company)*

setting, but also makes possible full use of the evaporator, improving the efficiency.

The equalizer connection normally penetrates the suction line 6 or 8 in downstream from the sensing bulb, unless the manufacturer's instructions advise differently.

Maximum Operating Pressure (MOP) Valves

The conventional thermostatic expansion valve uses a remote sensing bulb with some liquid refrigerant in it. This can cause operating problems during the OFF cycle as well as on startup. This type of charge can cause excess pressure buildup in the evaporator during the OFF cycle and high suction discharge temperatures which will overload the motor on compressor startup. This can be prevented by using a *maximum operating pressure* (MOP) valve, shown in Figure 5-5-14, which has a gas charged sensing bulb. Most all air conditioning systems using thermostatic expansion valves limit the amount of pressure that can develop in the evaporator.

An increase in temperature of a vapor produces only a slight increase in pressure. This solves the problem of liquid residual at compressor shutdown and startup.

MEASURING SUPERHEAT

It is important to accurately measure the superheat to determine whether the expansion valve is operating properly or if it needs adjustment. On close coupled installations, the operating suction pressure can be measured at the compressor service valve, as shown in Figure 5-5-15. This suction pressure can be converted to saturated evaporator temperature by referring to a pressure-temperature chart. The temperature of the suction vapor at the expansion valve bulb location can be read using an electronic thermometer. The difference between these two temperatures is the superheat.

If only a limited adjustment needs to be made, the valve stem or adjuster can be turned in small increments to change spring tension, which in turn will change the superheat setting. On many valves the adjustment is clockwise (cw) to increase superheat and counterclockwise (ccw) to decrease superheat. Valve instructions should be checked to be sure of correct adjustments.

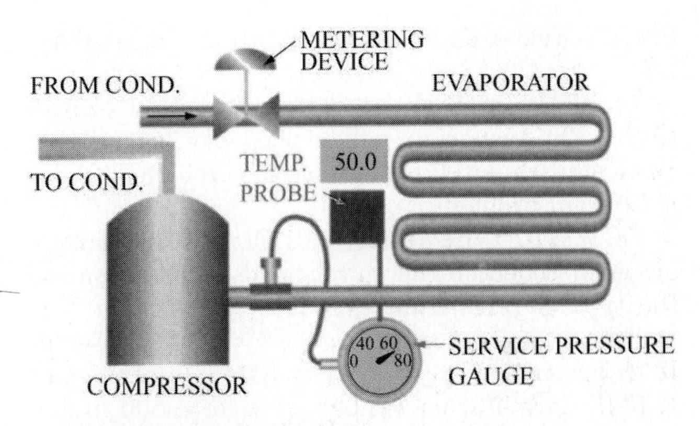

Figure 5-5-15 Method of measuring superheat on a system using a thermostatic expansion valve.

LIQUID DISTRIBUTORS

Liquid distributors are placed between the metering device and the evaporators with multiple circuit coils to equally distribute the refrigerant to each circuit, as shown in Figures 5-5-16a,b and 5-5-17. They are usually supplied by the coil manufacturer and are used where one metering device serves from 2 to 40 evaporator circuits, using connecting tubes ranging from 5⁄32 in OD to 3⁄8 in OD.

Some of these distributors have a thick flat washer inside, referred to as a nozzle, as shown in Figure 5-5-17. The purpose of this nozzle is to thoroughly mix the liquid and vapor refrigerant before the refrigerant enters the coil tubes. Some distributors have a special internal construction to accomplish this objective.

LOW LOAD LIMITS

Both distributor nozzles and thermostatic expansion valves have low load limitations. If the entire system is purchased as a package usually there is no problem. However, on field assembled units, difficulties can occur.

The load limit for most nozzles is usually 50% to 200% of its maximum capacity. The load limit for most thermostatic expansion valves is 35% to 50% of its maximum capacity.

When the flow through the nozzle drops below 50%, the gas and liquid refrigerant are not equally mixed. The superheat in the top circuits of the evaporator can be excessive due to liquid starving. The liquid floods back to the compressor from the lower circuits. This can easily be checked by measuring the temperature difference between the top and the bottom circuits of the coil.

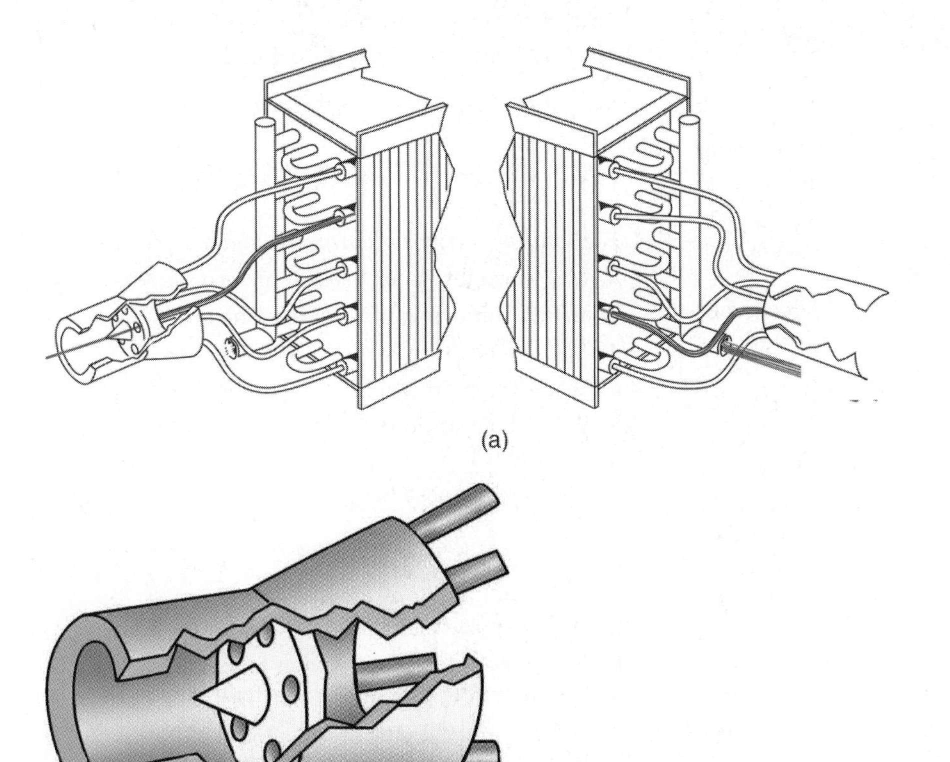

Figure 5-5-16 (a) Application of the distributor nozzle to both fin coils and plate evaporators; (b) Construction and use of a liquid distributor assembly. *(Courtesy Sporlan Valve Company)*

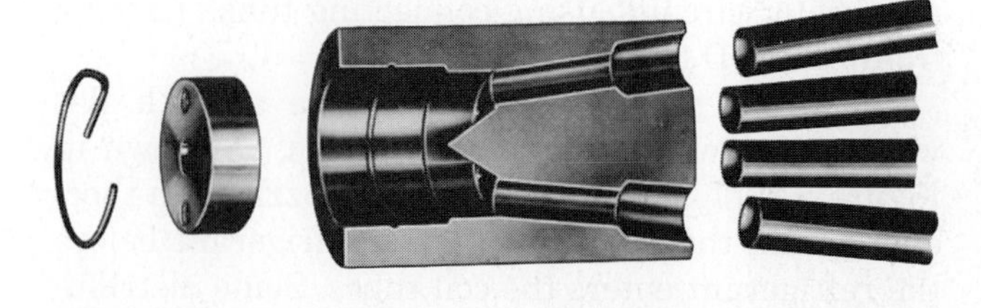

Figure 5-5-17 Detailed view of the nozzle used with a liquid distributor. *(Courtesy Sporlan Valve Company)*

This problem can be solved by replacing the nozzle with the next smaller size or arranging the control system to cut off one section of the evaporator when the low limit is exceeded.

When the flow through a thermostatic expansion valve drops below 30%, the valve will not hold constant settings. This condition is called hunting. Under these conditions superheat values fluctuate.

To correct this condition, a smaller expansion valve can be substituted or the control system can be modified to cut off a section of the evaporator when the capacity reaches the low limit.

On small capacity systems, it is sometimes advisable to use a double ported expansion valve. This valve can accommodate a two circuit evaporator without the need for a separate distributor nozzle on the coil.

COMMON METERING DEVICE PROBLEMS

Problems can be incurred with the use of metering devices. If they occur, it is important to find a solution as quickly as possible to prevent damage to the compressor. Most of these problems can be easily corrected if they are properly identified.

Below is a list of common metering device problems and the symptom that can be observed for each:

1. Stuck expansion valve If the valve is stuck open, the valve will flood the evaporator resulting in liquid refrigerant entering the compressor. There may be ice on the suction line at the compressor for freezer applications. If the valve is stuck closed, the evaporator will be starved, and the suction line will be too warm.

2. Oversized expansion valve The valve will overreact to changes in load, causing hunting. The suction pressure will continually rise and fall, never becoming stable.

3. Improper sensing bulb location On horizontal suction lines, the bulb should be located between 8 o'clock and 4 o'clock on a free draining system.

4. Valve superheat setting too low or too high The evaporator will be flooded or starved. If a valve adjustment does not solve the problem, look for a different problem.

5. Wrong type of valve TXV metering devices are color coded on the power head to correspond to the type of refrigerant that is in the system. The standard color codes are orange for R-11, yellow for R-12, green for R-22, purple for R-113, light blue for R-134a, dark blue for R-114, yellow for R-500, orchid for R-502, and white for R-717. Also, if the evaporator has a refrigerant distributor, the valve must be externally equalized.

6. Plugged equalizer line The end of the evaporator circuit being supplied by a plugged equalizer line may be warmer than the other circuits due to the restricted refrigerant flow.

7. Plugged distributor This will have the same symptoms as an expansion valve that is stuck closed.

8. Improperly sized distributor nozzle Some manufacturers use different size nozzles in the same expansion valve body. If the nozzle used is too small, the evaporator will be starved. If the nozzle is too large, the expansion valve will hunt, or may flood the evaporator.

9. Loss of thermostatic element power head charge The evaporator will be starved.

10. Sensing bulb loosely fastened to suction line The evaporator will be flooded.

UNIT 5—REVIEW QUESTIONS

1. What two actions occur in the metering device?
2. List the eight major types of metering device.
3. What are the two types of fixed metering device?
4. List one advantage of the capillary tube.
5. What are the six types of adjustable metering devices?
6. What is an obvious disadvantage of the hand expansion valve?
7. What is the most common application of the high side float?
8. The electronic expansion valve is activated by a(n) ________.
9. How do pulse modulated valves control flow?
10. Most air conditioning systems using the ________ limit the amount of pressure that can develop in the evaporator.
11. What is the load limit for most thermostatic expansion valves?
12. What will happen if you have a stuck expansion valve?

TECH TALK

Tech 1. I had a service call late yesterday afternoon and found something really odd. Someone had installed a solenoid in the liquid line up by the evaporator just before the metering device. Any idea why they would have done that?

Tech 2. Yes, as a matter of fact some manufacturers have found that putting a liquid line solenoid valve in just before a fixed bore metering device prevents the high side liquid from flowing into the evaporator during the off cycle. That simple addition for those manufacturers has provided them as much as a full SEER better efficiency.

Tech 1. Sounds interesting. How does it work?

Tech 2. Well, when a system with a fixed bore metering device first starts up you can watch your gauges. They start off at the same pressure and then they move around a bit, and slowly the high side pressure begins to rise.

Tech 1. Yeah, I see that all the time.

Tech 2. What happens with a fixed bore metering device is that when the system shuts down the liquid refrigerant is allowed to flow into the evaporator. It flows there because it is cold in the evaporator and the refrigerant collects where it is cold. You can actually hear it when the system shuts down as it gurgles through the device.

Tech 1. Yeah, I have heard that. I thought all systems did that.

Tech 2. No, most TXV used in air conditioning shut down as soon as the compressor goes off to keep all the liquid that is in the liquid line and condenser in place so when the system starts back up it is right there ready to cool. By putting the liquid line solenoid valve in, manufacturers are able to keep that liquid in the liquid line and condenser just like a TXV. So when the system starts up the valve opens and the liquid is there ready to start cooling immediately.

Tech 1. That sounds like a good idea. I think I will put it on my system.

Tech 2. No, you better not. That is something that engineers have worked out at some of the manufacturers. But it is not something that you just want to go out and stick on yourself. You could cause yourself some problems.

Tech 1. Thanks, I appreciate that.

UNIT 6

Refrigerant Piping

OBJECTIVES: In this unit the reader will learn about refrigerant piping including:

- Refrigerant Piping Requirements
- Pipe Sizing
- Insulation Considerations
- Piping Supports
- Vibration Dampeners
- Off Cycle Protection

REFRIGERANT PIPING REQUIREMENTS

The reliability and efficacy of a field assembled refrigeration or air conditioning system is greatly influenced by proper design and installation of the various parts of the piping system. Proper refrigeration piping is essential to the successful operation of the system. Improper layout or sizing can change the effectiveness of the various components, and thus alter the system's capacity, performance, and operating cost. It can also result in equipment damage and failure.

The piping layout is usually made by an application engineer, but the refrigeration technician who installs and services the system must also become involved with this layout because of the possibility of difficulties and system faults. Also, the application engineer's layout may be only a diagram, with little regard for the distances involved, either horizontal or vertical. Therefore, the technician is frequently in the position of needing to interpret the engineer's intent and then apply sound technical modifications to complete the installation properly.

Function of Refrigeration Piping

The piping that connects the four major components of the system shown in Figure 5-6-1 has two major functions: It provides a passageway for the circulation of refrigerant, either in liquid or vapor form depending on the portion of the system involved; and it provides a passageway through which lubricating oil which was carried over from the compressor is returned to the compressor. Each section of the piping should fulfill these two requirements with a minimum pressure drop of the refrigerant.

The second function, the return of oil to the compressor, is usually considered of secondary importance, but experience has proven that it is of equal importance to the function of carrying refrigerant. Reciprocating compressors, as well as rotary and centrifugal types, use various means to deliver oil to those bearings and surfaces requiring lubrication. These range from simple splash systems to pumped force feed systems. Regardless of the method, some oil is swept out of the compressor with the discharged refrigerant, resulting in the unavoidable loss of oil within the compressor.

For example, in the reciprocating type compressor, some of the oil in the crankcase gets on the cylinder walls during the downstroke or the intake stroke of the piston and is blown out with the compressed vapor refrigerant through the discharge valve parts on the up or compression stroke as shown in Figure 5-6-2.

Some compressors pump much less oil than others, depending on the design and manufacturing methods. There is no way, however, to design a compressor so that none of the oil escapes into the refrigerant piping. This oil serves no other useful purpose in the system except to lubricate the compressor.

The presence of oil in the heat exchangers (evaporator and condenser) can reduce the capacity of the heat exchange surfaces as much as 20%, because the excess oil forms an insulating coating on the interior surfaces of the refrigerant tubes. This, in turn, can cause thermal expansion valves to feed excess amounts of liquid refrigerant. Therefore, the presence of oil in the piping must be taken into consideration when installing piping. A piping system that is not correctly constructed to allow oil return can cause the following problems:

1. Seized compressor bearings due to insufficient oil returning to the compressor for lubrication.
2. Broken compressor valves, valve plates, pistons, and/or connecting rods due to liquid refrigerant and/or large quantities (slugs) of oil entering the

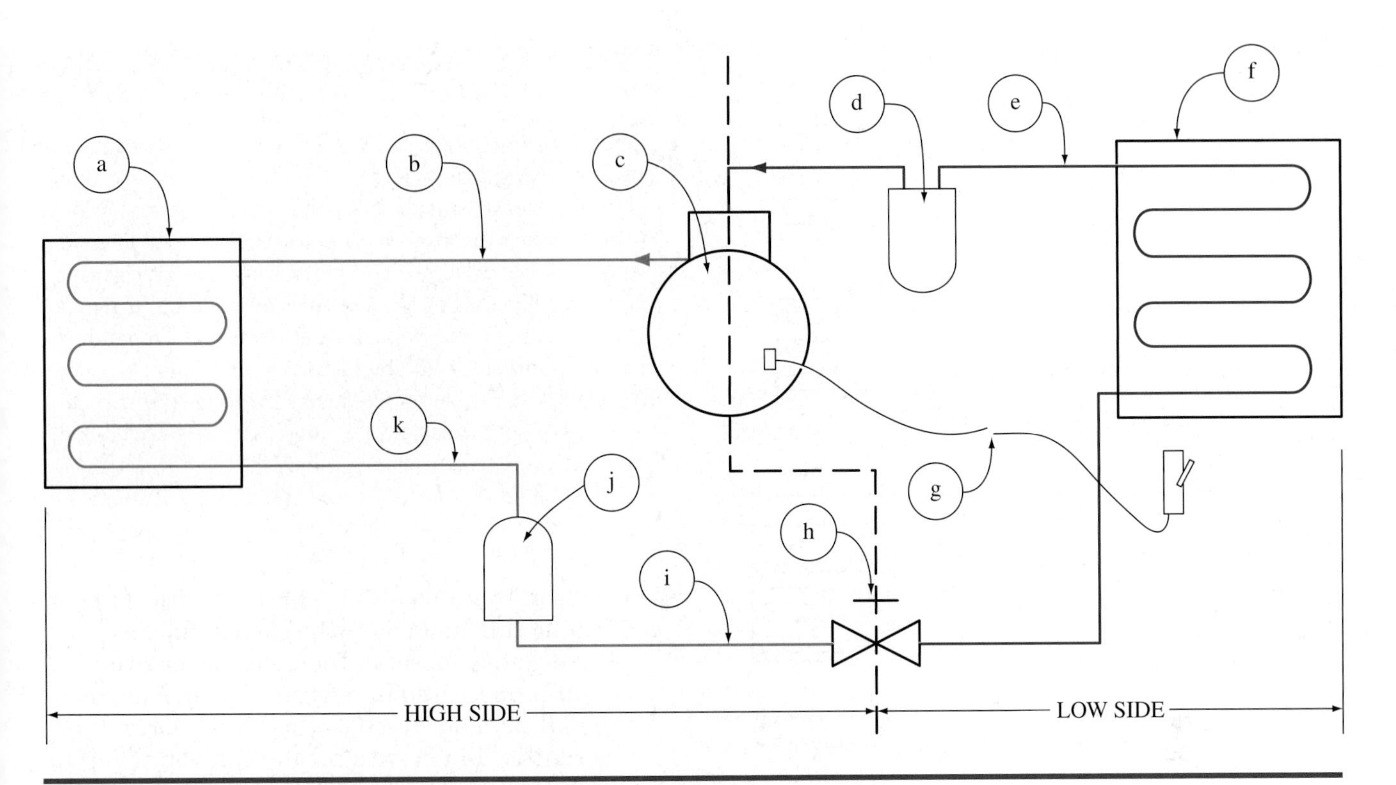

Figure 5-6-1 Major components of air conditioning and refrigeration systems: (a) Condenser; (b) Hot gas line; (c) Compressor*; (d) Accumulator; (e) Suctions (vapor) line*; (f) Evaporator*; (g) Operational controls and safeties*; (h) Metering device*; (i) Liquid line*; (j) Receiver; (k) Condensate line. (Components with asterisks appear on all air conditioning and refrigeration units.)

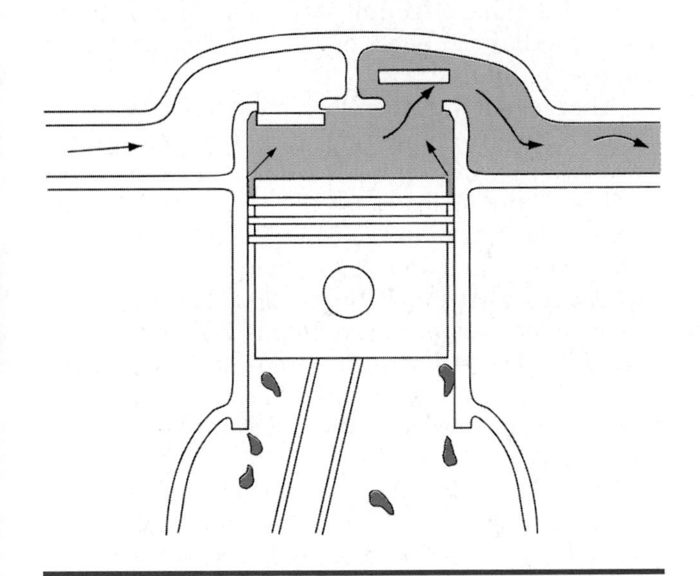

Figure 5-6-2 Compressor oil pumping.

compressor. It is important to understand that the compressor is designed to pump vapor and will not pump liquid.

3. Loss of capacity caused from the oil occupying portions of the evaporator, thus reducing the amount of effective surface and the overall system capacity.

BASIC PIPING PRECAUTIONS

A refrigeration technician must learn this subject well enough to avoid costly mistakes. There are five basic rules to keep in mind when piping a system:

1. Keep it clean. Cleanliness is a key factor in the actual installation. Dirt, metal filings, sludge, and moisture will cause breakdown in the system and must be avoided. Neat, clean work will avoid many service difficulties.

2. Proper sizing. Each section of the piping system must be properly sized to ensure proper oil return as well as to maximize system capacity and efficiency. In those installations where one of these functions is to be reduced, proper oil return takes precedence in being fully retained.

3. Use as few fittings as possible. See Figure 5-6-3. Fewer fittings mean less chance for leaks, and more importantly, less needless pressure drop.

4. Take special precautions in making every connection. Use the right material and follow the method recommended by the equipment manufacturer.

5. Pitch horizontal lines in the direction of refrigerant flow. See Figure 5-6-4. To aid in forcing oil to travel through lines that contain vapor (suction

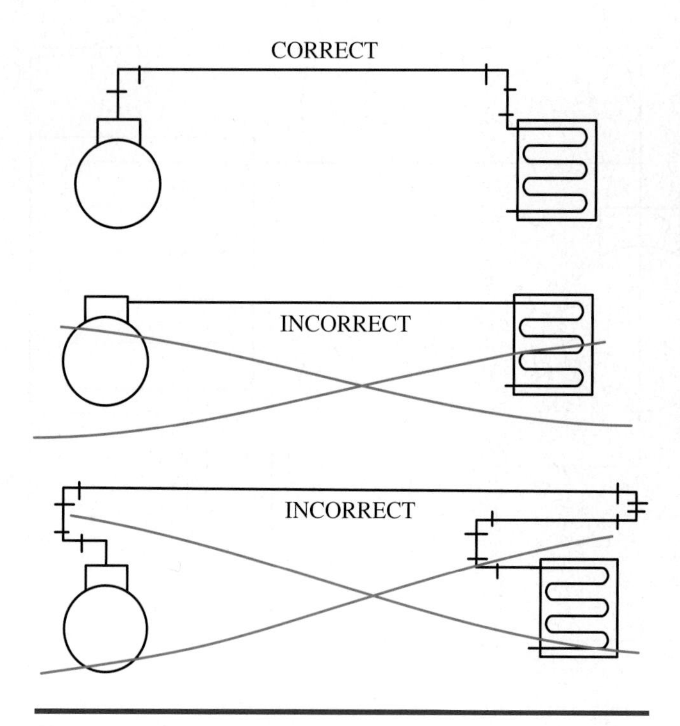

Figure 5-6-3 Proper and improper piping arrangement.

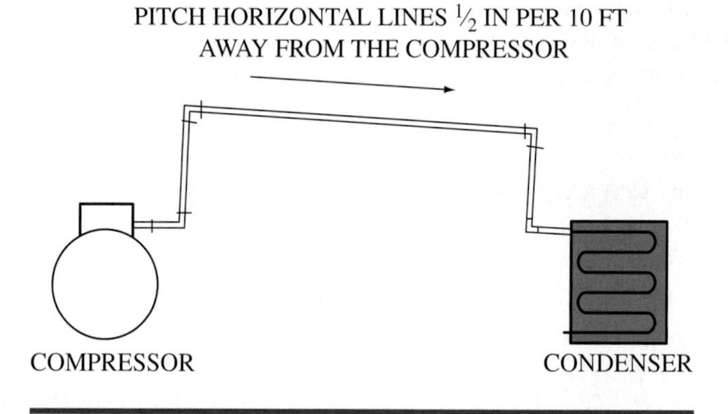

Figure 5-6-4 Proper pitch.

line, hot gas line), horizontal lines should be pitched in the direction of refrigerant flow. This pitch, which helps the oil flow in the right direction, should be a minimum of ½ in or more for each 10 ft of run. Pitch also helps to prevent backflow of the oil during shutdown.

In piping systems where sufficient return gas velocity can be ensured at all times, it is satisfactory to run the horizontal suction lines dead level. This may be desirable where headroom is at a premium or where a sloping run will interfere with other piping.

Suction Line

Oil circulates throughout the system and must be returned to the compressor to prevent damage to the compressor, as stated above. The most critical line in performing this function is the suction line.

For example, observe the behavior of two common refrigerants used in refrigeration and air conditioning, R-22 and R-410A. In liquid form these refrigerants will mix with oil and carry it along the piping with ease. Therefore, few, if any, oil problems exist in the liquid line. However, in their gaseous state the refrigerants are poor carriers of oil.

Oil in the suction line is at a lower temperature than the rest of the system, and therefore has higher viscosity, which slows down the flow over pipe surfaces. Also, the refrigerant is in vapor form and has only a mechanical effect on the oil. The vapor does not absorb the oil.

The suction line, therefore, must be carefully designed to ensure a uniform return of dry refrigerant gas as well as sufficient oil to the compressor.

The minimum load gas velocities within the suction line must be maintained at 500 ft/min (2.5 m/s) in horizontal runs and 1000 ft/min (5 m/s) through vertical rises with upward gas flow. The maximum recommended temperature drop is 2°F for R-22 refrigerants. The temperature drop can be converted into equivalent pressure drop if desirable; however, the equivalent pressure drop is different for various refrigerants.

The reason the suction pressure and temperature drop are so important is that they relate to the ratio between the discharge and suction pressures. Any increase in this ratio reduces the capacity of the compressor to pump refrigerant vapor and increases the power required.

TECH TIP

Selecting the correct refrigerant lines can be very complex involving many factors. One factor, however, that should be least considered is the cost of the materials. When cost is factored in, the system's copper line set tends to be undersized. Undersizing the refrigerant copper is like putting small wheels on your car. It significantly reduces the vehicle's performance. Depending on the number of 90° bends or other fittings in a refrigerant line set, undersized lines for long runs can reduce a system's capacity by as much as ½ ton on a 3-ton system.

Thermostatic Expansion Valve Systems

This system uses a thermostatic expansion valve (TXV) as a metering device to regulate the flow of refrigerant into the evaporator.

SERVICE TIP

When the suction line is installed during new construction, the air conditioning technician takes great care that the line does not get crimped. However, other trades are not as cautious and may damage the copper by bending it out of their way. For example, brick masons will often bend the copper up so it is easier for them to lay the first course or two of brick and then bend it back down. Every effort should be made to educate the other trades that will be working around your copper rough-in. If you suspect that the copper has been crimped, it may be possible for you to go into the residence and visually inspect the copper. A crimp in the suction line will significantly reduce the system's performance. Severe crimps can actually become a secondary metering device where ice will form from the crimp back to the compressor. This problem is extremely difficult to diagnose unless a temperature reading is taken on the suction line as it leaves the indoor coil and as it leaves the building. There should be no more than 2°F difference between these two readings. If the line is crimped, the temperature difference may be 10°–30° or more.

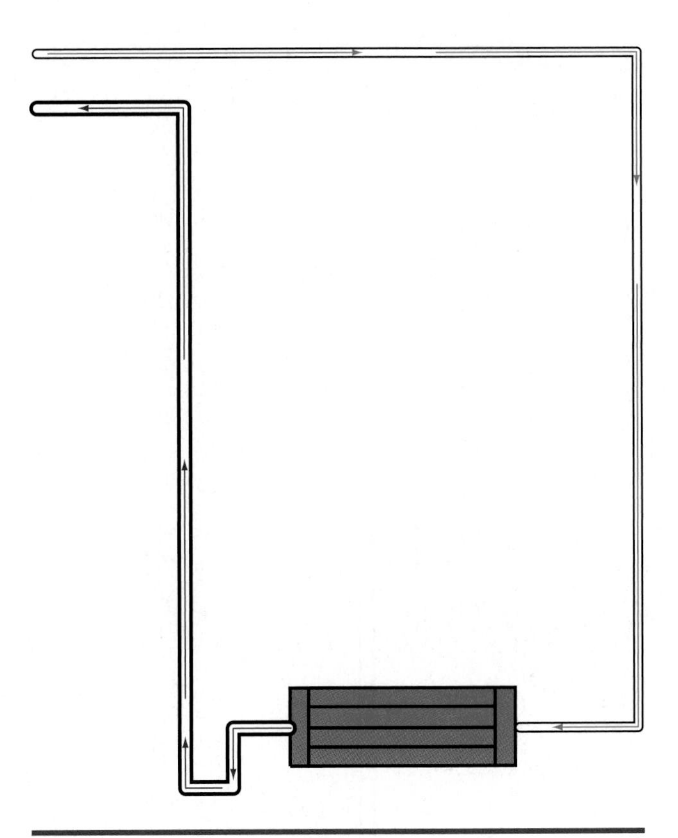

Figure 5-6-5 Suction line trap.

When the compressor is above the evaporator, liquid floodback to the compressor is not a problem. In all thermostatic expansion valve systems, especially in those where the compressor is at the same level or below the evaporator, a trap and riser to at least the top of the coil must be placed in the suction line, as shown in Figure 5-6-5.

This trap is to prevent liquid flowing from the direct expansion (DX) coil into the compressor during shutdown. The trap and riser combination also promotes free drainage of liquid refrigerant away from the thermostatic expansion valve bulb, thus permitting the bulb to sense suction gas superheat instead of evaporating liquid temperature.

Although it is important to prevent liquid refrigerant from draining from the evaporator to the compressor during shutdown, it is just as important to avoid unnecessary traps in the suction line near the compressor. Such traps would collect oil, which on startup might be carried to the compressor in the form of slugs, thereby causing serious damage.

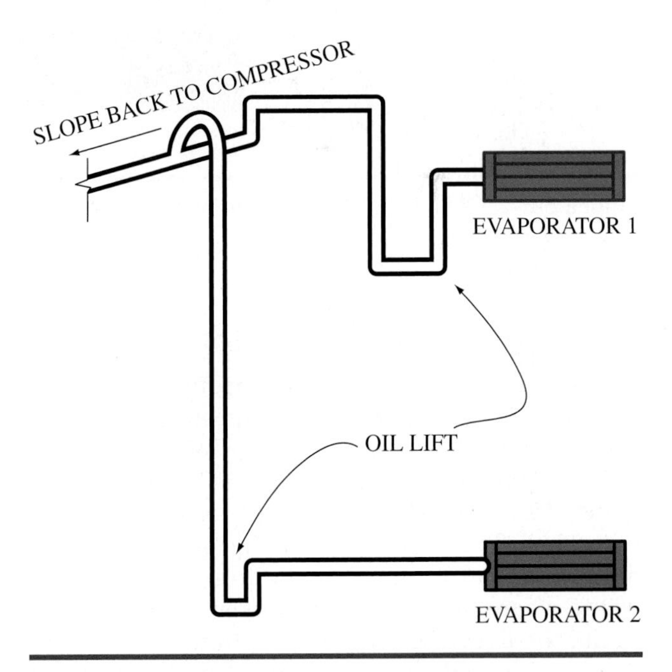

Figure 5-6-6 Oil lifts for multiple evaporators.

Piping for Multiple Evaporators

Figure 5-6-6 illustrates the use of multiple evaporators installed below the compressor. Notice that the piping is arranged so that refrigerant cannot flow from the upper evaporator into the lower evaporator.

Where the vertical rise is over 20 ft (6 m) on either suction or discharge lines, as illustrated in Figure 5-6-7, it is recommended that line traps be installed approximately every 20 ft (6 m) so that the storage and lifting of oil can be done in smaller stages.

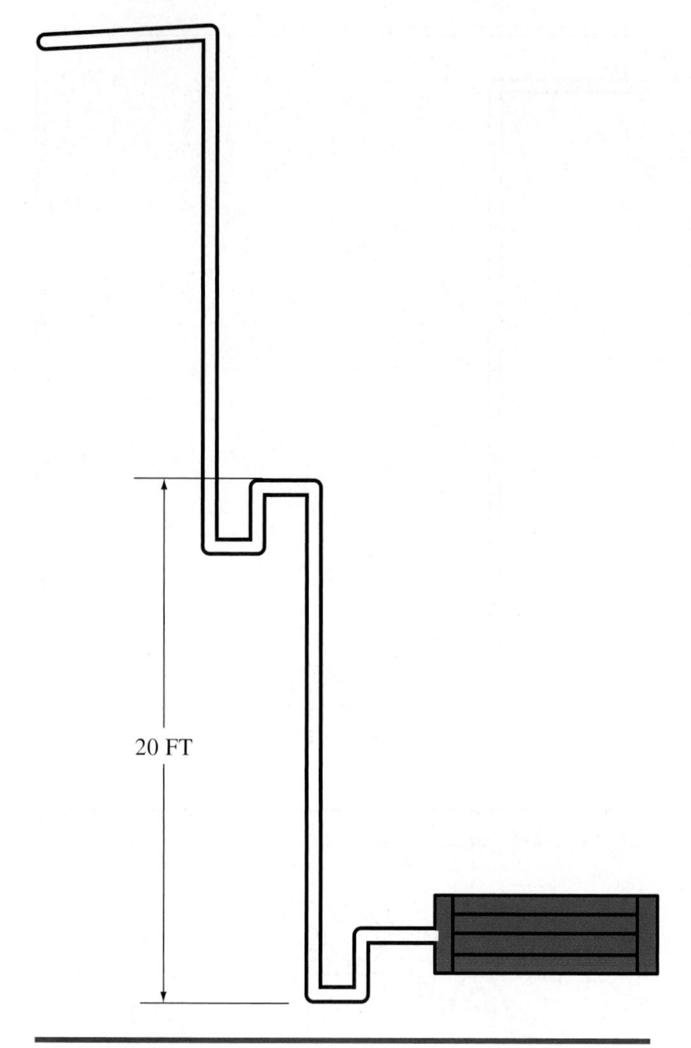

Figure 5-6-7 Oil lifts must be installed when the suction riser height is greater than 10 feet.

Preventing Liquid Slugging

To learn how to prevent liquid slugging, it is necessary to first understand what happens in a simple system when the compressor stops operating after its cooling requirements are satisfied. The evaporator is still filled with refrigerant—part liquid and part vapor. There will also be some oil present. The liquid refrigerant and oil may drain by gravity to points where, when the compressor starts running again, the liquid will be drawn into the compressor and cause liquid slugging. The piping design must prevent liquid refrigerant or oil from draining to the compressor during shutdown. When the compressor is above the evaporator, this is not a problem.

When the compressor is located in an area where the ambient temperature is cooler than the condenser and/or evaporator, OFF cycle refrigeration migration can be a problem. Traps in the discharge piping may be required as well as suction line accumulators and crankcase heaters. It is best

NOTE: Refrigerant copper comes in coils, and when properly unrolled it is relatively straight. However, if the copper is simply pulled from the center like a spring, it will have a series of loops. The loops are not significantly deep; however, each loop can collect a small amount of refrigerant oil. When the amount of oil in the loops gets large enough, it will begin moving from the first loop to the second where it picks up that oil and moves to the next. This moving slug of oil then can pass into the compressor and slug it. It moves much in the same way as the last bit of juice is sucked through a child's curlicue straw.

to avoid such problematic conditions where possible rather than to try to design fixes for them after the fact.

The greater the system refrigerant charge, the more likely it is that compressor failures may be experienced. Common sense dictates close coupled systems with clean, simple piping layouts and minimal refrigerant charges.

If the compressor is on the same level as or below the evaporator, as in Figure 5-6-8, a riser to at least the top of the evaporator must be placed in the suction line. This inverted loop is to prevent liquid draining from the evaporator into the compressor during shutdown. The sump at the bottom of the riser promotes free drainage of liquid refrigerant away from the thermostatic expansion valve bulb, thus permitting the bulb to sense suction gas superheat instead of evaporating liquid refrigerant.

PIPE SIZING

The technician must be able to spot and modify obvious field piping errors caused by improper design or installation. If piping sizes are incorrect, either undersized or oversized, they can cause poor system performance or even damage to the equipment. Figure 5-6-9 shows a diagram of a basic air conditioning system and has a key of symbols that will be used in the figures to come. Some of the common piping problems that are encountered for suction line sizing are shown in Figure 5-6-10a,b; hot gas (discharge) line sizing Figure 5-6-11a,b; liquid line sizing Figure 5-6-12a,b; and condensate line sizing Figure 5-6-13a,b. The suction line is the most critical, followed by the hot gas line, then the liquid line, and last, the condensate line.

The objective in sizing condensate lines is to provide ample size, allowing free drainage of the liquid refrigerant to the receiver while the gas refrigerant flows over the liquid in the opposite direction, thus serving as an equalizer line as well as a liquid drain.

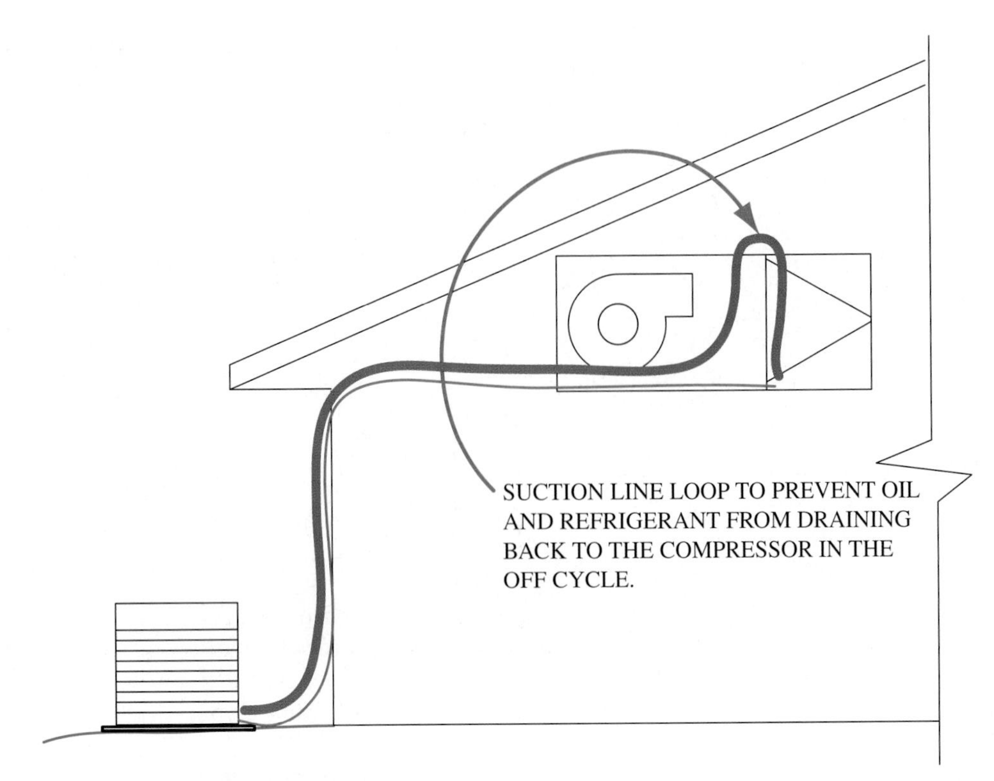

Figure 5-6-8 Many manufacturers require that the suction line rise above the height of the evaporator when the compressor is installed level with, or below, the evaporator coil.

Allowable Pressure Drop

In sizing refrigerant lines, it is desirable to keep the pressure drop within allowable limits. If the lines are too large the cost can be excessive; if the lines are too small the capacity of the equipment is reduced. A good compromise is to use a pressure drop roughly equal to a 2°F drop in saturation temperature. The equivalent change in pressure drop amounts to almost twice as much on the high side of the system as on the low side.

NOTE: An ideal refrigerant piping system has a pressure differential equivalent to less than 2°F temperature difference between the indoor coil and the outdoor unit. Refrigerant pressures and temperatures are not linear. There is greater difference between one pound of pressure at a lower temperature than there is at a higher temperature. For refrigeration applications caution must be taken so that the refrigerant lines are not sized using the same pressure drop calculations as for air conditioning, which operates at a higher temperature pressure range. Refer to the pressure temperature chart and the operating temperature range of the system you are working on to determine what pressure difference is related to a 2°F temperature drop at the system's normal operating temperature.

Where the system capacity is variable because of capacity control or some other arrangement, a short riser will usually be sized smaller than the remainder of the suction line, Figure 5-6-14, for a velocity of not less than 1000 ft/min (5 m/s) to ensure oil return up the riser. Although this smaller pipe has higher friction, its short length adds a relatively small amount to the overall suction line friction loss.

In general, the pressure drop for the total suction line should be a maximum equivalent of 2°F (1°C), or 2 psig (14 kPa), to avoid loss of system capacity. The compressor cannot pump or draw gas nearly as effectively as pushing or compressing the gas.

CAUTION: As the suction pressure drops, it has a greater effect on the compressor's compression ratio. The compression ratio is the absolute suction pressure divided into the absolute discharge pressure. If the suction pressure drops as a result of undersized refrigerant piping, the compression ratio will go up significantly. If the compression ratio gets too high, the refrigerant can decompose, forming acids; or the discharge temperatures leaving the compressor piston can be hot enough to actually burn away the discharge valves. Undersizing the suction line is extremely detrimental to the equipment's life.

Riser Sizes

Riser sizes for both suction and hot gas lines are critical to permit carrying the oil upward with the force of the flowing gas, as shown in Figure 5-6-15.

If the compressor is equipped with capacity control, the vertical runs must be sized properly to return the oil at minimum tonnage Figure 5-6-16. Vertical drops (flow down) will return the oil by gravity and are not critical as to size or velocity.

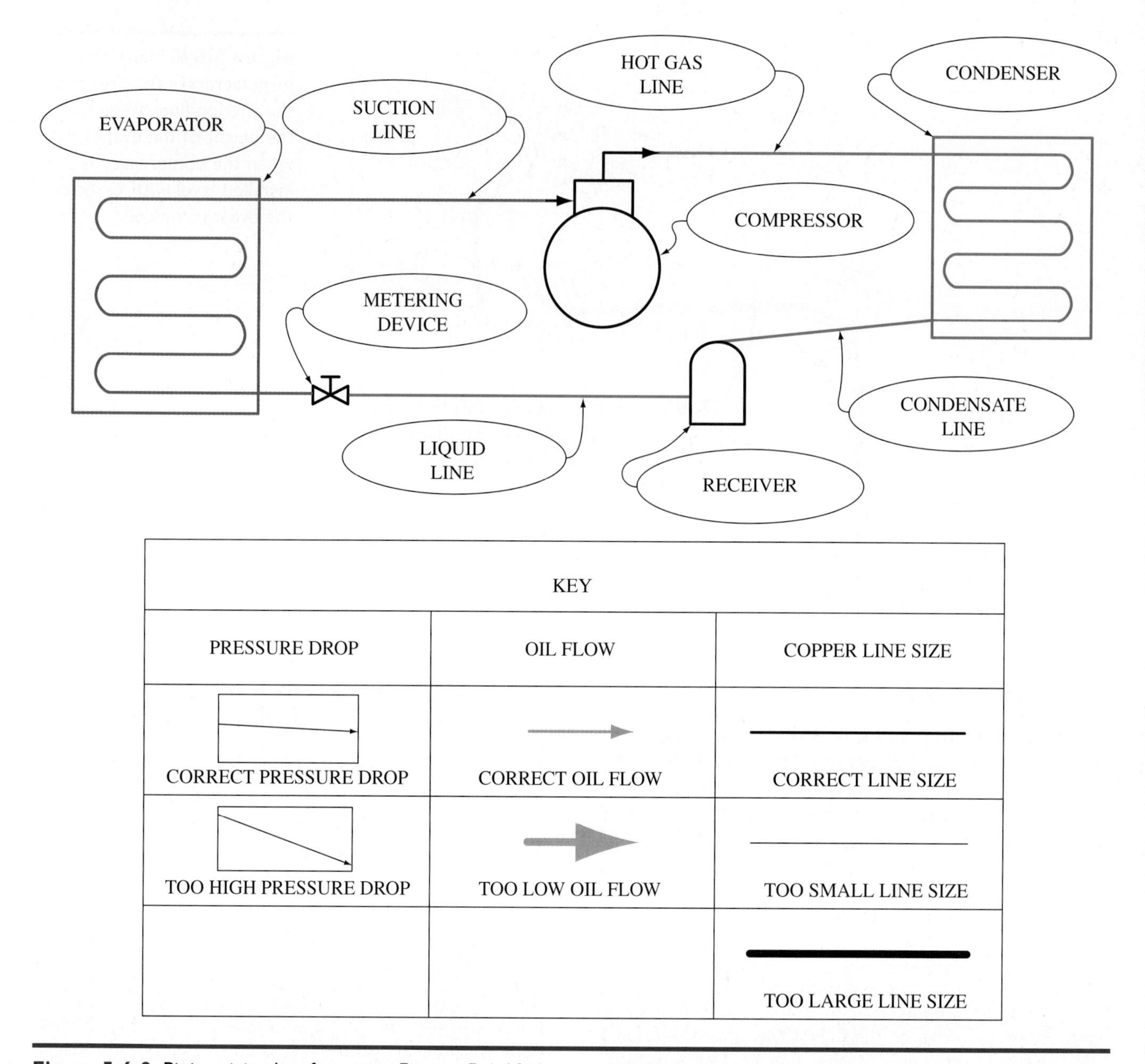

Figure 5-6-9 Piping sizing key for use in Figures 5-6-10 through 5-6-13.

An alternative to sizing the riser for minimum capacity is to provide a double riser in the suction line, as illustrated in Figure 5-6-17. If a double riser is used, additional oil is required in the system to fill the trap during periods when the overall capacity of the system is reduced, and only one riser is used. When full capacity is resumed this extra oil can return to the compressor and overload the oil capacity of the crankcase, causing oil slugging. It is, therefore, desirable to avoid the use of double risers wherever possible.

The operation of the double suction riser is similar to the hot gas double riser. When maximum cooling is required, the system will run at full capacity, and both risers will carry refrigerant and oil. On part load, as the amount of refrigerant being evaporated decreases, the gas velocity will also decrease to a point where it will not carry oil upward through the vertical risers.

The oil trap, which is located at the bottom of the large riser, will fill with oil. The entire refrigerant vapor will then pass up through the smaller riser, the

> **CAUTION:** Small high speed compressors have relatively small crankcase capacities. If double risers with oil seals are to be utilized, it may be necessary to add an auxiliary oil receiver. Be sure the system has sufficient oil to permit proper compression lubrication at full and minimum capacities.

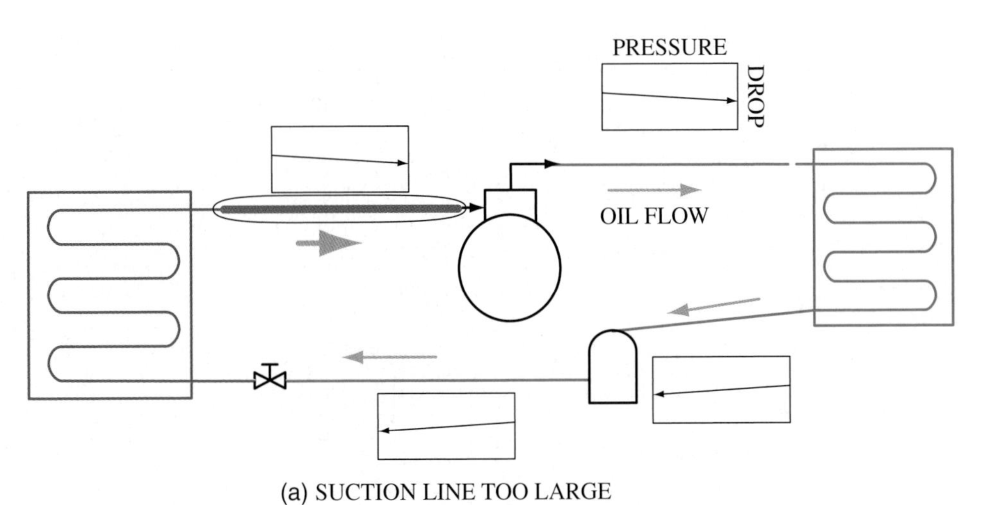

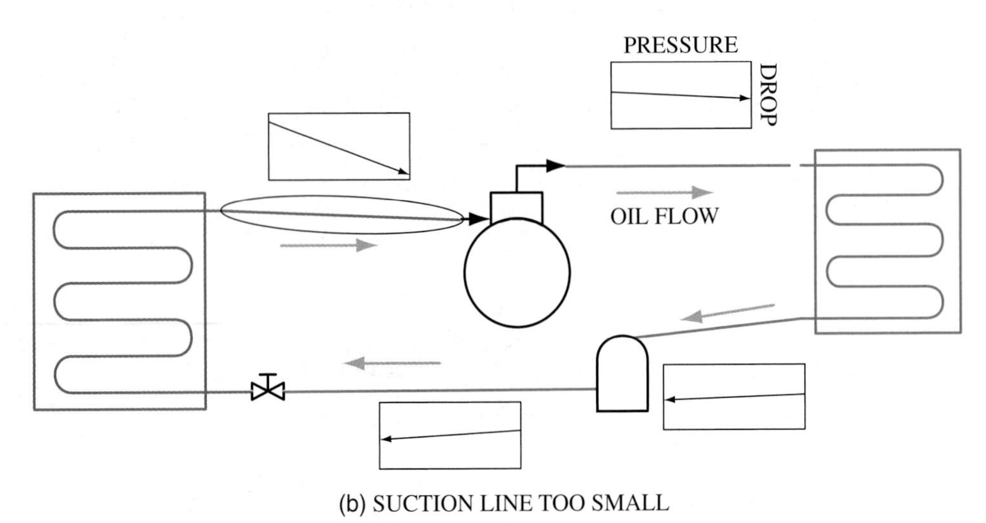

Figure 5-6-10 (a) Effect of oversized suction line; (b) Effect of undersized suction line.

pressure carrying oil with it. As the system load increases and more refrigerant is passed through the evaporator, this increased pressure will break the oil seal in the trap and carry oil upward through both risers.

Referring to Figure 5-6-17, riser 1 is used to carry the suction gas at minimum load. Riser 1 plus riser 2 are used to carry the suction gas at maximum load. The area of riser 1 plus the area of riser 2 is made equal to the area of the main suction line.

Capillary Tube Systems

Sizing of the suction line depends on the amount of suction vapor to be handled. Therefore, suction lines in capillary tube systems are sized the same way as in thermostatic expansion valve (TXV) systems.

The difference in suction line design between the two types of systems is in the use of traps. Traps are not recommended in capillary tube installations, as they only add resistance to the flow of refrigerant. The trap is composed of three 90° bends, and each bend is equal to 5 ft of straight pipe. Adding a trap is the same as adding 15 ft of straight pipe.

Practically all direct expansion (DX) valve coils using capillary tubes are bottom feed, which reduces the possibility of gravity drain during the OFF cycle of the system. Also, the refrigerant charge is limited to what the coil will hold during the OFF cycle.

When properly charged with refrigerant, the capillary tube system has less chance of liquid slugging. In addition, oil in the coil is lifted out into the suction header of the coil on each OFF cycle because the coil fills with liquid refrigerant when the pressures balance. This prevents oil accumulation in the coil.

Traps are also not required in vertical risers because of the high gas velocities on startup. Standing idle, the DX coil, full of liquid refrigerant, will acquire considerable thermal energy at the balance pressure. Immediately on startup of the compressor, when the suction pressure drops, a high quantity of vapor is produced in the DX coil, resulting in suction line velocities in excess of 6000 ft/min (30 m/s). This velocity is sufficient to lift oil in vertical riser suction lines up to 65 ft. It is not recommended that installations be made of capillary tube systems

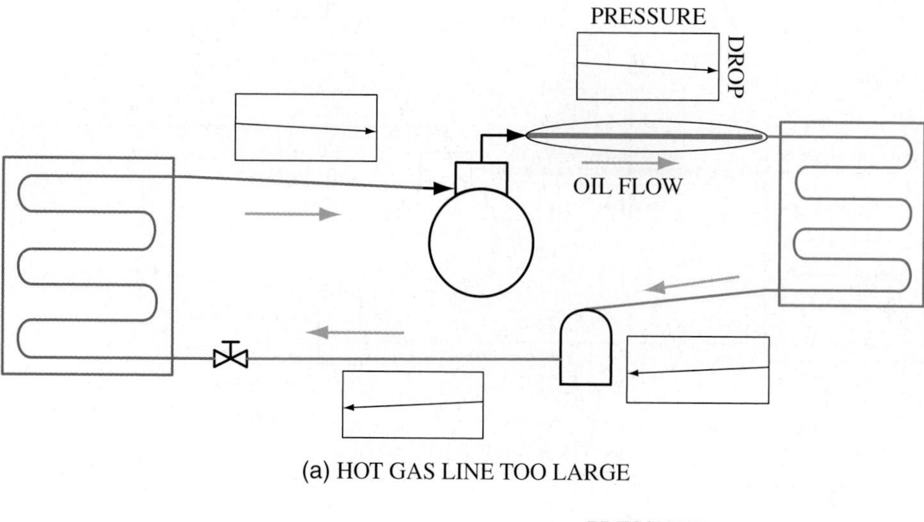

(a) HOT GAS LINE TOO LARGE

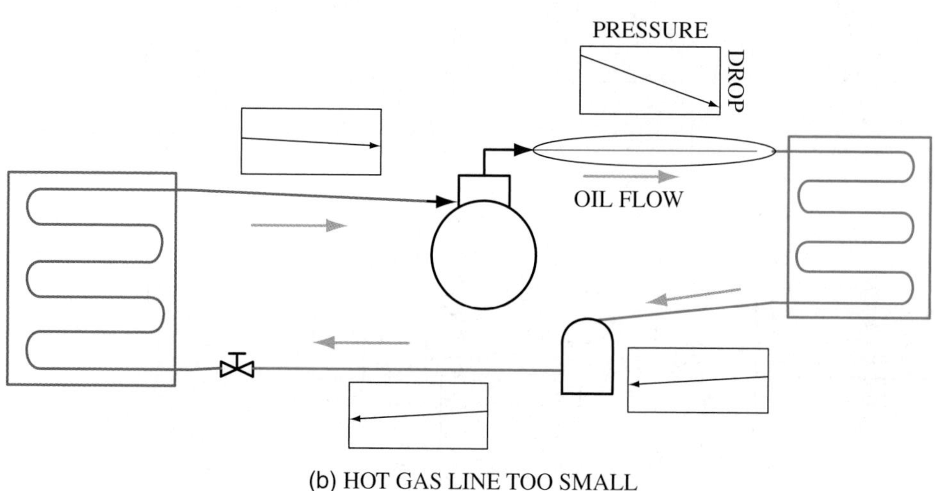

(b) HOT GAS LINE TOO SMALL

Figure 5-6-11 (a) Effect of oversized hot gas line; (b) Effect of undersized hot gas line.

where a rise in the suction line in excess of 65 ft is required. In these installations, TXV valves and suction line traps should be used. It is important to follow the manufacturer's recommendations where limitations of length of run, riser height, or charge are to be provided.

INSULATION CONSIDERATIONS

Suction Line Insulation

Insulation on the suction line is an absolute requirement. This eliminates the following:

1. Sweating of the suction line. Water condensing on the suction line can drip on occupants and can cause damage to ceilings, floors, furnishings, and electronics.

2. High suction temperature gain. Hermetic and semihermetic motor compressor assemblies are suction gas cooled. Therefore, the lower the returning suction gas temperature, the better the heat removal from the motor and the lower the motor operating temperature and compressor discharge temperature; also, a lower oil temperature results, enabling better lubrication and bearing heat removal. This promotes longer compressor life.

TECH TIP

The joints in the insulation on suction lines must be sealed. Often technicians have the misconception that the insulation is simply there to aid in system efficiency. That is only partially true. Because the refrigerant line is below the dew point, moisture can condense on the line. If the insulation vapor barrier is not sealed, water can collect around the copper and draw in through gaps and openings at the joints. The water can then run down the refrigerant line set and weep out, causing water stains and mold growth in the wall cavities. It is important that the seams of the refrigerant line insulation be sealed properly.

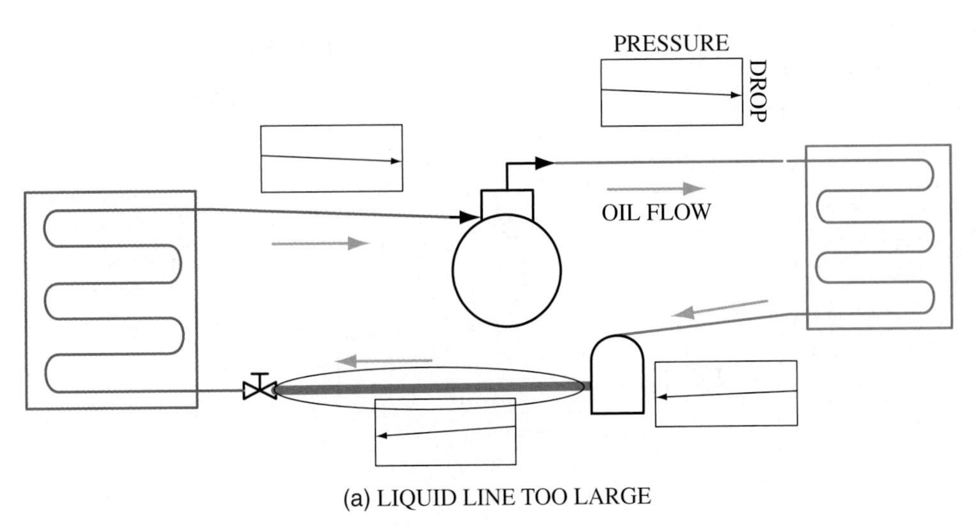

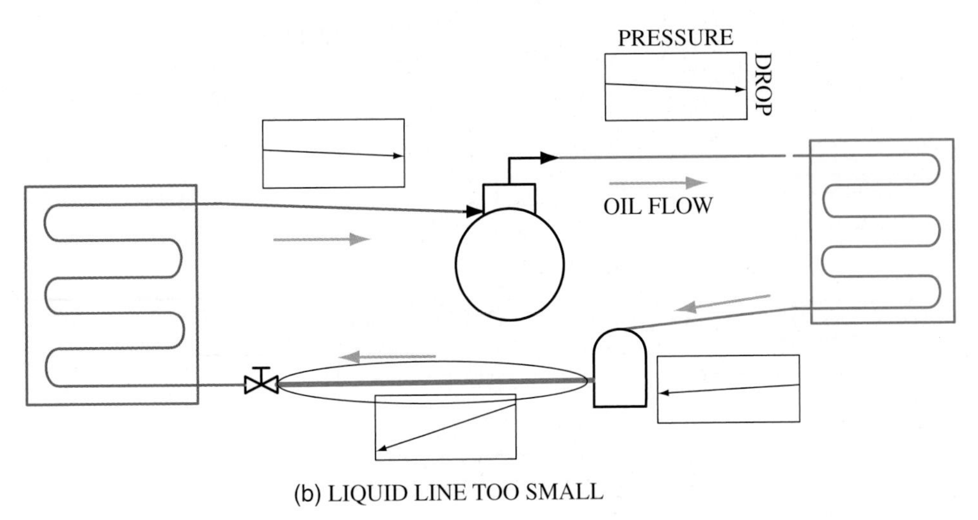

Figure 5-6-12 (a) Effect of oversized liquid line;
(b) Effect of undersized liquid line.

Hot Gas Line

In sizing and arranging hot gas lines, select tubing with a diameter small enough to provide the velocity to carry the hot vaporized oil to the condenser. On the other hand, the diameter must be large enough to prevent excessive pressure drop. In suction lines, the maximum allowable temperature drop is 2°F. In hot gas lines 2°F is recommended, but a greater loss can be accommodated if conditions demand it.

If a higher pressure drop is used, the velocity of gas flow through the line can be excessive, causing noise, vibration, and serious reduction in system capacity. There would also be an increase in operating cost due to the higher compressor discharge pressure required.

The installation of an oil separator, such as shown in Figure 5-6-18a,b, is an alternative to moving oil up the discharge riser. The oil from the separator is usually returned to the crankcase of the compressor. With this arrangement the hot gas riser can be sized for low pressure drop.

Hot Gas Line Insulation

In package units and condensing units with short hot gas lines between compressor and condenser, no insulation should be used on the hot gas line.

On remote condensers, insulating the hot gas line is advisable. If the unit is expected to operate in low outside temperatures, it is possible to reach the condensing temperature of the discharge refrigerant before the refrigerant reaches the condenser. This can cause liquid slugs to fall backward down the hot gas line into the superheated vapor from the compressor. Violent expansion of the slug vaporizing can cause "steam hammer," resulting in noise and vibration, even to the point of line breakage. Insulating the hot gas line would prevent this action.

When hot gas lines are run indoors in machinery rooms, they should be insulated or otherwise protected to prevent accidental burns to operating personnel.

Liquid Line

Liquid lines do not present a problem from an oil standpoint because liquid refrigerant and oil mix

Figure 5-6-13 (a) Effect of oversized condensate line; (b) Effect of undersized condensate line.

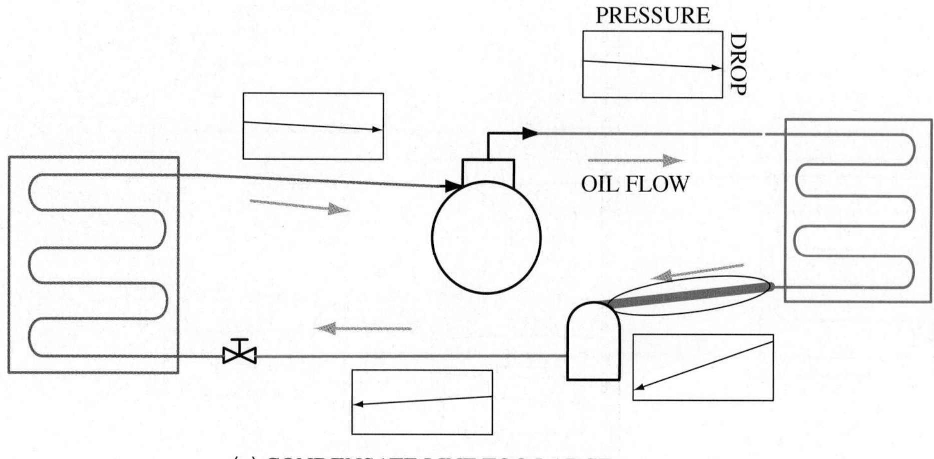

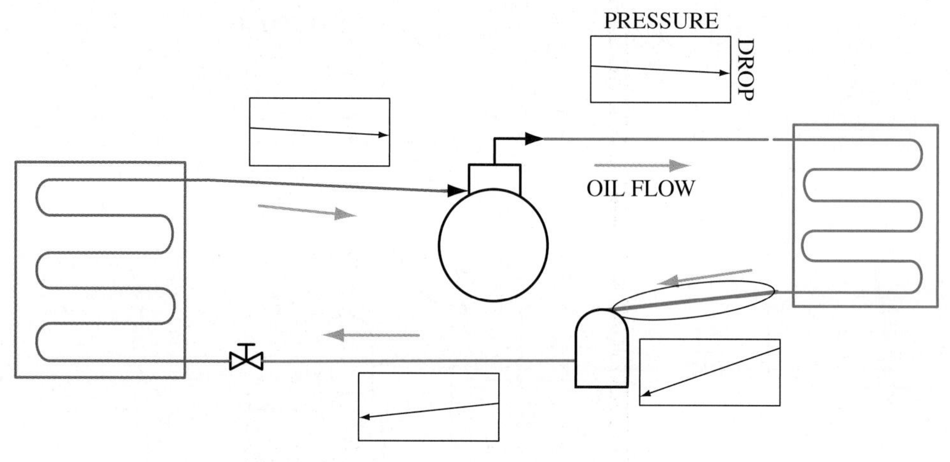

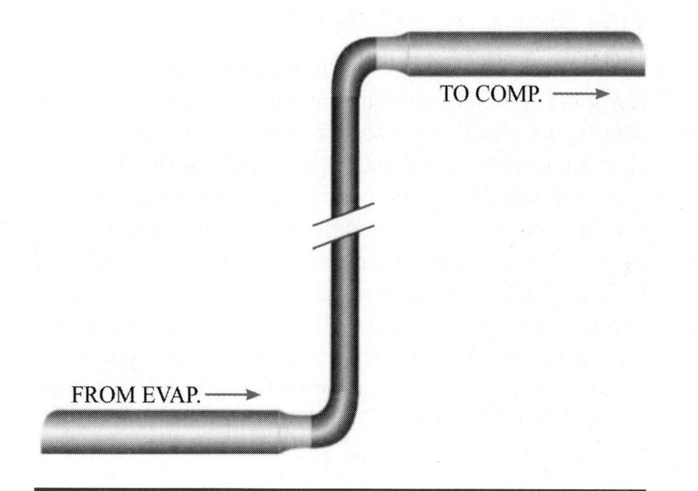

Figure 5-6-14 Reduced size suction gas pipe riser.

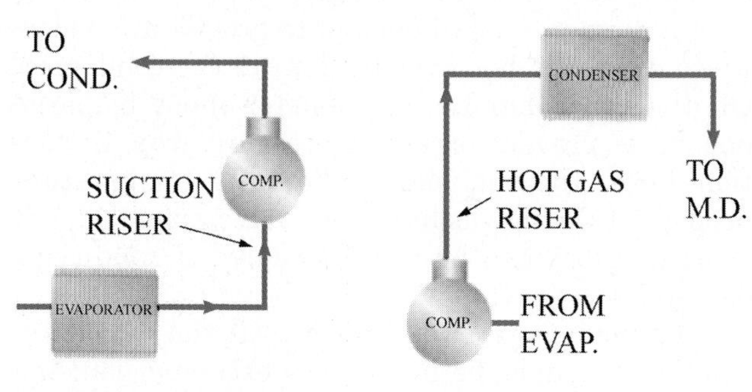

Figure 5-6-15 Piping risers for suction and hot gas lines. Sizing for both is critical.

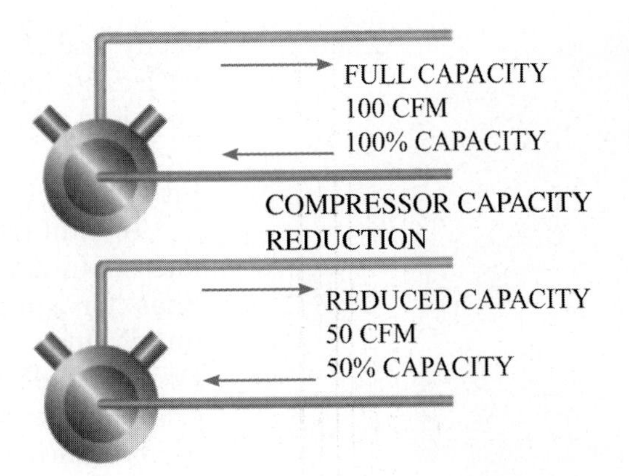

Figure 5-6-16 50% capacity reduction reduces the refrigerant flow 50%.

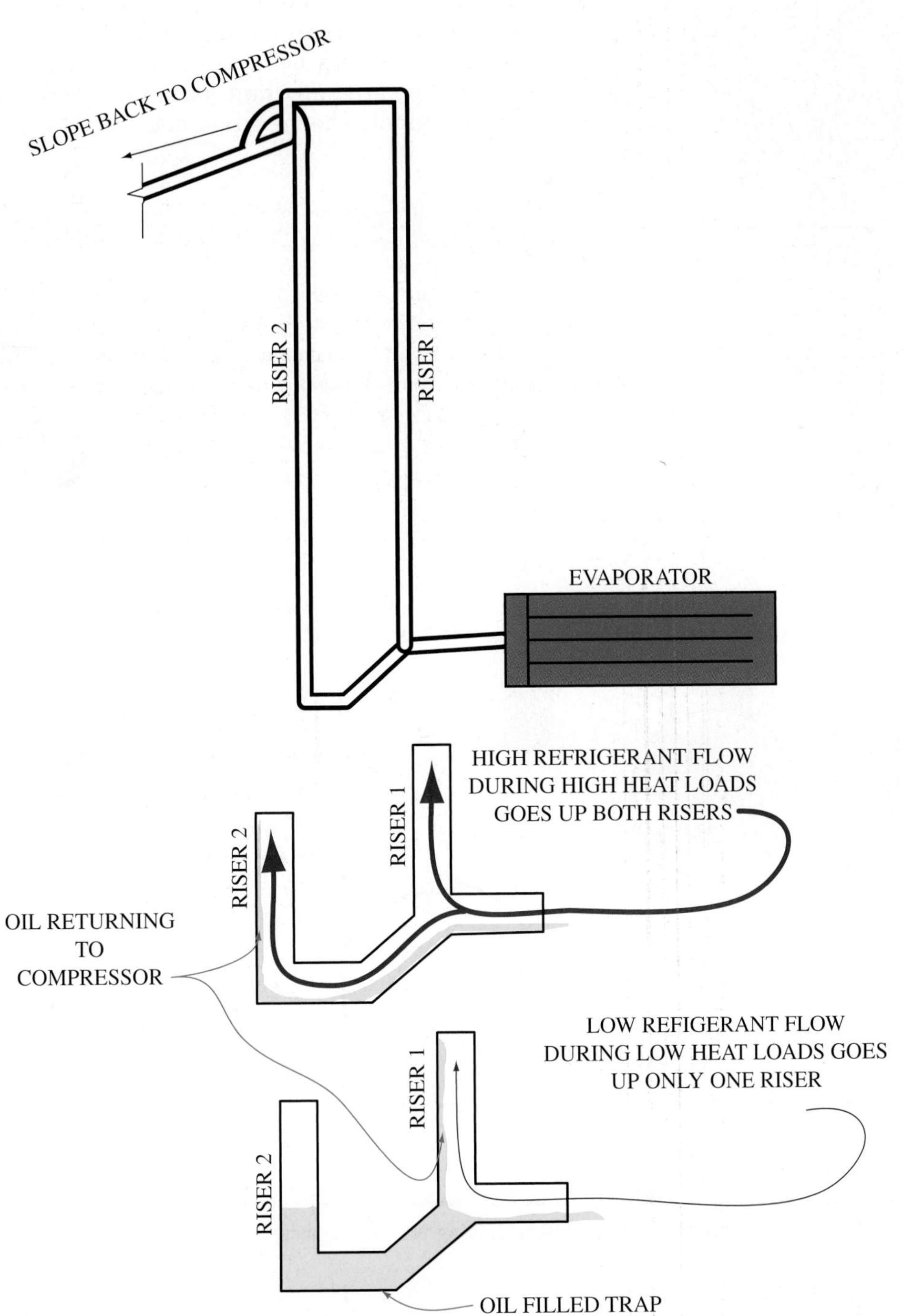

Figure 5-6-17 Double suction riser.

(a)

easily and the oil is carried through the liquid line with ease. Liquid lines, however, are critical for pressure loss both from a pressure drop due to pipe size and, in the case of the vertical upflow line, vertical lift of the refrigerant.

To limit the amount of refrigerant in the system, most manufacturers use ⅜ in and even ¼ in liquid lines for split systems using remote condensing units. This will vary with tonnage and line length. Do not attempt to change the liquid line size from the original specifications, as this will seriously affect performance and equipment life.

Some systems also depend on the pressure drop in the liquid line to add to the pressure drop when using a capillary tube for the pressure reduction device. In these cases, both the size and the length of the liquid line are important for peak performance. Do not attempt to change either one.

TECH TIP

The liquid line is typically not insulated; however, when it runs a great distance through a hot attic, it is advisable to insulate it from the extreme attic heat. In addition, insulating the liquid line will aid in performance of a heat pump. Finally, insulating the liquid line can reduce the noise caused by changeover of heat pumps.

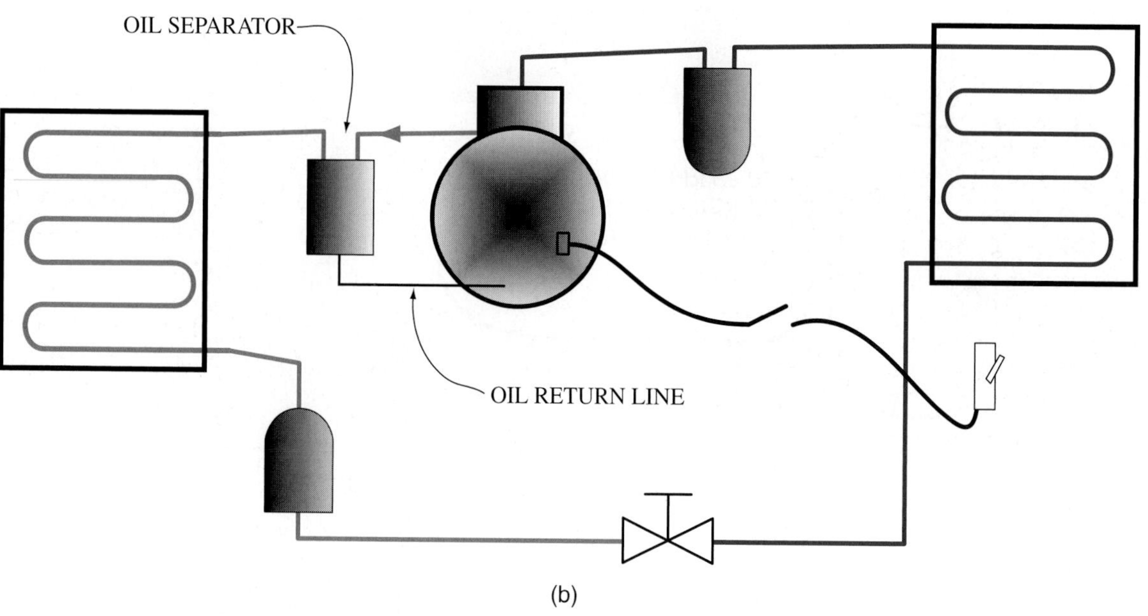

(b)

Figure 5-6-18 (a) Oil separator (*Courtesy Danfoss Inc.*); (b) Discharge piping with oil separator.

Preventing Flash Gas

Liquid lines can present a problem if the design causes a radical temperature change or pressure drop, since refrigerant leaving the condenser remains a liquid only as long as its boiling point (condensing temperature) is higher than its actual temperature. If a drop in pressure (drop in boiling point) or a rise in the liquid temperature were to reverse (boiling point fall below sensible temperature or sensible temperature rise above boiling point), vaporization of some of the liquid would occur before it passes through the pressure reducing device. The vapor formed is called pre-expansion flash gas.

TECH TIP

Liquid line sight glasses are used in some applications as an indicator of flash gas. However, the sight glass's usefulness is significantly diminished if it is in the liquid line near the condenser. At that location it simply tells you there is or is not flash gas there; however, flash gas can be forming in front of your metering device and go undetected. Do not rely on a sight glass as the sole way of determining whether or not flash gas is occurring in a system. Copper can work harden, meaning that if it is bent repeatedly, it will get significantly harder. It can get hard enough so that it is subject to brittle fractures. Vibration can cause copper to work harden. It is important that vibration be dampened not only for sound but to ensure that the copper does not become so work hardened that it will fracture.

An example of pre-expansion flash gas production is laying a liquid line on the roof of a building, where it is on tar that has been heated by the sun. Refrigerant lines should be raised at least 18 in above such a roof to minimize the effects of the sun. Flash gas within the liquid line is undesirable, since it displaces liquid at the port of the expansion valve, greatly reducing capacity. Known as gas binding, it also affects the capacity of capillary tubes, as the ability of the tubes to carry vapor is considerably less than the ability to carry liquid. Liquid leaving the condenser is subcooled although generally at a higher temperature than that of the surrounding air. Flash gas is therefore not likely to occur.

Static loss refers to the pressure difference that exists between the bottom and the top of a liquid filled pipe due to the weight of liquid in the line. Static loss is a frequent cause of the creation of pre-expansion flash gas. Friction and static losses do occur in properly sized lines.

For example, a standing column of liquid R-22 at 100°F and 210.6 psig, due to its weight, exerts a pressure of approximately 0.50 psi for every foot of height of the column (R-12 exerts 0.55 psi per foot of height). The pressure at the bottom of a 10 ft column of R-22 is 5 psi greater than the pressure at the top of the column. Conversely, for every 10 ft that R-22 is lifted in a vertical riser, the pressure at the top of the column is reduced by 5 psi.

Avoid, where possible, installations placing the condensing unit low and the evaporator high. An example is a condensing unit on grade and the air handler in the attic of a multistory dwelling. Flash gas is likely to be a problem.

If you cannot avoid this situation, provide liquid subcooling by stacking the liquid and suction lines together and insulating both lines inside a common wrap to promote heat exchange from the liquid to the suction line.

Figure 5-6-19 shows both a liquid lift and a liquid drop. The liquid lift is the condition where flash gas can exist at the top of the column, due to the drop in pressure.

Assume that the liquid R-22 leaves the condenser at standard conditions of 105°F condensing temperature with 3°F of subcooling. The pressure leaving the condenser would be 210.8 psig. Therefore, considering the 3°F of subcooling, the liquid leaves the condenser at 102°F. If the liquid line riser has a 30 ft lift, the pressure at the top of the riser would be 15 psi less due to the weight of the refrigerant. Therefore, the final pressure would be:

$$210.8 \text{ psig} - 15 \text{ psig} = 195.8 \text{ psig}$$
$$195.8 \text{ psig} - 6 \text{ psig line loss} = 189.8 \text{ psig}$$

This will produce a boiling point of 97.86°F. Since the liquid left the condenser at 102°F, enough liquid will vaporize to cool the remaining liquid to 97.9°F. To avoid the formation of flash gas in this riser, liquid subcooling leaving the condenser of 8°F or more is required instead of 3°F to compensate for lift and pressure drop in the line. This can be

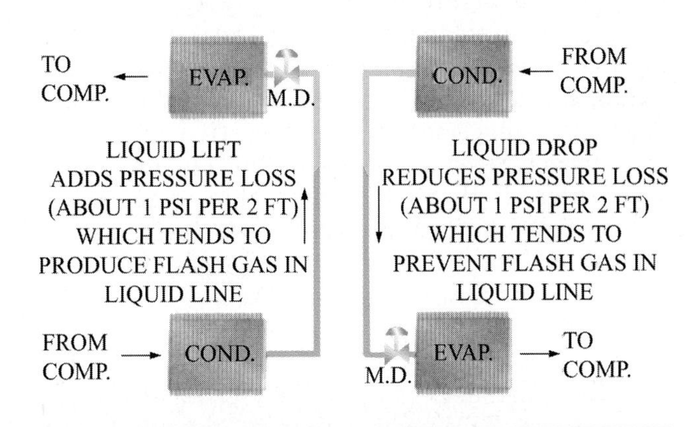

Figure 5-6-19 Liquid line static head with lift or drop conditions.

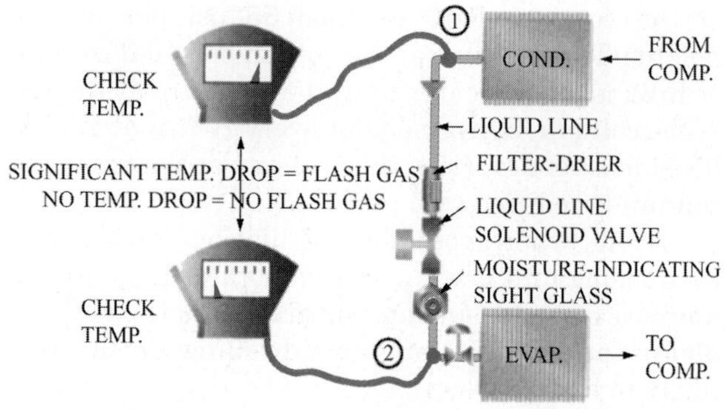

TEMPERATURE CHANGE = FLASHING IN LIQUID LINE

CHECK TEMP.
FROM COMP.
COND.
LIQUID LINE
SIGNIFICANT TEMP. DROP = FLASH GAS
NO TEMP. DROP = NO FLASH GAS
FILTER-DRIER
LIQUID LINE SOLENOID VALVE
MOISTURE-INDICATING SIGHT GLASS
CHECK TEMP.
EVAP.
TO COMP.

Figure 5-6-20 Liquid line piping showing test for flash gas.

accomplished by the use of a liquid-to-suction-line heat exchanger.

Again, assume that the same conditions are applied to a typical air conditioning unit that works at peak efficiency with 19°F of subcooling. Instead of leaving the condenser at 102°F the liquid would leave at 86°F. Again, assuming a 6 psig line loss for a properly sized line, the 30 ft of vertical lift would not produce flash gas. The vertical lift could increase to 105 ft before flash gas would occur.

The 19°F of subcooling assumes that a water cooled condenser, rather than an air cooled condenser, is used. It would not be practical on a hot day to obtain 86°F subcooled liquid using an air cooled condenser.

This demonstrates the value of a proper amount of subcooling. It not only provides latitude for the designer when laying out the liquid line, but the proper amount of subcooling provides the maximum system capacity at the lowest operating cost.

Locating Flash Gas

Referring to Figure 5-6-20, a method of spotting flash gas is to compare the temperature of the liquid line where a solid steam of liquid exists to the temperature of the liquid line at a point where flash gas is suspected. A typical location for the second thermometer would be at the entrance to a TXV valve. If there is a noticeable temperature drop at point 2, flash gas could be present. An exception would be where the presence of gas binding and low liquid flow would cause the liquid line to be warm at the TXV valve and mask the presence of the flash gas.

Some installers place a sight glass at the entrance to the TXV valve. The sight glass will show bubbles if there is flash gas in the line. This is a highly recommended procedure.

Liquid Line Insulation

Normally, no insulation is used on the liquid line because the greater the heat loss from the line, the lower the temperature of the liquid entering the pressure-reducing device, and the less system flash gas produced. However, when the liquid line runs through a hot space such as an attic, insulation of the liquid line may be required to prevent boiling from occurring prior to the expansion valve.

Insulation on the liquid line of a computer room air conditioning system is desirable. The liquid can get quite cold in winter and lines running through humidified space to the air conditioning unit will sweat and drip if not covered.

Condenser Drain Line

The line between a condenser and a liquid receiver, called the condenser drain line, must be carefully sized. Although it is almost impossible to oversize the line (0 psig pressure drop is desirable), undersizing or installing too long a line is to be avoided. An undersized or excessively long line can restrict the flow of refrigerant to the TXV valve to the extent that some refrigerant will be held back in the condenser. The residue reduces the effective condenser surface and reduces condenser capacity. This causes the head pressure to rise, decreasing the overall system capacity. At the same time, power requirements and operating costs increase.

For proper operation, the length of the line between the condenser and receiver must be kept as short as possible, preferably not more than 18 in. The top of the receiver should be level with or (preferably) below the bottom of the condenser. This eliminates the need for the liquid to be forced uphill to enter the receiver. Except in very special applications, the condenser drain line is not insulated.

Insulation Summary

The insulation needs of the piping are summarized in Table 5-6-1.

Generally the liquid line is not insulated. There are exceptions, however. If a liquid to suction line heat exchanger is used to avoid losing the subcooling, it is desirable to insulate it.

It is always good practice to insulate the suction line since it is usually lower in temperature than the surrounding air and can condense moisture. Insulation should be applied and thoroughly sealed with a good moisture barrier to prevent condensation on the outside of the pipe or in the insulation.

The hot gas line is seldom insulated since any heat in the hot gas that is rejected benefits the system, reducing the amount of heat the condenser must reject.

TABLE 5-6-1 Refrigerant Lines and Insulation

Refrigerant Line Type	Type of Insulation	
	Thermal*	Vapor Barrier
Liquid Line	Sometimes on heat pumps in cold climates and on AC's in very hot attics.	
Vapor Line	Always	Always
Hot Gas Line	Sometimes on heat pumps	
Condensate Line	Sometimes on rooftop installations	

*Sometimes refrigerant lines are insulated to reduce the transmission of sound from the lines to the occupied area.

The condenser drain line is usually not insulated since any loss of heat is desirable.

PIPING SUPPORTS

All piping must be properly supported (see Figure 5-6-21). The supports must allow for the expansion and contraction of the pipe. The recommended allowance is ¾ in movement per 100 ft of pipe. Hangers should be placed not more than 10 ft apart. They should always be placed near a bend in the piping, preferably on the longest straight connection to the bend. Where the pipe is insulated the hanger must have sufficient width not to crush the insulation. Sheet metal saddles or dense insulation block inserts (on large piping) will serve this purpose.

VIBRATION DAMPENERS

It is usually desirable to isolate vibrating equipment to reduce noise and to prevent damage to the piping or other equipment. With soft copper tubing, loops or a coil of tubing can be connected to the moving part. For hard copper tubing a similar dampening effect can be produced by running the suction and discharge lines 15 times the pipe diameter in each of two or three directions before securing the pipe hanger. This will provide some give to the piping without undue strain.

For larger equipment, rubber-in-sheaf or spring vibration eliminator mounts and flexible piping connections can be provided for the compressor to supply the necessary isolation. Concrete inertia blocks are sometimes incorporated into the base. Piping and electrical lines must be securely anchored beyond the isolators to be effective.

Figure 5-6-21 Proper support for piping.

The Hot Gas Muffler

The pulsations from a compressor, usually a reciprocating type, can cause serious vibrations and noise in the hot gas line. This can be most noticeable where a remote air cooled condenser requires a long vertical hot gas line. Usually the larger the compressor, the more noticeable the pulsations, although this is dependent on speed and number of cylinders.

The best way to solve this problem is to install a hot gas muffler in the compressor discharge line, as shown in Figure 5-6-22. The hot gas muffler should be placed in a vertical position, so that it does not trap oil, as close to the compressor as possible, and securely mounted to the compressor. This usually destroys the resonance that the compressor has built up in the hot gas line. If this does not help, it may be necessary to enlarge the discharge line and relocate it, to destroy the resonance pattern.

Off Cycle Protection

Using a vertical discharge riser directly from the compressor could permit oil and liquid refrigerant to

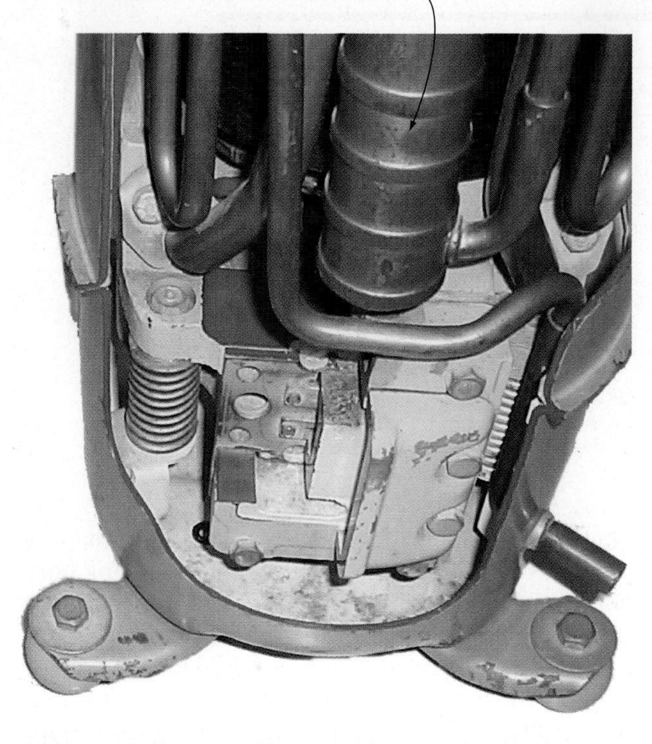

Figure 5-6-22 Using a hot-gas muffler to dampen discharge gas pulsations.

drain back down (or migrate) into the compressor head during the OFF cycle. This can break valves on a reciprocating compressor on startup. To prevent this, a discharge loop can be placed in the hot gas line with an optional check valve. Modest quantities of oil and liquid refrigerant can accumulate in this trap when the compressor shuts down and will be dissipated on startup without damage to the compressor.

UNIT 6—REVIEW QUESTIONS

1. What are the two major functions of the piping that connects the major components?
2. What problems can be caused by a piping system that is not correctly constructed to allow oil return?
3. List the five basic rules to keep in mind when piping a system.
4. The ________ must be carefully designed to ensure a uniform return of dry refrigerant gas as well as sufficient oil to the compressor.
5. What is the objective in sizing condensate lines?
6. What is an alternative to sizing the riser for minimum capacity?
7. What does sizing of the suction line depend on?
8. When properly charged with refrigerant, the capillary tube system has less chance of ________.
9. What does insulation on a suction line eliminate?
10. What does static loss refer to?
11. What is a method of spotting flash gas?
12. The best way to solve the problem of vibrations and noise from a compressor is to install a ________ in the discharge line.

TECH TALK

Tech 1. I picked up a tubing cutter the other day. A plumber threw it away, still works.

Tech 2. What did he throw it away for? Cutter wheel dull?

Tech 1. No. It's got a good sharp cutter, but when you start cutting, it kind of walks down the copper a little ways. But if you put enough pressure on it, you can get it to go ahead and track in and cut.

Tech 2. You're not using that on jobs are you?

Tech 1. Sure, I use it all the time. It's a nice cutter except for the fact that you have to make sure it doesn't spiral down the copper too far.

Tech 2. It could be bent or need oiling. That is why it walks. But did you know that copper is notch sensitive, like glass? You know, if you scratch it real deep and make a groove and then bend it, it will break right along that groove line just like when you scratch a piece of glass to break it.

Tech 1. You mean that those fittings I put together might someday come apart?

Tech 2. That's right. And when you're using R-410A, any deep groove on the outside of the copper can cause a catastrophic failure in that refrigerant line set, because R-410A actually causes the copper pipe to expand slightly at the high pressures. So every time it cycles off and on, that crack can open up just a little.

Tech 1. Gosh, I didn't know that. What should I do?

Tech 2. If you can go back and clean up those scratches with a little sanding using some sand cloth and they are not grooves and not very deep, you're OK. But if the grooves are deep at all, you are going to have to take that fitting apart and redo it to make it safe.

Tech 1. I guess that tubing cutter wasn't such a good deal after all, was it?

Tech 2. No it wasn't.

6

Refrigerants

UNIT 1

Refrigerants

OBJECTIVES: In this unit the reader will learn about refrigerants including:

- Definition of a Refrigerant
- Number Designation
- Chemical Composition
- Refrigerant Requirements
- Changing Refrigerants
- Guidelines for Selecting Refrigerants
- Properties of Common Refrigerants
- Pressure Temperature Relationships
- Contaminants in the Refrigerant
- Refrigerant Cylinders
- Safety

DEFINITION OF A REFRIGERANT

A refrigerant is a medium (fluid) for heat transfer, used in a refrigeration system to pick up heat by evaporating at a low temperature and pressure, and to give up heat upon condensing at a higher temperature and pressure.

Figure 6-1-1 shows a typical room cooler using the refrigerant R-22. Note that the evaporator of this unit is in the room being cooled, where the refrigerant picks up heat. The condenser is extended outside the building where the refrigerant gives up heat. The use of the compressor and the connecting piping is to form a closed system so that the refrigerant can be reused again and again.

The refrigerant characteristics for this application must be such that the evaporator boiling temperature is below room temperature and the condensing temperature is above the outside air temperature. These conditions are necessary so that proper heat transfer can take place to provide room cooling.

Many different substances can be used for refrigerants. Under certain conditions even water, R-718 can be used as a refrigerant. Referring to Figure 6-1-2, at standard conditions water boils at 212°F (100°C). In a vacuum of 29.75 in Hg, water boils at 40°F (4.4°C). Water is an undesirable refrigerant for most applications because of the expense in producing and maintaining such a low vacuum.

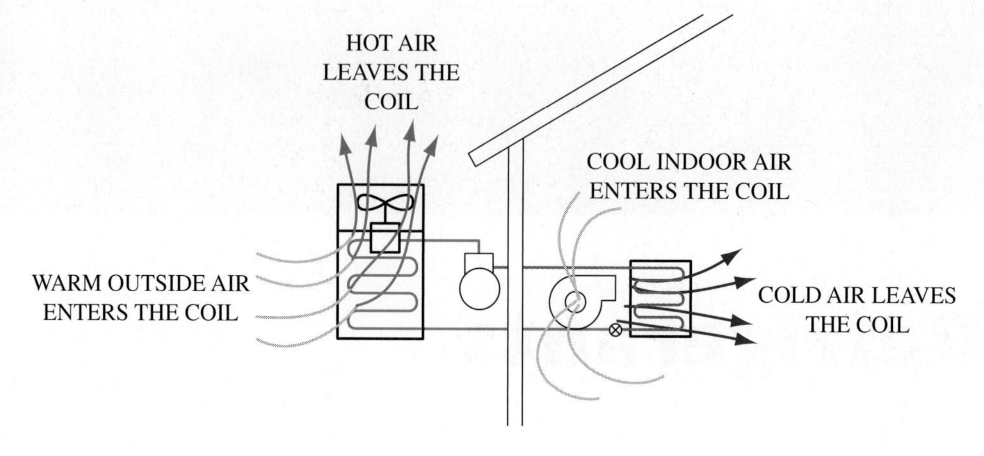

Figure 6-1-1 Typical refrigeration system showing the use of refrigerant to transfer heat from the inside to the outside.

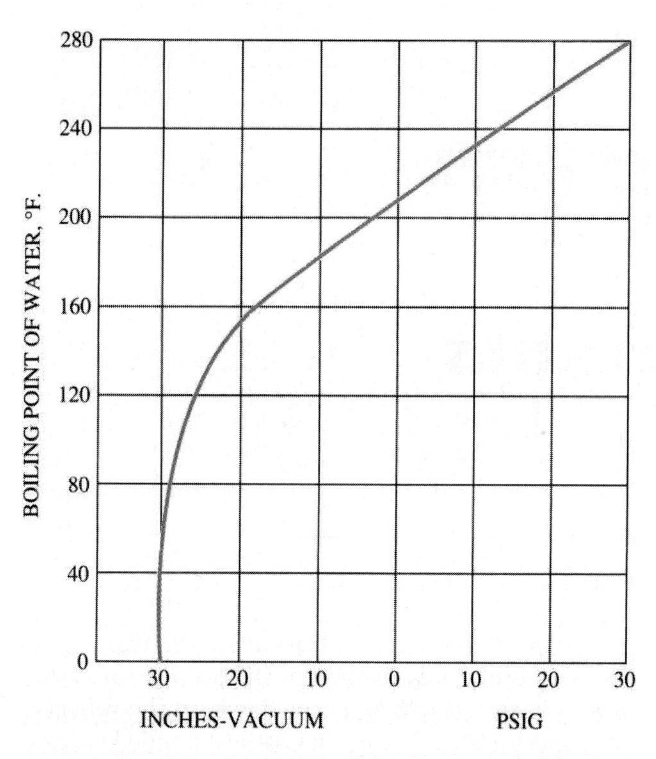

Figure 6-1-2 Pressure-temperature curve for water.

NUMBER DESIGNATION

Refrigerants are identified by number, preceded by the letter "R" (Refrigerant). This number designation has been established by the American Society of Heating, Refrigeration and Air Conditioning Engineers (ASHRAE) and is used throughout the industry.

Certain designations have been provided for the refrigerants in abbreviated form, to indicate the chemical composition, as well as to relate to the ozone depletion potential (ODP) for the refrigerant. ODP is a rating with the range of 1–0.

CHEMICAL COMPOSITION OF REFRIGERANTS

In Table 6-1-1, the common refrigerants are listed, along with their chemical name and ODP. The prefix "CFC" refers to the family of refrigerants containing chlorine, fluorine, and carbon. Compounds that also contain hydrogen precede the abbreviation with the letter "H" to signify an increased deterioration potential before reaching the stratosphere—for example, the prefix "HCFC." The "FC" family does not contain chlorine and can also be preceded with an "H," as in "HFC."

Note that the CFC refrigerants have a high ODP, the HCFC refrigerants have a low ODP, and the HFC refrigerants have a zero ODP.

Certain refrigerants, which are mixtures, use the R number and also have a designation that shows the constituents. For example, R-502 is composed of HCFC-22/115. The ODP given is a 0.28 rating.

NOTE: For decades there were a limited number of refrigerants available to the HVAC/R industry. However, following environmental concerns over the ozone depletion and greenhouse effect of many popular refrigerants, scientists began working on new refrigerant formulas. Today there are more types of refrigerants available than ever before in spite of the fact that some of the common refrigerants of the past have been phased out. Each of these new refrigerants offers the HVAC/R industry new opportunities to take advantage of the refrigerant's characteristics that make it most suitable for specific applications. For example, R-22 has a higher latent heat vaporization than R-410a which means that it is more efficient in heat removal for air conditioning applications.

TABLE 6-1-1 Common Refrigerants with Their Ozone Depletion Potential and Global Warming Potential

Refrigerant	Chemical Composition	Ozone Depletion Potential (ODP)	Global Warming Potential (GWP)
R-744	CO_2	0.0	1.0 (base)
R-11	CFC	1.0 (base)	1.30
R-12	CFC	0.93	3.70
R-22	HCFC	0.05	0.57
R-113	CFC	0.83	1.90
R-114	CFC	0.71	6.40
R-115	CFC	0.38	13.80
R-123	HCFC	0.02	0.28
R-125	HFC	0.0	
R-134a	HFC	0.0	0.40
R-401A	HCFC	0.03	
R-401B	HCFC	0.035	
R-402A	HCFC	0.03	
R-402B	HCFC	0.02	
R-507A	HFC	0.0	

Azeotropic, Zeotropic, and Near-Azeotropic Blends

A number of refrigerants, such as R-502, are made up of blends or chemically prepared mixtures of refrigerants.

The azeotropic blends consist of multiple components of different volumes that, when used in refrigeration cycles, do not change volumetric composition or saturation temperature as they evaporate or condense at constant pressure. These refrigerants have numbers in the 500 series, such as R-502.

The zeotropic blends consist of multiple components of different volumes that, when used in refrigeration cycles, change volumetric composition and saturation temperatures as they evaporate or condense at constant pressure. This process results in a slightly different temperature and pressure as the blend separates in the evaporator or combines in the condenser. This slight change is referred to as "glide."

The near-azeotropic blends consist of multiple components of different volumes that, when used in refrigeration cycles, change volumetric composition and saturation temperatures just like zeotropic blends but the glide is so small that it is not detectable in the field. Because the glide is so small, some manufacturers refer to R-410A as an azeotropic refrigerant blend.

REFRIGERANT REQUIREMENTS

There are many properties that are characteristic of a good refrigerant. An entire group of features needs to be considered. In selecting a refrigerant there are usually some compromises that need to be considered due to conflicts that arise between desirable characteristics. For example, ammonia refrigerant has a high performance factor in terms of net refrigerating effect (Btu/lb) but should not be selected for use in systems with a proximity to people due to its potential toxicity and flammability.

Performance

The refrigerant needs to operate within the temperature and pressure range required to perform the task assigned to it. For example, consider the requirements of a reach-in refrigerator. The box must be capable of maintaining temperatures in the range of 35° to 40°F. The refrigerant evaporating temperature needs to be in the range of 25° to 30°F (−3.9° to 1.1°C), which is 15°F (8.3°C) below box temperature. Using an air-cooled condenser, the condensing temperature typically will be in the range of 120° to 130°F (48.9° to 54.5°C.), which is 30°F (16.7°C) higher than ambient air. At these temperatures, the pressures must be in the range obtainable using a standard compressor.

Let us assume a compressor is available that can operate at pressures not to exceed 250 psig (1823.8 kPa). If R-134a is selected, the evaporating pressure would be 36.7 to 40.7 psig (261.4 to 281.3 kPa), with a condensing pressure of 185.8 to 199.2 psig (1281.6 to 1473.3 kPa). This is satisfactory.

Safety

The refrigerant should be safe to use. It should be nonpoisonous, nonexplosive, nonflammable, and nontoxic.

The ASHRAE Safety Code For Mechanical Refrigeration classifies the refrigerants with respect to their toxicity and flammability. The refrigerants are divided into groups: A for low toxicity and B for elevated toxicity. The number following the letter indicates the flammability of the refrigerant. The lower the number, the lower the flammability. A refrigerant designated as A1 has the lowest toxicity and the lowest flammability. Table 6-1-2 shows this information in diagram form.

Environmental Impact

The refrigerant should be free of any chemicals that when released to the atmosphere would damage the ozone layer (ozone depletion factor [ODP] = 0).

Efficiency

The refrigerant should have a high latent heat (liquid to vapor phase) and a low vapor volume per pound. Under these conditions, less refrigerant needs to be pumped, resulting in a smaller compressor and reduced piping sizes.

TABLE 6-1-2 Safety Classifications for Refrigerants

Flammability Classification	Toxicity Group: A, Lower Toxicity	Toxicity Group: B, Higher Toxicity
3 Higher Flammability	A3 R50 R290 R600	B3
2 Lower Flammability	A2 R42b R52b	B2 R717
1 No Flame Propagation	A1 R11 R12 R 22 R113 R114 R134a R500 R502	B1 R123

Stability

Both the liquid and the vapor should be stable so that the refrigerant will not decompose at normal operating temperatures and pressures.

Compatibility

The refrigerant must not react with or deteriorate materials it comes in contact with during the operation. These include metallic compressor parts, gaskets, O rings, seals, motor insulation and windings, piping, and condenser and evaporator heat transfer surfaces.

Leaks

Refrigerant leaks should be easy to detect and locate.

SERVICE TIP

Old style leak detectors can still be useful in detecting refrigerants that contain fluorocarbons. Newer leak detectors can identify leaks in any refrigerant. These leak detectors take advantage of the latest advances in electronics which allows them to be sensitive to the presence of the other refrigerants.

Price

The refrigerant should be reasonably priced in the quantities required for normal systems. Rating is in dollars per pound.

Lifetime of the Refrigerant

The lifetime of the refrigerant should be equal to or greater than the lifetime of the equipment it serves.

CHANGING REFRIGERANTS

In view of the restrictions placed on the use of many of the common refrigerants, the refrigerant manufacturers have been spending millions of dollars to find suitable alternates that meet the new requirements. Two of these new refrigerants now being used by some equipment manufacturers are R-123 (HCFC-123) and R-134a (HFC-134a). Table 6-1-3 gives a comparison of some of the characteristics of these alternate refrigerants with the common refrigerants they are designed to replace. R-123 is a replacement for R-11, and R-134a is a replacement for R-12.

R-22 is an HCFC type refrigerant, with an ODP of 0.05, and according to the present schedule, production will continue until the year 2030.

TABLE 6-1-3 Significant New Alternatives Policy (SNAP) Established by the EPA Lists the Existing Refrigerants and the Possible Alternative Replacement Refrigerants

Existing Refrigerants	Alternative Refrigerants
R-11	R-123
R-12	R-134a, 409A
R-22	R-410A
R-113	R-134a
R-500	R-401B
R-502	R-408A, R-404A

R-410A is an HFC near-azeotropic refrigerant that is a mixture of 60% HFC 32 and 40% HFC 125. It is a high pressure refrigerant designed to be used in place of R-22. It has an operating pressure nearly twice that of R-22 and is, therefore, not a drop-in replacement for R-22 and cannot be used in R-22 equipment. R-410A has an ODP of 0 and a GWP of 0.47.

SAFETY TIP

In the air conditioning refrigeration industry there are no distinctions between the service valve connections for the different refrigerants used. This differs from the automotive industry where different refrigerants have different sized access ports to prevent the accidental use of an improper gauge set on an unapproved refrigerant. In air conditioning it is strictly up to the technician to make certain before attaching the gauge set to a system that it is the proper gauge set. This is significant when working around systems that may contain R-410a because of its significantly higher pressure than most refrigerants used in the industry. Although the gauges will not explode, they can be overstressed to the point that they will not work properly when attached to the lower pressure refrigerants they were intended to be used with.

Ammonia is an inorganic refrigerant with an ODP of 0.0, which has limited use due to its toxic and flammable characteristics.

The absorption system refrigerant has an ODP of 0, and is limited to applications of absorption refrigeration. In view of the restrictions on the use of the formerly common refrigerants R-11 and R-12, it is believed that the market for ammonia and absorption systems will be increased.

R-123 has a satisfactory efficiency, a low global warming potential (GWP), a better ODP than R-22, and its cost is in the middle range. It has a B1 safety rating and, therefore, manufacturers that use it require the use of leak sensors and increased provision for ventilation.

R-134a has a favorable efficiency, an improved GWP compared to R-12, and a safety rating of A1. A sizable cost reduction is anticipated when larger quantities of R-134a are produced.

GUIDELINES FOR SELECTING REFRIGERANTS

For a new system, select equipment that uses a refrigerant that meets as many of the characteristics of good refrigerants as possible.

For example, a new reach-in refrigerator is being purchased. Previously this type of system used R-12 refrigerant, which is being phased out. Which one of the new refrigerants should it use?

From a performance standpoint the best selection would be HFC-134a. This refrigerant has an ODP of 0 and meets all safety requirements.

For an existing system, if a system is not leaking and is operating satisfactorily there is no technical reason to replace CFC refrigerants. In fact, it may void the U.L. listing of the unit.

For some types of service, such as supermarkets, the owner (or the manufacturer) is concerned about the availability of replacement refrigerant to maintain essential refrigeration. Under these conditions it may be advisable to replace the existing "CFC" refrigerant with a suitable substitute.

Whenever a replacement refrigerant is selected, either an interim HCFC refrigerant or a more permanent HFC refrigerant can be considered. The best selection is usually the HFC replacement, if it is available; however, the expected lifetime of the equipment is within the availability time of the interim refrigerant, the HCFC may be the best choice.

The lubrication requirements also need to be considered because they affect the cost of the changeover. If an HCFC refrigerant is selected, an alkyl benzone (AB) oil can be used.

If an HFC refrigerant is used, the lubricant must be a polyol ester (POE) type. They absorb moisture from the air so must be handled and packaged with much more care than conventional oils. As much of the original lubricant as possible must be removed before the new lubricant can be put in the system.

As an example, assume that the refrigerant in a commercial freezer using R-502 is being replaced with a new refrigerant.

If HCFC-402A is the new refrigerant, an AB lubricant can be used with one oil change. The ODP of 0.02, which is satisfactory for an interim refrigerant.

If HFC-404A is the new refrigerant, it would be a permanent replacement. A POE lubricant can be used with one oil change. The ODP of 0.

PROPERTIES OF COMMON REFRIGERANTS

The following information provides descriptions and tables for the properties of common refrigerants.

R-22

This refrigerant has been popular for use in domestic freezers and packaged air conditioning units. Fortunately, it has only a small effect on the ozone layer, with an OPF of 0.05. It has a low phase-out priority compared to R-11 or R-12.

At an evaporating temperature of 5°F (−15°C), the saturated vapor pressure is 28.19 psig (296 kPa) and at a condensing temperature of 86°F (30°C) the saturated vapor pressure is 158.17 psig (1190 kPa). The latent heat of vaporization at 5°F (−15°C) is 93.2 Btu/lb (217 J/g). It has a boiling temperature of −41°F (−41°C) at atmospheric pressure.

R-22 will absorb more water than R-12, and therefore a larger drier is needed. Oil will mix with the R-22 at most operating temperatures; however, at temperatures somewhat below −40°F the oil separates out.

R-123

This refrigerant was designed to replace R-11. The pressure-temperature curves show close performance characteristics between these two refrigerants.

At an evaporating temperature of 5°F (−15°C) the saturated vapor pressure is 2.30 psia (15.9 kPa) and at a condensing temperature of 86°F (30°C) saturated vapor pressure is 15.9 psia (109.60 kPa). The latent heat of vaporization at 5°F (−15°C) is 78.9 Btu/lb (83.5 J/g). It has a boiling temperature of 82.1°F (27.8°C) at atmospheric pressure.

R-123 has a safety group classification of B1, making it objectionable from the standpoint of toxicity. As a result some service companies refuse to use it. As yet no new universally acceptable refrigerant has been developed to replace R-11.

The ODP for R-123 is 0.02, which is low. Based on the need, therefore, it can serve as an interim replacement for R-11.

R-134a

This is a chlorine-free fluorinated refrigerant designed to replace R-12. It has an ODP of 0.0, and therefore meets the necessary environmental requirements. It is suitable for use in domestic refrigerators, automotive air conditioning, and medium- and high-temperature commercial applications. It is a suitable replacement for R-12 wherever the evaporation temperature is −10°F or higher.

At an evaporating temperature of 5°F (−15°C) the saturated vapor pressure is 9.1 psig (62.7 kPa) and at a condensing temperature of 86°F (30°C) saturated vapor pressure is 98.1 psig (676.4 kPa). The latent heat of vaporization at 5°F (−15°C) is 90.1 Btu/lb (209.6 J/g). It has a boiling temperature of −15.7°F (−26.5°C) at atmospheric pressure.

R-502

This refrigerant is an azeotropic mixture (blend of two or more component refrigerants whose equilibrium vapor phase and liquid phase compositions are the same at a given pressure). It is made up of 48.8% R-22 and 51.2% R-115. It is primarily used for low temperature refrigeration where the evaporator temperature is between 0 and 60°F (−18 to 51°C). It is a safe refrigerant, nontoxic and nonflammable. It has an ODP of 0.28 and, therefore, has a limited period of use.

At an evaporating temperature of 5°F (−15°C) the saturated vapor pressure is 35.8 psig (246.8 kPa) and at a condensing temperature of 86°F (30°C) saturated vapor pressure is 177 psig (1220 kPa). The latent heat of vaporization at 5°F (−15°C) is 67.3 Btu/lb (157.0 J/g). It has a boiling temperature of −50°F (−46°C) at atmospheric pressure.

R-717

Ammonia is one of the oldest refrigerants in use today. It has a much higher latent heat of evaporation than the other common refrigerants. At 5° the value is 565 Btu/lb (1314 J/g). This means that smaller piping can be used. It is corrosive to copper, but not to iron, steel, or aluminum. It therefore requires an all-iron or steel or aluminum system including the compressor, condenser, evaporator, controls, and piping. It is not destructive to the ozone layer. Its greatest defect is its toxicity and flammability. It has a safety rating of B2.

SAFETY TIP

When working around systems that contain ammonia as the refrigerant you must have a self contained breathing apparatus available for each worker in that area. It is not possible to simply hold your breath and run from an area if an ammonia leak occurs. Part of the problem is the reaction the lungs have when a small quantity of ammonia is inhaled. Even a small amount of concentrated ammonia can damage lungs. Exposure to concentrated ammonia can temporarily blind you. The only way to prevent these problems is to have the appropriate equipment ready and available for emergency use.

Ammonia does not absorb oil. It requires the use of an oil separator in the discharge line. Oil that is carried over to the evaporator must be drained off. It is extremely soluble in water.

Due to ammonia's safety hazards it is not found in appliances or normal comfort cooling applications. Typically it is used in large commercial or industrial applications where its operating efficiencies (lower horsepower per ton) are important and where plant engineers are available to operate the system. Examples would be dairies, ice cream plants, and large cold storage facilities.

Due to its toxicity, an operator must take special precautions to limit the quantity inhaled. Leaks are detected by use of litmus paper which changes color in the presence of ammonia, or a sulfur candle which creates smoke in contact with ammonia. Technicians must also take care to avoid ammonia contact with the skin.

At an evaporating temperature of 5°F (−15°C) the saturated vapor pressure is 19.6 psig (236 kPa) and a condensing temperature of 86°F (30°C) the saturated vapor pressure is 155 psig (1170 kPa). The latent heat of vaporization at 5°F (−15°C) is 565 Btu/lb (1314 J/g). It has a boiling temperature of −28°F (−33°C) at atmospheric pressure.

PRESSURE-TEMPERATURE RELATIONSHIPS

Figure 6-1-3 shows a series of pressure-temperature curves for some commonly used refrigerants. This chart shows the refrigerants in a saturated condition. One very good use of these curves is to observe the refrigerants that are closely matched to determine the suitability of substituting one for the other.

For example, note that R-502 and R-22 are closely matched. Since R-22 has only about 7% of the ozone depleting capability of R-502, it can be used as an interim replacement.

CONTAMINANTS IN THE REFRIGERANT

As the system operates, certain impurities tend to collect in the refrigerant, including water/vapor, acid, high-boiling-point residues, and particulates. When a refrigerant is recycled, many of these impurities can be removed. In order to resell the refrigerant it must be reclaimed by a certified source and brought back to the original level of purity indicated by ARI Standard 700-88 (or update) which gives maximum allowable contamination levels, shown in Table 6-1-4a and b.

SERVICE TIP

The two most common contaminants in refrigerant are air and water. Air cannot be condensed at refrigerant condensing temperatures and can be removed from the refrigeration system by recovering only liquid refrigerant from the system and then recovering the vapor from the system. The vapor recovery will contain the air since air does not dissolve in the liquid refrigerant. Once this has been done then the liquid, which will contain no air, can be recharged into the system with a small amount of new refrigerant added to top off the charge. Water and contaminant refrigerant can be removed with filter driers while the system is charged. Large quantities of moisture in the system may require several changes of the filter drier before the system is completely dry. Test kits are available to check the moisture level that remains in the system.

REFRIGERANT CYLINDERS

Figure 6-1-4 shows cylinders for various refrigerants. A disposable cylinder should never be refilled. A hand valve may be clamped to the top of the cylinder for shutoff.

Refillable (service) cylinders should only be used if adequate equipment is available for determining the amount of refrigerant transferred. No cylinders should be filled with liquid to more than 80% of its volume at 60°F (15°C). A fill limit device is often furnished with these cylinders, particularly when they are used for refrigerant recovery. This device automatically cuts out the flow of refrigerant into the cylinder when the fill level is reached. These cylinders may also be supplied with two valve stems to permit drawing either liquid or vapor from the cylinder.

Storage cylinders are used where large quantities of refrigerant are purchased for transfer to smaller cylinders. This should only be done when facilities are available for following the proper procedures.

Cylinders used for refrigerant must have a Department of Transportation (DOT) stamp of approval. Cylinders are made of steel or aluminum. DOT requires that cylinders that have contained a corrosive refrigerant must be tested every five years. Cylinders containing noncorrosive refrigerants must be tested every five years. Any cylinders over a 4½ in diameter or over 12 in long must have some type of pressure relief device. The maximum storage temperature of cylinders is 125°F

Refrigerant cylinders are color coded in accordance with the schedule in Table 6-1-5.

Refrigerant storage cylinders have a valve that uses two valve stems. One is used when liquid is drawn from the cylinder and the other for vapor. On the disposable cylinder the valve screws into threaded fittings that are welded to the top of the cylinder.

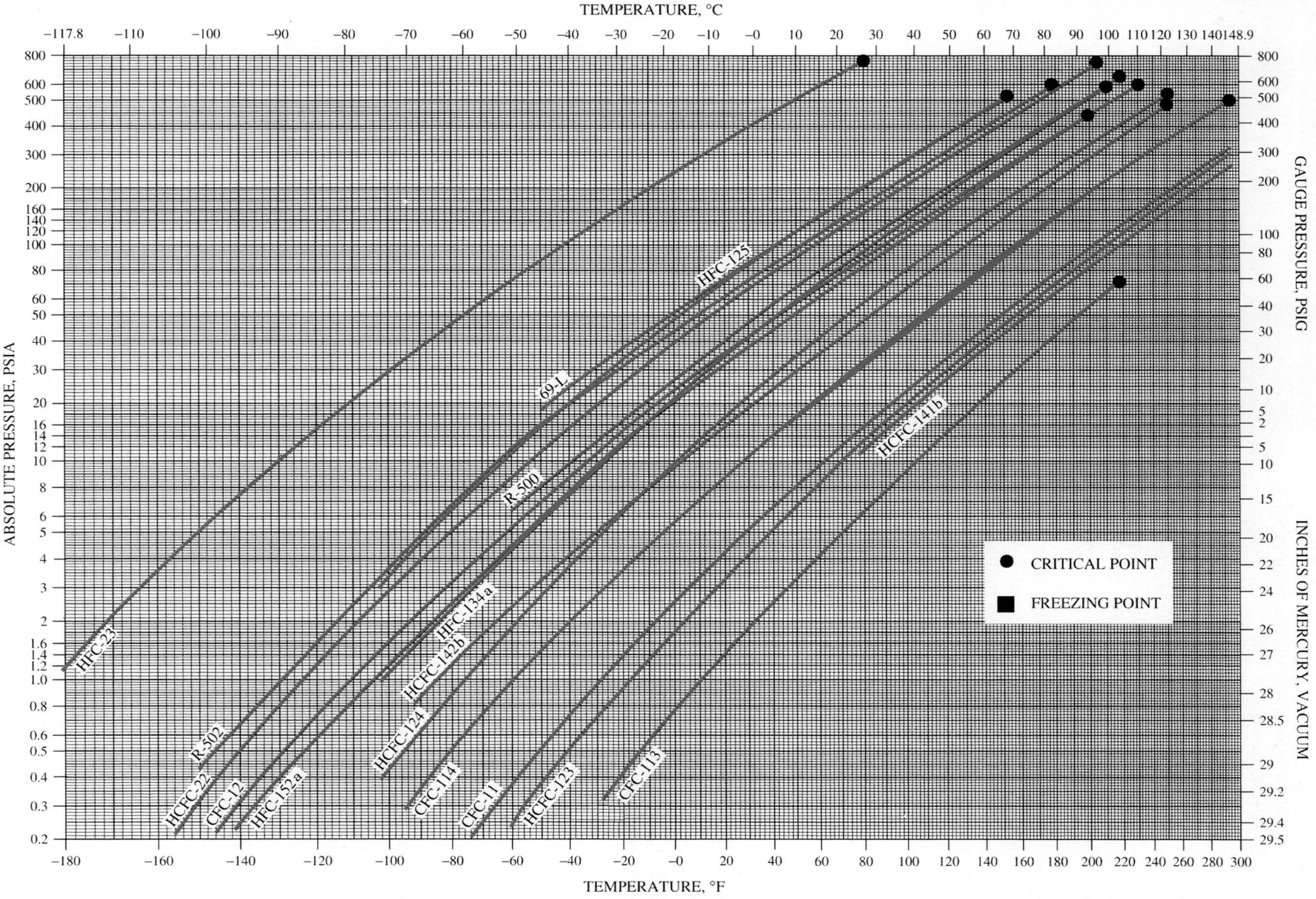

Figure 6-1-3 Pressure-temperature curves.

TABLE 6-1-4a ARI Refrigerant Standards #700-88 for Common Single Component Refrigerants. Maximum Contamination Levels

	Reporting Units	Reference (Subclause)	R-11	R-12	R-13	R-22	R-23	R-113	R-114	R-123	R-124	R-125	R-134a
Characteristics*													
Boiling Point*	°F @ 1.00 atm		74.9	−21.6	−114.6	−41.4	−115.7	117.6	38.8	82.6	12.2	−55.3	−15.1
	°C @ 1.00 atm	—	23.8	−29.8	−81.4	−40.8	−82.1	47.6	3.8	27.9	−11.0	−48.5	−26.2
Boiling Point Range*	K	—	0.3	0.3	0.5	0.3	0.5	0.3	0.3	0.3	0.3	0.3	0.3
Typical Isomer Content	by weight	—						0–1% R-113a	0–30% R-114a	0–8% R-123a	0–5% R-124a	N/A	0–5000 ppm R-134
Vapor Phase Contaminants													
Air and other non-condensibles	% by volume @ 25°C	5.9	N/A**	1.5	1.5	1.5	1.5	N/A**	1.5	N/A**	1.5	1.5	1.5
Liquid Phase Contaminants													
Water	ppm by weight	5.4	20	10	10	10	10	20	10	20	10	10	10
All other impurities including refrigerants	% by weight	5.10	0.50	0.50	0.50	0.50	0.50	0.50	0.50	0.50	0.50	0.50	0.50
High boiling residue	% by volume	5.7	0.01	0.01	0.05	0.01	0.01	0.01	0.01	0.01	0.01	0.01	0.01
Particulates/solids	Visually clean to pass	5.8	pass	pass	pass	pass	pass	pass	pass	pass	pass	pass	pass
Acidity	ppm by weight	5.6	1.0	1.0	1.0	1.0	1.0	1.0	1.0	1.0	1.0	1.0	1.0
Chlorides***	No visible turbidity	5.5	pass	pass	pass	pass	pass	pass	pass	pass	pass	pass	pass

*Boiling points and boiling point ranges, although not required, are provided for informational purposes.
**Since R-11, R-113 and R-123 have normal boiling points at or above room temperature, non-condensible determinations are not required for these refrigerants.
***Recognized chloride level for pass/fail is 3 ppm.

TABLE 6-1-4b ARI Refrigerant Standards #700-88 for Common Composition Refrigerants. Maximum Contamination Levels

	Reporting Units	Reference (Subclause)	R-401A	R-401B	R-402A	R-402B	R-500	R-502	R-503
Characteristics*									
Refrigerant Components			R-22/152a/124	R-22/152a/124	R-125/290/22	R-125/290/22	R-12/152a	R-22/115	R-23/13
Nominal Comp, weight %			53/13/34	61/11/28	60/2/38	38/2/60	73.8/26.2	48.8/51.2	40.1/59.9
Allowable comp, weight %			51–55/11.5–13.5/ 33–35	59–63/9.5–11.5/ 27–29	58–62/1–3/ 36–40	36–40/1–3/ 58–62	72.8–74.8/ 25.2–27.2	44.8–52.8/ 47.2–55.2	39–41/ 59–61
Boiling Point*	°F @ 1.00 atm		−27.6 to −16.0	−30.4 to −18.5	−56.5 to −52.9	−53.3 to −49.0			
	°C @ 1.00 atm		−33.4 to −26.6	−34.7 to −28.6	−49.1 to −47.2	−47.4 to −45.0	−33.5	−45.4	−88.7
Boiling Point Range*	K	—	—	—	—	—	0.5	0.5	0.5
Vapor Phase Contaminants									
Air and other non-condensibles	% by volume @ 25°C	5.9	1.5	1.5	1.5	1.5	1.5	1.5	1.5
Liquid Phase Contaminants									
Water	ppm by weight	5.4	10	10	10	10	10	10	10
All other impurities including refrigerants	% by weight	5.10	0.50	0.50	0.50	0.50	0.50	0.50	0.50
High boiling residue	% by volume	5.7	0.01	0.01	0.01	0.01	0.05	0.01	0.01
Particulates/solids	Visually clean to pass	5.8	pass	pass	pass	pass	pass	pass	pass
Acidity	ppm by weight	5.6	1.0	1.0	1.0	1.0	1.0	1.0	1.0
Chlorides**	No visible turbidity	5.5	pass	pass	pass	pass	pass	pass	pass

*Boiling points and boiling point ranges, although not required, are provided for informational purposes.
**Recognized chloride level for pass/fail is 3 ppm.

Figure 6-1-4 Refrigerant cylinders. (*Courtesy National Refrigerants, Inc.*)

SAFETY

Due to the potential dangers involved, it is extremely important that anyone handling refrigerants observe proper safety procedures.

Pressure

Most of the time the service person is dealing with confined liquids or vapors that have the capability of building up high pressures which have the potential of causing serious injury if carelessly handled. Precautions must be taken.

In storing refrigerant in approved cylinders, the liquid level must not exceed 80% of the volume of the container at 60°F (15°C). This allows space for the expansion of the refrigerant if heat is added.

TABLE 6-1-5 Characteristics of Some Common Refrigerants

Number	Chemical Composition	Cylinder Color	Refrigerant Name
R-11	CFC	Orange	Trichlorofluoromethane
R-12	CFC	White	Dichlorodifluoromethane
R-13	CFC	Light blue	Chlorotrifluoromethane
R-13B1	CFC	Coral	Bromotrifluoromethane
R-22	HCFC	Light green	Chlorodifluoromethane
R-23	HFC	Light gray	Trifluoromethane
R-113	CFC	Purple	Trichlorofluoromethane
R-114	CFC	Dark blue	Dichlorodifluoromethane
R-123	HCFC	Light gray	Dichlorotrifluoroethane
R-124	HCFC	Deep green	Chlorotrifluoromethane
R-125	HFC	Medium brown or tan	Pentafluoroethane
R-134a	HFC	Sky blue	Tetrafluoroethane
R-401A	Zeotropic (HCFC)	Coral red	R-22 + R-152a + R-124
R-401B	Zeotropic (HCFC)	Mustard yellow	R-22 + R-152a + R-124
R-401C	Zeotropic (HCFC)	Blue-green (aqua)	R-22 + R-152a + R-124
R-402A	Zeotropic (HCFC)	Pale brown	R-22 + R-125 + R-290
R-402B	Zeotropic (HCFC)	Green-brown	R-22 + R-125 + R-290
R-404A	Zeotropic (HCFC)	Orange	R-22 + R-143a + R-134a
R-406A	Zeotropic (HCFC)	Light gray-green	R-22 + R-142b + R-600a
R-407A	Zeotropic (HFC)	Bright green	R-32 + R-125 + R-134a
R-407B	Zeotropic (HFC)	Peach	R-32 + R-125 + R-134a
R-407C	Zeotropic (HFC)	Chocolate brown	R-32 + R-125 + R-134a
R-410A	Zeotropic (HFC)	Rose	R-32 + R-125
R-500	Azeotropic (CFC)	Yellow	Refrigerants 152a/12
R-502	Azeotropic (CFC)	Light purple	Refrigerants 22/115
R-503	Azeotropic (CFC)	Aquamarine	Refrigerants 23/13
R-507A	Azeotropic (HFC)	Teal	Refrigerants 125/143a
R-717	Inorganic Compound	Silver	Ammonia

Whenever a service person is called upon to examine a piece of refrigerating equipment, the pressures in the system should immediately be checked with suitable gauges. Depending on the R number of the refrigerant, an evaluation can be made of the condition of the system. In case there is any doubt of the type of refrigerant being used, test instruments are available to determine it. In no instance is it permissible to mix refrigerants.

Leaks

Refrigerants are heavier than air and, with the exception of ammonia and sulfur, have little or no odor, thus giving no warning of their presence. The air we breathe consists of about 21% oxygen and 78% nitrogen plus moisture vapor and traces of other gases such as carbon dioxide. If refrigeration vapors, from an accidental discharge or rupture, displace air in a confined space, pit, or low area, there can be great danger of asphyxiation in a short time to unsuspecting persons occupying or entering the space, causing loss of consciousness and possibly death. The risk is real and the technician must be aware of the hazard. The Safety Code for Mechanical Refrigeration requires oxygen depletion monitors for potentially hazardous conditions.

Charging

Whenever the refrigerant system is being serviced, especially when charging or evacuating, the service person should wear goggles and gloves. Contact with cold refrigerant liquid or spray can cause freezing or frostbite. Cold refrigerant in an eye can cause loss of sight. Proper precautions need to be taken.

Odor

Many of the common refrigerants have no odor and cannot be detected by this means. In case of doubt, assume that refrigerant is present in the atmosphere and protect yourself.

It is strongly recommended that a technician avoid working on systems or equipment using flammable or toxic refrigerants (other than those with an A1 safety rating) unless the technician has had special training and has proper equipment to deal safely with these materials.

Certain refrigerants have a pungent odor, such as ammonia (R-717) and sulfur dioxide (R-764). They can be dangerous if inhaled in sufficient quantities. Their odor even in small quantities provides a warning. If a service person must work in areas where these refrigerants are released, the person must wear a proper respirator.

Warning instruments are available to place in equipment rooms and rooms where refrigerant is stored or used. An alarm is sounded when a refrigerant is released. These instruments are also used in refrigerated areas of large systems to advise operators at an early stage when leaks occur.

Oil

It is important that oil be stored in airtight containers, since oil will absorb moisture if exposed to air. Moisture in the oil can be harmful to the equipment in which it is used. Unless moisture is removed it can cause hermetic compressor motor burnouts.

SAFETY TIP

When a hermetic compressor motor burns out, the oil can become acidic. If it comes in contact with the skin, it can produce a severe chemical burn. Precautions need to be taken.

SAFETY CODE REQUIREMENTS

The Safety Code of Mechanical Refrigeration gives certain safety requirements for the machinery room that must be followed:

For group A1 refrigerants, machinery rooms shall be equipped with an oxygen sensor to warn of oxygen levels below 19.5% volume since there is insufficient odor warning.

For other refrigerants, a refrigerant vapor detector shall be located in an area where refrigerant from a leak is likely to concentrate, and an alarm shall be employed.

The minimum mechanical ventilation required to exhaust a potential accumulation of refrigerant due to leaks or a rupture of the system shall be capable of moving air from the machinery room in the following quantity: (as specified in ANSI/ASHRAE 15-1992, Safety Code for Mechanical Refrigeration, Section 11.13.74).

$$Q = 100 \times G0.5 (Q = 70 \times G0.5)$$

where

Q = the airflow in CFM (l/s)

G = the mass of refrigerant in lb (kg) in the largest system, any part of which is located in the machinery room.

A sufficient part of the mechanical ventilation shall be: (a) operated, when occupied, at least 0.5 CFM/ft^2 (2.54 L/s/m^2) of machinery room area or 20

CFM per person (9.44 L/s) and (b) operable, if necessary for operator comfort, at a volume required to maintain a maximum temperature rise of 18°F (10°C) based on all of the heat producing machinery in the room.

SERVICE TIP

The minimum ventilation rates prescribed may not prevent temporary accumulations of flammable refrigerants above the lower flammability limit (LFL) in the case of catastrophic leaks or ruptures. The designer must consider the provisions of National Fire Protection Association (NFPA) code for such cases. Familiarity with the Safety Code is a must.

TECH TALK

Tech 1. How can you tell what type of refrigerant is in a system?

Tech 2. First of all you need to look for labels. There might a label on the unit itself, on the compressor, on the metering device, or in the paperwork that is fixed inside the system's blower door. If you are unable to find any label identifying the refrigerant it will be difficult but one way is to measure the system's pressure and temperature and then compare that with a pressure temperature chart. The refrigerant in the system should coincide with that same refrigerant on the chart. The other method of determining the type of refrigerant is by taking a sample of the refrigerant and having it analyzed. If you can determine it without incurring this expense for your customer obviously you are better off.

Tech 1. Are there any gizmos or gadgets that you can use that will tell you what kind of refrigerant you have?

Tech 2. Yes, there are some electronic sensors that can tell you what kind of refrigerant is in the system.

Tech 1. Those sound like a pretty good thing to have on the service van.

Tech 2. They certainly can save a lot of time and trouble when you can't determine the type of refrigerant.

Tech 1. Thanks, I appreciate that.

UNIT 1—REVIEW QUESTIONS

1. A refrigerant is a medium for ________.
2. What organization has established the number designation for refrigerants?
3. What does the prefix CFC refer to?
4. The ________ blends consist of multiple components of different volumes that when used in refrigeration cycles do not change volumetric composition or saturation temperature as they evaporate or condense at constant pressure.
5. What are the two groups refrigerants are divided into?
6. What are the characteristics of R-123?
7. R-134a is designed to replace ________.
8. ________ is one of the oldest refrigerants in use today.
9. What are the two most common contaminants in refrigerant?
10. Whose stamp of approval must a cylinder used for refrigerant have?
11. In storing refrigerant in approved cylinders, the liquid level must not exceed ________ of the volume of the container at 60°F.
12. What should the service person wear whenever the refrigerant system is being serviced?
13. Why is odor not a good detector of refrigerant?
14. How should oil be stored?
15. List the Safety Code of Mechanical Refrigeration safety requirements for the machinery room that must be followed.

UNIT 2

Accessing Sealed Systems

OBJECTIVES: In this unit the reader will learn about accessing sealed systems including:

- Install a Gauge Manifold
- Read Pressure and Temperature
- Preventive and Routine Maintenance
- Schrader Valve Access Tool
- Minimum Operating Head Pressure
- Pressure-Temperature Chart
- Operate Service Valves
- Install Access Valves

INSTALLATION OF GAUGE MANIFOLD TEST SET

The gauge manifold test set, often referred to as gauges, are a valuable service tool to the refrigeration service technician. It allows the technician a quick method of installing pressure gauges for checking system conditions, charging, and adding oil to the compressor, Figure 6-2-1.

SAFETY TIP

There are two major groupings of gauge manifold sets. One group is designed to be used with R-12, 22, 134A, R-408, R-407, and R-404. The other groups of manifold gauge sets are designed to be used with R-410A only. Do not use gauges designed for the lower pressured refrigerants on the very high pressure R-410A system.

Technicians use the gauge manifold to diagnose trouble in refrigeration systems. Gauges allow the operator to watch both gauges simultaneously during purging or charging operations, and save time on almost any work that must be done on the system. A cut-away view of a gauge manifold is shown in Figure 6-2-2.

The gauge manifold test set contains two shut-off valves, three external connections, and two pressure gauges. The gauges and the flexible hoses that connect to the manifold to connect it to the system are color-coded; blue is the low side of the system, red is the high side. The left hand gauge is called a compound or suction pressure gauge. The right hand gauge is called the high pressure or discharge pressure gauge. When both valves are closed (front seated) the center or utility port is isolated, Figure 6-2-3a. The parts above and below each valve are interconnected so the gauges will read at all times when connected to the system. When the valve is only slightly opened it's called cracking the valve to crack open the low-side valve to connect the low side and center hoses, (b). Some refrigerant can now flow from the center hose to the system. Fully opening the low side valve opens the low side port to the center port for full flow or refrigerant or for evacuating the system (c). Close the low side valve. Cracking open the high-side valve will allow some refrigerant to flow from the high side to the center hose, (d). Fully opening the high side valve opens the high side port to the center port to remove refrigerant from the system or for system evacuation (e). Opening both low and high side valves can allow the liquid in the high side line to be pulled back into the low side after the high side system access valve is closed or for system evacuation (f).

This is a test instrument and should be treated as such:

- Never drop or abuse the gauge manifold.
- Keep ports or charging lines capped when not in use.
- Never use any fluid other than clean oil and refrigerant.

Figure 6-2-4a shows the low pressure gauge. Figure 6-2-4b shows the high pressure gauge. Figure 6-2-4c shows gauges for R-410A.

The low side gauge is a compound gauge that measures both pressure and vacuum, Figure 6-2-5. It is usually calibrated from 0 to 30 in. of mercury and from 0 to 250 psi. Like the high pressure gauge, the compound gauge also has scales calibrated to read temperatures of various refrigerants such as R-12, R-22, and R-502. With these scales it is not necessary to refer to pressure-temperature tables or curves to calculate pressure-temperature relationships.

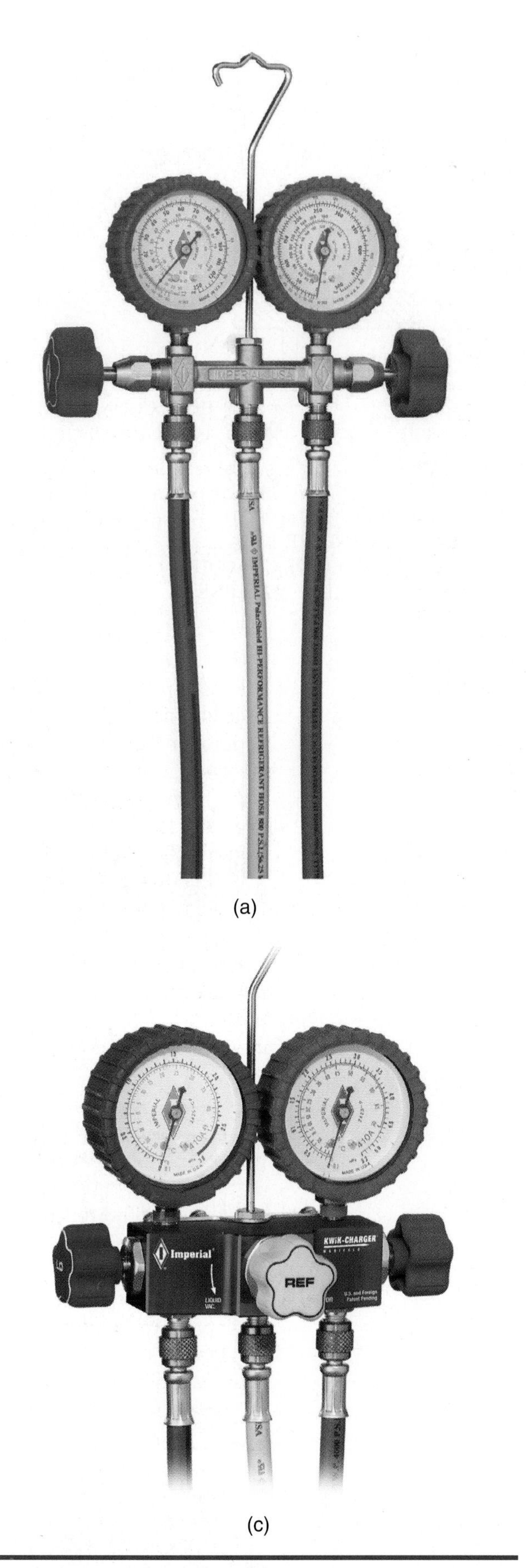

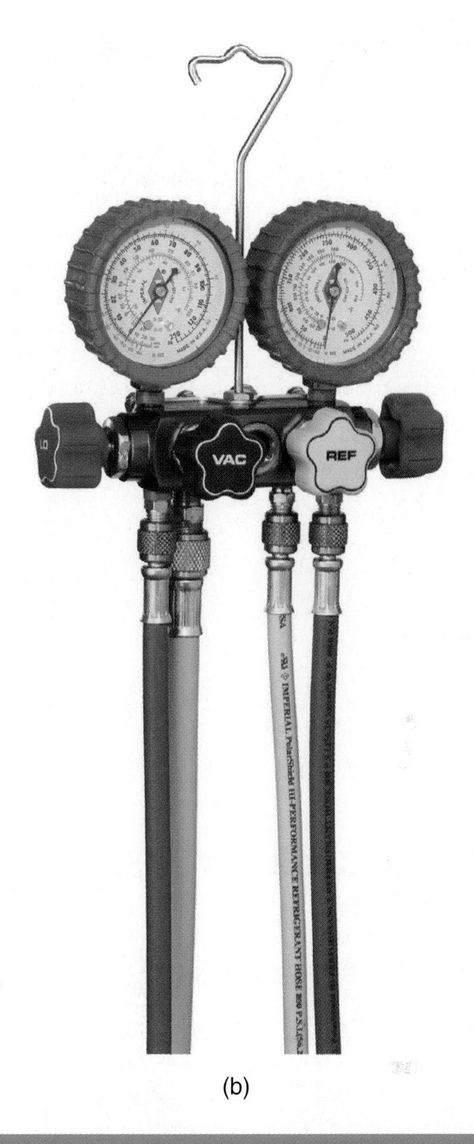

Figure 6-2-1 (a) Two valve gauge set; (b) Four valve gauge set; (c) Three valve R-410A gauge set. (*Courtesy Imperial*)

SERVICE TIP

Manifold gauges have a small adjustment screw that allows the gauge to be calibrated. From time to time the technician should attach the gauges to a new cylinder of refrigerant at a known temperature and compare the pressure reading to the temperature pressure chart so that the gauge can be calibrated. Some technicians adjust the gauge to the zero point with the hoses disconnected. However, if there is an error in the scaling of the gauge the adjustment to zero may result in a faulty reading at system operating pressures. It is far better to make your adjustments at a known temperature and pressure than simply to set it at zero.

The high pressure gauge, Figure 6-2-6, has a single continuous scale, usually calibrated (marked off) to read 0 to 500 psi. The scale may be marked in either 2 lb or 5 lb increments and is usually connected to the high side of the refrigeration system. The black scale is the pressure scale and the red scales indicate the

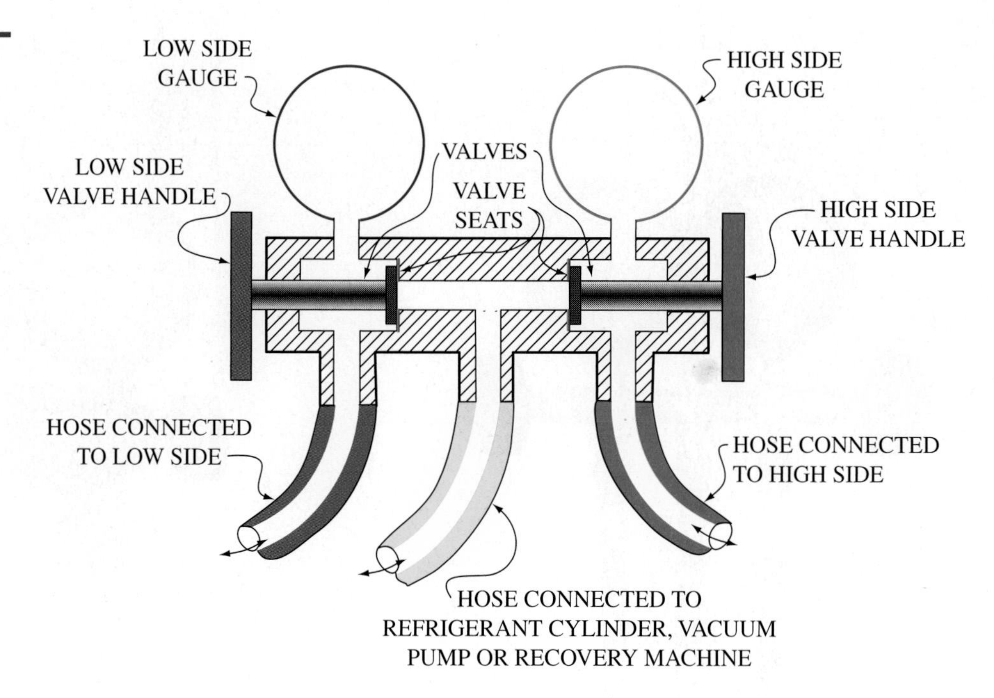

Figure 6-2-2 Cutaway of a gauge manifold.

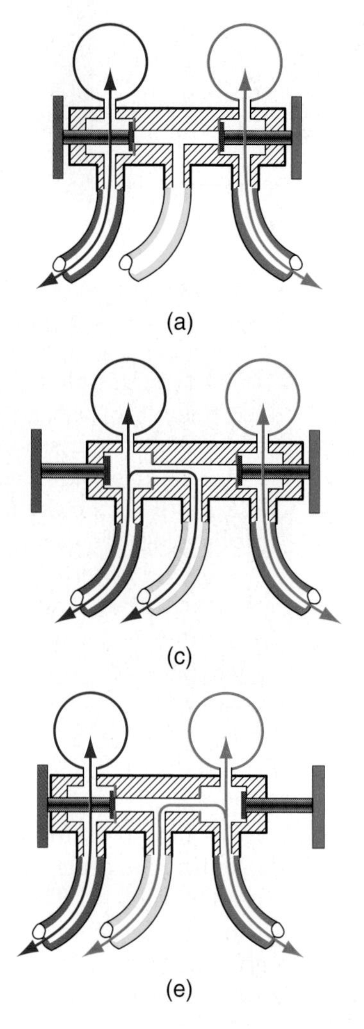

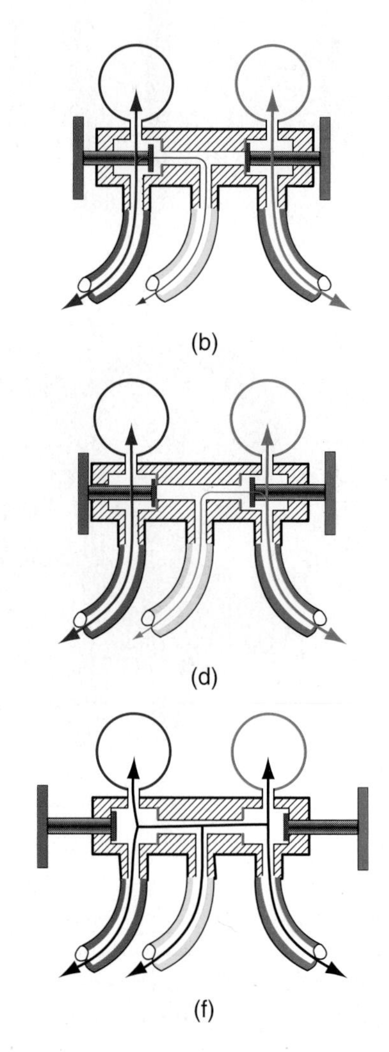

Figure 6-2-3 Gauge manifold valve positions.

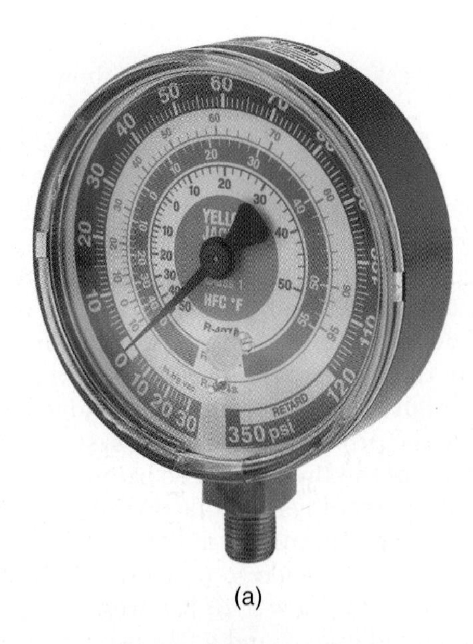

(a)

(b)

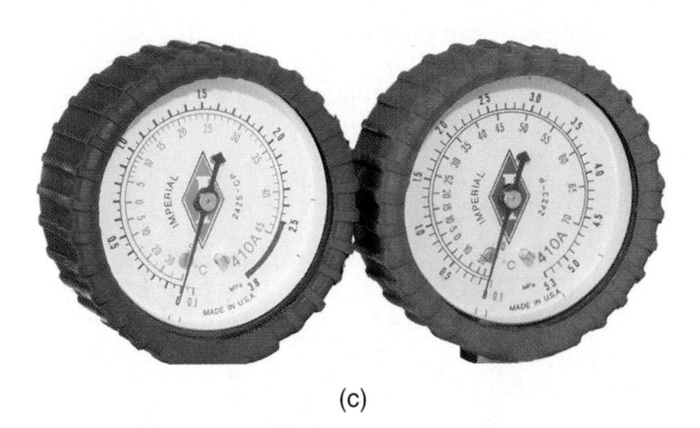

(c)

Figure 6-2-4 (a) Low side compound gauge; (b) High side gauge (*a, b Courtesy of Yellow Jacket Division, Ritchie Engineering Company*); (c) R-410A gauges. (*Courtesy Imperial*)

Figure 6-2-5 Low side gauge. (*Courtesy Imperial*)

Figure 6-2-6 High side gauge. (*Courtesy Imperial*)

temperature of the various refrigerants at their respective pressures. For example, if the gauge pointer indicated 200 psi pressure for R-22, the temperature of the refrigerant would be approximately 101°F.

Gauges use a Bourdon tube as the operating element, Figure 6-2-7. The Bourdon tube is a flattened metal tube sealed at one end, curved and soldered to the gauge fitting at the other end.

SERVICE TIP

The Bourdon tube inside of a gauge is made of a relatively thin gauge metal. These tubes are designed to withstand system pressure. However, if the valve is opened suddenly and pressure is allowed to pop the gauge, the Bourdon tube may be permanently deformed, damaging the gauge.

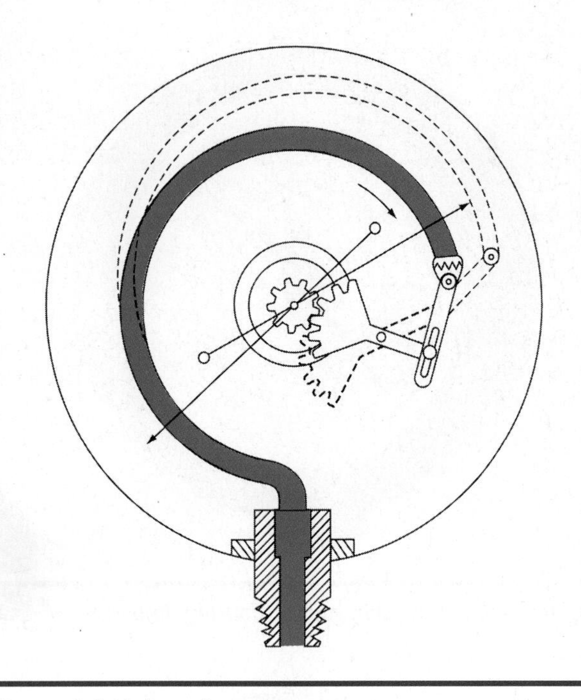

Figure 6-2-7 Bourdon tube.

A pressure rise in a Bourdon tube tends to make the mechanism straighten. This movement will pull on the link, which will turn the gear sector counterclockwise. The pointer shaft will move clockwise to move the needle. On a decrease in pressure, the Bourdon tube moves towards its original (clockwise) position and the pointer moves counterclockwise to indicate a decrease in pressure.

Many systems have a high pressure and low pressure service tap point for checking pressures and charging. They may be located on the compressor shutoff valve, liquid valves, or as independent points as shown in Figure 6-2-8. If the refrigeration system does not have installed service valves, the technician must tap the suction or discharge lines. Service access ports such as that shown in Figure 6-2-9 are sometimes used for this. Note that Schrader service valve ports require an adaptor fitting or a core remover between the service valve and the hose. These ports should always be leak-proof and capped or plugged when not in use.

The low side of the gauge manifold set is attached to the suction side of the system and air is purged out of the lines with refrigerant. You can also evacuate the lines up to the service valve.

The high side of the gauge manifold test set is attached to the discharge side of the system, and air is purged out of the lines again as before.

Pressure readings are taken while the system is operating. The compressor is running. Normal pressure readings of a medium temperature R-22 refrigeration unit may show a 76 psi pressure on the low side compound gauge, and a 210 psi pressure on the high side.

The temperature of the refrigerant may be calculated by observing the red scale for that particular refrigerant, in this case R-410A, and the corresponding pressure. The pointer will also point toward its corresponding temperature. Let's look at the compound gauge. This gauge indicates 150 psig, the evaporating temperature of R-410A at 45°F. The high-pressure gauge indicates 365 psig; the corresponding temperature for R-410A at this pressure is 110°F.

Now it is important to remove the gauge-manifold test set from the system. If the refrigeration system has two-way service valves installed or the technician has used a core removal tool, or has installed a valve to the flexible hoses connected to the suction and discharge service valves, shut off the service ports (suction and discharge), and purge the refrigerant remaining in the gauge lines, out through the center port of the gauge manifold. Remove the hoses and valves on the system and return to the storage position.

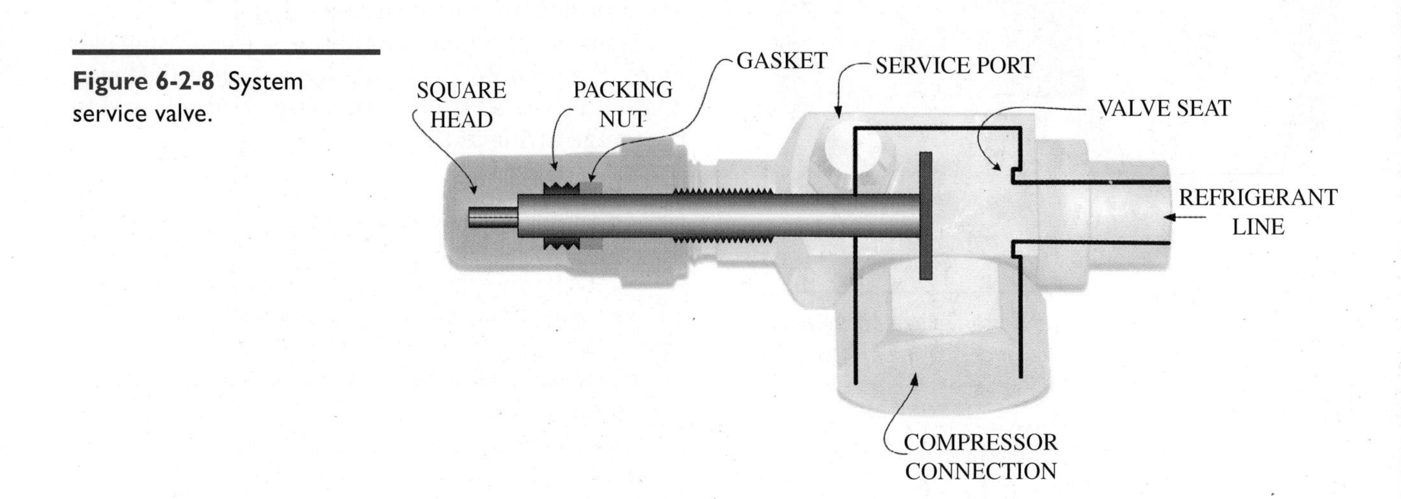

Figure 6-2-8 System service valve.

Figure 6-2-9 System service valve.

DETERMINING MINIMUM OPERATING HEAD PRESSURE

Every liquid has its own boiling (saturation) point at atmospheric pressure. The saturation temperature of any liquid may be changed by increasing or decreasing the pressure.

TECH TIP

Having an idea of the high pressure range on a system you are working on will allow you to quickly determine that there may be a problem with the system. A quick estimate of the proper range compared to the measured range is an excellent troubleshooting tool.

A pressure-temperature (P-T) relationship chart is available free of charge through most refrigeration wholesale supply outlets. One example can be seen in Table 6-2-1. This chart provides the technician with the pressure of various refrigerant liquids at any given temperature.

The P-T chart shown in Table 6-2-2 can be used by the technician, in conjunction with test manifold gauges, to diagnose a refrigeration system malfunction.

The first step is to determine the type of refrigerant used in the refrigeration system you are diagnosing. To accomplish this, look at the UL tag on the unit. If a compound gauge installed on the suction service port of a unit shows a pressure of 68 psig, the temperature of the cooling coil (evaporator) is 40°F, the saturation temperature that corresponds to 68 psi for R-22.

The same pressure-temperature relationship also applies to the high pressure side of the refrigeration system. With a high pressure gauge reading of 240 psig, the temperature of the refrigerant is 114°F.

A rule of thumb for checking the high side for proper operation is to take a temperature reading of the ambient air temperature. Ambient air temperature is the temperature of the air that flows across the condensing unit. Use an electronic temperature gauge or a superheat thermometer for an accurate reading.

Let's suppose that the ambient temperature is 80°F. In order to maintain a proper condensing temperature at the condenser, there must be approximately a 20–35°F difference in ambient temperature and condensing temperature. This means that the ambient temperature (°F) plus 20°F will give you the minimum condensing temperature for a low or medium temperature unit. The constant 35°F is used for high temperature units.

EXAMPLE

For an 80°F ambient temperature and a 20°F temperature difference using refrigerant R-410a, the condensing temperature is 100°F. Looking at the P-T chart, Table 6-2-1, this means that the high side gauge should read at least 320.4 psig pressure.

In analyzing the low side (compound gauge) pressure, suppose the unit is designed for an evaporator temperature of 30°F at 65°F ambient and 40°F at 85°F ambient. Using R-410A, the pressure of boiling refrigerant in the evaporator at 30°F would be 99.0 psig and at 40°F it would be 120.0 psig. ■

ENTERING THE SEALED SYSTEM

Refrigeration problems can be categorized as electrical failures or mechanical failures or a combination of both. Approximately 80% of the problems are electrical in nature, and of the remaining 20%, only a small portion requires extensive sealed-system service. One of the first things to make quite clear is to never enter the sealed system unless it is absolutely necessary.

SAFETY TIP

You must wear eye protection and gloves any time you access a refrigeration system. Liquid refrigerant accidentally released from a pressurized system can be at an extremely low temperature which can cause blindness and skin burns. Never position your face or head in a direct line with an access port as you are attaching or removing hoses.

If one or more of the electrical load devices is not functioning, the problem is electrical. If all components are functioning but the unit does not refrigerate, the problem is mechanical.

TABLE 6-2-1 Vapor Pressure/Temperature of Refrigerants

Temp (Deg F)	CFC R 11	CFC R 12	CFC R 113	CFC R 114	CFC R 500	CFC R 502	HCFC R 22	HCFC R 123	HCFC R 124	HFC R 125	HFC R 134a	HFC R 410a
-100	*29.8*	*27.0*			*26.4*	*23.3*	*25.0*	*29.9*	*29.2*	*24.4*	*27.8*	
-90	*29.7*	*25.7*			*24.9*	*20.6*	*23.0*	*29.8*	*28.8*	*21.7*	*26.9*	
-80	*29.6*	*24.1*			*22.9*	*17.2*	*20.2*	*29.7*	*28.2*	*18.1*	*25.6*	
-70	*29.4*	*21.8*			*20.3*	*12.8*	*16.6*	*29.6*	*27.4*	*13.3*	*23.8*	
-60	*29.2*	*19.0*			*17.0*	*7.2*	*12.0*	*29.5*	*26.3*	*7.1*	*21.5*	
-50	*28.9*	*15.4*			*12.8*	*0.2*	*6.2*	*29.2*	*24.8*	**0.3**	*18.5*	**5.8**
-40	*28.4*	*11.0*			*7.6*	**4.1**	**0.5**	*28.9*	*22.8*	**4.9**	*14.7*	**11.7**
-30	*27.8*	*5.4*			*1.2*	**9.2**	**4.9**	*28.5*	*20.2*	**10.6**	*9.8*	**18.9**
-20	*27.0*	**0.6**	*29.0*	*22.8*	**3.2**	**15.3**	**10.2**	*27.8*	*16.9*	**17.4**	*3.8*	**27.5**
-10	*26.0*	**4.4**	*28.6*	*20.5*	**7.8**	**22.6**	**16.4**	*27.0*	*12.7*	**25.6**	**1.8**	**38.8**
0	*24.7*	**9.2**	*28.1*	*17.7*	**13.3**	**31.1**	**24.0**	*26.0*	*7.6*	**35.1**	**6.3**	**49.8**
10	*23.1*	**14.6**	*27.5*	*14.3*	**19.7**	**41.0**	**32.8**	*24.7*	*1.4*	**46.3**	**11.6**	**63.9**
20	*21.1*	**21.0**	*26.7*	*10.1*	**27.2**	**52.4**	**43.0**	*23.0*	**3.0**	**59.2**	**18.0**	**80.2**
30	*18.6*	**28.4**	*25.7*	*5.1*	**36.0**	**65.6**	**54.9**	*20.8*	**7.5**	**74.1**	**25.6**	**99.0**
40	*15.6*	**37.0**	*24.4*	**0.4**	**46.0**	**80.5**	**68.5**	*18.2*	**12.7**	**91.2**	**34.5**	**120.0**
50	*12.0*	**46.7**	*22.9*	**3.9**	**57.5**	**97.4**	**84.0**	*15.0*	**18.8**	**110.6**	**44.9**	**144.9**
60	*7.8*	**57.7**	*20.9*	**7.9**	**70.6**	**116.4**	**101.6**	*11.2*	**25.9**	**132.8**	**56.9**	**172.5**
70	*2.8*	**70.2**	*18.6*	**12.6**	**85.3**	**137.6**	**121.4**	*6.6*	**34.1**	**157.8**	**70.7**	**203.6**
80	**1.5**	**84.2**	*15.8*	**18.0**	**101.9**	**161.2**	**143.6**	*1.1*	**43.5**	**186.0**	**86.4**	**238.4**
90	**4.9**	**99.8**	*12.4*	**24.2**	**120.4**	**187.4**	**168.4**	**2.6**	**54.1**	**217.5**	**104.2**	**277.3**
100	**8.8**	**117.2**	*8.5*	**31.2**	**141.1**	**216.2**	**195.9**	**6.3**	**66.2**	**252.7**	**124.3**	**320.4**
110	**13.1**	**136.4**	*3.8*	**39.1**	**164.0**	**247.9**	**226.4**	**10.5**	**79.7**	**291.6**	**146.8**	**368.2**
120	**18.3**	**157.7**	**0.8**	**48.0**	**189.2**	**282.7**	**259.9**	**15.4**	**94.9**	**334.3**	**171.9**	**420.9**
130	**24.0**	**181.0**	**3.8**	**58.0**	**217.0**	**320.8**	**296.8**	**21.0**	**111.7**	**380.3**	**199.8**	**478.9**
140	**30.4**	**206.6**	**7.3**	**69.1**	**247.4**	**362.6**	**337.2**	**27.3**	**130.4**	**430.2**	**230.5**	**542.5**
150	**37.7**	**234.6**	**11.2**	**81.4**	**280.7**	**408.4**	**381.5**	**34.5**	**151.0**	**482.1**	**264.4**	**612.1**
160								**42.5**	**173.6**		**301.5**	**684.0**
170								**51.5**	**198.4**		**342.0**	
180								**61.4**	**225.6**		**385.9**	
190								**72.5**	**255.1**		**433.6**	
200								**84.7**	**287.3**		**485.0**	
210								**98.1**	**322.1**		**540.3**	
220								**112.8**	**359.9**			
230								**128.9**	**400.6**			
240								**146.3**	**444.5**			
250								**165.3**	**491.8**			

italics are in inches of mercury vacuum (in/Hg or " Hg)

bold are in pounds per squair inch pressure (psig)

Mechanical problems include: poor capacity compressor (does not pump properly, bad valves), low side leak, low side restriction, high side leak, high side restriction, dirty condenser, and dirty evaporator.

Field Installed Service Valves

One of the easiest devices available for sealed system access is the saddle, piercing valve, or tap-a-line. Still other terms for this are used by some manufacturers. These piercing valves are clamped to the tubing, sealed by a bushing gasket, and then they pierce the tube with a tapered needle. Most contain some sort of shutoff control. The technician should keep in mind that these valves should be used to gain temporary access to a hermetically sealed system for checking system operating pressures or for pressurizing for leak testing. The piercing valve shown in Figure 6-2-10 allows quick access to system pressures to immediately start diagnosing the refrigeration problem.

SAFETY TIP

Before brazing a Schrader valve onto a existing refrigerant line, all of the refrigerant in the system must be recovered. Any refrigerant that is allowed to enter the torch flame will produce a noxious gas. Breathing this gas can be dangerous, and could even cause lung damage. Remove the Schrader core before brazing to keep it from becoming damaged.

TABLE 6-2-2 Pressure-Temperature Chart

SPORLAN PRESSURE-TEMPERATURE CHART

PSIG	TEMPERATURE °F PINK		SAND		ORANGE		GREEN		REDDISH PURPLE		BROWN	
	REFRIGERANT - (SPORLAN CODE) MP39 or 401A (X)		HP80 or 402A (L)		HP62 or 404A (S)		KLEA 60 or 407A		FX-10 or 408A		FX-56 or 409A	
5*	-23		-59		-57		-45		-54		-22	
4*	-22		-58		-56		-43		-52		-20	
3*	-20		-56		-54		-42		-51		-19	
2*	-19		-55		-53		-41		-49		-17	
1*	-17		-54		-52		-39		-48		-16	
0	-16		-53		-50		-38		-47		-15	
1	-13		-50		-48		-36		-44		-12	
2	-11		-48		-46		-33		-42		-9	
3	-9		-45		-43		-31		-39		-7	
4	-6		-43		-41		-29		-37		-5	
5	-4		-41		-39		-27		-35		-2	
6	-2		-39		-37		-25		-33		0	
7	0		-37		-35		-23		-31		2	
8	2		-36		-33		-21		-29		4	
9	4		-34		-32		-20		-27		6	
10	6		-32		-30		-18		-26		8	
11	8		-30		-28		-16		-24		9	
12	9		-29		-27		-15		-22		11	
13	11		-27		-25		-13		-21		13	
14	13		-26		-23		-12		-19		14	
15	14		-24		-22		-10		-18		16	
16	16		-23		-20		-9		-16		17	
17	17		-21		-19		-8		-15		19	
18	19		-20		-18		-6		-13		20	
19	20		-19		-16		-5		-12		22	
20	21		-17		-15		-4		-11		23	
21	23		-16		-14		-2		-9		25	
22	24		-15		-12		-1		-8		26	
23	25		-14		-11		0		-7		27	
24	27		-12		-10		1		-5		29	
25	28		-11		-9		2		-4		30	
26	29		-10		-8		4		-3		31	
27	30		-9		-7		5		-2		32	
28	32		-8		-5		6		-1		34	
29	33		-7		-4		7		0		35	
30	34		-6		-3		8		1		36	
31	35		-5		-2		9		3		37	
32	36		-4		-1		10		4		38	
33	37		-2		0		11		5		39	
34	38	BUBBLE POINT ↓	-1		1		12		6		40	BUBBLE POINT ↓
35	39		0		2		13		7		41	
36	40	30	0		3		14		8		43	
37	42	31	1		4		15		9		44	
38	43	32	2		5		16		10		45	30
39	44	33	3		6		17		11		46	31
40	45	34	4		7		18		12		47	32
42	46	36	6		9		19		13		48	34
44	48	38	8		10		21		15		50	36
46	50	40	10		12		23		17		DEW POINT ↑	38
48	DEW POINT ↑	42	11		14		24		19			39
50		44	13		15		26		20			41
52		45	14		17		28		22			43
54		47	16		19		29		24			45
56		49	18		20		31		25			46
58		50	19	BUBBLE POINT ↓	22	BUBBLE POINT ↓	32	BUBBLE POINT ↓	27	BUBBLE POINT ↓		48
60		52	20		23		33		28			50
62		53	22		25		35		30			51
64		55	23		26		36		31	30		53
66		56	25		27		38		32	32		54
68		58	26		29		39		34	33		56
70		59	27		30		40	30	35	34		57
72		61	29		31	30	41	31	37	36		58
74		62	30		33	32	43	32	38	37		60
76		64	31		34	33	44	34	39	38		61
78		65	32	30	35	34	45	35	40	40		63
80		66	34	31	36	35	46	36	42	41		64
85		69	37	34	39	38	49	39	45	44		67
90		73	40	37	42	41	DEW POINT ↑	42	48	47		70
95		76	42	40	45	44		45	50	50		73
100		78	45	43	48	47		47	DEW POINT ↑	52		76
105		81	48	45	50	50		50		55		79
110		84	50	48	DEW POINT ↑	52		53		57		82
115		87	DEW POINT ↑	50		55		55		60		84
120		89		53		57		57		62		87
125		92		55		59		60		65		89
130		94		57		61		62		67		92
135		96		60		64		64		69		94
140		99		62		66		66		71		96
145		101		64		68		68		73		99
150		103		66		70		70		76		101
155		106		68		72		72		78		103
160		108		70		74		74		80		105
165		110		72		76		76		81		107
170		112		74		78		78		83		109
175		114		75		80		80		85		111
180		116		77		81		81		87		113
185		117		79		83		83		89		115
190		119		81		85		85		91		117
195		121		82		87		87		92		119
200		123		84		88		88		94		121
205		125		86		90		90		96		123
210		127		87		92		91		97		124
220		130		91		95		94		100		128
230		133		94		98		97		104		131
240		136		97		101		100		107		134
250		140		99		104		103		109		137
260		143		102		107		106		112		141
275		147		106		111		110		116		145
290		151		110		114		114		120		149
305		155		114		118		117		124		153
320		159		118		122		121		127		157
335		163		121		125		124		131		161
350		167		125		129		128		135		165
365		170		128		132		131		138		169

* Inches mercury below one atmosphere

P-H DIAGRAM — BLENDS

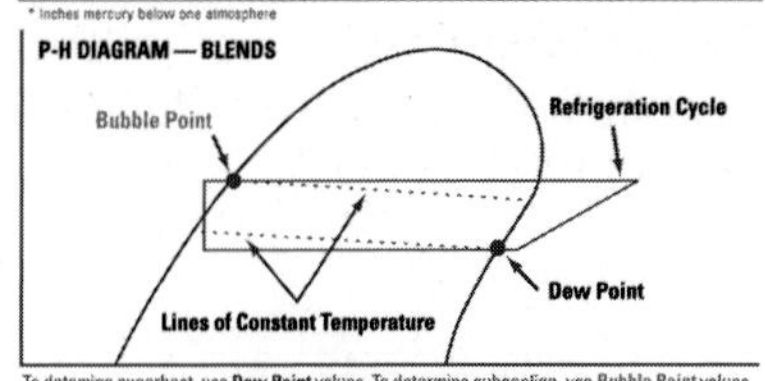

To detemine superheat, use Dew Point values. To determine subcooling, use Bubble Point values.

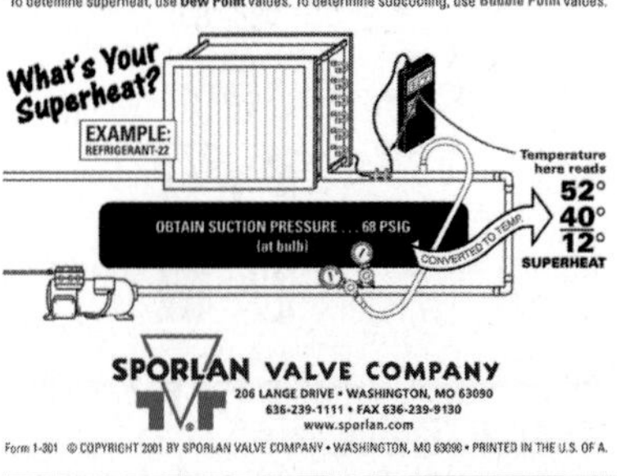

SPORLAN PRESSURE-TEMPERATURE CHART

PSIG	TEMPERATURE °F YELLOW	GREEN	GREEN	BLUE	PURPLE	TEAL	WHITE
	REFRIGERANT - (SPORLAN CODE) 12 (F)	22 (V)	124 (M)	134a (J)	502 (R)	AZ50 or 507 (P)	717 (A)
5*	-29	-48	3	-22	-57	-60	-34
4*	-28	-47	4	-21	-55	-58	-33
3*	-26	-45	6	-19	-54	-57	-32
2*	-25	-44	7	-18	-52	-55	-30
1*	-23	-43	9	-16	-51	-54	-29
0	-22	-41	10	-15	-50	-53	-28
1	-19	-39	13	-12	-47	-50	-26
2	-16	-37	16	-10	-45	-48	-23
3	-14	-34	18	-8	-42	-46	-21
4	-11	-32	21	-5	-40	-44	-19
5	-9	-30	23	-3	-38	-41	-17
6	-7	-28	26	-1	-36	-39	-15
7	-4	-26	28	1	-34	-38	-13
8	-2	-24	30	3	-32	-36	-12
9	0	-22	32	5	-30	-34	-10
10	2	-20	34	7	-29	-32	-8
11	4	-19	36	8	-27	-31	-7
12	5	-17	38	10	-25	-29	-5
13	7	-15	40	12	-24	-27	-4
14	9	-14	41	13	-22	-26	-2
15	11	-12	43	15	-20	-24	-1
16	12	-11	45	16	-19	-23	1
17	14	-9	46	18	-18	-21	2
18	15	-8	48	19	-16	-20	3
19	17	-7	49	21	-15	-19	4
20	18	-5	51	22	-13	-17	6
21	20	-4	52	24	-12	-16	7
22	21	-3	54	25	-11	-15	8
23	23	-1	55	26	-9	-14	9
24	24	0	57	27	-8	-12	11
25	25	1	58	29	-7	-11	12
26	27	2	59	30	-6	-10	13
27	28	4	61	31	-5	-9	14
28	29	5	62	32	-3	-8	15
29	31	6	63	33	-2	-7	16
30	32	7	65	35	-1	-6	17
31	33	8	66	36	0	-4	18
32	34	9	67	37	1	-3	19
33	35	10	68	38	2	-2	19
34	37	11	69	39	3	-1	20
35	38	12	71	40	4	0	21
36	39	13	72	41	5	1	22
37	40	14	73	42	6	2	23
38	41	15	74	43	7	3	24
39	42	16	75	44	8	4	25
40	43	17	76	45	9	4	26
42	45	19	78	47	11	6	28
44	47	21	80	49	13	8	29
46	49	23	82	51	15	10	31
48	51	24	84	52	16	11	32
50	53	26	86	54	18	13	34
52	55	28	88	56	20	15	35
54	57	29	90	57	21	16	37
56	58	31	91	59	23	18	38
58	60	32	93	60	24	19	40
60	62	34	95	62	26	21	41
62	64	35	97	64	27	22	42
64	65	37	98	65	29	24	44
66	67	38	100	66	30	25	45
68	68	40	101	68	32	26	46
70	70	41	103	69	33	28	47
72	71	42	104	71	34	29	49
74	73	44	106	72	36	30	50
76	74	45	107	73	37	32	51
78	76	46	109	75	38	33	52
80	77	48	110	76	40	34	53
85	81	51	114	79	43	37	56
90	84	54	117	82	46	40	58
95	87	56	120	85	49	43	61
100	90	59	123	88	51	45	63
105	93	62	126	90	54	48	66
110	96	64	129	93	57	51	68
115	99	67	132	96	59	53	70
120	102	69	135	98	62	55	73
125	104	72	138	100	64	58	75
130	107	74	140	103	67	60	77
135	109	76	143	105	69	62	79
140	112	78	145	107	71	64	81
145	114	81	148	109	73	66	82
150	117	83	150	112	75	68	84
155	119	85	152	114	77	70	86
160	121	87	154	116	80	72	88
165	123	89	157	118	82	74	90
170	126	91	159	120	83	76	91
175	128	92	161	122	85	78	93
180	130	94	163	123	87	80	95
185	132	96	165	125	89	82	96
190	134	98	167	127	91	83	98
195	136	100	169	129	93	85	99
200	138	101	171	131	95	87	101
205	140	103	173	132	96	88	102
210	142	105	175	134	98	90	104
220	145	108	178	137	101	93	107
230	149	111	182	140	105	96	109
240	152	114	185	143	108	99	112
250	156	117	188	146	111	102	115
260	159	120	192	149	114	105	117
275	163	124	196	153	118	109	121
290	168	128	201	157	122	113	124
305	172	132	205	161	126	117	128
320	177	136	209	165	130	120	131
335	181	139	213	169	133	124	134
350	185	143	217	172	137	127	137
365	188	146	221	176	140	130	140

* Inches mercury below one atmosphere

APPROXIMATE PRESSURE CONTROL SETTINGS

Pressure - Pounds Per Square Inch Gauge

APPLICATION	TEMPERATURE RANGE (°F)	EVAPORATOR TD (°F)	REFRIGERANT 22 Out	22 In	134a Out	134a In	404A Out	404A In	507 Out	507 In
Beverage Cooler Floral Cooler Produce Cooler	35 to 38	15	41	66	17	33	53	82	56	86
Smoked Meat Cooler Meat Reach Thru Service Deli Seafood	32 to 35	15	38	62	15	30	49	77	52	81
Multi-Deck Fresh Meat	26 to 29	15	32	54	11	25	42	68	45	72
Frozen Glass Door Frozen Walk-In	-10 to 0	10	9	24	-	-	15	33	16	35
Frozen Ice Cream Frozen Food - Open Type	-30 to -20	10	0	10	-	-	4	18	4	18

Pressure control settings assume a suction line loss equivalent to 2°F.

CARRYING CAPACITY OF REFRIGERATION LINES

Tons of Refrigeration - 100 Equivalent Feet of Pipe Length

Copper Tube Size O.D. Type L Inches	REFRIGERANT 22 Liquid Line	22 Suction Line 40°F Evap.	134a Liquid Line	134a Suction Line 20°F Evap.	404A / 507 Liquid Line	404A / 507 Suction Line -20°F Evap.	Iron Pipe Size Inches (Schedule 80)	REFRIGERANT 717 Liquid Line	717 Suction Line 5°F Evap.
3/8	1.5	0.3	1.2	0.1	1.0	0.1	3/8	14.4	-
1/2	3.8	0.6	2.8	0.3	2.4	0.2	1/2	28.4	-
5/8	7.0	1.1	5.3	0.4	4.4	0.3	3/4	64.1	-
7/8	18.6	2.9	14.0	1.1	11.6	0.8	1	126	3.8
1-1/8	37.7	5.9	28.4	2.3	23.5	1.6	1-1/4	271	8.2
1-3/8	66.0	10.2	49.6	4.1	41.0	2.8	1-1/2	413	12.5
1-5/8	104	16.1	78.5	6.4	64.8	4.5	2	811	24.5
2-1/8	217	33.2	164	13.2	135	9.2	2-1/2	1300	39.5
2-5/8	384	58.6	289	23.4	238	16.2	3	2330	70.8
3-1/8	614	93.3	462	37.3	379	25.9	4	4820	147
3-5/8	912	130	687	56.4	562	38.4	5	8790	268
4-1/8	1290	195	970	78.2	792	54.1	6	-	429

Refrigerant 717 (Ammonia) values are based on 86°F liquid temperature. Refrigerant 22, 134a, and 404A / 507 values are based on 100°F liquid temperature. Liquid line values based on pressure drop equivalent to 1°F charge in saturation temperatures. Suction line values based on pressure drop equivalent to 2°F charge in saturation temperature.

(Courtesy Sporlan Valve Company)

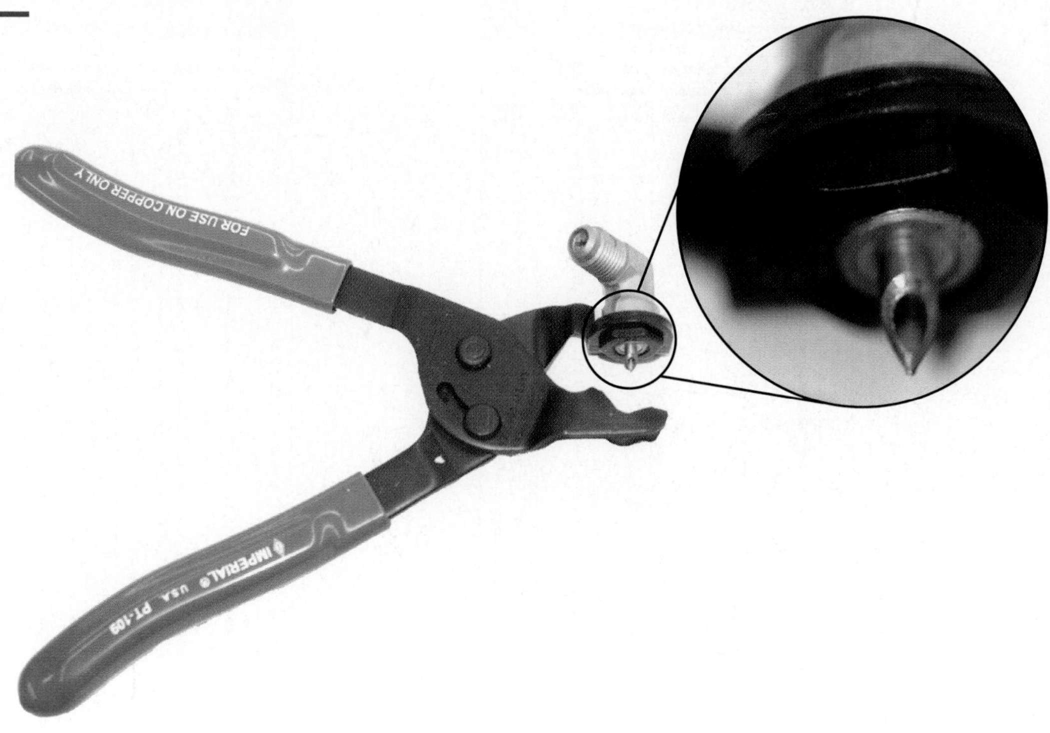

(a)

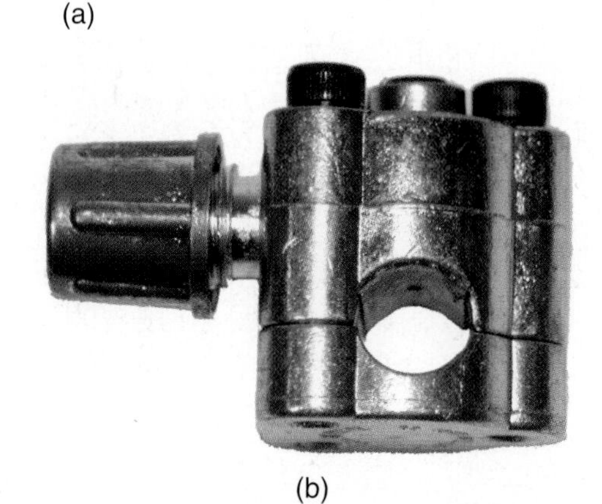

(b)

Figure 6-2-10 (a) Lever action line piercing tool; (b) Bolt on line piercing tool.

Access-piercing valves must be removed once the source of the sealed system malfunction has been located, Figure 6-2-10. Saddle valves, such as the one in Figure 6-2-11, have Schrader valve cores brazed on the tubing.

The Schrader valve core, shown in Figure 6-2-12, is a spring loaded device for position seating. The valve is like those used on automobile tires. The stem must be depressed to force the valve seat open against spring pressure. If a valve core leaks, it can be replaced by using a core-removal tool to unscrew it. Some tools, such as the one in Figure 6-2-13, allow this to be done while the system is under pressure.

Factory Installed Service Valves

With the implementation of the Clean Air Act, all manufacturers, with the exception of Type I equipment manufacturers, are required to install factory-installed service valves. Many commercial refrigeration and air conditioning systems have had factory-installed service valves for years.

Factory installed service valves may be a manually operated stem shutoff valve, as in Figure 6-2-14, or a Schrader type valve, as in Figure 6-2-15. Installed valves are usually located at the compressor as suction and discharge valves, and at the outlet of the

SERVICE TIP

You must put service valve caps back on all access ports. This is an EPA refrigerant management requirement. Failure to do so is a violation of EPA rules and regulations. Many service access valve caps have O rings to seal the system. Be sure these O rings are in place before installing the cap to ensure a proper seal. Caps that have a metal-to-metal seal must be tightened 1/8 turn with a wrench after they have been finger tightened.

Figure 6-2-11 Saddle valve.

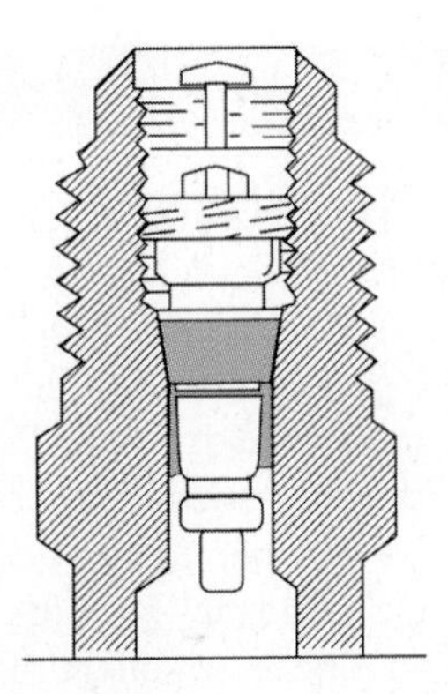

Figure 6-2-12 Schrader valve core.

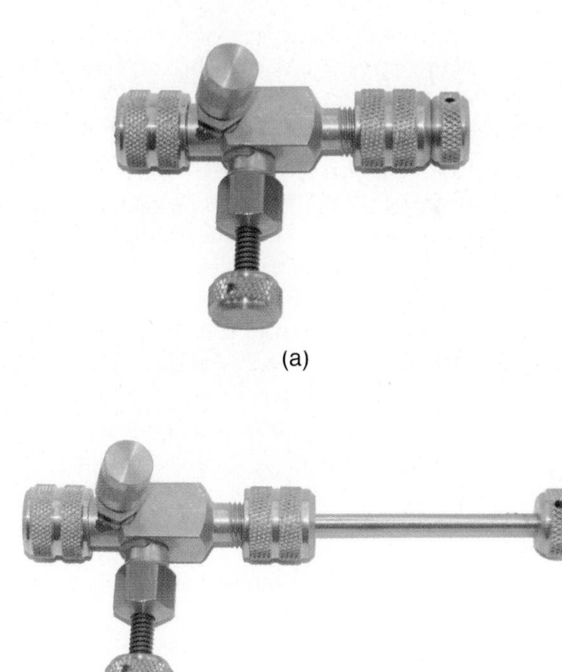

Figure 6-2-13 Manual shutoff valves.

receiver (the King valve). These service valves are equipped with a gauge service port. Operating refrigerant pressures may be observed on the service gauge manifold when hoses are connected to these ports.

The valve will be in the backseated (the stem turned all the way out, counterclockwise) position when you attach your gauges. This closes the gauge port, and the valve is open to the line connection. The valve is frontseated (the stem is turned all the way in, clockwise) to isolate the compressor from the system. The gauge port is open to the compressor but the line connection is closed. In order to read system pressures, the technician first checks to ensure that the service valve is backseated, then turns the stem in one or two turns (the service position, cracked), in

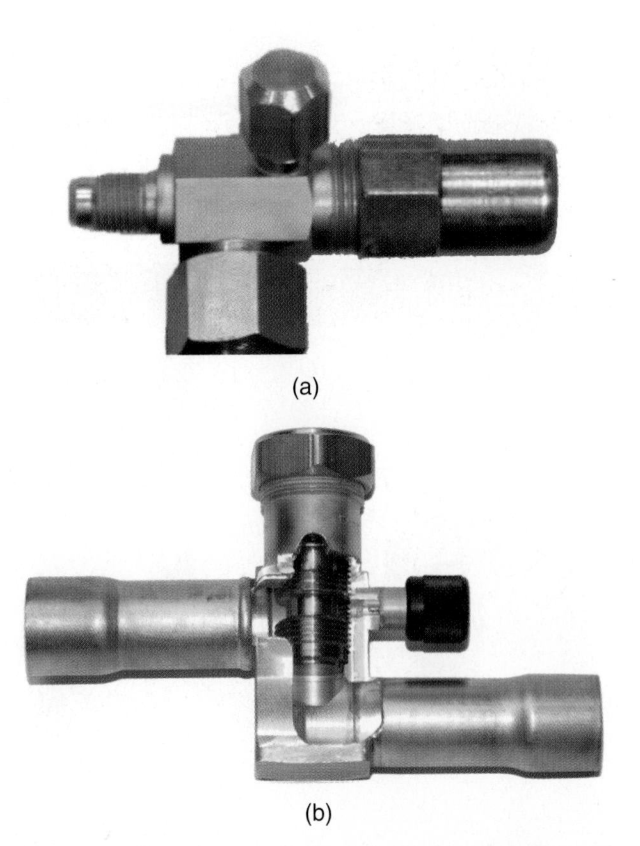

Figure 6-2-14 Service valves. (*b Courtesy Rheem Manufacturing*)

order to slightly open the connection to both line and gauge port. Service valves are returned to the backseated position after service and the gauges are removed. All caps are reinstalled on the system. Note that the compressor is always open to either the line connection or the gauge port, or both, if the valve is in the cracked position.

SAFETY TIP

Be sure that internal pressure in the compressor is relieved by recovery and vacuum procedures before attempting to remove an isolated compressor from the system.

Schrader valves provide a convenient method of checking system pressures or servicing the system, where it is not economical or convenient to use the compressor service valves with gauge ports. Some manufacturers provide Allen wrench stop valves in conjunction with the Schrader valve. The Schrader valve is similar to the air valves used on bicycle and automobile tires; however, they are not the same. The rubber used in tire valves is not compatible with refrigerants and would dissolve. The Schrader valve used in refrigeration and air conditioning systems

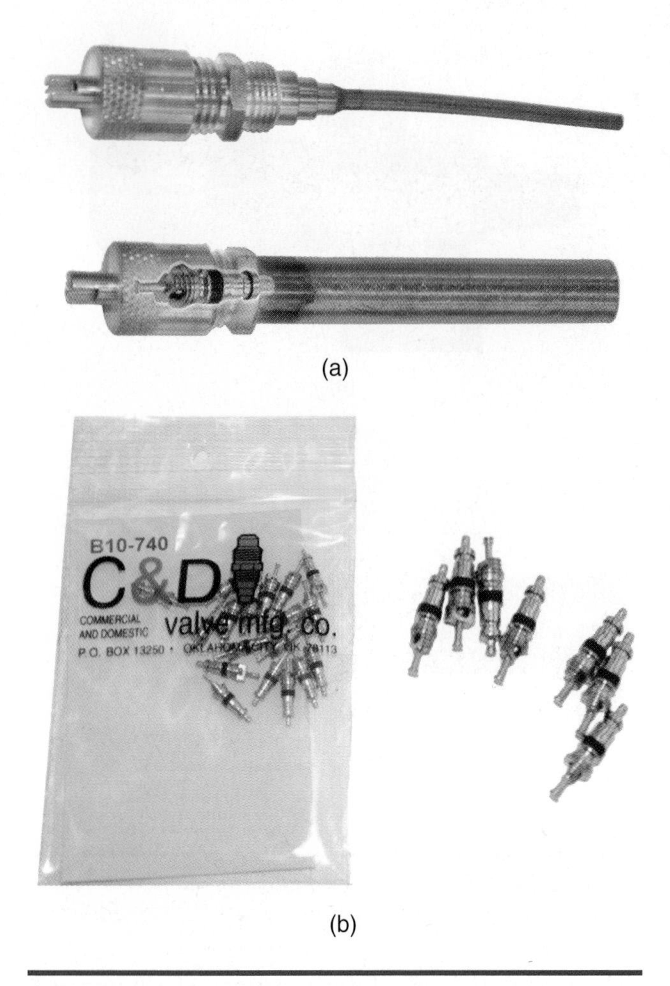

Figure 6-2-15 (a) Pigtail-type Schrader valve; (b) Schrader valve replacement cores.

must also have a cap for the fitting to ensure a leak-proof operation.

This type of service valve enables the technician to quickly check system operation without disrupting the unit's operation. Technicians should use quick-connect low-loss hose adapters to attach to the Schrader valves, greatly reducing the refrigerant loss, when connecting or disconnecting the service hoses.

TECH TALK

Tech 1. Every time I take my hoses off, the high and low pressure gauges still read the system pressure. Isn't that going to hurt the gauges?

Tech 2. No, the gauges are designed so that pressure is not going to affect them. Think for a minute about the gauges that are on a large system. They are at operating pressure year in and year out. They stay there for long periods of time and it does not affect them.

Tech 1. Oh, that's right, I have seen that.

Tech 2. It's just a myth that leaving pressure on the gauges for a long time is going to affect them. It is a good idea to get as much pressure out of the hoses as possible when you are removing the gauges.

Tech 1. How do you do that?

Tech 2. First of all you turn off the cylinder valve and then disconnect the high side gauge hose. The check valve will keep the refrigerant in the hose. Next, open the high side and low side hand wheels slowly. This will cause the liquid refrigerant in the high side hose to be drawn back into the unit while the system is running. That way you will only have vapor in the hoses and it will be at a lot lower pressure.

Tech 1. Sounds like a good idea, I'll try it. Thanks.

UNIT 2—REVIEW QUESTIONS

1. Technicians use the ________ to diagnose trouble in refrigeration systems.
2. What does a low side gauge measure?
3. What is a Bourdon tube?
4. How can the saturation temperature of any liquid be changed?
5. What does a pressure-temperature chart provide?
6. What is ambient air temperature?
7. Approximately ________ of refrigeration problems are electrical in nature.
8. List the mechanical problems that could cause refrigeration problems.
9. With the implementation of the Clean Air Act all manufacturers, with the exception of Type I equipment manufacturers, are required to ________.
10. ________ provide a convenient method of checking system pressures or servicing the system, where it is not economical or convenient to use the compressor service valves with the gauge ports.

UNIT 3

Refrigerant Management

OBJECTIVES: In this unit the reader will learn about refrigeration management including:

- EPA Regulations and Directives
- Procedures for Refrigeration Containment
- Effects of EPA Regulations
- Liquid and Vapor Refrigerant Recovery Methods
- Measuring and Testing Equipment
- Evacuate a System
- Refrigerant Leak Detection
- Liquid and Vapor Refrigerant Recharging
- Charging Charts and Calculators
- Subcooling While Charging
- Compressor Oil Recharging

PRACTICING REFRIGERATION CONTAINMENT

In 1990 the Congress of the United States passed a series of amendments to the Clean Air Act that greatly affected the refrigeration and air conditioning industry. The Act establishes a set of standards and requirements for the use and disposal of certain common refrigerants containing chlorine. These refrigerants, when allowed to escape into the atmosphere, are believed to damage the ozone layer. This stratospheric layer protects life forms from the harmful ultraviolet radiation from the sun.

SERVICE TIP

EPA regulations regarding the management and containment of refrigerants is not an optional program. Failure to follow the rules and regulations can result in significant fines and possible imprisonment of anyone caught in violation. It is essential that as a technician you not only learn about the EPA rules and regulations, but that you also become familiar with them and follow them in the field. Failure to do so can cause significant problems for you and the company you are working for.

EPA directives cover five main regulations:

- Refrigerant emissions to the atmosphere must be reduced. (The no-venting rules.)
- A certification program is established for technicians. Only certified technicians can purchase restricted refrigerants or service equipment that contains these refrigerants.
- Refrigerant reclaiming equipment must be certified.
- Substantial leaks must be repaired.
- Safe disposal regulations must be followed.

Refrigerant Leaks

Certain refrigerant leaks are difficult to prevent. EPA, therefore, will permit leaks under the following conditions:

- **De minimis releases.** These are releases that occur even when the approved procedures are followed, including the use of certified recovery and recycling machines.
- **Refrigerant released in the normal operation of equipment.** Regulations require that certain leaks must be repaired, however.
- **Leaks of R-22 mixed with nitrogen for leak testing.** Nitrogen cannot be added to a charged system for leak testing. The refrigerant in the system must first be recovered. Pure CFCs or HCFCs are not to be released for leak testing.
- **Purging hoses or hose connections during charging or service.** Low-loss hose fittings are required on all recovery and recycling equipment manufactured after November 15, 1993.

Effect of EPA Regulations

In evaluating the effects of the new regulations, two conditions need to be considered:

- New systems being evacuated and charged for the first time; and
- Existing systems that have been operated for some time and require service.

Each needs to be treated differently to comply with the regulations and is described below.

NICE TO KNOW

Some technicians with a long history of being in the HVAC/R field initially resented the extra effort required to follow the EPA regulations. However, the impact on the environment of the continued release of refrigerant is unacceptable. As technicians have become accustomed to the recovery, reclaim, and recycle procedures mandated by EPA, the amount of time and effort involved has decreased. Now the new methods are just as convenient as it was before the regulations came into effect. Technicians in the industry do not see following the EPA regulations as a burden, but as an asset to both our industry and the world's ecology.

New Systems

An example system has been constructed recently. Assume all of the components in the system are assembled and that the system is ready for leak testing, evacuation, and charging. The following procedures support EPA regulations.

- The gauge manifold is installed with hose connections to both the high and the low sides of the system in the regular manner.
- The system is then pressurized using nitrogen and a tracer of R-22 for leak testing. The pressures in the system should be comparable to the operating pressures unless local codes specify other testing pressures. Leak testing can be done with a halide torch or an electronic leak detector. If leaks are found, they must be repaired and the system retested.
- The system is then evacuated, venting the test charge to the atmosphere.
- It is advisable at this point to run a deep vacuum test. This means pulling down the system to at least 29 in Hg and closing off the valves on the gauge manifold, letting the system stand overnight in a deep vacuum. If the pressure in the system rises during the test period, there is a leak. If a leak is indicated, be certain that it is not in the gauge manifold hose connections.
- If the system is leak-tight, it is ready for charging. For most built-up systems using TXV metering valves, refrigerant is added to the system until the liquid line sight glass shows a solid stream.

Existing Systems

Existing systems that have been in operation may require repairs. The system may have a refrigerant leak or a defective system component that needs replacing. The refrigeration circuit needs to be opened up for service.

Under these conditions, EPA has very strict rules. If there is still refrigerant in the system, the refrigerant charge must be recovered before the system is opened. The kind of recovery operation is dependent on the type of system. The categories of systems are as follows:

- Small appliances, such as home refrigerators and freezers, having a refrigerant charge of 5 lb or less.
- High pressure appliances, such as residential air conditioners and heat pumps and most commercial air conditioning equipment.
- Low pressure appliances, such as centrifugal chillers.
- Very high pressure equipment, with a boiling point below −50°C (−58°F) at atmospheric pressure.

Evacuation Requirements

Each one of the appliance groups has its own requirement for level of evacuation, as follows:

The active method uses self contained recovery equipment. This method has its own compressor to pump out the system. It is required to reduce the system pressure to 4 to 10 in Hg, depending on the EPA requirement.

The passive method uses system dependent recovery equipment. This method uses only the small appliance compressor to recover the refrigerant. With this system, 90% of the refrigerant must be recovered if the compressor is running and 80% for a compressor that is not running. A recovery system that uses only a chilled recovery tank would qualify for this method.

SERVICE TIP

Use this tip when recovering refrigerant from an existing system. If the outdoor condenser fan and indoor blowers are operating the system will pick up a small amount of heat from the ambient air. This will speed up the recovery process significantly.

High Pressure Appliances

The recovery requirements depend on the refrigerant used and the amount of the charge, as shown in Table 6-3-1.

Low Pressure Appliances

To recover the refrigerant from centrifugal chillers and other low- pressure units, a combination of liquid- and vapor-recovery methods is used. The liquid-recovery procedure will recover about 70% of the charge, as shown in Figure 6-3-1. The balance of the

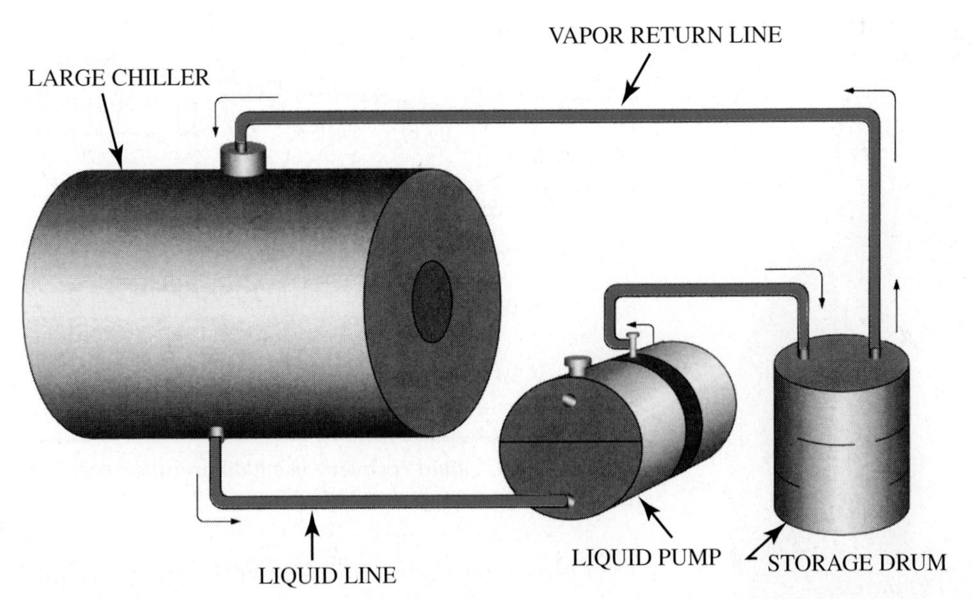

Figure 6-3-1 Liquid recovery using a liquid pump.

refrigerant is removed by the vapor method, shown in Figure 6-3-2.

The compressor in the vapor recovery machine creates a vacuum that draws out the vapor from the system. Some of these machines will also remove part of the liquid, which is separated out by an accumulator or another arrangement.

TECH TIP

Running the water through a low pressure refrigerant chiller will both increase the withdrawal rate and prevent possible freeze-up of the water.

TABLE 6-3-1 Required Levels of Evacuation for High Pressure Appliances

Recover Equipment Refrigerant and Charge	Maufacturing Date	
	Before 11/15/93 (in Hg)	After 11/15/93 (in Hg)
R-22 Appliance, <200 lb charge	0*	0
R-22 Appliance, >200 lb charge	4	10
Other high pressure appliance <200 lb charge	4	10
Other high pressure appliance >200 lb charge	4	15
Very high pressure equipment	0	0

*A zero (0) vacuum is atmospheric pressure. A perfect vacuum is 30 in Hg.

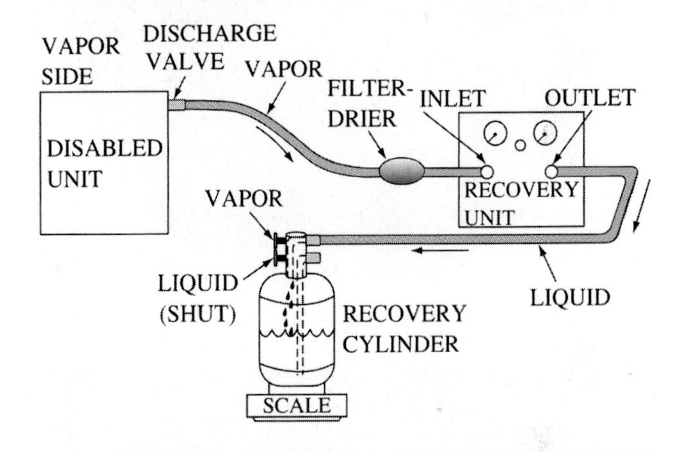

Figure 6-3-2 Vapor recovery.

The Recovery Cylinder

The recovery cylinder shown in Figure 6-3-3 is made especially for recovering refrigerants.

- It must have DOT approval.
- It should be color coded, gray with a yellow top, for all recovered refrigerants.

The recovery cylinder is usually supplied with the recovery machine. A fill-limit device is often part of the recovery equipment.

SAFETY TIP

Recovery cylinders must be hydrostatically tested every 5 years. They must have a valid date stamp indicating that they are within that time frame. The date stamped on the cylinder is the last date that cylinder can be used for recovering storing or handling refrigerants. It is your responsibility as the technician to verify this date. Using out of date cylinders is a violation of the DOT regulations.

(a)

(b)

Figure 6-3-3 (a) Recovery cylinders; (b) Refrigerant cylinders being refurbished. (*Courtesy National Refrigerants, Inc.*)

LIQUID REFRIGERANT RECOVERY

For liquid recovery either a liquid pump such as that shown in Figure 6-3-1 can be used or the differential pressure method shown in Figure 6-3-4.

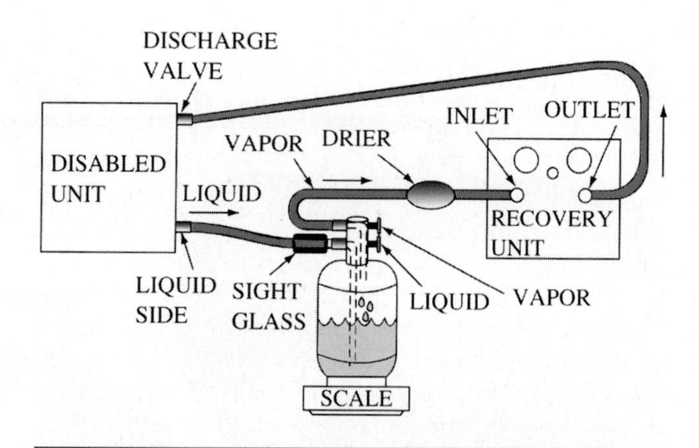

Figure 6-3-4 Liquid recovery using differential-pressure method.

This latter method is also known as the push pull technique and is the most common method of recovering liquid.

Referring to Figure 6-3-4, the liquid is forced out of the unit using the recovery machine to pull a vacuum on the recovery container. At the same time, the pressure is increased in the unit. This causes rapid movement of the liquid.

The advantage of the liquid method of transfer is that liquid recovery is much faster than the vapor recovery method. In order for the liquid transfer method to be feasible, suitable access fittings must be provided to access liquid as well as vapor in the equipment being serviced. The final evacuation must be done by the vapor transfer method.

VAPOR REFRIGERANT RECOVERY

A typical recovery procedure with a vapor recovery machine is as follows:

- First, it is important to be certain which refrigerant is being recovered and that the machine being used is certified for this service.
- The system must be turned off during this process and have all control valves open.
- The recovery cylinder must be evacuated prior to the recovery operation and be standing in an upright position.
- Connections are made to the system as shown in Figure 6-3-2. Whatever type of fill limit device is used must be in service.
- The valves accessing the gauges must be open.
- The vapor valve on the recovery cylinder is open.
- The recovery machine is turned on.

The vapor recovery machine is equipped with a low pressure cutout. The setting on this control is part of the certification. These controls are usually set below 29 in Hg for low pressure refrigerants and at 10 in Hg for high pressure refrigerants. The low pressure control setting is made at the factory and cannot be altered in the field. If the control setting is too low for a particular operation, the technician can shut off the machine manually and save operating time.

The machine also has a high pressure control. The setting is not adjustable and is based on the refrigerants for which the equipment is used. The high pressure control will shut off the machine in case there is an obstruction in the high pressure lines, loss of cooling, an overfilled cylinder, or air in the system. The setting of this control is part of the certification.

The recycle machine is usually equipped with filters and separators which are used primarily to protect the recovery equipment. The machine usually has two circuits for oil separation: one to remove system oil from incoming vapor, and the other to recover compressor oil coming from the recovery machine. The oil from the system is drained and properly disposed of, and the oil from the recovery unit flows from the separator back to the recovery compressor.

When the machine shuts off on the low pressure control, the evacuation is not necessarily completed. If the machine remains idle for a short period of time the pressure may creep up. The recovery machine should be run again. Usually, when the machine cuts off twice, the recovery is considered completed.

TECH TIP

Not all refrigerant recovery recycle machines are the same. Each manufacturer has specific advantages and disadvantages. Some machines have very high withdrawal rates for liquid or vapor. Other machines are very light. Some machines are known for their durability, while others are known for their economy. Before selecting a recovery recycle machine check the machine's features to be sure that they are the features that are most important to you.

MEASURING AND TESTING EQUIPMENT

In refrigeration (and air conditioning) work, many different kinds of instruments and test equipment are used. Previous chapters have detailed basic hand tools of the toolkit and certain test equipment for general use. This chapter covers measuring and testing equipment needed in the evacuation and charging operations.

Temperature Measurements

When analyzing a refrigeration system, accurate temperature readings are important. The most common temperature measuring device is the pocket glass thermometer, illustrated in Figure 6-3-5.

Note that it fits into a protective metal case. The thermometer head has a ring for attaching a string to suspend it if needed. The temperature ranges of glass thermometers vary, but −30 to +120°F is a common scale for refrigeration systems, and the thermometer

Figure 6-3-5 The metal cap on this pocket glass thermometer screws onto its metal tube carrying case.

is calibrated in 2°F marks. Some have a mercury fill, but others use a red fill that is easier to read.

To check the calibration of a pocket glass thermometer, insert it into a glass of ice water for several minutes. It should read 32°F, plus or minus 1°F. Should the fill separate, place the thermometer in a freezer, and the resulting contraction will probably rejoin the separated column of fluid. Another way to connect the separated fill is to carefully heat the stem, not the bulb. In most cases the liquid will coalesce, or gather together, as it expands.

Another form of pocket thermometer is the dial type, shown in Figure 6-3-6. It, too, has a carrying case with pocket clip. The dial thermometer is practical for measuring air temperatures in a duct. The stem is inserted into the duct, but the dial remains visible. Again, several ranges are available depending on the accuracy needed and the nature of the application. A common refrigeration range is −40 to 160°F.

A different type of thermometer is the superheat thermometer, illustrated in Figure 6-3-7. The highly accurate expansion-bulb thermometer is used to measure suction-line temperature(s) in order to calculate, check, and adjust superheat. A common range is −40 to 65°F. The sensing bulb is strapped or clamped to the refrigerant line and covered with insulating material (such as a foam rubber sheet) to prevent air circulation over the bulb while a reading

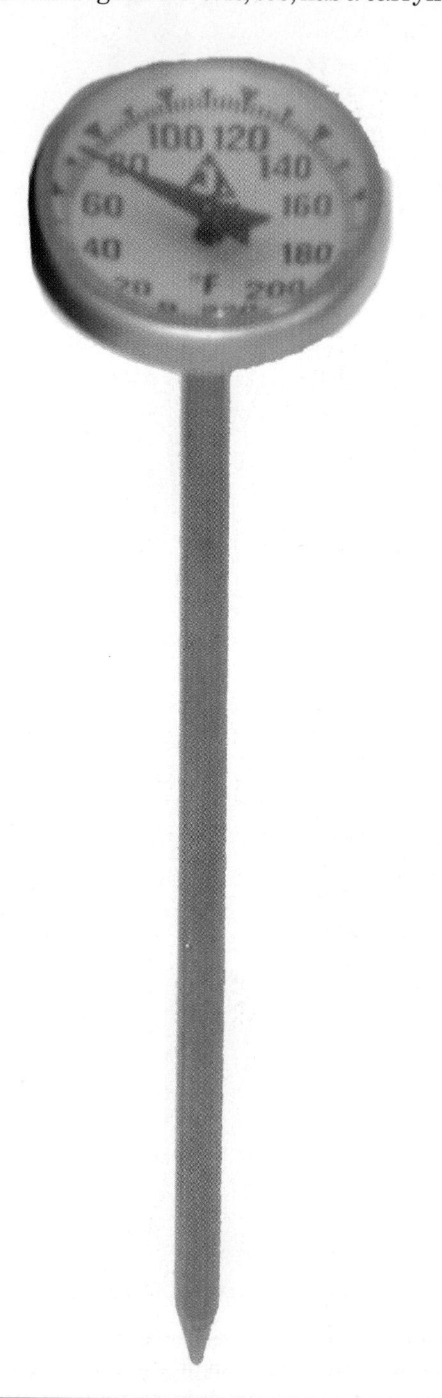

Figure 6-3-6 Dial-type pocket thermometer.

Figure 6-3-7 Superheat thermometer.

is taken. The superheat bulb thermometer can be used to measure air or water temperatures as well.

The thermometers just described have been the basic temperature measuring tools for many years; however, they do have certain limitations. For example, the operator is required to be present in the immediate area during the time of reading. An example of this limitation would be trying to measure the inside temperature of your home refrigerator without opening the door.

The thermometer illustrated in Figure 6-3-8 is also a hand held device, but it has a probe length of 30 in for "remote" temperature measurement.

Different probes are available for measuring surface temperatures. Some examples of these were seen in Figure 2-2-11a–c on page 56. In refrigeration use, the surface probe determines superheat settings of expansion valves, motor temperature, condensing temperature, and water temperature. This is a useful instrument for many applications but it is limited since only one reading at a time can be taken.

As a result of the rapid development of low cost electronic devices, the availability and use of electronic thermometers is now very common. The electronic thermometer as illustrated in Figure 6-3-8 consists of a tester with provision to attach one or several sensing leads.

The sensing lead tip is a thermistor element. Its resistance changes with temperature changes, varying the electrical current in the test circuit. The changes in electrical current are then converted to temperature readings. The sensing leads vary in length depending on the make of the unit, but extensions can be used for remote testing.

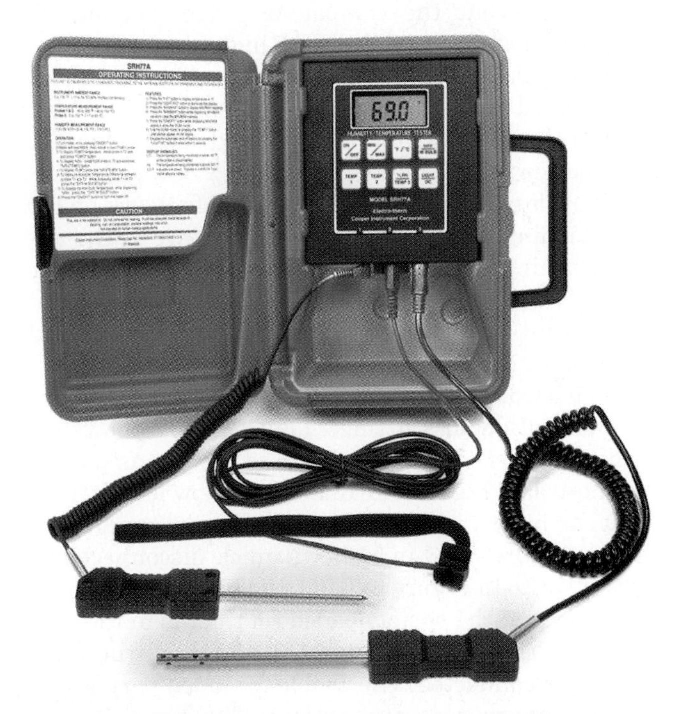

Figure 6-3-8 Thermometer capable of remote temperature measurement. (*Courtesy Cooper-Atkins Corporation*)

Once the sensing probes are positioned in the areas to be tested, the technician may record temperatures without actually going into each test area, such as the refrigerator, walk-in cooler, freezer, or air duct. The sensing probe can also be used to check superheat.

Sometimes it may be necessary to record temperatures over long periods of time such as a day or even a week, in order to examine the changes in system conditions. Recording thermometers and compact, portable recording thermometers, Figure 6-3-9, which consist of a hand wound chart driving unit, take the guesswork out of setting the system operating conditions or diagnosing and locating trouble areas—as well as providing permanent records of the results.

The final selection of temperature measuring instruments will depend on the scope of work with which a technician is involved. In servicing HVAC/R equipment, the technician will need a variety of thermometers. It is important to remember that these are sensitive devices and require consistent care and calibration to provide accuracy and reliability.

PRESSURE MEASUREMENTS

Temperature measurements are usually taken outside the operating system, but it is also necessary for the service technician to know what is going on inside the system. This must basically be learned from pressure measurements.

Two pressure gauges are necessary to determine the performance of the system as shown in Figure 6-3-10.

The right gauge of Figure 6-3-10 is the high pressure gauge, which measures high side or condensing pressures. It is normally graduated from 0 to 500 psi in 5 lb increments.

The compound gauge (left, in Figure 6-3-10) is used on the low side (suction pressures) and is normally graduated from a 30 in vacuum to 120 psi; thus, it can measure pressure above and below atmospheric pressure. This gauge is calibrated in 1-lb increments. Other pressure ranges are available for both gauges, but these two are the most common.

Some dial faces have an intrascale which gives the corresponding saturated refrigerant temperatures at a particular pressure. Gauges for specific refrigerants and for SI measurements are also available.

A service device that includes both the high pressure and compound gauges is called a gauge manifold. It enables the technician to check system operating pressures, add or remove refrigerant, add oil, purge noncondensibles, bypass the compressor,

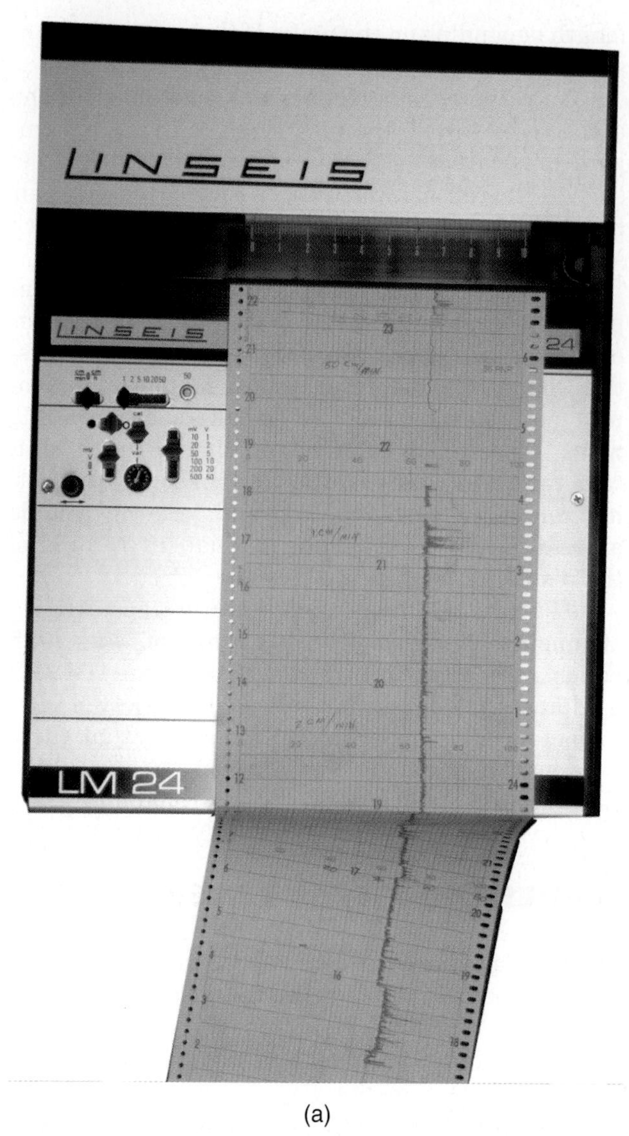

(a)

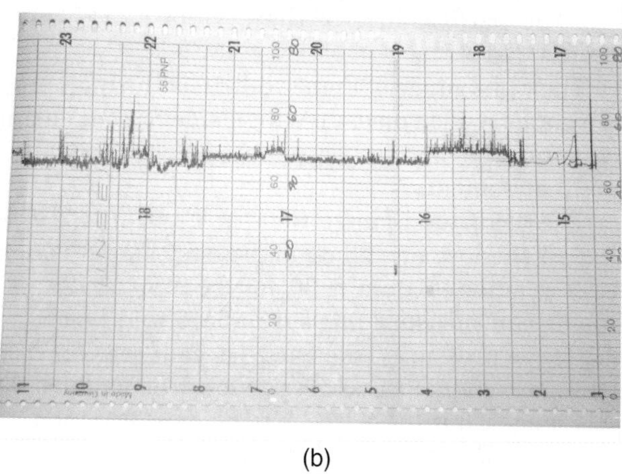

(b)

Figure 6-3-9 (a) Portable recording thermometer; (b) Recording thermometer chart.

Figure 6-3-10 Pressure gauge set. (*Courtesy Imperial*)

analyze system conditions, and perform many other operations without replacing gauges or trying to operate service connections in inaccessible places.

The testing manifold as illustrated in Figure 6-3-11 consists of a service manifold containing service valves.

On the left the compound gauge (suction) is mounted and on the right is the high pressure gauge (discharge). On the bottom of the manifold are hoses which lead to the equipment suction service valve (left), refrigerant drum (middle), and the equipment discharge, or liquid line valve (right).

Low loss fittings are not required for work on air conditioning and refrigeration as they are in the automotive industry. They are, however, a good idea for a number of reasons: first, they reduce the amount of refrigerant lost each time the system is accessed, and second, they prevent the refrigerant from possibly being released, causing skin burns. There are a number of types of low loss fittings. Figure 6-3-12 shows refrigerant hoses that have low loss fittings as part of the hose end. Figure 6-3-13a,b shows low loss fittings that can be screwed onto a standard set of refrigerant hoses. Figure 6-3-14a,b shows quick disconnect low loss fittings. In Figure 6-3-14a, the fittings are seated together and the spring loaded retainer is in the connector groove. But in Figure 6-3-14b, the fittings are shown disconnected. The visible threaded part of this fitting screws onto the system. The knurled part screws onto the hose. This allows for a snap on, snap off system for quick access. Figure 6-3-15 shows short

Figure 6-3-11 Testing manifold with glycerin filled gauges. (*Courtesy Imperial*)

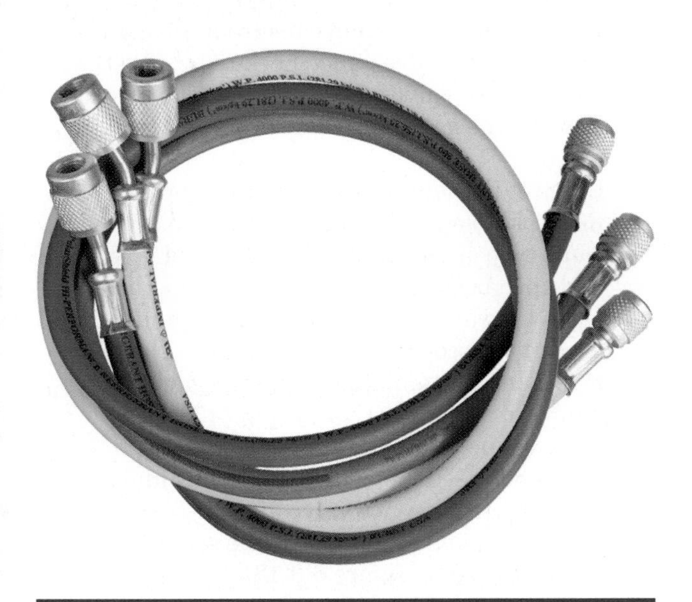

Figure 6-3-12 Hoses with low loss fittings.

stubby hoses that have a manual inline shutoff valve. These are a positive action low loss fitting that can be attached to the system. An advantage of these is that they are sometimes easier to fit onto access valves that are in very cramped locations.

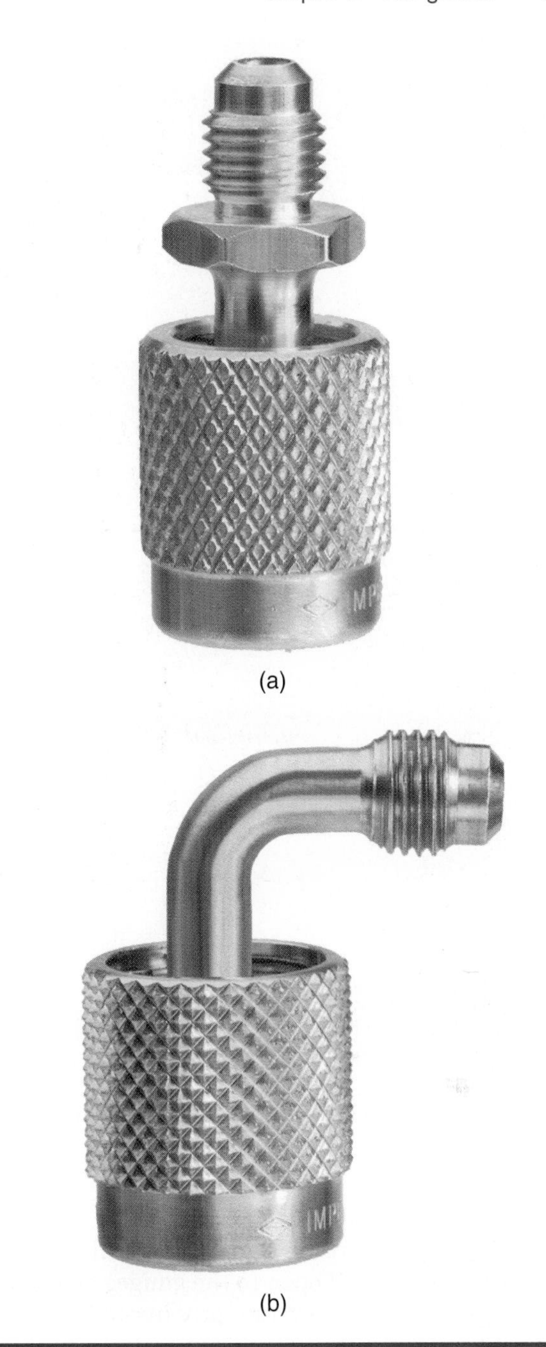

(a)

(b)

Figure 6-3-13 Low loss hose adaptor: (a) Straight; (b) 90°.

Many equipment manufacturers color code the low side gauge casing and hose blue and the high-side gauge and hose red. The center or refrigerant hose is yellow. This system is very helpful in avoiding crossing hoses and damaging gauges. A hook is provided on which to hang the assembly and free the operator from holding it.

By opening and the closing the refrigerant valves on gauge manifold a and b in Figure 6-3-16, different refrigerant flow patterns can be obtained.

The valve is so arranged that when the valves are closed (front seated), the center port on the manifold is closed to the gauges, as in Figure 6-3-16a.

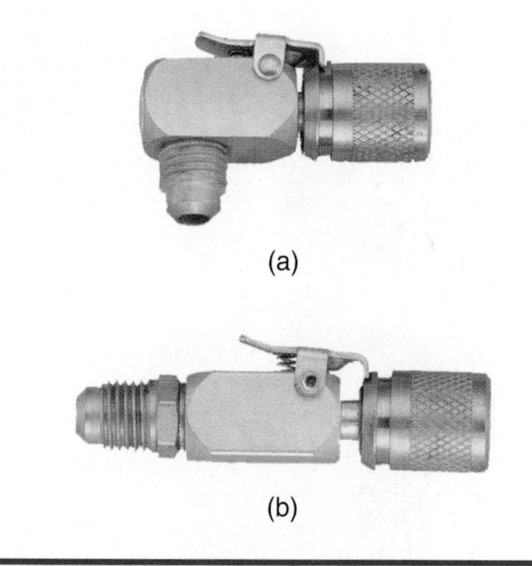

Figure 6-3-14 Low loss quick connects.

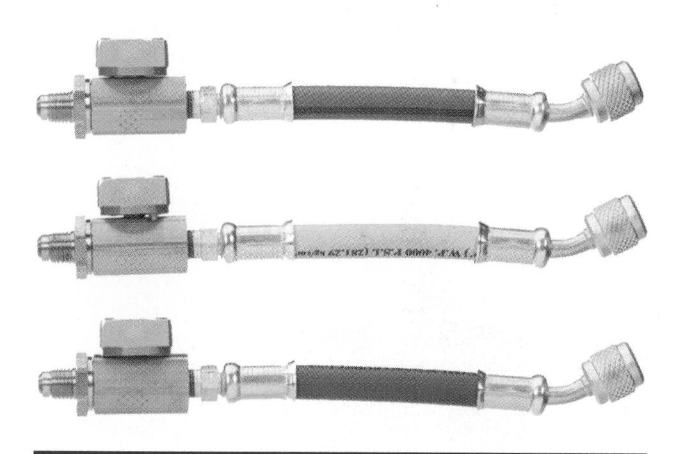

Figure 6-3-15 Short hose sections with manual shut off valves.

When the valves are in the closed position, gauge ports 1 and 2 are still open to the gauges, permitting the gauges to register system pressures.

With the low side valve open and the high side valve closed as in Figure 6-3-16b, the refrigerant is allowed to pass through the low side of the manifold and the center port connection. This arrangement might be used when refrigerant or oil is added to the system.

Figure 6-3-16c illustrates the procedure for bypassing refrigerant from the high side to the low side. Both valves are open and the center port is capped. Refrigerant will always flow from a high pressure area to a lower pressure area.

Figure 6-3-16d shows the valve arrangement for purging or removing refrigerant. The low side valve is closed. The center port is connected to an empty refrigerant drum. The high side valve is opened, permitting a flow of high pressure refrigerant out of the center port.

The method of connecting the gauge manifold to a refrigerant system depends on the state of the system—that is, whether the system is operating or just being installed. For example, let's assume that the system is operating and equipped with back seating inline service valves, as shown in Figure 6-3-17.

SERVICE TIP

Under EPA regulations, air conditioning technicians are allowed to purge the gauge hoses. This is done to remove air and other contaminants from the hose to protect the system. This is not a violation of regulations provided that it is a de mimimis release. Purging the line does not mean opening the valve and waiting for an extended period of time. Purging the hose is done by opening the valve and allowing some refrigerant to flow at a low rate through the hose as it is being attached to the system's access valve. A good rule of thumb is to not purge more than the count of 2.

The first step is to purge the gauge manifold of contaminants before connecting it to the system. Purging and connecting is done with the following procedure:

1. Remove the valve stem caps from the equipment service valves and check to be sure that both service valves are back seated.
2. Remove the gauge port caps from both service valves.
3. Connect the center hose from the gauge manifold to a refrigerant cylinder, using the same type of refrigerant that is in the system, and open both valves on the gauge manifold.
4. Open the valve on the refrigerant cylinder for about two seconds, and then close it. This will purge any contaminants from the gauge manifold and hoses.
5. Next, connect the gauge manifold hoses to the gauge ports—the low pressure compound gauge to the suction service valve and the high pressure gauge to the liquid line service valve, as illustrated in Figure 6-3-18.
6. Front seat or close both valves on the gauge manifold. Crack (turn clockwise) both service valves one turn off the back seat. The system is now allowed to register on each gauge. With the gauge manifold and hoses purged and connected to the system, we are free to perform whatever service function is necessary within the refrigeration cycle.

(a)

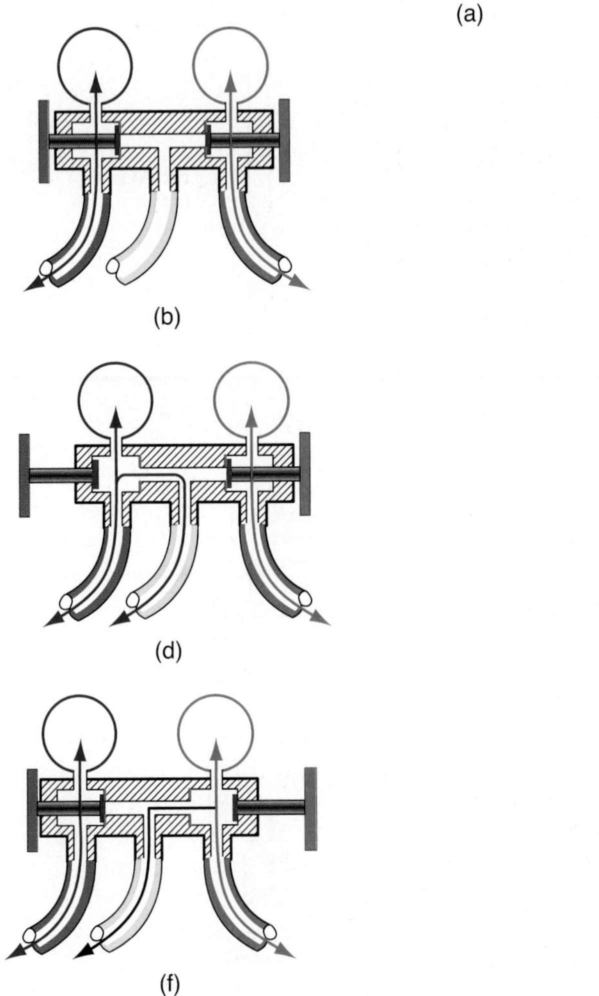

(b) (d) (f)

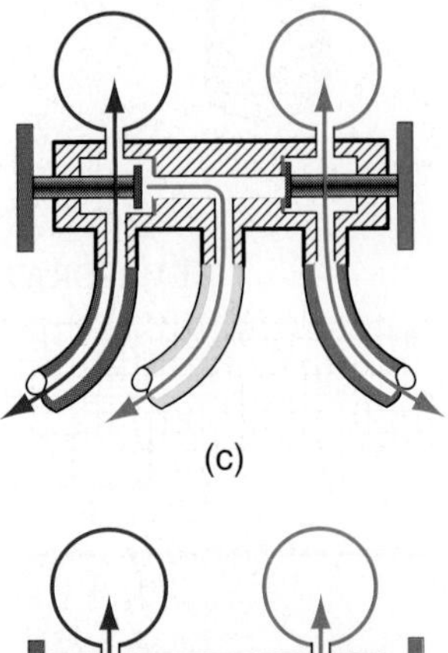

(c)

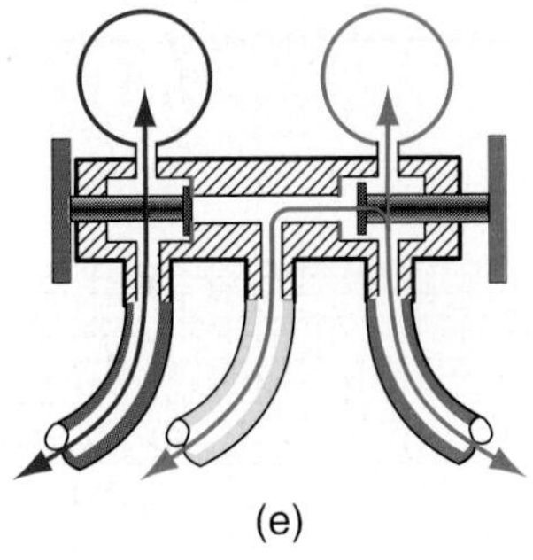

(e)

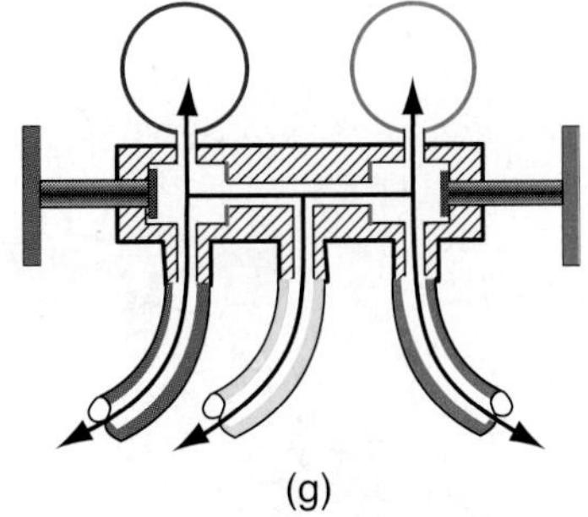

(g)

Figure 6-3-16 Operation of gauge manifold set.

To remove the gauge manifold from the system, follow this procedure:

1. Back seat the liquid line first (counterclockwise).
2. Mid seat the valves on the gauge manifold to allow the liquid refrigerant to be pulled into the suction side of the system.
3. After the pressures equalize, back seat the suction service valve on the compressor.
4. Remove hoses from gauge ports and seal the ends of the hoses by reattaching them to the back of the gauge manifold set.
5. Replace all gauge port and valve stem caps.
6. Leak test using a refrigerant leak detector or soap bubbles to guarantee that valves are fully closed and not leaking.

The manifold and gauges are necessary tools to perform many system operations. Once the system has been completed and cleaned of most of the air by purging, it must be tested for leaks. Also, whenever a component has been repaired or replaced, it is imperative that the entire system be checked for leaks. Leak testing is discussed further in Chapter 6, Unit 4.

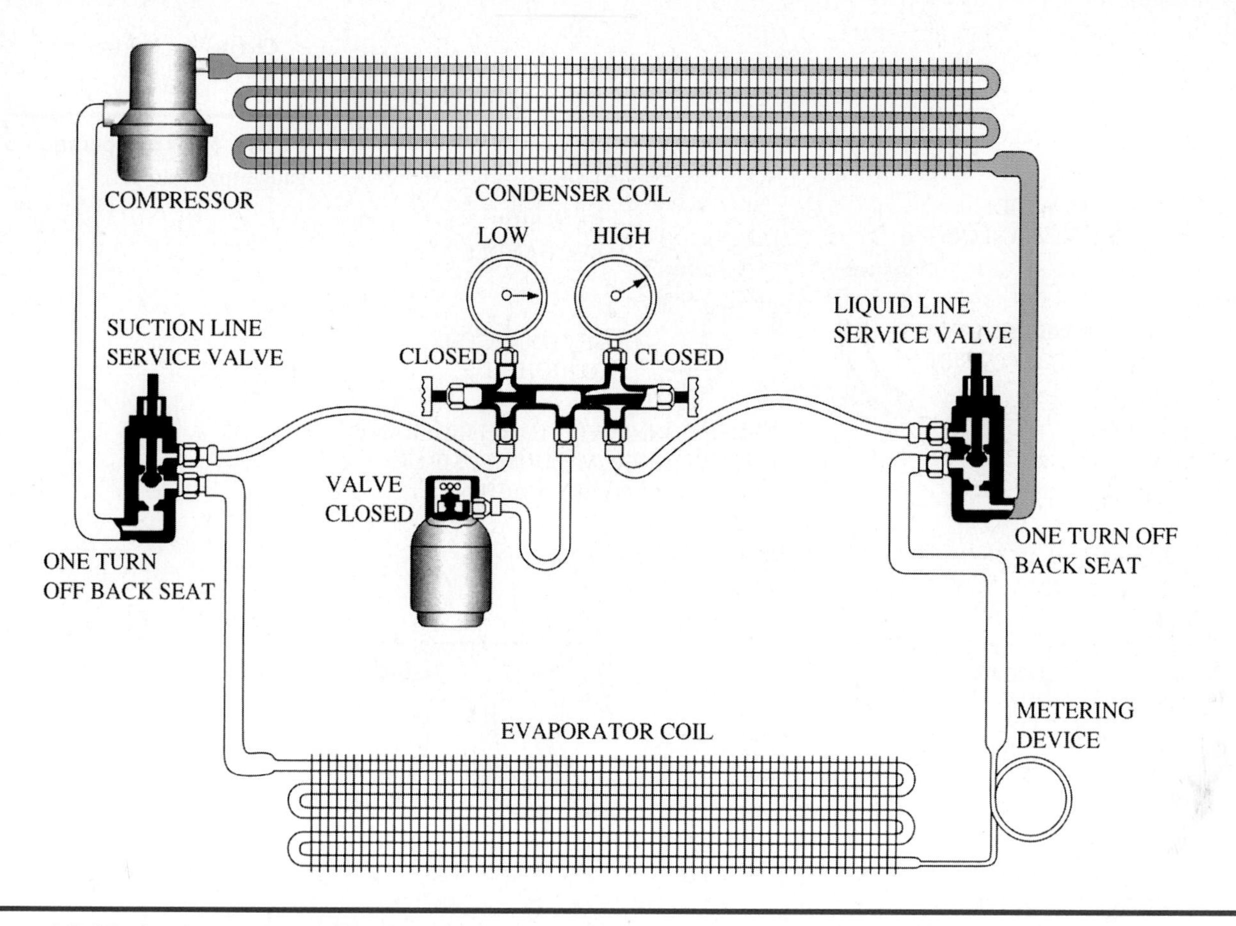

Figure 6-3-17 Connecting manifold.

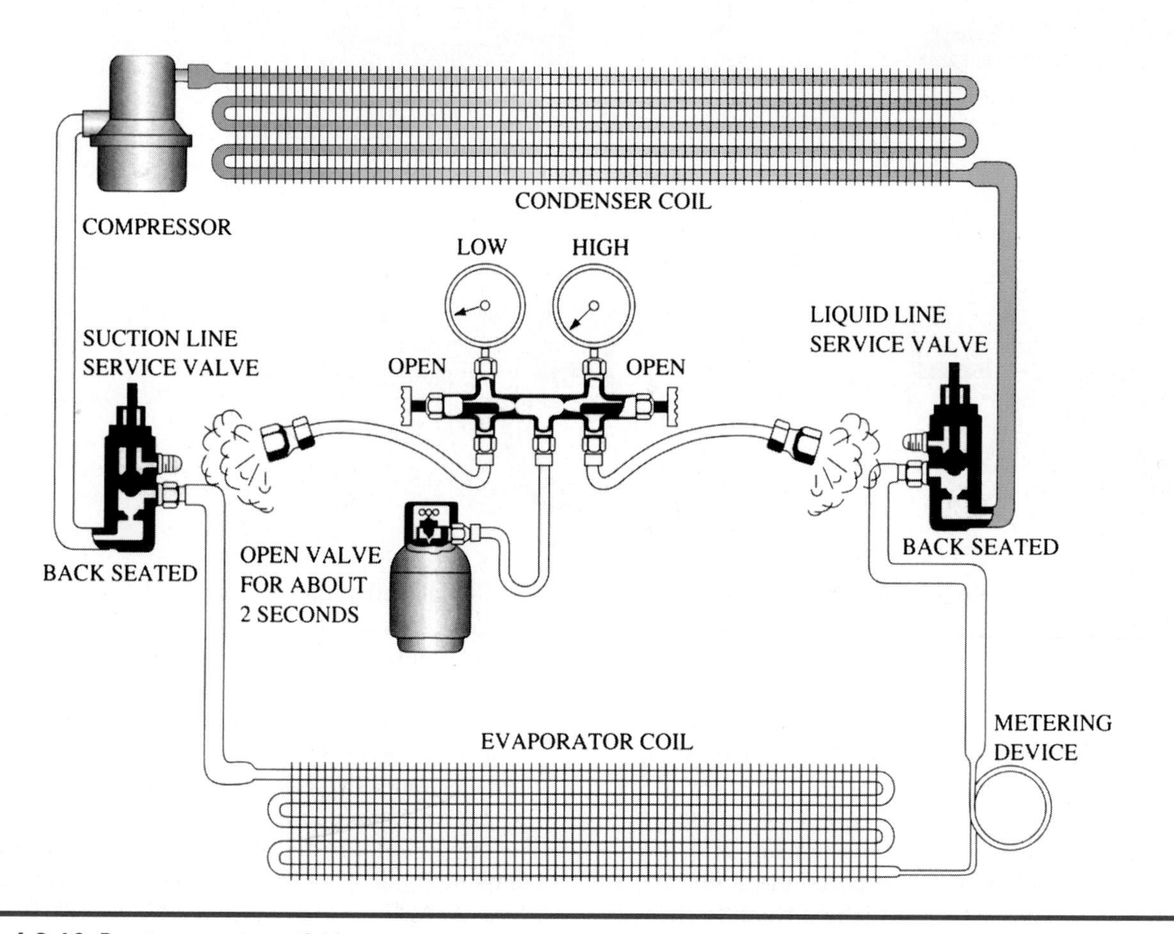

Figure 6-3-18 Purging gauge manifold.

Purging

Purging the entire system with pure CFC or HCFC refrigerants, following servicing the system, is not allowed under EPA regulations. If refrigerants are removed from the system, they must be recovered. If the system has been opened for any reason, and air and moisture permitted to enter, the system should be evacuated before recharging.

Whenever a defective component such as an expansion valve is to be removed, the system should be pumped down and the part of the system should be isolated by means of the service valves. Then when the new component is installed the lines should be purged from both sides. Pumping down means to store the refrigerant in the receiver or condenser.

EVACUATION

Proper evacuation of a unit will remove noncondensibles (mainly air, water, and inert gases) and assure a tight, dry system before charging. The tools needed to evacuate a system properly are a good vacuum pump and vacuum indicator.

Figure 6-3-19 shows a tool used to remove and replace the valve core from Schrader valves. By removing the core, a vacuum can be pulled much quicker. The valve core is a restriction to gas flow and should be removed. This can be done with this tool without opening the system to the atmosphere.

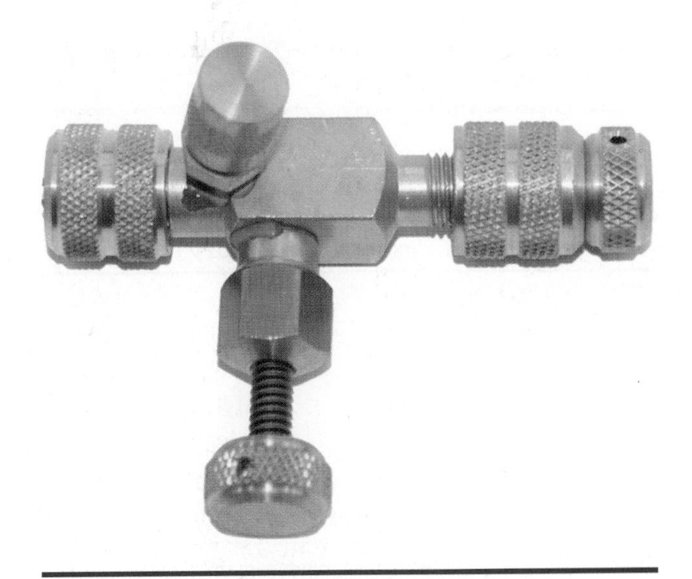

Figure 6-3-19 Valve core remover.

SERVICE TIP

Most manifold gauge sets have an internal opening and hoses with ¼ in diameter. Larger gauges and hoses are available that have a 3/8 in interior diameter. The larger the interior diameter and the shorter the hose, the faster a system can be evacuated. As the pressure in the system drops, the rate of withdrawal is significantly hampered when working with the smaller diameter hoses. For most residential and light commercial applications this time is not a major problem. However, for larger commercial installations it can become a significant factor in the overall job time. Many manufacturers of gauges provide these larger diameter hoses and frequently they are black in color and less flexible than the normal gauge hoses.

Vacuum Pump

A vacuum pump, illustrated in Figure 6-3-20, is somewhat like an air compressor in reverse. Most pumps are driven by a direct drive or belt driven electric motor, but gasoline engine driven pumps are also available. The pump may be single or two stage depending on the design. Most pumps for normal field service are portable. They have carrying handles or are mounted on dollies.

Figure 6-3-20 Vacuum pump. *(Courtesy Yellow Jacket Division, Ritchie Engineering Company)*

Vacuum pump sizes are rated according to the free air displacement in cubic feet per minute (CFM), or liters per minute (l/pm) in the SI system. Specifications may also include a statement as to the degree of vacuum the pump can achieve, expressed in terms of microns.

TABLE 6-3-2 Comparison of Three Different Pressure Measuring Systems

Boiling Point of Water		Unit of Absolute Pressure		Units of Vacuum (in Hg)
°F	°C	psia	Microns of Mercury	
212	100	14.7	—	0
79	26	0.5	25,400	28.9
72	22	0.4	20,080	29.1*
32	0	0.09	4,579	29.7*
−25	−31	0.005	250	29.8*
−40	−40	0.002	97	29.9*
−60	−51	0.0005	25	29.92*

*Too small a change to be seen on a gauge manifold set.

Microns

When the vacuum pressure approaches 29.5 to 30 in on the compound gauge, the gauge is working within the last half inch of pressure, and the readout beyond 29.5 in is not reliable for the single deep vacuum method. The industry has therefore adopted another measurement, called the micron. The micron is a unit of linear measure equal to 1/25,400 of an inch and is based on measurement above total absolute pressure, as opposed to gauge pressure, which can be affected by atmospheric pressure changes. Table 6-3-2 shows a comparison of measurements starting at standard atmospheric conditions and extending to a deep vacuum.

Table 6-3-2 not only demonstrates the comparison in units of measure but dramatically shows the changes in the boiling point of water as the evacuation approaches the perfect vacuum. This is the main purpose of evacuation—to reduce the pressure or vacuum enough to boil or vaporize the water and then pump it out of the system. It should be noted that the compound gauge could not possibly be read to such minute changes in units of in Hg.

HIGH VACUUM INDICATORS

To measure these high vacuums the industry developed electronic instruments, such as those shown in Figure 6-3-21.

In general, these are heat-sensing devices in which the sensing element, which is mechanically connected to the system being evacuated, generates heat. The rate at which heat is carried off changes as the surrounding gases and vapors are removed. Thus, the output of the sensing element (either thermocouple or thermistor) changes as the heat dissipation rate changes, and this change in output is indicated on a meter calibrated in microns of mercury.

The degree of accuracy of these instruments is approximately 10 microns, thereby approaching a perfect vacuum as shown in Table 6-3-1.

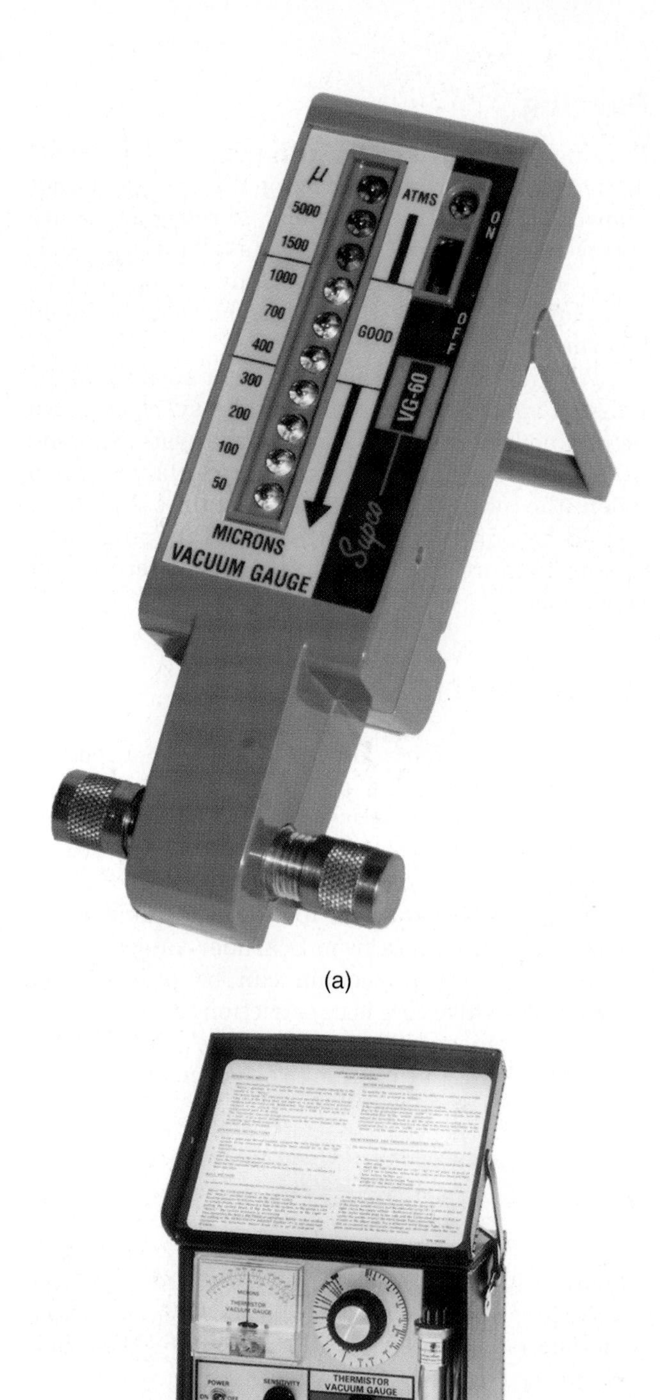

Figure 6-3-21 (a) Digital micron gauge; (b) Analog micron gauge. (*Courtesy Robinair SPX/OTC*)

DEEP METHOD OF EVACUATION

The single deep vacuum method is the most positive method of assuring a system free of air and water. It takes longer but the results are far more positive. A vacuum pump should be selected that is capable of pulling at least 500 microns and has a reliable electronic vacuum indicator. The procedure is illustrated in Figure 6-3-22 and described below.

SERVICE TIP

Evacuating the system is done to remove both air and moisture. In removing moisture it is often referred to as dehydrating the system. Moisture may exist in a system in one of two forms, as either liquid water or as water vapor. Liquid water is more difficult to remove because it requires 970 Btu per pound of heat input to convert it from a liquid to a vapor. Systems that contain molecular water require additional external heat to ensure that they are properly dehydrated.

1. Install the gauge manifold as described earlier.
2. Connect the center hose to the vacuum manifold assembly. This is simply a three valve operation for attaching the vacuum pump and vacuum indicator and a cylinder of refrigerant, each with a shutoff valve.
3. Open the valves to the pump and indicator. Close the refrigerant valve. Follow the pump manufacturer's instructions for pump suction line size, oil, indicator location, and calibration.
4. Open wide both valves on the gauge manifold and mid seat both equipment service valves.
5. Start the vacuum pump and evacuate the system until a vacuum of at least 500 microns is achieved.
6. Close the pump valve and isolate the system. Stop the pump for 5 minutes and observe the vacuum indicator to see if the system has actually reached 500 microns and is holding. If the system fails to hold, check all connections for tight fit and repeat evacuation until the system does hold.
7. Close the valve to the indicator.

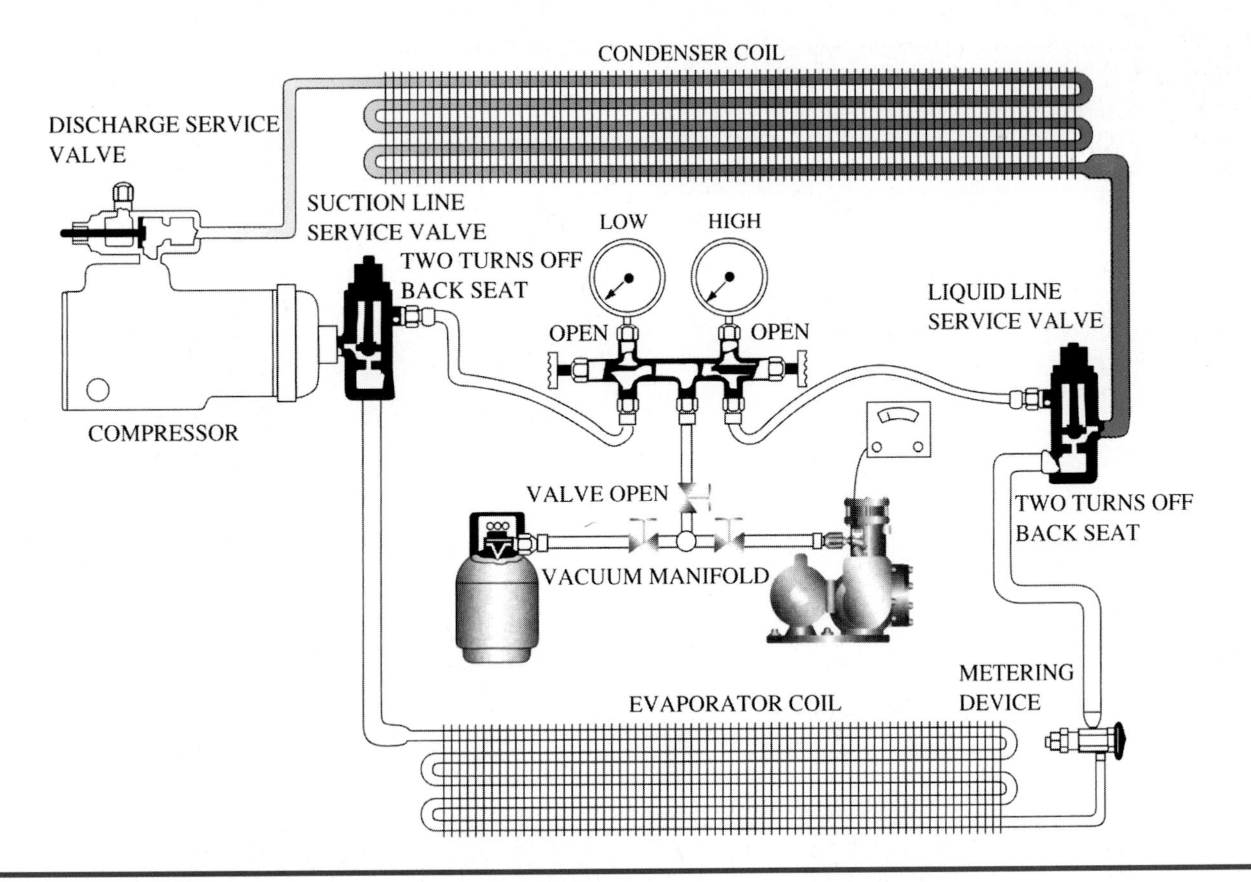

Figure 6-3-22 Deep vacuum method of evacuating the system.

TECH TALK

Tech 1. How would EPA know that I'm not recovering refrigerant?

Tech 2. You have to keep a refrigerant log showing where you have used refrigerant and where you have recovered it.

Tech 1. Yeah, but if I don't have the log they can't prove that I haven't been doing it right.

Tech 2. It doesn't work that way. If you don't have a log showing where you have recovered refrigerant then the EPA is going to assume you have not been doing it right.

Tech 1. Oh, my gosh! I didn't know that.

Tech 2. You need to keep a record of all the cylinders you buy and keep a log of where the refrigerant from each cylinder has gone. Write down the amount of the refrigerant, the date, the time, and the address. You could put it in a book or you can keep it on a list. But you have to keep it. You can pick up from the supply house a little paperback log book for refrigerant. That might be the easiest thing to do and it has all the information you need. In addition you need to keep a log on all the refrigerant you recover, how much you recovered and what type of system it was.

Tech 1. Gosh! I have not been doing any of that.

Tech 2. You really need to start and keep good records. That will keep you out of trouble.

Tech 1. Thanks. I better get busy.

8. Open the valve to the refrigerant cylinder and raise the pressure to at least 10 psig or charge the system to the proper level.
9. Disconnect the pump and indicator.

UNIT 3—REVIEW QUESTIONS

1. What does the Clean Air Act establish?
2. Under what conditions does the EPA permit refrigerant releases?
3. What is the advantage of the liquid method of transfer?
4. What is the most common temperature measuring device?
5. How do you check the calibration of a pocket glass thermometer?
6. What two pressure gauges are necessary to determine the performance of a system?
7. Why are low loss fittings on air conditioning and refrigeration a good idea?
8. Under EPA regulations are technicians allowed to purge the gauge hoses?
9. What tools are needed to evacuate a system properly?
10. How are vacuum pump sizes rated?
11. Define micron.
12. What is the most positive method of assuring a system free of air and water?

UNIT 4

Refrigerant Charging

OBJECTIVES: In this unit the reader will learn about refrigerant charging including:

- System Charging
- Leak Testing
- Oil Charging

SERVICE TIP

Too often technicians in the HVAC/R field are of the opinion that if a little refrigerant is good a lot is better and too much is just about right. Refrigerant is not a magical fluid that adds cooling capacity to a system. The cooling capacity of a system that is undercharged or overcharged is significantly decreased. The only time a system will operate with its designed peak performance is when the charge is at the manufacturer specified level. It is not an acceptable practice to add just a little more in case the system leaks.

SYSTEM CHARGING

Whether the system is a new one or an existing one that has been repaired, the final step in putting the system in operation is to charge it with refrigerant. In any event, the process is the same.

Some systems are more critical than others about the amount of the refrigerant charge. Systems that have a receiver—usually the larger systems—are not as critical since extra refrigerant can be stored in the receiver, and the expansion valve feeds the refrigerant into the evaporator as required to match the load.

In the smaller systems that do not have a receiver, any excess refrigerant will be stored in some part of the system where it reduces the effectiveness of that part and reduces the capacity of the system. If the system is short of refrigerant, the metering device is not supplied with a solid stream of refrigerant on full load and the evaporator will be starved for refrigerant. This will also reduce the capacity of the system. So, in the smaller systems it is critical that the proper charge be determined.

One way to determine the proper charge is to read it on the nameplate as shown in Figure 6-4-1. This is the charge specified by the manufacturer.

On some systems, the length of the refrigerant lines is determined by the conditions of the installation. For example, on a split system the amount of charge required is affected by the length of the refrigerant lines. Most manufacturers give specific information for determining the charge on this basis. For example, the manufacturer may state, "charge adequate for matched system including up to 25 ft of lines." Be sure to refer to the manufacturers' information.

A LIQUID OR VAPOR CHARGE

The unit can be charged with either liquid or vapor refrigerant. Whichever method is used, it is important to charge the system with the right amount of refrigerant and to protect the compressor from any damage that might be caused by liquid slugging the compressor.

Since the refrigerant cylinder is not filled over 80%, in an upright position the vapor is at the top and the liquid at the bottom. To charge with vapor, the refrigerant cylinder must be in the upright position. To charge with liquid, the cylinder is turned upside down, as shown in Figure 6-4-2.

It is usually considered good practice to charge with vapor, as in Figure 6-4-3, rather than liquid to prevent any danger of slugging the compressor with

F.L.A.		F.L.A.	
L.R.A.		L.R.A.	
H.P.		H.P.	
VOLTS		VOLTS	
HERTZ PHASE		HERTZ PHASE	
REFRIGERANT 22	5.0. LB.		KG.

Figure 6-4-1 Unit nameplate.

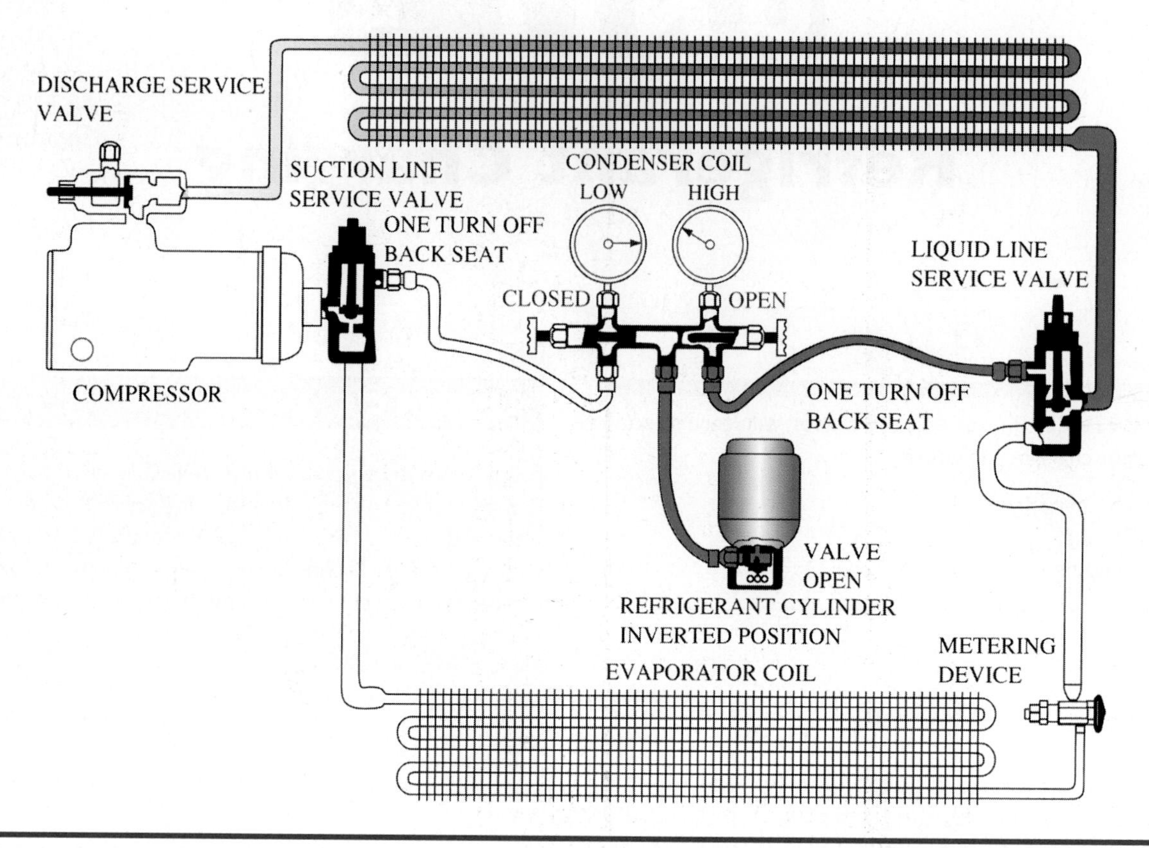

Figure 6-4-2 Liquid charging.

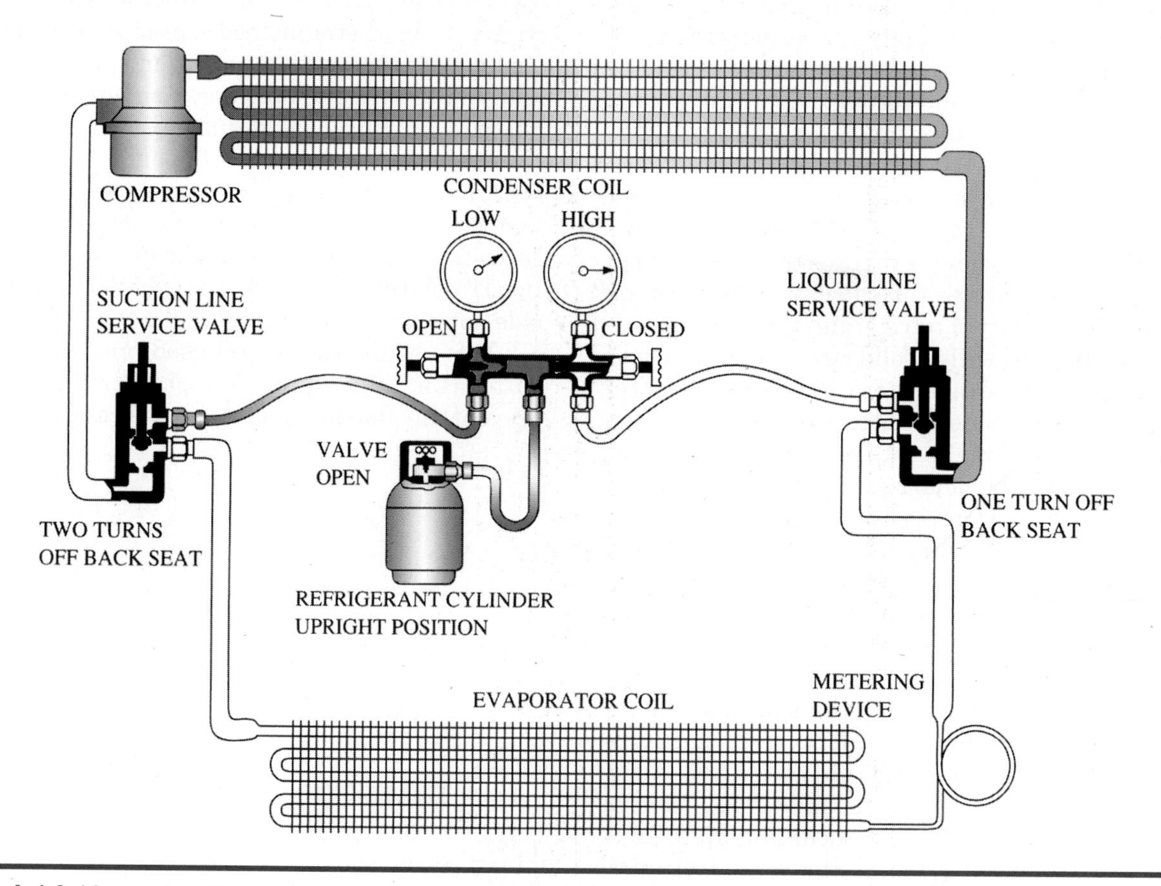

Figure 6-4-3 Vapor charging.

liquid. However, when using refrigerant blends, sometimes the tank will be constructed so that when the cylinder is right side up, it will discharge liquid. Refrigerant blends should not be charged as a vapor because the refrigerant will fractionate. Fractionation is the separation of the different gases that make up the refrigerant blend.

Vapor Charging

Prior to charging, the system must be leak tested and evacuated. When the charging is started, the system is under vacuum so that when the refrigerant enters the system it will be pushed into the unit due to the difference in pressure.

TECH TIP

When charging refrigerants that fractionate, such as R-401a, they must be charged as liquid to prevent the separation of the refrigerant as it enters the system. These refrigerants must come out of the cylinder as a liquid but must be a vapor when they enter the system. To accomplish this safely some gauge manifold sets have sight glasses so that you can use the low side gauge hand wheel as a metering device to allow the liquid refrigerant to vaporize at the gauge set. For gauges that do not have this, a liquid line sight glass that has 1/4 in flare fittings on both ends can be installed in the low side refrigerant hose.

To charge with vapor refrigerant, the system is connected to the gauge manifold in the usual way, as shown in Figure 6-4-4. Both valves on the gauge manifold are back seated. When the refrigerant is released from the cylinder, the vapor flows into both sides of the system.

The refrigerant will stop flowing when the pressures equalize and no more refrigerant will enter the system.

The next step is to close (front seat) the high side valve as shown in Figure 6-4-5 and to continue the balancing of the vapor charging through the low side of the system.

Figure 6-4-6 shows a charging sight glass that can be placed at the low side gauge port to see that only vapor is entering the system.

At this point check the scales or charging cylinder to see how much of the charge has entered the system and how much more is needed. Normally about 50 to 75% of the required charge is completed.

The balance of the vapor refrigerant will be charged with the compressor running. Even with the compressor running the process slows down. One way to speed up the process is to add heat to the cylinder. This is done by placing the cylinder in a water bath at a temperature not to exceed 125°F as shown in Figure 6-4-5.

The refrigerant measuring device should be watched closely. When the full charge has been added, turn off the compressor and disconnect the gauge manifold.

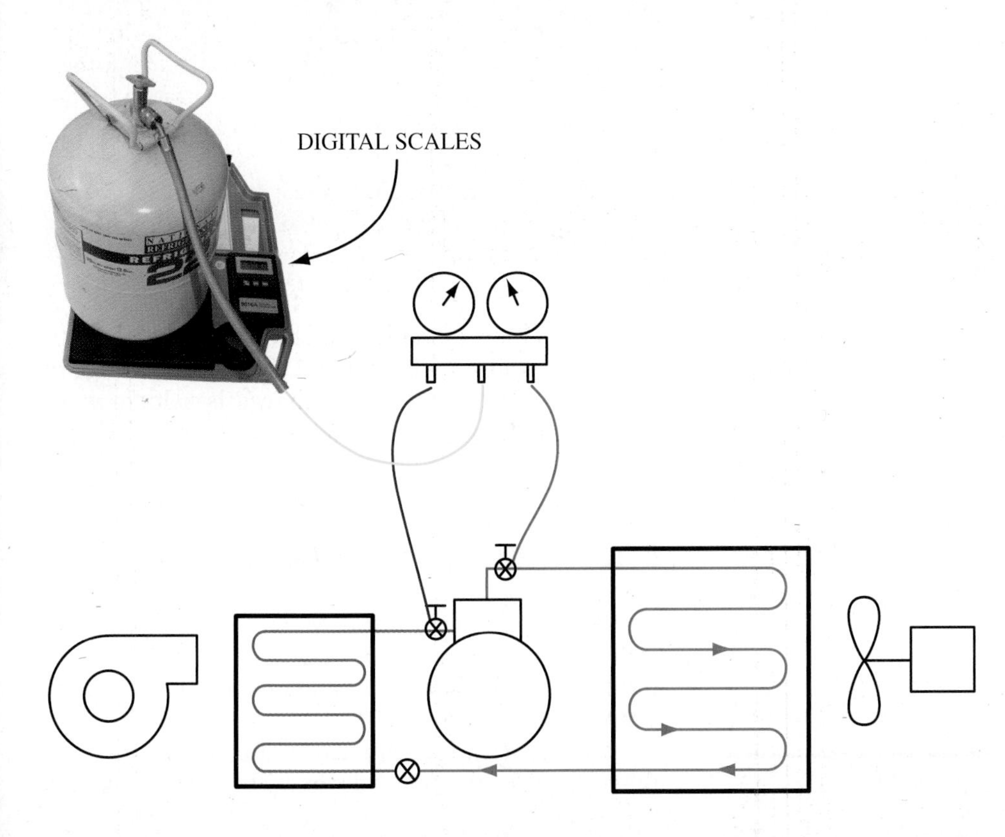

Figure 6-4-4 Refrigerant charging with vapor.

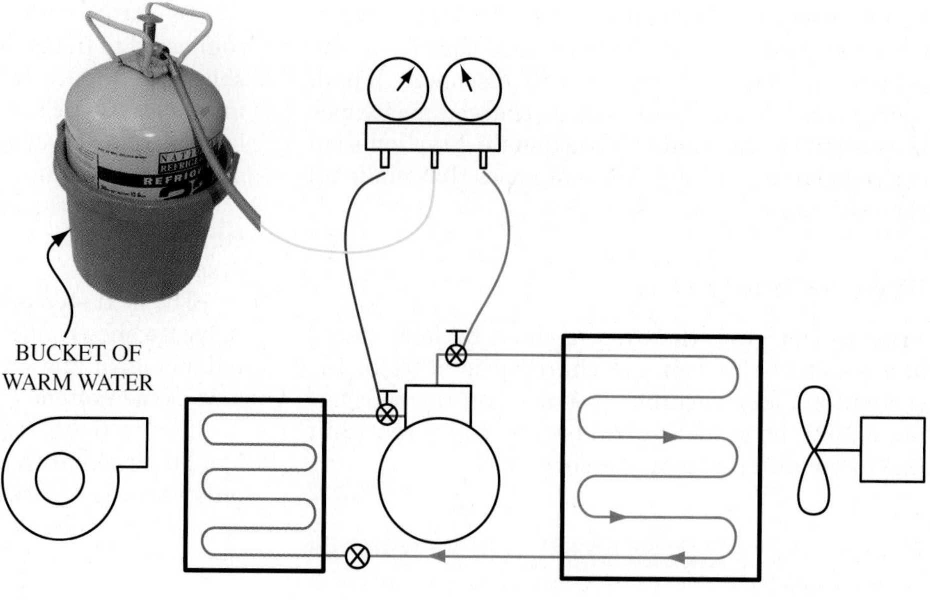

Figure 6-4-5 High side valve closed.

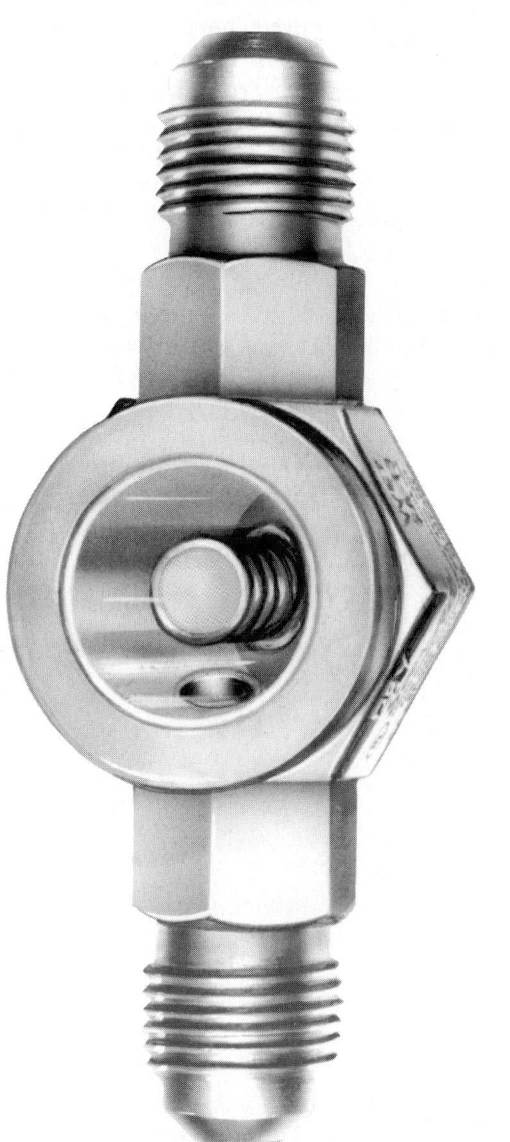

Figure 6-4-6 Low side vapor charging sight glass.

Liquid Charging

Liquid charging is always much faster than charging with vapor. Liquid is charged on the high side of the system. On larger systems, 20 tons or more, a king valve located between the condenser and the metering device offers a convenient means of charging the system on the high side, with the compressor running.

SERVICE TIP

On systems that require large quantities of liquid refrigerant to meet the basic charge requirement, liquid can be pumped in the system by using a cylinder heating band. These bands are thermostatically controlled and sense the cylinder pressure with a pressure sensitive switch that is connected to the refrigerant charging line. Cylinder heaters can significantly reduce the charging time.

On smaller systems the charging is done with the compressor off, Figure 6-4-7. The disadvantage of this method is that the full charge is seldom completed and the vapor method must be used to complete the process.

CORRECT CHARGING

Many problems occur if the system is overcharged or undercharged. The correct charge that is required and the charge that is installed should be carefully compared.

An undercharge can cause flash gas to form ahead of the metering device. This causes a low

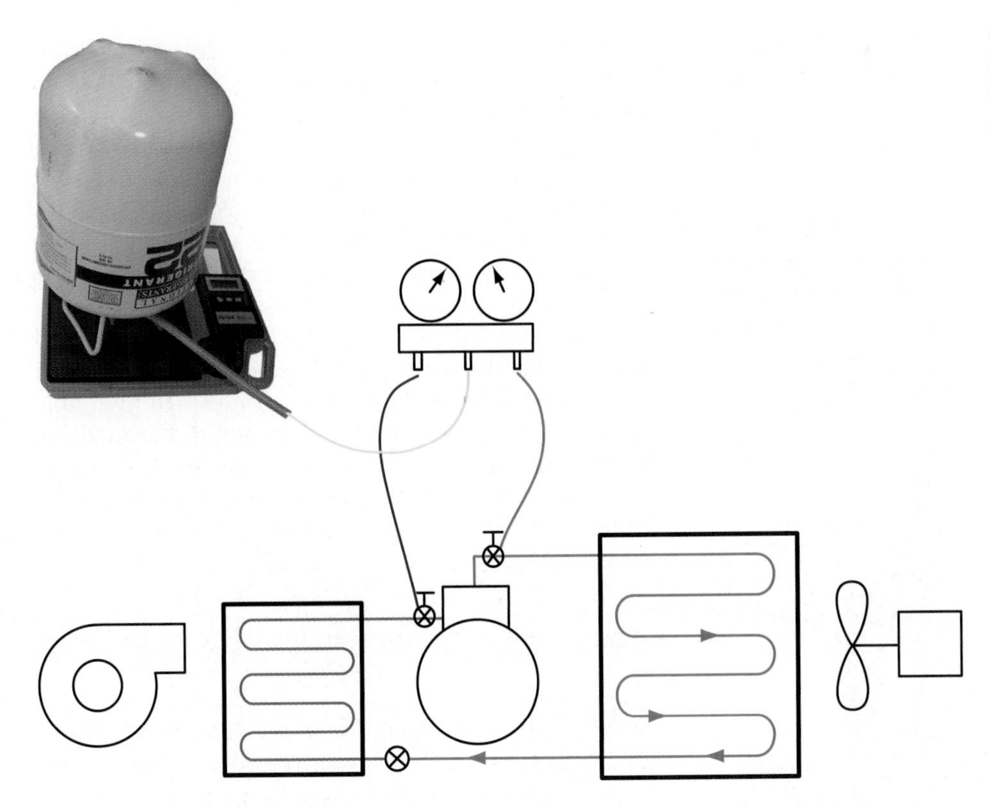

Figure 6-4-7 Liquid charging is preferred with the cylinder valve side down.

evaporator temperature and excess superheating of the suction gas. It can also cause poor cooling of a hermetically sealed gas-cooled motor. It can cause the discharge gas to be overheated. Sludge and harmful chemicals can form and the valves may carbonize.

An overcharge can cause the metering device to overfeed the evaporator. This causes liquid to flood back to the compressor. Liquid slugging will cause compressor damage. High side pressure will increase, causing loss of capacity.

Manufacturer Charging Charts and Calculators

Some manufacturers provide charging charts, charging calculators, or maximum performance charts for checking the charge while the system is in operation. It is important that the technician use whichever method the manufacturer provides to adjust the charge in factory charged units.

Some manufacturers' procedures require reading the outdoor dry bulb temperature, the indoor wet bulb temperature, and the system pressures. With this information, tables are available for making any necessary adjustments.

Charging with a Charging Cylinder

A charging cylinder, shown in Figure 6-4-8, offers an accurate method of charging a small system. The

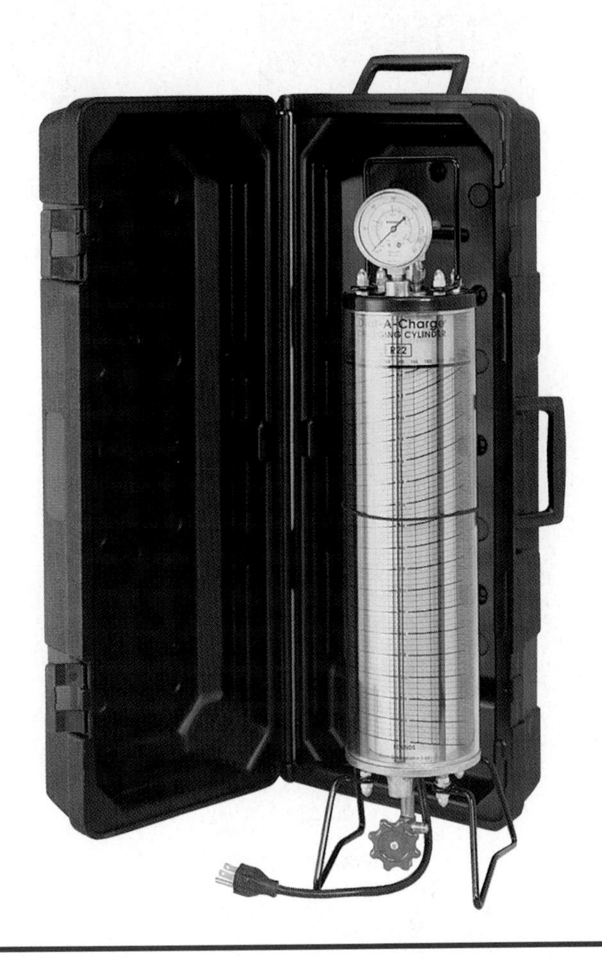

Figure 6-4-8 A charging cylinder. *(Courtesy of Robinair Division, SPX Corporation)*

scale on the side of the container is graduated in pounds and ounces. The amount of the charge can be accurately measured.

Charging by Weight

For a larger system, the best method is to charge by weight. Accurate scales like those shown in Figure 6-4-9 can be used to support the refrigerant cylinder and to weigh it and the refrigerant in it before and after the charge has been made. The difference between the two readings will give the weight of the refrigerant used. Some manufacturers require that systems be charged by the weighing in method when the outdoor temperature is 60°F or below.

For example, if the initial weight of the cylinder and refrigerant (before charging) in Figure 6-4-10 is 25 lb and the end weight (after charging) is 15 lb, 10 lb of refrigerant have been charged. The position of the hoses should not be changed during charging as this can cause inaccurate weighing in of refrigerant.

An electronic scale is preferred. These scales vary in size according to the amount of refrigerant being used. For small charges, scales are available that read down to fractions of an ounce. For medium sized charges, scales reading to the nearest ounce are satisfactory. For very large charges (50 tons of refrigeration or more), scales reading to the nearest pound are appropriate.

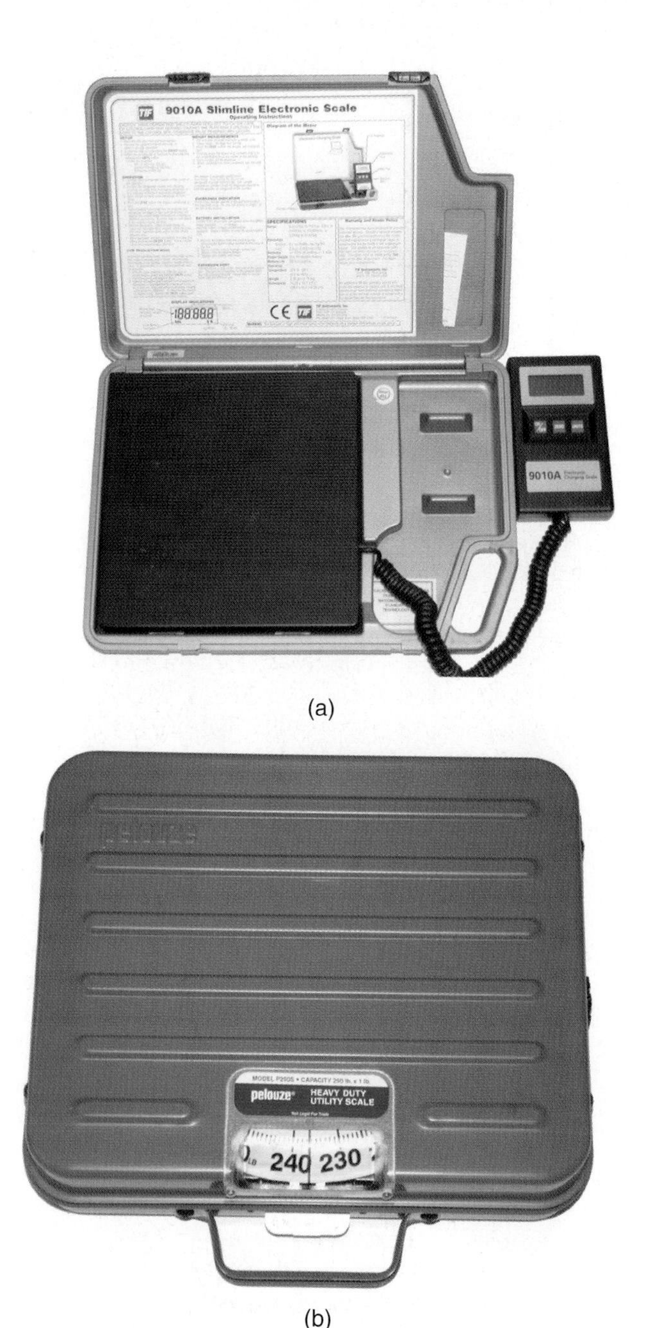

Figure 6-4-9 (a) Digital refrigerant scales; (b) Analog scale for weighing larger sized refrigerant drums.

TECH TIP

Digital scales are available in different weight ranges. Most scales will weigh cylinders weighing 50 lb or less. Heavy duty scales are available for the 125 lb tank.

Charging for Proper Superheat

Manufacturers may recommend checking the vapor superheat after the system has been put in operation,

Figure 6-4-10 Weighing refrigerant into a system.

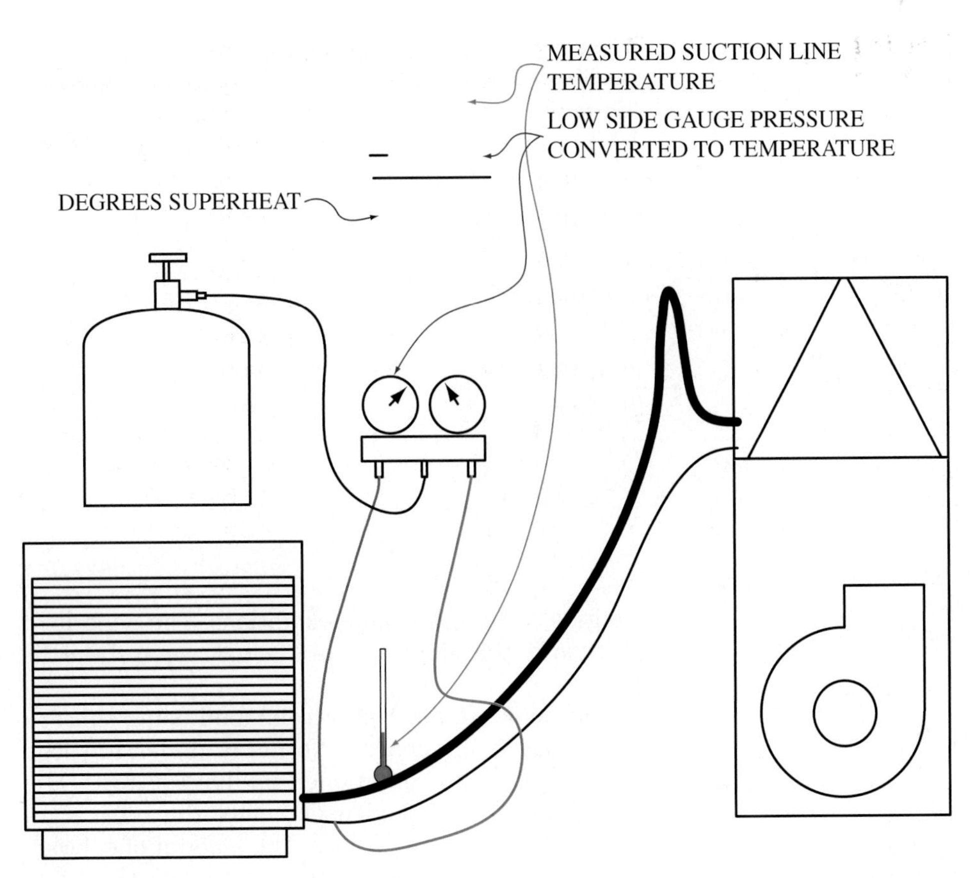

Figure 6-4-11 Calculating superheat.

to determine the proper refrigerant charge as shown in Figure 6-4-11.

The superheat charging method is used on fixed metering device systems. The full load superheat will vary depending on ambient temperatures.

The superheat on fixed metering device systems varies greatly under partial load conditions so it is important to follow the manufacturer's instructions, using the charts provided for the system being tested.

NICE TO KNOW

As the ambient temperature decreases, the superheat will increase. Conversely as the ambient temperature rises the superheat will decrease. A rule of thumb for determining superheat is to subtract the ambient temperature from 105°F. This will give you a ballpark value for most applications on setting the superheat. Under no circumstance should the superheat be set less than 5°F.

The superheat method is a very accurate means of checking the refrigerant charge. About a 1% change in refrigerant charge will change the superheat 3°F or more.

In making these tests it is important to use a fast reading resistance thermometer or an electronic temperature probe. Place the sensing element in good contact with a metal vapor line or suction line and insulate the element, along with the line.

Run the system for about 10 minutes to allow the temperatures and pressures to stabilize. Also, record the indoor and outdoor temperatures, since these are required in using the manufacturer's charts. Then read the pressure and temperature of the low temperature vapor line if it is a heat pump; or suction line, if it is a cooling-only line.

Use the pressure-temperature chart for the refrigerant being used and determine the saturation temperature that matches the pressure in the vapor line, or low side. Then subtract the saturation temperature from the actual vapor line temperature to obtain the superheat. Compare it with the manufacturer's recommendation. Make any adjustments necessary. Adding more charge gives less superheat; less charge gives more superheat.

For an example, refer to Figure 6-4-12. If the superheat is too high by more than about 5°F, correct this by adding vapor refrigerant through the low side port with the compressor running. If the superheat is too low by more than 5°F, remove vapor refrigerant by the proper recovery procedures.

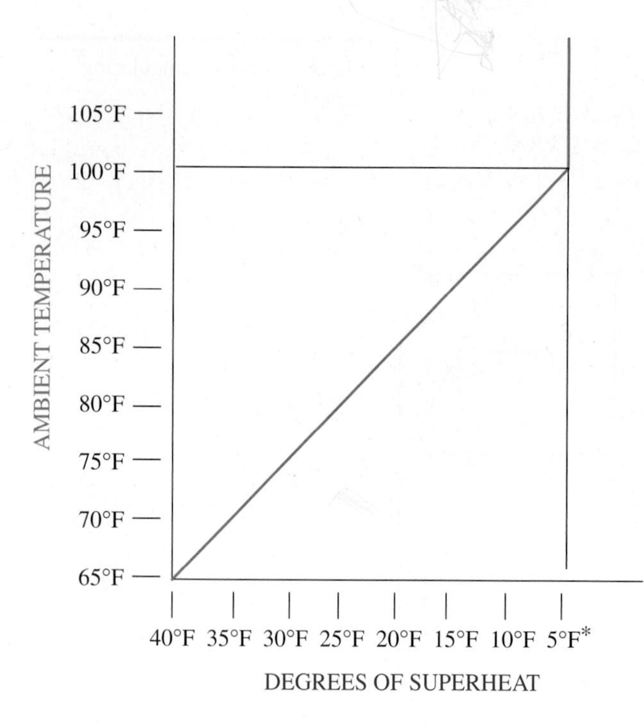

Figure 6-4-12 The degrees of superheat required for a correct charge change in direct proportion to the ambient temperature. The higher the temperature, the lower the degrees of superheat. The lower the temperature, the higher the degrees of superheat.

After any charge adjustment, repeat the test procedure to be certain that the charge is within the proper range.

Charging for Proper Subcooling

If the system uses a thermostatic expansion valve, the device will regulate the refrigerant flow over a wide range of load and charge conditions. The superheat method is therefore not satisfactory for testing the charge.

Some manufacturers recommend using subcooling to check the charge. Normally under full load conditions, an air conditioner will have 5° to 15°F of subcooling in the condenser before the refrigerant reaches the metering device. The amount of subcooling can be determined by measurements similar to those used in measuring superheat.

Compare the actual amount of subcooling with the manufacturer's recommendations. If the charge needs adjustment, more charge will increase the subcooling and less charge will decrease the subcooling.

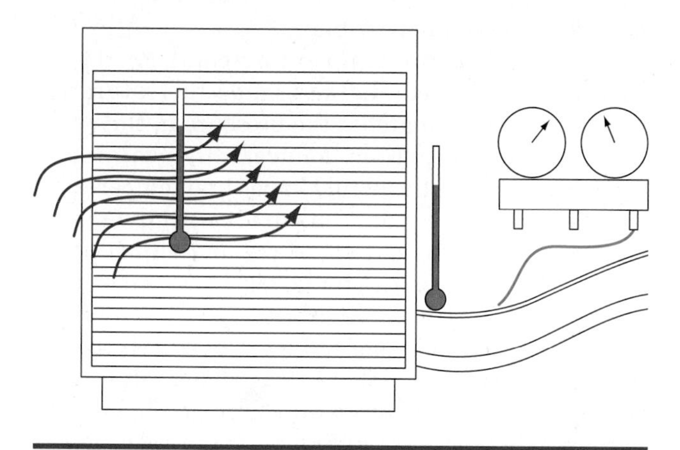

Figure 6-4-13 Checking charge using the approach method.

TABLE 6-4-1 Approach Method

Model No.	Approach Temperature liquid line–outdoor ambient °F(°C)
10ACB12	7 (3.9)
10ACB18	5 (2.8)
10ACB24	9 (5)
10ACB30	10 (5.6)
10ACB36	12 (6.7)
10ACB42	14 (8)
10ACB48	13 (7.2)
10ACB60	12 (6.7)
10ACB62	12 (6.7)

Approach Charging Method

In the approach charging method the difference between the liquid line temperature and outdoor ambient temperature are used, Figure 6-4-13. Once the temperature difference is obtained it is compared to the chart provide by the equipment manufacturer, Table 6-4-1.

LEAK TESTING

After a system has been repaired and before the final charge has been installed, the system needs to be leak tested. This is done by pressurizing the system with an inert gas such as nitrogen or carbon dioxide mixed with 5% or 10% R-22. Oxygen should not be used, since it can cause an explosion.

A halide leak detector and/or an electronic leak detector can be used to check for leaks. A bubble test can also be made. The bubble test is not affected by the type of refrigerant used.

The inert gas cylinder must be equipped with a suitable regulator. Install a pressure relief valve in the pressure feed line with a setting not to exceed the manufacturer's safe limit. Put the refrigerant gas in first. Turn the valve off and remove the refrigerant cylinder. Connect the inert gas cylinder and charge the system to the desired pressure.

> **NOTE:** Under current EPA regulations, the mixture of R-22 and inert gas are not considered to be refrigerant. Therefore they can be vented to the atmosphere following the leak check. However, the EPA does not look favorably on the addition of the inert gas to the refrigerant residue in the system as a means of leak checking. You must completely evacuate to the proper level all of the refrigerant in the system before introducing the trace gas.

OIL CHARGING

Sometimes oil needs to be added to the system.

In the compressor stopped method, the compressor is stopped, the crankcase is opened, and oil is poured into the compressor. This is done prior to evacuating and final charging, since air and moisture can get into the system during this operation.

In the evacuation method, the compressor suction service valve is closed and the compressor is evacuated to a slight vacuum. The oil hose is fitted to the compressor, making an airtight connection and the needed oil is drawn into the compressor. The oil line must have some means of starting and stopping the flow and the inlet must be below oil level.

Figure 6-4-14 Compressor oil level sight glass. (*Courtesy Hampden Engineering Corporation*)

On compressors that have a crankcase sight glass as in Figure 6-4-14, the correct oil level can be accurately observed and the quantity adjusted to the level indicated in the sight glass. It is important, however, to observe the oil level after the system has been in operation, since some oil remains in the system and does not get back to the compressor.

TECH TALK

Tech 1. There are a whole bunch of different charging methods. Which one is the best?

Tech 2. Some are more accurate than others, but the most accurate is to use the manufacturer's recommended weighed in charge. The charge level is listed in the material for the unit. They even have formulas for calculating the amount of additional charge you have to add if the line lengths are particularly long. The other charging methods using manufacturers' charging charts, the superheat method, the subcooling method, and the approach method are also accurate enough.

Tech 1. What about just using the sight glass to charge?

Tech 2. There have been a lot of studies to show that 60% of the systems that were charged just using the sight glass are overcharged.

Tech 1. Why are they overcharged? You just have to fill it up until the bubbles go away.

Tech 2. There are a lot of things that can cause flash gas bubbles to form in the sight glass. One, for example, is that sometimes the sight cavity is a lot larger than the refrigerant line. You'll remember as the volume increases the pressure decreases and that pressure can drop low enough to cause flash gas bubbles to be seen in the sight glass window. That does not mean that it is undercharged. That means there are bubbles being formed.

Tech 1. Oh, my goodness! I guess I've charged a lot of systems with way too much refrigerant then.

Tech 2. Yes, you probably have. So when you get a chance you may want to go by and check those system charges because a system that is overcharged won't function very well and it may cause the compressor to fail prematurely. It is just as bad as having the system undercharged as far as killing the compressor.

Tech 1. Thanks, I better get busy.

UNIT 4—REVIEW QUESTIONS

1. Why are systems that have a receiver not as critical about the amount of the refrigerant charge?
2. On some systems the length of the _______ is determined by the conditions of the installation.
3. Since the refrigerant cylinder is not filled over _______ in an upright position the vapor is at the top and the liquid at the bottom.
4. What is one way to determine the proper charge?
5. What must be done prior to charging the system?
6. What can happen if a system is undercharged?
7. The _______ method is a very accurate means of checking the refrigerant charge.
8. How should a system be leak tested?
9. When leak testing a system _______ should not be used.
10. What are the two methods for adding oil to the system?

SECTION III

Electricity

7

Basic Electricity

UNIT 1

Basic Electricity

OBJECTIVES: In this unit the reader will learn about basic electricity including:

- Static Electricity
- Current Electricity
- Direct Current
- Alternating Current
- Power Formula

TYPES OF ELECTRICITY

There are various ways to classify electricity. The following types relate to applications of this text:

1. Static electricity
2. Current electricity

Static Electricity

Static electricity is associated with the concept of electric charge, either negative or positive, possessed by bodies. If a body contains equal amounts of both, it will exert no force on other bodies and is considered uncharged. If an uncharged body acquires or loses charge, it is now charged and exerts a force on other, similarly charged, bodies. If both bodies have acquired the same kind of charge (both negative or both positive), they will repel each other. If the two charged bodies each have opposite charges, they will attract each other, Figure 7-1-1.

The amount of charge that a body has is measured by the *coulomb* and may be the amount either gained or lost by an uncharged body.

Probably the most common type of static electricity is lightning, occurring during a storm when cloud particles acquire large amounts of electric charge and discharge them to dissimilarly charged objects or bodies. The magnitude of the charge can cause serious damage because it is a natural, uncontrolled phenomenon. A smaller, more controlled charge is acquired in dry weather by a person by scuffing along a thick rug, collecting negative charges from the rug by friction. The negatively charged person exerts a force when touching a metal object, causing a spark to jump as the static electricity is discharged.

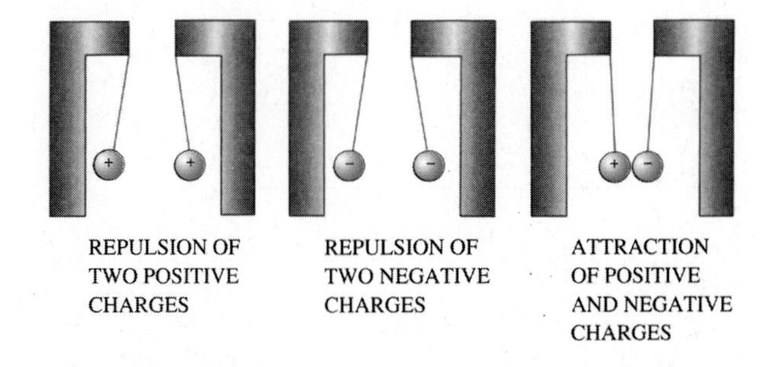

Figure 7-1-1 Static electricity negative and positive charges: similar charges repel and opposite charges attract.

SERVICE TIP

There are many computer printed circuit boards currently being used in the HVAC/R field. These circuit boards often contain electronic chips that are programmed to control the various functions of the system. These electronic chips can be damaged by an accidental discharge of static electricity. If you are working around these boards and notice that because of the low relative humidity that the static electricity is a problem, discharge yourself by touching the equipment frame before touching the microprocessor boards. To avoid possible damage to the program chips, aluminum wire is occasionally used as the primary service to residences and may be used for large amperage branch circuits such as electric heat. Aluminum wiring can form oxides that over time produce electrical resistance at the wire terminals, resulting in heat buildup and possible fires. To avoid this, any time aluminum wiring is used an anticorrosive cream must be used on the wire and terminal lugs.

This type of electricity has only a few uses; however, one rather common electrical HVAC/R device that uses static electricity is the *electrostatic air cleaner.* Air is passed through a "charging" section containing fine ionizer wires (charged with a very high DC voltage) alternating with grounded electrodes. This "high" field charges dust particles in the air. They are then passed through a collector section consisting of alternating charged and grounded plates which attract and hold the charged dust particles. Sometimes an oily adhesive coating is used to bind the dust to the collector plates until the unit is cleaned with a water spray wash cycle. The process is very effective in removing fine dust particles and pollen from the air. Maintenance is judged to be medium to high.

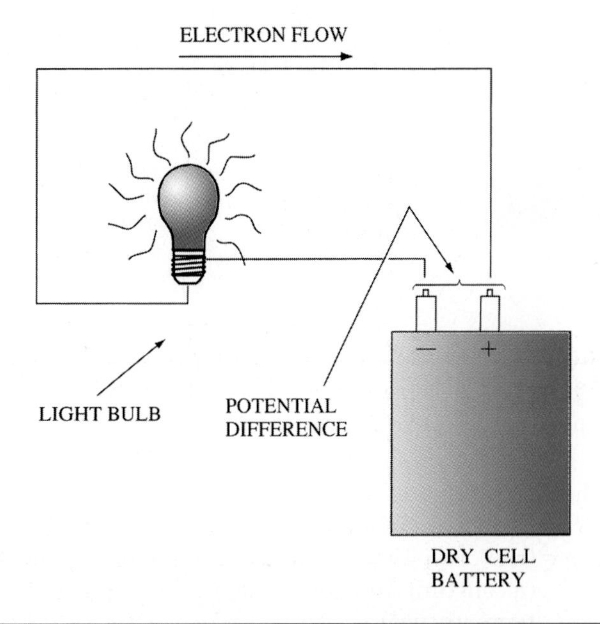

Figure 7-1-2 A simple electric circuit using a battery power source and a light bulb for the load.

ELECTRIC CURRENT

Current refers to the flow of electrons through a conductor such as a wire, or through a given space, or past a given point. The flow direction is from negatively to positively charged terminals and occurs because of the potential difference (in the charge) between terminals. The potential difference creates a force, called an *electromotive force* (EMF), as shown in Figure 7-1-2. This force is measured in volts (V). The letter symbol for voltage is E, referring to EMF, in circuits.

The unit for electron quantity, the coulomb, is rarely used; however, one definition for a volt is one coulomb of electrons passing a fixed point in the conductor per second.

The *current,* or rate of flow of the charge, is measured in units of amperes. The rate depends on the amount of voltage applied (the difference of potential between the two ends of the conductor), the size of the conductor, and the material of the conductor. The symbol for the current in amperes (A) is I. An electrical system illustrating current flow is shown in Figure 7-1-3.

The conductor may permit more or less current to flow because of resistance associated with its physical state. It is identified by applying a known value of voltage and measuring the resulting current. The ratio of voltage to current is the *resistance* and it is stated in ohms (Ω). For conductors, this depends on its dimensions and the material of which the conductor is made:

$$R = \rho l/A$$

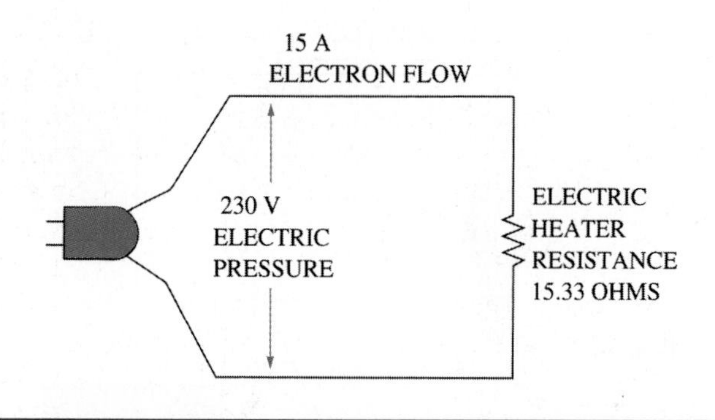

Figure 7-1-3 Simple electric circuit using an AC power source and an electric heater for the load.

where

R = resistance in ohms
ρ = (rho), resistivity of the material in ohm-meters (a constant for a given material at a given temperature)
l = length in meters
A = cross sectional area of conductor

The resistivities of commonly used conduction materials are:

Aluminum	2.62
Copper	1.72
Iron	9.71

From the above it can be seen why copper is the most frequently used for wires of electrical circuits. For example, a piece of copper wire 0.02 in (0.5 mm) in diameter and 39.37 in (1 m) long has a resistance of only about 0.09 ohm.

NICE TO KNOW

The most popular misconception is that voltage *E* and amperage *I* flow through all circuits equally. In reality voltage and amperage are only synchronized through resistance circuits. Voltage leads amperage in a inductive circuit and amperage leads voltage in a capacitance circuit. The shift between voltage and amperage is called phase shift. As the amount of phase shift increases, induction motors increase their heat and decrease their power output. Capacitors have the exact opposite effect on phase shift. They are used in conjunction with induction motors to increase their efficiency and decrease the motor's heat. That is the primary function of a run capacitor on a motor to correct the phase shift caused by the motor's windings.

As electric power wires travel great distances parallel to each other in rural areas, phase shift can occur as a result of the magnetic interference between the three power lines. For that reason you may see in rural areas three large gray cylinders

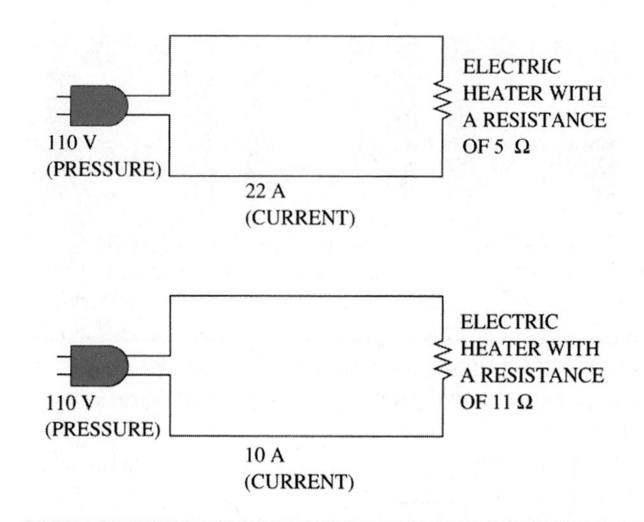

Figure 7-1-4 The effect of different loads on the current.

attached to the power lines. These are capacitor banks designed to bring the power back into phase.

In addition to the resistance of conductors, circuit elements called resistors are also used for various applications, Figure 7-1-4.

Putting it all together, current in a circuit is found to be directly proportional to the applied voltage and inversely proportional to the resistance of the circuit. This is known as *Ohm's law* and will be discussed in more detail later in the chapter.

The flow of electrons through a circuit dissipates energy in the form of heat in passing through the resistances that make up the circuit. Calculated amounts of this energy are called power, P, measured in watts (W). A watt of electricity is one ampere (A) of current, I, flowing with the force of one volt (V) of voltage, E. The power in watts in an electric circuit can be expressed by the equations for single-phase AC power:

$$P = EI\,(\mathrm{PF})$$
$$P = I^2R$$
$$P = E^2/R$$

where

PF (Power Factor) = phase angle between E and I.

Power factor equations vary for DC and AC power.

TECH TIP

Most motors used in HVAC are inductive type motors. In this type of motor, the rpm is determined by the number of poles in the motor and the power frequency. Therefore, a motor that is operating on 50 cycle power will spin at a slightly slower rpm than the same motor operating on 60 cycle power. The change in rpm can affect the system airflow. By reducing the rpm, the total cfm output by a blower will be reduced.

The amount of power required by electrical devices determines the design of the circuit. The electrical motor is the largest current-consuming device in most heating, cooling, and refrigeration systems. Other units that measure the output of such devices are the horsepower, equal to 746 W (power output), and the British thermal unit (Btu) where 1 W = 3.41 Btu/hour (heat output). A more abstract term commonly used by air conditioning manufacturers is the *energy efficiency rate* (EER) which identifies the amount of heat per watt of power consumed by equipment. It is becoming common to specify power ratings of many devices such as refrigeration systems, boilers, etc., in standard kilowatt (kW) terms.

Current electricity is produced as two types:

1. Direct current
2. Alternating current

Direct Current

Direct current (DC) is the type supplied by batteries. It is a continuous flow of electrical particles in one direction through a conductor. The source has a positive and a negative terminal. It is used mostly for transportation refrigeration equipment, electronic air cleaners, in computers, and for electronic circuits.

For example, a thermocouple used on a gas furnace as a safety device generates its own DC current for keeping the gas valve open. This device consists of a pair of wires made of different metals joined together at one end. When the joined end and the other end of the wires are at different temperatures (one end being heated by the gas flame), direct current is produced between the wires and flows through the circuit. When the heat from the pilot light goes off, electricity is no longer generated and the gas valve closes.

Transmission of large amounts of power involves unavoidable line losses due to the resistance of the wires. By utilizing high voltages, the power can be transmitted by using relatively low line current, thus minimizing the line losses (which may be calculated as I^2R). Power is produced from AC generators. It is expensive to convert AC voltages to DC and DC voltages to AC; therefore, alternating current (AC) is used for almost all ordinary power applications.

The most common method of obtaining AC voltage is by using a generator to create a magnetic field in which a conductor is rotated to induce voltage in the circuit. The amount of voltage produced is proportional to the rate at which the magnetic flux linking the circuit changes. The AC voltage in turn produces AC current.

Alternating Current

Figure 7-1-5 shows a typical cycle of alternating current. The AC waveform has a *sinusoidal* shape like that of a sine wave. Note that half the time the current is positive and half the time it is negative. Alternating current flows in one direction and then the other at regular intervals through the conductor. The frequency of AC current is the number of cycles per second. Frequency is measured in *Hertz* (Hz): one Hz is equivalent to one cycle per second. In the United States 60 Hz power is standard. In many other parts of the world including Canada the standard is 50 Hz.

Since most devices are designed to run on AC current, this unique source of power has universal usage. Alternating current will be discussed in detail later in this chapter.

Ohm's law, for DC circuits or AC single phase circuits with purely resistive loads is as follows:

The relationship between electrical potential, E, measured in volts (V); current flow, I, measured in amperes (A); and resistance, R, measured in ohms

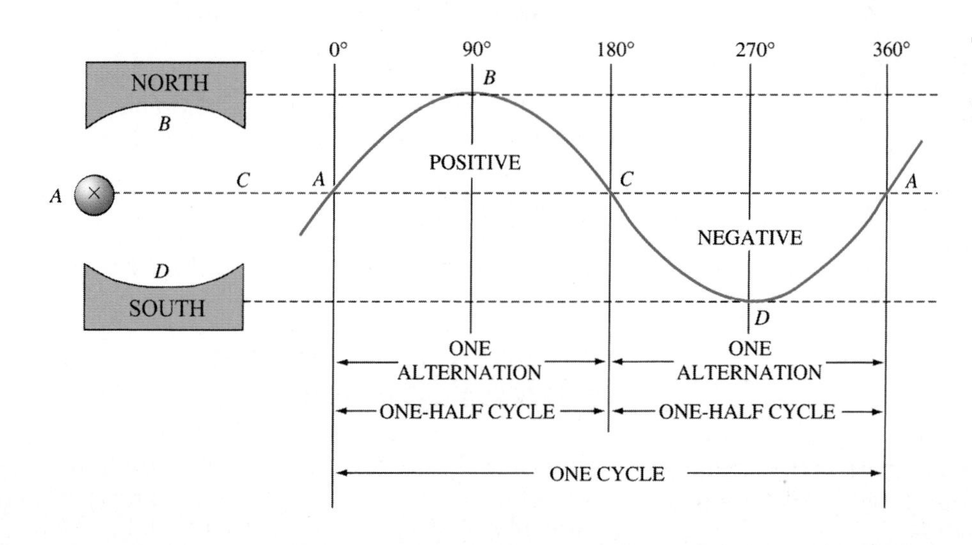

Figure 7-1-5 One cycle of alternating current.

(Ω), is expressed in Ohm's law. In simple terms it states that the greater the voltage, the greater the current; and the greater the resistance, the lesser the current flow. Ohm's law is expressed mathematically as "current is equal to electrical potential divided by resistance." Stated in symbols:

$$I = \frac{E}{R}$$

where

I = Current
E = Electrical potential
R = Resistance

The equation can be stated a number of ways. Use is made of whichever one applies. Other versions of Ohm's law are as follows:

$$E = IR \qquad R = E/I$$

This formula is very helpful in analyzing a circuit, since when any two of the terms are known or can be measured, the third value can be calculated using one of the above equations.

It should be noted that Ohm's law was first applied to circuits using direct current. It does not apply, without some modifications, to AC circuits since coils of wire produce different effects with alternating current. The flow of current in AC circuits can be influenced by such factors as *inductance* and *inductive reactance* (the "resistance" offered by inductance coil to the flow of alternating current), and *capacitance* and *capacitive reactance.* If modifications are made to resistance, it will be shown how the general principles contained in Ohm's law do apply to alternating current.

The following examples illustrate the relationships of voltage, current, and resistance in a simple electrical circuit. In the examples a DC circuit or a single phase AC circuit with purely resistive loads is used.

EXAMPLE

A simple electrical circuit, Figure 7-1-6, has a power source of 120 V and a resistance load of 10 Ω. How much is the current flow in amperes?

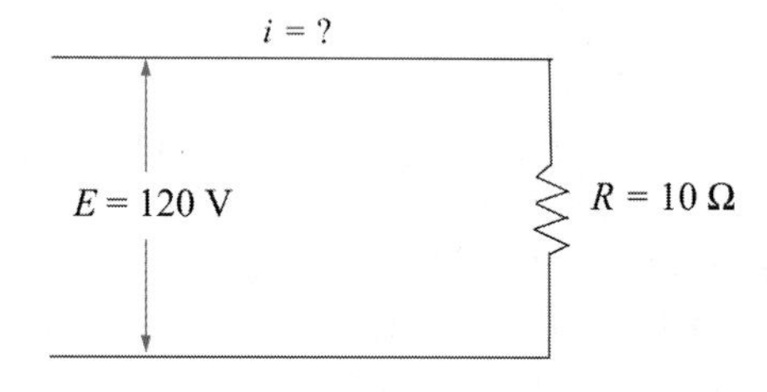

Figure 7-1-6 Calculating the current using Ohm's law, with voltage and resistance known.

Solution

$$I = E/R$$
$$I = 120\text{ V}/10\ \Omega$$
$$I = 12\text{ A}$$

■

EXAMPLE

Referring to Figure 7-1-7, what is the resistance of an electric heater, using a power source of 120 V and drawing a current of 10 A?

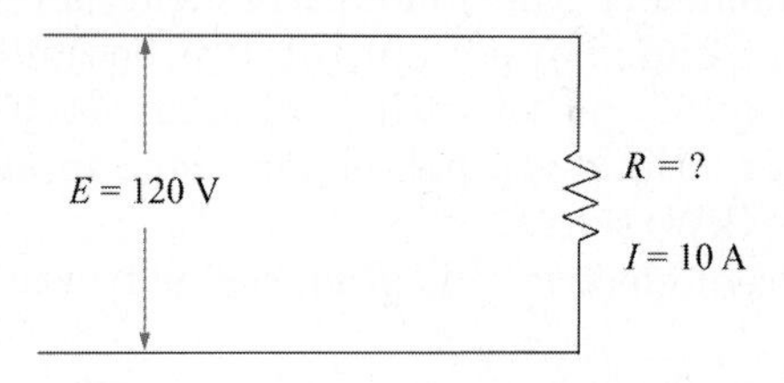

Figure 7-1-7 Calculating the resistance, using Ohm's law, with voltage and current known.

Solution

$$R = E/I$$
$$R = 120\text{ V}/10\text{ A}$$
$$R = 12\ \Omega$$

■

EXAMPLE

Assuming a circuit as shown in Figure 7-1-8, with a resistance of 23 Ω and a current flow of 10 A, what is the voltage of the power supply?

Solution

$$E = IR$$
$$E = (10\text{ A})(23\ \Omega)$$
$$E = 230\text{ V}$$

■

Note in the above examples that by using Ohm's law, if two factors are known, the third can be calculated.

Calculating electric power, for DC circuits or single phase AC systems with purely resistive loads can be done using the following formula:

$$P = EI$$

where E and I have the same meaning as used in calculations involving Ohm's law and P = power in watts.

In general, for single phase AC systems, $P = EI$ (PF), where PF = phase angle between E and I. For three phase AC systems, $P = \sqrt{3}\,EI$ (PF).

■ VARIATIONS OF THE POWER FORMULA

The power formula can be stated in three different ways with variations for DC, single phase AC, and three phase AC, as shown in Table 7-1-1.

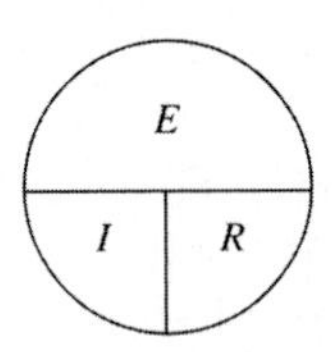

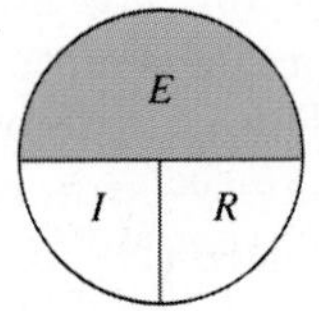

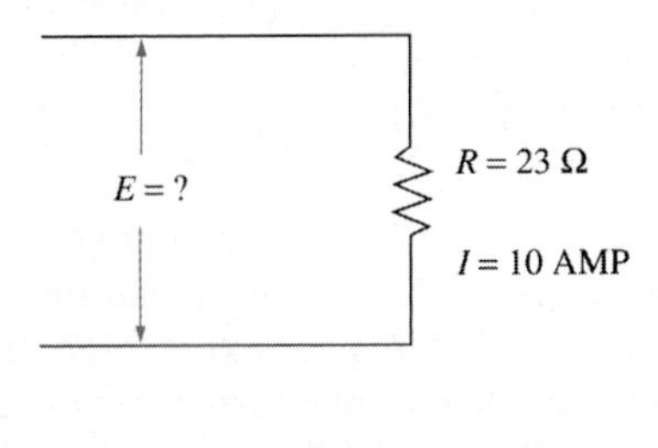

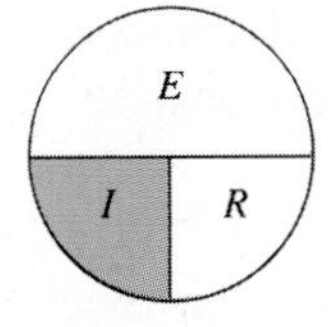

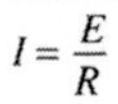

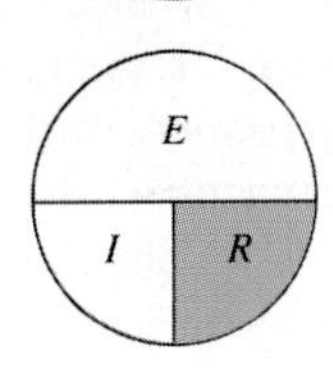

Figure 7-1-8 Various ways to use Ohm's law.

Let's work through some examples illustrating the power formula.

EXAMPLE

(Single phase AC, assuming PF = 1.0)
What is the power consumption in a circuit operating with 230 V and 5 A?

Solution

$$P = EI$$
$$P = (230\ \text{V})(5\ \text{A})$$
$$P = 1150\ \text{W}$$

■

EXAMPLE

(Single phase AC, assuming PF = 1.0)
What is the current draw of an 8000 W electric heater operating on a 230 V power supply?

Solution

$$I = P/E$$
$$I = 8000\ \text{W}/230\ \text{V}$$
$$I = 34.8\ \text{A}$$

■

EXAMPLE

(DC or single phase AC)
What is the power consumed by an electric resistance of 20 Ω, using a current of 6 A?

Solution

$$P = I^2R$$
$$P = (6\ \text{A})(6\ \text{A})(20\ \Omega)$$
$$P = 720\ \text{W}$$

■

EXAMPLE

(Three phase AC)
What is the power consumed by a three phase circuit with a voltage of 230 V, a current of 10 A, and a power factor of 0.80?

Solution

$$P = \sqrt{3}\,\text{EI(PF)}$$
$$P = (1.732)(230)(10)(.80)$$
$$P = 3186\ \text{W}$$

■

TABLE 7-1-1 Power Formula for DC, Single Phase AC, and Three Phase AC Current

	DC	Single Phase AC	Three Phase AC
Power Formula 1	P = EI	P = EI(PF)	$P = \sqrt{3}$EI(PF)*
Power Formula 2	$P = E^2/R$	$P = E^2/R$	$P = \sqrt{3}E^2IR$
Power Formula 3	$P = I^2R$	$P = I^2R$	$P = \sqrt{3}I^2R$

*$\sqrt{3} = 1.732$

ELECTRICAL CIRCUITS

There are three elements required for all electric circuits. They must contain a source, a path, and a load. The electrical power source can be provided by a battery or from the building's central electrical system, the path is provided by the interconnecting wiring and the load is any electrical consuming device such as motors, relay coils, lamps, resistance heaters, circuit boards transformers, etc.

There are three types of path arrangements for circuits, as follows:

1. The series circuit, which allows only one path for the current to flow;
2. The parallel circuit, which has more than one path;
3. The series parallel circuit, which is a combination of series and parallel circuits.

NOTE: An electrical circuit that contains a source and path without a load is called a short circuit. A circuit that contains a source and load without a path is called an open circuit. Sometimes people refer to any circuit that does not function as having a short. This is an improper use of the term. An open circuit is one that does not work. A short circuit is one that will trip the power breaker or cause something to overheat and possibly catch fire.

Series Circuits

In a series circuit, there is only one path for the current to follow. The power must pass through each electrical device in succession in that circuit to go from one side of the power supply to the other. An example of a series circuit is shown in Figure 7-1-9, where four resistance heaters are placed end-to-end in a single circuit.

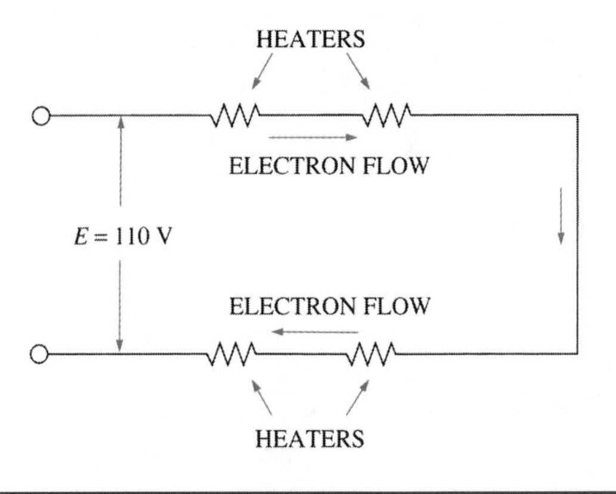

Figure 7-1-9 A series circuit with four resistances.

Series circuits are common on HVAC/R systems. Usually there is one load controlled by a series of switches, as shown in Figure 7-1-10. In this diagram the 208 V power supply terminals are indicated with the symbols L_1 and L_2. The one load is a compressor motor. The switches placed in series with the compressor motor are used to control its operation. These switches, shown in this diagram, are all safety switches and therefore are all normally closed (NC).

In a series circuit, all switches must be closed in order for current to flow through the circuit. Types of switches include the following:

1. The *high pressure switch,* which senses compressor discharge pressure, opens on a rise in pressure. It is set to cut out at a protective high limit pressure, but remain closed at normal operating pressures. It is also called a high pressure cutout.

2. The *low pressure switch,* which senses compressor suction pressure, opens on a drop in pressure. It is set to cut out at a protective low limit pressure, but remains closed at normal operating pressures. It is also known as a loss of charge or low pressure cutout switch.

3. The *compressor internal thermostat,* which senses compressor motor winding temperature, opens on a rise in temperature. It is set to cut out at a protective high temperature, but remain closed under normal operating conditions.

4. An *operating control* (not shown), is also placed in series with the compressor motor, starts and stops the compressor in response to temperature, pressure, humidity, or a time clock.

Calculations for a Series Circuit For the following calculations a DC circuit or a single phase AC circuit with purely resistive loads is used.

The current flowing through a series circuit is the same for each load in the circuit. For example, in the circuit shown in Figure 7-1-11, which has four loads, the current is equal through each load, and expressed in symbols:

$$I_1 = I_2 = I_3 = I_4$$

The total resistance of a series circuit is the sum of all the individual resistances that are placed in series between L_1 and L_2. To state this in symbol form:

$$R_T = R_1 + R_2 + R_3 + R_4$$

The voltage drop across each one of these resistances can be calculated using Ohm's law, and the total of the individual voltage drops should add up to the circuit voltage. Thus,

$$E_T = E_1 + E_2 + E_3 + E_4$$

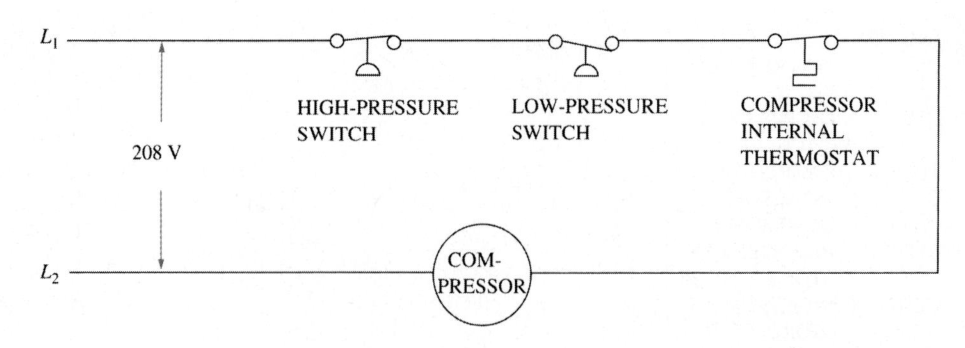

Figure 7-1-10 Three safety switches in series with a single load.

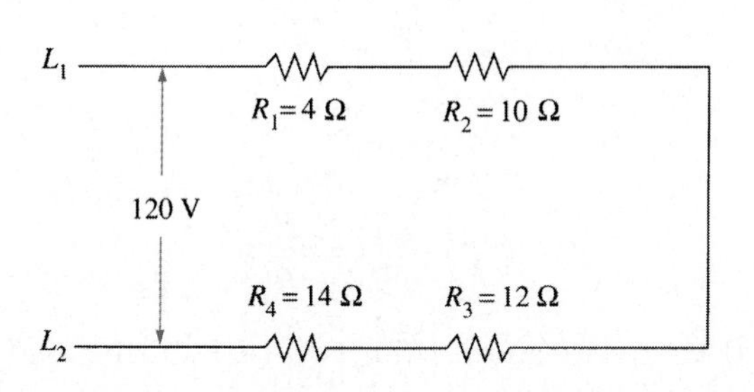

Figure 7-1-11 Calculating the current flow through a series circuit.

EXAMPLE

Using Figure 7-1-11, calculate the individual voltage drops to see if their total equals the circuit voltage.

Solution

1. Determine the total resistance of the circuit.

$$R_T = R_1 + R_2 + R_3 + R_4$$
$$R_T = 4\,\Omega + 10\,\Omega + 12\,\Omega + 14\,\Omega$$
$$R_T = 40\,\Omega$$

2. Determine the current flow in the circuit, using Ohm's law.

$$I = \frac{E}{R}$$
$$I = \frac{120\text{ V}}{40\,\Omega}$$
$$I = 3\text{ A}$$

3. Determine each voltage drop, using Ohm's law.

$$E_1 = I_1 \times R_1 \qquad E_2 = I_2 \times R_2$$
$$E_1 = 3 \times 4 \qquad E_2 = 3 \times 10$$
$$E_1 = 12\text{ V} \qquad E_2 = 30\text{ V}$$
$$E_3 = I_3 \times R_3 \qquad E_4 = I_4 \times R_4$$
$$E_3 = 3 \times 12 \qquad E_4 = 3 \times 14$$
$$E_3 = 36\text{ V} \qquad E_4 = 42\text{ V}$$

4. Check the accuracy of these calculations. Their total should equal the circuit voltage of 120 volts. The procedure is as follows:

$$E_T = E_1 + E_2 + E_3 + E_4$$
$$E_T = 12 + 30 + 36 + 42$$
$$E_T = 120\text{ V}$$

■

Parallel Circuits

Parallel circuits are used as power supplies for most HVAC/R equipment wiring. Each load has its own separate path for the current to flow. Most equipment loads are connected directly to the voltage source. The control circuits normally use a lower voltage; 24 V is common. The control system also has load devices, such as relays, which require separate circuits. Control voltage is usually provided by a stepdown transformer fed from the power source.

Figure 7-1-12 illustrates a parallel circuit with four loads. No controls are shown in this diagram, which would be necessary if the circuit were operational. In parallel circuits, each circuit has its own independent connection to the power source. If the switch is opened in one circuit, the other circuits will continue to operate.

Figure 7-1-13 shows a number of parallel load circuits, with switches or controls in series with the loads, that could be used for an air cooled condensing unit. There are three parallel circuits, C_1, C_2, and C_3. Going from top to bottom, they could be described as follows:

Circuit 1. The condenser fan motor #2 has a separate thermostat that turns it on and off.

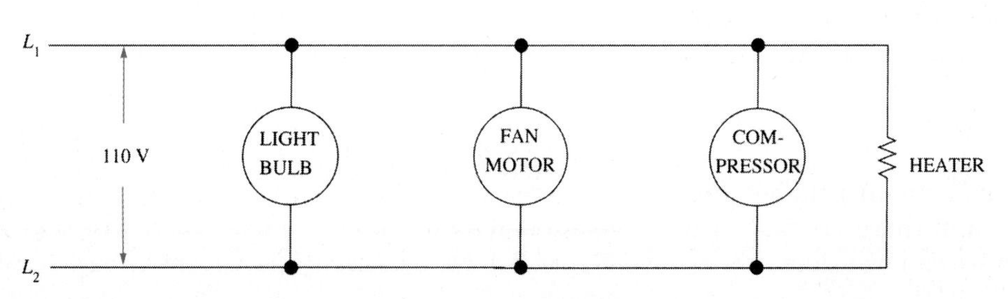

Figure 7-1-12 Parallel circuit with four different loads.

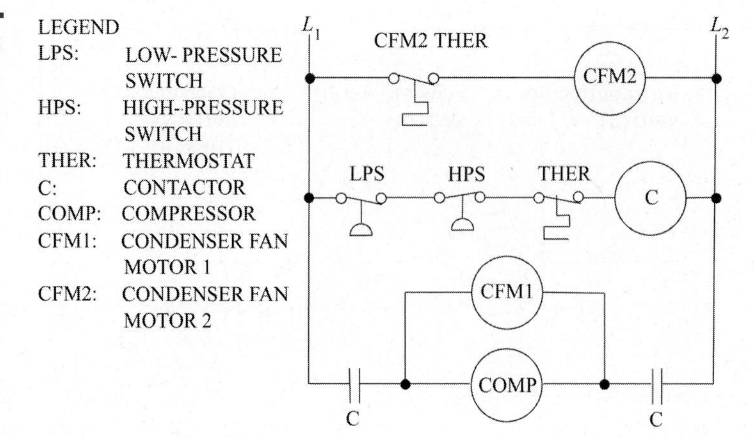

Figure 7-1-13 Condensing-unit wiring diagram showing three separate circuits.

Circuit 2. The compressor contactor coil (C) is energized when the primary thermostat calls for cooling, provided the two safety switches, LPS and HPS, are closed.

Circuit 3. The compressor contactor has two normally open switches, which are in series with the two loads. When the contactor coil in circuit 2 is energized, the two "C" switches in circuit 3 close, supplying power to the compressor motor and to condenser fan #1 at the same time. In effect the two loads, condenser fan motor #1 and the compressor motor, are themselves in parallel and both receive line voltage when the contactor switches close.

In actual practice, a unit may have many parallel circuits for individual loads, all operated in accordance with the design specification of the control system.

Calculations Using a Parallel Circuit

The following calculations use a DC circuit or a single phase AC circuit with purely resistive loads. The current draw for a parallel circuit is determined for each of its parts. The current consumed by the entire parallel system is the sum of the individual circuits. The calculation is made using Ohm's law. To obtain the current flowing in the circuit, both the voltage and the resistance of the load(s) must be known. Thus, the total current is calculated as follows:

$$I_{\mathrm{T}} = I_1 + I_2 + I_3 + I_4 + \ldots$$

The resistance of a parallel circuit gets smaller as more resistances are added. If there are only two resistances, the total resistance can be calculated as follows:

$$I_{\mathrm{T}} = I_1 + I_2 + I_3 + I_4 + \ldots$$

The resistance of a parallel circuit gets smaller as more resistances are added. If there are only two resistances, the total resistance can be calculated by the following formula:

$$R_{\mathrm{T}} = \frac{(R_1 \times R_2)}{(R_1 + R_2)}$$

If there are more than two resistances, use the following formula and solve for R_{T}.

$$\frac{1}{R_{\mathrm{T}}} = \frac{1}{R_1} + \frac{1}{R_2} + \frac{1}{R_3} + \frac{1}{R_4} + \ldots$$

TECH TIP

The total resistance of a parallel circuit is less than the resistance of the smallest single resistor in the circuit.

The voltage drop in a parallel circuit is the line voltage supplied to the loads, or simply stated:

$$E_{\mathrm{T}} = E_1 = E_2 = E_3 = E_4 = \ldots$$

Ohm's law can be used to calculate voltage, amperage, or resistance, if the other two values are known.

EXAMPLE

From the information given in Figure 7-1-14, calculate the current draw for each of the individual circuits and the total current draw.

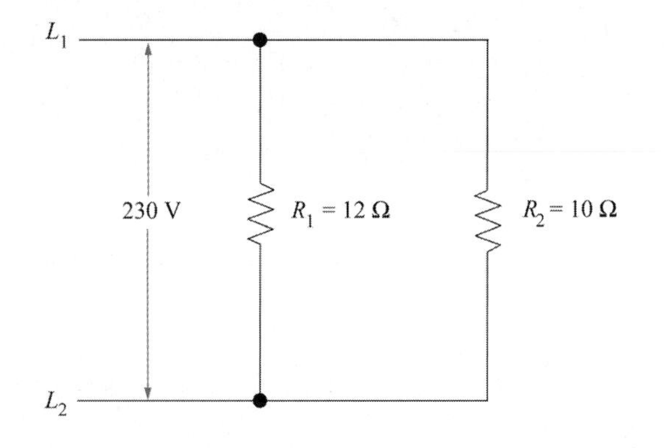

Figure 7-1-14 Calculating the current through parallel circuits.

Solution

$$I = E/R$$
$$I_1 = 120\text{ V}/12\ \Omega$$
$$I_1 = 10\text{ A}$$
$$I_2 = 120\text{ V}/10\ \Omega$$
$$I_2 = 12\text{ A}$$
$$I_T = I_1 + I_2$$
$$I_T = 12\text{ A} + 10\text{ A}$$
$$I_T = 22\text{ A}$$

■

EXAMPLE

Find the total resistance of the complete circuit shown in Figure 7-1-14.

Solution

$$R_T = \frac{(R_1 \times R_2)}{(R_1 + R_2)}$$
$$R_T = \frac{(12\ \Omega \times 10\ \Omega)}{(12\ \Omega + 10\ \Omega)}$$
$$R_T = \frac{120}{22}$$
$$R_T = 5.4\ \Omega$$

■

EXAMPLE

What is the total resistance with parallel resistances of 3 Ω, 4 Ω, and 5 Ω?

Solution

$$\frac{1}{R_T} = \frac{1}{R_1} + \frac{1}{R_2} + \frac{1}{R_3}$$
$$\frac{1}{R_T} = \frac{1}{3\ \Omega} + \frac{1}{4\ \Omega} + \frac{1}{5\ \Omega}$$

Converting to a common denominator:

$$\frac{1}{R_T} = \frac{20}{60} + \frac{15}{60} + \frac{12}{60}$$
$$\frac{1}{R_T} = \frac{47}{60}$$
$$R_T = \frac{60}{47}$$
$$R_T = 1.28\ \Omega$$

■

Series Parallel Circuits

A series parallel circuit, as the name implies, combines both a series and a parallel arrangement of electrical loads. A typical diagram is shown in Figure 7-1-15a.

Calculations As in previous calculations, a DC circuit or single phase AC circuit with purely resistive loads is used.

Both the voltage of the circuit and the values of all resistances are known (Figure 7-1-15a). With two of the factors provided, the third (current flow) can

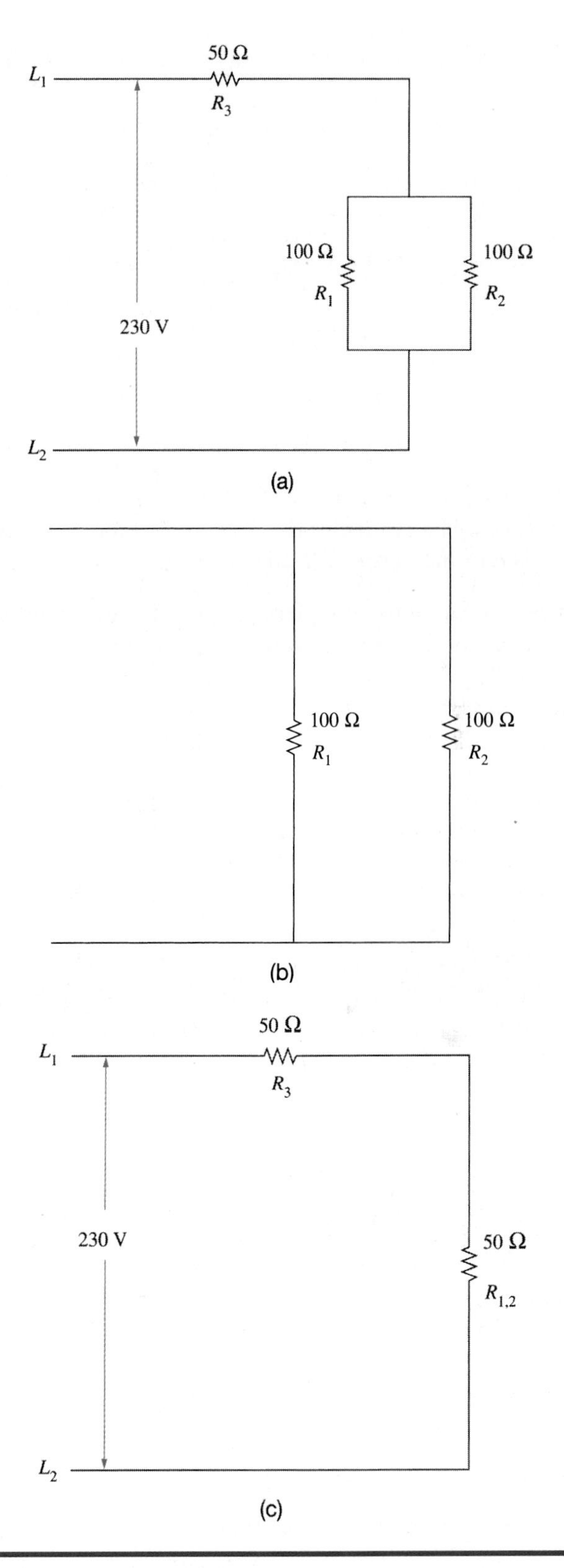

Figure 7-1-15 (a) Parallel series circuit showing values of three resistances; (b) Determine the combined resistance of the two parallel resistances; (c) Parallel-series circuit combining two parallel resistances to form a simple series circuit.

be determined using Ohm's law from the following steps:

Step 1 Calculate the resistance through the parallel circuit consisting of R_1 and R_2 (Figure 7-1-15b).

$$R_{1,2} = \frac{(R_1 \times R_2)}{(R_1 + R_2)}$$

$$R_{1,2} = \frac{(100 \times 100)}{(100 + 100)}$$

$$R_{1,2} = \frac{10{,}000}{200}$$

$$R_{1,2} = 50\ \Omega$$

Therefore, 50 Ω can be substituted for the parallel resistances. The main circuit has now become a strictly series circuit, Figure 7-1-15c.

Step 2 Calculate the current flow through the revised main circuit, Figure 7-1-15c.

$$I_T = \frac{E}{(R_{1,2} + R_3)}$$

$$I_T = \frac{230\ \text{V}}{(50\ \Omega + 50\ \Omega)}$$

$$I_T = \frac{230}{100}$$

$$I_T = 2.3\ \text{A}$$

Step 3 Calculate the current flow through R_1 and R_2. Since R_3 is in series with $R_{1,2}$ in Figure 7-1-15c, we calculate the voltage drop across R_3 and $R_{1,2}$ in Figure 7-1-15b.

$$E_3 = I \times R_3$$
$$E_3 = 2.3\ \text{A} \times 50\ \Omega$$
$$E_3 = 115\ \text{V}$$
$$E_{1,2} = E - E_3$$
$$E_{1,2} = 230 - 115$$
$$E_{1,2} = 115\ \text{V}$$

Step 4 Calculate the current through R_1 and R_2.

$$I_1 = \frac{E_{1,2}}{R_2}$$

$$I_1 = \frac{115\ \text{V}}{100\ \Omega}$$

$$I_1 = 1.15\ \text{A}$$
$$I_2 = I_1$$
$$I_2 = 1.15\ \text{A}$$
$$I_{1,2} = I_1 + I_2$$
$$I_{1,2} = 1.15\ \text{A} + 1.15\ \text{A}$$
$$I_{1,2} = 2.3\ \text{A}$$

Thus, the combined current through R_1 and R_2 is the same as through R_3, which is correct for a series circuit, and the answer is verified.

TECH TALK

Tech 1. Working with electricity really scares me.

Tech 2. It should. Electricity doesn't know whether you're a student, an experienced technician, or someone who accidentally touches it. If the jolt is large enough it can kill you.

Tech 1. That's why it scares me.

Tech 2. Everyone who works with electricity needs to have a healthy level of fear. It is when you become complacent and don't pay attention that you are most likely going to have a problem. You need to develop good working techniques so that you avoid the chance of being shocked.

Tech 1. How do you do that?

Tech 2. Well, number one, always check a circuit to make sure the electricity is off. Check it whether you're the one who turned it off or if someone tells you it is off. Never assume a circuit is off until you have checked and verified that it is off.

Tech 1. OK, always check that the power is really off. What else?

Tech 2. Whenever you are working on a system, federal regulations require that you lock the power off and tag it. It's called lock-out, tag-out and you have to do it any time you are working on a system.

Tech 1. Sure, I know about safety when the power is off. But sometimes you have a hot circuit and you have to do a test.

Tech 2. Well, you're right. From time to time we have to work on systems that are hot to get the voltage and amp draws. But I know lots of tips for testing safety.

Tech 1. Like what?

Tech 2. Well, when you're working on a system, make sure that you are not working in a wet or damp area. Also, always work on the equipment with dry tools with proper insulation. And when using a voltmeter, make sure that it is rated high enough for the voltages that you will be working with. And, importantly, don't lean against the equipment that you are working on. Follow all other safety practices.

Tech 1. I do all of that but it still makes me nervous.

Tech 2. Everybody who works with electricity is a little nervous. That is not bad, it makes you work safer.

UNIT 1—REVIEW QUESTIONS

1. Which type of electricity is associated with the concept of electric charge, either negative or positive, possessed by bodies?
2. What is the most common type of static electricity?
3. The current or rate of flow of charge is measured by ________.
4. Define resistance.
5. What determines the design of the circuit?
6. List the two types of current electricity.
7. What is the most common method of obtaining AC voltage?
8. How is Ohm's law expressed mathematically?
9. List the types of path arrangements for circuits.
10. In a series circuit, all switches must be ________ in order for current to flow through the circuit.
11. What is a series parallel circuit?

UNIT 2

Electrical Power

OBJECTIVES: In this unit the reader will learn about electrical power including:

- Power Sources
- Loads
- Electrical Circuit
- Control Devices and Switches
- Phase Shift
- Power Distribution

ELECTRICAL POWER SYSTEMS

The basic element that supplies power for all HVAC/R systems is the electrical power system. The electrical circuit is a means of delivering electrical power to the HVAC/R equipment. An electrical power system has three essential requirements and one optional requirement:

1. A source of power (could be a transformer)
2. An electrical load device
3. An electric circuit: a path for the current to flow
4. (Optional) A switch to control the flow

Power Sources

There are two types of power sources:

1. Direct current (DC)
2. Alternating current (AC)

The most common source of direct current is the battery. Batteries are used to supply power to many different types of electrical testing instruments, making them portable and convenient to use. DC power is used for the controls on automotive air conditioning systems. DC power obtained from AC power supplies is also used on certain solid state modules for defrost and overcurrent protection.

Alternating current is the most common source of power for most HVAC/R systems. AC power is generated by all the power companies. In residences, 120 V AC is used to power most small appliances. Larger appliances such as electric stoves and residential air conditioning units use 240 V. The power company supplies residential users with 240 V over the incoming lines. A portion of it is tapped to supply the 120 V requirement. Commercial and industrial customers normally use higher voltages in single phase and three phase systems.

Transformers increase or decrease incoming voltages to meet the requirements of the load. For control circuits, it is common to use a transformer to obtain 24 V from line voltages of 120 or 220 V. More detailed explanations of how transformers work and are used will be found in the next chapter.

NOTE: Most transformers that are used in the HVAC/R industry are 40 A. These transformers have a maximum current output capacity of 1.6 A. Larger transformers are available that can produce higher output amperages at 24 V. These transformers may be needed when additional controls are added to a standard HVAC system. Additional controls might include humidifiers, duct air booster fans, and economizers.

Loads

The second condition for an electrical power system is that it must have a load. A *load* is any electrical device that requires power to operate. The most common loads for HVAC/R systems are electric motors. Motors drive the compressors, fans, and pumps. Motors also drive damper motors and zone valves. Many other electrical components require power, such as resistance heaters and solenoid valves.

Electrical Circuit: Path for the Current

The third condition for a power system is that there must be a path for the current to travel, Figure 7-2-1. Every electric circuit has at least two wires, often indicated as line terminals, L_1 and L_2. In order for there to be a complete circuit, the path of the electrical service (wire connections) flows through one wire of the electric circuit, passes through one wire of the electric circuit, passes through the load, and returns through the other wire of the electric circuit. In AC systems, the direction of power flow reverses 60 times per second.

Control Device or Switches

The fourth (optional) condition for the power system is the *control device* or *switch*. The switch is a device to turn the load off and on. It may be manual or automatic, as in the case of a thermostat that turns a unit on and off in response to the surrounding temperature. The switch permits the circuit to be open, Figure 7-2-2, or closed, Figure 7-2-3. No current flows in the open circuit. When the circuit is closed the load receives power.

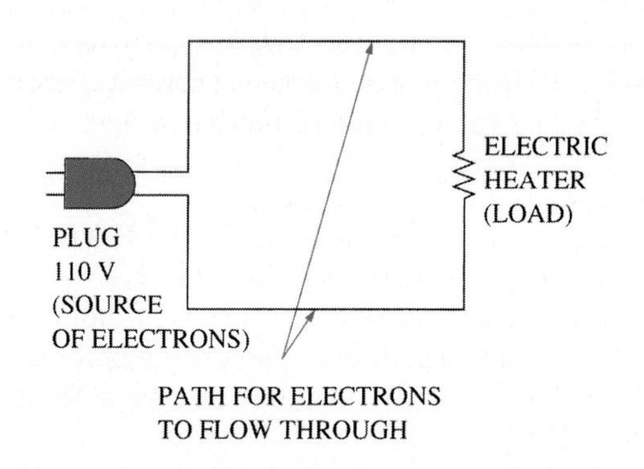

Figure 7-2-1 Principal components of a simple circuit.

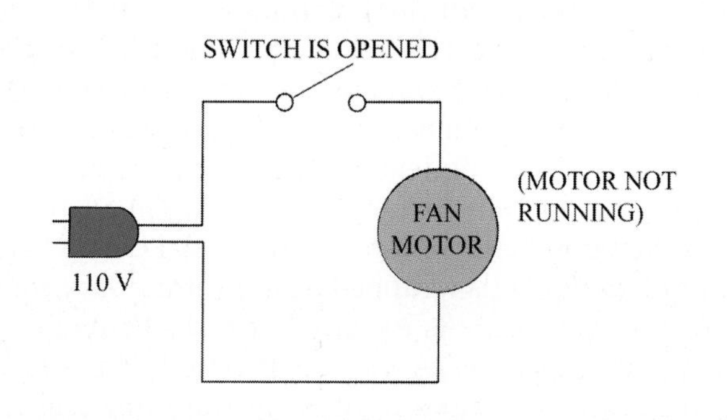

Figure 7-2-2 An open circuit: no current flows.

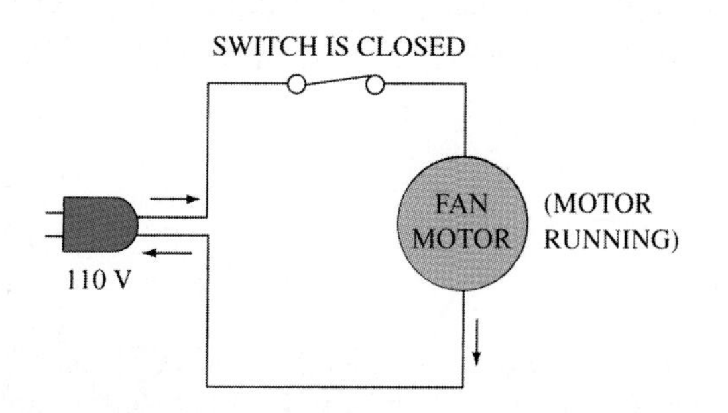

Figure 7-2-3 A closed circuit, with current flowing through the load.

ALTERNATING CURRENT

Alternating current is a potential difference with a sinusoidal waveform that alternates polarity continuously at a fixed rate or frequency. The number of times per second the polarity is reversed is the frequency expressed in Hertz (Hz), or cycles per second.

Magnetic Induction

Magnetic induction, demonstrated in Figure 7-2-4, is used to generate voltage for commercial and residential use. Induction occurs when a conductor is placed in a magnetic field. As the conductor is moved through the field, a potential difference is created in the conductor. The size of this potential difference is dependent on the strength of the field and the speed of the conductor through the field.

Alternating current is produced by rotating a coil within the magnetic field to produce an output voltage. Figure 7-2-5 is a diagram of a single cycle that is repeated 60 times per second (60 Hz), on a continuous basis. This is called single phase voltage.

Single phase power is commonly used for residential and small commercial systems. For larger motors, generally 5 Hp and above, three phase power is normally used. This requires the addition of one more conductor to the generator as shown in Figure 7-2-6. The three conductors are positioned 120° from each other.

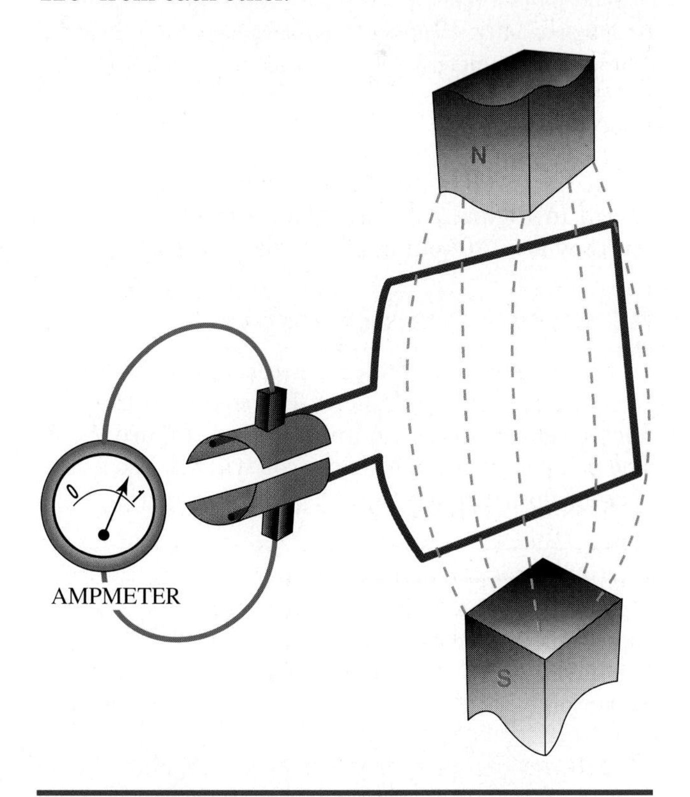

Figure 7-2-4 Current is induced in a coil of wire rotating in a magnetic field.

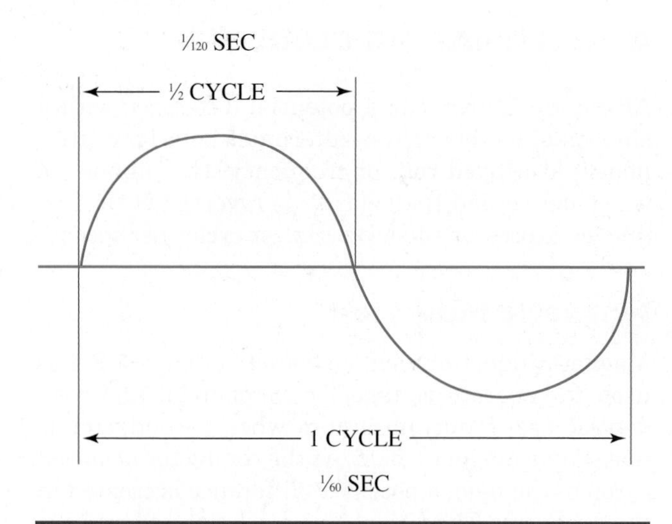

Figure 7-2-5 One cycle of alternating current.

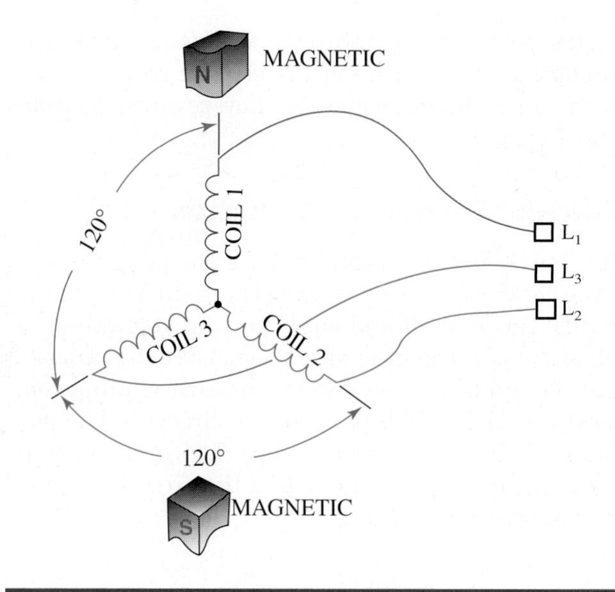

Figure 7-2-6 Three-phase alternating current is produced when three coils rotate inside of a magnetic field.

NICE TO KNOW

Sixty cycle current can create a low frequency hum in some motors and transformers. This hum is a result of the frequency of current that passes through the coil. Sometimes this noise can be very disturbing to the residents. Occasionally the sound is amplified because the motor or transformer is not tight on its mounting brackets. Secure the motor tightly to attempt to reduce the sound. If the sound can tracked to a vibrating sheet metal component, that component may be secured using a pop rivet or sheet metal screw to reduce the sound level.

A diagram of the three phase waveforms is shown in Figure 7-2-7. They have the same shape, but they are 120° out of phase with each other.

AC Circuit Characteristics

To review, when current passes through a conductor, a magnetic field is produced. The coiling of the conductor concentrates the lines of force, Figure 7-2-8. The polarity at the ends of the coil will reverse as the current flow through the coil changes direction.

If an iron bar is placed inside the coil as in Figure 7-2-9, current flow through the coil will cause the lines of force to magnetize the core with polarity opposite to that of the coil. Because unlike poles attract and like poles repel, the iron bar or core attempts to center itself within the coil.

This principle is applied to the construction of relays and contactors. In the resting position, gravity keeps the contacts separated. Referring to Figure 7-2-10, when current flows through the coil, the core is magnetized and tries to center itself in the coil. This raises the core and changes the position of the relay contacts, as shown in the diagram. A solenoid valve is another application of this principle.

This induction principle is also applied to the construction of AC motors, Figure 7-2-11. The basic motor has two coils wrapped around stationary cores called stator poles. As AC current flows through the coils, the stator poles are magnetized. The stator field induces an opposing field in the rotor and the principle of attraction and repulsion causes the motor to run.

Another application of the induction principle is used in the construction of a transformer, Figure

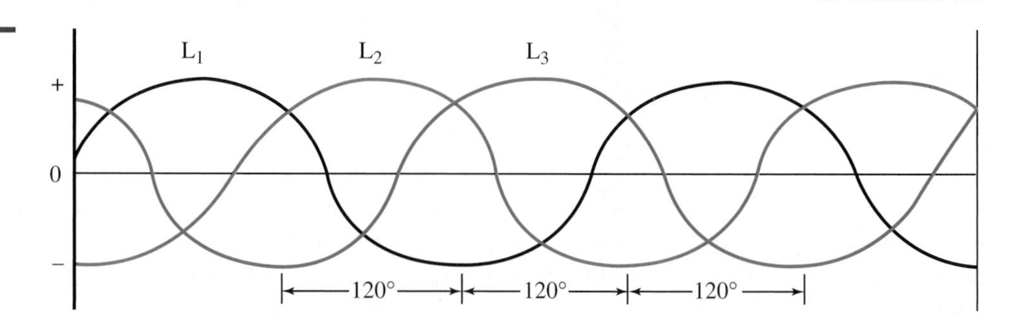

Figure 7-2-7 The three legs of power produced by a three phase A/C generator.

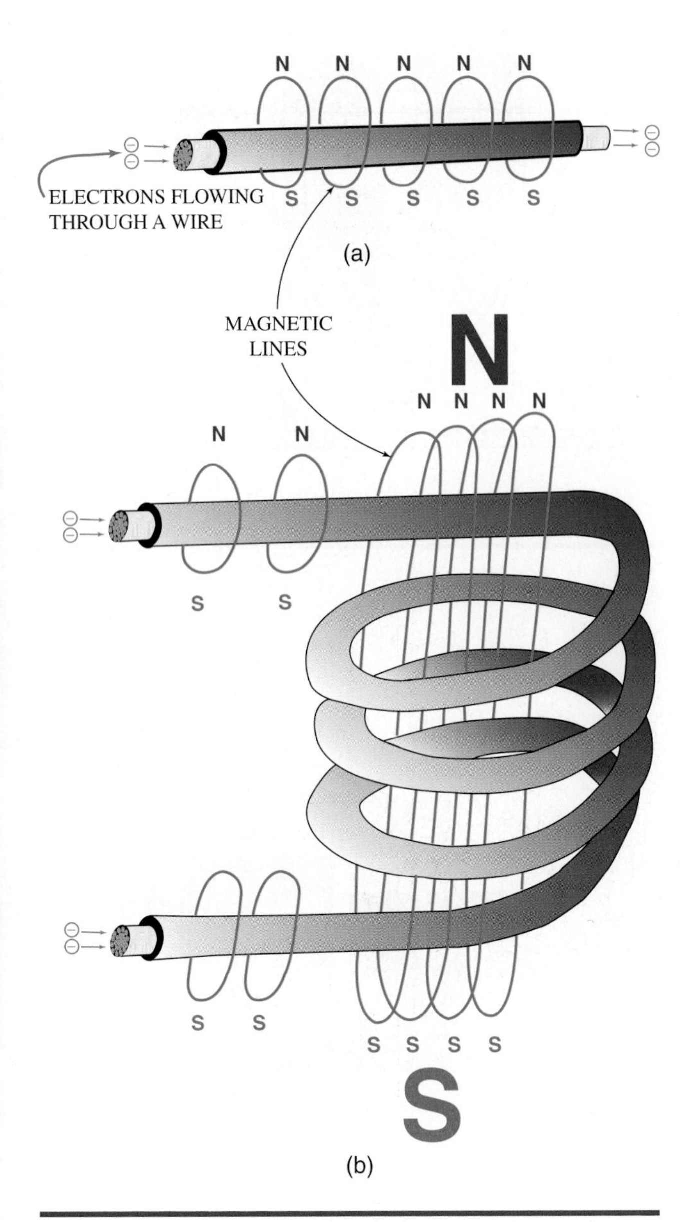

Figure 7-2-8 (a) Electrons flowing through a straight wire form a weak magnetic field around the wire; (b) When that same wire is coiled, the weak magnetic field around each wire combines to form a stronger magnetic field around the coil.

7-2-12. Transformers are used to step down the line voltage to 24 V for the controls used in many HVAC/R units. Transformers contain a single iron core that is wrapped with two separate coiled conductors known as primary and secondary windings. When AC voltage is applied to the primary winding, the resulting lines of force are carried through the core. These lines create a current flowing through the secondary winding, inducing a voltage in that winding.

There are two types of transformers: step down transformers and step up transformers, Figure 7-2-13. The amount of voltage induced in the secondary winding depends on the ratio of the number of

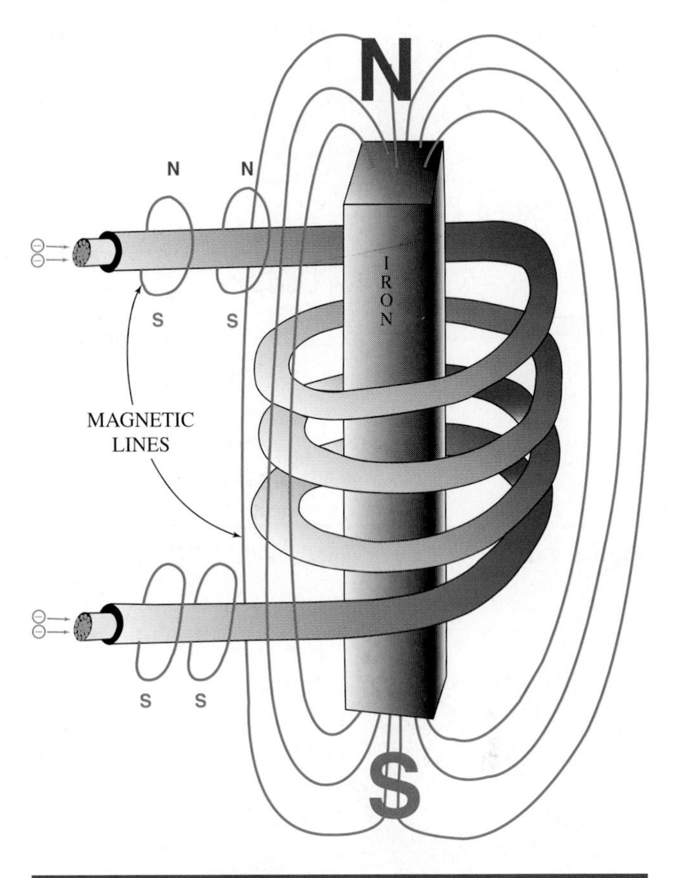

Figure 7-2-9 The addition of an iron core to the coil intensifies the magnetism and makes it stronger.

turns in the primary winding to the number of turns in the secondary winding.

Phase Shift

In working with AC power, certain characteristics affect the power calculation, $E \times I$, which are known as phase shift factors. The phase shift factors relate to:

1. Resistive circuits
2. Inductive circuits
3. Capacitive circuits

TECH TIP

A capacitor provides a phase shift in order to help an induction electric motor to start. A start capacitor added to an electric motor gives it greater phase shift for greater starting torque. Too much phase shift from an excessively large start capacitor can actually reduce some of the starting torque. Engineers have done extensive studies to determine the most appropriate size of start capacitor to optimize the starting torque. Refer to the manufacturer's literature any time you are applying a start capacitor to an induction motor.

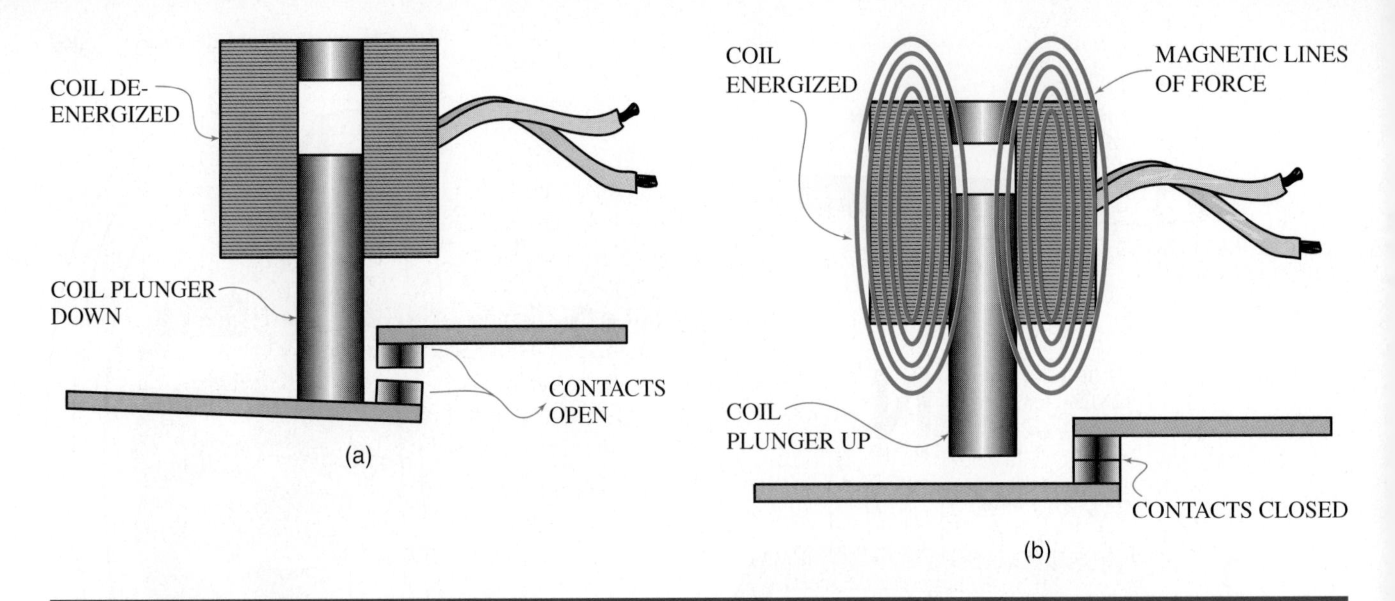

Figure 7-2-10 (a) Coil de-energized; (b) Coil energized.

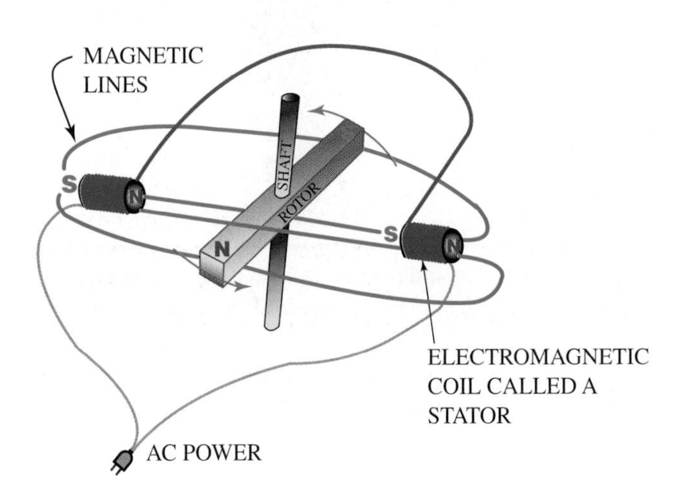

Figure 7-2-11 When AC power is applied to the coils of a motor, the rotor turns.

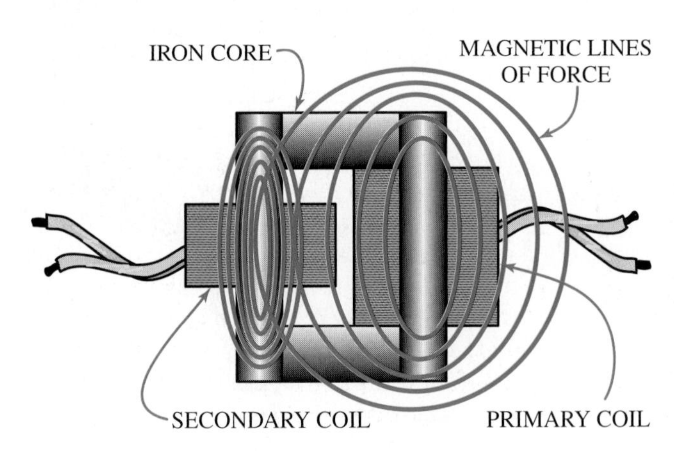

Figure 7-2-12 Power applied to the primary coil induces a current in the secondary coil of a transformer.

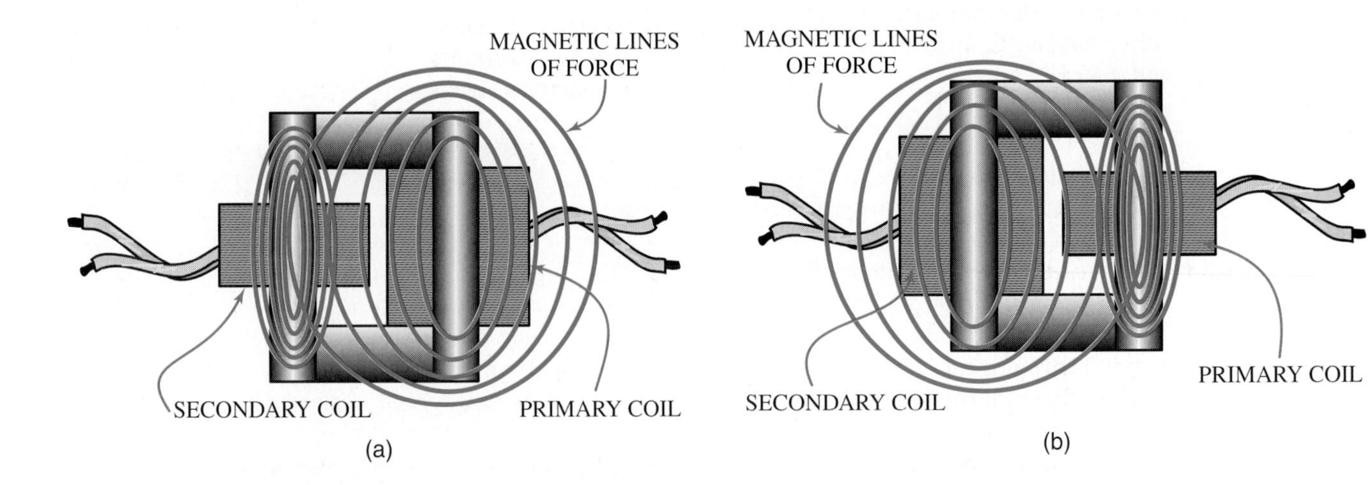

Figure 7-2-13 (a) Step down transformers have more windings on the primary coil; (b) Step up transformers have fewer windings on the primary coil.

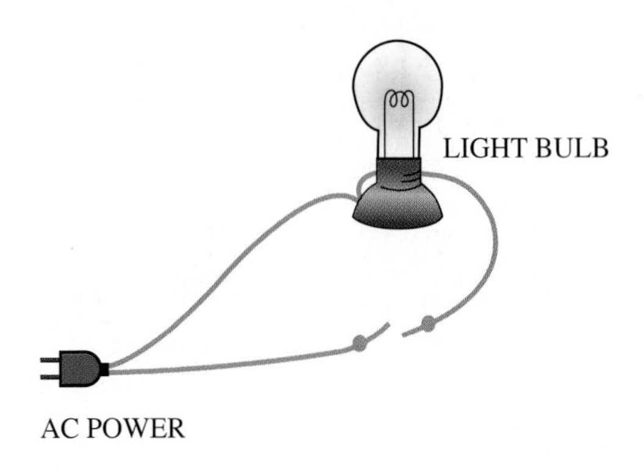

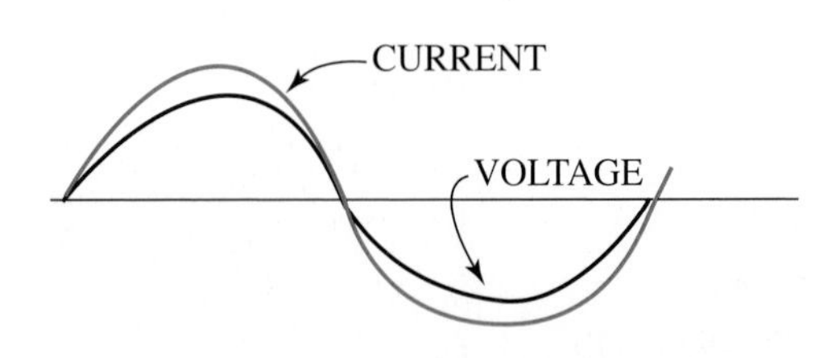

Figure 7-2-14 The AC voltage and current flow through resistance loads, like light bulbs, at the same time.

The *resistive circuit,* Figure 7-2-14, contains at least one resistive load, such as an electric heater or lamp. The current rises and falls with the voltage and the two are considered to be synchronized or "in phase." The maximum voltage occurs at the same time (or same phase angle) as the maximum amperage.

Motors, relays, transformers, and some other AC loads are constructed using coils of wire. These coils produce magnetism and are called inductive loads, Figure 7-2-15. The voltage and current in circuits containing inductive loads are phase shifted, or out of phase, sometimes as much as 90°.

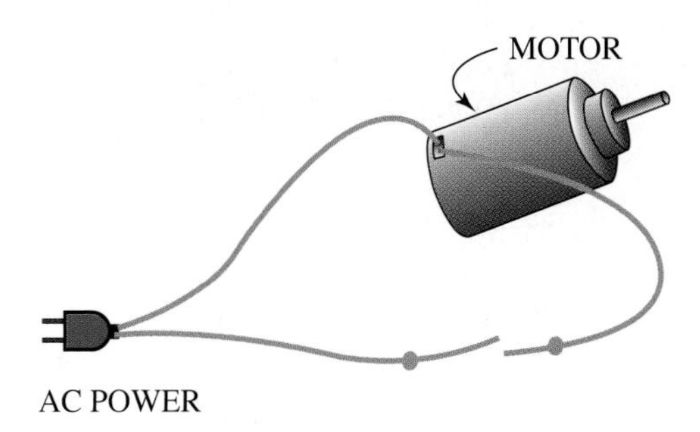

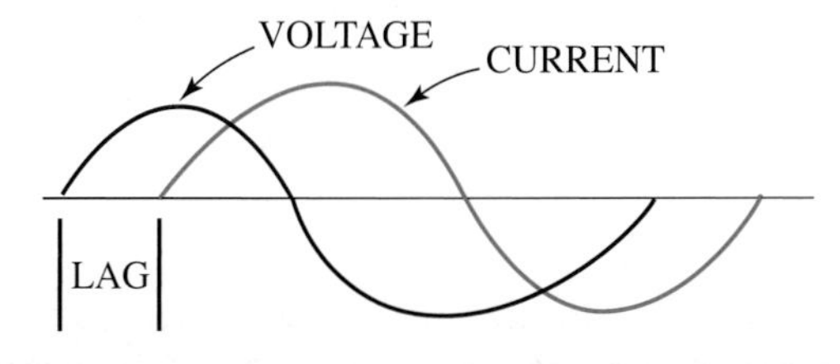

Figure 7-2-15 The AC voltage leads the current through inductive loads, like motors.

In the inductive circuit the current lags, or is out of phase with the voltage, as shown in the diagram. In this circuit, the current waveform peaks 90° after the voltage waveform.

Because of this current lag, the measured power in an inductive circuit will always be less than the calculated power ($E \times I$). This is because measured power is an instantaneous reading, and at any particular time, one or the other or both voltage and current readings are not at their peak.

The term power factor (PF) is used to indicate this difference.

$$\text{Power factor} = (\text{True power})/(\text{apparent power})$$

or

$$\text{Power factor} = (\text{Wattmeter reading})/(\text{E} \times \text{I})$$

A *capacitor* is an electrical device that is used to change the phase relationship between the current and the voltage. This effect can be used to increase the starting power (torque) of an electric motor. A capacitor consists of a layer of insulation (a dielectric) placed between two plates of highly conductive metal. A capacitor is usually connected in series with the load, Figure 7-2-16.

Obviously, no current can flow through the capacitor because of the dielectric. Initially, the current does flow through the series circuit, Figure 7-2-17. When the switch is closed, the supply voltage is applied across the capacitor. At that instant the electrons flow rapidly from the source to the right side of the capacitor and from the left side of the capacitor to the source, causing a current to flow through the load. The capacitor quickly reaches peak current. It is described as charging during this period.

Following the initial rapid flow of electrons, the rate of current flow reduces, as shown in the downward movement of the current waveform, Figure 7-2-18. As electrons leave one plate and accumulate on the other, a potential difference (voltage) begins to develop across the capacitor. This difference is created by current flow and therefore lags the current. In this circuit, which contains a resistive load, current leads the voltage by 45°, Figure 7-2-19.

As the supply voltage waveform crosses the baseline, the polarity across the capacitor changes and electrons leave the right-hand plate. The capacitor discharges and the current flows through the load in the opposite direction, Figure 7-2-20. The capacitor continues in this manner as long as the source voltage is being applied.

Impedance

Impedance is the opposition to the current flow in an AC circuit. Impedance is to an AC circuit what

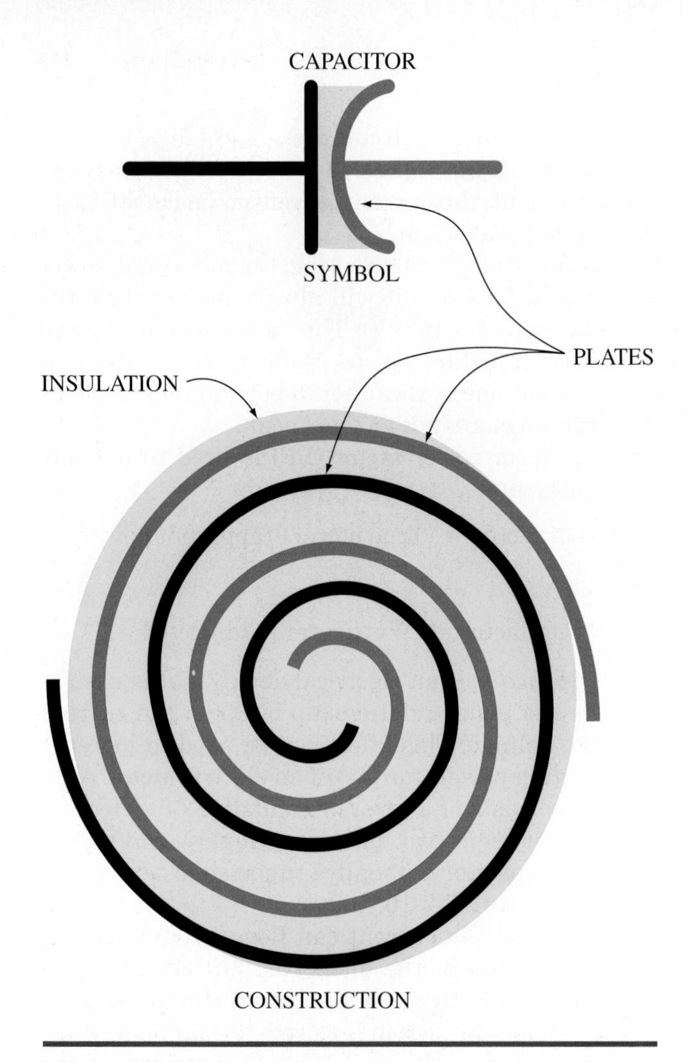

Figure 7-2-16 Capacitors are made up of two plates, thin metal foil, separated by an insulation paper and rolled up together.

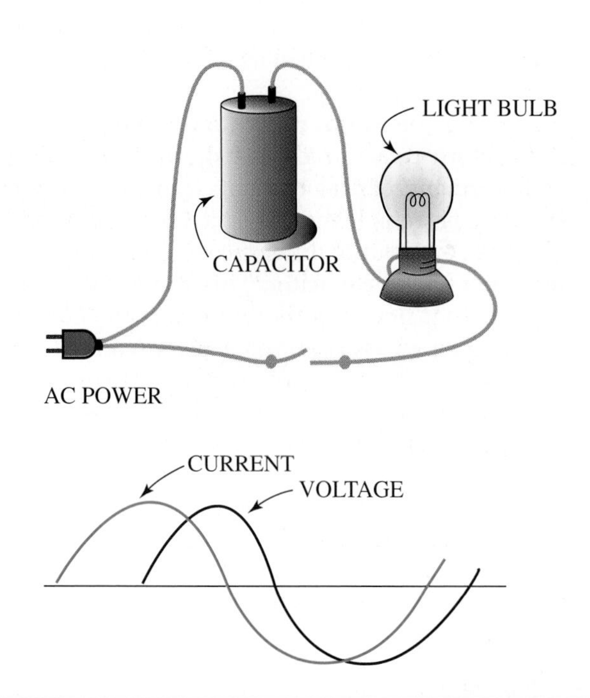

Figure 7-2-17 A capacitor causes a phase shift, so the current leads the voltage through a capacitance circuit.

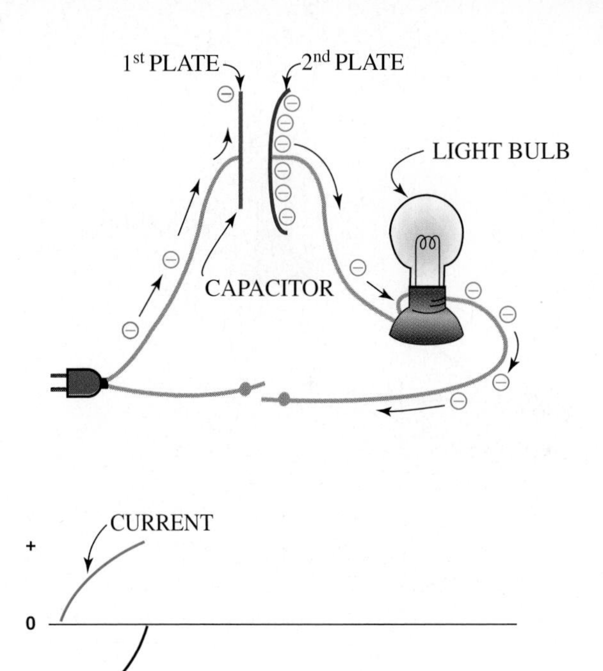

Figure 7-2-18 As the voltage starts to change from minus to plus, the electrons on the second of the capacitor's plates begin to flow out. This flow of electrons causes a surge of current before the voltage changes to positive, so the current is leading the voltage through the circuit.

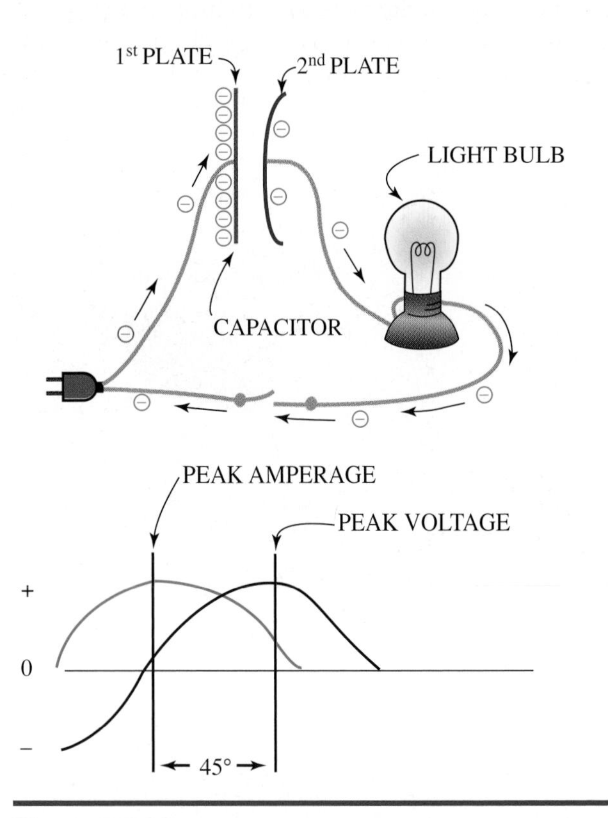

Figure 7-2-19 As the voltage on the first plate builds to its peak, most of the electrons have flowed off of the second plate.

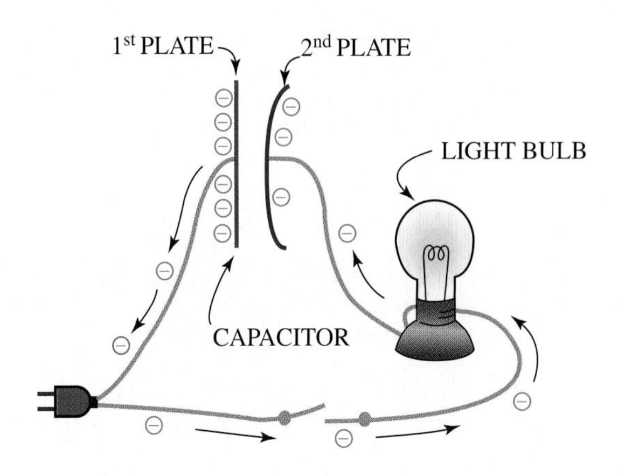

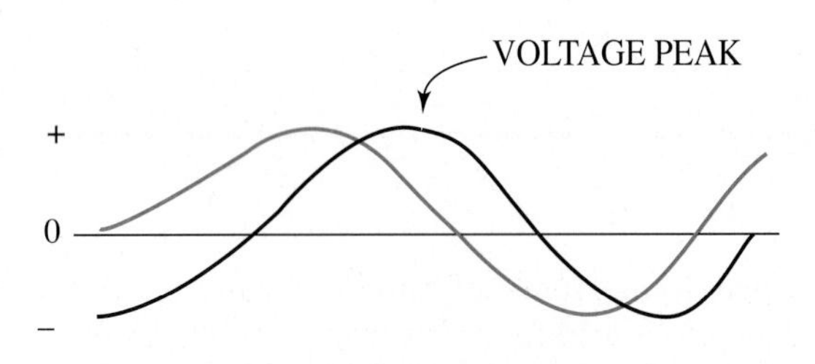

Figure 7-2-20 As the voltage peaks positive and starts back down toward zero, the electrons that were piled up on the first plate quickly flow off, forming a rush of current out of the capacitor.

resistance is to a DC circuit; however, multiple impedances in a circuit cannot be added like resistances are added in direct current because the currents through inductive, capacitive, and resistance loads in an AC circuit are out of phase with each other.

TECH TIP

The word inductance means to be induced. The electricity flowing through a coil of wire induces a magnetic field that cause another electric current to flow in a parallel coil. There is no electrical mechanical connection between the windings of an induction device such as a transformer. Each winding is completely separate. The only time the two windings are connected is when the transformer or motor has been overheated and the wiring insulation has broken down.

The following is an example of calculating impedance in an inductive circuit.

The formula that is used is:

$$Z = E/I$$

where

Z = impedance

EXAMPLE

What is the impedance of a single phase inductive circuit having a voltage of 240 V and a current flow of 10 A?

Solution

$$Z = \frac{240\text{ V}}{10\text{ A}} = 24\ \Omega$$

POWER DISTRIBUTION

Almost all of the electrical power used by consumers today is alternating current. It is supplied to substations at voltages as high as 120,000 V or more. There it is reduced to voltages between 4800 V and 34,000 V for distribution to areas of commercial or residential users. Most power companies are now using 23,000 V or 34,999 V. Local transformers reduce the voltage levels to suit user requirements as shown in Figure 7-2-21.

NICE TO KNOW

The primary reason that alternating current is used as the standard power source in commercial and residential buildings is that it is the most easily transformed from one voltage to another. Without the ability to transform power from the higher voltages used during transmission to the lower voltages used in our homes and businesses, cross country transmission lines would have to be made with extremely large diameter wires. Today, however, by using electronic switches, it is possible to more easily convert DC from one voltage to another and to convert power from DC to AC. A simple example of AC to DC conversion is a power inverter that uses an automobile's 12 volt battery to produce 110 volts of 60 cycle current. Power inverters are a great device to add to your service van because they allow you to run portable drill chargers and other electrical devices that normally must be plugged into a standard outlet.

Four common low voltage systems are available to consumers:

1. 230 V, single phase, 60 Hz systems
2. 230 V, three phase, 60 Hz systems
3. 208 V, three phase, 60 Hz systems
4. 480 V, three phase, 60 Hz systems

230 V, Single Phase, 60 Hz Systems

Single phase current is used for almost all residences. Any electrical appliance that operates on 120 V power is single phase equipment.

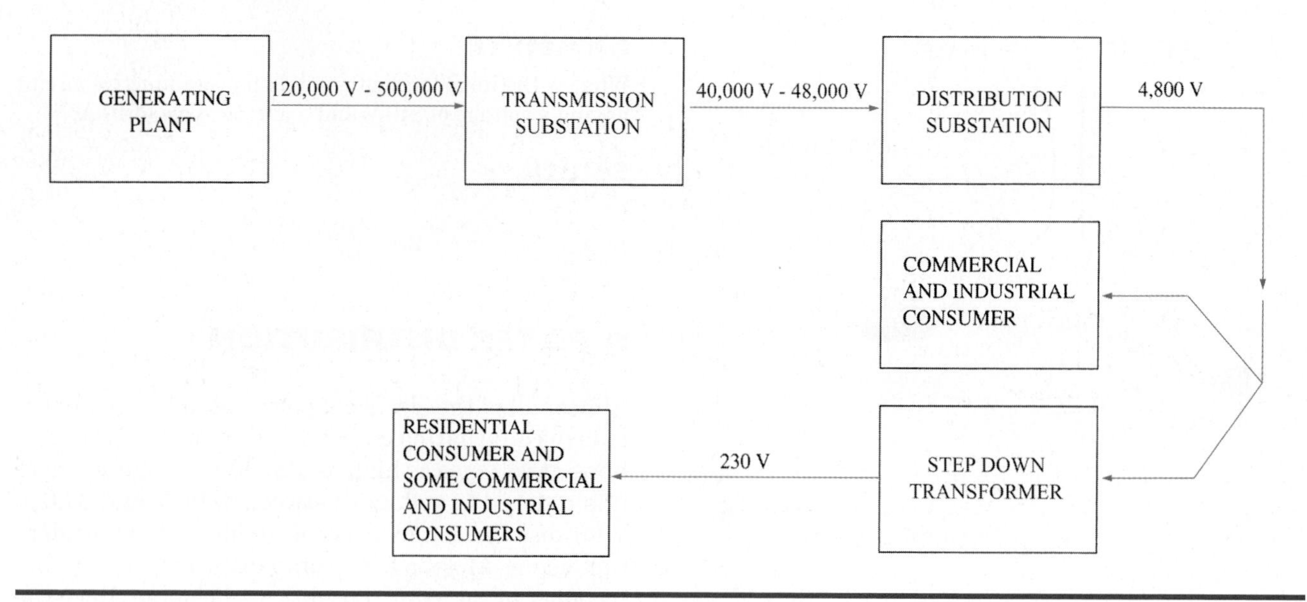

Figure 7-2-21 Voltage reduction in the power distribution system.

The most common service supplied to residential and small commercial users is the 230 V, single phase, 60 Hz system. The system uses three wires, two hot wires and one grounded neutral. A schematic diagram of this 230 V system is shown in Figure 7-2-22.

Electric utility companies use a transformer to produce this service, as shown in Figure 7-2-23.

All HVAC/R equipment is manufactured to operate satisfactorily on voltages of plus or minus 10% of the rated voltage unless otherwise specified. For example, if the equipment has a voltage rating of 230 V, the equipment should be able to operate at any voltage between 207 and 253 V. HVAC/R equipment has a tendency to operate more satisfactorily on maximum voltage than on minimum voltage.

The electric utility attempts to maintain a voltage at the load within this plus or minus 10% range. At peak load times, the line voltage may drop to near the permitted minimum. If your suspect any voltage problems, tests of the line voltage at the HVAC/R load should be measured at these times.

230 V, Three Phase, 60 Hz Systems

Three phase systems are commonly used for sizable commercial and industrial installations. These transformers have three hot legs of power and one neutral leg, as shown in Figure 7-2-24. This type of power supply is obtained from a delta transformer secondary hookup, as shown in Figure 7-2-25.

Three phase 230 V power is obtained by connecting to the three hot legs. Single phase 230 V power can be obtained by connecting to any two of the hot legs. Single phase 120 V power can be obtained by connecting to either of the adjacent hot legs and the midpoint neutral. Single phase 208 V power can be obtained by connecting to the non-adjacent hot leg and the ground. This is called the wild leg.

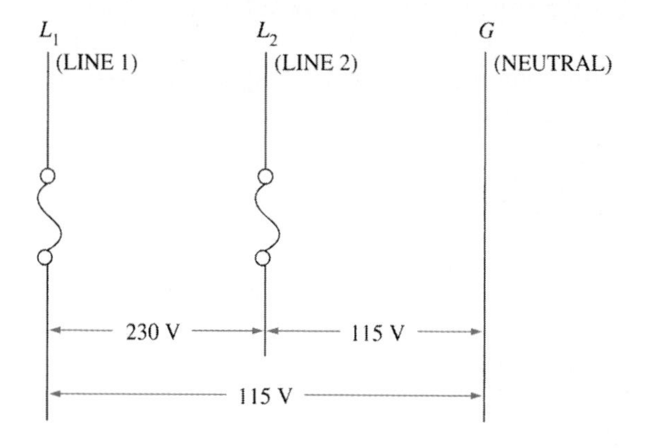

Figure 7-2-22 Line voltages for three wire 230 V single phase power supply.

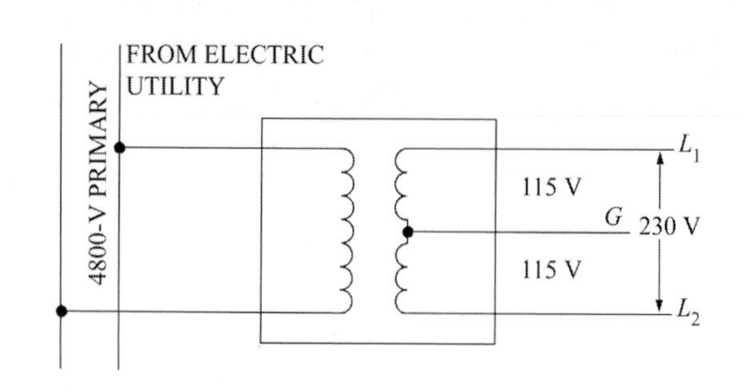

Figure 7-2-23 Step down transformer for a 230 V three wire single phase power supply.

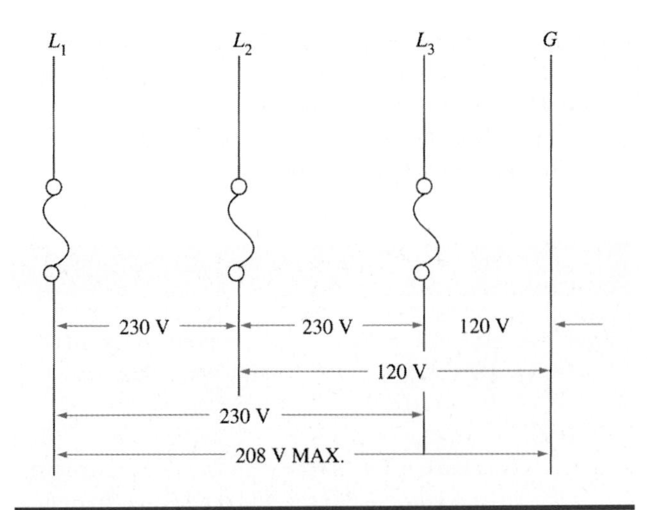

Figure 7-2-24 Line voltages for a four wire 230 V three phase power supply.

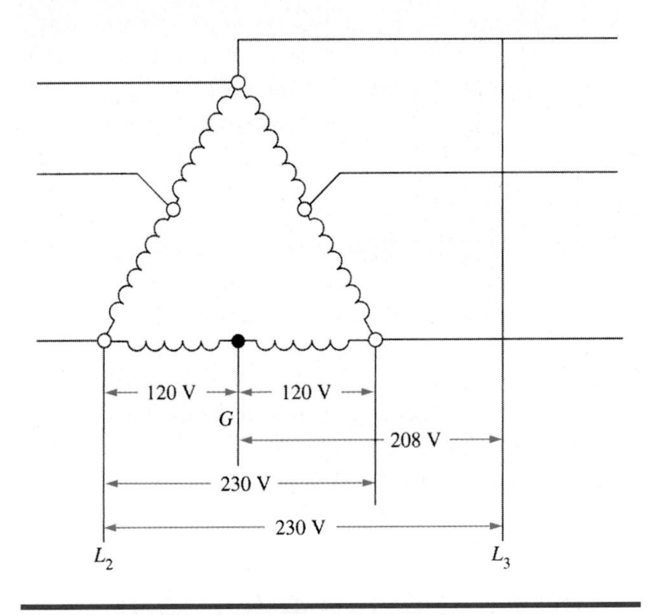

Figure 7-2-25 Line voltages for a four wire 230 V three phase system using a delta transformer secondary.

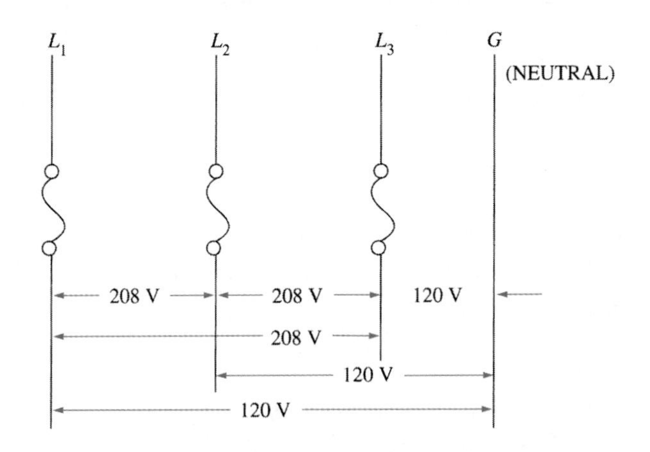

Figure 7-2-26 Line voltages for a four wire, 208 V three phase power supply.

208 V, Three Phase, 60 Hz Systems

These network systems are common in schools, hospitals and office buildings where 208 V three phase motors and 120 V single phase lighting and convenience circuits are required, as shown in Figure 7-2-26.

From this type of system, 208 V three phase, 208 V single phase, and 120 V single phase services are available. It is recommended that all motors used be rated for 208 V (not 230 V) for operation on 208 V systems.

The schematic for the transformer secondary hookup is shown in Figure 7-2-27.

480 V Systems

The 208 V/120 V three phase four wire wye connected network system, shown in Figure 7-2-26 and Figure 7-2-27, has been generally superseded in large buildings by the 480 V/277 V three phase four wire wye connected network, shown in Figure 7-2-28 and Figure 7-2-29. This improvement was made possible by the development of 277 V fluorescent lighting. Standard 460 V three phase motors can be used on 480 V systems. Convenience outlet circuits at 120 V are provided for by 480 V/208 V or 480 V/120 V stepdown transformers. The higher supply voltage permits larger loads to be serviced by smaller wires. In addition, voltage drops are reduced.

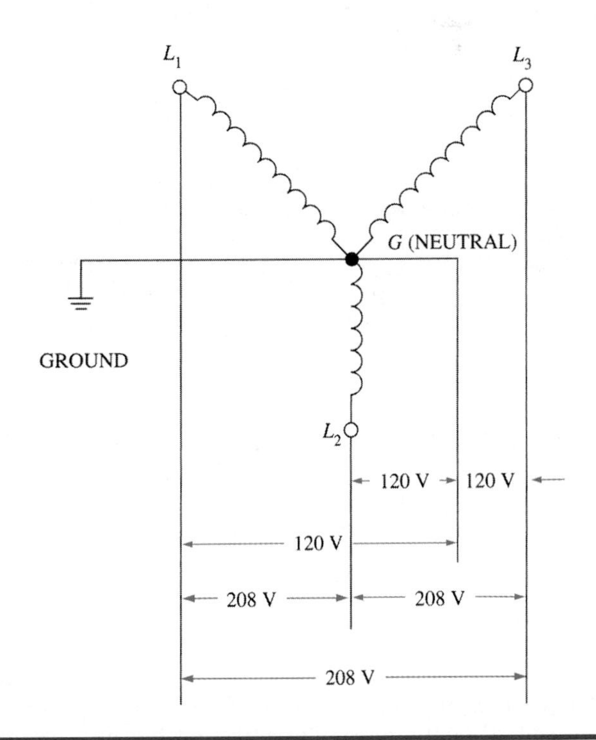

Figure 7-2-27 Line voltages for a four wire, 208 V three phase power supply using a wye transformer secondary.

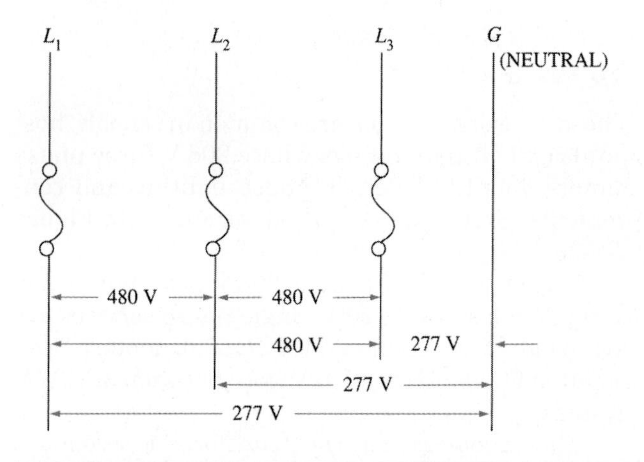

Figure 7-2-28 Line voltages for a four wire, 227 V/480 V three phase power supply.

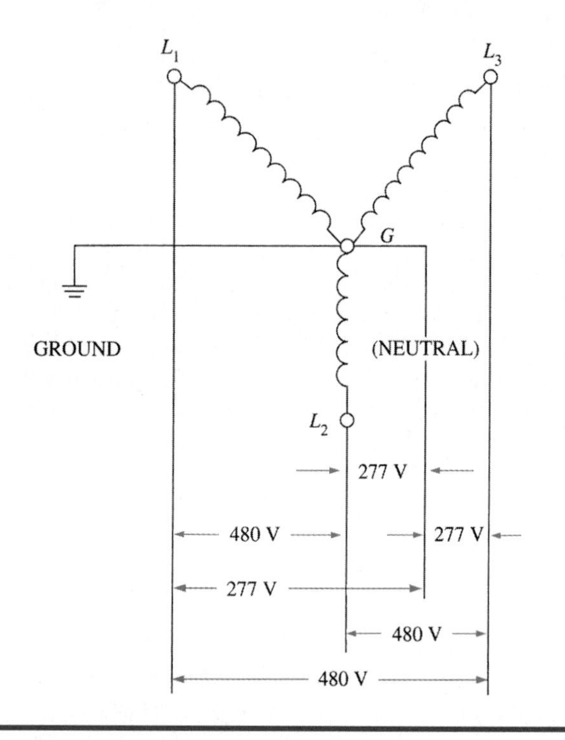

Figure 7-2-29 Line voltages for a four wire, 227 V/480 V three phase power supply, using a wye transformer secondary.

For example, a 50 Hp, 460 V, three phase, 60 Hz motor would have a full load of 57.5 A, while the same motor at 208 V would have a full load of 122 A.

Advantages of Three Phase Power

Most larger HVAC/R equipment operates on three phase power for three reasons. First, line current is reduced almost one-half for the same size motor load. This reduces the wire size and voltage drop. In addition, the line current is reduced by one-third. Second, there is less power loss to transformers in using it. Third, three phase motors are smaller, less expensive, more reliable, and more efficient than single phase. They do not require special capacitors to increase their starting torque.

TECH TALK

Tech 1. How come the wire on the poles in the street is so much smaller than the wires that run into the house?

Tech 2. Well, the voltage in the wires out in the street is a lot higher than those coming into the house. As the voltage goes up, the amperage goes down when the same current watts are flowing through the wire. It's the amperage that causes the wires to heat up, so by having a high voltage on the transmission wires the power company can use a smaller wire. You wouldn't want that very high voltage in your house; you wouldn't be safe. They use a transformer out on the pole to convert it down to 240 V. When you drop the voltage the amperage goes up and so the wire size has to be larger to carry the same amount of power.

Tech 1. That makes sense. As the amperage goes up the wire would get a lot hotter if it wasn't big enough.

Tech 2. That's right; so do you think you understand what's going on?

Tech 1. Yes. Thanks. I appreciate that.

UNIT 2—REVIEW QUESTIONS

1. What are the three essential and optional requirement for an electrical power system?
2. ________ power is used for the controls on automotive air conditioning systems.
3. A ________ is any electrical device that requires power to operate.
4. What is magnetic induction used to generate voltage for?
5. List the two types of transformers.
6. What is the term power factor used to indicate?
7. What is a capacitor?
8. Impedance is to an AC circuit what ________ is to a DC circuit.
9. What are the four common low voltage systems available to consumers?
10. What are three phase systems commonly used for?

UNIT 3

Electrical Meters

OBJECTIVES: In this unit the reader will learn about electrical meters including:

- Analog Meters
- Digital Meters
- Ammeters
- Ohmmeters

ELECTRIC METERS

The three electric meters that have the greatest use for installers and service personnel are the voltmeter, ammeter, and ohmmeter. They can be purchased as separate meters, or most commonly, they are all combined in a single meter called a multimeter.

There are two basic types of meters, the analog meter and the digital meter. These terms are familiar since they also apply to watches. The digital meter is solid state and gives a direct numerical readout of the measured value. The analog meter has a needle that points to the measured value. Each is shown in Figure 7-3-1.

CAUTION: Not all meters and leads have been independently certified and tested for safety. Look for markings on the meter to indicate that they have passed basic safety tests such as those provided by Underwriters Laboratories (UL). Meters that are not rated may be as safe. However, unless you have significant experience with these meters to know if they are capable of withstanding the types of service the HVAC/R industry requires, it is recommended that you avoid using unrated meters in the field.

Analog Meters

All analog meters operate on the same principle. When the current flows through a conductor, it produces a magnetic field around the conductor. If a magnetic needle is placed close to the current, the needle will attempt to line up with the field, Figure 7-3-2.

In order to conserve space, the current carrying wire is coiled and a scale is provided to indicate the position of the needle. The mechanism is so constructed that the greater the current flow, the greater the deflection of the needle on the scale.

NICE TO KNOW

Many people feel that digital meters are superior to analog meters in that they show you discrete values for voltage, amperage, and ohm readings. However, the analog meter will give you more of a relative reading that can be often understood more easily. For example, during the late 1980s and early 1990s, many car manufacturers provided digital speedometers for cars. These turned out not to be very popular because with digital displays it was no longer possible to simply glance at the speedometer needle to see your relative speed. In the same way analog meters allow the technician to glance at the meter and make a determination. Some digital meters have incorporated an analog sliding scale that appears across the digital meter face so that it is possible for the technician to have both the digital number and an analog reference.

Three important characteristics of analog meters need to be considered by those who use them:

1. The most accurate reading is at or near the midpoint of the scale. This is because the spring that opposes the deflection of the meter does not exert constant pressure across the scale. The meter may be inaccurate at either end of the scale. So whenever the operator has a choice of scales, the one selected should place the pointer in the most favorable (central) position.

2. Analog meters periodically need to recalibrated. Most meters include some type of adjustment and instructions for calibration.

3. The small coil of wire that forms part of the meter movement is sensitive to excessive current. The meter may be made completely inoperable if subjected to excessive current. In using a meter with multiple scales to choose from, always use the higher scale first and move down to the scale required.

Analog meter accuracy is normally specified as a percent of full scale reading, so readings should be

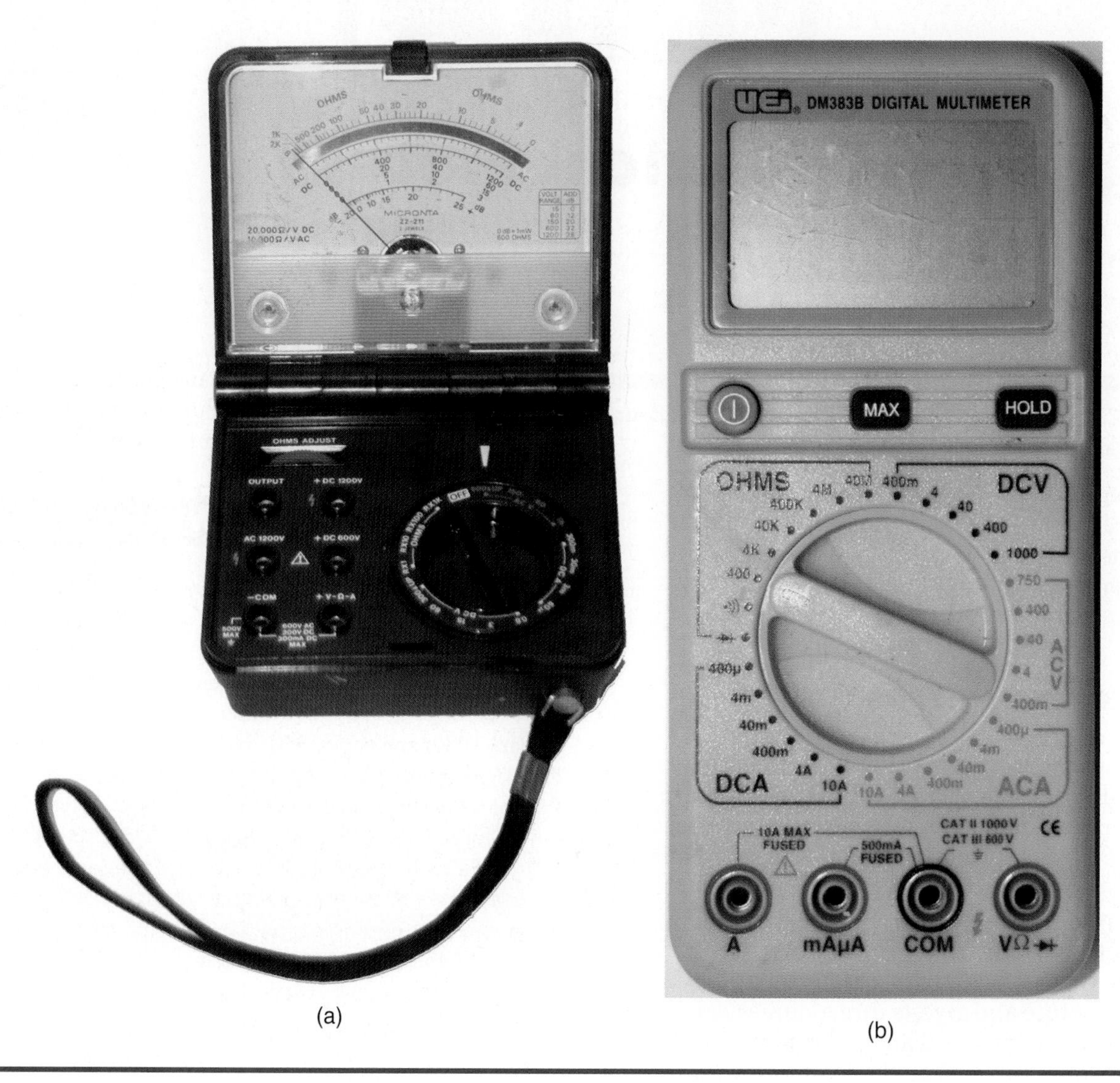

Figure 7-3-1 (a) Analog multimeter; (b) Digital multimeter.

taken in the upper two-thirds of the scale for testing accuracy.

More expensive (and more accurate) analog meters are often furnished with a mirror scale to enable more accurate readings. The mirror aids in minimizing parallax, an apparent difference in readings taken from different sight perspectives. When reading the meter correctly, the pointer and scale are aligned so the mirror image of the pointer disappears behind the pointer. This gives the most accurate reading.

TECH TIP

Because of the mechanical movements in an analog meter they tend to be more easily damaged as a result of rough treatment. Some analog meters have a cage for the meter movement which locks it down so that it does not bounce around and become damaged during transport.

SERVICE TIP

Many service technicians will want to have both analog and digital meters. Just like analog watches, analog meters have advantages such as the ease in reading changes and variations in the measurements.

Digital Meters

Digital meters offer a number of advantages, although they are usually more expensive. Rugged versions are available and recommended. Because of the expense of digital meters, it is important that the technician be thoroughly familiar with the use of both types. Some of the advantages of digital meters are:

1. They are direct reading. There is no need to interpret the scale.
2. Digital meters can be obtained that will give accurate readings to three decimal places.

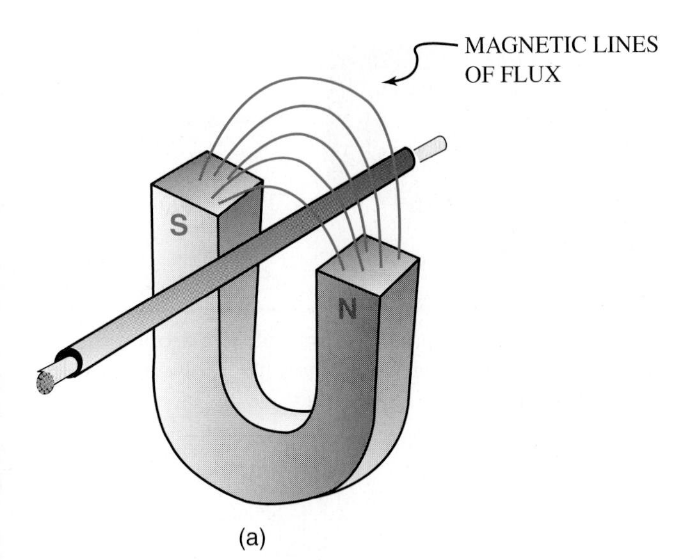

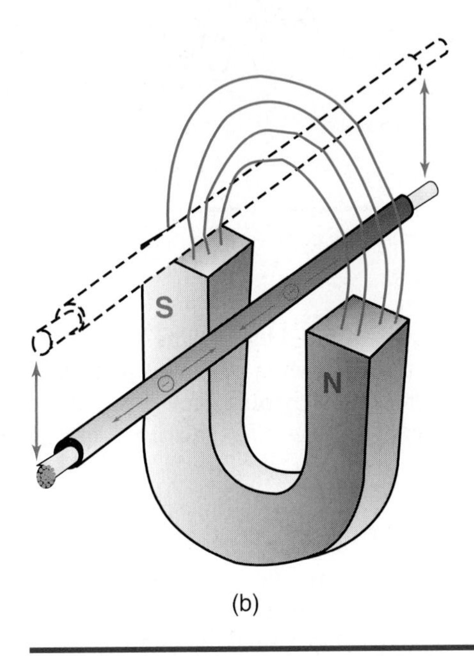

Figure 7-3-2 (a) If a wire is stationary in a magnetic field, no current is induced in the wire; (b) If the wire is moved up and down in the magnetic field, electrons in the wire flow forming an electric current.

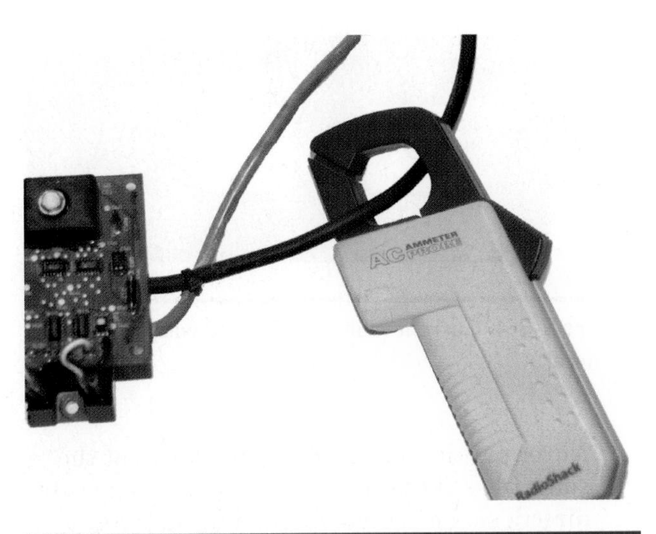

Figure 7-3-3 The most accurate amp measurement is when the wire passes through the center of the meter jaws.

SERVICE TIP

Some digital meter displays are susceptible to water or humidity damage. It is important that these meters be kept as dry as possible so that they do not become damaged. In some cases a digital display can be reconditioned by placing the meter in a warm location such as the cab of your service van to allow it to dry out and possibly begin working.

3. They have no moving parts and are less likely to fail or get out of calibration than analog meters.
4. They often have automatic scaling features.

AMMETERS

The clamp-on ammeter is one of the most useful of the electric meters for HVAC/R technicians. It is used to measure current flow through a single wire by enclosing the wire within the jaws of the instrument as shown in Figure 7-3-3.

This instrument functions like a transformer. The primary coil is the test wire encircled by the jaws of the instrument. The secondary is a coil of wire within the instrument that is connected to the current-indicating mechanism. The current in the primary wire induces a flow of current in the secondary winding, measuring the current flow. The greater the current flow through the test wire, the greater the induced current and the greater the deflection of the needle reading on the scale. It is not necessary to disconnect or make contact with any wires to obtain a reading. This is very convenient.

Care of the Instrument

More accurate tests and longer life for the clamp-on ammeter are obtained by proper care of the instrument. Some of the ways the instrument should be treated are:

1. Keep the jaws clean and aligned.
2. When taking a reading always start on the highest possible scale and then work down to the most appropriate scale.
3. Do not cycle a motor off and on while taking readings unless the meter is first set to the highest scale.

Figure 7-3-4 Current multiplier for use with a clamp-on ammeter.

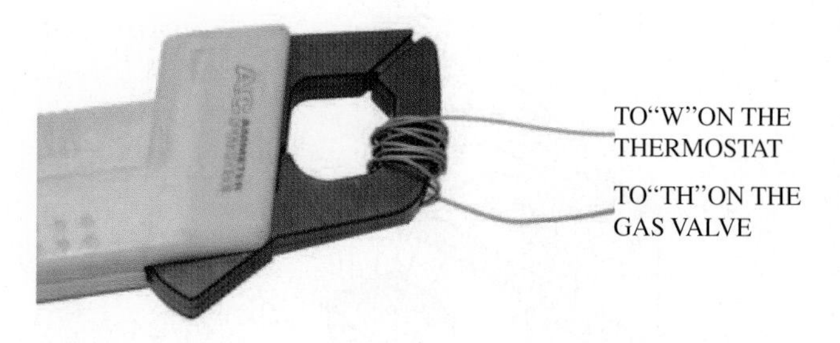

Figure 7-3-5 Wrapping ten turns of wire around the ammeter to read lower amp draws like those through the heat anticipator on a thermostat.

4. Never put the clamp around two wires at the same time. If the current is flowing in opposite directions, the meter will read the difference between the two. If the current is moving in the same direction in both wires, the meter will add the two.

Reading Small Amounts of Current The clamp-on ammeter is useful in reading small amounts of current. The procedure is to loop the wire around the jaws, as shown in Figure 7-3-4. Passing the wire twice through the jaws doubles the strength of the magnetic field. It is therefore necessary to divide the meter reading by two to determine the actual current. For two passes, the actual current is ½ the reading; for three passes the actual current is ⅓ the reading, etc.

SERVICE TIP

The most accurate place for a clamp-on ammeter to read amperage is for the conductor to pass near the center of the jaws' opening. For most amperage readings taken for motors, compressors, etc., whether or not the lead is in the center or not is relatively insignificant. But the lead placement can make a difference when taking very small amperage readings such as those for heating anticipators on thermostats.

One application of the clamp-on ammeter is to adjust the anticipator setting of a thermostat. The anticipator supplies false heat to the thermostat, causing it to shut off the heat to the space being heated before the temperature in the room reaches the set point. This prevents the residual heat in the furnace, supplied by the fan after the burners are off, from overheating the room.

For the anticipator adjustment, 10 passes of wire are wrapped around the jaws of the instrument. To obtain the actual current flow through the anticipator, the scale reading is divided by 10. The location of the ammeter with its coil of wire is shown in Figure 7-3-5. The wire that passes through the jaws of the ammeter is connected to the "W" terminal of the thermostat.

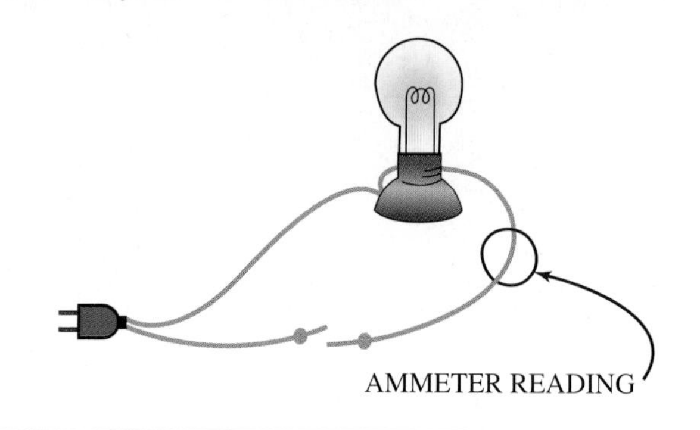

Figure 7-3-6 An ammeter is connected in series with the load.

The Inline Ammeter

Occasionally it is desirable to use an inline ammeter. The proper location for its connections in a circuit is shown in Figure 7-3-6. Note that it is connected in series with the circuit being tested. In DC circuits, verify the correct polarity of the ammeter used before energizing the circuit. Never connect an ammeter across a load. It will be destroyed by line current as there is no load to limit it!

VOLTMETERS

Two leads from the voltmeter are connected to the circuit being tested, Figure 7-3-7. Voltmeters are connected in parallel with the load to read the voltage drop or potential difference.

Figure 7-3-8 shows an analog voltmeter that has multiple scales. Each scale has a different resistance in the meter placed in series with the circuit being tested. A knob in the center face of the meter adjusts the meter to the scale being used. When using a multi-range meter, always start to measure voltage

Figure 7-3-7 Using a volt ohmmeter to check the voltage at a compressor disconnect box.

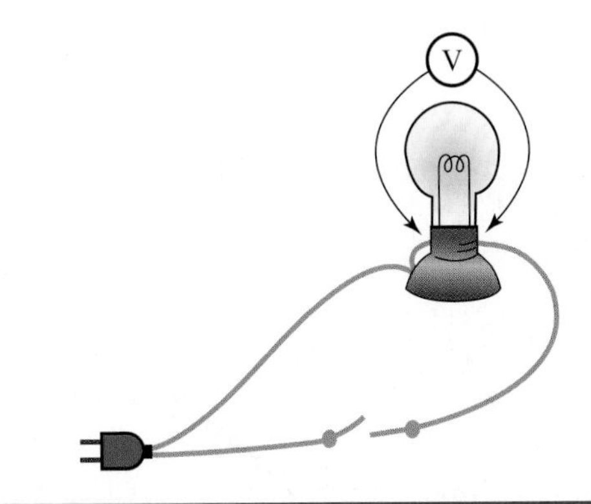

Figure 7-3-9 A voltmeter is connected parallel to the load.

Figure 7-3-8 Analog meters sometimes must have their needles zeroed by turning a small screw on the meter face.

Figure 7-3-10 The current through the very fine meter needle coil wire causes a magnetic field to form, which causes the meter needle to move.

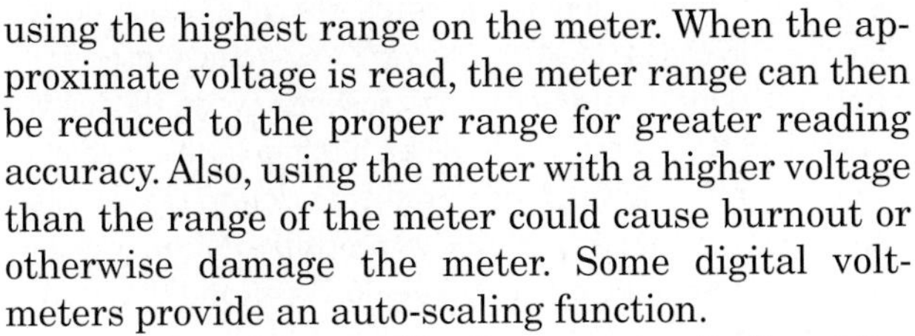

using the highest range on the meter. When the approximate voltage is read, the meter range can then be reduced to the proper range for greater reading accuracy. Also, using the meter with a higher voltage than the range of the meter could cause burnout or otherwise damage the meter. Some digital voltmeters provide an auto-scaling function.

In DC circuits, verify the correct polarity of the probes that are used before connecting the meter to the circuit.

In testing a motor to see that it has proper voltage, one lead goes on each side of the load, as shown in Figure 7-3-9. On a DC circuit the polarity of the leads must be observed.

The voltmeter can also be used to determine if a hidden switch is open or closed. This is very helpful in troubleshooting. If there is power in the circuit, and the leads of the voltmeter are placed on each side of the switch, a voltage reading indicates the switch is open and a zero reading indicates the switch is closed (if no other open switches are in the circuit).

OHMMETERS

The ohmmeter is different from an ammeter or a voltmeter in that it uses a battery as a power supply. The battery furnishes the current needed for resistance measurements.

The function of an ohmmeter is a direct application of Ohm's law. The higher the resistance, the lower the current flow. For analog meters, this produces reduced meter deflection, as shown in Figure 7-3-10. The resting place for the needle is on the left of the scale. For a high resistance the deflection is small. For a small resistance the deflection is large.

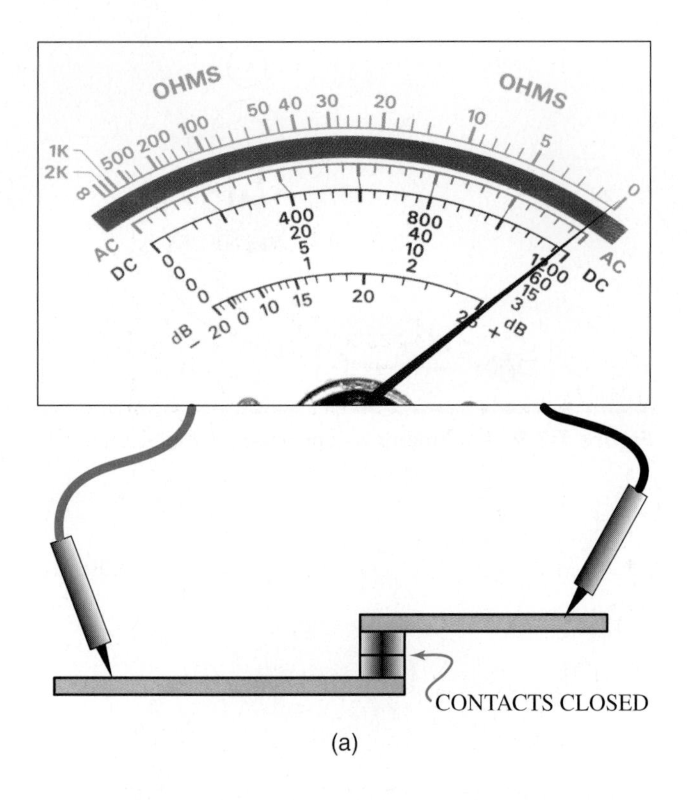

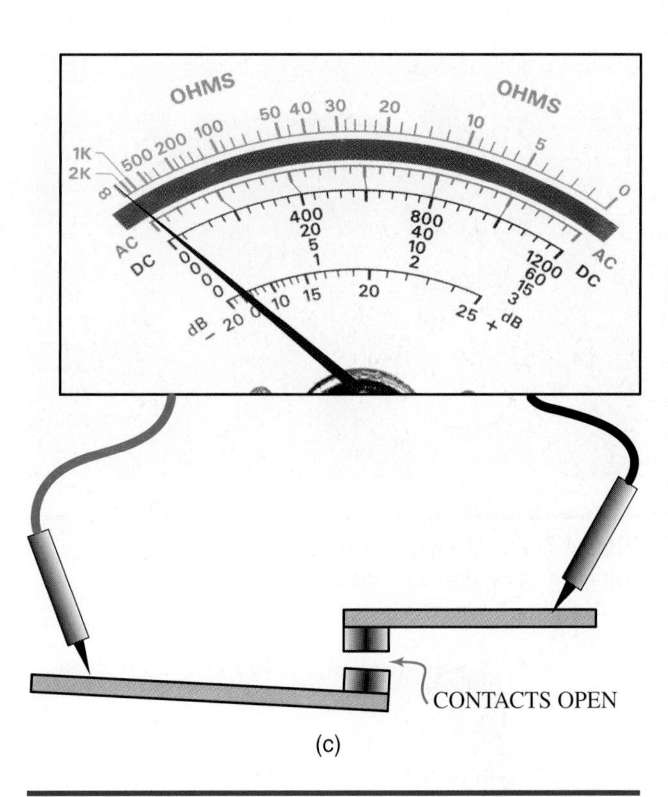

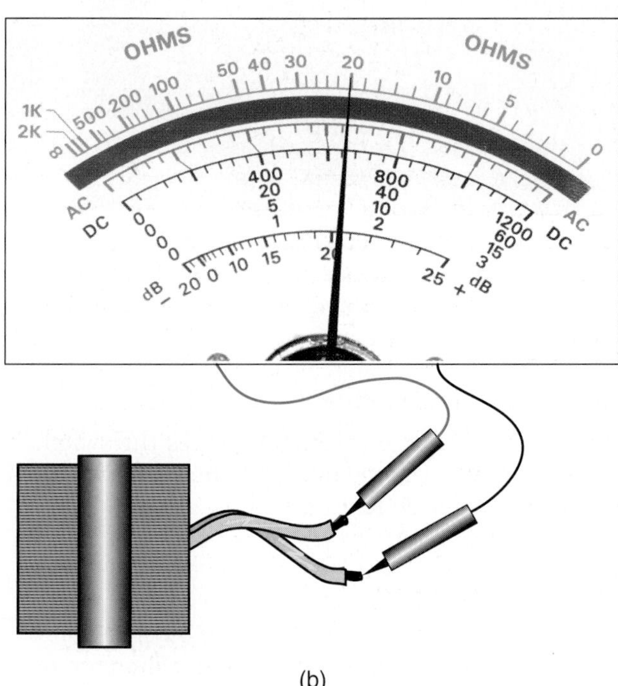

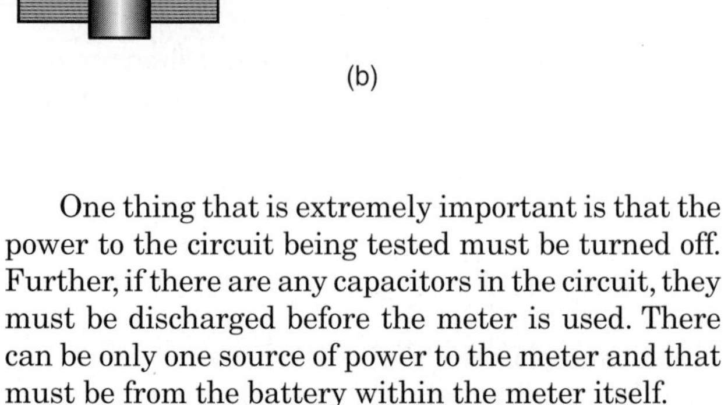

Figure 7-3-11 (a) Closed contacts or a short will read as 0 ohms; (b) A coil resistor will have a resistance reading between 0 and infinity; (c) An open contact or broken circuit will have an infinite (∞) resistance.

TECH TIP

An ohmmeter can be used as a quick check to test for a short in a capacitor. Analog meters are better for this because you can more easily see the meter movement. The test is performed by placing the two leads of the ohmmeter on each of the capacitor terminals for a few seconds and then swapping the leads. What this will do is charge the capacitor with the very low voltage used for ohm checking. In a functioning capacitor, when the leads are reversed, at first there will be a bump in the reading. Then, as the voltage bleeds off the capacitor, the needle will slowly swing back to the zero position. If the capacitor is shorted out, there will be no bump and the reading will stay at zero.

One thing that is extremely important is that the power to the circuit being tested must be turned off. Further, if there are any capacitors in the circuit, they must be discharged before the meter is used. There can be only one source of power to the meter and that must be from the battery within the meter itself.

Using the ohmmeter to check for open circuits is called continuity testing. Figure 7-3-11 shows three diagrams representing the three possible responses that the meter can give.

1. In Figure 7-3-11a, the meter is measuring the resistance through wire in a circuit and it registers zero. This indicates maximum current flow or 0 Ω resistance. This is the measurement of a short circuit.

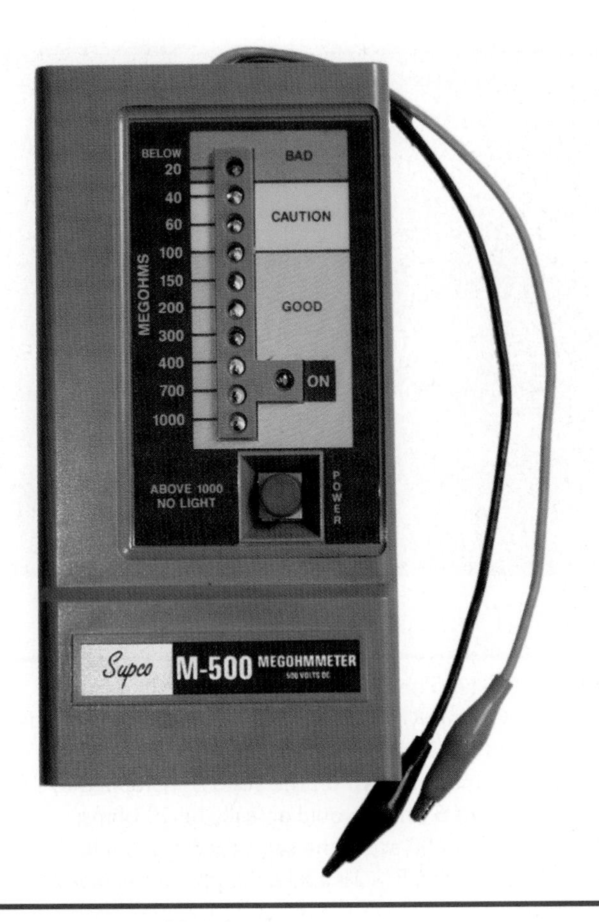

Figure 7-3-12 Megohmmeter.

2. In Figure 7-3-11b, the meter is measuring the resistance of a resistor, which has a measurable resistance that is read on the meter.
3. In Figure 7-3-11c, the meter is measuring the resistance of an open circuit, which is read on the meter as infinity. Infinity means that the resistance is so large that it cannot be measured. It means that at this point there is a lack of continuity or no current flow.

Some ohmmeters must be able to read resistances of tens of millions of ohms (megaohms), Figure 7-3-12. The higher amount of power is required for higher resistances.

One very handy feature on an analog ohmmeter is the zero ohm adjustment, shown in Figure 7-3-13. This knob makes possible a quick and easy method of calibrating the instrument each time it is used. To test the adjustment, the two leads are touched together and if the reading is not zero, the zero ohm adjustment knob is turned until the needle on the analog meter reads exactly zero.

Unlike the voltmeter, one scale of an analog meter is used to read all resistance ranges. To determine the resistance value, multiply the meter reading by the number shown next to the selector switch setting. For example, in Figure 7-3-14a, the reading would be multiplied by the selector setting (×1), giving a resistance of 5 Ω.

Figure 7-3-13 With the meter leads touching each other, zero the ohm needle.

Some ohmmeters have a selector position of R × 100,000 which is used in measuring very high resistances such as motor windings to ground.

Care must be taken to prevent errors in reading resistance when two or more circuits are connected in parallel, as shown in Figure 7-3-15. The meter in the illustration is actually reading the combined resistance of two parallel resistances, HTR1 and blower.

In order to read only one resistance, one side of the component being tested is disconnected, as shown in Figure 7-3-16.

One caution that needs to be followed: do not use an ohmmeter to test a solid state circuit unless the manufacturer specifically requires it. The internal battery voltage of the ohmmeter can damage an integrated circuit chip, as shown in Figure 7-3-17.

UNIT 3—REVIEW QUESTIONS

1. What are the two basic types of meters?
2. Describe a digital meter.
3. How is the accuracy of an analog meter normally specified?
4. List the advantages of digital meters.
5. What is an ammeter used to measure?

(a)

(d)

Figure 7-3-14 The range must be set to the proper scale. (a) On the R × 1 scale, the number on the scale is the same as the ohms being read; (b) On the R × 10 scale, the number shown on the scale is multiplied by 10, so a reading of 5 ohms would actually be 50 ohms; (c) On the R × 100 scale, the scale reading is multiplied by 100; (d) On the R × 1K scale, the number is multiplied by 1000.

(b)

(c)

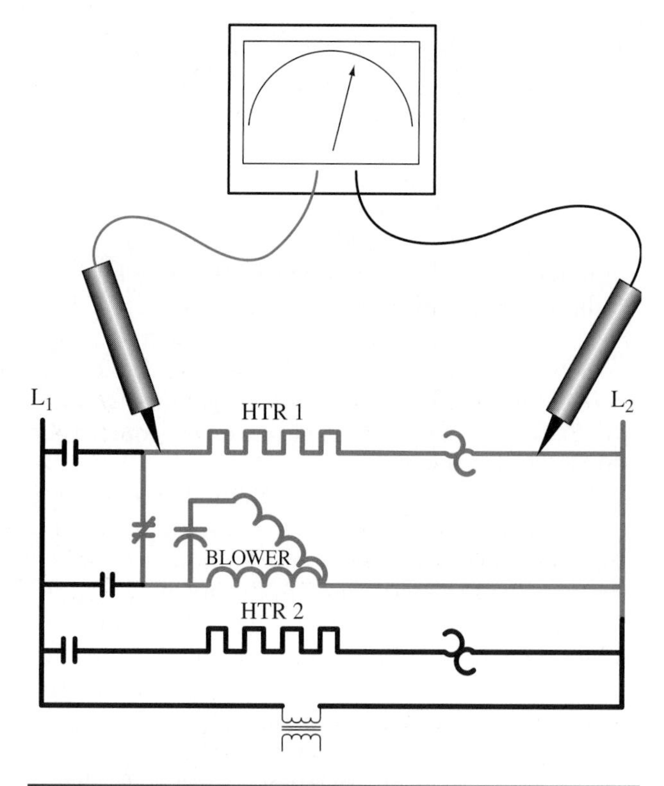

Figure 7-3-15 The ohm reading for Heater 1 would not be accurate because there are two paths, as shown in red.

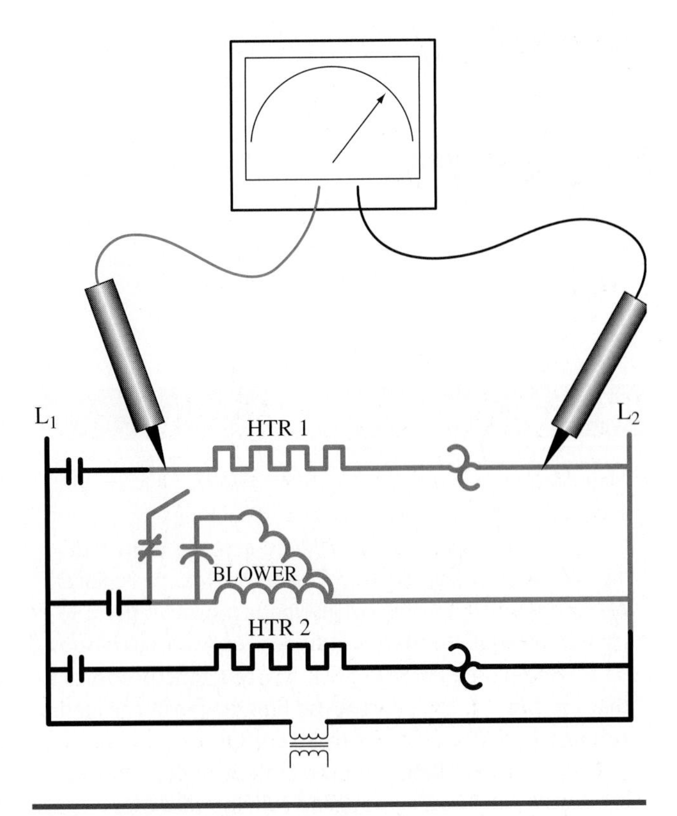

Figure 7-3-16 The ohm reading would be accurate because the second path is now broken at the normally closed contacts.

Figure 7-3-17 NEVER use a multimeter to check an integrated circuit board chip.

6. What is one application of the clamp-on ammeter?
7. How is the ohmmeter different from a voltmeter and an ammeter?
8. Using the meter to check for open circuits is called ________.
9. What does the zero adjustment knob on an analog ohmmeter do?
10. What is one caution that needs to be followed when using an ohmmeter?

TECH TALK

Tech 1. I was checking the circuit yesterday with my voltmeter and I wasn't getting any voltage. When I flipped one of the switches the fan started so I guess my meter wasn't working. What should I do?

Tech 2. Every time that you use your meter you need to do a quick meter check. The best one is to test a circuit that you know is electrically charged and then test the circuit that you think is not. Then recheck the circuit that you know is hot. This way you can see whether you have a reading or not.

Tech 1. I think I can do that. What else could the problem have been?

Tech 2. It is possible with some meters to have it on the wrong scale. You might check one voltage and the meter works and you check another voltage and it's too high and blows the fuse. That might fool you into thinking the circuit is dead when really you just blew out your meter.

Tech 1. Then the circuit could be live, but how would I check?

Tech 2. Just like you learned in class, you always start with the highest voltage scale and work your way down.

Tech 1. OK, I'll always do that method, highest scale first. But why do you have to go back and forth between a hot and a dead circuit?

Tech 2. By checking something that's hot, something you think is not hot and then something that is hot again, you know the meter is working that didn't show voltage and then shows voltage. That's a good practice any time you are going to be working on terminals to make sure they are not hot and going to shock you.

Tech 1. That sounds like an awful lot of work.

Tech 2. It just takes a second to do the test but it takes even less time to be electrocuted.

Tech 1. I guess you're right about that, I better start doing that.

Tech 2. It could save your life one day.

Tech 1. Thanks, I appreciate that.

UNIT 4

Electric Motors

OBJECTIVES: In this unit the reader will learn about electric motors including:

- AC Induction Motors
- Induction Motor Principles
- Capacitor Principles
- Single Phase Motors
- Three Phase Motors
- Motor Protection
- Testing Start Relays
- Principle of Motor Operation

NICE TO KNOW

Manufacturers can produce motors with high starting torque, high running torque, or good running efficiency. It is impossible, however, to produce a single motor that has all three characteristics. The performance required of motors will determine what type of motor to use. Motors that frequently start but do not run for long periods of time would benefit from high starting torque. Motors that operate for long periods of time under heavy loads would benefit from having high running torque. Motors that operate for long periods of time under relatively light loads will benefit from good running efficiency. Manufacturers of equipment take these three factors into consideration when specifying which motor is to be used. For that reason it is recommended that you select a replacement motor within the same category as the one originally supplied with the equipment.

AC INDUCTION MOTORS

Electric motors are the most important load device in the various types of HVAC/R units. They convert large portions of electrical energy to useful work. It is important, therefore, for the technician to understand how they operate and how they can be protected.

Motors are designed for various types of service. Important qualities are torque, speed, and power usage.

Torque is the twisting (or turning) force that must be developed by a motor to turn its load. The power required to power a load is directly related to the torque required and the speed. A greater amount of torque is required to start a motor than to run it. The starting torque requirements for a fan are low. The starting torque requirements for a reciprocating compressor are high. Providing extra starting torque is expensive. Therefore, to keep the cost down the motor is selected for as small a torque as will adequately perform the work for which it is intended.

INDUCTION MOTOR PRINCIPLES

The two principal parts of a motor are the stator and the rotor. The stator is the stationary part and the rotor is the rotating part or armature, as shown in Figure 7-4-1. The stator is a coil of wire wrapped around a magnetic casing. When alternating current is applied to the stator coil, a rotating magnetic field is produced.

The rotor is a series of aluminum (or copper) bars mounted on a soft iron core. The core provides a path for the magnetic field of the rotor. The conductor bars are shorted together by an end ring permitting current to flow, as shown in Figure 7-4-2.

The current passing through the stator creates a powerful magnetic field. The repulsion and attraction of the rotor and the stator parts cause the motor to turn, as shown in Figure 7-4-3.

No current is actually supplied to the rotor. The magnetic field of the rotor is produced by induction from the stator. The current induced in the rotor is in the opposite direction from the stator current. This opposing magnetic field of the rotor reacts with the stator field, producing rotation.

As shown in Figure 7-4-4, the like poles repel, causing rotation. In Figure 7-4-5, the rotor is shown as north and south poles relatively near each other. The two stator poles have wire coiled around them. Since like poles repel and opposite poles attract, with the polarity shown, the rotor will move toward the stator pole.

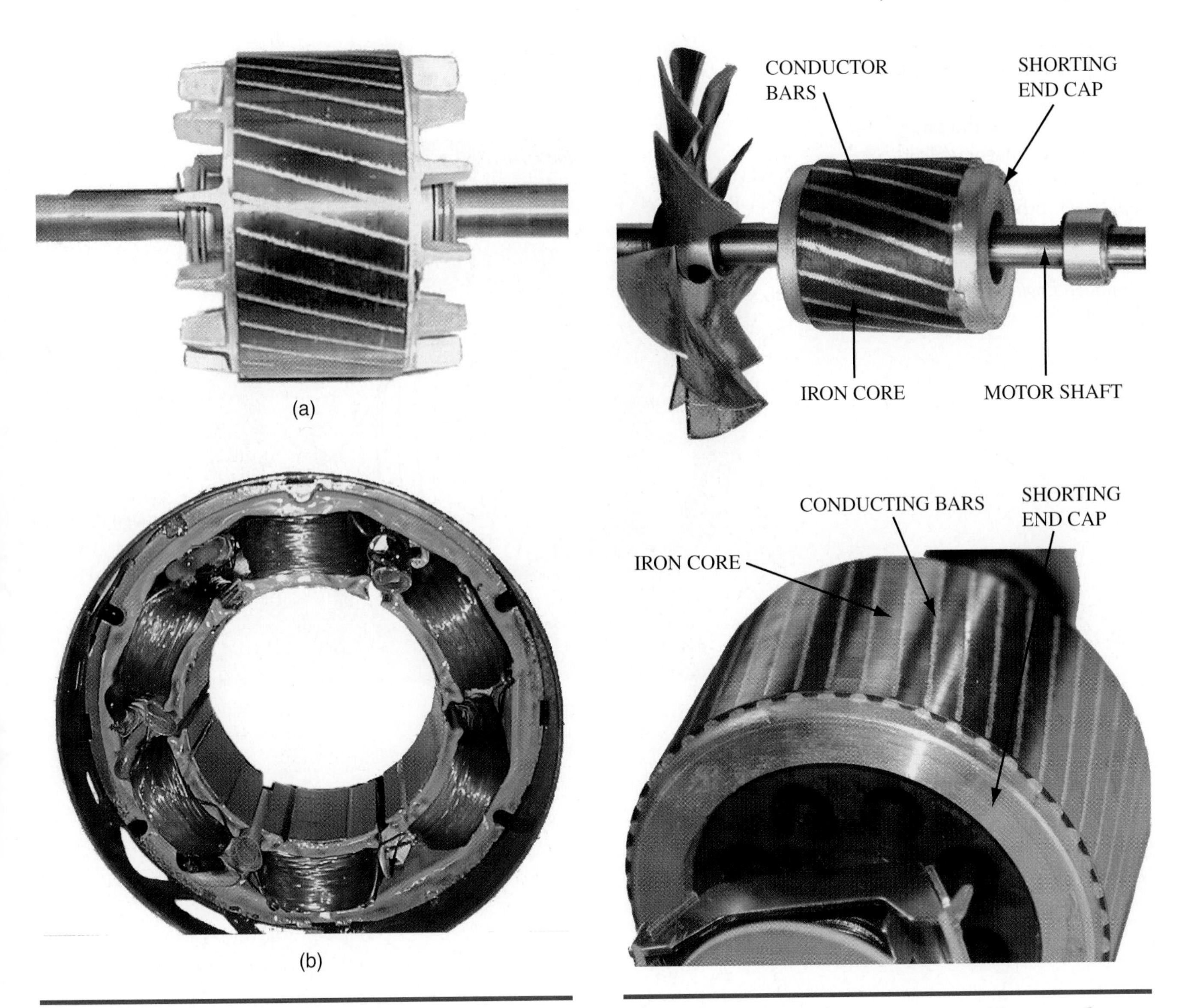

Figure 7-4-1 Parts of an electric motor: (a) Rotor; (b) Stator.

Figure 7-4-2 Squirrel cage type rotor. *(Courtesy of Hampden Engineering Corporation)*

As the alternating current in the stator coils changes direction, the polarity of the stator poles changes, as shown in Figure 7-4-6. The like poles are repelled and with further changes in the alternating current, continuous rotation is produced.

A problem can occur if the rotor is stopped in the position shown in Figure 7-4-7. In this position, regardless of polarity, no motion can occur. This position is sometimes described as dead center.

To correct this condition, a start winding is added, as shown in Figure 7-4-8. The original winding is called a run winding. The two windings—the run winding and the start winding—are out of phase with each other and enough torque is created to begin the turning. Motors of this type are called split phase motors. Because of their low starting torque they are used only on fractional horsepower applications.

Motor Speed

The speed of a motor is determined by the number of poles and the frequency (Hertz) of the alternating current. The greater the number of poles the slower the speed. The higher the frequency the faster the speed. Speed is measured in revolutions per minute (rpm). The maximum speed of a motor is known as the synchronous speed.

Motors are not 100% efficient. In actual performance there is some slippage, or inefficiency, in the motor operation. For motors used to power HVAC/R equipment, the actual speed is usually 95 to 97% of synchronous speed. Figure 7-4-9 shows two pole, four pole, six pole, and eight pole motors. The number of poles is used to calculate the synchronous motor speed.

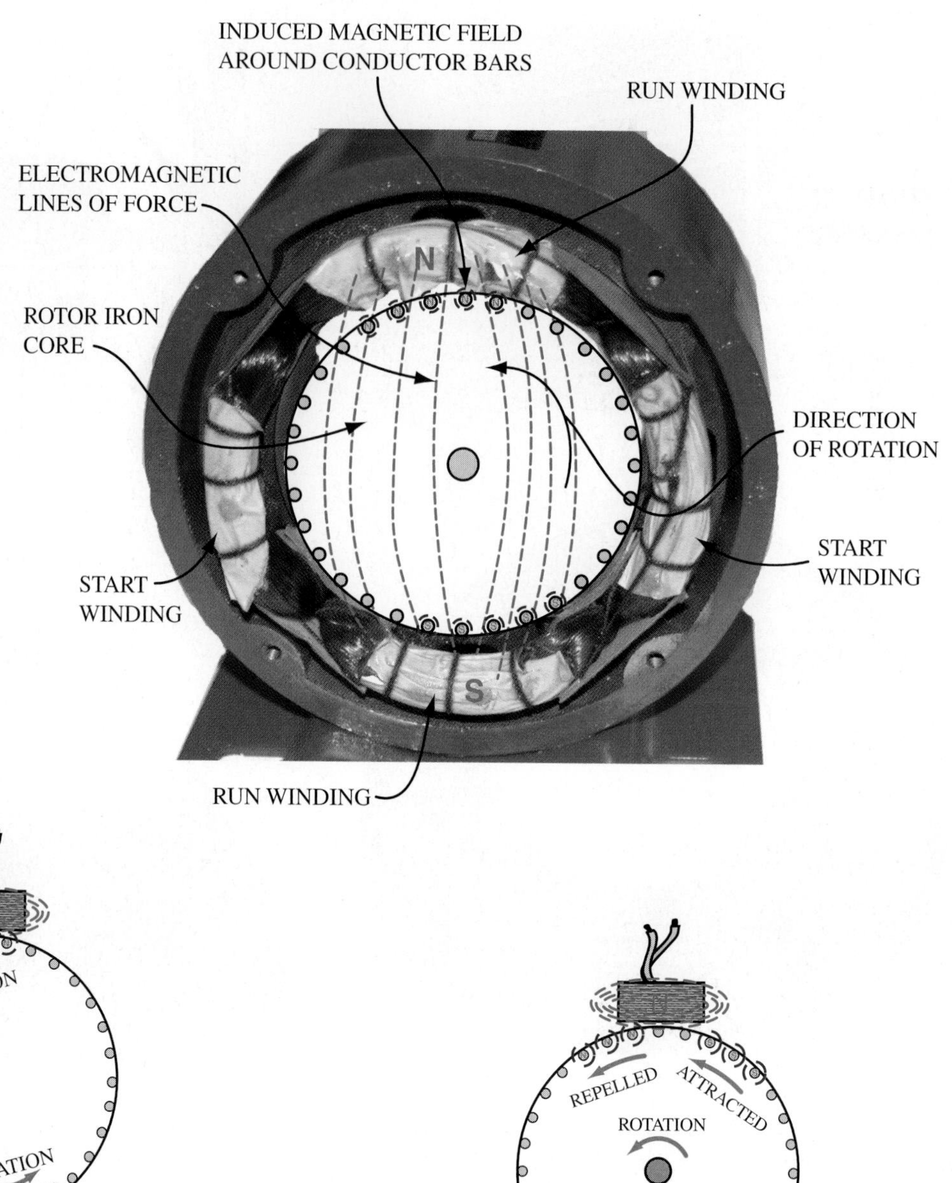

Figure 7-4-3 The induced magnetic field in the conductor bars is the same as the electromagnetic field. Because like magnetic poles repel each other, the rotor is forced to spin as the fields repel each other. (*Courtesy Hampden Engineering Corporation*)

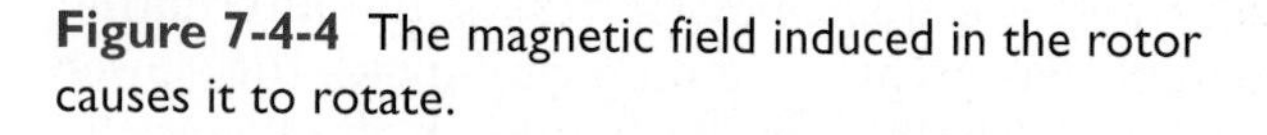

Figure 7-4-4 The magnetic field induced in the rotor causes it to rotate.

Figure 7-4-5 On one side the like magnetic poles are being repelled, and on the other side the opposite poles are being attracted.

CAUTION: The rpm of induction motors is totally dependent on the number of holes and frequency. An induction motor will turn at the same rpm even though the voltage supplied to the motor is lowered. However, as the voltage is decreased, the current the motor draws increases. Motors that are operating at low voltage will quickly overheat and may burn up. It is for that reason that rheostat motor controllers should not be used on induction motors.

The following is an example of calculating synchronous motor speed. The formula is:

$$\text{rpm} = \frac{\text{Hz} \times 60 \text{ sec/min}}{\frac{1}{2}p} \quad \text{or} \quad \frac{\text{Hz} \times 120}{p}$$

where

rpm = revolutions per minute
Hz = frequency in cycles/sec
p = number of poles

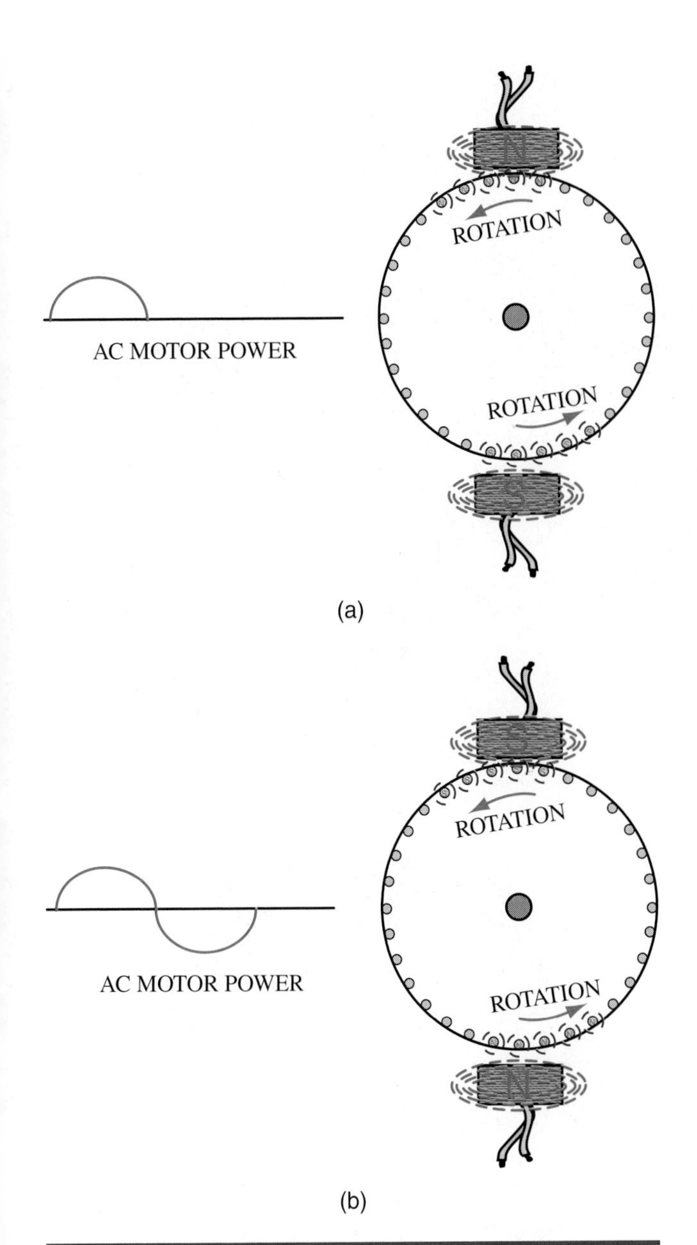

Figure 7-4-6 (a) In the first half of the AC power cycle, the rotor is pushed away from one side and attracted to the other; (b) In the second half of the AC power cycle, the current direction through the windings changes and the process is repeated. This is what keeps the motor spinning.

EXAMPLE

What is the speed of a four pole motor at 60 Hz?

Solution

$$\text{rpm} = \frac{60 \times 120}{4}$$

$$\text{rpm} = \frac{7200}{4}$$

$$= 1800 \text{ revolutions per minute}$$

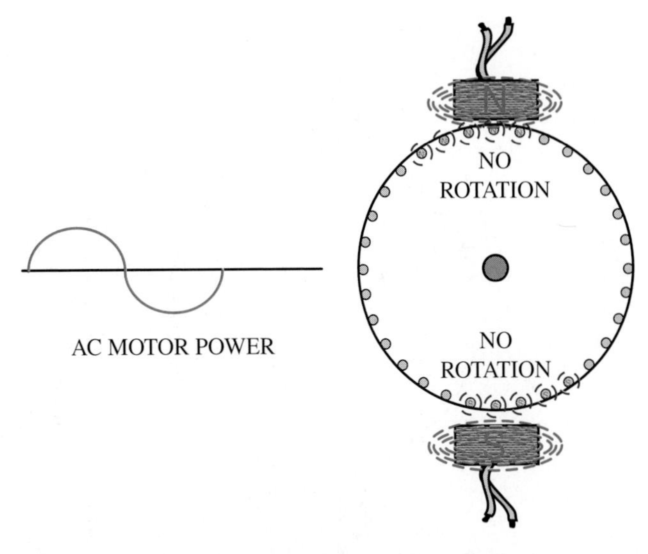

Figure 7-4-7 A motor cannot start rotating if the magnetic fields are at dead center.

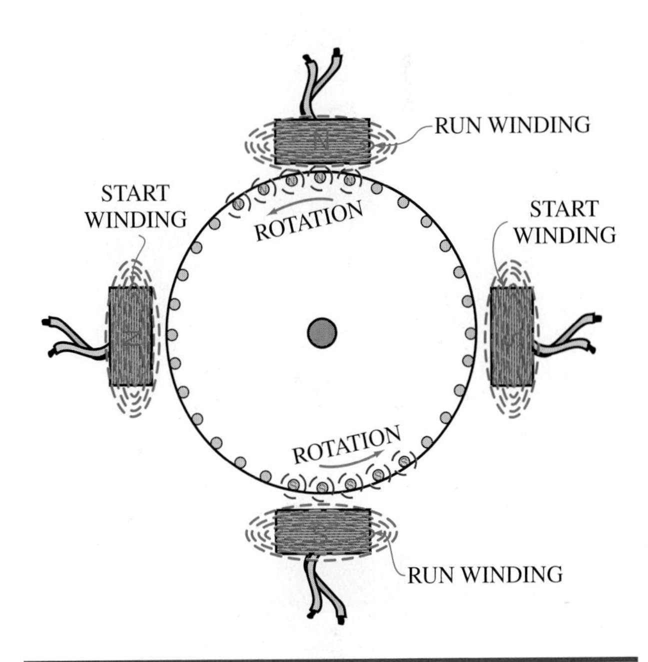

Figure 7-4-8 Start windings are offset from run windings to get the rotor out of dead center. Start windings are often made of finer wire than run windings.

CAPACITOR PRINCIPLES

In order to provide strong starting torque for a split phase motor, a start capacitor is placed in series with the start winding.

Referring to Figure 7-4-10, when voltage is applied, the current through the capacitor resistor circuit will lead the voltage.

Figure 7-4-11 shows a capacitor in series with the start winding. When voltage is applied, the magnetism

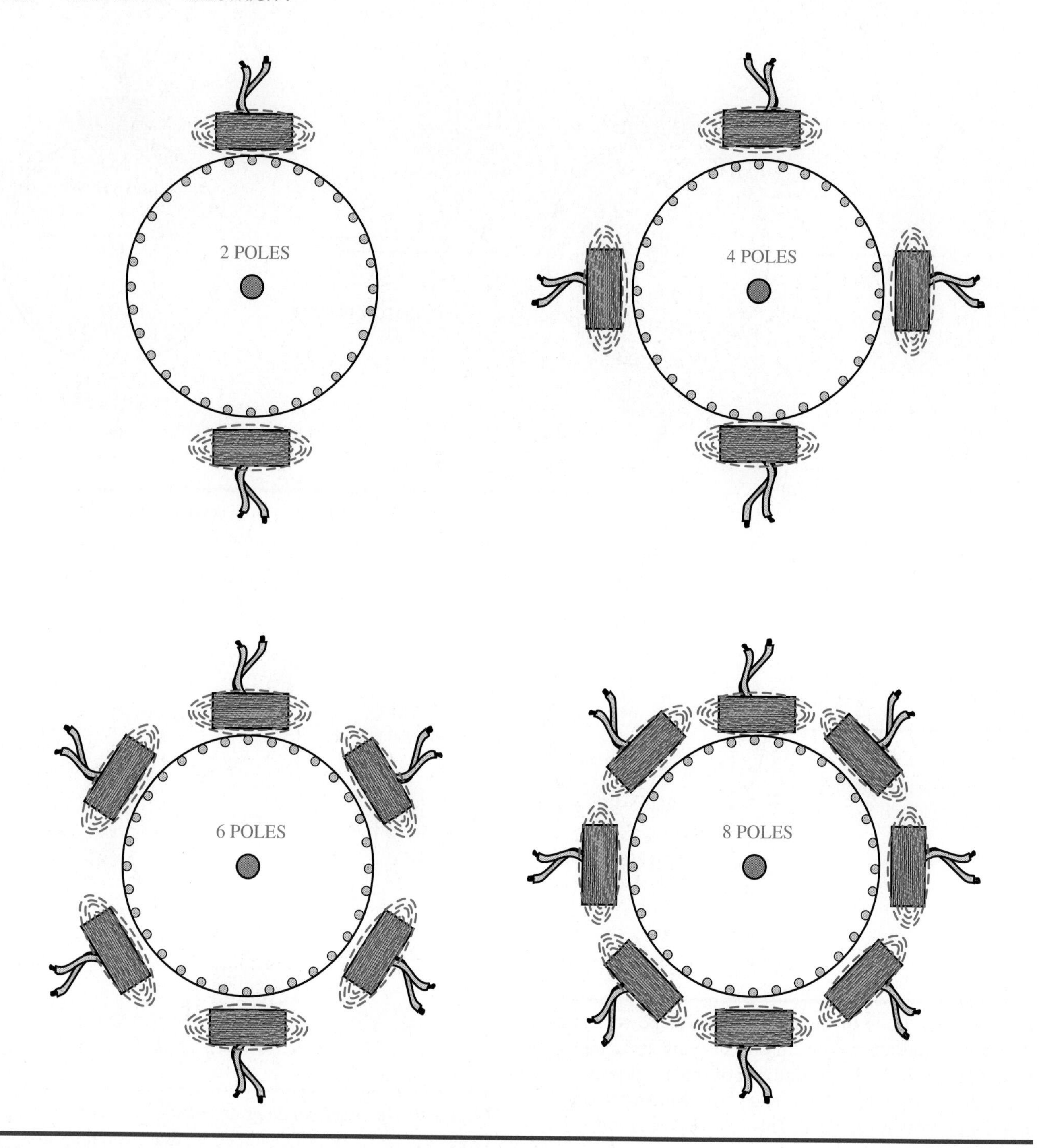

Figure 7-4-9 Synchronous speed depends on the number of poles.

in the start winding will occur earlier than in the run winding and will provide the push needed to start the motor. Figure 7-4-12 shows the use of the phase shift to advance the magnetic field in the start winding.

When the current direction in the start winding reverses, the polarity of the start winding also changes. By this time, the rotor south magnetic pole has rotated to the point where it is repelled by the new stator south pole, causing continuous rotation.

A capacitor is rated by its capacity and voltage limit. The unit of capacity is the microfarad (mfd and μF are commonly used abbreviations). A high mfd rating is obtained by either using large plates or a small amount of insulation. A low mfd is obtained by using smaller plates or more insulation. The voltage stamped on the outside of the capacitor is the maximum voltage that can be connected safely across the capacitor. If this voltage is exceeded the capacitor is likely to fail.

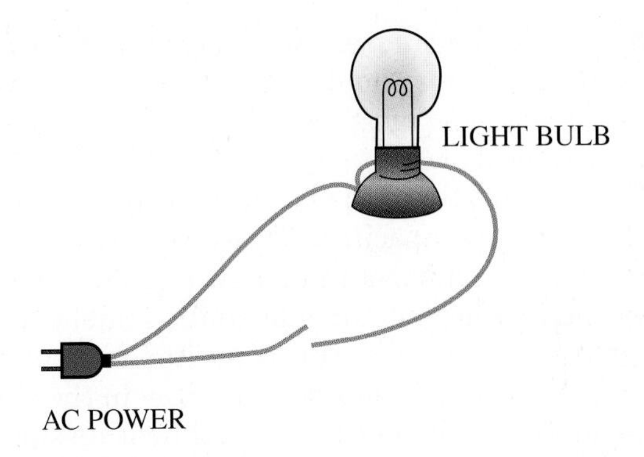

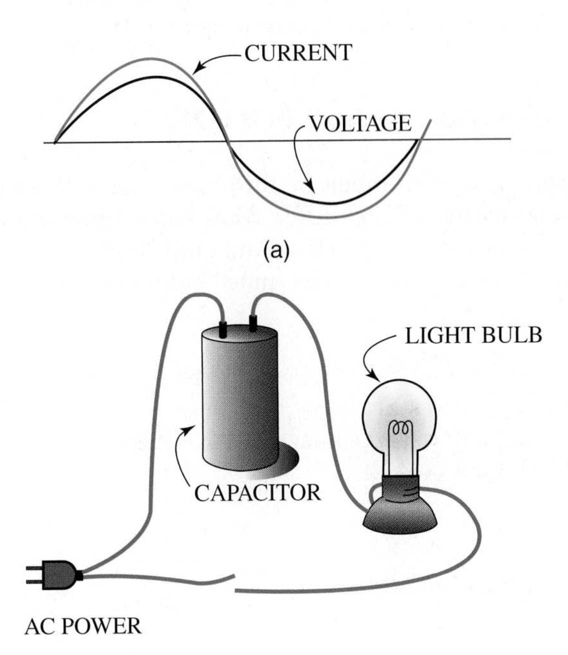

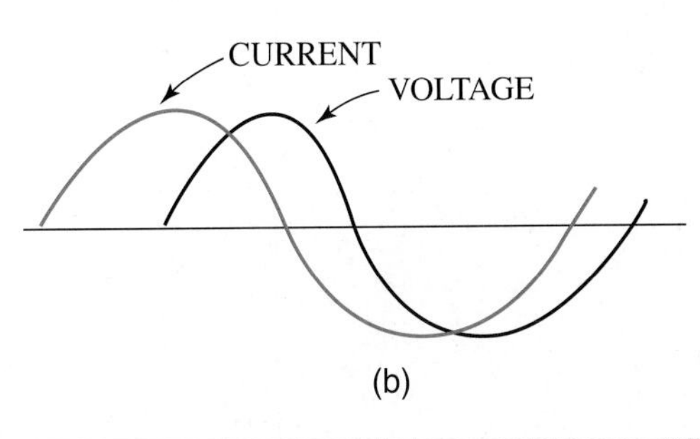

Figure 7-4-10 (a) The voltage and current flow through a resistor, like the light bulb load in this circuit, at the same time; (b) However, in the capacitor resistor circuit, the current leads the voltage through the circuit.

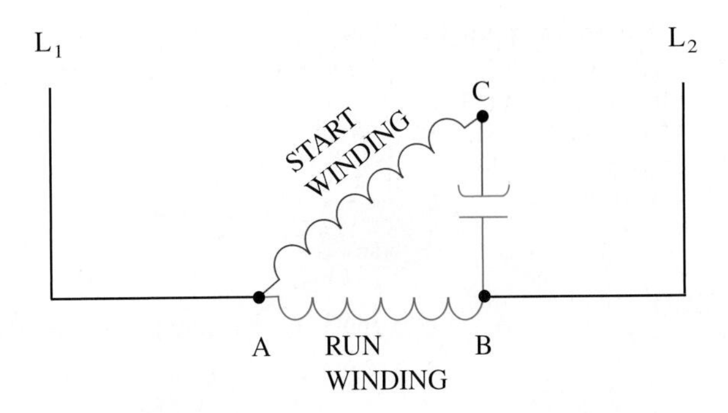

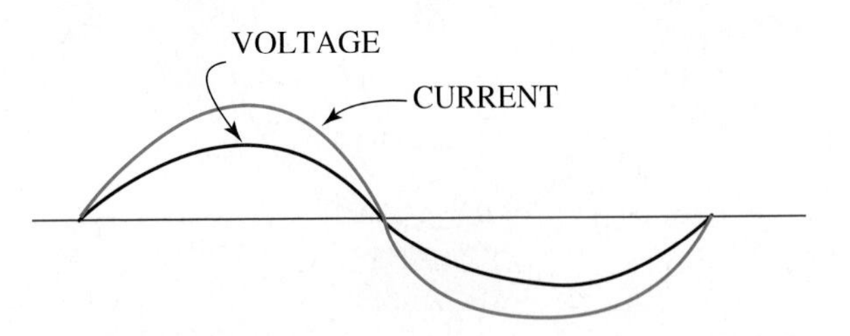

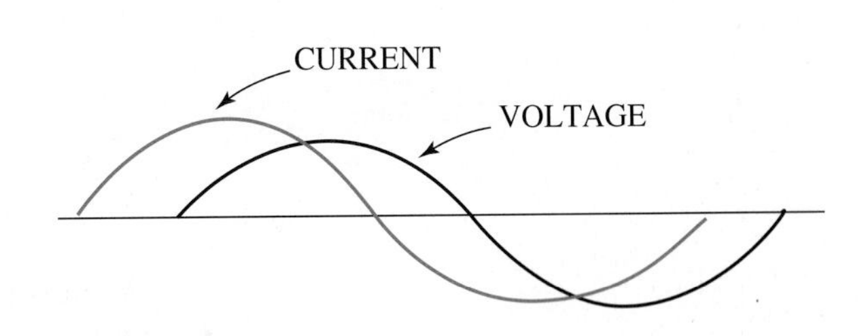

Figure 7-4-11 The phase shift caused by the capacitor causes the rotor to move, even if it is dead.

TECH TIP

Some people feel that a capacitor provides the motor with higher voltage. They make this assumption because often on the side of the capacitor there is a much higher voltage listed than the voltage the system operates at. For example, 377 is a common voltage listing for capacitors. Capacitors do not add voltage but merely shift the phase to increase the running efficiency or starting torque.

Capacitors are used to achieve the desired phase angle shift and to obtain the required current through the series load. Both of these qualities are obtained by selecting the proper mfd rating.

Figure 7-4-13 shows the wiring of a start capacitor and its relay, connected to a compressor motor. The start capacitor and the relay NC switch are in series with the start winding. The start relay coil is connected in parallel with the start winding. With this arrangement, the starting torque of the motor is increased. The relay removes the start capacitor from the circuit as soon as the motor is running.

Referring to Figure 7-4-14, a run capacitor has been connected to the power and start terminals of the fan motor. The run capacitor stays in the circuit

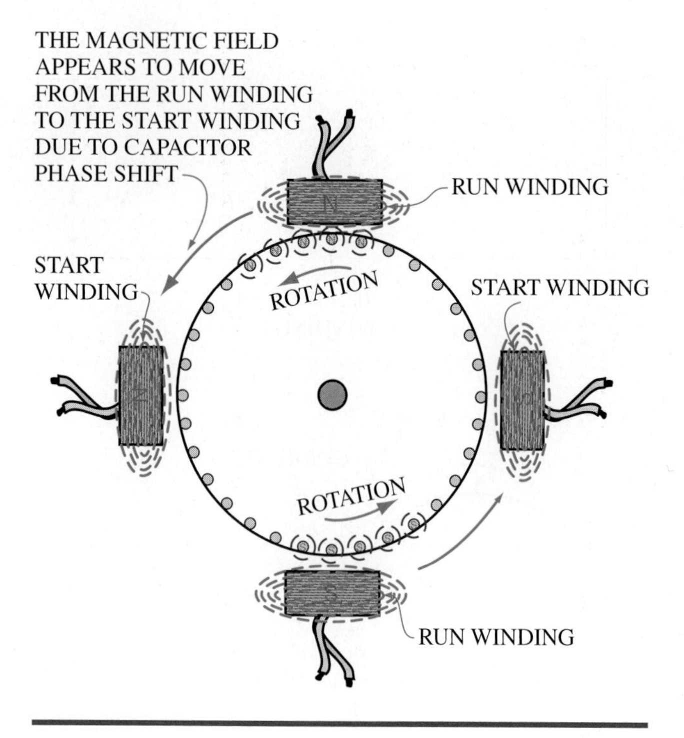

Figure 7-4-12 Rotating magnetic field due to capacitor phase shift.

all the time and increases the efficiency of the motor. If the run capacitor should fail, the current draw of the motor would be increased about 10% and the motor could overheat.

Figure 7-4-15 shows the normal appearance of the run and start capacitors. The run capacitor, which stays in the circuit continuously, is made of large plates and a large amount of insulation (dielectric) to dissipate the heat. The run capacitor may also be round. The start capacitor does not stay in the circuit long, and therefore does not have a heat dissipation problem. It is typically made in rolled form, sandwiching metal foil and insulating material.

SINGLE PHASE MOTORS

There are a number of different types of single phase motors. They differ from each other mainly by the amount of starting and running torque. The following types are the most commonly encountered:

1. Permanent split capacitor (PSC)
2. Capacitor start (CS)
3. Capacitor start/capacitor run (CSR)
4. Shaded pole

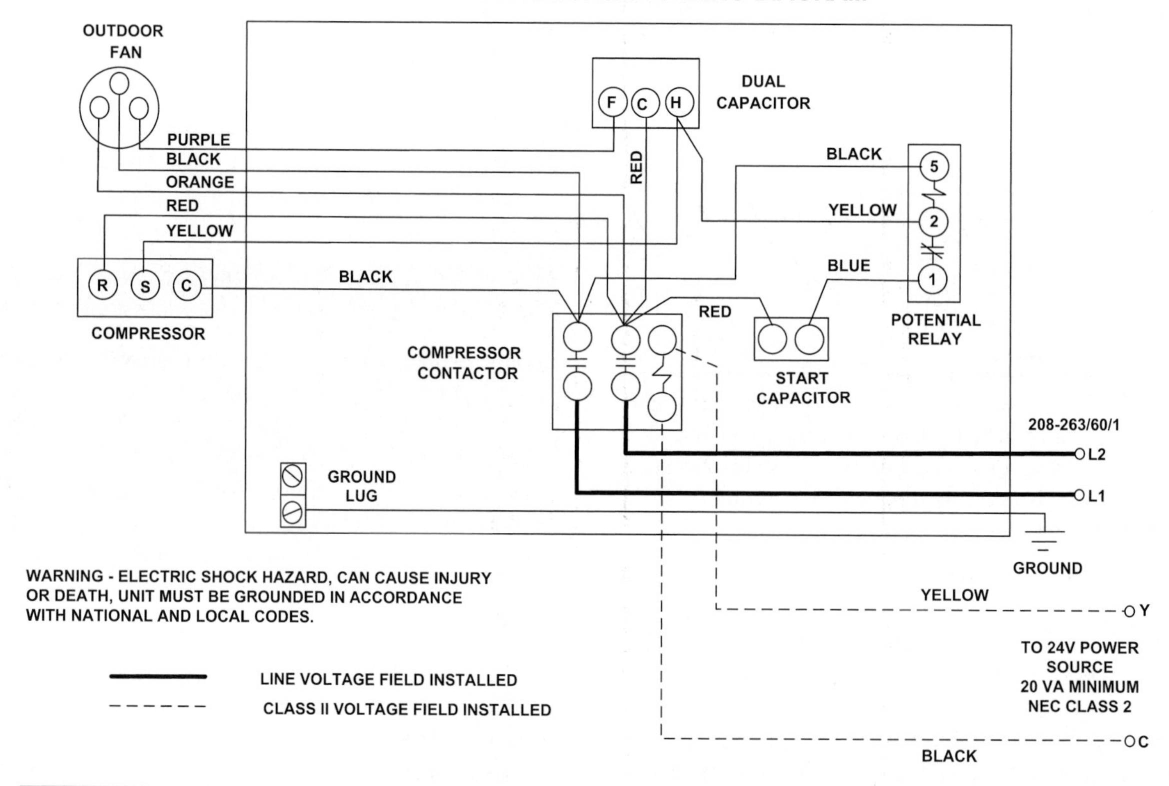

Figure 7-4-13 Start relay and start capacitor. *(Courtesy of Lennox Industries, Inc.)*

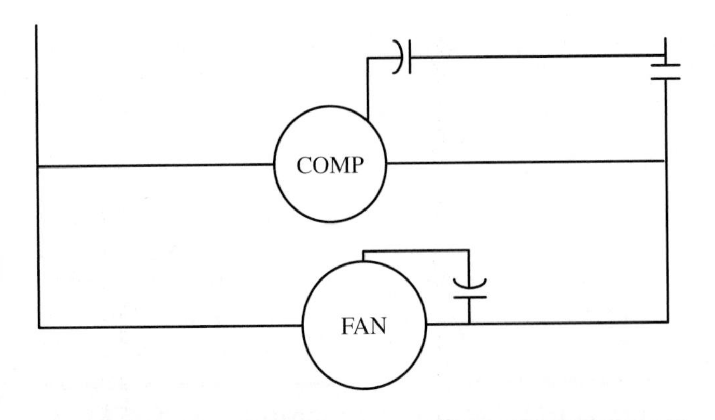

Figure 7-4-14 Run capacitor wired in to provide crankcase heat to the compressor.

The *permanent split capacitor (PSC) motor* has a run capacitor in series with the start winding as shown in Figure 7-4-16. This capacitor stays connected at all times. It starts the motor and then is left in the circuit to improve the efficiency of the motor after it is running.

The run winding has the number of turns of wire required to give the best motor performance at a given line voltage. The start winding has more turns of smaller wire, which gives it a higher resistance and lower current carrying capacity than the run winding.

The PSC motor will always have the common connection for the two windings attached to power. The run capacitor will be connected between the start winding and the run winding. The run capacitor therefore improves the performance of the motor both in starting and running.

TECH TIP

It is important that the proper size run capacitor be used with a PSC motor. If the capacitance changes more than 10% from the design capacitance, the motor windings will build up greater heat as they operate. If enough heat builds up, the motor can become damaged. It is therefore very important that the exact capacitor as specified by the manufacturer be installed on any PSC motor.

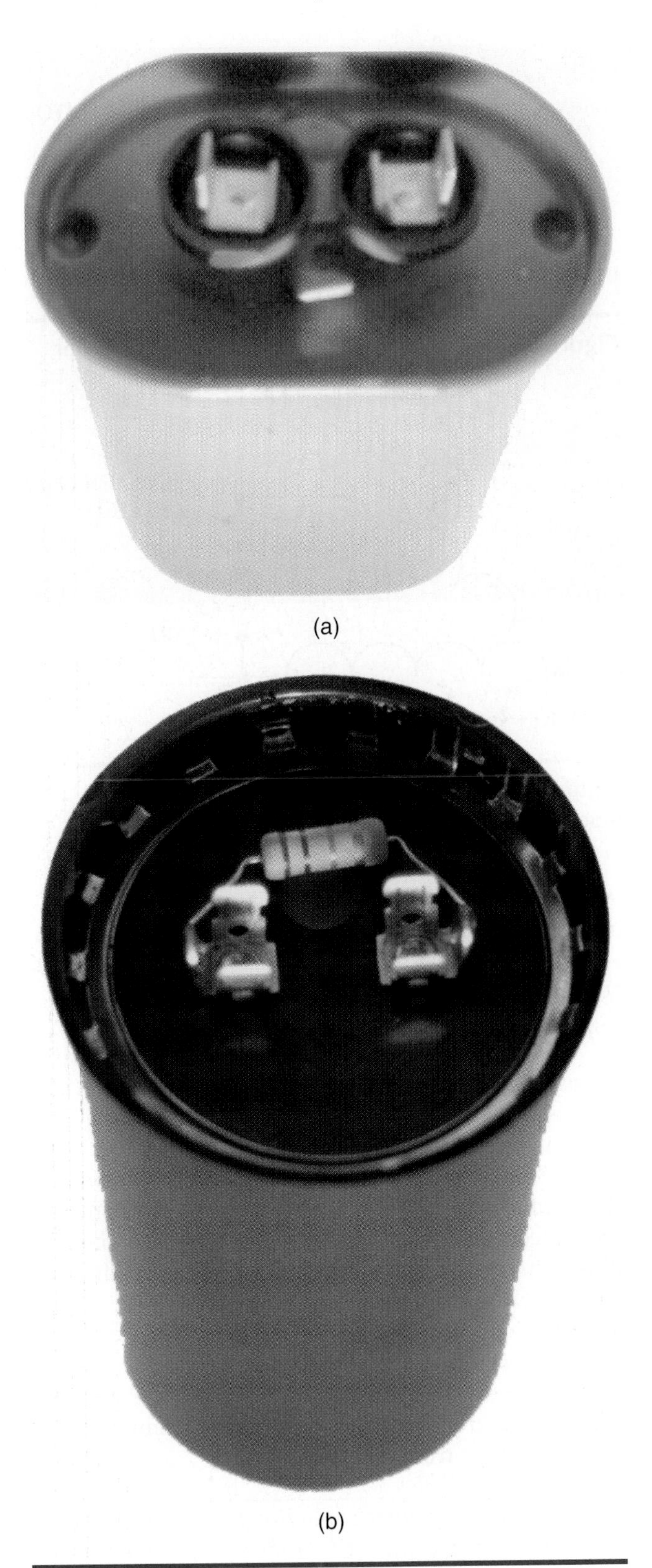

Figure 7-4-15 A comparison of the physical shape of start and run capacitors. (a) Run capacitor; (b) Start capacitor.

The PSC motor has moderate starting torque and good running efficiency. It is used to power fans and small compressors. The low mfd rating of the run capacitor results in a small phase angle shift, creating only a moderate starting torque. A diagram of the *capacitor start induction run (CSIR) motor* is shown in Figure 7-4-17. This motor has a high starting torque, but is not as efficient as the PSC motor. The reason for its lower efficiency is that the capacitor is

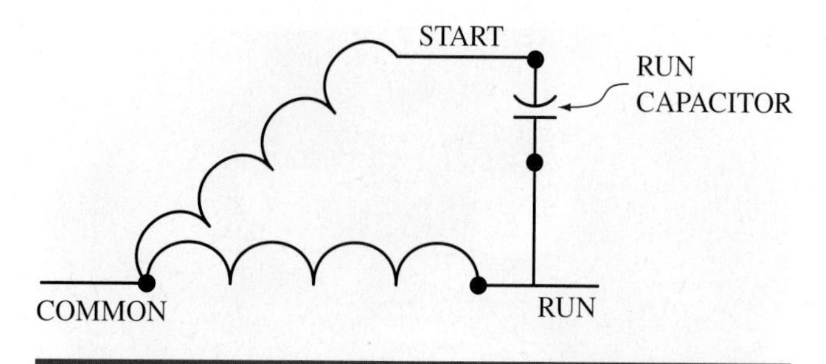

Figure 7-4-16 Permanent split capacitor (PSC) power.

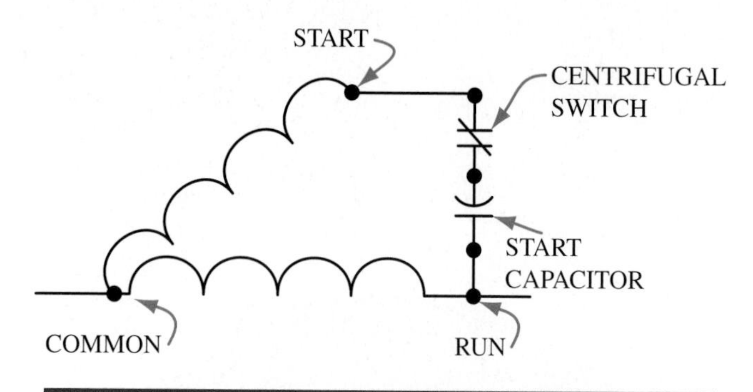

Figure 7-4-17 Capacitor start induction run (CSIR) motor.

switched out of the circuit immediately after starting. This motor has a high mfd start capacitor.

There are two ways that the start capacitor can be removed from the circuit:

1. Using a mechanical switch. This is a centrifugal switch attached to the motor shaft. When the motor reaches ⅔ or ¾ of its rated speed, centrifugal force opens the switch, shown in Figure 7-4-17.

2. The electromagnetic method. A potential relay is placed across the start winding. Its contacts are placed in series with the high mfd capacitor. When the motor is started, the capacitor produces a high starting torque. As the motor speed increases, the induced voltage across the start relay coil increases until it reaches a preset value. At this induced higher than line voltage value, the relay is energized and the NC switch is opened, removing the start capacitor from the circuit.

The *capacitor start capacitor run (CSCR) motor* has both a start capacitor and a run capacitor, as shown in Figures 7-4-18 and 7-4-19. It has excellent starting and running torque, but is not as efficient as the PSC motor. It is used to drive most compressors.

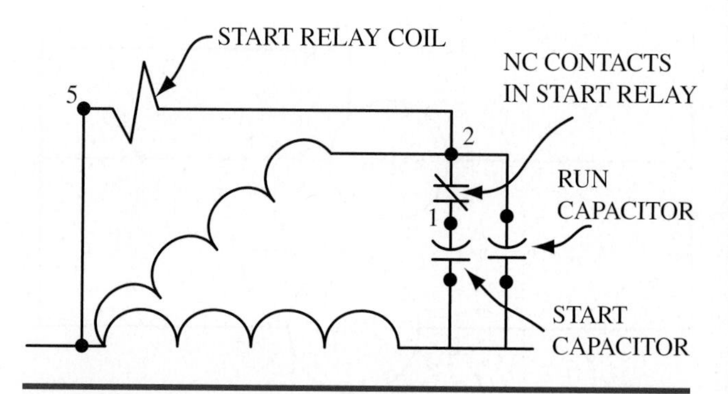

Figure 7-4-18 Capacitor start capacitor run (CSCR) motor.

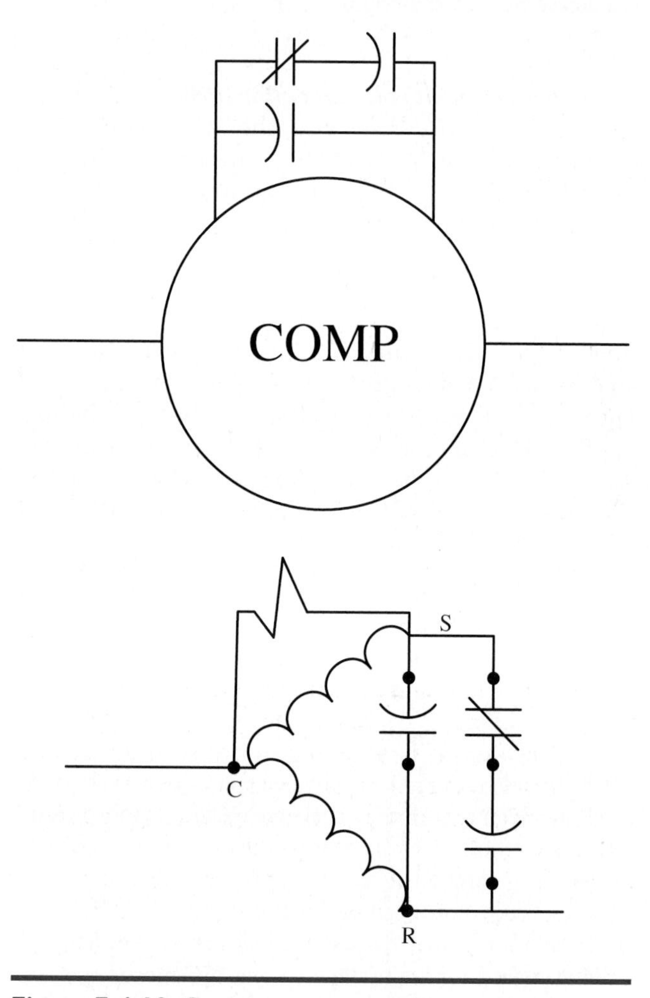

Figure 7-4-19 Capacitor start capacitor run motor diagrams.

Start Kits

Start kits are available for the technician's use whenever it is necessary to improve the starting torque of a motor. When a motor keeps tripping out on overload, the addition of a start kit may solve the problem.

Where a high starting torque is required, a hard-start kit can be installed. Where low voltage or a voltage lag is experienced, a soft-starting kit is used. Both kits contain a start capacitor and a start relay. The difference is in the size of the capacitor.

SERVICE TIP

Often start capacitors have bleed resistors connected between the two capacitor terminals. The purpose of this bleed resistor is to drain the capacitor's current during the motor's off cycle. This is done so that when the motor contactor starts the motor, there is not a capacitor discharge across the contactor's points. Such a discharge causes pitting of the contactor. Over a period of time this pitting will destroy the contacts. Bleed resistors are available from the supply house to be installed on start capacitors. If one is not installed, the contactor in the unit will become pitted and may actually weld or stick together.

A different soft-starting kit is available to be applied to PSC motors. This kit includes a *positive temperature coefficient (PTC) thermistor*. The PTC is a temperature sensitive device whose electrical resistance will increase as its temperature increases. This PTC is placed across the run and start terminals, parallel to the run capacitor of a PSC motor, as shown in Figure 7-4-20.

At room temperature, the PTC thermistor has a low resistance, about 25 or 50 Ω. When voltage is supplied, an initial surge of high current passes through the start winding, because the thermistor is effectively shorting out the capacitor. The surge causes an increased starting torque to start the motor. The temperature increase that results causes the PTC thermistor resistance to rise, removing the short from across the run capacitor. The motor then runs as a normal PSC motor.

The *shaded pole motor,* as shown in the diagram in Figure 7-4-21, has a modified stator pole. A groove separates a small portion of the stator pole from the rest of the pole. A bank of metal is placed around the smaller section of the pole, which provides a phase shift needed to start the motor. Shaded pole motors

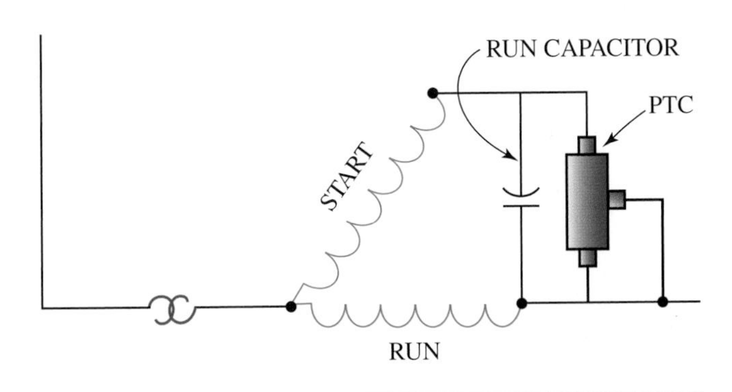

Figure 7-4-20 PTC start systems are sometimes referred to as soft start because they do not provide as much initial torque as a start capacitor and potential relay.

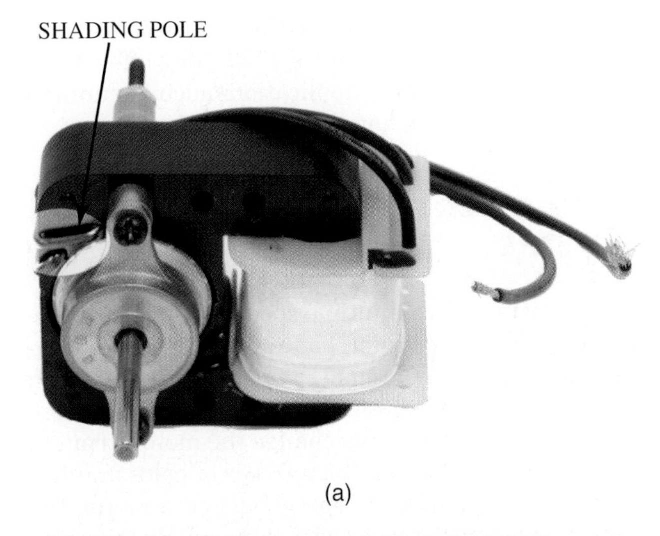

(a)

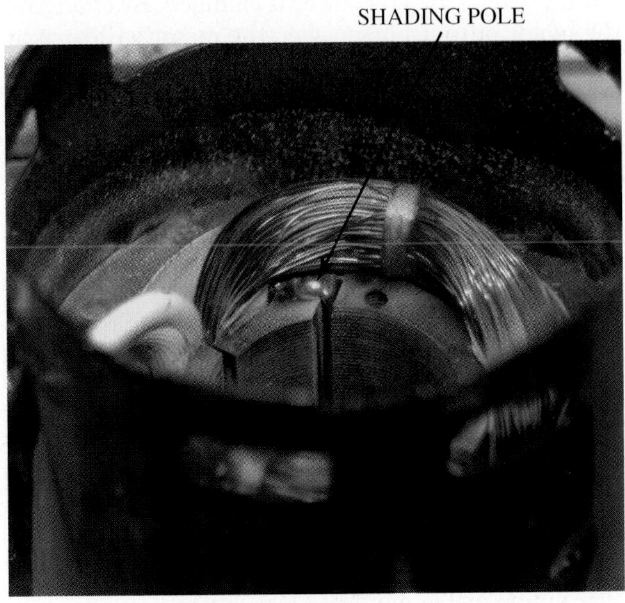

(b)

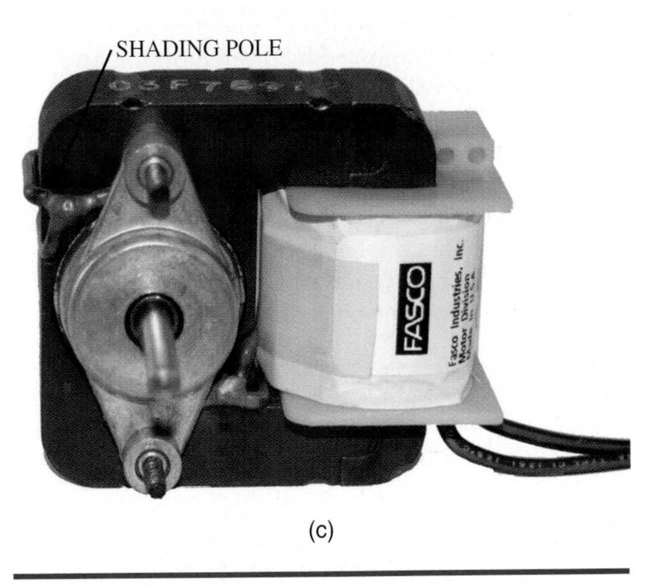

(c)

Figure 7-4-21 Shading pole on shaded pole motors.

have a low starting torque and their speed control under varying load conditions is poor. They offer a low cost motor for light duty applications such as running blowers on small air handling and heating units.

MULTIPLE SPEED BLOWER MOTORS

Blower motors have multiple speed settings often referred to as low, medium, and high; or low, medium-low, medium, medium-high, and high depending on the number of speed taps that the motor has. These speed taps do not actually change the motor's rpm. A motor's rpm is based on the number of poles and frequency. By changing the speed taps on a motor, the motor's effective horsepower is changed. By changing the motor output horsepower, the motor will turn at a different rate under different load conditions. For example, a ⅓ horsepower motor rated at 1750 rpm will turn at approximately 1750 rpm when the motor is on the high setting. However, when the motor is on the medium setting, it may produce only ¼ horsepower. Although it is attempting to turn at the same speed of 1750 rpm, it cannot because the load causes the motor to slow down, effectively reducing the rpm and CFM output of the blower. When the blower motor is on the low speed setting, it may produce only ⅛ horsepower. This would result in an even lower rpm and blower CFM output.

THREE PHASE MOTORS

Three phase motors have the following advantages over single phase motors:

1. They are easily reversible. The direction of rotation can be changed by interchanging any two supply voltage lines.
2. There is less running torque pulsation because at least one phase is always producing an induced rotational effect on the rotor.
3. They have higher starting torque because each winding is out of phase with the other windings and produces rotational torque.
4. They have higher efficiency.

CAUTION: Do not attempt to run a three phase motor on single phase power. Although the motor may start, it will quickly overheat and burn up the windings. The same thing will happen to a three phase motor if one of its legs of power is lost as the motor is being operated. For that reason, some large motors have single phase or lost leg motor protection systems, which will shut the motor down any time a single leg of power is lost to the motor.

The windings of a three phase motor are 120 degrees apart, which facilitates starting. Figure 7-4-22 shows a schematic drawing of a three phase motor. No capacitors, no auxiliary windings, and no switching circuits are required. Three phase motors create their own starting torque.

Three phase motors are wired in either a delta configuration as shown in Figure 7-4-23 or a wye configuration as shown in Figure 7-4-24. The delta configuration is named because the circuit resembles the triangle shape of the Greek letter delta (Δ). The wye configuration resembles a capital letter Y. In the delta configuration, the line voltages equal the phase voltages, and the line currents are 1.732 times the phase currents. In the wye

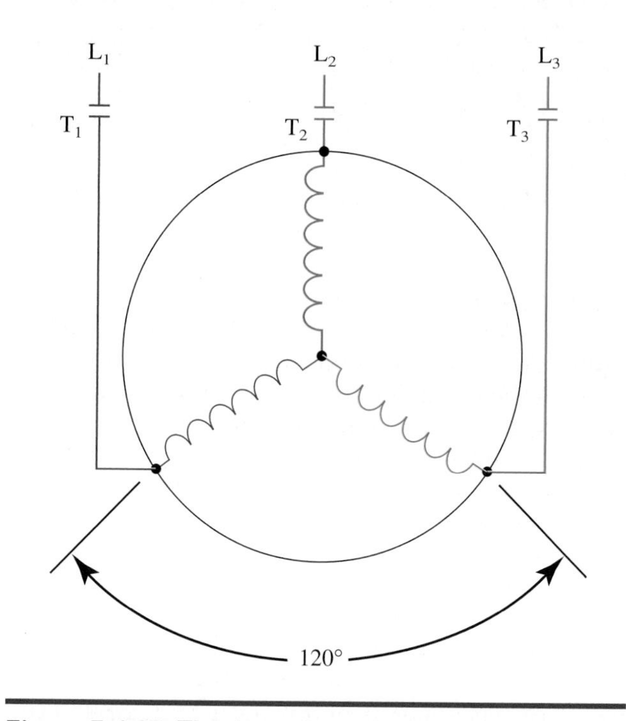

Figure 7-4-22 The three phase motor.

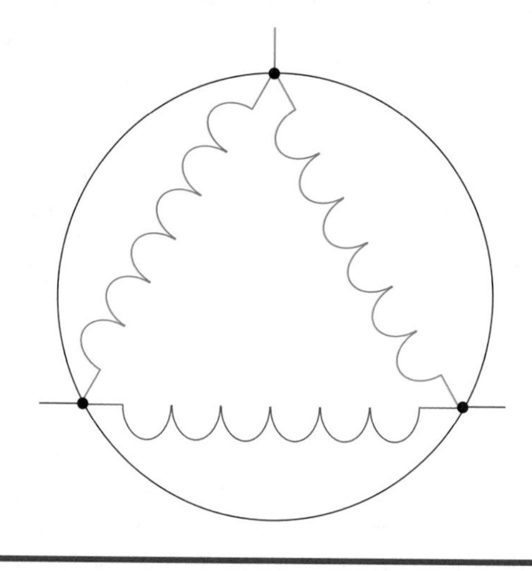

Figure 7-4-23 Delta three phase motor windings.

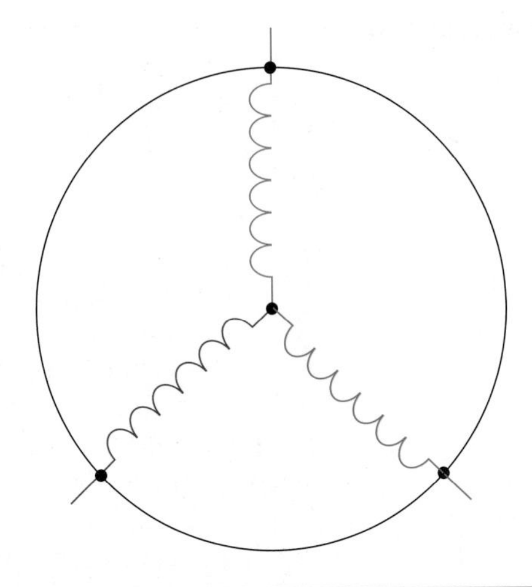

Figure 7-4-24 Wye three phase motor windings.

configuration, the line voltages equal 1.732 times the phase voltages, and the line currents equal the phase currents.

MOTOR PROTECTION

There are a number of causes of motor failure. The most common problem is excessive heating. Among the causes of excessive heating are:

1. Defective start relay (on a single phase motor)
2. Excessive load
3. Loss of refrigerant on a gas cooled motor
4. Excessive time on locked rotor current
5. Operation at too high or too low voltage
6. Single phasing (on a three phase motor)

All motor protective devices are designed to cut off the power to the motor before damage has occurred. Repeated cycling on the protective device can damage a motor.

The locked rotor current is the momentary starting current. This current can be three to five times the running current of the motor. If the motor is blocked from starting, this excessive current will be prolonged, causing heating and serious damage. A protective device is therefore required to disconnect power on either excessive temperature or excessive current. Fuses or circuit breakers are most often the devices that protect against excessive starting current.

All motors are designed to operate between certain voltage limits. If voltages outside these limits are applied, the life of the motor can be seriously reduced. Table 7-4-1 shows the limits of voltage that common motors are designed to handle.

Typically motors are designed to be applied within plus or minus 10% of nameplate voltage. Dual rated motors such as 208 V/230 V may be rated plus 10%, minus 5%. The manufacturer's information should be referenced.

TABLE 7-4-1 Voltages Handled by Common Motors

Nameplate	Upper Limit	Lower Limit
208	228	188
230	253	207
460	506	414
575	633	518

On three phase motors, if the voltage varies more than 2% between phases, the life of the motor will be reduced. This is caused by unequal currents and heating.

Single phasing is another serious problem common to only three phase motors. If the motor loses power in one of the three lines, the motor may continue to operate. This is called single phasing. This will cause sufficient unbalancing to increase the motor temperature and require motor protection.

Types of Motor Protection

There are three types of protective devices that are used to protect the motor against excessive heat:

- Temperature actuated
- Current actuated
- Combination of current and temperature

These devices can be either pilot duty or line duty arrangements, as shown in Figure 7-4-25.

SAFETY TIP

Motor protection for fractional horsepower compressors is often provided by an external snap disk (Klixon®) that clicks on to the compressor case. These devices are wired into the common power lead and are located inside of the terminal cover. They are supposed to be in direct contact with the compressor case so that they can send the compressor the sensed temperature. If the cover is not placed back on the terminal, or if the device is not in direct contact, it will not sense the temperature of the compressor. These devices are designed to open when the temperature and/or current exceeds the compressor's limits. They will not provide the proper protection for the compressor unless they are installed correctly.

Figure 7-4-25 Two types of motor protection.

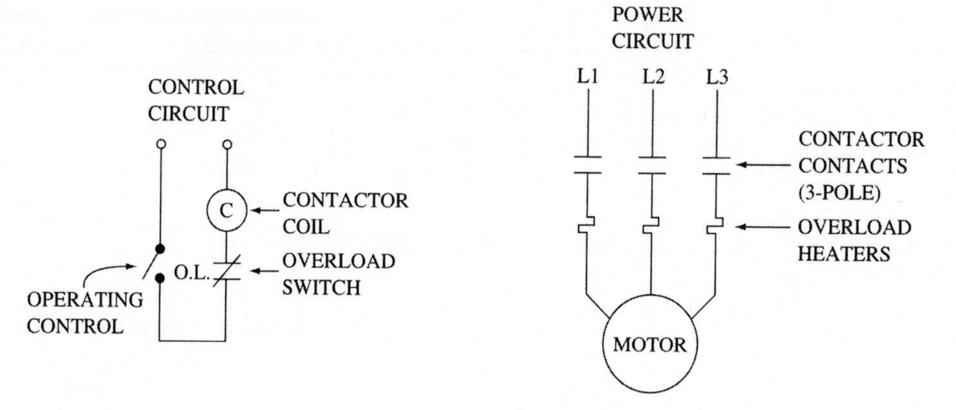

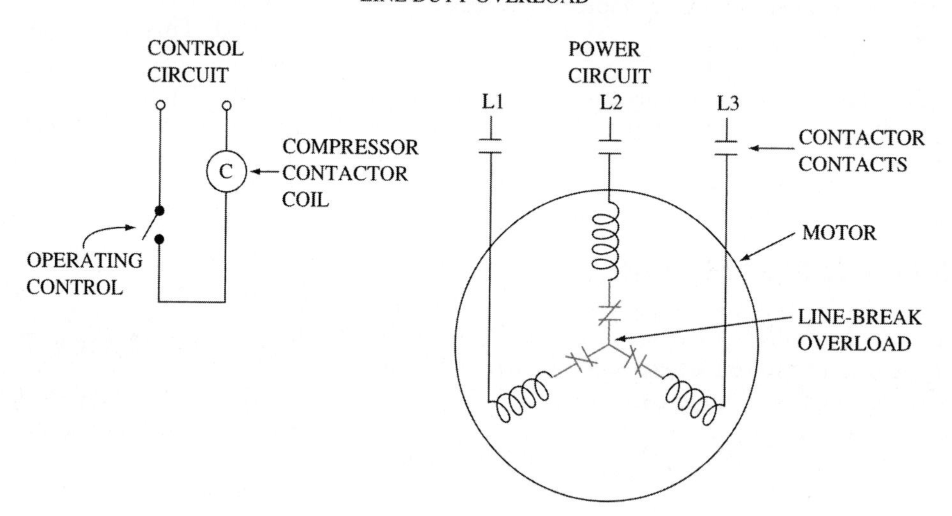

Motor Overload Devices (Line Break)

For a single phase motor, the overload device must interrupt one of the motor leads. For a three phase motor, the device must interrupt two or three of the motor leads. On a wye connected motor with a built-in overload device, this is often where the three windings are connected in common.

In a three phase motor, a pilot duty device senses current overload or excessive temperature within the motor, and opens the contactor circuit to remove power to the motor. A line duty device senses current and/or temperature in the motor winding, and if an overload occurs, will disconnect power by directly opening the motor winding circuit.

The line duty arrangement is commonly found on compressor motors used for domestic service. Most of these reset themselves automatically. However, since they are embedded in the motor winding (as shown in Figure 7-4-26), it takes some amount of time for the motor to cool down. During this waiting period it is easy to incorrectly diagnose the problem as compressor failure.

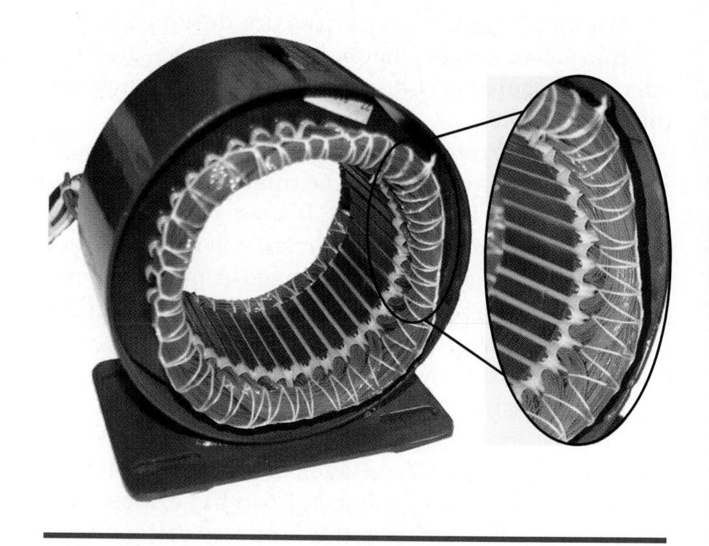

Figure 7-4-26 It's important to protect the motor windings because they are made of such small sized wire. *(Courtesy Hampden Engineering Corporation)*

Types of Reset

After a protective device has opened or tripped out, it needs to be reset. This can be done manually or automatically.

The advantage of manual reset is that a service technician has the opportunity to examine the cause of the problem before resetting the protective device.

The advantage of the automatic reset is that the unit can automatically go back into service in the case of a nuisance trip-out.

General automatic reset devices are employed only where the time to reset is sufficient to ensure the motor won't short cycle and be damaged by the protective device itself. Sometimes the protective device will reset automatically, but the control circuit will require resetting by switching the unit off then on from the thermostat or control switch.

Types of Overload Devices

The various types of overload devices that will be discussed are as follows:

1. Temperature overloads
2. Thermal overload relay
3. Heater element current overload
4. External supplemental overload
5. Magnetic overload (Heinemann)
6. Internal current and temperature overload
7. General Electric Thermotector®
8. Three phase overload

Figure 7-4-28 shows the external shell thermostat for pilot duty and Figure 7-4-29 shows the internal motor thermostat for line duty. Both of these operate by the principle that heating a bimetal disc or strip will cause it to warp and open a power switch. The external device is mounted on the motor shell and resets automatically when the temperature returns to normal. The internal device is wound into the motor windings and automatically resets when the motor cools down.

A motor starter is a contactor with motor overload protection added. Starters made for three-phase motors will have protection on each leg (for a total of three overloads). A motor starter is shown in Figure 7-4-27.

A thermal overload relay is current sensitive with automatic reset and is connected in series with the motor winding. On an increase in heat due to excessive motor current, the upper bimetal strip will warp, opening the pilot duty contacts wired in series with the contactor. Reset is automatic upon cooling. The lower bimetal strip prevents nuisance trip-outs caused by high ambient temperature.

A heater element current overload device for pilot duty uses a bimetal element (spiral or disc type) that responds to heat generated by a heater element.

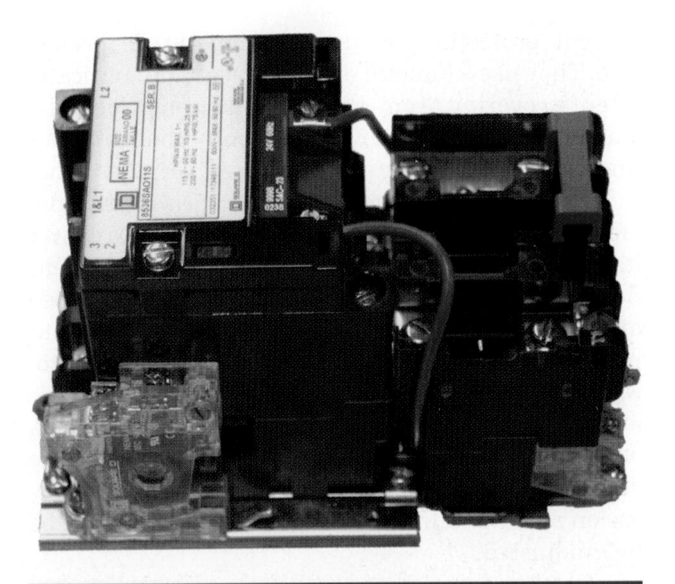

Figure 7-4-27 Motor starter. (*Courtesy of Hampden Engineering Corporation.*)

Figure 7-4-28 External type compressor overload.

When the bimetal heats sufficiently, the contacts open, disconnecting the contactor. It automatically resets on cooling. The heater elements are interchangeable and calculated for a specific range of current. They should be matched to the motor current.

Another version of this same device uses a solder pot relay instead of a bimetal sensing element. The solder pot relay has a ratchet and spindle construction and requires manual reset. It uses interchangeable heaters similar to the bimetal overload Heaters are not interchangeable between diffe makes of equipment.

External supplemental overload dev shown in Figure 7-4-28. These devices pro current protection or current temperature protection. They use a bimetal disc and reset automatically after recovering from the overload condition. The current temperature device is usually connected for line duty, but some versions are used for pilot duty.

The magnetic overload is a current sensitive device. A core placed inside a coil will attempt to center itself in the magnetic field created when current flows through the coil. The sensing coil is wrapped around a sealed tube filled with silicone oil containing the core. Excessive current will pull the core toward the armature, causing the contacts that are attached to the armature to open. The silicone has a dampening effect, which prevents operation of the device for temporary overload conditions such as occur on startup. Current rating is fixed and cannot be field-adjusted.

Another device that is sensitive to both internal current and temperature and is wound into the motor windings is the internal line duty overload device, shown in Figure 7-4-29. As the overload occurs, the bimetal strip within the device warps, opening the contacts. Reset is automatic.

The General Electric Thermotector® device is also an internally actuated temperature motor protector. On this device the case and the internal strip expand at different rates. The device will trip early if the temperature rises rapidly. If the temperature rise is gradual it will trip at its normal setting. This means it is a rate compensated device.

A three phase overload device is shown in Figure 7-4-30. These are current temperature line duty devices. They can be located internally or externally to the motor.

When an overload occurs, the bimetal disc warps, breaking the electrical circuit. They automatically reset when the overload no longer exists.

Figure 7-4-29 Internal motor overload.

Figure 7-4-30 Three phase overload.

Motor Circuit Testing

Since motors are so important in HVAC/R systems they are carefully protected and can be thoroughly tested to see that they are operating properly. The tests that follow begin with single phase motors. Since many single phase motors require capacitors, a word of warning needs to be added about handling capacitors. Capacitors can be dangerous. Safety precautions are required.

Capacitors can hold a high voltage charge even after the power is turned off. Always discharge capacitors before touching them.

TECH TIP

Capacitors should be discharged before you work on the system; however, the capacitor needs to be discharged with a bleed resistor. If the capacitor is discharged with a screwdriver or other metal device touching the two terminals, the spark that is generated externally can also occur internally, damaging the compressor.

Figure 7-4-31a–c shows the various types of instruments used for testing motors. The ohmmeter is used to check the resistance of the motor windings and the capacitor. The ammeter measures the current in the circuit. The voltmeter measures the operating voltages. A capacitor analyzer measures the actual mfd rating of the capacitor.

The first thing to do in testing is to check the motor windings, as shown in Figure 7-4-32. It may be necessary to remove a guard that encloses the terminals on a hermetic unit. The terminals should be marked C, S, and R.

CAUTION: Occasionally the terminal has been damaged and on a pressurized system the terminal can blow out. To avoid possible injury, unless the charge has been removed, use terminal points some distance away from the compressor. Be sure that all accessories such as capacitors and relays are disconnected and that the refrigerant charge has been removed.

Use an ohmmeter to read the resistance in both the run and start winding. Be sure good contact is made with the proper terminals.

If the terminals are not marked they can be identified by a simple test. First, measure the resistance between each pair of terminals. For example, assume that these readings are 5½, 4, and 1½ Ω. By diagramming them as shown in Figure 7-4-33, the terminals can be identified. In the example, the greatest reading is between 1 and 2. The common terminal is therefore 3, the one not being touched. So terminal 3 is C. Then, reading from C, the greatest resistance is to 2. Therefore, 2 is S. And, finally, since 2 is S and 3 is C, 1 must be R.

The motor is then tested for an open winding, a broken wire, or a shorted winding, as shown in Figure 7-4-34. To do this, a good low range ohmmeter (R × 1) is required. A zero resistance means a shorted winding. Low resistance means the winding is good. An infinity reading means it is an open winding. As a rule the start winding has a resistance three to five times the resistance of the run winding.

Large motors (5–10 Hp and higher) have heavy copper windings to carry the motor current. They may therefore indicate a reading very close to a short when winding resistance is read, depending on the ohmmeter.

Figure 7-4-35a,b shows the motor being tested for a grounded winding. The ohmmeter needs to be

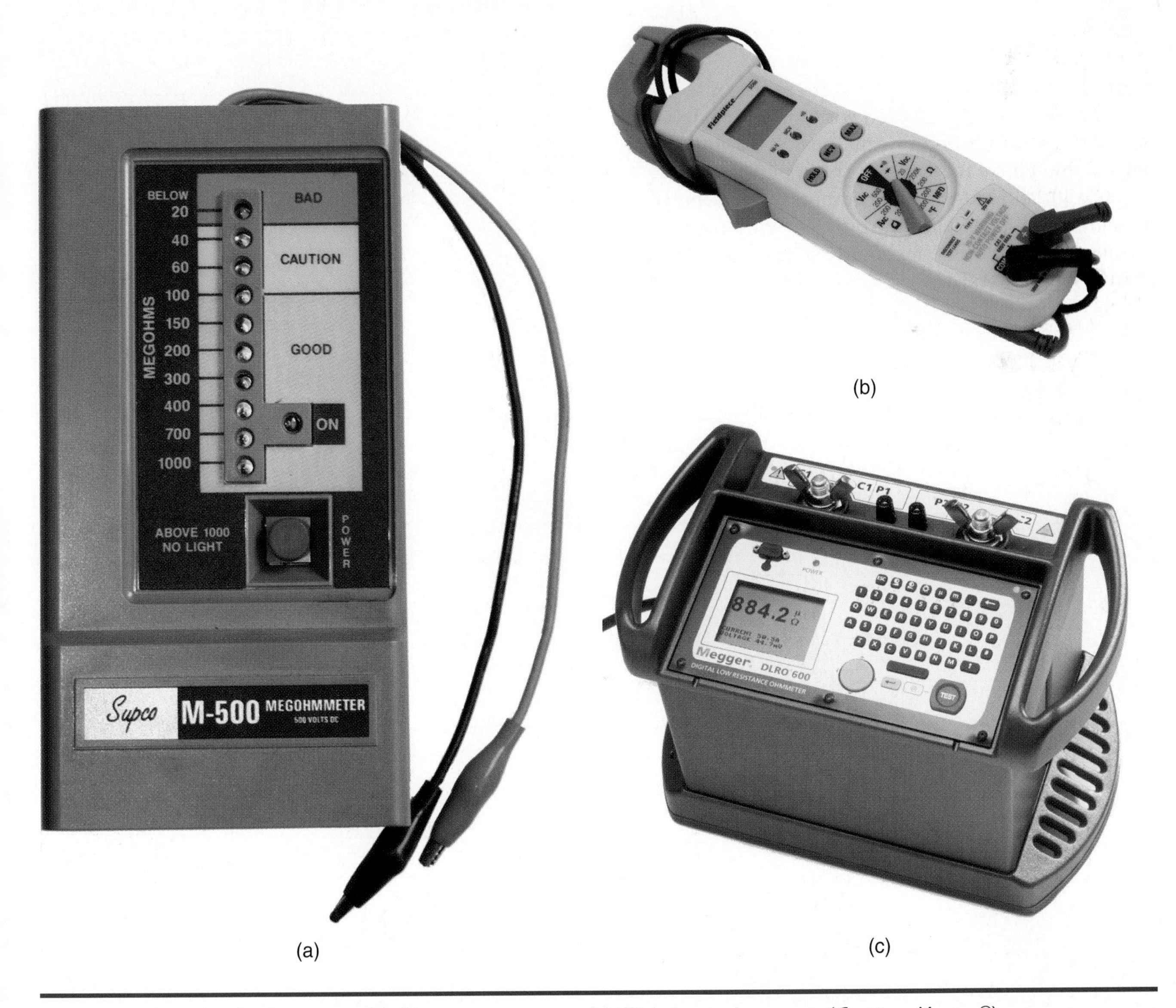

Figure 7-4-31 (a) Megohmmeter; (b) Clamp-on ammeter; (c) Digital megohmmeter. *(Courtesy Megger®)*

Figure 7-4-32 To test motor windings, first remove the leads.

capable of measuring very high resistance (R × 100,000). For an ungrounded winding the resistance is generally 1 to 3 MΩ (megaohms). This applies to both single phase motors and three phase motors.

The temperature of the compressor is important in testing for a partially grounded winding. If the compressor will run, it should be run for about 5 minutes before testing.

Insulation Resistance Testers

These are special testers that are invaluable for testing leakage resistance from motor windings to ground. They are often used to periodically test semihermetic motor insulation. The meters can test leakage at high voltage (500 V for 208–240 V motors and 1000 V for 480 V motors). They may be battery operated or use a hand cranked generator. They are often referred to as meggers. They may detect insulation faults, while an ordinary multimeter using a few volts DC would show a satisfactory reading.

SERVICE TIP

The wiring that is used for motor windings is insulated with a thin coat of varnish. As the motor is used and the windings are heated, this material will slowly carbonize. As it carbonizes, it will slightly change to a darker tan and eventually on to black. When it is completely carbonized, it does not provide any insulating capacity. As a motor heats up, this breakdown process can be accelerated to the extent that in extreme conditions it can carbonize almost immediately when overheated. A megger tests the degree to which carbonization has taken place within the motor winding insulation. The degree of breakdown in the insulation is an indication of the useful life that may remain with a motor or compressor.

Figure 7-4-33 Method of determining run, start, and common motor terminals.

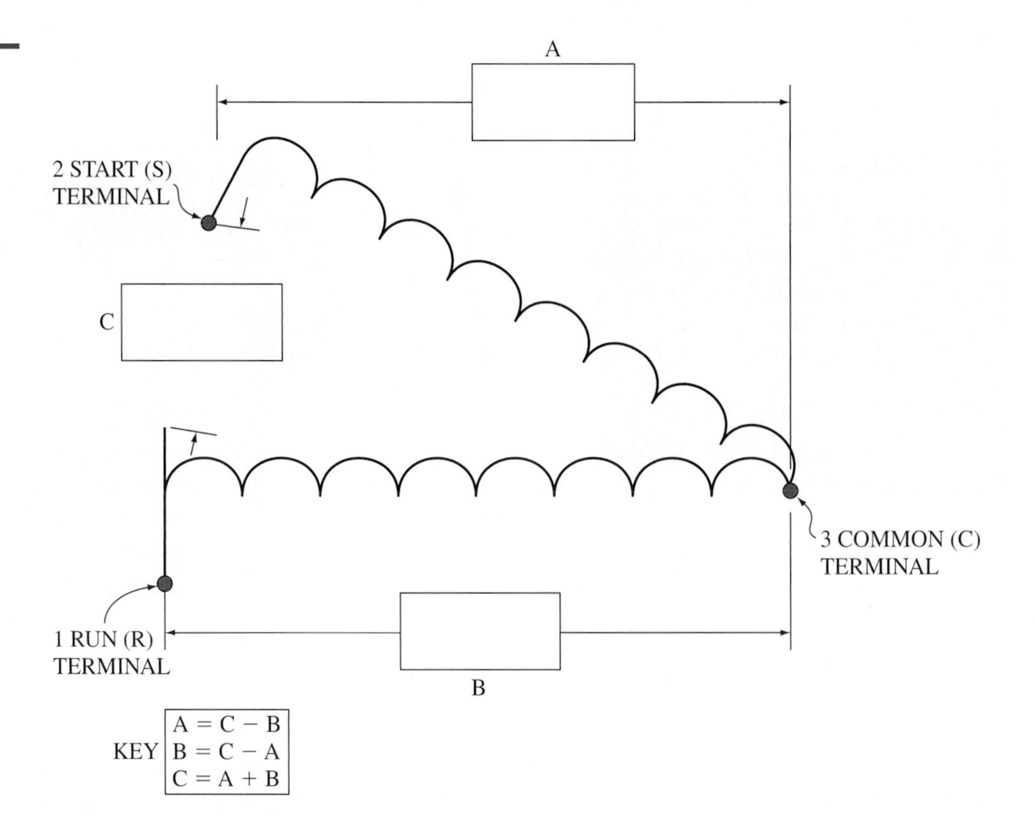

Figure 7-4-34 Method of testing open, shorted windings and broken wires.

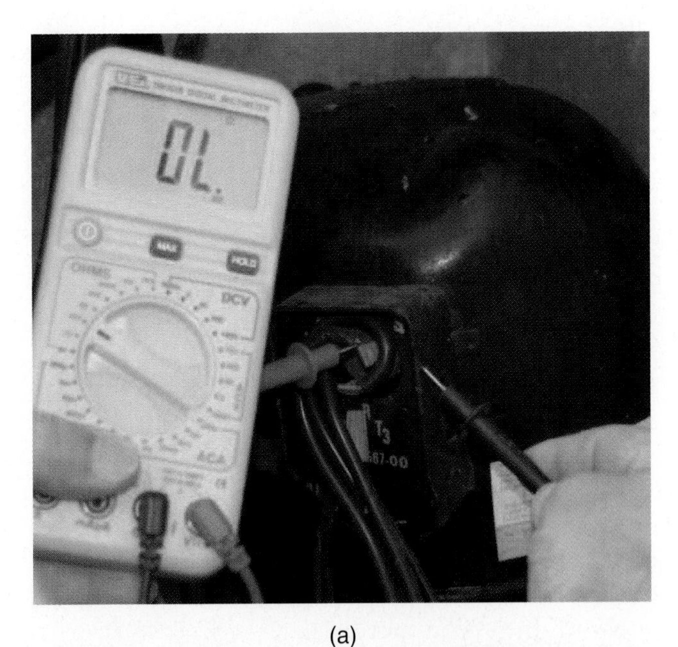

(a)

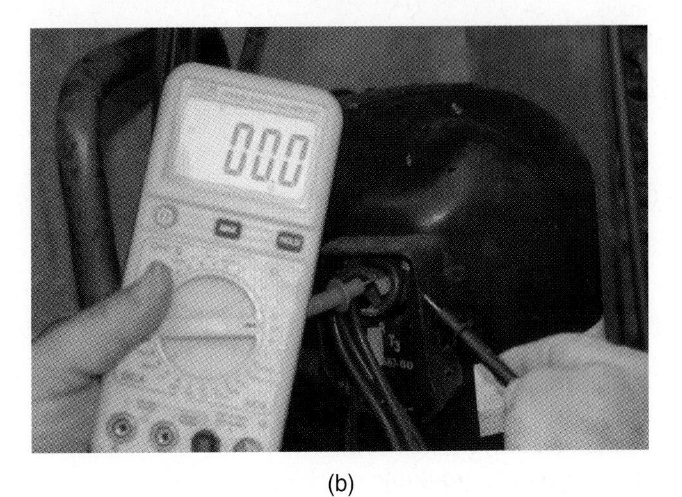

(b)

Figure 7-4-35 Testing for a grounded wiring: (a) Not grounded; (b) Grounded.

When testing for a grounded winding, one test lead is placed on the terminal and the other on the outer shell of the compressor. Be careful to make good contact with the motor shell. A coat of paint or a layer of dirt can hide a grounded winding. The copper refrigerant lines may also be used to check for grounds.

Figure 7-4-36 shows the terminals of the two types of three phase motors. The windings of a three phase motor all should have the same resistance. Remember when checking a three phase motor to be sure to reconnect it in the same manner as originally connected. Interchanging any two connections can reverse the rotation of the motor.

Testing Capacitors

The first operation in testing a capacitor is to discharge it. Do not discharge it by shorting out the terminals. This can damage the capacitor. To avoid electrical shock, the technician should never place fingers across the terminals before properly discharging the capacitor.

The proper way to discharge a capacitor is to put it in a protective case and connect a 20,000 Ω, 2 W resistor across the terminals, as shown in Figure 7-4-37. Most start capacitors have a bleed resistor; however, it is good practice to make sure the charge has been bled off.

Capacitors can be roughly checked by using an ohmmeter. The ohmmeter used in testing capacitors should have at least an R × 100 scale. To test the capacitor, disconnect it from the wiring and place the ohmmeter leads on the terminals, as shown in Figure 7-4-38.

If the capacitor is not shorted the needle will make a rapid swing toward zero and slowly return to infinity. If the capacitor has an internal short, the needle will stay at zero, indicating that the instrument will not take the charge. What you are actually doing is attempting to charge the capacitor using the battery in the ohmmeter (be sure the battery in the ohmmeter is good). An open capacitor will read high with no dip and no recovery.

The use of a capacitor analyzer as shown in Figure 7-4-31b is highly recommended. This instrument will

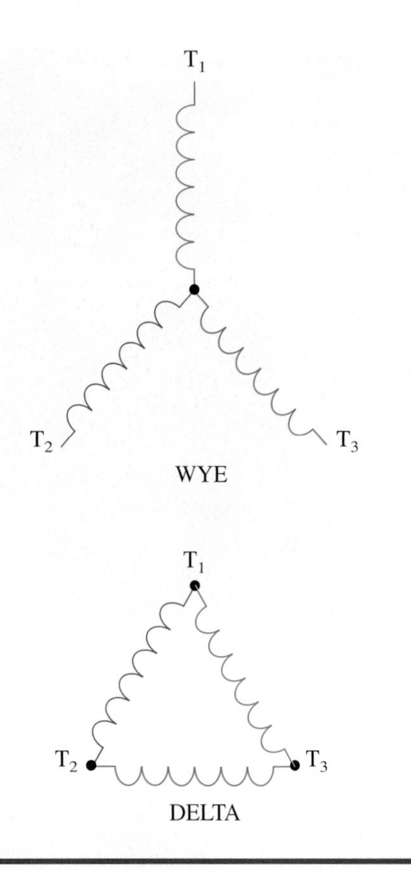

Figure 7-4-36 Two types of three phase motors wye and delta.

read the mfd rating and detect any breakdown in the insulation under load conditions. It will detect any capacitors that have failed to hold their ratings. It also is useful in measuring the rating of a capacitor that has an unreadable marking.

Some run capacitors have some sort of a mark, usually a red dot as shown in Figure 7-4-39, to indicate the terminal that should be connected to the run terminal.

With this arrangement, an internal short circuit to the capacitor case will blow the system fuses without passing the current through the motor start winding.

Where two run capacitors are used on the same equipment, multiple run capacitors with a common terminal are used. One of these is illustrated on the left side of Figure 7-4-40.

These two capacitors have different ratings as required by the equipment they serve. Testing is similar to the testing procedure for single run capacitors.

In Figure 7-4-40, the right device shows a start capacitor with a bleed resistor. The capacitor can be tested with this bleed resistor in place. The pop-out hole on the start capacitor allows insulation expansion if the capacitor is overheated. If the hole is ruptured, the capacitor must be replaced.

Figure 7-4-37 Using a bleed resistor on start capacitors.

In replacing a capacitor, it is desirable to use an exact replacement—a capacitor with the same mfd rating and voltage limit rating. When this is not possible, substitutions can be made as long as the rules for substitutions are carefully followed.

Do not interchange start and run capacitors. Start capacitors are high capacity (100–800 mfd) electrolytic units that are intended for momentary use in starting motors. They are normally encased in plastic.

Run capacitors have much lower capacitance ratings (2–40 mfd) but are made for continuous duty use. They are normally sealed in a metal can.

TESTING START RELAYS

In replacing start relays, it is important to do the following:

1. Always use the identical replacement. An improper substitution can damage the motor.

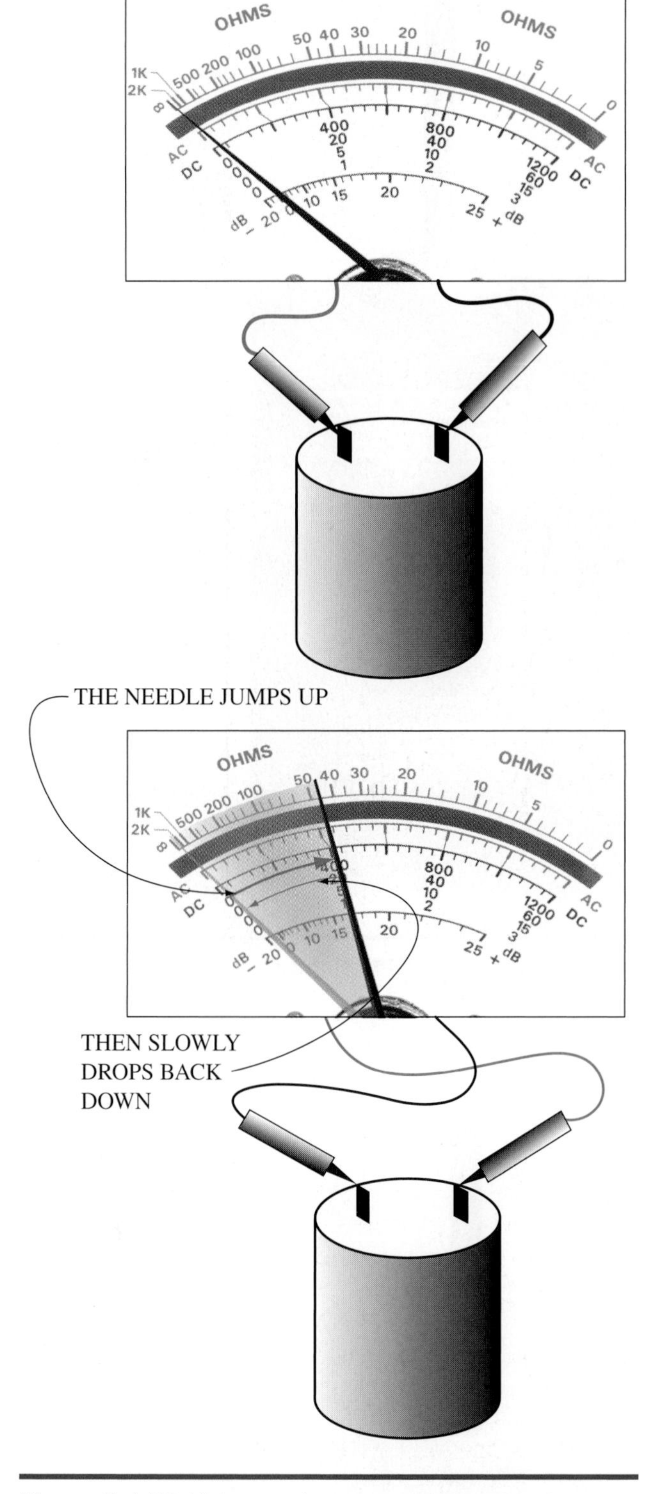

Figure 7-4-38 Using an ohmmeter to test capacitors.

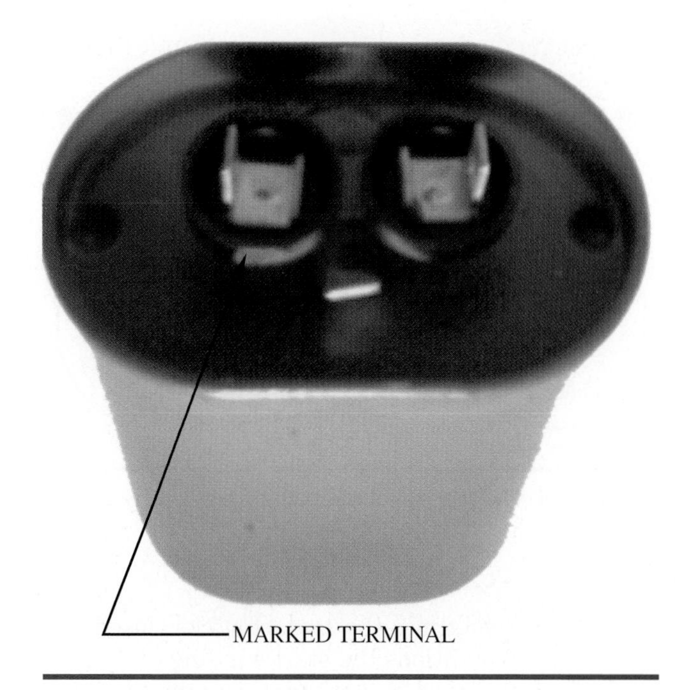

Figure 7-4-39 The marked terminal on the run capacitor should be connected to the run terminal of the compressor.

Figure 7-4-40 Multiple section or dual run capacitor and start capacitor.

2. The replacement must be mounted in the same position as the original and connected the same way. The method of testing the relay coil and contacts is shown in Figure 7-4-41.

When testing start relay contacts, the contacts should be closed and the ohmmeter will read zero

CAUTION: Many start relay coils have higher resistance than average control circuit relays. Be sure to test the coil on the R × 100 scale before deciding that the relay is defective. Note that the pull-in/drop-out voltage of the relay is unique. Do not attempt to replace it with an ordinary, similar voltage relay.

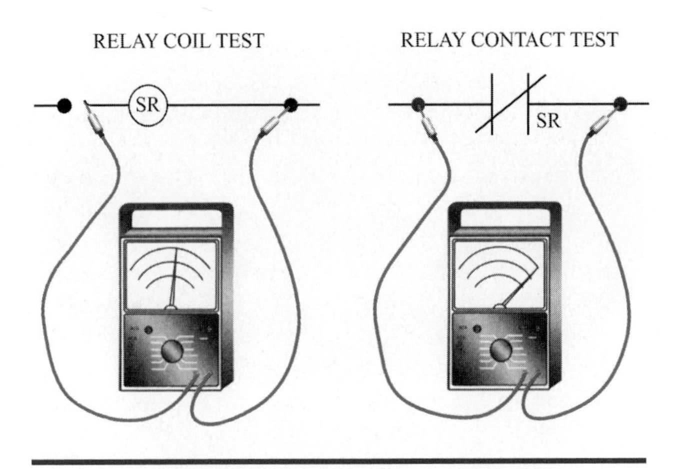

Figure 7-4-41 Using an ohmmeter to test a relay coil and its switch.

resistance. Sometimes the contact sticks closed and under these conditions the start winding of the motor is often damaged. Sometimes the normally closed contacts will be badly burned and make a poor connection. If such is the case, replace the relay. Do not attempt to clean the contacts.

GE ECM™ MOTORS

ECM™ motors were developed by the General Electric Corporation (GE) in the mid-1980s, Figure 7-4-42. ECM™ stands for electronically commutated motor. They were originally used as blower motors, but today they are used for every motor need including combustion air blower motors, condenser fan motors, and compressor motors. These motors are ultrahigh efficiency, programmable, brushless DC motors, 120 V or 240 V AC input. They utilize a permanent magnet rotor and a built-in AC to DC inverter to produce the DC that actually operates the motor, Figure 7-4-43. ECM™ motors have a number of advantages including:

- **High efficiency** ECM™ motors are approximately 70% efficient as compared to 45% or less efficiency for PSC motors.
- **Low maintenance** ECM™ motors use ball bearings as opposed to sleeves, which are used on PSC motors, Figure 7-4-44. Ball bearings are permanently lubricated as compared to sleeved motors, which must be routinely lubricated.
- **Speed control** ECM™ motors come preset for a specific cfm requirement. Because the rpm of DC motors is more easily adjusted, ECM™ motors can be field adjusted to a specific CFM and with the addition of a controller can be remotely controlled using a central energy management program, Figure 7-4-45.

Figure 7-4-42 GE ECM™ motor.

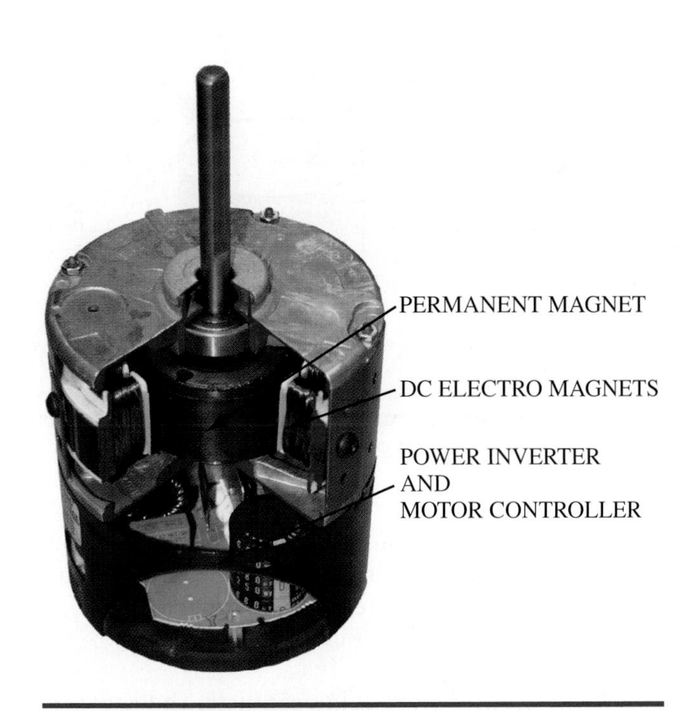

Figure 7-4-43 Cutaway of GE ECM™ motor.

Figure 7-4-44 Cutaway of GE ECM™ motor showing ball bearings.

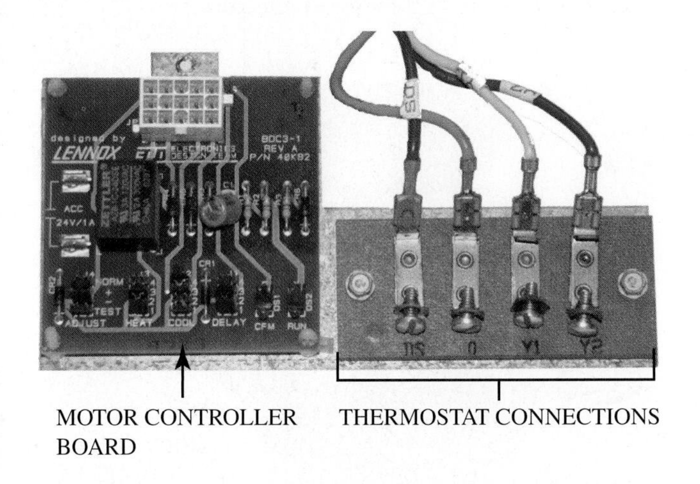

Figure 7-4-45 ECM™ motor controller PC board.

- **Constant CFM** ECM™ motors are designed to provide a constant CFM over a wide range of static pressures. The ECM™ motor rpm will increase automatically to provide the designed airflow.
- **Heat load** ECM™ motors operate at almost ambient temperatures, as compared to PSC motors, which typically operate from 90° to 150° above ambient.
- **Soft start** ECM™ motors are designed to start at a low rpm and gradually ramp up to the designed speed. This places less stress on the motor fan and other mechanical parts as compared to the almost instant starting of a PSC motor. Soft starting also decreases the sometimes-noticeable blast of air from conventional blowers associated with PSC motor starting.
- **Flexibility in size** Because the ECM™ motor's rpm can so easily be controlled without the loss of efficiency, it is possible for manufacturers to use a limited number of various horsepower rated ECM™ motors throughout their entire production line. This reduces inventory costs for the manufacturer and for the service company by reducing the number of replacement motors that they must carry on their service vehicles.

PRINCIPLE OF OPERATION

DC motors use a rotating magnetic field in the stator to create the rotating force in the permanent magnet armature. A microprocessor controls the speed of rotation of that magnetic field. An example of such a microprocessor embedded in the printed circuit board of the controller is shown in Figure 7-4-46. By increasing or decreasing the rate of rotation, the rpm of the motor is controlled. One advantage of DC motors is that they produce the same horsepower output, irrespective of the rpm.

There are a variety of motor controllers for ECM™ motors. One is shown in Figure 7-4-47. Some controllers permit the motor to be controlled directly at the motor. Other controllers allow the motor to be controlled remotely through a computer interlink, such as the one shown in Figure 7-4-48. Some of the features that motor controllers provide are constant

Figure 7-4-46 ECM™ motor control board showing speed jumpers.

Operation	Test Mode	Switch Selections				Expected Result of Typical System
		CONT FAN	HEAT	COOL	BK/PWM	
TSTAT	Fan-only	ON	OFF	OFF	OFF	Motor runs at fan airflow
	Heating	ON	ON	OFF	OFF	Motor runs at Heating airflow (higher than Fan only)
	Cooling	ON	OFF	ON	ON	Motor runs at Cooling airflow (higher than Fan only)
	Dehumidify	ON	OFF	ON	OFF	Motor runs at Dehumidification airflow (lower than cooling airflow)
VSPD	Variable Spd	ON	OFF	OFF	ON	Motor runs at 50% airflow
DSI	Digital Ser. Int.	ON	OFF	OFF	OFF	Motor runs at Fan airflow

INSTRUCTION ON BACK OF MOTOR TESTER

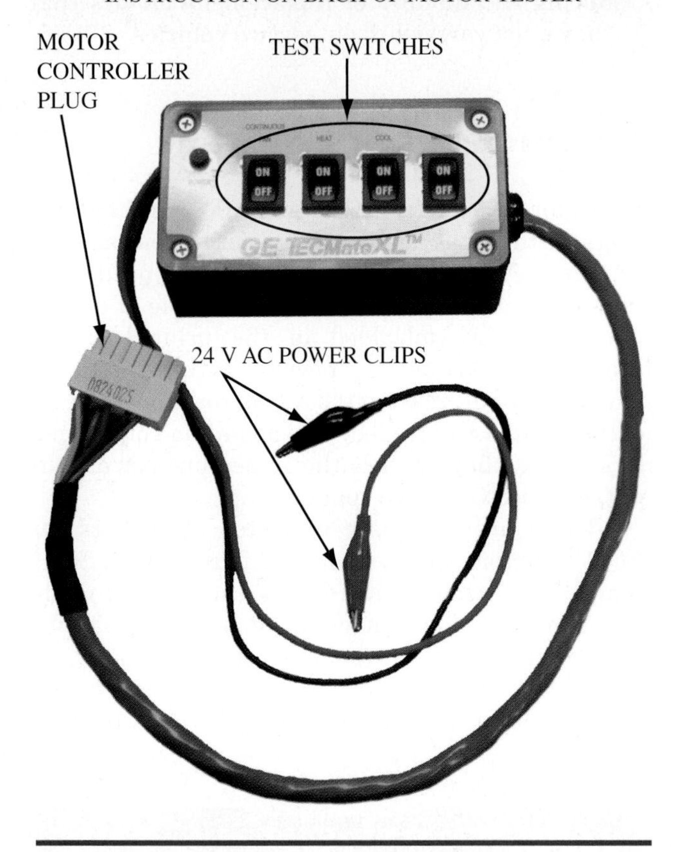

Figure 7-4-47 ECM™ motor tester.

Figure 7-4-48 ECM™ motor connectors.

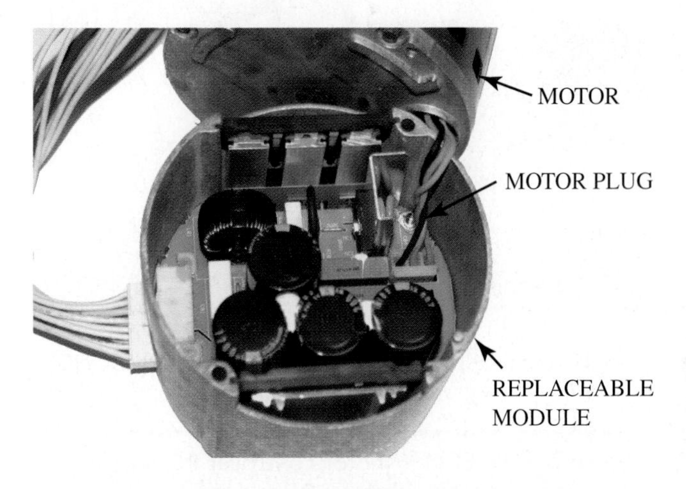

Figure 7-4-49 ECM™ motor replaceable module.

CFM, variable CFM, and constant rpm. Some have a replaceable module for control, as in Figure 7-4-49. Refer to the equipment installation guide or the manufacturer's data sheet to determine the appropriate way of adjusting ECM™ motors in HVAC equipment.

UNIT 4—REVIEW QUESTIONS

1. _______ is the twisting force that must be developed by a motor to turn its load.
2. What are two principal parts of a motor?
3. How is the speed of a motor determined?
4. What is a capacitor rated by?
5. _______ are used to achieve the desired phase angle shift and to obtain the required current through the series load.
6. List the most common types of single phase motors.
7. What are the two ways that the start capacitor can be removed from the circuit?
8. What advantages do three phase motors have over single phase motors?
9. What is the most common cause of motor failure?
10. According to Table 7-4-1, what is the upper and lower limit of a motor with the voltage of 460?

TECH TALK

Tech 1. I was out at a customer's house this morning and they keep complaining about the lights going dim every time the air conditioner starts. I've checked that thing over completely and it's all working perfectly. I mean, right down to the last thing. It's right on the manufacturer's specs. Any ideas?

Tech 2. Yes. A couple of years ago some of the power companies changed the way they calculate the line drops that come from the pole to the house. The new way they calculate them makes it just a little smaller than they used to be; and when the compressor starts, you get a voltage drop from the transformer on the pole to the house. And that's what causes the lights to dim.

Tech 1. They kept blaming the air conditioner. You mean it's not the air conditioner; it's the power company?

Tech 2. Actually, it's not anybody's fault, it's just one of those things that happen. What you can do is put on a hard start kit.

Tech 1. But the compressor starts okay. It doesn't need a hard start kit.

Tech 2. Well, if you put a hard start kit on, the compressor will start instantaneously. That way the lights don't dim out or flicker each time the compressor starts.

Tech 1. That sounds like a pretty easy fix.

Tech 2. It is. And since you know that's happening on that house, all the other houses in the neighborhood are going to be having the same problem. So next time you have a service call in that area, ask the people if they're having the problem. If they are, you can offer a fix.

Tech 1. That's a heck of a deal. Give them a little better service, make them appreciate what we do a little more, and besides that, the boss will be happy. It's one more item that I was able to fill.

Tech 2. Yes, and besides that, it's something they really need.

Tech 1. Thanks. I appreciate that, and so will the boss.

11. What happens if there is single phasing of a three phase motor?
12. List the three types of protective devices that are used to protect the motor against excessive heat.
13. Where is the line duty arrangement commonly found?
14. What is the advantage of manual reset?
15. What does an external supplemental overload device provide?
16. What is the first thing to do in testing?
17. What are insulation resistance testers?
18. What is the proper way to discharge a capacitor?
19. Why is the use of a capacitor analyzer highly recommended?
20. In the principle of operation, what is one advantage of DC motors?

8

Electrical Diagrams and Controls

UNIT 1

Electrical Components and Wiring Diagrams

OBJECTIVES: In this unit the reader will learn about electrical components and wiring diagrams including:

- Electrical Circuits
- Magnetism
- Transformers
- Electrical Load Devices
- Wiring Diagrams and Symbols
- Construction of Schematic Wiring Diagrams

ELECTRICAL CIRCUITS

All HVAC/R electrical systems are made up of electrical circuits. An electrical circuit, as shown in Figure 8-1-1, has three essential parts and one optional part as follows:

1. A source of power
2. A load
3. A path for the current to follow
4. A control (optional)

As a result of these electrical components, electrical current is transformed into heat, light, sound, or mechanical motion. Although the control is optional, meaning that the circuit will operate without it, most systems have controls to regulate the supply of power to, or remove the power from, the load.

Since many electrical components operate using an electrical force termed *magnetism,* the technician should be familiar with some of its characteristics.

MAGNETISM

Magnetism can be produced by electricity. Any time that current flows through a conductor it creates a magnetic field around it. To intensify the field, the conductor is coiled as shown in Figure 8-1-2. When

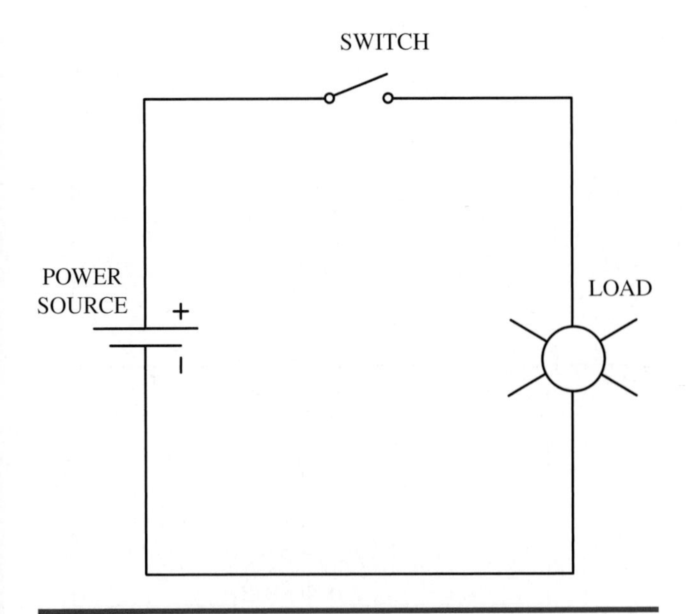

Figure 8-1-1 The basic components of an electric circuit.

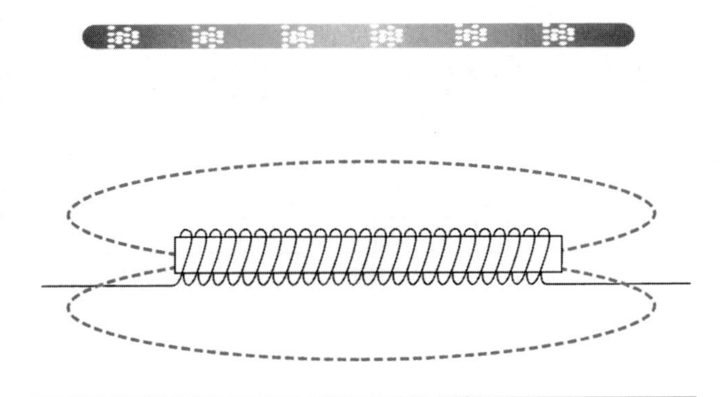

Figure 8-1-2 Electromagnetism produced by current flowing through a conductor.

this occurs, magnetic poles form at each end. The poles change polarity as the current alternates.

The magnetic force is further intensified by placing an iron rod in the center of the coil as shown in the figure. When the current flows through the coil the rod tends to center itself in the coil. This principle can be used to operate a switch as shown in Figure 8-1-3. This arrangement is useful in the construction of relays, contactors, solenoid valves, and motors.

Another characteristic of electromagnetism is its ability to induce a current flow in another conductor that passes through its magnetic field. This principle is known as induction. Induction is useful in the design of transformers, motors, and generators.

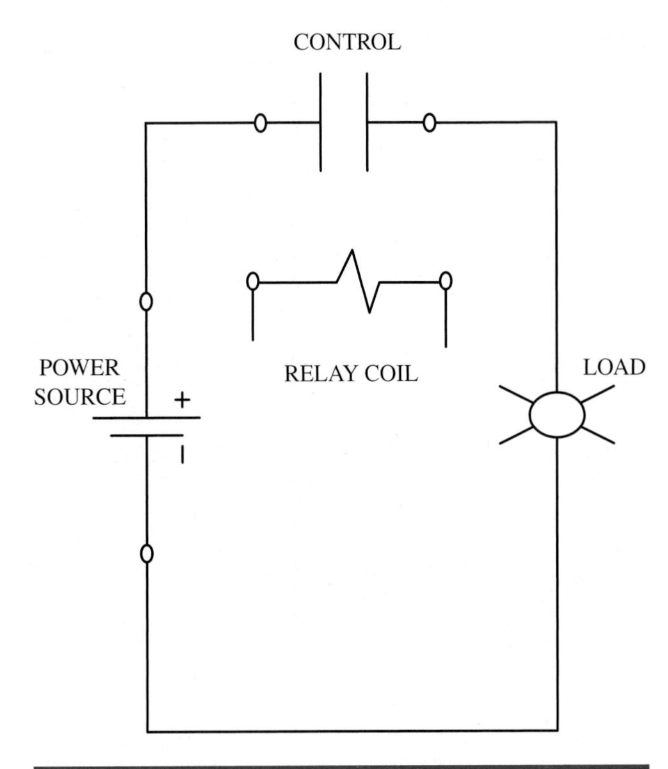

Figure 8-1-3 Electromechanical relay.

NICE TO KNOW

Motors, transformers, generators, and metal cores are all made of thin stamped sheet metal strips. The advantage of using thin sheet metal strips is twofold. (1) It is far easier to manufacture thin sheets because they can be stamped where the thicker material would have to be cut. (2) Sheet metal that is laminated together to form the metal core concentrates the magnetic field more intensely than a single solid piece of metal.

COMMON ELECTRICAL COMPONENTS AND THEIR SYMBOLS

In this section, some common electrical devices that are parts of standard units will be described to show how they are connected in the system to perform their proper function. As an example, a schematic for a simplified packaged air conditioning unit is shown in Figure 8-1-4.

Where the power comes into the building, it enters a service entrance panel for distribution to the various electrical loads. Each electrical circuit that comes from this panel is electrically protected by either

1. A fuse, or
2. A circuit breaker

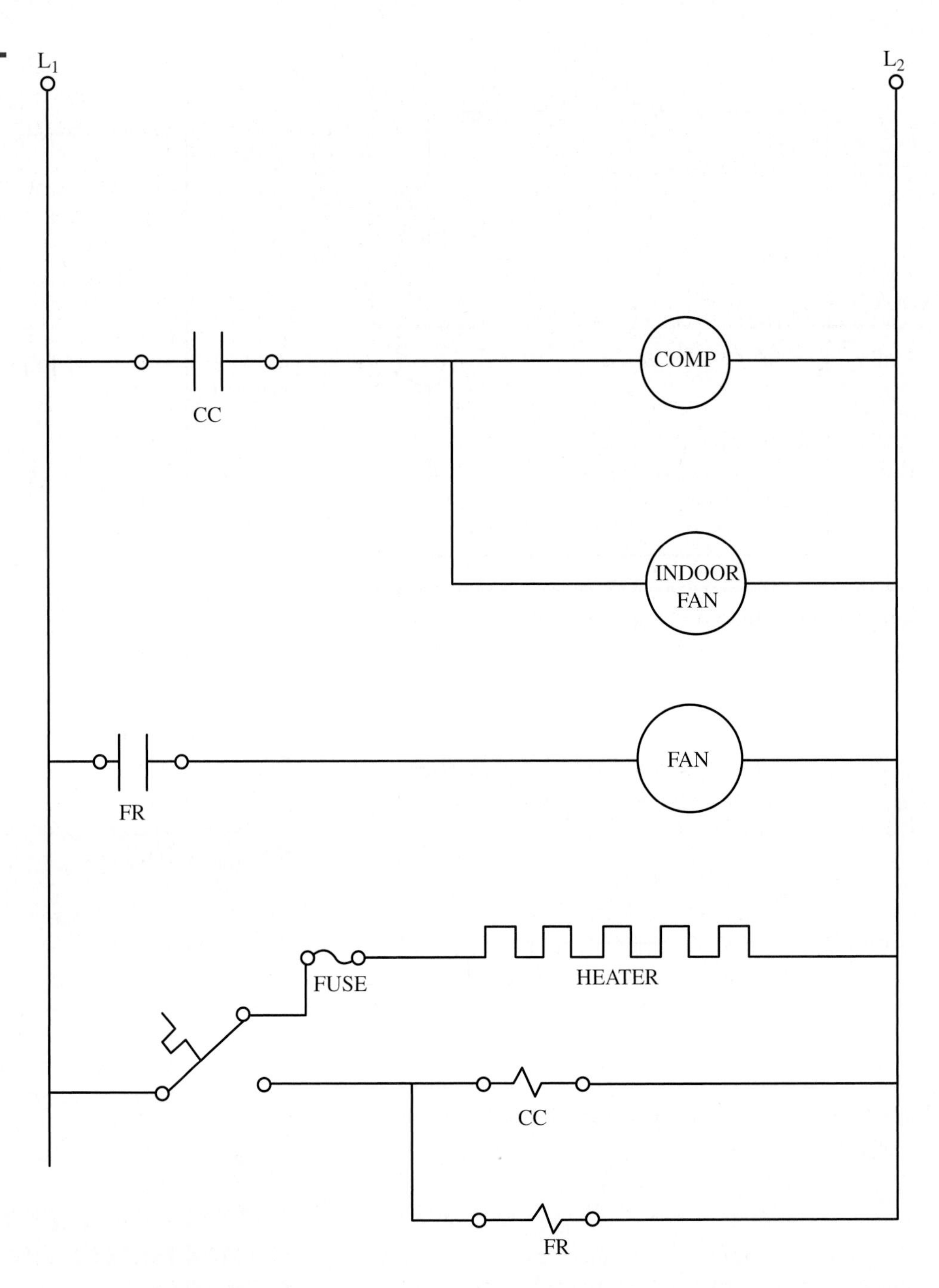

Figure 8-1-4 Packaged air conditioning wiring diagram.

SERVICE TIP

There are many standardized symbols used to represent components in air conditioning and refrigeration wiring diagrams. Although most manufacturers use the Institute of Electrical and Electronics Engineers (IEEE) standard symbols, some manufacturers choose to use their own symbols for components. Be sure and refer to the drawing legend to determine what each symbol means.

A sample fuse and two commonly used fuse symbols are shown in Figure 8-1-5. A fuse is a special electrical conductor that is placed in series with a load and melts when excessive current flows through it, breaking the circuit. Fuses are available in various types and sizes so that they can be selected to match the requirements of specific loads. If they are too small they melt before they should. If they are too large they do not offer the proper protection. Their selection follows the rules set forth in the electrical code or in the specifications accompanying the load.

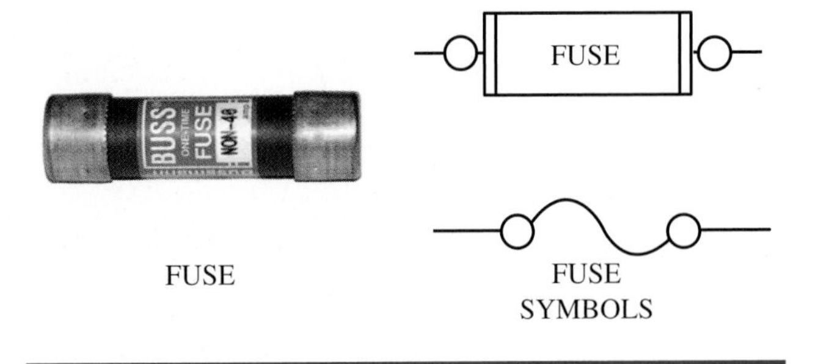

Figure 8-1-5 Symbol for fuses.

Where fuses are used to protect motors in the circuit, a special type of fuse is used called a *time delay fuse* or *dual element fuse*. This type of fuse has a built-in delayed action that will tolerate momentary heavy starting current on motor power-up, but functions the rest of the time to protect the motor against excessive running current.

All residences have some type of electrical panel, where the electrical service enters the building and is distributed to the various circuits in the building. Each circuit has some type of protective device to automatically disconnect the power in case the circuit is overloaded. This protection can either be a fuse as described above or a *circuit breaker*, as shown with its electrical symbols in Figure 8-1-6.

The advantage of a circuit breaker over a fuse is that it can be manually reset at the electrical service panel after an overload, rather than replaced. Also, the circuit can be manually opened in case there is a need to perform service on the circuit. The circuit breaker can either have a thermal or a magnetic trip mechanism. Note the symbols used to represent a circuit breaker in an electrical wiring diagram.

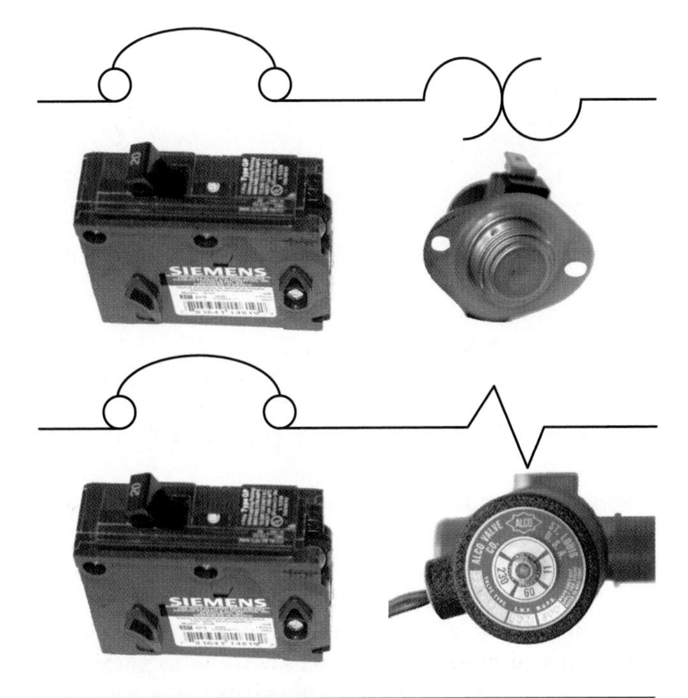

Figure 8-1-6 Circuit breaker symbols.

In addition to the protective equipment in the entrance panel, the electrical codes usually require that each circuit be protected at a distribution or subpanel. In addition, a service disconnect switch should be provided at the equipment being supplied current, as shown in Figure 8-1-7. This disconnect is conveniently located to provide an easy way to disconnect the unit for service.

Inside the unit itself the power supply is connected to a terminal strip where the power is distributed to the various circuits within the unit. Some circuits such as fan motors and compressors use the

Figure 8-1-7 Air conditioner disconnect.

Figure 8-1-8 Control voltage transformer.

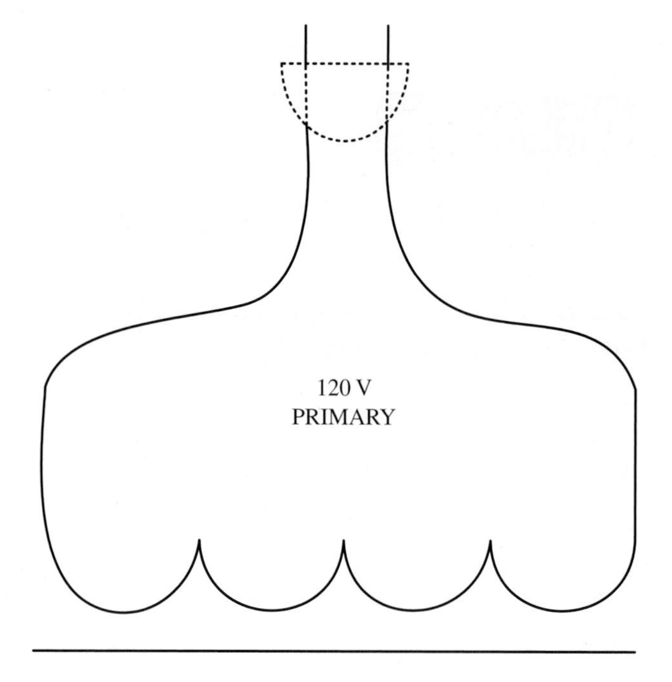

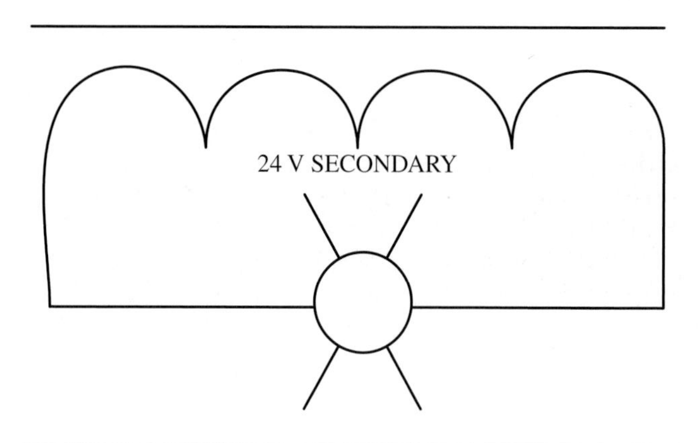

Figure 8-1-9 Primary and secondary voltages of a step down transformer.

full power source voltage. To produce the 24 V control voltage, a stepdown transformer is required as shown in Figure 8-1-8.

TRANSFORMERS

HVAC/R equipment often requires more than one voltage. One or more transformers are often used to step down the line voltage to supply load or control requirements. Occasionally a stepup transformer may be used.

> **NICE TO KNOW**
>
> There are several reasons that most HVAC/R equipment uses 24 V as the control voltage. First, under most state and local guidelines an electrician's license is not required to install and service these low voltage wires. Second, most codes do not require that these connections be made at electrical junction boxes. And third, under OSHA regulations, circuits that have less than 80 V fall under less stringent safety requirements.

Transformers are constructed using the induction characteristics of AC power. When current flows through a coil, a magnetic field is produced. When a second coil is placed in the field of the current carrying coil (primary), electric current can be transferred to the second coil (secondary), Figure 8-1-9. The process is made more efficient by wrapping the coils around a common metal core. The voltage transferred is directly in proportion to the ratio of the number of turns on the primary coil to the number of turns on the secondary coil. More than one secondary coil can be used if additional voltages or circuits are required.

ELECTRICAL LOAD DEVICES

The whole purpose of the electrical system is to supply power to the electrical devices in combination with a control sequence to produce a desired output. The proper operation of the load devices is of primary importance; therefore, the operation of some of the more common load devices will be examined to see how they fit into the wiring system.

The most useful type of diagram for observing the sequence of operation is the ladder diagram. In

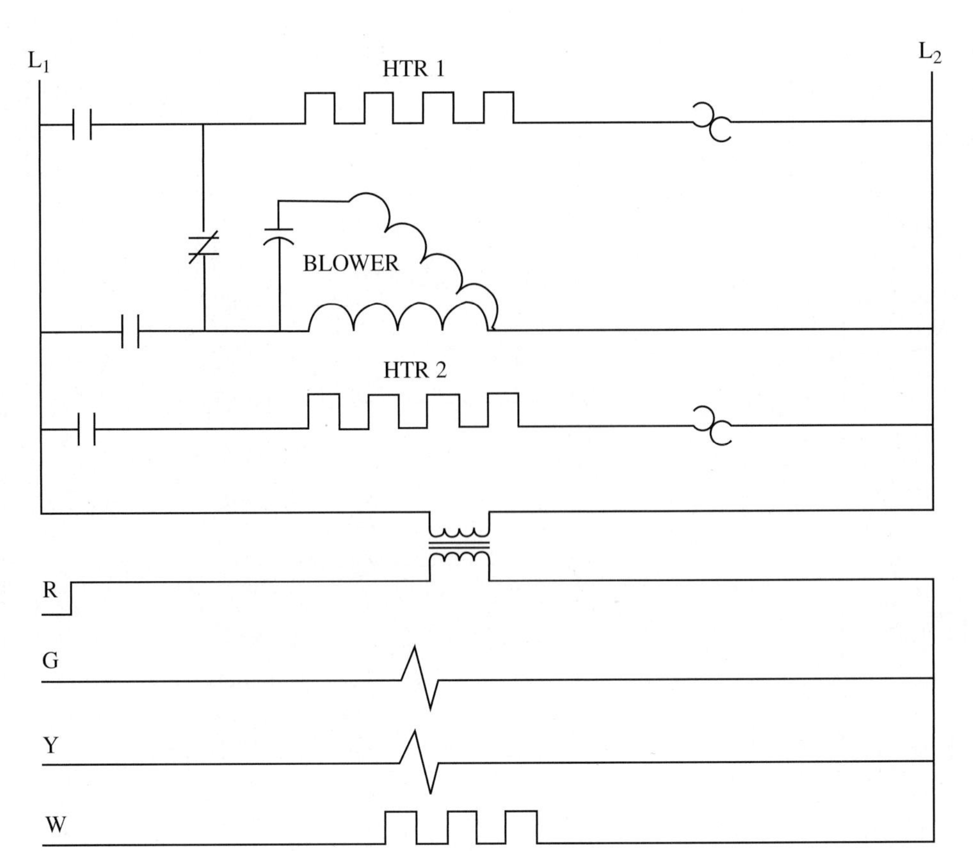

Figure 8-1-10 Basic ladder diagram.

this diagram, the power is represented by vertical lines shown as L_1 and L_2. Figure 8-1-10 shows a ladder diagram. The electrical circuits are represented by horizontal lines, stretching between L_1 and L_2, each circuit containing a load device. The controls (switches) are added in series with the loads to turn the loads on or off as the proper operation of the system requires.

The first major load that will be discussed is the *compressor*.

Compressors and Electric Motors

Some of the symbols that are used to represent compressors in power supply diagrams are shown in Figure 8-1-11. The symbol at the top, consisting of a circle between two horizontal lines, represents a motor with two power line connections. The caption COMP within the circle indicates that the motor is a compressor motor. The center diagram in Figure 8-1-11 indicates that the motor has both a start winding and run winding which need to be properly wired into the system. The bottom diagram in Figure 8-1-11 shows the three windings of a three phase compressor motor, and the three power connections required.

Figure 8-1-12 shows how the various motor symbols fit into the schematic wiring diagram. L_1 and L_2 represent the supply side of the power supply. Power is also being furnished to the primary connections of a 24 V control circuit transformer.

The circuits for three types of motor loads have been connected to L_2. The connections to L_1 will be added later when the control devices are known. Note that the fuse symbol has been placed in series with the secondary of the control transformer to protect this circuit against overload. Some transformer secondary windings are protected with circuit breakers.

SERVICE TIP

Some control voltage transformers have fuse protection. During troubleshooting for a short, you can clip a 24 V light into the fuse plug so that when the short is cleared the light will go out. In this manner you will not continually pop fuses as you try to clear the circuit. Another way of locating the short is to clip the leads of your continuity meter onto the fuse plug. When the short has been located and cleared, the meter's audible continuity tone will stop. Either process works well; however, the light can be less annoying when troubleshooting takes considerable time.

Heaters

Another type of load commonly used on air conditioning units and heat pumps is an electric heater, shown

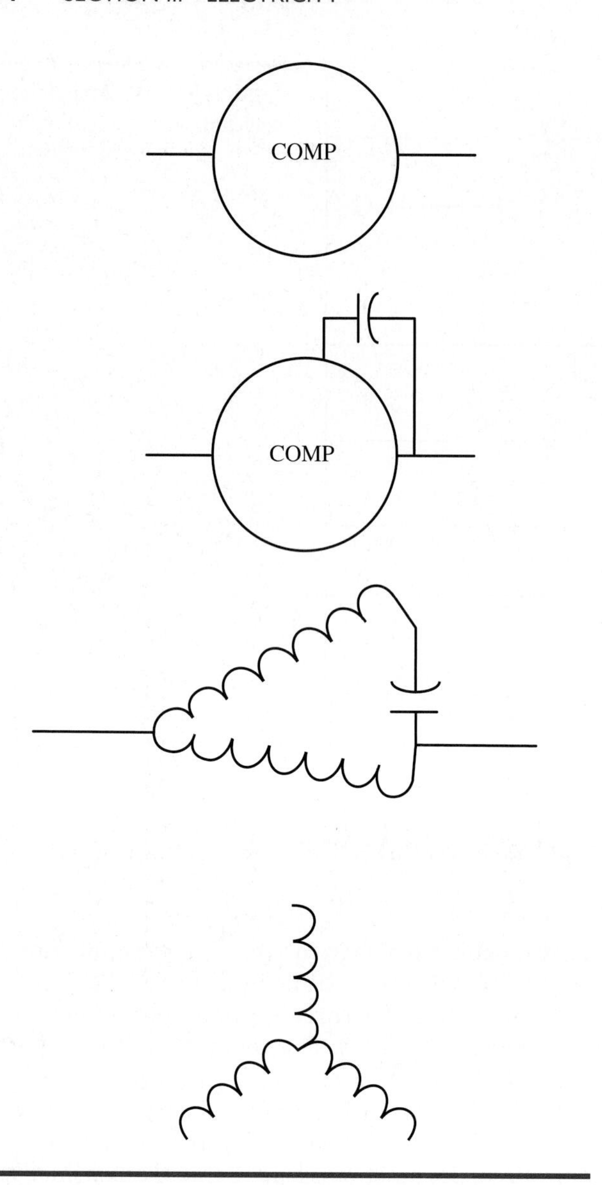

Figure 8-1-11 Compressor symbols.

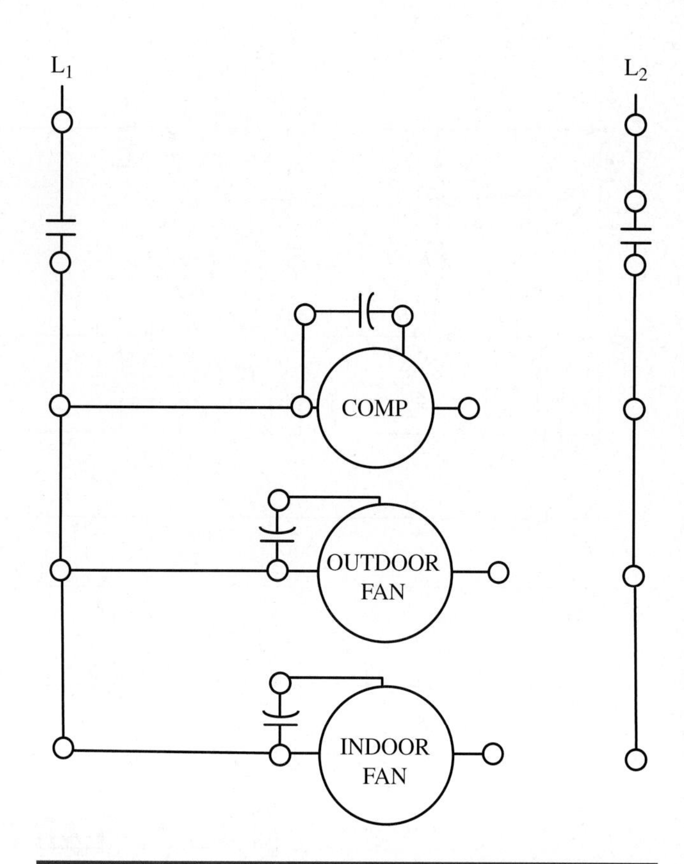

Figure 8-1-12 Basic ladder diagram with various motors shown.

in Figure 8-1-13. As indicated in the illustration, the heater symbol is a zigzag line with the letters HTR. The heater symbol is the same as the resistor symbol since these electric heaters are actually high wattage resistance units. As current flows through them they give off heat. The illustration shows the configuration for three different types of resistance heaters.

Figure 8-1-14 shows where the heater and its protective fuses are placed in the schematic diagram. Electric heaters may draw a high amount of current and should meet specific electrical requirements regarding fusing and overload protection in the heater circuit.

Load Devices in the Control Circuit

Two commonly used load devices are placed in the control circuit. These are magnetic coils used in relays and contactors, Figure 8-1-15. Relays and contactors have the same function except the contactor is larger, more rugged, and carries more current. A motor starter is a contactor with motor overload protection added. Actually these devices are combination units. They are composed of both a load device (coil) and one or more switches.

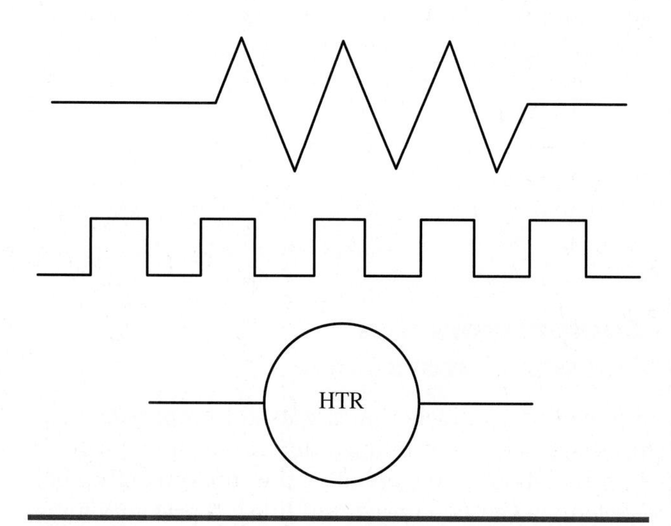

Figure 8-1-13 Symbols for resistance heating elements.

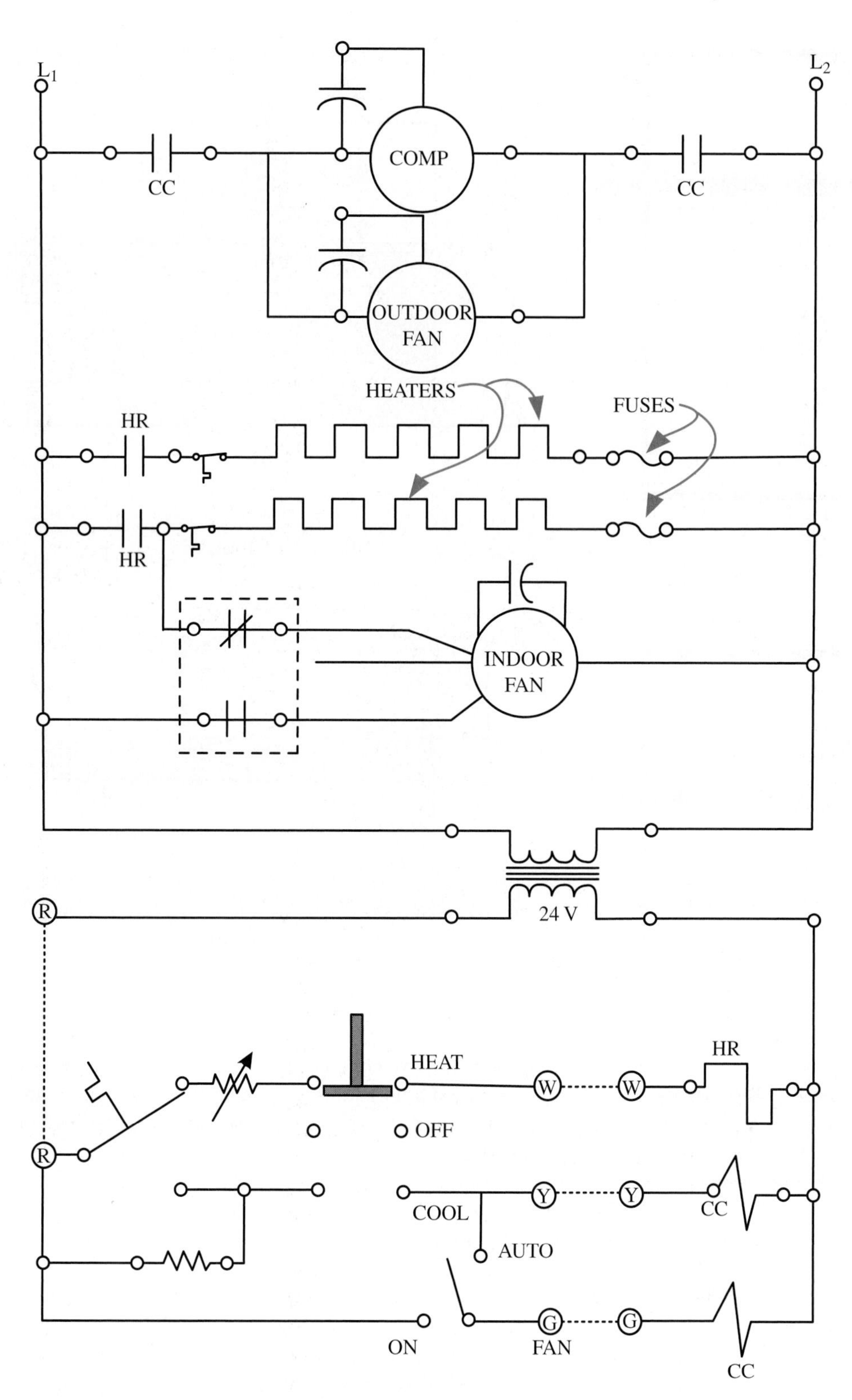

Figure 8-1-14 Over temperature and over current protection devices for resistance heating circuits.

SERVICE TIP

The heating element in a motor starter must be sized for the specific motor it is protecting. Refer to the motor manufacturer's technical data sheet to determine the size of the heater that must be installed for proper protection.

Figure 8-1-16 shows how contactors and relays operate. When the coil is energized (supplied with current), a magnetic field is set up which attracts a metallic armature. As the armature moves to the center of the coil it closes or opens the contacts (switches) which are attached to it. As long as the current flows through the coil the relay or contactor

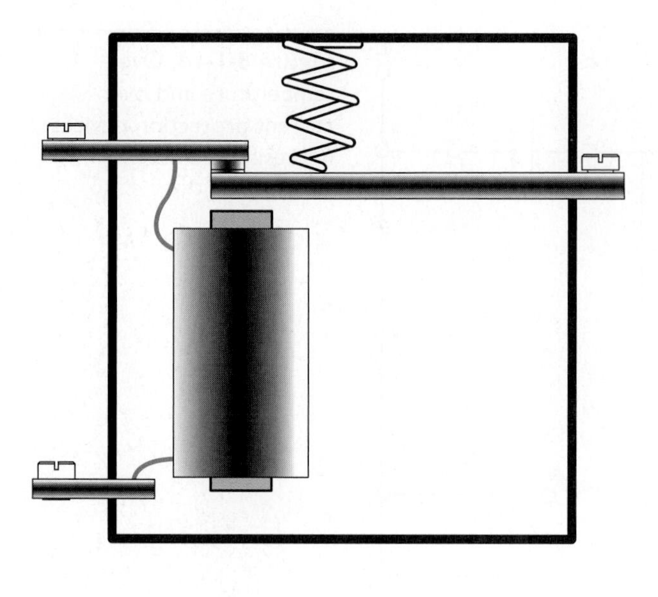

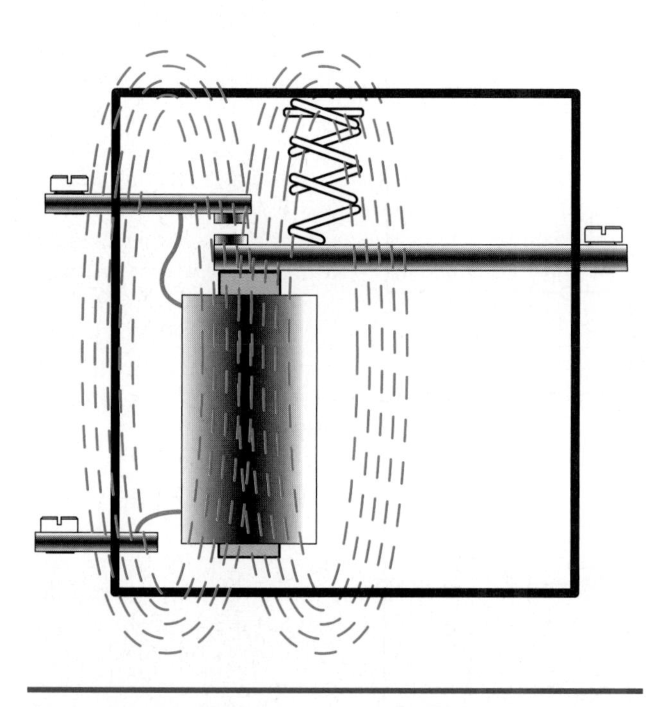

Figure 8-1-15 Magnetic field pulls in the relay contacts.

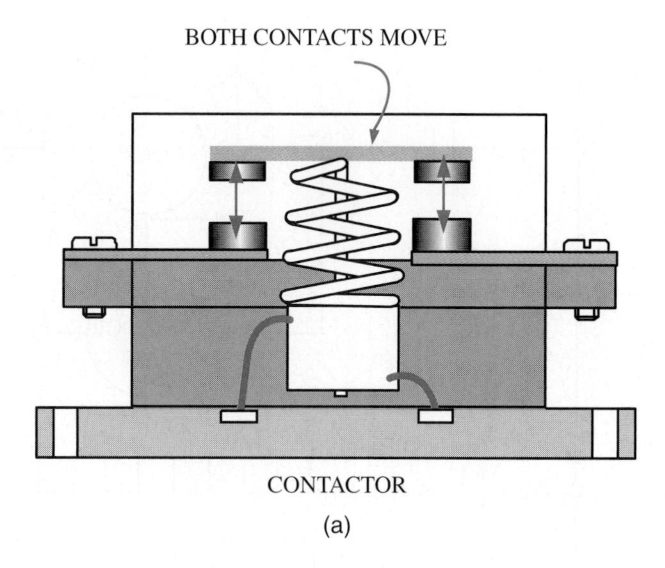

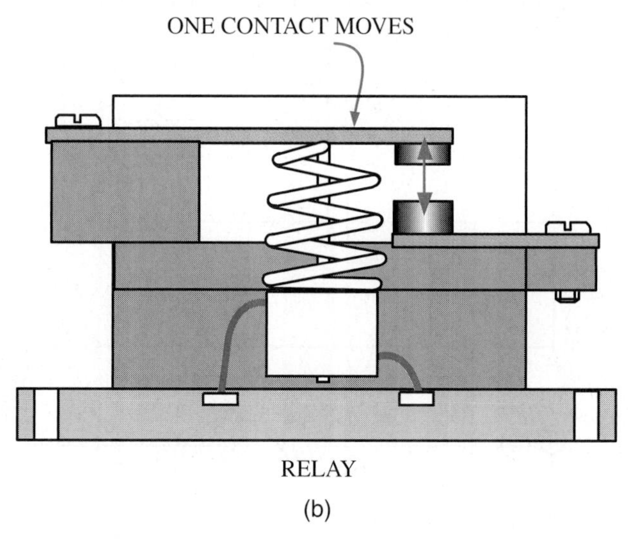

Figure 8-1-16 (a) Contactor. Note: contactors have two points on each circuit; (b) Relay. Note: relays only have one set of points for each circuit. Both contactors and relays can have circuit paths that cannot be opened or closed.

is energized and the switches are in the energized position.

These relay type devices are designed with both normally open (NO) switches and normally closed (NC) switches. The normal position of the switches is always the position of the switches when the relay coil is de-energized. The NC switch has a diagonal line through it as shown in Figure 8-1-17.

Note the letters that identify the relay, such as CC, appear both on the coil and on the switches. In this way switches that belong to each relay can be identified when they appear in different parts of a diagram. Very often the coil may operate at one voltage and the switches are located in another circuit operating at a different voltage.

A good example of the use of a relay in a schematic diagram is shown in Figure 8-1-18. Examine the diagram to determine the function of the control relay (CC). The coil is located in the low voltage part of the diagram and is controlled by the ON switch.

The two switches operated by the relay are located in the high voltage part of the diagram. Both of these switches are normally open and are in series with the compressor (COMP).

The control relay can also be diagrammed as shown in Figure 8-1-19. This view shows a separate coil circuit and three separated switches, two NO and the third NC. There is no electrical connection

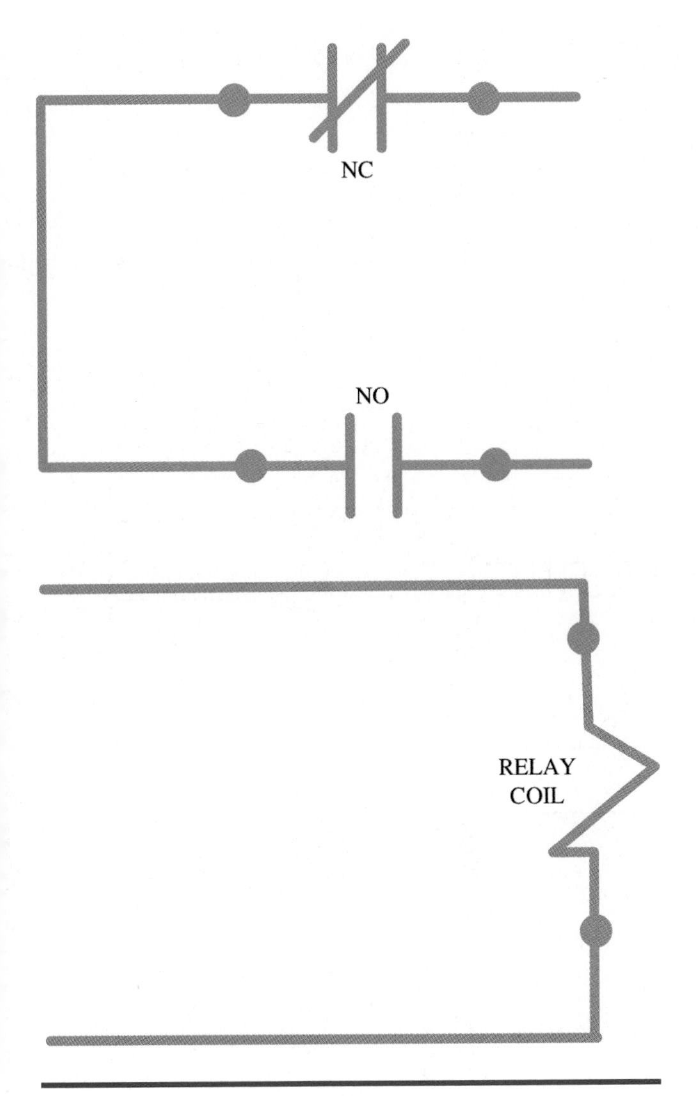

Figure 8-1-17 Normally open and normally closed contact symbols.

between any of these parts of the relay, only a mechanical connection.

WIRING DIAGRAMS

Wiring diagrams are a type of map, supplying complete information on how the electrical parts are connected to operate, control, and protect the unit. Certain standards have been set up relative to the use of electrical symbols for representing electrical components, but they are not always followed by every manufacturer. The technician, therefore, becomes more skillful with experience in interpreting an individual manufacturer's offerings.

Basically there are three types of wiring diagrams:

1. External
2. Connection (or panel)
3. Schematic

SERVICE TIP

Wiring diagrams for equipment, from time to time, change as the manufacturer of the equipment makes minor modifications in the equipment's components. Because wiring diagrams today are computer generated, it is easy for the manufacturer to make these minor changes without having to redraw the entire diagram. For this reason many of the manufacturer's diagrams may be slightly different from one another but look very similar. Make certain that you are using the correct diagram for the piece of equipment you are currently servicing by checking the model number on the diagram with the equipment model number.

The *external diagram* is supplied by many manufacturers with the installation instructions to show the type of electrical service required and how the unit is connected to it. A diagram of this type would show the location(s) of terminals on the equipment, the type of external fusing needed, and the type of power that must be furnished for the unit to operate properly. A typical external wiring diagram is shown in Figure 8-1-20.

It is common practice for these diagrams to show both the external wiring supplied by others as well as the external wiring that is field installed by the installation crew or their electrician. The drawing may also specify wire sizes and the type of insulation required, and may instruct the installer to obtain certain information from local codes.

This is an extremely important part of the job. If power is not properly applied, the unit will not function properly.

If the unit is factory assembled, the *connection diagram* is the one used at the factory for wiring the unit, Figure 8-1-21. The actual location of all electrical connections is shown on this drawing. If you were to examine the unit with a connection diagram in your hand, you would be able to locate all of the connections and be able to trace wires from one connection to the next. Regardless of what other wiring diagrams are available, this one would show as nearly as possible the way the wires are run on the actual unit. If you wanted to trace the wiring to locate a possible loose connection or a wire that possibly had been left out, the connection diagram would be the one to use.

To know how the unit is controlled or the sequence of operation of the various components, the diagram to use is the *schematic diagram*. If troubleshooting is needed, the schematic wiring diagram is practically a necessity. This diagram divides the system into a series of individual electrical circuits. Each circuit has a load (a component that requires

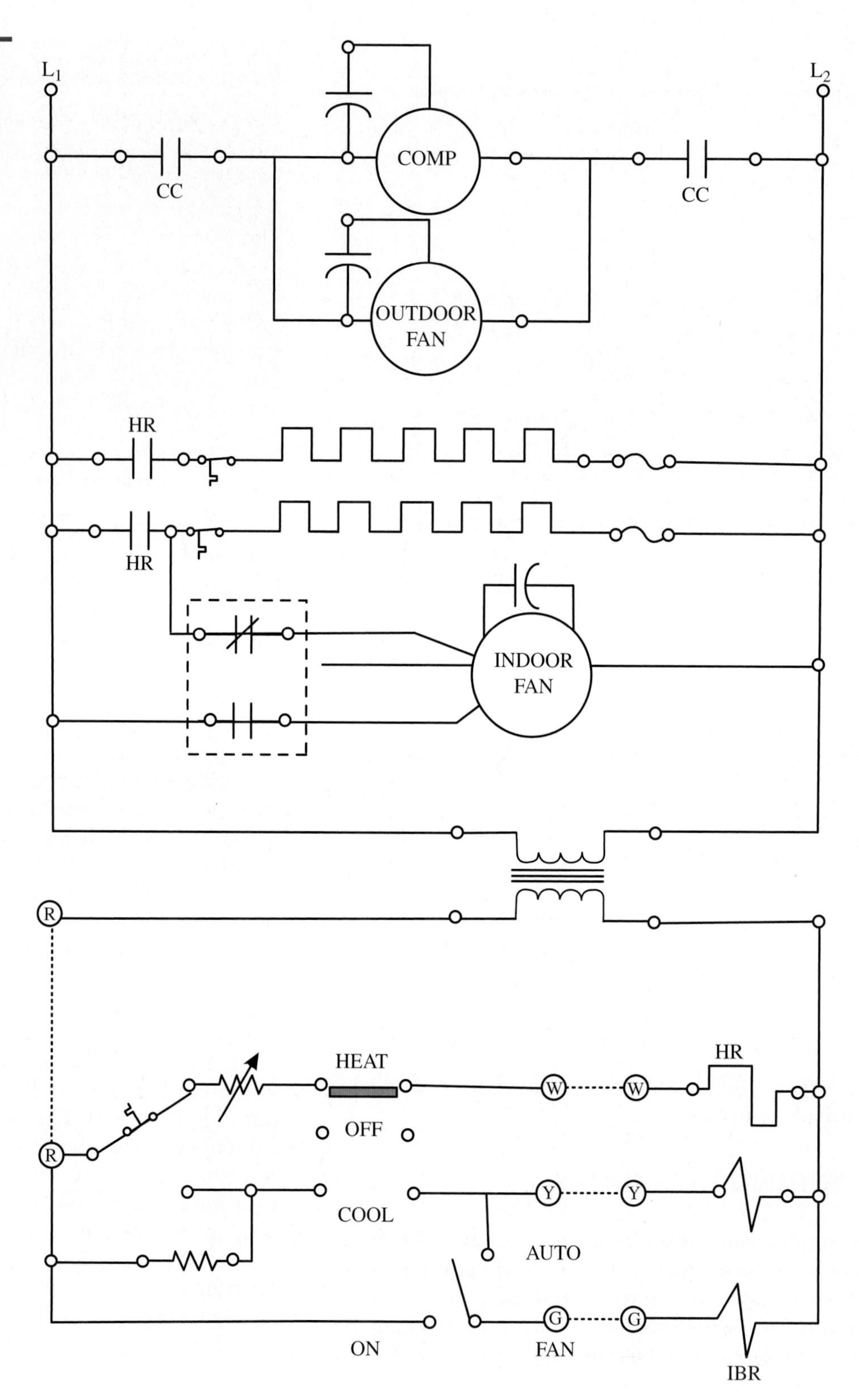

Figure 8-1-18 Ladder diagram showing the use of relays.

power), a path for the power to travel, and one or more switches (optional).

The schematic may be drawn as a ladder diagram. These are clear and easy to follow. Each circuit appears as a ladder rung between two power sources (L_1 and L_2) as shown in Figure 8-1-22. Due to the need for developing skill in using the schematic diagram, most of this chapter will be devoted to the use of ladder diagrams. Although a ladder drawing is usually made vertical, it can be made with the power lines horizontal (with the ladder lying on its side) if this is more convenient.

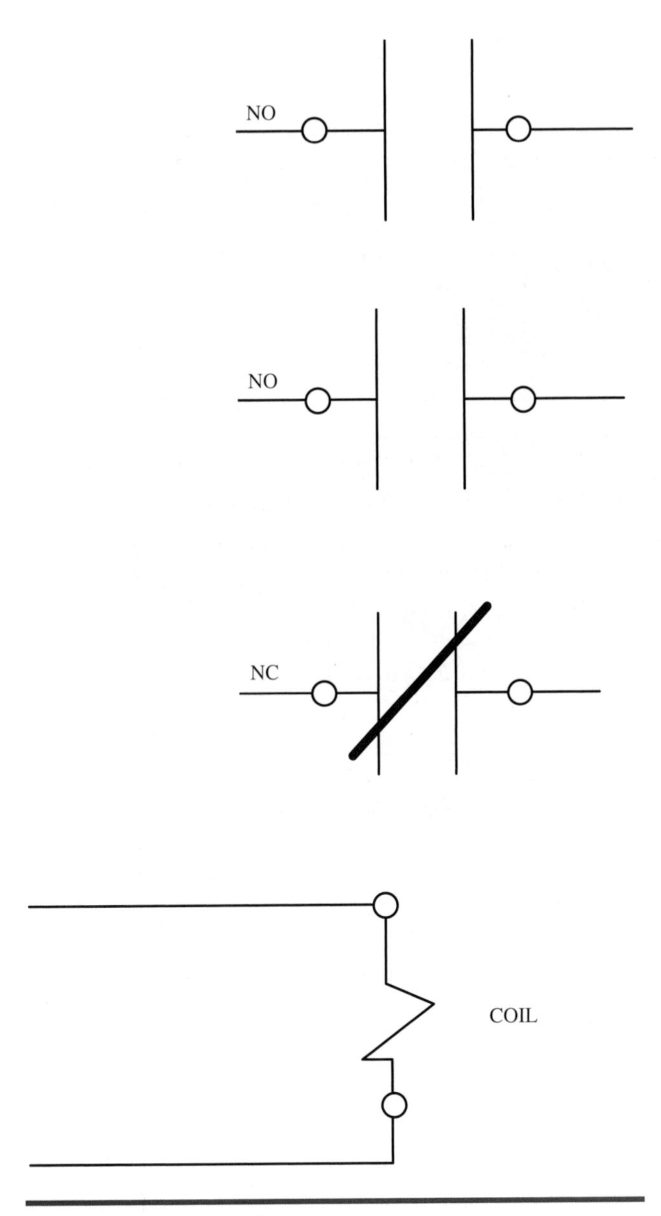

Figure 8-1-19 Control circuit relay diagram.

WIRING DIAGRAM SYMBOLS

In order to keep the diagram compact and meaningful, most of the electrical components are represented by symbols. A standard list of symbols is shown in Figure 8-1-23. These symbols are primarily used in making schematic wiring diagrams.

The wiring diagram symbols fall into the following categories:

1. Loads. These include any electrical device that uses power, including motors, transformers, resistance heaters, relays, lamps, and solenoid valves. Each electrical circuit must have some type of load.

2. Switches. A switch is any device placed in the electrical circuit that turns on or off the power supply to a load. There are many types of switches, as indicated in Figure 8-1-23. Some of them are operated mechanically, such as thermostats, pressurestats, humidistats, flow switches, and float switches. Some of them are operated by electrical power, such as relays, contactors, and starters. Some are protective in design, such as fuses and overloads.

All switches have a normally open (NO) or a normally closed (NC) position. These are usually the positions of the switch when the circuit is de-energized; however, there is a variation to this. A thermostat is usually shown in an open position. Other mechanical switches are usually shown in the position they would normally be if the unit were operating properly. For example, a high pressure cutout switch is shown in a closed position.

TECH TIP

The word "normal" as applied to contacts refers to the contact as it would be if it were in the box. It does not refer to the position of the contact most of the time. For example, the switch that turns the light on in a refrigerator is a normally closed switch, meaning that when it was in the box before it was installed it was in the closed position. However, when it is in the refrigerator, the contact is open almost 100% of the time so that the light in the refrigerator is off. So, although the switch is open most of the time, it is still considered to be normally closed. In reading diagrams you may have a normally open switch that is closed, or you may have a normally closed switch that is open; however, the designation of normally open or normally closed does not change, only the designation as to whether the switch is currently open or closed.

3. Combination load/switches. All switches that require electrical power to operate are combination load/switches. For example, a relay has a coil that requires power to operate it and it has one or more NO or NC switches.

4. Special electrical devices. This group includes all other electrical devices that do not fit in the above categories, such as capacitors and thermocouples. Capacitors were described in detail in the section covering single phase motors. A thermocouple is a unique device constructed by joining the ends of two dissimilar pieces of metal. When the two junction points are at different temperatures, current flows in the circuit. This device will be fully described in the warm air heating section. It is used as safety device to turn off the gas in case a standing gas pilot goes out.

5. Wiring conventions. The standard list of symbols shown in Figure 8-1-23 gives a number of drawing conventions that are useful, particularly the ones used for crossing wires on a diagram.

Figure 8-1-20 Typical external wiring diagram used for unit installation.

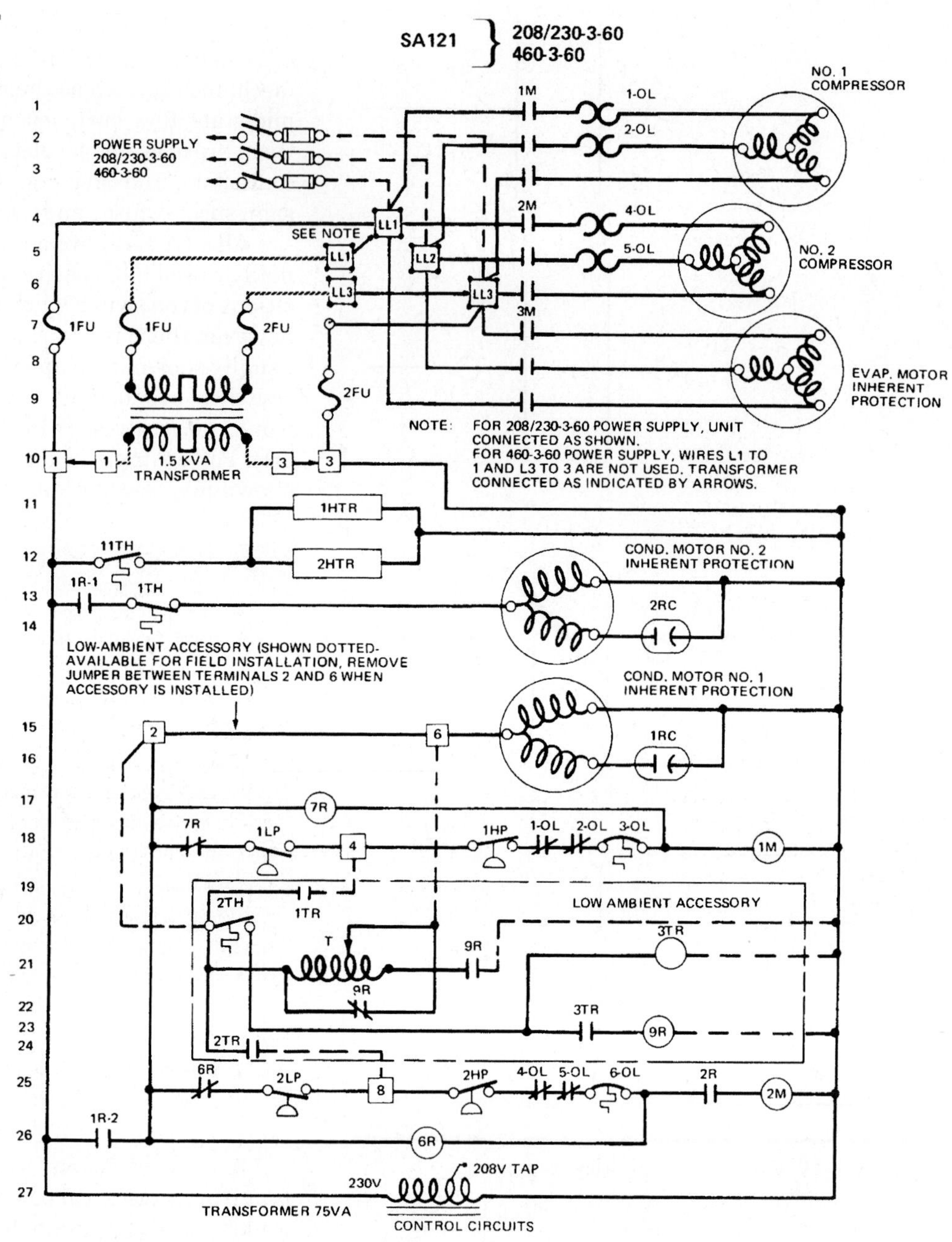

Unfortunately, all manufacturers do not adhere to the standard list of symbols shown in Figure 8-1-23. However, most manufacturers are consistent in their use of symbols. It is therefore recommended that the technician keep some type of notebook to record any new symbols encountered for future reference.

CONSTRUCTION OF SCHEMATIC WIRING DIAGRAMS

A schematic wiring diagram consists of a group of lines and electrical symbols arranged in ladder form to represent individual circuits controlling or operating a unit. The electrical symbols represent loads or switches. The rungs of the ladder represent individual electrical circuits. The unit can be an electrical-mechanical device such as an air conditioning system.

All schematic wiring diagrams are made up of one or more individual electrical circuits. Figure 8-1-24 shows a single circuit made up of a battery (power source), a light bulb (load), a switch (control), and the connecting wiring (path). The top diagram in the illustration is a pictorial diagram of the circuit and the diagram below it is the schematic.

Most circuits only have one load but as many switches as necessary to properly control the load. Since the power supply for the load shown in Figure 8-1-24 is a battery, the two sides of the power supply

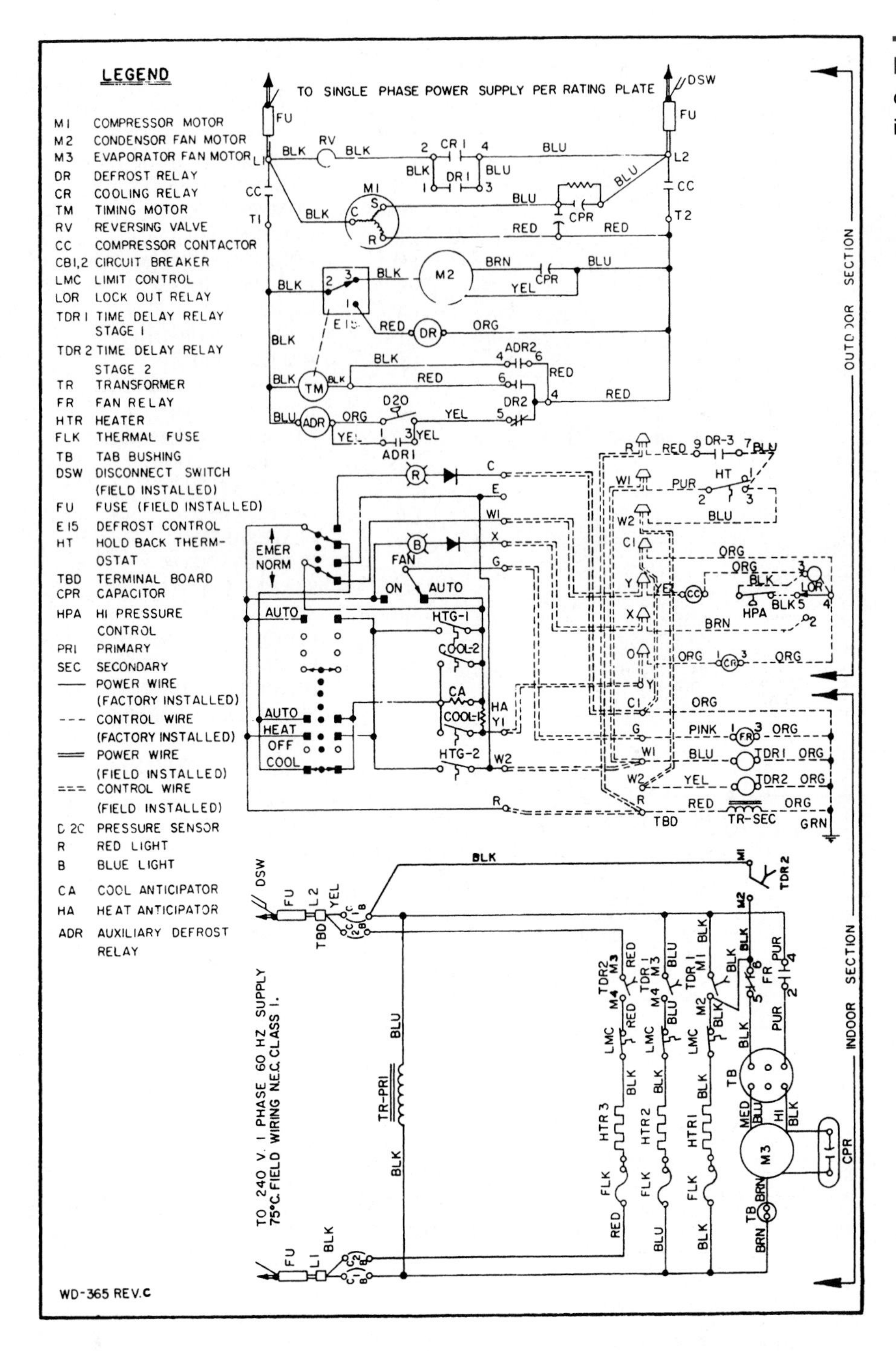

Figure 8-1-21 Typical connection diagram showing internal wiring of the unit.

are indicated as + and −. Note in the schematic diagram that symbols are used to represent the various electrical devices in the circuit. Since the simple circuit in Figure 8-1-24 shows the switch in the closed position, the circuit is completed and the lamp is lighted.

The Ladder Diagram

Figure 8-1-25 shows a preliminary view of a ladder type wiring diagram. Note that the power supply (symbols L_1 and L_2) are indicated by two vertical parallel lines. The three horizontal lines between

TECH TIP

Each type of wiring diagram has its advantage and purpose. A ladder diagram shows the electrical relativity of all of the components and is easiest to follow for troubleshooting. A schematic diagram shows the relative position of components and what wires are connected to each component. A schematic diagram is easiest to trace wires with. A pictorial diagram shows the appearance and location of the various parts. Pictorial diagrams are most advantageous for identifying the various parts in the circuit.

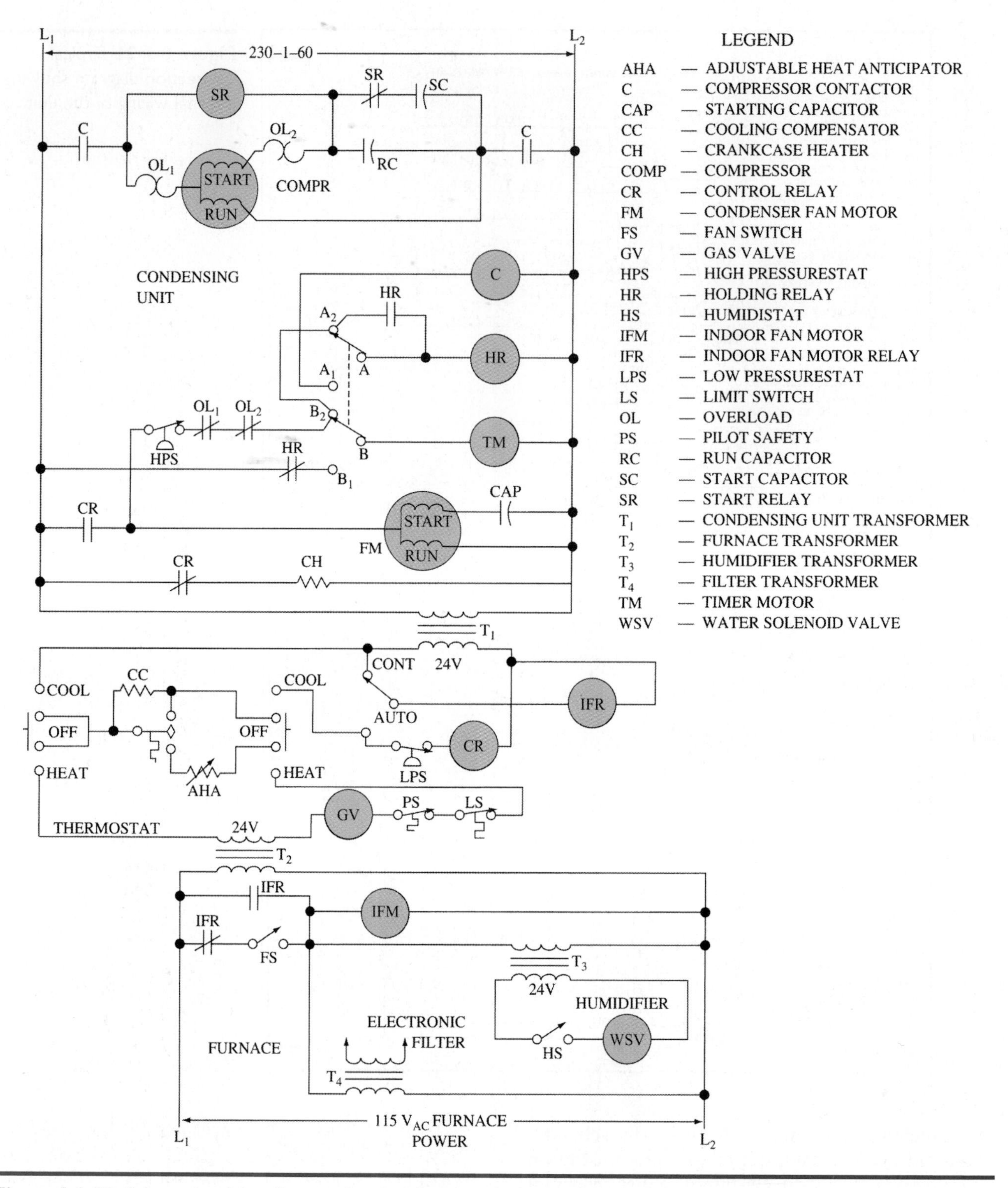

Figure 8-1-22 Schematic wiring diagram showing the various electrical circuits.

them and connecting them are electrical circuits. Each circuit has a load: the first circuit, a condenser fan motor (CFM); the second circuit, a control relay (CR) coil; and the third circuit, a compressor (COMP) motor. Note that the symbol for the thermostat shows that it is a cooling control since it closes on a rise in temperature. Note that the third circuit, in addition to the compressor, has two switches in series with the load, an operating switch (on-off) and a high pressure (HP) safety switch.

A few other electrical devices need to be added to make this an operating system. Referring to Figure 8-1-26, a circuit with a red light at the top of the diagram indicates when power is being supplied to the

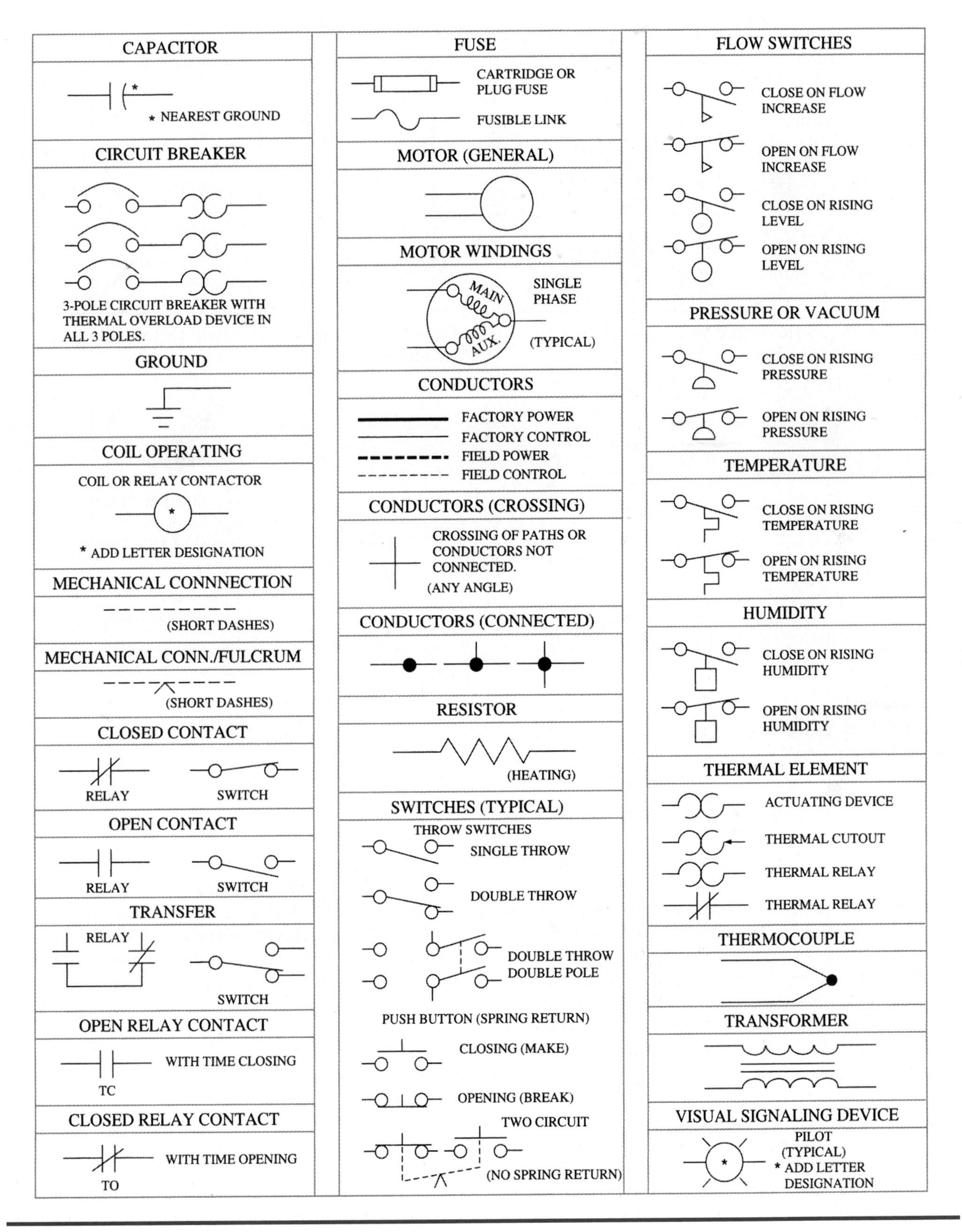

Figure 8-1-23 Recommended electric diagram symbols.

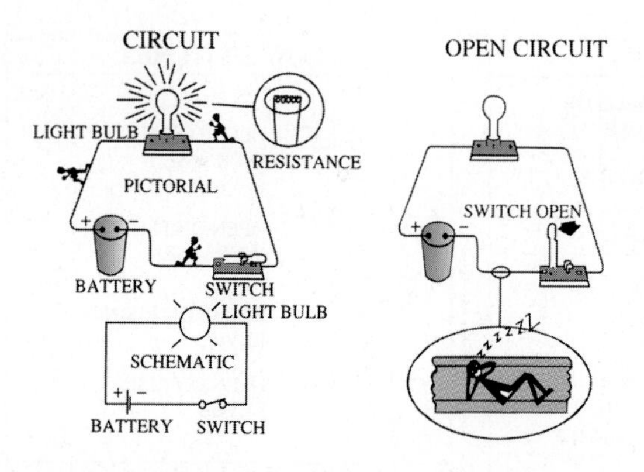

Figure 8-1-24 Simple electric circuit with battery power, a switch, and a lamp.

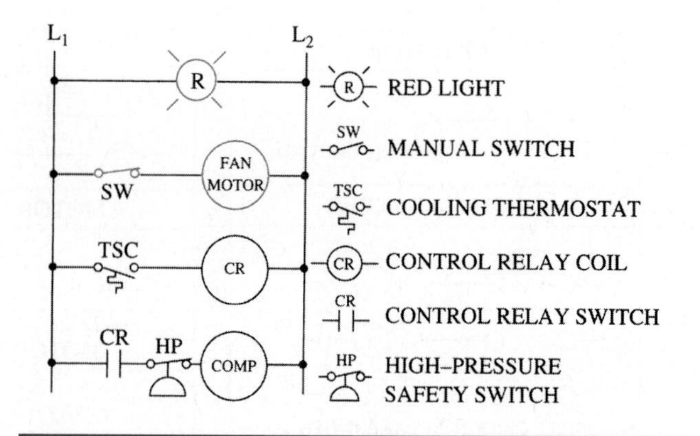

Figure 8-1-26 When the manual switch is closed, the fan motor runs.

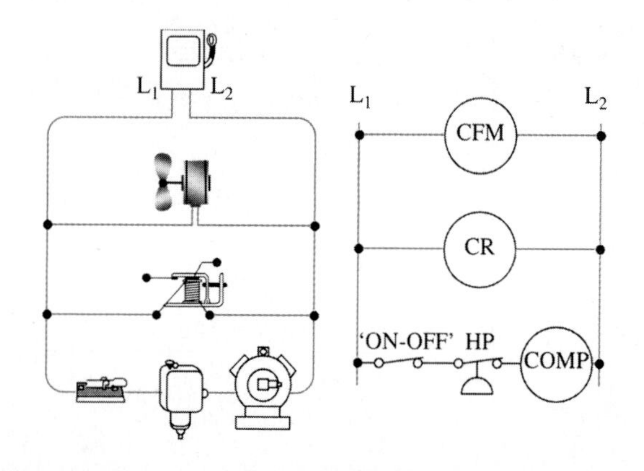

Figure 8-1-25 Ladder diagram showing a power supply and three loads.

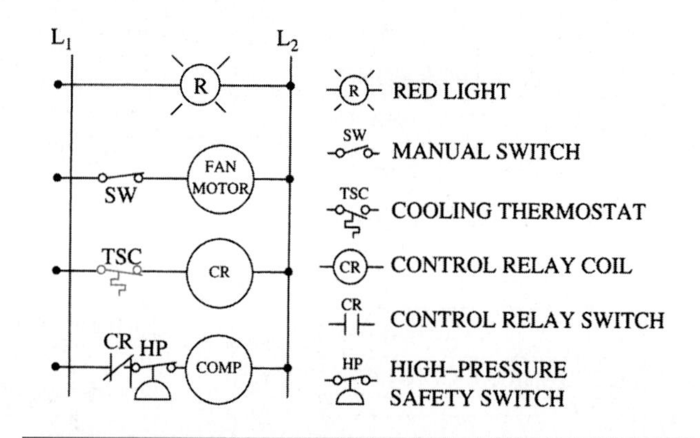

Figure 8-1-27 Schematic diagram showing thermostat calling for cooling.

system. A manual switch has been added in series with the fan to turn it on. A thermostat switch for cooling has been added in series with the control relay (CR).

Note that the symbol for the thermostat shows that it is a cooling control since it closes on a rise in temperature.

It is also interesting to note that the control relay (CR) with its coils in circuit 3 has one set of NO contacts in circuit 4. These contacts (also indicated as CR) when closed will start the compressor, provided that the on-off switch is closed.

The Sequence of Operation

From the ladder diagram in Figure 8-1-26, we can determine the sequence of operation of the system, that is, the order in which the loads are energized.

Referring to Figure 8-1-26, when power is supplied to L_1 and L_2 the red light comes on, verifying the power supply. Then the fan switch (SW) is closed manually, starting the fan motor.

In Figure 8-1-27, the cooling thermostat (temperature setting for cooling, or TSC) closes, indicating a call for cooling, energizing the control relay (CR) coil. When this occurs, the NO relay switch (also indicated as CR) in circuit 4 closes, starting the compressor (COMP) motor. The high pressure (HP) safety switch, also in series with the compressor motor, remains closed since it only opens on a malfunction.

Drawing a Schematic

We will now examine a more complex system and construct a schematic. Then using the information from the schematic we will determine the sequence of operations for the system.

Figure 8-1-28 shows a connection diagram for an air cooled condensing unit. A condensing unit is a packaged unit consisting primarily of a compressor, an air cooled condenser, and a control system. The control system consists of a wiring panel enclosing (1) the compressor starter (in the upper center of the diagram), (2) the start relay (SR in the

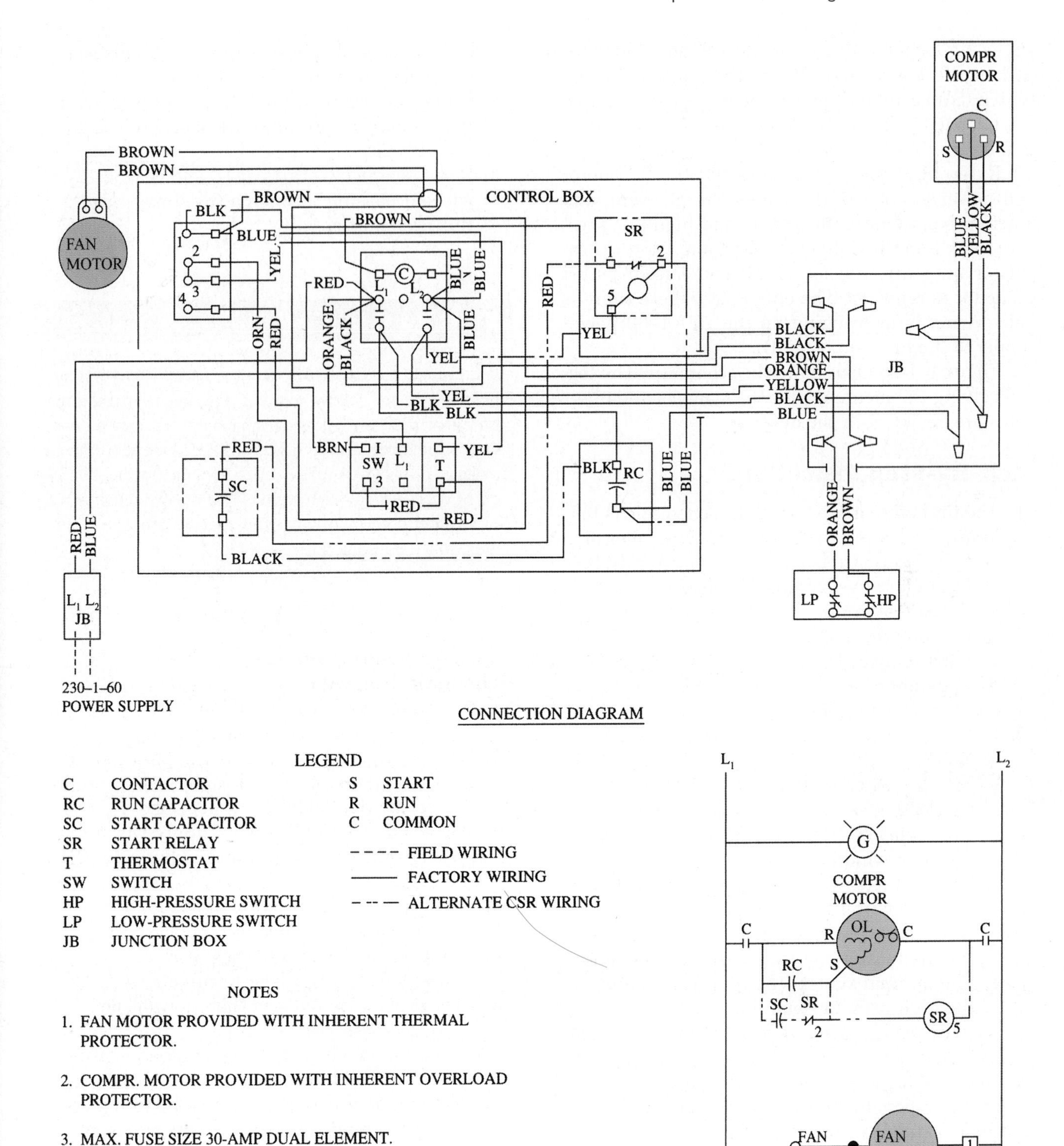

Figure 8-1-28 Diagrams for an air conditioning unit showing both the connection and schematic.

upper right of the diagram), (3) the run capacitor (RC in the lower right), (4) a thermostat (T) and switch (SW) combination (in the lower center), (5) a start capacitor (SC in the lower left), and (6) a junction box (in the lower left).

External to the wiring panel are the fan motor and power supply on the left of the diagram, and junction box, compressor motor, and high/low pressure control on the right of the diagram. The start capacitor and start relay are for alternate use if a start capacitor is required. The only place where there is field wiring is in bringing up the 230–1–60 power supply to the junction box (JB).

Figure 8-1-28 also shows the schematic. The legends giving the meanings for the symbols and other notes are shown with the diagrams.

Step-By-Step Procedure

1. The first step in preparing the schematic is to locate the loads and determine the number of circuits. There are five loads in Figure 8-1-28, and therefore five circuits. The loads are:
   a. Green test lamp
   b. Compressor motor
   c. Start relay coil
   d. Fan motor
   e. Contactor for compressor
2. Locate the switches in each circuit.
   a. None
   b. Two contactor switches and compressor overload in series with the run winding, and start relay (SR) switch (NC) in series with the start winding.
   c. Two contactor switches (same as in b above) in series with the start relay coil.
   d. One manual switch, or the manual switch and the thermostat, in series with the fan motor.
   e. The manual switch, the thermostat, and the high pressure and low pressure switches in series with the compressor contactor coil.
3. Draw each circuit, showing the position of each component. In the case of switches, show the normal position. The result of these three steps is shown in Figure 8-1-28 in the schematic.
4. Describe the sequence of operation for the system. This will be accomplished using the information shown in the schematic to trace the circuit, as described below.

Methods of Tracing Circuits

Three conditions can occur in tracing circuits:

1. Start at one side of the power supply (L_1), go through the resistance (load), and return to the other side of the power supply (L_2). This is a complete circuit.
2. If the technician starts at L_1 and goes through the resistance and cannot reach L_2, this is an open circuit.
3. If the technician starts at L_1 and reaches L_2 without passing through a resistance (load), this is a short circuit.

SERVICE TIP

It is sometimes easier for new technicians to physically hold a wire and follow it in the equipment at the same time that they are following the diagram. It may be necessary for more than one person to do this at a time to make it possible to follow both simultaneously. But by doing do, it will be much easier for the new technician to locate and determine what the specific problem is and what might be causing it.

Rules for Making Vertical Schematic Diagrams

1. Assign letters in the legend for each symbol to represent the names of the components.
2. When using a 120 V AC power supply, show the hot line (L_1) on the left side and the neutral line (L_2) on the right side of the diagram.
3. When using a 120 V power supply, the switches must be placed on the hot side (L_1) of the load.
4. Relay coils and their switches should be marked with the same (matched) symbol letters.
5. Numbers can be used to show wiring connections to controls or terminals.
6. Always show switches in their normal, de-energized position.
7. Thermostats with switching sub-bases, primary controls for oil burners, and other packaged electrical devices, can be shown by terminals only in the main diagram. Subdiagrams are used when necessary to show the internal control circuits.
8. It is common practice to start the diagram showing line voltage circuits first and low voltage (control) circuits second.

Equipment Wiring Diagrams

Most manufacturers supply comprehensive electrical circuit diagrams with the equipment when it is shipped. In many cases this is the only electrical information available to assist the technician

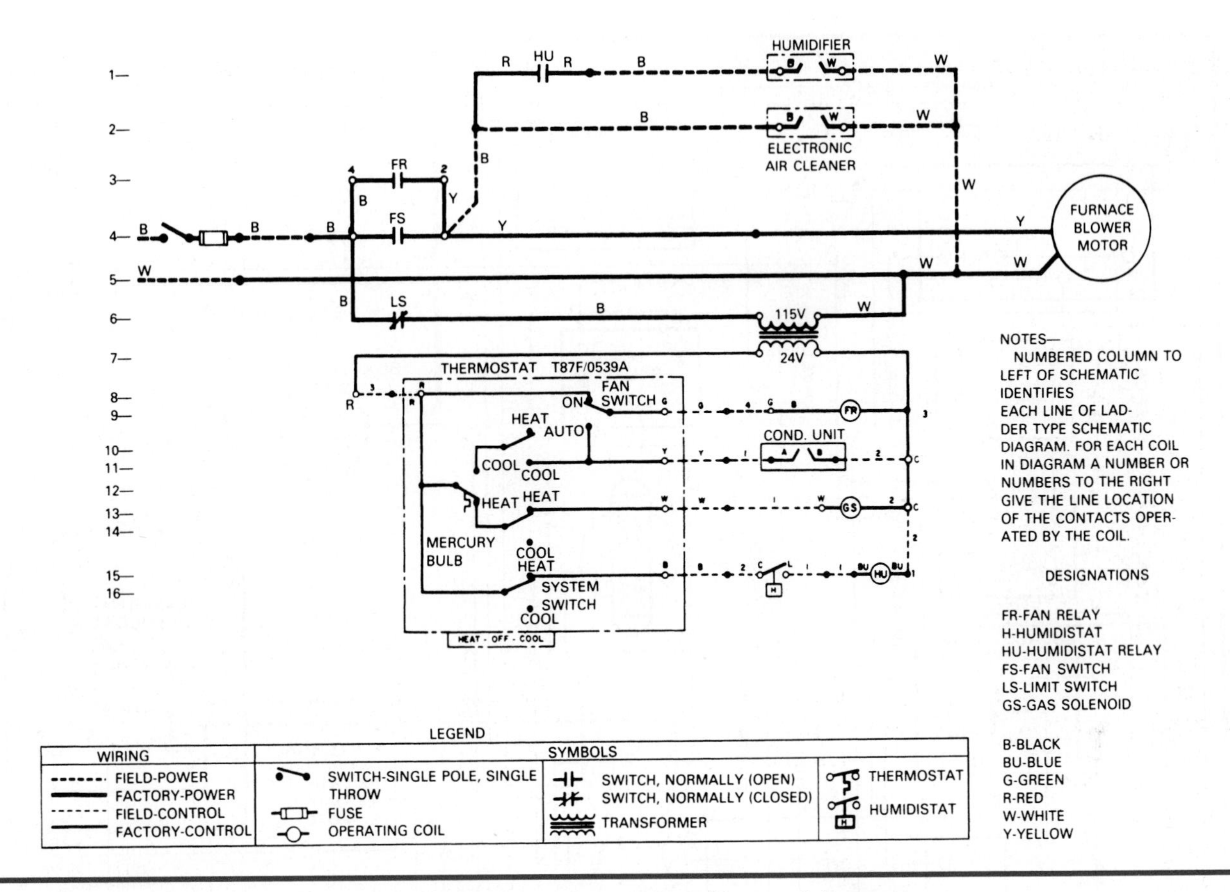

Figure 8-1-29 Label diagram comes with the equipment, pasted inside one of the panels.

installing and servicing the unit. It is important that this material be retained on the job for immediate and future reference.

Here are some to the types of diagrams that the technician is likely to find.

Label Diagrams A typical label diagram is shown in Figure 8-1-29. This is usually printed on peel-and-stick paper and attached to the unit in some convenient place, such as inside the control box cover or door of the equipment. It includes a component arrangement diagram, a wiring diagram, a legend, and notes.

The part of the drawing indicated as the component arrangement diagram has been highlighted and enlarged in Figure 8-1-30. This figure shows the actual location of the components inside the unit. In troubleshooting, a schematic drawing is helpful in determining the function(s) not performing. To test the circuits, however, the component arrangement diagram is absolutely essential for locating actual parts and connections within the unit itself. The component arrangement diagram is designed to help the technician in this respect.

The wiring diagram portion of the label diagram is shown in Figure 8-1-30. This drawing shows the actual internal wiring of the unit. It may show the color code for the wires used and the terminals where they are connected. It can be considered a schematic, however, since the components are not in their exact location. This liberty was taken in making the drawing to be able to adequately show all the necessary connections, some of which would be hidden from view on the actual unit.

The label wiring diagram is usually, but not always, organized showing the high voltage circuits in the upper portion of the diagram, as shown in Figure 8-1-21, and the control circuits (low voltage) in the lower portion of the diagram, as shown in Figure 8-1-22.

Referring to high voltage circuits (Figure 8-1-21) the primary power is shown in the upper left and the load circuits in the upper right.

Referring to the low voltage circuits (Figure 8-1-22) they almost always operate from a stepdown transformer. Note there are two elements of the control circuits. In the lower section are the contactor coils and switches that control them. In the high voltage section are the contacts for the contactors.

The legend and notes on label diagrams are important. The legend provides the meanings of

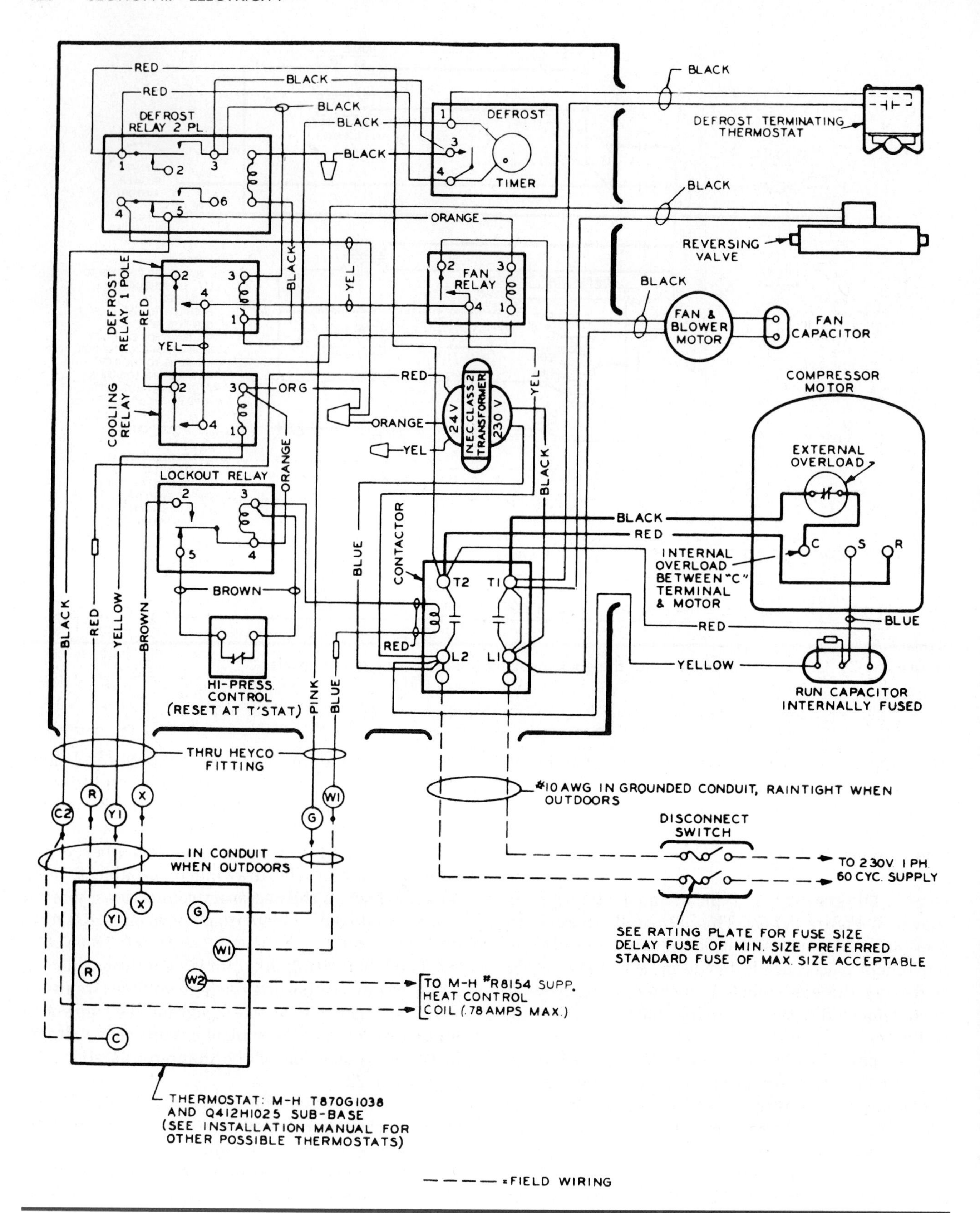

Figure 8-1-30 Component arrangement enlarged from label diagram.

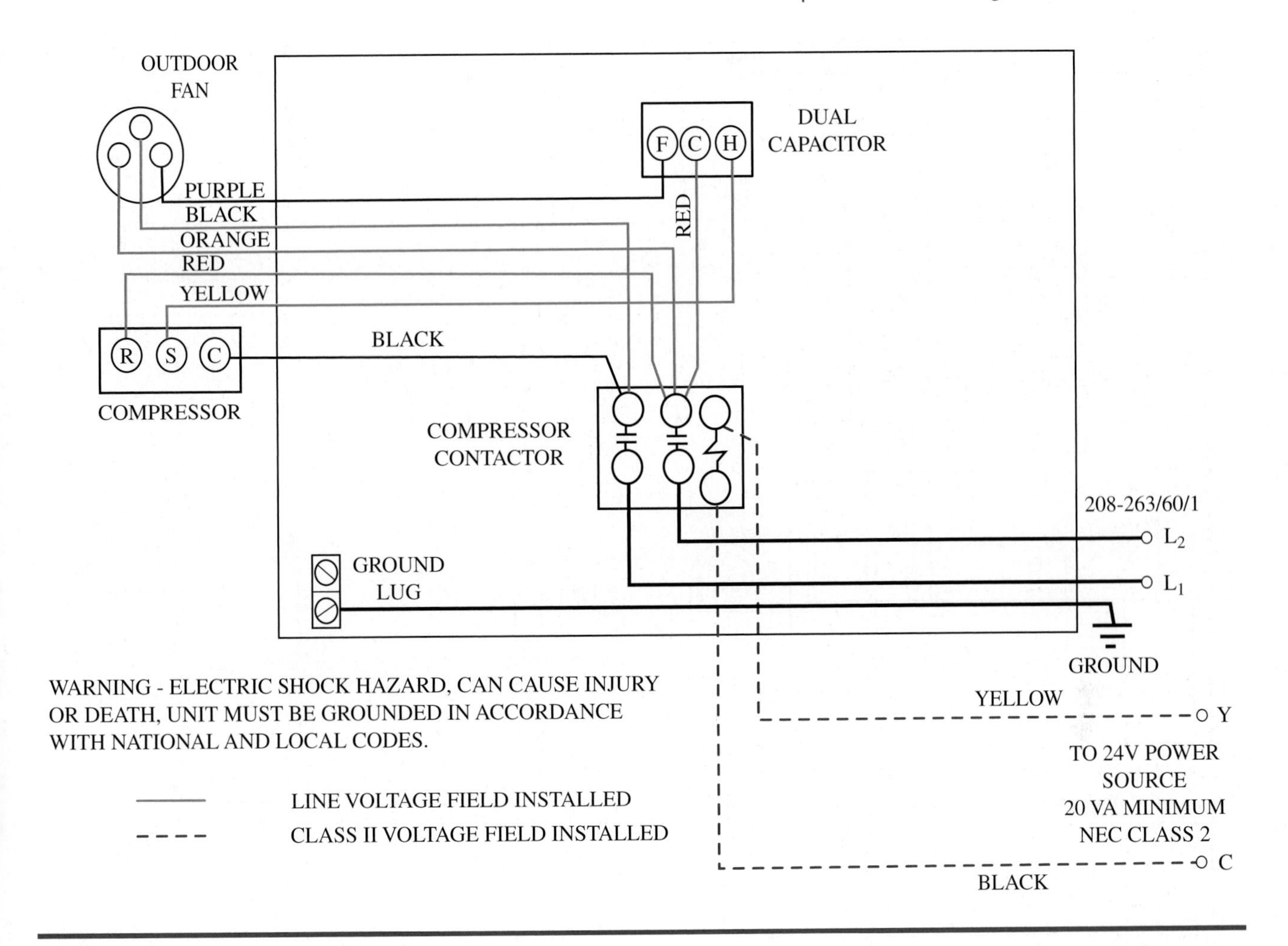

Figure 8-1-31 Installation wiring diagram.

abbreviations. The notes provide useful information that cannot be supplied elsewhere.

Installation Wiring Diagrams An installation wiring diagram (Figure 8-1-31) shows how to connect power from the building supply to the unit. It also shows unit-to-unit wiring, if applicable, as well as wiring to a remote-control device such as a thermostat. This drawing does not show the internal wiring unless it is necessary for the installation.

Connection Wiring Diagrams The wiring on connection diagrams is shown in a number of ways:

1. Pictorial. On these diagrams all of the components are shown in pictorial form in their actual location and the wiring is shown as it is actually used. A good illustration of this technique is shown in Figure 8-1-32.

2. Wiring Harness. Using this method, groups of wires are placed in harnesses, with individual wires connected to each component. This technique is illustrated in Figure 8-1-33.

3. Terminal Numbers. Using this method, each terminal is given a number. Few wires are shown. To find where wires are placed, the numbers are matched. This technique is illustrated in Figure 8-1-34. An increasing number of manufacturers are using this method. The absence of many wires on the diagram makes it simpler and easier to read.

For tracing the control circuit to understand how the circuit is wired and operates, the ladder diagram is still one of the easiest diagrams to follow. If the technician is having difficulty understanding a schematic or label diagram, it can be worthwhile redrawing it as a ladder diagram to more clearly understand the circuit.

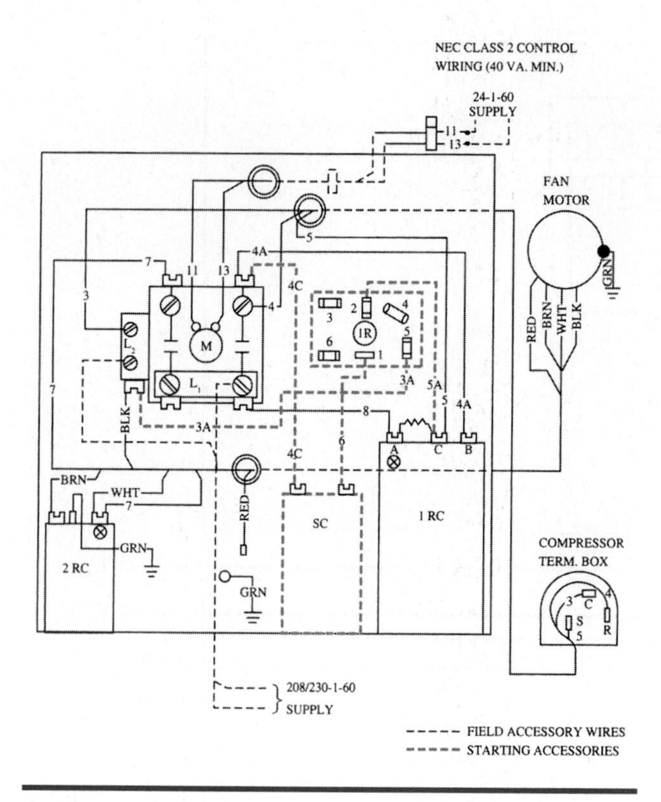

Figure 8-1-32 Connection or label wiring diagram. *(Courtesy of York International Corp.)*

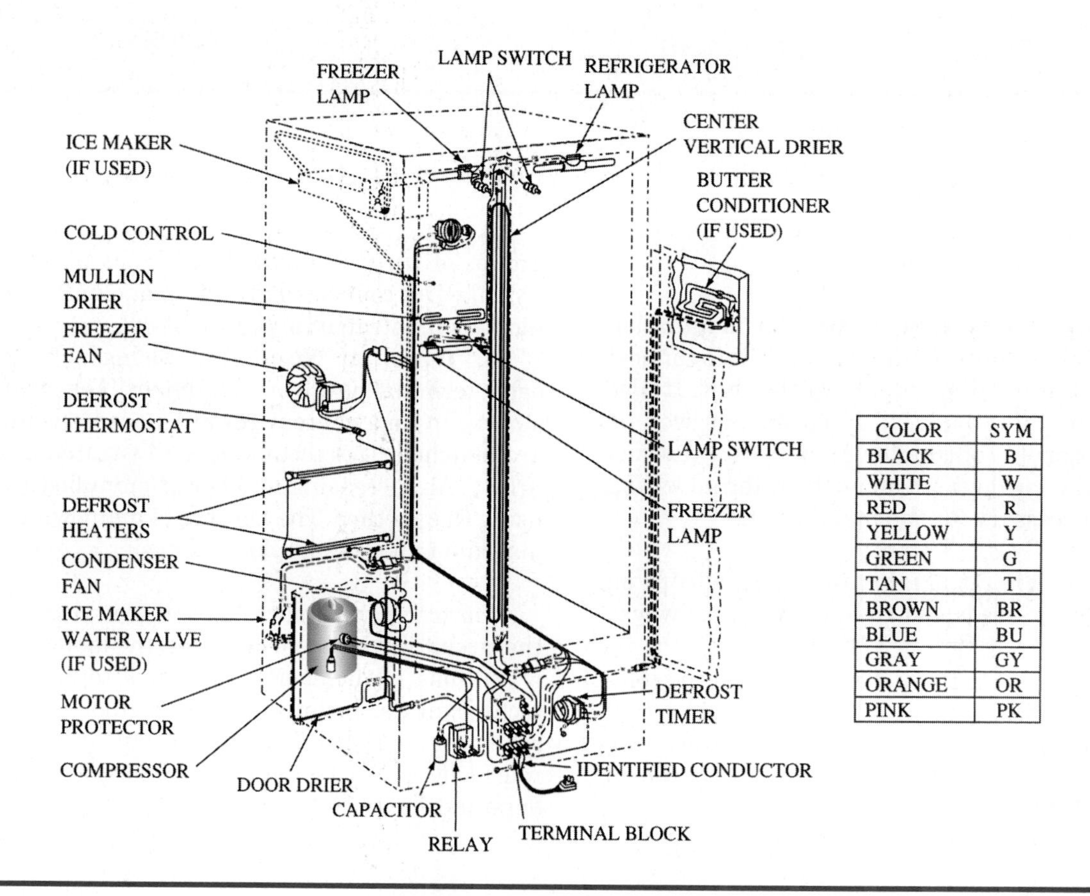

COLOR	SYM
BLACK	B
WHITE	W
RED	R
YELLOW	Y
GREEN	G
TAN	T
BROWN	BR
BLUE	BU
GRAY	GY
ORANGE	OR
PINK	PK

Figure 8-1-33 Wiring harness type wiring diagram.

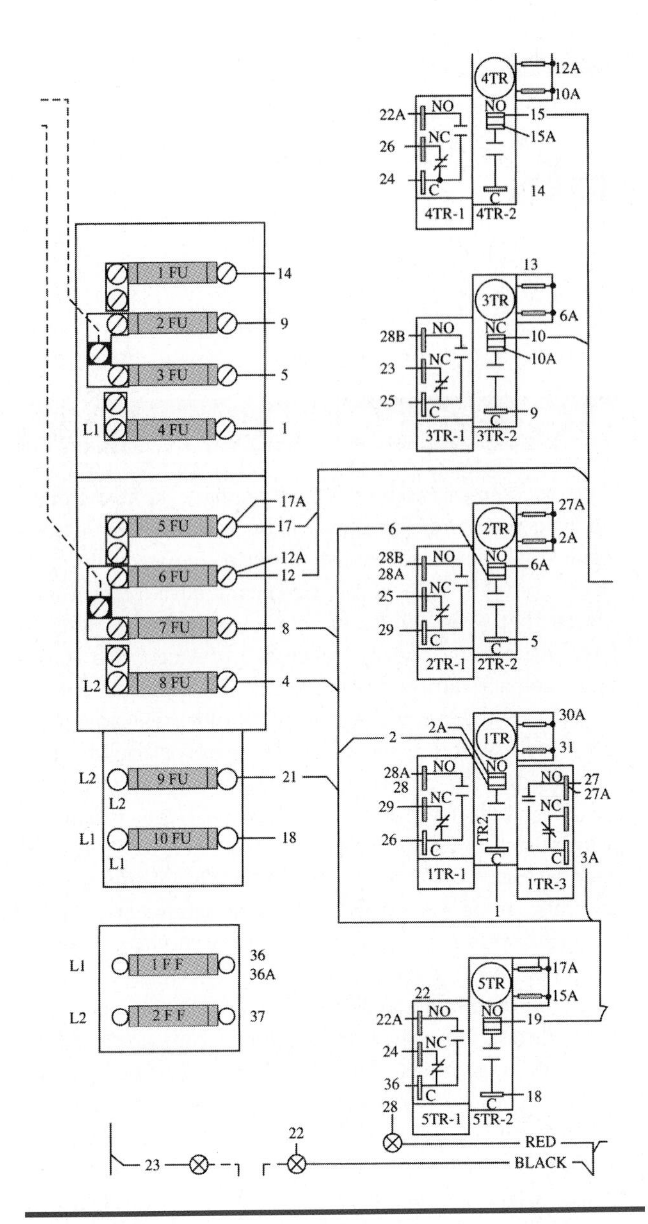

Figure 8-1-34 Terminal number type wiring diagram. *(Courtesy of York International Corp.)*

TECH TALK

Tech 1. I look at these wiring diagrams, and they just look like a bowl of spaghetti. I have no idea how you figure them out.

Tech 2. Well, the easiest thing to do when you first get started is to make a copy of the diagram. You can draw it on a piece of paper, or take it and have it photocopied, and then use colored pencils to follow the lines. Use one color of pencil on the L_1 side and the other colored pencil on L_2. This way you can see what the voltage is going to be between the different parts and where the voltage should be.

Tech 1. Doesn't that take a long time?

Tech 2. It does take a long time when you first start; but the more you practice, the easier it is; and the more you practice, the faster you will be at finding the problems.

Tech 1. Why don't I just draw on the diagram everything that's in the equipment so the next guy knows what's going on?

Tech 2. That's considered to be a poor practice because there are lots of lines and your color may be confusing to the next technician. They might be working on a completely different problem.

Tech 1. Oh, I see. I'll give it a try. Thanks for the help.

UNIT 1—REVIEW QUESTIONS

1. Define induction.
2. A ________ is a special electrical conductor that is placed in series with a load and melts when excessive current flows through it, breaking the circuit.
3. What is the advantage of a circuit breaker over a fuse?
4. The whole purpose of the ________ is to supply power to the electrical devices in combination with a control sequence to produce a desired output.
5. What is the difference between a relay and a contactor?
6. List the three types of wiring diagrams.
7. What are the categories that wiring diagram symbols fall into?
8. What does a schematic wiring diagram consist of?
9. What is the first step in preparing the schematic?
10. What is one of the three conditions that can occur in tracing circuits?
11. What is included on a typical label diagram?
12. List the three ways the wiring on connection diagrams is shown.
13. What is the easiest diagram to follow for tracing the control to understand how the circuit is wired and operates?

UNIT 2

Controls

OBJECTIVES: In this unit the reader will learn about controls including:

- Selecting the Control System
- Types of Basic Controls
- Operating Controls
- Metering Devices
- Relays, Contactors, and Starters
- Safety Controls

NICE TO KNOW

Many technicians assume that pneumatic controls are old fashioned, and that they will be replaced with electronic or electromechanical devices. However, pneumatic controls have some significant advantages that ensure that they will be in active use for many years. They are a zero energy consuming device once they have been activated or deactivated. Unless air pressure leaks from the controls they will hold their position indefinitely without the need for external power as compared to all of the other electronic and electromechanical devices that must be energized at some point during their operation and may need to be energized virtually 100% of the time. In addition, pneumatic controls can be more easily proportionately controlled ½ open, ¾ open, or full open, by simply modulating the air pressure. Electromechanical devices however are simply on or off and it takes a more sophisticated electronic device to provide proportional control for many mechanical functions. And the third major advantage pneumatic controls have is that by changing the building's central air pressure all of the thermostats can be set back for evening and weekend functions without having additional control wires to each thermostat as would be the case for the other control devices. It is these advantages that have kept pneumatic controls as a viable control system for commercial and light commercial applications.

DEFINITION OF CONTROLS

All of the controls described in this chapter can be termed *system controls*. This refers to their function, which is to start, stop, regulate, and protect the refrigeration cycle and its components. *Basic controls* are those that start and stop the equipment. *Operating controls* regulate processes, while those that protect the cycle are called *safety controls*.

The major components of the cycle can be modified and regulated to form a wide variety of applications. For example, if the application is a drinking-water cooler, a basic control (thermostat) starts and stops the refrigeration unit to maintain a desirable drinking water temperature. A safety control (motor overload) could be used to protect the compressor motor from damage from excess current conditions.

Electric and electronic controls are operated by electricity and are connected by wires. *Electronic controls* differ from *electrical controls* in that they use semiconductor materials and solid state controls. *Pneumatic controls* are nonelectric and are operated by air pressure.

There are three types of materials used in electrical control systems:

1. Conductors such as copper, aluminum, and iron.
2. Semiconductors, such as metal oxides and metal compounds.
3. Nonconductors, such as glass, plastic, rubber, and wood.

Conductors permit the free flow of electrons and offer the minimum amount of resistance to the flow of electric current.

Semiconductors restrict the flow of electrons and permit the limited flow of electricity under certain conditions. They perform unique electrical functions that are useful in solid state electronic systems. A separate section in this text describes their applications.

Nonconductors do not permit the flow of electrical energy and are used as insulators.

SELECTING THE CONTROL SYSTEM

The type of controls selected for any system depends on many factors: the age of the equipment, the size of the equipment, the location of the components, the relative cost, and even the designer's preference. Older units in the small and medium size class almost all use electrical controls. Larger HVAC/R systems use pneumatic controls when the cost of the air compressor can be justified. There is a preference for electronic controls due to their low cost, accuracy, and the extra performance features they are capable of executing.

Regardless of the type of controls, the function performed can be the same. A different type of control can perform the same function in a different way.

One of the big advantages of pneumatic control systems is their ability to modulate a damper or a valve to produce an infinite number of positions between fully open and fully closed. This makes it possible to provide close control of temperature and humidity where required.

The availability of microprocessor chips has greatly increased the popularity of solid state controls. A tiny chip contains a whole system of controls which replace complete circuits that formerly used electromechanical devices. Further, these chips are energy efficient and consume far less energy than the replaced devices.

A technician who is planning to service a wide variety of refrigeration and air conditioning equipment must be knowledgeable about electrical and electronic control systems. Both types of control are commonly used in the small and medium size jobs that constitute the largest part of the market. In this text the emphasis will be on these systems.

TYPES OF BASIC CONTROLS

There are four types of basic controls. The first is operated by temperature and is called a *thermostat*. The second is operated by pressure and is called a *pressurestat*. The third is operated by moisture or humidity and is called a *humidistat*. The fourth is operated by time and is called a *switch* or *time clock*. Each of these controls can be used to regulate cycle operation. For example, when a home refrigerator becomes too warm, that is, the temperature becomes too high for food storage, the thermostat senses this and starts the compressor.

Pressurestats Pressurestats are often used to control temperature conditions in a display case by controlling evaporator pressure. When the evaporator pressure and corresponding temperature become high, the pressurestat operates and starts the compressor. In storage rooms where humidity is all-important, the humidistat is designed to start the refrigeration cycle when the humidity rises to a predetermined level. Conversely, all the above controls will stop cycle operations when the conditions are satisfied.

Some systems are operated by a time clock, particularly where energy conservation is concerned.

Thermostats Thermostats respond to temperature. They can do this because of the warping effect of a bimetallic strip or because of fluid pressure.

Bimetallic Elements

The thermostat illustrated in Figure 8-2-1 is a bimetallic thermostat. The bimetallic element is composed of two different metals bonded together. As the temperature surrounding the element changes, the metals will expand or contract. As they are dissimilar metals, having different coefficients of expansion, one will expand or contract faster than the other. In the illustration, the open contracts are on the top. No current is flowing. If the temperature surrounding the bimetallic element is raised, metals A and B will both start to expand. However, metal A expands faster than metal B. This will cause the bimetallic strip to bend and close the contacts as shown on the bottom. As the temperature drops, A would contract faster than B, thereby straightening the element and opening the contacts.

This thermostat is widely used, especially in residential heating and cooling units and in refrigeration systems that work above freezing conditions, where the stat is actually located in the controlled

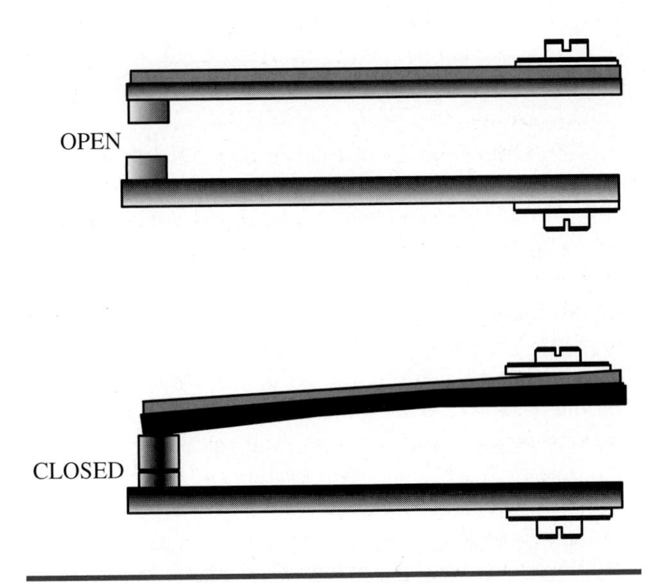

Figure 8-2-1 Bimetal thermostat.

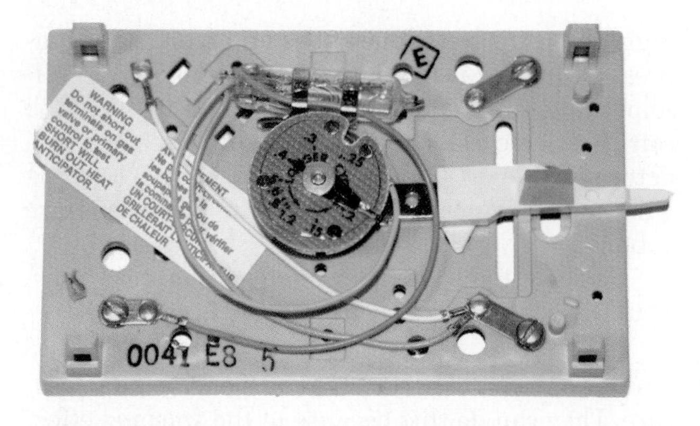

Figure 8-2-2 Mercury bulb thermostat.

space. It is simple and inexpensive to construct, yet dependable and easily serviced.

One of the common variations of the bimetallic thermostat is a mercury bulb thermostat, as shown in Figure 8-2-2.

Although the bimetallic principle is used, the contacts are enclosed in an airtight glass bulb containing a small amount of mercury as shown. The action of the metallic element tilts the bulb. By tilting the bulb to the left, mercury in the bulb will roll to the left and complete the electrical circuit. By tilting the bulb to the right, the mercury will flow to the end of the bulb and break the electrical circuit. Extra flexible insulated leads connect the mercury bulb to the circuit without interfering with the bulb's movement. This thermostat is used extensively in space heating and cooling. Although a little more expensive than the bimetallic open contact thermostat, its operation is more dependable because dirt cannot collect on the contacts.

TECH TIP

In some commercial and light commercial applications a placebo thermostat may be mounted in the occupied space. This thermostat can be adjusted by the occupants of the space without having an effect on the space temperature. Placebo thermostats have been used successfully to answer the need of some occupants who insist on having the control of their space. In some cases the placebo stat may activate the system fan without turning on the heating or cooling system. The actual thermostat in this case will be located in the return air plenum so that it monitors the return air temperature and maintains it at the preset space temperature. Another option customers use is an operating range thermostat that is used to limit the lowest and highest temperature that an occupant can control regardless of the settings on the thermostat.

Fluid Pressure

The second method of thermostat control is the fluid pressure type shown in Figure 8-2-3.

With a liquid and gas in the bulb, the pressure against the pressure diaphragm will increase or decrease as the temperature at the bulb varies. The thermostat illustrated is a heating thermostat. As the pressure in the bulb increases with the rise in bulb temperature, the pressure diaphragm expands and, through a mechanical linkage, opens the electrical contacts.

As the pressure at the diaphragm decreases with a lowering of bulb temperature, the diaphragm contracts and closes the electrical contacts. This thermostat is sometimes referred to as a remote bulb thermostat or temperature control thermostat. The controlling bulb can be placed in a location other than that used for the operating switch mechanism. For example, with a cooling thermostat in a refrigerated room, the bulb can be placed inside the room and the capillary tube run through the wall to the operating switch mechanism located outside the room. Thermostat adjustments can be made without entering the refrigerated room, and the switch mechanism is not exposed to the extreme conditions inside the room.

Another version of the pressure actuated thermostat is the diaphragm type, shown in Figure 8-2-4.

In this control the diaphragm is completely filled with liquid. The liquid expands and contracts with a change in temperature. The movement of the diaphragm is very slight, but the pressure that can be exerted is tremendous. Because the coefficient of expansion of liquid is small, relatively large volume bulbs are used on the control. This results in sufficient diaphragm movement and also ensures positive control from the bulb.

When installing a remote bulb temperature controller, make absolutely sure the maximum bulb temperature will not be exceeded during any part of the on or off cycle.

Using a subatmospheric fill control will let air in if a leak occurs. The rise in pressure will cause the control to shut off, thus it is a fail safe mechanism.

Set Point Adjustment

Figure 8-2-5 shows a dial arrangement for selecting the set point on a single bulb mercury tube thermostat. The dial is rotated until the indicator is placed at the setting. Note that this thermostat has other settings: a cool-off-heat system switch and a fan-on-auto switch. These switches are part of the subbase that is used in mounting the thermostat.

Figure 8-2-6 shows a heating-cooling thermostat, with two levers at the top for selecting the set

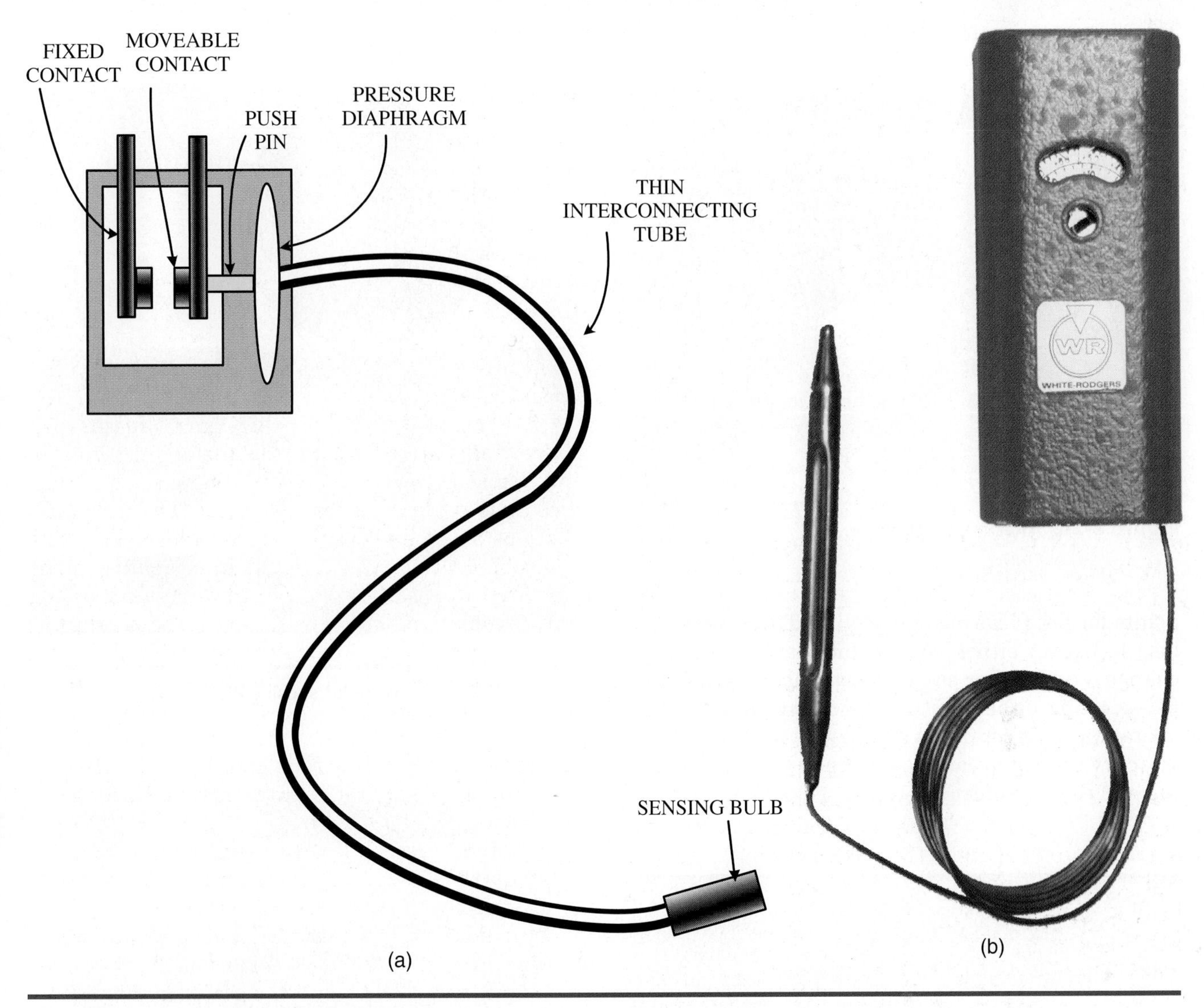

Figure 8-2-3 (a) Diagram of a fluid pressure thermostat; (b) Photo of a fluid pressure thermostat. (*Courtesy Hampden Engineering Corporation*)

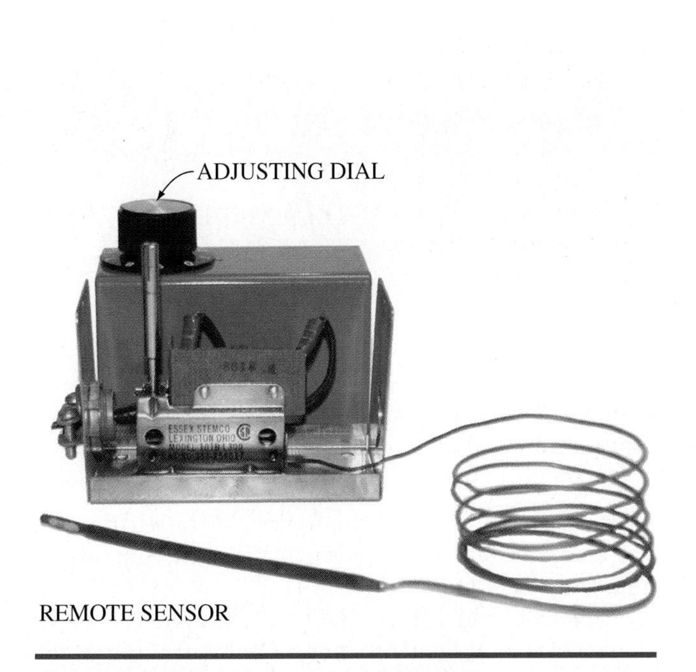

Figure 8-2-4 Diaphragm type thermostat.

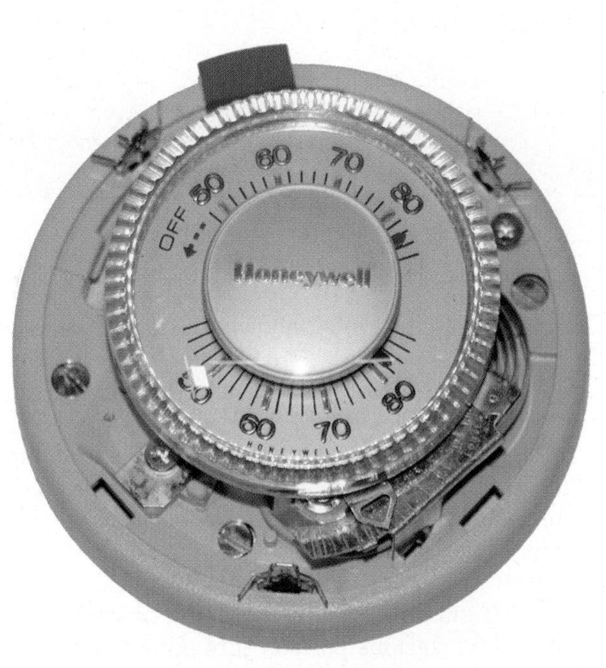

Figure 8-2-5 Thermostat.

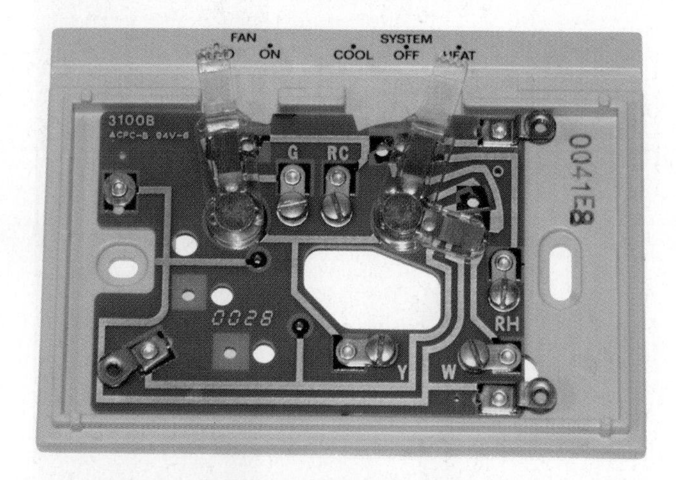

Figure 8-2-6 Thermostat subbase.

points for both heating and cooling. The system can also be set to either supply automatic or manual changeover. Another switch is available on the base for selecting either continuous or intermittent fan operation. This control is available with two to four mercury bulbs and can provide up to two stage heating and cooling controls.

> **CAUTION:** Mercury is considered to be a hazardous chemical and must be disposed of in an appropriate manner. You may not simply discard in the trash old mercury filled thermostat bulbs. They must be returned to an approved recycle center that is designated for that purpose. Failure to properly dispose of mercury filled thermostat bulbs can result in fines for you and your company. If a mercury thermostat bulb is dropped and broken on a job site the building owner must be notified so they can have the area decontaminated by an authorized hazardous materials cleanup crew. Mercury and mercury vapors have been associated with a number of serious health issues. It is not a material to be played with.

Electronic Thermostats

The inset in Figure 8-2-7 shows a common sensing element for an electronic thermostat called a thermistor. The thermistor has a resistance element that is affected by temperature. Using an electronic circuit, the thermostat can be set so that when the temperature set point is reached, a switch is opened, turning off the power to the fuel operating service such as a gas valve.

Electronic programmable thermostats such as shown in Figure 8-2-8a,b can be set for different temperatures for various periods throughout the week. These settings will take effect automatically. The great advantage of the electronic thermostat is its

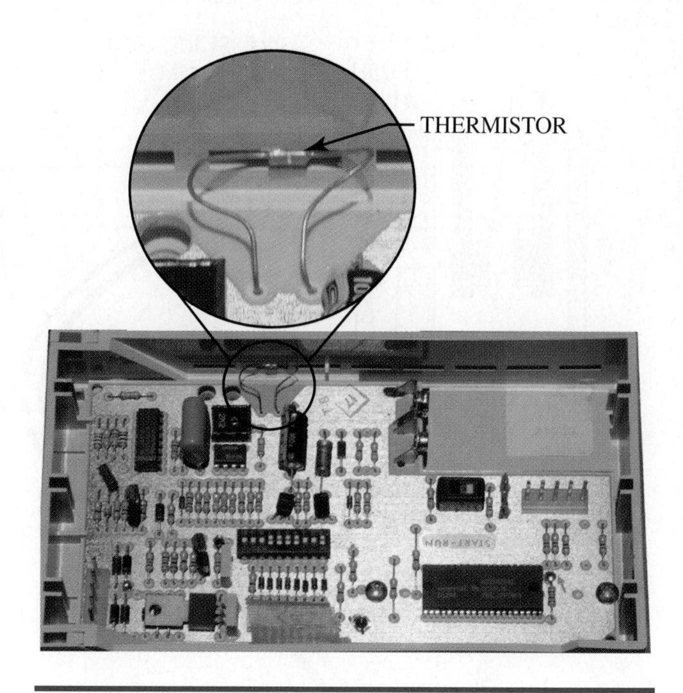

Figure 8-2-7 Thermostat sensing element.

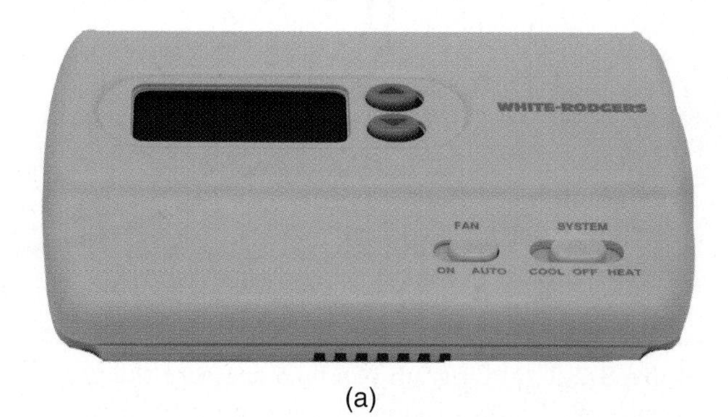

(a)

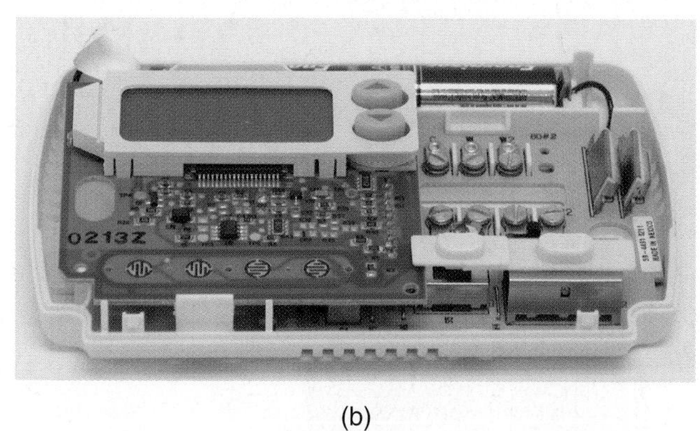

(b)

Figure 8-2-8 Programmable thermostat: (a) Cover appearance; (b) With cover removed.

accuracy. It will control the temperature within 1°F. For comparison, the bimetal thermostat controls to an accuracy of about 2°F.

Setback and *setup* are two common terms in thermostats. A heating thermostat with night (or unoccupied) setback will automatically reduce the

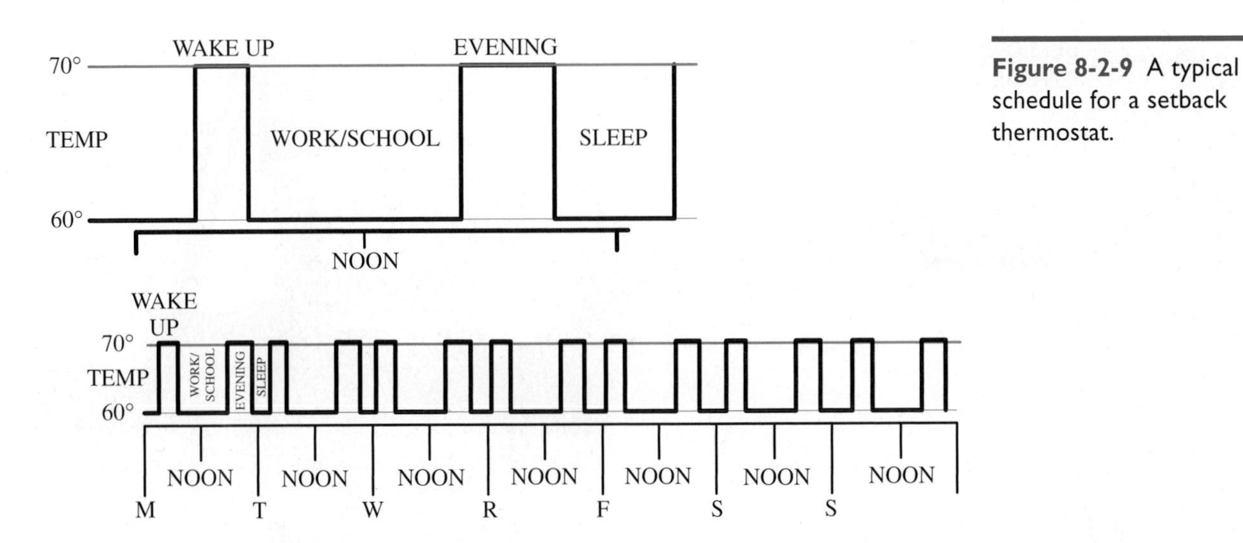

Figure 8-2-9 A typical schedule for a setback thermostat.

control setpoint during preset periods when lower than normal temperatures are acceptable. A cooling thermostat will have setup to raise the setpoint during scheduled hours. The intent of setup and setback is to reduce cooling and heating energy usage. A typical schedule for setback/setup thermostat times is shown in Figure 8-2-9.

Pressurestats

Pressure controls can also be divided into two categories: the bellows type and the Bourdon tube type. The bellows type is illustrated in Figure 8-2-10.

The bellows is connected directly into the refrigerant system through a capillary tube. As the pressure within the system changes, so does the pressure within the bellows, causing the bellows to move in and out in conjunction with pressure variation. As shown in Figure 8-2-11, the electrical connections are broken as the pressure falls.

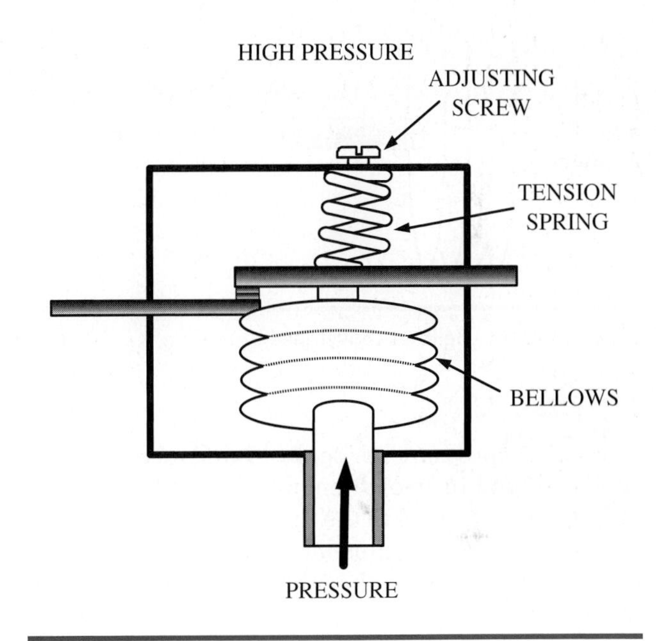

Figure 8-2-10 Typical components in a bellows type sensor.

SERVICE TIP

Many pressurestats have small orifices designed to sense the pressure differential. Unfortunately in many areas of the country there are various insects that build a mud nest in such openings. If the opening is not protected it may be subject to such insect activity. If you are having difficulty with pressure readings, use a thin wire to make certain the orifices are clean. If you find the orifices are covered, remove the insect nest. The end of the orifice can be protected with a fine wire screen to keep further insect activity from occurring.

These controls are available with normally open (NO) and normally closed (NC) and single pole double throw (SPDT) switching. Depending on the control pressure range, this type of control can be used as a low pressure control or a high pressure control. That is, it can be connected into the high or low side of the system. Because of its simplicity, dependability, and adaptability, this control is found on almost every air conditioning or refrigeration system. This type of control is found on many commercial air conditioning and refrigeration systems.

The low pressure control, shown in Figure 8-2-12a, is very similar in design to the high limit pressurestat shown in Figure 8-2-12b. The difference is that this control is used for low limit operation, and the contacts break on a drop in temperature below the setpoint.

The Bourdon tube pressure control control is ideally suited for mercury bulb operation and is frequently found in applications requiring enclosed

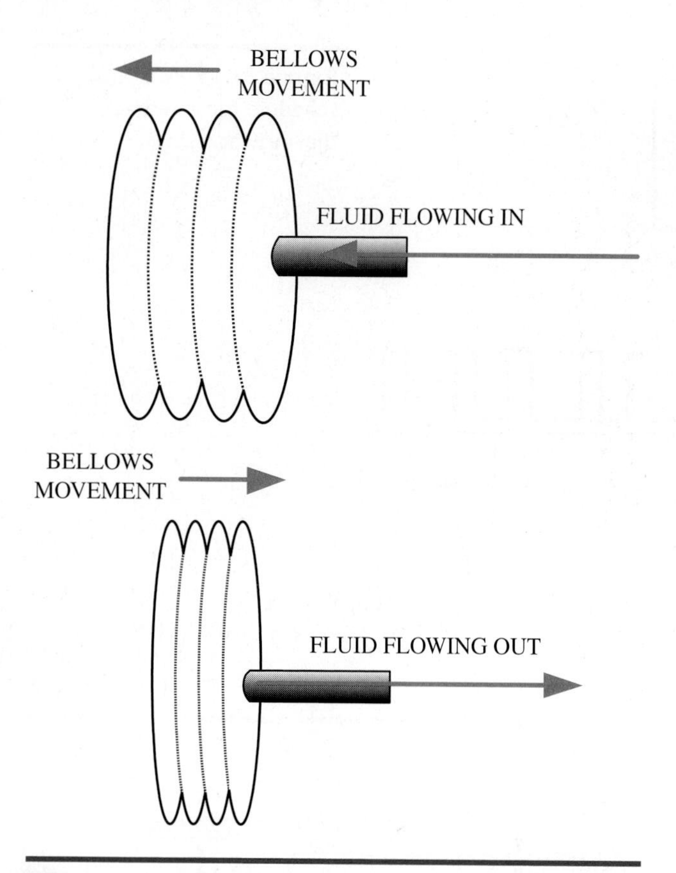

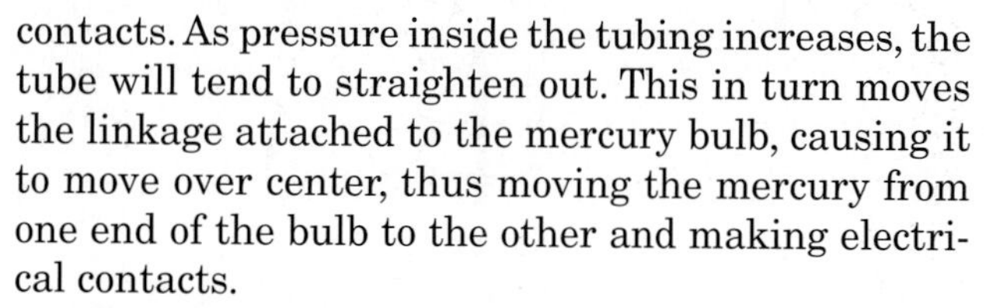

Figure 8-2-11 Bellows movement.

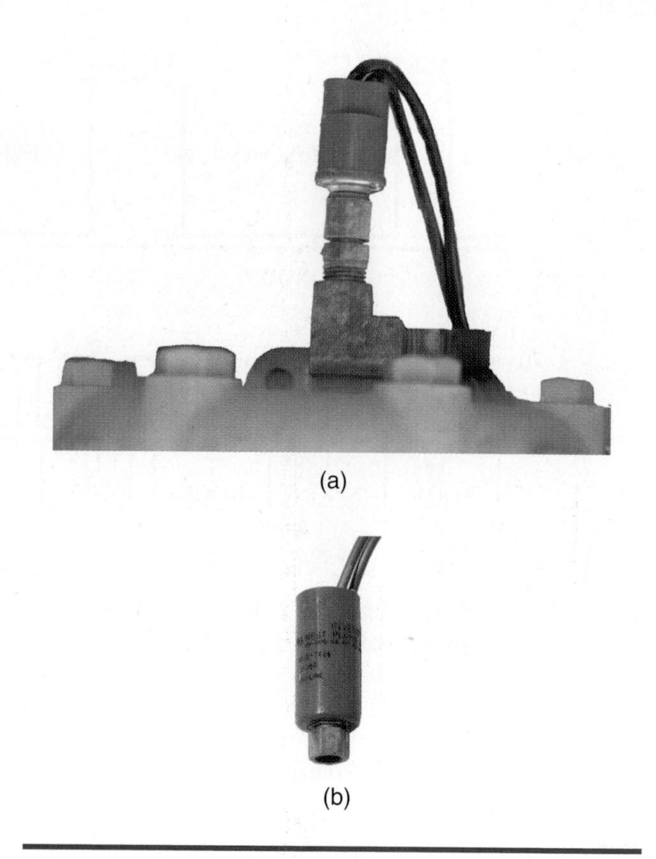

Figure 8-2-12 (a) Low pressure switch; (b) High pressure switch.

contacts. As pressure inside the tubing increases, the tube will tend to straighten out. This in turn moves the linkage attached to the mercury bulb, causing it to move over center, thus moving the mercury from one end of the bulb to the other and making electrical contacts.

The Bourdon tube design can be supplied either as a low pressure or high pressure range switch. A pressure rise of either type extends the tube. In Figure 8-2-13, a reduction of pressure in the tube causes the mercury tube switch to close. In Figure 8-2-13, the increase in pressure in the tube causes the mercury bulb switch to open. Mercury tube switches can be either single pole single throw (two wire) or single pole double throw (three wire) types.

Humidistats

The third type of basic control is the humidity control or humidistat, as shown in Figure 8-2-14a,b.

Hydroscopic elements are used on these controls, the most common being human hair. As the air becomes more moist and the humidity rises, the hair expands and allows the electrical contacts to close (or open). As the hair dries it contracts, thus transferring the electrical contacts. This type of control is susceptible to dirt and dust in the air. Although

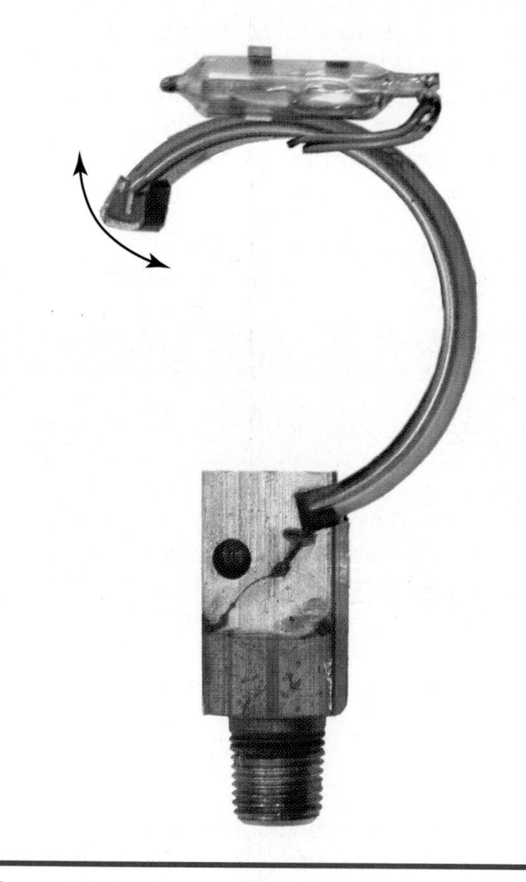

Figure 8-2-13 Bourdon tube switch and movement.

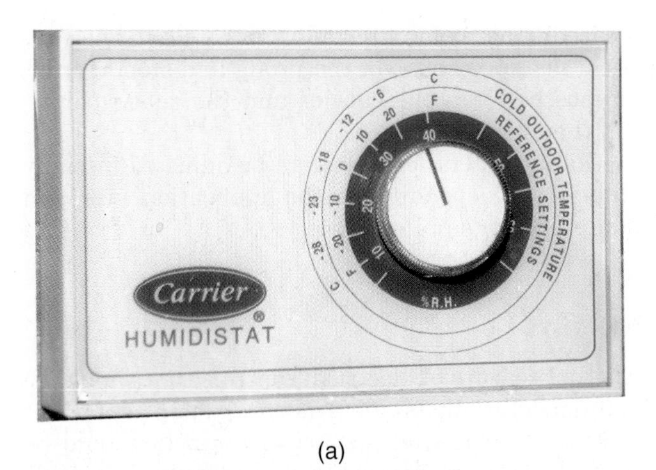

(a)

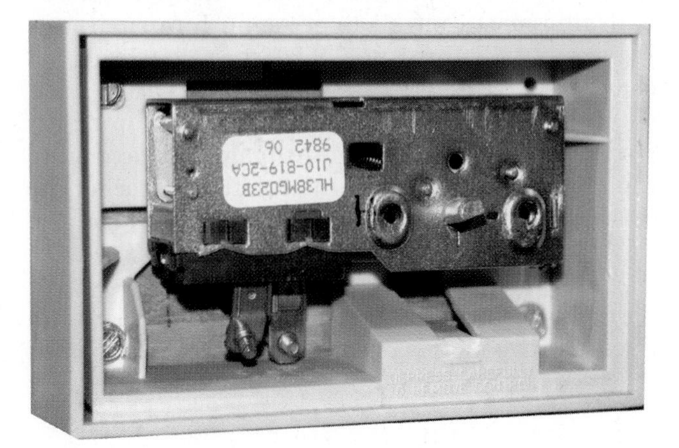

(b)

Figure 8-2-14 (a) Humidistat outer cover; (b) Humidistat, cover removed.

accurate, it must be carefully maintained and the calibration rechecked periodically.

Another form of humidistat uses a nylon element instead of a human hair. The nylon is bonded to a light metal in the shape of a coil spring. The expanding and contracting of the nylon creates the same effect as that found in the spiral bimetallic strip used in thermostats. Another type uses a thin, treated, nylon ribbon as a sensor.

An electronic circuit board can have hydroscopic properties and lithium salt is used for this purpose. Another arrangement uses carbon particles embedded in a hydroscopic material. In both cases, the sensing element acts like a thermistor. Changes in the humidity affect the resistance of the material and alter the current in the electronic circuit.

Time Switches

With the advent of microprocessors, time switches can be built into the electronic circuits and furnished at an affordable price.

Figure 8-2-15 Electronic setback thermostat.

Time clocks make possible automatic night setback functions in a programmable thermostat as well as such options as warmup cycles that precede early morning occupancy times. They permit programming a weekly scheduling of temperature setting, Figure 8-2-15. A commercial programmable time switch can turn the air conditioning off or on at preset times. Controllers of this type usually include batteries to operate the clock during power outages. A manual override permits the owner to change the schedule for special events.

SERVICE TIP

On some top end condensing units, time delay circuits prevent the short cycling or chattering of the contactor. These are used to protect the compressor. However, when you are testing a circuit, this time delay can be cumbersome. If by mistake you merely brush the thermostat wires together, you might find yourself waiting several minutes before you can continue with your test. If this becomes a problem you can temporarily bypass the time delay circuit so that you can continue actively pursuing the problem for which you were called to the job site. Be sure to remove the jumper when the job is complete.

RANGE ADJUSTMENT

There are almost as many possibilities for correct settings on controls as there are applications. Therefore, all controls have some way of being adjusted to compensate for the different conditions under which they may be required to operate. These may be field adjustable or fixed at the factory. The first such adjustment is range. *Range* is the difference between the minimum and the maximum operating points within which the control will function accurately.

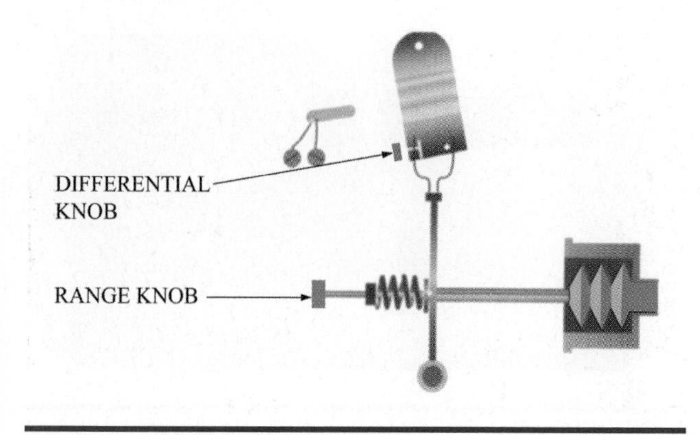

Figure 8-2-16 Control adjustment.

For example, a high pressure control may be used to stop the compressor when head pressure becomes too high. This control may have an adjustment that will allow a maximum cutout, or stopping point, of 300 psig and a minimum cutout of 100 psig. It may be set at any cutout point within this range. A control should never be set outside its range, as it will always be inaccurate and frequently will not even function.

Figure 8-2-16 shows one of the simpler methods of range control. As the pressure in the bellows increases, the lever is pushed to the left, thus opening the contacts. By varying the spring pressure, the bellows pressure can be increased or decreased as necessary to open the contacts—thus raising or lowering the system pressure at which the control will operate. Many controls use this principle and usually have some external adjustment for field use.

Differential Adjustment

If a control cuts out or breaks the circuit, it is just as important that it cuts in, or remakes the circuit. The cut-out and cut-in points cannot be the same or the control would chatter. The difference between these points is the differential. *Differential* can be defined as the difference between the cut-out and cut-in points of the control. For example, if the high pressure control discussed previously cuts out or breaks the electrical circuit at 250 psig and cuts in or remakes it at 200 psig, the pressurestat differential is 50 psig.

Referring to Figure 8-2-16, the control also has a differential adjustment. By opening or closing the effective distance between the prongs of the operating fork, the pressure at which the control will cut in can be varied. As the pressure on the bellows increases, the operating fork moves to the right, and the left prong of the fork will tilt the bulb and remake the electrical circuit. When the adjustable stop on the left prong of the fork is moved away from the prong on the right, the pressure in the bellows must decrease further before the bulb will be tilted. The electrical distance between the prong and stop, the difference between the cut-out and the cut-in points, can then be varied.

Range and differential can be adjusted in many other ways, depending on the application, size, and manufacturer's preference.

Detent

Another feature of electrical controls that should be mentioned briefly is called *detent* or snap action. For technical reasons, all electrical contacts should be opened and closed quickly and cleanly. Detent is built into most controls to accomplish this purpose. Figures 8-2-17 through 8-2-20 show four common examples.

Figure 8-2-17 shows the magnetic snap action switch that uses the pull of the magnet to accelerate the closing and opening rate of the contacts. The closer the magnet is to the bimetallic strip in the closed position, the more positive the snap action.

The bimetallic disk, as shown in Figure 8-2-18, is normally in a convex position. As the temperature rises, the two metals expand at different rates until

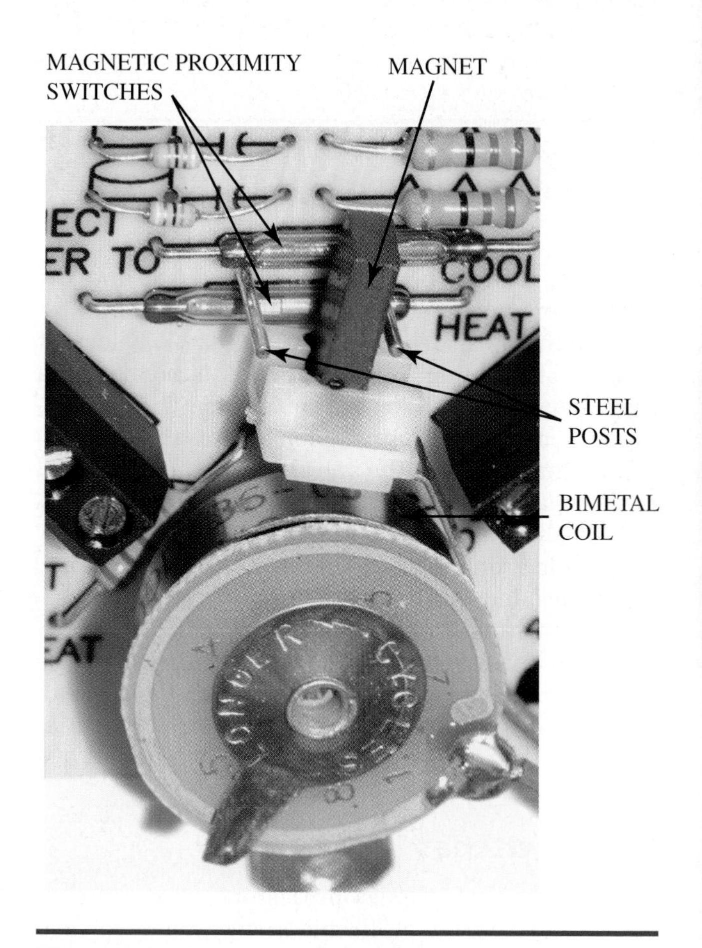

Figure 8-2-17 Snap action switch.

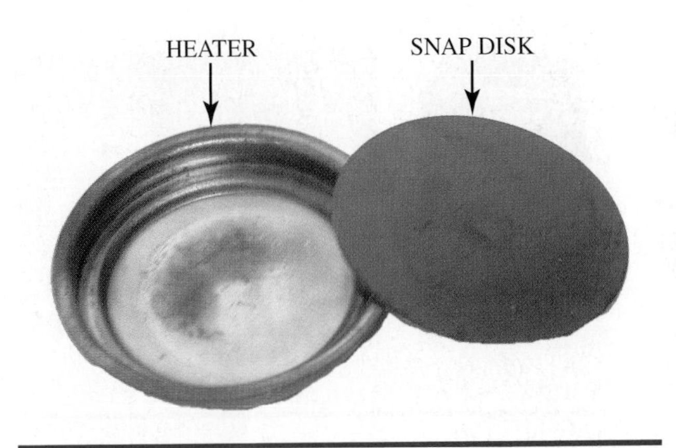

Figure 8-2-18 Bimetal disk.

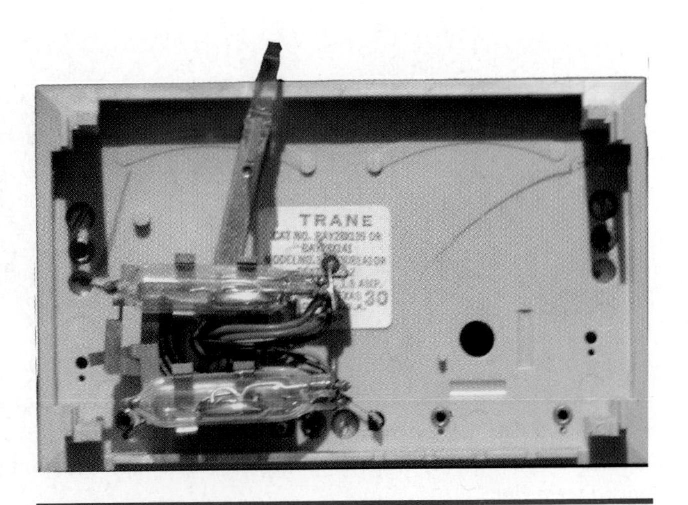

Figure 8-2-19 Mercury switch.

the disk snaps to a concave position, sharply breaking the electrical contacts.

The mercury bulb shown in Figure 8-2-19 is another method of obtaining detent action. As the element moves the bulb over the top center position, the heavy mercury runs from one end of the bulb to the other. This shifting of weight causes the bulb to move quickly from one side of the center to the other, thereby rapidly making or breaking contacts.

Figure 8-2-20a,b shows two examples of snap action induced by using a compressed spring. This type of device is commonly found in a household toggle switch. Force is applied to the operating arm. This rotates the toggle plate, which starts to compress the spring. As the switch approaches the center position the spring is at maximum compression. The moment the switch passes dead center, the compressed spring will force rapid completion of the switching action.

When the basic control sends a message to start or stop a unit, it usually acts through a relay, starter, or contactor, as shown in Figure 8-2-21a–e.

These devices consist of two parts: a coil that is energized when the primary control closes and one or more switches that change position when this occurs. The main difference between the three devices is mainly size. However, the starter is a contactor with the addition of overload protection.

Figure 8-2-22 shows a more detailed view of a contactor. The starter has one or more sets of contacts located on the armature. The armature moves up and down in the holding coil. When the coil is energized the armature moves and closes the contacts that start the motor. The motor contacts are considered normally open (NO) since they are open when no current is applied to the coil. Starters do some-

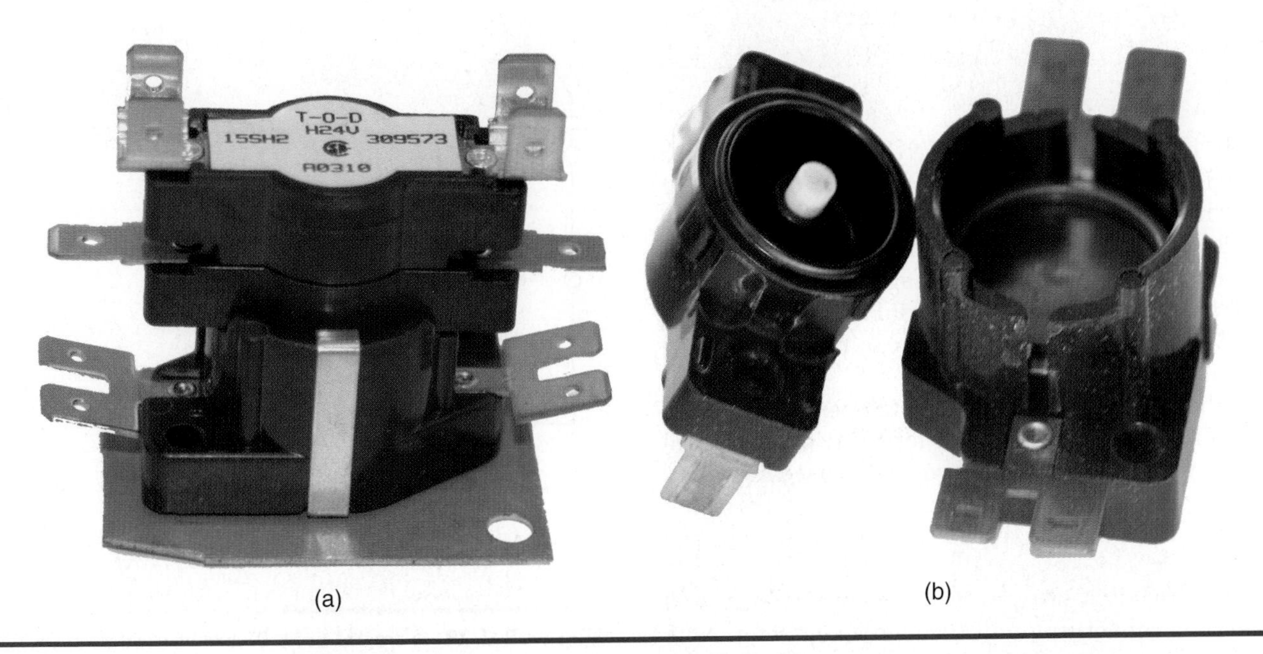

Figure 8-2-20 Snap action controls.

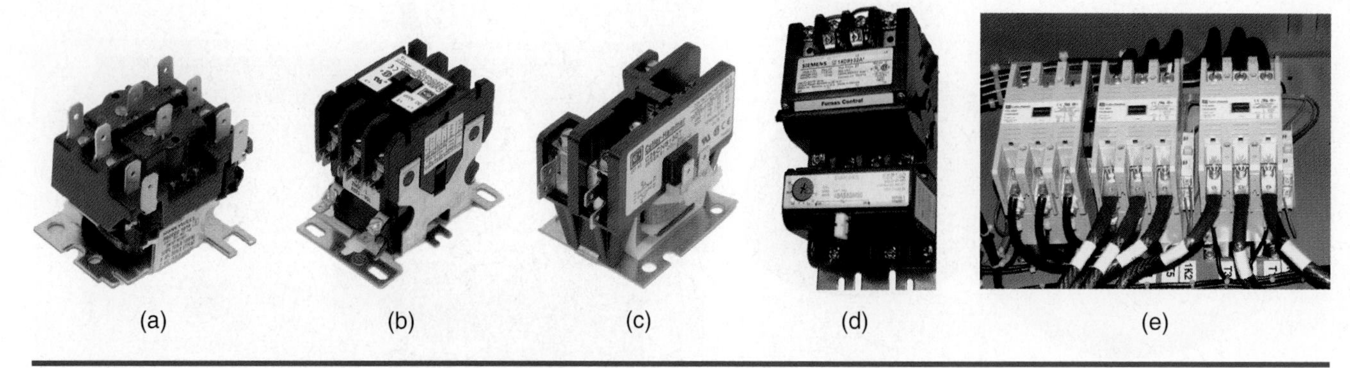

Figure 8-2-21 (a) Fan relay; (b) Three phase contactor; (c) Single pole contactor; (d) Motor starter; (e) Large amperage systems may use multiple contactors all wired together.

SERVICE TIP

The contact surface in relays and contactors is coated with a special alloy of very conductive material. This coating is very thin but effective in preventing the contacts from becoming quickly damaged from the momentary arc as they open and close. This arc would normally cause the contacts to weld themselves together. Over time this protective coating will wear away and the contactors may eventually begin to stick occasionally. When sticking occurs the contactor or contactor tips must be replaced. They cannot be reconditioned in the field with point files. The substructure of the contact is a thermally conductive material which helps carry the heat away from the coating. It is not resistant to arcing and welding itself together. A reconditioned contact surface will stick again very quickly. Do yourself and your customer a favor and replace sticking contacts when they are located.

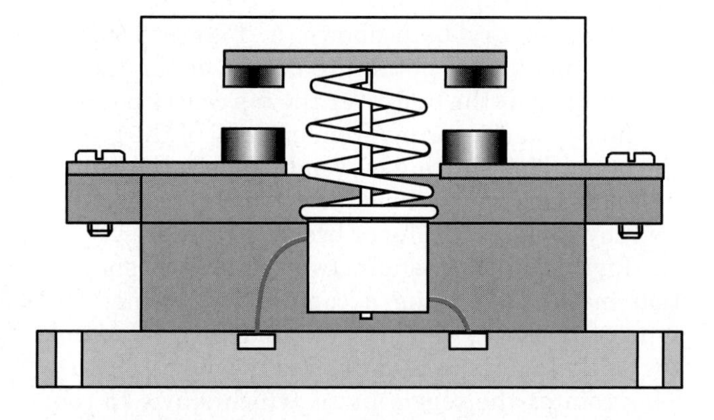

Figure 8-2-22 Contactor diagram.

times have normally closed (NC) contacts that are closed when the coil is de-energized.

The control voltage can be different for various sized jobs. Residential and light commercial control voltage is 24 V, while line voltages can be 115 V or 230 V. Commercial and industrial control voltage is 115 V, while line voltages there can be 240 V, 480 V, and 575 V.

The control voltage is used to operate the holding coil on a starter as well as for many other control functions. The lower control voltage is used for safety and to permit the use of more sensitive controls on these circuits as compared to the use of line voltage controls.

Figure 8-2-23 shows a typical control circuit used to start a condensing unit through a contactor. Note that a thermostat operating in a low voltage circuit

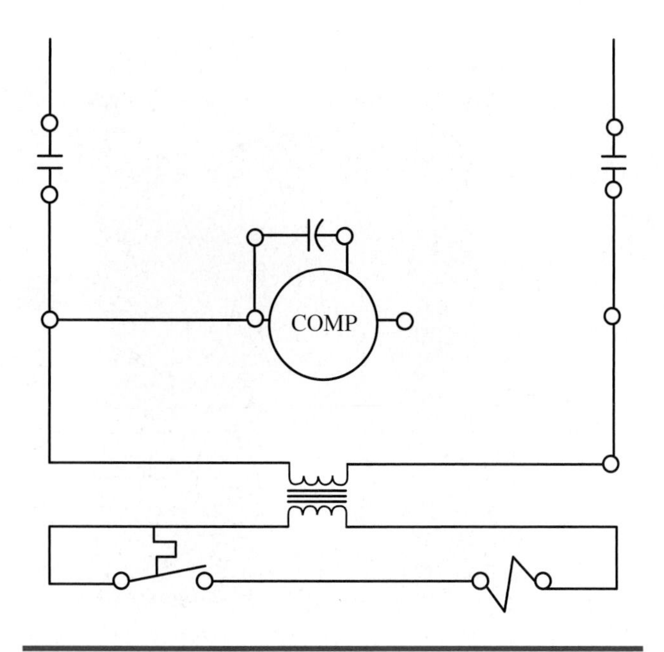

Figure 8-2-23 A typical control circuit used to start a condensing unit through a contactor.

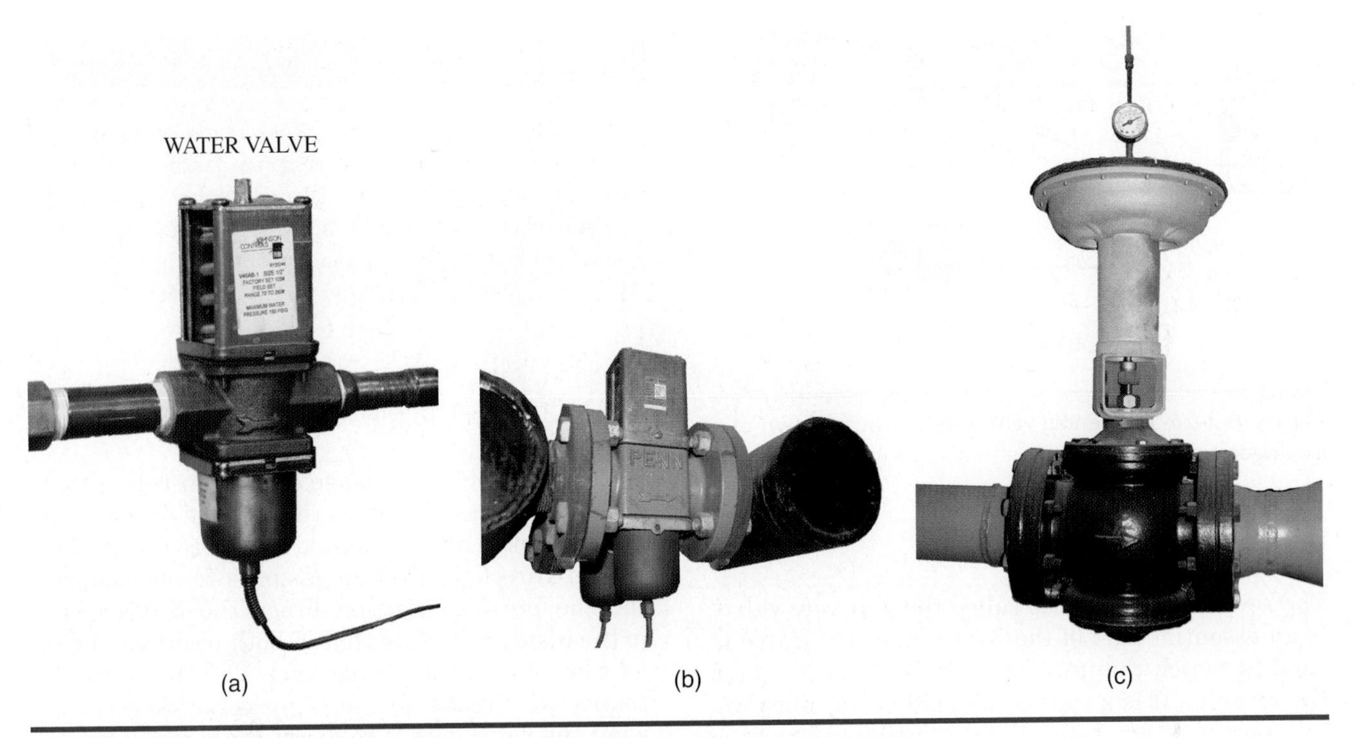

Figure 8-2-24 (a) Bellows type water valve that would be used on a small system (*Courtesy Hampden Engineering Corporation*); (b) Bellows type water valve that would be used on a large cooling tower; (c) Large bellows type water valve that can be pneumatically controlled.

closes the contacts in a high voltage circuit to operate the condensing unit. When the thermostat is satisfied, the contactor coil is de-energized and the condensing unit stops.

OPERATING CONTROLS

In this section, we will discuss electrical and mechanical controls which are used to operate the system, after it is started. We will refer to them as *operating controls.*

Metering Devices

Details on the various types of metering devices were covered in Chapter 5, Unit 5.

Condenser Water Valve

When a water cooled condenser is used, particularly where it uses city water, the valve regulates the quantity of water that flows through the condenser. It accomplishes two things. It regulates and conserves the flow to maintain a set condensing pressure and temperature. It also acts to conserve city water by shutting the condenser flow when the refrigeration cycle shuts off, Figure 8-2-24.

Refrigerant Solenoid Valve

Solenoid valves, as shown in Figure 8-2-25, are used for many purposes. Valves are available to operate on either line voltage or control circuit voltage. Some type of switch is operated to send power to the valve; for example, a thermostat. When the solenoid coil is energized the plunger is moved to control the flow of some fluid, such as refrigerant. In Figure 8-2-26, the solenoid valves are being used to start and stop the flow of refrigerant to the evaporator. The solenoid valve is placed in the liquid line ahead of the expansion valve. This type of control arrangement is often used where multiple evaporators operate from the same compressor and individual rooms require different temperatures.

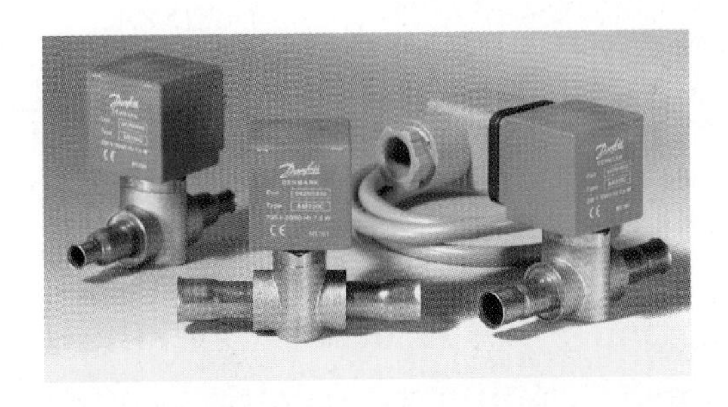

Figure 8-2-25 Solenoid valves. (*Courtesy Danfoss Inc.*)

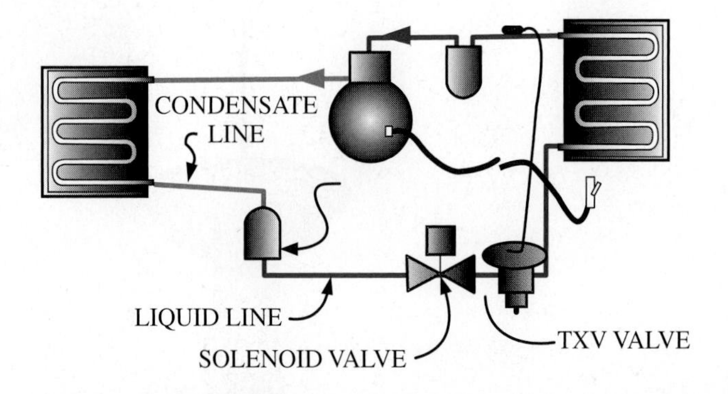

Figure 8-2-26 A solenoid valve used for pumpdown of a refrigeration system.

> **TECH TIP**
>
> Most solenoid valves on heat pump reversing valves are low voltage. But be careful, as some packaged systems use line voltage relay coils. Make certain when servicing these units that you verify the voltage for the reversing valve coil before changing the coil out.

Reversing Valve

The reversing valve, also called the four way valve, is an essential part of the heat pump. The valve is used to switch the unit from cooling to heating (or the reverse). It is a type of solenoid valve, since the two positions are controlled by a plunger that is moved by energizing or de-energizing a solenoid coil. Figure 8-2-27 shows the valve directing the flow of refrigerant through the unit to produce cooling. The refrigerant discharge from the compressor goes through the valve to the outside coil (condenser), through the indoor metering device to the inside coil (evaporator), through the valve again, and back to the suction side of the compressor. This is the same as a normal cycle for any cooling system, with the exception of the passage of the refrigerant vapors through the four way valve.

In Figure 8-2-28, the valve is positioned for heating. In this position the hot discharge gas from the compressor is directed to the inside coil to provide heating. After the vapor has condensed in the inside coil it goes through the outdoor metering device and to the outside coil. Here it vaporizes as it absorbs heat and returns to the suction side of the compressor as a gas.

Note the changed usage of the four connections to the valve. Starting from the left, the first connection is blocked off. The second and third connections (a loop) direct the suction gas back to the compressor. The fourth connection directs the discharge gas to the inside coil. Note that in both positions the inlet connection of the valve, shown on the bottom, is connected directly to the compressor. So basically what the valve does is to direct the compressor discharge gas either to the outside coil in the cooling cycle or to the inside coil in the heating cycle.

Back Pressure Valves

The back pressure valve has other names that may be more descriptive, such as the suction pressure regulator or the evaporator pressure regulator. Regardless of which name is used, the function of the back pressure valve, shown in Figure 8-2-29, is to provide a minimum suction pressure for a single coil or a group of coils, and thus maintain a constant evaporation temperature. The valve shown in Figure 8-2-29 is a self contained pressure activated valve.

The spring tension of the valve stem can be adjusted at the top for whatever suction pressure is

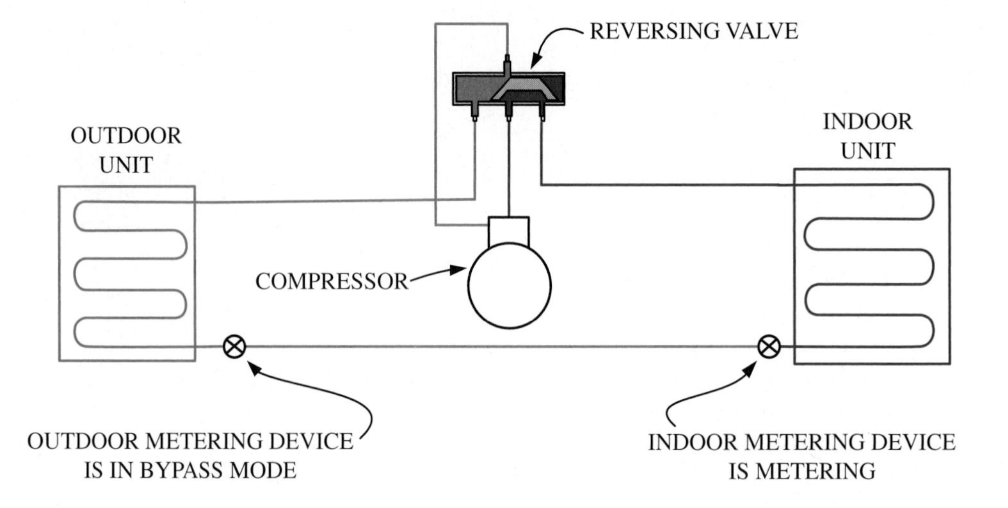

Figure 8-2-27 Reversing valve.

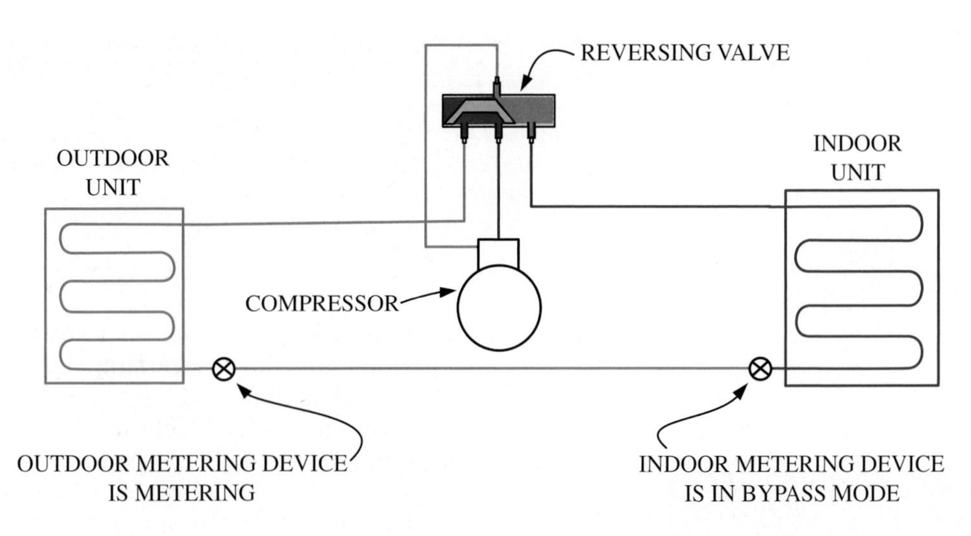

Figure 8-2-28 Reversing valve heating position.

desired (within the limits of the valve). It is usually placed in the suction line leaving the evaporator, as close to the coil(s) as possible. One variation of the valve is shown in Figure 8-2-30. This version of the valve has a pilot operated remote sensor that can be located at a more favorable control point. On larger jobs it is often desirable to control from the evaporator itself rather than from the suction line.

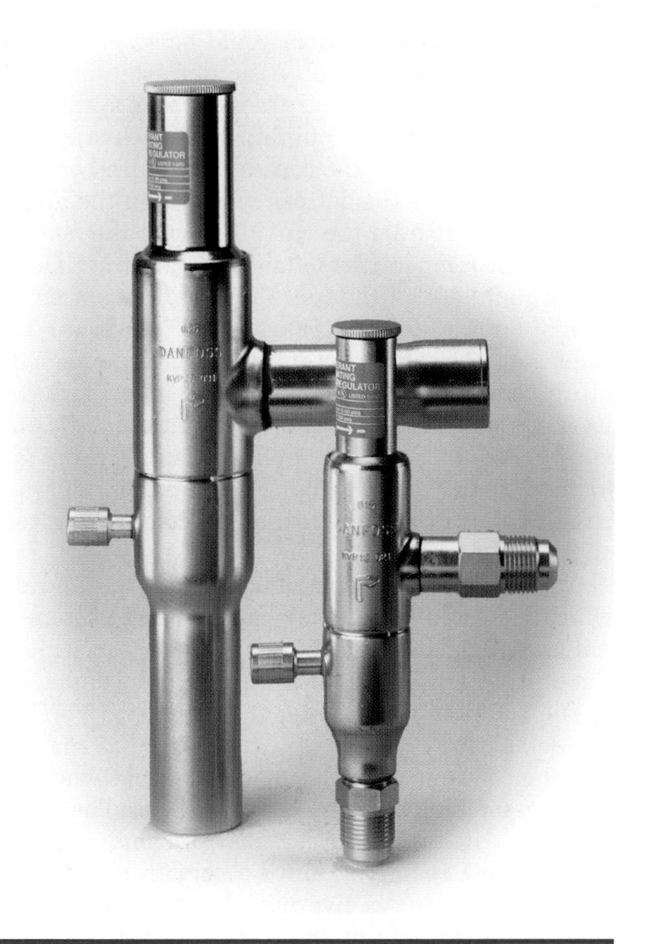

Figure 8-2-29 Pressure regulating valve. (*Courtesy Danfoss Inc.*)

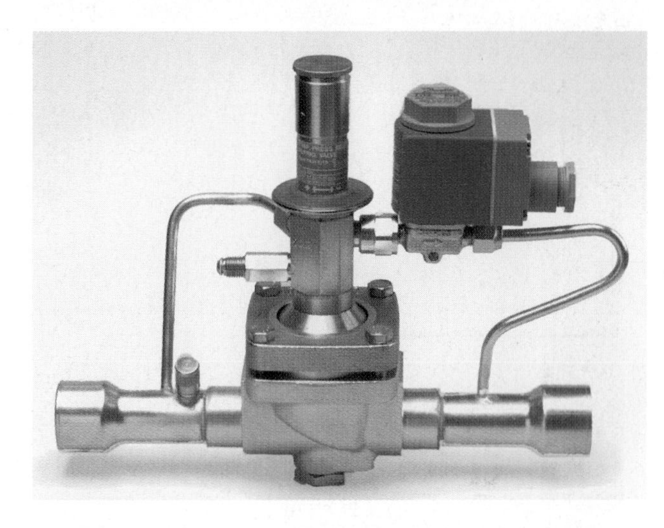

Figure 8-2-30 Pilot operated valve. (*Courtesy Danfoss Inc.*)

Figure 8-2-31 shows another variation of the evaporator pressure regulating valve (EPR), called the two temperature valve. This valve is used on applications where two evaporators with different evaporating temperatures are operated from the same compressor. The two temperature valve is placed in the suction line from the highest temperature evaporator. The two temperature valve is a snap action valve, in that it is either fully open or fully closed. It therefore cycles the evaporator as though it were directly connected to its own condensing unit and controlled by a pressurestat or thermostat.

Figure 8-2-31 Evaporator pressure regulator.

Check Valve

The check valve is a device that permits a flow of a fluid through piping in only one direction. Figure 8-2-32 shows the use of a swing type check valve.

When the flow moves from left to right, the flapper opens and the flow is free to move. If the flow is reversed, the flapper slams shut and the flow is blocked. Figure 8-2-33 shows an assortment of flapper type and ball type check valves.

The ball type check valve operates on the same principle as other types of check valves. When the flow is stopped, the ball falls into its seat. If the flow is in the direction of the arrow on the outside of the valve (see Figure 8-2-29), the ball moves away from its seat and permits the flow. Reversing the flow causes the ball to block the opening.

Check valves have many uses. They are required on heat pump systems and two temperature systems. They may be used in compressor discharge lines to prevent backflow of refrigerant and oil to the compressor during shutdown. This prevents damage that could occur to the compressor at the time of startup.

SERVICE TIP

Some ball type check valves will chatter as the system pressure drops during an off cycle. This chatter is normal and not an indication of a defective ball valve.

Timed Devices

Programmable thermostats incorporating time functions such as those shown in Figure 8-2-34 permit the system owner to set at least two temperatures for the day's system operation.

This resetting function would be considered an operating type of control. Also, defrost timers are used on some commercial freezer room systems to prevent excessive ice buildup on the evaporator. Figure 8-2-35 shows the control board used on heat pumps to initiate the defrost cycle. When the unit is absorbing heat from the outside during the heating cycle, if the temperature outside is below freezing, ice can accumulate on the coil. Two conditions are necessary for the defrost cycle to take place: (a) The timer must be calling for defrost and (b) the temperature sensor on the condenser coil must be below freezing. When the defrost cycle is initiated, the unit is returned to the cooling cycle, causing the hot gas to flow to the outside coil. Figure 8-2-36 shows a circuit using a timer to prevent the compressor from short cycling.

Short cycling means rapid turning on and off, which can occur on an air conditioning system in mild weather. It can also occur during servicing of the unit. In any event, such action can be damaging to the compressor. To prevent this, a short cycling control is used.

Some systems have a timer that starts when the compressor shuts down, Figure 8-2-37. The timer opens a switch in the circuit that prevents the compressor from restarting until a certain time period has elapsed, usually five minutes. On some of the solid state controls of this type, the timer is constructed to compensate for power outages.

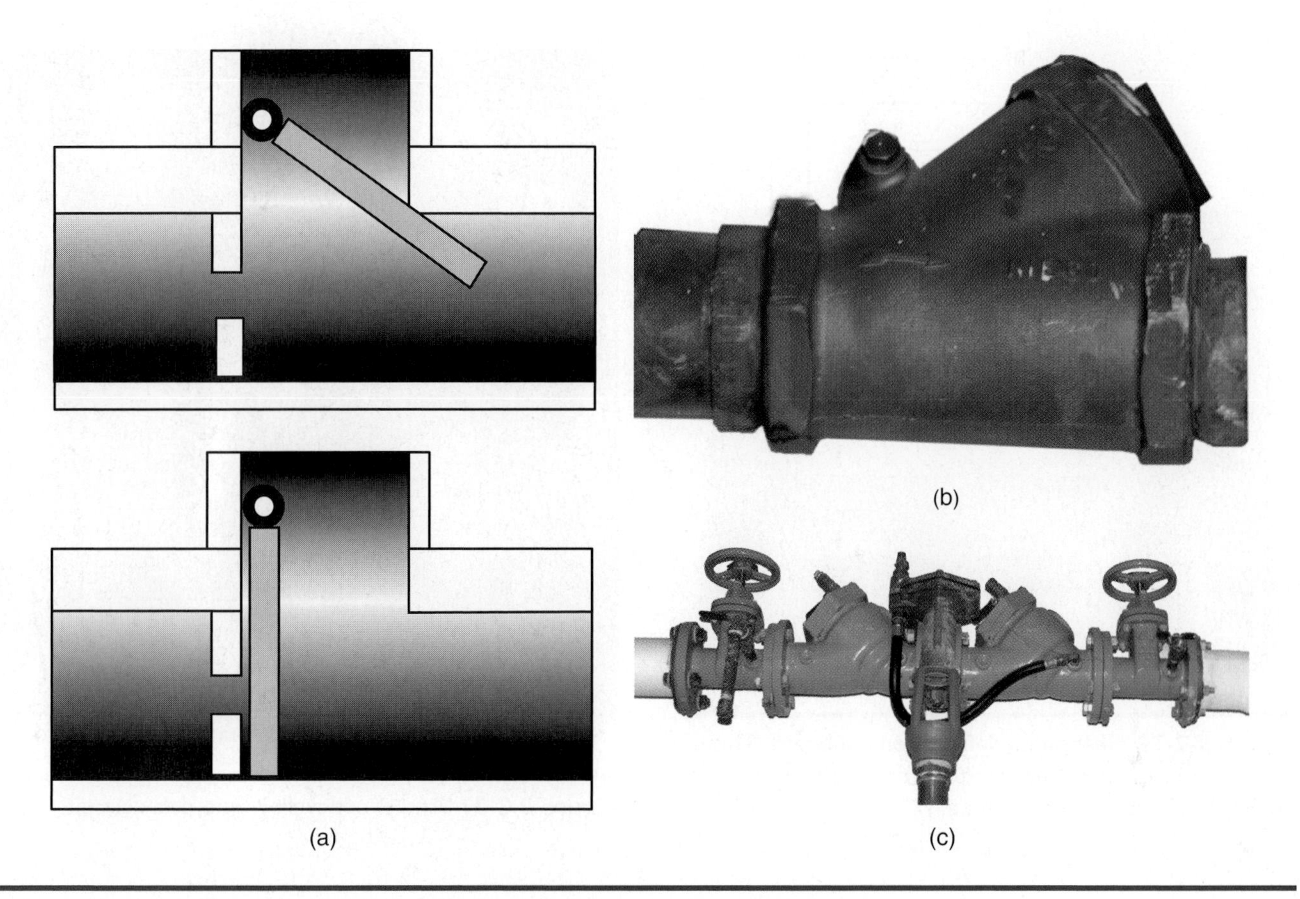

Figure 8-2-32 (a) Check valve internal operation; (b) Refrigerant check valve; (c) Double check valves, which are required for systems such as cooling towers that must be connected to the main water supply. These double check valves are designed to keep cooling tower water from contaminating the drinking water supply of a building.

Figure 8-2-33 Ball check valves. (*Courtesy Danfoss Inc.*)

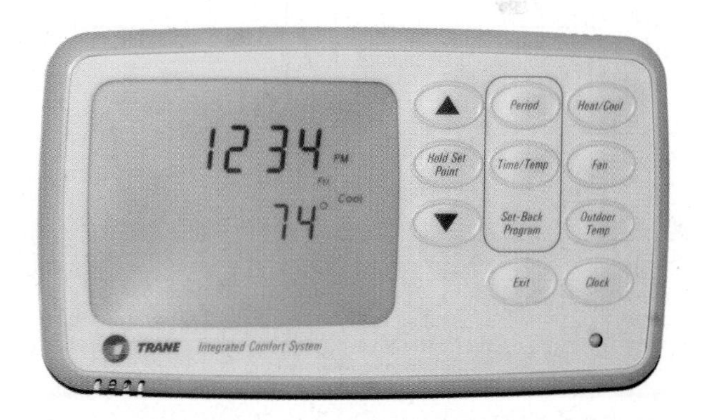

Figure 8-2-34 Programmable thermostat.

SAFETY CONTROLS

Safety controls protect the cycle and its components from damage. They monitor the operation, but do not change the operation unless something goes wrong. Some safety controls, however, prevent the unit from starting under adverse conditions.

If the safety controls are brought into action due to some problem, they must be reset after the necessary repair has been made and the unit is to resume normal operation. These resets can either be automatic or manual, depending on the nature of the problem and the particular design of the control. It is

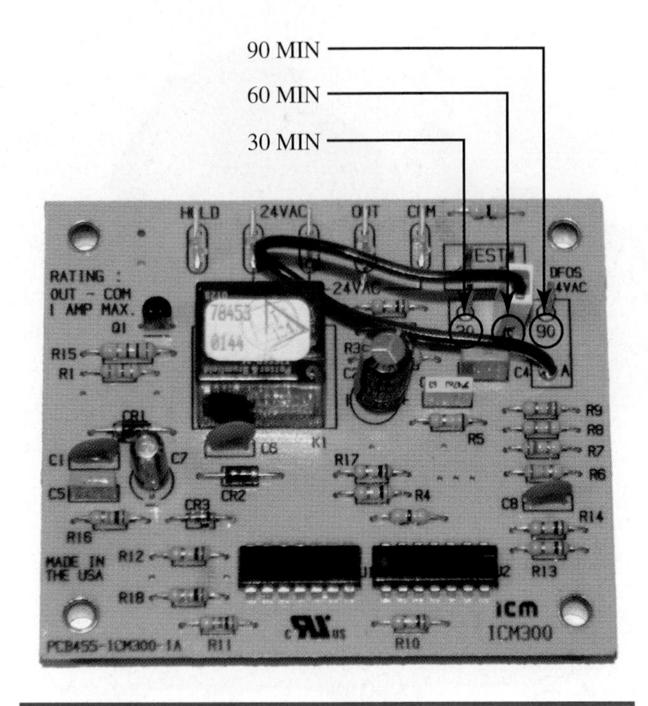

Figure 8-2-35 Heat pump defrost control board. This board can have the jumper set to initiate defrost every 30, 60, or 90 minutes of run time.

Figure 8-2-37 Anti short cycling time delay circuit.

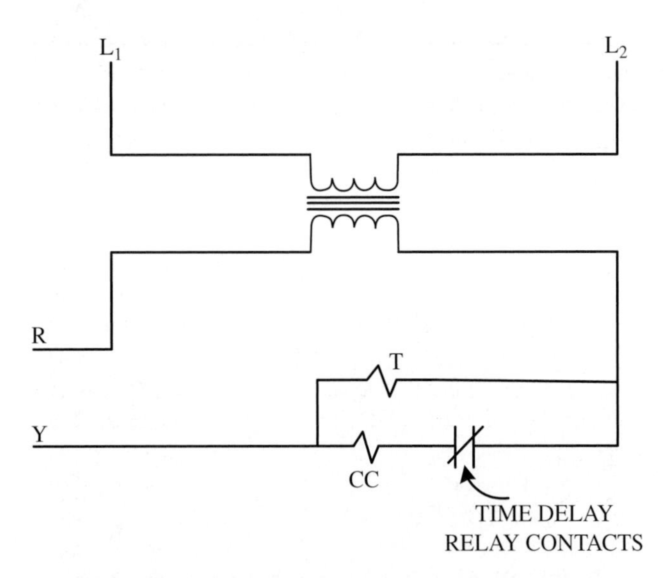

Figure 8-2-36 Time delay compressor circuit.

advisable to consult the manufacturer's information relative to the type of reset.

CAUTION: It is never acceptable to jumper across a safety device.

Reset

Temperature actuated safety controls can be designed with a wide differential setting so that automatic reset will not result in damaging short cycling. An example of this is a compressor motor winding thermostat.

Pressure actuated safety controls might recycle faster than desired. A manual reset high pressure switch could be used. Another option would be using a lockout relay so that the unit would not reset until the control circuit is interrupted by: (a) resetting the thermostat or (b) switching the unit's power off, then on. This permits the operator to try resetting the unit manually without gaining access to the equipment. The advantage: This procedure may avoid occasional nuisance service calls.

SERVICE TIP

Some equipment uses manual reset on oil pressure cutouts and low and high pressure switches. When these devices trip frequently it is often referred to as a nuisance call. There is no such thing as a nuisance call when it comes to safety devices; there is a problem that may not have been located causing the system to trip the switches. It is not a recommended practice to inform the building operator as to how they can reset these switches, because in doing so they may ultimately wind up causing more damage to the system.

Figure 8-2-38 Overload.

Types of Safety Controls

Electrical Overloads In all electrical circuits, some kind of protection against excessive current must be provided. In motor circuits a reset type of protector is frequently used. The overload is one method by which this is accomplished, Figure 8-2-38. It is used in conjunction with the contactor previously discussed. When a contactor is used, there is both a control circuit, in which the primary control is inserted, and a load circuit that is opened and closed by the contactor. When excessive current is drawn in the load circuit, this device will break the control circuit.

The control circuit to the contactor coil passes through the controls. If the load circuit passing through the bimetallic element becomes too high, the element bends to the side, forcing the contacts apart. This breaks the control circuit and allows the contactor to open, thereby interrupting the power to the load. The bimetallic element will now cool and return to its original position, but because of the slot in the bottom arm, the contacts will not be remade. The control circuit will remain broken until the reset button is pushed. This moves the arm to the right and closes the control-circuit contacts. Because this control must be reset by hand, it is usually referred to as a manual-reset overload. Some overload relays may be field adjusted for manual or auto reset function.

Current/Temperature Devices A second type of electrical overload protector is the bimetallic disk, Figure 8-2-39a–c. When the temperature is raised, the disk will warp and break the electrical circuit. The bimetallic disk overload protector is widely used in the protection of electrical motors.

The current relay is another type of electrical overload that can be manually or automatically reset. The greater advantage of this type of relay is that it is only slightly affected by ambient temperatures, thereby avoiding nuisance tripouts.

The current relay is made up of a sealed tube completely filled with a fluid and holding a movable iron core. When an overload occurs, the movable core is drawn into the magnetic field, but the fluid slows its travel. This provides a necessary time delay to allow for momentary high current during motor startup (locked rotor amps) without tripping the overload. When the core approaches the pole piece, the magnetic force increases and the armature is actuated, thus breaking the control circuit.

On short circuits, or extreme overloads, the movable core is not a factor because the strength of the magnetic field of the coil is sufficient to move the armature without waiting for the core to move. The time delay characteristics are built into the relay and are a function of the core design and fluid selection.

Many current overload devices sense both the current and temperature of the motor. These devices, like those located within the bakelite cover of small compressors, must be in their enclosure to function properly. If the bakelite cap is left off then the overload will only sense the current and not the temperature. This reduces their sensitivity and removes a large portion of the overload safety protection from a motor.

Thermostats The only difference between the thermostatic operating control previously discussed and the safety thermostat involves the application in which they are used; the controls themselves may be identical.

One thermostat used as a safety control can be found in most chill water refrigeration systems,

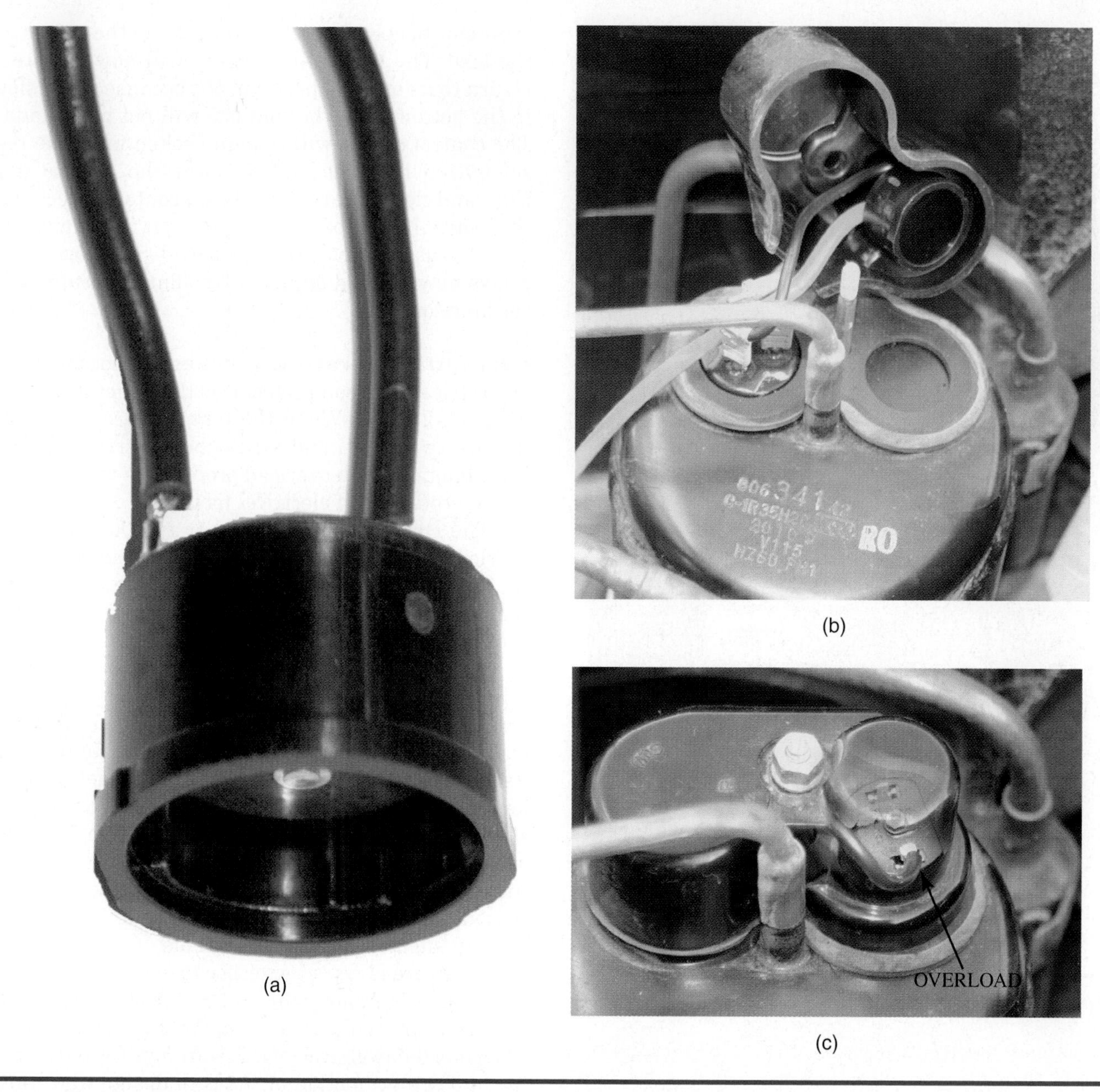

Figure 8-2-39 (a) Klixon current and temperature protector can be located inside or outside of the compressor; (b) When it is located outside of the compressor, it is under the motor terminal cover; (c) This motor cover holds it tightly against the compressor so it can sense the motor's temperature more accurately.

Figure 8-2-40. In chill water systems it is important that the water not be allowed to freeze, because it could then do physical damage to the equipment. In such systems a thermostat may be used with the temperature sensing element immersed in the water at the coldest point. The thermostat should be set so that it will break the control circuit at some temperature above freezing. This will stop the compressor and prevent a further lowering of water temperature and possible freezing. The temperature control used for safety purposes is physically the same as that used for operational purposes.

Pressurestats The high pressure cutout as shown in Figure 8-2-41 is probably the best example of a pressure control used for safety purposes.

It can be set to stop the compressor before excessive pressures are reached. Such conditions might occur because of a water supply failure in water cooled condensers or because of a fan motor stoppage on air cooled condensers.

Figure 8-2-40 Liquid sensing thermostat.

The operation of the low pressure limit control, Figure 8-2-42 is the same as discussed under the subject of basic controls. Mechanically, there is no difference between the low pressure control used as an operating control and the one used as a safety control except for settings. The low pressure control is used to stop the compressor at a predetermined minimum operating pressure. As a safety device, the low pressure control can protect against loss of charge, high compression ratios, evaporator freeze-ups, and entrance of air into the system through low side leaks.

Some pressurestats are connected to both the low and high pressure sides of the system. These combination pressurestats are actually two separate devices built into a single device, Figure 8-2-43. The low and high pressure setting can be adjusted individually without affecting the other setting.

Pressure Relief Valve Figure 8-2-44 shows a pressure relief valve that can be placed in the hot gas line, condenser, or liquid receiver as a high pressure safety control. These controls are required by the installation codes on most water cooled systems. This valve automatically resets when normal pressures are reestablished.

Fusible Pug and Rupture Disk Figure 8-2-45 shows two other safety fittings, the rupture disk and

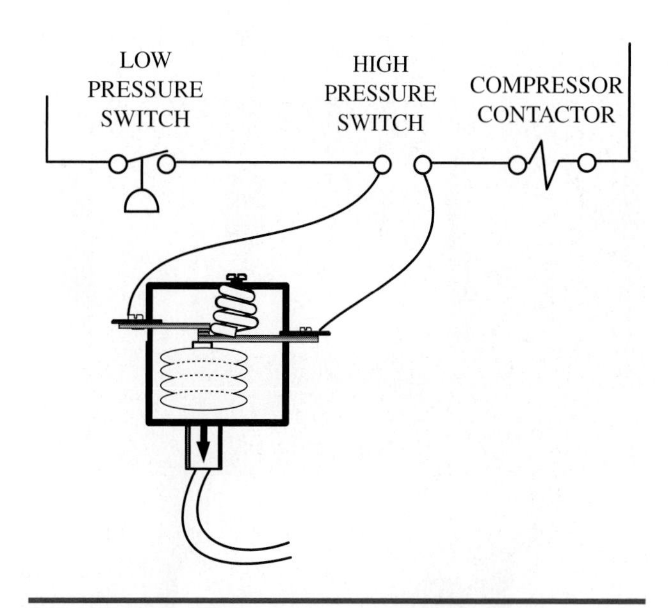

Figure 8-2-41 High pressure switch (auto reset).

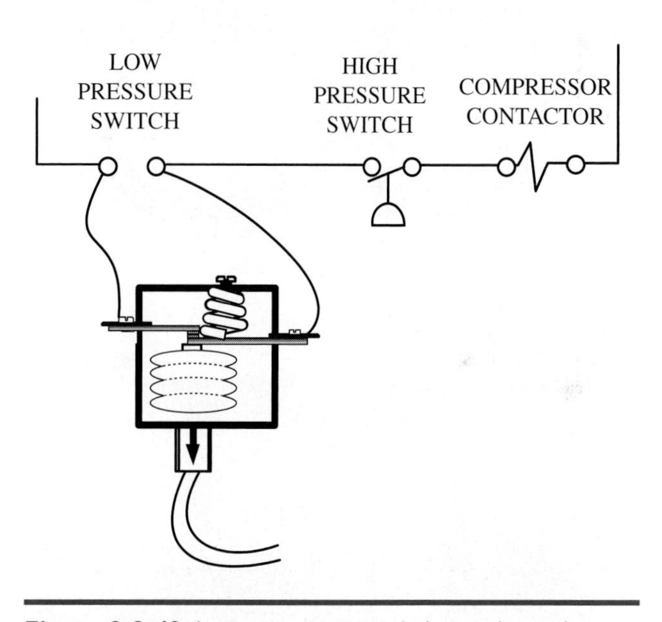

Figure 8-2-42 Low pressure switch (manual reset).

the fusible plug. The rupture disk opens on excessive pressure and the fusible plug melts open on excessive temperature.

It is advantageous to use the pressure relief valve in place of these fittings wherever possible since the relief valve automatically resets and does not permit losing the full refrigerant charge.

Oil Pressure Safety Switch The oil safety switch is activated by pressure; however, it is operated by a pressure *differential* rather than by straight system pressure. It is designed to protect against the loss of oil pressure. Inasmuch as the lubrication system is contained within the crankcase of the compressor, a pressure reading at the oil pump

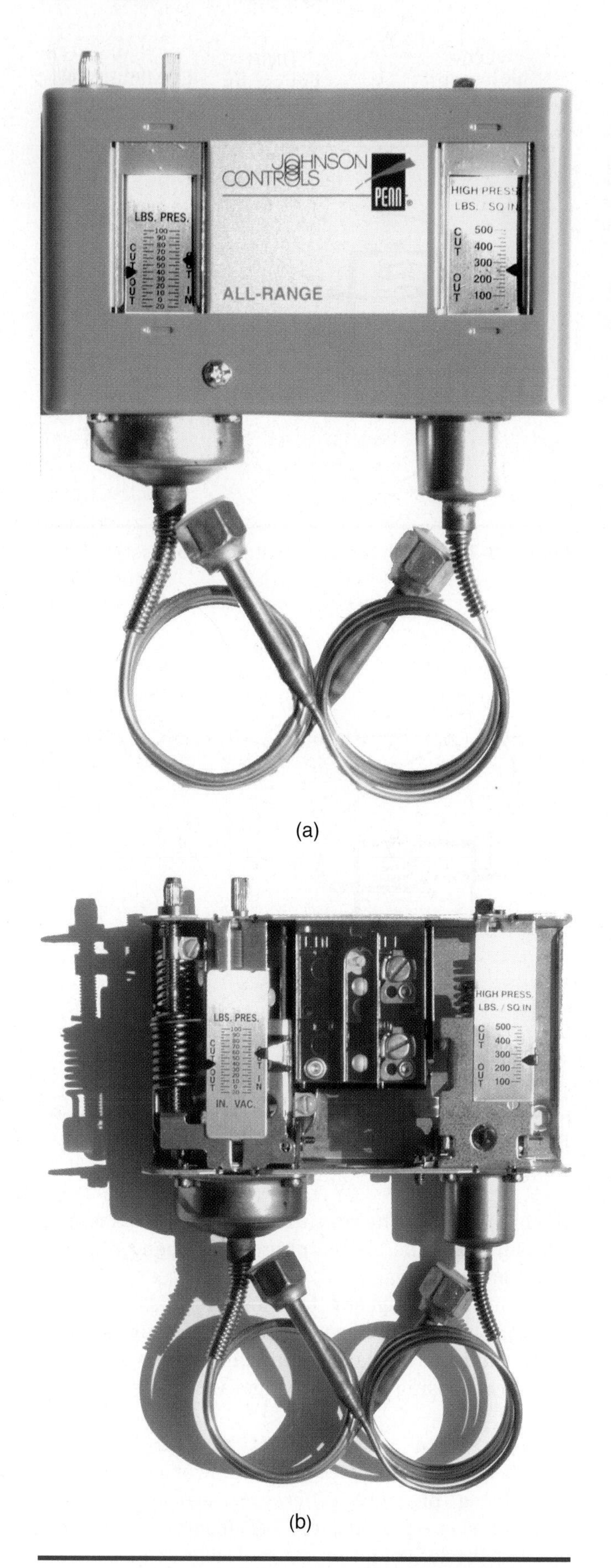

Figure 8-2-43 (a) Combination high and low pressure switch; (b) The high and low pressure sides of this control are separated inside the switch.

Figure 8-2-44 Small welded pressure relief valve used in a receiver.

discharge will be the sum of the actual oil pressure plus the suction pressure. The oil safety switch measures the pressure difference between the oil pump discharge pressure and compressor crankcase pressure and shuts down the compressor if the oil pump does not maintain an oil pressure as prescribed by the compressor manufacturer. For example, if a manufacturer indicated that a 15 lb net oil pressure was required, a compressor with a 40 lb back pressure must have at least a 55 lb pump discharge pressure. At any lower discharge pressure from the pump, this switch would stop the compressor. The control function depends on the new oil pressure overcoming the predetermined spring pressure. The spring pressure must equal the minimum oil pressure allowed by the manufacturer.

When a compressor starts, there is no oil pressure. Full oil pressure is not obtained until the compressor is up to speed. Therefore, a time delay device

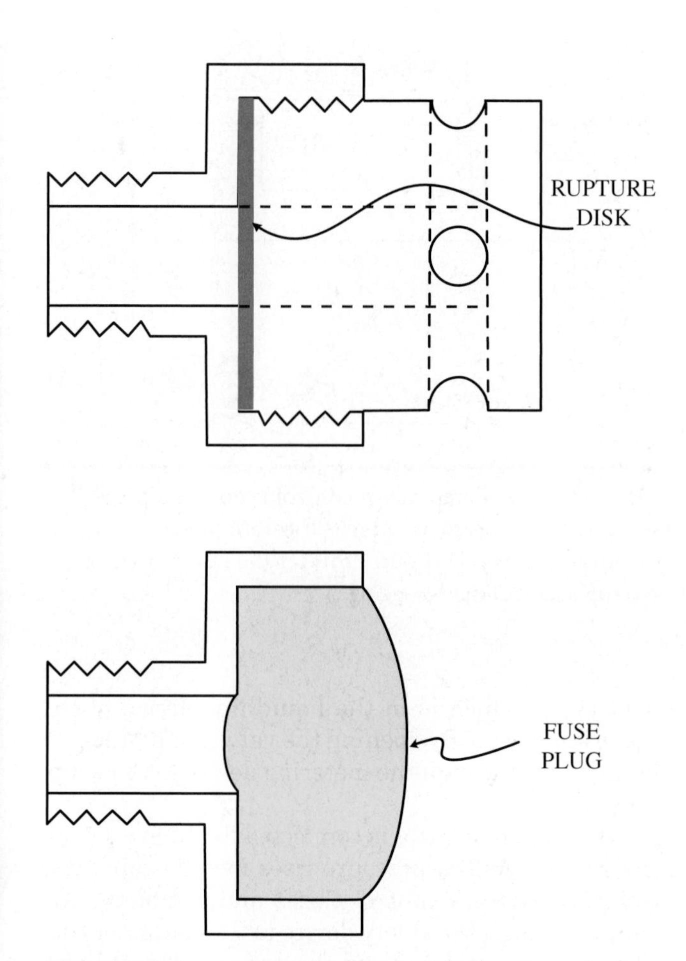

Figure 8-2-45 Fuse plug and rupture disk pressure relief valves.

SERVICE TIP

A popular misconception is that oil pumps produce pressure. Oil pumps only produce oil flow. It's the resistance to that oil flow provided by the proper bearing tolerances in a compressor that cause the oil pump to produce pressure. If a compressor is having low oil pressure problems in the oil pump, and the tolerances are correct, then the compressor may be developing a bad bearing. Do not attempt to increase the oil pressure as a way of resolving this problem. As the compressor bearing begins to wear, it will rapidly deteriorate from the constant wear and vibration in the compressor. Note that simply boosting the oil pressure will not stop the wear from occurring. If the compressor is taken out of service and the bearing is repaired before it completely destroys itself, a great deal of unnecessary cost can be avoided.

is built into the oil pressure safety switch to allow the compressor enough time to start. One of the methods used to create this time delay is a small resistance heater. When the compressor is started, because the pressure switch is normally closed, the resistance heater is energized. If the oil pressure does not build up to the cutout setting of the control, the resistance heater warps the bimetallic element, breaks the control circuit, and stops the compressor.

If the oil pressure rises to the cutout setting of the oil safety switch within the required time after the compressor starts, the control switch is opened and de-energizes the resistance heater, and the compressor continues to operate normally.

If oil pressure should drop below the cutin setting during the running cycle, the resistance heater is energized and, unless oil pressure returns to cutout pressure within the time delay period, the compressor will be shut down. The compressor can never be run longer than the predetermined time on subnormal oil pressure. The time delay setting on most of these controls is adjustable from approximately 60 to 120 seconds. Sometimes the timing and safety shutdown function will be accomplished by an electronic control module using a signal from an oil pressure differential switch.

Flow Switch The sail switch, shown in Figure 8-2-46, is a protective device to prevent the operation of a unit when there is inadequate fluid flow. In an air system, the sail switch is placed in the duct to

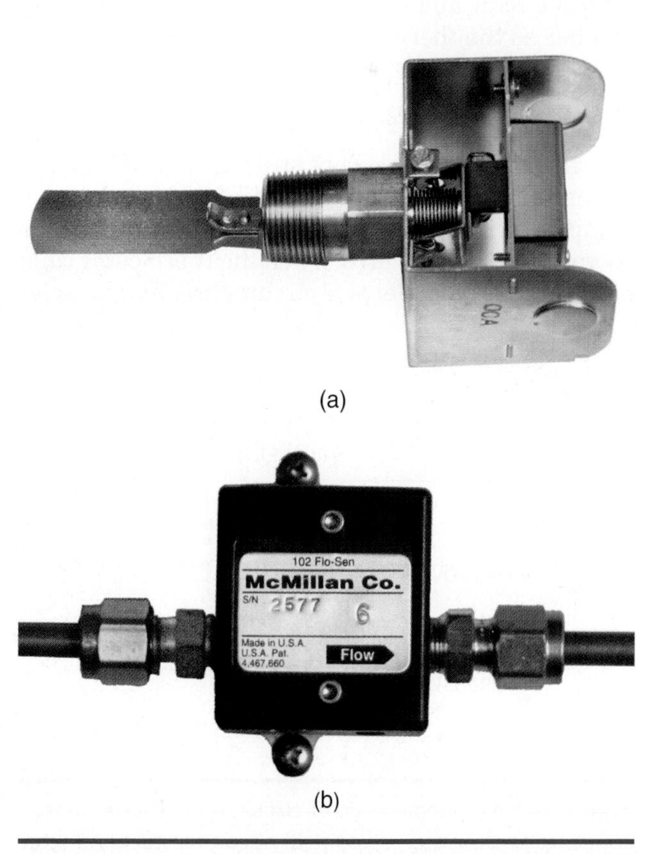

Figure 8-2-46 (a) Flow switch; (b) Electronic flow meter. *(Courtesy Hampden Engineering Corporation)*

sense the flow of air. Unless there is an adequate supply of air over the coil, the unit is either not started or shut down. Switches of this type can also be placed in a waterline feed of a water cooled condenser. If there is an inadequate supply of water, or no water, the unit is prevented from running.

OPERATING CONTROL SYSTEMS

There are as many control methods for operating refrigeration systems as there are engineers to design them, but there are two simple methods that are standard and will be discussed briefly.

Thermostats

The first control method, illustrated in Figure 8-2-47, is the simple thermostat start-stop method. As the thermostat calls for cooling and closes the contacts, the control circuit is completed through the contactor coil. This control circuit includes a Klixon, a pressure control, and a motor starter thermal overload relay as safety/limit controls. When the contactor coil is energized, it closes the contacts and completes the power circuit to the motor. The thermostat is the operating control and the motor cycles as the thermostat dictates. Any of the safety controls will break the control circuit and stop the compressor in the same manner as the thermostat.

Pumpdown Cycle

A slightly more complex but still widely used control method is shown in Figure 8-2-48. This is known as pumpdown control. In this method the thermostat operates a solenoid valve. When the thermostat calls for cooling, it completes a circuit through the solenoid valve, which is in the liquid line ahead of the metering device. By opening the valve, high pressure liquid flows through the metering device to the evaporator.

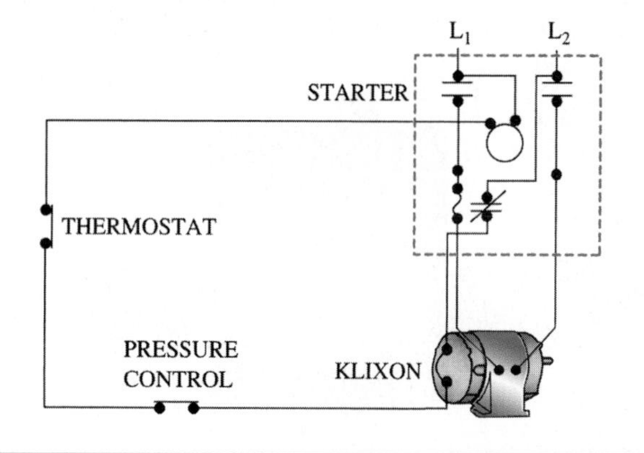

Figure 8-2-47 Simple control circuit, using thermostat connected to the started control circuit to start and stop the system.

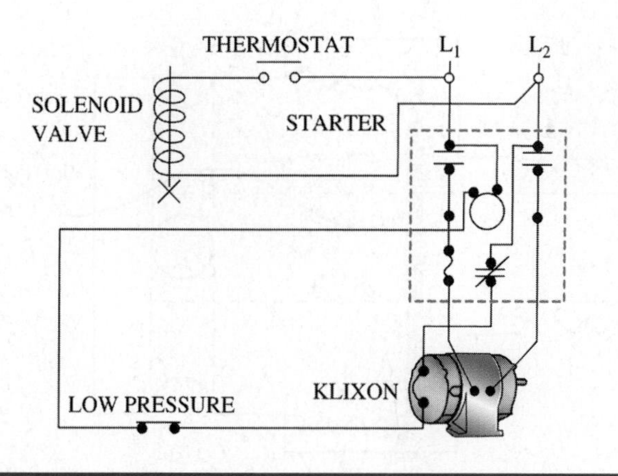

Figure 8-2-48 Pump down control used to stop the system on low pressure after refrigerant in the evaporator is pumped out. Protects the compressor on startup against liquid slugging.

A pressure control is connected to the low side of the system. As the pressure rises in the evaporator coil, the pressure control closes and completes the circuit through the safety device to the contactor coil. The compressor starts and the system operates normally. When the thermostat is satisfied, the solenoid coil is de-energized and the flow of liquid is cut off to the evaporator. Because the compressor is still running, the evaporator pressure is reduced to the cutout point of the low pressure control. The control contacts open and the compressor stops.

This system is designed to keep liquid refrigerant from filling the evaporator during shutdown. Regardless of the position of the thermostat, the low pressure control will operate the compressor upon sensing a rise in pressure in the evaporator.

UNIT 2—REVIEW QUESTIONS

1. What is the function of controls?
2. List the three types of materials used in electrical control systems.
3. What is one of the big advantages of the pneumatic control systems?
4. List the four types of basic controls.
5. What are the two categories that pressure controls can be divided into?
6. Where is the Bourdon tube control frequently found?
7. Define range.
8. Define differential.

TECH TALK

Tech 1. I was working on a system the other day and someone had installed a liquid line solenoid up in the attic just before the evaporator. What did they do that for?

Tech 2. What type of metering device did the system have?

Tech 1. It was a fixed bore, I think. One of those piston types.

Tech 2. Well, one thing that some companies do to increase the operating efficiency of their equipment when a piston type metering device is used is to put a liquid line solenoid that shuts down when the system goes off. That keeps the liquid in the condenser from flowing into the evaporator where it doesn't provide any cooling but drains off the pressure on the high side. If you can keep the liquid in the condenser then when the compressor starts up and the cooling cycle begins again you have immediate liquid coming into the evaporator and you have instantaneous cooling as compared to the system that might take a minute or two before it begins to equalize.

Tech 1. So they put it there to improve the efficiency of that type of system.

Tech 2. That's right. It's a pretty simple installation that some of the manufacturers have used to boost their equipment one SEER level without having to go to a TXV.

Tech 1. Now that makes sense. Thanks, I appreciate that.

9. What is the four way reversing valve used for?

10. What is a check valve?

11. What does short cycling mean?

12. What is the advantage of a current relay?

13. The ________ is a protective device to prevent the operation of a unit when there is inadequate fluid flow.

14. What are the two simple methods for operating refrigeration systems?

UNIT 3

Wiring Diagrams

OBJECTIVES: In this unit the reader will learn about wiring diagrams including:

- Circuits
- Total Comfort Systems
- Condenser Wiring

CIRCUITS

Wiring diagrams are made for different purposes, and it is important to recognize each type and know the intended purpose of each. Two common types are the *field connection diagram* and the *schematic diagram*.

The field connection diagram, Figure 8-3-1, identifies the various electrical controls in the control box and indicates the necessary field wiring by means of broken lines and shaded areas. The field connection diagram is designed to instruct the installing electrician or technician how to run the proper power supply to the unit and the correct wiring between sections if the unit consists of more than one section.

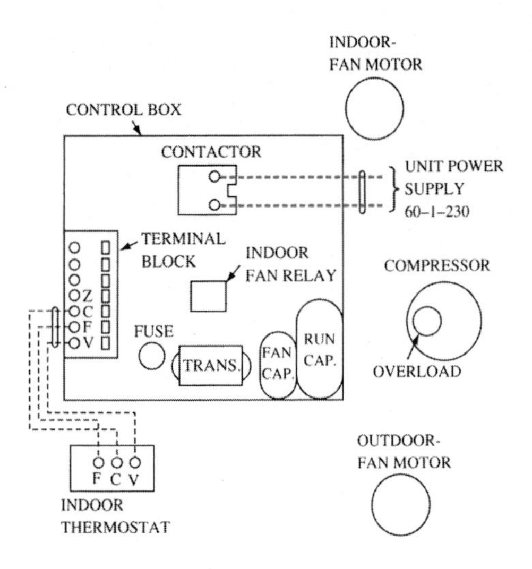

Figure 8-3-1 Field connection diagram for an air conditioning unit.

Internal wiring is not shown since the principal use of the field connection diagram is for installation.

A typical schematic diagram for an air cooled packaged air conditioning unit is shown in Figure 8-3-2. It is easy to see that this type of diagram contains more information than the field connection diagram. It is used by service personnel to see how the system works and why. Manufacturers include a copy of the diagram in the control cover or access panel of the unit. The load devices (motors) and the controls are represented by symbols, the circuits by lines. Note that some lines are heavy while others are light. The heavy lines represent wires in the power circuit, which supplies higher voltage current to the heavy electrical loads, such as motors. The light lines represent the control circuit which supplies low voltage through a stepdown transformer to the light load devices such as relay coils that are used only to actuate switches.

TECH TIP

In some cases the high voltage sealed wiring must be installed by a licensed electrician. Check with your local building official to determine the requirement in your area.

To build this cooling electrical circuit, component by component, start with the power supply to the unit as shown in Figure 8-3-3. Since these wires must be run by the installing electrician or the technician in the field, they are represented by heavy broken lines. For protection, the power supply must pass through a properly sized, fused disconnect switch. Proper fuse sizes are found in the equipment installation manual. If located outdoors, the fused switch must be installed in a weatherproof enclosure. The power supply shown is single phase, 60 Hz, 230 V current, coming in through wires L_1 and L_2.

The heaviest load, in this case the compressor, is placed in the diagram next and connected to the

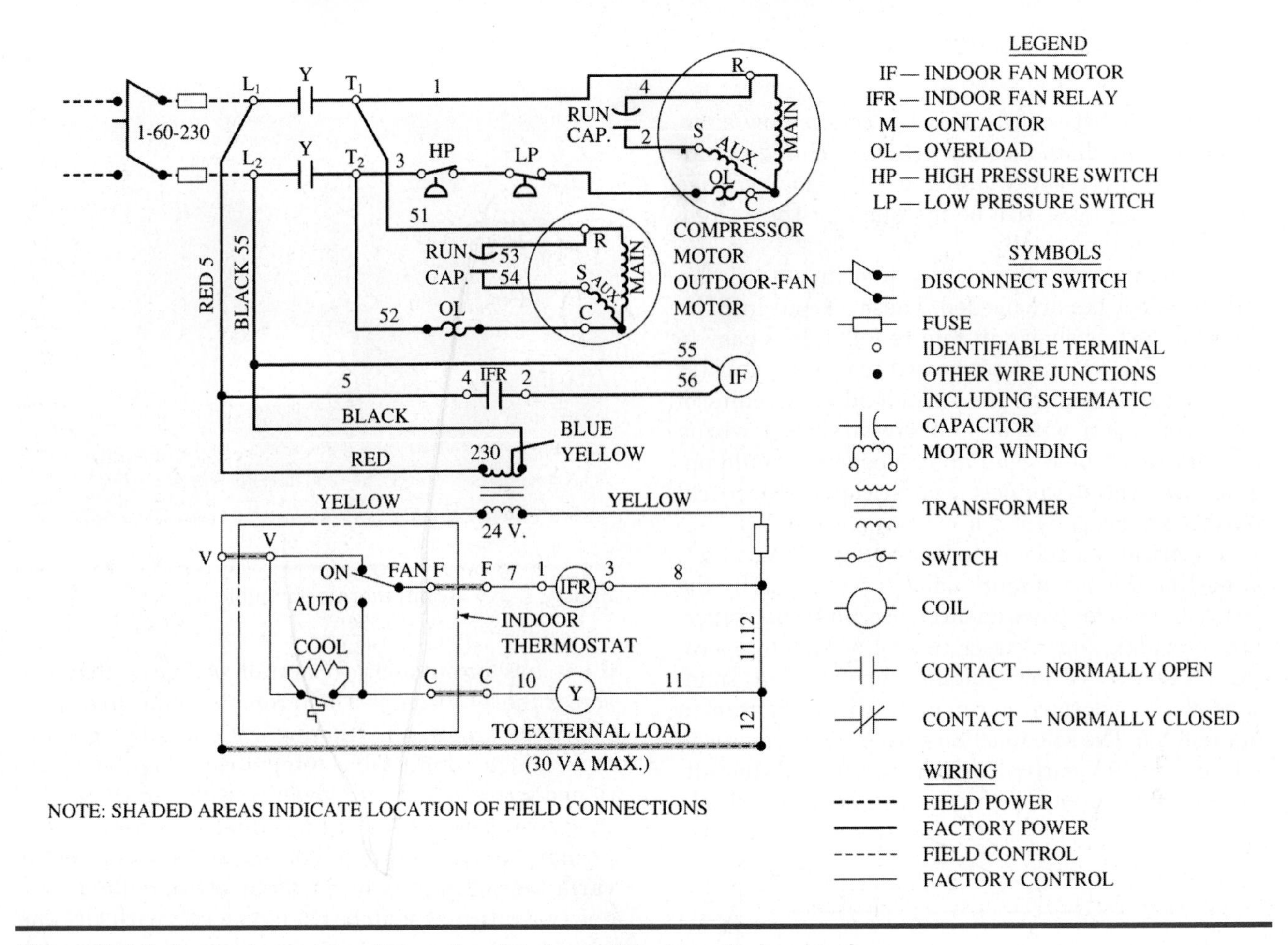

Figure 8-3-2 A complete wiring diagram for the air conditioning unit showing the connection of a FAN ON AUTO switch and cooling thermostat.

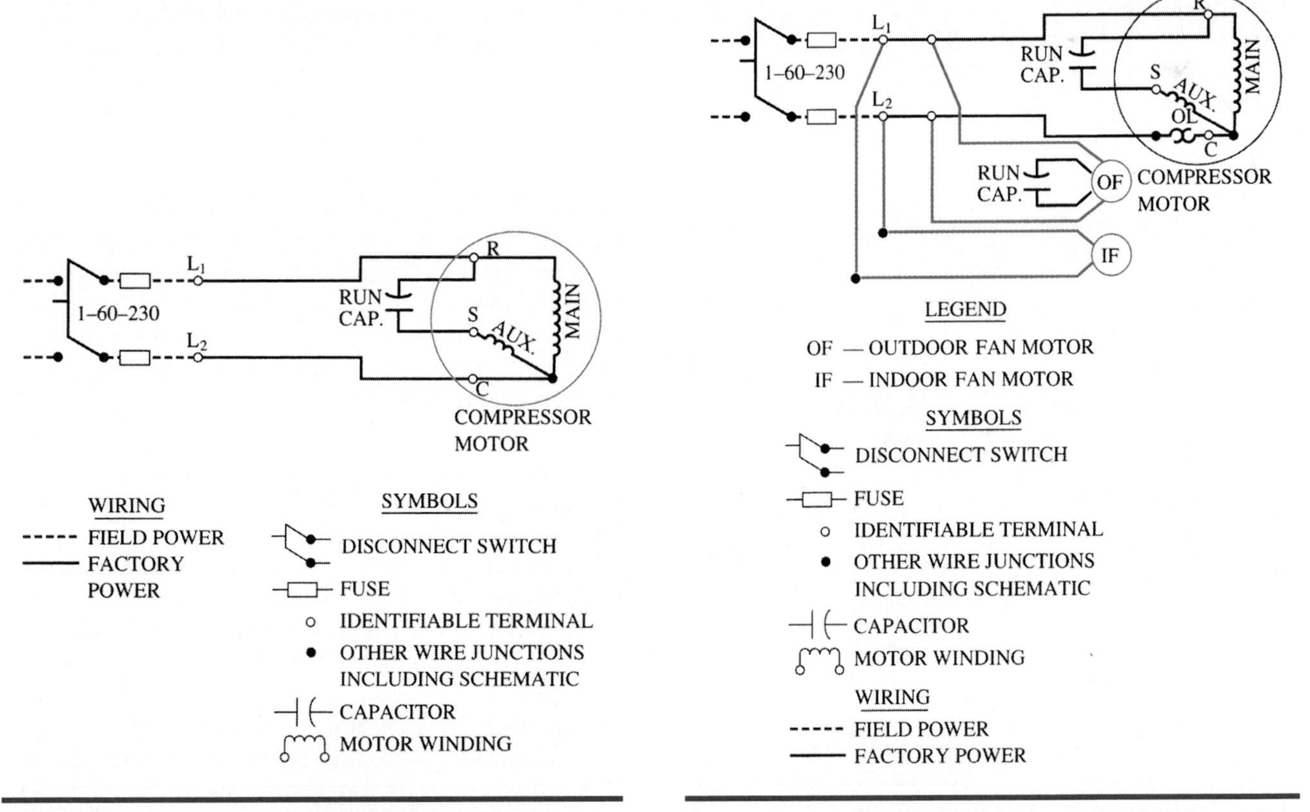

Figure 8-3-3 Diagram showing the wiring from the disconnect to the compressor motor.

Figure 8-3-4 Wiring diagram showing the addition of the outside and inside fan motors.

power supply between L_1 and L_2. This is now a circuit and if the disconnect switch were closed manually, the compressor would run. The low voltage control wiring must still be installed for the system to operate automatically.

Since this is an air cooled system, an outdoor fan and an indoor fan are needed. These are put into the diagram next as shown in Figure 8-3-4. It is easy to see that both fans are connected across L_1 and L_2, and all three motors are in individual circuits of their own. Again, with manual operation, the wiring diagram is complete, and all components would operate with the disconnect switch closed. Electrical controls are needed for automatic operation.

Electrical controls use a low voltage, 24 V circuit, permitting the use of much smaller wire, making the circuit much safer for residential use, and most important, providing much closer control of system operation. A source of low voltage current is a small stepdown transformer, shown in the wiring diagram in Figure 8-3-5. The sole function of the transformer is to reduce the 240 V current to 24 V, which is all the voltage needed to supply the necessary current to actuate

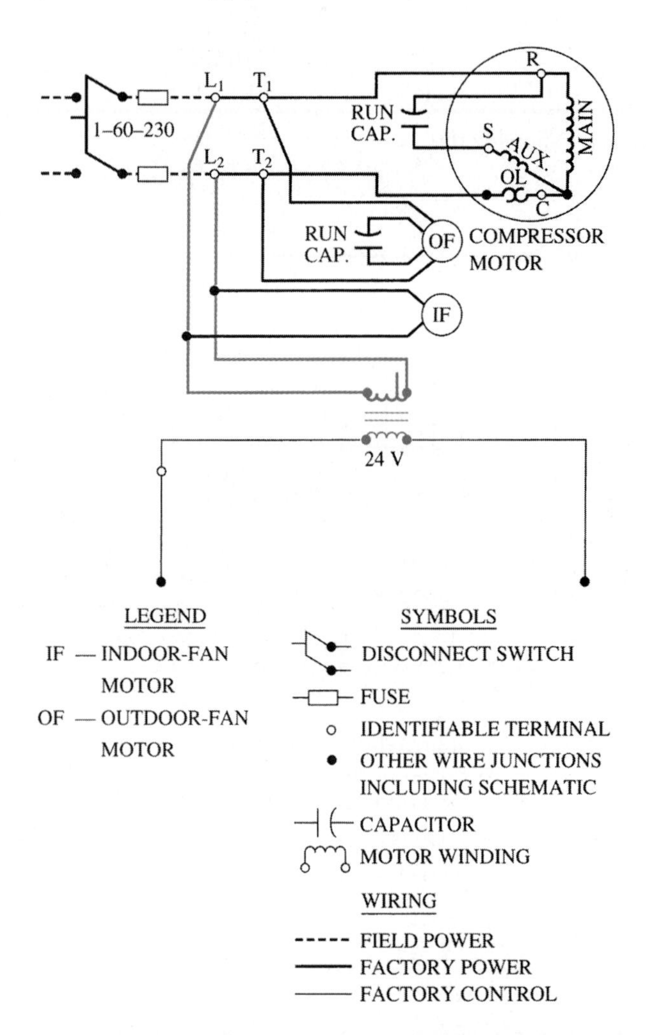

Figure 8-3-5 Wiring diagram showing the addition of a low voltage control transformer.

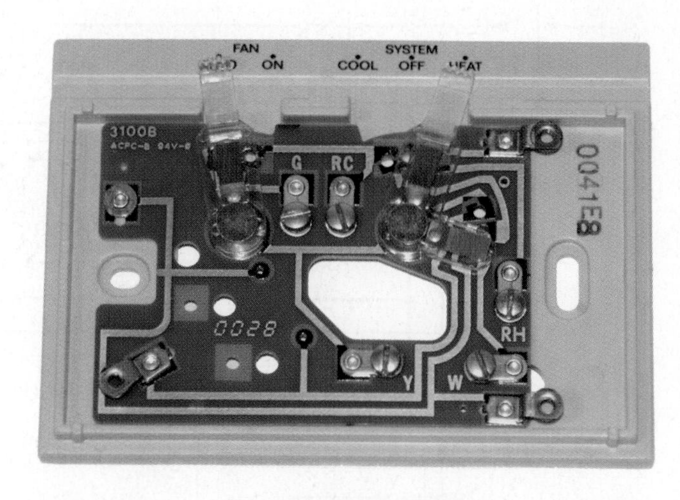

Figure 8-3-6 Room thermostat subbase.

the relays for automatic operation. A control installed across the 24 V source would complete the circuit.

Since an air conditioning system is designed to maintain a comfortable temperature in the conditioned space, a low voltage adjustable indoor cooling thermostat as shown in Figure 8-3-6 is used as the primary control system. Although there are many variations of indoor thermostats, the one illustrated has two internal switches: a fan switch with ON and AUTO positions, and a cooling control switch with positions marked OFF and COOL.

Most manufacturers of thermostats and central systems use a standard terminal identification such as shown in Table 8-3-1. Technicians will often find in the field that the thermostat wire that was provided does not contain a red, yellow, green, and white wire. Most often a four strand thermostat wire used for most heating and cooling applications contains a red, green, white, and black wire. Over time the yellow and white wires will fade and the colors look very similar. Most heat pumps use a six strand thermostat wire and the most common colors are red, white, green, black, blue, and brown. For systems requiring larger bundles of conductors in the thermostat wiring manufacturers provide wire bundles with nine and twelve conductors.

TECH TIP

Although the thermostats are often marked with color coded terminals, not all thermostat wiring has the same colors available. It is important that you pay attention when connecting a new thermostat that you reconnect the same terminal identification with the previously used thermostat wire for that purpose. For example, the former installer may have used the black conductor for the cooling wire marked Y in the thermostat. You must use the same black wire when replacing a thermostat to ensure that it will work correctly.

TABLE 8-3-1 Standard Terminal Identification for Thermostats

Terminal Identification*	Function
R	Power from transformer
RC	Power from cooling only transformer
RH	Power from heating only transformer
C	Common from transformer
X	Same as common from transformer
W	Heat valve or relay
W1	1st stage heat
W2	2nd stage heat
Y	Cooling contactor
Y1	1st stage cooling
Y2	2nd stage cooling
G	Fan relay
B	Heat pump changeover valve to heating
O	Heat pump changeover valve to cooling
E	Emergency heat relay
L	Emergency heat indicator lamp

* Some terminal identification letters are closely associated to the color of wire often used. For example R is often red, W is often white, G is often green, and Y may be yellow; however, not all thermostat control wires contain standard colored wires.

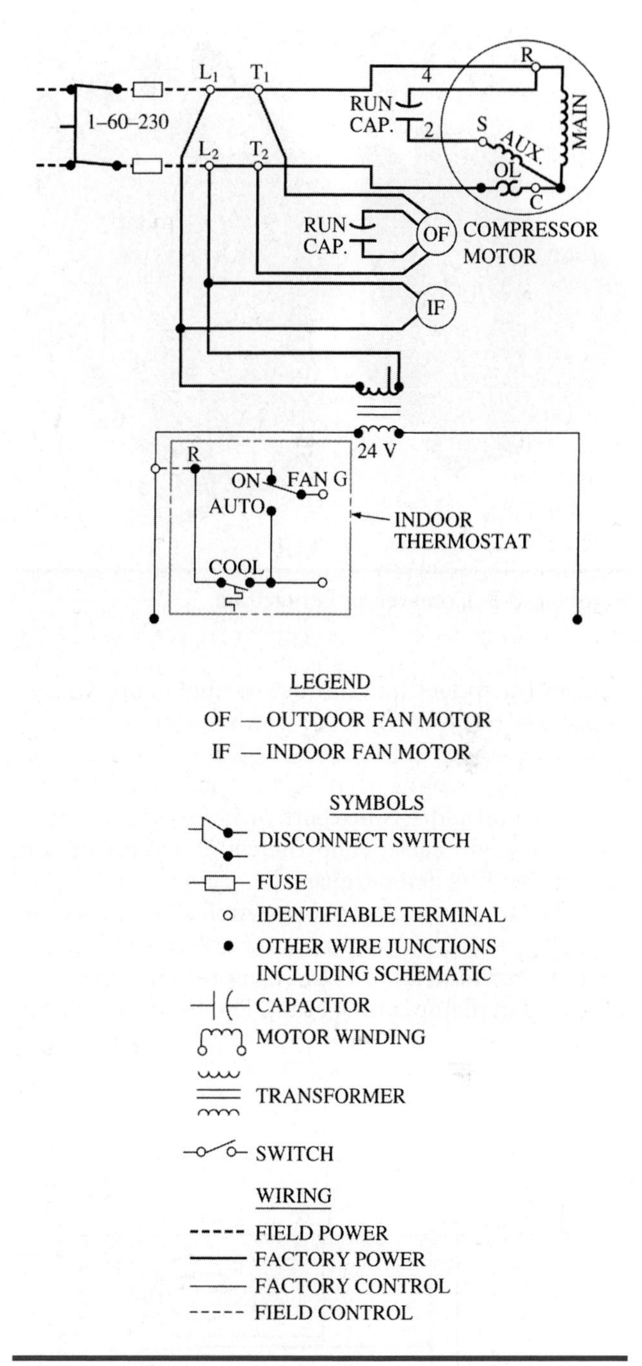

Figure 8-3-7 Wiring diagram showing the connection of the room thermostat to one side of the low voltage circuit.

The fan switch controls the indoor fan only. In the ON position, the fan will operate continuously; in AUTO, fan operation is actually controlled by the position of the cooling control switch. The contacts of the cooling switch are sealed inside a tilting bulb partially filled with mercury. When the bulb is tilted so that mercury covers only one contact, the switch is open. When the bulb is positioned so that mercury covers both contacts, the switch is closed. The bulb is tilted by means of a bimetal element that warps as it feels a change in temperature, to either close or open the mercury switch. The symbol R is the common low voltage terminal feeding through the fan switch to terminal G and through the mercury bulb to terminal Y.

When installed in the electrical circuit, the common terminal R of the thermostat is connected to the L_1 side of the low voltage supply as shown in Figure 8-3-7, but nothing will happen until a circuit is completed to the other side of the 24 V supply. A few more controls are needed for automatic operation.

To control the compressor motor automatically, a contactor is required as shown in Figure 8-3-8. This can be described as an oversized relay with a switch large enough to carry the heavy current drawn by motors, and a magnetic coil strong enough to actuate the larger switches.

In order for the cooling switch in the thermostat to govern the action of the contactor in starting and stopping the compressor motor, the contactor coil Y is wired in series with the cooling switch of the thermostat, Figure 8-3-9. The contactor switches must be connected in the power circuit so that they control both the compressor motor and outdoor fan motor,

Figure 8-3-8 Compressor contactors.

but not the indoor fan; it must be able to operate independently. Note that the contactor coil and both contactor switches are marked Y for identification. This simply means that coil Y actuates the two switches marked Y. Since both switches are normally open, they will close when the coil is energized and open when it is de-energized.

With the main disconnect switch closed now, current flows to the indoor fan motor and the 24 V transformer immediately, but the compressor motor and outdoor fan motor cannot start. The thermostat cooling switch is open so no current can flow to the contactor coil to close its switches to the compressor and outdoor fan motors.

When the thermostat cooling switch closes, calling for cooling, current immediately flows through contactor coil Y, and the two switches marked Y close at the same time. (See Figure 8-3-10.) This energizes and starts both the compressor motor and the outdoor fan motor at the same time. When the thermostat is satisfied, the cooling switch again moves to the open position. The compressor and outdoor fan motors stop, but the indoor fan motor continues to run. It will remain in operation as long as the disconnect is closed, regardless of the position of the fan switch.

Another control is needed for the indoor fan motor, IF (also referred to as the indoor blower motor). A relay with a coil and a normally open switch, both marked IFR for indoor fan relay (also referred to as the indoor blower relay), is used as shown in Figure 8-3-11. Note that the coil is wired in series with the fan switch in the thermostat, and its switch is wired into the indoor fan motor circuit.

With the fan switch in the ON position, low voltage current now flows through the fan switch to energize relay coil IFR and close its switch in the indoor blower power circuit. The indoor blower is now running, and it will continue to run as long as the fan switch remains in the ON position, regardless of the position of the cooling switch. The cooling system—the compressor and the outdoor fan—is free

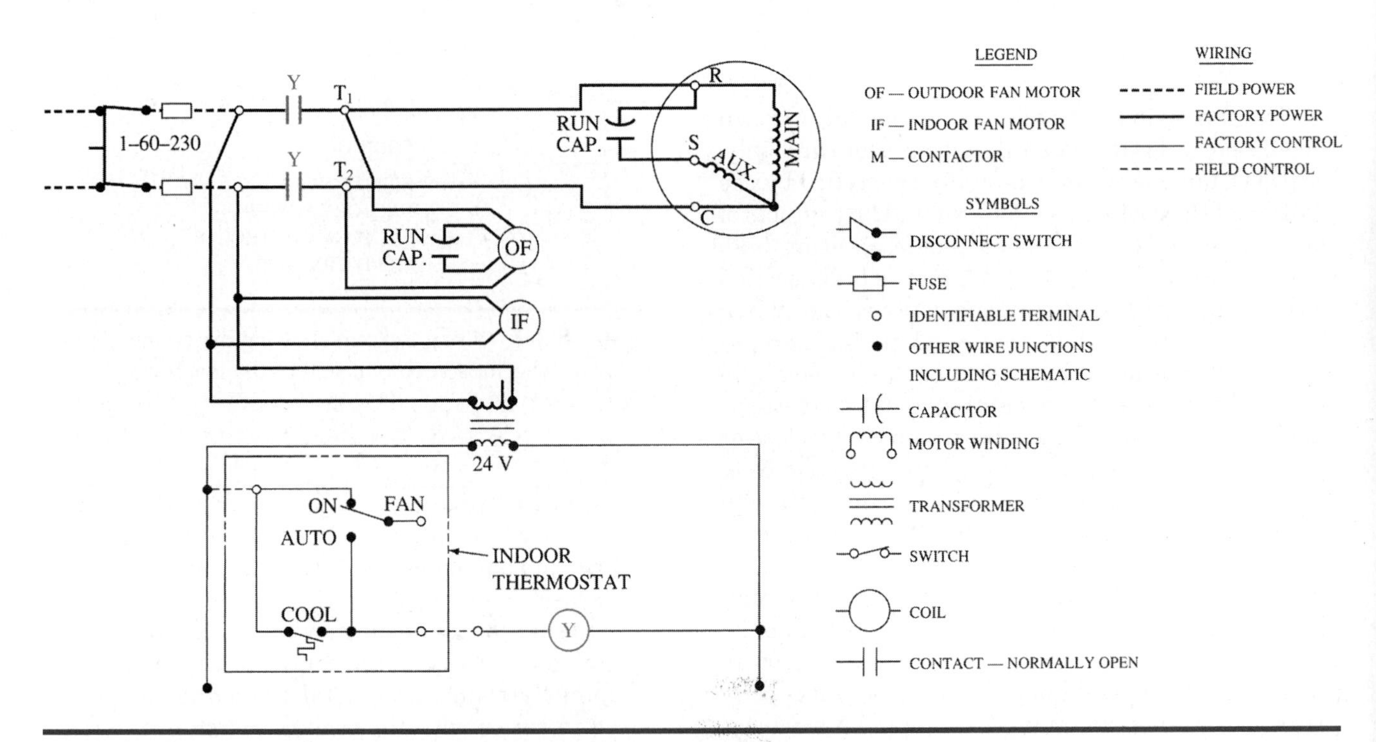

Figure 8-3-9 Wiring diagram showing the addition of the compressor contactor coil placed in the cooling circuit.

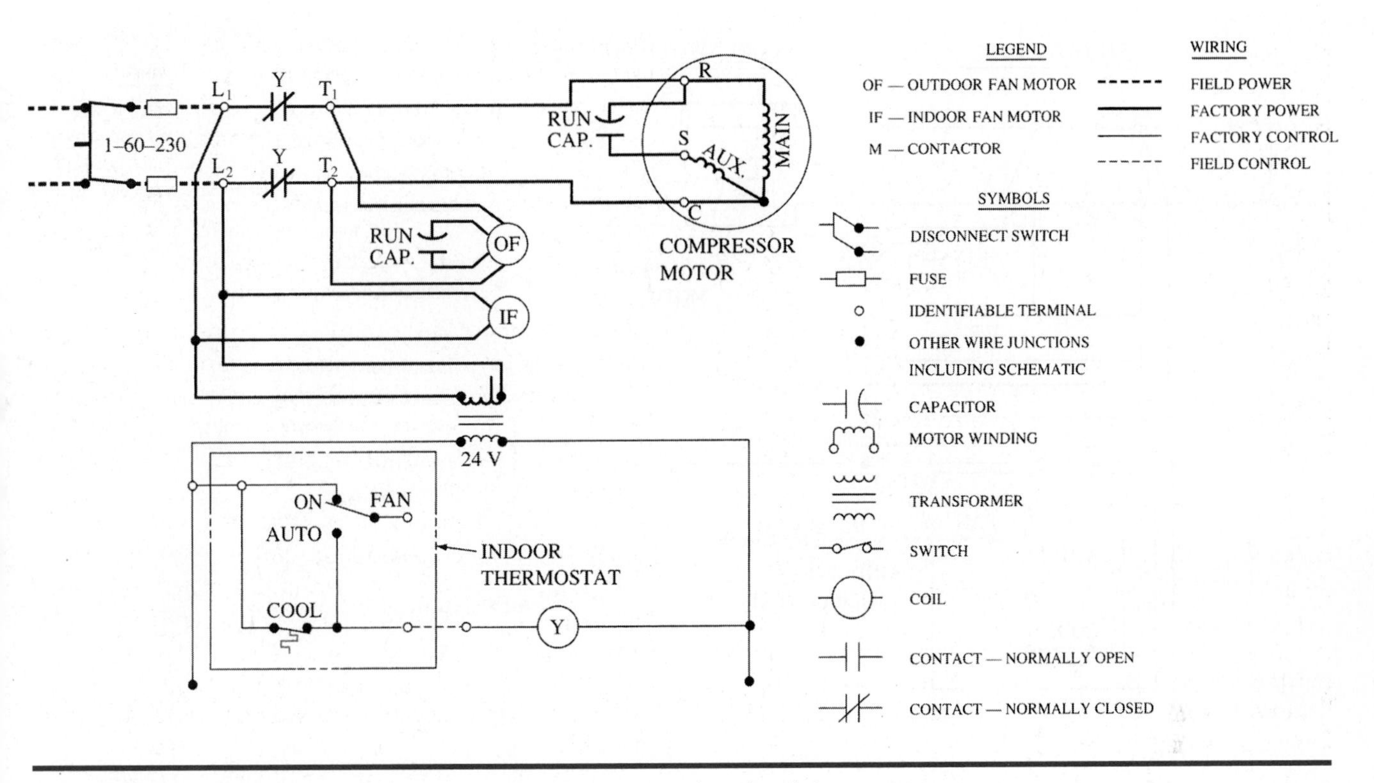

Figure 8-3-10 Wiring diagram with the cooling thermostat closed and the compressor contactor energized.

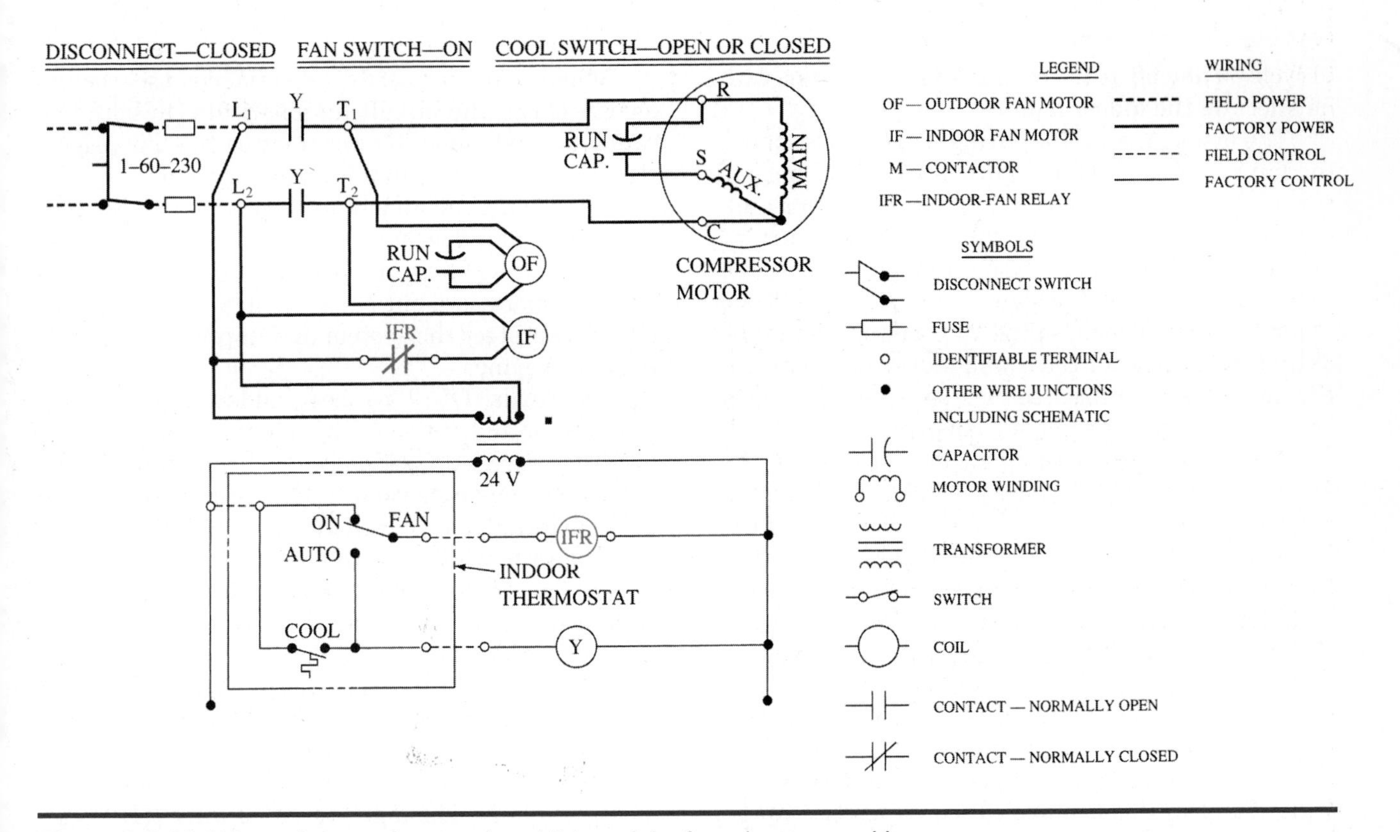

Figure 8-3-11 Wiring diagram showing the addition of the fan relay operated by the FAN ON switch.

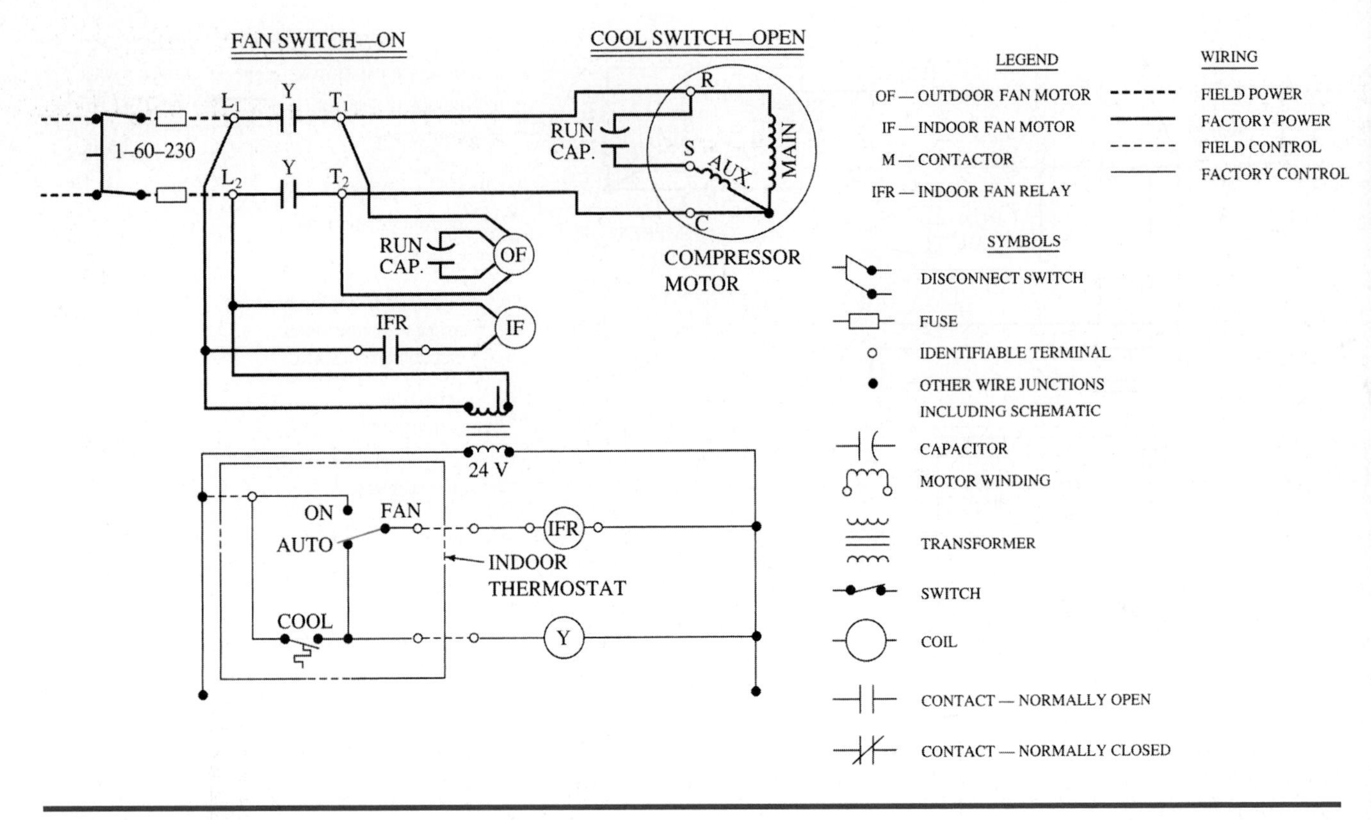

Figure 8-3-12 Fan switch in AUTO position.

to cycle on and off as the cooling switch dictates, with no effect on the indoor fan.

The last change to the wiring diagram is to move the fan switch to AUTO, Figure 8-3-12. The thermostat cooling switch is open; it is not calling for cooling. When the disconnect switch is closed, current immediately flows through the transformer into the 24 V supply, but the fan switch is in AUTO and the cooling switch is open, so the 24 V circuit cannot be completed. Contactor coil Y and indoor fan relay coil IFR cannot be energized, so their switches all remain open, and none of the motors can operate.

Assume that the cooling switch closes and calls for cooling without changing the position of the fan switch or the disconnect (disconnect closed) and the fan is on AUTO, as in Figure 8-3-13. Current immediately flows through the cooling switch and through both the contactor coil Y and the indoor fan relay coil IFR at the same time. This closes both switches marked Y in the power circuit to the compressor and outdoor fan motors, as well as the switch marked IFR in the power circuit to the indoor fan motor.

The compressor and outdoor fan motors are energized and started through closed switches Y. At the same time, the indoor blower motor is also started through closed switch IFR. The entire system is now energized and all motors are running automatically.

When the thermostat is satisfied, the cooling switch opens the circuit automatically to both the contactor and indoor blower relay coils, their respective switches immediately snap to their normally open positions, and all motors stop—compressor, outdoor fan, and indoor fan.

The wiring diagram is correct from the standpoint of operation, but safety controls must also be added to protect the system and improve its operating performance.

Overloads (OL) have been added to protect the compressor and outdoor fan motor windings against excessive current. Overloads may be installed inside the motor housing as shown for the compressor motor or externally as shown for the outdoor fan motor. Although the windings are not shown for the indoor blower motor, all small single phase motors are equipped with some type of protection.

As an added precaution against compressor operation under abnormal conditions such as excessive discharge pressure or dangerously low suction pressure, a high pressure switch and a low pressure switch have been wired in series with the compressor motor windings. The high pressure switch (HP) is actuated by system discharge pressure to open and stop the compressor if the discharge pressure exceeds the switch setting. The low pressure switch (LP) is

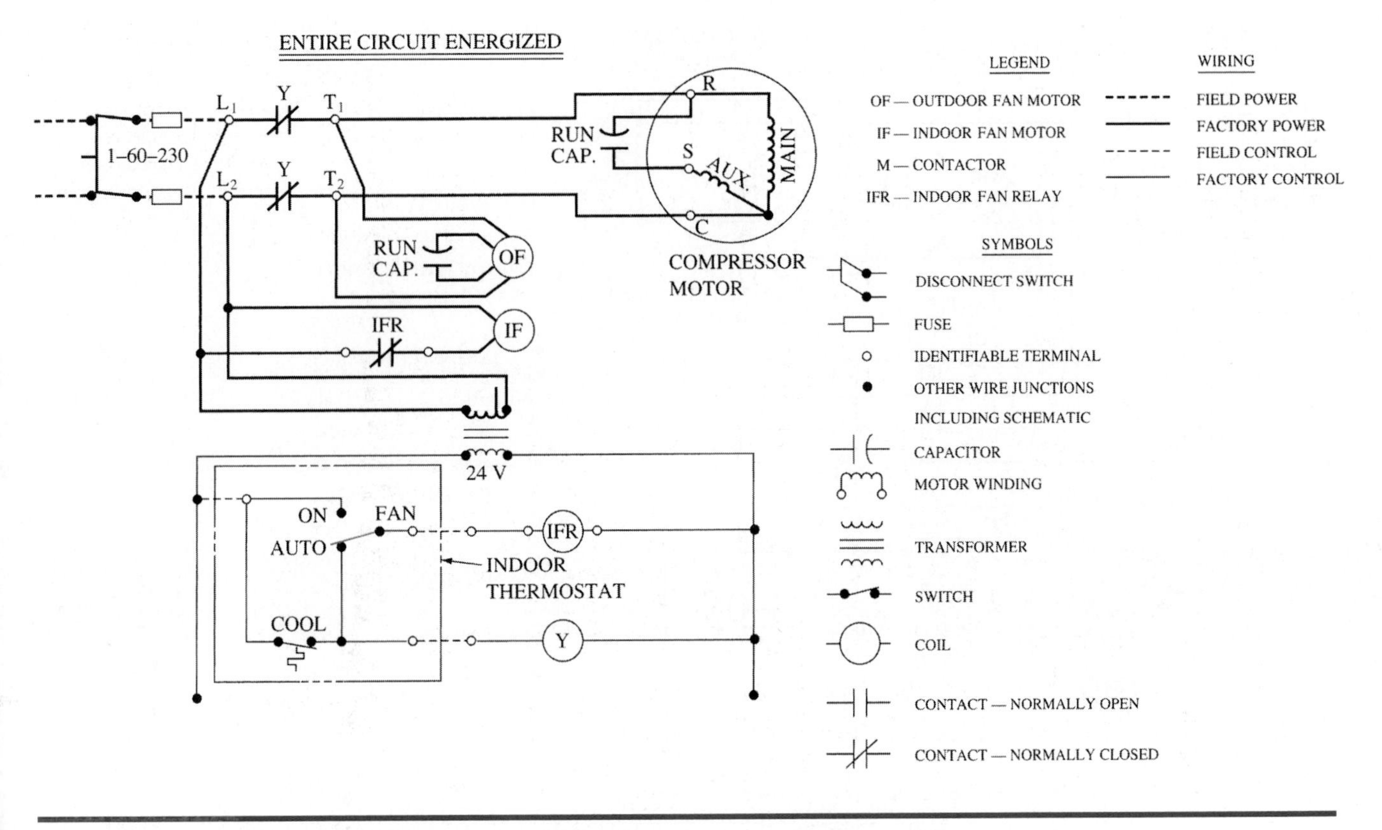

Figure 8-3-13 Wiring diagram showing fan in AUTO position and thermostat calling for cooling.

actuated by system suction pressure to open and stop the compressor if the suction pressure falls below the switch setting. To protect the relay coils and wiring in the low voltage control circuit, a fuse has been installed that is properly sized to open the 24 V circuit in case of excessive current.

For closer control of air temperature in the conditioned space, an anticipator has been wired across the terminals of the thermostat cooling switch. It is a fixed, nonadjustable type.

One relay coil may actuate many switches. Each switch may be located in a different section of the wiring diagram. As the diagram becomes larger and more relays are used, these switches become more difficult to locate. To make it easy to locate the switches of each relay, the ladder diagram is often used.

On the right side of the ladder diagram, Figure 8-3-14, opposite each relay coil, are shown the line numbers in which its switches are located. For example, the diagram shows that contactor coil Y has two switches located in lines 1 and 4, while relay coil IFR has switches located in lines 9 and 10.

The factory installed wiring has also been color coded or given a numerical designation by the manufacturer for ease of identification.

Total Comfort Systems

For simplicity, this text will review a total comfort system (TCS) using an upflow gas furnace with single speed fan operation, a split system air cooled condensing unit (with indoor coil), a combination heat/cool thermostat, a humidistat, and an electronic air cleaner.

First, the physical wiring and point to point connections are illustrated in Figure 8-3-15. This is what the electrical installer will use as a guide during the installation. Note the need for two junction boxes to make the required interlocks. The low voltage wiring is also detailed as to the number of conductors needed and the terminal or junction connections. Power wiring is black (B) and white (W), and ground conductors are sized according to the load and to meet local code requirements. The internal wiring of the furnace, condensing unit, and other components are not shown on the installation diagram.

The schematic ladder diagram for the same system is shown in Figure 8-3-16. First locate and relate the major components to the previous drawing: the furnace blower motor, the humidifier, the electronic air cleaner, and the condensing unit. Note that the combination thermostat has a HEAT-OFF-COOL

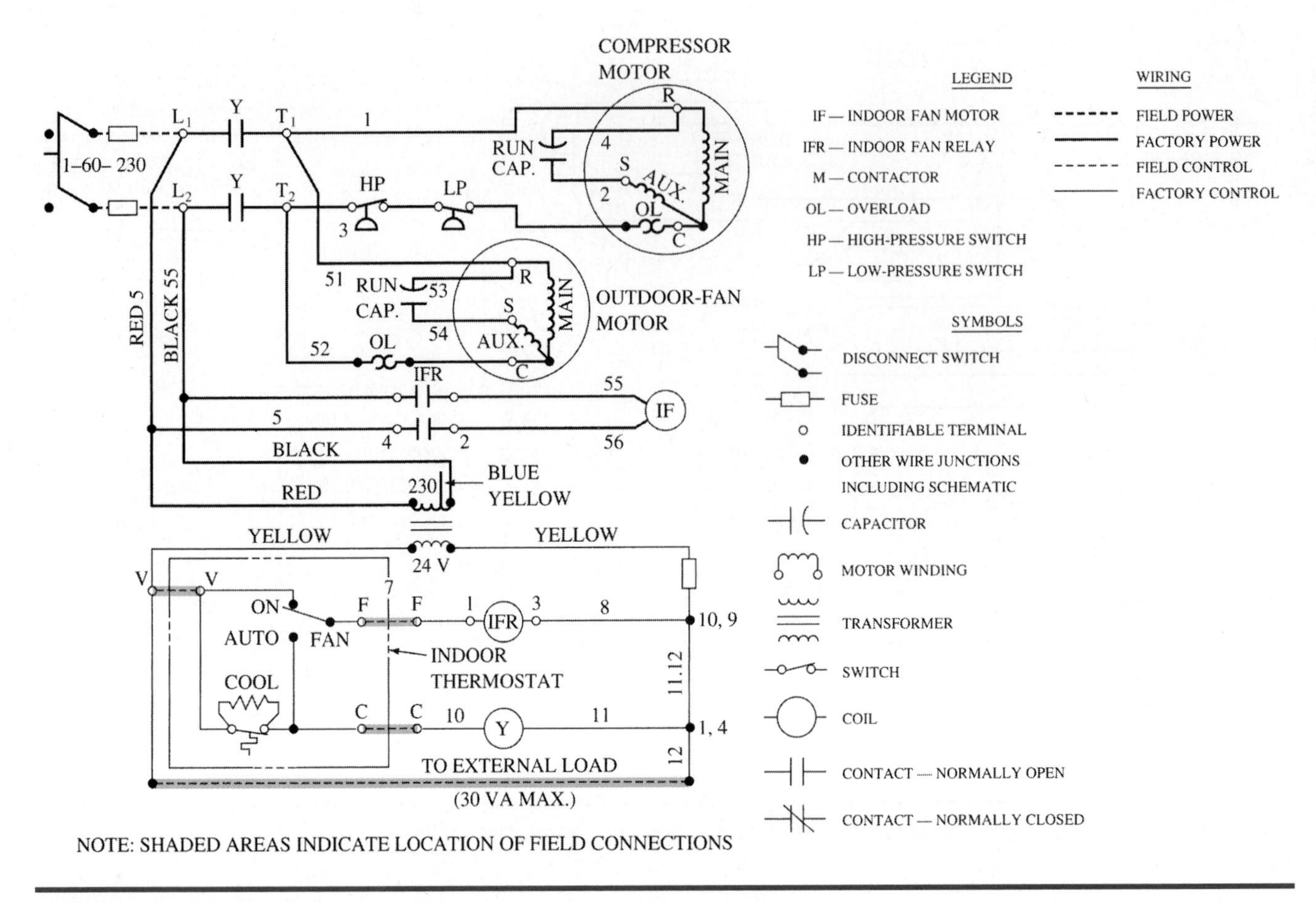

Figure 8-3-14 Complete ladder diagram for the air conditioning unit.

system selection and an ON-AUTO fan selection. Anticipators are not shown. The system diagram is shown on the heating mode.

Closing the disconnect switch (line 4) will produce a 120 V current to the transformer. Trace the black wire through the closed limit switch (LS) to the transformer primary coil and to the white power return (line 5). A low voltage potential is created.

Low voltage can now be established through the fan switch ON position and energize the FR (fan relay) coil in line 9. This closes the FR switch in line 3, allowing line voltage current to flow to the fan motor and to the humidifier relay (HU) in line 1. Thus the fan runs constantly. Note that the FS (furnace switch) in line 4 may or may not be closed, depending on whether there is sufficient heat to close its contacts. The gas valve (GS) operation in line 13 will depend on the activation of the mercury bulb by the room temperature. Note in line 15 that the HEAT position selector switch feeds low voltage to H (the humidistat), and if humidity is needed, it closes its contacts and energizes the humidistat relay coil (HU), also in line 15. The contacts of HU are located in line 1 and will activate the humidifier (fan motor or water solenoid valve). This is called permissive humidification by automatic demand from the humidistat on the heating cycle with the normal cycle of furnace fan. Command humidification would be a system where the humidistat can command fan-only operation, regardless of the heating/cooling mode. The electronic air cleaner simply parallels the fan operation and is energized any time the fan runs.

If the fan selection is AUTOMATIC, FR cannot be energized during heating, but the furnace fan cycles from its fan switch, line 4. On cooling, however, there is a contact to the automatic terminal and the fan cycles with the cooling thermostat. Also, on cooling there is a completed circuit on line 11 to the condensing unit contactor terminals A and B. This pulls in the compressor and the outdoor condenser fan. They then cycle off and on with the room temperature mercury switch. Note that the middle system switch on cool drops out the gas valve and the lower switch drops out the humidistat. Note that the middle system switch on cool drops out the gas valve and the lower switch drops out the humidistat. Note also that on cooling, the furnace fan can operate from the FR relay (ON or AUTO) and thus furnish power to the air cleaner whenever it runs.

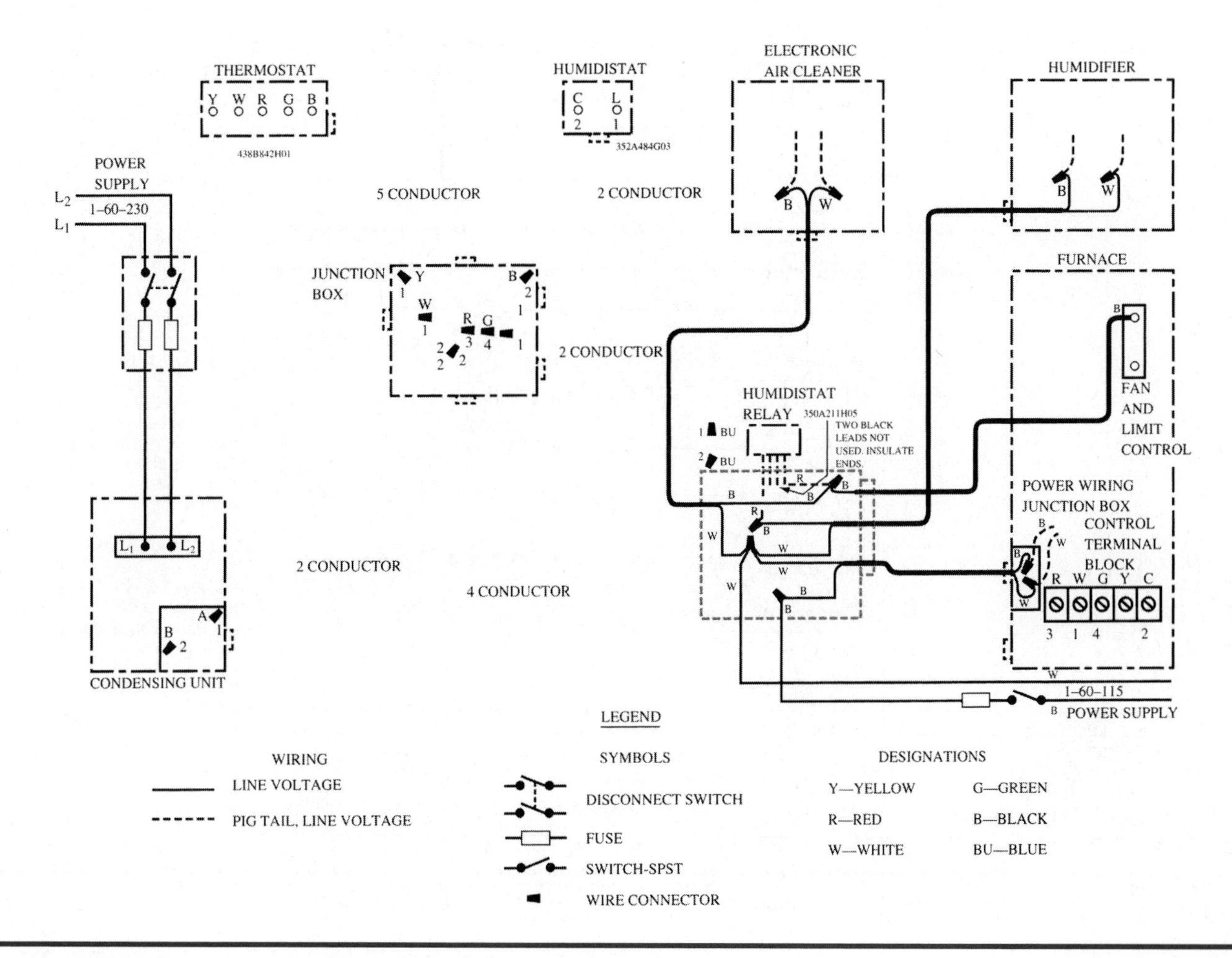

Figure 8-3-15 Connection diagram showing the wiring harnesses used to connect the terminals.

This is how a simple but common TCS is installed and wired schematically. The introduction of furnace fan speed control, two stage cooling and heating thermostats, and command humidification add sophistication and enlarge the diagram. These items do not necessarily add to its complexity if the schematic is analyzed step by step as to function and operation.

The types of wiring diagrams discussed in this unit are for air conditioning, gas heat, oil heat, heat pumps, and refrigeration systems. The various circuits on the diagrams included in this chapter have been color coded indicating line 1, line 2, or line 3 voltages. The drawing's color codes are done in such a way that if a voltmeter were to be used between any points of the same color, the voltage reading would be zero. If a voltage reading were to be taken between any two colors, a voltage would be indicated. The first simple example of this representation of the circuit is shown in Figure 8-3-17.

It is important to note that on many components both sides of the component are colored the same, Figure 8-3-18. This indicates that there is not a voltage difference or voltage drop across this component. A voltage difference or drop only occurs when current is flowing through a load. If current is not flowing, then most loads, such as transformer coils, motor windings, relay coils, and even a light, do not have a voltage drop; so both terminals are the same color.

Normally closed or closed contacts are not voltage consuming devices or loads; therefore, when current is flowing through these contacts, there is not a change in line color because there is no difference in voltage across the contacts. Normally open contacts or contacts that are open may have a color difference indicating that there is a voltage difference or voltage drop between the two terminals. A load that has a broken conductor that is powered up but not functioning will also show a color difference.

Primary voltage circuits are indicated with heavier colored lines than low voltage circuits. Lines that are not colored are not connected to the voltage circuits, Figure 8-3-19.

There are a number of control devices on the wiring diagrams. These devices can be grouped as

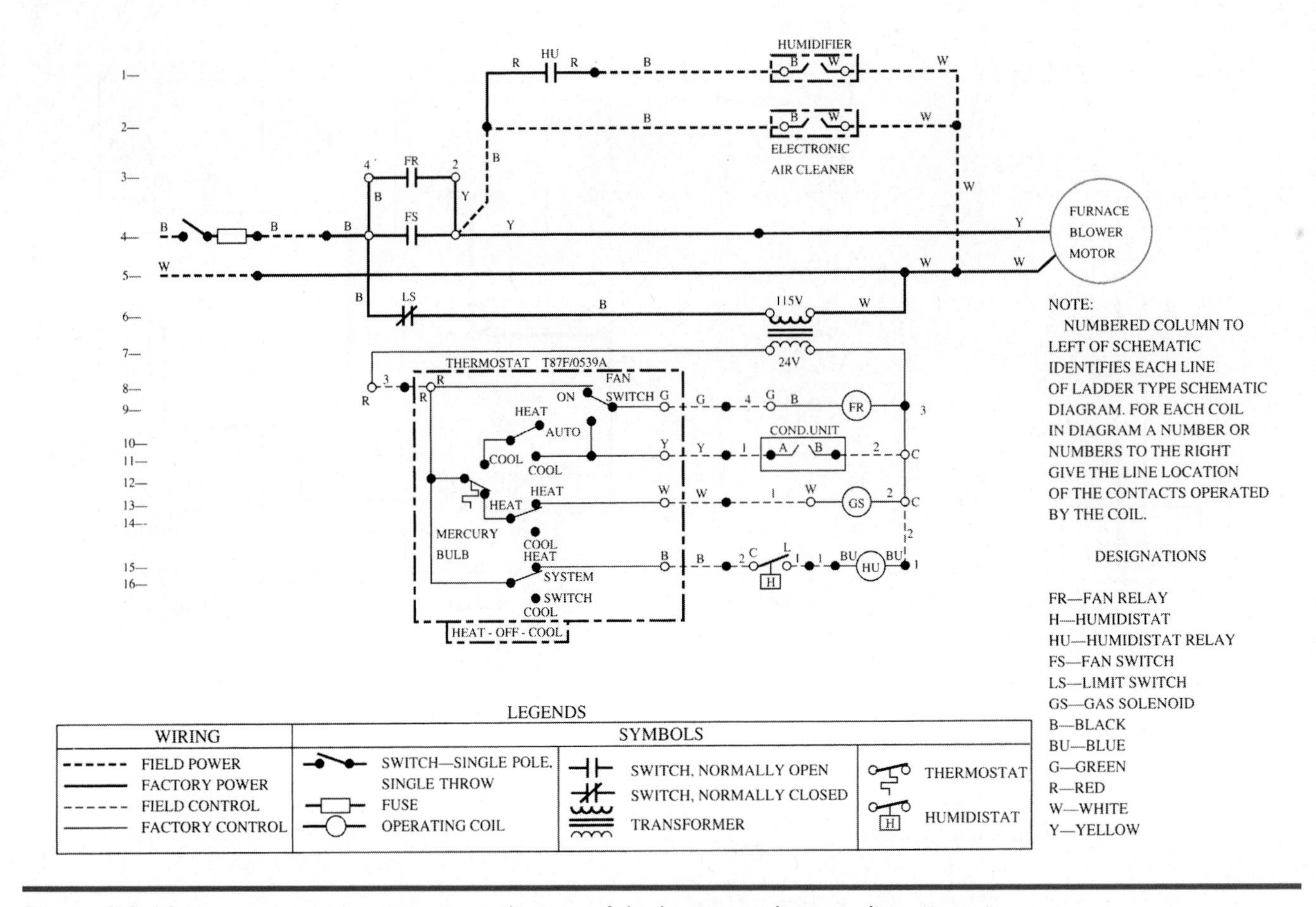

Figure 8-3-16 A complete schematic wiring diagram of the heating and air conditioning unit.

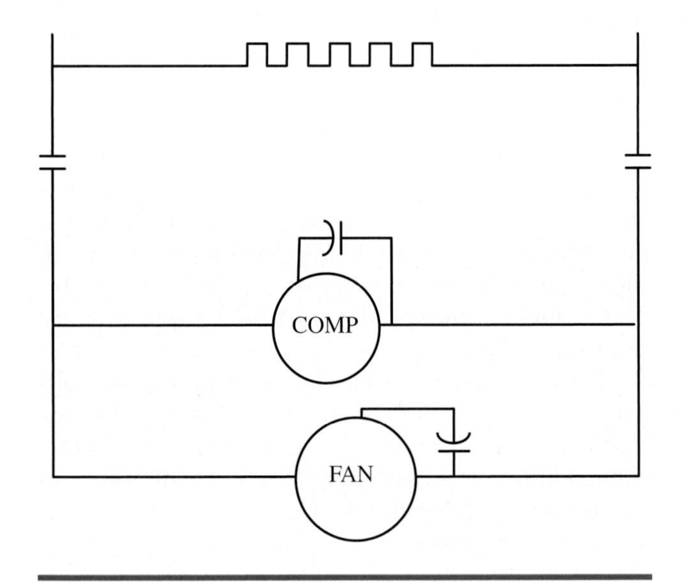

Figure 8-3-17 A ladder diagram for the complete system.

control devices or safety devices. Control devices are those components, such as pressure, temperature, current, or computer processors, that cause the system to start or stop as it provides heating, cooling, or ventilation. Safety devices are devices that prevent the system from operating with excessively high or low pressures, excessively high or low temperatures, excessively high current draws, or low voltage to prevent the system from operating within a dangerous or hazardous condition. The most common thermostat wire colors that are used to connect the outdoor unit to the indoor air handler contain a red and white wire. Some wires are available in 18 and 20 gauge sizes. The smaller 20 gauge wire may result in a significant voltage drop in the control circuit if the wire run is extremely long. It is therefore recommended that you use 18 gauge wire for any long run.

CAUTION: Never jumper across a safety device on any equipment.

CONDENSER WIRING

The basic condensing unit consists of a compressor fan and contactor, Figure 8-3-20a. When this unit is powered up the line voltage stops at the contactor. The secondary voltage, if any, that would be read from the common transformer lead would be read off of any point on the contactor coil wires.

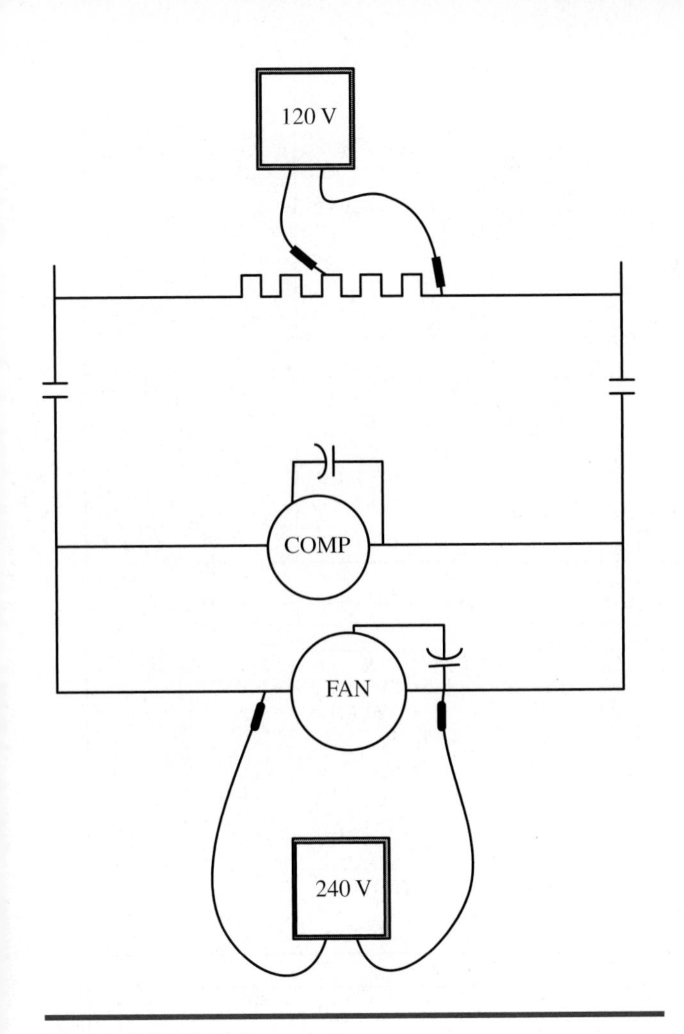

Figure 8-3-18 Voltage drop.

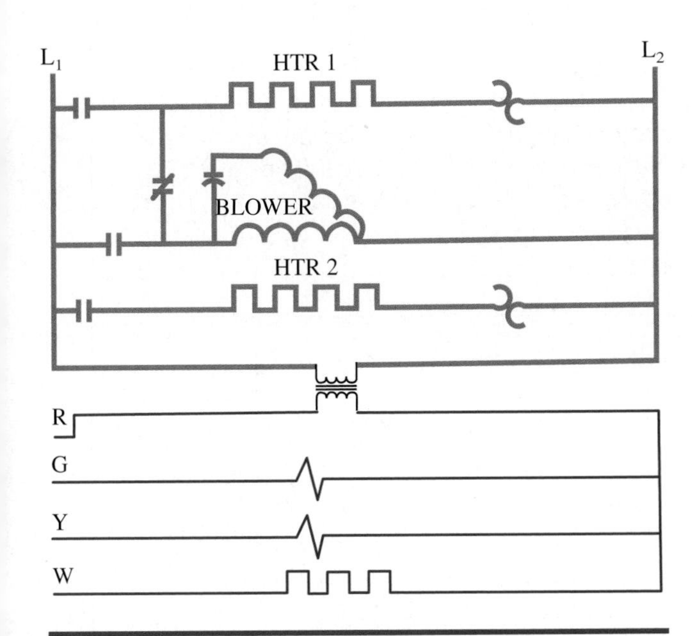

Figure 8-3-19 Primary (in red) and secondary (in blue) circuits.

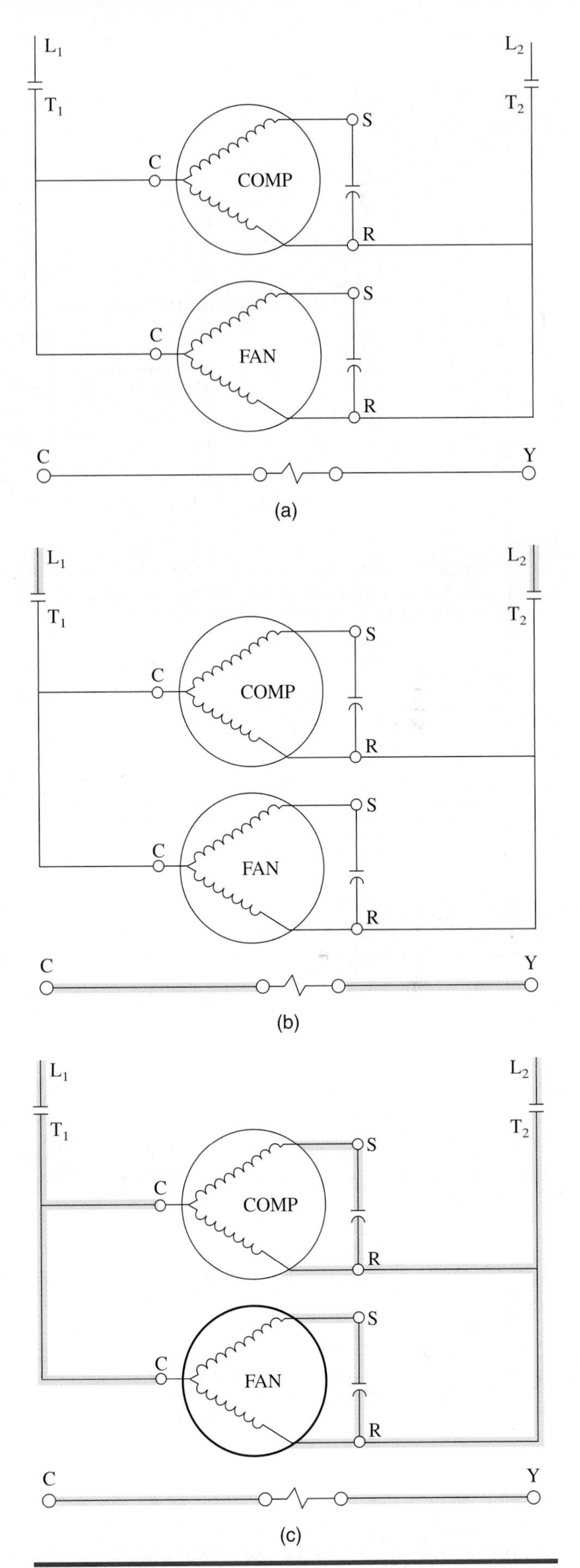

Figure 8-3-20 (a) Ladder diagram of a condensing unit; (b) Ladder diagram of a condensing unit, power on; (c) Ladder diagram of a condensing unit, system on.

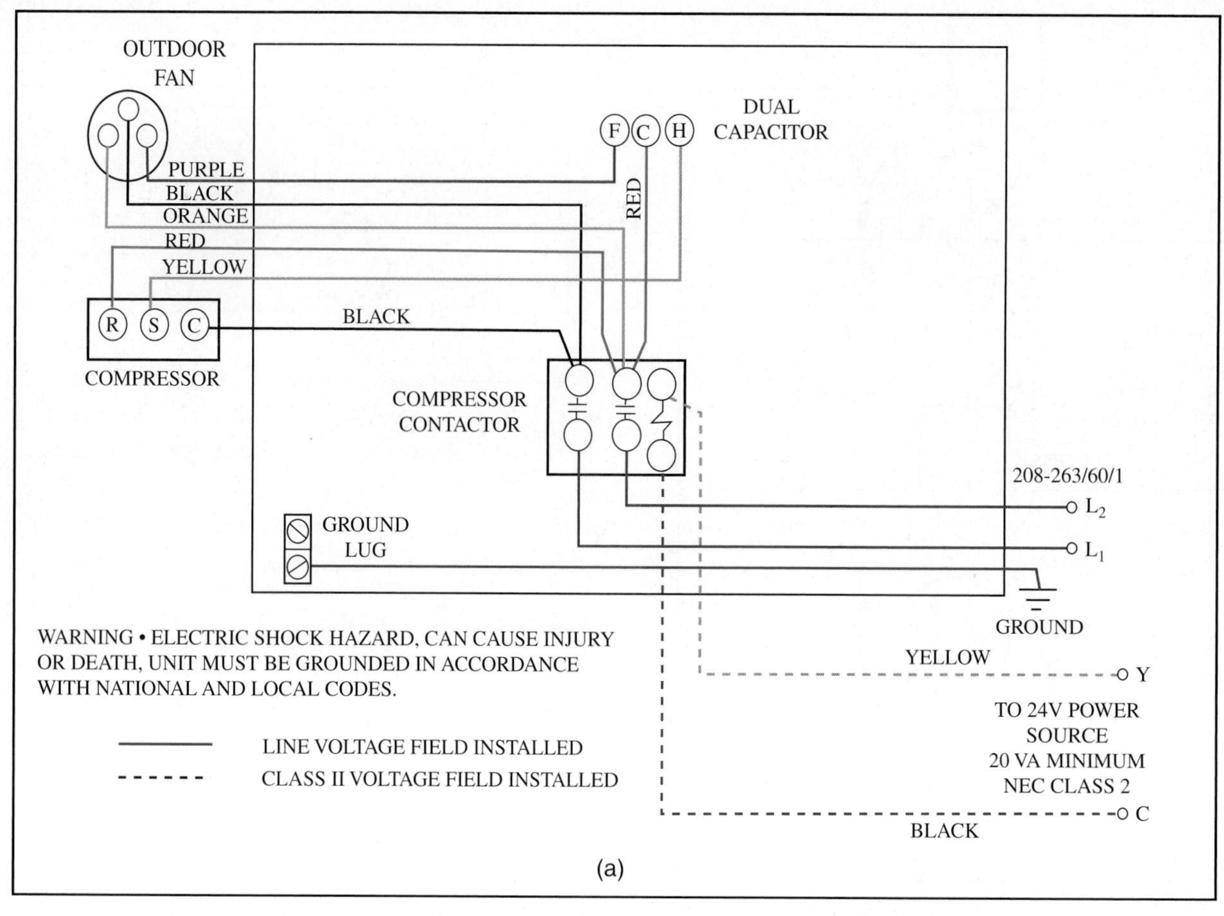

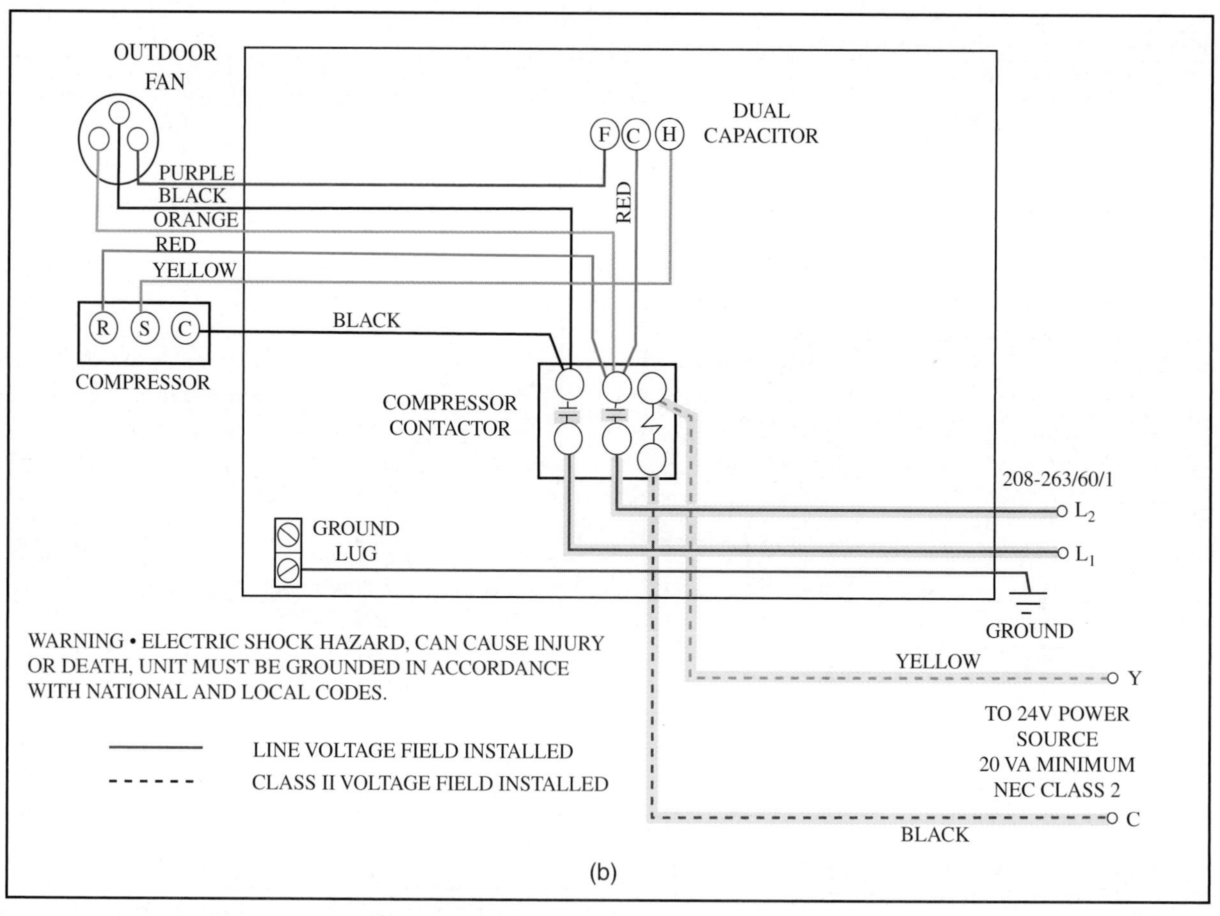

Figure 8-3-21 (a) Typical field wiring diagram; (b) Typical field wiring diagram, power on; (c) Typical field wiring diagram, system on. (*Courtesy Lennox Industries Inc.*)

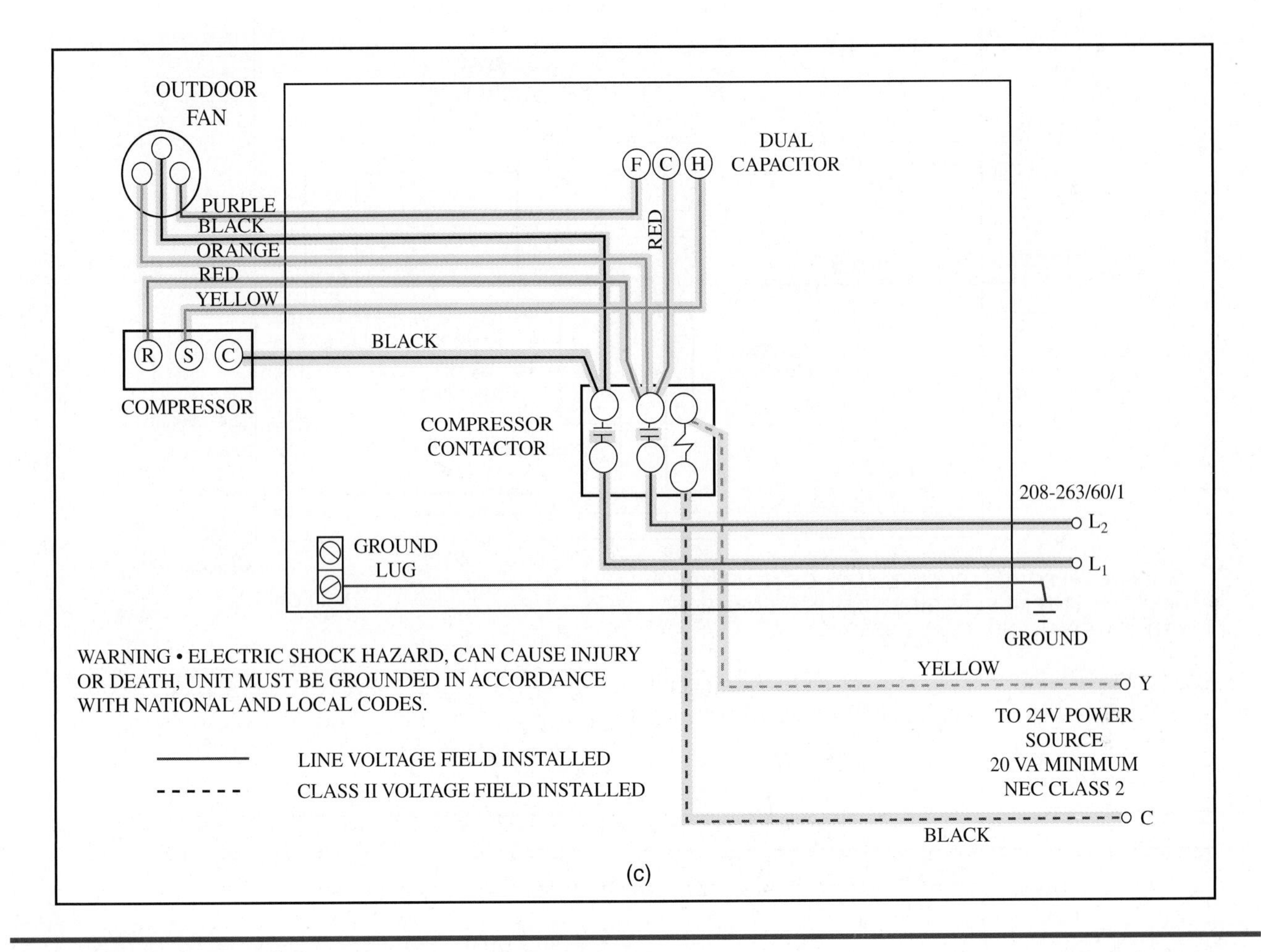

Figure 8-3-21 (continued) (*Courtesy Lennox Industries Inc.*)

When the system is on there is a voltage drop of 24 V that can be read across the contactor coil. This is indicated by the change in color. The voltage from line one passes across the closed contactor through terminal 1 to the common leads of both the compressor and fan. The voltage from line 2 passes across the closed contactor to terminal 2 and onto the run terminals of the compressor and fan. The voltage from line 2 can be read between any of the points shaded blue to any of the points shaded red.

Figure 8-3-21a is of a wiring diagram for a condensing unit. This condensing unit uses a dual (three terminal) run capacitor. When this unit is powered up the line voltage from line 1 can be measured up to the contactor. The line voltage from line 2 can be measured up to the contactor. If a voltage measurement is made between the two terminals on the contactor you would read a voltage from 208 to 240 V depending on the supply voltage. The control voltage would be the same as the control voltage of the common lead on the transformer. There is no voltage drop across the compressor contactor coil.

When the cooling contacts in the indoor thermostat close the circuit is completed through the outdoor compressor contact coil. This can be seen as a voltage drop or color change across the coil in Figure 8-3-21b. At this time the compressor contactor closes and line 1 voltage passes across the contacts to the common terminals on the compressor and fan as shown in Figure 8-3-21c. At the same line 2 voltage passes through the closed compressor contactor to the dual capacitor and the start and run terminals on the compressor and outdoor fan.

When this system is running correctly a voltage difference can only be measured between those terminals marked with red and those marked with blue.

Figure 8-3-22a is of a diagram similar to that shown in Figure 8-3-21a, however, in this diagram the outdoor unit has a start capacitor and a potential relay. When this system is powered on, Figure 8-3-22b, it has power in the same places as were shown in Figure 8-3-21b. When this system is on and running, there is a voltage drop across the coil on the potential relay between terminals 5 and 2, as shown in Figure 8-3-23. This measured voltage might be slightly higher than the actual measured line 1 and line 2 voltages. This higher voltage is the result of

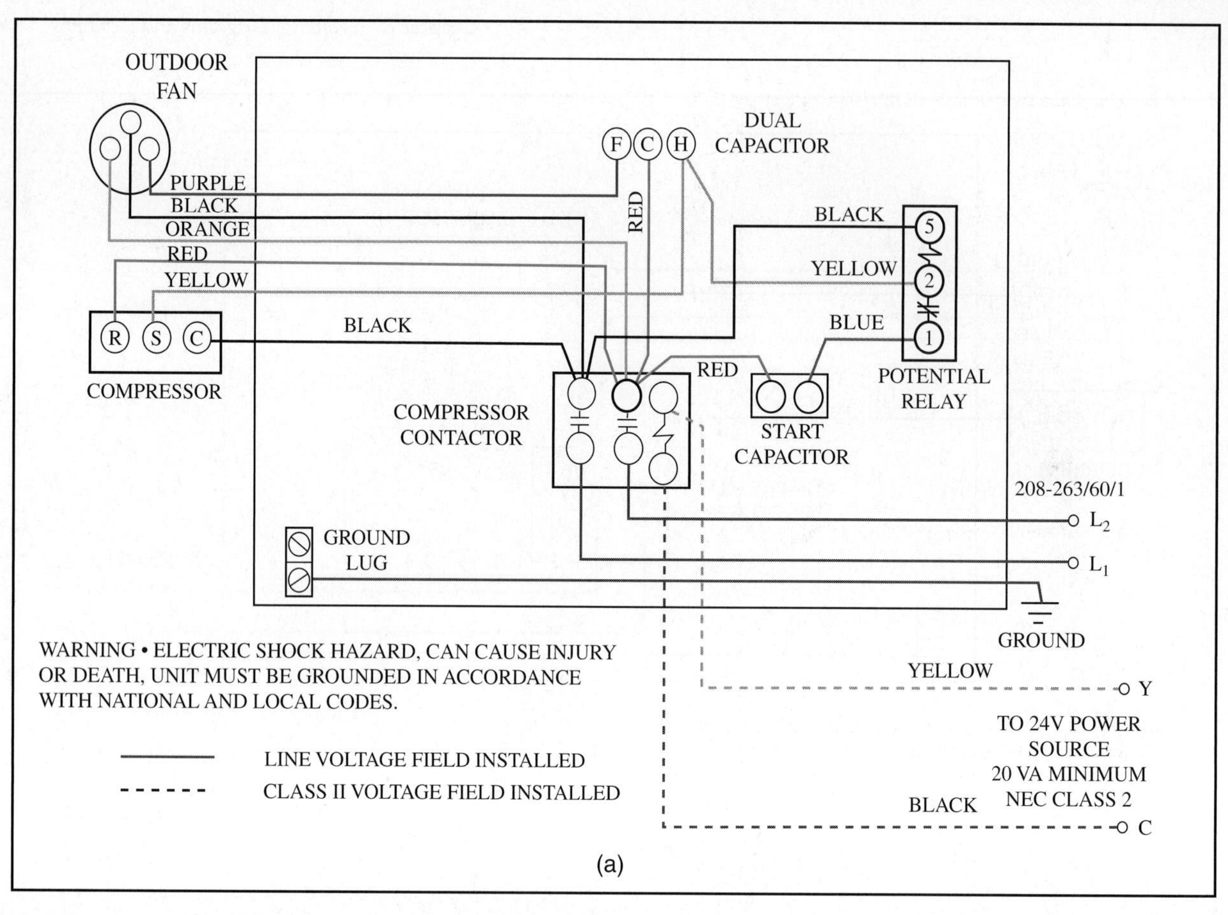

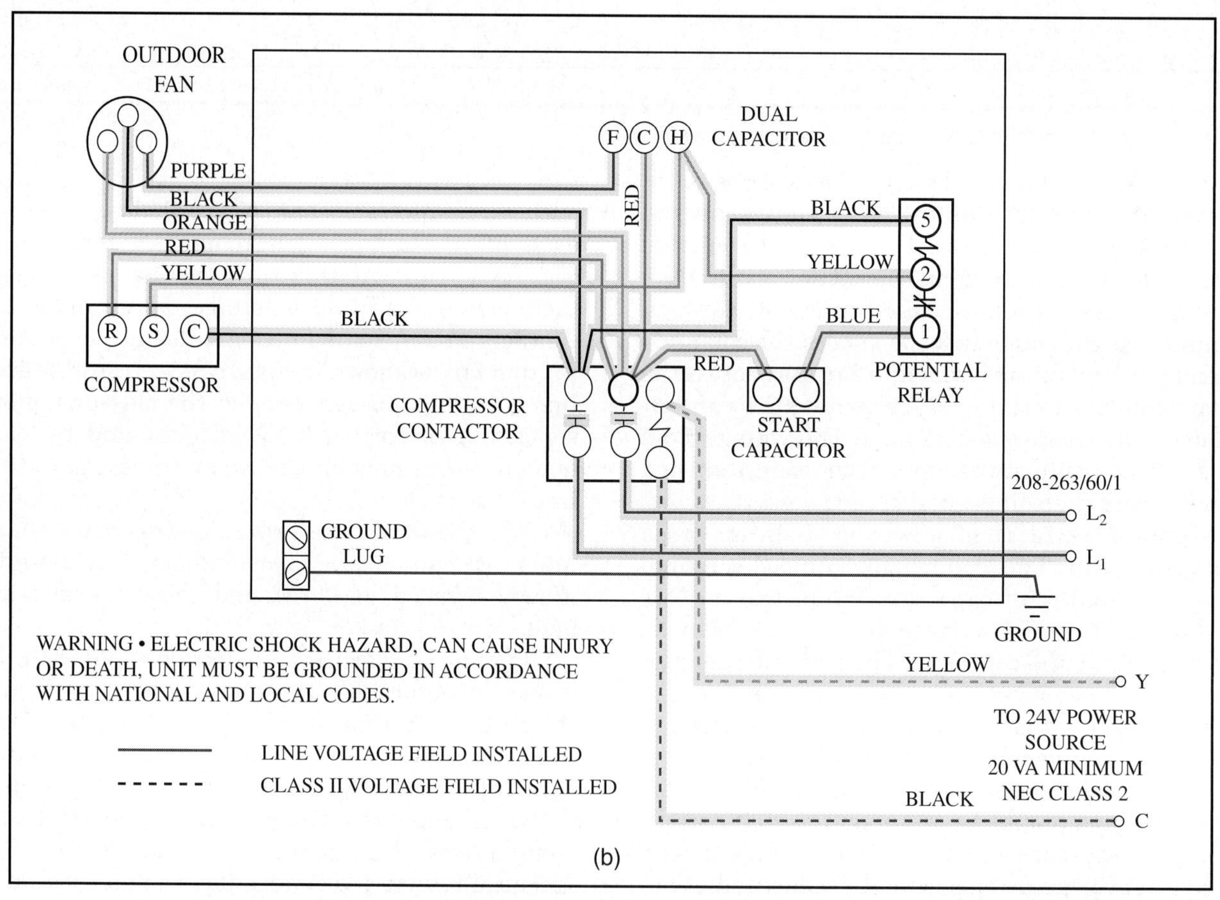

Figure 8-3-22 (a) Typical field wiring diagram with a hard start kit; (b) Typical field wiring diagram with a hard start kit, system running. (*Courtesy Lennox Industries Inc.*)

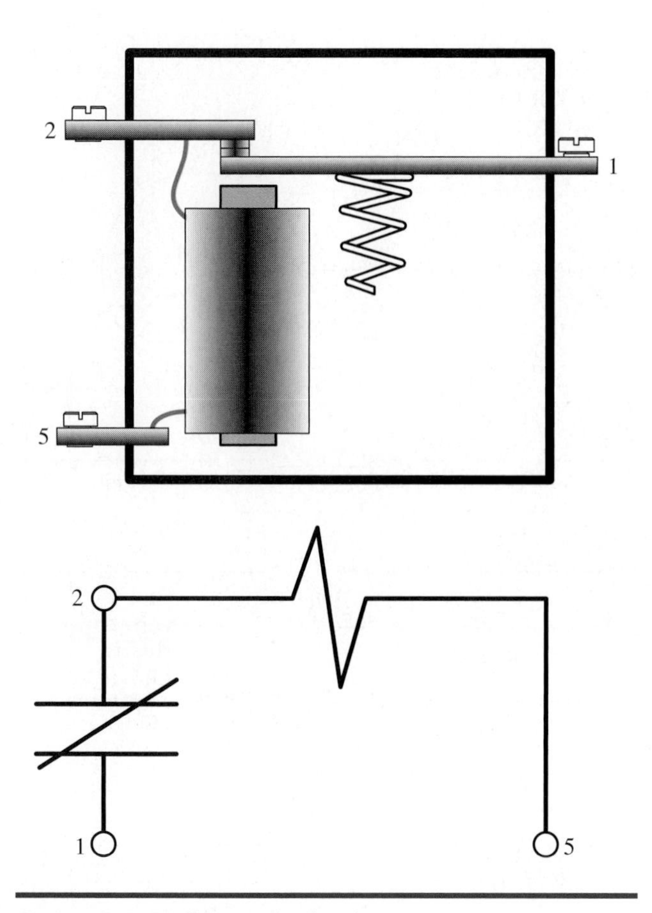

Figure 8-3-23 Potential relay diagram.

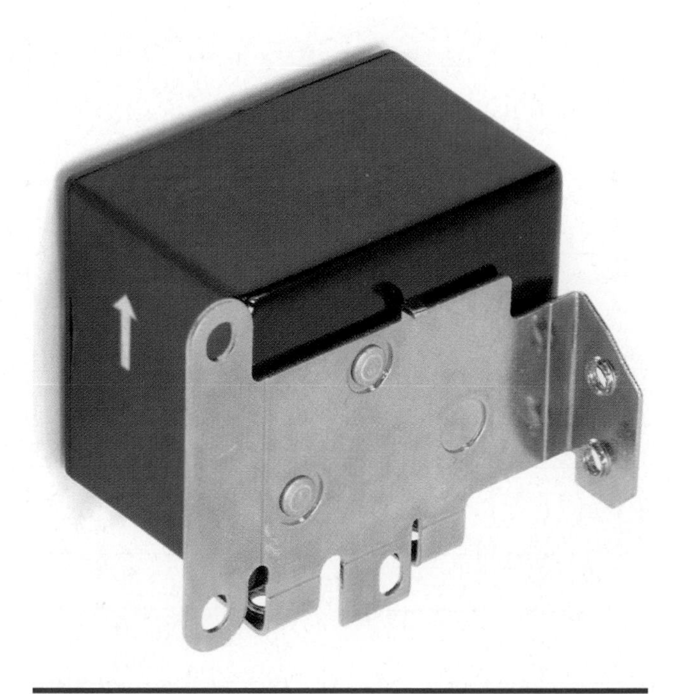

Figure 8-3-24 Potential relay.

the back EMF created as the compressor runs. It is this higher voltage or potential that causes the relay contacts located between terminals 2 and 1 to open. This occurs within a matter of a fraction of a second as the compressor starts.

If the contacts in the potential relay do not open then the current flowing through the start capacitor would quickly cause the compressor to overheat and shut down on internal overload. Many potential relays must be mounted in the equipment so that the arrow marking up is in the correct direction, as shown on the relay in Figure 8-3-24.

GAS FURNACE DIAGRAM

Figure 8-3-25a is a ladder diagram of a basic standing pilot gas furnace with central air conditioning. With the system power on and door switch closed line 1 voltage can be read at the fan switch and at the line 1 terminal of the transformer, Figure 8-3-25b. Line 2 voltage potential can be measured at all of the indoor blower motor terminals. It can be read at all of the terminals because without a current flow there is no voltage drop across the motor. Line 2 voltage can also be measured at the second power lead of the transformer. There is a voltage drop across the primary of the transformer because the transformer is energized any time the system is on and the blower compartment door switch is closed. The transformer provides power to the R terminal on the thermostat. Common on the transformer's potential voltage can be read on all sides of the cooling contactor, indoor fan relay, gas valves, and limit switch. This means that on the thermostat's sub-base a voltage potential could be read from R to Y, R to G, and R to W.

When the system is in heating as illustrated in Figure 8-3-25c line 1 voltage can be read on the low, medium, high, and start leads of the indoor blower motor. There is a voltage drop across the motor from the common terminal to all three speed terminals and the start terminal. A voltage drop can be measured across the primary windings of the control voltage transformer. A voltage drop can be measured across the gas valve but no voltage drop will occur across the cooling contactor or indoor blower relay because those circuits are not energized.

If the high limit switch opens, a voltage drop would occur across the high limit terminals and no voltage drop would be measured across the gas valve terminals. There would be no voltage drop on the gas valve because there would not be a current flow through the open high limit. It is important to note that the fan switch would remain closed even though the gas valve is off so that the fan can continue to blow out the excessive heat from the furnace.

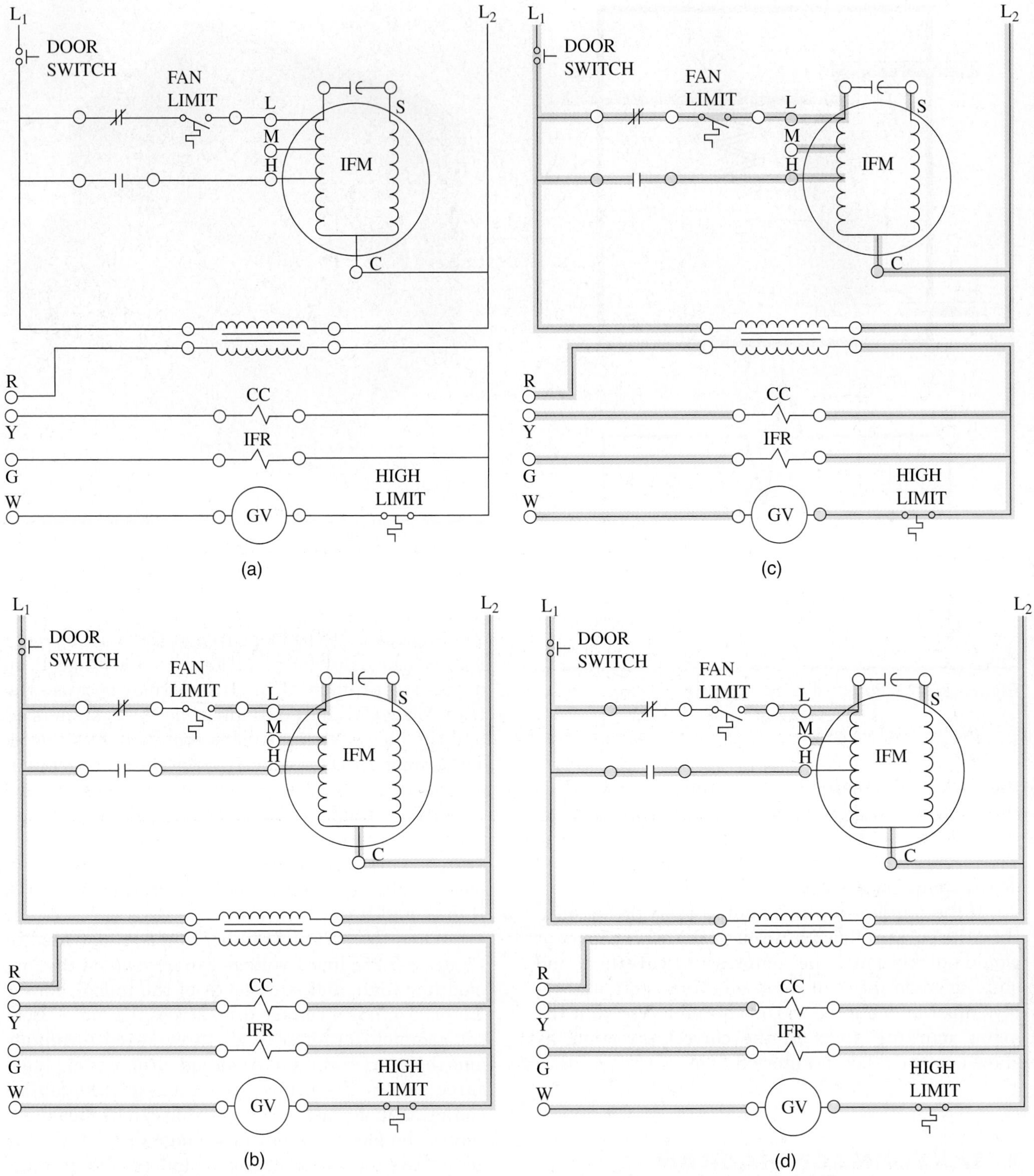

Figure 8-3-25 (a) Ladder diagram, gas furnace; (b) Ladder diagram, gas furnace power on; (c) Ladder diagram, gas furnace system on heating; (d) Ladder diagram, gas furnace system on cooling.

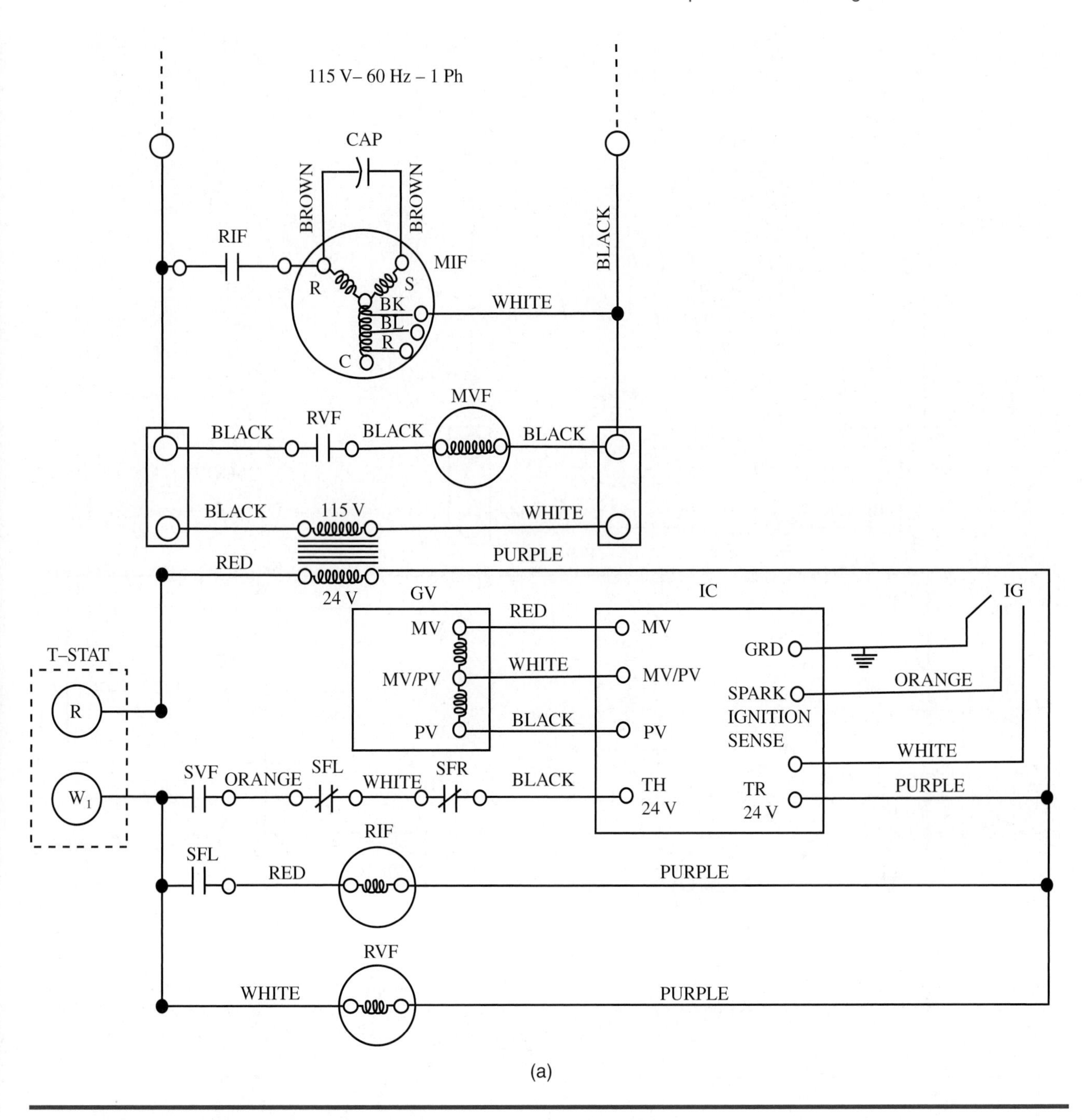

Figure 8-3-26 (a) Ladder diagram, gas furnace with spark ignition; (b) Power on; (c) Heat ignition; (d) System on heating.

When this system is in cooling, Figure 8-3-25d, the indoor blower relay contacts reverse so that the indoor blower motor is operating on high speed. The fan limit switch would remain open since there is no fire in the furnace to cause that temperature switch to close. There would be a voltage drop measurable across the cooling contactor and indoor blower relay. There would not be a voltage drop measurable across the gas valve because the gas valve would not be energized.

Figure 8-3-26a shows a ladder diagram for a central gas furnace with a spark ignition. Figure 8-3-26b illustrates the various parts of the circuit that would be at line 1 and line 2 voltage potential. Note that there is a voltage drop that would be measured across the blower relay terminals RIF and the induction fan relay terminals RVF and the primary leads on the transformer. Most of the low voltage components are at the same potential. The only areas that would measure a potential difference would be between the R and W_1 terminals in the thermostat.

Figure 8-3-26c shows those parts of the circuit that would have a potential voltage difference during

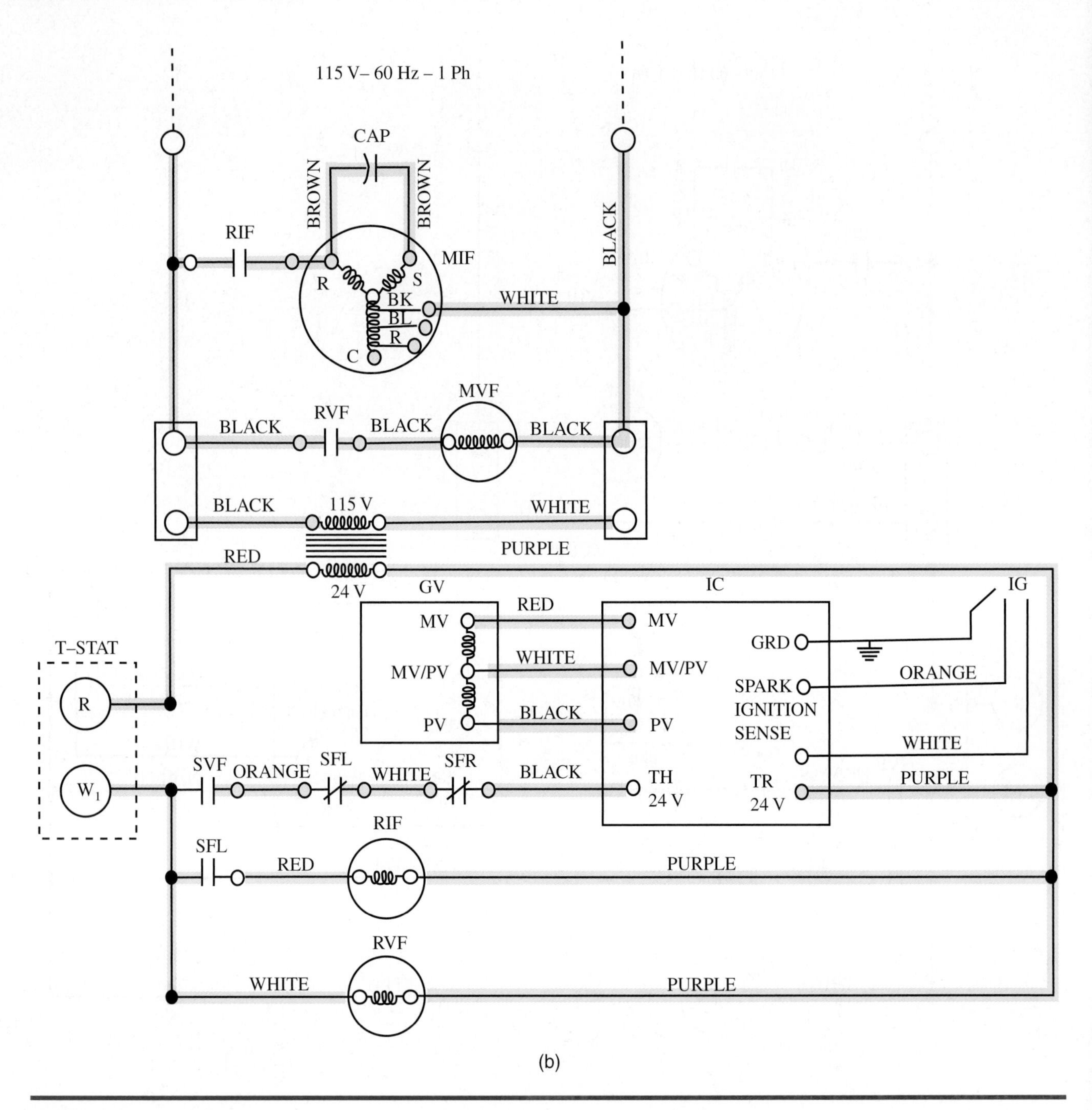

Figure 8-3-26 (continued)

the ignition phase of the furnace operation. This figure illustrates that only the induction fan motor and the transformer are energized on the line voltage side. The indoor blower motor does not become energized until after the furnace has had an opportunity to heat sufficiently. This delay in starting of the indoor blower motor is done so that cold air is not blown into the residence on startup.

On the low voltage circuit there is not a voltage difference across R and W_1 because the thermostat contacts would be closed. The gas valve terminals are energized across the PV and PV contacts and there is a voltage drop. There is also a voltage drop across the spark.

Figure 8-3-26d shows that the indoor blower motor is energized now and that the induction blower motor and transformer remain energized. On the secondary side of the circuit the indoor blower motor relay (RIF) is energized when the contacts FFL close following the time delay. The gas valve is now energized between the MV and MV/PV contacts. The voltage to the spark igniter has stopped once the flameproofing circuit of the IC controller has detected a flame.

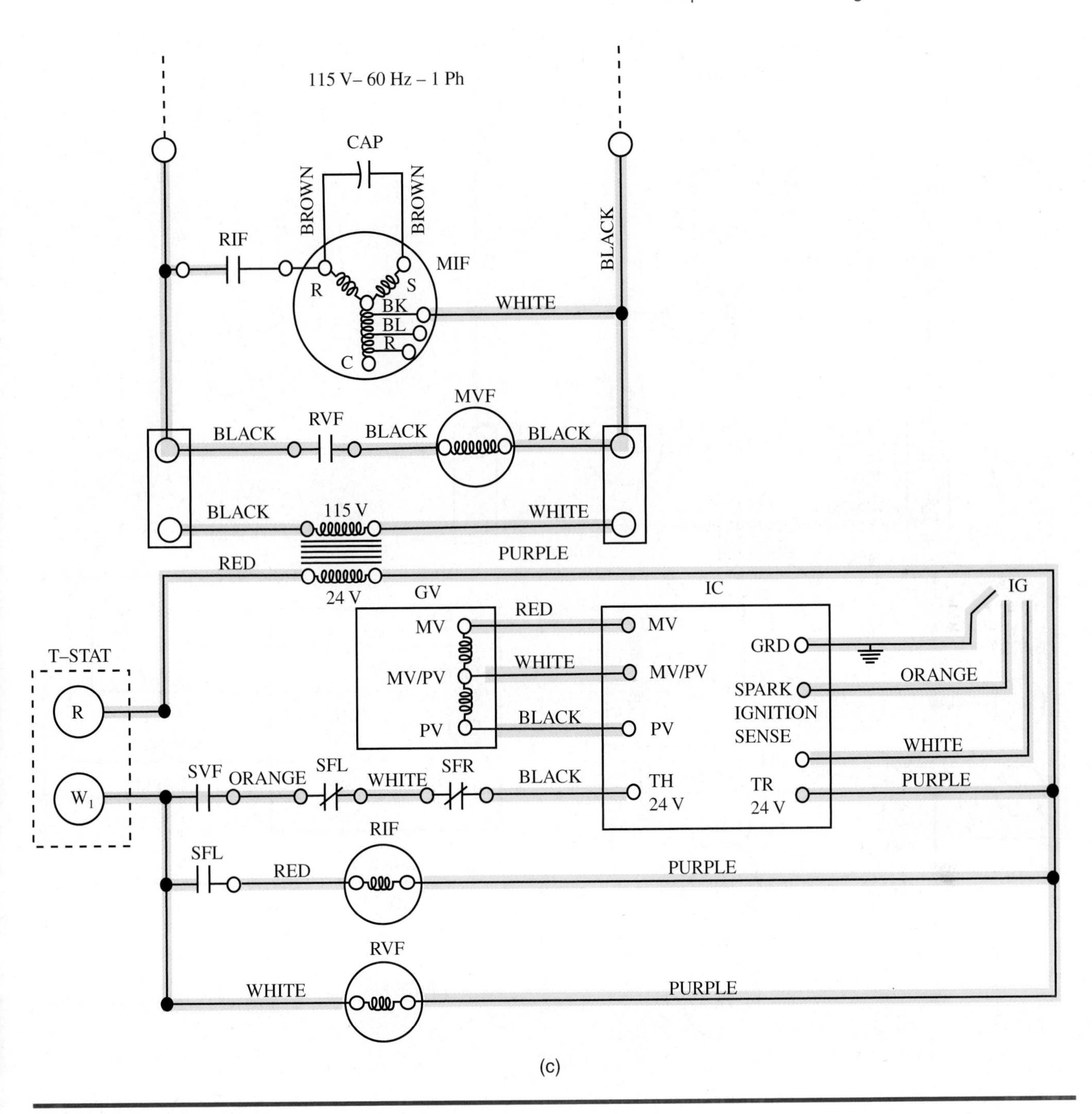

Figure 8-3-26 (continued)

ELECTRIC HEAT DIAGRAM

Figure 8-3-27a is a ladder diagram of a two element electric furnace diagram that has central air conditioning. Figure 8-3-27b shows the part of the circuit that is energized with line 1 and line 2 potential voltages.

Figure 8-3-27c illustrates all of the circuits that would be energized when the first stage heating element has come on. Note that there is not a voltage drop across the second stage heating element at this time. Only the first stage heat and fan are operating. On the secondary circuit the sequencer heat motor element has been energized. Figure 8-3-27d shows the parts of the circuit that would be energized when both the first and second stage heating element has been energized. When the second set of contacts close on the second stage heating circuit all of the elements are energized and show a voltage drop.

Figure 8-3-27e represents those circuits that would be energized when the system is in the cooling mode. Note that the heating elements are not energized and that the indoor blower relay has opened the contacts to the first stage heating element as the normally opened contacts are closed so that the indoor blower motor is operating. On the secondary

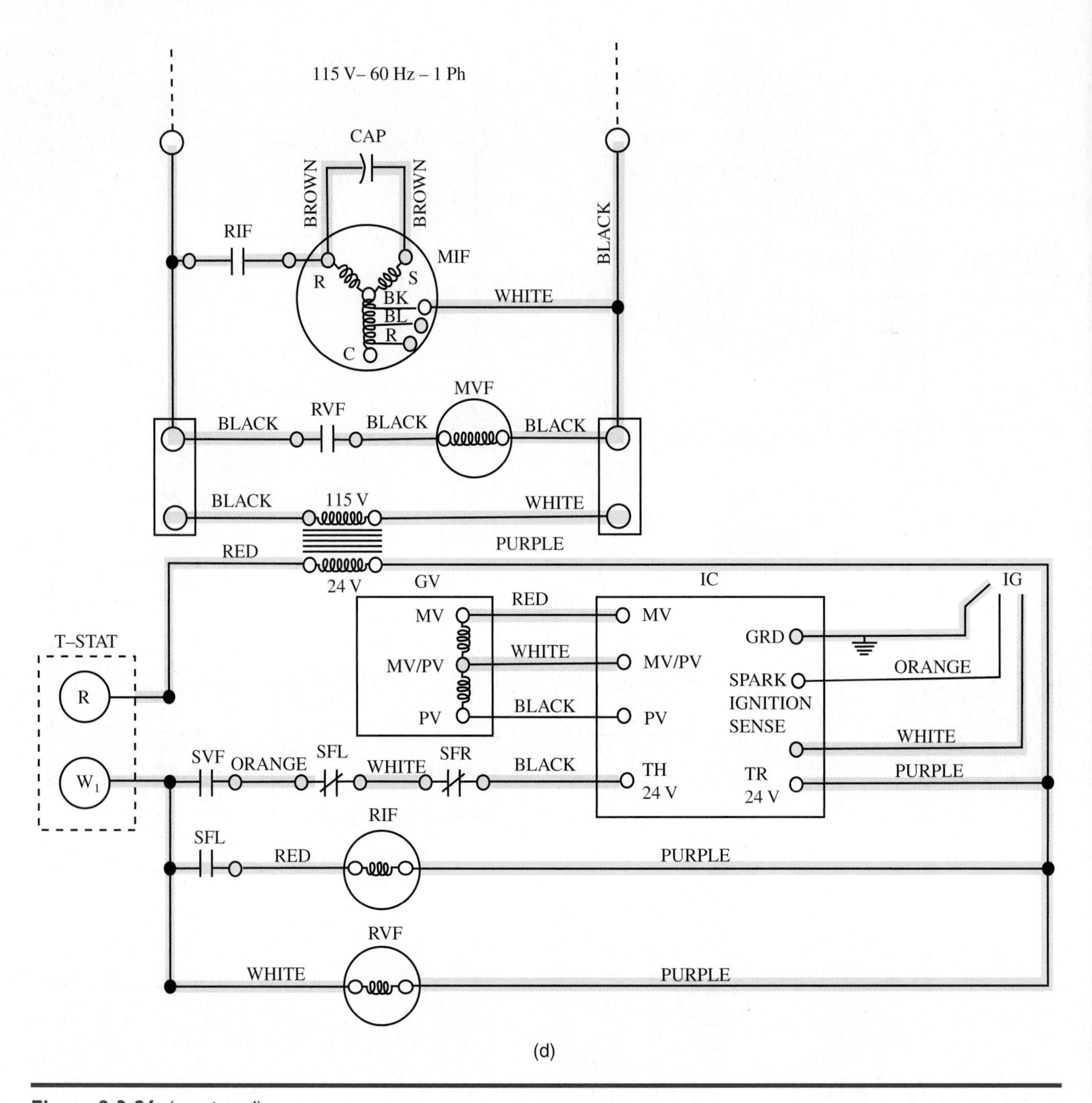

Figure 8-3-26 (continued)

there is a voltage drop across the indoor blower motor relay and outdoor cooling contactor.

HEAT PUMP WIRING DIAGRAMS

Figure 8-3-28a is a packaged heat pump diagram. The only difference between a packaged heat pump diagram and a split system diagram is that the indoor blower motor supplemental heat strips are located on a separate diagram.

Figure 8-3-28b shows the portions of the diagram that are energized with line 1 and line 2 voltage. A review of the diagram shows that the defrost relay is energized even though the system is not calling for heat or cooling. The primary voltage side of the transformer is energized. The low voltage control circuit is energized only to the red terminal on the indoor thermostat with 24 V. All the other components are at the common side of the transformer's potential voltage. Figure 8-3-28c shows the heat pump in the heating mode. Note that the compressor outdoor fan motor, indoor blower motor, and the compressor

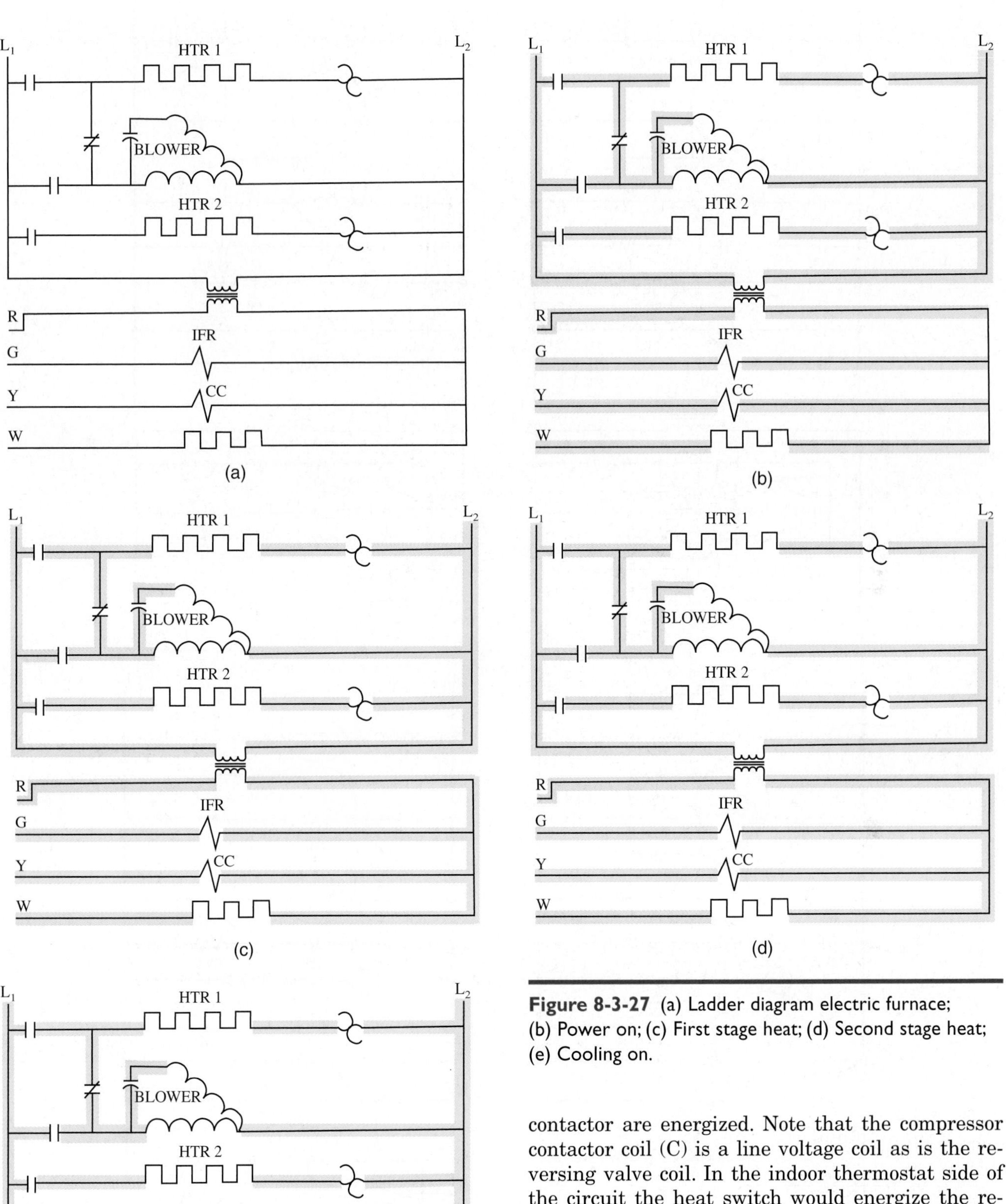

Figure 8-3-27 (a) Ladder diagram electric furnace; (b) Power on; (c) First stage heat; (d) Second stage heat; (e) Cooling on.

contactor are energized. Note that the compressor contactor coil (C) is a line voltage coil as is the reversing valve coil. In the indoor thermostat side of the circuit the heat switch would energize the reversing valve for W_1, but W_2, which is the supplemental heat relay (HC), will not be energized unless the heat pump is not able to maintain the residence indoor set temperature. The control relay is energized as a result of relay contacts RVR being closed. That also feeds current through the auto fan switch to energize the indoor blower relay through the G terminal on the indoor thermostat. Figure 8-3-28d represents the parts of the circuit that would be

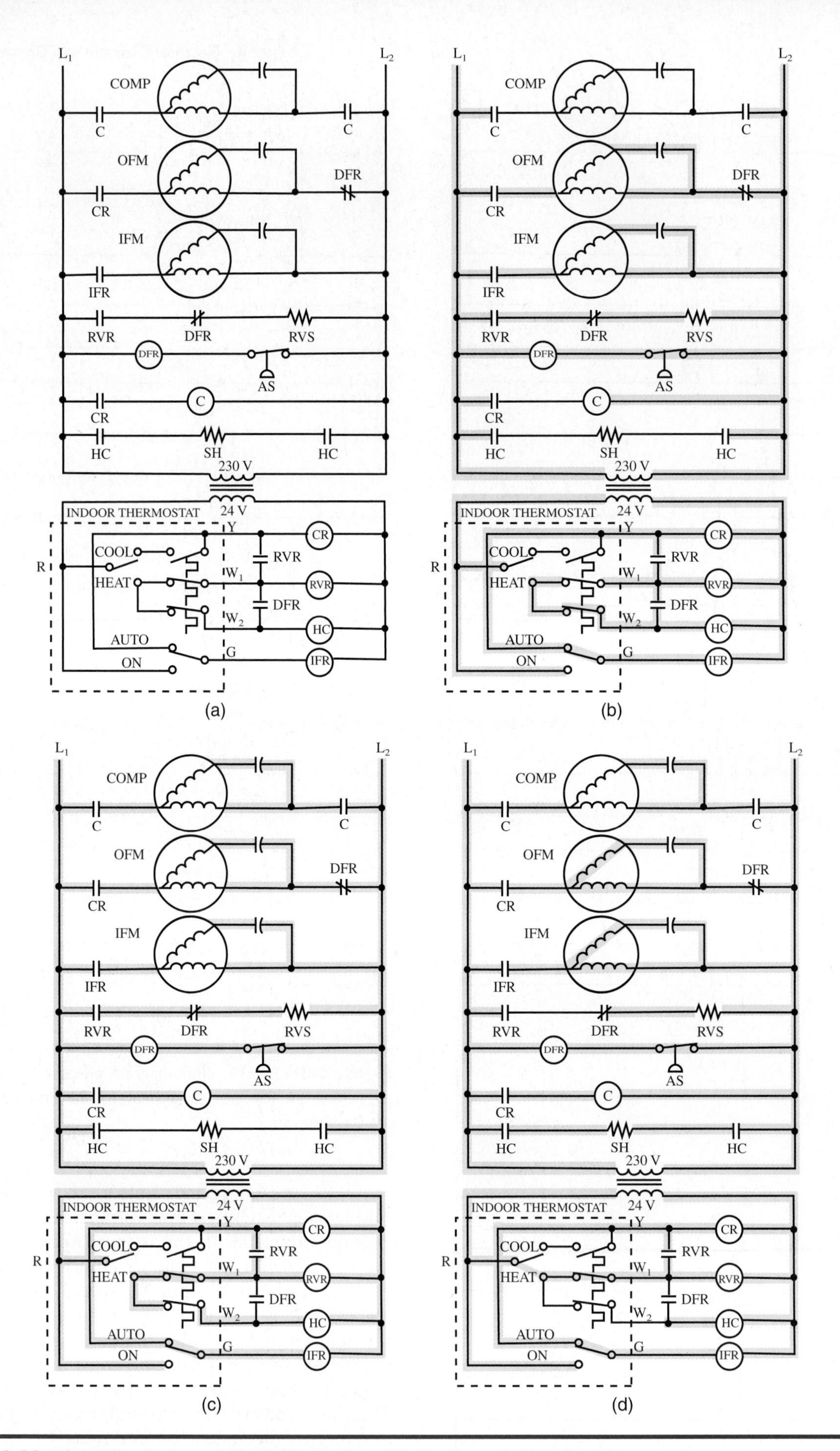

Figure 8-3-28 (a) Ladder diagram, packaged heat pump; (b) Power on ladder diagram, packaged heat pump; (c) Power on, heating; (d) Power on, defrost cycle.

TECH TALK

Tech 1. I get confused when I'm trying to figure out if a relay is open or closed when I'm checking voltage. It seems to me that if the relay is closed and the motor is running, then there should be voltage on the terminals. But it doesn't work that way. Why?

Tech 2. The only time you can read voltage on a circuit is when there is a voltage difference. When the contacts make a connection, they simply carry the current to the motor; so there is no voltage drop and no voltage difference. Therefore there's no voltage you can read. But there is voltage there. If you measure from that terminal on the relay to ground or to line 2, you'll read a voltage.

Tech 1. But what about the relay coil itself? Sometimes I have voltage and sometimes I don't.

Tech 2. Well, on a relay coil, when you have a voltage difference across the coil, that means the coil is energized. But you have to be careful. Just because there is energy to the coil doesn't mean the system is operating properly. If the coil windings are burnt out, it would still have voltage applied to it. And it would be a voltage difference because the coil is open. No current can flow through it, so you would have a voltage drop across it.

Tech 1. Then how do you tell the difference whether a coil is energized and working or energized and doesn't work?

Tech 2. Well, there are a couple of things you can do. One, you can turn the power off, pull the lead to the coil, and use your ohmmeter to see if it has continuity. The other thing you can do is use a clamp-on ammeter and check to see if there's any current flowing through the coil. But remember, on low voltage coils the current draw is going to be pretty small, and you may have to use a current multiplier in order for your meter to make a reading.

Tech 1. Okay. I think I understand it now. I appreciate that.

energized during a defrost cycle. Note that the outdoor fan is not running nor is the reversing valve. The supplemental heat (SH) is energized to provide warm air to the residence during the defrost cycle. On the control circuit side the defrost relay (DFR) has closed the contacts so that the supplemental heat relay coil (HC) has been energized. Note that the W_1 terminal is still supplying power to the reversing valve relay (RVR). However, the defrost relay has opened up the contact so that the reversing valve coil (RVS) is not energized. When the defrost cycle has terminated as a result of the pressure switch AS opening, the system returns to the standard heating mode.

UNIT 3—REVIEW QUESTIONS

1. What are two common types of wiring diagrams?
2. What is the field connection diagram designed for?
3. What is the schematic diagram used for?
4. To assemble the cooling electrical circuit, component by component, start with the ________.
5. Since an air conditioning system is designed to maintain a comfortable temperature in the conditioned space, a ________ is used as the primary system to control the wiring diagram.
6. What does the fan switch control?
7. To control the compressor automatically a ________ is required.
8. On the ladder diagram how are the horizontal lines numbered?
9. What is command humidification?
10. What does the basic condensing unit consist of?
11. When the system is running correctly a voltage difference can only be measured between ________ and ________.

SECTION **IV**

Residential Systems

9

Central Residential Air Conditioning

UNIT 1

Air Conditioning

OBJECTIVES: In this unit the reader will learn about air conditioning including:

- Adjusting Airflow
- Determining System Capacity
- Split System Conditioners
- Add-on Coils
- Air Cooled Condensing Unit
- Outdoor Installation
- Refrigerant Piping
- Condensing Unit and Evaporator

AIR CONDITIONING

To obtain the highest performance from the cooling unit it is necessary for the technician to give special attention to the following and make adjustments where necessary:

1. Airflow and temperature drop over the coil
2. Refrigerant charge
3. Capacity of the system
4. Adaptability to local conditions

It is important that the loads have been carefully calculated and that the equipment has been selected to match the loads.

Adjusting the Airflow

Residential and packaged commercial air conditioning units are commonly supplied with multispeed indoor blower motors. The wiring diagram will suggest the proper speed selection. For most applications, the factory setting will produce a supply air temperature of between 52°F and 58°F. If the supply airflow is low, the temperature may be below 52°F. On direct expansion systems, this can result in a suction pressure below 32°F, and this can cause frosting on the evaporator coil, as shown in Figure 9-1-1. First, check to see why the airflow is low. Check for obstructions in the ductwork, such as a piece of interior duct insulation that has come loose or a crushed flexible duct. On field assembled systems using air handlers, the plans will specify the required CFM. This should have been measured and verified prior to start-up (see Chapter 14, "Testing and Balancing"). But if the ductwork is clear and the duct system just has a higher than normal pressure drop, the indoor

Figure 9-1-1 Extremely iced slab AC coil due to low air flow.

SERVICE TIP

Most manufacturers recommend that a system's airflow be set so that there is approximately 400 CFM per ton of air conditioning. In areas that have high relative humidity a slightly slower fan speed can be used to aid in the humidification. In arid areas the fan speed can be increased to provide more sensible cooling. As a rule of thumb the fan speed should not be adjusted more than 10% above or below the manufacturer's recommendation. Excessively slow fan speeds can allow the evaporative coil to freeze up under light loads and excessively high fan speeds can put too large a load on the condenser under heavy load conditions. Table 9-1-1 shows blower performance at low, medium, and high load conditions.

blower speed may need to be increased. More serious airflow problems may require that the duct system be reworked to allow the proper airflow.

Most manufacturers supply operating data on their equipment such as shown in Table 9-1-2, indicating the total, sensible, and latent heat removal rating, at various outdoor dry bulb and indoor wet bulb temperatures at specific external static pressures. For these same conditions, they supply the operating suction and discharge pressures of the equipment. This information is given so that the technician can match the actual conditions on the job with the performance conditions shown on the manufacturer's chart. Notice that in Table 9-1-2 a system with a low external static pressure of .30 would have an airflow of 625 on low. But on a system with a higher external static pressure of .70, the blower would have to be set to high to obtain the same airflow.

TECH TIP

There is a great deal of information available to technicians on the manufacturer's technical data sheets. All of this information is valuable and if you learn how to properly use it you will become more valuable to your company and your customers. Most manufacturers provide explanations as to what each column of data can be used for. If you have difficulty understanding any of the material, most manufacturers or their distributors will provide you with an in-depth explanation of the data and how it can be used.

TABLE 9-1-1 Typical Cooling Unit Air Handler Performance Data

CB29M-21/26 BLOWER PERFORMANCE (208/230v)

External Static Pressure		Air Volume and Motor Watts at Specific Blower Taps								
		Low			Medium			High		
in. w.g.	Pa	cfm	L/s	Watts	cfm	L/s	Watts	cfm	L/s	Watts
.00	0	700	330	245	895	420	310	1030	485	375
.05	10	690	325	240	875	415	305	1010	475	370
.10	25	680	320	235	865	410	300	990	470	365
.15	35	665	315	230	850	400	290	970	460	355
.20	50	655	310	225	830	390	285	955	450	350
.25	60	640	300	220	810	385	280	925	440	345
.30	75	625	295	220	795	375	270	900	425	335
.40	100	595	280	210	750	355	255	850	400	320
.50	125	555	260	195	700	330	240	800	380	305
.60	150	510	240	185	640	300	225	725	340	290
.70	175	395	185	165	----	----	----	620	295	265
.75	185	----	----	----	----	----	----	570	270	255

NOTE - All air data is measured external to unit with air filter in place. Electric heaters have no appreciable air resistance.

TABLE 9-1-2 Typical Condenser Charging Data

RATINGS

NOTE - For Temperatures and Capacities not shown in tables, see bulletin - Cooling Unit Rating Table Correction Factor Data in Miscellaneous Engineering Data section.

Entering Wet Bulb Temperature	Total Air Volume		Outdoor Air Temperature Entering Outdoor Coil: 85°F (29°C)						95°F (35°C)						105°F (41°C)						115°F (46°C)					
			Total Cooling Capacity		Comp Motor kW	Sensible To Total Ratio (S/T) Dry Bulb			Total Cooling Capacity		Comp Motor kW	Sensible To Total Ratio (S/T) Dry Bulb			Total Cooling Capacity		Comp Motor kW	Sensible To Total Ratio (S/T) Dry Bulb			Total Cooling Capacity		Comp Motor kW	Sensible To Total Ratio (S/T) Dry Bulb		
	cfm	L/s	kBtuh	kW	Input	75°F 24°C	80°F 27°C	85°F 29°C	kBtuh	kW	Input	75°F 24°C	80°F 27°C	85°F 29°C	kBtuh	kW	Input	75°F 24°C	80°F 27°C	85°F 29°C	kBtuh	kW	Input	75°F 24°C	80°F 27°C	85°F 29°C
10ACC-018 — C33-18A COOLING CAPACITY																										
63°F (17°C)	450	210	16.8	4.9	1.30	.73	.84	.95	15.9	4.7	1.39	.74	.86	.98	15.0	4.4	1.47	.76	.89	1.00	14.1	4.1	1.55	.78	.92	1.00
	650	305	18.3	5.4	1.34	.79	.93	1.00	17.3	5.1	1.43	.81	.96	1.00	16.3	4.8	1.52	.83	.99	1.00	15.3	4.5	1.61	.86	1.00	1.00
	850	400	19.3	5.7	1.36	.85	1.00	1.00	18.3	5.4	1.45	.87	1.00	1.00	17.4	5.1	1.56	.91	1.00	1.00	16.4	4.8	1.65	.94	1.00	1.00
67°F (19°C)	450	210	17.7	5.2	1.32	.59	.70	.81	16.8	4.9	1.41	.60	.72	.83	15.8	4.6	1.50	.61	.73	.85	14.9	4.4	1.59	.62	.75	.88
	650	305	19.3	5.7	1.36	.63	.77	.90	18.2	5.3	1.45	.64	.79	.93	17.2	5.0	1.55	.66	.81	.96	16.1	4.7	1.64	.67	.84	.99
	850	400	20.2	5.9	1.38	.67	.83	.97	19.1	5.6	1.48	.68	.85	.99	17.9	5.2	1.57	.70	.88	1.00	16.8	4.9	1.67	.72	.92	1.00
71°F (22°C)	450	210	18.4	5.4	1.34	.47	.58	.68	17.5	5.1	1.43	.47	.58	.69	16.6	4.9	1.53	.48	.59	.71	15.7	4.6	1.62	.48	.61	.73
	650	305	20.2	5.9	1.38	.49	.62	.74	19.1	5.6	1.48	.49	.63	.76	18.1	5.3	1.58	.50	.64	.78	17.0	5.0	1.68	.51	.66	.81
	850	400	21.2	6.2	1.39	.50	.65	.80	20.0	5.9	1.50	.51	.67	.83	18.9	5.5	1.61	.52	.69	.86	17.8	5.2	1.71	.53	.71	.89

(Courtesy of Lennox Industries, Inc.)

Determining the Refrigerant Charge

The following is a summary of the various methods for determining refrigerant charge:

1. Observing the condition of the sight glass located on the inlet side of the evaporator metering device.
2. Measuring the superheat in the refrigerant leaving the evaporator (for capillary tube and fixed orifice systems only).
3. Measuring the subcooling in the refrigerant leaving the condenser (for thermal expansion systems only).
4. Measuring the wet bulb temperature drop across the evaporator.
5. Adding the required charges of each of the components.
6. Adjust the charge as needed to obtain the manufacturer's recommended levels of subcooling and/or superheat.

Determining the System Capacity

The actual cooling being done through the evaporator may be determined by measuring the entering and leaving air conditions, and the airflow across the coil. They are then used in the following formula previously discussed in Chapter 8:

$$Q_t = 4.45 \times \text{CFM} \times \Delta h$$

where

Q_t = the total (sensible and latent) cooling being done
CFM = airflow across the evaporator coil
Δh = the change of enthalpy of the air across the coil.

Other Useful Measurements

Other useful measurements can be made to indicate performance of specific systems. The technician must constantly be on the lookout for conditions that waste energy and reduce efficiency. Some of these additional tests that can be made are:

1. Measure the difference in temperature between the return air at the grill compared to return air temperature as it enters the unit. This difference should not exceed 2°F, as shown in Figure 9-1-2. If it does, the return duct needs to be insulated or there may be leaking openings in the duct that need to be sealed.

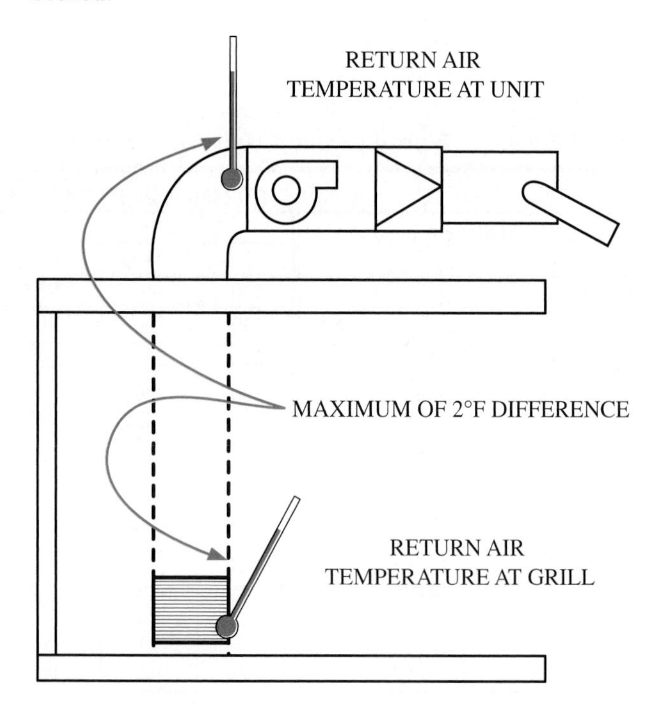

Figure 9-1-2 Checking return air system for leaks.

2. Measuring the refrigerant superheat in the evaporator may indicate that the expansion valve needs adjusting. The superheat should not exceed 10°F for TXV systems.

3. The compressor discharge pressure can be compared to the liquid line pressure. This pressure drop should not exceed 3 psig. An excessive drop indicates discharge piping that is too small or too high a pressure drop in the condenser.

4. Measure the degrees of subcooling in the liquid line. A safe value is 10°F. Less subcooling may permit flash gas to form and reduce the capacity of the expansion valve.

SPLIT SYSTEM CONDITIONERS

For many air conditioning systems, it is not practical to place all components in a single package, particularly those that involve the use of air cooled condensers. The air cooled condenser must have access to outside air so it is best placed outside. For this reason split systems have been developed with the inside unit consisting of a fan coil unit, with or without heating, an outside mounted air cooled condensing unit, and connecting refrigerant piping between the two, as shown in Figure 9-1-3.

Figure 9-1-3 Typical evaporator used on a split AC system.

TECH TIP

The final stage in manufacturing for split condensing systems is at the residence. Manufacturers have no control over the individuals who perform this last vital step in the production of their equipment. Manufacturers can't even insist that a matching system is installed. This final stage in manufacturing has a greater influence on the performance of a system than anything that the manufacturers can do. If the system is not installed with the correct size air duct systems, electrical wiring, and many other factors the result can be in the system providing poor performance or reduction of its operational life. It is therefore very important that all of the manufacturer's specifications and guidelines be followed during the installation so that your customer can have the most efficient and long life system possible.

Using a split system offers the opportunity to add cooling to an existing residential system heating unit where the necessary modifications are feasible.

Residential Split Systems

Where the furnace exists (gas, oil, or electric) and the size of the ductwork is adequate, a cooling coil may be added to the discharge side of the furnace. An air cooled condensing unit is located outdoors on a suitable base. The two are connected by properly sized liquid and suction refrigerant lines. A typical installation is illustrated in Figure 9-1-4.

Add-on Coils

Add-on coils are available in a number of configurations to fit various types of heating units, as shown in Figure 9-1-5. This illustration also shows the manner in which the position of the coils is related to the air handling unit. Coils are supplied for upflow, horizontal, and downflow furnace applications.

SAFETY TIP

Codes prohibit the installation of an evaporative coil before a gas or oil furnace heat exchanger. If the coil is placed before the heat exchanger, then condensate will form in the heat exchanger, causing it to quickly rust out.

Coil cases or cabinets are insulated to prevent sweating and all have pans for collecting condensate

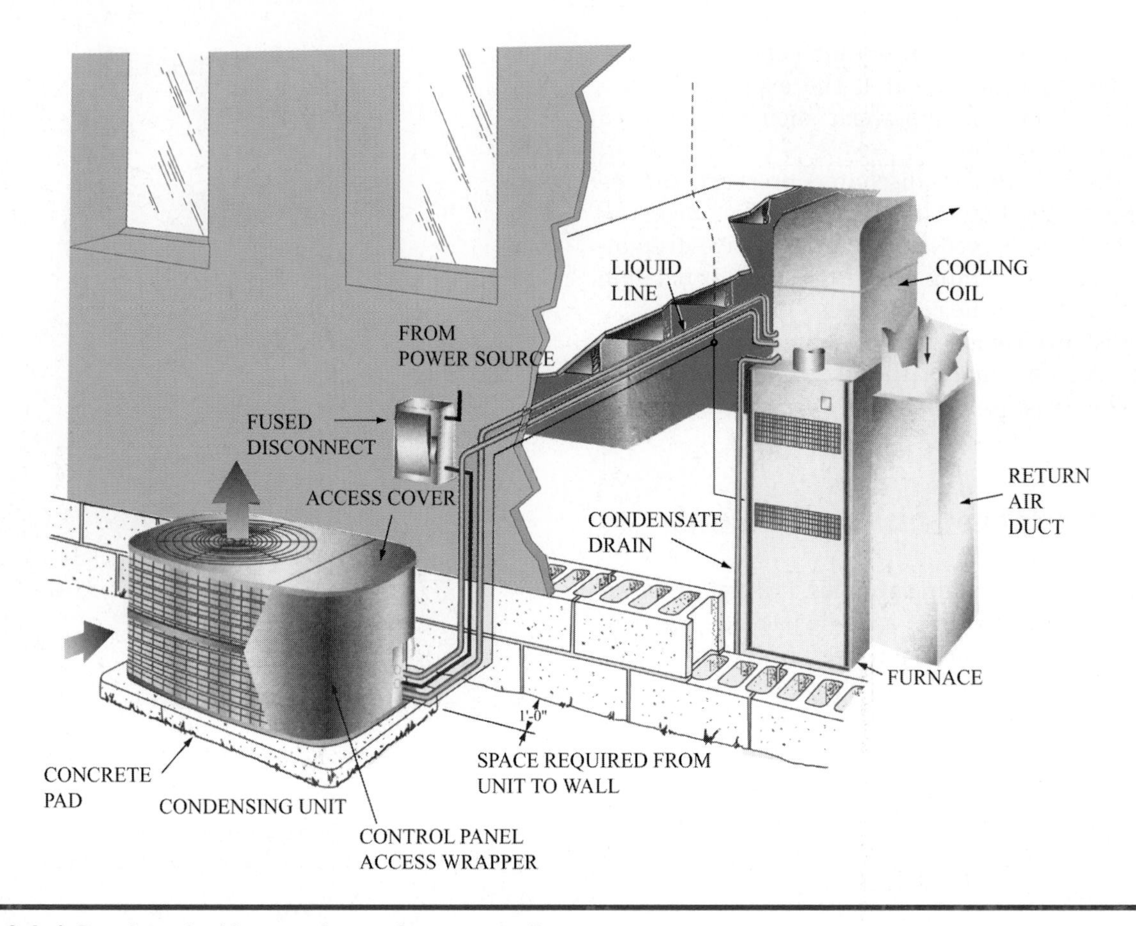

Figure 9-1-4 Residential add-on cooling to furnace installation.

water runoff. Plastic pipe may be used to connect condensate drain water to the nearest drain. If no nearby drain is available, a small condensate pump may be installed to pump the water to a drain or to an outdoor disposal arrangement.

An important consideration in applying an add-on cooling coil to an existing furnace is the air resistance it adds to the furnace blower. During the heating cycle the coil is inactive and dry. In summer, when the unit is cooling and dehumidifying, the coil is wet. The wet coil will add an average of 0.20–0.30 inches of water column (IWC) static pressure loss. This added resistance may be sufficient to require a change in furnace blower speed. Also, an increase in motor horsepower may be needed. If the installation is a new system, a furnace blower should be selected that is capable of producing sufficient external static pressure to deliver the required amount of air.

Air Cooled Condensing Unit

The outdoor air cooled condensing unit of an add-on system, Figure 9-1-6, consists of a compressor, condenser coil, a condenser fan, and the necessary electrical control box assembly. On residential condensing units, a fully hermetic compressor is used.

CONDENSER COIL

The condenser coil is a fin and tube arrangement that varies in design from manufacturer to manufacturer. A large surface area is desirable, and many

> **TECH TIP**
>
> Condenser fan motors turn at slower speeds than indoor blower motors. Not all condenser fan motors spin at the same rpm. Some operate around 1000 rpm while others may operate around 800 rpm. It is important when changing out a condenser fan that one of the exact same rpm be used. The fan blades are designed to move air at a certain speed. If the fan blade is spinning faster than it was designed for, it will not operate as efficiently. Increasing the rpm does not mean that you are going to increase the airflow. In addition, the height that the fan extends into the fan shroud is important. If the fan blade is too low or too high it will affect the airflow from the condenser. Mark the height of the blade before removing it to be sure it is put back in at the same height.

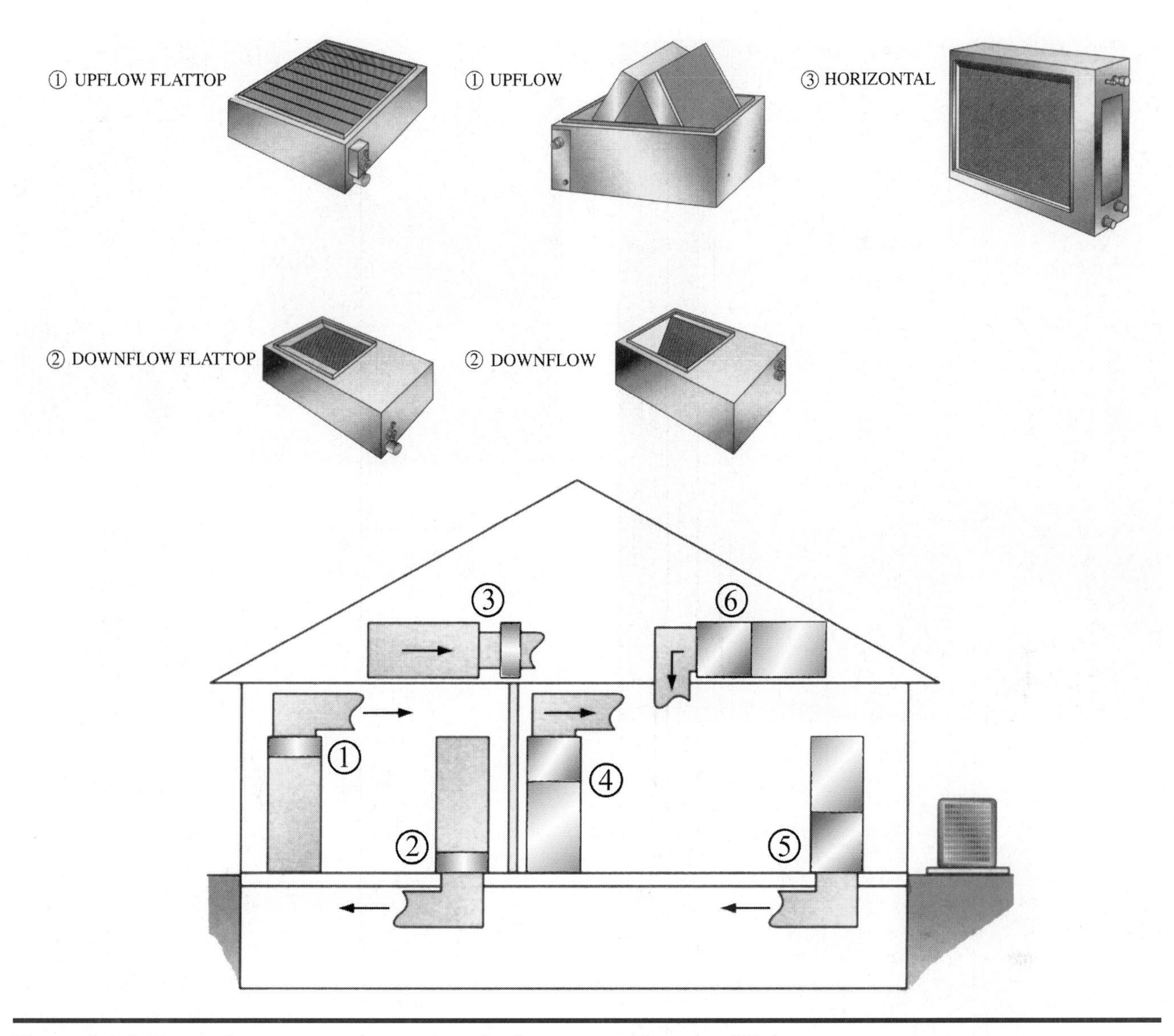

Figure 9-1-5 Types of evaporators for residential split systems. (*Courtesy of York International Corp.*)

units offer almost a complete wraparound coil to gain maximum coil surface area. The coil tube depth is limited in order to reduce resistance to airflow.

CONDENSER FAN

The condenser fan varies in design but is usually a propeller type fan, which can move large volumes of air through clean coils, which offer low resistance.

Airflow direction is a function of the cabinet and coil arrangement and there is no one best arrangement. Most units, however, use draw through operation over the condenser coil. Outlet air can have an effect on surrounding plant life. Top discharge is the most common arrangement.

Fan motors are sealed or covered with rain shields. Fan blades are shielded with a grill for the protection of hands and fingers.

TECH TIP

Axial fans, like those on condensers, are high volume and low static. If the coils on a condenser are dirty the fan will simply churn the air and throw it off the fan blade tips. You can actually take a very dirty condenser and place the sheet of paper in the center of the fan and it will suck it down while throwing the air off the fan tips. A clean condenser has a column of air that rises straight up from the fan. By feeling how the air is flowing a technician can determine to some extent whether the coil is clean or dirty. An easy example of this is to place a box fan flat on a table, point it up and turn it on. You will feel the air on the fan coming out the side and little or no air movement in the center. As the fan is lifted off the table the correct airflow will be established and the air will no longer come out the side of the blades.

(a)

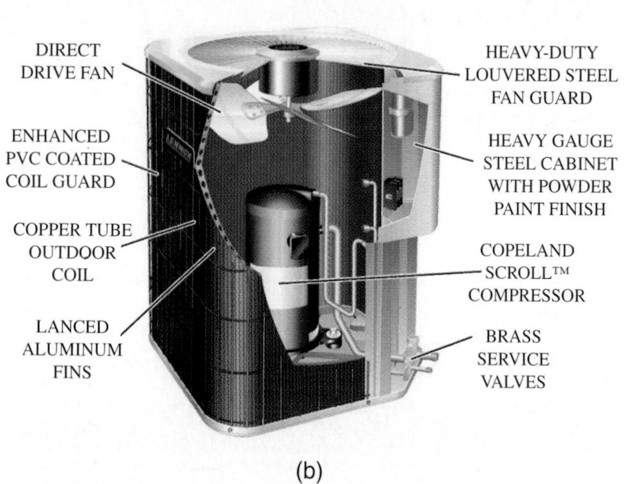

(b)

Figure 9-1-6 (a) Outdoor condensing unit for residential split system; (b) Cutaway of a typical residential split condenser. (*Courtesy Lennox Industries, Inc.*)

(a)

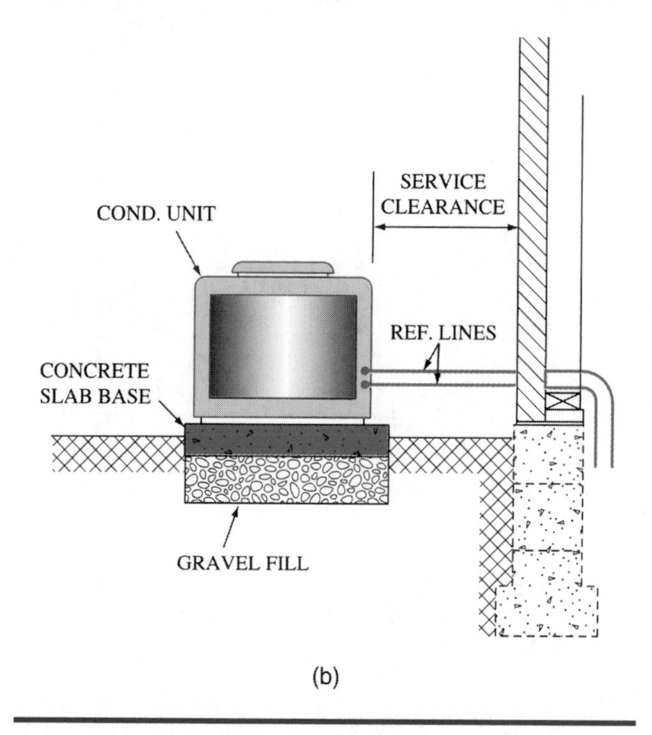

(b)

Figure 9-1-7 Outdoor condensing unit installed on a concrete slab: (a) Photo; (b) Illustration.

OUTDOOR INSTALLATION

When installing a condensing unit at ground level, it is very important to provide a solid foundation, Figure 9-1-7.

A concrete slab over a fill of gravel is recommended to minimize movement due to ground heaving. Where there is new construction, a foundation over unstable fill may settle sufficiently to place stress on or even break refrigerant lines.

The location of the outdoor condensing unit often is a compromise between several factors:

- The actual available space
- The length of run for refrigerant lines
- The aesthetic effects on the home or landscaping
- Noise/sound effects on the home and neighbors

In general, avoid placement of the unit directly underneath a window or immediately adjacent to a patio. Locate it where plants provide a visual as well as sound absorbing screen. Also, avoid placement of the unit where water runoff from the roof will enter the top of the condensing unit. During periods of heavy rain, the water runoff will place a heavy load on the condenser fan motor.

Noise

Noise is recognized as an environmental pollutant. Outdoor air cooled condensing units are sound producing mechanical devices, which some cities and towns regulate. Early attempts to create local ordinances prompted action on the part of the ARI to establish in 1971 an industry method of equipment sound rating and application standards (which local communities could adopt).

SERVICE TIP

Equipment is manufactured today to be as quiet as possible. There are noise standards that manufacturers follow. However, over time sheet metal parts and components may become loose and vibrate during operation. Refrigerant lines can rub or cause noise. Motors can become loose and vibrate. All of these factors increase the sound level. Check the unit to be sure that all nuts and bolts are in place and tight. If the holes have been enlarged, replace the sheet metal screws with one size larger to ensure a tight fit and support or brace any copper piping that is causing problems. Replace fans or motors that have excessive vibration that is causing noise.

ARI Standard 270 applies to the outdoor sections of factory made air conditioning and heat pump equipment (unitary air conditioners). Under the program, all participating manufacturers are required to rate the sound power levels of their equipment in accordance with the technical specifications contained in this standard. Test results by the manufacturers are submitted to ARI for review and evaluation. Units are sound rated with a single number, the sound rated number (SRN). Typical ratings are between 14 and 24. ARI anticipates that the sound rating program will encourage manufacturers to produce quieter units.

Along with equipment ratings, ARI Standard 270 contains recommended application procedures for using the sound rating to predict and control the sound level.

REFRIGERANT PIPING

Refrigerant piping is needed between the indoor coil and the outdoor condensing unit. On residential type equipment, three common methods of joining piping are:

- Brazed fittings,
- Compression fittings, and
- Flared fittings

The service port for checking refrigerant pressure is shown in Figure 9-1-8. The port is equipped with a valve similar to a tire valve (Schrader valve), that opens when depressed by gauge lines. When installing refrigerant lines, follow the manufacturer's installation directions on the radius of the bend for the size of tubing. Do not form vertical loops, which will create oil traps.

Figure 9-1-8 Service access valves.

Condensing Unit and Evaporator

Single compressor condensing units of 5 tons or less come for refrigerant lines up to 25 ft. When refrigerant line lengths used are longer or shorter than 25 ft, the refrigerant charge must be adjusted according to the equipment manufacturers' specifications.

Manufacturers limit both the length of line and height of refrigerant lift for split system piping, Figure 9-1-9.

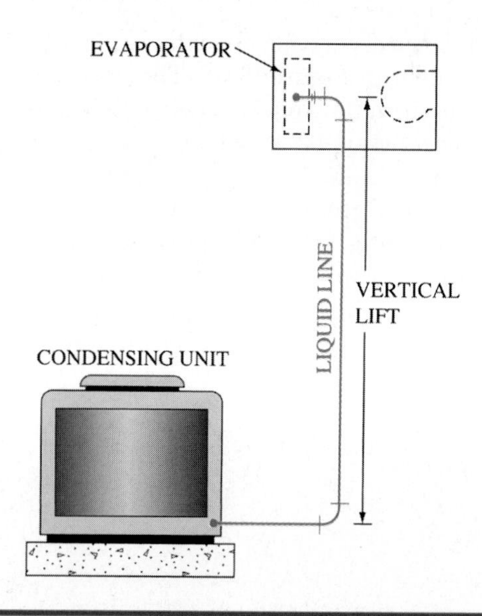

Figure 9-1-9 Piping from condensing unit to the evaporator.

SERVICE TIP

Manufacturers provide extensive testing with their condensers and evaporator coils. Some manufacturers may have 20, 30, 40 or more different air handler evaporator coil combinations for a single condenser. This research is carried out in scientific laboratories so that the most ideal combinations can be determined. Often, however, in the field technicians simply purchase from a supply house a coil that is most convenient and not one specifically designed for the condenser. Any change in design criteria done in the field will adversely affect performance, not enhance performance. Manufacturers are under tremendous pressure from regulatory agencies to produce the highest possible system efficiency. If a change in coil size could enhance efficiency, it would have been done in the lab and there would be test criteria and procedures set up. It is not recommended in any way that technicians undertake design changes from the manufacturer's recommendations. These matched systems are available from the manufacturer and are available from the ARI through publications and on their website.

TABLE 9-1-3 Refrigerant Pressure Loss for Vertical Lift

Vertical lift (ft)	5	10	15	20	25	30
Static pressure loss (psi)	$2^1/_2$	5	$7^1/_2$	10	$12^1/_2$	15

When the evaporator is located above the condensing unit, there is a pressure loss in the liquid line due to the weight of the column of liquid and friction of the liquid against the tubing walls. Table 9-1-3 shows the results of these losses in psi for R-22.

Thirty feet of vertical lift is considered the maximum height for most residential systems. If the pressure loss is great enough, vapor will form in the liquid line prior to normal vaporization in the evaporator, adversely affecting operation and capacity.

Where the evaporator is above the condenser, the suction line should be pitched toward the condenser a minimum of ⅛ in for every 10 ft of run to ensure proper gravity oil return.

Wiring and piping a typical single circuit add-on split system is shown in Figure 9-1-10.

Electrical

Electrical wiring consists of a line voltage of 208/240 V to the outdoor condensing unit through a fused outside disconnect switch. The 115 V line voltage is supplied to the furnace, which furnishes power to the fan motor. For the lower control voltage, a 115/24 V transformer is supplied. A 24 V combination heating and cooling room thermostat controls the on/off of cooling or heating through interlocking relays.

It is considered good electrical practice to use permanent split capacitor (PSC) motors in small hermetic compressors for residential work. The start winding remains energized at all times during motor operation. The run capacitor is added to provide additional torque both during starting and running. Where low voltage fluctuation exists, it may be necessary to add a starting capacitor for extra starting power. Hard start kits, which essentially consist of a starting capacitor, are available to overcome low voltage starting problems and minimize light flicker.

Equipment Sizes

The sizes of residential add-on split systems range from 1 ton to 5 tons, the most popular being 2 to 3 tons in size. Models are available in multiples of 6000 Btu/hr (i.e., 12,000, 18,000, etc. Btu/hr). Most manufacturers rate and certify their equipment in accordance with ARI Standard 210, which is based

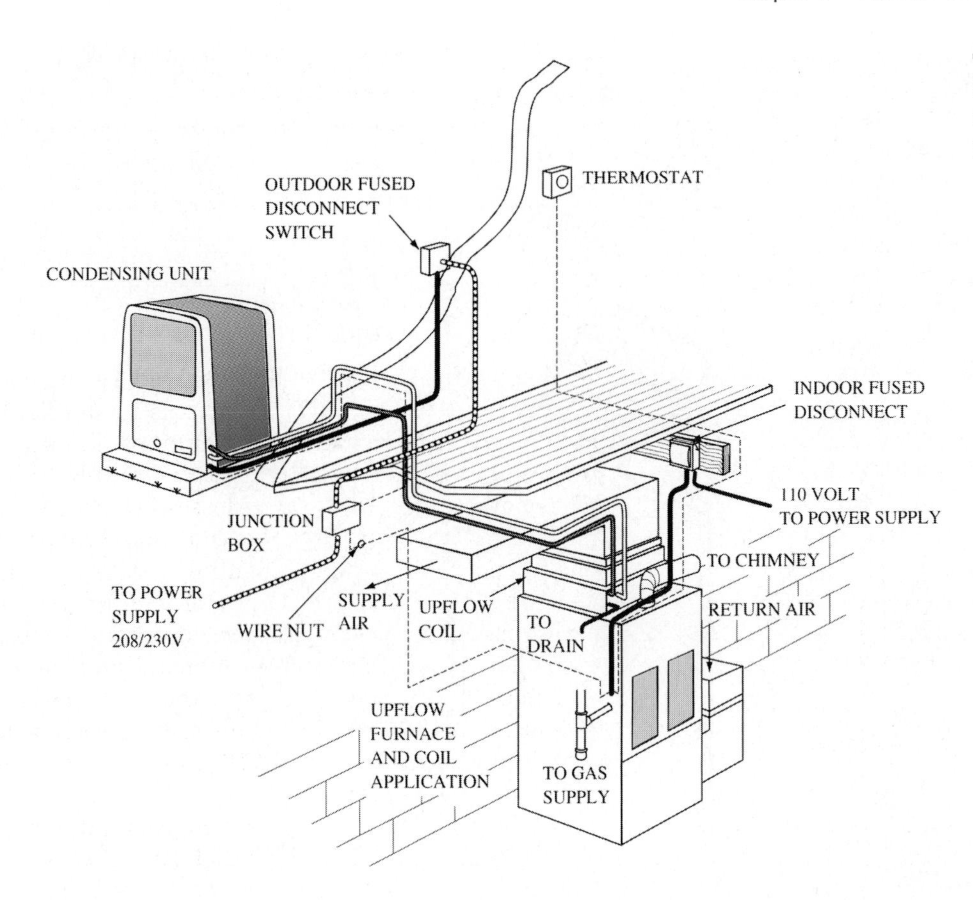

Figure 9-1-10 A single circuit add-on split system installation.

on matching specific size condensing units and cooling coils.

Split systems are also installed in new homes, apartments, and motels where the indoor equipment is a fan coil unit rather than a furnace. Figure 9-1-11 shows an evaporator blower installed in a dropped ceiling, in the furred down area above the closets, hallways, and bathroom. Room air is returned through a ceiling grill and conditioned air is discharged through high sidewall grills located in each room.

Fan coil units used in apartments and motels can also be supplied with electric resistance heaters for winter heating. Cooling sizes of ceiling evaporator blowers range from 12,000 to 30,000 Btu/hr. Condensate drain lines are carefully designed to prevent overflow that could result in damage to the ceiling.

COMMERCIAL SPLIT SYSTEMS

Split system equipment up to 5 tons in capacity may be classified as either commercial or residential. There is a wide range of applications in both markets using the same product; however, above 5 tons the application becomes distinctly commercial, and product designs use different components:

- Brazed refrigerant lines and expansion valves are used rather than fixed orifice type metering devices
- Multiple compressors and condenser fans are used for capacity reduction and low ambient operation
- More functional, heavier structural designs place less emphasis on appearance; and
- Compressors may use three phase electricity.

The air cooled condensing unit, located outside the building, can be designed to discharge air either vertically or horizontally. Vertical discharge has several advantages:

- These units expel large volumes of air, which are best discharged in an upward direction.
- The horizontal arrangement of fan blades is least affected by windmilling if the unit is off. This can be a problem if the fan tries to start with the blades rotating in the wrong direction.
- The use of direct drive multiple blade fans is almost universal in this type of equipment.
- Fan noise directed upward is less objectionable.

Capacity control is often necessary because commercial air conditioning loads are rarely constant. Capacity reduction can be accomplished in several ways. The most common are when multiple compressors are used for higher tonnages. Staging or sequencing the operation is accomplished by cutting compressors on and off to match the load; and with

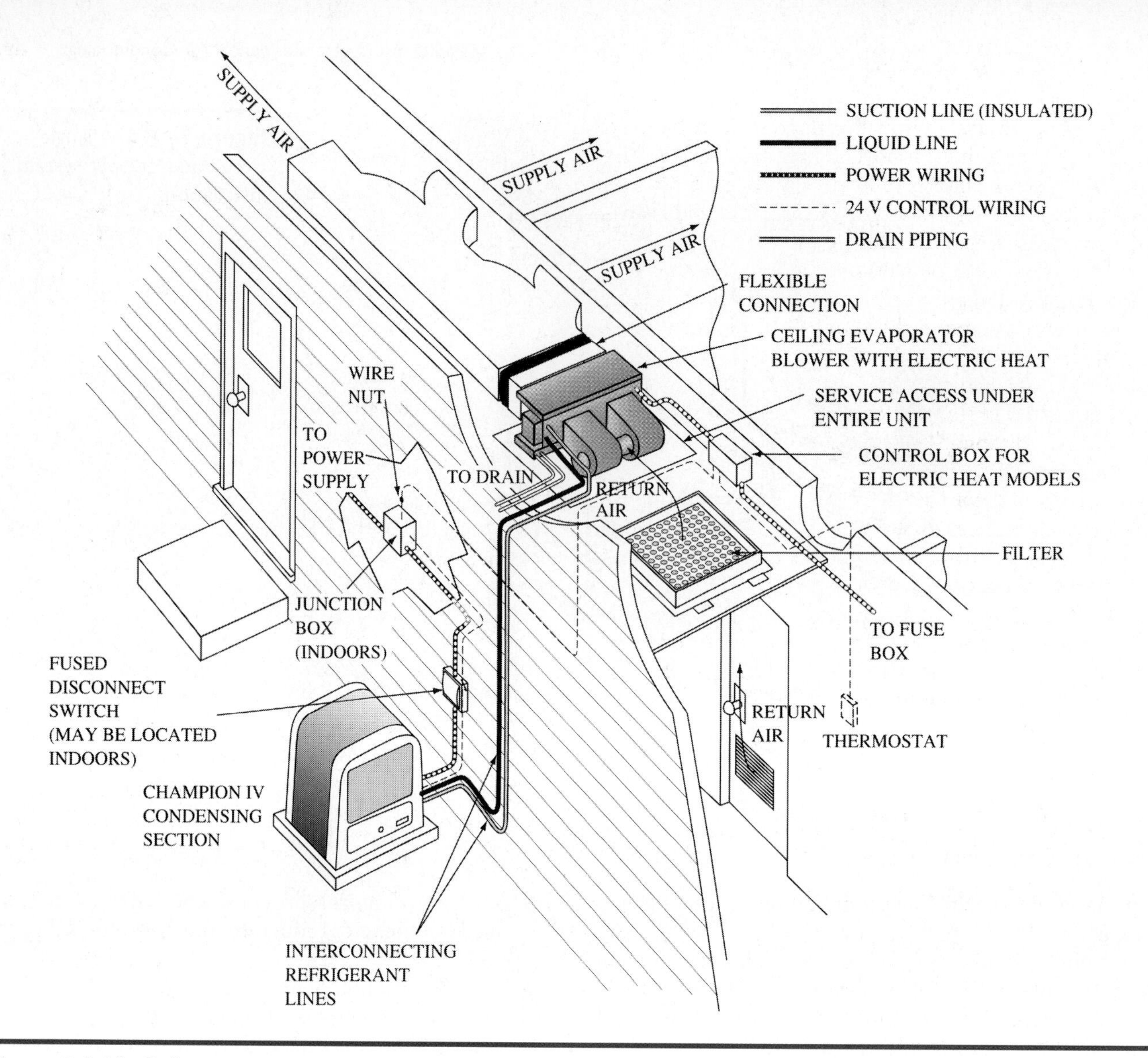

Figure 9-1-11 Ceiling evaporator blower installation.

TABLE 9-1-4 Performance Specifications for a Residential Cooling System

System		Air on Evaporator ft³/min	Air on Evaporator WB Temp. (°F)	Total Capacity (MBtu/hr)	WB Temp. of Evap. (°F)	Comp. and Cond. Fan Input (kW)	Sensible Capacity (MBtu/hr) Dry-Bulb Temp. on Evaporator (°F) 70	75	80	85	90
CA 91	(EB92-B)	3300	72	101	65.9	10.4	—	40	56	69	89
	(EBV92-B)		67	93	60.6	9.9	43	57	73	88	91
	(C90UX⁴)		62	86	55.4	9.4	57	73	84	86	86
	(92DX)		57	78	50.2	9.0	67	72	78	78	78
			72	—	—	—	—	—	—	—	—
	(EB122-B)	4400	67	98	60.6	10.2	43	59	77	93	98
	(EBV122-B)		62	91	55.4	9.7	61	79	87	91	91
	(122DX)		57	84	50.2	9.3	83	84	84	84	84

Cooling capacity–95°F DB air on condenser (MBtu/hr).

single compressors, cylinder unloaders can be used to vary the pumping capacity down 25% or less. The machine starts almost unloaded and the cylinders are cut in as the heat load demands. Unloaded starting reduces power demand.

The centrifugal blower used for commercial systems must deliver a higher airflow and at higher static pressure than for residential applications. A V-belt drive with an optional variable pitch motor pulley blower and motor is needed. The fan scroll can be rotated within the cabinet to obtain alternate discharge arrangements for greater application flexibility.

The evaporator coil(s) are circuited to match the number of compressors. Space is also provided for the inclusion of either freeze protected steam coils or hot water coils for winter heating.

Ratings of matched condensing units with specific air handlers are published by the manufacturer and are certified under ARI Standards. A sample specification is shown in Table 9-1-4.

UNIT 1—REVIEW QUESTIONS

1. How can the actual cooling being done through the evaporator be determined?
2. Why were split systems developed?
3. What does the outdoor air cooled condensing unit of an add-on system consist of?
4. In the condenser coil why is the coil tube depth limited?
5. List the factors that compromise the location of the outdoor condensing unit.
6. What does ARI Standard 270 apply to?
7. What are the three methods of piping on residential type equipment?
8. When the evaporator is located above the condensing unit why would there be a pressure loss in the liquid line?
9. Why is capacity control necessary in commercial split systems?
10. Who publishes the ratings of matched condensing units with specific air handlers?

TECH TALK

Tech 1. Why are we spending so much time getting this outdoor unit set level? I see them all the time on a slight angle.

Tech 2. Well, if the unit is not level the refrigerant that condenses to a liquid will collect in the coils on the low side. That significantly reduces the efficiency of the condenser.

Tech 1. I didn't realize it made a difference.

Tech 2. It does, and some units that are really unlevel can vibrate across their slabs putting a lot of strain on the wires and refrigerant lines. If that's allowed to happen, over time you can wind up with a refrigerant leak or the electric box pulled off the wall.

Tech 1. Sounds like I need to go back and level that unit.

Tech 2. That's right. Let me give you an easy trick to help level it. Put a little play sand underneath the pad. Play sand is sterile so you're not going to end up putting a lot of weeds in your customer's yard and the play sand is easier to work than some of this hard lumpy soil.

Tech 1. I guess I need to stop by the hardware store and pick up a bag of play sand on the way back out to that job.

Tech 2. Sounds like a good idea to me. I'll see you later.

UNIT 2

Troubleshooting Air Conditioning Systems

OBJECTIVES: In this unit the reader will learn about troubleshooting air conditioning systems including:

- Safety
- Problem Analysis
- Air System Problems
- Refrigeration System Problems
- Analyzing Problems

SAFETY

It is important in troubleshooting that the technician pay utmost attention to safety measures. The following general measures should be strictly followed on every job to provide for personal safety:

1. Wear safety glasses and gloves when handling refrigerants or when brazing.
2. Recover or recycle refrigerant using an approved device.
3. Shut off all power and lock it out and tag it out when working on electrical equipment.
4. If the work must be done while the electrical equipment is energized, remove all watches, rings, and other metal to reduce the risk of shock.
5. Always read and follow the specific safety recommendations in the manufacturer's installation and service literature.

GENERAL

Technicians servicing air conditioning equipment find that system performance can be greatly improved not only by good installation practices but also by good maintenance. Careful reference to the manufacturer's installation instructions is important when the system is installed. Good maintenance keeps the equipment running in its original efficient manner.

SAFETY TIP

OSHA regulations require that you follow safe work practice by locking out and tagging out any electrical device while you are working on or near it. You must carry with you the appropriate lock and tagging device to allow you to comply with this regulation.

In troubleshooting air conditioning systems, some of the most useful instruments are:

1. A gauge manifold for measuring suction and discharge pressures at the compressor.
2. A minimum of five thermometers or an electronic thermometer with provision for connecting to at least five remote temperature sensors (see below).
3. A sling psychrometer for measuring wet bulb and dry bulb temperatures to identify the conditions of the supply and return air.
4. A meter(s) with capability of reading amperes, voltages, and ohms.

The thermometers are used for measuring the following temperatures:

1. Supply air
2. Return air
3. Condenser entering and discharge air
4. Suction line at the coil outlet or at the compressor
5. Liquid line at the condenser outlet

By measuring the suction line temperature, the superheat can be determined. Since the suction pressure has been measured, the evaporating temperature can be read from a pressure-temperature chart. The superheat is the difference between the suction line temperature and the evaporating temperature.

By measuring the liquid line temperature, the subcooling can be determined. Since the discharge pressure has been measured, the condensing temperature can be read from the pressure-temperature chart. The subcooling is the difference between the liquid line temperature and the condensing temperature.

SERVICE TIP

We use superheat or subcooling as a charging method depending on the metering device. However, as part of the analyzing of the system you should take both the superheat and subcooling measurements. These temperatures will help resolve any refrigerant circuit problems that might exist.

PROBLEM ANALYSIS

In this unit the discussion is specifically directed to problems and solutions that apply to air conditioning systems. For those solutions that apply to the refrigeration system, please refer to Chapter 17.

A typical troubleshooting job begins with a unit that was originally checked out and put into proper operating condition; has been running satisfactorily for a period of time; and has developed a problem. Mechanical problems in an air conditioning system are classified in two categories: air side problems and refrigeration side problems.

Air System Problems

The primary problem that can occur in the air category is a reduction in quantity. Air handling systems do not suddenly increase in capacity, that is, increase the amount of air across the coil. On the other hand, the refrigeration system does not suddenly increase in heat transfer ability. The first check is therefore the temperature drop of the air through the DX coil. After measuring the return air and supply air temperatures and subtracting to get the temperature drop, is it higher or lower than it should be?

This means that "what it should be" has to be determined first. This is done by using the sling psychrometer to measure and determine the return air wet bulb temperature and relative humidity. From this the proper temperature drop across the coil can be determined from the chart in Figure 9-2-1.

SERVICE TIP

The most common air side problem in central residential air conditioning systems is low airflow. Too often the return duct or the supply duct system can not handle the necessary CFM for the system. An indication of a low airflow problem is a higher than normal temperature difference across the evaporator coil.

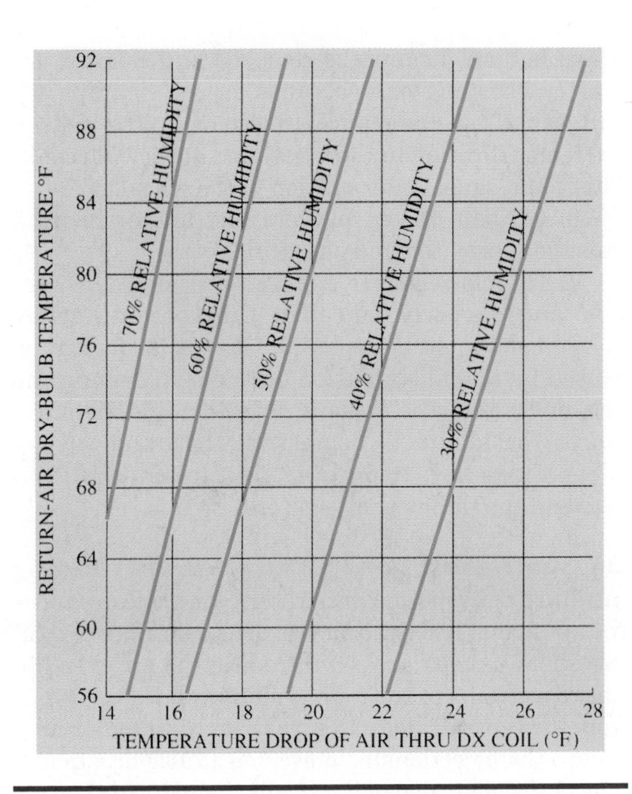

Figure 9-2-1 Air temperature drop for various return air conditions.

Using the required temperature drop as compared to the actual temperature drop, the problem can be classified as either an air problem or a refrigerant problem. If the actual temperature drop is greater than the required temperature drop, the air quantity has been reduced; look for problems in the air handling system. These could be:

1. Dirty air filters or evaporator coil
2. Blower motor and drive
3. Unusual restrictions in the duct system

Air Filters Air filters of the throwaway type should be replaced at least twice each year, at the beginning of both the cooling and heating seasons. In some areas of higher use or where dust is high, they may have to be replaced as often as every 30 days. In commercial and industrial applications, a regular schedule of maintenance must be worked out for best performance and longest equipment life. Because this is the most common problem of air failure, check the filtering system first.

Blower Motor and Drive Check the blower motor and drive in the case of belt-driven blowers to make sure that:

1. The blower motor is properly lubricated and operating freely.

2. The blower wheel is clean. The blades could be loaded with dust and dirt or other debris. If the wheel is dirty, it must be removed and cleaned. Do not try brushing only, because a poor cleaning job will cause an imbalance to occur in the wheel. Extreme vibration in the wheel and noise will result. This could cause deterioration of the wheel.

3. On belt driven blowers, the blower bearing must be lubricated and operating freely.

4. The blower drive belt must be in good condition and properly adjusted. Cracked or heavily glazed belts must be replaced. Heavy glazing can be caused by too much tension on the belt, driving the belt down into the pulleys. Proper adjustment requires the ability to depress the belt midway between the pulleys approximately 1 in for each 12 in between the pulley shaft centers.

Unusual Restrictions in Duct Systems Placing furniture or carpeting over return air grilles reduces the air available for the blower to handle. Shutting off the air to unused areas will reduce the air over the coil. Covering a return air grille to reduce the noise from the centrally located furnace or air handler may reduce the objectionable noise, but it also drastically affects the operation of the system by reducing the air quantity.

The collapse of the return air duct system will affect the entire duct system performance. Air leaks in the return duct will raise the return air temperature and reduce the temperature drop across the coil.

SERVICE TIP

Air distribution systems installed before the change in standards may have cloth or unapproved duct tape. These systems may need to be sealed properly in order for the central residential air conditioning system to function properly.

Refrigeration System Problems

When the temperature drop across the coil is less than required, this means that the heat removal capacity of the system has been reduced.

These problems can be simply divided into two categories: refrigerant quantity, and refrigerant flow rate. If the system has the correct amount of refrigerant charge and refrigerant is flowing at the desired rate, the system should work properly and deliver rated capacity. Any problems in either category will affect the temperatures and pressures that will occur in the unit when the correct amount of air is supplied over the DX coil for the capacity of the unit. Obviously, if the system is empty of refrigerant, a leak has occurred, and it must be found and repaired. The system must be evacuated and dehydrated thoroughly, down to 500 microns, before recharging with the correct amount of refrigerant. If the system will not operate at all, it is probably an electrical problem that must be found and corrected.

If the system will start and run but will not produce satisfactory cooling. This means that the amount of heat picked up in the coil plus the amount of motor heat added and the total rejected from the condenser is not the total heat quantity the unit is designed to handle. To determine the problem, all the information listed in Table 9-2-1 must be measured. These results compared to normal operating results will generally identify the problem. The use of the word "normal" does not imply a fixed set of pressures and temperatures. These will vary with each make and model of the system. A few temperatures are fairly consistent throughout the industry and can be used for comparison:

1. DX coil operating temperature
2. Condensing unit condensing temperature
3. Refrigerant subcooling

These items must also be modified according to the energy efficiency ratio (EER) of the unit. The reason for this is that the amount of evaporation and condensing surface designed into the unit are the main factors in efficiency rating. A larger condensing surface results in a lower condensing temperature and a higher EER. A larger evaporating surface results in a higher suction pressure and a higher EER. The energy efficiency ratio for the conditions is calculated by dividing the net capacity of the unit in Btu/hr by the watts input.

NICE TO KNOW

As an air conditioning system picks up heat the temperature and pressure of the system increases on the suction side. If the air conditioning system cannot pick up heat in the evaporator the suction pressure and temperature will go down. If the condenser is unable to reject the heat, the condenser temperature and pressure will go up. If the condenser is not receiving heat, its temperature and pressure will go down. Heat and temperature and pressure are all interrelated. If a system has lower than normal pressure and temperature, then heat is not being picked up. This may be caused by low airflow or low refrigerant charge. If a system has higher temperatures and pressures than normal it is either picking up more heat than can be rejected by the condenser or the condenser is unable to reject the heat normally. The bottom line is that if you follow the temperature and pressure you will follow the heat, which will enable you to diagnose refrigerant circuit problems.

TABLE 9-2-1 Troubleshooting Chart for Refrigeration and Air Conditioning Systems, Showing Symptoms and Probable Causes

Probable Cause	Low Side (Suction) Pressure (psig)	D.X. Coil Superheat (°F)	High Side (Hot Gas) Pressure (psig)	Condenser Liquid Subcooling (°F)	Cond. Unit Amperage Draw (A)
1 Insufficient or unbalanced load	Low	Low	Low	Normal	Low
2 Excessive load	High	High	High	Normal	High
3 Low ambient (cond. entering air °F)	Low	High	Low	Normal	Low
4 High ambient (cond. entering air °F)	High	High	High	Normal	High
5 Refrigerant undercharge	Low	High	Low	Low	Low
6 Refrigerant overcharge	High	Low	High	High	High
7 Liquid line restriction	Low	High	Low	High	Low
8 Plugged capillary tube	Low	High	Low	High	Low
9 Suction line restriction	Low	High	Low	Normal	Low
10 Hot gas line restriction	High	High	High	Normal	High
11 Inefficient compressor	High	High	Low	Low	Low

DX Coil Operating Temperature Normal coil operating temperatures can be found by subtracting the design coil split from the average air temperature going through the coil. The coil split will vary with the system design.

Systems in the EER range of 7.0 to 8.0 will have design splits in the range 25 to 30°F. Systems in the EER range of 8.0 to 9.0 will have design splits in the range 20 to 25°F. Systems with 9.0 + EER ratings will have design splits in the range 15 to 20°F. The formula used for determining coil operating temperatures is:

$$\text{COT} = \left(\frac{\text{EAT} + \text{LAT}}{2}\right) - \text{split}$$

where

COT = coil operating temperature
EAT = entering air temperature of the coil
LAT = leaving air temperature of the coil

The latter two temperatures added together and divided by 2 will give the average air temperature. This is also referred to as the mean temperature difference (MTD). It is also sometimes referred to as the coil TD or ΔT.

"Split" is the design split according to the EER rating. For example, a unit having an entering air condition of 80° DB and a 20°F temperature drop across the evaporator coil will have an operating coil temperature determined as follows:

EXAMPLE

For an EER rating of 7.0 to 8.0:

$$\text{COT} = \frac{80 + 60}{2} - 25 \text{ to } 30° = 40 \text{ to } 45°\text{F}$$

For an EER rating of 8.0 to 9.0:

$$\text{COT} = \frac{80 + 60}{2} - 20 \text{ to } 25° = 45 \text{ to } 50°\text{F}$$

For an EER rating of 9.0+:

$$\text{COT} = \frac{80 + 60}{2} - 15 \text{ to } 20° = 50 \text{ to } 55°\text{F}$$

This demonstrates that the operating coil temperature changes with the EER rating of the unit. ■

NICE TO KNOW

The current energy efficiency requirements mandate that new equipment have a minimum of a 10 SEER. It is important to note however that the EER will be lower and that there is not a specific standard requirement for the EER. As a rule of thumb the SEER will be somewhere between 80 and 90% of the SEER rating.

Condensing Unit Condensing Temperature The amount of surface in the condenser affects the condensing temperature the unit must develop to operate at rated capacity. The variation in the size of the condenser also affects the production cost and price of the unit. The smaller the condenser, the lower the price, but also the lower the efficiency (EER) rating. In the same EER ratings used for the DX coil, at 95°F outside ambient, the 7.0 to 8.0 EER category will operate in the 25 to 30° condenser split range, the 8.0 to 9.0 EER category in the 20 to 25° condenser split range, and the 9.0+ EER category in the 15 to 20° condenser split range.

This means that when the air entering the condenser is at 95°F, the formula for finding the condensing temperature would be:

$$RCT = EAT + split$$

where
RCT = refrigerant condensing temperature
EAT = entering air temperature of the condenser
split = design temperature difference between the entering air temperature and the condensing temperatures of the hot high pressure vapor from the compressor.

EXAMPLE

Using the formula with 95°F EAT, the split for the various EER systems would be:

For an EER rating of 7.0 to 8.0

$$RCT = 95° + 25 \text{ to } 30° = 120 \text{ to } 125°F$$

For an EER rating of 8.0 to 9.0

$$RCT = 95° + 20 \text{ to } 25° = 115 \text{ to } 120°F$$

For an EER rating of 9.0+

$$RCT = 95° + 15 \text{ to } 20° = 110 \text{ to } 115°F$$

This demonstrates that operating head pressures vary not only from changes in outdoor temperatures but with the different EER ratings. ■

Refrigerant Subcooling The amount of subcooling produced in the condenser is determined primarily by the quantity of refrigerant in the system. The temperature of the air entering the condenser and the load in the DX coil will have only a small effect on the amount of subcooling produced. The amount of refrigerant in the system has the predominant effect. Therefore, regardless of EER ratings, the unit should have, if properly charged, a liquid subcooled to 15 to 20°F. High outdoor temperatures will produce the lower subcooled liquid because of the reduced quantity of refrigerant in the liquid state in the system. More refrigerant will stay in the vapor state to produce the higher pressure and condensing temperatures needed to eject the required amount of heat.

ANALYZING PROBLEMS

Using the information obtained from the two pressure gauges, a minimum of five thermometers, the sling psychrometer, and a clamp-on ammeter, we can analyze the system problems by using the chart in Table 9-2-1.

The table shows that there are 11 probable causes of trouble in an air conditioning system. After each probable cause is the reaction that the cause

SERVICE TIP

If a refrigerant liquid line rises over 30 ft vertically, additional subcooling may be required to ensure that the metering device is receiving 100% liquid. The additional subcooling is required because of the pressure difference between the refrigerant at the condenser and pressure of the refrigerant at the metering device. This pressure drop is the result of the static head caused by the vertical lift of the refrigerant. Refer to the manufacturer's technical specifications to see what additional subcooling is required for unusually high vertical lifts.

would have on the refrigeration system low side or suction pressure, the DX coil superheat, the high side or discharge pressure, the amount of subcooling of the liquid leaving the condenser, and the amperage draw of the condensing unit.

Insufficient or Unbalanced Load

Insufficient air over the DX coil would be indicated by a greater than desired temperature drop through the coil. An unbalanced load on the DX coil would also give the opposite indication; some of the circuits of the DX coil would be overloaded, while others would be lightly loaded. This would result in a mixture of air off the coil that would cause a reduced temperature drop of the air mixture. The lightly loaded sections of the DX coil would allow liquid refrigerant to leave the coil and enter the suction manifold and suction line.

NICE TO KNOW

The final step in manufacturing air conditioning and refrigeration equipment takes place at the residence. This final step in manufacturing is done outside of the control of the manufacturer. Many manufacturers have found that significant problems can exist in system installation, including the mismatching of equipment sizes. If you suspect that your refrigerant circuit problems are the result of mismatched equipment, call the equipment supplier or manufacturer and provide them the system component model numbers so they can inform you as to whether the system is mismatched in size and make recommendations to resolve the problems if a mismatched system does exist.

In TXV systems, the liquid refrigerant passing the sensing bulb of the TXV would cause the valve to close down. This would reduce the operating temperature and capacity of the DX coil as well as lowering the suction pressure. This reduction would be very pronounced. The DX coil operating superheat would be very low, probably zero, because of the liquid leaving some of the sections of the DX coil.

High side or discharge pressure would be low due to the reduced load on the compressor, reduced amount of refrigerant vapor pumped, and reduced heat load on the condenser. Condenser liquid subcooling would be on the high side of the normal range because of the reduction in refrigerant demand by the TXV. Condensing unit amperage draw would be down due to the reduced load.

In systems using fixed metering devices, the unbalanced load would produce a lower temperature drop of the air through the DX coil because the amount of refrigerant supplied by the fixed metering device would not be reduced; therefore, the system pressure (boiling point) would be approximately the same.

The DX coil superheat would drop to zero with liquid refrigerant flooding into the suction line. Under extreme cases of imbalance, liquid returning to the compressor could cause compressor damage. The reduction in heat gathered in the DX coil and the lowering of the refrigerant vapor to the compressor will lower the load on the compressor. The compressor discharge pressure (hot gas pressure) will be reduced.

The flow rate of the refrigerant will be only slightly reduced because of the lower head pressure. The subcooling of the refrigerant will be in the normal range. The amperage draw of the condensing unit will be slightly lower because of the reduced load on the compressor and reduction in head pressure.

Excessive Load

In the case of excessive load the opposite effect exists. The temperature drop of the air through the coil will be less, because the unit cannot cool the air as much as it should. Air is moving through the coil at too high a velocity. There is also the possibility that the temperature of the air entering the coil is higher than the return air from the conditioned area. This could be from air leaks in the return duct system drawing hot air from unconditioned areas.

The excessive load raises the suction pressure. The refrigerant is evaporating at a rate faster than the pumping rate of the compressor. The superheat developed in the coil will be as follows:

1. If the system uses a TXV, the superheat will be normal to slightly high. The valve will operate at a higher flow rate to attempt to maintain superheat settings.

2. If the system uses fixed metering devices, the superheat will be high. The fixed metering devices cannot feed enough increase in refrigerant quantity to keep the DX coil fully active.

The high side or discharge pressure will be high. The compressor will pump more vapor because of the increase in suction pressure. The condenser must handle more heat and will develop a higher condensing temperature to eject the additional heat. A higher condensing temperature means a greater high side pressure. The quantity of liquid in the system has not changed, nor is the refrigerant flow restricted. The liquid subcooling will be in the normal range. The amperage draw of the unit will be high because of the additional load on the compressor.

Low Ambient (Condenser Entering Air) Temperature

In this case, the condenser heat transfer rate is excessive, producing an excessively low discharge pressure. As a result, the suction pressure will be low because the amount of refrigerant through the metering device will be reduced. This reduction will reduce the amount of liquid refrigerant supplied to the DX coil. The coil will produce less vapor and the suction pressure drops.

TECH TIP

Often air conditioning and refrigerant equipment die at night. This is frequently caused because of the lower ambient temperatures experienced in the evening. Low ambient operation of refrigeration and air conditioning compressors can result in a liquid floodback or slugging of the compressor. If a system is to be operated on a regular basis during low ambient conditions, it should be equipped with a low ambient kit to protect the compressor.

The decrease in the refrigerant flow rate into the coil reduces the amount of active coil, and a higher superheat results. In addition, the reduced system capacity will decrease the amount of heat removed from the air. There will be higher temperature and relative humidity in the conditioned area and the high side pressure will be low. This starts a reduction in system capacity. The amount of subcooling of the liquid will be in the normal range. The quantity of liquid in the condenser will be higher, but the heat transfer rate of the evaporator is less. The amperage draw of the condensing unit will be less because the compressor is doing less work.

The amount of drop in the condenser ambient air temperature that the air conditioning system will tolerate depends on the type of pressure reducing device in the system. Systems using fixed metering devices will have a gradual reduction in capacity as the outside ambient drops from 95°F. This gradual reduction occurs down to 65°F. Below this temperature the capacity loss is drastic, and some means of maintaining head pressure must be employed to prevent the evaporator temperature from dropping below freezing. The most reliable means is control of air through the condenser via dampers in the airstream of a variable speed condenser fan.

Systems that use TXV will maintain higher capacity down to an ambient temperature of 47°F. Below this temperature, controls must be used. The control of cfm through the condenser using dampers or the condenser fan speed control can also be used. In larger TXV systems, liquid quantity in the condenser is used to control head pressure.

High Ambient (Condenser Entering Air) Temperature

The higher the temperature of the air entering the condenser, the higher the condensing temperature of the refrigerant vapor to eject the heat in the vapor. The higher the condensing temperature, the higher the head pressure. The suction pressure will be high for two reasons: (1) the pumping efficiency of the compressor will be less; and (2) the higher temperature of the liquid will increase the amount of flash gas in the metering device, further reducing the system efficiency.

The amount of superheat produced in the coil will be different in a TXV system and a fixed metering device system. In the TXV system the valve will maintain superheat close to the limits of its adjustment range even though the actual temperatures involved will be higher. In a fixed metering device system, the amount of superheat produced in the coil is the reverse of the temperature of the air through the condenser. The flow rate through the fixed metering devices are directly affected by the head pressure. The higher the air temperature, the higher the head pressure and the higher the flow rate. As a result of the higher flow rate, the subcooling is lower.

Table 9-2-2 shows the superheat that will be developed in a properly charged air conditioning system using fixed metering devices. Do not attempt to charge a fixed metering system below 65°F, as system operating characteristics become very erratic.

The head pressure will be high at the higher ambient temperatures because of the higher condensing temperatures required. The condenser liquid subcooling will be in the lower portion of the normal range. The amount of liquid refrigerant in the condenser will be reduced slightly because more will stay in the vapor state to produce the higher pressure and condensing temperature. The amperage draw of the condensing unit will be high.

TABLE 9-2-2 The Effects of Outdoor (Ambient) Temperature on Superheat

Outdoor Air Temperature Entering Condenser Coil(°F)	Superheat (°F)
65	30
75	25
80	20
85	18
90	15
95	10
105 & above	5

Refrigerant Undercharge

A shortage of refrigerant in the system means less liquid refrigerant in the DX coil to pick up heat, and lower suction pressure. The smaller quantity of liquid supplied the DX coil means less active surface in the coil for vaporizing the liquid refrigerant, and more surface to raise vapor temperature. The superheat will be high. There will be less vapor for the compressor to handle and less head for the condenser to reject, lower high side pressure, and lower condensing temperature.

SERVICE TIP

The compressor in an air conditioning system is cooled primarily by the cool returning suction gas. Compressors that are low on charge can have a much higher operating temperature. The temperature can be high enough so that the motor windings begin to break down. As this occurs the motor can ultimately short out, resulting in a compressor change out. If an air conditioning system has a leak it must be located so that a low refrigerant charge can avoided.

The amount of subcooling will be below normal to none, depending on the amount of undercharge. The system operation is usually not affected very seriously until the subcooling is zero and hot gas starts to leave the condenser, together with the liquid refrigerant. The amperage draw of the condensing unit will be slightly less than normal.

Refrigerant Overcharge

An overcharge of refrigerant will affect the system in different ways, depending on the pressure reducing device used in the system and the amount of overcharge.

TECH TIP

It is unacceptable to add a little extra refrigerant to a system just in case it might have a leak. If you suspect that a refrigeration system has a leak don't overcharge the system but find the leak and fix it.

TXV Systems In systems using a TXV, the valve will attempt to control the refrigerant flow into the coil to maintain the superheat setting of the valve. However, the extra refrigerant will back up into the condenser, occupying some of the heat transfer area that would otherwise be available for condensing. As a result, the discharge pressure will be slightly higher than normal, the liquid subcooling will be high, and the unit amperage draw will be high. The suction pressure and DX coil superheat will be normal. Excessive overcharging will cause even higher head pressure, and hunting of the TXV.

For TXV systems with excessive overcharge:

1. The suction pressure will be high. Not only does the reduction in compressor capacity (due to higher head pressure) raise the suction pressure, but the higher pressure will cause the TXV valve to overfeed on its opening stroke. This will cause a wider range of hunting of the valve.
2. The DX coil superheat will be very erratic from the low normal range to liquid out of the coil.
3. The high side or discharge pressure will be extremely high.
4. Subcooling of the liquid will also be high because of the excessive liquid in the condenser.
5. The condensing unit amperage draw will be higher because of the extreme load on the compressor motor.

Fixed Bore and Capillary Tube Systems The amount of refrigerant in the fixed metering system has a direct effect on system performance. An overcharge has a greater effect than an undercharge, but both affect system performance, efficiency (EER), and operating cost.

Figures 9-2-2 through 9-2-4 show how the performance of an air conditioning system is affected by an incorrect amount of refrigerant charge.

Shown in Figure 9-2-2, at 100% of correct charge (55 oz), the unit developed a net capacity of 26,200 Btu/hr. When the amount of charge was varied 5% in either direction, the capacity dropped as the charge varied. Removing 5% (3 oz) of refrigerant reduced the net capacity to 25,000 Btu/hr. Another 5% (2.5 oz) reduced the capacity to 22,000 Btu/hr. From there on the reduction in capacity became very drastic: 85% (8 oz), 18,000 Btu/hr; 80% (11 oz), 13,000 Btu/hr; and 75% (14 oz), 8000 Btu/hr.

Addition of overcharge had the same effect but at a greater reduction rate. The addition of 3 oz of refrigerant (5%) reduced the net capacity to 24,600

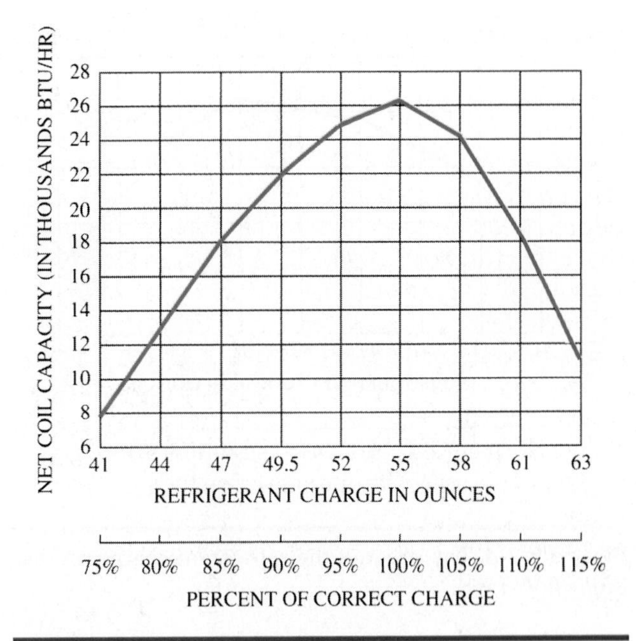

Figure 9-2-2 The effect of the refrigerant charge on the capacity of the unit.

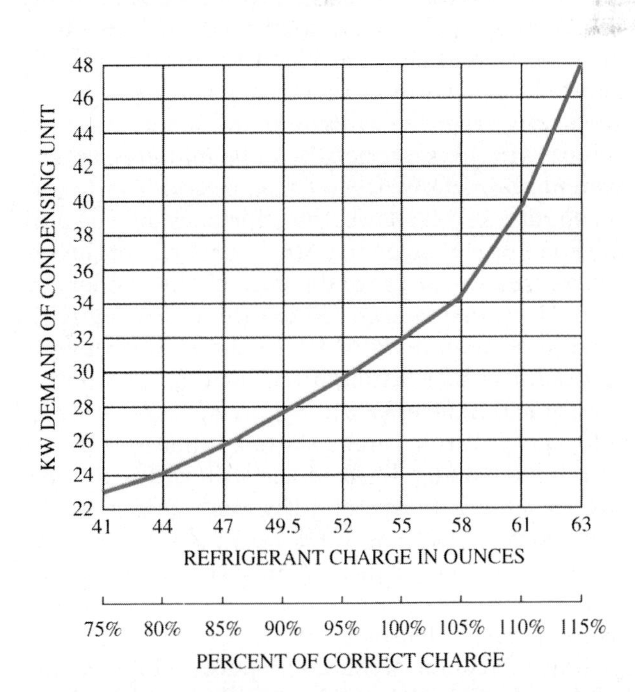

Figure 9-2-3 The effect of the refrigerant charge on the kW demand of the condensing unit.

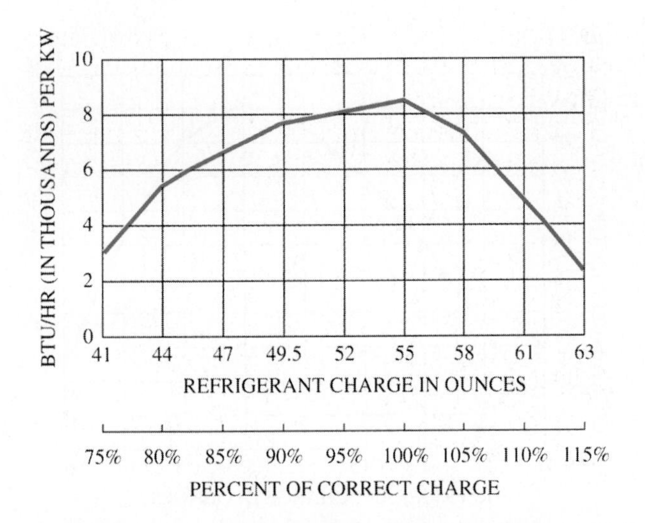

Figure 9-2-4 The effect of the refrigerant charge on the Btu/hr/kW ratio.

Btu/hr; 6 oz added (10%) reduced the capacity to 19,000 Btu/hr; and 8 oz added (15%) dropped the capacity to 11,000 Btu/hr. This shows that overcharging of a unit has a greater effect per ounce of refrigerant than does undercharging.

Figure 9-2-3 is a chart showing the amount of electrical energy the unit will demand because of pressure created by the amount of refrigerant in the system, with the only variable being the refrigerant charge. At 100% of charge (55 oz) the unit required 32 kW. As the charge was reduced, the wattage demand also dropped, 29.6 kW at 95% (3 oz), 27.6 kW at 90% (6.5 oz), 25.7 kW at 85% (8 oz), 25 kW at 80% (11 oz), and 22.4 kW at 75% (14 oz short of correct charge). When the unit was overcharged, the wattage required went up. At 3 oz (5% overcharge) the wattage required was 34.2 kW; at 6 oz (10% overcharge), 39.5 kW; and at 8 oz (15% overcharge), 48 kW.

Figure 9-2-4 shows the efficiency of the unit (EER rating) based on the Btu/hr capacity of the system versus the wattage demand of the condensing unit. At correct charge (55 oz) the efficiency (EER rating) of the unit was 8.49. As the refrigerant was reduced, the EER rating dropped to 8.22 at 95% of charge, 7.97 at 90%, 7.03 at 85%, 5.2 at 80%, and 3.57 at 75% of full refrigerant charge. When refrigerant was added, adding 5% (3 oz) the EER rating dropped to 7.19. At 10% (6 oz) the EER was 4.8, and at 15% overcharge (8 oz) the EER was 2.29. From these charts the only conclusion is that the capillary tube systems must be charged to the correct charge with only a +/−5% tolerance.

The effect of overcharge produces a high suction pressure because the refrigerant flow to the DX coil increases. Suction superheat will decrease because of the additional quantity to the DX coil. At approximately 8 to 10% of overcharge, the suction superheat becomes zero and liquid refrigerant will leave the DX coil. This will cause flooding of the compressor and greatly increases the chance of compressor failure. The high side or discharge pressure will be high because of the extra refrigerant in the condenser. Liquid subcooling will also be high for the same reason. The wattage draw will increase due to the greater amount of vapor pumped as well as the higher compressor discharge pressure.

SERVICE TIP

As the ambient temperature decreases the superheat will increase on fixed bore or capillary tube orifice systems. A rule of thumb to give a quick check on the charge is to subtract the ambient temperature from 105°F. The difference will be the approximate superheat that a normally functioning system would have. In no case however should the superheat be less than 5°F.

Liquid Line Restriction

Liquid line restriction reduces the amount of refrigerant to the pressure reducing device. Both TXV valve systems and fixed metering device systems will then operate with reduced refrigerant flow rate to the DX coil. The following observations can be made of liquid line restrictions:

1. The suction pressure will be low because of the reduced amount of refrigerant to the DX coil.

2. The suction superheat will be high because of the reduced active portion of the coil, allowing more coil surface for increasing the vapor temperature as well as reducing the refrigerant boiling point.

3. The high side or discharge pressure will be low because of the reduced load on the compressor.

4. Liquid subcooling will be high. The liquid refrigerant will accumulate in the condenser. It cannot flow out at the proper rate because of the restriction. As a result, the liquid will cool more than desired.

5. The amperage draw of the condensing unit will be low.

Plugged Fixed Bore or Capillary Tube Metering Devices

Either a plugged fixed metering device or plugged feeder tube between the TXV valve distributor and the coil will cause part of the coil to be inactive. The system will then be operating with an undersized coil, resulting in the following:

1. The suction pressure will be low because the coil capacity has been reduced.

2. The suction superheat will be high in the fixed metering device systems. The reduced amount of vapor produced in the coil and resultant reduction in suction pressure will reduce compressor capacity, head pressure, and the flow rate of the remaining active capillary tubes.

3. The high side or discharge pressure will be low.

4. Liquid subcooling will be high; the liquid refrigerant will accumulate in the condenser.

5. The unit amperage draw will be low.

In TXV systems, the following will result:

1. A plugged feeder tube reduces the capacity of the coil. The coil cannot provide enough vapor to satisfy the pumping capacity of the compressor and the suction pressure balances out at a low pressure.

2. The superheat, however, will be in the normal range because the valve will adjust to the lower operating conditions and maintain the setting superheat range.

3. The high side or discharge pressure will be low because of the reduced load on the compressor and condenser.

4. Liquid subcooling will be high because of the liquid refrigerant accumulating in the condenser.

5. The amperage draw of the condensing unit will be low.

Suction Line Restriction

Suction line restriction could be caused by a plugged suction line strainer, a kink in the suction line, or a solder joint filled with solder. It results in a high pressure drop between the DX coil and the compressor. The following symptoms will help diagnose a suction line restriction.

1. The suction pressure, if measured at the condensing unit end of the suction line, will be low.

2. The superheat, as measured by suction line temperature at the DX coil and suction pressure (boiling point) at the condensing unit, will be extremely high.

3. The high side or discharge pressure will be low because of reduced load on the compressor.

4. The low suction and discharge pressure usually indicate a refrigerant shortage. **Warning:** The liquid subcooling is normal to slightly above normal. This indicates a surplus of refrigerant in the condenser. Most of the refrigerant is in the coil, where the evaporation rate is low due to the higher operating pressure in the coil.

5. The amperage draw of the condensing unit would be low because of the light load on the compressor.

Hot Gas Line Restriction

The high side or compressor discharge pressure will be high if measured at the compressor outlet or low if measured at the condenser outlet or liquid line. In either case the compressor amperage draw will be high.

Therefore:

1. The suction pressure is high due to reduced pumping capacity of the compressor.

2. The DX coil superheat is high because the suction pressure is high.

3. The high side pressure is high when measured at the compressor discharge or low when measured at the liquid line.

4. Liquid subcooling is in the high end of the normal range.

5. Even with all of this, the compressor amperage draw is above normal. All symptoms point to an extreme restriction in the hot gas line. This problem is easily found when the discharge pressure is measured at the compressor discharge.

When the measuring point is the liquid line at the condenser outlet, the facts are easily misinterpreted. High suction pressure and low discharge pressure will usually be interpreted as an inefficient compressor. The amperage draw of the compressor must be measured. The high amperage draw indicates that the compressor is operating against a high discharge pressure. A restriction apparently exists between the outlet of the compressor and the pressure measuring point.

Inefficient Compressor

The problem of an inefficient compressor is last on the list because it is the least likely to be a problem. When the compressor will not pump the required amount of refrigerant vapor:

1. The suction pressure will balance out higher than normal.
2. The DX coil superheat will be high.
3. The high side or discharge pressure will be extremely low.
4. Liquid subcooling will be low because not much heat will be in the condenser. The condensing

temperature will therefore be close to the entering air temperature.

5. The amperage draw of the condensing unit will be extremely low, indicating that the compressor is doing very little work.

From this comparison of operational data, reference to a troubleshooting chart, such as shown in Table 9-2-1, will assist in locating the cause of the problem.

As soon as the cause is verified, in most cases the remedy is self evident. For example, if a wiring connection is loose, it needs to be tightened. When a motor is burned out, it needs to be replaced. Where replacement or service of specialized parts is required, usually the manufacturer provides detailed instructions for performing the work. When maintenance is required, such as cleaning a dirty coil or replacing worn out belts, helpful instructions will be found in the section of this manual on maintenance.

UNIT 2—REVIEW QUESTIONS

1. List the general measures that should be strictly followed on every job to provide for personal safety.
2. The ________ is the difference between the suction line temperature and the evaporating temperature.
3. What is the primary problem that can occur in the air category?
4. List the temperatures that are fairly consistent throughout the industry and that are used for comparison.
5. What is EER and how is it calculated?
6. What is the formula used for determining coil operating temperatures?
7. What would indicate insufficient air over the DX coil?
8. List the two reasons why high ambient temperature would make the suction pressure be high.
9. What does the shortage of refrigerant in the system mean?
10. What will happen when the compressor will not pump the required amount of refrigerant vapor?

TECH TALK

Tech 1. Why were you so concerned that the last unit we serviced was undercharged?

Tech 2. Well, you remember that the compression ratio of a system is calculated by dividing the absolute discharge pressure by the absolute suction pressure (psia discharge ÷ psia suction).

Tech 1. Yes, I remember that.

Tech 2. Well, just to make the math easier follow this example. If I have a 200 psia discharge pressure and a 20 psia suction pressure that would be 200 divided by 20 which equals 10 (200 psia ÷ 20 psia = 10). So that gives me a 10:1 ratio, because 200 psia discharge / 20 psia suction = 10:1.

Tech 1. Yes, I see that.

Tech 2. So if I have a dirty condenser and the discharge pressure doubles, that would be a 200 psia increase.

Tech 1. So you would have a 400 psia discharge.

Tech 2. That's correct; we would have a 400 psia discharge. If the suction remained at the same 20 psia that would give us 400 psia discharge divided by 20 psia suction or a compression ratio of 20 to 1, because 400 psia ÷ 20 psia = 20:1.

Tech 1. Yes, I see that.

Tech 2. So what if we have a system that is undercharged and the suction pressure drops by half?

Tech 1. That would be only a 10 psia drop in the suction pressure from 20 psia to 10 psia.

Tech 2. That's correct. So now let's calculate what would happen if the discharge pressure stayed the same at 200 and the suction pressure dropped to 10; that would give us a compression ratio of 20 to 1, because 200 psia discharge ÷ 10 psia suction = 20:1.

Tech 1. Wow, just a little drop in the suction pressure makes a whole lot more difference than a great big pressure increase on the discharge side for the compression ratio.

Tech 2. That's right. Remember that if the compression ratio gets too high then the refrigerant can break down and form acid. The acid then can heat up the compressor windings and bearings and everything else in the system and cause the oil to form sludge that will plug the filter dryer and metering device. So now you see why I was so concerned when the system showed an undercharge.

Tech 1. That makes sense. That's why you did an acid test to see if it had formed acid.

Tech 2. That's right. And we caught it soon enough. So we didn't have to put an acid neutralizer in the system.

Tech 1. I appreciate that. Next time I will pay more attention when I have a system with low charge.

10

Central Residential Gas Warm Air Heating

UNIT 1

Combustion and Fuels

OBJECTIVES: In this unit the reader will learn about combustion and fuel including:

- Gases
- Oils
- Electricity
- Types of Flames
- AFUE
- Practical Combustion Considerations
- Condensation and Corrosion

COMBUSTION

Combustion is the chemical process by which oxygen is combined rapidly with a fuel to release energy in the form of heat. Another form of energy released at the same time is light (electromagnetic energy), which gives visibility to the flame, as shown in Figure 10-1-1.

Conventional fuels are hydrocarbons. Hydrocarbons contain primarily hydrogen and carbon atoms in various compounds. Complete combustion products of all clean burning hydrocarbons are carbon dioxide (CO_2) and water (H_2O). However, not all flames burn clean. The main product of incomplete combustion is carbon monoxide (CO). The poorer the combustion the more CO produced. Incomplete combustion also will produce some small quantities of partially reacted fuel constituents (gases and liquid or solid aerosols). Most conventional fuels also contain small amounts of sulfur as a contaminant. In the flame sulfur reacts to form sulfur dioxide and sulfur trioxide. To a lesser amount other products of combustion include products such as ash and inert gases. Flue gas usually also contains excess air.

Combustion requires three things: (1) fuel, (2) heat, and (3) oxygen, as shown in Figure 10-1-2. The rate of combustion depends on:

1. The chemical reaction rate of the combustible fuel constituents with oxygen.
2. The rate at which oxygen is supplied to the fuel (mixing air and fuel).
3. The temperature in the combustion region.

The chemical reaction rate is fixed in the selection of the fuel. Increasing the mixing rate or the temperature increases the combustion rate.

Figure 10-1-1 The natural gas flame gives off both heat and light.

With complete combustion, all hydrogen and carbon are oxidized into H_2O and CO_2. Generally for complete combustion, excess air must be supplied beyond the amount required to oxidize the fuel.

Incomplete combustion occurs when the fuel is not completely oxidized in the combustion process.

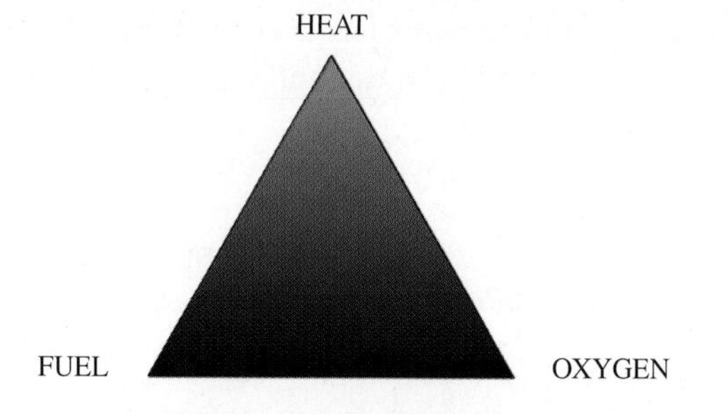

Figure 10-1-2 The three things that must be present for combustion to occur are fuel, heat, and oxygen.

TABLE 10-1-1 Ultimate CO_2 Values

Constituent	Chemical Formula	Ultimate CO_2 (%)
Methane	CH_4	11.73
Ethane	C_2H_6	13.18
Propane	C_3H_8	13.75
Butane	C_4H_{10}	14.05

Substances such as carbon monoxide and aldehydes are formed. This is caused by: (1) insufficient fuel and air mixing, (2) insufficient air supply to the flame, (3) flame impingement on a cold surface, and (4) insufficient flame temperature. Incomplete combustion is not only inefficient but hazardous because producing carbon monoxide is harmful to health.

The two most popular gaseous fuels are natural gas and liquid petroleum (LP). Natural gas consists primarily of methane and ethane. LP is usually propane or a mixture of propane and butane. The actual measured CO_2 needs to be compared to the ultimate (highest possible) CO_2 in order to be meaningful. Table 10-1-1 gives the ultimate CO_2 values that can be obtained for the chief constituents of these gaseous fuels.

Other interesting factors relating to the combustion of gaseous fuels are the flammability limits and the ignition temperatures. Fuels will burn in a continuous combustion process only when the volume percentages of fuel and air in a mixture are within their flammability limits and the mixture reaches ignition temperature. Table 10-1-2 gives upper and lower flammability limits and ignition temperatures for the chief constituents of gaseous fuels.

The heating values of fuels can be of two types: (1) higher heat values, which include the heat produced in condensing the water vapor in the products of combustion; and (2) lower heat values, which do not include the latent heat produced by condensing the water vapor. Table 10-1-3 lists the higher and lower heating values for the common constituents of gaseous fuels.

TABLE 10-1-2 Flammable Limits of Fuels

Constituent	Lower Flammability Limit (%)	Upper Flammability Limit (%)	Ignition Temperature (°F)
Methane	5.0	15.0	1301
Ethane	3.0	12.5	968 to 1166
Propane	2.1	10.1	871
Butane	1.86	8.41	761

TABLE 10-1-3 Heating Values of Fuels

Substance	Higher Heating Values (Btu/lb)	Lower Heating Values (Btu/lb)	Specific Volume (ft^3/lb)
Methane	23,875	21,495	23.6
Ethane	22,323	20,418	12.5
Propane	21,669	19,937	8.36
Butane	21,321	19,678	6.32

Heating requires the expenditure of some source of energy to raise the temperature of the air or water, depending on the type of equipment used. One can classify energy sources for heating as: (1) gases, both natural gas and from liquid petroleum, (2) oil, (3) electricity, and (4) other, including coal, wood, and solar. Natural gas and LP gas, oil, and electricity are the principal sources used for domestic heating.

GASES

Fuel gases are employed for various heating and air conditioning (cooling) processes and fall into three broad categories: *natural, manufactured,* and *liquefied petroleum.* Stories about the discovery and use of natural gas date back as far as 2000 B.C. History notes that the Chinese piped gas from shallow wells through bamboo poles and boiled seawater to obtain salt.

The great advantage of *natural gas* over other fuels is its relative simplicity of production, transportation, and use. When a sufficiently large quantity is discovered, it is pumped from drill wells to processing plants, and on to refineries and industrial centers where it has a large number of diverse uses in addition to heating.

Natural gas comes from sedimentation of trillions of tiny organisms at the bottom of the sea, buried and initially chemically converted into dense organic material. Over millions of years, pressure and heat gradually cracked this material into lighter hydrocarbon compounds. Liquids were liberated first, then the gases. Compaction squeezed them from the source rock and each migrated into the more porous reservoir rocks until stopped by impermeable barriers. Pools of gas and oil developed which are the targets of petroleum exploration.

Natural gas, Figure 10-1-3, is mostly methane, which consists of one carbon atom linked to four hydrogen atoms; and ethane, which consists of two carbon and six hydrogen atoms. Liquid petroleum (LP) gases are propane and butane or a mixture of the two, and these fuel gases are obtained from natural gas or as a by-product of refining oil. From Figure 10-1-3,

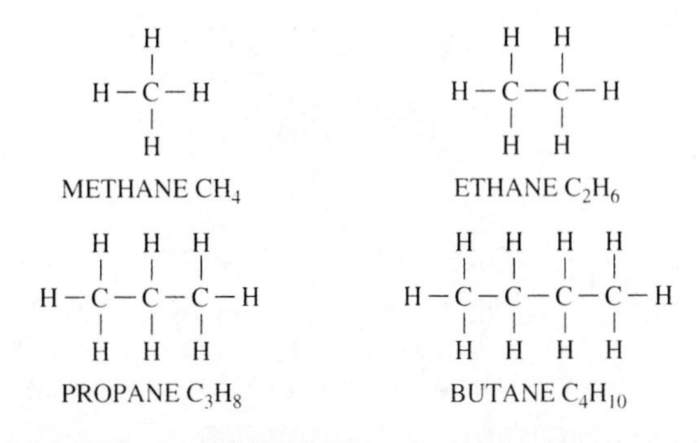

Figure 10-1-3 Chemical components of common fuels. Natural gas is mostly methane and ethane, and LP is mostly propane and/or butane.

you will note they contain more carbon and hydrogen atoms than natural gases, are thus heavier, and, as a result, have more heating value per cubic foot. Gas is a precious commodity and may someday not be available as a substitute for other fuels.

Manufactured gas, as the name states, is manmade as a by-product from other manufacturing operations. For example, in iron making, large amounts of gases are produced that can be used as fuel gases. The use of manufactured fuel gases has declined greatly in the United States. Today over 99% of sales by gas distribution and transmission companies is natural gas. Manmade gas is still a popular fuel in Europe.

Mixed gas, as the name implies, is a manmade mixture of gases such as natural gas and manufactured gas.

It is important to know the density of a gas as expressed by its specific gravity. Compared to the standard, air, which has a specific gravity of 1.0, natural gas ranges from 0.4 to 0.8 and thus is lighter than air, as shown in Figure 10-1-4. On the other hand, of

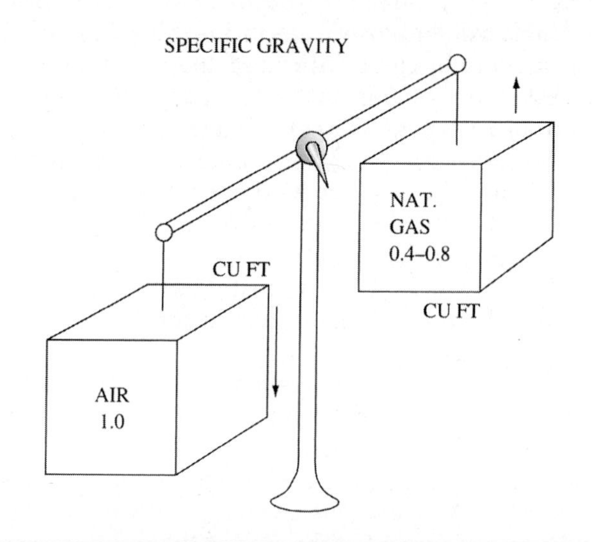

Figure 10-1-4 Weight of one cubic foot of air compared to one cubic foot of natural gas.

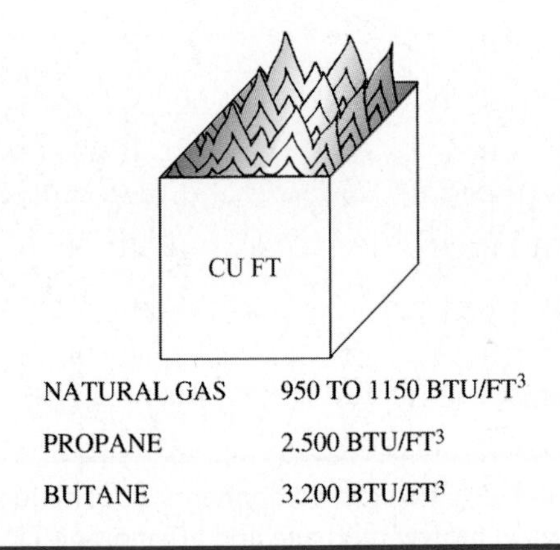

Figure 10-1-5 Heat delivered by gas fuels when burned.

the *liquid petroleum* gases, the specific gravity of propane is 1.5 and that of butane is 2.0, meaning they are heavier than air. The density of the gases is important because it affects the flow of the gas through orifices (small holes) and then to the burner. Should a leak develop in a gas pipe, natural gas will rise and probably dissipate while LP gases will drift to low spots and collect in pools, creating a greater hazard. Specific gravity also affects gas flow in supply pipes and the pressure needed to move the gas.

CAUTION: Because LP gases are heavier than air they can collect in low areas such as basements or in crawl spaces under houses.

The heat value of a gas is the amount of heat released when 1 ft^3 of the gas is completely burned, as shown in Figure 10-1-5. Natural gas has a heating value range of 950 to 1150 Btu/ft^3. The range of Btu's per cubic foot for natural gas is due to the fact that it is a mixture of gases with methane making up the largest share. 1000 Btu/ft^3 is the heat value most often used for estimating performance for natural gas furnaces. Propane has a heat value of approximately 2500 Btu/ft^3. The exact heating value of gases in your local area can be obtained from the gas company or LP gas distributor.

Propane is the LP gas mostly used for domestic heating. *Butane* has more agricultural and industrial applications.

A knowledge of combustion is closely related to understanding the importance of ventilation by the technician. Combustion, illustrated in Figure 10-1-6, takes place when fuel gases are burned in the presence of air. Methane gas combines with the oxygen and nitrogen present in the air and the resulting combustion reaction produces heat, with by-products of carbon dioxide, water vapor, and nitrogen. For every

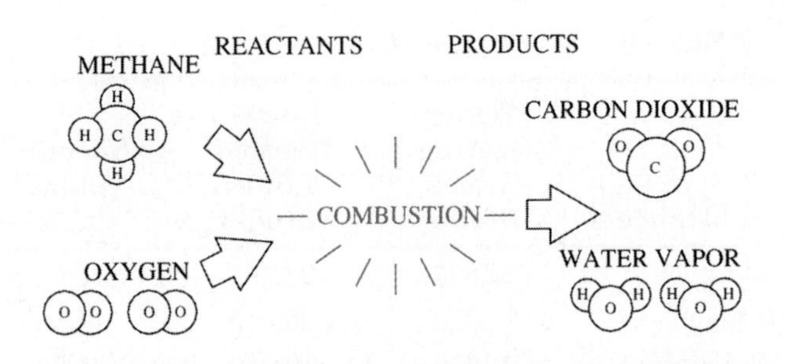

Figure 10-1-6 Combustion process when a fuel is burned.

1 ft^3 of methane gas, 10 ft^3 of air is needed for complete combustion, as shown in Figure 10-1-7. Although natural gas requires a 10:1 ratio of air, LP fuels require much more, due to the concentration (greater density) of carbon and hydrogen atoms. Liquid petroleum combustion must have more than 24 ft^3 of air per 1 ft^3 of gas to support proper combustion. When complete information on the fuel is not available, a frequently used value for estimating air requirements is that 0.9 ft^3 is required for 100 Btu of fuel. Excess air is also required to ensure complete combustion. Excess air is the free air around the flame that helps complete the combustion of the flame.

The by-products of combustion are called **flue gases** and are vented to the outside. In interior space, insufficient combustion air can produce hazardous conditions. If too little oxygen is supplied, part of the by-products will be dangerous carbon monoxide gas (CO) rather than harmless carbon dioxide gas (CO_2). The lack of combustion air can stop the flue gases from rising up the flue, which will cause the spillage of combustion products into the room. These flue gases in a living area cause serious health problems. Specific recommendations on venting and combustion air will be covered later, but remember, adequate combustion air is an essential element of the system's design, and central to indoor air quality.

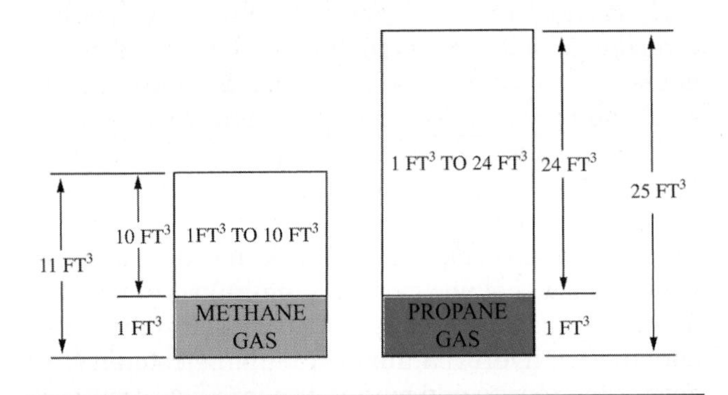

Figure 10-1-7 Amount of air required for combustion of methane compared to propane.

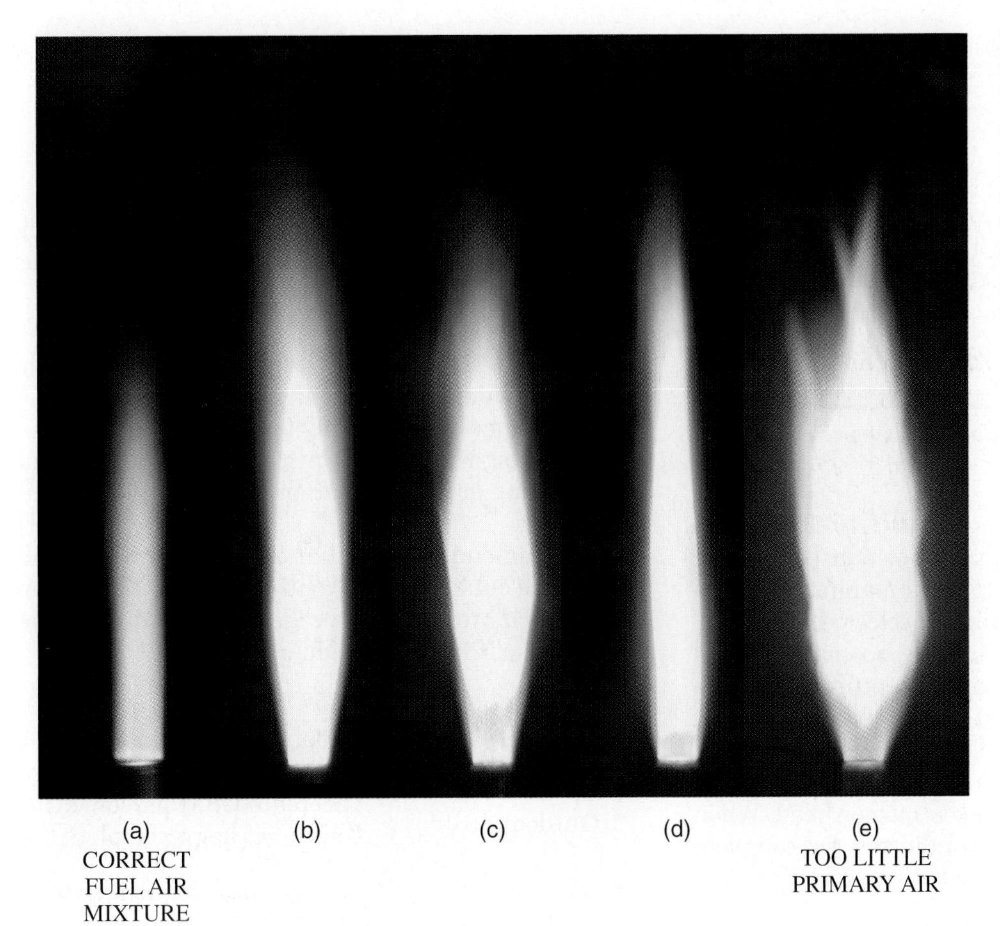

Figure 10-1-8 Natural gas flames: (a) Correct mixture of fuel and air; (b) Less primary air, cone reduced; (c) Less primary air, flame wider; (d) Less primary air, flame more irregular; (e) Too little primary air, irregular, wide flame.

SERVICE TIP

Combustion furnaces require a significant amount of combustion air. During new construction the quantity of air required is carefully checked and inspected; however, over time in older residences insulation or other debris can inadvertently fall over the makeup air vents for a furnace. When this happens it can adversely affect the furnace performance as well as create a potential safety problem. This problem is particularly a concern when the furnace is located inside of a dwelling space in an equipment closet. If the closet door is properly sealed the lack of proper makeup air will be negated when the technician has the door to the furnace closet open for service. Therefore, technicians must visually inspect makeup air vents to be sure they are free and open.

TYPES OF FLAMES

Basically there are two types of flames: yellow and blue, as shown in Figure 10-1-8a–e. The difference is mainly due to the manner in which the fuel is mixed with the air. A *yellow flame* is produced when gas is burned by igniting the gas gushing from an open end of a gas pipe, such as is common for ornamental lighting. Most modern gas burners burn with a blue flame. A *blue flame* is produced by a Bunsen burner such as those used in a laboratory, where 50% of the air requirement is mixed with the gas prior to ignition. This part of the air supply is called primary air. The balance of the air (secondary air) is supplied during combustion to the exterior of the flame for complete combustion.

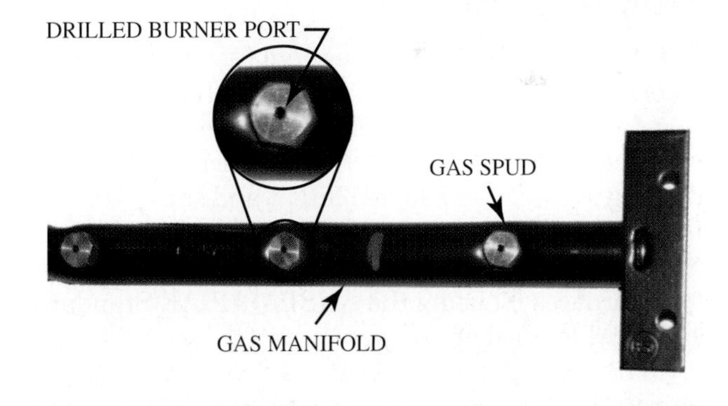

Figure 10-1-9 Burner bar.

The manifold and gas spud, shown in Figure 10-1-9, are used on nearly 80% of gas furnaces. A good example of the use of the use of primary and

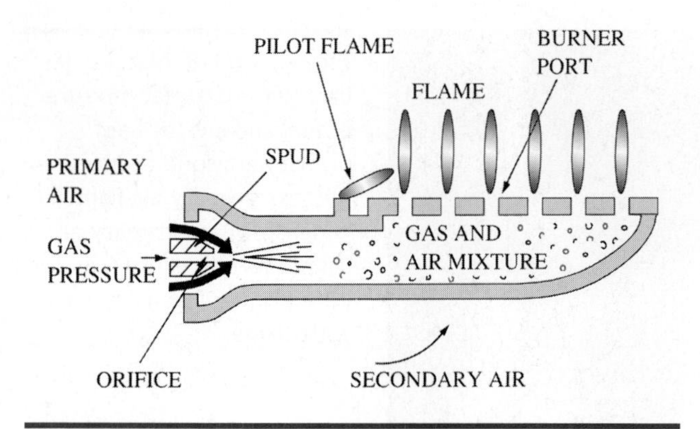

Figure 10-1-10 Cross section of a gas burner.

secondary air is shown in Figure 10-1-10. This figure shows the primary air adjustment on a drilled burner port. The gas is supplied through a manifold and metered into the burner by a properly selected gas orifice. The air enters through adjustable openings around the gas orifice. The purpose of the venturi is to create a vacuum as the pressurized gas is forced through it, sucking in the primary air. The air is mixed with the gas as they pass through the venturi tube. The secondary air is supplied above the burner head. When the air supplies are properly adjusted for complete combustion, the burner operates with a blue flame.

Annual Fuel Utilization Efficiency (AFUE)

Furnaces are rated for their annual fuel utilization efficiency. This rating is obtained by applying an equation developed by the National Institute of Standards and Technology (NIST) for 100% efficiency and deducting losses for exhausted latent and sensible heat, the effects of cycling, infiltration, and the pilot burner effect. The AFUE is determined for residential fan type furnaces by using the ANSI/ASHRAE Standard 103-1993 method of testing.

The federal law, effective January 1, 1992, requires that all new gas furnaces have a minimum AFUE efficiency of 80%.

SERVICE TIP

Gas furnaces are designed to last for many years. Many that were built before the change in efficiency laws have AFUE ratings of 60%. Over time these furnaces have lost some of that original efficiency and can be operating at a level well below 60% efficiency. Because of the higher cost of gas it is often in your customer's best interest to trade up to the newer higher efficiency model. Doing so will both save them money and reduce air pollution.

PRACTICAL COMBUSTION CONSIDERATIONS

Air Pollution

The combustion processes constitute the largest single source of air pollution in a home. Some of the ways that the heating technician can help to reduce air pollution are:

- Properly test installed furnaces and adjust the fuel burning device for highest efficiency.
- Encourage the use of high efficiency furnaces.
- Encourage the conservation of energy by recommending proper insulation and reduction of infiltration.
- Recommend the installation and operation of adequate outside air for ventilation and combustion.

  Calculate the percentage of air required to determine if it meets guidelines:

$$\text{Combustion air} = \frac{.5\ \text{CFM}}{1000\ \text{Btu}} \times \frac{\text{Total}}{\text{Btuh input}}$$

or

$$\%\ \text{Outdoor air} = \frac{\text{No. of people} \times \text{CFM/person} \times 100}{\text{Total flow}}$$

- Maintain clean plenum, air intakes, filters, ducts, and system components. Seal leaks.

CONDENSATION AND CORROSION

The heating technician must always be on the lookout for problems caused by condensation and corrosion produced by the flue gases. When the fuel burning system cycles on and off to meet the demand, the flue passages cool down during the off cycle. When the system starts up again, condensate forms briefly on the surfaces until they are heated above the dew point temperature. Low temperature corrosion occurs in system components (heat exchangers, flues, vents, chimneys). The condensate includes such corrosive substances such as sulfides, chlorides and fluorides.

Corrosion increases as the condensate dwell time increases. One of the common signs of this is the corrosion that takes place in the flue pipe between the furnace and the chimney. Even though a galvanized flue pipe is used, this piping needs to be replaced periodically due to the corrosion effect.

In the high efficiency furnace where condensation is allowed to occur inside the furnace on a continuous basis, the flue passages must be constructed of corrosion resistant materials such as stainless steel and PVC. The material used for condensate drains from

Figure 10-1-11 The CSA Blue Star Mark is for gas fueled products and indicates that a product complies with the requirements of applicable U.S. standards related to gas and electrical safety.

high efficiency furnaces is usually corrosion resistant PVC. Some cities require neutralizing the condensate before it enters the public sewers.

Soot

Soot is a form of carbon. It can be deposited on flue surfaces and acts as an insulating layer, reducing the heat transfer and lowering the efficiency. Soot can also clog flues, reduce draft and available air, and prevent proper combustion. Proper burner adjustment can minimize soot accumulation.

CERTIFICATION OF HEATING EQUIPMENT

Common to all types of heating apparatus sold domestically is the need to have these tested, certified, or listed by accredited testing agencies such as CSA International or Underwriters Laboratories, Inc. (UL).

CSA Group acquired the Certification and Testing business from the American Gas Association (AGA) in 1997. CSA America Inc., a division of CSA Group, is a standards development body in the United States for appliances and accessories fueled by natural gas, liquefied petroleum, and hydrogen gas. The CSA Technical Committees establish the minimum construction and performance standards for gas heating equipment.

CSA International, also a division of CSA Group, is a provider of product testing and certification services for electrical, mechanical, plumbing, gas-fueled, and a variety of other products. It is accredited as a Nationally Recognized Testing Laboratory (NRTL) by the US Department of Labor Occupational Health and Safety Administration (OSHA), and as such maintains testing facilities to certify oil heating

Figure 10-1-12 The CSA US Mark indicates that the product is in compliance with the requirements of the applicable U.S. standards.

equipment and electrical heating and air conditioning equipment to the applicable US requirements such as ANSI, UL, CSA America, NSF, and others.

Gas furnaces and other comfort heating gas appliances tested and found to be in compliance bear the CSA Blue Star Mark, Figure 10-1-11, and are listed in the CSA Certified Product Listings.

Electrical products certified by CSA International to the applicable US requirements are eligible to bear the CSA US Mark, as shown in Figure 10-1-12. This mark is accepted by regulatory authorities throughout the US and is the equivalent of the UL mark.

UL also gets involved in the approval and listing of heating and air conditioning equipment, even dealing with large centrifugal machinines of 100 tons and over. Local city codes and inspectors are guided by UL standards, and failure to comply with them may be costly to the manufacturer and installer. Most people are familiar with the UL mark, shown in Figure 10-1-13. UL maintains testing laboratories for certain types of products; however they often perform the necessary tests at the manufacturer's plant. CSA International also offers this service.

Figure 10-1-13 Underwriters' Laboratories (UL) seal. *(Courtesy Underwriters' Laboratories)*

TECH TALK

Tech 1. I worked on a really old gas furnace the other day. It had one of those cast iron heat exchangers. I told the customer that he would probably be able to keep that one going forever.

Tech 2. You are right, it probably will last a lot longer. However, those old cast iron furnaces were so inefficient.

Tech 1. I don't think that it looked that inefficient. It had a good blue flame.

Tech 2. That's true; you can have a clean burning flame, but the cast iron heat exchanger just really doesn't put all the heat in the house. A lot of it goes up the flue.

Tech 1. Does it make a that big of a difference?

Tech 2. Yes, it costs a lot more to operate and puts a lot more air pollution out in the community.

Tech 1. Then I probably should have recommended that he change that old thing out.

Tech 2. That would be a good idea. Good for the customer, good for the community.

Tech 1. Thanks, I will do that next time I am out there.

UNIT 1—REVIEW QUESTIONS

1. Define combustion.
2. List the three things required for combustion.
3. What are the two most popular gaseous fuels?
4. What are the two types of heating values of fuels?
5. Define mixed gas.
6. What is the difference between propane and butane?
7. What produces a blue flame?
8. How are furnaces rated?
9. What constitutes the largest single source of air pollution?
10. List some of the ways that the heating technician can help to reduce air pollution.
11. What is the function of the CSA?

UNIT 2

Warm Air Furnaces

OBJECTIVES: In this unit the reader will learn about warm air furnaces used in the HVAC industry including:

- Gas Forced Air Systems
- Gas High Efficiency Furnaces
- Electric Furnaces
- Air Handling Units and Duct Heaters

INTRODUCTION

Residential furnaces are available in a variety of configurations to facilitate installation locations and the type of heat source. Installation requirements have also been responsible for a variety of mounting arrangements and airflow designs. Styles Based on Installation Location.

Furnaces can be categorized into five styles: upflow, lowboy, downflow or counterflow, and horizontal.

The *upflow highboy* furnace, Figure 10-2-1, is most popular. Its narrow width and depth allow for location in first floor closets and/or utility rooms. It can still be used in most basement applications for heating only or with cooling coils where headroom space permits. Blowers are usually direct drive multispeed. Air intake can be either from the sides or from the bottom.

Lowboy furnaces, Figure 10-2-2, are built low in height to accommodate areas where headroom is minimal. These furnaces are approximately 4 ft high, providing for easy installation in a low ceiling height basement. Supply and return ducts are mounted on top for easy attachment. Blowers are commonly belt driven. Many of these furnaces are sold for retrofitting older homes.

The *counterflow* or *downflow* furnace, Figure 10-2-3, is similar in design and style to the highboy, except that the air intake and fan are at the top and the air discharge is at the bottom. These are widely used where duct systems are set in concrete or in a crawl space beneath the floor. A fireproof base is required when the furnace is installed on a combustible floor. An extra safety limit control is also used.

The fourth style, the *horizontal* furnace, Figure 10-2-4, is installed in low areas such as crawl spaces,

Figure 10-2-1 Upflow highboy furnace. (*Courtesy of Bard Manufacturing Company*)

attics, or partial basements. It requires no floor space. Intake air enters at one end and is discharged out the other end. Burners are usually field changeable for left-hand or right-hand application.

The fifth style, the multiposition furnace, can be configured by the installation technician on the job site to work as an upflow, horizontal, and/or downflow furnace. The introduction of the multiposition furnace has allowed manufacturers and contractors to reduce the number of furnaces they must keep in inventory to meet their local demand.

Typical installations of the horizontal, counterflow and upflow furnaces are shown in Figure 10-2-5.

TYPE OF HEAT SOURCE

Furnaces may also be described in relation to the type of heat source: *fuel burning* furnaces and *electric* furnaces.

Figure 10-2-2 Lowboy furnace. (*Courtesy of Bard Manufacturing Company*)

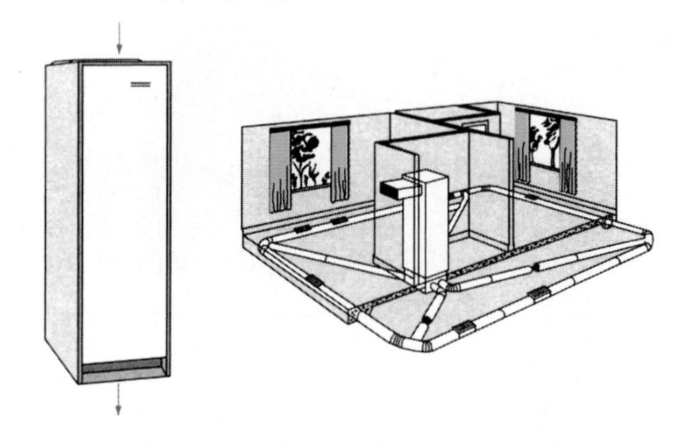

Figure 10-2-3 Counterflow furnace.

In fuel burning furnaces, fuel is burned in the combustion chamber. Circulating air passes over the outside surface of the heat exchanger. The products of combustion are vented to the atmosphere. Fuels used in these systems include:

1. gas, both natural and LP
2. oil

In an electric furnace, circulating air is directly heated by resistance type heating elements as the air passes through a metal sheath that encloses the resistance element.

GAS FORCED AIR SYSTEMS

Due to the need to conserve energy, two distinctly new types of furnaces have been developed:

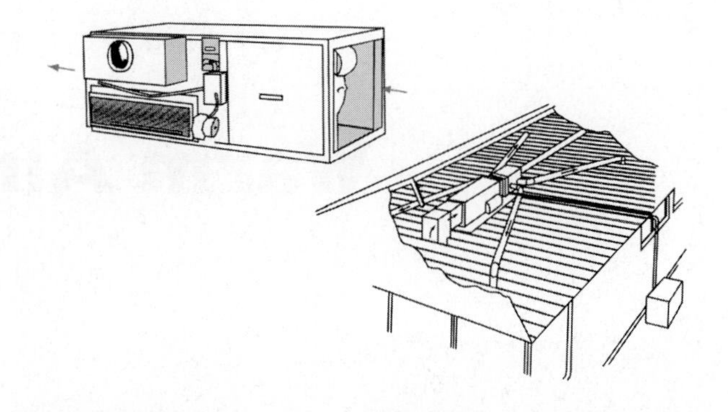

Figure 10-2-4 Horizontal furnace.

1. the mid-efficiency
2. the high efficiency condensing type furnace

The *mid-efficiency furnaces,* Figure 10-2-6, meet the minimum requirement of 80% AFUE (Annual Fuel Utilization Efficiency) and have a flue gas temperature above the dew point. The **high efficiency furnaces** have flue gas temperatures below the dew point and permit condensation within the furnace to pick up the extra latent heat. Since certain components are different, they will be described separately. Large numbers of the earlier **standard furnaces** are still in use and require service. They will also be described in this text, although they are no longer being manufactured.

Practicing Professional Services

During ignition startup of the unit, do not stand in front of the flame until it has ignited. Stand to the side of the unit and observe ignition. If there is flame rollout, this can prevent burning your face or skin. If there is a problem with ignition, shut down the unit at the disconnect and repair as necessary.

SAFETY TIP

Always inspect fittings for leaks by using soap bubbles or electronic leak detectors. If leaks are found, shut down the unit at the disconnect switch, turn off the fuel supply, and repair the leak. Safety should be emphasized when you work with warm air furnaces, which use combustible fuels.

CAUTION: When possible be sure that your head and body are below the height of the combustion chamber on gas furnaces during their ignition sequence. If flame rollout occurs for any reason the flames will generally rise safely above you.

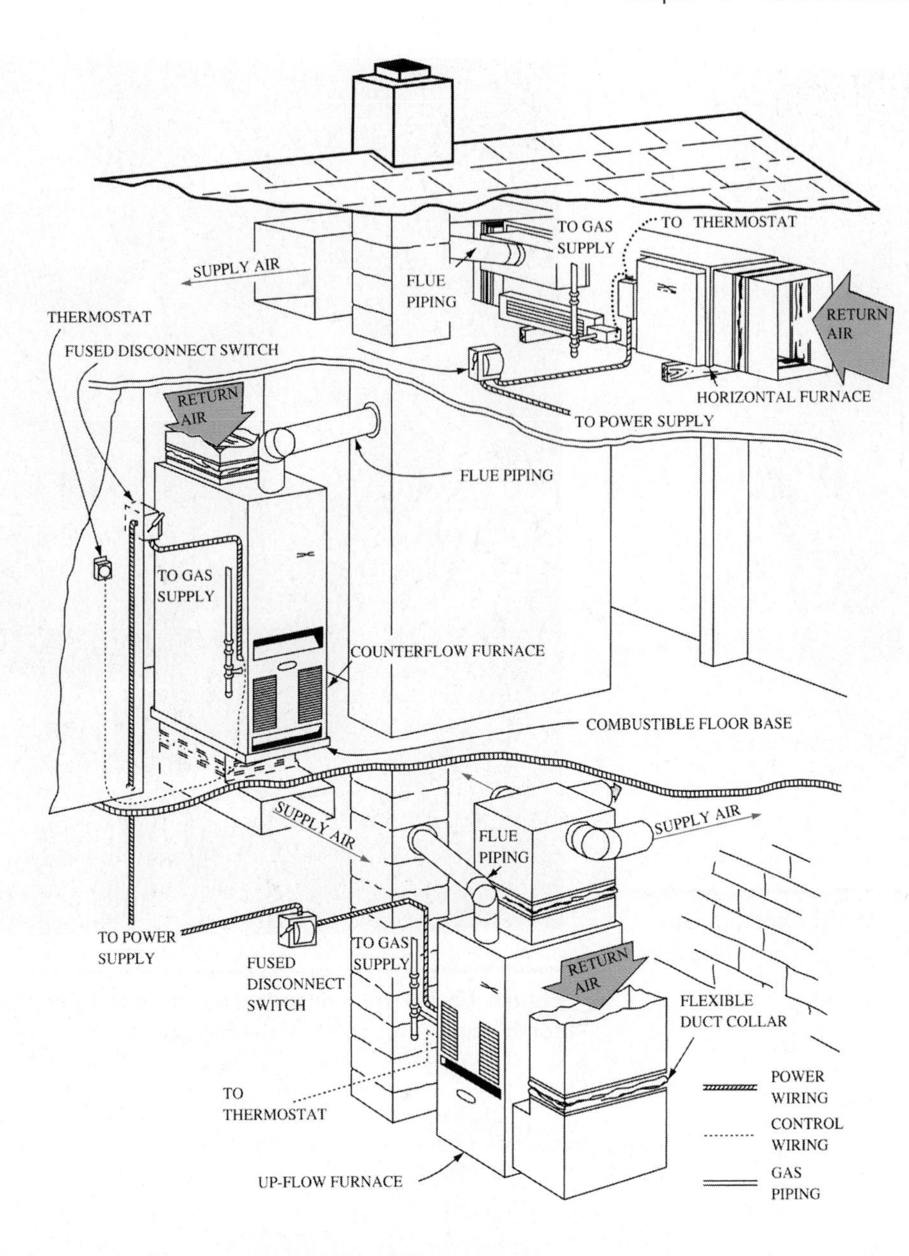

Figure 10-2-5 Composite diagram of three types of furnace installations: (1) Attic installation of a horizontal furnace; (2) First floor installation of a counterflow furnace; (3) Basement installation of a lowboy furnace.

SAFETY TIP

If during the ignition sequence a flame is not immediately visible after you hear the gas valve open, immediately shut down the furnace and the gas. If the ignition sequence is allowed to continue and a delayed ignition occurs, a large quantity of gas can be ignited in the combustion chamber. This will result in a large flame rolling out of the chamber. Severe delayed combustion can result in a large pop or minor explosion within the furnace combustion chamber. This condition needs to be avoided if at all possible and your prompt action can divert a potentially dangerous situation. Before reestablishing the ignition sequence, vent the combustion chamber of gas and determine the cause of the earlier failure before trying to relight the furnace.

Mid-efficiency Furnaces (80% AFUE)

A drawing of the internal components of a mid-efficiency furnace is shown in Figure 10-2-7. The internal components of a mid-efficiency furnace include:

- The *inducer blower assembly,* with a induced-draft blower for propelling the flue gases through the heat exchanger and into the vent, as pictured in Figure 10-2-8. The inducer motor forces the exact amount of air needed through the combustion exchanger. The assembly is resilient mounted to provide quiet operation.
- The *pressure switch,* which provides a positive indication that the inducer fan is operating before permitting the fuel to be ignited.

Figure 10-2-6 Internal view of the major components of a mid-efficiency gas furnace. (*Courtesy Lennox Industries, Inc.*)

Figure 10-2-8 An inducer-blower used to force the combustion gases through the furnace.

Figure 10-2-7 Internal view of the major components of a mid-efficiency gas furnace, with the component parts labeled. (*Courtesy Lennox Industries, Inc.*)

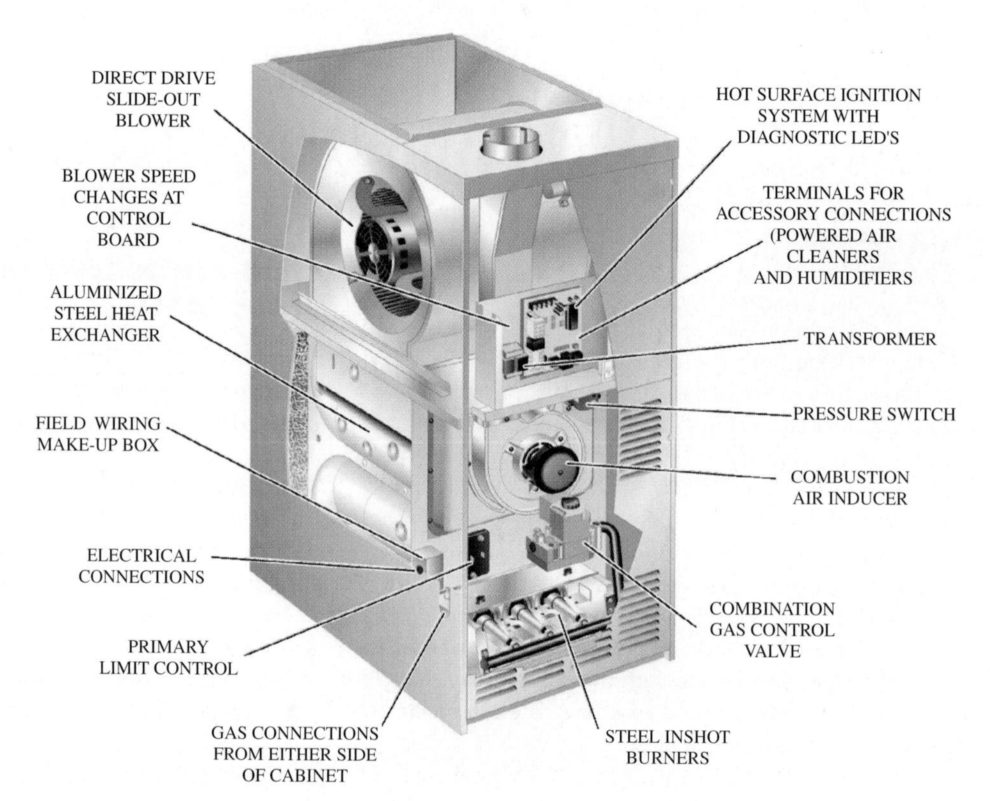

- The *gas control valve,* which meters the fuel gas into the burner. The gas valve opens slowly to provide a controlled ignition. It also provides for 100% shutoff to ensure safe operation.
- The *burner assembly,* which provides for proper mixing of fuel and air, and combustion of the fuel.
- The *blower door safety switch,* which disconnects the power supply to the unit whenever the front access panel is removed.
- The *control box,* which houses controls, including a microprocessor board that controls most operations and functions of the unit. It provides a blower delay on startup and shutdown, while monitoring furnace performance. The technician can use this self testing feature to identify a major component failure. The control board will check itself, then the inducer, silicon carbide ignition, low- and high-speed blower operation, and humidifier connections. Control boards often include a low amperage that protects the transformer and control board. The boards may also include an LED status indicator light.
- The *air filter retainer,* which holds the air filter.
- The *air filter,* which is either the replaceable type or an optional electronic filter that can be installed in the return air opening.
- The *wraparound casing,* which is a one piece seamless construction.
- The *heat exchanger,* which transfers the heat of combustion to the air distribution system.
- The *blower* and *blower motor,* which force the air over the external surface of the heat exchanger to pick up the heat for delivery to the space being conditioned. The motor is a direct drive multiple speed type.

SERVICE TIP

Some early Honeywell igniter controllers have an internal fuse. This fuse is not replaceable. If the leads are accidentally shorted it will blow and the entire control board must be replaced. When working with any electronic device be certain not to accidentally short any wires together for fear of damaging the component.

Product Data (80% AFUE Furnace)

Following the information supplied by the manufacturer about the product during installation is important. The technician needs to check the installation, either at the time of startup or whenever a service problem arises to determine that the product has been properly installed. Some of this information appears on the nameplate, but not all of it may be listed. This data can be very useful.

An external wiring diagram for a mid-efficiency furnace is shown in Figure 10-2-9. Although these units are primarily designed for heating, a number of accessories can be added, such as a cooling coil installed in the supply plenum, a humidifier installed on the supply duct, and an electronic air cleaner installed at the return air entrance to the unit.

CAUTION: It is against code to install an air conditioning coil in the return air plenum of a furnace. If a coil were to be installed in the return air plenum the heat exchanger would be cooled below the dew point during the air conditioning season. The resulting condensate would quickly rust out the furnace and heat exchanger.

The 80% AFUE units usually have output ranges between 35,000 and 124,000 Btu/hr. On larger jobs it may be necessary to install two of them linked together in a configuration called **twinning.** Twinning kits are available for field linking as shown in Figure 10-2-10. A number of parts are required, and the manufacturer's instructions must be carefully followed to coordinate the two systems.

SERVICE TIP

When working on a customer's furnace equipment, avoid tracking grease and/or dirt into their home after the work is complete. If you are in a basement and must exit the house across a clean floor or rug, remove your shoes or cover them with paper shoe covers. You must never leave any tracks behind. A little time spent ensuring your footwear is clean will help customer satisfaction.

GAS HIGH EFFICIENCY FURNACES (HEF)

High efficiency or condensing type gas furnaces are rated at 90% AFUE or better. They differ from the 80% AFUE mid-efficiency furnaces in that an extra heat exchanger (secondary heat exchanger) is added to extract more heat from the flue gases. This additional surface causes the moisture in the flue gas to condense. Additional heat is obtained by further lowering the temperature of the flue gas and picking up the heat rejected by the condensing moisture.

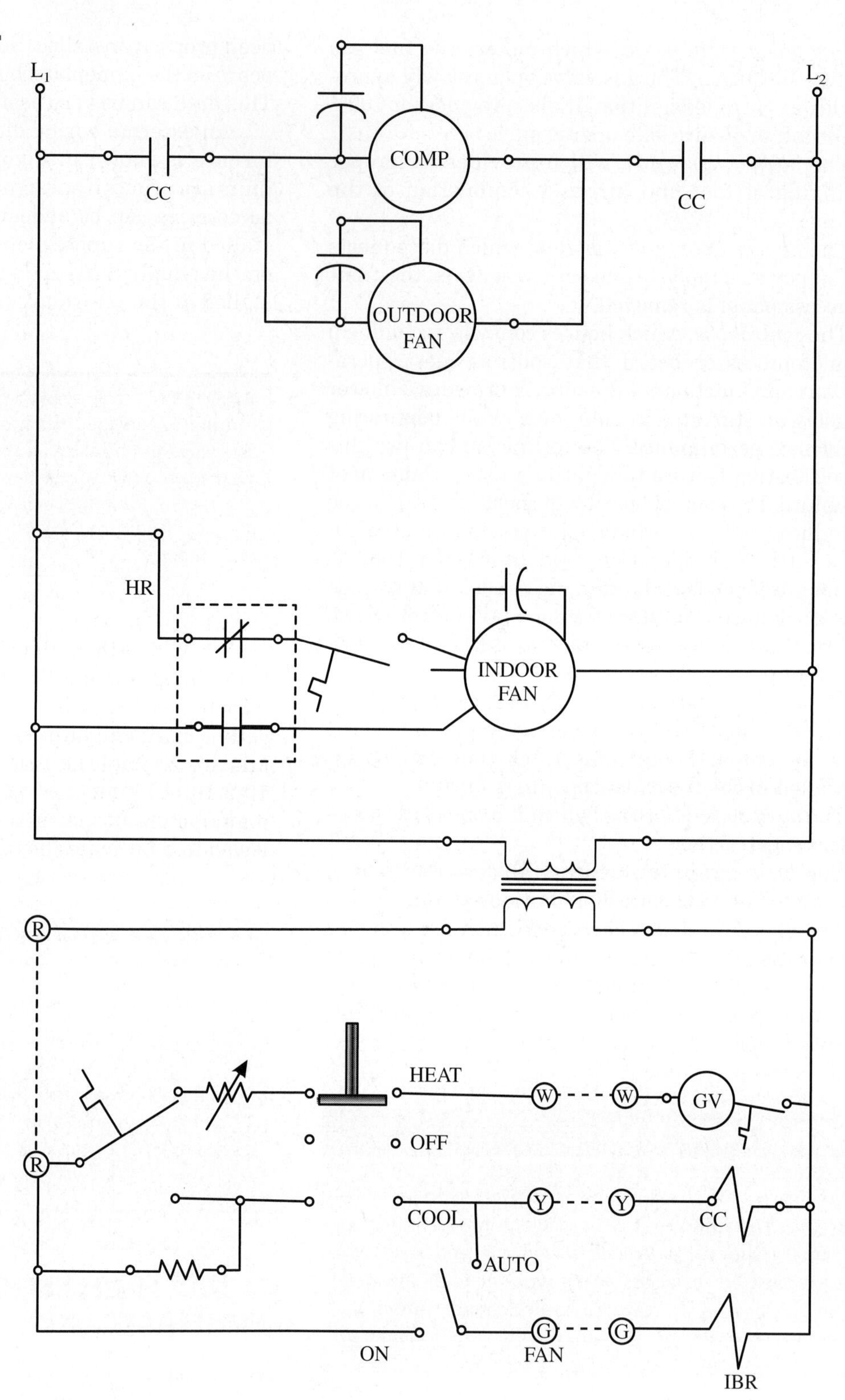

Figure 10-2-9 Typical schematic wiring diagram for a mid-efficiency furnace.

This construction reduces the volume and temperature of the flue gas, making it possible to use smaller vent pipes and simplify venting arrangements. Since the temperature of the flue gas is lower, PVC vent piping can be used and vent outlets can be run to the side of a building, if more convenient.

A condensate line is required to dispose of the condensed water. This needs to be run to the suitable drain. The condensate usually contains some contaminants. Local codes need to be examined to determine the possible need for a condensate neutralizer kit.

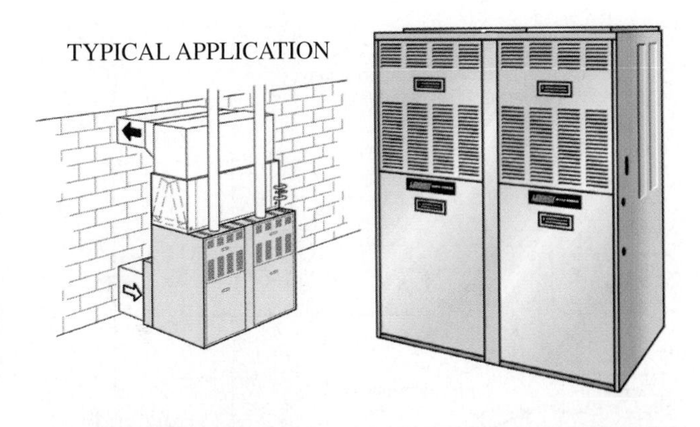

Figure 10-2-10 Typical twinning installation of upflow furnaces. (*Courtesy Lennox Industries, Inc.*)

Product Data (90% + AFUE Furnaces)

The product data for the 90%+ AFUE furnaces covers categories similar to that of the 80% AFUE models, with a few exceptions. Since these are condensing type furnaces, provision needs to be made for the combustion air and vent piping, as shown in Table 10-2-1.

The concentric vent, Figure 10-2-11a–c, is an interesting device. Since some manufacturers require 100% outside air for combustion on high efficiency furnaces, this device allows both vent type and combustion air pipes to terminate through a single exit in the roof or sidewall. One pipe runs inside the other, permitting venting through the inner pipe and combustion air to be drawn through the outer pipe.

There is also a limit on the length of an exposed insulated vent pipe (ft) in an unheated area, as shown in Table 10-2-2.

Typical installations of units and venting arrangements are shown in Figures 10-2-12 and 10-2-13.

SAFETY TIP

Because vent sizing is so important you must check with the equipment manufacturer and with local and state regulations before designing and installing furnace vent piping.

STANDARD GAS FURNACES

Lower efficiency gas furnaces refer to the older units now in the field, all of which were manufactured prior to 1992. Since so many are in use and will require service for some time, this section describes them and indicates differences from the newer designs.

A cross section of one of these units is shown in Figure 10-2-14. One distinctive characteristic of the older design is the convection flow of combustion gases from the burner, past the heat exchanger and into the vent. The newer units use an inducer fan. This convection action of the flue gases required a different type of venting.

Another distinguishing characteristic of the older units is that they use a standing, continuous flame gas pilot. (The newer units use electronic igniters.) Along with the pilot is a runner attached to the main burner. The purpose of the runner is to carry the gas flame from the pilot to all of the burners. The pilot flame impinges on the thermocouple about ½ in, which provides a safety arrangement for turning off the gas if the flame is ever extinguished.

The most common type of heat exchanger is made from two formed or stamped steel sheets welded together, as shown in Figure 10-2-15. These sections are headered at the top and bottom and welded into a fixed position. The number of sections depends on the capacity of the furnace. The heat exchanger has a protective coating to prevent rusting. The contours of the heat exchanger sections create turbulence to increase heat transfer.

Planned Maintenance (PM)

Set up the burner for the most efficient operation using a combustion analyzer.

The burner fits into the opening of the heat exchanger. Figure 10-2-16 shows a cross section of a typical atmospheric burner. Gas enters the burner cavity through the orifice, drilled precisely to meter the correct amount of gas. The gas is ejected into the venturi of the burner, drawing in the primary air. The air and gas are mixed and flow through the ports of the burner and are ignited by the pilot. The secondary air is added outside the ports and is controlled by the

TABLE 10-2-1 A Typical Table Giving Maximum Allowable Vent Pipe Length (ft). These Values Are for Altitudes of 0 to 2000 ft Above Sea Level, Using Either Two Separate Pipes or a 2 in Concentric Vent

		Number of 90° Elbows					
Unit Size	Pipe Diameter (in)	1	2	3	4	5	6
040-08	1½	70	70	65	60	60	55
	2	70	70	70	70	70	70

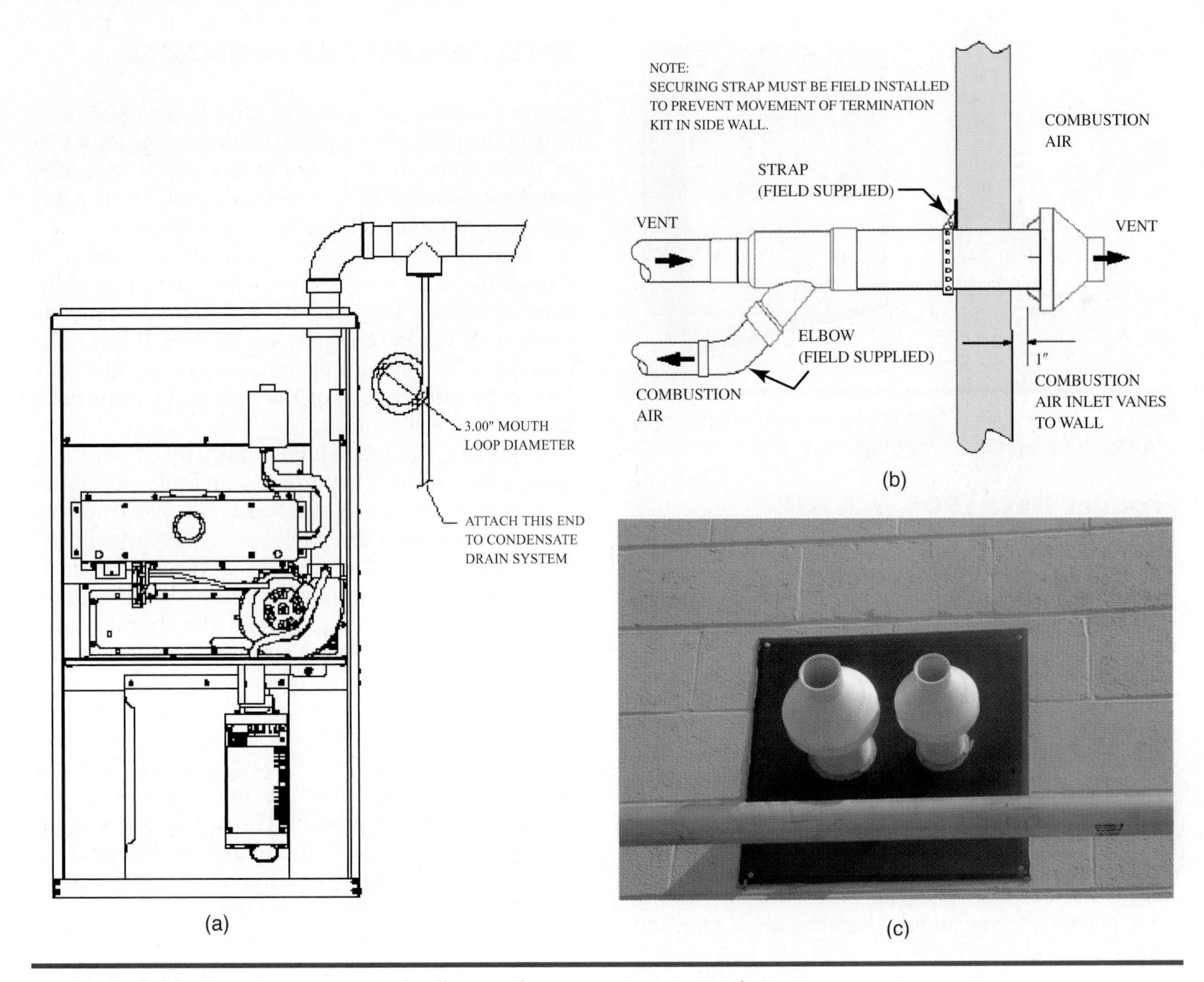

Figure 10-2-11 In order to have a high-efficiency furnace operate properly, it must have a correctly installed venting system. (a) It is possible to install the vent horizontally if the manufacturer's instructions are followed; (b) The vent pipe used is often made from an inner and outer tube. The outer tube is for combustion air for the furnace and the inner tube is for venting the flue gases; (c) The vent can be installed through the side wall of a building. Side venting reduces the possibility of roof leaks. *(Courtesy York International Corp., Unitary Products Group)*

TABLE 10-2-2 Limit on the Length of an Exposed Insulated Vent Pipe (ft) in an Unheated Area

Unit Size	Winter Temperature (°F)	Pipe Diameter (in)	Insulation Thickness (in) 0	3/8	1/2	3/4	1
040-08	20	1 1/2	31	56	63	70	70
	0	1 1/2	16	34	39	47	54
	−20	1 1/2	9	23	27	34	39

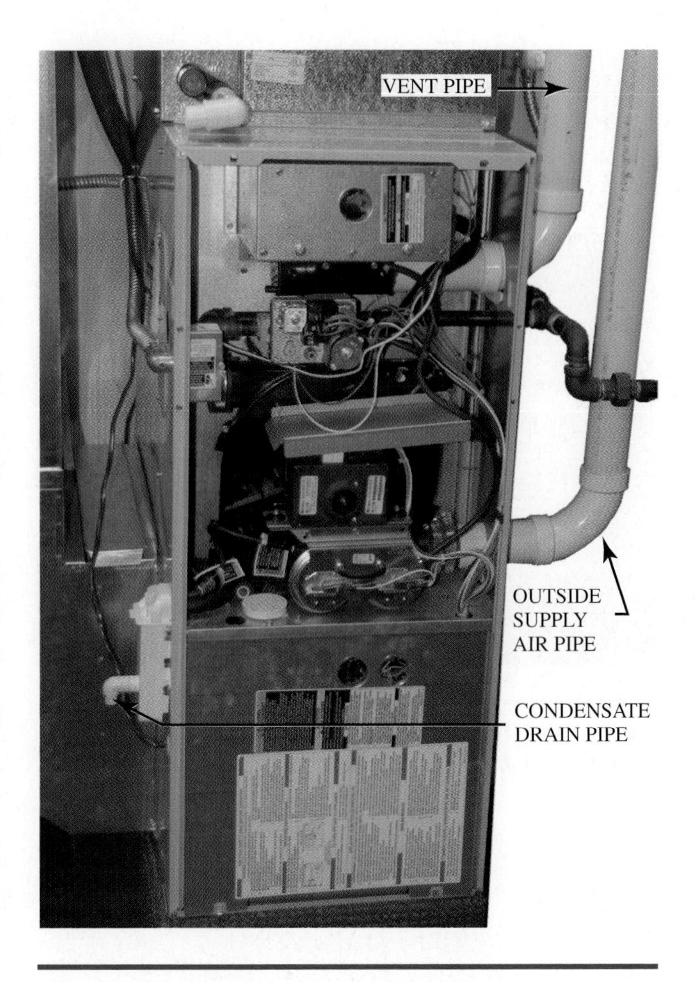

Figure 10-2-12 Side venting condensing gas furnace.

Figure 10-2-13 Top venting condensing gas furnace.

restrictor baffles designed into the heat exchanger. The primary air can be adjusted by positioning the air shutters as shown in Figure 10-2-17 to produce a blue flame for most efficient burning.

The main gas valve for the gas supply or flow performs many functions:

1. Manual on and off control of the gas supply.
2. Pilot supply, adjustment, and safety shutoff.
3. Pressure regulation of the gas feed.
4. On/off electric solenoid valve controlled by the thermostat.

The blower, Figure 10-2-18, delivers the heat from the furnace through the distribution system to the conditioned space. Most furnaces use a double inlet centrifugal blower. The blower can be either direct driven or belt driven. Direct drive blowers usually have multiple speed motors. The air quantity delivered by a belt driven blower can be regulated by adjusting the movable flange on the motor pulley, Figure 10-2-19, or by changing the diameter of the pulleys. The motor size is directly dependent on the required

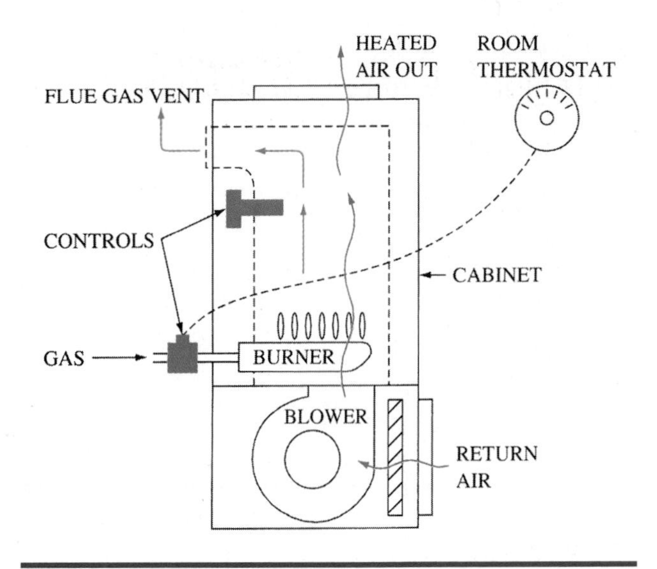

Figure 10-2-14 Cross section of a gas upflow warm air furnace.

Figure 10-2-15 Gas furnace heat exchanger. (*Courtesy York International Corp., Unitary Products Group*)

Figure 10-2-18 Blower assembly for a gas furnace. (*Courtesy York International Corp., Unitary Products Group*)

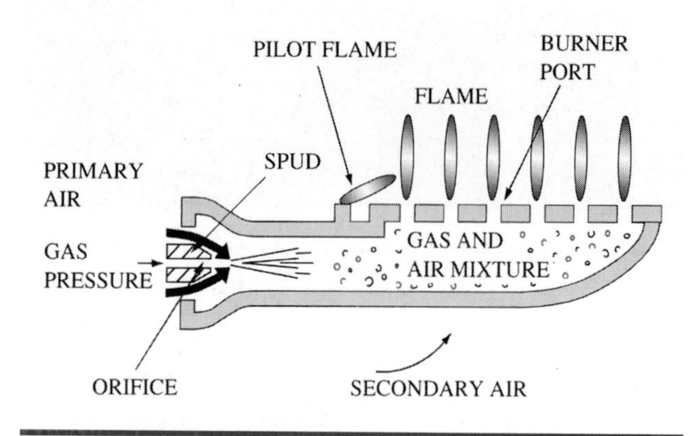

Figure 10-2-16 Cross section of a gas burner.

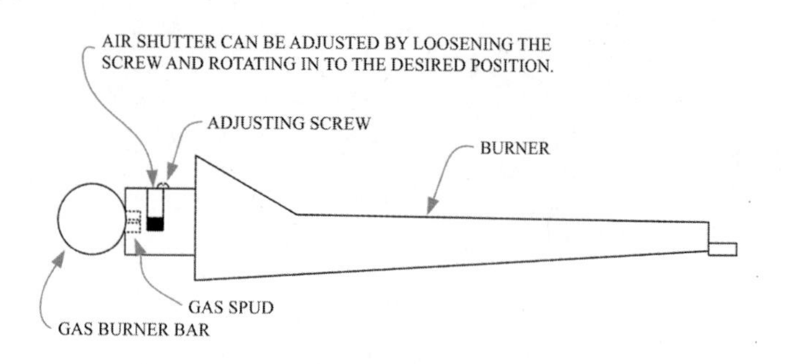

Figure 10-2-17 Adjusting the primary air on a gas furnace.

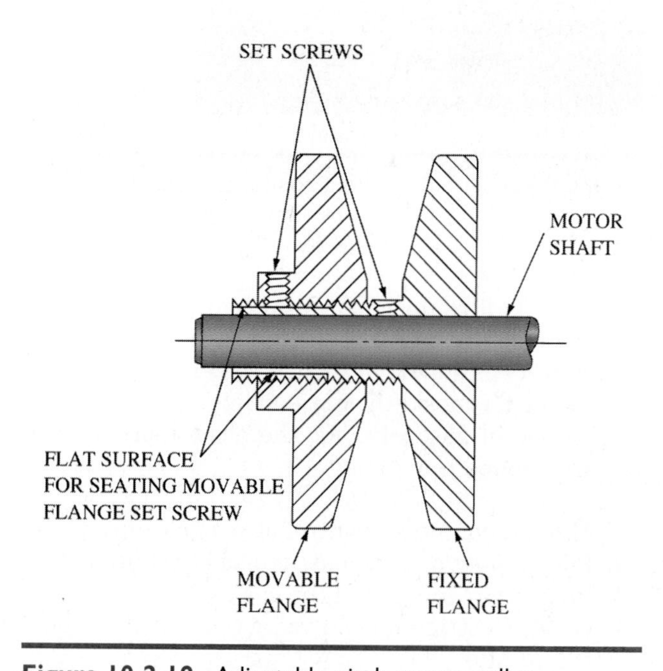

Figure 10-2-19 Adjustable pitch motor pulley.

air quantity and the resistance offered by the distribution system.

The cabinet provides support for all the internal components, as well as access to the controls, filters, blower drive components (if belts and pulleys are used), and includes the necessary openings for the combustion air supply. The cabinet is insulated where exposed to the warm parts of the furnace. It encloses and provides relief openings for the draft diverter. The draft diverter is required by the safety code to divert backdraft flue gases from extinguishing the burner flame.

UNIT 2—REVIEW QUESTIONS

1. List the four styles of furnaces.
2. Where is the fuel burned in a fuel burning furnace?
3. What is the difference between a mid-efficiency and a high efficiency furnace?
4. Why are furnaces twinned?
5. What are the AFUE ratings of high efficiency or condensing type gas furnaces?
6. List two types of high efficiency furnaces.
7. What is a concentric vent?
8. What is the purpose of a runner in standard gas furnaces?
9. What is the most common type of heat exchanger made from?
10. List the function performed by the main gas valve for the gas supply.

TECH TALK

Tech 1. I picked up a couple of furnaces at the supply house this morning but I think I got one that is not right. I want to make sure before I take it out to the job.

Tech 2. What makes you think that they gave you the wrong one?

Tech 1. Well one goes in the attic horizontal. The one that goes in the downstairs equipment closet is an upflow. Both the furnaces they gave me are exactly the same.

Tech 2. They didn't make a mistake. What you have are two multiposition furnaces.

Tech 1. What do you mean, multiposition?

Tech 2. It means that you can configure it for horizontal left to right, or vertical up.

Tech 1. How do I make it change?

Tech 2. The instructions are in the box. Just follow them. They are real simple.

Tech 1. That's neat, it means I don't have to take one back. Thanks.

UNIT 3

Installation, Startup, Checkout, and Operation

OBJECTIVES: In this unit the reader will learn about installation, startup, checkout, and operation used in the HVAC industry including:

- Heating System
- Checking the Gas Input
- Air Temperature Rise
- Efficiency Testing

THE HEATING SYSTEM

Heating startup, checkout, and operation will be discussed in separate sections:

1. natural gas
2. LP gas

NATURAL GAS

To check, test, and adjust a gas burning unit for the highest operating efficiency, the unit must have the proper gas input, the proper adjustment of the burners, the correct amount of combustion air, proper venting, and the correct amount of air for heat distribution.

To arrive at the correct gas input, two important factors must be known about the gas:

1. heat content in Btu/ft^3
2. specific gravity

Since natural gas is a mixture of methane and ethane, various sources have different heat contents ranging between 950 and 1150 Btu/ft^3. The specific gravity also varies between 0.56 and 0.72. These characteristics relate to sizing the piping and determining the amount of gas to be supplied to each burner.

Heating units are constructed to use gas pressure regulators set at an output of 3½ in WC (water column, a pressure measurement). Burner designs will allow operation between 3 and 4 in WC manifold pressure. These pressures must not be exceeded.

CAUTION: The diaphragm in a gas regulator is easily damaged with excessive pressure. Before pressure testing a gas piping system, the gas line should be removed from the gas valve and capped. This should be done even though there is a gas line shutoff so that the pipe joints from the shutoff to the gas valve can also be leak checked.

Checking the Gas Input

Figure 10-3-1 shows the index or dial of a typical domestic gas meter. Included are two test dials, one for ½ ft^3 per revolution and the other for 2 ft^3 per revolution. To determine if the correct amount of gas is being fed to the heating unit, it is only necessary to find the feed rate through the meter. For accuracy, all other appliances must be turned off. If the pilot lights of the other appliances including water heaters are turned off, be sure to relight before leaving. Usually, the requirements of pilot lights are so small that they are ignored.

Gas flow through the meter is determined by the time it takes the test dials to turn one revolution. To determine this time, the following formula can be used:

$$\text{Time (sec/ft}^3) = \frac{\text{seconds per hour}}{\text{ft}^3\text{/hr of gas}}$$

To determine how long it would take for 142.8 ft^3/hr of gas to flow, the formula would be set up as follows:

$$\text{sec/ft}^3 = \frac{60 \times 60}{142.8} = \frac{3600}{142.8}$$

It will require 25.2 sec for 1 ft^3 of gas to go through the meter. Thus, the ½ ft^3 dial would require 12.6 seconds per revolution, and the 2 ft^3 dial, 50.4 seconds per revolution. A stopwatch is recommended or a digital watch timing function.

Rather than use the formula, Table 10-3-1 shows revolution timing for various test dial sizes for various inputs. Using this table, the 142.8 ft^3/hr of gas would cause the ½ ft^3 dial to turn between 12 and 13 sec, the 1 ft^3 dial between 25 and 26 sec.

TABLE 10-3-1 Gas Input to the Meter (ft^3/hr)

Seconds for One Revolution	Size of Test Meter Dial				Seconds for One Revolution	Size of Test Meter Dial			
	One-Half Ft^3	One Ft^3	Two Ft^3	Five Ft^3		One-Half Ft^3	One Ft^3	Two Ft^3	Five Ft^3
10	180	360	720	1,800	52	35	69	138	346
11	164	327	655	1,636	53	34	68	136	340
12	150	300	600	1,500	54	33	67	133	333
13	138	277	555	1,385	55	33	65	131	327
14	129	257	514	1,286	56	32	64	129	321
15	120	240	480	1,200	57	32	63	126	316
16	112	225	450	1,125	58	31	62	124	310
17	106	212	424	1,059	59	30	61	122	305
18	100	200	400	1,000	60	30	60	120	300
19	95	189	379	947	62	29	58	116	290
20	90	180	360	900	64	29	56	112	281
21	86	171	343	857	66	29	54	109	273
22	82	164	327	818	68	28	53	106	265
23	78	157	313	783	70	26	51	103	257
24	75	150	300	750	72	25	50	100	250
25	72	144	288	720	74	24	48	97	243
26	69	138	277	692	76	24	47	95	237
27	67	133	267	667	78	23	46	92	231
28	64	129	257	643	80	22	45	90	225
29	62	124	248	621	82	22	44	88	220
30	60	120	240	600	84	21	43	86	214
31	58	116	232	581	86	21	42	84	209
32	56	113	225	563	88	20	41	82	205
33	55	109	218	545	90	20	40	80	200
34	53	106	212	529	94	19	38	76	192
35	51	103	206	514	98	18	37	74	184
36	50	100	200	500	100	18	36	72	180
37	49	97	195	486	104	17	35	69	173
38	47	95	189	474	108	17	33	67	167
39	46	92	185	462	112	16	32	64	161
40	45	90	180	450	116	15	31	62	155
41	44	88	176	440	120	15	30	60	150
42	43	86	172	430	130	14	28	55	138
43	42	84	167	420	140	13	26	51	129
44	41	82	164	410	150	12	24	48	120
45	40	80	160	400	160	11	22	45	112
46	39	78	157	391	170	11	21	42	106
47	38	77	153	383	180	10	20	40	100
48	37	75	150	375	190	9.5	19	38	95
49	37	73	147	367	200	9	18	36	90
50	36	72	144	360	210	8.5	17	34	86
51	35	71	141	353	220	8	16	33	82

Note: To convert to Btu per hour multiply by the Btu heating value of gas used.

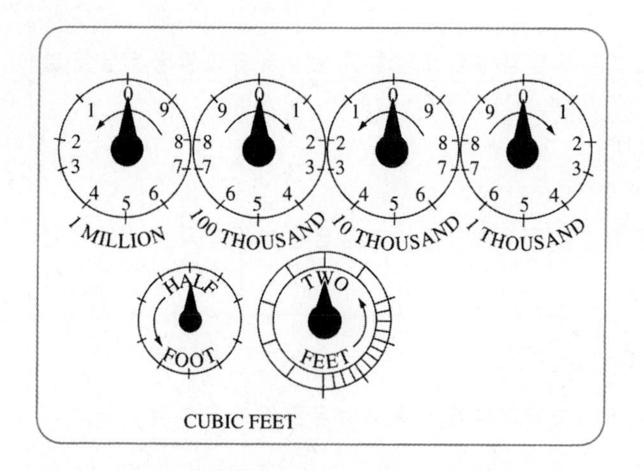

Figure 10-3-1 Typical domestic gas meter dials in ft^3.

If adjustments in the gas flow need to be made to produce the required input, it can be done by adjusting the gas pressure regulator, as long as the input pressure stays within the 3 to 4 in WC range. If the adjustment cannot be made within this range, the gas orifice size for the burners needs to be changed. The manufacturer's installation data supplies information for changing the orifice size.

LIQUID PETROLEUM (LP) GAS

Propane and butane are refined from crude oil. Although mixed, they are separable by condensing them at their respective boiling points. Propane produces 2522 Btu/ft^3 and butane yields 3261 Btu/ft^3.

The LP gas industry has established a set incoming line pressure for all LP gas burning appliances of 11 in WC. Therefore, it is only necessary to determine that the LP supply system is large enough to maintain 11 in WC at the units when the total connected load is operating. The LP supplier will provide the tank installation, pressure regulating devices, and piping to the furnace, but the service person should be familiar with the hookup procedure.

There are two basic systems, as illustrated in Figure 10-3-2. The single system on the right uses only one pressure regulator located at the tank. The two stage system shown on the left is used where the number of appliances and volume of gas must be greater. Therefore, the line pressure at the outlet of the tank regulator on the two stage system is 10 to 15 psig, whereas it is 11 in WC on the single stage system. The main supply line will carry more ft^3/hr at the higher pressure, which is smaller. Sometimes it is necessary to connect from a single-stage to a two-stage system if the supply lines are undersized.

With single stage systems, the pressure on the inlet of the regulator will be direct tank pressure and this will vary with fuel and temperature.

To ensure that the gas will flow from the tank into the supply system to the heating unit, the tank pressure must at all times be higher than the line pressure required to supply the system. To maintain a pressure of 11 in WC at the heating unit, allowing for a pressure loss in the line of ½ in WC, the minimum tank pressure would be 2 psig. This means that the minimum outside temperature must be considered when selecting the mixture of LP fuel.

Table 10-3-2 shows the tank pressure that will occur at temperatures of −30 to +110°F for mixtures of propane and butane. From this table it can be seen that butane is not usable below +40°F and even propane will not develop sufficient pressures below −30°F. In extremely cold climates, tank heaters are used to ensure adequate fuel supply.

TECH TIP

Some local and state governments require specific licenses to work on LP systems.

Combustion Air

Many burners have an adjustment for the primary air. Primary air is the air that is mixed with the fuel

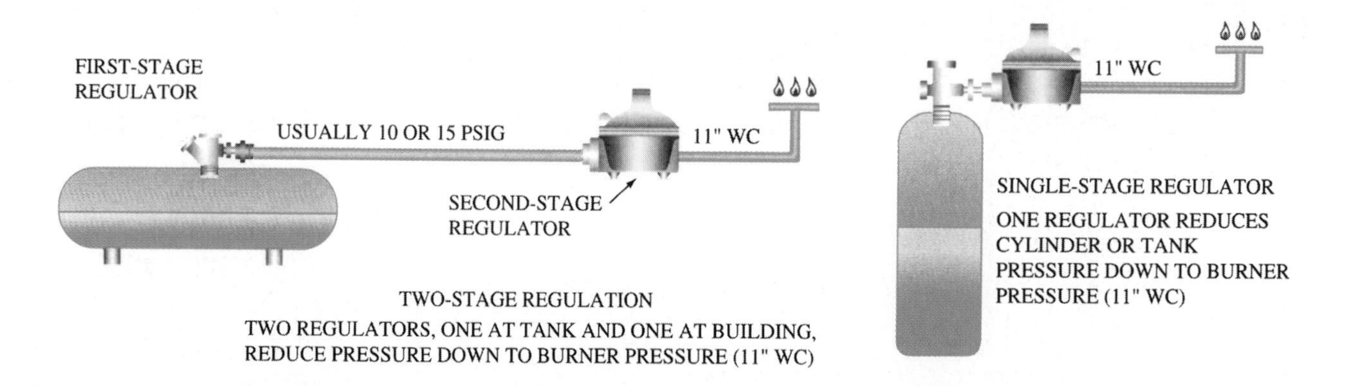

Figure 10-3-2 LP gas supply systems.

TABLE 10-3-2 LP Gas Tank Pressures (psig)

	Outside Temperature (°Fahrenheit)														
	−30	−20	−10	0	10	20	30	40	50	60	70	80	90	100	110
100% Propane	6.8	11.5	17.5	24.5	34	42	53	65	78	93	110	128	150	177	204
70% Propane 30% Butane	—	4.7	9	15	20.5	28	36.5	46	56	68	82	96	114	134	158
50% Propane 50% Butane	—	—	3.5	7.6	12.3	17.8	24.5	32.4	41	50	61	74	88	104	122
70% Butane 30% Propane	—	—	—	2.3	5.9	10.2	15.4	21.5	28.5	36.5	45	54	66	79	93
100% Butane	—	—	—	—	—	—	—	3.1	6.9	11.5	17	23	30	38	47

in the burner before combustion. The air adjustment should be set to produce a soft blue flame. Too much primary air will produce a hard blue flame that will waste fuel. Insufficient primary air will produce a flame with yellow tips. Insufficient secondary air causes incomplete combustion, which wastes fuel and forms carbon monoxide (CO) and soot. Figure 10-3-3 shows the proper technique for adjusting primary air on a gas burner.

Most high-efficiency furnaces use an in-shot gas burner, as shown in Figure 10-3-4, and an induced draft fan. On these burners normally the only adjustment that is made is on the gas pressure regulator, which should be set between 3.2 and 3.8 in WC for natural gas and 11 in WC for propane. These burners do not have a primary air adjustment. Regulation of the gas flow is by the gas pressure and the size of the orifice. The flame characteristic is clear blue.

After checking and setting the Btu/hr input and adjusting the primary air shutters, the vent draft condition should be checked.

Air Temperature Rise

The temperature rise of the air through the heating unit should be set as specified by the furnace manufacturer. Figure 10-3-5 shows the proper insertion of dial type thermometers into the supply and return plenums of the heating unit. Use a drill to make a hole large enough to take the 1/8 in diameter stem of the dial thermometer. Close the hole with a sheet metal screw after the test is completed.

The supply air thermometer should be located far enough away from the surface of the heat exchanger to prevent the effect of radiant heat on the thermometer from the heat exchanger. This is usually a minimum distance of 12 in. If the thermometer can be located in the main duct off the supply air plenum, the radiant effect will be practically eliminated.

TECH TIP

It is not always possible to get an ideal temperature location on furnaces that are located in equipment closets. Often the furnace plenum is not accessible. For these installations a temperature taken at the closest register to the furnace can be used.

Figure 10-3-3 Adjusting primary air on a gas burner. *(Courtesy of York International Corp.)*

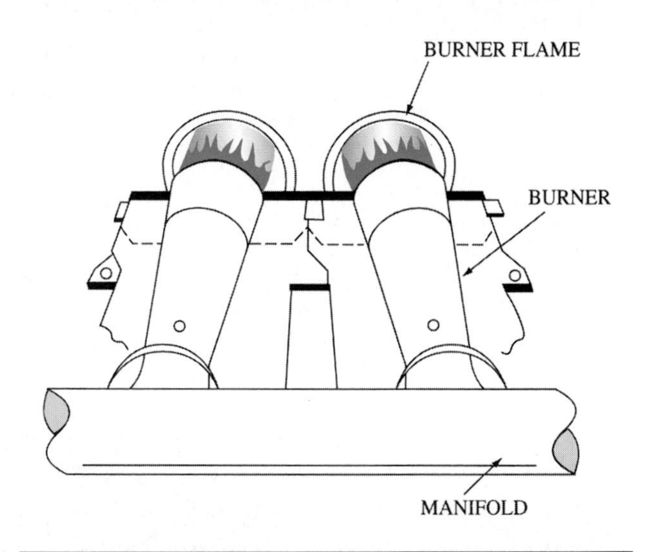

Figure 10-3-4 In-shot gas burner.

After the unit has operated long enough for the supply air thermometer to hold a steady reading (a stabilized reading), the return air temperature should be subtracted from the supply air temperature and the temperature rise recorded. If the temperature rise is below the specified temperature, the heating unit blower is moving too much air, and the speed should be reduced. If the temperature rise is above the specified temperature, the blower speed should be increased.

The procedure for this would depend on the type of drive used on the blower. Belt driven blowers are adjusted by changing the size of the motor or driving pulley, direct drive blowers by changing the electrical connections to the motor. Belt driven blowers use a combination of adjustable motor pulley (the driving pulley), blower pulley (the driven pulley), and drive belt. The speed of the blower is adjusted by changing the spread of the flanges of the driving pulley. Opening the pulley by spreading the flanges allows the belt to ride lower in the pulley, thus reducing the effective diameter of the pulley. This in turn reduces the pulley diameter rotation between the two flanges and reduces the blower speed.

Closing the pulley spread increases the drive pulley diameter, resulting in an increase in blower speed. The driven pulley usually has two setscrew flats to allow adjustment of the pulley in half turn increments. Use these flats—do not drive the setscrew into the pulley adjustment threads. This ruins the chance for future pulley adjustments.

After the blower speed has been set, the belt tension and alignment should be checked. See Figure 10-3-6. The belt tension should be tight enough to avoid slippage but not too tight to cause excessive wear. Approximately $^{3}/_{4}$ to 1 in of play should be allowed for each 12 in distance between the motor and blower shafts. Check the pulley alignment by using a straightedge.

On direct drive blowers, the choice of blower speeds is limited to the number of speeds built into the blower motor. On some heating only units, the blower has only one speed and no choice of temperature rise is available. If air conditioning is added to the unit, a change in blower motor or blower assembly is required to accommodate the additional pressure loss through the coil.

Figure 10-3-5 Measuring the air temperature rise though a furnace: (a) at the furnace; (b) at the return air supply.

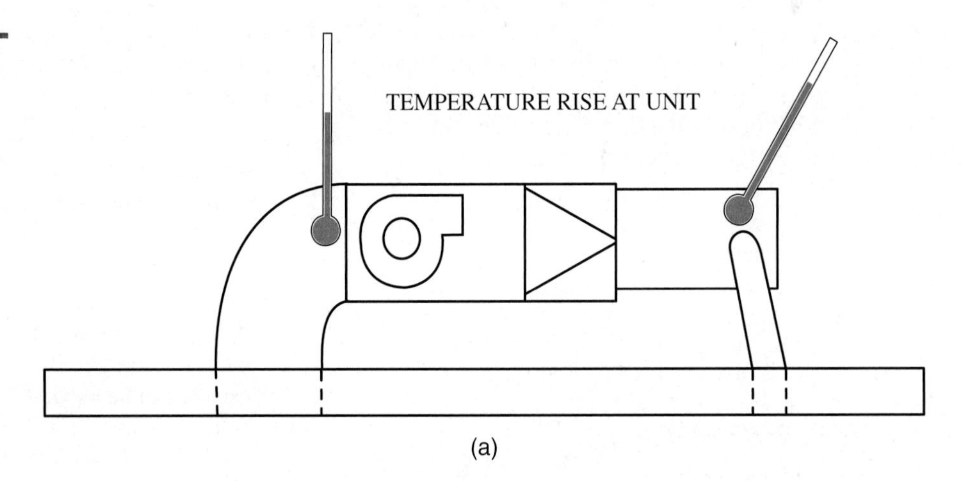

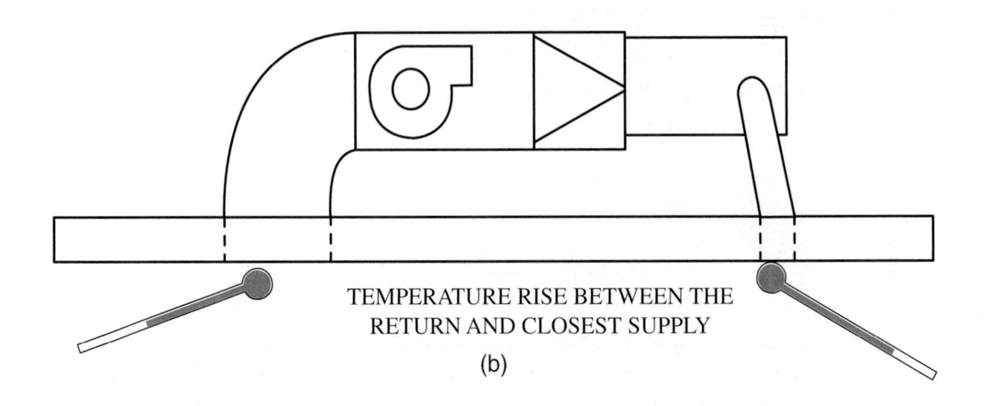

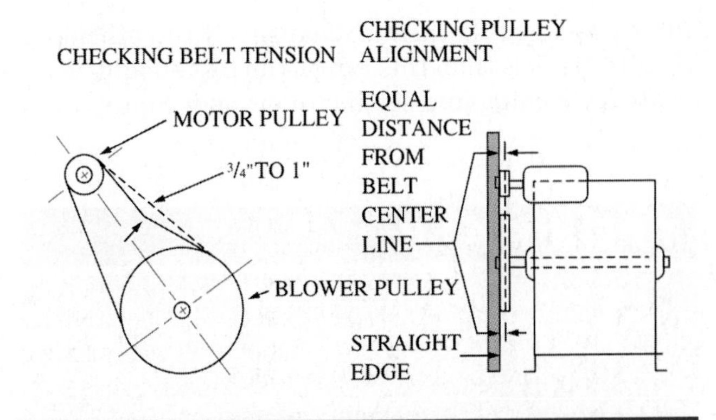

Figure 10-3-6 Aligning pulleys and tightening belts.

SERVICE TIP

After the pulley has been adjusted and tightened close the blower door and check the motor amperage. If the motor amperage is at or near the rated load amperage (RLA) for the motor it may be necessary to slow the blower down slightly to ensure that the motor does not overheat.

If the heating unit is equipped with a multispeed blower motor, observe which fan speed lead is connected: high, medium, or low. For the initial startup it is recommended that the fan be set for medium speed, subject to verification when the temperature rise is obtained.

In either case, increasing the blower speed will increase the CFM of air through the unit and lower the temperature rise. Conversely, lowering the blower speed will decrease the CFM through the unit and increase the temperature rise.

When increasing the blower speed, which increases the load in the motor, the amperage draw of the motor will also increase. A clamp type ammeter should be used to check the motor operating amperage. If the required CFM causes the motor to draw more than its amperage rating, a motor or blower assembly of larger capacity will have to be substituted.

Efficiency Testing

With the heating unit input set, the burners properly adjusted, and the unit operating at as close to desirable temperature rise as possible, the unit can be tested for operating efficiency. The objective of efficiency testing is to obtain as high an efficiency rating of the heating unit as possible, taking into account the operating cost, the equipment operating life, and the comfort obtained in the conditioned area.

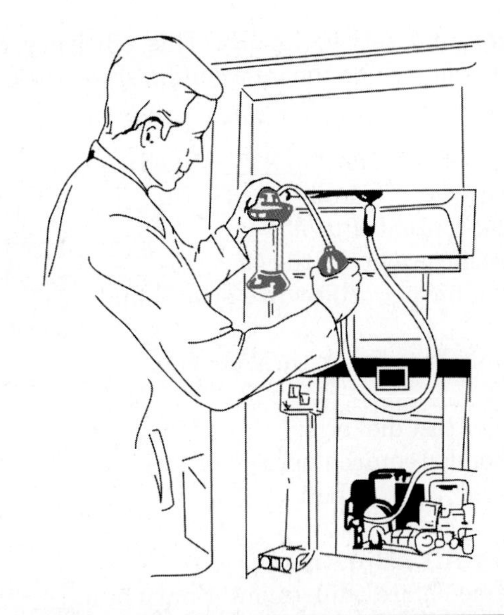

Figure 10-3-7 Testing the percent of CO_2 in the flue gas.

It is necessary to reach adjustments that achieve a balance between operating cost (efficiency) and comfort. Standards that are used for efficiency testing are:

1. *Input.* The unit must be supplied with the correct amount of fuel between 90 to 100% of its rated capacity.
2. *Burner primary air adjustment.* Soft blue fire without yellow color or flame lift.
3. *Air temperature rise.* Adjust the fan to obtain a temperature rise of 40° to 70°F for the air through the unit.
4. *Fan control settings.* According to manufacturer's specifications or with fan on, at 125 to 130°F; with fan off, at 115°F.
5. *CO_2.* The percentage CO_2 in the flue gas is an indicator of the overall combustion efficiency. It may be measured with CO_2 analyzer shown in Figure 10-3-7. Readings above 10% indicate incomplete combustion and possible production of carbon monoxide. Readings below 8% indicate too much combustion air.

TECH TIP

Checking a gas furnace fuel consumption as compared to heat output is a way of determining the actual operating efficiency of a particular system. This performance check is much like a driver confirming the fuel economy (miles per gallon) that their car is getting.

A means of recording such information is necessary so that it will become a permanent part of the unit operating and service history. This efficiency check sheet should include the information shown below.

Input

1. Type of gas: Nat. _______ Mixed _______
   Mfg. _______ Prop. _______ Bu. _______
2. Heat content (Btu/ft^3) _______
3. Specific gravity of the gas: _______
4. Main burner orifice drill size: Found _______
   Left _______
5. Manifold pressure (in WC): Found _______
   Left _______
6. Meter test dial size: _______ ft^3 per rev.
7. Seconds required per rev. of test dial:
   Found _______ Left _______

Primary Air Adjustment

1. Flame before adjustment: Sharp blue _______
   Soft blue _______ Yellow tips _______
2. Flame after adjustment: Soft blue _______

Air Temperature Rise

1. Supply air temperature:
   First test _______ Second test _______
   Left _______
2. Return air temperature:
   First test _______ Second test _______
   Left _______
3. Air temperature rise:
   First test _______ Second test _______
   Left _______

CO_2

1. First test _______%
2. Second test _______%
3. Left _______%

Stack Temperature Rise

1. Stack temperature
   First test _______ Second test _______
   Left _______
2. Combustion air temperature
   First test _______ Second test _______
   Left test _______
3. Stack temperature rise
   First test _______ Second test _______
   Left test _______

Combustion Efficiency

% Efficiency _______ _______ _______

Gas-burning equipment of standard design should always be capable of 75 to 80% efficiency. Unless the unit is of a higher efficiency design (when the manufacturer's instructions and settings must be followed), an efficiency above 80% could adversely affect the draft of the unit as well as cause condensation of moisture in the chimney or flue pipe and on the surfaces of the heat exchanger. If the efficiency results are less than this range, the test should be repeated, checking and setting the proper input.

TECH TALK

Tech 1. I was working on a furnace today and checked the temperature rise. It's got nearly 90°. That thing is working real efficient.

Tech 2. No, it's not working very efficient.

Tech 1. What do you mean? It has got the hottest air I have ever seen coming out of a furnace.

Tech 2. It's not how hot the air is that makes a furnace efficient. It's how much heat the furnace is putting in the air.

Tech 1. How do you tell the difference?

Tech 2. Well, it is fairly complex. You have to fill out all of the paperwork. But one of the things you are looking at is the temperature rise and the stack temperature.

Tech 1. What does the stack temperature have to do with it?

Tech 2. Well, the higher the stack temperature, the more heat that's going up and out of the house. You want the stack temperature to be within the proper range. If you had checked the stack temperature on that furnace you would have found that it was pretty high too.

Tech 1. Oh, so you mean I am losing a lot of heat. I didn't know that. Thanks

UNIT 3—REVIEW QUESTIONS

1. List the three leading domestic energy sources.
2. What two important factors must be known about the gas in order to arrive at the correct gas input?
3. Show the formula used to determine gas flow through the meter.
4. The two stage liquid petroleum gas system is mainly used in _______.
5. On direct drive blowers what is the choice of blower speeds limited to?
6. What is the recommended fan setting for the initial startup of a heating unit with a multispeed blower motor?
7. What is the objective of efficiency testing?
8. List the standards used for efficiency testing.
8. Gas burning equipment of standard design should always be capable of _______ efficiency.
10. What should be done if the efficiency results are less than the recommended range?

UNIT 4

Controls

OBJECTIVES: In the unit the reader will learn about controls used in the HVAC/R industry including:

- Control System Components
- Sensing Devices
- Operators
- Fuel Controls
- Gas Valves and Regulators

INTRODUCTION

Controls have been developed to answer the need for one or more factors, including operation, safety, personal convenience, and economy of the equipment.

Very simply, a control system checks or regulates, within prescribed limits, the functions of an HVAC/R system. Such a system consists of three major parts:

1. A source of power to operate the control system,
2. A load or loads to utilize the power, to obtain the desired results, and
3. Controllers to obtain the desired levels of the end results.

A typical control system for a year round air conditioning system using gas for winter heating and refrigerated air for summer conditions is shown in Figure 10-4-1. A power type humidifier is included for winter and an electronic air cleaner for year round air filtration. A multispeed blower motor in the heating (air handling) unit is used to provide the best results in both the heating and cooling phases.

The operation of the system is under the control of a room thermostat, which controls a gas valve in the heating phase and the condensing unit's operation in the cooling phase. The blower in the gas furnace must operate intermittently in the heating phase, depending on the furnace supply plenum temperature. A low voltage room thermostat is used to control a high voltage blower motor (120 V) and a higher voltage condensing unit (240 V).

When a device of one voltage is controlled by a device of another voltage, intermediate controls are required. These could be relays, contactors, or motor starters, depending on the load characteristics encountered.

In this control circuit, the power source is a transformer, which is part of a plate mounted relay/transformer assembly. The other component of the assembly is a relay containing two sets of single pole double throw contacts. One set of contacts controls the speed of the blower motor while the other set controls the operation of the blower motor—either under the control of the furnace fan switch or directly for cooling operation.

The room thermostat provides the control for both the heating and cooling operation. Intermediate controls consist of the fan switch and limit switch for heating operation and a circulator switch for continuous blower operation with humidification in the heating season.

The loads in this control circuit consist of the gas valve for field input control to the heating unit, the condensing unit contactor for cooling operation, the blower relay coil for control of the blower motor, and the blower motor for air circulation.

Servicing Electronic Control Boards

Much of the equipment comes from the factory with electronic boards to control specific functions. When working on these units, it is necessary to get the service manual or technical sheet from the manufacturer's supply house. To help simplify the troubleshooting at the board level, identify four areas on the board: the input, output, board ground, and power supply. The input is probably an on/off signal from a thermostat or variable signal from a thermistor, which indicates a change has occurred. The output is the signal coming out of the board to activate a relay or controlled device. Most solid state circuit boards rely on direct current to operate the

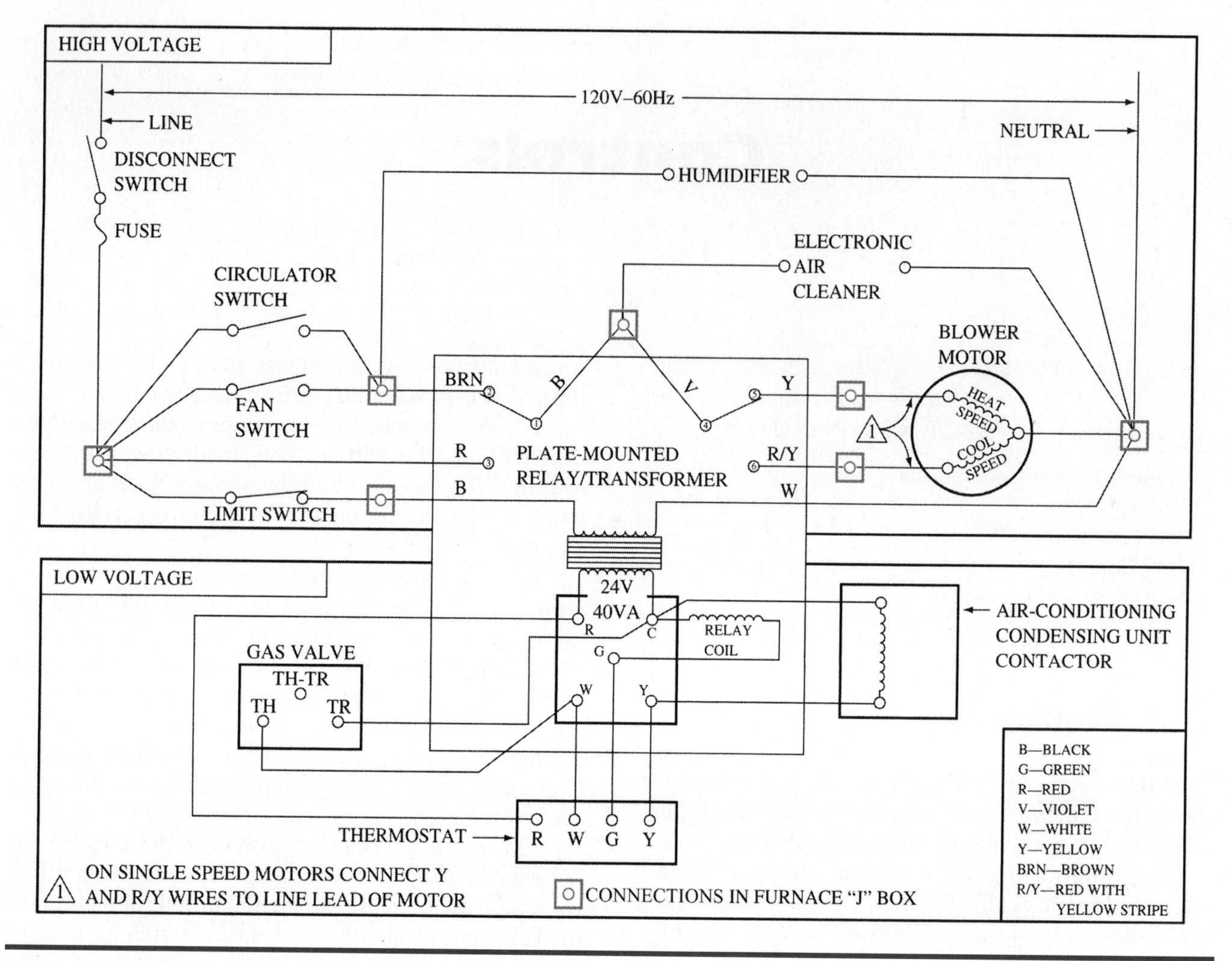

Figure 10-4-1 Typical control system for an air-conditioning system using gas heating and electric cooling.

electronic components, so they must be properly grounded. The power supply required is most likely low voltage, either 24 V AC or 5 V DC. If all four of these areas on the board are being provided as specified from the factory, there is a good chance the board has failed. It is not normally repaired in the field and should be replaced. Prior to replacement, however, verify there are no electrical shorts or wiring problems, which may have caused premature failure.

Planned Maintenance

Good routine maintenance includes annual startup and calibration on gas heat equipment prior to the heating season. The units should be tested and verified during normal operation to determine the operating and safety controls are working as designed. This includes checking air filters, oiling motors, and testing the plenum switches. On gas equipment, specific inspections should include pilot ignition, startup, and flame condition.

CONTROL SYSTEM COMPONENTS

Power Source

While the source of power to operate a control system is usually electricity, engineered systems also use electronic and pneumatic systems or a combination of all three. Here we will concentrate on electricity as the power source for residential and light commercial controls.

Heating control circuits can be designed to operate on either line voltage (115 to 120 V) or on low voltage, which is designated as a 24 V system. A low voltage control circuit is superior to a line voltage circuit because the wiring is simplified and safer, and low voltage thermostats provide closer temperature control than do line voltage thermostats.

The stepdown transformer or low voltage transformer that comes inside the furnace, Figure 10-4-2, is used in heating and air conditioning control systems.

Figure 10-4-2 AC voltage transformer for low voltage controls.

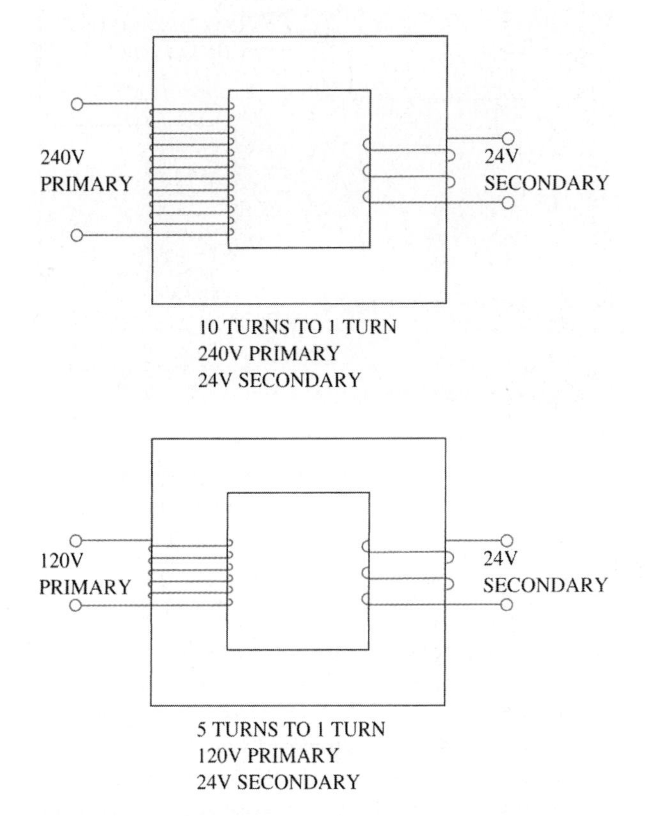

Figure 10-4-3 Construction of a step down transformer.

This transformer is used to reduce line voltage to operate the control components. Inside a simple step-down transformer are two unconnected coils of insulated wire wound around a common iron core, as shown in Figure 10-4-3. To go from 120 V (primary) to 24 V (secondary), there are five primary turns to one secondary turn. For a 240 V primary the ratio would be 10:1, etc. This, the induction ratio, is a direct proportion. Stepup transformers would be just the reverse.

Transformers used in the HVAC/R field are rated by their secondary power output. The most common size transformer found in residential gas furnaces is a 40 VA. The term VA (volt-amp) is used to express the watts of secondary power output. To convert VA to amps of output you divide the VA rating by the secondary voltage rating.

EXAMPLE

A given transformer has a primary voltage of 120 V and a secondary voltage of 24 V. Assuming that the transformer is rated at 40 VA, what is the secondary current?

Solution

Secondary (amps) = 40 VA/24 V = 1.67A ■

SERVICE TIP

A standard VA transformer has a very limited amperage capacity. If additional components such as humidifiers, duct boosters, and dampers are added to a system's control circuit, it may be necessary to increase the size of the transformer used.

Transformers are available in a variety of voltages and capacities. The capacity refers to the amount of electrical current expressed in volt-amperes. A transformer for a control circuit must have a capacity rating sufficient to handle the current (amperage) requirements of the loads connected to the secondary. 40 VA ratings are needed for air conditioning because electrical devices containing a coil and iron, such as solenoid valves and relays, have a power factor of approximately 50%. Thus, for secondary circuits with such controls, the capacity of a transformer must be equal to or greater than twice the total nameplate wattages of the connected loads.

The proper transformer rating for the electric control circuits will have been selected by the equipment manufacturer. If accessory equipment is added, however, the additional power draw must be considered. You may need to install a new larger VA transformer. This situation is common, for example, when cooling is added to an existing furnace and where the original transformer is too small. When replacing a defective transformer, make sure that the rating is equal to or greater than the original equipment.

THERMOSTATS

Snap action thermostats, Figure 10-4-4, are widely used. Low voltage thermostats, as shown in Figure 10-4-5a,b, are used in central system control circuits.

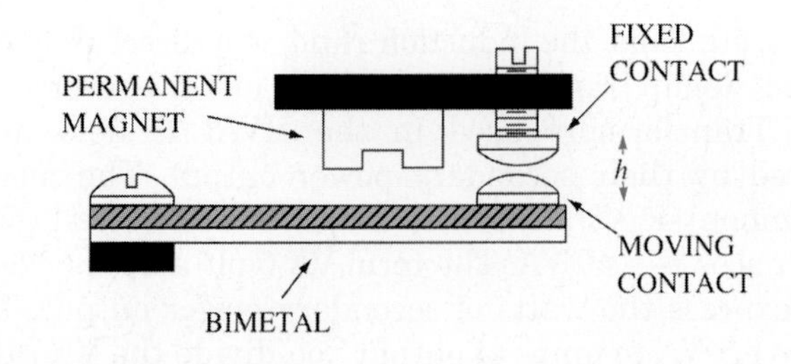

Figure 10-4-4 Snap action thermostat.

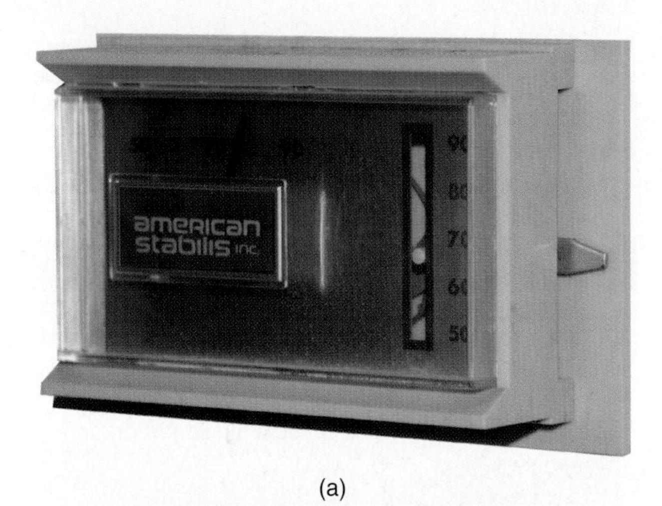

(a)

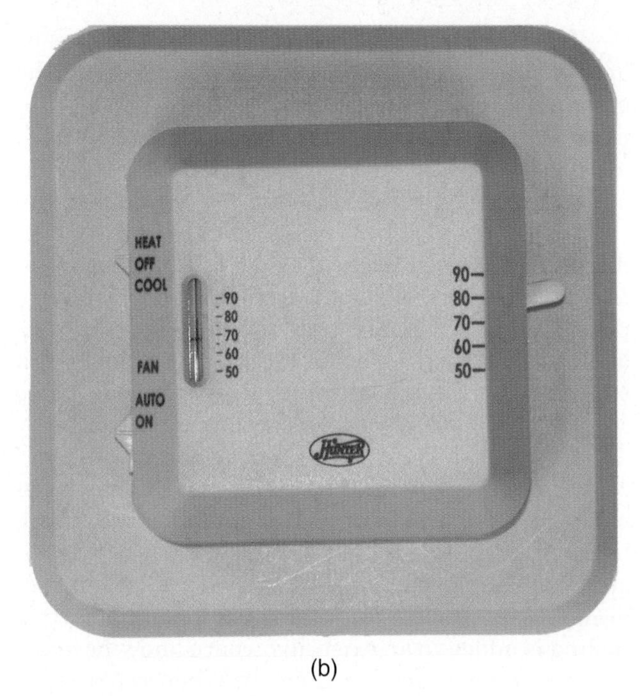

(b)

Figure 10-4-5 Low voltage thermostat used in a central system control circuits: (a) Heating only; (b) Heating and cooling.

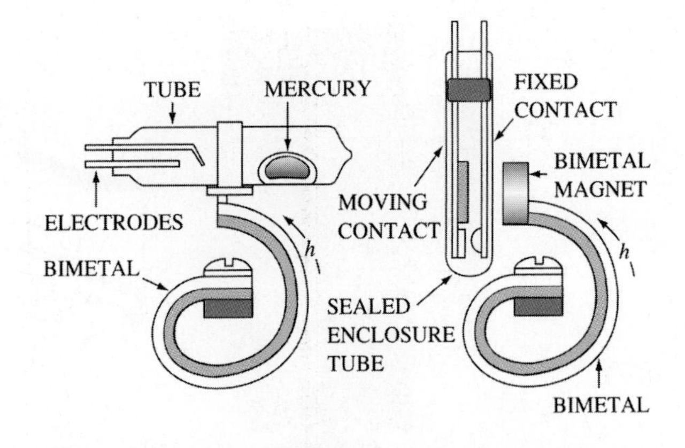

Figure 10-4-6 A single-action mercury-bulb thermometer on the left. A sealed tube with metal-to-metal contact on the right.

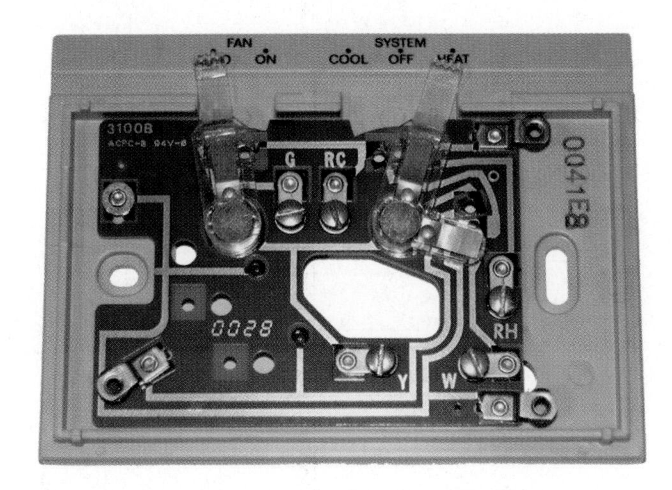

Figure 10-4-7 Subbase portion provides a mounting base for the thermostat and for system controls.

First, the use of spiral shaped, lightweight bimetallic elements increases the effective length and thus the sensitivity to temperature change. Second, the use of sealed contacts eliminates the problem of dirt and dust. Although there are minor variations among different manufacturers, the contacts are always sealed in a glass tube.

A single action mercury bulb design is shown in Figure 10-4-6. As the bimetallic strip expands and curves, the mercury fluid moves to the left, completing an electric circuit between the two electrodes which carry only 24 V. The differential gap between OFF and ON is very small: $^3/_4$ to 1°F from the set point. On the right is a sealed tube that has a metal to metal contact. The magnet provides the force that closes the contacts.

The subbase portion of the thermostat assembly, Figure 10-4-7, not only provides a mounting base for the thermostat, but is also used to control the system operation through a series of electrical switches. The system switch selects COOL, OFF, or HEAT. The blower operation is controlled by the fan switch, which

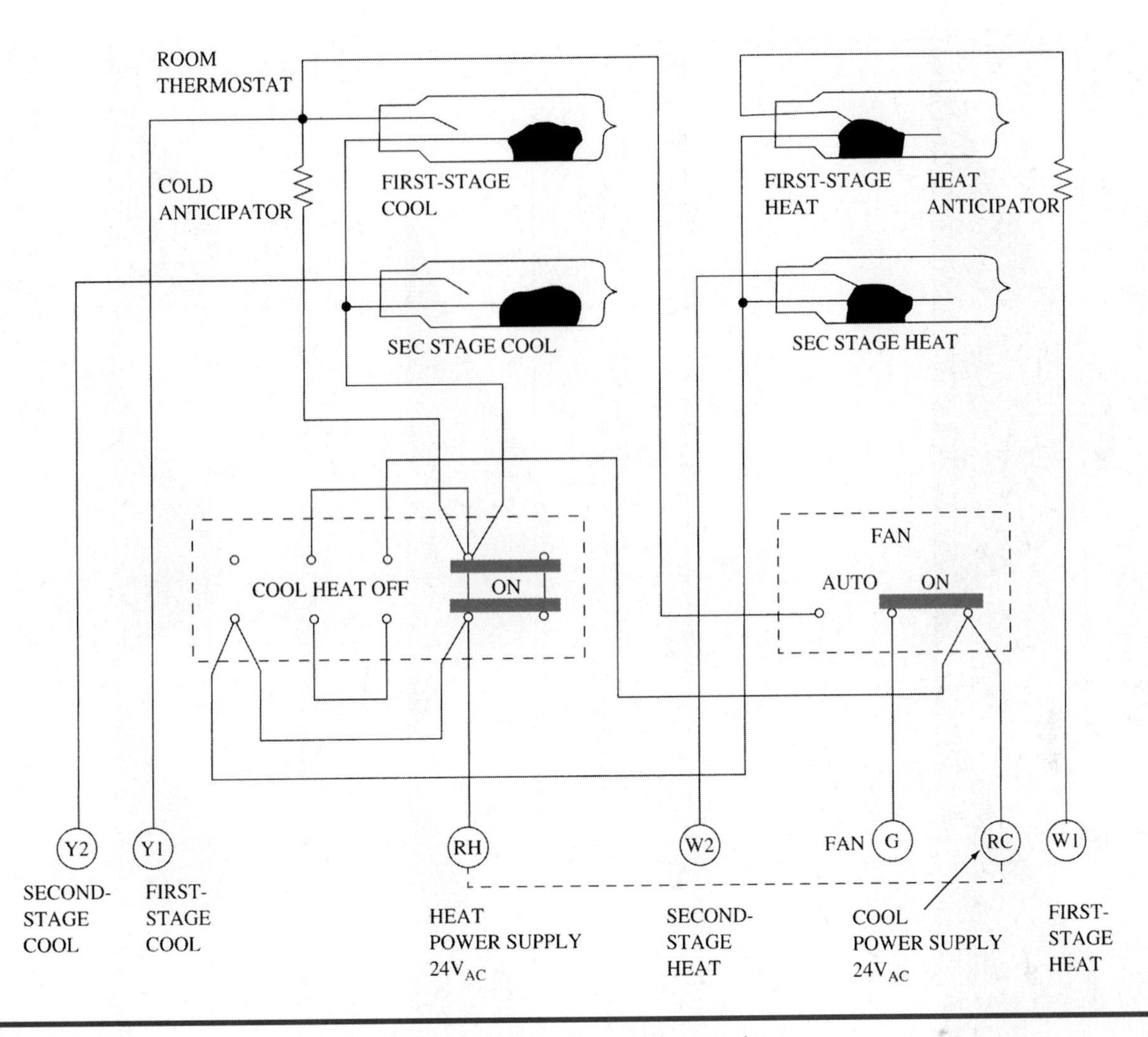

Figure 10-4-8 A two stage heating and two stage cooling schematic wiring diagram.

is a simple two position switch. When set on automatic, the fan will cycle with the furnace on heating. If in the ON position, the fan will run continuously.

More complex thermostats contain two adjustments that allow different control points for heating and cooling to be set. These thermostats may have an automatic changeover from heating to cooling. Figure 10-4-8 represents a two stage cooling schematic wiring diagram. Two stage thermostats are common in heat pumps or rooftop equipment, which use multiple compressors for cooling and two or more stages of heating. Note that seven electrical connections are required; however, RH and RC terminals are the same power source, so six wires are needed. With color coded low voltage wire, however, it is no problem to connect the thermostat to the mechanical equipment.

Heat Anticipation

The sensitivity of room thermostats is affected by both system lag and operating differential. System lag is the amount of time required for the heating or cooling system to produce a temperature change that is felt at the thermostat. The operating differential of a thermostat is the change in room air temperature needed to open or close the thermostat contacts.

Heat anticipators are used in low voltage thermostats to reduce the temperature swing caused by the systems. Heating anticipators are small electrical resistors that generate a little heat inside the thermostat when the heat is on. This "false heat" on the bimetal tricks the thermostat into thinking the room has reached the set temperature. The furnace burner goes off when the thermostat thinks the room is warm enough but the fan keeps running until the heat exchanger cools down. Without a heating anticipator the room temperature would overshoot the set temperature, Figure 10-4-9a. Temperature swing is the difference between the temperature when the furnace comes on and goes off.

Although heating anticipators result in closer room temperature control and less overshooting of heating, they can shorten the heating cycle time, as

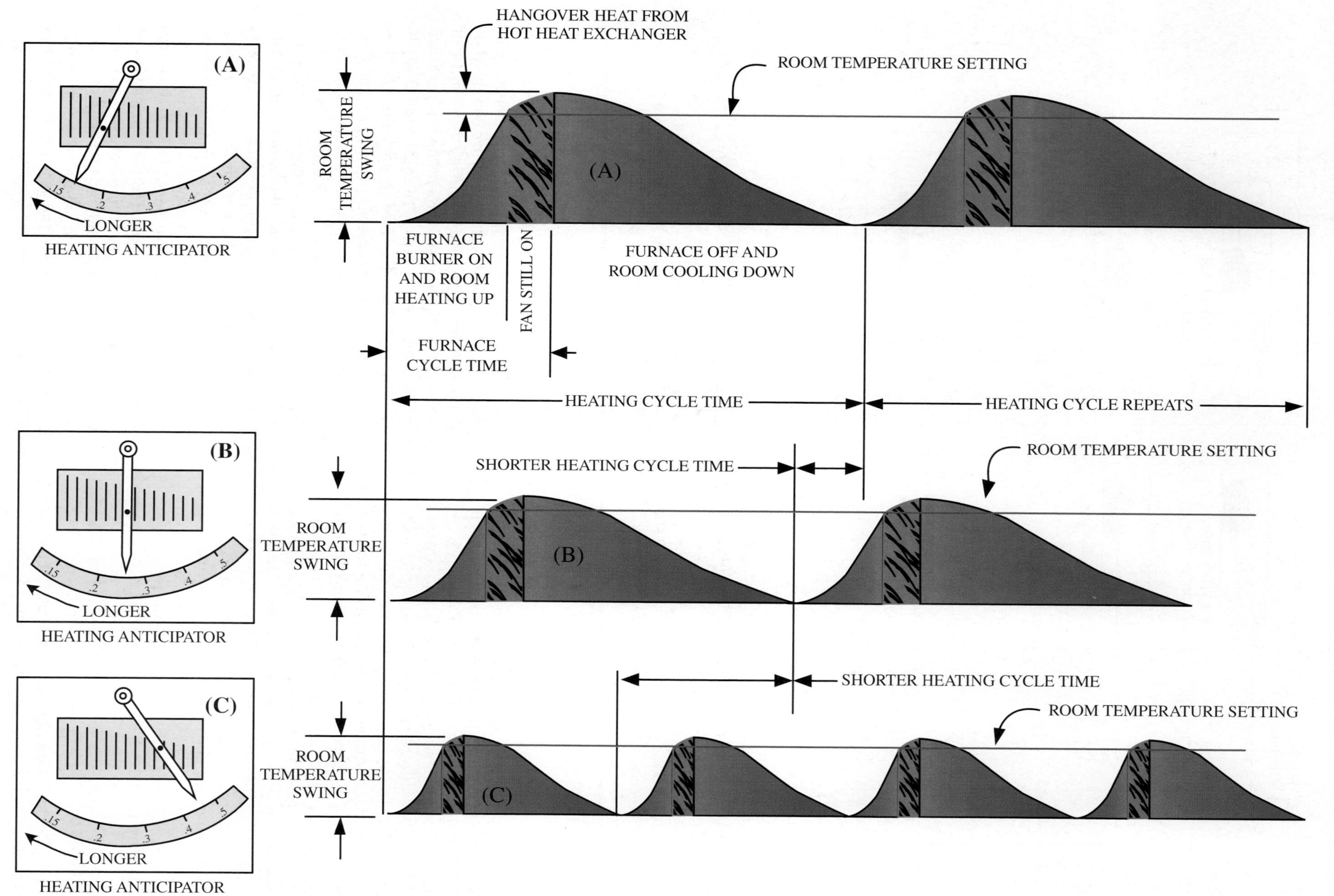

Figure 10-4-9 Effect of changing the heating anticipator on heating cycle times and temperature.

Figure 10-4-10 An adjustable heating anticipator.

shown in Figure 10-4-9b. If the heating anticipator is set too high the furnace will short cycle, as shown in Figure 10-4-9c.

Some heating anticipators are fixed, while others are wire wound variable resistors wired in series with the load. They are rated in fractions of amperes. On the adjustable type the installer will position the sliding arm to the proper load rating. A heating anticipator is shown in Figure 10-4-10. The initial setting should match the amp draw of the gas valve. But if on/off cycles are too long or too short, the system operation can be changed to give a faster or slower response.

SERVICE TIP

Some customers are more sensitive to temperature change than others. These customers might have a complaint that the room gets too cold before the heat comes on. They might also complain that this occurs even when the temperature has been turned up slightly. What they are feeling is the temperature swing resulting from the heating anticipator being set too low. By raising the current setting on the heating anticipator, the heating cycle will shorten, and the temperature swing will decrease.

TECH TIP

The longer a furnace runs during its heating cycle the more efficient it operates. Setting the heating anticipator so that a larger temperature swing occurs will result in energy savings. Conversely a short cycling heating system is less efficient to operate.

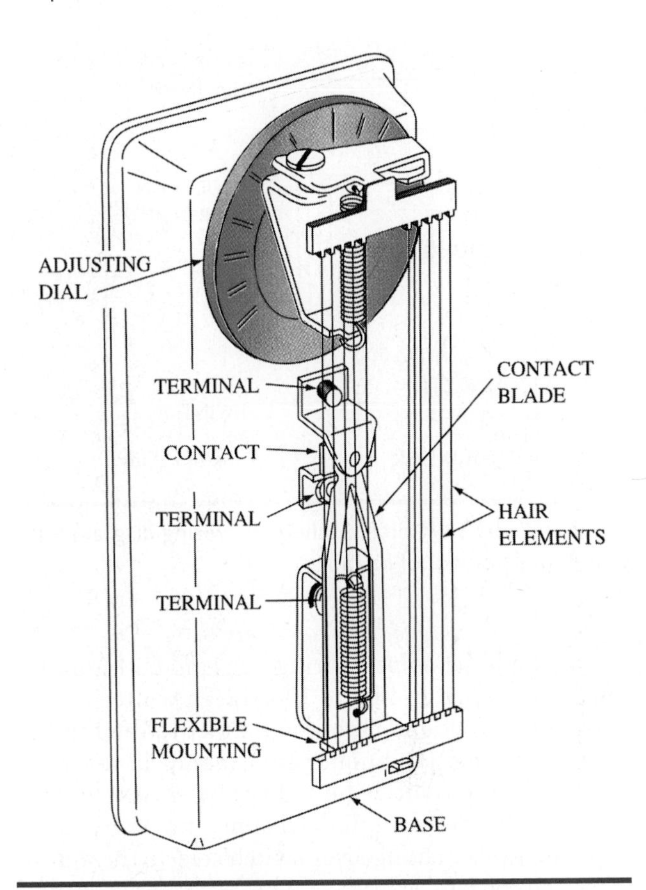

Figure 10-4-11 Construction of a wall mounted humidistat.

Cooling Anticipation

The cooling anticipator is a fixed electric heating resistor. It is on when the AC is off. The small amount of heat produced by the cooling anticipator will force the AC to come on from time to time under light loads. Humidity will build up in a house when the AC is on but cycling because of a light load, such as at night. The cooling anticipator can cause the AC system to cycle four or more time an hour to provide dehumidification. This is particularly important in warm humid climates.

Humidistats

Another type of controller found in residential comfort applications is the wall mounted humidistat, Figure 10-4-11. It is very similar to the low voltage thermostat, and it contains both a sensing element and a low voltage electric switch.

The sensing element consists of either an exceptionally thin moisture sensitive nylon ribbon or strands of human hair that react to changes in humidity. The movement of the sensing element is sufficient to make and break electrical contacts directly or when coupled with a mercury type switch. Humidistats for mounting on ductwork are also available.

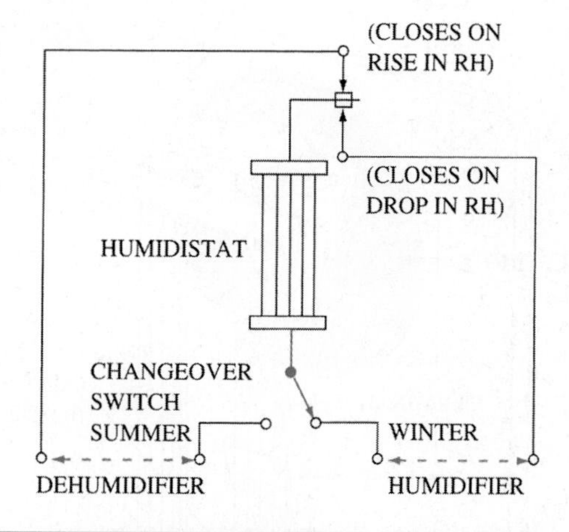

Figure 10-4-12 A simple schematic wiring diagram for year round control of humidity.

A simple schematic wiring diagram for humidity control is shown in Figure 10-4-12. In winter the elements contract and close the contact to the humidifier. In summer the humidistat expands in response to a rise in the relative humidity and closes the contacts to the dehumidifier (cooling unit) circuit. A summer/winter changeover switch selects the appropriate operation.

The humidistat is not a precise controller. It may allow a change of 5% relative humidity before switching into action. Because people cannot detect or react to changes in relative humidity with the same sensitivity as room temperature this swing is not noticed. More precise controls are available for specialized applications, such as computer rooms, libraries, and printing plants, where controllers such as wet bulb and dewpoint thermostats would be used.

SERVICE TIP

Adding humidity to a residence during the heating season will allow your customer to feel more comfortable at a lower thermostat setting. A large part of a person's comfort is the rate at which perspiration evaporates from their skin. As the relative humidity goes up that rate of evaporation decreases, thus they feel warmer even at a lower actual room temperature. A humidistat can provide that control which will increase your customer's comfort and save them heating costs.

Fan Limit Switch

The combination fan and limit switch, Figure 10-4-13a–c, is actually two switches in one. As a safety limit, some have fixed limit temperature settings; others are adjustable (approximately 180 to 200°F is the usual range). This allows a 50 to 60°F rise above normal operation before it opens. The fan control switch is also a temperature sensing device that is set to turn on the fan after the furnace has warmed up at least 15 to 20°F above room conditions, so that cold drafts are not experienced. It also stops the blower after the burner cuts off, so again there are no uncomfortable drafts. It is important to note that some systems employ constant fan operation and thus override this switch.

Other models may use spiral, flat bimetallic, or even liquid filled elements. Some forms of duct-mounted limit controls use a rod and tube or a liquid filled bulb to sense the air conditions.

GAS VALVES AND REGULATORS

At one time, the supplying of fuel gas to a heating unit was done by a combination of controls consisting of a gas-pressure regulator and a solenoid valve. These controls were combined into a single valve assembly, Figure 10-4-14, to meet the standards for proper ignition, input control, and quiet cutoff of gas unit operation.

CAUTION: Unless you have received specific manufacturer's training on servicing gas valves, do not try to repair one that is not working. It is not safe for technicians to disassemble and reassemble gas valves and gas regulators. If it is determined that the problem is in the gas valve it should be replaced and not repaired.

Valves have the following functions:

1. Manual control for ignition and normal operation;
2. Pilot supply, adjustment, and safety shutoff;
3. Pressure regulation of burner gas feed;
4. On/off electric solenoid valve controlled by the room thermostat.

Figure 10-4-15 shows a basic sketch of these functions for a natural gas valve. The schematic drawing of the valve shows the main diaphragm valve in the open condition that occurs during heat demand. When in this condition, the following assumptions can be made:

1. The schematic applies to a gas heating appliance with the pilot flame burning;
2. A thermocouple is connected to the automatic pilot magnet operator;
3. The lighting operation was previously performed to open the automatic pilot valve; and
4. The main gas cock has been turned to the ON position after the pilot is lit.

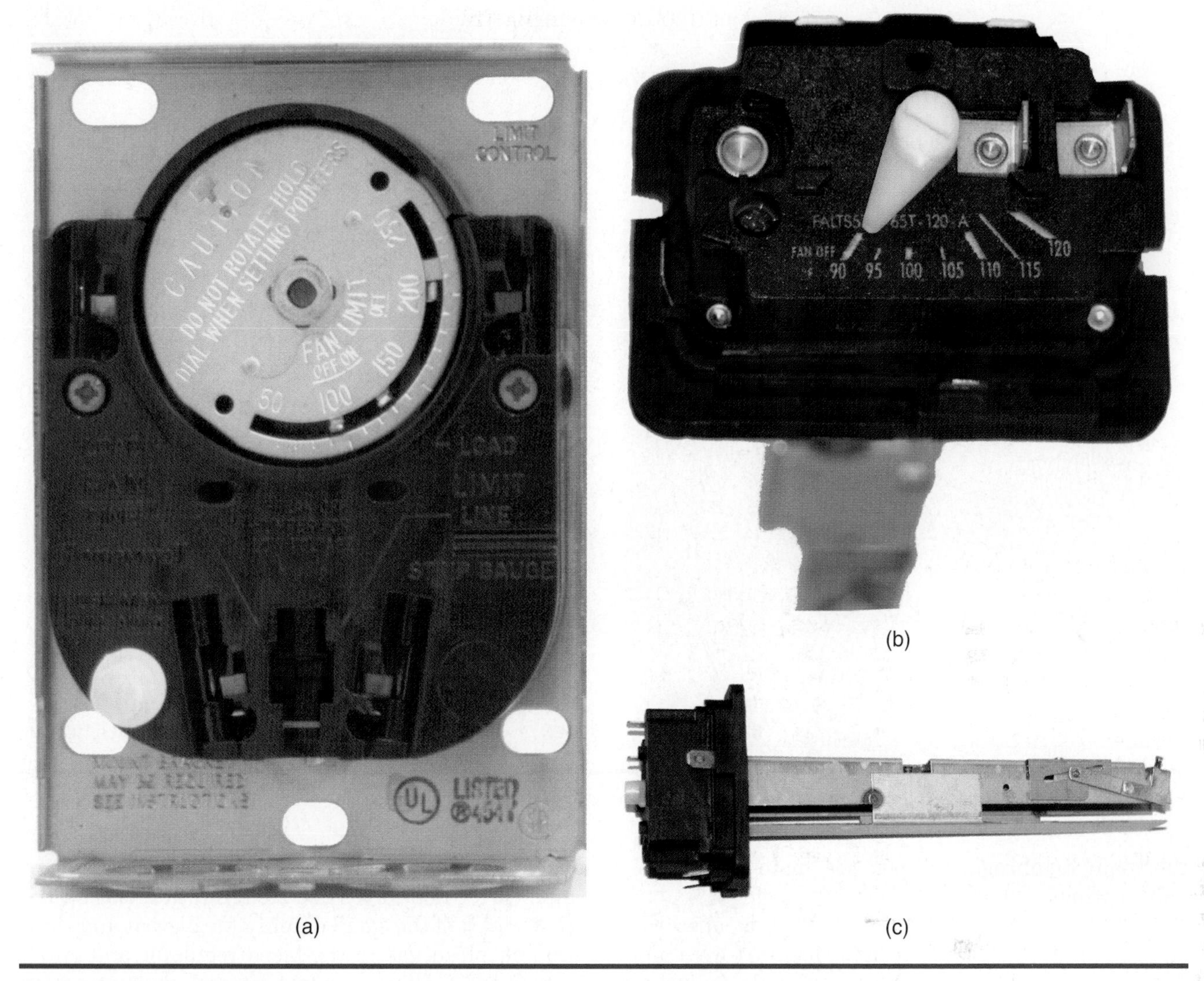

(a)

(b)

(c)

Figure 10-4-13 Combination fan and limit switch acts as a safety limit and temperature sensing device.

Figure 10-4-14 Component parts of a combination gas valve.

In the application, the 24 V operator and room thermostat are in series. Closure of the room thermostat switch on heat demand has energized the 24 V operator, which causes the armature to be attached to the pole face of the magnet, and results in a clockwise rotation of the armature as indicated by the arrows at the end of the armature. This rotation has overcome the valve spring and pulled the valve stem of the dual operator valve downward, allowing the diaphragm above it to seat on the valve seat. The seating of this diaphragm shuts off the bypass porting.

Bleed gas can enter the actuator cavity only through the bleed orifice. Bleed is allowed to flow from the actuator cavity through the dual operator valve and the regulator to the outlet through the outlet pressure sensing poet. (The valve stem has a square cross section in a circular guide, allowing gas to pass between the stem and the guide.) The resultant drop in pressure within the actuator cavity and through the main diaphragm allows the inlet pressure above

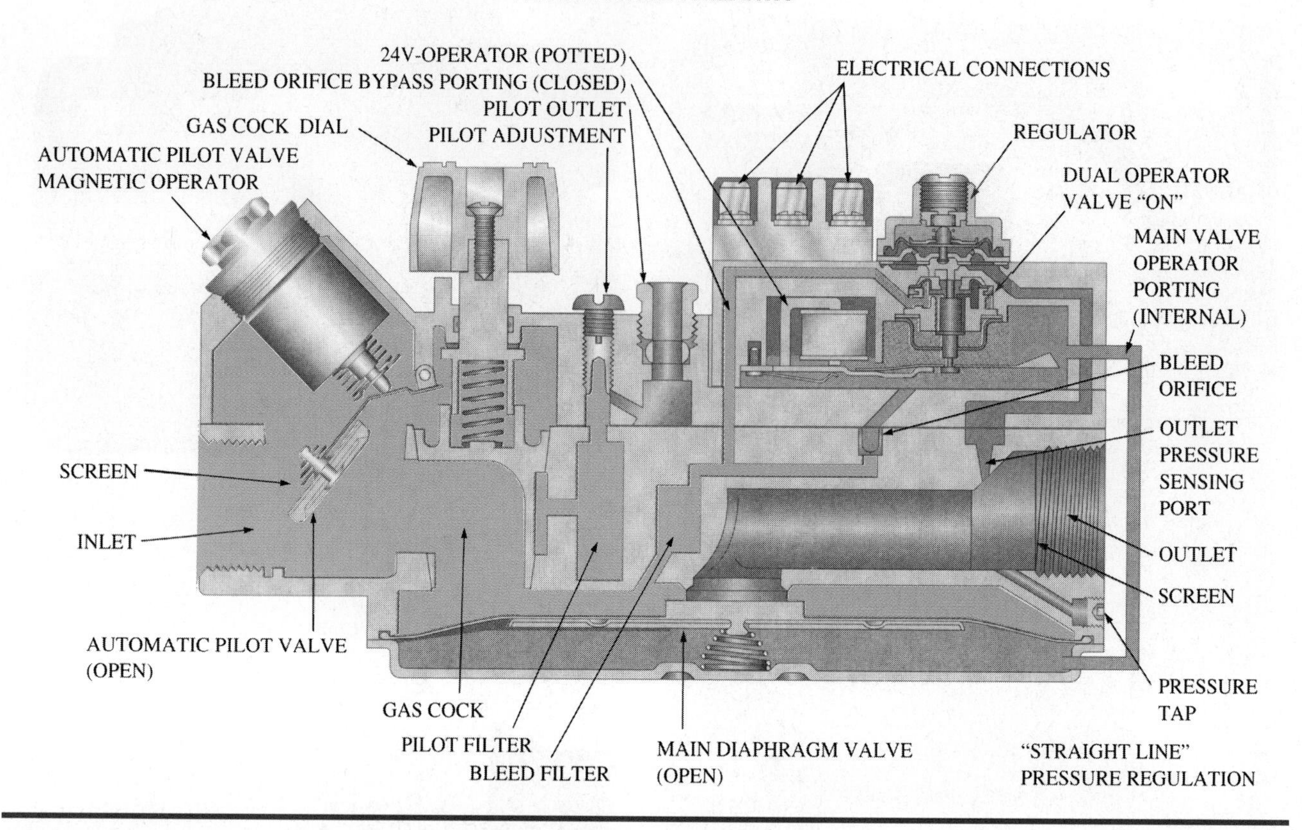

Figure 10-4-15 Cutaway of a combination gas valve in open position.

the main diaphragm to open the main diaphragm valve spring.

After the main diaphragm valve is open, as shown in Figure 10-4-15, straight-line pressure regulation is secured by the feedback of outlet pressure through the outlet sensing port to the pressure regulator in the bleed line. A rise in pressure at the control outlet above the set pressure causes a proportional closure of the regulator valve in the bleed line. The proportional closure of the regulator valve causes a corresponding rise in pressure in the bleed line ahead of the regulator. This rise in pressure in the actuator cavity increases the pressure beneath the main diaphragm and causes a partial closure of the main diaphragm valve, lowering the control outlet pressure to the pressure setting. Upon a drop in control outlet pressure, a like decrease in bleed line pressure transmitted to the underside of the main diaphragm through the action of the bleed line pressure regulator causes a proportional increase in the main valve opening to bring the delivered outlet pressure back up to the set pressure.

The schematic of the closed main diaphragm, Figure 10-4-16, shows the action of the bypass valve in the dual operator valve during a fast OFF response independent of the bleed orifice. Upon heating to the specified temperature, the thermostat switch opens and the 24 V operator is de-energized. The return spring on the valve stem of the dual operator valve then forces the stem upward, closing the center part of the small diaphragm above it and shutting off bleed gas to the bleed regulator and outlet sensing port. The resultant counterclockwise rotation of the armature is indicated by the arrows at the ends of the armature. As the small diaphragm above the valve stem is forced upward, the bleed orifice bypass porting is opened by the diaphragm, leaving the valve seat. The actuator cavity, main valve operator port, and cavity beneath the diaphragm are rapidly exposed to full inlet pressure, which acts to close the main diaphragm valve. The bleed orifice bypass port allows the pressure above and below the main diaphragm to be equalized rapidly, independently of the bleed orifice. This pressure is then equalized and the main valve is closed by the main valve spring.

The dual operator valve that provides the bleed orifice bypass porting control for the fast OFF response is an important feature. It helps to prevent flashback conditions. Flashback can cause pilot outage problems as well as burner and appliance sooting. Relatively slow closing has been a problem on larger capacity applications. The dual operator valve is an important factor in helping to eliminate this application problem.

Gas valves have built in protection against gas line contaminants. The inlets and outlets have

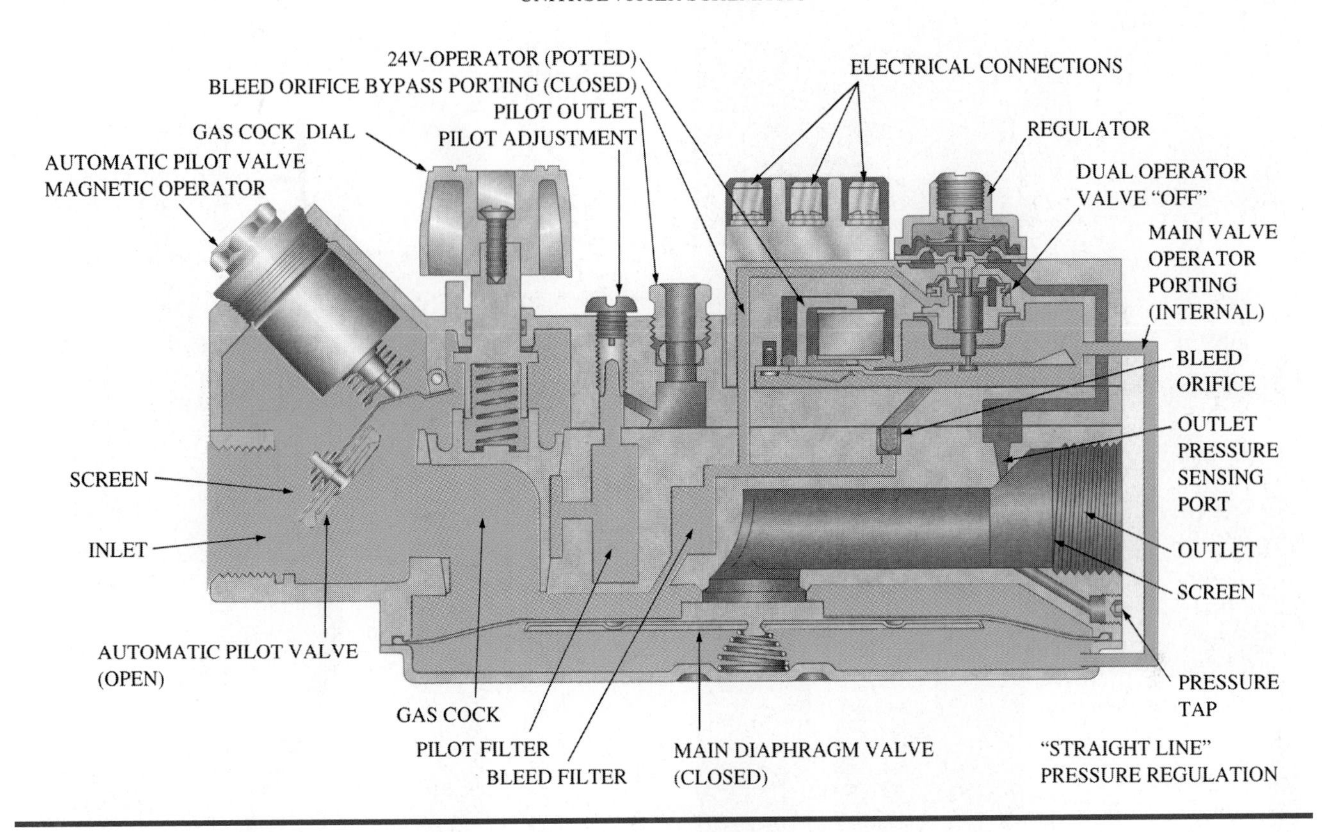

Figure 10-4-16 Cutaway of a combination gas valve in closed position.

screens for the main gas passages, a pilot filter for the pilot line, and a bleed filter for the bleed control line. These means protect the control from malfunction due to contaminants entering into control passages. In the highly unlikely case that the bleed orifice restrictor becomes clogged with the main valve in the open or ON position despite the bleed filter protection, the dual operator valve will enable the control to shut off the main gas valve when so signaled by the room thermostat or limit control.

LP gas is heavier than air, and over a prolonged period enough pilot gas could collect to be a hazard.

The methods of pilot ignition and gas valve control described above apply chiefly to the older residential furnaces, potentially requiring service. Electronic ignition systems, developed for commercial systems, have been adapted to the newer residential furnaces and require an entirely different approach to service.

An electronic-ignition system is shown in Figure 10-4-17. Pilot gas is fed to the assembly and burner orifice. The spark electrode is positioned to ignite the gas on a signal from the room thermostat. With the pilot ignited and burning, a sensing probe establishes an electric current sufficient to energize a relay that opens the main gas valve by closing normally open electrical contacts. At the same time the relay's normally closed electrical contacts in the spark ignition circuit open, terminating the spark. As long as the sensing probe recognizes the pilot flame, the relay coil will be energized, and normally closed contacts in the spark ignition system will remain open.

SAFETY TIP

Approximately 10,000 V are generated in the spark ignition circuit. Although the spark has very low amperage and is not considered a major safety hazard, it can cause a great deal of discomfort if it is energized while you are working in the area of the spark igniter or spark igniter wire.

The direct spark ignition system, Figure 10-4-18, uses an electricity carrying flame sensor rod mounted to have direct contact by the main burner flame. Because the gas flame will carry electrical energy by means of electrically charging the carbon atoms in the gas before combustion takes place, a current flow can be passed from the burner to a positive charged flame rod, Figure 10-4-19. When current is allowed to flow in only one direction, that current is said to be rectified. The flame rectifies the AC so it flows out of the flame as pulsed DC. This current flow controls a circuit in the solid state

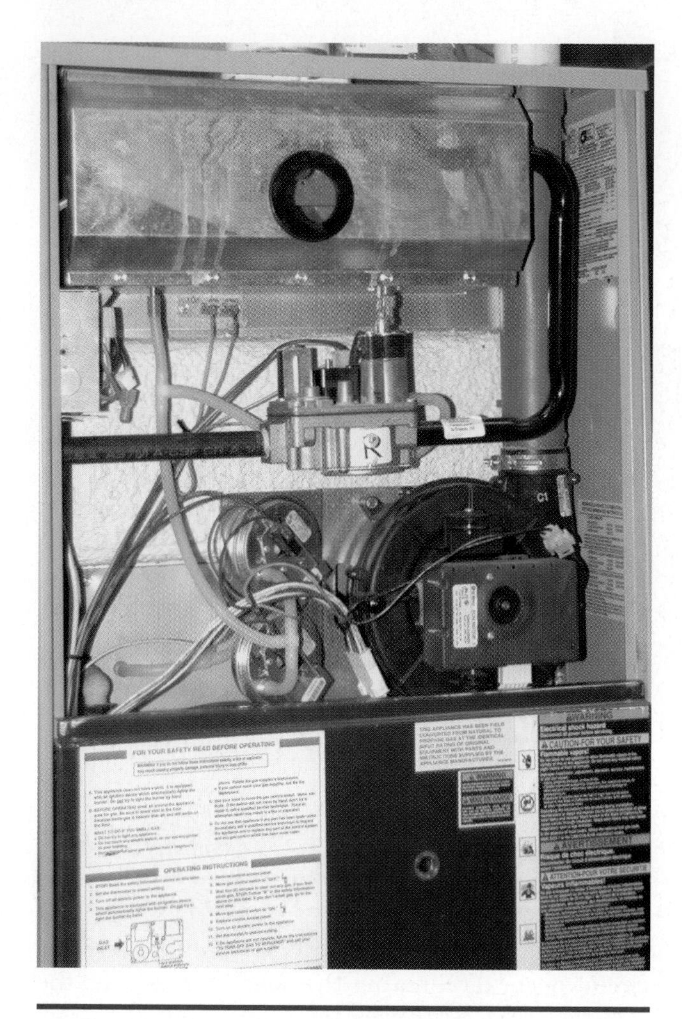

Figure 10-4-17 Gas furnaces that do have a standing pilot are often referred to as being "pilotless."

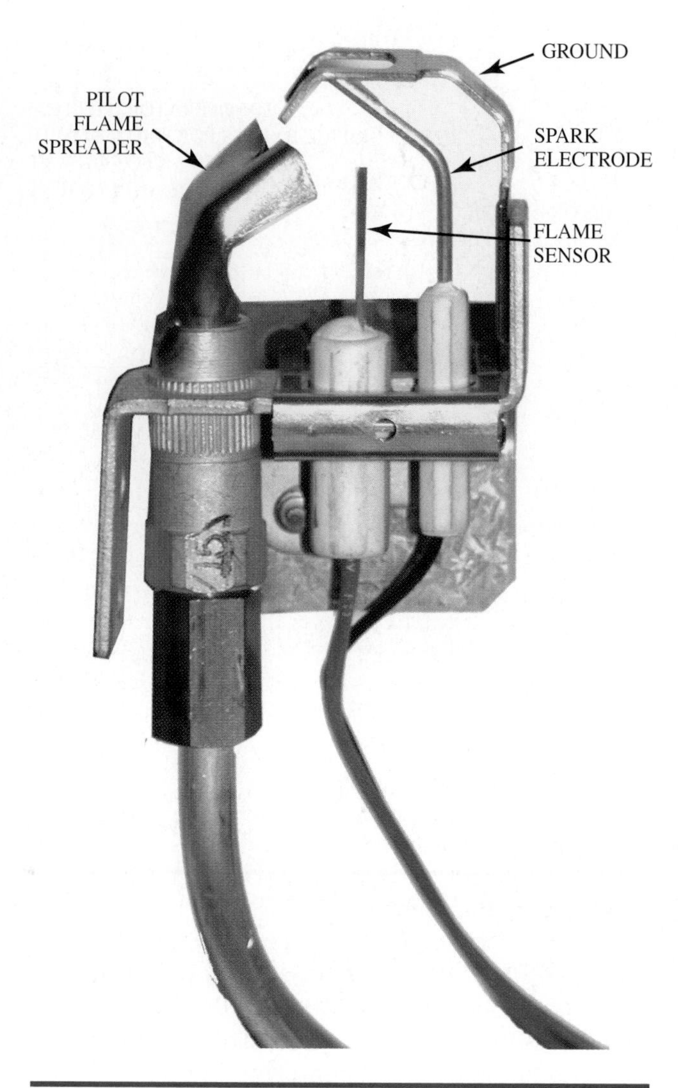

Figure 10-4-18 Spark ignition pilot assembly.

control module and keeps the gas valve energized, Figure 10-4-20. Flame rectification provides several unique features for a flame sensor. For example if the flame rod has shorted against the furnace it will pass the AC through it and the controller will not open the gas valve because pulsed DC is not present. Also if the flame does not light no current will flow and again the controller will not open the gas valve. And if the flame goes out, the controller will shut off the gas valve once the pulsed DC signal is lost.

Upon a call from the room thermostat, both the main gas valve and the spark ignitor are activated. Allowing a predetermined time for main flame ignition, the ignition control module will shut down the lockout circuit and maintain burner operation if main flame ignition occurs in that period. The period may be from 4 to 21 seconds, depending on the model of control module used. Usually, the higher the input to the gas unit, the shorter the proving time.

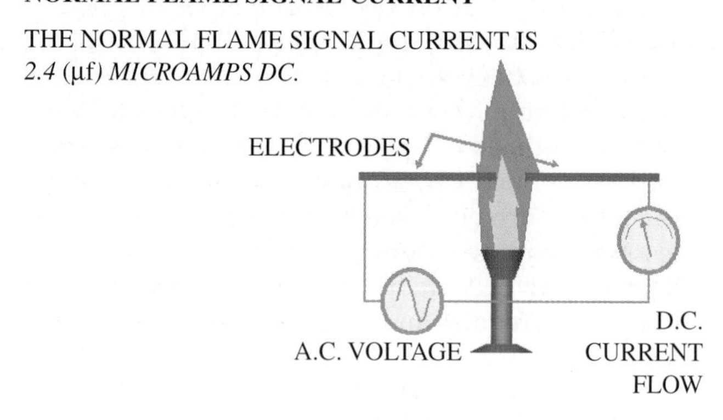

Figure 10-4-19 Flame rectification. (*Courtesy York International Corp., Unitary Products Group*)

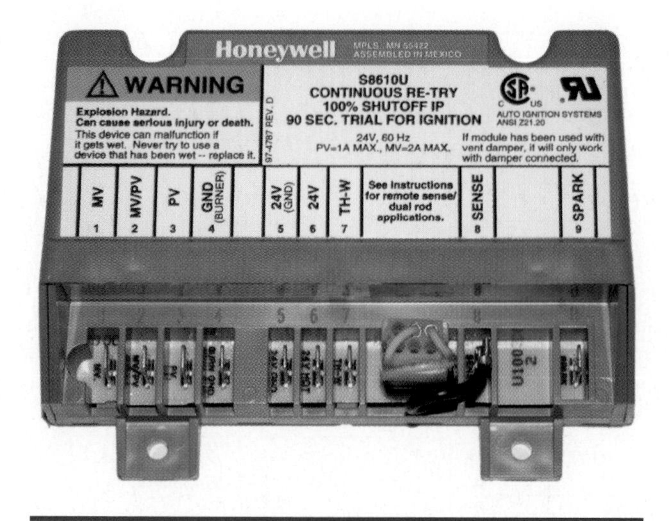

Figure 10-4-20 Pilot and gas valve controller.

If the main burner flame is not established in the set time, the control module automatically locks out. To reset the circuit, electrical power to the system must be cut off and then back on to start another cycle. Manual reset of such a system is used for maximum safety to the equipment and building.

Because this system uses the main burner assembly as the ground terminal of the spark system, it is absolutely necessary that the gas burning unit be thoroughly grounded to the electrical supply ground. It is wise and usually necessary to run a ground (green) wire from the power distribution panel to the unit to provide this positive ground. Because the white or neutral wire is a $^{120}/_{240}$ V supply system, it is a current carrying wire. It is not suitable for use as a unit ground.

Hot Surface Ignition

A hot surface ignition unit, Figure 10-4-21a,b, is used for igniting the gas burners on many of the latest furnaces. This unit uses a material called silicon carbide which has a very high resistance to current flow, and when energized reaches the ignition temperature of gas. It is very tough and will not burn up, something like a glow coil. Even though it will last a long time, it is also brittle and will break easily with rough treatment. The control system allows this material to reach the ignition temperature before the gas valve opens.

These units are powered by 120 V and draw a considerable amount of current when energized; however, they are used only a few minutes per day and therefore do not materially increase the electric bill. If the burner fails to light or the flame sensor fails to detect a flame after the gas is turned on for a few seconds, a safety lockout will occur to stop the flow of gas.

(a)

(b)

Figure 10-4-21 (a) Hot surface gas fuel igniter; (b) Hot surface gas fuel igniter glows red before gas valve opens.

RESIDENTIAL AND SMALL COMMERCIAL CONTROL SYSTEMS

Different manufacturers will have variations in the physical wiring, but the function remains the same in all systems. With a firm understanding of the construction and use of schematic wiring diagrams, the student should be able to wire any manufacturer's product. The industry uses standardized electrical symbols as shown in Figure 10-4-22, that help the technician identify the common controls. Also, the legend designations shown in Table 10-4-1 are related alphabetically to the devices they represent. For example, R is a general relay while CR is a cooling relay, HR is a heating relay, and DR is a defrost relay. Legends may vary with the particular wiring diagram or the manufacturer's method of labeling.

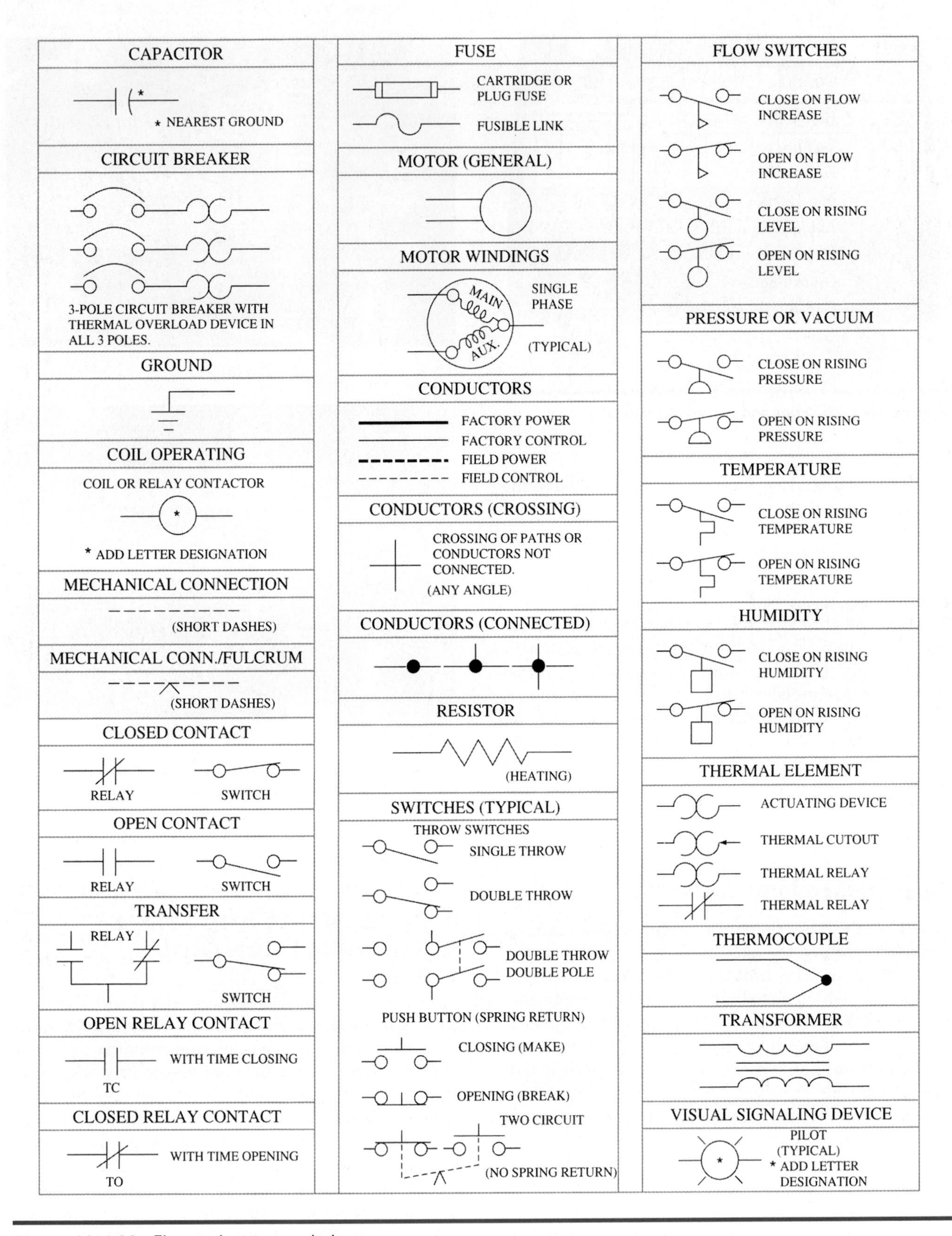

Figure 10-4-22 Electrical wiring symbols.

TABLE 10-4-1 Legend for the Wiring Diagrams

Relays		Switches		Miscellaneous	
R	Relay, General	DI	Defrost Initiation	C-HTR	Crankcase Heater
CR	Cooling Relay	DT	Defrost Termination	RES	Resistor
DR	Defrost Relay	DIT	Defrost Initiation Defrost Termination (dual function device)	HTR	Heater
FR	Fan Relay			PC	Program Control
IFR	Indoor Fan Relay			OL	Overload
OFR	Outdoor Fan Relay	GP	Gas Pressure	L	Indicating Lamp
GR	Guardistor Relay	HP	High Pressure	⊕	Manual Reset Device
HR	Heating Relay	LP	Low Pressure		
LR	Locking Relay (Lock-in or Lockout)	HLP	Combination High-Low Pressure	+	Automatic Reset Device
PR	Protection Relay (Relay in series with protective devices)	OP	Oil Pressure		
		RM	Reset, Manual		
		FS	Fan Switch		
VR	Voltage Relay	SS	System Switch		
TD	Time Delay Device	HS	Humidity Switch (Humidistat)		
THR	Thermal Relay (type)	TA	Thermostat, Ambient		
M	Contactor	TC	Thermostat, Cooling		
MA	Auxiliary Contact	TH	Thermostat, Heating		
Solenoids		TMA	Thermostat, Mixed Air		
S	Solenoid, General	CT	Thermostat, Compressor Motor		
CS	Capacity Solenoid				
GS	Gas Solenoid	HT	High Temperature		
RS	Reversing Solenoid	LT	Low Temperature		
		RT	Refrigerant Temperature		
		WT	Water Temperature		

SERVICE TIP

Manufacturers include wiring diagrams inside of the furnace cabinet. Diagrams are also available from the manufacturer and supply house for each piece of equipment. Be sure to study these diagrams carefully because manufacturers may make minor changes in the circuits within the furnace from year to year. Don't assume that the diagram that you are familiar with from previously working on a similar unit is exactly the same. Check it out carefully.

Wire Size

Wire sizes and fusing for independent appliances will be determined by the manufacturers and specified on their wiring diagrams. Local codes should be examined to determine whether or not conduit is required in bringing the wire from the main circuit breaker to the furnace or air conditioner. Some codes even require conduit for parts of the circuit on 120 V. Specification sheets will give the minimum wire and fuse sizes that will meet the requirements of the National Electric Code, and this wire size is based on 125% of the full load current rating. The length of the wire run will also determine the wire size, and Table 10-4-2 shows the maximum length of two wire runs for various wire sizes. This table is based on holding the voltage drop to 3% at 240 V. For other voltages, the multipliers at the bottom should be used. If the required run is nearly equal to or somewhat more than the maximum run shown by the calculation of the table, the next larger wire size should be used as a safety factor. It is always possible to use larger wire than called for on the specifications, but a smaller wire size should never be used, as this will cause nuisance trips, affect the efficiency of operation, and create a safety hazard. All replacement wire should be the same size and type as the original wire and should be rated at 90 to 105°C.

TABLE 10-4-2 Maximum Length of Two Wire Run Based on Length and Current Capacity*

Wire size	Amperes 5	10	15	20	25	30	35	40	45	50	55	70	80
14	274	137	91										
12		218	145	109									
10			230	173	138	115							
8					220	182	156	138					
6								219	193	175	159		
4									309	278	253	199	
3										350	319	250	219
2											402	316	276
1												399	349
0												502	439
00													560

*To limit voltage drop to 3% at 240 V. For other voltages use the following multipliers:

110 V	0.458	220 V	0.917
115 V	0.479	230 V	0.966
120 V	0.50	250 V	1.042
125 V	0.521		

For example, the maximum run for #10 wire carrying 30 amp at 120 V is 115 × 0.5 = 57 ft.

Note that if the required length of run is nearly equal to, or even somewhat more than, the maximum run shown for a given wire size—select the next larger wire. This will provide a margin of safety. The recommended limit on voltage drop is 3%. Something less than the maximum is preferable.

Heating Circuit

A simple schematic diagram for a typical gas fired upflow air furnace is shown in Figure 10-4-23. The wire conductors are represented by lines and the other components by symbols and letter designations. Note the very important legend that identifies these. Although other types of diagrams will be discussed later, the schematic is the type of diagram used by service personnel to analyze the system.

Only five major electrical devices are in the diagram:

1. The fan motor, which circulates the heated air;
2. The 24 V automatic gas solenoid valve to control the flow of gas to the burner;
3. The combination blower and limit control which controls fan operation and governs the flow of current to the low voltage control circuit;
4. The stepdown transformer, which provides 24 V current to operate the automatic gas valve; and
5. The room thermostat, which directly controls the opening and closing of the automatic gas valve.

To build this same schematic diagram, begin with the power supply. The wires for the power supply are usually run by the installing technician. They must carry line voltage (single phase, 60 Hz, 120 V) to operate the blower motor and are represented by heavy broken lines to indicate field wiring, as shown in Figure 10-4-24.

As a protection to the circuit, the power supply must be run through a fused disconnect switch. This is the main switch to the entire system.

In a gas fired heating system, the fan motor produces the heaviest load and is connected to the power supply, as shown in Figure 10-4-25. Since the fan motor is clearly identified on the diagram, it is not necessary to add it to the legend. Internal windings are not shown.

To test the system at this point, close the disconnect switch. The fan motor runs, since a simple circuit is completed from L_1 through the windings of the blower motor and back to neutral. Open the disconnect switch and continue the diagramming.

Since there must be automatic control of both the fan motor and the automatic gas solenoid valve, connect the combination fan and limit control, Figure 10-4-26. This control, as previously described, consists of a sensing element, which feels the temperature of the heated air inside the furnace, and two adjustable switches: a fan switch and a limit switch.

The fan switch controls the starting and stopping of the blower motor. When the burner starts, the

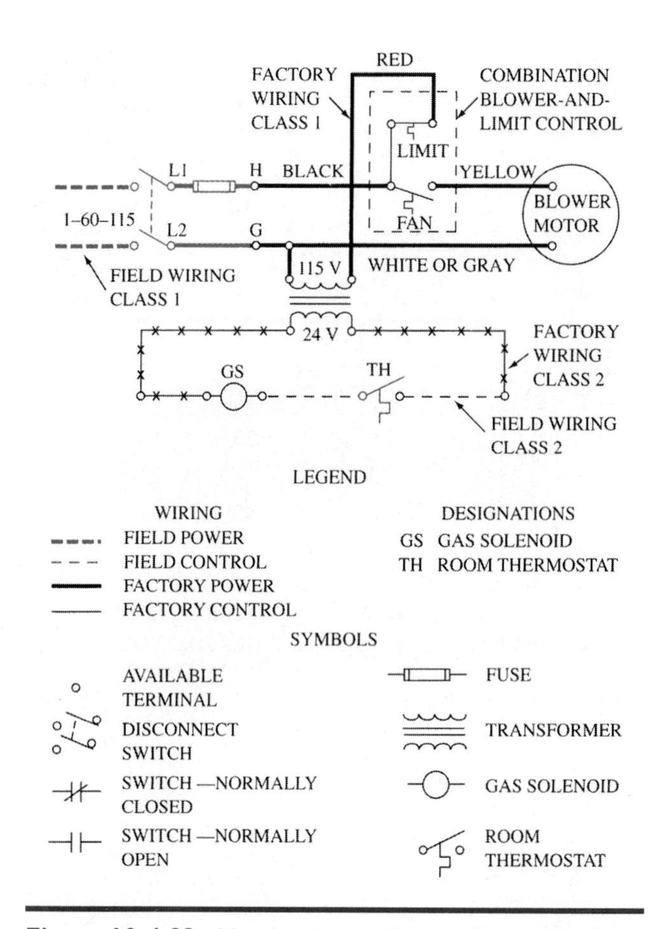

Figure 10-4-23 Heating circuit diagram for gas fired upflow furnace.

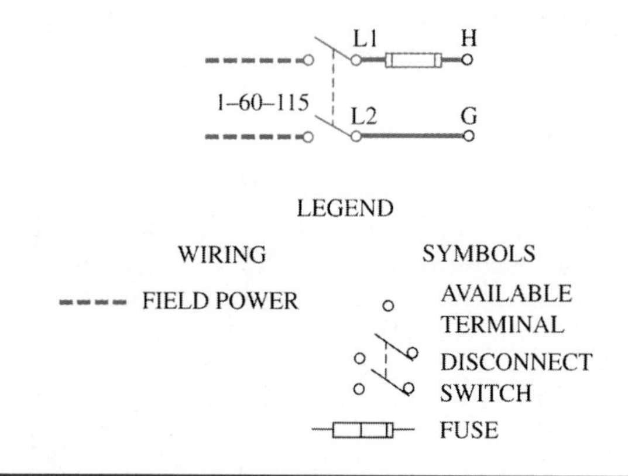

Figure 10-4-24 Diagram for a fused disconnect.

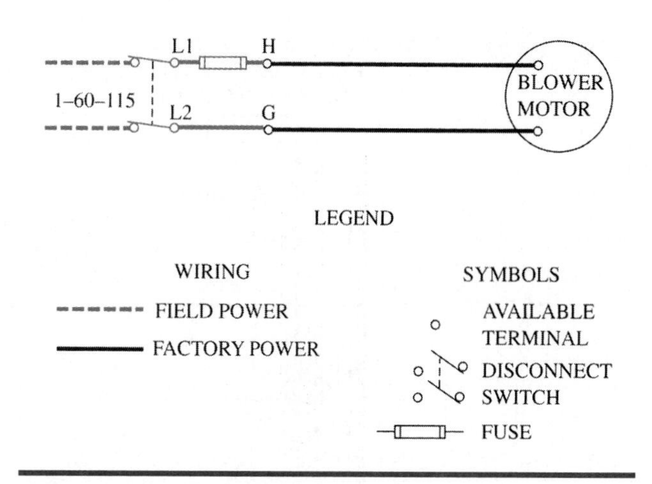

Figure 10-4-25 Diagram showing the fused disconnect wired to the blower motor.

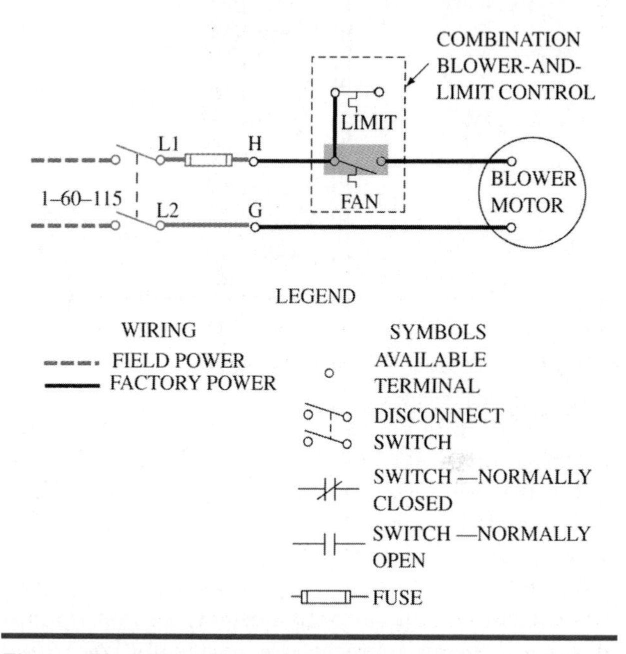

Figure 10-4-26 Wiring diagram showing the use of a fan limit control wired to a blower motor.

fan switch will not close to start the blower until the air temperature inside the furnace plenum chamber has warmed up to the FAN ON setting. When the burner stops, the fan switch remains closed, keeping the blower in operation until the air temperature inside the furnace cools down to the FAN OFF setting.

The limit switch is a safety control. It prevents the furnace from overheating. As long as the plenum chamber temperature is below the setting of the limit switch, the switch remains closed. If the plenum chamber temperature rises to the switch setting, it opens to deenergize the entire 24 V circuit and close the automatic gas valve.

As illustrated, the combination fan and limit control are partially connected. The fan switch is wired into L1 in series with the blower motor windings. If the disconnect switch is closed now, the fan motor will not run because the fan switch in the fan motor circuit is open and will remain open until the temperature in the furnace warms up to the switch setting.

Small electrical devices, such as relays and solenoid valves, do very little work and require little current for operation. A stepdown transformer to reduce

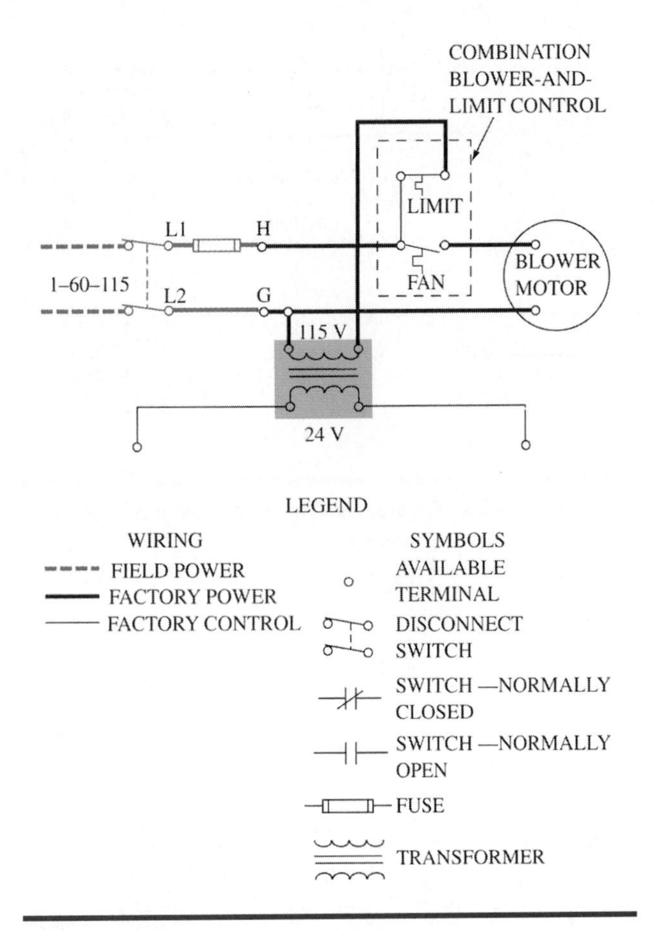

Figure 10-4-27 Schematic diagram showing the addition of a control circuit transformer.

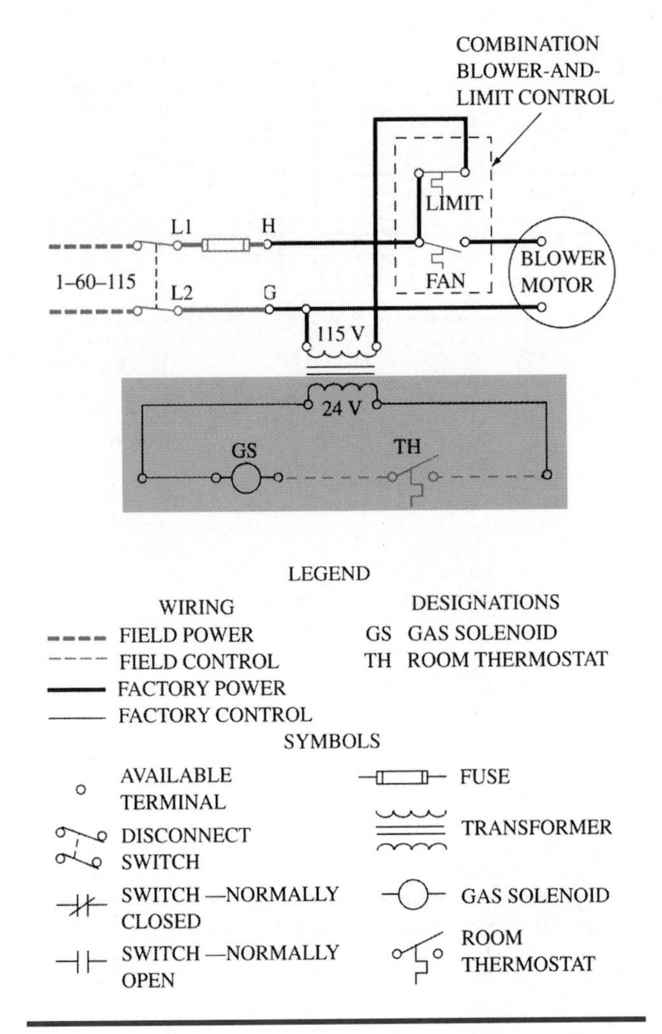

Figure 10-4-28 Schematic diagram of the low voltage circuit showing the thermostat operation of the gas valve.

line voltage to 24 V is required and properly installed in the wiring diagram, as shown in Figure 10-4-27. One side is connected to neutral and the other side is connected to L1 through the limit switch. This places the limit switch in control of the entire 24 V circuit.

If the disconnect switch is closed, the blower motor still cannot run since the fan switch must remain open until the air temperature warms up. The transformer and the 24 V power source, however, are energized immediately from L1 through the closed limit switch and back to L2. Although there is now a source of 24 V current, nothing can be accomplished until there is a circuit across the 24 V supply.

The automatic gas solenoid valve and room thermostat are wired in series across the 24 V circuit, Figure 10-4-28, since the opening and closing of the gas solenoid valve is controlled by the thermostat. The gas valve is factory wired, indicated on the diagram as a light, unbroken line between terminals. Normally, the thermostat is located in a room away from the furnace and must be field wired, as indicated by the light broken lines used to represent control circuit field wiring.

The circuit is now completely wired. Close the disconnect and limit switches and open the thermostat. The transformer and low voltage circuit are immediately energized from L1, through the closed limit switch back to neutral. With the thermostat open, the automatic gas valve remains closed. The fan motor cannot operate because the fan switch is open, and it will remain open until the air temperature in the furnace warms up to the FAN ON setting.

When the thermostat calls for heat, the automatic gas valve is energized, as shown in Figure 10-4-29, and the valve opens to admit gas to the burner. The fan switch would still be open so that the fan motor cannot run until the air temperature in the furnace plenum chamber warms up to the FAN ON setting.

Thermostat and gas valve circuits are reopened when the room temperature setting is reached. Fan and blower contacts remain closed until the temperature of the heated air cools to the FAN OFF setting.

If at any time during furnace operation the plenum chamber overheats, the air temperature soon reaches the setting of the limit switch. This is an important safety switch, which opens at the overheating setting to de-energize the entire 24 V circuit. When this happens, the automatic gas valve closes,

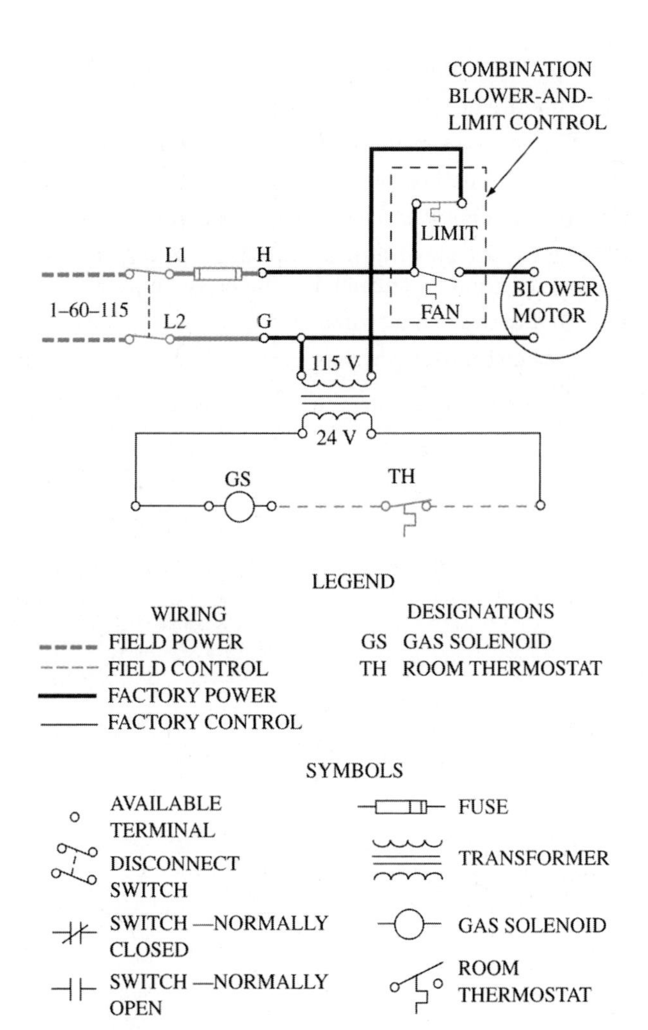

Figure 10-4-29 Schematic diagram showing the wiring of the limit control in series with the line voltage side of the transformer.

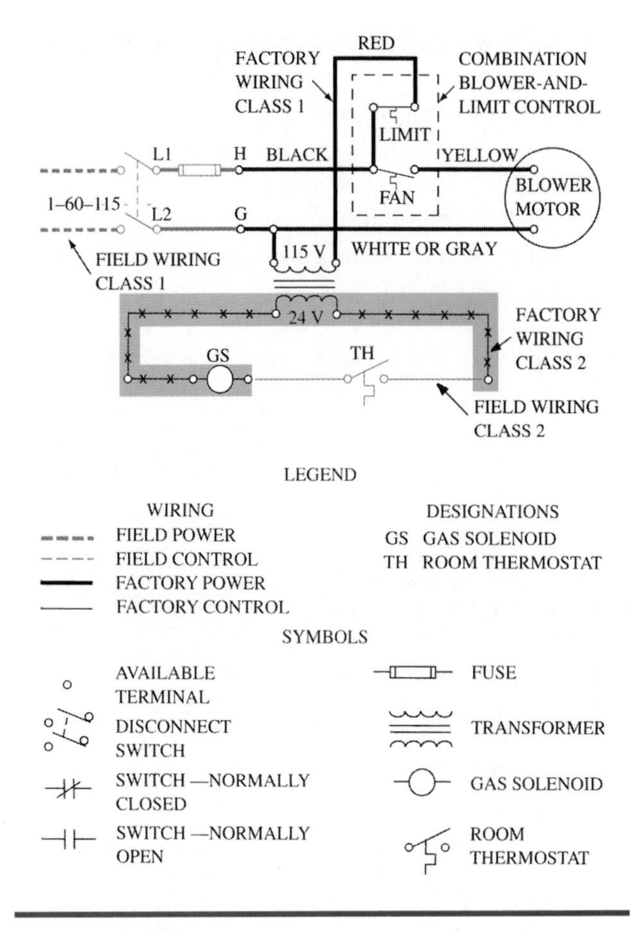

Figure 10-4-30 A diagram showing the gas valve circuit separating factory wiring from field wiring.

the burner is extinguished, and the thermostat is overridden. The fan switch remains closed and the blower continues to run as long as the heated air remains above the FAN OFF setting. The limit switch is manually reset by the service person after correcting the problem. Causes of overheating are discussed in the troubleshooting section of this text.

The operating sequence just completed proves that the wiring diagram is correct. The disconnect switch controls the entire electrical system; the blower motor operates independently of the burner, under the control of the fan switch. The automatic gas valve is cycled on and off by the room thermostat, under the control of the limit switch. The complete story of the electrical system, however, includes more than just the lines and symbols shown in Figure 10-4-30. The legend clearly identifies all symbol and letter designations used in the diagram, and the notes call attention to important facts relating to the circuit.

TECH TALK

Tech 1. There are just so many different parts in a furnace it is just really hard to know what they all do.

Tech 2. If you look at the wiring diagram in the corner there is a legend. The legend will tell you what each part's name is. From that if you have to you can look up and see what that part does. Because basically a gas valve does the same thing in every gas furnace, so does the fan switch, limit switch, and everything else. So once you learn where the parts are and how they are identified on the drawing it is easier to find them and easier to know what they do.

Tech 1. I see what you mean. Thanks.

UNIT 4—REVIEW QUESTIONS

1. List the three major parts of an HVAC/R system.
2. What are required when a device of one voltage is controlled by a device of another voltage?
3. What do intermediate controls consist of?
4. What is the most common size of transformer found in residential gas furnaces?
5. A transformer's capacity refers to _______.
6. What do the typical line voltage stats consist of?
7. What is system lag?
8. What are heating anticipators?
9. What does a sensing element consists of?
10. List the functions of valves.
11. What is a hot surface ignition unit used for?
12. In a gas fired heating system, what produces the heaviest load and is connected to the power supply?
13. What does the fan switch on a heating circuit control?
14. If the disconnect switch is closed can the blower motor run?

UNIT 5

Troubleshooting Gas Heating Systems

OBJECTIVES: In this unit the reader will learn about troubleshooting heating systems including how to:

- Identify Common Operating Problems
- Implement Systematic Troubleshooting of a Gas Furnace
- Identify Common Categories of Problems for Gas Furnaces
- Describe the Process of Troubleshooting Gas Furnace Solid State Controls
- State the Possible Cause of Gas System Problems
- Understand Troubleshooting Methods for Electrical or Mechanical Problems
- Analyze Malfunction Symptoms to Determine Cause of Problem
- Determine Remedial Action to Place System in Proper Operating Condition

INTRODUCTION

Any heating service organization is aware of the accumulation of service calls during the fall of the year. When homeowners first try to start their systems, the problems reveal themselves. The most serious of these is "no heat." Whenever a no-heat call is received, it receives top priority. It means that for some reason the furnace will not run. The technician must solve the problem as quickly as possible.

Preliminary information is essential. What type of furnace is being used? What type of fuel is being supplied? Has the unit operated properly in the past or is it a new installation? What has the customer done to get the unit started? Does any part of the unit run or is it completely "dead"?

A SYSTEMATIC ANALYSIS

The first thing to check is always the power supply. Is proper power being supplied to the unit? A voltmeter is a handy tool to use in checking the power supply. The second item to check is the thermostat. Is the thermostat calling for heat? If both of these conditions are satisfactory, the wiring diagram should be referenced.

Figure 10-5-1 shows a typical wiring diagram for a downflow gas furnace that meets the 80% AFUE requirement. The top portion of the diagram shows the low voltage wiring. The thermostat terminal connections to the furnace are shown on the top right. Toward the left in the diagram are the blower controls. Further to the left are the burner and ignition controls. At the extreme left is the control voltage transformer and the legend. In the lower part of the diagram are the line voltage connections to the circulating air blower and the combustion air blower.

Many technicians find it helpful to use a troubleshooting flowchart such as the one shown in Figure 10-5-2. The advantage of this procedure is the systematic process that it provides to help locate the problem.

TECH TIP

Manufacturers have done extensive research into the diagnostic testing of equipment. They provide this material in the form of flowcharts or troubleshooting tables. Using the manufacturer's troubleshooting information will make it much easier to locate and repair problems.

Gas Furnaces with Solid State Controls

Many of the newer furnaces use microprocessor controls and circuit boards as shown in the wiring diagram in Figure 10-5-3. The diagram on the left shows the component connections to the various parts of the printed circuit board. The diagram on the right shows the line voltage wiring at the top and the low voltage wiring, including the central processing unit (CPU), at the bottom.

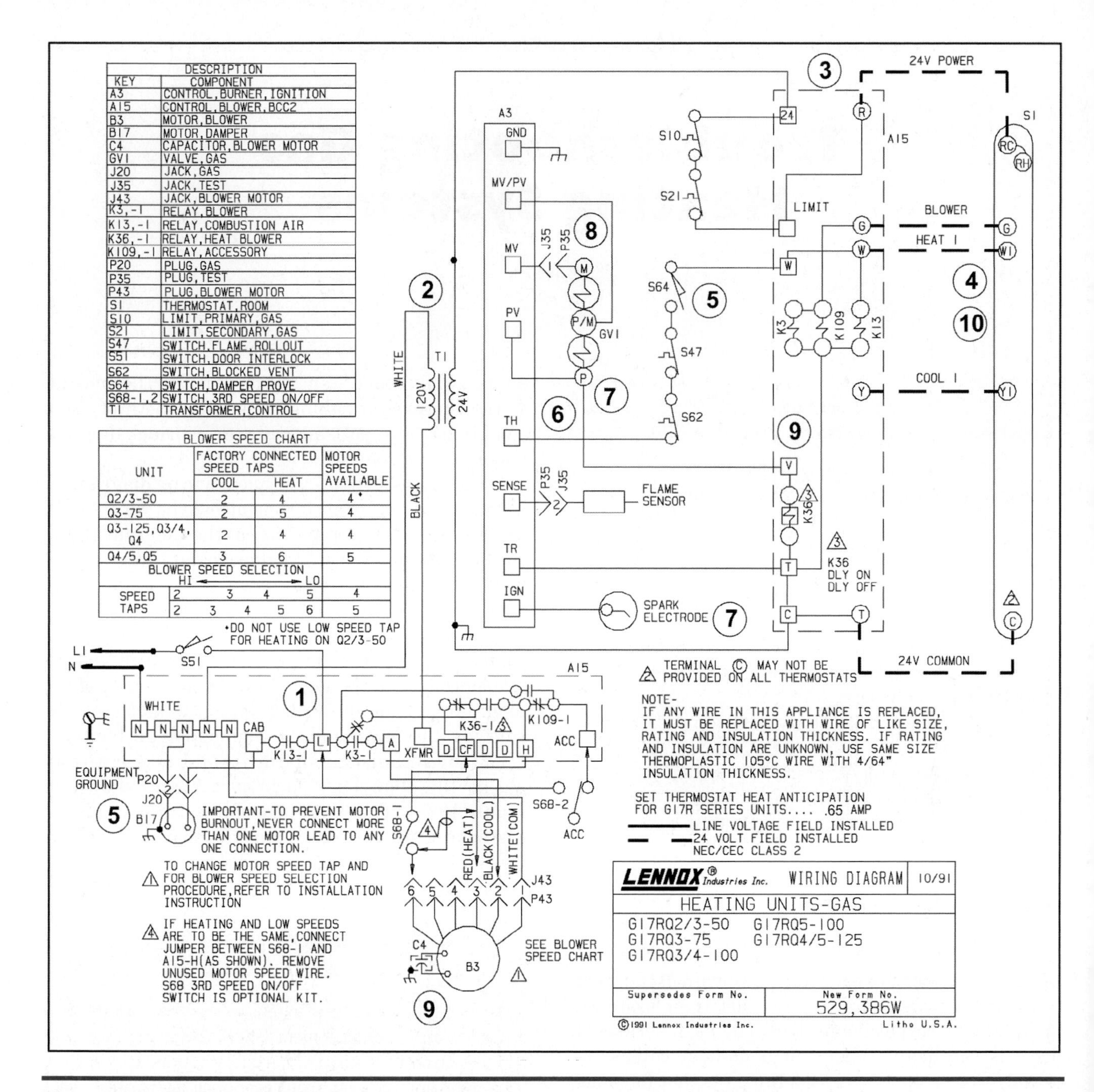

Figure 10-5-1 Wiring diagram for a down-flow gas furnace. (*Courtesy of Lennox Industries, Inc.*)

One helpful feature of the solid state control system is its ability to diagnose its own service problems. For example, in the system just described, a component test is available which allows all components, except the gas valve, to run for a short period of time to reveal any service problems or indicate a component failure.

As a preparation for troubleshooting, it is important for the technician to be completely familiar with the operating sequence of the furnace. This makes it possible to see what parts are operating, those that should be, and those not operating. The following is the sequence of operations for the heating mode of the furnace described in the wiring diagram in Figure 10-5-1:

Sequence of Operation

1. When disconnect is closed, 120 V feeds to line voltage side of the blower control (A15). Door interlock switch (S51) must be closed for A15 to receive voltage.
2. A15 supplies 120 V to transformer (T1).
3. T1 supplies 24 VAC to terminal "24" on A15. In turn, terminal "R" of A 15 supplies 24 VAC to terminal "RC" of the thermostat (S1).

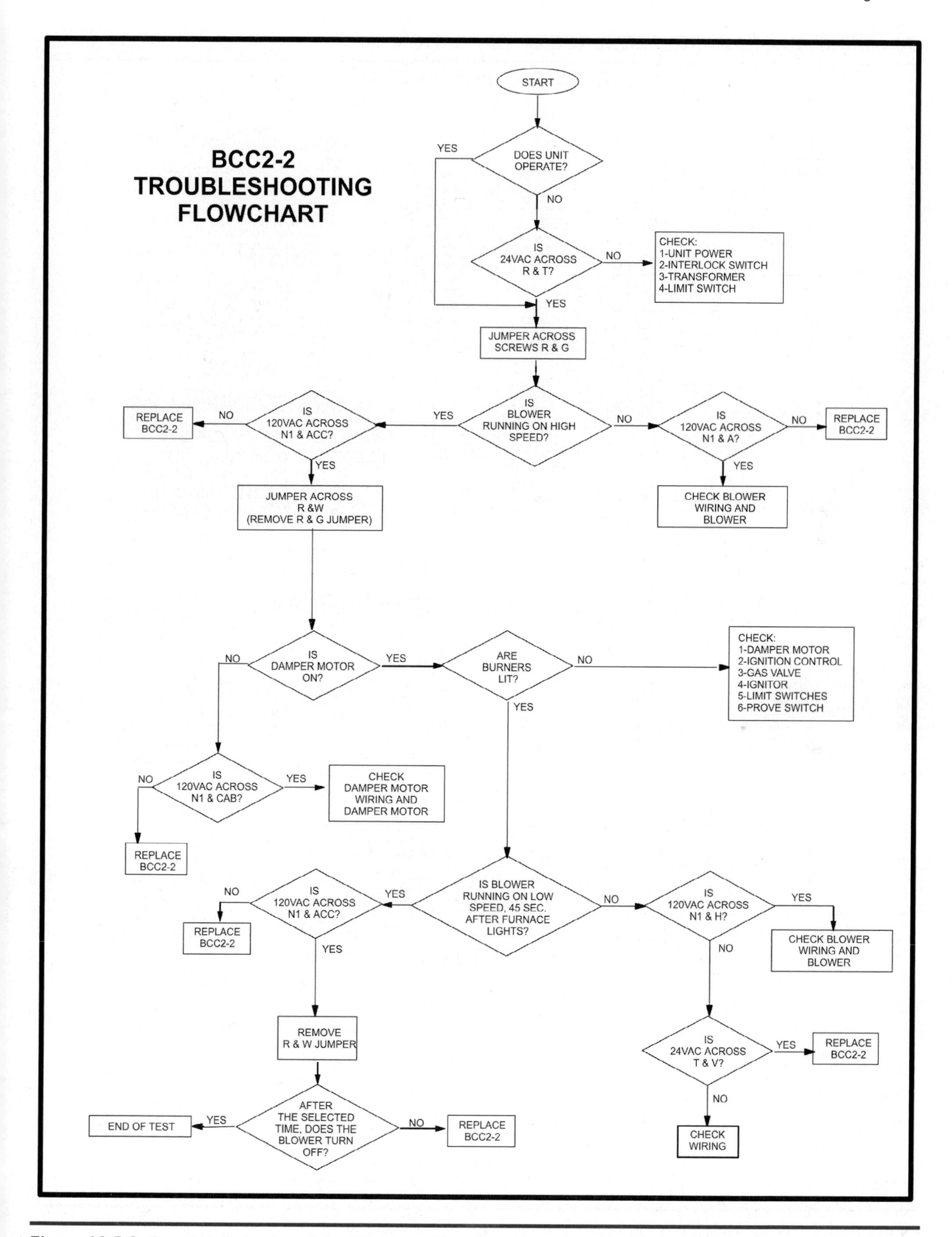

Figure 10-5-2 Troubleshooting flowchart. (*Courtesy of Lennox Industries, Inc.*)

VII-WIRING DIAGRAM AND SEQUENCE OF OPERATION

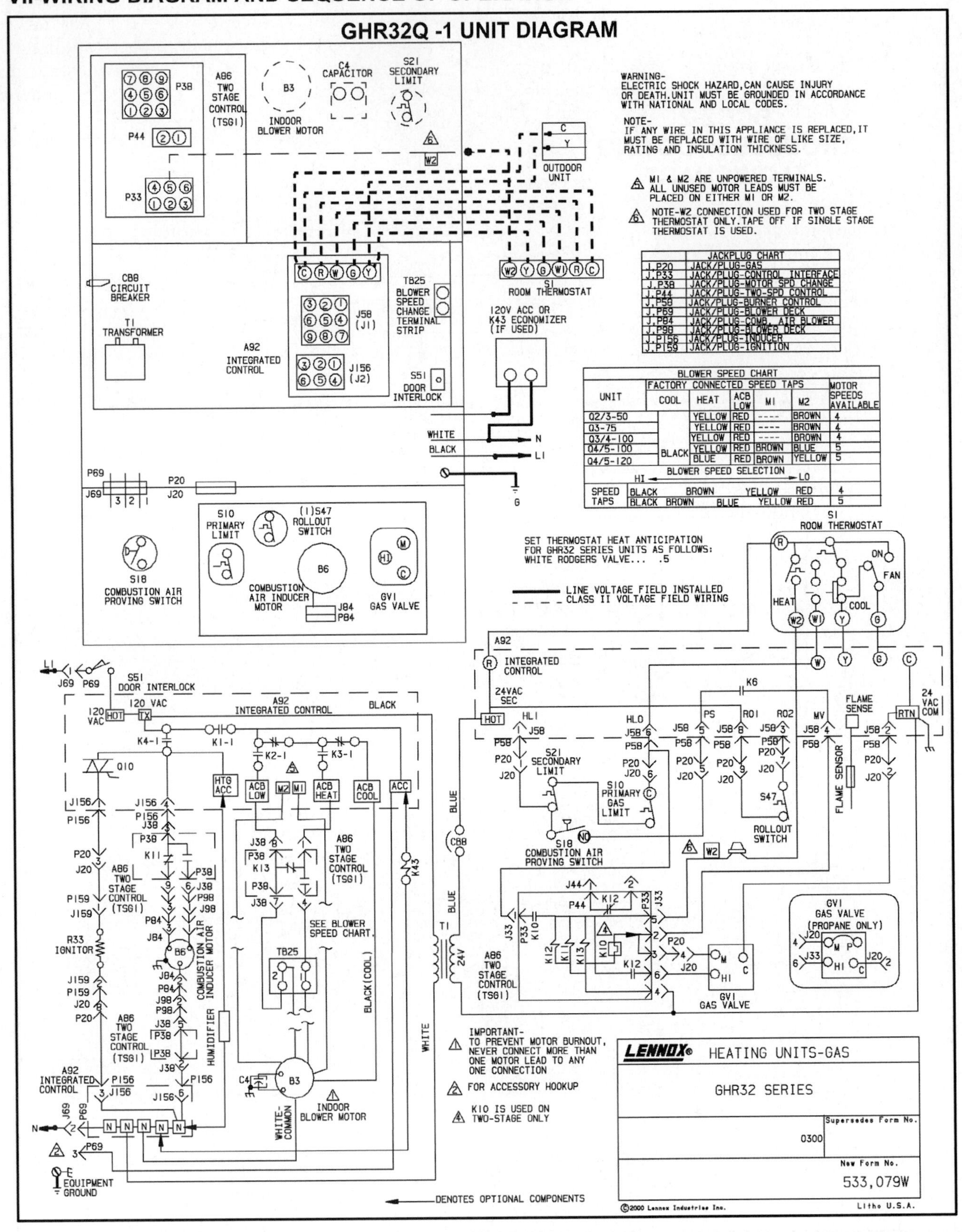

Figure 10-5-3 Wiring diagram of a gas furnace using microprocessor controls. *(Courtesy of Lennox Industries, Inc.)*

4. When there is a call for heat, W1 of the thermostat energizes W of the blower control with 24 VAC.
5. CAB of the blower control energizes the damper motor (B17) which opens the damper door. When door is in full open position, damper prove switch (S64) closes.
6. When S64 closes, assuming flame rollout switch (S47) and blocked vent shutoff switch (S62) are closed, 24 VAC is supplied to "TH" terminal of electronic ignition control (A3).
7. Through the electronic ignition control, the pilot valve "P" of the gas valve opens. The spark electrode ignites the pilot and the flame sensor senses the pilot.
8. When flame is sensed the main gas valve opens and supplies the burners with gas.
9. Terminal "V" (valve sense) of the blower control senses that the gas valve is energized and initiates a 45 second time delay. At the end of the 45 seconds the blower (B3) is energized.
10. When the heat demand has been satisfied, W1 of the thermostat deenergizes the gas valve and damper spring closes the damper door. As the damper door closes, the damper prove switch opens. The blower runs for a designated period (90-330 sec) as set by jumper on blower control.

NOTE: The ignition process will repeat several times before lockout occurs. Depending on the manufacturer's setup, lockout may automatically reset after an hour or more or it may have to be reset manually. Both the automatic and manual ignition lockouts can be reset manually by turning the power off for a few seconds, then powering up again.

SERVICE TIP

Many of the components found in medium and high efficiency furnaces contain computer chips. These chips are protected from accidental damage as a result of a minor static discharge. However, there may not withstand the current from an improperly installed jumper. Before jumpering terminals as part of board troubleshooting make certain that you are following the correct procedure.

Standard Furnaces

Since many thousands of standard furnaces are in residential use that were manufactured before the 80% AFUE minimum requirements, the service technician must be prepared to troubleshoot these units. Some important differences from the newer designs are natural draft venting and standing gas pilots. The troubleshooting information on standard furnaces is presented under the topics of (1) general procedures and (2) gas system problems.

Many of the procedures covered in this section also apply to the newer units.

GENERAL PROCEDURES

A suggested progression of analysis for troubleshooting starts with identifying the common categories of complaints or problems, then the possible problems in the system, followed by the symptoms and causes of specific problems. Causes can be either electrical or mechanical and related to the use of gas equipment.

TECH TIP

Before beginning the actual testing or troubleshooting of a unit take a few moments to check the system visually. Look for indications of arcing, burned or overheated wires, loose wires, or other easily visible mechanical signs of damage. Often the few minutes spent checking the system over can result in identifying the problem more quickly.

The categories of complaints or problems fall under the following headings:

1. Entire system operation.
2. Unit operation.
3. Burner operation.
4. Blower operation.
5. Heat exchanger complaints.
6. Cost of operation.
7. Noise.

GAS SYSTEM PROBLEMS

The possible causes for gas system problems follow.

Problems	Possible Causes*
Will not start	Season switch open Room thermostat set too low Disconnect switch open Blown fuse Limit control open Control transformer burned out (1)

(continued)

Problems	Possible Causes*
	Open circuit in thermostat Pilot outage (2) Gas valve stuck open/closed (3) Safety pilot burned out No fuel
Runs, but short cycles	Improper heat anticipator setting (4) Cycles on limit control (5) Gas input too low
Room temperature high	Thermostat setting too high Improper heat anticipator setting
Runs continuously	Short in thermostat circuit Gas valve stuck open
Blower cycles after thermostat is satisfied	Blower CFM adjustment Fan control setting
Blows cool air at start	Blower CFM adjustment Fan control setting (6)
Startup/cool down noise	Expansion noise in heat exchanger (11) Duct expansion Oil-can effect (12)
Noise from vibration	Blower wheel unbalanced or out of line Pulleys unbalanced (10) Defective blower belt Bearings burned out
Odor	Burning rust on start Improper venting Cracked heat exchanger
High fuel cost	Improper input Improper burner adjustment (7) Improper unit sizing
High electrical cost	Improper motor load Defective motor Incorrect motor speed
No fuel	No line pressure Defective regulator Supply valve closed
Gas valve won't open	Transformer burned out Open circuit in thermostat Gas valve stuck open/closed Defective safety pilot
Gas valve short cycles	Improper heat anticipator setting Improper limit control setting Dirty air filters Restriction in air supply
Delayed ignition	Improper burner adjustment Improper input Delayed valve opening Low line pressure
Pilot outage	Pilot orifice burned out Thermocouple burned out Safety pilot burned out Low line pressure Drafts
Extinction pop	Improper air adjustment Improper burner adjustment Improper orifice alignment Poor valve cutoff
Burns inside burner	Improper air adjustment Improper input Extinction pipe Leaking gas valve
Flame lift	Improper input Improper air adjustment
Flame roll out	Restriction in heat exchanger Improper venting Improper combustion air supply (8)
Yellow fire/carbon deposit	Improper air adjustment Improper input
Flashback	Improper input Improper air adjustment Improper venting (9) Restriction in heat exchanger
Intermittent blower operation	Improper CFM adjustment Improper fan control setting
Heat exchanger burn out	Improper input Chemical atmosphere Improper position of burners
Resonance (Pipe organ effect)	Resonance unit design

*See the number references for further information.

Many of the solutions to these problems are evident when the cause is determined. Below, however, is some further information regarding some of the remedies that can be used. Numbers refer to references above.

1. Control transformer burned out. An ohmmeter should be used to determine whether the primary or the secondary of the transformer is burned out. The cause is probably an overload in the secondary, which needs to be corrected before the transformer is replaced. If the overload cannot be reduced, a larger transformer must be used for the replacement.

2. Pilot outage. One of the most common causes of pilot failure is improper impingement of the flame on the thermocouple. See Figure 10-5-4. This figure illustrates the proper size and location of the flame. Sometimes the pilot is extinguished by the gas burner during lighting. It is actually blown out by the burner flame. To correct this it is often necessary to reposition the pilot to a more favorable location.

3. Gas valve stuck open or closed. A malfunctioning gas valve should be replaced.

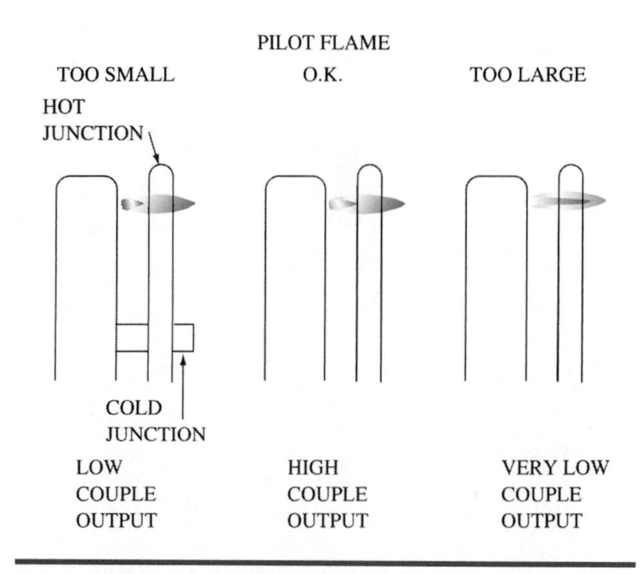

Figure 10-5-4 Proper size for a pilot flame.

4. Improper heat anticipator setting. The anticipator should be set at the amount of current traveling in the control circuit when the unit is operating. The current can be measured using a multiplier coil and clamp-on ammeter.

5. Cycling on the limit control. Occasionally a limit control will weaken and lower the operating range of the control. The normal range is to cut off between 140°F for a counterflow unit to 160 to 220°F on upright and horizontal units. If the control is cycling at a lower range, replace it.

6. Fan control settings. Almost all gas fired units operate best with fan control settings of 125 to 130°F fan on and 100 to 105°F fan off. If the unit blows cold air on startup, the fan on temperature can be changed to 145 to 150°F.

7. Improper burner adjustment. A properly adjusted burner will have approximately 40% of the combustion air mixing with 100% of the gas in the burner and 60% of the air mixing with the flame above the burner to complete the combustion process. Figure 10-5-5 shows a typical burner arrangement. As the gas is emitted from the orifice it expands and hits the proper place in the throat of the burner venturi. This produces maximum pull of primary air into the burner. Setting the burner for the correct flame condition will mean a minimum opening in the primary air control.

Figure 10-5-6 shows the size of the opening in an ordinary butterfly air control. When the burner and orifice are aligned properly and the burner is working correctly, a small opening in the primary air control will produce the soft blue flame that gives best overall unit performance.

An all blue sharp flame is receiving too much primary air. This means that there is less radiant heat

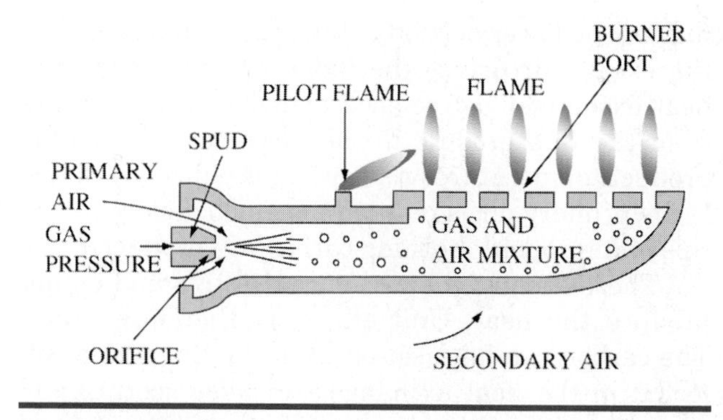

Figure 10-5-5 Primary and secondary air supply to the gas burner.

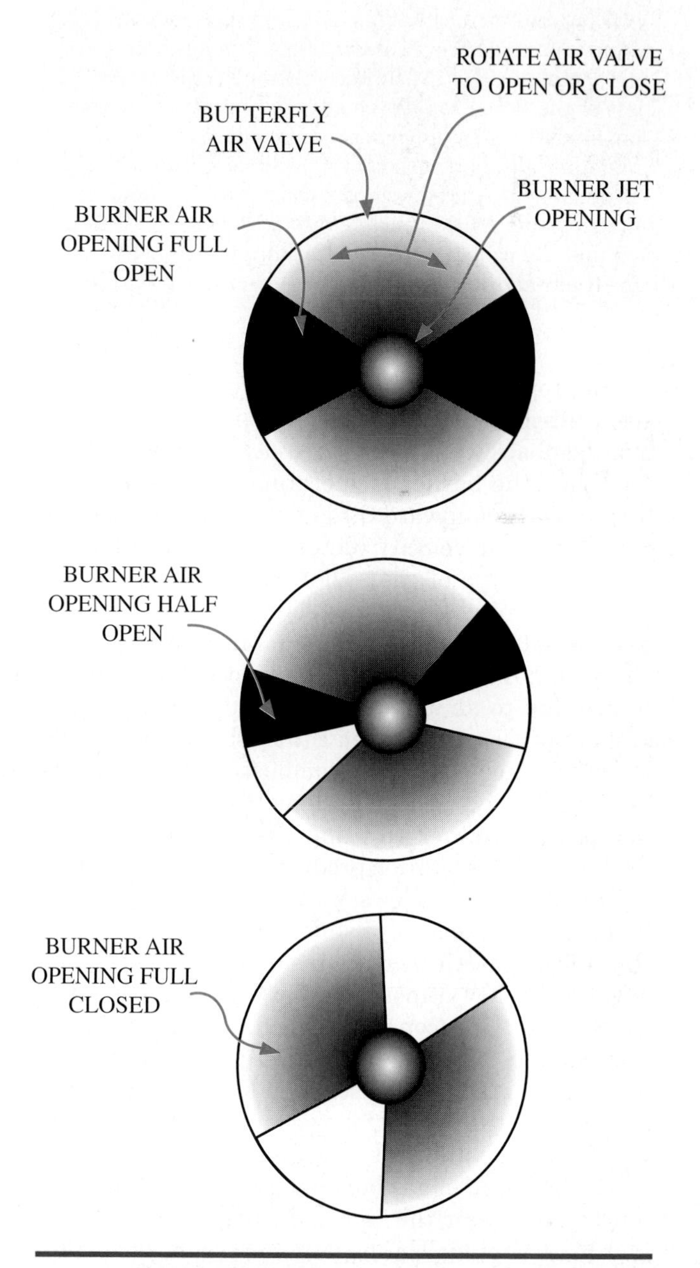

Figure 10-5-6 Primary air control opening size.

to heat the lower portion of the heat exchanger. Also, the excess air drives the flame products from the heat exchanger before good transfer of heat occurs from the flue product to the heat exchanger. Flue product temperatures rise and unit efficiency drops. If the primary air is reduced too much, heavy yellow tips of improperly burning carbon are produced.

These are much lower temperatures and do not produce the heat. Unit efficiency therefore drops. The carbon can be released from the flame and collected in the heat exchanger to cause sooting and plugging of the flue passages. The proper setting of the primary air quantity is the beginning step in producing high unit operating efficiency.

NOTE: Soot is most often caused by improper fuel air adjustments. However, if the flame impinges on internal metal parts of the furnace, then heat can be pulled out of the flame rapidly enough to retard proper combustion. Flame impingement can result in carbon forming. An example of carbon formed as the result of flame impingement can be seen if a copper wire is held in a match flame. Within moments black carbon will form around the copper wire as it withdraws the heat from the flame and retards the complete combustion process.

Improper setting of the primary air shutter can also contribute to pilot outage by producing extinction flashback, called *extinction pop*.

When the burner is operating, the gas/air mixture is blowing upward through the burner port at a given speed or velocity (determined by the burner design and type of gas). There is also a downward force or burning velocity, which is equalized by the gas/air outward velocity when burning is taking place. If, however, the burning velocity were to increase due to shutoff of the gas supply, the flame could approach the burner. Either the burner would absorb the heat below the combustion point and extinguish the flame or the flame would burn down through the burner port and ignite the mixture in the burner. This ignition produces the extinction pop.

Figure 10-5-7 shows what happens to the gas flame after the gas valve shuts off. At the moment of shutoff the gas/air mixture inside the burner is at a negative (below atmospheric) pressure. At full fire there is a full cone and a full tail of flame. Immediately after the gas valve closes, the burner pressure partially collapses. The full collapse of the fire down to the burner followed by extinction occurs when the burner absorbs the heat from the fire.

If the speed of gas burning is too high due to too much primary air, the flame does not collapse as rapidly as it should. The negative pressure is insufficient and there is an explosion of extinction pop within the burner when carry through occurs. This could cause a pressure wave over the pilot that blows the pilot out. A properly adjusted burner greatly reduces the chances of pilot outage.

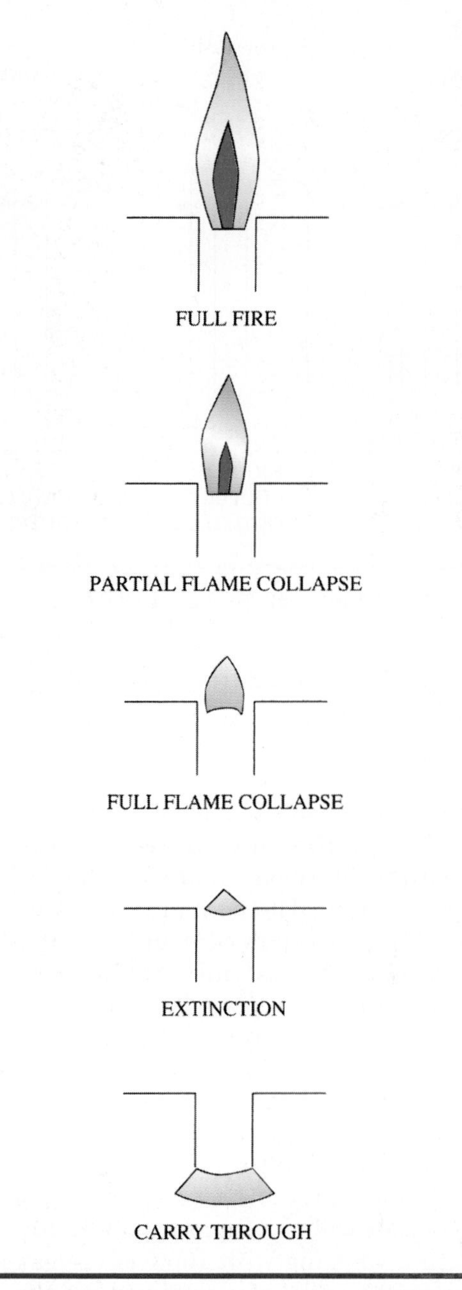

Figure 10-5-7 Types of flame action at the gas burner.

8. Improper combustion air supply. Excess air is needed for proper combustion, even though changes in gas pressure, heat content of the gas, and barometric pressure may occur. Draft conditions also change with barometric pressure and wind conditions.

When a unit encounters insufficient combustion air, the flame tends to become hazy and erratic and may even roll over the edge of the burner and out the burner pouch opening. The flame will seek air.

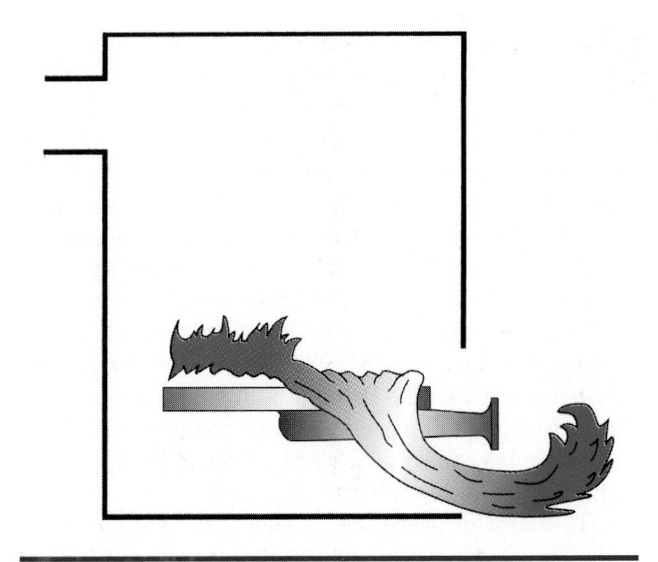

Figure 10-5-8 Rollout, due to insufficient secondary air.

Figure 10-5-8 shows the effect of insufficient secondary air causing a floating flame.

9. Improper venting. The mixture of gas and air produces a mixture of water, carbon dioxide, nitrogen, and excess air. All of this has to be removed from the heat exchanger. This removal process is called *venting.*

NICE TO KNOW

The inside lining of a double walled vent pipe is made of aluminum. Aluminum is used because it heats up very quickly once the furnace starts. If heavier metal were used that warmed up slower, the water vapor in the vent gases would condense inside of the flue pipe until it heated up. This condensed water would then run down and corrode the furnace and/or heat exchanger. Even with an aluminum lined vent pipe some condensation occurs and in extremely cold climates enough water can be condensed to still run down to the furnace. Fortunately, in cold climates the furnace will run long enough to evaporate most of that moisture before it has a chance to cause damage. Traces of the moisture can be found around the furnace vent. Technicians may assume that water is from a leaking roof stack, however often it is just condensation.

There are two types of venting: active (power) venting and atmospheric or passive (gravity) venting. *Active venting* uses a mechanical device such as a motor driven blower to either draw flue products from the heat exchanger or to force combustion air into the heat exchanger. The most popular type is the draw type, where the blower is mounted on the flue outlet and creates negative pressure in the outlet of the heat exchanger to get the desired combustion efficiency. Because the pressure difference is caused by mechanical power, wind and/or atmospheric conditions have little effect on the venting performance.

In *atmospheric* or *passive venting,* hot flue gases pass from the heat exchanger into a flue pipe, chimney, or vent stack. The driving force for a passive vent is obtained from the hot gases rising in the surrounding cooler air. The amount of force depends on the temperature of the hot gases and the height of the gravity vent. The hotter the gases and/or the higher the vent, the greater the amount of driving force or pull that is produced. Also, the greater the pull, the more secondary air is drawn through the heat exchanger.

Enough air must be drawn through to provide complete combustion as well as complete venting of the flue products. If too much air is drawn out of the heat exchanger before the correct amount of heat extraction is done, this results in higher flue temperatures and reduced unit efficiency.

If the passive vent pipe were connected directly to the flue outlet, the amount of air drawn through the heat exchanger would vary with factors like the pull of the vent stack, the wind effect on the vent stack, and outside temperature. Control of the venting rate on the heat exchanger would be impossible. Further, under some atmospheric conditions it may be possible to have a higher pressure at the outlet of the vent than the combustion process can overcome. This can produce poor combustion with the production of CO as well as the usual products of CO_2 and H_2O.

To overcome the effect of atmospheric conditions, all units use an opening in the venting system called a draft diverter. Figure 10-5-9 shows four typical heating unit draft diverters. These all consist of an opening from the flue outlet of the heating unit, an opening into the vent pipe, and a relief opening to the surrounding atmosphere.

Figure 10-5-10 shows the operation of a typical draft diverter under no-wind conditions and with updraft and downdraft conditions. The amount of flue products, the dilution air entering the relief opening and the amount of vent gases are indicated by the length of the arrows. With normal venting some air is pulled into the draft diverter by the pull of the passive vent. The mixture of flue products and surrounding air (called dilution air) that blows up the vent is called vent gas. The action of the draft diverter is to break the effect of the vent by introducing surrounding air and neutralizing the pull at the flue outlet. The heat exchanger then operates at approximately equal pressure from burner opening to flue outlet. The amount of air for combustion is then controlled by the flue restrictors.

If the conditions surrounding the vent stack increase the stack pull, additional air is drawn into

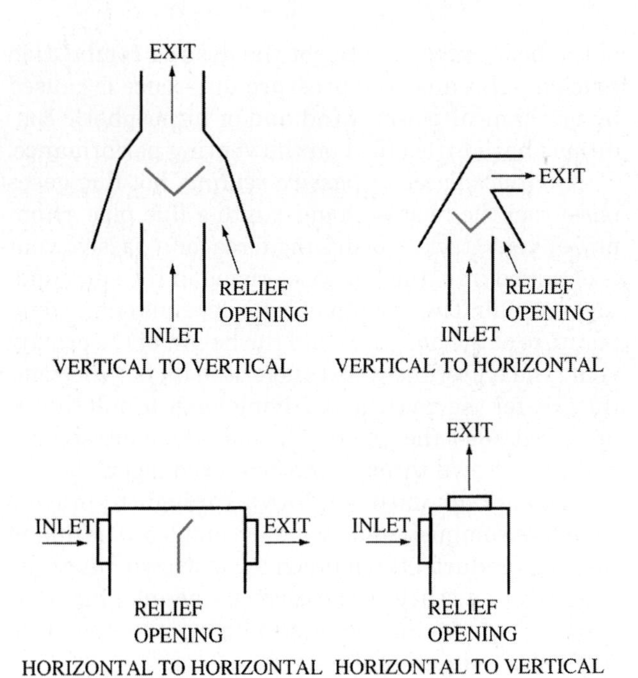

Figure 10-5-9 Typical gas appliance draft diverters.

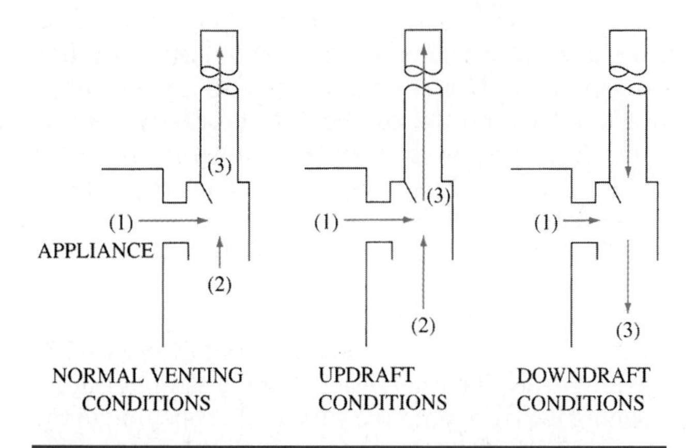

Figure 10-5-10 Operation of a draft diverter under various wind conditions.

the draft diverter to compensate for the increased pull. There is little effect on the heat exchanger performance.

Under conditions where the vent stack pull is reduced or even reversed, creating a downdraft, all combustion flue products are forced into the surrounding area. In addition, the increased pressure in the flue outlet will reduce the flow through the heat exchanger. This can cause incomplete combustion and produce odors carried by the gases and moisture produced by the combustion process. Even though no odors may result, the large amount of moisture produced in the combustion process can accumulate in the occupied area and create adverse living conditions or possible structural damage.

To check for proper operation of the vent system, use a candle placed below the bottom edge of the diverter opening. With the unit operating and up to temperature, the candle flame should bend in the direction of the opening in the diverter. If the flame is neutral, the draft is on the weak side. Possibly the vent stack is not high enough or large enough. If the candle flame bends outward, a draft problem definitely exists that must be corrected. If the vent stack cannot be lengthened or enlarged, a forced draft unit must be installed to overcome the problem.

10. Blower drive. Belt driven blowers have a higher probability of vibration problems than direct drive blowers due to the additional parts involved. The most common problem is due to belt tension. It is commonly believed that the tighter the belt, the better the performance, but the opposite is true. The tighter the belt, the harder the motor has to work to get the belt in and out of the pulleys. The belt should therefore be as loose as possible without slipping on startup.

Figure 10-5-11 gives the test for proper belt tension. It should be possible to easily depress the belt midway between the motor shaft and blower shafts $^3/_4$ to 1 in for each 12 in of distance between the

Figure 10-5-11 Aligning pulleys and checking belt tension.

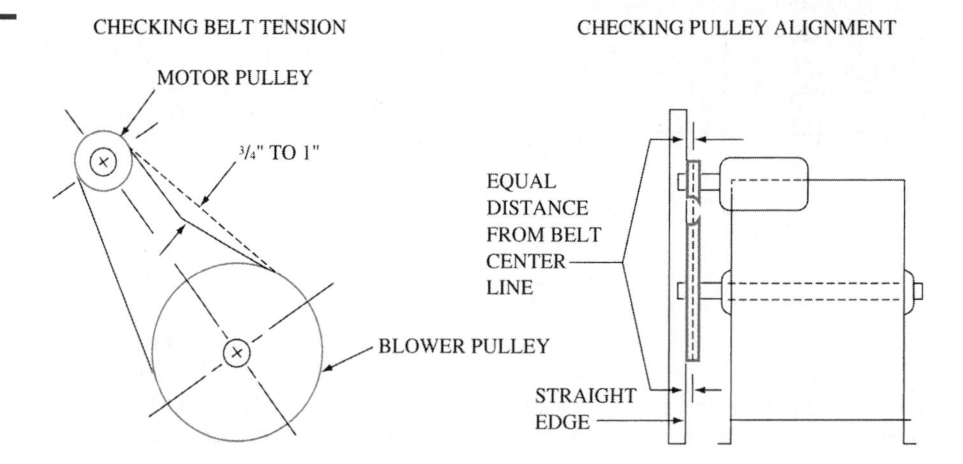

shafts. Alignment of the motor and blower pulleys is important to keep vibration to a minimum as well as to reduce wear on the sides of the belt.

Finally, each pulley, both motor and blower, should be checked for running true. Any warpage that creates wobble in the pulley requires replacement of the pulley.

11. Expansion noise in heat exchanger. Figure 10-5-12 shows a four section unit, each section composed of a right-hand and a left-hand drawn steel "clamshell" welded together. The sections are then welded into an assembly by fastening to the front mounting plate and rear retainer strap. Sometimes in the welding process stresses will be set up if the two metals are at different temperatures when the bond is made. This results in expansion noises, ticking, and popping as the heat exchanger heats and cools. Most of the time, these noises are muffled by the unit casing and duct system to a level where they are not objectionable.

In extreme cases, it is possible to reduce the noises by operating the unit with the blower disconnected, allowing the limit control to turn the unit off and on. This should be done through several cycles of the limit control. Cycling on this extreme heat will cause metal to stretch beyond the normal operation range and eliminate the sound. If this does not produce satisfactory results, the only cure is to change the heat exchanger.

12. Oil-can effect. This effect is caused by the sudden movement of a flat metal surface where a forming stress has been left in the surface. This stress causes the metal to have a slightly concave or convex position rather than a flat plane surface.

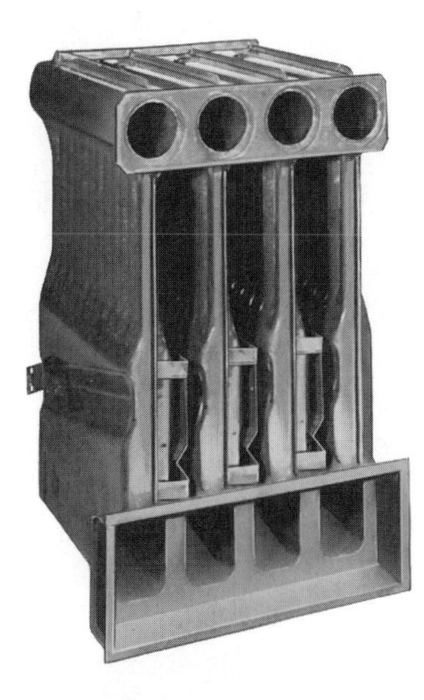

Figure 10-5-12 Clamshell gas heat exchanger. *(Rheem Manufacturing Air Conditioning Division)*

Temperature change will cause a stress increase in the material until the metal rapidly changes position to the opposite of its original position. This change will produce a loud bang. Ductwork is very prone to this action and must be cross broken over any large panel areas. Unless this surface is cross broken, it is subject to the oil-can effect. The best correction is removal of the panel and cross break it to relieve temperature stresses.

TECH TALK

Tech 1. I have no idea how you troubleshoot a furnace. There are just so many wires and so many controls it's totally baffling. I will never get this figured out.

Tech 2. Well, the first thing you need to do is get the diagram and take a look at it. Look at the legend so you know what the parts are and where they are shown on the diagram. Next take a look at the troubleshooting chart that the manufacturer supplies. See what the furnace is doing and what it is not doing and you can use that chart most of the time and tie the problem right off.

Tech 1. Yea, I have done that. But then when I try to find the part by chasing the wire I am just lost.

Tech 2. The easiest thing to do is take the diagram and see where the wire starts and where it goes. Once you have done that go to the equipment, find that wire where it starts and where it goes. You may even want to tag it. You can put a wire tie on it or a piece of tape, anything so you will know what it is when you come back to it. The next thing you do is find the next wire and follow it to where it goes. Do that to all the wires that come off the part that you think may be bad. That will let you see where they go and what they come off of.

Tech 1. Gee, that takes a whole lot of time.

Tech 2. It may take a whole lot of time the first time that you do it but after the first time you do it you will get a lot faster.

Tech 1. So, I guess you are saying practice makes perfect.

Tech 2. You shouldn't expect to be perfect right away, but practice makes troubleshooting go a lot faster.

Tech 1. Thanks, I appreciate that.

UNIT 5—REVIEW QUESTIONS

1. What is a voltmeter used for?
2. Describe the suggested progression of analysis for troubleshooting.
3. List the headings that complaints or problems fall under.
4. What could be the possible cause in a gas heating system if there is noise from vibration?
5. What could be the possible cause for in a gas system if there is a delayed ignition?
6. What should be used to determine whether the primary or the secondary of the transformer is burned out?
7. A properly adjusted burner will have approximately _______% of the combustion air mixing with _______% of the gas in the burner and _______% of the air mixing with the flame above the burner to complete the combustion process.
8. What happens if the speed of gas burning is too high due to too much primary air?
9. Describe the two types of venting.
10. What happens if too much air is drawn out of the heat exchanger before the correct amount of the heat extraction is done?
11. Why do belt driven blowers have a higher probability of vibration problems than direct drive blowers?
12. Describe the oil-can effect.

11

Central Residential Oil Warm Air Heating

UNIT 1

Oil Fired Forced Air Systems

OBJECTIVES: In this unit the reader will learn about oil fired forced air systems including:

- Types of Furnaces
- Ratings and Efficiencies
- Oil Storage
- Primary Oil Burner Controls
- Oil Valves

INTRODUCTION

Oil has been a popular fuel for heating in the northeast and northwest portion of the country. It is also widely used in rural areas where gas is not available and where the use of electricity is not practical or economical.

TYPES OF FURNACES

Like gas furnaces, oil forced air furnaces are designed in five basic configurations: upflow, Figure 11-1-1, lowboy, downflow (counterflow), horizontal, and multi-position models.

OIL

Fuel oils are mixtures of hydrocarbons derived from crude petroleum by various refining processes. They are divided into grades according to their characteristics and viscosity. Other properties like flash point, pour point, water and sediment content, carbon residue, and ash are all important. The *viscosity* determines whether the fuel oil can flow or be pumped through lines or if it can be atomized into small droplets.

For comfort heating applications, we are primarily interested in two grades of fuel oil: No. 1 and No. 2. Both contain high quantities of carbon and hydrogen with traces of sulfur. The carbon content for these fuel oils ranges between 84% to 86% with a maximum of 1% sulfur, Figure 11-1-2. The sulfur content of fuel oils is kept as low as possible to reduce air pollution.

No. 1 grade fuel oil is considered premium quality. It is used in room space heaters, which do not use

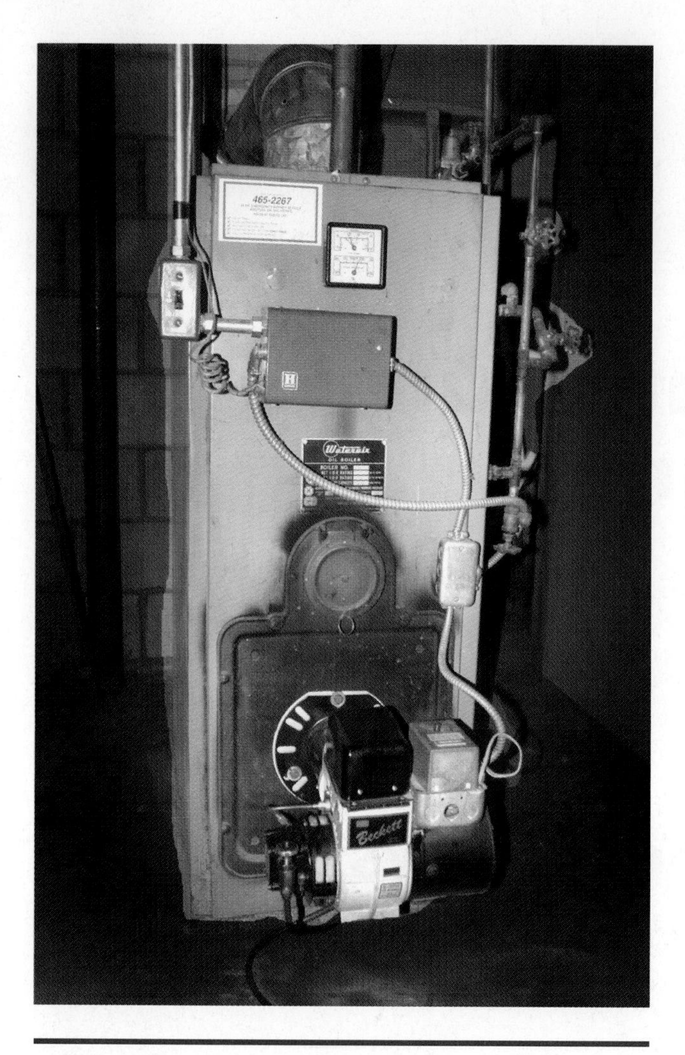

Figure 11-1-1 An upflow oil furnace.

SERVICE TIP

All heating oil has a very distinctive odor. You must be very careful when working in residences to not contaminate the residence property with oil. Careless work habits can result in the home smelling of oil for days after a technician leaves. Often oil fired furnaces are located in basements, some of which do not have outside access. If it is necessary for you to access the system through the residence, you should wear paper shoe covers as you enter and leave the residence and each time you enter and leave the basement. In addition, you must clean your hands thoroughly before entering the residence to adjust the thermostat, set dampers or vents and any other such service.

high pressure burners and depend on gravity flow, thus the need for the lower viscosity.

No. 2 grade is the standard heating oil sold. No. 2 oil is used in equipment that has pressure atomiz-

Figure 11-1-2 Carbon content of fuel oil and quantity of heat produced.

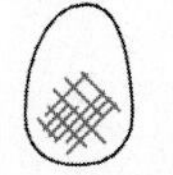

ing, which includes most forced air furnaces and boilers. The heating value is approximately 135,000 to 142,000 Btu/gal.

NICE TO KNOW

It's often assumed that because heating oils are thicker than water they are heavier than water. This is not true. Oil floats on water. Although it is thicker than water it is less dense. Most heating oils range between 6.8 and 7.2 lb per gallon as compared to water, which weighs 8.3 lb per gallon.

Heat Exchanger

The typical oil fired heat exchanger, Figure 11-1-3a,b, is a cylindrical shell of heavy gauge steel or cast iron in which combustion takes place. It offers additional surfaces for heat transfer from the products of combustion inside, to the air around the outside of the chamber. This type of heat exchanger is called a drum and radiator. The inside, containing the flame, is called the primary surface, and the outside is called the secondary surface.

Some manufacturers add baffles, flanges, fins, or ribs to the surfaces to provide faster heat transfer to the air passing over the surfaces.

The burner assembly is bolted to the heat exchanger. The firing assembly and blast tube extend into the primary surface in correct relationship to the combustion chamber or refractory. A flame inspection port is provided just above the upper edge of the refractory. This is used to observe proper ignition and the flame, and to measure over fire draft for startup and service operations.

SAFETY TIP

Never open the inspection port unless the power to the furnace is off, or there is a flame.

Refractory

High temperatures are required in the flame area of the heat exchanger to produce maximum burning

(a)

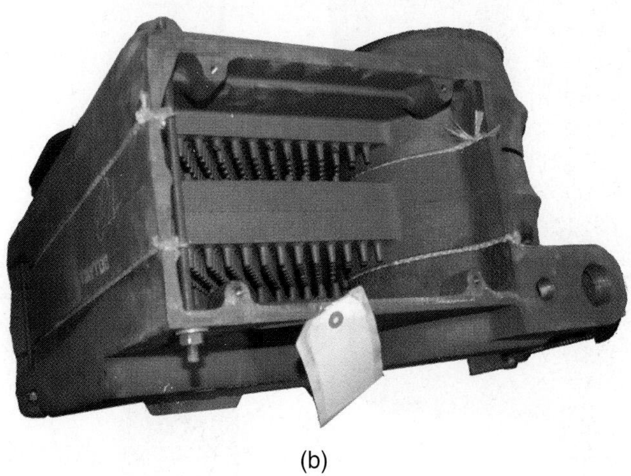

(b)

Figure 11-1-3 (a) Oil furnace cast iron firebox; (b) Oil furnace heat exchanger.

efficiency of the oil/air mixture. To obtain this high temperature, a reflective material, called the refractory, is installed around the combustion area. It is an insulation type of material designed to reach white hot surface temperatures quickly, with minimum deterioration. A refractory is shown in Figure 11-1-4. It should be noted that you should never use a vacuum cleaner on this type of heat exchanger.

Figure 11-1-4 White refractory insulation in burner chamber.

SERVICE TIP

Occasionally the refractory fire box on an oil furnace can be flooded with oil. When this happens the excess oil must be removed before the furnace can be restarted. With the oil burner assembly removed, use dry paper towels or rags without any quantity of liquid to absorb the remaining oil before attempting to relight the furnace. Never attempt to relight a furnace with liquid oil in the bottom of the refractory fire box. If the oil in the refractory fire box is ignited, it is very difficult to extinguish.

CAUTION: Only CO_2 type extinguishers should be used on oil burner fires because they leave no residue. Dry chemical fire extinguishers will put out an oil fire but the residue can cause an extensive and lengthy cleanup.

Burner

The high pressure atomizing gun burner shown Figure 11-1-5 is the type used on most residential and small commercial oil fired heating systems. Pressure oil burners burn with a yellow flame, as shown in Figure 11-1-6. Flame retention burners burn with a blue flame. Pressure oil burners have the following components:

1. Oil pump
2. Air blower
3. Electric motor

4. Ignition transformer
5. Blast tube, with nozzles and ignition system
6. Primary control

The burner illustrated in Figure 11-1-7 shows a detailed view of the operation. The pump and blower are driven from a common motor shaft. The oil pump is a gear arrangement. By changing a bypass plug in the housing, the pump can operate as a single stage or a two stage system. Single stage operation is used where the oil tank is above the burner and gravity oil feed to the burner is permitted. Two stage operation is needed where the oil tank is below the burner and the pump must lift the oil from the tank as well as

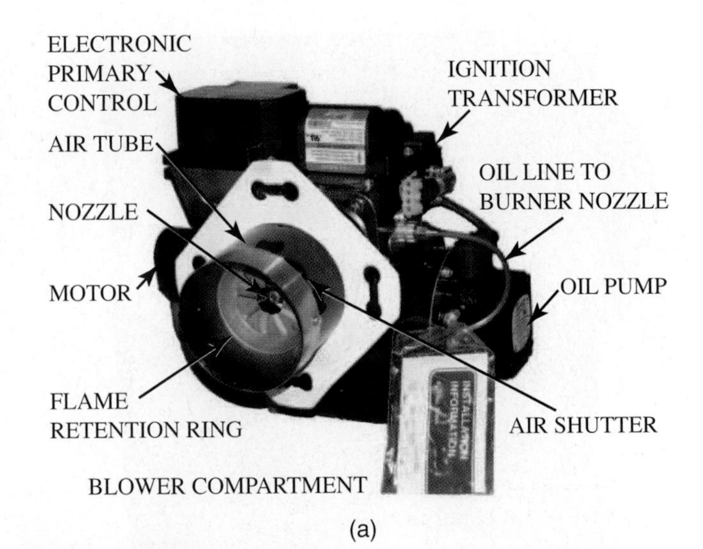

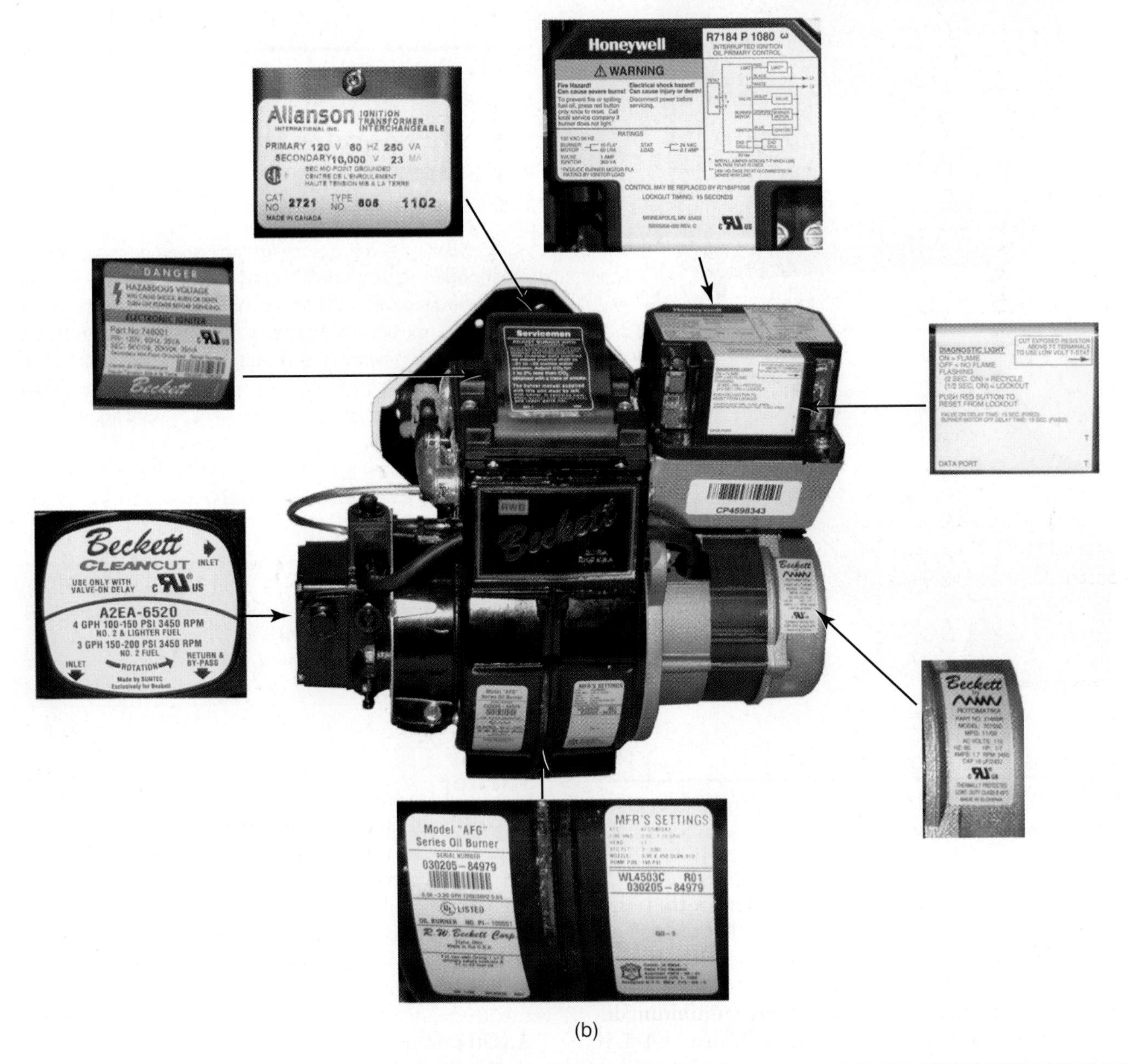

Figure 11-1-5 (a) High pressure atomizing gun type oil burner with parts labeled; (b) Typical labels on the high pressure atomizing gun type oil burner.

Figure 11-1-6 Oil burner flame.

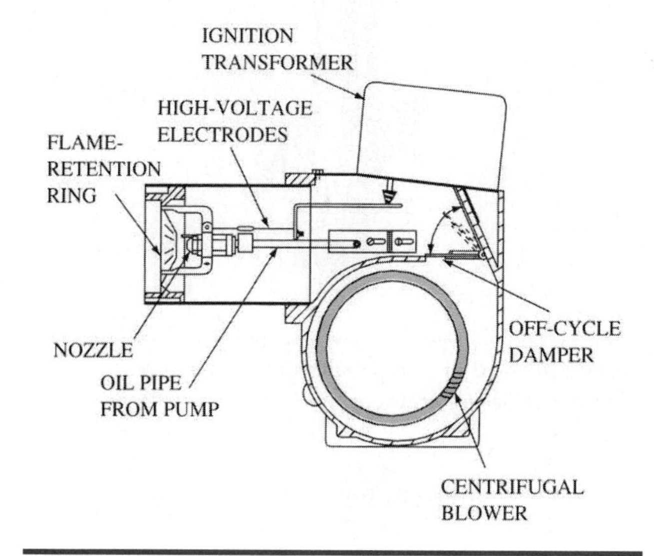

Figure 11-1-7 Cutaway view of the internal construction of an oil burner.

furnish pressure to the nozzles. The pump supplies oil to the nozzle at 100 to 300 psi.

The burner blower includes a centrifugal wheel also mounted on the common motor shaft. It furnishes air through the blast tube and produces air fuel turbulence for proper combustion. The amount of air is controlled by sliding a shutter band on the blower housing section, as shown in Figure 11-1-8.

Oil is pumped to the nozzle, which is mounted in an adapter, as shown in Figure 11-1-9. The orifice in the nozzle is factory bored to produce the correct firing rate, and must not be altered or changed in any way. The firing rate is approximately 0.8 gal/hr for each 100,000 Btu/hr output of the furnace. Under high pressure, the oil is atomized into fine droplets and mixes with the primary air.

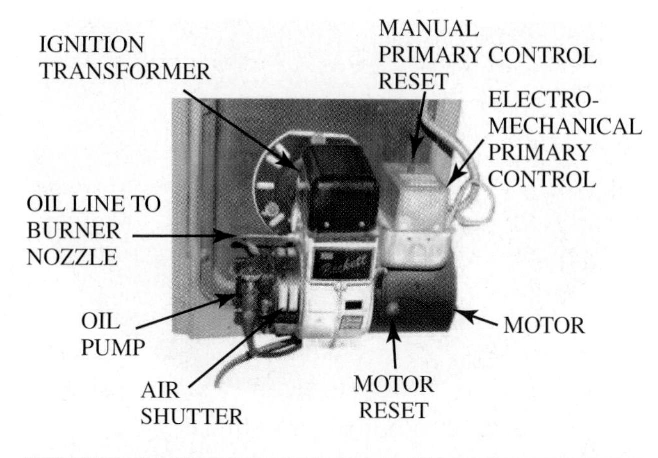

Figure 11-1-8 The air shutter shown at bottom left adjusts airflow for this old style oil burner.

The air deflector vane ring (turbulator) rotates or spins the atomized oil and air into the heat exchanger. Ignition by a high voltage electric spark is continuous when the motor is on or interrupted on start only combustion. The ignition transformer is located on the burner housing.

A safety device operates when there is an ignition failure. When there is no flame (light), a light sensitive cad cell will stop the burner motor, thus preventing oil from flowing into the heat exchanger. Some furnaces have a lockout system requiring the control to be reset manually before the burner can run again.

Flue Venting

Oil fired furnaces must have an ample supply of makeup air for combustion, and the methods of introducing makeup air for a gas furnace will also be adequate for oil equipment.

Masonry chimneys used for oil fired furnaces should be constructed as specified in the National Building Code of the National Board of Fire Underwriters. Prefabricated lightweight metal chimneys, Figure 11-1-10, are also available. These are rated for all fuels, Class A and Class B. Class A flues are used for solid and liquid fuels while Class B flues are made specifically for gas fired equipment.

The above roof dimension of the vent or chimney is the same for both gas and oil. Since oil fired furnaces operate on positive pressure from the burner blower, it is important to have a chimney that will develop a minimum draft of 0.01 to 0.02 in WC as measured at the burner flame inspection port. Consistency or stability of draft is also more critical, and the use of a barometric damper, Figure 11-1-11a–c, is required. The damper is usually installed

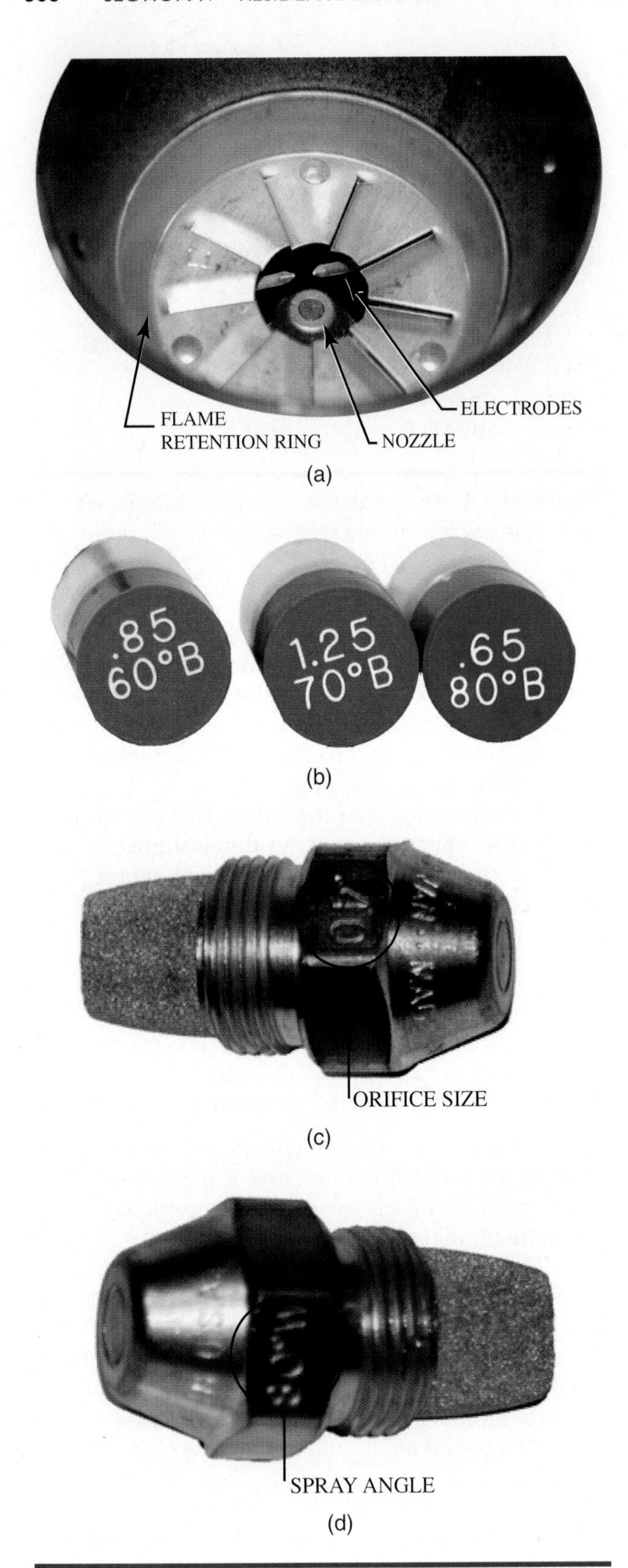

Figure 11-1-9 (a) Nozzle, electrodes, and air; (b) Various sizes and types of nozzles; (c) Nozzle size is stamped on the nozzle; (d) Nozzle angle is stamped on the nozzle.

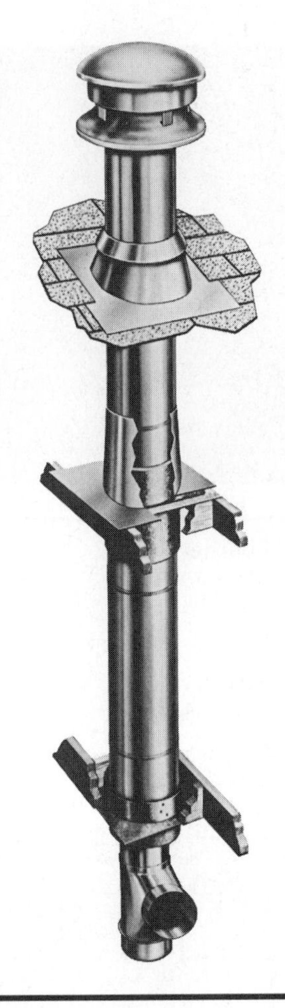

Figure 11-1-10 Factory built, all fuel chimney. *(Courtesy of Selkirk Metalbestos)*

in the horizontal vent pipe between the furnace and chimney. It may also be attached directly to the furnace flue outlet. The damper has a movable weight so that it can be set to counterbalance the suction and to maintain reasonably constant flue operation. It is adjusted while the furnace is in operation and the chimney is hot. The over fire draft at the burner should be in accordance with the equipment installation instructions. It is used to control draft, but will not help if the draft is insufficient.

Clearances

Cabinet temperatures and flue pipe temperatures run warmer for oil burning equipment, so clearances from combustible material should be adjusted accordingly. One inch clearances are common for the sides and rear of the cabinet, compared to zero clearances for many gas units. Flue pipe clearances of 9 in or more are needed, whereas only 6 in of clearance are needed for gas.

Front clearance is principally space needed to remove the burner assembly. Floor bases over

(a)

(b)

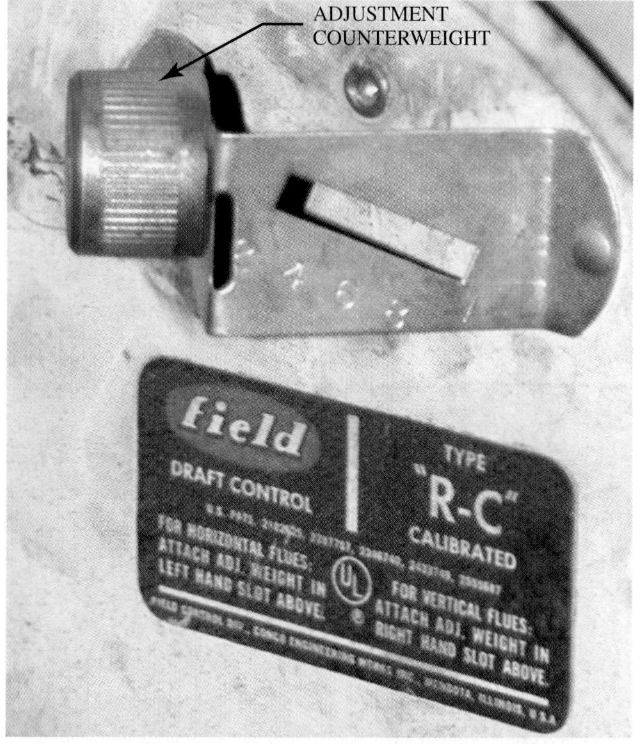

(c)

Figure 11-1-11 (a) Barometric damper, shut; (b) Barometric damper, open; (c) Damper adjustment counterweight.

combustible surfaces are generally increased for oil as compared to gas. For example, in horizontal oil furnaces installed in attics, the installer must pay particular attention to recommendations related to adjoining surfaces.

Clearances are an important part of the compliance and inspection procedures required by Underwriters' Laboratories (UL) for securing their seal of approval.

> **CAUTION:** Many oil fired furnaces are located in open basement areas where they are subject to having items stored on, against or near them. Such storage constitutes a major fire hazard and must be removed before you light the furnace. Caution the home owner or resident against such storage practices.

Ratings and Efficiencies

Oil burning furnaces are limited in the number of sizes available due to the restrictions in available oil burner nozzles. Table 11-1-1 lists some examples of models available from a typical manufacturer.

The efficiency ratings given in Table 11-1-1 are for the mid-efficiency furnaces. These new units are very similar in construction to the standard furnaces previously manufactured with certain improvements

TABLE 11-1-1 Examples of Mid-efficiency Ratings for Furnaces

Input (Btu/hr)	Output (Btu/hr)	AFUE% (ICS)	Application
70,000–154,000	56,000–125,000	80%	Upflow
105,000–210,000	85,000–166,000	Up to 80.6%	Basement (lowboy)
105,000–154,000	84,000–125,000	80%	Downflow or horizontal

to produce the new ratings. Some of the new features that have been added to increase the efficiency include:

1. A larger heat exchanger drum.
2. Electronic fan control, Figure 11-1-12, which turns blower on 60 to 90 seconds after thermostat demand.
3. Flame detector (cadmium sulfide).

Oil Storage

The installation of the fuel oil tank and connecting piping must conform to the standards of the National Fire Protection Association (NFPA) and to local code requirements. Regulations and space permitting, the oil tank can be located indoors as shown in Figure 11-1-13. Note the use of a shutoff valve and a filter to catch impurities.

If the oil tank is placed outdoors above ground, firm footings must be provided. The exposed tank and piping are subject to more condensation of water vapor and the possibility of freezing.

If the oil tank is installed below ground, Figure 11-1-14, it is important to keep it filled with oil during periods when the ground is saturated with water. Otherwise, high groundwater may force it to float upward. A heavy concrete cover over the tank may be advisable. Copper piping may provide flexibility if ground movement should occur.

TECH TIP

Because of environmental concerns many local and state governments have outlawed the installation of oil storage tanks below ground. Check with the local building official to see whether or not it is possible to install a tank underground. Another reason to check is to find out whether an underground tank can be replaced.

Note the use of suction and return lines in the illustration. It is normal practice to use a two stage fuel pump with a two pipe system whenever the oil supply is below the level of the burner.

The oilburner pump suction is measured in inches of mercury vacuum. A two stage pump should never exceed a 15 in vacuum. Generally, there is 1 in of vacuum for each 1 ft of vertical oil lift and 1 in of vacuum for each 10 ft of horizontal run of supply piping.

Typically most residential storage tanks, Figure 11-1-15, are located in the resident's basement. These storage tanks are typically made of steel and contain 275 to 1000 gallons of fuel oil. The 275 gallon tanks are the most commonly used in residential settings.

There are two pipes connected to the top of the oil storage tank. One pipe is used for filling the tank with oil, and the other pipe is a vent. Figure 11-1-16a,b shows the typical arrangement of these pipes outside the residence. The fill pipe has a cap that is easily unscrewed so that the fill nozzle from the truck can be attached to it, as shown in Figure 11-1-17. The vent pipe has an alarm whistle that stops sounding when the level of oil in the tank rises up to the fill level in the tank, as shown in Figure 11-1-18.

Some commercial installations may have more than one set of fill and alarm pipes, as shown in Figure 11-1-19. These fill and alarm pipes operate with the same system of indicating when the tank has been filled. An electronic alarm can be used to replace the mechanical air whistle on some systems, as shown in Figure 11-1-20.

Figure 11-1-12 Electronic fan control center for an oil fired furnace.

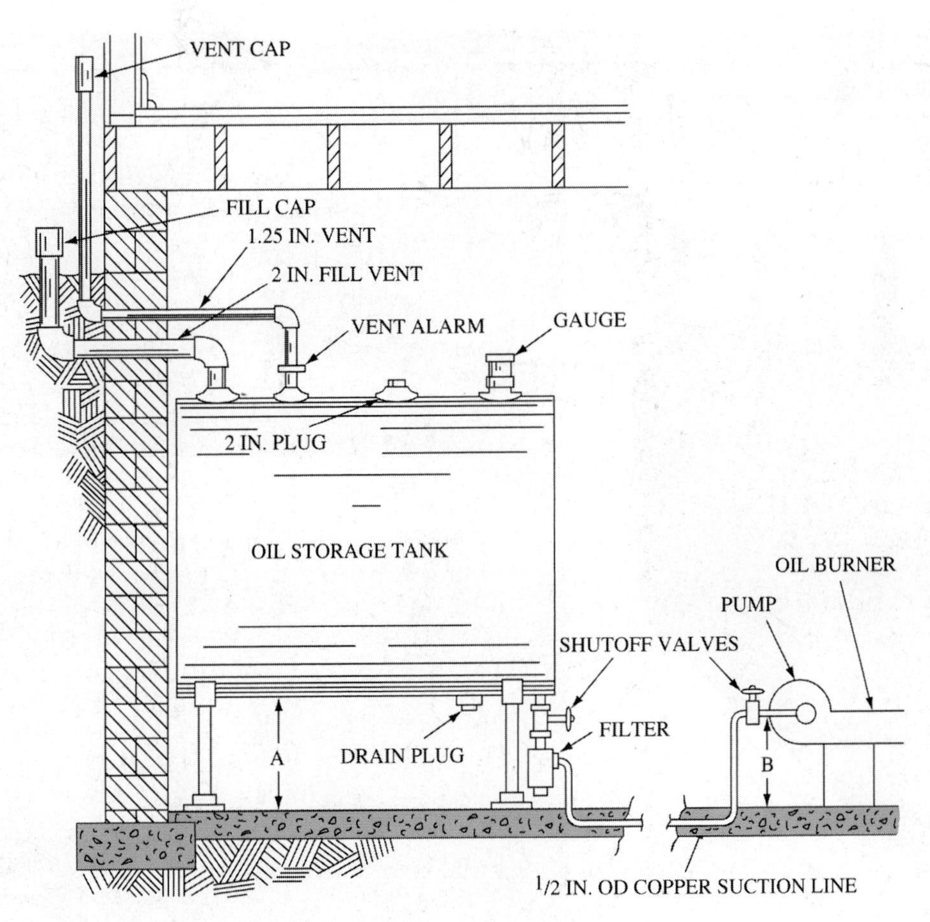

Figure 11-1-13 Indoor fuel oil storage tank.

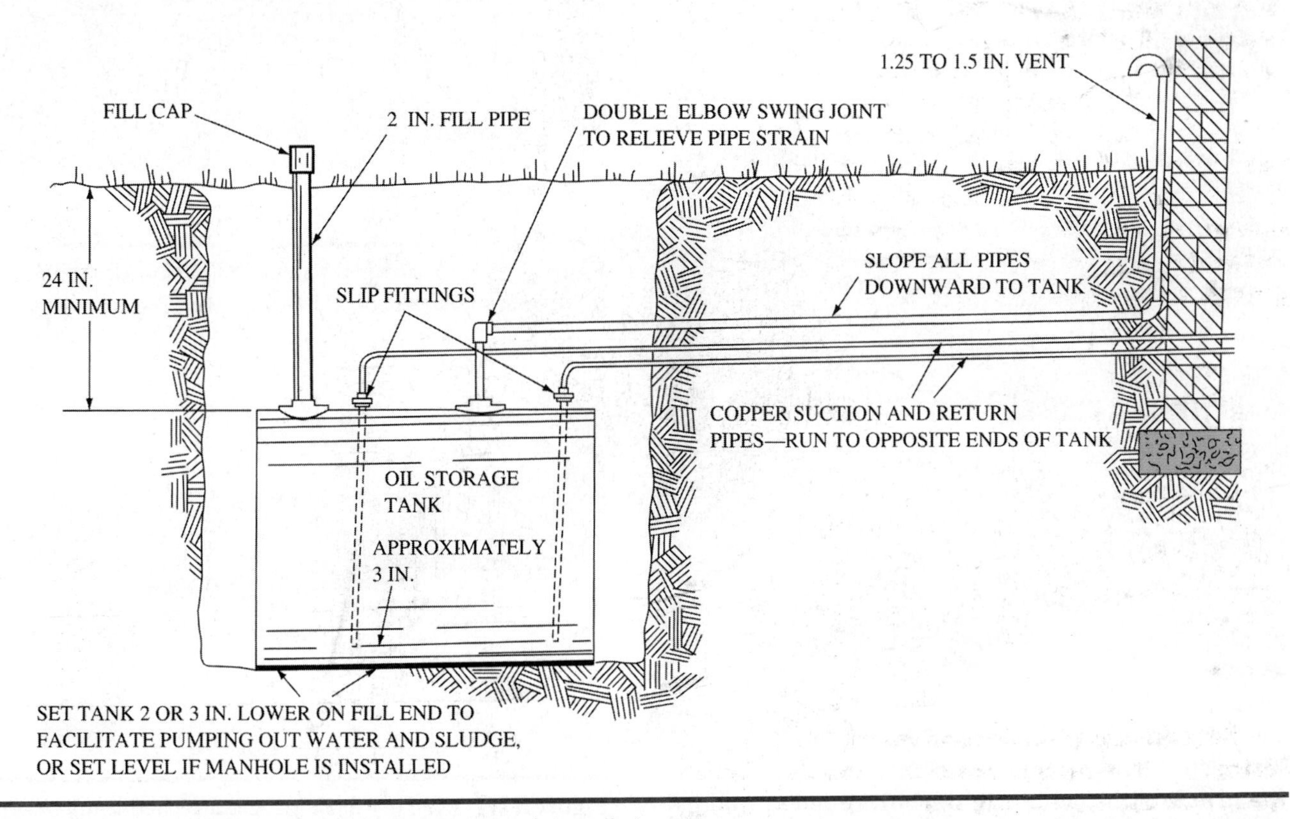

Figure 11-1-14 Underground fuel oil storage tank installation.

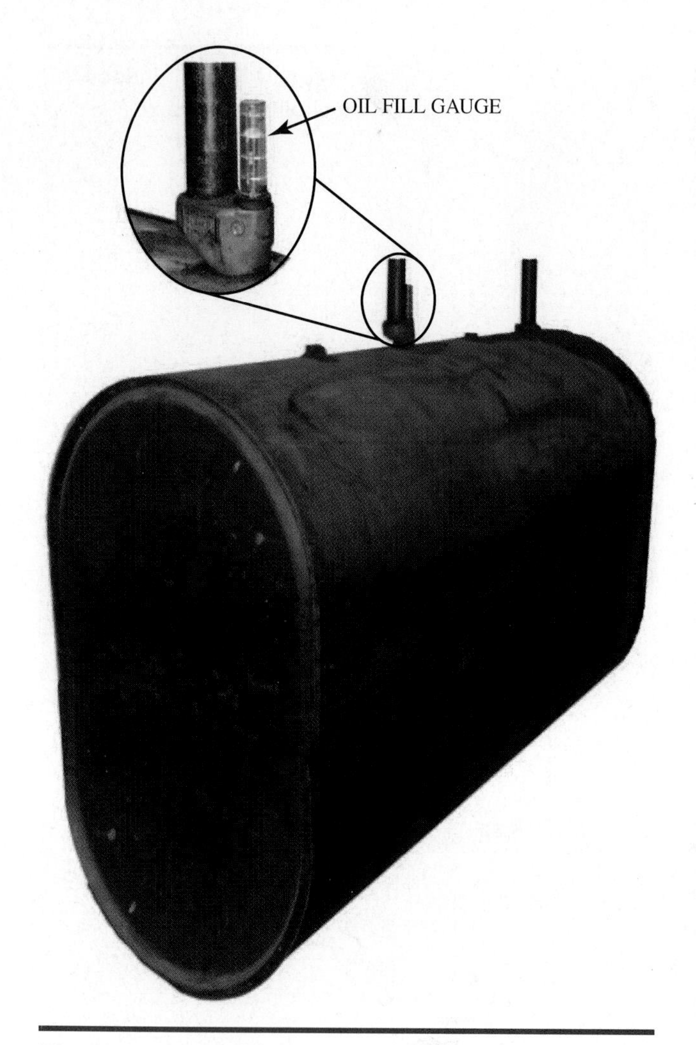

Figure 11-1-15 Oil storage tank and fill gauge.

Fuel oil is delivered to residences and small businesses by tank trucks similar to the one shown in Figure 11-1-21. These trucks are equipped with pumps and flowmeters that register the amount of oil delivered. A printed receipt is then left with the resident.

PRIMARY OIL SYSTEM CONTROLS

A typical oil fired forced air furnace control system is shown in Figure 11-1-22. The fan and limit control performs the same function as on a gas furnace. The safety limit, if open, Figure 11-1-23, will stop the electrical current to the burner before overheating. The fan switch is set to cycle the furnace blower as desired. The space thermostat (24 V) feeds directly to the burner low voltage secondary control circuit. This actuates the flame detector and relay, feeding the primary line voltage to the burner motor and ignition transformer.

(a)

(b)

Figure 11-1-16 Residential fuel oil fill and alarm pipes: (a) Typical oil fill cap; (b) Oil fill and vent pipes.

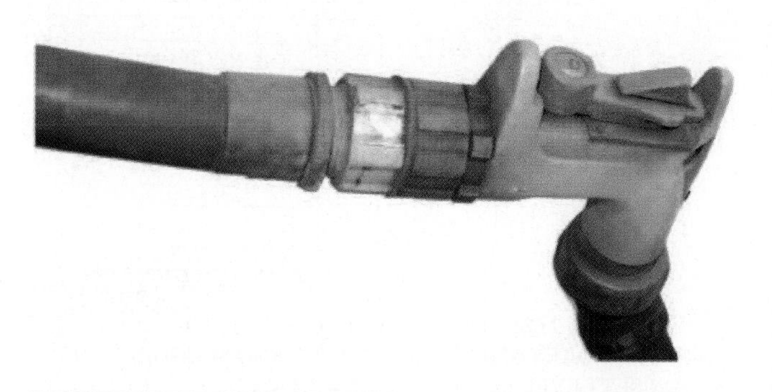

Figure 11-1-17 Fuel oil fill nozzle attached to oil fill pipe.

Figure 11-1-18 Fuel oil alarm cap showing internal screen.

Figure 11-1-19 Commercial fuel oil fill and alarm pipe sets.

> **CAUTION:** The ignition circuit on an oil fired furnace uses approximately a 10,000 to 15,000 V arc to continuously ignite the oil. Accidentally contacting this ignition transformer contacts can result in a severe electrical shock.

The original *primary oil burner control,* known as a protector relay, was mounted on the smoke pipe. The bimetallic element in the sensing tube would react to a rise in flue product temperature and cause the relay to keep the oil burner operating. Figure 11-1-24 shows

Figure 11-1-20 Electronic fill alarm.

an example of this control with the cover in place and the sensing (stack) element protruding from the rear.

Figure 11-1-25 shows a typical control for the primary oil burner. The control has its own power source transformer as well as temperature actuation contacts and manual reset safety switch. The operation of the control is by means of a slide clutch operating safety and holding contacts. The drive shaft at the top of the control provides the action for the slide clutch when moved by the expansion and contraction of the bimetallic element. Two relays are included, one to

Figure 11-1-21 Fuel oil delivery truck.

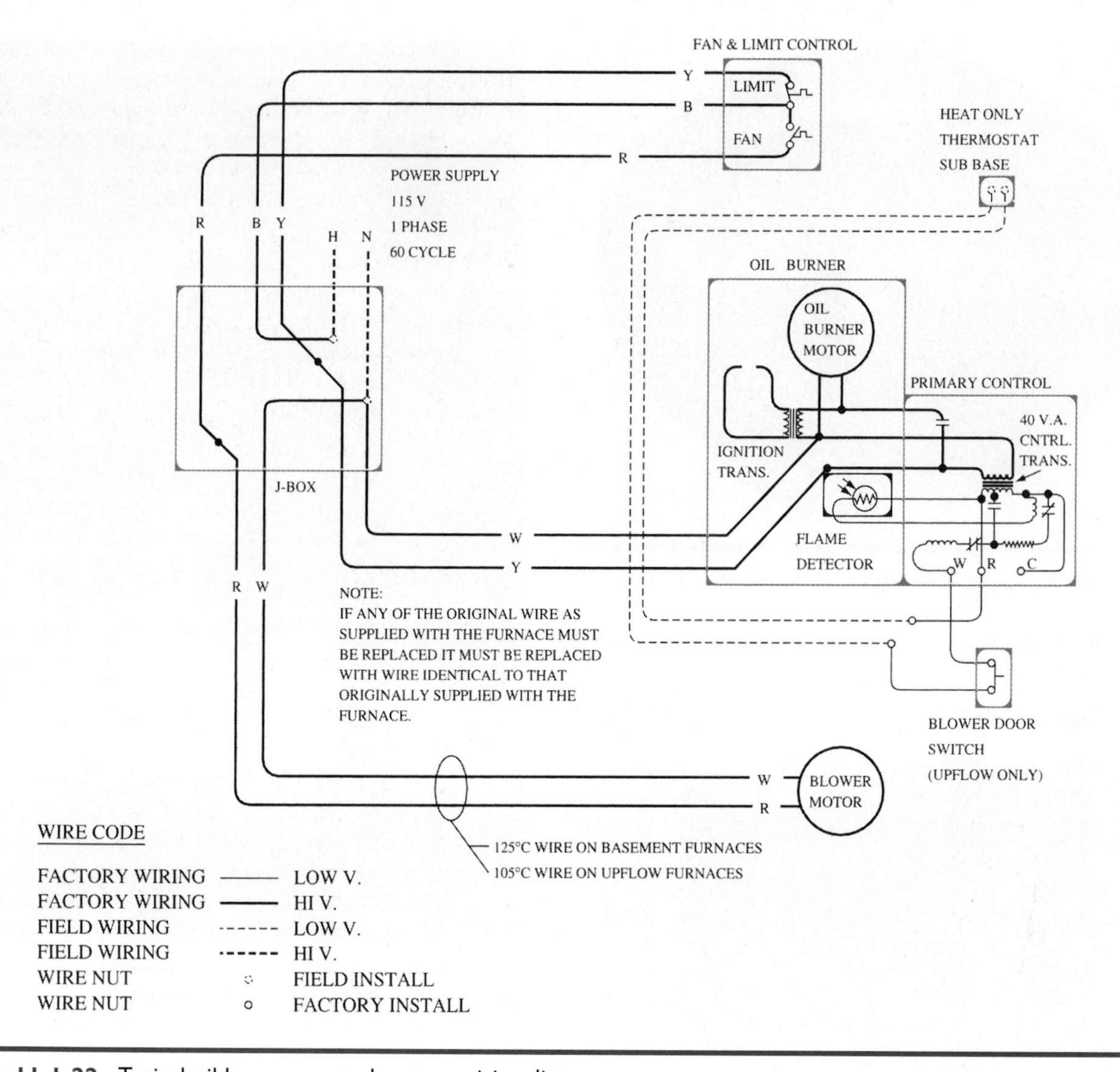

Figure 11-1-22 Typical oil burner control system wiring diagram.

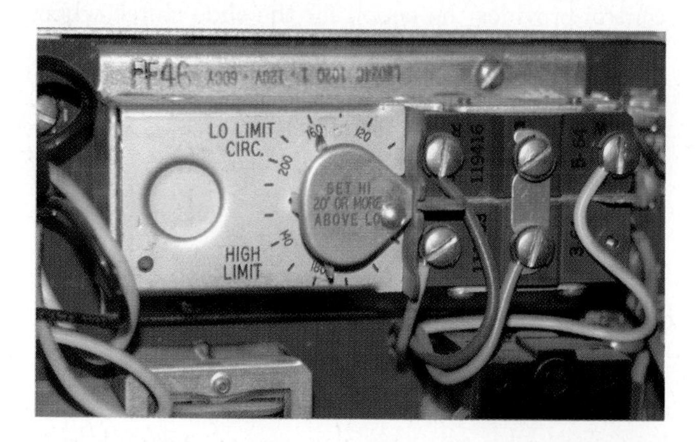

Figure 11-1-23 Oil burner temperature limit control.

Figure 11-1-24 Outer appearance of an oil burner stack mounted primary control.

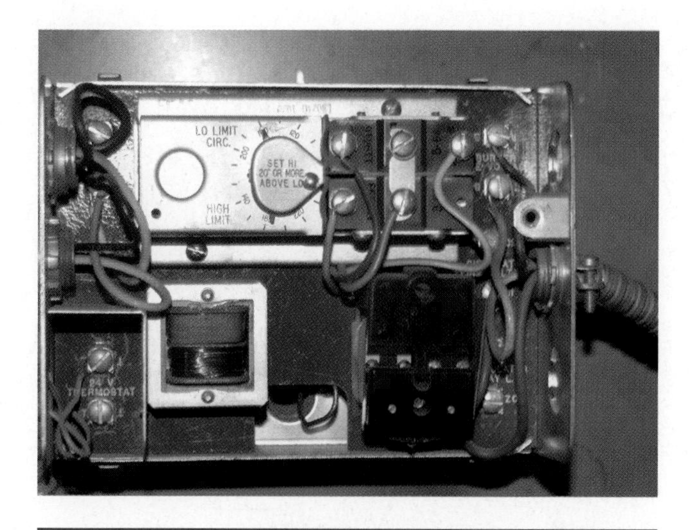

Figure 11-1-25 Oil burner stack mounted primary control.

control the oil burner ignition transformer (relay 1A) and the other to control the oil burner motor (relay 2A). Low voltage thermostat terminals on the lower left and high voltage (120 V) on the lower right are separated by an insulated barrier. Figure 11-1-26 shows the internal wiring arrangement of the control as well as the burner motor and ignition transformer circuits, 120 V power supply, and 24 V thermostat circuit.

Following through the control circuit, the action of the cycle would be as follows: The thermostat, calling for heat, closes the circuit from the transformer

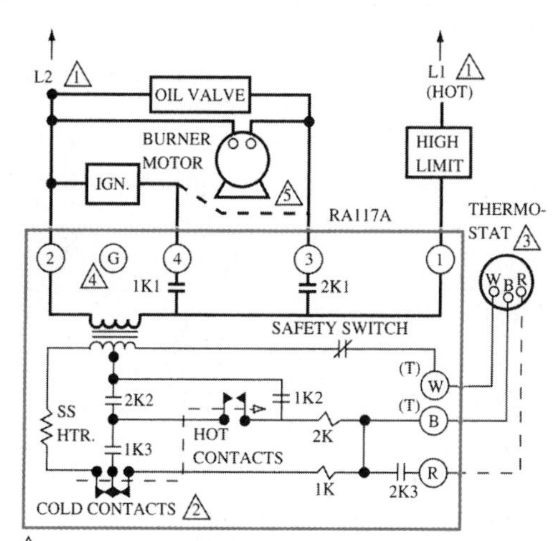

1 POWER SUPPLY. PROVIDE DISCONNECT MEANS AND OVERLOAD PROTECTION AS REQUIRED.

2 CONTACTS BREAK IN SEQUENCE ON RISE IN TEMP.

3 MAY BE CONTROLLED BY TWO-WIRE THERMOSTAT. CONNECT TO W AND B ONLY. TAPE LOOSE END OF RED WIRE, IF ANY.

4 CONTROL CASE MUST BE CONNECTED TO EARTH GROUND. USE GREEN GROUNDING SCREW PROVIDED.

5 TO REPLACE INTERMITTENT (FORMERLY CALLED CONSTANT) IGNITION DEVICE, WIRE IGNITION LEADWIRE TO TERMINAL 3 ON RA117A.

Figure 11-1-26 Wiring diagram of an oil burner primary control.

through the safety switch, through the thermostat, through relay coil 1K, through the right-hand cold contact, the left-hand cold contact, and the safety switch heater to the other side of the transformer.

Relay 1K pulls in closing contact 1K1, powering the ignition transformer; 1K2 energizes relay coil 2K and closing contact 1K3 to the center or common of the cold contacts.

When relay coil 2K is energized, it pulls in, closing contact 2K1, energizing the oil burner motor and oil valve (if used). Contact 2K2 removes the relays from the series circuit through the safety switch heater, reducing the amount of heat produced in the heater. If this action does not occur, the safety switch will open in a very short period of time, possibly 3 to 5 seconds. By removing the relay current from the safety switch circuit, the switch delay time is increased to 60 to 90 seconds.

The start of the oil burner produces hot flue products from the heat exchanger over the bimetallic helix in the sensing tube. This rise in temperature forces the drive shaft forward, opening the left-hand cold contact and removing the safety switch heater from the circuit. A further rise in temperature (forward action of the drive shaft) closes the hot contact; this allows the oil burner motor and oil valve (if used) to continue to operate by passing relay contact 1K2.

Finally, the drive shaft moves forward enough to open the right-hand cold contact, dropping out relay 1K. This stops the ignition and keeps the oil burner operating through the hot contacts. If a flameout occurs or the burner shuts down from the action of the thermostat, the hot contact opens immediately upon a drop in flue gas temperature.

Because contact 1K2 is open, the burner motor cannot operate until the cold contacts close, first the left side to energize the safety switch and then the right side to start the ignition oil burner motor cycle. By means of these lockout circuits, an explosive situation of spraying oil vapor into a white hot refractory can be prevented.

The stack mounted protector relay required extra wiring, as well as time and cost for manufacturing assembly or installation. To overcome this, a flame detection relay was developed. Figure 11-1-27 shows an oil burner protector relay using a cadmium cell flame detector instead of the bimetallic helix. It is mounted directly in the oil burner blast tube directly behind the turbulator plate. The cell sees the light of the burner the instant the flame is established. The interior arrangement of the relay using the light sensitive cadmium cell is shown in Figure 11-1-28. Two relays are used: 1K for control of the oil burner and ignition transformer, and 2K, the sensitive relay, controlled by the cadmium cell. Following the circuit action in the schematic in Figure 11-1-29, when the thermostat calls for heat and closes the

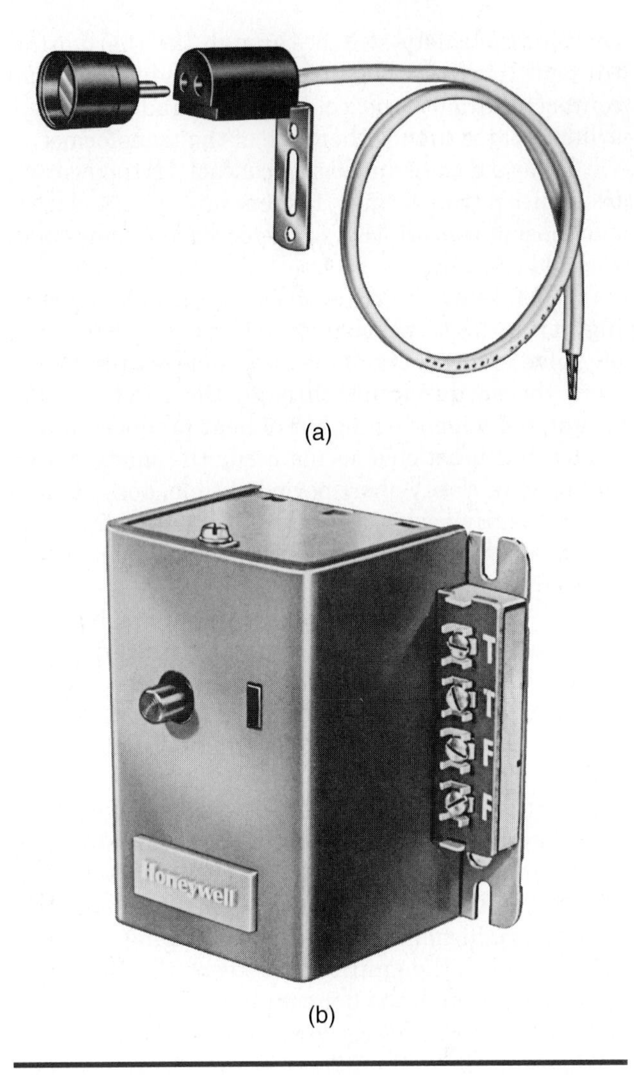

Figure 11-1-27 Oil burner relay using cad cell sensing element. (*Courtesy of Honeywell, Inc.*)

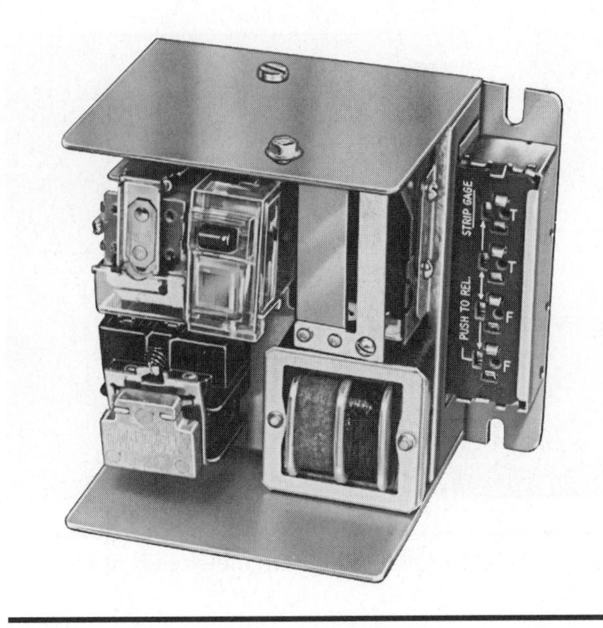

Figure 11-1-28 Oil burner relay using cad cell sensing element, with cover removed. (*Courtesy of Honeywell, Inc.*)

contact between T and T, current flows from the transformer through the thermostat through relay coil 1K, safety switch, timer contact 2, the safety switch heater, and contact 2K1 to the other side of the transformer. This puts a higher voltage through the coil and safety switch heater. If the relay 1K fails to act, the safety switch will open in only a few seconds.

When relay 1K pulls in, however, contact 1K1 pulls in, energizing the oil burner motor, oil valve (if used), and the ignition transformer. Also, contact 1K2 closes, bypassing the safety switch heater, reducing its current draw, and increasing the safety switch action time. If no flame is established, the heater will continue to receive voltage and heat until the safety

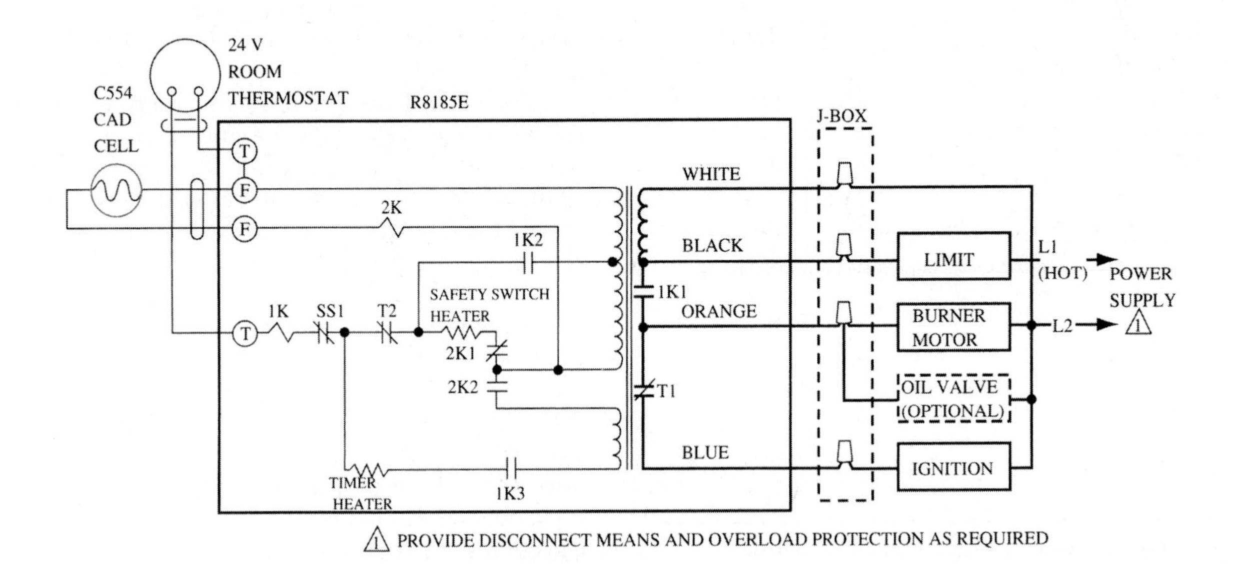

Figure 11-1-29 Oil burning wiring diagram, using a cad cell relay.

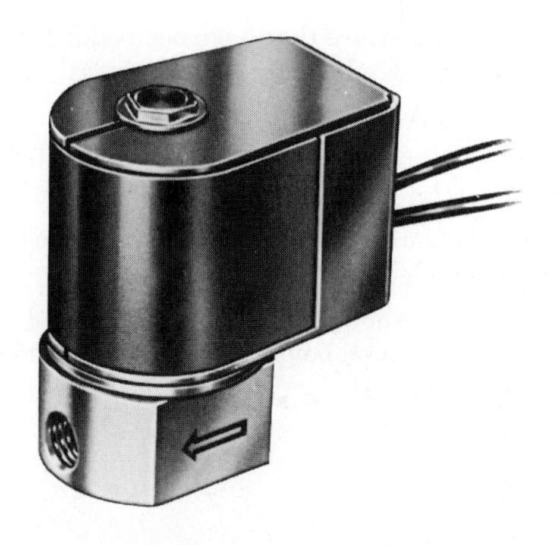

Figure 11-1-30 Delay type oil valve which is placed in the oil line to the burner. It is energized at the same time as the oil burner but has delayed opening. (*Courtesy of Honeywell, Inc.*)

switch contact (SS1) opens and shuts down the system. The switch is manually reset.

When a flame is established, the light strikes the cadmium cell, and immediately the resistance of the cell is reduced from over 10,000 Ω to about 1,500 Ω. The amount of current through relay coil 2K increases and the sensitive relay pulls in. This opens contact 2K1, which de-energizes the safety switch heater and energizes the timer heater through contact 1K3, which closed when relay 1K is pulled in. The timer heater now opens contact T1, shutting off the ignition transformer and the holding contact of the sensitive relay. This action will continue as long as burner operation is required.

If flame failure should occur, the cadmium cell resistance will increase, current flow through the sensitive relay will decrease, and the relay will drop out. This opens contact 2K2 which opens the circuit to relay 1K, and the burner shuts down.

The burner cannot come on until the timer heater has cooled sufficiently to close the ignition contact T1 and the relay 1K circuit contact T2. This assures the unit sufficient time to vent the furnace heat exchanger of unburned vapors as well as assuring ignition at the next startup.

These are only two of the various types of controls manufactured and used on oil fired equipment. The service person should collect and retain as much information as possible from all manufacturers and all types.

Oil Valves

The oil burner should reach operating speed and combustion air volume before oil is supplied and combustion is established. Also, at cutoff of the burner operation the best method of operation is to have instant cutoff of fuel and combustion.

Figure 11-1-30 shows a delay type oil valve that is installed between the oil pump and the burner fixing assembly. The valve, although wired to be energized at the same time as the oil burner motor (wired in parallel), has a delayed opening. This allows full operation of the burner blower and pump before oil flow is allowed. Using a thermistor in the valve coil circuit, the thermistor limits the current to the coil on start. As the thermistor heats, the resistance drops and the current increases. After 8 to 10 sec, the current increases sufficiently to cause the coil magnetic pull to open the valve. The valve acts the same as any other type of solenoid valve on cutoff. It closes immediately upon de-energizing the valve and motor. Thus, instant cutoff of combustion occurs.

TECH TALK

Tech 1. I was working on an oil furnace this morning and found a small oil leak at the bottom of their 500 gallon storage tank in the basement. The boss said it looked like it had rusted out. How would the tank rust out if it is full of oil? Doesn't oil stop rusting?

Tech 2. Oil will stop rusting. However if the oil has enough moisture in it, liquid water can form at the bottom of the tank where it can cause rusting.

Tech 1. How does the water get into the oil?

Tech 2. It can get into the oil from a number of sources. If the oil tank is left nearly empty all summer long, condensate can form in the tank and that condensate water will run to the bottom of the tank where it pools up. The other possibility is water can come in with a new supply of oil.

Tech 1. How do you get the water out?

Tech 2. There are oil treatments that should be put into the tank from time to time. The oil supply company can provide that to the homeowner. Some companies pretreat their oil; others have it as an add-on charge to the customer. When the customer doesn't purchase the service on a regular basis, water can build up and the rust can form. Rust has been such a problem that most new tanks are made from noncorrosive materials like fiberglass.

Tech 1. I guess it's too late for my customer to start treating their oil because their tank is already leaking.

Tech 2. That's right. The tank needs to be drained and either repaired or replaced.

Tech 1. That's what the boss told them. Thanks.

UNIT I—REVIEW QUESTIONS

1. List the four basic configurations of oil fired forced air systems.
2. The _______ determines whether the fuel oil can flow or be pumped through the lines or if it can be atomized into small droplets.
3. What are the two grades of fuel oil?
4. Where is No. 2 grade oil used?
5. List the components of pressure oil burners.
6. List the new feature that have been added to increase the efficiency of oil burning furnaces.
7. What are class A flues used for?
8. What would the action of the cycle be following the control circuit?
9. Why are oil burning furnaces limited in the number of sizes available?
10. The _______ should reach operating speed and combustion air volume before oil is supplied and combustion is established.

UNIT 2

Oil Service

OBJECTIVES: In this unit the reader will learn about oil including:

- Startup
- Operating Sequence
- Efficiency Testing
- Over Fire Draft
- Smoke Testing
- Calculation Efficiency
- Routine Maintenance
- Soot

STARTUP

Figure 11-2-1a,b show a typical wiring diagrams for an oil fired heating only installation using a thermostat controlling a cad cell flame sensor type of primary control with a blower door switch in the low voltage circuit as an air supply safety switch. The primary control is controlling the oil burner motor and ignition transformer through the limit control to prevent overheating the unit.

The final function is the operation of the system blower motor by the fan portion of the fan limit control. Properly set, this control will close at 125 to 130°F to start the blower motor and open at 95 to 100°F to stop the blower motor.

Operating Sequence

The step by step operation of the unit would be as follows:

1. With the blower compartment door closed and the blower door switch closed, the room thermostat, upon a call for heat in the conditioned area, will close. This causes current flow from the 40 VA transformer through the control circuits and the burner motor relay. The relay pulls the contact closed.

2. The closed contacts cause a current flow from the H or hot side of the supply line through the limit switch to the burner motor and ignition transformer. In addition, a current flow in the 24 V circuit exists through the dark cad cell and the safety switch heater.

3. Immediately, the ignition transformer establishes a spark across the electrodes located above the nozzle. The spark at this time is only 1/8 in long, the distance between the electrode tips.

4. The blower motor is energized at the same time as the ignition transformer. It reaches full load speed within 1 sec. When the full amount of air is delivered by the blower, the ignition spark is blown forward into the oil spray and the oil is ignited. This occurs directly in front of the nozzle and quickly expands into a full burst of fire in the combustion chamber.

5. As soon as the light from the fire reaches the cad cell located in the fire tube, the electrical resistance of the cell decreases. This decrease in resistance increases the current flow through the safety switch heater and prevents the safety switch from opening.

SERVICE TIP

On most oil fired furnaces the safety inter-lock with the cad cell does not respond instantly if the burner fails to light. This will result in some unburnt oil being sprayed into the combustion chamber. Repeated attempts to start the burner can result in a significant amount of oil collecting in the combustion chamber. This can occur if the homeowner repeatedly tries to restart the system by resetting the safeties. You must remove this oil before attempting to restart the furnace.

6. The flame efficiency increases as the refractory temperature rises. At a white-hot surface temperature of the refractory, the burner is operating at peak efficiency. At this point efficiency tests can be taken. To reach these conditions, the burner should be allowed to operate at least 5 min.

The initial system startup will require more than just turning on the burner. The first requirement will be to purge the fuel unit of air. This procedure would be different for the single pipe system where the fuel supply is above the burner and the two pipe system where the fuel supply is below the burner.

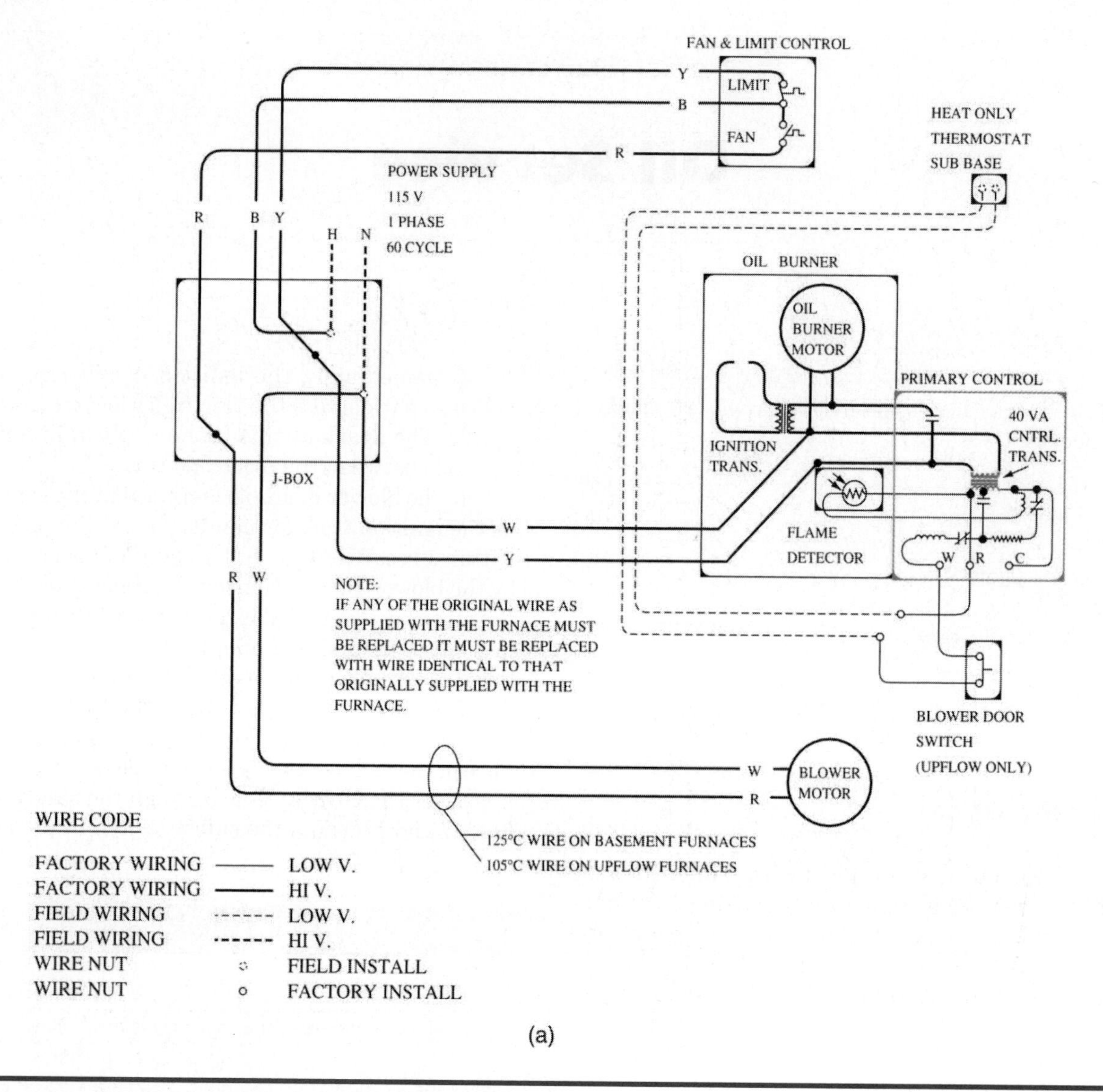

Figure 11-2-1 (a) Wiring diagram for typical oil burner control system; (b) Wiring diagram for the Lennox OF23 system. (*Courtesy of Lennox Industries Inc.*)

On the single pipe system with gravity feed (the supply is above the burner) it is only necessary to purge the air from the oil filter, fuel line, and fuel unit. The purge valve is located on the pump. It is only necessary to open this valve until the air in the system starts to flow out. Usually, this is one half to one turn of the valve. When the air is out and oil starts to flow out, close the valve to prevent oil leakage.

In a two pipe system, the pump will force the air down the oil return line to the tank and the system is self purging. It may take longer than the cycle time of the safety switch to purge the system and the burner may cut off before flame is established. It is then necessary to allow the safety switch to cool and be reset, and for the burner to be started again. This must be repeated until the oil lines are clean and the fire is established.

SERVICE TIP

Use this tip during purging. Place a plastic hose on the air purge valve and put the other end of the hose in a container. Run the pump until the hose is full of oil with no air bubbles.

EFFICIENCY TESTING

When the fire is established and the refractory obtains white heat, efficiency testing can be done on the unit. Figure 11-2-2 shows the label on all new models giving the expected efficiency. A test sheet is desirable to record the conditions to which the burner was

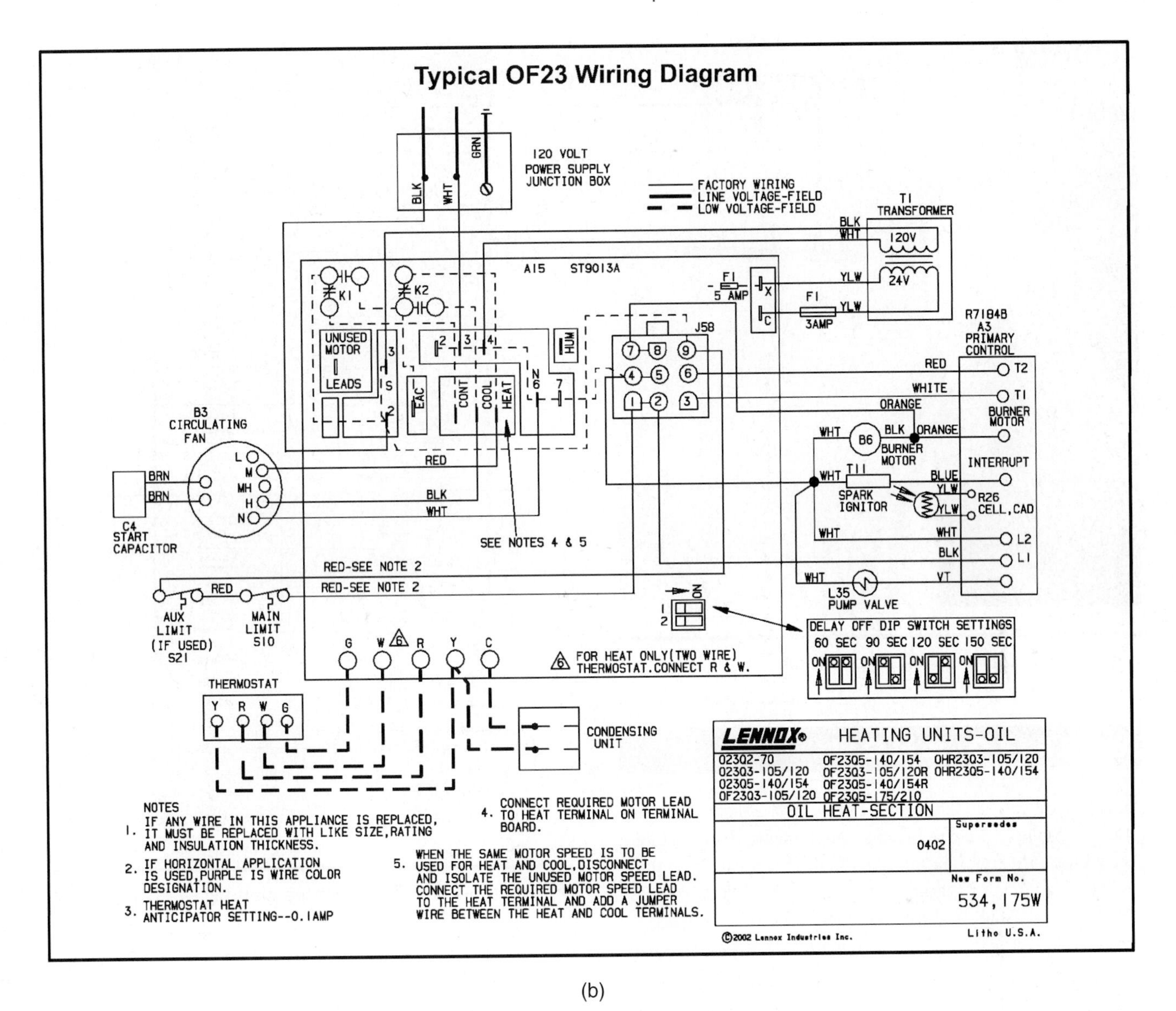

(b)

Figure 11-2-1 (continued)

adjusted, for future reference. The following information should be included on the test sheet.

1. Make of oil burner ______
2. Model No. ______ Serial No. ______
3. Nozzle: Size (gal/hr) ______ Type ______ Angle ______
4. Refractory: Shape ______ Operation
5. Over fire draft ______
6. Stack draft ______
7. Heat exchanger flow resistance ______
8. CO_2% ______
9. Net stack temperature ______
10. Efficiency ______
11. Smoke number ______
12. Air temperature (supply plenum) ______
13. Air temperature (return plenum) ______
14. Air temperature (rise) ______

It is a good idea to record on the test sheet the make, model number, and serial number of the burner for future situations. Recording the nozzle size, spray angle, and type forces the checking of the unit for proper input as well as spray angle for the shape of the refractory. Figure 11-2-3 shows a typical completed efficiency test sheet. Round or square refractories will take the 80 or 90° spray angle; a rectangular refractory, the 45 to 60° angle, depending on the length of the refractory. With the unit operating and refractory up to white heat, operating tests can be made.

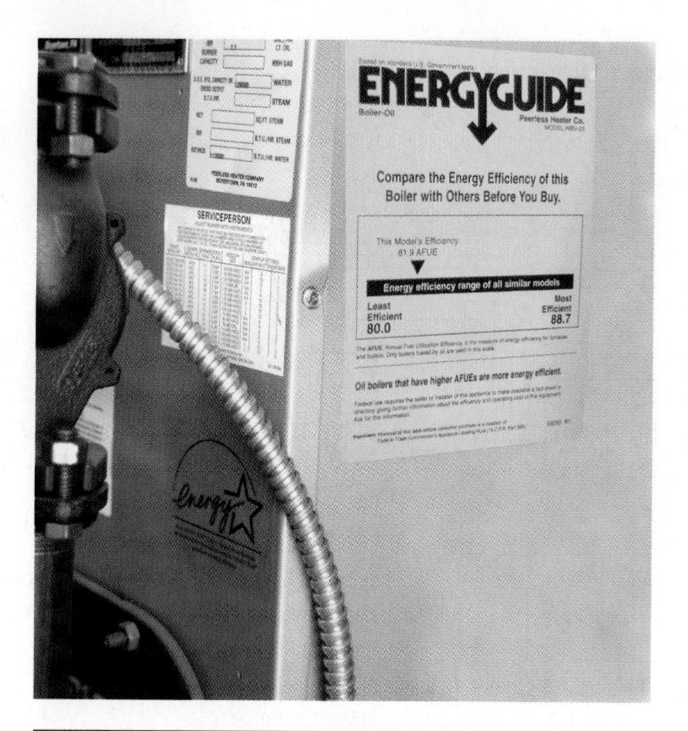

Figure 11-2-2 Energy efficiency sticker.

ACCURATE PRINTING CO • 104 PERIMETER ROAD • NASHUA, NH 03063

COMBUSTION EFFICIENCY TEST RESULTS

DATE 1-30-03

NAME

ADDRESS 38 Woodward Rd

TOWN Milford PHONE:

	BEFORE	AFTER
SMOKE		0
CO_2		8
DRAFT		.04
NET TEMP		450
EFFICIENCY		77¼%
DRAFT O.F.		.02
CO_2 O.F.		—
NOZZLE SIZE		100
NOZ. ANGLE		80
NOZ. PATTERN		h

OIL COMPANY, INC.

SER. TECH. Rob

DEALER COPY

Figure 11-2-3 Efficiency inspection tag.

NICE TO KNOW

Carbon or soot sublimes at a temperature slightly above 800°F. Subliming is the process of changing from a solid to a vapor without becoming a liquid. This means that the area of refractory material that is carbon free was at a temperature slightly above 800°F and the line where the carbon first begins is the beginning of the area that was below 800°F during system operation. If you observe a furnace firebox that has far less carbon free refractory material, this may be an indication of a low flame temperature.

Over Fire Draft

The first test should be *over fire draft*. Using a draft gauge such as those pictured in Figure 11-2-4 and Figure 11-2-5 and a ¼ in hole in the pressure relief door over the burner (sometimes called the observation door), insert the test pipe for the gauge through the hole until the end of the pipe is beyond the inner edge of the combustion chamber. To accomplish this, it is sometimes necessary to substitute a longer piece of ¼ in copper tubing for the gauge probe pipe. This test should result in a minimum pressure of 0.02 to −0.04 in of WC to ensure the proper draft for best operation. If the manufacturer of the unit does not specify the over fire draft, it is best to start at a draft control setting of −0.04 in of WC.

Figure 11-2-4 Draft gauge for measuring draft over the oil burner flame. (*Courtesy of Bacharach, Inc.*)

Figure 11-2-5 Stack draft gauge. *(Courtesy of Bachrach, Inc.)*

Some manufacturers specify stack draft for a setting. This means that the flow resistance of the heat exchanger must be taken into account for burner operation. By measuring both the over fire draft and stack draft on the new unit, the design draft resistance can be determined. To produce a negative pressure of a given amount over the fire, the stack draft must be a greater negative amount. The difference between the over fire draft and stack draft will be the heat exchanger draft flow resistance. For example, if it is necessary to have a stack draft of −0.06 in of WC to produce an over fire draft of −0.04 in of WC, the heat exchanger resistance would be 0.02 in of WC. When checking the unit performance after a period of operation, if the heat exchanger flow resistance has doubled, it is necessary to mechanically clean the soot from the heat exchanger flue passages.

A negative pressure must be maintained on the heat exchanger to prevent the products of combustion from being forced into the occupied area. Not only do they carry free carbon or soot, but they also contain a high percentage of CO. This is especially true at startup. Therefore, a minimum over fire draft of −0.02 in of WC is usually required for proper operation, with −0.04 in of WC to be on the safe side.

CAUTION: A carbon monoxide (CO) detector is a good recommendation for the homeowner to purchase and install near any combustion equipment, whether it is gas or oil. Carbon monoxide is often referred to as the silent killer because it can go unnoticed in a home for a long period of time if the home is not protected with a carbon monoxide monitor.

CO_2

The CO_2 sample is taken far enough ahead of the barometric draft control to obtain a good sample of flue gas without outside air mixture.

Again, this may mean a longer sampling tube on the analyzer. Use a CO_2 analyzer such as the ones pictured in Figure 11-2-6a,b, following the manufacturer's instructions. The following results should be obtained:

1. **Old style gun burners.** These burners have no special air handling parts other than an end cone and a stabilizer. A CO_2 reading of 7 to 9% should be obtained unless a CO_2 reading in this range results in more than a No. 2 smoke. If so, the CO_2 reading should be reduced until the smoke test results in below a No. 2 smoke.
2. **Newer style gun burners.** Burners with special air handling parts should be set in the range 9% to 11%. This should result in less than a No. 2 smoke (closer to a No. 1 smoke).
3. **Flame retention gun burners.** These burners should be set in the range of 10 to 12% with the 0 on the smoke test results.
4. **Rotary burners.** These burners should be set the same as old style gun burners.
5. **Rotary wall flame burners.** Higher CO_2 settings are available, up to 13.5%, with these older burners, but the maximum of No. 2 on the smoke test is required.

Smoke Testing

A smoke tester, such as the one shown in Figure 11-2-7, uses a pump piston to draw flue products through a filter inserted in the head of the pump between the sample tube connection and the piston body. This particular instrument requires ten slow strokes of the piston to draw the required amount of flue products through the filter. Some technicians want to make one more stroke on the piston to be sure they get a good test. This does not make a good test, it invalidates the test. Make only the exact number of strokes with the piston pump as specified by the manufacturer. The filtered sample is compared to numbered rings on the test card and should never be higher than the No. 1 ring. If higher, the

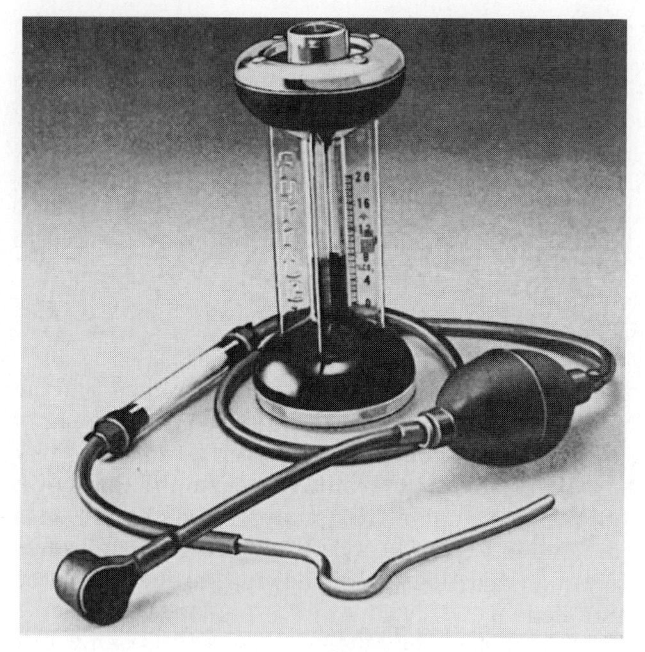

(a)

(b)

Figure 11-2-6 (a) Carbon dioxide (CO_2) analyzer. (*Courtesy of Bacharach, Inc.*); (b) Electronic CO_2 tester.

burner is receiving insufficient air and is not burning the carbon sufficiently. This will quickly produce carbon deposits and plug heat exchanger passes. On older units if less than the No. 1 ring, too much air is being allowed in the burner. The CO_2 content is too low and too much heat is being forced out of the heat exchanger. This results in considerably lower efficiency and much higher operating cost. On newer units the CO_2 reading should be as low as possible.

SERVICE TIP

Do not let the tip of the tube touch the far side of the flue pipe. You may pull soot off the pipe.

SERVICE TIP

Some furnaces are located in poorly lit basement areas. It may be necessary for you to use a drop light or flashlight to determine the difference between the shades of gray of the smoke spot test rings.

Stack Temperature

A high temperature thermometer, such as that shown in Figure 11-2-8, is inserted in the same opening through which the stack draft was taken. The temperature of the flue products will also determine the efficiency of the heating unit. The cleaner the heat exchanger, the more able it is to remove heat from the flue products as they pass through. If the CO_2 content is correct but the stack temperature is too high, this is usually a sign of excessive carbon deposit on the heat exchanger surfaces. To determine the net stack temperature, subtract the temperature of the air entering the burner from the stack temperature.

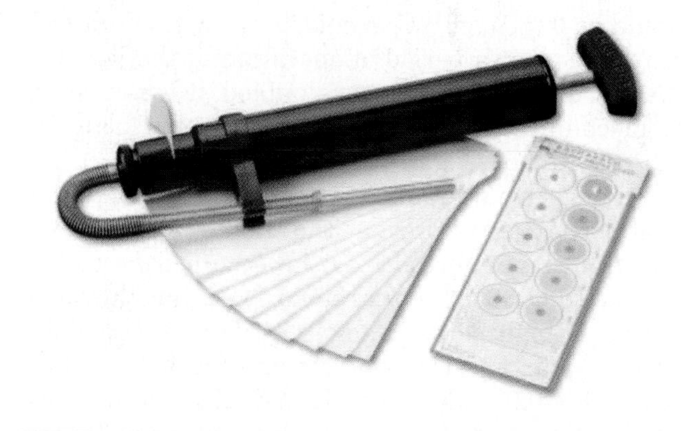

Figure 11-2-7 Smoke tester. (*Courtesy of Bachrach, Inc.*)

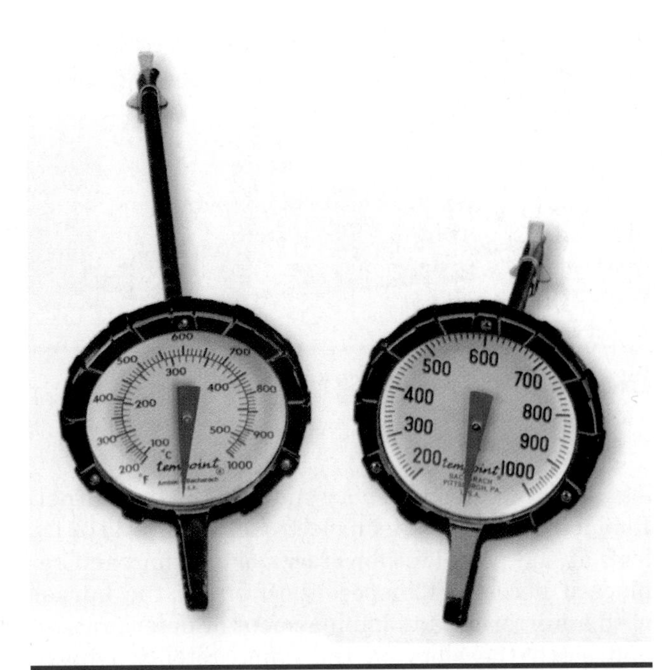

Figure 11-2-8 High temperature stack thermometer. *(Courtesy of Bacharach, Inc.)*

CALCULATION OF EFFICIENCY

Use an efficiency calculator such as that shown in Figure 11-2-9. The horizontal slider is set so that the net stack temperature to the nearest 50°F is shown in the net stack temperature window. The vertical slider is then adjusted so that the tip of the arrow is at the CO_2 determined by the analyzer test. The efficiency of the unit is then shown in the arrow in the vertical slider.

The efficiency results should be within the smoke test ratings as described above or the manufacturer's specifications. If not, the unit should be examined for the cause of the difference. Usually, a cleaning of the heat exchanger is required.

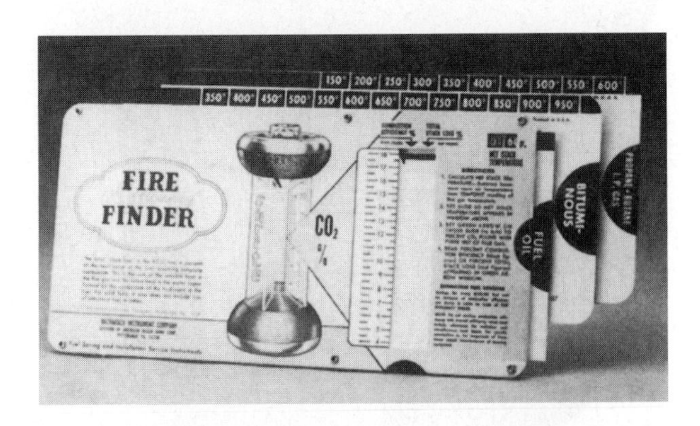

Figure 11-2-9 Oil burner efficiency calculator. *(Courtesy of Bacharach, Inc.)*

Soot

Soot deposited on flue surfaces acts as an insulating layer, reducing the heat transfer and lowering the efficiency. Soot can also clog flues, reduce draft and available air, and prevent proper combustion. Proper burner adjustment can minimize soot accumulation. An accumulation of 1/4 in of soot is equal to 1 in of insulation.

PLANNED MAINTENANCE

Good routine maintenance includes annual startup and calibration on oil heat equipment prior to the heating season. The units should be tested and verified during normal operation to determine the operating and safety controls are working as designed. This includes checking air filters, oiling motors, and testing the plenum switches. On oil equipment, inspections include replacing oil filters, strainers, checking electrodes, and cleaning the cad cell. It is also a good idea to open the cad cell circuit or put black electrical tape over it to verify the flame failure control will shut down the burner during a no flame condition.

One type of combination fan and limit control for a typical gas or oil warm air furnace, Figure 11-2-10,

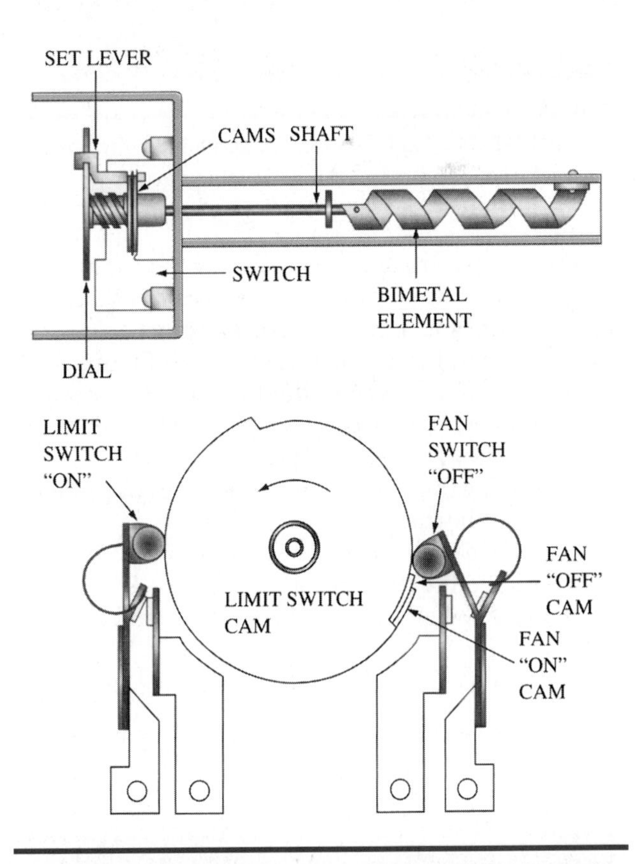

Figure 11-2-10 Construction of combination fan and limit control for gas or oil warm air furnaces. Shows how twisting motion of heated helix activates switches.

Figure 11-2-11 Combination fan and limit switch acts as a safety limit and temperature sensing device.

uses the power created by the rotary movement of a helix bimetal. The rotating cam makes or breaks the separate fan and limit electrical contacts. Another type of fan and limit switch is shown in Figure 11-2-11. It is the same type of dual switch discussed in Chapter 10, Unit 4, as this switch is used for both oil fired and gas warm air furnaces.

The following steps should be taken to properly maintain oil burners:

Shut off electrical power to the furnace.

Shut off oil to the burner.

Replace the oil line fuel filter, pictured in Figure 11-2-12.

Figure 11-2-12 Replacement fuel oil filter.

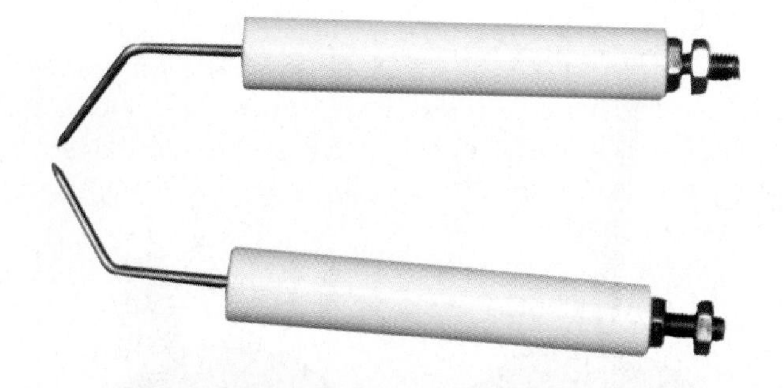

Figure 11-2-13 Oil furnace electrodes.

Remove the drawer assembly from the burner. Replace the nozzle. Clean electrodes, Figure 11-2-13, look at the porcelain for cracks or crazing, and replace if necessary. Inspect the front of the burner with a mirror for any impingement or deterioration, and clean the slots on the flame retention head. Adjust electrodes to factory specified settings, and then replace the drawer assembly.

SERVICE TIP

Replacing the nozzle as part of your routine maintenance is important because over a season the nozzle bore will be worn slightly larger as the result of oil flow. As this wear occurs, the furnace efficiency is affected. Replacing the nozzle each season assures that the furnace will operate properly.

Figure 11-2-14 Ignition transformer.

Clean the blower wheel, making sure it is tight on the motor shaft. Pay attention to the air intake slots on the air adjustment shutter and clean if required. While in the blower compartment check the pump coupling for alignment and tightness on the pump shaft.

Test ignition transformer, Figure 11-2-14, for the correct voltage output to the electrodes. Assure there is good contact from the transformer to the buss bar and that the buss bars are in good condition.

Test the oil pump outlet for pressure and inlet for proper vacuum. Replace the screen if the pump has one. Test the pump cutoff to verify a complete fuel shutoff. It should shut off at 80 psi.

Check the oil burner motor and the amperage draw. The amp reading should be under the nameplate rating.

Test the cad cell primary control and cad cell flame detector eye.

Turn the fuel on and start the burner.

TECH TALK

Tech 1. What causes carbon to form inside the chamber on an oil furnace? And is that carbon an indication of a problem?

Tech 2. Well first of all, some carbon will form in the combustion chamber even on a properly operating oil furnace. Second, the carbon is formed because of incomplete combustion. Any time that a fuel is burned and not burned 100% efficiently, there can be some carbon formed. It is only when the carbon gets to be excessive that it is an indication of a problem.

Tech 1. How can you have incomplete combustion in a firebox that is glowing white hot?

Tech 2. Well, the carbon doesn't form where it is white hot. It will form around the edges where the temperature is not as hot. And that carbon is okay because even a furnace operating at peak efficiency has a little carbon formed around the edges.

Tech 1. Okay, I think I understand.

UNIT 2—REVIEW QUESTIONS

1. The flame efficiency increases as the ________ temperature rises.
2. Properly set, the fan limit control will close at ________ to ________ °F.
3. What is the first requirement in the initial startup?
4. What should be recorded on the test sheet during the efficiency testing?
5. What is necessary if the heat exchanger flow resistance has doubled when checking the unit performance after a period of operation?
6. Why must a negative pressure be maintained on the heat exchanger?
7. The ________ of the flue products will also determine the efficiency of the heating unit.
8. What is included in good routine maintenance?
9. What is the function of the rotating cam in Figure 11-2-10?
10. What are the steps that should be taken to properly maintain oil burners?

UNIT 3

Troubleshooting Oil Heating Systems

OBJECTIVES: In this unit the reader will learn about troubleshooting heating systems including:

- General Troubleshooting Procedures
- Oil System Problems

GENERAL TROUBLESHOOTING PROCEDURES

As discussed in Chapter 10, Unit 5, the suggested progression of analysis for troubleshooting starts with identifying the common categories of complaints or problems, then the possible problems in system categories, followed by the symptoms and causes of specific problems. Causes can be either electrical or mechanical and related to the use of the oil equipment.

TECH TIP

No heat calls can require almost an emergency response when they occur during times of very low temperatures. Some residences can drop to near freezing temperatures inside when the heat goes off during a severe cold spell. Without prompt service water lines in the residence can begin to freeze and these low temperatures can become life threatening. For that reason you must move these no heat calls to the top of the service list during these critical low temperature periods.

As before when gas fired systems were covered, the categories of complaints or problems fall under the following headings:

- Entire system operation
- Unit operation
- Burner operation
- Blower operation
- Heat exchanger complaints
- Cost of operation
- Noise

OIL SYSTEM PROBLEMS

Principal causes of oil fired system problems follow.

Problems	Possible Causes (see the following number references for further information)
Will not start	Season switch open Room thermostat improperly set Safety switch open on protector relay Fuse blown Limit control open Protector relay transformer defective (1) Protector relay defective (2)
Starts, but will not continue to run	A. No fire established No fuel (6) No ignition Protector relay not functioning Nozzle defective Fuel pump defective B. Fire established Defective oil burner component Defective protector relay
Runs, but short cycles	Improper heat anticipator setting Cycling on limit control
Runs, but room temperature too high	High setting of thermostat Improper heat anticipator setting
Runs continuously	Short in thermostat circuit Stuck contacts in protector relay

Problems	Possible Causes (see the following number references for further information)
Blower cycles after thermostat is satisfied	Incorrect blower CFM Incorrect fan control setting Heat exchange heavy with soot
Blows cold air on start	Incorrect blower CFM Incorrect fan control setting
Startup/cooldown noise	Expansion noise in heat exchanger Duct expansion
Noise/vibration	Burner pulsation Blower wheel unbalanced Blower drive problems
Odor	Oil odor on unit startup Improper venting Cracked heat exchanger
High fuel cost	Improper input Improper burner adjustment
High electric cost	Improper load on blower motor
Smoke or odor from observation door	Improper draft or draft control setting Improper venting Improper input Delayed ignition
Burner pulses	Improper draft Improper venting
Burner cuts off on safety switch	A. No fire established No fuel No ignition Protector relay defective Components of burner defective B. Fire established Components of burner defective
Delayed ignition	Carbon deposit on firing head Improper electrode adjustment (3) Cracked electrode insulator (4) Ignition leads burned out Ignition transformer burned out (5)
Burns inside burner after cutoff	Fuel unit defective
Carbon deposits on refractory	Defective nozzle
Noisy flame	Improper air adjustment Improper nozzle
Fuel pump sings	Fuel unit defective (7)
Heat exchanger burnout	Input too high Defective refractory Defective nozzle
Blower short cycles	Input too low Blower CFM too high Fan control defective

Many of the solutions to these problems are evident when the cause is determined. Below, however, is further information regarding some of the remedies that can be used. Numbers refer to references above.

TECH TIP

When you are talking with a customer who has a problem with their system, listen to the customer's comment carefully. You may be able to ascertain some idea as to the problem. But more importantly if you act as if you are not interested you can alienate the customer. Alienating the customer can result in their loss of confidence in you and your company.

1. Protector relay transformer burned out. It is usually not possible to open the protector relay to reach the transformer to test it. It is not necessary since corrections to the transformer are done through the T-T terminals on the outside terminal board.

Figure 11-3-1 shows the inside wiring diagram of a typical protector relay circuit. With the control wires removed from terminals T-T, a circuit from the top terminal T, through the safety switch (SS), the transformer, normally closed contact (2K1), the safety switch heater, and relay coil (1K) should put 24 V across terminals T-T. With 120 V across the black and white leads to the relay, if no voltage is at T-T, replace the protector relay. If there is voltage at T-T and the thermostat will not operate the relay, check the thermostat and subbase on the control cable.

2. Protector relay defective. There are two types of protector relays: (a) a bimetallic control in the hot flue gases of a smoke pipe mount, and (b) a cad cell that operates the relay from the light of the fire.

a. Bimetallic controls. This type of control is clutch operated to move the hot and cold

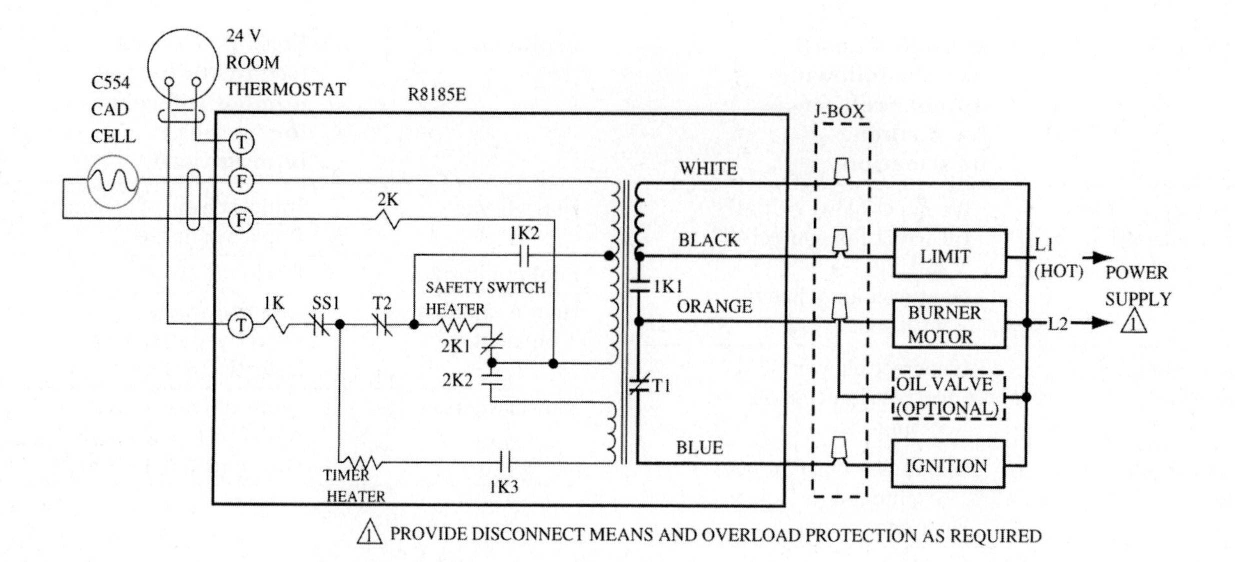

Figure 11-3-1 Oil burning wiring diagram, using a cad cell relay.

controls through their proper sequence. Any sharp blow to the control can release this clutch and throw the control out of sequence. The clutch also wears and can loosen the hold on the contact fingers.

If the control such as the one in Figure 11-3-2 refuses to operate, pull the drive shaft lever forward until the stop is reached. Slowly release the drive shaft lever to the cold position. This should close the contacts and the unit should operate. If not, the bimetallic element could be jammed or the contacts defective. To check the bimetallic element, remove the protector relay from the vent stack and check it. Usually carbon (soot) buildup through the bimetallic helix will be the cause of jamming the drive shaft lever. When cleaning the helix, be careful not to bend or break it.

Figure 11-3-2 Oil burner control type RA117A, cover removed. (*Courtesy of Honeywell, Inc.*)

SERVICE TIP

If a bimetal helix is not operating smoothly after cleaning it must be replaced. Any slight roughness in operation will only get worse and will result in a recall to the residence to fix the problem later.

With a clean helix, if the hot and cold contacts still do not hold, the clutch and contact leaves are worn and the entire control should be replaced.

b. Cad cell. This type of protector relay uses a light sensitive cadmium sulfide flame detector, as pictured in Figure 11-3-3a,b, mounted to the firing tube of the oil burner. When the cell is exposed to light, its resistance is very low, which allows current to flow through the cell. This current is sufficient to pull in the sensitive relay in the protector relay.

When the relay pulls in, it opens the circuit to the safety switch heater and prevents cutout of the burner. If the cell is not exposed to the light of the fire or if the cell becomes dirty or covered with soot then it will not allow the current to pull in the sensitive relay; the safety switch heater remains in the circuit until the safety switch opens. This breaks the thermostat circuit and the burner stops as shown in the circuit in Figure 11-3-4.

Figure 11-3-3 Light sensitive cadmium sulfide cell: (a) Front view; (b) Side view.

The safety switch is manually reset. When checking the protector relay for repeated burner cutoff even though the flame is established, make sure that the cad cell is clean. The cad cell can also be checked using an ohmmeter. Connect the ohmmeter leads across the leads of the cad cell. If the cad cell is exposed to light, the resistance will be 350 to 1,500 Ω. Placing a finger over the cell, cutting off the light will raise the resistance to 10,000 Ω or higher.

SERVICE TIP

When replacing a cad cell, be sure that it is completely and squarely seated. If the cad cell is not seated properly it might not be "looking" at the flame so it will not operate properly. An improperly seated cad cell can cause an intermittent problem.

If the cad cell checks out, check the protector relay by placing a jumper wire across the F-F terminals of the protector relay. The burner should start and continue to operate. If it cuts off, the timer contacts and heater circuit are defective and the relay should be replaced.

Before starting the burner, check for liquid oil in the bottom of the refractory. The customer knows that the burner should start if the reset button is pressed. If the reset button has been pressed a number of times, it is possible to have a considerable quantity of oil sprayed into the combustion chamber before the owner calls for help.

If any oil is pooling in the bottom of the refractory, it must be removed by soaking it up with sponges or rags. When reaching into the refractory, make sure that the power to the burner is off and your arm is covered. A fire extinguisher must be within reach and the observation door must be secured open. When the flame is established, considerable fire will develop until all the oil is burned out of the bottom where it has soaked into the refractory. If it is a large unit of 2.5 gal/hr input or larger, it is advisable to call the fire department for standby before lighting the burner.

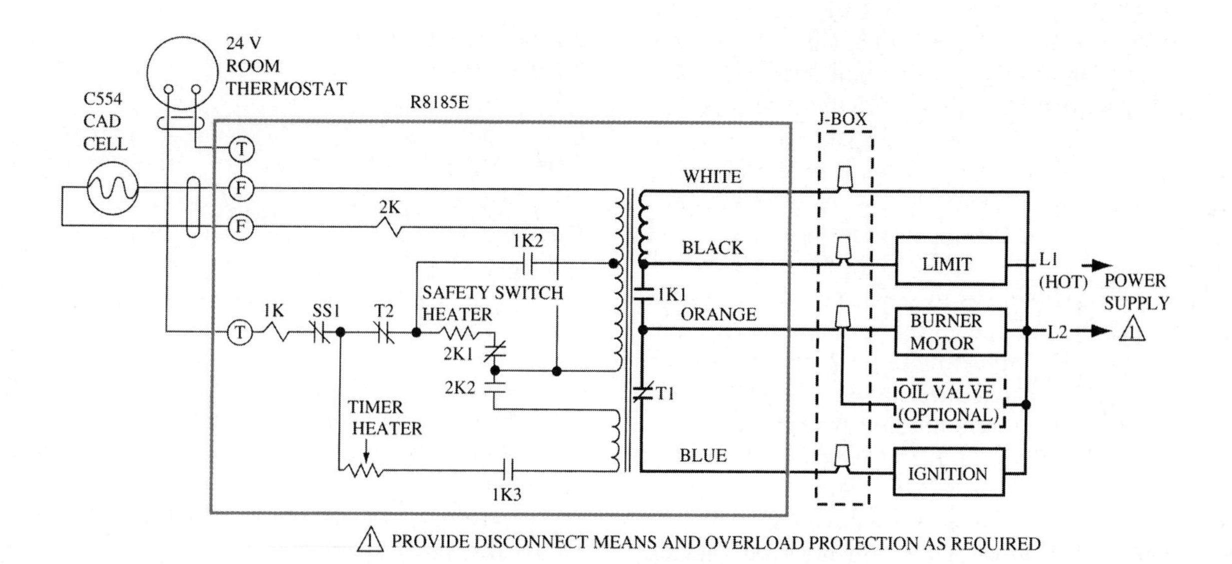

Figure 11-3-4 Oil burner wiring schematic diagram using cad cell protector relay.

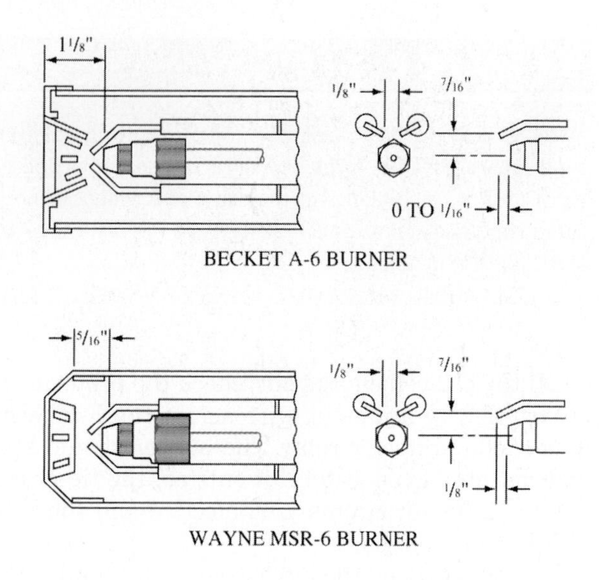

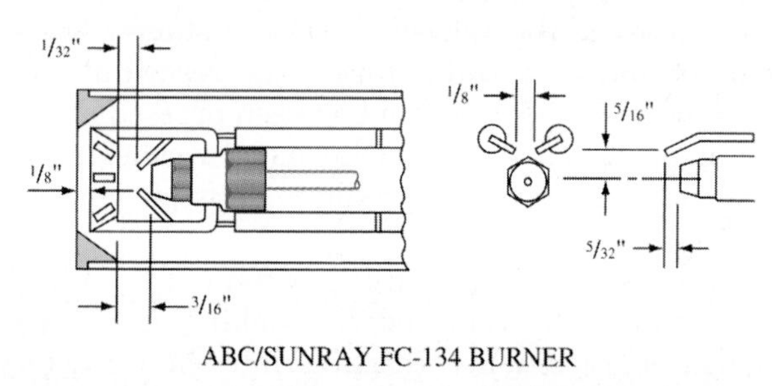

Figure 11-3-5 Oil burner electrode spacing dimensions. *(Courtesy of Rheem Manufacturing Air Conditioning Division)*

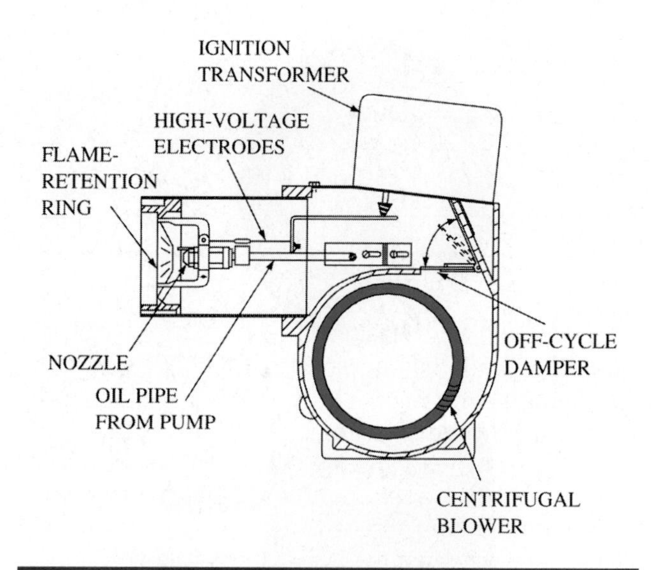

Figure 11-3-6 Oil burner cutaway view.

3. Improper electrode adjustment. To establish the spark necessary to ignite the oil, the electrodes have to be close enough together to present a minimum gap resistance to the 10,000 to 15,000 V supplied by the ignition transformer. To keep the electrodes out of the oil spray and still have the arc flame blown into the oil spray to ignite the oil spray, the electrodes must be high enough above the hole in the nozzle as well as the proper distance ahead of the nozzle end. These dimensions will vary with each manufacturer's unit. Figure 11-3-5 shows the burner setting used on a particular manufacturer's oil fired unit.

Another dimension is the distance of the nozzle end from the end of the firing tube air turbulator. This dimension is important to keep oil spray from impinging on the turbulator and causing carbon buildup and still have the nozzle back far enough to keep the effect of the heat of the fire to a minimum.

The burners have only one common dimension: the size of the gap between the electrodes, which is 1/8 in as shown in Figure 11-3-5. The electrode height and forward distance vary. When setting electrode positions, the manufacturer's specifications must be followed.

4. Cracked electrode insulators. Figure 11-3-6 shows the position of the high voltage electrodes in the firing assembly. The electrodes are held in a clamp device to ensure stability in the proper position. Clamped around the ceramic insulator, they hold the electrode and yet insulate the spark voltage from the grounded assembly. The electrode must insulate against 5,000 to 7,500 V (one-half the spark voltage) and still stand up against the heat of the burner and combustion changer when the burner shuts off.

The ceramic insulators are hard and brittle and crack easily. If any twisting or bending pressure is applied to change the position of the electrodes, loosen the electrode clamps. Do not attempt to bend the electrode wires; they are harder than the ceramic. Also, when tightening the clamps, they should not be overtightened.

SERVICE TIP

If any fire cracks or crazing are noted in the surface of the insulators, replace them. Do not take a chance on the old ones.

5. Ignition transformer defective. Ignition transformers are 10,000 to 15,000 V with a grounded center tap on the high voltage side. With 12,000 V between the terminals and 6,000 V from either terminal to ground, attempting to check the transformer by producing a spark with a wire, screwdriver, or other shorting means can be dangerous.

In time, the heat of operation dries the transformer, and cracks develop. Moisture enters the

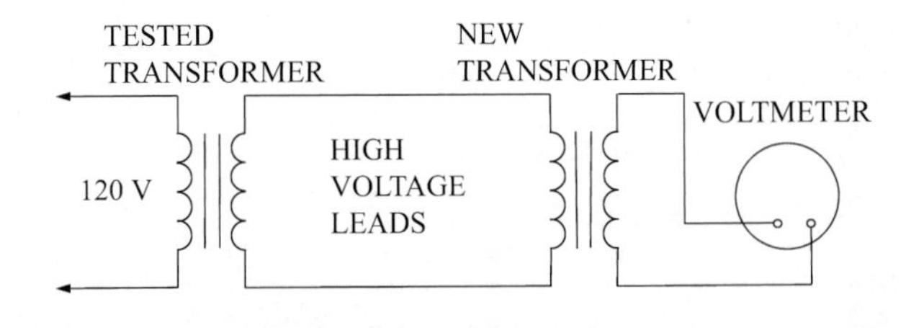

Figure 11-3-7 Transformer test wiring diagram.

assembly and is absorbed into the windings. Producing shorts between the windings, the output voltage is lowered to the point of failure of the spark across the ignition gap, and faulty ignition results.

To check a transformer, the best way is with a high voltage meter in the range of 10,000 to 15,000 V. If this meter is not available, two 120 V voltmeters and a new ignition transformer can be substituted. Figure 11-3-7 shows the wiring diagram for this test.

The secondary of each of the transformers (high voltage terminals) is connected together to make the new transformer a stepdown load of the test transformer. With 120 V applied to the test transformer (measured with one of the voltmeters), the output of the new or testing transformer should be within 10% of the applied voltage. In addition, the output voltage should hold steady. If there is more than a 10% difference or if the output voltage varies, the original transformer has internal shorts and should be replaced.

6. No fuel. Before taking the burner apart, first check the quantity of oil in the supply tank (the oil gauge on the inside tank or the dip rod for the outside buried tank). Second, make sure that the tank outlet valve is open. Third, close the tank outlet valve, open the filter cartridge case, and put in a new filter cartridge. Bleed the air from the filter cartridge after assembly.

If this procedure still does not supply oil to the burner and produce fire, remove the oil burner nozzle and check the nozzle filter. If this is plugged, replace the entire nozzle assembly with one of like capacity, spray angle, and cone type.

SERVICE TIP

If the oil filter or oil strainers are plugged with rust, the oil storage tank must be cleaned or replaced. A dirty oil storage tank will continually cause the filter and strainers to be plugged, resulting in numerous callbacks.

If these steps do not produce oil flow when the unit runs, check the inlet screen of the fuel unit. Figure 11-3-8 shows a single stage fuel unit cutaway.

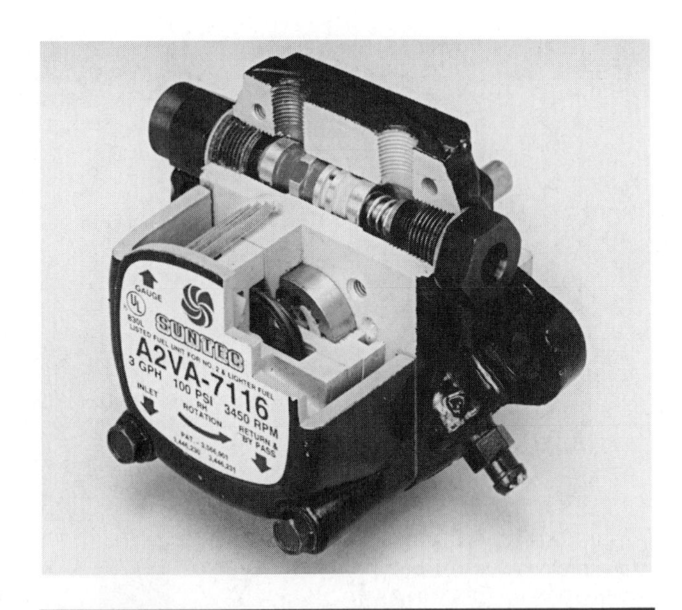

Figure 11-3-8 Fuel unit, cutaway view. *(Courtesy of Suntec Industries, Inc.)*

Within the unit, a fine mesh screen filters the oil supply before it enters the pump gear assembly. This screen can be removed and replaced. Do not leave this screen out of the pump.

After any of the portions of the fuel supply system have been opened, the system must be purged of air. A two pipe system will automatically purge itself of air. The single pipe gravity feed system does not have an automatic purge feature. Purging must therefore be done. With a short piece of plastic hose from the valve to a suitable container and the valve open, operate the unit until a clean stream of fuel oil is emitted from the hose. It may be necessary to reset the protector relay safety switch several times before the supply line is completely purged.

If it is not possible to obtain a flow of clear fuel oil, it is possible that the unit is receiving air through a line or fitting leak. This must be corrected to provide proper burner operation.

7. Fuel unit defective. The most common problem in fuel units is poor cutoff of the oil supply when the unit cycles off. Correct cutoff will provide instantaneous cutoff of the oil to the fuel pipe and nozzle when the pump pressure drops to 80% of the operating pressure. At 100 psig operating pressure the cutoff pressure is 80 psig. This is accomplished as the spring forces the control piston in the fuel unit against the fuel outlet seat and cuts off the flow of oil. The nozzle pressure drops immediately. Figure 11-3-9a,b shows the oil circuit of a single stage and a two stage fuel unit.

If particle buildup occurs on the face of the neoprene seat on the end of the piston, this prevents good full circle contact of the neoprene disk against

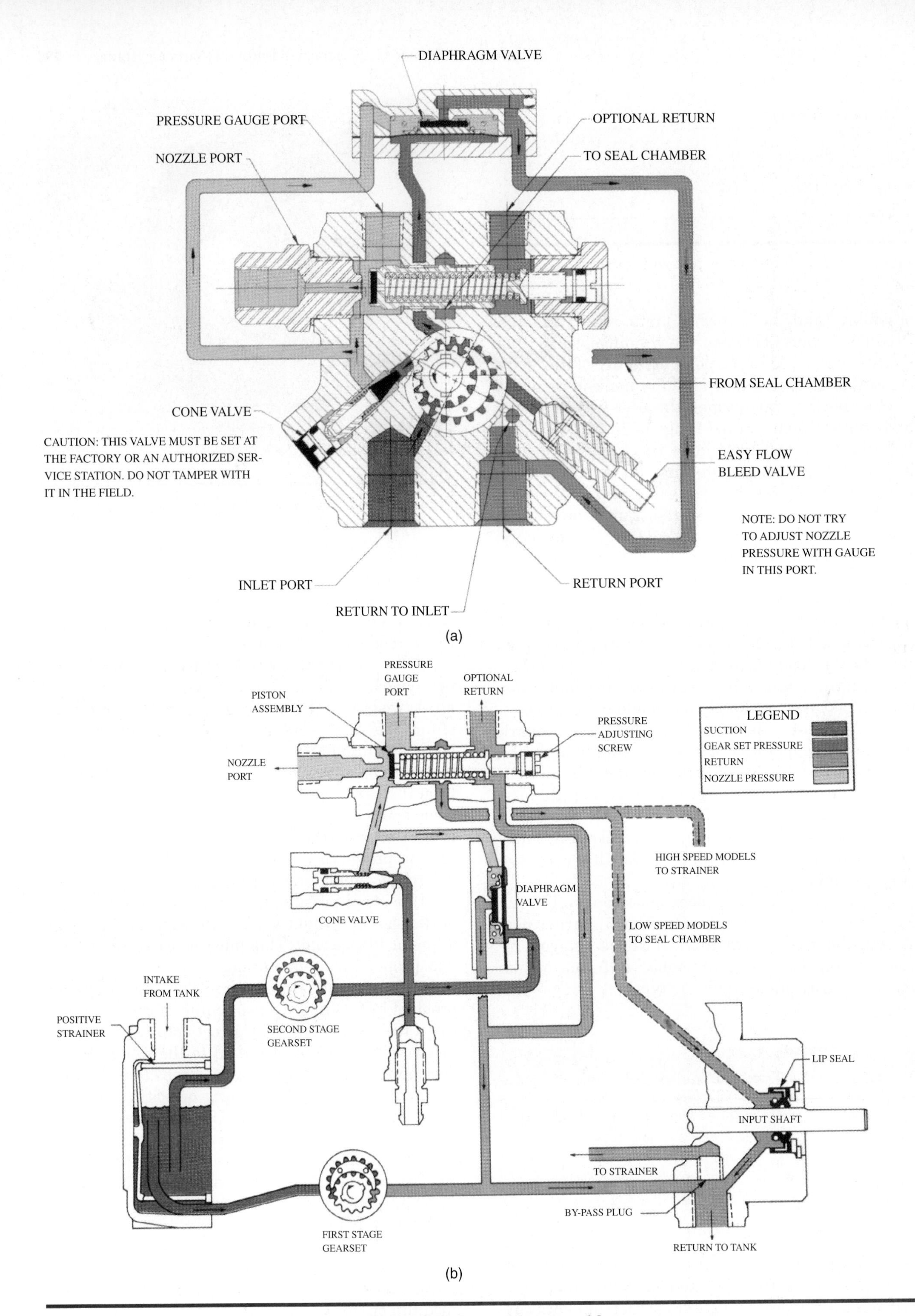

Figure 11-3-9 (a) Single stage fuel unit; (b) Two stage fuel unit. (*Courtesy of Suntec Industries, Inc.*)

Figure 11-3-10 Delayed action oil valve used with fuel unit. *(Courtesy of Suntec Industries, Inc.)*

the seat. Instead of positive cutoff, leakage occurs and the pressure gradually decreases in the nozzle. This gradual pressure reduction causes oil flow from the nozzle after the unit stops and the air supply disappears. The oil now burns with very little combustion air, producing a very smoky flame.

Carbon builds up on the turbulator end of the firing tube as well as the firing assembly. If the burner is not slanted at least 2° downward toward the combustion chamber, burning oil can flow back toward the blower. This can burn or smoke up the cad cell, with resulting cutoff of the burner on the safety switch. The correction for this is to clean the piston chamber of the fuel unit as well as the intake screen.

A persistent cause of this problem due to the quality of fuel oil supply can be reduced by double filtering of the oil before it reaches the unit and the use of a delayed action oil valve as shown in Figure 11-3-10.

Located in the fuel line between the outlet of the fuel unit and the firing head, this valve provides a time delay between the time the blower starts and the start of oil spray. This ensures that air for combustion as well as airflow through the heat exchanger is established before combustion starts. The valve also provides instant cutoff of oil pressure to the nozzle, regardless of the action of the fuel unit. This is a highly recommended accessory for any oil fired unit.

TECH TALK

Tech 1. How in the world do you stay clean when you are working on an oil fired furnace? I worked on one the other day and got soot all over me.

Tech 2. Unfortunately, any time you are working on an oil fired furnace there is a good chance you are going to get soot on you. You need to wear coveralls. Some guys wear the disposable ones that they can roll up in their disposable drop cloth when they are finished and throw the whole messy pile away.

Tech 1. I didn't know they had disposable coveralls.

Tech 2. Yeah. They are made of paper and they are made to be used one time and discarded. They are not that expensive but well worth having. You might want to get some disposable gloves, too.

Tech 1. Thanks, I appreciate that. I will get some.

UNIT 3—REVIEW QUESTIONS

1. What is the most serious problem call that is received?
2. What preliminary information should be gathered when a no heat call is received?
3. List the categories of complaints or problems.
4. What is the possible problem if an oil fired system runs continuously?
5. What is the possible problem if an oil fired system burner cuts off on safety switch?
6. What are the two types of protector relays?
7. What is the one common dimension of the three different burners?
8. To check a transformer the best way is with a high voltage meter in the range of ________ to ________.
9. When there is no fuel what are the three things that should be done when taking the burner apart?
10. What is the most common problem in fuel units?

12

Central Residential Electric Warm Air Heating

UNIT 1

Electric Furnace

OBJECTIVES: In this unit the reader will learn about electric furnaces including:

- Applications
- Air Handling Units and Duct Heaters
- System Components
- Safety
- Heating Elements

ELECTRIC HEATING

The growth of electric heating came in the 1950s and found its main use in areas not served by natural gas pipelines. In the late 1950s it became apparent that electric utility companies were going to be faced with heavy summer air conditioning power loads. To increase the sale of electricity in the winter the electric utilities and heating manufacturers began promoting electric heating. Special rates and promotional programs like "Live Better Electrically" and "Total Electric Home" were introduced in the 1960s, and the market developed rapidly. Over the past decade environmental concerns about air pollution produced at electric generating plants and the higher electric cost have reduced the number of straight electric furnaces being installed. The trend today is to have electric heat as supplemental heat for heat pumps.

The heating value of electric resistance heat is one watt of power produces 3.4 Btu/hr. Resistance heating is 100% efficient; there are no losses such as those experienced with oil and gas combustion processes. Electric power is measured in kilowatts (kW). One kilowatt is equivalent to 1000 W. The term kilowatt-hour (kW) is used to express the amount of energy used in an hour. Electric resistance furnace input is given in kWh and the output is given in Btuh (also referred to as Btu/h, but either way, it is Btu per hour). To find the heat output of an electric furnace from the input kW, multiply the kW input by 3,400. The input kWh can be calculated by dividing the furnace Btu/hr output by 3,400.

ELECTRIC FURNACES

The typical residential electric furnace, Figure 12-1-1, consists of a cabinet, blower compartment, filter, and resistance heating section. Most manufacturers provide a place for a cooling coil. An electrical furnace has many advantages:

1. It is more compact than the equivalent gas or oil furnace.

2. Due to cooler surface temperature, it has "zero" clearance requirements on all sides. It may, therefore, be located in small spaces such as closets.

3. Since there is no combustion process, there are no requirements for venting pipes, chimney, or makeup air. This reduces building costs and simplifies installation.

4. Units may be mounted for upflow, downflow, or horizontal airflow applications, Figure 12-1-2.

The blower compartment, Figure 12-1-3, usually houses a centrifugal multispeed direct drive fan. Larger units use a belt driven fan. An electric furnace has less resistance to airflow, and fan performance is more efficient. Mechanical air filters are usually used.

The heating section consists of banks of resistance heater coils of nickel chrome wire held in place by ceramic insulators, Figure 12-1-4. The heater

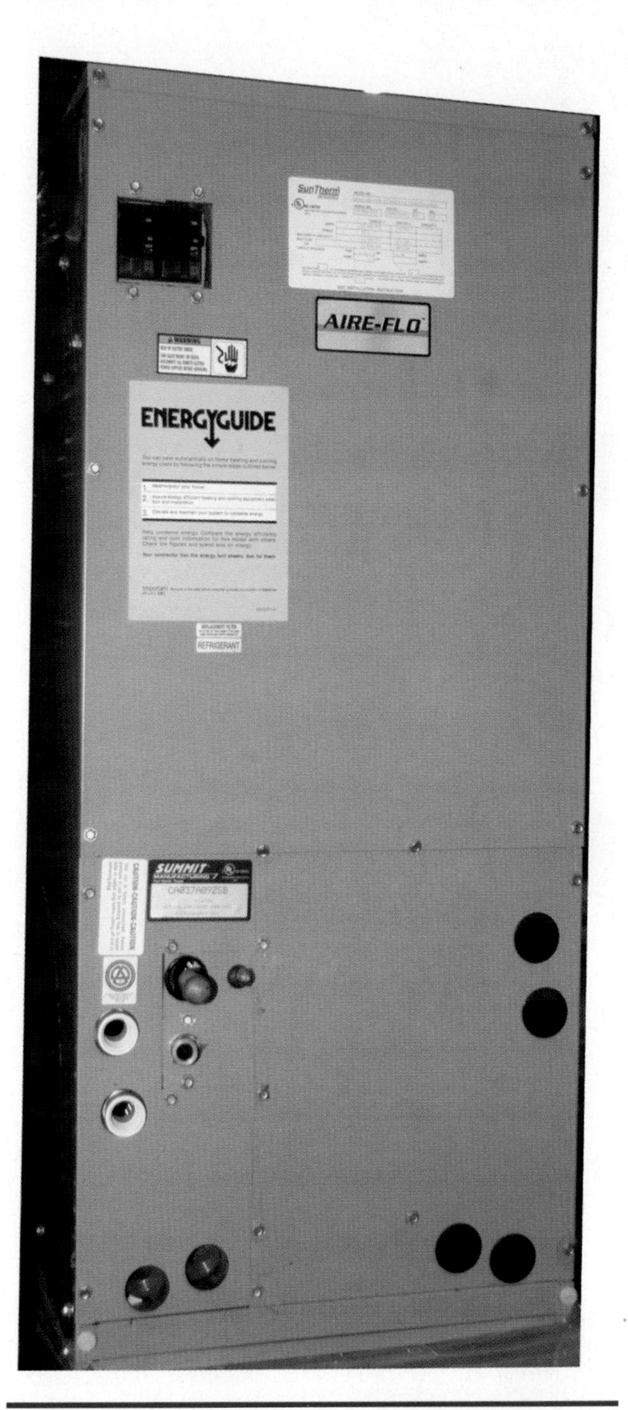

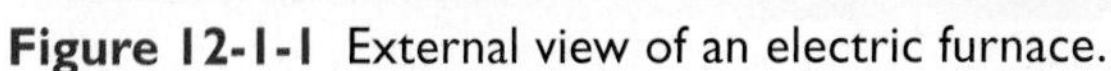

Figure 12-1-1 External view of an electric furnace.

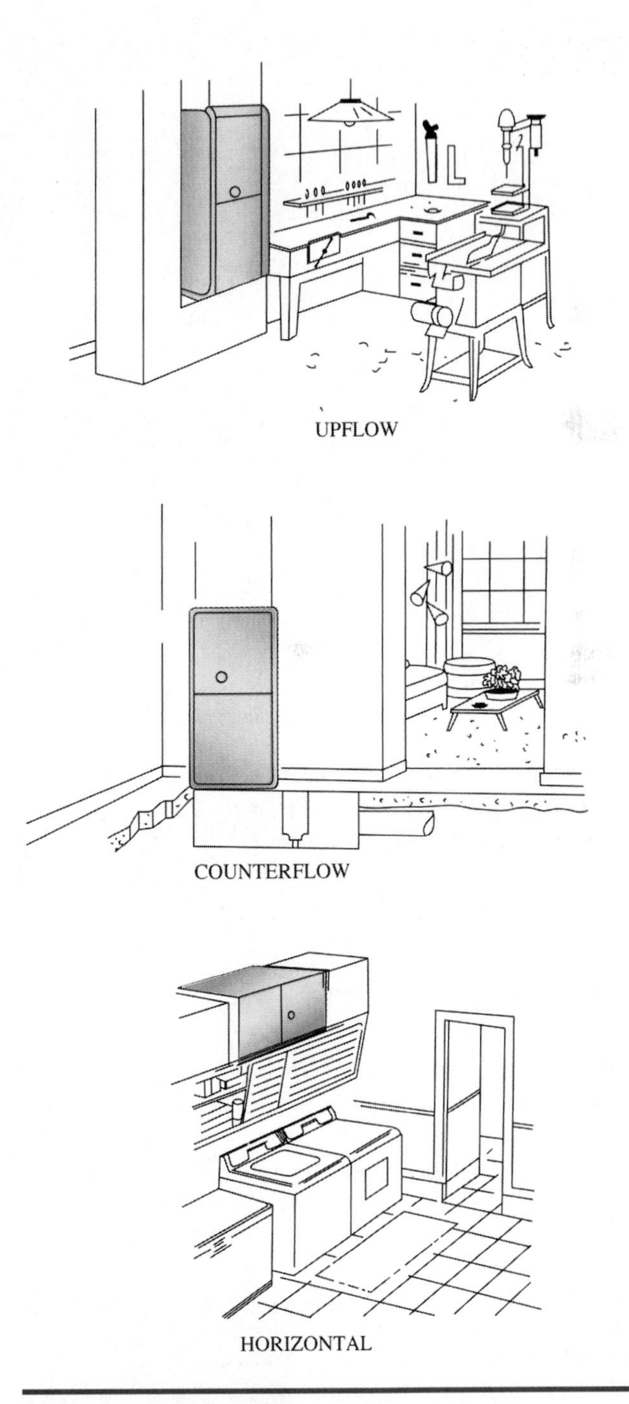

Figure 12-1-2 Various types of electric furnaces.

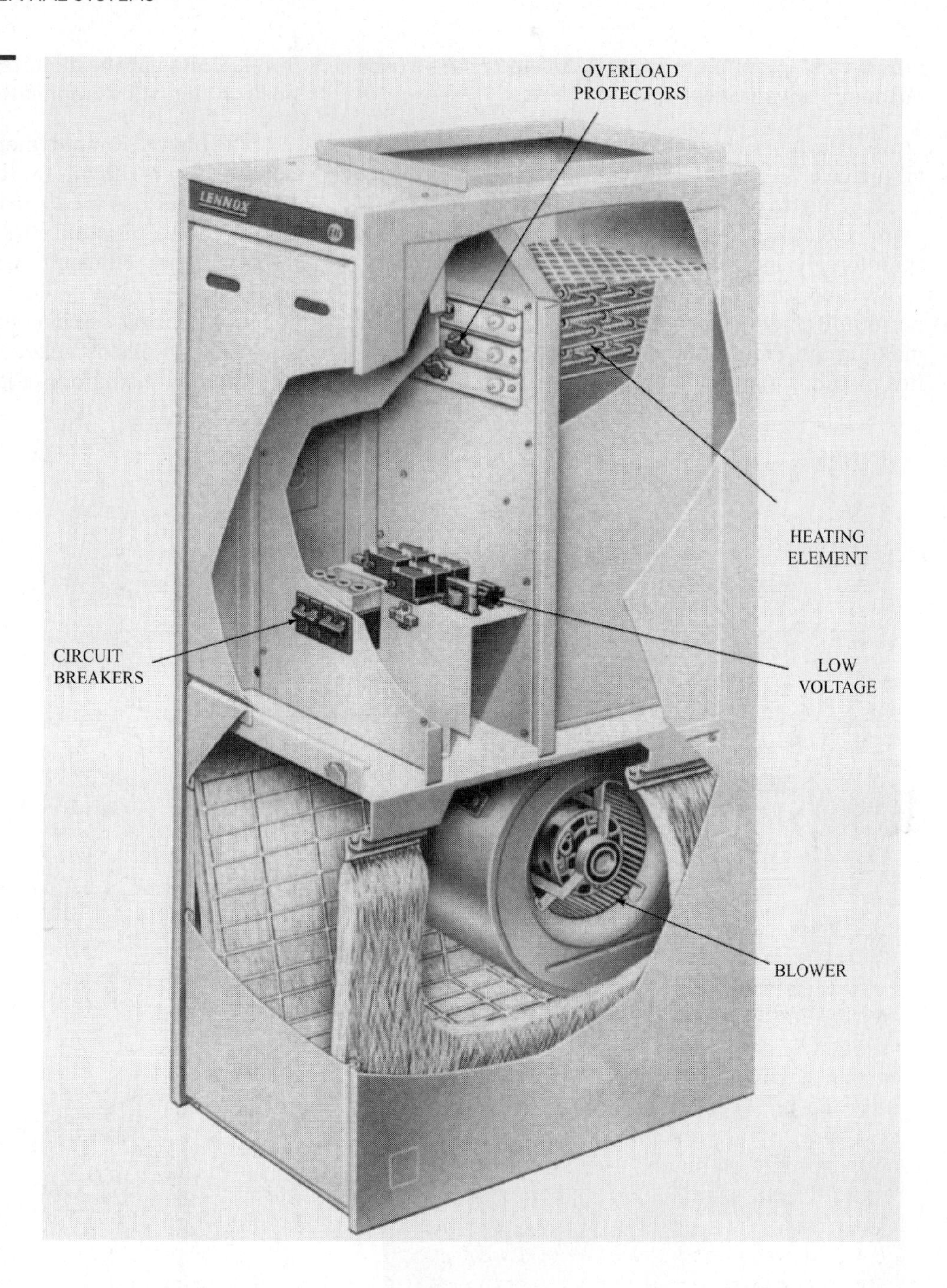

Figure 12-1-3 Internal diagram of an electric furnace. (*Courtesy of Lennox Industries, Inc.*)

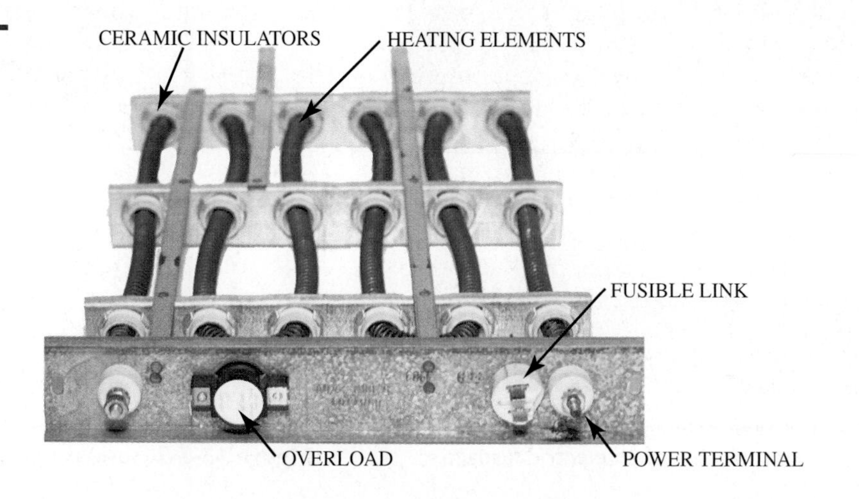

Figure 12-1-4 Electric heating element rack.

operates on a 208 V to 240 V power source. The amount of heat produced by each heating element is affected by the supply voltage.

The amount of heat per bank is a function of the current flow (amperage) draw. The amount of current that can be put on the line in one surge is limited by regulations, residences being limited to 200 A. Total furnace output capacities range from 5 kW (17,000 Btu) to 30 kW (102,360 Btu/hr). At 30 kW the amperage draw on 240 V approaches 125 A and, considering that 200 A is the total service to a residence, this only leaves 75 A for other electrical uses. Two electric furnaces would therefore be used, with sequencing and zoning.

SAFETY TIP

Many electric furnace installations use aluminum wire for the primary power. Aluminum wire is subject to oxidation at the terminals. Oxidation at the terminals can cause heat buildup to the point where arcing or a fire can occur. To prevent this from happening, use an anti-oxidation cream on the terminus before it is assembled. Follow the anti-oxidizing cream manufacturer instructions for use.

The sequenceer control is an important operation of the electrical furnace. There are single stage and two stage thermostat systems, depending on the size of the unit. When there is a call for heat, the electric sequencer closes its contacts, the circuit is energized and begins to heat up. The blower and first heater come on; there is a time delay before each additional heater stage is energized. This time delay (in seconds) is adequate to stagger the inrush current.

Larger furnaces and some heat pumps use multistage thermostats. For electric furnaces the first stage would bring on at least 50% of the total capacity. The second stage of the thermostat would respond only when full heating capacity is needed. With this added control, wide variations in indoor temperature are avoided. For heat pumps the first stage brings on the heat pump and the next stage brings on the electric strips as needed.

While these design features are feasible for modifying the load, code assumes all installations are used continuously, for safety purposes.

Electric heating elements are protected from any overheating that may be caused by fan failure or a blocked filter. High limit switches sense air temperature and open the electrical circuit when overheating occurs, Figure 12-1-5. As a backup, some furnaces have fusible links wired in series with the electric heater coil and melt at 300°F, opening the circuit.

Figure 12-1-5 Limit switch for electric heat elements.

Built-in fusing or circuit breakers are provided by some manufacturers in compliance with the National Electric Code K. This eliminates the need for external fuse boxes.

CAUTION: All replacement wiring used in an electric furnace must have a rating of 105°C with an AWM type appliance wire rating. This wire will stand a much higher environmental temperature which can be experienced within an electric furnace cabinet.

The air distribution system for an electric furnace requires the following special considerations:

1. Air temperatures coming off the furnace are lower, as compared to gas and oil equipment. Temperatures of 120°F and below can create drafts if improperly introduced into conditioned space. Additional air diffusers are recommended.

2. Duct loss through unconditioned spaces can be critical, so well insulated ducts are a must to maintain comfort and reduce operating cost.

Applications

One of the most useful applications of an electric furnace is its combination with a split system heat pump. In many geographical locations where heat pumps are installed, electric furnaces supply the additional heat necessary to keep the space at thermostat setpoint when the outdoor temperature drops below the point where the heat pump will not maintain indoor temperature. This type of installation is

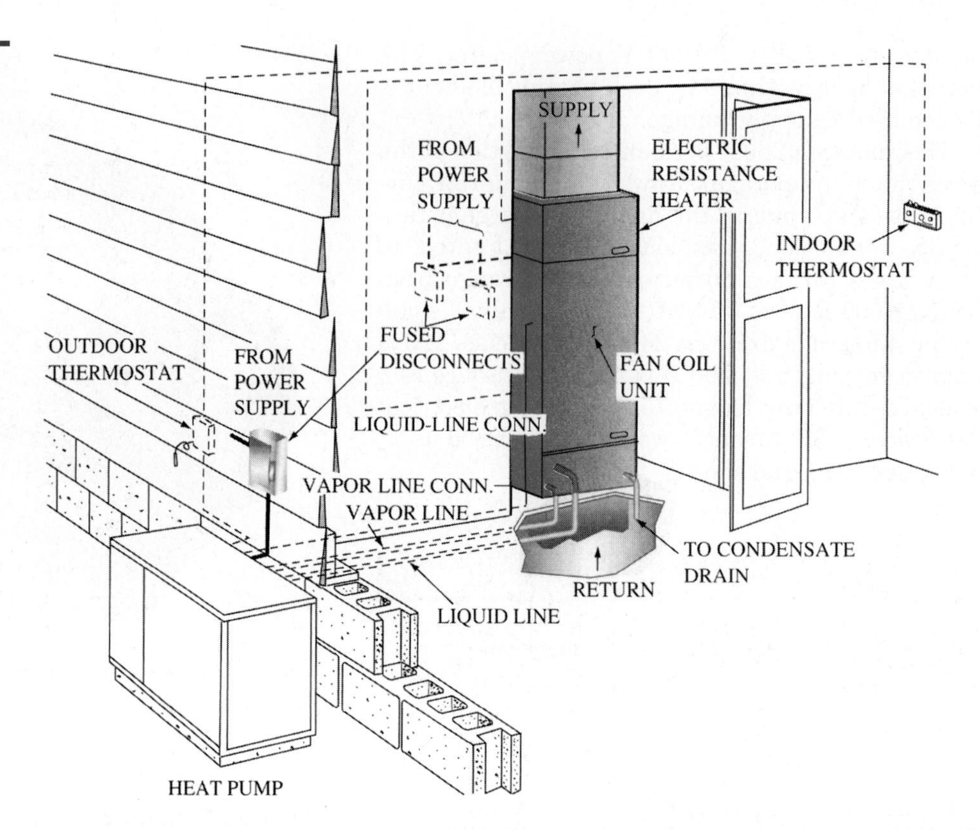

Figure 12-1-6 Split system heat pump.

shown in Figure 12-1-6. Some applications are required to have a low ambient cutout that will turn off the heat pump when the outdoor temperature drops too low.

For these installations the electric furnace acts as a fan coil unit. The refrigerated coil is installed in the furnace before the blower. The coil is connected by refrigerant tubing to the outdoor heat pump located on a slab outside the building. The thermostat is located in the conditioned space and controls the operation of the complete system to produce the heating or cooling as required.

Air handling capabilities of electric furnaces are similar to other forced air furnaces, using a multispeed direct drive blower. Multispeed direct drive blowers have the capability to adjust the blower speed to deliver the required air quantity depending on the static resistance of the supply and return air duct system.

AIR HANDLING UNITS AND DUCT HEATERS

A variation of the electric furnace is an air handling unit with duct type electric heaters. The air handler consists of a blower housed in an insulated cabinet with openings for connections to supply and return ducts. Electric resistance heaters are installed either in the primary supply trunk or in branch ducts leading from the main trunk to the rooms in the dwelling. When heaters are installed in branch runs, then there is comfort flexibility by zone control; for example, room temperature can be individually controlled. This system has not been as popular as the complete furnace package, primarily because it complicates installation requirements and adds costs.

Duct heaters are made to fit standard size ducts and are equipped with overheat protection devices. Electric duct heaters can also be used to supplement other types of ducted heating systems, to add heat in remote duct runs or to beef up the system if the house has been enlarged. They may be connected to come on with the furnace blower, and are thus controlled by a room thermostat. Additional safety may be achieved by installing a sail switch in the duct that must sense air movement before the duct heater will come on.

SYSTEM COMPONENTS

Sequencers

Electric heating elements have a very high inrush current draw as they initially come on. The current draw drops rapidly as the heating element gets hot. The initial high current draw is caused by the low temperature of the element. Electrical resistance increases as the temperature of the conductor increases. When the heating element begins to glow red, it draws less current.

Electric furnaces that have multiple heat strips are designed with a sequencer that allows each heat strip to come on individually or in small groups. Without a sequencer, the initial current draw for an electric furnace could be significantly higher than the

Figure 12-1-7 White center pin shown pushes upward on contact.

system's maximum amperage capacity. Most systems use sequencers to sequence or stage on elements. A sequencer uses a small heating element wrapped around a bimetal strip. When the bimetal strip warms up enough, it snaps or warps to turn a heat strip on. Some sequencers have multiple contacts that are staged on. Staging can be accomplished by using a center pin that pushes upward on each successive set of sequencer contacts, Figure 12-1-7. In some cases where several heating elements are sequenced on multiple sequencers are used, Figure 12-1-8.

To prevent the heating elements from overheating, the unit blower must come on with the first heating element and it must remain on throughout the heating cycle until the last stage heating element has cycled off. Some sequencers have specific terminals that are designated for the blower. This set of contacts is the first on and the last off in the sequence.

The time interval between when a sequencer is energized and when the sequencer's contacts make is predetermined, within a range, by the sequencer's manufacturer. Some manufacturers offer sequencers with different time cycles. Likewise, the time inter-

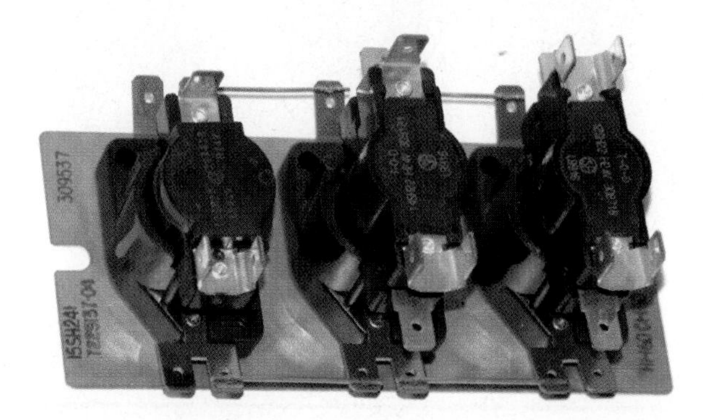

Figure 12-1-8 Multiple sequencers.

Figure 12-1-9 Fusible link.

SERVICE TIP

UL requires that the blower come on with or before the first heat element and must remain on until the last heating element has been de-energized.

val between when a sequencer is de-energized and each of the contacts drops out is preset by the manufacturer. In some cases sequencers stagger on and all go off at the same time. Check with your supply house or the manufacturer to see what time cycle is required for the unit you are working on.

Safety

All electric resistance heating circuits have one or more safety devices to protect the system from overheating and overcurrent. A fusible link, Figure 12-1-9, is a small one time use device that is wired in series with the heating element. Some manufacturers use fusible links that are mounted in ceramic. Others use fusible links that are spot welded to the heating element. Fusible links are made of an alloy that melts at a specific temperature. The advantage of a fusible link is that the melting temperature of the alloy remains constant over time. This provides the maximum level of safety because the fusible link can be counted on to open instantly any time an overcurrent situation occurs. Because the opening function of fusible links is based on the melting temperature of an alloy, they are highly reliable overload protectors.

An overload is another overcurrent protector for electric heating elements, Figure 12-1-10. Overloads

Figure 12-1-10 Overload protector.

sense both current and temperature so that either over temperature, overcurrent, or a combination of these two will cause the overload to open. Overloads will automatically reset once the bimetal strip in the overload has cooled sufficiently.

Inline cartridge (BUSS®) type fuses, Figure 12-1-11, are used to protect individual heating elements, groups of heating elements, or the entire furnace. The metal clips that hold cartridge fuses in place have sufficient spring tension when they are new to prevent heating of the connection between the fuse and the fuse holder. However, over time these connections can become loose, resulting in resistance that will cause them to heat up and lose their spring tension. If these clips have become discolored—even the slightest amount—they should be replaced before the resistance gets high enough that an arc occurs between the fuse and fuse clip. To reduce damage to the fuse spring clips and work more safely, you should remove the fuses with a fuse puller tool. Do not pry them out with a screwdriver.

(a)

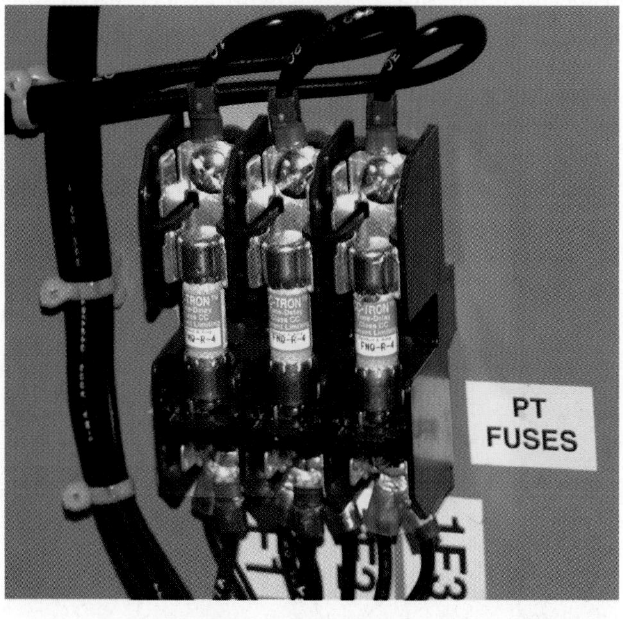

(b)

Figure 12-1-11 (a) Cartridge type fuse; (b) Fuses are held by metal clips and should be removed with a tube puller tool and not by a screwdriver to preserve the spring tension of the clips.

SERVICE TIP

Some inexpensive cartridge fuses will deteriorate over time when used in an electric furnace. These fuses can actually burn through the outer casing when they fail. This can result in damage to nearby wiring. Use only quality fuses such as those purchased from air conditioning and refrigeration supply houses. If a customer replaced fuses on their unit with the lower grade fuses, you should change them (or at least note it on your service ticket).

Circuit breakers, Figure 12-1-12, may be used as overcurrent protection. Circuit breakers used on the 208 V to 240 V systems have two breakers that are interlocked so that overcurrent in one leg causes both breakers to open, removing power from the load. Some furnaces use more than one set of breakers so that individual parts of the heating circuit can be protected separately. An advantage of circuit breakers is that they can be reset.

Heating Elements

The heating elements used in electric furnaces are made out of a nickel chrome alloy. The nickel chrome

Figure 12-1-12 Dual circuit breaker.

(a)

(b)

Figure 12-1-13 (a) Nameplate for 60 kW commercial heater; (b) Heating element for 60 kW commercial heater.

alloy is used because it has both a high electrical resistance and resistance to oxidation. Many metal alloys have an electrical resistance too low to be effective heating elements. Also, many alloys are susceptible to extreme oxidation at elevated temperatures. Heating elements appear to be long coil springs, and are bright in color until the coil is heated. As the elements are used, they will begin to discolor. Eventually, they will become dull black in color. The length of the heating element determines its kW capacity. Heating elements are available in 5 kW through 30 kW capacity for residential electric furnaces and may be as large as 60 kW for commercial electric furnaces. The typical kW options of a heating unit are shown in Figure 12-1-13a. Figure 12-1-13b shows the heating elements held in place by ceramic spacers.

Electric heating elements are rated by their kW capacity. The kW capacity can be converted to Btuh by multiplying the kW value by 3.412. Electric heating elements come in a pre-wired rack. Each electric furnace is designed to hold a specific type and size heating element rack. These racks are not interchangeable from manufacturer to manufacturer, or in many cases from one model furnace to another within the same manufacturer's product line. When replacing a heating element rack, be sure that you specify the model and serial number of the furnace the rack is fit into. Some heating element racks can be restrung with a replacement heating element, Figure 12-1-14. An advantage to restringing heating elements is that only a limited number of heating element sizes need to be stocked in your service van.

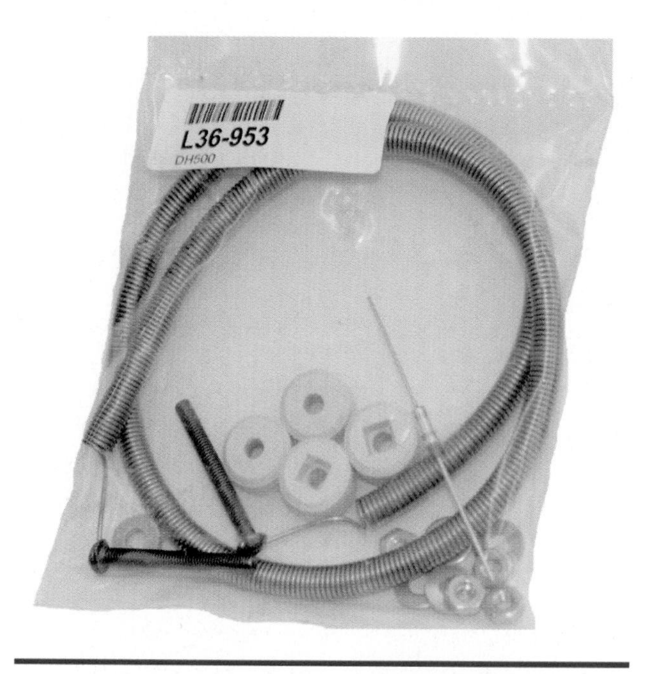

Figure 12-1-14 Electric heating element restringing kit.

An advantage to replacing the heating element rack is that it is a significant time savings over restringing, and you are providing the customer with a entirely new part. Sometimes the ceramic insulators can be damaged during restringing, such as the one shown in Figure 12-1-15. A damaged insulator may fail some time after the installation job is complete.

Because of energy conservation concerns, some cities, counties, and states have ordinances or laws that prevent straight electric furnaces from being installed in new construction. The electric furnace is often used in those areas in conjunction with a heat pump where it serves as supplemental heat. An advantage of using an electric furnace as a supplemental heat source in a heat pump application is that the heat pump indoor coil can be placed in the return air stream to the blower. This creates a draw through coil for the heat pump, which is the most efficient application of indoor heat pump coil. This is in contrast to gas or oil supplemental heating systems, where the coil must be located after the firebox to prevent condensation from forming on the heat exchanger.

TECH TALK

Tech 1. As expensive as electricity is, why would anyone want to put in an electric furnace?

Tech 2. You're right, electric heat can be a lot more expensive than oil, gas, or heat pump. But for some people who only use heat occasionally, it's worth it for them to have electric heat because the service required is less than for the other heating systems. In fact, unless an electric furnace cabinet has become damaged by rust or an electrical fire, there is very little reason it should ever be replaced. On an electric furnace the elements can be changed, the blower can be changed, and the controls can be changed so easily the owner probably will never want to pull the whole furnace out for a replacement.

Tech 1. What if the coil goes bad?

Tech 2. In most cases you can simply replace the coil. However, if you are going to a higher efficiency system, the new coil might be too large to fit in the original case. So I guess you are right—a bad coil would be a good reason to change out an electric furnace.

Tech 1. Thanks, I appreciate the information.

Figure 12-1-15 Damaged ceramic insulator.

UNIT 1—REVIEW QUESTIONS

1. What does the typical residential electrical furnace consist of?
2. What does the heating section consist of?
3. The amount of heat per bank is a function of the ________.
4. What special considerations are required for the air distribution system for an electric furnace?
5. What are electric duct heaters used for?
6. What is one of the most useful applications of an electric furnace?
7. What is a variation of the electric furnace?
8. What is the advantage of a fusible link?
9. What are inline cartridge (BUSS®) fuses used to protect?
10. What are the heating elements used in electric furnaces made of?
11. How are electric heating elements rated?
12. What is the advantage to using an electric furnace as a supplemental heat source to a heat pump?

UNIT 2

Troubleshooting Electric Furnaces

OBJECTIVES: In this unit the reader will learn about troubleshooting components of an electric furnace, including:

- Heating Element
- Fuse Line
- Overload
- Sequencers
- Transformers
- Indoor Blower Relay
- Determining CFM Through Furnace

Figure 12-2-1 Broken heating element.

HEATING ELEMENT SERVICE

From time to time the heating element in an electric furnace must be replaced because it breaks. When a heating element breaks, Figure 12-2-1, it cannot be repaired and it must be replaced.

There are several ways to identify a broken heating element. Broken elements can be identified by amp draw, ohm readings, or by visual inspection. To identify a broken heating element using the amp draw, first turn on the system and allow time for all of the sequencers to energize. Next, use a clamp-on ammeter to carefully check each of the circuit's power wires for amperage, Figure 12-2-2. Any broken elements will have a zero amp draw.

Figure 12-2-2 Clamp-on ammeter used to measure heating element amp draw.

SAFETY TIP

Some newer digital multimeters have internal protection to prevent meter damage if a live circuit is touched with the meter set to ohms. However, not all meters have this protection and accidentally touching your ohmmeter to a live circuit may blow an internal fuse or could actually destroy the meter. If your meter does not have automatic protection as described in the owner's manual, you should check the circuit for voltage before switching the meter over to ohm. This will help you prevent meter damage.

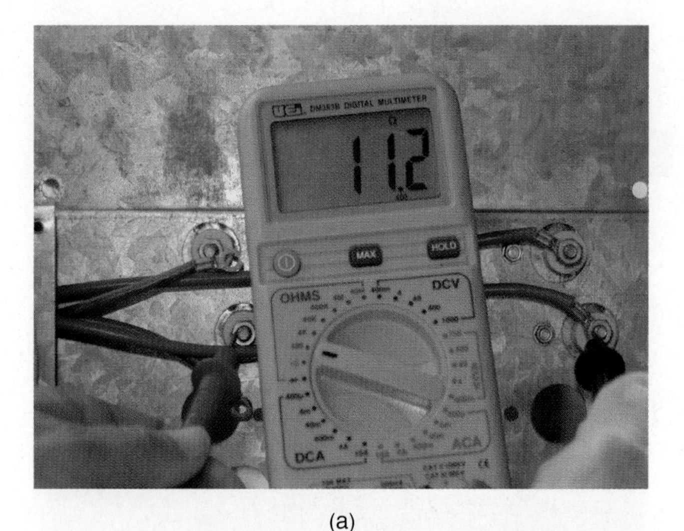

(a)

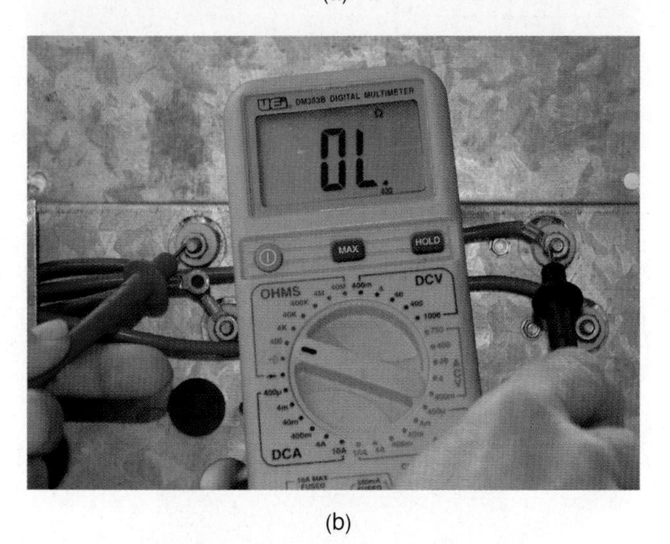

(b)

Figure 12-2-3 Using an ohmmeter to check heating element continuity: (a) A reasonable value of ohms indicates a working element; (b) A reading of ∞ or OL for overload indicates a broken element that must be replaced.

The power to the furnace must be turned off before doing an ohm check of the heating elements. To use an ohmmeter to check for broken heating elements, you must first isolate the heating element. Isolate the heating element by removing one of the power leads to the element. It is not necessary to remove both in order to isolate the circuit. Using your ohmmeter probes, check for continuity through each of the heating elements, Figure 12-2-3. The broken heating element will have infinite ohms, shown as OL (overload) in Figure 12-2-3b.

The heating elements must be removed from the furnace to do a visual inspection. Most furnace heating element racks can be removed once the power wires have been disconnected. Before disconnecting the wiring, make a sketch of the wire location so that it will be easier for you to reconnect them once the heating element racks have been replaced. There are typically 2, 4, or 6 sheet metal screws that hold the rack in place. Once the rack is out, it can be inspected for broken insulators, broken heating elements, or any other visible damage.

When reinstalling the heating element rack, you must make certain that the alignment pin or bracket on the heating element rack fits into the slot at the back of the furnace. In most cases, the opening at the back of the furnace is slightly indented so that the alignment pin will track into the center opening. If the pin is not seated properly, the heating element rack will either not fit completely in or be at a slight angle. Do not force the rack into place with the screws. You must reposition it until it fits properly in place.

Use the manufacturer's wiring diagram and the diagram you made as you disassembled the heating element rack as guides to reconnect the wiring. In some cases, components may have been changed in older furnaces so that the manufacturer's diagram is no longer exactly correct. Therefore you must rely heavily on your drawing when reconnecting the furnace wiring.

Replacement heating element coils are available either prestrung in a replacement rack or as individual coils that must be restrung by the technician, Figure 12-2-4. Make certain that you are replacing the heating element with the exact same capacity (kW) replacement element. If a larger element is used, you might exceed the amperage capacity of the main power circuit. If a smaller capacity element capacity is used, then the furnace may not produce sufficient heat to keep the residence warm.

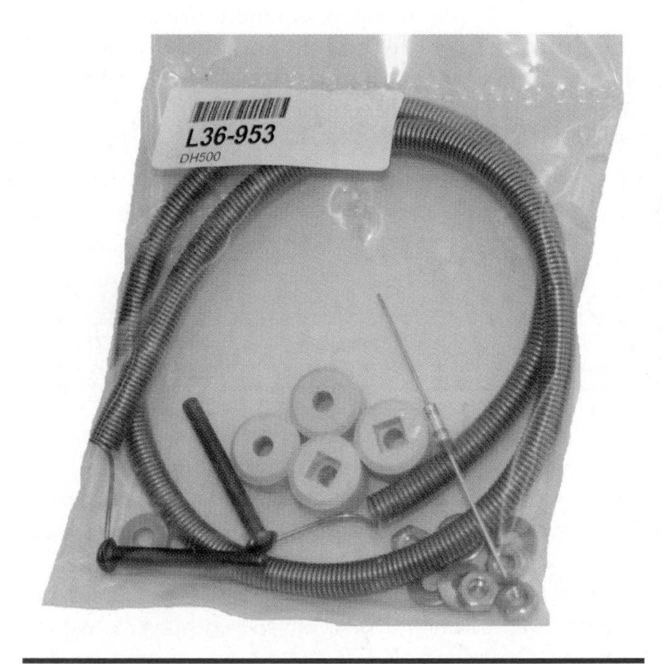

Figure 12-2-4 Electric heating element restringing kit.

Fusible Links

Fusible links are replaceable only if they are a separate component in the furnace heating circuit. Some fusible links are welded to the end of the heating element and are therefore not replaceable. Most replaceable fusible links are mounted on ceramic bases that attach in the sheet metal heating element rack plate, Figure 12-2-5a,b. Make certain that you order a replacement fusible link that is specifically designed for your particular furnace. There are a number of different designs and all are not interchangeable.

SERVICE TIP

Ceramic insulators are brittle so be sure that you do not force them as you are inserting them into their fitted location on the furnace heating rack. Do not attempt to hold a ceramic fitting down into place with the fastening screws because this will most often result in a cracked insulator. If the insulator is cracked during installation, the part must be replaced. Figure 12-2-6 shows a cracked ceramic insulator. It is not safe or acceptable to use damaged ceramic insulators.

(a)

(b)

Figure 12-2-5 Fusible links: (a) Installed in ceramic bases; (b) Replacement for fusible link.

Sequencers

There is a wide variety of different types of sequencers used in electric heating systems. It is important that you locate a matched sequencer when making a repair or replacement. Some sequencers have terminals that are marked A for auxiliary and/or F for fan/blower. Terminals such as the one pictured in Figure 12-2-7 are specifically designed for the blower.

When replacing a sequencer, make certain that you cycle the furnace on and off several times while making sure that the blower is going on and off in the correct time cycle.

TECH TIP

The contacts inside a sequencer cannot be visually inspected for wear. However, with an ohmmeter set to the R × 1 scale and the relay energized so that the contacts are made, you can test for resistance between the contact terminals. There should be 0 Ω or near 0 Ω read across these contact terminal points if the relay is good. A slightly higher resistance means that the contacts are worn and should be replaced; a high ohm reading indicates that the contacts are very pitted and must be replaced.

In order to test a sequencer to see that it is operating properly you must energize the relay's heat motor terminals with 24 V and test each set of relay contacts for continuity as they snap on. You should also test the sequencer as it is de-energized to see that each of the contacts opens as it should.

Figure 12-2-6 Broken ceramic insulator.

Figure 12-2-8 Overload protector.

Overload

The overload protector, Figure 12-2-8, is not easily tested to ensure that it is working properly. If you suspect that the overload is defective, it must be replaced. One way of testing overloads in an electric furnace is to use your clamp-on ammeter at the main power feed to the unit. Allow the unit to run for 15 to 20 minutes as you perform other checks. If any of the overloads are opening, the total current draw to the unit will drop. Some digital ammeters have data recording capacity so that the high and low amp readings remain in the instrument. These readings can be reviewed at the end of the test. This type of meter makes an overload protector test much easier.

Figure 12-2-7 Sequencer.

Another way of checking overloads is to place an electronic Tattle Tale®, Figure 12-2-9, across the overload. If the overload opens while the heating element is energized, the Tattle Tale® will indicate that the overload opened. After the heating element cools, the overload will close; however, the Tattle Tale® will continue to indicate the overload protector has opened.

Wires and Terminals

Electric furnace terminals carry large amperage loads. Any unusual resistance at the terminals will cause the terminal to discolor. If the resistance is significant enough, heat can be generated to both discolor the terminal and melt the wire insulation, as shown in Figure 12-2-10. Terminal resistance can be the result of a loose slip on connector or a set of contacts in a relay or a sequencer that are pitted and are heating up so that their heat is transmitted to the slip on connectors and connecting wire.

SAFETY TIP

Any wiring used in an area subject to high temperatures must have an appliance rating. These wires will have labels stating that they are suitable for internal wiring of appliances where the wire may be exposed to temperatures not to exceed 105°F, as shown in Figure 12-2-11. Any other non-approved wiring insulation may melt. This non-approved wire can cause a short if used inside an electric furnace.

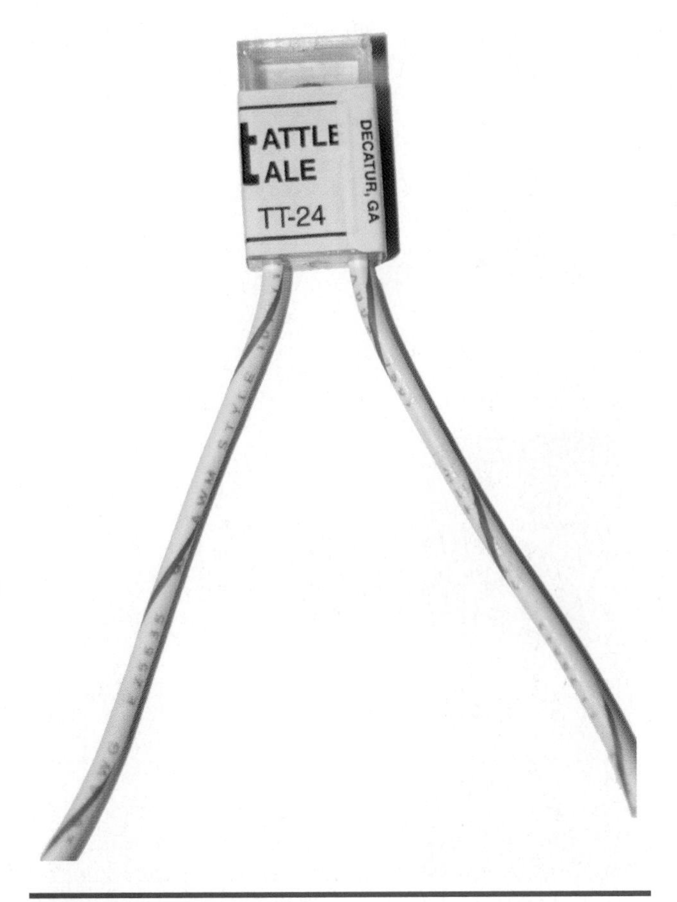

Figure 12-2-9 Tattle Tale® for testing overload protectors.

Figure 12-2-10 Overheating causes the terminal to discolor and could cause wire insulation to melt.

It is not recommended that you simply replace the terminal end if a wire and terminal have become overheated. The overheated copper conductor will have oxidized some distance back inside of the insulation; merely cutting off the end, even if you remove several inches, will not prevent this wire from overheating again. You must replace the entire wire with one of the same or larger gauge that has the proper AMW appliance rating. Any wires that are not properly rated will present a safety hazard. If the slip on terminal on a relay sequencer fuse or fuse block or other component has become discolored as a result of overheating, that component should be replaced. It is not recommended to simply sand or buff off the terminal's oxidization. Terminals have a thin coating of highly conductive material that may appear to be bright or buffed metallic. When the terminal is clean, the subbase material will be exposed; in some cases it is brass, copper, or steel. These metals are not as efficient a conductor as the coating that is applied to a new terminal. Cleaning these terminals will only serve as a temporary patch and not a permanent repair

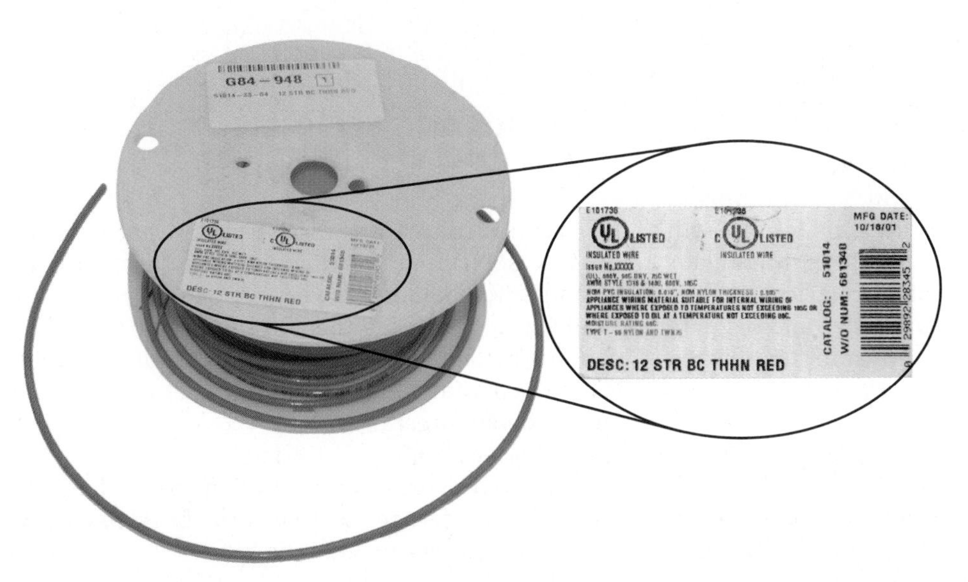

Figure 12-2-11 High temperature wire for appliance use.

because once a terminal has been overheated, it will never withstand the designated current load again.

TECH TIP

It is important that you use a quality crimping tool any time you are putting a new end on a wire that is going to be used in an electric furnace. Furnaces have fairly large amp draws and if the terminal is not extremely tight, the slightest resistance can cause the terminal and wire to overheat.

Blower Motor

Some electric furnaces come from the manufacturer with a single speed fan. A single speed fan is used because the typical airflow for an air conditioning system is the same as the typical airflow requirements for an electric furnace of 400 CFM per ton of air conditioning. For that reason there may be only a single speed fan motor in the unit. In most cases, a multispeed blower motor can be used to replace the single speed blower motor if a motor replacement is required.

Some new blower motors do not have oil ports. These motors are supplied with a "lifetime" of lubrication sealed in the motor bushings. If one of these sealed motors has developed a bearing problem, it cannot be lubricated and must be replaced. Motors that have oil ports should be lubricated no more than once annually for heating furnaces, and no more than twice annually for heating/cooling furnaces. The oil used to lubricate a blower motor must be specifically labeled as electric motor lubrication oil , such as the examples in Figure 12-2-12a,b. Any other oil, such as 3 in 1, WD-40, or motor oil, is not acceptable. Some of these oils have detergents or solvents that will actually damage the motor's windings over time.

Do not over oil blower motors. Most motors require 2 to 3 drops of lubricant in each end's oil ports. Excessive oil will run out of the bearing seals and be slung onto the motor and motor windings. Because of the dust present in an air handler, this oil film will collect a layer of dust. Dust on the motor and motor windings acts as an insulator and restricts airflow. In both cases, the motor will run hotter and begin to break down the motor winding insulation. Over a relatively short period of time, this can cause the motor windings to short out.

If the motor has to be removed in order to lubricate the inner bearing, make certain that it is reinstalled with the oil ports facing up. In addition, make certain that the oil filter port plugs have been replaced before the motor is put back in service. Failing to cap these ports will allow dirt and dust to collect in the port, which will be carried into the bearing the next time the motor is lubricated. This can cause damage to the motor bearings.

Figure 12-2-12 (a) Zip oiler; (b) Telescoping spout.

FURNACE EFFICIENCY AND CFM

Electric furnaces are the easiest furnaces to accurately determine the exact airflow through the air handler. This makes it very easy to properly set the fan speed to provide the 400 CFM of airflow per ton that is recommended for most air handlers. To calculate the actual CFM of an air handler, you must take the following readings:

1. Total amperage draw of the heating elements. This amperage must be of the heating elements only and not include the blower.

2. Closed circuit voltage at the heating elements as the unit is operating.
3. The return air temperature.
4. The supply air temperature. Make certain that this temperature is not taken with a probe inserted in the supply plenum so that it would be in a direct line of sight with the heating elements. The radiant heat that the elements produce would give you a false high reading.

Use the following formula:

$$\text{CFM} = \frac{V \times A \times 3.412}{\Delta T \times 1.08}$$

$$V = 220\text{ V}$$
$$A = 27.8\text{ A}$$
$$\Delta T = 47°\text{F}$$

$$\text{CFM} = \frac{220 \times 27.8 \times 3.412}{47 \times 1.08}$$
$$= \frac{20{,}868}{50.76}$$
$$\text{CFM} = 411$$

Transformer

The transformer, Figure 12-2-13, in an electric furnace is subject to a greater load than most air conditioning transformers because the system may have a number of sequencers, all of which have heat motors. You must check the secondary voltage supplied by a control voltage transformer while the system is operating to determine if there is sufficient voltage drop that will cause the system to be intermittent. The minimum operating voltage on the secondary is 18 V. If the transformer is not providing at least 18 V, it must be replaced.

SERVICE TIP

Electric furnaces use 208 to 240 V as their primary power. Most replacement transformers have multiple power taps. Refer to the transformer's specifications to make certain that you are connecting the proper leads for the voltage your unit operates on. It is not an acceptable practice to simply assume that you are on a 240 V tap. If you do this on a 208 V system, the secondary voltage will be lower than 24 V. The voltage will be low under some circumstances where there are a number of current drawing devices on the secondary, such as humidifiers, multiple sequencers, and dampener motors. For example, under these heavy loads there may not be sufficient voltage and current to operate all of these devices.

Figure 12-2-13 Control voltage transformer.

Blower Relay

The blower relay on electric furnaces is often wired into the number one heating element that comes on first and goes off last so that the blower is ensured to be functioning at the same time the strip heat is on. If the contact in the blower relay becomes stuck closed, current can flow through the circuit to the heating element whether or not the heat is on. This can occur during both air conditioning and fan on run cycles. An indication that this has occurred would be larger than normal utility bills during the cooling season. This can be checked by placing a clamp-on ammeter around the heating element power wires. If any element is energized during the cooling cycle the blower relay must be replaced.

TECH TIP

Occasionally during new construction metal debris can accidentally be left in a duct system. This debris can fall onto the heating elements because they are located at the top of the unit and are not protected by an evaporative coil the way that gas and oil burning systems are. Metal debris falling onto the heating elements will short a portion of the element to ground, turning it on.

UNIT 2—REVIEW QUESTIONS

1. What meter would you use to determine if a fuse link is open?
2. When one heating element burns out in a multielement electric furnace, do you replace the entire element rack?
3. One watt of electric energy will produce ________ Btu of heat.

TECH TALK

Tech 1. I have an electric furnace that's giving me fits! I've been out on the job several times and every time I'm there it works fine. I just can't seem to find the intermittent problem.

Tech 2. You need a Tattle Tale®.

Tech 1. I need a what? Sounds like something you'd have in elementary school.

Tech 2. No, a Tattle Tale® is a small tech tool that is clamped on to a safety overload so that if the overload opens while the circuit is energized the Tattle Tale® will trip. Then when you come back you can check every Tattle Tale® you put in the system and find the one that is causing problems.

Tech 1. Where do you get one of these things? Down at the supply house?

Tech 2. That's right—just check with the parts counter.

Tech 1. I will. Thanks.

4. What safety devices are used in electric furnaces to protect against overcurrent?
5. What is an advantage of electrical resistance heating?
6. What will happen if the furnace filter becomes restricted and air cannot pass through it?
7. Why is a high temperature limit control installed in an electric furnace?
8. If a high temperature limit control is installed in an electric furnace, why do you need a fuse link?
9. If a 5 kW electric heat strip draws 20.5 A on 240 V AC, what would a 30 kW heat strip package draw?
10. What is the function of a sequencer in an electric furnace?

13

Central Residential Heat Pump Systems

UNIT 1

Types of Heat Pump Systems

OBJECTIVES: In this unit the reader will learn about heat pumps including:

- History of Heat Pumps
- Heat Pump Cycles
- Air Source Systems
- Water Source Systems
- Air to Water Systems
- Heat Pump Efficiency Ratings

■ HEAT PUMPS

Heat pumps, like all refrigeration systems and air conditioners, transfer heat from one place to another by the change in state of a refrigerant. But the heat pump is able to reverse the action or direction of heat transfer. It can remove heat from inside for summer cooling and dispose of the heat into outside air, a water supply, or indirectly into the ground or other material. By reversing the action, it will also remove heat from the outside air source, the water supply, or from the earth or other material and supply it to the inside. Basically all refrigeration systems are heat pumps in that they transfer heat from a heat source at a low temperature to a heat sink at a higher temperature. Heat pumps are also called reverse cycle refrigeration systems.

Understanding the five basic laws of nature will aid in your understanding of the heat pump and other refrigeration systems.

- **Law 1** Heat exists in the air down to absolute zero, which is −460°F.
- **Law 2** Heat flows from a higher temperature to a lower temperature regardless of how small the temperature difference might be.
- **Law 3** Due to friction between molecules, all gases become warmer when compressed.
- **Law 4** Most matter can be in a solid (ice), liquid (water) or gas (steam) state.
- **Law 5** The temperature at which a material changes from a liquid to a gas or from a gas to a liquid depends on the pressure at which it is contained.

HISTORY OF HEAT PUMPS

The popularity of the heat pump rose with the increasing cost of electrical energy used for heating in residential and small commercial buildings. It generally cost two to three times less to heat with a heat pump than by using electric strip heat.

The heat pump was actually developed by Lord Kelvin in 1852. The first practically applied heat pump was installed in Scotland in 1927. Between 1927 and 1950, heat pumps were installed in hundreds of residential and small commercial applications throughout Europe and the southern part of the United States. Many of these installations were merely air conditioning units converted to heat pumps by the addition of reversing valves and applicable controls. Because these early heat pumps did not have defrost controls, crankcase heaters, and accumulators, they had high failure rates. As a result of this high failure rate, heat pumps acquired a bad reputation. Although improvements in design made heat pumps as reliable as any other HVAC system, the early high failure rates caused problems in the market for years. Further, the relatively low cost of energy for a period of time almost destroyed this market.

TECH TIP

Tell your customer about the vapor that will come off the outdoor unit during winter defrost cycles. Some owners may be startled when they see the vapor coming from the outdoor coil. The owner may think this vapor is smoke, and may believe the outdoor section is on fire. Prepare the owner so that when they see this, they will be assured the system is running properly.

The demand for heat pumps has grown over the past several decades. The reason for this new interest in heat pumps is both economic and environmental in nature. The cost of heating with electric heat became significant to home and small business owners. The rising concern over the pollution produced by power plants and the growing interest in energy conservation are other important issues.

HEAT PUMP CYCLES

In the conventional refrigeration cooling cycle, Figure 13-1-1, heat is absorbed in the indoor direct expansion (DX) coil or evaporator and discharged by the outside air cooled condenser. To reverse this process, the evaporator and condenser are not physically reversed. By means of a reversing valve, the refrigerant flow can be directed to make the process provide heating or cooling in the occupied area. The heat pump cycle is shown in Figures 13-1-2 and 13-1-3. The coils are relabeled as indoor and outdoor as they are now dual purpose. This reduces confusion because the outdoor coil is the condenser in the cooling cycle and the evaporator in the heating cycle. The indoor coil is the evaporator in the cooling mode and the condenser in the heating cycle.

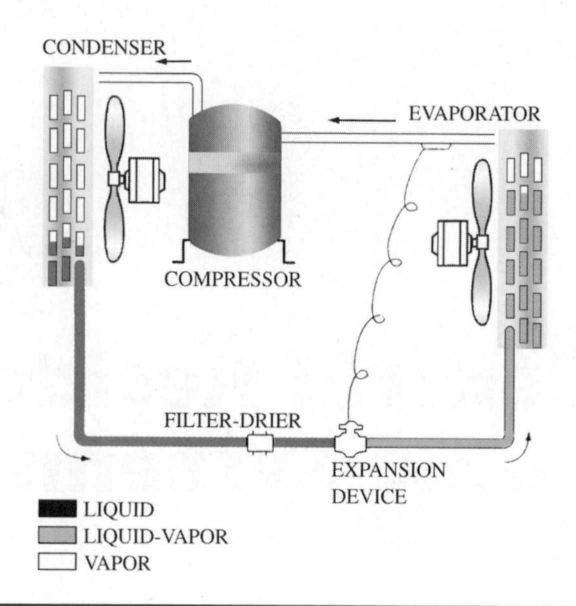

Figure 13-1-1 A conventional refrigeration cycle.

TECH TIP

Remember to tell the owner that the outdoor section will run during the winter heating mode. Some customers may make unnecessary service calls because they believe that the outdoor section only operates in the cooling mode.

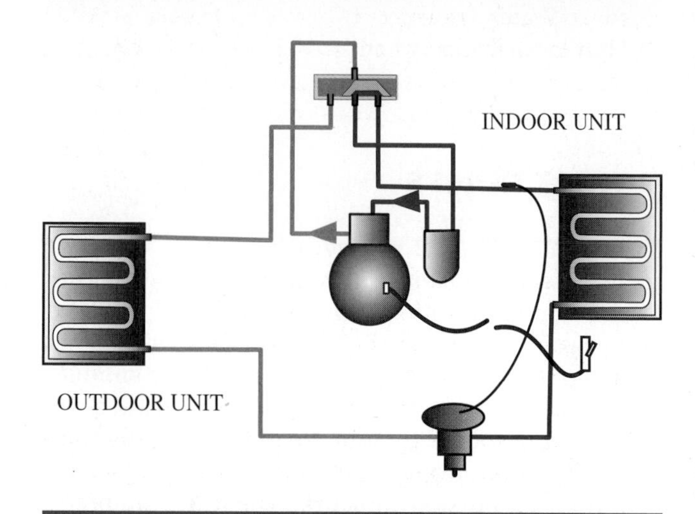

Figure 13-1-2 Heat pump system in cooling mode.

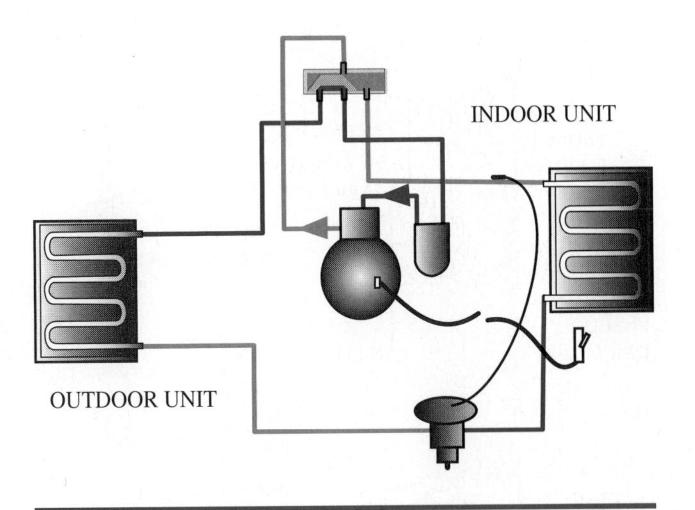

Figure 13-1-3 Heat pump system in heating mode.

To accomplish the reversing of the refrigerant flow, a reversing valve is used in the suction and discharge lines between the compressor and the vapor line to the coils. In some heat pump designs, check valves are connected in parallel, with the pressure reducing devices to permit removing the devices from the circuit when they are not to be used. A pressure reducing device is a generic term for a flow control or metering device. Referring to the circuit in the cooling phase the directional arrows show the high pressure/high temperature discharge gas being directed from the compressor to the outdoor coil (condenser), where it condenses to a high pressure, high temperature subcooled liquid. To avoid the restriction of the pressure reducing device connected to the outlet of the outdoor coil, a check valve is installed allowing the liquid to bypass the pressure reducing device. The liquid refrigerant travels to the indoor coil (evaporator) pressure reducing device. Forced to flow through the device because the check valve (connected in parallel with the pressure reducing device) has closed, the necessary pressure drop occurs to lower the refrigerant temperature so that heat can be absorbed in the indoor coil. The refrigerant vapor produced in the indoor coil now travels through the reversing valve and accumulator to the compressor. The cycle is complete.

TECH TIP

Tell the customer that even though heat pumps run longer than gas and oil heating systems, since they are more efficient, they are cheaper to operate.

In the heat cycle the reversing valve has changed position, changing the direction of gas flow. Remember, the compressor does not change the direction of refrigerant flow. The high pressure, high temperature

TABLE 13-1-1 Supply Air Temperature Ranges for Various Heating Systems

Heating Type	Discharge Temperature
Electric strip heat	110°F to 125°F
Gas heating	120°F to 140°F
Heat pumps	95°F to 115°F

gas from the compressor flows through the reversing valve to the indoor coil. The indoor coil now acts as a condenser, adding heat into the return air from the conditioned area. The hot vapor is condensed, and the temperature is reduced to produce a high pressure, high temperature subcooled liquid. To remove the cooling cycle pressure reducing valve from the circuit, the check valve opens and allows the liquid refrigerant to flow around the pressure reducing device. Continuing through the liquid line, the liquid refrigerant is forced through the outdoor coil pressure reducing device by the closing of the check valve connected in parallel with the pressure reducing device. When converted into a low pressure, low temperature liquid/vapor mixture, the refrigerant flows into the outdoor coil, which now acts as an evaporator. The refrigerant temperature will be lower than the outdoor temperature. Since the outdoor air is warmer than the refrigerant, heat is picked up from the outdoor air to evaporate the liquid refrigerant. This refrigerant vapor then flows through the vapor line, reversing valve, and accumulator to the compressor. The cycle is now complete. Accumulators are recommended for refrigeration and air conditioning systems; they are a necessity in heat pump systems. This will be covered more thoroughly under the subject of system defrost.

NOTE: Explain to the customer that discharge air temperature is lower with heat pumps. The lower air temperature can be a problem with customers who are accustomed to the high discharge air temperatures of gas and oil heating.

Heat pumps supply air temperature in a range from 95°F to 110°F. Higher discharge temperatures can even indicate an airflow problem. Electric heat has a discharge temperature of 110°F to 125°F and gas and oil heat supply air in the range of 120°F to 140°F, Table 13-1-1.

AIR SOURCE SYSTEMS

The basic cycle is an air to air heat pump where heat is taken from and given up to air. In this system two air type tube and fin heat exchangers are used. In

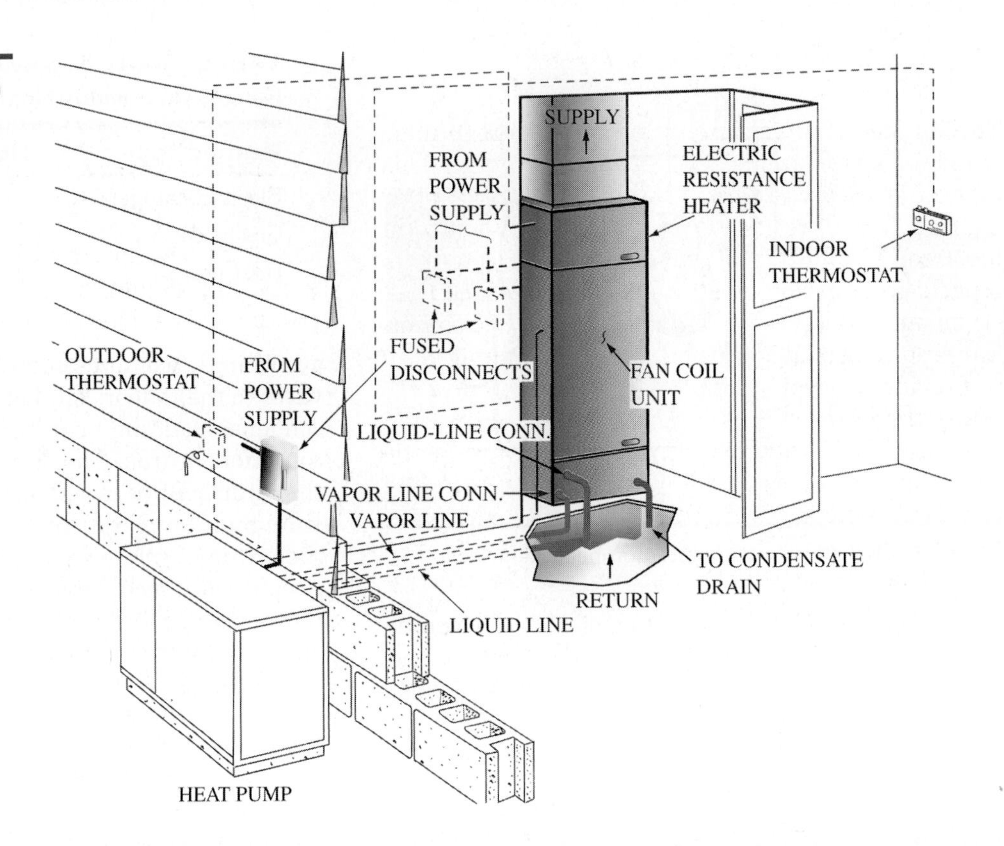

Figure 13-1-4 Split system air-to-air heat pump.

the cooling cycle, the indoor coil collects heat from the indoor air supply and the outdoor coil gives up heat to the outdoor air. In the heating cycle, the indoor coil gives up heat to the indoor air through the distribution system, and the outdoor coil collects heat. The blower assembly in the fan coil unit and the supply and return duct system circulates the air across the inside coil. The propeller fan assembly in the outdoor section provides the air through the outdoor coil, as shown in Figure 13-1-4.

WATER SOURCE SYSTEMS

Water to air heat pumps use a water to refrigerant heat exchanger in the high side or outdoor section of the heat pump assembly. This section should be located in the heated area to prevent possible freeze-up if the outdoor ambient temperature should drop below 32°F. If the water to refrigerant heat exchanger is part of a packaged unit, it would be located in the conditioned area.

A typical water to air unit installation is shown in Figure 13-1-5. In its lower compartment, the double tube water to refrigerant heat exchanger can be seen. Due to the variety of water conditions encountered, these heat exchangers are made of a copper nickel alloy or other corrosion resistant material. The construction is efficient and compact, called tube in tube or coaxial. The rest of the refrigeration system (compressor, pressure reducing devices, check valves, and reversing valve) is basically the same as in the air to air system. Since the operating temperature never drops below 32°F, this type of heat pump does not need a defrost cycle.

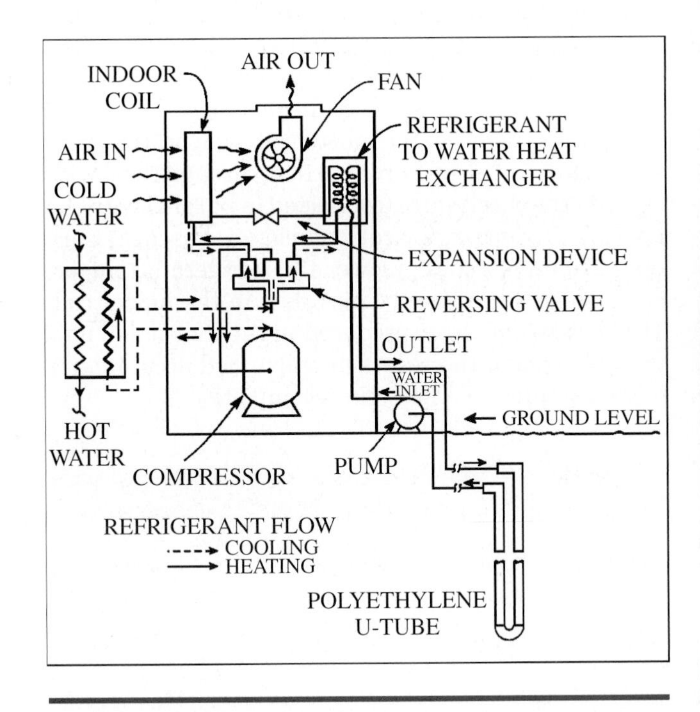

Figure 13-1-5 Typical installation diagram of an earth coupled water source heat pump. (*Copyright by the American Society of Heating, Refrigeration, and Air-Conditioning Engineers.*)

AIR TO WATER SYSTEMS

These types of heat pump systems have been marketed in small unit sizes for water heating. This system is used to replace expensive electric water heating. They are mostly in the capacity range of 6,000 to 12,000 Btu/hr and may be incorporated into a package unit along with a hot water storage tank or as an add-on unit for existing storage tanks. Larger units are available for commercial application to provide hot water and spot cooling. The unit becomes the primary heat source for domestic or small commercial hot water use, with the heat source of the water heater as backup. The higher efficiency rating or COP (coefficient of performance) of the air to water heat pump will usually supply hot water at a lower energy cost. Installation and service are covered in later units. There is also a water to water heat pump, which is the least common of all systems.

HEAT PUMP EFFICIENCY RATINGS

Efficiency ratings are easily defined, but often misused to make a point. This section will define the various efficiency ratings and how they are used to compare heating equipment.

This section will define the following terms:

- Energy efficiency ratio (EER)
- Seasonal energy efficiency ratio (SEER)
- Coefficient of performance (COP)
- Heating seasonal performance factor (HSPF)

Energy Efficiency Ratio—Cooling

Energy efficiency ratio or EER is the ratio of the cooling capacity divided by the watts for a given condition. For example, the EER of a 3 ton system at 75°F is calculated by dividing the measured total heat removing output of 38,000 btuh by the 3,000 W.

$$\text{EER} = \frac{\text{BTU}}{\text{W}} = \frac{38{,}000}{3{,}000} = 12.67\ \text{EER}$$

Normally, the cooler the outdoor ambient the greater the cooling capacity is. The compressor motor also draws fewer amps at these conditions. This air conditioning system will change the EER as the outdoor temperature changes, therefore using the EER method can be deceiving. The EER depends on the outdoor temperature. The old EER standard used 80°F dry bulb, 67°F wet bulb in indoor conditions as the test model. The problem with this model was that some systems were very efficient at 95°F but not at lower outdoor ambient temperatures. Therefore, the SEER test standard was developed to check the equipment operation across the normal operating outdoor conditions.

Seasonal Energy Efficiency Ratio—Cooling

SEER is the total amount of heat removed in a cooling season divided by total watts used during that cooling period. This is represented by the formula:

$$\text{SEER} = \frac{\text{Total cooling season BTU}}{\text{Total watts consumed}}$$

This is an elaborate test conducted by an independent laboratory. The test compresses time to simulate the full season operation of an air conditioning system. The BTU output and wattage is recorded. The final calculation is the SEER rating of the matched condenser and evaporator coil. The SEER rating is used to compare like capacity units for efficiency. The higher the SEER rating the lower the operating costs. The SEER rating is found on a temporary tag on the condensing unit. The tag has a short life. It is either lost in transit to the customer or separates from the unit after the first rain. The SEER rating can be found by contacting the equipment distributor or manufacturer. Unfortunately, this rating is not found on the nameplate of the equipment.

Coefficient of Performance—Heating in BTUH

COP is the ratio of total heat output divided by BTUH input. COP excludes any supplementary resistance (electric) heat. Since the fan motor adds heat to the air it is included in this rating. COP is represented by the following formula:

$$\text{COP} = \frac{\text{BTU heat output}}{\text{BTUH input}}$$

The COP for any electric heating system is near 1. In other words, the electric heat output in BTU is equal to the heat equivalent of the input wattage. Watts are converted into BTU by multiplying the watts by 3.4.

The COP rating for a heat pump is variable. The heat output will depend on the outdoor temperature and the power to move the heat from outdoors to indoors. The cooler the outdoor temperature the less heat in the air and the less heat absorbed by the heat pump. As the temperature drops, the COP drops. The COP of a heat pump is always higher than electric heat except at very low temperatures. The COP of a heat pump can be as low as 1 and may approach 4. The higher the COP the more efficient the operation and it costs less to operate. Most heat pumps operate in a COP range of 2 to 3.

Generally, heat pumps cost 2 to 2½ times less to operate compared to straight electric heat. Since COP can vary with the outdoor temperature it is not the best way to compare heat pump efficiency operation. Heat pumps with high SEER ratings will have high COP ratings. The SEER rating and COP rating track each other.

Heating Seasonal Performance Factor—Heating

HSPF is the total heat output of a heat pump, including supplementary electric heat during the heating season, divided by the total electric power in watts. It could be stated as the following formula:

$$\text{HSPF} = \frac{\text{Total BTU heating season output}}{\text{Total heating season watts}}$$

An independent laboratory determines the HSPF. The HSPF rating simulates the BTU values a heat pump will deliver during a heating season. This total BTU output is divided by the total watts used during the same time period. Unlike the COP rating, it is an effective way to compare heat pump efficiency ratings.

UNIT 1—REVIEW QUESTIONS

1. Describe the five laws of nature that will aid in understanding the heat pump and other refrigeration systems.
2. Why did the popularity of the heat pump rise with the increasing cost of electrical energy used for heating in residential and small commercial buildings?
3. How is the reversing of the refrigerant flow accomplished?
4. Why are accumulators recommended for refrigeration and air conditioning systems?
5. Heat pumps supply air temperature in a range from ________ to ________.
6. In the ________ system two, air type tube and fin heat exchangers are used.
7. Define EER.
8. Define SEER.
9. Define COP.
10. Generally heat pumps cost ________ to ________ less to operate compared to straight electric heat.

TECH TALK

Tech 1. I do not like working on heat pumps.

Tech 2. Why not? What's the problem?

Tech 1. It's so complex. You've got all those controls and wires and printed circuit boards. They're just really tough to figure out.

Tech 2. The system's not that complex if you stop and look at it. It's only got a heating and cooling system combined. No more complex than an air conditioner and a heater if you look at the wiring diagrams at the same time. What you need to do is isolate the problem. Is it in heating? Is it in cooling? Is it in defrost? Always look at that part of the circuit to find what you need.

Tech 1. But everything is all interconnected. It's hard to figure out which part does what.

Tech 2. They are complex, but if you take the time to follow the circuits they are a lot easier to work on.

Tech 1. So if I isolate the system that causes the problem then all I need to do is look at that part of the system. That should make it a lot easier. Thanks, I appreciate that.

UNIT 2

Heat Pump Components

OBJECTIVES: In this chapter the reader will learn the operation of following special components found on heat pumps:

- Heat Pump Thermostat
- Bi-flow Filter Drier
- Reversing Valve
- Two Metering Devices
- Bi-flow Pressure Reducing Devices
- Accumulator
- Check Valve
- Crankcase Heater
- Defrost Circuit
- Pressure Switches

INTRODUCTION

The heat pump uses all of the components found on a conventional air conditioning system with the addition of many specialty components. The specialty components are used to direct the refrigerant flow in the heating, cooling, and defrost modes. They are used to protect the equipment from refrigerant and electrical related damage. The components are designed to enhance system efficiency, reliability, and performance. Here is a list of components you might find on a conventional air to air heat pump system:

- Heat pump thermostat
- Bi-flow filter drier
- Reversing valve
- Two pressure reducing devices
- Bi-flow pressure reducing devices
- Accumulator
- Check valve
- Crankcase heater
- Defrost circuit
- Soft or hard start components
- Pressure switches
- Supplemental heating
- Outdoor ambient thermostat
- Holdback thermostat
- Miscellaneous—discharge temperature switch; equipment stands; wind baffles

HEAT PUMP THERMOSTAT

The heat pump thermostat is specially designed for the heat pump system. There are a variety of heat pump thermostats. The correct thermostat must be selected for the system to operate correctly. Most air to air heat pump systems use a single stage cooling and two stage heating thermostat. Figure 13-2-1 shows a common heat pump thermostat. The cooling side of the thermostat operates like a normal thermostat. A two stage thermostat is required for the heating operation. The first stage of heat energizes the refrigeration circuit of the heat pump to operate. If the heat pump is not able to maintain the space temperature within a degree or two, the second stage of heat will be energized. The second stage of heat is usually electric heat. The second stage of heat is called supplemental heat or auxiliary heat. Less commonly, gas or oil heat can also be used as the auxiliary or second stage heat.

Another important feature of the heat pump thermostat is that it has the capability of bypassing the heat pump operation and operating totally on the auxiliary heat. The second stage of heat is known as auxiliary heat or emergency heat. There is a difference in how the terms are used. When the heat pump and second stage heat are operating together, the second stage of heat is known as supplemental heat. To operate the system in the emergency mode, switch the thermostat to the emergency heat position and adjust the thermostat to the desired heat setting. The heat pump refrigeration circuit system is de-energized or locked out and only the auxiliary heat is energized.

Figure 13-2-1 Heat pump thermostat.

Most heat pump thermostats have two or more light indicators. On some models a light will indicate that the second stage of heat is energized. A different light will indicate that the emergency heat is operating. A third light may be used to indicate a system malfunction.

The customer should be instructed on how to use the heat pump thermostat. This will orient the customer on how to operate the system in the most efficient manner. This information will allow the customer to heat their building with emergency heat should a heat pump problem arise. Additionally, the operating instructions should be left with the homeowner.

BI-FLOW FILTER DRIER

The bi-flow filter drier, Figure 13-2-2, is used in the refrigerant line that has refrigerant flowing in both directions. Depending on the refrigeration circuit design, sometimes the refrigerant flows through the liquid line in both directions. If a conventional filter drier were installed in a flow reversing refrigerant line, most of the contaminants collected in one direction of flow would be flushed back into the system when the reverse flow starts. The design of the bi-flow drier prevents this from happening. The purpose of the filter drier is to remove suspended particles and moisture from the refrigeration system. The filter prevents contamination of the pressure reducing device. The bi-flow drier is similar to having two parallel filter driers with internal check valves that prevent the loss of trapped contaminants. The double sided arrow is used to indicate the bidirectional flow of the refrigerant through the drier. These driers should be changed every time the system is exposed to the atmosphere. The suction line filter drier is not considered to be a bi-flow drier because the refrigerant flow is always toward the compressor.

Some original equipment manufacturers (known as OEM) prefer to install two standard filter driers instead of the bi-flow drier. This design is shown in Figure 13-2-3. The drier is near the pressure reducing device. A check valve, piped across the pressure reducing device and the drier, prevents contaminations from flowing back into the refrigeration system. The advantage of this system over the bi-flow drier is that it provides more desiccant and filtering capacity, as well as less complicated drier parts and lower cost.

Figure 13-2-2 Heat pump bi-flow filter drier.

SERVICE TIP

Filter driers can be used to clean up a system after a compressor burnout. Once the new compressor or condenser is in place a less expensive model of filter drier can be temporarily installed to trap the major contaminations in the system. A severe burnout may require more than one filter drier changeout. Once the system has been cleaned up, a bi-flow filter drier can be installed permanently.

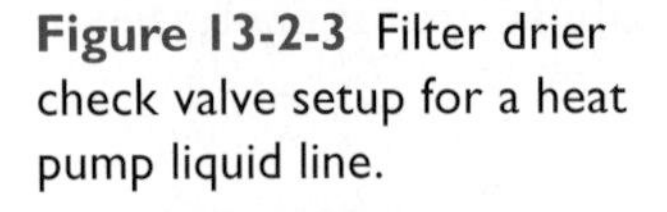

Figure 13-2-3 Filter drier check valve setup for a heat pump liquid line.

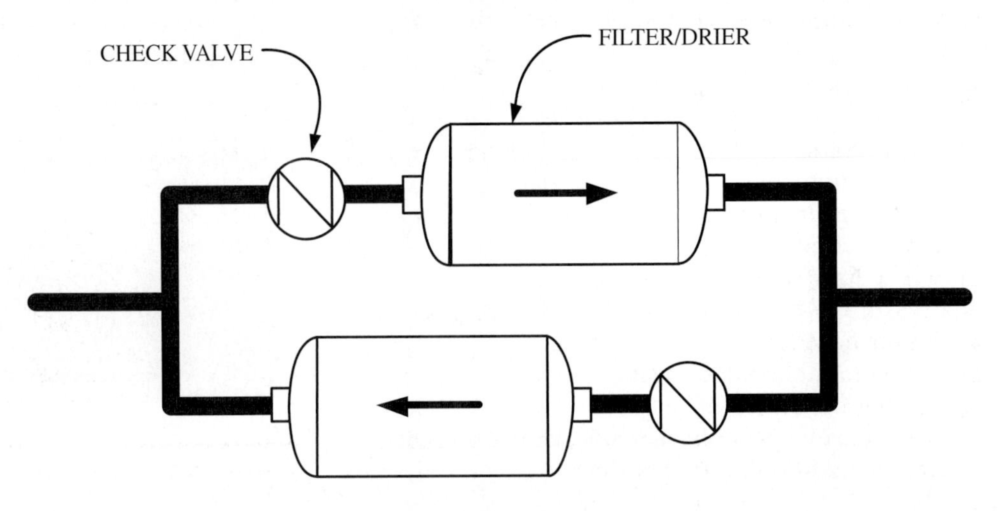

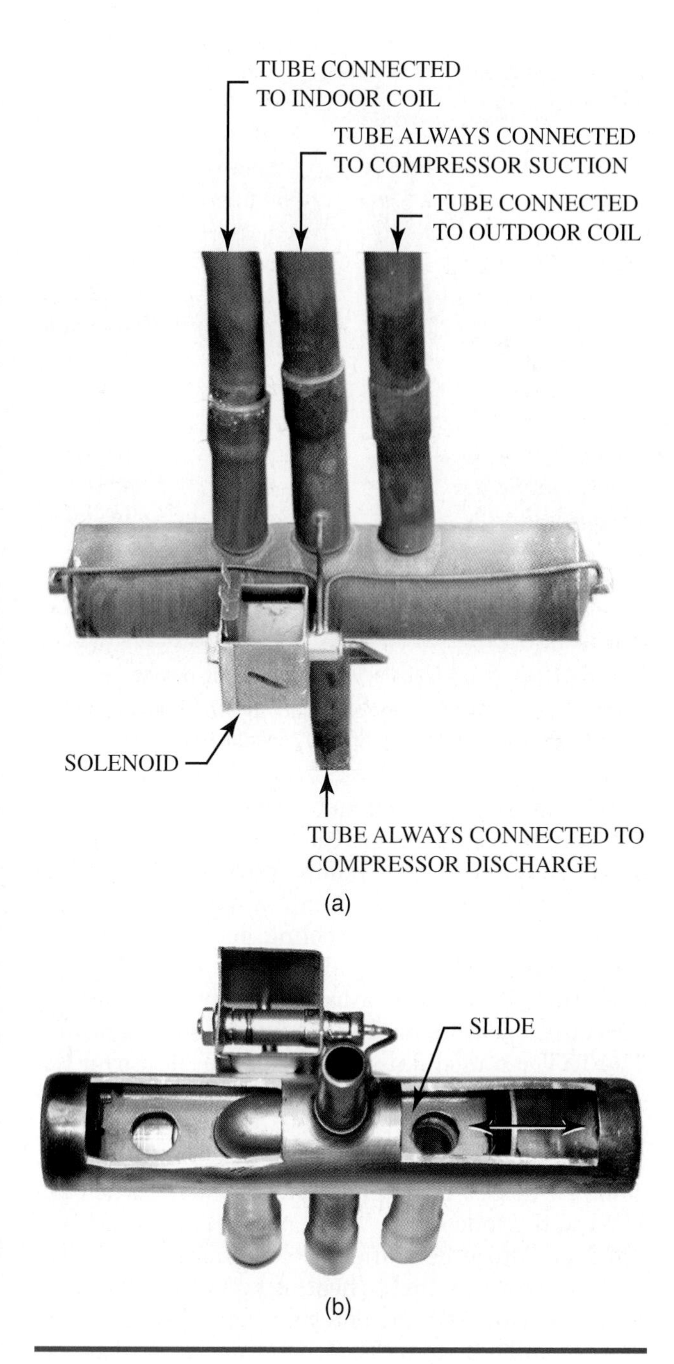

Figure 13-2-4 Heat pump reversing valve: (a) Outside view; (b) Cutaway view. *(Courtesy Rheem Manufacturing)*

Reversing Valve

The selection of the heat exchanger function to pick up heat (evaporator) or give up heat (condenser) is determined by the action of the reversing valve. A typical reversing valve is shown in Figure 13-2-4a,b. Internally, it is composed of two pistons connected to a sliding block or cylinder with two openings. The diagrams, Figures 13-2-5 and 13-2-6, show the position of the piston assembly in the heating mode and in the cooling mode.

The action of the piston is controlled by a solenoid valve that uses high pressure compressor discharge vapor to move the piston to the left or right, depending on whether the cooling or heating mode is needed. With the compressor discharge connected directly to the center of the piston chamber, equal pressure is exerted on the internal surfaces of each of the piston ends. To create the internal slider movement, a pressure difference is produced across the piston by bleeding the cylinder pressure into the suction side of the compressor. The bleeding action is controlled by the action of the two way solenoid valve. In Figure 13-2-5 the solenoid valve is de-energized and the control plunger is relaxed in the bottom position. The bottom port is closed and the top vent line is open to the equalizing line. The action that took

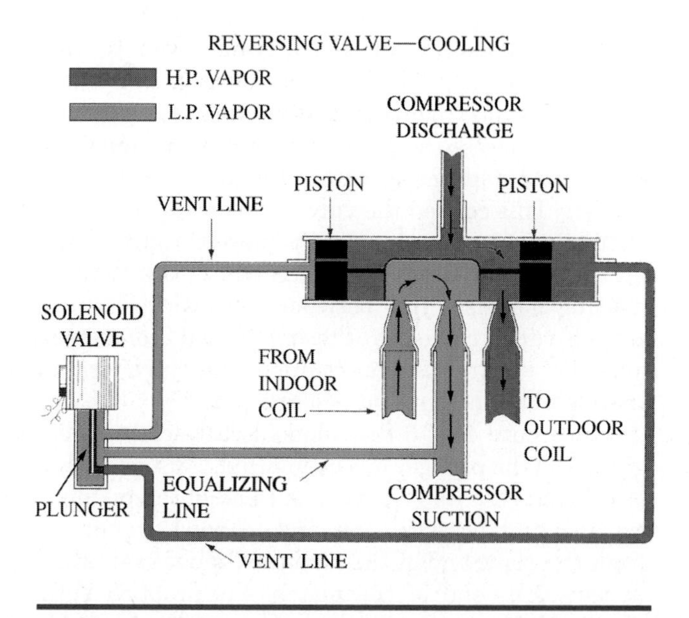

Figure 13-2-5 Reversing valve in cooling position.

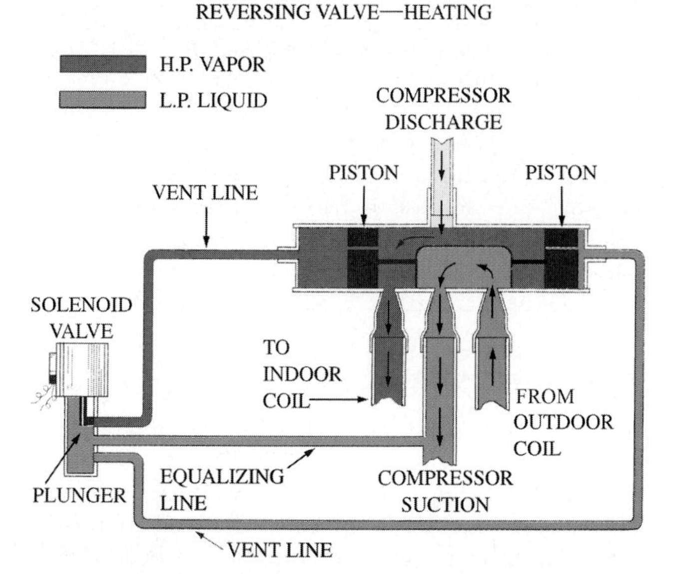

Figure 13-2-6 Reversing valve in heating position.

place was bleeding off the pressure from the left cylinder chamber. When the pressure in this chamber was reduced sufficiently to cause the pressure difference across the piston to move the piston (usually 75 to 100 psig or greater), the piston traveled to the left. This caused the valve slider to open the left control port to the center port and the right control port to the piston pressure chamber. This is now the heating mode. In this position the compressor discharge vapor travels to the indoor coil (condenser) and the vapor from the outdoor coil (evaporator) travels to the compressor suction.

In Figure 13-2-6 the solenoid valve coil is energized and the plunger has been lifted. This action has closed the vent line from the left end of the piston to create a high pressure drop and opened the vent line from the right end of the piston. When the pressure in the right end of the piston has dropped sufficiently, the pressure difference across the piston will force the piston to the extreme right end.

This action moves the slide valve, opening the outdoor coil to the compressor discharge and the indoor coil to the compressor suction. The unit is now performing like a standard air conditioning system. Not shown is a tip on the end of the piston that seals into a seat to prevent continuous bypass of the hot vapor from the cylinder through the piston bleed hole into the open vent line.

The piston bleed hole is a small opening through which the high pressure gas can slowly find its way to regulate the speed of the piston travel. In the crossover action, too rapid a change in pressure could result in system shock and excessive noise.

SERVICE TIP

A reversing valve has a relatively thin copper or brass outside skin. It can be easily damaged if struck by a hammer or wrench. Some technicians feel that reversing valve problems may be caused by a stuck reversing valve slide and strike the reversing valve tube in an attempt to dislodge the stuck slide. Striking a reversing valve with anything can only result in damage to the reversing valve and can never resolve a stuck slide. Never strike a reversing valve.

Two Metering Devices

The heat pump system needs two metering devices, one metering device for the cooling cycle and one for the heating cycle. There are several design options. The system can be designed with fixed metering, variable metering, or a combination of both. Fixed metering would be the capillary tube or piston device as shown in Figure 13-2-7a,b. Variable metering will be a thermostatic or electronic expansion valve as shown in Figure 13-2-8a,b. Electronic expansion valves are becoming more common in residential and small commercial heat pump systems. Some manufacturers design a heat pump system with a fixed metering device for the cooling operation and the thermostatic expansion valve (TXV) for the heating operation. The TXV is a better design choice in the heating mode compared to the fixed metering device. The TXV can modulate the refrigerant flow over the wide range of temperatures experienced in the heating mode.

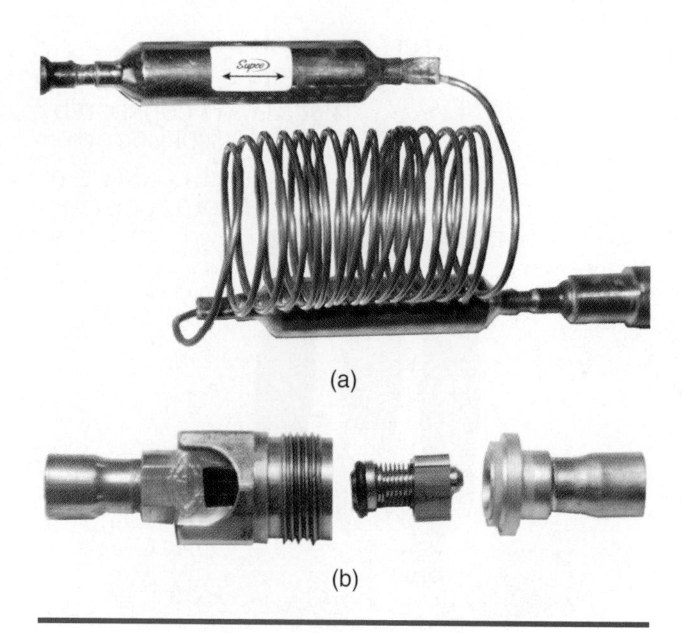

Figure 13-2-7 (a) Capillary tube metering device (*Courtesy Hampden Engineering Corporation*); (b) Piston type metering device. (*Courtesy Rheem Manufacturing*)

Capillary tubes in heat pump systems are the same as those in air conditioning systems when applied to the indoor coil. Those on the outdoor coil are sized for altogether different conditions of pressure and temperature. On the heating cycle, the heat pump operates with 70°F air entering the indoor coil (condenser) and −20 to 65°F air entering the outdoor coil (evaporator). Manufacturers' specifications should be used for the size and length of the capillary tubing. Also, the use of the fixed orifice (piston) metering device needs to be sized according to the capacity and operating conditions. The equipment manufacturer usually designs this.

Thermal expansion valves (TXV) used in heat pumps are not interchangeable with those used in standard air conditioning systems. The feeler bulb of the heat pump TXV valve is clamped to the cool suction line in one operating mode and to the hot gas line in the other mode. Thus, the valve must be able to operate when the coil is connected to the evaporator and yet stand the high temperatures of the hot gas line when the coil is connected to the condenser. The power element of the valves has a special pressure limiting charge to provide this range of operation.

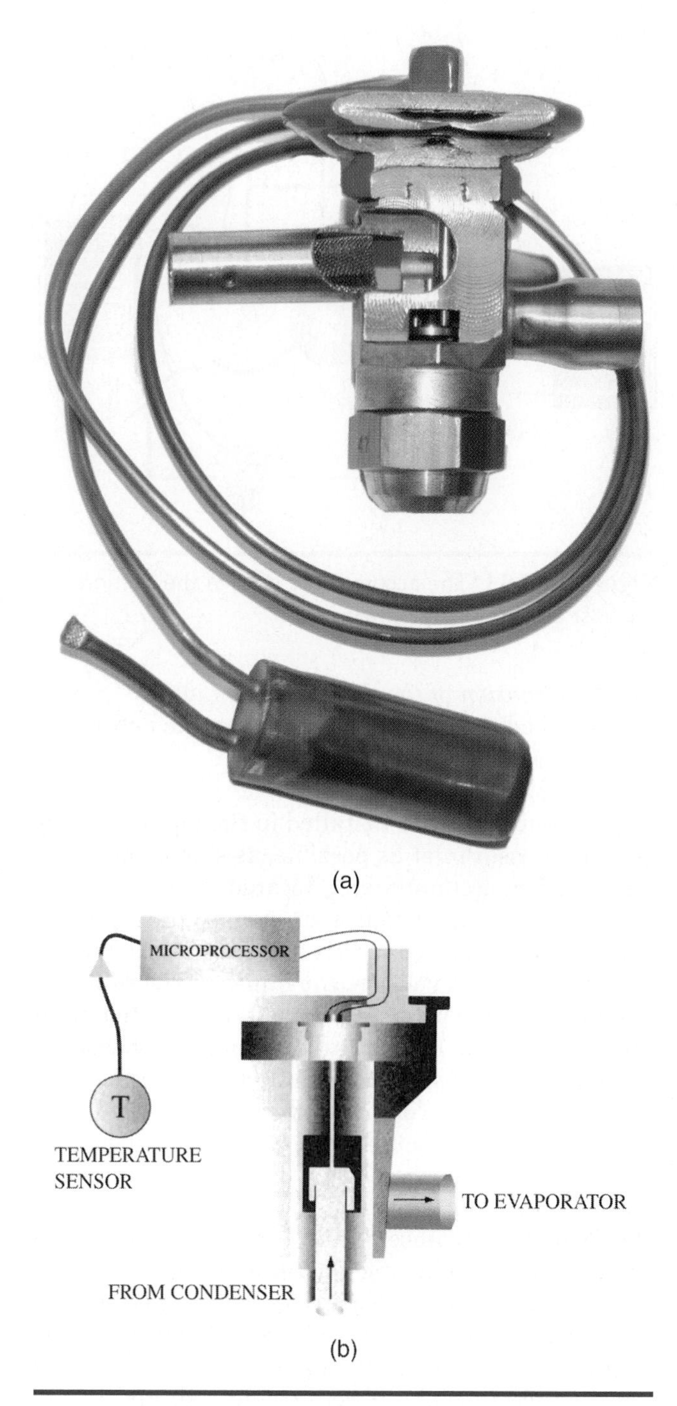

Figure 13-2-8 (a) Thermostatic expansion valve (*Courtesy Rheem Manufacturing*); (b) Electronic expansion valve.

Bi-flow Pressure Reducing Devices

The heat pump refrigeration cycle needs two pressure reducing devices. The single, bi-flow pressure reducing device is another way to obtain the pressure drop in the cooling and heating mode. The bi-flow device can be a fixed orifice device as shown in Figure 13-2-7 or a thermostatic expansion valve as shown in Figure 13-2-8.

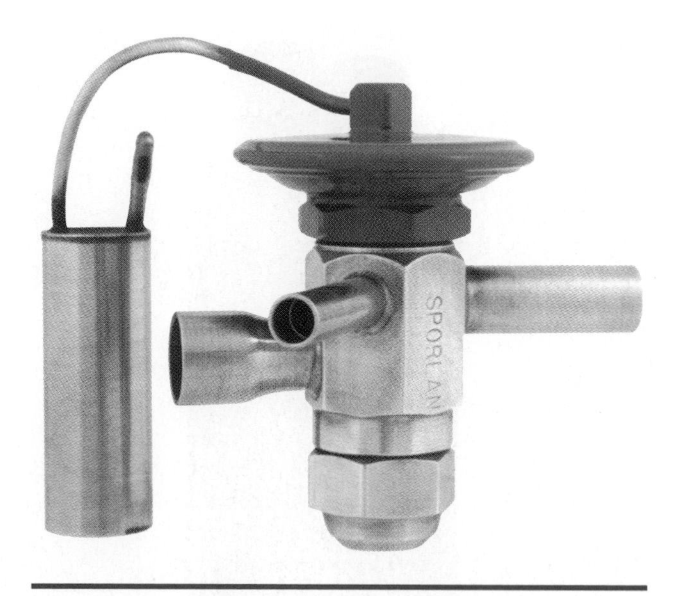

Figure 13-2-9 Bi-flow TXV. (*Courtesy Rheem Manufacturing*)

The fixed bi-flow device shifts the piston position between the cooling and heating modes. The piston must be sized properly for correct refrigerant metering.

The bi-flow TXV is designed to meter the refrigerant in either direction. Figure 13-2-9 shows a TXV with an external equalizer line. The newer bi-flow TXV has an internal equalizer that is not so obvious.

TECH TIP

Many new indoor coils come with piston metering devices. Often this device is located in a plastic or cloth bag tied to the coil liquid line. Check with the manufacturer before you leave the parts house to see that the piston supplied with the coil is the one required for the outdoor unit. Often a piston size must be changed to match the outdoor unit. Asking before you leave the parts house will save a great deal of trouble.

Accumulator

A suction line accumulator, Figure 13-2-10, is a cylinder placed in the suction line, ahead of the compressor, to catch any liquid refrigerant that has not evaporated before it reaches the compressor. It is a very simple device, but very important in certain types of installations. Some of the systems that routinely require suction line accumulators are those systems that:

1. Have wide or rapid load changes;
2. Utilize capacity control devices;
3. Are heat pump systems;

Figure 13-2-10 Suction line accumulator. *(Courtesy Rheem Manufacturing)*

4. Incorporate an evaporator defrost system, including a heat pump type defrost system;
5. Require compressor replacement due to liquid slugging.

Specifically, in the heat pump, the accumulator has three situations where compressor protection is needed. These are:

1. *Floodback on the cooling cycle.* If an air restriction causing light load and liquid runout occurs.
2. *Floodback on the heating cycle.* When excessive frost buildup on the outdoor coil occurs or if air troubles occur and cause liquid runout.

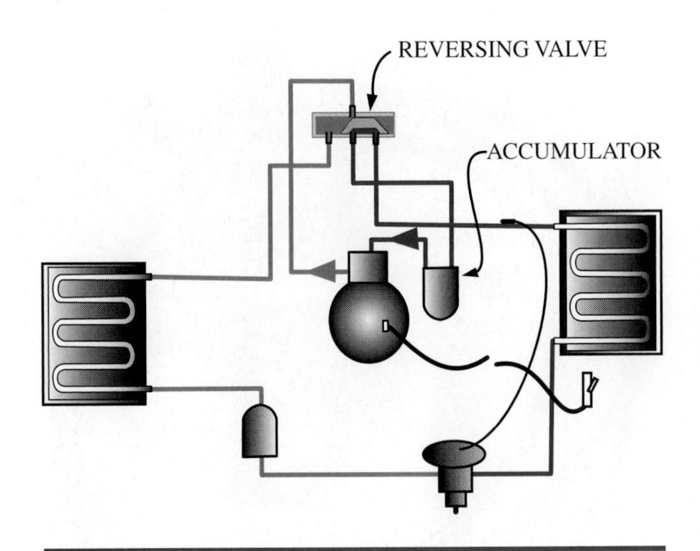

Figure 13-2-11 The accumulator goes in the suction line just after the reversing valve.

3. *Termination of the defrost cycle.* Liquid floodback will always occur when the defrost cycle is terminated.

Accumulators are installed in the piping as near the compressor inlet as possible, as shown in Figure 13-2-11. The accumulator is located between the reversing valve and compressor suction inlet. Due to the increased volume in this area and the reduction of the refrigerant velocity, the liquid refrigerant and oil drop to the bottom of the container rather than enter the suction opening to the compressor. The diagram shows the pipe at the bottom that permits the oil to return to the compressor crankcase. The liquid refrigerant in the accumulator gradually evaporates, entering the compressor as vapor. The accumulator shown in the illustration has a heater that speeds up the evaporation process. The heater and oil return line are not found on most accumulators.

Figure 13-2-12 shows a more common type of accumulator with a different type of construction. The suction gas enters at the upper right. Liquid slugs are directed toward the right-hand wall and run to the bottom. The metering orifice at the bottom of the return pipe slowly returns oil to the compressor through the suction line, plus a limited amount of liquid refrigerant.

Accumulators protect the compressor from liquid runout escaping the evaporator during light loads or air reduction problems. The heat pump employs the hot gas defrost method to melt the frost and ice off the outdoor coil. When the system action is reversed, heat is picked up in the inside coil and is raised in pressure and temperature by the compressor which forces it into the cold outside coil. With the outside blower or fan off, the coil rapidly heats up to melt the frost and/or ice on the outside surface of the coil. While

Figure 13-2-12 Accumulator. *(Courtesy Rheem Manufacturing)*

doing this, the outdoor coil fills with condensed liquid refrigerant. Before the defrost operation is completed, and the liquid out of the bottom of the outdoor coil reaches the proper termination temperature (around 55°F), pressure in the condenser can reach 350 psig and a temperature of 142°F.

When the defrost cycle terminates and the reversing valve switches over, this high pressure is relieved into the suction side of compressor. Immediately, vapor will form deep in the outdoor coil circuit and force liquid refrigerant out of the coil into the suction line. This action can be compared to removing the cap on the automobile radiator when the engine overheats.

Without the accumulator to catch and hold the liquid refrigerant, a liquid surge into the compressor can ruin valves. The protection of an accumulator in the heat pump system has encouraged variations in the amount of refrigerant used, which is not be to tolerated. The refrigerant charge of a heat pump system is very critical.

The accumulator must be changed in the case of a compressor burnout. The contamination in the accumulator cannot be removed. Purging or flushing may not clean the accumulator or clear the oil return orifice.

The accumulator acts as an oil trap. When an accumulator is installed or replaced, extra refrigeration oil must be added to the system. This oil can be directly poured into the accumulator or added in the suction line during the evacuation process. The specific amount of additional oil will vary between accumulator manufacturers.

The accumulator is sized for the heat pump system. Installing an undersized accumulator will not provide the required protection. Accumulators are designed to be installed either vertically or horizontally. Due to the confined space in the outdoor coil, vertical accumulator installations are most common.

SERVICE TIP

During refrigerant recovery it is sometimes difficult to get all of the refrigerant out of an accumulator. Some refrigerant can be trapped in the oil at the bottom of the accumulator. As the system is recovered, the accumulator may ice up around the base. If this occurs recovery can be helped by warming the accumulator. Never use a torch to warm any part of a system during a recovery. Warm water or a heat lamp can be used to warm the accumulator to speed the refrigerant recovery.

Check Valve

Checks valves are used in some heat pump designs to direct the flow of refrigerant around the metering device not being used. The check valves in the heat pump circuit are vital to ensure that proper pressure and refrigerant boiling points are maintained to provide the heat absorption and transfer at the rated capacity. This is easily accomplished, as these valves are either fully open or fully closed, depending on the direction of refrigerant flow. A swing type check valve with a directional flow arrow is shown in

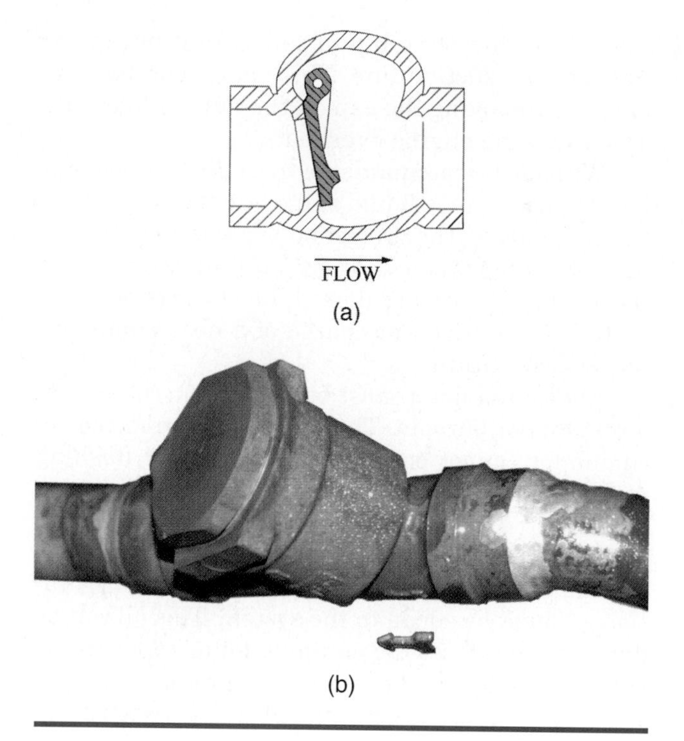

Figure 13-2-13 Check valve: (a) Diagram; (b) Typical appearance, with arrow showing flow direction.

Figure 13-2-13a,b. When the refrigerant flows in the direction of the arrow, the valve opens and allows full flow with minimal resistance. When the refrigerant is reversed, however, the check plate swings closed and stops the refrigerant flow. Thus, this action forces the refrigerant to flow through the pressure reducing device and proper operation is obtained.

The first check valves used in residential and small commercial heat pumps were of the disk type. This valve used a flat disk over a circular flat inset with a spring that had just a light spring constant to aid in closing and still cause very little resistance to refrigerant flow. In normal situations, where moderate pressure forced the refrigerant through the check valve and little reverse pressure was exerted when the system was idle, the check valve performed adequately. In heat pump systems, where radical pressure differences occur rapidly, the disk would be forced to one side and hang up. This prevented the shutoff of reverse refrigerant flow.

As an improvement to the original version, a check valve using a steel ball instead of a flat disk was developed. This type of check valve can withstand the heavy reversal of refrigerant flow pressure, especially when the unit completes the defrost cycle.

Usually, two check valves are incorporated in each capillary tube system, one connected in parallel with a pressure reducing device on each of the inside and outside coils. Newer TXVs used in heat pump application have a built-in check valve.

(a)

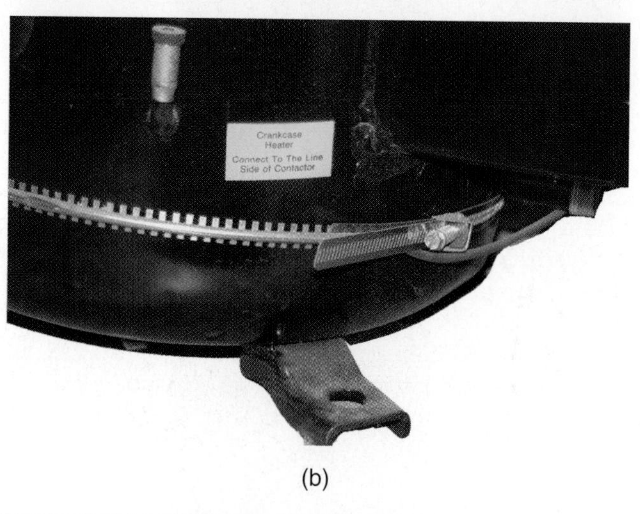

(b)

Figure 13-2-14 (a) Immersion type crankcase heaters must be installed with thermal mastic; (b) External type crankcase heater (sometimes referred to as "belly band.")

Crankcase Heater

The crankcase heater, Figure 13-2-14a,b, is an electrical heating device that is inserted in the compressor crankcase or around the lower part of a hermetically sealed compressor shell. It provides heat to evaporate any liquid refrigerant that reaches the crankcase during OFF-cycle. During shutdown, the compressor often becomes the coldest part of the system. The cold compressor creates a low pressure condition and condenses the vapor into liquid refrigerant. The process of attracting and condensing refrigerant is known as migration.

The crankcase heater protects the compressor when starting by vaporizing any liquid refrigerant that may enter the compressor during OFF-cycle. On many compressors the crankcase heater only operates during the OFF-cycle. On others it may be on all the time. The crankcase heater may be fastened to the bottom of the compressor crankcase or it may be inserted in a tube located in the crankcase. When it is wrapped around the compressor shell it is referred to as a bellyband crankcase heater.

Liquid refrigerant in the crankcase can be very damaging to the compressor. If refrigerant collects in the crankcase it mixes with the oil. This creates a condition known as flooded start. When the compressor cycles on, the crankcase pressure quickly falls to the level of the suction pressure. This causes the liquid refrigerant mixed with oil to boil violently, creating foam, which can result in compressor slugging and the oil being pumped from the compressor crankcase. At a minimum, the liquid refrigerant in the compressor crankcase dilutes the oil, causing improper lubrication and shortens the life of the compressor mechanical parts.

Defrost Circuit

The defrost circuit is required for all air to air heat pumps. Frost begins to form on the outdoor coil when the coil temperature falls below 32°F saturation temperature. The outdoor coil is defrosted by temporarily reversing the refrigerant flow to the cooling mode.

The defrost circuit is controlled by mechanical switches or solid state circuit boards depending on the defrost method. The defrost circuit is activated by at least two of the following conditions:

1. Low suction pressure
2. Low suction or low liquid line temperatures
3. High static pressure drop across the outdoor coil due to frost
4. Heat pump operating time (timer)

The defrost circuit operation will be described in detail in a later unit.

Start Components

These components are used to help start the compressor under low voltage or unequal pressure conditions. To obtain the maximum starting torque the phase shift between the voltage and current should be 90°. The soft start kit is a solid state device that improves starting torque but does meet the optimal starting phase angle of 90°. It is less expensive than the hard start kit and in many cases will do the job of starting the compressor motor. The soft start kit may include a start capacitor.

The hard start device is a matched start capacitor and potential relay, Figure 13-2-15. The compressor manufacturer determines the microfarad rating

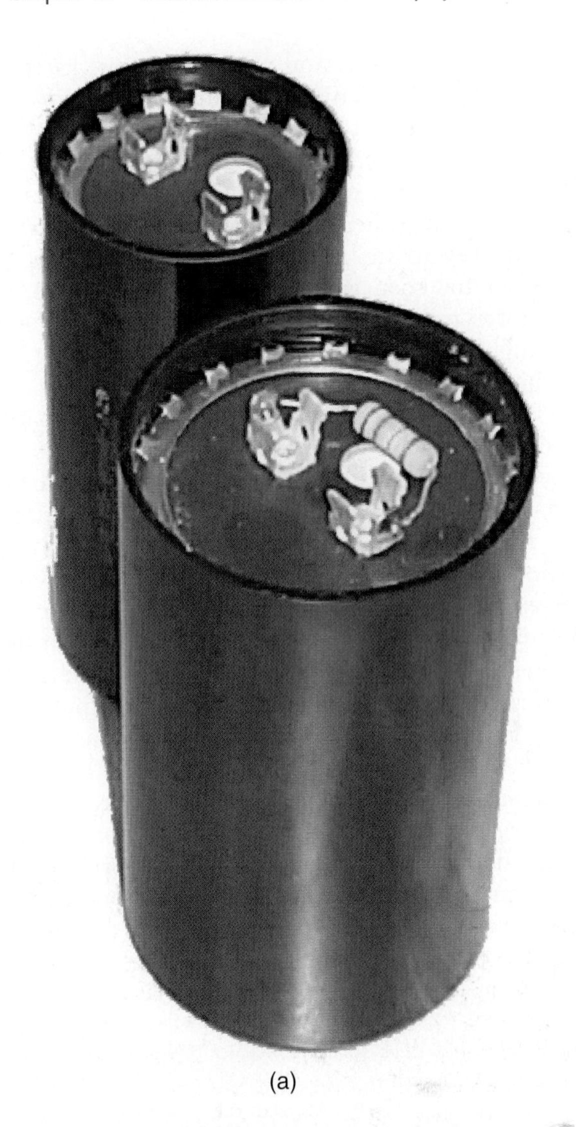

(a)

(b)

Figure 13-2-15 (a) Hard start kits consists of a hard start capacitor with a bleed resistor; (b) Another example of a hard start kit.

of the start capacitor and voltage dropout rating of the potential relay. This matched set will provide the maximum starting torque for the compressor motor. Using an undersized or oversized start capacitor will reduce starting torque. Using the wrong potential relay will cause the start capacitor to drop out too soon or stay in the circuit too long. Removing the capacitor too soon will reduce starting torque and the motor may draw locked rotor amps and stall. Leaving the capacitor in the circuit too long will overheat and burn out the capacitor.

There are a number of so called "hard start kits" marketed proclaiming that they will assist in starting compressor motors. There is no one hard start kit that fits all single phase compressor applications. The technician might be lucky if it helps at all. Again, the problem is with the sizing of the start capacitor and the amount of time the capacitor is in the circuit. This type of so called hard start kit uses various sizes of start capacitors in series with a solid state relay. If the technician is going to experiment with this type of start device it is recommend that several sizes be stocked for this activity. The size of the capacitor will determine the microfarad rating. The larger the capacitor, the greater the microfarad rating. If a hard starting condition is the problem, try each starting device with the hope that one of them will work. If this does not start the motor try the correctly rated start capacitor and potential relay.

TECH TIP

A common complaint in some regions of the country is that as the compressor starts the lights dim. This is caused by a general voltage drop to the residence. A hard start kit can be used to start the compressor more quickly, which reduces this brownout.

Pressure Switches

High pressure switches, Figure 13-2-16a, low pressure switches, Figure 13-2-16b, and loss of charge switches are used on some heat pumps. The high pressure switch can open in the cooling or heating mode of operation. Anything that creates a high pressure situation will cause the high pressure switch to open. Most high pressure switches will reset once the high pressure condition drops. This is known as automatic reset, which allows the heat pump to operate again. A high pressure switch that does not automatically close its contacts on a drop in pressure is known as a manual reset high pressure switch. A technician will need to push a button on the pressure switch to complete the electrical circuit and allow the heat pump to operate again. The technician will determine the reason for the high pressure condition.

The low pressure switch opens when the pressure in a heat pump system is too low. The heat pump can experience low-pressure conditions under cold weather operation. This would cause the heat pump to interrupt its operation. Most low pressure switches are automatic reset and will close the contacts again after the low side pressure rises.

Another type of low pressure switch is the loss of charge low pressure switch. The switch is designed to open at or near 0 psig. In very cold heating operation the heat pump low side pressure is low. A conventional R-22 low pressure switch opens at pressure of 38 psi or less. This would cause a heat pump to cycle off when operating in very low ambient conditions or on startup when the suction pressure temporarily drops to a low pressure condition.

Supplemental Heating

Supplemental or auxiliary heat is required for all air to air heat pump systems. Supplemental heat is used to provide additional heat as the outdoor temperature drops and the heat pump's refrigerant system capacity drops. Supplemental heat is also used when the heat pump goes into the defrost mode. Auxiliary heat is known as emergency heat when the heat pump operation is bypassed by switching the thermostat to the "emergency heat" mode.

Normally, electric strip heat is used as auxiliary or supplemental heat. If this is the case, there may be several sets or stages of electric heating elements. In a few instances, natural gas, propane, or oil heating is used as auxiliary heat.

Outdoor Ambient Thermostat

The outdoor ambient thermostat (OAT) controls the operations of the second and third stages of supplemental electric heat. The outdoor thermostat is set low enough to prevent the operation of all the electric heating elements when the outdoor temperature is mild. For example, a common setting is 40°F. The purpose of this control is to prevent unnecessary operation of electric heat should the thermostat operator turn the temperature setting high enough to operate the heat pump and the supplemental heat. The outdoor thermostat is bypassed when the heat pump thermostat is switched to the emergency heat mode.

Holdback Thermostat

The holdback thermostat is similar to the outdoor ambient thermostat, but it is used for a different purpose. The purpose of the holdback thermostat is to

Figure 13-2-16 (a) High pressure switch; (b) Low pressure switch.

TECH TIP

New energy codes that have been adopted in many cities and states require an outdoor thermostat on all installations. The outdoor thermostat must lock out the supplemental heat when the outdoor temperature is above 50°F. This allows the compressor to provide heat for the resident and keeps the more expensive energy consuming supplemental heat strip off.

prevent overheating of the space during the defrost mode. In some heat pump designs, overheating of the space during the defrost mode will terminate defrost early, thus not allowing a clean and frost-free outdoor coil. This incomplete defrost will prevent the most efficient operation of the heat pump. As this continues, each cycle will accumulate additional frost on the coil, further reducing heating capacity. In some cases, this repeated lack of total defrost condition will cause water refreezing to damage the lower fins and coil. Water expands, pushing apart the fins and collapsing the coil. It is important to have a complete defrost cycle to clear the frost and the water from the coil surfaces.

The customer must be notified when the holdback thermostat is used. The customer may experience cooler supply air conditions during the holdback thermostat operation.

Miscellaneous

Discharge Temperature Switches Some heat pumps have a variety of components not normally found on an air conditioning system. The discharge temperature switch can be used in two different ways. A discharge temperature switch can be attached to the discharge line of the compressor. A discharge line temperature in excess of 250°F indicates overheating of the compressor. If the discharge line temperature is 250°F, the internal temperature of the compressor is around 300°F. At this temperature the oil will begin to break down. Excessive discharge temperatures are caused by lack of compressor lubrication, low charge, high compression ratios, and high discharge pressures.

Another type of discharge temperature switch is used in the airstream between the supplemental heat provided by a gas furnace and the indoor coil as a component of a fossil fuel kit for some manufacturers. The temperature sensor is placed in the discharge of the gas furnace. This prevents the operation of the gas furnace and heat pump simultaneously. Discharging 130°F air from the gas furnace into the indoor coil will cause excessive discharge pressures in the heat pump system. This temperature switch prevents the heat pump from overheating and cutting out on high pressure or an overloaded condition.

Equipment Stands Heat pump equipment stands are available to raise the outdoor unit above the normal snow level and allow the defrost water to drain away from the base of the coil. Technicians like the stand because it elevates the outdoor section, making it easier to work on the equipment.

Wind Baffle Another option that is available is the wind baffle. Heat pumps that are exposed to strong prevailing winds have a difficult time clearing the coil in a defrost cycle. This is because the wind is blowing cold air across the coil and extends the length of time the unit remains in defrost. A 5 mph wind is equivalent to having the outdoor fan operating in the defrost mode.

UNIT 2—REVIEW QUESTIONS

1. Where is a bi-flow filter drier used?
2. What is the reversing valve composed of internally?
3. What controls the action of the piston in a reversing valve?
4. What is a piston bleed hole?
5. List the systems that routinely require suction line accumulators.
6. Why are check valves in the heat pump system vital?

TECH TALK

Tech 1. Drier heat pumps are much more efficient than other furnaces.

Tech 2. It's a lot easier to move heat than it is to generate it.

Tech 1. What do you mean by move heat? And why is that cheaper?

Tech 2. Well, the heat pump simply moves heat from outdoors to indoors or indoors to outdoors whether heating or cooling. It's a lot less expensive to move heat than it is to create heat. In the same way it's a lot less expensive to sail a sailboat across the lake than it is to drive a motorboat across the lake. Sailboats use the wind, which is free, while a motorboat uses gasoline.

Tech 1. So you're saying that the heat pump is a lot less expensive because it uses the free heat from outside to heat the house?

Tech 2. That's correct. I think you've got it.

Tech 1. Thanks.

7. What is a crankcase heater?
8. List the four conditions that can activate the defrost circuit.
9. _______ heat is used to provide additional heat as the outdoor temperature drops and the heat pump's capacity drops.
10. What does the outdoor ambient thermostat control?
11. What is the purpose of the holdback thermostat?
12. List the two ways that a discharge temperature switch can be used.

UNIT 3

Air and Water Source Applications

OBJECTIVES: In this unit the reader will learn about air and water source applications including:

- Air to Air
- Water to Air
- Water Source Systems
- Air to Water (Heat Only)

OVERVIEW

The appearance of a heat pump is exactly the same as that of its air conditioning counterpart. The difference is in the design of the two coil assemblies involved, as well as the extra refrigerant flow controls needed for the reverse action. The heat source used as well determines the general classification of the unit.

Heat pumps are classified by their operational systems as follows:

- Air to air
- Water to air
- Air to water

AIR TO AIR

Air to air heat pumps are the predominant type of unit sold. The capability of the air to air type to both heat and cool makes it a year-round application. Air to air heat pumps are available as both package and split systems. Package systems have all of the components in one preassembled unit, as shown in Figure 13-3-1. Split systems are fabricated at the jobsite. They are fabricated from several individual parts including an air handler, indoor coil, auxiliary heat source, an outdoor unit, and interconnecting wiring and copper tubing.

The basic cycle in an air to air heat pump consists of heat taken from air and given up to air. In this system two air type tube and fin heat exchangers are used. In the cooling cycle, the indoor coil collects heat from the indoor air supply and the outdoor coil gives up heat to the outdoor air. In the heating cycle, the indoor coil gives up heat to the indoor through the air distribution system, and the outdoor coil absorbs heat. The blower assembly in the fan coil unit and the supply and return duct system circulates the air through the inside coil. The propeller fan assembly in the outdoor section provides the airflow through the outdoor coil.

TECH TIP

The outdoor unit on a heat pump has to be mounted above the ground in areas where snowfall accumulates. It must be mounted at a height above the average annual snowfall for the region. See Figure 13-3-2.

Figure 13-3-3 shows a typical heat pump, including a complete cabinet view as well as an interior view of the refrigerant circuit and electrical components. Included in the package view are the optional electric heat strips located in the blower discharge area. Some type of auxiliary heat is required except in semitropical climates such as that in southern Florida. The easiest type of auxiliary heat system to incorporate into the heat pump assembly is the electric strip heater. When a gas or oil furnace is used for auxiliary heat, the combined system is referred to as being duel-fuel.

NICE TO KNOW

The auxiliary heat unit used in heat pumps has had a variety of names over the years. It used to be commonly referred to as emergency heat. However, having a system that required an emergency heating system was not very popular with the public. The next term that came into common usage was backup heat. Again this term carried some negative connotations. That is why today the most common names are auxiliary heat or supplemental heat.

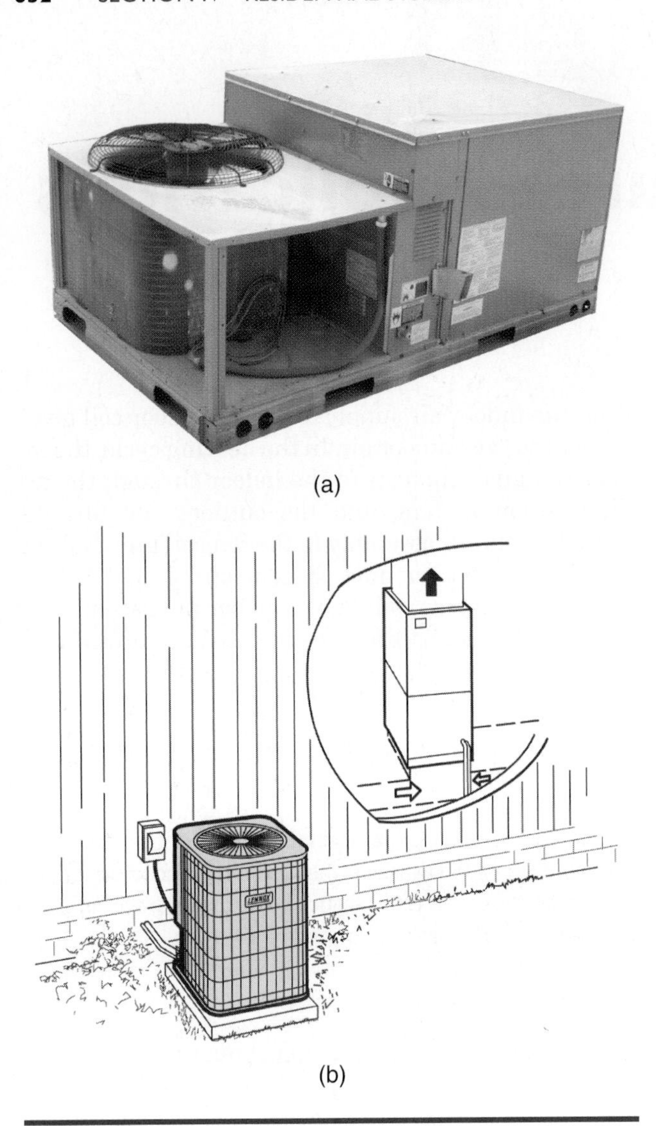

Figure 13-3-1 (a) Packaged heat pump; (b) Split system heat pump. (*Courtesy Lennox Industries Inc.*)

A split system includes both the indoor air handler with an electric heat strip and the outdoor section. The electric heat strip is located in the upper section of the blower discharge area, behind the control assembly and enclosing panel. To increase the indoor coil efficiency it is located in front of the blower so that air is drawn through the coil. By drawing the air through the coil all of the coil area is used, as compared to the central air blast stream effect for coils located on the discharge side of the blower, Figure 13-3-4.

Both types of equipment have their advantages and disadvantages. The package unit has no refrigerant line installation cost but is limited in application situations. The split system can be used in a wide range of vertical, horizontal, or downflow applications but installation cost is higher.

Figure 13-3-2 Split system heat pumps mounted above the ground for snow.

To calculate the net capacity on the cooling cycle, the same method is used for the heat pump as used for an air conditioning unit. This is possible since the reversing valve is in a position that permits the refrigeration cycle to operate in the normal

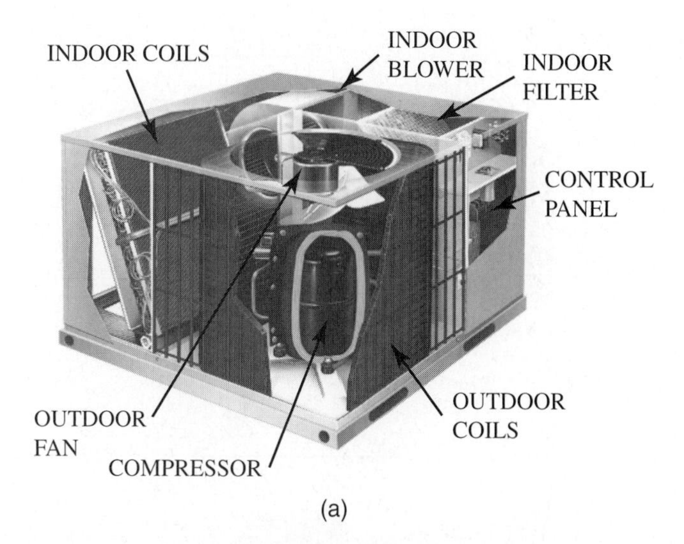

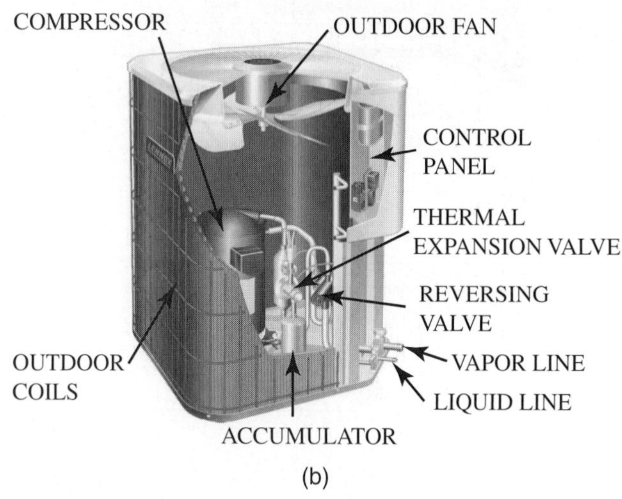

Figure 13-3-3 (a) Cutaway of a packaged system; (b) Cutaway of a split system. (*Courtesy Lennox Industries Inc.*)

way. Test methods can also be used for the heat pump similar to those described for the air conditioning unit.

To calculate the refrigerant system heating capacity, a number of quantities need to be added:

- The heat picked up by the outside coil
- The heat from the electrical energy of the compressor and blower motor (Watts × 3.415).

By testing, however, the gross heating capacity can be determined by knowing the CFM handled by the blower and the dry bulb temperature rise through the unit, using the standard formula:

$$H_S = \text{CFM} \times \text{TD} \times 1.08$$

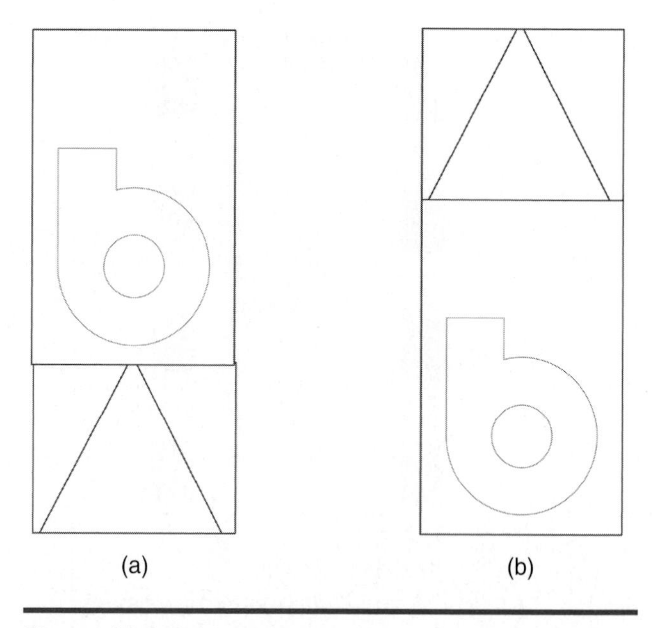

Figure 13-3-4 (a) Draw-through coil; (b) Blow-through coil.

WATER TO AIR

A number of applications allow the use of water to air heat pumps. These uses are illustrated in Figure 13-3-5, as follows:

- Water loop heat pump system
- Groundwater heat pump system
- Surface water heat pump system
- Closed loop surface water heat pump system
- Ground coupled heat pump system

WATER SOURCE SYSTEMS

Water to air heat pumps use a refrigerant to water heat exchanger instead of an outdoor coil. The heat exchanger should be located in an indoor area to prevent possible freezeup if the outdoor ambient temperature should drop below 32°F. A typical water to air unit, with the casing panels removed, is shown in Figure 13-3-6.

The water loop system uses water supplied by an evaporative cooler as a heat sink for heat pump cooling and heating. Supplemental heat is furnished by a boiler when needed. Where freezing temperatures may be encountered, the water loop must contain an antifreeze solution. Some installations use a flat plate heat exchanger between the outside and the inside water loop, to permit placing the antifreeze only in the outside loop.

The groundwater system uses well water as a heat sink and as a heat source for heating. Most systems of this type return the water to the ground through a reinjection well. Local codes must be

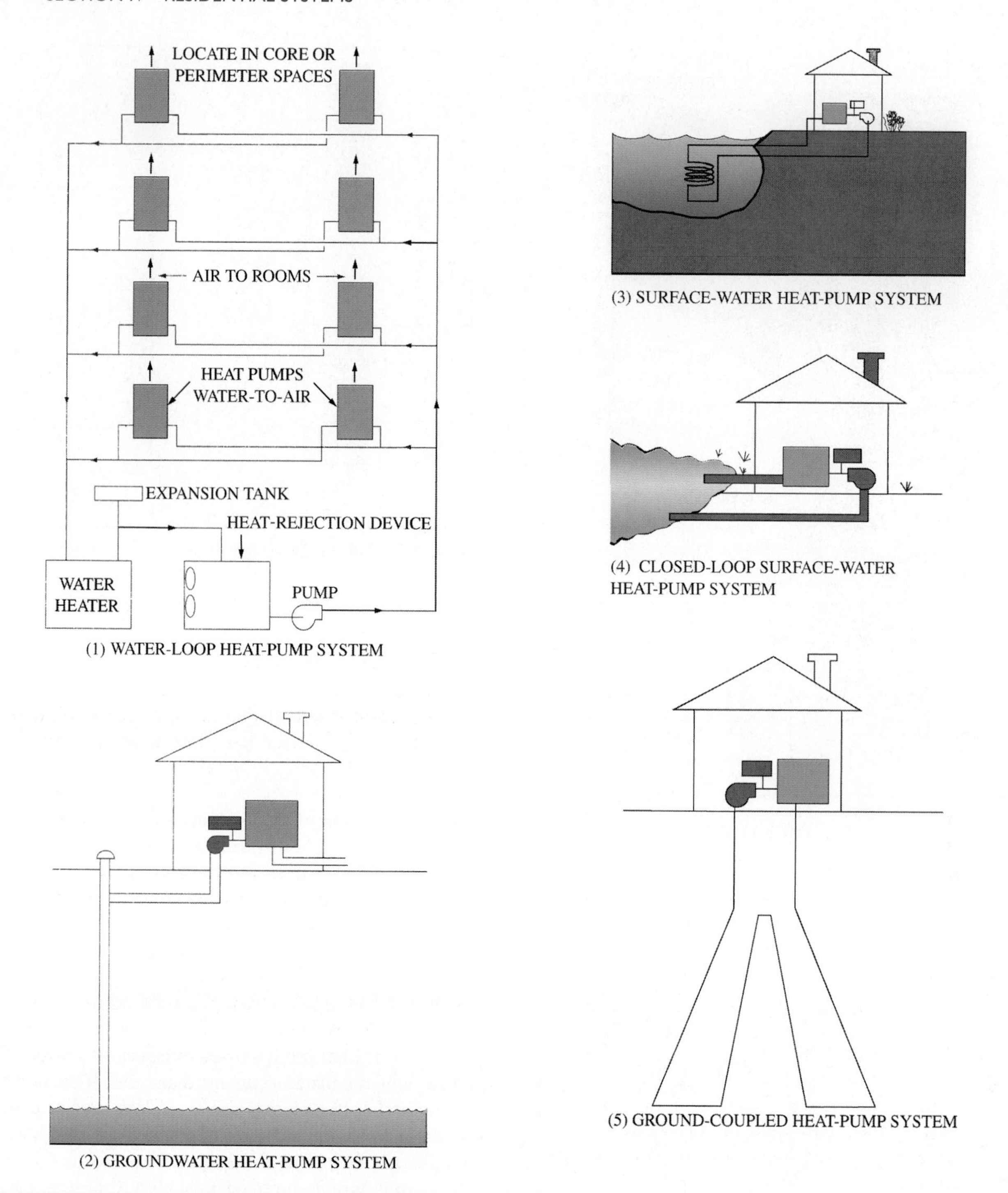

Figure 13-3-5 Various applications of the water source heat pump. *(Copyright by the American Society of Heating, Refrigeration, and Air-Conditioning Engineers.)*

examined relative to the permitted use and discharge of groundwater.

The surface water system uses and returns water to a nearby lake, stream, or canal. Some of these systems use closed loop arrangements where pipes or tubing are located on the surface of the water source to serve as a heat exchanger.

The ground coupled system takes advantage of the massive thermal capacity of the earth, which provides a temperature stabilizing effect on the circulating water loop. The success of these installations requires a detailed knowledge of the climate, the site, the soil heat transfer characteristics, and the performance of the heat exchanger.

Figure 13-3-6 Upright style water source heat pump with panels removed. *(Courtesy of Koldwave Division of Mestek, Inc.)*

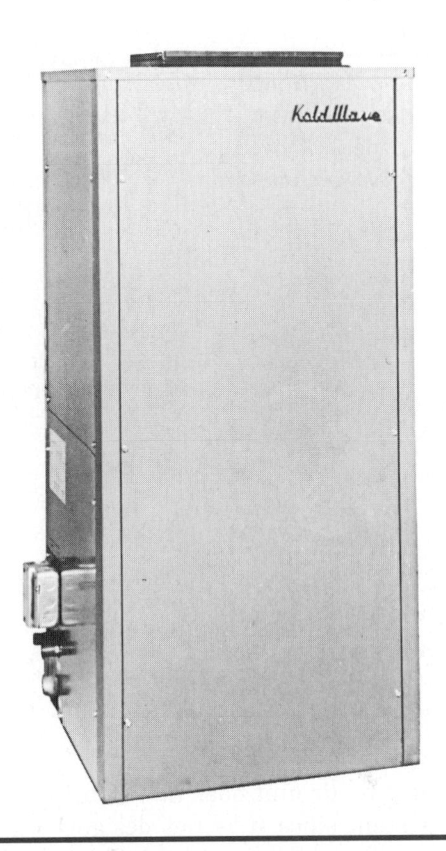

Figure 13-3-7 Vertical style water source heat pump. *(Courtesy of Koldwave Division of Mestek, Inc.)*

TECH TIP

Ground coupled heat pumps can be used in any climate because they are not subject to the low winter temperatures experienced in northern areas.

A vertical up self contained unit is shown in Figure 13-3-7 and a horizontal type in Figure 13-3-8. This type of unit is located totally within the conditioned area. The only connection to the outside area would be to the water supply and disposal.

Figure 13-3-9 shows the interior parts arrangement of the vertical model shown in Figure 13-3-7. The upper section of the unit contains the inside coil with the necessary refrigerant controls as well as the blower and motor assembly for the inside air movement. The bottom compartment contains the motor/compressor assembly, the water to refrigerant heat exchanger, reversing valve, and the operating controls.

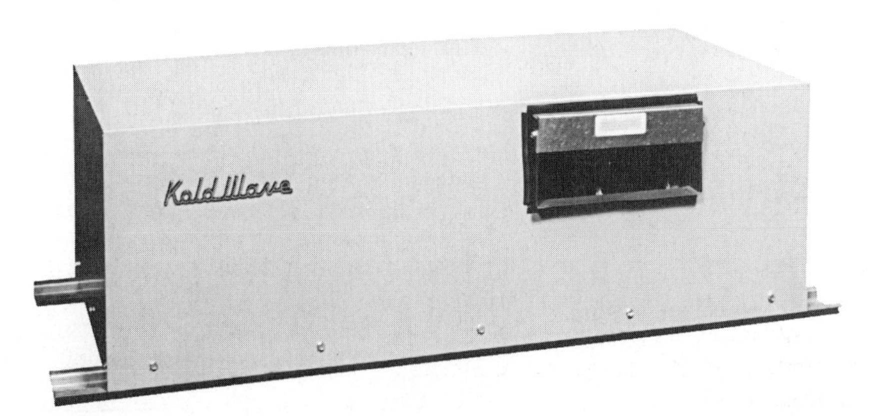

Figure 13-3-8 Horizontal style water source heat pump. *(Courtesy of Koldwave Division of Mestek, Inc.)*

Figure 13-3-9 Horizontal style water source heat pump with panels removed. *(Courtesy of Koldwave Division of Mestek, Inc.)*

A water to air unit does not contain the defrost control system since it is not designed to operate with any surfaces of the water to refrigerant heat exchanger below 32°F. This unit employs a tube in tube or coaxial type of heat exchanger as well as a water regulating valve to control the flow of water and a condensate pump for disposal of the inside coil condensate to drains elevated above the unit.

Figure 13-3-9 shows the interior parts arrangement of the horizontal unit shown in Figure 13-3-7.

The gross capacity on the cooling cycle can be calculated from the flow (gpm) through the water circuit and the temperature rise, using the following formula:

$$\text{HT} = \text{gpm} \times \text{TD} \times 500$$

To obtain the net cooling capacity, the compression motor heat must be deducted. The motor heat is calculated by

$$\text{HM} = \text{watts} \times 3.415 \times \text{PF}$$

To calculate the net cooling capacity,

$$\text{HC} = \text{HT} - \text{HM}$$

where
HT = Heat transfer (gross capacity)
HM = Motor heat
HC = Net cooling capacity
PF = Power factor

EXAMPLE

A water to air heat pump has a condenser using 4 gpm with a temperature rise of 10°F. What is its gross cooling capacity? The compressor draws 2784 watts with a power factor of 0.9. What is the net cooling capacity?

Solution

Using the above formulas:

$$\text{HT} = 4 \times 16 \times 500 = 32{,}000 \text{ Btu/hr}$$
$$\text{HM} = 2{,}784 \times 3.415 \times 0.9 = 8{,}556 \text{ Btu/hr}$$
$$\text{HC} = 32{,}000 - 8556 = 23{,}444 \text{ Btu/hr}$$

To obtain the heating capacity, the motor heat is added to the heat picked up by the outside coil. ■

EXAMPLE

A liquid to air heat pump has an outside coil that uses 5 gpm with a 9°F TD and a compressor that uses 2552 W at a power factor of 0.9. What is the heating capacity?

Solution

Using the standard formulas,

$$\text{HO} = 5 \times 9 \times 500 = 22{,}500 \text{ Btu/hr}$$
$$\text{HM} = 2552 \times 3.415 \times 0.9 = 7844 \text{ Btu/hr}$$
$$\text{HM} = 22{,}500 + 7844 = 30{,}344 \text{ Btu/hr}$$

where
HO = Heat absorbed from outside coil ■

NOTE: On water source heat pumps where the unit may be subject to freezing temperatures, antifreeze can be added to the circulating water. However, when antifreeze is added to water its specific heat is reduced. You can obtain from the refrigerant the antifreeze supplier tables that will let you calculate the new specific heat so that you can more accurately calculate the exact system performance using the above formulas.

AIR TO WATER (HEAT ONLY)

Air to water heat pump systems have been marketed in small unit sizes for water heating and larger sized units for swimming pool heaters. These systems are used to replace expensive electric water heating. Residential water heating heat pump units are mostly in the capacity range of 6,000 to 12,000 Btu/hr. They may be incorporated into a package unit along with a hot water storage tank or as an add-on unit for existing storage tanks. The unit can be the primary heat source for domestic or small commercial hot water use, with the heat source of the water heater as a backup.

Swimming pool heaters are available in capacities from 4 tons to over 10 tons. The higher efficiency rating or COP (coefficient of performance) of the air to water heat pump will supply hot water at a lower energy cost.

NOTE: Air to water heat pumps designed for swimming pool applications have corrosion resistant heat exchangers so that the chlorine swimming pool water does not damage them. For that reason they are more expensive than air to water heat pumps designed for hydronic heating systems.

Most air to water applications have been used for heating domestic hot water or where hot water and spot cooling are needed (kitchen, laundry, et. al.). The increase in the use of heated swimming pools has opened a market for large Btu/hr capacities in this type of heat pump. Figure 13-3-10 shows a smaller domestic water heating unit. This unit is designed to sit on top of a standard electric water heater. This provides hot water by taking heat from the surrounding air and heating the water with an COP factor of 2.9—thus more hot water for less energy cost.

Some units are designed to be connected into the water circulating system of the water heater. In this application, plastic hose (copper or galvanized pipe where required by code) is used to connect the heat pump to the hot water tank. The heat pump is set to maintain a higher water temperature than the settings of the thermostats of the gas or electric water heater. This means that the heat pump is the primary means of heating the water. The auxiliary electric or gas heat source cuts in only if the heat pump cannot provide sufficient heating capacity for the amount of water drawn at that particular time.

Usually, the thermostat settings are 130°F for the heat pump and 120°F for the auxiliary heat source. The exact settings will vary with each type of equipment, and the manufacturer's instructions should be followed.

Figure 13-3-10 Heat pump water heater. *(Courtesy of Applied Energy Recovery Systems)*

The air to water, single operation heat pump unit is used for water heating. It is designed to be directly connected to the domestic hot water system. The capacity of one of these units can be determined by measuring the time it takes to raise the temperature of a given quantity of water.

For example, if the unit can heat a 30 gal tank of water from 55°F to 75°F in one hour, the capacity is calculated using the formula

$$Q = 8.33 \times \text{gal} \times TD$$

where

Q = the heating capacity of the unit, Btu/hr
8.33 = conversion factor from gal to lb
TD = temperature rise of the water in one hour, °F

Using the data above,

$$Q = 8.33 \times 30 \times (75 - 55)$$
$$= 4998 \text{ Btu/hr}$$

UNIT 3—REVIEW QUESTIONS

1. List the classifications of heat pumps.
2. What is the predominant type of heat pump sold?
3. Where is the electric heat strip located?
4. List the applications that allow the use of water to air heat pumps.

TECH TALK

Tech 1. There are sure a lot of different types of heat pumps, aren't there?

Tech 2. Yes there are. They have become so popular because of their high energy efficiency.

Tech 1. How high of an efficiency can you get from one of those things?

Tech 2. Well, some of the real high efficiency heat pumps are a lot more efficient than using gas or oil. In fact several times more efficient if it is a ground source or water source heat pump compared to using gas or oil.

Tech 1. Why don't more people use them if they are so efficient?

Tech 2. Well, a lot of people are concerned that the heat pump is not going to provide enough heat or they are concerned that it is different technology and they are real familiar with gas. There are a lot of different reasons.

Tech 1. Sounds to me like in spite of all the reasons that they give for not using it, a heat pump might be the best choice for a lot of applications.

Tech 2. That's true.

5. Water to air heat pumps use a _______ instead of an outdoor coil.
6. What does the ground coupled system take advantage of?
7. What formula is used to calculate gross capacity of the cooling cycle?
8. What are most air to water applications used for?
9. What are the usual thermostat settings for the heat pump?
10. The _______ of a unit can be determined by measuring the time it takes to raise the temperature of a given quantity of water.

UNIT 4

Troubleshooting Heat Pump Systems

OBJECTIVES: In this unit the reader will learn about troubleshooting heat pump systems including:

- Air System Problems
- Refrigeration System Problems
- Problem Analysis

GENERAL

The heat pump is unique in design compared to other air conditioning apparatus in that it operates year-round, both in summer and winter, cooling in one season and heating in the other. The refrigeration cycle operates for a longer period of the year and can develop service problems more frequently.

Since heat pumps are a form of air conditioning, the technician's basic understanding of the fundamentals of air conditioning applies. Air is the principal medium used for transferring heat. The correct quantity must be delivered at the correct temperature over the variations of winter and summer to provide comfortable conditions.

In troubleshooting heat pumps, the technician must become thoroughly familiar with the controls and operating sequence of the particular design being serviced.

Each design potentially has its own selection of components and method of operation. To assist in finding a problem, many manufacturers describe helpful procedures in their installation and service bulletins.

It is important in troubleshooting heat pumps to find out as much as possible about the problem before starting a test procedure. Probably the most important question to ask first is: "Does the unit run or not run?" If the unit does not run, the first step is to check the power supply. If the unit does run, find out what components run and what components do not run. If all the components operate, then investigate whether the problem occurs on the heating cycle or the cooling cycle.

Basically, the service problems with heat pumps fall into two main categories: air circulation problems and refrigeration problems, each of which is discussed in the following sections.

SERVICE TIP

Manufacturers have produced troubleshooting charts for their equipment. These charts are extremely beneficial in resolving problems. However, they do not work as well when a system is mismatched. A mismatched system is one that has one manufacturer's outdoor unit and another manufacturer's indoor unit. In these cases you will need to follow the troubleshooting skills that you will develop from the following unit material.

AIR SYSTEM PROBLEMS

When in the cooling cycle, the inside coil must be able to remove the correct amount of sensible and latent capacity to produce the desired room conditions. Assuming that the system has previously been set to the required temperature drop, the first check is the temperature drop through the evaporator. If the temperature drop has increased, the amount of air through the coil has decreased. The capacity of the refrigeration system will not increase; therefore, a reduction in the CFM has occurred.

If the unit is in the heating cycle, an increase in the temperature rise of the air through the inside coil, now the condenser, indicates a reduction of air through the coil. The reduction can be severe enough to cause the unit to cut out on the high head pressure lockout relay system it equipped. Repetitive cutout bringing on the lockout light on the thermostat should be investigated by a competent service technician before damage to the compressor results.

The heat pump operates on a year-round basis, and the amount of air through the inside coil is more critical than in a heating or air conditioning system.

Throwaway air filters should be checked every 30 days. It is advisable to use electronic air cleaners with heat pumps because these filters have a low static resistance even when dirty.

Inspection of the blower and drive should be done at least once a year. Both blower motor and blower bearings should be lubricated with no more than ten drops of No. 20 electric motor oil. This oil is detergent-free oil. Use of automobile oil is discouraged because it has detergent (soap) which coats the outer surface of the sintered bronze bearing and prevents oil passage through the bearing.

Unusual restrictions in the duct system, such as closing off unused rooms and placement of furniture or carpeting over supply and/or return grills, will cause coil frosting in the cooling cycle and unit cutoff in the heating cycle. This practice has a greater effect on the heating cycle than on the cooling cycle.

Failure of the duct system also has a greater effect in the heating cycle than in the cooling cycle. With 70°F in the occupied area, the supply air temperature will only be in the range of 100 to 105°F when the outdoor temperature is 60 to 65°F and down to the 85°F range when it is −10°F outside. Any leakage in the duct system will therefore seriously reduce the capacity of the unit to handle the heating load. As a result, the operating cost will be higher than normal.

SERVICE TIP

Heat pumps have higher air flows per ton than other heating systems and their air supply is lower than other heating systems. This often results in customers complaining that heat pumps are cold and drafty. In order to prevent this problem from occurring the supply registers should have deep curved louvers so that the air is directed across the ceiling and not down into the room. By directing the air away from room occupants they are not as likely to feel a cold draft. Most register manufacturers have a specific designed series of registers for heat pumps.

REFRIGERATION SYSTEM PROBLEMS

When the temperature change across the inside coil is less than it should be on either the cooling cycle or the heating cycle, the refrigeration system should be suspected. As in an air conditioning system, it can be classified into (1) refrigerant quantity or (2) refrigerant flow rate. If the system has the proper amount of refrigerant and refrigerant is flowing at the desired rate, the system will work properly and deliver rated capacity. Any problem in either category will affect the temperatures and pressures that will occur in the unit when the correct amount of air is supplied over the inside coil for the capacity of the unit.

If the system is low on refrigerant, the problem may be a leak. It must be found and repaired. The system must then be evacuated thoroughly and recharged with the correct amount of refrigerant. If the system will simply not operate, the problem is probably electrical. The electrical fault must be located and corrected.

To compare the air conditioning system with the heat pump system, various systems are shown in Figures 13-4-1, 13-4-2, and 13-4-3. The only difference between the heat pump and the air conditioning system is the addition of a reversing valve, two check valves, and a second pressure reducing device.

In Figure 13-4-1 a conventional air conditioning system is shown. This system shows the refrigerant flow from the discharge of the compressor to the condenser, the outside coil. The refrigerant condenses and flows from the outside coil through the liquid line, through the filter drier and pressure reducing device to the evaporator, and finally through the inside coil. The refrigerant expands, picking up heat in the evaporator, and then flows as a vapor to the compressor.

In Figure 13-4-2 the same action takes place except that the following devices have been added.

A reversing valve has been added to the suction and discharge lines to enable the system to reverse the flow of refrigerant in the system. The position of

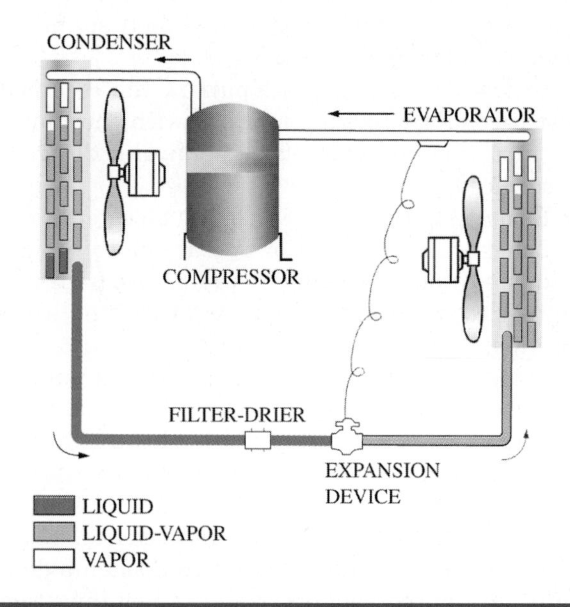

Figure 13-4-1 Schematic diagram of a conventional air conditioning/refrigeration system.

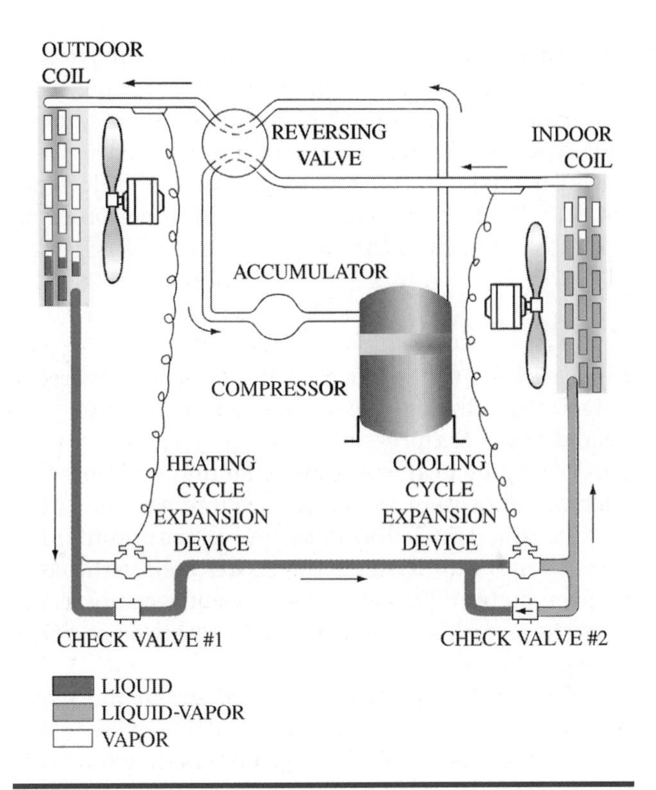

Figure 13-4-2 Schematic diagram of a heat pump system used for cooling, showing the position of the reversing valve.

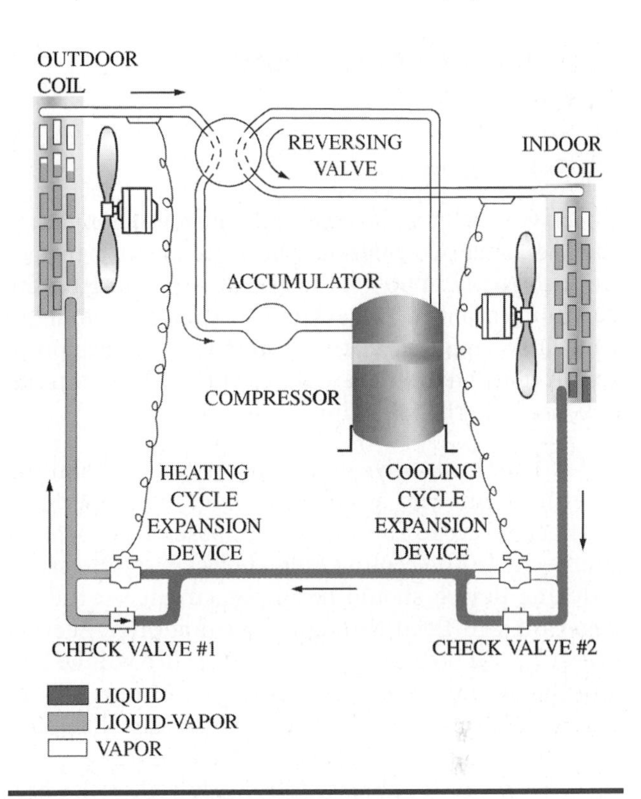

Figure 13-4-3 Schematic diagram of a heat pump system used for heating, showing the position of the reversing valve.

the reversing valve still directs the hot gas from the compressor to the outside coil and the vapor from the inside coil to the compressor.

CAUTION: During the heating mode, the vapor line on a heat pump is at the higher discharge pressure of the system. Make sure not to automatically attach the low pressure gauge to the vapor line. To do so can damage your low pressure gauge. There is a special cord that is connected between the reversing valve and compressor that is always in the low pressure side of the system.

In addition, a check valve has been installed around the pressure reducing device. It feeds refrigerant to the inside coil.

The check valve is connected to prevent flow around the pressure reducing device during cooling operations. It opens during the heating operation to eliminate the restriction of the pressure reducing device.

A pressure reducing device has also been added to the outside coil for this coil to operate as an evaporator during the heating cycle. A check valve is also connected around this pressure reducing device to remove its pressure drop during the cooling cycle. The refrigerant flow is the same as for the air conditioning unit, and this configuration is used for cooling.

In Figure 13-4-3, the reversing valve has changed position. The hot refrigerant vapor flows to the inside coil, the condenser. Giving up heat to the air supplied to the occupied area, the hot refrigerant vapor cools, condenses, and flows out of the bottom of the condenser. With the flow reversed, the check valve opens and allows the liquid refrigerant to flow around the pressure reducing device on the inside coil to eliminate any pressure loss. The liquid refrigerant flows in the reverse direction to the check valve on the outside coil. The refrigerant is forced to flow through the pressure reducing device on the outside coil (now the evaporator). Here the liquid refrigerant, at a lower pressure and boiling point, vaporizes and picks up heat from the outside air. The vaporized refrigerant flows through the reversing valve and the accumulator to the compressor.

NOTE: The only reversing of the refrigerant flow is from the reversing valve through the coils, pressure reducing devices, and check valves. The refrigerant vapor always flows from the reversing valve, accumulator, and compressor back to the reversing valve.

REFRIGERANT FLOW PROBLEMS

Check Valves

Check valves have the capability of sticking in either the open or closed position. The system will therefore work correctly in any of the cycles, depending on the valve's location in the system. If the valve is doing what it is supposed to, the system will operate properly. If not, the problem will show up. Common problems include the following.

1. The check valve on the inside coil sticks in the open position. The system will operate properly in the heating cycle. The valve is supposed to be open in this cycle. In the cooling cycle, however, no pressure reducing device should be in the circuit, so the refrigerant will flood through the evaporator, suction pressure will be high, discharge pressure will be low, and the accumulator will contain liquid refrigerant. It may be flooding back to the compressor if the liquid line is short or in a packaged heat pump.

2. The check valve on the inside coil sticks in a closed position. The system will operate properly in the cooling cycle, for which the valve is supposed to be closed. In the heating cycle, the suction pressure will be much lower than normal. In units using capillary tubes, the suction pressure will be the result of two capillary tubes in series. In systems using TXV valves, the valve will close and the compressor could pull the suction pressure into a vacuum. At the same time, the discharge pressure will be low due to little vapor for the compressor to pump. Before adding gas to the system, check the system by switching to the opposite cycle.

3. The check valve on the outside coil sticks in a closed position. The system will operate properly on heating. The valve is supposed to be closed. On cooling, low suction pressure and discharge pressure will result.

4. The check valve on the outside coil sticks open. The system will operate in cooling but not in heating. The outside check valve is supposed to be open on the cooling cycle.

CAUTION: Do not energize the reversing valve coil when it is not installed on the reversing valve. The coil can quickly overheat and burn up.

Usually, a check valve can be released from the stuck-open position by means of a magnet placed against the outlet end of the valve. Moving the magnet toward the center will force the ball or flapper to move to the seat. If in the stuck-closed position, place the magnet at the middle of the valve and move it to the outlet end. If this doesn't work, replace the valve. To reduce the possibility of future problems, use a ball-type check. Before installing the new valve, however, shake it. If it rattles, install it. If it does not rattle, it is already stuck and will not function properly. Take it back to the supplier.

Reversing Valves

Figure 13-4-4 shows the exterior view of a reversing valve. The single tube connection is always connected to the compressor discharge. The bottom middle connection is always connected to the compressor suction connection. When an accumulator is in the circuit, this connection is to the accumulator inlet. The accumulator outlet is then connected to the compressor suction. This puts the accumulator upstream from the compressor, and provides protection in either the heating or cooling mode.

The right and left connections are to the vapor line connection of the inside and outside coils. Which connection goes where depends on whether the operating coil is energized on heating or cooling. Most heat pumps are designed to operate in colder climates, where the heating operating hours are more than the cooling operating hours. Often the coil is energized during the cooling season and in a fail safe position for heating. In this case, the flow through the coil would be from the left connection to the middle and from the compressor discharge through the right-hand connection. The left-hand connection would be to the inside coil (the condenser) and the right-hand connection to the outside coil (the evaporator).

With the coil energized, the valve slide assembly would be in the opposite position. The hot gas from the compressor would flow through the left-hand

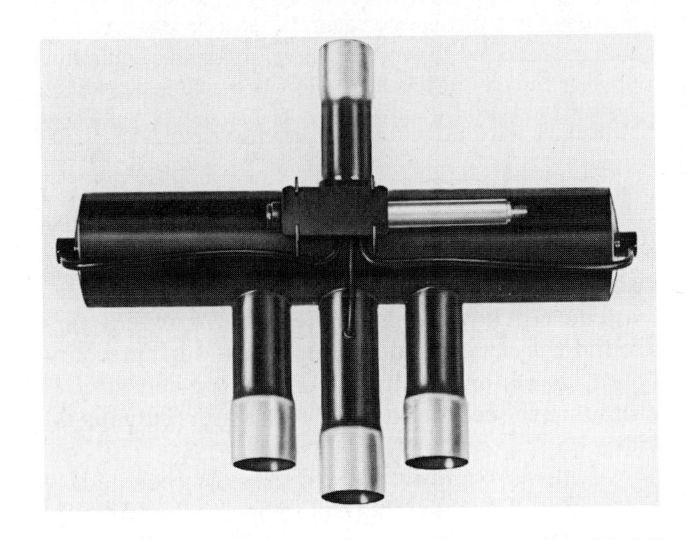

Figure 13-4-4 Reversing valve. (*Courtesy of Ranco North America*)

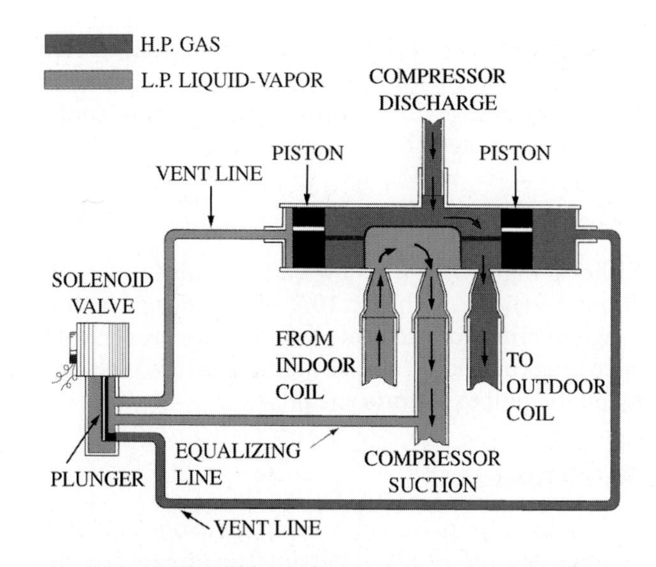

Figure 13-4-5 Reversing valve in the cooling position. *(Courtesy of Ranco North America)*

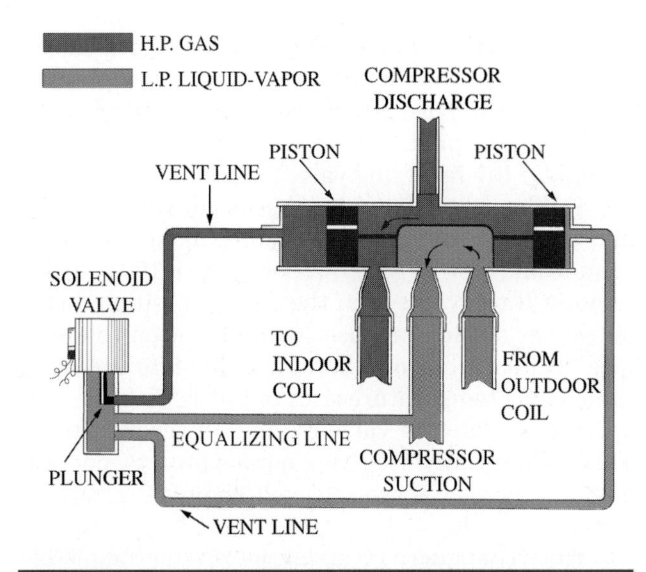

Figure 13-4-6 Reversing valve in the heating position. *(Courtesy of Ranco North America)*

connection to the outside coil, now the condenser. The cold gas from the inside coil (now the evaporator) flows through the right-hand connection to the middle connection to the accumulator and on to the compressor.

If the cycling of the valve is reversed, the coil is energized during the heating cycle and the outside connections would be reversed. Before checking the position of the valve operation, the electrical control system should be checked to determine the operating requirements of the valve.

Reviewing Figure 13-4-5, the solenoid valve is actually a single port double throw valve. The center connection is to the suction line and relieves pressure out of the valve. The bottom port has been closed. This port is normally closed. Because this port is closed, compressor discharge pressure builds up behind the main valve piston and in this line.

The top port is open and the pressure in this line bleeds into the suction line. Because discharge pressure is on the right side of the left piston in the main valve, and suction pressure on the left side, the piston has been forced to the left side of the main cylinder. This moves the bypass valve to cover the middle and left hand outlets. Not shown is the V point on the end of the piston that seats in the outlet to shut off gas flow through the piston port into the vent line.

When the pilot valve (solenoid valve) is energized, Figure 13-4-6, the ports in the pilot valve reverse; the lower valve opens and the upper valve closes. This lowers the pressure in the right end of the main cylinder to suction pressure. The pressure difference that develops across the right-hand piston forces the piston to the right end of the main cylinder. The bypass valve is moved to cover the center and right-hand outlets. When the piston stroke is completed, the gas flow is shut off by the V point entering the outlet valve seat.

This shows that the main valve piston is controlled by the action of the pilot valve to block the pressure off the ends of the main valve piston to the suction side of the compressor. The valve works on the compressor differential pressure. The minimum pressure required to operate the valve is 75 to 100 psig. This means that the refrigeration system must be fully charged with refrigerant and operating long enough to develop a 75 psig difference between suction and discharge pressure.

Problems in reversing valves are either electrical or mechanical. Electrical problems are confined to the solenoid coil on the pilot valve. When the solenoid coil is supposedly energized and nothing happens, test the coil as follows:

1. Feel the solenoid coil. If it feels hot, the coil is energized.

2. Make sure that voltage is applied to the coil. Some units have the coil in the 240 V portion of the system, while others use it in the 24 V portion. Check the wiring diagram for coil voltage before applying the leads of the voltmeter. Also, start the test with the voltmeter set to the higher range. This helps reduce meter burnout.

3. With voltage applied to the coil, remove the coil holding nut and attempt to pull the coil off the pilot valve plunger casing. If you feel a resistance to removing the coil, the coil is active. If no pull is felt, the coil is dead. Shut off the power, remove the coil heads, and check for continuity with the ohmmeter. If open, replace the coil. If a circuit exists, check the leads and the connections for continuity.

If the coil is active, when removing the coil a "click" should be heard when the pilot plunger returns to the normally closed position. If no clicks are heard, the pilot valve is stuck. The only repair is to replace the reversing valve.

4. When the pilot valve checks out satisfactorily and the main valve does not shift, make sure that suction and discharge show more than a 100 psig difference. If the valve is in the cooling position, block off the air to the condenser with plastic on the inlet face of the coil. Allow the unit to operate until the condensing temperature rises to 130°F. With the unit operating, cycle the valve on and off several times. This will usually free the main valve to operate again. If no results are achieved, change the valve.

When changing a reversing valve, after completely removing the refrigerant from the system, be sure to read the installation instructions supplied with the valve. Replace the valve with one of a comparable size. Always position the valve so that the main piston is in a horizontal position and the pilot valve is higher than the main valve. This is to keep oil from gathering in the pilot valve and affecting its operation.

The valve body must always be protected from heat when installing by wrapping the body with some type of thermoplastic material or wrapping with a water-saturated cloth. The maximum temperature the valve body will tolerate is 250°F.

SERVICE TIP

If the heat pump is operated for any period of time in the heating mode in warm weather the compressor can become overloaded. If it is necessary to run the heat pump in the heating mode for testing purposes when the weather is warmer than 60°F you should cover the coils to restrict the airflow. This will prevent the heat pump from picking up excessive amounts of heat and overloading the compressor while the system is being tested.

PROBLEM ANALYSIS

Various refrigeration, electrical, and air distribution problems relating to air conditioning systems have been discussed in previous chapters and should be referenced whenever applicable. In this chapter we will deal with only those problems uniquely related to heat pumps.

Refrigeration System

The primary refrigeration problem occurs when the unit will not change cycle, heating to cooling, or vice versa, regardless of thermostat switch settings. The changeover of the system is controlled by the action of the reversing valve. Malfunction of this control can be either electrical or mechanical.

Electrical

Either a high voltage or low voltage valve must have applied voltage within ±10% of the design voltage range in order for the control to operate properly. The operating coil should also be able to produce the proper magnetic pull to operate the pilot valve.

Mechanical

The main valve piston may be stuck in one end of the cylinder because it is not getting the proper pressure changes from the pilot valve. This can be determined by touching the coil connection tubes with the coil energized and de-energized. Any sticking of the pilot valve plugs in the small tubes between the pilot valve and the ends of the main valve, or seizure of the main valve, requires replacement of the valve. Repair of the valve is not justified because of the time cost and high percentage of failure repeat.

If the unit works fine on heating but poorly on cooling, the problem is usually found in the action of the check valves. If the suction pressure is higher than reasonable for the temperature of the air leaving the coil, the check valve on the inside coil may be stuck open. This removes the pressure reducing device from the circuit and refrigerant is flooding through the coil. Superheat will be low or nonexistent, and the temperature of the accumulator, reversing valve tubes, and possibly the compressor shell, will be very low.

If the suction pressure is lower than normal, the check valve on the outside coil is stuck closed. This keeps the pressure reducing device in the circuit. If it is a TXV valve, the valve will close and shut off the refrigerant flow to the liquid line. The suction pressure will drop extremely low, possibly into a vacuum.

If the pressure reducing devices are capillary tubes, refrigerant flow will take place but will be much less than normal. The unit will have two capillary tube sets in series with the higher pressure reduction of both sets. On some units, the high side pressure is measured on the liquid line between the check valve and the pressure reducing valve sets. With TXV valves, the head pressure gauge will indicate a loss of refrigerant. With capillary tubes, the head pressure will be low, along with a low suction pressure. In either case, do not add refrigerant until the subcooling has been checked. The subcooling will be found to be extremely high because the refrigerant has accumulated in the condenser.

SERVICE TIP

If it is necessary to replace the heat pump reversing valve the new valve must be protected from overheating during brazing. The valve contains plastic gaskets and seals that are easily damaged from excessive heat. Wrapping a wet rag around the valve body will help prevent overheating. In addition, an oxyacetylene torch with a large enough flame to heat the fitting quickly should be used.

Defrost System

To diagnose problems in the defrost system, the type of defrost system must be determined. The four most predominant systems are covered here. On those systems used by individual manufacturers, the manufacturer should be contacted for service information.

Temperature Differential Defrost System

1. *The unit will not go into defrost.* This is the most common complaint on this control. The control is operated by the temperature difference between the entering air temperature of the outside coil (the evaporator) and the coil operating temperature.

The predominant reason that the unit will not go into defrost is that the coil will not reach a low enough temperature to provide an increase in the control differential to activate the defrost cycle. The major reason for this is that the head pressure is too high, which forces the suction pressure and coil boiling point up. Only a 2°F rise in boiling point is necessary to require complete coverage of the coil with frost and ice before initiating the defrost cycle. Rather than correct the air problem—the duct system may be inadequate or the occupants dislike the low temperature discharge air—the service technician will often replace the control. Worse yet, the service person may try to adjust the control, which is not possible without closely controlled water baths.

The correct solution to this is to restore the correct flow of air over the indoor coil to design conditions. This will lower the head pressure and the suction pressure.

2. *The unit goes into defrost with very little frost on the bottom of the coil.* The location of the coil temperature sensing bulb is critical. If the bulb is located too far to the bottom of the coil and too close to the entering low pressure liquid refrigerant, the lower temperature of this portion of the coil will promote premature initiation of the defrost cycle. The cure is to move the coil temperature bulb up one return bend at a time until the correct location is obtained. Once the defrost cycle is initiated, the coil has to reach a preset temperature to terminate the defrost cycle. Usually, this temperature is 50 to 60°F.

3. *The unit does not defrost the entire coil; an ice ring builds up around the bottom of the coil.* The ice buildup indicates that the coil temperature bulb is located too high on the coil. The bulb is reaching termination temperature before the coil is completely defrosted. The bulb should be lowered one return bend at a time until complete defrosting takes place.

SERVICE TIP

Time defrost systems can be set for various defrost intervals. These are typically 30 min, 45 min, 1 hr, and 1 1/2 hr intervals. The higher the relative humidity in your area during the winter the more frequent defrost cycle times are needed . You should check with your local supplier for their recommendation on defrost time settings. The more frequent the defrost time the less efficiently the system will operate. Therefore, as long a time setting as possible should be used.

Pressure Temperature Defrost System

In this system, the pressure differential across the coil initiates the defrost cycle and the temperature of the liquid leaving the coil terminates it. The liquid line thermostat also controls the possibility of a defrost cycle.

A thermostat fastened to the liquid line at the bottom outlet of the coil is exposed to the expanding liquid refrigerant when the unit is operating in the heating cycle. If the evaporator coil is operating with an entering liquid of 26°F or higher, there is little chance that frost will form on the coil. If this temperature is below 26°F, the thermostat closes and completes the circuit through the defrost relay to the differential pressure switch.

When the frost buildup causes enough resistance to close the differential pressure switch, the defrost cycle is initiated. When the liquid leaving the coil reaches 55°F, the circuit is broken, the defrost relay circuit opens, and the relay drops out. This opens the holding circuit of the relay and completes the switchover to the heating cycle.

1. *The unit will not go into defrost.* Two possible causes can prevent the unit from going into defrost. The most probable cause is that the termination thermostat has come loose from the coil outlet pipe. The thermostat cannot reach 29°F or below to close and allow the defrost circuit to initiate. The other probable cause is plugged tubes to the differential pressure switch. Insects sometimes plug these pipes.

2. *The unit goes into defrost with very little frost on the bottom of the coil.* Because the pressure drop of the air through the coil is the determining factor for initiating the defrost cycle, both the cleanliness of the coil and the frost buildup make up this pressure drop. Therefore, the more dirt buildup on the coil, the less frost has to form to reach the required pressure drop to initiate the defrost cycle. When the unit starts defrosting with little frost buildup, clean the coil.

3. *The unit does not defrost the entire coil; an ice ring forms at the bottom of the coil.* The fact that the ice ring forms indicates that the defrost cycle is interrupted before the defrost cycle is completed. The most common cause of this is the time requirement set on the defrost control is too high.

When the coil is defrosted, the termination thermostat breaks the power to the defrost relay. The holding contact in the relay that keeps the relay energized is broken. The system must now develop the necessary frost buildup to close the pressure switch.

If the power supply to the defrost circuit is interrupted, the same action occurs. Therefore, if the reheat is great enough to warm the occupied area and cause the room thermostat to open before the defrost cycle is completed, the incomplete defrost cycle will leave an ice buildup at the bottom of the coil on the next heating demand. The unit usually starts in defrost and completes the defrosting cycle. The alternate thawing and freezing of the water held on the coil by the ice coating will cause collapse of the coil tubes and loss of refrigerant.

Time Temperature Defrost System

1. *The unit will not go into defrost.* This system uses a timer to control the defrost cycle under the control of the termination thermister. The power to the timer is supplied from the compressor circuit. The timer therefore operates even when the compressor is operating and the defrost thermostat is below 26°F.

The most common cause of this problem is a loose defrost thermostat. A poor contact between the thermostat and the coil tube prevents the thermostat temperature from dropping below 26°F. If contact is made, the thermostat should be checked to determine if it closes at a temperature below 26°F. Immersion in an ice and saltwater bath using an immersion thermometer will accomplish this. The timer motor should also be checked for proper motor operation.

2. *The unit goes into defrost with very little frost on the bottom of the coil.* The timer may be operating on too short a cycle time. Most timers have 30 min and 90 min cycle cams. Change the timer from the 30 min setting to the 90 min setting.

3. *The unit does not defrost the entire coil; an ice ring builds up around the bottom of the outside coil.* The correction to this problem is to reduce the amount of reheat used in the defrost cycle to an amount not more than the sensible capacity of the unit on the cooling cycle.

Pressure Time Temperature Defrost System

This system uses a differential air pressure switch to measure the amount of frost on the outside coil. When the pressure drop through the coil reaches a preset amount, the switch closes the timer circuit.

If the defrost thermostat is below 26°F, the timer will operate. This timer usually requires 5 min of pressure switch closure time to initiate the defrost cycle by energizing the defrost relay. When the termination thermostat reaches 55°F, the power to the defrost relay is interrupted and the unit changes over to the heating cycle.

1. *The unit will not go into defrost.* The most common cause of this is a loose or poorly connected termination bulb of the defrost control. It is very important that the bulb be securely fastened with heat transfer compound on the joint, insulated, and sealed from moisture. Ice can build up at this location and loosen the bulb fastening. The second cause can be failure of the pressure switch due to dirt and/or insects in the pressure switch tube connections. There is always the possibility of failure of the timer control, but this is remote. Check the previous two items before replacing any parts.

2. *The unit goes into defrost with very little frost on the bottom of the coil.* The major cause of this action is a dirty outside coil. Its frost-free resistance is so high that it takes very little increase in pressure drop to start the defrost cycle. A thorough coil cleaning is in order.

3. *The unit does not defrost the entire coil.* This system has an automatic defrost cycle termination anywhere from 5 to 12 min, depending on the frost accumulation rate of the outside weather conditions.

UNIT 4—REVIEW QUESTIONS

1. When troubleshooting heat pumps what is probably the most important question to ask first?
2. What are the two main categories of service problems with heat pumps?
3. How often should throwaway air filters be replaced?

TECH TALK

Tech 1. I've been trying to use this troubleshooting chart to find out what the problem is with that heat pump, but I just can't seem to get to the problem.

Tech 2. Let's take a look at the chart. Did you start here at the top and go through each of the steps and see how the system was responding?

Tech 1. No, I didn't. I know it's got to be a problem with the reversing valve, so I came down to this section where it talks about the reversing valve.

Tech 2. Well, it's not a good idea to jump ahead. There are a lot of things that can cause a problem that look like the reversing valve. Until you get more experience you don't want to do that. You want to go down each step on the list and see what the chart tells you should happen and if it doesn't happen then follow the procedures it recommends to replace the part that's been identified as the problem.

Tech 1. But that seems like it would take a whole lot more time than simply going down to where I know where the problem is.

Tech 2. Well, you've already spent a lot a time trying to figure out why the chart didn't work. You would have spent less time if you had followed the chart from the top.

Tech 1. Really? You're probably right. I guess I need to go back and start over from the beginning. Thanks.

4. List the classifications of a refrigeration system.
5. What is the only difference between the heat pump and the air conditioning system?
6. If a _______ is doing what it is supposed to, the system will operate properly.
7. What should be tested if the solenoid coil is supposedly energized and nothing happens?
8. What is the maximum temperature that the valve body will tolerate?
9. Where can the problem usually be found if the unit works fine on heating but poorly on cooling?
10. What is the most common complaint on the temperature differential defrost system?
11. The _______ system used a differential pressure switch to measure the amount of frost on the outside coil.

SECTION V

Indoor Air Systems

14

Air Distribution

UNIT 1

Duct Systems

OBJECTIVES: In this unit the reader will learn about duct systems including:

- Duct System Types
- Extended Plenum
- Air Distribution and Balancing
- Air Quantity

OVERVIEW

The major controlling factor affecting duct system design is the building's layout and space availability. The building's design affects the air handler location, which then controls the duct system layout. The building's designer may specify the depth, layout, and configuration, or they may leave it up to the air conditioning contracting company.

The most ideal duct system layout is one that provides for uniform temperature throughout the building without producing distracting sound levels or air movements and operates without notice. The system should be designed so that the location of lines, registers, and return grills are least obtrusive to the overall aesthetics of the building and its rooms. An ideal duct layout is one that allows for uniform air distribution and maximum system efficiency, and can do so without being noticed.

NICE TO KNOW

A number of studies have shown that traditional duct installation practices can result in systems that have excessive air leaks. New codes require the use of UL 181 approved duct tapes and UL 181 approved mastics to reduce air leaks and improve system efficiencies.

For maximum system efficiency the duct system should be located so that it is within the building's insulation envelope, Figure 14-1-1. Ducts located outside of the envelope, either in the attic space or unconditioned crawl space, even with R8 insulation and properly sealed, gain and lose heat. It is not

Figure 14-1-1 Duct installed in joist space so it is inside the building envelope.

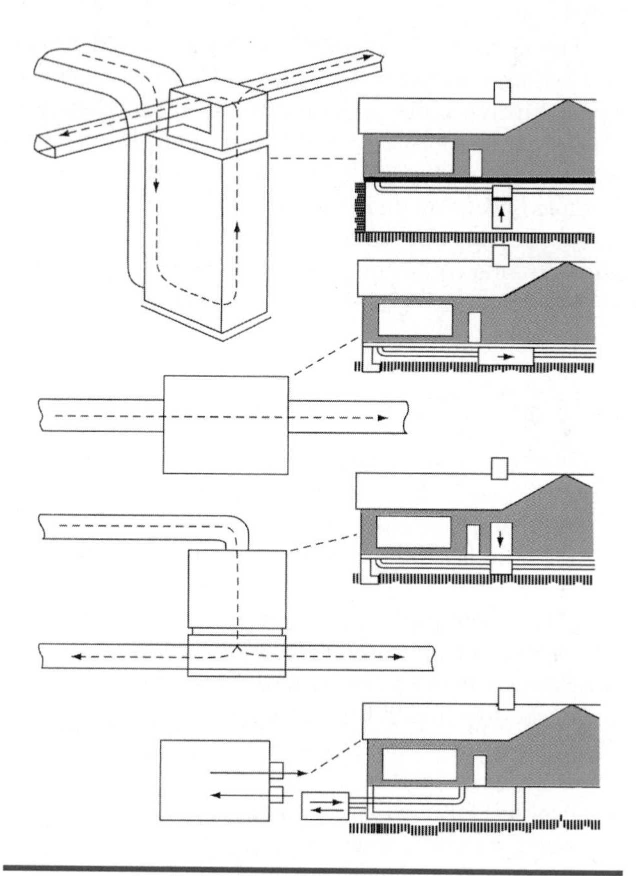

Figure 14-1-2 Various configurations of warm air furnaces.

always possible to locate the ducts within the building's envelope. The three most common areas to locate the duct system are in the attic, in the space between the joists in the ceiling or between floors, and under the building in either the basement or crawl space, Figure 14-1-2.

Energy efficiency guidelines require that ducts located outside of the building's envelope have either an R6 or R8 insulation value. Before beginning the layout and design for any building you are designing a duct system for, you must check with the local building department, county, or state code and/or regulatory agencies to determine the specific insulation requirements for your building's type and location. Under current Energy Star guidelines, a system's insulation value may be downgraded if the system energy efficiency ratings are upgraded. This is referred to as energy efficiency tradeoffs. Not all local or state codes allow for this trading of energy ratings.

SERVICE TIP

The new duct installation requirements result in a much thicker insulation layer around the ducts. This can make it far more difficult to pull these ducts through building structure members without snagging or tearing the flex duct's outer vapor barrier. If the outer vapor barrier is damaged it must be repaired with an approved tape. The tape must be spiraled around the duct with an approximately 75% overlap of each successive wrap. A required minimum of 3 wraps are required on all repairs. Large tears may require more than 3 wraps to completely seal the vapor barrier.

DUCT SYSTEM TYPES

The five most common duct configurations are the radial, the reducing radial, extended plenum, the reducing extended plenum with reducing trunks, and the perimeter loop. The central plenum duct system can easily be located in the attic or basement space. Some building floor joists, such as this open truss, Figure 14-1-3a, allow for the design and installation of central plenum duct systems within the floor space.

Radial duct systems are designed so that all or almost all of the duct runs originate at the central plenum. In some cases some of the duct runs may have wide or duct triangles as a means of joining additional ducts to an initial run, Figure 14-1-3b.

Extended Plenum

The extended plenum duct system uses a large trunk duct that travels out in two directions from the air handler. These systems are sometimes referred to as trunk duct systems, Figure 14-1-3c,d. Extended plenum systems, trunk ducts, do not reduce in size as they travel across the structure. Extended plenum systems can be located in the crawl space, attic, or basement.

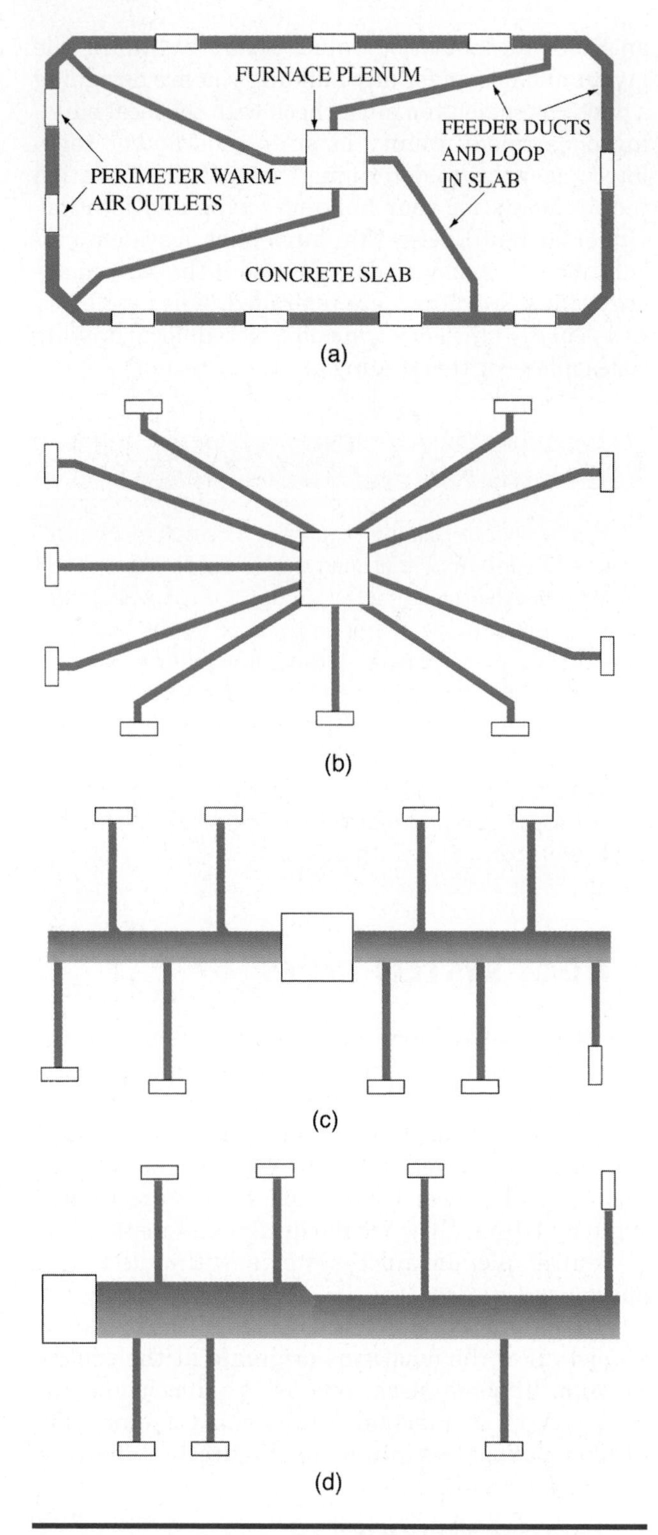

Figure 14-1-3 Most common duct systems:
(a) Perimeter loop system, with ductwork in concrete slab floor; (b) Central plenum duct system located in the basement for heating the space above; (c) Extended plenum or trunk duct system, located in basement;
(d) Extended plenum with reducing trunk system (located in the basement).

DUCT SYSTEMS

All ductwork designs should start with an accurate load calculation. The amount of air supplied to each area depends upon the space temperature, the supply air temperature, and the space heating and cooling loads.

The equipment used for forced air heating and cooling systems in residential and small commercial systems consists of furnaces, air conditioners, and heat pumps. Furnaces can be upflow, downflow, horizontalflow, or a packaged unit.

The most common air conditioning system uses a split system configuration with the evaporator located in the furnace plenum and a remote condensing unit. A packaged air conditioner contains all the components. The heat pump can be a split system or a packaged unit.

The accessories include humidifiers, air cleaners (filters), and economizer controls. There are several types of humidifiers, including self contained steam, atomizing, evaporative, and heated pan. Filters can be cleanable or throwaway types, with electronic air cleaners becoming increasingly popular. Economizers automatically use outside air for cooling or precooling when outside conditions permit it. All of these accessories affect the system airflow and pressure requirements.

System Design Procedure

The size and the performance of the system components Figure 14-1-4 are interrelated.

Following the calculation of heating and cooling loads, the following procedure should be followed:

1. Determine preliminary ductwork location.
2. Determine heating and cooling unit location.
3. Select accessory equipment. Provision may be necessary to add this equipment later.
4. Select control components.
5. Determine maximum airflow (heating and cooling) for each supply and return location.
6. Select heating and cooling equipment.
7. Finalize duct design and size system.
8. Select supply air outlets and return air grills.

CAUTION: An improperly designed and installed air conditioning air distribution system can significantly reduce the system performance. Unfortunately in many cases the duct systems in new construction are installed by subcontracting companies that have had little or no formal training of the proper installation techniques. When servicing one of these systems it may be necessary to do some reworking of the duct system to provide proper system performance.

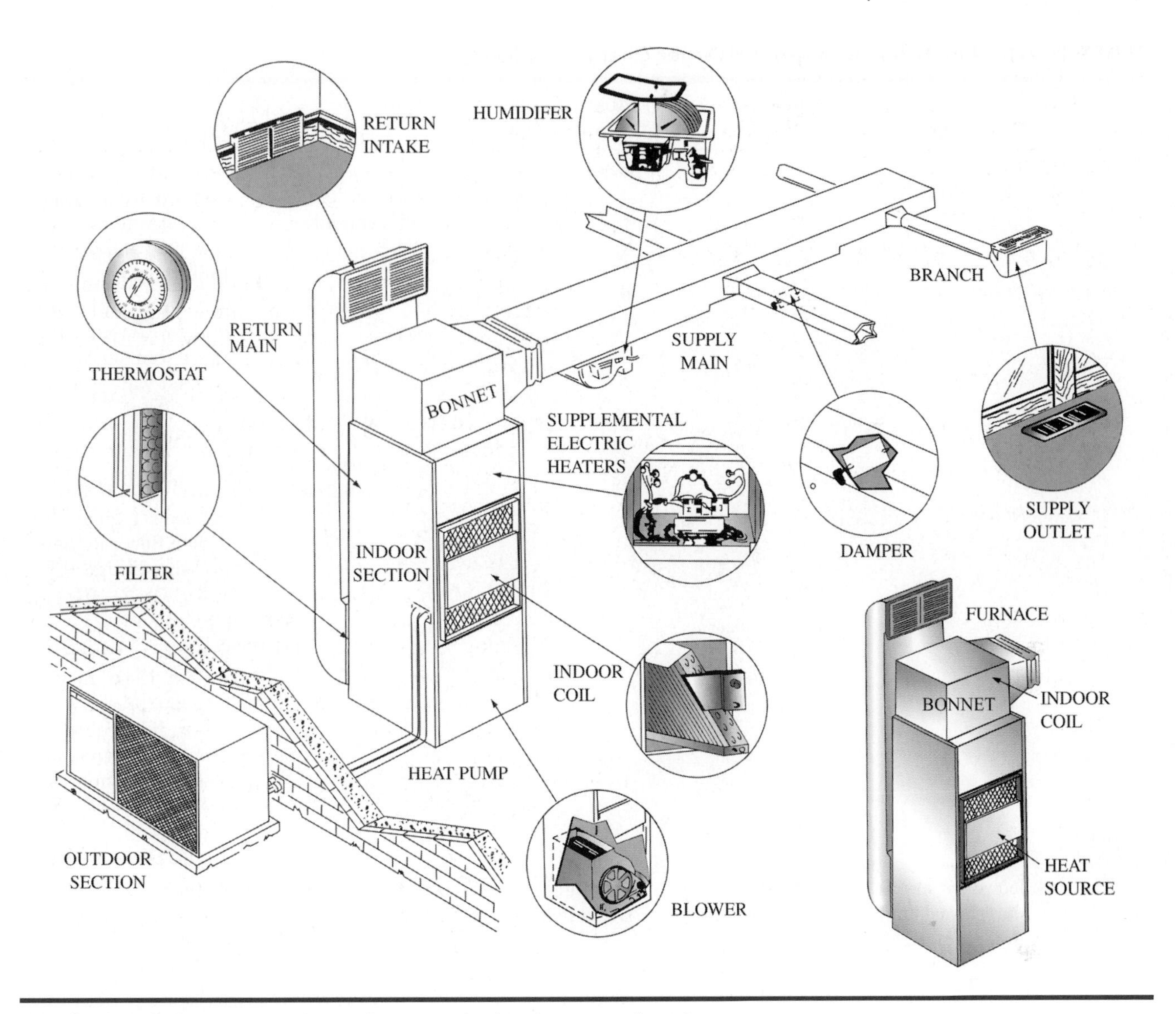

Figure 14-1-4 Components that make up a complete heating and cooling system. *(Copyright by the American Society of Heating, Refrigerating, and Air-Conditioning Engineers, Inc.)*

Locating Outlets, Ducts, and Equipment

The structure and layout of the building determine the available location for system components. In a residence, a full basement is an excellent location for the trunk ducts and conditioning equipment. Equipment can also be located in a closet space or utility room. All enclosures must meet local fire and safety codes.

For slab-type construction, ducts and even equipment can be placed in the attic.

Ducts located in unconditioned areas must be properly insulated, and allowance for heat loss included in the load calculation. Ducts for perimeter (heating only) systems can be located in the slab, as shown in Figure 14-1-3d.

Supply outlets can be classified in four different groups, defined by their air discharge patterns: horizontal high, vertical nonspreading, vertical spreading, and horizontal low. Table 14-1-1 lists the general characteristics of supply outlets. Note that there is no single outlet that is best for both heating and cooling.

Figure 14-1-5 shows the preferred return locations for different supply outlet positions. These return locations are based on a stagnant layer in the room that is beyond the influence of the supply outlet. The stagnant layer develops near the ceiling during cooling and near the floor on heating. Cooling returns should be placed high and heating returns low. For year-round systems a compromise should be made by placing returns where the largest stagnant area will occur.

TABLE 14-1-1 Tabulation of Supply Air Outlet Characteristics

Group	Outlet Type	Outlet Flow Pattern	Most Effective Application	Specific Use	Selection Criteria
1	Ceiling and high sidewall	Horizontal	Cooling	*Ceiling outlets* Full circle or widespread type	Select for throw equal to distance from outlet to nearest wall at design flow rate and pressure limitations.
				Narrow spread type	Select for throw equal to 0.75 to 1.2 times distance from outlet to nearest wall at design flow rate and pressure limitations.
				Two adjacent ceiling outlets	Select each so that throw is about 0.5 distance between them at design flow rate and pressure limits.
				High sidewall outlets	Select for throw equal to 0.75 to 1.2 times distance to nearest wall at design flow rate and pressure limits. If pressure drop is excessive, use several smaller outlets rather than one large one to reduce pressure drop.
2	Floor diffusers, baseboard, and low sidewall	Vertical, nonspreading	Cooling and heating	For cooling only	Select for 6 to 8 ft throw at design flow rate and pressure limitations.
3	Floor diffusers, baseboard, and low sidewall	Vertical, spreading	Heating and cooling	For heating only	Select for 4 to 6 ft throw at design flow rate and pressure limitations.
4	Baseboard and low sidewall	Horizontal	Heating only		Limit face velocity to 300 ft/min.

Selecting Equipment

A furnace selection should either match or be slightly larger than the load. A 40% limit on oversizing is recommended by the Air Conditioning Contractors Association (ACCA). This limit minimizes venting problems and improves part load performance. ASHRAE Handbook on Fundamentals recommends that cooling units not be oversized more than 25% of the sensible load.

Airflow Requirements

After selecting the equipment, the following determinations need to be made:

- The air quantity (CFM) for each area.
- The number and type of supply and return grills and registers for each area.

The next step is to proceed with the duct layout.

The design of the system must be based on airflow and the static pressure limitation of the fan. The CFM for each area is proportional to the load of each area. It is recommended that the system be designed using the medium speed of the fan to allow the flexibility to handle added accessories.

For systems designed for heating only, the air temperature rise must be within the manufacturer's recommendation (usually 40 to 80°F).

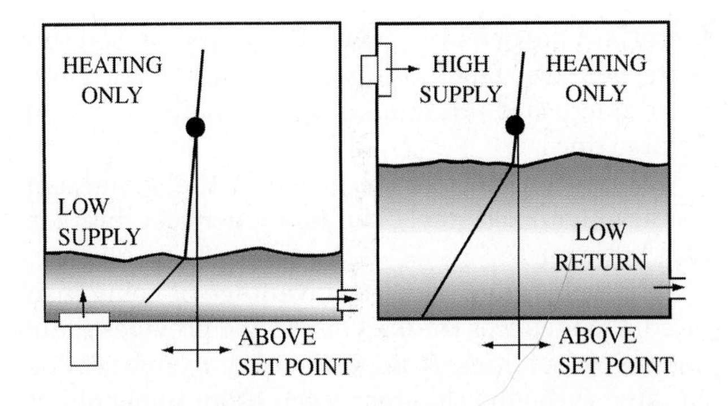

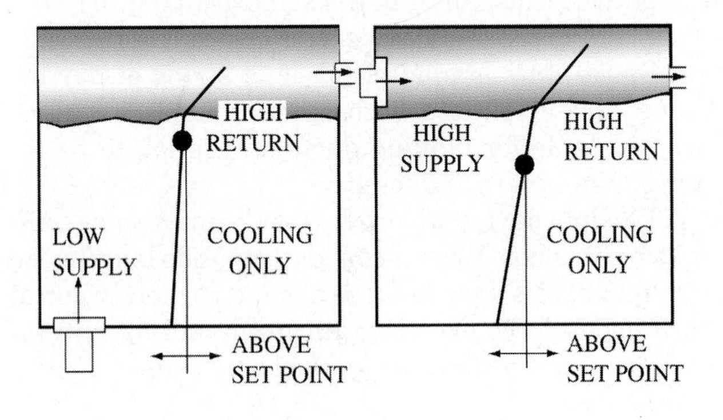

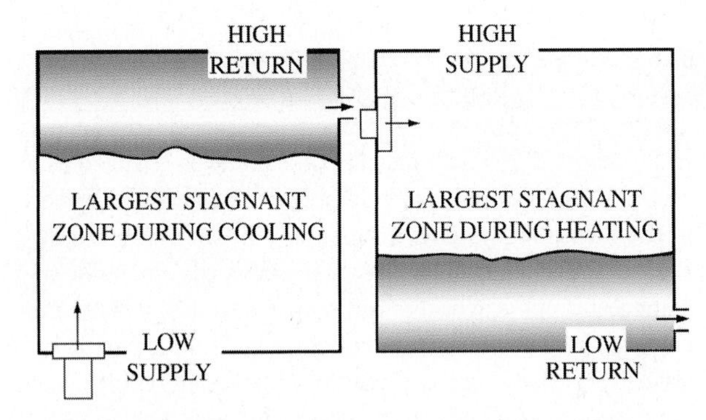

Figure 14-1-5 Preferred return air locations in relation to supply air outlets. *(Copyright by the American Society of Heating, Refrigerating, and Air-Conditioning Engineers, Inc.)*

For cooling only or heating and cooling systems, the flow rate can be determined by the following equation:

$$Q = Q_S(1.08 \times TD)$$

where

Q = flow rate in CFM
Q_S = sensible heat load in Btu/hr
TD = dry bulb temperature difference in °F

For preliminary design, an approximate TD is determined as follows:

Sensible Heat Ratio (SHR)	TD °F
0.75–0.79	21
0.80–0.85	19
0.85–0.90	17

TECH TIP

The constant of 1.08 in the heat formula may be seen as 1.1 in some equipment manufacturers' formulas. The difference between the two values is only two hundredths which is insignificant for most calculations unless the building is very large.

$$\text{SHR} = \frac{\text{calculated sensible load}}{\text{calculated total load}}$$

For example, if Q_S = 23,000 Btu/hr

Q_1 = 4900 Btu/hr

where

Q_1 = latent heat load

then

$$\text{SHR} = \frac{23{,}000}{(23{,}000 + 4900)} = 0.82$$

and

$$Q = \frac{23{,}000}{(1.08 \times 19)} = 1121 \text{ CFM}$$

The exact CFM value will be determined when the equipment is selected.

Duct Design

The duct design procedure shown here is a modified *equal friction method,* and should be used for systems requiring 2250 CFM or a 60,000 Btu/hr or less cooling load. This method has the advantage of simplifying the calculations and maintaining sufficient accuracy within this range.

The pressures available for the supply and return ducts are found by deducting coil, filter, grill, and accessories from the manufacturer's specified blower pressure. The remaining pressure is divided between the supply and return ducts as shown in Table 14-1-2.

General rules that apply to the design of ductwork are:

1. Keep main ducts as straight as possible.
2. Streamline transitions.
3. Design elbows with an inside radius of at least ⅓ the duct width, or use turning vanes.
4. Make ducts tight and sealed to limit air loss.

TABLE 14-1-2 Recommended Division of Duct Pressure Loss Between Supply and Return Air Systems

System Characteristics	Supply, %	Return, %
A Single return at blower	90	10
B Single return at or near equipment	80	20
C Single return with appreciable return duct run	70	30
D Multiple return with moderate return duct system	60	40
E Multiple return with extensive return duct system	50	50

TECH TIP

Manufacturers' technical data that is provided on equipment can be used to determine the exact pressure drop that each of the components in a system will have. It is important that you use this manufacturers' technical data when designing a system so that the system will function properly.

5. Insulate and/or line ducts, to conserve energy and limit noise.
6. Locate branch takeoffs at least 4 ft downstream from a fan or transition joint.
7. Separate air moving equipment from metal ducts using flexible connectors to isolate noise.
8. Install a volume damper in each branch for balancing air flow.

The noise level for a small system is controlled by limiting the duct velocities as follows:

Main ducts	700 to 900 fpm
Branch ducts	600 to 900 fpm
Branch risers	700 fpm

For combination heating and cooling systems calculations produce a separate CFM for heating and cooling. Since the cooling loads require greater CFM, they are usually used to size the ductwork. When two-speed fans are used, the air quantity can easily be reduced during the heating season.

Duct Materials and Standards

For many years galvanized sheet steel was used exclusively for air conditioning ductwork; however, the material is expensive and costly to install, so other materials are sometimes used. These include fiberglass ductboard, aluminum, spiral metal duct, and flexible duct.

Aluminum duct is fabricated in the same manner as galvanized steel duct. Its higher cost limits its use, however.

Fiberglass duct has the advantage of being insulated. This reduces the duct losses and provides sound absorbing qualities. It is particularly useful for duct running through cold areas, such as an unheated attic space. It is usually 1 in thick, fabricated into round ducts and made in flat sheets for custom fabricating.

Spiral duct is made from long strips of narrow metal and fabricated with spiral seams. Machines are available for making ducts on the job to fit required diameters and lengths.

Flexible duct, without the insulation, comes compressed in a box. When opened it expands lengthwise into ducts. It is easy to route around corners. Special care needs to be exercised during handling and installation not to flatten the duct and reduce the internal area.

SERVICE TIP

In the early 2000s duct installation standards changed as a result of increased energy efficiency demands. You may find in the field duct systems that are still in use that were designed and built before the new standards. In most instances it is not practical to remove these existing ducts and replace them with the higher more efficient duct materials unless local codes require it during remodeling. Check with your local building officials to determine whether or not it will be necessary to change out existing ducts.

Many localities adopt the metal duct standards developed by the Building Officials and Sheet Metal and Air Conditioning Contractor's National Association (SMACNA). A guideline for selecting metal thickness is shown in Table 14-1-3.

Sheet metal ducts are usually held together by the use of drive slips and S hooks, as shown in Figure 14-1-6. It is extremely important that air leakage from ductwork be held to a minimum. For this purpose, most can be sealed with mastic, Figure 14-1-7.

A problem, shown in Figure 14-1-8, illustrates the principles outlined above. The step by step procedure for making the calculations follows.

Duct Design Procedures Using the Equal Friction Method

1. Determine the heating and cooling load to be supplied by each outlet. Include duct losses or gains.

TABLE 14-1-3 Gauges Recommended for Sheet Metal Ductwork

	Comfort Heating or Cooling			Comfort Heating Only
	Galvanized Steel			
	Nominal Thickness (in Inches)	Equivalent Galvanized Sheet Gauge No.	Approximate Aluminum B & S Gauge	Minimum Weight Tin Plate Pounds Per Base Box
Round Ducts and Enclosed Rectangular Ducts				
14″ or less	0.016	30	26	135
Over 14″	0.019	28	24	—
Exposed Rectangular Ducts				
14″ or less	0.019	28	24	—
Over 14″	0.022	26	23	—

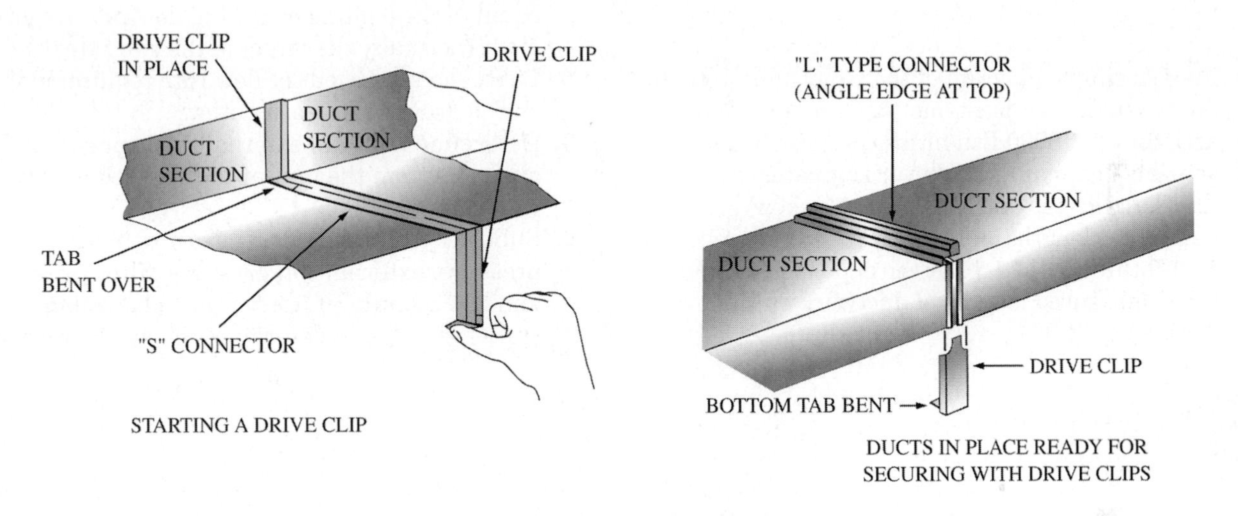

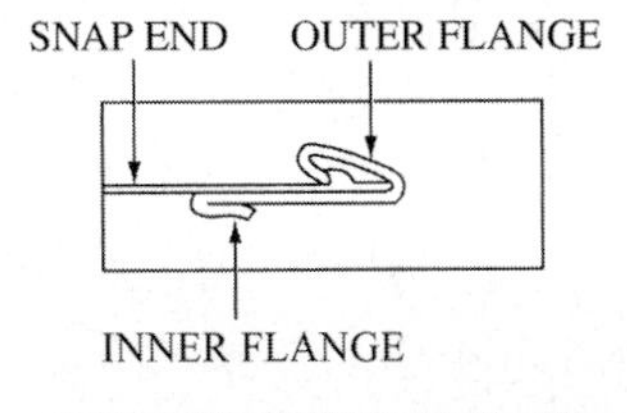

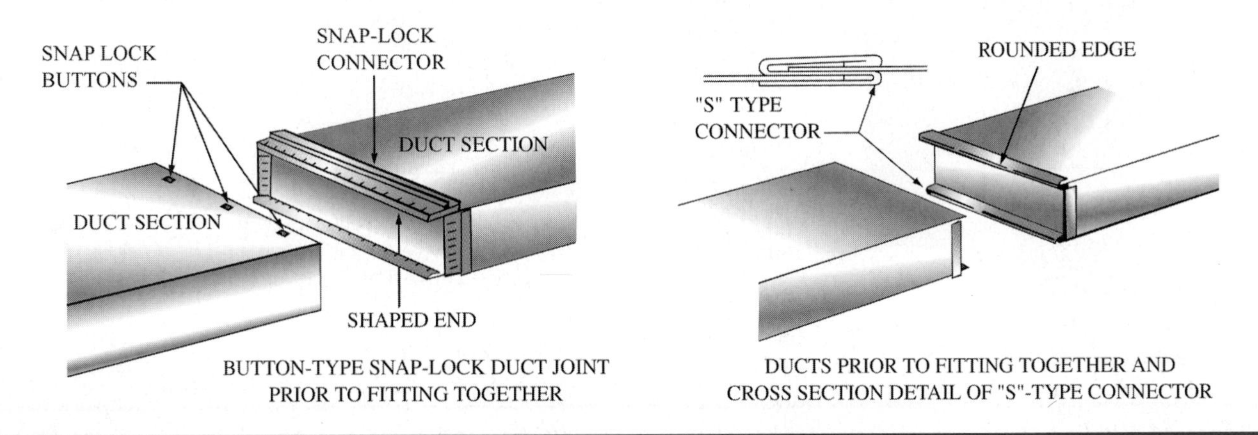

Figure 14-1-6 Sheet metal ductwork connections using drive clips and S hooks.

Figure 14-1-7 Applying mastic to sheet metal duct joints.

2. Make a simple diagram of the supply and return duct systems—at least one outlet in a room or area for each 8000 Btu/hr loss or 4000 Btu/hr sensible heat gain, whichever is greater.
3. Label all fittings and transitions to show equivalent lengths on the drawing. See Figures 14-1-9 through 14-1-17 at the end of the problem.
4. Show measured lengths of ductwork on the drawing.
5. Determine the total effective length of each branch supply. Beginning at the air handler,

$$\text{Effective length} = \text{equivalent length} + \text{measured length}$$

Branch R in the example, Figure 14-1-15, is 92 ft.
6. Proportion the total airflow rate to each supply outlet for both heating and cooling:

$$\text{Supply outlet flow rate} = \frac{\text{Outlet heat loss (gain)}}{\text{Total heat loss (gain)}} \times \text{Total system flow rate}$$

$$\text{Branch R, Figure 14-1-15} = \frac{5400}{60{,}000} \times 900 = 81 \text{ CFM}$$

7. The flow rate required for each supply outlet is equal to the heating or cooling flow rate requirement (normally the larger of the two rates).
8. Label the supply outlet flow rate requirement on the drawing for each outlet.
9. Determine the total external static pressure available from the unit at the selected airflow rate.
10. Subtract the supply and return register pressure, external coil pressure, filter pressure, and box plenum (if used) from the available static pressure to determine the static pressure

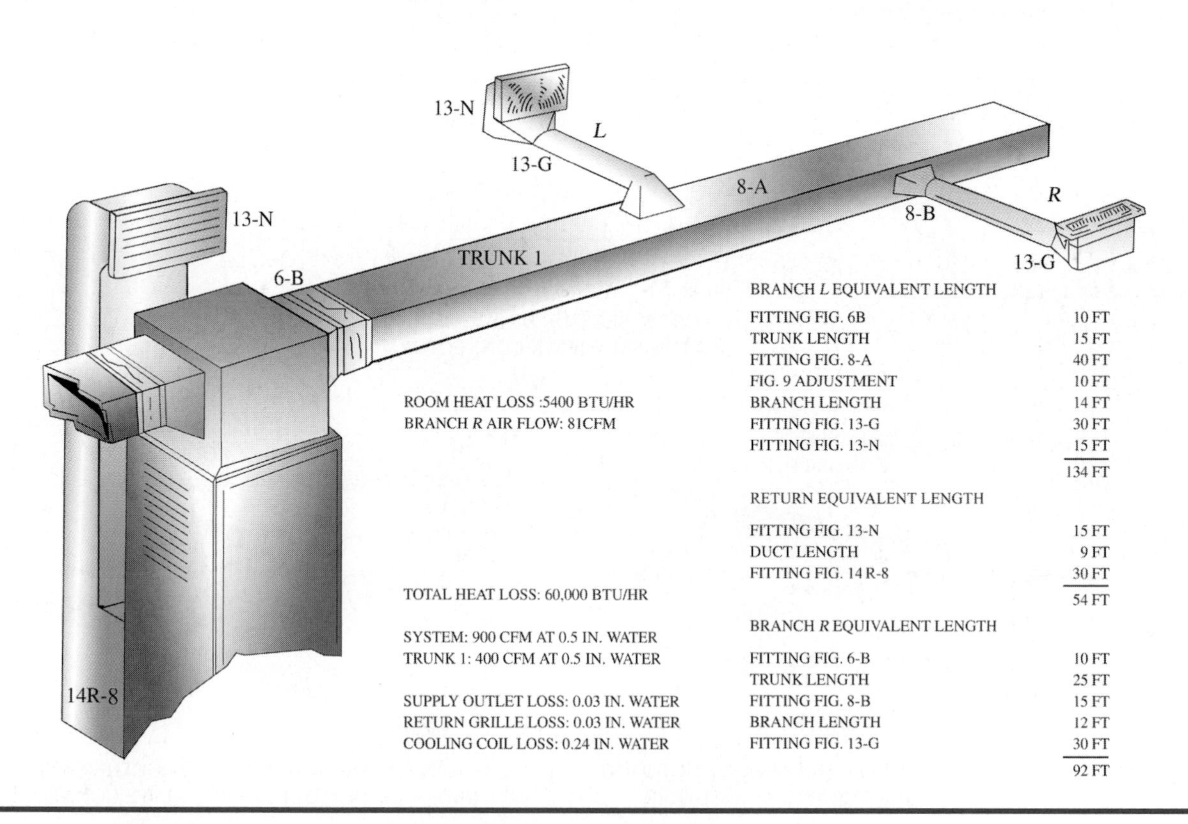

Figure 14-1-8 Equal friction duct design for sample calculations. (*Copyright by the American Society of Heating, Refrigerating, and Air-Conditioning Engineers, Inc.*)

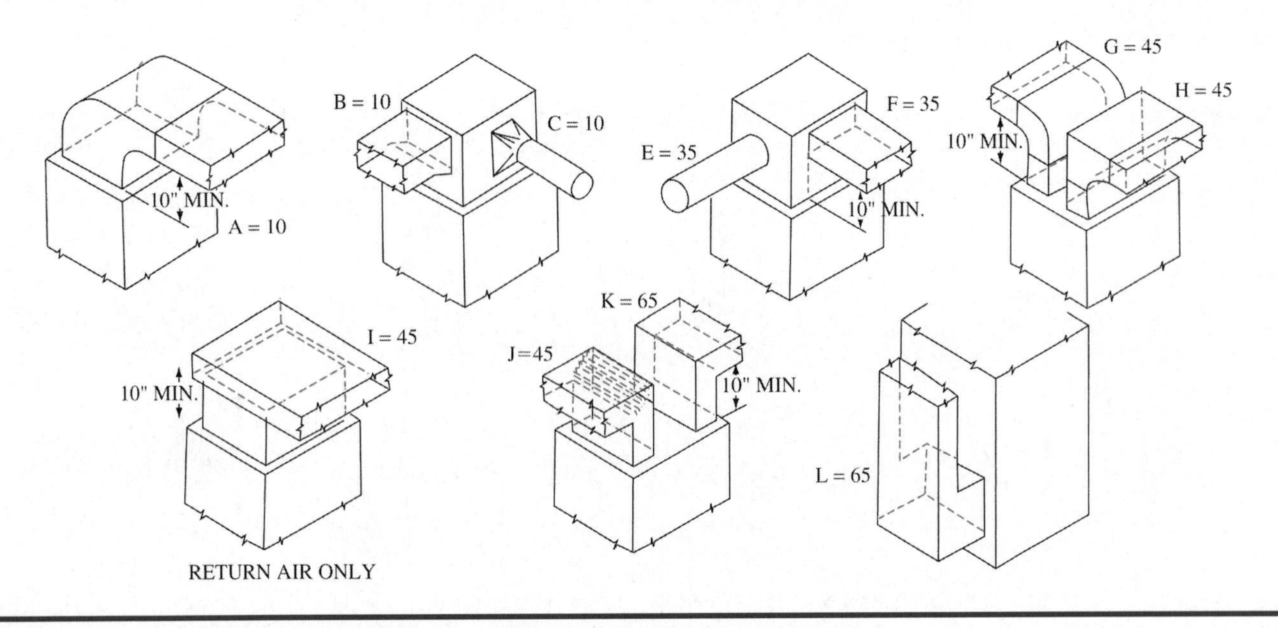

Figure 14-1-9 Equivalent lengths: supply and return air plenum fittings. *(Copyright by the American Society of Heating, Refrigerating, and Air-Conditioning Engineers, Inc.)*

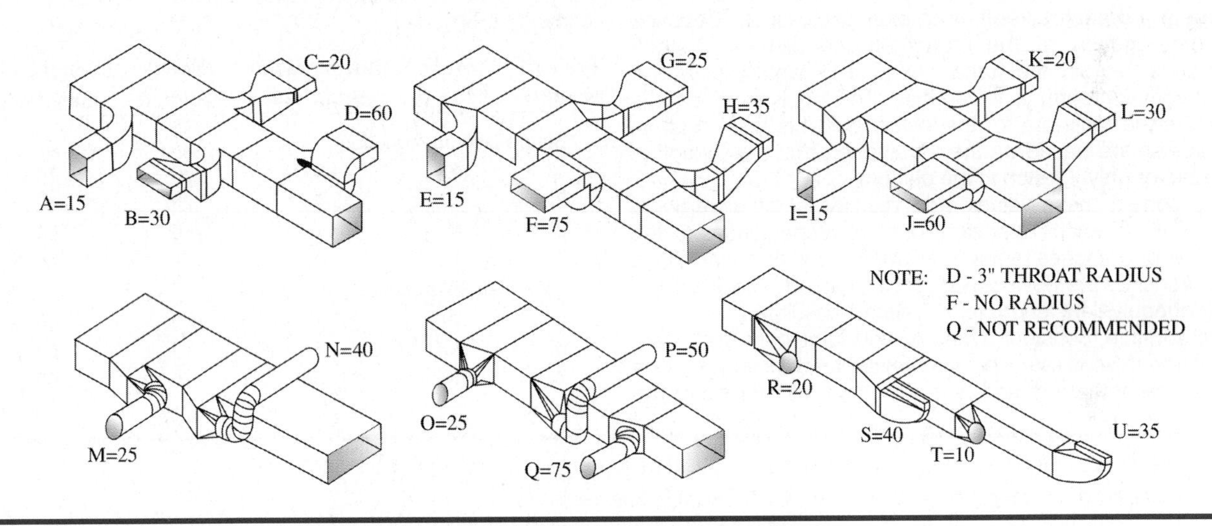

Figure 14-1-10 Equivalent lengths: reducing trunk duct fittings. *(Copyright by the American Society of Heating, Refrigerating, and Air-Conditioning Engineers, Inc.)*

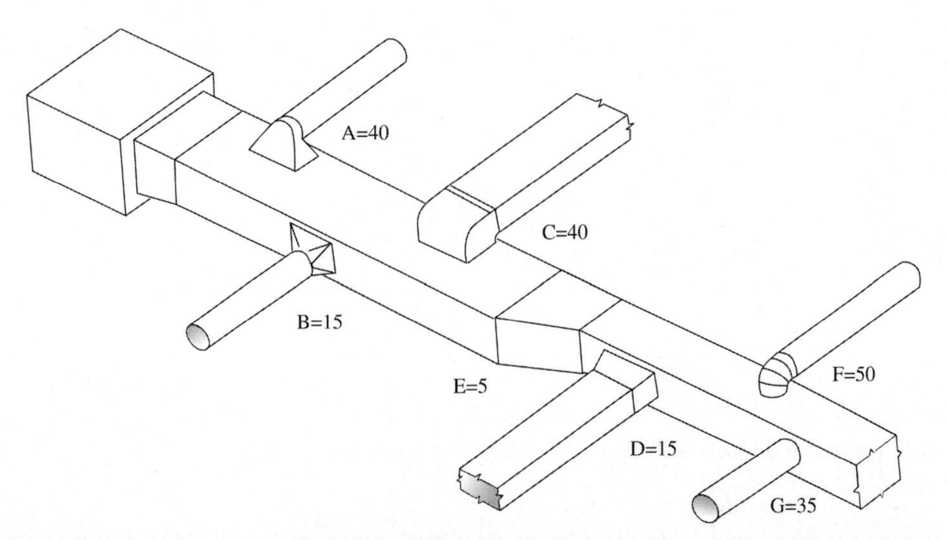

Figure 14-1-11 Equivalent lengths: extended plenum fittings. *(Copyright by the American Society of Heating, Refrigerating, and Air-Conditioning Engineers, Inc.)*

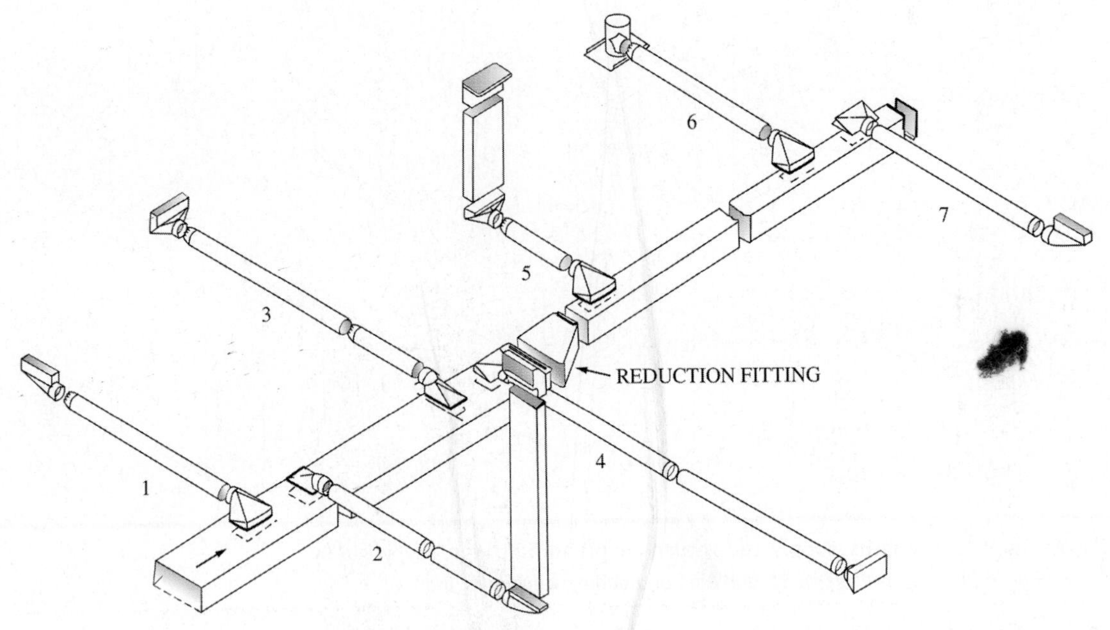

The pressure loss (expressed in equivalent length) of a fitting in a branch takeoff of an extended plenum depends on the location of the branch in the plenum. Takeoff branches nearest the furnace impose a higher loss, and those downstream, a lower loss. The equivalent lengths listed in the table are for the lowest possible loss—that is, the values are based on the assumption that the takeoff is the last (or only) branch in the plenum.

To correct these values, add velocity factor increments of 10 ft to each upstream branch, depending on the number of branches remaining downstream. For example, if two branches are downstream from a takeoff, add 2 × 10 or 20 ft of equivalent length to the takeoff loss listed.

In addition, consider a long extended plenum that has a reduction in width as two (or more) separate plenums—one before the reduction and one after. The loss for each top takeoff fitting in this example duct system is shown in the following table.

Branch Number	Takeoff Loss, ft	Downstream Branches	Velocity Factor, ft	Design Equiv. Length, ft
Before plenum reduction				
1	40	3	30	70
2	40	2	20	60
3	40	1	10	50
4	40	0	0	40
After plenum reduction				
5	40	2	20	60
6	40	1	10	50
7	40	0	0	40

Figure 14-1-12 Duct system example showing equivalent lengths. *(Copyright by the American Society of Heating, Refrigerating, and Air-Conditioning Engineers, Inc.)*

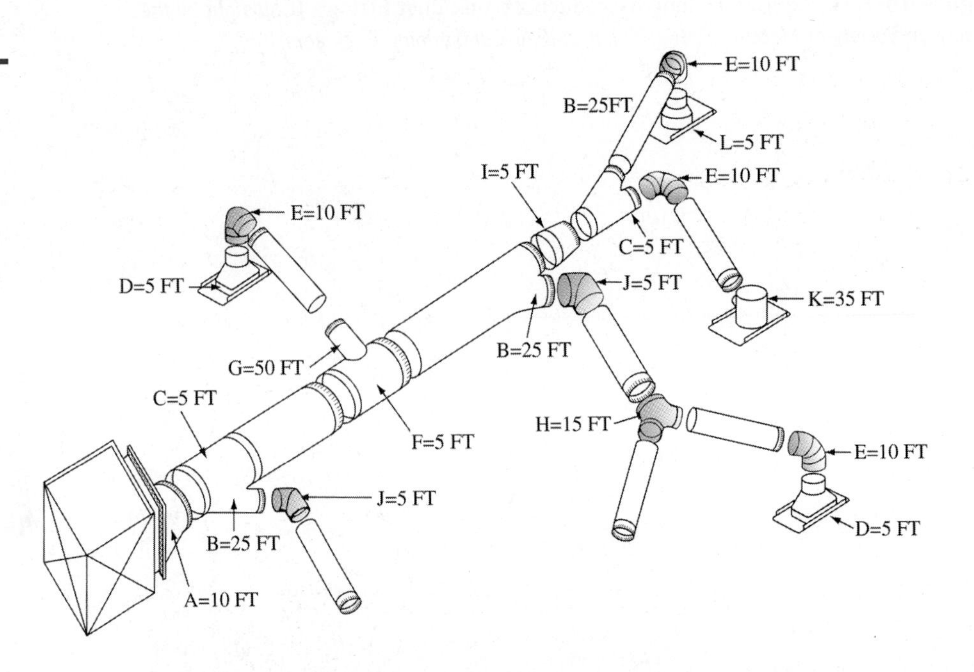

Figure 14-1-13 Equivalent length of round supply system fittings. *(Copyright by the American Society of Heating, Refrigerating, and Air-Conditioning Engineers, Inc.)*

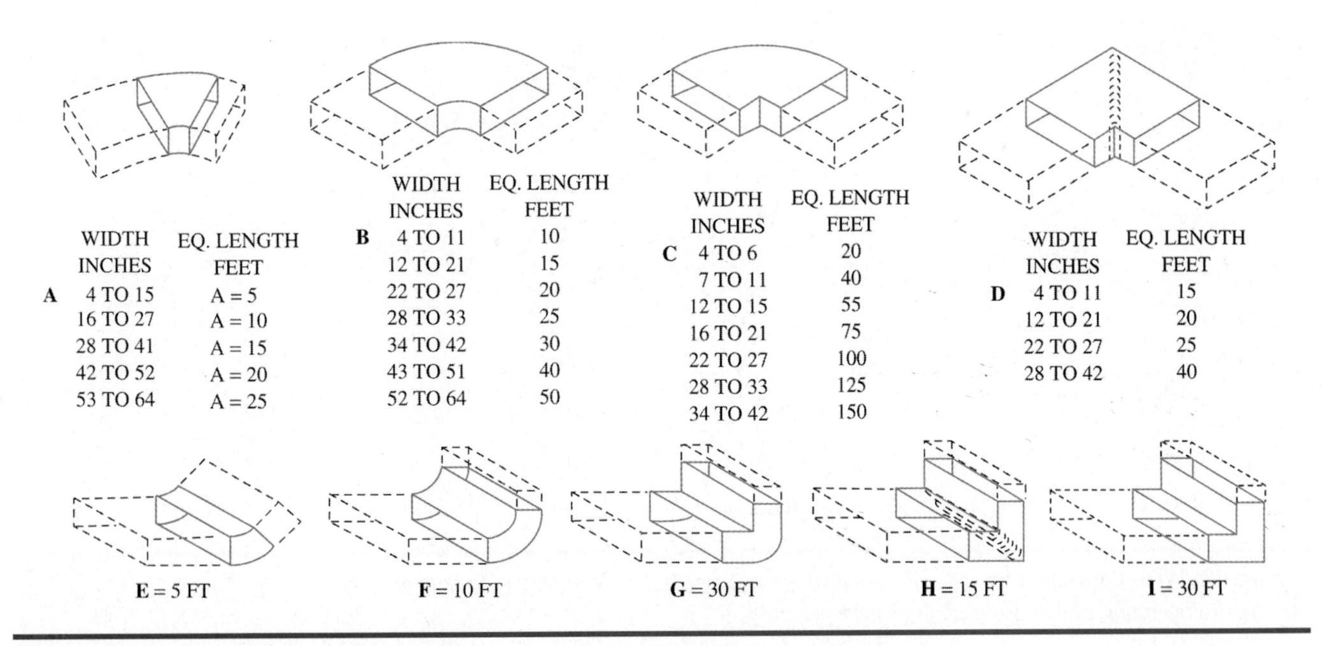

Figure 14-1-14 Equivalent lengths of angles and elbows for trunk ducts. *(Copyright by the American Society of Heating, Refrigerating, and Air-Conditioning Engineers, Inc.)*

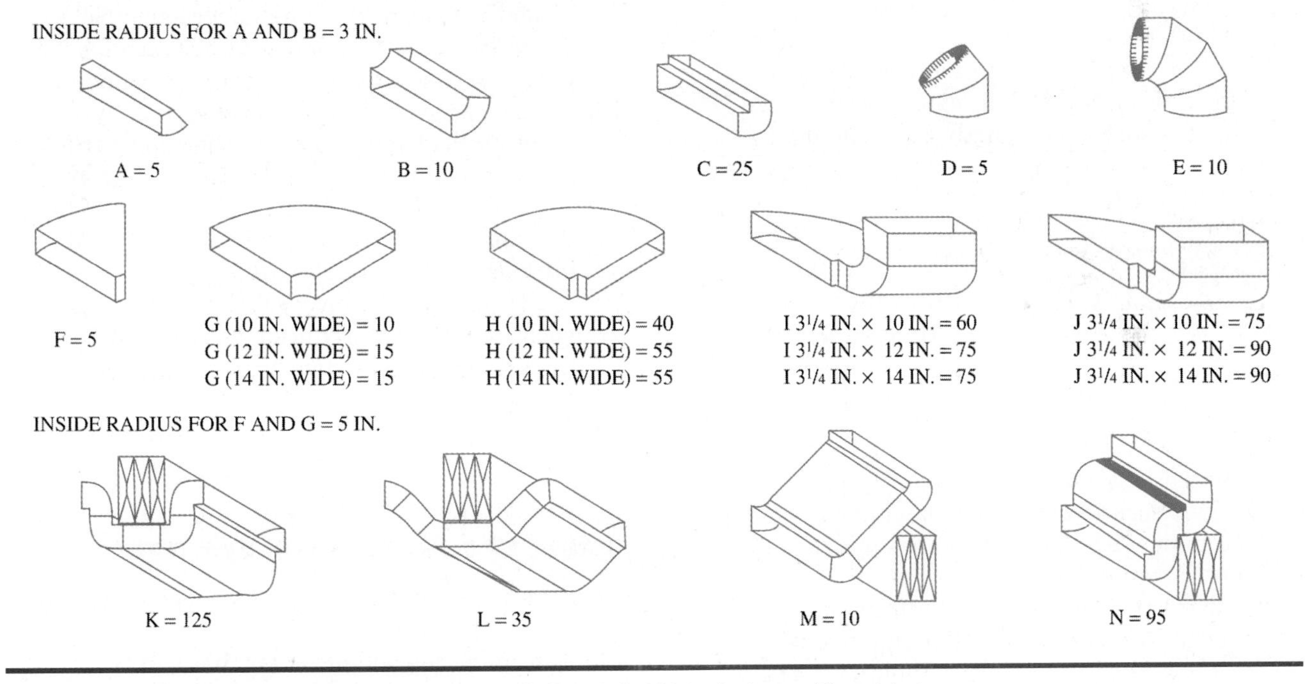

Figure 14-1-15 Equivalent lengths of angles and elbows for branch ducts. *(Copyright by the American Society of Heating, Refrigerating, and Air-Conditioning Engineers, Inc.)*

available for the duct design. Refer to Figure 14-1-15 for the following example.

Equipment		0.50 in water
Subtract:		
Cooling coil	0.24 in water	
Supply outlet	0.03 in water	
Return grill	0.03 in water	
	0.30 in water	

Total available static pressure = 0.20 in water

11. Proportion the available static pressure between the supply and return systems. Refer to Figure 14-1-15 for the following example.

Supply (75%)	0.15 in water
Return (25%)	0.05 in water
Total =	0.20 in water

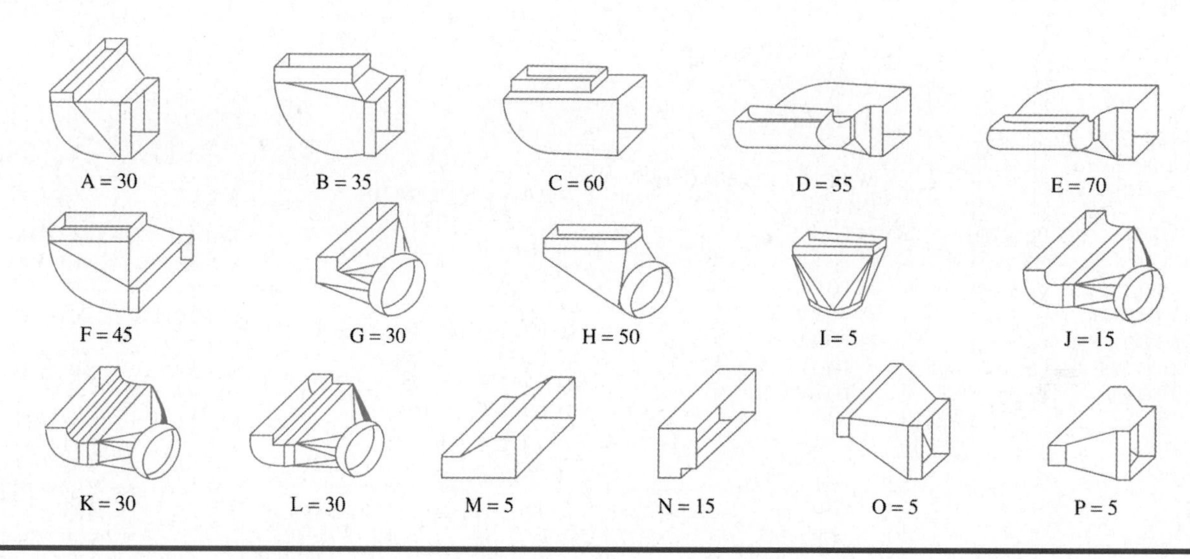

Figure 14-1-16 Equivalent lengths of boot fittings. (*Copyright by the American Society of Heating, Refrigerating, and Air-Conditioning Engineers, Inc.*)

Sizing the Branch Supply Air System

12. Use the supply static pressure available to calculate each branch design static pressure for 100 ft of equivalent length. For example:

$$\text{Branch R design static pressure rate} = \frac{100\ (\text{Supply static pressure available})}{\text{Effective length of each branch supply}}$$

$$100 \times \frac{0.15}{92} = \frac{0.16\ \text{in water}}{100\ \text{ft (Branch R)}}$$

13. Enter the friction chart, Figure 14-1-18, at the branch design static pressure (0.163) opposite the flow rate for each supply, and read the round duct size (5 in) and velocity (600 fpm).

14. If the velocity exceeds the maximum recommended values, increase the size and specify roundup damper.

15. Convert the round duct to rectangular, where needed by using the circular equivalents for rectangular ducts found in Table 14-1-4.

Sizing the Supply Trunk System

16. Determine the branch supply with the longest total effective length, and from this, determine the static pressure to size the supply trunk duct system. For example:

$$\text{Supply trunk design static pressure rate} = \frac{100\ (\text{Total supply static pressure available})}{\text{Longest effective length of branch duct supplies}}$$

$$0.15 \times \frac{100}{134\ (\text{Branch L})} = \frac{0.112\ \text{in water}}{100\ \text{ft}}$$

17. Total the heating airflow rate and the cooling airflow rate for each trunk duct section. Select the larger of the two flow rates for each section of duct between roundups or groups of roundups.

18. Design each supply trunk duct section by entering the friction chart at the supply trunk static pressure (0.112) and sizing each trunk section for the appropriate air volume handled by that section of duct. (Trunk 1—400 CFM; duct size = 9.5 in at 800 fpm)

Trunks should be checked for size after each roundup and reduced, as required, to maintain velocity above branch duct design velocity.

19. Convert round duct size to rectangular, where needed.

Sizing the Return Air System

20. Select the number of return air openings to be used.

21. Determine the volume of air that will be returned by each of the return air openings.

22. From step 11, select the return air static pressure.

23. Determine the static pressure available per 100 ft effective length for each return run, and design the same as for the supply trunk system. For example:

$$\text{Return trunk design static pressure} = \frac{100\ (\text{Total return static pressure available})}{\text{Longest effective length of return duct runs}}$$

$$0.05 \times \frac{100}{54} = \frac{0.093\ \text{in water}}{100\ \text{ft}}$$

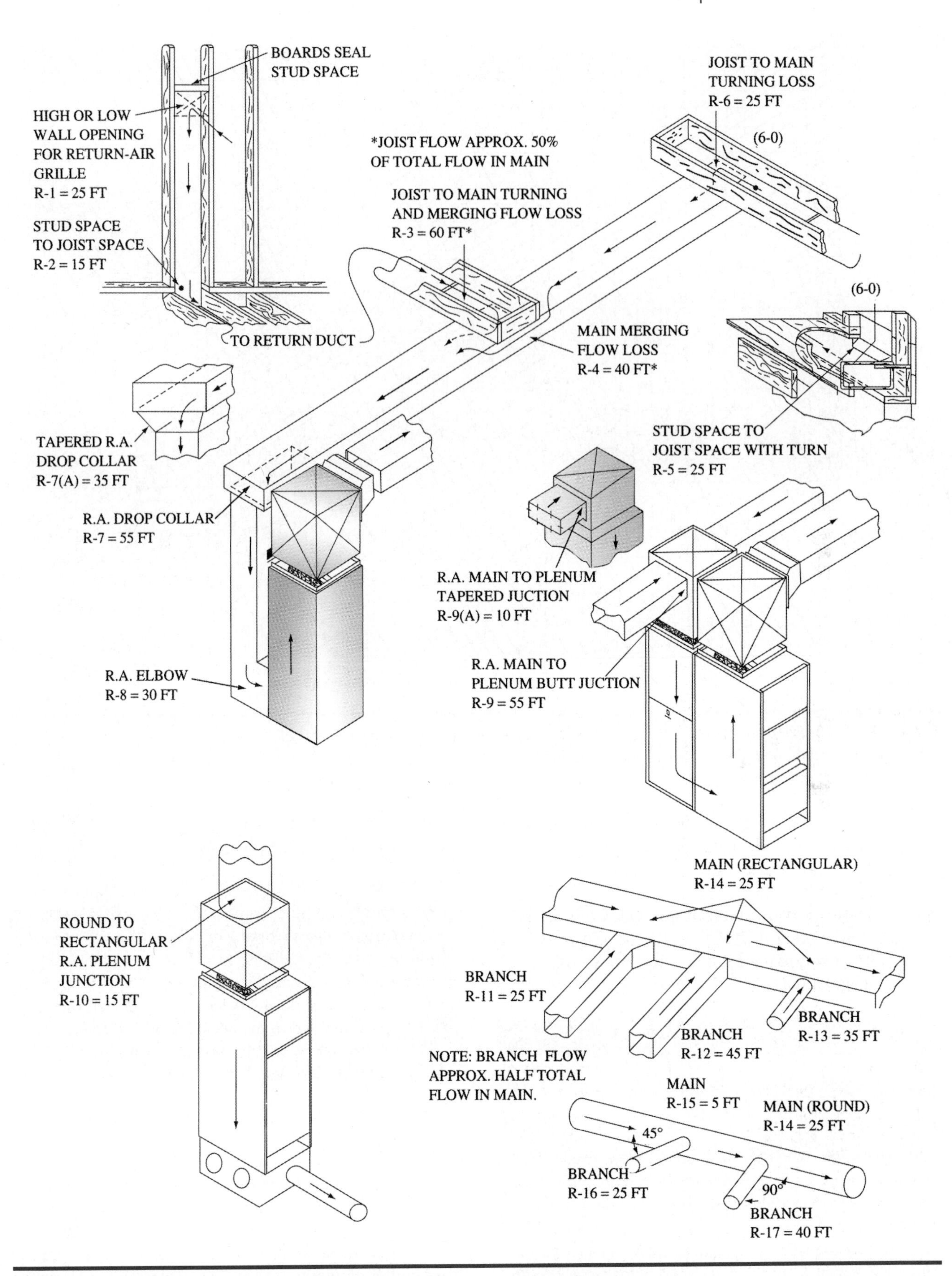

Figure 14-1-17 Equivalent lengths of special return fittings. (*Copyright by the American Society of Heating, Refrigerating, and Air-Conditioning Engineers, Inc.*)

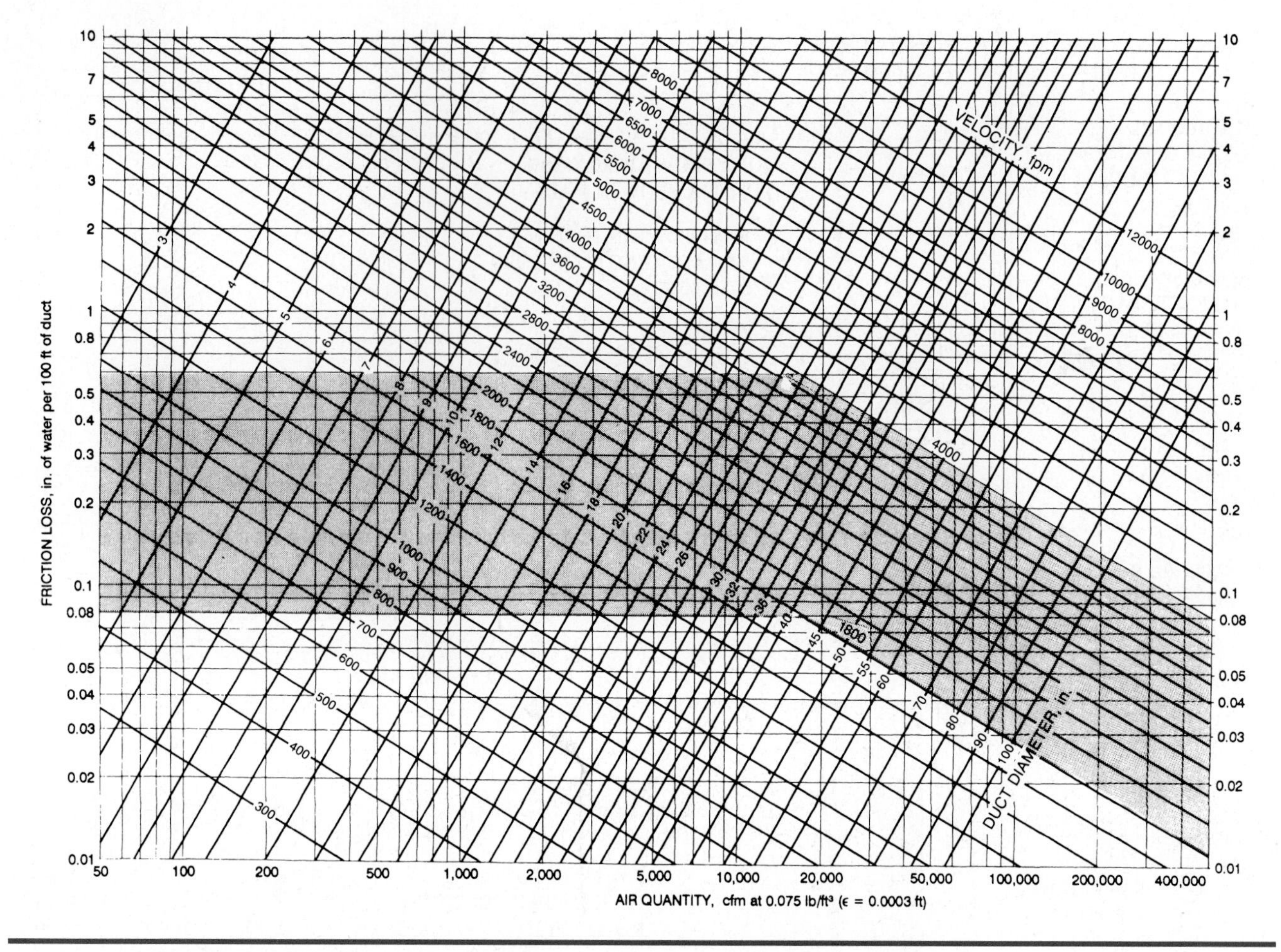

Figure 14-1-18 Friction chart for sizing ductwork. (*Copyright by the American Society of Heating, Refrigerating, and Air-Conditioning Engineers, Inc.*)

The trunk design for the return air is the same as the trunk design for the supply system. Example: Return duct in Figure 14-1-8, is as follows:

Enter friction chart at 900 CFM and 0.093 in of water. Duct size is 13.1 in. The velocity, however, is too high at 950 fpm, so use a larger duct.

SERVICE TIP

Some builders put return air grills in wall stud spaces. Typical construction with 16 in on center 2 × 4's have an approximately 14 in wide × 3½ in deep space between the studs. This is equivalent to about a 7 in round duct. However, in some cases filter grills are used. Filter grills are approximately 1½ in deep. This leaves only a 2 in × 14 in opening between the filter grill and the stud. This significantly reduces the system's performance. You may find that for these installations it will be necessary to install an additional return to provide adequate return air flow to the system for it to work properly.

SERVICE TIP

As airflow is restricted through a blower, the blower rpm will increase and motor amperage will decrease. Therefore, a motor that is working moving air through the system will have a much higher amp draw than a blower that is starved for air because the supply duct or return ducts are not allowing airflow. A quick check of the blower amperage as compared to the manufacturer's technical data can give you an idea as to the system's airflow.

The Blower

The blower must be capable of delivering the total air quantity against the external resistance of the conditioner, including the ductwork and grills. The air pressure required by a blower is relatively small. Instead of measuring this pressure in psi, it is measured in inches of water column (in WC).

Atmospheric pressure can be indicated as 14.7 psi, or in terms of an equivalent column of water it will

TABLE 14-1-4 Circular Equivalents of Rectangular Duct for Equal Friction and Capacity

Lgth Adj[b]	Length of One Side of Rectangular Duct (*a*), in.																
	4.0	4.5	5.0	5.5	6.0	6.5	7.0	7.5	8.0	9.0	10.0	11.0	12.0	13.0	14.0	15.0	16.0
3.0	3.8	4.0	4.2	4.4	4.6	4.7	4.9	5.1	5.2	5.5	5.7	6.0	6.2	6.4	6.6	6.8	7.0
3.5	4.1	4.2	4.6	4.8	5.0	5.2	5.3	5.5	5.7	6.0	6.3	6.5	6.8	7.0	7.2	7.5	7.7
4.0	4.4	4.6	4.9	5.1	5.3	5.5	5.7	5.9	6.1	6.4	6.7	7.0	7.3	7.6	7.8	8.0	8.3
4.5	4.6	4.9	5.2	5.4	5.7	5.9	6.1	6.3	6.5	6.9	7.2	7.5	7.8	8.1	8.4	8.6	8.8
5.0	4.9	5.2	5.5	5.7	6.0	6.2	6.4	6.7	6.9	7.3	7.6	8.0	8.3	8.6	8.9	9.1	9.4
5.5	5.1	5.4	5.7	6.0	6.3	6.5	6.8	7.0	7.2	7.6	8.0	8.4	8.7	9.0	9.3	9.6	9.9

Lgth Adj[b]	Length of One Side of Rectangular Duct (*a*), in.																			Lgth Adj[b]
	6	7	8	9	10	11	12	13	14	15	16	17	18	19	20	22	24	26	28	30	
6	6.6																				6
7	7.1	7.7																			7
8	7.6	8.2	8.7																		8
9	8.0	8.7	9.3	9.8																	9
10	8.4	9.1	9.8	10.4	10.9																10
11	8.8	9.5	10.2	10.9	11.5	12.0															11
12	9.1	9.9	10.7	11.3	12.0	12.6	13.1														12
13	9.5	10.3	11.1	11.8	12.4	13.1	13.7	14.2													13
14	9.8	10.8	11.4	12.2	12.9	13.5	14.2	14.7	15.3												14
15	10.1	11.0	11.8	12.6	13.3	14.0	14.6	15.3	15.8	16.4											15
16	10.4	11.3	12.2	13.0	13.7	14.4	15.1	15.7	16.4	16.9	17.5										16
17	10.7	11.6	12.5	13.4	14.1	14.9	15.6	16.2	16.8	17.4	18.0	18.6									17
18	11.0	11.9	12.9	13.7	14.5	15.3	16.0	16.7	17.3	17.9	18.5	19.1	19.7								18
19	11.2	12.2	13.2	14.1	14.9	15.7	16.4	17.1	17.8	18.4	19.0	19.6	20.2	20.8							19
20	11.5	12.6	13.5	14.4	15.2	16.0	16.8	17.5	18.2	18.9	19.5	20.1	20.7	21.3	21.9						20
22	12.0	13.0	14.1	15.0	15.9	16.8	17.6	18.3	19.1	19.8	20.4	21.1	21.7	22.3	22.9	24.0					22
24	12.4	13.5	14.6	15.6	16.5	17.4	18.3	19.1	19.9	20.6	21.3	22.0	22.7	23.3	23.9	25.1	26.2				24
26	12.8	14.0	15.1	16.2	17.1	18.1	19.0	19.8	20.6	21.4	22.1	22.9	23.5	24.2	24.9	26.1	27.3	28.4			26
28	13.2	14.5	15.6	16.7	17.7	18.7	19.6	20.5	21.3	22.1	22.9	23.7	24.4	25.1	25.8	27.1	28.3	29.5	30.6		28
30	13.6	14.9	16.1	17.2	18.3	19.3	20.2	21.1	22.0	22.9	23.7	24.4	25.2	25.9	26.6	28.0	29.3	30.5	31.7	32.8	30
32	14.0	15.3	16.5	17.7	18.8	19.8	20.8	21.8	22.7	23.5	24.4	25.2	26.0	26.7	27.5	28.9	30.2	31.5	32.7	33.9	32
34	14.4	15.7	17.0	18.2	19.3	20.4	21.4	22.4	23.3	24.2	25.1	25.9	26.7	27.5	28.3	29.7	31.0	32.4	33.7	34.9	34
36	14.7	16.1	17.4	18.6	19.8	20.9	21.9	22.9	23.9	24.8	25.7	26.6	27.4	28.2	29.0	30.5	32.0	33.3	34.6	35.9	36
38	15.0	16.5	17.8	19.0	20.2	21.4	22.4	23.5	24.5	25.4	26.4	27.2	28.1	28.9	29.8	31.3	32.8	34.2	35.6	36.8	38
40	15.3	16.8	18.2	19.5	20.7	21.8	22.9	24.0	25.0	26.0	27.0	27.9	28.8	29.6	30.5	32.1	33.6	35.1	36.4	37.8	40
42	15.6	17.1	18.5	19.9	21.1	22.3	23.4	24.5	25.6	26.6	27.6	28.5	29.4	30.3	31.2	32.8	34.4	35.9	37.3	38.7	42
44	15.9	17.5	18.9	20.3	21.5	22.7	23.9	25.0	26.1	27.1	28.1	29.1	30.0	30.9	31.8	33.5	35.1	36.7	38.1	39.5	44
46	16.2	17.8	19.3	20.6	21.9	23.2	24.4	25.5	26.6	27.7	28.7	29.7	30.6	31.6	32.5	34.2	35.9	37.4	38.9	40.4	46
48	16.5	18.1	19.6	21.0	22.3	23.6	24.8	26.0	27.1	28.2	29.2	30.2	31.2	32.2	33.1	34.9	36.6	38.2	39.7	41.2	48
50	16.8	18.4	19.9	21.4	22.7	24.0	25.2	26.4	27.6	28.7	29.8	30.8	31.8	32.8	33.7	35.5	37.2	38.9	40.5	42.0	50
52	17.1	18.7	20.2	21.7	23.1	24.4	25.7	26.9	28.0	29.2	30.3	31.3	32.3	33.3	34.3	36.2	37.9	39.6	41.2	42.8	52
54	17.3	19.0	20.6	22.0	23.5	24.8	26.1	27.3	28.5	29.7	30.8	31.8	32.9	33.9	34.9	36.8	38.6	40.3	41.9	43.5	54
56	17.6	19.3	20.9	22.4	23.8	25.2	26.5	27.7	28.9	30.1	31.2	32.3	33.4	34.4	35.4	37.4	39.2	41.0	42.7	44.3	56
58	17.8	19.5	21.2	22.7	24.2	25.5	26.9	28.2	29.4	30.6	31.7	32.8	33.9	35.0	36.0	38.0	39.8	41.6	43.3	45.0	58
60	18.1	19.8	21.5	23.0	24.5	25.9	27.3	28.6	29.8	31.0	32.2	33.3	34.4	35.5	36.5	38.5	40.4	42.3	44.0	45.7	60
62		20.1	21.7	23.3	24.8	26.3	27.6	28.9	30.2	31.5	32.6	33.8	34.9	36.0	37.1	39.1	41.0	42.9	44.7	46.4	62
64		20.3	22.0	23.6	25.1	26.6	28.0	29.3	30.6	31.9	33.1	34.3	35.4	36.5	37.6	39.6	41.6	43.5	45.3	47.1	64
66		20.6	22.3	23.9	25.5	26.9	28.4	29.7	31.0	32.3	33.5	34.7	35.9	37.0	38.1	40.2	42.2	44.1	46.0	47.7	66
68		20.8	22.6	24.2	25.8	27.3	28.7	30.1	31.4	32.7	33.9	35.2	36.3	37.5	38.6	40.7	42.8	44.7	46.6	48.4	68
70		21.1	22.8	24.5	26.1	27.6	29.1	30.4	31.8	33.1	34.4	35.6	36.8	37.9	39.1	41.2	43.3	45.3	47.2	49.0	70
72			23.1	24.8	26.4	27.9	29.4	30.8	32.2	33.5	34.8	36.0	37.2	38.4	39.5	41.7	43.8	45.8	47.8	49.6	72
74			23.3	25.1	26.7	28.2	29.7	31.2	32.5	33.9	35.2	36.4	37.7	38.8	40.0	42.2	44.4	46.4	48.4	50.3	74
76			23.6	25.3	27.0	28.5	30.0	31.5	32.9	34.3	35.6	36.8	38.1	39.3	40.5	42.7	44.9	47.0	48.9	50.9	76
78			23.8	25.6	27.3	28.8	30.4	31.8	33.3	34.6	36.0	37.2	38.5	39.7	40.9	43.2	45.4	47.5	49.5	51.4	78
80			24.1	25.8	27.5	29.1	30.7	32.2	33.6	35.0	36.3	37.6	38.9	40.2	41.4	43.7	45.9	48.0	50.1	52.0	80
82				26.1	27.8	29.4	31.0	32.5	34.0	35.4	36.7	33.0	39.3	40.6	41.8	44.1	46.4	48.5	50.6	52.6	82
84				26.4	28.1	29.7	31.3	32.8	34.3	35.7	37.1	38.4	39.7	41.0	42.2	44.6	46.9	49.0	51.1	53.2	84
86				26.6	28.3	30.0	31.6	33.1	34.6	36.1	37.4	38.8	40.1	41.4	42.6	45.0	47.3	49.6	51.7	53.7	86
88				26.9	28.6	30.3	31.9	33.4	34.9	36.4	37.8	39.2	40.5	41.8	43.1	45.5	47.8	50.0	52.2	54.3	88
90				27.1	28.9	30.6	32.2	33.8	35.3	36.7	38.2	39.5	40.9	42.2	43.5	45.9	48.3	50.5	52.7	54.8	90
92					29.1	30.8	32.5	34.1	35.6	37.1	38.5	39.9	41.3	42.6	43.9	46.4	48.7	51.0	53.2	55.3	92
96					29.6	31.4	33.0	34.7	36.2	37.7	39.2	40.6	42.0	43.3	44.7	47.2	49.6	52.0	54.2	56.4	96

TABLE 14-1-4 (continued)

Lgth Adj[b]	Length of One Side of Rectangular Duct (*a*), in.																				Lgth Adj[b]
	32	34	36	38	40	42	44	46	48	50	52	56	60	64	68	72	76	80	84	88	
32	35.0																				32
34	36.1	37.2																			34
36	37.1	38.2	39.4																		36
38	38.1	39.3	40.4																		38
40	39.0	40.3	41.5	42.6	43.7																40
42	40.0	41.3	42.5	43.7	44.8	45.9															42
44	40.9	42.2	43.5	44.7	45.8	47.0	48.1														44
46	41.8	43.1	44.4	45.7	46.9	48.0	49.2	50.3													46
48	42.6	44.0	45.3	46.6	47.9	49.1	50.2	51.4	52.5												48
50	43.6	44.9	46.2	47.5	48.8	50.0	51.2	52.4	53.6	54.7											50
52	44.3	45.7	47.1	48.4	49.7	51.0	52.2	53.4	54.6	55.7	56.8										52
54	45.1	46.5	48.0	49.3	50.7	52.0	53.2	54.4	55.6	56.8	57.9										54
56	45.8	47.3	48.8	50.2	51.6	52.9	54.2	55.4	56.6	57.8	59.0	61.2									56
58	46.6	48.1	49.6	51.0	52.4	53.8	55.1	56.4	57.6	58.8	60.0	62.3									58
60	47.3	48.9	50.4	51.9	53.3	54.7	60.0	57.3	58.6	59.8	61.0	63.4	65.6								60
62	48.0	49.6	51.2	52.7	54.1	55.5	56.9	58.2	59.5	60.8	62.0	64.4	66.7								62
64	48.7	50.4	51.9	53.5	54.9	56.4	57.8	59.1	60.4	61.7	63.0	65.4	67.7	70.0							64
66	49.4	51.1	52.7	54.2	55.7	57.2	58.6	60.0	61.3	62.6	63.9	66.4	68.8	71.0							66
68	50.1	51.8	53.4	55.0	56.5	58.0	59.4	60.8	62.2	63.6	64.9	67.4	69.8	72.1	74.3						68
70	50.8	52.5	54.1	55.7	57.3	58.8	60.3	61.7	63.1	64.4	65.8	68.3	70.8	73.2	75.4						70
72	51.4	53.2	54.8	56.5	58.0	59.6	61.1	62.5	63.9	65.3	66.7	69.3	71.8	74.2	76.5	78.7					72
74	52.1	53.8	55.5	57.2	58.8	60.3	61.9	63.3	64.8	66.2	67.5	70.2	72.7	75.2	77.5	79.8					74
76	52.7	54.5	56.2	57.9	59.5	61.1	62.6	64.1	65.6	67.0	68.4	71.1	73.7	76.2	78.6	80.9	83.1				76
78	53.3	55.1	56.9	58.6	60.2	61.8	63.4	64.9	66.4	67.9	69.3	72.0	74.6	77.1	79.6	81.9	84.2				78
80	53.9	55.8	57.5	59.3	60.9	62.6	64.1	65.7	67.2	68.7	70.1	72.9	75.4	78.1	80.6	82.9	85.2	87.5			80
82	54.6	56.4	58.2	59.9	61.6	63.3	64.9	66.5	68.0	69.5	70.9	73.7	76.4	79.0	81.5	84.0	86.3	88.5			82
84	55.1	57.0	58.8	60.6	62.3	64.0	65.6	67.2	68.7	70.3	71.7	74.6	77.3	80.0	82.5	85.0	87.3	89.6	91.8		84
86	55.7	57.6	59.4	61.2	63.0	64.7	66.3	67.9	69.5	71.0	72.5	75.4	78.2	80.9	83.5	85.9	88.3	90.7	92.9		86
88	56.3	58.2	60.1	61.9	63.6	65.4	67.0	68.7	70.2	71.8	73.3	76.3	79.1	81.8	84.4	86.9	89.3	91.7	94.0	96.2	88
90	56.8	58.8	60.7	62.5	64.3	66.0	67.7	69.4	71.0	72.6	74.1	77.1	79.9	82.7	85.3	87.9	90.3	92.7	95.0	97.3	90
92	57.4	59.3	61.3	63.1	64.9	66.7	68.4	70.1	71.7	73.3	74.9	77.9	80.8	83.5	86.2	88.8	91.3	93.7	96.1	98.4	92
94	57.9	59.9	61.9	63.7	65.6	67.3	69.1	70.8	72.4	74.0	75.6	78.7	81.6	84.4	87.1	89.7	92.3	94.7	97.1	99.4	94
96	58.4	60.5	62.4	64.3	66.2	68.0	69.7	71.5	73.1	74.8	76.3	79.4	82.4	85.3	88.0	90.7	93.2	95.7	98.1	100.5	96

[a]Table based on $D_e = 1.30(ab)^{0.625}/(a + b)^{0.25}$
[b]Length of adjacent side of rectangular duct (*b*), in.

support, 34 ft. For small pressure differences such as a blower would create, pressures are measured in inches WC. For example, an air conditioning unit for a small job, such as shown in Figure 14-1-19, could be capable of delivering 1200 CFM against an external static pressure of 0.12 in WC. These small pressures are measured using an inclined manometer, as shown in Figure 14-1-20.

Types of Pressures

In dealing with air distribution, and particularly duct systems, there are three types of pressure readings: static, velocity, and total. Static pressure is the force of the air against the sides of a vessel or duct, as shown in Figure 14-1-21.

Total pressure is read by placing the manometer so that air flows directly into the sensing tube opening, as shown in Figure 14-1-22. The velocity pressure is the difference between the total and static pressures and can be read as shown in Figure 14-1-23.

The velocity pressure is useful in measuring the velocity of air in a duct. Some manometers are calibrated to read duct velocity directly in feet per minute (fpm). A single instrument that can be inserted in the duct and connected to the manometer for reading velocity pressure, known as the *pitot* (pronounced Pea-Toe) *tube*, is shown in Figure 14-1-24.

Types of Fans

The terms *fans* and *blowers* are often used interchangeably; however, the accepted term for this group of equipment is *fan*. There are two distinctly different types of fans: centrifugal and axial, based on the direction of air flow through the impeller.

Figure 14-1-25 shows an exploded view of the centrifugal fan. These fans are sometimes called "squirrel cage" fans due to the shape of the impeller

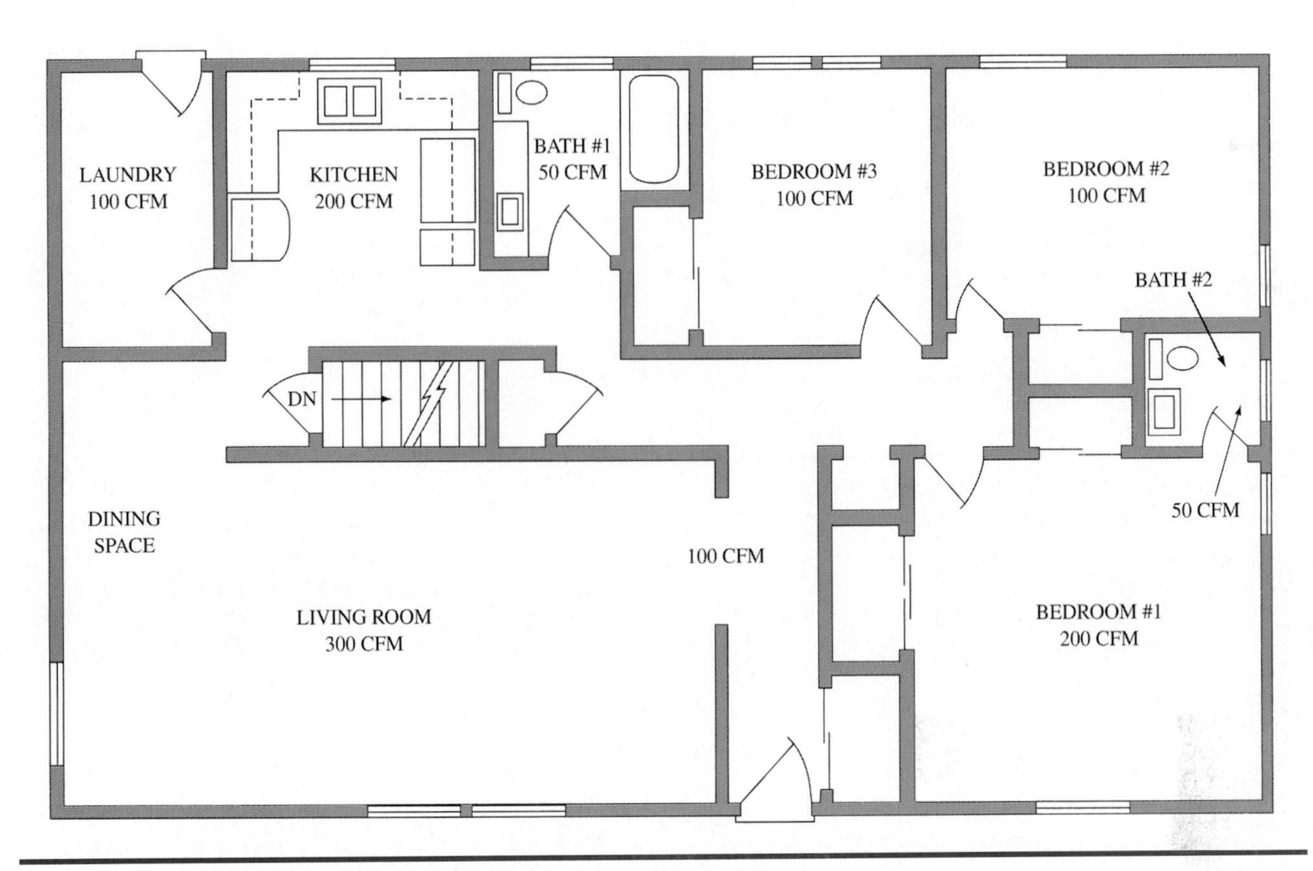

Figure 14-1-19 Floor plan showing air quantities (CFM) for each space.

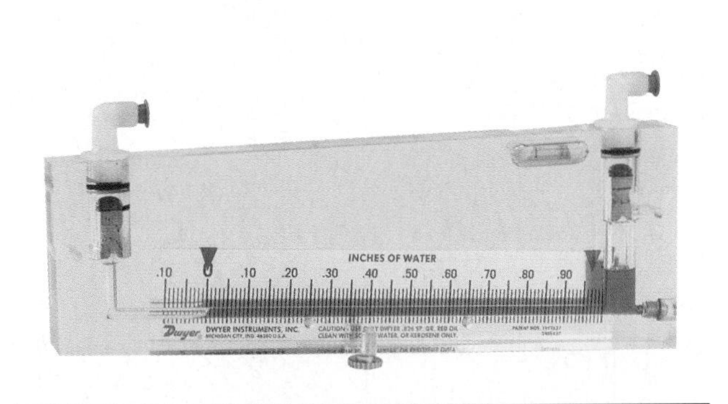

Figure 14-1-20 Inclined manometer for measuring air pressure. *(Courtesy Dwyer Instruments, Inc.)*

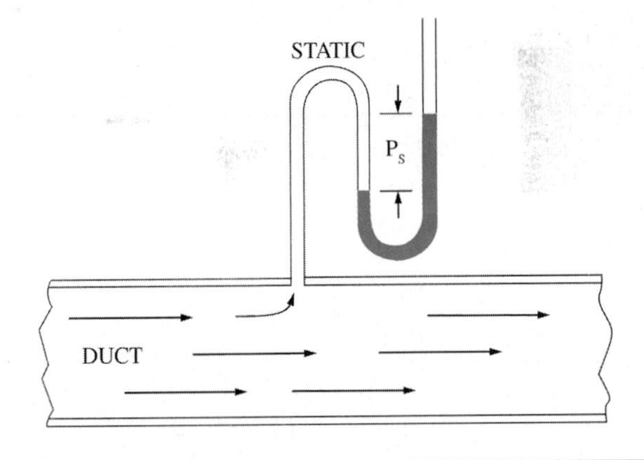

Figure 14-1-21 Manometer measuring static pressure, in inches of water column.

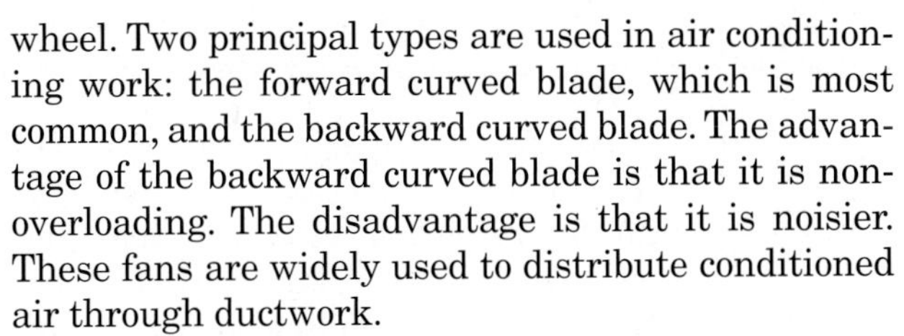

wheel. Two principal types are used in air conditioning work: the forward curved blade, which is most common, and the backward curved blade. The advantage of the backward curved blade is that it is nonoverloading. The disadvantage is that it is noisier. These fans are widely used to distribute conditioned air through ductwork.

Figure 14-1-26 shows an exploded view of the axial fan. There are two principal types of axial fans: the vane and the propeller. The propeller type will handle large volumes of air for low pressure applications. It has high usage for exhaust fans and condenser fans. Vane type fans are highly efficient but noisy. The blade pitch can be adjusted to control the amount of air they handle.

It is interesting to note that if the inlet to a fan is blocked, reducing the air quantity, the power required to drive it is also reduced. The power required is directly related to the quantity of air the fan pumps. If the outlet is blocked, the discharge pressure is increased, but the power required to operate the fan is reduced since it pumps fewer pounds of air. The greatest load on a fan is created when the fan operates in the open without ductwork restrictions, pumping the maximum quantity of air. There is always a danger under these conditions of overloading the fan motor.

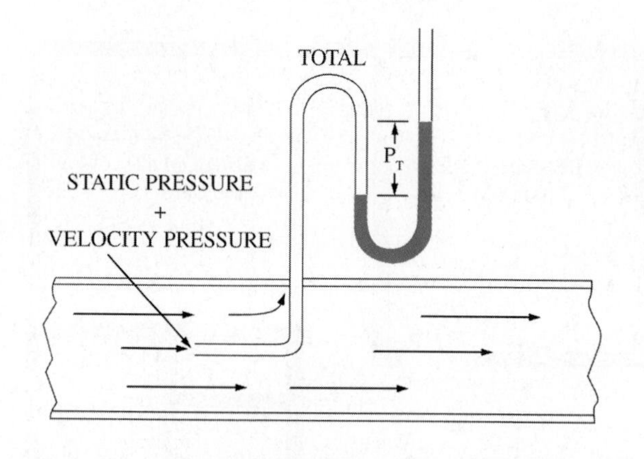

Figure 14-1-22 Manometer measuring total air pressure, in inches of water column.

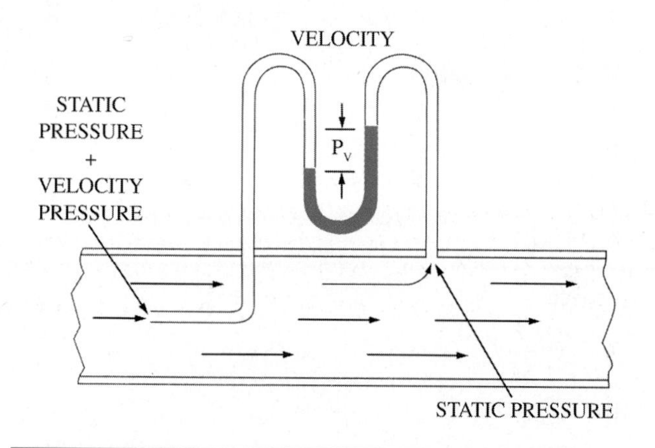

Figure 14-1-23 Manometer measuring velocity pressure in inches of water column.

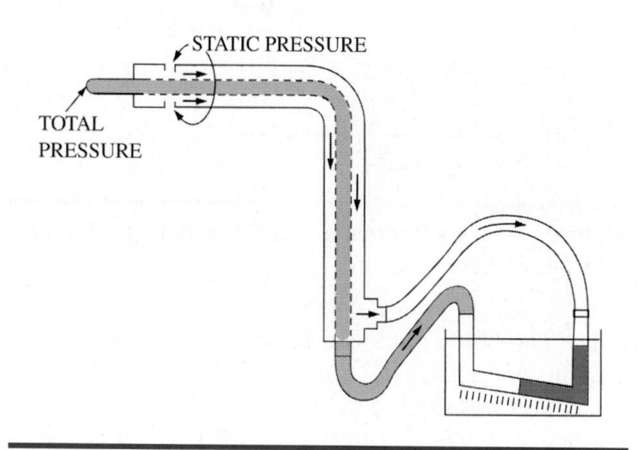

Figure 14-1-24 Pitot tube used in an air stream for measuring velocity pressure.

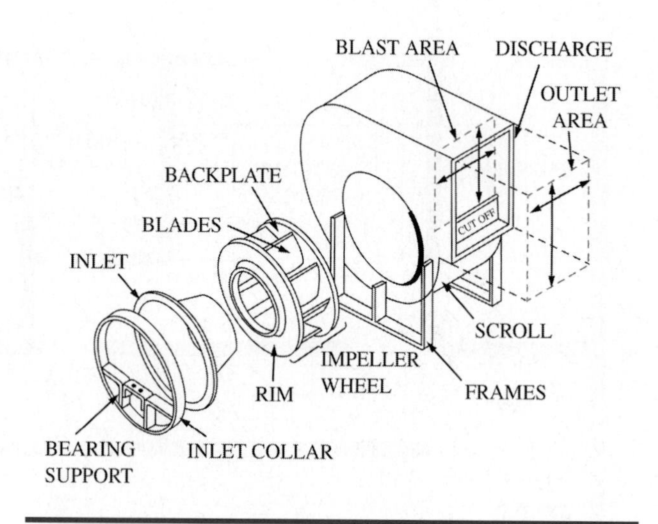

Figure 14-1-25 Expanded view of a centrifugal fan. *(Copyright by the American Society of Heating, Refrigerating, and Air-Conditioning Engineers, Inc.)*

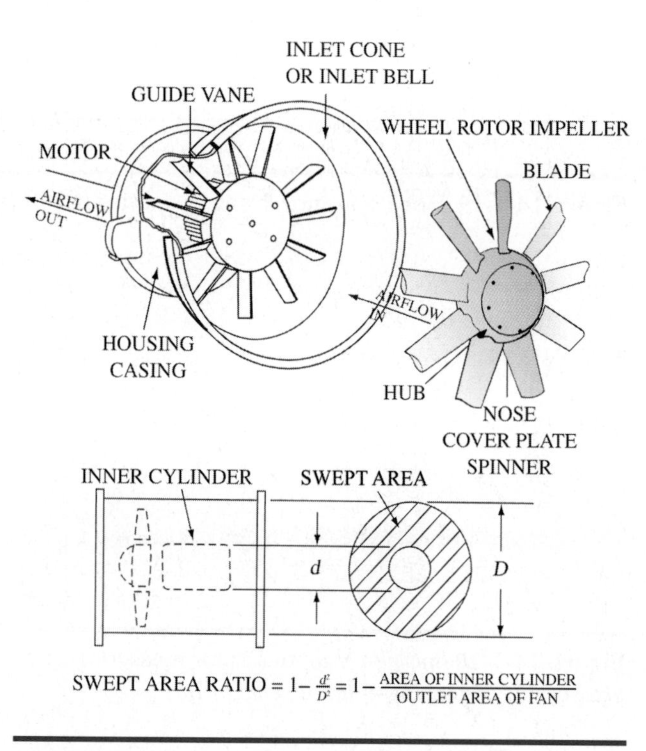

Figure 14-1-26 Cutaway view of an axial fan showing the component parts. *(Copyright by the American Society of Heating, Refrigerating, and Air-Conditioning Engineers, Inc.)*

Fans operate according to a set of formulas called fan laws. These laws show the relationship between changes in speed (rpm), air volume (CFM), static pressure (SP) and power in brake horsepower (Bhp). The fan laws are as follows:

$$\frac{\text{CFM}_2}{\text{CFM}_1} = \frac{\text{rpm}_2}{\text{rpm}_1}$$

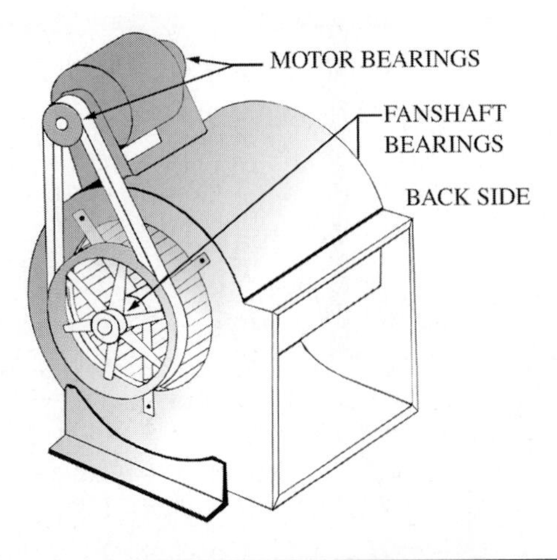

Figure 14-1-27 Belt driven blower with two pulley wheels for motor and blower.

$$\frac{SP_2}{SP_1} = \left(\frac{rpm_2}{rpm_1}\right)^2$$

$$\frac{Bhp_2}{Bhp_1} = \left(\frac{rpm_2}{rpm_1}\right)^3$$

These formulas are extremely useful in determining the effect of changing the speed of a fan. For example, if the speed is reduced 10%, the static pressure is reduced 19% and the brake horsepower is reduced 27%. A substantial amount of energy can be saved by slowing down a fan if it is delivering an excessive air volume.

Types of Fan Drives

There are two types of drive arrangements for air conditioning fans: belt and pulley drives, Figure 14-1-27, and direct drives, Figure 14-1-28.

Motors are available with synchronous speeds of 1800 rpm (actual speed about 1750 rpm) and 3600 rpm (actual speed about 3450 rpm).

For *belt driven motors* the speed of the fan is determined by the ratio of the motor pulley to the fan pulley. For example, if the motor pulley has a 3 in diameter, the fan pulley has a 6 in diameter and the motor speed is 1750 rpm, the speed of the fan is calculated as follows:

$$\text{Fan rpm} = \frac{\text{diameter of motor pulley} \times \text{motor rpm}}{\text{diameter of fan pulley}}$$

$$\text{Fan rpm} = \frac{3}{6} \times 1750$$

$$= 875 \text{ rpm}$$

Direct drive motors are available with multiple speed windings. The speed can be changed by switching wires in the motor terminal box. This can be done by the control system so that a different speed can be used for heating and cooling. Small fans use shaded pole motors. Direct drive fan motors are usually the PSC (permanent split capacitor) type.

SERVICE TIP

When adjusting the speeds on a belt driven motor check the motor amperage with the blower door closed. With the blower door open more free air is allowed to enter the blower and the amp draw will be higher than it would be with the door closed. It is important that you check the amp draw once the motor pulleys have been adjusted to make certain that the motor is not overamping which would result in its overheating and failing.

AIR DISTRIBUTION AND BALANCING

Air conditioning equipment is selected to control heating, cooling, ventilating, humidifying or dehumidifying, and cleaning the air in buildings. The effect is to produce comfortable conditions for the occupants. It also can serve to protect the contents of a building against damage due to improper temperature and humidity conditions.

System Components

Forced air systems are used to distribute heating and/or cooling in residential and small commercial buildings. An essential element in the process is the distribution of conditioned air.

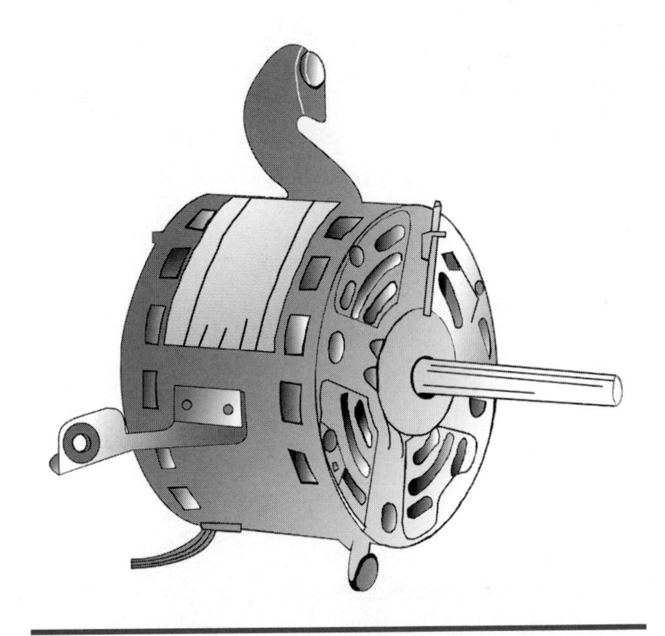

Figure 14-1-28 Direct drive fan motor for a blower.

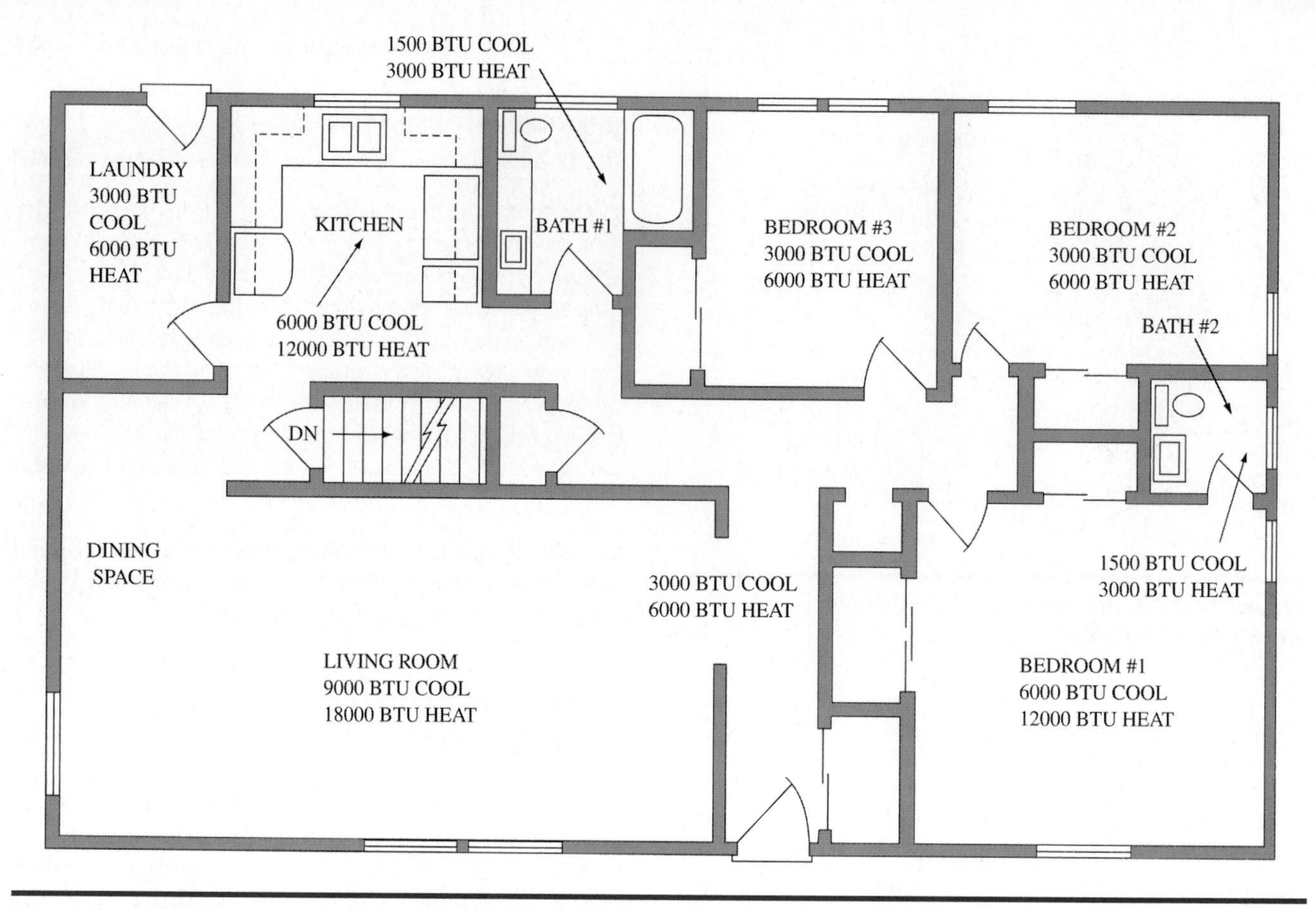

Figure 14-1-29 Floor plan showing heating and cooling load for each space.

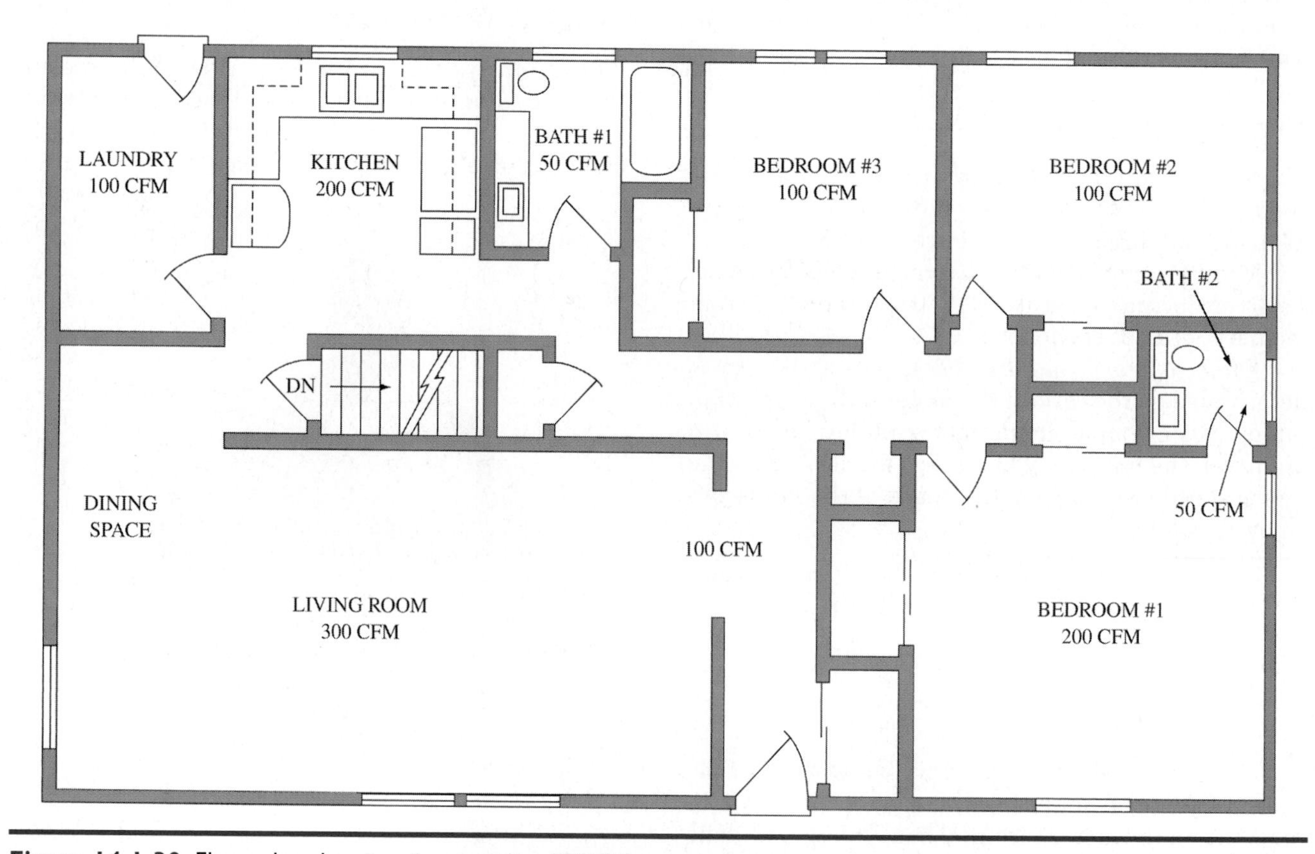

Figure 14-1-30 Floor plan showing air quantities (CFM) for each space.

For example, Figure 14-1-29 shows the required heating and cooling for various areas of a small building.

These are the loads under design conditions. The distribution system must be able to adequately supply conditioned air to each area during peak periods, as well as during partial load conditions. The distribution system is designed to handle peak requirements and the controls adjust the system to handle partial loads.

Assuming that a forced air system is used, the components consist of a blower, filters, humidifier, supply air ducts, return air ducts, ventilation air ducts, registers, and grills.

Air Quantity

In order for the system to operate properly, the correct air quantities must be delivered to each area in proportion to the load. In the example shown, the total cooling load is 35,000 Btu/hr and a 3 ton air conditioning unit is selected. Since a typical conditioner delivers 400 CFM/ton, assume that a fan is selected that will supply a total of 1200 CFM. Based on proportioning the air supply to each area according to the cooling loads shown, the CFM requirements for each area are selected and shown in Figure 14-1-30.

TECH TALK

Tech 1. Why do we have to put mastic and tape on all the fittings on the duct system? We didn't used to have to do that.

Tech 2. That's right, that is something new. They found out through studies that most of the duct systems we installed, if they didn't leak when we installed them, they would leak later. Seems like some of that old duct tape we used didn't stick as well as we thought.

Tech 1. I've noticed that some of that cloth duct tape that was put on years ago really dried out. Most of it just fell off.

Tech 2. That's right, and even some of the newer foil tapes we used didn't hold as well as we would have liked. In addition studies have shown that we really needed to put more tape on the joints. Now you have to put three complete wraps of foil tape around every round duct start collar. We used to put just one. Putting mastic on everything was only done in commercial work, but now we do it everywhere.

Tech 1. That stuff can sure be messy.

Tech 2. It can if you are not careful. Just carry a rag with you to wipe down your hands and your tools if you get any mastic on them. That certainly helps. I know it seems like using mastic and tape together on a lot of the joints seems like overkill, but if it keeps the duct system leak tight for many years it will certainly save our customers a lot of money and operating costs. It will also make the house a lot more comfortable.

Tech 1. Then I guess it's all worth it. Thanks.

UNIT 1—REVIEW QUESTIONS

1. What are the four most common duct configurations?
2. What should all ductwork designs start with?
3. In a residence, where is an excellent location for the trunk ducts and conditioning equipment?
4. What determinations need to be made after the equipment has been selected?
5. The design of the system must be based on _______ and _______.
6. List the general rules that apply to the design of ductwork.
7. What is spiral duct made from?
8. In dealing with air distribution and duct systems, what are the three types of pressure readings?
9. What are the two distinctly different types of fans?
10. Why are centrifugal fans sometimes called "squirrel cage" fans?
11. When is the greatest load on a fan created?
12. List the two different types of drive arrangements for air conditioning fans.
13. For belt driven motors the speed of the fan is determined by the _______.
14. If a forced air system is used what are the components?
15. In order for a system to operate properly, the _______ must be delivered to each area in proportion to the load.

UNIT 2

Zone Control Systems

OBJECTIVES: In this unit the reader will learn about zone control systems including:

- Principles of Zoning
- Types of Zone Dampers
- Types of Zoning Systems

PRINCIPLES OF ZONING

Many heating and air conditioning installations are installed in buildings where the floor plan is spread out, such as ranch style and split-level houses. On these jobs it is not practical to control comfort throughout the entire area using one thermostat. These layouts require zoning, because the loads differ from one area to another. Zoning is a method of controlling the supply of conditioned air to each unique area to match the load requirements. This is done by separating the air supply to each unique area by the use of a zone damper in the ductwork and controlling each zone damper with an individual thermostat.

Basically, there are two methods of zoning:

- Use zoning, and
- Building orientation zoning.

Use zoning is based on separating the areas depending on how they are used. For example, bedrooms can be on one zone and living areas on another. This is common for residential zoning.

TECH TIP

Occupancy sensors, devices that detect movement of people in an area, can be used to enhance the efficient operation of zoned systems. Occupancy sensors will allow a zone to be activated only when it is occupied. The popularity of occupancy sensors has resulted in a decrease in their cost so that they now can be used for many more applications. In fact, occupancy sensors have already found their way into residences for light switches.

Orientation zoning considers the directions the building faces. For example, the south side of a building would be on a separate zone from the north side. This type of zoning is common for commercial structures.

The following are some examples of zone controlled systems:

1. A three level split level system is shown in Figure 14-2-1. Each level of this residence has a separate trunk duct from the supply air plenum with a zone damper installed. Each of these dampers is controlled from an individual thermostat located in that zone.

2. Zone control for a bi-level house is shown in Figure 14-2-2. This system has three zones: living

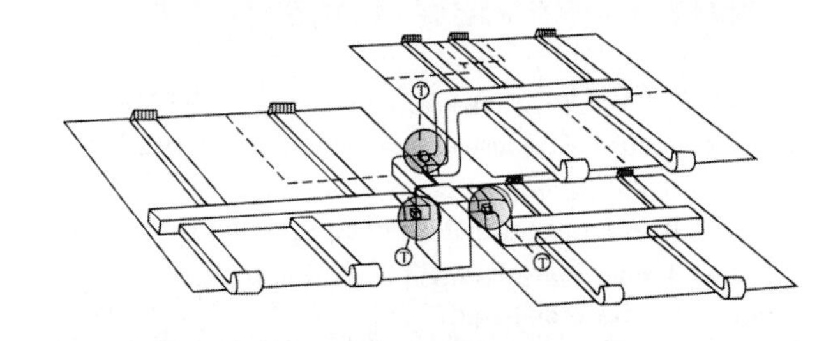

Figure 14-2-1 Three zone split level system.

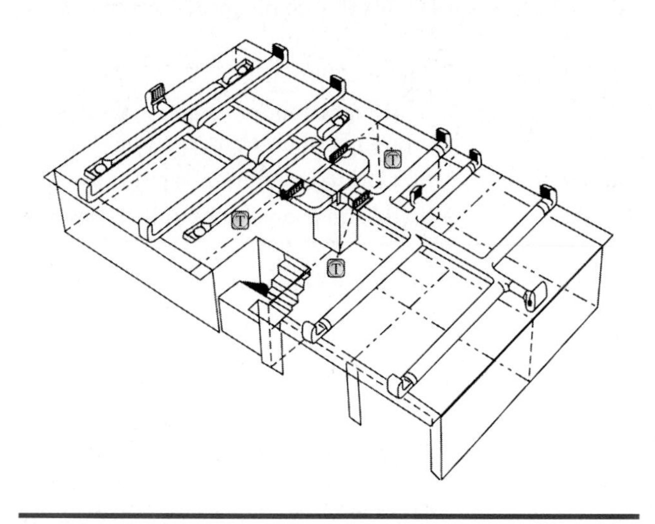

Figure 14-2-2 Zone control for a bi-level house.

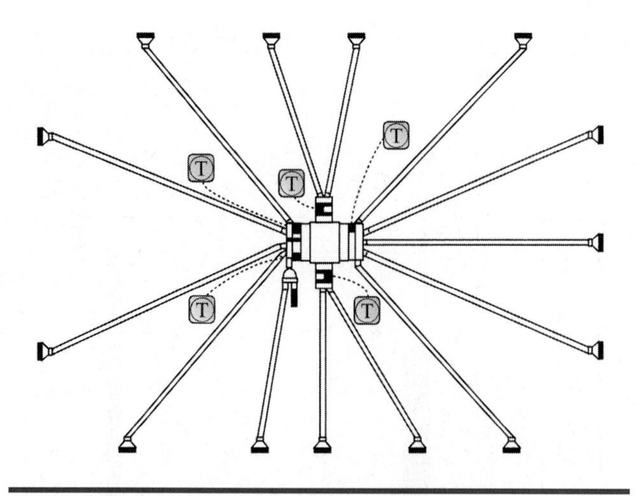

Figure 14-2-3 Five zone radial system.

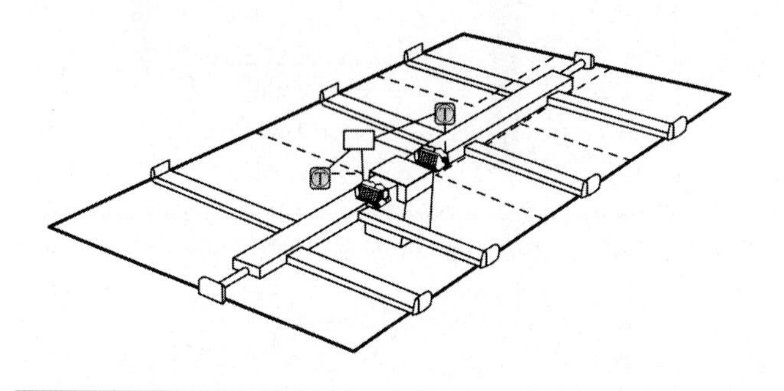

Figure 14-2-4 Two zone ranch style house.

area, bedrooms, and recreation room. The system uses a Honeywell Mastertrol, a 40 VA transformer, three zone dampers, and three thermostats.

3. A room by room temperature control is shown in Figure 14-2-3 for a five zone radial system. These ducts can be located overhead or in the slab floor. If located in the floor, the system must be used for heating only, since cold air could cause floor condensation.

4. A two zone ranch house is shown in Figure 14-2-4. Each zone has a thermostat controlling a zone damper. This system can be installed using a Honeywell Mastertrol unit arranged for two zones.

5. A four zone office building system is shown in Figure 14-2-5. Using Honeywell equipment, this system requires a four zone Mastertrol unit, four dampers and thermostats, and a 40 VA transformer.

6. A room by room comfort control layout is shown in Figure 14-2-6. In this illustration each outlet has an automatic square to round transition damper.

Relieving Excess Air Pressure

A problem that can occur on a system of thermostatically controlled zone dampers is the condition where some major zone dampers are closed off and the remaining zones are handling an excessive amount of

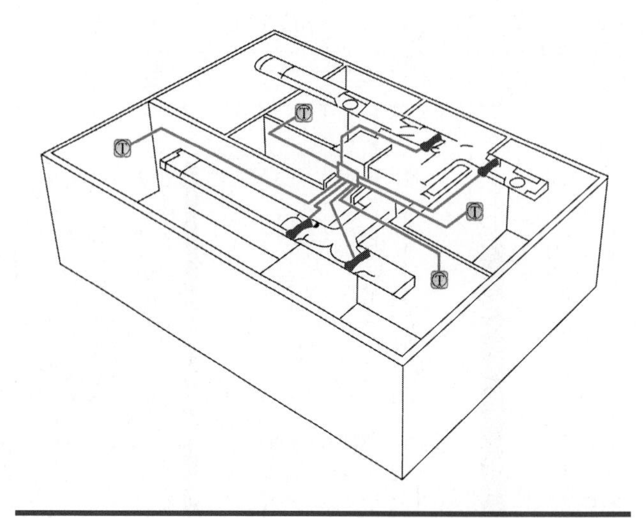

Figure 14-2-5 Four zone professional office system.

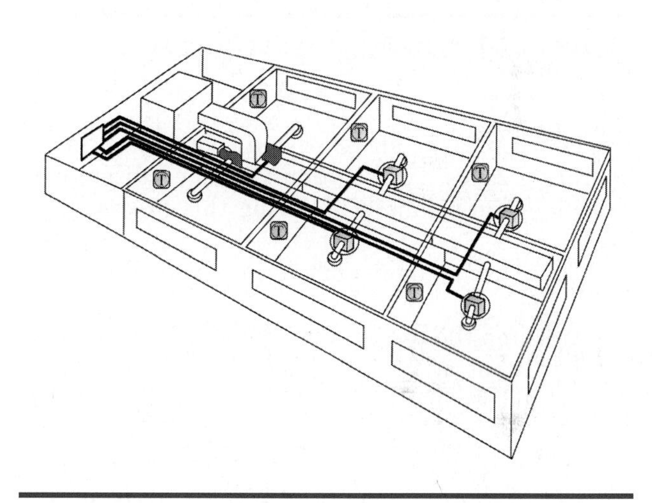

Figure 14-2-6 Individual room zone control system.

SERVICE TIP

Some homeowners will close off the vent to rooms that are not in use or commonly used such as spare bedrooms. If enough of the system's capacity is shut down as a result of this attempt to economize, it can actually cause major problems with the system itself. As the airflow is reduced below manufacturer's specifications, for example, the evaporator may not pick up enough heat and could result in refrigerant floodback to the compressor, which can shorten the compressor's life. Another example is when the airflow is restricted on a gas furnace to the point where it is constantly going off on high limit. An attempt to economize in this manner may ultimately result in higher utility bills and shorter equipment life for the owner.

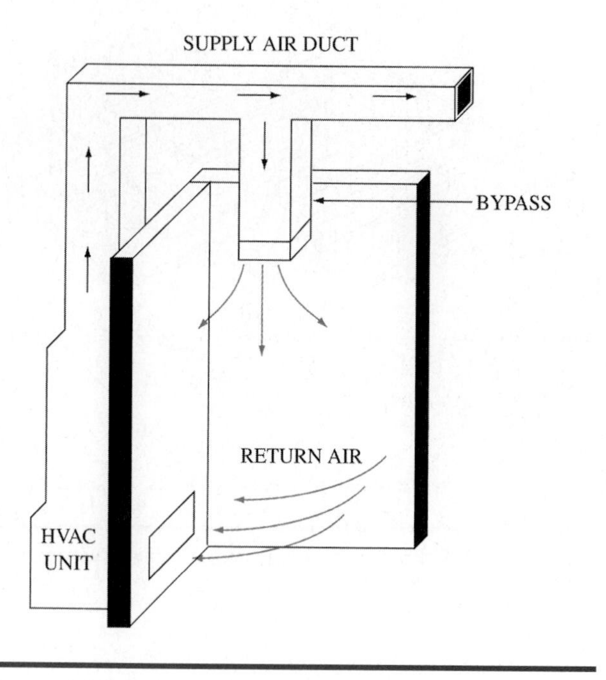

Figure 14-2-7 Bypassing air to a utility room.

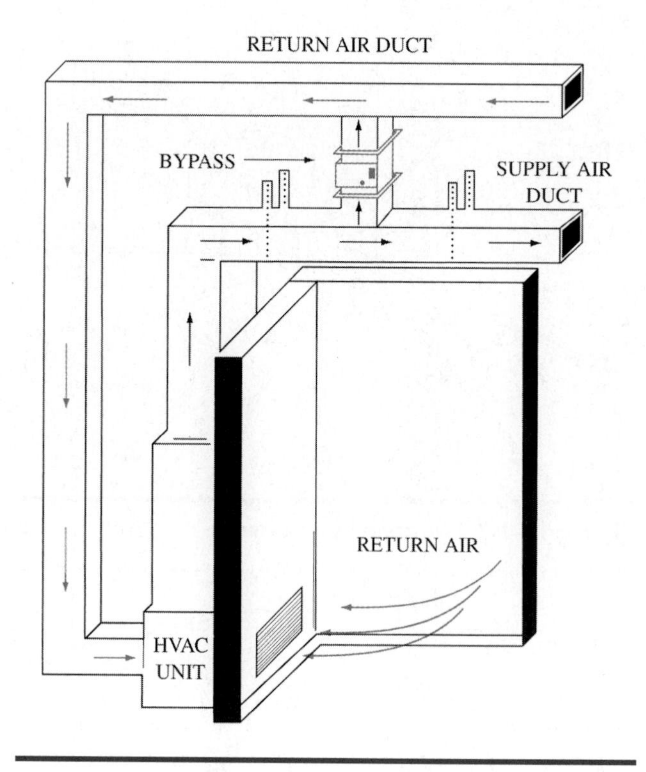

Figure 14-2-9 Relieving air pressure through bypassing air from the supply plenum back to the return.

air. This can cause air noise and poor control in the operating zones. The solution to this condition is to provide some means of relieving this excessive air pressure.

There are a number of ways to accomplish this:

1. Dump the excess air to another area such as the basement or utility room, as shown in Figure 14-2-7 and Figure 14-2-8. This is usually the preferred arrangement.

2. Excess air can be bypassed back into the return air duct as shown in Figure 14-2-9. The disadvantage of this arrangement is that with prolonged use on heating the high limit control may shut down the system, or on cooling the system may be shut down by the low limit control. Thus, short cycling can occur.

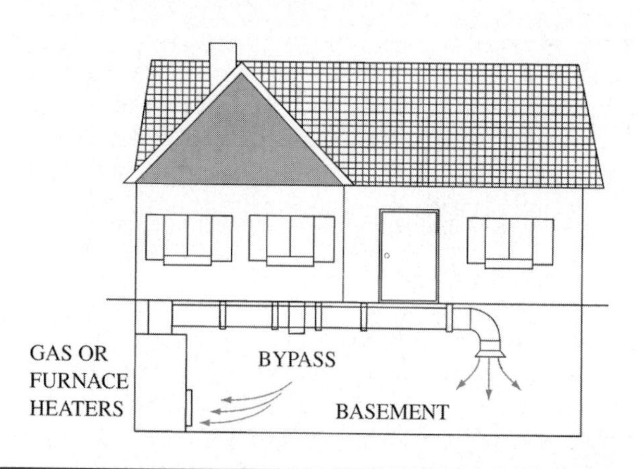

Figure 14-2-8 Relieving air pressure by dumping it into the basement.

3. Undersize the zone dampers so that some leakage will occur through the damper even when it is closed. For example, a 22 in × 8 in damper could be used in a 24 in × 8 in duct. The disadvantage of this arrangement is that the performance is difficult to predict.

4. Oversize each zone duct to handle 60 to 70% of the total air. This arrangement can only be used on small jobs.

Automatic Relief Dampers

In designing the control arrangement to relieve excess air pressure by dumping, described above, or bypassing, described above, automatic dampers can be used. There are two types:

1. Barometric static pressure relief dampers require no electrical connections. They are used on low pressure systems with static pressures usually below 0.5 in WC. They are activated by the pressure within the duct.

2. Motorized static pressure relief dampers are electrically or electronically operated in response to a signal from a remotely located duct pressurestat.

A single blade static pressure relief damper is shown in Figure 14-2-10.

During relief operations, it is important to maintain at least the minimum airflow through the air

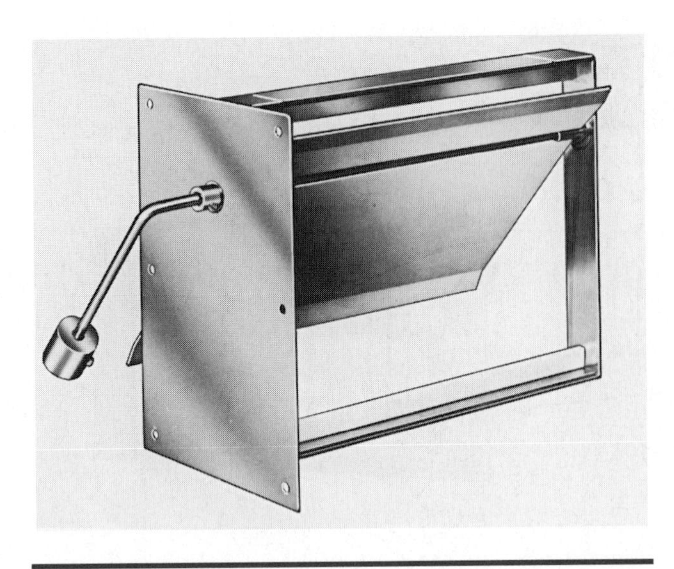

Figure 14-2-10 Static pressure relief damper. (*Courtesy of Honeywell, Inc.*)

handling unit; otherwise the coil performance will be unsatisfactory. The following formula can be used to determine the quantity of bypass air:

air handler (CFM) − smallest zone peak (CFM) − leakage of all closed dampers (CFM) = bypass air flow (CFM)

TYPES OF ZONE DAMPERS

There are many shapes, sizes, and styles of zone dampers designed to fit various requirements.

1. Opposed blade damper for rectangular ducts, Figure 14-2-11.
2. Round duct dampers, Figure 14-2-12.
3. Multi-valve supply register dampers, Figure 14-2-13.
4. Square ceiling diffuser dampers Figure 14-2-14.
5. Round ceiling diffuser dampers.
6. Floor diffuser dampers.

SERVICE TIP

Zone dampers must be sized as close as possible to the plenum or supply trunk dimensions. If they are too close to the register there is a possibility that air passing a partially opened damper will cause a whistling sound that can be heard in the room. If a sound can be heard and it is possible to move the damper further away one method of eliminating the sound is to put one or more 90° turns in the duct after the damper and before the room register.

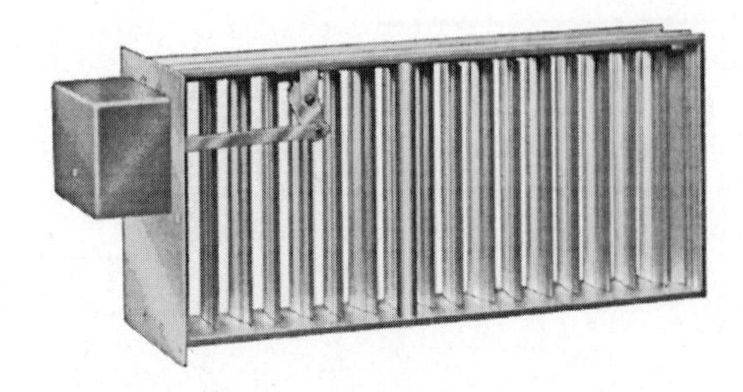

Figure 14-2-11 Zone damper for rectangular duct. (*Courtesy of Honeywell, Inc.*)

Each of these has a specific application. Dampers can either be modulating or two position type, depending on the type of control and thermostat selected.

TYPES OF ZONING SYSTEMS

A number of manufacturers make control systems for installations requiring zoning. These systems differ in number of zones handled, types of applications, methods of control, and types of adjustments provided. In this text, two commonly used systems will be described:

1. The Honeywell Trol-A-Temp zone control system.
2. The Lennox Harmony II system.

Honeywell Trol-A-Temp Zone Control System

The Honeywell Trol-A-Temp zone control system is an integrated set of devices for controlling a multizoned HVAC system. The control center is the Mastertrol automatic balancing system (MABS), shown in Figure 14-2-15. The basic controls are

Figure 14-2-12 Zone damper for a round duct. (*Courtesy of Honeywell, Inc.*)

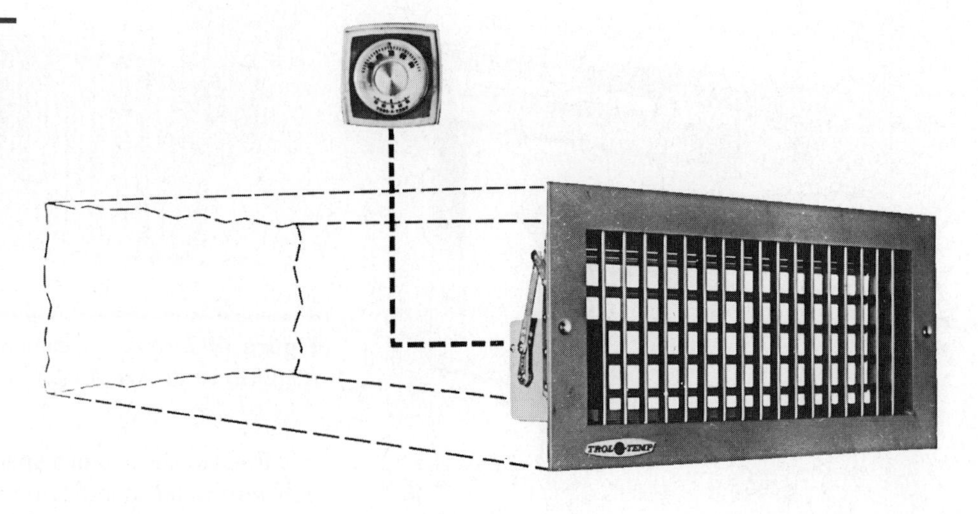

Figure 14-2-13 Register with a damper to regulate the air volume. *(Courtesy of Trol-A-Temp Division of Trolex Corporation)*

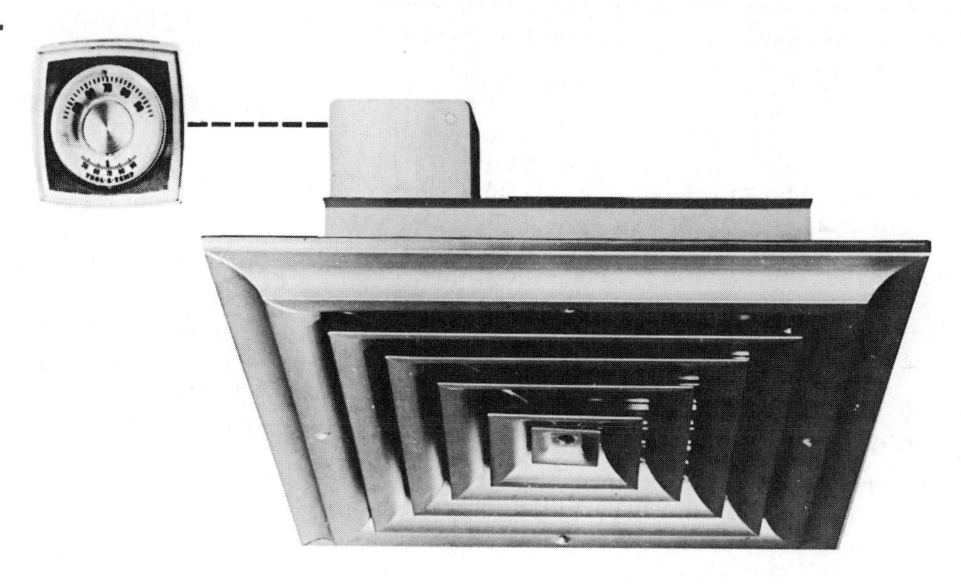

Figure 14-2-14 Square ceiling diffuser with damper. *(Courtesy of Trol-A-Temp Division of Trolex Corporation)*

capable of controlling a two zone or three zone system. If a greater number of zones are required, Mastertrol add-a-zone (MAZ) panels are used. Here are some of the features of the system:

1. Each zone damper is controlled by a room thermostat that is electrically connected to the control center.

2. When any zone calls for heating (or cooling), its zone damper opens and zones not called are closed.

3. Automatic changeover from heating to cooling is provided through the zone 1 thermostat. This thermostat is normally placed in the zone that has the greatest usage. In a residential installation, the best location would probably be the living room.

4. In the standard Mastertrol system, the zone damper operates in two positions. Dampers are either fully open or fully closed.

5. When all the zones are satisfied, all dampers go to a fully open position. This permits the fan to dissipate any residual heat that may be left in the furnace.

6. For the single stage system, the temperature reversing thermostat (TRT) is used with a Mastertrol changeover subbase (MCRS) in zone 1. The subbase also causes changeover at all other zone thermostats.

7. For a multi-stage system, a Mastertrol two stage (MCTS) thermostat is used in zone 1. For a heat pump system, a Mastertrol heat pump (MCHP) thermostat is used in zone 1. All other zones use standard two stage heating cooling thermostats (HCT-2S).

Mode Switches The MABS central control panel has three built-in mode switches: fan on in heat, two compressor, and one zone cooling.

The fan on in heat switch is used with heat pump equipment. When any zone calls for heat, the switch can be turned on to permit bringing the fan and the heating on at the same time. The two compressor

(a)

(b)

Figure 14-2-15 Honeywell Mastertrol Mark II zone control system. *(a. Courtesy of Honeywell, Inc.; b. Courtesy of Trol-A-Temp Division of Trolex Corporation)*

switch changes the thermostat operation in zones 1 and 2 to operate the first stage compressor and zone 3 thermostat to operate the second stage compressor. The one zone cooling switch permits individual control for cooling in the OFF position or only the zone 1 thermostat to control the cooling with all other zones locked open, in the ON position.

Accessory Equipment The Trol-A-Temp system uses the following accessory equipment:

1. A multiposition fresh air damper (MPFAD) and control can be provided, permitting the occupant to regulate the outside air intake.

2. A 7 day clock thermostat or a programmable thermostat with night setback is available for any zone.

SERVICE TIP

Many of the electronically controlled systems have unique programming features. Refer to the operating instructions provided by the manufacturer for the specific piece of equipment you are working on when making changes in settings. In some cases it may be necessary to contact the equipment manufacturer directly for assistance in some programming features. Many controller manufacturers have toll free phone numbers for technician's assistance with programming problems.

3. If only two zones are required, a Mastertrol mini-zone (MM-2) can be used in place of a MABS central control panel, at a savings in cost.

4. A Mastertrol junior (MABS-JR) panel is available for heating only or cooling only systems.

5. A slave driver control relay (SDCR) is available to control 2, 3, or 4 dampers simultaneously from a single thermostat.

Master Control Panel Wiring A typical wiring diagram for a Mastertrol panel, with T87F thermostats and single stage heating and cooling, is shown in Figure 14-2-16. The subbase on the zone 1 thermostat has provision for operating the changeover control.

Lennox Harmony II Zone Control System

The two principal components of the Lennox Harmony II zone control system are the control panel, Figure 14-2-17, and the control center, Figure 14-2-18. The control panel is used for owner access to operate the system along with the zone thermostats. The control center organizes the operation of the thermostats, dampers, and HVAC equipment to result in total comfort. The view shown includes the terminal connections and jumper blocks.

This control system manages the distribution of conditioned air to specific areas or zones in a house or small commercial building. The system is designed to operate with any of the following:

1. A Lennox high efficiency pulse type furnace used in combination with either a single speed or two speed condensing unit.

2. A Lennox variable speed blower coil unit used in combination with a single speed or two speed heat pump.

3. A Lennox high efficiency pulse type furnace used in combination with a Lennox heat pump and Lennox heat pump control.

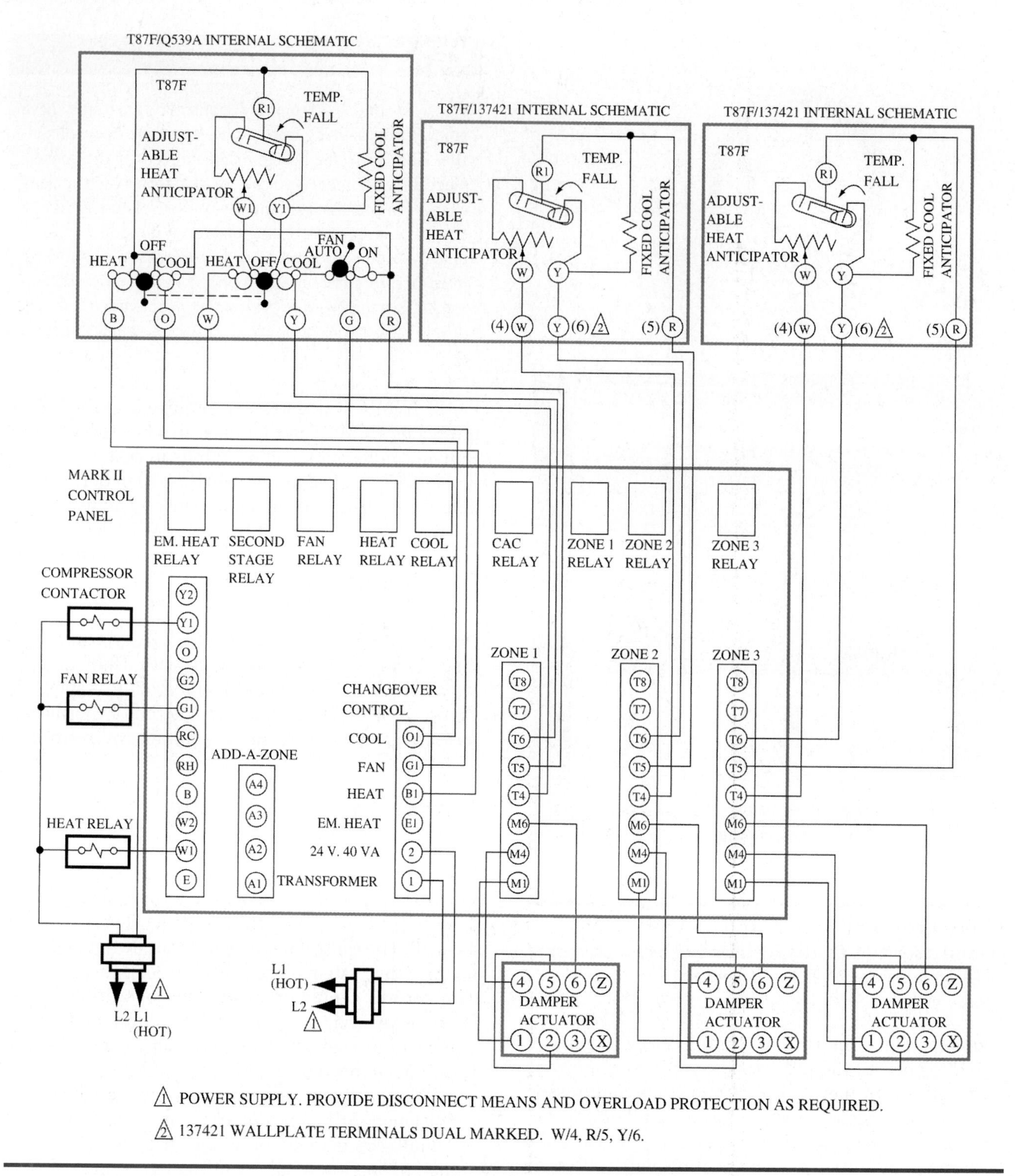

Figure 14-2-16 Wiring diagram for a Mastertrol panel. (*Courtesy of Honeywell, Inc.*)

The system operates in either of two modes: central control or zone control. Central control allows the heating and cooling to condition all zones. In the zone control mode, specific zones are conditioned only when a demand comes from that zone. LED lights on the panel indicate the operating mode.

The principal features of the system are:

1. The system controls the air volume of the blower, which eliminates the need for bypass dampers and discharge temperature limit switches.

2. The system does not require programming.

Figure 14-2-17 Control panel for the Lennox Harmony II zone-control system. (*Courtesy of Lennox Industries, Inc.*)

3. The system does not use sensors.

4. There is no overriding or reprogramming to condition a zone.

5. The standard components provide a balanced comfortable environment.

Some typical basement layouts showing damper locations for the reduced radial and extended plenum duct designs are shown in Figure 14-2-19 and Figure 14-2-20.

Accessory Components Thermostats are required for each zone and may be programmable or conventional. The zone 1 thermostat is designated as the master thermostat.

Dampers and damper transformers are required for each zone. Dampers are motorized, using 24 V AC to close and spring return to open.

A discharge air temperature probe monitors the supply air. It is used to gather temperature information across the airstream. It is not a limit switch.

A pressure switch is required for application with a heat pump. It acts as a safety switch in case of high head pressures during first stage and second stage heating.

System Operation—Zone/Central/System Off Modes In the Zone mode the system will respond to the demands of any zone. The only dampers that remain open are those supplying air to the demanding zones. The blower operates at the speed determined by the position of the CFM selection jumpers. In the Zone mode, it is recommended that the fan be placed in the Fan Auto position to minimize air mixing between zones.

In the Central mode, the system responds only to the master thermostat and all rooms receive conditioned air. When in the Central mode, the control panel should be set to Auto changeover so that the heat/cool selection can be made by the master thermostat.

In the System Off mode, the compressor and all heating equipment are turned off.

System Operation—Cool/Heat/Auto Changeover Modes When the Cool mode is selected, the system will respond to the cooling demands of the room thermostats. When the Heat mode is selected, the system will respond to the heating demands of the zone thermostats.

When the Auto mode is selected, the system will respond to both the heating and the cooling demands from the zone thermostats. For example, this will allow the system to heat in the morning and cool in the afternoon.

When in the Auto mode, there is a possibility that one zone may be calling for cooling at the same time that another zone is calling for heating. Opposing zones are satisfied on a first come first served basis. If opposing zones reach the control center at the same time, the heating will be supplied first. If opposing zones persist, the system will work to supply the current demand for a maximum of 20 min, then switch over and try to satisfy the opposing demand for a maximum of 20 min. When one or the other demand is satisfied, the system will work to satisfy the remaining demand.

System Operation—Fan Auto/Fan On Modes In the Fan Auto mode, the blower will cycle on and off with the demand, delivering air to the calling zones. During gas or electric strip heating, the blower will continue after the demand until the heater is cooled sufficiently.

In the Fan On mode with demands satisfied, all dampers open and air is circulated to all zones. When a cooling or heating demand is present, the system responds by closing all dampers except the demanding zone.

Time Delays Several different delay timers are used in the system as described in Table 14-2-1.

UNIT 2—REVIEW QUESTIONS

1. What are the two methods of zoning?
2. List the two types of automatic relief dampers.
3. During relief operations it is important to maintain ________ airflow through the air handling unit.

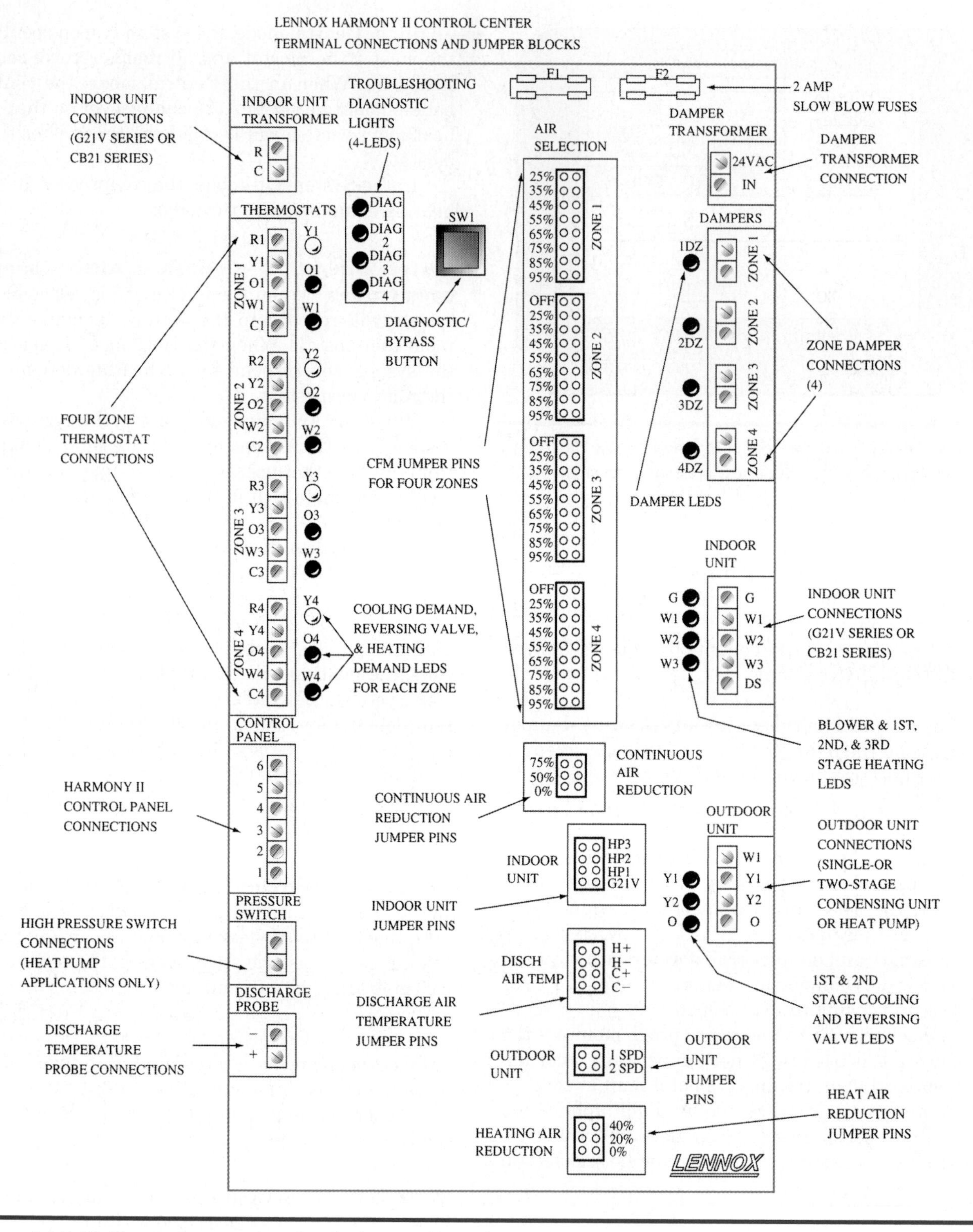

Figure 14-2-18 Terminal connections for the Lennox Harmony II control center. *(Courtesy of Lennox Industries, Inc.)*

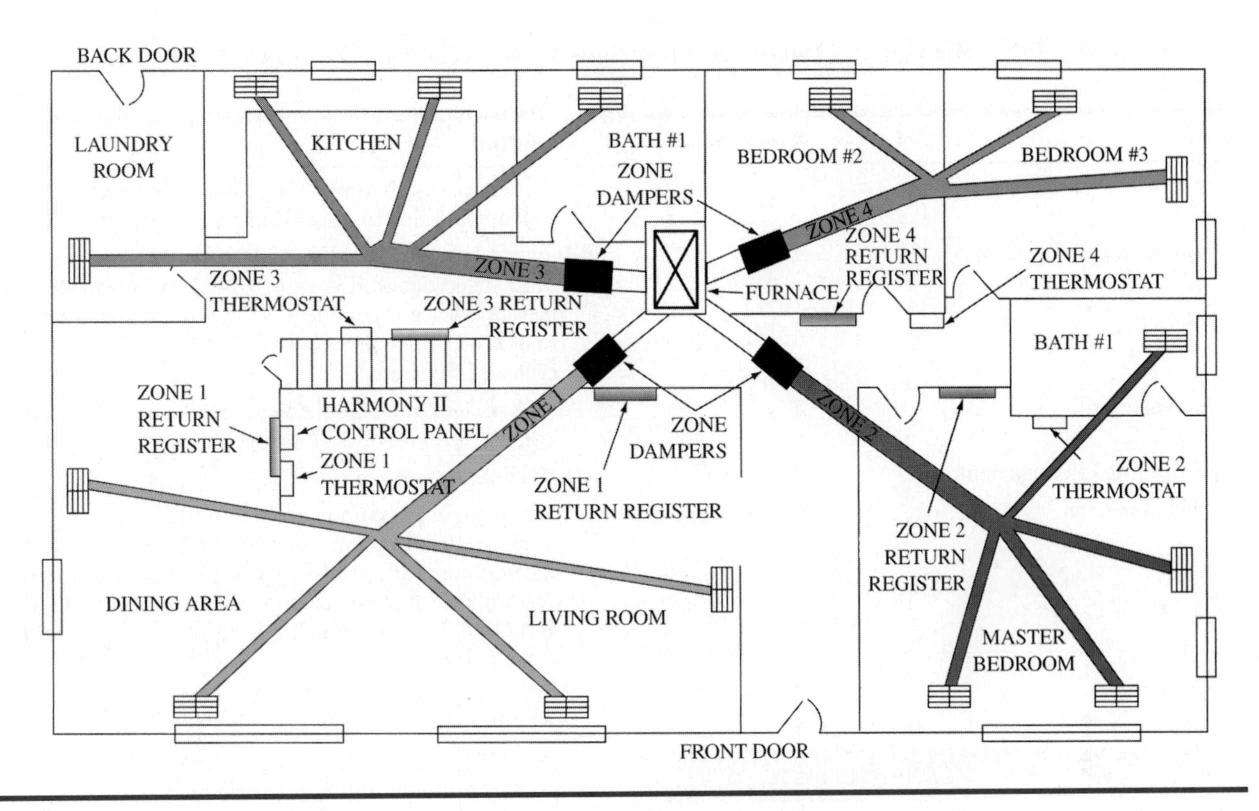

Figure 14-2-19 A typical radial zone design using the Harmony II control system. *(Courtesy of Lennox Industries, Inc.)*

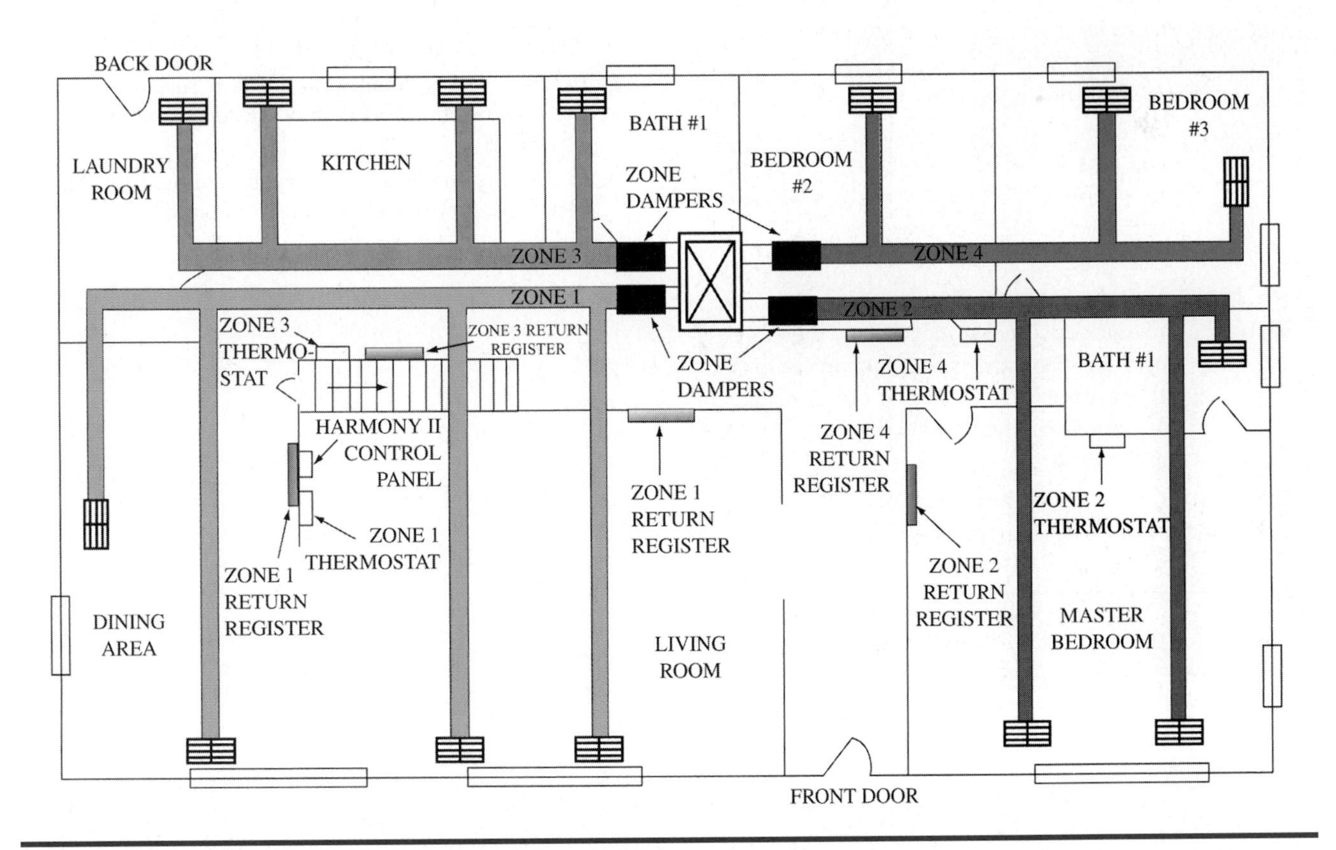

Figure 14-2-20 Typical extended plenum zone duct design using the Harmony II control system. *(Courtesy of Lennox Industries, Inc.)*

TABLE 14-2-1 Table Showing the Details of the Various Time Delay Relays Used in the Harmony II Control System

Delay	Time (min)	Function
Blower Delay (gas heat only)	5 ½	Gas furnace only. Dumps air into last zone called during cool down following heat demand.
Compressor Speed Change	4 + 1	Between low speed and high speed in order to make sure high speed demand is valid and to equalize refrigerant pressures. 1 min, due to TSC in outdoor unit. 4 min, due to delay in Harmony control center. Compressor can cycle off anytime.
Compressor Off	5	At end of demand. Equalizes pressure in refrigerant system and prevents short cycling.
Heat Staging (electric heat only)	2	Between stages up or down.
Auto-changeover	20	When opposing demands are present, Harmony II must work to satisfy current demand at least 20 min. If current demand is not satisfied after time has elapsed, system will changeover and satisfy opposing demand. On and Off delays above will also apply.

TECH TALK

Tech 1. What's the big deal about trying to put zone controls in all these offices?

Tech 2. Zone controls can save this building operator a lot of expense on utilities. The more efficient we can make any system operate the better it is for everybody.

Tech 1. What do you mean by that?

Tech 2. If everyone uses less electricity then there is going to be less pollution produced by the power plants. That's good for the environment, which is good for everyone.

Tech 1. I never thought of what we do as having an effect on the environment.

Tech 2. Since air conditioning accounts for the largest single use of electricity, anything we can do to cut the amount of electricity our customers use both saves them money and is good for the environment.

Tech 1. Oh, I see! So, putting zone controls in is good for the building operator and good for the environment.

Tech 2. That's right; I think you've got it!

Tech 1. Thanks.

4. The ________ system is an integrated set of devices for controlling a multi-zoned HVAC system.
5. What are the three built-in mode switches of the MABS central control panel?
6. What is the control panel of the Lennox Harmony II zone control system used for?
7. ________ are required for each zone and may be programmable or conventional.
8. In the ________ mode, the system responds only to the master thermostat and all rooms receive conditioned air.
9. In the Fan On mode when a cooling or heating demand is present how does the system respond?

UNIT 3

Testing and Balancing

OBJECTIVES: In this unit the reader will learn about testing and balancing including:

- General Requirements
- Report Forms
- Instruments for Testing
- Balancing the System

OVERVIEW

After the system is installed, an important function of the technician is to test its performance and make whatever adjustments are necessary to produce design requirements. For an HVAC system this usually means providing comfortable conditions at the lowest operating cost.

The equipment must produce its rated capacity. The air quantity handled by the blower must be adequate to handle the load. The temperature drop or rise over the heat exchanger or coil must be in the proper range. The air must be delivered to individual rooms in proportion to the requirement.

SERVICE TIP

A system that is not properly balanced will not be as efficient as one that is properly set up and balanced. It is to the building owner's advantage to invest in a total and complete air test and balance program prior to beginning operations. Often building operators are reluctant to spend the money to have a complete test and balance performed. It is your obligation as the installing technician to advise them of the advantages. On most large contracts, testing and balancing is included in the overall bid.

It is seldom found that a system installed in the normal manner is completely satisfactory without performing a proper test and balance (TAB) procedure. This process provides a systematic checkout of the critical factors that influence the performance of the job.

In existing buildings, changes take place in the use of the system that requires periodic TAB services. Many systems require rebalancing twice a year, when the system is changed over from heating to cooling and cooling to heating. Firms that specialize in this type of work also usually provide other maintenance services.

The technician who provides TAB services must be highly qualified. The person must be skillful and knowledgeable in the following:

1. Fundamentals of airflow
2. Fundamentals of hydronic flow
3. HVAC equipment
4. HVAC systems
5. Temperature control systems
6. Refrigeration systems
7. Temperature measurement
8. Pressure measurement
9. Flow measurement
10. Troubleshooting

In larger organizations doing much TAB work, a team of people is used to speed up the process. Training programs are available for technicians who wish to specialize in this work. Many sizable new air conditioning installations include a separate specification covering the TAB contract.

GENERAL REQUIREMENTS

In addition to the requirement for trained technicians, certain other needs must be supplied to perform a successful program of testing and balancing. These other requirements include:

1. The testing and balancing technician must have tools to measure airflow, air pressure, air temperature, air velocity, humidity, motor amps, and water pressure.
2. Report forms.
3. Specification for the work to be performed.

Instruments

Instruments for Air and Hydronic

Function/Measurement	Range	Minimum Accuracy
Rotation	0–5000 rpm	±2%
Temperature	–40°–120°F	Within ½ scale division
Temperature	0°–220°F	Within ½ scale division
Electrical	0–600 VAC	Within ½ scale division
	0–100 Amps	Within ½ scale division
	0–30 VDC	3% of full scale
	0–Infinite Ω	

Instruments for Air Balancing

Function/Measurement	Range	Minimum Accuracy
Air pressure	0–1.0 in WC	±.01 in WC
	0–10.0 in WC	
Pitot tube	18 in	N/A
	36 in	
Air velocity	100–3000 fpm	±10%
Humidity	10–90%	2% RH
Air volume	0–1400 CFM	±5%

Instruments for Hydronic Balancing

Function/ Measurement	Range	Minimum Accuracy
Hydraulic pressure	0–30 psi	±1% Full
	0–200 psi	
	30–60 psi	
Hydronic differential pressure	0 to 36 in WC	±1% Full

Report Forms

The use of report forms offers a convenient means of recording data and provides a reminder of the items that should be measured. The number and details of the forms used is dependent on the type of job. As a minimum, the forms needed for an air conditioning job include:

1. Apparatus description, Figure 14-3-1
2. Coil performance, Figure 14-3-2
3. Gas/oil heating apparatus, Figure 14-3-3
4. Fan performance, Figure 14-3-4
   If a duct traverse is made, a separate form should be used.
5. Duct traverse, Figure 14-3-5
   Other forms for the following may apply:
6. Air outlet test
7. Packaged chiller
8. Rooftop unit/heat pump
9. Compressor/condenser
10. Cooling tower
11. Pump
12. Boiler

Specifications Assessment (Minimum Requirements)

Technicians should visit a job during construction to verify that all fittings, damper control devices, and valves are properly located and installed. They should examine the distribution system to see that it is free from obstructions, moving parts are properly lubricated, and that valves and dampers are in open or in operating position. The technician should have access to proper, recently calibrated test instruments and adjust volume dampers, variable speed drives, balancing valves and control devices to meet the requirements of the job specifications. All air and hydronic volumes in the distribution system must be adjusted to within 10% of the specified quantities. The technician should document the tests and adjustments on suitable forms for

AIR APPARATUS TEST REPORT

PROJECT ______________________ SYSTEM/UNIT ______________________

LOCATION ______________________ DATE ______________________

Unit Data	
Make/Model No.	
Type/Size	
Serial Number	
Arr./Class	
Discharge	
Make Sheave	
Sheave Diam/Bore	
No. Belts/Make/Size	
No. Filters/Type/Size	

Motor Data	
Make/Frame	
Hp (W) rpm	
Volts/Phase/Hertz	
F.L. Amps/S.F.	
Make Sheave	
Make Sheave Diam/Bore	
Sheave ℄ Distance	

Test Data	Design	Actual
Total cfm (L/s)		
Total S.P.		
Fan rpm		
Motor Volts		
Motor Amps $T_1/T_2/T_3$		
Outside Air cfm (L/s)		
Return Air cfm (L/s)		

Test Data	Design	Actual
Discharge S.P.		
Suction S.P.		
Reheat Coil ΔS.P.		
Cooling Coil ΔS.P.		
Preheat Coil ΔS.P.		
Filters ΔS.P.		
Vortex Damp. Position		
Out. Air Damp. Position		
Ret. Air Damp. Position		

REMARKS:

TEST DATE ______________________ READINGS BY ______________________

Figure 14-3-1 Air apparatus test report. (*Courtesy of SMACNA*)

APPARATUS COIL TEST REPORT

PROJECT ______________________

Coil Data	Coil No.		Coil No.		Coil No.		Coil No.	
System Number								
Location								
Coil Type								
No. Rows-Fins/In. (mm)								
Manufacturer								
Model Number								
Face Area Ft.^{2} (m^2)								
	Design	**Actual**	**Design**	**Actual**	**Design**	**Actual**	**Design**	**Actual**
Air Qty., cfm (L/s)								
Air Vel., fpm (M/s)								
Press. Drop, In. (Pa)								
Out. Air DB/WB								
Ret. Air DB/WB								
Ent. Air DB/WB								
Lvg. Air DB/WB								
Air ΔT								
Water Flow, gpm (L/s)								
Press. Drop, psi (kPa)								
Ent. Water Temp.								
Lvg. Water Temp.								
Water ΔT								
Exp. Valve/Refrig.								
Refrig. Suction Press.								
Refrig. Suction Temp.								
Inlet Steam Press.								

REMARKS:

TEST DATE ______________________ READINGS BY ______________________

Figure 14-3-2 Coil performance test report. (*Courtesy of SMACNA*)

GAS/OIL FIRED HEAT APPARATUS TEST REPORT

PROJECT ______________________________

Unit Data	Unit No.		Unit No.		Unit No.		Unit No.	
System								
Location								
Make/Model								
Type/Size								
Serial Number								
Type Fuel/Input								
Output								
Ignition Type								
Burner Control								
Volts/Phase/Hertz								
HP (W)/rpm								
F.L. Amps/S.F.								
Drive Data								
Test Data	**Design**	**Actual**	**Design**	**Actual**	**Design**	**Actual**	**Design**	**Actual**
CFM (L/s)								
Ent./Lvg. Air Temp								
Air Temp ΔT								
Ent./Lvg. Air Press.								
Air Press ΔT								
Low Fire Input								
High Fire Input								
Manifold Press.								
High Limit Setting								
Operating Set Point								

REMARKS:

TEST DATE ______________________ READINGS BY ______________________

Figure 14-3-3 Gas/oil heating performance report. (*Courtesy of SMACNA*)

FAN TEST REPORT

PROJECT ____________________

Fan Data	Fan No.	Fan No.	Fan No.	Fan No.
Location				
Service				
Manufacturer				
Model Number				
Serial Number				
Class				
Motor Make/Frame				
Motor Hp (W)/rpm				
Volts/Phase/Hertz				
F.L. Amps/S.F.				
Motor Sheave Make				
Motor Sheave Diam/Bore				
Fan Sheave Make				
Fan Sheave Diam/Bore				
No. Belts/Make/Size				
Sheave ℄ Distance				

Test Data	Design	Actual	Design	Actual	Design	Actual	Design	Actual
CFM (L/s)								
Fan rpm								
Total S.P.								
Voltage								
Amperage $T_1/T_2/T_3$								

REMARKS:

TEST DATE ____________________ READINGS BY ____________________

Figure 14-3-4 Fan performance test report. (*Courtesy of SMACNA*)

Page____of____

INSTRUCTIONS

METHOD NO. 1
(EQUAL AREA)

TO DETERMINE THE AVERAGE AIR VELOCITY IN SQUARE OR RECTANGULAR DUCTS, A PITOT TUBE TRAVERSE MUST BE MADE TO MEASURE THE VELOCITIES AT THE CENTER POINTS OF EQUAL AREAS OVER THE CROSS SECTION OF THE DUCT. THE NUMBER OF EQUAL AREAS SHOULD NOT BE LESS THAN 15, BUT NEED NOT BE MORE THAN 64. THE MAXIMUM DISTANCE BETWEEN CENTER POINTS, FOR LESS THAN 64 READINGS, SHOULD NOT BE MORE THAN 6 INCHES (150 MM). THE READINGS CLOSEST TO THE DUCTWALLS SHOULD BE TAKEN AT ONE-HALF OF THIS DISTANCE. FOR MAXIMUM ACCURACY, THE VELOCITY CORRESPONDING TO EACH VELOCITY PRESSURE MEASURED MUST BE DETERMINED AND THEN AVERAGED.

METHOD NO. 2
(LOG)

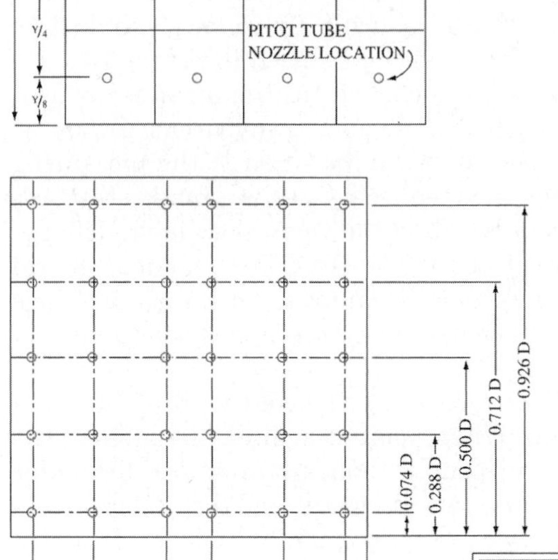

NO. OF POINTS OF TRAVERSE LINES	POSITION RELATIVE TO INNER WALL
5	0.074, 0.238, 0.500, 0.712, 0.926
6	0.061, 0.235, 0.437, 0.563, 0.765, 0.939
7	0.053, 0.203, 0.366, 0.500, 0.634, 0.797, 0.947

Figure 14-3-5 Rectangular duct traverse. *(Courtesy of SMACNA)*

review. A check will be made to assure that all national, state, and local codes have been followed.

NOTE: Verifying the proper installation of the air distribution system during construction is important. It may not be possible to balance a system that is installed significantly out of the design requirements. Even if a poorly installed system can be balanced it will be at the cost of overall operating efficiency. Do not rely on local building inspectors to catch problems with the air distribution system. It is not part of their job to check the engineering. They are primarily looking to see that the ducts are installed with the proper number of hangers so that the system is secure. It is your responsibility to see that the system is being installed according to the building prints and specifications.

INSTRUMENTS FOR TESTING

A complete set of calibrated test instruments is essential for quality TAB work. The following are some of the most common uses and a description of the test instruments used:

1. Air (gas) flow measuring
2. Pressure measuring
3. Rotation measuring
4. Temperature measuring
5. Electrical measuring
6. Hydronic flow measuring

Air (Gas) Flow Measuring

The U tube manometer, Figure 14-3-6, is used for measuring various types of gas flow. When not

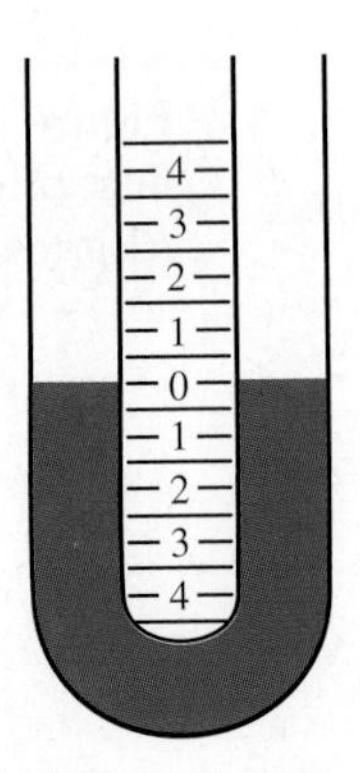

Figure 14-3-6 U tube manometer.

SERVICE TIP

The oil in an inclined manometer has a specific gravity of 0.8. Any oil that is used other than the specific inclined manometer oil will not provide you with the correct readings. Check with your local supplier if replacement oil is needed.

connected, both ends of the U tube are open to the atmosphere. The tube is partly filled with liquid, so that in the idle (open) condition, both tubes register zero. When one tube is attached to a gas under pressure, such as air in a duct, both liquid levels in the tubes change position. The pressure measurement (which is proportional to flow) is indicated by the difference in height of the two columns.

These gauges are recommended for measuring pressures above 1.0 in WC, across filters, coils, fans, terminal devices, and sections of ductwork. A U tube manometer is shown making these measurements in Figure 14-3-7a–c.

Instruments that use mercury should be equipped with safety reservoirs to prevent accidental blowing of mercury into the airstream. Tubes must be clean to give accurate results.

The standard manometer, Figure 14-3-8, or the inclined/vertical manometer, Figure 14-3-9, is useful in measuring static pressure, velocity pressure, and total pressure in a duct when connected to a pitot tube, Figure 14-3-10. This combination of instruments is used in measuring air volume in a duct by making a traverse such as shown in Figure 14-3-11.

The inclined and/or vertical manometer is usually constructed of molded transparent plastic so that it is easy to see the position of the measuring liquid. The inclined scale reads small pressures (below 1.0 in WC) and the vertical scale reads large pressures (up to 10.0 in WC). The inclined/vertical manometer is usually equipped with magnetic lugs so that it can be mounted conveniently on the side of the duct.

The *Pitot tube* is constructed with a tube in a tube design. When pointed against the airflow, the inner tube measures total air pressure, the outer tube measures static pressure, and the difference between the two pressures, connected as shown in Figure 14-3-8, is the velocity pressure. Convenient charts are available, shipped with the instrument, for converting velocity pressure to ft/min (fpm). Knowing the cross-sectional area of the duct in ft^2, the CFM can easily be calculated (CFM $= A \times$ fpm).

A traverse of the duct is important because a series of velocity readings can be taken and averaged to obtain a more accurate value.

(a)

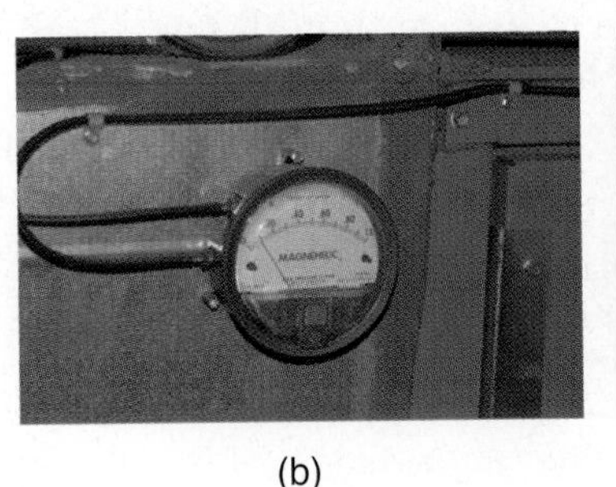

(b)

(c)

Figure 14-3-7 Magnehelic® gauges are used to measure inches of water: (a) This Magnehelic® has a range from 0 to 5 in; (b) This Magnehelic® has a range from 0 to 1 in, and is being used to measure the pressure drop across an air filter; (c) This is a Photohelic® gauge that is used to sound an alarm if the pressure is outside of the operating range as indicated by the two red pointers on the gauge face.

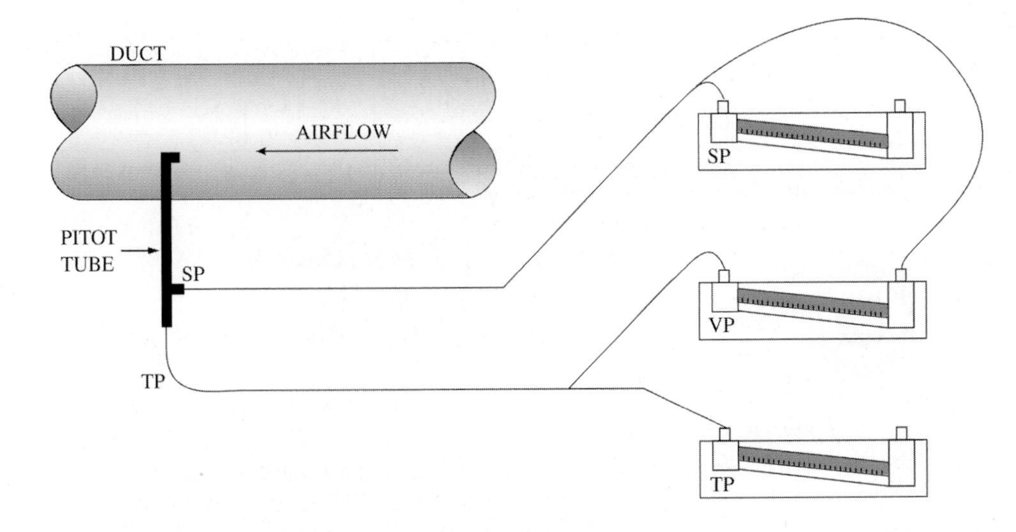

Figure 14-3-8 Standard manometer connection to a pitot tube. (*Courtesy of SMACNA*)

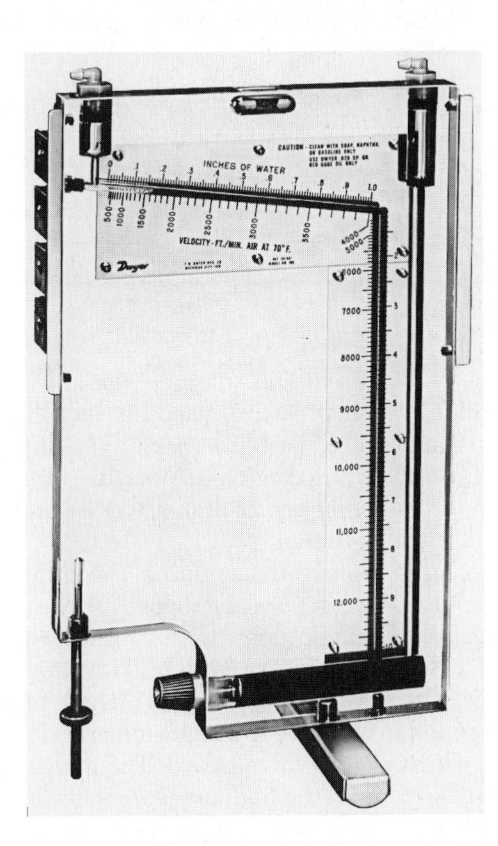

Figure 14-3-9 Inclined/vertical manometer. (*Courtesy of Dwyer Instruments, Inc.*)

The *rotating vane anemometer,* Figure 14-3-12a, consists of a lightweight air propelled wheel, geared to dials that record the linear feet of airflow passing through the instrument. The instrument is placed in various portions of the airstream and readings taken for a measured amount of time. These readings are averaged and converted to fpm.

The *electronic rotating vane anemometer,* Figure 14-3-12b, is much simpler to use. It is battery operated and the readout can be either digital or analog. The digital instrument will automatically average the readings and display the velocity readout in fpm. The analog instrument is direct reading with a choice of a number of different velocity scales.

The *Alnor Velometer,* Figure 14-3-13, meets the requirements of most TAB work. The instrument set consists of the meter, measuring probes, range selectors, and connecting hoses. Three velocity probes are provided: the low flow, diffuser, and Pitot tube. The low flow probe is used to measure airflows below 300 fpm, such as encountered in open spaces. The diffuser probe automatically averages the flow over the face of the supply or return grill. The Pitot tube probe is used to measure air velocities in ducts.

The thermal anemometer, Figure 14-3-14, operates on the principle that the resistance of a heated wire will change with its temperature. The probe is placed in the airstream and the velocity is indicated on the scale of the instrument. It is used for measuring very low velocities of air, such as found in a room, or it may be used in a duct to determine total flow.

SERVICE TIP

Air flows below 100 fpm are difficult or impossible to read with testing instruments other than the thermal anemometer.

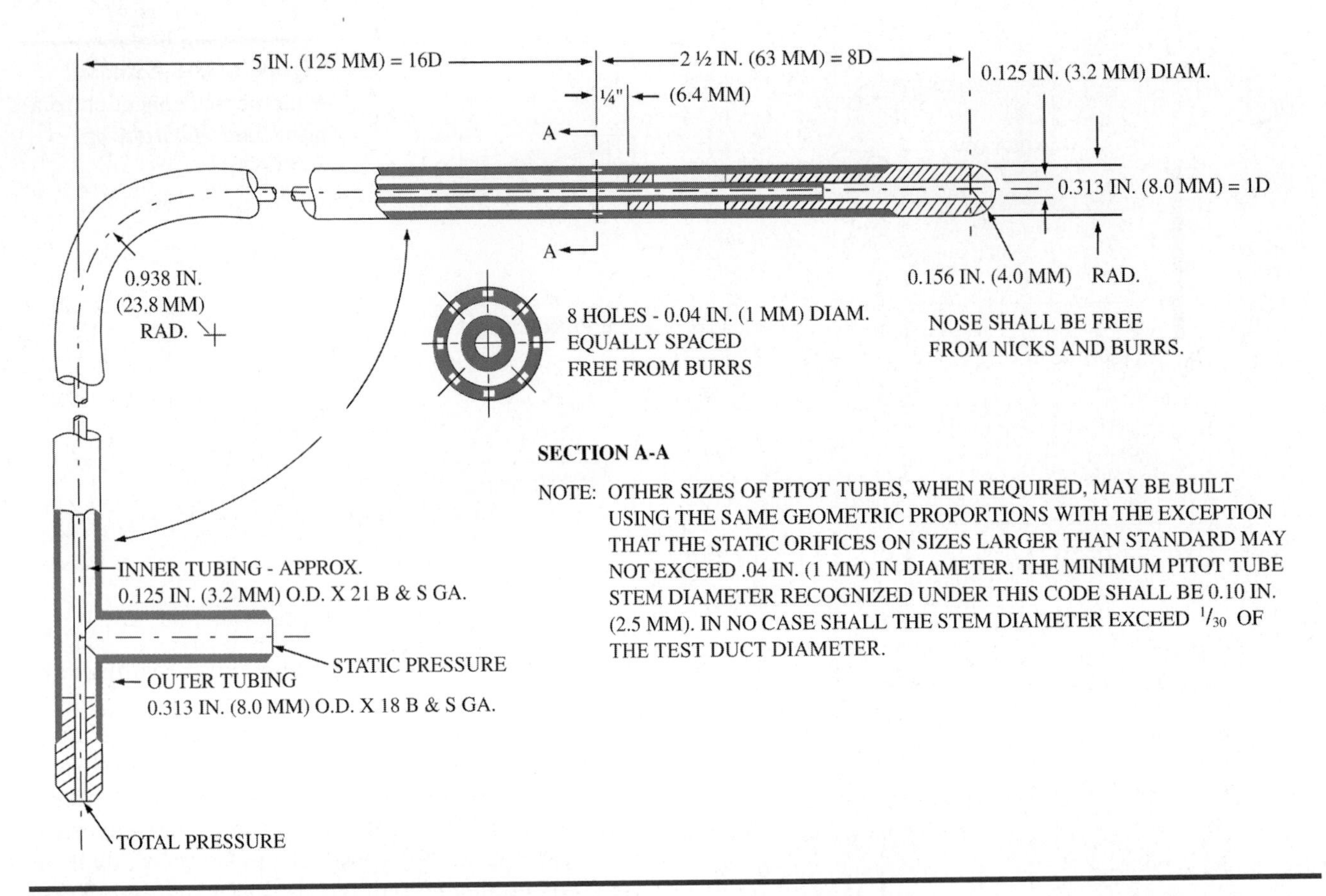

Figure 14-3-10 Detailed construction of a Pitot tube. (*Courtesy of SMACNA*)

The flow hood, Figure 14-3-15, is useful in accurately determining the total airflow in CFM from an outlet. It completely collects the air supply and directs it through a 1 ft^2 opening where a calibrated manometer provides a CFM readout. Hoods should be selected to as nearly as possible fit the diffuser being measured and have a tight seal around the opening.

Volume Air Balancer An air balancer, Figure 14-3-16a, makes direct measurements of CFM in residential air conditioning systems. It can be used on grills, registers, and diffusers. It averages the outlet conditions. It also reads air velocity in fpm. It requires no power or special maintenance.

Air Meter Kit A direct reading air meter kit is shown in Figure 14-3-16b, complete with meter, probes, and air velocity calculator.

Pressure Measurement

The calibrated pressure gauge, Figure 14-3-17, uses a Bourdon tube assembly for sensing pressure and may be constructed of stainless steel, monel metal, or bronze. The dials are 3½ in or 6 in diameter and are available with pressure, vacuum, or compound gauges. Gauges are used for checking pump pressures; coil, chiller, and condenser pressure drops; and pressure drops across orifice plates, valves, and other flow calibrated devices.

Most uses require determining the differential pressure drop. Two separate gauges can be used for this purpose or a single gauge with a rotating dial can be used as shown in Figure 14-3-18. The gauge is installed with two valve connections. The high pressure line is opened to measure the high side pressure, with the low pressure line valve closed. The gauge dial is rotated to zero. Then the high pressure valve is closed and the low pressure line valve opened. The dial then reads the differential pressure.

A differential pressure gauge, Figure 14-3-19, is available with two pressure connections and is used as shown in the illustration.

Rotation Measuring Instruments

A tachometer is an instrument for measuring the rotational speed of a shaft or wheel. The units of measurement are usually revolutions per minute (rpm). There are a number of types. Some have a digital readout. Others are mainly revolution counters and require a timing device.

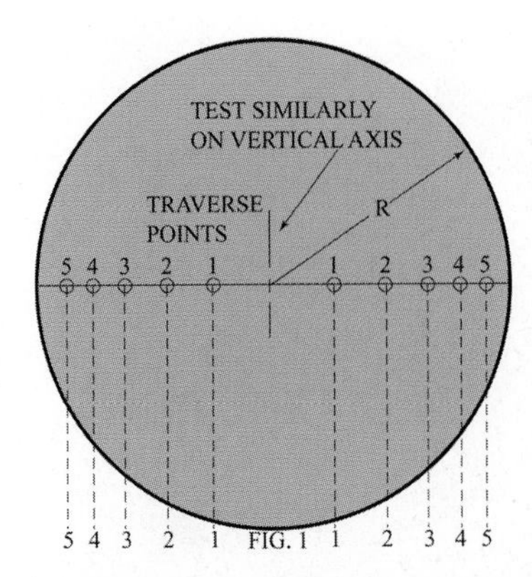

Fig.1: Locations for Pitot tube tip making a 10-point traverse across one circular pipe diameter. In making two traverses cross the pipe diameter, readings are taken at right angles to each other. The traverse points shown represent 5 annular zones of equal area.

Figure 14-3-11 Round duct traverse. *(Courtesy of SMACNA)*

Table 1
Distances of Pitot Tube Tip from Pipe Center

Duct diam.		Readings in one diam.	Point 1		Point 2		Point 3		Point 4		Point 5	
inches	(mm)		inches	(mm)	inches	(mm)	inches	(mm)	inches	(mm)	inches	(mm)
3	75	6	.612	15.3	1.061	26.5	1.369	34.2				
4	100	6	.812	20.4	1.414	35.4	1.826	45.6				
5	125	6	1.021	25.5	1.768	44.2	2.285	57.1				
6	150	6	1.225	30.6	2.121	53.0	2.738	68.5				
7	175	6	1.429	35.7	2.475	61.9	3.195	80.0				
8	200	6	1.633	40.8	2.828	70.7	3.651	91.3				
9	225	6	1.837	45.9	3.182	79.5	4.108	102.7				
10	250	8	1.768	44.2	3.062	76.6	3.950	98.8	4.677	116.9		
12	300	8	2.122	53.0	3.674	91.1	4.740	118.6	5.612	140.3		
14	350	10	2.214	55.3	3.834	95.8	4.950	123.7	5.857	146.4	6.641	166.0
16	400	10	2.530	63.2	4.382	109.5	6.567	141.4	6.693	167.3	7.589	189.7
18	450	10	2.846	71.1	4.929	123.2	6.364	159.1	7.530	188.2	8.538	213.4
20	500	10	3.162	79.1	5.477	136.9	7.077	176.8	8.367	209.1	9.487	237.2
22	550	10	3.479	87.0	6.025	150.6	7.778	194.4	9.213	230.1	10.435	260.9
24	600	10	3.798	94.9	6.573	164.3	8.485	212.1	10.040	251.0	11.384	284.6
26	650	10	4.111	102.8	7.120	178.0	9.192	229.8	10.877	171.9	12.222	308.3
28	700	10	4.427	110.7	7.668	191.7	9.900	247.5	11.713	292.8	13.282	332.0
30	750	10	4.743	118.6	8.216	205.4	10.607	265.1	12.550	313.7	14.230	355.7
32	800	10	5.060	126.5	8.764	219.0	11.314	282.8	13.387	334.6	15.179	379.4
34	850	10	5.376	134.4	9.311	232.7	12.021	300.5	14.233	355.6	16.128	403.2

For distance of traverse points from pipe center for pipe diameters other than those given in Table 1, use constants in Table 2.

Table 2
Constants To Be Multiplied By Pipe Diameter For Distances of Pitot Tube Tip From Pipe Center

Readings in One Diameter	Point 1	Point 2	Point 3	Point 4	Point 5
6	.2041	.3535	.4564		
8	.1768	.3062	.3953	.4677	
10	.1581	.2738	.3535	.4183	.4743

The chronometric tachometer, Figure 14-3-20, combines a revolution counter and a stopwatch in one instrument. The spindle is placed in contact with the rotating shaft. This sets the meter hand at zero and starts the stopwatch. After a fixed amount of time, usually 6 seconds, the counting mechanism is automatically uncoupled, and the instrument can be removed from the shaft and read.

The tachometer, Figure 14-3-21a,b, measures the rotation of a shaft by means of a rubber tip that is held against the shaft. The rpm measurement is shown on the analog scale in the figure. The optical tachometer uses a photocell to count the pulses as an object rotates. It includes a computerized circuit that produces a direct reading in rpm. It is completely portable since it is powered by batteries. It can be calibrated easily by directing it to a fluorescent light and comparing the reading against a 7200 rpm scale. A big advantage of this instrument is that it does not need to make direct contact with the rotating device and can be used where the rotating shaft is not accessible.

The stroboscope, Figure 14-3-22, is an electronic tachometer that uses an electronically flashing light.

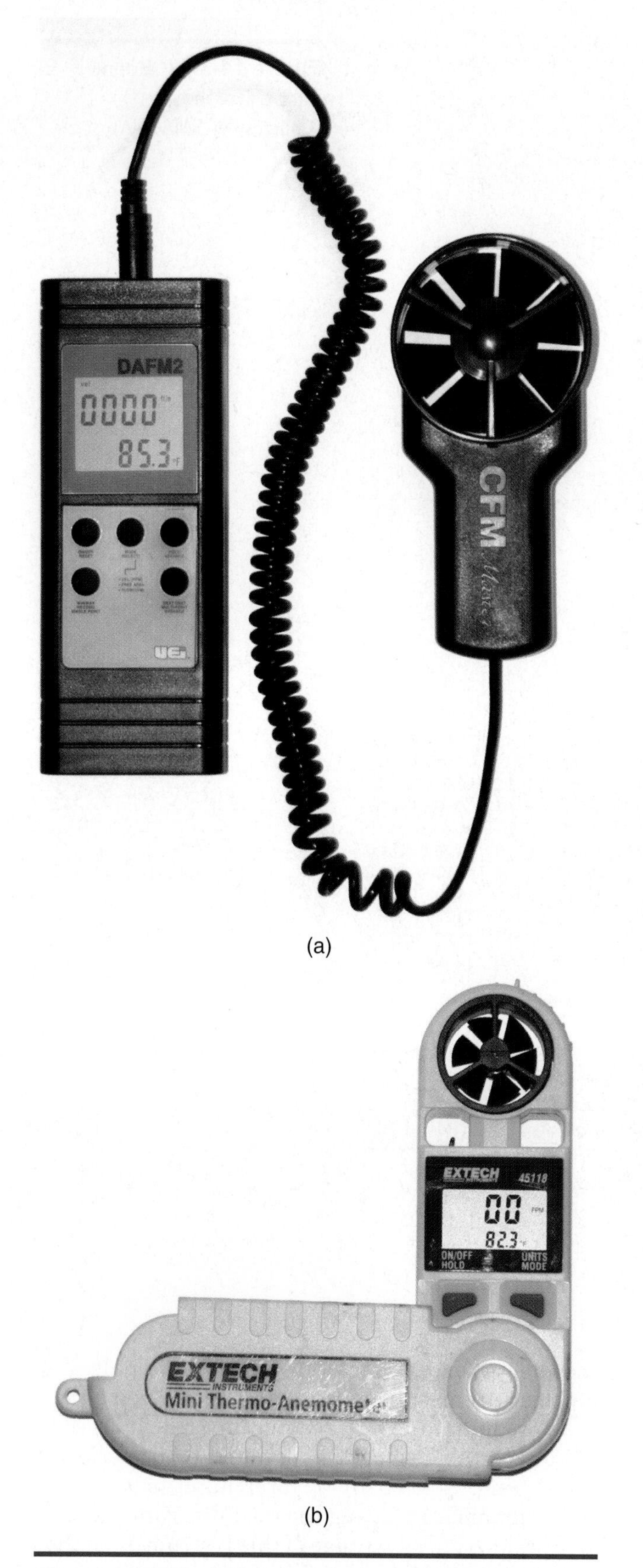

(a)

(b)

Figure 14-3-12 (a) Rotating vane anemometer; (b) Electronic rotating vane anemometer.

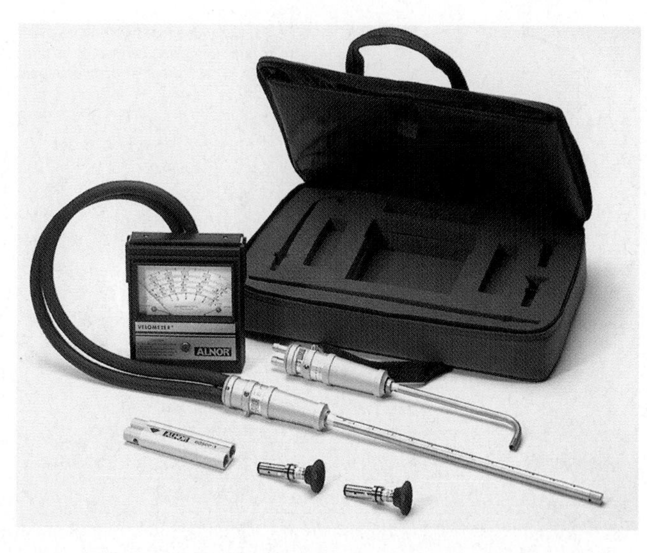

Figure 14-3-13 Velometer. (*Courtesy of Alnor Instrument Company*)

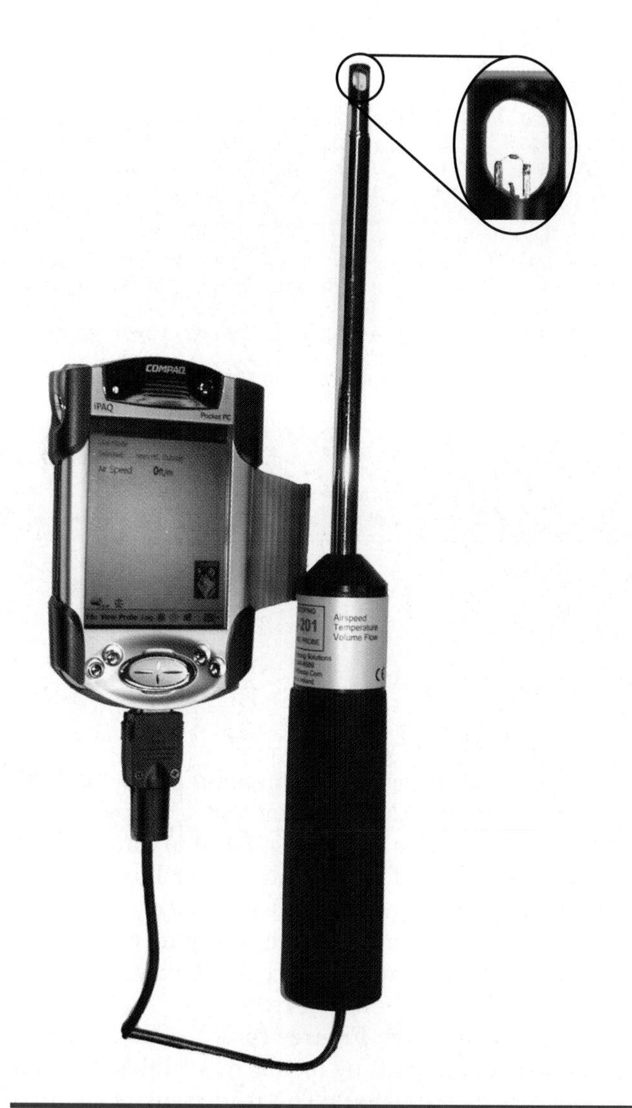

Figure 14-3-14 Thermal anemometer.

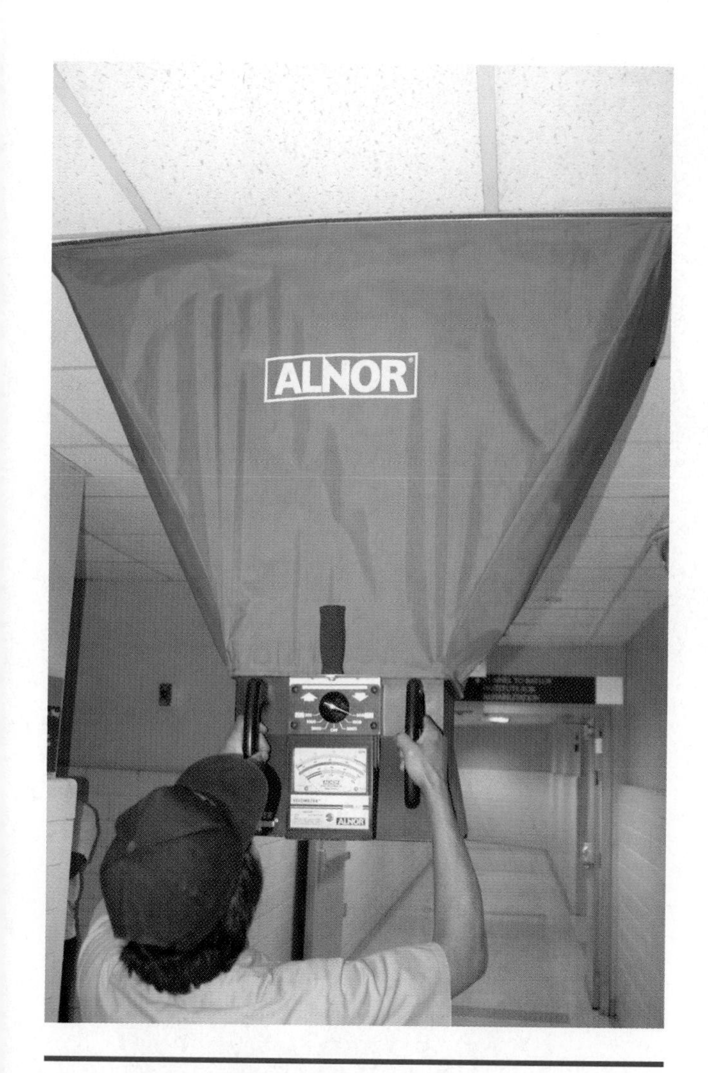

Figure 14-3-15 Flow measuring hood. *(Courtesy of Alnor Instrument Company)*

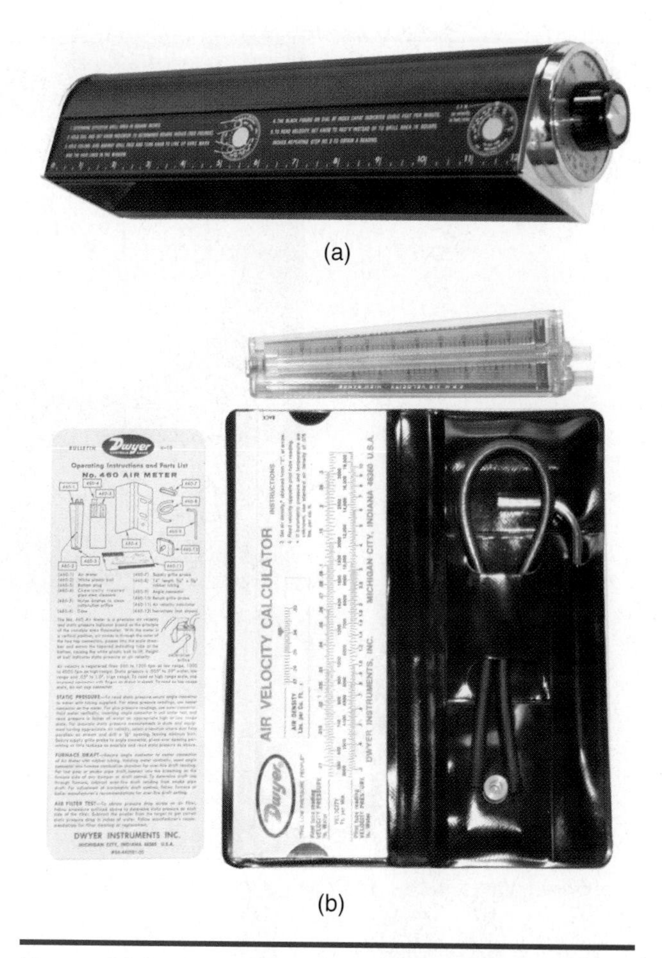

Figure 14-3-16 (a) Volume air balancer; (b) Air meter.

Figure 14-3-17 Calibrated pressure gauge. *(Courtesy of Robinair Division, SPX Corporation)*

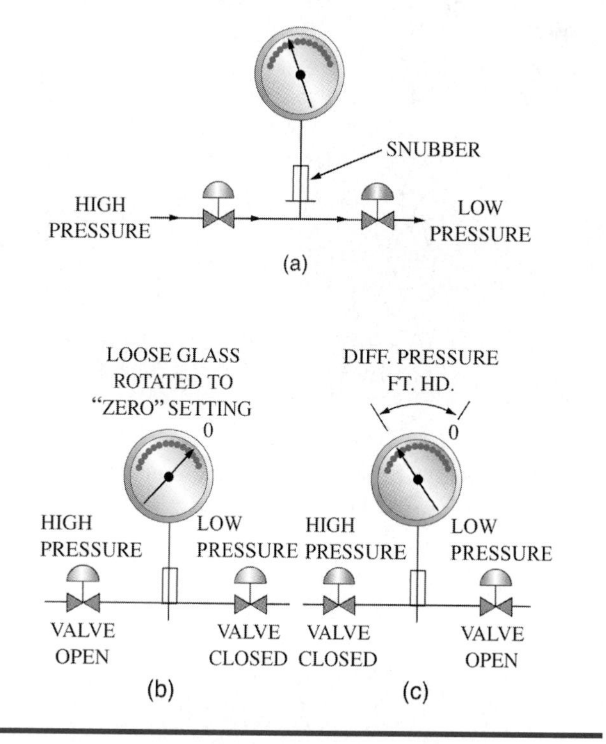

Figure 14-3-18 Single gauge for measuring differential pressures. *(Courtesy of SMACNA)*

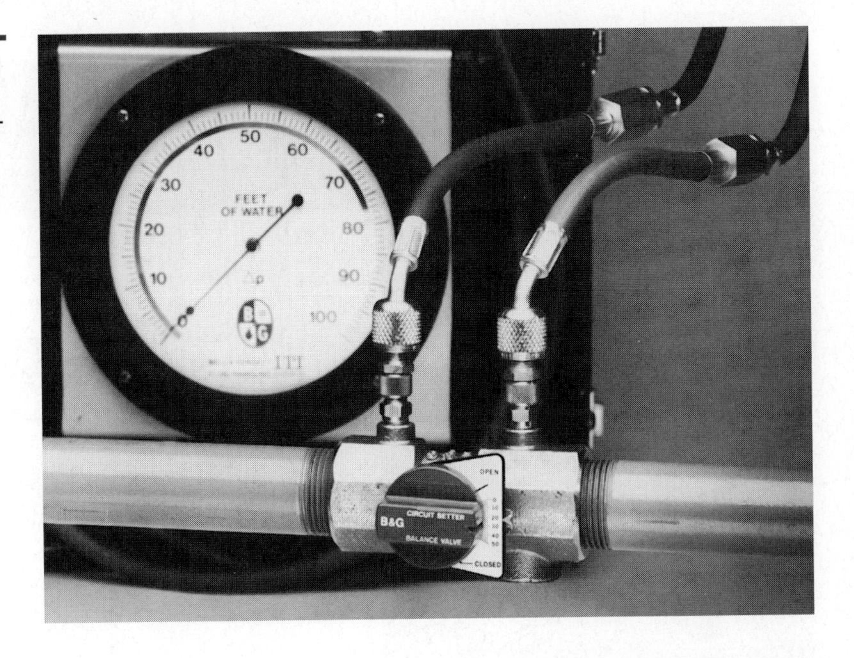

Figure 14-3-19 Differential pressure gauge with dual connections. (*Courtesy of ITT Bell & Gossett*)

Figure 14-3-20 Chronometric tachometer. (*Reprinted with permission by AVO International*)

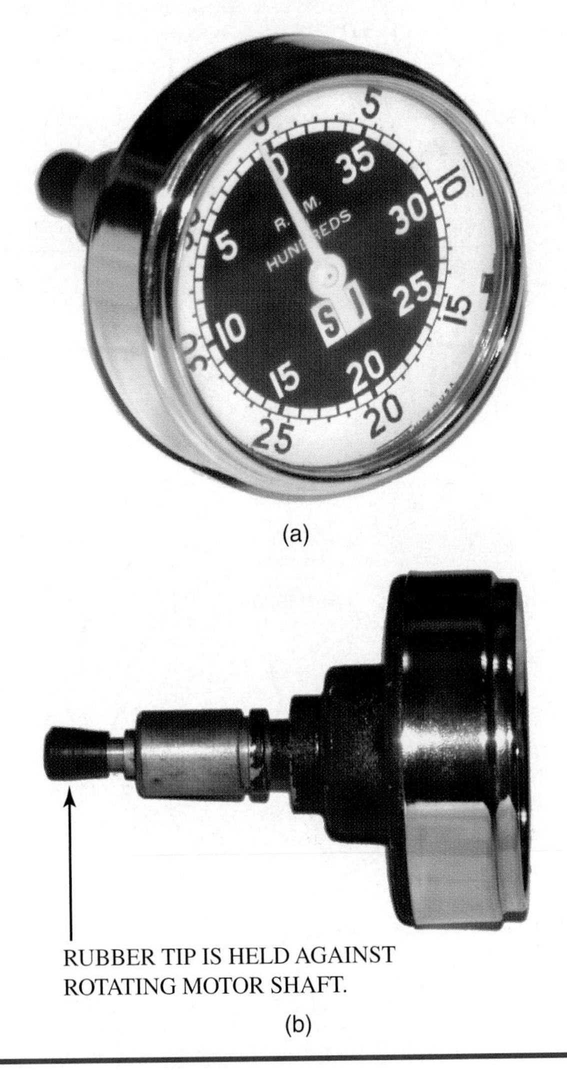

Figure 14-3-21 Tachometer: (a) Front view; (b) Side view.

Figure 14-3-22 Stroboscope for measuring rotational speed.

The frequency of the light flashes can be adjusted to equal the frequency of the rotating object. When the two are synchronized, the rotating object appears to be standing still. Care must be exercised in using it to prevent reading harmonics of the actual speed rather than the actual speed.

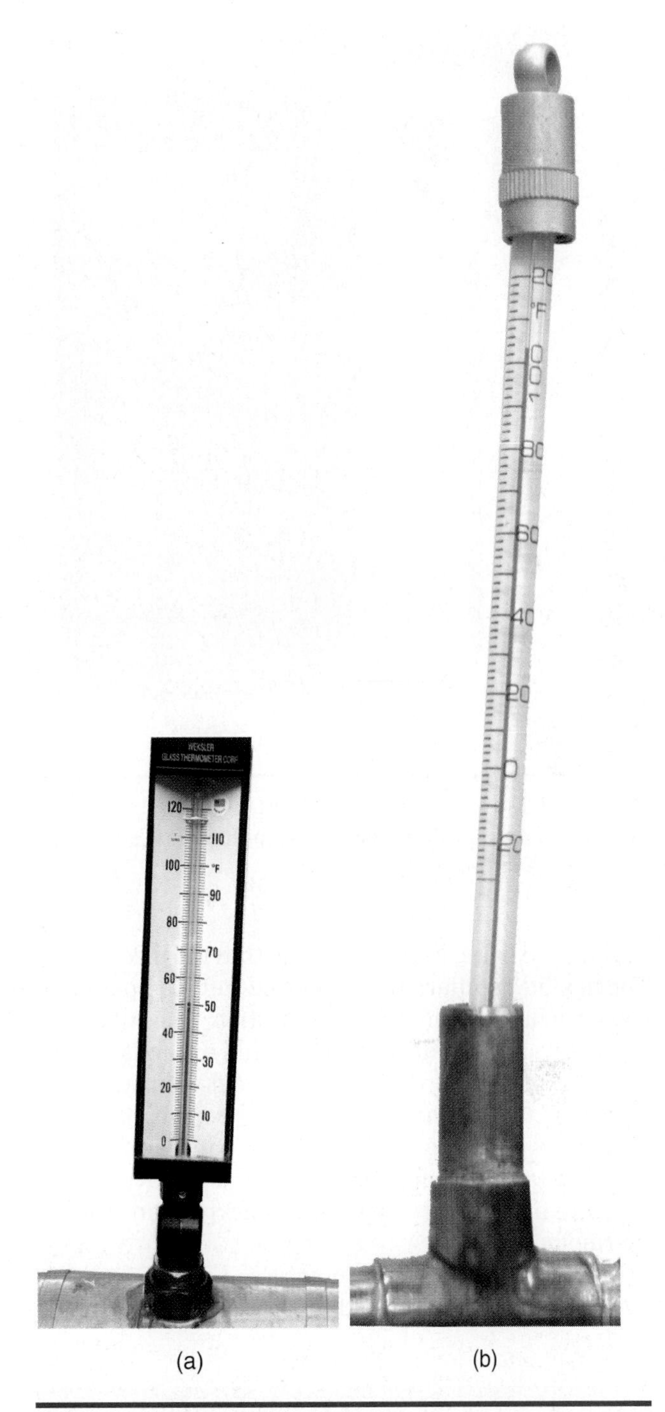

Figure 14-3-23 Glass tube thermometers: (a) In protective case; (b) With glass tube exposed.

Temperature Measurement

Glass tube thermometers, Figure 14-3-23a,b, are available in a number of ranges, scale graduations, and lengths. They have a useful range from −40° to over 220°F.

Dial thermometers are available in two general types: stem, Figure 14-3-24a, and flexible capillary tube. Dial thermometers are usually bimetal operated,

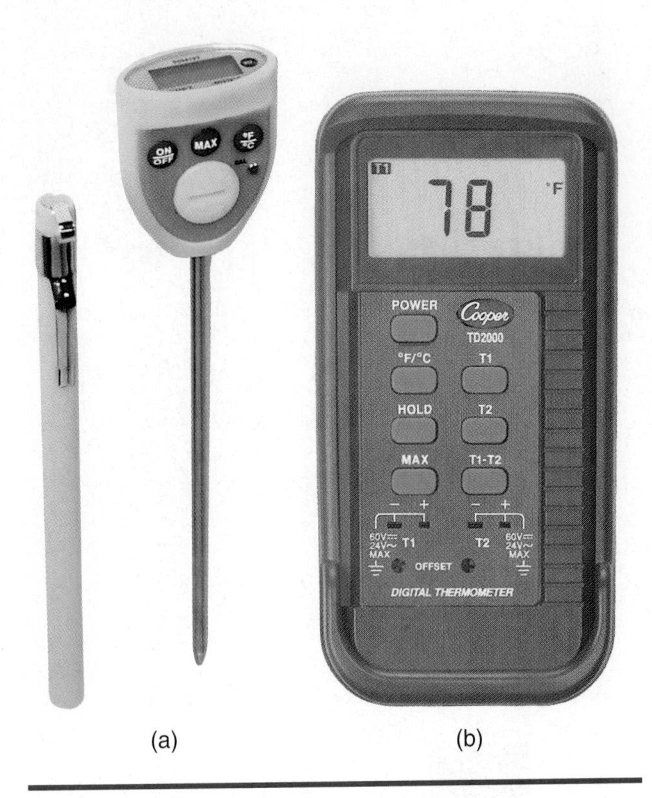

Figure 14-3-24 Dial thermometers: (a) Pocket; (b) Hand held superheat digital thermometer. (*Courtesy of Cooper-Atkins Corporation*)

the bimetal being in spiral form within the insert tube. The flexible capillary tube model permits temperature measurements from a remote location. The tubes are liquid filled or gas filled, and the fluid expands or contracts to operate a Bourdon tube held indicator. Dial thermometers usually have a longer reading time lag than glass tube thermometers. The hand held superheat digital thermometer shown in Figure 14-3-24b reads temperatures even more quickly than a glass thermometer.

Thermocouple thermometers, Figure 14-3-25, use a thermocouple sensing device and a millivolt-

SERVICE TIP

Dial thermometers may need to be calibrated from time to time because their needles may be bumped off slightly as they bounce around in a service vehicle. The best way of calibrating a dial thermometer is to place it in a glass of ice and water. While stirring the water the dial should read 32°F. If not, it could be adjusted. Next place the probe in a pot of boiling water and it should read very close to 212°F. If there is no difference between the low reading and the high reading, the dial thermometer may considered to be defective and should be discarded.

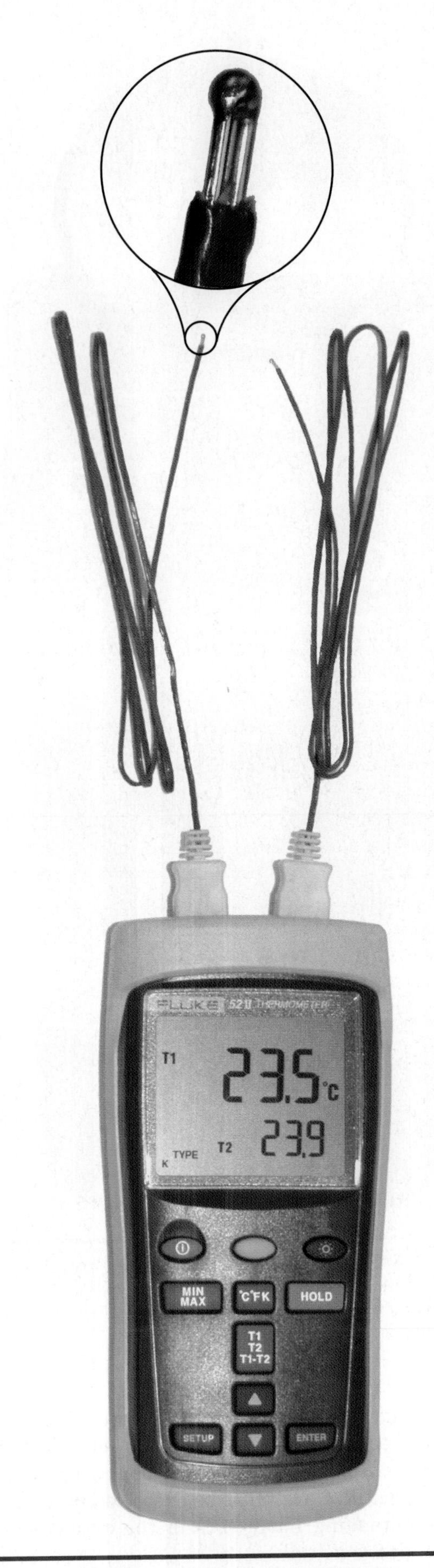

Figure 14-3-25 Thermocouple thermometers.

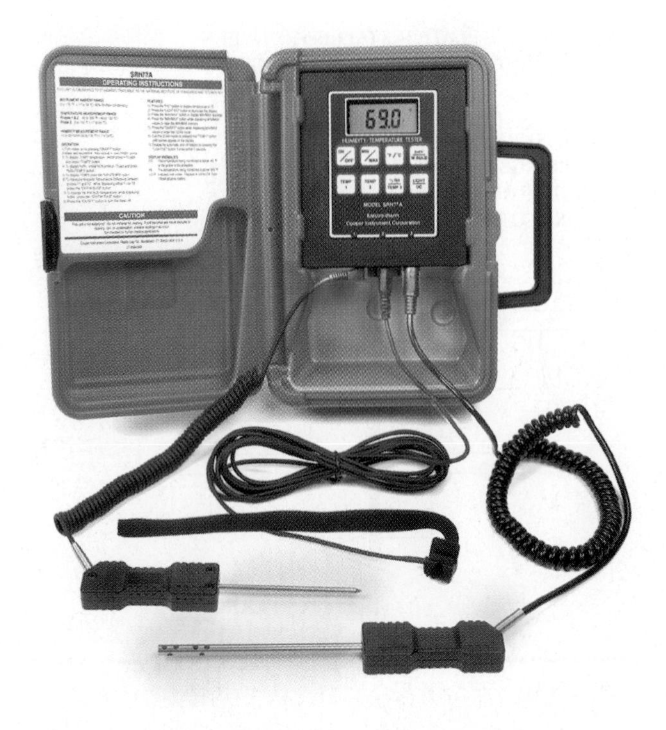

Figure 14-3-26 Electronic thermometer. *(Courtesy of Cooper-Atkins Corporation)*

meter with a scale calibrated to read temperatures directly. They are useful for remote reading and can be obtained with multiple connections for convenience in reading temperatures at a number of locations.

Electronic thermometers, Figure 14-3-26, have interchangeable probes to permit more accurate reading in selected ranges. There are a number of types: resistance temperature detectors (RTD), thermistors, thermocouple, and liquid crystal diode (LCD) or LED displays. The resistance type has a longer time lag than the thermocouple type.

Psychrometric Meters

The sling psychrometer consists of two mercury filled thermometers, one of which has a wick wetted by water surrounding the bulb. The frame that supports the two thermometers is hinged to permit revolving the wetted instrument in the air. The unit should be whirled at a rate of two revolutions per second for most accurate results. Readings are taken on both thermometers after the temperatures have stabilized. Readings indicate dry bulb and wet bulb temperatures, which can be plotted on a psychrometric chart to determine numerous properties of the air sampled.

The electronic thermohygrometer is usually constructed with a thin film capacitance sensor. As the moisture content and temperature change, the resistance of the sensor changes proportionately.

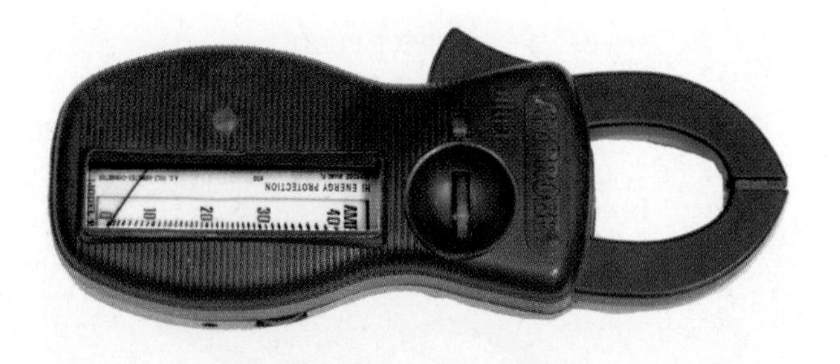

Figure 14-3-27 Clamp-on volt ammeter.

These instruments usually read directly in relative humidity. No wetted wick is necessary and the instrument can remain stationary when readings are taken.

Electrical Measuring Instruments

The clamp-on volt ammeter, Figure 14-3-27, is useful for making electrical measurements in the field. Most meters have several scales in amperes, voltages, and ohms. The clamp-on transformer jaws permit reading current flow without disconnecting the circuit. Care must be exercised when reading ampere flow to only clamp onto one wire. Enclosing two wires may result in a zero reading. Separate leads, with a DC battery in series, are required for resistance measurements.

Hydronic Flow Measurements

The orifice plate and venturi tube, Figure 14-3-28 and Figure 14-3-29, each have a specific area reduction in the path of the flow. The pressure differential across the restriction is related to the velocity of the fluid. By accurately measuring the pressure drops at a range of flow rates, a graph is set up to provide data for future measurements in the same range.

BALANCING THE SYSTEM

Balancing in its broadest meaning includes testing and adjusting an installed system to produce its design specifications. It is the final step in the HVAC project to prepare it for occupancy and use by the owner.

It is an important phase of the work, since many components, from many sources, have been assembled to perform a valuable service. Usually equipment adjustments and control settings are necessary to prepare for the desired operation.

Recommended procedures are described for balancing the following common systems:

1. Warm air heating installation.
2. Heating/cooling split system.

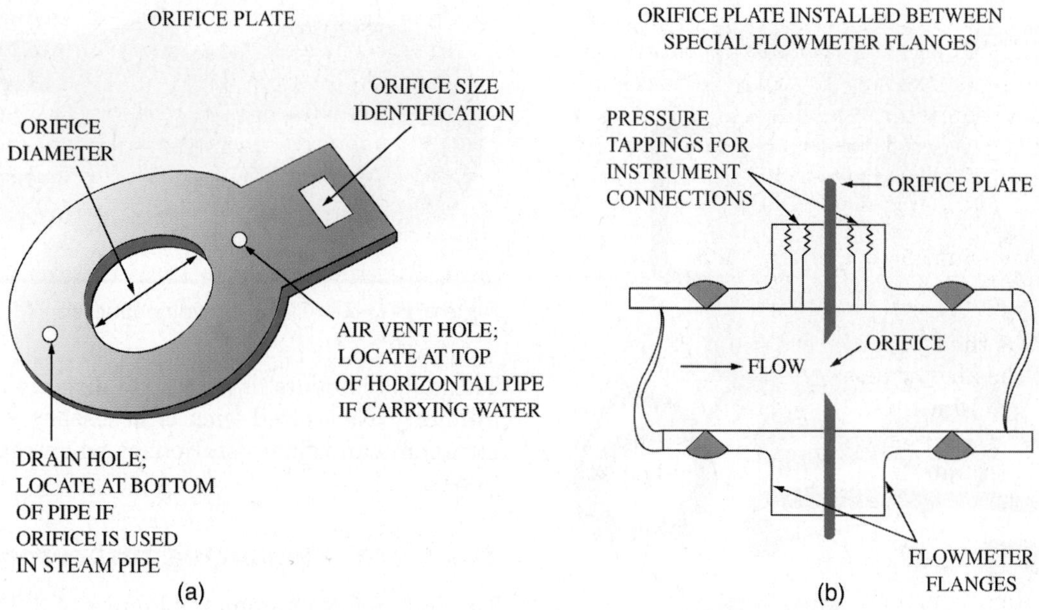

ORIFICE PLATE
ORIFICE SIZE IDENTIFICATION
ORIFICE DIAMETER
AIR VENT HOLE; LOCATE AT TOP OF HORIZONTAL PIPE IF CARRYING WATER
DRAIN HOLE; LOCATE AT BOTTOM OF PIPE IF ORIFICE IS USED IN STEAM PIPE
(a)
ORIFICE PLATE INSTALLED BETWEEN SPECIAL FLOWMETER FLANGES
PRESSURE TAPPINGS FOR INSTRUMENT CONNECTIONS
ORIFICE PLATE
ORIFICE
FLOW
FLOWMETER FLANGES
(b)

Figure 14-3-28 Orifice used as a measuring device. (*Courtesy of SMACNA*)

VENTURI TUBE
MODIFIED TUBE
FULL VENTURI TUBE
PIPE
THROAT
ENTRANCE CONE
RECOVERY CONE
ORIFICE PLATE
PIPE
ORIFICE PLATE
TURBULENCE
VENA CONTRACTA

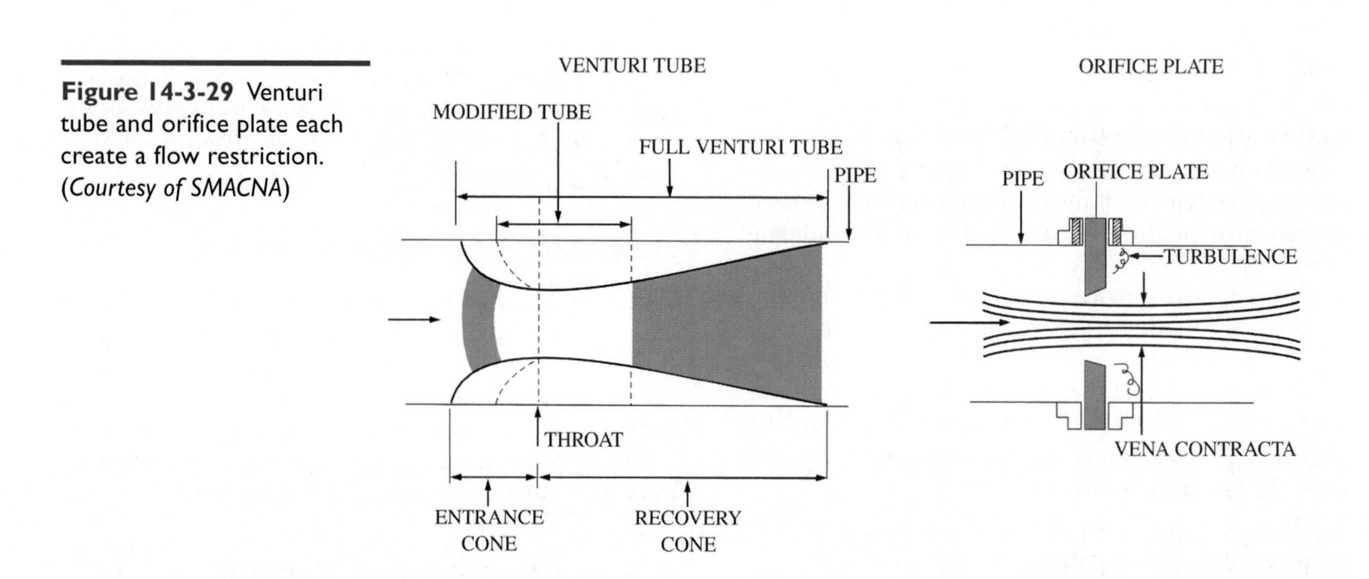

Figure 14-3-29 Venturi tube and orifice plate each create a flow restriction. (*Courtesy of SMACNA*)

3. Zoned air system installation.
4. Hydronic heating system.

Balancing a Warm Air Heating System

It is easier to balance a heating system during cold weather and a cooling system during warm weather; however, this is not always possible. It is therefore recommended that the first balance be made when the job is completed or before occupancy, and if necessary, the final balance made during a period as near design temperature conditions as possible.

The following is a suggested procedure for balancing a warm air heating system:

1. Collect complete information about the installation, including the specifications for the heating unit, details of the controls and electrical system, and a copy of the job layout, including the air quantities (CFM) required for each room.

2. Examine the entire installation. Look for problems that need to be corrected, such as leaky ductwork, missing parts, and incomplete wiring.

3. Open all dampers, turn the thermostat up so the unit will run continuously, and adjust the input to the furnace to meet the manufacturer's requirements.

4. Measure the temperature rise in the air passing through the furnace. Check the temperature rise with the manufacturer's specifications, to be certain it is within the required range. Based on the output rating of the furnace and the temperature rise, calculate the total CFM the unit is delivering, using the formula:

$$\text{CFM} = \frac{\text{furnace output in Btu/hr}}{(\text{TD} \times 1.08)}$$

where TD is the temperature rise in °F.

5. Adjust the airflow through each outlet to comply with the requirements, using either a flow hood measuring device or an anemometer to measure the face velocity in ft/min (fpm). The CFM is calculated by the formula,

$$\text{CFM} = \text{free area of grill (ft}^2) \times \text{fpm}$$

6. Place thermometers in a suitable location in each room. Set the room thermostat to a reasonable testing temperature. Let the system operate for an hour or more until it balances out. Read the thermometer temperatures in each room.

7. Make any airflow adjustments necessary to produce even temperatures. Pay particular attention to outlets farthest from the furnace that may have a higher supply air temperature drop.

8. Reset the thermostat to the proper room temperature. Instruct the owner on operating the system. Record the information used in balancing the system, including the weather conditions on the day the job was balanced.

Balancing a Heating/Cooling Split System

Balancing a split system consists of the following steps:

1. Collect complete information about the equipment, the controls, the electrical requirements, and the job layout.
2. Examine the entire installation. Look for problems that need to be corrected, such as leaky ductwork, missing parts, and incomplete wiring.
3. Measure the dry bulb temperature drop across the cooling coil with the entering air temperature held as near 80°F as possible. The TD should be about 20°F. TD is the entering air temperature minus the leaving air temperature.
4. Determine the quantity of air used for cooling, based on the sensible cooling (H_S) requirement and the TD across the coil, in accordance with the following formula:

$$\text{CFM} = H_S/(\text{TD} \times 1.08)$$

5. From the layout, record the CFM requirements for each outlet. If the total air required in the layout differs from the actual CFM determined in step 4 above, either adjust the unit fan or prorate the requirements, so that the two quantities match. Use the resulting CFM for balancing the job.
6. Adjust the airflow through each outlet to comply with the requirements, using either a flow hood measuring device or an anemometer to measure the face velocity in ft/min (fpm). The CFM is calculated by the following formula:

$$\text{CFM} = \text{free area of grill (ft}^2) \times \text{fpm}$$

7. Place thermometers in a suitable location in each room. Set the room thermostat to a reasonable testing temperature. Let the system operate for an hour or more until it balances out. Read the thermometer temperatures in each room.
8. Make any airflow adjustments necessary to produce even temperatures. Pay particular attention to outlets farthest from the unit that may have a higher supply air temperature rise.
9. Reset the thermostat to the proper room temperature. Instruct the owner on operating the system. Record the information used in balancing the system, including the weather conditions on the day the job was balanced.

Balancing a Zoned Air System Installation

The same procedure used for furnaces and split systems is followed for balancing a zoned air system, except:

1. By adjusting zone dampers, the total CFM of the fan must be proportioned for the requirements of each zone.

2. The CFM for each outlet on the zone is adjusted to meet individual requirements when the zone CFM has been properly set.

Balancing a Hydronic Heating System

This section applies to all types of piping layouts except the series loop. On the series loop system the only adjustment that can be made is in the total flow using a balancing valve in the main return before it enters the boiler. For all other systems the following procedure is recommended:

1. Collect complete information about the installation, including the specifications for the boiler, the terminal units, the controls, the electrical system, and the job layout, including the water quantities (gpm) required for each room.

2. Examine the entire installation. Look for problems that need to be corrected, such as air traps, missing parts or incomplete wiring.

3. Examine the piping. Be certain that the system has been installed with the necessary devices for balancing. A minimum should include pressure taps on each side of the pump, a pump flow volume balancing valve, and flow setters in the branch lines connecting each terminal unit.

4. Be sure all valves in the system are open. Start the system and measure the pressure drop across the pump. Refer to the manufacturer's pump data and determine the flow (gpm) under operating conditions.

5. Adjust the flow through each terminal unit to meet the requirements of the job. Recheck the pressure drop across the pump to verify the total flow requirements.

6. If the weather is suitable, check the temperatures in each room, with the thermostat set at design temperature. If a damper is provided, adjust the airflow in the terminal unit, or reset the branch water flow, to provide even room temperatures. Final adjustment may need to be made later during colder weather.

7. Instruct the owner in operating the system. Record the information used in balancing, including the weather conditions.

UNIT 3—REVIEW QUESTIONS

1. What is an important function of the technician after the system is installed?
2. What is a TAB procedure?
3. What are the four forms that are needed, as a minimum, for air conditioning jobs?
4. List the common test instruments used for TAB work.
5. What does a rotating vane anemometer consist of?
6. What is a tachometer used to measure?
7. What are the two general types of dial thermometers?
8. When is it recommended that the first balance be made on a warm air heating system?
9. When balancing a heating/cooling split system, what formula is used to determine the quantity of air used for cooling?
10. Record the information used in balancing the system, including ________.

TECH TALK

Tech 1. I don't understand why when I'm balancing the system I can't just turn the airflow down in this one room and get it to come up in the room down the hall.

Tech 2. When you shut the airflow down in one vent, that air is going to out the other vents, which means that all of the vent airflows will go up slightly.

Tech 1. Then how am I going to get more air to come out of the vent where I need it?

Tech 2. If you open up that vent then its air will come from the others, so their airflow will slightly go down. When you turn one damper down the other airflows will come up and as you open one damper to get more flow the air in the others will come down. You only have so many CFM of air to work with, and it will be divided among all of the vents.

Tech 1. It sounds like it could take a lot of time chasing the air all over the place.

Tech 2. The first time you do it will take a lot more time. After you have done it a few times you'll find that it is much easier and you can get it done fairly quickly.

Tech 1. Thanks, I'll give that a try on the next job.

15

Indoor Air Quality

UNIT 1

Fundamentals of Air Conditioning

OBJECTIVES: In this unit the reader will learn about the fundamentals of air conditioning including:

- Conditions for Comfort
- Psychrometrics
- Dew Point Temperature
- Calculating Performance

CONDITIONS FOR COMFORT

Comfort is the feeling of physical contentment with the environment. The air conditioning process is used to produce human comfort. Air conditioning provides control of temperature, relative humidity, air motion, radiant heat, removal of airborne particles, and contaminating gases.

The study of human comfort relates to:

1. How the body functions with respect to heat and
2. How the area around a person affects the feeling of comfort.

The body has a remarkable ability to adjust to temperature change. When a person goes from a warm house into the cold outdoors, some compensation needs to be made to prevent excessive heat loss. Involuntary shivering can occur, providing body heat. When a person goes from an air conditioned space into an outside temperature of 95°F, an adjustment in the circulatory and respiratory system takes place. The blood vessels dilate to bring the blood closer to the surface of the skin to provide better cooling. If this is not enough, sweating occurs, evaporating moisture, producing a body cooling.

The body behaves like a heating system. Fuel is consumed in the form of food, producing energy and heat. The body temperature is closely controlled at a temperature of 98.6°F, winter and summer. The proper functioning of the body is dependent on constantly maintaining this temperature. To be able to do this requires constantly losing heat to the surrounding area. Figure 15-1-1 illustrates a comfortable condition. With room conditions of 75°F and the proper humidity, the body loses heat at about the correct rate for a feeling of comfort.

The surrounding conditions must be cooler than the body temperature in order to dissipate heat.

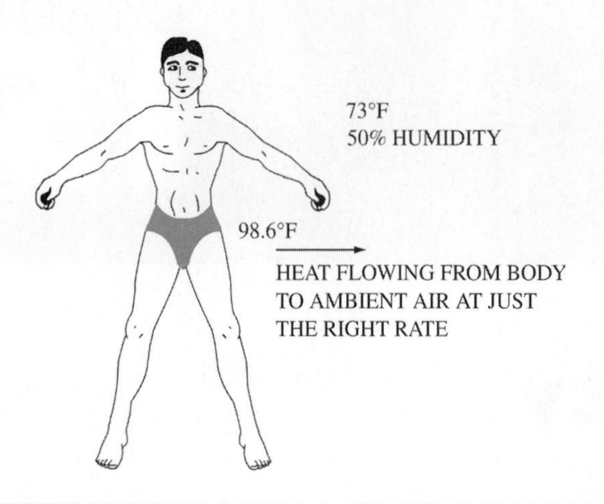

Figure 15-1-1 With ambient conditions of 73° and 50% RH, the body is able to give off the right amount of heat for comfort.

TECH TIP

As people age, their circulation tends to decrease. For that reason elderly residents will often complain about a chill even when the home's system is set well above the norm. One way to make the area more comfortable for seniors or anyone else with these complaints is to make sure that the room supply grills are directed so that there is the least amount of air movement in the occupied space. Air movement can cause a slight chill for these individuals even though the room is more than comfortably warm for the other occupants. It may be necessary to change out the supply grill faces with longer curved grill faces that are specifically designed for heat pump applications.

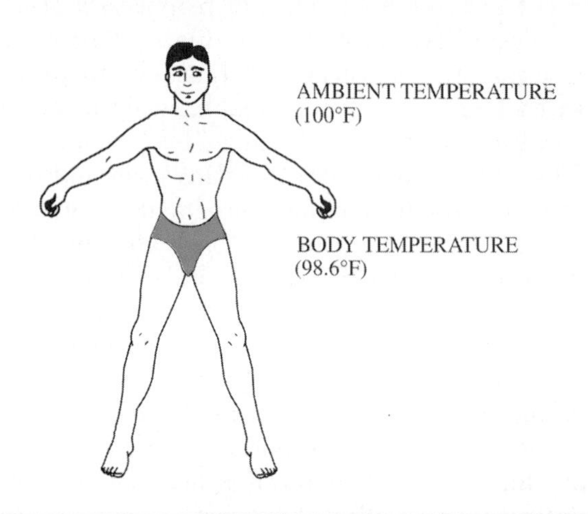

Figure 15-1-2 The body must dissipate heat or it will overheat in high ambient temperatures.

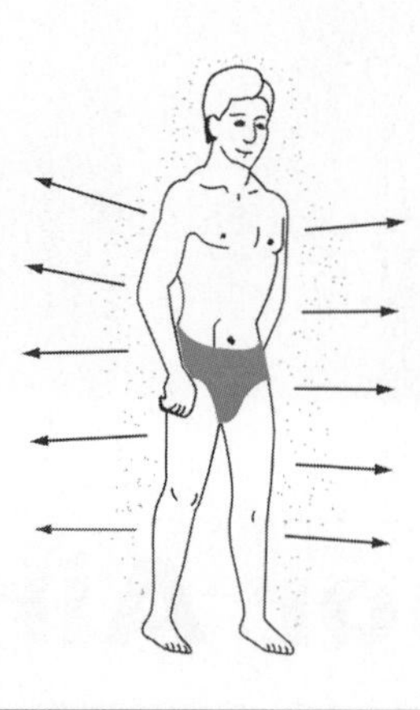

Figure 15-1-3 Heat is lost from the body by conduction to a film of air in contact with exposed skin.

Figure 15-1-2 shows a condition where the surroundings are warmer than body temperature. This condition cannot be tolerated for any length of time without some adjustment being made. There are four ways the body can lose heat to the environment:

1. Conduction, Figure 15-1-3
2. Convection of cool air, Figure 15-1-4
3. Radiation, Figure 15-1-5, and
4. Evaporation, Figure 15-1-6.

Many tests have been run to determine comfort conditions in winter as well as summer, due partly to the use of warmer clothing. The surrounding temperature, relative humidity, and air movement are influencing factors. The results of test data have been incorporated in comfort charts, one for summer conditions, Figure 15-1-7a, and one for winter conditions, Figure 15-1-7b.

The enclosed areas on both of these charts indicate the range of conditions at which people are comfortable.

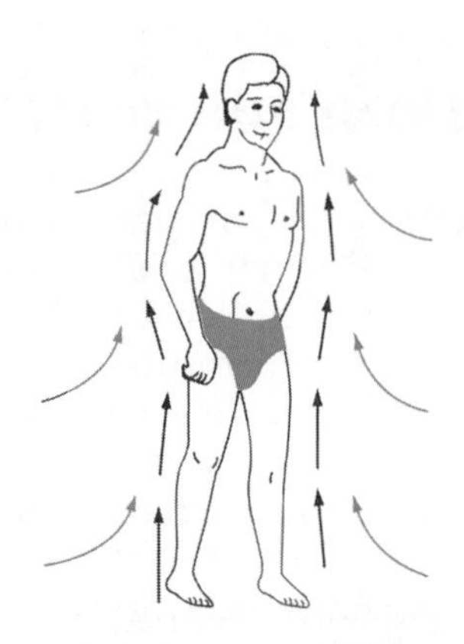

Figure 15-1-4 Heat is lost from the body by convection of the air around the body.

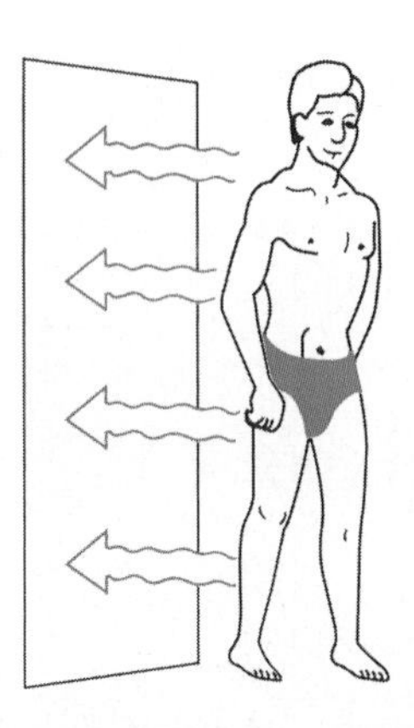

Figure 15-1-5 The body radiates heat to the surrounding surfaces/objects.

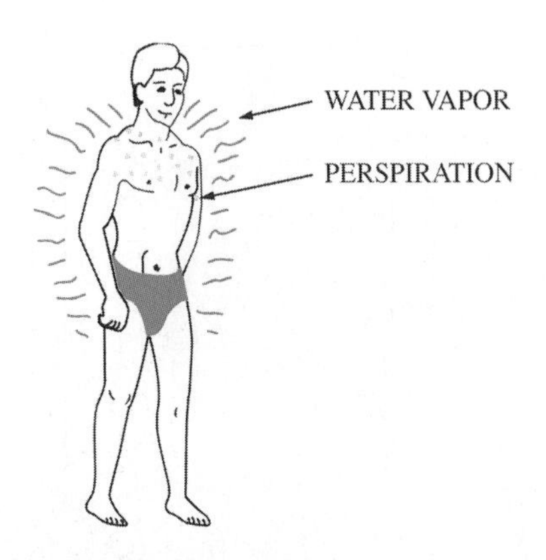

Figure 15-1-6 The body loses heat by evaporation of water from the skin.

PSYCHROMETRICS

The study of air and its properties is called psychrometrics. The psychrometric chart is used to show the relationship of these properties to each other.

Air has weight and occupies space. At standard conditions of 70°F and sea level atmospheric pressure (29.92 in Hg), dry air weighs 0.075 lb/ft^3 (density). The reciprocal of this figure is useful. One pound of air at standard conditions occupies 13.33 ft^3 (specific volume). These figures can be used effectively in air conditioning calculations.

NICE TO KNOW

Under standard conditions one pound of air occupies approximately 13.33 ft^3. A pound of air that size will fit in a box of approximately 2 ft 4 in. You can estimate the weight of air in your classroom by multiplying the length, width, and height of the classroom to determine the classroom's volume and then divide that total volume by 13.33. For example, a classroom that is 30 ft wide, 40 ft long, with a 10 ft ceiling would have 12,000 ft^3 of air; and that air would weigh approximately 900 pounds.

EXAMPLE

An air handling unit is capable of delivering air at the rate of 1000 ft^3/min (CFM). If a heating coil can raise the temperature of the air 10°F, how much heat is added?

Solution

$$Q_s = 1.08 \times \text{CFM} \times \text{TD}$$

where

Q_s = Sensible heat added in Btu/hr
= 1.08 conversion factor
CFM = 1000
TD = 10°F

Then

$Q_s = 1.08 \times 1{,}000 \times 10$
$= 10{,}800$ Btu/hr ■

All air contains some moisture (humidity). The amount is directly related to the temperature: the higher the temperature, the greater the amount of water will evaporate. The amount of moisture in the air can be measured in terms of relative humidity or percentage of saturation at a given temperature.

The process of vaporization takes two forms: evaporation, in which only part of the surface liquid turns into gas, and boiling, in which all parts of the liquid turn into gas. Boiling, which takes place at higher temperatures, can be considered an extreme case of evaporation.

The vapor formed above a liquid exerts pressure in direct proportion to the temperature. At standard conditions (70°F) 43.5°F WB the evaporating moisture from a pan of water would exert a vapor pressure of 0.28 in Hg. Evaporation will only exist if the vapor pressure in the air above the liquid water has a lower vapor pressure. For example, if the air above the pan is 72°F and at 40% relative humidity, the moisture in the air would have a vapor pressure of 0.45 in Hg. Under these conditions, evaporation from the pan would not take place.

When the temperature of the liquid is raised until its vapor pressure is higher than the prevailing atmospheric pressure, the vapor pressure is able to overcome the atmospheric pressure and the liquid changes into gas. This is known as the boiling point of the liquid. Water boils at 100°C (212°F), expanding more than 1500 times as it turns into saturated water vapor, or steam.

Partial pressures exerted by water vapor can be read from "Thermodynamic Properties of Water at Saturation," ASHRAE Handbook on Fundamentals.

In order to use the psychrometric chart, we need to review some of the fundamental properties of air.

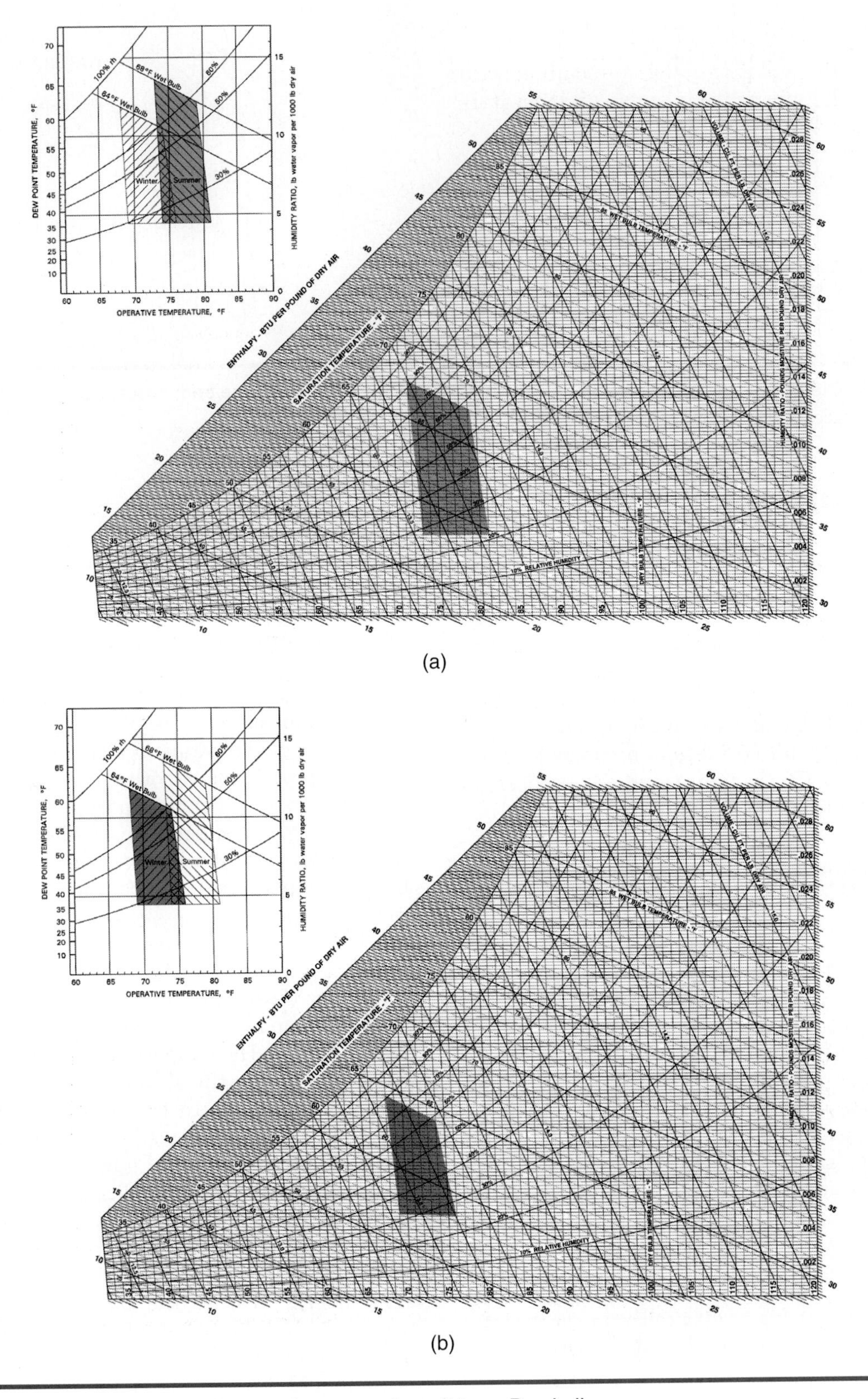

Figure 15-1-7 (a) Comfort chart showing the range of conditions. Dry bulb, wet bulb, and relative humidity ranges for comfortable summer conditions are shown in the rectangle. (b) Comfort chart for winter, showing range of dry bulb, wet bulb, and relative humidity for the most comfortable conditions.

Moisture in Air

Air can hold only a relatively small amount of water vapor or moisture at any given temperature. Moisture in air is measured in grains (gr) per pound of dry air. Grains is a unit of weight, and it takes 7000 grains to make one pound of moisture. Another way of saying this is that one pound equals 7000 grains. For example, a sample of air at 75°F and 50% relative humidity, at atmospheric pressure, holds 65 grains (0.00929 lb) of moisture per pound of air.

NICE TO KNOW

A popular misconception is that as air picks up moisture, it becomes denser. A water molecule is not as dense a molecule as the other molecules that make up air. It is much larger in size than the other molecules. As the water molecule is absorbed into the air, it is much like putting Styrofoam beads in with sand. Because the Styrofoam is much lighter, a box of Styrofoam and sand would be, therefore, lighter as the amount of Styrofoam mixed with the sand increases. One pound of moist air, therefore, occupies a greater space than one pound of dry air.

The percent of relative humidity is a convenient means of representing the amount of water vapor in the air. It is the ratio of the actual density of water vapor to the saturated density, at the same dry bulb temperature and barometric pressure.

Dry Bulb and Wet Bulb Temperatures

The dry bulb temperature of air is the temperature measured on an ordinary thermometer. The wet bulb temperature is measured by placing a wick soaked with water around the thermometer bulb and moving it rapidly through the air. Water evaporates from the wick, cooling the bulb and lowering the temperature.

The relation of the wet bulb temperature to the dry bulb temperature is a measure of the relative humidity. Instruments used to measure these conditions are called sling psychrometers, as shown in Figure 15-1-8a. There are a number of electronic psychrometers available, such as the one shown in Figure 15-1-8b.

The relative humidity value can be obtained by plotting the readings on a psychrometric chart. Tables and slide rules are available for the same purpose.

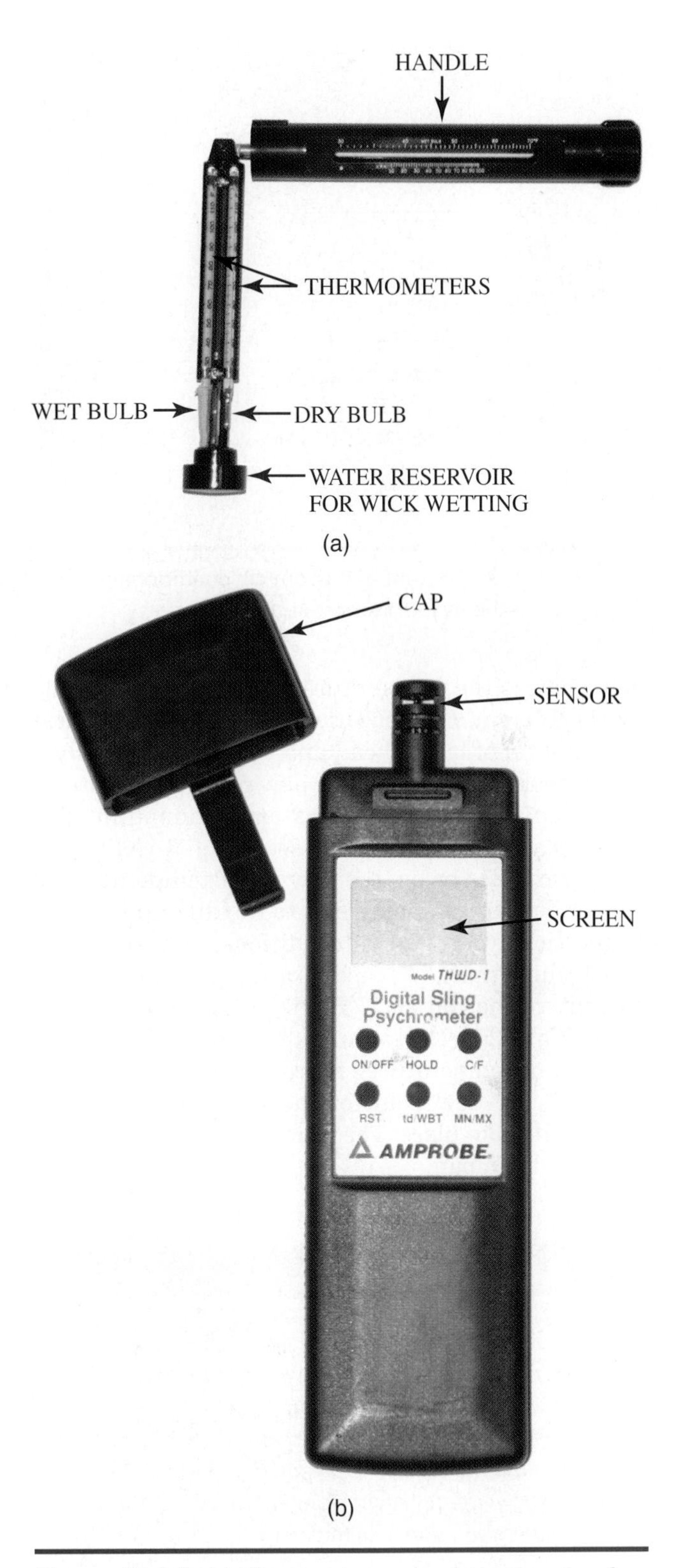

Figure 15-1-8 (a) Sling psychrometer for measuring dry bulb and wet bulb temperatures to determine relative humidity; (b) Electronic psychrometer.

Dew Point Temperature

The dew point temperature of a vapor is the temperature at which the vapor reaches the point of 100% humidity (saturation). It is not correct to state that the

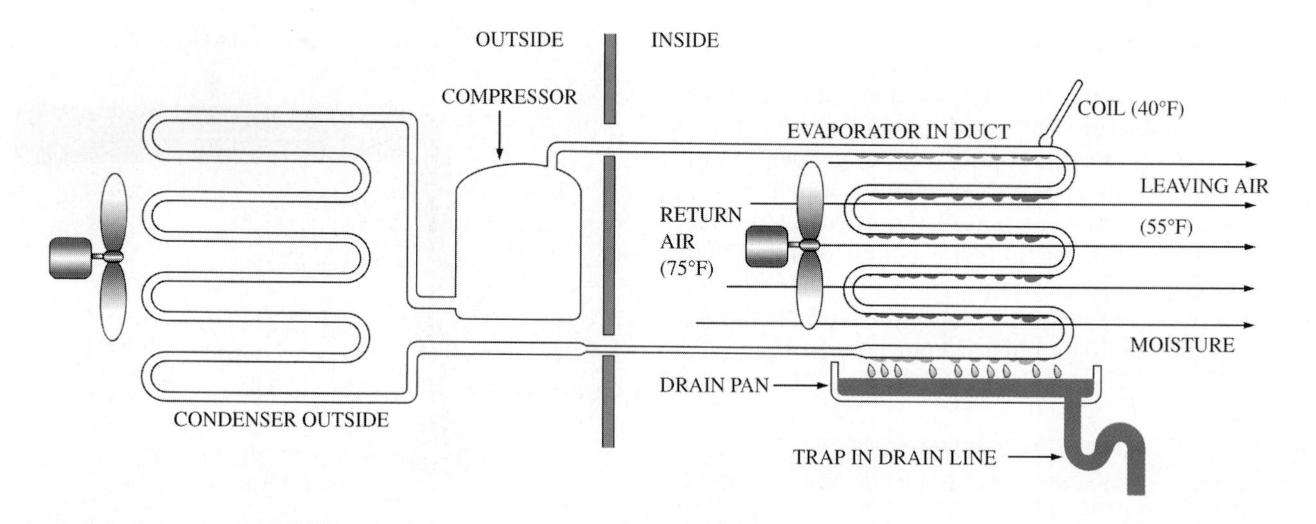

Figure 15-1-9 Diagram of a room air conditioning unit showing condensation of moisture on the evaporator, from the air.

dew point is the temperature at which condensation starts to occur. Condensation at dew point requires the removal of latent heat, and this can occur only if the vapor is cooled below the dew point temperature.

In an air conditioning system, dehumidification takes place when the air passes over a coil whose temperature is below the dew point temperature of the air. For an example, refer to the diagram of a refrigeration cycle of an air conditioner, Figure 15-1-9.

Assume that the evaporator temperature is 40°F. Air enters the coil at 75°F DBT (dry bulb temperature), 50% RH (relative humidity), and at 55°F DPT (dew point temperature). Since the coil temperature is below the dew point of the air, condensation of water will take place. The condensed water will drip into the drain pan.

TECH TIP

As the air velocity through an air conditioning coil increases, the amount of dehumidification decreases. Conversely, as the velocity of the air slows through a coil, greater dehumidification will occur. You can use slight changes in fan speed to help control an area's humidity. This can be very important along the coast or in areas of particularly high relative humidities.

PSYCHROMETRIC CHART

The psychrometric chart is useful in plotting the performance of air conditioning equipment. It provides a visual picture of the changes that take place in the properties of air passing through an air conditioning coil. It looks complicated because it incorporates so much information.

The simplest method of starting is to find the intersection of any two of the four principal air properties: dry bulb temperature, wet bulb temperature, dew point temperature, or relative humidity. From the intersection point, the value of the other two properties (and more) can be determined.

For example, assume that a sample of room air has a dry bulb temperature of 80°F and wet bulb temperature of 63.4°F. This represents a point on the psychrometric chart, Figure 15-1-10.

The following readings can then be taken directly from the chart:

1. Dry bulb temperature	80°F
2. Wet bulb temperature	64.4°F
3. Dew point temperature	53.5°F
4. Relative humidity	40%

and

5. Total heat	28.87 Btu/lb
6. Moisture	62 gr/lb
7. Specific volume	13.8 ft^3/lb

Figure 15-1-11 through 15-1-17 shows the direction of the lines on the chart where each of the above readings is taken.

For example, the vertical lines indicate dry bulb temperature values, Figure 15-1-11, and the horizontal lines indicate moisture content values.

Figures 15-1-18 through 15-1-23 show plots on the chart representing various air conditioning processes.

The changes that take place in the properties of the air during these processes are described as follows:

1. Heating Figure 15-1-18, adding sensible heat. The process is shown moving along a horizontal line to the right. The wet bulb temperature and total

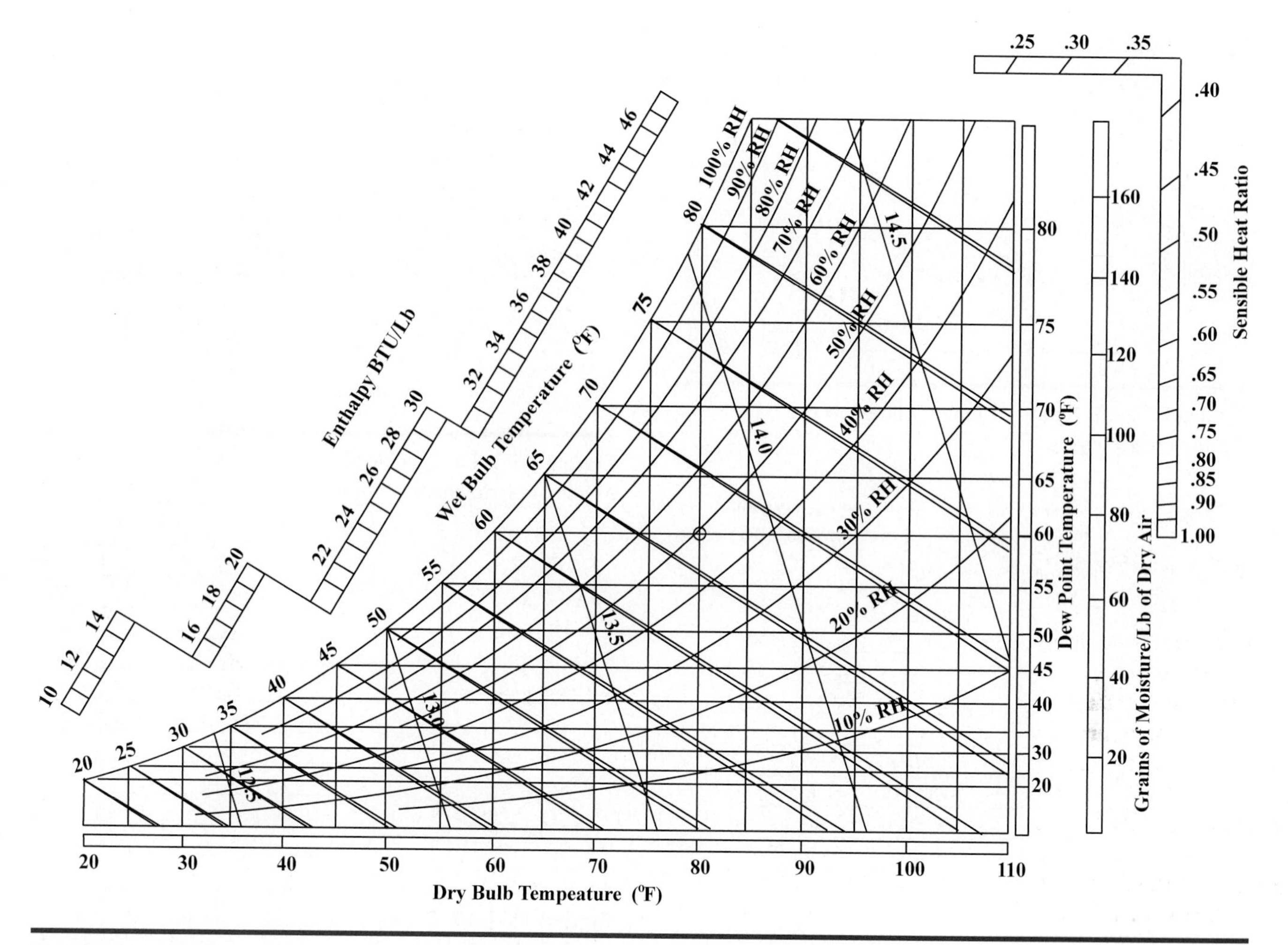

Figure 15-1-10 Psychrometric chart with 80°F dry bulb and 63.4°F wet bulb plotted, to identify other air properties.

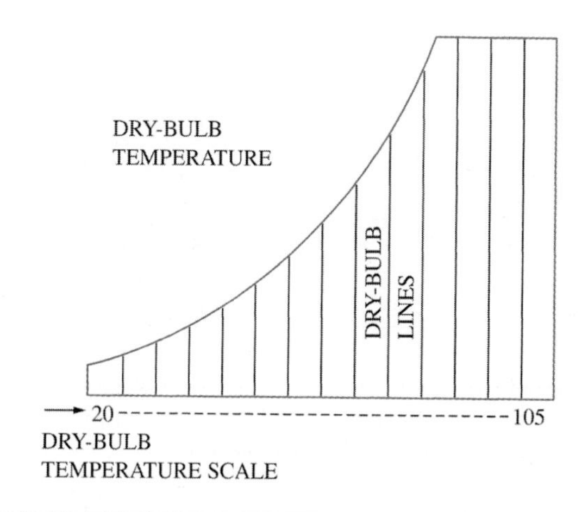

Figure 15-1-11 Dry bulb temperature lines on psychrometric chart.

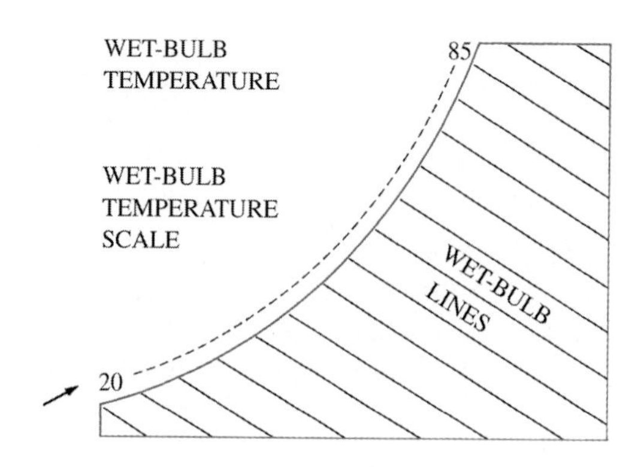

Figure 15-1-12 Wet bulb temperature lines on psychrometric chart.

heat content increase and the relative humidity decreases. The dew point temperature and the moisture content do not change.

2. Cooling with a dry evaporator Figure 15-1-19, reducing sensible heat. The process is also shown moving along a horizontal line, but in the opposite direction. The wet bulb temperature and total heat content decrease and the relative humidity increases. Again, there is no change in dew point or moisture content.

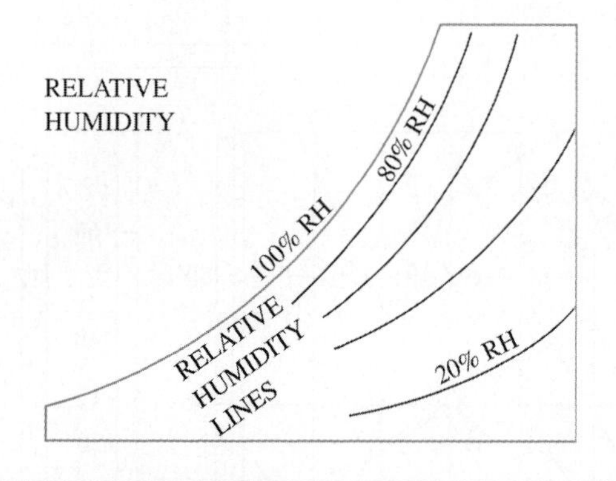

Figure 15-1-13 Relative humidity lines on the psychrometric chart.

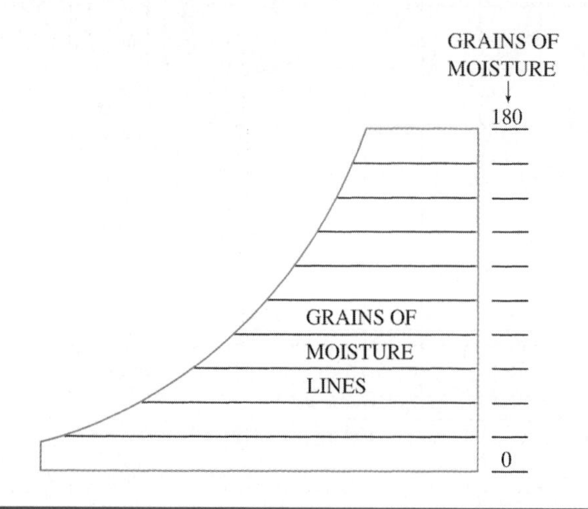

Figure 15-1-14 Grains of moisture line on the psychrometric chart.

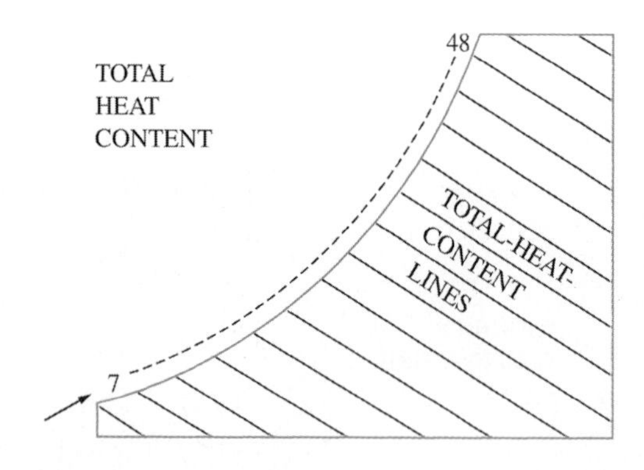

Figure 15-1-15 Total heat lines on the psychrometric chart.

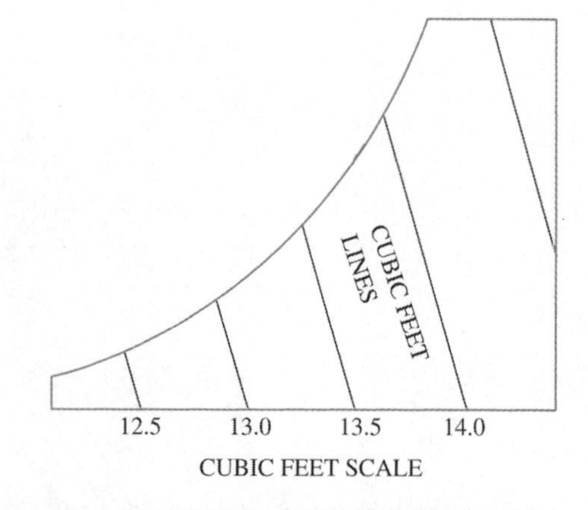

Figure 15-1-16 Specific volume lines on the psychrometric chart.

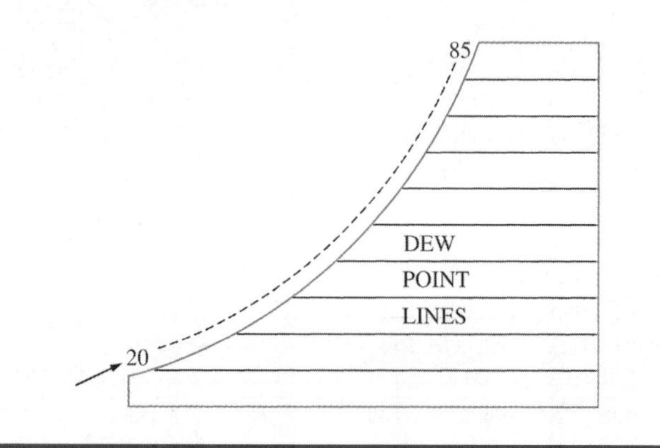

Figure 15-1-17 Dew point temperature lines on the psychrometric chart.

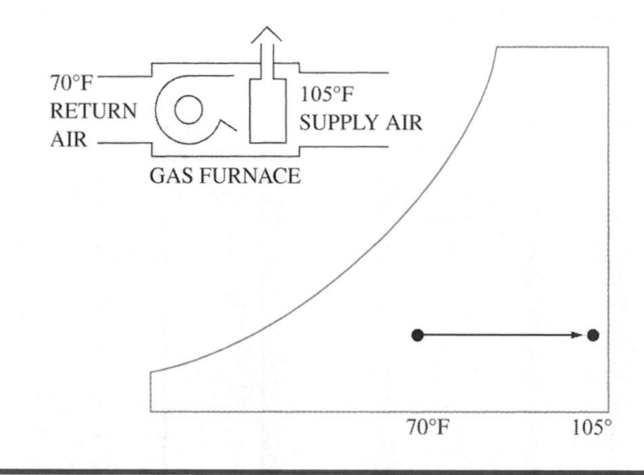

Figure 15-1-18 Plotting the addition of sensible heat by a furnace, on the psychrometric chart.

3. Humidifying Figure 15-1-20, adding moisture. This action is represented by a vertical line moving upward with moisture increase. The only property that does not change is the dry bulb temperature, which is held constant. Heat is added as evidenced by the increase in total heat, but the heat is used to evaporate the moisture. The wet bulb temperature, dew point, and relative humidity increase.

4. Heating and humidifying Figure 15-1-21, used for winter operation. The line representing the operation moves upward to the right. The air picks up both heat and moisture as it moves through the conditioner. The total heat and wet bulb and dew point temperatures all increase. The relative humidity could increase, remain the same, or decrease, depending on how much moisture is added. In actual

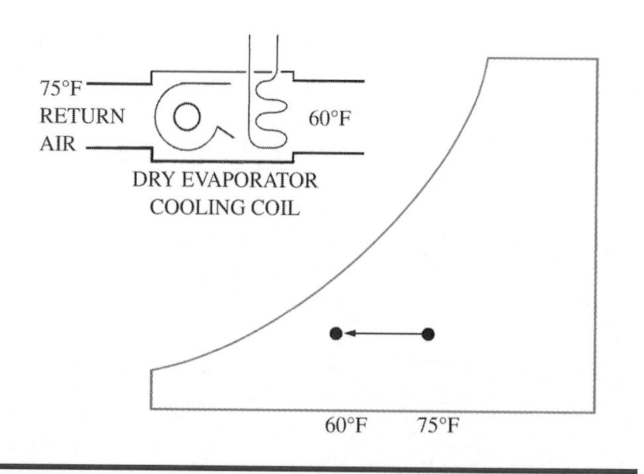

Figure 15-1-19 Plotting the temperature of air being cooled without dehumidification on the psychrometric chart.

operation the amount of heat and moisture added must represent the losses of the structure in order to maintain the entering air conditions (return air or room temperature).

5. Cooling and dehumidification Figure 15-1-22, used for summer operation. The line representing the operation moves downward to the left. The cooling coil both cools the air and condenses moisture as the air passes through it. The total heat and wet bulb and dew point temperatures all decrease. The relative humidity increases, since the air is nearly saturated when it leaves the coil. Again, the reduction of temperature and moisture in the process must be related to the load requirements.

6. Cooling and humidifying Figure 15-1-23. The process is called evaporative cooling. It uses evaporation of moisture to cool the air. The process follows the wet bulb line in the chart. The dry bulb temperature decreases as the moisture in the air and the relative humidity increase. Total heat is constant.

CALCULATING PERFORMANCE

The psychrometric chart can be useful in checking the performance of air conditioning equipment, as shown in the following example.

EXAMPLE

Assume that a 3 ton (1 ton = 12,000 Btu/hr) air conditioning system is installed in a residence. To check to see if the unit is producing its rated capacity, the following information is needed from the installation:

1. Dry bulb and wet bulb temperature, supply air
2. Dry bulb and wet bulb temperatures, inside air (A)
3. Dry bulb and wet bulb temperatures, outside air (B)
4. Air circulated by the blower, in CFM
5. Percent of ventilation air (or CFM)

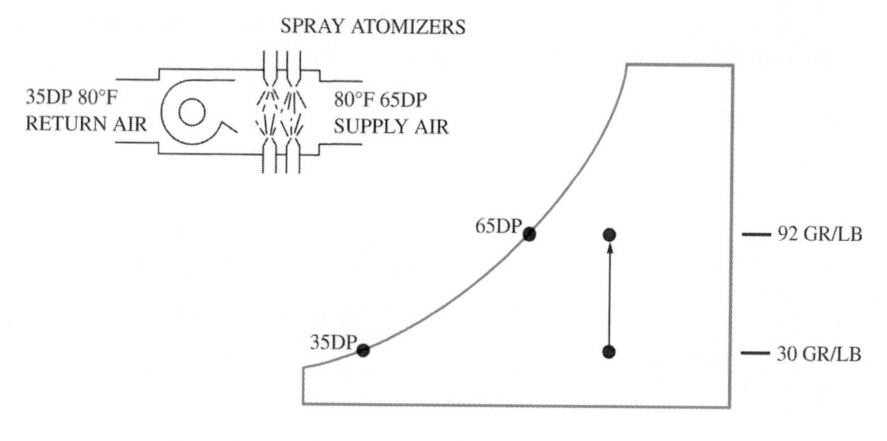

Figure 15-1-20 Plotting the addition of moisture to the air produced by a humidifier.

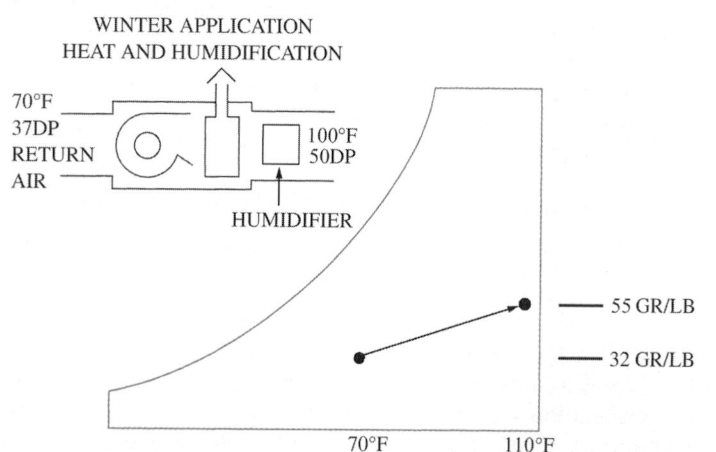

Figure 15-1-21 Heat and moisture added by a hot air furnace with a humidifier.

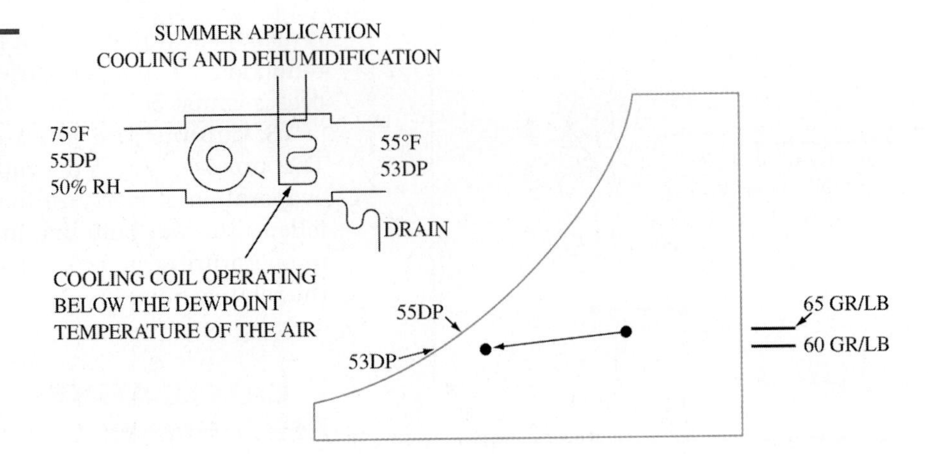

Figure 15-1-22 Cooling and dehumidification produced by an air conditioner, plotted on the psychrometric chart.

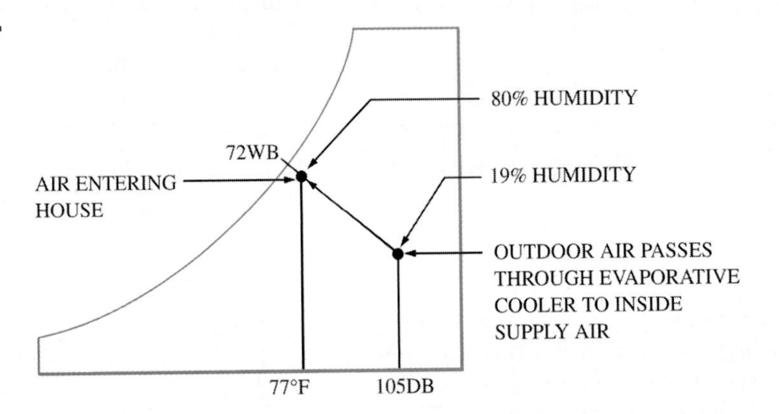

Figure 15-1-23 Cooling and humidification produced by an evaporative cooler plotted on the psychrometric chart.

For the example we will refer to Figure 15-1-24 and assume that the following readings were taken at the jobsite:

1. Inside temperatures: 75°F DBT, 62.5°F WBT
2. Outside temperatures: 95°F DBT, 67°F WBT
3. Total air circulated: 1100 CFM
4. Ventilation air: 25% (or 247 CFM)

To plot these conditions on the chart of Figure 15-1-24 and obtain further data, proceed as follows:

1. Locate the inside conditions. Mark this point A.
2. Locate the outside conditions. Mark this point B.
3. Connect points A and B with a straight line.
4. Determine the mixed air temperature. The dry bulb temperature of the mixture is found by taking 25% of the difference between inside and outside DBT (25% of 20° = 5°), adding this to the inside DBT (75 + 5 = 80°F), then locating this temperature on the line AB. Mark this point C. This is the condition of the air entering the unit.
5. For this example, we will assume that both cooling and dehumidification take place and that the condition of the air leaving the unit is measured to be 53°F DBT and 51°F WBT. Plot this on the chart and mark it point D.
6. Connect points C and D with a straight line. This represents the changes in the properties of air that take place when the air passes through the conditioning unit. The total amount of cooling and dehumidifying that takes place must conform to the load requirements.

A summary of information that can now be read from the chart is as follows:

Entering air conditions, point C
80°F DBT, 63.6°F WBT
Heat content of the air = 29.01 Btu/lb
Moisture content of the air = 62 gr/lb

Leaving air conditions, point D
53°F DBT, 51°F WBT
Heat content of the air = 20.86 Btu/lb
Moisture content of the air = 52 gr/lb ■

The following formulas are used in making the performance calculations:

Sensible heat removed

$$Q_s = 1.08 \times \text{CFM} \times \text{DBT difference} \qquad (1)$$

Latent heat removed

$$Q_1 = 0.68 \times \text{CFM} \times \text{gr moisture difference} \qquad (2)$$

Total heat removed

$$Q_t = Q_s + Q, \qquad (3a)$$

or

$$Q_t = 4.5 \times \text{CFM} \times \text{total heat difference} \qquad (3b)$$

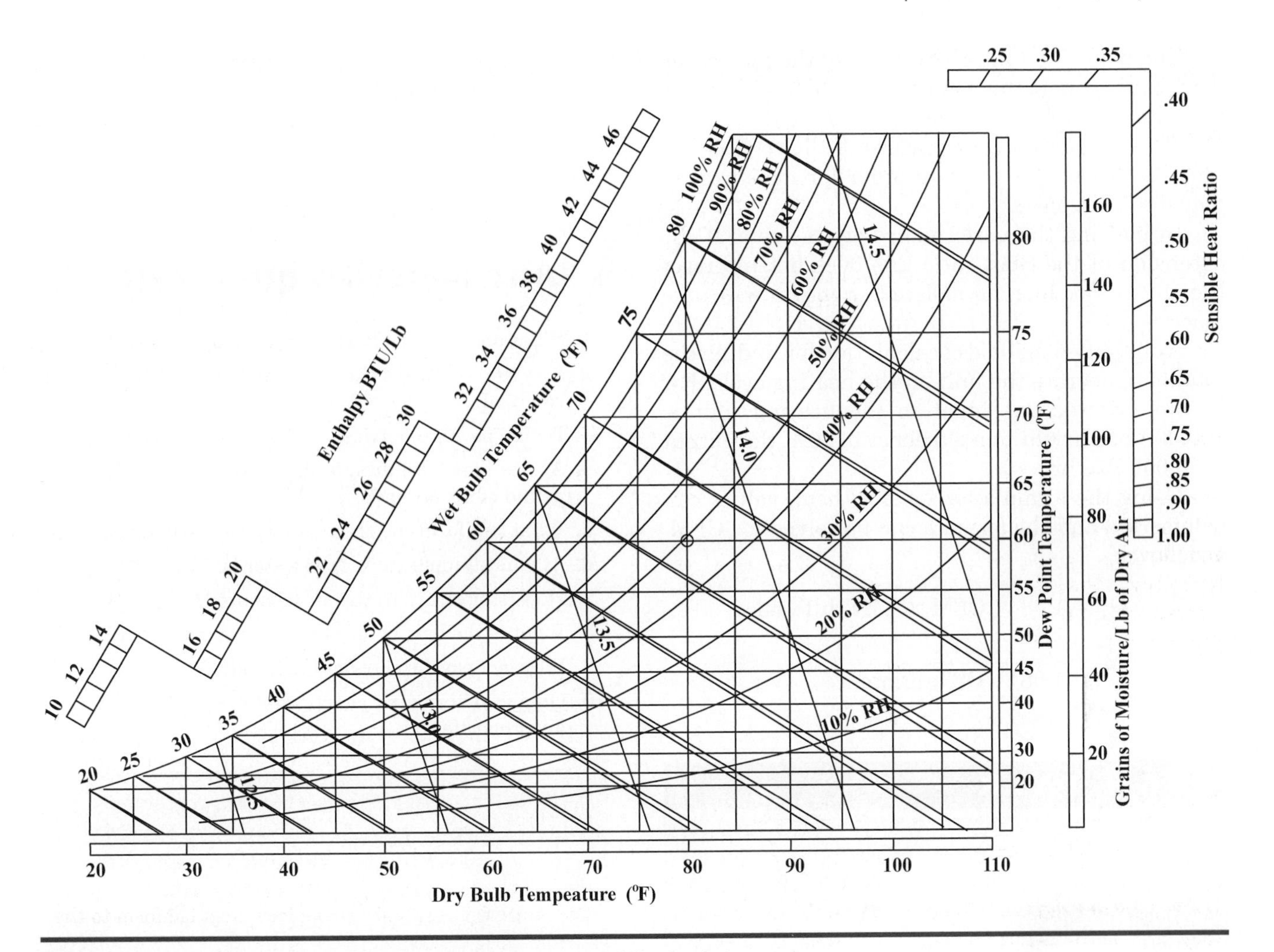

Figure 15-1-24 Plot of an air conditioning system using 25% outside air with temperature reduction and dehumidification.

If the chart is marked with a fine pen, the calculation of the total heat removed using formula (3a) should be nearly the same as using formula (3b). There will be some difference because there is a limit to how close the values can be read on the chart. Also formula (3b) is based on standard air.

EXAMPLE

Below are the calculations based on using the data indicated above.

Sensible heat removed

$$Q_s = 1.08 \times \text{CFM} \times \text{DBT difference}$$

$$Q_s = 1.08 \times 1100 \times 27 = 32{,}076 \text{ Btu/hr}$$

Latent heat removed

$$Q_l = 0.68 \times \text{CFM} \times \text{gr difference}$$

$$Q_l = 0.68 \times 1100 \times 10 = 7{,}480 \text{ Btu/hr}$$

Total heat removed

$$Q_t = 4.5 \times 1100 \times 8.15 = 40{,}343 \text{ Btu/hr}$$

The slope of the line CD represents the sensible heat ratio (SHR) of the air in passing through the cooling coil, Figure 15-1-24. In this example both the DBT and the moisture content are lowered. By definition:

$$\text{Sensible heat ratio} = \frac{\text{Sensible heat removed in Btu/lb of air}}{\text{Total heat removed in Btu/lb of air}}$$

In this example,

$$\text{SHR ratio} = \frac{32{,}076}{40{,}343} = 0.80$$ ■

The sensible heat ratio performed by the operating equipment must conform to the load requirements. Otherwise design conditions will not be maintained.

The point E on the chart is called the apparatus dew point. This is the point on the 100% saturation line intersected by an extension of the SHR ratio line CD. The evaporator temperature of the cooling coil must be below this point in order for moisture removal to take place.

In plotting the performance of a system, if the extension of the SHR ratio line does not intersect the saturation line, no moisture removal will take place.

A simplified method of calculation for finding the total is to operate the unit on the heating cycle. For example, if the gas input to the heating unit is 100,000 Btu/hr, with an efficiency of 80%, the output is 80,000 Btu/hr.

Using the formula for sensible heat, values can be inserted and the formula can be solved for CFM, as follows:

$$Q_s = 1.08 \times \text{CFM} \times \text{DBT difference}$$

$$\text{CFM} = \frac{Q_s}{1.08 \times \text{DBT difference}}$$

Assuming that the entering DBT = 70°F and the leaving DBT = 110°F

$$\text{CFM} = \frac{80{,}000}{1.08 \times 40}$$

$$\text{Total air} = 1852 \text{ CFM}$$

UNIT 1—REVIEW QUESTIONS

1. What does the study of human comfort relate to?
2. List the four ways the body can lose heat to the environment.
3. Define psychrometrics.
4. At standard conditions, how much space does one pound of air occupy?
5. What two forms does the process of vaporization take?
6. At what temperature does water boil?
7. How is moisture in the air measured?
8. What is the dew point temperature of a vapor?
9. Why is a psychrometric chart useful?
10. On a psychrometric chart, what are the four principal air properties?

TECH TALK

Tech 1. I don't understand why the boss is having me take all these wet bulb and dry bulb readings and airflow measurements on the system we are working on.

Tech 2. It's so that we can calculate the efficiency of the system.

Tech 1. You mean you can calculate how efficient the system is working?

Tech 2. That's right.

Tech 1. What difference does it make? We've done everything we need to do when we set it up.

Tech 2. That's true. But, you know how you calculate the gas mileage on your car to see how well the mechanic did the tuneup?

Tech 1. Sure. I take my gas mileage when I go on trips.

Tech 2. Well, in air conditioning that's not something we can do. We can't just take the gas mileage. But what we can do is calculate with the psychrometric chart how efficient the system is working.

Tech 1. So this is like getting the gas mileage for an air conditioning system?

Tech 2. That's right. When you calculate how efficient the system is, it will give you an idea of the type of job that was done. It's also a good way for determining when a system is starting to fail. As the compressor gets weaker over time, it doesn't pump as well and its efficiency will drop. As the evaporator condenser coil gets dirty or even the filter drier gets plugged up, it will change the operating efficiency. We can use our test results as a diagnostic tool much like a doctor taking your temperature and blood pressure when you come in.

Tech 1. It's an awful lot of work for us to do that.

Tech 2. That's true. But when you have a system that's causing problems, and you can't figure out what it is, it's a good way to see whether or not the problem is affecting the overall performance. It's also a great way to show a customer that it's time to change a unit out. If you can demonstrate to the customer that their system is rated by the manufacturer as a 12 EER, but today because of the condition of the system, it's only functioning at a 6 EER, they are more than likely to let you change it.

Tech 1. Won't they see that their utility bills are going up?

Tech 2. Often people do see that their utility bills are going up, but they may also simply say, "You know it was awfully hot last month," or "You know we had a lot of people over," or make an excuse for having a high utility bill. But in reality, it's the air conditioner.

Tech 1. Sounds like it's a pretty good deal for everybody if I go out and get those measurements so the boss can calculate that. Thanks. I appreciate that.

UNIT 2

Indoor Air Quality

OBJECTIVES: In this unit the reader will learn about indoor air quality including:

- Pollutants
- Particles
- Asbestos
- Vapors and Gases
- Energy Recovery Ventilator
- Services of the HVAC Technician

Air is composed of 21% oxygen, 78% nitrogen and 1% carbon dioxide, argon, and trace amounts of other rare gases. In addition, it contains varying amounts of water vapor and small quantities of microscopic solid matter called *permanent atmospheric particulates.*

The availability of oxygen is essential to human health. When we breathe, we inhale oxygen and exhale carbon dioxide. At atmospheric pressure, oxygen concentrations of less than 18% and carbon dioxide concentrations of greater than 5% are dangerous—even for short periods.

The permanent atmospheric particulates arise from natural processes such as wind erosion, sea spray evaporation, volcanic eruption, and biological processes. They usually create far less contamination than manmade activities. Some of the manmade activities that cause air contamination are power plant operation, industrial processing, and various types of transportation and agricultural activities. These suspended contaminants can be roughly classified as solids, liquids, and gases.

The size of the suspended particles is important, since it relates to the method of removal, usually through filtration. Figure 15-2-1 gives the size in microns of various common particulates.

A micron is a millionth of a meter, or one thousandth of a millimeter. To give some conception of the sizes shown, note that human hairs range in size from 25 to 300 microns. One micron is not visible to the naked eye. Smaller particles such as smoke or clouds are visible only in high concentrations. Health authorities are concerned with particles 2 microns or less since they can be retained in the lungs. Note the size range of a few common pollutants: pollen, 10 to

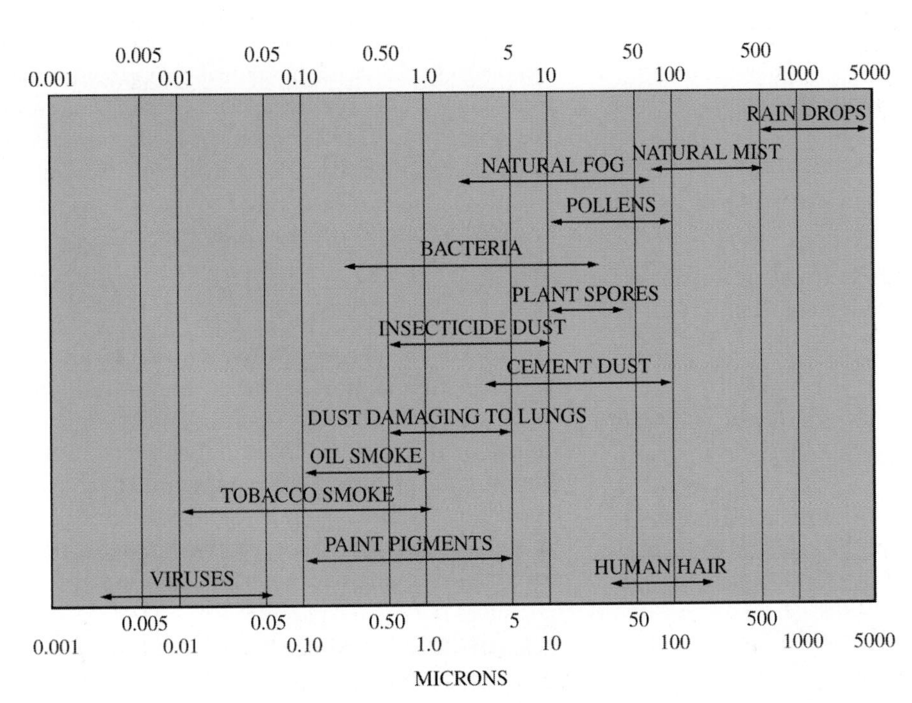

Figure 15-2-1 Sizes and characteristics of airborne solids and liquids.

100 microns; bacteria, 0.4 to 5.0 microns; tobacco smoke is a solid ranging in size from 0.1 to 0.3 microns; and viruses, 0.003 to 0.06 microns.

TECH TIP

Tobacco smoke is a fume. A fume is a solid that has condensed in the air from products of combustion. Because tobacco smoke is a solid it can be filtered and removed from the air efficiently with a filter capable of trapping particles as small as 0.1 microns. It should be noted however that some of the gases formed when cigarettes are smoked cannot be removed by filtration. These gases include carbon monoxide, formaldehydes, nitrogen oxide, sulfur dioxide, and ammonia. Concentrations of these gases from cigarette smoke can only be removed with ventilation.

A distinction needs to be made between a contaminant and a pollutant. A *contaminant* is any unwanted material that gets into the air. A contaminant may or may not be destructive to human health. A *pollutant* is a substance which has entered the air and can cause health problems. The study of air quality is particularly concerned with pollutants. The technician should know their source, the amount the body will tolerate, and how to control them.

Indoor air quality is the condition of air in an enclosed space, with respect to the presence of pollutants. It is desirable to maintain pollutant free air or at least an acceptable pollutant level.

The two general methods of pollutant control are:

1. Elimination at the source.
2. Reduction to an acceptable level.

THE POLLUTANTS

The nature of pollutants and their effect on health can be separated into three groups that have related characteristics:

A. Particles
   1. Bioaerosols
      a. Allergens: pollen, fungi, mold spores, insect parts, and feces
      b. Pathogens: bacteria and viruses (almost always carried in or on other particle matter)
   2. Respirable particles: 10 microns or less in size
   3. Asbestos fibers
B. Vapors and gases
   1. Formaldehyde (HCHO)
   2. Radon: Naturally occurring soil gas produced from the natural radioactive decay of radium and uranium, widely found
   3. Volatile organic compounds (VOCs)
C. Combination particles/vapors
   1. Environmental tobacco smoke
   2. Combustion products

Particles

Bioaerosols *Bioaerosols* are airborne biological agents know as allergens and pathogens. *Allergens* relate to those microorganisms that cause allergies and include pollen and fungi (yeasts and molds). *Pathogens* are microorganisms such as bacteria and viruses that cause diseases.

Biological growth sources of these microorganisms include wet insulation or moisture laden dirty ductwork, ceiling tile, furniture, and stagnant water in air conditioners, humidifiers, cooling towers, or cooling coils. Pets also bring disease bearing microorganisms into the house. Frequently the pollutants settle in the air distribution system of the house since they are too small to be removed by the ordinary filter.

These biological agents can cause sneezing, watery eyes, coughing, shortness of breath, lethargy, fever, and digestive problems. Prolonged exposure to mold spore allergens may wear down the immune system and increase an individual's susceptibility to infectious disease.

NICE TO KNOW

There is great concern over the effects of long term exposure to mold and mold spores. Mold can be controlled through filtration and by removing its growth sources. In order for mold to exist it needs both a source of food and moisture. Because mold can live on virtually any organic compound it is almost impossible to remove all of its food sources. Therefore it is much more easily controlled by removing the moisture.

Methods of control involve examining potential sources such as porous wet surfaces and water damaged material. Most remedies include good maintenance or preventive maintenance procedures, such as cleaning filters and wet areas located in the airstream, replacing water damaged carpets and insulation, maintaining relative humidities between 40 and 60%, and cleaning and disinfecting drain pans.

Respirable Particles These particles are usually considered to be 10 microns or smaller and are carried into the system through breathing. They can be either biological or nonbiological. The nonbiological particles often are carriers for bioaerosols and can deliver harmful substances to critical areas.

Common sources of respirable particles can be kerosene heaters, humidifiers, wood stoves, and tobacco smoke. Particles can be brought in the house on clothing, through the use of household sprays and cleaners, or by the deterioration of building materials and furnishings.

Particles and the substances they carry can cause nose, throat, and lung infections. They can impair breathing and increase the susceptibility to cancer.

Control is through introduction of outside air and use of good quality air filters. *High efficiency particulate arresting* (HEPA) filters and *electrostatic air cleaners* (EACs) can remove particulates very effectively. Appropriate filter maintenance, by cleaning and replacing at regular intervals, is essential.

SERVICE TIP

Not all residential air conditioning systems can handle the pressure drop created by a HEPA filter. In some cases if a HEPA filter is added to the system its airflow will drop below design levels. This can significantly reduce the system's overall performance to a level where it is impossible to maintain indoor air temperatures under heavy loads. Check with the equipment manufacturer to see if the pressure drop created by the HEPA filter is within their equipment design specifications.

Asbestos *Asbestos* is the name given to a group of naturally occurring silicates that occur in fiber bundles having unusual tensile strength and fire resistance. About 95% of the material is a substance called chrysolite. Once in place the material does not deteriorate rapidly. The danger occurs to the installer or to someone who damages the material when removing it. Its primary usages are for building materials, boiler insulation, sprayed-on fireproofing coatings, pipe insulation, and floor and ceiling tiles.

A number of diseases have been associated with the use of asbestos:

1. *Asbestosis,* or scarring of the lungs, leading to respiratory failure.
2. *Mesothelioma,* a damaging effect to the linings of the lungs and abdomen.
3. *Lung cancer,* a chronic, progressive, generally incurable disease.

There are no immediate symptoms of asbestos exposure. Most deaths have occurred as a result of high exposure in industrial locations. Studies of exposures in schools and homes have shown the exposure to be well below critical levels of 0.0001 fiber/ml.

The primary health risk is to operations and maintenance people who, in the course of their duties, disturb the asbestos containing materials. Therefore, unless you are specially trained and wearing appropriate personal protection equipment you must not disturb it.

CAUTION: Under most circumstances it is illegal for technicians to remove asbestos from the jobsite. Asbestos removal must be carried out by licensed asbestos abatement companies. Removing asbestos without the proper equipment can be hazardous to your health and could result in fines for you and your company. If you suspect that an older installation has asbestos you must request that an approved laboratory test the material to see whether or not asbestos is present.

Vapors and Gases

Formaldehyde Formaldehyde is a colorless gas at room temperatures and has a pungent odor. It is used in the manufacturing of a number of building products and usually emerges as a pollutant during construction or remodeling.

It is used in the manufacturing of plywood, fiberboard, paneling, particleboard, urea foam insulation, fiberglass, and wallboard. It also is a product of incomplete combustion.

The strength of the formaldehyde as well as the length of exposure influence the response to the material. Even low concentrations, besides the disagreeable odor, can cause eye and throat irritation and a biting sensation in the nose. A medium dose can cause a flow of tears in the eyes. A strong exposure can cause inflammation of the lungs and even danger of death.

The critical level is usually considered 0.1 parts per million (ppm). Homes built before 1990 without urea formaldehyde insulation have levels in the 0.02 to 0.05 ppm range. Special equipment is required to monitor the concentration accurately; however, the odor can indicate most critical levels.

Control is usually accomplished through careful selection of building materials and through an outgassing procedure. This is simply allowing time, after installation, for the undesirable vapors or gases to be emitted from the construction materials before the area is occupied. This process can be greatly speeded up by a "bake-out" procedure. A bake-out consists of three procedures: the indoor temperature is raised, maximum ventilation is provided, and adequate time is allowed for the process to do its work.

Materials that emit formaldehyde can also be coated to prevent leakage of gas. The National Aeronautics and Space Administration (NASA) found that ordinary houseplants can significantly reduce formaldehyde levels.

SERVICE TIP

Some individuals are far more susceptible to chemical contamination than others. They may experience respiratory difficulties, skin rashes, headaches, or other such symptoms. These chemicals can come from new construction materials, glue, and carpeting. It may be necessary to provide additional outside air ventilation for these residents so that the amount of indoor chemical air pollution can be maintained at a level for them to remain symptom free.

Radon *Radon* is a colorless odorless gas that is always present to some degree in the air supply. It is formed by the radioactive decay of radium and uranium. It produces radiation or radioactive elements, which can be a major health concern. These elements attach to particles in the air and are inhaled during breathing. About 30% of the charged particles are retained in the lungs, providing a source of injury to the tissues.

The principal source of radon is from soil. The gas enters a building through piping and other conduits in the foundation or through cracks and crevices in the foundation or through piping and other conduits in the foundation directly in contact with the soil. Tight construction and insulation capture the gases. This can be particularly significant if construction occurred on or near fill sites.

No immediate symptoms are noticed. Radon decay products increase the susceptibility to lung cancer. The critical level is very small, about 4 trillionth parts per liter.

The most effective means of controlling radon is to prevent it from entering the building. Cracks in the basement floor must be sealed. Foundations should provide some means of ventilating soil gases outside the structure. In residences, mechanical ventilation of the crawl space has proved satisfactory. In one location, increased ventilation at the first floor level and exhaust out of the basement took care of the problem.

Volatile Organic Compounds (VOCs) Some VOCs are always present in the indoor atmosphere. They constitute vapors which are emitted from many household products. Good examples are formaldehyde and products of tobacco smoke, such as benzene and phenols.

Some of the common sources are building materials, furniture materials, photocopying materials, disinfectants, gasoline, paints, and refrigerants. The ventilation system itself may bring in outside pollutants.

Some of the health effects include eye irritation, sore throat, nausea, drowsiness, fatigue and headaches. Studies have shown a relationship between certain VOCs and respiratory ailments, allergic reactions, heart disease, and cancer. The critical level seems to be in the area of $3mg/m^3$.

Where the source can be determined and removed or isolated from temperature building, this is the best method of control. Where this is not practical, increased ventilation is used to control the concentrations. Selective purchase of construction materials is always helpful. In new or remodelled construction, a bake-out procedure, previously described, can be effective.

VOC monitors are available with a cable for connecting the sensor to a datalogger or computer. The instruments will monitor VOCs and CO_2 at the same time. Additional temperature and humidity probes and specific gas monitors are available.

SAFETY TIP

Many of the adhesives used in sealing air conditioning duct systems are available as water soluble. These adhesives produce fewer VOCs than do mineral based adhesives. For most residential applications it is recommended that you use water soluble adhesives to eliminate or reduce this form of indoor air contamination for residences.

Combination Pollutants

Environmental Tobacco Smoke *Environmental tobacco smoke* is the emission from the burning end of a cigarette, cigar, or pipe. It is the secondhand smoke exhaled by the smoker. It is sometimes termed involuntary smoking. It contains irritating gases and tar products. Since tobacco does not burn completely, pollutants are given off such as nitrogen oxide, sulfur dioxide, ammonia, formaldehyde, benzene, and arsenic, to name a few.

Secondhand smoke has been shown to increase the risk of cancer in adults. Studies have indicated that involuntary smoking increases respiratory illness in children. The Environmental Protection Agency (EPA) has shown that inhaling tobacco smoke can cause permanent damage to the blood cells.

Control of involuntary smoke can take many forms:

1. Eliminate smoking entirely.
2. Confine it to certain areas where separate ventilation is provided.
3. Provide adequate ventilation to dilute the smoke to acceptable levels.
4. Filter the pollutants, using HEPA filters and electronic precipitators.

Since odor and irritation remain where the smoke is produced, ventilation and filtering do not completely solve the problem.

Combustion Products Combustion products include respirable products, carbon monoxide, nitrogen oxides, and VOCs. Some products are exceedingly small and lodge in the lungs. Carbon monoxide is a colorless, odorless, highly poisonous gas. Nitrous oxides are all irritating gases that can have an impact on human health. Other VOCs include hydrocarbons which act in an additive fashion with other pollutants.

Combustion products are released as a result of incomplete combustion that can occur during the burning of wood, gas, and coal stoves, unvented kerosene heaters, fireplaces during downdraft conditions, and environmental tobacco smoking.

Carbon monoxide reduces the ability of the blood to carry oxygen to the body tissues. Common symptoms are dizziness, dull headache, nausea, ringing in the ears, and pounding of the heart. An extreme dose can cause death. Nitrous oxide can cause irritation to the eyes, nose, and throat, respiratory infections, and some lung impairment. Combustion particles can affect lung function. The OSHA recommended limit is 50 ppm in an 8 hour period or about 40 mg/m^3 in a 1 hour period. Acute exposure can be fatal.

Control includes maintaining and properly adjusting fuel burning equipment. Houseplants can serve as a living air purifier for VOCs. When unusually high levels of combustion products are expected, additional ventilation can be used as a temporary measure. Do not use unvented heaters.

THE ENERGY RECOVERY VENTILATOR

The energy recovery ventilator (ERV) is an accessory for a heating and/or cooling system to supply and control the introduction of outside air and to exhaust stale air. It includes an air to air heat exchanger that is used to conserve energy, as shown in Figure 15-2-2.

The heat exchanger usually consists of a desiccant coated wheel that rotates between the two airstreams. In winter it recovers heat and moisture from the stale air and transfers them to the cold, dry, incoming air. A preheater adds extra heat when needed to the cold entering air. In summer, heat and moisture are removed from incoming air and transferred to the stale exhaust air. A standard air filter removes most airborne particles down to 5 microns in size before they enter the ERV unit. These filters are readily replaceable as required.

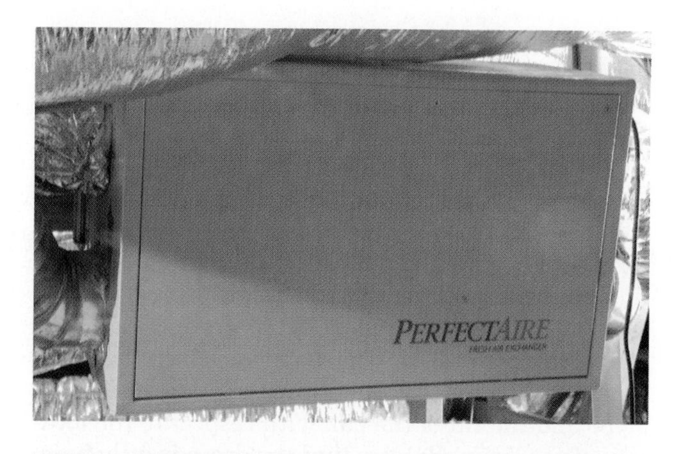

Figure 15-2-2 Energy recovering ventilator.

A *humidistat* is used to control the relative humidity in the occupied space and prevent condensation on windows in winter. An outside air control allows the occupants to adjust the speed of the fan to regulate the ventilation rate.

IAQ SERVICE OF HVAC SYSTEMS

There is great value in technician training to provide IAQ service.

> **TECH TIP**
>
> It is often the responsibility of the HVAC/R technician to monitor and maintain the quality of the air within a building. Failure to do so properly can result in significant problems for the building and building operator. Special training classes are available for individuals who are specifically involved in indoor air quality work.

Treatment for Pans and Coils

Condensate pans and cooling coils are two locations likely to contribute to IAQ problems. Cooling coils are usually cleaned with brushes and a proper biocide. When this material is removed it must be taken away from the area before the operation of the system can be resumed.

In addition to cleaning, an anti-foulant needs to be used to reduce temperature buildup of scale, dirt, and biological matter. The condensate which forms on the coil surface can then wash the surface during normal operation. A product such as First Strike, which is available through Indoor Environmental Quality Alliance (IEQA), serves this purpose. The products used on the coils should be checked for excessive alkalinity or acidity. Either extreme can cause coil pitting.

The condensate pan is fertile ground for many microbial pollutants. The condensate in the pan must drain properly. It is important to check the slope and the tilt of the pan with respect to the location of the drain. Be sure that the drain is fully functional. The drain pan must be properly trapped so that the fan will not pull air back through the drain and prevent the condensate from draining out.

The pan cleaning agent must be designed to function for a long period of time to maintain the scale in suspension so it will properly drain out. A pan guard is recommended by IEQA and operates to keep the pan clean. Condensate pan treatment products should also contain a wetting agent to reduce the possibility of contaminants sticking to the surface and building up. Look for a product with the lowest possible hazard rating on the material safety data sheet (MSDS).

If condensate is fed directly into the waste line, be sure the trap is kept full during long periods of disuse during the winter. This helps to reduce pollution that can migrate into the building from dry traps.

Measuring Other Sources of Pollution

Organizations like EPA, IEQA, and OSHA recognize carbon dioxide (CO_2) as a suitable measurement of IAQ. The CO_2 must be kept down to an acceptable level below 1000 ppm or 800 ppm as recommended by some groups. Basically, CO_2 is an indicator of outside air ventilation. Monitors are now readily available to measure CO_2.

As part of the inspection and service work on any HVAC/R system, the technician should determine that outside air intakes and the rest of the outside air supply system are working properly.

Since the presence of carbon monoxide (CO) is very serious, it also can be measured by monitors readily available. CO should be kept at levels of a few parts per million or zero levels.

Producing the Best IAQ

As a special service to valued customers, the technician can provide IAQ inspection and preventive maintenance service. Measurements can be made with an instrument such as shown in Figure 15-2-3, to alert the technician when critical elements are involved, and corrections can be made where needed. These services can help the client to provide a cleaner, healthier, and more productive environment.

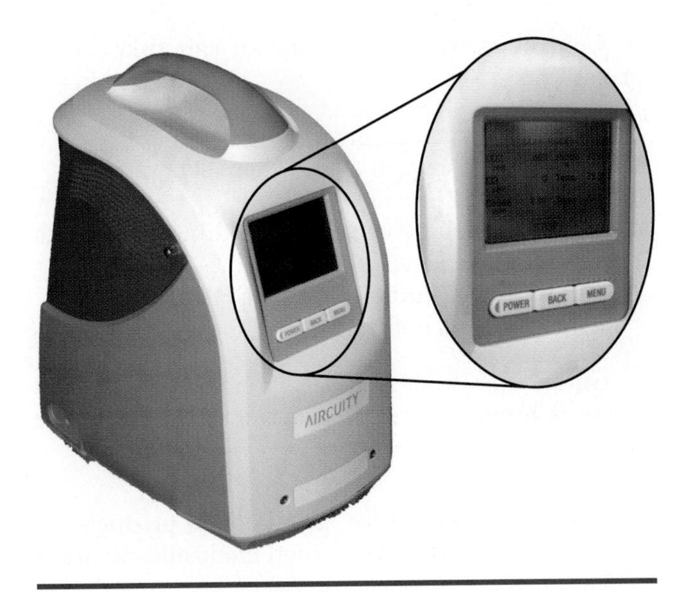

Figure 15-2-3 A comprehensive indoor air quality analyzer.

TECH TALK

Tech 1. I saw the news last night where a reporter was talking about a building downtown as being sick.

Tech 2. That's right. Sometimes when the indoor air quality in a building gets to be so bad a lot of people get sick in the building, then they refer to the building as having sick building syndrome.

Tech 1. I knew there had to be something like that. The building itself couldn't be sick.

Tech 2. That's true. In some severe cases the indoor air quality of a building can get to be so bad that almost everybody working in the building does get sick. When that happens the entire building could be shut down and require an extensive cleanup. It takes an extensive cleanup to get it back in condition so that it can be safely occupied.

Tech 1. That sounds expensive.

Tech 2. It is, and it takes a lot of time. So it is important that we make sure that the buildings we maintain are operating properly. We wouldn't want to get caught up in one of those sick building problems.

Tech 1. You are right about that. Thanks a lot.

UNIT 2—REVIEW QUESTIONS

1. What is the composition of air?
2. How small is a micron?
3. What are the two general methods of pollutant control?
4. Define bioaerosols.
5. What are some common sources of respirable particles?
6. What is asbestos?
7. What diseases have been associated with the use of asbestos?
8. How is radon formed?
9. What are some common sources of VOCs?
10. What is an energy recovery ventilator?

UNIT 3

Filters and Humidifiers

OBJECTIVES: In this unit the reader will learn about filters and humidifiers including:

- Mechanical Filtration
- Rating Air Filters
- Filter Installation
- Electronic Air Filters
- Evaporative Humidifiers
- Vaporizing Humidifiers
- Humidistats

SERVICE TIP

There are a wide variety of filter media available to the HVAC/R industry. Prices for these filters vary significantly. In some cases disposable air filters can be purchased at discount stores for less than a dollar. Unfortunately, not all air filters are of the same quality or have the same ability to trap airborne particles. Low quality filters can result in indoor air quality issues. Although high quality filters are far more expensive than those at discount stores, they are well worth the money to your customer because of the improvement they will provide in indoor air quality.

AIR FILTRATION

Air contaminants are of many types, including dust, fumes, smokes, mists, fogs, vapors, and gases. On a particle count basis, over 99% of the particles in a typical atmosphere are below 1 micron in size. The particle size for many of the airborne contaminants is shown in Figure 15-3-1.

A micron is 1 millionth part of a meter or $\frac{1}{25,400}$ part of an inch. A human hair is about 100 microns in thickness.

Referring to the chart in Figure 15-3-1, in order to remove airborne bacteria, the filter must be capable of removing particles down to 0.3 microns in size. To block out tobacco smoke, the filter must remove particles down to 0.01 microns. The chart indicates that this removal can be accomplished with either a high efficiency mechanical filter or an electrostatic filter. The common throwaway filters used in many furnaces are only capable of removing about 1 micron size particles.

A filter becomes more efficient as it fills with particles, although the disadvantages far outweigh the advantages. Air filters must be cleaned or replaced when they become dirty.

MECHANICAL FILTRATION

The term mechanical filter typically describes a fibrous material filter commonly called a throwaway or replacement type which comes as standard equipment on forced air heating units, Figure 15-3-2. It consists of a cardboard frame with a grill to hold the filter material in place. The filter media material consists of continuous glass fibers packed and loosely woven that face the entering air side and more dense packing and weave on the leaving air side. Air direction is clearly marked. Some makes also have the fiber media coated with an adhesive substance in order to attract and hold dust and dirt. Filter thickness is usually 1 in for residential equipment.

The recommended maximum air velocity across the face should be approximately 300 ft/min, with a maximum allowable pressure drop of about 0.5 in WC.

Many manufacturers and filter suppliers also offer a permanent or washable filter. It consists of a metal frame with a viscous impingement type of filter material supported by metal baffles with graduated openings or air passages. The filter material is coated with a thin spray of oil or adhesive to trap airborne particles. These filters may be removed, cleaned with detergent, dried, and recoated. Another variation of dry filter is reusable, supported by a metal frame, Figure 15-3-3.

In commercial and industrial installations there are variations of wet and dry filters for special uses where larger volumes of air or filtrations of chemical substances are required. Applications involving paint in spray booths, commercial laundries, hospital operating rooms, etc., usually require the services of experts.

CHARACTERISTICS OF PARTICLES AND PARTICLE DISPERSOIDS

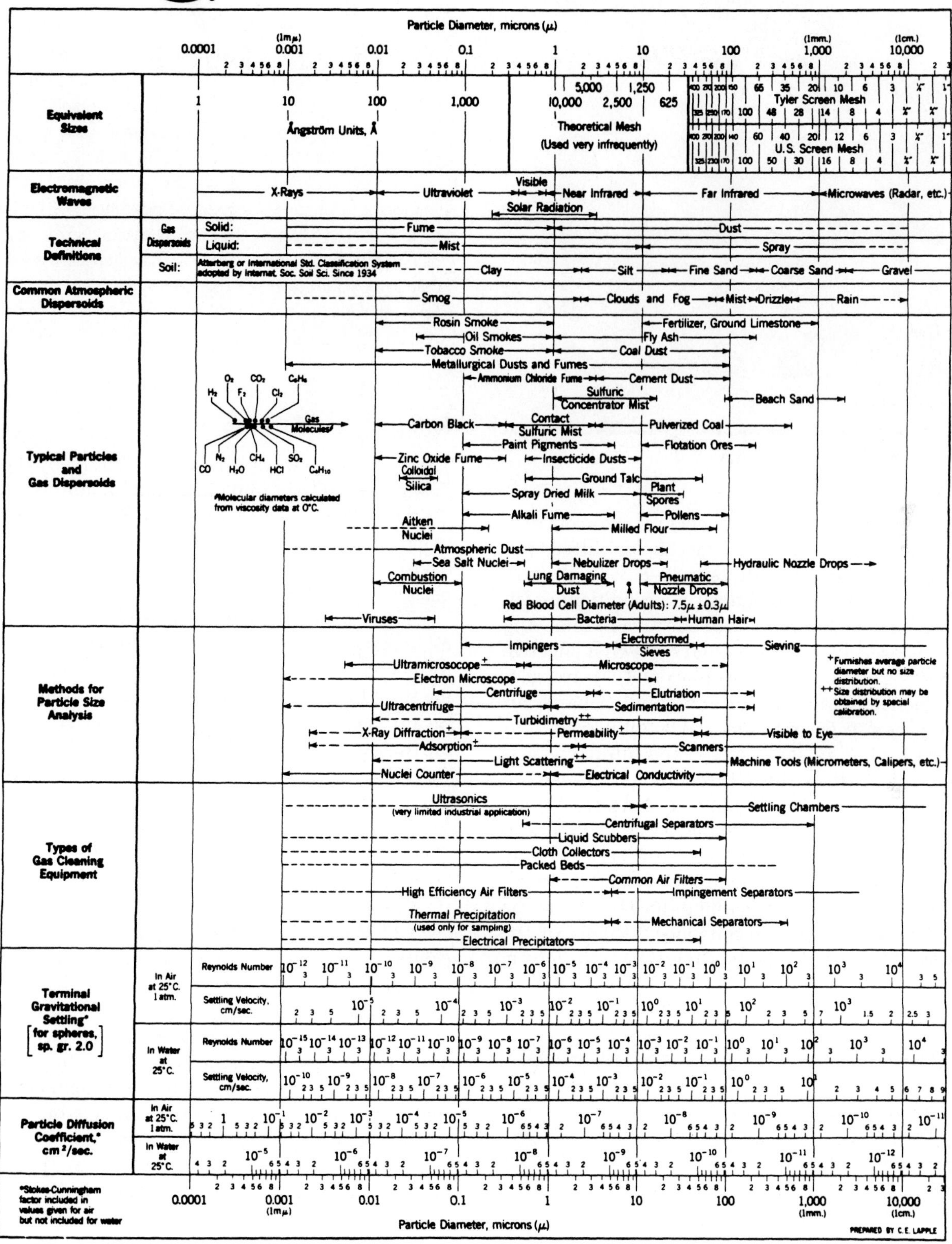

Reprinted from Stanford Research Institute Journal, Third Quarter, 1961. Single copies 8-1/2 by 11 inches free or $15 per hundred. 20 by 26 inch wall chart $15 each. Both charts available from Dept. 380.

SRI International • 333 RAVENSWOOD AVE. • MENLO PARK, CALIFORNIA 94025 • (415) 326-6200

Figure 15-3-1 Characteristics of particles carried by the airstream. (*C. E. Lapple, SRI JOURNAL, 5-94 [Third Quarter, 1961]. Reprinted with permission of SRI International.*)

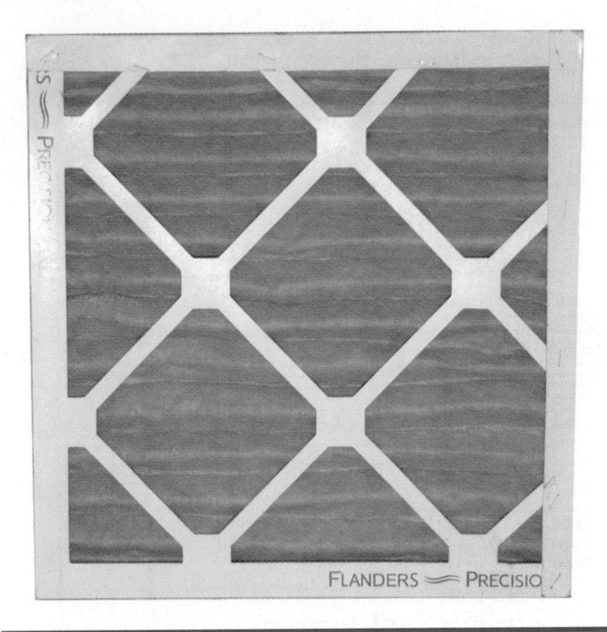

Figure 15-3-2 Throwaway filter.

The arrestance factor (efficiency) of standard residential and light commercial mechanical filters can vary from 25% for cheap filters to 80–85% for the better quality heating and central air conditioning filters. This means the filters are effective in removing lint, hair, large dust particles, and somewhat effective in removing common ragweed pollen. They are relatively ineffective for smoke and staining particles. For average use, clean filters do a creditable job.

Getting the homeowner to keep filters clean is not easy, and dirty filters are probably the main contributor to malfunctioning equipment. As the mechanical filter clogs with dirt, airflow is reduced to a point where the cooling coil will freeze, causing compressor failure. In a heating system, dirty filters can cause overheating and reduce the life of the heat exchanger or cause nuisance tripping of the limit switch. Operating costs increase with loss of efficiency.

Figure 15-3-3 Reusable filter.

SERVICE TIP

Better quality air filters capture more of the dust from the air than do low quality filters. Although the lower quality filter may cost significantly less than the more expensive high quality filters, the difference is quickly made up by the cost of indoor evaporator coil cleanings. Low quality filters can let enough dust and dirt through so that an annual cleaning is required on the evaporator coil. The evaporators become clogged quickly because the moisture on the coils during the cooling season entraps a great deal of dust and dirt when low quality filters are used. Encouraging your customer to invest in the higher quality filters will reduce their annual service coil cleaning cost.

Rating Air Filters

There are three ways of rating air filters: by their efficiency, resistance, and particulate holding capacity.

The *efficiency* of a filter can be determined by using procedures set up be ASHRAE Test Standard 52-76 to compare input with output air. Tests are performed by the filter manufacturer.

The *resistance* can be measured with an ordinary manometer. It is common practice on central systems to include a differential pressure measuring device for the filter section. This instrument compares the air pressure on the upstream and downstream sides of the filter assembly. The gauge on the instrument indicates when the filters should be changed.

Tests are available for measuring the *particulate holding capacity* of the filter. From this information the length of operating time before changing can be estimated for a given application. Filters made of pleated fabric, with depths of 6 to 12 in, have a large dust holding capacity and are often used on large systems.

Panel Type Filters

Viscous impingement panel filters are made of coarse fibers with a high porosity. The filter media is coated with a viscous substance, such as oil, which acts as an adhesive to the airborne particles coming into contact with it.

A number of materials have been used as the filtering media, including coarse glass fibers, animal hair, vegetable fibers, synthetic fibers, metallic wool, expanded metals and foils, crimped screens, random matted wire, and synthetic open cell foams.

Servicing these filters depends on their construction. Disposable filters are made of inexpensive materials and are discarded after one usage. Permanent filters have metal frames to stand continued usage.

Cleaning is done by washing with steam, water, or detergent.

Dry type extended surface filters are made of random fiber mats or blankets of varying thicknesses. Bonded glass fiber, wool felt, synthetics, and other materials are commonly used. Pleating of the media provides a larger filter surface area compared to the face area and a reasonable pressure drop. The efficiency of the dry filters is usually higher than the viscous coated type. In addition to their effectiveness, the dry filters have a larger dust holding capacity and, therefore, longer periods of use.

The *high efficiency particulate air* (HEPA) filters are the standard for cleanroom design. They have an extended surface configuration with deep folds of submicron glass asbestos fiber paper. This fiber will operate at 250 fpm face velocity with a pressure drop of 0.50 to 1.00 in of water or higher during their service life.

SERVICE TIP

In some commercial applications a prefilter is used in front of HEPA filters to extend the life of the more expensive HEPA filters. In some applications this prefilter media can be mounted on spools so that new filter material can be simply unwound from one spool as the dirty filter material is wound on the take-up spool. The dirty spool of filter material is significantly heavier than a clean spool.

Filter Installation

Filters should be placed ahead of heating and cooling coils and other mechanical equipment to protect the system from dust. The published performance of filters is based on straight through unrestricted airflow. Failure of filters to give satisfactory service is usually the result of improper installation and maintenance.

Renewable Media Filters

The *moving curtain viscous impingement filter* is an automatic moving curtain with the random fiber medium, treated with a viscous material, and furnished in roll form. The material rolls down from the top of the unit. As the exposed area becomes saturated with dirt, a clean section is automatically rolled into place. The used portion is collected in a roll at the bottom and thrown away. A fresh roll is again placed at the top to the unit and the filtering continues.

The *moving curtain dry media filter* is operated in a manner similar to the viscous impingement model. These filters are effectively used to remove the lint in the air in textile mills, dry cleaning establishments, and lint and ink mist in pressrooms.

ELECTROSTATIC AIR FILTERS

Where greater filtering efficiency is desired (above 80%), the use of electrostatic air cleaners is highly recommended.

Figure 15-3-4 is an example of a portable electrostatic room air cleaner. It contains a HEPA filter, electrostatic filter, and an ultraviolet light to provide clean, sterile air to improve indoor air quality.

The electrostatic air filter uses the precipitation principle to collect airborne particles. There are three types of units used for commercial service: ionizing plate, charged media nonionizing, and charged media ionizing.

With the *ionized plate unit,* the incoming air passes through a series of high potential ionized wires that generate positive ions. These ions adhere to the dust particles carried by the airstream. The charged dust particles then pass through an electric field, attracting the charged particles and removing them as shown in Figure 15-3-5. The direct current field is created by a 12,000 V DC current that maintains 6,000 V between the attracting plates. These plates offer little resistance to the airflow. For best results the airflow through the plates should be evenly distributed.

Conventional type prefilters are used ahead of the electrostatic filters to screen out the larger particles in the air. These filters offer some resistance to the airflow in the range of 0.14 to 0.26 in WC.

Figure 15-3-4 Room air purifier may have HEPA filters, activated charcoal, and ultraviolet lights to remove particles, odors, and biological impurities from the air.

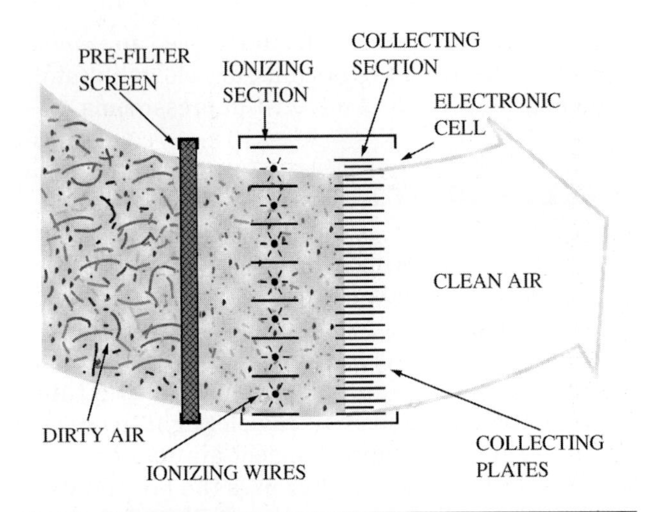

Figure 15-3-5 Ionizing air filter.

SERVICE TIP

The high voltage on an electrostatic air filter can cause discharges when the surface of the collector plate becomes excessively dirty or if a large particle or insect comes in contact with the collector plate. This electrical discharge or spark can be annoying to homeowners because of its sound and interference with television and radio reception. For installations where persistent arcing is a problem, the source of the arcing must be identified and controlled. It may mean more frequent cleaning of the collector plate or possibly the installation of a prefilter to catch or keep out larger particles or insects.

The operating efficiency of electrostatic air cleaners is a function of airflow. With a fixed number of ionizer wires and collector plates, a particular model may be used over a range of airflow quantities (ft^3/min or CFM). It will be rated at a nominal CFM, and then for other efficiency ratings for the recommended span of operation. The higher the CFM of air and the resultant velocity, the lower the efficiency. Residential electrostatic air cleaners come in sizes of 800 CFM to 2000 CFM. Static pressure drop will run about 0.20 in of WC at the nominal airflow rating point. Ratings are published in accordance with the National Bureau of Standards Dust Spot Test and are certified under ARI standards.

Collector plates are often used coated with oil to act as an adhesive. The filters are cleaned by use of water sprayed on the plates in place. Suitable drains are provided in the bottom of the filter compartment. Due to the high voltage used by the electrostatic units, safety switches are provided that will turn off the power when the filter access door is opened.

The *charged media nonionizing filters* are very different in construction. These filters consist of a dielectric filtering medium, usually arranged in pleats, as in typical dry filters. The dielectric medium consists of glass fiber mat, cellulose mat, or similar material, supported by a grid consisting of alternately grounded and charged members. The charged members are supplied with 12,000 V DC power. Airborne particles that approach the field are polarized and drawn to the filaments of the fibers of the media.

This type of filter offers about 0.10 in WC resistance to the flow of air when clean, with a face velocity of 250 fpm. This small resistance serves to equalize the flow of air over the exposed surface.

The *charged media ionizing electronic filters* combine the effects of the other two designs. Dust is charged in a corona discharge ionizer and collected on a charged media filter mat. This construction increases the effectiveness of the filter but is more critical to operate successfully.

In using electrostatic air filters, two conditions of operation can cause operating problems: (1) space charge and (2) ozone.

The unit needs to be carefully built so that charged dirt particles do not escape into the filtered space (space charge). If they do, they can darken the walls faster than if no cleaning arrangement was used.

All high voltage devices are capable of producing ozone or O_3. When the unit is operating correctly the amount of ozone produced is well within the recommended limits. If the unit is continuously arcing, it may yield levels of ozone, which are annoying and even poisonous. High levels of ozone are indicated by a strong odor.

Ultraviolet (UV) Lights

Ultraviolet lights (UV) are used to sterilize the air in duct systems. These lights have been popular for years in hospitals where 100% of the air entering an operating room has to be outside air. The air passes through a duct box containing a large number of ultraviolet lights where the air is sterilized before entering the operating room. Ultraviolet lights will kill mold, fungus, bacteria, and any other living organisms if the light is intense enough for the airflow.

In recent years ultraviolet light sterilizing units have become available for the residential and light commercial market. These lights are located in the plenum where they sterilize the air before it enters the occupied space. Lights are designed to come on during system operation. This is accomplished by tying their circuit into the fan blower relay circuit in the indoor equipment.

Ultraviolet light at the levels required for air sterilization is dangerous to technicians. The ultraviolet light can result in severe skin burns like those caused by too much exposure to the sun. In addition burns on the eyes can occur. The intensity of the sterilizing lights is such that it only requires a momentary exposure to cause damage. Ultraviolet light boxes must have an interlock circuit that prohibits the light from coming on when access panels are open.

TYPES OF HUMIDIFIERS

Humidifiers are used to add moisture to indoor air. The amount of moisture required depends on:

1. Outside temperatures
2. House construction
3. Amount of relative humidity that the interior of the house will withstand without condensation problems.

It is desirable to maintain 30 to 50% relative humidity. Too little humidity may cause furniture to crack. Too much humidity may cause mold problems. From a comfort and health standpoint, a range as wide as 20 to 60% relative humidity is acceptable.

The colder the outside temperature is, the greater the need for humidification. The amount of moisture the air will hold depends on its temperature. The cooler the outside air, the less moisture it contains. Air from the outside enters the house through infiltration. When outside air is warmed, its relative humidity is lowered unless moisture is added by humidification, Figure 15-3-6.

For example, air at 20°F and 60% relative humidity contains 8 grains of moisture per lb of air. When air is heated to 72°F, it can hold 118 grains of moisture per lb. Thus, to maintain 60% relative humidity in a 72°F house, the grains of moisture per lb must be increased to 71 ($118 \times 0.6 = 71$). One pound of air entering a house from the outside will require the addition of 63 grains of moisture ($71 - 8 = 63$). The amount of infiltration depends on the tightness of the windows and doors and other parts of the construction.

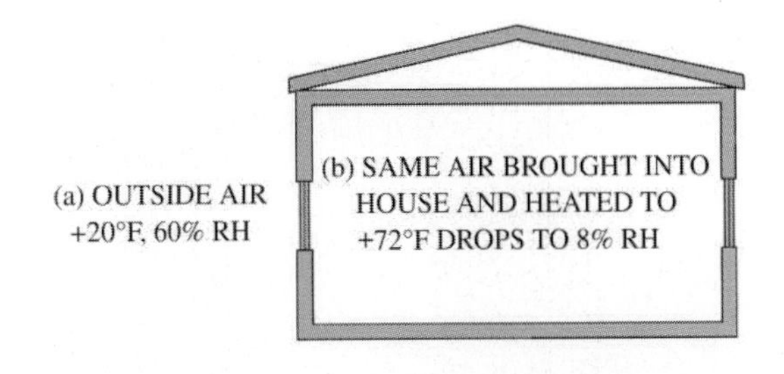

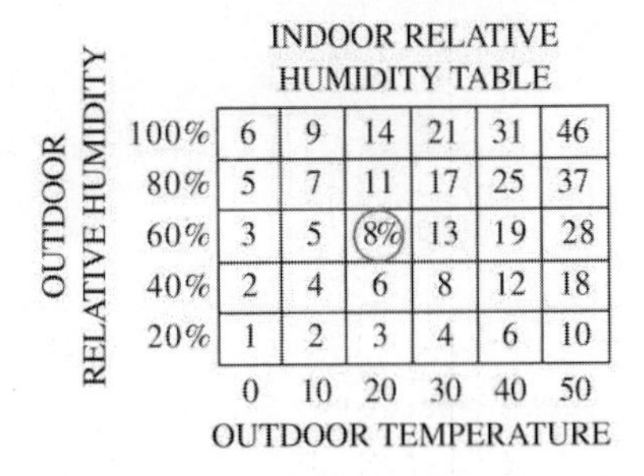

OUTDOOR RELATIVE HUMIDITY	0	10	20	30	40	50
100%	6	9	14	21	31	46
80%	5	7	11	17	25	37
60%	3	5	(8%)	13	19	28
40%	2	4	6	8	12	18
20%	1	2	3	4	6	10

Figure 15-3-6 Chart showing the effect of temperature on the relative humidity of air.

It is impractical in most buildings to maintain high relative humidity when the outside temperature is low. Condensation forms on the inside of the window when the surface temperature drops below the dew point of the air. (The dew point is the temperature at which water as vapor in the air has reached the saturation point, or 100% relative humidity.)

There are two types of humidifiers used with furnace installations:

1. Evaporative
2. Vaporizing
3. Steam

NICE TO KNOW

It is a popular misconception that electric heat dries the air out more than gas heat. The roots of this misconception can be traced back to a time when unvented gas heaters were in common use. These combustion byproducts of H_2O and CO_2 were released into the living space. Because of the looseness of the building construction during that time, excessive combustion contaminants did not collect in the residence; however, the moisture produced as a result of the combustion raised the relative humidity. Since electric space heaters did not provide additional moisture, it was assumed that the electric heat was, for some reason, burning out the moisture in the home. This misconception continues today when customers ask, "Isn't electric heat a lot drier?" And the answer to that question is "No."

Evaporative Humidifiers

Evaporative humidifiers provide a wetted surface, which adds moisture to the heated air. There are three types of evaporative humidifiers:

1. Plate
2. Rotating drum
3. Fan powered

The *plate evaporative humidifier* has a series of porous plates mounted in a rack, such as the one

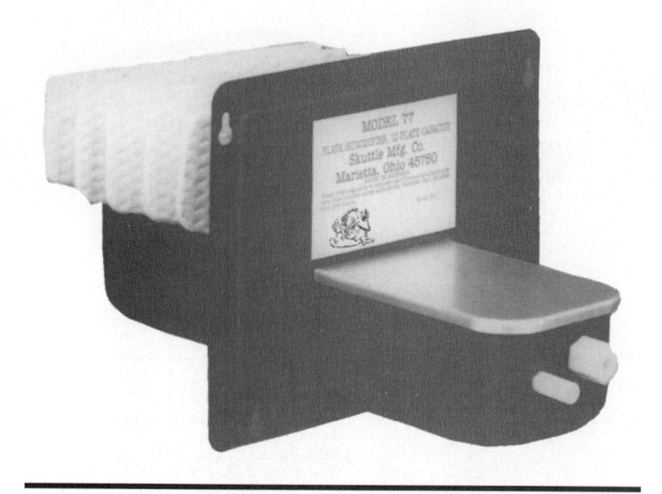

Figure 15-3-7 Plate type humidifier. (*Source: Skuttle Indoor Air Quality Products*)

shown in Figure 15-3-7. The lower section of the plates extends down into water contained in the pan. A float valve regulates the supply of water to maintain a constant level in the pan. The pan and plates are mounted in the warm air plenum.

The *rotating drum evaporative humidifier* has a slowly revolving drum covered with a polyurethane pad partially submerged in water, as shown in Figure 15-3-8. As the drum rotates, it absorbs water. The water level in the pan is maintained by a float valve. The humidifier is mounted on the side of the return air plenum. Air from the supply plenum is ducted into the side of the humidifier. The air passes over the wetted surface, absorbs moisture, and then goes into the return air plenum.

The *rotating plate evaporative humidifier* is similar to the drum type in that the water absorbing material revolves; however, this type is normally mounted on the underside of the main warm air supply duct, Figure 15-3-9.

The fan powered evaporative humidifier is mounted on the supply air plenum. Air is drawn in by the fan, forced over the wetted core, and delivered back into the supply air plenum. The water flow over the core is controlled by a water valve. A humidistat is used to turn the humidifier on and off, controlling both the fan and the water valve. The control system is set up so that the humidifier can operate only when the furnace fan is running. Wiring diagrams are shown in Figures 15-3-10 and 15-3-11.

Vaporizing Humidifiers

The *vaporizing humidifier* uses an electrical heating element immersed in a water reservoir to evaporate moisture into the furnace supply air plenum, as shown in Figure 15-3-12. A constant level of water is maintained in the reservoir. These humidifiers can operate even though the furnace is not supplying heat. The humidistat not only starts the water heater but also turns the furnace fan on if it is not running.

The *steam powered humidifier* is a type of vaporizing humidifier, pictured in Figure 15-3-13. It is

Figure 15-3-8 Rotating drum evaporative humidifier.

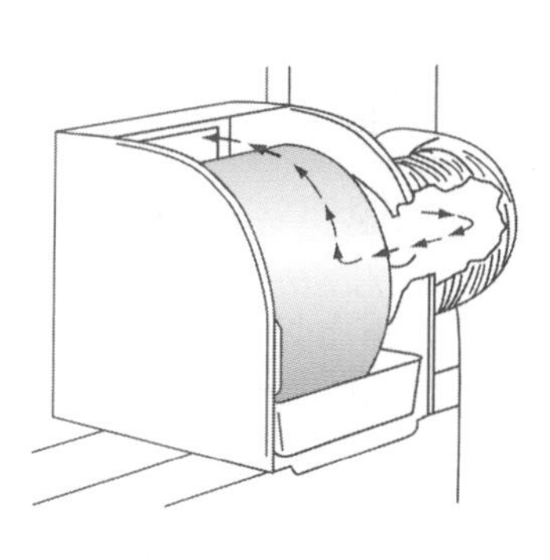

TYPICAL INSTALLATIONS

HIGHBOY FURNACE

HORIZONTAL FURNACE

LOWBOY FURNACE

COUNTER-FLOW FURNACE

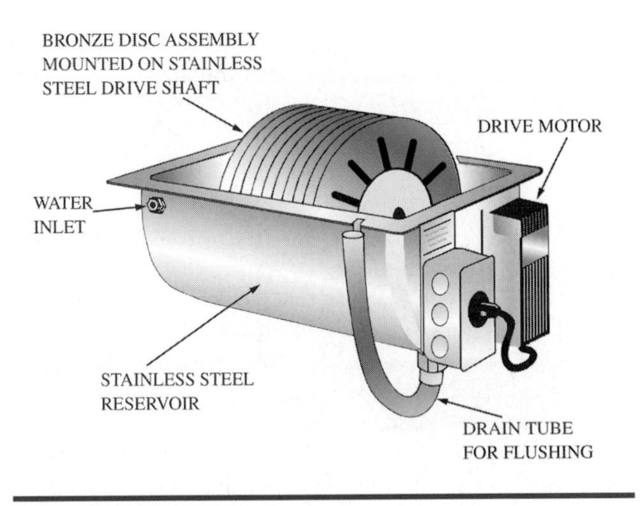

Figure 15-3-9 Rotating plate evaporative humidifier.

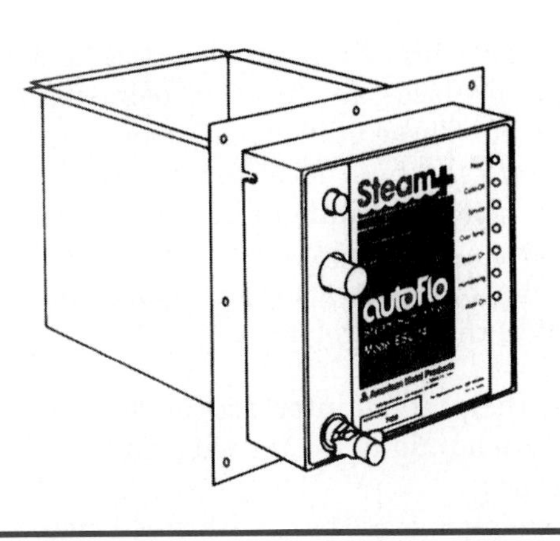

Figure 15-3-12 Vaporizing humidifier. (*Courtesy of American Metal Products*)

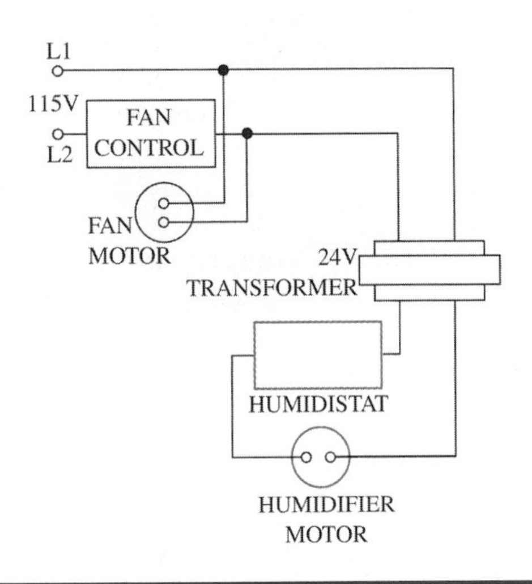

Figure 15-3-10 Wiring diagram for a humidifier with a single speed motor.

Figure 15-3-13 Steam humidifier. (*Source: Skuttle Indoor Air Quality Products*)

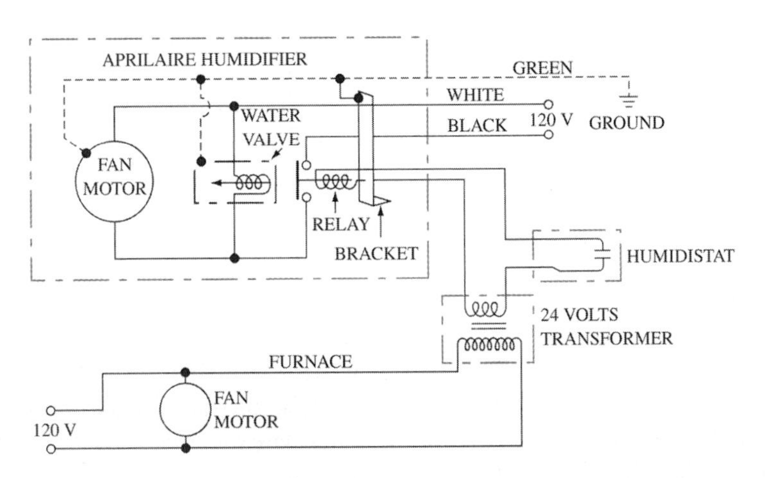

Figure 15-3-11 Wiring diagram for a fan powered evaporative humidifier.

ideal for use with heat pumps, electric furnaces, high efficiency furnaces, and furnaces using night setback thermostats. These units do not operate long enough or generate the high level temperatures necessary for evaporative humidifiers. The internal water temperature sensing device operates the fan independent of the cooling or heating mode. Some of the features of these units are:

1. The thermal fan interlock control allows the unit to humidify the air without heat from the furnace.
2. Cleaning is made easy with a one piece service drain petcock.
3. Low water safety cutoff switch is corrosion resistant and has a built-in overflow protection.
4. They use a minimum amount of water.

These units now feature a flushing timer and chlorine removal filters. The water in the unit is flushed out every 12 hr, removing any solid materials that may have accumulated. The carbon filter removes chlorine from the water supply, thus eliminating the corrosive effects of chlorine on the unit, and reduces mineral build up.

CAUTION: It is very important that humidifiers be checked for proper operation so that excessive moisture is not released into the house or duct system. Excessive moisture has been associated with mold growth, and mold growth is associated with poor indoor air quality.

Humidistats

It is desirable to control the amount of humidity in the house by the use of a humidistat. A humidistat is a device that regulates the ON and OFF periods of humidification, as shown in Figure 15-3-14a–c. The setting of a humidistat can be changed to comply with changing outside air temperatures. If the setting on the humidistat is too high, it may cause "sweating" on interior walls and windows. As it gets colder outside, the humidistat setting may need to be lowered.

(a)

(b)

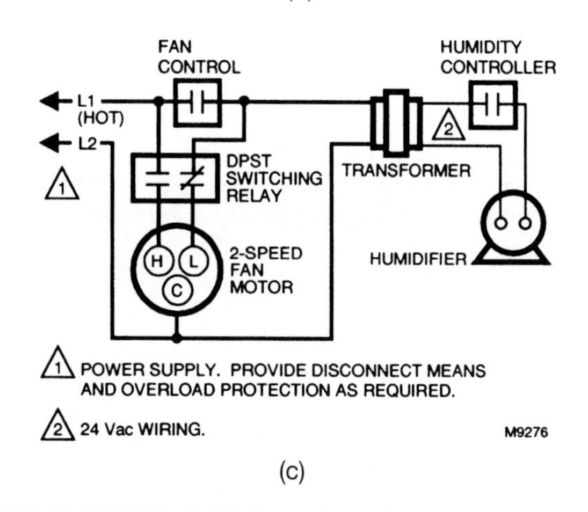

(c)

Figure 15-3-14 Two types of humidistat and a typical wiring diagram. (*Courtesy of Honeywell, Inc.*)

UNIT 3—REVIEW QUESTIONS

1. What does the term mechanical filtration typically describe?
2. What should the maximum air velocity across the face be?
3. What are the three ways of rating air filters?
4. Why is panel filter media coated with a viscous substance?
5. What is a HEPA filter?
6. What is the frequent cause of a filter's failure to give satisfactory service?

TECH TALK

Tech 1. I was in the filter room down at that big office building, and they had these great big filter bags.

Tech 2. Yes, those are common in commercial buildings because they can extend the filter life. They have a whole lot more area than a mere pleated filter.

Tech 1. If they're so good, why don't we put them in people's homes?

Tech 2. Well, bag filters have to have a constant airflow to keep the bag inflated. If the bags are allowed to inflate and deflate each time a system cycles off and on, the dirt and dust they catch will be knocked loose, much like shaking out a blanket. You wouldn't want all that dust to come flying back into the house each time a system started.

Tech 1. I didn't know that. So the blower has to stay on all the time when you have those bag filters?

Tech 2. It can be shut down, but not very often. Sometimes schools that are closed for winter break or summer vacation may shut their blowers down. But most everybody else that uses bag filters keeps them on 24 hours a day, 7 days a week.

Tech 1. Oh. Now I understand. I guess they wouldn't work in a house very well. Thanks. I appreciate that.

7. List the three types of electronic air filter units used for commercial service.
8. Residential electrostatic air cleaners come in sizes of _____ CFM to ______ CFM.
9. What is the desirable amount of relative humidity that should be maintained?
10. What are the two types of humidifiers used with furnace installations?
11. What are the three types of evaporative humidifiers?
12. A ________ is used to turn the humidifier on and off, controlling both the fan and the water valve.

16

Load Calculation

UNIT 1

Refrigeration Load Calculations

OBJECTIVES: In this unit the reader will learn about refrigeration load calculations including:

- Heat Transmission
- Sun Effect
- Design Temperatures
- Air Filtration
- Product Load
- Heat of Respiration
- Supplementary Loads
- Selection of Equipment

REFRIGERATION LOAD

The total refrigeration load of the system as expressed in Btu per hour (Btuh or Btu/hr) comes from many heat sources. For example, note the sources of heat in the refrigerated storage room of a supermarket, shown in Figure 16-1-1. The sources may be categorized as follows:

1. Heat transmission.
   a. A temperature difference of 60°F exists between the 95°F outside air and the room temperature of 35°F which causes much heat conduction.
   b. The effect of the sun on the roof and walls results in radiation heat gain.
2. Air infiltration.
   a. Air enters a room as a result of opening and closing the doors during normal working periods.
   b. Air enters the room through cracks in or around the structure and door seals.
   c. Air is purposely introduced for ventilation.
3. Product loads, or heat from the product(s) being stored:
   a. Dry or sensible heat, as from cooling a can of juice at room temperature to 35°F;
   b. A combination of dry and moist (latent) heat, as from produce;
   c. Heat from frozen products which have the latent heat of freezing;
   d. Some heat is also the result of chemical changes in the product, such as the ripening of fruit.
4. Supplementary loads are caused by heat emitting objects such as electric lights, motors, and tools—and people.

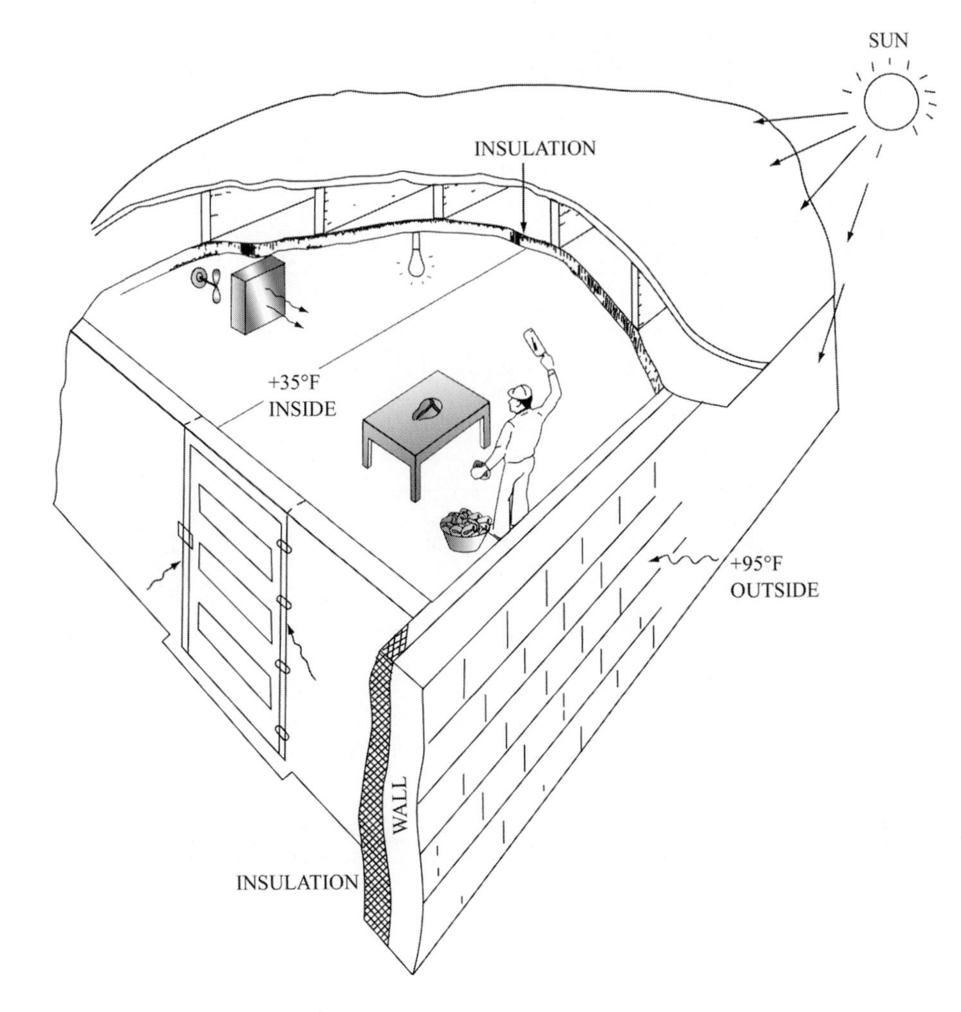

Figure 16-1-1 Refrigerated storeroom for supermarket.

Although refrigeration design engineers are primarily responsible for estimating the loads, refrigeration technicians should understand how these heat sources affect the operation of the system. They can then adjust the equipment to perform in a manner consistent with the design of the system.

TECH TIP

From time to time you may be asked to change a cold storage unit from a medium temperature application to a low temperature use. Unless the system was designed for low temperature application, this changeover may require a complete system changeout. A medium temperature application for example does not necessarily have a defrost cycle. Defrosting is required for all low temperature applications. In some cases, electric defrost elements can be added to the evaporator. But more importantly the compressor may not be designed for low temperature application, and it may also not have the capacity required to lower the box temperature to the new desired level.

Similar problems can be experienced when a low temperature system is being converted to a medium temperature application. The compressor intake valves are larger on low temperature compressors. For that reason low temperature compressors used in medium temperature applications may overheat and burn out quickly.

The subjects of refrigeration load calculation and selection of equipment are presented under the following topics:

- Heat transmission
- Sun effect
- Design temperatures
- Air infiltration
- Product load
- Supplementary loads
- Total hourly load
- Selection of equipment
- Sample calculations

Typical data sheets for recording the information are shown in Figure 16-1-2.

- Load Survey & Estimating Data -

DESIGN AMBIENT: _____ °F DB, _____ °F WB, _____ %RH, _____ °F SUMMER GROUND TEMP
(USE 55°F FOR INSULATED FREEZE FLOOR SLAB)

ROOM DESIGN: _____ °F DB, _____ °F WB, _____ % RH _____ °F WINTER DESIGN AMBIENT

ACCESS AREA: _____ °F DB, _____ °F WB, _____ %RH, (ANTE-RM/LOADING DOCK/OTHER)

ROOM DIM. OUTSIDE: _____ FT. W _____ FT. L _____ FT. H _____ TOTAL SQ. FT (OUTSIDE SURFACE)

	Insulation (Table 6)				Wall Thick-ness	Adj. Area °F	Effective Wall TD	°F Sun Effect (Table 8)	Total TD	Overall Wall Heat Gain BTU/24 hrs/sq. ft (Table 6)
	Type	Inches Thick	K Factor	U* Factor						
N. Wall								■		
S. Wall										
E. Wall										
W. Wall										
Ceiling										
Floor								■		

$$\text{* 'U' - Factor} = \frac{K}{\text{Insul. Thickness (in.)}}$$

REFRIG. DOOR(S): _______________ VENT. FAN(S): _______________

ROOM INT. VOL: _____ W x _____ L x _____ H = _____ CU. FT.
(INSIDE ROOM DIMENSION = OUTSIDE DIMENSION – WALL THICKNESSES)

FLOOR AREA _____ W x _____ L = _____ SQ. FT.

ELECTRICAL POWER _____ VOLTS, _____ PH. _____ HERTZ; CONTROL _____ VOLTS

TYPE CONTROL: _______________

PRODUCT DATA AND CLASS OF PRODUCT: _______________

Type Product	Amount of Product (Refer to Pg. 2 Item 3)			Product Temp. °F		Table 31				Table 42	
						Specific Heat					
	Amount Storage	Daily Turn-Over	Freezing or Cooling	Enter-ing	Final	Above Freeze	Below Freeze	Lat. Ht. Freeze Btu/lb.	Highest Product Freeze Temp.	Heat or Respir'n Btu/lb 24 Hr.	( ) Pull-Down ( ) Freezing Times Hrs.

EVAP. TD _____, TYPE DEFROST ☐AIR, ☐HOT GAS, ☐ELECTRIC,

CLASS PRODUCT _____,

NO. OF DEFROSTS & TOTAL TIME PER 24 HRS. _____ NO., _____ HRS.

COMPRESSOR RUNNING TIME _____ HRS.

BOX USAGE ☐AVERAGE, ☐HEAVY, ☐EXTRA HEAVY

PRODUCT LOAD AND ADDITIONAL INFORMATION: _______________

PACKAGING _____ CONTAINERS _____ WGT _____ SP. HT. _____ (CONTAINER)

PALLETS: NO. _____ SIZE _____ WGT. EA. _____ SP. HT. _____

PRODUCT RACKS: NO. _____ MAT'L _____ WGT. EA. _____ SP. HT. _____

ESTIMATING PRODUCT LOADING CAPACITY OF ROOM

ESTIMATED PRODUCT LOADING = 0.40 x _____ CU. FT x _____ LBS./CU. FT = _____ LBS:
(ROOM VOLUME) (LOADING DENSITY) (TABLE 45A OR B)

MISCELLANEOUS LOADS

PEOPLE NO. _____ HRS. _____

MOTORS (OTHER THAN EVAP. FAN)

USE: _____, _____ HP, _____ HRS.
_____, _____ HP, _____ HRS.

FORK LIFTS _____ NO., _____ HP, _____ HRS./DAY, OTHER
(REFER TO PG. 3, ITEM 4)
LIGHTS _____ WATTS/SQ, FT. (REFER TO PG. 3, ITEM 4)

Figure 16-1-2 Refrigeration load calculation sheet, blank form. (*Courtesy of Dunham-Bush, Inc.*)

Equipment Selection from Load Calculation Form

1. DETERMINE EVAP. TD REQUIRED FOR CLASS OF PRODUCT AND ROOM TEMP ______ °F (TD) (FROM LOAD SURVEY DATA)

2. DETERMINE COMPRESSOR RUNNING TIME BASED ON OPERATING TEMPERATURES AND DEFROST REQUIREMENTS ______ HRS. (FROM LOAD SURVEY DATA)

3. EVAPORATOR TEMP°F = ______ (ROOM TEMP) - ______ (EVAP. TD**) = ______ °F
**FROM LOAD SURVEY DATA

4. COMP, SUCT. TEMP.°F = ______ (EVAP. SUCT. TEMP) - ______ (SUCT. LINE LOSS) = ______ °F

BTU/24 HR. BASE REFRIGERATION LOAD WITH SAFETY FACTOR = ______
(NOT INCLUDING EVAPORATOR FAN OR DEFROST HEAT)

REFER TO PAGES 6 AND 7 "SELECTION OF REFRIGERATION EQUIPMENT TO DETERMINE PROCEDURE FOR EQUIPMENT SELECTION."

PRELIMINARY HOURLY LOAD = BTU/24 HR. (BASE LOAD) / HRS./DAY (COMP' RUNNING TIME) = ______ BTU/HR

FAN HEAT LOAD ESTIMATE BTU/HR. =

______ QTY. (MOTORS) x ______ WATTS EA. (INPUT) x 3.41 BTU/WATT x ______ HRS. = ______ BTU/24 HRS.

OR ______ QTY. (MOTORS) x ______ HP EA. x ______ BTU/HP/HR. (TABLE 13) x ______ HRS. = ______ BTU/24 HRS.

DEFROST HEAT LOAD ESTIMATE BTU/HR =

______ QTY. EXAPS. x ______ WATTS EA. x ______ HRS. x 3.41 BTU/WATT x ______ DEFROST LOAD FACTOR*

= ______ BTU/24 HRS.

*USE 0.50 FOR ELECTRIC DEFROST 0.40 FOR HOT GAS DEFROST

BTU/24 HR. TOTAL LOAD = ______ (BASE LOAD) + ______ (FAN HEAT) + ______ (DEFROST HEAT) = ______ BTU/24 HRS.

OR
(Refer to pg. 5) ______ (BASE LOAD) x ______ (BASE LOAD MULT.) = ______ BTU/24 HRS.

ACTUAL HOURLY LOAD = BTU/24 HRS. (TOTAL LOAD) / HRS/DAY (COMP. RUNNING TIME) = ______ BTU/HR.

Equipment Selection

	Compressor Units		Condensing Units		Evaporators		Condensers	
MODEL NO.								
QUANTITY								
BTU/ HR. CAPACITY (EA)								
CFM AIR VOLUME (EA.)								
	DESIGN	ACTUAL	DESIGN	ACTUAL	DESIGN	ACTUAL	DESIGN	ACTUAL
°F EVAPORATOR EMP.								
°F EVAP TD								
°F SUCTION TEMP.								
°F CONDENSING TEMP.								
°F DESIGN AMBIENT TEMP.								
°F MIN. OPER. AMBIENT TEMP								

Figure 16-1-2 (continued)

Heat Transmission

The heat gain through walls, floors, and ceilings varies with:

1. Type of construction,
2. Area that is exposed to different temperatures,
3. Type and thickness of insulation, and
4. Temperature difference between the refrigerated space and the ambient air.

Thermal conductivity (k) varies directly with time, area, and temperature difference. It is expressed in Btuh, per sq ft of area, per in of thickness, per °F of temperature difference.

Calculations

I WALL LOSS (TRANSMISSION LOAD)

SURFACE	TD	AREA OF SURFACE	ALL HEAT GAIN FACTOR TABLE 6	BTU/24 HRS
N. Wall		______ Ft. L x ______ Ft. H = ______ Sq. Ft. x	______ =	
S. Wall		______ Ft. L x ______ Ft. H = ______ Sq. Ft. x	______ =	
E. Wall		______ Ft. L x ______ Ft. H = ______ Sq. Ft. x	______ =	
W. Wall		______ Ft. L x ______ Ft. H = ______ Sq. Ft. x	______ =	
Ceiling		______ Ft. L x ______ Ft. W = ______ Sq. Ft. x	______ =	
Floor		______ Ft. L x ______ Ft. W = ______ Sq. Ft. x	______ =	
Box		(Table 14A, B, C) Total: Surface = ______ Sq. Ft. x	______ =	

I Total Wall Transmission Load BTU/24 HRS =

II_{SF} (SHORT FORM) USAGE HEAT GAIN ( ) AVG. ( ) HVY.( ) EX. HVY.
COOLERS ONLY

Refer to ii_{LF}=(Long Form) if application exceeds data shown in Table 15.

______ CU. FT. x ______ BTU/24 HR./CU. FT. (@______TD) =
(INT. BOX VOLUME) (TABLE 14) (USAGE HEAT GAIN) (TABLE 15)

NOTE: IF PRODUCT LOADS ARE UNUSUAL USE THE LONG FORM
EX. HVY. = 1½ x HVY. USAGE

TOTAL I + II_{SF} =

IF USAGE HEAT GAIN ABOVE IS USED DO NOT USE (II_{LF} III & IV)

II_{LF} (LONG FORM) INFILTRATION (AIR CHANGE LOAD)

______ CU. FT. x ______ AIR CHANGES/24 HRS. x ______ SRVC. FACTOR x ______ BTU /CU. FT =
(TABLE 14) (TABLE 9 OR 10) (FROM NOTES TABLE 9 OR 10) (TABLE 11)

II INFILTRATION LOAD BTU/24 HRS. =

III PRODUCT LOAD

PRODUCT TEMP. REDUCTION ABOVE FREEZING (SENSIBLE HEAT)

______ *LBS./DAY x ______ °F TEMP. REDUCTION x ______ SP. HT. =

PRODUCT FREEZING (LATENT HEAT LOAD)

______ *LBS./DAY x ______ BTU/LB. LATENT HEAT =

PRODUCT TEMP. REDUCTION BELOW FREEZING SENSIBLE HEAT)

______ *LBS./DAY x ______ °F TEMP. REDUCTION x ______ SP. HT. =

HEAT OF RESPIRATION

______ LBS. PRODUCT (STORAGE) x ______ BTU/LB./24 HRS. =

MISCELLANEOUS PRODUCT LOADS (1) CONTAINERS (2)PALLETS (3) OTHER

______ LBS./DAY x ______°F TEMP. REDUCTION x ______ SP. HT. =

______ LBS./DAY x ______°F TEMP. REDUCTION x ______ SP. HT. =

III TOTAL PRODUCT LOAD BTU/24 HRS. =

IV MISCELLANEOUS LOADS

(a) LIGHTS ______ Ft.² Floor Area x ______ Watts/Ft.² x 3.41 Btu/Watt x ______ HRS/24 HRS =
(1 TO 1½ WATTS/SQ. FT. IN STORAGE AREAS & 2 TO 3 FOR WORK AREAS)

(b) OCCUPANCY ______ NO. OF PEOPLE x ______ BTU/HR. x ______ HRS. ______ =

(c) MOTORS ______ BTU/HP/HR. x ______ HP x ______ HRS./24/HRS ______ =

______ BTU/HP/HR. x ______ HP x ______ HRS./24/HRS ______ =

(d) MATERIAL HANDLING
______ FORK LIFT(S) x ______ EQUIV. HP x 3100 BTU/HR./HP. x ______ HRS. OPERATION =
OTHER ______ =

IV TOTAL MISCEL. LOADS BTU/24 HRS. =

TOTAL BTU LOAD 1 TO IV BTU/24 HRS. =

ADD 10% SAFETY FACTOR =

* IF THE PRODUCT PULL-DOWN IS ACCOMPLISHED IN LESS THAN 24 HRS. THE DAILY PRODUCT WILL BE:

LBS. PRODUCT x $\frac{\text{24 HRS.}}{\text{PULLDOWN HRS.}}$

TOTAL BTU/24 HRS. WITH SAFETY FACTOR
(NOT INCLUDING EVAP. FAN OR DEFROST HEAT LOADS)
24 HR. BASE REFRIGERATION LOAD } =

Figure 16-1-2 (continued)

In order to reduce heat transfer, the thermal conductivity factor (based on material composition) should be as small as possible and the insulating material as thick as economically feasible.

Heat transfer through any material is also affected by surface resistance to heat flow caused by the type of surface, rough or smooth, and the surface position, vertical or horizontal, its reflective properties, and the rate of airflow over the surface.

Extensive testing has been done to determine accurate values for heat transfer through common building and structural materials. Certain materials have been found to have a high resistance to flow of heat and are good insulators while others are not.

Heat loss is measured by resistance (R), which is the resistance to heat flow of either 1 in of material, or a specified thickness, of an air space, an air film, or an entire assembly. R is the number of hours it takes one Btu to pass through one square foot of material with a 1°F TD. R11 means it took 11 hr for 1 Btu to pass through. R32 means it took 32 hr for the same 1 Btu to pass through. A high R value indicates low heat flow rates. The resistance of several components of a wall may be added together to obtain the total resistance:

TECH TIP

The higher the R value the longer it takes 1 Btu to pass through a material. It is however possible to spend more money on thicker insulation than the insulation will ever save on energy it will take to operate the cold storage facility. The amount of insulation is a balance between its cost and the cost to operate the refrigeration system.

$$R_t = R_1 + R_2 + R_3 + \ldots$$

Table 16-1-1 lists some R values for common building materials in order to illustrate the differences in heat flow characteristics.

The (U) value, overall heat transfer coefficient, is the reciprocal of the R value. The U value of a material is defined as the quantity of heat in Btu that will flow through 1 ft^2 of material in 1 hr with a 1°F TD. The U value is calculated by dividing R into 1 ($1/R$). For example if a material had an R value of 2 the U value would be 1 over 2 equals 0.5. So in 1 hour ½ a Btu would pass through an $R2$ material.

Both the R and U values are based on a one degree temperature difference between the inside and

TABLE 16-1-1 Typical Heat Transmission Coefficients

Material	Density (lb/ft^3)	Mean Temp. (°F)	Conductivity, k	Conductance, C	Resistance, R Per Inch	Resistance, R Overall
Insulating Materials						
Mineral wool blanket	0.5	75	0.32		3.12	
Fiberglass blanket	0.5	75	0.32		3.12	
Corkboard	6.5–8.0	0	0.25		4.0	
Glass fiberboard	9.5–11.0	−16	0.21		4.76	
Expanded urethane, R-11		0	0.17		5.88	
Expanded polystyrene	1.0	0	0.24		4.17	
Mineral wool board	15.0	0	0.25		4.0	
Insulating roof deck, 2 in		75		0.18		5.56
Mineral wool, loose fill	2.0–5.0	0	0.23		4.35	
Perlite, expanded	5.0–8.0	0	0.32		3.12	
Masonry Materials						
Concrete, sand and gravel	140		12.0		0.08	
Brick, common	120	75	5.0		0.20	
Brick, face	130	75	9.0		0.11	
Hollow tile, two-cell, 6 in		75		0.66		1.52
Concrete block, sand and gravel, 8 in		75		0.90		1.11
Concrete block, cinder, 8 in		75		0.58		1.72
Gypsum plaster, sand	105	75	5.6		0.18	

outside temperatures. As the temperature difference increases so does the rate of heat transfer. Heat loads are calculated in Btuh and the U value provides the Btuh for a square foot of material so to determine the quantity of heat transferred (Q) through a material you must know the U value, the area, and temperature difference:

$$Q = U \times A \times \text{TD}$$

where

Q = heat transfer, Btuh
U = overall heat transfer coefficient (Btuh/ft^2/°F)
A = area (ft^2)
TD = temperature difference between inside and outside design temperature and refrigerated space design temperature

EXAMPLE

Calculate the heat flow through an 8 in concrete block (cinder) wall (not including the air film effect), Figure 16-1-3, 100 ft^2 in area, having a 60°F temperature difference between the inside and the outside.

Solution

The R value of an 8 in concrete block wall, Table 16-1-1, is 1.72.
Therefore

$U = 1 \div$ R where R = total resistance of the individual components of the wall
$= 1 \div 1.72$
$= 0.58$ Btuh/°F/ft^2
$Q = 0.58 \times 100$ ft^2 $\times$ 60°F
$=$ 3,480 Btuh heat flow into the space

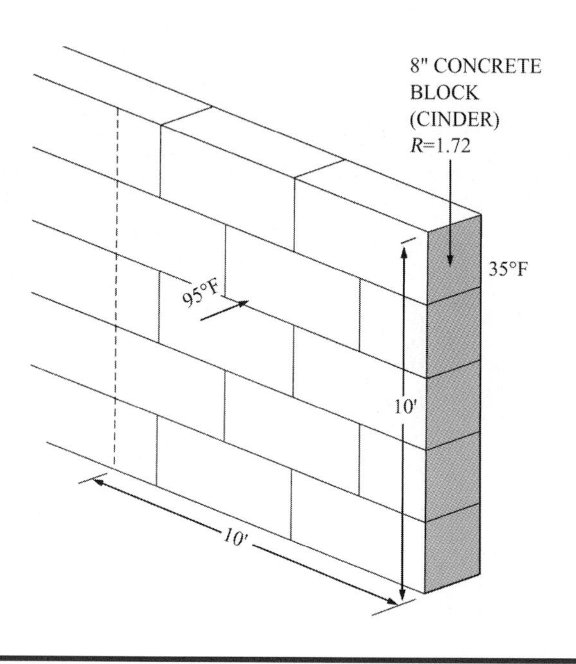

Figure 16-1-3 Heat flow through an 8 in cinder block wall.

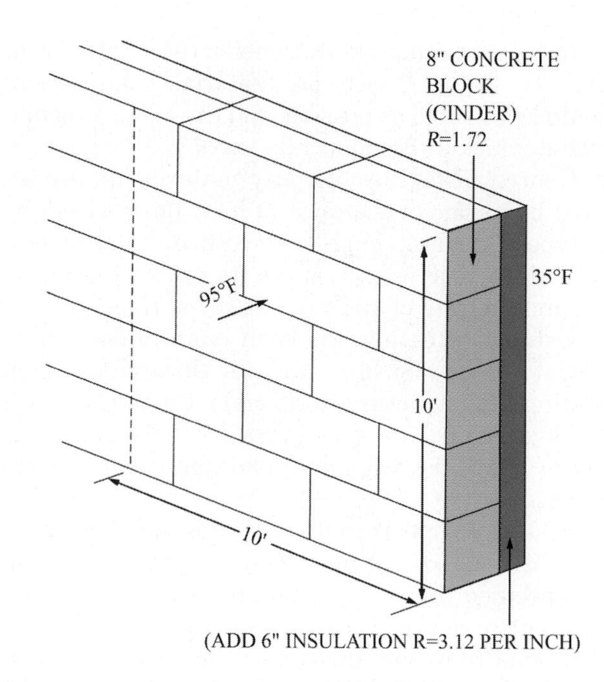

Figure 16-1-4 Heat flow through an 8 in insulated cinder block wall.

Add 6 in of fiberglass insulation to the wall, Figure 16-1-4, and recalculate the transmission load. Given that the R value for 1 in insulation is 3.12:

$R = (6 \times 3.12) = 18.72$
U factor $= 1 \div R$
$R_t = R_1$ (concrete block) $+ R_2$ (6 in insulation)
$= 1.72 + 18.72$
$= 20.44$

Therefore,

$U = 1 \div 20.44 = 0.049$
$Q = 0.049 \times 100$ ft^2 $\times$ 60°F $=$ 294 Btuh ■

The example above demonstrates the cumulative effect of R values on determining the total wall resistance. It also shows the heat reduction that can be achieved through greater insulation, from 3,480 Btuh to 294 Btuh. This would greatly reduce the size of the refrigeration equipment needed and the amount of the operating cost.

Insulation is the most effective method of reducing heat transmission. Types of insulation for various application are:

- loose fill
- rigid or semirigid
- reflective
- foamed in place

Loose fill or blown insulation, Figure 16-1-5, is used primarily in residential structures for reducing heat flow. Flexible insulation such as fiberglass, in

Figure 16-1-5 Loose fill insulation.

batts or rolls, Figure 16-1-6, are also common in new residential and commercial construction and come with a cover material such as Kraft paper, which acts as a vapor barrier. Flexible insulation may be used to wrap ducts or to place between wall studs or ceiling joists.

Rigid and semirigid insulations, Figure 16-1-7, are made of such materials as corkboard, polystyrene, foam glass, and polyurethane. These are available in boards and sheets, in various dimensions and forms. Some have a degree of structural strength.

The rigid and semirigid material is the type of insulation used in refrigeration equipment such as

NICE TO KNOW

Sawdust was used as an insulation material for both residences and refrigeration. Some homes and barns in New England that were built in the 1700s and 1800s still have sawdust insulated walls. Sawdust was used as an insulation material for medium and low temperature boxes through the 1950s. Some of the sawdust filled walls are still in service today.

walk-in coolers, freezers, and display cases. Because of its density and structure, it has a built-in vapor barrier to moisture.

Foamed in place insulation, Figure 16-1-8, is widely used for filling cavities that are hard to insulate. It is also used to cover drain pans and other places where temperature control and water seal are needed. Foamed insulation can be used in on-site, built-up refrigerated rooms in connection with rigid insulations.

TECH TIP

Foamed in place insulation provides the most effective vapor barrier of all of the insulation materials. Moisture infiltration to a cold storage box can sufficiently increase the box load and require more frequent defrosting. Foamed in place insulation should be recommended any time a cold storage box is to be constructed in a humid area.

(a)

(b)

Figure 16-1-6 Fiberglass insulation: (a) Rolls; (b) Batts.

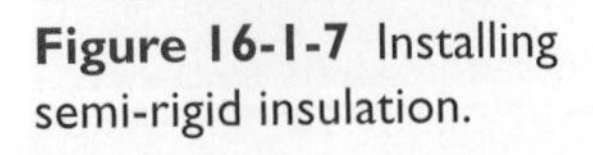

Figure 16-1-7 Installing semi-rigid insulation.

Figure 16-1-8 Application of foam type insulation.

INSULATION
83
START OF CONDENSATION
COLD SIDE
TEMP. 40° F
R.H. 80%
D.P. 34.5
TEMP. 90° F
R.H. 88%
D.P. 83°
60
70
50
80
WARM SIDE

Figure 16-1-9 Gradual temperature changes in an insulated wall from outside to inside.

For all types of insulation, moisture control is extremely important. Insulation should be dry when installed and be sealed perfectly so that it stays dry.

Figure 16-1-9 shows the gradual temperature changes that take place within an insulating material from 90°F on the warm outside to 40°F on the cold inside.

In the example, the 90°F warm air side has a dew point temperature of 83°F. (At dew point temperature air is saturated with moisture, and condensation from the vapor to a liquid occurs when the temperature is lowered.) As illustrated, when there is no effective vapor barrier or seal on the warm side, water will start to condense inside the insulation as the temperature drops below the dew point. Water is a good conductor of heat, about 15 times as fast as fiberglass. Thus, if water gets into the insulation, the material's insulating value is greatly reduced, and physical problems also occur.

Vapor seals can be formed from various materials, including metal casings, metal foil, plastic film, and asphalt coverings. The selection depends on the application.

The ability of a material to resist water vapor transmission is measured in perms, a term related to permeability. Vapor barriers of 1 perm or less have been found satisfactory for residential comfort heating and cooling work. In low temperature commercial refrigeration applications such as freezers, perm ratings of 0.10 and below may be accepted levels. As with insulation and heat flow, the resistance to vapor flow is a function of the composite of all the materials, as constructed, not just the rating of the vapor barrier itself.

One perm equals one grain (avoirdupois) of water vapor per hour flowing through one ft^2 of a material induced by a vapor pressure difference of one inch of mercury across the two surfaces. One perm is the generally accepted maximum allowable permeance value for the residential building envelope in a 5000 heating degree day climate.

The effectiveness of both insulation and vapor barriers are greatly reduced if any openings, however small, exist. Such openings may be caused by poor workmanship during construction and application, but they may also result from negligence in sealing around openings for piping and wiring. This is the responsibility of the service technician.

SERVICE TIP

An ultrasonic leak detector with an ultrasonic transmitter can be used to locate air leaks in refrigerated cabinets. Place the transmitter inside the cabinet and close the door. Follow all of the cabinet seams with the leak detector. Any opening that will allow air to pass through will allow the ultrasonic transmitter sound to be detected. This is one of the most effective ways of locating air leaks in refrigerated cabinets.

Sun Effect

The primary radiation factor in the refrigeration load is the heat gain from the sun's rays. If the walls of the refrigerated space are exposed to the sun, additional heat will be added to the heat load. A simplified method for estimating this effect is shown in Table 16-1-2.

The factors shown in °F for various conditions and locations are to be added to the normal temperature difference between indoor and outdoor design conditions.

Design Temperatures

Two design temperatures must be determined to proceed with the load calculation: outside and inside.

Recommended outside design conditions are the results of extensive studies by the National Weather Service. For air conditioning and refrigeration applications the maximum load occurs during the hottest weather. It is neither economical nor practical, however, to design equipment for the hottest temperature, since it might last for only a few hours over a span of several years. The design temperature chosen is therefore less than the peak temperature.

TABLE 16-1-2 Temperature Allowance for Sun Effect

Type of Structure	East Wall	South Wall	West Wall	Flat Roof
Dark Colored Surfaces (Slate, tar, black asphalt shingles, and black paint)	**8**	**5**	**8**	**20**
Medium Colored Surfaces (Brick, red tile, unpainted wood, dark cement)	6	4	6	15
Light Colored Surfaces (White stone, light colored cement, white asphalt shingles, white paint)	4	2	4	9

A segment of the ASHRAE outdoor design data chart is shown in Table 16-1-3. Recommended design dry bulb and wet bulb temperatures are given for American states and major cities. The inside design conditions are the recommended storage conditions for the product being stored.

Air Infiltration

Outside air entering the refrigerated space must be reduced to the storage temperature, thus increasing the refrigeration load. If the moisture content of the entering air is above that of the refrigerated space, the excess moisture will condense out of the air, also adding to the refrigeration load.

Traffic in and out of a refrigerator can vary with the size and volume of the refrigerator. Calculations for air infiltration from this source would therefore need to consider volume as well as the number of times refrigerator doors are opened.

Because of the many variables involved, it is difficult to calculate the total heat gain due to air infiltration. One method, the air change method, is based

TECH TIP

Some commercial refrigeration units that have high traffic use a combination of a strip plastic door enclosure and an air curtain. An air curtain is formed when a high velocity blast of air is blown straight down in front of a door opening. The combination of the strip curtain and air curtain can significantly reduce the load on a high traffic refrigeration system.

TABLE 16-1-3 Summer Outside Design Conditions: United States, Canada

Location	2½% Design Dry Bulb	Coincident Design Wet Bulb
Alabama		
Birmingham AP	94	75
Mobile AP	93	77
Montgomery AP	95	76
Alaska		
Anchorage AP	68	58
Barrow (S)	53	50
Juneau	70	58
Arizona		
Flagstaff AP	82	55
Phoenix AP (S)	107	71
Tucson AP (S)	102	66
Arkansas		
El Dorado AP	96	76
Fayetteville AP	94	73
Little Rock AP (S)	96	77
California		
Bakersfield AP	101	69
Eureka/Arcata AP	65	59
Los Angeles CO (S)	89	70
Sacramento AP	98	70
San Bernadino, Norton AFB	99	69
San Diego, AP	80	69
San Francisco CO	71	62
San Jose AP	81	65
Colorado		
Denver AP	91	59
Durango	87	59
Fort Collins	91	59
Connecticut		
Hartford, Brainard Field	88	73
New Haven AP	84	73
Delaware		
Wilmington AP	89	74
District of Columbia		
Washington National AP	91	74
Florida		
Gainesville AP (S)	93	77
Jacksonvlle AP	94	77
Key West AP	90	78
Miami AP (S)	90	77
Orlando AP	93	76
Tallahassee AP (S)	92	76
Tampa AP (S)	91	77
Georgia		
Atlanta AP (S)	92	74
Savannah-Travis AP	93	77
Valdosta-Moody AFB	94	77
Hawaii		
Hilo AP (S)	83	72
Honolulu AP	86	73
Idaho		
Boise AP (S)	94	64
Coeur D'Alene AP	86	61
Illinois		
Carbondale	93	77
Chicago CO	91	74
Springfield AP	92	74
Indiana		
Evansville AP	93	75
Indianapolis AP (S)	90	74
South Bend AP	89	73
Iowa		
Burlington AP	91	75
Des Moines AP	91	74
Mason City AP	88	74
Kansas		
Salina	100	74
Topeka AP	96	75
Wichita AP	98	73
Kentucky		
Bowling Green AP	92	75
Lexington AP (S)	91	73
Louisville AP	93	74
Louisiana		
Baton Rouge AP	93	77
New Orleans AP	92	78
Shreveport AP (S)	96	76
Maine		
Bangor, Dow AFB	83	68
Caribou AP (S)	81	67
Portland (S)	84	71
Maryland		
Baltimore CO	89	76
Cumberland	89	74
Salisbury (S)	91	75
Massachusetts		
Boston AP (S)	88	71
New Bedford	82	71
Springfield, Westover AFB	87	71
Worcester AP	84	70
Michigan		
Detroit	88	72
Grand Rapids AP	88	72
Marquette CO	81	69
Minnesota		
International Falls AP	83	68
Minneapolis/ St. Paul AP	89	73
Rochester AP	87	72
Mississippi		
Biloxi, Keesler AFB	92	79
Jackson AP	95	76
Tupelo	94	77
Missouri		
Kansas City AP	96	74
St. Louis CO	94	75
Springfield AP	93	74
Montana		
Billings AP	91	64
Butte AP	83	56
Great Falls AP (S)	88	60
Nebraska		
Lincoln CO (S)	95	74
Omaha AP	91	75
Scottsbluff AP	92	65
Nevada		
Las Vegas AP (S)	106	65
Reno CO	93	60
New Hampshire		
Berlin	84	69
Concord AP	87	70
Manchester, Grenier AFB	88	71

Location	2½% Design Dry Bulb	Coincident Design Wet Bulb
New Jersey		
Atlantic City CO	89	74
Newark AP	91	73
Trenton CO	88	74
New Mexico		
Albuquerque AP (S)	94	61
Las Cruces	96	64
Santa Fe CO	88	61
New York		
Albany CO	88	72
Buffalo AP	85	70
NYC-Central Park (S)	89	73
Rochester AP	88	71
Syracuse AP	87	71
North Carolina		
Charlotte AP	93	74
Raleigh/Durham AP (S)	92	75
Wilmington AP	91	78
North Dakota		
Bismark AP (S)	91	68
Fargo AP	89	71
Grands Forks AP	87	70
Ohio		
Cincinnati CO	90	72
Cleveland AP (S)	88	72
Columbus AP (S)	90	73
Oklahoma		
Muskogee AP	98	75
Oklahoma City AP	97	74
Tulsa AP	98	75
Oregon		
Eugene AP	89	66
Medford AP (S)	94	67
Portland AP	85	67
Pennsylvania		
Allentown AP	88	72
Philadelphia AP	90	74
Pittsburgh CO	88	71
Rhode Island		
Providence AP	86	72
South Carolina		
Charleston CO	92	78
Columbia AP	95	75
Greenville AP	91	74
South Dakota		
Aberdeen AP	91	72
Pierre AP	95	71
Sioux Falls AP	91	72
Tennessee		
Chattanooga AP	93	74
Memphis AP	95	76
Nashville AP (S)	94	74
Texas		
Corpus Christi AP	94	78
Dallas AP	100	75
Houston CO	95	77
Lubbock AP	96	69
San Antonio AP (S)	97	73
Utah		
Cedar City AP	91	60
Provo	96	62
Salt Lake City AP (S)	95	62
Vermont		
Barre	81	69
Burlington AP (S)	85	70
Rutland	84	70
Virginia		
Norfolk AP	91	76
Richmond AP	92	76
Roanoke AP	91	72
Washington		
Bellingham AP	77	65
Seattle-Tacoma AP (S)	80	64
Spokane AP (S)	90	63
West Virginia		
Charleston AP	90	73
Huntington CO	91	74
Wheeling	86	71
Wisconsin		
Green Bay AP	85	72
Madison AP (S)	88	73
Milwaukee AP	87	73
Wyoming		
Casper AP	90	57
Cheyenne AP	86	58
Sheridan AP	91	62
Alberta		
Calgary AP	81	61
Edmonton AP	82	65
British Columbia		
Vancouver AP (S)	77	66
Victoria CO	73	62
Manitoba		
Flin Flon	81	66
Winnipeg AP (S)	86	71
New Brunswick		
Fredericton AP (S)	85	69
Saint John AP	77	65
Newfoundland		
Gander AP	79	65
St. John's AP (S)	75	65
Northwest Terr.		
Fort Smith AP (S)	81	64
Yellowknife AP	77	61
Nova Scotia		
Halifax AP (S)	76	65
Yarmouth AP	72	64
Ontario		
Sudbury AP	83	67
Thunder Bay AP	83	68
Toronto AP (S)	87	72
Prince Edward Island		
Charlottetown AP (S)	78	68
Quebec		
Chicoutimi	83	68
Montreal AP (S)	85	72
Quebec AP	84	70
Saskatchewan		
Regina AP	88	68
Saskatoon AP (S)	86	66
Yukon Territory		
Whitehorse AP (S)	77	58

AP–Airport CO–City Office S–Solar Data Available

TABLE 16-1-4 Average Air Changes per 24 Hours, for Storage Rooms Due to Door Openings and Infiltration (Above 32°F)

Volume (cu ft)	Air Changes per 24 hr	Volume (cu ft)	Air Changes per 24 hr
200	44.0	6,000	6.5
300	34.5	8,000	5.5
400	29.5	10,000	4.9
500	26.0	15,000	3.9
600	23.0	20,000	3.5
800	20.0	25,000	3.0
1,000	17.5	30,000	2.7
1,500	14.0	40,000	2.3
2,000	12.0	50,000	2.0
3,000	9.5	75,000	1.6
4,000	8.2	100,000	1.4
5,000	7.2		

Note: For heavy usage multiply the above values by 2. For long storage multiply the above values by 0.6.

on the average number of air changes in a 24 hr period, compared to the refrigerator volume, as illustrated in Table 16-1-4.

This method is used for rooms where the temperature is above 32°F for average use. Values are increased by 2 where there is heavier use and higher temperatures. For storage at 0°F or below with decreased usage, the values are reduced.

Another means of computing infiltration is by the velocity of airflow through an open door. Charts are available that list average infiltration velocity based on door height and temperature difference. If the average time the door is opened each hour can be determined, the average hourly infiltration can be calculated.

Once the rate of infiltration in ft^3 per hr has been determined by either method, the heat load can be calculated from the heat removed per ft^3, given in Table 16-1-5.

EXAMPLE

Assume that the volume of a refrigerated room is 1,000 ft^3 and the storage temperature is 40°F, with an outside temperature of 95°F and 60% relative humidity. Calculate the heat load.

Solution

From Table 16-1-4, note that a 1,000 ft^3 volume would average 17.5 air changes per 24 hr. This would produce an infiltration of 17,500 ft^3 in 24 hr (1,000 × 17.5).

Referring to Table 16-1-5, for a room at 40°F with 95°F outside temperature and 60% RH, the Btu/ft^3 of heat is 2.62. Therefore, in 24 hr the load would be 2.62 × 17,500, or 45,850 Btu. In 1 hr it is 45,850 divided by 24, or 1,910 Btuh. ■

In systems where ventilation is provided by supply and/or exhaust fans, the ventilation load will replace the infiltration load if it is greater. The heat gain may be calculated on the basis of the ventilating equipment air volume.

Product Load

The product load is any heat gain from the product in the refrigerated space. The load may come from higher outside temperatures, from a chilling or freezing process, or from the heat of respiration of perishable products. It may also be the sum of various types of product loads.

TECH TIP

There is a significant difference in box design required for different types of product loads. For example a fresh flower storage unit must have a relatively high humidity to prevent dehydration of the stored flowers; these boxes would have evaporative coils designed for maximum sensible cooling and minimum latent cooling. Conversely a box designed to store wheat flour, rice flour, or corn meal must have a very low relative humidity to prevent mold growth. These storage units would have good sensible cooling and excellent latent cooling coils.

To calculate the refrigeration product load for food products (solids and liquids), it is necessary to know their freezing points, specific heats, percent water, etc.

Sensible Heat Load Above Freezing

Most products are at a higher temperature than the storage temperature when placed in the refrigerator. Since many foods have a high water content, their reaction to a loss of heat is quite different above and below the freezing point due to the change in the state of water. Specific heat of a product is defined as the Btu required to raise the temperature of 1 lb of the substance 1°F.

The heat to be removed from a product to lower its temperature (above freezing) may be calculated as follows:

$$Q = W \times C \times (T_1 - T_2)$$

TABLE 16-1-5 Heat Removed in Cooling Air to Storage-Room Conditions (Btu per cu ft)

Storage Room Temp °F	Temperature of Outside Air (°F) 85		90		95		100	
	Relative Humidity (Percent)							
	50	60	50	60	50	60	50	60
65	0.65	0.85	0.93	1.17	1.24	1.54	1.58	1.95
60	0.85	1.03	1.13	1.37	1.44	1.74	1.78	2.15
55	1.12	1.34	1.41	1.66	1.72	2.01	2.06	2.44
50	1.32	1.54	1.62	1.87	1.93	2.22	2.28	2.65
45	1.50	1.73	1.80	2.06	2.12	2.42	2.47	2.85
40	1.69	1.92	2.00	2.26	2.31	2.62	2.67	3.06
35	1.86	2.09	2.17	2.43	2.49	2.79	2.85	3.24
30	2.00	2.24	2.26	2.53	2.64	2.94	2.95	3.35

Storage Room Temp °F	Temperature of Outside Air (°F) 40		50		90		100	
	Relative Humidity (Percent)							
	70	80	70	80	50	60	50	60
30	0.24	0.29	0.58	0.66	2.26	2.53	2.95	3.35
25	0.41	0.45	0.75	0.83	2.44	2.71	3.14	3.54
20	0.56	0.61	0.91	0.99	2.62	2.90	3.33	3.73
15	0.71	0.75	1.06	1.14	2.80	3.07	3.51	3.92
10	0.85	0.89	1.19	1.27	2.93	3.20	3.64	4.04
5	0.98	1.03	1.34	1.42	3.12	3.40	3.84	4.27
0	1.12	1.17	1.48	1.56	3.28	3.56	4.01	4.43
−5	1.23	1.28	1.59	1.67	3.41	3.69	4.15	4.57
−10	1.35	1.41	1.73	1.81	3.56	3.85	4.31	4.74
−15	1.50	1.53	1.85	1.92	3.67	3.96	4.42	4.86
−20	1.63	1.68	2.01	2.09	3.88	4.18	4.66	5.10
−25	1.77	1.80	2.12	2.21	4.00	4.30	4.78	5.21
−30	1.90	1.95	2.29	2.38	4.21	4.51	4.90	5.44

(Copyright by the American Society of Heating, Refrigerating, and Air-Conditioning Engineers, Inc.)

where

Q = Btu to be removed
W = weight of product in lb
C = specific heat above freezing
T_1 = initial temperature, °F
T_2 = final temperature, °F (freezing or above)

For example, the heat to be removed in order to cool 1,000 lb of veal (whose freezing point is 29°F) from 42°F to 29°F can be calculated as follows:

$$Q = W \times C \times (T_1 - T_2)$$
$$= 1{,}000 \text{ lb} \times 0.71 \text{ specific heat (veal)} \times 13°\text{F}$$
$$= 9{,}230 \text{ Btu}$$

Heat of Respiration

Products such as fresh fruits and vegetables, even when stored at above freezing temperatures, give off some heat due to respiration. Respiration is the oxidation process of ripening, and carbon dioxide and heat are by-products. This load varies with the type and temperature of the product. Tabulated values given in Btu/lb/24 hr are shown in Table 16-1-6, and are applied to the total weight of the product stored, not just to the daily turnover. Use a value close to freezing (32°F).

The heat of respiration is a major factor for refrigerated storage of freshly picked fruit. The fruit

TABLE 16-1-6 Heat of Respiration

	Btu/lb/24 hrs.			
	Storage Temperature			
Product	**32°F**	**40°F**	**60°F**	**°F Other**
		Fruits		
Apples	.25–.450	.55–.80	1.5–3.4	
Apricots	0.55–.63	.70–1.0	2.33–3.74	
Avocados	–	–	6.6–15.35	
Bananas	–	–	2.3–2.75	@ 68° 4.2–4.6
Blackberries	1.70–2.52	5.91–5.0	7.71–15.97	
Blueberries	0.65–1.10	1.0–1.35	3.75–6.5	@ 70° 5.7–7.5
Cherries	0.65–0.90	1.4–1.45	5.5–6.6	
Cherries, Sour	0.63–1.44	1.41–1.45	3.0–5.49	
Cranberries	0.30–0.35	0.45–0.50	–	
Figs, Mission	–	1.18–1.45	2.37–3.52	
Gooseberries	0.74–0.96	1.33–1.48	2.37–3.52	
Grapefruit	0.20–0.50	0.35–0.65	1.1–2	
Grapes–American	0.30	0.60	1.75	
Grapes–European	0.15–0.20	0.35–0.65	1.10–1.30	
Lemons	0.25–0.45	0.30–0.95	1.15–2.50	
Limes	–	0.405	1.485	
Melons–Cantaloupes	0.55–0.63	0.96–1.11	3.70–4.22	
Melons–Honey Dew	–	0.45–0.55	1.2–1.65	
Oranges	0.20–0.50	0.65–0.8	1.85–2.6	
Peaches	0.45–0.70	0.70–1.0	3.65–4.65	
Pears	0.35–0.45	–	4.40–6.60	
Plums	0.20–0.35	0.45–0.75	1.20–1.40	
Raspberries	1.95–2.75	3.40–4.25	9.05–11.15	
Strawberries	1.35–1.90	1.80–3.40	7.80–10.15	
Tangerines	1.63	2.93	–	
		Vegetables		
Artichokes (Globe)	2.48–4.93	3.48–6.56	8.49–15.90	
Asparagus	2.95–6.60	5.85–11.55	11.0–25.75	
Beans, Green or Snap	–	4.60–5.7	16.05–22.05	
Beans, Lima	1.15–1.6	2.15–3.05	11.0–13.7	
Beets, Topped	1.35	2.05	3.60	
Broccoli	3.75	5.50–8.80	16.9–25.0	
Brussels Sprouts	1.65–4.15	3.30–5.50	6.60–13.75	
Cabbage	0.60	0.85	2.05	
Carrots, Topped	1.05	1.75	4.05	
Cauliflower	1.80–2.10	2.10–2.40	4.70–5.40	
Celery	0.80	1.20	4.10	
Corn, Sweet	3.60–5.65	5.30–6.60	19.20	
Cucumbers	–	–	1.65–3.65	
Garlic	0.33–1.19	0.63–1.08	1.18–3.0	
Horseradish	0.89	1.19	3.59	
Kohlrabi	1.11	1.78	5.37	

Btu/lb/24 hrs.				
	Storage Temperature			
Product	**32°F**	**40°F**	**60°F**	**°F Other**
		Vegetables		
Leeks	1.04–1.78	2.15–3.19	9.08–12.82	
Lettuce, Head	1.15	1.35	3.95	
Lettuce, Leaf	2.25	3.20	7.20	
Mushrooms	3.10–4.80	7.80	–	@ 50° 11.0
Okra	–	6.05	15.8	
Olives	–	–	2.37–4.26	
Onions, Dry	0.35–0.55	0.90	1.20	
Onions, Green	1.15–2.45	1.90–7.50	7.25–10.70	
Peas, Green	4.10–4.20	6.60–8.0	19.65–22.25	
Peppers, Sweet	1.35	2.35	4.25	
Potatoes, Immature	–	1.30	1.45–3.4	
Potatoes, Mature	–	0.65–0.90	0.75–1.30	
Potatoes, Sweet	–	0.85	2.15–3.15	
Radishes with Tops	1.59–1.89	2.11–2.30	7.67–8.5	
Radishes, Topped	0.59–0.63	0.85–0.89	3.04–3.59	
Rhubarb, Topped	0.89–1.44	1.19–2.0	3.41–4.97	
Spinach	2.10–2.45	3.95–5.60	18.45–19.0	
Squash, Yellow	1.3–1.41	1.55–2.04	8.23–9.97	
Tomatoes, Mature Green	–	0.55	3.10	
Tomatoes, Ripe	0.50	0.65	2.8	
Turnips	0.95	1.10	2.65	
Vegetables, Mixed	2.0	–	–	
		Miscellaneous		
Caviar, Tub	–	–	1.91	
Cheese, American	–	–	2.34	
Camembert	–	–	2.46	
Limburger	–	–	2.46	
Roquefort	–	–	–	@ 45° 2.0
Swiss	–	–	2.33	
Flowers, Cut	0.24 Btu/24 Hrs/Sq Ft Floor Area			
Honey	–	0.71	–	
Hops	–	–	–	@ 35° 0.75
Malt	–	–	–	@ 50° 0.75
Maple Sugar	–	–	–	@ 45° 0.71
Maple Syrup	–	–	–	@ 45° 0.71
Nuts	0.074	0.185	0.37	
Nuts, Dried	–	–	–	@ 35° 0.50

Notes:
All fruits and vegetables are living and give off heat in storage.
If heat of respiration not given, an approximate value or average value should be used.
For Btu/24 hrs./Ton/°F multiply by 2,000.

(Courtesy of Dunham-Bush, Inc.)

often comes in with a high sensible heat which increases its respiration rate. Rapid cooling of the fruit can reduce the respiration and slow the ripening process.

Load Due to Latent Heat of Freezing

To calculate the heat removal required to freeze food products having a high percentage of water, only the water needs to be considered. The latent heat of freezing is found by multiplying the latent heat of water (144 Btu/lb) by the percentage of water in the food product.

EXAMPLE

Veal is 63% water and its latent heat is 91 Btu/lb (0.63 × 144 Btu/lb = 91 Btu). Calculate the latent heat of freezing 1,000 lb of veal.

Solution

The heat to be removed from a product for the latent heat of freezing may be calculated as follows:

$$Q = W \times h_f$$

where

Q = Btu to be removed
W = weight of product, lb
h_f = latent heat of fusion, Btu/lb

The latent heat of freezing 1,000 lb of veal at 29°F is:

$$\begin{aligned} Q &= W \times h_f \\ &= 1{,}000 \text{ lb} \times 91 \text{ Btu/lb} \\ &= 91{,}000 \text{ Btu} \end{aligned}$$

■

Sensible Heat Load Below Freezing

Once the water content of a product has been frozen, sensible cooling can occur again in the same manner as freezing, with the exception that the ice in the product causes the specific heat to change. For example, note that the specific heat of veal above freezing is 0.71 while the specific heat of veal below freezing is 0.39.

The heat to be removed from a product to reduce its temperature below freezing is calculated as follows:

$$Q = W \times C_i \times (T_f - T_3)$$

where

Q = Btu to be removed
W = weight of product, lb
C_i = specific heat below freezing
T_f = freezing temperature
T_3 = final temperature

EXAMPLE

Calculate the heat to be removed in order to cool 1,000 lb of veal from 29°F to 0°F.

Solution

$$\begin{aligned} Q &= W \times C_i \times (T_f - T_3) \\ &= 1{,}000 \text{ lb} \times 0.39 \text{ specific heat} \times (29°\text{F} - 0°\text{F}) \\ &= 1{,}000 \times 0.39 \times 29 \\ &= 11{,}310 \text{ Btu} \end{aligned}$$

■

Total product load is the sum of the individual calculations for the sensible heat above freezing, the latent heat of freezing, and the sensible heat below freezing. In the preceding example of 1,000 lb of veal cooled and frozen from 42°F to 0°F, the total product load would be:

Sensible heat above freezing = 9,230 Btu
Latent heat of freezing = 91,000 Btu
Sensible heat below freezing = 11,310 Btu
Total product load = 111,540 Btu

If several different products are to be considered, separate calculations must be made for each item for an accurate estimate of the total product load. Note that in all the calculations above, the time factor is not considered, either for initial pull-down or storage life.

Supplementary Loads

Heat gain from other sources must also be included in the total cooling load estimate. Some example are:

1. Electric energy dissipated in the refrigerated space through lights and heaters (such as defrost) is converted into heat. One watt (W) of electric power equals 3.41 Btu.

2. For an electric motor in the refrigerated space, Table 16-1-7 gives the approximate Btuh.

3. For people working inside the refrigerated space, the following table gives the approximate heat generated. (Use Table 16-1-8 for Heat Equivalent/Person—Btuh.)

TECH TIP

Some supplemental box loads can be more significant than the product load. For example large refrigerated rooms are used to process fresh vegetables into salad mixes. These rooms contain large processing equipment including shredders, washers, baggers and sealers along with a substantial number of people. In addition, because this is a work area there is a substantial light load.

TABLE 16-1-7 Btu per Hour at Various Motor Hp Values

Motor Hp	Btuh	Motor Hp	Btuh
⅙	710	1	3,220
¼	1,000	1½	4,770
⅓	1,290	2	6,380
½	1,820	3	9,450
¾	2,680	5	15,600

TABLE 16-1-8 Heat Equivalent per Person per Btu per Hour at Various Cooler Temperatures

Cooler Temperature (°F)	Heat Equivalent/Person/Btuh
50	720
40	840
30	950
20	1,050
10	1,200
0	1,300
−10	1,400

The total supplementary load is the sum of the individual factors contributing to it. For example, the total supplementary load in a refrigerated storeroom maintained at 0°F in which there are 300 W of electric lights, a 3 Hp motor driving a fan, and two people working continuously would be as follows:

$$\begin{aligned} 300\text{ W} \times 3.41\text{ Btuh} &= 1{,}023\text{ Btuh} \\ 3\text{ Hp motor} &= 9{,}450\text{ Btuh} \\ 2\text{ people} \times 1{,}300\text{ Btuh} &= 2{,}600\text{ Btuh} \\ \text{Total supplementary load} &= 13{,}073\text{ Btuh} \end{aligned}$$

Total Hourly Load

For refrigeration appliances produced in quantity, the load is usually specified by the manufacturer. The refrigeration equipment is preselected and may be already installed in the fixture.

If it must be estimated, the expected load should be calculated by determining the heat gain due to each of the factors contributing to the total load. The most accurate methods use forms and data available from the manufacturer for such purposes, and each factor is considered separately as well.

Refrigeration equipment is designed to function continuously, and normally the compressor operating time is determined by the requirements of the defrost system. The load is calculated on a 24 hr basis and the required hourly compressor capacity is determined by dividing the 24 hr load by the desired number of hours of compressor operation during the 24 hr period. A reasonable safety factor must be provided to enable the unit to recover rapidly after a temperature rise, and to allow for any load that might be larger than originally estimated.

In cases where the refrigerant evaporating temperature does not drop below 30°F, frost will not accumulate on the evaporator, and no defrost period is necessary. The compressor for such applications is chosen on the basis of an 18 to 20 hr period of operation.

For applications with storage temperatures of 35°F or higher and refrigerant temperatures low enough to cause frosting, it is common practice to defrost by stopping the compressor and allowing the return air to melt the ice from the coil. Compressors for such applications should be selected for 16 to 18 hr operation periods.

On low temperature applications, some integrated means of defrosting must be provided. With normal defrost periods, 18 hr compressor operation is usually acceptable, although some systems are designed for continuous operation except during the defrost period.

TECH TIP

Hot gas defrosting is more efficient than electric resistance defrosting. However, electric resistance defrosting is a lot less expensive to install and maintain than a hot gas system. You must keep these factors in mind when making a recommendation or bid proposal to a customer.

An additional 5 to 10% safety factor is often added to load calculations as a conservative measure to be sure that the equipment will not be undersized. When data concerning the refrigeration load are uncertain, this practice may be desirable. Since, however, the compressor is sized on the basis of 16 or 18 hr operation, this provides a sizable safety factor. The load should be calculated on the basis of the peak demand at design conditions. Usually the design conditions are selected on the assumption that peak demand will occur no more than 1% of the hours during the summer months.

Selection of Equipment

The following information is important to know in the selection of equipment:

1. Daily cooling load
2. Running time for the compressors (hours per day)
3. Hourly cooling load in Btuh
4. Refrigerated room temperature

5. Refrigerated room relative humidity (%RH)
6. Coil operating temperature difference (TD = room temperature − evaporating temperature)
7. Evaporator temperature
8. Drop in suction line temperature due to pressure loss between coil and compressor
9. Condensing temperature of the compressor including discharge line temperature drop to remote condenser
10. Available coil and compressor sizes

The following tables are useful in determining the required compressor and evaporator operating conditions:

1. Storage temperature and humidity conditions for various foods, Table 16-1-9. Note the moisture classification for the foods.
2. Coil temperature differences for four classes of foods, Table 16-1-10.
3. Compressor running time, shown in Table 16-1-11.

TABLE 16-1-9 Storage Temperature and Humidity Conditions for Various Food Products

	Short-Time Storage		24–72 Hr. Storage	
Product	**°F**	**Moisture**	**°F**	**Moisture**
Vegetables	36–42	Class 2	32–36	Class 1
Fruits	36–42	Class 2	32–36	Class 2
Meats (cut)	34–38	Class 2	32–36	Class 2
Meats (carcass)	34–38	Class 3	32–36	Class 3
Poultry	32–36	Class 2	30–35	Class 1–2‡
Fish	35–40	Class 1	35–40	Class 1
Eggs	36–42	Class 2	31–35	Class 1
Butter, Cheese	38–45	Class 1*	35–40	Class 2–3†
Bottled Beverages	35–45	Class 4	40–45	Class 4
Frozen Foods			0	Class 2

ASRE 1959 Data Book and ASHRAE 1962. Guide and Data book. Reprinted by permission.
*If not packaged. †If packaged. ‡Freeze and hold at 0°F or below if held for more than 72 hours.

(Courtesy of Dunham-Bush, Inc.)

CLASSES OF FOOD

Class 1. Such products as eggs, unpackaged butter and cheese and most vegetables held for comparatively long periods of time. These products require very high relative humidity because it is necessary to effect a minimum of moisture evaporation during storage.

Class 2. Such foods as cut meats, fruits and similar products. These require high relative humidities but not as high as Class 1.

Class 3. Carcass meats and fruit such as melons which have tough skins. These products require only moderate relative humidities because they have surfaces whose rate of moisture evaporation is moderate.

Class 4. Canned goods, bottled goods, and other products which have a protective covering. These are products which need only low relative humidities or which are unaffected by humidity. Products from whose surfaces there is a very low rate of moisture evaporation, or none at all, fall into this class.

TABLE 16-1-10 Temperature Differences Between Evaporator Coil and Room for Various Types of Food Products

Type Coils	Class 1	Class 2	Class 3	Class 4
Forced Air Coils	6–9°F	9–12°F	12–20°F	Above 20°F
Gravity Coils	14–18°F	18–22°F	21–28°F	27–37°F

Temperature difference is defined as average fixture temperature minus average refrigerant temperature.

(Courtesy of Dunham-Bush, Inc.)

TABLE 16-1-11 Compressor Running Times for Various Food-Storage Room Temperatures

Room Temp. °F	Evap. Temp. °F	Defrosting		Compressor Running Time Hr	Rel. Hum. %	Evap. °F TD
		Type	No/24 Hrs			
Over 35	Over 30	None	–	18–20 & up	90	8–10
35 & up		Ambient	4	16	85	10–12
35 to 25	Below 30	Elec. or	4	18–(20*)	80	12–15
10 & less		Hot Gas	6	(18*)–20	75	16–20

* Preferred compressor running time.

(Courtesy of Dunham-Bush, Inc.)

Sample Calculations

EXAMPLE

Calculate the load and select the equipment for a walk-in cooler having the following specifications:

Location:	Atlanta, GA
Room use:	Storage mixed vegetables
Outside dimensions:	12 ft × 14 ft × 9 ft
Room temperature:	35°F
Insulation:	3 in molded polystyrene
Total room product load:	8,000 lb
Daily delivery:	2,000 lb precooled to 45°F
Containers:	Paper cartons (200 lbs)

Calculations are made as shown on the data sheet, Figure 16-1-10. For condensing unit performance see Table 16-1-12 and for evaporator performance see Table 16-1-13. ■

EXAMPLE

Calculate the load and select the equipment for a walk-in cooler having the following specifications:

Location:	Atlanta, GA
Room use:	Storage frozen beef
Outside dimension:	10 ft × 12 ft × 9 ft
Room temperature:	0°F
Insulation:	5 in molded polystyrene
Total room product load:	2,000 lb
Daily delivery:	300 lb precooled to 40°F
Containers:	Paper cartons (200 lb)
Type of defrost:	Electric

Calculations are made as shown on the data sheet, Figure 16-1-11. For condensing-unit performance, see Table 16-1-12; for evaporator performance see Figure 16-1-12. ■

TECH TALK

Tech 1. The boss had me working down at the supermarket all day yesterday starting a large walk-in cooler. He had me sit there and watch my gauges and clamp-on ammeter. Every time the amp draw went down a little he had me open up the low side service valve. That took almost the entire day. I don't understand why he had me do it; it seemed like a big waste of time.

Tech 2. Well, on a low temperature box like you were working on, if you simply turn it on and let all the heat in the box start going out to the compressor, then you would overwork the compressor. Low temperature compressors like that one work best when the suction pressure is around 20 psi. But on a hot refrigerator box that pressure could have gone up well over 100 psi. All of that would have overworked the compressor and it could have died before you ever got the box up and running.

Tech 1. Yeah, I noticed that the pressure on the suction was somewhere around 20 to 30 lb with amp draw where he wanted it.

Tech 2. Yes the amp draw and the suction pressure were both going to go down as the box got colder.

Tech 1. Thanks, I appreciate that.

Location: Atlanta, Ga.
Room use: Storage of mixed vegetables
Outside dimensions: 12 x 14 x 9
Room temperature: 35°F

Insulation: 3 in. molded polystyrene
Total room load: 8000 lb
Daily delivery: 2000 lb precooled to 45°F
Containers: Paper cartons, total weight 200 lb

- Load Survey & Estimating Data -

DESIGN AMBIENT: 95 °F DB, ____ °F WB, 50 %RH, ____ °F SUMMER GROUND TEMP (USE 55°F FOR INSULATED FREEZE FLOOR SLAB)

ROOM DESIGN: 35 °F DB, ____ °F WB, 80 % RH ____ °F WINTER DESIGN AMBIENT

ACCESS AREA: 95 °F DB, ____ °F WB, 50 %RH, (ANTE-RM/LOADING DOCK/OTHER)

ROOM DIM. OUTSIDE: 12 FT. W 14 FT. L 9 FT. H 804 TOTAL SQ. FT (OUTSIDE SURFACE)

	Type	Insulation (Table 6) inches Thick	K Factor	U* Factor	Wall Thickness	Adj. Area °F	Effective Wall TD	°F Sun Effect (Table 8)	Total TD	Overall Wall Heat Gain BTU/24 hrs/sq. ft (Table 6)
N. Wall	POLY.	3"	20	0.67	5"	95	60		60	96
S. Wall										
E. Wall										
W. Wall										
Ceiling										
Floor										

$$\text{* 'U' - Factor} = \frac{K}{\text{Insul. Thickness (in.)}}$$

REFRIG. DOOR(S): (1) 7 x 4 VENT. FAN(S): ____

ROOM INT. VOL: 11 W x 13 L x 8 H = 1144 CU. FT.
(INSIDE ROOM DIMENSION = OUTSIDE DIMENSION WALL THICKNESSES)

FLOOR AREA 11 W x 13 L = 143 SQ. FT.

ELECTRICAL POWER 240 VOLTS, 3 PH. 60 HERTZ; CONTROL 120 VOLTS

TYPE CONTROL: ____

PRODUCT DATA AND CLASS OF PRODUCT: ____

Type Product	Amount of Product (Refer to Pg. 2 Item 3) Amount Storage	Daily Turn-Over	Freezing or Cooling	Product Temp. °F Entering	Final	Table 31 Specific Heat Above Freeze	Below Freeze	Lat. Ht. Freeze Btu/lb.	Highest Product Freeze Temp.	Table 42 Heat or Respir'n Btu/lb 24 Hr.	( ) Pull-Down ( ) Freezing Times Hrs.
MIX. VEG.	8000	2000	COOLING	45	35	.9				2.0	

EVAP. TD 10, TYPE DEFROST ☐AIR, ☐HOT GAS, ☐ELECTRIC,

CLASS PRODUCT 2,

NO. OF DEFROSTS & TOTAL TIME PER 24 HRS. ____ NO., ____ HRS.

COMPRESSOR RUNNING TIME 16 HRS.

BOX USAGE ☐AVERAGE, ☐HEAVY, ☐ EXTRA HEAVY

PRODUCT LOAD AND ADDITIONAL INFORMATION: 2000 lb. mixed vegetables per day entering temperature 45°F

PACKAGING Paper cartons CONTAINERS ____ WGT 200 lbs. SP. HT. .32 (CONTAINER)

PALLETS: NO. ____ SIZE ____ WGT. EA. ____ SP. HT. ____

PRODUCT RACKS: NO. ____ MAT'L ____ WGT. EA. ____ SP. HT. ____

ESTIMATING PRODUCT LOADING CAPACITY OF ROOM

ESTIMATED PRODUCT LOADING = 0.40 x ____ (ROOM VOLUME) CU. FT x ____ (LOADING DENSITY) LBS./CU. FT = ____ LBS:

MISCELLANEOUS LOADS

PEOPLE NO. ____ HRS. ____

FORK LIFTS ____ NO., ____ HP, ____ HRS./DAY, OTHER

LIGHTS ____ WATTS/SQ. FT.

MOTORS (OTHER THAN EVAP. FAN)

USE: ____, ____ HP, ____ HRS.
____, ____ HP, ____ HRS.

Figure 16-1-10 Load calculation data sheet for food storage rooms above freezing.

Calculations

I WALL LOSS (TRANSMISSION LOAD)

SURFACE	TD	AREA OF SURFACE	WALL HEAT GAIN FACTOR	BTU/24 HR
N. Wall		____ Ft. L x ____ Ft. H = ____ Ft². x	____ =	
S. Wall		____ Ft. L x ____ Ft. H = ____ Ft². x	____ =	
E. Wall		____ Ft. L x ____ Ft. H = ____ Ft². x	____ =	
W. Wall		____ Ft. L x ____ Ft. H = ____ Ft². x	____ =	
Ceiling		____ Ft. L x ____ Ft. W = ____ Ft². x	____ =	
Floor		____ Ft. L x ____ Ft. W = 804 Ft². x	96 =	77,184
Box	60	Total: Surface = ____ Ft². x	____ =	77,184
		I TOTAL WALL TRANSMISSION LOAD	=	

II (LONG FORM) INFILTRATION (AIR CHANGE LOAD)

1144 CU. FT. x 17.5 AIR CHANGES/24 HR x 1 SERVICE FACTOR x 2.49 BTU /CU. FT³ = 49,850

II TOTAL INFILTRATION LOAD = 49,850

III PRODUCT LOAD

PRODUCT TEMP. REDUCTION ABOVE FREEZING (SENSIBLE HEAT)

2000 *LB/DAY x 10 °F TEMP. REDUCTION x .9 SP. HT. = 18,000

PRODUCT FREEZING (LATENT HEAT LOAD)

____ *LB/DAY x ____ BTU/LB/LATENT HEAT ____ =

PRODUCT TEMP. REDUCTION BELOW FREEZING (SENSIBLE HEAT)

____ *LB/DAY x ____ °F TEMP. REDUCTION x ____ SP. HT. =

HEAT OF RESPIRATION

8000 LB PRODUCT (STORAGE) x 2.0 BTU/LB./24 HR = 16,000

MISCELLANEOUS PRODUCT LOADS (1) CONTAINERS (2)PALLETS (3) OTHER

200 LB/DAY x 10 °F TEMP. REDUCTION x .32 SP. HT. = 640

____ LB/DAY x ____ °F TEMP. REDUCTION x ____ SP. HT. =

III TOTAL PRODUCT LOAD BTU/24 HRS. = 34,640

IV MISCELLANEOUS LOADS

(a) LIGHTS ____ Ft.² Floor Area x ____ W/Ft.² x 3.41 Btu/W x ____ HR/24 HR =
(1 TO 1½ W/FT² IN STORAGE AREAS & 2 TO 3 FOR WORK AREAS)

(b) OCCUPANCY ____ NO. OF PEOPLE x ____ BTU/HR x ____ HR =

(c) MOTORS ____ BTU/HP/HR x ____ HP x ____ HR/24 HR =

____ BTU/HP/HR x ____ HP x ____ HR/24 HR =

(d) MATERIAL HANDLING
____ FORK LIFT(S) x ____ EQUIV. HP x 3100 BTU/HR/HP x ____ HR OPERATION =
OTHER ____ =

IV TOTAL MISCELLANEOUS LOADS =

TOTAL BTU LOAD 1 TO IV = 161,674

ADD 10% SAFETY FACTOR = 16,167

TOTAL BTU/24 HR WITH SAFETY FACTOR (NOT INCLUDING EVAP. FAN OR DEFROST HEAT LOADS) 24 HR. BASE REFRIGERATION LOAD = 177,841

* IF THE PRODUCT PULL-DOWN IS ACCOMPLISHED IN LESS THAN 24 HR THE DAILY PRODUCT WILL BE:

POUNDS PRODUCT x $\frac{\text{24 HRS.}}{\text{PULL-DOWN HRS.}}$

Figure 16-1-10 (continued)

Equipment Selection from Load Calculation Form

1. DETERMINE EVAP. TD REQUIRED FOR CLASS
   OF PRODUCT AND ROOM TEMP __10__ °F (TD) (FROM LOAD SURVEY DATA)

2. DETERMINE COMPRESSOR RUNNING TIME BASED ON OPERATING TEMPERATURES
   AND DEFROST REQUIREMENTS __16__ HRS. (FROM LOAD SURVEY DATA)

3. EVAPORATOR TEMP°F = __35__ (ROOM TEMP) - __10__ (EVAP. TD) = __25__ °F (FROM LOAD SURVEY DATA)

4. COMP. SUCT. TEMP.°F = __25__ (EVAP. SUCT. TEMP) - __2__ (SUCT. LINE LOSS) = __23__ °F

BTU/24 HR. BASE REFRIGERATION LOAD WITH SAFETY FACTOR = __177,641__
(NOT INCLUDING EVAPORATOR FAN OR DEFROST HEAT)

PRELIMINARY HOURLY LOAD = BTU/24 HR. (BASE LOAD) / HR /DAY (COMP. RUNNING TIME) = ________ BTU/HR

FAN HEAT LOAD ESTIMATE BTU/HR. =

______ QTY. x ______ WATTS EA. x 3.41 BTU/W x ______ HR = ________ BTU/24 HR
(MOTORS) (INPUT)

OR __1__ QTY. x __1/6__ HP EA. x __4350__ BTU/HP/HR. x __24__ HR = __17,400__ BTU/24 HR
(MOTORS)

DEFROST HEAT LOAD ESTIMATE BTU/HR =

______ QTY. EVAPS. x ______ W EA. x ______ HR x 3.41 BTU/W x ______ DEFROST LOAD FACTOR*

= ______ BTU/ HR

*USE 0.50 FOR ELECTRIC DEFROST 0.40 FOR HOT GAS DEFROST

BTU/24 HR. TOTAL LOAD = __177,641__ (BASE LOAD) + __17,400__ (FAN HEAT) + ______ (DEFROST HEAT) = __195,041__ BTU/24 HR

OR ______ (BASE LOAD) x ______ (BASE LOAD MULT.) = ______ BTU/24 HR

ACTUAL HOURLY LOAD = __195,041__ BTU/24 HR (TOTAL LOAD) / __16__ HRS/DAY (COMP. RUNNING TIME) = __12,190__ BTU/HR

Equipment Selection

	Compressor Units		Condensing Units		Evaporators		Condensers	
MODEL NO.			15H		WJ 120			
QUANTITY			1		1			
BTU/ HR. CAPACITY (EA)			12,600		12,000			
CFM AIR VOLUME (EA.)								
	DESIGN	ACTUAL	DESIGN	ACTUAL	DESIGN	ACTUAL	DESIGN	ACTUAL
°F EVAPORATOR TEMP.					25			
°F EVAP TD								
°F SUCTION TEMP.			25					
°F CONDENSING TEMP.								
°F DESIGN AMBIENT TEMP.			100					
°F MIN. OPER. AMBIENT TEMP.								

Figure 16-1-10 (continued)

TABLE 16-1-12 Typical Condensing Unit Performance Data

	60 HZ.	50 HZ.		Low Temp.			Commercial Temp.					High Temp.		
				Saturated Suction Temperature °F										
				–40°F	–30°F	–20°F	–10°F	0°F	10°F	20°F	25°F	30°F	40°F	50°F
				Suction Pressure—Psig or *Vacuum Inches of Mercury										
HP	AH/AWH	AH/AWH	Ambient Air Temp. °F	10.9* Btu/hr	5.5* Btu/hr	0.6 Btu/hr	4.5 Btu/hr	9.2 Btu/hr	14.64 Btu/hr	21.04 Btu/hr	24.61 Btu/hr	28.45 Btu/hr	36.97 Btu/hr	46.70 Btu/hr
½	Δ5HCL		90	–	1,370	1,900	2,600	3,350	4,150	5,000	5,400	6,000	6,900	7,800
			100	–	1,250	1,750	2,350	3,050	3,800	4,650	5,100	5,800	6,700	7,600
¾	Δ8HCL		90	–	1,550	2,200	3,000	3,800	4,700	5,700	6,250	6,700	7,900	9,500
			100	–	1,400	2,000	2,700	3,500	4,400	5,400	5,950	6,500	7,700	9,000
	Δ9HCL		90	–	1,950	2,750	3,700	4,700	5,800	7,050	7,800	8,400	9,700	11,100
			100	–	1,750	2,500	3,400	4,350	5,400	6,500	7,200	7,700	8,900	10,200
1	Δ10HCL		90	–	2,250	3,200	4,350	5,550	6,900	8,300	9,150	9,800	11,400	13,000
			100	–	2,050	2,850	3,900	5,100	6,450	7,850	8,700	9,400	11,000	12,600
	Δ11HCL**		90	2,070	2,900	4,150	5,600	7,100	8,800	10,600	11,700	12,500	14,600	16,600
			100	1,850	2,650	3,800	5,100	6,600	8,400	10,000	11,100	12,000	14,100	16,100
	15H	15H5	90	–	–	–	–	–	11,800	12,200	13,400	14,600	16,800	18,900
			100	–	–	–	–	–	9,700	11,600	12,600	13,800	15,900	18,000
1½	15C	15C5	90	–	–	–	7,100	9,300	11,500	14,500	16,000	–	–	–
			100	–	–	–	6,500	8,700	10,900	13,600	15,300	–	–	–
	15L	15L5	90	2,800	4,200	5,800	7,600	9,600	–	–	–	–	–	–
			100	2,500	3,700	5,400	7,000	8,900	–	–	–	–	–	–
	20H	20H5	90	–	–	–	–	–	14,500	16,500	17,500	19,000	22,000	26,000
			100	–	–	–	–	–	12,000	14,500	15,500	17,000	20,000	23,000
2	20C	20C5	90	–	–	–	8,600	11,300	15,400	17,500	19,400	–	–	–
			100	–	–	–	8,100	10,600	14,000	16,200	18,000	–	–	–
	20L	22L5	90	3,600	5,500	7,500	9,800	12,200	–	–	–	–	–	–
			100	3,400	5,000	7,000	9,100	11,400	–	–	–	–	–	–
	30H	32H5	90	–	–	–	–	–	18,000	23,000	25,500	28,000	32,500	38,000
			100	–	–	–	–	–	17,000	21,500	24,000	26,000	30,500	35,400
	30C	30C5	90	–	–	–	–	18,600	23,000	28,700	32,000	–	–	–
			100	–	–	–	–	17,300	21,500	27,000	30,000	–	–	–
3	32C		90	–	–	–	12,500	16,200	19,900	25,400	28,200	–	–	–
			100	–	–	–	11,500	15,000	18,500	24,000	26,400	–	–	–
	30L	30L5	90	6,000	8,500	12,000	16,000	20,500	–	–	–	–	–	–
			100	5,400	8,000	11,000	15,000	19,000	–	–	–	–	–	–
		41PL5	90	6,560	9,700	13,900	18,900	24,100	–	–	–	–	–	–
			100	6,070	9,000	12,400	17,500	22,500	–	–	–	–	–	–

(continued)

TABLE 16-1-12 (continued)

HP	60 HZ.	50 HZ.		Low Temp.			Commercial Temp.					High Temp.		
				Saturated Suction Temperature °F										
				–40°F	–30°F	–20°F	–10°F	0°F	10°F	20°F	25°F	30°F	40°F	50°F
				Suction Pressure—Psig or *Vacuum Inches of Mercury										
HP	AH/AWH	AH/AWH	Ambient Air Temp. °F	10.9* Btu/hr	5.5* Btu/hr	0.6 Btu/hr	4.5 Btu/hr	9.2 Btu/hr	14.64 Btu/hr	21.04 Btu/hr	24.61 Btu/hr	28.45 Btu/hr	36.97 Btu/hr	46.70 Btu/hr
	50H	50H5	90	–	–	–	–	–	30,000	37,500	41,000	45,000	52,500	61,500
			100	–	–	–	–	–	29,300	36,000	39,000	42,700	50,000	58,500
	50C	51PC5	90	–	–	–	21,300	27,400	33,500	41,800	46,300	–	–	–
			100	–	–	–	18,600	24,600	30,600	38,300	43,000	–	–	–
	51PL		90	8,000	11,800	17,000	23,000	29,400	–	–	–	–	–	–
5			100	7,400	11,000	15,100	21,300	27,400	–	–	–	–	–	–
		51PH5	90	–	–	–	–	–	37,200	46,500	50,800	55,800	65,100	76,300
			100	–	–	–	–	–	36,300	44,600	48,400	53,000	62,000	72,500
		62PC5	90	–	–	–	26,200	33,700	41,200	51,400	57,000	–	–	–
			100	–	–	–	22,900	30,200	37,600	47,100	52,900	–	–	–
		61PL5	90	9,800	14,500	20,900	28,300	36,200	–	–	–	–	–	–
			100	9,100	13,500	18,600	26,200	33,700	–	–	–	–	–	–
	76PH/C		90	–	–	–	26,500	34,000	41,500	54,000	60,000	67,000	84,500	105,000
			100	–	–	–	23,900	32,400	38,000	50,500	56,000	62,500	78,500	96,000
	77PC	77PC5	90	–	–	–	32,000	41,000	54,000	67,000	75,000	–	–	–
7½			100	–	–	–	30,000	39,000	50,000	63,500	70,000	–	–	–
	76PL	75PL5	90	10,700	16,600	25,400	35,400	47,100	–	–	–	–	–	–
			100	9,700	15,500	23,300	32,600	43,500	–	–	–	–	–	–
		77PH5	90	–	–	–	–	–	50,400	66,400	73,800	82,400	104,000	129,000
			100	–	–	–	–	–	46,700	62,100	68,900	76,900	96,600	118,000
	101PH	101PH5	90	–	–	–	–	–	68,900	90,600	100,700	112,300	141,700	176,100
10			100	–	–	–	–	–	64,600	84,700	93,900	104,800	131,700	161,000
	101PC	101PC5	90	–	–	–	40,000	51,300	67,500	83,800	93,800	–	–	–
			100	–	–	–	37,500	48,800	62,500	79,400	87,500	–	–	–
	104PC		90	12,500	20,000	30,000	46,000	61,000	–	–	–	–	–	–
			100	11,750	18,800	28,200	43,300	57,400	–	–	–	–	–	–
	D154PH		90	–	–	–	–	–	74,800	101,200	106,400	119,000	147,600	182,000
			100	–	–	–	–	–	70,600	92,700	99,300	111,000	137,000	166,200
12	154PC		90	–	–	–	42,700	56,800	71,200	95,900	–	–	–	–
			100	–	–	–	39,400	52,500	67,200	89,600	–	–	–	–

† This unit will not operate at suction temperatures greater than 20°F. Δ AH units only. * Inches of mercury below one atmosphere.
** Not available for +30°F through +50°F SST 208v/1PH applications.

TABLE 16-1-13 Evaporator Performance Data Above Freezing for Wall Jet (WJ) Unit Coolers Series "C"

Wall Jet Unit Cooler Model	Unit Capacity BTU/HR		Fan		Motor HP 120/1/60†	Total Fan Mtr. Amps 120 V	Optional Heat Exchanger Recommended	Connections**			Approx. Shipping Weight lbs.
	@ 10°F TD	@ 15°F TD	CFM	Size Inch				Coil Inlet	Suction O.D.	Drain O.D.	
WJ35*	3,500	5,250	625	10	1/20	.82	B 25XS	1/2 FL.	1/2	1 1/8	40
WJ45*	4,500	6,750	725	10	1/20	.82	B 25 XS	1/2 FL.	1/2	1 1/8	46
WJ65	6,500	9,750	1050	12	1/10	1.70	B 50XS	1/2 FL.	1/2	1 1/8	52
WJ85	8,500	12,750	1300	12	1/10	1.70	B 75XS	1/2 FL.	3/4	1 1/8	60
WJ105	10,500	15,750	1600	14	1/6	1.70	B 75XS	1/2 FL.	3/4	1 1/8	66
WJ120	12,000	18,000	1800	14	1/6	1.70	B 75XS	1/2 FL.	3/4	1 1/8	72
WJ150	15,000	22,500	2375	16	1/6	1.70	B 120XS	1/2 FL.	7/8	1 1/8	82
WJ180	18,000	27,000	2900	16	1/6	1.70	B 120XS	1/2 FL.	7/8	1 1/8	88
WJ240	24,000	36,000	4000	16	1/2	3.00	B 200XS	1/2 FL.	1 1/8	1 1/8	108

Location: Atlanta, Ga.
Room use: Storage of frozen beef
Outside dimensions: 10 x 12 x 9
Room temperature: 0°F

Insulation: 5 in. molded polystyrene
Total room load: 2000 lb
Daily delivery: 300 lb precooled to 40°F
Containers: Paper cartons, total weight 200 lb
Type of defrost: electric

- Load Survey & Estimating Data -

DESIGN AMBIENT: 95 °F DB, ______ °F WB, 50 %RH, ______ °F SUMMER GROUND TEMP (USE 55°F FOR INSULATED FREEZE FLOOR SLAB)

ROOM DESIGN: ______ °F DB, 0 °F WB, ______ % RH ______ °F WINTER DESIGN AMBIENT

ACCESS AREA: 95 °F DB, 50 °F WB, ______ %RH, (ANTE-RM/LOADING DOCK/OTHER)

ROOM DIM. OUTSIDE: 10 FT. W 12 FT. L 9 FT. H 636 TOTAL FT^2 (OUTSIDE SURFACE)

	Insulation (Table 6)				Wall Thick-ness	Adj. Area	Effective Wall TD	Sun Effect (°F)	Total TD	Overall Wall Heat Gain
	Type	Thick (inches)	K Factor	U* Factor						BTU/24 hrs/ ft^2
N. Wall	Polystyrene	5"	.20	.04	6	95	95		95	91
S. Wall										
E. Wall										
W. Wall										
Ceiling										
Floor										

$$\text{* 'U' - Factor} = \frac{K}{\text{Insul. Thickness (in.)}}$$

REFRIG. DOOR(S): (1) - 7x4 VENT. FAN(S): ______

ROOM INT. VOL: 9 W x 11 L x 8 H = 792 CU. FT.
(INSIDE ROOM DIMENSION = OUTSIDE DIMENSION – WALL THICKNESSES)

FLOOR AREA 9 W x 11 L = 99 FT.^{3}

ELECTRICAL POWER 240 VOLTS, 3 PH. 60 HERTZ; CONTROL 120 V

TYPE CONTROL: ______

PRODUCT DATA AND CLASS OF PRODUCT: ______

Type Product	Amount of Product: Amount Storage	Daily Turn-Over	Freezing or Cooling	Product Temp. °F: Enter-ing	Final	Specific Heat: Above Freeze	Below Freeze	Lat. Ht. Freeze Btu/lb.	Highest Product Freeze Temp.	Heat or Respir'n Btu/lb 24 Hr.	( ) Pull-Down ( ) Freezing Times Hrs.
Beef	2000	300	Freezing	40	0	.80	.40	100	28		

EVAP. TD 10, TYPE DEFROST ☐AIR, ☐HOT GAS, ☐ELECTRIC,

CLASS PRODUCT II,

NO. OF DEFROSTS —TOTAL TIME PER 24 HR; 6 HR.

COMPRESSOR RUNNING TIME 18 HR.

BOX USAGE ☑AVERAGE, ☐ HEAVY, ☐ EXTRA HEAVY

PRODUCT LOAD AND ADDITIONAL INFORMATION: ______

PACKAGING ______ CONTAINERS ______ WGT 200 lbs SP. HT. .32 (CONTAINER)

PALLETS: NO. ______ SIZE ______ WGT. EA. ______ SP. HT. ______

PRODUCT RACKS: NO. ______ MAT'L ______ WGT. EA. ______ SP. HT. ______

ESTIMATING PRODUCT LOADING CAPACITY OF ROOM

ESTIMATED PRODUCT LOADING = 0.40 x ______ FT^3 x ______ LBS./FT^3 = ______ LBS:
(ROOM VOLUME) (LOADING DENSITY)

MISCELLANEOUS LOADS

PEOPLE NO. ______ HRS. ______

MOTORS (OTHER THAN EVAP. FAN)

USE: ______, ______ HP, ______ HR. ______
______, ______ HP, ______ HR. ______

FORK LIFTS ______ NO., ______ HP ______ HR /DAY, OTHER ______

LIGHTS ______ W/FT^2

Figure 16-1-11 Load calculation data sheet for storage rooms below freezing.

Calculations				
I WALL LOSS (TRANSMISSION LOAD)				
		WALL HEAT GAIN		
SURFACE	TD	AREA OF SURFACE	FACTOR	BTU/24 HR
N. Wall		____ Ft. L x ____ Ft. H = ____ Ft^2. x	____ =	
S. Wall		____ Ft. L x ____ Ft. H = ____ Ft^2. x	____ =	
E. Wall		____ Ft. L x ____ Ft. H = ____ Ft^2. x	____ =	
W. Wall		____ Ft. L x ____ Ft. H = ____ Ft^2. x	____ =	
Ceiling		____ Ft. L x ____ Ft. W = ____ Ft^2. x	____ =	
Floor		____ Ft. L x ____ Ft. W = 636 Ft^2. x	91 =	57,876
Box	95	(Table 14A, B, C) Total: Surface = ____ Ft^2. x	____ =	57,876
		I Total Wall Transmission Load	=	
II (LONG FORM) INFILTRATION (AIR CHANGE LOAD)				
792 CU. FT. x 15.3 AIR CHANGES/24 HR x 1 SRVC. FACTOR x 3.28 BTU /CU. FT			=	39,745
		II TOTAL INFILTRATION LOAD	=	39,745
III PRODUCT LOAD				
PRODUCT TEMP. REDUCTION ABOVE FREEZING (SENSIBLE HEAT)				
300 *LBS./DAY x 12 °F TEMP. REDUCTION x .80 SP. HT.			=	2880
PRODUCT FREEZING (LATENT HEAT LOAD)				
300 *LBS./DAY x 100 BTU/LB. LATENT HEAT			=	30,000
PRODUCT TEMP. REDUCTION BELOW FREEZING SENSIBLE HEAT)				
300 *LBS./DAY x 28 °F TEMP. REDUCTION x .40 SP. HT.			=	3660
HEAT OF RESPIRATION				
____ LBS. PRODUCT (STORAGE) x ____ BTU/LB./24 HRS.			=	
MISCELLANEOUS PRODUCT LOADS (1) CONTAINERS (2)PALLETS (3) OTHER				
200 LBS./DAY x 40 °F TEMP. REDUCTION x .32 SP. HT.			=	2560
____ LBS./DAY x ____ °F TEMP. REDUCTION x ____ SP. HT.			=	
		III TOTAL PRODUCT LOAD	=	38,800
IV MISCELLANEOUS LOADS				
(a) LIGHTS ____ Ft.^{2} Floor Area x ____ Watts/Ft.^{2} x 3.41 Btu/Watt x ____ HRS/24 HRS (1 TO 1½ WATTS/SQ. FT. IN STORAGE AREAS & 2 TO 3 FOR WORK AREAS)			=	
(b) OCCUPANCY ____ NO. OF PEOPLE x ____ BTU/HR. x ____ HR			=	
(c) MOTORS ____ BTU/HP/HR. x ____ HP x ____ HR/24/HR			=	
____ BTU/HP/HR. x ____ HP x ____ HR/24/HR			=	
(d) MATERIAL HANDLING ____ FORK LIFT(S) x ____ EQUIV. HP x 3100 BTU/HR./HP. x ____ HR OPERATION			=	
OTHER ____			=	
* IF THE PRODUCT PULL-DOWN IS ACCOMPLISHED IN LESS THAN 24 HRS. THE DAILY PRODUCT WILL BE:		IV TOTAL MISCELLANEOUS LOADS	=	
		TOTAL BTU LOAD 1 TO IV	=	136,421
LBS. PRODUCT x $\frac{\text{24 HRS.}}{\text{PULLDOWN HRS.}}$		ADD 10% SAFETY FACTOR	=	13642
		TOTAL BTU/24 HRS. WITH SAFETY FACTOR (NOT INCLUDING EVAP. FAN OR DEFROST HEAT LOADS) 24 HR. BASE REFRIGERATION LOAD	=	150,063

Figure 16-1-11 (continued)

Equipment Selection from Load Calculation Form

1. DETERMINE EVAP. TD REQUIRED FOR CLASS OF PRODUCT AND ROOM TEMP __10__ °F (TD) (FROM LOAD SURVEY DATA)

2. DETERMINE COMPRESSOR RUNNING TIME BASED ON OPERATING TEMPERATURES AND DEFROST REQUIREMENTS __18__ HRS. (FROM LOAD SURVEY DATA)

3. EVAPORATOR TEMP°F = __0__ (ROOM TEMP) - __10__ (EVAP. TD) = __-10__ °F (FROM LOAD SURVEY DATA)

4. COMP, SUCT. TEMP.°F = __-10__ (EVAP. SUCT. TEMP) - __3__ (SUCT. LINE LOSS) = __-13__ °F

BTU/24 HR. BASE REFRIGERATION LOAD WITH SAFETY FACTOR = ______
(NOT INCLUDING EVAPORATOR FAN OR DEFROST HEAT)

PRELIMINARY HOURLY LOAD = BTU/24 HR. (BASE LOAD) / HRS./DAY (COMP RUNNING TIME) = ______ BTU/HR

FAN HEAT LOAD ESTIMATE BTU/HR. =

__1__ QTY. (MOTORS) x __210__ WATTS EA. (INPUT) x 3.41 BTU/WATT x __24__ HR = __17,186__ BTU/24 HR

OR ______ QTY. (MOTORS) x ______ HP EA. x ______ BTU/HP/HR. x ______ HR = ______ BTU/24 HR

DEFROST HEAT LOAD ESTIMATE BTU/HR =

__1__ QTY. EVAPS. x __3380__ WATTS EA. x __1½__ HRS. x 3.41 BTU/WATT x __.5__ DEFROST LOAD FACTOR*

= ______ BTU/ HR

*USE 0.50 FOR ELECTRIC DEFROST 0.40 FOR HOT GAS DEFROST

BTU/24 HR. TOTAL LOAD = __150,063__ (BASE LOAD) + __17,186__ (FAN HEAT) + __8644__ (DEFROST HEAT) = __175,893__ BTU/24 HR

OR ______ (BASE LOAD) x ______ (BASE LOAD MULT.) = ______ BTU/24 HRS.

ACTUAL HOURLY LOAD = __175,893__ BTU/24 HRS. (TOTAL LOAD) / __18__ HRS/DAY (COMP. RUNNING TIME) = __9771__ BTU/HR.

Equipment Selection

	Compressor Units		Condensing Units		Evaporators		Condensers	
MODEL NO.			32 c		NDE 105			
QUANTITY			1		1			
BTU/ HR. CAPACITY (EA)			11,500		10,500			
CFM AIR VOLUME (EA.)					1420			
	DESIGN	ACTUAL	DESIGN	ACTUAL	DESIGN	ACTUAL	DESIGN	ACTUAL
°F EVAPORATOR EMP.					-10			
°F EVAP TD								
°F SUCTION TEMP.			-10					
°F CONDENSING TEMP.								
°F DESIGN AMBIENT TEMP.			100					
°F MIN. OPER. AMBIENT TEMP								

Figure 16-1-11 (continued)

Model	Capacity (Btu/hr)			Fan Data			Motor Data				Connections		
							Watts*	Standard 208-240/1/60		Optional 115/1/60 Amps* @ 115V			
	@ 10° F TD	@ 12° F TD	@ 15° F TD	cfm	Qty.	Size		Amps* @ 240V	Amps* @ 208V		Liquid Fl† (in.)	Suction O.D. (in.)	Drain O.D. (in.)
NDE 45	4,500	5,400	6,750	740	1	10"	105	.82	.76	1.5	1/2	1/2	1 1/8
NDE 65	6,500	7,800	9,750	880	1	10"	105	.82	.76	1.5	1/2	1/2	1 1/8
NDE 85	8,500	10,200	12,750	1,180	2	10"	210	1.64	1.52	3.0	1/2	5/8	1 1/8
NDE 105	10,500	12,600	15,750	1,420	2	10"	210	1.64	1.52	3.0	1/2	5/8	1 1/8
NDE 120	12,000	14,400	18,000	1,620	2	10"	210	1.64	1.52	3.0	1/2	5/8	1 1/8
NDE 150	15,000	18,000	22,500	2080	3	10"	315	2.46	2.28	4.5	1/2	5/8	1 1/8
NDE 180	18,000	21,600	27,000	2340	3	10"	315	2.46	2.28	4.5	5/8	7/8	1 1/8
NDE 280	28,000	33,600	42,000	3,580	4	10"	420	3.24	3.04	6.0	5/8	1 1/8	1 1/8

† Externally equalized expansion valves must be used for all models. * Watts or Amps are total for all fan motors.

Model	Approx. Refrig. Charge lbs.	Optional Heat Exchanger Recommended	Approx. Shipping Weight lbs.	Defrost Electrical Data							
				Standard 240/1/60		Optional 220/3/60		Optional 480/1/60		Optional 440/3/60	
				Amps	Watts	Amps	Watts	Amps	Watts	Amps	Watts
NDE 45	1.53	B 50 XS	80	6.6	1580	5.25	1450	3.30	1580	2.61	1450
NDE 65	2.24	B 75 XS	95	9.1	2180	7.15	2000	4.50	2180	3.41	2000
NDE 85	2.75	B 75 XS	110	11.6	2780	9.25	2540	5.80	2780	4.36	2540
NDE 105	3.56	B 120 XS	125	14.1	3380	11.30	3100	7.10	3380	5.30	3100
NDE 120	4.08	B 120 XS	140	16.6	3980	13.20	3640	8.30	3980	6.23	3640
NDE 150	5.10	B 200 XS	160	20.0	4780	15.90	4380	10.00	4780	7.50	4380
NDE 180	5.60	B 200 XS	175	21.5	5160	17.20	4720	10.80	5160	8.60	4720
NDE 280	8.60	B 500 XS	200	24.8	5960	19.70	5460	12.40	5960	9.85	5460

† Standard Stock Units

Electrical Defrost

Single Phase Defrost			Three-Phase Defrost		
NDE Units	No. of Units	Kit No.	NDE Units	No. of Units	Kit No.
45 THRU 280	1	1D	45 THRU 280	1	4D
45 THRU 105	2*	1D	45 THRU 120	2*	4D
120 THRU 280	2*	22D	150 THRU 280	2*	23D

* Any combination of units may be used

Figure 16-1-12 Evaporator performance data below freezing, electric defrost.
(Courtesy of Dunham-Bush, Inc.)

UNIT I—REVIEW QUESTIONS

1. List the sources that the total refrigeration load of a system may be categorized in.
2. _______ varies directly with time, area, and temperature difference.
3. What affects heat transfer through any material?
4. How is heat loss measured?
5. What is the formula for calculating the quantity of heat transmission through a substance or material?
6. List the types of insulation for various applications.
7. What is the primary radiation factor in the refrigeration load?
8. _______ is the oxidation process of ripening, and carbon dioxide and heat are by-products.
9. What needs to be considered to calculate the heat removal required to freeze food products having a high percentage of water?
10. What is the formula for calculating the heat to be removed from a product to reduce its temperature below freezing?
11. What is total product load the sum of?
12. List the information that is important to know in the selection of equipment.

UNIT 2

Residential Load Calculations

OBJECTIVES: In this unit the reader will learn about residential load calculations including:

- Residential Heating and Cooling Loads
- Heat Loss and Gain
- Heat Transmission Multiplier
- Daily Range
- Building Gains

TECH TIP

It is important to note that a properly sized heating and cooling systems will provide your customer with the best system for their home. Oversizing causes the system to start and stop too frequently. That's just like driving your car in stop and go traffic. When you drive a car in stop and go traffic it does not get as good gas mileage as it does when on the highway. A car driven mostly on the highway will last longer than one that is driven mostly in stop and go traffic. That is the same with all equipment, including heating and air conditioning systems. Therefore, longer periods of operation are better for both the efficiency and life expectancy of all HVAC equipment.

RESIDENTIAL HEATING AND COOLING LOADS

Residential and commercial load calculations have certain unique characteristics that are best served by using a procedure especially designed for each specific system type. This unit will concentrate on residential applications. Many procedures are available, including many for computer operation. One method which has industry-wide acceptance is Manual J: Load Calculation for Residential Winter and Summer Air Conditioning, published by the Air Conditioning Contractors of America (ACCA). The two divisions of the manual, one for heat losses and the other for heat gains, provide practical procedures and data for making load calculations.

Heat loss calculations are used to determine the quantity of heat in Btuh that will be needed to maintain the residence indoor temperature at the winter design temperature for the area. Heat gain calculations are used to determine the capacity of cooling in Btuh that will be needed to maintain the indoor temperature at the summer design temperature for the area. In both heat loss and heat gain calculations the systems design capacity in Btuh is for the design temperature and not for the higher maximum coldest or hottest temperatures. If the systems were designed for the maximum temperatures they would be oversized. It is important to avoid oversizing equipment because oversized equipment will cause the system to short cycle. Short cycling results in poor operating efficiency, poor humidity control, and shorter equipment life.

The following material from Manual J illustrates this procedure:

Heat Loss

1. Outside and inside design conditions: The manual describes selecting the outside design conditions from ASHRAE weather data and recommends an inside design temperature of 70°F.
2. Building losses, which includes those associated with the building envelope such as:
   a. Heat loss through glass windows and doors by conduction.
   b. Heat loss through solid doors by conduction.
   c. Heat loss through walls exposed to outdoor temperatures or through walls below grade.
   d. Heat loss through partitions which separate spaces within the structure that are at different temperatures.
   e. Heat loss through ceilings to a colder room or to an attic.
   f. Heat loss through a roof-ceiling combination.
   g. Heat loss through floors to a colder basement, crawl space, or to the outside.
   h. Heat loss through on-grade slab floors or through basement floors.

i. Heat loss due to infiltration through windows and doors or through cracks and penetrations in the building envelope.

3. System losses:
   a. Heat loss through ducts located in an unheated space.
   b. Ventilation air which must be heated before it is introduced into the space. (In older structures infiltration provided enough fresh air to the space making ventilation unnecessary. In newer structures, tighter construction may require ventilation.)
   c. Bathroom and kitchen exhaust systems tend to increase infiltration, but they are not a design factor because they are used intermittently.
   d. Combustion air for gas or oil fired furnaces must be provided. In older homes, infiltration will meet the combustion air requirement. In newer, tighter homes it may be necessary to introduce combustion air to the burner or the furnace room. Generally, codes require 1 in^2 of opening to a source of outside air for each 1,000 Btu/hr of fuel input. Local codes must be consulted.

For heating garages, the manual recommends a separate heater and control.

Heat Transmission Multiplier The heat transmission multiplier (HTM) for various exposures is the product of the heat transmission factor (U) multiplied by the temperature difference (TD). The infiltration HTM is calculated by dividing the winter infiltration Btu/hr by the total window and door area.

Heat Gain

The calculation of heat gain includes consideration of the following conditions:

1. The summer outside design temperature.
2. Radiation from the sun.
3. Heat and moisture given off by equipment and appliances.
4. Heat and moisture given off by people.
5. Heat and moisture gained by infiltration.
6. Heat from conduction through the building envelope.

These conditions may or may not occur simultaneously.

Outside Design Conditions

The outside design conditions are selected from ASHRAE 97.5% weather data. These conditions are only exceeded 50 to 100 hr from June through September. Using the 97.5% conditions prevents excessive oversizing of the equipment to handle a temporary high load.

NICE TO KNOW

The design temperature for any locale is based on a 10 yr average. Over time a design temperature for a locale may vary if there was a very cold winter or hot summer. These minor fluctuations will be reflected in the 10 yr average as provided by the U.S. Weather Service.

Daily Range

The difference between the average high and low temperatures is the daily range given for U.S. and Canada locations. The high usually occurs late in the afternoon and the low at about daybreak. The daily range affects the cooling load since a low night temperature can reduce the daytime load due to the storage factor. The daily range factors are given for three ranges: the L range below 15°F, the M range for temperatures between 16° and 25°F, and the H group which refers to a range above 25°F.

Storage

When the sun shines in a window, its radiant heat is stored when it reaches the interior surfaces of the house and later gradually heats room air. The same kind of action occurs on the exterior of structures. Radiant heat from the sun warms the surfaces and is stored. Gradually, at a later time, the heat reaches the interior where it warms the interior air. The net effect of storage is to delay and smooth out solar loads. The delayed effect of the sun's load is incorporated in the procedures used for calculating heat gain.

Building Gains

The building gains included in the calculations are as follows:

1. Heat gained by solar radiation through glass.
2. Heat transmitted through glass by conduction.
3. Heat transmitted through walls exposed to outside air.
4. Heat gained through partitions which separate conditioned and unconditioned spaces.
5. Heat gained through ceilings from the attic.
6. Heat gained through roofs and roof/ceiling combinations.

7. Heat gained through ceilings and floors that separate conditioned and unconditioned spaces.
8. Heat gained by infiltration through doors and windows or building envelope.
9. Heat produced by people.
10. Heat produced by lights.
11. Heat produced by appliances and equipment.

The system gains included in the calculations are as follows:

1. Heat gained through ducts located in unconditioned spaces.
2. Heat gains associated with ventilation air mechanically introduced through the system.
3. Bathroom and kitchen exhaust tend to increase infiltration. These devices only operate a small part of the time.

Walls below grade and slabs are not included in the heat gain calculations. Moisture loads due to the introduction of outside air or infiltration, people and appliances are included in the load.

Data Sheets

Winter heating loads are summarized in Figure 16-2-1. A sample data sheet filled in with typical calculations is shown in Figure 16-2-2.

Summer cooling loads are summarized in Figure 16-2-3. A sample data sheet filled in with typical calculations is shown in Figure 16-2-4.

House Orientation and Solar Gain

No house is symmetrical, having the exact same number of windows on all sides. For that reason the direction the house is facing will significantly affect the heat gain for cooling load calculations. West and east facing windows have the highest heat gain per square foot of glass. North facing windows have the least heat gain per square foot of glass. Using the house floor plan shown in Figure 16-2-5, we are going to observe the effect that house orientation has on the heat load. The amount of heat gained from the floor and ceiling will not change as the house is rotated to face north, south, east, and west. In addition the latent and sensible load of the house does not change either. For that reason the ceiling and floor sensible and latent interior loads have been omitted from this calculation so the solar gain from the windows can be more easily observed. Table 16-2-1 list the wall and window areas for each of the house sides.

Using the basic heat transfer formula in from the footnote of Table 16-2-1, we can calculate the heat gain for each of the walls. Using Table 16-2-1, we can determine the amount of solar heat gain each

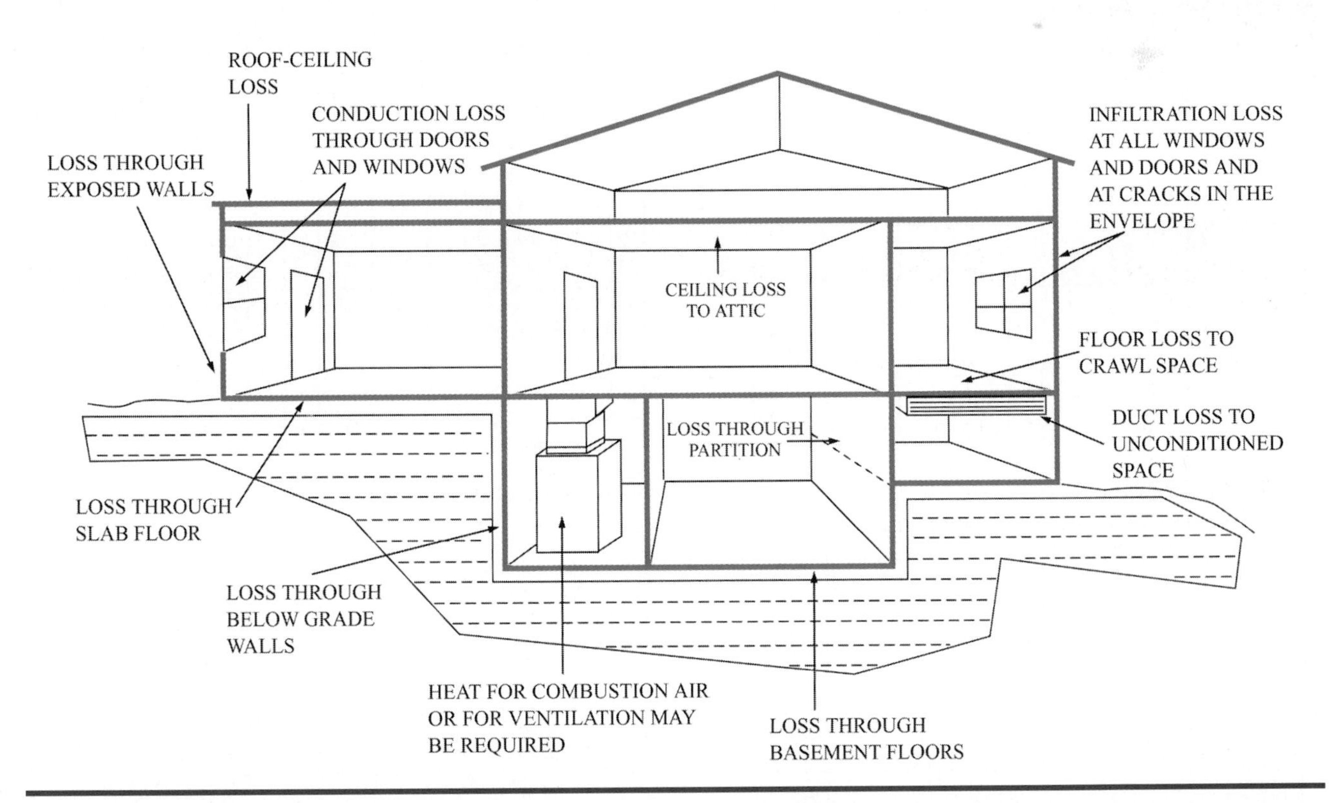

Figure 16-2-1 Winter heating loads. (*Copyright by Air Conditioning Contractors of America [ACCA]*)

1	Name of Room					Entire House			1 Living			2 Dining			3 Laundry			4 Kitchen			5 Bath - 1		
2	Running Ft. Exposed Wall					160			21			25			18			11			9		
3	Room Dimensions Ft.					51 x 29			21 x 14			7 x 18			7 x 11			11 x 11			9 x 11		
4	Ceiling Ht Ft.	Directions Room Faces				8			8	West		8	North		8			8	East		8	East	
	Type of Exposure		Const. No.	HTM Htg.	HTM Clg.	Area or Length	Btu/hr Htg.	Btu/hr Clg.	Area or Length	Btu/hr Htg.	Btu/hr Clg.	Area or Length	Btu/hr Htg.	Btu/hr Clg.	Area or Length	Btu/hr Htg.	Btu/hr Clg.	Area or Length	Btu/hr Htg.	Btu/hr Clg.	Area or Length	Btu/hr Htg.	Btu/hr Clg.
5	Gross	a	12-d			1280			168			200			144			88			72		
	Exposed	b	14-b			480																	
	Walls &	c	15-b			800																	
	Partitions	d																					
6	Windows	a	3-A	41.3		60	2478		40	1652		20	826										
	& Glass	b	2-C	48.8		20	976																
	Doors Clg.	c	2-A	35.6		105	3738											11	392		8	285	
		d																					
7	Windows	North																					
	& Glass	E & W																					
	Doors Clg.	South																					
		Basement																					
8	Other Doors		11-E	14.3		37	529								17	243							
9	Net	a	12-d	6.0		1078	6468		128	768		180	1080		127	762		77	462		64	384	
	Exposed	b	14-b	10.8		460	4968																
	Walls &	c	15-b	5.5		800	4400																
	Partitions	d																					
10	Ceilings	a	16-d	4.0		1479	5916		294	1176		126	504		77	308		121	484		99	396	
		b																					
11	Floors	a	21-a	1.8		1479	2662																
		b																					
12	Infiltration HTM			78.8		222	17490		40	3152		20	1576		17	1340		11	867		8	630	
13	Sub Total Btu/hr Loss = 6 + 8 + 9 + 10 + 11 + 12						49625			6748			3896			2653			2205			1695	
14	Duct Btu/hr Loss			0%		—			—			—			—			—					
15	Total Btu/hr Loss = 13 + 14						49625			6748			3896			2653			2205			1695	
16	People @ 300 & Appliances 1200																						
17	Sensible Btu/hr Gain = 7 + 8 + 9 + 10 + 11 + 12+16																						
18	Duct Btu/hr Gain																						
19	Total Sensible Gain = 17 + 18																						

Figure 16-2-2 Sample heat loss calculation sheet.

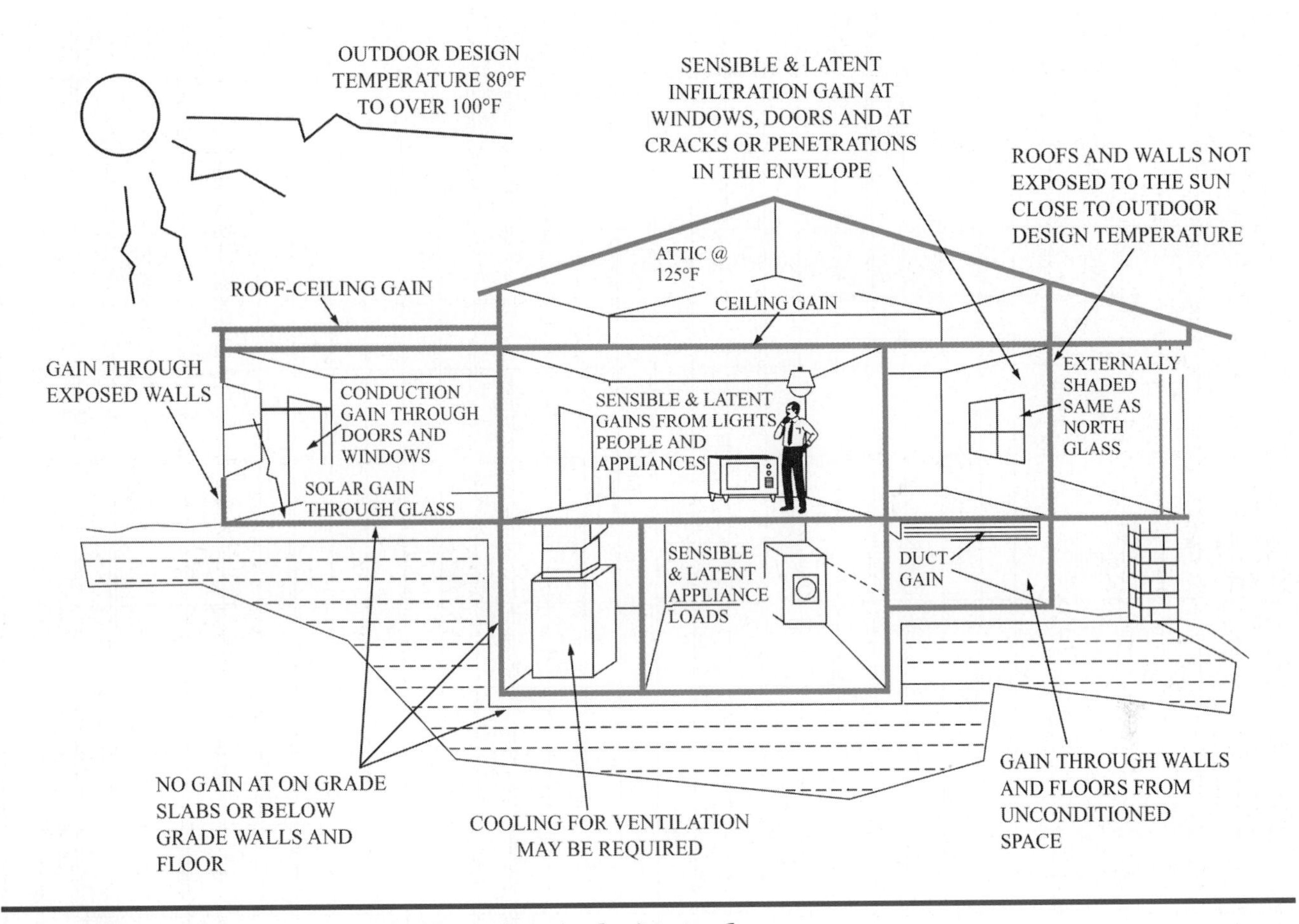

Figure 16-2-3 Summer cooling loads. (*Copyright by Air Conditioning Contractors of America [ACCA]*)

wall's windows will receive when the front of the house is facing north, east, south, west.

Table 16-2-1 shows the total heat gain from the walls, windows, and doors. You will note that when the house is facing east its heat gain is 38% below the heat gain when the house is facing south. It is important to note that the direction the house is facing has a significant impact on the heat load.

In addition to using the heat load to determine the size of the heating and cooling system, a room by room load should be used to determine the duct system layout and design.

TECH TIP

It is important to note that when you are calculating the heat load on a wall that you subtract the area of the doors and windows from the load.

TABLE 16-2-1 Total Heat Load for a Residence's Four Sides, Front, Left Side, Back, and Right Side

	Total Wall	Total Glass	Door	Net Wall	Wall (Btuh)	Door (Btuh)	Glass (Btuh)	Total
Front	221	40	20	161	466	122	2,488	3,076
Left side	253	20	0	233	674		1,244	1,918
Back	221	73		148	428		4,540	4,968
Right side	253	13		240	694		808	1,502
		146	20	782		122	9,080	11,464

#2 Btuh wall = $\frac{1}{9}$ × wall area × 55°
Btuh window = 62.2 × window area
Btuh door = 6.1 × door area

1	Name of Room					Entire House			1 Living			2 Dining			3 Laundry			4 Kitchen			5 Bath - 1		
2	Running Ft. Exposed Wall					160			21			25			18			11			9		
3	Room Dimensions Ft.					51 x 29			21 x 14			7 x 18			7 x 11			11 x 11			9 x 11		
4	Ceiling Ht Ft.	Directions Room Faces				8			8	West		8	North		8			8	East		8	East	
Type of Exposure			Const. No.	HTM		Area or Length	Btu/hr		Area or Length	Btu/hr		Area or Length	Btu/hr		Area or Length	Btu/hr		Area or Length	Btu/hr		Area or Length	Btu/hr	
				Htg.	Clg.		Htg.	Clg.		Htg.	Clg.		Htg.	Clg.		Htg.	Clg.		Htg.	Clg.		Htg.	Clg.
5	Gross	a	12-d			1280			168			200			144			88			72		
	Exposed	b	14-b			480																	
	Walls &	c	15-b			800																	
	Partitions	d	13N			232																	
6	Windows	a																					
	& Glass	b																					
	Doors Htg.	c																					
		d																					
7	Windows	North			14	20		280				20		280									
	& Glass	E & W			44	115		5060	40		1760							11		484	8		352
	Doors Clg.	South			23	30		690															
		Basement			70/36	8/8		848															
8	Other Doors		10-e		3.5	37		130							17		60						
9	Net	a	12-d		1.5	1078		1617	128		192	180		270	127		191	77		116	64		96
	Exposed	b	14-b		1.6	233		373															
	Walls &	c	15-b		0																		
	Partitions	d	13-n		0																		
10	Ceilings	a	16-d		2.1	1479		3106	294		617	126		265	77		162	121		254	99		208
		b																					
11	Floors	a	21-a		0																		
		b	19-f		0																		
12	Infiltration HTM				9.0	218		1962	40		360	20		180	17		153	11		99	8		72
13	Sub Total Btu/hr Loss = 6 + 8 + 9 + 10 + 11 + 12																						
14	Duct Btu/hr Loss																						
15	Total Btu/hr Loss = 13 + 14																						
16	People @ 300 & Appliances 1200							3000	3		900	3		900			—			1200			—
17	Sensible Btu/hr Gain = 7 + 8 + 9 + 10 + 11 + 12+16							17066			3829			1895			566			2153			728
18	Duct Btu/hr Gain										—			—			—			—			—
19	Total Sensible Gain = 17 + 18							17066			3829			1895			566			2153			728

Figure 16-2-4 Sample heat gain calculation sheet.

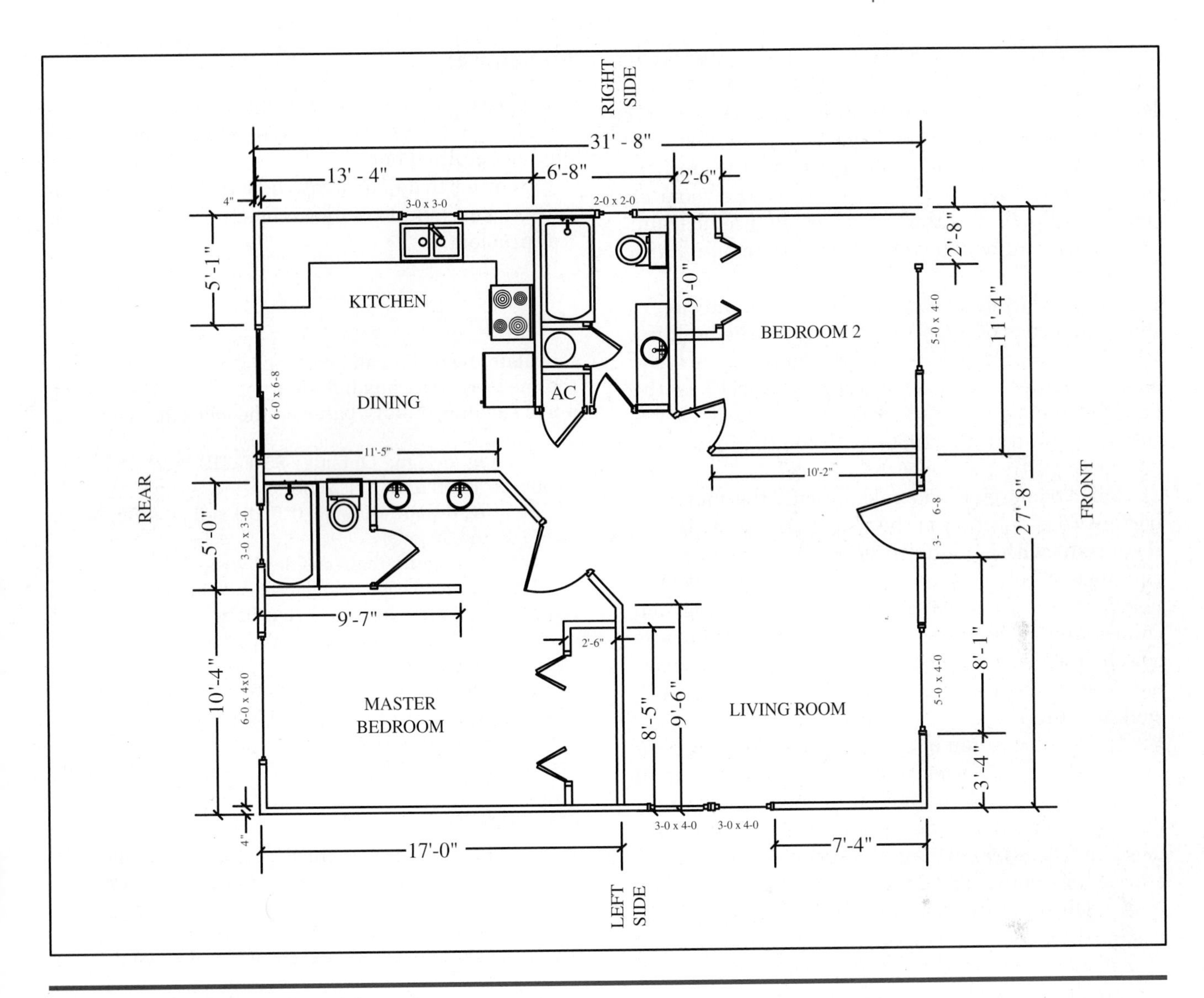

Figure 16-2-5 Basic residential floor plan.

TECH TIP

The effect that glass has from the total heat load relative to the direction that it faces is significant. It is possible for an architect or building designer to use this information in the design process. On one home, there is a 2½ ton difference in heat load depending on the floor plan orientation. Some computerized heat gain and loss programs will automatically give you the best and worst case scenarios for the house you are designing the load for.

COMMERCIAL HEATING AND COOLING LOAD

There are a number of similarities in load calculations for refrigerated rooms, residential buildings, and commercial establishments. Each requires the calculation of heat transmission through the enclosure, gain or loss due to infiltration or ventilation of outside air, and gains due to internal loads such as people or product.

TECH TIP

Commercial and residential load calculating programs are not interchangeable. If you use a residential load calculating program for a commercial building the system will be significantly oversized. The oversizing occurs because many commercial buildings use zone control so that a smaller system can provide the necessary cooling or heating since only those areas requiring heat or cool air are supplied.

The load calculations for each of these applications differs in respect to the major factor that comprises the load. For example, for refrigerated loads, a large factor in the total load is the internal load. To correctly calculate this load, the cooled product must be analyzed from a number of standpoints: What condition is the product in when it is brought into the room? Is it frozen in the room or just cooled? What is the nature of the product and what is the timing in bringing various amounts into the room?

In calculating residential loads, the condition of the structure is of major importance. Is the building low and spread out like a ranch house, or are the spaces compact as in an apartment house? Is the house well insulated and tightly constructed, or was the house built in the 1930s with limited insulation and loosely fitted windows and doors?

For commercial load calculations, the most important consideration is the use of the space. Is it a restaurant with a high concentration of people consuming hot food or is it an office with many pieces of heat producing equipment such as duplicating machines and computers? Is it a laboratory where closely controlling temperature and humidity are necessary or is it a clean room requiring highly efficient air filters?

In the discussion of commercial loads, one general application is selected that fits many small and medium sized commercial spaces where unitary equipment can be applied. These units are used in restaurants, small offices, banks, retail stores, and many other commercial buildings. The load calculation process can use a simplified form. A sample of such a form, consisting of seven tables and two data sheets for figuring both heating and cooling loads, is shown in Tables 16-2-2 through 16-2-10. The step-by-step procedure, using this simplified form, is described under the following topics:

For calculating cooling loads:

1. Design conditions
2. Solar heat gain (Table 16-2-2)
3. Heat transmission (Table 16-2-3)
4. Heat gain from occupants (Table 16-2-4)
5. Heat gain from appliances (Table 16-2-5)
6. Infiltration/ventilation (Tables 16-2-6 and 16-2-7)
7. Duct heat gain (Table 16-2-8)

For calculating heating loads:

1. Heat transmission (Table 16-2-3)
2. Infiltration/ventilation (Tables 16-2-6 and 16-2-7)
3. Duct heat loss (Table 16-2-8)

Procedure

The instructions for calculating cooling load, shown in Table 16-2-9, can be followed, and the cooling load data sheet filled out.

In order to demonstrate how the simplified form is used, an example is shown using the following typical problem.

EXAMPLE

Calculate the cooling and heating load (Table 16-2-10) for the one story office building shown in Figure 16-2-6, using a free standing unitary, based on the following conditions:

Cooling design: Outside, 95°F DB and 78°F WB, Inside, 80°F DB and 67°F WB

Heating design: Outside, 0°F DB and 20 gr/lb, Inside, 70°F DB and 32 gr/lb

Solar heat gain: Light inside shades

Heat transmission: Single glass, floors over unconditioned room, roof insulated with ceiling

Heat from occupants: five people, seated, light work, 320 W of fluorescent lights

Heat gain from appliances: 200 W from computers

Ventilation/infiltration: 20 CFM per person

Duct heat gain: None

Solution

The cooling load form is filled out using the above information and the tables that apply, shown in Table 16-2-11.

Note that loads are recorded in two separate columns, one for sensible heat gains and one for latent heat gains for occupants and ventilation.

The following are a few explanations that will assist in filling out the form:

Design conditions: For specific humidity values in gr/lb, refer to the psychrometric chart in Figure 16-2-7.

Solar heat gain: Glass areas must be separated by the direction they are facing.

Heat transmission: The temperature difference comes from the determination in design conditions.

Heat gain from occupants: Type of activity influences the load factor.

Heat gain from appliances: The wattage used by the appliance is the basis for the load.

Infiltration/ventilation: Using 20 CFM per person makes ventilation a larger quantity than obtained by determining the air entering by infiltration and therefore is used in the calculations.

Duct heat gain: Not a factor on this job since there is no ductwork. ■

Now fill out the heating load form in Table 16-2-12 using the above example and the tables that apply.

TABLE 16-2-2 Solar Heat Gain for Commercial Building Using Unitary Equipment (Btu/hr/sq ft Sash Area)

Direction Windows Face	N	NE & NW	E & W	SE & SW	S	Horiz.
Clear glass (single or double), no protection	40	130	200	160	100	
Shaded completely by awnings	12	48	56	45	29	265
Light inside shades or Venetian blinds	24	77	122	95	58	
Glass brick, no protection	16	52	80	64	40	

(The Trane Company)

TABLE 16-2-3 Average Transmission Factors for Commercial Buildings Using Unitary Equipment

	Factor [Btu per (hr) (sq ft) (F)]			
Item	Cooling Load		Heating Load	
Windows				
Single glass	1.06		1.18	
Double glass	0.61		0.65	
Doors	*		*	
Walls and interior partitions	0.32		0.32	
Ceiling under unconditioned room	0.24		0.29	
Floors				
Over unconditioned room	0.29		0.24	
Over basement	0.20		0.23	
On ground	0.00		†	
	No Ceiling	Ceiling	No Ceiling	Ceiling
Roofs:‡				
Uninsulated frame	§1.20	§1.07	0.48	0.33
Uninsulated light masonry	§1.87	§1.60	0.70	0.41
Insulated	§0.80	§0.67	0.18	0.14

*For glass doors, use window factors; for all other doors, use wall factors.
†For slab floors on ground, use 50 Btuh per lineal foot of exposed building perimeter.
‡Factors based on flat roof construction.
§These factors include an average allowance for increased temperature caused by sun load.
This table was extracted from ARI Standard.

NOTE: Factors can be adjusted to actual U factors, if known.

(The Trane Company)

TABLE 16-2-4 Heat Gain from Occupants in a Commercial Building Using Unitary Equipment*

Degree of Activity	Typical Application	Sensible Heat (Btuh)	Latent Heat (Btuh)
Seated at rest	Theater—Matinee	225	105
	Theater—Evening	245	105
Seated, very light office work	Offices, Hotels, Apartments	245	155
Moderately active office work	Offices, Hotels, Apartments	250	200
Standing; light work, walking slowly	Dept. Store, Retail Store	250	200
Walking; seated Standing; walking slowly	Drug Store Bank	250	250
Sedentary work	Restaurants§	275	275
Light bench work	Factory	275	475
Moderate dancing	Dance Hall	305	545
Walking, 3 mph Moderately heavy work	Factory	375	625
Bowling‡ Heavy Work	Bowling Alley Factory	580	870

*This table was extracted with permission from the 1993 ASHRAE Handbook: *Fundamentals*.

‡For *bowling*, figure one person per alley actually bowling and all others as sitting (400 Btuh) or standing (550 Btuh).

§The adjusted total heat value for *sedentary work, restaurant* includes 60 Btu per hour for food per individual (30 Btuh sensible and 30 Btuh latent).

NOTE: The above values are based on 75°F room dry-bulb temperature. For 80°F room dry-bulb temperature, the total heat gain remains the same, but the sensible heat values should be decreased by approximately 20 percent and the latent heat values increased accordingly.

(The Trane Company)

TABLE 16-2-5 Recommended Rate of Heat Gain from Restaurant Equipment Located in Air-Conditioned Area

		Input Rating, Btu/hr		Recommended Rate of Heat Gain, Btu/hr			
				Without Hood			With Hood
Appliance	Size	Maximum	Standby	Sensible	Latent	Total	Sensible
Electric, no hood required							
Barbeque (pit), per pound of food capacity	80 to 300 lb	136	—	86	50	136	42
Barbeque (pressurized), per pound of food capacity	44 lb	327	—	109	54	163	50
Blender, per quart of capacity	1 to 4 qt	1,550	—	1,000	520	1,520	480
Braising pan, per quart of capacity	108 to 140 qt	360	—	180	95	275	132
Cabinet (large hot holding)	16.2 to 17.3 ft^3	7,100	—	610	340	960	290
Cabinet (large hot serving)	37.4 to 406 ft^3	6,820	—	610	310	920	280
Cabinet (large proofing)	16 to 17 ft^3	693	—	610	310	920	280
Cabinet (small hot holding)	3.2 to 6.4 ft^3	3,070	—	270	140	410	130
Cabinet (very hot holding)	17.3 ft^3	21,000	—	1,880	960	2,830	850
Can opener		580	—	580	—	580	0
Coffee brewer	12 cup/2 brnrs	5,660	—	3,750	1,910	5,660	1,810
Coffee heater, per boiling burner	1 to 2 brnrs	2,290	—	1,500	790	2,290	720
Coffee heater, per warming burner	1 to 2 brnrs	340	—	230	110	340	110
Coffee/hot water boiling urn, per quart of capacity	11.6 qt	390	—	256	132	388	123
Coffee brewing urn (large), per quart of capacity	23 to 40 qt	2,130	—	1,420	710	2,130	680
Coffee brewing urn (small), per quart of capacity	10.6 qt	1,350	—	908	445	1,353	416
Cutter (large)	18 in. bowl	2,560	—	2,560	—	2,560	0
Cutter (small)	14 in. bowl	1,260	—	1,260	—	1,260	0
Cutter and mixer (large)	30 to 48 qt	12,730	—	12,730	—	12,730	0
Dishwasher (hood type, chemical sanitizing), per 100 dishes/h	950 to 2,000 dishes/h	1,300	—	170	370	540	170

NOTE: This table was extracted with permission from the 1993 ASHRAE *Handbook—Fundamentals*. (The Trane Company)

TABLE 16-2-6 Infiltration for Commercial Buildings Using Unitary Equipment

Type of Window	Description of Window	Wind Velocity (mph) Summer 7½	Winter 15
Double hung	Total for average window, nonweatherstripped, 1/16 in crack and 3/64 in clearance	14	39
Wood sash	Ditto, weatherstripped	8	24
(Unlocked)	Total for poorly fitted window, nonweatherstripped, 3/32 in crack and 3/32 in clearance	48	111
	Ditto, weatherstripped	13	34
Double hung	Nonweatherstripped locked	33	70
Metal sash	Nonweatherstripped unlocked	34	74
	Weatherstripped unlocked	13	32
Rolled	Industrial pivoted, 1/16 in crack	80	176
Section	Architectural projected, 3/64 in crack	36	88
Steel	Residential casement, 1/32 in crack	23	52
Sash	Heavy casement section, Projected 1/32 in crack	16	38
	Description of Door		
Poorly fitted		96	222
Well fitted		48	111
Weatherstripped		24	56

More current information is available in the 1993 ASHRAE *Handbook*.
(The Trane Company)

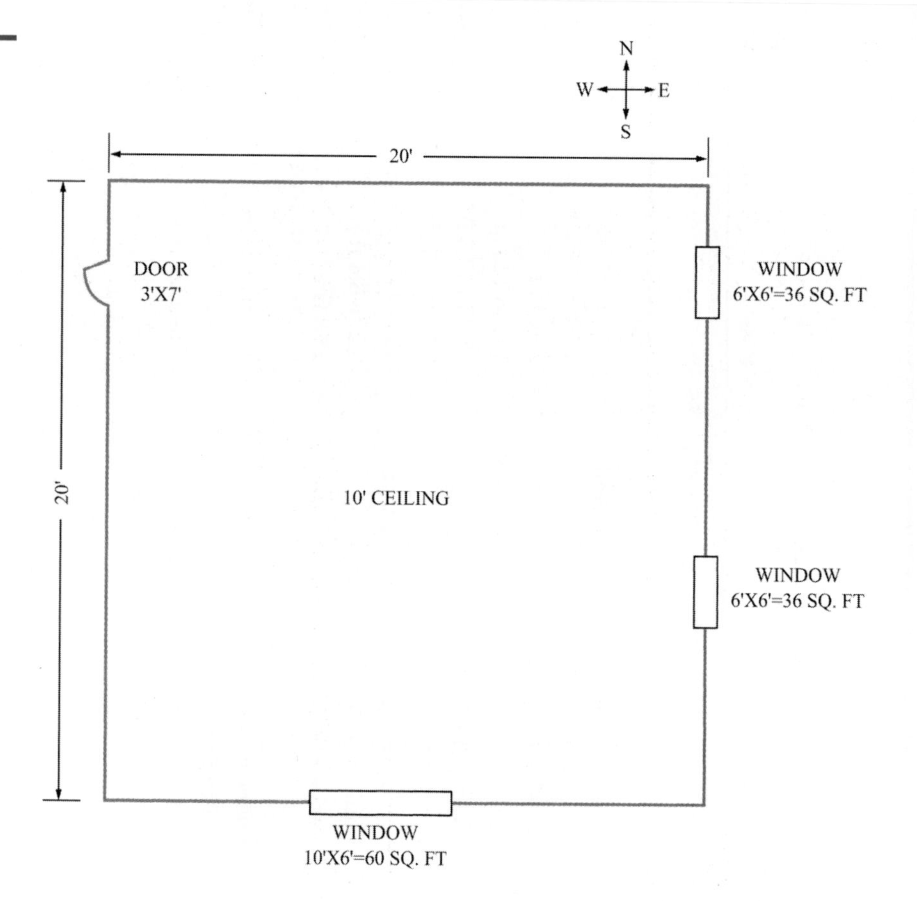

Figure 16-2-6 Sample commercial building layout.

TABLE 16-2-7 Ventilation for Commercial Buildings Using Unitary Equipment*

Application	Est. Max. Occupancy p/1,000 ft²	Outdoor Air Requirements	
		CFM/person	CFM/ft²
Food and beverage service			
Dining rooms	70	20	
Cafeteria, fast food	100	20	
Kitchens (cooking)	20	15	
Garages, service stations			
Enclosed parking garage			1.50
Auto repair rooms			1.50
Hotels, Motels, Resorts			***Room***
Bedrooms			30
Living rooms			30
Baths			35
Lobbies	30	15	
Conference rooms	50	20	
Assembly rooms	120	15	
Gambling casinos	120	30	
Offices			
Office space	7	20	
Reception area	60	15	
Data entry rooms	60	20	
Conference rooms	50	20	
Retail stores			***ft²***
Basement, street floors	30		.30
Upper floors	20		.20
Malls	20		.20
Elevators			1.00
Supermarkets	8	15	
Hardware, drugs, fabric	8	15	
Beauty	25	25	
Barber	25	15	
Sports			
Spectator areas	150	15	
Game rooms	70	25	
Ice arenas (playing areas)			.50
Swimming pools			.50
Gymnasium (playing floors)	30	20	
Bowling alleys (seating areas)	70	25	
Theaters			
Lobbies	150	20	
Auditorium	150	14	

This table was extracted with permission from Table 2, pp. 8–9, from ASHRAE STANDARD 62-1989.

*The ventilation air should be calculated as follows:

_____ CFM per person × _____ persons = _____CFM

or

_____ CFM per sq ft × _____ sq ft = _____CFM

(The Trane Company)

TABLE 16-2-8 Duct Heat Transfer vs. Duct Air Flow at Various Temperature Differences for Ducts 100 Feet Long

Temp. Diff. (°F)	Air Flow (CFM)							
	1,000	**2,000**	**3,000**	**4,000**	**5,000**	**6,000**	**7,000**	**8,000**
	Heat Transfer—Gain or Loss (Btu/hr)*							
20	900	1,200	1,480	1,725	1,975	2,190	2,300	2,400
40	1,800	2,450	3,000	3,500	3,900	4,300	4,625	4,900
60	2,575	3,500	4,375	5,125	5,800	6,375	6,900	7,325
80	3,525	4,775	5,900	6,900	7,800	8,525	9,225	9,800
100	4,375	5,925	7,325	8,575	9,650	10,625	11,500	12,300
120	5,375	7,275	8,950	10,400	11,675	12,800	13,775	14,675

*For 2 in insulation, use Btuh values shown.
For 1 in insulation, multiply by 1.8.
For no insulation, multiply by 7.6.

(The Trane Company)

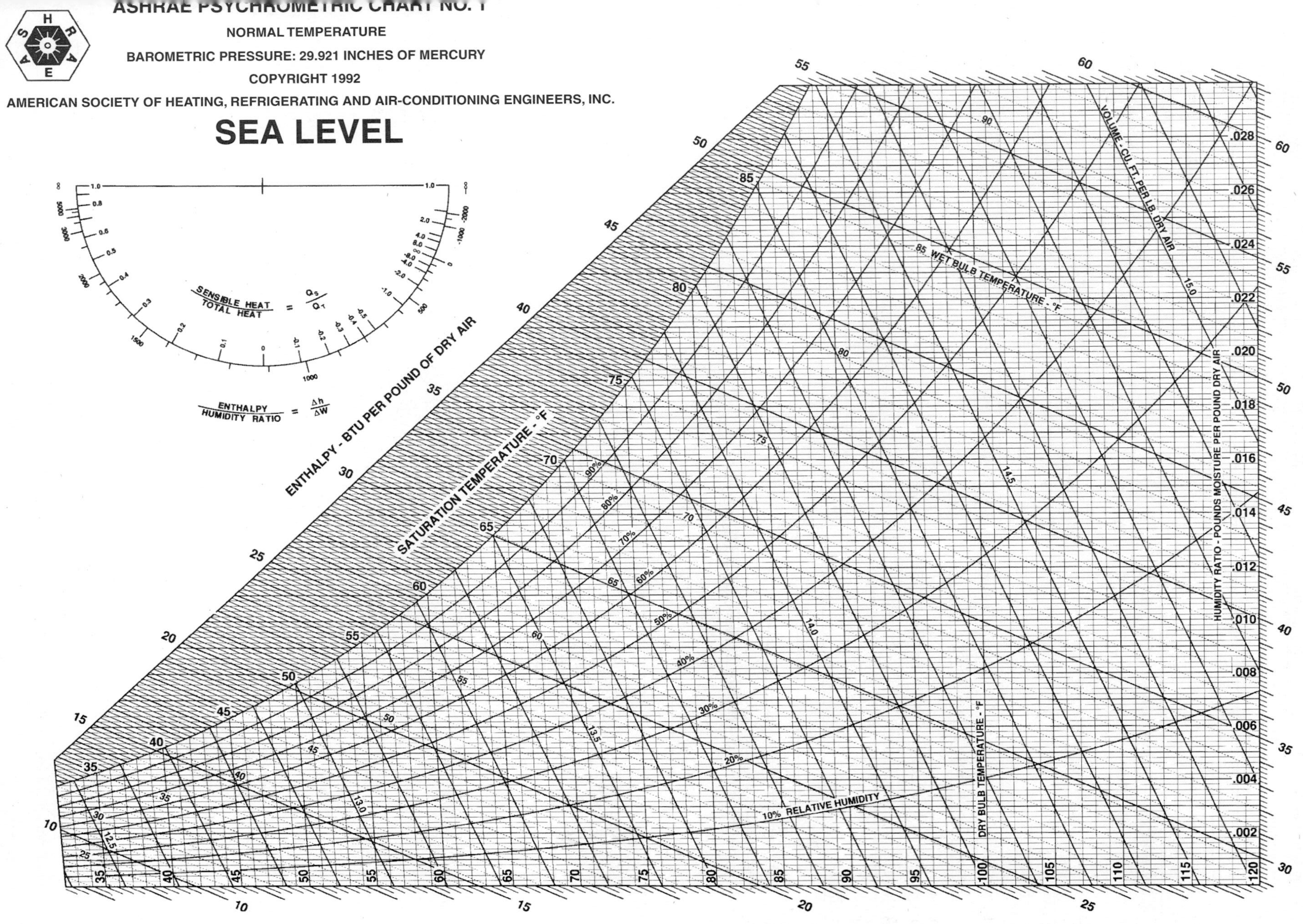

Figure 16-2-7 Psychrometric chart. (*Copyright by the American Society of Heating, Refrigerating, and Air-Conditioning Engineers, Inc.*)

TABLE 16-2-9 Cooling Load Calculation Sheet

1. Design Conditions	Dry Bulb (°F)	Wet Bulb (°F)	Specific Humidity (gr/lb)
Outside			
Inside			
Difference		–	

ITEMS				COOLING LOAD (Btuh) Sensible	Latent
2. Sensible Heat Gain Through Glass (Table 16-2-2)					
Sq. Ft.	×	**Factor**			
			=		
			=		
			=		
			=		
Use largest single load from one side only. Transfer sum of all glass areas to window portion of transmission gain calculation (Step 3).					
3. Transmission Gain (Table 16-2-3)					
	Sq. Ft. ×	**Factor** ×	**Dry Bulb Temp. Diff.**		
Windows			=		
Walls			=		
Partitions			=		
Roof			=		
Ceiling			=		
Floor			=		
4. Internal Heat Gains—People, Lights (Table 16-2-4)					
	Number ×	**Sensible Factor** ×	**Latent Factor**		
People			– =		–
		–	=	–	
	Watts ×	**Factor**			
Lights		3.4	=		–
		*4.25	=		–
5. Internal Heat Gains—Other					
	Horsepower ×	**Factor**			
Motors		3,393	=		–
Appliances (Use Table 16-2-5)			=		
			=		
			=	–	
6. Ventilation or Infiltration (Tables 16-2-6 and 16-2-7, use larger quantity)					
CFM ×	**Dry Bulb Temp. Diff.** ×	**Factor**			
		1.08	=		
CFM ×	**Specific Humidity Diff.** ×	**Factor**			
		0.67	=		
7. Duct Heat Gain (Table 16-2-8)					
Factor for Insulation Thickness ________ × Heat Gain ________ × Duct Length (ft.) ________ ÷100			=		–
8. Total Sensible And Latent Heat Gain			=		
9. Total Cooling Load			=		

* For use with fluorescent lamps with ballast within conditioned space.

(The Trane Company)

TABLE 16-2-10 Heating Load Calculation Sheet

1. Design Conditions	**Dry Bulb (°F)**	**Specify Humidity gr/lb**
Outside		
Inside		
Difference		

2. Transmission Gain Items (Table 16-2-3)	**Calculations**			**Heat Load (Btu/hr)**
Windows	**Sq. Ft.** ×	**Factor** ×	**Dry Bulb Temp. Diff.**	
			=	
			=	
Walls			=	
Partitions			=	
Roof			=	
Floor			=	
Other			=	

3. Ventilation or Infiltration Items (Tables 16-2-6 and 16-2-7, use larger quantity)	**Calculations**			**Heat Load (Btu/hr)**
Sensible Load	**CFM** ×	**Dry Bulb Temp. Diff.** ×	**Factor**	
			1.08 =	
Humidification Load	**CFM** ×	**Specific Humidity Diff.** ×		
			0.67 =	

4. Duct Heat Loss Items (Table 16-2-8)	**Calculations**			**Heat Load (Btu/hr)**
Heat Loss ×	Factor for Insulation Thickness ×	Duct Length (ft)		
			÷ 100 =	
5. Total Heating Load			=	

(The Trane Company)

TABLE 16-2-11 Cooling Load Form Filled Out

1. Design Conditions	Dry Bulb (°F)	Wet Bulb (°F)	Specific Humidity (gr/lb)
Outside	95	78	118
Inside	80	67	78
Difference	15	–	40

ITEMS				COOLING LOAD Btuh	
				Sensible	**Latent**
2. Sensible Heat Gain Through Glass (Table 16-2-2)					
	Sq. Ft.	× **Factor**			
E	72	122	=	8,784	
S	60	58	=	3,480	
			=		
	132		=		

Use largest single load from one side only. Transfer sum of all glass areas to window portion of transmission gain calculation (Step 3).

3. Transmission Gain (Table 16-2-3)					Sensible	Latent
	Sq. Ft.	× **Factor**	× **Dry Bulb Temp. Diff.**			
Windows	132	1.06	15	=	2,099	
Walls	668	.32	15	=	3,206	
Partitions				=		
Roof	400	.67	15	=	4,020	
Ceiling				=		
Floor	400	.29	15	=	1,740	

4. Internal Heat Gains—People, Lights (Table 16-2-4)						Sensible	Latent
People	**Number**	× **Sensible Factor**	× **Latent Factor**				
	5	245	–	=		1,225	
	5	–	155	=	–		775
Lights	**Watts**	× **Factor**					
		3.4		=		–	
	320	4.25		=		1,360	

5. Internal Heat Gains—Other				Sensible	Latent
Motors	**Horsepower**	× **Factor**			
		3,393	=	–	
Appliances (Use Table 16-2-5)			=		
			=		
Computer 200 × 3.4			=	680	

6. Ventilation or Infiltration (Tables 16-2-6 and 16-2-7, use larger quantity)					Sensible	Latent
CFM	× **Dry Bulb Temp. Diff.**	× **Factor**				
100	15	1.08	=		1,620	
CFM	× **Specific Humidity Diff.**	× **Factor**				
100	40	0.67	=	–		3,149

7. Duct Heat Gain (Table 16-2-8)		Sensible	Latent
Factor for Insulation Thickness ______ × Heat Gain ______ × Duct Length (ft) ______ ÷100	=	–	
8. Total Sensible and Latent Heat Gain	=	28,214	3,455
9. Total Cooling Load	=	31,669	

(The Trane Company)

TABLE 16-2-12 Heating Load Form Filled Out

1. Design Conditions		Dry Bulb (°F)		Specify Humidity gr/lb
Outside		0°F		20
Inside		70°F		32
Difference	TD =	70°		12
2. Transmission Gain Items (Table 16-2-3)		**Calculations**		**Heat Load (Btu/hr)**
Windows	**Sq. Ft. ×**	**Factor ×**	**Dry Bulb Temp. Diff.**	
	132	1.13	70 =	10,903
			=	
Walls	668	.32	70 =	14,963
Partitions			=	
Roof	400	.14	70 =	3,920
Floor	400	.24	35 =	3,360
Other			=	
3. Ventilation or Infiltration Items (Tables 16-2-6 and 16-2-7, use larger quantity)		**Calculations**		**Heat Load (Btu/hr)**
Sensible Load	**CFM ×**	**Dry Bulb Temp. Diff. ×**	**Factor**	
	100	70	1.08 =	7,560
Humidification Load	**CFM ×**	**Specific Humidity Diff. ×**		
	100	12	0.67 =	804
4. Duct Heat Loss Items (Table 16-2-8)		**Calculations**		**Heat Load (Btu/hr)**
Heat Loss ×	**Factor for Insulation Thickness ×**	**Duct Length (ft)**		
			÷ 100 =	
5. Total Heating Load			=	41,510

(The Trane Company)

TECH TALK

Tech 1. I want to put a new air conditioner in my house. I am going to put a new system in so it doesn't have to work as hard.

Tech 2. Why would you want to put a bigger system in with the other one working?

Tech 1. Yes, but it runs all the time.

Tech 2. Well, there are a couple of things you need to do. Number one you need to bring in all the information so the boss can run a heat load and find out exactly what size system you need. But more importantly, a properly designed and sized system will run almost continuously throughout the heat of the day and on hot days it will run continuously.

Tech 1. But won't that wear it out faster? And doesn't it cost me more money running all the time?

Tech 2. No, it doesn't cost you more money and it won't wear out faster. In fact the hardest thing on any piece of equipment, whether it is your car, boat, lawnmower, or air conditioning and heating system is stopping and starting. The longer it runs continuously the more efficiently it runs. Just like stopping and starting a car doesn't get near the gas mileage as out running on the highway.

Tech 1. I never thought of it that way. I appreciate it, thanks.

UNIT 2—REVIEW QUESTIONS

1. What is Manual J?
2. What are heat loss calculations used to determine?
3. _______ cycling results in poor operating efficiency and shorter equipment life.
4. How is the HTM calculated?
5. List the considerations in the calculation of heat gain.
6. The difference between the average high and low temperature is the _______.
7. Where are the outside design conditions selected from?
8. When the sun shines in a window when is its radiant heat released?
9. In calculating residential loads what is of major importance?
10. What is the most important consideration for commercial load calculations?

SECTION **VI**

Commercial Systems

17

Packaged Heating/Cooling Systems

UNIT 1

Air Conditioning Systems

OBJECTIVES: In this unit the reader will learn about air conditioning systems including:

- Types of Unitary Equipment
- Room Air Conditioners
- Construction and Installation
- Performance and Operation
- Performance
- Controls
- Dehumidifier Units
- Single Package Conditioners
- Horizontal Conditioner
- Vertical Conditioner
- Rooftop Conditioners
- Desiccant Cooling Systems

DEFINITIONS

Unitary air conditioning equipment has been developed to provide factory built and tested systems, complete as much as possible, with piping, controls, wiring, and refrigerant. These packages are usually simple to install, requiring only service connections and in some cases ductwork for field applications.

Unitary air conditioning equipment consists of one or more factory made assemblies, which normally include an evaporator or cooling coil, a compressor and condenser combination, and possibly include a heating unit. When the air conditioner is connected to a remote condensing unit, such as in residential applications, the system is often referred to as a split system. A packaged air conditioner, where all components are included in one assembly, such as a room cooler, is referred to as a self contained system.

The sizes of unitary equipment range from small fractional tonnage room coolers to large packaged rooftop units in the 100 ton category.

Types of Unitary Equipment

The various types of units described as unitary equipment include the following:

- Room air conditioners
- Console type through the wall conditioners
- Dehumidifier units

- Single package conditioners
- Split system conditioners
- Rooftop conditioners
- Desiccant cooling systems

Heat pumps are also classified as unitary equipment. However, due to their unique characteristics, they will be discussed in a separate chapter.

ROOM AIR CONDITIONERS

Room air conditioners were primarily developed to provide a simplified means of adding air conditioning to an existing room. These units are considered semiportable in that they can easily be moved from one room to another or from one building to another. They provide cooling, dehumidifying, filtering, and ventilation, and some units provide supplementary heating.

Figure 17-1-1 Window air conditioning unit for cooling.

Figure 17-1-2 Window air conditioning unit on Harry Elkins Widner building at Harvard University.

In numbers sold, room air conditioners such as the one shown in Figure 17-1-1 outsell all other types of unitary equipment. They are relatively low in cost, easy to install, and can be used in almost any type of structure, as shown in Figure 17-1-2.

The disadvantage of room air conditioners is that they may either block part of the window area and prevent the window from being opened or require a special hole through the wall. Some people object to operating noise that they produce close to the occupants. They are best used to condition a single room; however, the spillover can supply some conditioning to adjacent areas.

TECH TIP

A common problem with window air conditioning units is the nuisance trips of the residence circuit breakers. Window units are often used in older homes where the electrical service is not adequate for the air conditioner load. It may be possible to rewire the residence for a single circuit to the air conditioner or it may be best to downsize the system to one that will not trip the circuit breaker. Downsizing the unit may provide less than desired levels of cooling on extremely hot days. However the constant shutting down of the oversized unit electrical service can result in the same lack of cooling. Additionally it may be possible to use more than one window unit in a room if the electrical outlets in that room are serviced by more than one circuit breaker.

Construction and Installation

There are basically two parts to the unit, as shown in Figure 17-1-3. One section goes inside the room where the evaporator fan draws in room air through the filter and cooling coil, delivering conditioned air back into the room. The other section extends outside the room where the condenser fan forces outside air through the condenser, exhausting the heat absorbed by the evaporator. One motor operates both the indoor blower and outdoor fan. The motor shaft extending through the separating partition and drives both fans. Condensate from the evaporator coil flows into the drain pan, which extends below the condenser fan. The condenser fan tip dips into condensate, splashing it onto the hot condenser where it evaporates and is blown into the outside air.

The window mounted units are supplied with a kit of parts for installation. Sill brackets, window mounting strips, and sealing strips are set in place for installations in double hung windows, as shown in Figure 17-1-4. Side curtains fold out to fill up the extra window space, as shown in Figure 17-1-5. A sponge rubber seal is provided for the opening where the sash

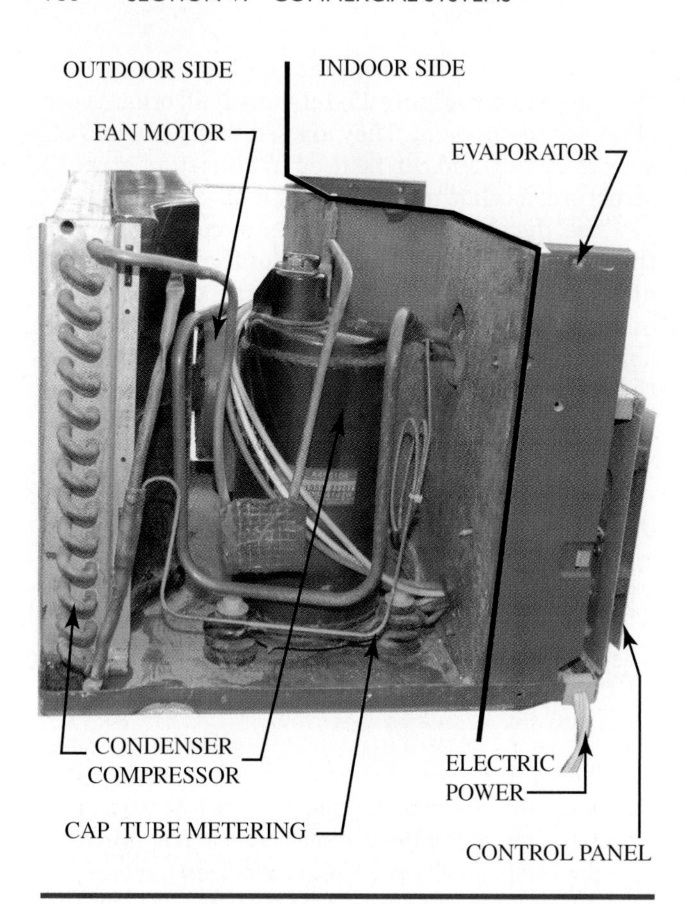

Figure 17-1-3 Two basic parts of a window unit.

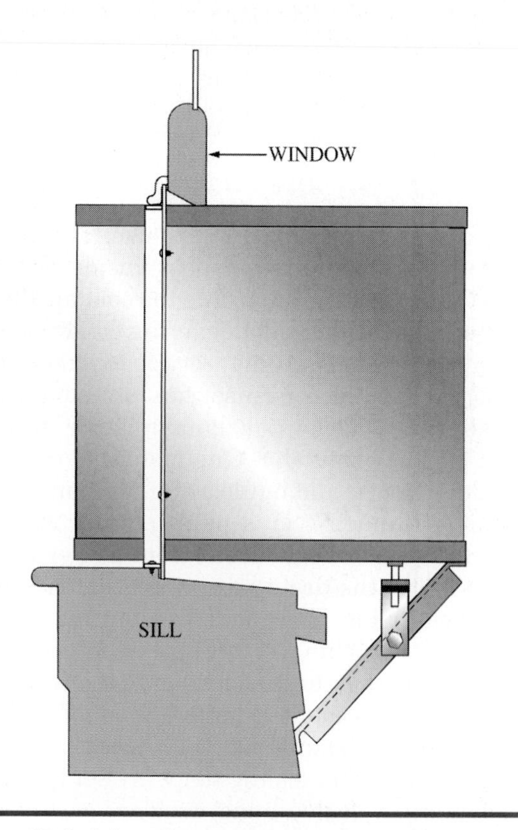

Figure 17-1-4 Installation diagram of a room cooler showing the supporting bracket, from the outside.

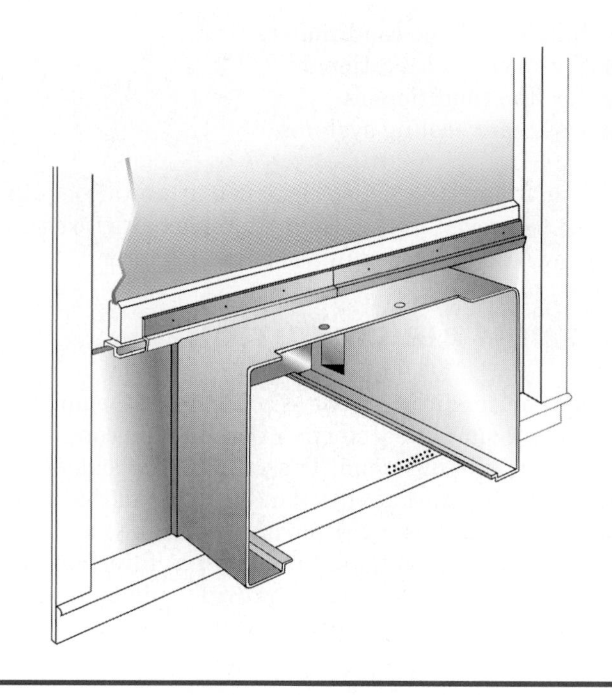

Figure 17-1-5 Installation of a diagram of a room cooler showing the window sealing strips and filler boards.

overlaps, and a sash bracket is installed to lock the lower sash in place, as shown in Figure 17-1-6.

Room air conditioners are available in a vertical configuration for mounting in sliding window openings, as shown in Figure 17-1-7. These vertical conditioners are held in place by the sides as opposed to the conventional shaped room air conditioners that are secured in place on their tops and bottoms.

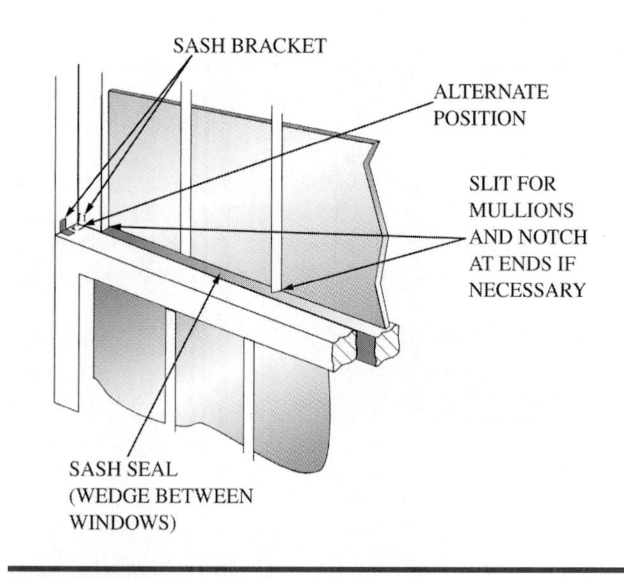

Figure 17-1-6 Installation diagram of a room cooler showing the sponge rubber seal between the upper edge of the lower sash and the upper sash of a double hung window.

Figure 17-1-7 Vertically constructed window unit for horizontal siding window.

Performance and Operation

A schematic diagram of the refrigeration cycle is shown in Figure 17-1-8. The system uses a capillary tube metering device. A typical wiring diagram for a cooling unit is shown in Figure 17-1-9.

A selection switch offers the following modes of operation: FAN ONLY (for ventilation), LOW COOL (using the low evaporator fan speed), HIGH COOL (using the high speed of the evaporator fan), and OFF.

An electric heater strip is sometimes provided to supply heat during mild weather. Some units use a heat pump for heating (heat pumps are described in a separate chapter).

The specifications for a typical series of room coolers are shown in Table 17-1-1. The size range is from a 5,000 Btu/hr to 20,000 Btu/hr. These ratings are based on inside air at 80°F dry bulb temperature (DBT), 67°F wet bulb temperature (WBT), and outside air at 95°F DBT. All of these units will operate on 115 V current, except the three large models, which operate on 230/208 V.

Console Through the Wall Conditioners

A console through the wall conditioner is a type of room cooler that is designed for permanent installation. It was developed to provide individual room conditioning for hotels, motels, and offices where it is impractical or uneconomical to install a central plant system. An opening needs to be made in the outside wall adjacent to the unit for condenser air and ventilation.

NOTE: Package terminal air conditioners (PTAC) are most commonly used in hotels, nursing homes, apartment complexes, and other such areas where the area being cooled is limited. PTAC units may be straight air conditioning, heat pumps, heat pumps with electric resistance supplemental heating, or straight air conditioning with electric resistance heating. All of these units are similar in appearance and most fit in a standard wall sleeve. These units are accessible from inside the occupied space and in most cases the condensate is simply allowed to drain outside the building. Some cities and municipalities have ordinances controlling condensate from PTAC but many do not.

These units are also known as package terminal air conditioners (PTAC), as shown in Figure 17-1-10.

Units are efficient, quiet, and easy to install. Temperature efficiency is usually stated in terms of energy efficient ratio (EER). The EER is equal to the cooling output in Btu/hr divided by the power input in watts under standard rating conditions. Standard rating conditions set up by ARI are based on 80°F DBT, 67°F WBT indoor entering air and 95°F DBT outdoor ambient air. For example, a unit with an output of 6,600 Btu/hr and 660 watts input, under standard conditions, would have an EER of 10 (6,600/660), which is considered a good rating.

These units should comply with standards set up by the following associations:

- Canadian Standards Association (performance standards).
- Underwriters Laboratory (electrical and safety standards).
- Air-Conditioning and Refrigeration Institute (ARI): Standard #310 (for package terminal air conditioners and heat pumps), and Standard #380 (for refrigerant cooled liquid coolers, remote type).

PTAC units should also meet ASTME (American Society of Testing and Materials Engineers) wind and rain infiltration standards.

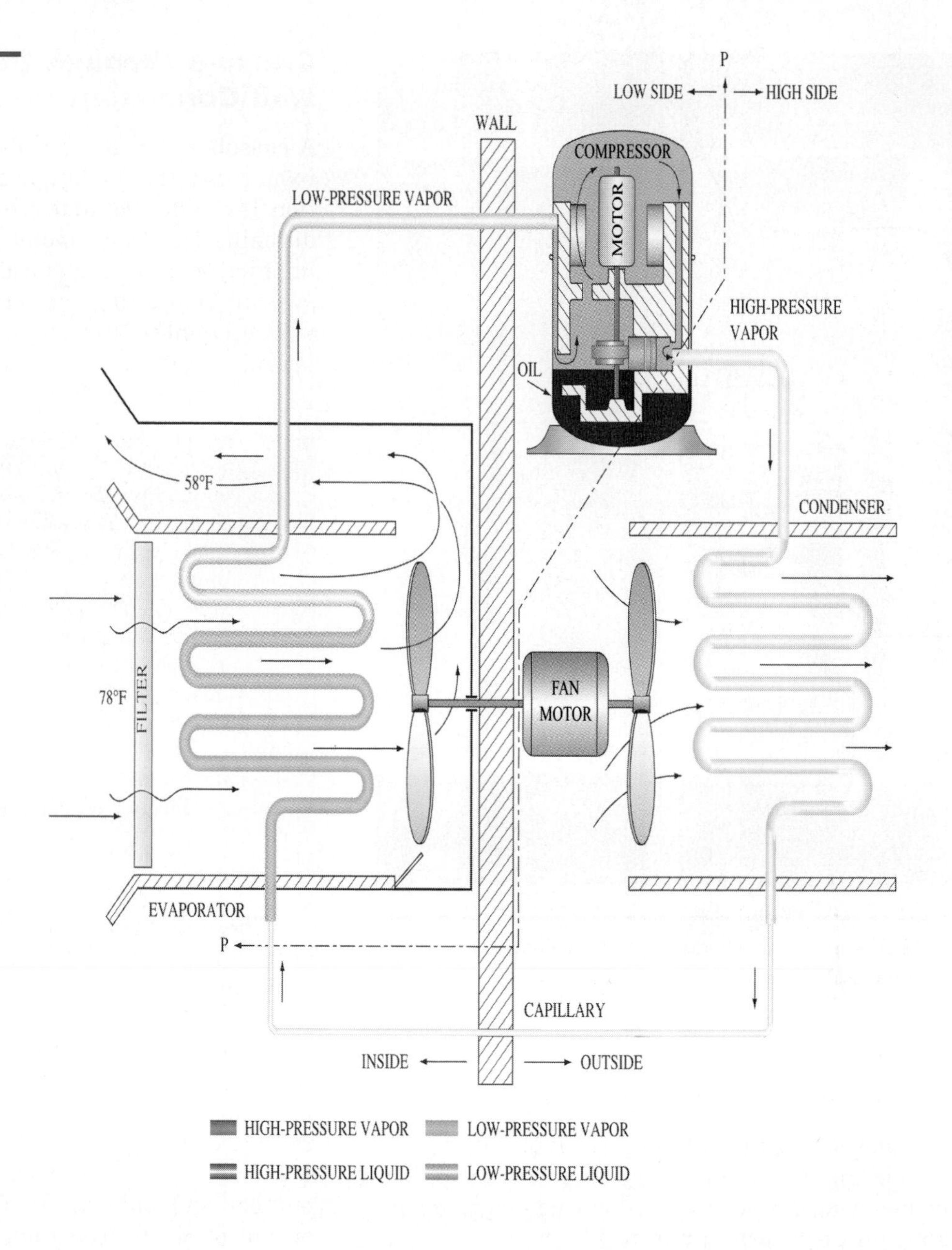

Figure 17-1-8
Refrigeration cycle of a room cooler.

The information in this section applies to PTAC console conditioners with electric heat. PTAC units are also manufactured using heat pumps, which are described in a separate chapter.

Performance

The performance data for PTAC units are shown in Table 17-1-2. The sizes range from 6,800 to 14,900 Btu/hr at standard rating conditions. The quantity of outside air that the units can admit ranges from 40 to 55 CFM depending on the size of the unit. Units are available for 115 and 208/230 V, single phase, AC power.

Electric heat capacity from 1.5 to 5.0 kW can be installed in any size unit. Power receptacle configurations depend on the amperage drawn.

Components and Installation

Figure 17-1-11 shows the location of the following essential parts: (1) indoor fan cover; (2) temperature control; (3) heat/cool/fan switch; (4) evaporator; (5) temperature sensor; (6) compressor; (7) condenser; (8) metering device; (9) condensate level float/drain; (10) condensate pan. The air filter is located behind the front return air panel and can be easily changed.

An optional duct package that can be supplied for a PTAC unit is shown in Figure 17-1-12. This makes it possible to condition several rooms with one unit.

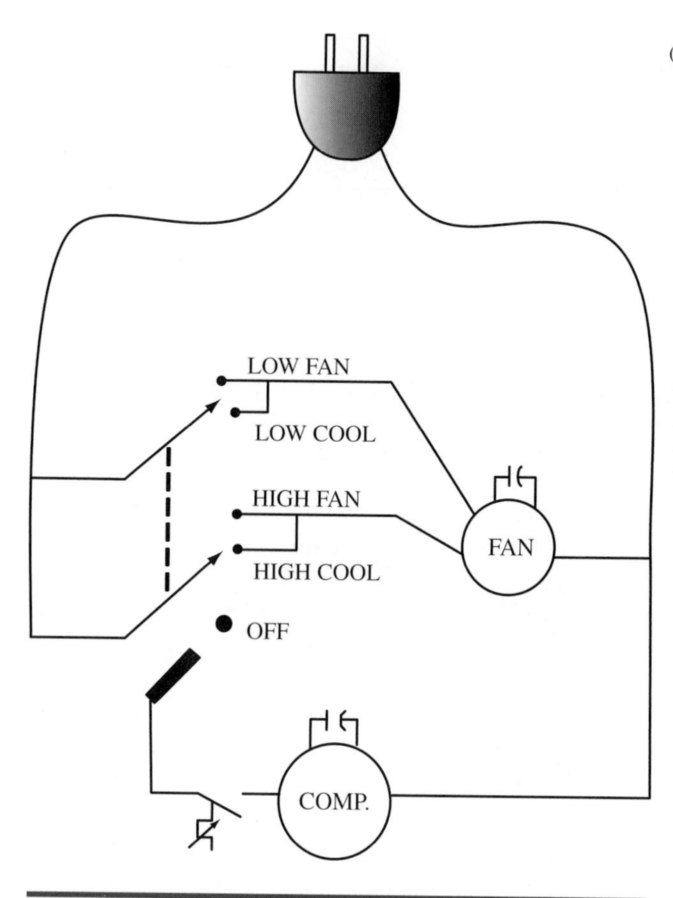

Figure 17-1-9 Wiring diagram for a room cooler.

Figure 17-1-10 (a) Exterior view of a package terminal air conditioner (PTAC); (b) Interior view.

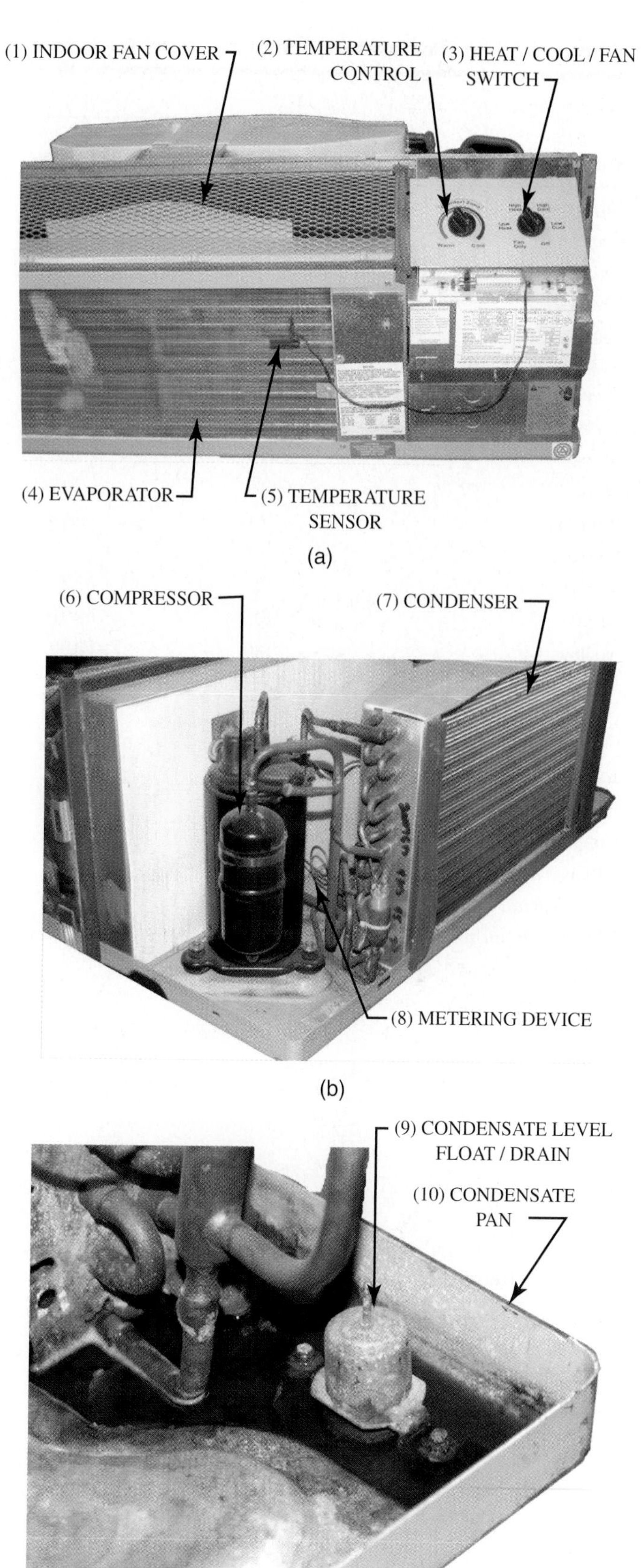

Figure 17-1-11 Expanded views of a PTAC unit listing major parts.

TABLE 17-1-1 Performance Specifications for Room Air Conditioners

	Portable				
Models	**5P2MY**	**7P2MY**	**5P2MC**	**7P2MC**	**9P2MC**
Capacity†					
Cooling (Btu)	5,000	6,500	5,300	6,600	8,600
Energy Efficiency Ratio					
EER	9.0	9.2	9.6	10.0	9.0
Dehumidification (Pints Per Hour)	1.1	2.0	1.4	2.0	3.0
Electrical					
Voltage	115	115	115	115	115
Amps	5.2	6.5	5.1	5.9	8.3
Watts	555	705	550	660	955
Plug Type	PAR	PAR	PAR	PAR	PAR
Features					
Fan Speeds	2	2	3	3	3
RPM	750/1000	750/1000	750/875/1000	750/875/1000	1000/1125/1250
Airflow (CFM)††	180	180	180	180	190
Ventilation	—	—	Exhaust	Exhaust	Exhaust
Controls	Rotary	Rotary	Touch Cooling™	Touch Cooling™	Touch Cooling™
Easy Filter Access	Yes	Yes	Yes	Yes	Yes
Rotary Compressor	Yes	Yes	Yes	Yes	Yes
Copper Tubing	Yes	Yes	Yes	Yes	Yes
24 Hour Timer	—	—	—	—	—
Installation					
Window Mounting	Instamount	Instamount	Instamount	Instamount	Instamount
Slide Out Chassis	—	—	—	—	—
Fits Window Widths	22½–40 in	22½–40 in	22½–40 in	22½–40 in	22½–40 in
(Min Max. Inches) 76.2–111.8 cm	57.2–101.6 cm	57.2–101.6 cm	57.2–101.6 cm	57.2–101.6 cm	57.2–101.6 cm
Thru-Wall (Max. Wall Thickness)	—	—	—	—	—
Dimensions					
Height	13⅜ in/34 cm	13⅜ in/34 cm	13⅜ in/34 cm	13⅜ in/34 cm	13⅜ in/34cm
Width	19 in/48.3 cm	19 in/48.3 cm	19 in/48.3 cm	19 in/48.3 cm	19 in/48.3 cm
Depth (with front on)	20¹⁄₁₆ in/51 cm	20¹⁄₁₆ in/51 cm	20¹⁄₁₆ in/51 cm	20¹⁄₁₆ in/51 cm	20¹⁄₁₆ in/51 cm
Shipping Weight	71 lb/32.2 kg	71 lb/32.2 kg	71 lb/32.2 kg	71 lb/32.2 kg	84 lb/38.1 kg

† Rating conditions are 80°F/27°C db, 67°F/19°C wb indoor air and 95°F/35°C db.
75°F/24°C wb outdoor air.
†† Wet coil with fan on high speed.
* To mount in storm or mobile home window use adapter kit RISAK.
** With optional mounting kit (RSMK1).
Temperature control has approximately 4° differential on cooling.
Dual voltage units based on operation from 253 to 197 V.
*** Fold out side curtains require some assembly.

Compact					High Capacity		
10C2MA	**12C2MB**	**12C3V**	**10C2MT**	**12C2MT**	**14C2MA**	**18C3MA**	**21C3MS**
10,000	11,800	11,800/11,700	10,000	11,800	13,700	17,900/17,400	20,500/20,300
9.6	9.7	10.0	9.6	9.7	10.4	9.0	8.5
2.0	3.3	3.3/3.3	2.0	3.3	3.5	5.1/5.1	6.8/6.8
115	115	230/208	115	115	115	230/208	230/208
9.2	10.5	5.2/5.6	9.2	10.5	12.0	9.0/9.6	10.0/11.6
1040	1215	1180/1170	1040	1215	1315	1990/1940	2400/2390
PAR	PAR	TAN	PAR	PAR	PAR	TAN	TAN
3	3	3	3	3	3	3	3
850/795/1100	850/975/1100	850/1100	850/975/1100	850/975/1100	800/950/1115	800/950/1115	800/950/1115
340	340	340	340	340	420	450	450
Exhaust Touch	Exhaust Touch	Exhaust	Exhaust	Exhaust	Exhaust Touch	Exhaust Touch	Exhaust
Cooling™	Cooling™	Rotary	Rotary	Rotary	Cooling™	Cooling™	Rotary
Yes	Yes	Yes	Yes	Yes	Yes	Yes	Yes
Yes	Yes	Yes	Yes	Yes	Yes	Yes	Yes
Yes	Yes	Yes	Yes	Yes	Yes	Yes	Yes
—	—	—	Yes	Yes	—	—	—
Instamount*	Instamount*	Optional	Instamount*	Instamount*	Instamount***	Instamount***	Instamount***
Yes	Yes	Yes	Yes	Yes	Yes	Yes	Yes
28–42 in	28–42 in	28–42 in	28–42 in	28–42 in	30–44 in	30–44 in	30–44 in
71.1–106.7 cm	71.1–106.7 cm	71.1–106.7 cm	71.1–106.7 cm	71.1–106.7 cm	76.2–111.8 cm	76.2–111.8 cm	76.2–111.8 cm
10¼ in/26 cm	10¼ in/26 cm	10¼ in/26 cm	10¼ in/26 cm	10¼ in/26 cm	10½ in/26.7 cm	10½ in/26.7 cm	10½ in/26.7 cm
15⅞ in/40.3 cm	15⅞ in/40.3 cm	15⅞ in/40.3 cm	15⅞ in/40.3 cm	15⅞ in/40.3 cm	17⅛ in/43.5 cm	17⅛ in/43.5 cm	17⅛ in/43.5 cm
24½ in/62.2 cm	24½ in/62.2 cm	24½ in/62.2 cm	24½ in/62.2 cm	24½ in/62.2 cm	26½ in/67.3 cm	26½ in/67.3 cm	26½ in/67.3 cm
23⅞ in/60.6 cm	23⅞ in/60.6 cm	23⅞ in/60.6 cm	23⅞ in/60.6 cm	23⅞ in/60.6 cm	28¼ in/71.8 cm	28¼ in/71.8 cm	28¼ in/71.8 cm
118 lb/53.5 kg	118 lb/53.5 kg	117 lb/53.1 kg	118 lb/53.5 kg	118 lb/53.5 kg	165 lb/74.8 kg	165 lb/74.8 kg	165 lb/74.8 kg

TABLE 17-1-2 General Performance Data—Air Conditioner with Electric Heat Models

Model Type	Air Conditioner											
Model No.	**PTEB 07**			**PTEB 09**			**PTEB 12**			**PTEB 15**		
Voltage[1]	230	208	265	230	208	265	230	208	265	230	208	265
Capacity[2] (Btuh)	6800	6800	6800	9200	9100	9200	11900	11600	11800	14900	14600	14900
Indoor Fan CFM												
High/Low (Wet coil)	240/210	215/185	240/210	240/210	215/185	240/210	280/260	260/240	280/260	280/260	260/240	280/260
High/Low (Dry coil)	300/260	280/240	300/260	300/260	280/240	300/260	350/325	325/300	350/325	350/325	325/300	350/325
Fresh Air CFM (Dry coil)[3]		40			40			55			55	
Approx. Ship Wt (lb)		129			133			148			157	
Refrig. Charge (oz)		20.0			30.0			29.0			36.0	
Oil Charge (oz)		8.8			7.4			10.8			13.9	

Minimum operating ambient temperature for cooling is 45 degrees.
[1]Minimum voltage on 230/280 V units is 197 V; maximum is 253 V.
[2]All capacities based on approved ARI rating point 80/67 entering air temperature, 95°F outdoor ambient.
[3]CFM rating is with the unit in the "fan only" setting.

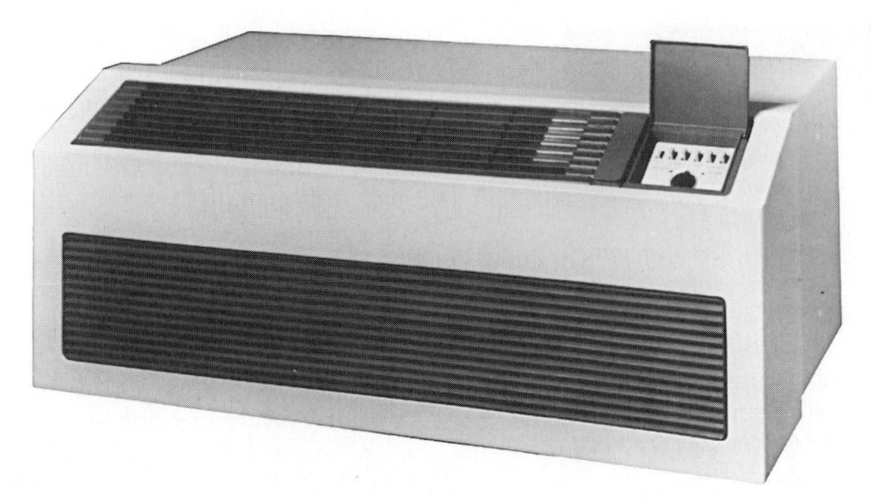

Figure 17-1-12 Duct package for PTAC units. (*The Trane Company*)

SERVICE TIP

Most PTAC units use rotary compressors and capillary tube metering devices. Rotary compressors have an excellent service life for this application provided that the condenser is kept clean. High compression ratios will cause refrigerant to break down and rotary compressors are unusually susceptible to damage from excessive heat because the motor is on the high side of the system.

Controls

A rotary switch is provided with the following choices:

OFF	Turns unit off.
FAN ONLY	Indoor fan operates.
COOL	Provides cooling with indoor fan.
HEAT	Provides heating with indoor fan.

A rocker switch provides the following choices:

HI	High fan speed
LOW	Low fan speed

The following additional controls are provided:

Adjustable temperature limiting device Limits the range of the room thermostat.

Outside air damper Control lever can be positioned to permit zero for fully open supply of outside air.

Fan cycle switch Allows continuous or intermittent fan operation.

Remote thermostat (optional) Unit can be wired to use a remote wall mounted thermostat rather than the one normally supplied in the unit.

Front desk control interface (optional) Units may be individually started and stopped with an energy management panel from a central location.

Room freeze protection (optional) Overrides OFF signal when room thermostat goes below 40°F and turns off equipment automatically.

DEHUMIDIFIER UNITS

Dehumidifier units are small, portable, self contained refrigeration systems designed to extract moisture from the air. They are used in localities where high humidity can cause damage to stored materials. They are usually installed where cooling is not required. They consist of a motor compressor unit, a condenser, an evaporator, an air circulating fan, and a means of collecting and/or disposing of the condensate, and a cabinet, as shown in Figure 17-1-13.

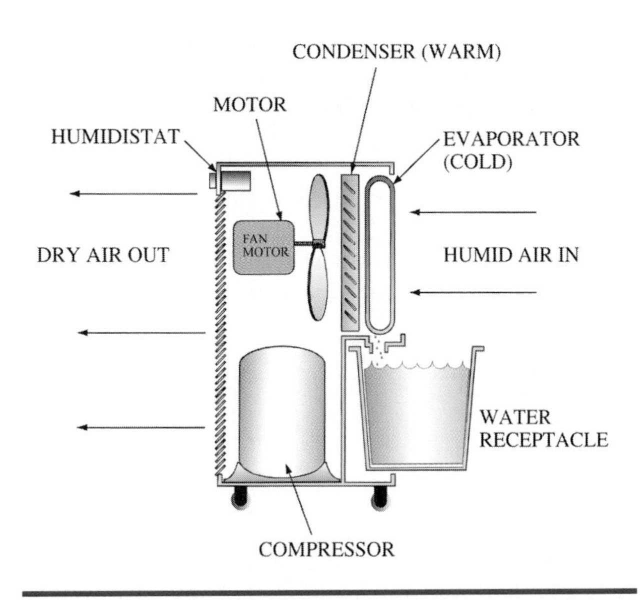

Figure 17-1-13 Diagrammatic view of a dehumidifier. (*Copyright by the American Society of Heating, Refrigerating, and Air-Conditioning Engineers, Inc.*)

NICE TO KNOW

Dehumidifiers are very common in areas with basements particularly when the basement is in older homes where the basement walls were not vapor filled before construction. Many homes in the New England areas have stone walls in their basement and humidity is a significant problem. Any area where humidity can be excessive is subject to mold growth. Dehumidifiers therefore are an excellent addition to a residence as a way of controlling mold in damp basement areas. Mold needs moisture and food for growth. It can grow on virtually any organic material so it is difficult to remove the food supply for mold. It is much easier to remove the moisture.

The fan draws moist room air over the evaporator and cools it below dew point temperature. The removed moisture drains into a collecting pan or drops into an open drain. The cooled air then passes over the condenser coil where it is heated and discharged into the room at a higher dry bulb temperature and at a lower relative humidity. Continuous circulation gradually reduces the relative humidity in the room.

The compressor is a fractional horsepower hermetically sealed unit, typically requiring from 200 to 700 W of power input. Most refrigeration systems use a capillary metering device, although some larger units use thermal expansion valves. The airflow rate is usually 125 to 250 CFM. The capacity ranges from 11 to 50 pints per day at standard test conditions of 80°F DBT and 60% RH.

The controls supplied with the unit vary depending on the manufacturer and the model. The following controls are desirable and come as either standard equipment or options:

- On-off switch.
- Humidity sensing control to cycle the unit automatically.
- Automatic sensing switch to turn the unit off when the water receptacle is full and requires emptying.
- Defrost controls, which cycle the compressor off under frosting conditions.

Servicing Dehumidifiers

A room dehumidifier is a portable device for removing moisture from an area where cooling is not required. Due to its portability and relatively light weight it can be taken into the shop when service is required. It consists of a hermetically sealed refrigeration unit that extracts moisture from room air and reheats it by passing the air through the condenser coil. The discharged air is therefore about the temperature of the room air, and no cooling is taking place.

The proper functioning of the following controls on dehumidifiers is important.

- A manual switch turns the unit on and off.
- A humidistat permits the unit to run only when the humidity level is above the setting of the control.
- A frost control stops the compressor before the coil temperature reaches freezing temperature.

The following are types of service problems that can occur:

- **Water leakage** The customer reports that water is leaking from the unit onto the floor. This may be caused by not emptying the water receptacle often enough or leakage in the water runoff system.
- **Continuous running** This may be due to dirty coils that cause the unit to run inefficiently. If the unit is operating properly, it may be that the unit is too small to handle the load. Resetting the humidistat may be necessary. Dirty coils can also cause the compressor to cycle on overload.
- **Evaporator collects ice** This may point to a defective frost control or a frost control that is not making proper contact with the coil. It could also indicate a restriction in the refrigerant line, requiring hermetically sealed system service.
- **Unit will not operate** This may be due to a blown fuse, low voltage, a broken wire, or a defective on/off switch. The electrical system needs to be thoroughly checked.

SINGLE PACKAGE CONDITIONERS

A single package conditioner, often called a Unitaire, is a complete self contained factory built unit, for permanent installation, to condition larger spaces than practical using room coolers.

In this category there are two variations of available equipment:

- **Horizontal conditioner** Horizontal packaged heating and cooling units, with integral air cooled condensers, in the size range of 1½ to 5 tons.
- **Vertical conditioner** Vertical self contained air conditioners, with water cooled or remote air cooled condensers, in the size range of 3 to 15 tons.

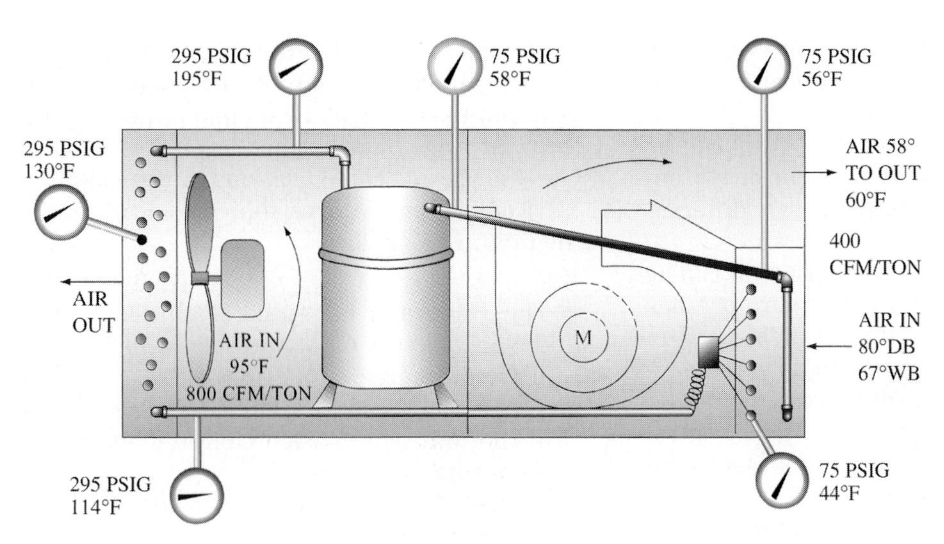

Figure 17-1-14 Component arrangement for a single-package conditioner.

HORIZONTAL CONDITIONER

The horizontal unit usually uses ductwork for air distribution. In residences it can be used to supply cooling, humidification, and ventilation for a radiation heated (hydronic) house. These systems are sometimes called split systems.

The horizontal unit is completely self contained, including the air cooled condenser. It therefore must be placed either entirely outside or at least with the condenser section outside. A schematic diagram of the arrangement of parts is shown in Figure 17-1-14.

The unit is primarily a cooling unit although electric heaters can be installed in the unit as shown in Figure 17-1-15.

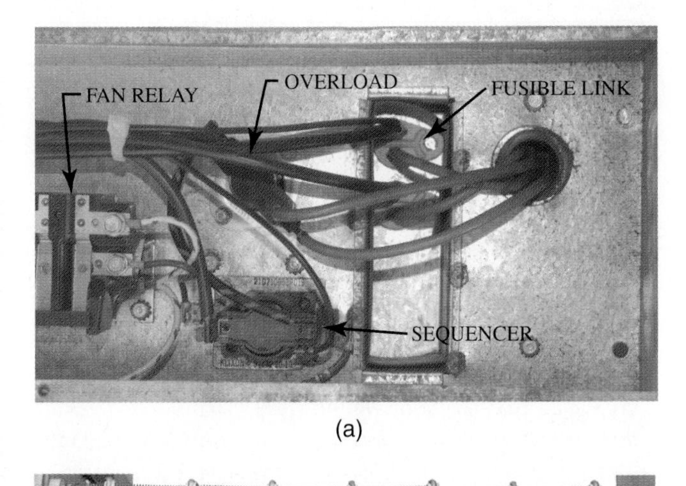

Figure 17-1-15 (a) Electric heaters for a package air conditioner; (b) Nichrome resistance heating wire.

The unit can be equipped with supply and return air duct connections. Ducts can be entered from the side or bottom, depending on the application.

For residential use, the horizontal unit can be arranged for a ground level installation.

For commercial use the unit can be installed on the roof, as shown in Figure 17-1-16. This type of application is used for shopping malls, factories, and other commercial buildings.

The horizontal unit comes equipped with the following features:

- **Water protection** A weather resistant cabinet along with a water shedding base pan with elevated downflow openings and a perimeter channel prevent water from draining into the ductwork.
- **Low ambient control (optional)** Kits are available that control the condenser head pressure to permit the unit to cool the ambient temperature.
- **Economizer (optional)** An economizer and dry bulb temperature sensor can be supplied for

Figure 17-1-16 Roof installation of a single package conditioner.

downflow installations. This makes it possible to use outside air for cooling when outdoor temperatures and humidity permit.

- **Enthalpy control kit (optional)** This can be supplied in place of the dry bulb sensor, or two enthalpy controls can be paired to provide differential enthalpy control.
- **Fresh air (25%) kit (optional)** This kit can be mounted over the horizontal return air openings for downflow requirements. It also can be used on horizontal applications by cutting a hole in the return air duct or in the unit filter access panel.
- **Fan delay relay kit (optional)** This control keeps the indoor blower on for about 90 sec to improve the EER.
- **Anti short cycle timer** A time-off device ensures a minimum of 5 min off between compressor cycles.

Performance

Units range in capacity from 18,000 to 60,000 Btu/hr and range in air quantity from 600 to 2,000 CFM at standard rating conditions. All units are available for 208/230 V, single phase, 60 Hz power. Efficiencies range from 10.0 EER and higher depending on the size. Optional electric heaters range in size from 3.74 to 29.80 kW.

VERTICAL CONDENSER

This is a commercial packaged unit for installation, usually inside the space being conditioned. The condenser is water cooled. Water from a remotely located cooling tower needs to be piped in and out of the shell type condenser(s), as shown in Figure 17-1-17.

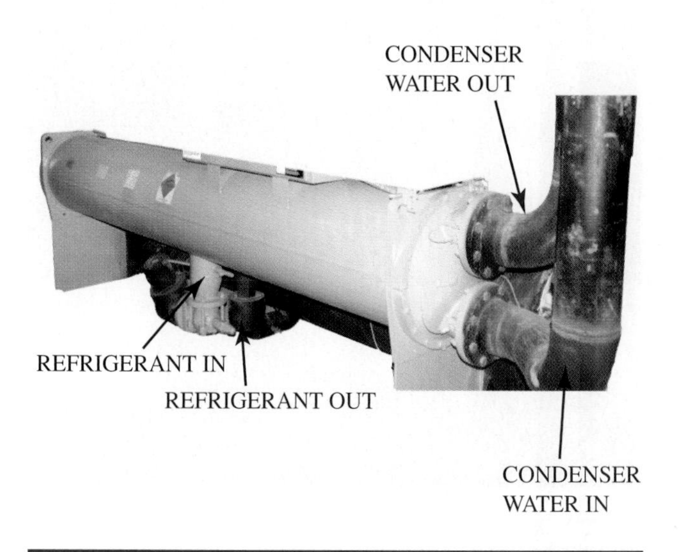

Figure 17-1-17 Shell type condenser.

Discharge air can either be free throw or ducted horizontally or vertically. An accessory plenum and grill that fits on top of the unit is supplied for the free throw arrangement. When ductwork is used, a number of discharge configurations can be supplied.

Application features that are incorporated into the unit are as follows:

- The evaporator fan speed is adjustable, affecting the air delivery CFM and the available ductwork static pressure.
- Thermostats can be supplied as an integral part of the unit or provision can be made for remote mounting.
- An anti short cycle timer is provided to protect the compressor from excess cycling.
- If an air cooled condenser is used, a low ambient control can be provided to permit the unit to cool at low temperatures.

ROOFTOP CONDITIONERS

Rooftop conditioners are similar to single package conditioners except that they are thoroughly weatherproofed and provide for duct access at the bottom of the unit. They are popular for air conditioning low story commercial buildings because they offer a substantial savings of space within the building.

From the standpoint of the service technician, they are desirable because they offer plenty of access space around the unit for servicing. On many jobs, however,

SERVICE TIP

One of the prime concerns and problems with all rooftop installations is water leakage. Most commercial buildings use flat roofs and water in a rainstorm can build up to several inches in depth. If the equipment panels are not properly reinstalled to provide an adequate water seal, rainwater can easily enter the building. In addition, if the proper curb is not provided for the unit, then the system will leak. Many commercial roofs are "bonded" which means that there is a specific insurance policy taken out on the roof by the roofing company at the time of installation. This provides the building operator with insurance that the roof will not leak. That roofing contractor is the only one who can do any work on the roof that would affect the integrity of that roof without violating that bond. If an air conditioning contractor were to violate the bond on the roof by putting in a vent pipe that contractor could be ultimately liable for any water damage resulting from leaks. Check with the building operator before any work is done on bonded roofs.

access to the roof is only by ladder and in bad weather the units may be difficult to reach and offer physical restraints in supplying needed tools and parts.

Rooftop self contained air conditioning units are commonly used on commercial installations. The sizes range from 3 tons to 130 tons of cooling capacity under standard rating conditions. Besides the difference in size of the components, individual units differ in the type of heating supplied with the package.

For the units in the 3 to 25 ton range, gas fired or electric heat can be supplied. Larger units can also be equipped with hot water or steam coils. When the units are located outside, adequate freeze protection needs to be provided where the unit contains water. Condensate drain pans must be free draining.

Since the most frequently used units are in the 3 to 25 ton range, this equipment will be described in the following sections.

Dual compressor models are usually available starting at 7½ tons and higher. When dual compressors are furnished, dual refrigeration circuits are also supplied. This arrangement makes possible better performance ratings and increased energy savings at partial loads.

CABINET CONSTRUCTION

Cabinets are constructed of zinc coated heavy gauge steel and are weather tight. All services can be performed through access panels on one side. Supply and return ductwork connections can be made at the bottom or side of the unit. Roof curb frames are available for roof mounting.

AIR FILTERS

Air filters are available for systems that are disposable (one time use), reusable, and permanent. Filters should be selected to provide the optimal performance for the environment. Some filters have a very low static which will allow the maximum free air flow through the filter while providing a minimal amount of filtration. These filters are acceptable in open areas, commercial buildings, or retail stores where there is no concern about dust, pollen, or other contaminants. For occupied spaces where allergens, pollen, and mold are a concern, filters with a finer mesh (HEPA filters) are recommended. However, some of these filters have exceedingly high static pressure drops. These pressure drops may exceed the acceptable range for the equipment. If the filter has an excessively high static pressure drop it will adversely affect the total CFM. This can significantly reduce the system performance. Permanent filters have limited use in most applications. They do, however, work well as prefilters for electronic filtration systems.

Check with the filter manufacturer for specifications on pressure drop and particle size as well as the equipment manufacturer to make certain that you are selecting the most appropriate filter medium for your customer.

COMPRESSORS

Reciprocating or scroll type, direct drive, hermetic compressors are used. Compressor motors are suction gas cooled, and are protected with temperature and current sensitive overloads. Crankcase heaters are standard equipment.

GAS HEATING

The heat exchanger is made of corrosion resistant steel. One manufacturer uses a drum and tube design, as shown in Figure 17-1-18. Units use a forced

Figure 17-1-18 Drum and tube heat exchanger.

combustion blower and hot surface ignition. Gas will not ignite unless the combustion air blower is operating.

On an initial call for heat, the combustion air blower will purge the heat exchanger with fresh air. If the attempt to light the main burner flame is unsuccessful, the system will purge the heat exchanger again and start a second trial for ignition. If there are three unsuccessful trials for ignition, the entire heating system will lock out until it is manually reset.

CONDENSER AND EVAPORATOR FAN

The condenser fan is a propeller type with a permanently lubricated, overload protected motor. The evaporator fan is a centrifugal type, belt driven in most sizes, with adjustable sheaves. Units are capable of delivering nominal airflows at 1 in external static pressure (ESP).

The unit can be equipped to handle a supply duct system using variable air volume (VAV) control.

CONTROLS

The newer units are provided with microprocessor controls for all 24 V functions. Control decisions are automatically made in response to the input from indoor and outdoor temperature sensors. Anti short cycle compressor controls are provided. The control system can be interfaced with a central direct digital control (DDC) system.

Outside Air Dampers

Manually positioned outside air dampers that can be adjusted to provide up to 25% outside air are supplied with a rain screen and hood for field installation.

ECONOMIZER

An economizer is an optional arrangement for using outside air for cooling if conditions are feasible, as sensed by the control system. The alterations to the unit are shown in Figure 17-1-19. The assembly includes a fully modulated 0–100% motor and dampers. The barometric relief damper provides automatic closing of the outside air opening when the equipment is not operating.

Desiccant Cooling Systems

A desiccant is a material that absorbs moisture without causing a chemical change in the material. The

TECH TIP

There are local, state and national codes that affect the amount of outdoor air that must enter an area depending on its use. In some cases higher ventilation requirements exist for buildings such as schools or theaters, where large numbers of people gather, than there are for warehouses or areas where minimal numbers of people gather. Each area has its specifications for makeup air. Outside dampers provide this essential replacement of air for building interiors. Make certain that your dampers are set according to the specific needs. If the dampers are set to provide less than the required amount of fresh air then indoor air quality will be significantly affected. If the dampers are set to provide excessive outdoor air then the building operating costs will be excessive.

Figure 17-1-19 An economizer installation in a package rooftop conditioner.

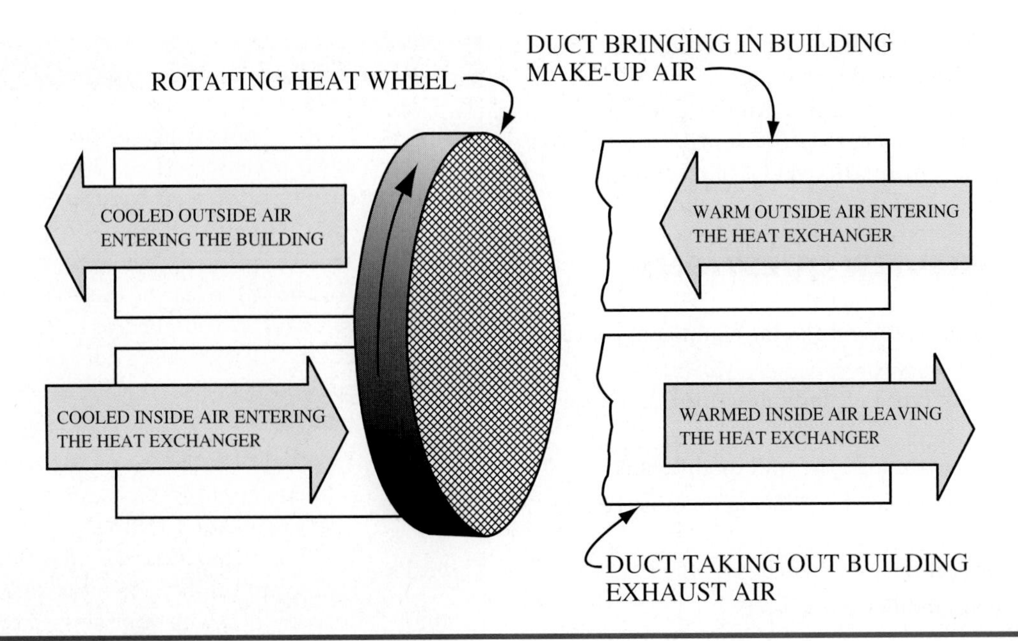

Figure 17-1-20 Energy recovery wheel.

material can thus be reactivated and reused by applying heat to the saturated product. Systems with desiccant are used for dehumidifying air. The term desiccant cooling is applied to these systems since dehumidification is a part of the air conditioning (cooling) process. These systems can only be justified in relation to their ability to save air conditioning energy.

Some applications employ a rotating wheel impregnated with a porous desiccant. The air passages through the wheel are divided into two parts, one for low temperature outside air entering the building, and the other for hot moist air that contacts the desiccant. The moisture is absorbed and at the same time produces heat (970 Btu/lb of moisture absorbed). As the wheel rotates into the outside airstream the heat held by the desiccant is given off, preheating the cool outside air before it enters the building. This is an efficient use of the desiccant in winter.

In summer the desiccant wheel can be used to dry the incoming air and eliminate the need to chill the air deeply to condense the moisture. This reduces the amount of energy required to produce cooling. This is an efficient use of the desiccant.

ENERGY RECOVERY VENTILATORS

The unitary equipment that uses the desiccant wheel is the energy recovery ventilator (ERV). An illustration of one of these units that has been developed for residential application is shown in Figure 17-1-20. The desiccant wheel is designed to transfer water vapor from one airstream to another.

The wheel will also transfer heat. It will recover enough heat from exhaust air to warm incoming air to 60°F when the outside air temperature is as low as 5°F. On subzero days, an electric frost control operates an electric preheater to warm incoming air a few more degrees. Different heater sizes are available for different design temperature ranges.

Another type of heat recovery ventilator unit is shown in Figure 17-1-21. This unit is primarily used to transfer heat from one airstream to the other. In winter the transfer core uses the heat of indoor air to warm the incoming cool fresh air. In the warm months air conditioned indoor air cools the incoming fresh air. The core recovers about 75% of the energy during this process and at the same time removes a substantial

Figure 17-1-21 Energy recovery ventilator (ERV).

amount of humidity from the warm incoming air to reduce the air conditioning requirement.

Both of these recovery units are designed to provide a continuous supply of outside (ventilation) air to improve the indoor air quality (IAQ) of the building.

UNIT 1—REVIEW QUESTIONS

1. What does unitary air conditioning equipment consist of?
2. List the various types of units described as unitary equipment.
3. What is a console through the wall conditioner?
4. List the standards that the PTAC units must comply with.
5. List the choices of a rocker switch.
6. Where are dehumidifier units used?
7. What is a single packaged conditioner?
8. For commercial use where can a horizontal unit be installed?
9. What is the difference between a condenser and evaporator fan?
10. What is a desiccant?

TECH TALK

Tech 1. I don't like working on commercial units very much because you are always on top of the building or on top of a tower or high on a catwalk. Heights scare me.

Tech 2. You are right—heights should scare you. Falls are one of the leading causes of injuries in our industry. So you need to take precautions. It doesn't mean you are not going to be scared but you need to take all the necessary precautions if you are going to be up on heights.

Tech 1. You mean like wearing a safety harness and helmet and all those other things?

Tech 2. That's right. You have to every time you go up. You have to tie the ladder off when you are going up and you have to make sure that you're securely tied with your lanyard to something that is substantial. If you are on a catwalk or if you are near the edge you need to take precautions that you don't accidentally fall because it is your life that is in your hands.

Tech 1. Even though I do all that, it still makes me nervous.

Tech 2. Yes, and you probably will be nervous for the rest of your life working on heights. But that is still good. It makes you pay attention and be careful. It is when you get so comfortable working on heights that you can get careless that it really becomes a danger to yourself and others.

Tech 1. So you are telling me that it is not going to get any better.

Tech 2. It may get a little better. But you hope that it doesn't get so comfortable that you are up jumping around or that you feel that you don't have to take the precautions this time because you are not that high. Any time you are working at any height greater than 6 ft you need to take precautions. Certainly almost all commercial work is going to put you at that height or higher.

Tech 1. Well, I guess I will just have to work at getting more comfortable but staying alert. Thanks. I appreciate that.

UNIT 2

Air Handling Units and Accessories

OBJECTIVES: In this unit the reader will learn about air handling units and accessories including:

- Types of Air Handling Units
- Fan Coil Units
- Central Station Air Handling Units

TYPES OF AIR HANDLING UNITS

The air handling unit is that portion of the air conditioning system that conveys the conditioned air to and from the conditioned space. Historically, these units distributed air by means of gravity action like space heaters, which are still used for certain applications. Most units, however, now use a fan (or blower) to force the air through the conditioning equipment and its distribution system.

Air handling units for these forced air systems are separated from the primary heating or cooling apparatus. These units form a part of a central plant, a built-up system that is flexible in makeup to fit the application.

There are two general types of air handling units that are used for central plant systems: ones with remote air handling units, called fan coil units, and ones with central station air handling units.

Fan coil unit systems consist of room units that are supplied with hot water or cold water directly from central plant boilers and chillers. Each one of these units has fans, water coils, filters, a fresh air supply, and a control system. They were primarily designed to supply conditioned air to individual rooms.

Central station systems use one or more large air handling units equipped with fans, heating and cooling coils and other accessories, and duct work to convey the conditioned air to remote areas.

FAN COIL UNITS

Fan coil units for combination heating and cooling come in a variety of designs, the most familiar being the individual room conditioner, Figure 17-2-1.

This type of unit consists of a filter, direct driven centrifugal fan(s), and a coil suitable for handling chilled or hot water. The size of the unit is based on its cooling ability; it is usually more than adequate

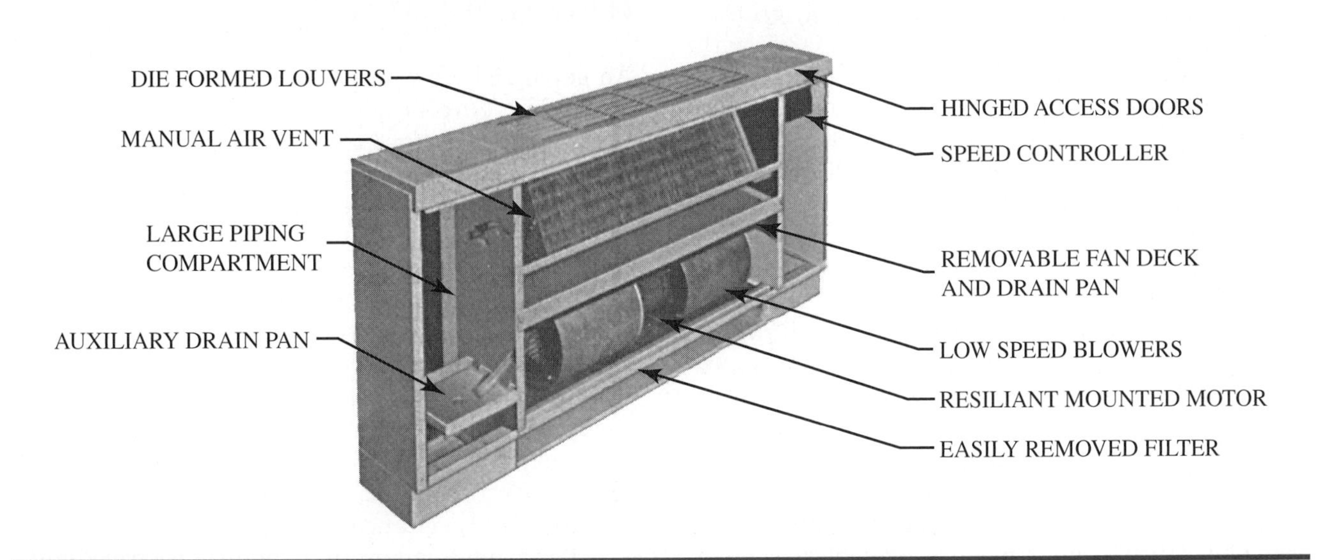

Figure 17-2-1 Typical fan coil unit, showing component parts. (*Courtesy of Dunham-Bush, Inc.*)

for heating. A variety of water flow control packages are available for manual, semi-automatic, or fully automatic motorized or solenoid valve operation.

TECH TIP

Often around fluorescent lights in a drop ceiling there is a narrow slit. That opening is the return air. The advantage of having a return air around the fluorescents is that the heat generated by the light is pulled back up into the return and is not radiated as much into the occupied space. Because of the high artificial light levels in most commercial buildings fluorescents can represent a significant load. This removes the load from the occupied space to make it feel more comfortable.

Airflow is controlled by fan speed adjustment manual or automatic. Outside air is introduced through a dampered opening to the outside wall. The size of the cabinet is designated in ft^3 of airflow and ranges from 200 to 1200 ft^3/min. These units may be installed on two, three, or four pipe system designs as illustrated in Figure 17-2-2.

With modifications to the cabinet, the same components are assembled in a horizontal ceiling mounted version, Figure 17-2-3. Where hot water heating is not desirable for heating, electric resistance heaters are installed in the cabinet on the leaving air side of the cooling coil.

Another version of the room fan coil unit is a vertical column design, Figure 17-2-4, which may be installed exposed or concealed in the wall. These are placed in common walls between two apartments, motel rooms, etc. Water piping risers are also included in the same wall cavity. They are not designed for ducting but can serve two rooms by adding air supply grills.

Small ducted fan coil units, Figure 17-2-5, with water coils may be installed in a drop ceiling or a closet, with ducts running to individual rooms in an apartment. These range in sizes from 800 to 2,000 ft^3/min. The unit's cooling capacity is selected based on the desired ft^3/min airflow.

Larger fan coil units, Figure 17-2-6, used to condition offices, stores, etc., where common air distribution is feasible, resemble cabinets of self contained store conditioners. They can be equipped with supply and return air grills for in-spacing application, or they may be remotely located with ducting. Sizes range from 800 to 15,000 ft^3/min. In larger sizes, cabinet and fan discharge arrangements permit flexible installations. These units approach the next category of equipment called air handlers, but, in general, they do not have the size and functions available in central station air handlers.

Fan coil units have a wide variety of applications. One advantage they have is the ability to provide individual room conditioning. This is valuable in applications for hospital rooms, motel rooms, and individual private offices. An accessory fresh air duct can be provided in the bottom rear of the unit to permit the entrance of outside ventilation air. Each room unit can be individually controlled without mixing air from an adjacent room.

One very interesting application of fan coil units is in small church air conditioning. Units are placed around the perimeter of the sanctuary, heating or cooling the exterior exposure of the building and gently supplying conditioned air to the occupants. The units are quiet, effective, and can be used year round.

For the installation of fan coil units that include cooling, it is important to provide a method of disposal for the dehumidification condensate from the cooling coil. This condensate falls into a condensate pan below the coil and is removed by gravity flow to an open drain.

Larger sized fan coil units, sometimes called unit ventilators, are used for school classroom conditioning. A single unit ventilator is installed below window level in each classroom with extension supply air ducts and grills running along the sill the full length of the room. An opening in the back supplies outside air from a through-the-wall opening. The units include air filters to clean the air. Units are individually controlled and thus allow for differences in sun load, depending on the orientation of the room. They also meet the ventilation requirements of local codes for classrooms.

CENTRAL STATION AIR HANDLING UNITS

An expanded view of a typical large air handler is shown in Figure 17-2-7. Often these units are made in modules or sections that can be joined together on the job. This arrangement offers flexibility in the selection of components required for a specific installation.

A schematic diagram of a two fan central station air handling system is shown in Figure 17-2-8. On a large system, the reason for using the return air fan is to be able to control the building pressurization. With a single fan, the suction pressure in the return system can create a negative pressure in many parts of the building. This can cause undesirable infiltration, affect the operation of doors, and put an extra resistance on the operation of the fan. It is desirable to maintain a slightly positive building pressure.

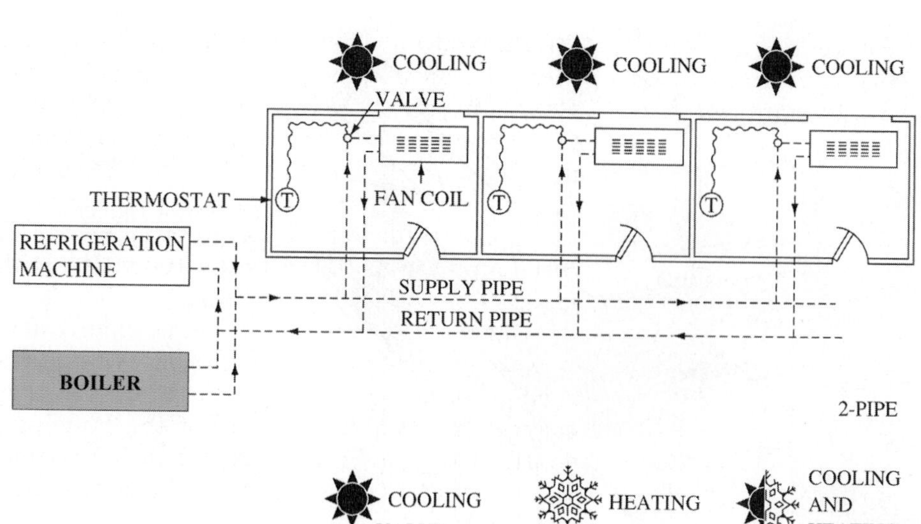

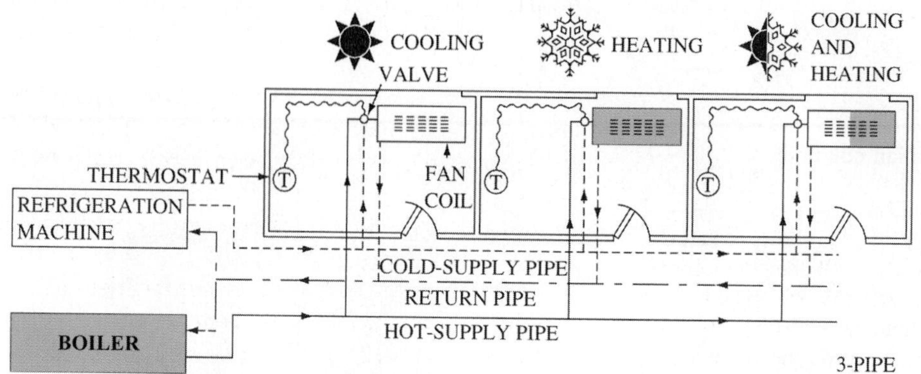

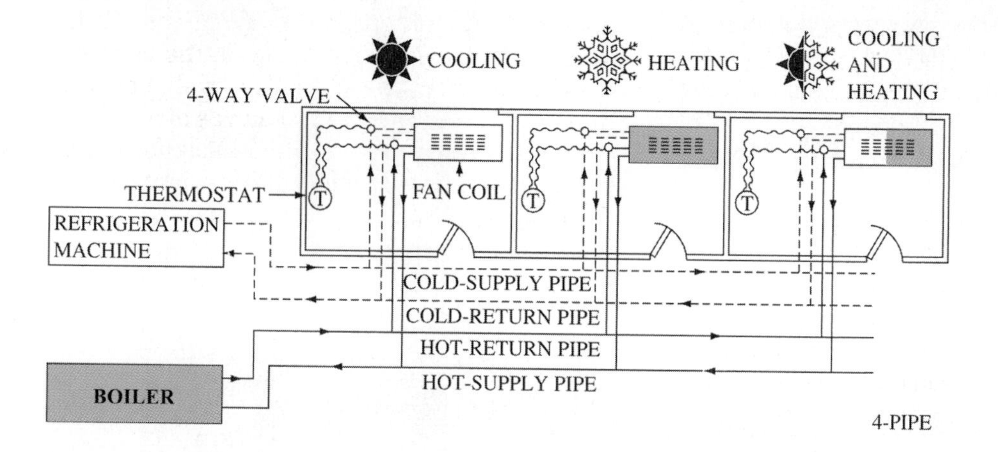

2-PIPE SYSTEM
EITHER HOT OR CHILLED WATER IS PIPED THROUGHOUT THE BUILDING TO A NUMBER OF FAN-COIL UNITS. ONE PIPE SUPPLIES WATER AND THE OTHER RETURNS IT. COOLING OPERATION IS ILLUSTRATED HERE.

3-PIPE SYSTEM
TWO SUPPLY PIPES, ONE CARRYING HOT AND THE OTHER CHILLED WATER, MAKE BOTH HEATING AND COOLING AVAILABLE AT ANY TIME NEEDED. ONE COMMON RETURN PIPE SERVES ALL FAN-COIL UNITS.

4-PIPE SYSTEM
TWO SEPARATE PIPING CIRCUITS – ONE FOR HOT AND ONE FOR CHILLED WATER. MODIFIED FAN COIL UNIT HAS A DOUBLE OR SPLIT COIL. PART OF THIS HEATS ONLY, PART COOLS ONLY.

Figure 17-2-2 Three types of piping arrangements for fan coil units.

The return air fan is installed in the ductwork and is not part of the air handling unit. Likewise the preheating coil is usually installed in the ductwork near the entrance of the outside air into the building. The preheating coil is provided with some type of freeze protection. The air handling unit itself consists of:

- Heating and cooling coils
- A humidifier
- Mechanical air filters
- Electronic air filters (optional)
- Mixing dampers and (optional) exhaust dampers
- Fans and motors

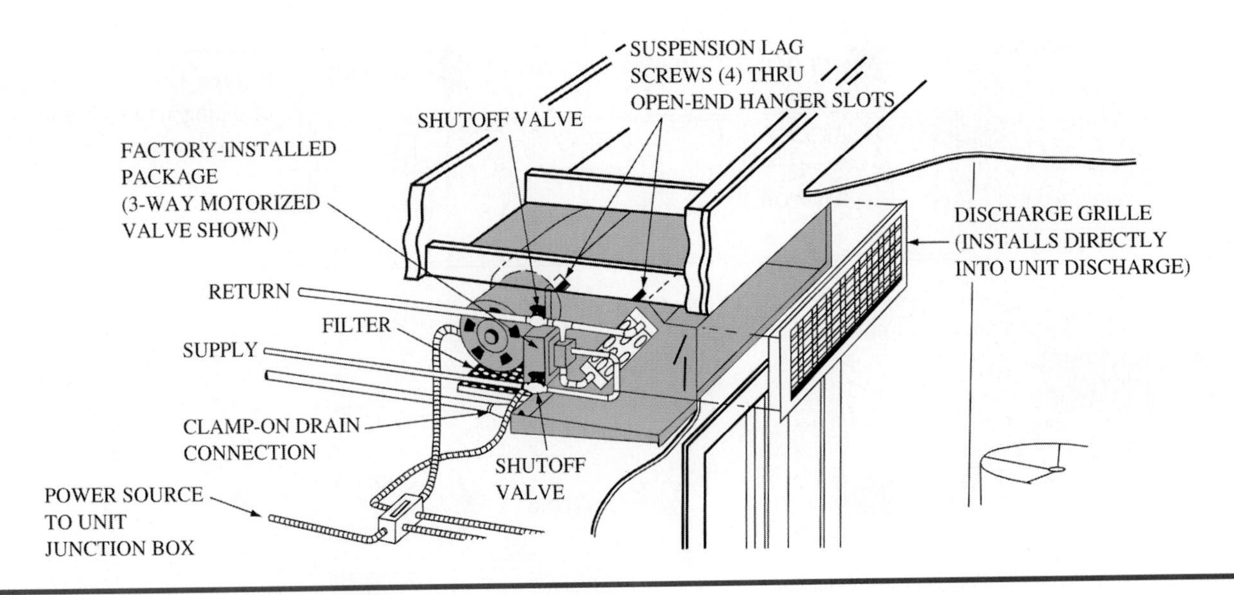

Figure 17-2-3 Ceiling-mounted fan coil unit.

NOTE: Central station air handlers can be very large. Some blower wheels are more than 20 feet in height and are powered by very large variable speed motors. These blowers handle tens of thousands of ft^3 of air per minute. Blower area is actually a large room that may be located in a subbasement or in high rises on a separate floor in the building. When you enter this area you are actually walking into the blower chamber. Be careful when opening any door in this area or accessing the area because there may be a substantial pressure difference and the door can lunge open or closed.

Heating and Cooling Coils

With central air handling units of this type it is not possible to operate both the heating and cooling coils at the same time, for two reasons:

1. The energy conservation codes do not permit it.
2. There would be no advantage since a single zone unit can only provide heating or cooling at one time.

The control system includes a provision for changing over from heating to cooling (and the reverse) when conditions warrant it.

The heating coil is usually selected for hot water heating, although occasionally steam is used. The advantage of hot water is that it is easier to control. There can be parts of the building that never require heat, such as an interior zone. For those units, the heating coil may be optional depending on the application.

The chilled water cooling coil is similar in construction to the hot water heating coil. The cooling coil, however, will probably be deeper (more rows deep), since more surface is necessary for dehumidification.

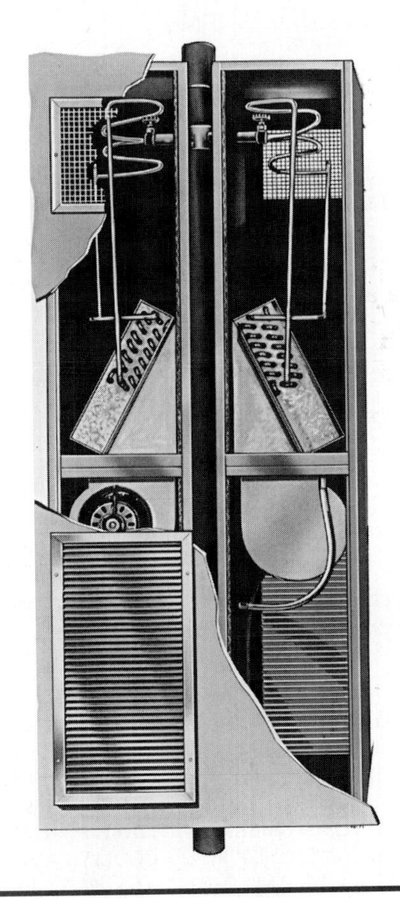

Figure 17-2-4 Vertical-column fan coil unit. *(Courtesy of York International Corp.)*

A typical finned tube water coil for heating and/or cooling is shown in Figure 17-2-9. Water coils are available in depths of 1, 2, 3, 4, 5, 6, and 8 rows deep with various types of circuiting. They are free draining and are provided with ¼ in NPT fittings in the header for draining and venting. Tubes are staggered to improve thermal efficiency.

Figure 17-2-5 Fan coil unit arranged for the ductwork. *(Courtesy of York International Corp.)*

Humidifiers

Humidity is the amount of water vapor within a given air space. *Absolute humidity* is the weight of the water vapor per pound of dry air. Relative humidity is the ratio of the actual amount of water vapor in the air to the amount that it would hold if 100% saturated (the dew point).

For example, referring to the psychometric chart in Figure 17-2-10, if the dry bulb temperature is 72° and the wet bulb temperature is 56°F, the absolute humidity is 41 gr/lb, the relative humidity is 36%, and the dew point is 43°F.

There are four benefits in maintaining the proper humidity in the occupied space:

1. Comfort
2. Preservation
3. Health
4. Energy conservation

TECH TIP

Humidifiers that are not properly maintained are subject to mold and mildew growth. Because of concerns with indoor air quality it is very important that any humidifier be properly maintained to prevent these growths. If a growth occurs and the system is allowed to dry, mold and mildew will release spores and the spores are the primary irritants in a building. Before a system is restarted you must remove all of the growth. Follow the equipment manufacturer's procedures and all local health code requirements when removing mold and mildew from humidifiers or any other part of a building.

Comfort is greatly affected by the relative humidity of air. At a temperature of 72°F, with 75% RH you would feel very uncomfortable. At the same temperature of 72°F with 35% RH you would feel cozy. At 72°F and 10% RH you would probably feel cool. Too much or too little humidity is unpleasant. It is important to maintain the proper humidity.

For the *preservation* of hydroscopic materials such as wood, leather, paper, and cloth, a fixed amount of moisture is necessary to preserve their proper condition. These materials shrink when they dry out. If moisture loss is rapid, warping and cracking can take place. To retain their good quality, proper humidity must be maintained.

From a *health* standpoint, medical science reveals that the nasal mucus contains some 96% water. Doctors indicate that the drying out of the nasal tissues in winter helps to initiate the common cold. Maintaining proper humidity can greatly affect one's susceptibility to colds.

Figure 17-2-6 Large size air handling unit. *(Courtesy of Rheem Manufacturing Air Conditioning Division)*

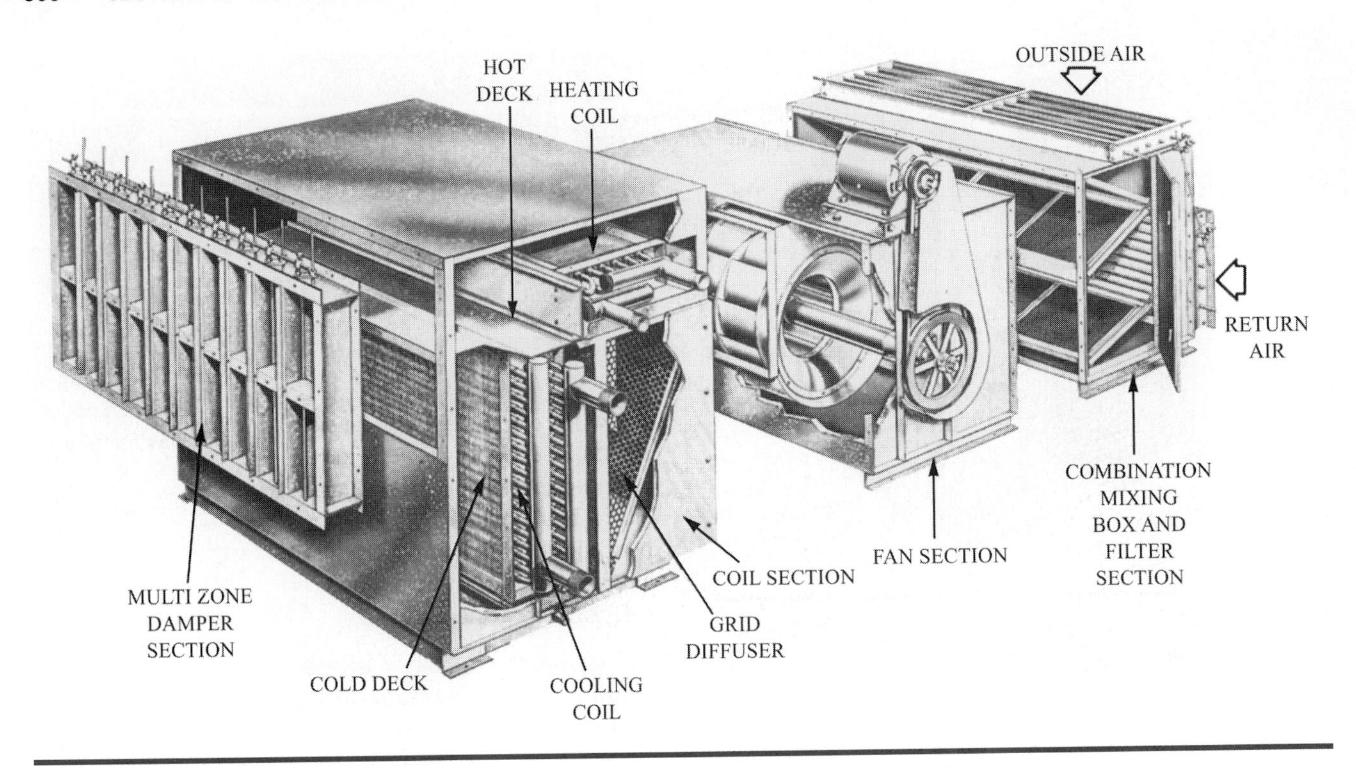

Figure 17-2-7 Central multizone air handling unit, modular construction. *(Courtesy of York International Corp.)*

Low humidities can cause shrinkage of the framing around doors and windows, increasing the infiltration of outside air, causing increased energy usage. Also, proper humidity permits lower comfort temperatures, providing *energy conservation*.

One of the limiting factors in maintaining proper humidity is the *condensation* that occurs on windows in winter. Table 17-2-1 shows the outside temperature at which condensation will occur for various types of glass.

Not shown in the air handler illustrated, Figure 17-2-7, are humidifiers that can be used to add moisture to the air. This would be done in the hot deck depending on the need and application. Several types

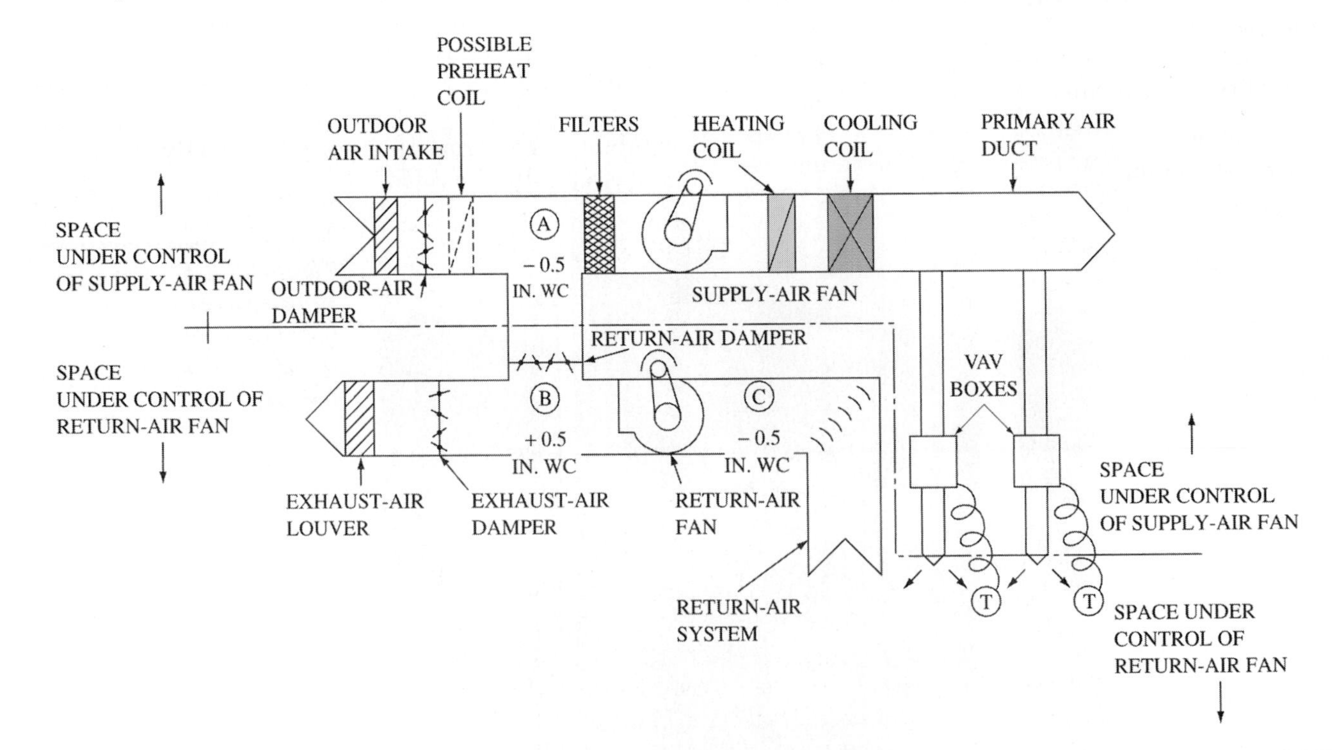

Figure 17-2-8 Schematic view of a central station air handling unit.

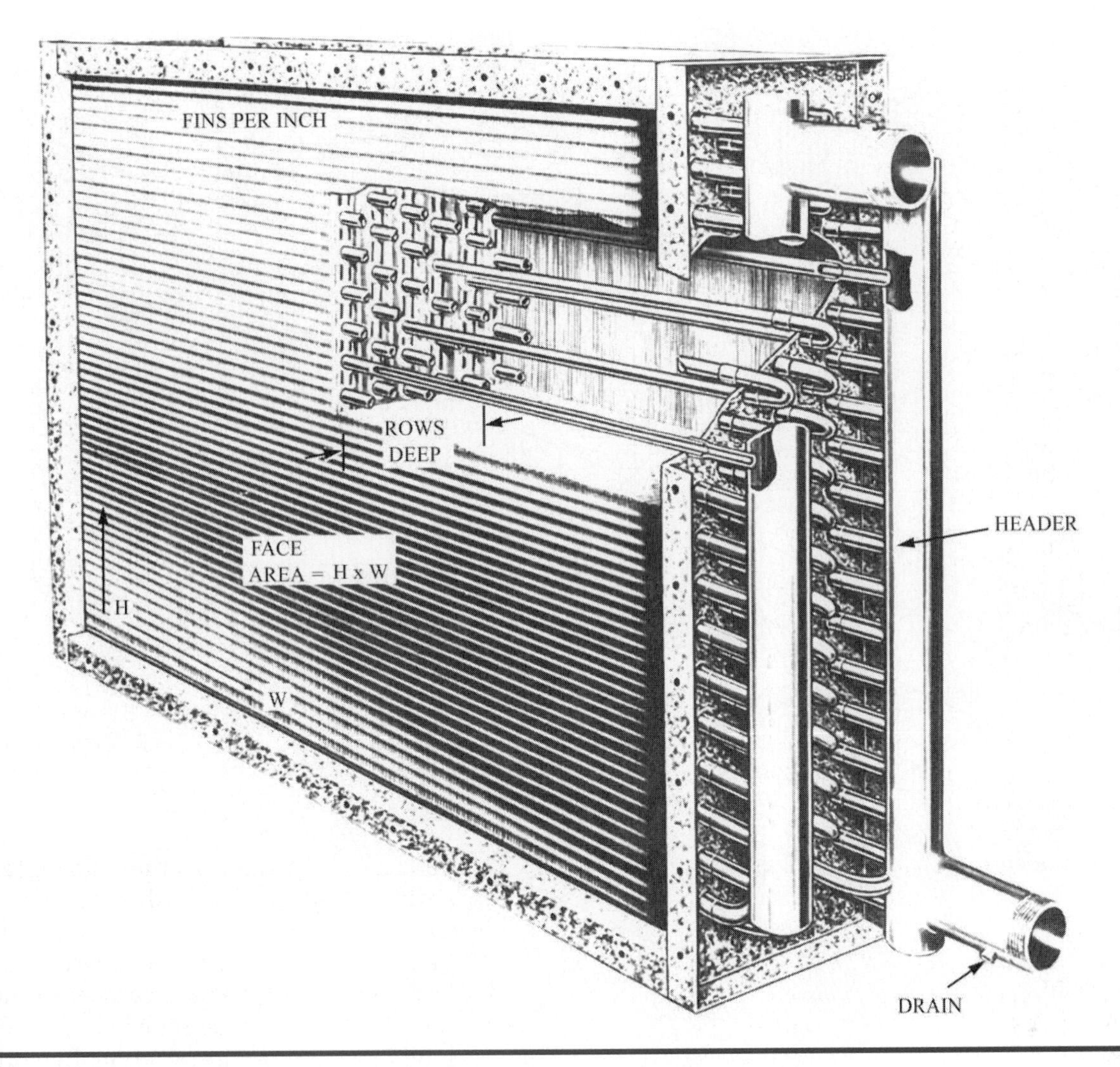

Figure 17-2-9 Typical water coil for a central plant air handling unit. (*Courtesy of York International Corp.*)

TABLE 17-2-1 The Effect of Single, Double, and Triple Pane Windows on Condensation Temperatures

Inside Temp. 70°F	Outside Temperature at which Condensation Will Probably Occur °F*		
Inside RH	**Single Pane**	**Double Pane**	**Triple Pane**
50%	43°	18°	10°
45%	38°	11°	0°
40%	34°	2°	−10°
35%	28°	−8°	−25°
30%	22°	−20°	−35°
25%	15°	−30°	
20%	8°		
15%	0°		
10%	−11°		

* With constant air circulation over the windows, a higher inside relative humidity can be maintained. Heavy drapes, closed blinds, etc., will have an adverse effect on the load.

(Courtesy of Research Products Corporation)

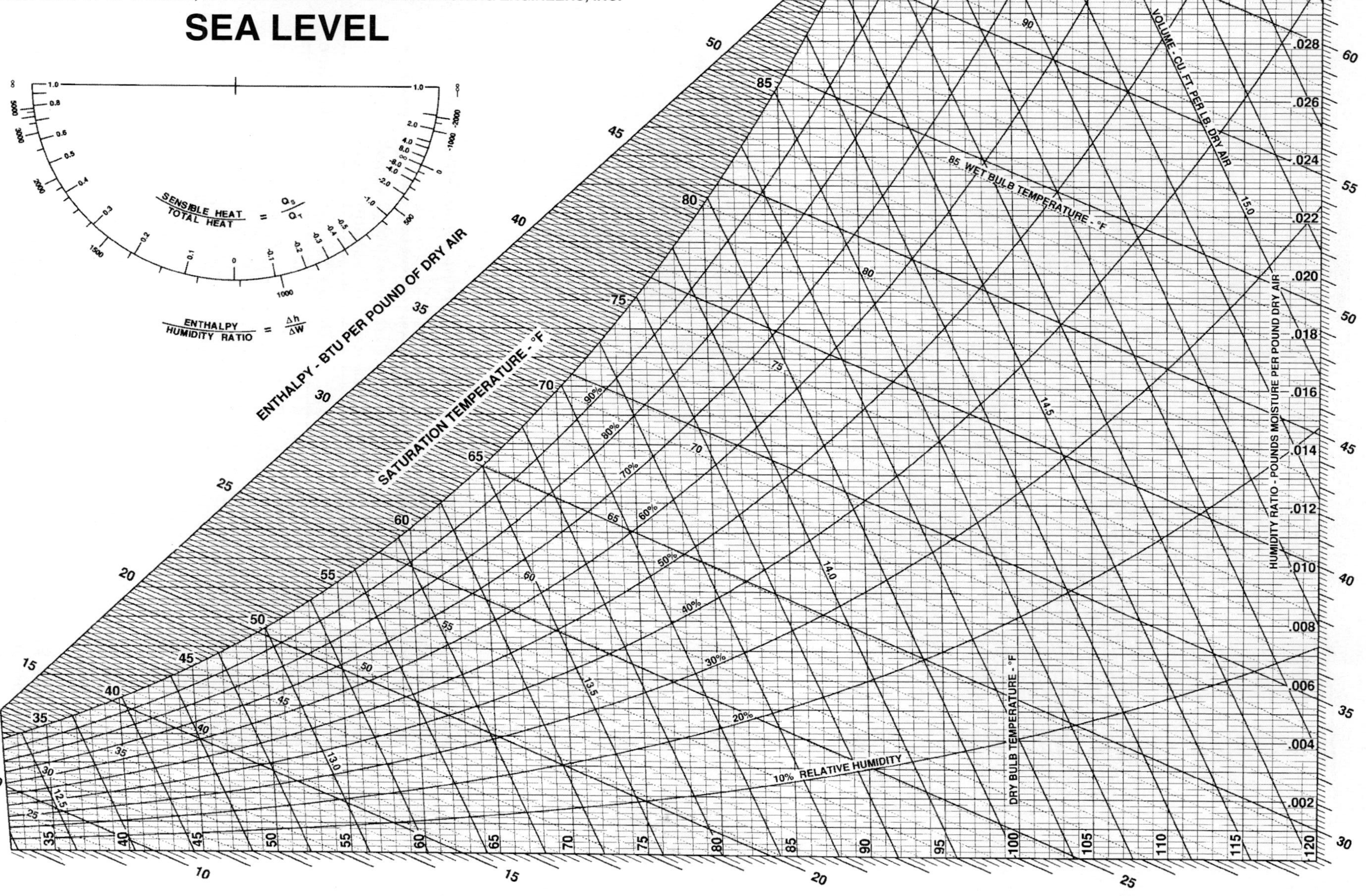

Figure 17-2-10 Psychrometric chart. (*Copyright by the American Society of Heating, Refrigerating and Air-Conditioning Engineers, Inc.*)

Figure 17-2-11 Spray type humidifier.

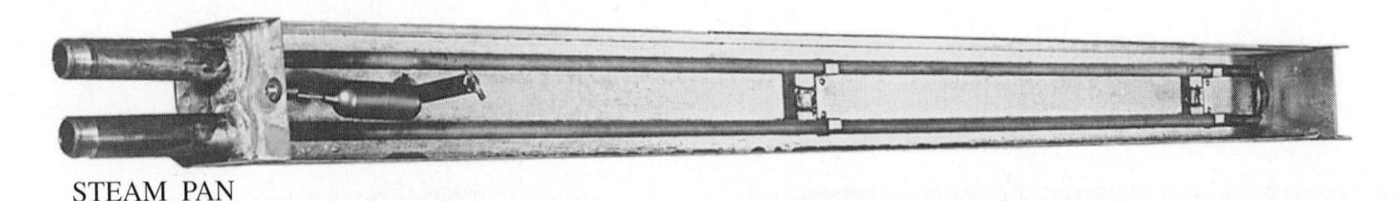

Figure 17-2-12 Pan type humidifier.

Figure 17-2-13 Steam grid humidifier.

are used, as shown in Figures 17-2-11, 17-2-12, and 17-2-13.

Water spray types are used with hot water heating and provide optimum performance in applications where the humidity level is fairly low and precise control is not required.

The steam pan type is used when the introduction of steam directly into the air stream is undesirable. The vaporization of water from the pan provides moisture to the conditioned air.

The steam grid type is highly recommended because it offers simplicity in construction and operation, and humidification can be closely controlled. A source of steam must be available.

UNIT 2—REVIEW QUESTIONS

1. What is the air handling unit of an air conditioner?
2. What are the two general types of air handling units that are used for central plant systems?
3. What is one advantage of fan coil units?
4. What are larger sized fan coil units used for?
5. What does an air handling unit consist of?
6. Define relative humidity.
7. List the four benefits of maintaining proper humidity.
8. At a temperature of 72°F with ________ RH you would feel cozy.
9. What is one of the limiting factors in maintaining proper humidity?
10. Why is the steam grid type of humidifier recommended?

TECH TALK

Tech 1. I was out at a building the other day and when I raised the ceiling tile there wasn't any duct at all on that fan coil.

Tech 2. Yes, that is pretty common in commercial. They use the space above the drop ceiling as the return. This is sometimes called a wild return.

Tech 1. Well, aren't there problems with dust and dirt?

Tech 2. No, the air handler has its air filter just before the air goes into the coil. But that part of the building is sealed so it's not open to the outside. It is not like it is blowing air from the attic in a residence which may have attic vents. In a commercial building that space is sealed and it is just part of the building envelope but it is above the drop ceiling.

Tech 1. What is the advantage of that?

Tech 2. Well, it is certainly a whole lot easier to put in returns. But secondly a lot of the heat in a commercial building comes in through the roof and with the return air pulled across the roof it picks that heat up and puts it through the coil and then puts it out into the occupied space. If you can stop the heat from getting into the occupied space you can make it more comfortable.

Tech 1. I see. So it is easier to put the returns in and it makes the occupied space a little more comfortable. I guess having a wild return is not such a bad idea. Thanks.

UNIT 3

Package Unit Conditioned Air Control Systems

OBJECTIVES: In this unit the reader will learn about package unit electrical systems including:

- Mixing Dampers
- Mixed Air Control
- Face and Bypass Control
- Variable Air Volume Control System
- Multizone Units
- Fans and Motors

MIXING DAMPERS

Dampers installed on air handling units control the flow of air through various parts of the system. The function of dampers depends on the design of the system. There are two general types of multiple leaf dampers: parallel blade and opposing blade. Parallel blade dampers tend to direct the air as they open, whereas opposing blade dampers are usually constructed to more readily offer a positive close-off. Most of the dampers now supplied are of the opposing blade type.

TECH TIP

The proper adjustment of mixing dampers is critical to maintaining an efficiently operating system which provides the proper air distribution within a building. Mixing dampers blend the air so that you have the proper temperature and are able to control humidity within the building. If these dampers are not set properly the building operating cost will go up dramatically.

Dampers can operate either in two positions, open or closed, or they can be modulated so that they can be positioned anywhere between fully open and fully closed, depending on the requirements of the control system.

Figure 17-3-1 Return air and outside air mixing box with dampers. *(Courtesy of Dunham-Bush, Inc.)*

Dampers can be linked together, so that when one opens the other closes. For example, in a mixing box such as shown in Figure 17-3-1, where the return air is mixed with outside air, the dampers can be linked and controlled so that when the outside air is increased, the return air is decreased (and the reverse).

Most dampers on central station air handlers are of the modulating type. This provides the control of the airflow needed for the wide variety of conditioning requirements. The following are some of the basic applications.

Mixed Air Control

To illustrate some of the many uses of dampers, refer to Figure 17-3-2. Of primary importance is the mixing of return air and outside air. During normal operation only the return air and the minimum outside air dampers are involved. The minimum outside air damper is a two position damper. Normally it is fully open to provide ventilation air to meet code

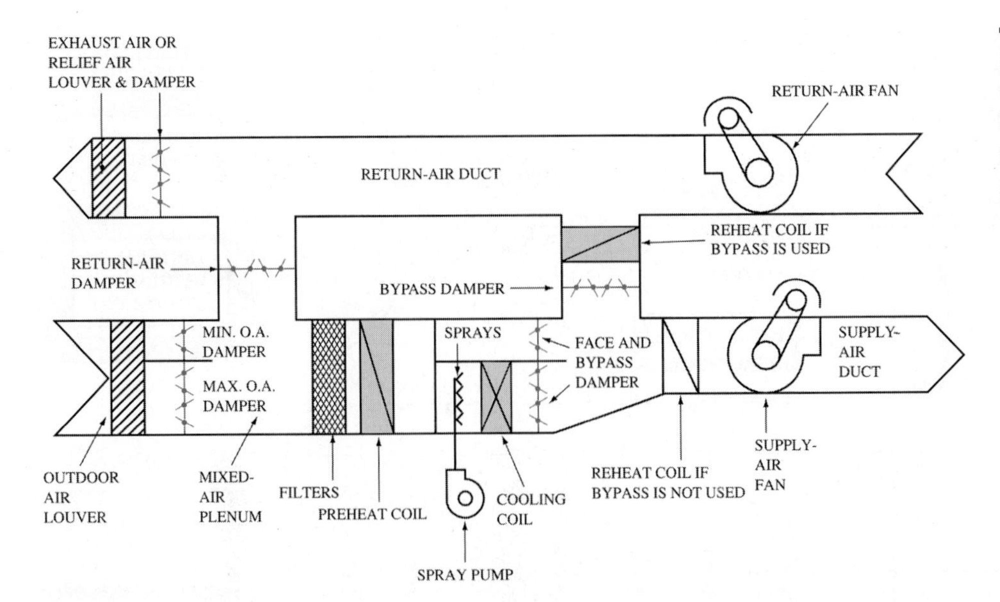

Figure 17-3-2 Schematic diagram of a typical air handling unit showing the various types of damper arrangements.

requirements. The only time it closes is when the building is unoccupied or the system is shut down.

Provision is made in the damper system and the control system to use additional outside air for free cooling (economizer cycle) when practical. A maximum outside air damper is modulated along with the return air and exhaust air damper to maintain a slightly positive pressure in the building. The total air quantity through the air handler does not change but the proportions of outside air to return air do change to meet the control requirements.

Face and Bypass Control

One way to control the amount of cooling or heating is to use a face damper in series with the coil surface and a bypass damper in a duct connection to the return air, as shown in Figure 17-3-2. With this arrangement the total air volume remains constant and the air entering the fan is a mixture of cooled of heated air and return air. The damper control modulates the air quantity from each source to match the load. The two sets of dampers are linked together either mechanically or electrically so that when one modulates toward the open position, the other modulates toward the closed position (and the reverse).

Variable Air Volume Control System

This system differs from the constant volume/variable temperature system since basically it maintains a nearly constant air temperature but matches the load by changing the air volume. This system became popular when attention was given to conserving energy without decreasing the comfort level during partial load conditions.

One of the features of these systems is the use of a variable frequency fan speed controller that is applied to the air handler fan. This electrical device, along with the necessary sensors, controls the speed of the central fan to match the total air volume required by the individual zone terminal units. Since the power required to drive the fan is proportional to the cube of the fan speed, tremendous power savings can result from slowing down the fan when the extra air is not needed.

In addition to the arrangement for controlling the air volume at the central unit, the VAV systems use a terminal unit in each zone. A typical system of this type is shown in Figure 17-3-3, along with a solid state control system. Each terminal unit has an automatic air volume controlled damper to supply the needs of the zone, as shown in Figure 17-3-4.

Multizone Units

A multizone unit is a specially designed air handling unit that provides separate zone ducts for the distribution of conditioned air, as shown in Figure 17-3-5. In addition, each zone is provided with mixing dampers to supply conditioned air to meet the requirements of the zone. These units operate on the basis of using dampers to mix the air from a heat source and a cold source to produce the required temperature.

FANS AND MOTORS

The proper operation of the air handling unit's fan and motor is essential to a good air conditioning job. They not only need to be sized properly to begin with, but most jobs require field adjustment of these components to fit the actual conditions of the installation.

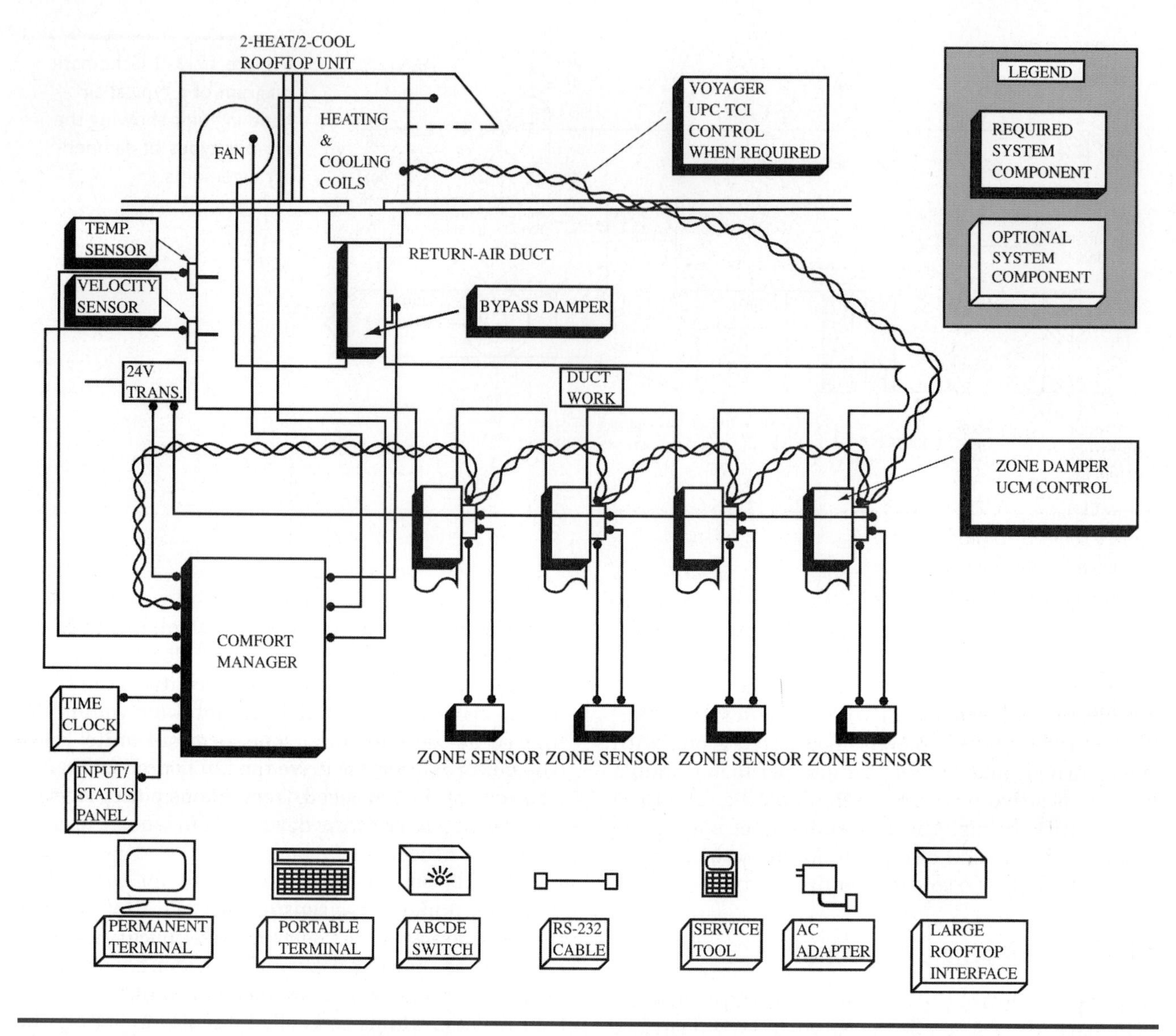

Figure 17-3-3 Schematic view of a typical variable air volume (VAV) system.

Figure 17-3-4 VAV terminal unit. (*Courtesy of the Trane Company*)

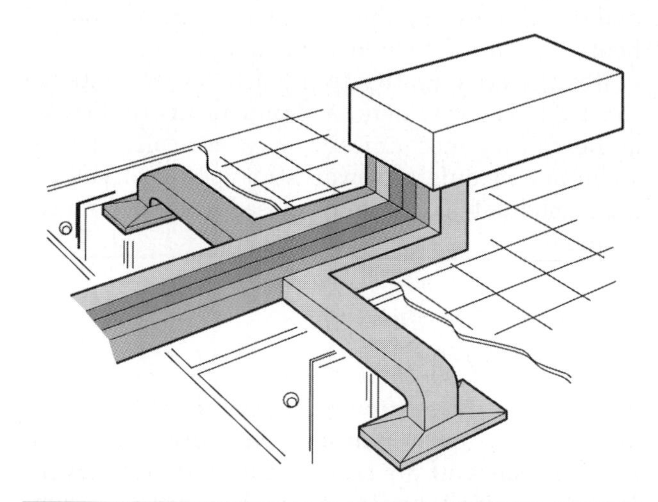

Figure 17-3-5 Constant volume multizone heating and cooling unit.

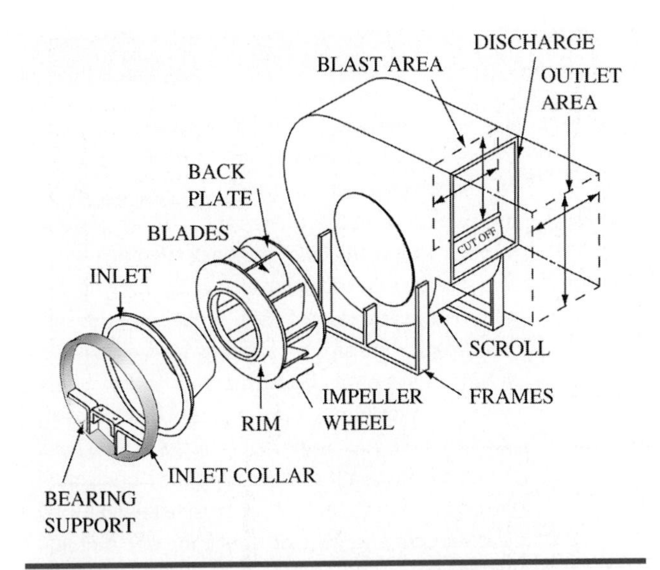

Figure 17-3-6 Exploded view of a centrifugal fan. *(Copyright by the American Society of Heating, Refrigerating, and Air-Conditioning Engineers, Inc.)*

There are two basic type of fans used in central station air handlers: (1) centrifugal, Figure 17-3-6 and (2) axial, Figure 17-3-7.

TECH TIP

Centrifugal fans are designed to produce high static pressures and axial fans provide high volumes of air. Centrifugal fans work very well on duct air distribution systems. Axial fans are not capable of producing enough static to move air through long ducts; however, they are very effective in moving air in large openings such as building ventilation. Each fan type has its purpose and application and they are not interchangeable.

The centrifugal fan is much like a centrifugal pump, except that it handles air instead of water. It is used to "pump" small or large quantities of air through equipment that usually offers a relatively high resistance to the airflow. The fan is belt driven with variable sized pulley wheels on the fan shaft and motor shaft, so that the speed can be adjusted and the motor sized to meet the needs of the job.

There are two types of centrifugal fans: (a) backward curved blades and (b) forward curved blades, Figure 17-3-8. The backward curved blade fans usually run faster than the forward curved type and therefore are noisier, but have the advantage of being non-overloading. The forward curved blade fans are preferred because of their flexibility for adjustment and quieter operation.

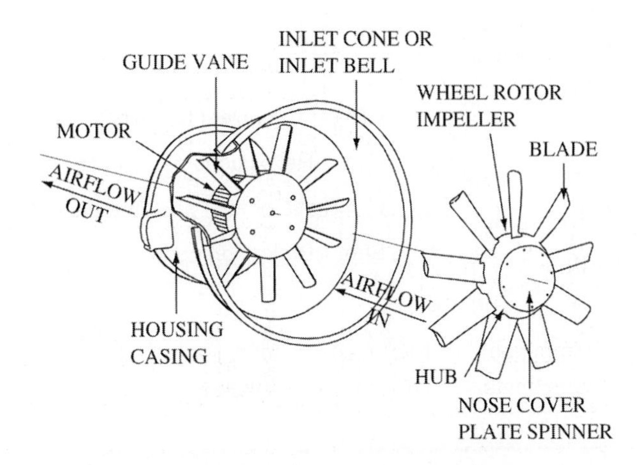

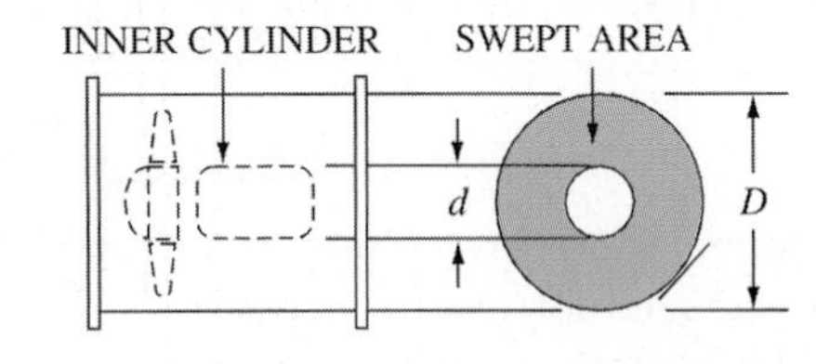

$$\text{SWEPT AREA RATIO} = 1 - \frac{d^2}{D^2} = 1 - \frac{\text{AREA OF INNER CYLINDER}}{\text{OUTLET AREA OF FAN}}$$

NOTE: THE SWEPT AREA RATIO IN AXIAL FANS IS EQUIVALENT TO THE BLAST AREA RATIO IN CENTRIFUGAL FANS.

Figure 17-3-7 Schematic drawings of axial fans. *(Copyright by the American Society of Heating, Refrigerating, and Air-Conditioning Engineers, Inc.)*

The axial fan is used where large volumes of air are required at a relatively low static pressure. This fan also operates at high speeds and can be noisy. It therefore needs to be strategically located.

In adjusting the speed of centrifugal fans, the technician needs to be knowledgeable of the fan laws, Figure 17-3-9, that govern the relation between speed, static pressure and horsepower requirements. A great deal of electrical energy can usually be saved by properly selecting the fan motor size.

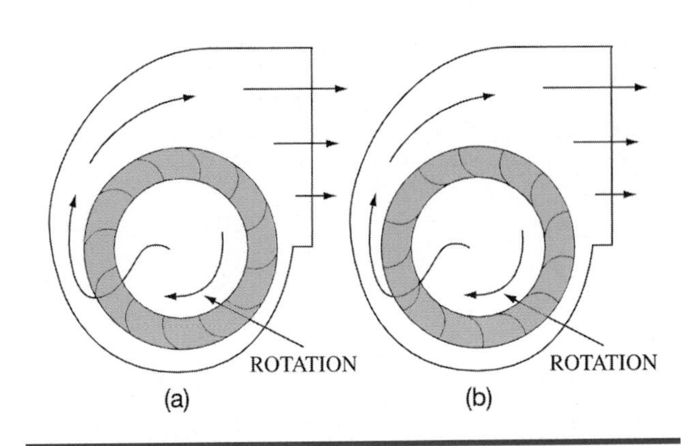

Figure 17-3-8 Schematic view of the construction of (a) Backward curved fan and (b) Forward curved fan.

Fan Equations

a) $\frac{CFM_2}{CFM_1} = \frac{rpm_2}{rpm_1}$

b) $\frac{P_2}{P_1} = \left(\frac{rpm_2}{rpm_1}\right)^2$

c) $\frac{Bhp_2}{Bhp_1} = \left(\frac{rpm_2}{rpm_1}\right)^3$

d) $\frac{\text{rpm (fan)}}{\text{rpm (motor)}} = \frac{\text{Pitch diam. motor pulley}}{\text{Pitch diam. fan pulley}}$

CFM = Cubic feet per minute

rpm = Revolutions per minute

P = Static or total pressure (in WC)

Bhp = Brake horsepower

Figure 17-3-9 Fan laws showing the relationships between speed, air quantity, and motor requirements.

Squirrel cage three phase motors with laminated rotors are used for most central station air handling units. These motors have starting torques of 125 to 275% of full load torque, depending on the design. The application voltage and ampere draw of the motor should always be measured on the job, under load conditions, to be certain the use is compatible with the nameplate rating.

UNIT 3—REVIEW QUESTIONS

1. What are the two general types of multiple leaf dampers?
2. ________ dampers tend to direct the air as they open.
3. What are the positions in which dampers can operate?
4. What is one way to control the amount of cooling or heating?
5. How does the variable air volume control system differ from the constant volume/variable temperature system?
6. What is a multizone unit?
7. List the two basic types of fans used in central station air handlers.
8. What is a centrifugal fan?
9. Why are the forward curved blade fans preferred?
10. ________ motors with laminated rotors are used for most central station air handling units.

TECH TALK

Tech 1. I don't understand why these companies have these variable air volume systems. If the building needs so much air, why don't you just put it in? Looks to me like there's an awful lot of technology and awful lot of expense for nothing.

Tech 2. No, you are absolutely wrong. By varying the air volume you can vary the capacity of the system to provide air conditioning to the building. If you didn't have the variable air volume system, you would put the same amount of cooling in when you only need a little of it. That's going to waste a lot of energy. So by being able to adjust the system to meet the building load you can cut the cost of operating the building significantly.

Tech 1. I thought variable air volume was just to make the building air movement more comfortable so it doesn't blow people around.

Tech 2. No. The air distribution system is designed so it does not blow things around. Variable air volume is just for economics and building operation. It makes the building more economical, it also makes the building more comfortable.

Tech 1. That explains why they spend so much time and money putting those things in.

Tech 2. You're right, and they're worth it!

UNIT 4

Troubleshooting Refrigeration Systems

OBJECTIVES: In this unit the reader will learn about troubleshooting refrigeration systems including:

- Identifying Problems of a Refrigeration System
- Performing Electrical Troubleshooting
- Interpreting Manufacturer's Wiring Diagrams
- Diagnostic Tests to Isolate Component Failures
- Troubleshooting Mechanical Problems
- Power On and Power Off Inspections
- Factors Causing Shortened Compressor Life
- Causes of a Failed Compressor
- Troubleshooting a Dead Compressor
- Troubleshooting Evaporators and Condensers
- Metering Devices
- Hermetically Sealed Systems

OVERVIEW

Any operating mechanical/electrical equipment will at some time require service. The repair work necessary to place the equipment back on line is one of the important functions of the HVAC/R technician. The basic term that describes this type of work is troubleshooting. Troubleshooting is the process of determining the cause of an equipment malfunction and performing corrective measures. Depending on the problem, this may require a high degree of knowledge, experience, and skill.

Basically there are two types of problems, electrical and mechanical, although there is much overlap. Whatever the nature of the problem, it is good practice to follow a logical, structured, systematic approach. In this manner the correct solution is usually found in the shortest possible time.

Since the greatest number of malfunction problems are electrical, it is common practice to perform electrical troubleshooting (including controls) first. If the problem is mechanical, the electrical analysis will usually point the technician in that direction.

TECH TIP

Before beginning any troubleshooting you must locate the manufacturer's troubleshooting guide for the equipment you are working on. Manufacturers have developed troubleshooting techniques that when followed will result in accurate and rapid location of the problem. These charts may seem complex and difficult to follow if you look at the entire chart. In order to properly use one of these charts start at the beginning and follow each and every step. Do not simply jump ahead assuming that you know what the answer is. Do the test and report the results and then move to the next test as indicated by the flowchart.

ELECTRICAL TROUBLESHOOTING

In preparing to analyze electrical problems, four items are important to know or recognize:

1. The operating sequence of the unit.
2. The functions of the equipment that are working and those that are not working.
3. The electrical test instruments that are needed to analyze the problem.
4. The power circuit driving the system.

The operating sequence is usually supplied by the manufacturer in the service instructions or it can be determined by the technician by studying the schematic wiring diagram. The functions of the operating and nonoperating equipment are determined by examination and testing. Necessary test instruments include: the volt ohm meter (VOM), the clamp-on ammeter, the capacitor tester, and the temperature analyzer. The technician must be proficient in the use of these instruments.

The power circuit is the first to be examined, because power must be available to operate the loads. For example, on a refrigeration system with an air

TECH TIP

Wiring diagrams are often included inside the equipment panels. If the diagrams are not there contact the manufacturer. They are available from local suppliers and are often available from the manufacturers themselves through their websites. If you don't have a diagram troubleshooting electrical circuits can be lengthy, time consuming, and inaccurate.

cooled condenser, the two principal loads that must be energized are the compressor motor and two fan motors. Before proceeding with anything else, the technician must be certain that the proper power can be supplied to loads.

Each electrical circuit has one or more switches that start or stop the operation of a load. This switching operation is called the control function. In troubleshooting, when a load is not working, the technician must determine whether the problem is in the load itself or in the switches that control the load.

To assist in analyzing an operational problem of the unit, the manufacturer furnishes one or more of the following:

1. Wiring diagrams
2. Installation and service instructions
3. Troubleshooting tables
4. Fault isolation diagrams
5. Diagnostic tests

The *wiring diagrams* usually consist of connection diagrams and schematic diagrams. The *connection diagram* shows the wires to the various electrical component terminals in their approximate location on the unit. This is the diagram that the technician must use to locate the test points. The *schematic diagram* separates each circuit to clearly indicate the function of switches that control each load. This is the diagram that the technician uses to determine the sequence of operation for the system.

The installation and service instructions supply a wide variety of information that the manufacturer believes is necessary to properly install and service the unit. This bulletin includes the wiring diagram, the sequence of operation, and any notes or cautions that need to be observed in using them.

The troubleshooting guide shown in Figure 17-4-1 is helpful as a guide to corrective action. By a process of elimination, this guide offers a quick way to solve a service problem. The process of elimination permits the technician to examine each suggested remedy and disregard ones that do not apply or are impractical, leaving only the solution(s) that fits the problem.

A *fault isolation diagram,* Figure 17-4-2, starts with a failure symptom and goes through a logical decision action process to isolate the failure.

⚠ WARNING

DISCONNECT ALL POWER TO UNIT BEFORE SERVICING. CONTACTOR MAY BREAK ONLY ONE SIDE. FAILURE TO SHUT OFF POWER CAN CAUSE ELECTRICAL SHOCK RESULTING IN PERSONAL INJURY OR DEATH.

SYMPTOM	POSSIBLE CAUSE	REMEDY
Unit will not run	• Power off or loose electrical connection • Thermostat out of calibration - set too high • Defective contactor • Blown fuses • Transformer defective • High pressure control open (if provided)	• Check for correct voltage at contactor in condensing unit • Reset • Check for 24 volts at contactor coil - replace if contacts are open • Replace fuses • Check wiring-replace transformer • Reset-also see high head pressure remedy-The high pressure control opens at 450 PSIG
Outdoor fan runs, compressor doesn't	• Run or start capacitor defective • Start relay defective • Loose connection • Compressor stuck, grounded or open motor winding. Open internal overload • Low voltage condition	• Replace • Replace • Check for correct voltage at compressor check & tighten all connections • Wait at least 2 hours for overload to reset. • If still open, replace the compressor. • Add start kit components
Insufficient cooling	• Improperly sized unit • Improper indoor airflow • Incorrect refrigerant charge • Air, non-condensibles or moisture in system	• Recalculate load • Check - should be approximately 400 CFM per ton. • Charge per procedure attached to unit service panel • Recover refrigerant, evacuate & recharge, add filter drier
Compressor short cycles	• Incorrent voltage	• At compressor terminals, voltage must be ± 10% of

Figure 17-4-1 Typical troubleshooting table. (*Courtesy of Lennox Industries, Inc.*)

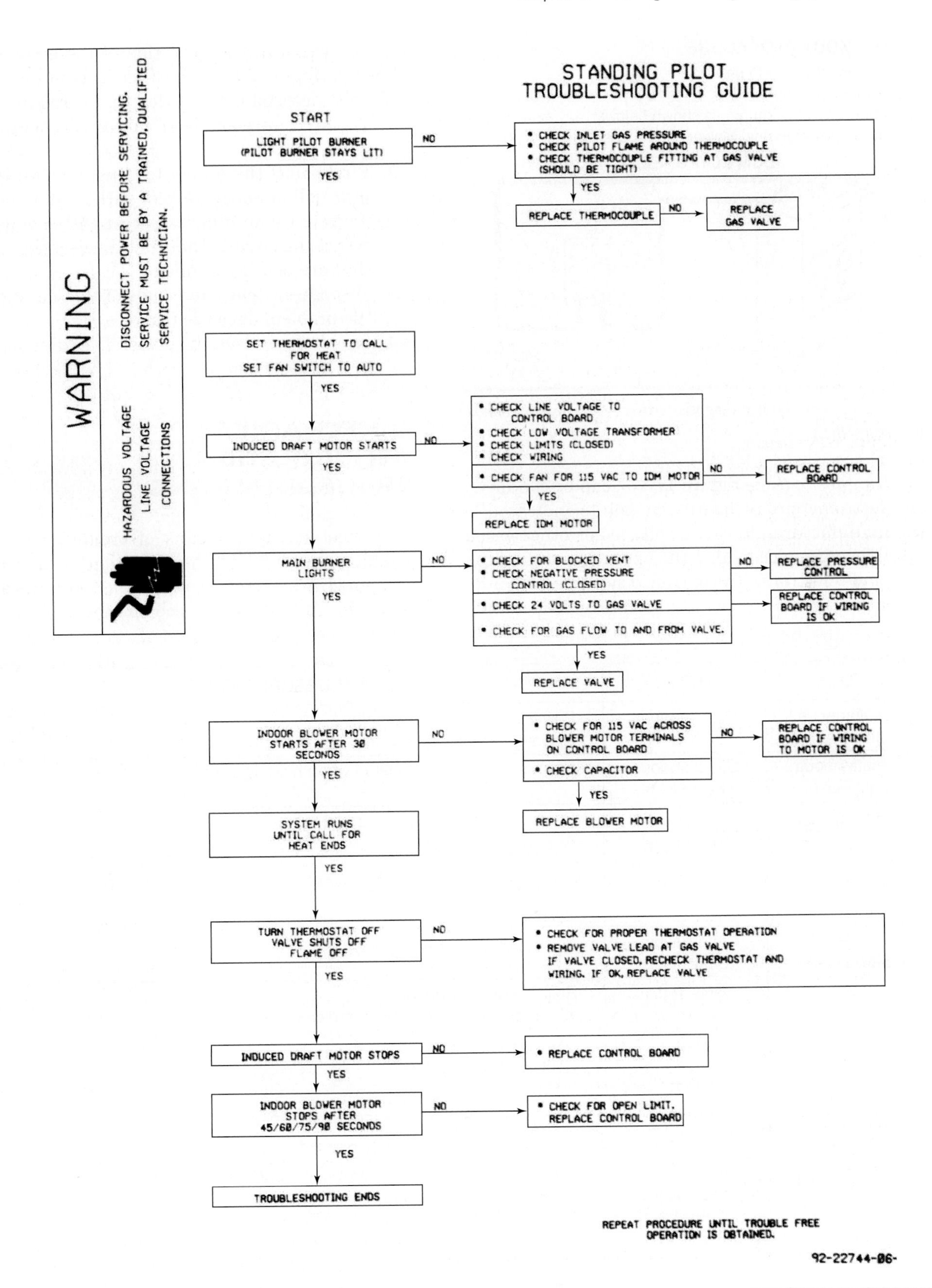

Figure 17-4-2 Typical fault isolation diagram. (*Courtesy of Rheem Manufacturing*)

CHECKOUT PROCEDURE

1- Disconnect power to unit.
2- Disconnect P49 from J49.
3- Connect voltage source as shown below.
4- Turn on power to unit. Blower should operate at low speed.

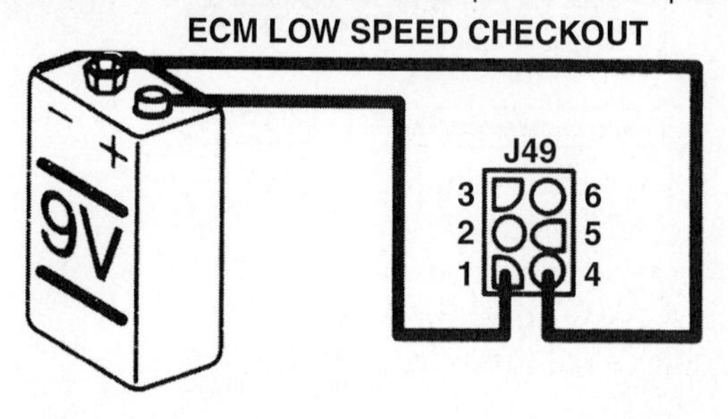

Figure 17-4-3 Typical diagnostic test. (*Courtesy of Lennox Industries, Inc.*)

Diagnostic tests, Figure 17-4-3, can be conducted on electronic circuit boards, at points indicated by the manufacturer, to check voltages or other essential information critical to the operation of the unit.

Some electronically controlled systems have automatic testing features, which indicate by code number a malfunction in the operation of the equipment as shown in Figure 17-4-4. Further tests are usually required to determine the action that is required.

The procedure for electrical troubleshooting is to:

1. First select the proper test instrument. If the unit will not operate, test with an ohmmeter. If parts of the unit operate, test with a voltmeter.
2. Select the circuits that contain electrical devices that are not functioning.
3. Test the switches and loads in that circuit until the problem device is found.
4. Repair or replace the defective equipment.

MECHANICAL REFRIGERATION TROUBLESHOOTING

The most efficient means of troubleshooting mechanical problems in the operation of refrigeration systems is a systematic approach. Shortcuts are possible, depending on the problem, the type of system,

LED #1	LED #2	DESCRIPTION
Simultaneous Slow Flash	Simultaneous Slow Flash	Power on – Normal operation. Also signaled during cooling and continuous fan.
Simultaneous Fast Flash	Simultaneous Fast Flash	Normal operation – signaled when heating demand initiated at thermostat.
Slow Flash	On	Primary or secondary limit switch open.
Off	Slow Flash	Pressure switch open or has opened 5 times during a single call for heat; or Blocked inlet/exhaust vent; or Condensate line blocked; or Pressure switch closed prior to activation of combustion air blower.
Alternating Slow Flash	Alternating Slow Flash	Watchguard – burner failed to ignite.
On	Slow Flash	Rollout switch open; or 12-pin connector improperly attached.
On On Off	On Off On	Circuit board failure; or Control wired incorrectly.
Fast Flash	Slow Flash	Main power polarity reversed. Switch line and natural.
Slow Flash	Fast Flash	Low flame signal. Measures below 0.7 microAmps. Replace flame sensor rod.
Alternating Fast Flash	Alternating Fast Flash	The following conditions are sensed during the ignitor warm-up period only: 1. Improper main ground; 2. Broken ignitor; or Open ignitor circuit; 3. Line voltage below 75 volts. (If voltage lower than 75 volts prior to ignitor warm-up, control will signal waiting on call for thermostat, and will not respond.)
NOTE – Slow flash rate equals 1 Hz (one flash per second). Fast flash rate equals 3 Hz (three flashes per second). Minimum flame sensor current = 0.15 microAmps.		

Figure 17-4-4 Fault message. (*Courtesy of Lennox Industries, Inc.*)

and the experience of the technician, but it is usually helpful to follow a step by step procedure. Here are the steps:

SERVICE TIP

When troubleshooting the mechanical refrigeration system follow the heat. As heat is picked up, pressure is increased. If heat is not picked up pressures do not increase. If heat is being rejected on the condenser side the pressure will decrease. If it is not being rejected the pressure will increase. If you keep in mind that the system's primary function is to pick up heat on the evaporator side and release it on the condenser side and the amount of heat entering affects the temperatures and pressures, then you can quickly determine where the problem in the refrigeration system is.

1. Collect information about the problem.
   a. A description of the problem when the service call was received.
   b. Direct information about the problem by discussion with the customer.
   c. Conduct a preliminary power off visual system inspection.
   d. Conduct a preliminary power on system inspection.
2. Read and calculate the system's vital signs.
   a. Read and record vital signs, including suction and discharge pressures for type of refrigerant being used.
   b. Calculate the refrigerant liquid subcooling at the metering device.
   c. Calculate the refrigerant gas superheat at the compressor.
3. Compare typical versus actual values.
   a. Determine typical values for the conditions and system.
   b. Compare typical with actual conditions.
4. Consult troubleshooting aids.
   a. Perform basic system analysis. Using a basic analysis guide, Table 17-4-1, select possible system problems, based on a comparison of the five actual to typical vital values shown in the guide.
   b. Using the manufacturer's troubleshooting information, Table 17-4-2, perform a detailed analysis. Eliminate possible causes of the problem by test or observation, and select the cause that fits the condition.

In using these steps to diagnose the cause of the problem, the answer may be found in the first two steps, eliminating the need to go further. A difficult problem may require completing all four steps. Proceed through the steps only so far as necessary to find the cause of the problem.

For example, in step 1c, preliminary power off inspection, one of the following causes for low capacity may be found:

1. Dirty or missing filters
2. Dirty or loose fan
3. Loose belts
4. Dirty or corroded coil or fins
5. Loose or uninsulated TXV bulb
6. Damaged interconnecting piping

In step 1d, preliminary power on inspection, one of the following conditions may indicate the source of the problem:

1. Incorrect fan rotation direction
2. Insufficient air circulation
3. Noise from a loose pulley wheel
4. Odor from an overheated transformer
5. Hot spots on a bearing

In order to illustrate the use of the step by step procedure to locate the problem, follow the example below:

EXAMPLE

A customer using a split system air conditioner reports that the compressor is running but that the cooling is inadequate. What should you do?

Solution

Proceed with step by step analysis and use the basic symptom analysis (step 4a) to solve the problem. Tests and calculations indicate the following:

1. Discharge pressure is high.
2. Suction pressure is high.
3. Superheat is low.
4. Subcooling is high.
5. Amps are high.

Referring to the chart, this condition could be caused by only one condition: an overcharge of refrigerant. Now that the problem is identified, the manufacturer's charging charts can be used to adjust the charge for the present indoor and outdoor conditions. ■

TROUBLESHOOTING THE COMPRESSOR

Compressors built today are expected to provide many years of constant, troublefree, quiet operation. In many applications the compressor is required to run 24 hours per day, 365 days a year. Such continuous

Table 17-4-1 Basic Refrigerant System Analysis

System Problem	Cooling System Operating Conditions							
	Superheat	Subcooling	Low Side Pressure	High Side Pressure	Compressor Amperage	Indoor Blower Amperage	Condenser Air TD	Evaporator Air TD
Overcharged TXV Metering	≡	↑	≡	↑	↑	—	↑	≡
Overcharged Fixed Metering	↓	↑	↑	↑	↑	—	↑	↑
Undercharged TXV Metering	↑	↓	↓	↓	↓	—	↓	↓
Undercharged Fixed Metering	↑	↓	↓	↓	↓	—	↓	↓
Restricted Liquid Line	↑	↑	↓	↓	↓	—	↓	↓
Restricted Vapor Line	↑	↑	↓	↓	↓	—	↓	↓
Dirty Condenser Coil	↑	↓	↑	↑	↑	—	↑	↓
Dirty Evaporator Coil	↓	↑	↓	↓	↓	↓	≡	↑
Dirty Air Filter	↓	↑	↓	↓	↓	↓	≡	↑
Higher than Design Load	↑	↓	↑	↑	↑	—	↓	↑
Lower than Design Load	↓	↑	↓	↓	↓	—	↓	↓
Worn Compressor	↑	NA	↑	↓	↓	—	↓	↓

Key
— Normal Conditions
≡ Near Normal Conditions
↓ Below Normal Conditions
↑ Higher than Normal Conditions
NA Non-applicable

TABLE 17-4-2 Troubleshooting Refrigerant System Components

Problem Reported	Symptoms Observed	Possible Causes
Moisture Indication in Sight Glass	Sight glass indicator changed from dry to wet	System not dehydrated before charging Leak in water to refrigerant heat exchanger Leak in low side below atmospheric pressure
Oversized Filter/Drier	Fixed metering device systems slugging during startup System undercharged due to too large a refrigerant volume of filter/dryer	Too large a filter/dryer installed
Undersized Filter/Drier	Low system pressures when properly charged Low current draw Low system capacity	Too small a filter/dryer installed
Restricted Liquid Line Filter/Dryer	Dry evaporator Lower than normal low side pressure Temperature drop across the filter/dryer	Moisture in refrigerant Copper oxides from improper brazing techniques Oil sludge or wax
No Crankcase Heat	System shuts down on low oil during startup High current on startup Compressor stalls during startup and draws lock rotor amperage Compressor starts slow with a lot of noise	No crankcase heater installed Crankcase heat not on long enough before startup Defective heater element Defective control circuit
Reversing Valve Stuck in Cooling	Little or no heating	Damaged reversing valve tube Sludge plugging reversing valve bleed tubes Brazing heat damage to slide seals Non-working solenoid coil Stuck solenoid coil plunger No power to solenoid coil Low refrigerant charge
Reversing Valve Stuck in Heating	No cooling	Damaged reversing valve tube Sludge plugging reversing valve bleed tubes Brazing heat damage to slide seals Non-working solenoid coil Stuck solenoid coil plunger No power to solenoid coil Low refrigerant charge
Oil Separator Not Working	Low oil levels in compressor Oil logged evaporator Low system capacity Noisy compressor Oil slugging on startup	Sludge in oil plugging oil return pipe Damaged float or float assembly

operation, however, is often not as hard on a compressor as is a cycling operation, where temperatures constantly change and oil is not maintained at a constant viscosity.

SERVICE TIP

Compressors are vapor pumps. They are designed to move only vapor. If liquid refrigerant or oil slugs the compressor it will damage the compressor and may actually destroy it. Some procedures used in the past for testing compressors recommended that the suction line service valve be closed to see how deep a vacuum the compressor would pull. This procedure can result in deep vacuum which can allow internal arcing to occur in the electrical windings of the compressor. This practice is not recommended by any refrigerant compressor manufacturer and must not be performed. It can result in compressor failure.

The compressor must not only be designed to withstand normal operating conditions, but also occasional abnormal conditions such as liquid slugging and excessive discharge pressure. Compressors have been designed to take extra punishment and yet function properly. Most compressor failures are caused by system faults and not from operating fatigue. The degree of skill technicians use to install, operate, and maintain the equipment will ultimately determine the actual life expectancy of the system, particularly the compressors. It is therefore helpful to review some of the factors that shorten the life of a compressor.

Loss of Efficiency

The loss of efficiency of a compressor is usually an indication that the compressor is being subjected to system problems that are wearing some of the component parts. For a reciprocating machine this can result from a number of conditions:

1. If liquid enters the compressor, the efficiency and resulting capacity will be seriously affected. The physical damage reduces the effectiveness of the internal parts.
2. Leaking discharge valves reduce the pumping efficiency and cause the crankcase pressure to rise, increasing the load on the machine.
3. Leaking suction valves seriously affect the compressor efficiency (and capacity) especially at lower temperature applications.
4. Loose pistons cause excessive blow-by and lack of compression.
5. Worn bearings, especially loose connecting rods and wrist pins, prevent the pistons from rising as far as they should on the compression stroke. This has the effect of increasing the clearance volume and results in excessive reexpansion.
6. Belt slippage on belt driven units.

Motor Overloading

When the compressor is not performing satisfactorily, the motor load sometimes provides a clue to the trouble. Either an exceptionally high or exceptionally low motor load is an indication of improper operation. Here are some of the causes of motor overloading:

1. Mechanical problems such as loose pistons, improper suction valve operation, or excessive clearance volume usually lead to reduction in motor load.
2. Another common problem is a restricted suction chamber or inlet screen (caused by system contaminants). The result is much lower actual pressure in the cylinders at the end of the suction stroke than the pressure in the suction line as registered on the suction gauge. If so, an abnormally low motor load will result.
3. Improper discharge valve operation, partially restricting ports in the valve plate (which do not show up on the discharge pressure gauge), and tight pistons will usually be accompanied by high motor load.
4. Abnormally high suction temperatures created by an excess load will cause a high motor load.
5. Abnormally high condensing temperatures, created by problems associated with the condenser, will also lead to high motor load.
6. Low voltage at the compressor, whether the source is the power supply or excessive line loss, will contribute to high motor loading.

Noisy Operation

Noisy operation usually indicates that something is wrong. There may be some noisy condition outside the compressor or something defective or badly worn in the compressor itself. Before changing the compressor, a check should be made to determine the cause of the noise. Here are some possible causes outside the compressor:

1. *Liquid slugging* Make sure that only superheated vapor enters the compressor.
2. *Oil slugging* Possibly oil is being trapped in the evaporator or suction line and is intermittently coming back in slugs to the compressor.
3. *Loose flywheel (on belt driven units)*
4. *Improperly adjusted compressor mountings* In externally mounted hermetic type compressors, the feet of the compressor may be bumping the studs. The hold-down nuts may not be backed off

sufficiently, or springs may be too weak, thus allowing the compressor to bump against the base.

Compressor Noises

Noises coming from the inside of the compressor may be one of the following:

1. *Insufficient lubrication* The oil level may be too low for adequate lubrication of all bearings. If an oil pump is incorporated, it may not be operating properly, or it may have failed entirely. Oil ports may be plugged by foreign matter or oil sludge from moisture and acid in the system.
2. *Excessive oil level* The oil level may be high enough to cause excessive oil pumping or slugging.
3. *Tight piston or bearing* A tight piston or bearing can cause another bearing to knock—even though it has proper clearance. Sometimes in a new compressor such a condition will "wear in" after a few hours of running. In a compressor that has been in operation for some time, a tight piston or bearing may be due to copper plating, resulting from moisture in the system.
4. *Defective internal mounting* In an internally spring mounted compressor, the mountings may be bent, causing the compressor body to bump against the shell.
5. *Loose bearings* A loose connecting rod, wrist pin, or main bearing will naturally create excessive noise. Misalignment of main bearings, shaft to crankpins or eccentrics, main bearings to cylinder walls, etc., can also cause noise and rapid wear.
6. *Broken valves* A broken suction or discharge valve may lodge in the top of a piston and hit the valve plate at the end of each compressor stroke. Chips, scale, or any foreign material lying on a piston head can cause the same result.
7. *Loose rotor or eccentric* In hermetic compressors a loose rotor on the shaft can cause play between the key and the keyway, resulting in noisy operation. If the shaft and eccentric are not integral, a loose locking device can be the cause of knocking.
8. *Vibrating discharge valves* Some compressors, under certain conditions, especially at low suction pressure, have inherent noise which is due to vibration of the discharge reed or disc on the compression stroke. No damage will result, but if the noise is objectionable, some modification of the discharge valve may be available from the compressor manufacturer.
9. *Gas pulsation* Under certain conditions noise may be emitted from the evaporator, a condenser, or suction line. It might appear that a knock and/or a whistling noise is being transmitted and amplified through the suction line or discharge tube. Actually, there may be no mechanical knock, but merely a pulsation caused by the intermittent suction and compression stroke, coupled with certain phenomena associated with the size and length of refrigerant lines, the number of bends, and other factors.

ANALYSIS OF A DEAD COMPRESSOR

This exercise assumes that the power system has been checked and that the thermostat is calling for cooling, so that even though the proper power is available at the compressor, it will not start. The compressor may not even "hum" when power is applied to it or it may "hum" and cycle on the compressor overload. In any event, it is the duty of the technician to analyze the problem, locate the cause, and provide a remedy. Here are some of the causes of dead compressors:

1. Open control contacts
2. Overload contacts tripped
3. Improper wiring
4. Overload cutout destroyed
5. Shortage of refrigerant
6. Low voltage
7. Start capacitor defective or wrong one
8. Run capacitor defective or wrong one
9. Start relay defective or wrong one
10. High head pressure
11. Compressor winding burnt out
12. Piping restriction

To locate the cause, the electrical system and, if necessary, the refrigeration system must be thoroughly tested. The best procedure usually is to check the electrical system first. The procedure is outlined below.

Prepare for Testing

Disconnect the power supply and remove the refrigerant from the system. All electrical devices such as relays, capacitors, and external overloads must be disconnected.

Locate the Compressor Motor Terminals

Find the two terminals that provide the highest resistance reading. That reading represents the combined resistance of two windings, from R to S. The remaining terminal is the common (C) terminal. Put one ohmmeter lead on the C terminal and find which of the remaining terminals gives the highest resistance reading. That will be the start winding and therefore the S terminal. The remaining terminal is the R terminal, as shown in Figure 17-4-5.

Figure 17-4-5 Measuring resistances of compressor windings. Mathematically, the numbers should add up.

S-R	5 Ω	4
S-C	4 Ω	−1
R-C	1 Ω	5

R-C	1 Ω	1
S-C	4 Ω	+4
S-R	5 Ω	5

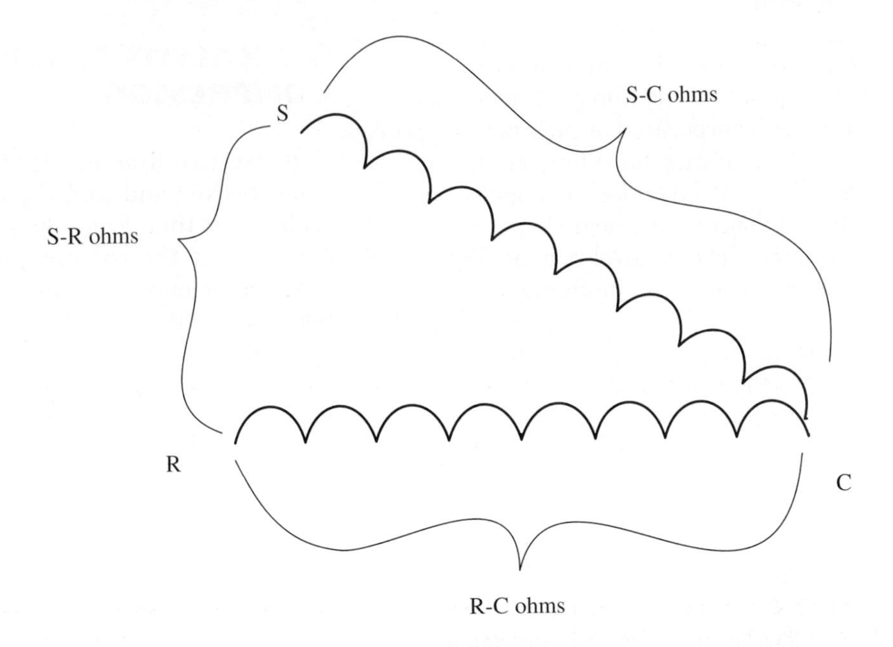

Check Motor Windings with an Ohmmeter

Read the resistance across each pair of terminals. For example, a typical set of readings for a good single phase compressor could be as follows:

$$\text{C to S} = 4\ \Omega$$
$$\text{C to R} = 1\ \Omega$$
$$\text{S to R} = 5\ \Omega$$

Note that the S to R reading is the sum of the other two and that the C to S resistance is always higher than C to R. A good rule of thumb is that the resistance of the start winding is three to five times that of the run winding.

If "zero" is read for any one of these pairs, there is a shorted winding and a defective compressor, as shown in Figure 17-4-6. If "infinity" is read for any one of these pairs, there is an open winding and a defective compressor.

To test for a grounded winding, the ohmmeter must be capable of measuring a very high resistance. It needs to have a scale that can be set to R × 100,000. One lead of the ohmmeter is placed on a compressor terminal and the other on the housing, Figure 17-4-7. The resistance between the winding and the housing ranges from 1 to 3 million Ω for an ungrounded winding. When placing the meter lead on the housing, be sure that the meter is making good contact. A coat of paint, a layer of dirt, or corrosion can hide a grounded winding.

In testing three phase compressor motors, a good motor will show equal resistances between each of the terminals, as shown in Figure 17-4-8. Testing the windings is done in the same way as for single phase motors. One caution: after the testing is done, be sure to reconnect the windings exactly as they were. Interchanging any two windings of a three phase motor will reverse the direction of rotation. Also, when checking windings of a compressor that is under pressure, make the meter connections "upstream" from the terminals.

Test the Capacitors and Relays

Capacitors If the windings of the compressor motor are satisfactory, the problem may be in one of the capacitors or the relay. There are two types of capacitors, the start capacitor and the run capacitor, as shown in Figure 17-4-9. The procedure for testing the capacitors is as follows:

1. Discharge the capacitor. Never place your fingers across the terminals of a capacitor. Do not place a direct short across the terminals of a capacitor—this can damage the capacitor. Place it in a protective case and connect a 20,000 Ω, 2 W resistor across the terminals and gradually bleed off the charge.

OPEN START WINDING

S-R	∞ Ω
S-C	∞ Ω
R-C	1 Ω

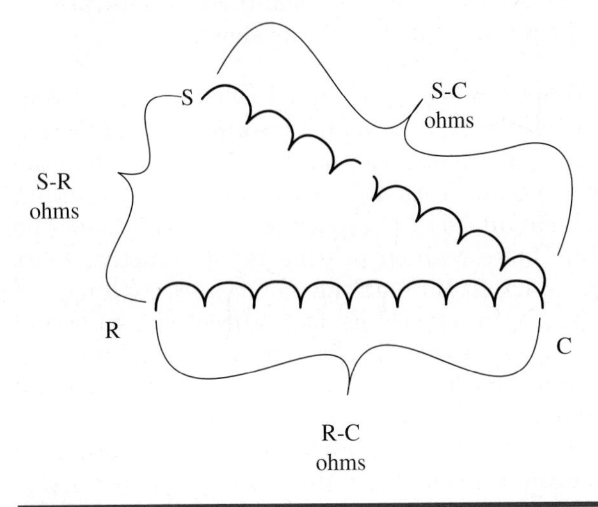

SHORTED START WINDING

S-R	0 Ω
S-C	0 Ω
R-C	1 Ω

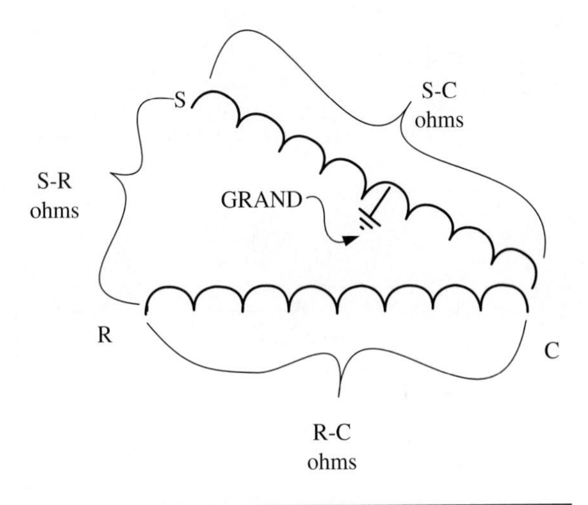

Figure 17-4-6 Testing for shorted or open windings.

S-R	5 Ω
S-C	4 Ω
R-C	1 Ω

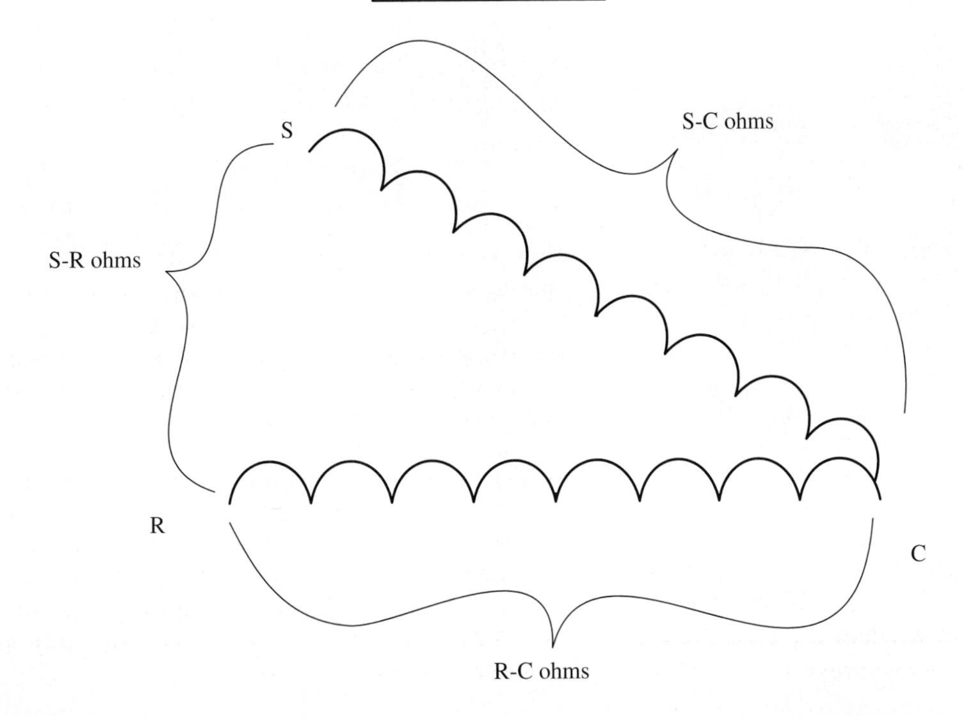

Figure 17-4-7 Testing for grounded motor windings.

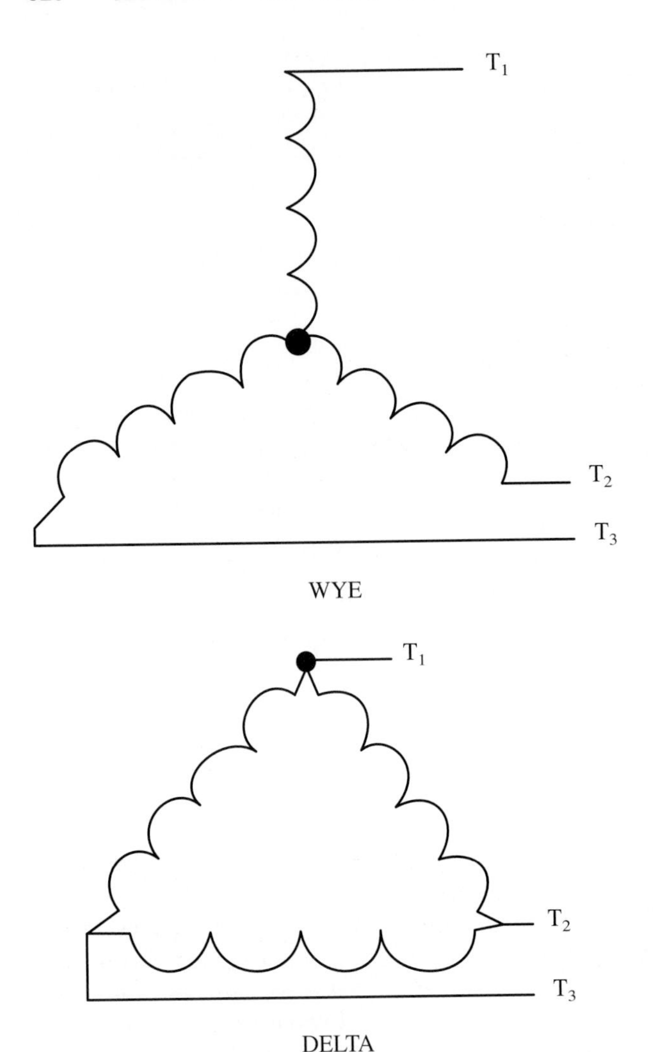

Figure 17-4-8 Three phase motors.

Figure 17-4-9 Various types of capacitors.

2. Disconnect the circuit wiring and connect the leads of an analog ohmmeter across the terminals and observe the meter needle. If the capacitor is OK, the needle will make a rapid swing toward zero and slowly return to infinity. If the capacitor has an internal short, the needle will stay at zero.
3. A capacitor analyzer is used to read the microfarad value of the capacitor and detect insulation breakdown under load conditions. In this case, the meter indicates a "dead short."

Some run capacitors may have a marked terminal as in Figure 17-4-9. This terminal should be connected to the motor run terminal and one leg of the power circuit. With this arrangement, an internal short circuit to the capacitor case will blow the system fuses without passing the destructive short circuit current through the motor start winding.

Some run capacitors have three terminals and are used in the circuits of two motors. They are actually two capacitors in the same container. They are tested in the same way as single capacitors.

Many start capacitors are built with a bleed resistor connected between the terminals. This resistor discharges the capacitor after it is switched out of the circuit.

A "pop out" hole on the start capacitor allows insulation to expand if the capacitor is overheated. If the hole is ruptured, the capacitor must be replaced.

Testing Start Relays The start relay removes the start capacitor from the circuit when the motor reaches a certain rpm. If it is necessary to replace the relay, an identical replacement must be used, because these relays are built with unique characteristics to match the compressor motor with which they are used. When replaced they must be located in the same position, and wired exactly the same as the original relay.

SERVICE TIP

Many potential start relays use gravity as part of the force opening and closing the contacts. These relays have an arrow that indicates up. Other potential or start relays are not marked and do not use gravity as part of the force for opening and closing. However, they are position sensitive. You should make sure that any relay that is marked is installed with the side up and that during testing of the system the relay not be improperly positioned if it is being temporarily placed as part of a testing procedure.

The coil and the switch are tested separately, as shown in Figure 17-4-10. Disconnect the relay from the circuit. In testing the coil it should register a

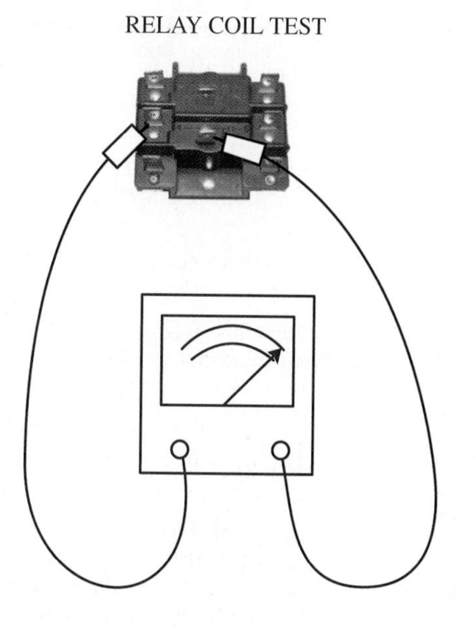

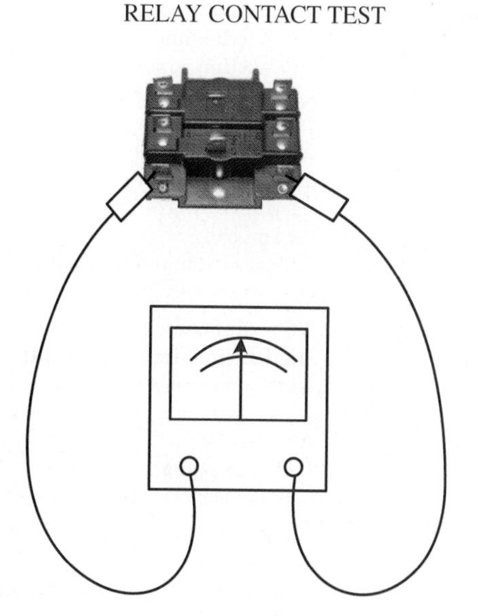

Figure 17-4-10 Testing relays.

substantial resistance. Use the R × 100 scales on the ohmmeter. In testing the contacts, the ohmmeter should read zero, indicating that the switch is closed. Relay switches usually fail in the open position. When they fail by sticking closed, the start capacitor or the start winding of the compressor motor will usually be damaged.

TROUBLESHOOTING EVAPORATORS AND CONDENSERS

Evaporators

The following is a list of some of the refrigeration system problems associated with evaporators:

1. Low airflow
2. Excessive airflow
3. Uneven airflow over coils
4. Low refrigerant supply
5. Uneven refrigerant distribution to coil circuits
6. Low water flow in cooler (water cooling evaporator for chillers)
7. Uneven water flow through cooler
8. Low refrigerant supply to cooler

The symptoms and possible causes for these conditions are shown in Table 17-4-3. The technician, by process of elimination, uses this information to arrive at the cause of a particular problem. The following information will be helpful in analyzing some of the common problems.

A dirty filter is probably the number one cause of low airflow, particularly if the owner has not provided periodic maintenance of air filters. On larger systems a differential pressure sensing meter can be used to indicate the pressure drop through the filters when filter changing is necessary. A responsible person is required to perform the cleaning or replacement necessary periodically. Filters provide a valuable function but do require maintenance.

Excessive airflow can lower the efficiency, interfere with the comfort level, and produce noisy operation. Excessive airflow is indicated by a low air temperature drop over the evaporator coil. For proper airflow the temperature drop should normally be between 16° and 18°F. Adjusting the airflow should be a remedial measure in the checkout procedure. Since the power (watts) supplied to the blower motor varies as the cube of the blower rpm, a large savings can be affected on most jobs by supplying the proper air quantity. Too high an airflow reduces the dehumidifying function of an air conditioning system and therefore interferes with the comfort level. The humidity can be tested by measuring the entering dry bulb and wet bulb temperatures. The noise factor should be greatly improved by reducing the fan speed.

A low refrigerant supply is most commonly the result of a leak in the refrigerant lines. This may be checked by using an electronic refrigerant leak detector and locating the leak. Should a leak be found, it must be repaired.

A low refrigerant supply can also be traced to a plugged filter drier. This is a comparatively easy item to check. Due to the restriction in the filter drier, some vaporization of the refrigerant can occur, lowering the temperature at the outlet. By sensing a temperature drop through the drier, the restriction is located.

A low water flow through a cooler will be evidenced by a low water temperature leaving the chiller. This is usually caused by a restriction in the water

TABLE 17-4-3 Troubleshooting Chart for Evaporators

Problem Reported	Symptoms Observed	Possible Causes
Low Airflow	Low suction line temperature Low suction superheat Low condensing temperature Low compressor amp draw Low supply-air temperature Low system capacity Lack of cooling Iced on coil Refrigerant liquid floodback Compressor slugging	Dirty coil low air flow Dirty filters
Excessive Elevated Airflow	High supply-air temperature High suction line temperature High compressor amp draw Ducts whistle Air handler Water dripping and supply ductwork close to air handler	Too high fan speed Wrong fan drive package and/or setting Coil too small Condensate blown into fan and supply ductwork
Uneven Refrigerant Distribution to Coil Circuits or to Cooler Circuits	Inadequate system capacity Low suction temperature Low suction-gas superheat non existent TXV Liquid refrigerant to compressor Compressor slugging Irregular coil surface temperature Irregular condensate formation on evaporator Evaporator frosted in places but not on others	Plugged evaporator capillary tubes Kinked or crushed capillary tube(s) Distributor has partial blockage Distributor too large Applied air distribution nozzle too large Distributor installed wrong Evaporator tube (especially crushed or kinked return bends) Plugged evaporator (or cooler) circuit
Low Water Flow in Cooler (Water Cooling Evaporator for Chillers)	Low suction line temperature Low suction line superheat Saturated condensing too low temperature Low compressor amperage Low chilled water temperature from evaporator Inadequate system capacity Warm rooms High temperature drop between entering and leaving chilled water Low water thermostat shuts chiller down intermittently	Too small chilled water pump Defective pump Damaged or blocked pump impeller Chilled water line or valve restricted Water baffle(s) in D-X cooler in wrong position blocking flow Water side of evaporator scaled up Chilled water piping has blockage Constriction in water flow valve
Uneven Water Flow Through Cooler	Low system performance Low suction temperature Compressor floodback Compressor slugging Leaving chilled-water temperature Low temperature drop between entering and leaving chilled water	(D-X Cooler): Misaligned or broken baffle(s) Large amount of air in water system Debris inside shell of cooler (Flooded Cooler): Water tube(s) scaled up Kinked or crushed water tube(s) Plugged water tube(s) or water box
Low Refrigerant Supply to Cooler	Low system performance Leaving chilled-water temperature too high Low suction temperature High suction gas superheat Low compressor amperage draw Low saturated condensing temperature Conditioned space temperature too warm Chiller compressor cycles off frequently on low-pressure switch	System undercharged Low bleed pressure at low outdoor temperatures Refrigerant leaking from system Flooded Cooler: Refrigerant flow from condenser to cooler blocked or restricted Cooler refrigerant supply valve stuck Refrigerant distributor or nozzle plugged or restricted Electronic expansion valve malfunction or microprocessor problem

line, a water pump with the wrong impeller rotation, or a defective pump. Many times a shutoff valve is found that has been improperly opened and is creating the problem. As soon as the problem is found, a relatively simple solution can often solve the problem.

Condensers

The following is a list of some of the refrigeration problems associated with condensers:

1. High head pressure
2. Refrigerant charge incorrect
3. Low head pressure

The symptoms and possible causes for these conditions are shown in Table 17-4-4. The technician, by process of elimination, uses this information to arrive at the cause of a particular problem. The following information will be helpful in analyzing some of the common problems.

The most common cause of high head pressure can be traced to a dirty condenser coil. Dirt, lint, and grass clippings can restrict the airflow through the condenser coil causing high head pressure. This accumulation of dirt and debris can be removed with the use of a foaming coil cleaner. Open the unit disconnect and spray the coil. Follow the cleaner manufacturer instructions as to the correct mixture of

TABLE 17-4-4 Troubleshooting Chart for Condensers

Problem Reported	Symptom Observed	Possible Causes
High Head Pressure (Saturated Condensing Temperature or Saturated Discharge Temperature)	Compressor cycles off frequently on high-pressure switch with system calling for cooling Compressor cycles off frequently on compressor-motor protection switch High condensing temperature High discharge-gas superheat Compressor overheats Compressor locks up Compressor burnout High compressor power Amp draw Low system capacity Low system capacity Saturated suction temperature too high Too large condenser water-flow rate	Head pressure control device malfunction Condenser coil dirty or stopped up Condenser fan motor inoperative Large area of fin damage Hot condenser air recirculating over condenser coils Condenser fan blades dirty Airflow over condenser coils blocked Condenser fan rotating backwards Condenser fan belt broken or slipping Damaged condenser fan blade(s) Water cooled condensing tubes blocked or restricted Condenser water valve closed Cooling tower water low Condenser vapor locked by restricted condensate line System overcharged Air in condenser
Refrigerant Charge Incorrect	High head pressure High liquid subcooling Low system capacity High saturated suction temperature High compressor amperage draw Low head pressure Low saturated suction temperature Low system capacity Low or no liquid subcooling Flash gas at metering-device inlet bubbles in sight glass	System overcharged System undercharged
Low Head Pressure	Low saturated condensing temperature Low system capacity Low saturated suction temperature Low compressor power draw (kW)	Head-pressure control device operating improperly Refrigerant system leaking refrigerant System has low refrigerant charge Too high water flow through condenser water valve

cleaner and water. Be sure to rinse the cleaner from the coil when the cleaning is completed.

A high head pressure can often be traced to non-condensable gases present in the refrigerant. The saturated condensing refrigerant has an equivalent condensing pressure, as shown in the charts. Reading a higher condensing pressure on the equipment is an indication that non-condensables are present. These non-condensables, usually air in the system, need to be removed for the system to operate efficiently.

A high head pressure can also be due to a dirty condenser surface. Occasionally an operator will spray an air cooled condenser coil with water during extremely hot weather to increase its capacity. As a result a deposit is left on the coil that fills the fin space and reduces the capacity of the condenser. This deposit is hard to remove, usually removed with acid and a stiff brush. It can also be so destructive that the coil must be replaced.

If the system has excess refrigerant, depending on the system, too much of the condenser can be filled with liquid, not leaving enough room for condensation. This can cause high head pressure. The overcharge needs to be removed to have the condenser operate normally.

Low head pressure is often an indication of lack of refrigerant due to a leak. If so, the leak needs to be located with a suitable leak detector (usually electronic), the leak repaired, and the system recharged with the correct amount of refrigerant.

TROUBLESHOOTING METERING DEVICES

The following is a list of some of the refrigeration system problems associated with metering devices:

1. Evaporator overfeeding ("flooding")
2. Evaporator underfeeding ("starving")
3. Thermal expansion valve (TXV) hunting
4. Distributor nozzles unevenly feeding

The symptoms and possible causes for these conditions are shown in Table 17-4-5. The technician, by process of elimination, uses this information to arrive at the cause of a particular problem. The following information will be helpful in analyzing some of the common problems.

Evaporator flooding can be caused by a low superheat setting or loose sensing bulb. When this occurs, the application and installation of the valve needs to be carefully checked. Most valves are set at the factory for 10°F superheat. The coil may have a high pressure drop and require an external equalizer to operate properly. The sensing bulb must be tightly held to the suction line and the connection insulated if it is picking up stray heat from other sources.

A starved evaporator may be caused by an undercharged system. If this is the condition, look for a refrigerant leak, particularly if the system has been operating satisfactorily for some time.

A starved evaporator may also be caused by an expansion valve sensing bulb that has lost some of its charge. As the sensing bulb cannot exert as much pressure to the expansion valve, it feeds less refrigerant to the evaporator, starving the evaporator.

A hunting TXV is usually a sign that the expansion valve is too large for the application. Most valves operate best for loads down to 50% of their capacity. A valve that is too large will hunt too much of the time. Hunting is the condition where the valve continually opens and closes rather than reaching a stabilized condition.

When the distributor nozzle is furnished with the TXV it must be sized properly to fit the load and to comply with the installation requirements. The tubes of the coil can be inspected to see that all circuits are being fed equally. If not, a properly sized nozzle needs to be applied.

SERVICE TIP

Most thermostatic expansion valves have an adjustment for superheat. This adjustment virtually never needs adjusting in the field once the TXV is installed. Technicians frequently turn this adjustment, thinking it is going to resolve a refrigerant problem. In reality the adjustment should be the last option considered. However, inside of TXV's and other metering devices there are strainers. These strainers can become clogged with debris from the system or wax from the refrigerant oils. It may be necessary from time to time to clean these strainers out. Refer to the metering device manufacturer's recommendation on procedures for cleaning out the strainers.

Other Areas for Troubleshooting

In the above descriptions, troubleshooting certain major components of the system has been suggested. If additional areas need to be examined to find the problem, here are some further areas to troubleshoot:

1. Refrigeration cycle accessories
2. Refrigeration piping
3. The quality of the installation process
4. The quality of the evacuation/dehydration process

TABLE 17-4-5 Troubleshooting Chart for Metering Devices

General Problem Category	Symptoms	Possible Causes
Evaporator Overfeed ("Flooding")	High saturated suction temperature Low suction line superheat Liquid refrigeration back to compressor Compressor slugging Compressor too cool Low compressor amperage draw Compressor fails Compressor pumps improperly TXV cycles excessively	Refrigerant overcharge (fixed metering device) Metering device too large TXV doesn't cycle properly TXV superheat setting too low TXV type doesn't match system refrigeration Uninsulated TXV sensing bulb pushing up ambient heat Sensing bulb loose TXV sensing bulb installed in wrong area Metering device wrong low load conditioner Too much oil circulating in system In system with fixed metering device, head pressure too high
Evaporator Underfeed ("Starvation")	Low system capacity Low suction line temperature High suction line superheat Low compressor amperage draw Low condensing temperature High discharge line superheat Supply air too warm Evaporator blocked or restricted by ice or frost	System undercharged Metering device too small Metering device restricted Distributor or nozzle restricted TXV distributor or nozzle too small Kinked or crushed capillary tube TXV stuck closed TXV power element has lost charge TXV does not match refrigerant in system TXV external equalizer line stopped up or kinked TXV superheat setting too high TXV sensing bulb in wrong area Ice in system blocks refrigerant flow Low head pressure (fixed metering device) Stacking liquid refrigerant in condenser Head pressure control device not working
TXV Hunting	Saturated suction temperature cycles High then low cycles Suction gas superheat cycles high and low Compressor amperage cycles high and low Intermittent compressor slugging Supply air temperature changing Rapidly higher and lower Unstable evaporator surface Temperature	TXV too large TXV too lightly loaded under partial load conditions Cooling load too small Cooling load changes rapidly High side pressure cycles high and low Flashing in liquid line Poor evaporator circuiting (applied air handlers)
Distributor Nozzles (TXV Applications)	Evaporator not receiving enough refrigeration (see symptoms above) Evaporator unevenly fed by refrigerant (see symptoms on evaporator sheet)	Distributor nozzle too small (quite unlikely on comfort work) Nozzle too large Nozzle wrong low load Part load control sequence for evaporator Sections not working correctly

SEALED SYSTEM TROUBLESHOOTING

A hermetically sealed refrigeration system is one that is completely enclosed by welding or soldering that prevents the escape of refrigerant or the entrance of air. Further, these systems use a hermetic compressor and a capillary tube metering device, requiring no external adjustments. The units are assembled under controlled conditions, evacuated, charged, and tested at the factory. Everything possible is done by the manufacturer to provide units with long, troublefree service. With this assurance many manufacturers offer a 5 yr or 10 yr limited warranty on the sealed system. A warranty of this type applies only if the equipment is given normal and proper use. The unit must be correctly installed and certain items of maintenance need to be performed, such as keeping the condenser cleaned, and changing the air handler filter regularly.

Like all mechanical equipment, with continuous use there will come a time when the system must be opened for service. The following topics pertain to information for troubleshooting these systems when they fail:

1. Refrigerant leaks
2. Restrictions in the refrigerant system
3. Installation problems

Refrigerant Leaks

Refrigerant leaks on a refrigeration system can be either inward or outward. If the system operates below atmospheric pressure, an inward leak will cause air and moisture to enter the cycle. This will cause an increase in discharge pressure and temperature, which increases corrosion and can stop the operation quite quickly. An inward leak can be more serious than an outward leak.

NOTE: Current EPA regulations do not mandate that leaks in systems containing less than 50 lb be repaired. However, good environmental practice and economic practice should prevail. All leaks should be addressed and those that can be repaired should be repaired. It will save the customer money and any reduction in refrigerant leakage to the atmosphere is good for the environment.

Several methods of outward leak detection can be used with any halocarbon refrigerant. The best method is usually the electronic leak detector such as those shown in Figure 17-4-11. It is usually a time saver over some of the former methods.

The halide torch, Figure 17-4-12, is a fast and reliable method for use with hydrocarbon refrigerants. The fuel that operates the torch is either methyl alcohol or a hydrocarbon such as butane or propane. The flame heats a copper element. Air to support combustion is drawn through the tube. The free end of the tube is passed over any suspected leak areas. When a halocarbon vapor passes over the hot copper element, the flame changes from normal color to bright green or purple.

A halide torch should only be used in well ventilated areas. When working in direct sunlight, it is sometimes hard to see the color change.

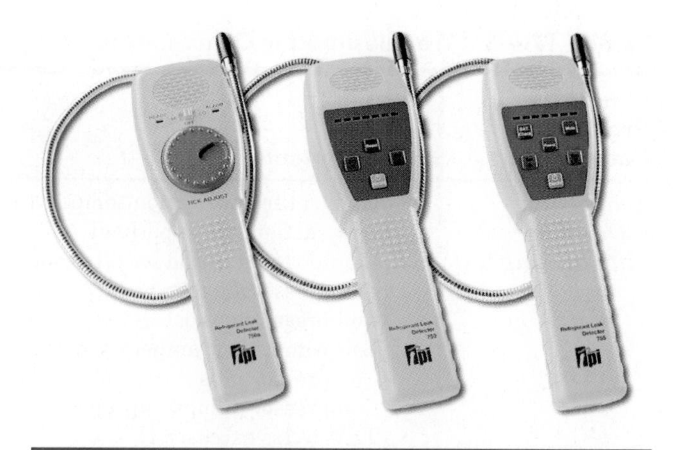

Figure 17-4-11 Electronic refrigerant leak detectors. *(Courtesy Test Products International, Inc.)*

Figure 17-4-12 Halide refrigerant leak detector.

Another reliable method of leak detection is soap bubble testing as shown in Figure 17-4-13. One method is to paint the suspected leak area with the soap solution and then, using a bright light, watch for the formation of bubbles.

Another method of testing is to pressurize the system with dry nitrogen and submerge the suspected

(a)

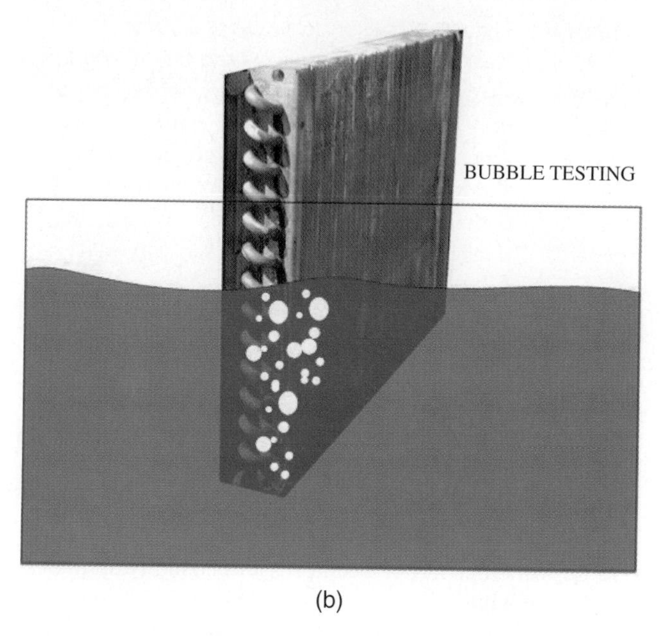

(b)

Figure 17-4-13 The use of soap bubbles as a leak detector: (a) Spraying; (b) Immersing.

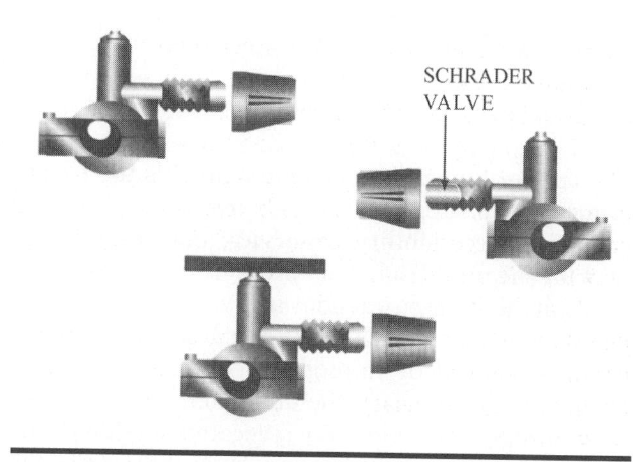

Figure 17-4-14 The use of piercing valves for access to a hermetic system.

leak area in water, and watch for bubbles. The system pressure should be at least 50 psig.

To repair a leak in a hermetic system, some type of access valve(s) needs to be installed. Piercing valves, Figure 17-4-14, are handy for penetrating refrigerant lines that have no service valve. The valve is bolted or brazed to the line, depending on the type of valve. A needle in the valve penetrates the tubing when the stem is turned down. This permits installing a gauge manifold and proceeding with recovery of the remaining refrigerant, and repairing, evacuating, and recharging the system.

Restrictions in the System

In a hermetically sealed system, a restriction in the refrigerant circuit sometimes develops. This can be caused by kinked or blocked tubing, a moisture restriction, or a blocked filter drier. Occasionally a capillary tube is bent and kinked, or contaminants may be in the system that block the opening of the tube, preventing refrigerant flow.

If moisture is in the system and the evaporator is operating below freezing temperatures, ice can form at the metering device, stopping the flow of refrigerant. If the filter drier is full of waste materials collected from the system, it can cause a restriction in the flow of refrigerant.

A restriction is usually easy to diagnose, since it prevents the normal flow of refrigerant. The discharge pressure, the suction pressure, and the amperage all drop. The superheat and the subcooling both increase.

Normally, restrictions take place in the liquid line. When they do there is a decided temperature drop across the restriction.

Whenever a restriction does occur, its location and cause must be found. Remedial action depends on the nature of the restriction. Kits are available for cleaning out a plugged capillary tube; alternatively, the exact replacement can be installed. A plugged filter drier can be replaced.

Installation Problems

Problems in the installation of hermetically sealed systems can be minimized by following specific installation instructions supplied by the manufacturer. Since there are many types and uses of hermetic systems, only general information can be given here. For example, there is a considerable difference between the proper installation of a room cooler and a domestic refrigerator, although they both contain hermetic systems.

Generally, installation instructions for units with hermetic refrigeration systems include the following topics:

1. Location of equipment
2. Leveling the equipment

3. Electrical and plumbing connections
4. Adding accessories
5. Necessary maintenance

The equipment location must provide the proper clearances to supply ventilation air for an air cooled condenser, accessibility for service, and suitable support for the unit. High head pressure would be an indication that the outside air was not properly circulating through the air cooled condenser. This could be caused by improper clearances or a dirty condenser. If necessary, the unit should be relocated. Occasionally the condenser is located outside of the building where shrubbery has grown to restrict the air into the condenser. Vegetation should be trimmed away from the unit and this noted for future maintenance.

Units need to be installed in a level position, so that condensate will flow to the drain opening for proper disposal. If the drain pan becomes dirty or the opening for condensate removal is blocked, these areas must be cleaned to permit proper flow after leveling the unit.

The unit needs to be located so that electrical and plumbing connections can be properly applied. Usually a drain trap is required and must be provided. Otherwise, under certain conditions the drainage will flow in the reverse direction, causing the unit to flood. Electrical connections must be kept dry and accessible.

Sometimes a unit can be improved by adding certain accessories. In some cases the manufacturer has available an air deflector to direct the moving air over the condenser. Without the accessory, the prevailing wind may pass through the condenser in the wrong direction, not cooling the unit and causing high head pressure.

Very often, the installation instructions will indicate certain maintenance that must be provided for the equipment to work properly. For example, an air cooled condenser coil must be cleaned periodically. Maintenance records should be examined to see whether this has been done, what history the problem may have, and what previous actions have been taken.

TECH TALK

Tech 1. I've looked at all the troubleshooting charts and talked to a lot of other guys and no one seems to agree on what is the first thing I should do when I start troubleshooting a system.

Tech 2. The first thing you should do on any troubleshooting call is to take some time to look the system over. You can often see a loose connection, burned wire, and sometimes heater refrigerant leaks. In addition a refrigerant leak over time will leave a small residue of oil. So check the system thoroughly for any visual indication of what the problem is before you simply start doing tests.

Tech 1. That sounds like the simplest thing to do.

Tech 2. You're right, it takes very little time if you can visually find a burned wire or loose connection or leaky refrigerant and get that fixed. Once you have done that you need to check and see what may have caused that wire to come off, burn up, or for a leak to occur so it does not happen again just after you leave.

Tech 1. Thanks, I appreciate that.

UNIT 4—REVIEW QUESTIONS

1. Define troubleshooting.
2. What four items are important to know in preparing to analyze electrical problems?
3. In troubleshooting when the load is not working, the technician must determine what?
4. What is the procedure for electrical troubleshooting?
5. What is the most efficient means of troubleshooting mechanical problems in the operation of refrigeration systems?
6. What is the cause of most compressor failures?
7. Loss of efficiency of a compressor is usually an indication of what?
8. List the possible causes of noises coming from inside of the compressor.
9. When testing three phase compressor motors what will a good motor show?
10. What are the two types of capacitors?
11. A ________ is probably the number one cause of low airflow.
12. What is an indication of low airflow?
13. List the refrigeration problems associated with condensers.
14. The most common cause of high head pressure can be traced to a ________.
15. A ________ is usually a sign that the expansion valve is too large for the application.
16. What is a hermetically sealed refrigeration system?
17. What is one reliable method of refrigerant leak detection?
18. Why do units need to be installed in a level position?

18

Commercial Refrigeration

UNIT 1

Refrigeration Systems

OBJECTIVES: In this unit the reader will learn about refrigeration systems including:

- Types of Systems
- Industrial Refrigeration Units

SYSTEMS

Both air conditioning and refrigeration are processes of moving heat from a substance or area where it is not wanted to an area where it is not objectionable. Although they both use the same process to remove heat their purpose for removing the heat differ. Air conditioning is used for creature comfort, to improve our living environment. Refrigeration is used for almost everything else that needs to be cooled. Refrigeration can be used for food storage or preparation. It can also be used to make ice for many different needs from chilling a drink to slowing concrete's curing rate on hot days. When refrigeration is used in manufacturing it is called process cooling.

Refrigeration is divided into four broad areas based on the evaporator coil temperature. High temperature refrigeration coil temperatures are no cooler than 32°F. Medium temperature refrigeration coil temperatures range from 32°F down to 0°F. Low temperature refrigeration coils are 0°F and below. Most low temperature refrigeration uses coils no colder than −40°F. Cryogenics is a very low temperature process with temperatures of −250°F and below.

High temperature refrigeration is used for product storage above freezing. When we think of high temperature refrigeration we usually think of food storage. Food items like milk, cheese, and fresh fruit and vegetables are stored at temperatures below 40°F and above 32°F. Storage temperatures within this range are also used for things like flowers or medicine.

Medium temperature refrigeration is used for storing frozen foods. The most common example of medium temperature refrigeration is a home refrigerator's freezer compartment.

Low temperature refrigeration is used to quickly freeze and store foods. Temperatures below 0°F freeze foods quickly to preserve their quality. Foods at these temperatures can be stored for long periods of time as compared to foods stored in the medium temperature range. The most common example of low temperature refrigeration is a deep freeze.

Cryogenic temperatures are used for the preservation of materials for an extremely long period of time. Materials stored at these temperatures are almost suspended in time. Maintaining cryogenic temperatures requires a lot of refrigeration and is very expensive, so this type of storage is mainly used to store medical and scientific materials.

TECH TIP

Generally, the colder a food product is stored the longer its shelf life. For example, milk stored at 40°F in a refrigerator may only last a few days before souring, as compared to milk stored at 32°F, which may last two or more weeks before souring. Even foods that are frozen will remain at a higher quality longer the lower the storage temperature. Food in a refrigerator freezer can be stored and used longer when the temperature in the freezer is closest to 0°F. A deep freeze temperature is below 0 and foods at these temperatures can be stored for much longer periods of time while maintaining their quality.

There are many types of refrigeration systems. Each type of system is designed for specific applications. The current equipment being produced must meet the following criteria:

1. The quality of the product being cooled must remain satisfactory.
2. A minimum amount of energy must be used to perform the operation.
3. The process must be operated to comply with the laws relating to protecting the environment.

TYPES OF SYSTEMS

The following are various types of systems used in commercial refrigeration, and are described in this section:

1. Single component systems
2. Multiple evaporator systems
3. Multiple compressor systems
4. Multistage compressor systems
   a. Compound
   b. Cascade
5. Evaporative cooling systems
6. Secondary refrigerants
7. Thermoelectric refrigerators
8. Expendable refrigerants
9. Absorption refrigeration systems

Single Component Systems The essential elements of a *single component system* as shown in Figure 18-1-1 are the: compressor, condenser, metering device, and evaporator. Also shown are some of the important accessories common to commercial refrigeration systems: a high/low pressure control, condenser fan, receiver tank, suction to liquid line heat exchanger, filter drier, evaporator fan, thermostat, and suction line accumulator. Each of these accessories helps maintain an efficient system.

There are many variations of these basic elements that are useful in a wide variety of applications. For example, Figure 18-1-2 shows a water cooled condenser used in place of the air cooled condenser and fan coil unit evaporator of an air conditioning system. Note also the additional accessories shown: compressor and receiver service valves, a liquid line solenoid, and a water regulating valve for the condenser.

SERVICE TIP

Refrigeration systems require the addition of components not normally found on air conditioning systems. Each of these devices that are added create resistance to the flow of liquid or vapor refrigerant. Such resistance reduces the overall operating efficiency of a system. For that reason technicians must exercise good judgment when deciding whether to add additional accessories to an operating refrigeration system. They must weigh the advantages and necessity for such accessories to the proper operation of the system as compared to any detrimental effect it has on system capacity. This is of prime concern with systems that are marginally meeting current demand.

Multiple Evaporator Systems This system is common in supermarkets. It makes possible using a single compressor to control a number of different case or fixture temperatures. Figure 18-1-3 shows a three temperature system.

Evaporator pressure regulators (EPRs) are placed in the suction lines to the two higher temperature evaporators. These are adjusted to maintain the desired evaporator temperature.

A check valve, Figure 18-1-4, is installed in the suction line from the lowest temperature coil. This prevents migration of the refrigerant from the higher temperature coils to the low temperature coil when the compressor is producing cooling for the higher temperature coils.

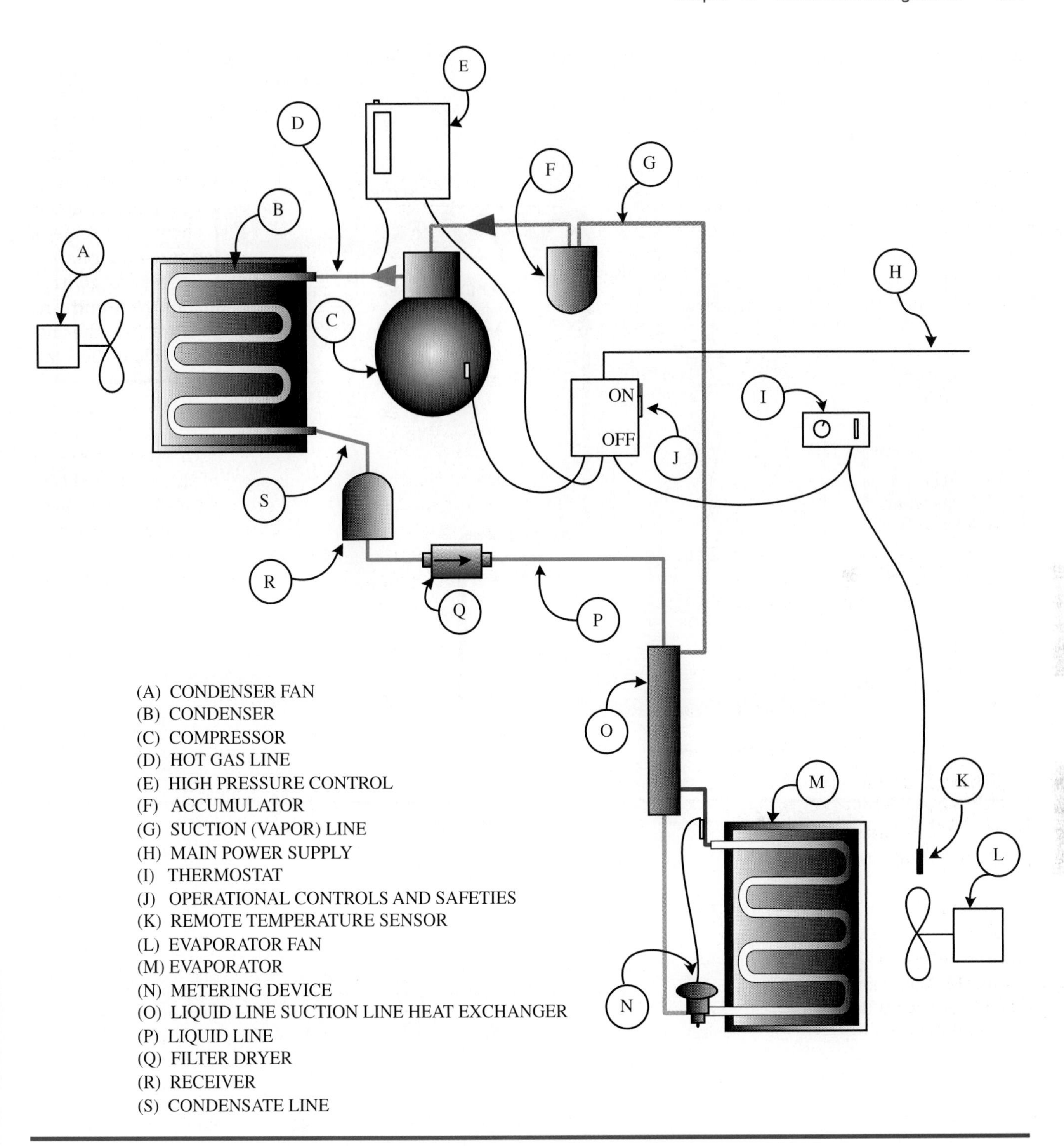

Figure 18-1-1 A simple refrigeration system with a single compressor and air cooled condenser.

When all of the coils require refrigeration, such as on startup, the compressor will operate at the suction pressure required to cool the highest temperature coil first, then the middle temperature coil, and last, the low temperature coil. The compressor must be sized to produce the entire cooling load at the evaporator pressure of the lowest temperature coil.

The low temperature coil will receive very little refrigeration until the higher temperature coils are satisfied. For this reason the load on the low temperature coil must account for 60% of the total load, otherwise it may not receive adequate refrigeration.

On a small system it is general practice to install a surge tank in the suction line of the low temperature evaporator to prevent short cycling.

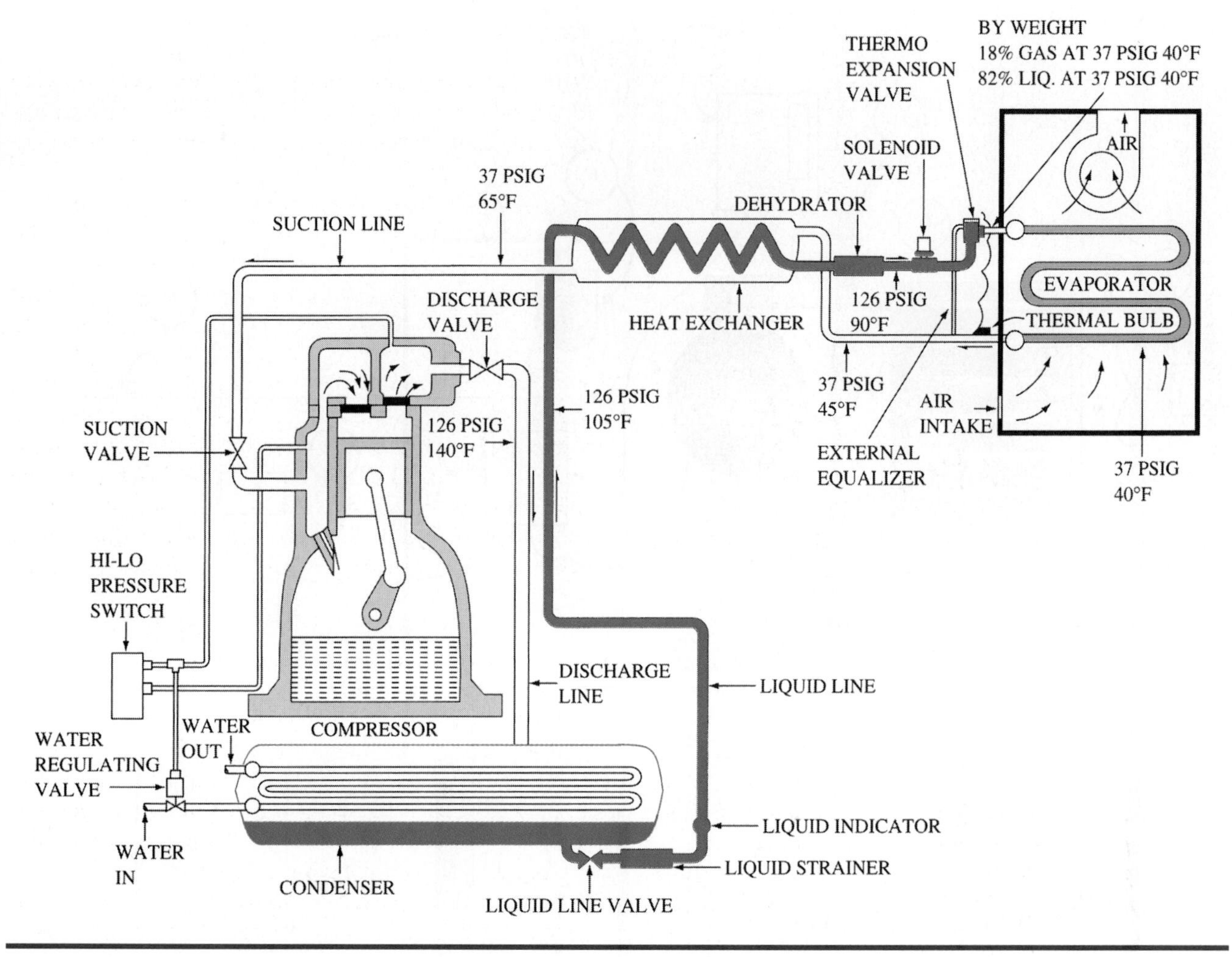

Figure 18-1-2 A simple refrigeration system with a single compressor and water cooled condenser.

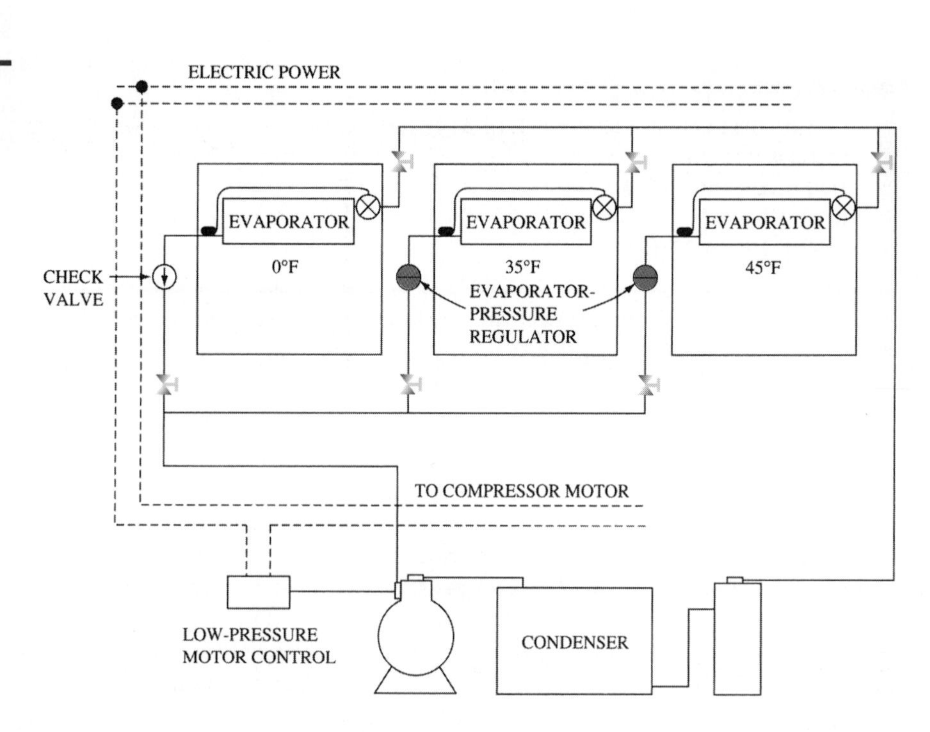

Figure 18-1-3 Multiple evaporator system using evaporator pressure regulators to maintain different conditions in each box.

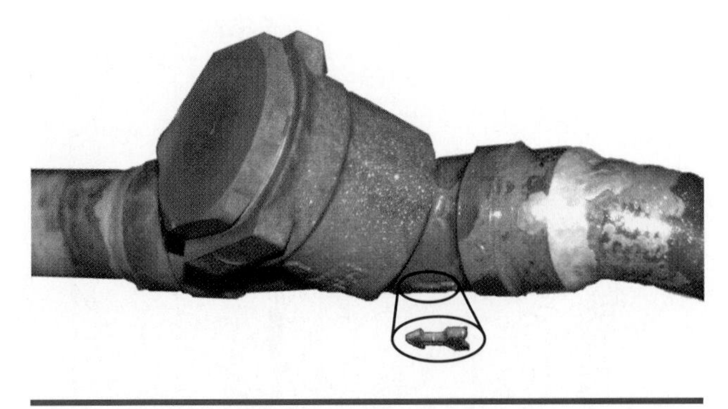

Figure 18-1-4 Refrigerant check valves have arrows showing the direction of flow.

SERVICE TIP

Restaurants come and go, but the refrigeration equipment stays behind. Frequently refrigeration equipment designed for the original restaurant may not meet the needs of the new operation. When the primary loads change it is not possible simply to make adjustments to multiple evaporator systems to accommodate these changes. It is not possible to simply turn down or up the low temperature side of a system without adversely affecting the medium or high temperature side of a system. Restaurant managers are not always aware of this issue. It is in your best interest to take the time to explain to the manager what the difficulties can be for changing the load for refrigeration systems.

Multiple Compressor Systems

Where the refrigeration load varies over a wide range, such as in supermarkets, it is desirable to use multiple compressors connected in parallel, as shown in Figure 18-1-5.

A suction pressure control is used to turn on and off individual compressors as required to match the load. These controls also have the ability to change the lead compressor to obtain approximately equal running time on each compressor. A single condenser and receiver is used for all units.

The refrigerant piping must be done properly, as shown in Figure 18-1-6 and Figure 18-1-7. Referring to Figure 18-1-6, the suction piping should be brought in above the level of the compressor. With multiple compressors, a common suction header should be used and the piping should be designed so that the oil return to several compressors is as nearly equal as possible.

If an oil level control is not used, the discharge piping as shown in Figure 18-1-7 should be used with the discharge piping running to a header near to floor level. With this arrangement, a discharge line trap is not required since the header serves this purpose.

It is also important to provide equalization of the oil level in all compressors using an arrangement such as shown in Figure 18-1-8. Oil is pumped by the compressor to a common discharge header and into an oil separator. Since the oil separator has a large holding capacity, oil is then transferred to an oil reservoir.

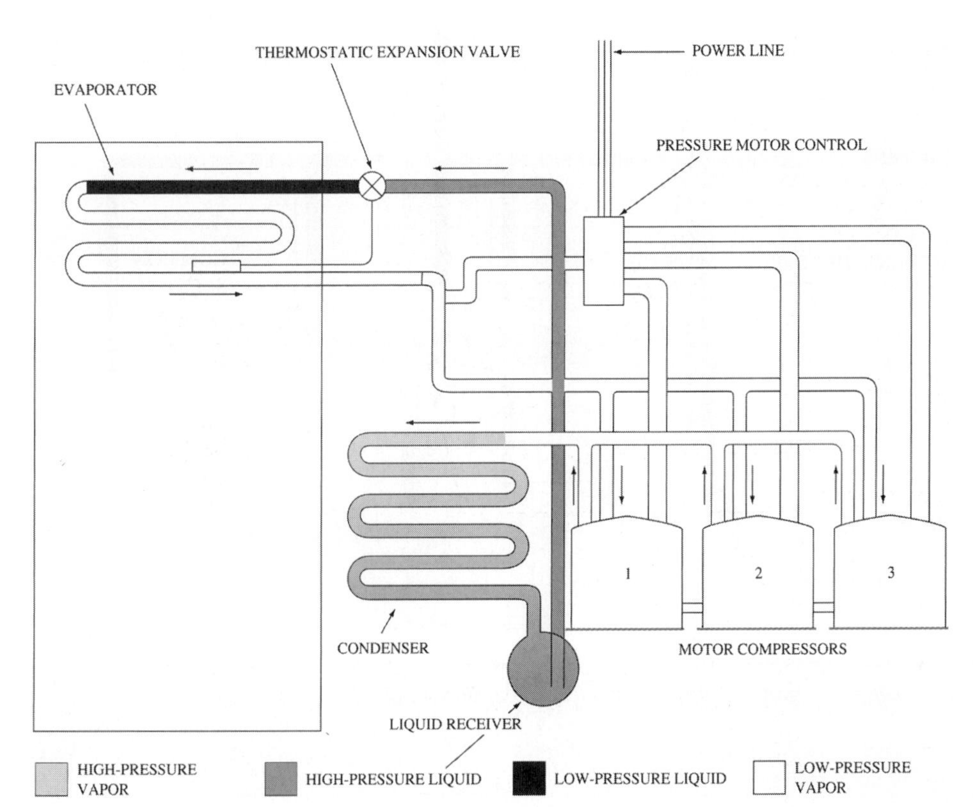

Figure 18-1-5 Multiple compressor system using three compressors connected in parallel.

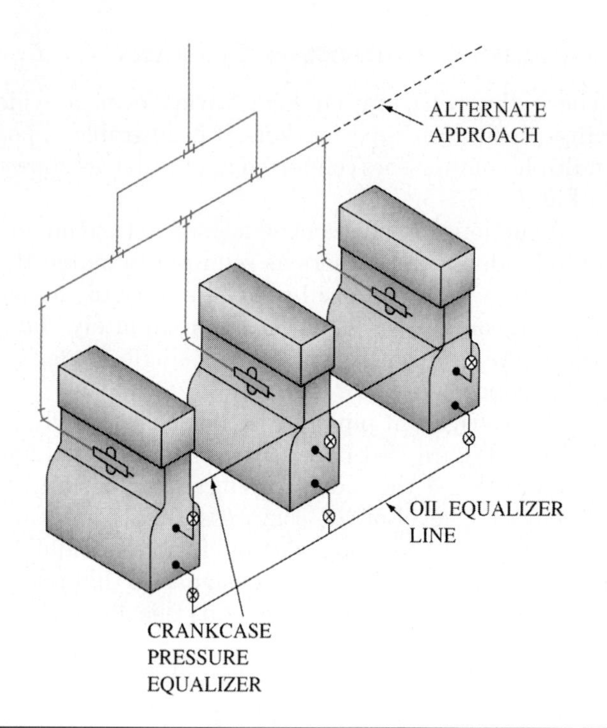

Figure 18-1-6 Suction line piping for parallel compressors.

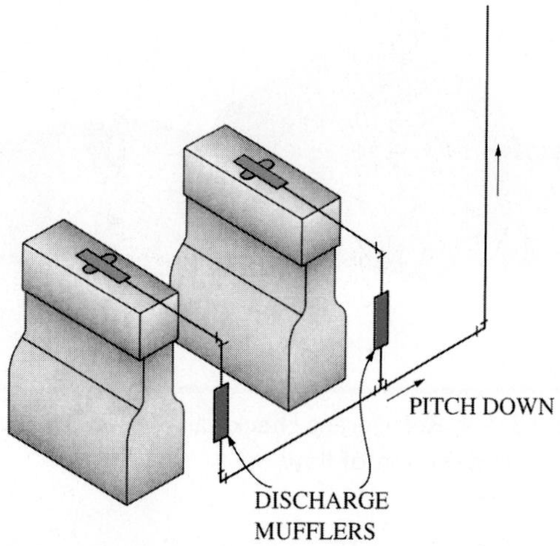

Figure 18-1-7 Discharge line piping for parallel compressors.

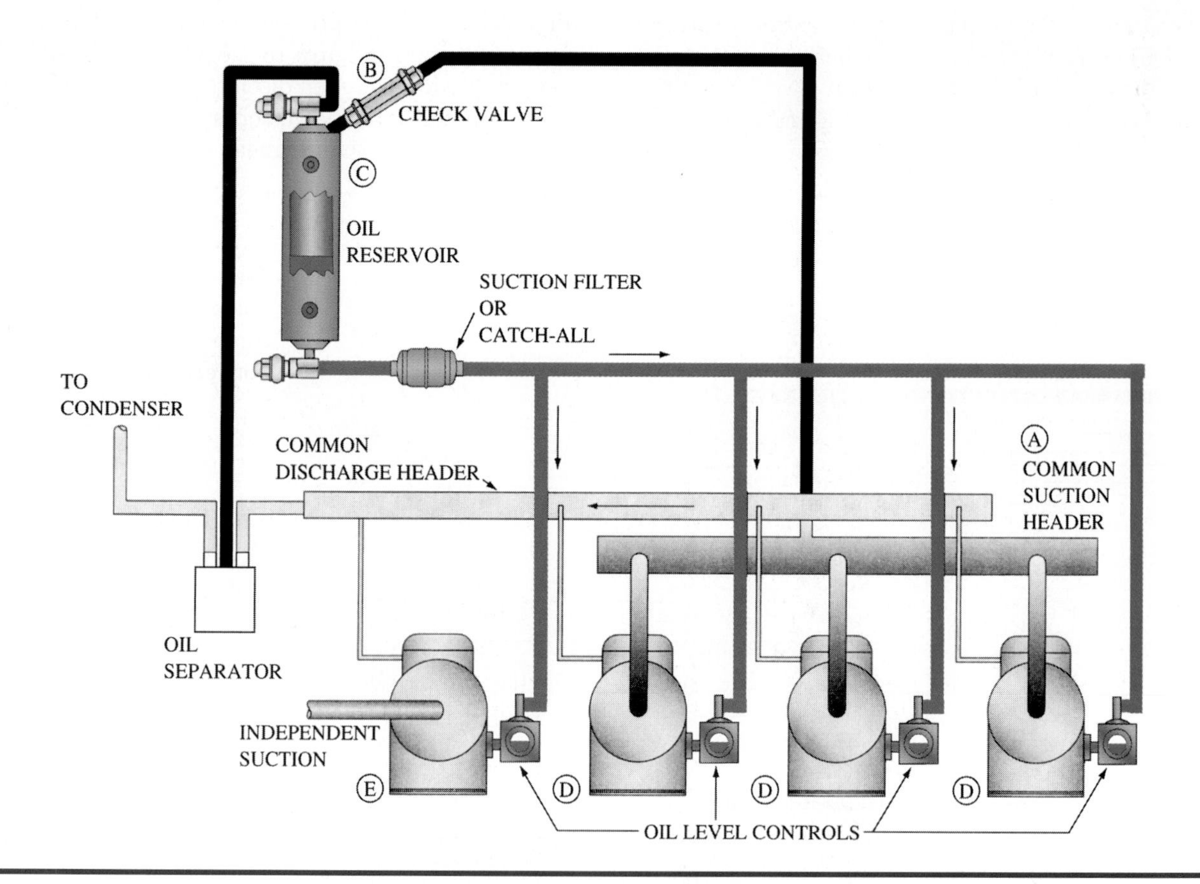

Figure 18-1-8 Oil level control system for multiple compressors. (*Courtesy of Sporlan Valve Company*)

CAUTION: Maintaining proper oil level between multiple compressors is critical. If the compressor oil level drops, the system can be severely damaged or destroyed. Compressors have oil safety switches that are designed to trip and take the compressor offline when the oil pressure drops below a critical level. However, system owners and managers frequently see oil trips as a "nuisance" call and will push the reset button themselves. It is important that you stress to your customers that you be notified each and every time a system shuts down as a result of an oil problem so that you are able to identify the root cause of the problem and correct it.

Pressure in the reservoir is reduced by boiling the refrigerant contained in the oil and relieving the pressure above the oil through a vent line to the suction header. An oil level control meters the oil to the compressors equal to the pumping rate and thereby maintains the oil level specified by the manufacturer.

SERVICE TIP

Oil is returned to a compressor as a result of the velocity of refrigerant vapor in the suction line. If an evaporator defrost system is not working correctly, then the evaporator will ice over, resulting in a slower flow of refrigerant vapor. This can result in oil slugging of the evaporator. If the oil level in an entire system appears to be low, do not simply add oil. Look for the primary cause of the oil shortage. Refrigerant oil does not simply evaporate. It stays in the system somewhere and it is important that you locate it and determine what is necessary to have it return to the compressor during system operation.

Multistage Compressor Systems

There are two general methods of multistage compressors, compound and cascade.

A *compound system* offers a method for producing low temperature refrigeration. In order to achieve a low suction pressure, two compressors are connected in series as shown in Figure 18-1-9.

The suction line from the evaporator feeds the first compressor. Then the discharge from the first compressor enters the suction of the second compressor. The discharge from the second compressor goes to the condenser. Although both compressors handle about equal loads, the first compressor handles the greatest volume of gas, since the density of the gas in the first stage is less.

An important additional device in this system is the *intercooler* between the two stages that reduces the superheat in the discharge gas from the first compressor. This superheat, produced by the work done by the first compressor, must be removed to keep the gas temperature of the second compressor within limits. A single temperature control operates both compressor motors. On startup, both motors are started at the same time.

It would be possible to use a single compressor to achieve the multistage effect; however, the disadvantage of a single multistage compressor is that there is a fixed ratio between the high side and the low side volumes which only applies to applications where this ratio produces a satisfactory evaporating temperature. Compound systems using two compressors give the added flexibility of producing a wider range of low evaporating temperatures. Compound arrangements of compressors are seldom considered unless the desired evaporating temperature is below −20°F.

A single refrigerant is used throughout the entire compound system, in contrast to a cascade system which may use separate refrigerants for each compressor.

A *cascade system* is also used for low temperature refrigeration, but has the capability of reaching much lower temperatures than the compound system. The cascade system shown in Figure 18-1-10 uses two compressors; however, for lower temperatures, three or more stages (compressors) are used.

CAUTION: Because of the extreme load placed on low temperature compressors during startup, it may be necessary to use the suction line valve as a metering device to throttle the system capacity back so that it comes online more slowly. You can monitor the compressor amperage as the service valve is slowly opened so that it does not exceed the rated load by more than 10%. Failure to do this on large low temperature refrigeration systems, especially cascade system second stage compressors, can result in almost immediate failure of the compressor.

The system has an interesting configuration. Each stage has a separate refrigeration circuit. Each stage can use a different refrigerant if there is an advantage in doing so. The interconnection is through a common heat exchanger. In the system shown in Figure 18-1-10, the evaporator of the second stage machine cools the condenser of the first stage machine.

When the system is started up, the high stage compressor is started first. After it lowers the evaporating temperature to the required level, the low stage compressor is started. To facilitate shutdown, an expansion tank is added to the low stage circuit, as shown in Figure 18-1-11.

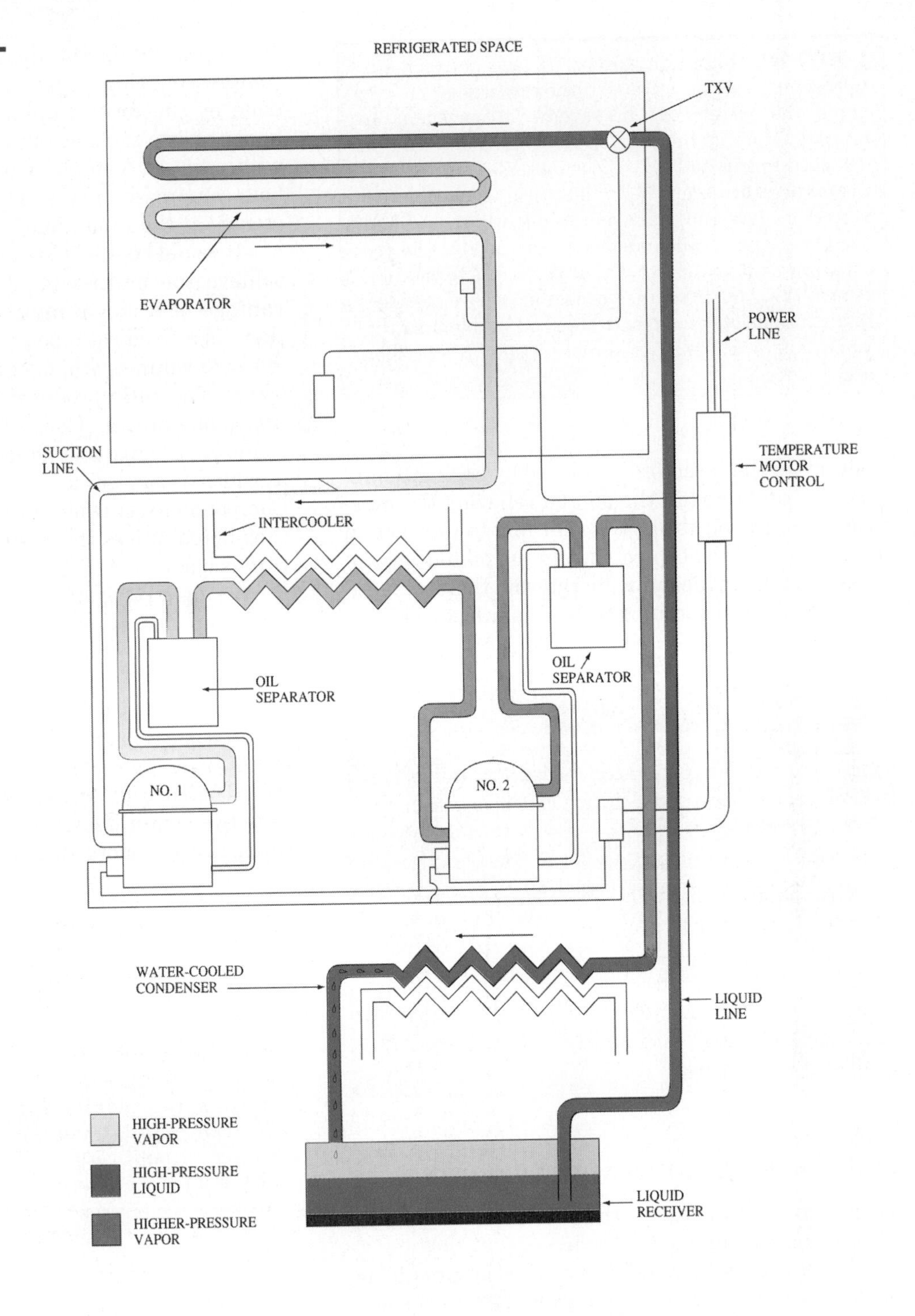

Figure 18-1-9 Compound refrigeration system using two compressors.

This is done to prevent the pressures from becoming excessive when the refrigerant is subjected to room temperatures. A relief valve on the high side circuit discharges into the expansion tank. Otherwise it would be necessary to transfer refrigerant to storage cylinders during shutdown.

Evaporative Cooling Systems

Evaporative cooling is the process of absorbing heat by the evaporation of water. Approximately 970 Btu of heat is required to evaporate a pound of water. Two general types of evaporative cooling are:

1. An open system
2. A closed circuit system.

A diagram of an open system is shown in Figure 18-1-12.

A cooling tower such as shown in Figure 18-1-13 uses evaporative cooling to reduce the temperature of the hot water from the condenser so that it can be reused at a lower temperature.

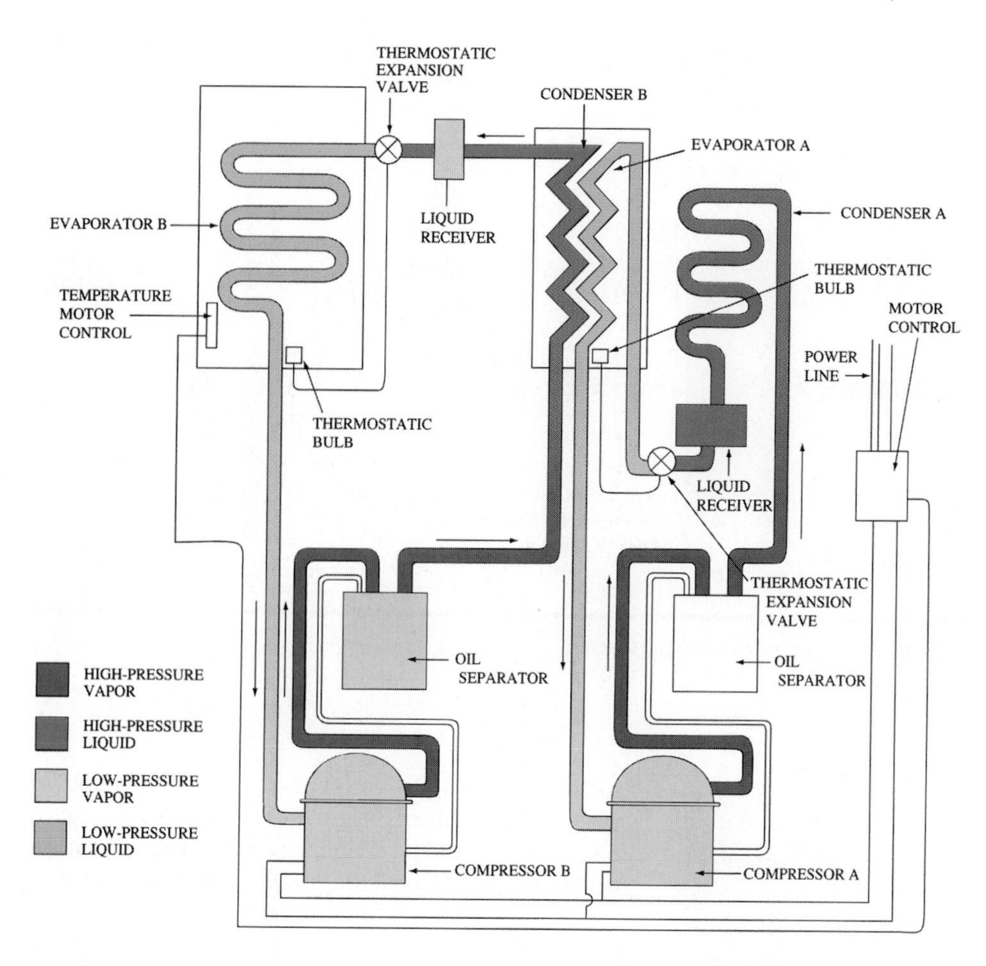

Figure 18-1-10 Cascade refrigeration system using two compressors.

Greenhouses, as shown in Figure 18-1-14, use evaporative cooling to provide a safe temperature for plants. Roof sprays are used to reduce roof temperatures to lower the building heat load. All of these systems are effective as long as the water temperature provided for evaporation is higher than the wet bulb temperature of the air.

TECH TIP

Because as water evaporates the dissolved minerals remain behind, evaporative cooling systems can be adversely affected by mineral deposits. An easy way to reduce the mineral deposit buildup in areas where water contains large quantities of dissolved minerals is to increase the water load to the system so that approximately 10% of the water in the pump flows into the waste water drain. This excessive water will keep the concentration of minerals in the basin low enough so that it does not pose as great a problem to system performance over time.

A *closed circuit* evaporative cooler is built like an evaporative condenser, as shown in Figure 18-1-15. Instead of refrigerant circulating through the coil, the

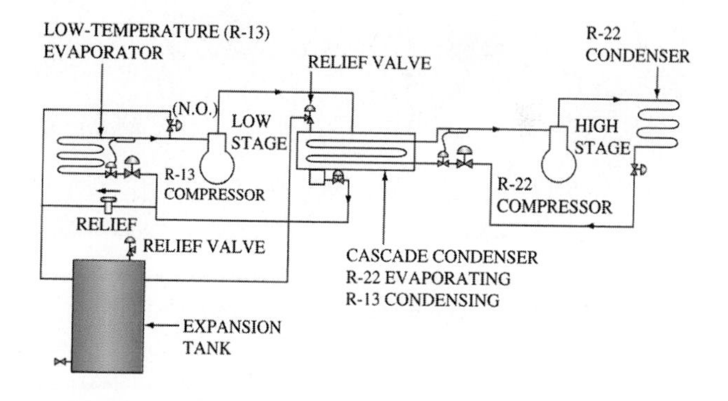

Figure 18-1-11 Cascade refrigeration system showing expansion tank added to the low stage circuit.

coil is supplied with hot water from the condensers. The hot water entering the cooler is cooled by the evaporative action of the water on the outside of the tubes. This condensed water is pumped back to the condensers for reuse. The evaporative cooler coil, the circulating pump, the water cooled condensers, and an expansion tank are piped in series to form a closed loop.

Secondary Refrigerants A secondary refrigerant is a fluid cooled by direct refrigeration and used to transfer cooling to a distant area where cooling is

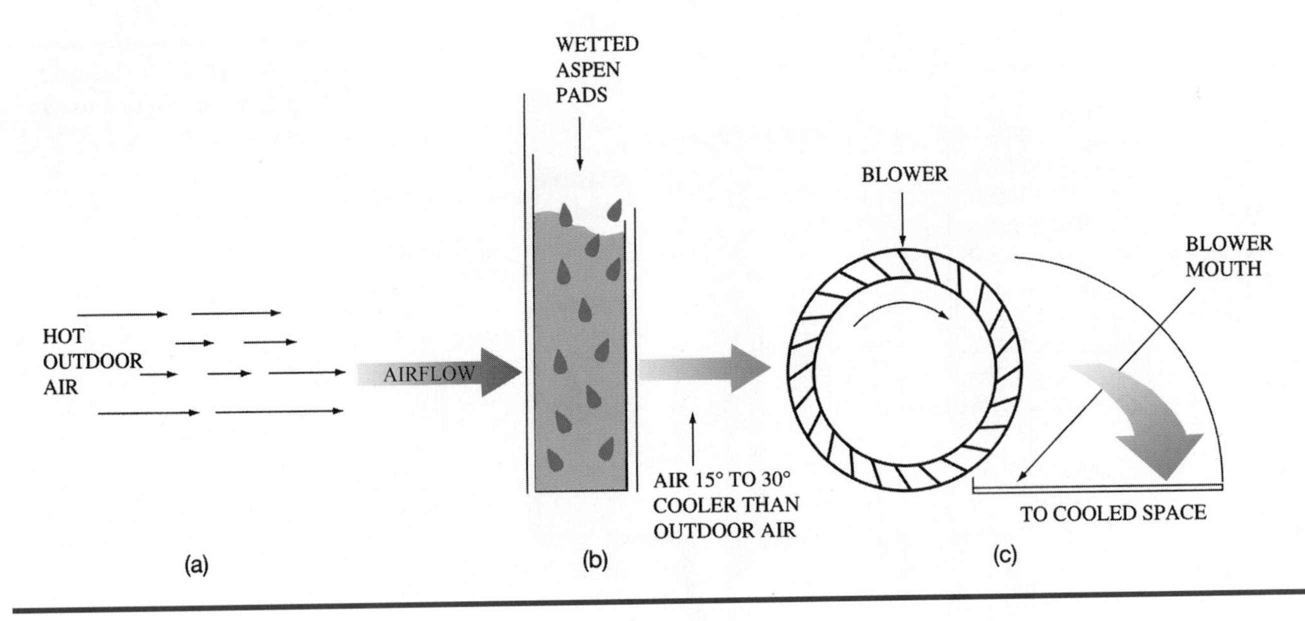

Figure 18-1-12 Schematic drawing of an evaporative cooling system.

Figure 18-1-13 Cooling tower.

needed and long direct expansion lines are not practical or economical. A good example of the common use of a secondary refrigerant is in the application of *water chillers* to provide cooling for a large building.

A very effective use of water as the refrigerant for cooling is the *ice storage* application, as shown in Figure 18-1-16a,b.

These systems are used for comfort cooling installations where cooling is required only during the day. The refrigeration system builds up icy slurry during the night when electric rates are low. During the day, water is circulated through the icy slurry and through coils in the air conditioning system to provide cooling.

Where the secondary coolants are needed below the freezing temperature of water, a brine solution is used. *Brine* is the name given to a solution of a

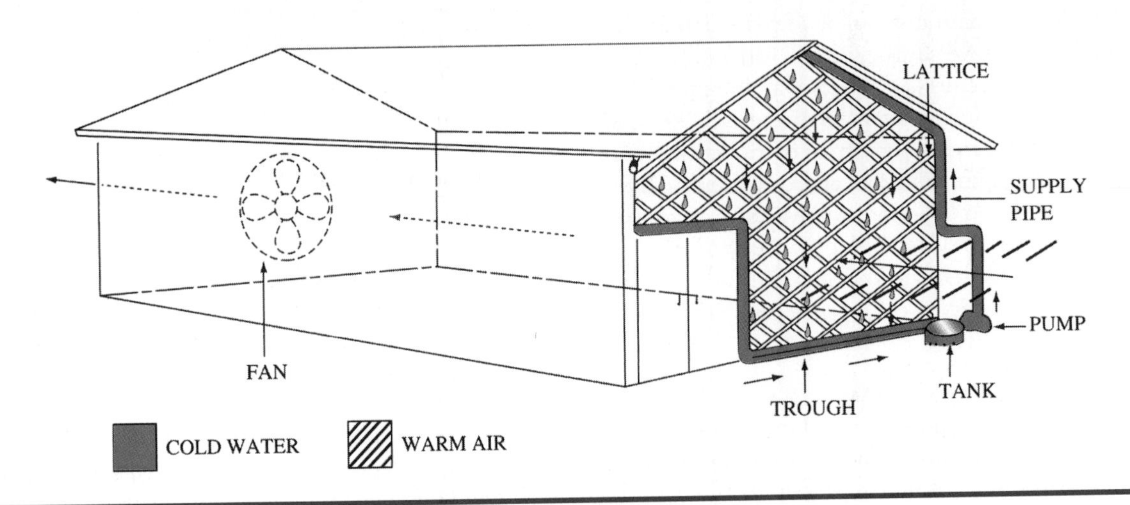

Figure 18-1-14 Application of evaporative cooling to a greenhouse.

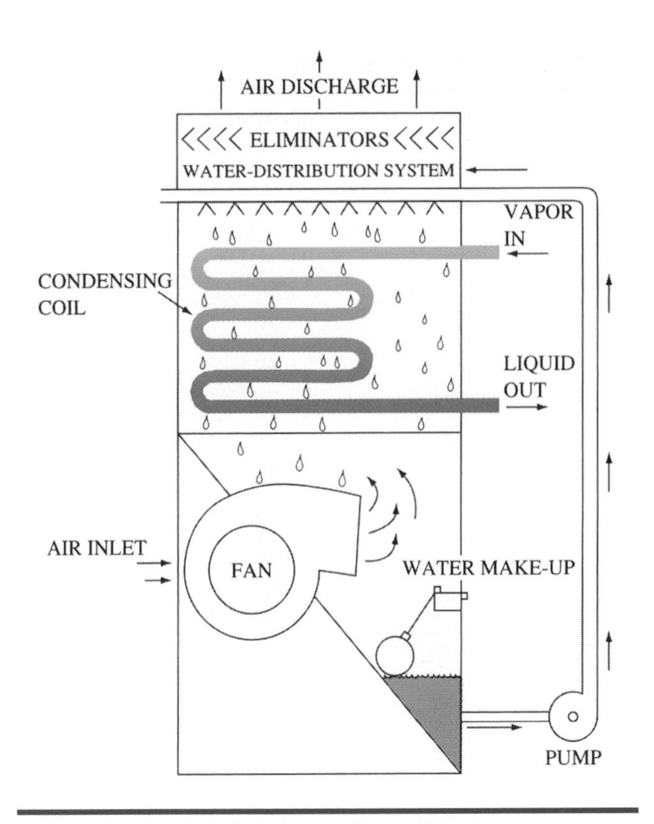

Figure 18-1-15 Diagram of a closed circuit evaporative cooler.

substance in water that lowers the freezing point of the water. Common additives are calcium chloride, sodium chloride, ethylene and propylene glycols, methyl alcohol, and glycerin. The amount of additive used affects the freezing point of the solution as shown in Table 18-1-1.

NICE TO KNOW

The inventor of the mercury thermometer, Daniel Fahrenheit, used the lowest temperature that he could create using water, ice, and salt as the zero point on his newly invented thermometer. The zero Fahrenheit point was the base point for his thermometer. The middle point was 32 degrees or the freezing point of water, and the high point was 96 degrees for the approximate temperature for the human body. These three fixed points were then used to calibrate and mark the scale for all of the thermometers Fahrenheit produced. During the 1720s it was important for scientists to be able to reproduce these scaled points so that thermometers manufactured in different parts of the world would have a uniform scale that could easily be related to temperature readings taken by other scientists in different parts of the globe. The 0°F point on the Fahrenheit scale therefore coincides with the lowest temperature a mixture of brine can be cooled to without the formation of ice crystals.

(a)

(b)

Figure 18-1-16 (a) Ice storage used for cooling storage; (b) The pneumatic motor turns the paddles to "feel" when the ice storage tank is full so the compressor can be stopped.

A good example of the use of brine for secondary cooling is the ice rink application. The general plan for the piping is shown in Figure 18-1-17.

The brine used in modern rinks is a glycol solution. The layout includes supply and return headers with piping loops under the ice to carry the brine. The brine is pumped through the piping. It enters at a temperature of 12°F (−11°C) and returns to the chiller at 14°F (−10°C). Chillers are usually in the 100 to 125 Hp range.

Thermoelectric Refrigeration The thermoelectric refrigerator uses the *Peltier effect,* a physical principle discovered by Jean Peltier in 1834. He found that if direct current was passed through the

TABLE 18-1-1 Freezing Points of Aqueous Solution Used for Secondary Cooling

Alcohol		Glycerine		Ethylene Glycol		Propylene Glycol	
% by Wt	°F	% by Wt	°F	% by Vol	°F	% by Vol	°F
5	28.0	10	29.1	15	22.4	5	29.0
10	23.6	20	23.4	20	16.2	10	26.0
15	19.7	30	14.9	25	10.0	15	22.5
20	13.2	40	4.3	30	3.5	20	19.0
25	5.5	50	−9.4	35	−4.0	25	14.5
30	−2.5	60	−30.5	40	−12.5	30	9.0
35	−13.2	70	−38.0	45	−22.0	35	2.5
40	−21.0	80	−5.5	50	−32.5	40	−5.5
45	−27.5	90	+29.1			45	−15.0
50	−34.0	100	+62.6			50	−25.5
55	−40.5					55	−39.5
						59*	−57.0

*Above 60% fails to crystalize at −99.4°F.

(Copyright by the American Society of Heating, Refrigerating, and Air-Conditioning Engineers, Inc.)

junction of two dissimilar metals, the junction point would become hot or cold depending on the direction of current flow.

Figure 18-1-18 illustrates the construction of a thermoelectric refrigerator using semiconductors.

There are two types of semiconductors, the *N-type* which conducts electricity by the flow of negatively charged particles, and the *P-type* that conducts electricity by the flow of positively charged particles. When current flows from the N-type into the P-type, the junction absorbs heat. When current flows from the P-type to the N-type, the junction becomes warm and gives off heat. A single thermocouple produces a small amount of cooling, so a number of N-P junctions put in series produce a greater cooling effect. The diagram in Figure 18-1-18 shows a series of thermoelectric couples that pick up heat from the inside of the box and transfer the heat to the outside of the box.

Thermoelectric refrigerators are available in small portable units that can operate on 12 V automotive systems. Small electrically operated refrigerators can be found in hotel guest rooms and offices. Submarines and space ships are often equipped with thermoelectric refrigerators.

One advantage of these units is that they have no moving parts requiring service. A disadvantage is that they are not as efficient as other types of electric refrigerators.

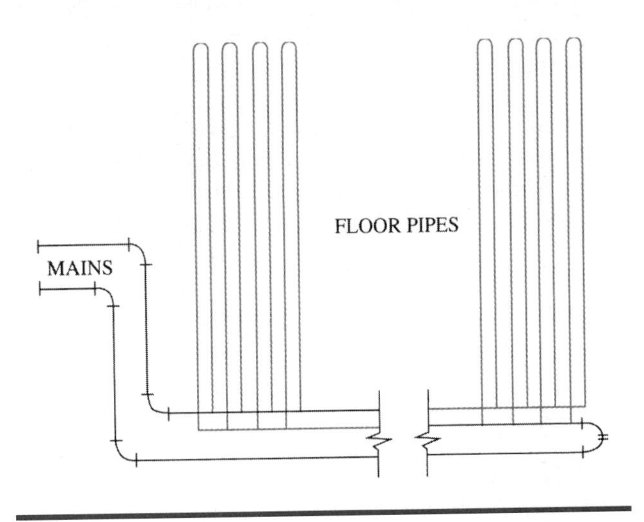

Figure 18-1-17 Piping schematic for an ice rink.

Expendable Refrigerants An expendable refrigerant is one that performs cooling by permitting the liquid refrigerant to boil at atmospheric pressures and is released to the atmosphere after the cooling operation is completed. Liquid nitrogen, with a boiling temperature of −320°F, and carbon dioxide, which has a temperature of −108°F, are the most common refrigerants used. Liquid carbon dioxide cannot exist at atmospheric pressures. It boils so quickly that it turns itself into a solid. Because there is no liquid carbon dioxide, the solid is called dry ice. Dry ice *sublimes,* which means it changes from a solid to a gas without a liquid phase. These refrigerants, when released to the atmosphere, are not considered destructive since they are both normally present in the environment.

There are two basic ways to apply the refrigerants: (1) using cold plates, Figure 18-1-19, and (2) spray cooling, Figure 18-1-20.

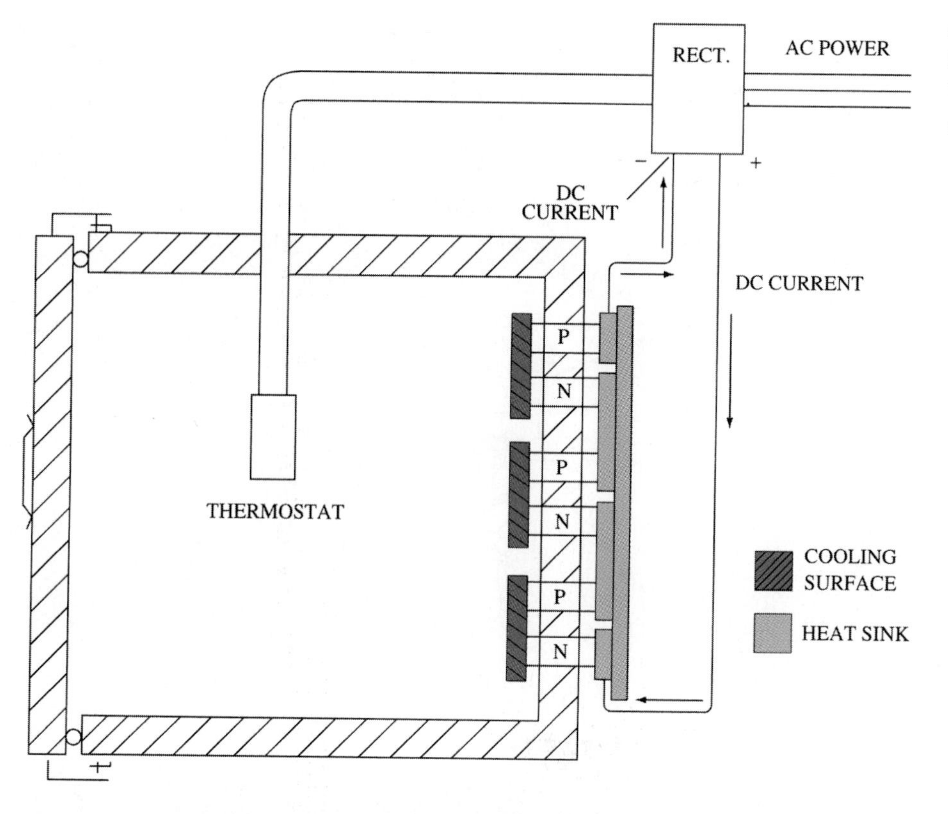

Figure 18-1-18 Thermoelectric module used for cooling.

TECH TIP

Dry ice is commonly available; it can be used in an emergency to keep frozen foods from thawing. It can serve as a temporary stopgap while you are servicing a system to prevent product from thawing and being lost.

The most common usages are for transportation vehicles, truck bodies, and railroad cars, and for shipping containers used to transport perishable items.

A diagram showing a typical spray type system for a truck body is shown in Figure 18-1-21. The storage cylinder is filled through the fill valve. The refrigerant is released at a rate to satisfy the thermostat that regulates a liquid control valve in the refrigerant line. Temperatures are maintained in the range between −20 and 60°F depending on the product being cooled. Gauges record the tank pressure and the liquid level in the storage cylinder. A safety relief valve and a vent valve are provided for the storage tank. A safety vent is installed in the truck body that will automatically open if the refrigerant pressure in the truck increases.

CAUTION: Safety precautions must be observed. Physical contact with the low temperature refrigerants can cause immediate freezing of body tissue.

Figure 18-1-19 Expendable refrigeration system using cold plate evaporator.

Absorption Systems Absorption systems have certain unique characteristics that give them a special use:

1. They use a heat source for power to operate the cycle rather than mechanical energy as required for compression refrigerating systems.
2. The absorption systems have very few moving parts. On the smaller systems the moving parts

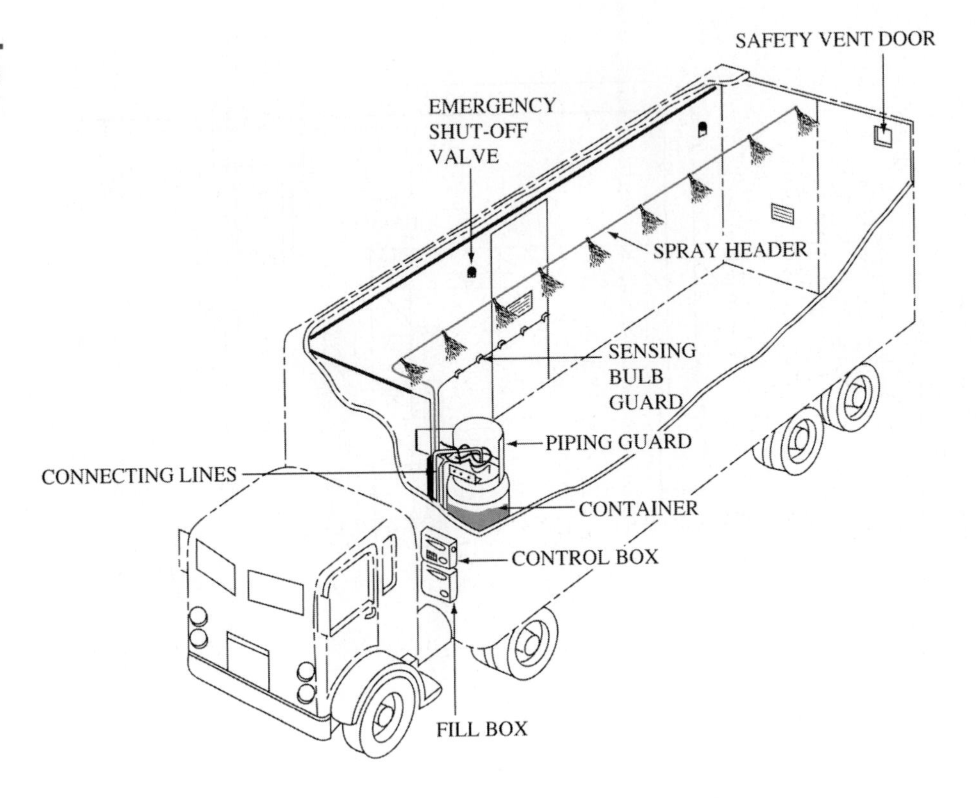

Figure 18-1-20 Expendable refrigeration system using a spray header.

are in the controls. The larger systems may use fans or pumps.

3. Most systems are sealed at the factory and the refrigeration cycle is not normally serviced in the field.

Many of these systems are used in recreation vehicle (RV) units and for refrigerators where electricity is not available. The systems are also used for comfort air conditioning, although they are less popular than mechanical systems. The heat source can be liquid petroleum, kerosene, steam, solar heat, or an electrical heater.

Figure 18-1-22 shows a diagram of the operation of a typical absorption unit using ammonia as the refrigerant and an aqueous ammonia solution as the absorbent.

A pump is used to maintain the pressure difference between the high and low sides of the system. The same pump transfers the strong solution of water and ammonia from the absorber to the generator.

In the generator the solution of water and ammonia is heated, boiling off the ammonia vapor. The ammonia vapor then goes to the condenser (water cooled or air cooled) where the ammonia is condensed. The liquid ammonia then enters the evaporator where it boils and picks up heat from the product load, which in this case is water. The ammonia vapor that has boiled off then goes to the absorber where it enters into solution with the water. From there, the pump forces the strong solution into the generator and the cycle is repeated.

Larger units use water as a refrigerant and a lithium bromide solution for an absorber. A diagram of this absorption circuit is shown in Figure 18-1-23.

This system uses a solution pump, a refrigerant pump and a solution heat exchanger in addition to the standard parts. The generator condenser section is separated from the evaporator absorber section, as shown.

INDUSTRIAL REFRIGERATION UNITS

Certain manufacturers specialize in producing refrigeration for industry. Many of these units are highly specialized in design to supply a specific need in a production process. Often these units are individually tailored, assembled, tested at the factory, and shipped as a package.

Examples of industrial refrigeration units are shown in the next three figures. Figure 18-1-24 shows an industrial quality compressor unit for process use. Figure 18-1-25 shows a custom built 300 ton water chiller that is used in the manufacture of new refrigerant. Figure 18-1-26 shows a 70 ton ammonia system designed to contain only about 25% as much refrigerant ammonia as a conventional system.

UNIT 1—REVIEW QUESTIONS

1. What is the difference between air conditioning and refrigeration?
2. What are the temperatures of cryogenics?

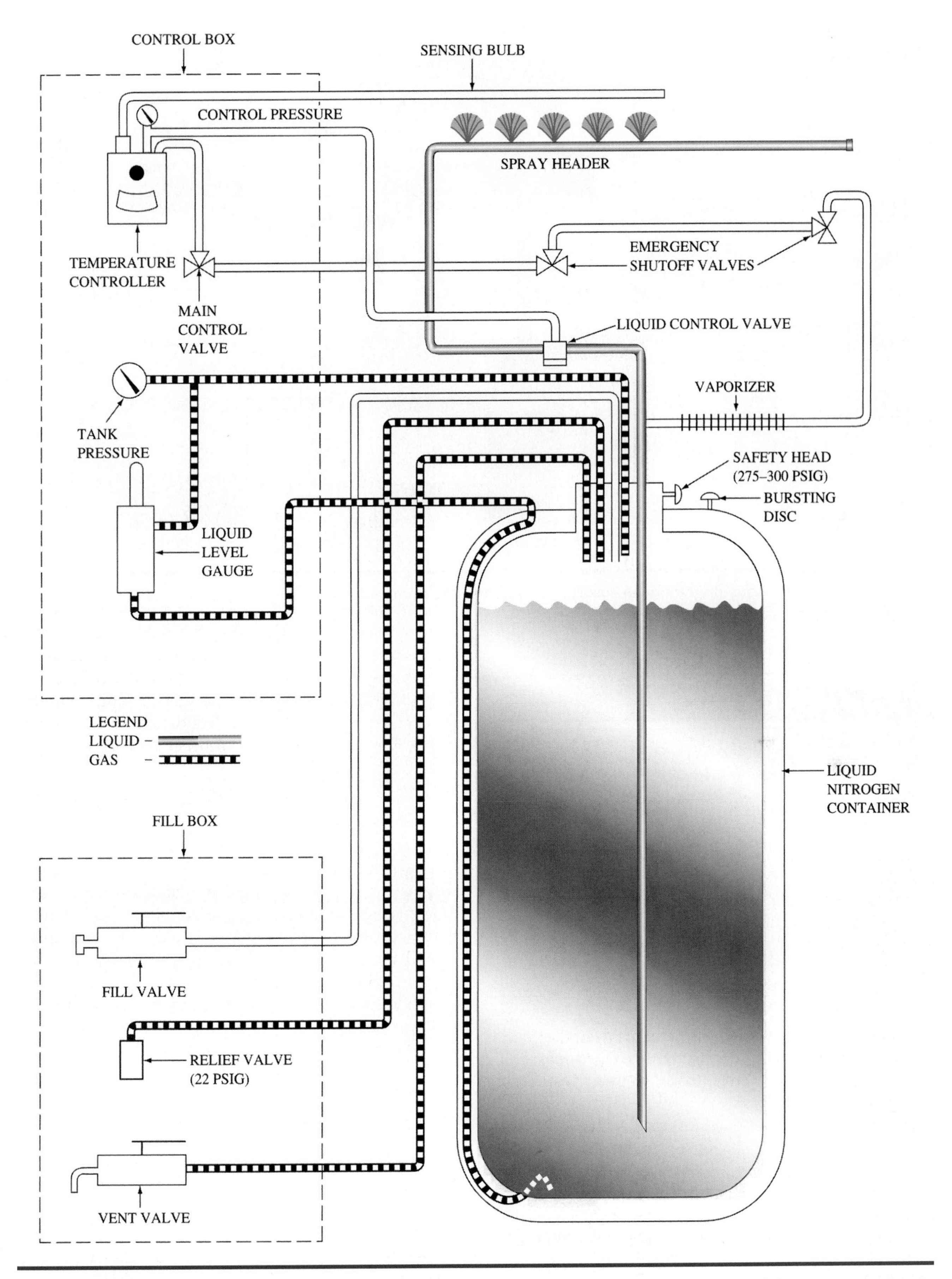

Figure 18-1-21 Piping schematic for an expendable refrigeration system.

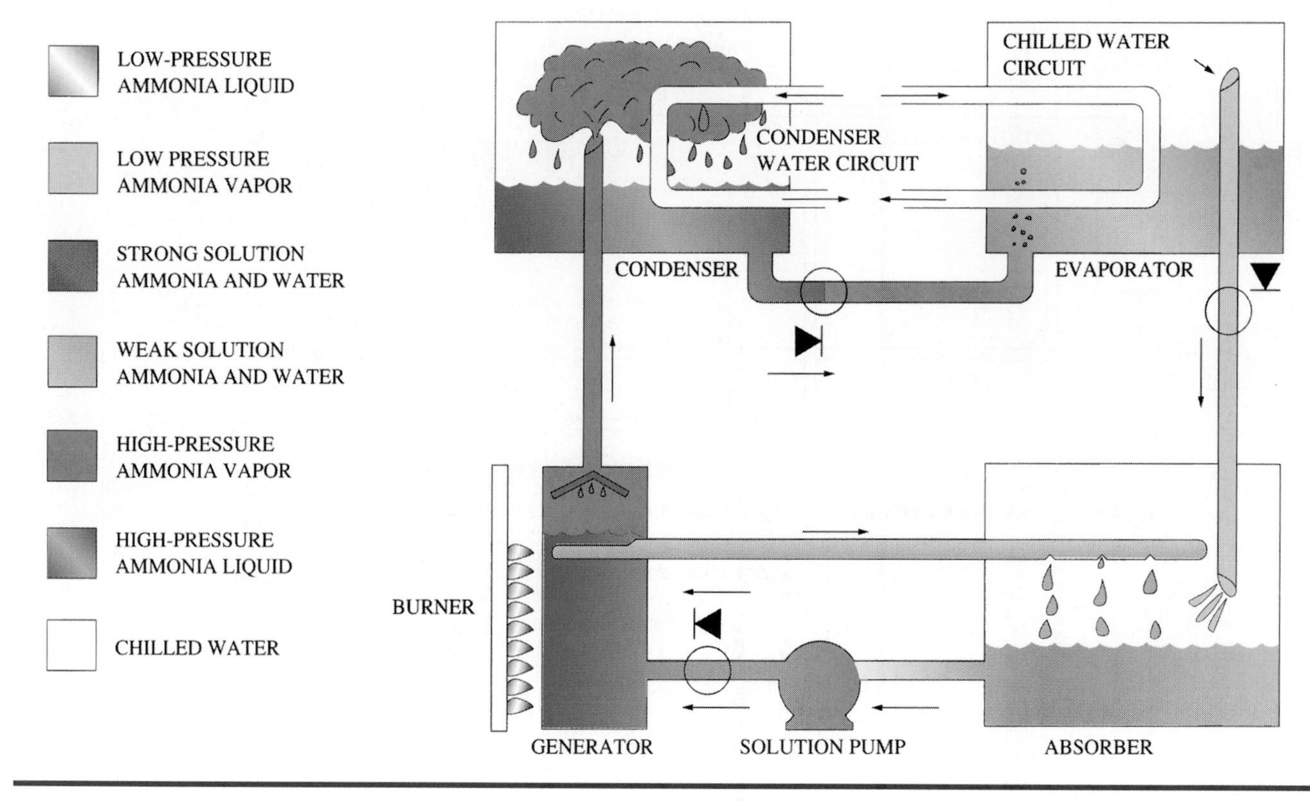

Figure 18-1-22 Schematic diagram of an absorption system using ammonia as a refrigerant and water as an absorbent.

TECH TALK

Tech 1. I just got finished working on a medium temperature box. It seems to be doing fine. There must be a problem but I could not figure out what it is.

Tech 2. What do you mean by that?

Tech 1. Well there was ice all over the suction line. It went all the way to the compressor. It must be liquid flowback. But the box seems to be doing OK. I don't understand what is going on.

Tech 2. Just because there is ice on the suction line doesn't mean that there is liquid flowback to the compressor.

Tech 1. Yes it does. Every time that I have worked on an air conditioning system when there was ice back to the compressor there was a problem with the charge.

Tech 2. Well, that is because air conditioning is high temperature refrigeration. The coil is supposed to be above freezing so if there is ice on that there is a problem with the system. But on a medium temperature box the coil is operating below freezing, causing the vapor in the suction line to be so cold that ice will form.

Tech 1. I didn't know that.

Tech 2. Your assumption unfortunately is real common with technicians used to working in the air conditioning field. If they see ice on the suction line then they think there is a problem. In refrigeration if you *don't* see ice there may be a problem.

Tech 1. Thanks. I appreciate that.

3. What is the most common example of medium temperature refrigeration?
4. List the criteria that current refrigeration equipment must meet.
5. What are the essential elements of a single component system?
6. Where are multiple evaporator systems commonly used?
7. List the two general methods of multistage compressors.
8. An important additional device in the multistage compressor system is the _______ between the two stages that reduces the superheat in the discharge gas from the first compressor.
9. How many compressors does the cascade system use?
10. Define evaporative cooling.

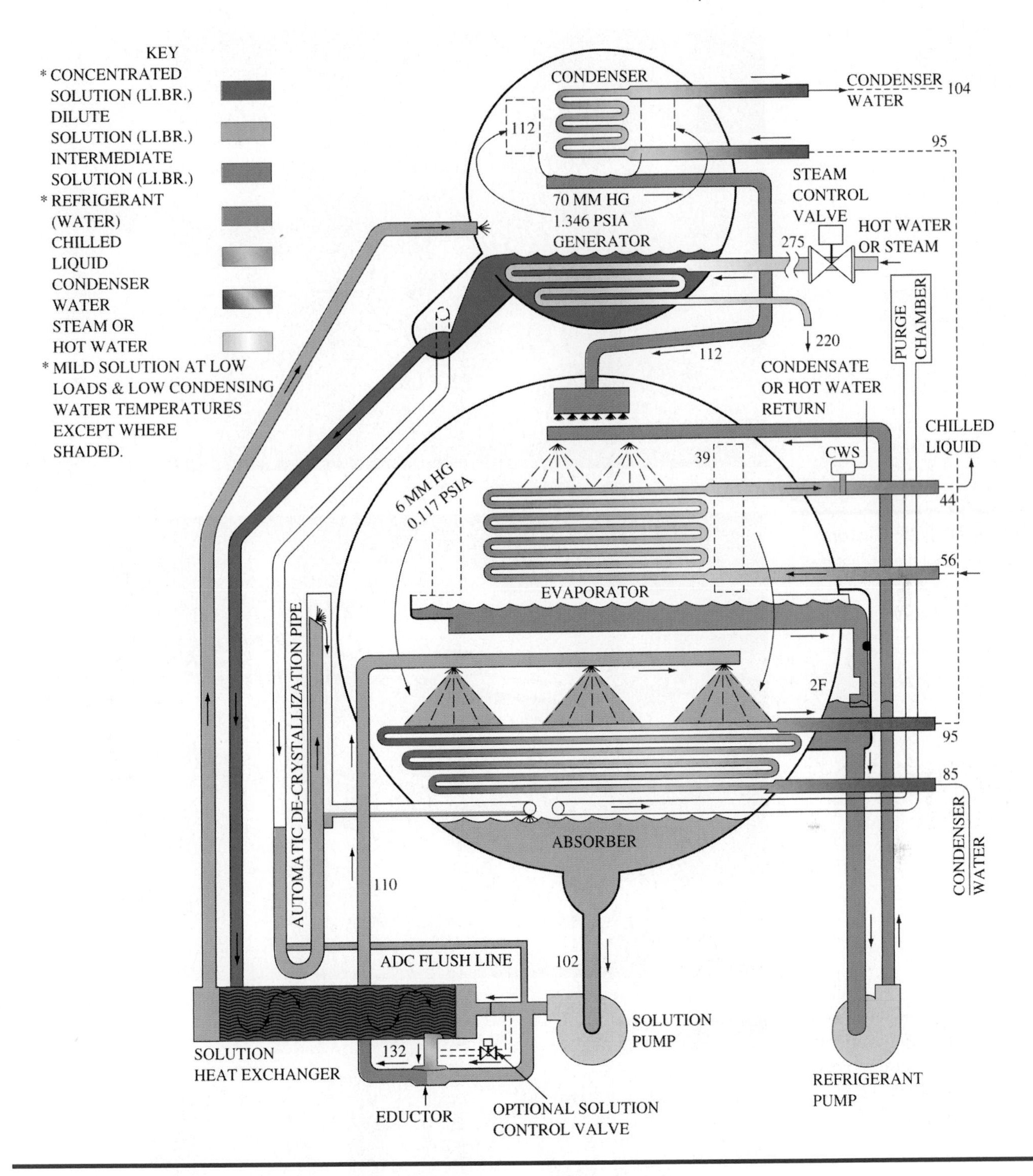

Figure 18-1-23 Schematic diagram of an absorption system using water as the refrigerant and lithium bromide solution as the absorbent.

11. How is a closed circuit evaporative cooler different from an evaporative condenser?
12. What is a good example of a common use of a secondary refrigerant?
13. _______ is the name given a solution of a substance in water that lowers the freezing point of the water.
14. What are the two types of semiconductors?
15. Where can thermoelectric refrigerators often be found?
16. What are the two basic ways that refrigerants can be applied?
17. _______ systems use a heat source for power to operate the cycle rather than mechanical energy as required for compression refrigeration systems.
18. Where are absorption systems used?

Figure 18-1-24 Industrial quality compressor unit. *(Photo provided courtesy of Vilter Manufacturing Corp.)*

Figure 18-1-25 Custom built water chiller. *(Photo provided courtesy of Vilter Manufacturing Corp.)*

Figure 18-1-26 Ammonia refrigeration system. *(Photo provided courtesy of Vilter Manufacturing Corp.)*

UNIT 2

Refrigeration Equipment

OBJECTIVES: In this unit the reader will learn about refrigeration equipment including:

- Classification of Equipment
- Refrigerated Food Storage Units

CLASSIFICATION OF EQUIPMENT

There are many ways to classify refrigeration equipment. One of the simplest ways is to separate packaged equipment from field assembled equipment since installation requirements are different. Another method is to separate all food refrigeration equipment from industrial processing equipment. This, however, results in considerable overlapping. Equipment can also be separated by the temperature it maintains, such as medium temperature for fresh food storage, low temperature for frozen foods and ultra low temperature for industrial processing. In each category the same equipment can be used for various applications and the applications govern the modifications of the equipment to perform the assigned task.

NICE TO KNOW

Most refrigeration and freezer equipment used in food handling, storage, and preparation is stainless steel. Stainless steel has its advantages to the owner in that it is more easily sterilized using harsh chemicals than painted boxes. Some manufacturers offer product lines with both painted and stainless finishes as options. For some applications stainless steel is required by local or state codes.

The discussion that follows presents mainly factory packaged units and the applications to which they apply. Some of these packages perform the entire task, such as a drinking water cooler. Other packages are connected to components on the job, such as a cooling brine for an ice rink. The important element is that the technician view the equipment in terms of the entire system when performing his/her work, so that the system operates to accomplish its design requirements. The following equipment is presented in this context:

1. Refrigerated food storage units
   - Reach in cabinets
   - Walk in coolers
2. Ice making units
   - Cube ice machines
   - Flake ice machines
3. Supermarket equipment
   - Display cases
   - Rack mounted condensing units
   - Parallel compressor units
   - Protocol system
4. Packaged liquid chillers
   - Air cooled
   - Water cooled
5. Industrial refrigeration units
6. Drinking water coolers
7. Restaurant/bar equipment
8. Dispensing freezers
   - Soft serve ice cream
   - Slush machine
   - Shake machine
   - Hard serve ice cream

REFRIGERATED FOOD STORAGE UNITS

The use of refrigerated storage of both fresh and frozen food products constitutes a large part of the refrigeration industry. This was the first use of refrigeration and this segment of the business now offers many variations.

In commercial refrigeration these uses can be divided into the following product groups: (1) reach in refrigerators and freezers, (2) walk in refrigerators and freezers, (3) storage warehouses.

Reach In Refrigerators and Freezers

In this type of unit the refrigeration is contained in the top section. With the single-door arrangement the door can be installed to open either to the left or to the right. Care must be exercised to locate the unit where the floor offers adequate support, since some models can weigh as much as 3000 lb.

Adjustable legs are furnished to permit leveling the unit. Leveling is important to permit proper draining of the evaporator condensate. A floor drain connection is needed or a condensate vaporizer can be used. The unit needs to be wired with a plug in connector located on the electrical box. The unit must be connected to a separately fused circuit on the electrical panel. A drain line heater is used on freezer models.

The storage capacity typically ranges from 200 ft^3 to 700 ft^3. The compressors for the refrigerators are ¼, ⅓, and ½ Hp and for the freezers ½, ¾, and 1 Hp, respectively. All units can run on 115 V/60 Hz/1 phase power. The 1 Hp size is also available for 208 230 V/60 Hz/1 phase power.

The thermal sensing bulb for the refrigerator is located on the evaporator coil and thermostat for the freezer is in the air. The refrigerator is controlled to maintain 38°F and the freezer 0°F. The refrigerator uses air defrost during the OFF cycle. The freezer uses electrical defrost. A high pressure cut off is standard on all models with a limit setting of 440 psi.

The wiring diagrams are as follows: refrigerator cooling cycle Figure 18-2-1; freezer cooling cycle Figure 18-2-2; freezer defrost cycle Figure 18-2-3.

The sequence of operation is as follows.

Refrigerator Cooling With the main switch in the "on" position, the current flows through the high pressure cut out to the relay. When the relay is energized the NO contacts close and with the thermostat calling for cooling, the compressor and condenser fan start. The evaporator fan stays on unless a door is opened. When a door opens the fan goes off and the interior light goes on.

When the thermostat is satisfied, the compressor and condenser fan are stopped. They restart when the thermostat reaches 38°F.

Freezer Cooling With the main switch in the "on" position, the current energizes the automatic defrost time clock, as shown in Figure 18-2-4. At the same time the current flows through the high pressure cut out, energizing the relay. With contact "4" on the time clock closed and the thermostat calling for cooling, the refrigeration will start. With contact "N" closed on the time clock, current will flow through the NO contacts of the light/fan switch (with the door closed). When the evaporator coil temperature reaches 30–35°F the defrost end and fan delay thermostat close, energizing the evaporator fan.

Freezer Defrosting At preset times the time clock contacts will switch to defrost the evaporator. Contact "N" opens, stopping the evaporator fan. Contact "4" opens on the time clock, stopping the refrigeration system. Contact "1" closes, energizing the defrost heater. As the evaporator temperature rises, the defrost end and fan delay thermostat will open, terminating the defrost cycle. Contacts "N" and "4" will close. Contact "1" will open, de-energizing the defrost heater and starting the refrigeration system. The system will then be under the control of the thermostat.

SERVICE TIP

Refrigerator coolers are designed to operate at temperatures above freezing. These systems do not typically have mechanisms for defrost. If these boxes are operated as freezers ice will build up on the coils. In addition compressors are specifically designed to work under certain load conditions and they may not function properly if the box is changed from a refrigeration system to a freezer. Although many refrigeration and freezer compressors look similar a big difference between the two is the size of the suction port inside the compressor. Freezers require larger suction ports because they operate at lower suction pressures. The larger port allows more of the low pressure refrigerant in the piston. If one of these compressors is used in a refrigeration application the compressor can overheat due to the excessively large load. Before making such a change refer to the manufacturer's literature or the compressor to make sure it can operate in the new temperature range.

Walk In Refrigerators and Freezers

Walk in coolers and freezers are made with modular insulated metal clad panels. Using corner sections and standard wall and ceiling panels, various sized boxes can be made. Panels are assembled on the job using special eccentric cam fasteners. Insulated panels are used for the floors. The interior surfaces of wall and ceiling panels are usually clad with aluminum, floors with galvanized iron.

An assembled walk in cooler is shown in Figure 18-2-5. The sectional panels are shown in Figure 18-2-6. The installation of a partition panel is shown in Figure 18-2-7. This makes it possible to include a cooler and a freezer in the same structure. By entering the cooler first and then the freezer, more economical operation is possible since a common wall is used.

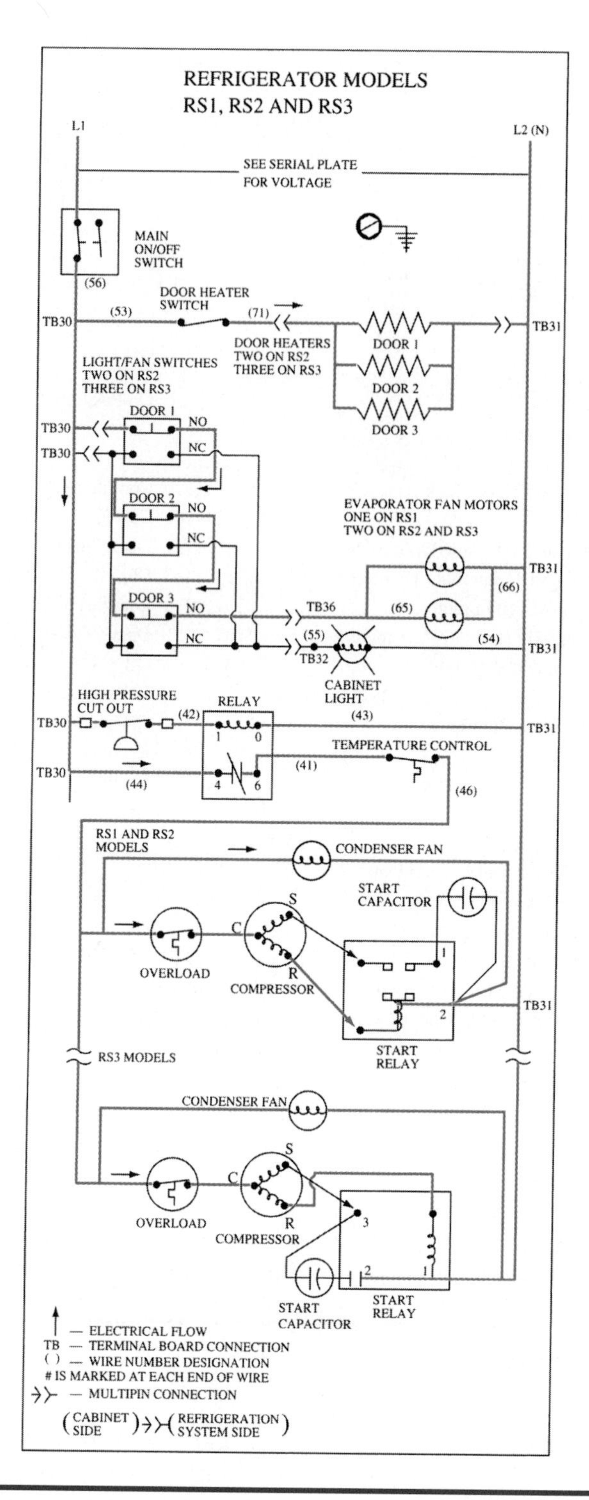

1. Main ON/OFF Switch .... On Position
2. Door Heater Switch (Optional On/Off) .... On
   A. Door Heater .... On
3. Light/Fan Switch(es) (Door Switches) .... Door Closed
   A. Normally Open (N.O.) Contact .... Closed
      1. Evaporator Fan Motor .... Energized
   B. Normally Closed (N.C.) Contact .... Open
      1. Interior Cabinet Light .... Off
4. High Pressure Cut-Out .... Closed
5. Relay .... Energized
   A. Normally Open (N.O.) Contacts .... Closed
6. Temperature Control .... Closed
   A. Compressor .... Energized
   B. Condenser Fan Motor .... Energized

Figure 18-2-1 Wiring diagram for reach in refrigerator cooling cycle. *(Courtesy of The Manitowoc Company)*

CAUTION: Local, state, and federal laws require that all walk in refrigerators and coolers have safety latches that can be opened from the inside. It's a good practice when working alone to confirm that the latch is operating properly so that the door can be opened easily from the inside. There may also be a safety indicator visible from the outside as shown in Figure 18-2-8.

Storage Warehouses

Most modern refrigerated warehouses are one story structures. A typical floor plan is shown in Figure 18-2-9. The building construction is shown in Figure 18-2-10. Note that the walls are constructed so that as few structural members as possible penetrate the insulated envelope. Insulated panels applied to the outside structural frame prevent conduction through

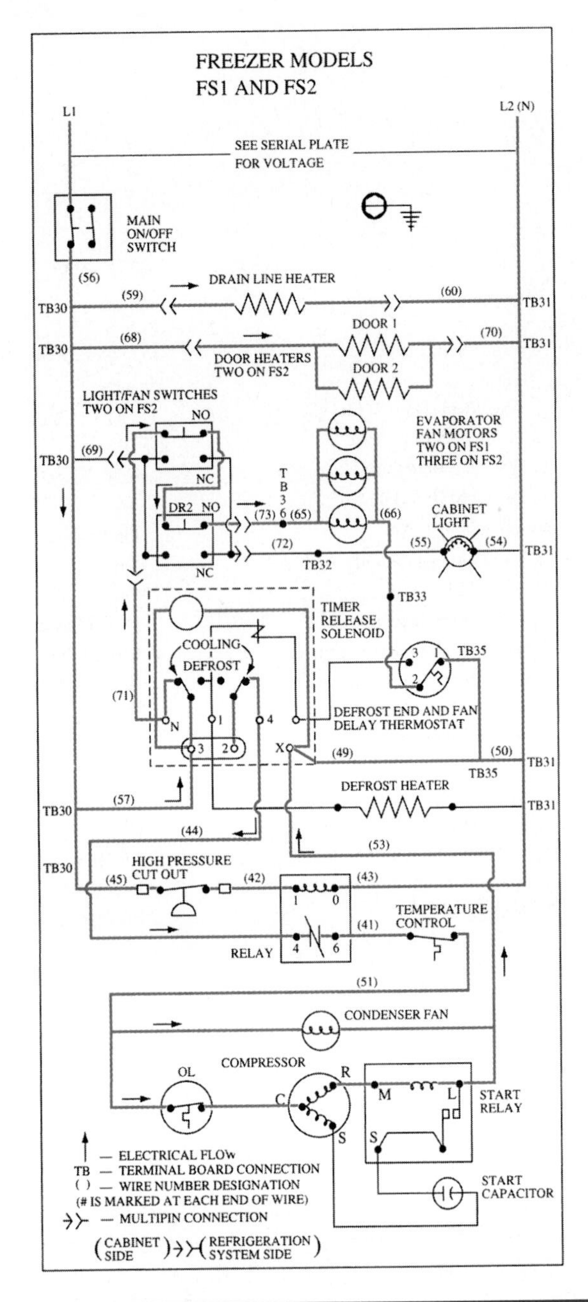

1. Main ON/OFF Switch	On Position
A. Drain Line Heater	On
B. Door Heater(s)	On
2. Automatic Defrost Time Clock	Cooling cycle
A. Contact "N"	Closed
1. Light/Fan Switch(es) (Door Switch(es))	Door Closed
a. Normally open (N.O.) contact	Closed
Evaporator fan motor(s)	Energized
b. Normally closed (N.C.) contact	Open
Interior cabinet light	Off
B. Contact #4	Closed
1. Temperature Control	Closed
a. Compressor	Energized
b. Condenser fan motor(s)	Energized
C. Contact #1	Open
1. Defrost Heater	Off
3. Defrost End and Fan Delay Thermostat	
A. Normally Closed (N.C.) Contacts	Closed
B. Normally Open (N.O.) Contacts	Open
4. High Pressure Cut-out	Closed
5. Relay	Energized
A. Normally Open (N.O.) Contact	Closed

Figure 18-2-2 Wiring diagram for reach in freezer cooling cycle. (*Courtesy of The Manitowoc Company*)

the framing. It is usually convenient to locate the refrigeration equipment in a penthouse room, as shown in Figure 18-2-11.

UNIT 2—REVIEW QUESTIONS

1. What is one of the simplest ways to classify refrigeration equipment?
2. What is the storage capacity of reach in refrigerators and freezers?
3. Why must care be exercised to locate the unit where the floor offers adequate support?
4. Where is the thermal sensing bulb in a reach in refrigerator and freezer located?
5. What temperature must be maintained in a reach in refrigerator and freezer?
6. At what temperature does the compressor and condenser fan restart?
7. What is the interior surface of the walls and ceiling of a walk in refrigerator and freezer made of?
8. Why is a partition panel added to walk in refrigerators?

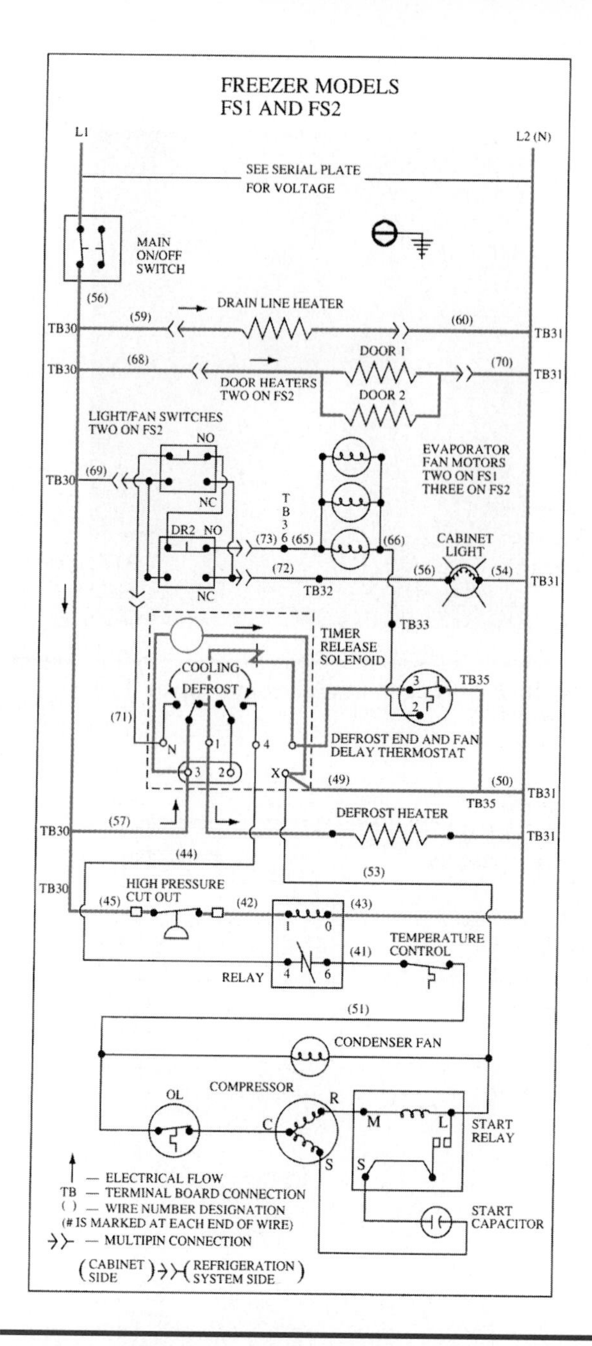

1. Main ON/OFF Switch .... On Position
   - A. Drain Line Heater .... On
   - B. Door Heater(s) .... On
2. Automatic Defrost Time Clock .... Defrost Cycle
   - A. Contact "N" .... Open
     - 1. Light/Fan Switch(es) (Door Switch(es)) .... Door Closed
       - a. Normally open (N.O.) contacts .... Closed
         - Evaporator fan motor(s) .... De-Energized
       - b. Normally closed (N.C.) contacts .... Open
         - Interior cabinet light .... Off
   - B. Contact #4 .... Open
     - 1. Temperature Control .... Closed
       - a. Compressor .... De-Energized
       - b. Condenser fan motor(s) .... De-Energized
   - C. Contact #1 .... Closed
     - 1. Defrost heater .... On
3. Defrost End and Fan Delay Thermostat (end of defrost)
   - A. Normally Closed (N.C.) Contacts .... Open
   - B. Normally Open (N.O.) Contacts .... Closed
4. High Pressure Cut-out .... Closed
5. Relay .... Energized
   - A. Normally Open (N.O.) Contact .... Closed

Figure 18-2-3 Wiring diagram for reach in freezer defrost cycle. *(Courtesy of The Manitowoc Company)*

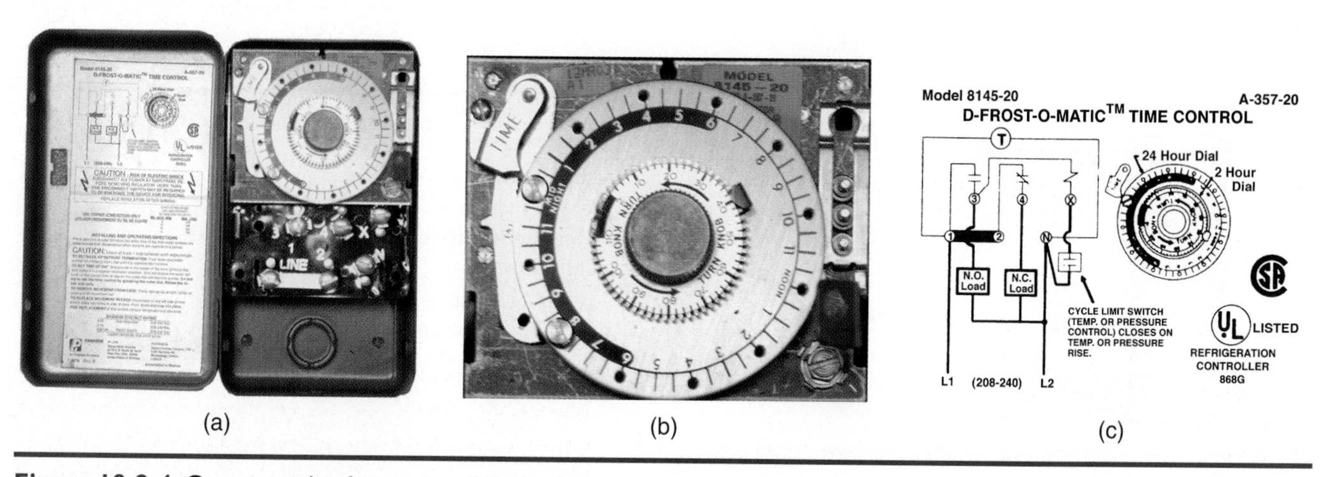

Figure 18-2-4 Commercial refrigeration defrost timer.

Figure 18-2-5 Walk in cooler/freezer.

Figure 18-2-8 Walk-in freezer/cooler alarm.

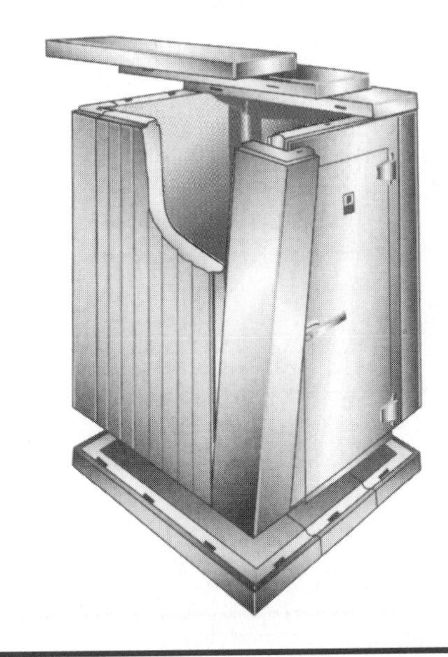

Figure 18-2-6 Sectional panels for a walk in cooler.

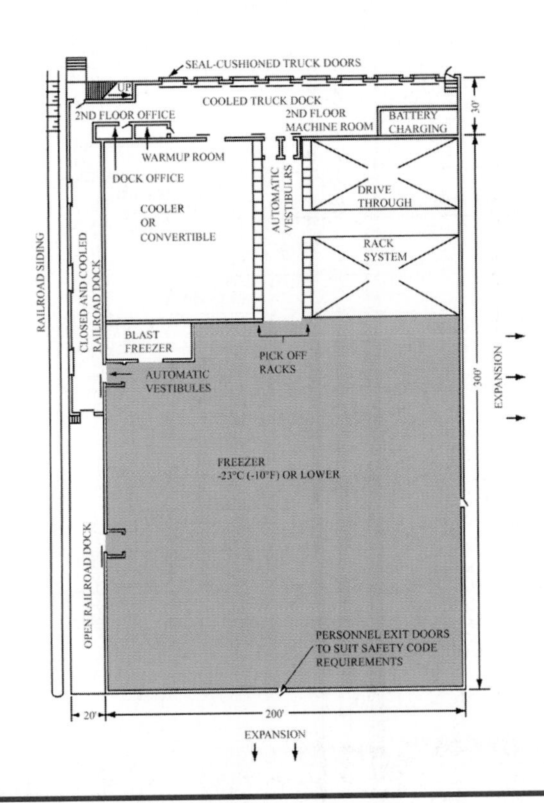

Figure 18-2-9 One story refrigerated warehouse.

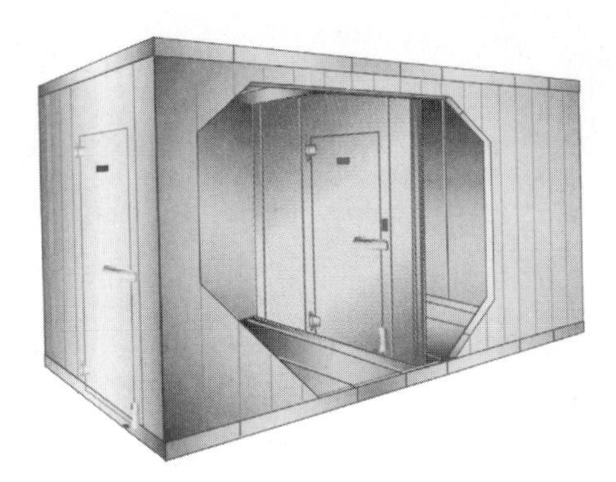

Figure 18-2-7 Partition location in a walk in cooler.

9. Most modern refrigerated warehouses are ______ structures.
10. What is applied to the outside structural frame of a storage warehouse to prevent conduction through the framing?

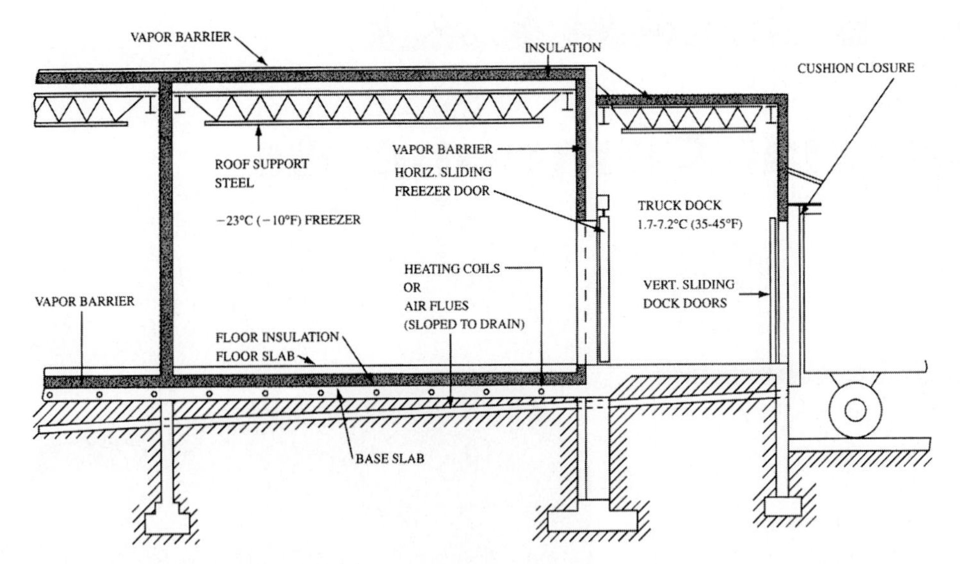

Figure 18-2-10 Refrigerated warehouse construction.

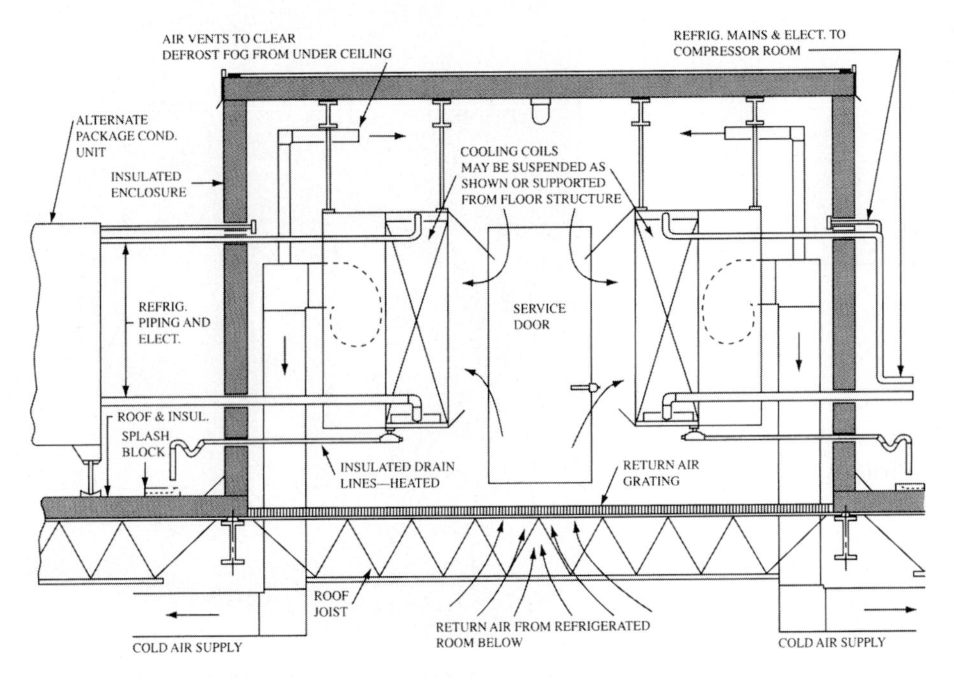

Figure 18-2-11 Penthouse equipment room for a refrigerated warehouse.

TECH TALK

Tech 1. Why does the fan run continuously in all the walk in coolers?

Tech 2. Well, the fan runs so that the temperature in the box stays consistent from one side to the other and from floor to ceiling. If the fan didn't run the cooler might develop warm spots. That wouldn't be good for product storage.

Tech 1. Oh, is that the only reason?

Tech 2. No, actually the other reason that it runs is so that the temperature sensor in the coil can tell when the box temperature has begun to rise. That way the compressor can come on to maintain temperature. If you didn't keep the air moving it would be real difficult to decide where to put the thermostat to keep the box running at the right temperature.

Tech 1. Thanks, I always wondered why they run all the time. I thought there may have been a problem at first until I saw all the boxes doing that, then I knew that there had to be a reason.

UNIT 3

Special Components

OBJECTIVES: In this unit the reader will learn about special components including:

- Defrost Systems
- Condenser Pressure Temperature Control
- Evaporator Pressure Temperature Control
- Low Ambient Controls
- High/Low Pressure Controls
- Suction Accumulator
- Suction to Liquid Heat Exchanger
- Oil Separator
- Receivers and Valves
- Hot Gas Mufflers
- Heat Recovery Units
- Crankcase Pressure Regulator

DEFROST SYSTEMS

The most popular types of defrost systems are air, electric, and hot gas. Other defrost systems include glycol, water, and reverse cycle.

The glycol type requires special apparatus to heat the glycol solution, remove the excess water that is absorbed, and maintain the proper concentration of glycol.

The water defrost system uses a relatively large quantity of water, relying on the water to quickly move through the unit, melting the ice and reaching the drain before it freezes.

Both the glycol and water defrost arrangements have gradually been replaced by the electric and hot gas methods.

The reverse cycle defrost operates using a four way valve to divert discharge gas to the evaporator for defrosting. The hot gas method is preferred over the reverse cycle since it offers a closer control of defrost temperatures.

A typical electric defrost unit cooler is shown in Figure 18-3-1.

Figure 18-3-1 Electric defrost unit cooler. (*Courtesy of Hampden Engineering Corporation*)

CONDENSER PRESSURE TEMPERATURE CONTROL

For *air cooled condensers* some of the options for controlling condenser pressure are: fan cycling, air volume control, fan speed control, and refrigerant flooding. For *water cooled condensers* a water regulating valve can be used. Figure 18-3-2a is a picture

SERVICE TIP

Defrosting the evaporator coil can accomplish two things: (1) Remove the ice buildup to improve the efficiency, and (2) aid in the return of refrigerant oil. As ice builds up on the evaporator, the coil temperature decreases because the ice acts as an insulator. As the coil temperature decreases so does the velocity of the returning refrigerant vapor. The returning vapor velocity can decrease enough so that oil is not being returned to the compressor. When this happens oil builds up in the evaporator. When oil builds up in the evaporator it is called oil logging. The defrost cycle warms the evaporator and increases the velocity of the returning refrigerant vapor which will bring along with it the oil. In some cases oil buildup in the evaporator can occur under normal conditions. If this happens it may be necessary to schedule more frequent defrost cycles that are required to keep the coil ice free. These defrosting cycles are used to harvest the oil from the evaporator.

of a water valve, and Figure 18-3-2b is a cutaway view of the valve.

On a cooling tower system the water temperature is controlled by cycling the fan or water bypass. On evaporative condensers head pressure is controlled by air supply dampers.

EVAPORATOR PRESSURE TEMPERATURE CONTROL

Evaporator pressure temperature control usually requires an evaporator pressure regulator (EPR) valve such as shown in Figure 18-3-3. It is used primarily on multiple evaporator systems where different temperatures are maintained in each unit. A two temperature control can be applied where only two box temperatures are needed.

CRANKCASE PRESSURE REGULATOR

The installation of a crankcase pressure regulator (CPR) valve in the suction line to the compressor is shown in Figure 18-3-4b. Examples of this valve are shown in Figure 18-3-4b–g. This type of valve offers an automatic means of preventing the overloading of the compressor on startup. The function of this valve is similar to the practice of manually throttling the suction service valve until the machine can handle the load. Either of these arrangements will increase the pulldown time, but they are important for protecting the compressor.

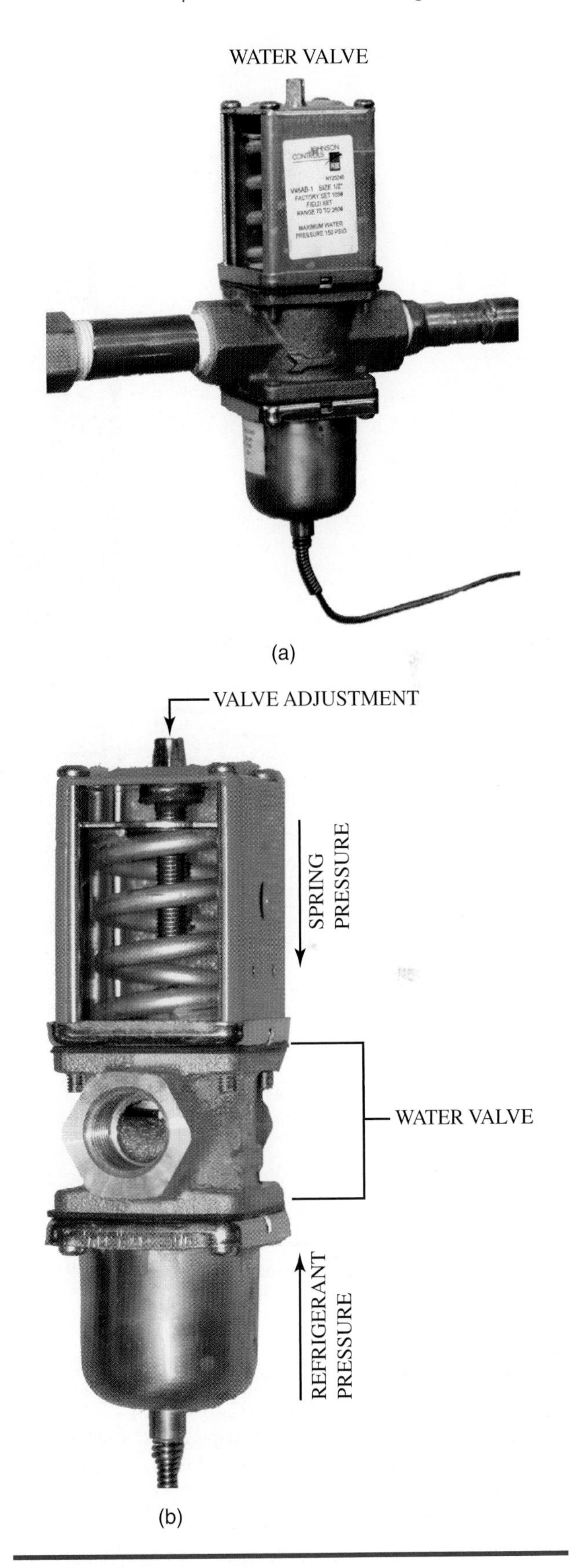

Figure 18-3-2 Condenser water regulating valve used for head pressure control. (*Courtesy of Hampden Engineering Corporation*)

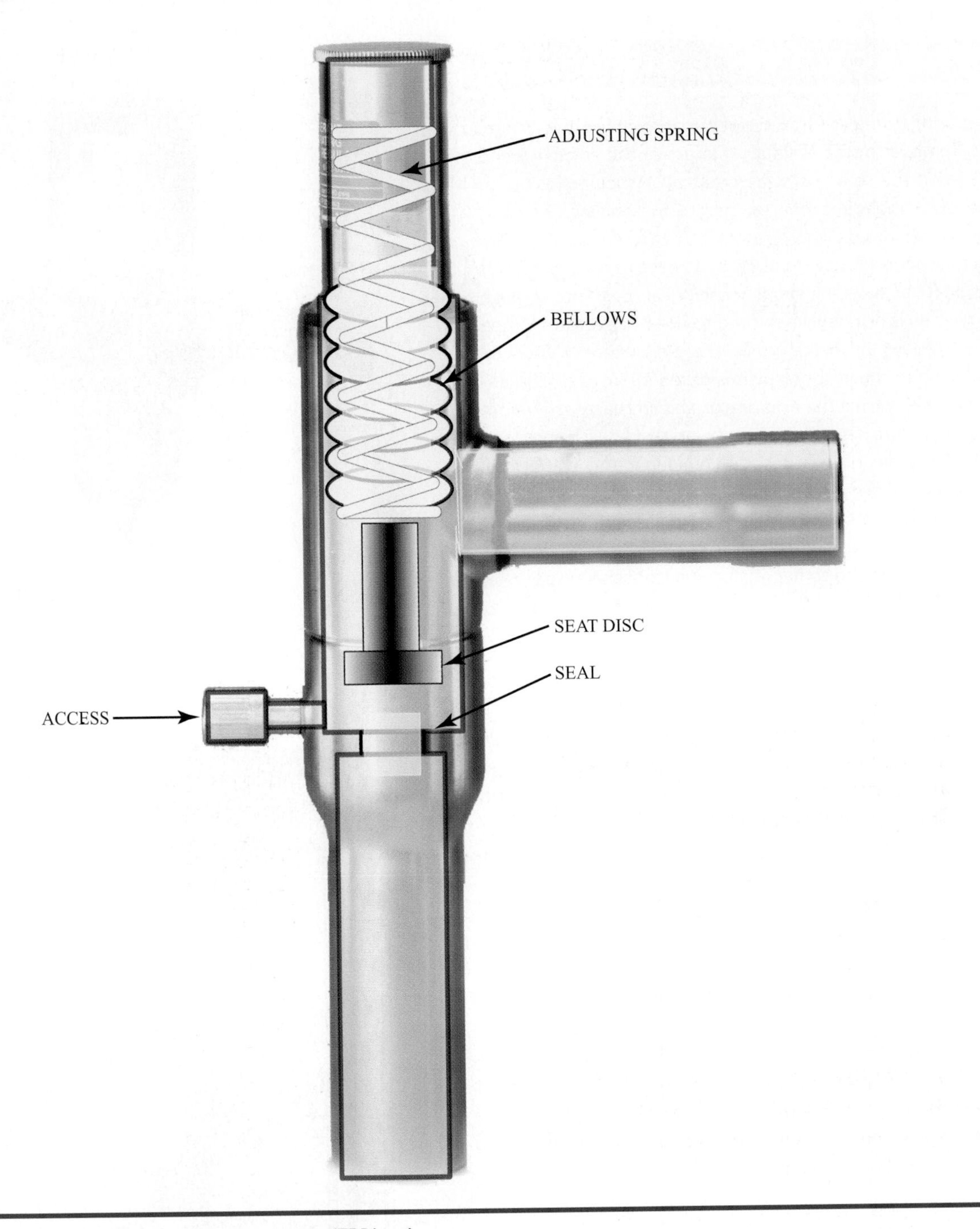

Figure 18-3-3 Evaporator pressure regulating (EPR) valve.

SERVICE TIP

Although many evaporator pressure regulators and crankcase pressure regulators look exactly alike they are not. They both have significantly different operating functions and cannot be interchanged. When doing a new installation that has both EPR and CPR scheduled for installation you must make certain that they are being placed in the correct location prior to brazing them in place.

LOW AMBIENT CONTROLS

Where air cooled condensers are subjected to low outside temperatures, the condensing pressure can drop below the setting of the low limit control and prevent starting. To eliminate this problem, some type of head pressure control needs to be employed. Under extremely cold conditions the partially flooded evaporator needs to be used to maintain high enough condensing pressure to permit the compressor to start.

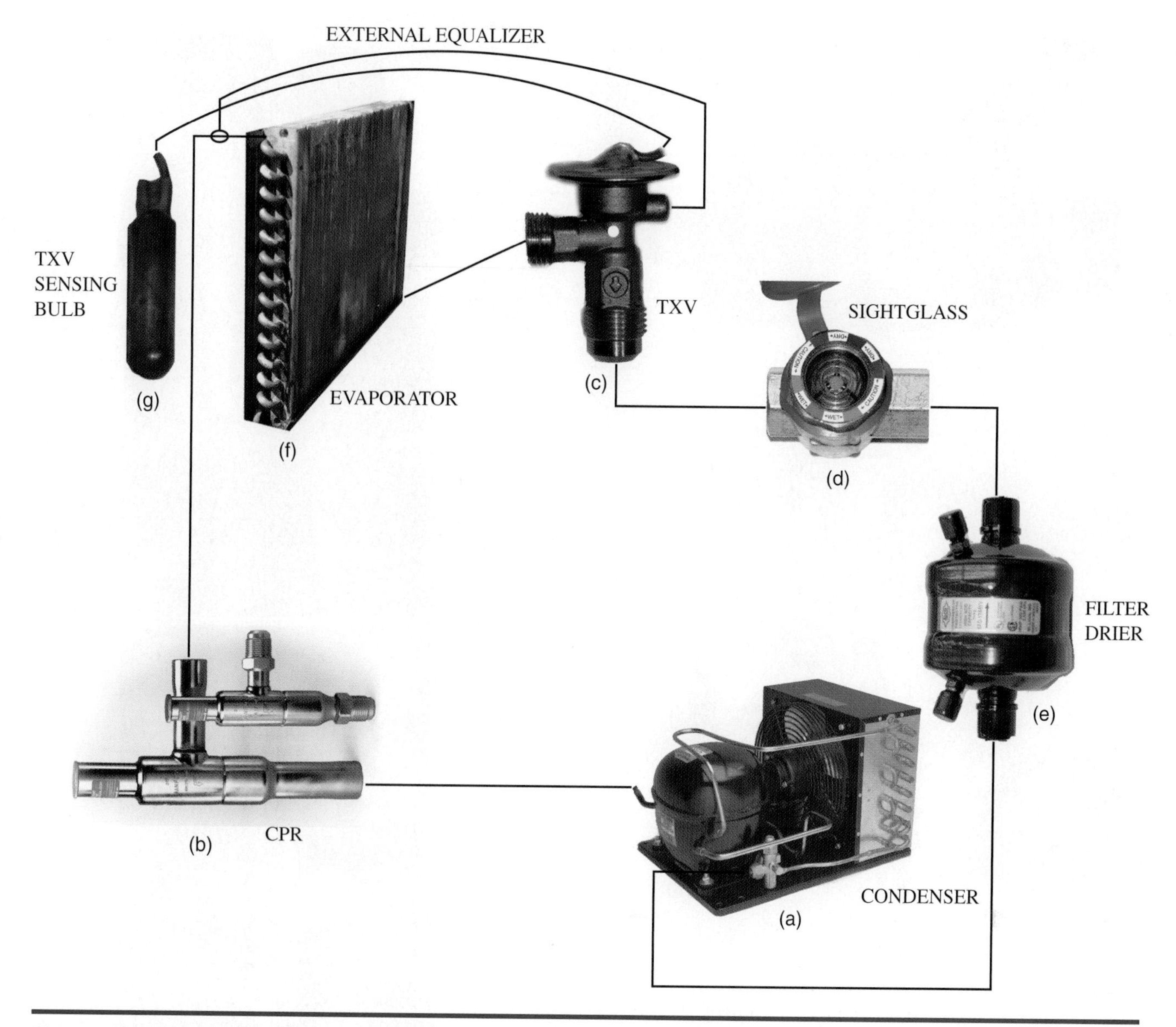

Figure 18-3-4 Crankcase pressure regulator used to limit capacity on start-up: (a) Condenser; (b) CPR; (c) TXV; (d) Sightglass; (e) Filter drier; (f) Evaporator; (g) TXV sensing bulb.

HIGH/LOW PRESSURE CONTROLS

High/low pressure controls, Figure 18-3-5a,b, are to be used as a safety control and not as an operating control. It is usually considered to be standard equipment on most commercial compressor systems. It stops the compressor on excessive pressures or extremely low pressures. For example, it would stop the compressor on low pressure if there were a loss of refrigerant. It would stop the compressor on high pressure if the discharge of the compressor were blocked for any reason.

Many small refrigerator units use low pressure controls as operating controls, shutting down when the evaporator temperature drops below its minimum setpoint.

SUCTION ACCUMULATOR

The suction accumulator, Figure 18-3-6, is placed in the suction line close to the compressor suction inlet to catch any liquid refrigerant that would cause slugging in the compressor. This equipment is important particularly on startup since liquid may accumulate in the evaporator during the OFF cycle. Using a suction accumulator not only protects the compressor but also permits low superheat, thus increasing the capacity of the evaporator.

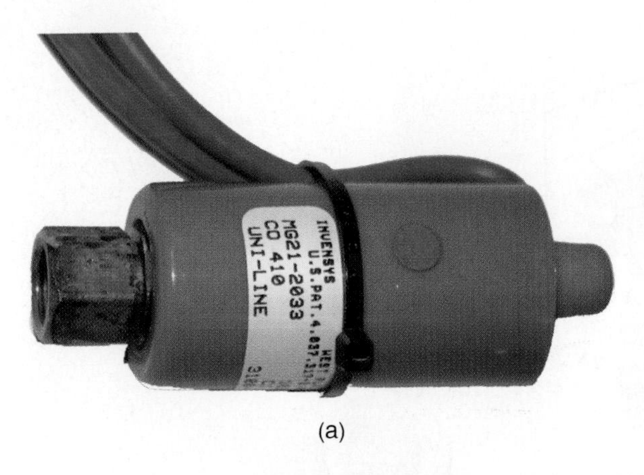

(a)

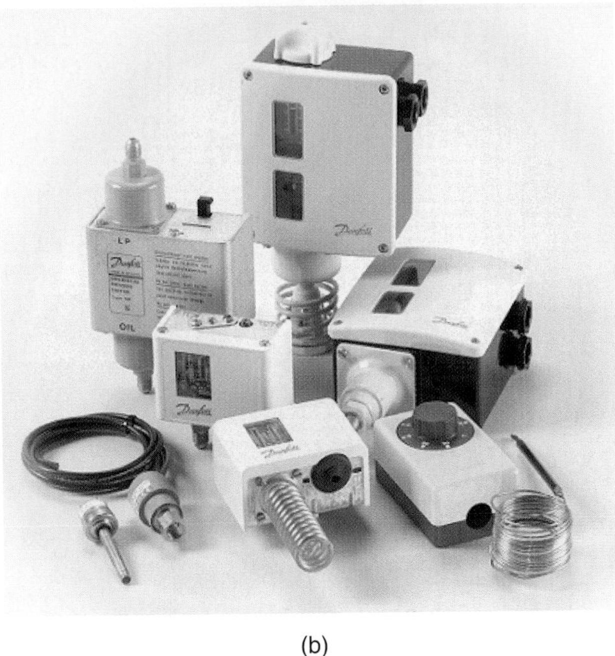

(b)

Figure 18-3-5 High/low pressure limit control.

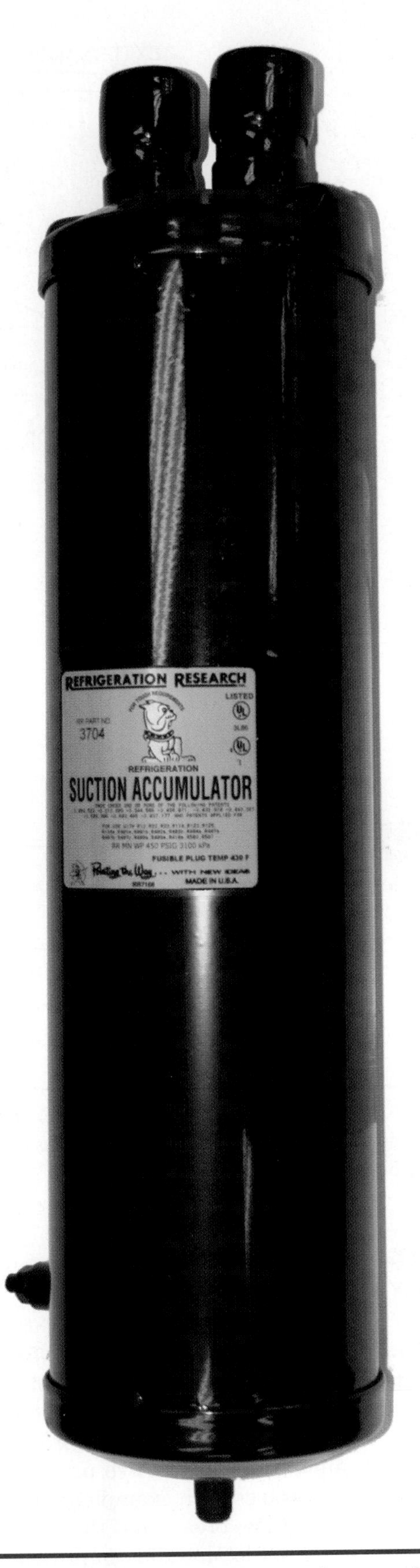

Figure 18-3-6 Suction pressure accumulator used to prevent liquid refrigerant from entering the suction of the compressor. (*Courtesy of Rheem Manufacturing*)

SUCTION TO LIQUID HEAT EXCHANGER

The suction to liquid heat exchanger, Figure 18-3-7a,b, serves to subcool the liquid refrigerant and superheat the suction gas, as well as increase the capacity of the system. It is commonly used in commercial refrigeration evaporators.

OIL SEPARATOR

Oil separators, Figure 18-3-8, are placed on the discharge of the compressor to remove excess oil being carried with the discharge gas. The oil that is removed is automatically returned to the crankcase of the compressor. Certain compressors, such as the screw type particularly, require their use.

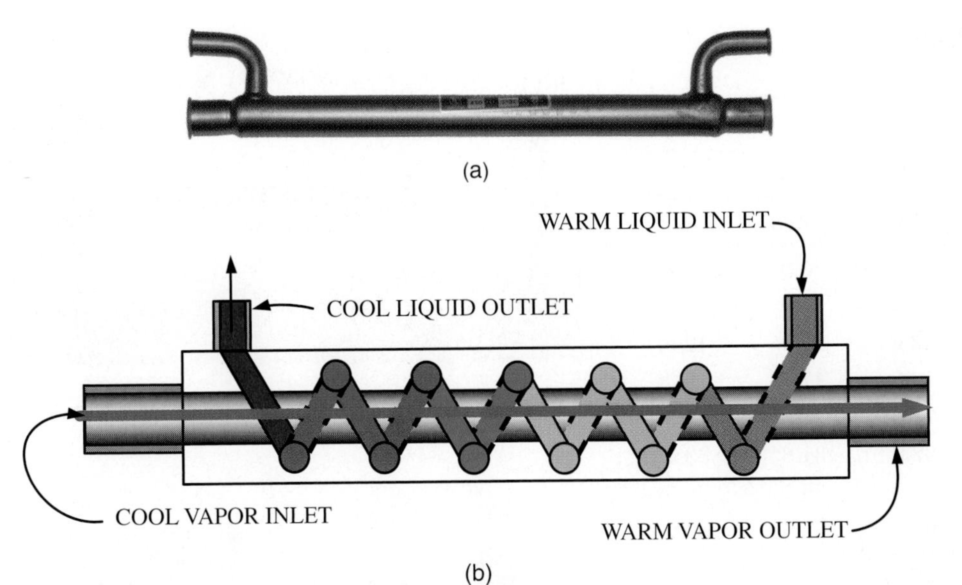

Figure 18-3-7 Suction to liquid line heat exchanger used to subcool the liquid and superheat the suction gas.

TECH TIP

An oil separator may be required in refrigeration systems that have extremely long refrigerant lines. When a compressor starts, much of its oil is pumped out during the first moments it is operating. The oil pressure switch has a time delay that allows the oil to flow out to the evaporator and back to the compressor. If the oil is not returned, the compressor will shut down on oil safety. This can frequently occur and is often deemed by the equipment owner as a nuisance call. An oil separator can capture the oil as it leaves the compressor and return it to the crankcase rapidly enough to prevent an oil safety lockout from occurring.

Figure 18-3-8 Oil separators used to remove oil from the discharge gas. (*Courtesy Danfoss Inc.*)

RECEIVERS AND VALVES

Liquid receivers, Figure 18-3-9, provide a storage tank for the refrigerant in the system.

The receiver needs to have a sufficient volume to contain the refrigerant when the system needs to be opened up for service. Table 18-3-1 shows the recommended receiver sizes for various systems. The receiver outlet pipe inside the receiver runs to the bottom of the tank, as shown in Figure 18-3-10. This opening has a fine mesh screen to prevent contamination from entering.

Receivers usually are equipped with both an inlet and an outlet valve. The outlet valve is called a king valve. A system can be charged with liquid refrigerant through the gauge port on the king valve. When this is done the king valve is closed or front seated to permit the system to draw refrigerant from the attached cylinder.

Receivers are equipped with safety devices. The minimum requirement is a fusible plug, although receivers may also have a pressure relief valve. If the receiver is located inside the building, a pipe to the outside is connected to the relief valve.

TABLE 18-3-1 Recommended Liquid Receiver Volumes

		Weight (lb)			
		Refrigerant			
Hp	Volume (in^2)	R-12	R-22	R-500	R-502
½	150	6.8	6.2	5.9	6.3
¾	225	10.3	9.3	8.9	9.4
1	300	13.7	12.4	11.9	12.9
1½	450	20.5	18.6	17.9	19.3
2	600	27.4	24.8	23.8	25.8
3	750	35.0	32.0	31.8	33.0
5	900	41.0	37.0	35.5	38.5
7½	1500	70.0	64.0	61.6	66.0

(Courtesy of Standard Refrigeration Co.)

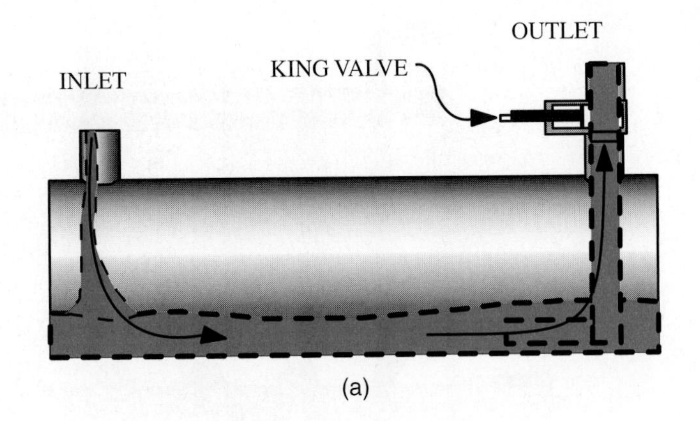

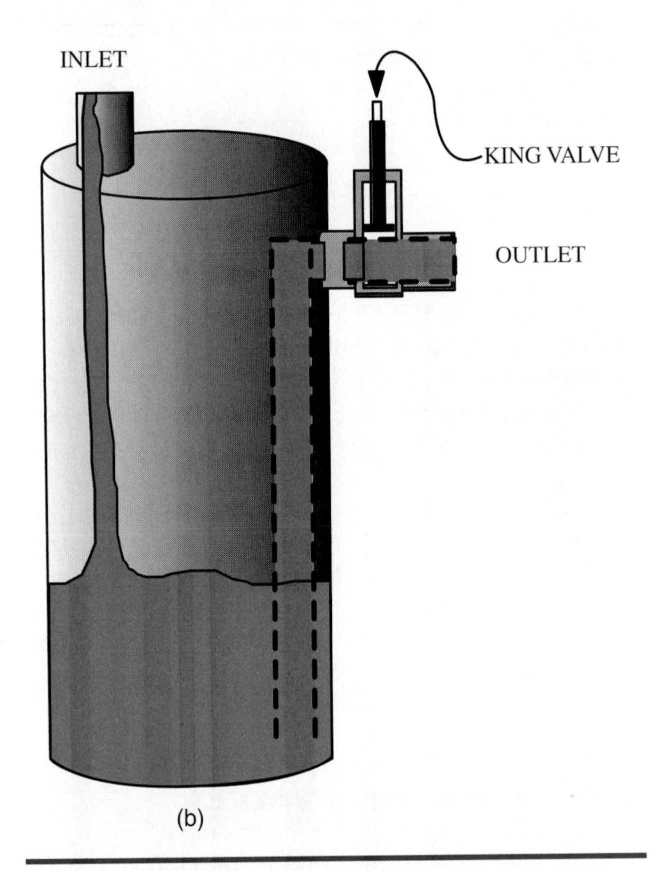

Figure 18-3-10 Common types of receivers showing the outlet connection going to the bottom of the tank: (a) Horizontal liquid receiver; (b) Vertical liquid receiver.

Figure 18-3-9 Liquid receiver showing inlet and outlet valves and safety relief.

On systems with air cooled condensers, using a flooded condenser arrangement for low ambient control, the receiver may be located under the condenser. It has sufficient capacity to hold the added refrigerant charge.

HOT GAS MUFFLERS

Mufflers, Figure 18-3-11, are used in the discharge piping from compressors with remote air cooled condensers located outside. To some extent they are valuable for reducing sound, but their greatest value is in dampening vibration from the compressor that is carried along the discharge line.

HEAT RECOVERY UNITS

A typical heat recovery system for a supermarket is shown in Figure 18-3-12. Since the refrigeration equipment is used all year, excess heat that would normally be dissipated by the condenser can be diverted to heat the market in the winter.

Figure 18-3-11 Hot gas line muffler used primarily with remote air cooled condensers. (*Courtesy of AC & R Components, Inc.*)

The diagram in Figure 18-3-12 shows the hot gas from the compressor bypassed to the heat recovery air handling unit. A pressure regulator is used to maintain a sufficient discharge gas temperature for effective heating.

UNIT 3—REVIEW QUESTIONS

1. What are the most popular types of defrost systems?
2. List the options for controlling condenser pressure.
3. What controls the water temperature on a cooling tower system?
4. An evaporator pressure regulator valve is primarily used on _______.
5. Why does a partially flooded evaporator need to be used in extremely cold conditions?
6. _______ is a safety control and is not to be used as an operating control.
7. What is the purpose of a suction to liquid exchanger?
8. _______ are placed on the discharge of the compressor to remove excess oil being carried with the discharge gas.
9. Where may the receiver be located on systems with air cooled condensers, using a flooded condenser arrangement for low ambient control?
10. A _______ is used to maintain a sufficient discharge gas temperature for effective heating.

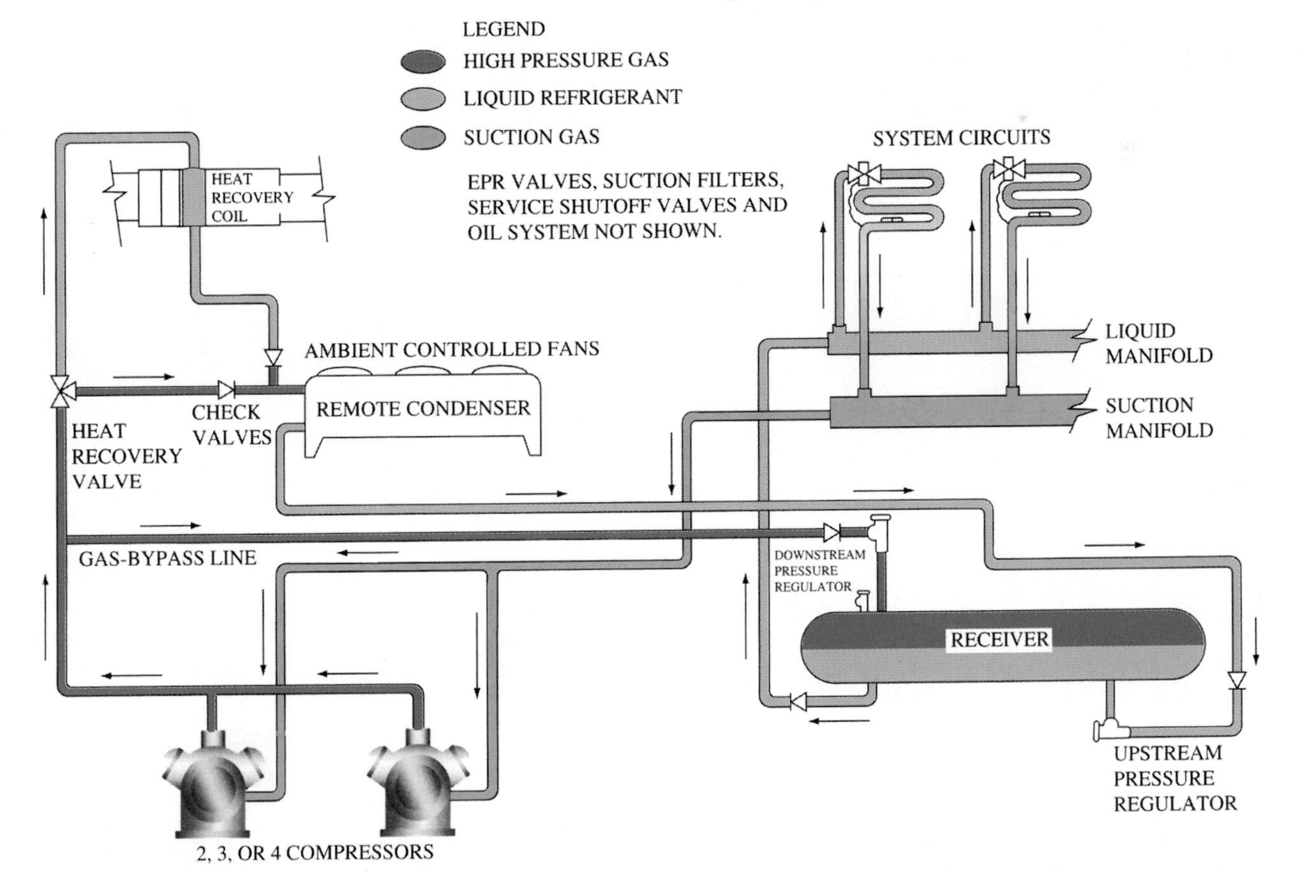

Figure 18-3-12 Heat recovery system for supermarket installation. (*Copyright by the American Society of Heating, Refrigerating, and Air-Conditioning Engineers, Inc.*)

TECH TALK

Tech 1. I noticed that a lot of the refrigeration systems that I work on have very few accessories. Aren't accessories a good idea? Don't they help the system?

Tech 2. Well, first of all, every time you add an accessory to a refrigeration system it acts as a slight resistance to refrigerant flow, which can slightly decrease the system operational efficiency. If you put a lot of unnecessary accessories in a system you can significantly decrease the overall operating efficiency of the system enough so that marginal systems may not meet their load.

Tech 1. But how do you know what accessories a system needs?

Tech 2. First of all you have to establish whether an accessory is needed. Perhaps a system is not functioning properly. It may be too noisy and need a gas muffler. It may have floodback and need a suction accumulator. Before you decide, you need to define the problem.

Tech 1. So you have to evaluate the system carefully before making a recommendation that it needs an accessory to work.

Tech 2. That's correct. Proper evaluation of the system is the most important in making any suggestion to add a part to an operating system.

Tech 1. Thanks, I appreciate that.

UNIT 4

Restaurant and Supermarket Equipment

OBJECTIVES: In this unit the reader will learn about restaurant and supermarket equipment including:

- Dispensing Freezers
- Functions of the Refrigeration System
- Function of the Control System
- General Guide to Service
- The Protocol System
- Air Cooled Chiller
- Water Cooled Chiller
- Flake Ice Machines
- Packaged Liquid Chillers
- Ice Making Units

RESTAURANT EQUIPMENT

Normally restaurants use reach in refrigerators and freezers, and depending on the size, walk in coolers. Where the restaurant is also a bar, serving draft beer, special refrigeration equipment needs to be provided.

Bottled beer is pasteurized and can be stored in beverage coolers, which maintain a temperature of 38 to 40°F. Draft beer does not have preservatives and is a perishable product that requires refrigeration for storing and dispensing.

There are a number of kinds of draft beer dispensers. The three fundamentals common to each dispenser are the ability to satisfactory clean the equipment, temperature control, and product pressure control.

It is important that the beer be maintained at the correct temperature. It should be drawn at 38°F, to assure that when the customer receives the glass, the temperature is between 40 and 42°F.

The pressure controls the speed of the draw at the faucet. Barrels are delivered at a pressure of 12 to 15 psi. Some carbon dioxide gas needs to be added to most direct draw systems to maintain a pressure between 14 and 16 psi.

Dispensing Equipment

Bottle coolers are usually self contained with the refrigeration equipment located in a lower compartment at one end. A back bar cooler, Figure 18-4-1, with front door access is available for storing bottles or cans of beer and other items that need refrigeration.

A direct draw beer cooler is used for small bars that have space available under the counter. The refrigeration equipment is placed in a separate compartment at the end of the cabinet. Forced circulation of cold air keeps the beer cold right up to the faucet shanks.

There are two types of remote system air cooled beer lines and coolant cooled beer lines.

The air cooled remote system is shown in Figure 18-4-2. Two separate insulated ducts run side by side from the storage cooler to the dispensing cabinet. One carries the beer lines and the cold air to the tapping cabinet. The other carries the cold air back to the storage cooler. This type of system requires a sealed dispenser cabinet to prevent cold air leakage.

Figure 18-4-1 Back bar for storing beverages and food products. (*Courtesy of Nor-Lake, Inc.*)

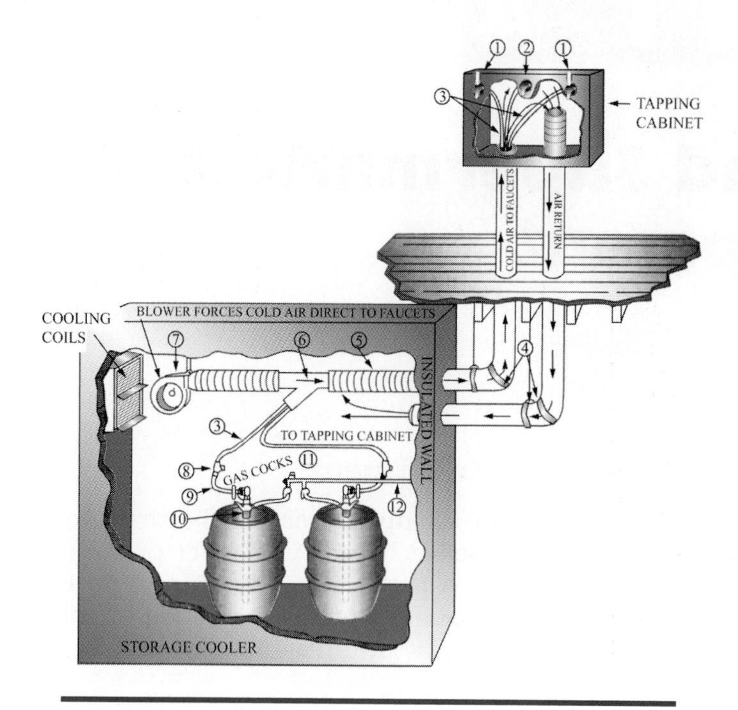

Figure 18-4-2 Air cooled remote draft beer system. *(Courtesy of Master-Bilt Products, a Division of Standex International Corporation)*

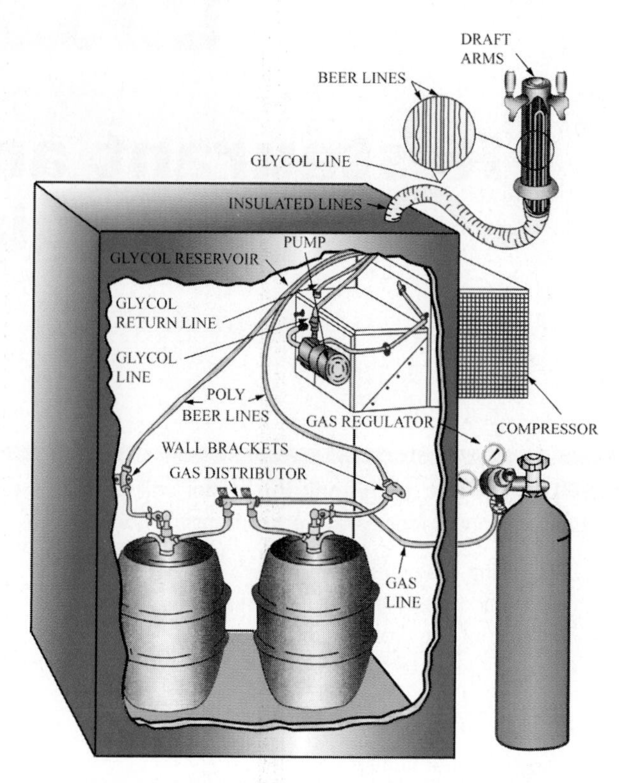

Figure 18-4-3 Coolant cooled draft beer system. *(Courtesy of Master-Bilt Products, a Division of Standex International Corporation)*

The coolant cooled remote system is shown in Figure 18-4-3. This system uses a single, well insulated, hard rubber tube to carry the beer lines, supply, and return coolant lines. Beer lines to the dispenser are cooled by contact with the cold coolant lines.

TECH TIP

Some states have extremely restrictive laws relative to underage individuals being in or near the sale of wine, beer, and liquor. The age and restriction vary from municipality to municipality and state to state. Before you engage in any work during regular business hours of a bar you should check with your local officials or the bar owner to see if there are any such restrictions in your area.

DISPENSING FREEZERS

This group of products includes soft serve ice cream, slush, and milkshake machines. All of these products are made by cooling prepared mixes. Soft serve ice cream and milkshake machines use mixes that include dairy products, and were developed first. Slush machines came later, using nondairy products, and principally involved freezing carbonated beverages. Both the dairy and nondairy product freezing processes are essentially the same. They differ primarily in the constituents of the mix and serving temperature of the product.

The proper serving temperature is an essential condition of the product when it is to be consumed. The different serving temperatures are listed in Table 18-4-1.

If there is a problem with the appearance of the product, the freezing time and the product temperature should be checked. If these conditions are correct, the problem is in the mix.

TABLE 18-4-1

Mix Description	Serving Temperature
Dairy	
Soft ice cream	20 to 22°F
Milkshake	26 to 28°F
Hard ice cream	5 to 9°F
Non-dairy	
Slush or FCB*	24 to 26°F

*Frozen Carbonated Beverages

Dairy Product Mixes

Ice cream and milkshake basic mixes consist of the following ingredients:

1. Butterfat
2. Milk solids, nonfat
3. Sugar
4. Emulsifier stabilizers (substances having a high water holding capacity such as gelatin or vegetable gums)
5. Water

Because of the standardization of the mixes, problems that affect the quality of the product may be operational; however, the service person must not assume that the appearance of the finished product is due to mechanical problems of the refrigeration system unless product temperature is not being maintained. Figure 18-4-4 lists some common defects in ice cream.

SERVICE TIP

Nothing is more repugnant to customers than the smell of sour milk around a frozen dairy product machine. To prevent this it is essential that the machine be cleaned thoroughly and that all covers and seals be replaced following service so that milk cannot seep into areas the owner cannot easily clean.

Nondairy Product Mixes

Slush mix, as it is now dispensed, consists of carbonated beverages which have variations in flavor, stabilizers, sugar, and water.

Overrun To the mix ingredients, air must be added, called overrun, which is essential to the consistency of the product. Most machines have a fixed overrun, set at the factory for the particular mix being handled.

Function of the Operator The draw rates for the machines are limited by their individual capacities. If the designed draw rate is exceeded by the operator, the product will be too soft to stand up, and therefore unsatisfactory. This condition can be checked by the technician during a peak operating time.

Proper daily machine cleaning is important to ensure the safety and quality of the product being dispensed by dairy machines. When delivering a new machine be sure to take time to train the owner in the proper care and maintenance of the new piece of equipment. Follow the manufacturer's guidelines for daily cleanings. Do not use unapproved cleaning solvents in dairy equipment. Some unapproved products can present health concerns to customers.

- *Sandy or grainy texture.* A quick test of the quality of the mix is to take a "lick" of the ice cream then rub your tongue against the roof of your mouth. If the ice cream is grainy (ice crystals), the mix is poor quality. It should be creamy and smooth.
- *Butter churning,* which imparts a greasy taste to the ice cream, is a result of improper homogenization of the mix. Homogenizers are high-pressure pumps for breaking down the fat globules of the mix into tiny particles. For soft-serve mixes, because of repeated whipping, pressures of 3,000 to 3,560 psig need to be maintained, while 2,000 psig is satisfactory for hard ice creams. The supplier of the mix should be contacted and the operator should empty the freezing cylinder and wash all traces of butter clinging to the cylinder walls or dasher assembly.
- *Wet, sloppy product.* Serum solids (milk solids non-fat) are important for improving firmness or "body" of the finished product. If the product does not "stand up" at the desired temperature or is wet and sloppy, or does not absorb air or overrun during churning, the quality of the serum solids in the mix can be questioned. Variations in this ingredient occur naturally and the mix manufacturer should have procedures in place for identifying and compensating for these variations. However, the service person must first determine that the product temperature is being satisfactorily maintained.
- *Sugar content of the mix may vary.* The higher the content, the lower the temperature required to obtain the same firmness of the product. Small variations can often be compensated for by adjusting the temperature control either higher or lower. A large amount of sugar, requiring two or three degrees below normal range, could result in a considerable loss of capacity for the soft-serve dispenser. The service person can use a thermometer to check product temperature and quality over a period of time to know whether the problem is in the mix or in the manner in which the freezer is operated.

Figure 18-4-4 Common problems with soft serve ice cream.

Functions of the Refrigeration System There are two unique parts to this machine: (1) the freezing cylinder evaporator, and (2) the control system.

The evaporator is usually a cylinder within a cylinder. The fully flooded refrigerant occupies the

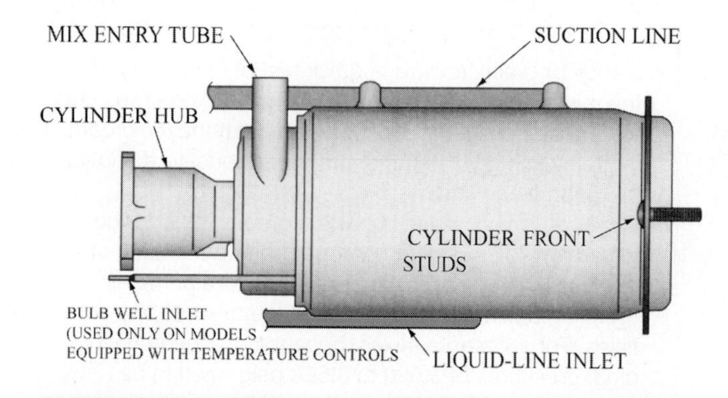

Figure 18-4-5 Evaporator of freezing cylinder.

space between the two cylinders as shown in Figure 18-4-5. Typical refrigeration systems are shown in Figures 18-4-6, 18-4-7, and 18-4-8.

Refrigeration cylinders can be damaged by using water that is too hot during cleaning. Cleaning should be done with cold or cool water.

Measure the temperature of the product. If the product temperature is higher than normal for the mix being used, it is likely that a mechanical problem exists.

Function of the Control System Temperature, pressure, and torque controls (consistency controls) are used. The temperature sensor is located as close to the product as possible. The torque control requires proper adjustment of the belt tension.

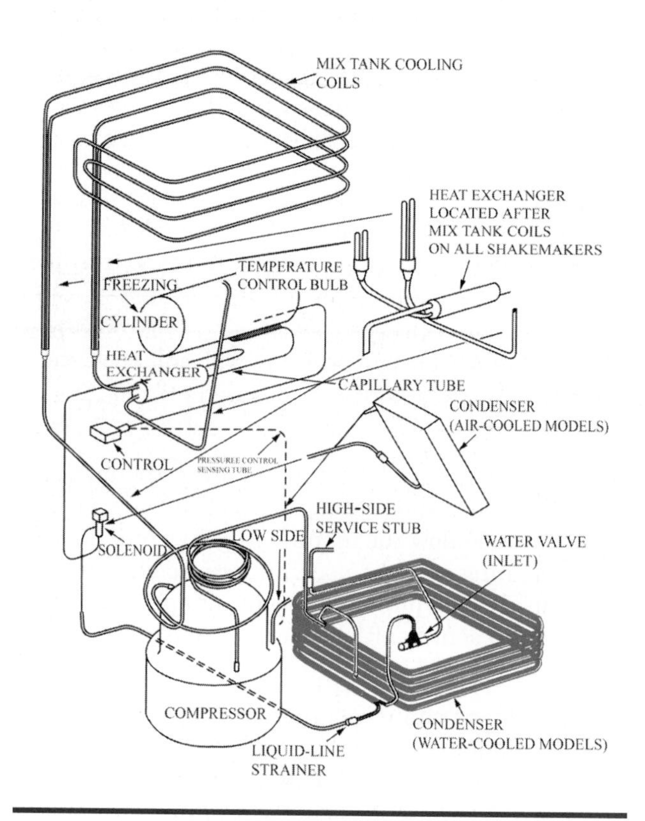

Figure 18-4-6 Refrigeration system of ice cream freezer using single capillary tube flow control.

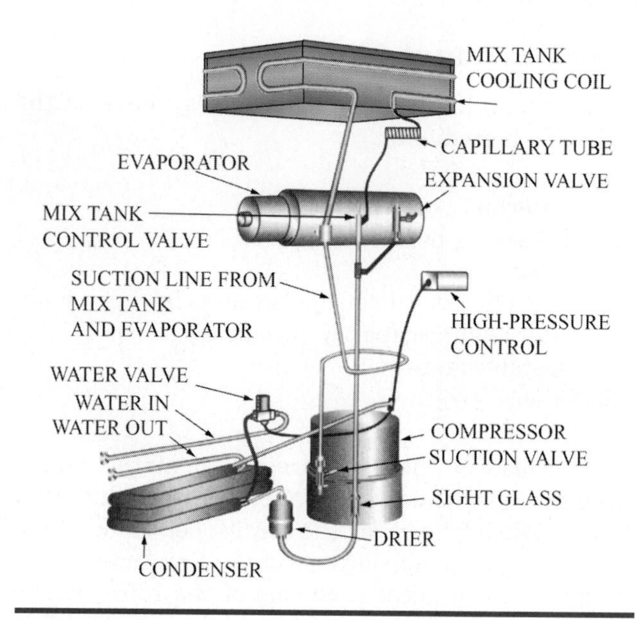

Figure 18-4-7 Ice cream freezer refrigeration system using capillary tube.

> **NOTE:** As a product begins to freeze it requires greater force for the paddles or internal mixing blades to move through the product. This additional force, called torque, is measured by the control system so that as the product begins to freeze and thicken the refrigerant cycle can be terminated. These paddles or blades continue moving through the product to keep its constancy. In addition the torque control will sense the decrease in torque required to move the blades or paddles as the product softens so that the refrigerant system can be initiated.

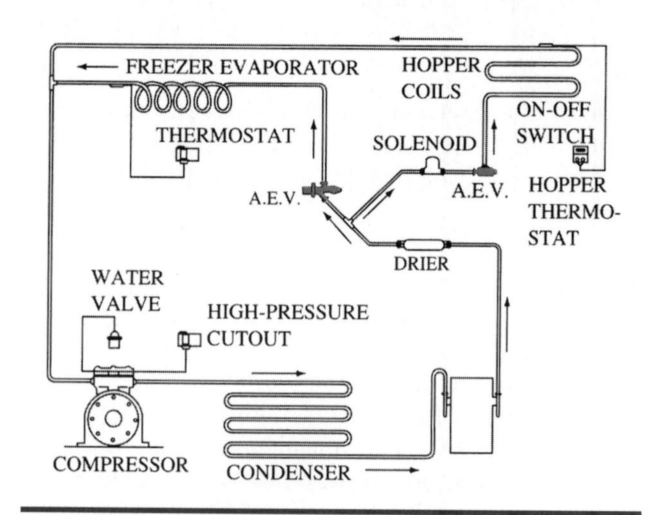

Figure 18-4-8 Refrigeration circuit using two automatic expansion valves for refrigerant control.

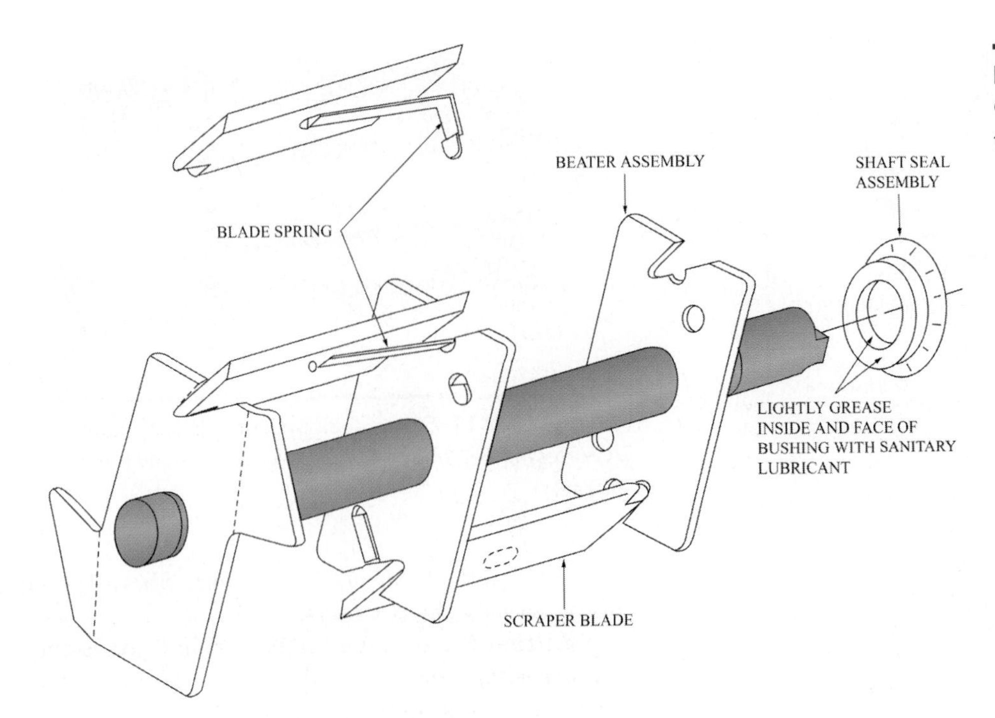

Figure 18-4-9
Construction of the dasher that beats the mix.

Many systems use an anticipator circuit that functions every time a serving of the product is made. The anticipator functions to start the machine, even though it has just turned off. The basic functions of the controls for all types of machines are as follows:

1. The refrigeration system must cut in as quickly as possible after the product has been drawn from the machine. This is necessary since fresh mix is automatically supplied to the freezer to replace product that has been drawn out. If the refrigeration does not come on after three or four servings, the cut-in point is too high.
2. The refrigeration must remain on during heavy draw conditions. If the compressor cycles off during heavy loads, the cut-out point is set too high.
3. The product temperature must be as even as possible between cut-in and cut-out points.
4. ON cycles should be as short as possible and OFF cycles should be as long as possible, to avoid unnecessary whipping of the product.

Function of the Dasher Assembly The dasher assembly has three functions: (1) to scrape the frozen product from the cylinder, (2) to whip the product, and (3) to eject the product.

The dasher blades must be kept in good condition and not be damaged during the cleaning operation. Most modern machines use dashers with plastic blades as shown in Figure 18-4-9. The whipping action puts the necessary air into the product. The CO_2 or air pressure within the freezer also assists in the ejection process.

Care must be taken to remove and replace the same dasher assembly in the machine it was taken out of during the cleaning operation. These units "wear in" and should not be interchanged.

Function of the Mix Feed System Mix feed systems automatically meter liquid mix into the machine to replace that which has been drawn off as shown in Figure 18-4-10. Most mix feed systems require some adjustment at the time of original startup and whenever a different mix is used. Some of the basic points relating to mix feed systems are as follows:

1. Mechanical feed systems must be adjusted to meter an equivalent amount of liquid mix by weight into the cylinder to replace the product that has been drawn off.
2. The gravity feed systems do not require such precise settings. An orifice size is selected that permits the proper flow of mix.
3. Some models have different valve assemblies for soft ice cream and for milkshake makers. Milkshake machines require larger quantities of mix. These valve assemblies should not be interchanged.
4. The freezer cylinder must be preloaded with the correct amount of mix at the start of the day's operation. Consult the service manual for the machine being used for the initial loading and valve adjustment.

General Guide to Service Always separate the problem into one of these three categories: (1) mix, (2) operational, and (3) mechanical.

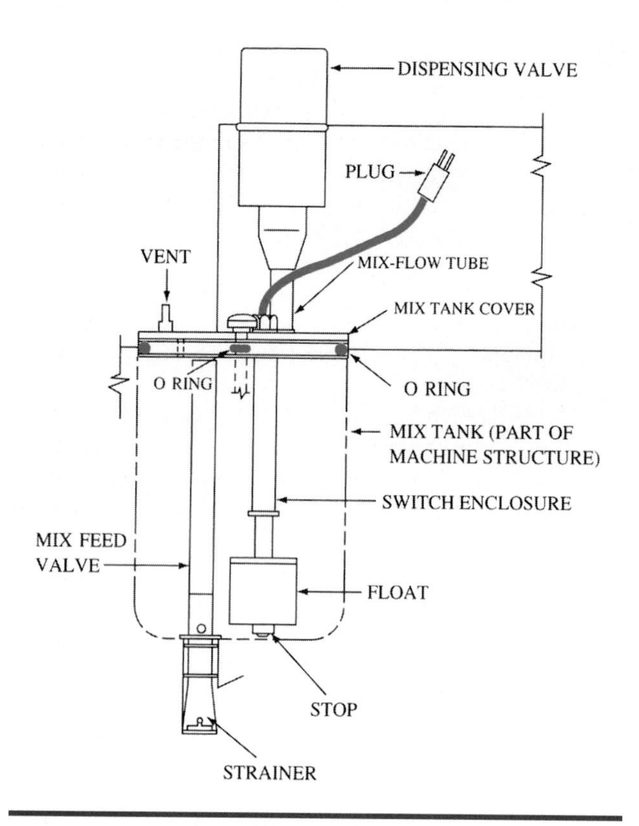

Figure 18-4-10 Typical mix feed system designed to operate the electric dispensing valves as well as automatically meter liquid mix into freezing cylinder.

Mix problems relate to the appearance and taste of the product. Solving these problems requires jointly working with the operator and mix supplier.

Operational problems require actually observing the use and cleaning of the equipment. It is important to explain to the operator the necessity of following accepted procedures.

Solving *mechanical* problems requires a thorough knowledge of the equipment being used. There are many different models—each having its own construction and service requirements. A careful study of the manufacturer's service manuals for the equipment being used makes the analysis of problems much easier.

SERVICE TIP

When possible refer to the manufacturer's troubleshooting chart to isolate and resolve problems with equipment.

SUPERMARKET EQUIPMENT

Food store systems vary depending on the size of the store, the food products being refrigerated, and the area in which the equipment is located. Generally speaking, there are two types of fixtures: (1) *self service equipment,* where the customers select their own products; and (2) *service equipment,* where the employee selects the products.

Figure 18-4-11 Single deck produce display cabinet. *(Courtesy of Hussmann Corporation, an Ingersoll Rand business)*

Within these categories are found fixtures made in open single deck, closed single deck, open multideck, and closed multideck styles.

For produce, the most generally used style is the open single deck. A commonly used produce case is shown in Figure 18-4-11. It is sometimes referred to as a "vision type" case because mirrors are used to enhance the appearance of the product.

The produce sections in supermarkets also use a walk in cooler, not accessible to the public, maintained at about 38°F and 80% RH for storage, and a preparation room maintained at about 55°F.

For meat and dairy products, single deck, multideck, and island self service cases are used. A typical meat and deli merchandiser is shown in Figure 18-4-12. A plan view and an elevation view of the cabinet's construction are shown in Figure 18-4-13.

An evaporator using a gravity coil is located in the top of the cabinet to provide a gentle circulation of cool air without drying out the product. Meat markets also usually have a walk in storage cooler and a meat preparation room.

Figure 18-4-12 Meat and deli merchandiser cabinet. *(Courtesy of Hussmann Corporation, an Ingersoll Rand business)*

For dairy products a multideck refrigerator as shown in Figure 18-4-14 is commonly used.

Some of these have a rear access for loading the cabinet from racks. An illustration of the typical construction of the dairy merchandiser is shown in Figure 18-4-15. A forced air refrigerated evaporator coil is located below the lower deck.

For frozen foods and ice cream, single deck island type, multideck open, and multideck reach in cabinets are used. Figure 18-4-16 shows an island frozen food cabinet and Figure 18-4-17a–c shows its construction. The forced air evaporator coil is located below the product space.

SERVICE TIP

The condensate drain on most frozen food cases has a heater to prevent ice blockage of the drain during the defrost cycle. These heaters are very small wattage, designed to provide only enough heat so that condensate can flow freely from the case during the defrost cycle. When these heaters go out, condensate will collect and freeze at the bottom of the case. Severe ice buildup can cause damage to the coil and to the cabinet and may result in tripping the safety. When the box begins to warm the ice will melt and the water will run off. This can make it very difficult to troubleshoot drain heater problems.

Refrigeration Systems for Food Stores

The refrigeration systems for food stores range from single condensing units to multiple parallel compressor banks with remote condensers. These options include the following systems.

Single Condensing Unit per Fixture This usually consists of a remote condensing unit as shown in Figure 18-4-18.

Where a number of these are used in a single installation, they are placed on racks in an equipment room. Controlled ventilation is provided to maintain a satisfactory ambient air temperature for the air cooled condensers.

Each individual refrigeration circuit has its own temperature and defrost control. This arrangement offers the greatest flexibility, but requires the greatest amount of compressors, piping, and electrical connections.

Single Condensing Unit for Multiple Cases (Multiplexing) This arrangement is used where a number of cases are all maintained at the same conditions. Individual cases are controlled by thermostats operating liquid line solenoid valves and evaporator pressure regulators (EPRs). Installations

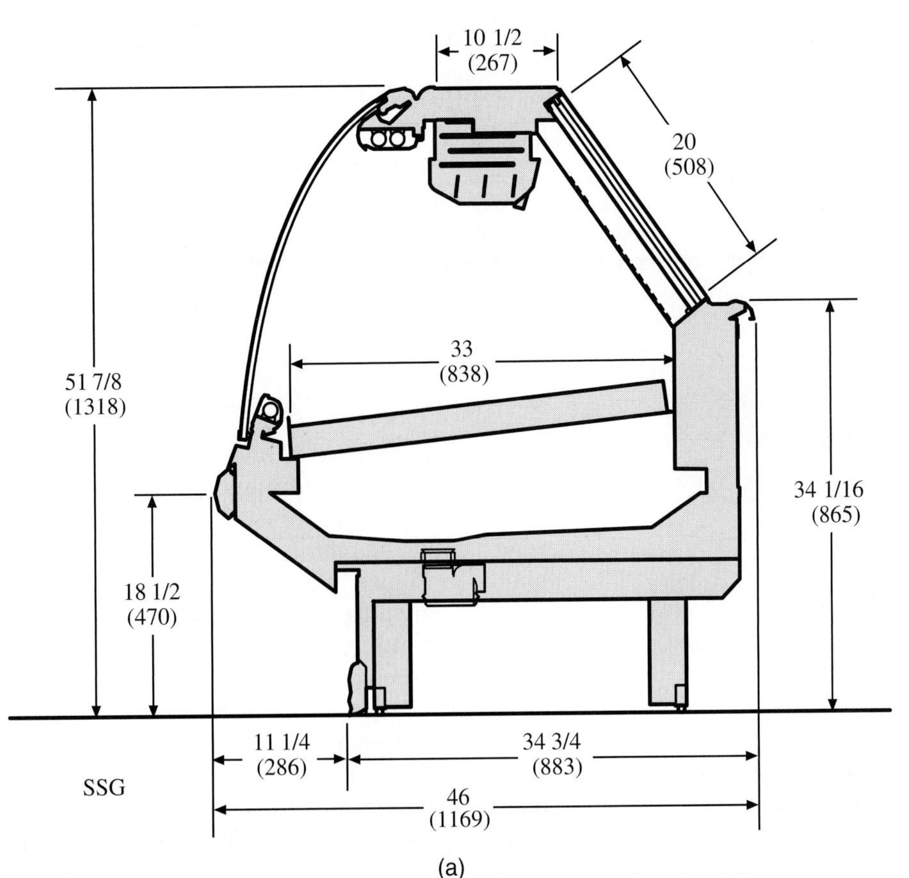

Figure 18-4-13 Schematic diagram of meat and deli cabinet showing the gravity airflow evaporator coil. *(Courtesy of Hussmann Corporation, an Ingersoll Rand business)*

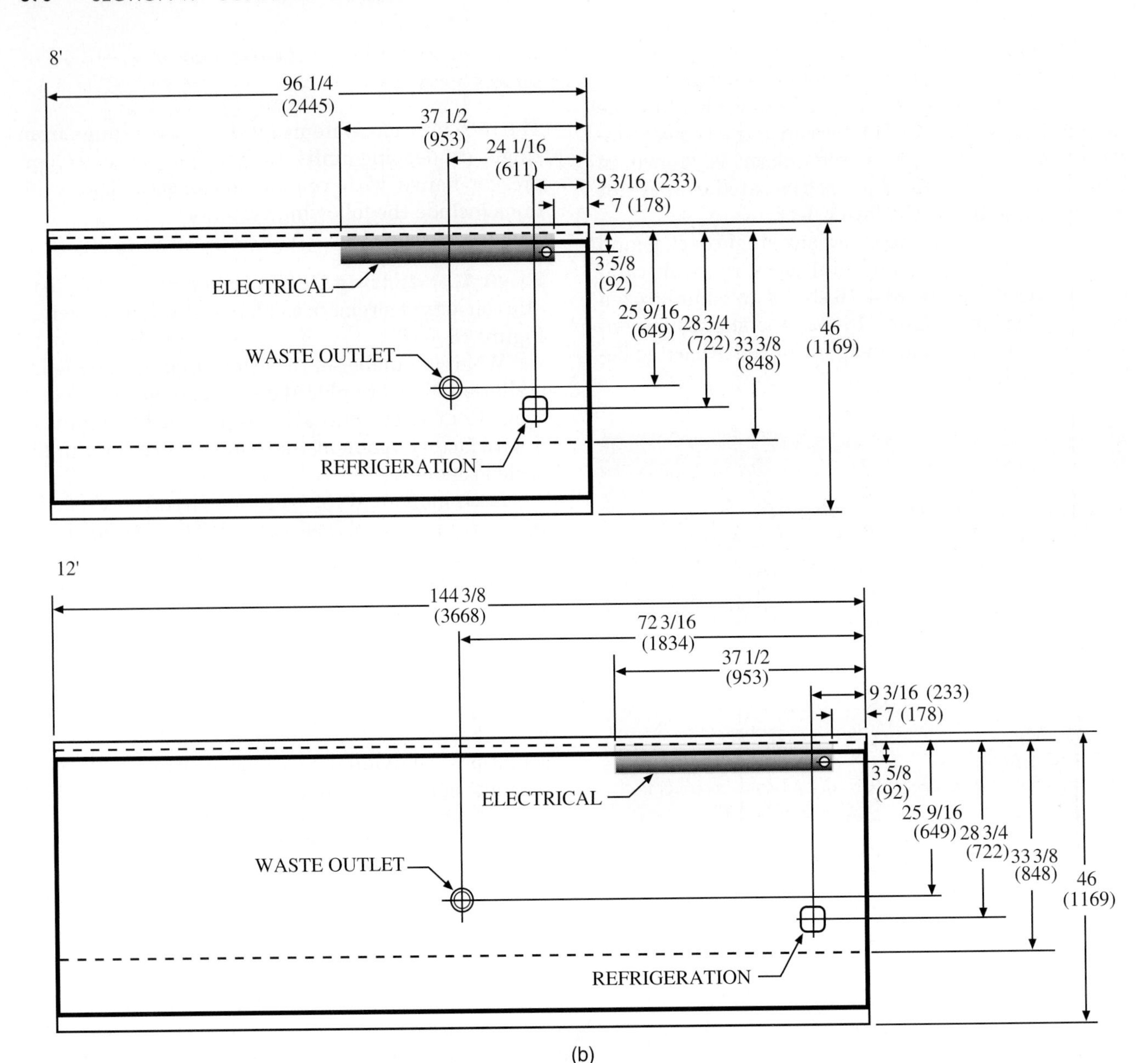

Figure 18-4-13 (continued)

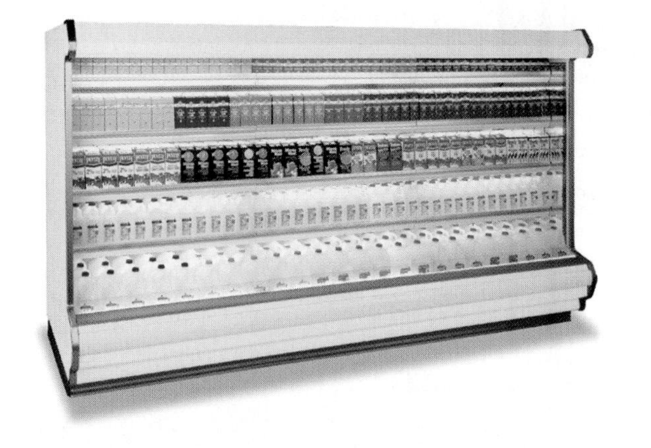

Figure 18-4-14 Multi deck dairy merchandiser. *(Courtesy of Hussmann Corporation, an Ingersoll Rand business)*

are more complex; however, less energy is used and the overall cost is reduced.

Parallel Systems These consist of a bank of parallel connected compressors and accessories mounted on a common base, piped and wired at the factory, ready for field connection to individual cases as shown in Figure 18-4-19.

The advantage of this type of arrangement is that individual compressors can be cycled on and off to provide the required capacity. In case of compressor failure, other compressors carry the load.

This equipment configuration is advantageous for latent heat defrosting, heat recovery, and electrically operated total control systems. Using this arrangement, defrosting can be selective, with some units

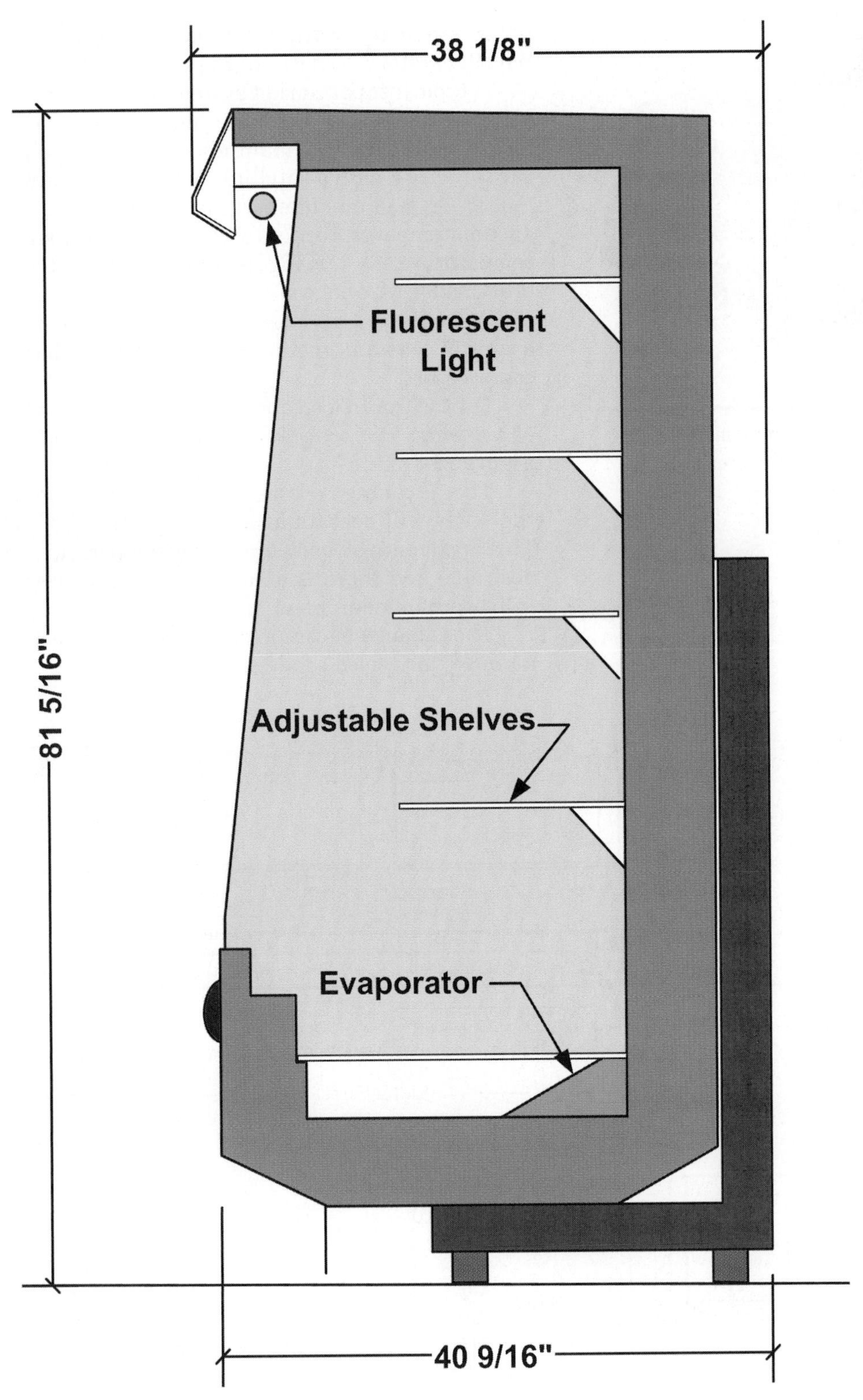

Figure 18-4-15 Schematic diagram of the dairy merchandiser cabinet showing the location of the forced air evaporator. *(Courtesy of Hussmann Corporation, an Ingersoll Rand business)*

operating while others defrost. The operating units can provide the necessary heat for the defrosting units.

For maximum efficiency a bank of compressors should be connected to evaporators operating at similar temperatures. From an operational standpoint, fixtures of different evaporating temperatures can be connected to the bank; however, this causes a substantial energy loss.

SERVICE TIP

A disadvantage of parallel systems is that a single leak can shut down a sizable number of cases. When possible a loss charge refrigerant monitor should be used to alert the owner or service company of the problem.

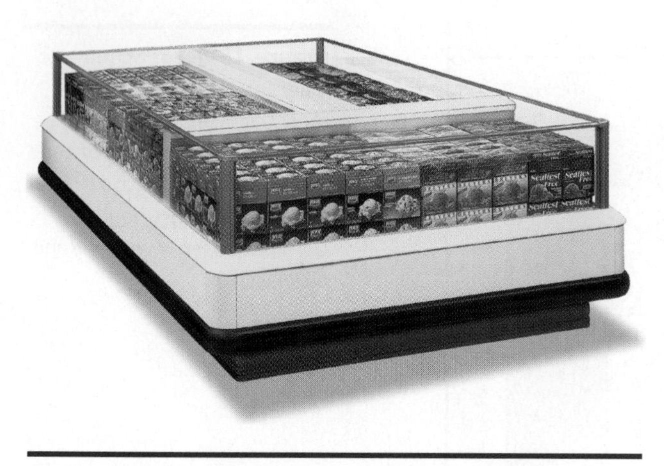

Figure 18-4-16 Frozen food/ice cream island merchandiser. *(Courtesy of Hussmann Corporation, an Ingersoll Rand business)*

A schematic refrigerant piping diagram for a parallel system is shown in Figure 18-4-20.

A typical supermarket system produces refrigeration in the low or medium temperature ranges. A low temperature rack maintains a suction temperature of −25°F with a satellite suction temperature of −33°F. A medium temperature rack maintains a suction temperature of +16°F and a satellite suction temperature of +7°F. Figure 18-4-21 lists system operations for parallel compressors.

A *suction pressure controller* is used to maintain a nearly constant suction pressure for the bank of compressors.

For the frozen food and ice cream cabinets a bank of *two stage compressors* can be used, as shown in Figure 18-4-22.

The advantage of the two stage compressors is that they will operate at a lower suction pressure. This is advantageous where lower temperatures are desirable and there is a sufficient load to economically warrant equipment of this type.

These banks of compressors can be connected to a remote air cooled condenser.

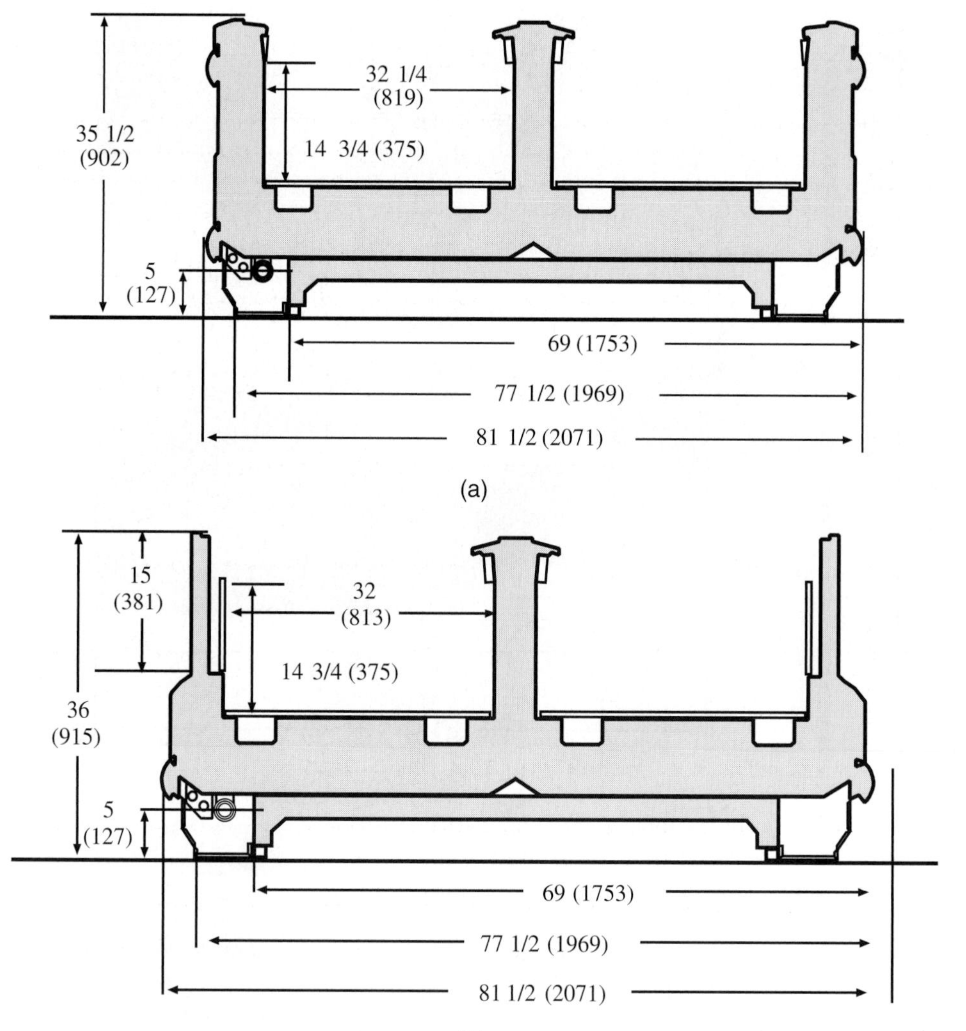

Figure 18-4-17 Schematic diagram of the island-type merchandiser cabinet. *(Courtesy of Hussmann Corporation, an Ingersoll Rand business)*

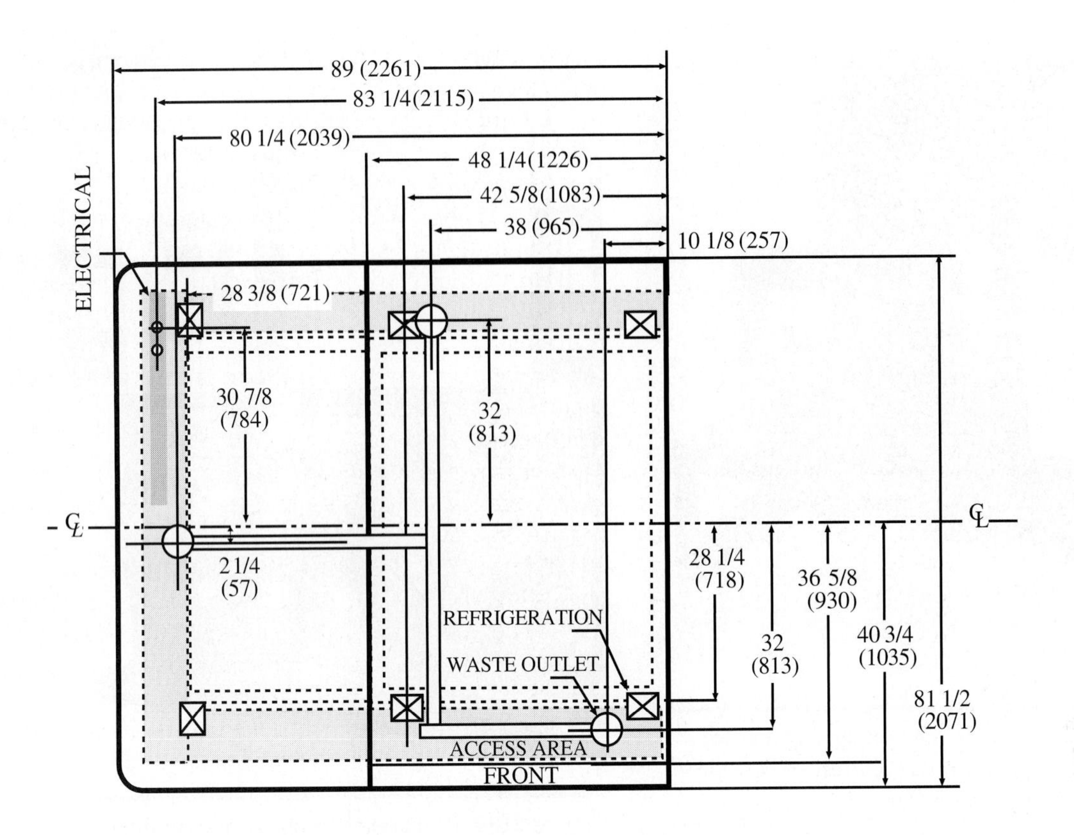

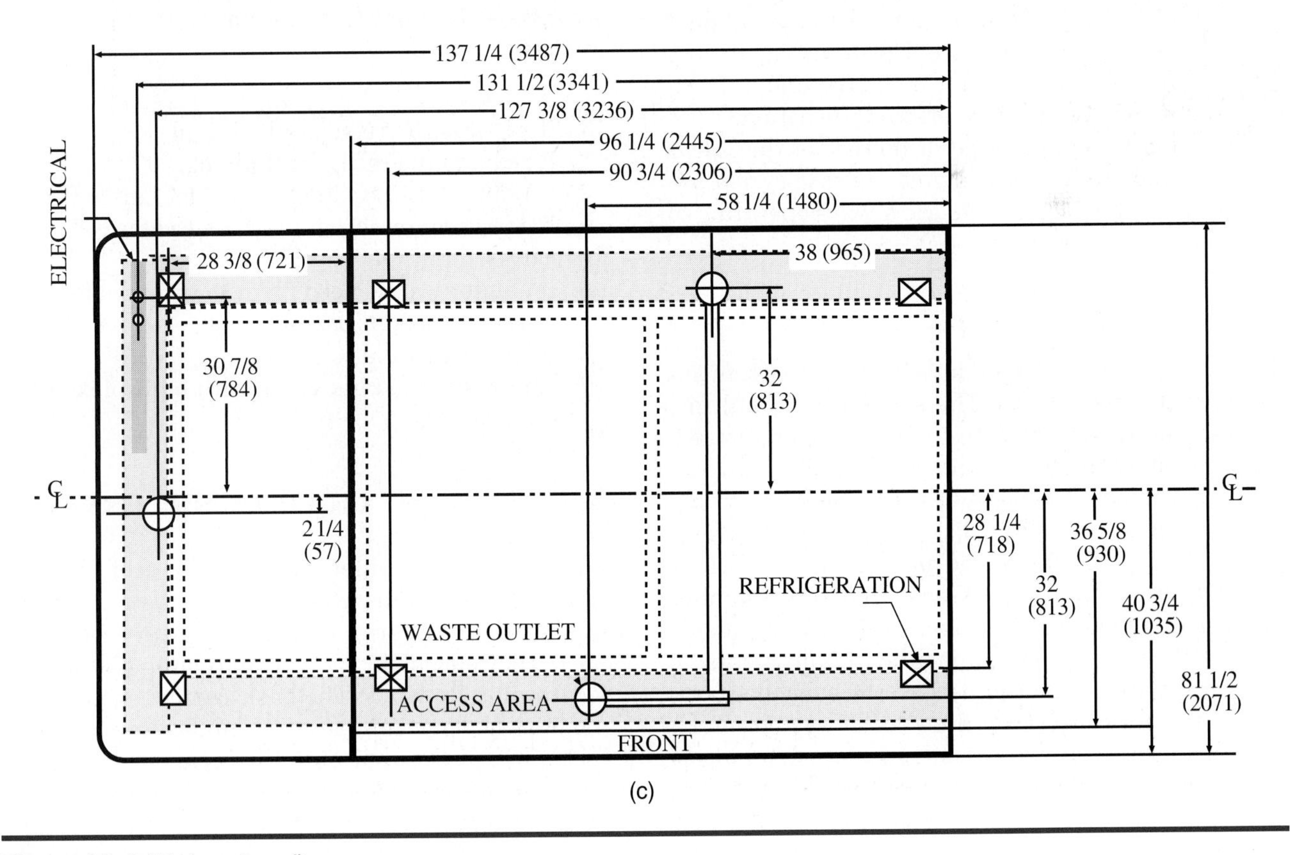

(c)

Figure 18-4-17 (continued)

Figure 18-4-18 Indoor air cooled condensing unit for high, medium, and low applications. *(Courtesy of Hussmann Corporation, an Ingersoll Rand business)*

Figure 18-4-19 Parallel compressors mounted on a single base complete with wiring, piping, and controls. *(Courtesy of Hussmann Corporation, an Ingersoll Rand business)*

Piping from the central unit to the individual cases is usually carried in trenches underneath the floor. Piping must be properly insulated and isolated to prevent the possibility of electrolytic action.

The most common type of defrost is the *latent heat method*. In larger systems, defrost cycles are staggered so that some units are always operating to produce heat.

The Protocol System

The protocol system is an advanced approach to supermarket refrigeration. The system uses multiple compressor units located near the cases being cooled. A typical store might have as many as 10 to 15 multiplexed compressors located throughout the store, as shown in Figure 18-4-23.

Each unit, such as the one shown in Figure 18-4-24, uses scroll compressors that provide quiet operation, and an HFC refrigerant R-507 that has a zero ODF to meet environmental requirements. A compact plate condenser is part of each package, cooled from a closed loop fluid cooler system. Each unit includes an electronic controller which manages both compressor cycling and scheduled defrosts. Compressors are cycled to match the load requirements.

A central pedestal mounted power distribution panel furnishes electrical power to each unit by means of a four wire drop cord that supplies power for the compressors, fans, lights, and anti-sweat heaters. The advantages of this system are as follows:

1. It reduces the refrigerant charge.
2. It reduces the refrigerant piping.
3. It reduces the possibility of refrigerant leaks.
4. It decreases or eliminates the need for EPR valves.
5. It eliminates the need for a central area for refrigeration equipment.
6. It lowers installation costs.
7. It provides load matching with multiplexed compressors.

PACKAGED LIQUID CHILLERS

Packaged water chillers are used to cool water or brine, as a secondary refrigerant, for both air conditioning and refrigeration applications. A good example of the refrigeration use is in cooling glycol brine for freezing ice rinks. The basic components include a motor driven compressor, a liquid cooler (evaporator), a water cooled or air cooled condenser, a refrigerant metering device, and a control system.

A schematic view of a basic refrigeration cycle is shown in Figure 18-4-25. Chilled water enters the cooler at 54°F and leaves at 44°F. Condenser water leaves a cooling tower at 85°F and returns to the tower at 95°F. Air cooled condensers can be used in place of the water cooled type. Liquid chillers are available in capacity sizes ranging from 25 to 3000 tons.

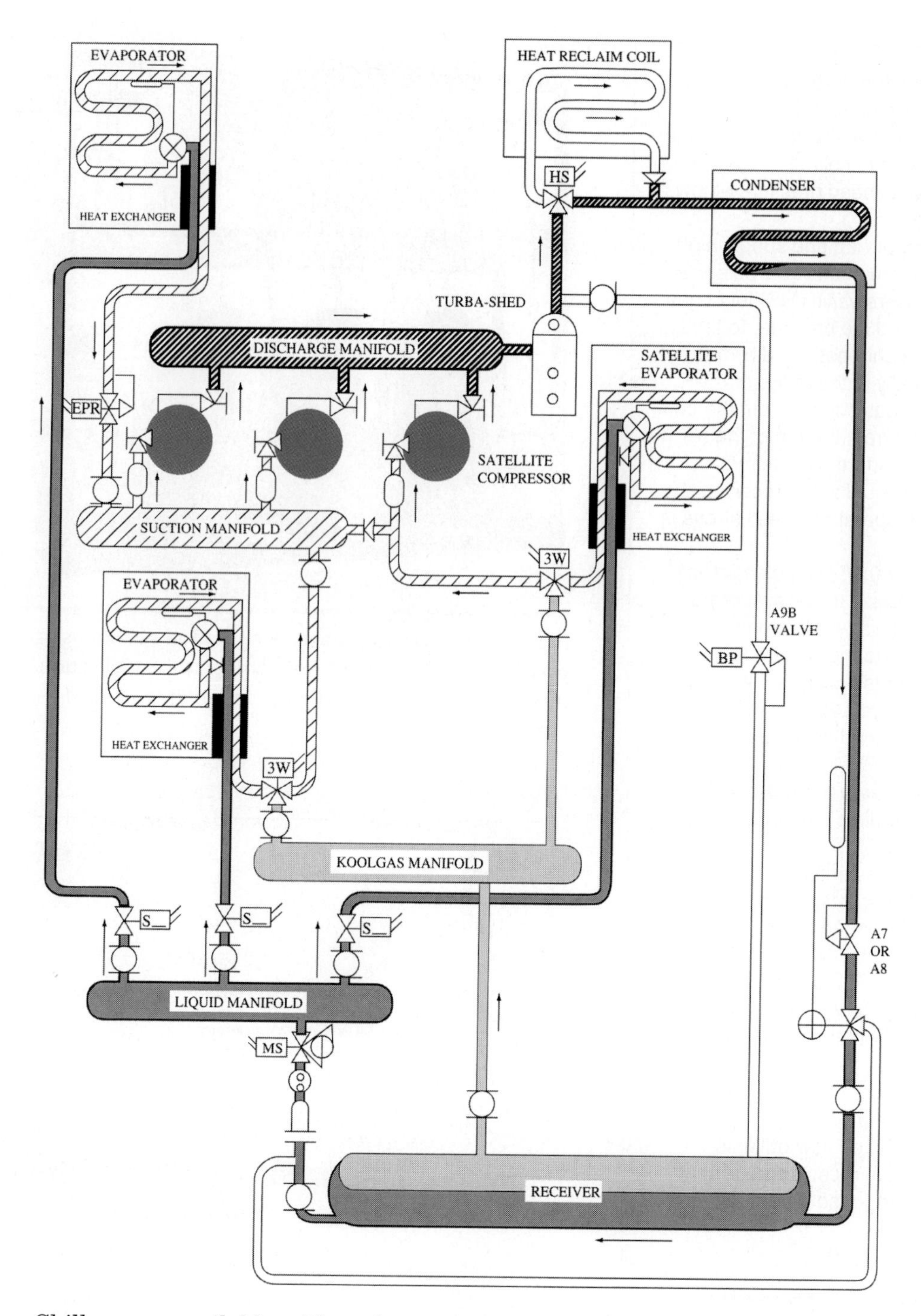

Figure 18-4-20 A typical refrigeration system using parallel compressors for food store application. *(Courtesy of Hussmann Corporation, an Ingersoll Rand business)*

Chillers are available with reciprocating, scroll, screw, or centrifugal compressors. Chillers are also constructed using the absorption refrigeration cycle.

The type of control system varies somewhat depending on the type of compressor used. All chillers have a leaving liquid temperature sensor that can be adjusted to maintain the desired liquid temperature. Most chillers have some arrangement for changing the capacity of the chiller to match the load. For example, a reciprocating compressor may have unloaders to reduce its capacity by 25, 50, and 75%.

Two popular types of modern chillers are the air cooled chiller with multiple reciprocating compressors and the water cooled chiller with a rotary screw compressor.

Air Cooled Chiller

An air cooled chiller can range in size from 75 to 200 tons. It uses *multiple reciprocating compressors* and multiple condenser fans to achieve capacity reduction and maintain efficient operation at partial loads.

A diagram of an electronic expansion valve is shown in Figure 18-4-26. The valve uses a stepper motor controlled by a microprocessor to regulate the refrigerant flow. It is capable of operating at lower condensing temperatures than a normal TXV valve system, thus improving the efficiency.

A two circuit shell and tube chiller is shown in Figure 18-4-27. At half the load, only half the chiller

1. Compressed gas flows into the *Turba-shed,* which separates the vapor refrigerant from the liquid oil.
2. The three-way *heat-reclaim valve* directs the refrigerant to the condenser or reclaim coil.
3. The *flooding valve* maintains head pressure at low ambient conditions.
4. *Twin receivers* act as a vapor trap and supply high-quality liquid refrigerant to the main liquid solenoid.
5. The *main differential-pressure valve* (MS) functions during *Koolgas* defrost to reduce pressure to the liquid manifold. *Koolgas* is the hot gas used by this system for defrost and is usually in the range of 85°F. This is cooler than the hot gas coming from the compressor since it is removed from the top of the receiver. When the *Koolgas* performs its defrost function it is returned as a liquid to the receiver. By using lower defrost gas temperatures, less stress is placed on the piping.
6. For the defrost cycle, the *branch line solenoid valve* (S) closes off refrigerant supply to the evaporator.
7. The *heat reclaim three-way valve* (HS) routes the discharge heat-laden vapor to the remote reclaim coil. The check valve prevents flooding of the reclaim coil when the heat reclaim cycle is off.
8. Receiver pressure is maintained by the *back-pressure regulating valve* (BP).
9. The receiver vapor flows directly to the *Koolgas manifold* and in the reverse direction through the evaporator to defrost it.
10. The *Koolgas vapor* condenses and flows into the reduced-pressure liquid line through a *bypass check valve* around the TXV, returning to the liquid-line manifold.
11. When the defrost is called for, the suction-line valve closes and a two-way *Koolgas valve* opens to permit the flow of *Koolgas* from the manifold to the evaporator.
12. The oil level in the compressor crankcases is maintained by the *oil level regulators.*
13. The *autosurge valve* directs the flow of liquid refrigerant either through the receiver or around the receiver to obtain the required amount of subcooling.

Figure 18-4-21 Systems operations for parallel compressors.

is used. Since units are usually running at reduced capacity, this feature also helps to improve partial load efficiency.

The control system is self diagnostic. In case there is a service problem the electronic circuit will flash a code number on the control panel to indicate the nature of the problem. During normal operation, the operator can read eight different system temperatures to indicate the performance of the unit.

Figure 18-4-22 Two stage refrigeration system using parallel compressors for low temperature application. (*Courtesy of Hussmann Corporation, an Ingersoll Rand business*)

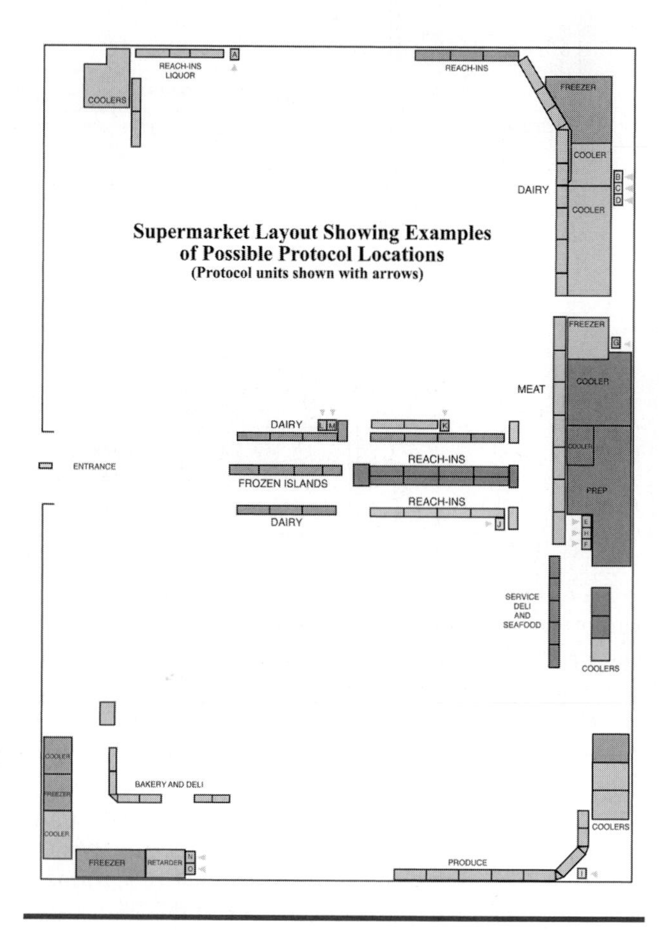

Figure 18-4-23 Store layout using protocol refrigeration units. (*Courtesy of Hussmann Corporation, an Ingersoll Rand business*)

(a)

(b)

Figure 18-4-24 (a) Interior view of a typical condensing unit for a protocol refrigeration system; (b) Exterior view. *(Courtesy of Hussmann Corporation, an Ingersoll Rand business)*

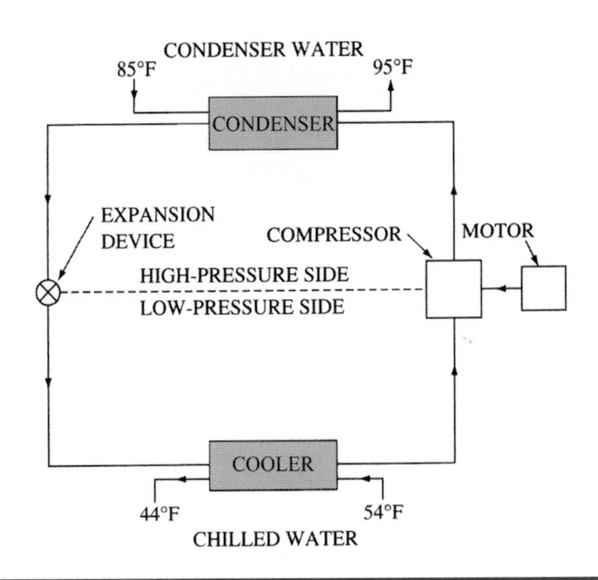

Figure 18-4-25 Cycle diagram for a water chiller. *(Copyright by the American Society of Heating, Refrigerating, and Air-Conditioning Engineers, Inc.)*

Water Cooled Chiller

A typical water cooled chiller using multiple rotary screw compressors is shown in Figure 18-4-28. The advantages of using screw compressors are as follows:

1. The efficiency is higher than reciprocating units. Full load efficiencies of approximately 0.80 kW/ton are offered.
2. They have about one tenth the number of moving parts and therefore require less maintenance.
3. They are quieter in operation.
4. They have infinite capacity modulation, using a slide valve operation. Leaving water temperature can be controlled to within 0.5°F from full load to 30%.

The size range for the series illustrated is from 50 to 300 tons. The number of condensing units varies from one to four depending on the total capacity required. Multiple condensing units offer increased efficiency at partial load and simplify capacity reduction.

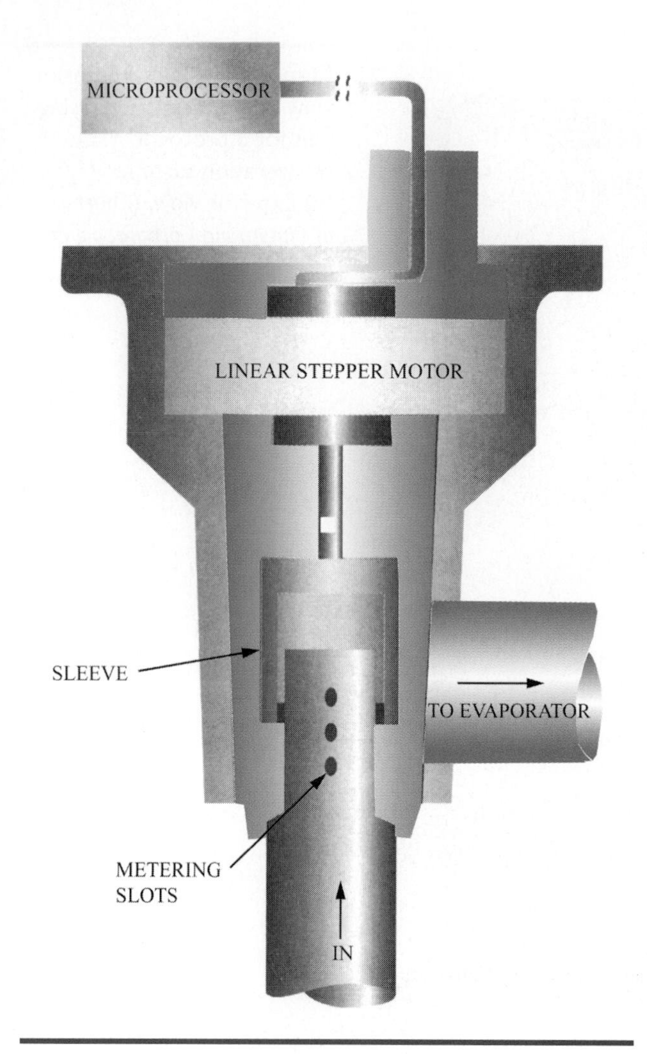

Figure 18-4-26 Stepper motor metering device.

Figure 18-4-27 Cycle diagram for air cooled chiller showing two refrigeration circuits.

(a)

(b)

Figure 18-4-28 (a) Water cooled chiller using multiple rotary screw compressors; (b) Screw compressor.

ICE MAKING UNITS

There are two general types of ice making machines: those that make clear ice cubes used to cool beverages and those that make flake ice used to cool products like fish.

Ice Cube Machines

Ice cube machines such as the one shown in Figure 18-4-29 are usually self contained refrigeration systems that produce and store clear ice cubes, which are preferred by bars, restaurants, and hotels.

The cubes are frozen using running water. This process removes the air to create a clear cube. If there is air in the ice, the carbonation in a drink can be destroyed. Various machines produce different

NICE TO KNOW

There are three major types of ice produced by ice machines. They are cube ice, flake ice, and nugget ice. Cube ice machines represent the largest single group of machines sold. They can produce cubes that are in a variety of shapes such as rectangular, cylindrical, pillow shaped, and square. Cube ice, of various shapes, is used primarily for beverage service. Flake ice is easily packed around a product for display or transit. Flake ice is frequently used in supermarket display cases or as a means of preserving fresh seafood as it is brought to market in the storage holds of fishing vessels. Flake ice may contain up to 20% liquid water which can be a disadvantage for use in soft drinks. Nugget ice consists of small cylindrical pieces that are extruded through the freeze mold and broken off. Nugget ice can be used for product display or beverage service. Nugget ice provides a great deal of surface area to more rapidly cool warm beverages when served.

sizes and shapes of the cubes, depending on the shape and construction of the evaporator.

Cubers are available having capacities to produce 450 to 1800 pounds of ice per day. Bins that collect the ice cubes fit under the cuber and can be selected in various sizes to fit the needs of the user.

An automatic self cleaning system is available as an accessory which cleans the water passages of the machine on a selected schedule. Periodic cleaning is essential in maintaining a supply of clear ice as shown in Figure 18-4-30.

To reduce the maintenance, the air to the condenser is filtered to remove lint, dust, and grease. The sequence of operations of the evaporator in producing ice cubes is as follows:

1. *Initial startup.* Prior to the time that the refrigeration system is started, the water pump and the water dump solenoid are energized to flush the water passages for 45 seconds. At the completion of this cycle, the compressor and condenser fan (on air cooled models) are energized.
2. *Freeze cycle.* After a 30 sec delay, the water pump is restarted and water is supplied to each freeze cell, where it freezes. When the required amount of ice is formed, the ice thickness probe is contacted, and after 7 sec the harvest cycle begins.
3. *Harvest cycle.* The hot gas valve opens, directing hot gas into the evaporator, loosening the cubes. The water dump solenoid is opened for 45 seconds, purging the excess water into the trough. The water pump and dump solenoid are

(a)

(b)

Figure 18-4-29 Ice cube machines with ice bins.

Figure 18-4-30 Switch used to operate the water pump with the refrigeration system not running. (*Courtesy of The Manitowoc Company*)

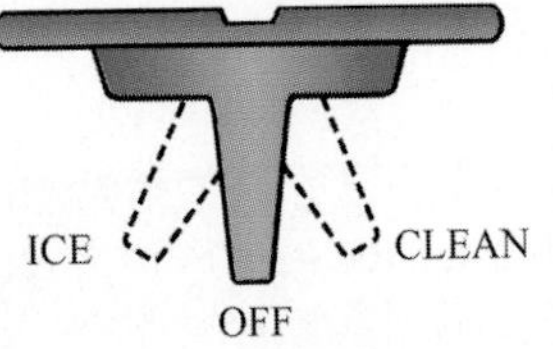

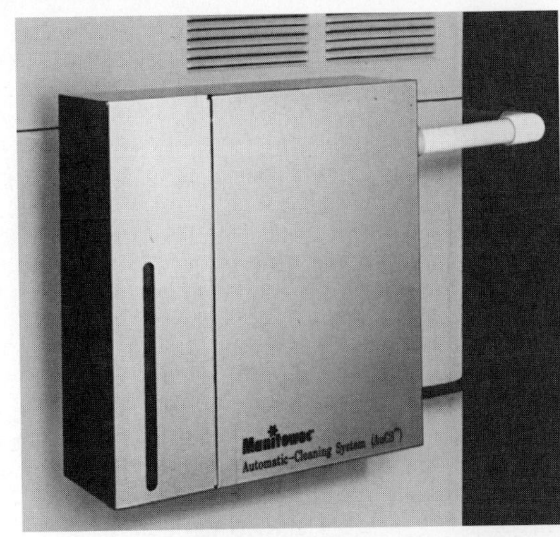

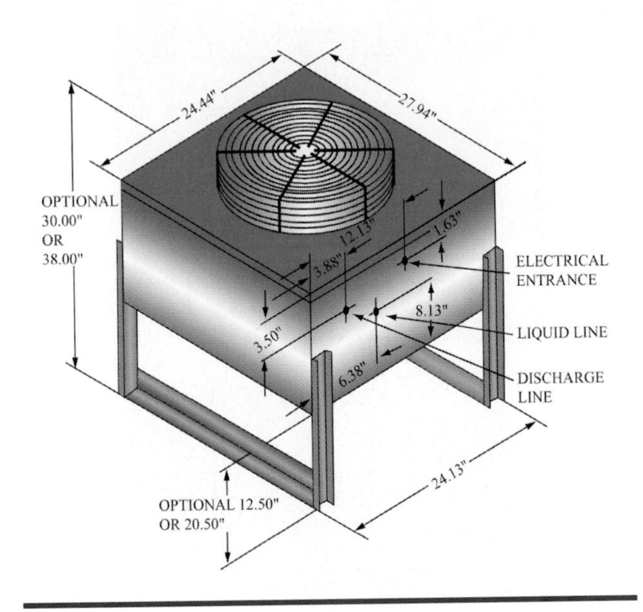

Figure 18-4-31 Remote air cooled condenser used with ice cube machine. (*Courtesy of The Manitowoc Company*)

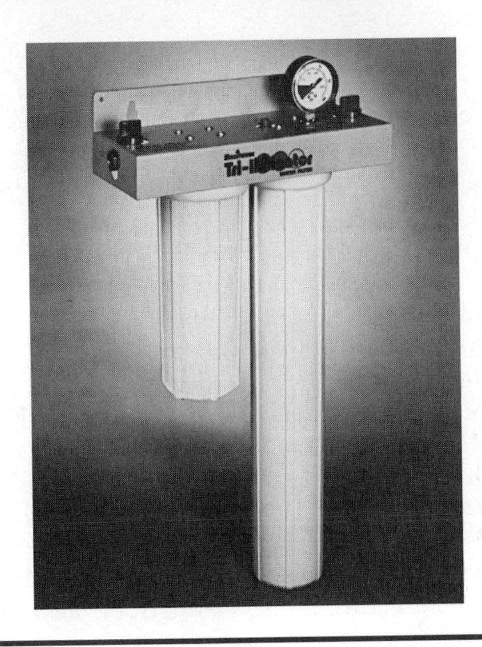

Figure 18-4-33 Water filter for an ice cube machine. (*Courtesy of The Manitowoc Company*)

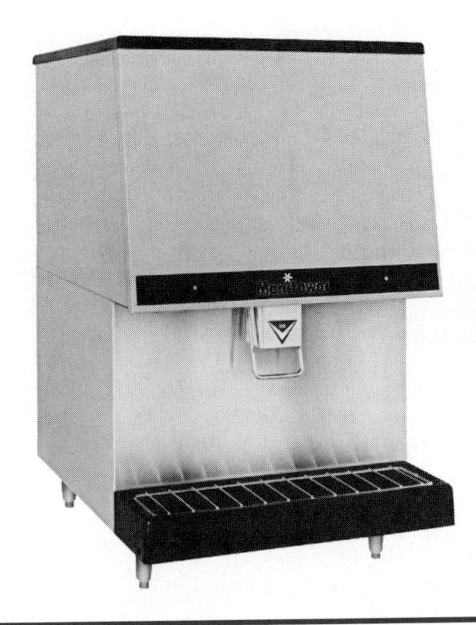

Figure 18-4-32 Ice dispenser for self serve beverage machine. (*Courtesy of The Manitowoc Company*)

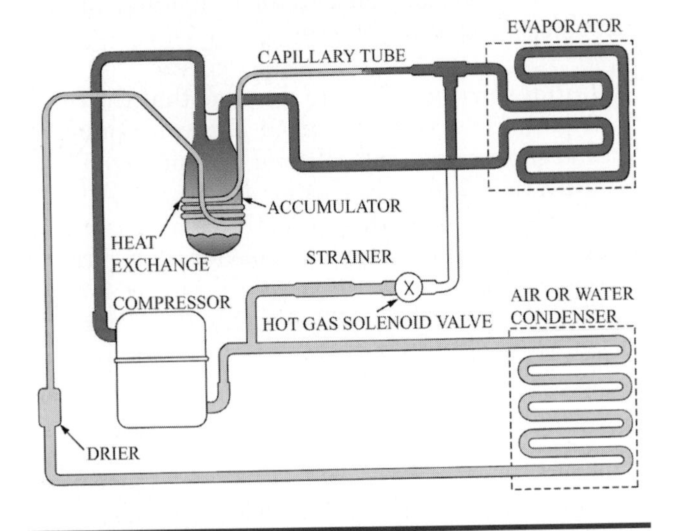

Figure 18-4-34 Refrigeration cycle for an ice cube machine. (*Courtesy of The Manitowoc Company*)

then de-energized. The cubes slide into the bin. The sliding cubes move the water curtain out, activating the bin switch. This terminates the harvest cycle, returning the machine to the freeze cycle.

4. *Automatic shutoff.* When the bin is full of ice, no more cubes can slide into the bin and the water curtain is held open. If this occurs for more than 7 sec, the machine automatically shuts off. As ice is used from the bin, space develops for accepting more ice. The water curtain slips back into operating position and the machine restarts.

The machine can use a remote air cooled condenser as shown in Figure 18-4-31.

The ice dispenser shown in Figure 18-4-32 is a variation of the cuber, which is commonly used in self service restaurants.

Water filters, Figure 18-4-33, can be effectively used in the incoming water lines. These units offer a means of reducing scale formation, filtering sediment, and removing chlorine taste and odor.

Figure 18-4-34 shows the refrigeration system components and piping for a typical unit. Figure 18-4-35 shows a typical wiring diagram.

For other servicing information, consult the manufacturer's service manual.

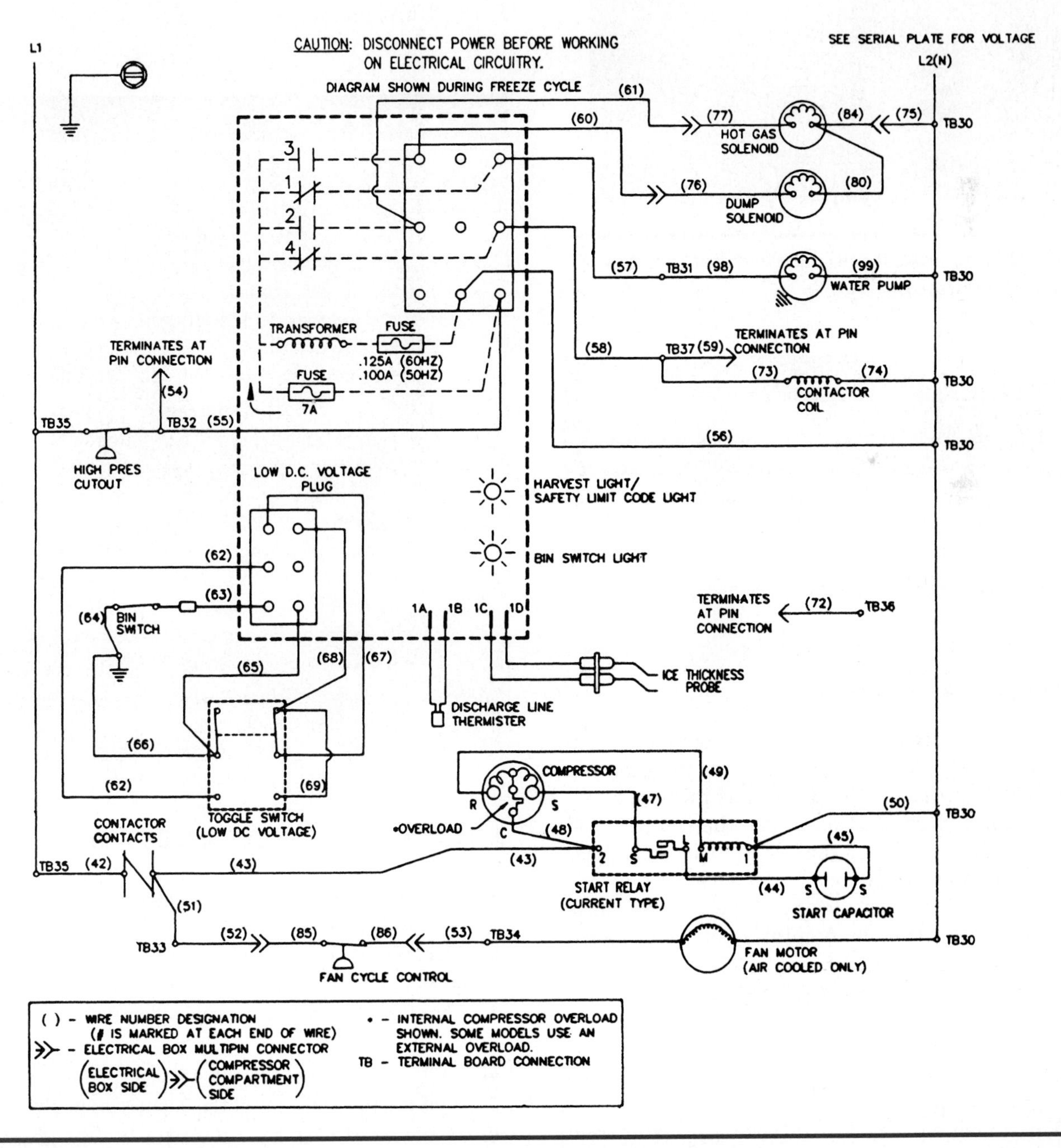

Figure 18-4-35 Wiring diagram for an ice cube machine. (*Courtesy of The Manitowoc Company*)

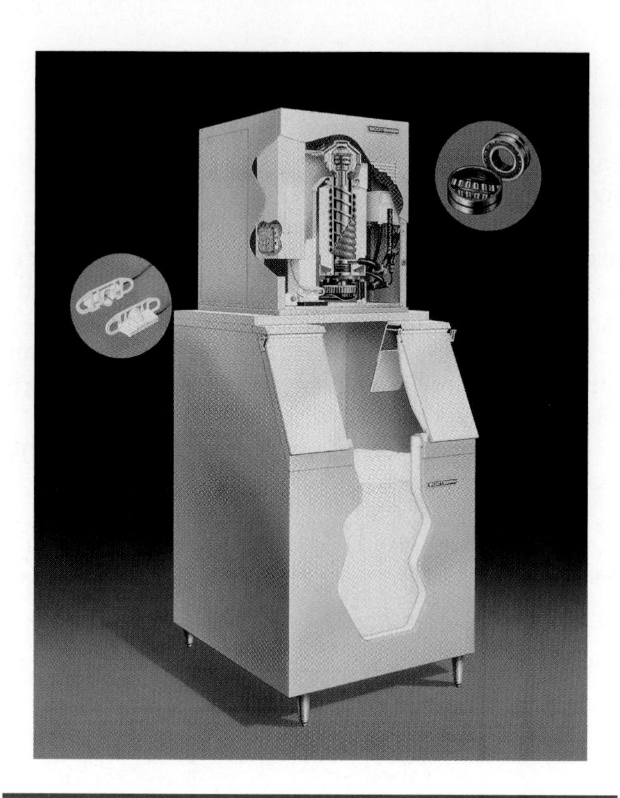

Figure 18-4-36 Flake ice machine with bin. (*Courtesy of Scotsman Ice Systems*)

SERVICE TIP

Note that the ice in ice maker storage bins is not refrigerated. It is allowed to slowly melt away. This prevents ice from developing a "freezer" taste. It is important that ice have a very neutral flavor so that it does not adversely affect the beverage or product it is used in. The amount of melting that occurs in a storage bin is minimal.

Flake Ice Machines

Flake ice machines have always been popular for contact cooling, such as cooling fish, where maintaining high humidity is required. These machines are available in small sizes such as under the counter and floor mounted dispenser models, and in industrial sizes for making 1 to 10 tons of flake ice per day.

An innovation in the design of these machines is shown in Figure 18-4-36. A cutaway of the evaporator is shown in Figure 18-4-37.

Referring to the illustrated evaporator in Figure 18-4-37, the water distribution system at the top evenly distributes water over the subzero evaporator surface. Below this cast aluminum tray is the squeegee and ice blade.

The rubber squeegee removes excess water from the ice. The excess water is recirculated to be used for further ice making. No water is wasted. The stainless steel ice blade wedges the ice off the evaporator without touching the surface.

The shaft holds the major components and is the only moving part. It rotates at only 2 rpm. The shaft is supported by two bronze sleeve bearings. Ice falls by gravity into the bin. The flaker operates continuously and there is no need for hot gas defrost.

The refrigeration cycle component piping is illustrated in Figure 18-4-38. A typical wiring diagram is shown in Figure 18-4-39.

Industrial flake ice machines are available using a remote condensing unit that will freeze either freshwater or seawater. A number of different refrigerants are available including R-22, R-404A, R-134A, and ammonia.

TECH TALK

Tech 1. The boss made me go home and change clothes yesterday before he would send me out on that service call at the supermarket.

Tech 2. That was probably a great idea. People don't like to see technicians in dirty clothes working around food and food handling equipment. In some cases the health code will not allow you to work in an area unless your uniform is clean.

Tech 1. I don't see what difference it would make. I was going to be in the back anyway.

Tech 2. Whether you are working in the back or out in the front with the customers, having a clean uniform is part of good work habits. This is especially true when you are dealing with food and the public. You never want someone to say that they got sick from something and have someone point a finger at you and say he was here and not very clean.

Tech 1. I never thought of it that way. I guess it was a good idea. Thanks.

UNIT 4—REVIEW QUESTIONS

1. List the three fundamentals of satisfactory equipment for draft beer dispensers.
2. What is the proper serving temperature for a milkshake?
3. What ingredients do ice cream and milkshakes consist of?
4. List the two unique parts to a refrigeration system.
5. What are the functions of the dasher assembly?
6. What type of food storage system is most generally used in a supermarket for produce?

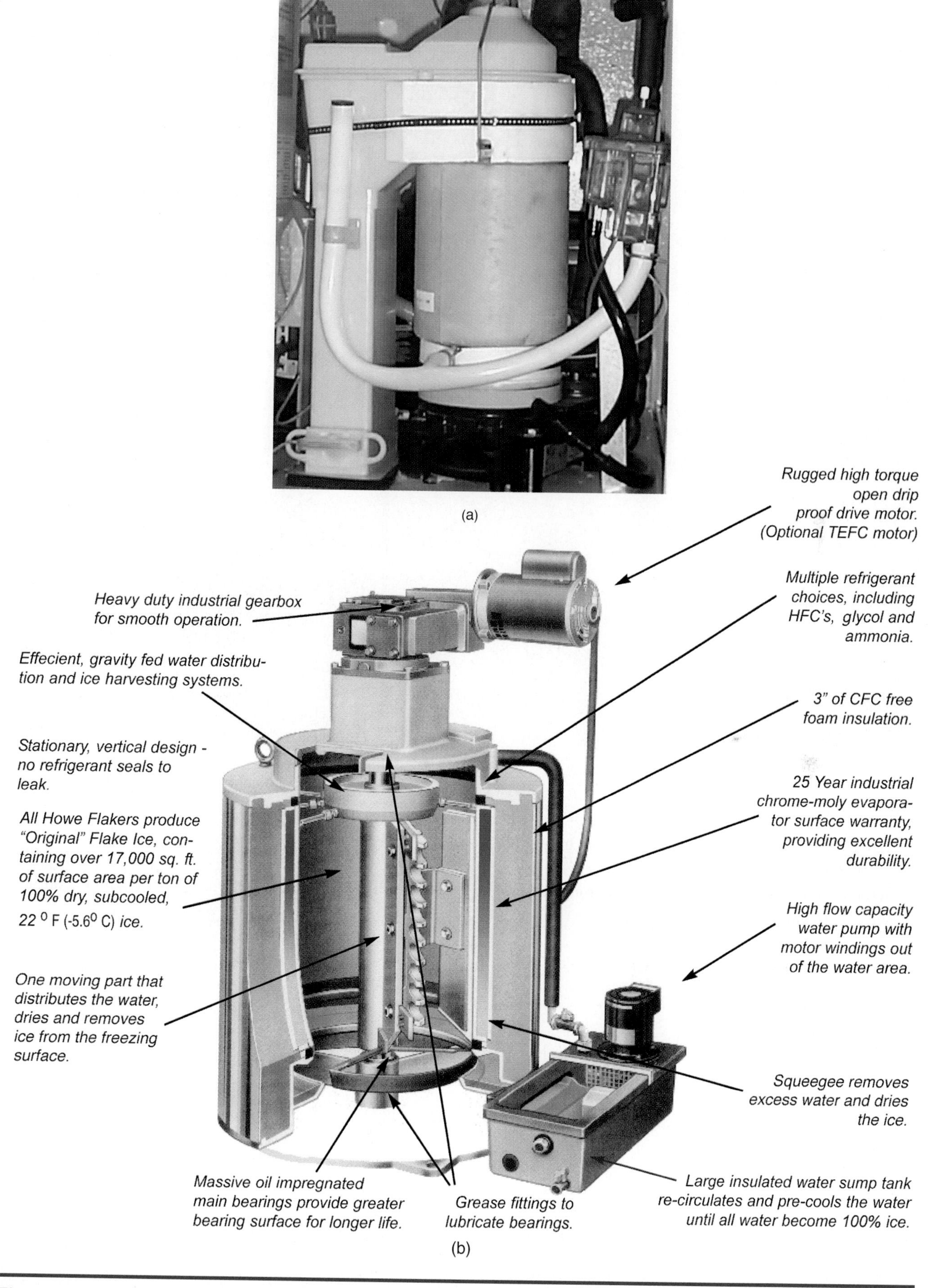

Figure 18-4-37 (a) Evaporator of a flake ice machine; (b) Cutaway view of the evaporator of a flake ice machine. (*Courtesy of Howe Corporation*)

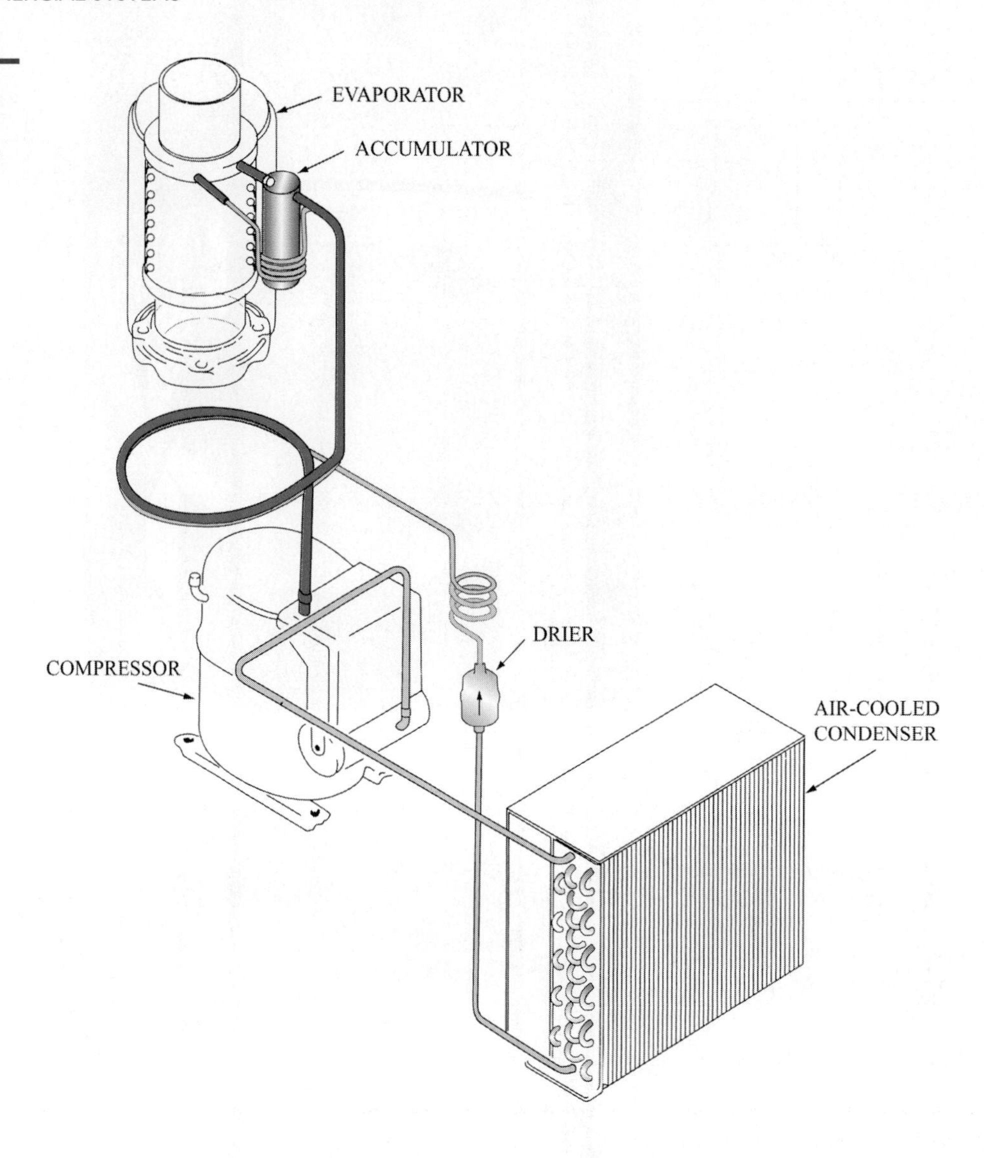

Figure 18-4-38
Refrigeration cycle for a flake ice machine. (*Courtesy of Scotsman Ice Systems*)

7. What type of food storage system is used in a supermarket for dairy products?
8. What is the advantage of the two stage compressors?
9. What is the most common type of defrost?
10. List the two types of modern chillers.
11. What are the advantages of using the screw compressors?
12. During the sequence of operations of the evaporator what happens in the harvest cycle?

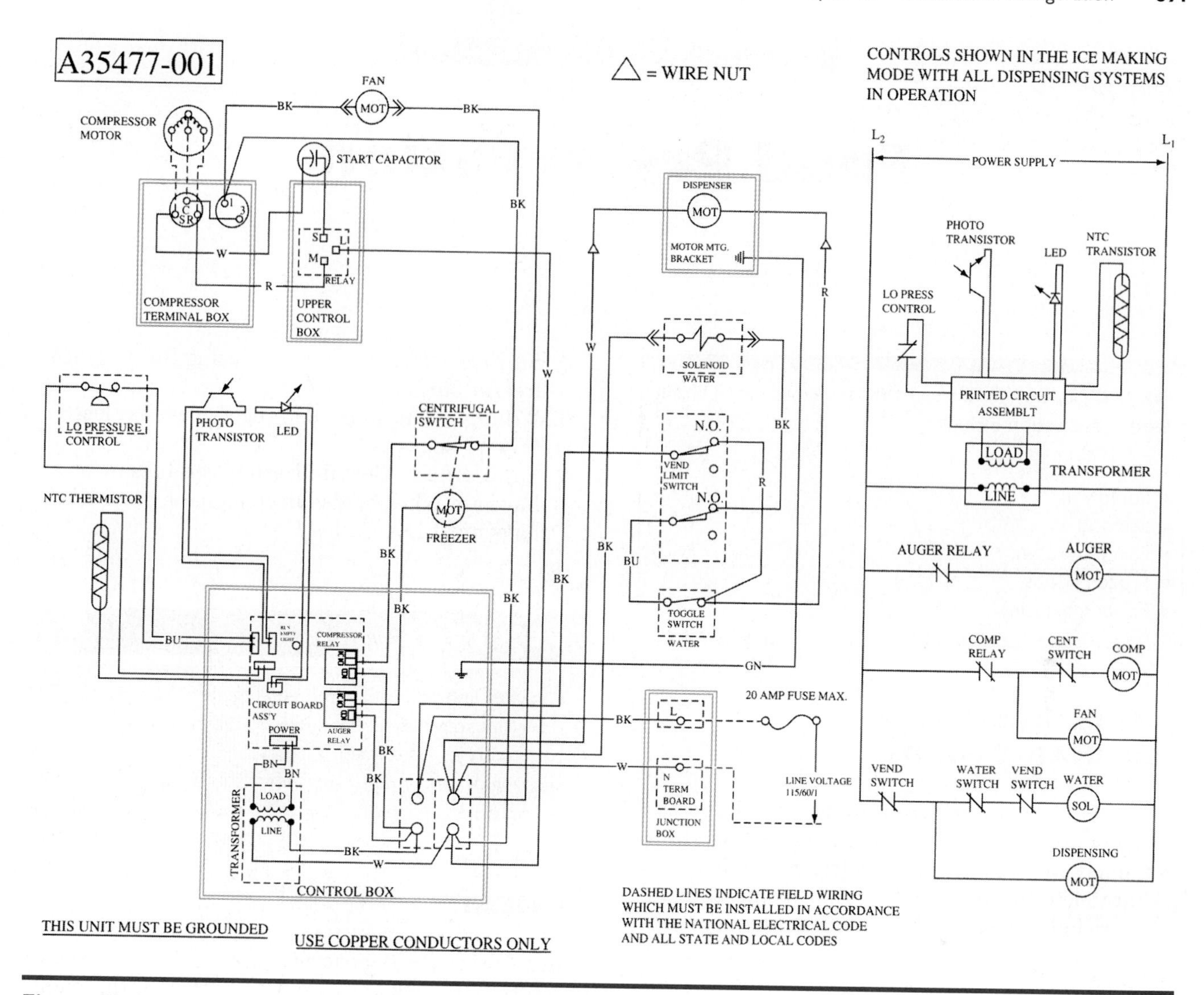

Figure 18-4-39 Wiring diagram for a flake ice machine. (*Courtesy of Scotsman Ice Systems*)

UNIT 5

Food Preservation

OBJECTIVES: In this unit the reader will learn about food preservation including:

- Meat
- Poultry
- Seafood
- Dairy and Eggs
- Frozen Foods
- Food Processing

PRESERVATION OF PERISHABLE FOODS

The perishable food industry is one of the largest industries in the country. An industry of this size is extremely important. Proper refrigeration is an important factor in the success of this business.

Perishable foods can be classified into six groups:

1. Meats
2. Poultry
3. Seafood
4. Fruits
5. Vegetables
6. Dairy products

These perishable foods can be divided into three groups: animal, vegetable, and dairy products. Each group requires separate treatment to preserve the products and to keep them palatable. The storage of animal products requires the prevention of deterioration of the nonliving products. Fruits and vegetables, however, are as much alive while they are being transported as they were while growing and require an entirely different set of preservation conditions. The principal causes of food spoilage are as follows:

1. *Microbiological.* These include bacteria, molds, and fungi.
2. *Enzymes.* These are chemical in nature and do not deteriorate.
3. *Oxidation changes.* These are caused by atmospheric oxygen coming in contact with the food, producing discoloration and rancidity.
4. *Surface dehydration.* In freezing this is called freezer burn.
5. *Wilting.* This applies to vegetables that lose their crispness.
6. *Suffocation.* Certain fresh vegetables must have air. When sealed in cellophane bags, the bags must have holes.

SERVICE TIP

It is very important that the equipment used to process, store, or serve food products be cleaned. Food borne contaminants are a major health issue and only stringent cleaning of equipment can control this problem.

MEATS

These products deteriorate through the action of bacteria. The enzymes serve to tenderize the meat. Aging is the process of utilizing the good effects of the enzymes without the harmful effects of the bacteria. Sanitation is the most important factor in controlling bacteria. Air has many forms of bacteria present. One of the best ways of controlling this infection is through the use of germicidal or ultraviolet lamps. Oxidation is detrimental to meats, causing undesirable appearance and deterioration of the flavor. Dehydration can be controlled to a large extent by maintaining high humidity in the storage room. High humidity also protects against moisture loss, which lowers the weight of the meat.

Good practice requires that pork be rapidly cooled after it is cut. This prevents destructive enzymatic action that causes discoloration, rancidity, and poor flavor. Figure 18-5-1 shows a chill cabinet in a truck loaded with pork trimmings that can be unloaded into the chiller room through side opening doors. Fans circulate the chilled air over the meat at low velocity. Air temperature is in the 0°F range with a TD (temperature difference) of 10°F. A continuous trimming chiller is shown in Figure 18-5-2. Spray coolers are used in this type of chiller.

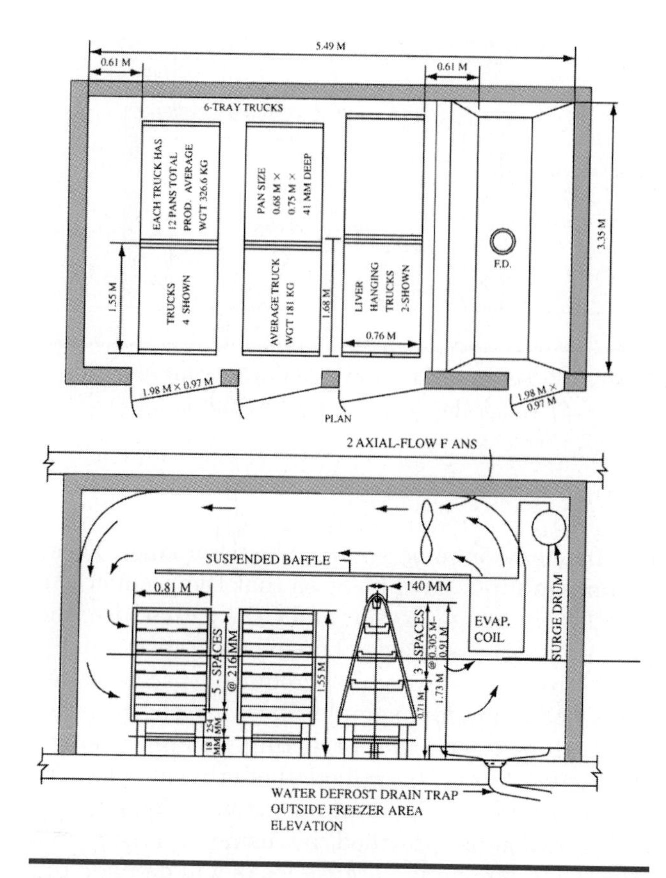

Figure 18-5-1 Chilling room for variety meats. *(Copyright by the American Society of Heating, Refrigerating, and Air-Conditioning Engineers, Inc.)*

The layout for a sausage dry room is shown in Figure 18-5-3. The keeping qualities of a variety of dry sausages produced depend on curing ingredients, spices, and removal of moisture from the product by drying. The purpose of this room is to remove about 30% of the moisture, to a point where the sausage will keep for a long time virtually without refrigeration. This process is used as an alternative to the smoking process. The U.S. Department of Agriculture (USDA) requires that this room be maintained at temperatures above 45°F, and the length of time in the room depends on the diameter of the sausage after stuffing and method of preparation.

POULTRY

Problems associated with the preservation of poultry are similar to those of meat in many respects except that poultry spoils much faster. Poultry, however, can be precooled by the use of cold water without detrimental effects. This is a relatively simple and effective process and therefore quite generally used. Bacteria and enzyme action is useful only in preserving game birds, as such action has a tendency to enhance the "game flavor."

CAUTION: All meat products have some level of bacteria. However, poultry has a particularly high level of salmonella. If you cut yourself while working on equipment that processes or stores any raw meat, especially poultry, you must take particular precautions to ensure that the cut does not become infected. If it does become infected you should seek medical advice.

SEAFOOD

This product is the most perishable of all the animal foods, yet there is a vast difference in the keeping quality of different kinds of fish. For example, swordfish can be kept refrigerated for 24 days and be in a more palatable condition than mackerel refrigerated for 24 hours. Commercial fish are usually refrigerated with ice.

FRUITS AND VEGETABLES

The unique situation with fruits and vegetables is that they are still alive after they are picked. They grow, breathe, and ripen. Most fruits and vegetables are picked in an unripened condition. The best tasting products are ripened before they are picked. The purpose of refrigeration is to slow down the ripening process so that these products can reach consumers before spoiling.

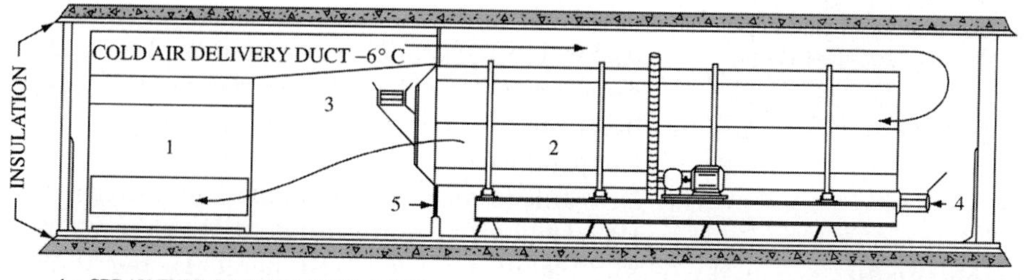

Figure 18-5-2 Continuous trimming chiller. *(Copyright by the American Society of Heating, Refrigerating, and Air-Conditioning Engineers, Inc.)*

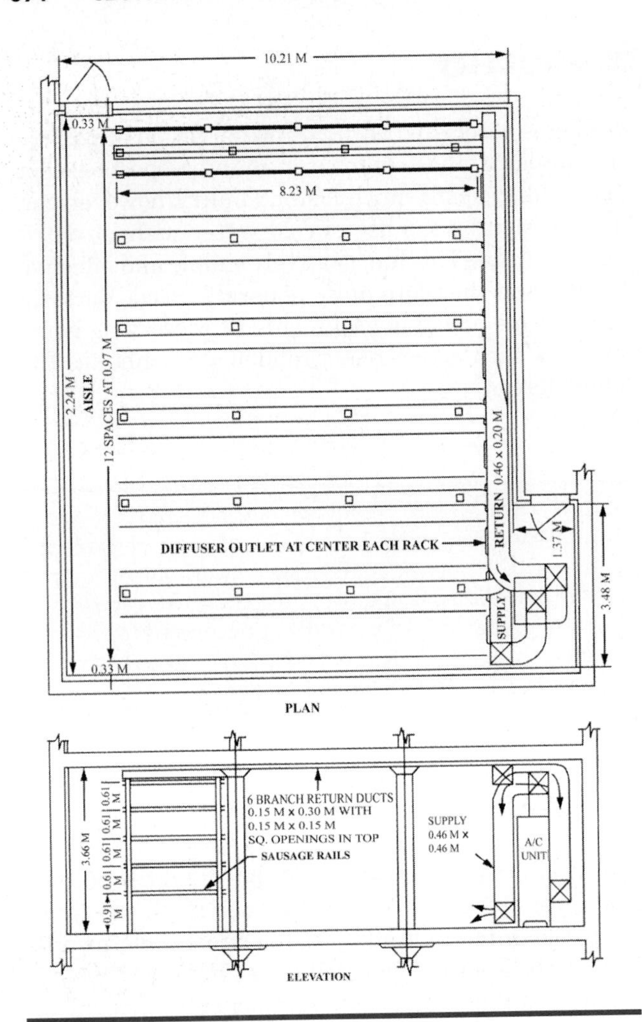

Figure 18-5-3 Sausage dry room. (*Copyright by the American Society of Heating, Refrigerating, and Air-Conditioning Engineers, Inc.*)

Vegetables quickly lose their vitamin content when surface drying takes place. It is interesting to note that products shipped from California to Chicago that have been properly iced after harvest will be fresher than produce supplied from Illinois farms and shipped to a Chicago market without being iced.

SERVICE TIP

All fruits and vegetables give off some quantity of moisture. This moisture release is called respiration. The ability to cool a product can be significantly affected by its rate of respiration due to the added latent load it represents.

Another way to improve the product when it reaches the user is to package the product. This cuts down on surface drying. Packages are usually made of cellophane or some similar plastic product. These containers must have holes so that the product can breathe (exchange oxygen and CO_2). Otherwise, the product will die, and a dead product will spoil rapidly.

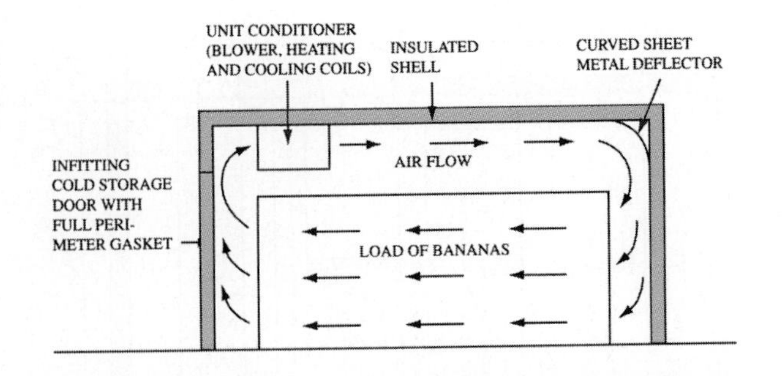

Figure 18-5-4 Banana room. (*Copyright by the American Society of Heating, Refrigerating, and Air-Conditioning Engineers, Inc.*)

A number of products require special treatment—bananas, for example. These are picked green and must be ripened for marketing. Banana ripening is initiated by the introduction of ethylene gas. For this to be effective, banana rooms must be airtight. Refrigeration is provided by using a refrigerant other than ammonia because leaks will damage the fruit. Rooms such as those shown in Figure 18-5-4 are cooled using 45 to 65°F air. A design temperature difference of 15°F and a refrigerant temperature of 40°F are considered good practice.

An interesting application of refrigeration is the processing of fruit juice concentrate as shown in Figure 18-5-5. Hot gas discharge from the compressor is used to supply heat for the evaporation of the juices. The water vapor is condensed by evaporating liquid refrigerant in a shell in tube condenser. Water vapor is used to superheat the suction gas. This type of apparatus provides a continuous process.

DAIRY PRODUCTS AND EGGS

Sanitation is extremely important in all stages of handling milk. The bacteria content of milk must be controlled. Mechanical refrigeration begins to cool it even during milking, from 90 to 50°F within the first hour and from 50 to 40°F within the next hour. As more milk is added, the blended liquid must not rise above 45°F. Limits are set for the number of bacteria (the bacteria count) for milk supplied by the producer.

Milk is stored in insulated or refrigerated silo type tanks that maintain a 40°F temperature. After milk is pasteurized and homogenized it is again cooled in a heat exchanger (a plate or tubular unit) to 40°F or lower, and packaged.

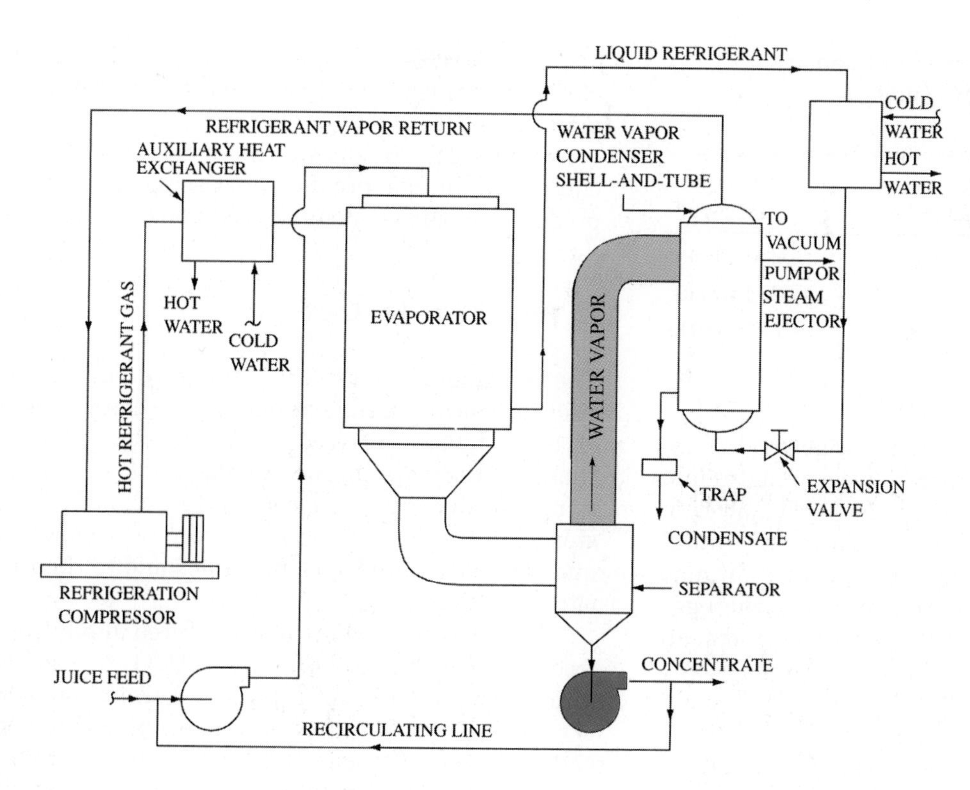

Figure 18-5-5 Equipment for fruit juice concentrates. *(Copyright by the American Society of Heating, Refrigerating, and Air-Conditioning Engineers, Inc.)*

Butter

Butter is manufactured from 30–40% cream obtained from the separation of warm, acidified milk. It is cooled to 46–55°F, then churned to remove excess water. The average composition of butter on the market has these ranges:

Fat:	80–81.2%
Salt:	1.0–2.5
Moisture:	16–18
Curd, etc:	0.5–1.5

Butter keeps better if stored in bulk. If kept for several months, the temperature should not be above 0°F and preferably below −20°F. For short periods, 32–40°F is satisfactory. If stored improperly, the quality of butter deteriorates from absorption of atmospheric odors, loss of weight through evaporation, surface oxidation, growth of microorganisms and resulting activity of enzymes, and low pH (high acid) of salted butter. Low temperatures, a clean environment, use of a good quality cream, avoidance of light, copper, and iron, and adjustment of the pH to 6.8–7.0 eliminates most of these problems.

Cheeses

Cheeses are refrigerated to prevent too rapid mold growth. The surface must be kept moist or the cheese will become hard and brittle. Moisture, meanwhile, facilitates mold. While some mold enhances the flavor, too much mold creates waste because it must be removed before sale. The ideal storage temperature range for various types of cheese is in the range of 30–34°F for natural cheeses and 40–45°F for processed cheeses. Maximum temperatures range from 45–60°F for the natural cheeses while the processed cheeses may be kept on open shelves at 75°F.

NICE TO KNOW

Many cheeses contain active organisms that provide the cheese with its unique flavor. Improper storage of these cheeses can damage these organisms and the flavors that they provide the cheese. It is very important that the specific temperature and humidity range for a particular cheese be obtained from the customer before a storage system is established. Some cheeses have very broad temperature and humidity tolerances while others are extremely narrow and sensitive. The active organisms in cheese can also continue to generate heat after the cheese is in the cooler. This contributes additional load and extends the pulldown time.

Eggs

Eggs should be refrigerated at all stages of handling and storage. Shell eggs account for about 75% of all eggs used. Table 18-5-1 lists the temperatures, humidity, and time they can be stored.

Research has shown that microbial growth associated with Salmonella can be controlled by holding eggs at less than 40°F. This has led to major changes

TABLE 18-5-1 Storage Period of Food at Various Temperature and Humidity Conditions

°F Temperature	Relative Humidity (%)	Storage Period
50–60	75–80	2–3 weeks
45	75–80	2–4 weeks
29–31	85–92	5–6 months

in storage and display areas not refrigerated or inefficiently refrigerated. All egg storage areas should maintain ambient temperatures at 45°F; however, with mechanized processing and packaging procedures which insulate the eggs within cartons, it may require up to one week of storage before the eggs reach the temperature of the storage room. If shipped earlier to sell a "fresh" product, eggs are only partially cooled. Methods are being developed to improve cooling in the processing plant.

Shipment in refrigerated trucks is mandatory. Problems that arise are associated with the truck design, manner of loading, size of the shipment, and distance. A 1993 USDA survey found that over 80% of the trucks used to deliver eggs were unsuitable to maintain 45°F.

Figure 18-5-6 shows the use of mechanized equipment in egg product processing. The various products are shown along the right edge of the figure.

FROZEN FOODS

The freezing of food, although highly product specific, is basically a time temperature related process of three phases: (1) cooling to freezing point, (2) changing the water in the product to ice, and (3) lowering the freezing temperature to optimum frozen storage temperature. Product differences as well as quality relate to the specific values and to the rates of these stages.

The following factors are considered in selecting a freezing system for a specific product: (1) special handling requirements, (2) capacity, (3) freezing times, (4) quality consideration, (5) yield, (6) appearance, (7) first cost, (8) operating costs, (9) automation, (10) space availability, and (11) upstream/downstream processes.

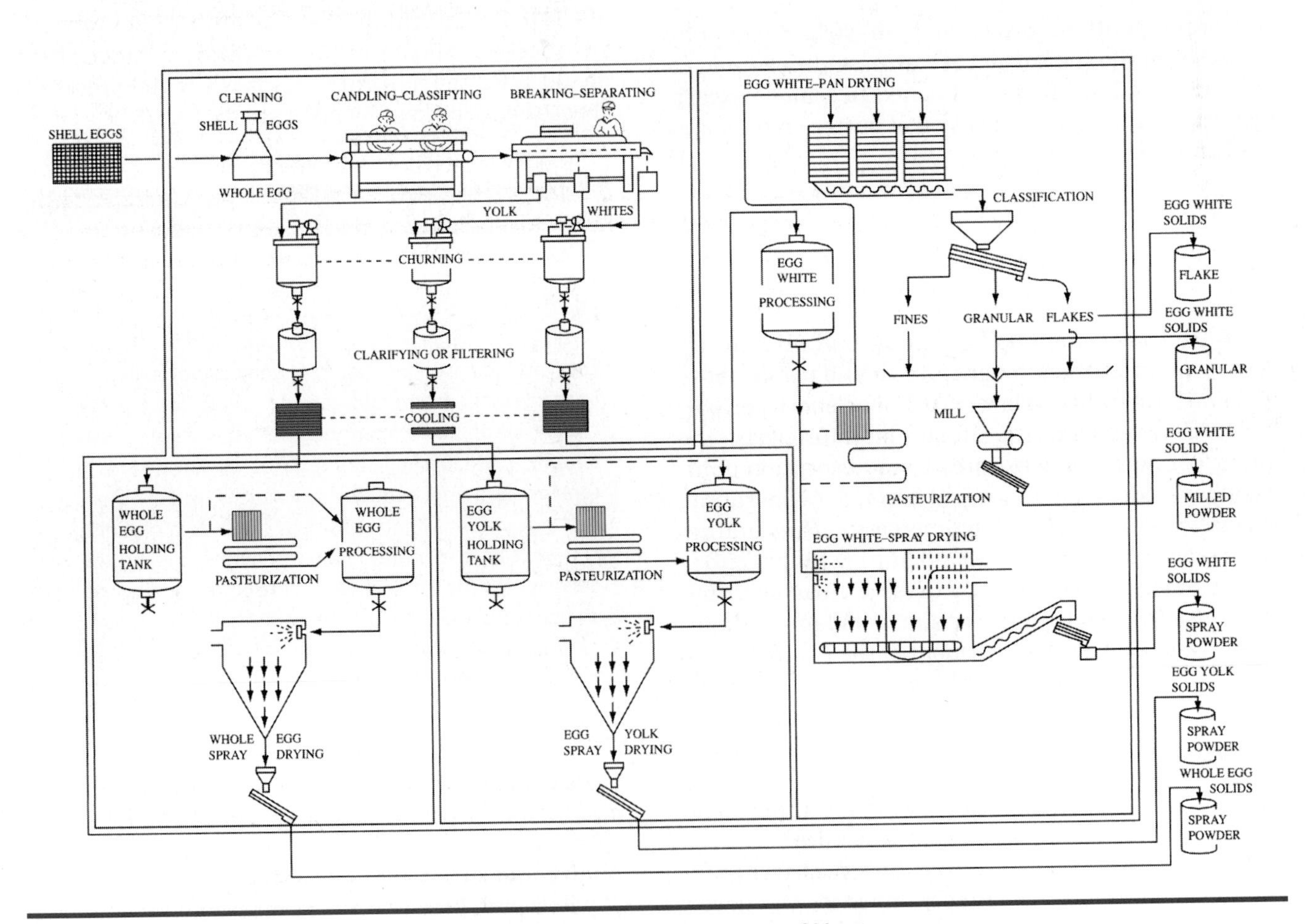

Figure 18-5-6 Egg product processing. *(Copyright by the American Society of Heating, Refrigerating, and Air-Conditioning Engineers, Inc.)*

Quick Freezing

This essential process produces small ice crystals which are less damaging to the product. Ideally, the ripe produce should be frozen immediately after harvest—the sooner the better. Small packages are better to freeze than large packages because the interior freezes more quickly.

Small packages may be frozen on or between refrigerated plates or in a "blast" freezer. Foods are frozen at temperatures between −5 and −20°F. Freezer burn should be avoided. This is the condition of surface oxidation that causes discoloration of the product. It is prevented by packaging in airtight containers or by waxing or glazing the product. Ice glazing is used to prevent surface drying of fish. Fruits are often glazed with a sugar syrup to prevent oxidation.

NICE TO KNOW

As the water in products begins to freeze it forms very long sharp crystals. The slower the freezing process the larger these crystals become. If the crystal size is large enough it will puncture the cell walls of the product being frozen. When the product thaws and the ice crystals melt, fluids in the individual cells of the product can be lost. This loss of fluid results in a significant decrease in the quality of the product. To reduce this problem many products are fast frozen. This process produces the smallest possible crystal form thus reducing the loss of special product fluids and maintaining product quality.

Vegetables must be *blanched* before freezing. This consists of placing the product in boiling water or steam to kill bacteria and to stop enzyme action. Air is removed from citrus juice before freezing.

Commercial freezing systems can be divided into four groups:

1. Air blast freezers
2. Contact freezers
3. Immersion freezers
4. Cryogenic freezers

Air blast freezing can best be described as a convection system, where cold air at high velocities is circulated over the product. The air removes heat from the product and releases it to an air refrigerant heat exchanger before being recirculated.

The air blast freezer using a stationary tunnel, Figure 18-5-7, produces satisfactory results for practically all products, in or out of packages. Products are placed in trays that are held in racks, placed so

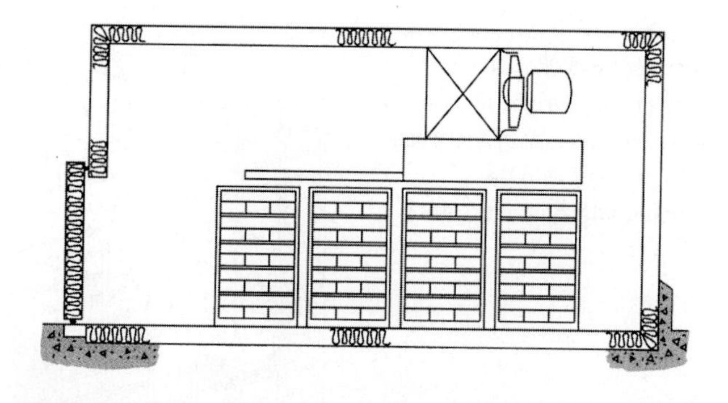

Figure 18-5-7 Stationary freezing tunnel. *(Copyright by the American Society of Heating, Refrigerating, and Air-Conditioning Engineers, Inc.)*

that air bypass is minimized. Air blast freezers can be mechanized to provide a continuous process, as shown in Figure 18-5-8 and 18-5-9. Two stage belt freezers permit precooling at 15–25°F before transferring to the second belt for freezing at temperatures of −25 to −40°F.

Belt type freezers use vertical airflow and greatly improve the contact between air and product. Two types are shown in Figures 18-5-10 and 18-5-11.

The fluidization principle is illustrated in Figure 18-5-12. Solid particulate products such as peas, sliced and diced carrots, and shredded potatoes or cheese are floated upward and through the freezer by streams of air. The product is frozen in 3 to 11 minutes by refrigerant temperatures of −40°F. These freezers are packaged, factory assembled units. Figure 18-5-13 shows a typical fluidized bed freezer.

Contact freezers are conduction type freezers. Products are placed on or between horizontal or vertical refrigerated plates that provide efficient heat transfer and short freezing time. Relatively thin packages of food or products, 2 to 3 in thick, are used in this system. A typical freezer is shown in Figure 18-5-14.

Cryogenic freezers utilize both convection and/or conduction by exposing food to temperatures below −76°F in the presence of liquid nitrogen or liquid carbon dioxide refrigerants. Liquid nitrogen boils at

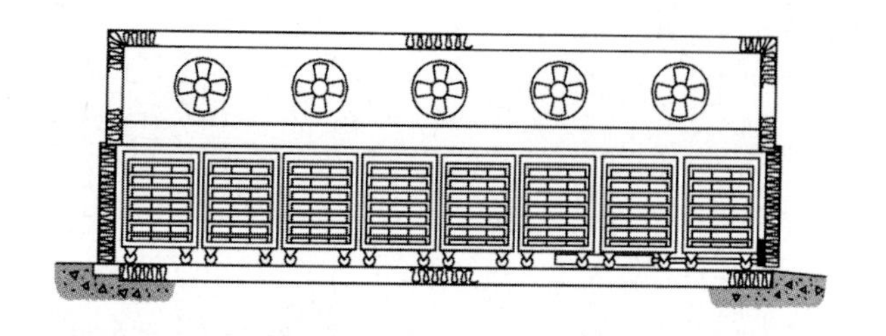

Figure 18-5-8 Push through tunnel. *(Copyright by the American Society of Heating, Refrigerating, and Air-Conditioning Engineers, Inc.)*

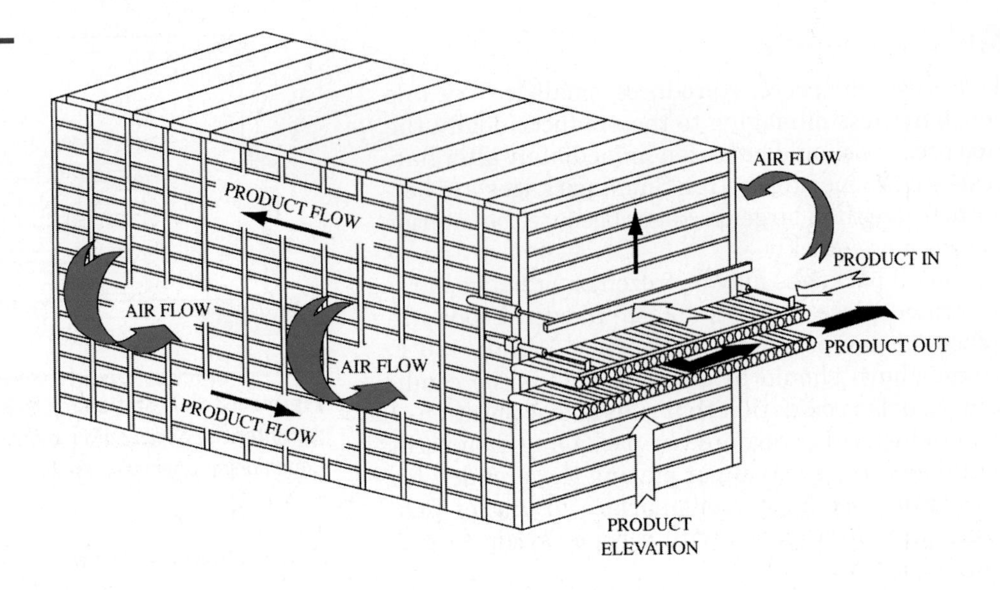

Figure 18-5-9 Carrier freezer. *(Copyright by the American Society of Heating, Refrigerating, and Air-Conditioning Engineers, Inc.)*

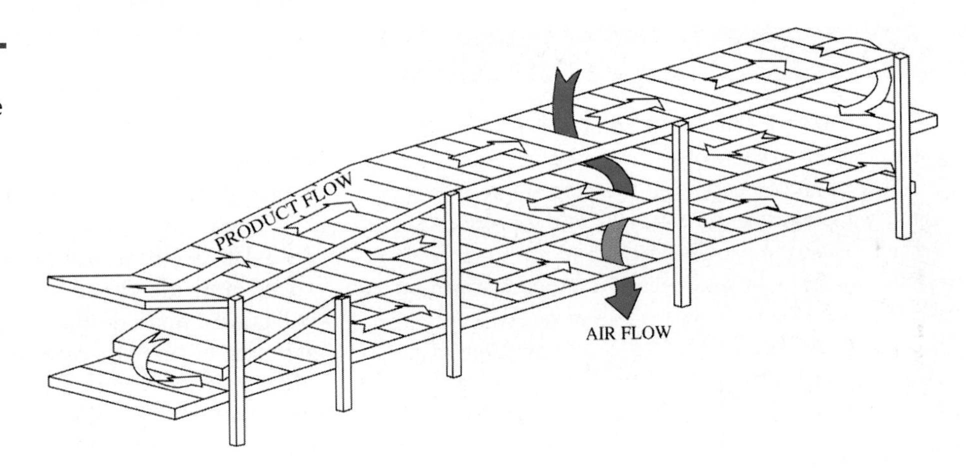

Figure 18-5-10 Multiple belt freezer. *(Copyright by the American Society of Heating, Refrigerating, and Air-Conditioning Engineers, Inc.)*

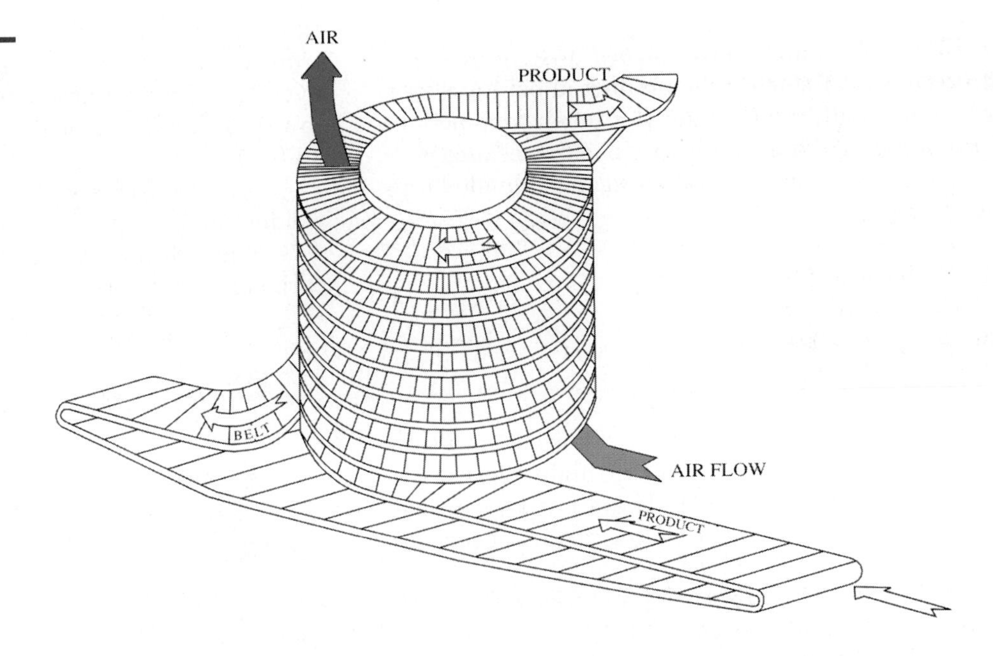

Figure 18-5-11 Spiral belt freezer. *(Copyright by the American Society of Heating, Refrigerating, and Air-Conditioning Engineers, Inc.)*

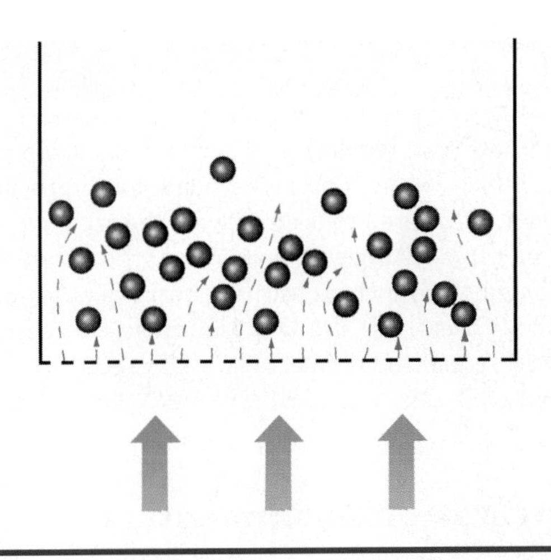

Figure 18-5-12 Fluidization principle. *(Copyright by the American Society of Heating, Refrigerating, and Air-Conditioning Engineers, Inc.)*

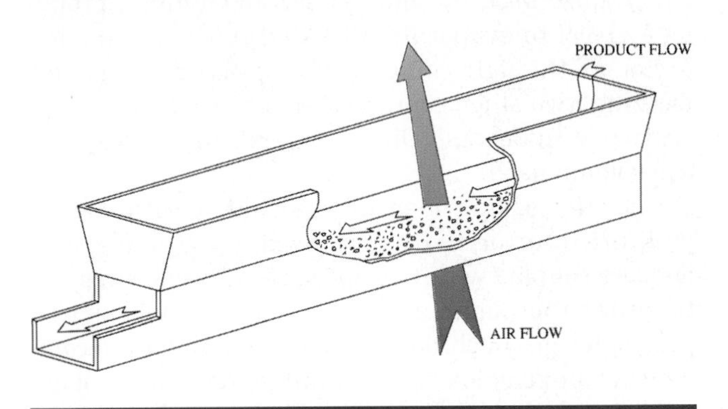

Figure 18-5-13 Fluidized bed freezer. *(Copyright by the American Society of Heating, Refrigerating, and Air-Conditioning Engineers, Inc.)*

−320°F and carbon dioxide boils at approximately −110°F. The freezers may be cabinets, straight belt freezers, spiral conveyors, or liquid immersion freezers. The boiling liquid comes in direct contact with the product. After use, the refrigerant is wasted to the atmosphere. A freezer of this type is shown in Figure 18-5-15. While operating costs are high, the small initial investment makes this economical for certain foods.

Some products, such as shrimp, are frozen by *immersion* in a boiling, highly purified refrigerant.

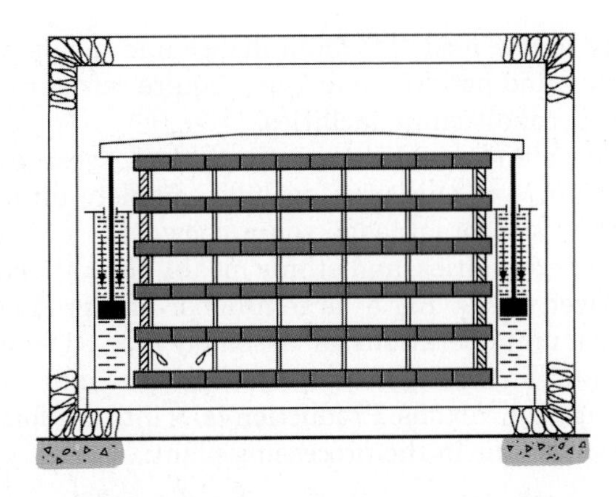

Figure 18-5-14 Plate freezer.

The surface of a sticky or delicate product is "set" by this rapid freezing, reducing dehydration and improving the handling characteristics of the product. The product is then removed and the freezing process completed in a mechanical freezer. In these systems the refrigerant is recovered by condensing on the surface of refrigerated coils.

Refreezing

When a vegetable product is in a frozen state, for all practical purposes it is considered "dead"; however, microbes and enzymes may remain there in an inactive state. When thawed, a large amount of water is present from the ruptured tissues to provide a favorable environment for the growth of microbes and deterioration due to the enzymes. All of these processes serve to lower the quality of the product and could continue after refreezing. The product should be heated sufficiently to kill these destructive agents before refreezing. Some canneries freeze products to prevent spoilage until they can schedule the final canning process.

FOOD PROCESSING

The production of precooked and prepared foods developed into an important industry the last half of the twentieth century. These foods, which include

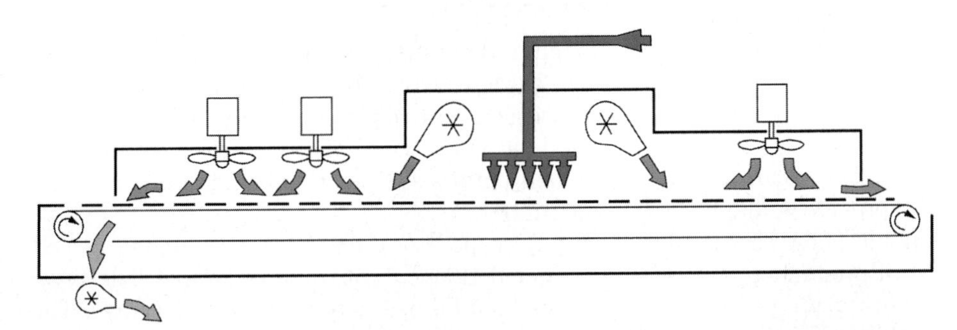

Figure 18-5-15 Cryogenic freezer. *(Copyright by the American Society of Heating, Refrigerating, and Air-Conditioning Engineers, Inc.)*

ready to use foods for main dishes and meals, vegetables, and potato production, require refrigeration and air conditioning facilities.

Main dishes, which constitute the largest group of products in this area, include complete dinners, lunches and breakfasts, soups/chowders, low calorie/diet specialties, and ethnic meals. They are characterized by having a large number of ingredients, several unit operations, an assembly type packaging line, and final refrigerating or freezing of individual packages or cartons. Production falls into the following operations in the processing plant:

1. *Preparation, processing, and unit operations.* This involves the initial preparation of all ingredients to be assembled, including refrigeration and/or freezing needs. These require specific attention to individual requirements for selecting processes, equipment, space, controls, and safeguards.
2. *Assembly, filling, and packaging.* This includes all handling of components for putting into containers or packages, packaging, and placing the containers and packages in refrigerators or freezers. It is considered good practice to air condition filling and packaging bins to control bacteria and increase worker productivity.
3. *Cooling, freezing, and casing.* There is a constant effort to improve the economy and efficiency of production, often altering original design conditions. Space and equipment capacity allowances should be 25–50% for increasing requirements. Maintenance should include checking temperatures and other specific conditions of the particular product involved. Defrosting on conveyor systems should be checked when there are hangups or stoppages. When the position or place of packages is changed in the storage processes, the quality of the product should be closely monitored.
4. *Finishing: storage and shipping.* Infiltration and product pulldown loads occur when there is negative air pressure due to exhausting more air than is supplied by ventilation. This causes a serious load on the refrigeration, making it difficult to maintain proper storage temperatures.

Refrigeration Loads

Records need to be kept of operating conditions to identify poor performance and to provide guidance for new systems. These records should show conditions for time of day, season, on/off shift production, evaporator temperature, and equipment type/function.

SERVICE TIP

Many plants have requirements for hairnets and protective clothing while working around food processing equipment. You are required to wear the same type and level of personal hygienic materials as other workers. These items may include hairnets, beard nets, lab coats, and shoe covers. It is also important that you keep your tools very clean and disinfected.

Refrigeration Equipment

The refrigerant used for many large refrigeration systems is ammonia. To avoid the hazard of a potential ammonia spill to workers in the plant, glycol chillers are used by some plants to circulate propylene glycol to evaporators located in the production areas. With high compression ratios required for freezing, two stage compressors are used with evaporative condensers. Direct expansion evaporators are seldom used.

Energy saving measures include floating head pressure controls with oversized evaporative condensers coupled with two speed fans, single stage refrigeration for small areas and loads, variable speed pumps for glycol chiller systems, ice builders to compensate for peak loads, door infiltration protection devices, insulation, and computerized control systems.

Vegetables

All prepared vegetables are precooked and cooled before freezing as discussed in the previous section. Refrigeration is used for raw product cooling and storage, product cooling after blanching, freezing, and process equipment located in freezer storage facilities, and freezer storage warehouses. Loads vary widely depending on the product.

In vegetable facilities that operate only for short periods at peak capacity—1500 to 2500 hr/yr—spare equipment cannot be economically justified and good maintenance is important to avoid downtime losses.

Potatoes

Prepared potato products in various forms dominate the frozen ready to use vegetable group and are processed year round. Products include French fries, hash browns, twice baked potatoes, potato skins, and boiled potatoes. French fries are probably the most popular.

Raw potatoes for fries are steam peeled and trimmed, then cut into desired shapes. The slivers are graded out for use as puffs, tots, and wedges. The fries are blanched, and then partially dried and oil fried. They are frozen on a straight belt freezer system with three separate conveyors for precooling and totally freezing the fries to 5–10°F. Sorting is done at 15°F and packaged in an air conditioned area.

Storage

Freezing and thawing temperatures of animal, vegetable, dairy, and egg products vary widely. Temperatures must be maintained that preserve the quality and safety of products over the time periods required. As a result of ongoing research, lower temperatures are being recommended.

Most modern refrigerated warehouses are one story structures. A typical floor plan is shown in Figure 18-5-16. The building construction is shown in Figure 18-5-17. Note that the outer walls are independent of the rest of the building. It is usually convenient to use penthouse refrigeration equipment rooms, as shown in Figure 18-5-18. Consideration must be given to:

1. Entering temperatures
2. Duration of storage
3. Required product temperature for maximum/minimum protection
4. Uniformity of temperatures
5. Air movement and ventilation
6. Humidity
7. Traffic in and out of storage space
8. Sanitation
9. Light

Freezer storage at vegetable processing plants must also consider the following potential additional loads: (1) extra reserve capacity needed for product pulldown during peak processing, (2) negative pressure that can increase infiltration by direct flow through, and (3) the process machinery load (particularly pneumatic conveyors) associated with repack operations.

Regulations and guidelines for the refrigerated storage of foods have been established by the following

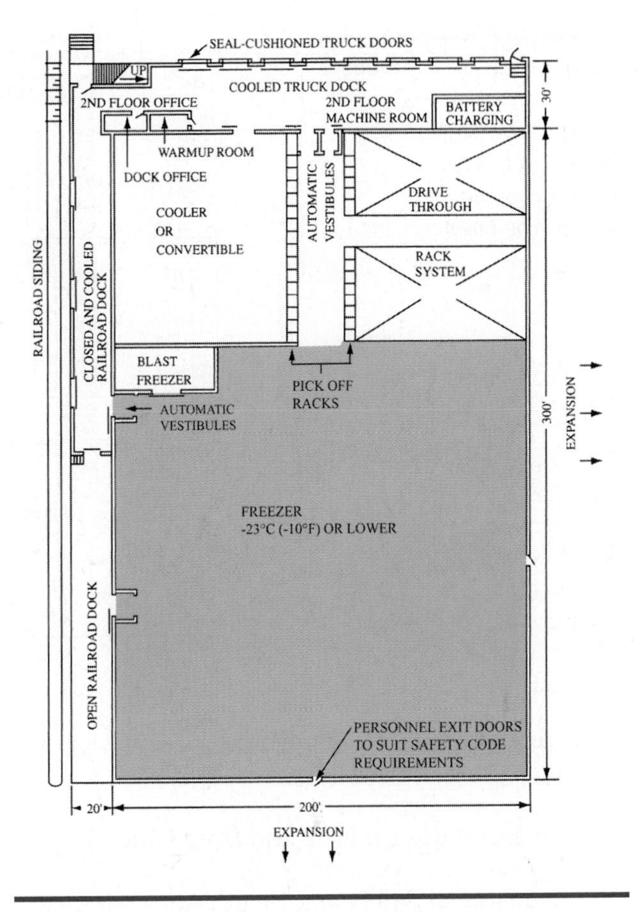

Figure 18-5-16 One story warehouse. *(Copyright by the American Society of Heating, Refrigerating, and Air-Conditioning Engineers, Inc.)*

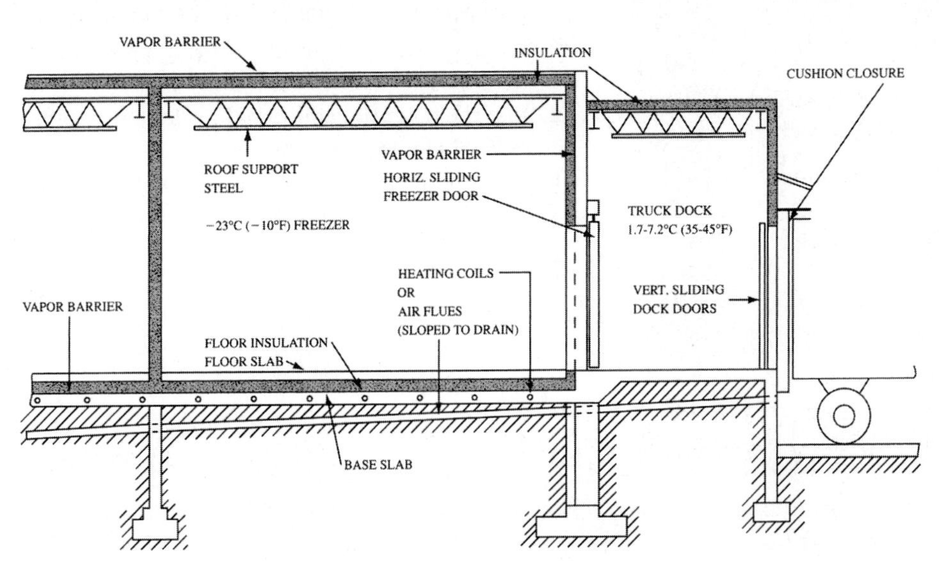

Figure 18-5-17 Typical one story construction with hung insulated ceiling and underfloor warming pipes. *(Copyright by the American Society of Heating, Refrigerating, and Air-Conditioning Engineers, Inc.)*

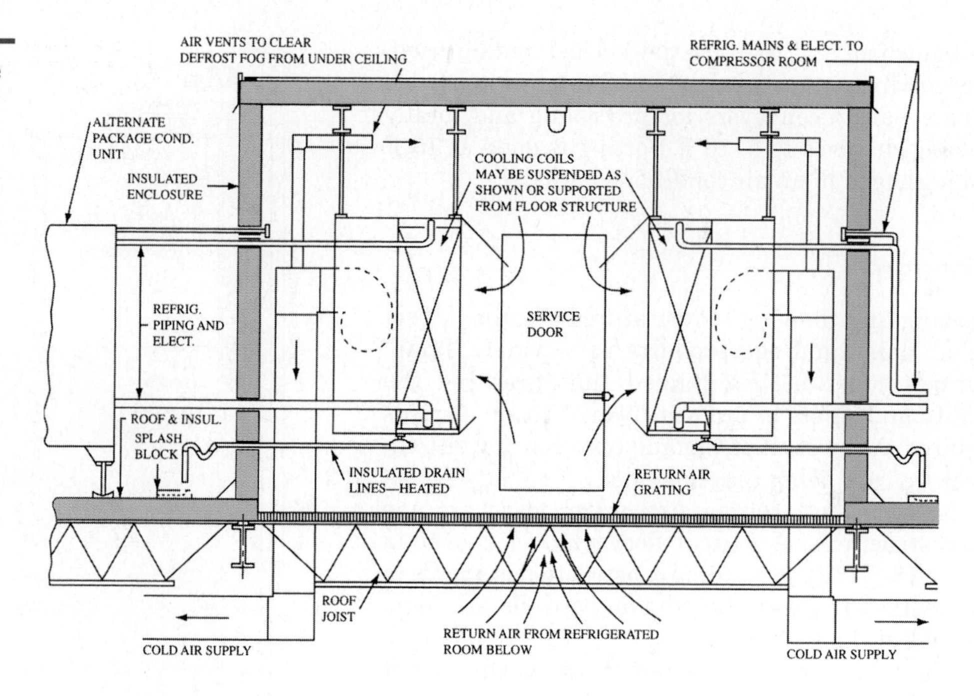

Figure 18-5-18 Penthouse application of cooling units. *(Copyright by the American Society of Heating, Refrigerating, and Air-Conditioning Engineers, Inc.)*

agencies and should be familiar to the serviceperson working in these areas:

1. The Association of Food and Drug Officials (AFDO)
2. Occupational Safety and Health Act (OSHA)
3. U.S. Department of Agriculture (USDA)
4. Environmental Protection Agency (EPA)

TECH TALK

Tech 1. I like working around food. There are always free samples. Especially those cheeses.

Tech 2. You can't simply pick up free samples while you are working around that stuff for a number of reasons but primarily it is theft.

Tech 1. What do you mean theft? Everybody else is doing it.

Tech 2. That may be true and it may be true for the people that work there but we are part of this service company and we have no right or business eating any of their product without paying for it.

Tech 1. I guess I could get in big trouble.

Tech 2. Yeah and besides that some of those cheeses are extremely expensive. You don't want to eat something that is going to take your whole paycheck to buy.

Tech 1. Thanks I am not going to do any of that anymore.

UNIT 5—REVIEW QUESTIONS

1. List the six categories of perishable foods.
2. What are the principal causes of spoilage in foods?
3. _______ is the most perishable of all the animal foods.
4. What is the ideal storage temperature range for various types of cheese?
5. When should eggs be refrigerated?
6. List the three phases of freezing food.
7. What are the factors that are considered when selecting a freezing system?
8. Vegetables must be _______ before freezing.
9. Describe the contact freezing system.
10. How are products like shrimp frozen?
11. List the four groups of commercial freezing systems.
12. What is included in the assembly, filling and packaging operation of foods in a processing plant?
13. What is the refrigerant used for most refrigeration systems?
14. List the potential additional loads that vegetable processing plants must consider.
15. List the agencies that establish the regulations and guidelines for the refrigerated storage of foods.

19

Central Plant Hydronic Systems

UNIT 1

Air Handling Units

OBJECTIVES: In this unit the reader will learn about air handling units including:

- Air Distribution Systems
- Control Systems
- Direct Digital Control
- Pneumatic Control

AIR HANDLING UNITS

As shown in Figure 19-1-1 the air handling unit is the focal point of a number of subsystems that make up a typical central HVAC layout. Referring to the diagram, the subsystems are as follows.

Heating

The heating circuit consists of the boiler, a pump, the heating coil, and the necessary piping.

Cooling

The cooling circuit consists of a chiller, a pump, the cooling coil, and the necessary piping.

Condenser Water

This circuit consists of a cooling tower, a pump, a water cooled condenser (part of the chiller), and the necessary piping. This is actually a subcircuit to the cooling circuit.

Air Handling

This circuit has many parts. In tracing this circuit, we will start at the entrance to the air handling unit. Air that enters the unit is made up of both return air from the conditioned space and outside ventilation air. In the air handling unit, air passes through the filter, humidifier (not shown), and various heat transfer coils and into the fan. The fan delivers air through the distribution system to supply conditioned air to meet the space requirements.

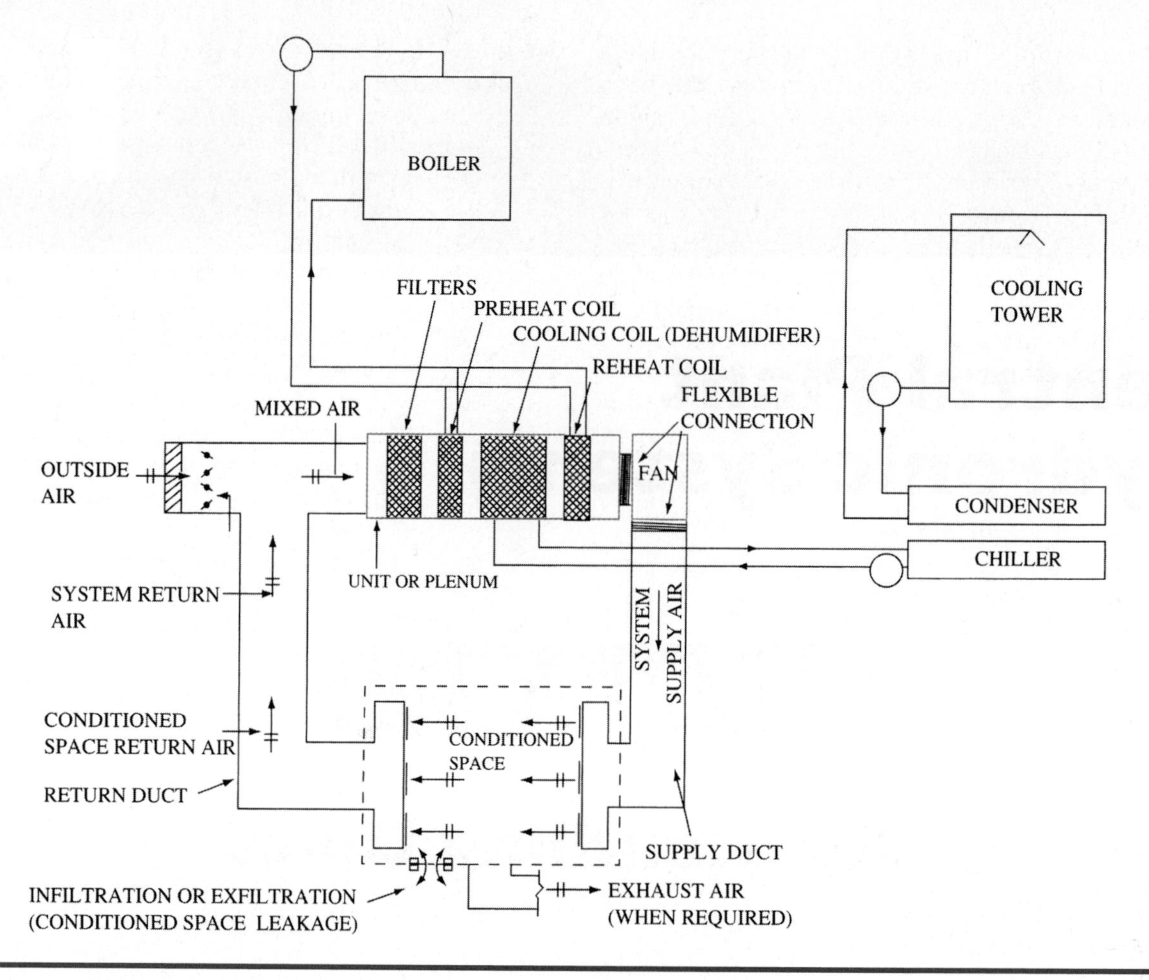

Figure 19-1-1 Typical central HVAC system showing air handler unit.

Note that provision is made in the air system for the use of 100% outside air during an economizer cycle. The excess air is exhausted after it has served its purpose, through the air relief opening.

AIR DISTRIBUTION SYSTEMS

For simplicity, types of commercial air distribution may be classified as low, medium, or high velocity systems.

Low velocity systems are those associated with the application of smaller unitary package and split system units. The use of limited ductwork or none at all (free blow) is typical of the classification. Where ductwork is used, the external static pressure is held down to the range 0.25 to 0.50 in WC. The use of concentric supply and return ducts is a common application. The type of duct design is the same as that of the equal friction method, which permits the prediction or control of total static.

TECH TIP

The higher the air distribution system's operating static the lower the system efficiency. This is so because the blower has to work harder to move enough air. But to reduce the static pressure, larger ducts are required. So the savings from having a lower static pressure must be balanced against the higher cost of the larger duct system.

Air Distribution Applications

Air distribution applications in central station systems fall into three broad areas: interior or core areas, exterior or perimeter areas or zones, and entire building applications.

Core areas have conditioning loads subject only to interior loads, such as lighting, people, and equipment. Consequently, they are basically cooling only loads, except for top floors and/or warmup cycles in extremely cold climates.

Perimeter zones are exposed to outer building skin variables, such as wall and window loads, wind effects and exposure effects, as well as the items in the core areas. The perimeter zones have to handle a wide range of conditions, including such variances as high solar gain on a winter day requiring cooling, while shaded parts of the building require heating.

Entire building applications are a combination of these systems, the selection of which depends on building size and economics. Small buildings frequently cannot justify one system for the core and another for the perimeter. In these instances, any one of the perimeter systems can be used for the entire building.

The systems are all of the type that use primary (supply) air from the air conditioner, and the maximum quantity needed is based on the maximum load conditions of the area.

Core Area Applications

If the interior core of a building is a large open area, a *single zone constant volume system* Figure 19-1-2 can be used at a reasonable initial cost. A single heat cool thermostat with automatic changeover provides year round control of the coil temperatures. Top floor and ground floor systems will have heating for morning warm up. Intermediate floors usually only have provisions for cooling. The duct design follows conventional practice, with room air distribution from the ceiling.

For core areas divided into smaller spaces, with variations in lighting and people loads, a *variable air volume* (VAV) *single zone system* has typically been used. There are several approaches to this application. In commercial buildings air circulation and fresh air requirements are easily met with VAV systems, because the fans can run continuously without providing excessive heating or cooling. In addition as the fan slows down under light loads less energy is used.

Figure 19-1-3 shows a constant volume air handler connected to individual VAV terminals. The terminal may be an individual control box, a system air powered VAV slot diffuser, or a self contained temperature actuated VAV diffuser. In all these cases, the variable air volume is accomplished by throttling the airflow at the individual duct run, causing the constant volume air handler to "back up" the fan curve to a new balance point of lower CFM at a higher static pressure. The controls for this type of system are self contained within the air distribution system, and do not require any more control than the thermostat noted above.

Another approach for the core area is to use VAV terminals, with the capability of handling several duct runs from each terminal, Figure 19-1-4. In this case, the air handler is also constant volume, but the VAV terminal throttles the air supply to the ducts, and dumps the remainder back into the ceiling space return air plenum. The space is controlled with variable air volume, but the air handler is operating at constant volume. The control for this subsystem requires that the space thermostat controls an actuator on the VAV terminal, as well as controls the cooling and heating demand.

In either of these cases, a single air handler is sufficient in the core zone and can provide proper comfort conditions. The air handler is typically constant volume because the loads are relatively steady, regardless of the season or ambient air temperature.

An approach often seen on older systems, but prohibited by current energy codes, is commonly called *zone reheat*. As seen in Figure 19-1-5, an air handler delivers *constant volume, constant temperature* air to

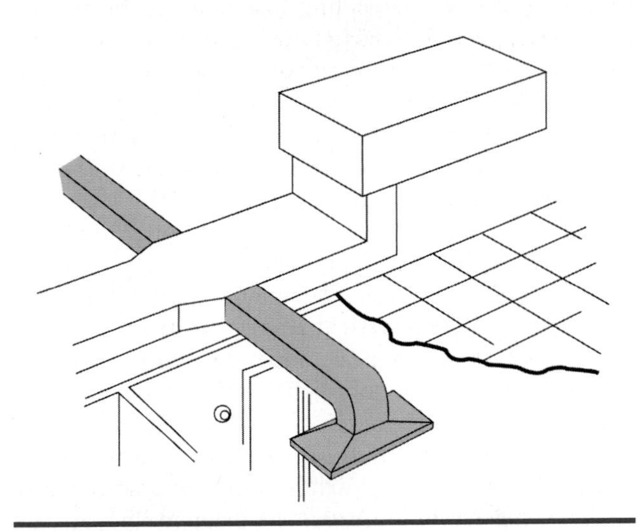

Figure 19-1-2 Single zone system using constant volume.

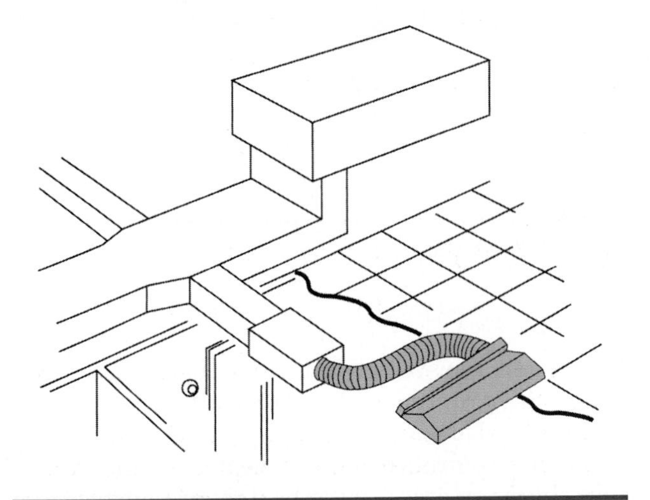

Figure 19-1-3 Single zone system using variable air volume (VAV).

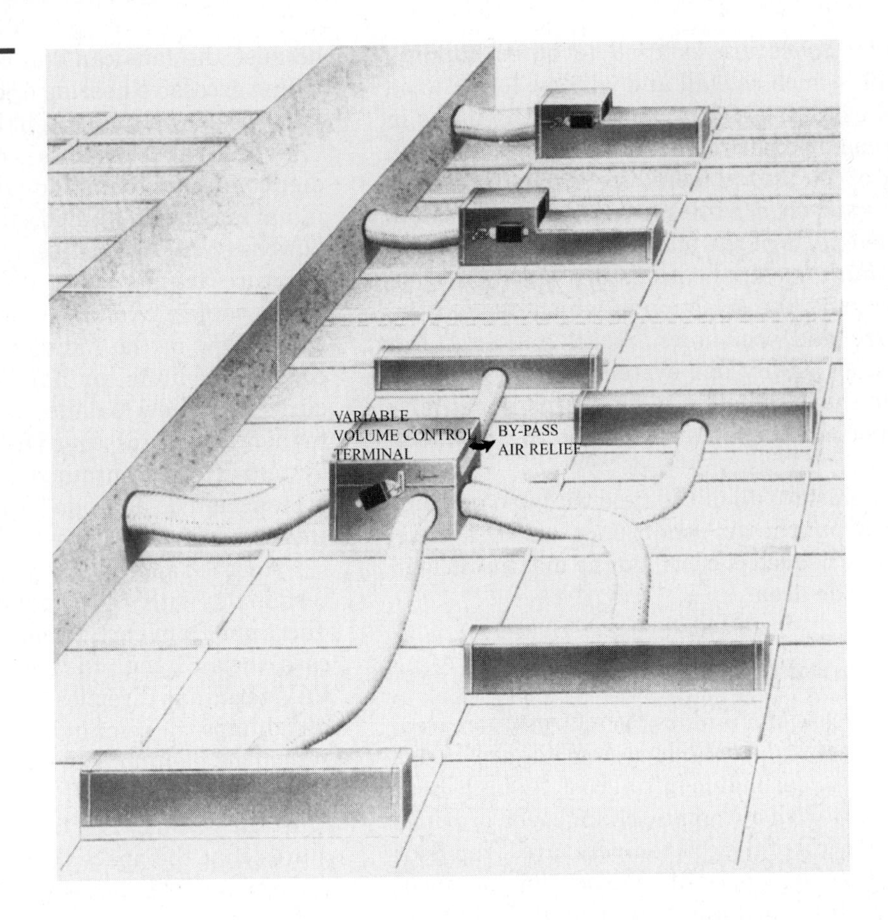

Figure 19-1-4 Variable air volume system using terminal units. (*Courtesy of York International Corp.*)

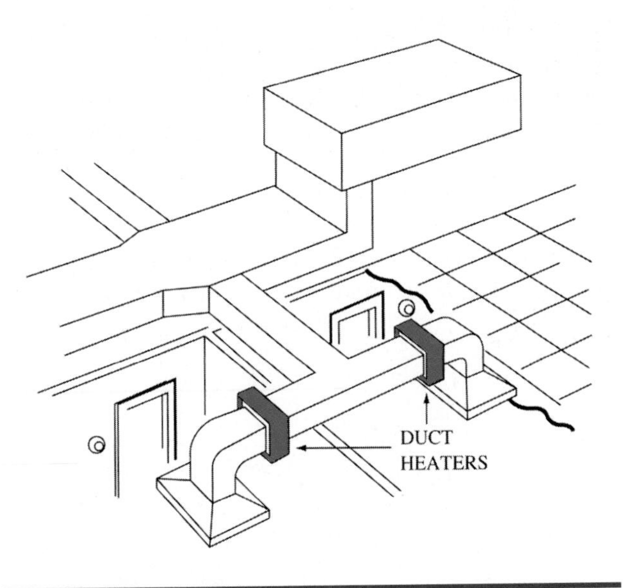

Figure 19-1-5 Constant volume system using zone duct heaters.

the distribution system. Individual zones or branches have heaters in them, which reheat the cool air to a comfort level for each space. These heaters may be individually controlled electric heaters or hot water coils. In many cases, these heaters are applied to a problem area in a system, to provide adequate heat where the original design was insufficient.

Perimeter Zone Applications

Perimeter zones require *terminal systems* which can handle the wide range of conditions, from the coldest morning warm up to the hottest solar gain cooling load. In many systems, this range was handled by providing perimeter radiant or forced air heat systems on the outer wall and under windows to furnish the necessary heat for the cold loads. A separate air system provided the necessary air movement, outside air requirements, and the cooling load responsibility. The air systems could be either constant or variable volume, but it has been found that only VAV really can provide the necessary control for comfort. In milder climates, the perimeter heating system is eliminated, without much loss of comfort. In some cases, the entire perimeter heating and cooling load is handled by console units on the outside wall, with individual outside air sources and temperature controls.

The *VAV systems* for perimeter applications are generally provided with primary air from either a central air handler for the entire building, or an entire floor, depending on the size of the project. A single air handler may handle up to 40,000 ft^2 of occupied space, but in many cases the ductwork becomes unmanageable. Variable volume systems may be low velocity or high velocity, depending on the type of controller and the terminal units used. The volume of primary air is automatically adjusted to

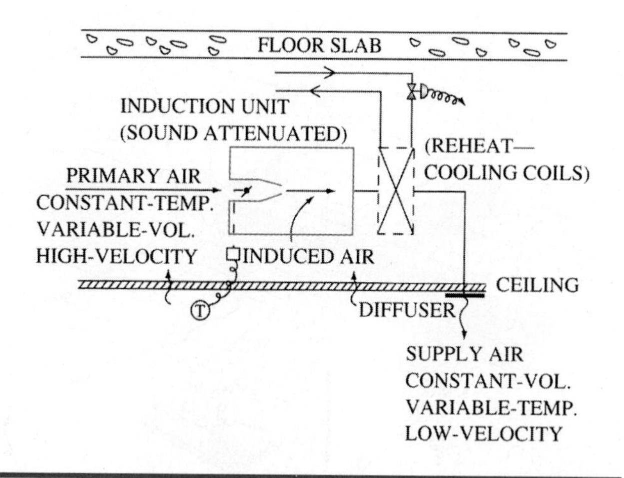

Figure 19-1-6 Induction terminal.

the total cooling demand by duct sensors controlling fan volume controls.

The terminal units for VAV systems have gone through an evolution since the advent of VAV. Some earlier systems were *induction air distribution systems*. Induction systems use a primary air system delivering a relatively small quantity of cooled high velocity, high pressure air to induction room terminals. This saves space, since the ductwork is much smaller than conventional ducting. The terminal unit takes the high velocity air through a nozzle arrangement, as shown in Figure 19-1-6, that induces room air into the unit through a heating coil. The temperature control is both from the VAV of the primary air and the reheat and/or cooling coils in the unit.

Some of these systems were not VAV. Figure 19-1-7 shows an induction room terminal which is constant volume, with temperature control achieved by reheat. Note that these terminal units do not rely on fans; the induced airflow provides the necessary secondary air.

Induction air systems have some trouble spots. First, high pressure air requires very tight and sealed ductwork. High pressure air leaks are very noisy. Second, these systems tend to use much energy, since the fan pressure requires considerable power; the VAV aspects are limited because the induction effect falls off quickly with a reduction of airflow.

Other VAV terminals for low pressure systems are the system powered boxes or diffusers, such as the one shown previously in Figure 19-1-4, that use system air pressure to control inflation of bladders that, in turn, control the air volume in the unit. These are simple in concept and effective at a reasonable cost. They provide VAV space conditioning, but usually without the benefit of variable air volume central fan systems. Some of these diffuser units use self contained, temperature powered actuators to modulate the diffuser openings. These can provide fairly simple, reasonable cost VAV to smaller systems.

Other systems provide cost effective zone control with a combination of VAV duct dampers and a sophisticated temperature control system. These allow

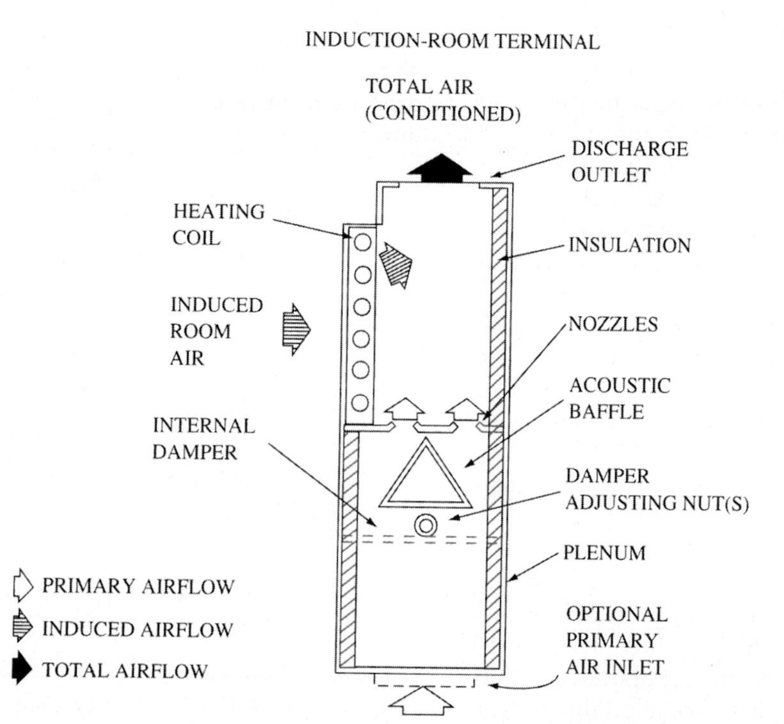

Figure 19-1-7 Sectional view of an induction type room terminal unit.

the system to provide only heating or cooling at any given time. The demands of a number of zones are programmed to switch the system to that mode, for example, cooling. Any zones requiring heat are temporarily closed off. With a heating demand in place, the system controls will then switch the system back to heating, closing off the zones requiring cooling. The individual zone dampers are modulating, so there are no abrupt changes. A fan speed controller is part of the system, reducing airflow to a practical minimum, as allowable. These systems are generally designed for unitary equipment, but they can be used as part of a central system.

VAV systems can be used to provide minimum airflows to meet ventilation requirements. The terminal units are fan powered, providing constant volume in the conditioned zone, with primary air operating from 100% down to about 20% of total air requirements. The low powered terminal fan provides terminal reheat at reduced air conditions for the zone. The primary air handler is variable volume, constant temperature, with the cooling plant demand proportional to the airflow. The individual zone controls can be stand alone systems, but usually are integrated through a DDC system controlling the overall system.

Entire Building Applications

In the chapter on central station air handlers, mention is made of multizone and double duct applications. Before energy codes were enacted, these were the ultimate in zoned comfort systems. By providing a blend of always available heated and cooled air to any zone, with the addition of good pneumatic control systems, these systems supplied excellent building comfort. The major limitation of them was the cost of energy to provide simultaneous heating and cooling. Later, the heat rejected from the chiller was used to provide warm water for heating. Other heat recovery approaches were tried, but the fully energy efficient approaches were not always the simplest or most economical to install. The primary differences in these two types of systems are noted below.

Multizone systems provide one trunk duct to each zone from the multizone air handler, Figure 19-1-8. The air handler is a blow through design with a hot and a cold deck. The airflow from each of these decks is fed through a set of dampers, 90 degrees opposed to each other on the same shaft. When the cooling damper is fully closed, the heating damper is fully open, and vice versa. Most of the time, the dampers are modulating in response to a temperature signal from the zone. The overall system is low to medium pressure, constant volume. The multizone's limitations, from a design standpoint, are that only a limited number of zones are available on the

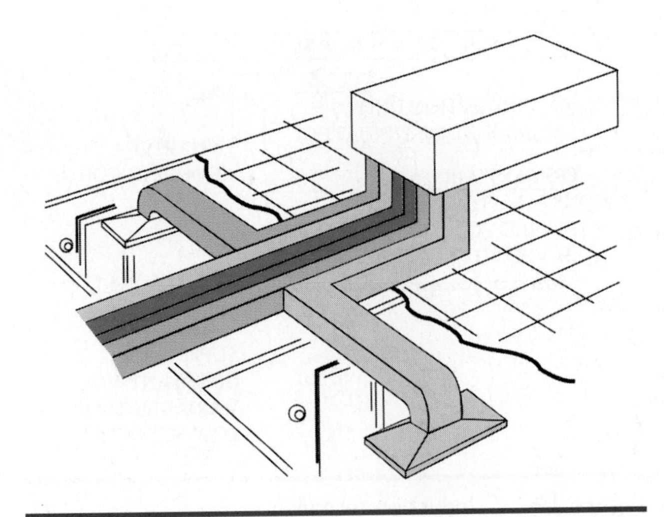

Figure 19-1-8 Constant volume multizone unit.

air handler. Consequently, adjustments are always needed to get an even balance. The smallest zone is typically 8–10% of the full load.

Double duct systems were the ultimate in design flexibility, but are very high in initial cost. Double duct systems are high velocity, high pressure systems. Some are variable volume, but most are constant volume. Two full sized supply ducts are required for the system, one for cooling and one for heating, Figure 19-1-9. The major benefit of these systems is that the two ducts each serve one terminal mixing box, which controls the air temperature for each zone. The mixing boxes are available in very small sizes, down to 200 CFM, allowing for zones as small as ½ ton. Zoning is limited by this small increment and the overall size of the system. The high pressure air in the main supply ducts (4 to 5 in WC) is reduced in the mixing boxes, so that the distribution ducting is normally low pressure design, with low noise levels.

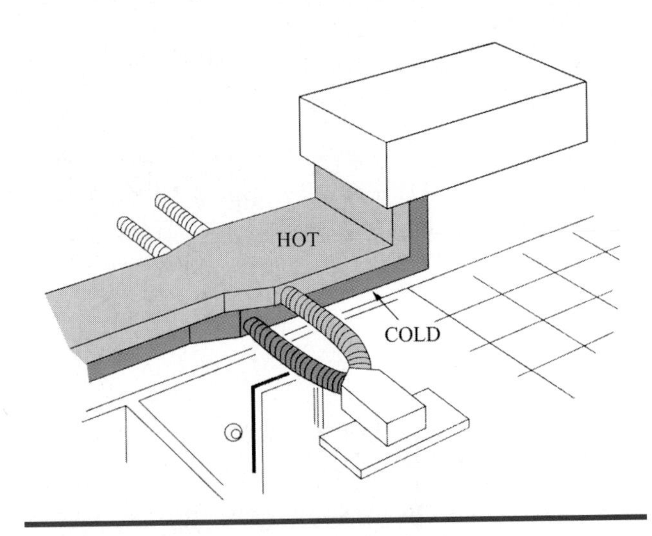

Figure 19-1-9 Double duct system.

CONTROL SYSTEMS

The *control system* must direct the operation of all elements automatically. In large installations, the control system is usually a separate installation from the air conditioning equipment. Controls may be electric, electronic, or pneumatic (air), or a combination of all three. The controls must be included in the initial construction stages to provide the total integration of the system.

> **CAUTION:** Because these systems can be started and stopped automatically, you must make sure the power to the unit is off before beginning service to prevent the unit from starting while you are working.

In a basic control system for cooling, a thermostat operates the chiller to maintain a set leaving water temperature. The condenser pump operates when the chiller is running. The cooling tower fan is controlled by the condenser water temperature, as is the water bypass control.

For the heating system, boiler firing is activated by the hot water thermostat. Space thermostats and humidistats control the functions of the air handler.

Electric Control Circuits

An example of a commercial and engineered system is a rooftop air conditioning unit with gas heating, low ambient control and an economizer cycle. It is common for the manufacturer to present the control information in three parts:

1. Basic refrigeration circuit controls, Figure 19-1-10
2. Natural gas heating circuit controls, Figure 19-1-11
3. Legend, to identify the control components, Table 19-1-1

The basic control circuit is line voltage, with wiring for the dual compressors and fan motors. The heating circuit is low voltage and includes the selector switch and thermostats. The legend defines the various control abbreviations.

Refrigeration Circuit, High Voltage

Referring to Figure 19-1-10, a power supply of 208/230 V, three phase, 60 Hz has been selected for sizing fuses and the disconnect switch. From the manufacturer's Table 19-1-2, this model will require a 100 A disconnect and a maximum of 70 A dual element fuses. Assuming that the distance of the switch to the panel is 125 ft, the minimum wire size is 4 AWG 60°C wire.

In Figure 19-1-10, compressor motors are three phase, with external overloads (OL) in the motor winding circuit. The evaporator motor is also three phase, with inherent motor protection. All three motors are started by the contactors, 1M, 2M, and 3M, respectively.

> **CAUTION:** Never work on any outdoor electrical system in the rain or when the system is wet.

> **CAUTION:** Never work on any electrical system when there is lightning in the area because a high voltage spike could come into the building through the electrical wiring.

The balance of the line voltage controls is single phase to permit series circuit controls with ordinary control devices. The condenser fans in this illustration are single phase, as well. Some units may use three phase motors; they will require relays or contactors for control, as the compressors do. Note that the single phase power is connected at terminals 1 and 3 on line 10. A transformer is shown, with a note showing its use in this unit for only 460 V power, to convert it to the control voltage of 240 V. Since the example is shown as 240 V main power, a transformer is not required. Other manufacturers use 120 V for control voltage, and will require a transformer (or a separate power source) for all primary power voltages.

SERVICE TIP

Control voltage transformers are available that can operate on multiple primary voltages. This type of transformer is preferred for service since it can be used on 120 V, 208 V, and 240 V power. This reduces the number of types of transformers that must be carried on your service vehicle.

Note that fuses 1FU and 2FU are provided to protect the control circuits whether they are from direct or transformed power. On 120 V control circuits there is no fuse on the secondary or neutral side of the circuit. This maintains the protection of a grounded neutral for safety.

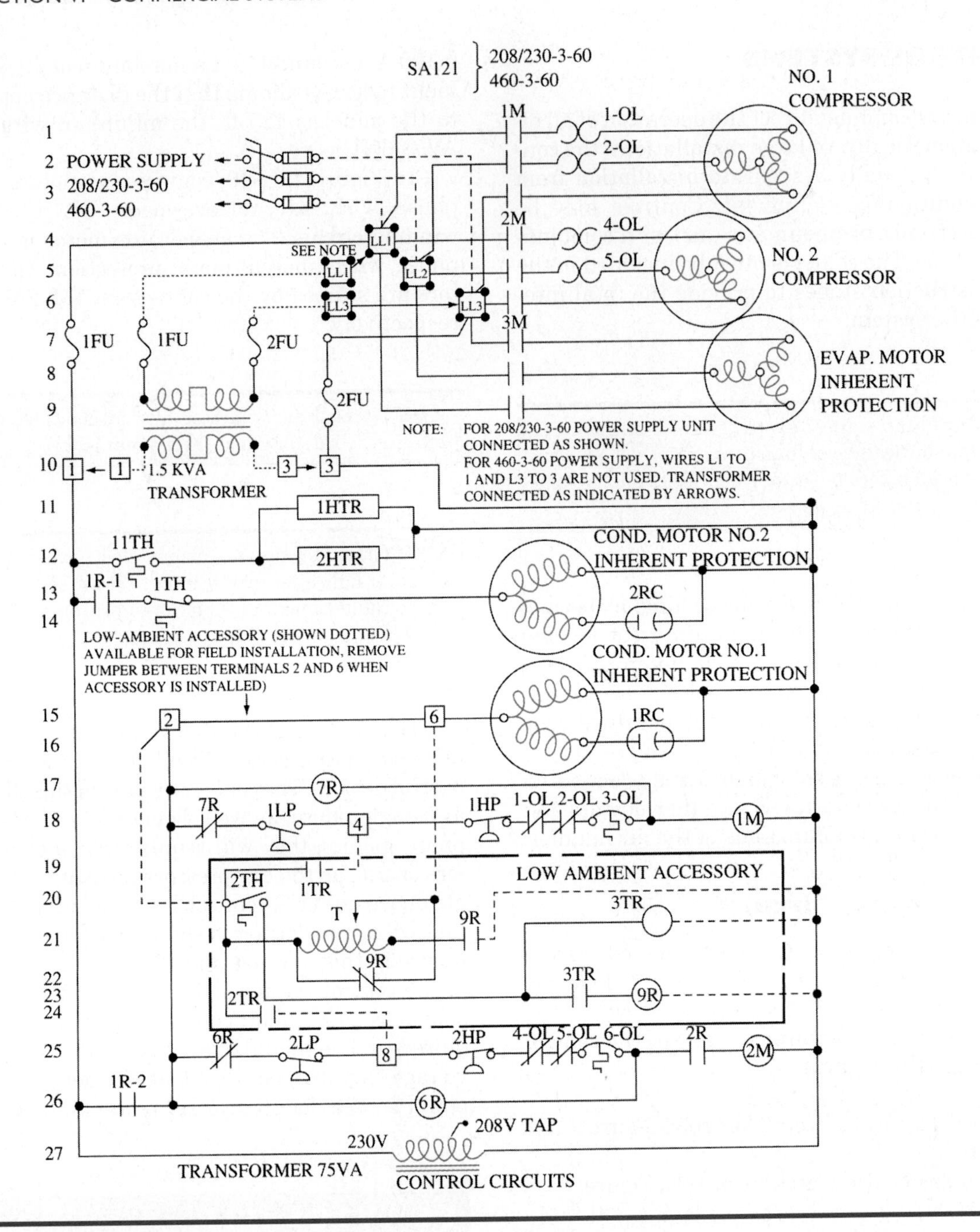

Figure 19-1-10 Line voltage wiring diagram for a central station air conditioning system. (*Courtesy of York International Corp.*)

Dropping down to lines 11 and 12 of Figure 19-1-10, note that the crankcase heaters (1HTR and 2HTR) come on when the crankcase thermostat closes in response to the oil temperature. The heaters can operate as long as the power is on, regardless of the system operation. It is therefore important that the power not be turned off in cold weather.

Both condenser fans have single phase motors with external running capacitors (1RC and 2RC) and inherent motor protection. Since this unit is equipped with low ambient control, note that in line 13 the #2 condenser fan motor is energized by the fan relay (1R-1) to start and run. It can also be dropped out by the first low ambient thermostat (1TH).

If it were not for the economizer cycle, the #1 condenser fan motor would be wired the same as the #2 fan. But the economizer system works within the same temperature range as the first stage low ambient control, and the two must be electrically interlocked as shown in lines 15 and 26.

1R-2 in line 26 is the #2 cooling relay and the key to energizing the circuit that feeds terminals 2 and

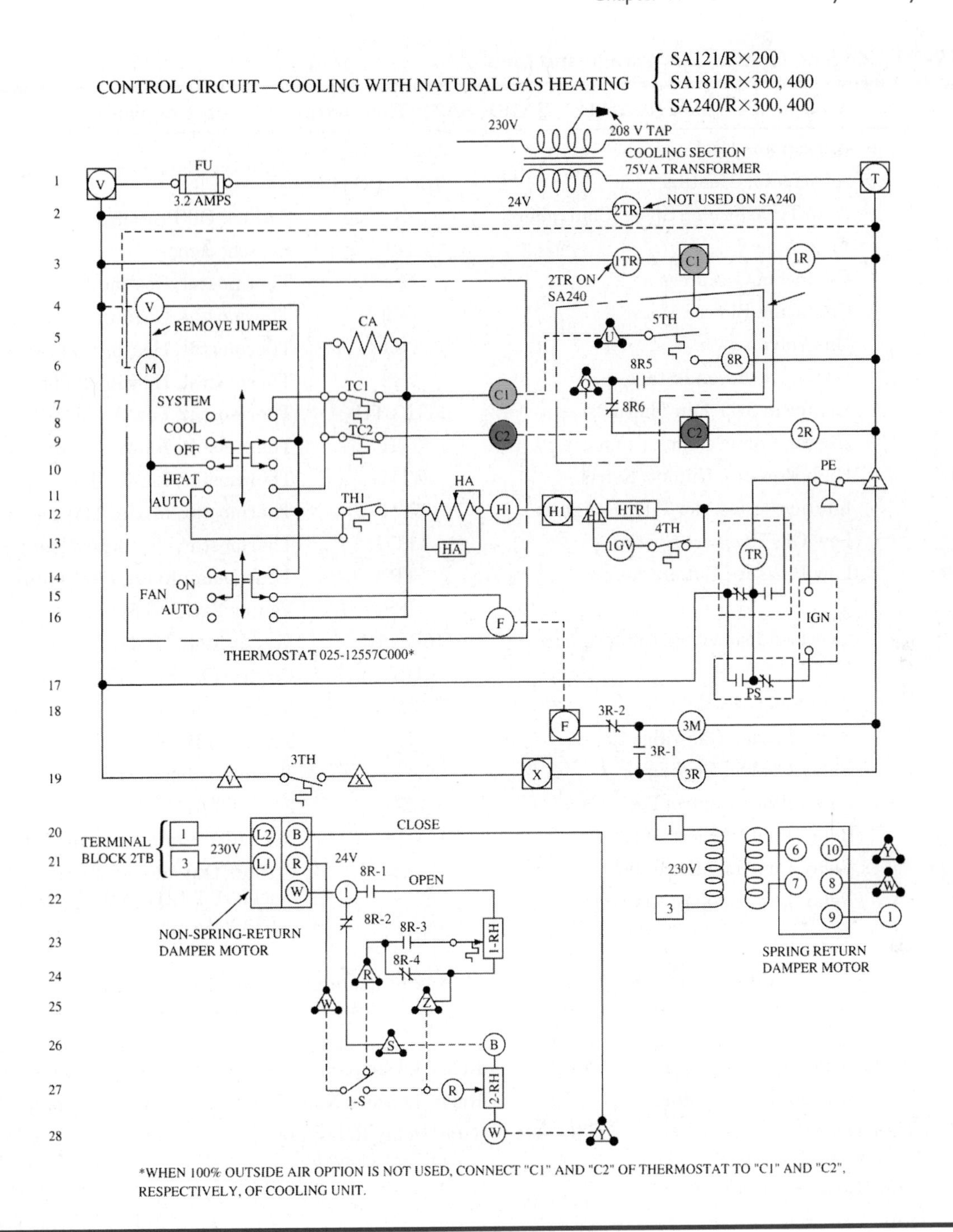

Figure 19-1-11 Low voltage wiring diagram for a central station heating and air conditioning system. *(Courtesy of York International Corp.)*

6 and the #1 condenser fan motor. When the low ambient accessory kit is used, however, the jumper between terminals 2 and 6 is removed, and the circuit shown in dashed lines is connected. Following line 26, the relay switch 1R-2 releases power to coil 6R and on to 2R (cooling contacts) and to contactor coil 2M, closing the #2 compressor contactor. 2R functions from the second stage of the cooling thermostat. The 6R coil is the key to furnishing power to line 25 through contact 6R. Line 25 has a low pressure cutout (2LP), high pressure cutout (2HP), electrical overload switches 4-0L and 5-0L, and thermal cutout 6-0L, all of which protect #2 compressor.

Note that line 18 is a similar protective circuit for compressor #1. The two lockout relays 6R and 7R are normally closed.

Without a complete explanation of the low ambient accessory, it can be noted that 2TH, line 20 (low ambient thermostat), is the key to controlling the #1 condenser fan motor through a system of time delay

TABLE 19-1-1 Legend for the Low Voltage and Line

Common Legend for SA121, SA181, SA240 Elementary Wiring Diagrams

CA	Anticipator, Cooling	1RC, 2RC, 3RC, 4RC	Running Capacitor
HA	Anticipator, Heating	S	Switch, Oil Pressure
T	Auto Transformer, Speed Controller	1-S	Switch, Bypass
FU, 1FU, 2FU	Fuse	TC1	Thermostat, Cooling 1st Stage
1M, 2M	Contactor, Compressor	TC2	Thermostat, Cooling 2nd Stage
3M	Contactor, Blower Motor	TH1	Thermostat, Heating 1st Stage
1GV	Gas Valve	TH2	Thermostat, Heating 2nd Stage
2GV	Gas Valve, Second Stage	1TH, 2TH	Thermostat, Low Ambient Control
HTR	Heater in 3HT Fan Thermostat	3TH	Thermostat, Blower (Heat)
1HTR, 2HTR	Heater, Compressor Crankcase	4TH	Thermostat, Limit (Heat)
1HP, 2HP	High Pressure Cutout, Refrig.	5TH	Thermostat, Mixing Box
IGN	Ignition Trans. (for Pilot Relighter)	11TH	Thermostat, Crankcase Heater
PE	Low Gas Pressure Switch	TR	Time Delay Relay, Pilot Ignition
1LP, 2LP	Low Pressure Cutout, Refrig.	VFS	Venter Fan Sail Switch
1-OL, 2-OL, 3-OL, 4-OL, 5-OL, 6-OL	Overload Protectors, Compressor	10R, 11R	Venter Motor Relays
		1RH	Control Damper, Mixed Air
PS	Pilot Safety Switch	2RH	Control Damper Controller, Min. Position
R	Second Stage Gas Valve Relay	□	Terminal Block, 1TB
1R, 2R	Relay, Control Cooling	□	Terminal Block, 2TB
3R	Low Voltage Control Relay	▣	24 Volt Terminal Block, 3TB
4R, 5R	Relay, Control Electric Heat	△	Identified Connection in Heating Section
8R	Relay, Control, Mixing Box	△	0-100% Outside Air Terminal Block, 6TB (SA 121) 7TB (SA 181): 4TB (SA 240)
9R	Relay, Low Ambient Control		

Specific Legend for SA121, SA181, SA240 Elementary Wiring Diagram

	SA121		SA181		SA240
6R	Relay, Lockout No. 2 System	6R	Relay, Lockout No. 2 System	6R	Relay, Lockout
7R	Relay, Lockout No. 1 System	7R	Relay, Lockout No. 1 System	MP	Compr. Protection
1TR, 2TR	Time Delay Relay (Low Ambient Accessory)	1TR, 2TR	Time Delay Relay, Low Ambient Control	1TR	Time Delay Relay, Part Winding Start
3TR	Time Delay Relay, Cond. Fan				
T	Auto Transformer (Low Ambient Accessory)			2TR	Time Delay Relay, Low Ambient Control
				3TR	Time Delay Relay, Oil Pressure Switch
				1-SOL	Solenoid, Compr. Unloader
				2-SOL	Solenoid, Evap. Unloader

(Courtesy of York International Corp.)

TABLE 19-1-2 Wire, Fuse, and Disconnect Sizes for Several Sizes of Central Station Air Conditioners

Model	Power Supply	Length Circuit One Way, Ft Up To	Min. Wire Size Copper AWG 60 C 2% Voltage Drop	Max. Fuse Size Dual Element	Disconnect Switch Size, Amps
		100	6		
SA91-25A	A	175	4	55	60
		200	3		
		250	2		
SA91-45A	A	175	10	30	30
		250	8		
208/230 V		125	4		
		175	3		
SA121-258	A	200	2	70	100
		250	1		
SA121-46B	A	200	8	40	60
		250	6		
		150	1		
SA181-25A	A	200	0	100	100
		250	000		
SA181-45A	A	250	4	60	60
		150	00		
SA240-25C	A	200	000	175	200
		250	250MCM		

(Courtesy of York International Corp.)

relays 1TR, 2TR, and 3TR, eventually feeding to terminal 6 and the condenser fan motor. T (line 20) is an auto transformer fan speed control that modulates the fan motor rpm instead of providing a straight on-off. Note that it is put in the circuit when 9R switches reverse in response to the need for low ambient control. Otherwise, the fan speed is full rpm when the current goes around T, because 9R in line 21 is open. Line 27 is the primary line voltage side of the low voltage control transformer. It is rated at 75 VA (volt amps).

Heating Circuit, Low Voltage

Referring to Figure 19-1-11, line 1 is 24 V secondary, with fused protection based on amperage draw. The space thermostat has a COOL OFF HEAT AUTO system selection. In automatic mode, the system will bring on heating, cooling, or the economizer system without manual selection. TC1 and TC2 are the first and second stage cooling stats. TH1 is the first stage heating stat. The indoor fan selection on ON (continuous) or AUTO to cycle with the 3TH blower control.

The economizer control thermostat, 5TH in the mixing boxes, is wired to the C1 first stage cooling circuit, so that #1 compressor cannot operate as long as it is open. This means that the temperature of the outside air is sufficient to provide cool air without refrigeration. But when the outside ambient is high enough to call for cooling, 5TH closes; furnishing power to relay 8R and, through contacts 8R5 and 8R6, allowing both cooling and relays 1R and 2R to close on command of the cooling stats. 1R and 2R are time delay relays in the cooling circuit that prevent both compressors from coming on at the same time.

In the heating mode, the economizer system is bypassed. Current from the TH1 stat (line 12) goes through the heat anticipator HA to terminal H1. From H1 it feeds to the gas valve 1GV, provided that the limit control contact 4TH is closed. Another circuit on line 17 provides an alternate flow of current to the pilot safety switch, PS, and through its contacts feeds the IGN pilot ignition transformer. Ignition of the pilot cannot take place unless there is sufficient gas pressure to close PE (pressure/electric) switch (line 11). At the same time the TR (time delay relay) is also energized, the contacts below it are reversed, and the gas valve opens.

If the fan selector is in the ON position, current from terminal F goes through normally closed 3R-2 switch and energizes 3M (the contactor coil) for the indoor or evaporator fan. In the AUTO position the circuit is essentially bypassed. Current then (on line 19) flows through 3TH, the blower control thermostat (if heat is present to close the switch), and to coil 3R, which then closes 3R-1 and opens 3R-2. Thus, 3M is again energized and the indoor blower comes on. It

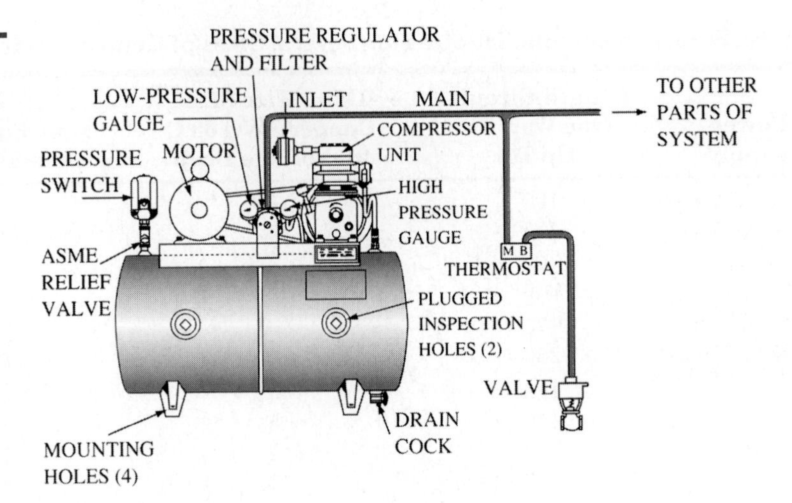

Figure 19-1-12 Air compressor for a pneumatic control system.

will remain on until 3TH opens to indicate heat in the plenum.

The power (240 V on lines 20 and 21) to operate the economizer outside air damper is picked up at terminal blocks 1 and 3 from line 10 in Figure 19-1-10. Two diagrams are shown. One is a damper motor with a nonspring return; the other has a spring return. With the spring return the motor will close completely in the event of power failure. If there is no spring return, the damper will close to a minimum position based on a minimum amount of outside air needed. Note that relay 8R, coil line 6, is the key to all switch actions: 8R-1, 8R-2, 8R-3, and 8R-4 position the control dampers 1-RH and 2-RH.

Pneumatic Control Systems

Pneumatic control systems use compressed air to supply energy for the operation of valves, motors, relays and other pneumatic control equipment. Consequently the circuits consist of air piping, valves, orifices, and similar mechanical devices.

Pneumatic control systems offer some distinct advantages:

- They provide an excellent means of modulating control operation.
- They provide a wide variety of control sequences with relatively simple equipment.
- They are relatively free of operational problems.
- They cost less than electrical controls if the codes require electrical conduit.

Pneumatic controls are made up of the following elements:

1. A constant supply of clean, dry compressed air.
2. Air lines consisting of mains and branches, usually copper or plastic, to connect the control devices.
3. A series of controllers including thermostats, humidistats, humidity controllers, relays, and switches.
4. A series of controlled devices including motors and valves called operators or actuators.

The air source is usually an electrically driven compressor Figure 19-1-12, which is connected to a storage tank. The air pressure is maintained between fixed limits (usually between 20 and 35 psi for low pressure systems). Air leaving the tank is filtered to remove the oil and dust. Many installations use a small refrigeration system to dehumidify the air. Pressure reducing valves control the air pressure.

How the Controls Operate

The controller function is to regulate the position of the controlled device. It does this by taking air from the supply main at a constant pressure and adjusting the delivered pressure according to the measured conditions.

One type of thermostat is the bleed type, shown in Figure 19-1-13. The bimetal element reacts to the temperature and controls the branch line bleed off pressure. These thermostats do not have a wide range of control, therefore the branch line is often run to a relay that controls the action. Bleed controls cause a constant drain on the compressed air source.

Non-bleed controllers use air only when the branch line pressure is being increased. The air pressure is regulated by a system of valves, Figure 19-1-14, which eliminates the constant bleeding characteristic of the bleed type unit. Valves C and D are controlled by the action of the bellows (A) resulting from the changes in room temperature. Although the exhaust is a bleeding action, it is relatively small and occurs only on a pressure increase.

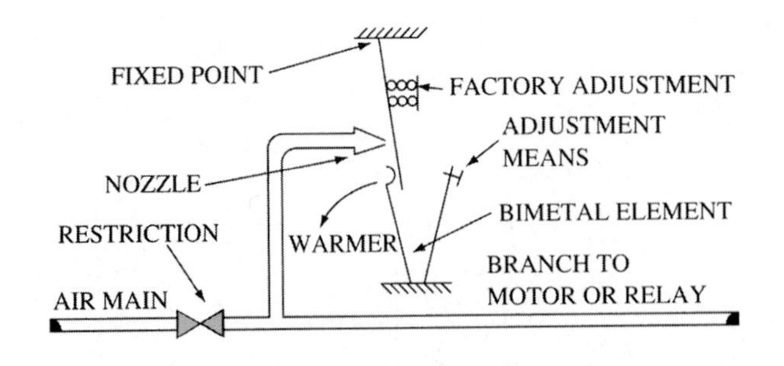

Figure 19-1-13 Diagram of a bleed pneumatic thermostat.

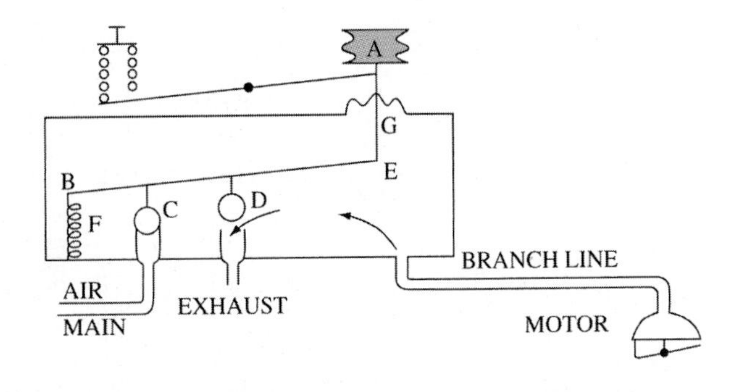

Figure 19-1-14 Diagram of a nonbleed pneumatic thermostat.

Controlled devices, operators or actuators, are mostly pneumatic damper motors or valves. The principle of operation is the same for both. Figure 19-1-15 is a diagram of a typical motor. The movement of the bellows as the branch line changes activates the lever arm or valve stem. The spring exerts an opposing force so that a balanced, controlled position can be stabilized. The motor arm L can be linked to a number of functions.

Figure 19-1-16 shows a pictorial review of some of the functions in a pneumatic control system. There is always some crossover between the air devices and the electrical system. The device most widely used is the pneumatic/electric relay.

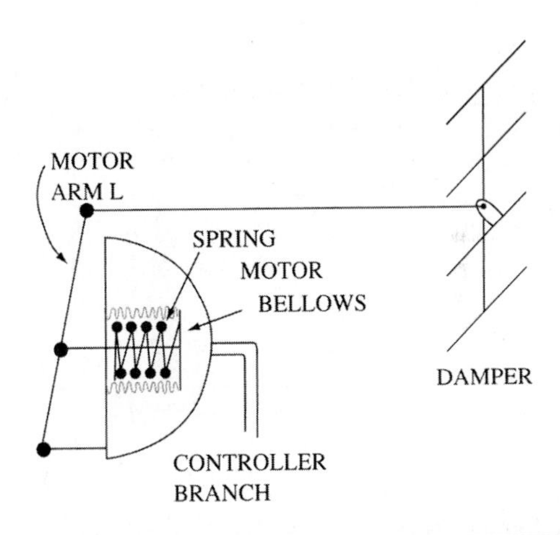

Figure 19-1-15 Pneumatic actuator with normally open damper.

NICE TO KNOW

Unlike electronic and electrical control systems, pneumatic systems do not require any energy when they are in a "hold position." Some electrical and electronic devices consume power even when they are off. This power consumption is referred to as the background power. Although the background power for any single device can be very small the total power consumed by a large building's control system can be significant. The federal government feels that background power consumption has become so large nationally that the EPA has established guidelines for manufacturers regarding background power consumption. Because pneumatic systems do not have any background power they are exempt from these regulations.

Electronic Control Systems

Electronic control may also be used effectively for central station equipment. Due to a number of advantages it is rapidly gaining in popularity:

1. It has no moving parts.
2. The response is fast.
3. The regulatory element can easily be a reasonable distance from the sensing element, permitting:
   a. Adjustments made at a central location.
   b. Cleaner conditions at a central location than at the location of the sensing element.
   c. Easier temperature averaging.

In the Fundamentals of Electronic Devices and Circuits chapter, a basic overview of electronic circuits is presented. In this section, the emphasis is on the specific control functions of electronic controls. With a comprehension of these fundamentals, most electronic control circuits can be reduced to simple components. The interaction of these controls and the logic behind them is the heart of electronic control of engineered air conditioning systems.

Only simple low voltage connections are needed between the sensing element and the electric circuit. Flexibility is important since electronic circuits can be combined with both electric and pneumatic circuits to provide results that could not usually

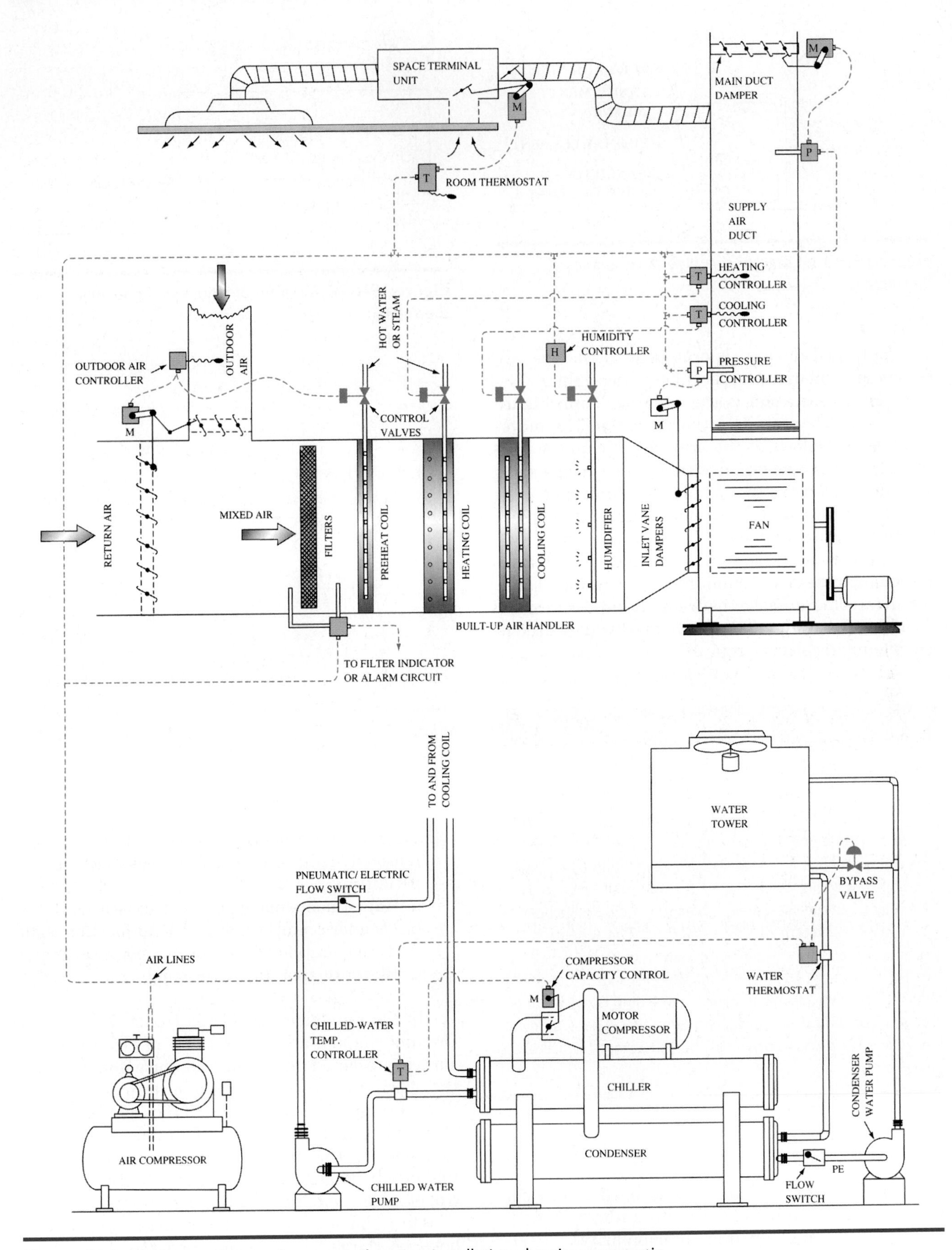

Figure 19-1-16 System diagram for a central station installation showing pneumatic controls.

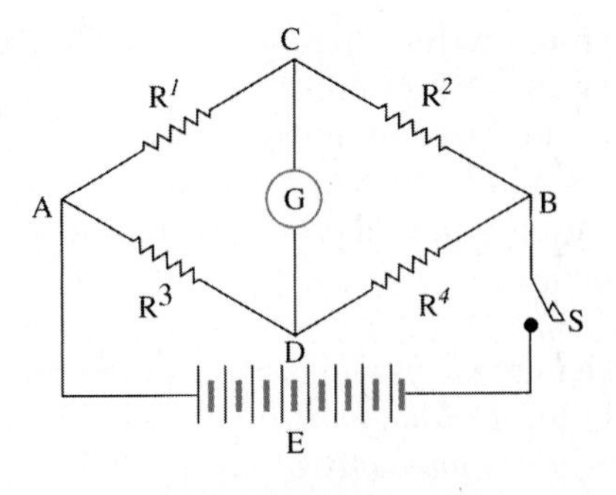

Figure 19-1-17 Wheatstone bridge circuit.

be achieved separately. Electronic circuits can coordinate temperature changes from several sources such as room, outdoor air, and fan discharge air, and program action accordingly.

Electronic controls are based on the principle of the Wheatstone bridge, Figure 19-1-17, which is composed of two series resistors (R_1 and R_2, R_3 and R_4), connected in parallel across a DC voltage source. A galvanometer G (a sensitive indicator of electrical current) is connected across the parallel branches at junctions C and D between the series resistors. If switch S is closed, voltage E (DC battery) energizes both branches. If the potential at C equals the potential at D, the net potential difference is zero. When this condition exists, the bridge is in balance.

If the resistance of any one leg is changed, the galvanometer will register a flow of current. The bridge is now unbalanced.

If that resistance is changed as a result of a temperature reaction, we now have an electronic method of measuring current in relation to temperature change. With a few changes we can create an electronic main circuit such as in Figure 19-1-18. A 15 V AC circuit replaces the DC battery. The galvanometer is replaced by a combination voltage amplifier phase discriminator switching relay unit. Resistor R_2 is replaced by sensing element T of an electronic controller.

The purpose of the *voltage amplifier* is to take the small voltage from the bridge and increase its magnitude by stage amplification to do the work. *Phase discrimination* means determining the sensor action. In an electric bimetal thermostat, the mechanical movement is directly related to the temperature changes; however, the electronic sensing element is a non-moving part and the phase discriminator determines whether the signal will indicate a rise or fall in temperature. The *relay* then operates the final element action. Phase discrimination can be two position or, with certain modifications, can be converted to a modular system.

The crossover point from the electronic to electric occurs at the output of the amplifier and relay signal. The motor that it operates is a conventional ON OFF motor, or a proportional (modulating) electric motorized valve or damper actuator.

Electronic temperature sensing elements are room thermostats, outdoor thermostats, insertion thermostats for ducts or insertion thermostats for liquids. The typical room thermostat is a coil of wire wound on a bobbin. The resistance of the wire varies directly with the temperature changes.

A *sensor* is any device that converts a nonelectrical impulse such as sound, heat, light, or pressure into an electrical signal. Sensors have been developed to provide the necessary input for controlling the pressure in a duct or the relative humidity in a conditioned area. The sensor provides an input to a solid state controller. The logic in the controller sends an output to some mechanical device to produce the required action.

For example, to control humidity, a pair of electronic sensors read wet and dry bulb temperatures. The logic in the controller converts these readings to a relative humidity value. Based on the limits set up, the connected mechanical device is programmed to add or deduct moisture from the air to meet the requirements. The use of electronic equipment makes possible more accurate control of the space conditions than can be accomplished by pneumatic or electromechanical control equipment.

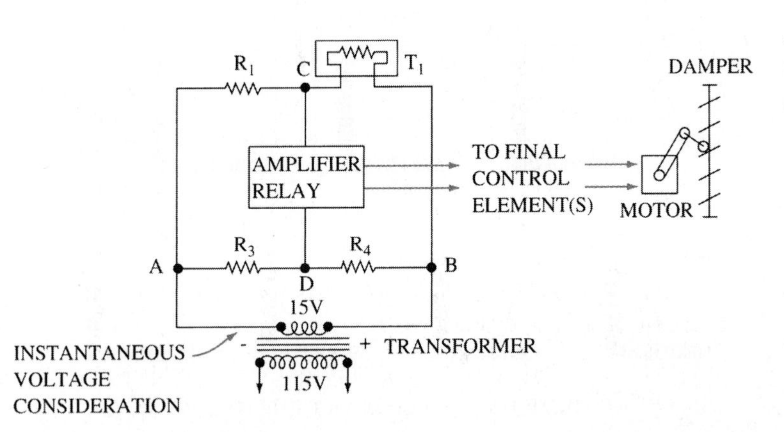

Figure 19-1-18 Electronic main bridge circuit.

SERVICE TIP

Make sure you follow the manufacturer procedures when replacing electronic sensors because some can be damaged by improper handling. Because they may be easily damaged, some supply houses will not allow electronic sensors to be returned once they are removed from their packing. Make sure you are ready to use the sensor right away before you open the package.

Direct Digital Control Systems

In a direct digital control (DDC) system, the computer acts as the primary control for all HVAC/R functions. Valves, dampers, fan speeds, etc., are all controlled by the computer without the use of conventional control devices such as thermostats, humidistats, timers, etc.

The computer directly senses the building environmental conditions, and based on a user defined programmed set of instructions, initiates the proper control actions in the HVAC/R system. Direct digital control of HVAC/R components gives more accurate control and greater flexibility than other commonly used mechanical and electrical control devices. It also has the capability of coordinating inputs from a number of sensing devices and arriving at an output that takes into consideration numerous influencing factors. The following are some examples of the capabilities of the DDC control systems:

1. The DDC system can control a VAV terminal box to discharge the proper air supply based on a variety of inputs such as dry bulb temperature, relative humidity and mean radiant temperature. In this way, considering the total environmental conditions, a greater feeling of comfort is produced for the occupants in the space.
2. In a central station system with a large supply fan, the microprocessor can regulate the speed of the fan to produce the required airflow using a minimum amount of power.
3. Computers are currently used to turn on chillers and boilers at an optimum time to recover from a period when the building is unconditioned.

A typical DDC system is shown in Figure 19-1-19. A central *stand alone controller* (SAC) is connected to a series of *remote control units* (RCUs). The SAC is wired to a computer terminal, a printer and modem, Figure 19-1-20. The terminal is used for input communications by the operator. The printer is used to record any information concerning the operation of the system. The modem is an electronic device that permits the computer to communicate through the telephone lines.

The individual RCUs are located in the building near the equipment being controlled. These units have a series of input and output wiring connections that go to sensors and controls on the HVAC/R system that permit the SAC unit to control the operations of the system. Both the input and output functions are of two types: analog and digital. The *analog functions* supply or deliver modulated information. For example, an analog temperature sensor may be capable of reading temperatures between 0°F and 100°F. This input could be converted by the computer to produce an analog output signal to control a damper to any position between fully open to fully closed.

The *digital function* is a binary or two position function. For an input, the signal could monitor whether a switch is open or closed. For output, the binary signal could position the switch in either an ON or OFF position.

The SAC unit, RCU units, computer terminal, printer, and modem are all considered *hardware*. The control program, which is programmed for an individual system, is called *software*. The software is installed in the SAC unit at the time that the DDC takes over the operation of the system. The operator periodically monitors the operation of the system

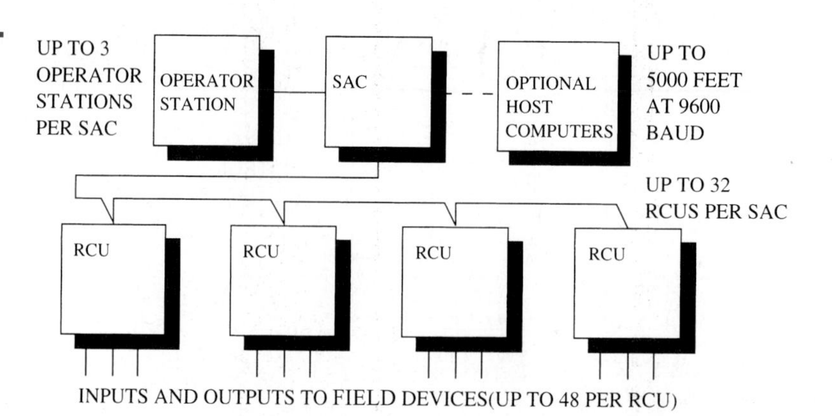

Figure 19-1-19 Data control center for a central station system.

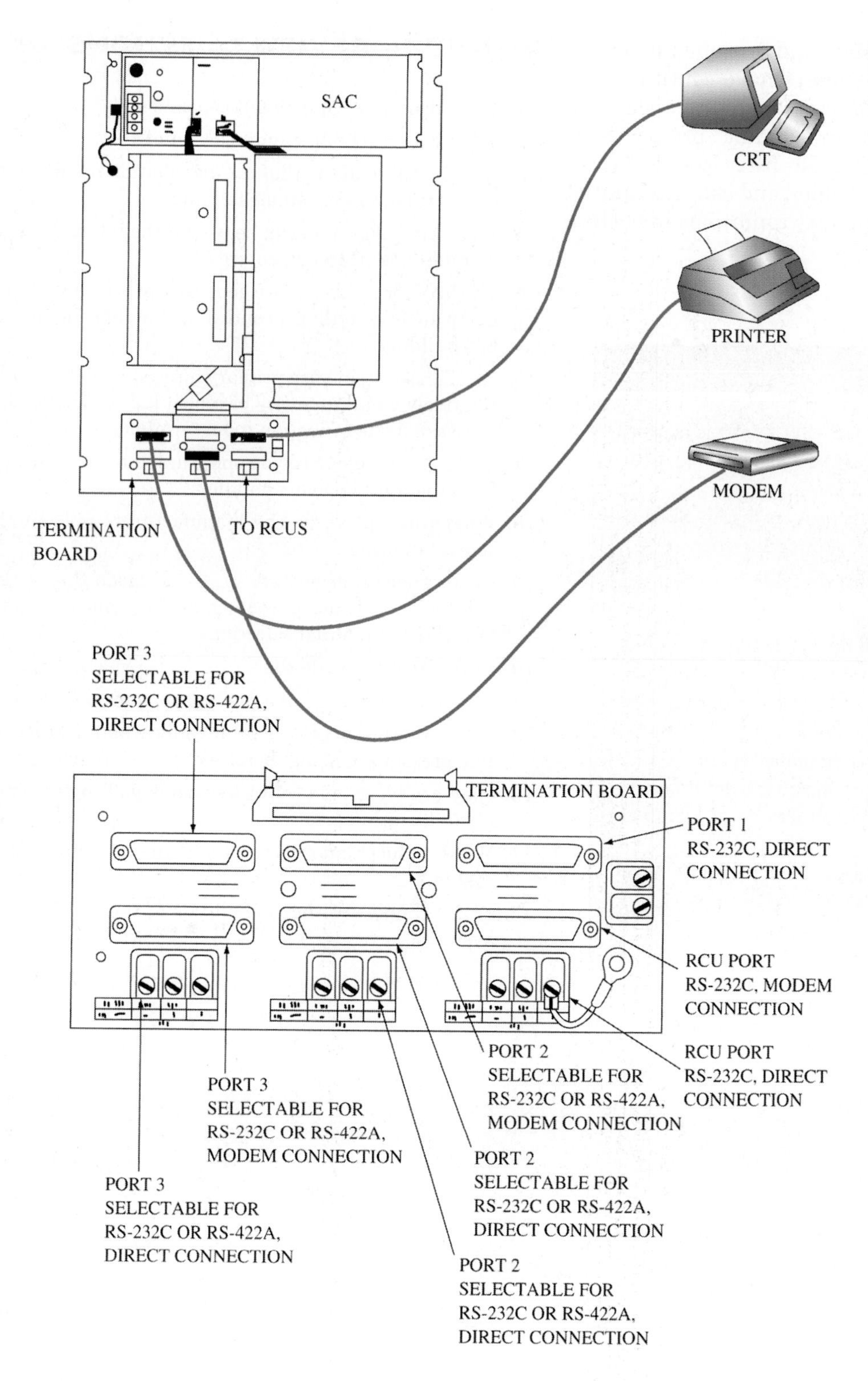

Figure 19-1-20 Stand alone controller (SAC).

through the computer, notes the reports and alarms, and makes any changes in the program that are considered advantageous.

The DDC systems are often called *energy management systems* since one of their main functions, and usually justification for their adoption, is saving energy. Special provision has been made in the selection of sensors to make possible continuous monitoring of the energy usage. The control system is set up to energize loads only when necessary and to use such features as free cooling (economizer operation) whenever possible.

The control center collects key operating data from the HVAC/R system and incorporates remote control devices to supervise the system's operation. The elaborateness of the data center is related to the type and size of the system and to economic considerations. Some control centers have continuous scanners with alarm indicators to monitor refrigeration machines, oil and refrigerant pressures, chilled water temperatures, air filter conditions, low water conditions in the boiler, and conventional space temperature and humidity conditions in each zone.

TECH TALK

Tech 1. Troubleshooting electronic controls is usually a hit or miss process so I just change out the most likely parts to fail.

Tech 2. It's not a hit or miss process if you follow the manufacturer's troubleshooting chart.

Tech 1. Do those things really work?

Tech 2. Yes, they do.

Tech 1. Well, what do you do if they are not with the equipment? Use any one that you have?

Tech 2. No wonder you are having problems troubleshooting. You have to use the one specifically for the piece of equipment you're working on. If you don't have it, you can get it faxed to you from the manufacturer or get it off the world wide web.

Tech 1. Thanks. I did not know that.

UNIT 1—REVIEW QUESTIONS

1. What does a heating circuit consist of?
2. What does a cooling circuit consist of?
3. List the three areas that air distribution applications in central station systems fall into.
4. What kind of system can be used if the interior core of a building is a large open area?
5. The VAV for ______ are generally provided with primary air from either a central air handler for the entire building.
6. ______ systems use primary air system delivering a relatively small quantity of cooled high velocity, high pressure air to induction room terminals.
7. ______ systems provide one trunk duct to each zone from the multizone air handler.
8. What three parts do the manufacturers present the control information in?
9. In the space thermostat's ______ mode the system will bring on heating, cooling, or the economizer system without manual selection.
10. In the heating mode the ______ is bypassed.
11. List the advantages of a pneumatic control system.
12. ______ thermostats use air only when the branch line pressure is being increased.
13. Why are electronic control systems gaining in popularity?
14. What is the primary control in a direct digital control system?
15. List examples of the capabilities of the DDC control systems.

UNIT 2

Water Chillers

OBJECTIVES: In this unit the reader will learn about water chillers used in the HVAC industry including:

- Water Chiller Types and Configurations
- Boiler Classification
- Circulating Pumps
- Chiller Components

WATER CHILLERS

The water in a chiller is cooled to approximately 43 to 45°F. The chilled water pump circulates the chilled water supply (CHWS) to the cooling coil in the air handler. The cooling coil normally has piping isolation valves and a control valve. The heat absorbed by the cooling coil warms the water about 10°F at full load, with a chilled water return (CHWR) temperature of 53 to 55°F.

There are various configurations of water chillers. Basically they differ in: (1) type of condenser, (2) type of compressor, and (3) size. Based on consideration for these parameters, we will be discussing the following types:

1. Water cooled chillers
2. Air cooled chillers
3. Centrifugal chillers
4. Screw compressor chillers
5. Absorption chillers

Water Cooled Chillers

The design and range of water cooled chillers is related to the compressor and condensing medium (air or water). All chillers have compressors, liquid chillers, compressor starters, controls, and refrigerant and oil pressure gauges neatly assembled into a package.

Starting at the lower capacities, package water chillers use one or more reciprocating compressors. Figure 19-2-1 shows an example of a small water cooled chiller in the 20 ton range.

The larger package chillers, also water cooled, with reciprocating compressors, range upward from 40 to 200 tons as shown in Figure 19-2-2. These have multiple compressors that permit close control of capacity. They may also have standby compressors in the event of a malfunction. Where single compressors are used, these are equipped with cylinder unloading to allow capacity reduction and minimum starting power requirements.

In addition to reciprocating compressors, the scroll type compressor is being utilized increasingly in package chillers. These compressors use a pair of mating scroll shaped surfaces to compress the refrigerant with pure rotary motion, as shown in the cutaway diagram in Figure 19-2-3.

The individual scroll compressors are very efficient and are used in sizes from 5 to 20 tons. Since they do not have any capacity control, they are usually used in multiples of 2, 3, or 4 compressors in a chiller unit for efficient operation.

Heat rejection takes place in a shell and tube condenser such as the one shown in Figure 19-2-4. Water flows through the tubes, and refrigerant vapor fills the shell. The refrigerant vapor condenses to a liquid as heat is transferred from the refrigerant to the water. The liquid collects in the bottom where it is subcooled an additional 10 to 15°F for greater cooling capacity. In the water circuit, the coldest condenser water enters the lower part of the shell and circulates through the tubes. The water will make two or three passes through the shell before it is discharged. This is arranged by circuit baffles in the condenser heads. The higher the number of passes, the greater the pressure drop and pressure required to produce the required flow rate. There are also cross baffles within the shell which serve to hold the tube bundles but also help to spread the refrigerant gas over the entire length of the shell. Condenser capacity is normally based on 85°F entering water temperature. There is normally a 10°F rise between the exiting and entering water temperatures.

The chiller (evaporator) shown in Figure 19-2-5 is a direct expansion type, associated with R-22 reciprocating compressors. Refrigerant flows through the tubes and will generally make two passes for standard operation, which gives a counterflow arrangement, versus the typical water flow pattern.

Figure 19-2-1 Small package chiller. *(Courtesy of York International Corp.)*

Figure 19-2-2 Large package chiller with reciprocating compressor.

NICE TO KNOW

A special sea water condenser is used for marine duty where salt water is the cooling medium. These tubes are made of cupronickel steel to withstand the salt corrosion effects.

The cooler shell and suction lines must be properly insulated to prevent sweating. This is done at the factory with a layer of closed cell foam insulation prior to painting.

Both condenser and cooler shells must comply with ANSI B9.1 and the applicable ASME safety codes for pressure vessels.

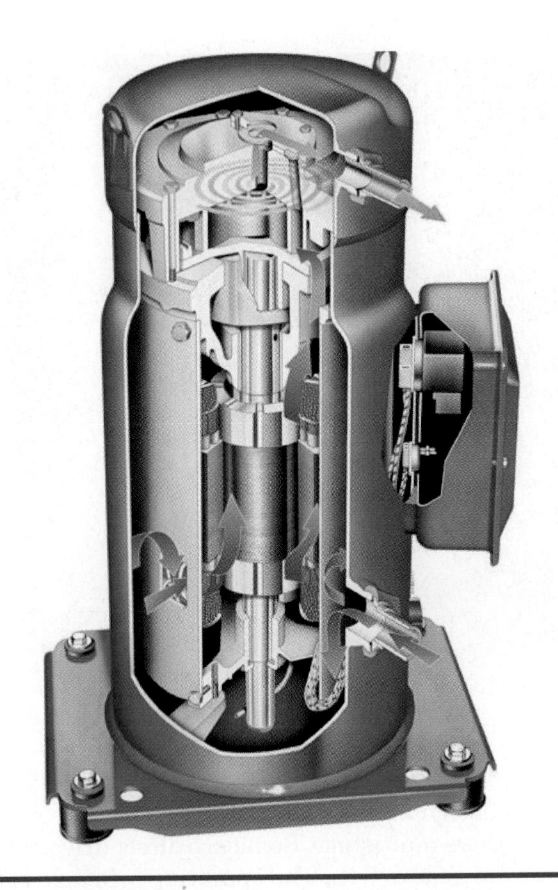

Figure 19-2-3 Cutaway view of a scroll compressor. (*Courtesy Danfoss Inc.*)

Standard chiller ratings are based on ARI Standard 590: 44°F leaving water temperature off the cooler at 105°F and 120°F condensing temperatures; and 95°F leaving water off the condenser with a 10°F rise. The 95°F condenser water rating will

CAUTION: Never weld, cut, or braze on any pressure vessel without first obtaining written approval from a licensed engineer and proper welding certification that meet any and all other local, state or federal regulations. Any such work can damage the pressure vessel, which can cause it to fail.

produce a condensing temperature near 105°F. The rating point of 120°F condensing temperature is used for applications where remote air cooled condensers, such as the one in Figure 19-2-6, are used instead of water cooled types.

Air Cooled Chillers

These chillers, as shown in Figure 19-2-7, are complete packages that can be mounted on the roof or outside the building. They are similar to large air cooled condensing units, except the cooler shell is suspended beneath the condenser coil and fan section. The chiller shell (evaporator) must be protected from freezing. Electric heating elements are wrapped around the shell and then covered with a thick layer of insulation. Some manufacturers also add a final protective metal jacket that doubles as a good vapor barrier.

The sizes of reciprocating compressor air cooled package water chillers range from 10 to over 100 tons. These are also rated in accordance with ARI Standard 590 at 44°F leaving chilled water temperature and 95°F DB condenser entering air temperature.

Centrifugal Chillers

For very large installations the industry offers a range of hermetic centrifugal compressor water

Figure 19-2-4 Shell and tube condenser.

Figure 19-2-5 Direct expansion chiller.

chillers of up to 2000 tons in a single assembly. Very large buildings or complexes such as college and school campuses, sports arenas, airports, and high-rise office buildings may use more than one chiller to meet their cooling needs.

Hermetic centrifugal compressors, Figure 19-2-8, vary in design and refrigerant use. Some are single stage, others multistage. Some are direct drive, while others are gear driven. The operating principle of all centrifugal compressors is the same: a rotating im-

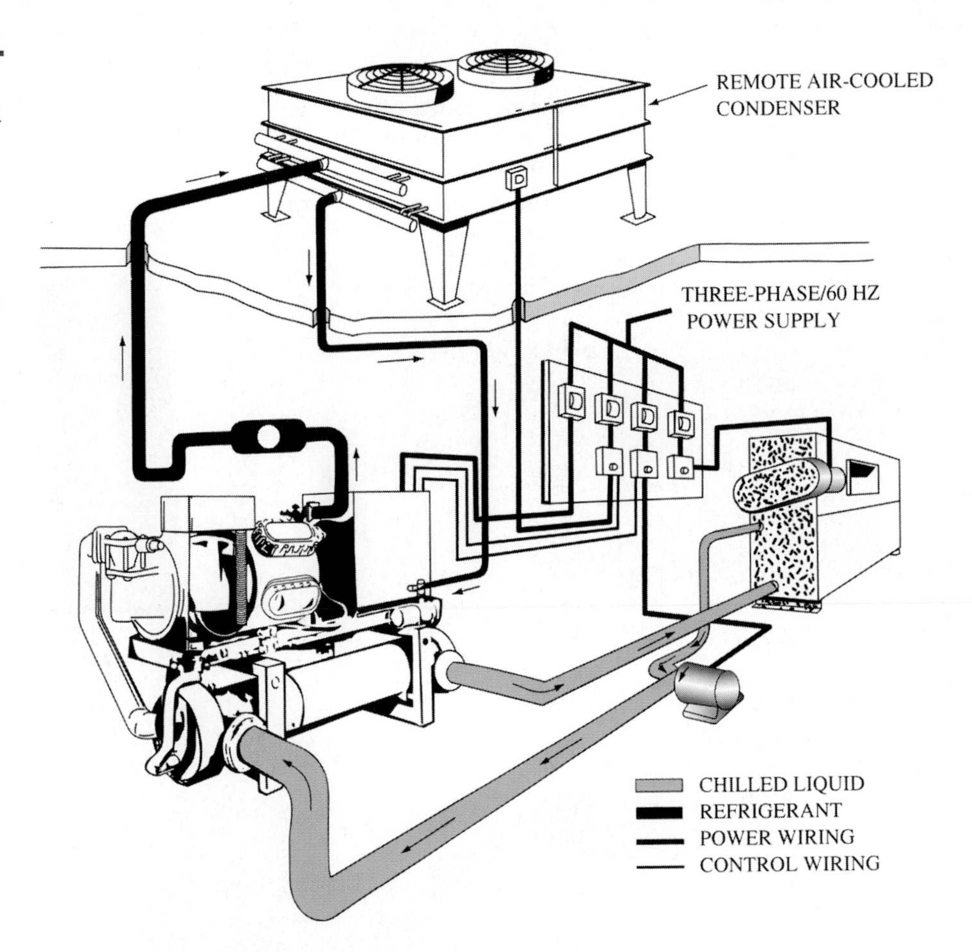

Figure 19-2-6 Air cooled condenser installation with a package chiller.

Figure 19-2-7 Package air cooled water chiller.

Figure 19-2-8 Hermetic centrifugal compressor and drive.

peller is used to draw suction gas from the chiller (cooler) and compress it through a spiral discharge passage into a condenser. The speed of the impeller is a function of the design. Some gear driven impellers reach speeds of 20,000 to 25,000 rpm. Capacity control is accomplished by a set of inlet vanes that throttle the suction gas to load or unload the impeller.

A complete water cooled hermetic centrifugal chiller assembly is illustrated in Figure 19-2-9. This unit uses a combination chiller and condenser in one shell although they are separated internally. Chillers of this size and design do not use expansion valves. They use either a float control or a metering device to flood the chiller shell with liquid refrigerant.

Some centrifugal chillers have compressors with an open drive. Units of this type produce up to 5000 tons. The choice of drive can be gas, steam, diesel, or electric motor. These units are somewhat specialized and not as easily designed, selected, and installed as are the complete package units.

Screw Compressor Chillers

An important type of package water chiller uses the helical rotary compressor, commonly known as a screw compressor because of the appearance of the rotors. The twin rotor screw compressor, illustrated in Figure 19-2-10, uses a mating pair of rotors with lobes that rotate much like a pair of gears.

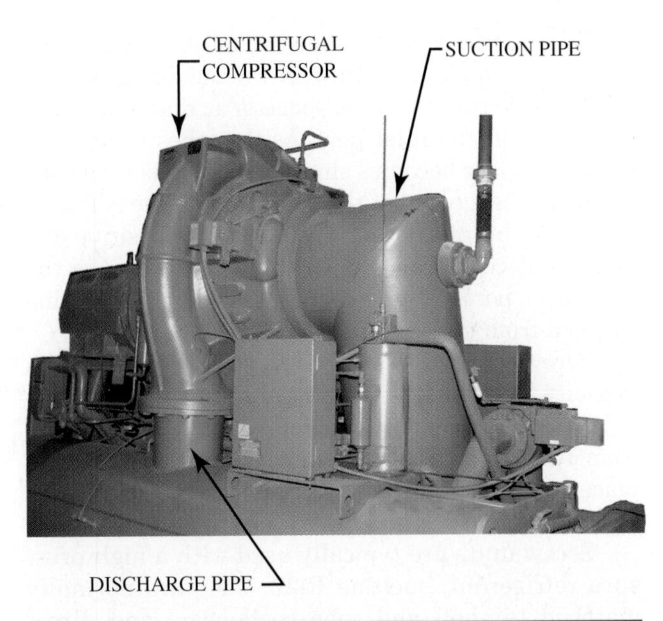

Figure 19-2-9 Hermetic centrifugal liquid chiller.

TECH TIP

Some chiller systems use heat recovery units to provide hot water for the building. Heat recovery units pick up waste heat from the condenser to provide "free" hot water.

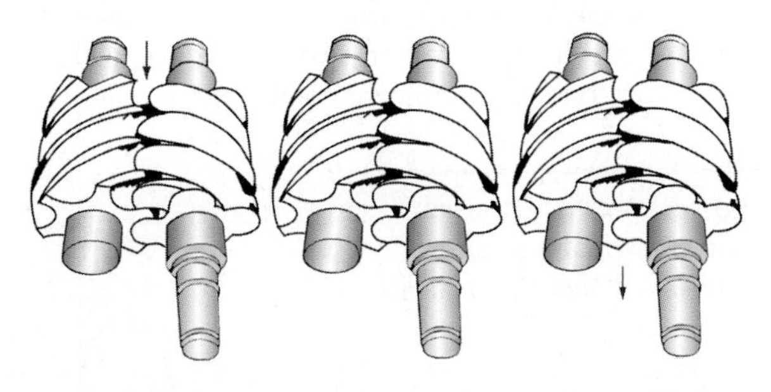

Figure 19-2-10 Diagram of the rotors of a screw compressor.

During rotation the space or mesh between the lobes first expands to draw in the suction gas. At a point where the interlobe space is at maximum, the lobes seal off the inlet port. As the lobes rotate, the interlobe space becomes smaller as the gas is carried to the discharge end of the compressor. The refrigerant gas is internally compressed by this positive displacement compressor until the rotors uncover the discharge port, where the compressed gas is discharged from the compressor.

Screw compressors are used because of their high capacity for a small unit, and their continuously variable (stepless) capacity control, typically modulating from 100% to 10% of full capacity. Being positive displacement compressors, they can also be used with remote condensers, and have other piping flexibility.

Screw units are typically used with a high pressure refrigerant, such as R-22. They are normally matched to shell and tube condensers and direct expansion chillers, similar to those used on the reciprocating package chillers already reviewed.

Screw compressors and their chiller packages were originally developed as effective units for the refrigeration needs of the food and chemical industries. They have been effectively adapted to the needs of the comfort air conditioning market, with industrial based technology.

Twin rotor screw compressors come in a variety of configurations, depending on the manufacturer and application. The most common type is the horizontal open drive unit, typically driven at 3500 rpm by an external motor, Figure 19-2-11. Other units are of vertical shaft construction, for minimum floor space usage. Many package chiller units use semihermetic construction on the compressor, eliminating the problems of shaft alignment and mechanical shaft seal leakage.

Screw compressors also come in single rotor designs, using gate rotors to seal the low pressure from the high pressure side of the compressor. An example of a semihermetic single rotor unit is shown in Figure 19-2-12a,b.

Absorption Chillers

Unlike the conventional mechanical compression refrigeration cycle used in all the other equipment discussed, an absorption chiller, Figure 19-2-13, uses steam, hot water, or direct firing by natural gas as an energy source to produce a pressure differential in a generator section. Some absorption units generate both chilled water and hot water from the same unit. Since they can operate on waste steam or direct fired natural gas the actual costs of operation may be less

SERVICE TIP

Most power companies use a building peak load as the basis for establishing the billing rate for electrical power. The peak load for electrical billing rates is based on the highest electricity usage during the prime electrical usage time of the day. Because chillers require larger amounts of energy during startup than during operation, they must be started before the start of the prime cooling time. In most parts of the country the prime electrical usage time is between 9 am and 6 pm.

Once the peak load rate is established, it will remain in effect for all electrical usage for the next 12 months. Starting a chiller late and establishing a new higher peak rate can cost a building operator tens of thousand of dollars in higher utility bills.

Figure 19-2-11 External drive screw compressor. *(Courtesy of York International Corp.)*

(a)

(b)

Figure 19-2-12 (a) Semihermetic screw compressor; (b) Screw compressor.

than electrically driven equipment, depending on the relative energy costs.

Absorption chillers operate most efficiently when they run at a steady state for long periods of time. For this reason they may be one unit as part of a large building multiple chiller system. In a multiple chiller instillation the absorption chiller can provide the base cooling needed to maintain the building over 24 hours and other chillers can be started and stopped during the day as needed to meet peak cooling loads.

CIRCULATING WATER PUMPS

The major uses of pumps in an HVAC/R system are for: (1) pumping chilled or heated water, (2) pumping condenser water from the cooling tower, and (3) circulating water in a cooling tower or evaporative condenser water circuit. These pumps are generally centrifugal types with configurations as shown in Figure 19-2-14. It is common practice in central station systems to provide identical pumps operating in parallel, one of which is a spare for critical uses.

Pumps are selected to perform a specific purpose, providing the proper flow at a pressure to overcome the resistance of the circuit in which they are placed. They are selected from pump curves provided by the manufacturer, such as the sample in Figure 19-2-15. These curves show the flow in gpm for various head pressures measured in feet (1 foot = 2.31 psi). The curves also show the performance of the pump with different diameter impellers, as well as the required motor size, based on the desired performance. For example, using the pump curve in Figure 19-2-15, if 320 gpm is required at a 42 ft head, a 7.75 in impeller could be used with a 7.5 Hp motor.

It is important to understand the difference between open and closed piping systems in working with pumps. An open system has some part of the circuit open to the atmosphere, such as a cooling pump water circuit. In this type of circuit the height that

Figure 19-2-13 Large absorption chiller. *(Courtesy of York International Corp.)*

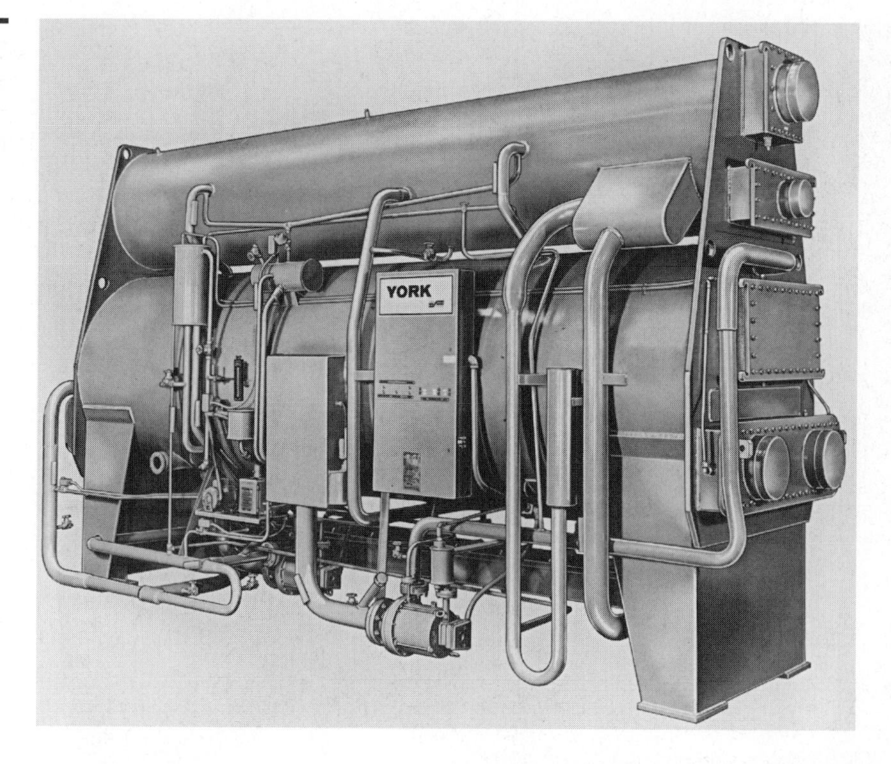

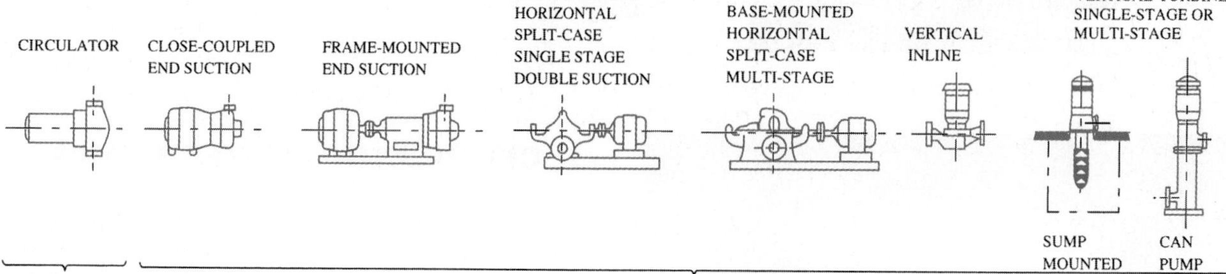

Figure 19-2-14 Typical centrifugal pumps. *(Copyright by the American Society of Heating, Refrigerating, and Air-Conditioning Engineers, Inc.)*

the water must be lifted must be added to the friction loss of the piping to determine the pump head. The suction lift of the pump is also limited.

In a closed system all piping is in series with the pump, such as in a circulating hot water heating system. The fluid that is pumped out returns to the suction side of the pump. An expansion tank is required in the circuit to allow for the volume change of the fluid due to variations in temperature. The location or the expansion tank and pump are important, as shown in Figure 19-2-16. In a closed system the system pressure can be regulated or limited by a pressure relief valve and an automatic makeup valve.

Table 19-2-1 describes common pumping problems and presents solutions.

UNIT 2—REVIEW QUESTIONS

1. List the three ways that water chillers differ.
2. Describe a scroll compressor.
3. What is the function of a shell and tube condenser?
4. What is the range of sizes of reciprocating compressor air cooled package water chillers?
5. ________ are used because of their high capacity for a small unit, and their continuously variable capacity control.
6. What is the most common type of twin rotor screw compressors?
7. Unlike the conventional mechanical compression refrigeration cycle used in all the other equipment, an absorption chiller uses ________.

TABLE 19-2-1 Pumping System Trouble Analysis Guide

Complaint	Possible Cause	Recommended Action
Pump or system noise	Shaft misalignment	■ Check and realign.
	Worn coupling	■ Replace and realign.
	Worn pump/motor bearings	■ Replace, check manufacturer's lubrication recommendations. ■ Check and realign shafts.
	Improper foundation or installation	■ Check foundation bolting or proper grouting. ■ Check possible shifting because of piping expansion/contraction. ■ Realign shafts.
	Pipe vibration and/or strain caused by pipe expansion/contraction	■ Inspect, alter, or add hangers and expansion provision to eliminate strain on pump(s).
	Water velocity	■ Check actual pump performance against specified, and reduce impeller diameter as required. ■ Check for excessive throttling by balance valves or control valves.
	Pump operating close to or beyond end point of performance curve	■ Check actual pump performance against specified, and reduce impeller diameter as required.
	Entrained air or low suction pressure	■ Check expansion tank connection to system relative to pump suction. ■ If pumping from cooling tower sump or reservoir, check line size. ■ Check actual ability of pump against installation requirements. ■ Check for vortex entraining air into suction line.
	Pump running backward (three-phase)	■ Reverse any two-motor leads.
	Broken pump coupling	■ Replace and realign.
	Improper motor speed	■ Check motor nameplate wiring and voltage.
	Pump (or impeller diameter) too small	■ Check pump selection (impeller diameter) against specified system requirements.
	Clogged strainer(s)	■ Inspect and clean screen.
	Clogged impeller	■ Inspect and clean.
	System not completely filled	■ Check setting of PRV fill valve. ■ Vent terminal units and piping high points.
Inadequate or no circulation	Balance valves or isolating valves improperly set	■ Check settings and adjust as required.
	Air-bound system	■ Vent piping and terminal units. ■ Check location of expansion tank connection line relative to pump suction. ■ Review provision for air elimination.
	Air entrainment	■ Check pump suction inlet conditions to determine if air is being entrained from suction tanks or sumps.
	Insufficient NPSHR	■ Check NPSHR of pump. ■ Inspect strainers and check pipe sizing and water temperature.

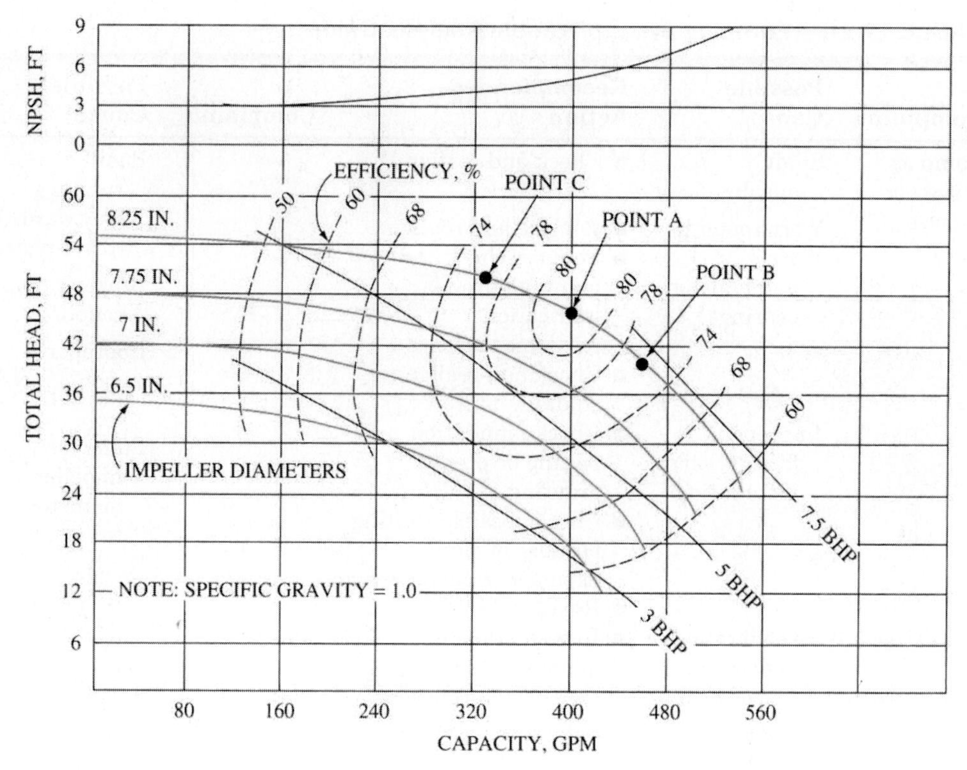

Figure 19-2-15 Typical pump performance curves provided by manufacturers. *(Copyright by the American Society of Heating, Refrigerating, and Air-Conditioning Engineers, Inc.)*

8. What is meant by the term peak power rate?

9. A ________ is a pressure vessel designed to transfer heat to a fluid, usually water.

10. List the three ways used to classify boilers.

11. What material are most boilers made of?

What do boiler controls regulate?

12. List the major uses of pumps in the HVAC/R system.

13. Describe the difference between open and closed piping systems in working with pumps.

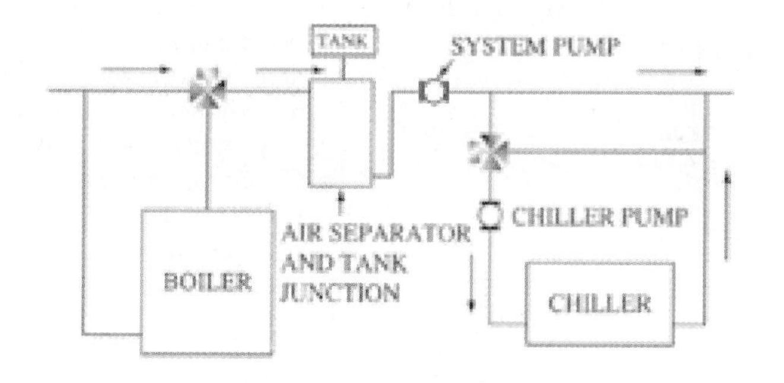

Figure 19-2-16 Correct expansion tank connection.

TECH TALK

Tech 1. Here it is the hottest part of August and yesterday when I got the number 2 chiller fixed down at the mall, the building manager would not let me start it up.

Tech 2. That's right. He did not want you setting a new peak load for the building.

Tech 1. What do you mean set a new peak load for the building?

Tech 2. Large electric users like the mall pay for electricity based on their highest electrical use during the heat of the day. If you had started up that chiller in the middle of the day, you could have set a new higher peak power level for the building.

Tech 1. But the mall was getting warm. They could have used the chiller.

Tech 2. That may be true. But if you had set a new higher utility rate, they would have been paying for your mistake for the next 12 months. Turning on a chiller during peak hours would have cost them more than the sales lost due to the mall being a little warm for one afternoon. You know on a mall, school, or large building, increasing the peak just a little can cost them more than you make a year in higher utility cost.

Tech 1. Oh, I see. Starting a chiller is not like flipping on the lights at home. It can cost the mall big time if I had started that thing up in the heat of the afternoon.

Tech 2. Now you get it.

Tech 1. Thanks.

UNIT 3

Hydronic Heating

OBJECTIVES: In this unit the reader will learn about hydronic heating including:

- Hydronic Heating Systems
- Heat Transfer
- Limitations of Hot Water Heat
- Safety Measures
- Circulator Pumps
- Properties of Water
- Components and Accessories
- Piping Layouts

OVERVIEW

Hydronics can be defined as a science that utilizes water or steam to transfer heat from the source where it is produced, to an area where it can be used, through a closed system of piping. Hydronics is applied also to cooling systems; however, this unit is confined to heating applications.

The amount of heat transfer that takes place is dependent on three conditions:

1. The temperature difference between the water and the surrounding medium.
2. The quantity of water flowing through the piping, expressed in gpm or gph.
3. The design characteristics of the heat exchanger.

Although water can be circulated by gravity, modern heating systems use forced circulation for two reasons:

1. Smaller pipe sizes can be used.
2. Systems are easier to control.

Advantages of hot water heating systems are based on overall comfort for the following reasons:

1. Heat is supplied at the base of the outside wall, warming the exterior exposure, by convection, to near room temperature. This prevents the uncomfortable feeling of radiating body heat to the cold outside walls.
2. The gravity circulation of warm air from the convectors supplies heat to the room by a gentle movement of air. This prevents uncomfortable drafts and air velocities that absorb moisture from the skin, creating a cooling effect.
3. The hot water heats and cools slowly. This prevents sudden changes in temperature.

Disadvantages of hot water heating are as follows:

1. They cost more than an equivalent warm air installation.
2. If humidification and air filtration are desired, a separate air system needs to be installed.
3. If cooling is desired as well as heating, dual systems are required.

PROPERTIES OF WATER

In order to deal with hydronic heating systems, it is advantageous to review some of the properties of water. Fortunately, most tap water used in these closed systems is free from contamination and requires no special treatment. Certain properties, however, are critical and must be considered during application and installation, as follows:

1. Water expands or contracts as its temperature is changed. In a liquid state the change in volume with temperature is relatively small (but significant). When its temperature increases to the steam temperature, the change in volume is enormous. In closed hot water heating systems, to allow for the expansion characteristics, certain safety measures are provided as follows:
   - **a.** Both temperature and pressure limiting controls are used to maintain the water within safe conditions.
   - **b.** A pressure relief device is supplied in the piping to relieve excess pressure should it occur.

c. An expansion tank is provided in the piping, partly filled with air, to permit normal expansion and contraction of the water.

2. Water will freeze into ice at temperatures around 32°F, depending on the pressure. As water freezes it expands and can burst pipes or containers. Provisions need to be made to:
   a. Prevent the water from reaching the freezing point, or
   b. Use an antifreeze solution in the system in place of pure water, to lower the freezing point to a safe value.
3. Water has considerable weight (62.3 lb/ft^3), and allowances often need to be made in the supporting structure for equipment containing water.
4. Water has friction as it flows through piping, fittings, and equipment. This frictional force must be considered when selecting the flow of a circulating pump.

CAUTION: It is not easy to drain all of the fluid out of hydronic system pipes for service. Sometimes water can be trapped and be released suddenly as a result of vibration from shaking the pipe or when air breaks the siphon. It is important when working on electrical parts such as pumps that you make certain that both the power to the pump is off and that all water in the system has been removed. A sudden gush of water can create a major safety hazard from electric shock if these precautions are not taken.

EXAMPLE

Determine the total pump head required for selecting a cooling tower pump for the installation shown in Figure 19-3-1, based on the following conditions:

Piping—75 ft of steel pipe
Fittings—10 standard elbows
4 gate valves
Static lift—5 ft
Water flow—15 gpm
Condenser pressure drop—30 ft

Solution

From Table 19-3-1, a 1¼ in steel pipe is selected, with a head loss of 6.35 ft/100 ft of pipe. ■

EXAMPLE

From Table 19-3-2, the equivalent length of the piping and fittings is:

Item	Equivalent Length
75 ft of 1¼ in steel pipe	75.0 ft
10 1¼ in standard elbows @ 3.5	35.0 ft
4 1¼ in open gate valves @ 0.74	2.96 ft
Total Equivalent Length	112.96 ft

$$\text{Head loss in piping and fittings} = (112.96/100) \times 6.35 = 7.17 \text{ ft}$$

Pressure Loss	Total Pump Head
Due to pipe and fittings	7.17 ft
Due to condenser (13 × 2.31)	30.00 ft
Due to static lift	5.00 ft
Total pump head	42.00 ft

■

PIPING, PUMPS, AND ACCESSORIES

Piping

The supply piping conveys heat from the source to the terminal units. A proportion of its heat is released

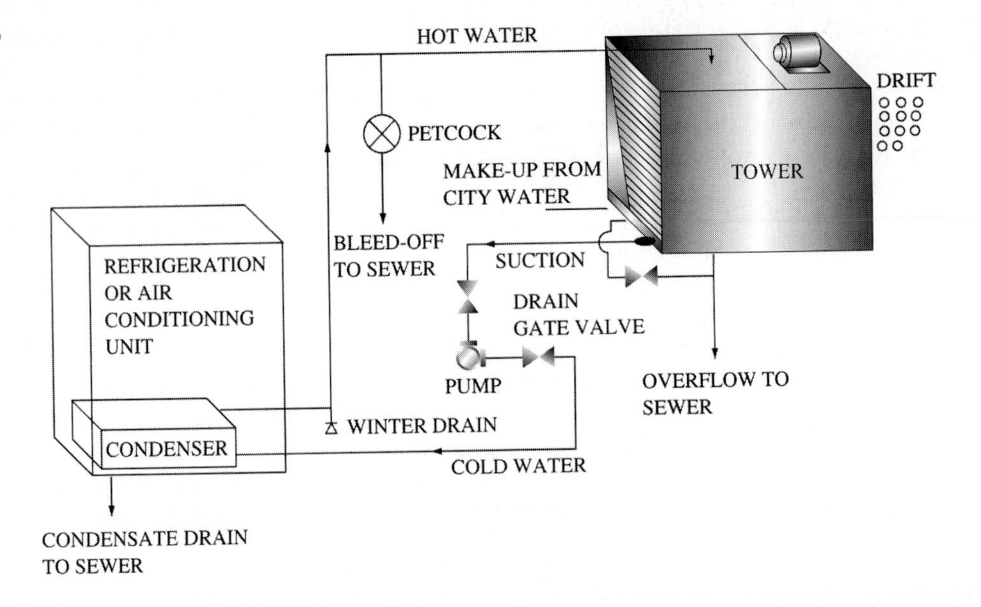

Figure 19-3-1 Diagram of a hydronic heating system's piping arrangement.

TABLE 19-3-1 Friction Losses of Standard Steel and Type L Copper Tubing in Feet of Head per 100 Feet of Pipe

		¾ in		1 in		1¼ in		1½ in		2 in	
Water Flow (gal/min)	Type of Pipe or Tubing	Velocity (ft/sec)	Head Loss (ft/100 ft)	Velocity (ft/sec)	Head Loss (ft/100 ft)	Velocity (ft/sec)	Head Loss (ft/100 ft)	Velocity (ft/sec)	Head Loss (ft/100 ft)	Velocity (ft/sec)	Head Loss (ft/100 ft)
6	Std. steel	3.61	14.7	2.23	4.54						
	Copper type L	3.98	11.5	2.34	3.13						
9	Std. steel	5.42	31.1	3.34	9.72	1.93	2.75				
	Copper type L	5.96	24.2	3.50	6.63	2.30	2.38				
12	Std. steel			4.46	16.4	2.57	4.31	1.89	2.04		
	Copper type L			4.67	11.3	3.06	4.04	2.16	1.73		
15	Std. steel			5.57	24.9	3.22	6.35	2.36	3.22		
	Copper type L			5.84	17.1	3.83	6.12	2.70	2.62		
22	Std. steel					4.72	13.2	3.47	6.25	2.10	1.85
	Copper type L					5.21	12.5	3.96	5.57	2.28	1.40
30	Std. steel							4.73	11.1	2.87	3.29
	Copper type L							5.41	9.44	3.11	2.45
45	Std. steel									4.30	6.96
	Copper type L									4.66	5.20

Note: Data on friction losses based on information published in *Cameron Hydraulic Data* by Ingersoll Rand Company. Data based on clear water and reasonable corrosion and scaling.

(Data source: Ingersoll Rand Company)

TABLE 19-3-2 Friction Loss of Fittings and Valves in Equivalent Feet of Pipe

Pipe Size (in)	Gate Valve Fully Open	45° Elbow	Long Sweep Elbow or Run of Std. Tee	Std. Elbow or Run of Tee Reduced One-Half	Std. Tee through Side Outlet	Close Return Bend	Swing Check Valve Fully Open	Angle Valve Fully Open	Globe Valve Fully Open
¾	0.44	0.97	1.4	2.1	4.2	5.1	5.3	11.5	23.1
1	0.56	1.23	1.8	2.6	5.3	6.5	6.8	14.7	29.4
1¼	0.74	1.6	2.3	3.5	7.0	8.5	8.9	19.3	38.8
1½	0.86	1.9	2.7	4.1	8.1	9.9	10.4	22.6	45.2
2	1.10	2.4	3.5	5.2	10.4	12.8	13.4	29.0	58.0

(Data source: Crane Company)

and the return piping carries water back to the boiler for reheating to its original discharge temperature.

SERVICE TIP

Water pumps cannot pump air. Therefore it may be necessary to bleed the air out of pumps on new systems after pump replacement. Most pumps have a bleed port, which is a small plug that can be opened on the pump housing. The bleed port allows the trapped air to escape. On systems that may have air trapped throughout the piping, more than one pump priming may be necessary as additional air is forced back into the pump along with the system circulating water. A pump that has lost its prime because of trapped air can overheat the impeller if it is operated for any significant period of time.

A good piping system has the following qualities:

1. The piping is the correct size.
2. It is well supported, with hangers that permit adequate pipe expansion and contraction.
3. The piping is properly assembled with leak tight joints.

Pipe sizing is usually performed by using a friction chart, as shown in Figure 19-3-2. The water velocity should not exceed 5 ft/sec (5 fps). Figure 19-3-2 gives the flow rates for water in copper tubing. This chart shows head loss in ft/100 ft on the vertical axis, flow rate in gpm on the horizontal axis. Pipe velocity lines run upward to the left, and pipe size lines run upward to the right. Having given the flow rate and the pipe velocity, the pipe size and the loss in head per 100 ft can be read from the chart.

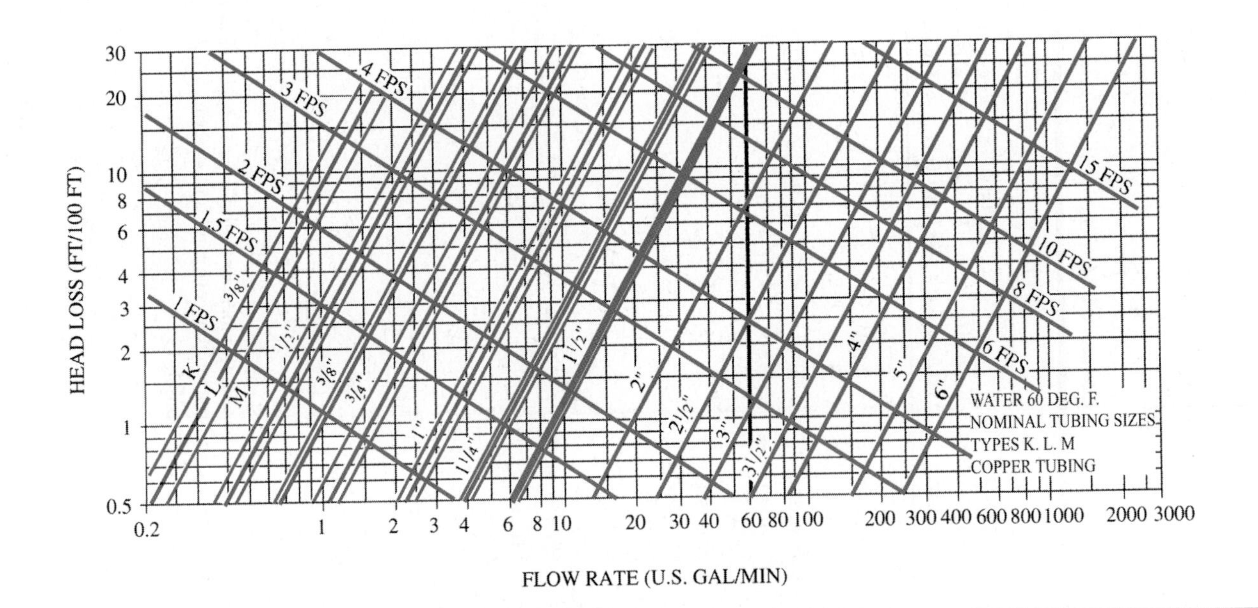

Figure 19-3-2 Friction losses of water flowing in copper tube showing pipe sizes and velocities. *(Copyright by the American Society of Heating, Refrigerating, and Air-Conditioning Engineers, Inc.)*

EXAMPLE

Given a boiler with an output capacity of 100,000 Btu/hr, a supply water temperature of 210°F, a return water temperature of 190°F, and a piping velocity requirement of 4 fps, what size copper pipe main should be used?

In order to convert the boiler output to gpm, the following formula is used:

$$\text{gpm} = \text{boiler output in Btu/hr}/(\text{TD} \times 500)$$

where

$$\text{TD} = 210 - 190°\text{F} = 20°\text{F}$$

Substituting in the formula,

$$\text{gpm} = 100{,}000\ \text{Btu/hr}\ (20°\text{F} \times 500) = 10\ \text{gpm}$$

Solution

Plot the intersection of the 10 gpm line and the 4 fps line. At this point the chart indicates a 1 in copper pipe. ■

Commonly used pipe supports consist of hanger rods which extend down from the ceiling, with loosely fitting circular openings for the pipe. This construction provides support, but permits the pipe to move during expansion or contraction.

Pipe anchors need to be provided at regular intervals to control and contain piping movements. These anchor supports are clamped tightly to the pipe. Where pipes are insulated, the hanger must surround the insulation without compressing it.

Piping materials for heating systems are normally steel or copper. Joints can be flanged, screwed or welded for steel pipes and brazed for copper. The joints must be tight and verified by approved tests.

Valves Valves are an important part of the system. They can be two position or adjustable to regulate flow. Some of the types of valves and their uses are shown as follows:

Type	Function
Gate valve	Stops and starts fluid flow
Globe valve	Throttles or controls fluid flow
Ball valve	Same as globe valve
Butterfly valve	Same as globe valve
Check valve	Prevents backward flow in pipes
Safety or relief valve	Relieves the excessive pressure and/or temperature

Valves must be selected to operate satisfactorily with the pressure in the system in which they are located.

Circulator Pumps

Pumps are used to force the flow of a certain quantity of liquid against the pressure drop of the system. The pumps used for hydronic systems are usually the centrifugal type.

> **SERVICE TIP**
>
> It is important to get all of the air out of the piping system before leaving the job. Air trapped in the system can cause noise and may result in the water pump losing its prime. The noise is a nuisance and the vibration can over time cause the pipe to fatigue and break. If the pump loses its prime the impeller can overheat and become so damaged that the pump will not work. On large systems it may be necessary to connect an auxiliary pump in order to have sufficient flow velocity to remove all of the trapped air.

Some small circulating pumps used in hot water systems have replaceable cartridges, such as the one shown in Figure 19-3-3. The replacement part contains all the moving parts and allows the pump to be serviced instead of replacing the entire unit. It is self lubricating and contains no mechanical seal.

An example of a pump performance chart is shown in Figure 19-3-4. Six curves are shown for the various pumps that can be applied. For example, with a 0010 pump 18 gpm can be obtained with an 8 ft total head.

Terminal Units *Terminal units* transfer heat from hot water or steam to the various building

Figure 19-3-3 Centrifugal pump used for hydronic systems. (*Courtesy of Taco, Inc.*)

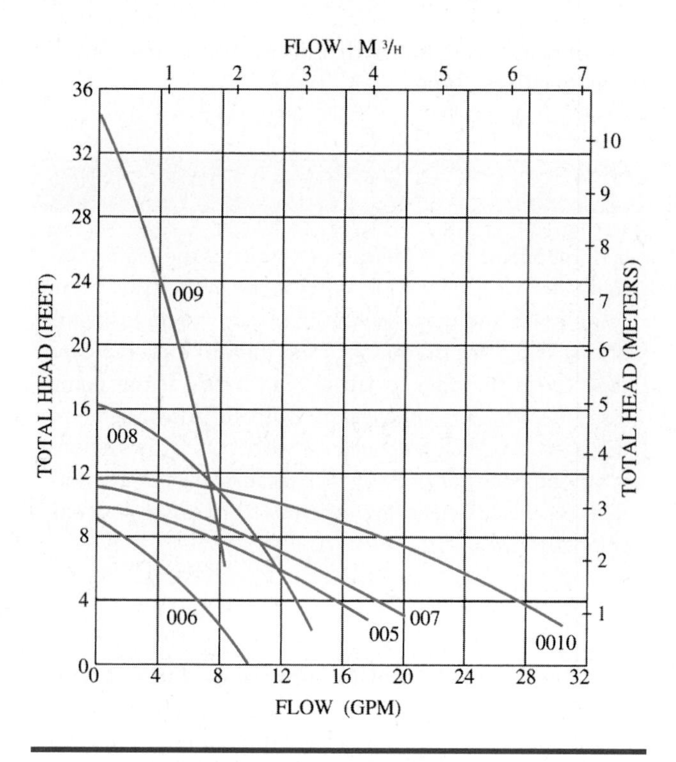

Figure 19-3-4 Typical performance of Taco "00" series centrifugal pumps. (*Courtesy of Taco, Inc.*)

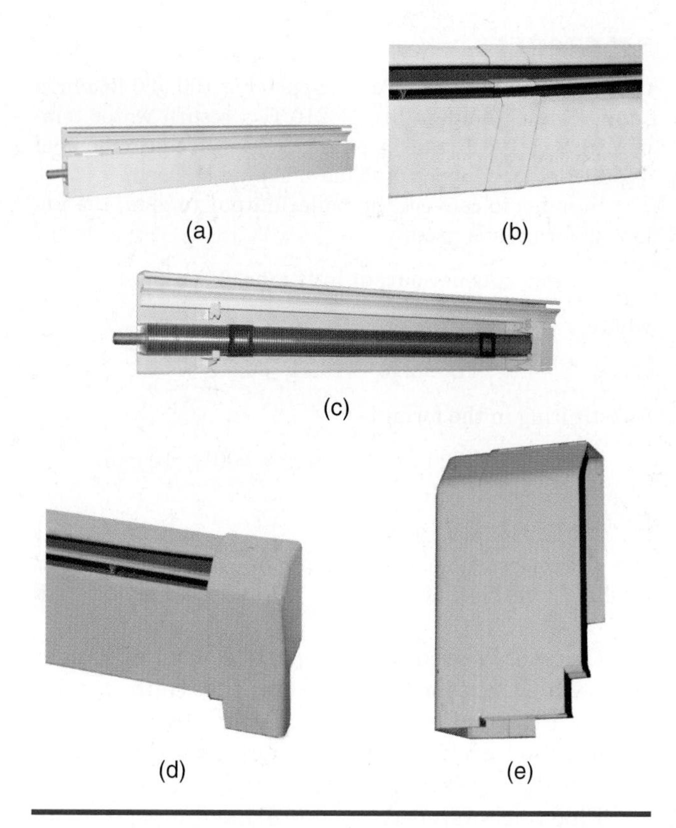

Figure 19-3-5 Trim pieces for baseboard enclosures.

TECH TIP

Circulating pumps on closed loop systems only work against the vertical lift head when the system is initially being filled. Once the system is completely filled with water, the pump effective head is only the flow resistance of the piping system itself. As the water returns to the pump, its weight counteracts the water being lifted on the supply side.

areas. The heat supplied by these units is controlled to provide comfortable conditions.

Normally this equipment is rated in Btu/hr or MBtu/hr; however, these ratings can be stated in terms of equivalent direct radiation (EDR). For steam systems, 1 ft^2 EDR = 240 Btu, with 1 psig steam condensing in the unit. For hot water systems 1 ft^2 EDR = 150 Btu, based on an average water temperature of 170°F. Hot water units are rated based on the water temperature drop through the unit which can vary between 10 and 30°F.

A number of types of terminal heating units are available, including: baseboard units, convectors, fan coil units, and radiators.

Baseboard units are installed near the floor on the outside walls of each room. The heating element is finned tube construction or cast iron. Heat is supplied by radiation and convection.

A typical finned tube along with the accessory pieces for corners and ends are shown in Figure 19-3-5a–e. Typical ratings are shown in Table 19-3-3. Dampers can be provided to regulate the airflow. The most common units have elements using copper tube and aluminum fins.

Convector units such as the one shown in Figure 19-3-6 can be free standing or recessed. These units heat the room mainly by convection, with the air entering the bottom and leaving from the upper front. The free standing model is supplied either with a flat or slanting top. An adjustable damper can be supplied, if desired.

A typical rating table for units with a 20°F hot water temperature drop is shown in Table 19-3-4. A typical rating table for a steam unit using 215°F steam is shown in Table 19-3-5.

Fan coil units as shown in Figure 19-3-7 can serve the same function as convectors, but have additional features that increase their efficiency and usefulness. Some of these features are as follows:

1. The unit has a centrifugal fan, with manual speed control, for circulating the room air through the unit.
2. Air filters are standard equipment.
3. The unit can be used for both heating and cooling.

TABLE 19-3-3 Rating Table for Slope-Top Convectors with Steel Elements

Tube Dia. (in)	Fin Style	Fins per Ft	Tiers	Encl. Type	Average Water Temperature (°F) 240	230	220	215	210	200	190	180	170	160	Installed Height (in)
1¼	41S025	32	1	S512	1530	1390	1280	1220	1160	1050	950	840	740	650	16½
			1	S520	1650	1500	1390	1320	1250	1140	1030	910	800	700	24
			1	S524	1680	1530	1410	1340	1270	1150	1050	930	820	710	28
			2	S520	2460	2250	2070	1970	1870	1690	1540	1360	1200	1040	28
			2	S524	2590	2360	2170	2070	1970	1780	1610	1430	1260	1100	28
			3	S524	3180	2900	2670	2540	2410	2180	1980	1750	1550	1350	28
1¼	41S025	40	1	S512	1690	1540	1420	1350	1280	1160	1050	930	820	720	16½
			1	S520	1840	1680	1540	1470	1400	1260	1150	1010	900	780	24
			1	S524	1875	1710	1580	1500	1420	1290	1170	1040	920	800	28
			2	S520	2710	2470	2280	2170	2060	1870	1690	1500	1320	1150	24
			2	S524	2790	2540	2340	2230	2120	1920	1740	1540	1360	1180	28
			3	S524	3210	2930	2700	2570	2440	2210	2000	1770	1570	1360	28
2	42S025	24	1	S512	1290	1170	1080	1030	980	890	800	710	630	550	16½
			1	S520	1350	1230	1130	1080	1030	930	840	750	660	570	24
			1	S524	1450	1320	1220	1160	1100	1000	900	800	710	620	28
			2	S520	2060	1880	1730	1650	1570	1420	1290	1140	1010	870	24
			2	S524	2210	2020	1860	1770	1680	1520	1380	1220	1080	940	28
			3	S524	2810	2570	2360	2250	2140	1940	1760	1550	1370	1190	28
2	42S025	32	1	S512	1510	1380	1270	1210	1130	1040	940	830	740	640	16½
			1	S520	1640	1490	1380	1310	1240	1130	1020	900	800	690	24
			1	S524	1750	1600	1470	1400	1330	1200	1090	970	850	740	28
			2	S520	2350	2140	1970	1880	1790	1620	1460	1300	1150	1000	24
			2	S524	2550	2910	2140	2040	1940	1750	1590	1410	1240	1080	28
			3	S524	2960	2700	2490	2370	2250	2040	1850	1640	1450	1260	28
1¼	41S032	32	1	S512	1630	1480	1370	1300	1240	1120	1010	900	790	690	16½
			1	S520	1750	1600	1470	1400	1330	1200	1090	970	850	740	24
			1	S524	1780	1620	1490	1420	1350	1220	1110	980	870	750	28
			2	S520	2630	2400	2200	2100	2000	1800	1640	1450	1280	1110	24
			2	S524	2750	2510	2310	2200	2090	1890	1720	1520	1340	1170	28
			3	S524	3380	3080	2840	2700	2570	2320	2110	1860	1650	1430	28
1¼	41S032	40	1	S512	1790	1630	1500	1430	1360	1230	1120	990	870	760	16½
			1	S520	1950	1780	1640	1560	1480	1340	1220	1080	950	830	24
			1	S524	2000	1824	1680	1600	1520	1380	1250	1100	980	850	28
			2	S520	2890	2630	2430	2310	2190	1990	1800	1600	1410	1220	24
			2	S524	2960	2700	2490	2370	2250	2040	1850	1630	1450	1260	28
			3	S524	3410	3110	2870	2730	2590	2350	2130	1880	1670	1450	28
2	42S032	24	1	S512	1360	1240	1140	1090	1040	940	850	750	670	580	16½
			1	S520	1440	1310	1210	1150	1090	990	900	790	700	610	24
			1	S524	1540	1400	1290	1230	1170	1060	960	850	750	650	28
			2	S520	2190	2000	1840	1750	1660	1510	1370	1210	1070	930	24
			2	S524	2350	2140	1970	1880	1790	1620	1470	1300	1150	1000	28
			3	S524	2990	2720	2510	2390	2270	2060	1860	1650	1460	1270	28
2	42S032	32	1	S512	1610	1470	1350	1290	1230	1110	1010	890	790	680	16½
			1	S520	1740	1580	1460	1390	1320	1200	1080	960	850	740	24
			1	S524	1860	1700	1560	1490	1420	1280	1160	1030	910	790	28
			2	S520	2500	2280	2100	2000	1900	1720	1570	1380	1220	1060	24
			2	S524	2710	2470	2280	2170	2060	1860	1690	1500	1320	1150	28
			3	S524	3150	2860	2650	2520	2390	2170	1970	1740	1540	1340	28

All ratings are Btuh/linear ft and are based on 65°F entering air and water velocity of 3 ft/sec. Elements are located with longest dimension in horizontal position.

(Courtesy of Dunham-Bush, Inc.)

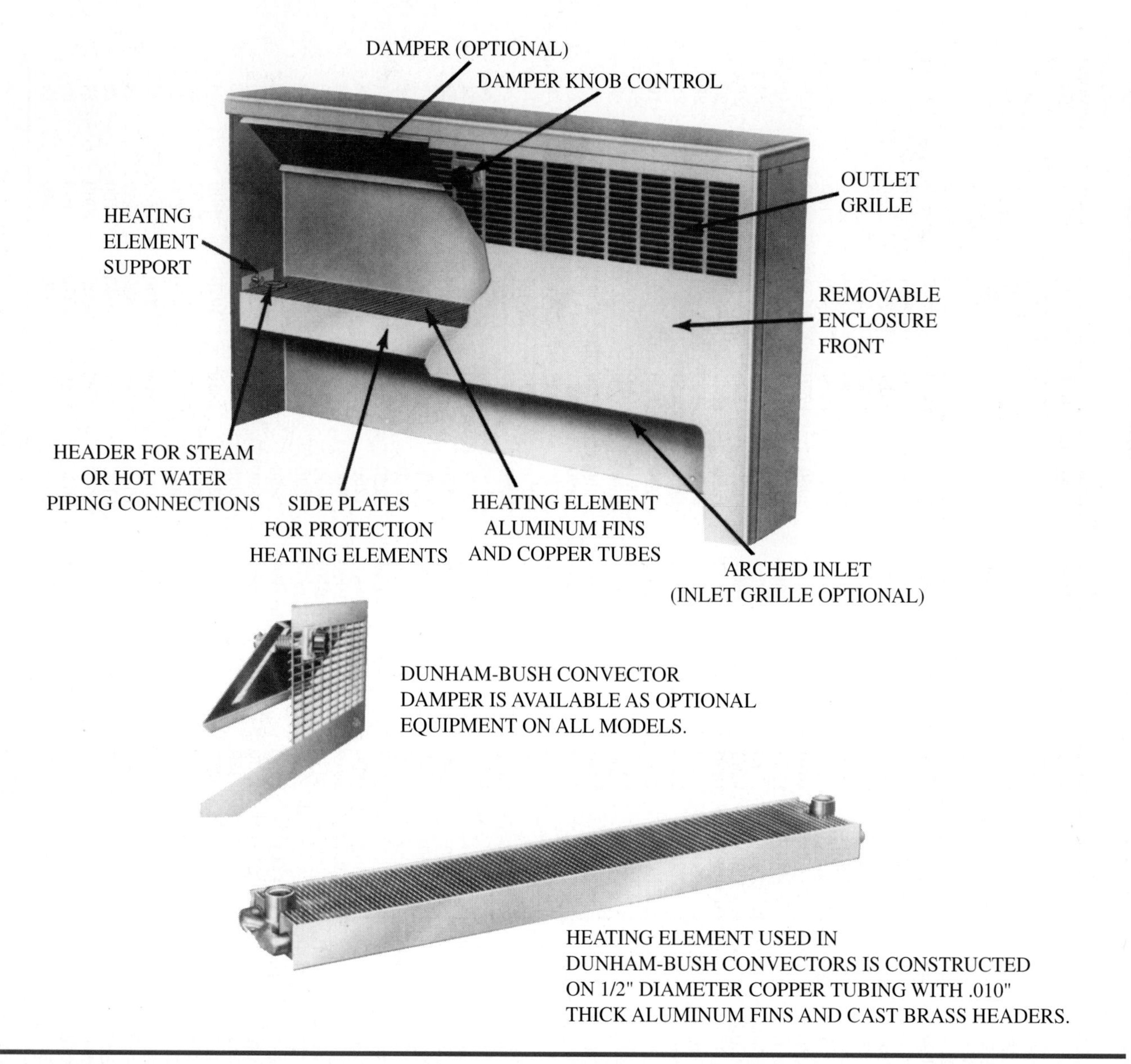

Figure 19-3-6 Typical convector.

4. Various piping arrangements can be used for heating/cooling applications to facilitate operation.
5. Various automatic control arrangements can be used.
6. Outside air can be supplied through the back of the unit.

Units are available with air volume ratings from 200 to 1200 CFM and hot water heating capacities from 7.6 to 112.0 MBtu/hr.

Radiators are primarily used for steam installations. They were used more frequently in past years. Table 19-3-6 gives the dimensions and ratings for small tube cast iron radiators. These have been replaced on new installations with more efficient heat transfer units.

Radiant heat panels using hot water can be used for heating from floors or ceilings. Typical installations use floor panel temperatures of 80 to 85°F and ceiling panel temperatures from 120 to 130°F. If water pipes are placed in concrete or plaster, the expansion coefficients of both must be the same. Pipes must be treated to prevent corrosion.

TYPES OF PIPING SYSTEMS

Hot water heating systems can be installed with a variety of piping systems. It is important that the technician be able to recognize each of the common systems, because only in so doing is he able to make adjustments where necessary and/or provide service.

TABLE 19-3-4 Typical Hot Water Capacities for Front Outlet Convector

Height (in)	Depth (in)	Length (in) 16	20	24	28	32	36	40	44	48	56	64	72	80	88	96	104	112
Type FH 18	4	1.0	1.1	1.4	1.7	2.0	2.3	2.5	2.8	3.1	3.6	4.2	4.7	5.3	5.8	6.3	6.9	7.4
	6	1.4	1.7	2.1	2.5	2.9	3.3	3.7	4.2	4.6	5.4	6.3	6.9	7.7	8.5	9.3	10.1	10.9
	8	1.7	2.1	2.6	3.3	3.9	4.5	5.1	5.7	6.3	7.5	8.6	9.0	10.0	11.4	12.4	13.7	14.8
	10	2.1	2.4	2.9	3.5	4.1	4.8	5.5	6.3	7.0	8.5	10.0	10.4	11.6	12.9	14.1	16.5	17.1
Type FH 20	4	1.1	1.2	1.6	2.0	2.3	2.6	3.0	3.3	3.7	4.3	5.0	5.6	6.3	6.9	7.6	8.3	9.0
	6	1.5	1.9	2.3	2.8	3.3	3.8	4.3	4.8	5.3	6.2	7.2	7.7	8.7	9.6	10.5	11.4	12.3
Type WH 14	8	1.9	2.3	3.0	3.6	4.2	4.8	5.4	6.1	6.7	7.9	9.2	10.0	11.2	12.5	13.7	14.8	16.0
	10	2.4	2.7	3.4	4.1	4.8	5.5	6.2	6.9	7.6	9.1	10.6	11.5	12.9	14.2	15.6	17.0	18.3
Type FH 24	4	1.2	1.5	1.8	2.2	2.6	3.0	3.4	3.8	4.2	4.9	5.7	6.2	7.0	7.8	8.5	9.2	9.9
	6	1.8	2.2	2.8	3.4	3.9	4.7	5.0	5.6	6.2	7.3	8.4	9.4	10.4	11.4	12.4	13.4	14.5
Type WH 18	8	2.2	2.6	3.3	4.0	4.6	5.3	6.0	6.7	7.4	8.7	10.1	11.1	12.3	13.6	14.8	16.1	17.4
	10	2.7	3.0	3.7	4.5	5.3	6.0	6.8	7.6	8.3	9.9	11.4	12.6	14.1	15.5	17.0	18.5	20.0
Type FH 26	4	1.3	1.5	1.9	2.3	2.6	3.1	3.4	3.9	4.3	5.0	5.7	6.6	7.3	8.1	8.9	9.6	10.4
	6	1.9	2.3	2.8	3.4	4.0	4.6	5.2	5.8	6.4	7.5	8.7	9.8	10.8	12.0	13.1	13.6	14.6
Type WH 20	8	2.4	2.6	3.3	3.7	4.8	5.4	6.2	6.9	7.6	9.0	10.4	11.7	13.0	14.4	15.6	17.1	18.5
	10	2.8	3.0	3.8	4.6	5.4	6.2	7.0	7.8	8.6	10.1	11.7	13.0	14.6	16.1	17.6	19.1	20.6
Type FH 32	4	1.4	1.6	2.0	2.5	2.9	3.3	3.7	4.2	4.6	5.4	6.3	6.9	7.8	8.5	9.2	10.0	10.8
	6	2.1	2.5	3.1	3.8	4.4	5.1	5.7	6.4	7.0	8.3	9.6	10.5	11.6	12.7	14.0	14.9	16.0
Type WH 26	8	2.7	2.9	3.6	4.4	5.1	5.9	6.6	7.4	8.2	9.7	11.2	12.3	13.9	15.3	16.7	18.1	19.6
	10	3.1	3.3	4.1	5.0	5.9	6.7	7.6	8.4	9.3	11.0	12.7	14.5	15.5	17.1	19.5	20.4	21.4
Type FH 38	4	1.5	1.7	2.1	2.3	3.0	3.5	3.9	4.4	4.8	5.8	6.7	7.4	8.3	9.1	9.9	10.8	11.6
	6	2.3	2.6	3.3	4.0	4.6	5.3	5.9	6.6	7.3	8.6	10.0	11.1	12.3	13.4	14.5	15.7	16.8
Type WH 32	8	2.8	3.0	3.8	4.6	5.4	6.2	6.9	7.7	8.5	10.1	11.7	12.9	14.5	15.9	17.4	18.9	20.3
	10	3.2	3.4	4.3	5.2	6.1	7.0	7.9	8.8	9.7	11.4	13.2	14.6	16.3	18.0	19.7	21.4	22.5
Type WH 38	4	1.6	1.7	2.2	2.7	3.2	3.6	4.1	4.6	5.0	6.1	7.0	8.0	9.0	9.7	10.7	11.5	12.2
	6	2.4	2.7	3.4	4.1	4.7	5.4	6.1	6.8	7.5	8.9	10.4	11.9	13.0	14.2	15.3	16.5	17.7
	8	3.0	3.1	3.9	4.7	5.5	6.4	7.2	8.0	8.8	10.4	12.0	13.6	15.2	16.6	18.1	19.6	21.1
	10	3.4	3.5	4.4	5.4	6.3	7.2	8.1	9.0	10.0	11.8	13.7	15.3	17.0	17.5	20.5	22.2	23.8

TABLE 19-3-5 Typical Steam Capacities for Front Outlet Convector

Height (in)	Depth (in)	Length (in)																
		16	20	24	28	32	36	40	44	48	56	64	72	80	88	96	104	112
18	4	8.4	9.3	11.6	13.8	16.1	18.3	20.6	22.8	25.0	29.6	34.2	38.5	42.7	47.3	51.4	55.7	60.3
	6	11.7	13.5	16.8	20.2	23.6	27.0	30.4	33.8	37.2	44.1	51.0	55.7	62.3	58.9	75.4	82.0	88.9
	8	13.5	16.8	21.5	26.6	31.7	36.6	41.4	46.3	51.1	62.1	70.2	73.4	81.4	92.3	101.1	111.7	120.4
	10	17.1	19.7	23.3	27.5	33.3	39.2	44.8	51.2	57.0	69.1	81.3	84.6	94.4	104.5	114.8	134.4	139.3
20	4	8.8	10.0	13.1	15.9	18.7	21.5	24.3	27.0	29.7	35.3	40.8	45.6	51.1	56.5	62.1	67.5	72.9
	6	12.4	15.1	19.0	23.0	27.0	30.8	34.9	38.8	42.7	50.6	58.3	63.0	70.5	78.0	85.5	93.0	100.3
	8	15.8	18.9	24.0	29.0	34.0	39.1	44.2	49.2	54.3	64.4	74.4	81.1	90.9	101.3	111.6	120.2	129.9
	10	19.2	22.0	27.7	33.4	39.2	45.0	50.8	56.5	62.1	73.6	86.1	93.8	104.7	115.8	126.7	137.9	148.8
24	4	9.6	11.8	15.0	18.1	21.3	24.4	27.5	30.7	33.8	40.0	46.3	50.7	57.1	63.5	68.9	74.8	80.4
	6	14.5	17.7	22.4	27.0	31.6	36.2	40.8	45.4	50.0	59.2	68.4	76.4	84.5	92.9	101.0	109.2	118.0
	8	18.2	21.0	26.5	32.0	37.5	43.1	48.7	54.2	59.8	70.9	82.0	90.9	100.1	110.2	120.4	131.0	141.3
	10	21.7	24.0	30.2	36.5	42.8	49.0	55.3	61.5	67.7	80.2	92.7	102.2	114.4	126.4	138.4	150.4	162.4
26	4	10.8	12.3	15.4	18.5	21.8	25.0	28.2	31.4	34.6	41.0	47.4	53.5	59.7	66.0	72.3	78.4	84.8
	6	15.8	18.3	23.1	27.9	32.6	37.4	42.2	47.0	56.9	61.2	70.8	79.3	88.3	97.2	106.1	110.7	118.8
	8	19.9	21.4	27.2	32.4	38.6	44.3	50.0	55.7	61.4	72.8	84.3	95.1	105.9	116.7	127.2	138.8	152.3
	10	23.1	24.6	31.0	37.5	43.9	50.4	56.8	63.2	69.6	82.4	95.3	105.7	118.3	130.6	142.9	155.3	167.6
32	4	11.4	13.0	16.5	20.0	23.4	26.9	30.3	33.8	37.2	44.1	51.0	56.3	63.5	68.9	75.0	81.2	87.8
	6	17.1	20.2	25.5	30.6	36.0	41.2	46.5	51.7	58.1	67.4	77.9	85.0	95.4	103.1	113.7	121.2	130.1
	8	21.6	23.2	29.4	35.5	41.6	47.8	54.0	60.2	66.4	78.8	91.0	100.3	112.9	124.2	135.5	147.4	159.6
	10	24.9	26.6	33.5	40.6	47.6	54.5	61.5	68.4	75.4	89.3	103.1	113.2	126.3	139.4	158.5	165.8	178.6
38	4	12.0	13.6	17.3	21.0	24.7	28.4	32.0	35.8	39.4	46.8	54.2	60.0	67.2	74.3	81.6	88.8	96.0
	6	18.6	21.0	26.5	32.0	37.4	42.8	48.3	53.8	59.2	70.1	81.2	89.9	110.7	111.3	122.3	132.9	143.6
	8	23.0	24.2	30.8	37.0	43.6	50.0	56.4	62.8	69.2	82.0	94.8	105.1	117.9	130.3	142.8	155.3	167.9
	10	26.2	27.7	34.9	42.3	49.6	56.8	64.0	71.2	78.6	92.9	107.4	118.8	132.6	146.5	160.4	174.3	183.2

Figure 19-3-7 Typical fan coil unit.

Each of these systems controls the supply of hot water from the boiler to the terminal units (and the return) in such a way that comfortable conditions are produced.

The common systems of piping for hot water heating applications in residential and small commercial installations are as follows:

1. Series loop, single circuit
2. Series loop, double circuit
3. One pipe venturi system, single circuit
4. One pipe venturi system, double circuit
5. Two pipe system, direct return
6. Two pipe system, reverse return
7. A zoned system, using pumps
8. A zoned system, using motorized valves.

Series Loop, Single Circuit The series loop system is probably the simplest of all the piping arrangements. A schematic diagram of the single circuit piping layout is shown in Figure 19-3-8. In this piping system the main supply pipe from the boiler enters the first baseboard unit. The outlet from the first unit is connected directly to the inlet of the second unit. This arrangement is continued until all baseboard sections are connected in series. The outlet from the last section is connected to the circulator pump, which connects to the boiler. Thus, all the water flowing through the system passes through each unit.

In many cases the bathroom or kitchen does not have sufficient baseboard space and a convector is used in place of baseboard. The convector can be tied into the system using a venturi fitting as shown in the diagram.

When these systems are installed, the entire perimeter of the building has baseboard enclosures.

TABLE 19-3-6 Ratings for Small Tube Cast Iron Radiators

Number of Tubes per Section	Catalog Rating per Section[a]		A Height[c]	Section Dimensions B Width		C Spacing[b]	D Leg Height[c]
				Min.	Max.		
	ft²	Btu/hr (W)	in (mm)	in (mm)	in (mm)	in (mm)	in (mm)
3	1.6	384 (113)	25 (635)	3.25 (83)	3.50 (89)	1.75 (44)	2.50 (64)
4	1.6	384 (113)	19 (483)	4.44 (113)	4.81 (122)	1.75 (44)	2.50 (64)
	1.8	432 (127)	22 (559)	4.44 (113)	4.81 (122)	1.75 (44)	2.50 (64)
	2.0	480 (141)	25 (635)	4.44 (113)	4.81 (122)	1.75 (44)	2.50 (64)
5	2.1	504 (148)	22 (559)	5.63 (143)	6.31 (160)	1.75 (44)	2.50 (64)
	2.4	576 (169)	25 (635)	5.63 (143)	6.31 (160)	1.75 (44)	2.50 (64)
6	2.3	552 (162)	19 (483)	6.81 (173)	8 (203)	1.75 (44)	2.50 (64)
	3.0	720 (211)	25 (635)	6.81 (173)	8 (203)	1.75 (44)	2.50 (64)
	3.7	888 (260)	32 (813)	6.81 (173)	8 (203)	1.75 (44)	2.50 (64)

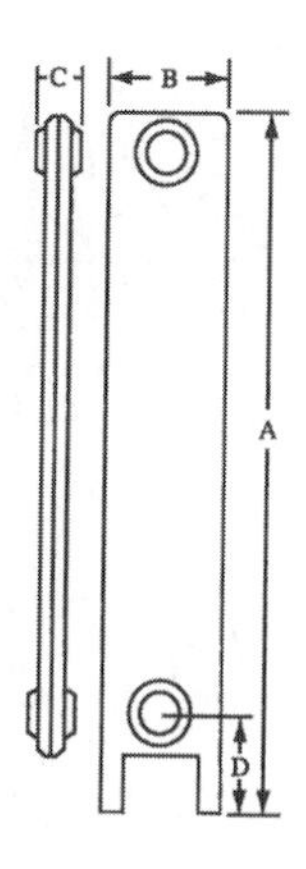

[a]These ratings are based on steam at 215°F (101.7°C) and air at 70°F (21.1°C). They apply only to installed radiators exposed in a normal manner, not to radiators installed behind enclosures, grills, or under shelves.
[b]Length equals number of sections multiplied by 1.75 in (44 mm).
[c]Overall height and leg height, as produced by some manufacturers, are 1 in (25 mm) greater than shown in Columns A and D. Radiators may be furnished without legs. Where greater than standard leg heights are required, this dimension shall be 4.5 in (114 mm).

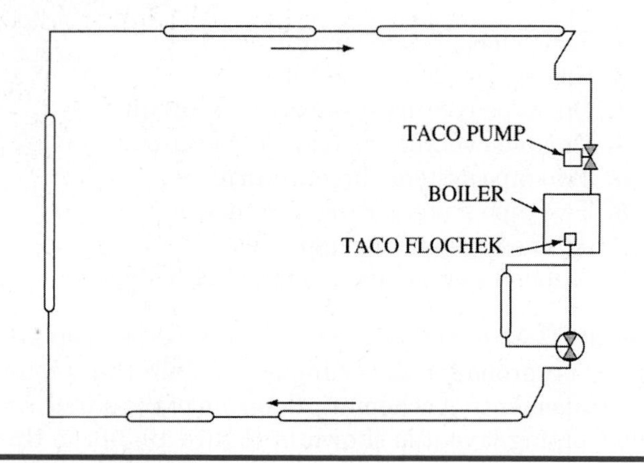

Figure 19-3-8 Series loop piping layout, single circuit. *(Courtesy of Taco, Inc.)*

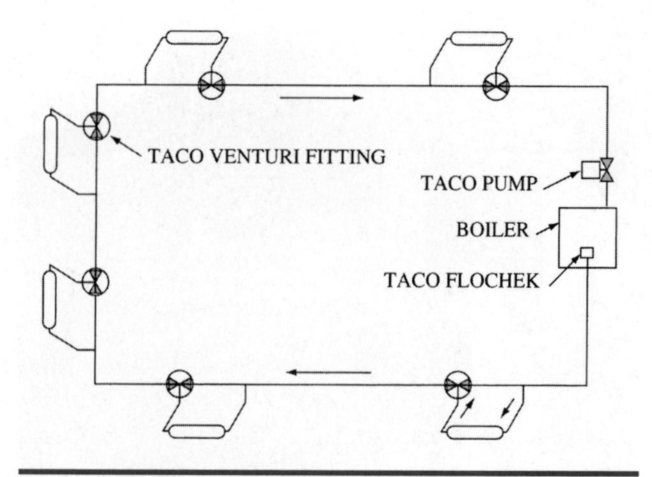

Figure 19-3-10 One pipe venturi system, single circuit. *(Courtesy of Taco, Inc.)*

The amount of finned tube element placed in these enclosures is dependent on the heat loss requirements. Where the main pipe passes a doorway, an offset is made in the piping to run the main below the floor until it passes the door opening.

With this piping arrangement each downstream unit is supplied with cooler water than the preceding one. The actual hot water temperature drop between terminal units is about 2°F. To compensate for this, each successive unit is oversized about 2%.

The use of a series loop system reduces the cost of the installation to a minimum.

Series Loop, Double Circuit On larger systems, better performance can be obtained by dividing the system into two (or more) approximately equal circuits, such as the example in Figure 19-3-9. When this is done, balancing valves are installed in each branch to permit adjusting the flow.

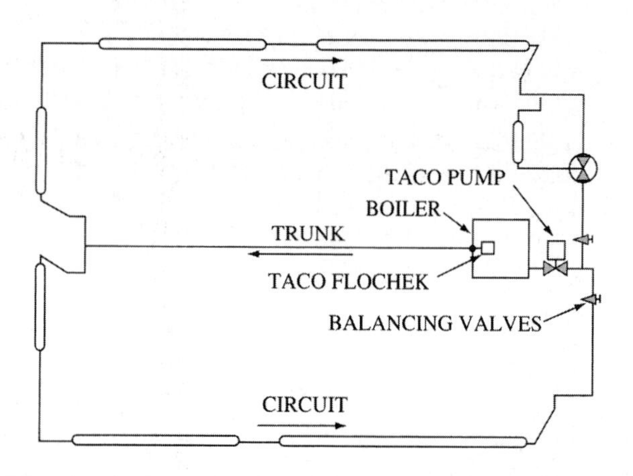

Figure 19-3-9 Series loop piping layout, double circuit. *(Courtesy of Taco, Inc.)*

One Pipe Venturi System, Single Circuit The one pipe venturi fitting system has a single main extending from the supply to the return connection of the boiler. The terminal units are fed by a supply and return branch connected to the main as shown in Figure 19-3-10. The supply connection uses a standard tee and the return connection uses a special venturi fitting. The venturi creates a negative pressure at the main connection, drawing the necessary flow through the branch.

This system differs from the series loop in that only a portion of the total flow enters the terminal unit. There is still a temperature drop in the main as it reaches the location of successive units. The reduced flow, however, improves the performance of the terminal units.

One Pipe Venturi Unit, Double Circuit On larger systems using this piping arrangement, better performance can be obtained by dividing the system into two (or more) approximately equal circuits, as shown in Figure 19-3-11. When this is done balancing valves are installed in each branch to permit adjusting the flow.

Two Pipe System, Direct Return In the two pipe system, separate supply and return mains are used. The return water from the terminal units is collected and returned to the boiler. With this piping arrangement, each unit will receive the same supply water temperature.

With the direct return, the two mains are run side by side with a supply branch run to each unit from the supply main and a return branch run from each unit to the return main. The first unit taken off the supply main is the first unit on the return main before it reaches the boiler, Figure 19-3-12. Obviously, the shortest piping is used on the unit nearest the

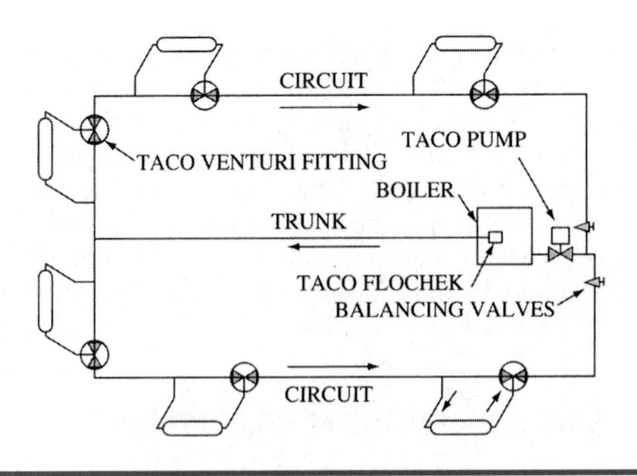

Figure 19-3-11 One pipe venturi system, double circuit. *(Courtesy of Taco, Inc.)*

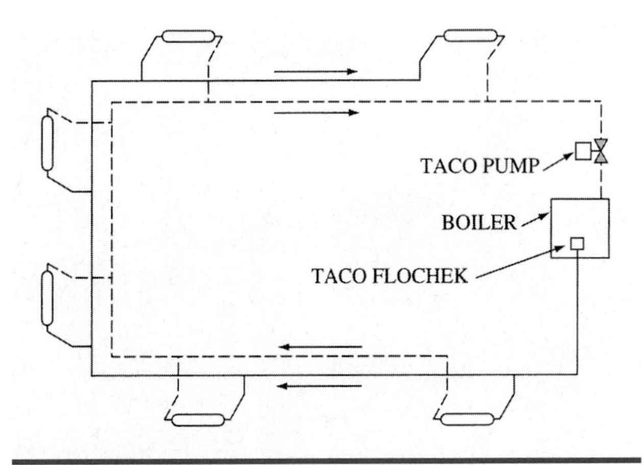

Figure 19-3-13 Two pipe reverse return system. *(Courtesy of Taco, Inc.)*

boiler. The longest piping is used on the unit farthest from the boiler. This difference in piping pressure drop must be equalized by using a square head cock in either the supply or return branch to each unit.

The two pipe direct return system uses more pipe than the one pipe system, however; it is considered to be more efficient by providing better water distribution and better control. This system can also be designed using two or more circuits.

Two Pipe System, Reverse Return The two pipe reverse return system is similar to the direct return system except the piping is arranged so that the same length of pipe (supply + return) is used for each terminal unit. This is accomplished, as shown in Figure 19-3-13, by taking the unit nearest the boiler off the supply main first, connecting it to the return main last and proceeding on this basis to equalize the piping lengths for all units. This equalization of the piping loss to each unit provides a balanced condition so that each unit receives its proper share of water.

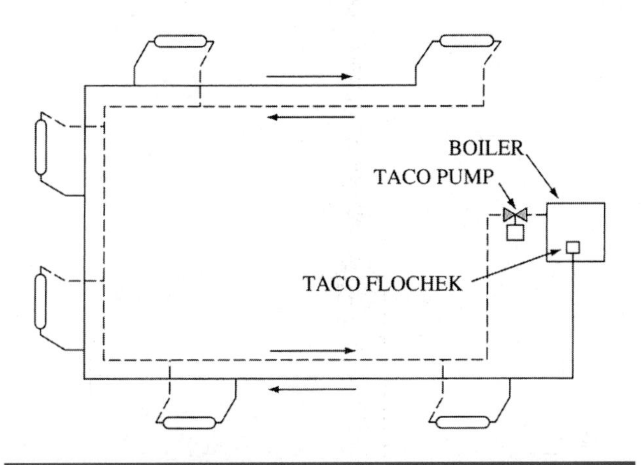

Figure 19-3-12 Two pipe direct return system. *(Courtesy of Taco, Inc.)*

Zoned System, Using Pumps Zoning is provided where it is desirable to control the supply of heat (hot water) to different areas of the building by separate thermostats. The areas selected can be determined by usage or building orientation. For example, to zone by usage, the living areas of a residence can be placed in one zone and the sleeping areas in another. To illustrate orientation zoning, a small commercial building can place all the rooms with southern exposure in one zone and the rooms with northern exposure in another. Whatever separation is made, allowance must be made in the piping and controls to provide this feature.

The piping needs to provide a separate hot water supply circuit to the terminal units in each zone. The number of supply water circuits is the same as the number of zones. One way to control the flow of hot water in each zone is to provide a separate circulator for each zone as shown in Figure 19-3-14. Any one of the systems described can be provided with zoning. The circulator in each zone responds to the requirement of the thermostat in that zone.

TECH TIP

Zoning hydronic systems can have significant energy savings by both providing heat only where it is needed and the thermal mass of the water within the system will allow the heating unit to operate for longer periods of time even during light loads. As the water in the system is heated and the heat is used at a terminal unit located in a zoned area the furnace may cut off and remain off for a period of time while the circulating pump still provides hot water to the zone. The furnace will not relight until the water temperature has dropped below the upper setpoint. This allows for longer furnace run times which increases the overall operating efficiency.

Figure 19-3-14 Zoned system using separate circulators for each zone.

TECH TALK

Tech 1. It was so cold last night that you did not even have to touch the wall to feel the cold. Just hold your hand up there and you could feel that cold coming right at you.

Tech 2. Actually you were not feeling the cold coming to you.

Tech 1. Yes, you can feel it.

Tech 2. No, actually heat flows from your warm hand to that cold wall so fast that it makes your hand feel cold as the heat is leaving.

Tech 1. It certainly feels like it's a cold wall without touching it.

Tech 2. That's true. It's called the cold wall effect. That is one of the advantages I have in my house with baseboard hydronic heating.

Tech 1. What do you mean?

Tech 2. Well, the heat comes up along the outside wall so it doesn't feel like that wall is as cold. In fact it doesn't feel as if it is cold at all.

Tech 1. Sounds like a big advantage of baseboard hydronic heating.

Tech 2. Yes, it's why can keep our house at 70°F and feel real comfortable but without it you would have to have the heat higher to feel as warm as we do.

Tech 1. Keeping the heat down as low as you do saves a lot of money. That's a good idea. Thanks.

Zoned System, Using Motorized Valves The zoning arrangement using motorized valves is similar to the system with pump zoning, with the exception that the motorized valves control the flow in each zone. Each supply circuit has a motorized valve controlled by the thermostat in that zone. A motorized valve is also used in a bypass around the circulator pump to relieve the pressure when only one zone valve is open.

UNIT 3—REVIEW QUESTIONS

1. Define hydronics.
2. Why do modern heating systems use forced circulation?
3. Why should water be prevented from freezing in pipes?
4. List the qualities of a good piping system.
5. What is the function of a glove valve?
6. What is the function of a check valve?
7. What is the function of a terminal unit?
8. List the types of terminal heating units that are available.
9. What are radiators primarily used for?
10. Describe the one pipe venturi fitting system.

UNIT 4

Boilers and Related Equipment

OBJECTIVES: In this unit the reader will learn about boilers including:

- Types of Boilers
- Boiler Controls

BOILERS

For heating, a hot water generator, commonly called a hot water boiler, or just a boiler, produces water at 180 to 200°F. The hot water pump delivers this hot water supply (HWS) to the heating coil in the air handler. As with the cooling coil, the heating coil has isolation and control valves. The hot water normally has a greater temperature change than the chilled water, giving up 20 to 40°F with a hot water return (HWR) temperature as low as 140°F.

A boiler is a pressure vessel designed to transfer heat (produced by combustion or by electrical resistance) to a fluid, usually water. If the fluid being heated is air, the unit is called a furnace.

The heating surface of a boiler is the area of the fluid backed surface exposed to the products of combustion, or fire side surface. Boiler design provides for connections to a piping system which delivers heated fluid to the point of use and returns the fluid to the boiler.

Boiler Classification

Boilers are classified in three ways: (1) working temperature/pressure, (2) fuel used, and (3) materials of construction.

Working Temperature/Pressure Low pressure boilers have a working pressure of up to 15 psi steam and/or up to 160 psi water pressure at a maximum of 250°F operating temperature. If a package boiler is used, the temperature and pressure limiting devices are supplied with the package; otherwise they must be applied during installation.

Medium and high pressure boilers operate above 15 psi steam and/or over 160 psi water pressure at temperatures above 250°F.

Steam boilers are available for up to 50,000 lb/hr of steam. Many of them are used in central station systems for heating medium and large commercial buildings that are beyond the range of a furnace. They are also used for industrial heating and industrial processing.

Water boilers are available from outputs of 50,000 Btu to 50,000,000 Btu (50 MBH). Many of these are low pressure boilers and are used for space heating. Some are equipped with internal or external heat exchangers to supply domestic hot water.

Every steam and hot water boiler is rated at a maximum working pressure determined by the ASME code under which it is constructed and tested.

CAUTION: Boiler operators are licensed and regulated by local, state, and federal agencies. Only individuals holding appropriate licenses may be operating engineers in charge of boilers.

SAFETY TIP

All boilers must be routinely inspected and certified for operation. The boiler certification must be posted in plain view in the boiler room.

Fuel Used Boilers may be designed to burn coal, wood, various grades of fuel oil, various types of fuel gas, or operate as electric boilers. Each fuel has its own special firing arrangement depending on the type of application.

Materials of Construction Most boilers are made of cast iron or steel. Some small boilers are made of copper or copper clad steel.

Cast iron boilers are constructed of individually cast sections assembled in groups or sections. Push or screw nipples or an external header join the section pressure tight and provide passages for water, steam, and products of combustion.

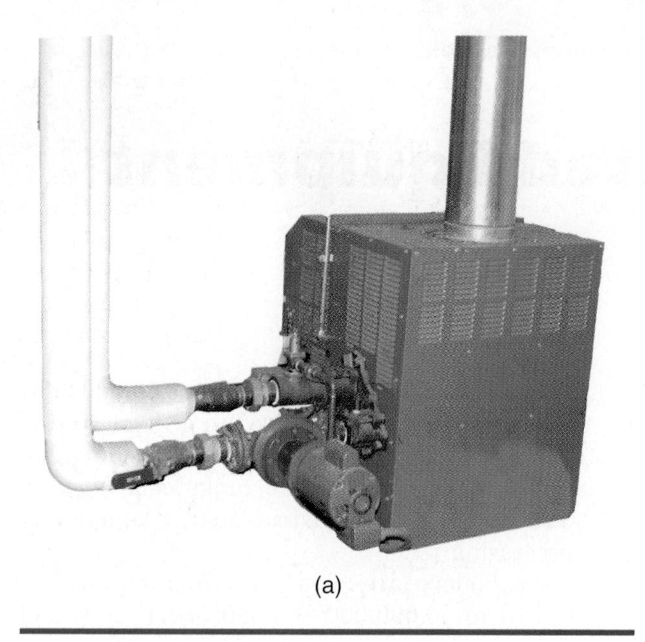

(a)

(b)

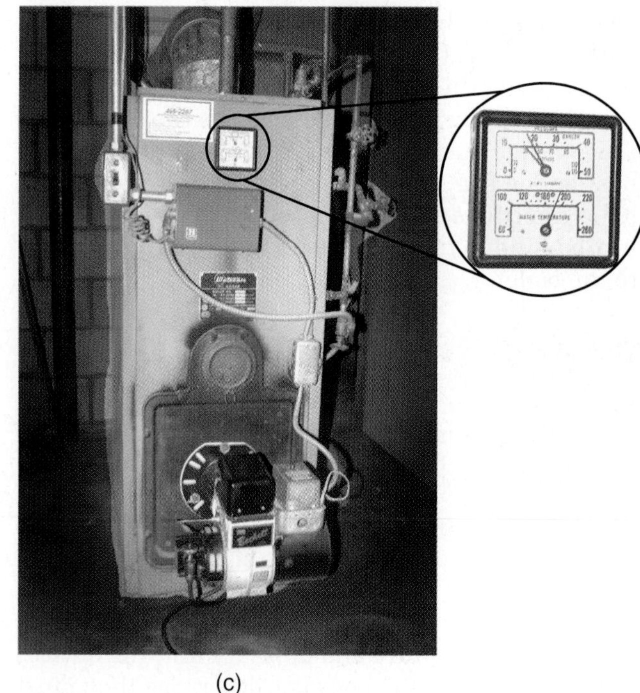

(c)

Figure 19-4-1 Types of residential boilers.

Steel boilers are fabricated into one assembly of a given size, usually by welding. The heat exchangers for steel boilers can be fire tube or water tube design. The fire tube construction, which is most common, has flue gas passage space between the water holding sections. The water tube construction uses water filled tubes for the heat exchanger with the flue gases in contact with the external surface of the tubes.

Copper boilers are usually some variation of the water tube type. The two most common tube types are parallel finned copper tube coils with headers, and serpentine copper tube units. Some are offered as residential wall hung boilers.

Condensing boilers have recently been developed to improve the efficiency of boilers. Previously the flue gases were not allowed to condense in the boiler due to the corrosion it could cause. These new condensing boilers are constructed with heat exchangers made of materials to resist corrosion. Depending on the constituents of the fuel and the application, overall efficiencies as high as 97% have been achieved using this new construction.

Electric boilers are in a separate class. No combustion takes place and no flue passages are needed. Electric elements can be immersed in the boiler water.

Figure 19-4-1 shows the configuration of a number of types of residential boilers.

BOILERS AND RELATED EQUIPMENT

All boilers are used to heat water. Where hot water is used to transfer heat, the output of the boiler maintains a temperature below the steaming

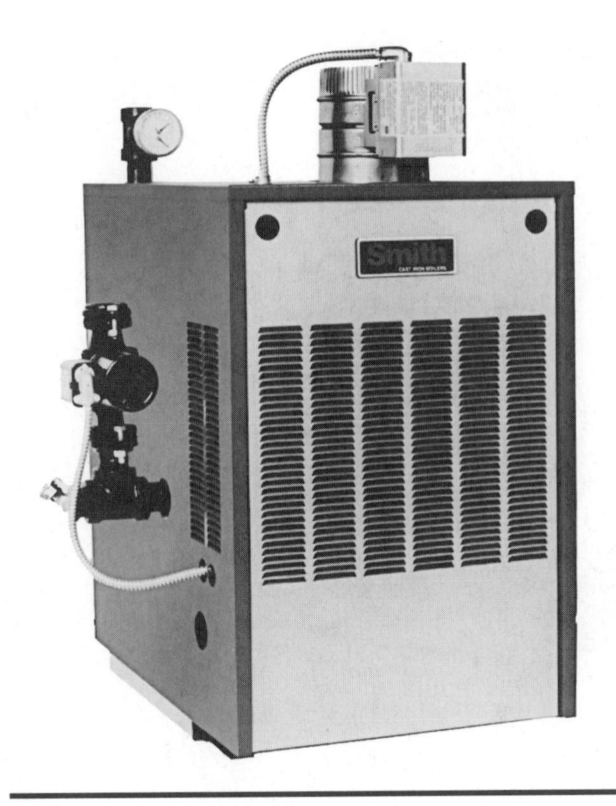

Figure 19-4-2 Typical cast iron gas boiler. (*Courtesy of Mestek, Inc.*)

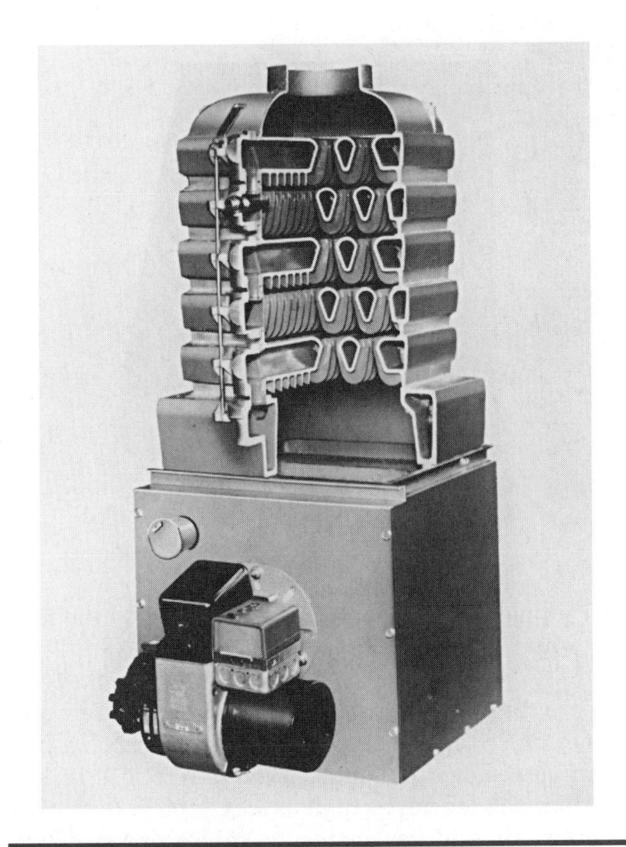

Figure 19-4-3 Typical cast iron oil boiler. (*Courtesy of Mestek, Inc.*)

temperature. At sea level atmospheric pressure, water boils at 212°F; at higher pressures the boiling point increases.

Where large amounts of heat need to be transferred, the boiler is used to produce steam. This is due to the fact that about 970 Btu/lb of heat can be added to water at 212°F and atmospheric pressure to change it into steam at the same temperature.

In general, for residential and small commercial installations, hot water boilers are used rather than steam due to the size of the installations and accuracy of control that can be provided.

Hot water boilers are constructed using cast iron, steel or stainless steel. A typical gas fired cast iron boiler is shown in Figure 19-4-2, and an oil fired cast iron boiler is shown in Figure 19-4-3. The surface of steel boilers can be electroplated or clad with nickel or other corrosion resistant material. Although this adds significant cost to the boiler, it significantly extends the unit operational life. Corrosion protection can also be provided by placing a sacrificial zinc rod in the tank. Zinc is more reactive than steel so it corrodes away first and can be easily replaced when it is gone.

Both of these boilers are assembled and shipped as package units. The package usually includes the circulator pump and complete controls. The wall thermostat is shipped separately for field mounting and wiring. Some packages also include a diaphragm type expansion tank and a check valve piped into the unit.

A steel boiler is shown in Figure 19-4-4. This boiler is also shipped as a package with accessories and controls. The design illustrated is a high efficiency unit with a condensing type heat exchanger producing efficiencies in the AFUE 90+% range.

A unique integrated home comfort system, using a boiler, is shown in Figure 19-4-5. It supplies

Figure 19-4-4 Steel boiler, gas fired.

Figure 19-4-5 Integrated home comfort system. (*Courtesy of GlowCore A.C. Inc.*)

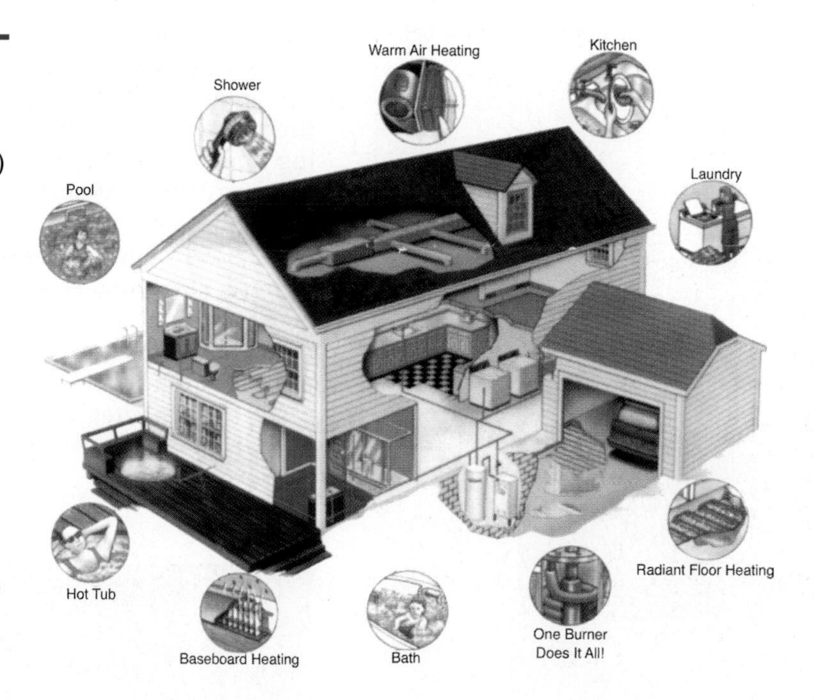

Figure 19-4-6 Household functions supported by integrated heating system. (*Courtesy of GlowCore A.C. Inc.*)

many household heating functions as shown in Figure 19-4-6. There are three connected units in the system:

1. A package gas fired boiler.
2. A storage tank for domestic hot water.
3. An air handling unit that can supply either heating or cooling. The cooling coil is arranged for connection to a remote condensing unit for operation as a split system air conditioner or heat pump. When operated as a heat pump, supplementary heat can be supplied by the hot water heating coil.

For larger hot water installations, a high efficiency pulse type steel boiler is available, such as the examples shown in Figure 19-4-7a,b. These boilers produce efficiencies as high as AFUE 96%. Both combustion air and gas metering valves are standard equipment. A small fan is used to deliver combustion air on startup, but shuts off after ignition. The unit is primarily controlled by a microprocessor.

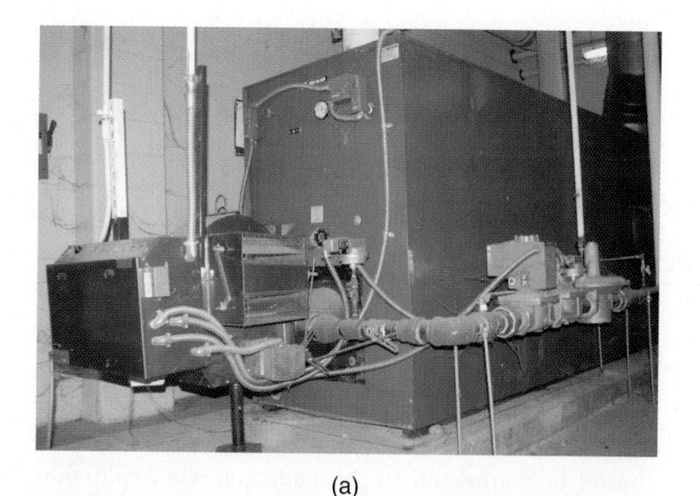

(a)

(b)

Figure 19-4-7 (a) Forced draft large high efficiency steel boiler; (b) Pulse high efficiency boiler.

PIPING DETAILS

A number of accessory devices are used in piping hot water systems, such as shutoff valves, air purging fittings, automatic fill valves, flow control valves, expansion tanks, safety controls, and gauges. Each of these devices has a special function in the proper operation of the system.

One of the main concerns in the selection and use of these accessories is to maintain a system of piping that contains only water and is free from air pockets. Air in the system can prevent proper operation of the pump and interfere with the transfer of heat in the terminal units.

Any air that forms in the system must be either directed to the expansion tank or vented. The system must be properly filled with water and the pump pressure maintained to produce the required flow.

TECH TIP

To reduce the expense in piping for large water systems steel pipe is used. However, steel pipe has the potential problem of rusting. To reduce the rust or corrosion problem steel pipes are treated with a "pickling" solution. This treatment forms a barrier to rust formation in the pipe. In addition, chemicals are added to the water circulating through the system that further retard rust. Check with the chemical manufacturer for the proper mixing ratio.

Air Purging Arrangement

Air must be eliminated from the piping system. Air vents need to be placed at high points of the system and on terminal units. Even if the system is free from air when filled, air can enter by the heating of water during operation. Depending on the selection of accessories, air that collects during operation can be vented at the expansion tank. Some air remains in the expansion tank to act as a cushion for the expansion and contraction of water during temperature change.

Water Volume Control

An expansion tank is placed in the piping to permit the expansion and contraction of the water volume. The tank holds both air and water. The air is compressed when the water volume expands and the air expands when the water volume decreases.

There are two types of expansion tanks: the diaphragm type, Figure 19-4-8, and the air cushion compression type, Figure 19-4-9.

The diaphragm type has a flexible separator between the air space and the water space that moves during operation to allow for the changing volume of water. As shown in the illustration, the air removed from the system is released through a float vent.

The air cushion compression type feeds air into the tank. This fitting permits the air bubbles to rise

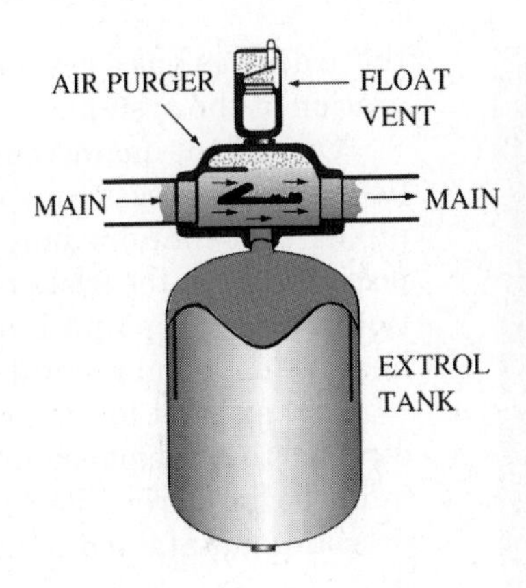

Figure 19-4-8 Diaphragm type expansion tank.

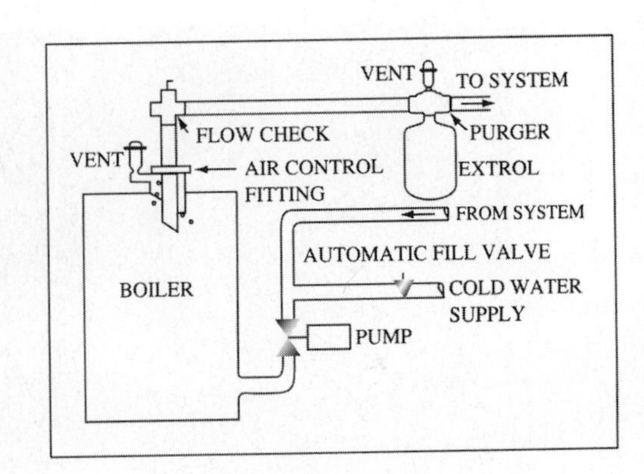

directly into the tank, but restricts the flow of water back into the tank. The air cushion provides adequate pressurization for all fluctuations of water volume.

CAUTION: It is extremely important that boilers not be allowed to run low or out of water. If a boiler runs low of water it can overheat, causing damage to the boiler. If a boiler runs out of water it can explode as a result of steam pressure. A boiler that is extremely low of water can shut down or can implode as violently as if it exploded. Implosion comes from the rapid condensing of steam into water, creating a negative pressure in the boiler.

Location of the Expansion Tank

The expansion tank can be located on the suction side of the pump, as shown in Figure 19-4-10, or the discharge side of the boiler, as shown in Figure 19-4-11. Both the relief valve and the reducing valve should be connected to the boiler on the expansion tank side, as shown in the illustrations.

Filling the System

It is important in filling the system to supply a continuous flow of water to replace the air in the piping. A good method of doing this can be illustrated by referring to the diagram in Figure 19-4-12. The drain valve is opened, valve "B" is closed, and valve "A" is opened. This arrangement generally uses city water pressure to fill the piping. Care must be exercised to return the system to 12 psi when the automatic feed is placed in control.

Safety Controls and Gauges

The location of the pressure relief valve and the temperature pressure gauge is shown in Figure 19-4-12. The pressure relief valve is usually set for 30 psig on a domestic or small commercial system and piping is arranged to discharge into an approved drain. The

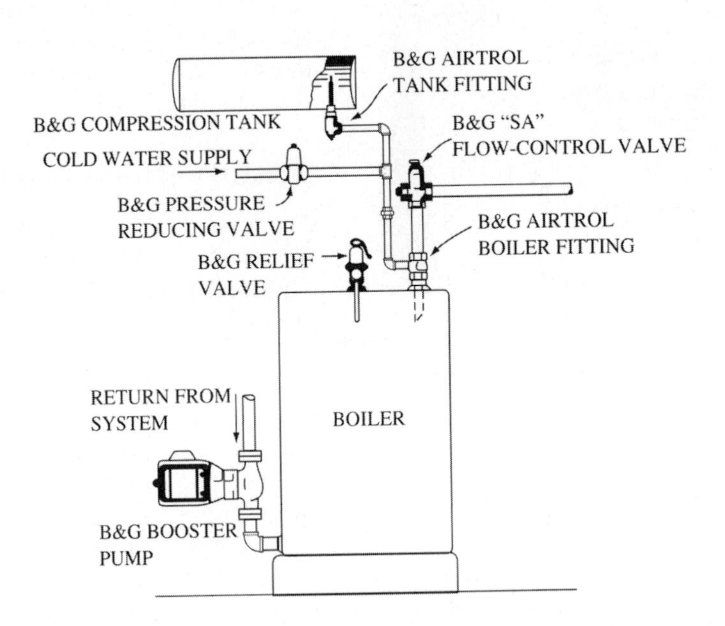

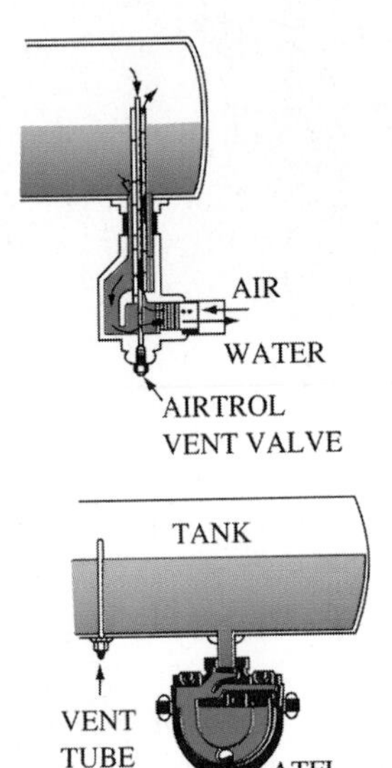

Figure 19-4-9 Compression type expansion tank.

Figure 19-4-10 Recommended expansion tank hookup connected to suction side of pump.

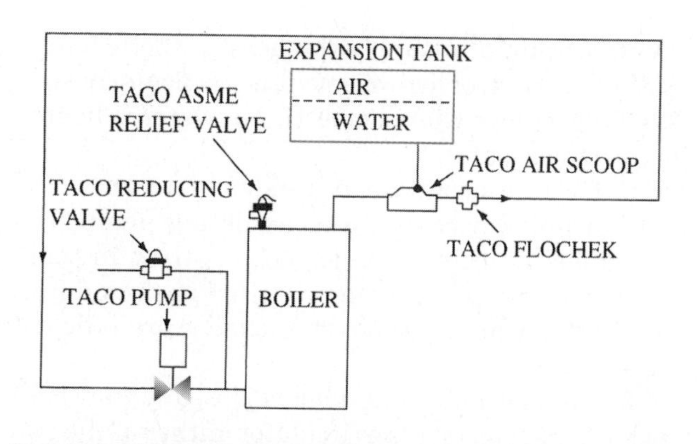

Figure 19-4-11 Recommended expansion tank hookup connected to discharge side of boiler. (*Courtesy of Taco, Inc.*)

gauge measures water pressure (in psig or ft of head) and boiler water temperature.

Controls

There are a number of ways residential or small commercial hot water heating systems can be controlled:

1. An aquastat can be used to maintain boiler water temperature by controlling the fuel burning device. A room thermostat starts and stops the

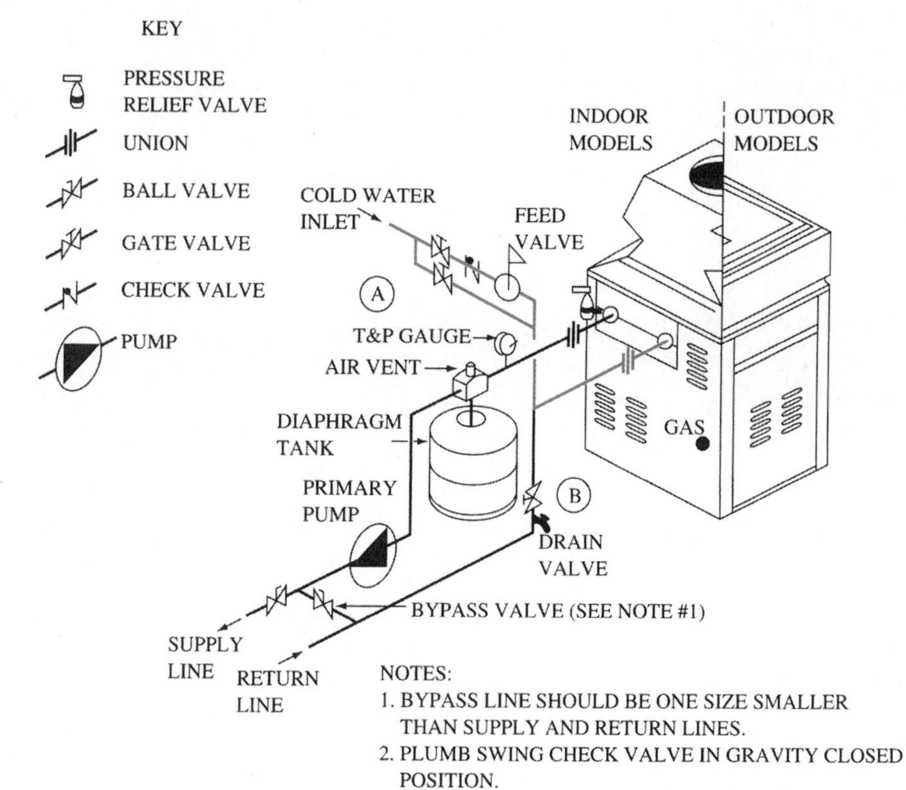

Figure 19-4-12 Diagram showing piping arrangement for filling the system.

circulator pump to supply the heat to the terminal units. This control arrangement is particularly suitable for systems where domestic hot water is heated by the boiler water.

2. The thermostat can be used to control both the fuel burning device and the circulating pump. The advantage of this arrangement is that in mild weather the boiler water temperature is lower due to the shorter running time, producing more efficient operation.

3. For a system using zone valves, any zone calling for heat can start the circulator. Either of the two above arrangements can be used to control the boiler water temperature.

TECH TALK

Tech 1. I was just out in the chiller room and noticed that a lot of the pumps are leaking water. Do I need to order new pumps or can we replace the packing?

Tech 2. No, a little leak is okay. In fact most of the pumps will have small leak that will develop over time and it is no problem. That is why we have the automatic fill on the water system.

Tech 1. Is that why there are so many drains in the floor around the pump station?

Tech 2. That's right. The system is expected to have some leaks and they actually put floor drains in so that the water doesn't become a safety problem.

Tech 1. How do you know when to change the pump if they all have leaks?

Tech 2. Watch the leak. If it gets more than just a drop every now and then you may want to look at changing the pump packing or at least tightening it.

Tech 1. Okay, I will keep an eye on them. Thanks.

UNIT 4—REVIEW QUESTIONS

1. At what temperature does water boil at sea level atmospheric pressure?
2. What is included and shipped in a boiler package?
3. What kind of boiler is available for larger hot water installations?
4. List the accessory devices used in piping hot water systems.
5. Where is an expansion tank located?
6. Where should air vents be placed on the piping system?
7. What are the two types of expansion tanks?
8. The ________ type feeds air into the tank through an airtrol tank fitting.
9. Why is it important in filling the system to supply a continuous flow of water?
10. A ________ starts and stops the circulator pump to supply heat to the terminal unit.

UNIT 5

Cooling Towers

OBJECTIVES: In this unit the reader will learn about cooling towers used in the HVAC industry including:

- Cooling Tower Purpose
- Evaporative Condensers

COOLING TOWERS

The function of the water tower is to pick up the heat rejected by the condenser and discharge it into the atmosphere by the process of evaporation. The cooling tower delivers the condenser water supply (CWS) at about 85°F, pumped by the condenser pump, to the water cooled condenser on the water chiller. The condenser water is warmed by the rejected heat from the chiller to about 95°F, and returned through the condenser water return (CWR) to the cooling tower, where it is cooled by evaporation in an air stream.

Cooling towers are an essential part of most central station air conditioning systems. For the smaller systems, even up to 100 tons, there has been a trend toward the use of remote air cooled condensers. Their use is limited, however, to installations where the length of the refrigerant piping can be short enough to be practical. Most all large condensing units and large packaged chiller units use water cooled condensers with cooling towers.

Figure 19-5-1 is a schematic drawing showing the principle of operation used by a cooling tower. The water from the heat source is distributed over the wet deck surface by spray nozzles. Air is simultaneously blown upward over the wet deck surface, causing a small portion of the water to evaporate. This evaporation removes the heat from the remaining water. The cooled water is collected in the tower sump and returned to the heat source.

EFFECT OF WET BULB TEMPERATURE ON TOWER OPERATION

The evaporation of water from any surface requires the removal of a certain amount of heat from the water in order to bring about this change of state. This heat is called the *latent heat of vaporization*. When absorbing heat from water in this manner, air is capable of cooling water below the ambient dry bulb temperature. It takes approximately 1000 Btu to evaporate 1 lb of water. This removal of latent heat by air is the cooling effect that makes it possible to cool the water in a cooling tower. *Relative humidity* is the ratio of the quantity of water vapor actually present in a cubic foot of air to the amount of water vapor that air could hold if it were saturated. When the relative humidity is 100% , the air cannot hold any more water and, therefore, evaporation does not take place. But when the relative humidity of the air is less than 100%, water will evaporate from the drops of falling water in a cooling tower. The lower the relative humidity, the greater the evaporation and cooling.

The drier the air, the more water will evaporate and the greater the difference between the dry bulb and wet bulb temperatures. It follows, then, that cooling tower operation does not depend on the dry bulb temperature. The ability of a cooling tower to cool water is a measure of how close the tower can bring the water temperature to the wet bulb temperature of the surrounding air. The lower the wet bulb temperature the lower the tower can cool the water. It is important to remember that no cooling tower can ever cool water below the wet bulb temperature of the incoming air. In actual practice the final water temperature will always be at least a few degrees above the wet bulb temperature selected in designing cooling towers for refrigeration and air conditioning service. It is usually close to the average maximum wet bulb for the summer months at the given location.

There are two main types of cooling towers: mechanical draft, Figure 19-5-2a, and atmospheric draft, Figure 19-5-2b. A mechanical draft tower utilizes a motor driven fan to move air through the tower, the fan being an integral part of the tower. They are available in many configurations and arrangements, including forced draft, induced draft, vertical discharge, and horizontal discharge.

Typically, the water enters the tower at the top of upper distribution basin. It then flows through holes

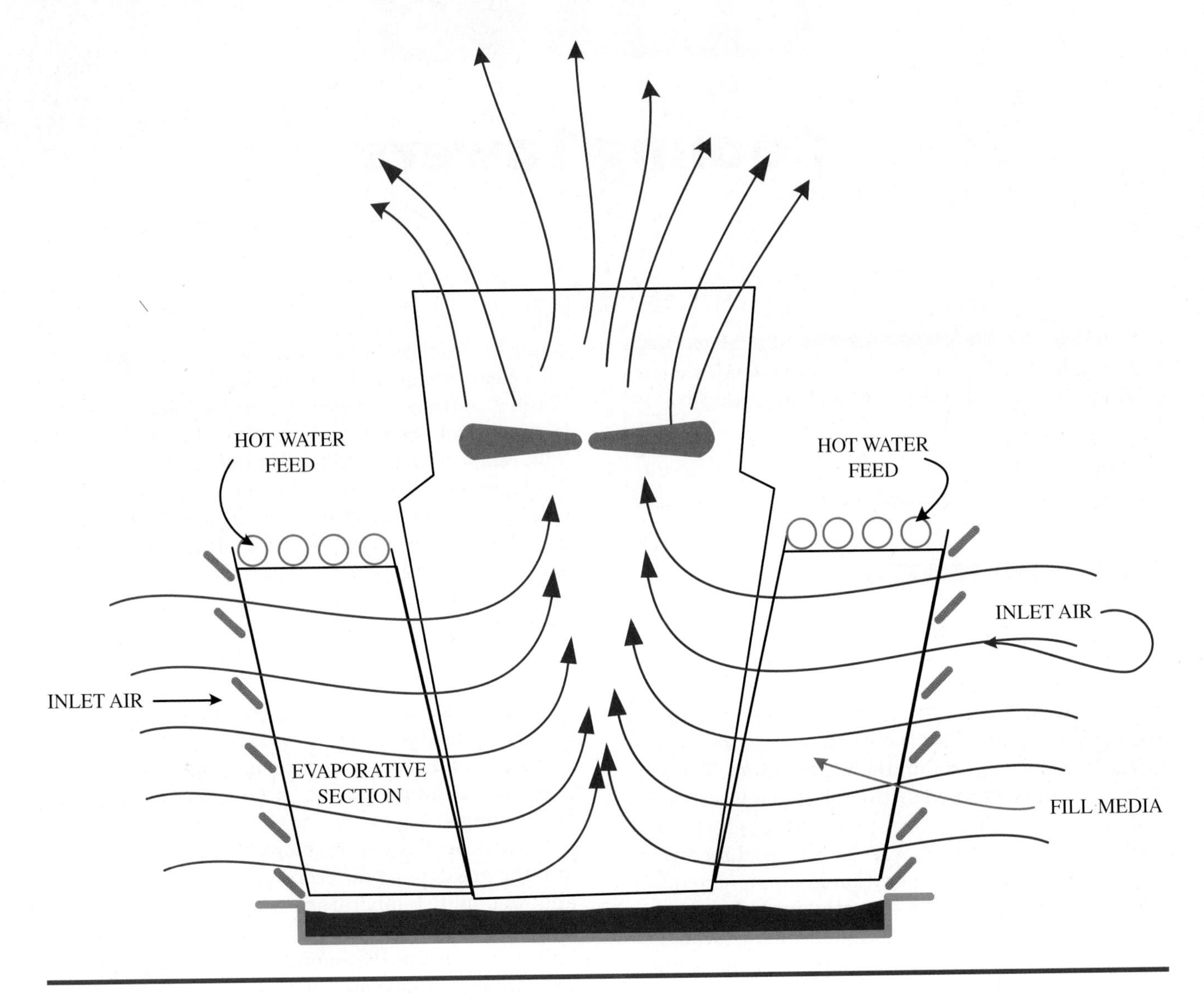

Figure 19-5-1 Schematic view of cooling tower construction.

NICE TO KNOW

The visible cloud that exits cooling towers during cool damp weather can be a problem in some circumstances. Airports, for example, cannot have large plumes of water vapor rising that might interfere with control tower visibility at the airfield. Plume suppression kits are available. These kits draw in additional air and disperse the plume in a turbulent draft caused by the suppression fans. Even though some plume may be visible a few feet above the tower, the plume does not extend hundreds of feet in the air as is the case with an unsuppressed plume.

in the distribution basin and into the tower filling, which retards the fall of the water and increases its surface exposure. The concept of splash type filling is shown in Figure 19-5-1. Meanwhile, the fan is pulling air through the filling. This air passes over and intimately contacts the water, and the resulting evaporation transfers heat from the warm water into the air. Finally, the falling water is cooled and collects in the lower (cold water) basin of the tower. It is then pumped back to the water cooled condenser to pick up more heat.

The atmospheric draft tower depends on the spray nozzles to break up the water and affect air movement. This tower has no filling or fan, and its size, weight, and location requirements (compared to mechanical draft towers) reduce its use considerably. They are seldom encountered; however, in normal refrigeration service work there may be occasions when the technician is called upon for service or maintenance on such units, and thus it is important to be familiar with the operation of these natural draft towers.

The following terms and definitions apply to all cooling towers and are illustrated in Figure 19-5-3.

(a)

(b)

Figure 19-5-2 Two types of cooling towers: (a) Mechanical draft; (b) Atmospheric draft.

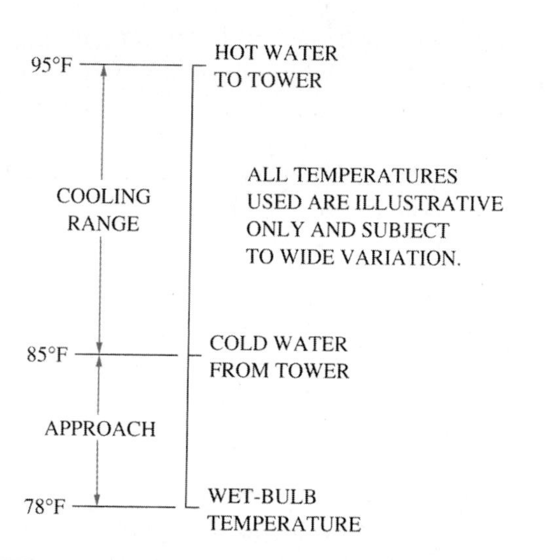

Figure 19-5-3 Range and approach as applied to cooling tower selection.

Cooling Range The number of degrees (Fahrenheit or Celsius) through which the water is cooled in the tower. It is the temperature difference between the hot water entering the tower and the cold water leaving the tower.

Approach The difference in degrees (Fahrenheit or Celsius) between the temperature of the cold water leaving the cooling tower and the wet bulb temperature of the air entering the tower.

Heat load The amount of heat "thrown away" by the cooling tower in Btu per hour (or per minute). It is equal to the pounds of water circulated multiplied by the cooling range.

EXAMPLE

Given a tower circulating 18 gpm with a 10°F cooling range, what would be the capacity?

Solution

Use the following formula:

$$Q = \text{gpm} \times 500 \times \text{TD}$$

where:

Q = heat rejection in Btu/hr
gpm = flow in gallons per minute
500 = factor derived from specific heat of water (1) × 60 min/hr × 8.33 lb/gal
TD = temperature difference in °F

Therefore,

$$Q = 18 \text{ gpm} \times 500 \times 10°\text{F}$$
$$= 90{,}000 \text{ btu/hr}$$

Or, based on 1 ton tower capacity, equivalent to 15,000 Btu/hr,

$$Q = 6.0 \text{ cooling tower tons}$$ ■

The above formula is useful for calculating heat rejection of hot water, condenser water, or chiller water systems.

■ PIPING HOOKUP FOR COOLING TOWER

Figure 19-5-4 represents a typical mechanical draft tower piping arrangement to the condenser of a package refrigeration or air conditioning unit.

Cooling tower pump head refers to the pressure required to lift the returning warm water from the cold water basin operating level to the top of the tower and force it through the distribution system. This data is found in the manufacturer's specifications and is usually expressed in feet of head (1 lb of pressure = 2.31 ft of head). *Drift* is the small amount of water lost in the form of fine droplets carried away

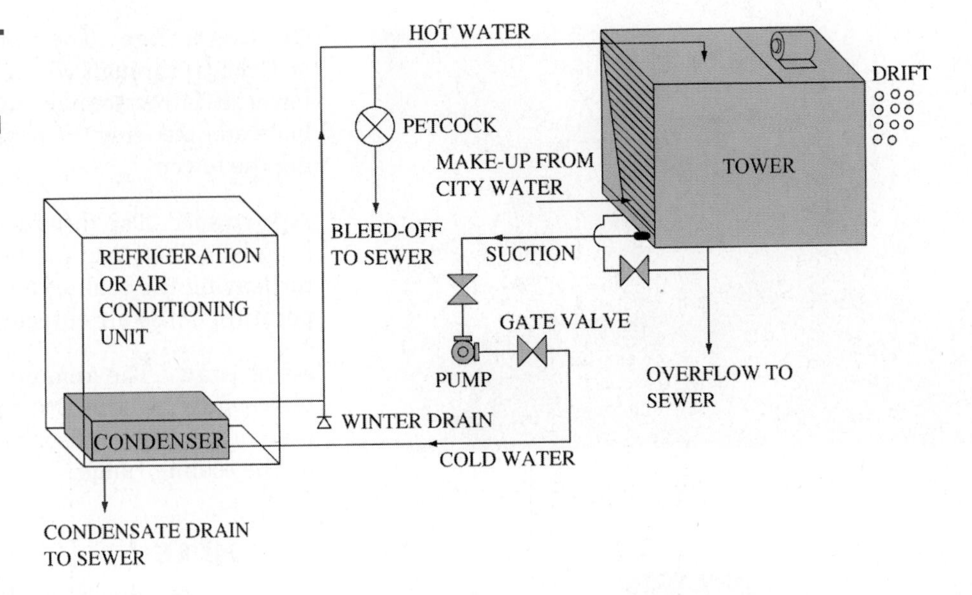

Figure 19-5-4 Piping diagram for a system using water cooled condenser and cooling tower.

by the circulating air, as shown in Figure 19-5-5. It is independent of, and in addition to, evaporation loss. *Bleed off* (often called blow down) is the continuous or intermittent wasting of a small fraction of circulating water to prevent the buildup and concentration of scale forming minerals and other nonvolatile impurities in the water. *Makeup* is the water required to replace water lost by evaporation, drift, and bleed off.

The first step in the design of the piping system is to determine the water flow to be circulated, based on the heat load given above. Normally, towers run between 3.0 and 4.0 gpm/ton.

Figure 19-5-5 Drift can be seen rising out of this cooling tower on a hot, humid summer day.

SERVICE TIP

In areas that have a great deal of dissolved minerals in the water extensive buildup of scale can be controlled by putting approximately 10% additional makeup water in the cooling tower so that the extra water is flushed down through the basin overflow. This will significantly increase the time intervals between descaling.

Water supply lines should be as short as conditions permit. Standard weight steel pipe (galvanized), type L copper tubing, and CPVC plastic pipe are among the satisfactory materials, subject to job conditions and local codes.

Piping should be sized so that water velocity does not exceed 8 ft/sec. Table 19-5-1 lists approximate friction losses in standard steel pipe and type L copper tubing. Plastic pipe will have the same general friction loss as copper. The data are based on clear water, reasonable corrosion and scaling, and velocity flow at or below the 5 ft/sec range.

EXAMPLE

Based on Table 19-5-1, 100 feet of 1¼ in standard steel pipe would have a pressure loss of 4.31 ft at 12 gpm. Determine the pressure losses for 50 and 200 ft lengths.

Solution

50 ft would have a pressure loss of

$$4.31 \times 50/100 = 2.15 \text{ ft}$$

and 200 ft would have a pressure loss of

$$4.31 \times 200/100 = 8.62 \text{ ft}$$

■

TABLE 19-5-1 Friction Losses for Steel Pipe and Copper Tubing

Water Flow (gal/min)	Type of Pipe or Tubing	¾ in		1 in		1¼ in		1½ in		2 in	
		Velocity (ft/sec)	Head Loss (ft/100 ft)	Velocity (ft/sec)	Head Loss (ft/100 ft)	Velocity (ft/sec)	Head Loss (ft/100 ft)	Velocity (ft/sec)	Head Loss (ft/100 ft)	Velocity (ft/sec)	Head Loss (ft/100 ft)
6	Std. steel	3.61	14.7	2.23	4.54						
	Copper type L	3.98	11.5	2.34	3.13						
9	Std. steel	5.42	31.1	3.34	9.72	1.93	2.75				
	Copper type L	5.96	24.2	3.50	6.63	2.30	2.38				
12	Std. steel			4.46	16.4	2.57	4.31	1.89	2.04		
	Copper type L			4.67	11.3	3.06	4.04	2.16	1.73		
15	Std. steel			5.57	24.9	3.22	6.35	2.36	3.22		
	Copper type L			5.84	17.1	3.83	6.12	2.70	2.62		
22	Std. steel					4.72	13.2	3.47	6.25	2.10	1.85
	Copper type L					5.21	12.5	3.96	5.57	2.28	1.40
30	Std. steel							4.73	11.1	2.87	3.29
	Copper type L							5.41	9.44	3.11	2.45
45	Std. steel									4.30	6.96
	Copper type L									4.66	5.20

Note: Data on friction losses based on information published in *Cameron Hydraulic Data* by Ingersoll Rand Company. Data based on clear water and reasonable corrosion and scaling.

TABLE 19-5-2 Friction Losses of Valves and Fittings Used for Cooling Tower Piping

Pipe Size (in)	Gate Valve Full Open	45° Elbow	Long Sweep Elbow or Run of Std. Tee	Std. Elbow or Run of Tee Reduced One-Half	Std. Tee Through Side Outlet	Close Return Bend	Swing Check Valve Full Open	Angle Valve Full Open	Globe Valve Full Open
¾	0.44	0.97	1.4	2.1	4.2	5.1	5.3	11.5	23.1
1	0.56	1.23	1.8	2.6	5.3	6.5	6.8	14.7	29.4
1¼	0.74	1.6	2.3	3.5	7.0	8.5	8.9	19.3	38.6
1½	0.86	1.9	2.7	4.1	8.1	9.9	10.4	22.6	45.2
2	1.10	2.4	3.5	5.2	10.4	12.8	13.4	29.0	58.0

Note: Data on fittings and valves based on information published by Crane Company.

In using Table 19-5-1, the smallest pipe size should be selected that will provide proper flow and velocity, to keep installation costs to a minimum. Friction loss is expressed in feet of head per 100 ft of straight pipe length.

The entire piping circuit should be analyzed to establish any need for proper valves for operation and maintenance of the system. A means of adjusting water flow is desirable; shutoff valves should be placed so that each piece of equipment can be isolated for maintenance.

Valves and fittings (elbows, tees, etc.) create added friction loss and pumping head, Table 19-5-2 lists the approximate friction loss expressed in equivalent feet of pipe.

The following is an example of calculating pipe sizing:

EXAMPLE

Determine the total pump head required for a 5 ton installation requiring 75 ft of steel pipe, 10 standard elbows, 4 gate valves, and a net tower static lift of 60 in. Water circulation will be 15 gpm and the pressure drop across the condenser is 13 psi (data obtained from the manufacturer).

Solution

1 ¼ in pipe, since the velocity of 15 gpm is less than 5 ft/sec.

Quantity of 1¼ in Pipe	Equivalent Pipe Length
75 ft of 1¼ in standard steel pipe =	75.0 ft
(10) 1¼ in standard elbows × 3.5 =	35.0 ft*
(4) 1¼ in open gate valves × 0.74 =	2.96 ft*
Total	112.96 ft

From Table 19-5-1, we find that for 100 ft of 1¼ in pipe, the loss is 6.35 ft; and for 112.96 ft:

$$\text{loss} = \frac{112.96 \times 6.35}{100} = 7.17 \text{ ft}$$

Pressure loss due to piping and fittings =	7.17 ft
Pressure loss due to condenser** = 13 × 2.31 =	30.00 ft
Pressure loss due to static lift minus cooling tower pump head =	5.00 ft
Total Head	42.17 ft

Pipe size is adequate since velocity is less than 5 ft/sec.

*See Table 19-5-2.

**To convert pounds per square inch pressure to feet of head multiply by 2.31. ■

In the above example, a tower pump may be selected based on the water flow (e.g., 15 gpm) and the total head (42.17 ft, or 43 ft). The pump manufacturer's catalog will rate the pump capacity in gpm versus feet of heat and the horsepower size needed to do the job.

There are many types of pumps from which to choose. For most air conditioning applications an iron body, bronze fitted, end suction, centrifugal pump with mechanical seals will do the job. Close coupled, 3500 rpm pumps are economical and do not have to be aligned. If continuity of service is important, install a standby pump. By locating the condenser pump outdoors below the tower, a leaking seal will be less of a problem. For some applications, 1750 rpm base mounted pumps are specified. Motor replacements are easier and the motors run quieter.

The installation of the pumps includes the following:

1. The pump should be located between the tower and the refrigeration or air conditioning unit so that the water is "pulled" from the tower and "pushed" through the condenser. See Figure 19-5-4 for a typical piping diagram. It is good practice to place a flow control valve (a gate valve is satisfactory) in the pump discharge line.

2. The pump should be installed so that the pump suction level is lower than the water level in the cold water basin of the tower. This assures pump priming.

3. If the pump is located indoors, consideration should be given to noise and water leakage should the seal fail.

4. If an open, drip proof motor is used outdoors, a rain cover will provide additional protection. Make sure ventilation is adequate.

5. The pump should be accessible for maintenance and installed to permit complete drainage for winter shutdown.

SERVICE TIP

When replacing motors that are frame mounted it is very important that the alignment between motor and pump be accurate. Misalignment will cause noise and excessive bearing wear. Shims are available to make the alignment process easier.

Cooling Tower Wiring

The most desirable wiring and control arrangement varies, depending on the size of the equipment being installed. In every case, the objective is to provide the specified results with optimum operating economy and protection to the equipment involved.

For small refrigeration and air conditioning equipment, the ideal arrangement is based on a sequence beginning with the cooling tower pump. The starter controlling the fan and pump would then activate the compressor motor starter through an interlock. This method, illustrated in Figure 19-5-6, assures sufficient condenser water flow so that compressor short cycling is eliminated in the event of pump motor failure. In other words, the compressor cannot run unless the tower is operating. The tower fan is wired to allow cycling by a tower thermostat to maintain condenser water temperature.

There are other, more economical methods based on using the compressor starter to activate the pump and fan, but water temperature or flow sensing devices should be incorporated as protection for the compressor. Where multiple refrigeration units are used on a common tower, the first unit that is turned on activates the cooling tower.

Protective Measures for Cooling Towers in Cold Weather

Winter operation or low ambient operation of a cooling tower is subject to special treatment for temperatures near freezing. Water that exits the tower too cold may cause thermal shock to the condenser and result in a very low condensing temperature. One method of preventing the temperature of the water leaving the tower from falling too low is to turn off or reduce the speed of the tower fan. This will reduce the tower's thermal capability, causing the water temperature to rise. If this is not sufficient, a bypass valve can be placed in the tower piping to permit dumping warm water directly into the tower base, thus bypassing the spray or water distribution system. A combination of both fan and total bypass may be needed to assure leaving water temperatures during cold weather, but reducing the water flow will contribute to tower freezeup. Prolonged below freezing conditions usually require total shutdown and draining of the tower sump and all exposed piping.

SAFETY TIP

For a number of reasons including environmental, expense, and maintenance, some cooling tower operators do not put antifreeze in their cooling towers. These towers are typically protected by either draining the towers in prolonged cold spells or running the circulator pumps. A significant disadvantage in this is that a sudden cold spell may freeze the system before it can be drained. Also, power failure resulting from cold weather can stop the circulating pumps, freezing the system. The owner of towers that are unprotected with antifreeze need to be precautioned regarding these two potential problems.

A piping and control arrangement to provide winter operation is shown in Figure 19-5-7. This arrangement provides an inside sump. During normal operation, when the ambient temperature is above freezing, the water flows from the inside sump, through the condenser into the tower and back to the sump. When the thermostat senses near freezing water temperatures, the three way valve is repositioned to direct the flow of water from the condenser directly to the inside sump, bypassing the tower. When there is no flow to the tower, the tower fan is cycled off.

Condensing Unit Performance

The foregoing discussion on water cooled condensers was presented as if most refrigeration and/or air conditioning equipment had separate condensers. While separate components are sold for built up systems, these are relatively few. Most condensers are assembled at the factory as part of a water cooled condensing unit, Figure 19-5-8 or as part of a complete water chilling unit. The performance rating of

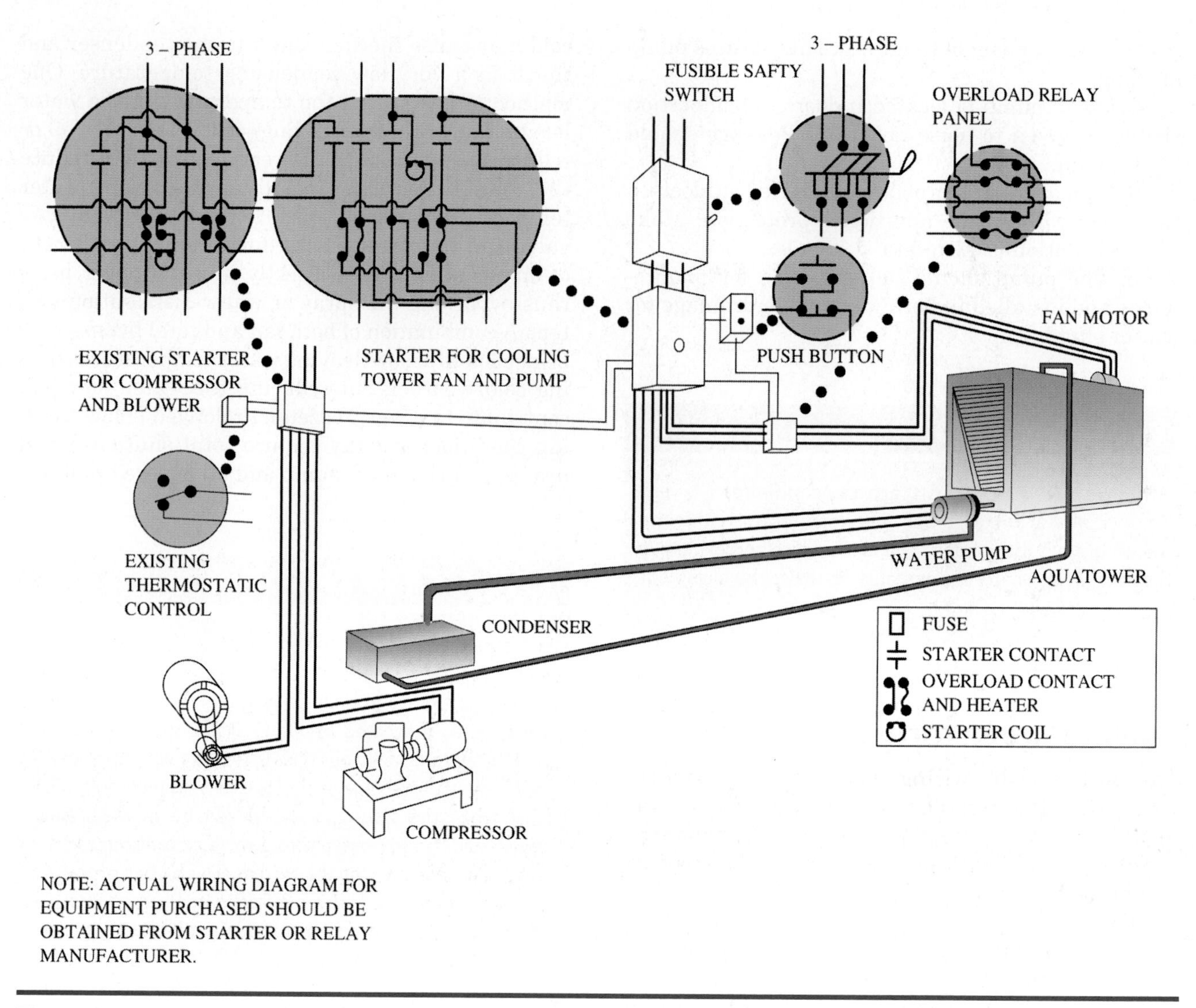

Figure 19-5-6 Typical wiring diagram for water cooled condensing unit and cooling tower.

the equipment therefore becomes a system rating similar to the capacity rating, Table 19-5-3, wherein the temperature of the water leaving the condenser and the saturated suction temperature to the compressor determine the unit tonnage, kilowatt power consumption, resulting condensing temperature, and condenser flow rate.

Additionally, the manufacturer's data will list the water pressure drop across the condenser (in psi or ft of head) for various flow rates and numbers of passes; the refrigerant charge in lb (maximum and minimum) when the condenser is used as a receiver or with an external receiver; and the pumpdown capacity, which is usually equal to about 80% of the net condenser volume but does not exceed a level above the top row of tubes.

Pumpdown is used to store or contain the entire system refrigerant in the condenser and receiver, so as to be able to perform service or maintenance operations on other components and not lose the refrigerant. Shutoff valves on the condenser permit this operation.

Maintenance for Water Cooled Equipment

Peak performance of a water cooled condenser and cooling tower system depends heavily on regular maintenance. Growths of slime or algae, which reduce heat transfer and clog the system, should be prevented.

The technician should become familiar with the companies, chemicals, and cleaning techniques recommended locally. The success of any water treatment lies in starting it early and using it regularly. Once scale deposits have formed it can be costly to remove them.

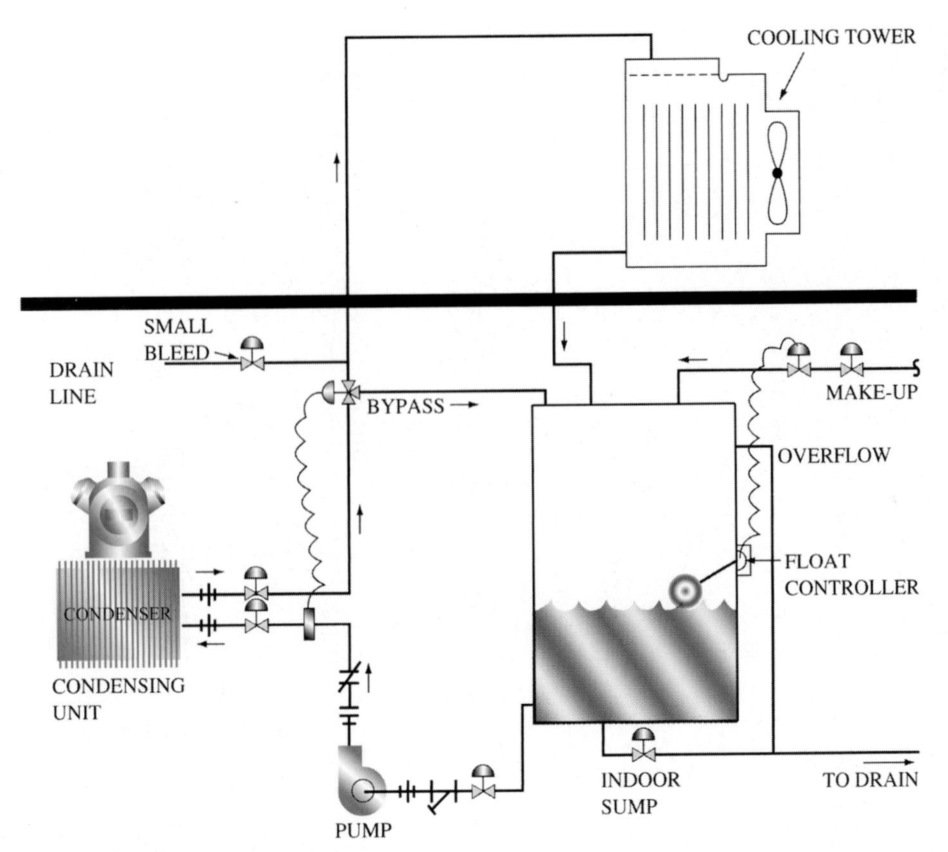

Figure 19-5-7 Winter operation using a cooling tower.

Figure 19-5-8 Typical water-cooled condensing unit used for commercial installations.

In addition to preventing scale, water treatment should protect the system components against corrosion. This is critical on systems using steel pipe. The water tower continuously aerates the water, adding oxygen. Open circuit cooling tower loops are much more subject to corrosion than closed circuit chilled or hot water systems.

In addition to chemical treatment, regular draining, cleaning and flushing of the tower basin is recommended. Also, a bleed line valve should be adjusted to cause a small amount of overflow that trickles down the drain; this is called bleed off, or sometimes blow down. It is the continuous or intermittent removal of a small amount of water (1% or less) from the system. Dissolved concentrates are continually diluted and flushed away.

SERVICE TIP

Cooling towers are protected with screen wires to prevent birds such as pigeons from roosting in the towers. However even the best screening allows some birds access to the tower. Often towers become roost for large flocks of pigeons during evening off hours. Unfortunately when the tower is brought on line in the morning many of these birds drown. Their skeletons are washed into the basin. This organic material can cause the mineral deposits normally found in the basin during cleanup to become extremely nauseating. Better screening and more frequent basin cleanout will help reduce this problem.

Water treatment and maintenance can be expensive and time consuming. Once commonly used compounds, chromates are no longer used. Chemicals used for treatment are being increasingly regulated. A treatment should be selected that complies with local regulations.

TABLE 19-5-3 Typical Rating Data for Commercial Condensing Units

		Condensing Unit Model Number											
		JS43L-KW413						JS53M-12W523					
Condenser Leaving Water Temp. (F)	Saturated Suction Temp. (F)	Tons Cap.	KW	Cond. Temp. (F)	Heat Rej. MBH	Cond. GPM	Cond. Δ P	Tons Cap.	KW	Cond. Temp. (F)	Heat Rej. MBH	Cond. GPM	Cond. Δ P*
80	20	27.0	25.0	86.8	409	81.8	9.5	31.6	30.9	86.4	485	97.0	8.6
	30	33.5	26.6	88.3	493	98.6	13.3	39.5	32.8	87.8	586	117.2	12.2
	40	41.0	28.0	90.2	588	117.6	18.3	48.1	34.5	89.6	695	139.0	16.6
	50	—	—	—	—	—	—	58.3	36.2	91.8	824	164.8	22.7
85	20	26.0	26.0	91.6	401	80.2	9.1	30.5	32.1	91.2	476	95.2	8.3
	30	32.4	27.8	93.1	484	96.8	12.9	38.1	34.4	92.7	574	114.8	11.7
	40	39.8	29.4	94.9	578	115.6	17.8	46.7	36.3	94.4	684	136.8	16.1
	50	48.1	31.1	97.1	683	136.6	24.3	56.7	38.2	96.4	810	162.0	22.0
90	20	25.1	26.8	96.4	393	78.6	8.8	29.5	33.2	96.0	467	93.4	8.0
	30	31.5	29.0	97.9	477	95.4	12.6	37.0	35.9	97.5	567	113.4	11.4
	40	38.5	30.9	99.6	568	113.6	17.2	45.3	38.1	99.1	674	134.8	15.7
	50	46.8	32.8	101.7	673	134.6	23.6	55.0	40.2	101.1	797	159.4	21.4
95	20	24.2	27.6	101.1	384	76.8	8.5	28.5	34.2	100.9	459	91.8	7.8
	30	30.4	30.1	102.5	468	93.6	12.2	35.7	37.2	102.2	555	111.0	11.0
	40	37.4	32.3	104.3	559	111.8	16.7	44.0	39.9	103.8	664	132.8	15.3
	50	45.5	34.5	106.4	664	132.8	23.0	53.5	42.3	105.8	786	157.2	20.8
100	20	23.3	28.3	106.0	377	75.4	8.2	27.4	35.1	105.7	449	89.8	7.5
	30	29.4	31.2	107.3	460	92.0	11.8	34.5	37.6	107.0	542	108.4	10.5
	40	36.0	33.7	109.0	547	109.4	16.1	42.5	41.6	108.6	652	130.4	14.8
	50	44.0	36.1	111.0	651	130.2	22.1	51.9	44.5	110.4	775	155.0	20.3
105	20	22.3	28.9	110.8	367	73.4	7.8	26.3	35.8	110.5	438	87.6	7.1
	30	28.3	32.2	112.2	450	90.0	11.3	33.2	39.8	111.8	534	106.8	10.3
	40	35.0	35.0	113.7	540	108.0	15.7	41.2	43.3	113.3	642	128.4	14.4
	50	42.7	37.8	115.7	641	128.2	21.5	50.5	46.7	115.1	765	153.0	19.8

(Courtesy of York International Corp.)

*Δ P = Pressure Drop (ft. H_2O)

EVAPORATIVE CONDENSERS

In an evaporative condenser the hot gas piping from the compressor passes through the condenser like it does on an air cooled condenser, except the tubes are bare, and no fins are needed. In addition, the evaporative condenser has water nozzles that spray water over the tubes. As water flows over the hot tubes, it evaporates and picks up heat from the refrigerant, causing it to condense. At the same time, the blower in the evaporative condenser exhausts the humid heat laden air to the outside. The water absorbs approximately 1000 Btu/lb of moisture evaporated.

The evaporative condenser has a sump in the bottom with a float valve. As the water is evaporated, makeup water is added to the sump to replace the water that has evaporated. Water from the sump is pumped to the spray nozzles at the top of the condenser and recirculated to provide continuous flow.

When the water evaporates, it leaves a mineral residue in the sump that can accumulate. To prevent this, a continuous bleed off is provided. This residue also will adhere to the tubes and needs to be periodically cleaned off. If it accumulates on the tubes it will reduce the transfer rate. Water treatment is also recommended.

Evaporative condensers can be located inside or outside the building. If they are located inside, outside air must be supplied to them. The heat laden air they exhaust needs to be ducted to the outside of the building.

They are more efficient than either the air cooled or water cooled condenser. But of the three types, they are the least popular. Internal corrosion causes them to have the highest maintenance requirements. Their use is more frequent on industrial jobs. They range in capacity from about 10 to 1000 tons.

One desirable feature of the evaporative condenser is that under standard conditions it can operate at a low 105°F condensing temperature, the same as the water cooled units.

The capacity of cooling towers and evaporative condensers is governed by the lower wet bulb temperatures, whereas air cooled condensers are related to the higher ambient air dry bulb temperatures.

An interesting application of the evaporative condenser principle is included in the design of a room cooler. The condensed water vapor from the evaporator is directed to a small sump in the bottom of the condenser fan housing. A slinger ring on the fan splashes this condensate onto the surface of the condenser coil. When this moisture evaporates from the surface of the hot condenser coil, it removes heat, similar to the action that takes place using an evaporative condenser. When the condensation is light or nonexistent, the condenser acts strictly as an air cooled condenser.

TECH TALK

Tech 1. Cooling towers seem to be a real hassle. They are a constant problem. What good are they?

Tech 2. Cooling towers are the most efficient way of getting rid of a building's heat.

Tech 1. What makes them so efficient?

Tech 2. Well, it is because of the evaporating water taking with it so much heat. Every pound of water that evaporates takes almost 1000 Btu of heat out of the building.

Tech 1. If they are so good, why don't people put them on their homes?

Tech 2. Years ago a lot of houses had cooling towers. They were mostly redwood and that is probably one of the reasons why they weren't successful. Those old wooden towers require a great deal more maintenance and service than our new ceramic ones.

Tech 1. Oh, now I understand why they are so good for big buildings and are not user friendly for homeowners.

UNIT 5—REVIEW QUESTIONS

1. What is the function of the water tower?
2. What is relative humidity?
3. List the two main types of cooling towers.
4. Define heat load.
5. ________ is the small amount of water lost in the form of fine droplets carried away by the circulating air.
6. What is the first step in the design of the piping system?
7. How is friction loss expressed?
8. With the installation of the pump, what should be considered if a pump is placed indoors?

   Water cooled condensers are usually equipped with a ________ to vent noncondensable gases, and permit connection of a water regulating valve or pressure stat.
9. What does peak performance of a water cooled condenser and cooling tower system depend heavily on?
10. What governs the capacity of cooling towers and evaporative condensers?
11. List the factors that can cause the condensing temperature to be too high.
12. Excessive head pressure is a major cause of ________.
13. What is control floodback?
14. What happens when warm humid air from the condenser discharge is mixed with outdoor air by the use of bypass dampers and duct work?

UNIT 6

Thermal Storage Systems

OBJECTIVES: In this unit the reader will learn about thermal storage systems including:

- Modes of Operation
- Types of Storage Equipment

THERMAL STORAGE SYSTEMS

Thermal storage is the temporary storage or removal of heat for later use. An example of thermal storage is the storage of solar heat energy for night heating. An example of the storage of "cold" heat removal for later use is ice made during the cooler night hours for use during the hot daylight hours. Most thermal storage systems are used in air conditioning and refrigeration applications. The advantages of thermal storage are:

- Commercial electrical rates are lower at night.
- It takes less energy to make ice when it is cool at night.
- A smaller, more efficient system can do the job of a much larger unit by running for more hours.

Most thermal storage applications involve a 24 hour storage cycle, although weekly and seasonal cycles are also used. In the context of the energy efficient system applications, storage of cooling is of prime interest.

For HVAC/R purposes, water and phase change materials (PCMs)—particularly ice—constitute the principal storage media. Water has the advantage of universal availability, low cost, and transportability through other system components. Ice has the advantage of approximately 80% less volume than that of water storage, for a temperature range of 18°F, which is the difference in temperature between a fully charged and a fully discharged storage vessel.

Modes of Operation

There are five modes of operation of a cooling storage system that stores and releases thermal energy. These are illustrated in Figure 19-6-1, and are listed below.

1. *Charging storage* The refrigeration system is extracting heat from the storage vessel.
2. *Simultaneous recharging storage and live load chilling* Some of the refrigeration capacity is being used by the load at the same time the storage is being recharged.
3. *Live load chilling* Identical to normal water chiller operation, providing cooling as needed.
4. *Discharging and live load chilling* The refrigeration unit operates and the storage vessel is being discharged at the same time to satisfy the cooling load requirement.
5. *Discharging* Cooling needs are met only from the storage, with no refrigeration operating.

Selecting the Method of Operation

Thermal storage, particularly ice storage, has its roots in applications involving short duration loads, with relatively long times between loads. The classic examples are churches, sports facilities, and older movie theaters. Current applications are driven by increasing electric utility rates, time of day rate schedules, and large demand charges for electric power.

The essential purpose of ice storage is to make ice while the power rates are lower. During periods when the rates are higher, either the power use is minimized by using ice and less refrigeration (load leveling, strategy 4 above), or using only the stored ice for cooling (full storage, strategy 5 above). The optimum combinations are derived from an analysis of all the elements involved, including:

1. Total cooling load.
2. Total daily ton hours of load (the sum of each of the hourly loads in a day's use).
3. Available ice recharging time.
4. Steady loads and temporary loads.
5. Energy cost, peak and off peak, and hours of billing.
6. Space for storage.
7. Relative costs of type of systems.

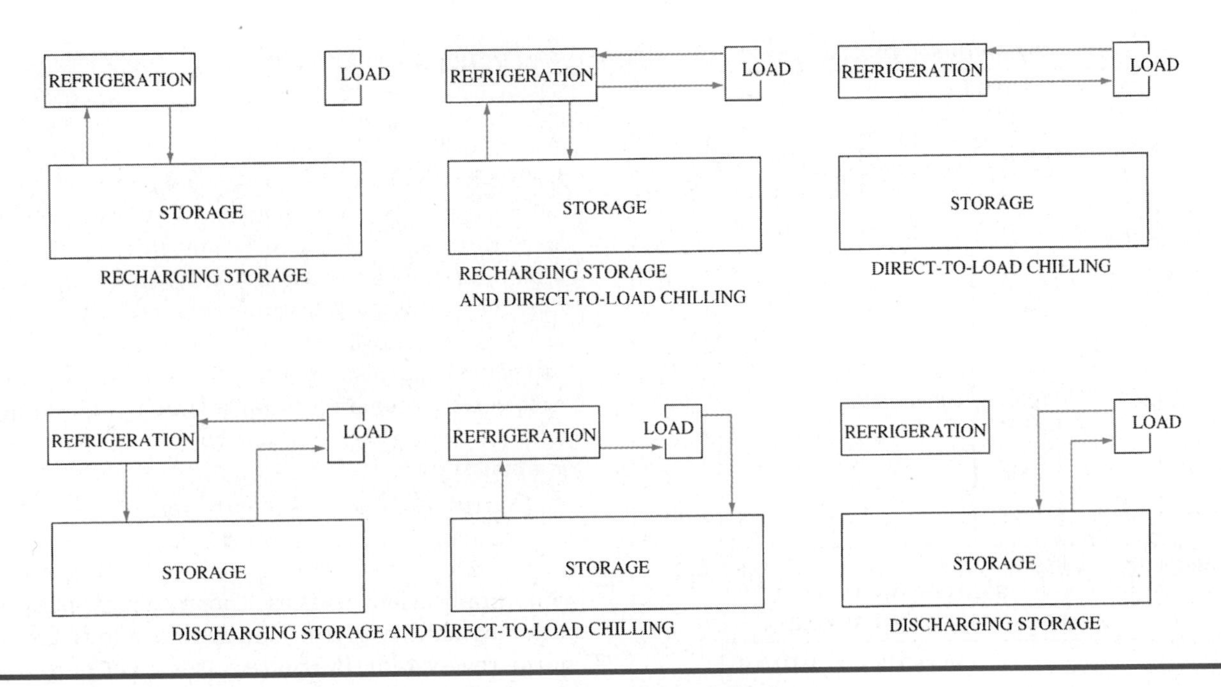

Figure 19-6-1 Operating strategies for ice storage systems. (*Copyright by the American Society of Heating, Refrigerating, and Air-Conditioning Engineers, Inc.*)

Typical HVAC/R cooling load profiles of a building are shown in Figure 19-6-2.

Water based thermal storage has certain specialized applications, and has been successfully applied many times. Problems are: (1) providing stratified storage temperatures, so that the recirculated, warmer water does not disturb the stored chilled water; (2) space and structural limitations; and (3) open systems requiring water treatment.

Ice storage systems are practical on systems as low as 5 tons. Ice provides cold (33–34°F) water for cooling and has the following advantages:

1. Colder water allows a greater temperature rise, requiring less flow for the capacity.
2. Lower flows mean smaller pipes and less pumping cost.
3. Colder water allows lower leaving air temperatures from cooling coils.
4. Colder supply air allows greater temperature rise, requiring less cfm.
5. Less CFM (airflow) allows smaller fans and ductwork.

NICE TO KNOW

Because of the high energy efficiency obtainable through thermal storage systems, many manufacturers are working on smaller residential systems. As these become more commonly available, they can provide significant energy savings to homeowners.

Applications utilizing ice storage systems require modifications of conventional system designs.

Types of Ice Storage Equipment

Ice on Coil Systems Ice on coil systems represent the oldest technology for ice storage currently in use. This type of unit is normally called an *ice builder,* since it forms ice on the outside of a pipe bundle, with refrigerant in the pipes. Figure 19-6-3 shows an ice builder. Ice builders are sized for tons of refrigeration as a chiller, and for total capacity of ice (usually in tons of ice) formed on the pipes.

These are very efficient units, since the ice is formed on the outside of the pipes and is then melted from the outside of the ice surface back to the bare pipe. This keeps the melting ice in contact with the chilled water until the ice is gone. These units can be charged with a variety of direct refrigerants, including ammonia.

An ice thickness control is required for these systems so that the ice builder does not bridge ice between pipes, or worse, freeze solid. Ice sensors are used to monitor ice thickness, and shut down the refrigeration when the total ice build is completed. The ice builder has water flow baffling to create even water flow over the pipes to ensure uniform ice formation and melting.

Solid Ice Brine Coil Systems Another method, using brine as a secondary refrigerant, is less efficient than the direct refrigeration system since a double heat exchange is required. The most common

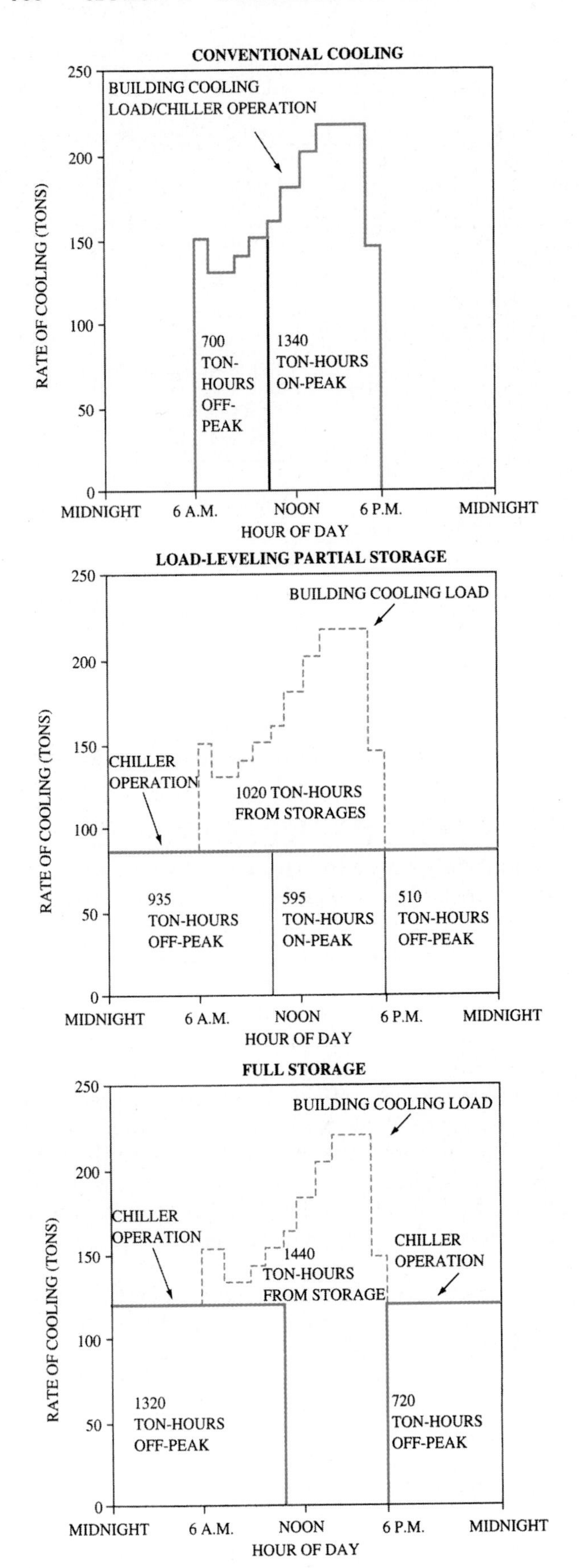

Figure 19-6-2 Cooling load profiles for ice storage systems. (*Copyright by the American Society of Heating, Refrigerating, and Air-Conditioning Engineers, Inc.*)

TECH TIP

The clearer the ice that forms on a coil the more efficient the process. As ice forms, small pockets of air in the ice will cause it to have a milky color. These frozen bubbles act as small insulators and reduce the freezing efficiency. If the water in the freezing tank is agitated even slightly these tiny bubbles will float to the surface and not form voids in the ice. Agitation of the freeze tank can be accomplished with a small pump or by introducing larger bubbles from a compressor. These larger bubbles float to the surface and bring the smaller bubbles up behind them in a bubble train.

brine application uses a conventional packaged chiller to chill a glycol solution to about 25°F, to charge the system. It also requires a glycol to water system to interface with the chilled water system. The most effective ice melt system is the direct melt to a chilled water system. This system provides simple operation, and some of the benefits of a cold chilled water system noted above.

In this glycol system, plastic mats containing brine coils are tightly rolled and placed inside a cylindrical water tank, Figures 19-6-4 and 19-6-5.

The mats occupy approximately 10% of the tank volume; another 10% of the volume is left empty to allow for the expansion of the water when it freezes, and the rest is filled with water. A brine solution, typically 25% ethylene glycol, is cooled by a packaged chiller and circulates through the coils and freezes water in the tank. Ice is built up to about ½ in thickness on the coils. During discharge, the cool brine solution circulates through the system cooling loop and returns to the tank to be cooled again.

Control is very simple, since the water expands 9% upon freezing. The level in the vessels is monitored, and the refrigeration is turned off when the frozen level is reached.

The offsetting factors for these units are that they use conventional package chillers for their cooling, and the ice vessels are very economical to purchase, and very lightweight to install. The weight of the water/ice still needs to be considered when supporting the units, but the overall weight is significantly less than the ice builder units. These units are currently the most common units in use for ice storage because of their overall cost of installation, simplicity of operation, and competitive energy savings.

Ice in Containers In this type of system, small water filled containers are placed inside a tank, with the brine filling the tank, and circulating over the containers. Some containers are spherical, with dimples to allow for expansion on freezing, as shown in

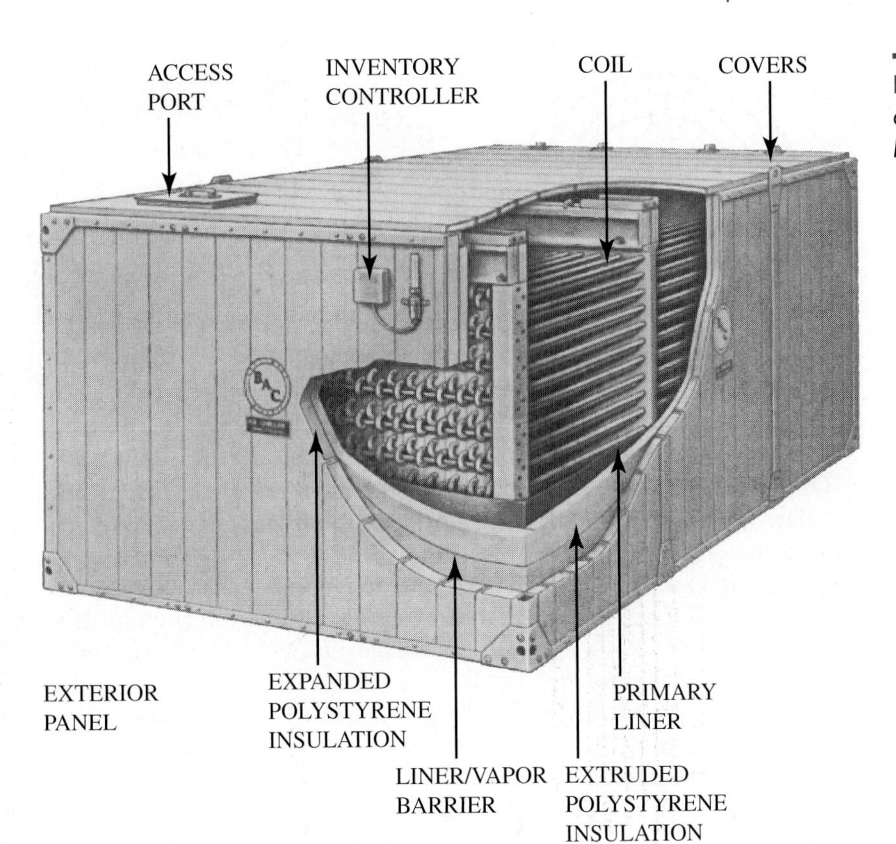

Figure 19-6-3 Ice builder construction. (*Courtesy of Baltimore Aircoil Company*)

Figure 19-6-4 Installation of a series of tanks using the solid ice brine coil system. (*Courtesy of Calmac Manufacturing Corp.*)

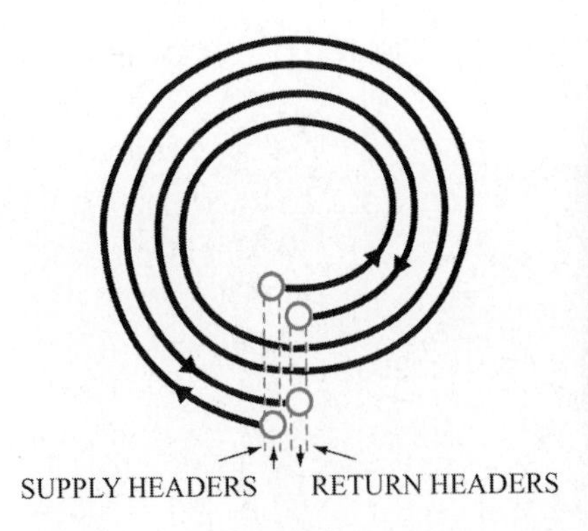

Figure 19-6-5 Internal construction of a solid ice brine coil tank. (*Courtesy of Calmac Manufacturing Corp.*)

the diagram in Figure 19-6-6. Other units are rectangular, with flexible sides to allow for expansion. These containers are stacked in storage tanks that can vary in shape and design. The piping in and out of the tank must create a flow path that is effective for both charging and discharging.

This container type of system operates much like the solid ice brine coil system above, in that the ice is formed from contact with a brine solution, and the ice is first melted away from the contact surface with the brine, requiring the ice to transfer its heat through water to the brine.

The economic factors for this type of system include storage tank shape and configuration, and simple installation. The water filled containers are heavier than the ice tanks of the solid ice brine coils, but their use is more flexible. The control monitors the liquid level of the tanks to determine the amount of ice formed. In the case of the ice in containers, the containers expand, raising the glycol level in the tank, which is monitored for control.

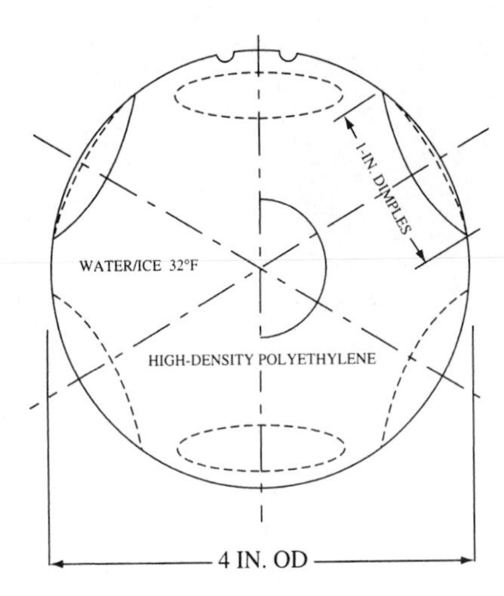

Figure 19-6-6 Spherical container for ice thermal storage.

Ice Harvesting Systems *Ice harvesting systems* are another application of industrial ice makers/water chillers. These units build ice in sheets on the surface of vertical refrigerated plates. The ice is harvested by slightly warming the plates with refrigerant discharge gas, which causes the ice sheets to break away from the plates and fall into a water filled storage tank. The sheets of ice break up when they hit the water, providing many exposed ice surfaces for heat transfer to the water. The thin sheets of ice provide a very even discharge cycle, with consistently low water temperatures. Monitoring of the tank level determines refrigeration operation.

The storage tank must be constructed with weirs to prevent the ice from getting into the circulation system. Since the ice floats and distributes itself in the tank, water flow distribution is less critical than with some of the other systems.

SERVICE TIP

Ice harvesting equipment is the most complex of the various thermal storage systems. However, it is often the most efficient method of ice production when properly set up. A major factor in the efficiency of ice harvesters is the actual harvesting cycle. Harvesting is typically performed by diverting the hot condenser gas back into the evaporator plate where the ice has formed. This momentary heating causes the ice to release, clearing the evaporator surface for more ice to form. Manufacturer guidelines list recommended harvest times to optimize ice production.

The ice harvesting systems are designed to operate on direct refrigerant applications only, which allows them to have highly efficient operation. A disadvantage is that the equipment is large and heavy, with limited configurations. They usually require field piped refrigeration systems.

TECH TALK

Tech 1. How many pounds of ice does it take to produce 1 ton of air conditioning?

Tech 2. Two thousand pounds. It takes a ton, that is why it is called a ton of air conditioning.

Tech 1. Really? I thought air conditioning had been around for a long time but thermal storage hadn't been.

Tech 2. Actually early air conditioning was done with ice. They didn't necessarily make it often. They harvested it from lakes and stored it until the summer when it was needed.

Tech 1. Seems like that would have been a lot of work.

Tech 2. It was. At the time they didn't have economical ways of making ice but they still needed to stay cool.

Tech 1. How long have they been doing that?

Tech 2. Actually different cultures have been harvesting ice and storing it for centuries to help preserve food.

Tech 1. Wow, I didn't know that. Thanks.

UNIT 6—REVIEW QUESTIONS

1. List the examples of thermal storage.
2. What are the five modes of operation of a cooling storage system that stores and releases thermal energy?
3. What are three problems with water based thermal storage?
4. List the advantages of cold water for cooling.
5. What is the oldest technology for ice storage?
6. ________ are used to monitor ice thickness and shut down the refrigeration when the total ice build is completed.
7. Why is using brine as a secondary refrigerant less efficient than a direct refrigeration system?
8. In a solid ice brine coil system, the most common brine application uses a ________ to chill a glycol solution to about 25°F, to charge the system.
9. Control with the solid ice brine coil system is simple, since water expands ________ upon freezing.
10. What is the disadvantage of ice harvesting systems?

SECTION VII

Unitary Systems

20

Appliances

UNIT 1

Space Heaters

OBJECTIVES: In this unit the reader will learn about space heaters including:

- Gas Space Heaters
- Infrared Heaters
- Vaporizing Oil Pot Heaters
- Electric Space Heaters

INTRODUCTION

Space heater is a general term applied to "in-space" heating equipment where the fuel is converted to heat in the space to be heated. Such heaters may be permanently installed or portable and employ a combination of radiation, natural convection, and forced convection to transfer the heat produced. The energy source may be liquid, solid, gaseous, or electric.

GAS SPACE HEATERS

Gas space heaters are available in a variety of styles—room heaters, wall furnaces and floor furnaces. Each has its own unique construction and application.

Room Heaters

Room heaters are self contained free standing units that transfer the heat produced from fuel combustion to room air through a heat exchanger system. The heat is transferred by radiation and convection without mixing the flue gas with the circulating room air. Since air for combustion is supplied from the room air, these units must be installed in a space that has sufficient infiltration to supply the necessary combustion air. Room heaters are made in a number of different constructions: vented, unvented, and catalytic.

Vented room heaters have an opening that is permanently connected to the chimney to convey the flue gas to the outdoors. An example of a vented room heater is shown in Figure 20-1-1. The combustion gases pass through a heat exchanger which transfers the heat to the room air by radiation and convection. Cool room air enters the grill at the bottom of the unit and the heated air is distributed from the top. These heaters often have a glass panel on the front to supply radiant heat. Room air is completely separated from the combustion gases.

Some room heaters are operated entirely by gravity. Others use a fan to circulate the room air

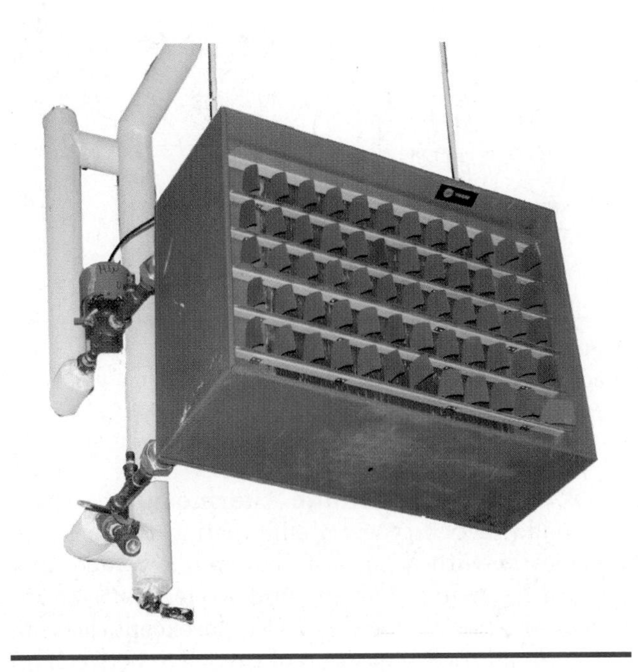

Figure 20-1-1 Room space heater.

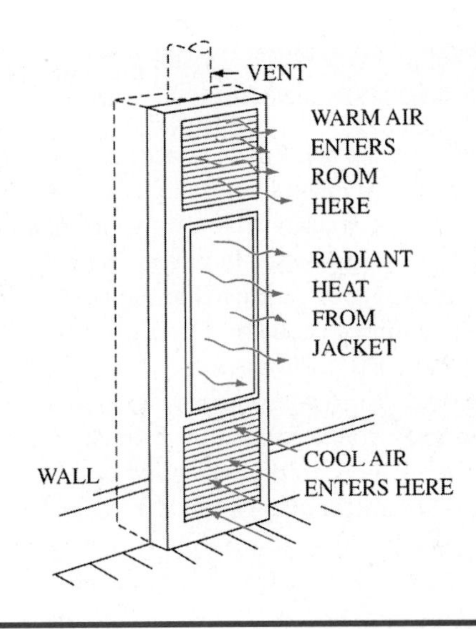

Figure 20-1-2 Wall furnace.
(Copyright by the American Society of Heating, Refrigerating, and Air-Conditioning Engineers, Inc.)

TECH TIP

Many local codes do not allow unvented space heaters to be used in many occupied areas. This may mean that when replacing an unvented space heater with new equipment that a vented space heater may need to be used. The primary concern for unvented space heaters is the oxygen deprivation and possible carbon monoxide buildup. For that reason it is strongly recommended and may be mandated by ordinance that carbon monoxide detectors be located in any area where unvented space heaters are being used.

through the unit. The fan increases the efficiency of the heater and provides better distribution of the heated air.

Since the air for combustion is taken from the room, applications are limited to spaces where there is sufficient infiltrated air to supply the required combustion air. These units are available in sizes ranging from 10,000 to 75,000 Btu/hr.

Unvented room heaters discharge the products of combustion into the room. They are limited in application to commercial projects where the area is relatively open. They are often used during building construction to supply temporary heat. One type commonly used is called a "Salamander." It is portable and can be located where needed.

The *catalytic heater* transfers heat from a glowing heat exchanger. It has no flame. The heat exchanger is constructed of fibrous material impregnated with a catalytic substance that accelerates the oxidation of a gaseous fuel. Catalytic heaters transfer heat by radiation and convection. The surface temperature is below red heat, usually about 1200°F.

Wall Furnaces

Wall furnaces are designed in a vertical configuration to fit into the stud space Figure 20-1-2. They heat a single room or have a rear boot to also supply heat to an adjacent room. The units are usually 6 or 8 in. deep, so that part of the cabinet protrudes into the room. The units are completely self contained and operate from a room thermostat.

Some units have gravity air circulation, others use a small blower. Some units have conventional venting arrangements with a flue extending above the roof. Others, located on the outside wall, have direct vents.

These units pull the air for combustion from the outside and therefore can be placed in a tight room. Heating capacities are available from 6,000 to 65,000 Btu/hr.

Floor Furnaces

Floor furnaces, Figure 20-1-3, are constructed for suspension from a floor. The unit is constructed for supplying heated air through the center of the grill, with the return air entering through the outside corners, as shown in the illustration.

Combustion air is drawn from the outside of the building. The common application of these units is in a central room of a small house, where often a single unit is used to circulate heated air through the entire house.

SERVICE TIP

Some wall furnaces and floor furnaces use micro voltage thermostat and gas valves. These units are typically ones that do not have circulating blowers so there is no need to supply line voltage to the furnaces. The voltage to operate the system comes from a thermopile. A thermopile is much like a thermal couple except that it generates between 600 and 800 millivolts which is significantly more than the 20 to 30 millivolts produced by a thermocouple. The thermopile's voltage is enough to operate the gas valve. Because the operative voltage is so slight, any loose connections in the control wiring can cause the system to fail to operate.

Minimum Efficiency Requirements

The National Appliance Energy Conservation Act of 1987 established minimum efficiency requirements for all gas fired direct heating equipment, shown in Table 20-1-1. The AFUE values are obtained by the test methods set up by the Department of Energy.

Controls

The thermostats used by space heaters are of two types:

1. Wall thermostats. These thermostats can be either 24 V or use millivolt power. They are selected to operate whatever gas valve is being used. A suitable source of power must be supplied.

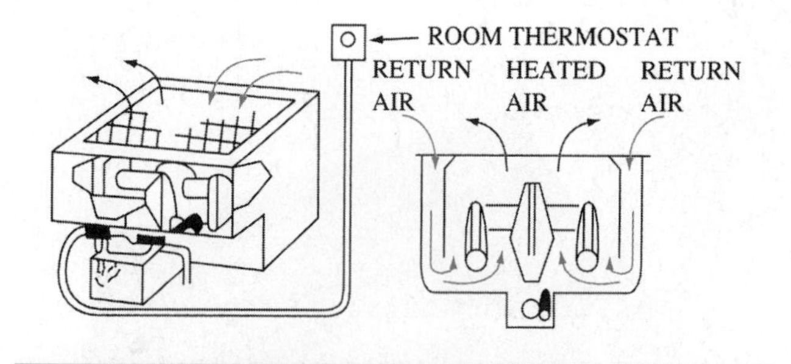

Figure 20-1-3 Floor furnace. *(Copyright by the American Society of Heating, Refrigerating, and Air-Conditioning Engineers, Inc.)*

2. Built-in hydraulic thermostats. These are made in two types: a snap action two position thermostat with a liquid filled capillary tube temperature sensing element, and a modulating type thermostat, similar to the first type, except the temperature alters the position of the valve between off and fully open.

INFRARED HEATERS

A *multimount electric infrared heater,* Figure 20-1-4, has many applications, including total area heat, spot heat, and snow and ice control. They can be mounted horizontally, or at a 30°, 60°, or 90° angle. The double reflectors can be placed to provide a wide variety of patterns as shown in Figure 20-1-5. Two elements are used per fixture. Three different lengths are available and units can be mounted end to end for additional lengths. Units are available in sizes 3200, 5000 and 7300 W.

TABLE 20-1-1 Gas Fired Direct Heating Equipment Efficiency Requirements

Input, 1000 Btu/hr	Minimum AFUE,%	Input, 1000 Btu/hr	Minimum AFUE,%
Wall Furnace (with fan)		***Floor Furnace***	
<42	73	<37	56
>42	74	>37	57
Wall Furnace (gravity type)			
<10	59	***Room Heaters***	
>10–<12	60	<18	57
>12–<15	61	>18–<20	58
>15–<19	62	>20–<27	63
>19–<27	63	>27–<46	64
>27–<46	64	>46	65
>46	65		

(Copyright by the American Society of Heating, Refrigerating, and Air-Conditioning Engineers, Inc.)

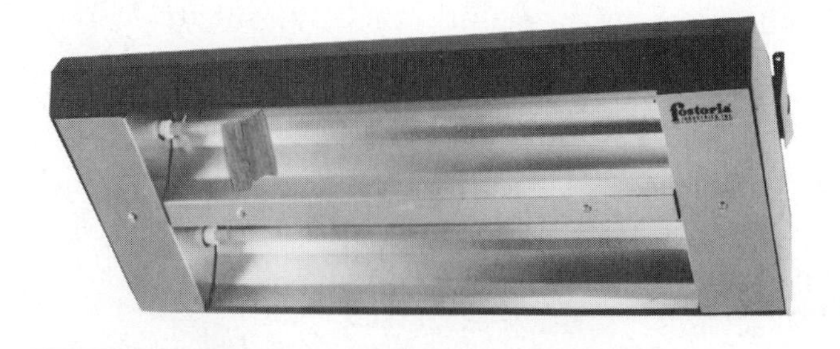

Figure 20-1-4 Multimount infrared heater. *(Courtesy of Fostoria Industries, Inc.)*

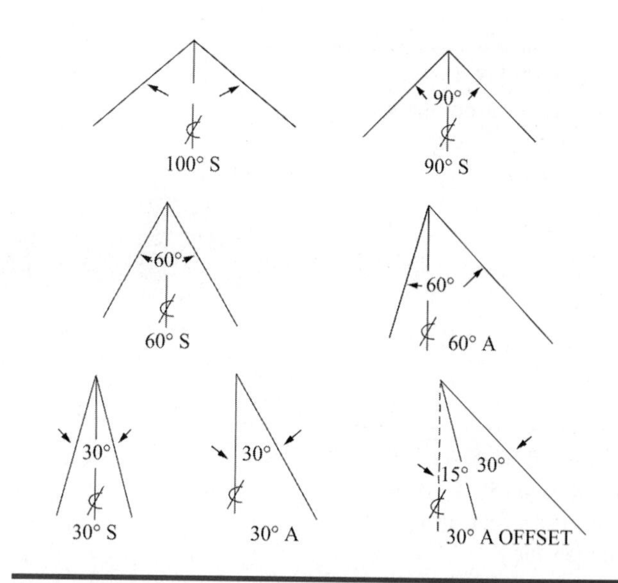

Figure 20-1-5 Reflector patterns for an infrared heater. *(Courtesy of Fostoria Industries, Inc.)*

Any surface or object should be 24 in away from direct radiation from the unit. For surface mounting, UL requires fixtures to be at least 3 in from the ceiling and 24 in from a vertical surface. Rows of fixtures should be separated by a minimum of 36 in.

A typical gas fired infrared heater installation for a warehouse or large open area is shown in Figure 20-1-6. Heaters are located overhead, and radiate heat to the area below. By using radiation, heat is directed to the people and objects in the space that require warming, rather than wasting it to the air or other nonessential surfaces. This makes possible maintaining lower room temperatures, yet creating comfortable conditions and saving energy.

A more detailed view of these heaters is shown in Figure 20-1-7. This system consists of individual burners ranging from 20,000 through 120,000 Btu/hr, fired in series, with branches connected to one vacuum pump. These systems are custom engineered to meet specific floor plans as well as heating requirements. They can be operated in a condensing mode for the ultimate in heating efficiency. Direct spark ignition is used. Combustion chambers can be either fabricated

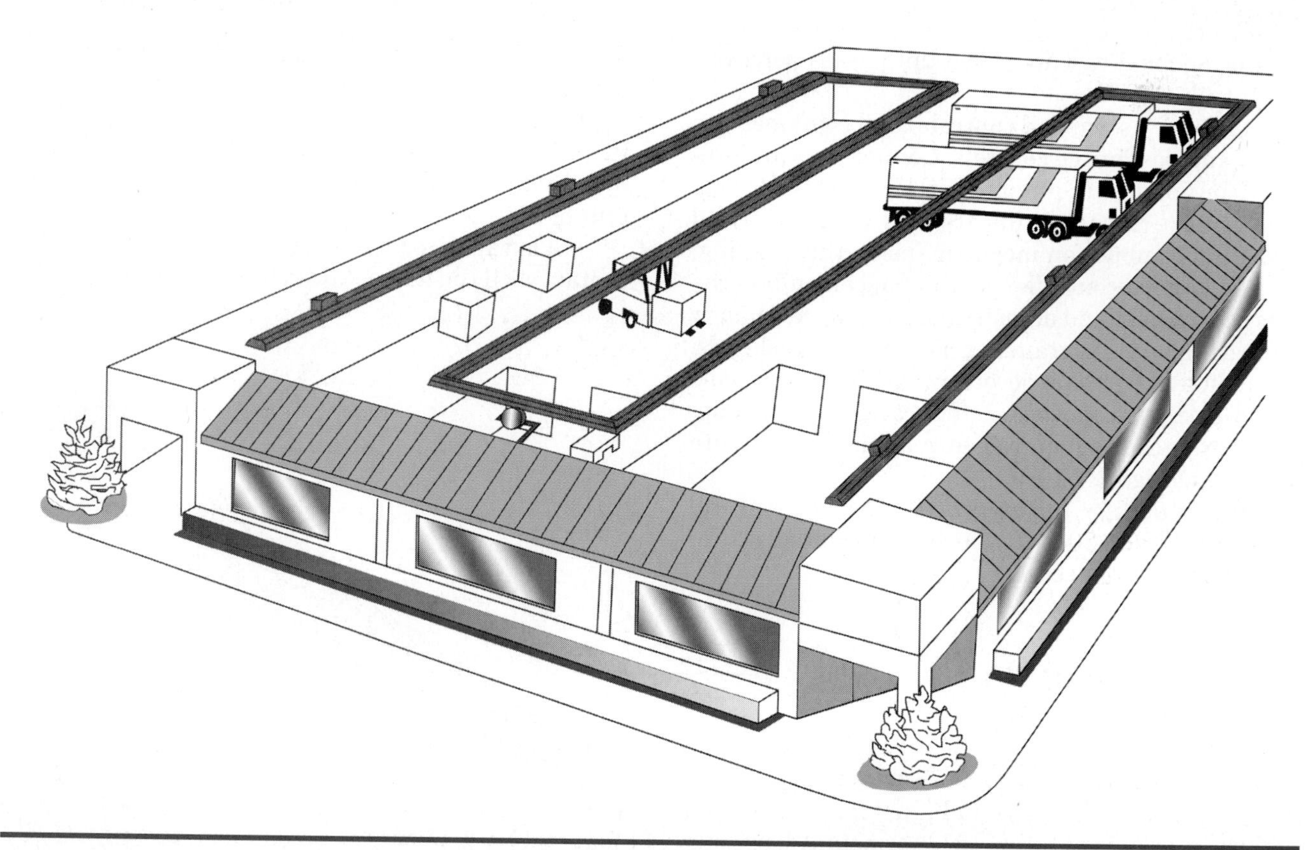

Figure 20-1-6 Infrared heating equipment installation. *(Courtesy of Roberts Gordon, Inc.)*

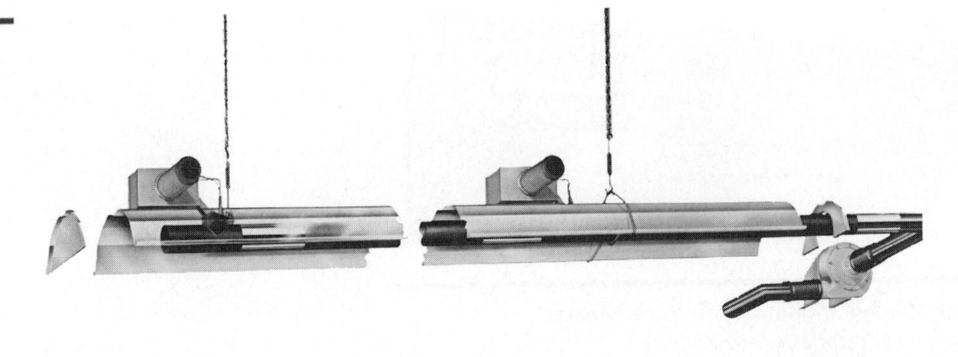

Figure 20-1-7 Low intensity infrared heater. *(Courtesy of Roberts Gordon, Inc.)*

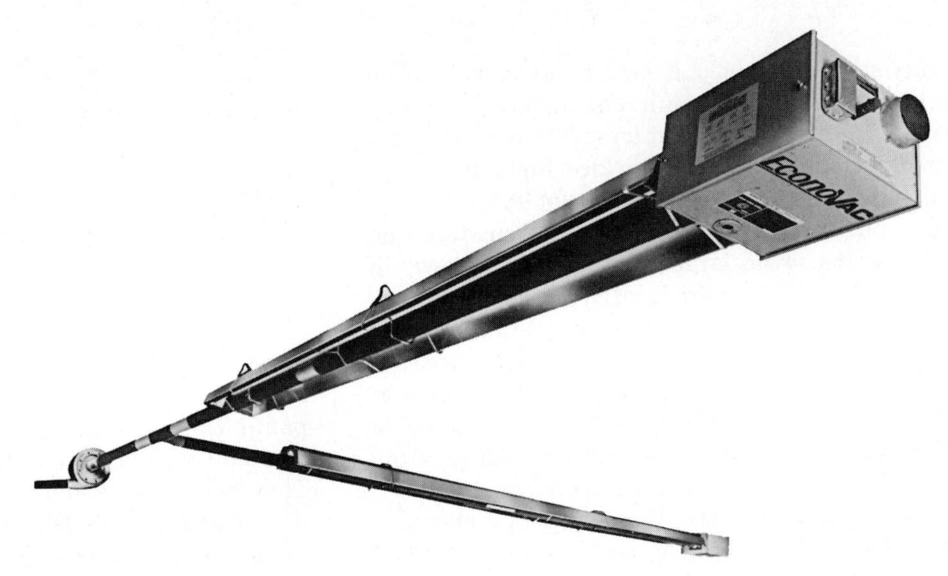

Figure 20-1-8 Vacuum assisted infrared heater. *(Courtesy of Roberts Gordon, Inc.)*

steel or optional cast iron. The tail pipe is porcelain lined.

A factory assembled control panel combines pre purge and post purge controls. Connections are provided for up to four low voltage thermostats for zone temperature control. Units are fully vented to avoid release of combustion moisture inside the building.

An optional controller continuously monitors the demand for heat and adjusts firing cycles. Automatic setback can be programmed for nights, weekends, and holidays. Optional decorative grills are available for drop ceilings. Various configurations can be supplied to match the requirements, as shown in Figure 20-1-8 and Figure 20-1-9.

High intensity infrared heaters, using natural gas or LP, Figure 20-1-10, are available for direct radiant heat where needed.

VAPORIZING OIL POT HEATERS

Vaporizing type oil burning units use No. 1 grade fuel oil. A typical unit is shown in Figure 20-1-11. The oil is burned in a bowl or pot (A). A constant level metering device, with an adjustable needle, controls the flow of oil into the combustion changer and pilot burner (B), through a burner oil supply pipe (C). The adjustable needle can be manually placed in an OFF, PILOT, or VARIABLE FLOW setting. The oil flows by gravity from a 2 or 3 gallon tank attached to the unit or a larger tank located outside.

Oil is vaporized by the heat of combustion and by contact with the hot metal surfaces of the combustion

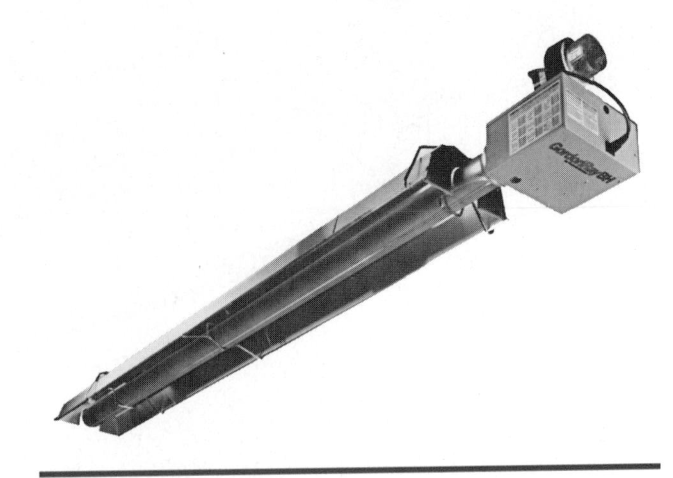

Figure 20-1-9 Unitary infrared heater. *(Courtesy of Roberts Gordon, Inc.)*

Figure 20-1-10 High intensity infrared heater.

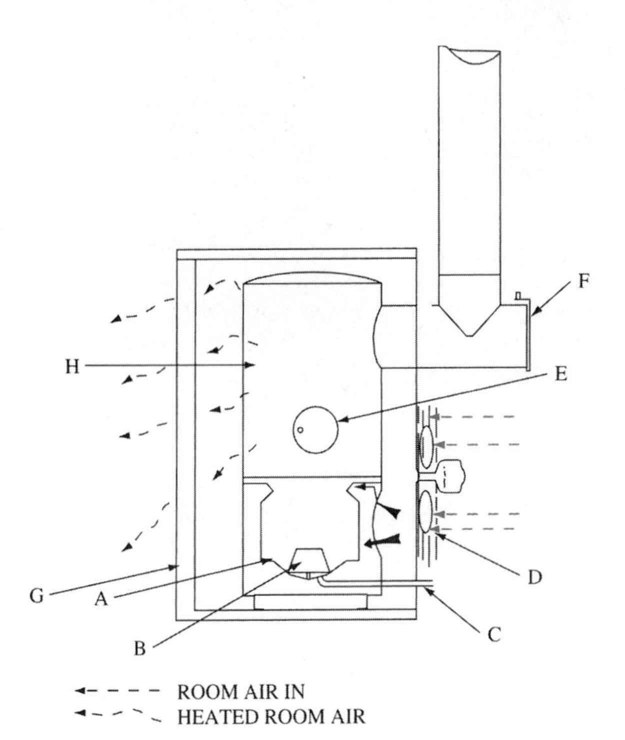

Figure 20-1-11 Oil fired space heater. *(Copyright by the American Society of Heating, Refrigerating, and Air-Conditioning Engineers, Inc.)*

chamber. The combustion air enters from the room and mixes with the vaporized oil.

Units can be supplied for gravity air circulation or with an air circulator (D). A port opening for cleaning the unit and for access for lighting the pilot (E) is provided on the front of the drum type heat exchanger. An automatic draft controller (F) is used on the flue pipe to maintain the required draft. Flue gases are vented to the outside. A perforated metal grill (G) is located on the lower front of the cabinet to admit return air for heating. The supply of heated air enters the room from the top of the unit. The steel heat exchanger (H) furnishes radiant heat to the room by heating the side walls of the cabinet. A thermostat can be used to operate the unit by turning on and off the unit at a selected firing rate.

ELECTRIC IN-SPACE HEATERS

There are various types of electrical in-space heaters. Due to the fact that there are no products of combustion and no flue, their location is very flexible. These types are described:

1. Forced air heaters
2. Suspended heaters
3. Cabinet type heaters
4. Baseboard heaters
5. Ceiling radiant heating cable
6. Floor panels

NICE TO KNOW

Although electric heat is the most expensive heat for heating it is the least expensive for equipment and maintenance. Electric space heaters are ideal for applications such as sunrooms, vacation homes and other areas where heating would only be used on limited bases. Electric space heaters can sit for long periods of time unused and still function reliably.

Forced air heaters add a measure of comfort to electric space heating. In homes and offices the recessed models such as the one shown in Figure 20-1-12 combine style with the forced air circulation, resulting in more uniform room temperature control. Most are wall mounted but models are also available for recessing into the ceiling, and even under kitchen counters. Wall mounted units provide up to 3000 W (10,240 Btu/hr), which can warm larger rooms or offices.

Suspended force fan unit heaters, Figure 20-1-13, are most functional, being used in homes (garages, workrooms, playrooms, etc.) where large capacity and

Figure 20-1-12 Wall mounted recessed space heater. (*Source: Federal Pacific Electric*)

Figure 20-1-13 Suspended space heater.

positive air circulation are important as well as in commercial and industrial establishments of all kinds. They may be suspended in vertical or horizontal fashion with a number of control options such as wall switch only, thermostats, timers, and night setback operation. Capacities generally range from 3 to 12 kW and above. Discharge louvers may be set to regulate air motion patterns. The heavy duty construction of this heater category is geared for commercial application.

Cabinet type space heaters, Figure 20-1-14, are found in classrooms, corridors, foyers, and similar

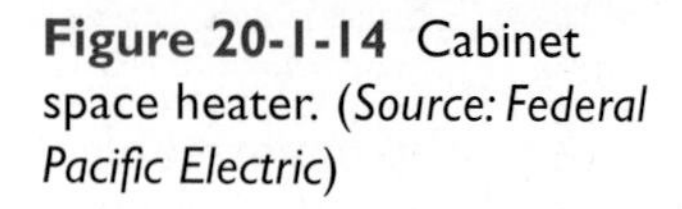

Figure 20-1-14 Cabinet space heater. (*Source: Federal Pacific Electric*)

areas of commercial and institutional buildings where cooling is not a consideration. Heating elements go up to 24 kW, and the fans are usually the centrifugal type for quiet but high volume airflow. Cabinets have provision for introducing outside air where use and/or local codes require minimum ventilation air. Outdoor air intake damper operation can be manual or automatic. Control systems can be simple one unit operation or multizone control from a central control station.

The most frequently used electric space heater is the *convection baseboard model,* Figure 20-1-15, installed in residential and commercial buildings. A cross section through the heater shows the convective airflow across the finned tube heating element. The contour of the casing provides warm air motion away from the walls, thus keeping them cooler and cleaner.

It is important that drapes do not block the air flow and cause overheating, or carpeting block inlet air. Most baseboard units have a built-linear type of thermal protection that prevents overheating. If blockage occurs, the safety limit stops the flow of electrical current. It cycles off and on until the blockage is removed. Baseboard heaters come in lengths of 2 to 10 ft in the nominal 120, 208, 240, and 277 V rating. Standard wattage per foot is 250 at the rated voltage. But lower heat output can be obtained by applying heaters on lower voltage.

For example, a standard 4 ft heater rated at 250 W/ft at 240 V will produce 1000 W of heat. Most manufacturers offer these reduced output (low density) models by changing the heating element to 187 W/ft. Low density is generally preferred by engineers and utilities and is recommended for greatest comfort. Controls may be incorporated in the baseboard or through the use of wall mounted thermostats. Accessories such as corner boxes, electrical outlets, and plug in outlets for room air conditioners are available.

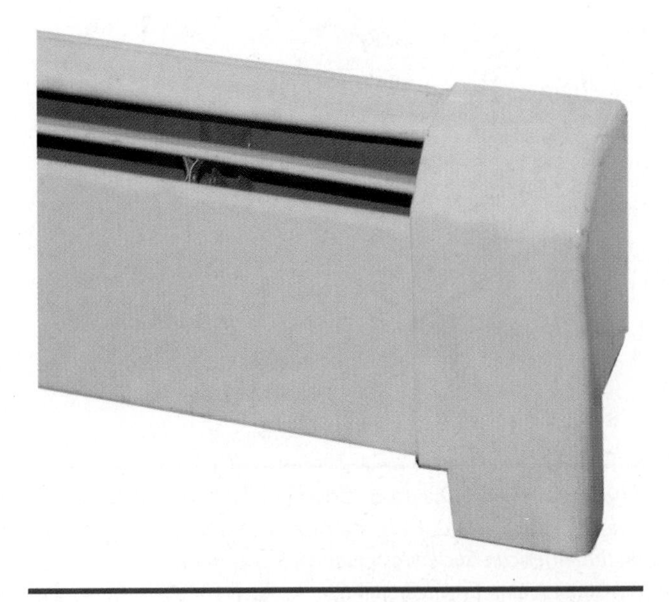

Figure 20-1-15 Baseboard space heater.

Electric radiant heating cable, Figure 20-1-16, was one of the most popular methods of heating early in the development of this technology. It is an invisible source of heat and does not interfere with the

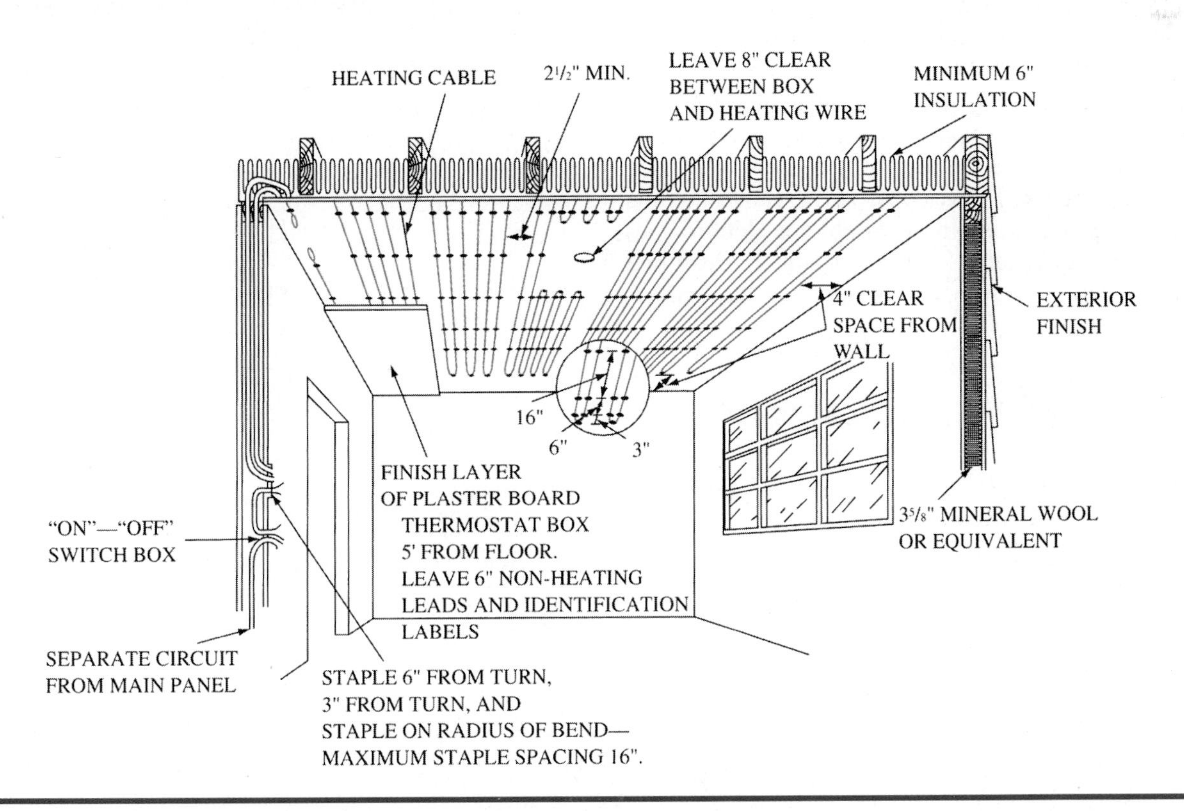

Figure 20-1-16 Diagram showing installation of ceiling cable for radiant heating.

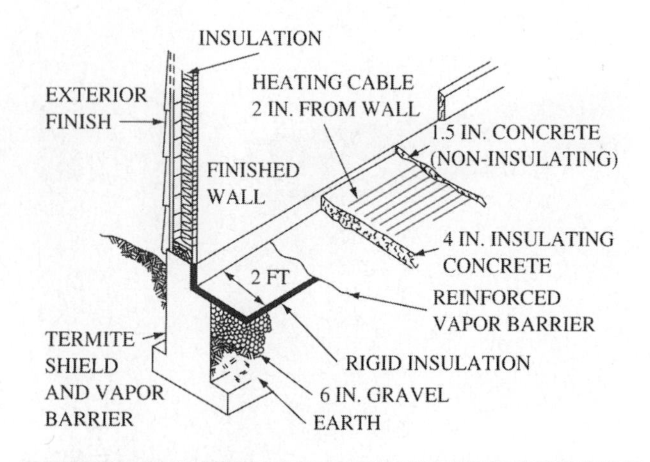

Figure 20-1-17 Diagram showing installation of electric heating cable in the floor for radiant heating. (*Copyright by the American Society of Heating, Refrigerating, and Air-Conditioning Engineers, Inc.*)

placement of furnishings. The source of warmth is spread evenly over the area of the room. It can be installed within either plaster or drywall ceilings. Staples hold the wire in place until the finish layer of plaster or drywall is applied. Wall mounted thermostats control space temperature.

Floor panels, Figure 20-1-17, can be constructed using perforated mats. These mats consist of PVC insulated heating cable, woven in, or attached to metal or glass fiber mesh. Such assemblies are available in sizes from 2 to 100 ft^2, with various watt densities ranging from 15 to 25 W/ft^2.

Another effective method of slab heating uses mineral insulated (MI) heating cable. MI cable is small diameter, highly durable, solid, electrical resistance heating wire surrounded by compressed magnesium oxide electrical insulation and enclosed in a metal sheath. Several MI cable constructions are available, such as single conductor, double conductor and double cable, as well as custom designed cables. Either of these constructions can be embedded in concrete.

TECH TALK

Tech 1. I saw an advertisement for an electric space heater that was oil filled. What's the oil for if it is an electric space heater?

Tech 2. One of the disadvantages of resistance electric heat is that it is really warm when it is on so it feels really cold when it is off. Heaters, electric heaters that are designed so that there is an oil reservoir around the heating element, dampen the temperature swing so that they don't feel as hot and cold as they cycle.

Tech 1. Isn't it more expensive to have all that oil being heated if all it does is dampen the amount of heat you get out of the heater?

Tech 2. No actually the oil filled heaters are more efficient because they are not cycling off and on as often. You know any time the heat stays on the longer cycles it's more efficient than heat that cycles off and on.

Tech 1. Sounds like they are kind of a good deal all the way around.

Tech 2. Yes; and the other advantage that they have is that sometimes electric heaters can be hazardous because they get so hot if something gets close to them they can catch fire but with one of these oil filled heaters they don't get nearly as hot. You also will have the smell of burning dust whenever one of them starts up for the first time like you do with some electric space heaters.

Tech 1. So they are more efficient, give you better heat, and are safer. Sounds like they have a lot of advantages.

Tech 2. You're right, they do.

UNIT 1—REVIEW QUESTIONS

1. Define space heater.
2. How do room heaters operate?
3. Where are wall furnaces designed to fit?
4. What act established the minimum efficiency requirements for all gas fired heating equipment?
5. What are the two thermostats used by space heaters?
6. List the types of electrical in-space heaters.
7. Which of the in-space heaters are the most functional and are used in homes?
8. The most frequently used electric space heater is the ________.
9. What happens if a blockage occurs in a baseboard heating unit?
10. What is electric radiant heating cable?

UNIT 2

Special Refrigeration Applications

OBJECTIVES: In this unit the reader will learn about special refrigeration applications including:

- Ice Cream
- Bakery Refrigeration
- Ice Rinks
- Industrial Processing
- Transportation Refrigeration
- Mechanical Refrigeration
- Drinking Water Coolers

SERVICE TIP

When working on equipment used in the manufacturing or sale of food and food products, you must wear the same types of hygienic protection required of other workers in the area. In some cases you must wear hair nets, beard nets, boot covers, jumpsuits, or other such items as a precaution to avoid possible contamination of the manufacturer's product as you are servicing their equipment.

ICE CREAM

Hard ice cream is defined legally by its butterfat content. The minimum, set by federal standards, is 8% for bulk flavored ice cream mixes (chocolate) and 10% or above for other flavors (vanilla). For the specialty trade, the richer ice creams are 16–18% fat, but the average is 10–12%. Butterfat accounts for the rich flavor of the product. Refrigeration is a major cost in the making of ice cream, and its effectiveness depends on how well it transfers heat throughout the production processes. The making of ice cream begins with making the mix, followed by freezing and storing the mix, and then using the mix to produce various kinds of ice cream. The service person should know the processes involved in order to recognize the source of problems which may arise, maintain efficient operations, avoid extra power use, and finish with a satisfactory product.

Creating the Mix

Pasteurizing The basic ingredients for a typical ice cream are milk fat (12.5%), serum solids (nonfat solids of milk, 10.5%), sugar (15.0%), a stabilizer/emulsifier (0.3%), and (optionally) egg solids. The ingredients are placed in a vat, mixed, and heated (pasteurized) at 155°F for 0.5 hours to dissolve the solids and to destroy any pathogenic organisms. They are then precooled in a plate section using cool water. The final cooling to just above the freezing point is done in a second plate section using chilled water or glycol.

In the large plants, having automated and computerized processes, all liquid ingredients are used. These are blended, preheated, homogenized, then heated with plate equipment, a heat exchanger, or a direct steam injector or infuser. This is followed by vacuum chamber treatment for precooling, which improves the flavor of the mix. It is then cooled through a regenerative plate section and additionally cooled indirectly to 40°F or less with chilled water.

Homogenizing The heated mix is blended to reduce the size of the fat globules so the fat will not churn out during freezing. Should this happen, the ice cream has a greasy taste.

Cooling and Holding the Mix The mix should be as cold as possible. In smaller plants the vats are equipped with both precooling and final cooling sections. The mix is precooled to about 10°F warmer than the entering water temperature, using city, well, or cooling tower water. The final cooler using chilled water reduces the mix to about 40°F, and brine, or direct expansion refrigerant, reduces the mix temperature to 30–33°F.

Larger plants usually use separate equipment for the final cooling. An ammonia jacketed scraped surface chiller is often used.

Where the mix can be held overnight, part of the final mix cooling is done by means of a refrigerated surface built into the tanks. The rate of cooling is about 1°F per hour.

Freezing Ice Cream

Two kinds of freezers, for making either batch or continuous ice cream, use a cylinder with annular space or coils around the cylinder for direct refrigerant cooling. The space is either flooded, with an accumulator, or the cooling controlled by a thermostatic expansion valve. A dasher inside the cylinder has sharp metal blades which scrape the ice cream off the sides as it freezes. For efficient operation the blades should be kept sharpened.

Small operations use batch freezers which have an average maximum output of eight batches per hour depending on: the sharpness of the blades, the refrigeration supplied, and the overrun desired. At optimum operation the ice cream is drawn from the freezer at about 24°F, with refrigerant temperature around the freezer cylinder of about −15°F.

Large operations use one or more continuous freezers. The ice cream is discharged from several machines connected together to facilitate packaging. The mix is continuously pumped into the freezer cylinder. Air pressure for the overrun is drawn into the cylinder at 20 to over 100 psig, either with the mix or from a separate air compressor. A dasher with freezer blades agitates the mix and air as it moves through the cylinder. The ice cream is discharged at the other end of the cylinder at an average temperature of 22°F. The flooded system surrounding the cylinder, when operating with ammonia, is −25°F. Ice cream temperatures as low as 16°F can be obtained with some mixes by regulating the evaporator temperature around the freezer cylinder using a suction pressure regulating valve.

Hardening Ice Cream

The semi solid ice cream leaving the cylinder is further refrigerated to a solid condition at 0°F for storage and distribution. Rapid freezing to obtain a smooth texture is achieved in hardening rooms kept at −20 to −30°F. Forced air is circulated from a unit cooler or a remote bank of coils. Ice cream in containers up to 5 gallons will harden in about 10 hours when spaced to allow air circulation. Systems using overhead coils or coil shelves with gravity circulation take twice this time.

Some larger plants use air blast type ice cream hardening tunnels with operating temperatures ranging from −30 to −50°F to shorten hardening time to 4 hours.

> **CAUTION:** Extremely low temperatures used in some refrigeration plants can cause skin burns and possible lung damage if you fail to wear proper protective clothing when working in these areas. Skin exposed to temperatures below −40° can quickly freeze especially when working in an area that uses air blast type cooling.

Horizontal continuous plate hardening systems automatically load and unload ice cream packages, synchronized with the filler machines. They save space and power and eliminate package bulging, but can only be used for rectangular packages.

Refrigeration Equipment

Most of the larger commercial ice cream plants use ammonia systems while smaller operations use single stage halogen refrigerant compressors. These compressors are usually operated at conditions above the maximum compression ratio recommended by the manufacturer.

For maintaining −20°F freezer rooms, multistage compression is the most economical. One or more booster compressors are used at the same suction pressure, discharging into second stage compressors. Where a hardening tunnel is used, at least two booster compressors should be used for the tunnel and for the freezer/storage room. Both units discharge into the second stage compressor system. For plants with larger volumes, a three stage system may be the most economical. A low temperature booster is used for the tunnel and a second stage booster discharges into the third stage compressor system.

Maintenance of Operating Efficiencies

Conditions which lower the efficiency and effectiveness of the refrigeration system in ice cream plants are related to the following:

1. *Heat transfer:* Air films, frost and ice, scale, noncondensable gases, abnormal temperature differentials, clogged sprays, slow liquid circulation, poor air circulation, and foreign particles.
2. *Ice cream mix:* Viscous mix, low overrun percentage, high mix temperature, low ice cream discharge temperature, and high sugar content in the mix.
3. *Dull scraper blades*
4. *Refrigeration equipment:* Low evaporator efficiency caused by rapid frost and ice development and an oil film.

Automatic defrost is recommended. Regular oil purges and an oil separator in the discharge line of the compressor will minimize oil film.

Neglected condensers lead to high head pressures and higher electrical requirements. Condenser surfaces should be kept clean of mineral deposits and other forms of scale or debris.

Noncondensable gases in compressor operation should be purged in order to avoid extra power use. Automatic purgers are available for this.

Door and conveyor openings are the cause of significant refrigeration losses. Insulation in storage rooms should be sufficient.

Cold rooms and ice banks should be serviced when the ambient air temperature is 26°F or lower.

BAKERY REFRIGERATION

Proper refrigeration is an essential part of bakery production. Refrigeration is used primarily in three areas:

1. Storage of the ingredients prior to use.
2. Controlling the temperature of the dough during the mixing process.
3. Storing and freezing the products.

Storing Ingredients

Bulk quantities require temperature protection during shipment or storage. Certain items, such as corn syrup, liquid sugar, lard, and vegetable oil, are held at 125°F to prevent crystallization or congealing. Fructose syrups are often used because they can be stored at a lower temperature, 84°F, to maintain fluidity and require less refrigeration during mixing. Smaller amounts and specialized sugars and shortenings in drums, bags, or cartons are stored at conditions for preventing mold and rancidity. Yeast should always be stored below 45°F. Cocoa, milk products, spices, and other raw materials subject to insect infestation should be stored at the same temperature as yeast.

NICE TO KNOW

Bulk wheat flour, rice flour, and corn meal along with other similar products are stored under refrigeration. These products, like all natural food products, contain insect eggs that when not stored under refrigerated conditions will hatch. If the eggs hatch, the insects will contaminate the product. Insect eggs are present in most raw food products as a result of the products being grown in a natural setting. There is no way of growing these products without some insect eggs being present. As the products are cooked they are sterilized.

Some food service programs that are only operated periodically such as those at summer camps should be encouraged to store their products in their walk in coolers during the off seasons as a way of ensuring the products remain insect free.

Total plant air conditioning is often used in new plant construction and includes filters to protect the equipment from airborne flour dust.

Mixing Process

During mixing, heat is produced in two ways: heat of friction from the electrical energy of the mixer and heat of hydration, produced when the material in the mixer absorbs water. The temperature rise in the dough is also dependent on the specific heat of each ingredient.

The temperature at which yeast is most active is quite critical. It is extremely active when mixed with water and fermentable sugars in a range of 80 to 100°F. The cells are killed at temperatures around 140°F and below 26°F, the freezing point of yeast. To maintain the proper mixing temperature, excess heat is removed by chilled water in contact with the dough.

For many years the principal refrigerant used in bakeries was ammonia. Other refrigerants are now in use for cooling the water to the desired temperature.

Fermentation Two methods are used for processing the dough: the straight dough process and the sponge dough process.

In the straight dough process all the ingredients are mixed at once (yeast, flour, malt, and liquid) in an unjacketed tank at a temperature of 69 to 80°F and allowed to ferment for 1 hour. The second hour fermentation takes place in a jacketed tank at about 84°F. A continuous process allows the dough to be drawn from one tank while the other is fermenting.

In the sponge dough process, only part of the flour and water are mixed with the yeast for the first fermentation period. After fermentation takes place, the remaining flour and water are added. The total process takes 3½ to 5 hours.

Bread Cooling and Processing

Bread is cooled to 90 to 95°F for handling and slicing. This prevents condensation within the wrapper, which can cause mold to develop.

Refrigeration is required for the bread wrapping machine. Evaporator temperature is usually around 10°F. The plate surface temperature must be held to about 16°F. Refrigeration is not required if bags are used.

Bread is frozen at temperatures between 16 and 20°F and stored at temperatures below 0°F to prevent moisture loss and crystallization of the starch. Freezing rooms maintain storage temperatures between 0 and −30°F. Two stage refrigeration equipment is generally used.

ICE RINKS

The category of "ice rinks" includes all types of ice sheets, indoors or outdoors, created for recreation such as skating, hockey, and curling.

The rinks are usually made by laying down a series of pipe coils below the level of the design surface of the ice. A secondary refrigerant (brine), such as glycol, methanol, or calcium chloride solution, circulates through the coils from a central chilling system. The system usually uses R-22, although other refrigerants have been used.

NICE TO KNOW

The ice on skating rinks is produced in layers. The first layer is formed and then its top service sprayed with white paint. This obscures the system's piping from view. Next, a thin layer of ice is formed and any markings such as hockey boundaries or logos are then painted on this ice layer's surface. Finally, the top layer is frozen. This is the layer the skaters are actually skating on. By producing the ice in multiple layers it makes it possible for rink operators to change the markings without having to allow the entire ice slab to melt.

The amount of refrigeration is usually dependent on the location of the rink (indoors or outdoors), the use of the rink, and the conditions of the building for indoor rinks, Table 20-2-1.

Ice Rink Conditions

Indoor rinks are operated from 6 to 11 months a year, depending mainly on profitability. Rooms for heated rinks are usually maintained at 50 to 60°F with relative humidity ranging from 60 to 90%.

The temperature of the ice is controlled closely either by the brine temperature or the ice surface temperature. The temperature of ice for hockey is 22°F, for figure skating 26°F, and for recreational skating 26 to 28°F. A 1 in thickness of ice is usually maintained.

Average brine temperatures are 10°F lower than the required ice temperature. Usually two compressors are used for pulldown and one for normal operation. The temperature differential between supply and return brine is approximately 2°F.

Construction and Equipment

Brine systems generally use 1 in steel or polyethylene pipe 4 in on center. Pipes are set dead level, with sand fill around and over the pipes. A balanced flow distribution system is shown in Figure 20-2-1. An expansion tank must be installed in the piping.

The construction of the space below the rink is very important, Figure 20-2-2. There must be proper drainage to prevent "heaving" caused by water freezing below the rink. Many new rinks install heater cables below the floor of the rink or circulate warm brine to prevent heaving. A "header trench" is located

TABLE 20-2-1 Ice Rink Refrigeration Requirements for a Variety of Applications

Four to Five Winter Months, above 37° Latitude	ft²/ton	m²/kW
Outdoors, unshaded	85 to 300	2.2 to 7.8
Outdoors, covered	125 to 200	3.3 to 5.2
Indoors, uncontrolled atmosphere	175 to 300	4.6 to 7.8
Indoors, controlled atmosphere	150 to 350	4.0 to 9.1
Curling rinks, indoors	200 to 400	5.2 to 10.4
Year Round (Indoors), Controlled Atmosphere	**ft²/ton**	**m²/kW**
Sports arena	100 to 150	2.6 to 4.0
Sports arena, accelerated ice making	50 to 100	1.3 to 2.6
Ice recreation center	130 to 175	3.4 to 4.6
Figure skating clubs and studios	135 to 185	3.5 to 4.8
Curling rinks	150 to 225	4.0 to 5.9
Ice shows	75 to 130	2.0 to 3.4

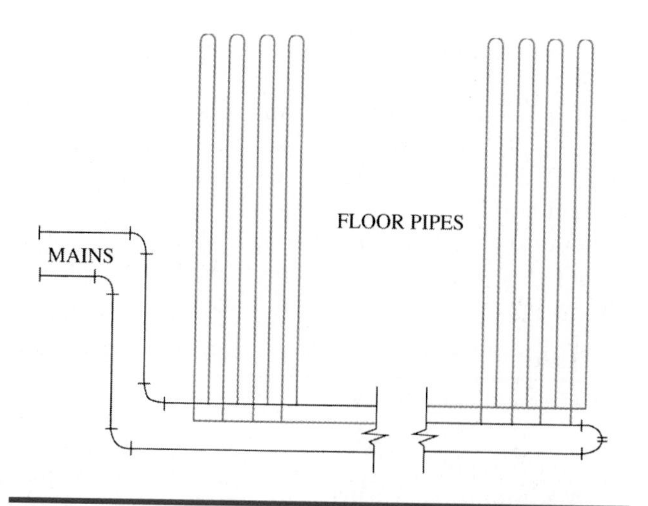

Figure 20-2-1 Piping diagram for ice rink to provide balanced flow.

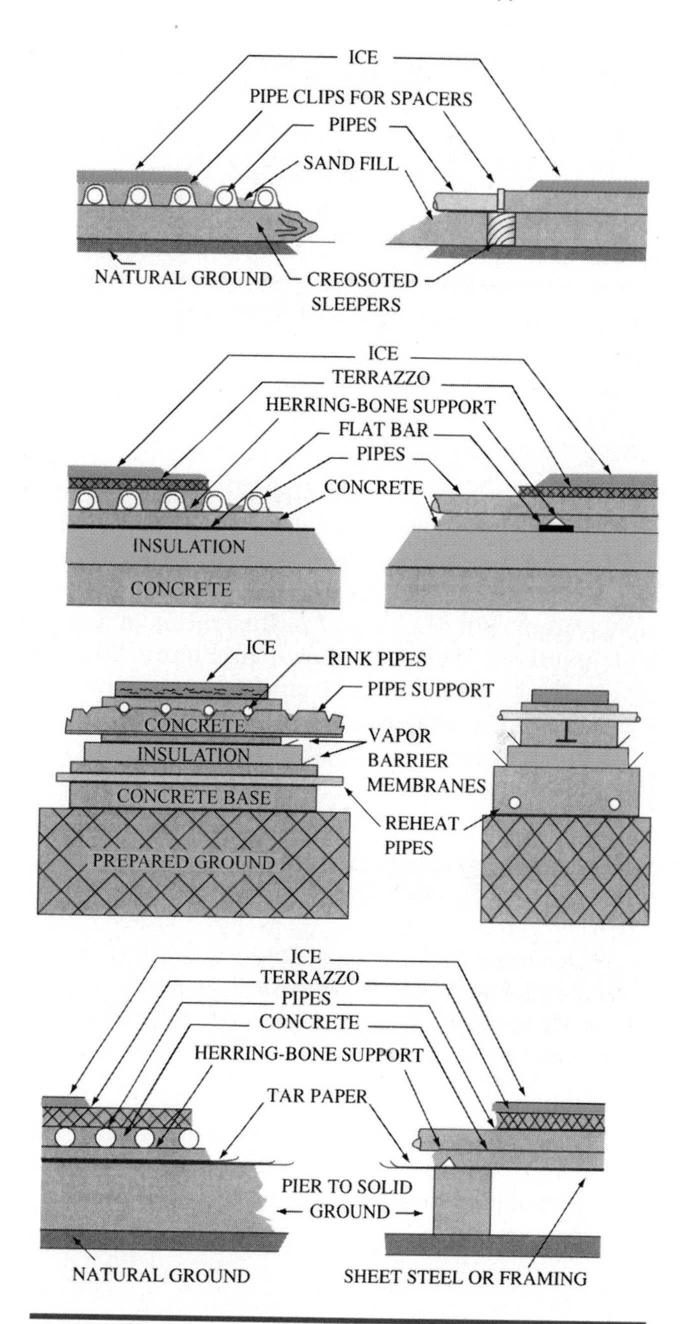

Figure 20-2-2 Types of ice rink surface floors.

at one end of the rink to house the piping header. A snow melting pipe is usually located at the opposite end of the rink for melting the scraped ice, removed by scraping the surface of the rink.

INDUSTRIAL PROCESSING

Some of the industries where refrigeration is most prevalent are: commercial ice plants that produce large quantities of flake ice for refrigerating fresh fish and other perishable products, soil stabilization systems which prevent problems associated with "frost heave" encountered in cold climates like Alaska, chemical processes that require refrigeration, usually for the removal of unwanted heat generated during manufacturing processes, and environmental test facilities that use refrigeration to simulate environments where their products will be used. For example, one of the large automobile factories tests car operation in a −40°F room.

TECH TIP

When a commercial facility is air conditioned for the purpose of aiding in the manufacturing of the company's products it is referred to as a process cooling as opposed to an air conditioning system. Air conditioning is used for creature comfort. The EPA has different regulatory requirements relative to the allowable leak rate for air conditioning and process cooling. If you suspect the system is used for process cooling, verify that with the building owner before beginning work that would require recovering or recycling the refrigerant.

TRANSPORTATION REFRIGERATION

The transportation of refrigerated products is a large and important aspect of commercial refrigeration. Many truck bodies are well insulated, and reliable refrigeration units are available to maintain the required temperatures. There are a number of types of refrigerated trucks and trailers. There is some difference between those used for long and short hauls, as well as for the different types of products being transported.

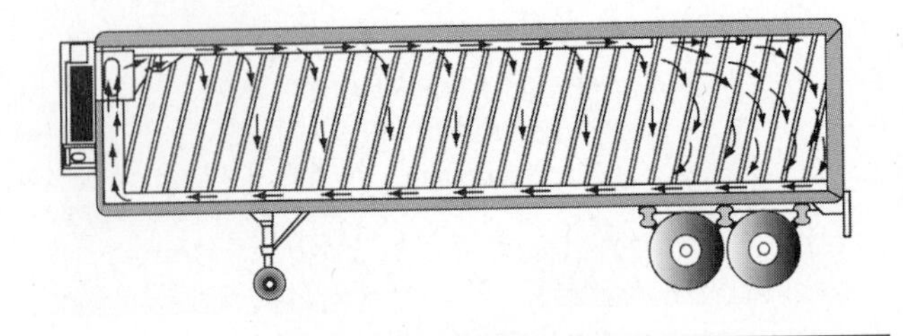

Figure 20-2-3 Section of a trailer showing air circulation.

Long Haul Systems

Long distance hauls are usually handled by trailer trucks. These trailers can be detached from the engine or tractor portion of the assembly. The trailer is usually equipped with a stand alone refrigeration system called a reefer. An illustration of a typical insulated trailer is shown in Figure 20-2-3. Notice the bulkhead in the front for return air coming back to the evaporator through a raised floor. This illustration shows a "piggyback" or clip on refrigeration unit.

SERVICE TIP

It is relatively easy to adjust the temperature of a semi trailer refrigeration unit. This feature allows the trucking company to haul both frozen and refrigerated products.

Some trailers use refrigeration consisting of the evaporation of liquid nitrogen (N_2) or liquid carbon dioxide (CO_2), Figure 20-2-4. In these systems the evaporated refrigerant is wasted to the outdoors.

The semi trailer with stand alone refrigeration can be placed on a railroad flat car and transported. Local delivery is accomplished by connecting a tractor and making the local delivery.

Four different temperature conditions are maintained in refrigerated transport, depending on the type of product being handled. These are:

Type of Service	Temperature (°F)
Air conditioning for floral products, candy, etc.	55 to 70
Medium temperature for perishable foods	32 to 40
Fresh meats	28 to 32
Frozen foods	0 to −5

Some systems that are used both winter and summer have heating equipment as well as cooling equipment. Reverse cycle refrigeration is popular for these applications. Reverse cycle refrigeration can also be used for defrosting evaporators where conditions cause the accumulation of ice on the coils.

There are basically two different types of transport refrigeration units: one type for trucks, Figure 20-2-5, and the other type for trailers, Figure 20-2-6.

The most common type of trailer unit is mounted at the top front of the trailer, Figure 20-2-7. Note that the complete unit is factory assembled, ready for installation. All units are available with diesel engine drive, a complete refrigeration system for cooling and heating, a control system, and accessories. Optional electric motors are available for stationary operation.

The net cooling capacity depends on the temperature maintained. Capacities range from 4,000 to 19,500 Btu/hr at 0°F and 100°F ambient. The fuel consumption varies from 0.13 to 5 gph, depending on the type of service.

Figure 20-2-4 Liquid nitrogen cylinders used for charging truck mounted refrigerant tanks.

Figure 20-2-5 Front mount diesel powered cooling/heating unit for large trucks. (*Courtesy of Thermo King Corporation*)

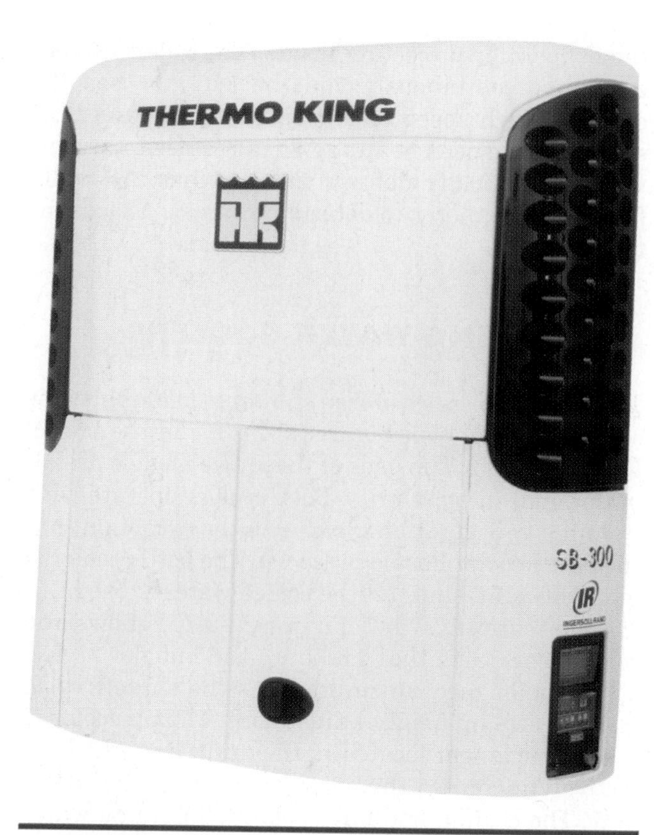

Figure 20-2-6 Fuel saver heating and cooling, high speed and low speed. Runs continuously. Automatic or manual defrost. *(Courtesy of Thermo King Corporation)*

Figure 20-2-7 Front mount cooling/heating unit. *(Courtesy of Thermo King Corporation)*

Selection of the proper unit is based on the length of the truck body, the temperature maintained inside the body, the amount of insulation, and the number of door closings per day.

Local Delivery Equipment

When smaller trucks are used for local deliveries, refrigeration systems with eutectic plates are advantageously used. Eutectic plates are constructed with a provision for refrigeration storage capacity. Plates

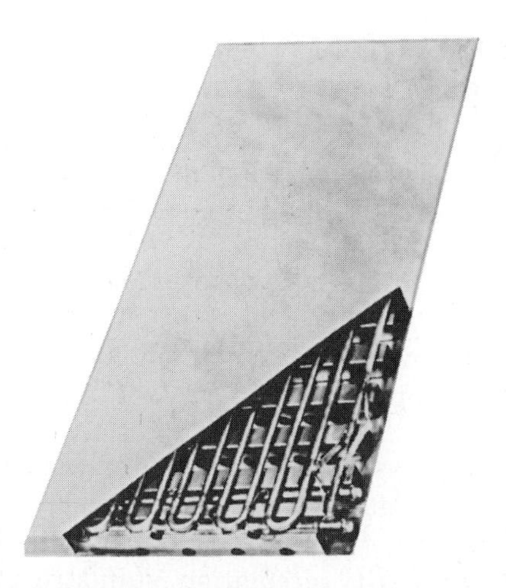

Figure 20-2-8 Plate evaporator with eutectic solution for holdover cooling capacity. *(Courtesy of Dole Refrigerating Company)*

are made with an interior volume that holds a special liquid called a eutectic solution. This liquid stores the cold by changing the state of the solution. Just as ice freezes at 32°F, these eutectic solutions freeze at various temperatures depending on the desired conditions in the truck. Plates are available for operating temperatures of −58, −29, −14, −12, −10, −8, −6, +18, +23, and +26°F.

The plates are usually connected to the refrigeration compressor at night, when the truck is not being used. The refrigeration capacity of the plates is sufficient to carry the load during the truck's next day use. Some trucks carry their own compressor so that they can be easily plugged into electric service wherever they may be. Illustrations of the eutectic plates are shown in Figure 20-2-8.

SERVICE TIP

The solution in delivery trucks' thermal banks does not lose its storage capacity over time. However, as a truck's refrigeration equipment wears it may lose enough capacity so that the thermal storage plates are not completely frozen during the overnight recharging cycle. If a vehicle's refrigerated compartment no longer maintains its temperature throughout the delivery day it may be necessary to do a complete compressor analysis to see if the compressor's capacity has diminished to the point where it can no longer provide enough refrigeration to keep the compartment cool.

Another arrangement for ice cream or frozen food service consists of equipping the truck with auxiliary drive or an electric motor to operate the compressor while the truck is in service.

MECHANICAL REFRIGERATION

Many types of systems are available for furnishing the power to drive the compressor, including:

1. Independent gasoline (or propane or butane) engines
2. Power takeoffs from vehicle engines
3. Electric motors

The compressor may be either on board the vehicle or at the garage for out of service use only. Evaporators are of many types: finned and pipe coils for forced or gravity circulation, standard plates, or eutectic plates.

Equipment must be designed to withstand severe motion, shock, and vibration. Short, rigid lines are subject to cold working. They become hard and brittle through continual motion. Thus, flexible hose connections are often used. Air cooled condensers need to be designed so that normal airflow due to movement of the truck does not prevent adequate air circulation. Units with independent engines require batteries for cranking, fuel tanks and pumps.

The typical refrigeration cycle is shown in Figure 20-2-9 and Figure 20-2-10, and the typical heating and defrost cycle is shown in Figure 20-2-11 and Figure 20-2-12.

The units are factory assembled with engine compressor and condenser in the outside enclosure and the evaporator projecting inside the trailer. The engine runs continuously and the operator has the option of high speed or low speed cooling and high speed or low speed heating. When required, the unit will automatically defrost with hot gas. All units have an hour meter to determine service periods for the engine.

DRINKING WATER COOLERS

There are two styles of free standing drinking water coolers: the bubbler, Figure 20-2-13, and the bottle, Figure 20-2-14. The uses of these two models are almost equal in popularity. Both coolers operate with a similar refrigeration cycle, with the exception that the bottle cooler has no precooler. The bottle cooler is more efficient, since all the cooled water is used.

The cutaway view, Figure 20-2-15, shows the component parts that make up the bubbler styles. The bubbler mechanism includes the shutoff valve and the stream height adjustment. The precooler is a heat exchanger located in the drain line which effectively increases the capacity of the cooler about 40%. The cooling not only cools the incoming water but also can store cooling by creating an ice bank.

Many of these "coolers" also provide a source of hot water for making tea or coffee, using a resistance heater accessory.

The capacities of water coolers are based on ARI standard conditions of 90°F room temperature, 90°F

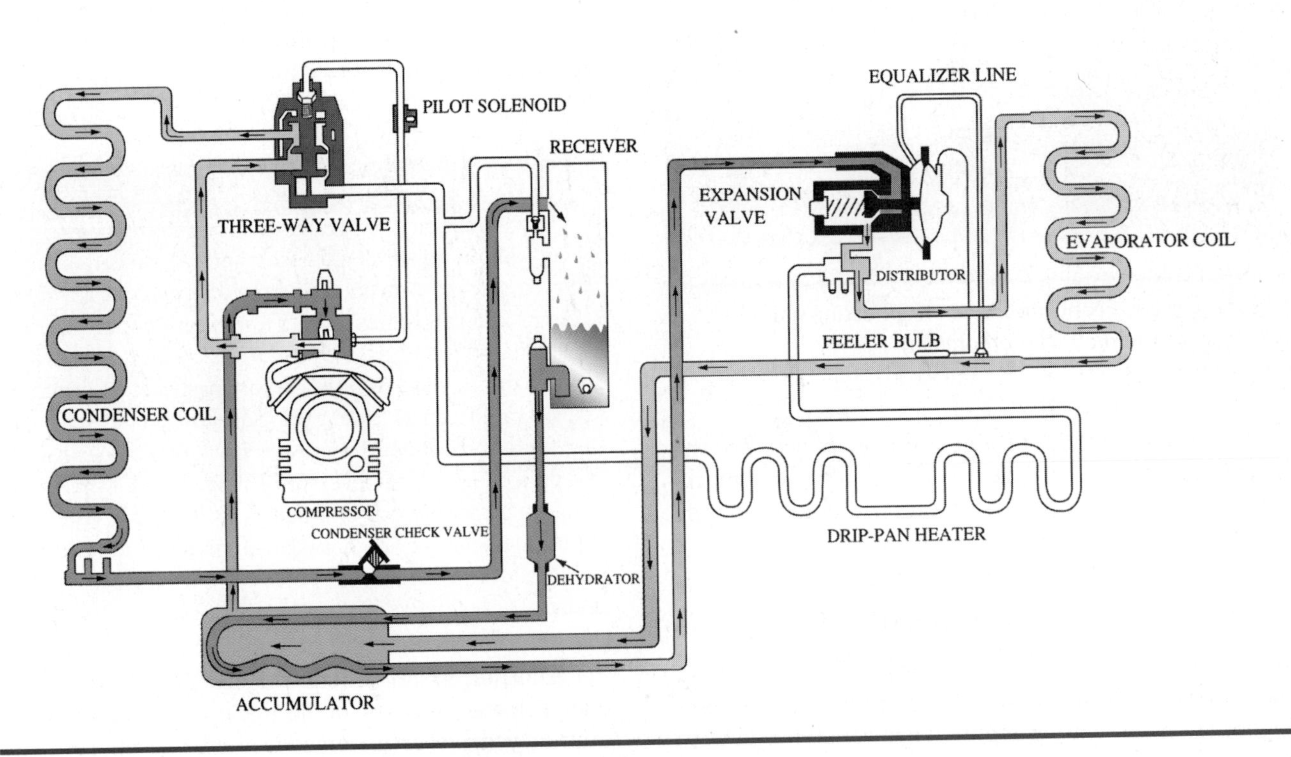

Figure 20-2-9 Cooling cycle for a truck refrigeration unit.

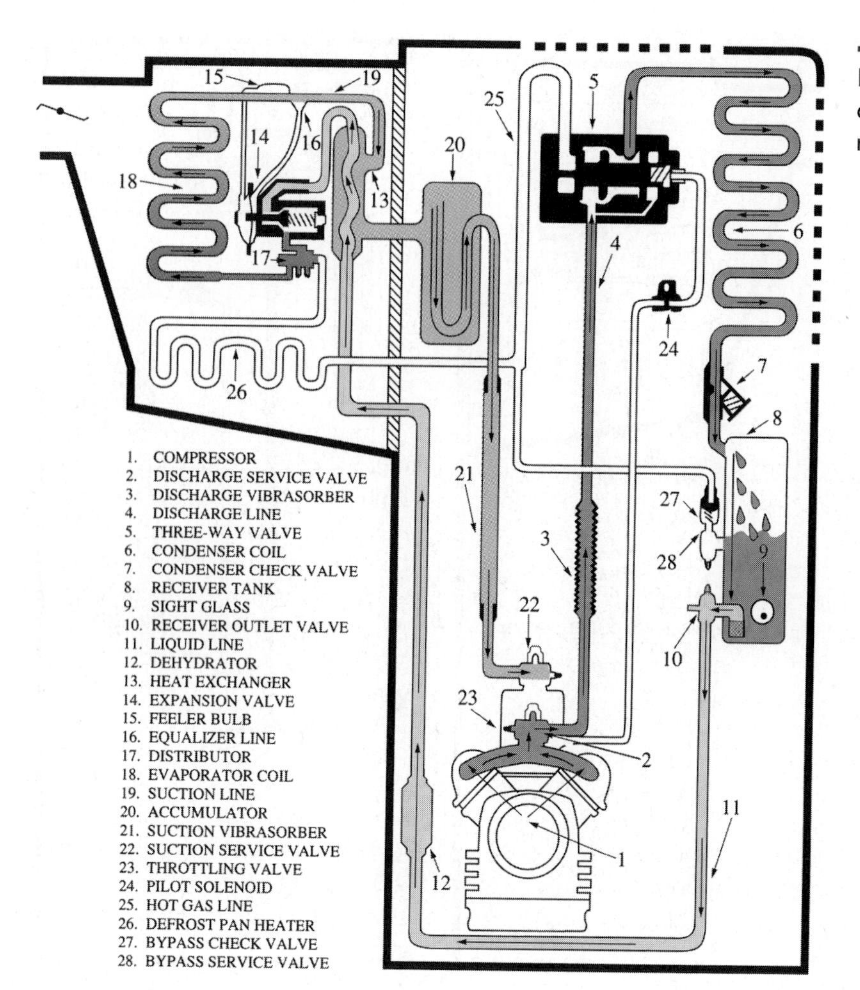

Figure 20-2-10 Cooling cycle for an engine driven refrigeration unit.

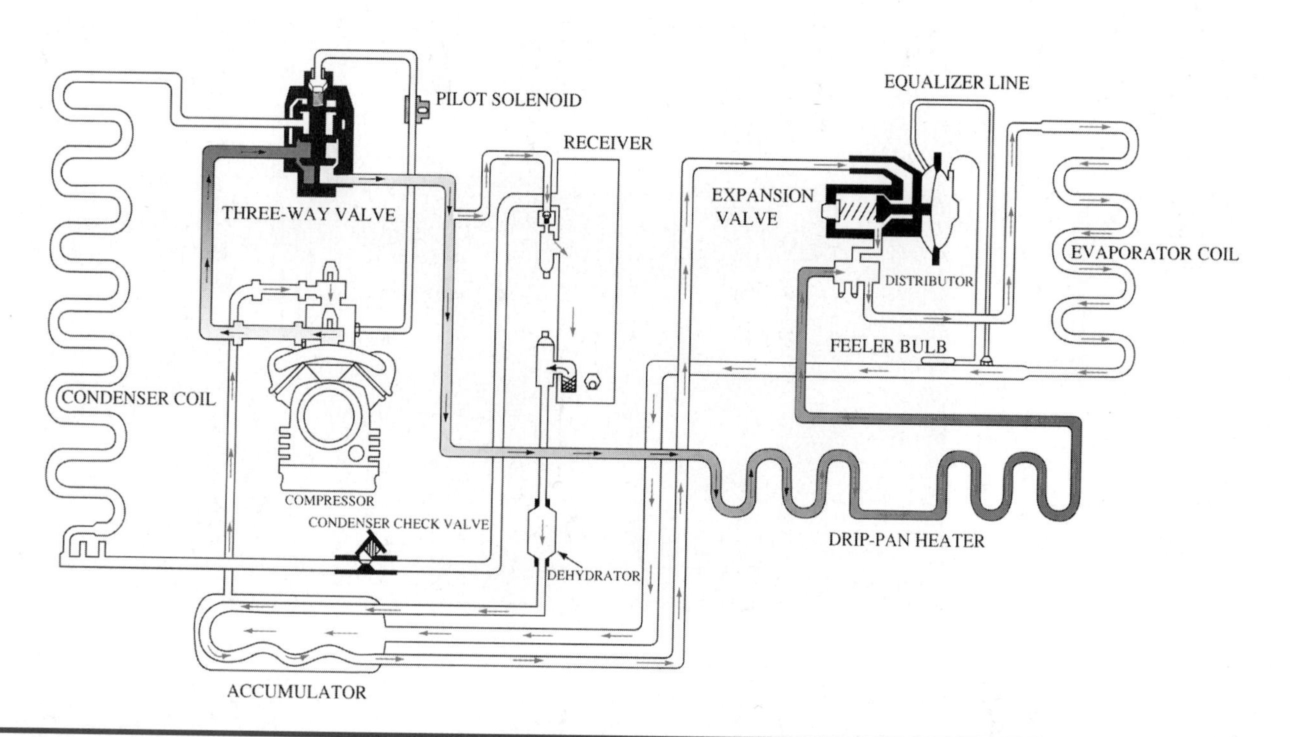

Figure 20-2-11 Heating and defrost cycle. Heating is by reverse cycle principle using hot gas discharge from compressor.

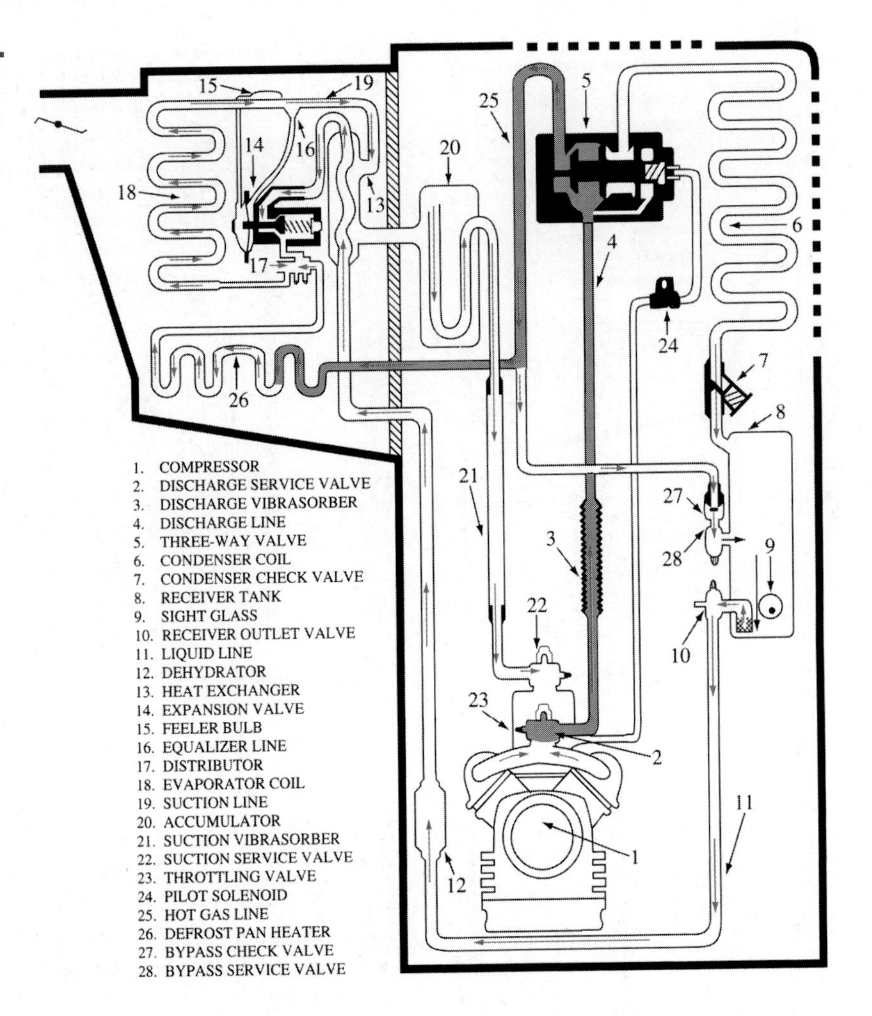

Figure 20-2-12 Heating/defrost cycle for an engine driven refrigeration unit.

Figure 20-2-13 Bubbler drinking water cooler.

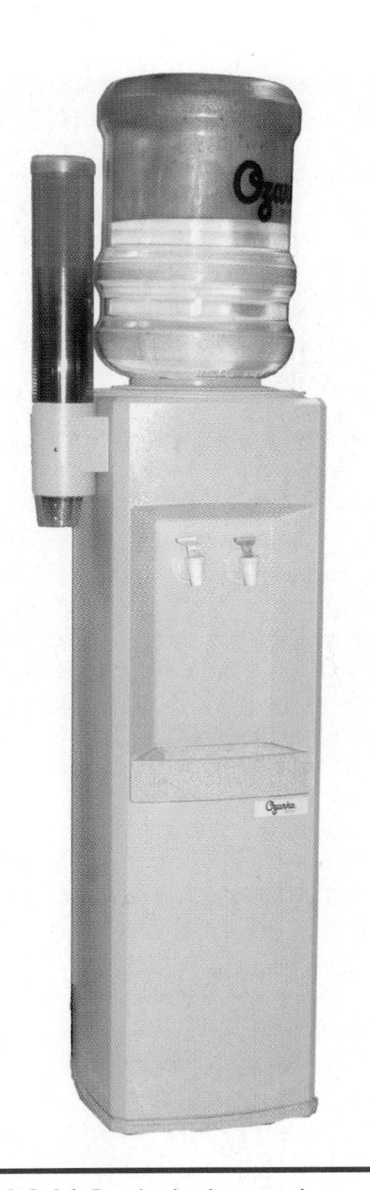

Figure 20-2-14 Bottle drinking cooler.

Figure 20-2-15 Cutaway view showing internal parts of a bubbler drinking water cooler. (*Courtesy of Ebco Manufacturing Company*)

supply water temperature, and 50°F drinking water temperature.

The refrigerant systems are hermetically sealed using a capillary tube metering device and are normally not serviced in the field. The compressor motors are fractional horsepower, split phase, with a relay to cut out the start winding when the compressor is near full speed, as shown in Figure 20-2-16.

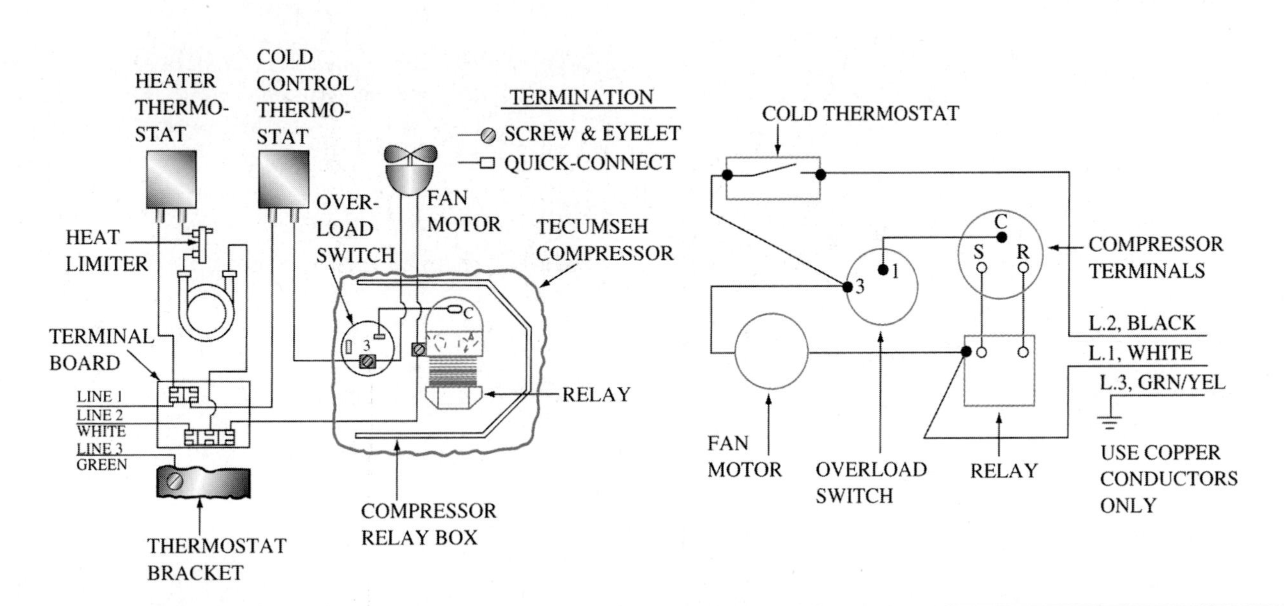

Figure 20-2-16 Wiring diagram for a drinking water cooler. (*Courtesy of Ebco Manufacturing Company*)

The system is started and stopped by a water temperature thermostat.

The refrigerants used have been R-12 and R-500; however, most manufacturers now use an HFC type refrigerant, such as R-134a. The amount of the refrigerant charge is usually less than one pound. To meet environmental regulations, EPA requires the recovery of any CFC refrigerant if the unit is to be discarded or destroyed.

UNIT 2—REVIEW QUESTIONS

1. How is ice cream frozen?
2. List the conditions that can lower the efficiency and effectiveness of the refrigeration system in ice cream plants.
3. _______ lead to high head pressures and higher electrical requirements.
4. What should the ambient temperature be when cold rooms and ice banks are serviced?
5. What are the two ways heat is produced during mixing?
6. Yeast is extremely active when mixed with water and fermentable sugars in a range of _______ to _______.
7. Why is bread cooled to 90 to 95°F for handling and slicing?
8. What is included in the category of "ice rinks?"
9. List the four different temperature conditions that are maintained in refrigerated transport.
10. What are two different types of transport refrigeration units?
11. What are the systems available for furnishing the power to drive the compressor?
12. What are the two styles of free standing drinking water coolers?

TECH TALK

Tech 1. Hey, I've got a question. The boss won't let me work on that refrigerator unit out back, but he turned around and let the mechanic work on it. I know the mechanic's not certified by the EPA. Why is he able to work on it?

Tech 2. Actually, you're only certified for Type I, which means small appliances.

Tech 1. Yeah, that's right, but what about that mechanic? He doesn't have any certification.

Tech 2. He's certified under Section 609 for MVA, which stands for motor vehicle air conditioning.

Tech 1. OK, he does have that certification, but he's not certified to work on that refrigerator on the truck.

Tech 2. Well, you are not right there either. Under Section 609 a mechanic who is certified to work on the truck air conditioner can work on what is called motor vehicle like appliances. Those are things that are on vehicles like that refrigerator. He could also work on the air conditioners on a camp trailer or offroad vehicles and things like that. He is certified to work on it because it is on the truck and considered to be a motor vehicle like piece of equipment.

Tech 1. That sounds real confusing to me.

Tech 2. Well, it does at first. If all the mechanic shops had to have both 608 and 609 certification every time they wanted to work on a piece of air conditioning equipment on a truck, bus, or trailer, that would be a hardship on everybody. So the EPA made provisions for them to work on it. They can only work on it if it's part of that mobile system.

Tech 1. Well, then I guess it does make sense.

Tech 2. Yes, and what you need to do is go back and study the EPA rules and regulations and get your Type II, Type III, or maybe the Universal certification so the boss can send you out on more jobs.

Tech 1. Yeah, I think I'll look into that. Thanks.

SECTION **VIII**

Employment Skills

21

Job Skills

UNIT 1

Installation Techniques

OBJECTIVES: In this unit the reader will learn about installation techniques including:

- Codes, Ordinances, and Standards
- International Residential Code
- Equipment Placement
- Charging the System with Refrigerant
- Charging the System with Oil

CODES, ORDINANCES, AND STANDARDS

The statement is often made that, "All work should be performed in accordance with applicable national and local codes and standards." These regulations may apply to the design and performance of a product, its application, its installation, or safety considerations. Ordinances are set up locally to require conformity to national standards. The national standards that apply to the HVAC industry usually originate from ARI, ASHRAE, ASME, UL, or NFPA. The symbols that designate these organizations are shown in Figure 21-1-1. The following descriptions indicate the scope of the regulations they produce.

ARI

The Air-Conditioning and Refrigeration Institute (ARI) has its headquarters in Arlington, Virginia. It is an association of manufacturers of refrigeration and air conditioning equipment and allied products. Although it is a public relations and information center for industry data, one of the institute's most important functions is to establish product or application standards by which the associated members can design, rate, and apply to their hardware. In some cases, the products are submitted for test and are subject to certification and are listing in nationally published directories. The intent is to provide the user with equipment that meets a recognized standard.

ASHRAE

The American Society of Heating, Refrigeration, and Air-Conditioning Engineers (ASHRAE) started in 1904 as the American Society of Refrigeration Engineers. Today its membership includes thousands of

Figure 21-1-1 Symbols of industry organizations that produce national standards for HVAC equipment.

engineers and technicians from all areas of the industry. ASHRAE creates standards, but its most important contribution probably has been the publication of a series of handbooks that have become the reference bibles of the industry. These include Fundamentals, Applications, HVAC Systems and Equipment, and Refrigeration.

ASME

The American Society of Mechanical Engineers (ASME) is concerned primarily with codes and standards related to the safety aspects of pressure vessels.

UL

The Underwriters' Laboratories (UL) is a testing and code agency which specializes in the safety aspects of electrical products, while also including an overall review of some products. The UL seal on household appliances is a familiar one and it is also applied to the approval of refrigeration and air conditioning equipment. Its scope of activity has expanded into large centrifugal refrigeration machines.

UL approval for certain types of refrigeration and air conditioning products may be mandatory by local electrical inspectors. Compliance with UL is the responsibility of the manufacturer, and the approved products are listed in a directory. Installation in accordance with the approved standards is the responsibility of the installer. Violation of these standards can cause a safety hazard as well as possibly voiding the user's insurance coverage should an accident or fire result. Installation procedures should therefore always conform to UL approval standards.

NFPA

Closely associated with the work of UL in terms of electrical safety is the National Electrical Code sponsored by the National Fire Protection Association (NFPA). The original code was developed in 1897 as a united effort of various insurance, electrical, architectural and allied interests. Although it is called the National Electrical Code, its intent is to guide local parties in the proper application and installation of electrical devices. It is the backbone of most state or local electrical codes and ordinances. It is essential, therefore, that those involved with installation procedures be thoroughly familiar with the scope and content of the National Electrical Code Book and be able to find any information it contains.

INTERNATIONAL RESIDENTIAL CODE™

The International Residential Building Code is a comprehensive, stand alone residential construction code that has been adopted by many states and municipalities. It is replacing the uniform mechanical code for most one and two family dwellings. This code encompasses all aspects of residential construction including all aspects of HVAC mechanical systems.

Local codes are usually divided into: electrical, plumbing and air conditioning/refrigeration, and other codes, such as sound control. Local inspectors enforce these codes based on the permits issued.

TECH TIP

Although there are many up to date codes available in some locations only those codes approved by the local or state governing agents are in effect. As a result often the most current code may not have been adopted at the time your project's construction begins. In these cases you must follow the earlier adopted code unless you receive a waiver from the city or state allowing you to use the more current codes. Failure to get prior approval, even for a newer code, can result in your job failing an inspection and receiving a red tag.

EQUIPMENT PLACEMENT

Despite what may seem to be a great many possibilities for positioning major components during installations, three factors must be considered in the

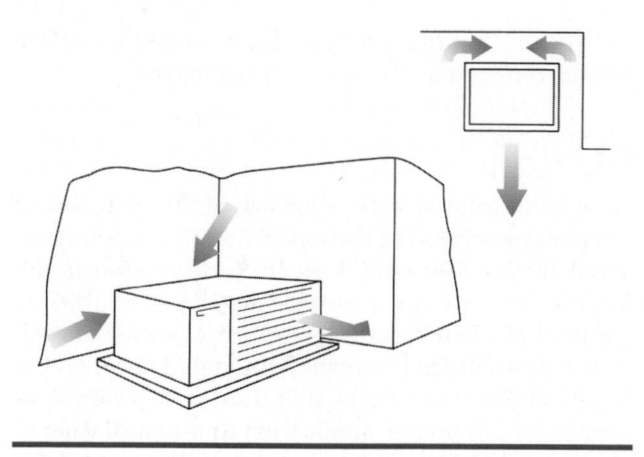

Figure 21-1-2 A properly placed air cooled condensing unit.

placement of equipment if satisfactory installation and proper operation are to be assured:

1. Ample space must be provided for air movement around air cooled condensing equipment to and from the condenser.
2. All major components must be installed so that they may be serviced readily. When an assembly is not easily accessible for service, the cost of service becomes excessive.
3. Vibration isolation must always be considered, not only in regard to the equipment itself, but also in relation to the interconnecting piping and sheet metal ductwork. All manufacturers supply recommendations of the space required; these recommendations should be followed.

Figure 21-1-2 shows a properly placed air cooled condensing unit. Although the unit has been installed in an inside corner, sufficient room has been left for the passage of incoming air around and over the unit.

An example of a common error in positioning major system components is shown in Figure 21-1-3. A shell and tube water cooled condenser has been placed in such a position that the entire condensing unit must be moved to replace a single condenser tube. Be sure to allow room for replacement of items such as compressors, fan motors, fans, and filters. High service costs are often attributable to the poor placement of system components.

Figure 21-1-3 Service accessibility is important in positioning major system components.

Noise is also an important factor in the placement of air cooled condensing equipment. Noise generated within the unit will be carried out by the discharge air. It is poor practice to "aim" the condenser discharge air in a direction where noise may be disturbing, such as a neighboring window.

SERVICE TIP

Some metropolitan areas have maximum noise level ordinances affecting HVAC/R equipment. These ordinances may affect the equipment selected and equipment placement in order to keep the sound level within the ordinance guidelines. Check with your local building office regarding noise level ordinances.

Vibrations set up by rotating assemblies such as compressors, fans, and fan motors can break refrigerant lines, cause structural building damage, and create noise. Vibration isolation is required on all refrigeration and air conditioning equipment where noise or vibration may be disturbing. Almost all manufacturers use some form of vibration isolation in the production of their equipment. This is usually enough for the average installation; however, individual conditions must be investigated to determine whether more stringent measures need to be taken.

The compressor is considered the chief source of system vibration. Since it is good practice to isolate vibration at its source, compressor vibration isolation is essential during installation, as shown in Figure 21-1-4a–d.

When the compressor or condensing unit is to be installed on the roof or in the upper stories of a multistory building, vibration isolators as shown in Figure 21-1-4 may be used. This type of isolator is usually available from the manufacturer of the unit, and in many cases, it is standard equipment.

An isolation pad designed specifically for vibration dampening is shown in Figure 21-1-4. This type of material is designed to dampen the vibration from a given amount of weight per square inch of area. As the right amount and type of this material can be properly selected only when both weight and vibration cycles are taken into consideration, a competent engineer should be consulted.

A final method of vibration isolation, used with small hermetic compressors, is shown in Figure 21-1-4.

In this case the compressor is spring mounted within the hermetic shell.

Figure 21-1-5a–c illustrates a vibration eliminator inserted in the discharge line of the compressor. This eliminator is designed to absorb compressor discharge line pulsation before it creates noise or breaks refrigerant lines. The eliminator consists of a flexible corrugated metal hose core with an overall metal braid. The braid will allow some linear movement of the flexible material, but no expansion or contraction. This type of eliminator is normally placed in the compressor discharge line as close to the compressor as practical. It is particularly effective on installations where the compressor and condenser are on different bases, yet quite close together.

The vibration isolator should be placed in the line so that the movement it absorbs is in a plane at

(a)

(b)

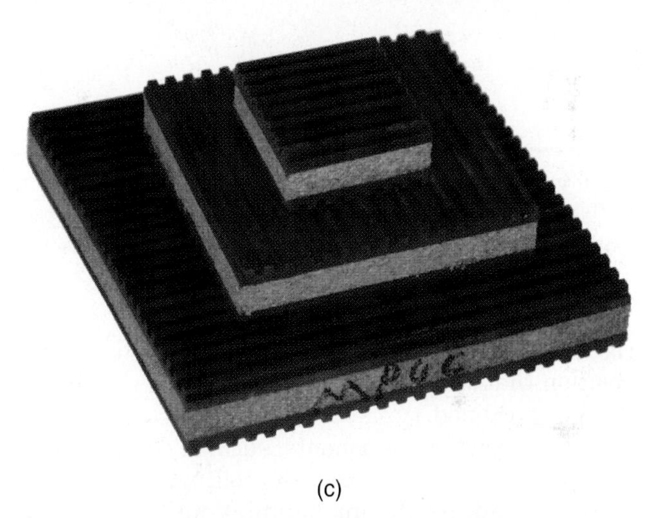

(c)

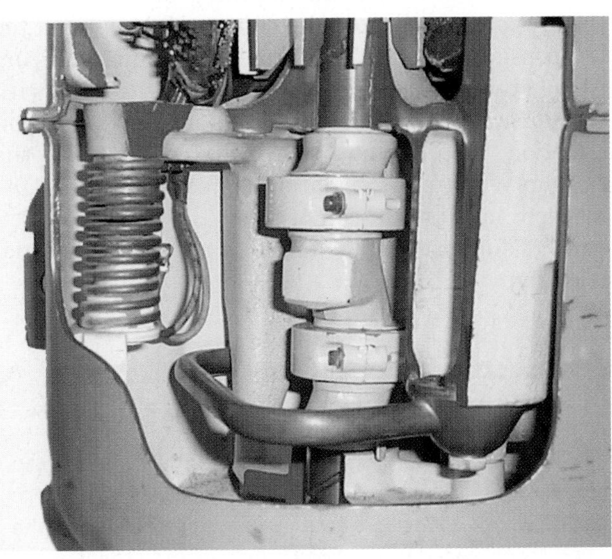

(d)

Figure 21-1-4 Types of vibration isolators used for compressors: (a) Rubber isolation bushings; (b) Hanging spring mountings; (c) Isolation pads in various sizes; (d) Internal isolation springs.

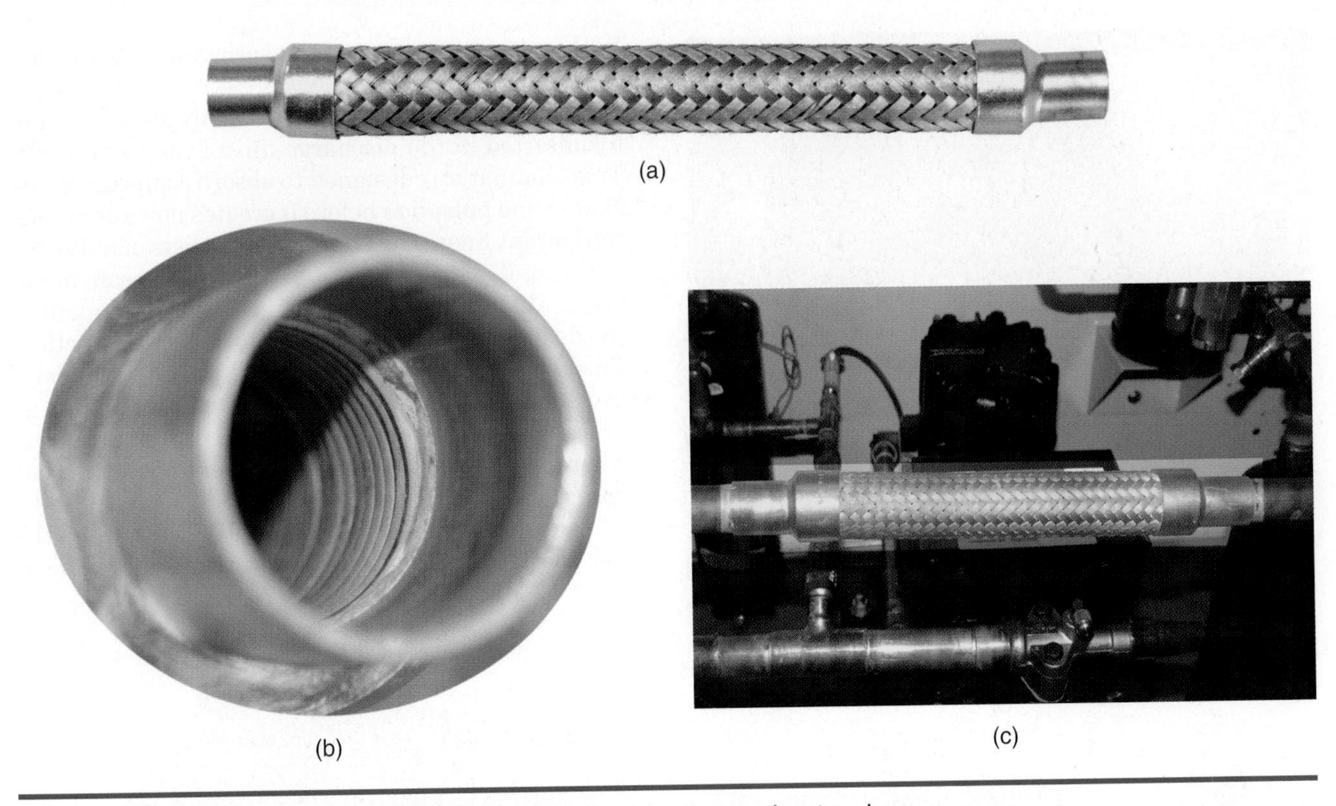

Figure 21-1-5 (a) Vibration eliminator; (b) Internal ribs that allow a vibration damper to flex; (c)Typical installation of a vibration damper as it is installed on a refrigeration system. *(Courtesy of Hampden Engineering Corporation)*

right angles to the device. Do not place this unit in a position that will put tension on it, as its useful life will be shortened.

Some system components, such as the evaporator, may be suspended from the ceiling, as shown in Figure 21-1-6. As the air handling unit containing the evaporator will contain a fan and fan motor, it too is a possible source of vibration. Most manufacturers isolate the fan and fan motor inside the unit with rubber mounts. If this supplies sufficient vibration isolation, the unit may be directly connected to the ceiling. When more isolation is required, the same method used with compressors is effective. Coil springs support the air handling unit for additional isolation.

SERVICE TIP

When hanging an air handler from the attic rafters it can be set on wooden blocks first so it can be leveled before attaching the hanging straps. This will make leveling the unit much easier. In addition to the common sheet metal straps there are other hanging systems including cable and threaded rods which make hanging and leveling much easier and faster.

OTHER MAJOR PHASES OF INSTALLATION

In addition to the placement of major system components, most refrigeration system installations have three other major phases:

1. Erection of piping (both refrigerant and water)
2. Making electrical connections
3. Erection of ductwork

The erection of refrigerant piping is a primary responsibility of the refrigeration installer, along with the placement of major system components. Although not always the responsibility of the refrigeration installer, electrical and ductwork are important parts of most installations. The refrigeration installer should be familiar with good electrical and air duct installation techniques, since quite often, particularly with small equipment, the refrigeration installer is called upon to do the electrical wiring and make connections to ductwork.

Piping

The art of making flared, soldered and brazed connections in copper tubing has been discussed, as well as the procedures for sizing lines, installing traps,

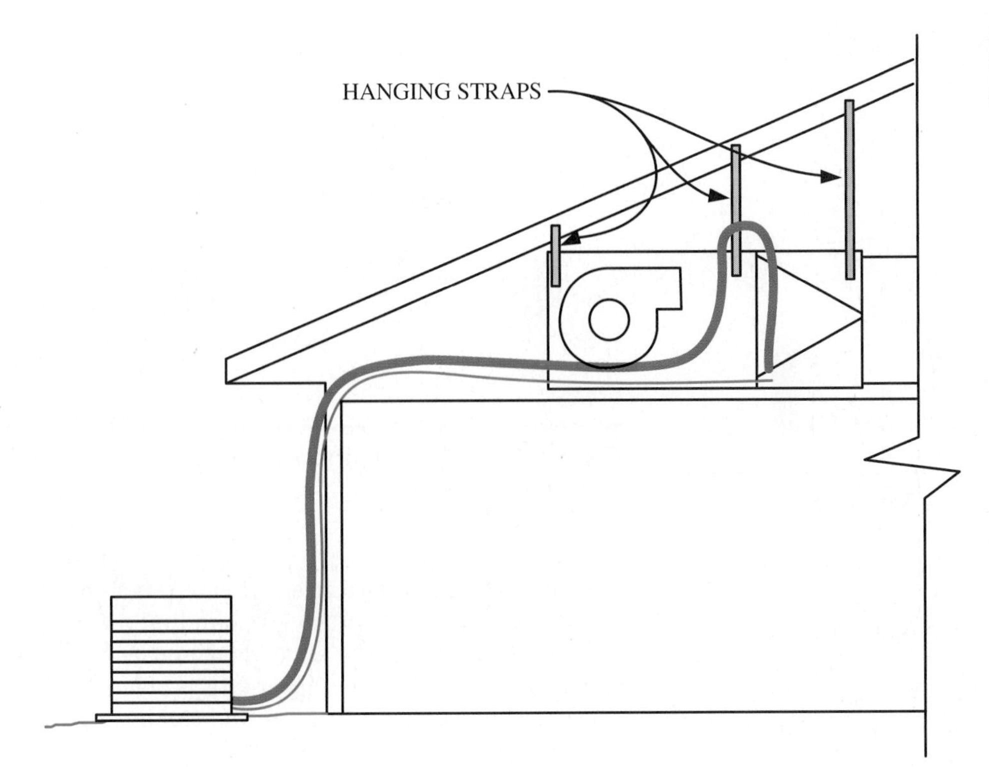

Figure 21-1-6 Suspended air handling unit.

etc., but several important points should be remembered while actually erecting the system piping.

When hard copper tubing is selected for refrigerant piping, it is recommended that low temperature silver alloy brazing materials be used. These alloys have flow points ranging from 1,100 to 1,400°F.

To attain these temperatures, oxyacetylene welding equipment is required. Figure 21-1-7 shows the equipment necessary for this type of brazing. Both oxygen and acetylene tanks with gauges and pressure regulating valves are needed. Also shown is the torch used with the tanks. At the right is a bottle of dry nitrogen also equipped with gauges and pressure regulating valves. The use of nitrogen is recommended, since it serves to keep the interior of the pipe clean during the brazing process. During low temperature brazing, the surface of the copper will reach a temperature at which the metal will react with oxygen in the air to form a copper oxide scale. If this scale forms on the interior surface of the pipe, it might be washed off by the refrigerant and oil circulating in the system. The scale can clog strainers or capillary tubes and will plug orifices.

This scaling can be prevented by replacing the air in the pipe with nitrogen or carbon dioxide. As nitrogen and carbon dioxide are inert gases and will not combine with copper even under high temperature conditions, the pipe interior will remain clean during brazing and no scale is formed, though discoloration may occur with overheating.

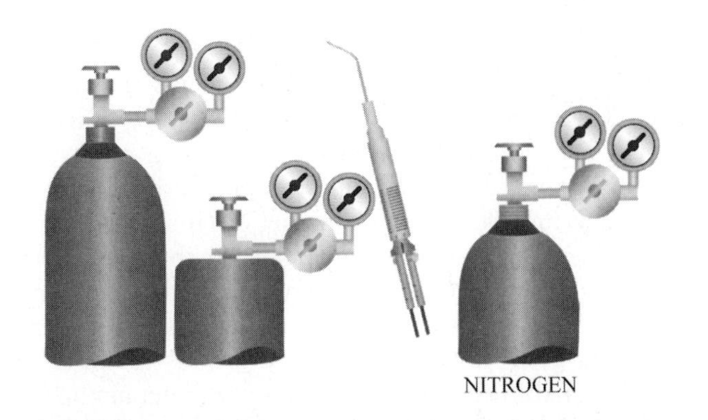

Figure 21-1-7 Equipment necessary for oxyacetylene welding or brazing.

SERVICE TIP

When using an inert gas as a purge to prevent oxide formation only a very slight flow of gas is required. Excessively high flow rates may create enough internal pressure to cause pinholes during the brazing process as the gas is forced out through the joint.

Figure 21-1-8 shows a method by which nitrogen may be introduced into the refrigerant tubing during the brazing operation. Connect the nitrogen bottle to the liquid line service valve Schrader valve with a refrigeration hose. By use of the pressure regulating

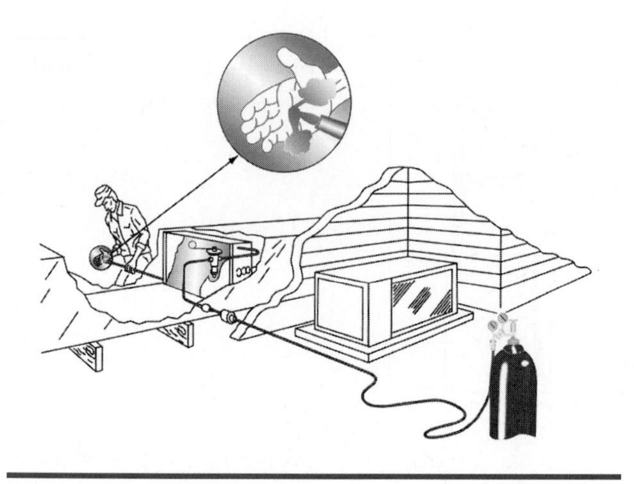

Figure 21-1-8 Brazing operation using nitrogen in the tube to eliminate scaling.

valve on the nitrogen bottle, a slight pressure is admitted to the tubing. This is just enough pressure to assure that air will be forced from the pipe. As shown in the inset, if the nitrogen flow can be felt on the palm of the hand, the flow is sufficient. This nitrogen pressure is kept in the tubing throughout the entire brazing operation, thus assuring an oxygen free pipe.

CAUTION: Never use oxygen or acetylene to develop pressure when checking for leaks. Oxygen will cause an explosion in the presence of oil. Acetylene will decompose and explode if the regulator pressure is over 15 psig to 30 psig (30 psia to 45 psia or 210 kPa to 310 kPa).

Clean pipe is essential in refrigeration installation. Therefore, the use of nitrogen and carbon dioxide is extremely important in the brazing operation as it assures scale free interior pipe walls.

The temperatures during brazing operations can warp metals and burn or distort plastic valve seats. It is important that metal and plastic components are not overheated during the brazing process. Figure 21-1-9 shows the result of brazing heat damage and also one method of assuring this will not happen. The valve at the upper left was not protected from brazing heat; the plastic seat has been damaged. Also, the seat holder has been warped, as shown in the inset. The valve obviously will not operate properly and would require immediate replacement. The valve at the bottom right, however, has a wet rag wrapped around the body. The water absorbs the heat that flows to the valve body during the brazing operation. By keeping the rag wet, the valve and its component parts are protected from heat damage.

In many cases, the suction line of the refrigerant system will run through a nonconditioned area. The

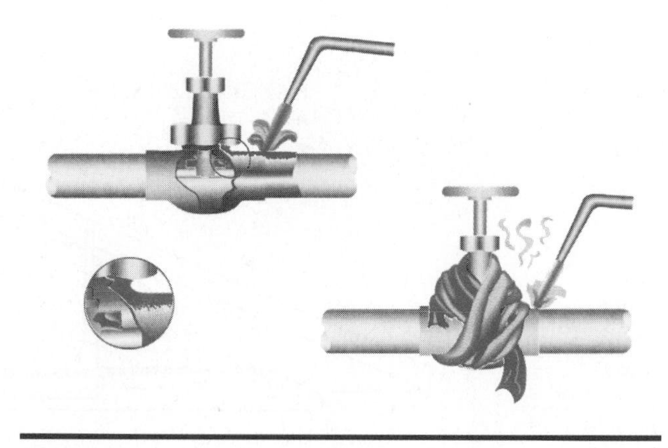

Figure 21-1-9 Protecting valves from overheating during brazing operation.

outside surface temperature of this pipe is frequently below the dew point of the surrounding air. In this case the moisture in the air will condense on the exterior of the pipe. This will create problems where this continuous moisture drips and can be annoying. The pipe must be insulated to avoid condensation and excessive heat from being absorbed by the refrigerant.

SERVICE TIP

Moisture has been associated with the formation of mold and mildew in buildings. Condensate can be a significant source of moisture. It is very important that tubing insulation on refrigerant lines be sealed at the joints so that condensate can not form.

The insulation must be of good quality so that the temperature of its exterior surface will never drop below the dew point of the surrounding area. It must also be well sealed, so that air and the moisture it contains cannot reach the pipe, thus causing condensation underneath the insulation.

At the right in Figure 21-1-10, is an example of a typical pipe hanger that might be found on a small commercial refrigeration installation. This hanger serves as a vibration isolator as well as a pipe supporter. A short length of light gauge rustproof metal is used to form a cradle to support the insulated pipe. A metal strap has been attached to this length of metal. In some cases it is merely wrapped around the length of sheet metal. The free end of the strap is then fastened to the joist or ceiling.

The insulation in such a hanger will act as the vibration isolator. The purpose of the short length of metal is to prevent the thin strap from cutting the insulation.

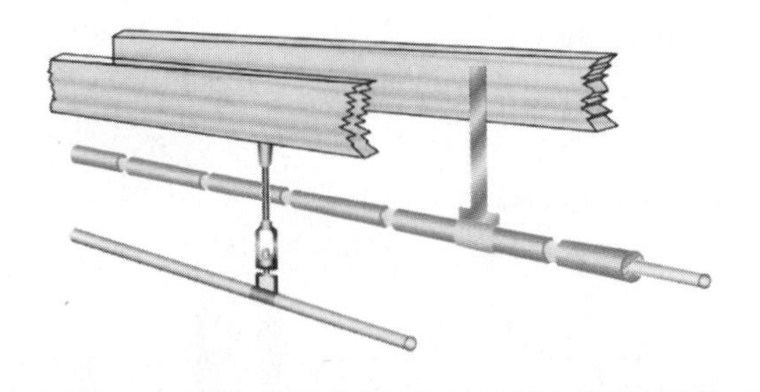

Figure 21-1-10 Pipe hangers used to support insulated pipes.

CAUTION: Although codes require that all electricity to equipment be able to be disconnected at a single disconnect point, this is not always the case. Occasionally, host installation wiring may bring voltage into equipment that does not go through the disconnect. Before you begin working inside electrical panels on equipment, make certain that there is no voltage. Use your voltmeter to test all connections between each power connection and each power connection and a known ground.

The hanger shown on the left in Figure 21-1-10 has a height adjustment feature for leveling or pitching the pipe if required for oil return. This type of hanger can be used with or without insulation.

Electrical Connections

The service technician who installs the refrigeration equipment is sometimes responsible for the final wiring connections between the installed unit and the fused disconnect switch shown in Figure 21-1-11. All electrical power to the refrigeration unit must pass through this switch. When this switch is pulled, or opened, all electrical power to the unit must be disconnected. This same disconnect switch also contains fuses, which will interrupt the flow of current whenever a severe electrical overload occurs. This mechanism is a protection against fires and explosions and also against electrical shocks to people.

Electrical codes, both national and local, are made to protect property and life, and should always be followed. All refrigeration equipment installations having electrical connections are governed by national or local electrical codes. For example, electrical codes require that the fused disconnect switch in Figure 21-1-11 always be placed within sight of the unit that receives the power passing through the switch.

When electrical circuits are to be connected by the refrigeration installer, wiring should be made to assure good electrical contacts. Lug type connectors are recommended.

When stranded wire is used, a single wire may separate and create a potential electrical hazard. Loose wires might contact other wires or "ground," causing electrical problems. Make sure there are no loose strands and the connection is tight to ensure a good mechanical connection. This will ensure good contact and eliminate the hazard of wire separation.

Ductwork

The refrigeration system installer is frequently required to make the final connection between the evaporator or air handling unit and the ductwork. This final connection for metal duct is made with a flexible connector, Figure 21-1-12, that will eliminate vibration transmission from the air handler to the ductwork. It must be installed correctly. If, as shown in the upper left in the illustration, it is too loose, it will drop into the airstream and obstruct normal

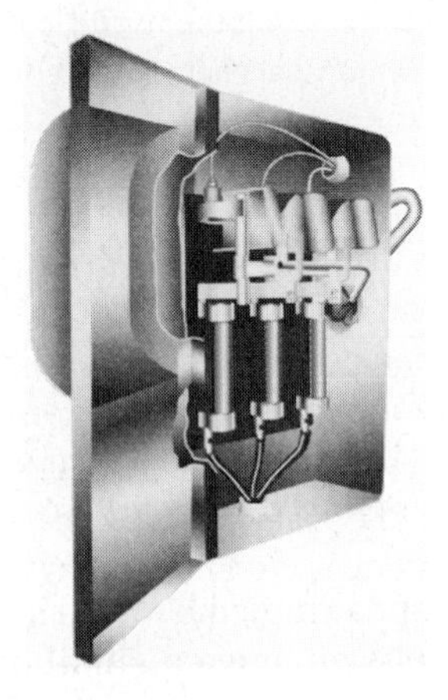

Figure 21-1-11 Fused disconnect switch for the power supply.

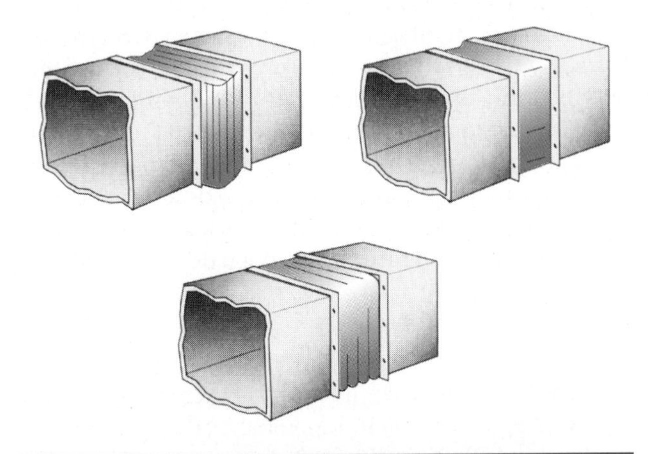

Figure 21-1-12 Flexible duct connector to eliminate vibration.

airflow. If too tight it will stretch, harden, and disintegrate over a period of time, causing leaks. If the canvas is damp and installed too tightly, the ductwork might be pulled out of alignment. The center of Figure 21-1-12 shows the proper application: the canvas is loose enough to absorb vibration and not so loose as to interfere with airflow. Blanket insulation should be installed over the flexible connection and properly sealed.

SERVICE TIP

Current codes and standards require that UL 181 mastic and pressure sensitive tapes be used to seal all ductwork to prevent air leaks.

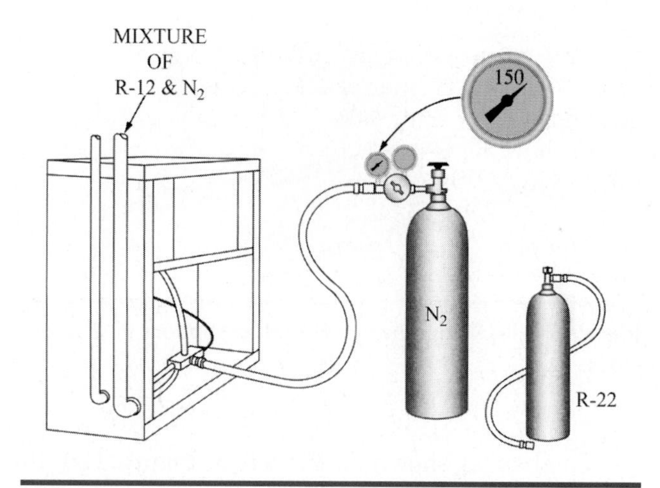

Figure 21-1-13 Leak testing using nitrogen to pressurize the piping.

Final Checking

When the refrigeration system has been completely assembled and all electrical and duct connections completed, a number of important steps must still be taken before the equipment is started:

1. The unit must be leak tested and charged.
2. Field leak detection of the refrigerant is generally done using an electronic leak detector.
3. All belts must be checked for tension and alignment.
4. There must be an electrical motor check of the direction of rotation.
5. The bearings must be oiled if needed.
6. The power source must be checked to be absolutely sure that the correct power will be applied to the unit.

After all interconnecting tubing has been assembled, some refrigerant is introduced into the system as a gas. Although a leak test could be made at this time, some refrigerants do not exert sufficient pressure at room temperature to assure trustworthy results. By using nitrogen, the pressure in the system may be built up to approximately 150 psig, at which pressure a true leak test can be made, Figure 21-1-13. The mixture of refrigerant and nitrogen inside the unit will cause a reaction on the detector if a leak is present. Since nitrogen is an inert gas, the system must be evacuated after it has been determined to be leak free.

TECH TIP

Venting di minimus mixtures of R-22 and nitrogen that are used for leak testing permitted by the EPA.

Following the leak test, the system is ready for evacuation, dehydration and charging.

CHARGING THE SYSTEM WITH REFRIGERANT

The system may be charged with refrigerant as a liquid or gas. Although a number of factors may affect the method of charging, the most important is the quantity of refrigerant involved.

SERVICE TIP

Some studies have shown that as many as 50% of the systems that are in service today are overcharged. There is a feeling among some technicians that a little extra refrigerant should be added to the system just in case there is a leak. If you suspect a leak, fix the leak do not overcharge the system. Excessive refrigerant charges can be as detrimental to the equipment and to the equipment performance as undercharged equipment.

Because it is difficult to remove excess refrigerant from a system you should be very cautious as you approach the proper charge level as to not overcharge the system. Any excess refrigerant would have to be transferred from the system to a recovery drum which would be time consuming and wasteful of a valued company product.

Relatively speaking, most refrigerant containers, regardless of size, have a small outlet. In small units, sufficient gas vapor may be passed through this outlet to complete the full charging of the unit in a reasonable length of time. On large units, however, the time required for the proper amount of vapor to pass through this small outlet may take so long that gas charging is impractical. When this is the case, liquid charging is used. The small orifice will pass a much greater weight of liquid refrigerant in any given amount of time.

The point at which the refrigerant will actually enter the system is determined by one basic consideration: whether the unit is to be charged with refrigerant in a gas or liquid form.

Under normal charging conditions the refrigerant cylinder will be at ambient temperature and corresponding pressure. As most field refrigerant charging is done with the unit operating, the refrigerant cylinder pressure will usually be below the head pressure of the condenser and above the back pressure of the evaporator.

SERVICE TIP

Liquid must be used to charge a system when the refrigerant is a zeotrope or near zeotrope because a vapor will fractionate. These refrigerants are blends and as vapor refrigerant is drawn from the charging cylinder more of one gas would be withdrawn than others, leaving an imbalance in the refrigerant charge. Many of these refrigerant cylinders have a dip tube that allows liquid to be withdrawn from the top of the cylinder in the upright position. Read the cylinder labels carefully to determine if the particular cylinder you are using has such a tube.

CAUTION: Caution must be taken when charging a system with liquid refrigerant so that the compressor is not overloaded or slugged with liquid. Slugging or overloading can cause compressor damage. To prevent this from happening you must meter the liquid refrigerant between the charging gauge set and the vapor port on the unit. Liquid can be changed to a vapor by either using the low side gauge hand wheel as a metering device or by using a liquid charging device installed in the suction line hose.

If the refrigerant is to be charged as a liquid, the charging should take place ahead of the metering device to protect the compressor from damage due to liquid flooding.

With open type and serviceable hermetic systems that are small enough to make gas charging practical, or where only a small amount of refrigerant is required, charging is usually done through the suction service valve on the compressor. Welded hermetic systems are normally small enough for gas charging.

SERVICE TIP

The last thing a technician should consider doing when servicing a system is adding refrigerant. Unless there is a known leak in the system the technician should first check to see what problem may be causing low pressures that indicate low refrigerant charge. If the evaporator is dirty it will have a low side pressure and little or no superheat. If the condenser is dirty there will be a high side pressure with little or no subcooling. A plugged liquid line filter drier or metering device will result in both a low high side pressure and a low low side pressure which mimics a low charge; however, there would be a very high degree of subcooling because most of the liquid refrigerant would have backed up into the condenser. Check all aspects of the refrigerant cycle to determine where the refrigerant is before connecting a refrigerant cylinder to simply add refrigerant.

Pinch off tubes are made available at the compressor for this purpose. These pinch off tubes are connected into the suction side of the compressor and are designed to allow the compressor to pump gas directly from a refrigerant drum.

In units of this type, the original charge is generally done at the factory, but occasionally the charge must be replaced in the field. When this is necessary, the pinch off tube may be cut and a connection placed on the tube. The unit is then charged through the tube, which is repinched and brazed closed. Occasionally, a valve is placed permanently on this pinch off line. This is practical where the pinch off tube is either very short or inaccessible.

There are also devices that are designed to puncture a refrigerant line for charging purposes. The device then remains on the line and serves as a valve for either future charging or as a gauge connection.

Another device used for charging the hermetic circuit is a type of valve called a Schrader fitting. This valve contains a plunger which, when depressed, opens the circuit. A special adapter on the charging hose will depress the plunger when the hose is firmly attached to the valve. Refrigerant will then flow through the valve into the system. When the charging hose is removed, the plunger returns to its original position and reseals the refrigerant circuit. The valve cap must always be replaced after servicing.

The most accurate method of charging any air conditioning or refrigeration system is according to the manufacturers charging specifications. The second most acceptable method of charging is a weighed or measured in charge using a digital scale or graduated charging cylinder. The next two most common accepted practices use superheat for fixed bore orifice metering devices and subcooling for TXV metering devices. All other charging methods are less accurate and therefore less acceptable.

CHARGING THE SYSTEM WITH OIL

The proper amount of oil can be measured into the system in several ways:

1. In a new system, it can be measured or weighed in. Unit installation instructions include the compressor oil requirements in either weight or liquid measurements. This method is also applicable following a compressor overhaul, when all of the oil has been removed from the compressor; however, it should be used only when the system has no oil in it.

2. The dipstick method is used primarily with small, vertical shafted, hermetic compressors. Some larger, open types of compressors may also have openings designed for the use of a dip stick. The manufacturer's recommendations of the correct level should always be followed (if provided).

3. The compressor crankcase sight glass is used after the system has operated for a period of time under normal conditions to determine the proper oil level. This procedure will assure proper oil return to the crankcase. It will also allow the oil lines and reservoirs to fill and give the refrigerant an opportunity to absorb its normal operating oil content, if applicable.

SERVICE TIP

Refrigerant oil should be periodically tested to see whether or not it contains any acids. Acid formed in the refrigerant oil as a result of the breakdown of the refrigerant. Refrigerant breakdown is often the result of moisture in the system or excessively high compression ratios. Moisture can be removed with a new liquid line filter drier and compression ratios can be corrected by cleaning the evaporator and or condensers. The acidity in the oil can be neutralized by using an acid neutralizing additive in the oil or by changing the liquid line filter drier. Failure to remedy the acid or problem causing the acid will result in the premature failure of the refrigeration compressor.

When a compressor is replaced, the new unit should be charged with the same amount of oil as the old unit.

Oil is normally introduced into a refrigerant system by one of two methods, as shown in Figure 21-1-14. It may be poured in as shown on the left, providing the compressor crankcase is at atmospheric pressure. This method is normally used prior to dehydration since it will expose the compressor crankcase interior to air and the moisture the air contains.

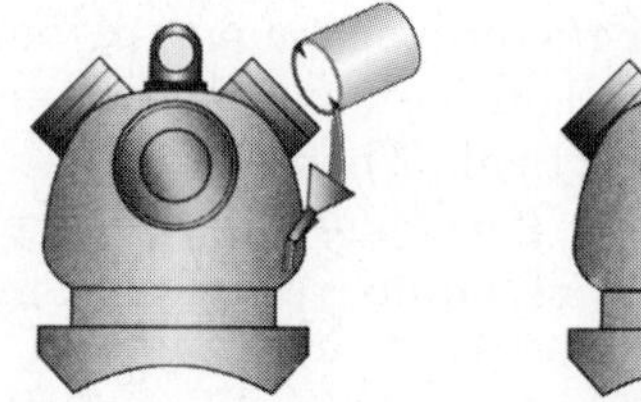
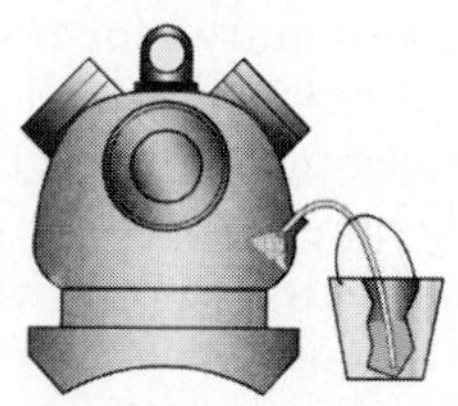

Figure 21-1-14 Two methods of charging oil into the compressor.

On the right in Figure 21-1-14 the method normally used with an operating unit is shown. In this case the crankcase is pumped down below atmospheric pressure and the oil is drawn in. When this method is used, the tube in the container subjected to air pressure should never be allowed to get close enough to the surface of the oil to draw air. As shown, the tube is well below the level of the oil in the container.

Oil charging of replacement welded hermetic compressors should be done according to the manufacturer's recommendations, and depends on whether the replacement compressor has been shipped with or without an oil charge.

There are three precautions to take in charging or removing oil:

1. Use clean, dry oil. Hermetically sealed oil containers are available and should be used.

2. Pressure must be controlled when the crankcase is opened to the atmosphere. Too much pressure can force oil out through the opening rapidly and create quite a mess.

3. System overcharging should be avoided. Not only will this create the possibility of oil slugs damaging the compressor, but it also can hinder the performance of the refrigerant in the evaporator. Oil overcharging may also cause liquid refrigerant to return to the compressor from the evaporator.

UNIT 1—REVIEW QUESTIONS

1. List the associations that set up the national standards that apply to the HVAC industry.
2. What is the function of ARI?
3. What does the UL specialize in?
4. List the three categories that local codes are usually divided into.

TECH TALK

Tech 1. I was working on a job the other day with the new guy. Before we installed that replacement compressor he had me drain all the oil out of the old compressor, measure it, and then drain all the oil out of the new compressor and measure it. He then had me put a little of the oil back in the new compressor before we installed it. That didn't make any sense at all to me.

Tech 2. Actually, he did exactly right. Compressor manufacturers have as much as 10 times more oil in the compressor than the compressor actually needs. They put the extra oil in there so that as it circulates through the system. The compressor does not run short of oil, but when a compressor dies and you are doing a changeout, a lot of that oil is out of the system. If you put a new compressor in the system, it could have 10 times more oil than it needs. If all that oil ends up back in the compressor at the same time, the excessive oil will cause high current draws as the crankshaft splashes through all that oil at the bottom of the crankcase. If that happens, the compressor is going to burn up. Some studies have shown if you don't do what that technician had you do there is a 50% chance that your new compressor is going to die prematurely and if you replace that second compressor and you don't do it that third compressor is more likely going to die 90% of the time prematurely. You need to make sure you do that each time. Drain the oil out of the new compressor, drain it out of the old compressor and put only enough oil back in the new compressor to match the amount of oil that was in the crankcase in that old compressor.

Tech 1. Have people always done that?

Tech 2. No. Years ago they used to let technicians purge refrigerant through the system to blow it out. When you purge through the system you flush out all that old oil so you don't have to when you put the compressor in. Today, purging the system is very difficult. There are some solutions you can pump through to clean it out, but unless you do that you've got to fix the oil level or you are going to have a problem pretty quick with the new compressor.

Tech 1. Thanks, I'll remember that.

5. What three factors must be considered in the placement of equipment?
6. When hard copper tubing is selected for refrigerant piping what type of brazing material is recommended?
7. If the refrigeration system installer is required to make the final connection between the evaporator or air handling unit and ductwork, what is the final connection made with?
8. After all interconnecting tubing has been assembled, what is introduced into the system as a gas?
9. What determines the point at which the refrigerant will actually enter the system?
10. A _______ will result in both a low high side pressure and a low low side pressure, which mimics a low charge.
11. What is a Schrader filling?
12. What is the most accurate method of charging any air conditioning or refrigeration system?
13. List two methods by which oil is normally introduced into a refrigerant system.
14. List the three precautions to take in charging or removing oil.

UNIT 2

Planned Maintenance

OBJECTIVES: In this unit the reader will learn about planned maintenance including:

- Heating Systems Maintenance
- Electric Heating
- Gas Heating
- Oil Fired Heating
- Hydronic Heating
- Heat Pumps
- Coil Maintenance
- Filter Replacement
- Sound Level Monitoring
- Visual Inspection
- Calibration
- Safety Checks

INTRODUCTION

In this section we have reached the point where the components of the system have been identified, the system has been operated, and vital measurements have been made, and a list of planned maintenance tasks have been set up that need to be performed as listed in Figure 21-2-1.

PLANNED MAINTENANCE

Refrigeration/Air Conditioning Maintenance

TECH TIP

Planned maintenance is designed to increase the operating efficiency and maintain the equipment so that its life expectancy can be extended as much as possible. Planned maintenance can also be used to detect components that are near the end of their useful life so that they can be replaced before they fail, leaving their owner without heat or cooling.

Every maintenance contractor should perform a series of planned maintenance tasks at the time of each inspection as listed in Figure 21-2-2. The following list of maintenance items used by one contractor is an example:

1. Clean and flush condensate drain pans and drain lines. Clean, lubricate, and test condensate pumps.
2. Check fan belts. Replace worn or cracked belts with new ones. Adjust belt tension.
3. Lubricate bearings. For oil type bearings, use nondetergent oil (not automotive engine oil) of weight (viscosity) specified by bearing or motor manufacturers. Technicians find that plastic containers with extended spouts are useful for reaching out of the way oil cups.

SERVICE TIP

Do not over oil motors. Usually one or two drops in each bearing is adequate. Over oiling will result in the excess oil draining out of the bearing and running across the motor face plate where it will collect dust. As the dust collects it both blocks the cooling air flow and acts as an insulator, both of which will cause the motor to overheat. Overheating is the primary cause of motor failure.

For grease type bearings, use grade and type of grease specified. Do not over grease. Forcing excessive grease into bearings can damage the grease seals and lead to eventual failure. Where a drain or vent plug is provided, follow instructions and allow excess lubricant to escape.

4. Clean up the equipment. Wipe up any oil or grease spills. Clean equipment makes it easy to spot oil leaks, which sometimes indicate refrigerant leaks and bearing problems. Be sure to tighten packing glands on refrigerant valves and replace valve caps.
5. Check crankcase oil heaters. Follow the instructions for turning on the heater for 12 to 24 hrs before attempting to start the compressor.

Acetylene Regulator Hose & Tips (Turbo)	25 Foot Steel Tape
Flare/Swage Set	Tubing Hand Formers
Nitrogen or Carbon Dioxide Tank	220 V Quick Start Kit
Pressure Regulator and Relief Valve	110 V Quick Start Kit
First Aid Kit	Leak Detector
Dial Indicator	20 Gauge Capacity Lockformer
Schrader Valve Core Removal Tool	Wooden Mallet
Low Loss Fittings	Pop Rivet Gun
Industrial Flashlight	Reamers
Fuse Pullers	Service Valve Kit
Hacksaw	Tin Snips
Ballpeen Hammer	Right
Hand Tool Set—Refrigeration	Left
Nut Driver Set	Straight
Pinch Off Tool	Soldering Gun
Pliers	Squares—Combination
Slip Joint	Tap and Die Set
Needle Nose	Pipe Threading Dies
Linesman	Pipe Vise
Locking	Three Foot Metal Rule
Scratch Awl	Vacuum Cleaner
Screwdriver Set	Oxyacetylene Welding Unit
Straight	Wheel Puller Set
Philips	Pipe Wrench Set
Sockets & Ratchet Set	Torque Wrench
½ in and ⅜ in Drive, ¼ in Drive	Chisel Set
Torch Tip—Propane	Combination Wrench Set
Tubing Bender Set	Diagonal Cutters
Tubing Cutter Kit	Metal File
Tubing/Wrench Set	Flat
Wire Strippers	Round
Allen Wrenches	Vacuum Pump
Wire End Crimpers	Micron Gauge
Safety Glasses	Refrigerant Gauge Set
Gloves	Recovery Machine
Fire Extinguisher	Recovery Drum
Bar Holder	VOM
Bench/Wood Top with Vise	Temperature Probe
C Clamps	Clamp-on Ammeter
Conduit Tubing Bender	Pocket Thermometer
Drills ⅜ in, ½ Elect.	

Figure 21-2-1 Suggested tool list for HVAC/R service and maintenance.

6. Check and tighten all power connections. Loose connections will overheat, damaging wire insulation and could cause motor failures.
7. Examine starter contacts for burning or excessive pitting. Check for free operation of the armature without binding. After getting approval from the owner, rebuild or replace defective units.
8. Examine the air filters. Replace dirty disposable filters or replace with washable filters. Ordinary disposable air filters only have an efficiency rating of 5 to 10%. Often it is possible to install a better filter with 30% cleaning efficiency (same size), at only a modest increase in cost. Most owners appreciate an improvement in indoor air quality.
9. Examine the water pumps. Mechanical seals do not require service. Packed stuffing boxes usually need repacking periodically. Do not tighten the packing gland so tight that no dripping occurs. This can score the shaft or burn the packing rings. Some dripping is desired and normal.
10. Check the refrigeration circuit. Attach a gauge manifold and record operating pressures. Convert refrigerant pressures to their corresponding refrigerant temperatures. Check the refrigerant sight glass (if provided) to see that the charge is adequate. Check compressor oil sight glass (if provided) to verify that the oil charge is adequate.
11. Check the cooling towers or evaporative condensers. Clean the sumps before filling.

AIR CONDITIONING/HEATING SERVICE REPORT

WORK ORDER No. ____________________

OUTDOOR	INDOOR
MAKE	MAKE
MODEL	MODEL
SERIAL NUMBER	SERIAL NUMBER

FACILITY	
ADDRESS	DATE
APPLIANCE	MODEL S/N
TECHNICIAN NAME and ID No.	
COMPLAINT	

QTY.	MATERIALS & SERVICES	UNIT PRICE	AMOUNT
	REFRIGERANT R _____ # ____		
	FILTERS X X		
	FILTERS X X		
	BELTS		
	TOTAL MATERIALS		

HRS.	LABOR	RATE	AMOUNT
		TOTAL LABOR	

TECHNICIAN SIGNATURE	DATE
CUSTOMER NAME	DATE
CUSTOMER SIGNATURE	

ENVIRONMENTAL CHECKLIST

WORK PERFORMED	QTY.	TYPE/DISPOSITION
☐ RECOVERED		
☐ RECYCLED		
☐ RECLAIMED		
☐ RETURNED		
☐ DISPOSAL		
TOTAL		

☐ **LEAK TEST**: __ELECTRONIC DETECTOR, __FLOURESENT DYE, __SOAP BUBBLES, __DEEP VACUUM, __ULTRASONIC DETECTOR

☐ **LEAK FOUND**, LEAK RATE ________%

☐ **LEAK ISOLATED**

☐ **LEAK REPAIRED**

☐ **LEAK AUDIT DATE**:______________

__ELECTRONIC DETECTOR, __FLOURESENT DYE, __SOAP BUBBLES, __DEEP VACUUM, __ULTRASONIC DETECTOR

DESCRIPTION OF WORK PERFORMED

TOTAL SUMMARY

TOTAL MATERIALS	
TOTAL LABOR	
ENVIRONMENTAL CHARGES	
TRAVEL CHARGE	
TAX	
TOTAL	

WORK PERFORMED

CONDENSED UNIT

- CLEANED COIL
- CHECKED CHARGE
- _____ PSIG @ _____°F AMBER
- SUPER HEAT __________°F
- SUBCOOLING __________°F
- REPAIRED LEAK IN COIL
- REPAIRED LEAK IN COPPER
- _______ # REF. R _______
- CHECK MOTOR
- VOLTAGE _____ RATED _____
- AMPERAGE _____ FLA _____
- CHANGED MOTOR
- REPLACED BELT
- ADJUSTED BELT
- REPLACED CONTACTOR
- REPLACED START RELAY
- CHECKED START CAP
- mf/V RATING __________
- REPLACED START CAP
- CHECKED RUN CAP
- mf/V RATING __________
- REPLACED RUN CAP
- CHECKED CONTACTOR
- CONTACTOR RATING __________
- REPAIRED WIRING
- REPLACED FUSE
- CHECKED COMPRESSOR
- VOLTAGE _____ RATED _____
- AMPERAGE _____ FLA _____
- CHANGED COMPRESSOR

EVAPORATION COIL

- TYPE __________
- METERING DEVICE TYPE
- FIXED_______ TXV _______
- REPLACED EXP. VALVE
- ADJUSTED EXP. VALVE
- REPLACED FIXED DEVICE
- CLEANED FIXED DEVICE
- REPAIRED COIL LEAK
- REPAIRED COPPER
- CLEANED COIL
- WET BULB __________°F
- RETURNED AIR __________°F
- SUPPLY AIR __________°F
- DELTA T __________°F

CONDENSATE DRAINS

- CLEANED MAIN DRAIN
- PREPARED MAIN DRAIN
- CHECKED AXL. DRAIN
- REPLACED AXL. DRAIN

FURN. OR FAN COIL

- REPLACED BELT
- ADJUSTED BELT
- REPLACED PULLEY
- ADJUSTED PULLEY
- CLEANED BLOWER
- REPLACED BEARINGS
- OILED MOTOR
- OILED BEARINGS
- CLEANED HEAT EXCH.
- CLEANED OR ADJ. PILOT
- REPLACED THERMOCOUPLE
- CHECKED GV
- GAS PRESSURE _______ "H2O
- REPLACED GV
- CLEANED BURNERS

DUCT

- REPAIRED
- ADJUSTED

THERMOSTAT

- CHECKED THERMOSTAT
- ADJUSTED
- REPLACED

REFRIGERANT LINES

- VAPOR LINE DIA. __________
- LIQUID LINE DIA. __________
- CHECK VAPOR LINE F/D
- CHANGE VAPOR LINE F/D
- CHECK LIQUID LINE F/D
- CHANGE LIQUID LINE F/D

ELECT. HTR.

- REPLACED LINK
- REPLACED OL
- REPAIRED WIRE
- REPLACED CONT.
- CHECKED HTR. ELEMENTS
- VOLTAGE __________
- AMPERAGE __________
- CFM __________
- REPLACED HTR. ELEMENT

FILTER ☐ CLEANED ☐ REPLACED

ADDITIONAL INFORMATION ON THE BACK OF THIS REPORT. ☐

Figure 21-2-2 Air conditioning/heating service report.

Scrape and paint rusty spots. Make sure all spray nozzles or openings are clear and functioning properly.

12. Check the operating temperatures of the system. Many experienced technicians can make a preliminary check by feeling various parts of the system to determine any abnormal temperatures.
13. Check the condition of the evaporator and air cooled condenser coils. They must be clean for the system to operate properly.
14. Always shut off the power before servicing equipment. Technicians tend to become complacent and forget the hazard of electric shock. Be safe—always shut off the power before servicing.
15. Cleanup. A dirty littered workplace can be hazardous and certainly makes a poor impression on the owner or manager. Old belts, dirty filters, broken controls, empty refrigerant drums and oil cans should be removed from the premises or placed in the dumpster. A good cleanup lets people know the technician is professional.

Heating Systems Maintenance

Heating system planned maintenance is much like air conditioning insofar as common elements such as filters, fans, motors, and belt drives are concerned. In addition to these common air side components, the other necessary maintenance is associated with the type of fuel used or the manner in which the heat is supplied. Airside maintenance is discussed under air conditioning maintenance (above). The unique features related to the fuels that require maintenance are as follows:

Electric Heating Electric heating can be panel type, baseboard type or forced air. For the system to maintain its original efficiency, the heat exchanger must be kept clean. Periodic cleaning is particularly important on finned tube baseboards. Due to the gentle airflow over the heating elements, no air filter can be used. Dust and lint can accumulate in the fins and restrict heat transfer.

Special attention also is necessary wherever electrical contacts are exposed. This condition exists on certain types of thermostats. They need to be inspected and cleaned when necessary. Dirty contacts can prevent the thermostat and other controls from working properly.

The fuses on individual heating elements sometimes fail, keeping the circuit open. This problem may appear to be a burned out element, when simply replacing the fuse will put the element back in service.

Gas Heating The gas should burn with a clear blue flame. Most furnaces should not have any yellow tips on the flame. If there are, and they cannot be cleared up by adjustment, they are probably due to rust, dust, or scale on the burners or manifold. Occasionally this material can be removed by a brush and a vacuum cleaner. If the material adheres tightly, however, it may be necessary to remove the burner and use a brush and/or an acetylene torch tip cleaner.

Standing gas pilots can be a problem. The pilot provides several functions: A flame to ignite the main burner, a support to hold the thermocouple in proper position, and a supply of heat for the thermocouple to produce the millivoltage necessary to hold in the safety cutoff. A steady flame should impinge on the thermocouple about ½ in to ⅔ in on the thermocouple. Occasionally these units need to be cleaned. Compressed air usually removes unwanted material that interferes with the flame.

CAUTION: If a leak is suspected in a heat exchanger you must completely disassemble the heat exchanger if necessary to confirm that there is or is not a leak. Never allow a possibly leaking heat exchanger to remain in service.

If the furnace uses a natural draft vent, the draft action can be checked by holding a draft gauge or a lighted match near the vent opening. The draft gauge is the preferred method but if a match is used then the draft should draw the flame toward the opening. Any obstructions need to be removed.

On high efficiency furnaces, the induced draft action can be checked by holding a lighted match near the vent opening. The draft should draw the flame toward the opening. Any obstructions need to be removed.

On high efficiency furnaces, the induced draft blower may need periodic lubrication. The condensing portion of the secondary heat exchanger may be a finned surface that should be inspected for possible obstructions. The condensate drain must be kept free from any blockage.

It is usually desirable to observe the operational sequence of the furnace from the call for heat by the thermostat to the actual supply of heat; the same holds true for the reverse action, when the thermostat is satisfied. The burner should light properly on startup, with delayed starting of the fan. On shut down, the burner flame is extinguished and the fan continues to run until the residual heat is removed. On high efficiency furnaces there is a purge cycle operated by the induced draft fan which precedes the lighting of the burner.

The limit control setting can be checked by operating the burner with the fan disconnected. This is a good safety procedure.

Oil Fired Heating This equipment requires regular maintenance in order to function properly. Burning oil is a more involved process than burning gas. The liquid oil is sprayed into the burner and mixed with the correct amount of air to burn.

Special equipment has been designed for burning oil. Oil is delivered to a spray nozzle through tubing under a pressure of 100–300 psi. A forced draft blower supplies combustion air to the oil spray. The oil is ignited by a high voltage spark positioned close to the spray. The flame retention ring controls the shape of the flame to fit the insulated combustion chamber. The draft over the fire is held constant at a negative pressure of −0.04 to −0.01 in WC using a counterbalanced draft regulator. Safety controls are provided to shut off the flow of oil in case the flame is extinguished. The burner must operate efficiently or soot can accumulate and interfere with proper operation.

The service technician must constantly be on the lookout for leaks in the oil lines. Oil can leak out of pressurized oil lines.

SERVICE TIP

Fuel oil leaks are considered to be hazardous chemical spills. Small areas of soil contamination from fuel oil leaks can typically be easily cleaned up. Larger areas of contamination will need to be addressed by someone trained in the proper cleanup procedures for fuel oil spills. Report any spills to the owner or property manager.

If the system uses a tank submerged in the ground with an oil line operating in a vacuum, air or water can leak in, mix with the oil and create a fuel burning problem. It may be desirable to eliminate the outside tank and replace it with an inside tank.

Water can accumulate in an oil tank due to condensation, particularly when the tank is nearly empty. Each year the technician should check to see if any accumulation of water is in a submerged tank, requiring corrective action. For this reason, it is considered good practice to keep the tank filled with oil when the system is not in use.

It is also good practice to provide planned maintenance in the spring after the heating system is taken out of service. This prevents the soot from hardening throughout the off season. Planned service requires checking the entire combustion process and making adjustments or replacing parts where necessary. Annual replacement of the oil filter is standard procedure.

The annual maintenance for an oil burner should include the following:

1. Burner assembly
   a. Clean fan blades, fan housing and screen.
   b. Oil motor with a few drops of SAE No. 10 oil.
   c. Clean pump strainer.
   d. Adjust oil pressure to manufacturer's specification.
   e. Check oil pressure cutoff.
   f. Conduct combustion test and adjust air to burner for best efficiency.
2. Nozzle assembly
   a. Replace nozzle.
   b. Clean nozzle assembly.
   c. Check ceramic insulators for hairline cracks and replace if necessary.
   d. Check location or electrodes and adjust if necessary.
   e. Replace cartridges in oil line strainers.
3. Ignition system and controls
   a. Test transformer spark.
   b. Clean thermostat contacts.
   c. Clean control elements that may be contaminated with soot, especially those that protrude into the furnace or flue pipe.
   d. Check system electrically.
4. Furnace
   a. Vacuum combustion chamber and clean flue passages.
   b. Clean furnace fan blades.
   c. Oil fan motor.
   d. Replace air filter.

After this work is completed, run the furnace through a complete cycle and check all safety controls. Clean up the exterior of the furnace and the area around it.

CAUTION: If the unit runs out of oil or has a leak in the oil lines, the fuel pump can become air bound and not pump oil. To correct this, the air must be bled from the pump and replaced with oil. On a one pipe system, this is done by loosening the plug on the port side of the intake. Put a hose on the port to direct the oil into a container. Start the furnace and run until oil flows out the opening, then turn it off and replace the plug. The system can then be put back into operation. A two pipe system is considered to be self priming; however, if it fails to prime, follow the procedure just described for a one pipe system.

Hydronic Heating The planned maintenance for hydronic systems involves both air and water components. Terminal units, where the heat is transferred to the space, must be kept clean. Fan motors must be properly lubricated.

One of the problems encountered in hydronic systems is air collecting in the top of a room convector, interfering with the flow of water in the system. Normally it is standard practice to install manual or automatic air vents at the highest points of the system. These vents can be used to exhaust the air from the system.

Occasionally, depending on the system and the accessories, the expansion tank will become waterlogged and lose its cushion of air. The tank needs to be replaced with a diaphragm type design where the air cushion is maintained.

SERVICE TIP

Most large hydronic systems leak some water during their normal operations. Small leaks around pump seals and valves are not uncommon. Because of this there must be a way of providing makeup water to replace that loss. Makeup water cannot be plumbed into the system without proper check or reverse flow valves. These valves are required to prevent the siphoning of system water back into the city water system if the city system were to lose pressure. In some cases it is much easier to use a separate fill tank and pump than it is to connect directly to the city water system because of the regulations involved when a direct connect is made. A fill tank and pump will also allow the additional stabilizing and antifreeze chemicals that are often required in hydronic systems.

If the system overheats, it may be due to maintaining too high a boiler water temperature. Either the operating control needs to be reset or a control arrangement selected that will vary the boiler water temperature with the changes in outside temperature.

Circulator pumps may require maintenance. Periodic lubrication is needed on some types. Others have couplings connecting the motor to the pump shaft, where the alignment needs to be checked.

If the water being circulated in the system is contaminated, the system should be drained and refilled with treated fresh water. Water treatment must comply with local requirements.

Heat Pumps Airflow maintenance on a heat pump is similar to that of an air conditioning unit. Reference can therefore be made to the information supplied above.

Where the heat pump involves a split system, the technician should give special attention to the condition of the outside coil. Any blockage can interfere with airflow and cause a reduction in capacity. The outside coil must be kept clean.

At least once a year the compressor contactor should be examined for pitting and loose connections. This device is used more frequently on a heat pump than on an air conditioning system and shows wear much quicker. Damaged contactors or wiring with poor insulation should be replaced.

Another critical condition that should be examined is the frost build up on the outdoor coil during the heating cycle. Unless the frost is removed, the capacity of the unit will be reduced and the safety controls can shut the unit down completely. The defrost control on most systems permit the technician to "test" the defrost cycle for proper operation. The defrost cycle should begin when the coil temperature reaches 26°F and should terminate when the coil temperature reaches 55°F.

SERVICE TIP

The indoor coil on a heat pump becomes the condenser during the heating season. If the air flow across the coil is restricted due to a dirty filter, dirty coil, improper duct system sizing, low blower speed, etc., the high side pressure can increase resulting in a high discharge gas temperature above normal. When this occurs the refrigerant will begin to break down, forming acid in the system. This acid will have an adverse effect on the motor and motor windings, causing them to fail prematurely. Make certain that during planned service of heat pump systems that the indoor coil is clean and that the air flow is within manufacturer's specified rates.

COIL MAINTENANCE

Although the filters remove much of the dust and dirt carried by an air conditioning system, foreign materials do collect on finned tube coils and need to be removed. Moisture on specific coils can increase the holding capacity for these materials that gradually block the flow of air and reduce the heat transmission rate.

One of the best ways to clean a finned coil on the job, if allowed, is to use a sprayer filled with an appropriate cleaning material to loosen dirt, lint, oil and other grime and remove it from the area. Let set according to the manufacturer's instructions; then rinse coil thoroughly. The following points need to be considered when using this procedure.

1. Some portable means of pressure needs to be provided. This can be an air compressor or a small CO_2 with a pressure reducing valve.

2. The cleaning material must be nonflammable and nontoxic.

3. The cleaning material must not have an objectionable odor and the room must be well ventilated.

4. The operator of the hose or spray must be protected from over spray of the chemicals. The use of rubber gloves, goggles, boots, and special clothing may be advisable under certain conditions where such risks exist.

5. If electrical machinery is to be cleaned, the cleaning material must not contain water. Both sides of the electrical line must be disconnected before the cleaning takes place.

6. Where dust and lint are loose, the use of a vacuum cleaner is recommended.

MAINTENANCE OF ELECTRONIC AIR FILTERS

Electronic air filter screens and cells must be cleaned periodically with detergent and hot water. Some designs include an automatic washing arrangement with the filter screens and cells in place. Other designs require the removal of the screens and cells for cleaning. The frequency of cleaning depends upon the application. After initial startup, the filters should be inspected frequently. Once the rate of buildup has been observed, the appropriate frequency for regular cleaning is easily determined. The filter washing procedure for unit without an automatic built-in washer is as follows:

1. Turn the electronic filter, blower, and furnace power off.
2. Remove lint screens and ionizing collecting cells.
3. Clean using hot soapy water, and rinse thoroughly.
4. Replace lint screens and ionizing collecting cells after they have dried thoroughly.

A properly operating unit will be indicated by black water when the cell is washed. If a cheese cloth placed over the air outlet grille becomes discolored, the electronic air filter is not working properly.

FILTER REPLACEMENT

One of the most neglected items of maintenance on air conditioning systems is the periodic cleaning or replacement of air filters. As the mechanical filter clogs with dirt, it can reduce airflow to the point where the cooling coil will freeze and possibly lead to a compressor failure. In a heating system, dirty filters can cause overheating and reduce the life of the heat exchanger, or it can cause nuisance tripping or the limit switch. Dirty filters increase operating cost as the efficiency goes down.

The filter(s) furnished on new HVAC units are often the throwaway type. They are easily accessible by removing a panel on the unit. They can then slide out and be replaced by a similar filter. The filter material consists of continuous glass fibers loosely woven to face the entering air side. Air direction is clearly marked. Filters are usually 1 in. thick for residential equipment. Some filters have fiber material coated with an adhesive substance in order to attract and hold dust and dirt.

Many technicians replace these with a permanent or washable filter which has longer life. It consists of a metal frame with a washable viscous impingement type of material supported by metal baffles and graduated openings or air passages. When these filters are removed and cleaned with detergent, they are dried and recoated before reinstallation.

BELT DRIVE ASSEMBLY CHECKING

When the installed equipment is belt driven both the alignment and the tension of the belt must be checked. Improperly aligned belts show excessive wear and have an extremely short life. Loose belts wear rapidly, slap, and frequently slip. Belts that are too tight may cause excessive motor bearing and fan bearing wear. Belt alignment is simple to check; a straightedge laid along the side of the pulley and flywheel will show any misalignment immediately.

SERVICE TIP

Many air handler units in commercial and light commercial applications run 24 hours a day, seven days a week to provide both conditioned air and ventilation. Most of these units are belt driven. In addition to the wear that occurs on the belt over time the pulleys themselves will begin to wear out. The sides of the pulley provide the necessary friction to drive the system. As the pulley sides wear they become slick and the gap becomes excessively wide. When this occurs even with a new belt there will be excessive belt slippage which will more rapidly wear out the new belt and cause a loss in overall system efficiency. Check the pulley surfaces for wear and replace as necessary as part of planned service.

Belts should be set at the right tension. The correct tension allows 1 in. deflection on each side of the belt.

SOUND LEVEL MONITORING

Unacceptable levels of noise and vibrations from HVAC/R systems can be avoided once the technician becomes familiar with some fundamental principles about sound and vibration. This includes knowing the characteristics of sound, the pathways sound takes, the effect of sound on people, and the factors to apply in achieving satisfactory installation and operation in any given location.

Sound is generally considered a longitudinal wave traveling in a fluid medium such as air or water. It is identified by its speed, frequency, and wave length, which are interrelated. Sound is generated by a vibrating body or a turbulent air or water stream. HVAC/R systems produce both. The number of cycles per second, expressed in Hz, is the frequency, and defines the audible levels of sound as ranging between 20 Hz to 29 kHz. Within any given range, the frequency is also characterized by its sound power output, expressed as octave or ⅓ octave bands (ANSI standards for rating the acoustics, for example, of an operating room).

Noise is any unwanted sound or a broadband of sound having no distinguishable frequency characteristics. Broadbands are often used to mask other sounds.

The intensity of sound is measured in a number of ways. In acoustics, where the quality of sounds in a room is being analyzed, the strength of sound sources, levels and attenuation are measured by the decibel (dB) unit. Since the human ear is pressure sensitive, a decibel scale for sound pressure is based on pressure squared being proportional to intensity. The unit used for this measurement is the micropascal (μPa) and the threshold of hearing is about 20 μPa. The human ear can tolerate a broad range of sound pressures before reaching the level of pain. Figure 21-2-3 shows some typical sound sources, their sound pressure levels and the human reactions to them.

The quality of sound, as well as the loudness, is also important. Diffusers are often used to fill in the frequencies of equipment sounds which have unpleasant low or high tones. This becomes complex when a number of sources are present.

Measuring sound levels is accomplished with a battery operated, hand-held sound level meter with an attached filter set. The meter has a microphone, internal electronic circuits, and a readout display to measure the pressure at any given location.

In measuring sound, a type "A" filter is placed in the microphone circuit to reduce the intensity of the low frequencies. A typical instrument for measuring sound is shown in Figure 21-2-4. This is known as a Noise Dosimeter. It measures the sound level in dBA units (decibels, using an "A" filter). It is supplied with an integrated data logger. The stored data can be viewed on a computer monitor, or a hard copy of the data can be produced on a printer attached to the unit. It records up to 31 hrs at 1 min. intervals of average and maximum readings for each period.

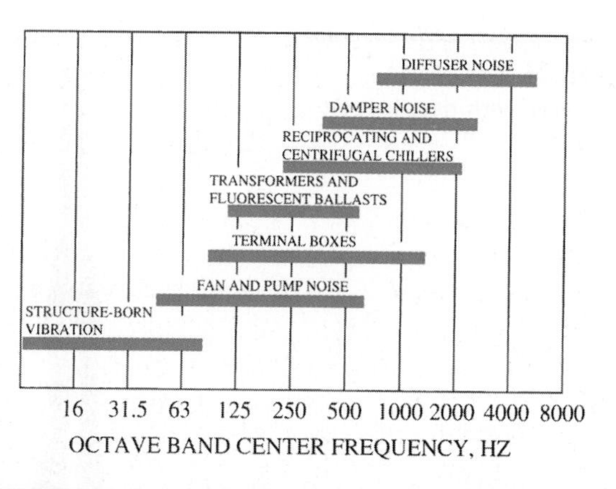

Figure 21-2-3 Sound frequencies of various types of mechanical and electrical equipment. *(Copyright by the American Society of Heating, Refrigerating, and Air-Conditioning Engineers, Inc.)*

The technician is concerned with reducing noise and vibration from HVAC equipment wherever possible. Noise can be disturbing to the customer. Vibration can create noise, but in many cases the most serious effect is that it can be destructive.

LUBRICATION

Electric motors that require periodic lubrication should be checked and oiled with care. Motors can be damaged by over oiling.

When a motor is over oiled, oil runs out of the bearing into the housing and is sprayed off the rotor throughout the interior of the motor. This coats the windings, terminals, and starting switch, causing gumming, carbonizing the moving parts, and often causing motor failure. It is better to oil the motor every two years than to do it more often than required.

The type of oil is also very important. The oil should be heavy enough to provide proper lubrication. It should not contain soaps, detergents, wax, or any other additives. Three-in-one, WD-40™ or any other rust routine oil should never be used in blower or fan motors. Also, ordinary automotive engine oil should never be used, regardless of the make or grade. Use electric motor oil of a weight or grade specified on the motor. Generally, this is No. 10 weight.

(a) (b)

Figure 21-2-4 Noise dosimeter with data logger.

> **CAUTION:** Some spray lubricants have volatile solvents that will break down motor winding insulation. If these spray lubricants are used on electric motors they will result in the motor failing as a result of shorts that will occur in the motor windings. Never use unapproved lubricants on any HVAC/R equipment.

VISUAL INSPECTION

Most service and maintenance technicians make a visual inspection of the entire system before applying instruments or starting their work. Many problems are evident through visual indicators, and considerable time can be saved by this procedure.

Here are some of the system conditions that point to problems or the need for maintenance:

1. Water leaks. A condensate pan may have a clogged drain causing water to overflow onto the floor.

2. A dirty or obstructed condenser coil. Outside condensing units are subject to air blockage due to the growth of shrubbery. There also may have been construction work in the area and airborne dirt has collected on the coil.

3. A motor or transformer may be discolored due to excessive heat. This can be an indication of an overload.

4. An excessive operating noise. This could be a broken valve on the compressor or a loose pulley or blower wheel, requiring further investigation.

5. Reduced airflow. Dirty filters reduce the airflow and the capacity of the system.

Here are some of the refrigeration conditions that point to problems or the need for maintenance:

1. Short cycling on the low pressure control. This may indicate a restriction in the liquid line that needs further investigation.

2. Oil on a refrigerant line. This may indicate a refrigerant leak.

3. Bubbles in the refrigerant sight glass. This is usually a sign that the system is short of refrigerant.

4. Low oil level in the compressor sight glass. Oil probably needs to be added to the compressor crankcase.

5. A kink in a refrigerant line or capillary tube. This usually creates a restriction and may indicate a vibration problem that may later cause a break in the tubing.

CALIBRATION

It is important in using test instruments to be certain that they are indicating accurate readings. Some means needs to be provided for periodically checking most instruments.

Instruments do vary widely in their need for and method of calibration. Some instruments, such as a liquid manometer, do not require recalibration unless fluid is lost. Under normal usage they retain their original accuracy.

Some instruments, such as an analog ohmmeter, require recalibration every time they are used. Each ohmmeter has a calibrating wheel that is turned as required to "zero in" the instrument before it is used.

Some instruments, such as a pressure gauge, have a calibration adjustment on the face of the instrument which is used occasionally, particularly after the gauge has been subjected to pressures beyond the scale reading.

Instruments, such as a refrigerant monitor, can be calibrated in the field using a special calibration kit furnished by the manufacturer.

Other instruments, such as an electronic thermometer, need to be sent into the factory for calibration.

Any time an instrument is used or purchased, the technician should determine the type of calibration required for that particular instrument. By comparison with another instrument of the same kind, it may be feasible to determine whether or not the instrument is reading correctly. The instructions for recalibration usually are supplied with printed material that comes with the instrument or can be obtained from the manufacturer.

SCHEDULING PARTS CHANGES

During a planned maintenance call, the technician may find certain parts that need changing. Some of these are normal parts changes, such as replacing the oil filter and oil burner nozzle in the fall before the seasonal startup. Other parts changes may be needed due to wear or to reduce certain problems from recurring.

For example, the technician may find on a semi-hermetic compressor that the valves are leaking and should be replaced. This may be caused by the liquid refrigerant slugging the compressor. As a result the service person may conclude that a suction line accumulator needs to be installed. The problem is not solved by just replacing the valve plate, but the cause must be corrected to reduce recurrence.

Properly scheduling the work requires securing the approval of the owner for extra expense, securing the necessary parts and tools, and arranging for a return date.

One condition to be avoided is making a parts change without proper analysis (trail and error method). For example, if a unit is stopped by the high limit control, the problem is not usually solved by changing the limit control. The cause may be an excessive load. When the load is reduced, the unit is found to run in a satisfactory manner. A well trained technician is confident that the part is needed before replacing it.

The planned maintenance technician should carry with him all the parts and tools required for normal maintenance. He or she should have a record of the size required for such items as replacement air filters. If a return call is necessary to complete the work, the technician should be well equipped with materials and tools to efficiently complete the job.

SAFETY CHECKS

Too often accidents occur because safety controls fail to function. An important part of the technician's job is to be certain that everything possible is done to provide for safety to the occupants, the workers, and the equipment.

Each system has a series of devices and work practices designed to provide safety. These range from overload protection to devices that stop the system if a dangerous condition exists.

Here is an example of a protective device that needs to be checked by the planned maintenance technician during planned maintenance. Every boiler installation has at least one pressure relief device that relieves the excess pressure if it builds up to limiting levels. Since these devices are seldom (if ever) used, corrosion can form. Inspect them for any signs of leaks and replace if necessary. They must function if ever needed.

Another example of an important safety device that must be checked during planned maintenance is the "no-flame" control on a furnace. If the burner fails to light when fuel and ignition are supplied, the fuel must shut off. This can be tested both on an oil burner and a gas burner. If the safety equipment is defective, it should be replaced.

The monitoring of environmental conditions involving HVAC systems is increasing to reduce the occurrence of health problems. Detecting operational problems at an early stage is a key to reducing health problems.

TECH TALK

Tech 1. The last company I worked at did preventative maintenance work and you guys do planned maintenance work. What's the difference?

Tech 2. Actually, there is not any difference, just a different term.

Tech 1. Why the different term?

Tech 2. Actually, the term "preventive maintenance" has been used for years for what actually has been planned maintenance. Although you are trying to prolong the life of the equipment as much as possible there is no way you could realistically prevent it from dying.

Tech 1. Yeah, I guess you're right. The equipment is going to wear out no matter what you do.

Tech 2. That's correct. No matter what you do the equipment is going to eventually fail; it can't last forever. So what you are doing is extending the life as much as possible, keeping the equipment's efficiency working as well as possible through planned maintenance and service.

Tech 1. Then we're actually doing the same thing we've always done; we're just calling it by the new name, but it is still often referred to by the initials PM.

Tech 2. That's right.

UNIT 2—REVIEW QUESTIONS

1. What planned maintenance operations should every maintenance contractor perform at each inspection?
2. What functions do standing gas pilots provide?
3. On high efficiency furnaces how can the induced draft action be checked?
4. What is required in planned service of a heating system?
5. What is one of the problems encountered in hydronic systems?
6. What is one of the best ways to clean a finned coil on the job?
7. What is the filter washing procedure for units without an automatic built-in washer?
8. How is sound generated?
9. _______ are often used to fill in the frequencies of equipment sounds which have unpleasant low or high tones.
10. What happens if a motor is over oiled?
11. Ohmmeters and other instruments require _______ every time they are used.
12. What is an example of a protective device that needs to be checked by the planned maintenance technician?

UNIT 3

Being a Professional HVAC/R Technician

OBJECTIVES: In this unit the reader will learn about being a professional HVAC/R technician including:

- Publications
- Professional Certifications
- Industry Competency Exams
- North American Technical Excellence (NATE)
- Professional Appearance
- Technician and Customer Communication

INTRODUCTION

The air conditioning and refrigeration industry has more professional organizations, trade associations, publications, and other related organizations than any other technical field. These groups provide the HVAC/R industry with the most up to date and current information. As a student, you should consider becoming involved with a student organization such as ACCA's student club, RSES student club, or ASHRAE's student club. These will give you an opportunity to begin developing extremely important business and technical contacts in the local HVAC/R industry. These contacts will serve you very well as you enter this profession.

As a professional in the trade it is to your advantage to maintain a close relationship with one or more of these professional organizations. Each group provides its members with the latest trends and most current technical information that give members a significant competitive edge. Most of the organizations provide ongoing technical and business training classes. Many of the classes cover the latest trends in equipment, regulations, codes and standards, local building regulations, and business practices. Being able as a member to participate in these ongoing educational opportunities will keep you at the leading edge of your new profession.

Each of these professional organizations has publications and many provide the industry with codes and standards. These publications would be an excellent addition to your technical library. Having a current and up to date technical library will help you provide your employer and customers with the best possible service while making you significantly more valuable to your employer. HVAC/R is an ongoing learning process for even the most skilled technician.

Figure 21-3-1 lists a number of these professional associations. Most have websites that are accessible through the world wide web and many have local, regional, and state chapters that you can become affiliated with.

TECH TIP

Once you enter the profession as an HVAC/R technician, all costs associated with taking additional classes, purchasing books, and membership dues may be tax deductible.

Many of the HVAC/R professional organizations have industrial trade shows. These shows are an excellent opportunity for you to see the various manufacturers' latest equipment, tools, supplies, and services. Some of the trade shows are local, others may be regional, some are national, and a few international, Figure 21-3-2.

PUBLICATIONS

Another excellent way of keeping up with the latest information in the HVAC/R field is by subscribing to one or more of the HVAC/R publications, Figure 21-3-3. Some of these publications are weekly, while others are monthly. They all contain well written articles specifically addressing HVAC/R industry concerns.

Some of the professional organizations have their own newsletters that are published and provided to their members. Some local and state chapters of these organizations have additional newsletters that are provided to their members.

Figure 21-3-1 United States organizations.

ACCA—Air Conditioning Contractors of America
AFEAS—Alternative Fluorocarbons Environmental Acceptability Study
AGA—American Gas Association
AHAM—Association of Home Appliance Manufacturers
AMCA—Air Movement & Control Association
ANSI—American National Standards Institute
ARI—Air-Conditioning & Refrigeration Institute
ARWI—Air-Conditioning & Refrigeration Wholesalers International
ASAE—American Society of Association Executives
ASHRAE—American Society of Heating, Refrigerating and Air-Conditioning Engineers
ABC—Associated Builders & Contractors
Boiler and Refrigeration Engineers
BOMA International—Building Owners and Managers Association
COBRA—The Association of Cogeneration
CDA—Copper Development Association
EEI—Edison Electric Institute
EPRI—Electrical Power Research Institute
EHCC—Eastern Heating & Cooling Council
Envirosense Consortium Inc.
EPEE—European Partnership for Energy and the Environment
FMI—The Food Marketing Institute
GAMA—Gas Appliance Manufacturers Association
Geothermal Heat Pump Consortium
GMA—Grocery Manufacturers of America
HARDI—Heating, Airconditioning and Refrigeration Distributors International
HPBA—Hearth, Patio and Barbeque Association
HI—Hydraulic Institute
HRAI—Heating, Refrigerating, & Air-Conditioning Institute of Canada
IDDBA—International Dairy, Deli, Bakery Association
IFPA—International Fresh-Cut Produce Association
IIAR—International Institute of Ammonia Refrigeration
IHACI—Institute of Heating and Air Conditioning Industries
ISA—The Instrumentation, Systems, and Automation Society
MCAA—Mechanical Contractors Association of America
MSCA—Mechanical Service Contractors of America
NACS—National Association of Convenience Stores
NADCA—National Air Duct Cleaners Association
NAHB—National Association of Home Builders
NAFEM—National Association of Food Equipment Manufacturers
NAM—National Association of Manufacturers
NATE—North American Technician Excellence Program
National Restaurant Association
NEMA—National Electrical Manufacturers Association
NFFS—Non-Ferrous Founders' Society
NIPC—National Inhalant Prevention Coalition
PHCC—Plumbing Heating Cooling Contractors Association
PIMA—Polyisocyanurate Insulation Manufacturers Association
PMA—Produce Marketing Association
RACCA—Refrigeration & Air Conditioning Contractors Association
RSES—Refrigeration Service Engineers Society
SMACNA—Sheet Metal and Air Conditioning Contractors' National Association
UL—Underwriters Laboratories Inc.

PROFESSIONAL CERTIFICATION

Every HVAC/R technician must become certified under the EPA Section 608 regulations. Compliance with these regulations regarding the management of refrigerants is mandatory for everyone in the trade. Following your successful completion of any and all of the appropriate levels, it remains your responsibility to stay current with any changes in these regulations. As unfair as you might think it is, you can be fined significantly for violating an EPA regulation pertaining

CANMET Energy Technology Centre
EPEE—European Partnership for Energy and the Environment
EUROVENT-CECOMAF—European Committee of Air Handling & Refrigerating Equipment Mfrs.
EUROVENT Certification
ICARMA—International Council of Air-Conditioning and Refrigeration Manufacturers' Associations
globalEDGE International Business Resources Desk
International Energy Agency (IEA) Heat Pump Centre
IIAR—International Institute of Ammonia Refrigeration
IIR—International Institute of Refrigeration
JRAIA—Japan Refrigeration and Air-Conditioning Industry Association
LATCO's Tools of the Trade (Latin American International Trade Sites)
Strategic Information for Trade Efficiency (UN-ETO)
Trade Compass
United Nations Environment Programme (UNEP)
USA*Engage
The World Bank
World Trade Point Federation
Worldclass

Figure 21-3-2 International organizations.

The Air-Conditioning, Heating & Refrigeration (ACHR) News
American School & University (AS&U)
APPLIANCE Magazine
Appliance Manufacturer
ASHRAE Journal
Buildings
Building Design and Construction
Consulting-Specifying Engineer
Contracting Business
Contractor Magazine
Energy User News
Engineered Systems
Facilities Net
Heating/Piping/AirConditioning (HPAC)
Japan Air Conditioning, Heating & Refrigeration News (JARN)
Japan Refrigeration & Air Conditioning News
Plant Engineering
RSES Journal
SchoolDesigns.com
Skylines
Supply House Times
Western HVAC/R News

Figure 21-3-3 Publications.

to refrigerants even if that regulation took effect after your successful completion of the certification. In addition, it is your sole responsibility to remember and follow all of the EPA regulations pertaining to refrigerant management. For that reason, it would be a good business practice to occasionally take a refresher course in EPA rules and regulations.

Industry Competency Exam (ICE)

The Air Conditioning and Refrigeration Institute (ARI) in conjunction with the Gas Appliance Manufacturers Association (GAMA); the Heating, Air Conditioning, and Refrigeration Distributors International (HARDI); Plumbing, Heating, Cooling Contractors Association (PHCC); and Refrigeration Service Engineers Society (RSES) have established a competency examination that is designed for students who have completed or nearly completed a technical training program. This examination is voluntary; but it does provide students leaving a training program, whether from high school, trade school, or community college, with an opportunity to evaluate their knowledge with an industry standardized test.

The ICE exam has been developed over the years with input from manufacturers, trade associations, instructors, and other industry experts. This exam can also provide your institution with an overall evaluation of its training program. Upon your successful completion of the ICE exam, your name, along with your school's, is published and made available to area contractors who might be looking for new skilled employees. In short, the successful completion of the ICE exam can put you well ahead of other graduates from programs not participating in the ICE examination.

ARI and its affiliates provide training institutions with incentives as encouragement to participate in the ICE examination by directing many of its manufacturing members' equipment donation programs toward the schools, institutes, and colleges that participate in the ICE program. These equipment donations can become an excellent source of the latest equipment you will be seeing in the field.

Skills USA VICA

Skills USA VICA is a vocational industrial club for students in high schools, trade schools, and community colleges. VICA clubs are open to students in all areas of specialty including HVAC/R. The national organization, Vocational Industrial Clubs of America, provides local chapters and students with many opportunities to develop leadership, citizenship, and interpersonal skills that are invaluable to the success of individuals in any profession. The Skills USA VICA logo is shown in Figure 21-3-4a.

In addition to the opportunities of individual professional growth, VICA sponsors regional, state, national, and international skills competitions. These skills competitions are available on the secondary and post-secondary levels. By participating in the competition students are given the opportunity to demonstrate their troubleshooting skills in diagnosing real world problems under the supervision of highly skilled individuals serving as judges.

The winners of local, state, and national competitions move on to the international skills competition held in a different country each year. At the international competition the best and brightest students from around the world compete to see who has the greatest knowledge and expertise in each of the technical areas including HVAC/R. Every student who participates in these skills Olympics at any level receives recognition. Students involved in this program are shown in Figure 21-3-4b,c. This recognition is invaluable to the students because such recognition is valued by prospective employers.

NICE TO KNOW

My name is Larry Jeffus, and I am the author of this book. As a high school student back in the early 1960s I joined my school's VICA club. In the VICA club I was given the opportunity to develop my leadership skills along with a level of self confidence. Both of these have served me well throughout my education and professional career. Much of what I have done I owe to the strong basis of business and professional skills I developed as a member and officer in my high school VICA club. Being an active member of VICA had such a dramatic impact on me that nearly 40 years later I still have fond memories of friends, advisors, and businesspeople I had the opportunity to work with as part of my club activities.

(a)

(b)

(c)

Figure 21-3-4 (a) Skills USA VICA logo; (b, c) Students competing in the national skills Olympics.

Care

The Council of Air Conditioning and Refrigeration Educators (CARE) is an organization that was founded in late 1990s by a group of air conditioning educators, counselors, and administrators responsible

Figure 21-3-5 NATE certified air conditioning technician checking the refrigerant charge in a residential system.

for the varies aspects of HVAC/R training. This group comprised individuals from secondary schools, post-secondary schools, and colleges representing institutions from various regions of the country.

The purpose of HVAC/R is to educate in order to meet or exceed the needs of the industry.

CARE membership is open to instructors, counselors, and administrators who are involved in some aspects of HVAC/R training. Through this organization individuals can come together to learn and share experiences for the betterment of the HVAC/R students and program.

North American Technical Excellence (NATE)

NATE was formed in the 1997 as a result of a concern expressed by many in the industry that there was not a way of distinguishing quality, highly skilled HVAC/R technicians from every other person working in the field. NATE was formed and still enjoys the support of all of the major equipment manufacturers, major component manufacturers, professional and trade associations, and the National Skills Standard Board. Figure 21-3-5 shows a technician with a patch showing NATE certification working on a system.

The NATE certification program has two levels for each of the five specialty areas. Installation certification is primarily designed for the technician who is involved with the installation or removal and installation of HVAC equipment. Installers assemble the system and fabricate the necessary connections to complete an efficient system. They also set up the operational controls under the supervision of a service technician. After the system is started, the installation technician records the readings for temperature, pressure, voltage, current, and any other measurements required by the manufacturer or service company for the completion of the warranty paperwork. There are no educational requirements for anyone taking the installation examination but it is recommended that you have at least one year of field experience.

The service technician must have all of the skills of the installation technician plus be able to work independently. Service technicians must be able to perform field diagnostics to determine the cause of system failures and to make the needed repairs. There are currently five specialty areas. They are:

1. Air conditioning
2. Air distribution
3. Gas heating
4. Heat pumps
5. Oil heating

NATE is currently working on additional areas of specialty certification. Areas currently being developed include certifications for indoor air quality, ground source heat pumps, and refrigeration.

In addition to the technical skills required to pass the various areas of specialization, each technician must take as part of the exam process a core exam. The core exam covers safety; tools; soft skills; principles of heat, heat transfer, and total comfort; as well as electrical questions. A well trained, experienced technician should be able to take and pass both the core and area specialty exams. A number of

International Association of Plumbing and Mechanical Officials (IAPMO)
International Building Officials and Code Administrators International, Inc. (BOCA)
International Code Council (ICC)
International Conference of Building Officials(ICBO)
International Fire Code Institute (IFCI)
National Conference of States on Building Codes & Standards (NCS/BCS)
National Fire Protection Association(NFPA)
Southern Building Code Congress International Inc. (SBCCI)

Figure 21-3-6 Code groups.

organizations provide NATE pretest tutorial classes to bring technicians' skill levels up to pass the exam. The exam passing score is 70% or better.

NICE TO KNOW

I have been involved with the NATE certification program from its early conceptual years through to the current program. It has been my experience in working with the people who represent the various industries by serving on the NATE test preparation committees are extremely knowledgeable and experienced in the HVAC/R skill areas. There have been tens of thousands of hours of volunteer time and travel invested by these individuals into the development of the NATE test and the continued updating of NATE test materials. The passion for the HVAC/R industry exhibited by the committee members is certainly commendable.

NATE has provided technicians with Knowledge Areas of Technician Expertise (KATEs) which give detailed outlines of all of the material that a technician can expect to be questioned about on an exam. These KATEs represent all of the knowledge and skills that a quality HVAC/R technician should have. The KATEs also help the prospective test candidate to focus on those areas that the industry feels are most important.

CODES AND STANDARDS

There are many organizations and agencies that provide codes that are used throughout the HVAC/R industry. Figure 21-3-6 lists many of these organizations. In addition to codes there are standards that are provided. The difference between a code and a standard is that codes often carry with them the force of law and standards do not. Organizations providing standards to the HVAC/R industry are listed in Figure 21-3-7.

The HVAC/R industry has many consumer groups that over the years have worked with the industry and industry leaders to help us provide consumers with the most efficient and effective service. Some of these consumer information groups are listed in Figure 21-3-8.

THE PROFESSIONAL TECHNICIAN'S APPEARANCE

As a service technician you are seen by the customer as a representative of your company. Customers often assume the appearance and professionalism of

Figure 21-3-7 Standards.

AFNOR—Association Francaise de Normalisation
ANSI—American National Standards Institute
BSI—British Standards Institution
CSA International—Canadian Standards Association
DIN—Deutches Institut für Normunge. V.
CEN— The European Committee for Standardization
CENELEC—The European Committee for Electrotechnical Standardization
EU—European Union
IEC—International Electrotechnical Commission
ISO—International Organization for Standardization
JIS—Japanese Industrial Standards
NSSN—National Standards Systems Network
SASO—Saudi Arabian Standards Organization
SES—Standards Engineering Society

Buying Energy Efficient Products (DOE Federal Energy Management Program)
Consumer Information Center (GSA)
North American Technician Excellence Program (NATE)
Consortium for Energy Efficiency (CEE)

Figure 21-3-8 Consumer information.

the technician is a reflection on the technician and company's technical skills. It is, therefore, important that you present yourself in a very clean and neat professional manner to the customer.

In some cases you might find that during the course of the day that your uniform becomes soiled. It is, therefore, a good idea to carry at least an extra clean shirt in the service van so that you can change if necessary.

If your company does not provide uniforms, you should dress appropriately. In some cases blue jeans and a jersey or denim shirt are acceptable. However, in other cases where you might be working in an office building, slacks and a shirt would be appropriate. Check with your employer to see what the company dress code is.

In addition to your clothing, you must have a clean and neat professional appearance. That means clean well kept hair, for men, clean shaven or well kept beard, and clean hands.

You must keep your service vehicle clean and neat. It is a rolling billboard for your company and it is important that it look sharp. A clean service van provides a better, more efficient work area making it easier to find tools and supplies. A clean and neat service vehicle also makes a better impression on the customer. Unless you have permission from residential customers, you should not park your service vehicle in their driveway. You do not want to be responsible for cleaning up an oil spill; and your vehicle can obstruct their access, which to some customers is very aggravating.

When you present yourself to a customer's door, you should have your hands in plain sight either at your side or holding your tools and clipboard. When the customer opens the door, if you take one step back, it will give the customer a greater sense of comfort. Many companies provide photo ID badges, and you must have yours clearly displayed. If it is necessary for you to enter a dwelling for service, ask permission from the homeowner before entering.

Working Neat

Some companies provide technicians with paper shoe covers to prevent tracking dirt into residences. If you suspect your shoes are dirty, either remove them or use the covers. You and your company can be responsible for cleaning the carpet if it becomes soiled. When working on an indoor furnace, place a drop cloth on the floor in front of the furnace so that any debris will be contained. When you are finished working in the furnace area, a small battery powered vacuum cleaner can be used to pick up loose dirt and debris in the area. Use a damp rag to wipe down any fingerprints that are on the equipment or that may be on the door or woodwork.

SERVICE TIP

On rare occasions you may be asked to perform service or replace a refrigeration or air conditioning system that is completely empty of refrigerant. You should make note of that on your service ticket. In addition at some point during your professional career, you will be servicing a refrigeration or air conditioning system when it or your equipment fails, resulting in an accidental release of refrigerant. On a rare occasion you may not be able to stem the refrigerant release in spite of every effort you may put forth. If such an accident occurs, you should note that on the service ticket. Include a full statement as to what happened and how you might prevent it from occurring in the future. Include the type and approximate quantity of refrigerant that was accidentally released.

TECH TIP

Cleaning up the equipment area in a residence is an excellent PR move. You should vacuum up all of the debris in the area including all the burnt matches that have been left there by the customer over the years as they may have lit the pilot. Many customers judge your work by appearance. All they may understand is neat and clean.

In many residences the air handler is located in an attic. If access to the unit is through a pulldown stair casing, make certain that any dirt and debris that falls from the stairs when it is pulled down is cleaned up. If you are going to be going in and out of the attic a number of times during the service, place a drop cloth on the floor below the stairs to catch any insulation or other debris that might fall on the floor. Do not attempt to carry a large number of tools up and down the stairs. This can cause you to be caught off balance and possibly drop your toolbox. In addition, most attic stairs have a weight limit that could be exceeded by you and a large toolbox. Attic stair treads, like any ladder, are strongest next to the side rail. For that reason, when you ascend and descend the stairs, place your foot as close as possible to the side next to the side rail to reduce the possibility of breaking the stairs.

Many customers have extensively landscaped their homes including the area around an outdoor condensing unit. Even though their landscaping may encroach within the manufacturer's recommended free air space around the unit, do not remove this landscaping; but notify the homeowner as to how it should be trimmed so that they or their landscaper can remove the vegetation. When working around vegetation, it is important that you be as careful as possible so as not to damage any plants. On very soft, wet ground your repeated trips to the service van can wear a path. To avoid this, each time you have to cross soft, wet ground, take a slightly different route when possible.

It is not considered to be a good practice to leave packaging and boxes left over from the installation of new equipment in the customer's trash receptacles. This material should be taken with you so as not to overload their receptacle.

TECH TIP

Working in air conditioning and refrigeration is often a very hot job. From time to time customers may offer you a glass of water, tea, or cold drink. Use your discretion on accepting these offers. However, it is never appropriate for you to accept a beer or other alcoholic beverage from a customer. Even if you do not open the drink but take it with you, you have significantly damaged your credibility with the customer and your boss. If you take a beer, even though you don't drink it, and there is a problem with the service you provided the customer, the first thing the customer is going to tell your boss and everyone else is that you were drinking on the job, in spite of the fact they gave you the beer. Do not take any alcoholic beverages that are offered by customers.

Technician and Customer Communication

A large part of the technician's job is to educate the customer as to the problem that was found and the available options for its repair. Tell the customer what failed, why it failed, and the options to fix the problem. Do not simply tell the customer that they need a part, and you are going to replace it.

SERVICE TIP

You should make a note on the service ticket that a system was leak checked with a nitrogen R-22 charge, so that if someone were to question whether or not you were venting refrigerant from the system, you would have supportive documentation.

Many customers would like an estimate of the job's cost. In some municipalities you are obligated under consumer protection laws to provide customers with such a quote. In some cases the quote must be in writing. If however, as part of your service you uncover a situation that could not have been foreseen or was not visible in your initial evaluation that will require additional work, immediately stop and inform the customer of the new problem. Do not simply make the repairs and expect the customer to accept the higher charge at the conclusion of the job.

TECH TIP

If you locate an additional problem or pending problem with a customer's system and after notifying the customer of your concerns they choose not to have you provide that repair, you should note that on the customer's invoice as part of your service call record keeping.

Under no circumstance should you tell a customer that the previous technician messed up their system when they did the last service. All that will result from such statements is the loss of faith the customer may have in you and your professionalism; but more importantly, you might find out that it was one of your colleagues who works for your company that did the last service job.

TECH TIP

Good professional technicians never knock the competition. Their skill and knowledge will set them apart from everyone else without the need to brag.

One of the primary complaints customers have on any service call is punctuality. It is not always possible to be at a job exactly when you anticipated being there. Earlier service calls may take more time than you initially estimated. However, as soon as you realize you are not going to make the schedule and as early as possible, let the customer know and give them an opportunity if they so choose, to reschedule. If you do reschedule, you must be at the next scheduled appointment on time. It is extremely important that you provide a very clear, clean, and legible invoice or bill to the customer. If it is necessary for you to get prices, look up information, or check your spelling, you should do all of that in your service vehicle and not in front of the customer. Even though you may not be a literary expert, customers do expect you to provide them with clear, concise, well written statements. If you have a problem writing clear and concise statements, you may want to invest in a PDA (Personal Data Assistant) device. There are many of these devices on the market and many can have spell check and writing programs installed.

TECH TIP

Some air conditioning and refrigeration work in the summer requires that the technician work in relatively hot environments such as attics or buildings without working air conditioning systems. It is important that you guard against dehydration by drinking large quantities of water or sports drinks. You must drink enough so that you have to use the restroom at least once every couple of hours throughout the day. In addition to dehydration, you may also develop kidney and bladder problems if you do not drink enough fluids while working in hot environments.

Carbonated beverages, milk, fruit juice, beer, and other beverages do not replace the body's electrolytes. Without replacing the electrolytes you can feel fatigue and may experience cramps. For that reason, only sports drinks with the essential electrolytes and water are recommended as your primary drinks when working in the heat.

UNIT 3—REVIEW QUESTIONS

1. What are three student organizations that students should consider becoming involved with?
2. Every HVAC/R technician must become certified under the ________ regulations.
3. Why is it beneficial to take the ICE exam?

TECH TALK

Tech 1. Hey, I got that part out of your service van that you wanted. It looked like you have a library back there.

Tech 2. I do. A pretty good one.

Tech 1. Why on earth would you want a library in your service van?

Tech 2. Well, there are a lot of times out on the job where you are not really sure that you remember exactly what a setting is supposed to be or what a code requirement might be. So rather than take on doing something that might not be right I can go out to my van and in just a minute pull our a code book or manufacturers spec, even pull out my textbook and look something up. Then I know I am right. I would rather know I am right than risk doing something wrong.

Tech 1. That sounds like a good idea. Maybe I should start building a library. What sort of books should I have?

Tech 2. A good textbook, a code book, as many of the manufacturer's equipment specification books as you can get, two or three supply catalogs from different parts houses, and a math book. In addition to those books I have a subscription to a weekly air conditioning heating and refrigeration newspaper to keep current.

Tech 1. I think I will start collecting those books today for my service van.

4. What is Skills USA VICA?
5. Why was NATE formed?
6. What are the five specialty areas of the NATE certification program?
7. What is KATE?
8. Why is a service technician's appearance and professionalism important?
9. A large part of the technician's job is to ________ the customer as to the problem that you found and the reason for your recommendation for its repair.
10. What is one of the primary complaints customers have on any service call?

APPENDIX A

Temperature Conversions

Temperature Conversion

°F	Temp. to be Converted	°C
-112.0	-80	-62.2
-110.2	-79	-61.7
-108.4	-78	-61.1
-106.6	-77	-60.6
-104.8	-76	-60.0
-103.0	-75	-59.4
-101.2	-74	-58.9
-99.4	-73	-58.3
-97.6	-72	-57.8
-95.8	-71	-57.2
-94.0	-70	-56.7
-92.2	-69	-56.1
-90.4	-68	-55.6
-88.6	-67	-55.0
-86.8	-66	-54.4
-85.0	-65	-53.9
-83.2	-64	-53.3
-81.4	-63	-52.8
-79.6	-62	-52.2
-77.8	-61	-51.7
-76.0	-60	-51.1
-74.2	-59	-50.6
-72.4	-58	-50.0
-70.6	-57	-49.4
-68.8	-56	-48.9
-67.0	-55	-48.3
-65.2	-54	-47.8
-63.4	-53	-47.2
-61.6	-52	-46.7
-59.8	-51	-46.1
-58.0	-50	-45.6
-56.2	-49	-45.0
-54.4	-48	-44.4
-52.6	-47	-43.9
-50.8	-46	-43.3
-49.0	-45	-42.8
-47.2	-44	-42.2
-45.4	-43	-41.7
-43.6	-42	-41.1
-41.8	-41	-40.6
-40.0	-40	-40.0
-38.2	-39	-39.4
-36.4	-38	-38.9
-34.6	-37	-38.3
-32.8	-36	-37.8
-31.0	-35	-37.2
-29.2	-34	-36.7
-27.4	-33	-36.1
-25.6	-32	-35.6
-23.8	-31	-35.0
-22.0	-30	-34.4
-20.2	-29	-33.9
-18.4	-28	-33.3
-16.6	-27	-32.8
-14.8	-26	-32.2
-13.0	-25	-31.7
-11.2	-24	-31.1
-9.4	-23	-30.6
-7.6	-22	-30.0
-5.8	-21	-29.4
-4.0	-20	-28.9
-2.2	-19	-28.3
-0.4	-18	-27.8
1.4	-17	-27.2
3.2	-16	-26.7
5.0	-15	-26.1
6.8	-14	-25.6
8.6	-13	-25.0
10.4	-12	-24.4
12.2	-11	-23.9
14.0	-10	-23.3
15.8	-9	-22.8
17.6	-8	-22.2
19.4	-7	-21.7
21.2	-6	-21.1
23.0	-5	-20.6
24.8	-4	-20.0
26.6	-3	-19.4
28.4	-2	-18.9
30.2	-1	-18.3
32.0	0	-17.8
33.8	1	-17.2
35.6	2	-16.7
37.4	3	-16.1
39.2	4	-15.6
41.0	5	-15.0
42.8	6	-14.4
44.6	7	-13.9
46.4	8	-13.3
48.2	9	-12.8
50.0	10	-12.2
51.8	11	-11.7
53.6	12	-11.1
55.4	13	-10.6
57.2	14	-10.0
59.0	15	-9.4
60.8	16	-8.9
62.6	17	-8.3
64.4	18	-7.8
66.2	19	-7.2
68.0	20	-6.7
69.8	21	-6.1
71.6	22	-5.6
73.4	23	-5.0
75.2	24	-4.4
77.0	25	-3.9
78.8	26	-3.3
80.6	27	-2.8
82.4	28	-2.2
84.2	29	-1.7
86.0	30	-1.1
87.8	31	-0.6
89.6	32	0.0
91.4	33	0.6
93.2	34	1.1
95.0	35	1.7
96.8	36	2.2
98.6	37	2.8
100.4	38	3.3
102.2	39	3.9
104.0	40	4.4
105.8	41	5.0
107.6	42	5.6
109.4	43	6.1
111.2	44	6.7
113.0	45	7.2
114.8	46	7.8
116.6	47	8.3
118.4	48	8.9
120.2	49	9.4
122.0	50	10.0
123.8	51	10.6
125.6	52	11.1
127.4	53	11.7
129.2	54	12.2
131.0	55	12.8
132.8	56	13.3
134.6	57	13.9
136.4	58	14.4
138.2	59	15.0
140.0	60	15.6
141.8	61	16.1
143.6	62	16.7
145.4	63	17.2
147.2	64	17.8
149.0	65	18.3
150.8	66	18.9
152.6	67	19.4
154.4	68	20.0
156.2	69	20.6
158.0	70	21.1
159.8	71	21.7
161.8	72	22.2
163.4	73	22.8
165.2	74	23.3
167.0	75	23.9
168.8	76	24.4
170.6	77	25.0
172.4	78	25.6
174.2	79	26.1
176.0	80	26.7
177.8	81	27.2
179.6	82	27.8
181.4	83	28.3
183.2	84	28.9
185.0	85	29.4
186.8	86	30.0
188.6	87	30.6
190.4	88	31.1
192.2	89	31.7
194.0	90	32.2
195.8	91	32.8
197.6	92	33.3
199.4	93	33.9
201.2	94	34.4
203.0	95	35.0
204.8	96	35.6
206.6	97	36.1
208.4	98	36.7
210.2	99	37.2
212.0	100	37.8
213.8	101	38.3
215.6	102	38.9
217.4	103	39.4
219.2	104	40.0
221.0	105	40.6
222.8	106	41.1
224.6	107	41.7
226.4	108	42.2
228.2	109	42.8
230.0	110	43.3
231.8	111	43.9
233.6	112	44.4
235.4	113	45.0
237.2	114	45.6

APPENDIX B

Properties of Refrigerants

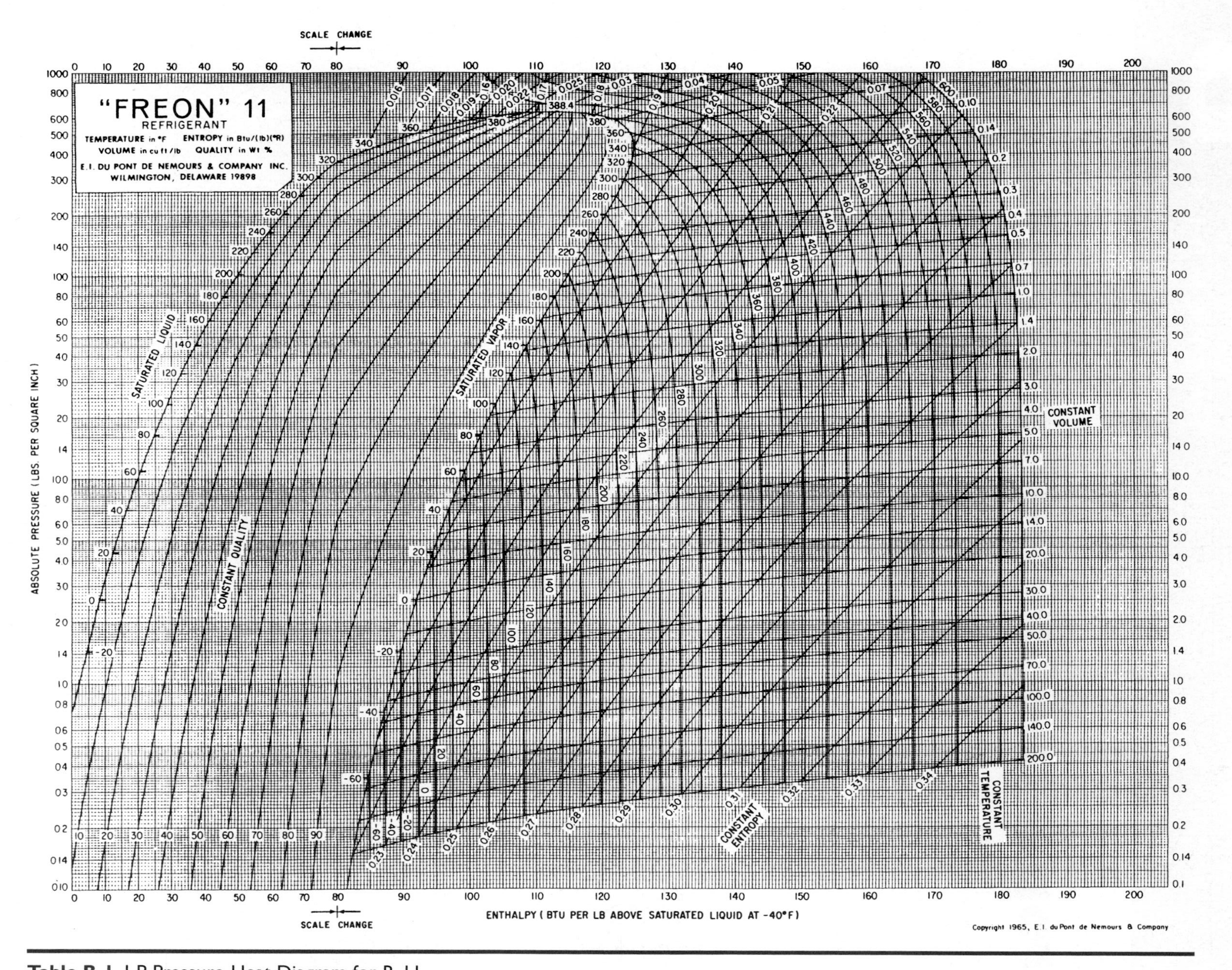

Table B-1 I-P Pressure-Heat Diagram for R-11

Tables B-1 through B-12 courtesy of DuPont Fluoroproducts.

TEMP.	PRESSURE		VOLUME cu ft/lb		DENSITY lb/cu ft		ENTHALPY Btu/lb			ENTROPY Btu/(lb)(°R)		TEMP.
°F	PSIA	PSIG	LIQUID v_f	VAPOR v_g	LIQUID $1/v_f$	VAPOR $1/v_g$	LIQUID h_f	LATENT h_{fg}	VAPOR h_g	LIQUID s_f	VAPOR s_g	°F
− 58	0.38715	29.13297*	0.0097455	80.831	102.61	0.012371	− 3.634	88.714	85.080	−0.008848	0.21201	− 58
− 57	0.40215	29.10241*	0.0097524	78.002	102.54	0.012820	− 3.432	88.628	85.197	−0.008346	0.21176	− 57
− 56	0.41765	29.07086*	0.0097594	75.288	102.46	0.013282	− 3.230	88.543	85.313	−0.007845	0.21150	− 56
− 55	0.43365	29.03829*	0.0097664	72.684	102.39	0.013758	− 3.028	88.458	85.430	−0.007346	0.21125	− 55
− 54	0.45016	29.00466*	0.0097735	70.185	102.32	0.014248	− 2.826	88.373	85.547	−0.006848	0.21100	− 54
− 53	0.46720	28.96997*	0.0097805	67.785	102.24	0.014752	− 2.624	88.288	85.664	−0.006351	0.21075	− 53
− 52	0.48479	28.93417*	0.0097875	65.482	102.17	0.015271	− 2.423	88.203	85.781	−0.005855	0.21050	− 52
− 51	0.50292	28.89724*	0.0097946	63.269	102.10	0.015806	− 2.221	88.119	85.898	−0.005361	0.21026	− 51
− 50	0.52163	28.85916*	0.0098017	61.144	102.02	0.016355	− 2.019	88.034	86.015	−0.004868	0.21002	− 50
− 49	0.54092	28.81988*	0.0098088	59.102	101.95	0.016920	− 1.817	87.950	86.133	−0.004376	0.20979	− 49
− 48	0.56080	28.77940*	0.0098159	57.139	101.88	0.017501	− 1.615	87.865	86.250	−0.003885	0.20955	− 48
− 47	0.58130	28.73766*	0.0098230	55.253	101.80	0.018099	− 1.413	87.781	86.368	−0.003395	0.20932	− 47
− 46	0.60243	28.69466*	0.0098301	53.439	101.73	0.018713	− 1.211	87.697	86.485	−0.002907	0.20909	− 46
− 45	0.62419	28.65034*	0.0098373	51.695	101.65	0.019344	− 1.010	87.612	86.603	−0.002419	0.20886	− 45
− 44	0.64661	28.60469*	0.0098444	50.017	101.58	0.019993	− 0.808	87.528	86.721	−0.001933	0.20864	− 44
− 43	0.66970	28.55768*	0.0098516	48.404	101.51	0.020660	− 0.606	87.444	86.839	−0.001448	0.20842	− 43
− 42	0.69348	28.50926*	0.0098588	46.851	101.43	0.021344	− 0.404	87.360	86.957	−0.000964	0.20820	− 42
− 41	0.71797	28.45941*	0.0098660	45.356	101.36	0.022048	− 0.202	87.276	87.075	−0.000482	0.20798	− 41
− 40	0.74317	28.40809*	0.0098733	43.917	101.28	0.022770	0.000	87.193	87.193	0.000000	0.20776	− 40
− 39	0.76911	28.35527*	0.0098805	42.532	101.21	0.023511	0.202	87.109	87.311	0.000481	0.20755	− 39
− 38	0.79581	28.30092*	0.0098878	41.198	101.14	0.024273	0.404	87.025	87.429	0.000960	0.20734	− 38
− 37	0.82327	28.24500*	0.0098951	39.914	101.06	0.025054	0.606	86.942	87.548	0.001438	0.20713	− 37
− 36	0.85153	28.18748*	0.0099023	38.676	100.99	0.025856	0.808	86.858	87.666	0.001915	0.20693	− 36
− 35	0.88059	28.12832*	0.0099097	37.483	100.91	0.026679	1.010	86.775	87.784	0.002392	0.20673	− 35
− 34	0.91047	28.06748*	0.0099170	36.333	100.84	0.027523	1.212	86.691	87.903	0.002867	0.20652	− 34
− 33	0.94119	28.00493*	0.0099243	35.225	100.76	0.028389	1.414	86.608	88.022	0.003341	0.20633	− 33
− 32	0.97277	27.94063*	0.0099317	34.157	100.69	0.029277	1.616	86.524	88.140	0.003814	0.20613	− 32
− 31	1.0052	27.8745*	0.0099391	33.127	100.61	0.030187	1.818	86.441	88.259	0.004285	0.20593	− 31
− 30	1.0386	27.8066*	0.0099464	32.133	100.54	0.031120	2.020	86.358	88.378	0.004756	0.20574	− 30
− 29	1.0729	27.7369*	0.0099539	31.175	100.46	0.032077	2.222	86.275	88.497	0.005226	0.20555	− 29
− 28	1.1081	27.6652*	0.0099613	30.250	100.39	0.033058	2.425	86.192	88.616	0.005695	0.20536	− 28
− 27	1.1442	27.5915*	0.0099687	29.357	100.31	0.034063	2.627	86.108	88.735	0.006162	0.20518	− 27
− 26	1.1814	27.5159*	0.0099762	28.496	100.24	0.035093	2.829	86.025	88.854	0.006629	0.20499	− 26
− 25	1.2195	27.4383*	0.0099837	27.664	100.16	0.036148	3.031	85.942	88.974	0.007095	0.20481	− 25
− 24	1.2586	27.3586*	0.0099911	26.861	100.09	0.037228	3.233	85.859	89.093	0.007559	0.20463	− 24
− 23	1.2988	27.2768*	0.0099987	26.086	100.01	0.038335	3.436	85.776	89.212	0.008023	0.20446	− 23
− 22	1.3401	27.1928*	0.010006	25.337	99.938	0.039468	3.638	85.694	89.331	0.008485	0.20428	− 22
− 21	1.3824	27.1067*	0.010014	24.613	99.863	0.040628	3.840	85.611	89.451	0.008947	0.20411	− 21
− 20	1.4258	27.0183*	0.010021	23.914	99.788	0.041816	4.043	85.528	89.570	0.009408	0.20393	− 20
− 19	1.4703	26.9276*	0.010029	23.239	99.712	0.043032	4.245	85.445	89.690	0.009867	0.20376	− 19
− 18	1.5160	26.8346*	0.010036	22.586	99.637	0.044276	4.448	85.362	89.810	0.010326	0.20360	− 18
− 17	1.5629	26.7392*	0.010044	21.954	99.561	0.045549	4.650	85.279	89.929	0.010783	0.20343	− 17
− 16	1.6109	26.6414*	0.010052	21.344	99.486	0.046852	4.852	85.197	90.049	0.011240	0.20327	− 16
− 15	1.6602	26.5411*	0.010059	20.754	99.410	0.048184	5.055	85.114	90.169	0.011696	0.20310	− 15
− 14	1.7106	26.4383*	0.010067	20.183	99.334	0.049547	5.258	85.031	90.289	0.012151	0.20294	− 14
− 13	1.7624	26.3330*	0.010075	19.630	99.259	0.050941	5.460	84.948	90.408	0.012605	0.20279	− 13
− 12	1.8154	26.2250*	0.010082	19.096	99.183	0.052367	5.663	84.866	90.528	0.013058	0.20263	− 12
− 11	1.8697	26.1144*	0.010090	18.579	99.107	0.053824	5.865	84.783	90.648	0.013510	0.20247	− 11
− 10	1.9254	26.0011*	0.010098	18.079	99.031	0.055314	6.068	84.700	90.768	0.013961	0.20232	− 10
− 9	1.9824	25.8850*	0.010106	17.594	98.955	0.056837	6.271	84.617	90.888	0.014411	0.20217	− 9
− 8	2.0408	25.7661*	0.010113	17.125	98.879	0.058393	6.474	84.535	91.008	0.014860	0.20202	− 8
− 7	2.1006	25.6443*	0.010121	16.671	98.803	0.059984	6.677	84.452	91.129	0.015308	0.20187	− 7
− 6	2.1618	25.5197*	0.010129	16.231	98.726	0.061609	6.879	84.369	91.249	0.015756	0.20173	− 6
− 5	2.2245	25.3920*	0.010137	15.805	98.650	0.063270	7.082	84.287	91.369	0.016202	0.20158	− 5
− 4	2.2887	25.2614*	0.010145	15.393	98.574	0.064966	7.285	84.204	91.489	0.016648	0.20144	− 4

*Inches of mercury below one atmosphere

Table B-2 I-P Properties of Saturated Liquid and Vapor for R-11

TEMP.	PRESSURE		VOLUME cu ft/lb		DENSITY lb/cu ft		ENTHALPY Btu/lb			ENTROPY Btu/(lb)(°R)		TEMP.
°F	PSIA	PSIG	LIQUID v_f	VAPOR v_g	LIQUID $1/v_f$	VAPOR $1/v_g$	LIQUID h_f	LATENT h_{fg}	VAPOR h_g	LIQUID s_f	VAPOR s_g	°F
−3	2.3544	25.1277*	0.010153	14.993	98.497	0.066699	7.488	84.121	91.609	0.017093	0.20130	−3
−2	2.4216	24.9908*	0.010160	14.605	98.421	0.068468	7.691	84.038	91.730	0.017536	0.20116	−2
−1	2.4904	24.8508*	0.010168	14.230	98.345	0.070275	7.894	83.956	91.850	0.017979	0.20102	−1
0	2.5607	24.7076*	0.010176	13.866	98.268	0.072119	8.098	83.873	91.970	0.018421	0.20088	0
1	2.6327	24.5610*	0.010184	13.513	98.191	0.074003	8.301	83.790	92.091	0.018863	0.20075	1
2	2.7063	24.4112*	0.010192	13.171	98.115	0.075925	8.504	83.707	92.211	0.019303	0.20062	2
3	2.7816	24.2578*	0.010200	12.839	98.038	0.077887	8.707	83.624	92.332	0.019743	0.20049	3
4	2.8586	24.1011*	0.010208	12.517	97.961	0.079890	8.910	83.542	92.452	0.020181	0.20036	4
5	2.9373	23.9408*	0.010216	12.205	97.884	0.081933	9.114	83.459	92.572	0.020619	0.20023	5
6	3.0178	23.7770*	0.010224	11.902	97.807	0.084018	9.317	83.376	92.693	0.021056	0.20010	6
7	3.1000	23.6095*	0.010232	11.608	97.730	0.086145	9.521	83.293	92.814	0.021492	0.19998	7
8	3.1841	23.4383*	0.010240	11.323	97.653	0.088315	9.724	83.210	92.934	0.021927	0.19985	8
9	3.2700	23.2634*	0.010248	11.046	97.576	0.090528	9.928	83.127	93.055	0.022362	0.19973	9
10	3.3578	23.0847*	0.010257	10.778	97.498	0.092785	10.131	83.044	93.175	0.022795	0.19961	10
11	3.4475	22.9020*	0.010265	10.517	97.421	0.095087	10.335	82.961	93.296	0.023228	0.19949	11
12	3.5391	22.7155*	0.010273	10.263	97.344	0.097433	10.539	82.878	93.416	0.023660	0.19937	12
13	3.6327	22.5249*	0.010281	10.017	97.266	0.099826	10.742	82.795	93.537	0.024091	0.19925	13
14	3.7283	22.3303*	0.010289	9.7785	97.189	0.10226	10.946	82.712	93.658	0.024521	0.19914	14
15	3.8259	22.1316*	0.010297	9.5465	97.111	0.10475	11.150	82.628	93.778	0.024951	0.19903	15
16	3.9256	21.9286*	0.010306	9.3210	97.033	0.10728	11.354	82.545	93.899	0.025379	0.19891	16
17	4.0274	21.7214*	0.010314	9.1019	96.956	0.10987	11.558	82.462	94.020	0.025807	0.19880	17
18	4.1313	21.5099*	0.010322	8.8890	96.878	0.11250	11.762	82.379	94.140	0.026234	0.19869	18
19	4.2373	21.2940*	0.010331	8.6821	96.800	0.11518	11.966	82.295	94.261	0.026661	0.19859	19
20	4.3456	21.0736*	0.010339	8.4810	96.722	0.11791	12.170	82.212	94.382	0.027086	0.19848	20
21	4.4560	20.8487*	0.010347	8.2855	96.644	0.12069	12.374	82.128	94.502	0.027511	0.19837	21
22	4.5687	20.6192*	0.010356	8.0954	96.566	0.12353	12.578	82.045	94.623	0.027935	0.19827	22
23	4.6837	20.3851*	0.010364	7.9106	96.488	0.12641	12.783	81.961	94.744	0.028358	0.19817	23
24	4.8010	20.1462*	0.010372	7.7308	96.409	0.12935	12.987	81.878	94.864	0.028781	0.19806	24
25	4.9207	19.9026*	0.010381	7.5560	96.331	0.13235	13.191	81.794	94.985	0.029202	0.19796	25
26	5.0428	19.6541*	0.010389	7.3859	96.253	0.13539	13.396	81.710	95.106	0.029623	0.19787	26
27	5.1672	19.4006*	0.010398	7.2205	96.174	0.13850	13.600	81.626	95.227	0.030043	0.19777	27
28	5.2942	19.1422*	0.010406	7.0595	96.096	0.14165	13.805	81.542	95.347	0.030463	0.19767	28
29	5.4236	18.8786*	0.010415	6.9028	96.017	0.14487	14.009	81.458	95.468	0.030881	0.19758	29
30	5.5556	18.6100*	0.010423	6.7503	95.938	0.14814	14.214	81.374	95.588	0.031299	0.19748	30
31	5.6901	18.3361*	0.010432	6.6019	95.860	0.15147	14.419	81.290	95.709	0.031716	0.19739	31
32	5.8273	18.0569*	0.010441	6.4574	95.781	0.15486	14.624	81.206	95.830	0.032133	0.19730	32
33	5.9670	17.7723*	0.010449	6.3168	95.702	0.15831	14.828	81.122	95.950	0.032548	0.19721	33
34	6.1094	17.4823*	0.010458	6.1798	95.623	0.16182	15.033	81.038	96.071	0.032963	0.19712	34
35	6.2546	17.1868*	0.010466	6.0465	95.544	0.16539	15.238	80.953	96.192	0.033377	0.19703	35
36	6.4025	16.8857*	0.010475	5.9166	95.464	0.16902	15.443	80.869	96.312	0.033791	0.19694	36
37	6.5532	16.5789*	0.010484	5.7901	95.385	0.17271	15.648	80.784	96.433	0.034204	0.19686	37
38	6.7067	16.2664*	0.010493	5.6668	95.306	0.17647	15.853	80.700	96.553	0.034616	0.19677	38
39	6.8630	15.9480*	0.010501	5.5468	95.226	0.18029	16.059	80.615	96.674	0.035027	0.19669	39
40	7.0223	15.6238*	0.010510	5.4298	95.147	0.18417	16.264	80.530	96.794	0.035437	0.19660	40
41	7.1844	15.2936*	0.010519	5.3158	95.067	0.18812	16.469	80.445	96.915	0.035847	0.19652	41
42	7.3496	14.9574*	0.010528	5.2047	94.988	0.19214	16.675	80.361	97.035	0.036256	0.19644	42
43	7.5178	14.6150*	0.010537	5.0964	94.908	0.19622	16.880	80.276	97.156	0.036665	0.19636	43
44	7.6890	14.2664*	0.010545	4.9908	94.828	0.20037	17.086	80.190	97.276	0.037073	0.19628	44
45	7.8633	13.9115*	0.010554	4.8879	94.748	0.20459	17.291	80.105	97.396	0.037480	0.19621	45
46	8.0407	13.5502*	0.010563	4.7876	94.668	0.20887	17.497	80.020	97.517	0.037886	0.19613	46
47	8.2213	13.1825*	0.010572	4.6897	94.588	0.21323	17.702	79.935	97.637	0.038292	0.19606	47
48	8.4051	12.8083*	0.010581	4.5943	94.508	0.21766	17.908	79.849	97.757	0.038697	0.19598	48
49	8.5922	12.4274*	0.010590	4.5012	94.428	0.22216	18.114	79.764	97.877	0.039101	0.19591	49
50	8.7825	12.0399*	0.010599	4.4105	94.347	0.22673	18.320	79.678	97.998	0.039505	0.19584	50
51	8.9762	11.6456*	0.010608	4.3219	94.267	0.23138	18.526	79.592	98.118	0.039907	0.19577	51

*Inches of mercury below one atmosphere

Table B-2 (continued)

TEMP.	PRESSURE		VOLUME cu ft/lb		DENSITY lb/cu ft		ENTHALPY Btu/lb			ENTROPY Btu/(lb)(°R)		TEMP.
°F	PSIA	PSIG	LIQUID v_f	VAPOR v_g	LIQUID $1/v_f$	VAPOR $1/v_g$	LIQUID h_f	LATENT h_{fg}	VAPOR h_g	LIQUID s_f	VAPOR s_g	°F
52	9.1733	11.2444*	0.010617	4.2355	94.186	0.23610	18.732	79.506	98.238	0.040310	0.19570	52
53	9.3737	10.8362*	0.010626	4.1512	94.106	0.24089	18.938	79.420	98.358	0.040711	0.19563	53
54	9.5776	10.4211*	0.010635	4.0690	94.025	0.24576	19.144	79.334	98.478	0.041112	0.19556	54
55	9.7850	9.9988*	0.010645	3.9887	93.944	0.25071	19.350	79.248	98.598	0.041512	0.19549	55
56	9.9960	9.5693*	0.010654	3.9103	93.864	0.25573	19.556	79.162	98.718	0.041912	0.19542	56
57	10.210	9.133*	0.010663	3.8338	93.783	0.26083	19.763	79.075	98.838	0.042311	0.19536	57
58	10.429	8.688*	0.010672	3.7592	93.702	0.26602	19.969	78.989	98.958	0.042709	0.19529	58
59	10.650	8.237*	0.010681	3.6863	93.620	0.27128	20.176	78.902	99.078	0.043107	0.19523	59
60	10.876	7.778*	0.010691	3.6151	93.539	0.27662	20.382	78.816	99.198	0.043504	0.19517	60
61	11.105	7.311*	0.010700	3.5456	93.458	0.28204	20.589	78.729	99.317	0.043900	0.19511	61
62	11.338	6.836*	0.010709	3.4777	93.377	0.28754	20.795	78.642	99.437	0.044295	0.19505	62
63	11.575	6.354*	0.010719	3.4114	93.295	0.29313	21.002	78.555	99.557	0.044690	0.19499	63
64	11.816	5.864*	0.010728	3.3467	93.213	0.29880	21.209	78.468	99.676	0.045085	0.19493	64
65	12.061	5.366*	0.010737	3.2834	93.132	0.30456	21.416	78.380	99.796	0.045478	0.19487	65
66	12.309	4.859*	0.010747	3.2216	93.050	0.31040	21.623	78.293	99.916	0.045872	0.19481	66
67	12.562	4.345*	0.010756	3.1612	92.968	0.31633	21.830	78.205	100.035	0.046264	0.19475	67
68	12.819	3.822*	0.010766	3.1022	92.886	0.32235	22.037	78.118	100.154	0.046656	0.19470	68
69	13.080	3.291*	0.010775	3.0445	92.804	0.32846	22.244	78.030	100.274	0.047047	0.19464	69
70	13.345	2.752*	0.010785	2.9882	92.722	0.33465	22.451	77.942	100.393	0.047437	0.19459	70
71	13.614	2.204*	0.010795	2.9331	92.640	0.34094	22.658	77.854	100.512	0.047827	0.19454	71
72	13.887	1.647*	0.010804	2.8792	92.557	0.34731	22.866	77.766	100.631	0.048217	0.19448	72
73	14.165	1.081*	0.010814	2.8266	92.475	0.35378	23.073	77.678	100.751	0.048605	0.19443	73
74	14.447	0.507*	0.010823	2.7751	92.392	0.36034	23.280	77.589	100.870	0.048993	0.19438	74
75	14.733	0.037	0.010833	2.7248	92.310	0.36700	23.488	77.501	100.989	0.049381	0.19433	75
76	15.024	0.328	0.010843	2.6756	92.227	0.37375	23.696	77.412	101.108	0.049768	0.19428	76
77	15.319	0.623	0.010853	2.6275	92.144	0.38059	23.903	77.323	101.226	0.050154	0.19423	77
78	15.619	0.923	0.010862	2.5804	92.061	0.38754	24.111	77.234	101.345	0.050540	0.19419	78
79	15.924	1.228	0.010872	2.5344	91.978	0.39458	24.319	77.145	101.464	0.050925	0.19414	79
80	16.233	1.537	0.010882	2.4893	91.895	0.40172	24.527	77.056	101.583	0.051309	0.19409	80
81	16.546	1.850	0.010892	2.4453	91.812	0.40895	24.735	76.967	101.701	0.051693	0.19405	81
82	16.865	2.169	0.010902	2.4021	91.728	0.41629	24.943	76.877	101.820	0.052076	0.19400	82
83	17.188	2.492	0.010912	2.3600	91.645	0.42374	25.151	76.787	101.938	0.052459	0.19396	83
84	17.516	2.820	0.010922	2.3187	91.561	0.43128	25.359	76.698	102.057	0.052841	0.19391	84
85	17.848	3.152	0.010932	2.2783	91.477	0.43893	25.567	76.608	102.175	0.053222	0.19387	85
86	18.186	3.490	0.010942	2.2388	91.394	0.44668	25.776	76.518	102.293	0.053603	0.19383	86
87	18.529	3.833	0.010952	2.2001	91.310	0.45453	25.984	76.427	102.411	0.053983	0.19379	87
88	18.876	4.180	0.010962	2.1622	91.226	0.46250	26.192	76.337	102.529	0.054363	0.19375	88
89	19.229	4.533	0.010972	2.1251	91.142	0.47057	26.401	76.246	102.647	0.054742	0.19371	89
90	19.587	4.891	0.010982	2.0888	91.057	0.47874	26.610	76.156	102.765	0.055121	0.19367	90
91	19.950	5.254	0.010992	2.0532	90.973	0.48703	26.818	76.065	102.883	0.055499	0.19363	91
92	20.318	5.622	0.011002	2.0184	90.889	0.49543	27.027	75.974	103.001	0.055876	0.19359	92
93	20.691	5.995	0.011013	1.9844	90.804	0.50394	27.236	75.883	103.119	0.056253	0.19355	93
94	21.070	6.374	0.011023	1.9510	90.719	0.51256	27.445	75.791	103.236	0.056629	0.19352	94
95	21.454	6.758	0.011033	1.9183	90.635	0.52130	27.654	75.700	103.354	0.057005	0.19348	95
96	21.843	7.147	0.011044	1.8863	90.550	0.53015	27.863	75.608	103.471	0.057380	0.19345	96
97	22.238	7.542	0.011054	1.8549	90.465	0.53912	28.072	75.516	103.588	0.057755	0.19341	97
98	22.638	7.942	0.011064	1.8241	90.379	0.54820	28.281	75.424	103.706	0.058129	0.19338	98
99	23.044	8.348	0.011075	1.7940	90.294	0.55740	28.491	75.332	103.823	0.058502	0.19334	99
100	23.456	8.760	0.011085	1.7645	90.209	0.56672	28.700	75.240	103.940	0.058875	0.19331	100
101	23.873	9.177	0.011096	1.7356	90.123	0.57616	28.909	75.147	104.057	0.059248	0.19328	101
102	24.296	9.600	0.011106	1.7073	90.038	0.58572	29.119	75.055	104.174	0.059620	0.19325	102
103	24.724	10.028	0.011117	1.6795	89.952	0.59540	29.329	74.962	104.290	0.059991	0.19322	103
104	25.159	10.463	0.011128	1.6523	89.866	0.60521	29.538	74.869	104.407	0.060362	0.19319	104
105	25.599	10.903	0.011138	1.6256	89.780	0.61514	29.748	74.776	104.524	0.060732	0.19316	105
106	26.045	11.349	0.011149	1.5995	89.694	0.62519	29.958	74.682	104.640	0.061102	0.19313	106

*Inches of mercury below one atmosphere

Table B-2 (continued)

TEMP.	PRESSURE		VOLUME cu ft/lb		DENSITY lb/cu ft		ENTHALPY Btu/lb			ENTROPY Btu/(lb)(°R)		TEMP.
°F	PSIA	PSIG	LIQUID v_f	VAPOR v_g	LIQUID $1/v_f$	VAPOR $1/v_g$	LIQUID h_f	LATENT h_{fg}	VAPOR h_g	LIQUID s_f	VAPOR s_g	°F
107	26.497	11.801	0.011160	1.5739	89.608	0.63538	30.168	74.589	104.757	0.061471	0.19310	107
108	26.956	12.260	0.011170	1.5487	89.522	0.64569	30.378	74.495	104.873	0.061840	0.19307	108
109	27.420	12.724	0.011181	1.5241	89.435	0.65612	30.588	74.401	104.989	0.062208	0.19304	109
110	27.890	13.194	0.011192	1.4999	89.349	0.66669	30.798	74.307	105.105	0.062575	0.19301	110
111	28.367	13.671	0.011203	1.4763	89.262	0.67739	31.009	74.213	105.221	0.062942	0.19299	111
112	28.850	14.154	0.011214	1.4530	89.175	0.68822	31.219	74.118	105.337	0.063309	0.19296	112
113	29.339	14.643	0.011225	1.4302	89.088	0.69919	31.429	74.023	105.453	0.063675	0.19294	113
114	29.834	15.138	0.011236	1.4079	89.001	0.71029	31.640	73.929	105.569	0.064041	0.19291	114
115	30.336	15.640	0.011247	1.3860	88.914	0.72152	31.851	73.833	105.684	0.064406	0.19289	115
116	30.844	16.148	0.011258	1.3645	88.826	0.73289	32.061	73.738	105.800	0.064770	0.19286	116
117	31.359	16.663	0.011269	1.3434	88.739	0.74440	32.272	73.643	105.915	0.065134	0.19284	117
118	31.880	17.184	0.011280	1.3227	88.651	0.75605	32.483	73.547	106.030	0.065498	0.19281	118
119	32.408	17.712	0.011291	1.3024	88.564	0.76784	32.694	73.451	106.145	0.065861	0.19279	119
120	32.943	18.247	0.011303	1.2824	88.476	0.77977	32.905	73.355	106.260	0.066223	0.19277	120
121	33.484	18.788	0.011314	1.2629	88.388	0.79184	33.116	73.259	106.375	0.066585	0.19275	121
122	34.032	19.336	0.011325	1.2437	88.300	0.80406	33.328	73.162	106.490	0.066947	0.19273	122
123	34.587	19.891	0.011336	1.2249	88.211	0.81642	33.539	73.066	106.605	0.067308	0.19271	123
124	35.149	20.453	0.011348	1.2064	88.123	0.82893	33.751	72.969	106.719	0.067669	0.19269	124
125	35.718	21.022	0.011359	1.1882	88.034	0.84159	33.962	72.872	106.834	0.068029	0.19267	125
126	36.294	21.598	0.011371	1.1704	87.946	0.85440	34.174	72.774	106.948	0.068388	0.19265	126
127	36.877	22.181	0.011382	1.1529	87.857	0.86735	34.385	72.677	107.062	0.068747	0.19263	127
128	37.467	22.771	0.011394	1.1358	87.768	0.88046	34.597	72.579	107.176	0.069106	0.19261	128
129	38.064	23.368	0.011405	1.1189	87.679	0.89372	34.809	72.481	107.290	0.069464	0.19259	129
130	38.668	23.972	0.011417	1.1024	87.589	0.90714	35.021	72.383	107.404	0.069822	0.19257	130
131	39.280	24.584	0.011429	1.0861	87.500	0.92071	35.233	72.284	107.518	0.070179	0.19256	131
132	39.899	25.203	0.011440	1.0702	87.410	0.93443	35.446	72.186	107.631	0.070536	0.19254	132
133	40.525	25.829	0.011452	1.0545	87.321	0.94832	35.658	72.087	107.745	0.070892	0.19252	133
134	41.159	26.463	0.011464	1.0391	87.231	0.96237	35.870	71.988	107.858	0.071248	0.19251	134
135	41.801	27.105	0.011476	1.0240	87.141	0.97657	36.083	71.888	107.971	0.071604	0.19249	135
136	42.450	27.754	0.011488	1.0091	87.051	0.99094	36.296	71.789	108.084	0.071959	0.19248	136
137	43.106	28.410	0.011499	0.99456	86.960	1.0055	36.508	71.689	108.197	0.072313	0.19246	137
138	43.771	29.075	0.011511	0.98023	86.870	1.0202	36.721	71.589	108.310	0.072667	0.19245	138
139	44.443	29.747	0.011523	0.96615	86.779	1.0350	36.934	71.489	108.423	0.073021	0.19243	139
140	45.123	30.427	0.011536	0.95232	86.689	1.0501	37.147	71.388	108.535	0.073374	0.19242	140
141	45.810	31.114	0.011548	0.93873	86.598	1.0653	37.360	71.287	108.648	0.073727	0.19241	141
142	46.506	31.810	0.011560	0.92538	86.507	1.0806	37.574	71.186	108.760	0.074079	0.19239	142
143	47.210	32.514	0.011572	0.91226	86.415	1.0962	37.787	71.085	108.872	0.074431	0.19238	143
144	47.921	33.225	0.011584	0.89937	86.324	1.1119	38.001	70.983	108.984	0.074783	0.19237	144
145	48.641	33.945	0.011597	0.88670	86.232	1.1278	38.214	70.881	109.096	0.075134	0.19236	145
146	49.369	34.673	0.011609	0.87424	86.141	1.1439	38.428	70.779	109.207	0.075484	0.19235	146
147	50.105	35.409	0.011621	0.86200	86.049	1.1601	38.642	70.677	109.319	0.075835	0.19233	147
148	50.850	36.154	0.011634	0.84996	85.957	1.1765	38.856	70.574	109.430	0.076184	0.19232	148
149	51.603	36.907	0.011646	0.83813	85.864	1.1931	39.070	70.472	109.542	0.076534	0.19231	149
150	52.364	37.668	0.011659	0.82650	85.772	1.2099	39.284	70.368	109.653	0.076883	0.19230	150
151	53.133	38.437	0.011671	0.81507	85.680	1.2269	39.499	70.265	109.764	0.077231	0.19229	151
152	53.912	39.216	0.011684	0.80383	85.587	1.2440	39.713	70.161	109.874	0.077579	0.19228	152
153	54.698	40.002	0.011697	0.79278	85.494	1.2614	39.928	70.057	109.985	0.077927	0.19227	153
154	55.494	40.798	0.011709	0.78191	85.401	1.2789	40.142	69.953	110.096	0.078275	0.19227	154
155	56.298	41.602	0.011722	0.77122	85.308	1.2966	40.357	69.849	110.206	0.078622	0.19226	155
156	57.111	42.415	0.011735	0.76071	85.214	1.3146	40.572	69.744	110.316	0.078968	0.19225	156
157	57.933	43.237	0.011748	0.75037	85.121	1.3327	40.787	69.639	110.426	0.079314	0.19224	157
158	58.763	44.067	0.011761	0.74021	85.027	1.3510	41.003	69.533	110.536	0.079660	0.19223	158
159	59.603	44.907	0.011774	0.73021	84.933	1.3695	41.218	69.428	110.646	0.080006	0.19223	159
160	60.451	45.755	0.011787	0.72037	84.839	1.3882	41.433	69.322	110.755	0.080351	0.19222	160
161	61.309	46.613	0.011800	0.71070	84.745	1.4071	41.649	69.215	110.865	0.080695	0.19221	161

Table B-2 (continued)

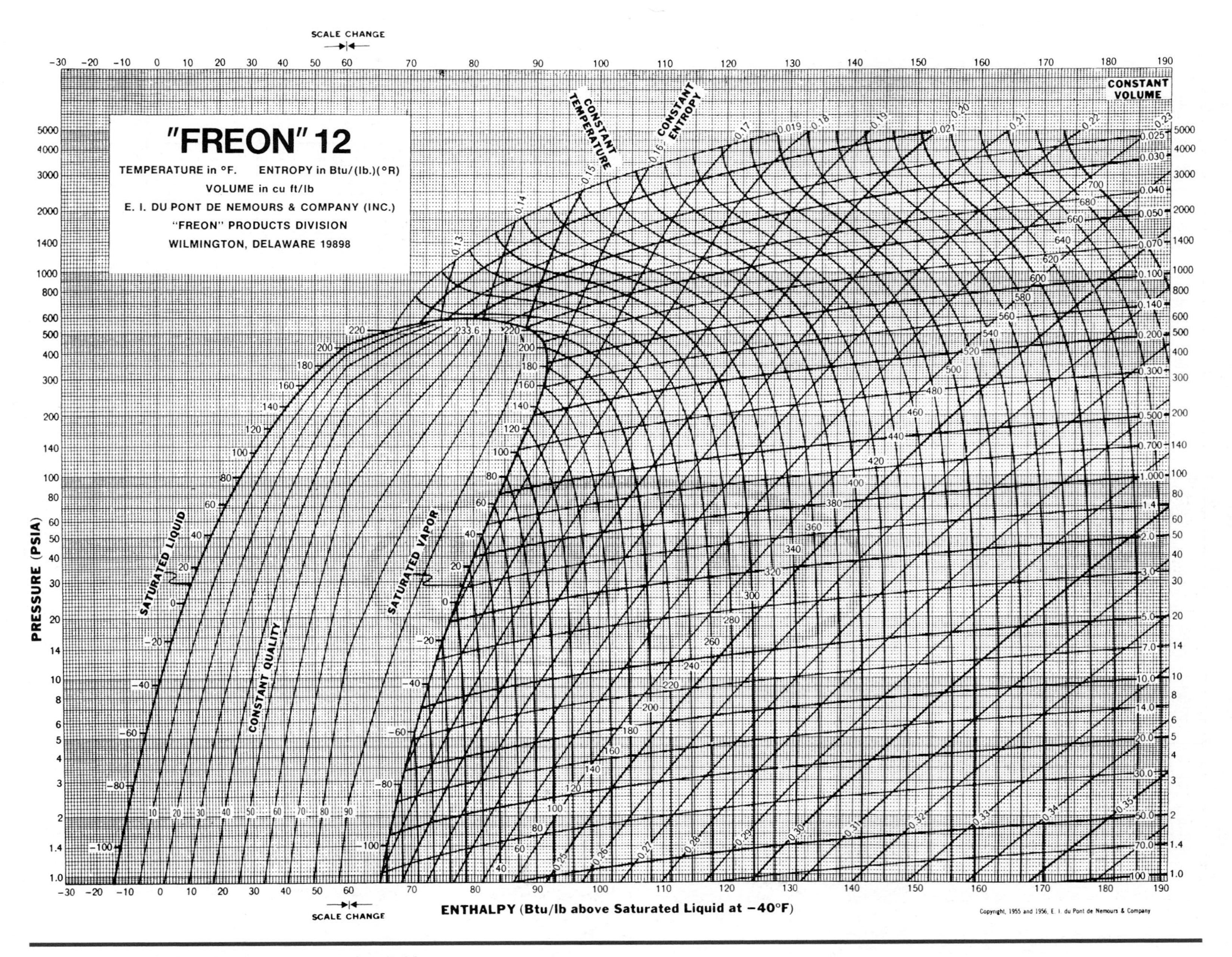

Table B-3 I-P Pressure-Heat Diagram for R-12

TEMP.	PRESSURE		VOLUME cu ft/lb		DENSITY lb/cu ft		ENTHALPY Btu/lb			ENTROPY Btu/(lb)(° R)		TEMP.
°F	PSIA	PSIG	LIQUID v_f	VAPOR v_g	LIQUID $1/v_f$	VAPOR $1/v_g$	LIQUID h_f	LATENT h_{fg}	VAPOR h_g	LIQUID s_f	VAPOR s_g	°F
−40	9.3076	10.9709*	0.010564	3.8750	94.661	0.25806	0	72.913	72.913	0	0.17373	−40
−39	9.5530	10.4712*	0.010575	3.7823	94.565	0.26439	0.2107	72.812	73.023	0.000500	0.17357	−39
−38	9.8035	9.9611*	0.010586	3.6922	94.469	0.27084	0.4215	72.712	73.134	0.001000	0.17343	−38
−37	10.059	9.441*	0.010596	3.6047	94.372	0.27741	0.6324	72.611	73.243	0.001498	0.17328	−37
−36	10.320	8.909*	0.010607	3.5198	94.275	0.28411	0.8434	72.511	73.354	0.001995	0.17313	−36
−35	10.586	8.367*	0.010618	3.4373	94.178	0.29093	1.0546	72.409	73.464	0.002492	0.17299	−35
−34	10.858	7.814*	0.010629	3.3571	94.081	0.29788	1.2659	72.309	73.575	0.002988	0.17285	−34
−33	11.135	7.250*	0.010640	3.2792	93.983	0.30495	1.4772	72.208	73.685	0.003482	0.17271	−33
−32	11.417	6.675*	0.010651	3.2035	93.886	0.31216	1.6887	72.106	73.795	0.003976	0.17257	−32
−31	11.706	6.088*	0.010662	3.1300	93.788	0.31949	1.9003	72.004	73.904	0.004469	0.17243	−31
−30	11.999	5.490*	0.010674	3.0585	93.690	0.32696	2.1120	71.903	74.015	0.004961	0.17229	−30
−29	12.299	4.880*	0.010685	2.9890	93.592	0.33457	2.3239	71.801	74.125	0.005452	0.17216	−29
−28	12.604	4.259*	0.010696	2.9214	93.493	0.34231	2.5358	71.698	74.234	0.005942	0.17203	−28
−27	12.916	3.625*	0.010707	2.8556	93.395	0.35018	2.7479	71.596	74.344	0.006431	0.17189	−27
−26	13.233	2.979*	0.010719	2.7917	93.296	0.35820	2.9601	71.494	74.454	0.006919	0.17177	−26
−25	13.556	2.320*	0.010730	2.7295	93.197	0.36636	3.1724	71.391	74.563	0.007407	0.17164	−25
−24	13.886	1.649*	0.010741	2.6691	93.098	0.37466	3.3848	71.288	74.673	0.007894	0.17151	−24
−23	14.222	0.966*	0.010753	2.6102	92.999	0.38311	3.5973	71.185	74.782	0.008379	0.17139	−23
−22	14.564	0.270*	0.010764	2.5529	92.899	0.39171	3.8100	71.081	74.891	0.008864	0.17126	−22
−21	14.912	0.216	0.010776	2.4972	92.799	0.40045	4.0228	70.978	75.001	0.009348	0.17114	−21
−20	15.267	0.571	0.010788	2.4429	92.699	0.40934	4.2357	70.874	75.110	0.009831	0.17102	−20
−19	15.628	0.932	0.010799	2.3901	92.599	0.41839	4.4487	70.770	75.219	0.010314	0.17090	−19
−18	15.996	1.300	0.010811	2.3387	92.499	0.42758	4.6618	70.666	75.328	0.010795	0.17078	−18
−17	16.371	1.675	0.010823	2.2886	92.399	0.43694	4.8751	70.561	75.436	0.011276	0.17066	−17
−16	16.753	2.057	0.010834	2.2399	92.298	0.44645	5.0885	70.456	75.545	0.011755	0.17055	−16
−15	17.141	2.445	0.010846	2.1924	92.197	0.45612	5.3020	70.352	75.654	0.012234	0.17043	−15
−14	17.536	2.840	0.010858	2.1461	92.096	0.46595	5.5157	70.246	75.762	0.012712	0.17032	−14
−13	17.939	3.243	0.010870	2.1011	91.995	0.47595	5.7295	70.141	75.871	0.013190	0.17021	−13
−12	18.348	3.652	0.010882	2.0572	91.893	0.48611	5.9434	70.036	75.979	0.013666	0.17010	−12
−11	18.765	4.069	0.010894	2.0144	91.791	0.49643	6.1574	69.930	76.087	0.014142	0.16999	−11
−10	19.189	4.493	0.010906	1.9727	91.689	0.50693	6.3716	69.824	76.196	0.014617	0.16989	−10
−9	19.621	4.925	0.010919	1.9320	91.587	0.51759	6.5859	69.718	76.304	0.015091	0.16978	−9
−8	20.059	5.363	0.010931	1.8924	91.485	0.52843	6.8003	69.611	76.411	0.015564	0.16967	−8
−7	20.506	5.810	0.010943	1.8538	91.382	0.53944	7.0149	69.505	76.520	0.016037	0.16957	−7
−6	20.960	6.264	0.010955	1.8161	91.280	0.55063	7.2296	69.397	76.627	0.016508	0.16947	−6
−5	21.422	6.726	0.010968	1.7794	91.177	0.56199	7.4444	69.291	76.735	0.016979	0.16937	−5
−4	21.891	7.195	0.010980	1.7436	91.074	0.57354	7.6594	69.183	76.842	0.017449	0.16927	−4
−3	22.369	7.673	0.010993	1.7086	90.970	0.58526	7.8745	69.075	76.950	0.017919	0.16917	−3
−2	22.854	8.158	0.011005	1.6745	90.867	0.59718	8.0898	68.967	77.057	0.018388	0.16907	−2
−1	23.348	8.652	0.011018	1.6413	90.763	0.60927	8.3052	68.859	77.164	0.018855	0.16897	−1
0	23.849	9.153	0.011030	1.6089	90.659	0.62156	8.5207	68.750	77.271	0.019323	0.16888	0
1	24.359	9.663	0.011043	1.5772	90.554	0.63404	8.7364	68.642	77.378	0.019789	0.16878	1
2	24.878	10.182	0.011056	1.5463	90.450	0.64670	8.9522	68.533	77.485	0.020255	0.16869	2
3	25.404	10.708	0.011069	1.5161	90.345	0.65957	9.1682	68.424	77.592	0.020719	0.16860	3
4	25.939	11.243	0.011082	1.4867	90.240	0.67263	9.3843	68.314	77.698	0.021184	0.16851	4
5	26.483	11.787	0.011094	1.4580	90.135	0.68588	9.6005	68.204	77.805	0.021647	0.16842	**5**
6	27.036	12.340	0.011107	1.4299	90.030	0.69934	9.8169	68.094	77.911	0.022110	0.16833	6
7	27.597	12.901	0.011121	1.4025	89.924	0.71300	10.033	67.984	78.017	0.022572	0.16824	7
8	28.167	13.471	0.011134	1.3758	89.818	0.72687	10.250	67.873	78.123	0.023033	0.16815	8
9	28.747	14.051	0.011147	1.3496	89.712	0.74094	10.467	67.762	78.229	0.023494	0.16807	9
10	29.335	14.639	0.011160	1.3241	89.606	0.75523	10.684	67.651	78.335	0.023954	0.16798	10
11	29.932	15.236	0.011173	1.2992	89.499	0.76972	10.901	67.539	78.440	0.024413	0.16790	11
12	30.539	15.843	0.011187	1.2748	89.392	0.78443	11.118	67.428	78.546	0.024871	0.16782	12
13	31.155	16.459	0.011200	1.2510	89.285	0.79935	11.336	67.315	78.651	0.025329	0.16774	13
14	31.780	17.084	0.011214	1.2278	89.178	0.81449	11.554	67.203	78.757	0.025786	0.16765	14
15	32.415	17.719	0.011227	1.2050	89.070	0.82986	11.771	67.090	78.861	0.026243	0.16758	15

* Inches of mercury below one atmosphere

Table B-4 I-P Properties of Saturated Liquid and Vapor for R-12

TEMP. °F	PRESSURE		VOLUME cu ft/lb		DENSITY lb/cu ft		ENTHALPY Btu/lb			ENTROPY Btu/(lb)(° R)		TEMP. °F
	PSIA	PSIG	LIQUID v_f	VAPOR v_g	LIQUID $1/v_f$	VAPOR $1/v_g$	LIQUID h_f	LATENT h_{fg}	VAPOR h_g	LIQUID s_f	VAPOR s_g	
15	32.415	17.719	0.011227	1.2050	89.070	0.82986	11.771	67.090	78.861	0.026243	0.16758	15
16	33.060	18.364	0.011241	1.1828	88.962	0.84544	11.989	66.977	78.966	0.026699	0.16750	16
17	33.714	19.018	0.011254	1.1611	88.854	0.86125	12.207	66.864	79.071	0.027154	0.16742	17
18	34.378	19.682	0.011268	1.1399	88.746	0.87729	12.426	66.750	79.176	0.027608	0.16734	18
19	35.052	20.356	0.011282	1.1191	88.637	0.89356	12.644	66.636	79.280	0.028062	0.16727	19
20	35.736	21.040	0.011296	1.0988	88.529	0.91006	12.863	66.522	79.385	0.028515	0.16719	20
21	36.430	21.734	0.011310	1.0790	88.419	0.92679	13.081	66.407	79.488	0.028968	0.16712	21
22	37.135	22.439	0.011324	1.0596	88.310	0.94377	13.300	66.293	79.593	0.029420	0.16704	22
23	37.849	23.153	0.011338	1.0406	88.201	0.96098	13.520	66.177	79.697	0.029871	0.16697	23
24	38.574	23.878	0.011352	1.0220	88.091	0.97843	13.739	66.061	79.800	0.030322	0.16690	24
25	39.310	24.614	0.011366	1.0039	87.981	0.99613	13.958	65.946	79.904	0.030772	0.16683	25
26	40.056	25.360	0.011380	0.98612	87.870	1.0141	14.178	65.829	80.007	0.031221	0.16676	26
27	40.813	26.117	0.011395	0.96874	87.760	1.0323	14.398	65.713	80.111	0.031670	0.16669	27
28	41.580	26.884	0.011409	0.95173	87.649	1.0507	14.618	65.596	80.214	0.032118	0.16662	28
29	42.359	27.663	0.011424	0.93509	87.537	1.0694	14.838	65.478	80.316	0.032566	0.16655	29
30	43.148	28.452	0.011438	0.91880	87.426	1.0884	15.058	65.361	80.419	0.033013	0.16648	30
31	43.948	29.252	0.011453	0.90286	87.314	1.1076	15.279	65.243	80.522	0.033460	0.16642	31
32	44.760	30.064	0.011468	0.88725	87.202	1.1271	15.500	65.124	80.624	0.033905	0.16635	32
33	45.583	30.887	0.011482	0.87197	87.090	1.1468	15.720	65.006	80.726	0.034351	0.16629	33
34	46.417	31.721	0.011497	0.85702	86.977	1.1668	15.942	64.886	80.828	0.034796	0.16622	34
35	47.263	32.567	0.011512	0.84237	86.865	1.1871	16.163	64.767	80.930	0.035240	0.16616	35
36	48.120	33.424	0.011527	0.82803	86.751	1.2077	16.384	64.647	81.031	0.035683	0.16610	36
37	48.989	34.293	0.011542	0.81399	86.638	1.2285	16.606	64.527	81.133	0.036126	0.16604	37
38	49.870	35.174	0.011557	0.80023	86.524	1.2496	16.828	64.406	81.234	0.036569	0.16598	38
39	50.763	36.067	0.011573	0.78676	86.410	1.2710	17.050	64.285	81.335	0.037011	0.16592	39
40	51.667	36.971	0.011588	0.77357	86.296	1.2927	17.273	64.163	81.436	0.037453	0.16586	40
41	52.584	37.888	0.011603	0.76064	86.181	1.3147	17.495	64.042	81.537	0.037893	0.16580	41
42	53.513	38.817	0.011619	0.74798	86.066	1 3369	17.718	63.919	81.637	0.038334	0.16574	42
43	54.454	39.758	0.011635	0.73557	85.951	1.3595	17.941	63.796	81.737	0.038774	0.16568	43
44	55.407	40.711	0.011650	0.72341	85.836	1.3823	18.164	63.673	81.837	0.039213	0.16562	44
45	56.373	41.677	0.011666	0.71149	85.720	1.4055	18.387	63.550	81.937	0.039652	0.16557	45
46	57.352	42.656	0.011682	0.69982	85.604	1.4289	18.611	63.426	82.037	0.040091	0.16551	46
47	58.343	43.647	0.011698	0.68837	85.487	1.4527	18.835	63.301	82.136	0.040529	0.16546	47
48	59.347	44.651	0.011714	0.67715	85.371	1.4768	19.059	63.177	82.236	0.040966	0.16540	48
49	60.364	45.668	0.011730	0.66616	85.254	1.5012	19.283	63.051	82.334	0.041403	0.16535	49
50	61.394	46.698	0.011746	0.65537	85.136	1.5258	19.507	62.926	82.433	0.041839	0.16530	50
51	62.437	47.741	0.011762	0.64480	85.018	1.5509	19.732	62.800	82.532	0.042276	0.16524	51
52	63.494	48.798	0.011779	0.63444	84.900	1.5762	19.957	62.673	82.630	0.042711	0.16519	52
53	64.563	49.867	0.011795	0.62428	84.782	1.6019	20.182	62.546	82.728	0.043146	0.16514	53
54	65.646	50.950	0.011811	0.61431	84.663	1.6278	20.408	62.418	82.826	0.043581	0.16509	54
55	66.743	52.047	0.011828	0.60453	84.544	1.6542	20.634	62.290	82.924	0.044015	0.16504	55
56	67.853	53.157	0.011845	0.59495	84.425	1.6808	20.859	62.162	83.021	0.044449	0.16499	56
57	68.977	54.281	0.011862	0.58554	84.305	1.7078	21.086	62.033	83.119	0.044883	0.16494	57
58	70.115	55.419	0.011879	0.57632	84.185	1.7352	21.312	61.903	83.215	0.045316	0.16489	58
59	71.267	56.571	0.011896	0.56727	84.065	1.7628	21.539	61.773	83.312	0.045748	0.16484	59
60	72.433	57.737	0.011913	0.55839	83.944	1.7909	21.766	61.643	83.409	0.046180	0.16479	60
61	73.613	58.917	0.011930	0.54967	83.823	1.8193	21.993	61.512	83.505	0.046612	0.16474	61
62	74.807	60.111	0.011947	0.54112	83.701	1.8480	22.221	61.380	83.601	0.047044	0.16470	62
63	76.016	61.320	0.011965	0.53273	83.580	1.8771	22.448	61.248	83.696	0.047475	0.16465	63
64	77.239	62.543	0.011982	0.52450	83.457	1.9066	22.676	61.116	83.792	0.047905	0.16460	64
65	78.477	63.781	0.012000	0.51642	83.335	1.9364	22.905	60.982	83.887	0.048336	0.16456	65
66	79.729	65.033	0.012017	0.50848	83.212	1.9666	23.133	60.849	83.982	0.048765	0.16451	66
67	80.996	66.300	0.012035	0.50070	83.089	1.9972	23.362	60.715	84.077	0.049195	0.16447	67
68	82.279	67.583	0.012053	0.49305	82.965	2.0282	23.591	60.580	84.171	0.049624	0.16442	68
69	83.576	68.880	0.012071	0.48555	82.841	2.0595	23.821	60.445	84.266	0.050053	0.16438	69
70	84.888	70.192	0.012089	0.47818	82.717	2.0913	24.050	60.309	84.359	0.050482	0.16434	70

Table B-4 (continued)

TEMP. °F	PRESSURE		VOLUME cu ft/lb		DENSITY lb/cu ft		ENTHALPY Btu/lb			ENTROPY Btu/(lb)(°R)		TEMP. °F
	PSIA	PSIG	LIQUID v_f	VAPOR v_g	LIQUID $1/v_f$	VAPOR $1/v_g$	LIQUID h_f	LATENT h_{fg}	VAPOR h_g	LIQUID s_f	VAPOR s_g	
70	84.888	70.192	0.012089	0.47818	82.717	2.0913	24.050	60.309	84.359	0.050482	0.16434	70
71	86.216	71.520	0.012108	0.47094	82.592	2.1234	24.281	60.172	84.453	0.050910	0.16429	71
72	87.559	72.863	0.012126	0.46383	82.467	2.1559	24.511	60.035	84.546	0.051338	0.16425	72
73	88.918	74.222	0.012145	0.45686	82.341	2.1889	24.741	59.898	84.639	0.051766	0.16421	73
74	90.292	75.596	0.012163	0.45000	82.215	2.2222	24.973	59.759	84.732	0.052193	0.16417	74
75	91.682	76.986	0.012182	0.44327	82.089	2.2560	25.204	59.621	84.825	0.052620	0.16412	75
76	93.087	78.391	0.012201	0.43666	81.962	2.2901	25.435	59.481	84.916	0.053047	0.16408	76
77	94.509	79.813	0.012220	0.43016	81.835	2.3247	25.667	59.341	85.008	0.053473	0.16404	77
78	95.946	81.250	0.012239	0.42378	81.707	2.3597	25.899	59.201	85.100	0.053900	0.16400	78
79	97.400	82.704	0.012258	0.41751	81.579	2.3951	26.132	59.059	85.191	0.054326	0.16396	79
80	98.870	84.174	0.012277	0.41135	81.450	2.4310	26.365	58.917	85.282	0.054751	0.16392	80
81	100.36	85.66	0.012297	0.40530	81.322	2.4673	26.598	58.775	85.373	0.055177	0.16388	81
82	101.86	87.16	0.012316	0.39935	81.192	2.5041	26.832	58.631	85.463	0.055602	0.16384	82
83	103.38	88.68	0.012336	0.39351	81.063	2.5413	27.065	58.488	85.553	0.056027	0.16380	83
84	104.92	90.22	0.012356	0.38776	80.932	2.5789	27.300	58.343	85.643	0.056452	0.16376	84
85	106.47	91.77	0.012376	0.38212	80.802	2.6170	27.534	58.198	85.732	0.056877	0.16372	85
86	108.04	93.34	0.012396	0.37657	80.671	2.6556	27.769	58.052	85.821	0.057301	0.16368	**86**
87	109.63	94.93	0.012416	0.37111	80.539	2.6946	28.005	57.905	85.910	0.057725	0.16364	87
88	111.23	96.53	0.012437	0.36575	80.407	2.7341	28.241	57.757	85.998	0.058149	0.16360	88
89	112.85	98.15	0.012457	0.36047	80.275	2.7741	28.477	57.609	86.086	0.058573	0.16357	89
90	114.49	99.79	0.012478	0.35529	80.142	2.8146	28.713	57.461	86.174	0.058997	0.16353	90
91	116.15	101.45	0.012499	0.35019	80.008	2.8556	28.950	57.311	86.261	0.059420	0.16349	91
92	117.82	103.12	0.012520	0.34518	79.874	2.8970	29.187	57.161	86.348	0.059844	0.16345	92
93	119.51	104.81	0.012541	0.34025	79.740	2.9390	29.425	57.009	86.434	0.060267	0.16341	93
94	121.22	106.52	0.012562	0.33540	79.605	2.9815	29.663	56.858	86.521	0.060690	0.16338	94
95	122.95	108.25	0.012583	0.33063	79.470	3.0245	29.901	56.705	86.606	0.061113	0.16334	95
96	124.70	110.00	0.012605	0.32594	79.334	3.0680	30.140	56.551	86.691	0.061536	0.16330	96
97	126.46	111.76	0.012627	0.32133	79.198	3.1120	30.380	56.397	86.777	0.061959	0.16326	97
98	128.24	113.54	0.012649	0.31679	79.061	3.1566	30.619	56.242	86.861	0.062381	0.16323	98
99	130.04	115.34	0.012671	0.31233	78.923	3.2017	30.859	56.086	86.945	0.062804	0.16319	99
100	131.86	117.16	0.012693	0.30794	78.785	3.2474	31.100	55.929	87.029	0.063227	0.16315	100
101	133.70	119.00	0.012715	0.30362	78.647	3.2936	31.341	55.772	87.113	0.063649	0.16312	101
102	135.56	120.86	0.012738	0.29937	78.508	3.3404	31.583	55.613	87.196	0.064072	0.16308	102
103	137.44	122.74	0.012760	0.29518	78.368	3.3877	31.824	55.454	87.278	0.064494	0.16304	103
104	139.33	124.63	0.012783	0.29106	78.228	3.4357	32.067	55.293	87.360	0.064916	0.16301	104
105	141.25	126.55	0.012806	0.28701	78.088	3.4842	32.310	55.132	87.442	0.065339	0.16297	105
106	143.18	128.48	0.012829	0.28303	77.946	3.5333	32.553	54.970	87.523	0.065761	0.16293	106
107	145.13	130.43	0.012853	0.27910	77.804	3.5829	32.797	54.807	87.604	0.066184	0.16290	107
108	147.11	132.41	0.012876	0.27524	77.662	3.6332	33.041	54.643	87.684	0.066606	0.16286	108
109	149.10	134.40	0.012900	0.27143	77.519	3.6841	33.286	54.478	87.764	0.067028	0.16282	109
110	151.11	136.41	0.012924	0.26769	77.376	3.7357	33.531	54.313	87.844	0.067451	0.16279	110
111	153.14	138.44	0.012948	0.26400	77.231	3.7878	33.777	54.146	87.923	0.067873	0.16275	111
112	155.19	140.49	0.012972	0.26037	77.087	3.8406	34.023	53.978	88.001	0.068296	0.16271	112
113	157.27	142.57	0.012997	0.25680	76.941	3.8941	34.270	53.809	88.079	0.068719	0.16268	113
114	159.36	144.66	0.013022	0.25328	76.795	3.9482	34.517	53.639	88.156	0.069141	0.16264	114
115	161.47	146.77	0.013047	0.24982	76.649	4.0029	34.765	53.468	88.233	0.069564	0.16260	115
116	163.61	148.91	0.013072	0.24641	76.501	4.0584	35.014	53.296	88.310	0.069987	0.16256	116
117	165.76	151.06	0.013097	0.24304	76.353	4.1145	35.263	53.123	88.386	0.070410	0.16253	117
118	167.94	153.24	0.013123	0.23974	76.205	4.1713	35.512	52.949	88.461	0.070833	0.16249	118
119	170.13	155.43	0.013148	0.23647	76.056	4.2288	35.762	52.774	88.536	0.071257	0.16245	119
120	172.35	157.65	0.013174	0.23326	75.906	4.2870	36.013	52.597	88.610	0.071680	0.16241	120
121	174.59	159.89	0.013200	0.23010	75.755	4.3459	36.264	52.420	88.684	0.072104	0.16237	121
122	176.85	162.15	0.013227	0.22698	75.604	4.4056	36.516	52.241	88.757	0.072528	0.16234	122
123	179.13	164.43	0.013254	0.22391	75.452	4.4660	36.768	52.062	88.830	0.072952	0.16230	123
124	181.43	166.73	0.013280	0.22089	75.299	4.5272	37.021	51.881	88.902	0.073376	0.16226	124
125	183.76	169.06	0.013308	0.21791	75.145	4.5891	37.275	51.698	88.973	0.073800	0.16222	125

Table B-4 (continued)

TEMP. °F	PRESSURE PSIA	PRESSURE PSIG	VOLUME cu ft/lb LIQUID v_f	VOLUME cu ft/lb VAPOR v_g	DENSITY lb/cu ft LIQUID $1/v_f$	DENSITY lb/cu ft VAPOR $1/v_g$	ENTHALPY Btu/lb LIQUID h_f	ENTHALPY Btu/lb LATENT h_{fg}	ENTHALPY Btu/lb VAPOR h_g	ENTROPY Btu/(lb)(R) LIQUID s_f	ENTROPY Btu/(lb)(R) VAPOR s_g	TEMP. °F
125	183.76	169.06	0.013308	0.21791	75.145	4.5891	37.275	51.698	88.973	0.073800	0.16222	125
126	186.10	171.40	0.013335	0.21497	74.991	4.6518	37.529	51.515	89.044	0.074225	0.16218	126
127	188.47	173.77	0.013363	0.21207	74.836	4.7153	37.785	51.330	89.115	0.074650	0.16214	127
128	190.86	176.16	0.013390	0.20922	74.680	4.7796	38.040	51.144	89.184	0.075075	0.16210	128
129	193.27	178.57	0.013419	0.20641	74.524	4.8448	38.296	50.957	89.253	0.075501	0.16206	129
130	195.71	181.01	0.013447	0.20364	74.367	4.9107	38.553	50.768	89.321	0.075927	0.16202	130
131	198.16	183.46	0.013476	0.20091	74.209	4.9775	38.811	50.578	89.389	0.076353	0.16198	131
132	200.64	185.94	0.013504	0.19821	74.050	5.0451	39.069	50.387	89.456	0.076779	0.16194	132
133	203.15	188.45	0.013534	0.19556	73.890	5.1136	39.328	50.194	89.522	0.077206	0.16189	133
134	205.67	190.97	0.013563	0.19294	73.729	5.1829	39.588	50.000	89.588	0.077633	0.16185	134
135	208.22	193.52	0.013593	0.19036	73.568	5.2532	39.848	49.805	89.653	0.078061	0.16181	135
136	210.79	196.09	0.013623	0.18782	73.406	5.3244	40.110	49.608	89.718	0.078489	0.16177	136
137	213.39	198.69	0.013653	0.18531	73.243	5.3965	40.372	49.409	89.781	0.078917	0.16172	137
138	216.01	201.31	0.013684	0.18283	73.079	5.4695	40.634	49.210	89.844	0.079346	0.16168	138
139	218.65	203.95	0.013715	0.18039	72.914	5.5435	40.898	49.008	89.906	0.079775	0.16163	139
140	221.32	206.62	0.013746	0.17799	72.748	5.6184	41.162	48.805	89.967	0.080205	0.16159	140
141	224.00	209.30	0.013778	0.17561	72.581	5.6944	41.427	48.601	90.028	0.080635	0.16154	141
142	226.72	212.02	0.013810	0.17327	72.413	5.7713	41.693	48.394	90.087	0.081065	0.16150	142
143	229.46	214.76	0.013842	0.17096	72.244	5.8493	41.959	48.187	90.146	0.081497	0.16145	143
144	232.22	217.52	0.013874	0.16868	72.075	5.9283	42.227	47.977	90.204	0.081928	0.16140	144
145	235.00	220.30	0.013907	0.16644	71.904	6.0083	42.495	47.766	90.261	0.082361	0.16135	145
146	237.82	223.12	0.013941	0.16422	71.732	6.0895	42.765	47.553	90.318	0.082794	0.16130	146
147	240.65	225.95	0.013974	0.16203	71.559	6.1717	43.035	47.338	90.373	0.083227	0.16125	147
148	243.51	228.81	0.014008	0.15987	71.386	6.2551	43.306	47.122	90.428	0.083661	0.16120	148
149	246.40	231.70	0.014043	0.15774	71.211	6.3395	43.578	46.904	90.482	0.084096	0.16115	149
150	249.31	234.61	0.014078	0.15564	71.035	6.4252	43.850	46.684	90.534	0.084531	0.16110	150
151	252.24	237.54	0.014113	0.15356	70.857	6.5120	44.124	46.462	90.586	0.084967	0.16105	151
152	255.20	240.50	0.014148	0.15151	70.679	6.6001	44.399	46.238	90.637	0.085404	0.16099	152
153	258.19	243.49	0.014184	0.14949	70.500	6.6893	44.675	46.012	90.687	0.085842	0.16094	153
154	261.20	246.50	0.014221	0.14750	70.319	6.7799	44.951	45.784	90.735	0.086280	0.16088	154
155	264.24	249.54	0.014258	0.14552	70.137	6.8717	45.229	45.554	90.783	0.086719	0.16083	155
156	267.30	252.60	0.014295	0.14358	69.954	6.9648	45.508	45.322	90.830	0.087159	0.16077	156
157	270.39	255.69	0.014333	0.14166	69.770	7.0592	45.787	45.088	90.875	0.087600	0.16071	157
158	273.51	258.81	0.014371	0.13976	69.584	7.1551	46.068	44.852	90.920	0.088041	0.16065	158
159	276.65	261.95	0.014410	0.13789	69.397	7.2523	46.350	44.614	90.964	0.088484	0.16059	159
160	279.82	265.12	0.014449	0.13604	69.209	7.3509	46.633	44.373	91.006	0.088927	0.16053	160
161	283.02	268.32	0.014489	0.13421	69.019	7.4510	46.917	44.130	91.047	0.089371	0.16047	161
162	286.24	271.54	0.014529	0.13241	68.828	7.5525	47.202	43.885	91.087	0.089817	0.16040	162
163	289.49	274.79	0.014570	0.13062	68.635	7.6556	47.489	43.637	91.126	0.090263	0.16034	163
164	292.77	278.07	0.014611	0.12886	68.441	7.7602	47.777	43.386	91.163	0.090710	0.16027	164
165	296.07	281.37	0.014653	0.12712	68.245	7.8665	48.065	43.134	91.199	0.091159	0.16021	165
166	299.40	284.70	0.014695	0.12540	68.048	7.9743	48.355	42.879	91.234	0.091608	0.16014	166
167	302.76	288.06	0.014738	0.12370	67.850	8.0838	48.647	42.620	91.267	0.092059	0.16007	167
168	306.15	291.45	0.014782	0.12202	67.649	8.1950	48.939	42.360	91.299	0.092511	0.16000	168
169	309.56	294.86	0.014826	0.12037	67.447	8.3080	49.233	42.097	91.330	0.092964	0.15992	169
170	313.00	298.30	0.014871	0.11873	67.244	8.4228	49.529	41.830	91.359	0.093418	0.15985	170
171	316.47	301.77	0.014917	0.11710	67.038	8.5394	49.825	41.562	91.387	0.093874	0.15977	171
172	319.97	305.27	0.014963	0.11550	66.831	8.6579	50.123	41.290	91.413	0.094330	0.15969	172
173	323.50	308.80	0.015010	0.11392	66.622	8.7783	50.423	41.015	91.438	0.094789	0.15961	173
174	327.06	312.36	0.015058	0.11235	66.411	8.9007	50.724	40.736	91.460	0.095248	0.15953	174
175	330.64	315.94	0.015106	0.11080	66.198	9.0252	51.026	40.455	91.481	0.095709	0.15945	175
176	334.25	319.55	0.015155	0.10927	65.983	9.1518	51.330	40.171	91.501	0.096172	0.15936	176
177	337.90	323.20	0.015205	0.10775	65.766	9.2805	51.636	39.883	91.519	0.096636	0.15928	177
178	341.57	326.87	0.015256	0.10625	65.547	9.4114	51.943	39.592	91.535	0.097102	0.15919	178
179	345.27	330.57	0.015308	0.10477	65.326	9.5446	52.252	39.297	91.549	0.097569	0.15910	179
180	349.00	334.30	0.015360	0.10330	65.102	9.6802	52.562	38.999	91.561	0.098039	0.15900	180

Table B-4 (continued)

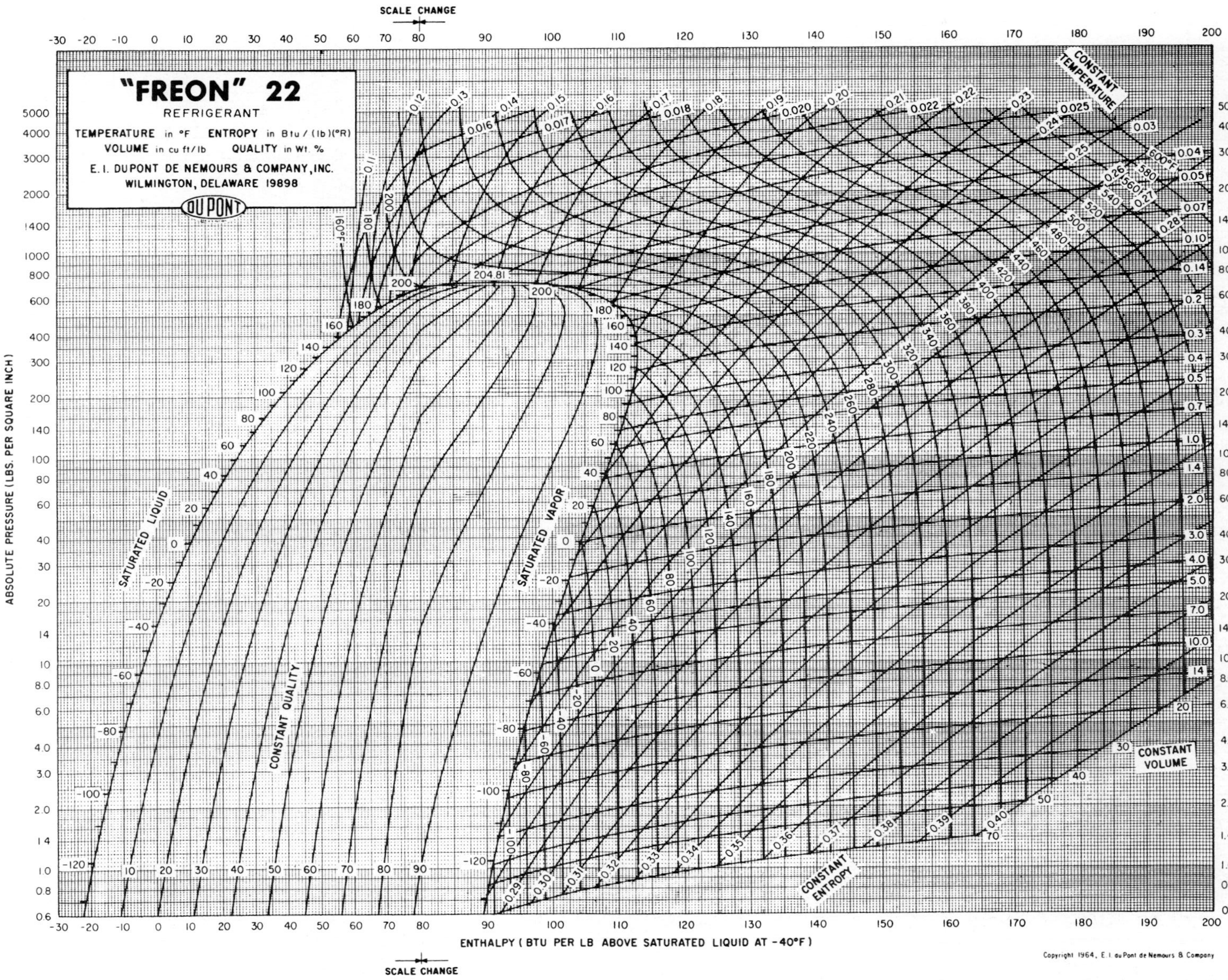

"FREON" 22

REFRIGERANT

TEMPERATURE in °F ENTROPY in Btu / (lb)(°R)
VOLUME in cu ft/lb QUALITY in Wt. %

E. I. DUPONT DE NEMOURS & COMPANY, INC.
WILMINGTON, DELAWARE 19898

DU PONT

SCALE CHANGE

ABSOLUTE PRESSURE (LBS. PER SQUARE INCH)

CONSTANT TEMPERATURE

SATURATED LIQUID

SATURATED VAPOR

CONSTANT QUALITY

CONSTANT ENTROPY

CONSTANT VOLUME

ENTHALPY (BTU PER LB ABOVE SATURATED LIQUID AT -40°F)

SCALE CHANGE

Table B-5 I-P Pressure-Heat Diagram for R-22

TEMP. °F	PRESSURE PSIA	PSIG	VOLUME cu ft/lb LIQUID v_f	VAPOR v_g	DENSITY lb/cu ft LIQUID $1/v_f$	VAPOR $1/v_g$	ENTHALPY Btu/lb LIQUID h_f	LATENT h_{fg}	VAPOR h_g	ENTROPY Btu/(lb)(°R) LIQUID s_f	VAPOR s_g	TEMP. °F
−45	13.354	2.732*	0.011298	3.7243	88.507	0.26851	−1.260	100.963	99.703	−0.00301	0.24046	−45
−44	13.712	2.002*	0.011311	3.6334	88.407	0.27523	−1.009	100.823	99.814	−0.00241	0.24014	−44
−43	14.078	1.258*	0.011324	3.5452	88.307	0.28207	−0.757	100.683	99.925	−0.00181	0.23982	−43
−42	14.451	0.498*	0.011337	3.4596	88.207	0.28905	−0.505	100.541	100.036	−0.00120	0.23951	−42
−41	14.833	0.137	0.011350	3.3764	88.107	0.29617	−0.253	100.399	100.147	−0.00060	0.23919	−41
−40	15.222	0.526	0.011363	3.2957	88.006	0.30342	0.000	100.257	100.257	0.00000	0.23888	−40
−39	15.619	0.923	0.011376	3.2173	87.905	0.31082	0.253	100.114	100.367	0.00060	0.23858	−39
−38	16.024	1.328	0.011389	3.1412	87.805	0.31835	0.506	99.971	100.477	0.00120	0.23827	−38
−37	16.437	1.741	0.011402	3.0673	87.703	0.32602	0.760	99.826	100.587	0.00180	0.23797	−37
−36	16.859	2.163	0.011415	2.9954	87.602	0.33384	1.014	99.682	100.696	0.00240	0.23767	−36
−35	17.290	2.594	0.011428	2.9256	87.501	0.34181	1.269	99.536	100.805	0.00300	0.23737	−35
−34	17.728	3.032	0.011442	2.8578	87.399	0.34992	1.524	99.391	100.914	0.00359	0.23707	−34
−33	18.176	3.480	0.011455	2.7919	87.297	0.35818	1.779	99.244	101.023	0.00419	0.23678	−33
−32	18.633	3.937	0.011469	2.7278	87.195	0.36660	2.035	99.097	101.132	0.00479	0.23649	−32
−31	19.098	4.402	0.011482	2.6655	87.093	0.37517	2.291	98.949	101.240	0.00538	0.23620	−31
−30	19.573	4.877	0.011495	2.6049	86.991	0.38389	2.547	98.801	101.348	0.00598	0.23591	−30
−29	20.056	5.360	0.011509	2.5460	86.888	0.39278	2.804	98.652	101.456	0.00657	0.23563	−29
−28	20.549	5.853	0.011523	2.4887	86.785	0.40182	3.061	98.503	101.564	0.00716	0.23534	−28
−27	21.052	6.536	0.011536	2.4329	86.682	0.41103	3.318	98.353	101.671	0.00776	0.23506	−27
−26	21.564	6.868	0.011550	2.3787	86.579	0.42040	3.576	98.202	101.778	0.00835	0.23478	−26
−25	22.086	7.390	0.011564	2.3260	86.476	0.42993	3.834	98.051	101.885	0.00894	0.23451	−25
−24	22.617	7.921	0.011578	2.2746	86.372	0.43964	4.093	97.899	101.992	0.00953	0.23423	−24
−23	23.159	8.463	0.011592	2.2246	86.269	0.44951	4.352	97.746	102.098	0.01013	0.23396	−23
−22	23.711	9.015	0.011606	2.1760	86.165	0.45956	4.611	97.593	102.204	0.01072	0.23369	−22
−21	24.272	9.576	0.011620	2.1287	86.061	0.46978	4.871	97.439	102.310	0.01131	0.23342	−21
−20	24.845	10.149	0.011634	2.0826	85.956	0.48018	5.131	97.285	102.415	0.01189	0.23315	−20
−19	25.427	10.731	0.011648	2.0377	85.852	0.49075	5.391	97.129	102.521	0.01248	0.23289	−19
−18	26.020	11.324	0.011662	1.9940	85.747	0.50151	5.652	96.974	102.626	0.01307	0.23262	−18
−17	26.624	11.928	0.011677	1.9514	85.642	0.51245	5.913	96.817	102.730	0.01366	0.23236	−17
−16	27.239	12.543	0.011691	1.9099	85.537	0.52358	6.175	96.660	102.835	0.01425	0.23210	−16
−15	27.865	13.169	0.011705	1.8695	85.431	0.53489	6.436	96.502	102.939	0.01483	0.23184	−15
−14	28.501	13.805	0.011720	1.8302	85.326	0.54640	6.699	96.344	103.043	0.01542	0.23159	−14
−13	29.149	14.453	0.011734	1.7918	85.220	0.55810	6.961	96.185	103.146	0.01600	0.23133	−13
−12	29.809	15.113	0.011749	1.7544	85.114	0.56999	7.224	96.025	103.250	0.01659	0.23108	−12
−11	30.480	15.784	0.011764	1.7180	85.008	0.58207	7.488	95.865	103.353	0.01717	0.23083	−11
−10	31.162	16.466	0.011778	1.6825	84.901	0.59436	7.751	95.704	103.455	0.01776	0.23058	−10
− 9	31.856	17.160	0.011793	1.6479	84.795	0.60685	8.015	95.542	103.558	0.01834	0.23033	− 9
− 8	32.563	17.867	0.011808	1.6141	84.688	0.61954	8.280	95.380	103.660	0.01892	0.23008	− 8
− 7	33.281	18.585	0.011823	1.5812	84.581	0.63244	8.545	95.217	103.762	0.01950	0.22984	− 7
− 6	34.011	19.315	0.011838	1.5491	84.473	0.64555	8.810	95.053	103.863	0.02009	0.22960	− 6
− 5	34.754	20.058	0.011853	1.5177	84.366	0.65887	9.075	94.889	103.964	0.02067	0.22936	− 5
− 4	35.509	20.813	0.011868	1.4872	84.258	0.67240	9.341	94.724	104.065	0.02125	0.22912	− 4
− 3	36.277	21.581	0.011884	1.4574	84.150	0.68615	9.608	94.558	104.166	0.02183	0.22888	− 3
− 2	37.057	22.361	0.011899	1.4283	84.042	0.70012	9.874	94.391	104.266	0.02241	0.22864	− 2
− 1	37.850	23.154	0.011914	1.4000	83.933	0.71431	10.142	94.224	104.366	0.02299	0.22841	− 1
0	38.657	23.961	0.011930	1.3723	83.825	0.72872	10.409	94.056	104.465	0.02357	0.22817	0
1	39.476	24.780	0.011945	1.3453	83.716	0.74336	10.677	93.888	104.565	0.02414	0.22794	1
2	40.309	25.613	0.011961	1.3189	83.606	0.75822	10.945	93.718	104.663	0.02472	0.22771	2
3	41.155	26.459	0.011976	1.2931	83.497	0.77332	11.214	93.548	104.762	0.02530	0.22748	3
4	42.014	27.318	0.011992	1.2680	83.387	0.78865	11.483	93.378	104.860	0.02587	0.22725	4
5	42.888	28.192	0.012008	1.2434	83.277	0.80422	11.752	93.206	104.958	0.02645	0.22703	5
6	43.775	29.079	0.012024	1.2195	83.167	0.82003	12.022	93.034	105.056	0.02703	0.22680	6
7	44.676	29.980	0.012040	1.1961	83.057	0.83608	12.292	92.861	105.153	0.02760	0.22658	7
8	45.591	30.895	0.012056	1.1732	82.946	0.85237	12.562	92.688	105.250	0.02818	0.22636	8
9	46.521	31.825	0.012072	1.1509	82.835	0.86892	12.833	92.513	105.346	0.02875	0.22614	9

*Inches of mercury below one atmosphere

Table B-6 I-P Properties of Saturated Liquid and Vapor for R-22

TEMP.	PRESSURE		VOLUME cu ft/lb		DENSITY lb/cu ft		ENTHALPY Btu/lb			ENTROPY Btu/(lb)(°R)		TEMP.
°F	PSIA	PSIG	LIQUID v_f	VAPOR v_g	LIQUID $1/v_f$	VAPOR $1/v_g$	LIQUID h_f	LATENT h_{fg}	VAPOR h_g	LIQUID s_f	VAPOR s_g	°F
10	47.464	32.768	0.012088	1.1290	82.724	0.88571	13.104	92.338	105.442	0.02932	0.22592	10
11	48.423	33.727	0.012105	1.1077	82.612	0.90275	13.376	92.162	105.538	0.02990	0.22570	11
12	49.396	34.700	0.012121	1.0869	82.501	0.92005	13.648	91.986	105.633	0.03047	0.22548	12
13	50.384	35.688	0.012138	1.0665	82.389	0.93761	13.920	91.808	105.728	0.03104	0.22527	13
14	51.387	36.691	0.012154	1.0466	82.276	0.95544	14.193	91.630	105.823	0.03161	0.22505	14
15	52.405	37.709	0.012171	1.0272	82.164	0.97352	14.466	91.451	105.917	0.03218	0.22484	15
16	53.438	38.742	0.012188	1.0082	82.051	0.99188	14.739	91.272	106.011	0.03275	0.22463	16
17	54.487	39.791	0.012204	0.98961	81.938	1.0105	15.013	91.091	106.105	0.03332	0.22442	17
18	55.551	40.855	0.012221	0.97144	81.825	1.0294	15.288	90.910	106.198	0.03389	0.22421	18
19	56.631	41.935	0.012238	0.95368	81.711	1.0486	15.562	90.728	106.290	0.03446	0.22400	19
20	57.727	43.031	0.012255	0.93631	81.597	1.0680	15.837	90.545	106.383	0.03503	0.22379	20
21	58.839	44.143	0.012273	0.91932	81.483	1.0878	16.113	90.362	106.475	0.03560	0.22358	21
22	59.967	45.271	0.012290	0.90270	81.368	1.1078	16.389	90.178	106.566	0.03617	0.22338	22
23	61.111	46.415	0.012307	0.88645	81.253	1.1281	16.665	89.993	106.657	0.03674	0.22318	23
24	62.272	47.576	0.012325	0.87055	81.138	1.1487	16.942	89.807	106.748	0.03730	0.22297	24
25	63.450	48.754	0.012342	0.85500	81.023	1.1696	17.219	89.620	106.839	0.03787	0.22277	25
26	64.644	49.948	0.012360	0.83978	80.907	1.1908	17.496	89.433	106.928	0.03844	0.22257	26
27	65.855	51.159	0.012378	0.82488	80.791	1.2123	17.774	89.244	107.018	0.03900	0.22237	27
28	67.083	52.387	0.012395	0.81031	80.675	1.2341	18.052	89.055	107.107	0.03958	0.22217	28
29	68.328	53.632	0.012413	0.79604	80.558	1.2562	18.330	88.865	107.196	0.04013	0.22198	29
30	69.591	54.895	0.012431	0.78208	80.441	1.2786	18.609	88.674	107.284	0.04070	0.22178	30
31	70.871	56.175	0.012450	0.76842	80.324	1.3014	18.889	88.483	107.372	0.04126	0.22158	31
32	72.169	57.473	0.012468	0.75503	80.207	1.3244	19.169	88.290	107.459	0.04182	0.22139	32
33	73.485	58.789	0.012486	0.74194	80.089	1.3478	19.449	88.097	107.546	0.04239	0.22119	33
34	74.818	60.122	0.012505	0.72911	79.971	1.3715	19.729	87.903	107.632	0.04295	0.22100	34
35	76.170	61.474	0.012523	0.71655	79.852	1.3956	20.010	87.708	107.719	0.04351	0.22081	35
36	77.540	62.844	0.012542	0.70425	79.733	1.4199	20.292	87.512	107.804	0.04407	0.22062	36
37	78.929	64.233	0.012561	0.69221	79.614	1.4447	20.574	87.316	107.889	0.04464	0.22043	37
38	80.336	65.640	0.012579	0.68041	79.495	1.4697	20.856	87.118	107.974	0.04520	0.22024	38
39	81.761	67.065	0.012598	0.66885	79.375	1.4951	21.138	86.920	108.058	0.04576	0.22005	39
40	83.206	68.510	0.012618	0.65753	79.255	1.5208	21.422	86.720	108.142	0.04632	0.21986	40
41	84.670	69.974	0.012637	0.64643	79.134	1.5469	21.705	86.520	108.225	0.04688	0.21968	41
42	86.153	71.457	0.012656	0.63557	79.013	1.5734	21.989	86.319	108.308	0.04744	0.21949	42
43	87.655	72.959	0.012676	0.62492	78.892	1.6002	22.273	86.117	108.390	0.04800	0.21931	43
44	89.177	74.481	0.012695	0.61448	78.770	1.6274	22.558	85.914	108.472	0.04855	0.21912	44
45	90.719	76.023	0.012715	0.60425	78.648	1.6549	22.843	85.710	108.553	0.04911	0.21894	45
46	92.280	77.584	0.012735	0.59422	78.526	1.6829	23.129	85.506	108.634	0.04967	0.21876	46
47	93.861	79.165	0.012755	0.58440	78.403	1.7112	23.415	85.300	108.715	0.05023	0.21858	47
48	95.463	80.767	0.012775	0.57476	78.280	1.7398	23.701	85.094	108.795	0.05079	0.21839	48
49	97.085	82.389	0.012795	0.56532	78.157	1.7689	23.988	84.886	108.874	0.05134	0.21821	49
50	98.727	84.031	0.012815	0.55606	78.033	1.7984	24.275	84.678	108.953	0.05190	0.21803	50
51	100.39	85.69	0.012836	0.54698	77.909	1.8282	24.563	84.468	109.031	0.05245	0.21785	51
52	102.07	87.38	0.012856	0.53808	77.784	1.8585	24.851	84.258	109.109	0.05301	0.21768	52
53	103.78	89.08	0.012877	0.52934	77.659	1.8891	25.139	84.047	109.186	0.05357	0.21750	53
54	105.50	90.81	0.012898	0.52078	77.534	1.9202	25.429	83.834	109.263	0.05412	0.21732	54
55	107.25	92.56	0.012919	0.51238	77.408	1.9517	25.718	83.621	109.339	0.05468	0.21714	55
56	109.02	94.32	0.012940	0.50414	77.282	1.9836	26.008	83.407	109.415	0.05523	0.21697	56
57	110.81	96.11	0.012961	0.49606	77.155	2.0159	26.298	83.191	109.490	0.05579	0.21679	57
58	112.62	97.93	0.012982	0.48813	77.028	2.0486	26.589	82.975	109.564	0.05634	0.21662	58
59	114.46	99.76	0.013004	0.48035	76.900	2.0818	26.880	82.758	109.638	0.05689	0.21644	59
60	116.31	101.62	0.013025	0.47272	76.773	2.1154	27.172	82.540	109.712	0.05745	0.21627	60
61	118.19	103.49	0.013047	0.46523	76.644	2.1495	27.464	82.320	109.785	0.05800	0.21610	61
62	120.09	105.39	0.013069	0.45788	76.515	2.1840	27.757	82.100	109.857	0.05855	0.21592	62
63	122.01	107.32	0.013091	0.45066	76.386	2.2190	28.050	81.878	109.929	0.05910	0.21575	63
64	123.96	109.26	0.013114	0.44358	76.257	2.2544	28.344	81.656	110.000	0.05966	0.21558	64

Table B-6 (continued)

TEMP.	PRESSURE		VOLUME cu ft/lb		DENSITY lb/cu ft		ENTHALPY Btu/lb			ENTROPY Btu/(lb)(°R)		TEMP.
°F	PSIA	PSIG	LIQUID v_f	VAPOR v_g	LIQUID $1/v_f$	VAPOR $1/v_g$	LIQUID h_f	LATENT h_{fg}	VAPOR h_g	LIQUID s_f	VAPOR s_g	°F
65	125.93	111.23	0.013136	0.43663	76.126	2.2903	28.638	81.432	110.070	0.06021	0.21541	65
66	127.92	113.22	0.013159	0.42981	75.996	2.3266	28.932	81.208	110.140	0.06076	0.21524	66
67	129.94	115.24	0.013181	0.42311	75.865	2.3635	29.228	80.982	110.209	0.06131	0.21507	67
68	131.97	117.28	0.013204	0.41653	75.733	2.4008	29.523	80.755	110.278	0.06186	0.21490	68
69	134.04	119.34	0.013227	0.41007	75.601	2.4386	29.819	80.527	110.346	0.06241	0.21473	69
70	136.12	121.43	0.013251	0.40373	75.469	2.4769	30.116	80.298	110.414	0.06296	0.21456	70
71	138.23	123.54	0.013274	0.39751	75.336	2.5157	30.413	80.068	110.480	0.06351	0.21439	71
72	140.37	125.67	0.013297	0.39139	75.202	2.5550	30.710	79.836	110.547	0.06406	0.21422	72
73	142.52	127.83	0.013321	0.38539	75.068	2.5948	31.008	79.604	110.612	0.06461	0.21405	73
74	144.71	130.01	0.013345	0.37949	74.934	2.6351	31.307	79.370	110.677	0.06516	0.21388	74
75	146.91	132.22	0.013369	0.37369	74.799	2.6760	31.606	79.135	110.741	0.06571	0.21372	75
76	149.15	134.45	0.013393	0.36800	74.664	2.7174	31.906	78.899	110.805	0.06626	0.21355	76
77	151.40	136.71	0.013418	0.36241	74.528	2.7593	32.206	78.662	110.868	0.06681	0.21338	77
78	153.69	138.99	0.013442	0.35691	74.391	2.8018	32.506	78.423	110.930	0.06736	0.21321	78
79	155.99	141.30	0.013467	0.35151	74.254	2.8449	32.808	78.184	110.991	0.06791	0.21305	79
80	158.33	143.63	0.013492	0.34621	74.116	2.8885	33.109	77.943	111.052	0.06846	0.21288	80
81	160.68	145.99	0.013518	0.34099	73.978	2.9326	33.412	77.701	111.112	0.06901	0.21271	81
82	163.07	148.37	0.013543	0.33587	73.839	2.9774	33.714	77.457	111.171	0.06956	0.21255	82
83	165.48	150.78	0.013569	0.33083	73.700	3.0227	34.018	77.212	111.230	0.07011	0.21238	83
84	167.92	153.22	0.013594	0.32588	73.560	3.0686	34.322	76.966	111.288	0.07065	0.21222	84
85	170.38	155.68	0.013620	0.32101	73.420	3.1151	34.626	76.719	111.345	0.07120	0.21205	85
86	172.87	158.17	0.013647	0.31623	73.278	3.1622	34.931	76.470	111.401	0.07175	0.21188	86
87	175.38	160.69	0.013673	0.31153	73.137	3.2100	35.237	76.220	111.457	0.07230	0.21172	87
88	177.93	163.23	0.013700	0.30690	72.994	3.2583	35.543	75.968	111.512	0.07285	0.21155	88
89	180.50	165.80	0.013727	0.30236	72.851	3.3073	35.850	75.716	111.566	0.07339	0.21139	89
90	183.09	168.40	0.013754	0.29789	72.708	3.3570	36.158	75.461	111.619	0.07394	0.21122	90
91	185.72	171.02	0.013781	0.29349	72.564	3.4073	36.466	75.206	111.671	0.07449	0.21106	91
92	188.37	173.67	0.013809	0.28917	72.419	3.4582	36.774	74.949	111.723	0.07504	0.21089	92
93	191.05	176.35	0.013836	0.28491	72.273	3.5098	37.084	74.690	111.774	0.07559	0.21072	93
94	193.76	179.06	0.013864	0.28073	72.127	3.5621	37.394	74.430	111.824	0.07613	0.21056	94
95	196.50	181.80	0.013893	0.27662	71.980	3.6151	37.704	74.168	111.873	0.07668	0.21039	95
96	199.26	184.56	0.013921	0.27257	71.833	3.6688	38.016	73.905	111.921	0.07723	0.21023	96
97	202.05	187.36	0.013950	0.26859	71.685	3.7232	38.328	73.641	111.968	0.07778	0.21006	97
98	204.87	190.18	0.013979	0.26467	71.536	3.7783	38.640	73.375	112.015	0.07832	0.20989	98
99	207.72	193.03	0.014008	0.26081	71.386	3.8341	38.953	73.107	112.060	0.07887	0.20973	99
100	210.60	195.91	0.014038	0.25702	71.236	3.8907	39.267	72.838	112.105	0.07942	0.20956	100
101	213.51	198.82	0.014068	0.25329	71.084	3.9481	39.582	72.567	112.149	0.07997	0.20939	101
102	216.45	201.76	0.014098	0.24962	70.933	4.0062	39.897	72.294	112.192	0.08052	0.20923	102
103	219.42	204.72	0.014128	0.24600	70.780	4.0651	40.213	72.020	112.233	0.08107	0.20906	103
104	222.42	207.72	0.014159	0.24244	70.626	4.1247	40.530	71.744	112.274	0.08161	0.20889	104
105	225.45	210.75	0.014190	0.23894	70.472	4.1852	40.847	71.467	112.314	0.08216	0.20872	105
106	228.50	213.81	0.014221	0.23549	70.317	4.2465	41.166	71.187	112.353	0.08271	0.20855	106
107	231.59	216.90	0.014253	0.23209	70.161	4.3086	41.485	70.906	112.391	0.08326	0.20838	107
108	234.71	220.02	0.014285	0.22875	70.005	4.3715	41.804	70.623	112.427	0.08381	0.20821	108
109	237.86	223.17	0.014317	0.22546	69.847	4.4354	42.125	70.338	112.463	0.08436	0.20804	109
110	241.04	226.35	0.014350	0.22222	69.689	4.5000	42.446	70.052	112.498	0.08491	0.20787	110
111	244.25	229.56	0.014382	0.21903	69.529	4.5656	42.768	69.763	112.531	0.08546	0.20770	111
112	247.50	232.80	0.014416	0.21589	69.369	4.6321	43.091	69.473	112.564	0.08601	0.20753	112
113	250.77	236.08	0.014449	0.21279	69.208	4.6994	43.415	69.180	112.595	0.08656	0.20736	113
114	254.08	239.38	0.014483	0.20974	69.046	4.7677	43.739	68.886	112.626	0.08711	0.20718	114
115	257.42	242.72	0.014517	0.20674	68.883	4.8370	44.065	68.590	112.655	0.08766	0.20701	115
116	260.79	246.10	0.014552	0.20378	68.719	4.9072	44.391	68.291	112.682	0.08821	0.20684	116
117	264.20	249.50	0.014587	0.20087	68.554	4.9784	44.718	67.991	112.709	0.08876	0.20666	117
118	267.63	252.94	0.014622	0.19800	68.388	5.0506	45.046	67.688	112.735	0.08932	0.20649	118
119	271.10	256.41	0.014658	0.19517	68.221	5.1238	45.375	67.384	112.759	0.08987	0.20631	119

Table B-6 (continued)

TEMP. °F	PRESSURE PSIA	PSIG	VOLUME cu ft/lb LIQUID v_f	VAPOR v_g	DENSITY lb/cu ft LIQUID $1/v_f$	VAPOR $1/v_g$	ENTHALPY Btu/lb LIQUID h_f	LATENT h_{fg}	VAPOR h_g	ENTROPY Btu/(lb)(°R) LIQUID s_f	VAPOR s_g	TEMP. °F
120	274.60	259.91	0.014694	0.19238	68.054	5.1981	45.705	67.077	112.782	0.09042	0.20613	120
121	278.14	263.44	0.014731	0.18963	67.885	5.2734	46.036	66.767	112.803	0.09098	0.20595	121
122	281.71	267.01	0.014768	0.18692	67.714	5.3498	46.368	66.456	112.824	0.09153	0.20578	122
123	285.31	270.62	0.014805	0.18426	67.543	5.4272	46.701	66.142	112.843	0.09208	0.20560	123
124	288.95	274.25	0.014843	0.18163	67.371	5.5058	47.034	65.826	112.860	0.09264	0.20542	124
125	292.62	277.92	0.014882	0.17903	67.197	5.5856	47.369	65.507	112.877	0.09320	0.20523	125
126	296.33	281.63	0.014920	0.17648	67.023	5.6665	47.705	65.186	112.891	0.09375	0.20505	126
127	300.07	285.37	0.014960	0.17396	66.847	5.7486	48.042	64.863	112.905	0.09431	0.20487	127
128	303.84	289.14	0.014999	0.17147	66.670	5.8319	48.380	64.537	112.917	0.09487	0.20468	128
129	307.65	292.95	0.015039	0.16902	66.492	5.9164	48.719	64.208	112.927	0.09543	0.20449	129
130	311.50	296.80	0.015080	0.16661	66.312	6.0022	49.059	63.877	112.936	0.09598	0.20431	130
131	315.38	300.68	0.015121	0.16422	66.131	6.0893	49.400	63.543	112.943	0.09654	0.20412	131
132	319.29	304.60	0.015163	0.16187	65.949	6.1777	49.743	63.206	112.949	0.09711	0.20393	132
133	323.25	308.55	0.015206	0.15956	65.766	6.2674	50.087	62.866	112.953	0.09767	0.20374	133
134	327.23	312.54	0.015248	0.15727	65.581	6.3585	50.432	62.523	112.955	0.09823	0.20354	134
135	331.26	316.56	0.015292	0.15501	65.394	6.4510	50.778	62.178	112.956	0.09879	0.20335	135
136	335.32	320.63	0.015336	0.15279	65.207	6.5450	51.125	61.829	112.954	0.09936	0.20315	136
137	339.42	324.73	0.015381	0.15059	65.017	6.6405	51.474	61.477	112.951	0.09992	0.20295	137
138	343.56	328.86	0.015426	0.14843	64.826	6.7374	51.824	61.123	112.947	0.10049	0.20275	138
139	347.73	333.04	0.015472	0.14629	64.634	6.8359	52.175	60.764	112.940	0.10106	0.20255	139
140	351.94	337.25	0.015518	0.14418	64.440	6.9360	52.528	60.403	112.931	0.10163	0.20235	140
141	356.19	341.50	0.015566	0.14209	64.244	7.0377	52.883	60.038	112.921	0.10220	0.20214	141
142	360.48	345.79	0.015613	0.14004	64.047	7.1410	53.238	59.670	112.908	0.10277	0.20194	142
143	364.81	350.11	0.015662	0.13801	63.848	7.2461	53.596	59.298	112.893	0.10334	0.20173	143
144	369.17	354.48	0.015712	0.13600	63.647	7.3529	53.955	58.922	112.877	0.10391	0.20152	144
145	373.58	358.88	0.015762	0.13402	63.445	7.4615	54.315	58.543	112.858	0.10449	0.20130	145
146	378.02	363.32	0.015813	0.13207	63.240	7.5719	54.677	58.159	112.836	0.10507	0.20109	146
147	382.50	367.81	0.015865	0.13014	63.034	7.6842	55.041	57.772	112.813	0.10564	0.20087	147
148	387.03	372.33	0.015917	0.12823	62.825	7.7985	55.406	57.380	112.787	0.10622	0.20065	148
149	391.59	376.89	0.015971	0.12635	62.615	7.9148	55.774	56.985	112.758	0.10681	0.20042	149
150	396.19	381.50	0.016025	0.12448	62.402	8.0331	56.143	56.585	112.728	0.10739	0.20020	150
151	400.84	386.14	0.016080	0.12265	62.187	8.1536	56.514	56.180	112.694	0.10797	0.19997	151
152	405.52	390.83	0.016137	0.12083	61.970	8.2763	56.887	55.771	112.658	0.10856	0.19974	152
153	410.25	395.56	0.016194	0.11903	61.751	8.4011	57.261	55.358	112.619	0.10915	0.19950	153
154	415.02	400.32	0.016252	0.11726	61.529	8.5284	57.638	54.939	112.577	0.10974	0.19926	154
155	419.83	405.13	0.016312	0.11550	61.305	8.6580	58.017	54.515	112.533	0.11034	0.19902	155
156	424.68	409.99	0.016372	0.11376	61.079	8.7901	58.399	54.087	112.485	0.11093	0.19878	156
157	429.58	414.88	0.016434	0.11205	60.849	8.9247	58.782	53.652	112.435	0.11153	0.19853	157
158	434.52	419.82	0.016497	0.11035	60.617	9.0620	59.168	53.213	112.381	0.11213	0.19828	158
159	439.50	424.80	0.016561	0.10867	60.383	9.2020	59.557	52.767	112.324	0.11273	0.19802	159
160	444.53	429.83	0.016627	0.10701	60.145	9.3449	59.948	52.316	112.263	0.11334	0.19776	160
161	449.59	434.90	0.016693	0.10537	59.904	9.4907	60.341	51.858	112.199	0.11395	0.19750	161
162	454.71	440.01	0.016762	0.10374	59.660	9.6395	60.737	51.394	112.131	0.11456	0.19723	162
163	459.87	445.17	0.016831	0.10213	59.413	9.7915	61.136	50.923	112.060	0.11518	0.19696	163
164	465.07	450.37	0.016902	0.10054	59.163	9.9467	61.538	50.446	111.984	0.11580	0.19668	164
165	470.32	455.62	0.016975	0.098956	58.909	10.106	61.943	49.961	111.904	0.11642	0.19640	165
166	475.61	460.92	0.017050	0.097393	58.651	10.268	62.351	49.469	111.820	0.11705	0.19611	166
167	480.95	466.26	0.017126	0.095844	58.390	10.434	62.763	48.969	111.732	0.11768	0.19581	167
168	486.34	471.65	0.017204	0.094309	58.125	10.603	63.178	48.461	111.639	0.11831	0.19552	168
169	491.78	477.08	0.017285	0.092787	57.855	10.777	63.596	47.945	111.541	0.11895	0.19521	169
170	497.26	482.56	0.017367	0.091279	57.581	10.955	64.019	47.419	111.438	0.11959	0.19490	170
171	502.79	488.09	0.017451	0.089783	57.303	11.138	64.445	46.885	111.330	0.12024	0.19458	171
172	508.37	493.67	0.017538	0.088299	57.019	11.325	64.875	46.340	111.216	0.12089	0.19425	172
173	513.99	499.30	0.017627	0.086827	56.731	11.517	65.310	45.786	111.096	0.12155	0.19392	173
174	519.67	504.97	0.017719	0.085365	56.438	11.714	65.750	45.221	110.970	0.12222	0.19358	174

Table B-6 (continued)

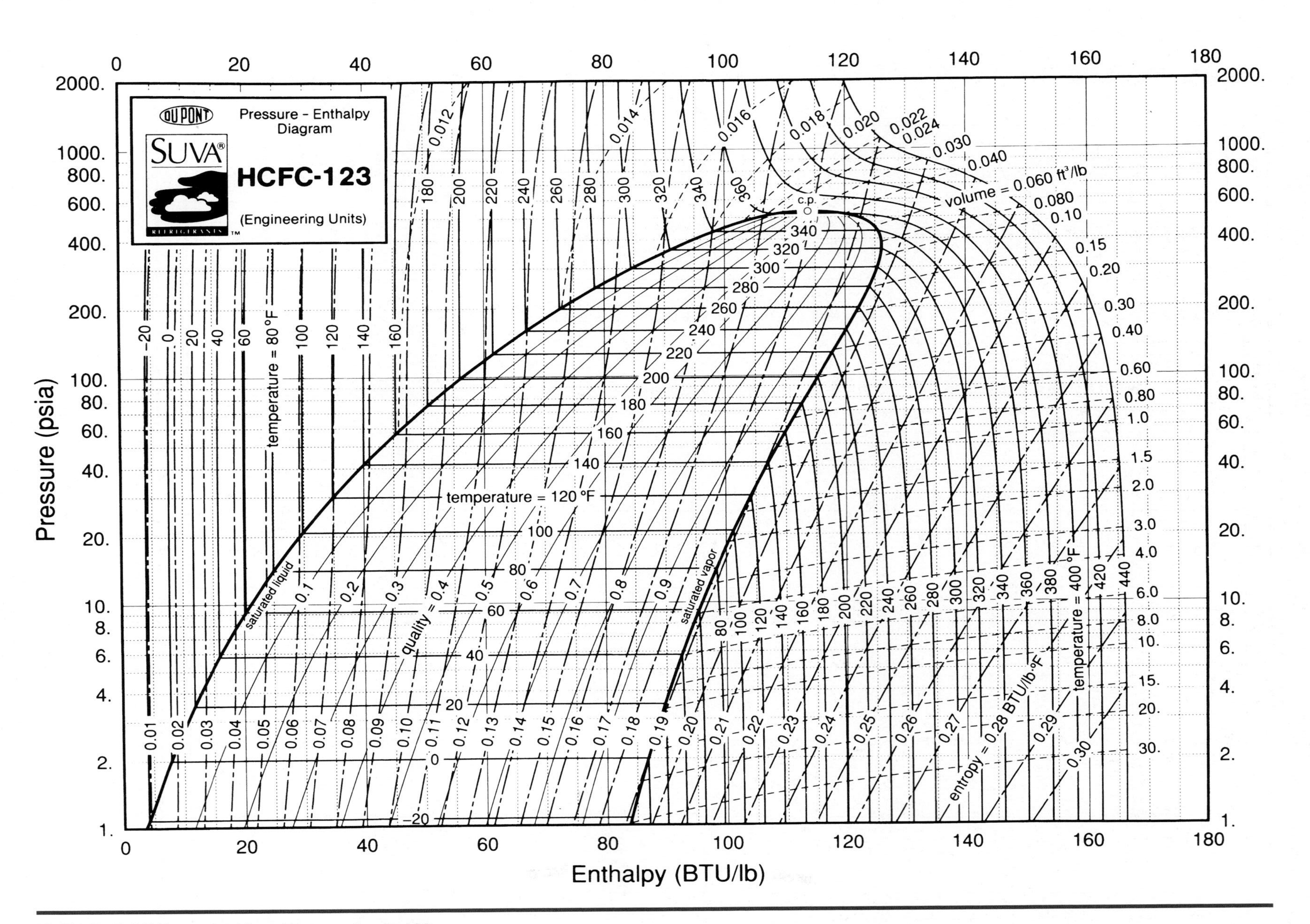

Table B-7 I-P Pressure-Heat Diagram for R-123

TEMP.	PRESSURE	VOLUME ft³/lb		DENSITY lb/ft³		ENTHALPY Btu/lb			ENTROPY Btu/(lb)(°R)		TEMP.
°F	psia	LIQUID v_f	VAPOR v_g	LIQUID $1/v_f$	VAPOR $1/v_g$	LIQUID h_f	LATENT h_{fg}	VAPOR h_g	LIQUID s_f	VAPOR s_g	°F
−30	0.776	0.0100	38.6100	100.0	0.026	2.0	80.7	82.7	0.0047	0.1925	−30
−29	0.803	0.0100	37.4532	99.9	0.027	2.2	80.6	82.8	0.0052	0.1924	−29
−28	0.830	0.0100	36.2319	99.8	0.028	2.4	80.6	83.0	0.0056	0.1923	−28
−27	0.859	0.0100	35.0877	99.7	0.029	2.6	80.5	83.1	0.0061	0.1922	−27
−26	0.888	0.0100	34.0136	99.7	0.029	2.8	80.5	83.3	0.0065	0.1921	−26
−25	0.918	0.0100	33.0033	99.6	0.030	3.0	80.4	83.4	0.0070	0.1920	−25
−24	0.948	0.0100	32.0513	99.5	0.031	3.2	80.4	83.5	0.0074	0.1919	−24
−23	0.980	0.0101	31.0559	99.4	0.032	3.4	80.3	83.7	0.0078	0.1918	−23
−22	1.012	0.0101	30.1205	99.4	0.033	3.5	80.3	83.8	0.0083	0.1917	−22
−21	1.046	0.0101	29.2398	99.3	0.034	3.7	80.2	83.9	0.0087	0.1916	−21
−20	1.080	0.0101	28.3286	99.2	0.035	3.9	80.2	84.1	0.0091	0.1915	−20
−19	1.115	0.0101	27.4725	99.1	0.036	4.1	80.1	84.2	0.0096	0.1914	−19
−18	1.151	0.0101	26.6667	99.1	0.038	4.3	80.1	84.4	0.0100	0.1913	−18
−17	1.188	0.0101	25.9067	99.0	0.039	4.5	80.0	84.5	0.0104	0.1912	−17
−16	1.227	0.0101	25.1889	98.9	0.040	4.7	80.0	84.6	0.0108	0.1911	−16
−15	1.266	0.0101	24.4499	98.8	0.041	4.9	79.9	84.8	0.0113	0.1910	−15
−14	1.306	0.0101	23.7530	98.8	0.042	5.1	79.9	84.9	0.0117	0.1909	−14
−13	1.347	0.0101	23.0415	98.7	0.043	5.2	79.8	85.1	0.0121	0.1908	−13
−12	1.390	0.0101	22.3714	98.6	0.045	5.4	79.8	85.2	0.0125	0.1907	−12
−11	1.433	0.0102	21.7391	98.5	0.046	5.6	79.7	85.3	0.0130	0.1906	−11
−10	1.478	0.0102	21.1416	98.4	0.047	5.8	79.7	85.5	0.0134	0.1905	−10
−9	1.524	0.0102	20.5339	98.4	0.049	6.0	79.6	85.6	0.0138	0.1905	−9
−8	1.571	0.0102	19.9601	98.3	0.050	6.2	79.6	85.8	0.0142	0.1904	−8
−7	1.619	0.0102	19.4175	98.2	0.052	6.4	79.5	85.9	0.0146	0.1903	−7
−6	1.669	0.0102	18.8679	98.1	0.053	6.6	79.5	86.0	0.0151	0.1902	−6
−5	1.720	0.0102	18.3486	98.1	0.055	6.8	79.4	86.2	0.0155	0.1901	−5
−4	1.772	0.0102	17.8571	98.0	0.056	6.9	79.4	86.3	0.0159	0.1901	−4
−3	1.825	0.0102	17.3611	97.9	0.058	7.1	79.3	86.5	0.0163	0.1900	−3
−2	1.880	0.0102	16.8919	97.8	0.059	7.3	79.3	86.6	0.0167	0.1899	−2
−1	1.936	0.0102	16.4474	97.8	0.061	7.5	79.2	86.7	0.0171	0.1898	−1
0	1.993	0.0102	16.0000	97.7	0.063	7.7	79.2	86.9	0.0176	0.1898	0
1	2.052	0.0102	15.5521	97.6	0.064	7.9	79.1	87.0	0.0180	0.1897	1
2	2.113	0.0103	15.1515	97.5	0.066	8.1	79.1	87.2	0.0184	0.1896	2
3	2.174	0.0103	14.7493	97.4	0.068	8.3	79.0	87.3	0.0188	0.1896	3
4	2.238	0.0103	14.3472	97.4	0.070	8.5	79.0	87.4	0.0192	0.1895	4
5	2.303	0.0103	13.9860	97.3	0.072	8.7	78.9	87.6	0.0196	0.1894	5
6	2.369	0.0103	13.6054	97.2	0.074	8.9	78.9	87.7	0.0200	0.1894	6
7	2.437	0.0103	13.2626	97.1	0.075	9.1	78.8	87.9	0.0205	0.1893	7
8	2.507	0.0103	12.9199	97.0	0.077	9.3	78.7	88.0	0.0209	0.1893	8
9	2.578	0.0103	12.5786	97.0	0.080	9.4	78.7	88.1	0.0213	0.1892	9
10	2.651	0.0103	12.2549	96.9	0.082	9.6	78.6	88.3	0.0217	0.1892	10
11	2.726	0.0103	11.9474	96.8	0.084	9.8	78.6	88.4	0.0221	0.1891	11
12	2.802	0.0103	11.6414	96.7	0.086	10.0	78.5	88.6	0.0225	0.1890	12
13	2.880	0.0103	11.3507	96.7	0.088	10.2	78.5	88.7	0.0230	0.1890	13
14	2.960	0.0104	11.0619	96.6	0.090	10.4	78.4	88.9	0.0234	0.1889	14
15	3.042	0.0104	10.7759	96.5	0.093	10.6	78.4	89.0	0.0238	0.1889	15
16	3.126	0.0104	10.5152	96.4	0.095	10.8	78.3	89.1	0.0242	0.1888	16
17	3.212	0.0104	10.2459	96.3	0.098	11.0	78.3	89.3	0.0246	0.1888	17
18	3.299	0.0104	10.0000	96.3	0.100	11.2	78.2	89.4	0.0251	0.1888	18
19	3.389	0.0104	9.7466	96.2	0.103	11.4	78.1	89.6	0.0255	0.1887	19
20	3.480	0.0104	9.5147	96.1	0.105	11.6	78.1	89.7	0.0259	0.1887	20
21	3.574	0.0104	9.2764	96.0	0.108	11.8	78.0	89.8	0.0263	0.1886	21
22	3.669	0.0104	9.0498	95.9	0.111	12.0	78.0	90.0	0.0267	0.1886	22
23	3.767	0.0104	8.8339	95.9	0.113	12.2	77.9	90.1	0.0272	0.1885	23
24	3.867	0.0104	8.6207	95.8	0.116	12.4	77.8	90.3	0.0276	0.1885	24
25	3.969	0.0105	8.4104	95.7	0.119	12.6	77.8	90.4	0.0280	0.1885	25
26	4.073	0.0105	8.2102	95.6	0.122	12.9	77.7	90.6	0.0284	0.1884	26
27	4.180	0.0105	8.0192	95.5	0.125	13.1	77.6	90.7	0.0288	0.1884	27
28	4.288	0.0105	7.8309	95.5	0.128	13.3	77.6	90.8	0.0293	0.1884	28
29	4.400	0.0105	7.6453	95.4	0.131	13.5	77.5	91.0	0.0297	0.1883	29

Table B-8 I-P Properties of Saturated Liquid and Vapor for R-123

TEMP.	PRESSURE	VOLUME ft³/lb		DENSITY lb/ft³		ENTHALPY Btu/lb			ENTROPY Btu/(lb)(°R)		TEMP.
°F	psia	LIQUID v_f	VAPOR v_g	LIQUID $1/v_f$	VAPOR $1/v_g$	LIQUID h_f	LATENT h_{fg}	VAPOR h_g	LIQUID s_f	VAPOR s_g	°F
30	4.513	0.0105	7.4627	95.3	0.134	13.7	77.5	91.1	0.0301	0.1883	30
31	4.629	0.0105	7.2886	95.2	0.137	13.9	77.4	91.3	0.0305	0.1883	31
32	4.747	0.0105	7.1225	95.1	0.140	14.1	77.3	91.4	0.0310	0.1882	32
33	4.868	0.0105	6.9541	95.0	0.144	14.3	77.3	91.6	0.0314	0.1882	33
34	4.991	0.0105	6.7935	95.0	0.147	14.5	77.2	91.7	0.0318	0.1882	34
35	5.117	0.0105	6.6401	94.9	0.151	14.7	77.1	91.9	0.0322	0.1882	35
36	5.245	0.0105	6.4893	94.8	0.154	14.9	77.1	92.0	0.0327	0.1881	36
37	5.376	0.0106	6.3412	94.7	0.158	15.1	77.0	92.1	0.0331	0.1881	37
38	5.509	0.0106	6.1958	94.6	0.161	15.4	76.9	92.3	0.0335	0.1881	38
39	5.646	0.0106	6.0569	94.6	0.165	15.6	76.9	92.4	0.0339	0.1881	39
40	5.785	0.0106	5.9207	94.5	0.169	15.8	76.8	92.6	0.0344	0.1881	40
41	5.927	0.0106	5.7904	94.4	0.173	16.0	76.7	92.7	0.0348	0.1880	41
42	6.071	0.0106	5.6593	94.3	0.177	16.2	76.6	92.9	0.0352	0.1880	42
43	6.219	0.0106	5.5340	94.2	0.181	16.4	76.6	93.0	0.0357	0.1880	43
44	6.369	0.0106	5.4142	94.1	0.185	16.7	76.5	93.2	0.0361	0.1880	44
45	6.522	0.0106	5.2938	94.1	0.189	16.9	76.4	93.3	0.0365	0.1880	45
46	6.679	0.0106	5.1787	94.0	0.193	17.1	76.4	93.4	0.0370	0.1880	46
47	6.838	0.0107	5.0659	93.9	0.197	17.3	76.3	93.6	0.0374	0.1879	47
48	7.000	0.0107	4.9554	93.8	0.202	17.5	76.2	93.7	0.0378	0.1879	48
49	7.166	0.0107	4.8497	93.7	0.206	17.7	76.1	93.9	0.0383	0.1879	49
50	7.334	0.0107	4.7461	93.6	0.211	18.0	76.1	94.0	0.0387	0.1879	50
51	7.506	0.0107	4.6425	93.6	0.215	18.2	76.0	94.2	0.0391	0.1879	51
52	7.681	0.0107	4.5455	93.5	0.220	18.4	75.9	94.3	0.0396	0.1879	52
53	7.860	0.0107	4.4484	93.4	0.225	18.6	75.8	94.5	0.0400	0.1879	53
54	8.041	0.0107	4.3535	93.3	0.230	18.9	75.7	94.6	0.0404	0.1879	54
55	8.226	0.0107	4.2626	93.2	0.235	19.1	75.7	94.7	0.0409	0.1879	55
56	8.415	0.0107	4.1736	93.1	0.240	19.3	75.6	94.9	0.0413	0.1879	56
57	8.607	0.0107	4.0866	93.1	0.245	19.5	75.5	95.0	0.0417	0.1879	57
58	8.802	0.0108	4.0016	93.0	0.250	19.8	75.4	95.2	0.0422	0.1879	58
59	9.001	0.0108	3.9185	92.9	0.255	20.0	75.3	95.3	0.0426	0.1879	59
60	9.203	0.0108	3.8388	92.8	0.261	20.2	75.3	95.5	0.0430	0.1879	60
61	9.410	0.0108	3.7594	92.7	0.266	20.4	75.2	95.6	0.0435	0.1879	61
62	9.619	0.0108	3.6832	92.6	0.272	20.7	75.1	95.8	0.0439	0.1879	62
63	9.833	0.0108	3.6075	92.6	0.277	20.9	75.0	95.9	0.0443	0.1879	63
64	10.050	0.0108	3.5348	92.5	0.283	21.1	74.9	96.1	0.0448	0.1879	64
65	10.272	0.0108	3.4638	92.4	0.289	21.4	74.8	96.2	0.0452	0.1879	65
66	10.497	0.0108	3.3944	92.3	0.295	21.6	74.8	96.3	0.0457	0.1879	66
67	10.726	0.0108	3.3267	92.2	0.301	21.8	74.7	96.5	0.0461	0.1879	67
68	10.958	0.0109	3.2605	92.1	0.307	22.0	74.6	96.6	0.0465	0.1879	68
69	11.195	0.0109	3.1959	92.0	0.313	22.3	74.5	96.8	0.0470	0.1879	69
70	11.436	0.0109	3.1328	92.0	0.319	22.5	74.4	96.9	0.0474	0.1879	70
71	11.682	0.0109	3.0713	91.9	0.326	22.7	74.3	97.1	0.0479	0.1879	71
72	11.931	0.0109	3.0111	91.8	0.332	23.0	74.2	97.2	0.0483	0.1879	72
73	12.184	0.0109	2.9525	91.7	0.339	23.2	74.2	97.4	0.0487	0.1879	73
74	12.442	0.0109	2.8952	91.6	0.345	23.4	74.1	97.5	0.0492	0.1880	74
75	12.704	0.0109	2.8393	91.5	0.352	23.7	74.0	97.7	0.0496	0.1880	75
76	12.970	0.0109	2.7840	91.4	0.359	23.9	73.9	97.8	0.0501	0.1880	76
77	13.241	0.0109	2.7307	91.3	0.366	24.2	73.8	98.0	0.0505	0.1880	77
78	13.517	0.0110	2.6788	91.3	0.373	24.4	73.7	98.1	0.0509	0.1880	78
79	13.796	0.0110	2.6274	91.2	0.381	24.6	73.6	98.2	0.0514	0.1880	79
80	14.081	0.0110	2.5780	91.1	0.388	24.9	73.5	98.4	0.0518	0.1880	80
81	14.369	0.0110	2.5291	91.0	0.395	25.1	73.4	98.5	0.0523	0.1881	81
82	14.663	0.0110	2.4814	90.9	0.403	25.3	73.3	98.7	0.0527	0.1881	82
83	14.961	0.0110	2.4355	90.8	0.411	25.6	73.2	98.8	0.0531	0.1881	83
84	15.264	0.0110	2.3901	90.7	0.418	25.8	73.1	99.0	0.0536	0.1881	84
85	15.572	0.0110	2.3452	90.6	0.426	26.1	73.1	99.1	0.0540	0.1881	85
86	15.885	0.0110	2.3020	90.6	0.434	26.3	73.0	99.3	0.0545	0.1882	86
87	16.203	0.0111	2.2594	90.5	0.443	26.6	72.9	99.4	0.0549	0.1882	87
88	16.525	0.0111	2.2183	90.4	0.451	26.8	72.8	99.6	0.0553	0.1882	88
89	16.853	0.0111	2.1777	90.3	0.459	27.0	72.7	99.7	0.0558	0.1882	89

Table B-8 (continued)

TEMP. °F	PRESSURE psia	VOLUME ft³/lb LIQUID v_f	VOLUME ft³/lb VAPOR v_g	DENSITY lb/ft³ LIQUID $1/v_f$	DENSITY lb/ft³ VAPOR $1/v_g$	ENTHALPY Btu/lb LIQUID h_f	ENTHALPY Btu/lb LATENT h_{fg}	ENTHALPY Btu/lb VAPOR h_g	ENTROPY Btu/(lb)(°R) LIQUID s_f	ENTROPY Btu/(lb)(°R) VAPOR s_g	TEMP. °F
90	17.186	0.0111	2.1381	90.2	0.468	27.3	72.6	99.9	0.0562	0.1883	90
91	17.523	0.0111	2.0991	90.1	0.476	27.5	72.5	100.0	0.0567	0.1883	91
92	17.866	0.0111	2.0610	90.0	0.485	27.8	72.4	100.1	0.0571	0.1883	92
93	18.215	0.0111	2.0239	89.9	0.494	28.0	72.3	100.3	0.0576	0.1883	93
94	18.568	0.0111	1.9877	89.8	0.503	28.3	72.2	100.4	0.0580	0.1884	94
95	18.927	0.0111	1.9524	89.8	0.512	28.5	72.1	100.6	0.0584	0.1884	95
96	19.291	0.0112	1.9175	89.7	0.522	28.7	72.0	100.7	0.0589	0.1884	96
97	19.661	0.0112	1.8836	89.6	0.531	29.0	71.9	100.9	0.0593	0.1884	97
98	20.036	0.0112	1.8501	89.5	0.541	29.2	71.8	101.0	0.0598	0.1885	98
99	20.417	0.0112	1.8179	89.4	0.550	29.5	71.7	101.2	0.0602	0.1885	99
100	20.803	0.0112	1.7857	89.3	0.560	29.7	71.6	101.3	0.0606	0.1885	100
101	21.195	0.0112	1.7547	89.2	0.570	30.0	71.5	101.5	0.0611	0.1886	101
102	21.593	0.0112	1.7241	89.1	0.580	30.2	71.4	101.6	0.0615	0.1886	102
103	21.997	0.0112	1.6943	89.0	0.590	30.5	71.3	101.8	0.0620	0.1886	103
104	22.406	0.0112	1.6650	88.9	0.601	30.7	71.2	101.9	0.0624	0.1887	104
105	22.821	0.0113	1.6364	88.8	0.611	31.0	71.1	102.0	0.0628	0.1887	105
106	23.242	0.0113	1.6082	88.8	0.622	31.2	71.0	102.2	0.0633	0.1887	106
107	23.670	0.0113	1.5808	88.7	0.633	31.5	70.9	102.3	0.0637	0.1888	107
108	24.103	0.0113	1.5540	88.6	0.644	31.7	70.8	102.5	0.0642	0.1888	108
109	24.542	0.0113	1.5277	88.5	0.655	32.0	70.6	102.6	0.0646	0.1888	109
110	24.988	0.0113	1.5017	88.4	0.666	32.2	70.5	102.8	0.0650	0.1889	110
111	25.440	0.0113	1.4765	88.3	0.677	32.5	70.4	102.9	0.0655	0.1889	111
112	25.898	0.0113	1.4518	88.2	0.689	32.7	70.3	103.1	0.0659	0.1890	112
113	26.362	0.0113	1.4276	88.1	0.701	33.0	70.2	103.2	0.0664	0.1890	113
114	26.833	0.0114	1.4037	88.0	0.712	33.2	70.1	103.4	0.0668	0.1890	114
115	27.310	0.0114	1.3805	87.9	0.724	33.5	70.0	103.5	0.0672	0.1891	115
116	27.794	0.0114	1.3576	87.8	0.737	33.7	69.9	103.6	0.0677	0.1891	116
117	28.284	0.0114	1.3353	87.7	0.749	34.0	69.8	103.8	0.0681	0.1891	117
118	28.781	0.0114	1.3134	87.6	0.761	34.3	69.7	103.9	0.0686	0.1892	118
119	29.285	0.0114	1.2918	87.6	0.774	34.5	69.6	104.1	0.0690	0.1892	119
120	29.796	0.0114	1.2708	87.5	0.787	34.8	69.5	104.2	0.0694	0.1893	120
121	30.313	0.0114	1.2502	87.4	0.800	35.0	69.4	104.4	0.0699	0.1893	121
122	30.837	0.0115	1.2300	87.3	0.813	35.3	69.2	104.5	0.0703	0.1894	122
123	31.368	0.0115	1.2102	87.2	0.826	35.5	69.1	104.7	0.0707	0.1894	123
124	31.906	0.0115	1.1908	87.1	0.840	35.8	69.0	104.8	0.0712	0.1894	124
125	32.451	0.0115	1.1716	87.0	0.854	36.0	68.9	105.0	0.0716	0.1895	125
126	33.004	0.0115	1.1529	86.9	0.867	36.3	68.8	105.1	0.0721	0.1895	126
127	33.563	0.0115	1.1346	86.8	0.881	36.6	68.7	105.2	0.0725	0.1896	127
128	34.130	0.0115	1.1166	86.7	0.896	36.8	68.6	105.4	0.0729	0.1896	128
129	34.703	0.0115	1.0989	86.6	0.910	37.1	68.5	105.5	0.0734	0.1897	129
130	35.285	0.0116	1.0817	86.5	0.925	37.3	68.3	105.7	0.0738	0.1897	130
131	35.873	0.0116	1.0646	86.4	0.939	37.6	68.2	105.8	0.0742	0.1898	131
132	36.470	0.0116	1.0480	86.3	0.954	37.8	68.1	106.0	0.0747	0.1898	132
133	37.073	0.0116	1.0317	86.2	0.969	38.1	68.0	106.1	0.0751	0.1899	133
134	37.684	0.0116	1.0156	86.1	0.985	38.4	67.9	106.3	0.0755	0.1899	134
135	38.303	0.0116	0.9999	86.0	1.000	38.6	67.8	106.4	0.0760	0.1899	135
136	38.930	0.0116	0.9844	85.9	1.016	38.9	67.7	106.5	0.0764	0.1900	136
137	39.565	0.0117	0.9693	85.8	1.032	39.1	67.5	106.7	0.0768	0.1900	137
138	40.207	0.0117	0.9545	85.7	1.048	39.4	67.4	106.8	0.0773	0.1901	138
139	40.857	0.0117	0.9398	85.6	1.064	39.7	67.3	107.0	0.0777	0.1901	139
140	41.515	0.0117	0.9255	85.5	1.081	39.9	67.2	107.1	0.0781	0.1902	140
141	42.181	0.0117	0.9115	85.4	1.097	40.2	67.1	107.3	0.0786	0.1902	141
142	42.856	0.0117	0.8977	85.3	1.114	40.4	67.0	107.4	0.0790	0.1903	142
143	43.538	0.0117	0.8841	85.2	1.131	40.7	66.8	107.5	0.0794	0.1903	143
144	44.229	0.0117	0.8708	85.2	1.148	41.0	66.7	107.7	0.0799	0.1904	144
145	44.928	0.0118	0.8578	85.1	1.166	41.2	66.6	107.8	0.0803	0.1904	145
146	45.635	0.0118	0.8449	85.0	1.184	41.5	66.5	108.0	0.0807	0.1905	146
147	46.351	0.0118	0.8323	84.9	1.202	41.7	66.4	108.1	0.0812	0.1905	147
148	47.075	0.0118	0.8199	84.8	1.220	42.0	66.2	108.3	0.0816	0.1906	148
149	47.808	0.0118	0.8078	84.7	1.238	42.3	66.1	108.4	0.0820	0.1906	149

Table B-8 (continued)

TEMP. °F	PRESSURE psia	VOLUME ft³/lb LIQUID v_f	VAPOR v_g	DENSITY lb/ft³ LIQUID $1/v_f$	VAPOR $1/v_g$	ENTHALPY Btu/lb LIQUID h_f	LATENT h_{fg}	VAPOR h_g	ENTROPY Btu/(lb)(°R) LIQUID s_f	VAPOR s_g	TEMP. °F
150	48.549	0.0118	0.7959	84.6	1.257	42.5	66.0	108.5	0.0824	0.1907	150
151	49.299	0.0118	0.7841	84.5	1.275	42.8	65.9	108.7	0.0829	0.1907	151
152	50.058	0.0119	0.7726	84.4	1.294	43.1	65.8	108.8	0.0833	0.1908	152
153	50.825	0.0119	0.7613	84.3	1.314	43.3	65.6	109.0	0.0837	0.1909	153
154	51.602	0.0119	0.7502	84.2	1.333	43.6	65.5	109.1	0.0842	0.1909	154
155	52.387	0.0119	0.7393	84.1	1.353	43.9	65.4	109.2	0.0846	0.1910	155
156	53.181	0.0119	0.7285	84.0	1.373	44.1	65.3	109.4	0.0850	0.1910	156
157	53.985	0.0119	0.7180	83.8	1.393	44.4	65.1	109.5	0.0854	0.1911	157
158	54.797	0.0119	0.7077	83.7	1.413	44.6	65.0	109.7	0.0859	0.1911	158
159	55.619	0.0120	0.6974	83.6	1.434	44.9	64.9	109.8	0.0863	0.1912	159
160	56.450	0.0120	0.6875	83.5	1.455	45.2	64.8	109.9	0.0867	0.1912	160
161	57.290	0.0120	0.6776	83.4	1.476	45.4	64.6	110.1	0.0871	0.1913	161
162	58.140	0.0120	0.6680	83.3	1.497	45.7	64.5	110.2	0.0876	0.1913	162
163	58.999	0.0120	0.6585	83.2	1.519	46.0	64.4	110.4	0.0880	0.1914	163
164	59.868	0.0120	0.6491	83.1	1.541	46.2	64.3	110.5	0.0884	0.1914	164
165	60.746	0.0120	0.6399	83.0	1.563	46.5	64.1	110.6	0.0888	0.1915	165
166	61.634	0.0121	0.6309	82.9	1.585	46.8	64.0	110.8	0.0892	0.1916	166
167	62.532	0.0121	0.6220	82.8	1.608	47.0	63.9	110.9	0.0897	0.1916	167
168	63.439	0.0121	0.6133	82.7	1.631	47.3	63.8	111.1	0.0901	0.1917	168
169	64.357	0.0121	0.6047	82.6	1.654	47.6	63.6	111.2	0.0905	0.1917	169
170	65.284	0.0121	0.5963	82.5	1.677	47.8	63.5	111.3	0.0909	0.1918	170
171	66.221	0.0121	0.5880	82.4	1.701	48.1	63.4	111.5	0.0914	0.1918	171
172	67.169	0.0122	0.5798	82.3	1.725	48.4	63.2	111.6	0.0918	0.1919	172
173	68.126	0.0122	0.5718	82.2	1.749	48.6	63.1	111.7	0.0922	0.1919	173
174	69.094	0.0122	0.5639	82.1	1.773	48.9	63.0	111.9	0.0926	0.1920	174
175	70.072	0.0122	0.5561	82.0	1.798	49.2	62.8	112.0	0.0930	0.1921	175
176	71.060	0.0122	0.5485	81.9	1.823	49.4	62.7	112.2	0.0934	0.1921	176
177	72.059	0.0122	0.5410	81.8	1.849	49.7	62.6	112.3	0.0939	0.1922	177
178	73.068	0.0122	0.5336	81.7	1.874	50.0	62.5	112.4	0.0943	0.1922	178
179	74.088	0.0123	0.5263	81.6	1.900	50.2	62.3	112.6	0.0947	0.1923	179
180	75.118	0.0123	0.5192	81.4	1.926	50.5	62.2	112.7	0.0951	0.1923	180
181	76.159	0.0123	0.5121	81.3	1.953	50.8	62.1	112.8	0.0955	0.1924	181
182	77.211	0.0123	0.5052	81.2	1.980	51.0	61.9	113.0	0.0959	0.1924	182
183	78.274	0.0123	0.4984	81.1	2.007	51.3	61.8	113.1	0.0964	0.1925	183
184	79.348	0.0123	0.4916	81.0	2.034	51.6	61.6	113.2	0.0968	0.1926	184
185	80.432	0.0124	0.4850	80.9	2.062	51.9	61.5	113.4	0.0972	0.1926	185
186	81.528	0.0124	0.4785	80.8	2.090	52.1	61.4	113.5	0.0976	0.1927	186
187	82.635	0.0124	0.4721	80.7	2.118	52.4	61.2	113.6	0.0980	0.1927	187
188	83.753	0.0124	0.4659	80.6	2.147	52.7	61.1	113.8	0.0984	0.1928	188
189	84.882	0.0124	0.4596	80.5	2.176	52.9	61.0	113.9	0.0988	0.1928	189
190	86.023	0.0124	0.4536	80.4	2.205	53.2	60.8	114.0	0.0993	0.1929	190
191	87.175	0.0125	0.4475	80.2	2.234	53.5	60.7	114.2	0.0997	0.1929	191
192	88.339	0.0125	0.4416	80.1	2.264	53.7	60.6	114.3	0.1001	0.1930	192
193	89.514	0.0125	0.4358	80.0	2.295	54.0	60.4	114.4	0.1005	0.1931	193
194	90.700	0.0125	0.4301	79.9	2.325	54.3	60.3	114.6	0.1009	0.1931	194
195	91.899	0.0125	0.4244	79.8	2.356	54.6	60.1	114.7	0.1013	0.1932	195
196	93.109	0.0126	0.4189	79.7	2.387	54.8	60.0	114.8	0.1017	0.1932	196
197	94.331	0.0126	0.4134	79.6	2.419	55.1	59.9	115.0	0.1021	0.1933	197
198	95.565	0.0126	0.4080	79.5	2.451	55.4	59.7	115.1	0.1025	0.1933	198
199	96.811	0.0126	0.4027	79.3	2.483	55.6	59.6	115.2	0.1029	0.1934	199
200	98.069	0.0126	0.3975	79.2	2.516	55.9	59.4	115.3	0.1034	0.1934	200
201	99.340	0.0126	0.3923	79.1	2.549	56.2	59.3	115.5	0.1038	0.1935	201
202	100.622	0.0127	0.3872	79.0	2.583	56.5	59.1	115.6	0.1042	0.1935	202
203	101.917	0.0127	0.3822	78.9	2.616	56.7	59.0	115.7	0.1046	0.1936	203
204	103.224	0.0127	0.3773	78.8	2.650	57.0	58.8	115.9	0.1050	0.1937	204
205	104.544	0.0127	0.3725	78.7	2.685	57.3	58.7	116.0	0.1054	0.1937	205
206	105.876	0.0127	0.3677	78.5	2.720	57.6	58.6	116.1	0.1058	0.1938	206
207	107.221	0.0128	0.3630	78.4	2.755	57.8	58.4	116.2	0.1062	0.1938	207
208	108.578	0.0128	0.3583	78.3	2.791	58.1	58.3	116.4	0.1066	0.1939	208
209	109.948	0.0128	0.3538	78.2	2.827	58.4	58.1	116.5	0.1070	0.1939	209

Table B-8 (continued)

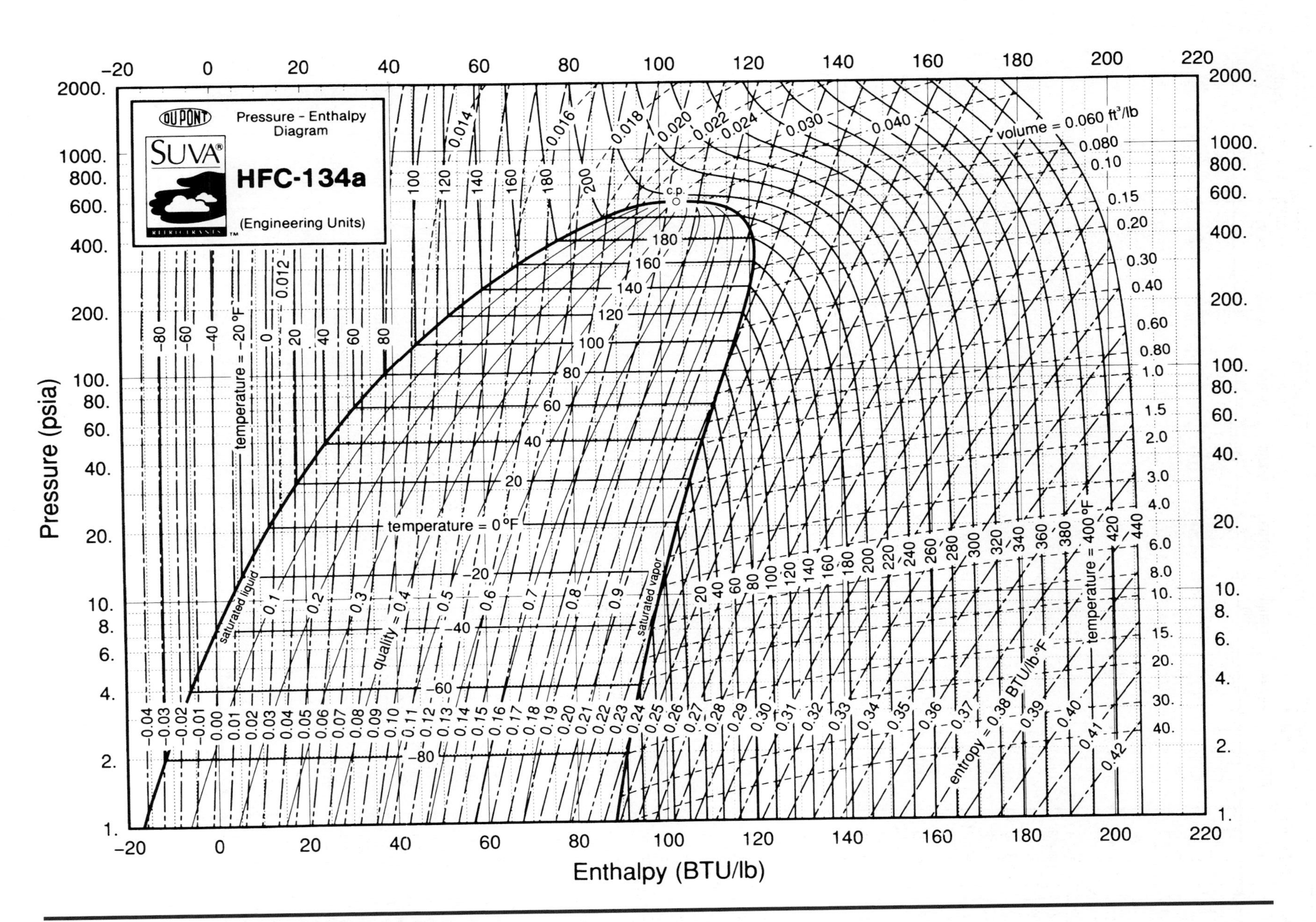

Table B-9 I-P Pressure-Heat Diagram for R-134a

TEMP. °F	PRESSURE psia	VOLUME ft³/lb LIQUID v_f	VOLUME ft³/lb VAPOR v_g	DENSITY lb/ft³ LIQUID $1/v_f$	DENSITY lb/ft³ VAPOR $1/v_g$	ENTHALPY Btu/lb LIQUID h_f	ENTHALPY Btu/lb LATENT h_{fg}	ENTHALPY Btu/lb VAPOR h_g	ENTROPY Btu/(lb)(°R) LIQUID s_f	ENTROPY Btu/(lb)(°R) VAPOR s_g	TEMP. °F
−30	9.851	0.0115	4.4366	87.31	0.2254	3.0	95.7	98.7	0.0070	0.2297	−30
−29	10.126	0.0115	4.3234	87.21	0.2313	3.3	95.5	98.8	0.0077	0.2296	−29
−28	10.407	0.0115	4.2141	87.11	0.2373	3.6	95.4	99.0	0.0084	0.2294	−28
−27	10.694	0.0115	4.1068	87.01	0.2435	3.9	95.2	99.1	0.0091	0.2292	−27
−26	10.987	0.0115	4.0032	86.91	0.2498	4.2	95.1	99.3	0.0098	0.2291	−26
−25	11.287	0.0115	3.9032	86.81	0.2562	4.5	94.9	99.4	0.0105	0.2289	−25
−24	11.594	0.0115	3.8066	86.71	0.2627	4.8	94.8	99.6	0.0112	0.2288	−24
−23	11.906	0.0115	3.7120	86.61	0.2694	5.1	94.6	99.7	0.0119	0.2286	−23
−22	12.226	0.0116	3.6206	86.51	0.2762	5.4	94.5	99.9	0.0126	0.2284	−22
−21	12.553	0.0116	3.5323	86.41	0.2831	5.7	94.3	100.0	0.0132	0.2283	−21
−20	12.885	0.0116	3.4471	86.30	0.2901	6.0	94.2	100.2	0.0139	0.2281	−20
−19	13.225	0.0116	3.3625	86.20	0.2974	6.3	94.0	100.3	0.0146	0.2280	−19
−18	13.572	0.0116	3.2819	86.10	0.3047	6.6	93.9	100.5	0.0153	0.2278	−18
−17	13.927	0.0116	3.2031	86.00	0.3122	6.9	93.7	100.6	0.0160	0.2277	−17
−16	14.289	0.0116	3.1270	85.89	0.3198	7.2	93.6	100.8	0.0167	0.2276	−16
−15	14.659	0.0117	3.0525	85.79	0.3276	7.5	93.4	100.9	0.0174	0.2274	−15
−14	15.035	0.0117	2.9806	85.69	0.3355	7.8	93.3	101.1	0.0180	0.2273	−14
−13	15.420	0.0117	2.9104	85.59	0.3436	8.1	93.1	101.2	0.0187	0.2271	−13
−12	15.812	0.0117	2.8417	85.48	0.3519	8.4	92.9	101.4	0.0194	0.2270	−12
−11	16.212	0.0117	2.7762	85.38	0.3602	8.7	92.8	101.5	0.0201	0.2269	−11
−10	16.620	0.0117	2.7115	85.28	0.3688	9.0	92.6	101.7	0.0208	0.2267	−10
−9	17.037	0.0117	2.6490	85.17	0.3775	9.3	92.5	101.8	0.0214	0.2266	−9
−8	17.461	0.0118	2.5880	85.07	0.3864	9.7	92.3	102.0	0.0221	0.2265	−8
−7	17.893	0.0118	2.5291	84.97	0.3954	10.0	92.1	102.1	0.0228	0.2264	−7
−6	18.334	0.0118	2.4716	84.86	0.4046	10.3	92.0	102.3	0.0235	0.2262	−6
−5	18.784	0.0118	2.4155	84.76	0.4140	10.6	91.8	102.4	0.0241	0.2261	−5
−4	19.242	0.0118	2.3613	84.65	0.4235	10.9	91.7	102.5	0.0248	0.2260	−4
−3	19.709	0.0118	2.3084	84.55	0.4332	11.2	91.5	102.7	0.0255	0.2259	−3
−2	20.184	0.0118	2.2568	84.44	0.4431	11.5	91.3	102.8	0.0262	0.2258	−2
−1	20.669	0.0119	2.2065	84.34	0.4532	11.8	91.2	103.0	0.0268	0.2256	−1
0	21.163	0.0119	2.1580	84.23	0.4634	12.1	91.0	103.1	0.0275	0.2255	0
1	21.666	0.0119	2.1106	84.13	0.4738	12.4	90.9	103.3	0.0282	0.2254	1
2	22.178	0.0119	2.0644	84.02	0.4844	12.7	90.7	103.4	0.0288	0.2253	2
3	22.700	0.0119	2.0194	83.91	0.4952	13.0	90.5	103.6	0.0295	0.2252	3
4	23.231	0.0119	1.9755	83.81	0.5062	13.4	90.4	103.7	0.0302	0.2251	4
5	23.772	0.0119	1.9327	83.70	0.5174	13.7	90.2	103.9	0.0308	0.2250	5
6	24.322	0.0120	1.8911	83.60	0.5288	14.0	90.0	104.0	0.0315	0.2249	6
7	24.883	0.0120	1.8508	83.49	0.5403	14.3	89.9	104.2	0.0322	0.2248	7
8	25.454	0.0120	1.8113	83.38	0.5521	14.6	89.7	104.3	0.0328	0.2247	8
9	26.034	0.0120	1.7727	83.27	0.5641	14.9	89.5	104.4	0.0335	0.2245	9
10	26.625	0.0120	1.7355	83.17	0.5762	15.2	89.4	104.6	0.0342	0.2244	10
11	27.227	0.0120	1.6989	83.06	0.5886	15.5	89.2	104.7	0.0348	0.2244	11
12	27.839	0.0121	1.6633	82.95	0.6012	15.9	89.0	104.9	0.0355	0.2243	12
13	28.462	0.0121	1.6287	82.84	0.6140	16.2	88.9	105.0	0.0362	0.2242	13
14	29.095	0.0121	1.5949	82.73	0.6270	16.5	88.7	105.2	0.0368	0.2241	14
15	29.739	0.0121	1.5620	82.63	0.6402	16.8	88.5	105.3	0.0375	0.2240	15
16	30.395	0.0121	1.5298	82.52	0.6537	17.1	88.3	105.5	0.0381	0.2239	16
17	31.061	0.0121	1.4984	82.41	0.6674	17.4	88.2	105.6	0.0388	0.2238	17
18	31.739	0.0122	1.4678	82.30	0.6813	17.7	88.0	105.7	0.0395	0.2237	18
19	32.428	0.0122	1.4380	82.19	0.6954	18.1	87.8	105.9	0.0401	0.2236	19
20	33.129	0.0122	1.4090	82.08	0.7097	18.4	87.7	106.0	0.0408	0.2235	20
21	33.841	0.0122	1.3806	81.97	0.7243	18.7	87.5	106.2	0.0414	0.2234	21
22	34.566	0.0122	1.3528	81.86	0.7392	19.0	87.3	106.3	0.0421	0.2233	22
23	35.302	0.0122	1.3259	81.75	0.7542	19.3	87.1	106.5	0.0427	0.2233	23
24	36.050	0.0122	1.2995	81.64	0.7695	19.6	87.0	106.6	0.0434	0.2232	24
25	36.810	0.0123	1.2737	81.52	0.7851	20.0	86.8	106.7	0.0440	0.2231	25
26	37.583	0.0123	1.2486	81.41	0.8009	20.3	86.6	106.9	0.0447	0.2230	26
27	38.368	0.0123	1.2240	81.30	0.8170	20.6	86.4	107.0	0.0453	0.2229	27
28	39.166	0.0123	1.2000	81.19	0.8333	20.9	86.2	107.2	0.0460	0.2229	28
29	39.977	0.0123	1.1766	81.08	0.8499	21.2	86.1	107.3	0.0467	0.2228	29

Table B-10 I-P Properties of Saturated Liquid and Vapor for R-134a

TEMP. °F	PRESSURE psia	VOLUME ft³/lb LIQUID v_f	VOLUME ft³/lb VAPOR v_g	DENSITY lb/ft³ LIQUID $1/v_f$	DENSITY lb/ft³ VAPOR $1/v_g$	ENTHALPY Btu/lb LIQUID h_f	ENTHALPY Btu/lb LATENT h_{fg}	ENTHALPY Btu/lb VAPOR h_g	ENTROPY Btu/(lb)(°R) LIQUID s_f	ENTROPY Btu/(lb)(°R) VAPOR s_g	TEMP. °F
30	40.800	0.0124	1.1538	80.96	0.8667	21.6	85.9	107.4	0.0473	0.2227	30
31	41.636	0.0124	1.1315	80.85	0.8838	21.9	85.7	107.6	0.0480	0.2226	31
32	42.486	0.0124	1.1098	80.74	0.9011	22.2	85.5	107.7	0.0486	0.2226	32
33	43.349	0.0124	1.0884	80.62	0.9188	22.5	85.3	107.9	0.0492	0.2225	33
34	44.225	0.0124	1.0676	80.51	0.9367	22.8	85.2	108.0	0.0499	0.2224	34
35	45.115	0.0124	1.0472	80.40	0.9549	23.2	85.0	108.1	0.0505	0.2223	35
36	46.018	0.0125	1.0274	80.28	0.9733	23.5	84.8	108.3	0.0512	0.2223	36
37	46.935	0.0125	1.0080	80.17	0.9921	23.8	84.6	108.4	0.0518	0.2222	37
38	47.866	0.0125	0.9890	80.05	1.0111	24.1	84.4	108.6	0.0525	0.2221	38
39	48.812	0.0125	0.9705	79.94	1.0304	24.5	84.2	108.7	0.0531	0.2221	39
40	49.771	0.0125	0.9523	79.82	1.0501	24.8	84.1	108.8	0.0538	0.2220	40
41	50.745	0.0125	0.9346	79.70	1.0700	25.1	83.9	109.0	0.0544	0.2219	41
42	51.733	0.0126	0.9173	79.59	1.0902	25.4	83.7	109.1	0.0551	0.2219	42
43	52.736	0.0126	0.9003	79.47	1.1107	25.8	83.5	109.2	0.0557	0.2218	43
44	53.754	0.0126	0.8837	79.35	1.1316	26.1	83.3	109.4	0.0564	0.2217	44
45	54.787	0.0126	0.8675	79.24	1.1527	26.4	83.1	109.5	0.0570	0.2217	45
46	55.835	0.0126	0.8516	79.12	1.1742	26.7	82.9	109.7	0.0576	0.2216	46
47	56.898	0.0127	0.8361	79.00	1.1960	27.1	82.7	109.8	0.0583	0.2216	47
48	57.976	0.0127	0.8210	78.88	1.2181	27.4	82.5	109.9	0.0589	0.2215	48
49	59.070	0.0127	0.8061	78.76	1.2405	27.7	82.3	110.1	0.0596	0.2214	49
50	60.180	0.0127	0.7916	78.64	1.2633	28.0	82.1	110.2	0.0602	0.2214	50
51	61.305	0.0127	0.7774	78.53	1.2864	28.4	81.9	110.3	0.0608	0.2213	51
52	62.447	0.0128	0.7634	78.41	1.3099	28.7	81.8	110.5	0.0615	0.2213	52
53	63.604	0.0128	0.7498	78.29	1.3337	29.0	81.6	110.6	0.0621	0.2212	53
54	64.778	0.0128	0.7365	78.16	1.3578	29.4	81.4	110.7	0.0628	0.2211	54
55	65.963	0.0128	0.7234	78.04	1.3823	29.7	81.2	110.9	0.0634	0.2211	55
56	67.170	0.0128	0.7106	77.92	1.4072	30.0	81.0	111.0	0.0640	0.2210	56
57	68.394	0.0129	0.6981	77.80	1.4324	30.4	80.8	111.1	0.0647	0.2210	57
58	69.635	0.0129	0.6859	77.68	1.4579	30.7	80.6	111.3	0.0653	0.2209	58
59	70.892	0.0129	0.6739	77.56	1.4839	31.0	80.4	111.4	0.0659	0.2209	59
60	72.167	0.0129	0.6622	77.43	1.5102	31.4	80.2	111.5	0.0666	0.2208	60
61	73.459	0.0129	0.6507	77.31	1.5369	31.7	80.0	111.6	0.0672	0.2208	61
62	74.769	0.0130	0.6394	77.19	1.5640	32.0	79.7	111.8	0.0678	0.2207	62
63	76.096	0.0130	0.6283	77.06	1.5915	32.4	79.5	111.9	0.0685	0.2207	63
64	77.440	0.0130	0.6175	76.94	1.6194	32.7	79.3	112.0	0.0691	0.2206	64
65	78.803	0.0130	0.6069	76.81	1.6477	33.0	79.1	112.2	0.0698	0.2206	65
66	80.184	0.0130	0.5965	76.69	1.6764	33.4	78.9	112.3	0.0704	0.2205	66
67	81.582	0.0131	0.5863	76.56	1.7055	33.7	78.7	112.4	0.0710	0.2205	67
68	83.000	0.0131	0.5764	76.44	1.7350	34.0	78.5	112.5	0.0717	0.2204	68
69	84.435	0.0131	0.5666	76.31	1.7649	34.4	78.3	112.7	0.0723	0.2204	69
70	85.890	0.0131	0.5570	76.18	1.7952	34.7	78.1	112.8	0.0729	0.2203	70
71	87.363	0.0131	0.5476	76.05	1.8260	35.1	77.9	112.9	0.0735	0.2203	71
72	88.855	0.0132	0.5384	75.93	1.8573	35.4	77.6	113.0	0.0742	0.2202	72
73	90.366	0.0132	0.5294	75.80	1.8889	35.7	77.4	113.2	0.0748	0.2202	73
74	91.897	0.0132	0.5206	75.67	1.9210	36.1	77.2	113.3	0.0754	0.2201	74
75	93.447	0.0132	0.5119	75.54	1.9536	36.4	77.0	113.4	0.0761	0.2201	75
76	95.016	0.0133	0.5034	75.41	1.9866	36.8	76.8	113.5	0.0767	0.2200	76
77	96.606	0.0133	0.4950	75.28	2.0201	37.1	76.6	113.7	0.0773	0.2200	77
78	98.215	0.0133	0.4868	75.15	2.0541	37.4	76.3	113.8	0.0780	0.2200	78
79	99.844	0.0133	0.4788	75.02	2.0885	37.8	76.1	113.9	0.0786	0.2199	79
80	101.494	0.0134	0.4709	74.89	2.1234	38.1	75.9	114.0	0.0792	0.2199	80
81	103.164	0.0134	0.4632	74.75	2.1589	38.5	75.7	114.1	0.0799	0.2198	81
82	104.855	0.0134	0.4556	74.62	2.1948	38.8	75.4	114.3	0.0805	0.2198	82
83	106.566	0.0134	0.4482	74.49	2.2312	39.2	75.2	114.4	0.0811	0.2197	83
84	108.290	0.0134	0.4409	74.35	2.2681	39.5	75.0	114.5	0.0817	0.2197	84
85	110.050	0.0135	0.4337	74.22	2.3056	39.9	74.8	114.6	0.0824	0.2196	85
86	111.828	0.0135	0.4267	74.08	2.3436	40.2	74.5	114.7	0.0830	0.2196	86
87	113.626	0.0135	0.4198	73.95	2.3821	40.5	74.3	114.9	0.0836	0.2196	87
88	115.444	0.0135	0.4130	73.81	2.4211	40.9	74.1	115.0	0.0843	0.2195	88
89	117.281	0.0136	0.4064	73.67	2.4607	41.2	73.8	115.1	0.0849	0.2195	89

Table B-10 (continued)

TEMP.	PRESSURE	VOLUME ft³/lb		DENSITY lb/ft³		ENTHALPY Btu/lb			ENTROPY Btu/(lb)(°R)		TEMP.
°F	psia	LIQUID v_f	VAPOR v_g	LIQUID $1/v_f$	VAPOR $1/v_g$	LIQUID h_f	LATENT h_{fg}	VAPOR h_g	LIQUID s_f	VAPOR s_g	°F
90	119.138	0.0136	0.3999	73.54	2.5009	41.6	73.6	115.2	0.0855	0.2194	90
91	121.024	0.0136	0.3935	73.40	2.5416	41.9	73.4	115.3	0.0861	0.2194	91
92	122.930	0.0137	0.3872	73.26	2.5829	42.3	73.1	115.4	0.0868	0.2193	92
93	124.858	0.0137	0.3810	73.12	2.6247	42.6	72.9	115.5	0.0874	0.2193	93
94	126.809	0.0137	0.3749	72.98	2.6672	43.0	72.7	115.7	0.0880	0.2193	94
95	128.782	0.0137	0.3690	72.84	2.7102	43.4	72.4	115.8	0.0886	0.2192	95
96	130.778	0.0138	0.3631	72.70	2.7539	43.7	72.2	115.9	0.0893	0.2192	96
97	132.798	0.0138	0.3574	72.56	2.7981	44.1	71.9	116.0	0.0899	0.2191	97
98	134.840	0.0138	0.3517	72.42	2.8430	44.4	71.7	116.1	0.0905	0.2191	98
99	136.906	0.0138	0.3462	72.27	2.8885	44.8	71.4	116.2	0.0912	0.2190	99
100	138.996	0.0139	0.3408	72.13	2.9347	45.1	71.2	116.3	0.0918	0.2190	100
101	141.109	0.0139	0.3354	71.99	2.9815	45.5	70.9	116.4	0.0924	0.2190	101
102	143.247	0.0139	0.3302	71.84	3.0289	45.8	70.7	116.5	0.0930	0.2189	102
103	145.408	0.0139	0.3250	71.70	3.0771	46.2	70.4	116.6	0.0937	0.2189	103
104	147.594	0.0140	0.3199	71.55	3.1259	46.6	70.2	116.7	0.0943	0.2188	104
105	149.804	0.0140	0.3149	71.40	3.1754	46.9	69.9	116.9	0.0949	0.2188	105
106	152.039	0.0140	0.3100	71.25	3.2256	47.3	69.7	117.0	0.0955	0.2187	106
107	154.298	0.0141	0.3052	71.11	3.2765	47.6	69.4	117.1	0.0962	0.2187	107
108	156.583	0.0141	0.3005	70.96	3.3282	48.0	69.2	117.2	0.0968	0.2186	108
109	158.893	0.0141	0.2958	70.81	3.3806	48.4	68.9	117.3	0.0974	0.2186	109
110	161.227	0.0142	0.2912	70.66	3.4337	48.7	68.6	117.4	0.0981	0.2185	110
111	163.588	0.0142	0.2867	70.51	3.4876	49.1	68.4	117.5	0.0987	0.2185	111
112	165.974	0.0142	0.2823	70.35	3.5423	49.5	68.1	117.6	0.0993	0.2185	112
113	168.393	0.0142	0.2780	70.20	3.5977	49.8	67.8	117.7	0.0999	0.2184	113
114	170.833	0.0143	0.2737	70.05	3.6539	50.2	67.6	117.8	0.1006	0.2184	114
115	173.298	0.0143	0.2695	69.89	3.7110	50.5	67.3	117.9	0.1012	0.2183	115
116	175.790	0.0143	0.2653	69.74	3.7689	50.9	67.0	117.9	0.1018	0.2183	116
117	178.297	0.0144	0.2613	69.58	3.8276	51.3	66.8	118.0	0.1024	0.2182	117
118	180.846	0.0144	0.2573	69.42	3.8872	51.7	66.5	118.1	0.1031	0.2182	118
119	183.421	0.0144	0.2533	69.26	3.9476	52.0	66.2	118.2	0.1037	0.2181	119
120	186.023	0.0145	0.2494	69.10	4.0089	52.4	65.9	118.3	0.1043	0.2181	120
121	188.652	0.0145	0.2456	68.94	4.0712	52.8	65.6	118.4	0.1050	0.2180	121
122	191.308	0.0145	0.2419	68.78	4.1343	53.1	65.4	118.5	0.1056	0.2180	122
123	193.992	0.0146	0.2382	68.62	4.1984	53.5	65.1	118.6	0.1062	0.2179	123
124	196.703	0.0146	0.2346	68.46	4.2634	53.9	64.8	118.7	0.1068	0.2178	124
125	199.443	0.0146	0.2310	68.29	4.3294	54.3	64.5	118.8	0.1075	0.2178	125
126	202.211	0.0147	0.2275	68.13	4.3964	54.6	64.2	118.8	0.1081	0.2177	126
127	205.008	0.0147	0.2240	67.96	4.4644	55.0	63.9	118.9	0.1087	0.2177	127
128	207.834	0.0147	0.2206	67.80	4.5334	55.4	63.6	119.0	0.1094	0.2176	128
129	210.688	0.0148	0.2172	67.63	4.6034	55.8	63.3	119.1	0.1100	0.2176	129
130	213.572	0.0148	0.2139	67.46	4.6745	56.2	63.0	119.2	0.1106	0.2175	130
131	216.485	0.0149	0.2107	67.29	4.7467	56.5	62.7	119.2	0.1113	0.2174	131
132	219.429	0.0149	0.2075	67.12	4.8200	56.9	62.4	119.3	0.1119	0.2174	132
133	222.402	0.0149	0.2043	66.95	4.8945	57.3	62.1	119.4	0.1125	0.2173	133
134	225.405	0.0150	0.2012	66.77	4.9700	57.7	61.8	119.5	0.1132	0.2173	134
135	228.438	0.0150	0.1981	66.60	5.0468	58.1	61.5	119.6	0.1138	0.2172	135
136	231.502	0.0151	0.1951	66.42	5.1248	58.5	61.2	119.6	0.1144	0.2171	136
137	234.597	0.0151	0.1922	66.24	5.2040	58.8	60.8	119.7	0.1151	0.2171	137
138	237.723	0.0151	0.1892	66.06	5.2844	59.2	60.5	119.8	0.1157	0.2170	138
139	240.880	0.0152	0.1864	65.88	5.3661	59.6	60.2	119.8	0.1163	0.2169	139
140	244.068	0.0152	0.1835	65.70	5.4491	60.0	59.9	119.9	0.1170	0.2168	140
141	247.288	0.0153	0.1807	65.52	5.5335	60.4	59.6	120.0	0.1176	0.2168	141
142	250.540	0.0153	0.1780	65.34	5.6192	60.8	59.2	120.0	0.1183	0.2167	142
143	253.824	0.0153	0.1752	65.15	5.7064	61.2	58.9	120.1	0.1189	0.2166	143
144	257.140	0.0154	0.1726	64.96	5.7949	61.6	58.6	120.1	0.1195	0.2165	144
145	260.489	0.0154	0.1699	64.78	5.8849	62.0	58.2	120.2	0.1202	0.2165	145
146	263.871	0.0155	0.1673	64.59	5.9765	62.4	57.9	120.3	0.1208	0.2164	146
147	267.270	0.0155	0.1648	64.39	6.0695	62.8	57.5	120.3	0.1215	0.2163	147
148	270.721	0.0156	0.1622	64.20	6.1642	63.2	57.2	120.4	0.1221	0.2162	148
149	274.204	0.0156	0.1597	64.01	6.2604	63.6	56.8	120.4	0.1228	0.2161	149

Table B-10 (continued)

TEMP.	PRESSURE	VOLUME ft³/lb		DENSITY lb/ft³		ENTHALPY Btu/lb			ENTROPY Btu/(lb)(°R)		TEMP.
°F	psia	LIQUID v_f	VAPOR v_g	LIQUID $1/v_f$	VAPOR $1/v_g$	LIQUID h_f	LATENT h_{fg}	VAPOR h_g	LIQUID s_f	VAPOR s_g	°F
150	277.721	0.0157	0.1573	63.81	6.3584	64.0	56.5	120.5	0.1234	0.2160	150
151	281.272	0.0157	0.1548	63.61	6.4580	64.4	56.1	120.5	0.1240	0.2159	151
152	284.857	0.0158	0.1525	63.41	6.5593	64.8	55.7	120.6	0.1247	0.2158	152
153	288.477	0.0158	0.1501	63.21	6.6625	65.2	55.4	120.6	0.1253	0.2157	153
154	292.131	0.0159	0.1478	63.01	6.7675	65.6	55.0	120.6	0.1260	0.2156	154
155	295.820	0.0159	0.1455	62.80	6.8743	66.0	54.6	120.7	0.1266	0.2155	155
156	299.544	0.0160	0.1432	62.59	6.9831	66.4	54.3	120.7	0.1273	0.2154	156
157	303.304	0.0160	0.1410	62.38	7.0940	66.9	53.9	120.7	0.1279	0.2153	157
158	307.100	0.0161	0.1388	62.17	7.2068	67.3	53.5	120.8	0.1286	0.2152	158
159	310.931	0.0161	0.1366	61.96	7.3218	67.7	53.1	120.8	0.1293	0.2151	159
160	314.800	0.0162	0.1344	61.74	7.4390	68.1	52.7	120.8	0.1299	0.2150	160
161	318.704	0.0163	0.1323	61.52	7.5584	68.5	52.3	120.9	0.1306	0.2149	161
162	322.646	0.0163	0.1302	61.30	7.6801	69.0	51.9	120.9	0.1312	0.2148	162
163	326.625	0.0164	0.1281	61.08	7.8042	69.4	51.5	120.9	0.1319	0.2146	163
164	330.641	0.0164	0.1261	60.86	7.9308	69.8	51.1	120.9	0.1326	0.2145	164
165	334.696	0.0165	0.1241	60.63	8.0600	70.2	50.7	120.9	0.1332	0.2144	165
166	338.788	0.0166	0.1221	60.40	8.1917	70.7	50.3	120.9	0.1339	0.2142	166
167	342.919	0.0166	0.1201	60.16	8.3262	71.1	49.8	120.9	0.1346	0.2141	167
168	347.089	0.0167	0.1182	59.93	8.4635	71.5	49.4	120.9	0.1352	0.2140	168
169	351.298	0.0168	0.1162	59.69	8.6037	72.0	49.0	120.9	0.1359	0.2138	169
170	355.547	0.0168	0.1143	59.45	8.7470	72.4	48.5	120.9	0.1366	0.2137	170
171	359.835	0.0169	0.1124	59.20	8.8934	72.8	48.1	120.9	0.1373	0.2135	171
172	364.164	0.0170	0.1106	58.95	9.0431	73.3	47.6	120.9	0.1380	0.2133	172
173	368.533	0.0170	0.1087	58.70	9.1961	73.7	47.2	120.9	0.1386	0.2132	173
174	372.942	0.0171	0.1069	58.45	9.3527	74.2	46.7	120.9	0.1393	0.2130	174
175	377.393	0.0172	0.1051	58.19	9.5129	74.6	46.2	120.8	0.1400	0.2128	175
176	381.886	0.0173	0.1033	57.92	9.6770	75.1	45.7	120.8	0.1407	0.2126	176
177	386.421	0.0173	0.1016	57.66	9.8451	75.5	45.2	120.8	0.1414	0.2125	177
178	390.998	0.0174	0.0998	57.39	10.0173	76.0	44.7	120.7	0.1421	0.2123	178
179	395.617	0.0175	0.0981	57.11	10.1939	76.5	44.2	120.7	0.1428	0.2121	179
180	400.280	0.0176	0.0964	56.83	10.3750	76.9	43.7	120.7	0.1435	0.2119	180
181	404.987	0.0177	0.0947	56.55	10.5609	77.4	43.2	120.6	0.1442	0.2116	181
182	409.738	0.0178	0.0930	56.26	10.7518	77.9	42.7	120.5	0.1449	0.2114	182
183	414.533	0.0179	0.0913	55.96	10.9481	78.4	42.1	120.5	0.1456	0.2112	183
184	419.373	0.0180	0.0897	55.66	11.1498	78.8	41.6	120.4	0.1464	0.2109	184
185	424.258	0.0181	0.0880	55.35	11.3575	79.3	41.0	120.3	0.1471	0.2107	185
186	429.189	0.0182	0.0864	55.04	11.5713	79.8	40.4	120.2	0.1478	0.2104	186
187	434.167	0.0183	0.0848	54.72	11.7916	80.3	39.8	120.1	0.1486	0.2102	187
188	439.192	0.0184	0.0832	54.40	12.0189	80.8	39.2	120.0	0.1493	0.2099	188
189	444.264	0.0185	0.0816	54.07	12.2536	81.3	38.6	119.9	0.1501	0.2096	189
190	449.384	0.0186	0.0800	53.73	12.4962	81.8	38.0	119.8	0.1508	0.2093	190
191	454.552	0.0187	0.0784	53.38	12.7472	82.3	37.4	119.7	0.1516	0.2090	191
192	459.757	0.0189	0.0769	53.02	13.0072	82.8	36.7	119.5	0.1523	0.2087	192
193	465.026	0.0190	0.0753	52.65	13.2769	83.4	36.0	119.4	0.1531	0.2083	193
194	470.346	0.0191	0.0738	52.27	13.5570	83.9	35.3	119.2	0.1539	0.2080	194
195	475.717	0.0193	0.0722	51.88	13.8484	84.4	34.6	119.1	0.1547	0.2076	195
196	481.139	0.0194	0.0707	51.48	14.1522	85.0	33.9	118.9	0.1555	0.2072	196
197	486.614	0.0196	0.0691	51.07	14.4693	85.5	33.1	118.7	0.1563	0.2068	197
198	492.142	0.0197	0.0676	50.64	14.8012	86.1	32.3	118.4	0.1572	0.2063	198
199	497.724	0.0199	0.0660	50.20	15.1493	86.7	31.5	118.2	0.1580	0.2059	199
200	503.361	0.0201	0.0645	49.73	15.5155	87.3	30.7	118.0	0.1589	0.2054	200
201	509.054	0.0203	0.0629	49.25	15.9020	87.9	29.8	117.7	0.1597	0.2049	201
202	514.805	0.0205	0.0613	48.75	16.3113	88.5	28.9	117.4	0.1606	0.2043	202
203	520.613	0.0207	0.0597	48.22	16.7466	89.1	27.9	117.0	0.1616	0.2037	203
204	526.481	0.0210	0.0581	47.66	17.2121	89.8	26.9	116.7	0.1625	0.2031	204
205	532.410	0.0212	0.0565	47.06	17.7129	90.5	25.8	116.3	0.1635	0.2024	205
206	538.402	0.0215	0.0548	46.42	18.2558	91.2	24.7	115.8	0.1645	0.2016	206
207	544.458	0.0219	0.0531	45.73	18.8499	91.9	23.4	115.3	0.1656	0.2008	207
208	550.581	0.0222	0.0513	44.98	19.5084	92.7	22.1	114.8	0.1667	0.1998	208
209	556.773	0.0227	0.0494	44.14	20.2504	93.5	20.6	114.1	0.1680	0.1988	209

Table B-10 (continued)

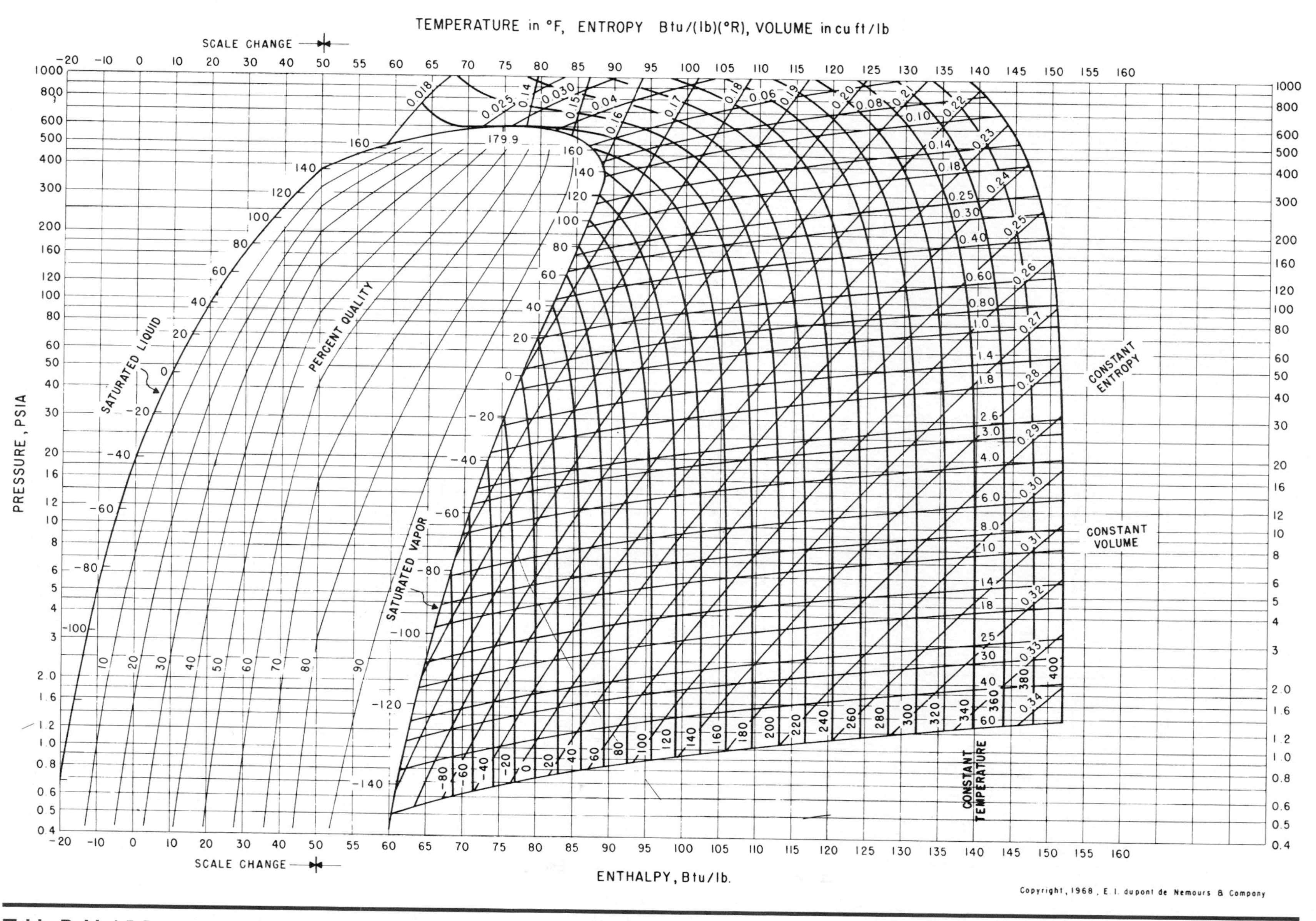

Table B-11 I-P Pressure-Heat Diagram for R-502

TEMP. °F	PRESSURE		VOLUME cu ft/lb		DENSITY lb/cu ft		ENTHALPY Btu/lb			ENTROPY Btu/(lb)(°R)		TEMP. °F
	PSIA	PSIG	LIQUID	VAPOR	LIQUID	VAPOR	LIQUID	LATENT	VAPOR	LIQUID	VAPOR	
−40	18.802	4.106	0.010941	2.0453	91.393	0.48891	0.000	73.114	73.114	0.00000	0.17421	−40
−39	19.269	4.573	0.010955	1.9987	91.280	0.50030	0.228	73.003	73.232	0.00054	0.17408	−39
−38	19.746	5.050	0.010968	1.9535	91.167	0.51189	0.457	72.892	73.350	0.00108	0.17395	−38
−37	20.231	5.535	0.010982	1.9094	91.053	0.52370	0.687	72.780	73.468	0.00163	0.17381	−37
−36	20.726	6.030	0.010996	1.8666	90.939	0.53571	0.918	72.667	73.585	0.00217	0.17368	−36
−35	21.230	6.534	0.011010	1.8249	90.825	0.54795	1.149	72.553	73.702	0.00271	0.17355	−35
−34	21.744	7.048	0.011023	1.7844	90.711	0.56040	1.381	72.438	73.819	0.00325	0.17342	−34
−33	22.267	7.571	0.011037	1.7449	90.597	0.57307	1.613	72.323	73.936	0.00380	0.17330	−33
−32	22.800	8.104	0.011051	1.7065	90.482	0.58597	1.846	72.206	74.053	0.00434	0.17317	−32
−31	23.343	8.647	0.011065	1.6691	90.367	0.59909	2.080	72.089	74.170	0.00488	0.17305	−31
−30	23.897	9.201	0.011080	1.6328	90.252	0.61244	2.315	71.971	74.286	0.00543	0.17293	−30
−29	24.460	9.764	0.011094	1.5973	90.136	0.62602	2.550	71.853	74.403	0.00597	0.17281	−29
−28	25.033	10.337	0.011108	1.5628	90.021	0.63984	2.785	71.733	74.519	0.00651	0.17269	−28
−27	25.617	10.921	0.011122	1.5292	89.905	0.65390	3.021	71.613	74.635	0.00706	0.17257	−27
−26	26.212	11.516	0.011137	1.4965	89.789	0.66819	3.258	71.492	74.751	0.00760	0.17245	−26
−25	26.817	12.121	0.011151	1.4646	89.673	0.68273	3.496	71.370	74.866	0.00815	0.17234	−25
−24	27.433	12.737	0.011166	1.4336	89.556	0.69752	3.734	71.247	74.982	0.00869	0.17222	−24
−23	28.059	13.363	0.011180	1.4033	89.439	0.71256	3.973	71.124	75.097	0.00923	0.17211	−23
−22	28.697	14.001	0.011195	1.3739	89.322	0.72785	4.212	71.000	75.212	0.00978	0.17200	−22
−21	29.346	14.650	0.011210	1.3451	89.205	0.74340	4.452	70.875	75.327	0.01032	0.17189	−21
−20	30.006	15.310	0.011224	1.3171	89.088	0.75920	4.693	70.749	75.442	0.01087	0.17178	−20
−19	30.678	15.982	0.011239	1.2898	88.970	0.77527	4.934	70.622	75.556	0.01141	0.17167	−19
−18	31.361	16.665	0.011254	1.2632	88.852	0.79160	5.176	70.494	75.671	0.01196	0.17156	−18
−17	32.056	17.360	0.011269	1.2373	88.734	0.80820	5.418	70.366	75.785	0.01250	0.17146	−17
−16	32.762	18.066	0.011284	1.2120	88.615	0.82507	5.661	70.237	75.899	0.01305	0.17135	−16
−15	33.480	18.784	0.011299	1.1873	88.496	0.84222	5.905	70.107	76.012	0.01359	0.17125	−15
−14	34.211	19.515	0.011315	1.1632	88.377	0.85964	6.149	69.976	76.126	0.01414	0.17115	−14
−13	34.954	20.258	0.011330	1.1397	88.258	0.87734	6.394	69.844	76.239	0.01469	0.17105	−13
−12	35.709	21.013	0.011345	1.1168	88.138	0.89533	6.640	69.712	76.352	0.01523	0.17095	−12
−11	36.476	21.780	0.011361	1.0945	88.018	0.91361	6.886	69.578	76.465	0.01578	0.17085	−11
−10	37.256	22.560	0.011376	1.0727	87.898	0.93218	7.133	69.444	76.577	0.01632	0.17075	−10
− 9	38.049	23.353	0.011392	1.0514	87.778	0.95104	7.380	69.309	76.690	0.01687	0.17066	− 9
− 8	38.854	24.158	0.011408	1.0307	87.657	0.97020	7.628	69.173	76.802	0.01741	0.17056	− 8
− 7	39.673	24.977	0.011423	1.0104	87.536	0.98966	7.877	69.037	76.914	0.01796	0.17047	− 7
− 6	40.504	25.808	0.011439	0.99065	87.415	1.0094	8.126	68.899	77.025	0.01851	0.17037	− 6
− 5	41.349	26.653	0.011455	0.97134	87.293	1.0295	8.376	68.761	77.137	0.01905	0.17028	− 5
− 4	42.207	27.511	0.011471	0.95247	87.172	1.0498	8.626	68.622	77.248	0.01960	0.17019	− 4
− 3	43.078	28.382	0.011487	0.93406	87.050	1.0705	8.877	68.482	77.359	0.02014	0.17010	− 3
− 2	43.964	29.268	0.011503	0.91607	86.927	1.0916	9.129	68.341	77.470	0.02069	0.17001	− 2
− 1	44.863	30.167	0.011520	0.89851	86.804	1.1129	9.381	68.199	77.580	0.02124	0.16992	− 1
0	45.775	31.079	0.011536	0.88135	86.681	1.1346	9.633	68.056	77.690	0.02178	0.16983	0
1	46.702	32.006	0.011552	0.86458	86.558	1.1566	9.887	67.913	77.800	0.02233	0.16975	1
2	47.643	32.947	0.011569	0.84821	86.434	1.1789	10.141	67.769	77.910	0.02287	0.16966	2
3	48.599	33.903	0.011586	0.83220	86.310	1.2016	10.395	67.624	78.019	0.02342	0.16958	3
4	49.568	34.872	0.011602	0.81657	86.186	1.2246	10.650	67.478	78.128	0.02397	0.16949	4
5	50.553	35.857	0.011619	0.80129	86.062	1.2479	10.906	67.331	78.237	0.02451	0.16941	5
6	51.552	36.856	0.011636	0.78635	85.937	1.2716	11.162	67.183	78.346	0.02506	0.16933	6
7	52.565	37.869	0.011653	0.77175	85.811	1.2957	11.419	67.035	78.454	0.02561	0.16925	7
8	53.594	38.898	0.011670	0.75748	85.686	1.3201	11.676	66.885	78.562	0.02615	0.16917	8
9	54.638	39.942	0.011687	0.74353	85.560	1.3449	11.934	66.735	78.670	0.02670	0.16909	9
10	55.697	41.001	0.011704	0.72988	85.434	1.3700	12.193	66.584	78.777	0.02724	0.16901	10
11	56.771	42.075	0.011722	0.71654	85.307	1.3955	12.452	66.432	78.885	0.02779	0.16893	11
12	57.861	43.165	0.011739	0.70349	85.180	1.4214	12.711	66.280	78.991	0.02834	0.16885	12
13	58.967	44.271	0.011757	0.69073	85.053	1.4477	12.972	66.126	79.098	0.02888	0.16878	13
14	60.088	45.392	0.011774	0.67825	84.925	1.4743	13.232	65.971	79.204	0.02943	0.16870	14

Table B-12 I-P Properties of Saturated Liquid and Vapor for R-502

TEMP. °F	PRESSURE		VOLUME cu ft/lb		DENSITY lb/cu ft		ENTHALPY Btu/lb			ENTROPY Btu/(lb)(°R)		TEMP. °F
	PSIA	PSIG	LIQUID	VAPOR	LIQUID	VAPOR	LIQUID	LATENT	VAPOR	LIQUID	VAPOR	
15	61.225	46.529	0.011792	0.66604	84.797	1.5013	13.494	65.816	79.310	0.02997	0.16863	15
16	62.379	47.683	0.011810	0.65410	84.669	1.5288	13.756	65.660	79.416	0.03052	0.16855	16
17	63.548	48.852	0.011828	0.64241	84.540	1.5566	14.018	65.503	79.521	0.03107	0.16848	17
18	64.734	50.038	0.011846	0.63098	84.411	1.5848	14.281	65.345	79.626	0.03161	0.16841	18
19	65.936	51.240	0.011864	0.61979	84.282	1.6134	14.545	65.186	79.731	0.03216	0.16833	19
20	67.155	52.459	0.011883	0.60884	84.152	1.6424	14.809	65.026	79.836	0.03270	0.16826	20
21	68.391	53.695	0.011901	0.59812	84.022	1.6719	15.073	64.866	79.940	0.03325	0.16819	21
22	69.643	54.947	0.011920	0.58762	83.891	1.7017	15.339	64.704	80.043	0.03379	0.16812	22
23	70.913	56.217	0.011938	0.57735	83.760	1.7320	15.604	64.542	80.147	0.03434	0.16805	23
24	72.199	57.503	0.011957	0.56730	83.629	1.7627	15.871	64.379	80.250	0.03488	0.16799	24
25	73.503	58.807	0.011976	0.55745	83.497	1.7938	16.138	64.215	80.353	0.03543	0.16792	25
26	74.824	60.128	0.011995	0.54781	83.365	1.8254	16.405	64.050	80.455	0.03597	0.16785	26
27	76.163	61.467	0.012014	0.53837	83.232	1.8574	16.673	63.884	80.557	0.03652	0.16778	27
28	77.520	62.824	0.012033	0.52912	83.099	1.8898	16.941	63.717	80.659	0.03706	0.16772	28
29	78.894	64.198	0.012053	0.52007	82.966	1.9228	17.210	63.550	80.760	0.03761	0.16765	29
30	80.286	65.590	0.012072	0.51120	82.832	1.9561	17.480	63.381	80.861	0.03815	0.16759	30
31	81.697	67.001	0.012092	0.50251	82.698	1.9900	17.750	63.212	80.962	0.03870	0.16752	31
32	83.126	68.430	0.012111	0.49399	82.563	2.0242	18.020	63.042	81.062	0.03924	0.16746	32
33	84.573	69.877	0.012131	0.48565	82.428	2.0590	18.291	62.871	81.162	0.03979	0.16739	33
34	86.038	71.342	0.012151	0.47748	82.292	2.0943	18.563	62.698	81.262	0.04033	0.16733	34
35	87.522	72.826	0.012171	0.45947	82.156	2.1300	18.835	62.526	81.361	0.04087	0.16727	35
36	89.026	74.330	0.012192	0.46162	82.020	2.1662	19.107	62.352	81.460	0.04142	0.16721	36
37	90.548	75.852	0.012212	0.45393	81.883	2.2029	19.381	62.177	81.558	0.04196	0.16715	37
38	92.089	77.393	0.012233	0.44638	81.746	2.2401	19.654	62.001	81.656	0.04250	0.16708	38
39	93.649	78.953	0.012253	0.43899	81.608	2.2779	19.928	61.825	81.754	0.04305	0.16702	39
40	95.229	80.533	0.012274	0.43175	81.469	2.3161	20.203	61.647	81.851	0.04359	0.16696	40
41	96.828	82.132	0.012295	0.42464	81.330	2.3549	20.478	61.469	81.948	0.04413	0.16690	41
42	98.447	83.751	0.012316	0.41767	81.191	2.3941	20.754	61.290	82.044	0.04468	0.16684	42
43	100.08	85.38	0.012337	0.41084	81.051	2.4339	21.030	61.110	82.140	0.04522	0.16678	43
44	101.74	87.04	0.012359	0.40414	80.911	2.4743	21.307	60.928	82.235	0.04576	0.16673	44
45	103.42	88.72	0.012380	0.39757	80.770	2.5152	21.584	60.746	82.331	0.04630	0.16667	45
46	105.12	90.42	0.012402	0.39113	80.629	2.5566	21.861	60.563	82.425	0.04685	0.16661	46
47	106.84	92.14	0.012424	0.38481	80.487	2.5986	22.140	60.380	82.520	0.04739	0.16655	47
48	108.58	93.88	0.012446	0.37861	80.345	2.6412	22.418	60.195	82.613	0.04793	0.16649	48
49	110.34	95.64	0.012468	0.37252	80.202	2.6843	22.697	60.009	82.707	0.04847	0.16644	49
50	112.12	97.42	0.012490	0.36655	80.058	2.7280	22.977	59.822	82.800	0.04901	0.16638	50
51	113.92	99.22	0.012513	0.36070	79.914	2.7723	23.257	59.635	82.892	0.04955	0.16632	51
52	115.74	101.05	0.012536	0.35495	79.769	2.8172	23.538	59.446	82.984	0.05009	0.16627	52
53	117.59	102.89	0.012558	0.34931	79.624	2.8627	23.819	59.257	83.076	0.05063	0.16621	53
54	119.45	104.75	0.012581	0.34377	79.479	2.9088	24.100	59.066	83.167	0.05117	0.16616	54
55	121.34	106.64	0.012605	0.33834	79.332	2.9555	24.382	58.875	83.257	0.05171	0.16610	55
56	123.25	108.55	0.012628	0.33301	79.185	3.0028	24.665	58.682	83.347	0.05225	0.16605	56
57	125.18	110.48	0.012652	0.32777	79.038	3.0508	24.948	58.489	83.437	0.05279	0.16599	57
58	127.13	112.43	0.012675	0.32263	78.890	3.0994	25.231	58.294	83.526	0.05333	0.16594	58
59	129.10	114.41	0.012699	0.31759	78.741	3.1486	25.515	58.099	83.615	0.05387	0.16588	59
60	131.10	116.40	0.012723	0.31263	78.592	3.1985	25.799	57.903	83.703	0.05441	0.16583	60
61	133.12	118.42	0.012748	0.30777	78.441	3.2491	26.084	57.705	83.790	0.05495	0.16577	61
62	135.16	120.46	0.012772	0.30299	78.291	3.3004	26.370	57.507	83.877	0.05549	0.16572	62
63	137.22	122.53	0.012797	0.29830	78.140	3.3523	26.656	57.308	83.964	0.05602	0.16566	63
64	139.31	124.61	0.012822	0.29369	77.988	3.4049	26.942	57.107	84.050	0.05656	0.16561	64
65	141.42	126.72	0.012847	0.28916	77.835	3.4582	27.229	56.906	84.135	0.05710	0.16556	65
66	143.55	128.86	0.012872	0.28471	77.682	3.5122	27.516	56.704	84.220	0.05764	0.16550	66
67	145.71	131.01	0.012898	0.28034	77.528	3.5669	27.804	56.500	84.304	0.05817	0.16545	67
68	147.89	133.19	0.012924	0.27605	77.373	3.6224	28.092	56.296	84.388	0.05871	0.16539	68
69	150.09	135.39	0.012950	0.27183	77.217	3.6786	28.380	56.091	84.471	0.05925	0.16534	69

Table B-12 (continued)

TEMP. °F	PRESSURE		VOLUME cu ft/lb		DENSITY lb/cu ft		ENTHALPY Btu/lb			ENTROPY Btu/(lb)(°R)		TEMP. °F
	PSIA	PSIG	LIQUID	VAPOR	LIQUID	VAPOR	LIQUID	LATENT	VAPOR	LIQUID	VAPOR	
70	152.32	137.62	0.012976	0.26769	77.061	3.7356	28.669	55.884	84.554	0.05978	0.16529	70
71	154.57	139.87	0.013003	0.26362	76.904	3.7933	28.959	55.676	84.636	0.06032	0.16523	71
72	156.84	142.15	0.013029	0.25961	76.747	3.8517	29.249	55.468	84.717	0.06085	0.16518	72
73	159.14	144.44	0.013056	0.25568	76.588	3.9110	29.539	55.258	84.798	0.06139	0.16512	73
74	161.46	146.77	0.013083	0.25181	76.429	3.9711	29.830	55.047	84.878	0.06193	0.16507	74
75	163.81	149.11	0.013111	0.24801	76.269	4.0319	30.122	54.835	84.958	0.06246	0.16502	75
76	166.18	151.49	0.013139	0.24428	76.108	4.0936	30.414	54.622	85.037	0.06299	0.16496	76
77	168.58	153.88	0.013167	0.24060	75.947	4.1561	30.706	54.408	85.115	0.06353	0.16491	77
78	171.00	156.30	0.013195	0.23699	75.784	4.2194	30.999	54.193	85.193	0.06406	0.16485	78
79	173.45	158.75	0.013223	0.23344	75.621	4.2836	31.292	53.977	85.270	0.06460	0.16480	79
80	175.92	161.22	0.013252	0.22995	75.457	4.3487	31.586	53.759	85.346	0.06513	0.16474	80
81	178.41	163.72	0.013281	0.22651	75.292	4.4146	31.880	53.541	85.421	0.06566	0.16469	81
82	180.94	166.24	0.013310	0.22313	75.126	4.4815	32.175	53.321	85.496	0.06620	0.16463	82
83	183.48	168.79	0.013340	0.21981	74.959	4.5492	32.470	53.100	85.570	0.06673	0.16458	83
84	186.06	171.36	0.013370	0.21654	74.791	4.6179	32.766	52.878	85.644	0.06726	0.16452	84
85	188.66	173.96	0.013400	0.21333	74.622	4.6874	33.062	52.654	85.717	0.06780	0.16446	85
86	191.28	176.59	0.013431	0.21017	74.453	4.7580	33.359	52.430	85.789	0.06833	0.16441	86
87	193.94	179.24	0.013462	0.20705	74.282	4.8295	33.656	52.204	85.860	0.06886	0.16435	87
88	196.62	181.92	0.013493	0.20399	74.111	4.9020	33.953	51.977	85.930	0.06939	0.16429	88
89	199.32	184.62	0.013524	0.20098	73.938	4.9755	34.251	51.748	86.000	0.06992	0.16423	89
90	202.05	187.36	0.013556	0.19801	73.764	5.0500	34.550	51.519	86.069	0.07045	0.16418	90
91	204.81	190.12	0.013588	0.19509	73.590	5.1255	34.849	51.288	86.137	0.07098	0.16412	91
92	207.60	192.90	0.013621	0.19222	73.414	5.2021	35.148	51.055	86.204	0.07151	0.16406	92
93	210.41	195.72	0.013654	0.18939	73.237	5.2798	35.448	50.822	86.271	0.07205	0.16400	93
94	213.25	198.56	0.013687	0.18661	73.059	5.3585	35.749	50.587	86.336	0.07258	0.16394	94
95	216.12	201.43	0.013721	0.18387	72.880	5.4384	36.050	50.350	86.401	0.07311	0.16388	95
96	219.02	204.32	0.013755	0.18117	72.700	5.5194	36.352	50.113	86.465	0.07364	0.16382	96
97	221.94	207.25	0.013789	0.17852	72.519	5.6015	36.654	49.874	86.528	0.07417	0.16375	97
98	224.90	210.20	0.013824	0.17590	72.336	5.6848	36.956	49.633	86.590	0.07470	0.16369	98
99	227.88	213.18	0.013859	0.17333	72.152	5.7693	37.259	49.391	86.651	0.07522	0.16363	99
100	230.89	216.19	0.013895	0.17079	71.967	5.8550	37.563	49.147	86.711	0.07575	0.16356	100
101	233.93	219.23	0.013931	0.16829	71.781	5.9419	37.867	48.902	86.770	0.07628	0.16350	101
102	237.00	222.30	0.013967	0.16583	71.593	6.0301	38.172	48.656	86.828	0.07681	0.16343	102
103	240.09	225.40	0.014004	0.16340	71.405	6.1196	38.477	48.407	86.885	0.07734	0.16337	103
104	243.22	228.52	0.014042	0.16101	71.214	6.2104	38.783	48.158	86.941	0.07787	0.16330	104
105	246.38	231.68	0.014079	0.15866	71.023	6.3026	39.090	47.906	86.997	0.07840	0.16323	105
106	249.56	234.87	0.014118	0.15634	70.829	6.3961	39.398	47.653	87.051	0.07893	0.16316	106
107	252.78	238.08	0.014157	0.15405	70.635	6.4910	39.705	47.398	87.103	0.07946	0.16309	107
108	256.02	241.33	0.014196	0.15180	70.439	6.5873	40.013	47.142	87.155	0.07998	0.16302	108
109	259.30	244.60	0.014236	0.14958	70.241	6.6851	40.322	46.883	87.206	0.08051	0.16295	109
110	262.61	247.91	0.014277	0.14739	70.042	6.7843	40.631	46.623	87.255	0.08104	0.16288	110
111	265.94	251.25	0.014318	0.14523	69.841	6.8851	40.942	46.361	87.303	0.08157	0.16281	111
112	269.31	254.62	0.014359	0.14311	69.639	6.9875	41.252	46.098	87.350	0.08210	0.16273	112
113	272.71	258.02	0.014401	0.14101	69.435	7.0914	41.564	45.832	87.396	0.08263	0.16265	113
114	276.15	261.45	0.014444	0.13894	69.229	7.1970	41.876	45.564	87.441	0.08316	0.16258	114
115	279.61	264.91	0.014488	0.13690	69.022	7.3042	42.189	45.294	87.484	0.08368	0.16250	115
116	283.10	268.41	0.014532	0.13489	68.812	7.4132	42.503	45.022	87.525	0.08421	0.16242	116
117	286.63	271.94	0.014576	0.13290	68.601	7.5239	42.817	44.748	87.566	0.08474	0.16234	117
118	290.19	275.50	0.014622	0.13095	68.388	7.6364	43.132	44.472	87.605	0.08527	0.16225	118
119	293.78	279.09	0.014668	0.12902	68.173	7.7507	43.448	44.194	87.642	0.08580	0.16217	119
120	297.41	282.71	0.014715	0.12711	67.956	7.8669	43.765	43.913	87.678	0.08633	0.16208	120
121	301.07	286.37	0.014762	0.12523	67.737	7.9850	44.082	43.630	87.713	0.08686	0.16199	121
122	304.76	290.06	0.014811	0.12337	67.515	8.1052	44.401	43.344	87.746	0.08739	0.16190	122
123	308.49	293.79	0.014860	0.12154	67.292	8.2274	44.720	43.056	87.777	0.08792	0.16181	123
124	312.25	297.55	0.014910	0.11973	67.066	8.3516	45.040	42.766	87.807	0.08845	0.16172	124

Table B-12 (continued)

TEMP. °F	PRESSURE		VOLUME cu ft/lb		DENSITY lb/cu ft		ENTHALPY Btu/lb			ENTROPY Btu/(lb)(°R)		TEMP. °F
	PSIA	PSIG	LIQUID	VAPOR	LIQUID	VAPOR	LIQUID	LATENT	VAPOR	LIQUID	VAPOR	
125	316.04	301.35	0.014961	0.11795	66.838	8.4781	45.361	42.472	87.834	0.08899	0.16163	125
126	319.87	305.17	0.015013	0.11618	66.608	8.6067	45.684	42.176	87.860	0.08952	0.16153	126
127	323.73	309.04	0.015065	0.11444	66.375	8.7377	46.007	41.878	87.885	0.09005	0.16143	127
128	327.63	312.94	0.015119	0.11272	66.140	8.8710	46.331	41.576	87.907	0.09058	0.16133	128
129	331.57	316.87	0.015174	0.11102	65.901	9.0067	46.656	41.271	87.928	0.09112	0.16122	129
130	335.54	320.84	0.015229	0.10934	65.661	9.1449	46.983	40.963	87.946	0.09165	0.16112	130
131	339.55	324.85	0.015286	0.10769	65.417	9.2858	47.310	40.652	87.962	0.09219	0.16101	131
132	343.59	328.89	0.015344	0.10605	65.171	9.4293	47.639	40.337	87.977	0.09273	0.16090	132
133	347.67	332.97	0.015403	0.10443	64.921	9.5755	47.969	40.019	87.989	0.09326	0.16078	133
134	351.79	337.09	0.015463	0.10283	64.669	9.7246	48.300	39.697	87.998	0.09380	0.16067	134
135	355.94	341.24	0.015524	0.10124	64.413	9.8767	48.633	39.372	88.006	0.09434	0.16055	135
136	360.13	345.44	0.015587	0.099682	64.154	10.031	48.967	39.043	88.011	0.09488	0.16042	136
137	364.36	349.67	0.015651	0.098133	63.892	10.190	49.303	38.709	88.013	0.09543	0.16030	137
138	368.63	353.94	0.015716	0.096601	63.626	10.351	49.641	38.372	88.013	0.09597	0.16017	138
139	372.94	358.25	0.015783	0.095086	63.356	10.516	49.980	38.030	88.010	0.09652	0.16004	139
140	377.29	362.60	0.015852	0.093586	63.082	10.685	50.320	37.683	88.004	0.09706	0.15990	140
141	381.68	366.98	0.015922	0.092101	62.805	10.857	50.663	37.331	87.995	0.09761	0.15976	141
142	386.11	371.41	0.015993	0.090630	62.523	11.033	51.008	36.975	87.983	0.09817	0.15962	142
143	390.58	375.88	0.016067	0.089174	62.237	11.213	51.354	36.613	87.968	0.09877	0.15947	143
144	395.09	380.39	0.016142	0.087732	61.946	11.398	51.703	36.245	87.949	0.09928	0.15931	144
145	399.64	384.95	0.016220	0.086303	61.650	11.587	52.054	35.872	87.927	0.09983	0.15916	145
146	404.24	389.54	0.016299	0.084886	61.350	11.780	52.408	35.493	87.901	0.10039	0.15899	146
147	408.88	394.18	0.016381	0.083481	61.044	11.978	52.764	35.107	87.871	0.10096	0.15882	147
148	413.56	398.86	0.016465	0.082088	60.732	12.181	53.123	34.714	87.838	0.10152	0.15865	148
149	418.29	403.59	0.016551	0.080706	60.415	12.390	53.485	34.314	87.800	0.10209	0.15847	149
150	423.06	408.35	0.016641	0.079335	60.092	12.604	53.850	33.907	87.757	0.10267	0.15828	150
151	427.87	413.18	0.016732	0.077973	59.762	12.824	54.218	33.491	87.710	0.10325	0.15809	151
152	432.74	418.04	0.016827	0.076620	59.425	13.051	54.590	33.067	87.657	0.10383	0.15789	152
153	437.65	422.95	0.016925	0.075276	59.082	13.284	54.966	32.633	87.599	0.10442	0.15769	153
154	442.60	427.91	0.017026	0.073939	58.730	13.524	55.346	32.190	87.536	0.10501	0.15747	154
155	447.61	432.91	0.017131	0.072610	58.371	13.772	55.730	31.736	87.467	0.10561	0.15725	155
156	452.66	437.97	0.017240	0.071286	58.002	14.027	56.119	31.271	87.391	0.10622	0.15701	156
157	457.77	443.07	0.017353	0.069968	57.625	14.292	56.513	30.794	87.308	0.10683	0.15677	157
158	462.92	448.22	0.017470	0.068655	57.237	14.565	56.913	30.304	87.218	0.10746	0.15652	158
159	468.13	453.43	0.017593	0.067345	56.839	14.848	57.320	29.800	87.120	0.10809	0.15625	159
160	473.38	458.69	0.017721	0.066037	56.429	15.142	57.732	29.281	87.013	0.10872	0.15598	160
161	478.70	464.00	0.017854	0.064730	56.007	15.448	58.153	28.744	86.898	0.10937	0.15569	161
162	484.06	469.37	0.017994	0.063423	55.571	15.767	58.581	28.190	86.772	0.11004	0.15538	162
163	489.48	474.79	0.018141	0.062114	55.121	16.099	59.019	27.616	86.635	0.11071	0.15506	163
164	494.96	480.27	0.018296	0.060802	54.654	16.446	59.466	27.019	86.486	0.11140	0.15472	164
165	500.50	485.80	0.018460	0.059484	54.169	16.811	59.925	26.398	86.324	0.11210	0.15436	165
166	506.10	491.40	0.018634	0.058158	53.665	17.194	60.397	25.749	86.146	0.11283	0.15398	166
167	511.75	497.06	0.018818	0.056822	53.138	17.598	60.883	25.069	85.953	0.11357	0.15358	167
168	517.47	502.78	0.019016	0.055472	52.586	18.026	61.385	24.355	85.740	0.11434	0.15314	168
169	523.26	508.56	0.019228	0.054104	52.005	18.482	61.906	23.599	85.506	0.11514	0.15268	169
170	529.11	514.41	0.019458	0.052714	51.392	18.970	62.449	22.798	85.248	0.11597	0.15217	170
171	535.03	520.33	0.019708	0.051295	50.739	19.495	63.019	21.941	84.960	0.11684	0.15163	171
172	541.01	526.32	0.019983	0.049839	50.041	20.064	63.619	21.018	84.638	0.11775	0.15103	172
173	547.07	532.38	0.020288	0.048335	49.288	20.688	64.258	20.016	84.274	0.11873	0.15037	173
174	553.20	538.51	0.020633	0.046769	48.465	21.381	64.944	18.913	83.858	0.11978	0.14962	174
175	559.41	544.72	0.021029	0.045118	47.552	22.163	65.693	17.681	83.374	0.12092	0.14878	175
176	565.70	551.01	0.021496	0.043348	46.518	23.069	66.525	16.273	82.798	0.12219	0.14779	176
177	572.08	557.38	0.022071	0.041396	45.307	24.156	67.480	14.607	82.087	0.12365	0.14659	177
178	578.54	563.84	0.022829	0.039137	43.803	25.551	68.635	12.515	81.151	0.12542	0.14505	178
179	585.09	570.39	0.023993	0.036203	41.677	27.621	70.210	9.518	79.729	0.12784	0.14275	179
179.889	591.00	576.30	0.028571	0.028571	35.000	35.000	74.654	0.000	74.654	0.13476	0.13476	179.889

Table B-12 (continued)

APPENDIX C

Commonly Used HVAC/R Formulas

1. Formula to convert degrees Fahrenheit to degrees Celsius (Chapter 1, Unit 4)

$$°F = 9/5(°C) + 32$$

or

$$°C = 5/9(°F - 32)$$

2. Formula to measure power (Chapter 2, Unit 2)

$$\text{watts} = \text{volts} \times \text{amps} \times \text{power factor}$$

3. Formula to measure area (Chapter 3, Unit 6)

$$A = L \times W$$

4. Formula to measure quantity of heat (Chapter 4, Unit 1)

$$\text{Btu} = W \times \Delta T$$

5. Formula to measure quantity of specific heat (Chapter 4, Unit 1)

$$\text{Btu} = W \times c \times \Delta T$$

6. Conversion factors for heat transfer (Chapter 4, Unit 1)

$$1\ \text{J/g} = 0.4299\ \text{Btu/lb}$$
$$1\ \text{Btu/lb} = 2.326\ \text{J/g}$$

7. Formula to calculate sensible heat added or removed from a substance (Chapter 4, Unit 1)

$$Q = W \times \text{SH} \times \Delta T$$

where
Q = quantity of heat
W = weight of the material
SH = specific heat
ΔT = change in temperature

8. Formula to determine the latent heat added or removed from a substance (Chapter 4, Unit 1)

$$Q = W \times \text{LH}$$

9. Formula to measure pressure (Chapter 4, Unit 3)

$$\text{Pressure} = \frac{\text{Force}}{\text{Area}}$$

or

$$P = \frac{F}{A}$$

10. Formula to measure head pressure (Chapter 4, Unit 3)

$$p = 0.433 \times h$$

where
p = pressure, psi
h = head, feet of water

11. Formula to measure density (Chapter 4, Unit 3)

$$D = \frac{W}{V}$$

where
D = density
W = weight
V = volume

12. Formula to measure specific gravity (Chapter 4, Unit 3)

$$\text{Pressure} = \text{head} \times \text{density}$$
$$p = h \times D$$

where
p = pressure, lb/ft^2
h = head or depth below the surface, ft
D = density, lb/ft^3

or where
p = pressure, psi
h = head or depth below the surface, in
D = density, lb/in^3

13. Formula to measure gas pressure (Boyle's Law) (Chapter 4, Unit 3)

$$p_1V_1 = p_2V_2$$

where
p_1 = original pressure
V_1 = original volume
p_2 = new pressure
V_2 = new volume

14. Formula to measure gas expansion (Charles' Law) (Chapter 4, Unit 3)

$$\frac{V_1}{V_2} = \frac{T_1}{T_2}$$

and

$$\frac{p_1}{p_2} = \frac{T_1}{T_2}$$

where
T = absolute temperature
p = absolute pressure

To clear the fractions, these may also be expressed as

$$V_1T_2 = V_2T_1 \quad \text{and} \quad p_1T_2 = p_2T_1$$

15. Formula to determine refrigeration effect (Chapter 4, Unit 4)

$$W = \frac{200}{\text{NRE}}$$

where
W = weight of refrigerant circulated per minute, lb/min
200 = 200 Btu/min—the equivalent of 1 ton of refrigeration
NRE = net refrigerating effect, Btu/lb of refrigerant

16. Formula to measure coefficient of performance (COP) (Chapter 4, Unit 4)

$$\text{COP} = \frac{\text{refrigerating effect}}{\text{heat of compression}}$$

17. Formula to measure electric current (Chapter 7, Unit 1)

$$R = pl/A$$

where
R = resistance in ohms
p = *(rho)*, resistivity of the material in ohm-meters (a constant for a given material at a given material at a given temperature)
l = length in meters
A = cross sectional area of conductor

18. Formula to measure power (Ohm's Law) (Chapter 7, Unit 1)

$$P = EI\,(\text{PF})$$
$$P = I^2R$$
$$P = E^2/R$$

where
PF (Power Factor) = phase angle between E and I

19. Formula to measure alternating current (also Ohm's Law) (Chapter 7, Unit 1)

$$I = \frac{E}{R}$$

where
I = Current
E = Electrical potential
R = Resistance

20. Formulate to measure/calculate Ohm's Law (another Ohm's Law version) (Chapter 7, Unit 1)

$$E = IR \quad R = E/I$$

21. Formula to calculate electric power (Chapter 7, Unit 1)

$$P = EI$$

22. Power Formula for DC, Single Phase AC, and Three Phase AC Current (Chapter 7, Unit 1)

	DC	Single-Phase AC	Three-Phase AC
Power Formula 1	$P = EI$	$P = EI(\text{PF})$	$P = \sqrt{3EI(\text{PF})}$*
Power Formula 2	$P = E^2/R$	$P = E^2/R$	$P = \sqrt{3E^2IR}$
Power Formula 3	$P = I^2R$	$P = I^2R$	$P = \sqrt{3I^2R}$

*$\sqrt{3} = 1.732$

23. Formulas to calculate series circuits (Chapter 7, Unit 1)

$$I_1 = I_2 = I_3 = I_4$$
$$R_T = R_1 + R_2 + R_3 + R_4$$
$$E_T = E_1 + E_2 + E_3 + E_4$$

24. Formulas to calculate parallel circuits (Chapter 7, Unit 1)

$$I_T = I_1 + I_2 + I_3 + I_4 + \cdots$$
$$R_T = \frac{R_1 \times R_2}{R_1 + R_2}$$
$$\frac{1}{R_T} = \frac{1}{R_1} + \frac{1}{R_2} + \frac{1}{R_3} + \frac{1}{R_4} + \cdots$$
$$E_T = E_1 = E_2 = E_3 = E_4 = \cdots$$

25. Formulas to measure power factor (Chapter 7, Unit 2)

Power factor = (True power)/(apparent power)
or
Power factor = (Wattmeter reading)/(E × I)

26. Formula to calculate impedance in an inductive circuit (Chapter 7, Unit 2)

$$Z = E/I$$

where
Z = impedance

27. Formula to Calculate Synchronous Motor Speed (Chapter 7, Unit 4)

$$\text{rpm} = \frac{\text{Hz} \times 60 \text{ sec/min}}{\frac{1}{2}p} \quad \text{or} \quad \frac{\text{Hz} \times 120}{p}$$

where
rpm = revolutions per minute
Hz = frequency in cycles/sec
p = number of poles

28. Formula to determine the system capacity (Chapter 9, Unit 1)

$$Q_t = 4.45 \times \text{CFM} \times \Delta\text{h}$$

where
Q_t = the total (sensible and latent) cooling being done
CFM = airflow across the evaporator coil
Δh = the change of enthalpy of the air across the coil

29. Formula to determine coil operating temperatures (Chapter 9, Unit 2)

$$COT = \left(\frac{EAT + LAT}{2}\right) - split$$

where
COT = coil operating temperature
EAT = temperature of air entering the coil
LAT = temperature of air leaving the coil

30. Formula to find the condensing temperature (Chapter 9, Unit 2)

$$\text{RCT} = \text{EAT} + \text{split}$$

where
RCT = refrigerant condensing temperature
EAT = temperature of the air entering the condenser
split = design temperature difference between the entering air temperature and the condensing temperatures of the hot high-pressure vapor from the compressor.

31. Formula to measure gas flow through the meter (Chapter 10, Unit 3)

$$\text{Time (sec/ft}^3) = \frac{\text{seconds per hour}}{\text{ft}^3\text{/hr of gas}}$$

32. Formula to measure airflow through the furnace (Cubic Feet Per Minute) (Chapter 12, Unit 2)

$$\text{CFM} = \frac{V \times A \times 3.412}{\Delta T \times 1.08}$$

33. Formula to measure energy efficiency ratio in a heat pump (Chapter 13, Unit 1)

$$\text{SEER} = \frac{\text{Total cooling season BTU}}{\text{Total watts consumed}}$$

34. Formula to measure coefficient of performance (Chapter 13, Unit 1)

$$\text{COP} = \frac{\text{BTU heat output}}{\text{Electric heat equivalent}}$$

35. Formula to measure total heat output of a heat pump (Chapter 13, Unit 1)

$$\text{HSPF} = \frac{\text{Total BTU heating season output}}{\text{Total heating season watts}}$$

36. Formula to calculate gross heating capacity (Chapter 13, Unit 3)

$$\text{H}_s = \text{CFM} \times \text{TD} \times 1.08$$

37. Formula to calculate gross cooling capacity (Chapter 13, Unit 3)

$$\text{HT} = \text{gpm} \times \text{TD} \times 500$$
$$\text{HM} = \text{watts} \times 3.415 \times \text{PF}$$

38. Formula to calculate net cooling capacity (Chapter 13, Unit 3)

$$\text{HC} = \text{HT} - \text{HM}$$

where
HT = Heat transfer (gross capacity)
HM = Motor heat
HC = Net cooling capacity
PF = Power factor

39. Formula to calculate heating capacity of heat pump (Chapter 13, Unit 3)

$$\text{Q} = 8.33 \times \text{gal} \times \text{TD}$$

where
Q = the heating capacity of the unit, Btu/hr
8.33 = conversion factor from gal to lb
TD = the temperature rise of the water in one hour, °F

40. Formula to determine airflow rate of a system (Chapter 14, Unit 1)

$$Q = Q_s(1.08 \times \text{TD})$$

where
Q = flow rate in CFM
Q_s = sensible-heat load in But/hr
TD = dry bulb temperature difference in°F

41. Formulas to calculate for fan operation ("Fan Laws") (Chapter 14, Unit 1)

a. $$\frac{\text{CFM}_2}{\text{CFM}_1} = \frac{\text{rpm}_2}{\text{rpm}_1}$$

b. $$\frac{SP_2}{SP_1} = \left(\frac{rpm_2}{rpm_1}\right)^2$$

c. $$\frac{Bhp_2}{Bhp_1} = \left(\frac{rpm_2}{rpm_1}\right)^3$$

42. Formula to calculate total airflow through a unit (Chapter 14, Unit 3)

$$\text{CFM} = \frac{\text{furnace output in Btu/hr}}{(\text{TD} \times 1.08)}$$

where
TD is the temperature rise in °F

43. Formula to determine CFM (Chapter 14, Unit 3)

$$\text{CFM} = \text{free area of grill (ft}^2) \times \text{fpm}$$

44. Formula to determine quantity of air used for cooling, based on the sensible cooling (Chapter 14, Unit 3)

$$\text{CFM} = H_s/(\text{TD} \times 1.08)$$

45. Performance Calculation Formulas (Chapter 15, Unit 1)

a. Sensible heat removed

$$Q_s = 1.08 \times \text{CFM} \times \text{DBT difference}$$

b. Latent heat removed

$$Q_1 = 0.68 \times \text{CFM} \times \text{gr moisture difference}$$

c. Total heat removed

$$Q_t = Q_s + Q_l$$

or

$$Q_t = 4.5 \times \text{CFM} \times \text{total heat difference}$$

46. Formula to Calculate Rate of Heat Transfer (Chapter 16, Unit 1)

$$Q = U \times A \times \text{TD}$$

Where

Q = heat transfer, Btuh
U = overall heat transfer coefficeint (Btuh/ft^2/°F)
A = area(ft^2)
TD = temperature difference between inside and outside design temperature and refrigerated space design temperature

47. Formula to calculate the heat to be removed from a product to lower its temperature (above freezing) (Chapter 16, Unit 1)

$$Q = W \times C \times (T_1 - T_2)$$

where

Q = Btu to be removed
W = weight of product in lb
C = specific heat above freezing
T_1 = initial temperature,°F
T_2 = final temperature, °F (freezing or above)

48. Formula to calculate the heat to be removed from a product for the latent heat of freezing (Chapter 16, Unit 1)

$$Q = W \times h_f$$

where

Q = Btu to be removed
W = weight of product, lb
h_f = latent heat of fusion, Btu/lb

49. Formula to calculate the heat to be removed from a product to lower is temperature below freezing (Chapter 16, Unit 1)

$$Q = W C_i \times (T_f - T_3)$$

where

Q = Btu to be removed
W = weight of product, lb
C_i = specific heat below freezing
T_f = freezing temperature
T_3 = final temperature

50. Formula to convert the boiler output to gpm (Chapter 19, Unit 3)

gpm = boiler output in Btu/hr/(TD × 500)

where

TD = 210 − 190°f = 20°F

Substituting in the formula,

gpm = 100,000 Btu/hr (20°F × 500) = 10 gpm

51. Formula to calculate heat (Chapter 19, Unit 5)

A tower circulating 18 gpm with a 10°F cooling range:

$$Q = \text{gpm} \times 500 \times \text{TD}$$

where

Q = heat rejection in Btu/hr
gpm = flow in gallons per minute
500 = factor derived form, specific heat of water (1) × 60 min/hr × 8.33 lb/gal
TD = temperature difference in °F

APPENDIX D

Tables and Figures with Useful HVAC/R Data

English Language Glossary

Absolute Humidity The amount of moisture actually in a given volume of air.

Absolute Pressure Gauge pressure plus atmospheric pressure.

Absolute Temperature Temperature measured from absolute zero.

Absolute Zero Temperature at which all molecular motion ceases (−460°F and −273°C).

Absorbent Substance which has the ability to take up or absorb another substance.

Absorber A device containing liquid for absorbing refrigerant vapor or other vapors.

Accumulator A storage vessel located in the suction line ahead of the compressor. Used to limit liquid refrigerant return to the compressor and store excess refrigerant in the heating mode.

Activated Carbon Specially processed carbon commonly used to clean air.

Adiabatic Compression Compressing refrigerant gas without removing or adding heat.

Adsorbent Substance which has property to hold molecules of fluids without causing a chemical or physical change.

Air Binding A condition in which a bubble or other pocket of air is present in a pipeline that prevents the desired flow in the pipeline.

Air Changes The amount of air leakage through a building in terms of the number of building volumes exchanged.

Air Conditioner Device used to control temperature, humidity, cleanliness, and movement of air in conditioned space.

Air Cushion Tank (See Expansion Tank)

Air Shutter An adjustable shutter on the primary air openings of a burner, which is used to control the amount of combustion air.

Air Source Heat Pump Heat pump that transfers heat from outdoor air to an indoor air circulation system.

Air Vent A valve installed at the high points in a hot water system to eliminate air from the system.

Aldehyde A class of compounds that can be produced during incomplete combustion of a fuel gas.

Allen Head Screw Screw with recessed head designed to be turned with a hex shaped wrench.

Alternating Current Abbreviated AC. Current that reverses polarity or direction periodically. It rises from zero to maximum strength and returns to zero in one direction then goes to similar variation in the opposite direction. This is a cycle which is repeated at a fixed frequency. It can be single phase, two phase, three phase, and polyphase. Its advantage over direct or undirectional current is that DC voltage can be stepped up by transformers to a high value which reduces transmission costs.

Alternator A machine that converts mechanical energy into alternating current.

Ambient Temperature Temperature of the air that surrounds an object.

American Wire Gauge Abbreviated AWG. A system of numbers that designate cross-sectional area of wire. As the diameter gets smaller, the number gets larger, e.g., AWG #14 = 0.0641 in, AWG #12 = 0.0808 in.

Ammeter An electric meter used to measure current.

Ampere Unit of electric current equivalent to flow of one coulomb per second.

Ampere Turn Abbreviated AT or NI. Unit of magnetizing force produced by a current flow of one ampere through one turn of wire in a coil.

Amplitude The maximum instantaneous value of alternating current or voltage. It can be in either a positive or negative direction.

Anemometer Instrument for measuring the velocity of air.

Angle of Lag or Lead Phase angle difference between two sinusoidal wave forms having the same frequency.

Annealing Process of heat treating metal to obtain desired properties of softness and ductility.

Anticipator A heater used to adjust thermostat operation to produce a closer temperature differential than the mechanical capability of the control.

Armature The moving or rotating component of a motor, generator, relay or other electromagnetic device.

ASME American Society of Mechanical Engineers.

Aspect Ratio Ratio of length to width of rectangular duct.

Aspirating Psychrometer A device that draws in a sample of air to measure for humidity.

Atom Smallest particle of an element.

Atomic Weight The number of protons in an atom of a material.

Atomize Process of changing a liquid to a fine spray.

Attenuate Decrease or lessen in intensity.

Authorized Dealer A dealer authorized to install heat pumps by the manufacturer.

Automatic Defrost System of removing ice and frost from evaporators automatically.

Automatic Expansion Valve (AEV) Pressure controlled valve used as a metering device.

Autotransformer A transformer in which both primary and secondary coils have turns in common. Stepup or stepdown of voltage is accomplished by taps on common winding.

Back Pressure Pressure in the low side of the refrigeration system. Also called low side pressure or suction pressure.

Back Seat The position of a service valve that is all the way counterclockwise.

Balance Fitting A pipe fitting or valve designed so that its resistance to flow may be varied. This type of fitting is used to obtain the desired flow rate through parallel circuits.

Balance Point The outdoor temperature at which the heating capacity of a heat pump in a particular installation is equal to the heat loss of the conditioned area. An outdoor temperature, usually between 30–45° Fahrenheit, at which the heat pump's output exactly equals the heating needs of the house. Below the balance point, supplementary heat is needed to maintain indoor comfort.

Balance Point, Initial The outdoor ambient temperature at which the heating capacity of the heat pump only balances the heat loss of the conditioned area.

Balance Point, Second The outdoor temperature at which the heating capacity of the heat pump plus the auxiliary heat capacity balances the heat loss of the conditioned area.

Ball Check Valve A check valve that uses a ball against a seat as a shutoff means.

Ball Valve A valve in which modulation or shutoff is accomplished by a one quarter turn of a ball that has an opening through it.

Barometer Instrument for measuring atmospheric pressure.

Baseboard A terminal unit resembling the base trim of a house.

Battery Two or more primary or secondary electrically interconnected cells.

Baudelot Cooler Heat exchanger in which water flows by gravity over the outside of the tubes or plates.

Bellows Corrugated cylindrical container which moves as pressures change.

Bending Spring Coil spring that is used to keep a tube from collapsing while being bent.

Bimetal Strip Two dissimilar metals with unequal expansion rates welded together.

Bleed Off The continuous or intermittent wasting of a small fraction of the water to a cooling tower to prevent the buildup of scale-forming minerals.

Bleed Valve Valve that permits a minimum fluid flow when valve is closed.

Blower A centrifugal fan.

Boiler Horsepower The equivalent evaporation of 34.5 lb of water per hr at 212°F. This is equal to a heat output of $970.3 \times 34.5 = 33{,}475$ Btu/h.

Boiling Temperature Temperature at which a fluid changes from a liquid to a gas.

Bore Inside diameter of a cylindrical hole.

Bourdon Tube Thin walled tube of circular shape that tends to straighten as pressure inside is increased.

Brazing Soldering with a filler material whose melting temperature is higher than 840°F.

Break Electrical discontinuity in the circuit generally resulting from the operation of a switch or circuit breaker.

Breaker Strip Strip of plastic used to cover the joint between the outside case and the inside liner of a refrigerator.

Brine Water saturated with a chemical such as salt.

British Thermal Unit (Btu) The amount of heat required to raise the temperature of one pound of water (about one pint) by one degree Fahrenheit.

Bull Head The installation of a pipe tee in such a way that water enters (or leaves) the tee at both ends of the run (the straight through section of the tee) and leaves (or enters) through the side connection only.

Bunsen Burner A gas burner in which combustion air is premixed with the gas supply within the burner body before the gas burns at the burner port.

Burner A device for the final conveyance of gas, or a mixture of gas and air, to the combustion zone.

Burnout Accidental passage of high voltage through an electrical circuit or device that causes damage.

Cadmium Cell A device whose resistance changes according to the amount of light sensed.

Calibrate To adjust an indicator so that it correctly indicates the variable sensed.

Calorie Heat required to raise the temperature of one gram of water one degree Celsius.

Capacitor Type of electrical storage device used in starting and/or running circuits on many electric motors.

Capillary Tube A fixed restriction pressure reducing device. Usually consists of lengths of small inside diameter tubing. The flow restriction produces the necessary reduction in pressure and boiling point of the liquid refrigerant before entering the evaporator.

Carbon Dioxide (CO_2) A nontoxic product of combustion.

Carbon Monoxide (CO) A chemical resulting from incomplete combustion. It is odorless, colorless, and toxic.

Cascade System One having two or more refrigerant circuits, each with a compressor, condenser, and evaporator, where the evaporator of one circuit cools the condenser of the other (lower temperature) circuit.

Celsius The metric system temperature scale.

Centimeter Metric unit of linear measurement, equals 0.3937 in.

Centrifugal Compressor Compressor that compresses gaseous refrigerants by centrifugal force.

CFM Cubic feet per minute (ft^3/min).

Charge The amount of refrigerant in a system.

Charging Cylinder A device for charging a predetermined weight of refrigerant into a system.

Charging the Loop Filling the earth loop with the correct mixture and purging all the air from the loop.

Check Valve A flow control valve that permits flow in one direction only.

Chimney Effect The tendency of air or gas in a duct or other vertical passage to rise when heated.

Circuit A tubing, piping, or electrical wire installation that permits flow from the energy source back to energy source.

Circuit Breaker A device that senses current flow and opens when its rated current flow is exceeded.

Circulator A pump.

Clearance Space in cylinder not occupied by piston at end of compression stroke.

Closed Loop Any cycle in which the primary medium is always enclosed and repeats the same sequence of events.

Coaxial Fitting A fitting that connects to a single port of a hot water storage tank. Or, tube in a fitting. The hot water goes through the tube into the tank, and the cold water comes out of the fitting surrounding the tube.

Coefficient of Performance (COP) A ratio calculated by dividing the total heating capacity provided by the refrigeration system, including circulating fan heat but excluding supplementary resistance heat (Btu/h), by the total electrical input (watts) $\times$ 3.412.

Coil A wound conductor that creates a strong magnetic field when current passes through it.

Coil, Device Subcooler A section in the outdoor evaporator coil used to increase the liquid subcooling during the cooling mode as well as act as an extra defrost surface during the defrost mode.

Coil, Inside: The coil located in the inside portion of the heat pump system. Performs as the evaporator in the cooling mode and the condenser in the heating mode.

Coil, Outside The coil located in the outside portion of an air to air heat pump system. Performs as an evaporator in the heating mode and as the condenser in the cooling mode.

Cold The absence of heat.

Cold Junction That part of the thermoelectric system that absorbs heat.

Combustion The rapid oxidation of fuel gases accompanied by the production of heat.

Combustion Air Air supplied for the combustion of a fuel.

Combustion Products Constituents resulting from the combustion of a fuel gas with the oxygen in air, including the inerts, but excluding excess air.

Comfort Chart Chart used in air conditioning to show the dry bulb temperature and humidity for human comfort.

Commercial Buildings Such buildings as stores, shops, restaurants, motels, and large apartment buildings.

Commutator A ring of copper segments insulated from each other and connecting the armature and brushes of a motor or generator. It passes power into or from the brushes.

Compound Gauge Pressure gauge that has scales both above and below atmospheric pressure.

Compound Refrigerating System System that has several compressors in series.

Compression Ratio The absolute discharge pressure divided by the absolute suction pressure for a compressor.

Compression Tank (See Air Cushion Tank)

Compressor The pump of a refrigerating mechanism that draws a vacuum or low pressure on the cooling side of refrigerant cycle and squeezes or compresses the gas into the high pressure or condensing side of the cycle.

Compressor Displacement The volume discharged by a compressor in one rotation of the crankshaft.

Compressor Seal Leakproof seal between crankshaft and compressor body.

Condensate Drain Trap A pipe arrangement that provides a water seal in the drain line to prevent airflow through the drain line.

Condensate Pan A pan located under an evaporator to collect condensate from the coil and carry it to the drain line.

Condensate Pump Device used to remove fluid condensate.

Condensation Liquid that forms when a vapor is cooled below its condensing temperature or dew point.

Condense The action of changing a saturated vapor to a saturated liquid.

Condenser The part of the refrigeration system that receives the high temperature, high temperature vapor from the compressor and extracts the heat in the vapor reducing it to a high pressure, medium temperature liquid.

Condenser, Air Cooled A heat exchanger that transfers heat energy from the refrigerant vapor to air.

Condenser, Liquid Cooled A heat exchanger that transfers heat energy from the refrigerant vapor to the liquid heat sink.

Condenser Water Pump A device used to force the liquid through a liquid cooled condenser.

Condensible A gas that can be easily converted to liquid form.

Condensing Pressure The refrigerant pressure inside the condenser coil.

Condensing Temperature The temperature at which a vapor changes to a liquid at a given pressure.

Condensing Unit The portion of the refrigeration system that converts low pressure, low temperature vapor to high pressure, medium temperature liquid. Commonly called the high side.

Conductance (Thermal) The C factor. The time rate of heat flow per unit area through a material.

Conduction (Thermal) Particle to particle transmission of heat.

Conductivity The ability of a substance to allow the flow of heat or electricity.

Conductor Material or substance which readily passes electricity.

Connected Load The sum of the capacities or continuous ratings of the load-consuming apparatus connected to a supplying system.

Connecting Rod That part of compressor that connects the piston to the crankshaft.

Constrictor Tube or orifice used to restrict flow.

Contact The part of a switch or relay that carries current.

Contactor A device for making or breaking load carrying contacts by a pilot circuit through a magnetic coil.

Contaminant A substance (dirt, moisture, etc.) foreign to refrigerant or refrigerant oil in system.

Control, Low Pressure A pressure operated control connected to the suction side of the refrigeration system to prevent unit operation below a set pressure or coil operating boiling point.

Control, Refrigerant A device used to provide the necessary pressure reduction of the liquid refrigerant to obtain the proper boiling point of the refrigerant in the evaporator.

Control, Temperature A device that uses changes in temperature to operate contacts in an electrical circuit.

Controller Measures the difference between sensed output and desired output and initiates a response to correct the difference.

Controls Devices designed to regulate the gas, air, water, or electricity.

Convection Transfer of heat by means of movement of a fluid, either liquid or air.

Cooling Anticipator A resistor in a room thermostat that causes the cooling cycle to begin prematurely.

Cooling Tower Device that cools water by evaporation in air. Water is cooled to wet bulb temperature of air.

Coulomb An electrical unit of charge; one coulomb per second equals one ampere or 6.25×10^{23} electrons past a given point in one second.

Counter EMF Counterelectromotive force. The EMF induced in a coil that opposes applied voltage.

Counterflow Two liquids and/or vapors flowing in directions opposite to each other.

Cracking a Valve Opening a valve a small amount.

Crankcase Heater A heating device fastened to the crankcase or lower portion of the compressor housing intended to keep the oil in the compressor at a higher temperature than the rest of the system to reduce the migration of the refrigerant.

Crankshaft Seal Leakproof joint between crankshaft and compressor body.

Current The flow of electrons through a conductor.

Cut-In The temperature or pressure at which an automatic control switch closes.

Cut-Out The temperature or pressure at which an automatic control switch opens.

Cylinder Head Part that encloses compression end of compressor cylinder.

Damper Valve for controlling airflow.

Dead Band A range of temperature in a heating/cooling system in which no heating or cooling is supplied.

Decibel Unit used for measuring relative loudness of sounds.

Dedicated Geothermal Well Water is drawn from the top of the well and returned to the bottom of the well. The well supplies only the heat pump.

Defrost Control A control system used to detect frost or ice buildup on the outside coil of an air to air heat pump during the heating mode and causes the system to reverse in order to remove the frost or ice from the coil.

Defrost Cycle Refrigerating cycle in which evaporator frost and ice accumulation is melted.

Defrost Mode The portion of the operation of the system in which the evaporator frost or ice is removed.

Defrost Timer A device connected into the electrical control system that controls the frequency and duration of the defrost operating mode.

Degree Day Unit that represents one degree of difference from given point in average outdoor temperature of one day.

Dehydrate To remove water in all forms from a material or system.

Delta Connection The connection in a three phase system in which terminal connections are triangular, similar to the Greek letter delta (Δ).

Demand The size of any load generally averaged over a specified interval of time.

Demand (Billing) The demand upon which billing to a customer is based.

Demand Meter An instrument used to measure peak kilowatt hour consumption.

Density Weight per unit volume.

Desiccant Substance used to collect and hold moisture.

Design Load The amount of heating or cooling required to maintain inside conditions when the outdoor conditions are at design temperature.

Design Temperature Difference The difference between the design indoor and outdoor temperatures.

Design Water Temperature Difference The difference between the temperature of the water leaving the unit and entering the unit when the system is operating at design conditions.

Desuperheater A heat exchanger in the hot gas line between the compressor discharge and reversing valve. Used to heat domestic water by removing the superheat from the hot refrigerant vapor.

Detector, Leak A test instrument used to detect and locate refrigerant leaks.

Dew Point Temperature at which water vapor begins to condense out of air. Air at the dew point—100% relative humidity—is referred to as saturated vapor.

Diaphragm Flexible membrane.

Dielectric An insulator or nonconductor.

Dies (Thread) Tool used to cut external threads.

Differential As applied to controls, the differential between the cut-in and cut-out temperature or pressure setpoints of a control.

Diffuser Air distribution outlet designed to direct airflow into a room.

Dilution Air Air that enters a draft hood and mixes with the flue gases.

Diode A device that will carry current in one direction but not the reverse direction.

Direct Current Abbreviated DC. Electric current that flows only in one direction.

Direct Expansion Evaporator An evaporator coil using a pressure reducing device to supply liquid refrigerant at the correct pressure and boiling point to obtain the desired rate of heat absorption into the liquid refrigerant.

Direct Return A two pipe system in which the first terminal unit taken off the supply main is the first unit connected to the return main.

Disconnecting Switch A knife switch that opens a circuit.

Disturbed Earth Effect Denotes the heat content change that affects the earth temperature by the addition to or removal of heat energy by the heat pump system.

Domestic Hot Water Heated water used for cooking, washing, etc.

Domestic Geothermal Well Water is drawn from the top of the well. Only the water from the heat pump is returned to the bottom of the well. The difference is used for building requirements.

Double Pole Switch Simultaneously opens and closes two wires of a circuit.

Double Thickness Flare Tubing end that has been formed into two-wall thickness.

Dowel Pin Pin pressed through two assembled parts to ensure accurate alignment.

Downdraft Downward flow of flue gas.

Draft Gauge Instrument used to measure air pressure.

Drier A substance or a device used to hold moisture from circulating through a refrigeration system.

Drift Entrained water carried from a cooling tower by wind.

Drilled Port Burner A burner in which the ports have been formed by drilled holes.

Drip Pan Pan shaped panel or trough used to collect condensate from evaporator coil.

Dry Bulb Temperature The actual (physical) temperature of a substance. Usually refers to air.

Dry Ice Solid carbon dioxide.

Duct Round or rectangular sheet metal or fiberglass pipe that carries the air between the conditioning unit and the conditioned area.

Dynamometer Device for measuring power.

E Symbol for volts. From the term "electromotive force."

Earth Temperature, Minimum (T) The lowest temperature the earth will become in the winter season. Used for heat pump design. Will vary with location and weather conditions.

Eccentric A circle or disk mounted off center on a shaft.

Effective Area The actual opening in a grill or register through which air can pass. The effective area is the gross (overall) area of the grill face minus the area of the deflector vanes or bars.

Efficiency (Electrical) A percentage value denoting the ratio of power output to power input.

Electric Field A magnetic region in space.

Electric Heating A heating system that uses electric resistance elements as the heat energy source.

Electric Heating Element A unit consisting of resistance wire, insulated supports, and connection terminals for connecting wire to a source of electrical power.

Electrolyte A solution of a substance (liquid or paste) that is capable of conducting electricity.

Electromagnet A magnet created by the flow of electricity through a coil of wire.

Electromotive Force Abbreviated EMF. The difference in potential or voltage of electrical energy between two points.

Electronic Leak Detector An electronic instrument that measures the changes in electron flow across an electrically charged gap that indicates the presence of refrigerant vapor molecules.

Electronics Field of science dealing with electronic devices and their uses.

Electrostatic Filter Type of filter that gives particles of dust electric charge.

End Bell End structure of electric motor which usually holds motor bearings.

Endplay Slight movement of shaft along a center line.

Energy Efficiency Ratio (EER) The comparison of the heat transfer ability of a refrigeration system and the electrical energy used as expressed in Btuh per watt. A ratio calculated by dividing the cooling capacity in Btu per hour (Btu/hr) by the power input in watts at any given set of rating conditions, expressed in Btu/hr per watt.

Enthalpy Heat content of a refrigerant, usually with respect to a reference point (Btu/lb).

Epoxy (Resins) A synthetic plastic adhesive usually comprised of two parts, the rosin and hardener.

EPR (Evaporation Pressure Regulator) An automatic pressure regulating valve used to maintain a predetermined pressure in the evaporator.

Equalizer Tube Device used to maintain equal pressure or equal liquid levels between two containers.

Evacuate To remove water in all forms from a material or system.

Evaporation A term used to describe the changing of a liquid to a vapor by the addition of heat energy. Change from liquid to vapor at a relatively slow rate.

Evaporative Condenser A device that uses a water spray to cool a condenser.

Evaporator Part of a refrigerating mechanism in which the refrigerant evaporizes and absorbs heat.

Evaporator Coil A piece of equipment made of tubing (liquid) or tubing and fin (air) used to transfer heat energy from the material into refrigerant.

Evaporator, Dry Type An evaporator into which refrigerant is fed from a pressure reducing device.

Evaporator Fan Fan that moves air through the evaporator.

Evaporator, Flooded An evaporator containing liquid refrigerant at all times.

Excess Air Air that is in excess of the required amount for complete combustion of a fuel gas.

Exfiltration Outward flow of air from an occupied area through openings in the building structure.

Exhaust Opening Any opening through which air is removed from a space.

Expansion Joint A fitting or piping arrangement designed to relieve stress caused by expansion of piping.

Expansion Tank A closed tank that allows for water expansion without creating excessive pressure.

Expansion Valve A device in refrigerating system that maintains a pressure difference between the high side and low side.

Expendable Refrigerant System System that discards the refrigerant after it has evaporated.

Extended Surface Heat transfer surface, one side of which is increased in area by the use of fins.

External Equalizer A pressure connection from an area below the diaphragm of a thermostatic expansion valve and the suction outlet of an evaporator.

Fahrenheit The common scale of temperature measurement in the English system of units.

False Defrost The system goes into the defrost mode even if the outdoor has no frost or ice buildup.

Fan A radial or axial flow device using blades in a venturi for moving air.
Fan Coil A terminal unit consisting of a finned tube coil and a fan in a single enclosure.
Farad The base unit of measurement for capacitance. Since a farad is a very large unit, a much smaller unit is usually used. (See Microfarad)
Feedback The transfer of energy output back to input.
Feet of Head (ft HD) The term used to designate the flow resistance the pump must overcome to deliver the required amount of liquid. 2.307 ft HD = 1 psig.
Female Thread A thread on the inside of a pipe or fitting.
Ferrous Objects made of iron or steel.
Field The space involving the magnetic lines of force.
Filter A device for removing foreign particles from vapor or liquid.
Fin Comb Comb-like device used to straighten the metal fins on coils.
Fire Tube Boiler A steel boiler in which the hot gases from combustion are circulated through tubes that are surrounded by water.
Flame Detector The components of a flame detection system that detects the presence or absence of a flame.
Flame Sensing Element A device that uses heat, infrared, optical, rectification, rectifying photo cell, or ultraviolet light to detect a flame.
Flare An angle formed at the end of a tube.
Flare Nut Fitting used to clamp the tubing flare against another fitting.
Flashback The movement of the gas flame down through the burner port upon shutdown of the gas supply.
Flash Gas Vapor that is produced by the boiling of liquid refrigerant to reduce the temperature of the liquid refrigerant from the condensing temperature minus the subcooling to the boiling temperature of the evaporator.
Flash Point Temperature at which an oil will give off sufficient vapor to support a flash flame.
Float Valve Type of valve that is operated by a sphere that floats on the liquid surface and controls level of liquid.
Floc Point The temperature at which the wax in an oil will start to separate from the oil.
Flooded System Type of refrigerating system in which liquid refrigerant fills the evaporator.
Flow Hood A direct reading instrument that measures air quantity (cubic feet per minute [CFM]) out of or into a duct system and directing the air over a sensing element.
Flow Meter Instrument used to measure the volume of a liquid flowing through a pipe or tube in quantity per time (gallons per minute, pounds per hour, liters per minute, etc.).
Flow Switch An automatic switch that senses fluid flow.
Flue A passage in the chimney to carry flue gas.
Flue Gases Products of combustion and excess air.
Fluid Substance in a liquid or gaseous state.
Flushing the Loop Forcing liquid through the loop at sufficient velocity to remove all forms of foreign material and air.
Flux Brazing, Soldering Substance applied to surfaces to be joined to free them from oxides.
Flux, Electrical The electric or magnetic lines of force in a region.
Flux, Magnetic Lines of force of a magnet.
Foaming Formation of a foam in an oil refrigerant mixture due to rapid evaporation of refrigerant dissolved in the oil.
Foot of Water A measure of pressure. One foot of water is the pressure created by a column of water 1 ft high. This is equal to 0.433 psig.
Foot Pound Unit of work.
Forced Convection Movement of a liquid or vapor by a mechanical means.
Forced Draft Burner A burner in which combustion air is supplied by a blower.
Free Area The total area of an opening in a grill or register through which air passes.
Freeze Up Frost or ice formation on an evaporator that restricts air or liquid flow through the evaporator.
Freezing Change of state from liquid to solid.
Freezing Point The temperature at which a liquid will solidify upon removal of heat.
Freon Trade name for a family of synthetic chemical refrigerants manufactured by DuPont, Inc.
Frequency The number of complete cycles in a unit of time.
Friction Head In a hydronic system, the loss in pressure resulting from the flow of water in the piping system.
Frost Back A condition in which liquid refrigerant boils in the suction line and produces a frosted surface on the line.
Frost Control, Automatic A control that automatically reverses a refrigeration system to remove frost/ice from the evaporator.
Frost Free Refrigerator A refrigerated cabinet which operates with an automatic defrost.
Frost Line The depth to which the ground freezes over a winter season. The frost line is usually set at 12 in below the deepest recorded frost penetration.
Full Load Amperage (FLA) The amount of amperage an inductive load (motor) will draw at its full design load.
Furnace That part of a warm air heating system in which combustion takes place.
Fuse Electrical safety device consisting of strip of fusible metal in circuit that melts when the circuit is overloaded.
Fusible Plug A plug or a mode with a metal of low melting temperature to release pressures in case of fire.

Galvanic Current generated chemically that results when two dissimilar conductors are immersed in an electrolyte.
Galvanic Action The destruction of one metal by the generation of electrical energy when two different metals are joined.
Gang Mechanical connection of two or more switches.
Gasket A resilient material used between mating surfaces to provide a leakproof seal.
Gas A substance in the vapor state.
Gas, Noncondensible A gas that will not form into a liquid under available pressure temperature conditions.

Gauge, Compound A pressure indicating instrument that measures pressures both above and below atmospheric pressure.

Gauge, Gas Manifold Pressure A direct reading gauge for measuring the gas manifold pressure of a vapor burning heating unit. A more convenient substitute for a U tube manometer. Measures in units of inches of water.

Gauge, High Pressure An instrument used to measure pressures above atmospheric pressure in the 0 to 500 psig range.

Gauge, Low Pressure An instrument used to measure pressures above atmospheric up to 50 psig.

Gauge, Manifold A device that contains a combination of gauges and control valves to control the flow of vapors through the device.

Gauge, Standard An instrument designed to measure pressures above atmospheric.

Gauge, Vacuum An instrument used to measure pressures below atmospheric.

Generator A machine that converts mechanical energy into electrical energy. (Also see Alternator)

Geothermal Heat Pump (See Ground Coupled Heat Pump)

Geothermal Well A drilled well in which the returned water is returned to the same well from which the supply water is taken.

Grade Level The top surface of the earth.

Gravity System A heating system in which the distribution of the warm air or water relies upon the buoyancy of that air or water. There is no fan or pump.

Grill An ornamental or louvered opening through which air enters or leaves a duct system or area.

Gross Capacity, Heating The gross capacity of a heat pump in the heating mode is the total amount of heat energy transferred to the inside air or material. This is made up of the heat energy picked up from a heat source (air or liquid) plus the heat energy equivalent of the electrical energy required to operate the heat pump.

Gross Output A rating applied to boilers. It is the total quantity of heat which the boiler will deliver and at the same time meet all limitations of applicable testing and rating codes.

Ground Connection between an electrical circuit and earth.

Ground Coil (Earth Coil) A heat exchanger coil buried in the earth that is used to either remove heat energy from the earth or add heat energy to the earth.

Ground Coupled Heat Pump A heat pump that uses a ground coil to exchange heat.

Ground Wire An electrical wire that will safely conduct electricity from a structure into the ground.

Halide Refrigerants Family of refrigerants containing halogen chemicals.

Halide Torch Type of torch used to detect halogen refrigerant leaks.

Head As used in this course, head refers to a pressure difference. (See Pressure Head, Pump Head, Available Head)

Head (Total) In flowing fluid, the sum of the static and velocity pressures at the point of measurement.

Header A piping arrangement for interconnecting two or more supply or return tappings of a boiler.

Head Pressure The pressure in the condensing or high side of the refrigeration system.

Head, Pressure Control A pressure operated control that opens an electrical circuit if the pressure exceeds the cutout point of the control.

Head, Static Pressure of a liquid or vapor in terms of the height of a column of a liquid or mercury.

Head, Velocity In flowing fluid, height of fluid equivalent to its velocity pressure.

Heat A form of energy. Heat affects the molecular activity of a substance. This is reflected in the temperature of the substance. Addition of heat energy increases the molecular activity and temperature. Removal of heat energy lowers the molecular activity and temperature.

Heat Anticipator A resistor in a room thermostat that shuts the heating cycle off prematurely.

Heat Exchanger A device used to transfer heat energy from a higher temperature to a lower temperature. Evaporators, condensers, and earth coils are examples.

Heat Lag When a substance is heated on one side, it takes time for the heat to travel through the substance.

Heat, Latent Heat energy used to change the state of a substance; solid to liquid and liquid to vapor by heat addition, or vapor to liquid and liquid to solid by heat removal, without a change in the temperature of the material at the time of change.

Heat Load The amount of heat energy, measured in Btu/hr, that is required to maintain a given temperature in a conditioned area at design conditions both inside and outside the conditioned area.

Heat Loss The rate of heat transfer from a heated building to the outdoors.

Heat Motor An electrical device that produces motion by means of the temperature change of bimetal elements. Used to produce time delay in the sequence of operation of a device. (See Sequencer)

Heat of Compression The heat added to the refrigerant by the work being done by the compressor.

Heat of Fusion The heat energy needed to accomplish the change in state of a material between a liquid and solid; addition for solid to liquid, removal for liquid to solid.

Heat of Respiration When carbon dioxide and water are given off by foods in storage.

Heat Pump A name given to an air conditioning system that is capable of either heating or cooling an area on demand.

Heat Pump, Air to Air A device that transfers heat between two different air quantities in either direction on demand.

Heat Pump, Air to Liquid A device that transfers heat from an air source to a liquid by means of a refrigeration system. Units in this category are nonreversible.

Heat Pump, Liquid to Air A device that transfers heat between liquid and air in either direction on demand.

Heat Pump, Liquid to Liquid A device that transfers heat between two different liquids in either direction on demand.

Heat Pump, Water Heater An air to liquid refrigeration system used to heat domestic water with air as the heat source.

Heat Pump; Water Heater, Remove The refrigeration system is a complete unit designed to be connected to an existing water tank.

Heat Pump; Water Heater, Self Contained The component parts of the refrigeration system (condenser) are an integral part of the hot water tank.

Heat, Sensible The heat energy used to change the temperature of a material—solid, liquid, or vapor—without a change in state.

Heat Sink A place or material into which heat energy is placed. A body of air or liquid that accepts the heat removed from the house. In a heat pump, the air outside the house is used as a heat sink during the cooling cycle.

Heat Source A place or material from which heat energy is obtained. A body of air or liquid from which heat is collected. In an air source heat pump, the air outside the house is used as a heat source during the heating cycle.

Heat, Total The sum of both the sensible and latent heat energy in air. Expressed as Btu/lb of air.

Heat Transfer Movement of heat energy from one body or substance to another. Heat may be transferred by any combination of or all of the three methods (radiation, conduction, and convection). The process of transferring heat from one location to another.

Heating Coil A heat transfer device designed to add heat to liquid or vapor.

Heating Mode The operating phase of a system that is adding heat energy to an occupied area.

Heating Seasonal Performance Factor (HSPF) The total heating output of a heat pump during its normal annual usage period for heating divided by the total electric power input in watt-hours during the same period.

Heating Value The number of Btu produced by the combustion of one cubic foot of gas.

Henry Abbreviated H. The unit of inductance.

Hermetic Completely sealed.

Hermetic Motor Compressor drive motor sealed within same casing that contains the compressor.

High Pressure Cutout An electrical pressure control switch operated by the high side pressure set to stop compressor action operation at maximum pressure safety limits.

High Side Parts of a refrigerating system which are under condensing or high side pressure.

High Velocity System Usually large commercial or industrial air distribution systems designed to operate with static pressures of 6 to 9 in WG.

Horsepower A unit of power equal to 33,000 ft-lb of work per minute.

Hot Gas Defrost A defrosting system in which hot refrigerant gas from the compressor is directed through the evaporator to remove frost or ice from the evaporator.

Hot Gas Line The refrigerant tube that carries the high pressure, high temperature refrigerant vapor from the compressor to the condenser.

Hot Junction That part of thermoelectric circuit that releases heat.

Humidifiers Device used to add humidity.

Humidistat A control that is affected by changes in the relative humidity in air. Used to control the operation of a humidifier.

Humidity Moisture in air.

Hydrocarbon Any of a number of compounds composed of carbon and hydrogen.

Hydrometer Floating instrument used to measure specific gravity of a liquid.

Hydronics Pertaining to heating or cooling with water or vapor.

Hygrometer An instrument used to measure the percentage of moisture (relative humidity) in air.

Hygroscopic Ability of a substance to absorb and retain moisture.

Ice Bank A thermal accumulator in which ice is formed or charged during off-peak periods of refrigeration demand, and used during peak demand for refrigeration to supplement the compressor capacity by melting the ice.

Ice Harvest Switch A device located at the end of the water plate. It resets the timer on the status indicator control card after each ice harvest.

Ice Maker A cyclic-type automatic ice making machine that has separate and sequential water fill, freezing, and ice harvesting phases for the production of ice.

Ice Ring An accumulation of ice at the bottom of the outdoor coil due to incomplete defrost operation.

Idler A pulley used on some belt drives to provide the proper belt tension.

Ignition Temperature The minimum temperature at which combustion can be started.

Ignition Transformer A transformer designed to provide a high voltage current. Used in many heating systems to ignite fuel.

Impedence A type of electrical resistance that is only present in alternating current circuits.

Impeller Rotating part of a centrifugal pump.

Indoor Coil The portion of a heat pump that is located in the house and functions as the heat transfer point for warming or cooling indoor air.

Induced Draft Burner A burner that depends on draft induced by a fan or blower at the flue outlet to draw in combustion air and vent flue gases.

Inductance The characteristic of an alternating current circuit to oppose a change in current flow.

Induction The act that produces induced voltage in an object by exposure to a magnetic field.

Induction Motor An AC motor that operates on the principle of the rotating magnetic field. The rotor has no electrical connection, but receives electrical energy by transformer action from field windings.

Inductive Load A device that uses electrical energy to produce motion.

Inductive Reactance Opposition, measured in ohms, to an alternating or pulsating current.

Inerts Noncombustible substances in a fuel, or in flue gases, such as nitrogen or carbon dioxide.

Infiltration The air that leaks into the refrigerated or air conditioned space.

Infrared Lamp An electrical device which emits infrared rays; invisible rays just beyond red in the visible spectrum.

In Phase The condition existing when two waves of the same frequency have their maximum and minimum values of like polarity at the same instant.

Insulation, Thermal Substance used to retard or slow the flow of heat through a wall or partition.

Interlock A safety device that allows power to a circuit only after a predetermined function has taken place.

IR Drop Voltage drop resulting from current flow through a resistor.

Isothermal At constant temperature.

Joints, Brazed A solder type connection or joint obtained by the joining of the metal parts with metallic mixtures or alloys that have a melting temperature above 1,000°F up to 1,500°F.

Joint, Soldered A solder type connection or joint obtained by the joining of the metal parts with metallic mixtures or alloys that have a melting temperature below 1,000°F.

Joint, Welded A solder type pipe connection of joint obtained by the joining of the metal parts with metallic mixtures or alloys that have a melting temperature above 1,500°F.

Joule A measure of heat in the metric system.

Journal, Crankshaft Part of the shaft that contacts the bearing.

Junction Box Group of electrical terminals housed in protective box or container.

Kelvin Scale (K) Thermometer scale on which unit of measurement equals the centigrade degree and according to which absolute zero is 0°, the equivalent of −273.16°C.

Kilopascal A measure of pressure in the metric system.

Kilowatt A unit of electrical energy equal to 1,000 watts.

Kilowatt Hour A unit for measuring electrical energy equal to 3,413 Btu.

King Valve The valve located at the outlet of the receiver.

Latent Heat Heat characterized by a change of state of the substance concerned.

Latent Heat of Fusion Amount of heat required to change from solid to liquid at the same temperature, Btu/lb.

Latent Heat of Vaporization Amount of heat required to change from liquid to vapor at the same temperature, Btu/lb.

Leak Detector A device or instrument used to detect and locate leaks of gases or vapors.

Limit Control A device used to open or close electrical circuits if temperature or pressure reaches preset limits.

Line Voltage Voltage existing at wall outlets or terminals of a power line system above 30 V.

Liquefied Petroleum Gases The terms liquefied petroleum gases, LPG, and LP gas mean and include any fuel gas that is composed predominantly of any of the following hydrocarbons, or mixtures of them: propane, propylene, normal butane or isobutene, and butylenes.

Liquid Line The tube that carries liquid refrigerant from the condenser or liquid receiver to the pressure reducing device.

Liquid Nitrogen Nitrogen in liquid form.

Liquid Receiver Cylinder connected to condenser outlet for storage of liquid refrigerant in a system.

Liter Metric unit of volume.

LNG Liquefied natural gas. Natural gas that has been cooled until it becomes a liquid.

Load The amount of heat per hour that the refrigeration system is required to supply at design conditions. The amount of electrical energy expected to be connected to an electrical power supply.

Locked Rotor Amperage (LRA) The amount of energy a motor will draw under stalled conditions.

Lock Out Relay A control scheme that prevents the restarting of a compressor after a safety control opens, even after the safety control resets itself.

Louvers Sloping, overlapping boards or metal plates intended to permit ventilation and shed falling water.

Low Pressure Control A device used to start or stop the system operation when the low pressure (low side pressure, suction pressure) drops to preset limits.

Low Pressure Cutout Device used to keep low side evaporating pressure from dropping below a certain pressure.

Low Side The portion of the refrigeration system that operates at the evaporator pressure. Consists of the pressure reducing device, evaporator, and suction line.

Low Side Pressure Pressure in the low side of the refrigeration system. Also called low side pressure or suction pressure.

Low Voltage Control Controls designed to operate at voltages of 20 to 30 V.

Magnetic Field The space in which a magnetic force exists.

Makeup Air Fresh air. The air supplied to a building to replace air exhausted from the building.

Makeup Water Line The line connected to the loop system for filling or adding liquid as necessary.

Male Thread A pipe or fitting with a thread on the outside.

Manifold Gauge A device constructed to hold compound and high pressure gauges and valves to control flow of fluids through it.

Manometer An instrument used to measure the pressure of gases or vapors. The gas pressure is balanced against a column of liquid, such as water or mercury, in a U-shaped tube open to atmospheric pressure.

Manometer, Inclined A manometer on which the liquid tube is inclined from horizontal to produce wider, more accurate readings over a smaller range of readings.

Mass A quantity of matter cohering together to make one body which is usually of indefinite shape.

Mean Annual Earth Temperature The mean average temperature of the earth as it changes throughout the year. The largest factor for earth temperature change is sunshine.

Mean Temperature Difference The average temperature between the temperature before a process begins and the temperature after the process is completed.

Mechanical Cycle Cycle that is a repetitive series of mechanical events.

Mega Prefix for one million.

Megohm 1,000,000 ohms of electrical resistance.
Melting Point Temperature at atmospheric pressure, at which a substance will melt.
Mercaptan An odorant additive that gives natural gas its characteristic odor.
Mercury Bulb An electrical circuit switch which uses a small quantity of mercury in a sealed glass tube to make or break electrical contact with terminals within the tube.
Meter Metric unit of linear measurement.
Methane A hydrocarbon gas with the formula CH_4, the principal component of natural gas.
MFD Abbreviation for microfarad.
Micro A combining form denoting one millionth.
Microfarad The unit of measurement for capacitance, equal to a millionth of a farad.
Micron A metric unit. For length it is one millionth (1/1,000,000) of a meter. For pressure it is 1/254,000 of an inch of mercury (1/254,000 of 1 in Hg.).
Micron Gauge An instrument used to measure pressure below atmospheric (vacuum) close to perfect vacuum.
Milli A term used to denote one thousandth (1/1,000) of a unit. For example, a milliampere is 1/1,000 of an ampere.
Modulating A type of device or control that tends to adjust by increments (minute changes) rather than by either full on or full off operation.
Moisture Indicator Instrument used to measure moisture content of a refrigerant.
Molecule The smallest portion of a compound that retains the identity and characteristics of the compound.
Mollier Diagram Graph of refrigerant pressure, heat, and temperature properties.
Motor Burnout A condition in a motor where the insulation has deteriorated due to overheating.
Motor, Capacitor Type A single phase induction motor that uses both run and start windings in the running mode. The start winding is connected in series with a capacitor to change the electrical characteristics of the winding.
Motor Control A control to start and stop a motor at preset pressures or temperatures.
Motor Starter A high capacity electrical contact operated by an electromagnetic coil and containing properly sized overload cutouts.
Mullion Stationary part of a structure between two doors.
Multimeter A volt ohm meter.

National Electrical Code Abbreviated NEC. A code of electrical rules based on fire underwriters' requirements for interior electric wiring.
Natural Convection Circulation of a gas or liquid due to difference in density resulting from temperature differences.
Natural Gas Any gas found in the earth, as opposed to gases that are manufactured.
Needle Valve A valve used to accurately control very low flow rates of fluids.
Neoprene A synthetic rubber that is resistant to hydrocarbon oil and gas.
Net Refrigeration A measure of the amount of cooling that takes place in evaporator. It is the enthalpy of gas leaving the evaporator minus enthalpy of gas entering evaporator. The units are Btu/lb.
Nominal Size Tubing measurement that has an inside diameter the same as iron pipe of the same stated size.
Noncondensible Gas Gas that does not change into a liquid at operating temperatures and pressures.
Nonferrous Group of metals and metal alloys that contain no iron; also metals other than iron or steel. In heating systems the principal nonferrous metals are copper and aluminum.
Normally Closed (NC) Switch contacts closed with the circuit de-energized.
Normally Open (NO) Switch contacts open with the circuit de-energized.
Nozzle The device on the end of the oil fuel pipe used to form the oil into fine droplets by forcing the oil through a small hole to cause the oil to break up.

Odorant A substance added to an otherwise odorless, colorless, and tasteless gas to give warning of gas leakage and to aid in leak detection.
Off Mode Cycle That part of the operating mode when the system has been shut down by the controls.
Ohm (Ω) A unit of resistance.
Oil Binding Physical condition when an oil layer on top of refrigerant liquid hinders it from evaporating at its normal pressure-temperature condition.
Oil Rings Expanding rings mounted in grooves and piston; designed to prevent oil from moving into compression chamber.
Oil Separator A device used to separate refrigerant oil from refrigerant gas and return the oil to crankcase of compressor.
One Pipe Fitting A specially designed tee for use in a one-pipe system to connect the supply or return branch into a circuit.
One Pipe System A forced hot water system using one continuous pipe or main from the boiler supply to the boiler return.
Open Circuit An interrupted electrical circuit that prevents electrical flow.
Open Compressor One with a separate motor or drive.
Orifice An accurately sized opening that controls the flow of vapor or liquid at given pressures across the opening.
Orifice, Oil Metering A small hole in the pickup tube in the accumulator used to ensure oil return to the compressor in small easily handled quantities.
Orifice Spud A removable plug containing an orifice that determines the quantity of gas that will flow.
Oscilloscope A fluorescent coated tube that visually shows an electrical wave.
Outdoor Coil The portion of a heat pump that is located outside the home and functions as a heat transfer point for collecting heat from or dispelling heat to the outside air.
Outside Air Atmosphere exterior to the conditioned area.
Overload Load greater than load for which system or mechanism was intended.
Overload Protector A device that will stop operation of unit if dangerous conditions arise.
Ozone A gaseous form of oxygen (O_3).

Package Boiler A boiler having all components assembled as a unit.

Package Unit A complete refrigeration or air conditioning system in which all the components are factory assembled into a single unit.

Parallel Circuit connected so current has two or more paths to follow.

Parallel Circuit An electrical circuit in which there is more than one different path across the power supply.

Partial Pressures Condition where two or more gases occupy a space and each one creates part of the total pressure.

Pascal's Law A pressure imposed upon a fluid is transmitted equally in all directions.

Permeance The ratio of water vapor flow to the vapor pressure difference.

pH A term based on the hydrogen ion concentration in water, which denotes whether the water is acid, alkaline, or neutral.

Photoelectricity A physical action wherein an electrical flow is generated by light waves.

Pilot A small flame that is used to ignite the gas at the main burner.

Pilot Generator A device used to generate voltage in a millivolt heating system.

Pilot Safety Valve A valve that will shut off main gas if the pilot flame is not proved.

Pilot Switch A control used in conjunction with gas burners. Its function is to prevent operation of the burner in the event of pilot failure.

Pinch Off Tool Device used to press walls of a tubing together until fluid flow ceases.

Piston Close fitting part which moves up and down in a cylinder.

Piston Displacement Volume obtained by multiplying area of cylinder bore by length of piston stroke.

Pitch Pipe slope.

Pitot Tube A device that senses static and total pressure.

Plenum A cube shaped duct or box on the supply or return side of an air handling unit to connect the supply or return duct system to the unit.

Point of Vaporization The location of the position in the evaporator coils where the last bit of liquid refrigerant is vaporized.

Polarity The condition denoting direction of current flow.

Polybutylene A plastic used for pipe material that has excellent creep resistance as well as high resistance to stress cracking. Recommended for earth loops.

Polyethylene A plastic used for tubing for ground loop systems, cold water lines, and heat pump piping.

Polystyrene Plastic used as an insulation in some refrigerator cabinet structures.

Ponded Roof Flat roof designed to hold quantity of water.

Potable Water Water that is suitable for human consumption.

Potential The amount of voltage or electrical pressure between two points of an electrical circuit.

Potentiometer A variable resistor.

Pour Point (Oil) Lowest temperature at which a particular oil will pour.

Power Time rate at which work is done or energy consumed. Source or means of supplying energy.

Power Element Sensitive element of a temperature operated control.

Power Factor The rate of actual power as measured by a wattmeter in an alternating circuit to the apparent power determined by multiplying amperes by volts.

Pressure Force per unit area.

Pressure, Back The pressure the pump must overcome to be able to force the liquid into the pressure tank. The pressure in the tank.

Pressure Drop The pressure difference between two locations in a circuit, the difference being a result of flow resistance in the circuit.

Pressure Limiter Device that remains closed until a certain pressure is reached and then opens and releases fluid to another part of system.

Pressure Motor Control A device which opens and closes on electrical circuit as pressures change to desired pressures.

Pressure Operated Altitude (POA) Valve Device that maintains a constant low side pressure independent of altitude of operation.

Pressure Reducing Device The device used to reduce the pressure on the liquid refrigerant and thus the boiling point before entering the evaporator.

Pressure Reducing Valve A diaphragm operated valve installed in the makeup water line of a hot water heating system.

Pressure Regulator A device for controlling a uniform outlet gas pressure.

Pressure Relief Valve A device for protecting a tank from excessive pressure by opening at a predetermined pressure.

Pressure, Static The pressure in the system when the pump is idle.

Pressure, Suction The pressure in the low pressure or evaporator section of the refrigeration system.

Primary Air The combustion air introduced into a burner which mixes with the gas before it reaches the port.

Primary Control One type of operating controller for an oil burner.

Primary Voltage The voltage of the circuit supplying power to a transformer.

Process Tube Length of tubing fastened to hermetic unit dome, used for servicing unit.

Propane A hydrocarbon fuel.

psi Unit of pressure measurement; pounds per square inch.

psia Pressure measured in pounds per square inch absolute.

psig A symbol or initials used to indicate pressure in pounds per square inch gauge.

Psychrometer An instrument used to measure the dry bulb and wet bulb temperatures of air.

Psychrometric Chart A chart that graphs the relationship of the temperature, pressure, and moisture content of air.

P/T Plugs Pressure temperature plugs that allow entrance into the system by the gauge stems of thermometers and pressure gauges without draining or removing the pressure of the system.

PTC Positive temperature coefficient; a thermistor, working like a temperature sensitive resistor, commonly used as a component for measuring temperature with electronic controls. Sometimes used as an electronic start relay in the start circuits for compressors. Commonly used to measure temperature of outdoor coils on heat pumps to initiate and terminate defrost cycles.

Pump A motor driven device used to mechanically circulate water in the system.

Pumpdown A service procedure where the refrigerant is pumped into the receiver.

Pump Head The difference in pressure on the supply and intake sides of the pump created by the operation of the pump.

Purging Releasing compressed gas to atmosphere through some part or parts for the purpose of removing contaminants from that part or parts.

Pyrometer Instrument for measuring temperatures.

Quenching Submerging hot solid object in cooling fluid.

Quick Connect Coupling A device that permits an easy and fast means of connecting two fluid lines or fittings together without the use of solder.

Radiant Heating A heating system in which only the heat radiated from panels is effective.

Radiation Transfer of heat by heat rays.

Radiator A heating unit exposed to view within the room or space to be heated.

Radiator Valve A valve installed on a terminal unit to manually control the flow of water through the unit.

Range Pressure or temperature settings of a control.

Rankine Scale Absolute Fahrenheit scale.

Reactance Opposition to alternating current by either inductance or capacitance, or both.

Reciprocating Action in which the motion is back and forth in a straight line.

Recirculated Air Return air passed through the conditioner before being again supplied to the conditioned space.

Rectifier, Electric An electrical device for converting AC into DC.

Reducing Fitting A pipe fitting designed to change from one pipe size to another.

Reed Valve Thin flat tempered steel plate fastened at one end.

Refractory A material that can withstand very high temperature.

Refrigerant Substance used in a refrigerating mechanism to absorb heat in an evaporator coil and to release its heat in a condenser.

Refrigerant Charge Quantity of refrigerant in a system.

Refrigerating Effect The amount of heat in Btu/hr the system is capable of transferring.

Refrigeration The process of transferring heat from one place to another.

Refrigeration Cycle A process during which a refrigerant absorbs heat at a relatively low temperature and reflects heat at a relatively higher temperature.

Refrigeration Oil Specially prepared oil used in refrigerator mechanism. It circulates with refrigerant.

Refrigeration System The combination of interconnecting devices, tubes, and/or pipes in which the refrigerant is circulated for the purpose of exchanging heat to produce cooling.

Register Combination grill and damper assembly on an air opening or at the end of an air duct.

Relative Humidity The ratio of the weight of moisture that air actually contains at a certain temperature as compared to the amount that it could contain if it were saturated.

Relay Electrical mechanism that uses small current in the control circuit to operate a valve switch in the operating circuit.

Relief Opening The opening in a draft hood to permit ready escape to the atmosphere of flue products.

Relief Valve Safety device designed to open before dangerous pressure is reached.

Repulsion Start Induction Motor Type of motor that has an electrical winding on the rotor for starting purposes.

Resistance The opposition to current flow by a physical conductor.

Resonance The pipe organ effect produced by a gas furnace when the frequency of the burner flame combustion and the pressure wave distance in the burner pouch are in exact synchronization.

Return Air Air returned from conditioned or refrigerated space.

Return Piping That portion of the piping system that carries water from the terminal units back to the boiler.

Reverse Acting Control A switch controlled by temperature and designed to open on temperature drop and close on temperature rise.

Reverse Cycle Defrost A method of defrosting the evaporator by means of flow valve(s) to move hot vapor from the compressor into the evaporator.

Reverse Cycle Refrigeration System Commonly called a heat pump. A refrigeration system capable of reversing its operation and direction of heat transfer.

Reverse Return A two pipe system in which the return connections from the terminal units into the return main are made in the reverse order from that in which the supply connections are made.

Reversing Valve A device used to change the direction of refrigerant vapor flow between the evaporator to compressor and compressor to condenser depending on the heating or cooling effect desired.

Rheostat An adjustable or variable resistor.

Rich Mixture A mixture of gas and air containing too much fuel or too little air for complete combustion of the gas.

Rollout A condition where flame rolls out of a combustion chamber when the burner is turned on.

Roof Mounted The unit is mounted on a platform designed to distribute the weight of the unit over as wide an area of the roof as possible.

Rotary Compressor Mechanism that pumps fluid by using a rotating motion.

Rotor Rotating part of a mechanism.

Running Winding Electrical winding of motor which has current flowing through it during normal operation of motor.

Runout This term generally applies to the horizontal portion of branch duct between the main trunk and the diffuser.

Saddle Valve Valve body shaped so it may be silver brazed to refrigerant tubing surface.

Safety Control Device that will stop the refrigerating unit if unsafe pressures and/or temperatures are reached.

Safety Plug Device that will release the contents of a container above normal pressure conditions and before rupture pressures are reached.

Saturated Vapor A vapor condition that will result in condensation into droplets of liquid as vapor temperature is reduced.

Saturation Condition when both liquid and vapor are present or either phase is just about to appear or disappear.

Scaling The formation of lime and other deposits on the water side surfaces of heat exchangers.

Schrader Valve Spring loaded device that permits fluid flow when a center pin is depressed.

Scotch Yoke Mechanism used to change reciprocating motion into rotary motion or vice versa.

SCR A solid state device (silicon controlled rectifier) used to modulate the capacity of an electric heating element.

Seasonal Energy Efficiency Ratio (SEER) The total cooling of a central air conditioner in BTU during its normal annual usage period for cooling divided by the total electric input in watt-hours during the same period.

Secondary Air Combustion air externally supplied to a burner flame at the point of combustion.

Secondary Voltage The output, or load supply voltage, of a transformer.

Second Law of Thermodynamics Heat will flow only from material at a certain temperature to material at a lower temperature.

Semihermetic Compressor A serviceable hermetic compressor.

Sensible Heat Heat energy added to or removed from a material that causes a change in temperature of the material without a change in state of the material.

Sensor A material or device that changes characteristics (electrical or mechanical) with a change in temperature and pressure.

Sequencer A control device used to control electrical circuits that use a heat motor for time delay of the operation of the device. (See Heat Motor)

Series A circuit with one continuous path for current flow.

Serviceable Hermetic Hermetic unit housing containing motor and compressor assembled by use of bolts or threads.

Service Valve Device used by service technicians to check pressures and change refrigerating units.

Shaded Pole Motor A small AC motor used for light start loads.

Shaft Seal A device used to prevent leakage between shaft and housing.

Short Circuit A low resistance connection (usually accidental and undesirable) between two parts of an electrical circuit.

Short Cycling Refrigerating system that starts and stops more frequently than it should.

Sight Glass Glass tube or glass window in refrigerating mechanism that shows amount of refrigerant or oil in system.

Silica Gel Chemical compound used as a drier, that has ability to absorb moisture.

Silver Brazing Brazing process in which brazing alloy contains some silver as part of joining alloy.

Single Package Heat Pump A system that has all components completely contained in one unit.

Sling Psychrometer Humidity measuring device with wet and dry bulb thermometers.

Slugging A condition where liquid refrigerant is entering an operating compressor.

Smoke Test Test made to determine completeness of combustion.

Soft Flame A flame partially deprived of primary air such that the combustion zone is extended and inner cone is ill defined.

Solar Heat Heat from visible and invisible energy waves from the sun.

Soldering Joining two metals by adhesion of a low melting temperature metal (less than 840°F).

Solenoid A movable plunger activated by an electromagnetic coil.

Solenoid Valve Valve actuated by magnetic action by means of an electrically energized coil.

Soot A black substance, mostly consisting of small particles of carbon, that can result from incomplete combustion.

Specific Gravity For a liquid or solid, the ratio of its density compared to water. For a vapor, the ratio of its density to air.

Specific Heat The amount of heat energy needed to change the temperature of a material at a given pressure as compared to an equal quantity of water or air at the same pressure.

Specific Volume The volume of a substance per unit mass; the reciprocal of density units; cubic feet per pound, cubic centimeters per gram, etc.

Split Phase Motor Motor with two stator windings.

Split System Refrigeration or air conditioning installation that places condensing unit remote from evaporator.

Split System Heat Pump A heat pump with components located both inside and outside of a building—the most common type of heat pump installed in a home.

Squirrel Cage Fan that has blades parallel to fan axis and moves air at right angle to fan axis.

Standard Conditions Temperature of 68°F, pressure of 29.92 in Hg and relative humidity of 30%.

Start Relay An electrical device that connects and/or disconnects the start winding of electric motor.

Starting Winding Winding in electric motor used for only the brief period when the motor is starting.

Static Pressure The pressure exerted against the inside surfaces of a container or duct. Sometimes defined as bursting pressure.

Stator Stationary part of electric motor.

Steam Jet Refrigeration Refrigerating system which uses a steam venturi to create high vacuum (low pressure) on a water container causing water to evaporate at low temperature.

Steam Trap A device that will prevent the flow of steam, but will allow the flow of condensate.
Strainer A screen used to retain solid particles while liquid passes through.
Stratification Condition in which air lies in temperature layers.
Subcooled Liquid Liquid at a temperature lower than is possible when it is in equilibrium with its vapor. The pressure is higher than the vapor pressure.
Subcooling The reduction of the temperature of a liquid below its condensing temperature.
Sublimation Condition where a substance changes from a solid to a gas without becoming a liquid.
Suction Line Tubing or pipe used to carry refrigerant vapor from the evaporator to the compressor in single function systems or from the reversing valve to the compressor in dual function systems.
Suction Pressure Pressure in the low side of the refrigeration system. Also called low side pressure or suction pressure.
Suction Service Valve A two way manually operated valve located at the inlet to the compressor.
Superheat Heat energy added to a gas so that the enthalpy is higher than for saturated vapor at same pressure. The heat added to a vapor to raise the sensible temperature of the vapor above its boiling point.
Supplementary Heat The auxiliary heat provided at temperatures below the heat pump balance point. In most cases this is done with electric heating elements that are part of the heat pump system installation. A gas or oil furnace also can be used to provide supplementary heat when a heat pump is added to an existing fossil fuel heating system.
Surge Tank Container connected to a refrigerating system that increases gas volume and reduces rate of pressure change.
Swaging Enlarging one tube end so end of other tube of same size will fit within.
Swash Plate Wobble plate. Device used to change rotary motion to reciprocating motion.
Sweating This term has two definitions in air conditioning or heat pump work: (1) Formation of moisture on the outside of cold pipes or ducts; and (2) joining of two metals by the adhesive action of a third metal.
Sweet Water Tap water.

Tankless Water Heater An indirect water heater designed to operate without a hot water storage tank.
Tap (Screw Thread) Tool used to cut internal threads.
Temperature Degree of hotness or coldness as measured by a thermometer.
Terminal Units Radiators, convectors, baseboard, unit heaters, finned tube, etc.
Therm A unit of heat having a value of 100,000 Btu.
Thermal Conductivity The ability of a material to transmit heat.
Thermal Conductivity, Earth The rate at which heat energy flows through an earth material. Expressed in Btu/ft^2 of the material surface times the temperature difference per thickness in feet of the material.
Thermal Cutout An overcurrent protection device that contains a heater element that affects a bimetal element designed to open a circuit in the event of electrical current flow above the rated amount of the device.
Thermal Resistance The resistance a material offers to the transmission of heat.
Thermistor An electrical device that changes electrical resistance with a change in temperature of the device.
Thermocouple Device that generates electricity using the principle that if two dissimilar metals are welded together and the junction is heated, a voltage will develop across the open ends.
Thermoelectric Refrigeration A refrigerator mechanism that depends on the Peletier effect.
Thermometer Device for measuring temperatures.
Thermopile A pilot generator.
Thermostat A control device that responds to surrounding air temperatures.
Thermostat, Outdoor Ambient A control used to limit the amount of auxiliary electric heat according to the outdoor ambient temperature to reduce electrical surge and cost.
Thermostat, Termination A thermostat mounted on the outdoor coil that interrupts the defrost mode when the temperature of the coil reaches the cutout setpoint of the control.
Thermostatic Control A device that controls the operation of equipment according to the temperature of the air surrounding the control.
Thermostatic Expansion Valve A valve operated by temperature and pressure within the evaporator coil.
Thermostatic Valve Valve controlled by thermostatic elements.
Throttling Expansion of gas through an orifice.
Timers Mechanism used to control the time cycling of an electrical circuit.
Timer, Defrost A timer that operates at the same time as the refrigeration system. After a set period of operating time, the timer trips to initiate a defrost operation if the termination thermostat is closed.
Ton Refrigerating effect equal to the melting of one ton of ice in 24 hours.
Torque Turning or twisting force.
Torque Wrenches Wrenches which may be used to measure torque applied.
Total Pressure The sum of the static pressure and the velocity pressure at the point of measurement.
Transformer A device designed to change voltage.
Tube in a Tube Coaxial. A heat exchanger constructed of a tube inside a tube sealed off from each other. Usually liquid through the inner tube and refrigerant through the outer tube.
Turbulent Flow The movement of a liquid or vapor in a pipe in a constantly churning and mixing fashion.
Turndown The ratio of maximum to minimum input rates.
Two Pipe System A hot water heating system using one pipe from the boiler to supply heated water to the terminal units, and a second pipe to return the water from the terminal units back to the boiler.

U Factor Unit of measure of thermal conductivity.
Ultraviolet Invisible radiation waves with frequencies shorter than wavelengths of visible light and longer than X-ray.
Underwriters Laboratories Abbreviated UL. Underwriters Laboratories, Inc. maintains and operates laboratories for the examination and testing of devices.
Unit Heater A fan and motor, a heating element, and an enclosure hung from a ceiling or wall.
Unit Ventilator A terminal unit in which a fan is used to mechanically circulate air over the heating coil.
Urethane Foam Type of insulation that is foamed in between inner and outer walls of a display case.

Vacuum Reduction in pressure below atmospheric.
Vacuum Pump A high efficiency vapor pump used for creating deep vacuum in refrigeration systems for testing and/or drying purposes.
Valve Device used for controlling fluid flow.
Valve, Check A valve that will permit fluid flow on only one direction. Sometimes called a one way valve.
Valve Plate Part of compressor located between top of compressor body and head that contains compressor valves.
Valve, Reversing A valve used to change the direction of refrigerant flow in a heat pump system. Because there are four pipe connections, it is also called a four way valve.
Valve, Service A device used by service technicians to connect pressure gauges into the refrigeration system.
Valve, Slide The slide valve portion of the reversing valve that shifts the refrigerant flow.
Valve, Solenoid A flow control valve controlled by an electromagnetic coil actuating a plunger off a seat.
Valve, TX, Bi-Flow A thermostatic expansion valve that is designed to provide pressure reduction and refrigerant flow control in either direction.
Vapor Barrier Thin plastic or metal foil sheet used to prevent water vapor from penetrating insulating material.
Vapor Line Found only in dual action heat pumps. It is the suction line in the cooling mode and the hot gas line in the heating mode.
Vapor, Saturated A vapor whose temperature has been reduced to the point of condensation but condensation has not started.
Variable Pitch Pulley Pulley that can be adjusted to provide different pulley ratios.
V Belt Type of drive belt.
Velocimeter Instrument used to measure air velocities.
Velocity Pressure The pressure exerted in direction of flow.
Vent Gases Products of combustion.
Ventilation The introduction of outdoor air into a building by mechanical means.
Venturi A section in a pipe or a burner body that narrows down and then flares out again.
Viscosity Measure of a fluid's ability to flow.
Volt The unit of electrical potential or pressure.
Voltage Relay One that functions at a predetermined voltage value
Volumetric Efficiency Ratio of the actual performance of a compressor and calculated performance.
VOM A meter that measures voltage and resistance (ohms).

Walk In Cooler Large commercial refrigerated room.
Water Column Abbreviated as WC. A unit used for expressing pressure.
Water Cooled Condenser A heat exchanger that uses water to remove the heat from the high temperature compressor discharge vapor.
Water Cooled Condensing Unit A condensing unit (high side) that is cooled by the use of water.
Water Tube Boiler A hot water boiler in which the water is circulated through the tubes and the hot gases from combustion of the fuel are circulated around the tubes.
Watt A unit of electrical power.
Well Water Source, Closed Loop A system that removes water from the earth by means of a drilled well and returns it to the earth by means of a separate drilled well.
Well Water Source, Open Loop A system that removes water from the earth by means of a drilled or bored well and returns the water to the earth through a separate disposal system.
Wet Bulb A thermometer that uses a wet sac on the testing tip or bulb to measure the evaporation rate of the air sample. Evaporation of the moisture lowers the temperature of the wet bulb thermometer as compared to the dry bulb thermometer. The two readings and a psychrometric chart are used to determine the characteristics of the air.
Wind Effect The increase in evaporation rate due to air travel over a water surface.

Zone That portion of a building whose temperature is controlled by a single thermostat.

Spanish Language Glossary

Absolute Humidity Humedad Absoluta. La cantidad de vapor de agua en un cierto volumen de aire.

Absolute Pressure Presión Absoluta. Presión de manómetro más presión atmosférica.

Absolute Temperature Temperatura Absoluta. Temperatura medida a partir del cero absoluto.

Absolute Zero Cero Absoluto. Temperatura a la cual cesa todo movimiento molecular (equivalente a −460° Farenheit y −273° Celsius).

Absorbent Absorbente. Substancia que tiene la capacidad de absorber otra substancia.

Absorber Tanque de absorción. Aparato que contiene un líquido para absorber vapor refrigerante u otro tipo de vapor.

Accumulator Acumulador. Depósito que recibe líquido refrigerante del evaporador e impide que fluya hacia el tubo de succión.

Activated Carbon Carbón activado. Carbón procesado industrialmente que se usa comúnmente para limpiar aire.

Adiabatic Compression Compresión adiabática. Compresión de gas refrigerante sin añadir o eliminar calor.

Adsorbent Adsorbente. Substancia que tiene la propiedad de retener moléculas de un fluído sin causar una reacción química o un cambio físico.

Air Binding Atascamiento de aire. Condición en la cual una burbuja o una bolsa de aire en una tubería impide el flujo deseado.

Air Changes Cambios de aire. La cantidad de aire que se escapa de un edificio en términos del número de volúmenes del edificio intercambiados.

Air Conditioner Aire Acondicionado. Aparato utilizado para controlar la temperatura, humedad, limpieza y movimiento de aire en el espacio acondicionado.

Air Cushion Tank Depósito de cojín de aire. Recipiente cerrado que permite la expansión del agua sin que la presión aumente de manera excesiva.

Air Shutter Obturador de aire. Obturador ajustable situado en los orificios primarios de aire de una caldera que se utiliza para controlar la cantidad de aire de la combustión.

Air Source Heat Pump Bomba térmica de alimentación de aire. Bomba térmica que transfiere calor del aire exterior a un sistema de circulación de aire interior.

Air Vent Respiradero. Válvula instalada en los puntos altos de un sistema de agua caliente para eliminar aire de dicho sistema.

Aldehyde Aldehído. Una clase de compuestos que se pueden producir durante la combustión incompleta de un carburante.

Allen Head Screw Tornillo Allen. Tornillo de cabeza empotrada, diseñado para girar con una llave de sección hexagonal.

Alternating Current Corriente Alterna (Abreviatura: AC). Corriente eléctrica que cambia de polaridad o dirección de manera periódica: aumenta de cero hasta un máximo, vuelve a cero y aumenta de nuevo hasta dicho máximo pero en la dirección opuesta, en un ciclo que se repite a una frecuencia fija. La corriente alterna puede ser de una sola fase, dos fases, tres fases y polifásica. La ventaja de la corriente alterna sobre la corriente continua es que el voltaje puede aumentarse a valores muy altos con la ayuda de transformadores, lo cual reduce los costes de transmisión.

Alternator Alternador. Una máquina que convierte la energía mecánica en corriente alterna.

Ambient Temperature Temperatura Ambiente. Temperatura del fluído, generalmente aire, que rodea a un objeto.

American Wire Gauge Sistema Americano de Clasificación de Hilo Conductor (Abreviatura: AWG). Un sistema de numeración utilizado para designar la sección de un hilo conductor. Este número es mayor conforme disminuye el diámetro del hilo; por ejemplo, AWG #14 = 0.0641 pulgadas; AWG #12 = 0.0808 pulgadas)

Ammeter Amperímetro. Un aparato de medición que se utiliza para medir la corriente eléctrica.

Ampere Amperio. Unidad de corriente eléctrica equivalente al flujo de un culombio por segundo.

Ampere Turn Amperio Vuelta (Abreviatura: AT o NI). Unidad de fuerza magnética que se produce cuando una corriente eléctrica de un Amperio circula por una vuelta de una bobina.

Amplitude Amplitud. Valor máximo instantáneo de una corriente o un voltaje alterno, tanto en la dirección positiva como en la negativa.

Anemometer Anemómetro. Instrumento para medir el coeficiente de flujo de aire.

Angle of Lag or Lead Ángulo de Desfase. Diferencia de fase entre dos formas sinusoidales de la misma frecuencia.

Annealing Revenir o destemplar. Proceso de tratamiento térmico de metales para obtener las características deseadas de dureza y ductilidad.

Anticipator Anticipador. Un calentador que se utiliza para ajustar la operación de un termostato con el fin de reducir el diferencial de temperatura por encima de la capacidad mecánica del controlador.

Armature Armadura. El componente móvil o rotatorio de un motor, generador, relé u otro tipo de dispositivo electromagnético.

ASME American Society of Mechanical Engineers. Asociación Americana de Ingenieros Mecánicos.

Aspect Ratio Relación de aspecto o exposición. Relación entre la longitud y la anchura de un conducto rectangular.

Aspirating Psychrometer Higrómetro aspirante. Dispositivo que aspira una muestra de aire para determinar su humedad.

Atom Átomo. La particular más pequeña de un elemento.

Atomic Weight Peso atómico. El número de protones en un átomo de un material.

Atomize Atomizar o pulverizar. Proceso por el cual un líquido se transforma en un polvo fino.

Attenuate Atenuar. Reducir intensidad.

Authorized Dealer Concesionario autorizado. Representante comercial autorizado por el fabricante a instalar sus bombas térmicas.

Automatic Defrost Descongelador automático. Sistema para eliminar automáticamente hielo y escarcha de un evaporador.

Automatic Expansion Valve (AEV) Válvula de Expansión Automática.Válvula controlada a presión que se utiliza como dispositivo de dosificación.

Autotransformer Autotransformador. Un transformador en el que el lado primario y el secundario comparten parte del bobinado. Se puede obtener un mayor o menor voltaje de salida a través de diferentes tomas en el bobinado común.

Back Pressure Contrapresión. Presión en el lado inferior del sistema de refrigeración; también llamada presión de succión o presión del lado inferior.

Back Seat Posición de reposo. Posición de una válvula de servicio cuando se ha movido en el sentido contrario de las agujas del reloj hasta su tope.

Balance Fitting Conexión compensadora. Conexión de tubería con resistencia variable al flujo de un fluído. Este tipo de conexión es utilizada para obtener el caudal deseado por circuitos paralelos.

Balance Point Punto de equilibrio. Temperatura exterior a la cual la capacidad calorífica de una bomba de calor en una instalación de calefacción es idéntica a la pérdida de calor de la zona acondicionada. Temperatura exterior, generalmente entre 30 y 45 grados Fahrenheit, a la cual el rendimiento de la bomba térmica y los requerimientos de calefacción de una casa se igualan. Por debajo del punto de equilibrio, es preciso tener una fuente de calor adicional para mantener confort en el interior de la casa.

Balance Point, Initial Punto de equilibrio inicial. Temperatura ambiental exterior a la cual el rendimiento de la bomba térmica solamente reemplaza la pérdida de calor en el área acondicionada.

Balance Point, Second Punto de equilibrio posterior. Temperatura ambiental exterior a la cual el rendimiento de la bomba térmica más la fuente de calor auxiliar reemplazan la pérdida de calor en el área acondicionada.

Ball Check Valve Válvula unidireccional de bola. Conjunto de válvula que permite el flujo de un fluído en una sola sección.

Ball Valve Válvula de bola. Una válvula que modula o impide el flujo de un fluído al girar un cuarto de vuelta una bola con un orificio que la atraviesa.

Barometer Barómetro. Instrumento usado para medir la presión atmosférica.

Baseboard Tablero. Unidad terminal de calefacción que tiene el aspecto del zócalo de una casa.

Battery Batería o pila. Dos o más células primarias o secundarias interconectadas eléctricamente.

Baudelot Cooler Refrigerador Baudelot. Intercambiador de calor en el cual el agua fluye por la parte exterior de tubos o placas por la acción de la fuerza de la gravedad.

Bellows Fuelle. Contenedor cilíndrico corrugado que se mueve con los cambios de presión.

Bending Spring Muelle de flexion. Muelle helicoidal que se utiliza para impedir que un tubo se pliegue al flexionarlo.

Bimetal Strip Pletina bimetal: Dos metales distintos con diferentes coeficientes de expansión soldados juntos.

Bleed Off Purga. La eliminación continua o intermitente de una pequeña fracción del agua de una torre de refrigeración con el fin de prevenir la acumulación de sedimentos químicos.

Bleed Valve Válvula de purga. Válvula que permite el flujo mínimo de un fluído cuando está cerrada.

Blower Soplador. Un ventilador centrífugo.

Boiler Horsepower Potencia de una caldera. La evaporación equivalente de 34.5 libras de agua sin cambiar la temperatura de 212° Fahrenheit. Esta potencia equivale a una liberación de calor de 970,3 × 34,5 = 33.475 Btu/h (Unidades Térmicas Británicas por hora)

Boiling Temperature de ebullición. Temperatura a la cual un fluído cambia del estado líquido al estado gaseoso.

Bore Diámetro interior. El diámetro interior de un agujero cilíndrico.

Bourdon Tube Tubo de bordón. Tubo de pared delgada y forma circular, que tiende a enderezarse cuando aumenta la presión interior.

Brazing Estañar. Soldar con un material de relleno (estaño, bronce o latón) cuya temperatura de fusión es superior a los 800° Fahrenheit.

Break Corte eléctrico. Discontinuidad eléctrica en un circuito, generalmente causada al operar un interruptor o un cortacircuitos.

Breaker Strip Tira de refuerzo. Tira de plástico que se utiliza para cubrir la junta entre la tapa exterior y el forro interior de un refrigrerador.

Brine Salmuera. Agua saturada con un producto químico como la sal.

British Thermal Unit (BTU) Unidad Térmica Británica (Btu). Cantidad de calor requerida para cambiar la temperatura de una libra de agua un grado Fahrenheit.

Bull Head Conexión en forma de astas. Una conexión de fontanería en forma de "t" en la que el agua entra (o sale) por ambos lados de la conexión horizontal y sale (o entra) por el conducto perpendicular.

Bunsen Burner Quemador tipo Bunsen. Un calentador de gas en el cual el aire de combustión se mezcla con gas en el interior del calentador antes de que éste se queme en la lumbrera.

Burner Quemador. Un dispositivo para el transporte final de gas, o una mezcla de gas y aire, a la zona de combustión.

Burnout Fundir. Paso accidental de una corriente eléctrica muy alta por un circuito o dispositivo eléctrico que lo daña.

Cadmium Cell Célula o pila de Cadmio. Un dispositivo cuya resistencia eléctrica varía de acuerdo con la cantidad de luz que percibe.

Calibrate Calibrar. Ajustar un indicador para que determine correctamente la variable a medir.

Calorie Caloría. Cantidad de calor requerida para cambiar la temperatura de un gramo de agua un grado Celsius.

Capacitor Condensador. Dispositivo que almacena electricidad, usado para arrancar o mantener circuitos de motores eléctricos.

Capillary Tube De tubos capilares. Una clase de control de refrigeración que consiste en una gran longitud de varios tubos de diámetro interior muy pequeño.

Carbon Dioxide (CO_2) Dióxido de carbono. Gas no tóxico que se produce en la combustión.

Carbon Monoxide (CO) Monóxido de carbono. Un producto químico que se produce en la combustión incompleta. El monóxido de carbono no tiene color ni olor y es muy tóxico.

Cascade System Sistema en cascada. Sistema que contiene dos o más circuitos de refrigeración, cada uno de ellos con un compresor, condensador y evaporador, y donde el evaporador de uno de los circuitos enfría el condensador de otro circuito (de menor temperatura).

Celsius Celsius o Centígrado. La escala de medición de temperatura del sistema métrico.

Centimeter Centímetro. Unidad de longitud del sistema métrico que equivale a 0.3937 pulgadas.

Centrifugal Compressor Compresor centrífugo. Compresor que utiliza fuerza centrífuga para comprimir refrigerantes gaseosos.

CFM Pies cúbicos por minuto. Unidad de medida del flujo de un fluído.

Charge Carga. La cantidad de refrigerante de un sistema.

Charging Cylinder Cilindro de carga. Un dispositivo diseñado para cargar una cantidad predeterminada de refrigerante en un sistema.

Charging the Loop Cargar el circuito cerrado. Rellenar el circuito con la mezcla correcta y purgar el aire atrapado.

Check Valve Válvula unidireccional. Válvula que permite el flujo de un fluído en una sola dirección.

Chimney Effect Efecto chimenea. La tendencia a subir al calentarse del aire o gas en una tubería.

Circuit Circuito. Una instalación que puede ser de cableado eléctrico, tuberías o tubos flexibles que permite el flujo desde una fuente de energía hasta volver a dicha fuente.

Circuit Breaker Cortocircuito. Un dispositivo que percibe el flujo de una corriente eléctrica y abre el circuito cuando dicha corriente excede el valor nominal del dispositivo.

Circulator Circulador. Bomba.

Clearance Holgura. Espacio en un cilindro que no es ocupado por el pistón al final de su carrera.

Closed Loop Circuito cerrado. Un ciclo en el cual el medio primario se encuentra siempre encerrado y repite la misma secuencia.

Coaxial Fitting Conexión coaxial. Una conexión a un orificio único de un depósito de agua caliente. Un tubo en una conexión. El agua caliente que entra en el depósito por el tubo. El agua fría que sale por la conexión alrededor del tubo.

Coefficient of Performance (COP) Coeficiente de Rendimiento. Energía útil a la salida de un dispositivo dividida por la energía a la entrada.

Coil Bobina. Un hilo conductor enrollado, que crea un campo magnético muy fuerte al ser atravesado por una corriente eléctrica.

Coil, Deice Subcooler Serpentín de sobreenfriamiento y descongelación. Una sección en el Serpentín exterior de evaporación que se utiliza para sobreenfriar líquido refrigerante y, durante la fase de descongelación, como superficie adicional.

Coil, Inside Serpentín interior. El serpentín situado en la parte interior del sistema de la bomba de calor. Dicho serpentín funciona como evaporador en la fase de enfriamiento y como condensador en la fase de calentamiento.

Coil, Outside Serpentín exterior. El serpentín situado en la parte exterior de una bomba de calor de aire. Dicho serpentín funciona como evaporador en la fase de calentamiento y como condensador en la fase de enfriamiento.

Cold Frío. Es la ausencia de calor.

Cold Junction Junta en frío. Aquella parte del sistema termoeléctrico que absorbe calor.

Combustion Combustión. La oxidación rápida de los gases de un carburante, acompañada de la producción de calor.

Combustion Air Aire de combustion. Aire proporcionado para la combustión de un carburante.

Combustion Products Productos de la combustion. Componentes que resultan de la combustión de un gas carburante con el oxígeno en el aire, incluyendo algunos compuestos inertes, pero excluyendo el exceso de aire.

Comfort Chart Tabla de comfort. Tabla que se utiliza en proyectos de aire acondicionado para mostrar la temperatura y humedad necesarias para la comodidad del ser humano.

Commercial Buildings Edificios comerciales. Construcciones tales como tiendas, grandes almacenes, restaurantes, moteles y edificios de apartamentos de gran tamaño.

Commutator Conmutador. Un anillo de segmentos de cobre aislados entre sí que conectan la armadura y las escobillas de un motor o de un generador. El conmutador dirige la corriente eléctrica en la dirección de las escobillas.

Compound Gauge Manómetro compuesto. Indicador de presión que tiene escala de medición por encima y por debajo del nivel de presión atmosférica.

Compound Refrigerating System Sistema de refrigeración compuesto. Sistema que tiene varios compresores en serie.

Compression Ratio Relación de compresión. Presión absoluta de descarga dividida por la presión absoluta de succión de un compresor.

Compression Tank Tanque o depósito de compresión. (Ver depósito de cojín de aire).

Compressor Compresor. La bomba de un mecanismo de refrigeración que crea un vacío o baja presión en el lado de enfriamiento del ciclo de refrigeración y comprime el gas en el lado de condensación de dicho ciclo.

Compressor Displacement Desplazamiento o cilindrada de un compressor. El volumen que descarga un compresor en una rotación del cigüeñal.

Compressor Seal Retén o sello del compressor. Junta a prueba de fugas entre el cigüeñal y el cuerpo del compresor.

Condensate Drain Trap Sifón de desagüe del condensador. Forma de tubería que crea estanqueidad en la cañería del desagüe con el fin de impedir el flujo de aire.

Condensate Pan Bandeja de condensación. Contenedor situado debajo de un evaporador para contener el condensado que se escapa del serpentín y llevarlo a la tubería de desagüe.

Condensate Pump Bomba de condensado. Dispositivo utilizado para eliminar fluído condensado.

Condensation Condensación. Líquido que se forma cuando un gas se enfría por debajo de su temperatura de condensación.

Condense Condensar. El cambio de vapor saturado a líquido saturado.

Condenser Condensador. La parte del mecanismo de refrigeración que recibe gas refrigerante a alta presión y temperatura del compresor y lo enfría hasta que pasa al estado líquido.

Condenser, Air Cooled Condensador enfriado por aire. Un intercambiador de calor que transfiere la energía calorífica del vapor refrigerante al aire.

Condenser, Liquid Cooled Condensador con refrigeración líquida. Un intercambiador de calor que transfiere la energía calorífica del vapor refrigerante a un elemento absorbente de calor en estado líquido.

Condenser Water Pump Bomba de agua del condensador. Dispositivo utilizado para forzar el paso de agua a través del condensador.

Condensible Condensable. Un gas que se puede convertir fácilmente en líquido.

Condensing Pressure Presión de condensación. La presión del refrigerante dentro del serpentín del condensador.

Condensing Temperature Temperatura de condensación. La temperatura a la cual tiene lugar la condensación del refrigerante dentro del condensador.

Condensing Unit Unidad o sistema de condensación. Unidad de refrigeración que consta del compresor, el condensador, y los mecanismos de control.

Conductance (Thermal) Conductancia (termal). El coeficiente de paso de calor a través de un material por unidad de área.

Conduction (Thermal) Conducción (termal). Transmisión de calor de una partícula a otra partícula.

Conductivity Conductividad. La capacidad de una substancia de permitir el paso de calor o electricidad.

Conductor Conductor. Material o substancia que permite el paso de electricidad con gran facilidad.

Connected Load Carga de conexión. La suma de las capacidades de los aparatos conectados a una fuente de alimentación eléctrica.

Connecting Rod Biela. La parte del compresor que conecta el pistón con el cigüeñal.

Constrictor Constrictor. Tubo u orificio que se utiliza para restringir el flujo de un fluído.

Contact Contacto. La parte de un interruptor o relé que permite el paso de la corriente eléctrica.

Contactor Interruptor automático. Un dispositivo que abre o cierra el contacto con circuitos de carga mediante un bobinado magnético.

Contaminant Contaminante. Una substancia (humedad, polvo, suciedad, etc.) ajena al refrigerante o al aceite de engrase de un sistema de refrigeración.

Control, Low Pressure Controlador de baja presión. Control a presión conectado al lado de succión del sistema de refrigeración para impedir que éste opere por debajo de una cierta presión o punto de ebullición.

Control, Refrigerant Controlador del refrigerante. Dispositivo que se utiliza para reducir la presión del líquido refrigerante con el fin de obtener el punto de ebullición adecuado del refrigerante en el evaporador.

Control, Temperature Control por temperature. Dispositivo que utiliza cambios de temperatura para operar contactos en un circuito eléctrico.

Controller Combinador. Dispositivo que calcula la diferencia entre la variable real de salida y la variable deseada e inicia una respuesta con el fin de corregir esa diferencia.

Controls Controles. Dispositivos diseñados para regular gas, aire, agua o electricidad.

Convection Convección. Transferencia de calor mediante el movimiento de un fluído.

Cooling Anticipator Anticipador de enfriamiento. Resistencia en el termostato de una habitación que hace que el ciclo de enfriamiento comience de manera prematura.

Cooling Tower Torre de enfriamiento. Aparato que enfría agua mediante el proceso de evaporación de agua en el aire. El agua se enfría a la temperatura ambiente.

Coulomb Culombio. Unidad de carga eléctrica; un culombio por segundo equivale a un amperio o al paso de $6{,}25 = 10^{18}$ electrones por un punto en un segundo.

Counter EMF Fuerza contraelectromotriz. La fuerza que se induce en un bobinado al aplicársele una diferencia de potencial a sus bornes. Dicha fuerza tiene el signo opuesto al voltaje aplicado.

Counterflow Contraflujo. Flujo en el sentido opuesto.

Cracking a Valve Abrir una válvula. Permitir el paso de una pequeña cantidad de fluído.

Crankcase Heater Calentador del carter. Dispositivo calefactor fijado al cárter o parte inferior de la carcasa del compresor para mantener el aceite en el compresor a una temperature superior a la del resto del sistema de refrigeración con el fin de reducir pérdidas del refrigerante.

Crankshaft Seal Sello del cigüeñal. Junta a prueba de fugas entre el cigüeñal y el cuerpo del compresor.

Current Corriente. El flujo de electrones a través de un conductor.

Cut-In Salto de conexión. La temperatura o presión a la cual un interruptor automático cierra el circuito.

Cut-Out Salto de desconexión. La temperatura o presión a la cual un interruptor automático abre el circuito.

Cylinder Head Culata. Pieza que cubre y protege el lado de compresión del cilindro compresor.

Damper Registro. Válvula para controlar el flujo de aire.

Dead Band Banda inoperative. El rango de temperaturas en un sistema de calefacción o de refrigeración en el cual el sistema no entra en funcionamiento.

Decibel Decibelio. Unidad de medición de la intensidad relativa del sonido.

Dedicated Geothermal Well Pozo geotérmico dedicado. Se obtiene agua de la parte superior del pozo y se devuelve a la parte inferior. El pozo funciona como bomba de calor.

Defrost Control Control de descongelación. Un sistema de control situado en una bomba de calefacción usado para detectar la acumulación de hielo en el bobinado exterior durante el ciclo de calentamiento y que causa la inversión del sistema con el fin de eliminar el hielo mediante gases calientes.

Defrost Cycle Ciclo de descongelación. Ciclo de refrigeración en el cual se derrite el hielo acumulado en el evaporador.

Defrost Mode Descongelación. La parte de la operación del sistema refrigerante en la cual se remueve la escarcha o el hielo del evaporador.

Defrost Timer Ruptor de descongelación. Dispositivo conectado al circuito eléctrico que desconecta la unidad el tiempo suficiente para permitir que el hielo o escarcha acumulados en el evaporador se derritan.

Degree Day Grado día.Unidad que representa un grado de temperatura de diferencia en la temperatura media de un día con respecto a otro.

Dehydrate Deshidratar. Eliminar agua.

Delta Connection Conexión Delta. La conexión en un sistema eléctrico a tres fases en el cual las terminales son triangulares, de una manera similar a la letra griega delta (Δ).

Demand Demanda o requerimiento. La cantidad media de carga en un período de tiempo determinado.

Demand (Billing) Demanda (facturación). La demanda en que se basa la facturación a un cliente.

Demand Meter Medidor de demanda. Un instrumento utilizado para medir el uso de Kwh. en condiciones punta.

Density Densidad. Masa por unidad de volume.

Desiccant Desecativo. Substancia usada para retener y eliminar humedad.

Design Load Carga de diseño. La cantidad de calefacción (o refrigeración) requerida para mantener las condiciones internas cuando las condiciones exteriores se encuentran a la temperatura de diseño.

Design Temperature Difference Diferencia en la temperatura de diseño: La diferencia entre las temperaturas interior y exterior de diseño.

Design Water Temperature Difference Diferencia de temperatura de diseño del agua. La diferencia entre las temperaturas de salida y entrada del agua por un dispositivo cuando dicho dispositivo opera en las condiciones de diseño.

Desuperheater Eliminador de sobrecalentamiento. Un intercambiador de calor en la línea de gas caliente situado entre la zona de descarga del compresor y la válvula de inversión. Se utiliza para calentar agua para uso doméstico al eliminar el sobrecalentamiento del vapor refrigerante.

Detector, Leak Detector de fugas. Instrumento de ensayo utilizado para detectar y localizar fugas de refrigerante.

Dew Point Temperatura de condensación. La temperatura a la cual el vapor (en condiciones de 100% humedad) comienza a pasar al estado líquido.

Diaphragm Diafragma. Membrana flexible.

Dielectric Dieléctrico. Un aislante (no conductor).

Dies (Thread) Matriz (Rosca). Herramental utilizado para cortar roscas exteriores.

Differential Diferencial. En el caso de su aplicación a sistemas de refrigeración y calefacción, se refiere a la diferencia entre las temperaturas o presiones de conexión y desconexión.

Diffuser Difusor. Boca de distribución diseñada para dirigir el flujo de aire en una habitación.

Dilution Air Aire de dilución. Aire que entra en una campana de extracción y se mezcla con los gases de la chimenea.

Diode Diodo. Un dispositivo que permite el paso de corriente eléctrica en un sentido pero no en el sentido opuesto.

Direct Current Corriente directa (Abreviatura: DC). Corriente eléctrica que fluye solamente en un sentido.

Direct Expansion Evaporator Evaporador de expansión directa. Evaporador que contiene líquido y vapor refrigerante.

Direct Return Retorno directo. Sistema de cañería doble en el que la primera unidad terminal a partir de la fuente de alimentación es la primera unidad conectada a la cañería de retorno.

Disconnecting Switch Interruptor de desconexión. Interruptor tipo cuchillo que abre un circuito eléctrico.

Disturbed Earth Effect Efecto de perturbación de la tierra. Cambio de la cantidad de calor que afecta a la temperatura de la tierra cuando el sistema de la bomba de calor añade o elimina energía calorífica.

Domestic Hot Water Agua caliente residencial. El agua caliente que se utiliza para cocinar, lavar, etc.

Domestic Geothermal Well Pozo geotérmico doméstico. Se obtiene agua de la parte superior del pozo. Solamente el agua que proviene de la bomba de calor vuelve a la parte inferior del pozo. La diferencia se utiliza para las necesidades de consumo del edificio.

Double Pole Switch Interruptor de doble polaridad. Interruptor que abre o cierra dos cables de un circuito simultáneamente.

Double Thickness Flare Ensanchamiento de doble espesor. Extremo de un tubo que ha sido formado con un espesor doble del resto del tubo.

Dowel Pin Perno o chaveta. Bulón ensamblado a presión en dos piezas de un conjunto para asegurar que se encuentran alineadas correctamente.

Downdraft Flujo descendente. Flujo de un gas hacia abajo.

Draft Gauge Indicador de corriente de aire. Instrumento utilizado para medir la presión del aire.

Drier Secador. Una substancia o un dispositivo utilizado para eliminar humedad de un sistema de refrigeración.

Drift Agua desplazada. Agua estancada en una torre de refrigeración que ha sido transportada por el viento.

Drilled Port Burner Quemador de orificios barrenados. Quemador en el que las lumbreras fueron formadas por agujeros taladrados.

Drip Pan Colector. Panel o cubeta en forma de bandeja que se usa para recoger refrigerante condensado del serpentín de evaporación.

Dry Bulb Temperature Temperatura ambiente. Temperatura del aire, tal y como la indica un termómetro corriente.

Dry Ice Hielo seco. Dióxido de carbono en estado sólido.

Duct Conducto. Tubería de chapa metálica o de fibra de vidrio de sección circular o rectangular que transporta aire entre el sistema de acondicionamiento de aire y la zona acondicionada.

Dynamometer Dinamómetro. Dispositivo para medir fuerza.

E E. Símbolo de diferencia de potencial (voltaje) eléctrico.

Earth Temperature, Minimum (T) Mínima temperatura de la tierrra (T). La temperatura más baja que puede alcanzar la tierra durante el invierno. Varía según zonas y las condiciones metereológicas.

Eccentric Excéntrico. Círculo o disco cuyo centro no coincide con el eje de un vástago.

Effective Area Área efectiva. Área neta de flujo de un fluído.

Efficiency (Electrical) Eficiencia (eléctrica). Valor en porcentaje que representa la relación entre el potencial de salida y el potencial de entrada.

Electric Field Campo eléctrico. Una región magnética en el espacio.

Electric Heating Calefacción eléctrica. Sistema de calefacción en el cual resistencias eléctricas producen calor.

Electric Heating Element Calentador eléctrico: Dispositivo que consiste de una resistencia eléctrica, soportes aislados y terminales que conectan dicha resistencia a una fuente de alimentación eléctrica.

Electrolyte Electrolito. Una solución (en forma líquida o pastosa) de una substancia que tiene la capacidad de conducir la electricidad.

Electromagnet Electroimán. Imán creado debido al flujo de electricidad por una bobina conductora.

Electromotive Force Fuerza electromotriz (Abreviatura: EMF). Diferencia de potencial eléctrico entre dos puntos.

Electronic Leak Detector Detector electrónico de fugas. Instrumento electrónico que mide los cambios en el flujo de electrones por un espacio cargado eléctricamente, lo que indica la presencia de moléculas de un vapor refrigerante.

Electronics Electrónica. Campo de las ciencias que estudia dispositivos basados en electrones y sus posibles usos.

Electrostatic Filter Filtro electrostático. Tipo de filtro que proporciona carga eléctrica a partículas de polvo o suciedad.

End Bell Campana terminal. Elemento estructural final de un motor eléctrico que generalmente contiene los cojinetes de dicho motor.

Endplay Juego longitudinal. Movimiento ligero del vástago sobre su eje.

Energy Efficiency Ratio (EER) Relación de eficiencia energética. Btu por hora de refrigeración que se produce por cada vatio de potencia eléctrica de alimentación.

Enthalpy Entalpía. Cantidad total de calor en una libra de una substancia.

Epoxy (Resins) Resina sintética. Un adhesivo sintético de plástico.

EPR (Evaporation Pressure Regulator) Regulador de presión de evaporación. Válvula automática de regulación de presión que se utiliza para mantener una presión predeterminada en el evaporador.

Equalizer Tube Tubo equilibrador. Dispositivo utilizado para mantener la misma presión o el mismo nivel en el líquido presente en dos recipientes.

Evacuate Vaciar. Eliminar agua en todas las formas posibles de un material o sistema.

Evaporation Evaporación. Término que se aplica al cambio de un líquido al estado gaseoso.

Evaporative Condenser Condensador de evaporación. Un dispositivo que usa un pulverizador de agua para enfriar un condensador.

Evaporator Evaporador. Parte de un mecanismo de refrigeración en el cual el refrigerante se evapora y absorbe calor.

Evaporator Coil Serpentín del evaporador. Dispositivo fabricado con tubos (líquido) o tubos con aletas (aire) que se utiliza para transferir energía calorífica del material al refrigerante.

Evaporator, Dry Type Evaporador en seco. Un evaporador en el cual el refrigerante es alimentado a través de un dispositivo de reducción de presión.

Evaporator Fan Ventilador del evaporador. Ventilador que mueve aire a través del evaporador.

Evaporator, Flooded Evaporador inundado. Un evaporador que contiene líquido refrigerante en todo momento.

Excess Air Exceso de aire. Aire por encima de la cantidad requerida para la combustión completa de un gas.

Exfiltration Exfiltración. Flujo de aire a través de una habitación hacia el exterior.

Exhaust Opening Apertura de escape. Cualquier apertura a través de la cual se elimina aire de un espacio.

Expansion Joint Junta de expansion. Una conexión de fontanería diseñada para mitigar la tensión causada por la expansión de la cañería.

Expansion Tank Tanque o depósito de expansion. (Ver Depósito de cojín de aire).

Expansion Valve Válvula de expansion. Dispositivo en un sistema de refrigeración que mantiene el diferencial de presión entre el lado alto y bajo de dicho sistema.

Expendable Refrigerant System Sistema de eliminación de refrigerante. Sistema que elimina el refrigerante después de su evaporación.

Extended Surface Superficie extendida (o de radiación). Superficie diseñada para la transmisión de calor. Uno de sus lados tiene más superficie que el otro gracias a la adición de aletas.

External Equalizer Equilibrador externo. Conexión a presión entre una zona por debajo del diafragma de una válvula termostática y el conducto de descarga de un evaporador.

Fahrenheit Fahrenheit. La escala de medición de temperatura en el sistema de unidades británico.

False Defrost Falsa descongelación. El sistema entra en la zona de descongelación aún cuando no hay escarcha o hielo en su exterior.

Fan Ventilador. Dispositivo que crea movimiento radial o axial usando aspas para mover aire.

Fan Coil Sistema de serpentín y ventilador. Una unidad terminal que consiste de un tubo en serpentín con aletas y de un ventilador en un solo contenedor.

Farad Faradio. Una unidad de medida de capacitancia grande.

Feedback Realimentación. La transferencia de energía de la salida de un sistema a su entrada.

Feet of Heat (FT HD) Pies de calor. Término usado para designar la resistencia que una bomba debe vencer para suministrar la cantidad de líquido requerida [2.307 FT HD = 1 psig.]

Female Thread Rosca hembra. Una rosca en el diámetro interior de un tubo.

Ferrous Ferroso. Objetos fabricados en acero o fundición.

Field Campo. El espacio que ocupan las líneas de fuerza magnética.

Filter Filtro. Dispositivo utilizado para eliminar pequeñas partículas de un fluído.

Fin Comb Peine de aletas. Dispositivo en forma de peine usado para enderezar las aletas metálicas de un serpentín.

Fire Tube Boiler Caldera de tubos de humo: Una caldera de acero en la cual los gases calientes de combustión circulan por tubos rodeados de agua.

Flare Abocinamiento o ensanche. Ángulo formado en el extremo de un tubo.

Flare Nut Tuerca abocinada o de union. Conexión utilizada para sujetar un tubo abocinado con otro tubo.

Flashback Retroceso de la llama. El movimiento de la llama hacia el orificio del quemador cuando se cierra la alimentación de gas.

Flash Gas Fogonazo. Gas que se forma debido a una reducción repentina de presión.

Flash Point Punto de inflamación. Temperatura a la cual un aceite emana suficientes vapores para sostener una llama por fogonazo.

Float Valve Válvula de flotador. Tipo de válvula operada por una esfera que flota en una superficie líquida para controlar su nivel.

Floc Point Temperatura de flóculo. Temperatura a la cual la cera en un aceite empieza a separarse del aceite.

Flooded System Sistema inundado. Tipo de sistema de refrigeración en el cual el evaporador se encuentra lleno de líquido refrigerante.

Flow Hood Campana de flujo. Instrumento de medición directa que calcula la cantidad de aire (en pies cúbicos por metro) que sale de o entra a un conducto mediante el paso del aire por un sensor.

Flow Meter Medidor de caudal. Instrumento utilizado para medir el volumen de líquido que fluye por una tubería por unidad de tiempo (galones por minuto, libras por hora, litros por minuto, etc.).

Flow Switch Interruptor de flujo. Interruptor automático que percibe el flujo de un fluído.

Flue Tragante o escape. Zona de paso en una chimenea para eliminar los gases de escape.

Flue Gases Gases de escape. Productos de la combustión y del exceso de aire.

Fluid Fluído. Substancia en estado líquido o gaseoso.

Flushing the Loop Lavado a presión del circuito cerrado. Forzar un líquido por un circuito cerrado a una velocidad tal que elimina materiales extraños y aire atrapado que se encuentran en él.

Flux Brazing, Soldering Fundente de soldadura de latón. Substancia que se aplica a superficies que se van a unir por soldadura para eliminar depósitos de óxido.

Flux, Electrical Flujo eléctrico. Las líneas de fuerza eléctrica o magnética en una región.

Flux, Magnetic Flujo magnético. Las líneas de fuerza creadas por un imán.

Foaming Espumar. Formación de espuma en una mezcla de aceite y refrigerante debido a la rápida evaporación del refrigerante disuelto en el aceite.

Foot of Water Pie de agua. Una medida de presión.

Foot Pound Pie-libra. Una unidad de trabajo.

Forced Convection Convección forzada. Transferencia de calor que resulta del movimiento forzado de un líquido o un gas por medio de una bomba o un ventilador.

Forced Draft Burner Caldera de tiro forzado. Una caldera en la cual un soplador proporciona el aire de combustión.

Free Area Superficie libre. La superficie mínima total de las aperturas de una rejilla.

Freeze Up Atascamiento por congelación. 1 La formación de hielo en el dispositivo de control de refrigeración que detiene el flujo de refrigerante al evaporador. 2 Formación de escarcha en un serpentín que detiene el flujo de aire en su interior.

Freezing Congelación. Cambio del estado líquido al estado sólido.

Freezing Point Temperatura de congelación. Temperatura a la cual un líquido se solidifica debido a la eliminación de calor.

Freon Freón. Marca comercial para una familia de refrigerantes sintéticos manufacturados por DuPont, Inc.

Frequency Frecuencia: El número de ciclos completos en una unidad de tiempo.

Friction Head Caída de presión. En un sistema hidráulico, la pérdida de presión que se produce al fluír agua por las tuberías.

Frost Back Retorno del refrigerante frío. Situación que tiene lugar cuando líquido refrigerante fluye desde el evaporador a la línea de succión.

Frost Control, Automatic Control automático de descongelación. Un control que invierte automáticamente el sistema de refrigeración para eliminar hielo y escarcha del evaporador.

Frost Free Refrigerator Nevera libre de escarcha. Armario de refrigeración que opera con un dispositivo automático que previene la formación de escarcha.

Frost Line Línea de congelación. El espesor de congelación de la tierra durante el invierno. La línea de congelación generalmente se sitúa 12 pulgadas por debajo de la penetración real registrada.

Full Load Amperage (FLA) Corriente de plena carga. La cantidad de corriente eléctrica que pasa por un motor de inducción a su carga total de diseño.

Furnace Horno. La parte de un sistema de calefacción por aire donde tiene lugar la combustión.

Fuse Fusible. Dispositivo eléctrico de seguridad que consiste en una tira o hilo de un metal que se funde cuando el circuito está sobrecargado.

Fusible Plug Tapón fusible. Un tapón con un metal de baja temperatura de fusión para permitir la reducción de presión en caso de incendio.

Galvanic Galvánico(a). Corriente eléctrica generada químicamente, al sumergir dos conductores diferentes en un electrolito.

Galvanic Action Acción galvánica. Acción corrosiva que ocurre entre dos metales con diferente actividad electrónica.

Gang En equipo. Conexión mecánica de dos o más interruptores.

Gasket Junta. Material elástico que se coloca entre dos superficies para sellar a prueba de fugas.

Gas, Noncondensible Gas incondensable. Gas que no pasa al estado líquido bajo ninguna combinación posible de presión y temperatura.

Gas Gas. Una substancia en estado de vapor.

Gauge, Compound Manómetro compuesto. Indicador de presión que tiene escala de medición por encima y por debajo del nivel de presión atmosférica.

Gauge, Gas Manifold Pressure Manómetro de presión de collector. Manómetro de medición directa para determinar la presión de colector en una calefacción de vapor. Suele utilizarse en lugar de manómetros de tubo en forma de "U". Mide en incrementos de pulgadas de agua.

Gauge, High Pressure Manómetro de alta presión. Instrumento que se utiliza para medir presiones por encima de la presión atmosférica con valores entre 0 y 500 libras por pulgada cuadrada.

Gauge, Low Pressure Manómetro de baja presión. Instrumento que se utiliza para medir presiones por encima de la presión atmosférica hasta 50 libras por pulgada cuadrada.

Gauge, Manifold Colector de manómetros. Dispositivo que contiene una combinación de manómetros y válvulas para controlar el flujo de vapor por dicho dispositivo.

Gauge, Standard Manómetro estándar. Dispositivo diseñado para medir presiones por encima de la presión atmosférica.

Gauge, Vacuum Manómetro de vacío. Instrumento utilizado para medir presiones por debajo de la presión atmosférica.

Generator Generador. Una máquina que transforma la energía mecánica en energía eléctrica (ver también "alternador").

Geothermal Well Pozo geotérmico. Pozo del cual se obtiene el agua de entrada a un sistema de refrigeración y al cual retorna el agua después de circular por dicho sistema.

Grade Level Nivel de tierra. La superficie de la tierra.

Gravity System Sistema de calefacción por gravedad. Un sistema de calefacción en el cual el aire caliente es distribuído gracias al efecto flotador de ese aire sobre agua. Dicho sistema no utiliza bomba o ventilador.

Grill Rejilla. Elemento ornamental que se coloca al final de una salida de aire.

Gross Capacity, Heating Capacidad térmica bruta. La capacidad bruta de una bomba térmica es la cantidad total de energía en forma de calor que puede transmitir al aire interior. Esta capacidad incluye la energía térmica producida por una fuente de calor (sea aire o líquida) más la energía térmica equivalente requerida para operar la bomba.

Gross Output Potencial bruto. Una clasificación que se aplica a calderas. Es la cantidad total de calor que la caldera puede proporcionar cumpliendo las limitaciones aplicables de ensayo y códigos legales.

Ground Toma de tierra. Conexión entre un circuito eléctrico y la Tierra (potencial cero).

Ground Coil (Earth Coil) Serpentín enterrado. Un intercambiador de calor colocado bajo tierra que se utiliza bien para obtener calor de la tierra o para desviar calor hacia ella.

Ground Wire Conexión de tierra. Cableado eléctrico que conduce la electricidad de una estructura a tierra por razones de seguridad.

Halide Refrigerants Refrigerantes halógenos. Familia de refrigerantes que contienen compuestos químicos halógenos.

Halide Torch Antorcha halógena. Tipo de antorcha usada para detectar fugas de refrigerantes halógenos.

Head Diferencial o delta. En este curso, se utiliza este término para referirse a la diferencia de presión.

Head (Total) Presión total. La suma de la presión estática y dinámica de un fluído en movimiento.

Header Colector. Una forma de cañería utilizada para interconectar dos o más entradas o salidas de una caldera.

Head Pressure Presión principal. Presión existente en el lado de condensación de un sistema de refrigeración.

Head, Pressure Control Control de la presión principal. Un sistema de control activado a presión que impide que la presión principal caiga por debajo del valor deseado.

Head, Static Presión estática. Presión de un fluído expresada en términos de la altura de una columna de un fluído como agua o mercurio.

Head, Velocity Presión mecánica. En un fluído en movimiento, altura del fluído equivalente a su presión de velocidad.

Heat Calor. Forma de energía que provoca un aumento de temperatura; energía asociada al movimiento aleatorio de las moléculas.

Heat, Anticipator Anticipador de calor. Resistor en el termostato situado en una habitación que corta el ciclo de calefacción prematuramente.

Heat Exchanger Intercambiador de calor. Dispositivo utlizado para transferir calor de una superficie caliente a otra fría (o menos caliente).

Heat Lag Atraso en la transmisión de calor. Cuando una substancia se calienta por un lado, el tiempo necesario para que el calor pase a través de dicha substancia.

Heat, Latent Calor latente. Energía térmica utilizada para cambiar el estado de una substancia; del estado sólido al estado líquido y del estado líquido al estado gaseoso al añadir calor, o del estado gaseoso al líquido y de éste al estado sólido al eliminar calor. Dicho cambio de estado tiene lugar sin un cambio de temperatura del material.

Heat Load Carga de calor. La cantidad de calor que debe sustraerse de un espacio refrigerado o acondicionado para mantener la temperatura de diseño.

Heat Loss Pérdida de calor. La tasa de transferencia de calor de un edificio al exterior.

Heat Motor Motor térmico. Dispositivo eléctrico que crea movimiento gracias al cambio de temperatura de elementos bimetales. Generalmente se utiliza para retrasar la operación de un dispositivo.

Heat of Compression Calor de compresión: El calor añadido a un refrigerante debido al trabajo ejercido por el compresor.

Heat of Fusion Calor de fusión: Calor liberado por una substancia al pasar del estado líquido al sólido.

Heat of Respiration Calor de respiración: Dióxido de carbono y agua liberados por alimentos almacenados.

Heat Pump Bomba de calor. Sistema reversible de acondicionamiento de aire, capaz de eliminar o proporcionar calor a un espacio definido.

Heat Pump, Air to Air Bomba térmica, aire a aire. Dispositivo utilizado para transferir calor en una y otra dirección entre dos zonas de aire separadas.

Heat Pump, Air to Liquid Bomba térmica, aire a líquido. Dispositivo utilizado para transferir calor de una zona de aire a un líquido a través de un sistema de refrigeración. Unidades de este tipo no son reversibles.

Heat Pump, Liquid to Air Bomba térmica, líquido a aire. Dispositivo utilizado para transferir calor entre un líquido y aire en una y otra dirección.

Heat Pump, Liquid to Liquid Bomba térmica, líquido a líquido. Dispositivo utilizado para transferir calor entre dos líquidos diferentes en una y otra dirección.

Heat Pump, Water Heater Bomba térmica, calentador de agua. Sistema de refrigeración, aire a líquido, utilizado para calentar agua para uso doméstico.

Heat Pump; Water Heater, Remove Bomba térmica, calentador de agua desmontable. El sistema de refrigeración es una unidad completa diseñada para ser conectada directamente a un depósito de agua.

Heat Pump; Water Heater, Self Contained Bomba térmica, calentador de agua integral. Los components del sistema de refrigeración (condensador) son una parte integral del depósito de agua caliente.

Heat Sensible Calor sensible. La energía térmica utilizada para cambiar la temperatura de un material (sólido, líquido o gaseoso) sin modificar su estado.

Heat Sink Absorbente de calor. Material en el cual el sistema de refigeración elimina el calor no deseado.

Heat Source Fuente térmica. Lugar o material del cual se obtiene energía térmica. Aire o líquido del cual se puede obtener calor. En una bomba térmica de aire, el aire exterior es utilizado como fuente térmica en el ciclo calefactor.

Heat, Total Calor total. La suma del calor sensible y calor latente en el aire. Se expresa en unidades térmicas británicas [Btu/libra de aire].

Heat Transfer Transferencia de calor. Movimiento de calor de una substancia a otra.

Heating Coil Serpentín térmico. Dispositivo de transmisión térmica diseñado para añadir calor a un líquido o gas.

Heating Mode Ciclo calefactor. La fase de la operación de un sistema calefactor en la cual se añade calor a una zona ocupada.

Heating Seasonal Performance Factor (HSPF) Factor de rendimiento térmico estacional. El rendimiento total de una bomba térmica durante el período de tiempo que está en funcionamiento en un año dividido por la potencia eléctrica en vatios-hora utilizada durante el mismo tiempo.

Heating Value Valor calorífico. El número de Btu producido por la combustión de un pie cúbico de un gas.

Henry Henrio (Abreviatura: H). Unidad de inductancia.

Hermetic Hermético: Completamente sellado.

Hermetic Motor Motor hermético. Motor de propulsión de un compresor que se encuentra sellado en la misma carcasa que contiene al compresor.

High Pressure Cutout Interruptor de alta presión. Un interruptor eléctrico accionado a presión que se utiliza para cortar el suministro eléctrico a un compresor cuando éste alcanza el límite máximo de presión de funcionamiento.

High Side Lado alto (de un sistema de refrigeración). La parte de un sistema de refrigeración en el lado de condensación (o alta presión).

High Velocity System Sistema de alta velocidad. Sistema de distribución de aire, generalmente utilizado en proyectos industriales y comerciales, diseñado para operar bajo presiones estáticas de 7 a 9 pulgadas de agua.

Horsepower Caballo de vapor. Unidad de potencial equivalente a un trabajo 33.000 libras-pie por minuto.

Hot Gas Defrost Descongelador de aire caliente. Un sistema de descongelación en el cual gas refrigerante caliente es enviado a través del evaporador.

Hot Gas Line Línea de aire caliente. La línea que transporta el aire caliente de descarga del compresor al condensador.

Hot Junction Empalme o junta en caliente. Aquella parte del circuito termoeléctrico que libera calor.

Humidifiers Humidificadores. Dispositivos usados para añadir humedad.

Humidistat Humidistat. Control eléctrico que opera en base a cambios de los niveles de humedad.

Humidity Humedad. Nivel de vapor de agua presente en el aire.

Hydrocarbon Hidrocarbono. Uno de los muchos compuestos químicos que contienen carbono e hidrógeno.

Hydrometer Densímetro: Instrumento flotante que se utiliza para medir el peso específico de un líquido.

Hydronics Hidrónica. Término que se aplica al calentamiento o enfriamiento con agua o vapor.

Hygrometer Higrómetro. Instrumento utilizado para medir el grado de humedad en la atmósfera.

Hygroscopic Higroscópico(a). Habilidad de una substancia de absorber y retener humedad.

Ice Ring Anillo de hielo Hielo acumulado en la parte inferior de un bobinado o serpentín exterior debido a una descongelación incompleta.

Idler Rueda guía. Polea utilizada en algunos sistemas de transmisión por correa para proporcionar la tensión requerida.

Ignition Temperature Temperatura de ignición o encendido. La temperatura mínima a la cual puede empezar la combustión.

Ignition Transformer Transformador de ignición Transformador diseñado para proporcionar una corriente de alto voltaje, que es utilizado en muchos sistemas de calefacción para causar la inflamación del carburante.

Impedance Impedancia. Tipo de resistencia eléctrica que sólo se encuentra en circuitos de corriente alterna.

Impeller Impulsor. Pieza giratoria de una bomba centrífuga.

Indoor Coil Serpentín interior. La parte de una bomba térmica situada dentro de una casa, que funciona como punto de transmisión de calor para calentar o enfriar aire interior.

Induced Draft Burner Caldera de tiro inducido o aspirado. Una caldera que depende de una corriente de aire inducida por un ventilador o un soplador a la salida del tragante para obtener aire de combustión y eliminar gases de escape.

Inductance Inductancia. Característica de un circuito de corriente alterna que se opone al cambio de dirección de la corriente.

Induction Inducción. El acto de producir voltaje eléctrico en un objeto expuesto a la acción de un campo magnético.

Induction Motor Motor de inducción. Motor de corriente alterna que opera bajo el principio del campo magnético rotante. El rotor no está conectado a un circuito eléctrico pero recibe energía eléctrica gracias a la acción transformadora de un bobinado.

Inductive Load Carga inductive. Dispositivo que utiliza energía eléctrica para producir movimiento.

Inductive Reactance Reactancia inductive. La oposición, medida en ohmios, a una corriente alterna.

Inerts Inertes. Substancias no combustibles en un carburante, o en gases de escape, como el nitrógeno o el dióxido de carbono.

Infiltration Infiltración. El aire que penetra en un espacio refrigerado o acondicionado.

Infrared Lamp Lámpara de rayos infrarrojos. Dispositivo eléctrico que emite rayos infrarrojos (rayos invisibles más allá del rojo en el espectro visible).

In Phase En fase. Condición que ocurre cuando dos ondas de la misma frecuencia tienen sus valores de máxima y mínima polaridad en el mismo instante.

Insulation, Thermal Aislamiento térmico. Substancia que se utiliza para reducir o detener la transferencia de calor a través de una pared.

Interlock Mecanismo de interbloqueo. Dispositivo de seguridad que solamente permite el paso de electricidad a un circuito cuando una función predeterminada ha tenido lugar.

IR Drop Caída de voltaje resistive. Diferencia de potencial resultante del paso de corriente eléctrica a través de una resistencia.

Isothermal Isotérmico(a). A temperatura constante.

Joints, Brazed Soldadura de latón. Conexión o unión de soldadura que se obtiene al unir piezas metálicas con aleaciones o mezclas de metales con temperaturas de fusión entre 1.000 y 1.500 grados Fahrenheit.

Joint, Soldered Junta estañada. Conexión o unión de soldadura que se obtiene al unir piezas metálicas con aleaciones o mezclas de metales con temperaturas de fusión por debajo de 1.000 grados Fahrenheit.

Joint, Welded Junta soldada. Conexión o unión de soldadura que se obtiene al unir piezas metálicas con aleaciones o mezclas de metales con temperaturas de fusión por encima de 1.500 grados Fahrenheit.

Joule Julio. Unidad de calor en el sistema métrico decimal.

Journal, Crankshaft Pivote del cigüeñal. Parte del vástago del cigüeñal en contacto con el cojinete.

Junction Box Caja de conexión. Grupo de terminales eléctricos en una caja protectora.

Kelvin Scale (K) Escala Kelvin. Escala de temperatura en la cual la unidad de medida es el grado centígrado y en la cual el cero absoluto equivale a –273.16°C.

Kilopascal Kilopascal. Unidad de presión en el sistema métrico.

Kilowatt Kilovatio. Unidad de potencial eléctrico, equivalente a 1.000 vatios.

Kilowatt Hour Kilovatio hora. Unidad de medida de energía eléctrica, equivalente a 3.413 Btu.

King Valve Válvula maestro. La válvula situada a la salida del receptor.

Latent Heat Calor latente. Calor caracterizado por el cambio de estado de la substancia en cuestión.

Latent Heat of Fusion Calor latente de fusion. Cantidad de calor [Btu/libra] que se requiere para transformar una substancia del estado sólido al líquido sin variar la temperatura.

Latent Heat of Vaporization Calor latente de vaporización. Cantidad de calor [Btu/libra] que se requiere para transformar una substancia del estado líquido al gaseoso sin variar la temperatura.

Leak Detector Detector de fugas. Dispositivo o instrumento usado para la detección de fugas de gases o vapores.

Limit Control Control límite. Control utilizado para abrir o cerrar circuitos eléctricos cuando se alcanzan ciertos límites de temperatura y presión.

Line Voltage Voltaje de línea. Voltaje disponible en enchufes de suministro o terminales de un sistema eléctrico por encima de los 30 voltios.

Liquefied Petroleum Gases Gases licuados de petróleo. Este término se aplica a todos los gases combustibles compuestos principalmente de uno de los hidrocarbonos que a continuación se relacionan, o de mezclas de ellos: propane, propileno, butano o isobutano y butilenos.

Liquid Line Línea líquida. Se aplica al tubo que transporta líquido refrigerante del condensador o cilindro de almacenaje a un dispositivo de reducción de presión.

Liquid Nitrogen Nitrógeno líquido. Nitrógeno en estado líquido.

Liquid Receiver Cilindro de almacenaje. Cilindro conectado a la salida del condensador para almacenar líquido refrigerante en un sistema.

Liter Litro. Unidad métrica de volumen.

LNG Gas natural licuado. Gas natural que ha sido enfriado hasta pasar al estado líquido.

Load Carga. Aplícase al dispositivo que convierte energía eléctrica en otra forma de energía utilizable.

Locked Rotor Amperage (LRA) Corriente de rotor trabado. La cantidad de corriente eléctrica que atraviesa un motor calado.

Lock Out Relay Relé interruptor. Sistema de control que impide el arranque automático de un compresor después de que un control de seguridad ha saltado, incluso después de que dicho control se ha repuesto por sí mismo.

Louvers Lumbreras tipo persiana. Placas de metal o tableros superpuestos en ángulo que permiten el paso de aire pero no el de agua de lluvia.

Low Pressure Control Control de baja presión. Dispositivo utilizado para arrancar o parar un sistema cuando la presión en el lado inferior (presión de succión) cae por debajo de un límite establecido.

Low Pressure Cutout Interruptor de baja presión. Dispositivo utilizado para impedir que la presión del lado de evaporación caiga por debajo de un cierto valor.

Low Side Zona baja. La parte del sistema de refrigeración que se encuentra bajo la presión de evaporación más baja.

Low Side Pressure Presión del lado inferior o de la zona baja. Presión en el lado inferior de un sistema de refrigeración. También llamada presión de succión.

Low Voltage Control Control de baja tension. Controles diseñados para operar entre voltajes de 20 a 30 voltios.

Magnetic Field Campo magnético. El espacio en el cual existe una fuerza magnética.

Makeup Air Aire renovado. El aire que se introduce a un edificio para reemplazar al aire eliminado por un sistema extractor.

Makeup Water Agua de relleno. Agua necesaria para reemplazar la pérdida de agua de una torre de refrigeración por evaporación, arrastrada por el viento o por fugas.

Makeup Water Line Línea de agua de ajuste. Línea conectada al circuito cerrado para rellenar o añadir líquido conforme a las necesidades del sistema.

Male Thread Rosca macho. Una rosca en el diámetro exterior de un tubo.

Manifold Gauge Colector de indicadores. Dispositivo construído para contener indicadores compuestos y de alta presión y con válvulas que permiten controlar el flujo de fluídos por su interior.

Manometer Manómetro. Instrumento para medir la presión de gases y vapores.

Manometer, Inclined Manómetro inclinado. Manómetro en el cual el tubo que contiene el líquido se encuentra a un ángulo de la posición horizontal con el fin de poder proporcionar mediciones más precisas dentro de un cierto rango de presiones.

Mass Masa. Una cierta cantidad de materia que forma un cuerpo individual, generalmente de forma indefinida.

Mean Annual Earth Temperature Temperatura media anual de la tierra. La media de todas las temperaturas registradas a lo largo de un año. El factor de mayor influencia en dicha medida es el sol.

Mean Temperature Difference Media del diferencial de temperature. El punto intermedio entre la temperatura antes del proceso y la temperatura una vez que el proceso ha terminado.

Mechanical Cycle Ciclo mecánico. Ciclo compuesto de una serie de eventos mecánicos.

Mega Mega. Prefijo que indica un millón; por ejemplo, megohmio es un millón de ohmios.

Megohm Megohmio. 1.000.000 ohmios de resistencia eléctrica.

Melting Point Temperatura de fusion. Temperatura a la cual una substancia expuesta a la presión atmosférica pasa al estado líquido.

Mercaptan Mercaptan. Substancia que proporciona al gas natural su olor característico.

Mercury Bulb Interruptor de mercurio. Interruptor de circuitos eléctricos formado por una pequeña cantidad de mercurio en un tubo de cristal hermético. El mercurio hace contacto o lo abre con terminales dentro del tubo.

Meter Metro. Unidad de medida linear del sistema métrico.

Methane Metano. Hidrocarbono gaseoso (CH_4); componente principal de los gases naturales.

MFD mfd. Abreviatura de microfaradio.

Micro Micro. Prefijo que indica una millonésima parte.

Microfarad Microfaradio. Unidad de medida de capacitancia, equivalente a una millonésima de un faradio.

Micron Micra. Unidad de medida en el sistema métrico; una milésima parte de un milímetro.

Micron Gauge Indicador de vacío. Instrumento para medir presiones muy cercanas al vacío completo.

Milli Mili. Prefijo que indica una milésima parte; por ejemplo, un milivoltio es una milésima de un voltio.

Modulating Modulador. Un tipo de dispositivo o control que ajusta por incrementos pequeños y no abriendo y cerrando el paso de una variable.

Moisture Indicator Indicador de humedad. Instrumento que se usa para medir el contenido de humedad de un refrigerante.

Molecule Molécula. Es la unidad más pequeña de un elemento o un compuesto químico que retiene la identidad y las características de dicho elemento o compuesto.

Mollier Diagram Diagrama de Mollier. Gráfico en el que se expresan la características de presión, calor y temperatura de un refrigerante.

Motor Burnout Motor quemado. Condición que se da cuando el aislamiento de un motor eléctrico se ha deteriorado debido a un sobrecalentamiento.

Motor, Capacitor Type Motor tipo condensador. Motor de inducción de una sola fase que usa tanto el bobinado de arranque como el de marcha durante su funcionamiento normal. El bobinado de arranque está conectado en serie a un condensador para modificar las características eléctricas del bobinado.

Motor Control Controlador de un motor. Un control que arranca y para un motor en determinadas condiciones de presión y temperatura.

Motor Control Control de un motor. Dispositivo para arrancar o parar un motor a ciertas condiciones de temperatura o presión.

Motor Starter Motor de arranque. Un contacto eléctrico de alta capacidad operado por un bobinado electromagnético que contiene saltos de desconexión del tamaño apropiado.

Mullion Parteluz. Parte estacionaria de la estructura entre dos puertas.

Multimeter Multímetro. Medidor de voltaje y resistencia.

National Electrical Code Código Nacional de Electricidad. El código de reglamentación eléctrica basado en los requerimientos de las aseguradoras para instalaciones interiores.

Natural Convection Convección natural. Circulación de un gas o un líquido debido a la diferencia de densidad a diferentes temperaturas.

Natural Gas Gas natural. Cualquier gas obtenido directamente de la tierra, que no ha sido manufacturado.

Needle Valve Válvula de aguja. Una válvula que se utiliza para controlar con gran precisión velocidades muy pequeñas de la corriente de un fluído.

Neoprene Caucho sintético (neopreno). Goma sintética resistente al aceite de hidrocarbono y a los gases.

Net Refrigeration Refrigeración neta. Una medida de la cantidad de refrigeración que tiene lugar en un evaporador. Es equivalente a la entalpía del gas que sale del evaporador menos la entalpía del gas que entra en el evaporador y se mide en Btu/libra.

Nominal Size Tubing Tubo de diámetro nominal. Clase de tubería que tiene un diámetro interior idéntico al de cañería de fundición de una clase predeterminada.

Noncondensible Gas Gas incondensable. Gas que no pasa al estado líquido bajo ninguna combinación de presión y temperatura operativas.

Nonferrous No ferroso. Grupo de metales y aleaciones de metal que no contienen hierro; también se refiere a metales que no son aceros ni fundiciones. El cobre y el aluminio son los metales no ferrosos más utilizados en sistemas de calefacción.

Normally Closed (NC) Cerrado(a) en condiciones normales. Contacto interruptor que se encuentra cerrado cuando no hay corriente en el circuito.

Normally Open (NO) Abierto(a) en condiciones normales. Contacto interruptor que se encuentra abierto cuando no hay corriente en el circuito.

Nozzle Boquilla. Dispositivo al final de la tubería para pulverizar el aceite combustible al hacerlo pasar por un orificio minúsculo.

Odorant Aromatizante. Substancia añadida a un gas inodoro, incoloro e insípido para avisar de una fuga de gas y para que ésta pueda ser detectada.

Off Mode Cycle Ciclo de parade. La parte de la operación en la que el sistema ha sido detenido por sus controles.

Off Cycle Ciclo de parade. La parte del ciclo de refrigeración en la que el sistema no está en operación.

Ohm Ohmio. Unidad de medida de resistencia eléctrica.

Oil Binding Obstrucción o taponamiento de aceite. Condición física que tiene lugar cuando una capa de aceite sobre el líquido refrigerante le impide la evaporación en condiciones normales de presión y temperatura.

Oil Rings Anillos de control de aceite. Anillos dilatantes colocados en ranuras y en el pistón, con el fin de impedir que entre aceite a la cámara de compresión.

Oil Separator Separador de aceite. Dispositivo utilizado para mantener el aceite separado del gas refrigerante y hacer que regrese al cigüeñal del compresor.

One Pipe Fitting Conexión para sistema de una tubería. Una conexión en forma de "t" diseñada especialmente para sistemas de una tubería que se usa para conectar el ramal de entrada o salida a un circuito.

One Pipe System Sistema de una tubería. Un sistema de agua caliente a presión que utiliza una tubería continua desde la salida de la caldera de suministro al retorno a la caldera.

Open Circuit Circuito abierto. Un circuito eléctrico interrumpido con el fin de impedir el flujo de corriente eléctrica.

Open Compressor Compresor abierto. Compresor cuyo motor se encuentra separado.

Orifice Orificio. Abertura de dimensiones definidas para controlar el flujo de un fluído.

Orifice, Oil Metering Orificio dosificador de aceite. Un agujero pequeño en el tubo del acumulador que se utiliza para asegurar que el aceite regresa al compresor en pequeñas cantidades fáciles de absorber.

Orifice Spud Escoplo horadado. Un tapón de quita y pón que tiene un orificio que determina la cantidad de gas que puede pasar por él.

Oscilloscope Osciloscopio. Tubo revestido de un material fluorescente que permite observar la forma de una onda eléctrica.

Outdoor Coil Serpentín exterior. La parte de una bomba térmica situada fuera de una casa, que funciona como punto de transmisión de calor para calentar o enfriar aire exterior.

Outside Air Aire exterior. Aire que se encuentra fuera del espacio refrigerado o acondicionado.

Overload Sobrecarga. Carga superior a la carga para la cual el sistema o mecanismo ha sido diseñado.

Overload Protector Protector de sobrecarga. Dispositivo que detiene la operación de un mecanismo en condiciones peligrosas.

Ozone Ozono. Una forma gaseosa de oxígeno (O_3).

Package Boiler Caldera unitaria. Caldera que forma una unidad con todos sus componentes ensamblados.

Package Unit Unidad ensamblada. Un sistema completo de refrigeración o acondicionamiento de aire en el que todos sus componentes has sido ensamblados en fábrica en una unidad independiente.

Parallel En paralelo. Circuito conectado en una forma tal que la corriente eléctrica puede circular en dos o más direcciones.

Parallel Circuit Circuito en paralelo. Un circuito eléctrico en el cual hay más de una salida de la corriente desde la fuente de alimentación.

Partial Pressures Presiones parciales. Condición que se da cuando dos o más gases ocupan el mismo espacio y cada uno crea parte de la presión total.

Pascal's Law Ley de Pascal. La presión ejercida sobre un líquido se transmite igualmente en todas las direcciones.

Permeance Permeabilidad de vapor de agua. Relación entre la transmisión de humedad a través de una pared en función del gradiente o diferencial de presión de vapor entre las dos caras.

pH pH. Término basado en la concentración de iones de hidrógeno en el agua, que denota la acidez o alcalinidad del agua.

Photoelectricity Fotoelectricidad. Acción física que se produce cuando ondas de luz generan flujo eléctrico.

Pilot Llama piloto. Pequeña llama usada para iniciar el encendido en la caldera principal.

Pilot Generator Generador piloto. Dispositivo utilizado para generar tensión eléctrica en un sistema de calefacción de minivoltaje.

Pilot Safety Valve Válvula piloto de seguridad. Válvula que cierra el suministro principal de gas si la llama piloto no se enciende o se apaga.

Pilot Switch Interruptor piloto. Un control que se utiliza en calderas de gas. Su función es impedir que la caldera opere si existe un problema con la llama piloto.

Pinch Off Tool Aplasta-tubos. Herramienta utilizada para aplastar las paredes de un tubo con el fin de impedir el paso de un fluído.

Piston Pistón. Pieza cilíndrica de dimensiones muy similares a un tubo interior que circula a lo largo de dicho tubo por su interior.

Piston Displacement Desplazamiento del pistón o cilindrada. Volumen que se obtiene al multiplicar el área interior del tubo por la carrera del pistón.

Pitch Pendiente. Grado de inclinación de una tubería.

Pitot Tube Tubo pitoto. Dispositivo que detecta la presión estática y la presión total.

Plenum Espacio. Un conducto en forma de cubo en el lado de entrada o salida de un sistema de tratamiento de aire.

Point of Vaporization Punto de vaporización. El lugar específico en el serpentín de un evaporador donde se vaporizan las últimas moléculas de líquido refrigerante.

Polarity Polaridad. Condición que denota la dirección de la corriente eléctrica.

Polybutylene Polibutileno. Plástico utilizado en la fabricación de tuberías y conductos que tiene gran resistencia a la deformación y a la ruptura. Recomendado para circuitos cerrados de tierra.

Polyethylene Polietileno. Plástico utilizado para tuberías para circuitos cerrados, líneas de agua fría, y conducciones en bombas térmicas.

Polystyrene Poliestireno. Plástico que se utiliza como aislante en algunos armarios de refrigeración.

Ponded Roof Tejado de retención. Tejado construído en forma horizontal para retener una cierta cantidad de agua.

Potable Water Agua potable. Agua adecuada para el consumo humano.

Potential Tensión o diferencia de potencial. El voltaje entre puntos de un circuito eléctrico.

Potentiometer Potenciómetro. Un resistor variable.

Pour Point (Oil) Punto de fluidez (Aceite). La temperatura más baja a la cual el aceite se puede verter.

Power Potencial. Trabajo o energía por unidad de tiempo.

Power Element Elemento de potencia. Elemento sensible de un sistema que opera por temperatura.

Power Factor Factor de potencia (o coseno de Π). Relación de potencia actual, es decir, la potencia de un circuito de corriente alterna medida en vatios dividida por la potencia aparente que resulta de multiplicar voltios por amperios.

Pressure Presión. Fuerza por unidad de superficie.

Pressure, Back Contrapresión. Presión que la bomba debe vencer para enviar líquido a un depósito a presión. La presión en el depósito.

Pressure Drop Caída de presión. Diferencia de presión entre dos puntos de un circuito como consecuencia de la resistencia al flujo de un líquido en dicho circuito.

Pressure Limiter Limitador de presión. Dispositivo que se mantiene cerrado hasta que se alcanza una presión determinada y al abrirse permite el paso de un fluído a otra parte del sistema.

Pressure Motor Control Control de motor a presión. Dispositivo que abre y cierra un circuito eléctrico conforme la presión adquiere valores determinados.

Pressure Operated Altitude (POA) Valve Válvula de altitud operada a presión. Dispositivo que mantiene una presión constante en el lado inferior del sistema independientemente de la altura de la operación.

Pressure Reducing Device Dispositivo de reducción de presión. El dispositivo utilizado para producir una reducción de presión y de la correspondiente temperatura de ebullición antes de introducir el refrigerante en el evaporador.

Pressure Reducing Valve Válvula de reducción de presión. Una válvula con diafragma instalada en la línea del agua de relleno en un sistema de calefacción por agua caliente.

Pressure Regulator Regulador de presión. Un dispositivo para mantener una presión de salida constante.

Pressure Relief Valve Válvula de descarga de presión. Dispositivo que protege un contenedor de gas abriéndose y dejando escapar parte del gas cuando la presión es demasiado alta.

Pressure, Static Presión estática. La presión en un sistema cuando la bomba no está en funcionamiento.

Pressure, Suction Presión de succión. La presión en el lado inferior (zona del evaporador) de un sistema de refrigeración.

Primary Air Aire primario. El aire de combustión que entra en una caldera y se mezcla con el gas combustible antes de llegar a la lumbrera.

Primary Control Control primario. Tipo de control de la operación de una caldera de aceite.

Primary Voltage Voltaje primario. El voltaje del circuito que proporciona la tensión a un transformador.

Process Tube Tubo de servicio. La longitud del tubo sujeto a la unidad terminal hermética, que se usa para labores de reparación y mantenimiento.

Propane Propano. Un hidrocarbono combustible.

psi psi. Presión medida en libras por pulgada cuadrada.

psia psia. Presión absoluta medida en libras por pulgada cuadrada.

psig psig. Símbolo o iniciales que se aplican a presion de manómetro en libras por pulgada cuadrada.

Psychrometer or Wet Bulb Hygrometer Psycrómetro o higrómetro de bulbo seco y empapado. Instrumento usado comúnmente para medir la cantidad de humedad en el aire mediante la determinación de las temperaturas en seco y del bulbo empapado.

Psychometric Chart Gráfico psycrométrico. Un diagrama que muestra la relación entre la humedad presente en el aire y su temperatura.

P/T Plugs Tapones a presión y temperature. Tapones que permiten el acceso de material al sistema a través de un conjunto de manómetros y termómetros sin que se altere la presión de dicho sistema.

PTC Coeficiente de temperatura positive. Termistor, que funciona como un resistor variable con la temperatura, que se usa comúnmente como un componente de medición de la temperatura con componentes electrónicos. Se utiliza a veces como un relé electrónico en el circuito de arranque de un compresor. Usado a menudo para medir la temperatura de bobinas exteriores en bombas de calefacción con el fin de iniciar o terminar ciclos de descongelación.

Pump Bomba. Dispositivo impulsado por motor que se utiliza para hacer circular agua en un sistema.

Pumpdown Vaciado. Un procedimiento de mantenimiento o reparación en el que el refrigerante es bombeado y enviado al receptor.

Pump Head Diferencial de presión de una bomba. La diferencia de presión entre el suministro y la salida de una bomba que se origina cuando ésta entra en funcionamiento.

Purging Purgar. Liberar gas comprimido a la atmósfera a través de una pieza o conjunto de piezas con el fin de eliminar contaminantes en la pieza o conjunto.

Pyrometer Pirómetro. Instrumento para medir temperaturas.

Quenching Templado. La inmersión de un objeto sólido caliente en un líquido con el fin de enfriarlo.

Quick Connect Coupling Manguito de acoplamiento rápido. Un dispositivo que permite la conexión rápida de dos líneas hidráulicas.

Radiant Heating Calefacción por radiación. Sistema de calefacción en el que solamente el calor irradiado por paneles es efectivo.

Radiation Radiación. Transmisión de calor a través de rayos.

Radiator Radiador. Un aparato de calefacción por radiación que se coloca dentro del espacio que se desea calentar.

Radiator Valve Válvula de radiador. Una válvula instalada en una unidad de radiación para controlar manualmente el flujo de agua en su interior.

Range Escala. Amplitud de regulación de presión y temperatura de un control.

Rankine Scale Escala Rankine. Escala Fahrenheit absoluta.

Reactance Reactancia. Oposición a la corriente alterna por inductancia o por capacitancia, o por ambas.

Reciprocating Alternativo. Acción en la cual el movimiento alterna de un sentido al opuesto, pero siempre en línea recta.

Recirculated Air Aire recirculado. Aire de salida que vuelve a pasar por el acondicionador antes de ser enviado de nuevo al espacio acondicionado.

Rectifier, Electric Rectificador eléctrico. Dispositivo eléctrico que transforma la corriente alterna en corriente continua.

Reducing Fitting Conexión reductora. Conexión de tubería para unir dos tubos de diferente tamaño.

Reed Valve Válvula de lengüeta. Placa delgada y plana de acero templado sujeta en un lado.

Refractory Refractario(a). Un material o substancia que puede soportar temperaturas muy altas.

Refrigerant Refrigerante. Substancia usada en un sistema de refrigeración para absorber calor en el serpentín del evaporador y para liberarlo en el condensador.

Refrigerant Charge Carga de refrigeración. Cantidad de refrigerante en el sistema.

Refrigerating Effect Efecto refrigerante. La cantidad de calor que el sistema es capaz de transferir en Btu por hora.

Refrigeration Refrigeración. El proceso de transferir calor de un lugar a otro.

Refrigeration Oil Aceite de refrigeración. Aceite formulado para uso en mecanismos de refrigeración. Este aceite circula por el sistema con el refrigerante.

Register Registro. Combinación rejilla y difusor que cubre una salida de aire o el final de un conducto de ventilación.

Relative Humidity Humedad relative. Relación entre el peso del vapor de agua en el aire a una cierta temperatura con la cantidad máxima que puede contener si estuviera saturado.

Relay Relé. Mecanismo eléctrico que usa una pequeña corriente en el circuito de control para operar un interruptor en el circuito de operación.

Relief Opening Apertura de descarga o escape. La apertura en una campana de aire que permite la salida de gases de escape a la atmósfera.

Relief Valve Válvula de decompression. Dispositivo de seguridad diseñado para abrirse antes de que se alcance una presión peligrosa.

Repulsion Start Induction Motor Motor de inducción con arranque a repulsion. Tipo de motor que tiene un bobinado eléctrico en el rotor para arrancar.

Resistance Resistencia. La oposición al paso de la corriente eléctrica por un conductor físico.

Resonance Resonancia. El efecto de sonido que se produce en un horno de gas cuando la frecuencia de la llama de combustión está en perfecta sincronización con la longitud de la onda de presión en el quemador.

Return Air Aire de retorno. Aire extraído del espacio acondicionado o refrigerado.

Return Piping Tuberías de retorno. La parte del sistema de fontanería que lleva agua de las unidades terminales a la caldera.

Reverse Acting Control Control inversor. Un interruptor controlado por temperatura diseñado para abrirse cuando cae la temperatura y para cerrarse cuando sube.

Reverse Cycle Defrost Descongelación de ciclo opuesto. Método de calentar el evaporador para eliminar escarcha usando válvulas que envían gases calientes desde el compresor.

Reverse Cycle Refrigeration System Sistema de refrigeración reversible. Conocido comúnmente como "bomba térmica". Sistema de refrigeración en el que la dirección en que se transmite el calor se puede invertir.

Reverse Return Retorno invertido. Un sistema de doble cañería en el cual las conexiones de retorno desde las unidades terminales a la tubería de retorno principal están en el orden opuesto al de las conexiones en el lado de suministro.

Reversing Valve Válvula de inversion. El control usado para regular el flujo de refrigerante en una bomba de calefacción.

Rheostat Reostato. Resistencia variable o ajustable.

Rich Mixture Mezcla rica. Mezcla de gas y aire que contiene demasiado carburante o insuficiente aire para la combustión completa del gas.

Rollout Llamarada. Situación que ocurre cuando la llama sale de la cámara de combustión cuando la caldera está encendida.

Roof Mounted Instalada sobre el tejado. Dícese de la unidad de refrigeración colocada sobre una plataforma diseñada para distribuir el peso de la unidad sobre la superficie de un tejado lo más grande posible.

Rotary Compressor Compresor rotativo. Mecanismo que bombea fluído como consecuencia del movimiento giratorio.

Rotor Rotor. Parte giratoria de un mecanismo.

Running Winding Bobinado encendido. Bobinado eléctrico de un motor por el que fluye la corriente eléctrica durante la operación normal del motor.

Runout Conducto horizontal. Término que se aplica a la porción horizontal de un ramal de un circuito.

Saddle Valve Válvula en forma de silla. Válvula cuyo cuerpo tiene la forma adecuada para que pueda ser soldado a la superficie exterior de un tubo de refrigeración.

Safety Control Control de seguridad. Dispositivo que detiene el funcionamiento de un aparato de refrigeración cuando éste alcanza temperaturas o presiones peligrosas.

Safety Plug Tapón de seguridad. Dispositivo que libera el contenido de un recipiente por encima de la presión normal de funcionamiento antes de alcanzar una presión que pueda ocasionar su ruptura.

Saturated Vapor Vapor saturado. Condición del vapor que causa condensación en gotas líquidas al bajar la temperatura.

Saturation Saturación. Condición que se da cuando una substancia contiene la máxima cantidad posible de otra substancia a esas condiciones de presión y temperatura.

Scaling Decapado. La formación de cal y otros depósitos en las líneas de contacto con la superficie del agua en un intercambiador de calor.

Schrader Valve Válvula Schrader. Dispositivo con un muelle que permite el flujo de un fluído al apretar una clavija en su centro.

Scotch Yoke Brida escocesa. Mecanismo que se utiliza para transformar movimiento linear alternativo en movimiento giratorio o viceversa.

SCR Rectificador de silicio. Un dispositivo de estado sólido utilizado para modular la capacidad de un elemento calefactor eléctrico.

Seasonal Energy Efficiency Ratio (SEER) Tasa de rendimiento energético estacional. El rendimiento total de un sistema de aire acondicionado central durante el período de tiempo que está en funcionamiento en un año dividido por la potencia eléctrica en vatios-hora utilizada durante el mismo tiempo.

Secondary Air Aire secundario. Aire suministrado del exterior a la llama de una caldera en el lugar donde tiene lugar la combustión.

Secondary Voltage Voltaje o tensión secundario. El voltaje de salida de un transformador.

Second Law of Thermodynamics Segunda ley de la termodinámica. El calor sólo puede fluír de un material a una cierta temperatura a otro material a una temperatura más baja.

Semihermetic Compressor Compresor semihermético. Un compresor hermético que puede ser reparado o se puede abrir por razones de mantenimiento o reparación.

Sensible Heat Calor sensible. Término utilizado en el mundo de la refrigeración y calefacción para indicar la porción del calor que cambia solamente la temperatura de la substancia en cuestión. También se usa para describir el calor que cambia la temperatura de un objeto sin modificar su forma.

Sensor Sensor. Material o dispositivo cuyas características intrínsecas (eléctricas o mecánicas) sufren cambios debido a un cambio de presión y temperatura.

Sequencer Secuenciador. Un controlador usado en sistemas de calefacción eléctricos para escalonar la operación de los elementos calefactores.

Series En serie. Un circuito en el que la corriente eléctrica solamente puede circular en una dirección.

Serviceable Hermetic Hermético reparable. Dícese de la carcasa de la unidad de refrigeración ensamblada con tornillos que contiene un motor y un compresor.

Service Valve Válvula de reparación. Mecanismo utilizado por los técnicos de reparación para comprobar presiones y cambiar las unidades de refrigeración.

Shaded Pole Motor Motor de polaridad libre. Un motor pequeño de corriente alterna que se usa para cargas ligeras de arranque.

Shaft Seal Sello del vástago o eje. Dispositivo utilizado para impedir el escape de fluído entre el eje y la carcasa.

Short Circuit Cortocircuito. Una conexión de baja resistencia (a menudo accidental y no deseada) entre dos partes de un circuito eléctrico.

Short Cycling Ciclado corto. Sistema de refrigeración que arranca y para con más frecuencia de la deseada.

Sight Glass Indicador transparente. Tubo o ventana de cristal en un sistema de refrigeración que permite ver la cantidad de refrigerante o aceite en el sistema.

Silica Gel Polvo de sílice. Compuesto químico que se usa frecuentemente como desecante, porque absorbe la humedad.

Silver Brazing Soldadura con plata. Proceso de soldadura en el cual la aleación utilizada para soldar contiene plata.

Single Package Heat Pump Bomba térmica autónoma. Sistema que tiene todos sus componentes en una unidad independiente.

Sling Psychrometer Psycrómetro tipo matraca. Aparato utilizado para medir la cantidad de humedad en el aire que consta de dos termómetros, uno para medir la temperatura del aire y otro para medir la temperatura del bulbo empapado.

Slugging Inundación del compressor. Situación en la que el líquido refrigerante entra en un compresor en funcionamiento.

Smoke Test Prueba de humo. Ensayo realizado para determinar si la combustión es completa.

Soft Flame Llama débil. Llama que recibe insuficiente aire primario, lo que aumenta la zona de combustión y no permite definir su cono interior.

Solar Heat Calor solar. Calor producido por las ondas visibles e invisibles del sol.

Soldering Soldadura. Unión de dos metales por adhesión de un metal con bajo punto de fusión (por debajo de 800° Fahrenheit).

Solenoid Solenoide. Émbolo móvil activado por un bobinado electromagnético.

Solenoid Valve Válvula de solenoide. Válvula actuada por la acción magnética creada por un bobinado eléctrico.

Soot Carbonilla. Substancia negra, generalmente formada por partículas pequeñas de carbón, que resulta como consecuencia de la combustión incompleta.

Specific Gravity Peso específico. Para un líquido o para un sólido, la relación entre su densidad y la del agua. Para un gas, la relación entre su peso y el del aire.

Specific Heat Calor específico. El calor absorbido (o liberado) por una unidad de masa de una substancia cuando su temperatura aumenta (o desciende) un grado. Unidades usadas con más frecuencia: Btu por libra por grado Fahrenheit y calorías por gramo por grado centígrado. Para gases, se describen tanto el calor específico a presión constante (C_p) y calor específico a volumen constante (C_v). En sistemas de aire acondicionado, se utiliza C_p.

Specific Volume Volumen específico. El volumen de una substancia por unidad de masa. Es el recíproco de la unidad de densidad: pies cúbicos por libra, centímetros cúbicos por gramo, etc.

Split Phase Motor Motor de fase dividida. Motor con dos bobinas en el estator.

Split System Sistema dividido. Instalación de refrigeración o de aire acondicionado en la que el condensador se encuentra alejado del evaporador.

Split System Heat Pump Bomba térmica dividida. Una bomba térmica cuyos componentes se encuentran tanto dentro como fuera de un edificio. Es el tipo de bomba térmica que se usa más frecuentemente en viviendas.

Squirrel Cage Ventilador de jaula de ardilla. Ventilador cuyas aspas son paralelas a su eje, que mueve el aire en ángulo recto a dicho eje.

Standard Conditions Condiciones de ambiente. Temperatura de 68 grados Fahrenheit, presión de 29,92 pulgadas de mercurio y humedad relativa del 30 por ciento.

Starting Relay Relé de arranque. Dispositivo eléctrico que conecta y desconecta el bobinado de arranque de un motor.

Starting Winding Bobinado de arranque. Bobina en el motor eléctrico que se usa durante un corto período de tiempo hasta que el motor arranca.

Static Pressure Presión estática. Presión ejercida contra el interior de un conducto o una tubería en todas las direcciones.

Stator Estator. La parte estacionaria de un motor eléctrico.

Steam Jet Refrigeration Refrigeración por chorro de vapor. Sistema de refrigeración que usa un difusor de vapor para crear un vacío (baja presión) en un recipiente de agua, lo cual hace que el agua se evapore a bajas temperaturas.

Steam Trap Bolsa de vapor. Dispositivo que impide el flujo de vapor, pero permite el flujo de condensado.

Strainer Colador o filtro. Tamiz usado para retener partículas sólidas pero permitir el paso de líquidos.

Stratification Estratificación. Condición del aire cuando forma capas de diferentes temperaturas.

Subcooled Liquid Líquido sobreenfriado. Líquido a una temperatura inferior a la temperatura de equilibrio con su vapor (la presión es superior a la presión de vapor).

Subcooling Subenfriamiento. Enfriamiento del líquido refrigerante por debajo de su temperatura de condensación.

Sublimation Sublimación. El paso de una substancia del estado sólido al estado gaseoso sin pasar por el estado líquido.

Suction Line Línea de succión. Tubo o cañería que sirve para transportar gas refrigerante del evaporador al compresor.

Suction Pressure Presión de succión. La presión en el lado inferior (zona del evaporador) de un sistema de refrigeración.

Suction Service Valve Válvula de succión de servicio. Una válvula de dos posiciones que se opera manualmente y se encuentra a la entrada al compresor.

Superheat Sobrecalentamiento. Temperatura del vapor por encima de la temperatura de ebullición de su líquido a esa presión.

Supplementary Heat Calor supletorio. Calor adicional que se requiere a temperaturas inferiores a la temperatura de equilibrio de la bomba; en la mayoría de los casos es proporcionado por una resistencia eléctrica que forma parte de la instalación de la bomba térmica. También se puede usar una caldera de gas o aceite para proporcionar calor supletorio a un sistema calefactor por combustible de origen fósil.

Surge Tank Tanque de compensación. Recipiente conectado a un sistema de refrigeración que aumenta el volumen de gas y reduce la velocidad de cambio de presión.

Swaging Abocardar. Aumentar el diámetro de un tubo en un extremo para que otro tubo del mismo diámetro se pueda encajar en él.

Swash Plate Wobble Plate. Placa balanceante. Dispositivo usado para transformar movimiento giratorio en movimiento linear alternativo.

Sweating Sudar. Dícese de la formación de agua en una superficie fría debido a la condensación de vapor presente en el aire.

Sweet Water Agua dulce. Agua del grifo.

Tankless Water Heater Calentador de agua sin depósito. Calentador indirecto de agua diseñado para operar sin un depósito de agua caliente.

Tap (Screw Thread) Macho de roscar. Herramienta utilizada para formar roscas interiores.

Temperature Temperatura. Grado de frío o calor que se mide con un termómetro.

Terminal Units Unidad terminal. Término en el que se incluyen radiadores, convectores, zócalos calefactores, calentadores, tubos con aletas, etc.

Therm Termos. Unidad de calor equivalente a 100.000 Btu.

Thermal Conductivity Conductividad térmica. Capacidad de un material de transmitir calor.

Thermal Conductivity, Earth Conductividad térmica de la tierra. La velocidad a la que energía térmica fluye a través de un material; se calcula multiplicando Btu. por pie cuadrado de la superficie del material por el espesor en pies de dicho material.

Thermal Cutout Salto de desconexión térmica. Dispositivo de seguridad; contiene un elemento calefactor que actúa sobre un dispositivo bimetal diseñado para abrir un circuito cuando el flujo de la corriente eléctrica excede el valor de diseño del dispositivo.

Thermal Resistance Resistencia térmica. Resistencia al transmisión de calor que ofrece un material.

Thermistor Termistor. Semiconductor cuya resistencia eléctrica varía con la temperatura.

Thermocouple Par termoeléctrico. Dispositivo que genera electricidad, usando el principio que dice que si dos metales diferentes son soldados juntos y el conjunto se calienta, se genera una diferencia de potencial entre los dos metales.

Thermoelectric Refrigeration Refrigeración termoeléctrica. Mecanismo de refrigeración basado en el efecto de Peletier.

Thermometer Termómetro. Dispositivo para medir temperaturas.

Thermopile Termopila. Generador piloto.

Thermostat Termostato. Interruptor que se acciona por temperatura.

Thermostat, Outdoor Ambient Termostato de temperatura exterior. Control que se utiliza para limitar la cantidad de calor supletorio de acuerdo a la temperatura exterior, con el fin de reducir sobretensión, consumo y el coste asociado.

Thermostat, Termination Termostato de corte. Un termostato situado sobre el serpentín exterior que interrumpe el ciclo de descongelación cuando la temperatura del serpentín alcanza el calor de desconexión establecido.

Thermostatic Control Control termostático. Dispositivo que controla la operación de un sistema según la temperatura del aire que lo rodea.

Thermostatic Expansion Valve Válvula de expansión termostática. Válvula que opera a presión y temperatura dentro del serpentín de evaporación.

Thermostatic Valve Válvula termostática. Válvula controlada por elementos termostáticos.

Throttling Estrangulador. Expansión de gas a través de un orificio.

Timer Temporizador. Mecanismo utilizado para controlar el tiempo de ciclado de un circuito eléctrico.

Timer, Defrost Temporizador de descongelación. Un temporizador que opera en conjunción con un sistema de refrigeración. Tras un período específico de funcionamiento, el temporizador salta e inicia el ciclo de descongelación si el termostato final está cerrado.

Ton Tonelada (de refrigeración). Efecto refrigerador que equivale a la fusión de una tonelada de hielo en 24 horas.

Torque Momento de torsion. Fuerza de giro.

Torque Wrenches Medidor del momento de torsion. Llave que se utiliza para medir el momento de torsión aplicado.

Total Pressure Presión total. También recibe el nombre de presión de impacto. Es la suma de la presión estática y la presión dinámica.

Transformer Transformador. Dispositivo diseñado para cambiar el voltaje.

Tube in a Tube (Coaxial) Intercambiador de calor coaxial. Un intercambiador de calor formado por dos tubos sellados, uno de los cuales se encuentra dentro del otro. En general, el líquido a refrigerar pasa por el tubo interior y el refrigerante se encuentra en el tubo exterior.

Turbulent Flow Flujo turbulento. El movimiento de un líquido o vapor en un tubo cuando cambia constantemente de dirección. Es lo opuesto al flujo linear.

Turndown Capacidad multiplicative. El ratio entre las relaciones máxima y mínima de entrada.

Two Pipe System Sistema de doble cañería. Un sistema de calefacción por agua caliente que utiliza una tubería para suministrar el agua caliente desde la caldera a las unidades terminales, y otra tubería para devolver el agua de las terminales a la caldera.

U factor Factor U. Conductividad térmica.

Ultraviolet Ultravioleta. Ondas invisibles de radiación de frecuencia más corta que la luz visible pero más larga que los rayos "X".

Underwriters Laboratories Underwriters Laboratories, Inc. (Abreviatura: UL). Corporación americana dedicada a la operación de laboratorios para ensayos de dispositivos eléctricos, de refrigeración, etc.

Unit Heater Unidad de calefacción. Ventilador con su motor, elemento calefactor y el recinto en que se encuentran, que generalmente va colgado del techo o de una pared.

Unit Ventilator Unidad de ventilación. Unidad terminal en la cual un ventilador hace circular aire sobre una bobina de calefacción.

Urethane Foam Goma-espuma. Tipo de aislamiento en forma de espuma que se coloca entre las paredes exteriores e interiores de un estuche de exposición.

Vacuum Vacío. Presión por debajo de la atmosférica.

Vacuum Pump Bomba de vacío. Compresor especial de gran eficiencia usado para crear presiones muy bajas.

Valve Válvula. Dispositivo utilizado para controlar el flujo de un fluído.

Valve, Check Válvula unidireccional. Válvula que permite el paso de un fluído en un solo sentido.

Valve Plate Placa de valvulería. Parte del compresor situada entre la parte superior del cuerpo y la cabeza, que contiene las válvulas de compresión.

Valve, Reversing Válvula de inversión de flujo. Válvula que se utiliza para modificar el sentido del movimiento de un refrigerante en un sistema de bomba térmica. Como tiene cuatro conexiones de tubería, también recibe el nombre de válvula de cruz.

Valve, Service Válvula de reparación. Dispositivo utilizado por técnicos de reparación para conectar manómetros al sistema de refrigeración.

Valve, Slide Válvula deslizable o de corredera. La parte deslizante de la válvula de inversión de flujo que modifica la dirección en que circula el rerigerante.

Valve, Solenoid Válvula de solenoide. Válvula de control de flujo que funciona al excitar un bobinado electromagnético que acciona un pulsador normalmente asentado en una superficie plana.

Valve, TX, Bi-Flow Válvula termostática bidireccional. Válvula termostática de expansión diseñada para reducir presiones y controlar el flujo de refrigerante en ambos sentidos.

Vapor Barrier Barrera de vapor. Plástico delgado u hoja de papel metálico que se usa para impedir que el vapor de agua penetre un material aislante.

Vapor Line Línea de vapor. Es la línea de succión en el ciclo de enfriamiento y la línea de gas caliente en el ciclo de calentamiento que se observa en bombas de calor de doble acción.

Vapor, Saturated Vapor saturado. Vapor cuya temperatura ha descendido hasta el punto de condensación sin que ésta haya empezado.

Variable Pitch Pulley Polea de paso variable. Polea que se puede ajustar para proporcionar diferentes ratios de polea.

V Belt Correa V. Tipo de correa de tranmisión.

Velocimeter Velocímetro. Instrumento que se utiliza para medir la velocidad del aire.

Velocity Pressure Presión mecánica. La presión ejercida en la dirección del flujo de un fluído.

Vent Gases Gases de escape. Productos de la combustión.

Ventilation Ventilación. La introducción de aire del exterior en un edificio por un medio mecánico.

Venturi Cono de difusión o tubo venture. Una sección de cañería o del cuerpo de una caldera que primero se estrecha y después se ensancha.

Viscosity Viscosidad. Medida de la capacidad de fluír de un fluído.

Volt Voltio. Unidad de potencial eléctrico.

Voltage Relay Relé de potencial. Un relé que entra en funcionamiento a un determinado voltaje.

Volumetric Efficiency Eficiencia volumétrica. Relación entre el rendimiento real de un compresor y el rendimiento teórico calculado.

VOM VOM. Medidor de voltaje y resistencia.

Walk In Cooler Cámara de refrigeración industrial. Habitación refrigerada para uso industrial a la que pueden acceder personas.

Water Column Columna de agua (Abreviatura: WC). Unidad para expresar presión.

Water Cooled Compressor Compresor enfriado por agua. Un intercambiador de calor que usa agua para eliminar el calor del vapor de descarga del compresor.

Water Cooled Condensing Unit Unidad de condensación enfriada por agua. Un condensador (lado alto) enfriado con agua.

Water Tube Boiler Caldera con cañería de agua. Una caldera de agua caliente en la cual el agua circula por las tuberías y los gases calientes procedentes de la combustión del carburante circulan por el exterior de dichas tuberías.

Watt Vatio. Unidad de potencia eléctrica.

Well Water Source, Closed Loop Fuente natural de agua en circuito cerrado. Sistema que obtiene agua de la tierra gracias a un pozo perforado y que la devuelve a través de otro pozo independiente del primero.

Well Water Source, Open Loop Fuente natural de agua en circuito abierto. Sistema que obtiene agua de la tierra gracias a un pozo perforado y que la devuelve a través de un sistema de desagüe.

Wet Bulb Bulbo húmedo. Termómetro que usa una bolsa húmeda en la punta de ensayo (bulbo) para medir la velocidad de evaporación de una muestra de aire. La evaporación de la humedad del aire hace descender la temperatura del bulbo húmedo en una relación diferente al termómetro de bulbo seco. Las dos mediciones ayudan a determinar las características del aire al evaluarlas en un gráfico psycrométrico.

Wind Effect Efecto del viento. El aumento de la tasa de evaporación debido al movimiento de aire sobre una superficie acuática.

Zone Zon. La porción de un edificio cuya temperatura es controlada por un termostato.

Index